MECHANICAL DESIGN HANDBOOK

Harold A. Rothbart Editor

Dean Emeritus
College of Science and Engineering
Fairleigh Dickinson University
Teaneck, N.J.

McGraw-Hill

New York   San Francisco   Washington, D.C.   Auckland   Bogotá
Caracas   Lisbon   London   Madrid   Mexico City   Milan
Montreal   New Delhi   San Juan   Singapore
Sydney   Tokyo   Toronto

Library of Congress Cataloging-in-Publication Data

Mechanical design handbook / Harold A. Rothbart, editor.
    p.   cm.
    Rev. ed. of: Mechanical design and systems handbook. c1985.
    Includes index.
    ISBN 0-07-054038-1 (acid-free paper)
    1. Machine design—Handbooks, manuals, etc. I. Rothbart, Harold
A. II. Title: Mechanical design and systems handbook.
TJ230.M433   1996
612.8′15—dc20   95-11164
                                     CIP

McGraw-Hill

A Division of The McGraw·Hill Companies

ISBN 0-07-054038-1

*The sponsoring editor for this book was Robert W. Hauserman, the editing supervisor was
Stephen M. Smith, and the production supervisor was Donald F. Schmidt. It was set in
Times Roman by Terry Leaden of McGraw-Hill's Professional Book Group composition
unit.*

Printed and bound by R. R. Donnelley & Sons Company.

Previously published as *Mechanical Design and Systems Handbook,* copyright © 1985,
1964 by McGraw-Hill, Inc.

To Florence, Ellen, Dan, and Jane

CONTENTS

v

CONTRIBUTORS

William J. Anderson *Vice President, NASTEC Inc., Cleveland, Ohio* (Sec. 19, Rolling-Element Bearings)

William H. Baier *Director of Engineering, The Fitzpatrick Co., Elmhurst, Ill.* (Sec. 24, Belts)

Philip Barkan *Professor Emeritus of Mechanical Engineering, Stanford University, Stanford, Calif.* (Sec. 31, Impact Design)

Stanley J. Becker *Adjunct Professor of Mechanical Engineering, Villanova University, Philadelphia, Pa.* (Sec. 14, Shrink- and Press-Fitted Assemblies)

Stephen B. Bennett *Manager of Research and Product Development, Delaval Turbine Division, Imo Industries, Inc., Trenton, N.J.* (Sec. 2, Mechanics of Materials)

Phyllis Zych Budka *President, Technical Communications Unlimited, Schenectady, N.Y.* (Sec. 16, Welding, Brazing, Soldering, and Adhesive Bonding)

George A. Costello *Professor of Theoretical and Applied Mechanics, University of Illinois, Urbana, Ill.* (Sec. 26, Wire Rope)

John J. Coy *Chief of Mechanical System Technology Branch, NASA Lewis Research Center, Cleveland, Ohio* (Sec. 27, Gearing)

Thomas A. Dow *Professor of Mechanical and Aerospace Engineering, North Carolina State University, Raleigh, N.C.* (Sec. 22, Friction Clutches, and Sec. 23, Friction Brakes)

Izhak Etsion *Professor of Mechanical Engineering, Technion—Israel Institute of Technology, Haifa, Israel* (Sec. 17, Seals)

David Fleming *Senior Research Engineer, NASA Lewis Research Center, Cleveland, Ohio* (Sec. 18, Fluid-Film Bearings)

Earlwood T. Fortini *Director of Engineering, Cordell Engineering, Inc., Peabody, Mass.* (Sec. 10, Dimensions and Tolerances)

Ferdinand Freudenstein *Stevens Professor of Mechanical Engineering, Columbia University, New York, N.Y.* (Sec. 3, Kinematics of Mechanisms)

Theodore Gela *Professor Emeritus of Metallurgy, Stevens Institute of Technology, Hoboken, N.J.* (Sec. 9, Properties of Engineering Materials)

Bernard J. Hamrock *Professor of Mechanical Engineering, Ohio State University, Columbus, Ohio* (Sec. 19, Rolling-Element Bearings)

Glenn L. Haugen *Programmer Analyst, City of Los Angeles, Los Angeles, Calif.* (Sec. 13, Rivets and Riveted Joints)

Anthony J. Healey *Consultant, Marine Division, Brown and Root, Inc., Houston, Tex.* (Sec. 30, Hydraulic Components)

John E. Johnson *Manager, Mechanical Model Shops, T.R.W. Corp., Redondo Beach, Calif.* (Sec. 20, Power Screws)

Kailash C. Kapur *Professor and Director of Industrial Engineering, University of Washington, Seattle, Wash.* (Sec. 8, Reliability)

Keith T. Kedward *Professor of Mechanical Engineering, University of California—Santa Barbara, Santa Barbara, Calif.* (Sec. 15, Composites)

Robert P. Kolb *Manager of Engineering (Retired), Delaval Turbine Division, Imo Industries, Inc., Trenton, N.J.* (Sec. 2, Mechanics of Materials)

Leonard R. Lamberson *Professor and Dean, College of Engineering and Applied Sciences, West Michigan University, Kalamazoo, Mich.* (Sec. 8, Reliability)

Stuart H. Loewenthal *Staff Technical Consultant, Lockheed Missile and Space Corp., Sunnyvale, Calif.* (Sec. 21, Shafts, Couplings, Keys, Etc.)

Richard F. Lytle *Technical Consultant, Fisher Controls Company, Marshalltown, Iowa* (Sec. 29, Pneumatic Components)

Ernest F. Nippes *Professor Emeritus of Metallurgical Engineering, Rensselaer Polytechnic Institute, Troy, N.Y.* (Sec. 16, Welding, Brazing, Soldering, and Adhesive Bonding)

M. O. M. Osman *Professor of Mechanical Engineering, Concordia University, Montreal, Canada* (Sec. 12, Bolts and Bolted Joints)

Burton Paul *Asa Whitney Professor of Dynamical Engineering, Department of Mechanical Engineering and Applied Mechanics, University of Pennsylvania, Philadelphia, Pa.* (Sec. 6, Machine Systems)

Abilio A. Relvas *Manager—Technical Assistance, Associated Spring, Barnes Group Inc., Bristol, Conn.* (Sec. 28, Springs)

Harold A. Rothbart *Dean Emeritus, College of Science and Engineering, Fairleigh Dickinson University, Teaneck, N.J.* (Sec. 4, Cam Mechanisms)

George N. Sandor *Research Professor Emeritus of Mechanical Engineering, Center for Intelligent Machines, University of Florida, Gainesville, Fla.* (Sec. 3, Kinematics of Mechanisms)

Bruce M. Steinetz *Senior Research Engineer, NASA Lewis Research Center, Cleveland, Ohio* (Sec. 17, Seals)

David Tabor *Professor Emeritus, Laboratory for the Physics and Chemistry of Solids, Department of Physics, Cambridge University, Cambridge, England* (Sec. 1, Friction, Lubrication, and Wear)

Steven M. Tipton *Associate Professor of Mechanical Engineering, University of Tulsa, Tulsa, Okla.* (Sec. 7, Static and Fatigue Design)

George V. Tordion *Professor of Mechanical Engineering, Université Laval, Quebec, Canada* (Sec. 25, Chains)

Dennis P. Townsend *Senior Research Engineer, NASA Lewis Research Center, Cleveland, Ohio* (Sec. 27, Gearing)

Eric E. Ungar *Chief Consulting Engineer, Bolt, Beranek and Newman, Inc., Cambridge, Mass.* (Sec. 5, Mechanical Vibrations)

John J. Viegas *Consulting Engineer, Saddle River, N.J.* (Sec. 11, Standards for Mechanical Elements)

Erwin V. Zaretsky *Chief Engineer of Structures, NASA Lewis Research Center, Cleveland, Ohio* (Sec. 27, Gearing)

PREFACE

This book, the *Mechanical Design Handbook*, is the successor to the two editions of the *Mechanical Design and Systems Handbook*. The new title reflects a reduction of the emphasis on systems. The broad subject of systems is well-covered elsewhere in the literature. The *Mechanical Design Handbook* concentrates on the basic principles and data that optimize the mechanical design or machine design of structures, machines, and components.

This book includes conventional practical design data plus the latest theoretical design concepts and information for designers and engineers. It is intended also to be of value to physicists, scientists, and students as a source reference.

The theoretical fundamentals of this handbook are presented in a classical manner with the subjects of friction, mechanics of materials, and mechanical vibrations. The traditional applied machine design elements of materials, factors of safety, shafts, rivets, friction brakes, clutches, belts, wire rope, springs, screws, chains, bolted joints, gearing, bearings, couplings, and keys are treated with the latest data available.

Included in this book is information on mechanical phenomena in modern machinery not shown in conventional machine design books. These topics are kinematics of mechanisms, cam mechanisms, reliability, shrink- and press-fitted assemblies, impact design, and wear. Also included is a section on the dynamics of machine systems based on analytical mechanics and digital computation. And a new section has been written on the important subject of composites, which are used in lightweight structures and high-speed machinery. Finite-element numerical analysis techniques for machine elements of complex shapes such as shaft couplings have been added, as well as new material on lubrication, rolling-element bearings, welding, adhesive bonding, seals, static design, and fatigue design.

The subjects of pneumatic and hydraulic components are found in this book; they provide the mathematical definitions for their specific system, whether a control system or otherwise. Many more examples and references have been added, and the elaborate index will assist the practitioner in selecting the material contained in this book.

Now I will reflect on the complex field of mechanical design. It is both a creative discipline and a decision-making process in which the mechanical designer/engineer works in the realm of ideas and at the same time deals in tangibles. Design is a constantly synthesizing and innovating process, and the designer must define, analyze, and test. The designer also uses imagination and invention to produce something new, novel, and original. Accordingly, the designer should possess engineering maturity and experience in the state of the art.

The purpose of any design is to produce a new product for a customer at an acceptable overall cost; the three controlling factors are cost of the components that go into the product, time, and quality of the product. Overall cost determines all other things and should be established early in the design. The customer always desires that the product be the least expensive possible and have the best quality for the time frame involved. The planning stage of the design requires that the designer understand the design problem and the customer's essential requirements. Redesign or corrections later in the design become quite involved and expensive.

A design plan should include the tasks needed to fulfull the project and the objective of each task with the personnel, time, and costs delineated. The operation's strategy will best be served by a concurrent engineering structure of personnel. In concurrent engineering a team of designers, engineers, manufacturers, quality control staff, materials specialists, servicing personnel, industrial designers, marketing staff, vendors, and customers determines the design criteria.

At present, design is aided by sophisticated analytical and computer methods including reliability analysis, probabilistic analysis, finite-element techniques, stress analysis, computer-aided design (CAD/CAM), stereolithography, vibration analysis, and machine dynamic analysis. All of these and other sophisticated tools should be employed to produce the initial design.

The other stages of the design include the establishment of both the function and the structure of the machine. The structure of the machine is optimized by the proper choices of physical construction, materials, size, lubrication, and other pertinent factors; the customer generally desires a machine that is small and attractive.

A significant decision in the design is the determination of when to produce a model or prototype, whether analytical, physical, or simulation. The model is studied to eliminate the ambiguities and unknowns that were not revealed or understood in the design analysis. Concern should be shown for the deviation of the model from actual operating conditions. Also, particular attention should be paid to the machinists or toolmakers who make the parts and assemble the machine. These skilled artisans are talented, and may apply innovation and invention working with all kinds of machinery.

Last, and as soon as feasible, the machine should be forwarded to the customer for installation and appraisal. The designer should be in direct contact with the customer, in order to maximize the flow of ideas. The designer should be sensitized to the customer's suggestions on the performance of the machine, i.e., adaptability, strength, life, wear, corrosion, maintenance, and safety. Furthermore, the customer's machine operators may be strategically helpful in making the machine perform properly.

In the end, frequent communication between all involved will produce the best design, with the customer being the final judge.

I would like to express my appreciation to Loretta Miller for typing the manuscript. In addition, I am indebted to Ellen Haugen for assisting in editing. Grateful acknowledgment is due the section contributors for their commitment to this handbook. Also, my thanks go to Stephen Bennett, Sheldon Kaminsky, Stuart Loewenthal, and Irwin Mortman for their invaluable ideas. And I wish to note a special thanks to Erwin Zaretsky for the many creative hours and thoughts shared in this endeavor.

Harold A. Rothbart

P · A · R · T · 1

FUNDAMENTALS

SECTION 1
FRICTION, LUBRICATION, AND WEAR

David Tabor, Sc.D.

Professor Emeritus
Laboratory for the Physics and Chemistry of Solids
Department of Physics
Cambridge University
Cambridge, England

1.1   INTRODUCTION

Sliding friction is primarily a surface phenomenon. Consequently it depends very markedly on surface conditions, such as roughness, degree of work hardening, type of oxide film, and surface cleanliness.[4,6,11] In general, in unlubricated sliding the roughness has only a secondary effect, but surface contamination can have a profound influence on friction (and wear), particularly with surfaces that are nominally clean. Because of this the account given here concentrates mainly on the mechanisms involved in friction.[4,11,20a] In this way the reader may be better able to assess the main factors involved in any particular situation. Tables of friction values are given, but they must be used with caution. Very wide differences in friction may be obtained under apparently similar conditions, especially with unlubricated surfaces.

1.2 DEFINITIONS AND LAWS OF FRICTION

1.2.1 Definition

The friction between two bodies is generally defined as the force at their surface of contact which resists their sliding on one another. The friction force F is the force required to initiate or maintain motion. If W is the normal reaction of one body on the other, the coefficient of friction

$$\mu = F/W \tag{1.1}$$

1.2.2 Static and Kinetic Friction

If the force to initiate motion of one of the bodies is F_s and the force to maintain its motion at a given speed is F_k, there is a corresponding coefficient of static friction $\mu_s = F_s/W$ and a coefficient of kinetic friction $\mu_k = F_k/W$. In some cases, these coefficients are approximately equal; in most cases $\mu_s > \mu_k$.

1.2.3 Basic Laws of Friction

The two basic laws of friction, which are valid over a wide range of experimental conditions, state that:[4]

1. The frictional force F between solid bodies is proportional to the normal force between the surfaces, i.e., μ is independent of W.
2. The frictional force F is independent of the apparent area of contact.

These two laws of friction are reasonably well obeyed for sliding metals whether clean or lubricated. With polymeric solids (plastics) the laws are not so well obeyed: in particular, the coefficient of friction usually decreases with increasing load as a result of the detailed way in which polymers deform.

1.3 SURFACE TOPOGRAPHY AND AREA OF REAL CONTACT

1.3.1 Profilometry and Asperity Slopes

When metal surfaces are placed in contact they do not usually touch over the whole of their apparent area of contact.[4,11] In general, they are supported by the surface irregularities which are present even on the most carefully prepared surfaces. Such roughnesses are usually characterized by means of a profilometer in which a fine stylus runs over the surface and moves up and down with the surface contour. The movement is measured electrically and may be recorded digitally for future detailed analysis by appropriate interfaced display units.[28] These units can provide information (see below) concerning the mean asperity heights, the distribution of peaks, valleys, slopes, asperity-tip curvatures, correlation lengths, and other features. Some commercial units display not only the essential parameters but also some which are redundant or even pointless. For visualization of the surface topography it is convenient to display the stylus movements on a chart. Since changes in height are generally very small compared with the horizontal distance traveled by the stylus, it is usual to compress the horizontal movement on the chart by a factor of 100 or more. As a result the chart record appears to suggest that the surface is covered with sharp jagged peaks.[30] In fact, when allowance is made for the difference in vertical and horizontal scales, the average slopes are rarely more than a few degrees (see Fig. 1.1).[30]

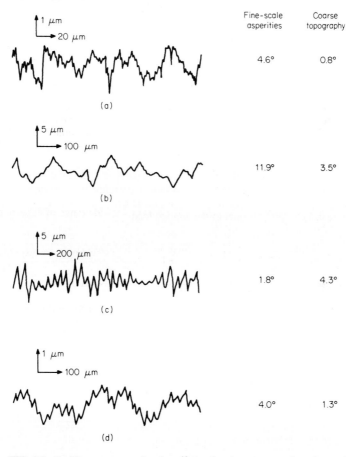

FIG. 1.1 Profilometry traces of surfaces[16] showing the average surface slopes of the fine-scale asperities and of the coarser topography. Surface treatments are (a) ground; (b) shot peened; (c) turned; (d) diamond turned.

1.3.2 Elastic and Plastic Deformation of Conical Indenters

The characterization of surface topographies and the detailed way in which the asperities deform under contact have become the subject of a number of specialized studies of varying degrees of sophistication.[28,30] We consider here the simplest case, in which the individual asperity is represented by a right circular cone of a slope θ (semiapical angle $90° - \theta$). If the cone is pressed against a smooth, flat, nondeformable surface and if it deforms elastically, the mean contact pressure p is independent of the load and is given by

$$p = (E \tan \theta)/2(1 - v^2) \qquad (1.2)$$

where E is Young's modulus and v is Poisson's ratio of the cone material.[29] Ignoring the problem of infinite stresses at the cone tip we may postulate that plastic deformation of the cone will occur when p equals the indentation hardness H of the cone,* that is, when

$$(E \tan \theta)/H = 2(1 - v^2) \approx 2 \qquad (1.3)$$

Thus the factors favoring elastic deformation are (1) smooth surfaces, that is, low θ, and (2) high hardness compared with modulus, that is, a low value of E/H.* We may at once apply this to various materials to show the conditions of surface roughness under which the asperities will deform elastically (Table 1.1). The results show that with pure metals the surfaces must be extremely smooth if plastic deformation is to be avoided. By contrast, ceramics and polymers can tolerate far greater roughnesses and still remain in the elastic regime. From the point of view of low friction and wear, elastic deformation is generally desirable, particularly if interfacial adhesion is weak (see below). Further, on this model the contact pressure is constant, either (E tan θ)/2(1 − v^2) for elastic deformation or H for plastic deformation. Thus the area of contact will be directly proportional to the applied load.

TABLE 1.1   Mean Asperity Slope for Conical Asperity Marking Transition from Elastic to Plastic Deformation

Material	Ratio E/H	Critical asperity slope for plastic deformation
Pure metals (Annealed)	200–400	$> \frac{1}{2}°$
Pure metals (Work-hardened)	70–100	1°
Alloys in hardened state	30–50	2°
Ceramics	20–30	5°
Diamond	8–10	10°
Thermoplastics (Below T_g)	5	> 20°
Cross-linked plastics	3–5	> 20°

1.3.3   Elastic and Plastic Deformation of Real Surfaces

Real surfaces are conveniently described by two main classes of representation, each of which requires two parameters.

Random or Stochastic Process.   Here the profile is treated as a two-dimensional random process and is described by the root-mean roughness (see below) and the distance

---

*The indentation hardness for a conical asperity depends on the cone angle, but for shallow asperities (80° < θ < 90°) the variation in H is small compared with E tan θ in Eq. (1.3).

over which the autocorrelation function decays to a certain fraction of its initial value. This is sometimes referred to as the "correlation distance," though it is not quite the same thing. This representation is extremely powerful but not so convenient conceptually for the purposes of this section.

Statistical Height Description.   Here the two most important parameters are:

1. The radius of curvature ß of the tip of the asperities (sharp-pointed conical asperities are unrealistic). This quantity may be treated as approximately constant for a fixed surface or it may be given a distribution of values.

2. The center-line average R_a, or alternatively the root-mean-square σ of the asperity heights (see Fig. 1.2). If in any length L of the surface the distance of any point from the mean is y,

$$R_a = \frac{1}{L} \int_0^L y \, dL \qquad (1.4)$$

$$\text{and } \sigma = \left(\frac{1}{L} \int_0^L y^2 \, dL \right)^{1/2} \qquad (1.5)$$

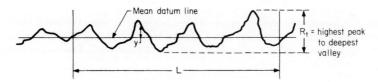

FIG. 1.2   Typical surface profile indicating the main parameters used to describe the heights of the surface roughnesses.[28]

For roughnesses resembling sine functions $\sigma \simeq 1.1R_a$ and for gaussian roughnesses $\sigma \approx 1.25R_a$. In engineering practice it is usual to use the center-line average R_a to specify surface roughness. However, in topographical theories it is more useful to use the rms value σ. (Note that in some conventions the symbol for the rms value is R_q.) However, it is clear that a measure of the mean surface height does not include such features as the spacing between significant peaks. This is particularly important for surfaces which have been milled or planed or turned, for here the topography is very different along or across the direction of machining. The random-process analyses specifically include a correlation distance between asperity peaks as well as a distribution of radii of curvature of the asperity tips. The sampling length is also of great importance.[28]

If an individual asperity is pressed against a hard smooth surface under a load ω and if it deforms elastically, the area of contact is[29]

$$A = \pi\{\tfrac{3}{4}\, \omega\beta[(1 - \nu^2)/E]\}^{2/3} \qquad (1.6)$$

Thus for each asperity A is proportional to $\omega^{2/3}$ and the contact pressure increases as $\omega^{1/3}$. The overall behavior of the real surface then depends on the way in which the asperities deform as the total load is increased. Clearly, existing asperity contacts will grow in size while new asperities will come into contact. Some of the initial asperities may reach a contact pressure exceeding the elastic limit and plastic flow will occur.

Greenwood and Williamson[8] showed that, with surfaces for which the asperity heights followed an exponential distribution, the total area of contact, even if some asperities undergo elastic and others plastic deformation, will be directly proportional to the applied load.[26,28] Physically this means that the distribution of asperity-contact areas remains almost constant so that increasing the load merely increases the number of contacts proportionately. However, real surfaces do not show an exponential distribution. The distribution is more nearly gaussian.* The detailed behavior now depends in the Greenwood-Williamson model on (1) the *surface topography* which can be described by the square root of α/β; (2) the deformation properties of the material as represented by the ratio E'/H, where $E' = E/(1 - v^2)$ and H is the contact pressure at which plastic deformation occurs. For a spherical asperity, plastic deformation is initiated when the contact pressure is about $H_v/3$, where H_v is the Vickers indentation hardness, and gradually increases with further deformation.[23] Thus H is not a crisp constant. However, to a good approximation the situation is described analytically in terms of the plasticity index ψ where

$$\psi = \frac{E'H}{(\sigma/\beta)^{1/2}} \tag{1.7}$$

The analysis shows that if $\psi < 0.6$, the deformation will be elastic over an enormous load range. The mean asperity contact pressure increases somewhat as the load is increased, but the change is not large: it is of order $0.3E(\sigma/\beta)^{1/2}$ over a large load range, so that the area of contact is very nearly proportional to the load. For extremely smooth surfaces the true contact pressure turns out to be between 0.1 and $0.3H$.[25]

For most engineering surfaces $\psi > 1$: the deformation is now plastic over an enormous range of loads and the true contact pressure is close to H. The area of contact is again proportional to the load.

The elastic and plastic regimes are shown in Fig. 1.3 for a gaussian distribution of roughnesses on a series of solids of different hardnesses. For a given surface finish the deformation is elastic if the nominal pressure is below each line. If it passes across the line, a fraction of the asperities will begin to deform plastically. This fraction will increase with increasing load until the major part of the contact becomes plastic. It will be noted that with aluminum ($H = 40$ kg/mm²), a nominal pressure of only 4 kg/mm² will give predominantly plastic deformation for even the smoothest surface. Only with ball-bearing steel ($H = 900$ kg/mm²) is the contact predominantly elastic even for relatively rough surfaces.[25] Similar results have been obtained with a stochastic treatment of surface asperities.

We conclude that for plastic deformation the true contact pressure will be equal to H. For elastic contact, over an extremely wide range it will lie between $0.1H$ and $0.3H$; indeed, it is difficult to envisage any type of asperity distribution involving elastic deformation for which contact pressure is less than about $0.1H$. This provides limiting values to the true area of contact A for the most diverse situations; for metals it will lie between W/H (plastic) and $10W/H$ (elastic), where W is the applied load.

Finally, we may note that $(\sigma/\beta)^{1/2}$ is a direct measure of the average slope of the asperity. In fact, for sine-wave asperities, it is roughly equal to θ. Thus the results obtained in the topographical model assuming spherical asperities [Eq. (1.7)] merge with the results deduced for conical asperities [Eq. (1.3)]. We may also note that if the surface roughness is characterized by a correlation length ℓ and rms asperity height σ (stochastic treatment), the average asperity slope is $2.3\sigma/\ell$.

---

*Note that the tail of a gaussian is approximately exponential, so that the highest asperities would first deform in the way described below.

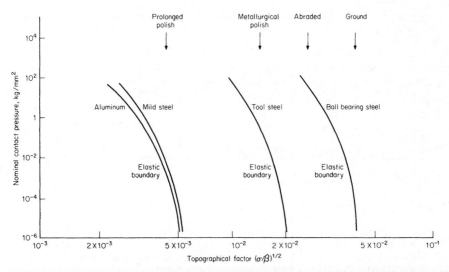

FIG. 1.3 Graph showing the nominal pressure at which the transition from elastic to the onset of plastic deformation occurs for a flat metal of specified roughness pressed on to a flat ideally smooth hard surface.[12] The materials are aluminium ($H = 40$ kg·mm^{-2}); mild steel ($H = 120$ kg·mm^{-2}); tool steel ($H = 400$ kg·mm^{-2}); ball-bearing steel ($H = 900$ kg·mm^{-2}).

1.4 FRICTION OF CLEAN METALS

1.4.1 Theory of Metallic Friction

Friction involves three major factors: (1) the area of true contact A between the surfaces, (2) the nature of the adhesion or bonding at the regions of real contact, and (3) the way in which the junctions so formed are sheared during sliding.

We have already seen that over a wide range of experimental conditions, A is proportional to the applied load and independent of the size of the bodies. For clean surfaces the adhesion that occurs at these regions is a process resembling the cold welding of metals. Consequently strong junctions are formed at the interface. Then if s is the specific shear strength of the interface, the force to produce sliding is

$$F = As + P$$

where P is a deformation or ploughing term which arises if a harder surface slides over a softer one. In general, for unlubricated surfaces, the adhesion term As is very much larger than the deformation term P, so that $F = As$. Thus the friction is proportional to the load and independent of the size of the bodies. This mechanism also explains the type of adhesive wear and surface damage which occurs between unlubricated surfaces. If the surfaces are contaminated, the adhesion is weaker and the amount of surface plucking and transfer—that is, the wear—is much less. The friction will also be smaller, but the two laws of friction still apply. However, the deformation term P may become more important relative to the adhesion term.

If metal surfaces are thoroughly cleaned in a vacuum, it is almost impossible to slide them over one another.[4,6] An attempt to do so causes further deformation at the regions of contact. The surfaces, being clean, adhere strongly wherever they touch, so that marked junction growth occurs (see Sec. 1.4.2 on microdisplacements before slid-

TABLE 1.2 Static Friction of Metals (Spectroscopically Pure) in Vacuum (Outgassed) and in Air (Unlubricated)

	Metals											
Conditions	Ag	Al	Co	Cr	Cu	Fe	In	Mg	Mo	Ni	Pb	Pt
μ, metal on itself in vacuo	S*	S	0.6	1.5	S	1.5	S	0.8	1.1	2.4	S	4
μ, metal on itself in air	1.4	1.3	0.3	0.4	1.3	1.0	2	0.5	1.9	1.7	1.5	1.3

*S signifies gross seizure ($\mu = 10$).

ing). The resistance to motion increases the harder the surfaces are pulled, and complete seizure may readily occur (top row, Table 1.2). Hydrogen and nitrogen generally have little effect, but the smallest trace of oxygen or water vapor produces a profound reduction in friction by inhibiting the formation and growth of strong metallic junctions. With most metals in air the surface oxide film serves a similar role, and the friction μ is in the range 0.5 to 1.3 (second row, Table 1.2).

The results in Table 1.2 are for spectroscopically pure metals.[4] The friction values depend crucially on the state of surface cleanliness. Thus, measurements carried out in even better vacuum will give higher values than those quoted in the table.[6]

Small amounts of impurities do not have a marked effect on the friction if (1) they do not produce a second phase, (2) they do not diffuse to the surface and dominate the

TABLE 1.3 Static Friction of Unlubricated Metals and Alloys Sliding on Steel in Air*

Metal or alloy	μ_s
Aluminum (pure)	0.6
Aluminum bronze	0.45
Brass (Cu 70%, Zn 30%)	0.5
Cast iron	0.4
Chromium (pure)	0.5
Constantan	0.4
Copper (pure)	0.8
Copper-lead (dendritic: Pb 20%)	0.2
Copper-lead (nondendritic: Pb 27%)	0.28
Indium (pure)	2
Lead (pure)	1.5
Molybdenum (pure)	0.5
Nickel (pure)	0.5
Phosphor-bronze	0.35
Silver	0.5
Steel (C 0.13%, Ni 3.42%)	0.8
Tin (pure)	0.9
White metal (tin-base): (Sb 6.4%, Cu 4.2%, Ni 0.1%, Sn 89.2%)	0.8
White metal (lead base); (Sb 15%, Cu 0.5%, Sn 6%, Pb 78.5%)	0.5
Wood's alloy	0.7

*The values are for sliders of pure metals and alloys sliding over 0.13 percent C, 3.42 percent Ni normalized steel. The results on mild steel are essentially the same.

surface properties, and (3) they do not appreciably modify the nature of the oxide film. The friction values obtained in air will depend on the extent to which the surface oxide is ruptured by the sliding process itself. Indeed, with metals in air, especially in the absence of lubricant films, the removal of the surface oxide either by an adhesion or ploughing (abrasion) mechanism—and its reformation—often play an important part in both the friction and wear mechanisms.

For softer metals, pickup occurs onto the harder surface and with repeated traversals of the same track, the sliding becomes characteristic of the softer metal sliding on itself.

The data of Tables 1.3–1.5 show that the friction does not vary monotonically with the hardness and is not greatly dependent on it. The reason is that with softer metals the area of contact A is large for a given load, but the interface is weak and therefore s is small. Conversely with hard metals A is small but s is large. Consequently the product $F = As$ is scarcely affected by the hardness. However, the friction of hard metals is, on the whole, somewhat less than that of softer metals. This is partly because of the reduced ductility of the metal junctions, which restricts junction growth, but mainly because the harder substrate provides greater support to the surface oxide film (see results for copper-beryllium alloy in Table 1.5). The friction also depends on the nature and strength of the oxide film itself. With ferrous alloys the homogeneity of the alloy is at least as important a factor, since heterogeneous materials will give weakened junctions.

TABLE 1.4 Static Friction of Unlubricated Ferrous Alloys Sliding on Themselves in Air*

Alloy	VPN kg/mm^2	μ_s
Pure iron	150	1–1.2
Normalized steel (C 0.13%, Ni 3.42%)	170	0.7–0.8
Austenitic steel (Cr 18%, Ni 8%)	200	1
Cast iron (pearlitic)	200	0.3–0.4
Ball race steel (Hoffman)	900	0.7–0.8
Tool steel (C 0.8%, containing carbides)	900	0.3–0.4
Chromium plate (hard bright)	1000	0.6

*The values indicate the effect of structure and the lack of correlation with hardness.[4]

TABLE 1.5 Static Friction of Hard Steel on Beryllium Alloy in Air*

Condition	Approximate VPN, kg/mm^2	μ_s
As quenched	120	1
Fully aged	410	0.4
Overaged	200	0.9–1

*The values are for a hard-steel slider (VPN 464 kg/mm^2) on beryllium alloy (Be 2%, Co 0.25%). They show the effect of heat treatment.

1.4.2 Microdisplacements before Sliding[4,11]

When surfaces are placed in contact under a normal load and a tangential force F is applied, the combined stresses produce further flow in the junctions long before gross sliding occurs. On a microscopic scale the surfaces sink together, increasing the area of contact; at the same time a minute tangential displacement occurs. As the tangential force is increased this process continues until a stage is reached at which the applied shear stress is greater than the strength of the interface. Junction growth comes to an end and gross sliding takes place. The tangential force has its critical value F_s. The tangential displacements before sliding are always very small. The values given below correspond to the stage where F has reached 90 percent of the value necessary to produce gross sliding ($F = 0.9F_s$). The results are for a hemispherically tipped conical slider on a flat, finely abraded surface of the same metal. As a crude approximation these tangential displacements are proportional to the square root of the normal load (Table 1.6).

TABLE 1.6   Microdisplacements before Gross Sliding[4]

| Surfaces | Vicker's hardness kg/mm^2 | Tangential displacements at $F = 0.9F_s$, 10^{-4} cm | | |
		1-g load	100-g load	10,000-g load
Indium	1	30	100	
Tin	7	1	20	200
Gold	19	0.5	5	
Platinum	117		1	60
Mild steel	280		1	8

1.4.3 Breakdown of Oxide Films

If the surface deformation produced during sliding is sufficiently small, the surface oxide may not be ruptured so that all the sliding may occur within the oxide film itself. The junctions formed in the oxide film are often weaker than purely metallic junctions, so that the friction may be appreciably less than when the oxide is ruptured. Since the shearing process occurs within the oxide film, the surface damage and wear are always considerably reduced. The criterion for "survival" of the oxide film is that it should be sufficiently soft or ductile compared with the substrate metal itself, so that it deforms with it and is not easily ruptured or fractured. Thus the oxide normally present on copper is not easily penetrated, whereas aluminum oxide, being a hard oxide on a soft substrate, is readily shattered during sliding, and even at the smallest loads there is some metallic interaction (Table 1.7). Thicker oxide films often provide more effective protection to the surfaces. Thus with anodically oxidized surfaces of aluminum or aluminum alloys, the sliding may be entirely restricted to the oxide layer.

TABLE 1.7   Breakdown of Oxide Films Produced during Sliding

| Metal | Vickers hardness, kg/mm | | Load, g, at which appreciable metallic contact occurs, g |
	Metal	Oxide	
Gold	20		0
Silver	26		0.003
Tin	5	1650	0.02
Aluminum	15	1800	0.2
Zinc	35	200	0.5
Copper	40	130	1
Iron	120	150	10
Chromium plate	800	2000	> 1000

Similarly, with very hard metal substrates, such as chromium, the surface deformation may be so small that the oxide is never ruptured.[4]

In Table 1.7 the breakdown is detected by electrical conductance measurements. The results are for a spherical slider on a flat, electrolytically polished surface. The actual breakdown loads will depend on the geometry of the surfaces and on the thickness of the oxide film. The values given in this table provide a relative measure of the protective properties of the oxide film normally present on metals prepared by electrolytic polishing.

1.4.4 Friction of Metals after Repeated Sliding

Much of the earlier basic work on the friction of metals dealt with single traversals of one body over the other. This emphasized the initial deformation and shearing at the interface and tended to give prominence to plastic deformation even if the surfaces were covered with a thin protective film of oxide (see Sec. 1.4.3). In practical systems where repeated traversals take place, work hardening and gradual surface conformity may occur; the asperities may achieve a state of plastic-elastic shakedown and the deformation may be quasi-elastic. Such a condition can never be achieved with thoroughly clean surfaces but with engineering surfaces, in air, where oxide films form this may be the more common mode.

Friction is then primarily due to shearing of oxide layers which can reform if worn away. A small part of the friction may also arise from deformation losses. In the presence of effective lubrication the friction is due to the viscous shearing of the lubricant (see Sec. 1.5). In both cases the asperities are subjected to repeated loading-unloading cycles and, even in the state of plastic-elastic shakedown, they will gradually fail by fatigue. At this stage the surfaces are worn out (see Sec. 1.6).

1.4.5 Friction of Hard Solids

The friction of hard solids in air is generally small, partly because they lack ductility and partly because of the presence of surface films. Further, there is some evidence that with covalent solids the adhesion at the interface will generally be weaker than the cohesion within the solid itself. Higher frictions are observed if the surface films are removed and/or if the sliding occurs at elevated temperatures; with covalent solids, this will favor the formation of strong interfacial bonds (see Table 1.8).

1.4.6 Friction of Thin Metallic Films

If soft metal films of suitable thickness are plated onto a hard metal, the substrate supports the load, while sliding occurs within the soft film. This can give very low coefficients of friction which persist up to the melting point of the surface film. Copper-lead–bearing alloys function in this way. Note, however, that the shear properties of the metallic film may depend on the contact pressure.

Table 1.9 gives typical friction values for thin films of indium, lead, and copper (10^{-3} to 10^{-4} cm thick) deposited on various metal substrates. The other sliding member is a steel sphere 6 mm in diameter. In air, lead films have been found to show remarkable viability.

A very effective low-friction, low-wear combination for sliding electrical contacts consists of a thin flash of gold on plated rhodium.

TABLE 1.8 Friction of Very Hard Solids

	Unbonded "hard metals"*		
	Coefficient of friction, μ_s		
	In air	Outgassed and measured in vacuo	
Material	at 20°C	20–1000°	Comments
Boron carbide	0.2	0.9	Rises rapidly above 1800°C
Silicon carbide	0.2	0.6	
Silicon nitride	0.2		
Titanium carbide	0.15	1.0	Rises rapidly above 1200°C
Titanium monoxide	0.2	0.6	
Titanium sesquioxide	0.3		
Tungsten carbide	0.15	0.6	Rises rapidly above 1000°C

Diamond and sapphire†	
Surfaces and conditions	μ_s
Diamond on self:	
In air at 20°C, clean and lubricated	0.05–0.15
Outgassed at 1000°C and measured in vacuo at 20°C	0.5
Surface films worn away be repeated sliding in high vacuum[6]	0.9
Diamond on steel:	
In air at 20°C, clean and lubricated	0.1–0.15
Sapphire on self:	
In air at 20°C, clean	0.2
Outgassed at 1000°C and measured in vacuo at 20°C	0.9
Sapphire on steel	
In air at 20°C, clean and lubricated	0.15–0.2

*Carbides, oxides, and nitrides sliding on themselves in air.

†Diamond shows marked frictional anisotropy. On the cube face {100}, μ_s = 0.05 along the edge direction <100>. As with sapphire, lubrication has little effect in air.[4]

TABLE 1.9 Static Friction of Thin Metallic Films, Unlubricated Room Temperature, in Air, Spherical Steel Slider Diameter 6 mm[4]

	Coefficient of static friction, μ_s			
Load, g	Indium film on steel	Indium film on silver	Lead film on copper	Copper film on steel
4000	0.07	0.1	0.18	0.3
8000	0.04	0.07	0.12	0.2

1.4.7 Friction of Polymers

The friction of polymers is fairly adequately explained in terms of the adhesion theory of friction. There are, however, three main differences from the behavior of metals. First, Amontons' laws are not accurately obeyed; the coefficient of friction tends to decrease with increasing load; it also tends to decrease if the geometric contact area is decreased. Second, if the surfaces are left in contact under load, the area of true contact may increase with time because of creep and the starting friction may be correspondingly larger. Third, the friction may show changes with speed which reflect the viscoelastic properties of the polymer, but the most marked changes occur as a result of frictional heating.[27] Even at speeds of only a few meters per second, the friction of unlubricated polymers can, as a result of thermal softening, rise to very high values.

TABLE 1.10 Friction of Steel on Polymers, Room
Temperature, Low Sliding Speeds*

Material	Condition	μ
Nylon	Dry	0.4
Nylon	Wet (water)	0.15
Perspex (Plexiglas)	Dry	0.5
PVC	Dry	0.5
Polystyrene	Dry	0.5
Low-density polythene (No plasticizer)	Dry or wet (water)	0.4
Low-density polythene (With plasticizer)	Dry or wet (water)	0.1
High-density polythene (No plasticizer)	Dry or wet (water)	0.15
Soft wood	Natural	0.25
Lignum vitae	Natural	0.1
PTFE† (low speeds)	Dry or wet (water)	0.06
PTFE (high speeds)	Dry or wet (water)	0.3
Filled PTFE (15% glass fiber)	Dry	0.12
Filled PTFE (15% graphite)	Dry	0.09
Filled PTFE (60% bronze)	Dry	0.09
Rubber (polyurethane)	Dry	1.6
Rubber (isoprene)	Dry	3–10
Rubber (isoprene)	Wet (water-alcohol solution)	2–4

*The slider is 0.13% carbon, 3.42 percent Ni normalized steel.
Mild steels give essentially the same result.
†Polytetrafluoroethylene.

On the other hand, at extremely high speeds the friction may fall again because of the formation of a molten lubricating film (see Table 1.10). With very soft rubbers, sliding may occur by a type of ruck moving through the interface so that motion resembles the movement of a caterpillar.

1.4.8 Shear Properties of Thin Polymer Films

The shear strengths of thin films of polymer trapped between hard surfaces have been studied experimentally.[5] (See Table 1.11.) It is found that s depends to some extent on speed and temperature, but most markedly on contact pressure p. To a first approximation

$$s = s_0 + \alpha p \tag{1.8}$$

where s_0 is the shear strength at negligibly small pressure and α is a coefficient which is approximately constant for a given material. This has an interesting relation to the friction coefficient of polymers sliding on themselves or on harder solids.[1]

Under a load W, the true area of contact A is given by $A = W/p$, where p is the true contact pressure acting on the polymer. The frictional force, ignoring the deformation term, is $F = As = (W/p)s = (W/p)(s_0 + \alpha p)$. Consequently,

TABLE 1.11   Shear Properties of Solid Polymer Films at Room Temperature (Low Speeds)[5]

Polymer	s_0, 10^7 N/m$_2$	α	$\mu_s{}^*$
Polytetrafluoroethylene (PTFE)	0.1	0.08	0.06
High-density polyethylene (HDPE)	0.25	0.10	0.12
Low-density polyethylene (LDPE)	0.6	0.14	0.4†
Polystyrene (PS)	1.4	0.17	0.5†
Polyvinyl choride (PVC)	0.45	0.18	0.5†
Polymethyl methacrylate	2.00	0.77	0.5
Plexiglas, Perspex			
Stearic acid	0.15	0.07	0.07
Calcium stearate	0.10	0.08	0.08

*See Table 1.9.
†s_0 is too large to be neglected in Eq. (1.9).

$$\mu_s = F/W = s_{0/p} + \alpha \qquad (1.9)$$

The first term is usually small compared with the second, so that

$$\mu_s \approx \alpha \qquad (1.10)$$

1.4.9   Kinetic Friction

Kinetic friction is usually smaller than static. The behavior is complicated by frictional heating which may produce structural changes near the surface or influence oxide formation. At speeds of a few meters per second, these effects are not as marked as at very high speeds (compare Tables 1.12 and 1.13), but they may be significant. The results in Table 1.12 are for stationary sliders rubbing on a mild steel disk in air. The materials are grouped in descending order of friction.

At very high sliding speeds the friction generally falls off because of the formation of a very thin molten surface layer which acts as a lubricant film. Other factors may

TABLE 1.12   Kinetic Friction of Unlubricated Metals at Speeds of a Few Meters per Second

Slider	μ_k
Nickel, mild steel	0.55–0.65
Aluminum, brass (70:30), cadmium, magnesium	0.4–0.5
Chromium (hard plate), steel (hard)	0.4
Copper, copper-cadmium alloy	0.3–0.35
Bearing alloys:	
Tin-base	0.46
Lead-base	0.34
Phosphor-bronze	0.34
Copper-lead (Pb 20)	0.18
Nonmetals:	
Brake materials (resin-bonded asbestos)	0.4
Garnet	0.4
Carbon	0.2
Bakelite	0.13
Diamond	0.08

TABLE 1.13   Kinetic Friction of Unlubricated Metals at Very High Sliding Speeds

Surface	Duration of experiment, s	Coefficient of friction, μ_k			
		9 m/s	45 m/s	225 m/s	450 m/s
Bismuth	1–10	0.25	0.1	0.05	—
Lead	1–10	0.8	0.6	0.2	0.12
Cadmium	1–10	0.3	0.25	0.15	0.1
Copper	1–10	>1.5	1.5	0.7	0.25
Molybdenum	1–10	1	0.8	0.3	0.2
Tungsten	1–10	0.5	0.4	0.2	0.2
Diamond	1–10	0.06	0.05	0.1	~ 0.1
Bismuth	10^{-3}	0.25	0.1	0.05	0.3
Lead	10^{-3}	0.7	0.5	0.2	0.3
Copper	10^{-3}	—	0.25	0.15	0.1
Steel (mild)	10^{-3}	—	0.2	0.1	0.08
Nylon	10^{-3}	0.4	0.25	0.1	0.08

*Up to 600 m/s.

also be involved. For example, with steel sliding on diamond, the friction first diminishes and then increases, because at higher speeds steel is transferred to the diamond so that the sliding resembles that of steel on steel.[4,6] In some cases the solids may fragment at these very high speeds, particularly if they are of limited ductility. Again, if appreciable melting occurs, the friction may increase at high speeds because of the viscous resistance of the liquid interface: this occurs with bismuth.

The results in Table 1.13 are for a rapidly rotating sphere of ball-bearing steel rubbing against another surface in a moderate vacuum.[4] The friction is roughly independent of load over the load range examined (10 to 500 g). However, it may depend critically on the duration of sliding, since cumulative frictional heating may greatly change the sliding conditions. The duration in the upper part of the table is about 1 to 10 s; in the lower part where a special rebound technique was used[5] the duration was about 10^{-3} s.

1.4.10   New Tribological Materials: Composites, Ceramics

Because most lubricating oils oxidize and form gums at temperatures above 250°C, considerable effort has been expended in developing high-temperature materials which are self-lubricating over a wide temperature range. One very promising approach is the formation of composite surfaces by plasma spraying, electrodeposition, or by "ion-plating." It is now possible to deposit almost any required material onto any substrate and to achieve strong adhesion to the substrate. Ceramics are particularly effective as low-friction, low-wear surfaces for operation at elevated temperatures. They may be made less brittle and tougher by incorporating other constituents during deposition. In some cases the additives act as binders, as in cermets, or as solid lubricants or structural modifiers. Ceramics are often covalent solids. As a result, when ceramics slide on one another, the interfacial adhesion is weaker than the cohesion in the bulk, so that sliding should occur at the interface itself. If they are extremely brittle, interfacial sliding may be regarded as a Type II fracture. The friction will be small compared with metals. The wear will also be smaller unless shear produces interfacial fragmentation. In general some ductility is desirable. However, at very high temperatures interfacial covalent bonding may be activated, the solids may become more ductile, and the behavior will begin to resemble that of metals.[21]

1.5 LUBRICATION

1.5.1 Hydrodynamic or Fluid Lubrication

If a convergent wedge of fluid can be established between surfaces in relative motion it will, because of its viscosity Z, generate a hydrodynamic pressure in the fluid film.[7,18] If there is no solid-solid contact, the whole of the resistance to motion is due to the viscous shear of the fluid. In journal bearings the bearing has a radius of curvature a little bigger than that of the journal (by about one part in 1000). The journal acquires a position which is slightly eccentric relative to the bearing so that lubricant is squeezed through the converging gap between the surfaces. Under properly designed conditions, the hydrodynamic pressure built up in the lubricant film is sufficient to support the normal load W. The friction is very low ($\mu_k \approx 0.001$) and there is, in principle, no wear of the solid surfaces.

Journal bearings operate under average pressures P of order 10^6 to 10^7 N/m^2 and speeds of revolution N of the order of 100 r/min. For a journal of radius R, diameter D, length L, radial clearance c, the torque G to overcome the viscous resistance of the lubricant in a full bearing may be calculated fairly reliably, simply by assuming that the journal and bearing run concentrically:

$$G = (4\pi^2 R^3 L/60c)ZN \tag{1.11}$$

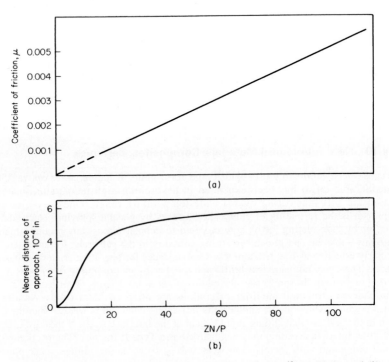

FIG. 1.4 Hydrodynamic lubrication between journal and bearing[18]: (*a*) friction and (*b*) distance of nearest approach. The dimensionless parameter ZN/P is in mixed units, with Z in centipoise, N in r/min, and P in lb/in^2.

Since the nominal pressure $P = W/2RL$, and the couple G may be written $G = \mu WR$, we obtain

$$\mu = (2\pi^2/60)(R/c)(ZN/P) \qquad (1.12)$$

where N is in revolutions per minute and all the other parameters are in consistent units.

Results for a typical full oil bearing are shown in Fig. 1.4a for $R = 15$ mm and $R/c = 1000$. Evidently μ is very small and can be reduced by working at very low values of ZN/P. However, there is a limit to this. As ZN/P diminishes, the distance h_{min} of nearest approach (Fig. 1.4b) diminishes in order to maintain adequate convergence of the lubricant film. The film may then become smaller than the surface roughness and penetration of the film may occur. In engineering practice in oil bearings, the average film thickness is of order 10^{-3} cm, and in air bearings perhaps 10 times smaller. Figure 1.5 shows the distance of nearest approach (h_{min}) for a full bearing (of infinite length) and for a full bearing of length $L = 2R$. The quantities are dimensionless and may therefore be used for both oil and air bearings if self-consistent units are used. This implies that in English units, all lengths should be in inches, forces (and loads) in pound-force, viscosity in reyns, where 1 reyn = 6.9×10^6 cP. In SI units, all forces should be in newtons, viscosity in pascals per second, where 1 Pa/s = 10^3 cP. In both systems N should be in revolutions per second. For further design charts, see Ref. 18.

In hydrodynamic lubrication (HL) it is essential to maintain an adequate value of h_{min} relative to surface roughness. The most important properties of the oil are its vis-

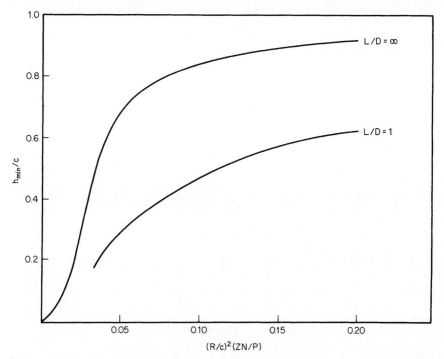

FIG. 1.5 Nondimensional graphs showing the distance of nearest approach (h_{min}) for a full journal and bearing.

cosity, its viscosity temperature dependence, and its chemical stability, especially stability against oxidation.[17]

1.5.2 Elastohydrodynamic Lubrication

In normal hydrodynamic lubrication, the hydrodynamic pressures developed in the oil film are too small to produce appreciable elastic deformation of the bearing. However, if rubber bearings are used, appreciable elastic deformation may occur and there may be a significant change in the geometry of the convergent film.[7] The hydrodynamic equations must then be combined with the equations for elastic deformation. If both surfaces are metallic and the contact pressures are high (as in rolling-element bearings or in the contact between gear teeth), there may again be sufficient elastic deformation to produce a significant change in the geometry of the contacting surfaces. A new feature is that with most lubricating oils the high pressures produce a prodigious increase in the viscosity of the oil. Thus at contact pressures of 30, 60, and 100 kg/mm^2 (such as may occur between gear teeth), the viscosity of a simple mineral oil is increased 200-, 400-, and 1000-fold, respectively. The harder the surfaces are pressed together, the harder it is to extrude the lubricant. As a result, effective lubrication may be achieved under conditions where it would normally be expected to break down. The film thickness in elastohydrodynamic lubrication (EHL) is of order 10^{-5} cm (0.1 μm), so that for safe operation, surface finish and alignment are of great importance. A full and detailed account of EHL is given in Sec. 19. It will be observed that the EHL film can show elastic, viscous, and viscoelastic properties. At sufficiently high contact pressures where the lubricant solidifies (that is, below its glass transition temperature), the oil behaves as a solid wax and its shear behavior is essentially that of a plastic solid.

1.5.3 Boundary Lubrication

Under severe conditions the EHL film may prove inadequate. Metallic contact and surface damage may occur, particularly if the oil-film thickness is too small relative to surface roughness. It is then found that the addition of a few percent of a fatty acid, alcohol, or ester may significantly improve the lubrication even if the thickness of the lubricant film is no larger than 100 Å (10^{-2} μm) or so. This is the regime of boundary lubrication (BL).[9] Radioactive tracer experiments show that while a good boundary lubricant may reduce the friction by a factor of about 20 (from $\mu \approx 1$ to $\mu \approx 0.05$), it may reduce the metallic transfer by a factor of 20,000 or more. Under these conditions the metallic junctions contribute very little to the frictional resistance. The friction is due almost entirely to the force required to shear the lubricant film itself. For this reason two good boundary lubricants may give indistinguishable coefficients of friction, but one may easily give 50 times as much metallic transfer (i.e., wear) as the other. Thus with good boundary lubricants the friction may be an inadequate indication of the effectiveness of the lubricant.[4]

In boundary lubrication the film behaves in a manner resembling EHL. There is, however, one marked difference. Because most boundary additives are adsorbed at the surface to form a condensed film or react with the surface to form a metallic soap, they are virtually solid; they do not depend on high contact pressures to achieve the load-bearing capacity of an EHL film. They are able to resist penetration by surface asperities (and here their protective properties may be enhanced by the high contact pressures), and thus they provide protection which cannot be achieved with ordinary EHL films. If, however, the temperature is raised, the boundary film may melt or it may dissolve in the superincumbent bulk fluid, and lubrication may then become far less effective (see Fig. 1.6).

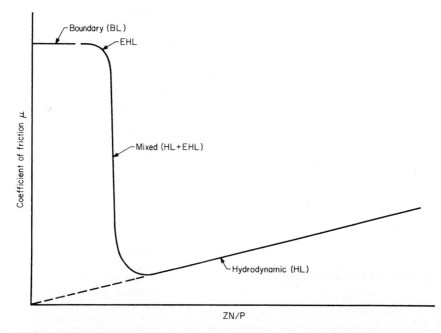

FIG. 1.6 Lubrication of a journal bearing showing the regimes of hydrodynamic (HL), elastohydrodynamic (EHL), and boundary lubrication (BL).

In the older literature the transition from HL to BL was referred to as the regime of "mixed" lubrication. We now recognize that HL gradually merges into EHL and that this then merges into BL. The coefficient of friction with good boundary lubricants ($\mu = 0.05$ to $\mu = 0.1$) is indeed similar to that observed in EHL.

In view of the nature of boundary films it is not surprising to find that their shear behavior resembles that of thin polymeric films.[5] The shear strength s per unit area of film again depends to some extent on speed and temperature, but so long as it is solid it is affected most by the contact pressure p. We find

$$s = s_0 + \alpha p$$

where α for long-chain fatty acids or esters is of order 0.05 to 0.08 and s_0 is very small (see Tables 1.14 to 1.16).

Another approach is to form a protective film by chemical attack, a small quantity of a suitable reactive compound being added to the lubricating oil. The most common materials are additives containing sulfur or chlorine or both. Phosphates are also used. The additive must not be too reactive, otherwise excessive corrosion will occur; only when there is danger of incipient seizure should chemical reaction take place. The earlier work suggested that metal sulfides and chlorides were formed and the results in Table 1.17 are based on idealized laboratory experiments in which metal surfaces were exposed to H_2S or HCl vapor and the frictional properties of the surface examined. The results show that the films formed by H_2S give a higher friction than those formed by HCl. However, in the latter case the films decompose in the presence of water to liberate HCl, and for this reason chlorine additives are less commonly used than sulfur additives.

The detailed behavior of commercial additives depends not only on the reactivity

TABLE 1.14 Static Friction of Pure Metals Sliding on Themselves in Air and When Lubricated with 1 Percent Fatty Acid in Mineral Oil (Room Temperature)[4]

	Coefficient of friction, μ_s	
Metal	Unlubricated	Lubricated
Aluminum	1.3	0.3
Cadmium	0.5	0.05
Chromium	0.4	0.34
Copper	1.4	0.10
Iron	1.0	0.12
Magnesium	0.5	0.10
Nickel	0.7	0.3
Platinum	1.3	0.25
Silver	1.4	0.55

TABLE 1.15 Lubrication of Mild Steel Surfaces by Various Lubricants (Room Temperature)[4]

Lubricant	μ_s
None	0.6
Vegetable oils	0.08–0.10
Animal oils	0.09–0.10
Mineral oils:	
Light machine	0.16
Heavy motor	0.2
Paraffin	0.18
Extreme pressure	0.10
Oleic acid	0.08
Trichloroethylene	0.3
Ethyl alcohol	0.4
Benzene	0.5
Glycerine	0.2

TABLE 1.16 Friction of Metals Lubricated with Certain Protective Films[4]

Protective film	Coefficient of friction μ_s	Temperature up to which lubrication is effective
PTFE (Teflon)	0.05	320°C
Graphite	0.07–0.13	600°C
Molybdenum disulfide	0.07–0.1	800°C

TABLE 1.17 Effect of Sulfide and Chloride Films on Friction of Metals[4]

	Coefficient of friction, μ_s				
		Sulfide films		Chloride films	
Metal	Clean	Dry	Covered with lubricating oil	Dry	Covered with lubricating oil
Cadmium on cadmium	0.5	—	—	0.3	0.15
Copper on copper	1.4	0.3	0.2	0.3	0.25
Silver on silver	1.4	0.4	0.2	—	—
Steel on steel (0.13% C, 3.42% Ni)	0.8	0.2	0.05	0.15	0.05

of the metal and the chemical nature of the additive but also on the type of carrier fluid used (e.g., aromatic, naphthenic, paraffinic). Further, the chemical reactions which occur are far more complicated than originally supposed. With sulfurized additives, oxide formation appears to be at least as important as sulfide formation. With phosphates the surface reaction is still the subject of dispute.[2] The most widespread phosphate is zinc dialkyldithio phosphate (ZDP), and in most applications it provides a low coefficient of friction ($\mu < 0.1$) up to elevated temperatures.

1.6   WEAR

1.6.1   Laws of Wear

Although the laws of friction are fairly well substantiated, there are no satisfactory laws of wear. In general, it is safe to say that wear increases with time of running and that with hard surfaces the wear is less than with softer surfaces, but there are many exceptions, and the dependence of wear on load, nominal area of contact, speed, etc., is even less generally agreed upon. This is because there are many factors involved in wear and a slight change in conditions may completely alter the importance of individual factors or change their mode of interaction.[26,31]

1.6.2   Mild and Severe Wear[10]

One of the most general characteristics of metallic wear, both for clean and for lubricated surfaces, is that below a certain load the wear is small (mild wear); above this load it rises catastrophically to values that may be 1000 or 10,000 times greater ("severe wear"). In severe wear, which occurs most readily with unlubricated surfaces, the wear is mainly due to adhesion and the shearing of the intermetallic junctions so formed.[4,26] If the junctions are very strong, shearing takes place a short distance from the interface; lumps of metal are torn out of one or both of the surfaces and these later appear as wear fragments. This often occurs in the sliding of similar metals since the junctions at the interface are highly work-hardened. If the junctions are weaker than one surface but stronger than the other, fragments will be torn out of the softer metal and the wear will generally be lower. This often occurs in the sliding of dissimilar metals. In this regime of severe wear the wear of various metallic pairs may vary by a factor of say 100 to 1, although the friction may be substantially the same.

Mild wear occurs with metals in the presence of suitable oxide films.[4,10] If the surface deformation is below a critical value, the oxide retains its integrity and the shearing occurs in the oxide film itself. The wear rate is very small and the oxide is able to reform. Mild wear also occurs with lubricated surfaces.

If wear is due to interfacial adhesion and the shearing of junctions, it may be shown on a simple model that wear is proportional to the load, is not greatly dependent on the nominal area, and is little affected by the sliding speed if frictional heating is not excessive.[4] The wear volume Z per unit distance of sliding may be written as

$$Z = K(W/3p) \qquad (1.14)$$

where Z is in cubic millimeters per millimeter (or cubic centimeters per meter), the load W is in kilogram-force, and p, the yield pressure or indentation hardness of the softer of the two bodies, is in kilograms per square millimeter. In the earlier work the quantity K was regarded as the fraction of friction junctions which produce a wear fragment. More recent work suggests that it may be more meaningful to regard it as a measure of the rate at which subsurface fatigue causes cracking and the release of a wear fragment, often in the form of a flake (delamination).[22] Table 1.18 shows wear rates of different materials in combination.

There are some combinations of friction pairs in which the interfacial adhesion is weak and sliding appears to occur truly at the interface (e.g., polyethylene on steel). Minute wear rates may then be regarded as long-term fatigue of surface asperities. A similar situation appears to occur in the wear of lubricated metals.

TABLE 1.18 Wear Rates of Various Combinations of Materials

Surfaces	Hardness p, kg/mm^2	K (calculated)
Similar metals, 0.01 cm/s:		
Cadmium	20	10^{-2}
Zinc	38	10^{-1}
Silver	43	10^{-2}
Mild Steel	160	10^{-2}
Dissimilar metals, 0.01 cm/s:		
Cadmium on mild steel	—	10^{-4}
Copper on mild steel	—	10^{-3}
Platinum on mild steel	—	10^{-3}
Mild steel on copper	—	4×10^{-4}
Similar metals, 180 cm/s:		
Mild steel	190	7×10^{-3}
Hardened tool steel	850	10^{-4}
Sintered tungsten carbide	1300	10^{-6}
On hard tool steel, 180 cm/s:		
Brass 60 : 40	90	6×10^{-4}
Silver steel	320	6×10^{-5}
Beryllium copper	210	4×10^{-5}
Stellite 1	690	5×10^{-5}
Nonmetals on hard steel, 180 cm/s:		
PTFE (Teflon, Fluon)	5	2×10^{-6}
Perspex (Plexiglas)	20	7×10^{-6}
Bakelite (poor)	5	7×10^{-6}
Bakelite (good)	30	7×10^{-7}
Polythene	2	1×10^{-7}

1.6.3 Effect of Environment

The surrounding atmosphere can have a marked effect on friction, and in many cases air or oxygen or water vapor reduce the wear rate. However, this is not always the case. If, for example, the metal oxide is hard and the conditions favor abrasive wear, the continuous formation of oxidized wear fragments may lead to a large increase in wear rate. With ferrous materials in air, the atmospheric nitrogen may play an important part. Frictional heating and rapid cooling can produce martensite, but with low-carbon steels, surface hardening can still occur by reaction with nitrogen. When these hard surface films are formed the wear generally decreases.[10]

1.6.4 Effect of Speed

The main effect of speed arises from increased surface temperatures.[4] Four of the most important consequences are:

1. High hot-spot temperatures increase reactivity of the surfaces and the wear fragments with the environment.
2. Rapid heating and cooling of asperity contacts can lead to metallurgical changes which can change the wear process.
3. High temperatures may greatly increase interdiffusion and alloy formation.
4. Surface melting may occur. In some cases, if melting is restricted to the outermost surface layers, the friction and wear may become very low.

1.6.5   Wear by Abrasives

Wear by hard abrasive particles is very common in running machinery. Measurements of the wear rate Z on abrasive papers show that the abrasion resistance $1/Z$ increases almost proportionally with the hardness of the metal.[12] This is shown in Fig. 1.7. The surfaces are grooved by the abrasive particles, but the wear is mainly in the form of fine shavings. It has been estimated that the amount of wear corresponds to about 10 percent of the volume of the material displaced in the grooves.[3]

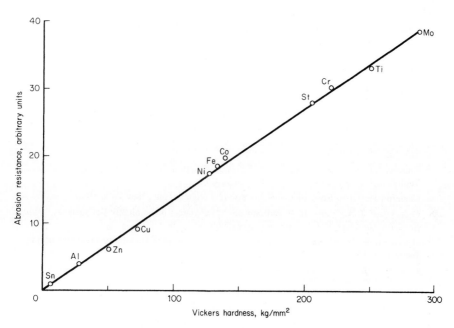

FIG. 1.7   Variation of abrasion resistance (reciprocal of wear rate) as a function of hardness for metals rubbed under standard conditions on dry abrasive paper. (The symbol St refers to 1.2 percent carbon steel.)

1.6.6   Wear Behavior of Specific Materials

There are many current publications, proceedings of conferences, etc., which deal with the wear of specific materials.[13a,13b,21,26,27,31] In addition, there are individual papers which deal, for example, with the wear of aluminum, carbon brushes,[14] bronze bearings,[15] steel,[20] and polymers.[13]

1.6.7   Identification of Wear Mechanisms

The following are possible ways of identifying wear mechanisms in a particular piece of machinery.

1. Examination of the wear debris (collected, for example, from the lubricating oil): large lumps imply adhesive wear; fine particles, oxidative wear; chiplike particles, abrasive wear; flakelike particles, delamination wear.

2. Examination of the worn surfaces: heavy tearing implies adhesive wear; scratches imply abrasive wear; burnishing indicates nonadhesive wear.

3. Metallographic examination of the surface and subsurface structure. This may reveal the type of deformation produced by the sliding process, the generation of subsurface cracks, incipient delamination, etc.

REFERENCES

1. Amuzu, J. K. A., B. J. Briscoe, and M. M. Chaudhri: "Frictional Properties of Explosives," *J. Phys. D. Appl. Phys.,* vol. 9, pp. 133–143, 1976.

2. Barcroft, F. T., K. J. Bird, J. F. Hutton, and D. Parks: "The Mechanism of Action of Zinc Thiophosphates as Extreme Pressure Additives," *Wear,* vol. 77, pp. 355–382, 1982.

3. Battacharya, S.: "Wear and Friction of Aluminum and Magnesium Alloys and Brasses," "Wear of Materials," ASME, New York, p. 40, 1981.

4. Bowden, F. P., and D. Tabor: "Friction and Lubrication of Solids," Clarendon Press, Oxford, England, part 1, 1954; part 2, 1964.

5. Briscoe, B. J., and D. Tabor: "Shear Properties of Polymeric Films," *J. Adhesion,* vol. 9, pp. 145–155, 1978.

6. Buckley, D. H.: "Surface Effects in Adhesion, Friction, Wear and Lubrication," Elsevier Publishing Co., Amsterdam, The Netherlands, 1981.

7. Cameron, A. (ed.): "Principles of Lubrication," Longman, Inc., New York, 1966.

8. Greenwood, J. A., and J. B. P. Williamson: "The contact of nominally flat surfaces," *Proc. Roy. Soc. London,* vol. A295, pp. 300–319, 1966.

9. Hardy, Sir W. B.: "Collected Scientific Papers," Cambridge University Press, Cambridge, England, 1936.

10. Hirst, W.: "Basic Mechanisms of Wear," *Proc. Lubrication and Wear, Proc. Inst. Mech. Eng.,* vol. 182 (3A), pp. 281–292, 1968.

11. Kragelsky, I. V., M. N. Dobychin, and V. S. Kombalov: "Friction and Wear-Calculation Methods" (in English), Pergamon Press, Oxford, England, 1982.

12. Kruschov, M. M.: "Principles of Abrasive Wear," *Wear,* vol. 28, pp. 69–88, 1974.

13. Lancaster, J. K.: "Friction and Wear," in "Polymer Science, a Materials Science Handbook," A. D. Jenkins ed., North Holland Publishing Co., Amsterdam, The Netherlands, pp. 960–1046, 1972.

13a. "Selecting Materials for Wear Resistance," Third International Conference on Wear of Materials, ASME, San Francisco, 1981.

13b. Ludema, K. C.: The Biennial Wear Conferences organized by ASME.

14. McNab, I. R., and J. L. Johnson: "Brush Wear," in "Wear Control Handbook," ASME, New York, p. 1053, 1980.

15. Murray, S. F., M. B., Peterson, and F. Kennedy: "Wear of Cast Bronze Bearings," INCRA Project 210, International Copper Research Association, New York, 1975.

16. Neale, M. J. (ed.): "Tribology Handbook," Butterworth & Company, London, 1973.

17. Norton, A. E.: "Lubrication," McGraw-Hill Book Company, Inc., New York, p. 24, 1942.

18. O'Connor, J. J., and J. Boyd: "Standard Handbook of Lubrication Engineering," McGraw-Hill Book Company, Inc., New York, 1978.

19. Rabinowicz, E.: "Friction and Wear of Materials," John Wiley & Sons, Inc., New York, 1965.

20. Salesky, W. J.: "Design of Medium Carbon Steels for Wear Applications," "Wear of Materials," ASME, New York, p. 298, 1981.

20a. Singer, I. L., and H. M. Pollock: "Fundamentals of Friction: Mocroscopic and Microscopic

Processes," Proceedings of a NATO Conference, 1991, Kluwer Academic Publishers, Dordrecht, The Netherlands, 1992.

21. Suh, N. P., and N. Saka (eds.): "International Conference on Fundamentals of Tribology," MIT Press, Cambridge, Mass., 1978.

22. Suh, N. P.: "The Delamination Theory of Wear," *Wear,* vol. 25, pp. 111–124, 1973.

23. Tabor, D.: "Hardness of Metals," Cambridge University Press, Cambridge, England, 1951.

24. Tabor, D.: "A Simplified Account of Surface Topography and the Contact between Solids," *Wear,* vol. 32, pp. 269–271, 1975.

25. Tabor, D.: "Interaction between Surfaces: Adhesion and Friction," chap. 10 in "Surface Physics of Materials," vol. II, J. M. Blakely (ed.), Academic Press, New York, 1975.

26. Tabor, D.: "Wear—A Critical Synoptic View," *J. Lub. Technol.,* vol. 99, pp. 387–395, 1977.

27. Tanaka, K., and Y. Uchiyama: "Friction and Surface Melting of Crystalline Polymers," in "Advances in Polymer Friction and Wear, Polymer Science and Technology," 5A and 5B, pp. 499–532, H. H. Lee (ed.), Plenum Press, Inc., New York, 1974.

28. Thomas, T. R. (ed.): "Rough Surfaces," Longman, Inc., New York, 1982.

29. Timoshenko, S., and J. N. Goodier: "Theory of Elasticity," McGraw-Hill Book Company, Inc., New York, 1951.

30. Whitehouse, D. J., and M. J. Phillips: "Discrete Properties of Random Surfaces," *Phil. Trans. Roy. Soc.* (London), vol. 290, pp. 267–298, 1976.

31. Winer, W. O. (ed.): "Wear Control Handbook," ASME, New York, 1980.

SECTION 2
MECHANICS OF MATERIALS

Stephen B. Bennett, Ph.D.

Manager of Research and Product Development
Delaval Turbine Division
Imo Industries, Inc.
Trenton, N.J.

Robert P. Kolb, P.E.

Manager of Engineering (Retired)
Delaval Turbine Division
Imo Industries, Inc.
Trenton, N.J.

2.1 INTRODUCTION

The fundamental problem of structural analysis is the prediction of the ability of machine components to provide reliable service under its applied loads and temperature. The basis of the solution is the calculation of certain performance indices, such as stress (force per unit area), strain (deformation per unit length), or gross deformation, which can then be compared to allowable values of these parameters. The allowable values of the parameters are determined by the component function (deformation constraints) or by the material limitations (yield strength, ultimate strength, fatigue strength, etc.). Further constraints on the allowable values of the performance indices are often imposed through the application of factors of safety.

This section, "Mechanics of Materials," deals with the calculation of performance indices under statically applied loads and temperature distributions. The extension of the theory to dynamically loaded structures, i.e., to the response of structures to shock and vibration loading, is treated elsewhere in this handbook.

The calculations of "Mechanics of Materials" are based on the concepts of force equilibrium (which relates the applied load to the internal reactions, or stress, in the body), material observation (which relates the stress at a point to the internal deformation, or strain, at the point), and kinematics (which relates the strain to the gross deformation of the body). In its simplest form, the solution assumes linear relationships between the components of stress and the components of strain (hookean material models) and that the deformations of the body are sufficiently small that linear relationships exist between the components of strain and the components of deformation. This linear elastic model of structural behavior remains the predominant tool used today for the design analysis of machine components, and is the principal subject of this section.

It must be noted that many materials retain considerable load-carrying ability when stressed beyond the level at which stress and strain remain proportional. The modification of the material model to allow for nonlinear relationships between stress and strain is the principal feature of the theory of plasticity. Plastic design allows more effective material utilization at the expense of an acceptable permanent deformation of the structure and smaller (but still controlled) design margins. Plastic design is often used in the design of civil structures, and in the analysis of machine structures under emergency load conditions. Practical introductions to the subject are presented in Refs. 6, 7, and 8.

Another important and practical extension of elastic theory includes a material model in which the stress-strain relationship is a function of time and temperature. This "creep" of components is an important consideration in the design of machines for use in a high-temperature environment. Reference 11 discusses the theory of creep design. The set of equations which comprise the linear elastic structural model do not have a comprehensive, exact solution for a general geometric shape. Two approaches are used to yield solutions:

The geometry of the structure is simplified to a form for which an exact solution is available. Such simplified structures are generally characterized as being a level surface in the solution coordinate system. Examples of such simplified structures include rods, beams, rectangular plates, circular plates, cylindrical shells, and spherical shells. Since these shapes are all level surfaces in different coordinate systems, e.g., a sphere is the surface $r =$ constant in spherical coordinates, it is a great convenience to express the equations of linear elastic theory in a coordinate invariant form. General tensor notation is used to accomplish this task.

The governing equations are solved through numerical analysis on a case-by-case basis. This method is used when the component geometry is such that none of the available beam, rectangular plate, etc., simplifications are appropriate. Although several classes of numerical procedures are widely used, the predominant procedure for the solution of problems in the "Mechanics of Materials" is the finite-element method.

2.2 STRESS

2.2.1 Definition[2]

"Stress" is defined as the force per unit area acting on an "elemental" plane in the body. Engineering units of stress are generally pounds per square inch. If the force is normal to the plane the stress is termed "tensile" or "compressive," depending upon whether the force tends to extend or shorten the element. If the force acts parallel to the elemental plane, the stress is termed "shear." Shear tends to deform by causing neighboring elements to slide relative to one another.

2.2.2 Components of Stress[2]

A complete description of the internal forces (stress distributions) requires that stress be defined on three perpendicular faces of an interior element of a structure. In Fig. 2.1 a small element is shown, and, omitting higher-order effects, the stress resultant on any face can be considered as acting at the center of the area.

The direction and type of stress at a point are described by subscripts to the stress symbol σ or τ. The first subscript defines the plane on which the stress acts and the second indicates the direction in which it acts. The plane on which the stress acts is indicated by the normal axis to that plane; e.g., the x plane is normal to the x axis. Conventional notation omits the second subscript for the normal stress and replaces the σ by a τ for the shear stresses. The "stress components" can thus be represented as follows.

Normal stress:

$$\sigma_{xx} \equiv \sigma_x$$

$$\sigma_{yy} \equiv \sigma_y \qquad (2.1)$$

$$\sigma_{zz} \equiv \sigma_z$$

Shear stress:

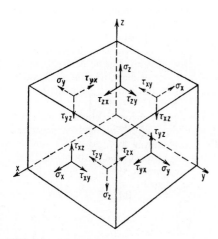

FIG. 2.1 Stress components.

$$\sigma_{xy} \equiv \tau_{xy} \qquad \sigma_{yz} \equiv \tau_{yz}$$

$$\sigma_{xz} \equiv \tau_{xz} \qquad \sigma_{zx} \equiv \tau_{zx} \tag{2.2}$$

$$\sigma_{yx} \equiv \tau_{yz} \qquad \sigma_{zy} \equiv \tau_{zy}$$

In tensor notation, the stress components are

$$\sigma_{ij} = \begin{pmatrix} \sigma_x & \tau_{xy} & \tau_{xz} \\ \tau_{yx} & \sigma_y & \tau_{yz} \\ \tau_{zx} & \tau_{zy} & \sigma_z \end{pmatrix} \tag{2.3}$$

Stress is "positive" if it acts in the "positive-coordinate direction" on those element faces farthest from the origin, and in the "negative-coordinate direction" on those faces closest to the origin. Figure 2.1 indicates the direction of all positive stresses, wherein it is seen that tensile stresses are positive and compressive stresses negative.

The total load acting on the element of Fig. 2.1 can be completely defined by the stress components shown, subject only to the restriction that the coordinate axes are mutually orthogonal. Thus the three normal stress symbols σ_x, σ_y, σ_z and six shear-stress symbols τ_{xy}, τ_{xz}, τ_{yx}, τ_{yz}, τ_{zx}, τ_{zy} define the stresses of the element. However, from equilibrium considerations, $\tau_{xy} = \tau_{yx}$, $\tau_{yz} = \tau_{zy}$, $\tau_{xz} = \tau_{zx}$. This reduces the necessary number of symbols required to define the stress state to σ_x, σ_y, σ_z, τ_{xy}, τ_{xz}, τ_{yz}.

2.2.3 Simple Uniaxial States of Stress[1]

Consider a simple bar subjected to axial loads only. The forces acting at a transverse section are all directed normal to the section. The uniaxial normal stress at the section is obtained from

$$\sigma = P/A \tag{2.4}$$

where P = total force and A = cross-sectional area.

"Uniaxial shear" occurs in a circular cylinder, loaded as in Fig. 2.2a, with a radius which is large compared to the wall thickness. This member is subjected to a torque distributed about the upper edge:

$$T = \Sigma Pr \tag{2.5}$$

Now consider a surface element (assumed plane) and examine the stresses acting. The

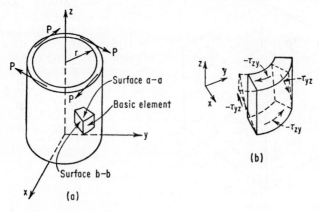

FIG. 2.2 Uniaxial shear basic element.

stresses τ which act on surfaces *a-a* and *b-b* in Fig. 2.2*b* tend to distort the original rectangular shape of the element into the parallelogram shown (dotted shape). This type of action of a force along or tangent to a surface produces shear within the element, the intensity of which is the "shear stress."

2.2.4 Nonuniform States of Stress[1]

In considering elements of differential size, it is permissible to assume that the force acts on any side of the element concentrated at the center of the area of that side, and that the stress is the average force divided by the side area. Hence it has been implied thus far that the stress is uniform. In members of finite size, however, a variable stress intensity usually exists across any given surface of the member. An example of a body which develops a distributed stress pattern across a transverse cross section is a simple beam subjected to a bending load as shown in Fig. 2.3*a*. If a section is then taken at *a-a*, F'_1 must be the internal force acting along *a-a* to maintain equilibrium. Forces F_1 and F'_1 constitute a couple which tends to rotate the element in a clockwise direction, and therefore a resisting couple must be developed at *a-a* (see Fig. 2.3*b*). The internal effect at *a-a* is a stress distribution with the upper portion of the beam in tension and the lower portion in compression, as in Fig. 2.3*c*. The line of zero stress on the transverse cross section is the "neutral axis" and passes through the centroid of the area.

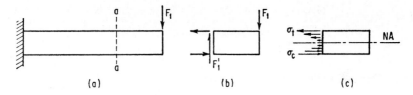

FIG. 2.3 Distributed stress on a simple beam subjected to a bending load.

2.2.5 Combined States of Stress

Tension-Torsion. A body loaded simultaneously in direct tension and torsion, such as a rotating vertical shaft, is subject to a combined state of stress. Figure 2.4*a* depicts such a shaft with end load *W*, and constant torque *T* applied to maintain uniform rotational velocity. With reference to *a-a*, considering each load separately, a force system

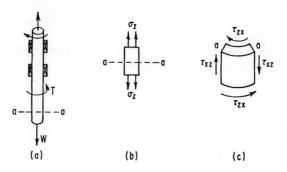

FIG. 2.4 Body loaded in direct tension and torsion.

as shown in Fig. 2.2*b* and *c* is developed at the internal surface *a-a* for the weight load and torque, respectively. These two stress patterns may be superposed to determine the "combined" stress situation for a shaft element.

Flexure-Torsion. If in the above case the load *W* were horizontal instead of vertical, the combined stress picture would be altered. From previous considerations of a simple beam, the stress distribution varies linearly across section *a-a* of the shaft of Fig. 2.5*a*. The stress pattern due to flexure then depends upon the location of the element in question; e.g., if the element is at the outside (element *x*) then it is undergoing maximum tensile stress (Fig. 2.5*b*), and the tensile stress is zero if the element is located on the horizontal center line (element *y*) (Fig. 2.5*c*). The shearing stress is still constant at a given element, as before (Fig. 2.5*d*). Thus the "combined" or "superposed" stress state for this condition of loading varies across the entire transverse cross section.

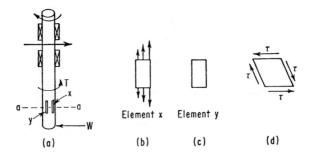

Element x   Element y

(a)   (b)   (c)   (d)

FIG. 2.5   Body loaded in flexure and torsion.

2.2.6   Stress Equilibrium

"Equilibrium" relations must be satisfied by each element in a structure. These are satisfied if the resultant of all forces acting on each element equals zero in each of three mutually orthogonal directions on that element. The above applies to all situations of "static equilibrium." In the event that some elements are in motion an inertia term must be added to the equilibrium equation. The inertia term is the elemental mass multiplied by the absolute acceleration taken along each of the mutually perpendicular axes. The equations which specify this latter case are called "dynamic-equilibrium equations" (see Sec. 5).

Three-Dimensional Case.[5,13]   The equilibrium equations can be derived by separately summing all *x*, *y*, and *z* forces acting on a differential element accounting for the incremental variation of stress (see Fig. 2.6). Thus the normal forces acting on areas *dz dy* are $\sigma_x\, dz\, dy$ and $[\sigma_x + (\partial\sigma_x/\partial x)\, dx]\, dz\, dy$. Writing *x* force-equilibrium equations, and by a similar process *y* and *z* force-equilibrium equations, and canceling *higher-order* terms, the following three "cartesian equilibrium equations" result:

$$\partial\sigma_x/\partial x + \partial\tau_{xy}/\partial y + \partial\tau_{xz}/\partial z = 0 \tag{2.6}$$

$$\partial\sigma_y/\partial y + \partial\tau_{yz}/\partial z + \partial\tau_{yx}/\partial x = 0 \tag{2.7}$$

$$\partial\sigma_z/\partial z + \partial\tau_{zx}/\partial x + \partial\tau_{zy}/\partial y = 0 \tag{2.8}$$

or, in cartesian stress-tensor notation,

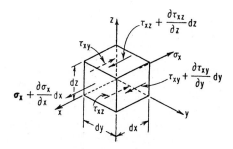

FIG. 2.6 Incremental element (dx, dy, dz) with incremental variation of stress.

$$\sigma_{ij,j} = 0 \qquad i,j = x,y,z \qquad (2.9)$$

and, in general tensor form,

$$g^{1k}\sigma_{ij,k} = 0 \qquad (2.10)$$

where g^{ik} is the contravariant metric tensor.

"Cylindrical-coordinate" equilibrium considerations lead to the following set of equations (see Fig. 2.7):

$$\partial\sigma_r/\partial r + (1/r)(\partial\tau_{r\theta}/\partial\theta) + \partial\tau_{rz}/\partial z + (\sigma_r - \sigma_\theta)/r = 0 \qquad (2.11)$$

$$\partial\tau_{r\theta}/\partial r + (1/r)(\partial\sigma_\theta/\partial\theta) + \partial\tau_{\theta z}/\partial z + 2\tau_{r\theta}/r = 0 \qquad (2.12)$$

$$\partial\tau_{rz}/\partial r + (1/r)(\partial\tau_{\theta z}/\partial\theta) + \partial_{\sigma z}/\partial z + \tau_{rz}/r = 0 \qquad (2.13)$$

The corresponding "spherical polar-coordinate" equilibrium equations are (see Fig. 2.8)

$$\frac{\partial\sigma_r}{\partial r} + \frac{1}{r}\frac{\partial\tau_{r\theta}}{\partial\theta} + \left(\frac{1}{r\sin\theta}\right)\frac{\partial\tau_{r\phi}}{\partial\phi} + \frac{1}{r}(2\sigma_r - \sigma_\theta - \sigma_\phi + \tau_{r\theta}\cot\theta) = 0 \qquad (2.14)$$

$$\frac{\partial\tau_{r\theta}}{\partial r} + \frac{1}{r}\frac{\partial\sigma_\theta}{\partial\theta} + \left(\frac{1}{r\sin\theta}\right)\frac{\partial\tau_{\theta\phi}}{\partial\phi} + \frac{1}{r}[(\sigma_\theta - \sigma_\phi)\cot\theta + 3\tau_{r\theta}] = 0 \qquad (2.15)$$

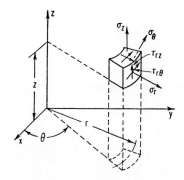

FIG. 2.7 Stresses on a cylindrical element.

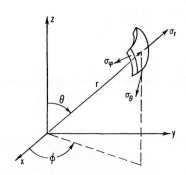

FIG. 2.8 Stresses on a spherical element.

$$\frac{\partial \tau_{r\phi}}{\partial r} + \frac{1}{r}\frac{\partial \tau_{\theta\phi}}{\partial \theta} + \left(\frac{1}{r \sin\theta}\right)\frac{\partial \sigma_\phi}{\partial \phi} + \frac{1}{r}(3\tau_{r\phi} + 2\tau_{\theta\phi}\cot\theta) = 0 \qquad (2.16)$$

The general orthogonal curvilinear-coordinate equilibrium equations are

$$h_1 h_2 h_3\left(\frac{\partial}{\partial \alpha}\frac{\sigma_\alpha}{h_2 h_3} + \frac{\partial}{\partial \beta}\frac{\tau_{\alpha\beta}}{h_3 h_1} + \frac{\partial}{\partial \gamma}\frac{\tau_{\gamma\alpha}}{h_1 h_2}\right) + \tau_{\alpha\beta}h_1 h_2 \frac{\partial}{\partial \beta}\frac{1}{h_1}$$

$$+ \tau_{\gamma\alpha}h_1 h_3 \frac{\partial}{\partial \gamma}\frac{1}{h_1} - \sigma_\beta h_1 h_2 \frac{\partial}{\partial \alpha}\frac{1}{h_2} - \sigma_\gamma h_1 h_3 \frac{\partial}{\partial \alpha}\frac{1}{h_3} = 0 \qquad (2.17)$$

$$h_1 h_2 h_3\left(\frac{\partial}{\partial \beta}\frac{\sigma_\beta}{h_3 h_1} + \frac{\partial}{\partial \gamma}\frac{\tau_{\beta\gamma}}{h_1 h_2} + \frac{\partial}{\partial \alpha}\frac{\tau_{\alpha\beta}}{h_2 h_3}\right) + \tau_{\beta\gamma}h_2 h_3 \frac{\partial}{\partial \gamma}\frac{1}{h_2}$$

$$+ \tau_{\alpha\beta}h_2 h_1 \frac{\partial}{\partial \alpha}\frac{1}{h_2} - \sigma_\gamma h_2 h_3 \frac{\partial}{\partial \beta}\frac{1}{h_3} - \sigma_\alpha h_2 h_1 \frac{\partial}{\partial \beta}\frac{1}{h_1} = 0 \qquad (2.18)$$

$$h_1 h_2 h_3\left(\frac{\partial}{\partial \gamma}\frac{\sigma_\gamma}{h_1 h_2} + \frac{\partial}{\partial \alpha}\frac{\tau_{\gamma\alpha}}{h_2 h_3} + \frac{\partial}{\partial \beta}\frac{\tau_{\beta\gamma}}{h_3 h_1}\right) + \tau_{\gamma\alpha}h_3 h_1 \frac{\partial}{\partial \alpha}\frac{1}{h_3}$$

$$+ \tau_{\beta\gamma}h_3 h_2 \frac{\partial}{\partial \beta}\frac{1}{h_3} - \sigma_\alpha h_3 h_1 \frac{\partial}{\partial \gamma}\frac{1}{h_1} - \sigma_\beta h_3 h_2 \frac{\partial}{\partial \gamma}\frac{1}{h_2} = 0 \qquad (2.19)$$

where the α, β, γ specify the coordinates of a point and the distance between two coordinate points ds is specified by

$$(ds)^2 = (d\alpha/h_1)^2 + (d\beta/h_2)^2 + (d\gamma/h_3)^2 \qquad (2.20)$$

which allows the determination of h_1, h_2, and h_3 in any specific case.

Thus, in cylindrical coordinates,

$$(ds)^2 = (dr)^2 + (r\,d\theta)^2 + (dz)^2 \qquad (2.21)$$

so that
$$\alpha = r \qquad h_1 = 1$$
$$\beta = \theta \qquad h_2 = 1/r$$
$$\gamma = z \qquad h_3 = 1$$

In spherical polar coordinates

$$(ds)^2 = (dr)^2 + (r\,d\theta)^2 + (r\sin\theta\,d\phi)^2 \qquad (2.22)$$

so that
$$\alpha = r \qquad h_1 = 1$$
$$\beta = \theta \qquad h_2 = 1/r$$
$$\gamma = \phi \qquad h_3 = 1/(r\sin\theta)$$

All the above equilibrium equations define the conditions which must be satisfied by each interior element of a body. In addition, these stresses must satisfy all surface-stress-boundary conditions. In addition to the cartesian-, cylindrical-, and spherical-coordinate systems, others may be found in the current literature or obtained by reduction from the general curvilinear-coordinate equations given above.

In many applications it is useful to integrate the stresses over a finite thickness and express the resultant in terms of zero or nonzero force or moment resultants as in the beam, plate, or shell theories.

Two-Dimensional Case—Plane Stress.[2] In the special but useful case where the stresses in one of the coordinate directions are negligibly small ($\sigma_z = \tau_{xz} = \tau_{yz} = 0$) the general cartesian-coordinate equilibrium equations reduce to

$$\partial\sigma_x/\partial x + \partial\tau_{xy}/\partial y = 0 \tag{2.23}$$

$$\partial\sigma_y/\partial y + \partial\tau_{yx}/\partial x = 0 \tag{2.24}$$

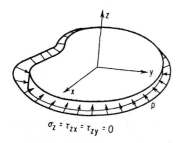

$\sigma_z = \tau_{zx} = \tau_{zy} = 0$

FIG. 2.9 Plane stress on a thin slab.

The corresponding cylindrical-coordinate equilibrium equations become

$$\partial\sigma_r/\partial r + (1/r)(\partial\tau_{r\theta}/\partial\theta) + (\sigma_r - \sigma_\theta)/r = 0 \tag{2.25}$$

$$\partial\tau_{r\theta}/\partial r + (1/r)(\partial\sigma_\theta/\partial\theta) + 2(\tau_{r\theta}/r) = 0 \tag{2.26}$$

This situation arises in "thin slabs," as indicated in Fig. 2.9, which are essentially two-dimensional problems. Because these equations are used in formulations which allow only stresses in the "plane" of the slab, they are classified as "plane-stress" equations.

2.2.7 Stress Transformation: Three-Dimensional Case[4,5]

It is frequently necessary to determine the stresses at a point in an element which is rotated with respect to the *x, y, z* coordinate system; i.e., in an orthogonal *x´, y´, z´* system. Using equilibrium concepts and measuring the angle between any specific original and rotated coordinate by the direction cosines (cosine of the angle between the two axes) the following transformation equations result:

$$\begin{aligned}
\sigma_{x'} = &[\sigma_x \cos(x'x) + \tau_{xy}\cos(x'y) + \tau_{zx}\cos(x'z)]\cos(x'x) \\
&+ [\tau_{xy}\cos(x'x) + \sigma_y\cos(x'y) + \tau_{yz}\cos(x'z)]\cos(x'y) \\
&+ [\tau_{zx}\cos(x'x) + \tau_{yz}\cos(x'y) + \sigma_z\cos(x'z)]\cos(x'z)
\end{aligned} \tag{2.27}$$

$$\begin{aligned}
\sigma_{y'} = &[\sigma_x \cos(y'x) + \tau_{xy}\cos(y'y) + \tau_{zx}\cos(y'z)]\cos(y'x) \\
&+ [\tau_{xy}\cos(y'x) + \sigma_y\cos(y'y) + \tau_{yz}\cos(y'z)]\cos(y'y) \\
&+ [\tau_{zx}\cos(y'x) + \tau_{yz}\cos(y'y) + \sigma_z\cos(y'z)]\cos(y'z)
\end{aligned} \tag{2.28}$$

$$\begin{aligned}
\sigma_{z'} = &[\sigma_x \cos(z'x) + \tau_{xy}\cos(z'y) + \tau_{zx}\cos(z'z)]\cos(z'x) \\
&+ [\tau_{xy}\cos(z'x) + \sigma_y\cos(z'y) + \tau_{yz}\cos(z'z)]\cos(z'y) \\
&+ [\tau_{zx}\cos(z'x) + \tau_{yz}\cos(z'y) + \sigma_z\cos(z'z)]\cos(z'z)
\end{aligned} \tag{2.29}$$

$$\begin{aligned}
\tau_{x'y'} = &[\sigma_x \cos(y'x) + \tau_{xy}\cos(y'y) + \tau_{zx}\cos(y'z)]\cos(x'x) \\
&+ [\tau_{xy}\cos(y'x) + \sigma_y\cos(y'y) + \tau_{yz}\cos(y'z)]\cos(x'y) \\
&+ [\tau_{zx}\cos(y'x) + \tau_{yz}\cos(y'y) + \sigma_z\cos(y'z)]\cos(x'z)
\end{aligned} \tag{2.30}$$

$$\tau_{y'z'} = [\sigma_x \cos(z'x) + \tau_{xy} \cos(z'y) + \tau_{zx} \cos(z'z)] \cos(y'x)$$

$$+ [\tau_{xy} \cos(z'x) + \sigma_y \cos(z'y) + \tau_{yz} \cos(z'z)] \cos(y'y)$$

$$+ [\tau_{zx} \cos(z'x) + \tau_{yz} \cos(z'y) + \sigma_z \cos(z'z)] \cos(y'z) \qquad (2.31)$$

$$\tau_{z'x'} = [\sigma_x \cos(x'x) + \tau_{xy} \cos(x'y) + \tau_{zx} \cos(x'z)] \cos(z'x)$$

$$+ [\tau_{xy} \cos(x'x) + \sigma_y \cos(x'y) + \tau_{yz} \cos(x'z)] \cos(z'y)$$

$$+ [\tau_{zx} \cos(x'x) + \tau_{yz} \cos(x'y) + \sigma_z \cos(x'z)] \cos(z'z) \qquad (2.32)$$

In tensor notation these can be abbreviated:

$$\tau_{k'l'} = A_{l'n}A_{k'm}\tau_{mn} \qquad (2.33)$$

where $\quad A_{ij} = \cos(ij) \qquad m,n \to x,y,z \qquad k',l' \to x',y',z'$

A special but very useful coordinate rotation occurs when the direction cosines are so selected that all the shear stresses vanish. The remaining mutually perpendicular "normal stresses" are called "principal stresses."

The magnitudes of the principal stresses σ_x, σ_y, σ_z are the three roots of the cubic equations associated with the determinant:

$$\begin{vmatrix} \sigma_x - \sigma & \tau_{xy} & \tau_{zx} \\ \tau_{xy} & \sigma_y - \sigma & \tau_{yz} \\ \tau_{zx} & \tau_{yz} & \sigma_z - \sigma \end{vmatrix} = 0 \qquad (2.34)$$

where $\sigma_x, \ldots, \tau_{xy}, \ldots$ are the general nonprincipal stresses which exist on an element.

The direction cosines of the principal axes x', y' z' with respect to the x, y, z axes are obtained from the simultaneous solution of the following three equations considering separately the cases where $n = x', y' z'$:

$$\tau_{xy} \cos(xn) + (\sigma_y - \sigma_n) \cos(yn) + \tau_{yz} \cos(zn) = 0 \qquad (2.35)$$

$$\tau_{zx} \cos(xn) + \tau_{yz} \cos(yn) + (\sigma_z - \sigma_n) \cos(zn) = 0 \qquad (2.36)$$

$$\cos^2(xn) + \cos^2(yn) + \cos^2(zn) = 1 \qquad (2.37)$$

2.2.8 Stress Transformation: Two-Dimensional Case[2,4]

Selecting an arbitrary coordinate direction in which the stress components vanish, it can be shown, either by equilibrium considerations or by general transformation formulas, that the two-dimensional stress-transformation equations become

$$\sigma_n = [(\sigma_x + \sigma_y)/2] + [(\sigma_x - \sigma_y)/2] \cos 2\alpha + \tau_{xy} \sin 2\alpha \qquad (2.38)$$

$$\tau_{nt} = [(\sigma_x - \sigma_y)/2] \sin 2\alpha - \tau_{xy} \cos 2\alpha \qquad (2.39)$$

where the directions are defined in Figs. 2.10 and 2.11 ($\tau_{xy} = -\tau_{nt}$, $\alpha = 0$).

The principal directions are obtained from the condition that

$$\tau_{nt} = 0 \qquad \text{or} \qquad \tan 2\alpha = 2\tau_{xy}/(\sigma_x - \sigma_y) \qquad (2.40)$$

where the two lowest roots of (first and second quadrants) are taken. It can be easily seen that the first and second principal directions differ by 90°. It can be shown that the *principal* stresses are also the "maximum" or "minimum normal stresses." The "plane of maximum shear" is defined by

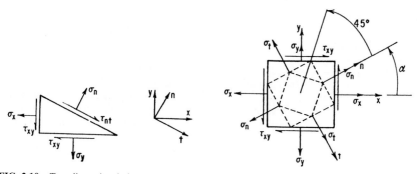

FIG. 2.10 Two-dimensional plane stress.

FIG. 2.11 Plane of maximum shear.

$$\tan 2\alpha = -(\sigma_x - \sigma_y)/2\tau_{xy} \tag{2.41}$$

These are also represented by planes which are 90° apart and are displaced from the principal stress planes by 45° (Fig. 2.11).

2.2.9 Mohr's Circle

Mohr's circle is a convenient representation of the previously indicated transformation equations. Considering the x, y directions as positive in Fig. 2.11, the stress condition on any elemental plane can be represented as a point in the "Mohr diagram" (clockwise shear taken positive). The Mohr's circle is constructed by connecting the two stress points and drawing a circle through them with center on the σ axis. The stress state of any basic element can be represented by the stress coordinates at the intersection of the circle with an arbitrarily directed line through the circle center. Note that point x for positive τ_{xy} is below the σ axis and vice versa. The element is taken as rotated counterclockwise by an angle α with respect to the x-y element when the line is rotated counterclockwise an angle 2α with respect to the x-y line, and vice versa (Fig. 2.12).

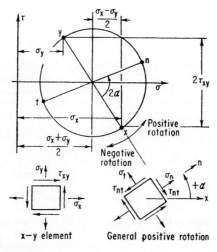

FIG. 2.12 Stress state of basic element.

2.3 STRAIN

2.3.1 Definition[2]

Extensional strain ϵ is defined as the extensional deformation of an element divided by the basic elemental length, $\epsilon = u/l_0$.

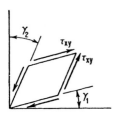

In large-strain considerations, l_0 must represent the instantaneous elemental length and the definitions of strain must be given in incremental fashion. In small strain considerations, to which the following discussion is limited, it is only necessary to consider the original elemental length l_0 and its change of length u. Extensional strain is taken positive or negative depending on whether the element increases or decreases in extent. The units of strain are dimensionless (inches/inch).

"Shear strain" γ is defined as the angular distortion of an original right-angle element. The direction of positive shear strain is taken to correspond to that produced by a positive shear stress (and vice versa) (see Fig. 2.13). Shear strain γ is equal to $\gamma_1 + \gamma_2$. The "units" of shear strain are dimensionless (radians).

FIG. 2.13 Shear-strain-deformed element.

2.3.2 Components of Strain[2]

A complete description of strain requires the establishment of three orthogonal extensional and shear strains. In cartesian stress nomenclature, the strain components are:

Extensional strain:

$$\epsilon_{xx} \equiv \epsilon_x$$

$$\epsilon_{yy} \equiv \epsilon_y \qquad (2.42)$$

$$\epsilon_{zz} \equiv \epsilon_z$$

Shear strain:

$$\epsilon_{xy} = \epsilon_{yx} \equiv \tfrac{1}{2}\gamma_{xy}$$

$$\epsilon_{yz} = \epsilon_{zy} \equiv \tfrac{1}{2}\gamma_{yz} \qquad (2.43)$$

$$\epsilon_{zx} = \epsilon_{xz} \equiv \tfrac{1}{2}\gamma_{zx}$$

where positive ϵ_x, ϵ_y, or ϵ_z corresponds to a positive stretching in the x, y, z directions and positive γ_{xy}, γ_{yz}, γ_{zx} refers to positive shearing displacements in the xy, yz, and zx planes. In tensor notation the strain components are

$$\epsilon_{ij} = \begin{pmatrix} \epsilon_x & \tfrac{1}{2}\gamma_{xy} & \tfrac{1}{2}\gamma_{zx} \\ \tfrac{1}{2}\gamma_{xy} & \epsilon_y & \tfrac{1}{2}\gamma_{yz} \\ \tfrac{1}{2}\gamma_{zx} & \tfrac{1}{2}\gamma_{yz} & \epsilon_z \end{pmatrix} \qquad (2.44)$$

2.3.3 Simple and Nonuniform States of Strain[2]

Corresponding to each of the stress states previously illustrated there exists either a simple or nonuniform strain state.

In addition to these, a state of "uniform dilatation" exists when the shear strain vanishes and all the extensional strains are equal in sign and magnitude. Dilatation is defined as

$$\Delta = \epsilon_x + \epsilon_y + \epsilon_z \qquad (2.45)$$

and represents the change of volume per increment volume.

In uniform dilatation,

$$\Delta = 3\epsilon_x = 3\epsilon_y = 3\epsilon_z \qquad (2.46)$$

2.3.4 Strain-Displacement Relationships[4,5,13]

Considering only small strain, and the previous definitions, it is possible to express the strain components at a point in terms of the associated displacements and their derivatives in the coordinate directions (e.g., u, v, w are displacements in the x, y, z coordinate system).

Thus in a "cartesian system" (x, y, z),

$$\epsilon_x = \partial u/\partial x \qquad \gamma_{xy} = \partial v/\partial x + \partial u/\partial y$$

$$\epsilon_y = \partial v/\partial y \qquad \gamma_{yz} = \partial w/\partial y + \partial v/\partial z \qquad (2.47)$$

$$\epsilon_z = \partial w/\partial z \qquad \gamma_{zx} = \partial u/\partial z + \partial w/\partial x$$

or, in stress-tensor notation,

$$2\epsilon_{ij} = u_{i,j} + u_{j,i} \qquad i,j \rightarrow x,y,z \qquad (2.48)$$

In addition the dilatation

$$\Delta = \partial u/\partial x + \partial v/\partial y + \partial w/\partial z \qquad (2.49)$$

or in tensor form

$$\Delta = u_{i,j} \qquad i \rightarrow x,y,z \qquad (2.50)$$

Finally, all incremental displacements can be composed of a "pure strain" involving all the above components, plus "rigid-body" rotational components. That is, in general

$$U = \epsilon_x X + \tfrac{1}{2}\gamma_{xy}Y + \tfrac{1}{2}\gamma_{zx}Z - \overline{\omega}_z Y + \overline{\omega}_y Z \qquad (2.51)$$

$$V = \tfrac{1}{2}\gamma_{xy}X + \epsilon_y Y + \tfrac{1}{2}\gamma_{yz}Z - \overline{\omega}_x Z + \overline{\omega}_z X \qquad (2.52)$$

$$W = \tfrac{1}{2}\gamma_{zx}X + \tfrac{1}{2}\gamma_{yz}Y + \epsilon_z Z - \overline{\omega}_y X + \overline{\omega}_x Y \qquad (2.53)$$

where U, V, W represent the incremental displacement of the point $x + X$, $y + Y$, $z + Z$ in excess of that of the point x, y, z where X, Y, Z are taken as the sides of the incremental element. The rotational components are given by

$$2\overline{\omega}_x = \partial w/\partial y - \partial v/\partial z$$

$$2\overline{\omega}_y = \partial u/\partial z - \partial w/\partial x \qquad (2.54)$$

$$2\overline{\omega}_z = \partial v/\partial x - \partial u/\partial y$$

or, in tensor notation,

$$2\overline{\omega}_{ij} = u_{i,j} - u_{j,i} \qquad i,j = x,y,z \qquad (2.55)$$

$$\overline{\omega}_{zy} \equiv \overline{\omega}_x, \, \overline{\omega}_{xz} \equiv \overline{\omega}_y, \, \overline{\omega}_{yx} \equiv \overline{\omega}_z$$

In *cylindrical coordinates,*

$$\epsilon_r = \partial u_r/\partial r \qquad\qquad \gamma_{\theta z} = (1/r)(\partial u_z/\partial\theta + \partial u_\theta/\partial z$$

$$\epsilon_\theta = (1/r)(\partial u_\theta/\partial\theta) + u_r/r \qquad \gamma_{zr} = \partial u_r/\partial z + \partial u_z/\partial r \qquad (2.56)$$

$$\epsilon_z = \partial u_z/\partial z \qquad\qquad \gamma_{r\theta} = \partial u_\theta/\partial r - u_\theta/r + (1/r)(\partial u_r/\partial\theta)$$

The dilatation is

$$\Delta = (1/r)(\partial/\partial r)(ru_r) + (1/r)(\partial u_\theta/\partial\theta) + \partial u_z/\partial z \qquad (2.57)$$

and the rotation components are

$$2\overline{\omega}_r = (1/r)(\partial u_z/\partial\theta) - \partial u_\theta/\partial z$$

$$2\overline{\omega}_\theta = \partial u_r/\partial z - \partial u_z/\partial r \qquad (2.58)$$

$$2\overline{\omega}_z = (1/r)(\partial/\partial r)(ru_\theta) - (1/r)(\partial u_r/\partial\theta)$$

In *spherical polar coordinates,*

$$\epsilon_r = \frac{\partial u_r}{\partial r} \qquad\qquad \gamma_{\theta\phi} = \frac{1}{r}\left[\frac{\partial u_\phi}{\partial\theta} - u_\phi \cot\theta\right] + \frac{1}{r\sin\theta}\frac{\partial u_\theta}{\partial\phi}$$

$$\epsilon_\theta = \frac{1}{r}\frac{\partial u_\theta}{\partial\theta} + \frac{u_r}{r} \qquad \gamma_{\phi r} = \frac{1}{r\sin\theta}\frac{\partial u_r}{\partial\phi} + \frac{\partial u_\phi}{\partial r} - \frac{u_\phi}{r} \qquad (2.59)$$

$$\epsilon_\phi = \frac{1}{r\sin\theta}\frac{\partial u_\phi}{\partial\phi} + \frac{u_\theta}{r}\cot\theta + \frac{u_r}{r} \qquad \gamma_{r\theta} = \frac{\partial u_\theta}{\partial r} - \frac{u_\theta}{r} + \frac{1}{r}\frac{\partial u_r}{\partial\theta}$$

The dilatation is

$$\Delta = (1/r^2 \sin\theta)[(\partial/\partial r)(r^2 u_r \sin\theta) + (\partial/\partial\theta)(ru_\theta \sin\theta) + (\partial/\partial\phi)(ru_\phi)] \qquad (2.60)$$

The rotation components are

$$2\overline{\omega}_r = (1/r^2 \sin\theta)[(\partial/\partial\theta)(ru_\phi \sin\theta) - (\partial/\partial\phi)(ru_\theta)]$$

$$2\overline{\omega}_\theta = (1/r\sin\theta)[\partial u_r/\partial\phi - (\partial/\partial r)(ru_\phi \sin\theta)] \qquad (2.61)$$

$$2\overline{\omega}_\phi = (1/r)[(\partial/\partial r)(ru_\theta) - \partial u_r/\partial\theta]$$

In general *orthogonal curvilinear coordinates,*

$$\epsilon_\alpha = h_1(\partial u_\alpha/\partial\alpha) + h_1 h_2 u_\beta(\partial/\partial\beta)(1/h_1) + h_3 h_1 u_\gamma(\partial/\partial\gamma)(1/h_1)$$

$$\epsilon_\beta = h_2(\partial u_\beta/\partial\beta) + h_2 h_3 u_\gamma(\partial/\partial\gamma)(1/h_2) + h_1 h_2 u_\alpha(\partial/\partial\alpha)(1/h_2)$$

$$\epsilon_\gamma = h_3(\partial u_\gamma/\partial\gamma) + h_3 h_1 u_\alpha(\partial/\partial\alpha)(1/h_3) + h_2 h_3 u_\beta(\partial/\partial\beta)(1/h_3) \qquad (2.62)$$

$$\gamma_{\beta\gamma} = (h_2/h_3)(\partial/\partial\beta)(h_3 u_\gamma) + (h_3/h_2)(\partial/\partial\gamma)(h_2 u_\beta)$$

$$\gamma_{\gamma\alpha} = (h_3/h_1)(\partial/\partial\gamma)(h_1 u_\alpha) + (h_1/h_3)(\partial/\partial\alpha)(h_3 u_\gamma)$$

$$\gamma_{\alpha\beta} = (h_1/h_2)(\partial/\partial\alpha)(h_2 u_\beta) + (h_2/h_1)(\partial/\partial\beta)(h_1 u_\alpha)$$

$$\Delta = h_1 h_2 h_3 [(\partial/\partial\alpha)(u_\alpha/h_2 h_3) + (\partial/\partial\beta)(u_\beta/h_3 h_1) + (\partial/\partial\gamma)(u_\gamma/h_1 h_2)] \quad (2.63)$$

$$2\bar{\omega}_\alpha = h_2 h_3 [(\partial/\partial\beta)(u_\gamma/h_3) - (\partial/\partial\gamma)(u_\beta/h_2)]$$

$$2\bar{\omega}_\beta = h_3 h_1 [(\partial/\partial\gamma)(u_\alpha/h_1) - (\partial/\partial\alpha)(u_\gamma/h_3)] \quad (2.64)$$

$$2\bar{\omega}_\gamma = h_1 h_2 [(\partial/\partial\alpha)(u_\beta/h_2) - (\partial/\partial\beta)(u_\alpha/h_1)]$$

where the quantities h_1, h_2, h_3 have been discussed with reference to the equilibrium equations.

In the event that one deflection (i.e., w) is constant or zero and the displacements are a function of x, y only, a special and useful class of problems arises termed "plane strain," which are analogous to the "plane-stress" problems. A typical case of plane strain occurs in slabs rigidly clamped on their faces so as to restrict all axial deformation. Although all the stresses may be nonzero, and the general equilibrium equations apply, it can be shown that, after combining all the necessary stress and strain relationships, both classes of *plane* problems yield the same form of equations. From this, one solution suffices for both the related *plane-stress* and *plane-strain* problems, provided that the elasticity constants are suitably modified. In particular the applicable strain-displacement relationships reduce in *cartesian coordinates* to

$$\epsilon_x = \partial u/\partial x$$

$$\epsilon_y = \partial v/\partial y \quad (2.65)$$

$$\gamma_{xy} = \partial v/\partial x + \partial u/\partial y$$

and in *cylindrical coordinates* to

$$\epsilon_r = \partial u_r/\partial r$$

$$\epsilon_\theta = (1/r)(\partial u_\theta/\partial\theta) + u_r/r \quad (2.66)$$

$$\gamma_{r\theta} = \partial u_\theta/\partial r - u_\theta/r + (1/r)(\partial u_r/\partial\theta)$$

2.3.5 Compatibility Relationships[2,4,5]

In the event that a single-valued continuous-displacement field (u, v, w) is not explicitly specified, it becomes necessary to ensure its existence in solution of the stress, strain, and stress-strain relationships. By writing the strain-displacement relationships and manipulating them to eliminate displacements, it can be shown that the following six equations are both necessary and sufficient to ensure compatibility:

$$\partial^2\epsilon_y/\partial z^2 + \partial^2\epsilon_z/\partial y^2 = \partial^2\gamma_{yz}/\partial y\,\partial z$$

$$2(\partial^2\epsilon_x/\partial y\,\partial z) = (\partial/\partial x)(-\partial\gamma_{yz}/\partial x + \partial\gamma_{zx}/\partial y + \partial\gamma_{xy}/\partial z) \quad (2.67)$$

$$\partial^2\epsilon_z/\partial x^2 + \partial^2\epsilon_x/\partial z^2 = \partial^2\gamma_{zx}/\partial x\,\partial z$$

$$2(\partial^2\epsilon_y/\partial z\,\partial x) = (\partial/\partial y)(+\partial\gamma_{yz}/\partial x - \partial\gamma_{zx}/\partial y + \partial y_{xy}/\partial z) \quad (2.68)$$

$$\partial^2\epsilon_x/\partial y^2 + \partial^2\epsilon_y/\partial x^2 = \partial^2\gamma_{xy}/\partial x\,\partial y$$

$$2(\partial^2\epsilon_z/\partial x\,\partial y) = (\partial/\partial z)(+\partial\gamma_{yz}/\partial x + \partial\gamma_{zx}/\partial y - \partial\gamma_{xy}/\partial z) \quad (2.69)$$

In tensor notation the most general compatibility equations are

$$\epsilon_{ij,kl} + \epsilon_{kl,ij} - \epsilon_{ik,jl} - \epsilon_{jl,ik} = 0 \qquad i,j,k,l = x,y,z \qquad (2.70)$$

which represents 81 equations. Only the above six equations are essential.

In addition to satisfying these conditions everywhere in the body under consideration, it is also necessary that all *surface strain* or *displacement boundary conditions* be satisfied.

2.3.6 Strain Transformation[4,5]

As with stress, it is frequently necessary to refer strains to a rotated orthogonal coordinate system (x', y', z'). In this event it can be shown that the stress and strain tensors transform in an identical manner.

$$\sigma_{x'} \to \epsilon_{x'} \qquad \sigma_x \to \epsilon_x \qquad \tau_{x'y'} \to \tfrac{1}{2}\gamma_{x'y'} \qquad \tau_{xy} \to \tfrac{1}{2}\gamma_{xy}$$

$$\sigma_{y'} \to \epsilon_{y'} \qquad \sigma_y \to \epsilon_y \qquad \tau_{y'z'} \to \tfrac{1}{2}\gamma_{y'z'} \qquad \tau_{yz} \to \tfrac{1}{2}\gamma_{yz}$$

$$\sigma_{z'} \to \epsilon_{z'} \qquad \sigma_z \to \epsilon_z \qquad \tau_{z'x'} \to \tfrac{1}{2}\gamma_{z'x'} \qquad \tau_{zx} \to \tfrac{1}{2}\gamma_{zx}$$

In tensor notation the strain transformation can be written

$$e_{k'l'} = A_{l'n}A_{k'm}\epsilon_{mn} \qquad \begin{array}{l} m, n \to x, y, z \\ l'k' \to x', y' \end{array} \qquad (2.71)$$

As a result the stress and strain principal directions are coincident, so that all remarks

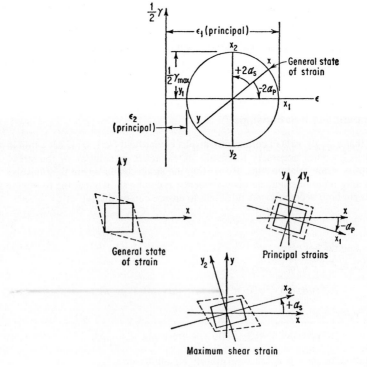

FIG. 2.14   Strain transformation.

made for the principal stress and maximum shear components and their directions apply equally well to strain tensor components. Note that in the use of Mohr's circle in the two-dimensional case one must be careful to substitute $\frac{1}{2}\gamma$ for τ in the *ordinate* and ϵ for σ in the abscissa (Fig. 2.14).

2.4 STRESS-STRAIN RELATIONSHIPS

2.4.1 Introduction[2]

It can be experimentally demonstrated that a one-to-one relationship exists between uniaxial stress and strain during a single loading. Further, if the material is always loaded within its *elastic* or *reversible* range, a one-to-one relationship exists for all loading and unloading cycles.

For stresses below a certain characteristic value termed the "proportional limit," the stress-strain relationship is very nearly linear. The stress beyond which the stress-strain relationship is no longer reversible is called the "elastic limit." In most materials the proportional and elastic limits are identical. Because the departure from linearity is very gradual it is often necessary to prescribe arbitrarily an "apparent" or "offset elastic limit." This is obtained as the intersection of the stress-strain curve with a line parallel to the linear stress-strain curve, but offset by a prescribed amount, e.g., 0.02 percent (see Fig. 2.15a). The "yield point" is the value of stress at which continued deformation of the bar takes place with little or no further increase in load, and the "ultimate limit" is the maximum stress that the specimen can withstand.

Note that some materials may show no clear difference between the apparent elastic, inelastic, and proportional limits or may not show clearly defined yield points (Fig. 2.15b).

The concept that a useful linear range exists for most materials and that a simple mathematical law can be formulated to describe the relationship between stress and strain in this range is termed "Hooke's law." It is an essential starting point in the "small-strain theory of elasticity" and the associated mechanics of materials. In the above-described tensile specimen, the law is expressed as

$$\sigma = E\epsilon \tag{2.72}$$

as in the analogous torsional specimen

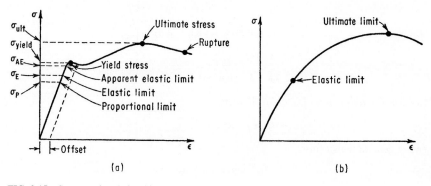

FIG. 2.15   Stress-strain relationship.

$$\tau = G\gamma \tag{2.73}$$

where E and G are the slope of the appropriate stress-strain diagrams and are called the "Young's modulus" and the "shear modulus" of elasticity, respectively.

2.4.2 General Stress-Strain Relationship[2,4,5]

The one-dimensional concepts discussed above can be generalized for both small and large strain and elastic and nonelastic materials. The following discussion will be limited to small-strain elastic materials consistent with much engineering design. Based upon the above, Hooke's law is expressed as

$$\epsilon_x = (1/E)[\sigma_x - \nu(\sigma_y + \sigma_z)] \qquad \gamma_{xy} = \tau_{xy}/G$$

$$\epsilon_y = (1/E)[\sigma_y - \nu(\sigma_z + \sigma_x)] \qquad \gamma_{yz} = \tau_{yz}/G \tag{2.74}$$

$$\epsilon_z = (1/E)[\sigma_z - \nu(\sigma_x + \sigma_y)] \qquad \gamma_{zx} = \tau_{zx}/G$$

where ν is "Poisson's ratio," the ratio between longitudinal strain and lateral contraction in a simple tensile test.

In *cartesian tension form* Eq. (2.74) is expressed

$$\epsilon_{ij} = [(1 + \nu)/E]\sigma_{ij} - (\nu/E)\delta_{ij}\sigma_{kk} \qquad i,j,k = x,y,z \tag{2.75}$$

where
$$\delta_{ij} = \begin{cases} 0 & i \neq j \\ 1 & i = j \end{cases}$$

The stress-strain laws appear in inverted form as

$$\sigma_x = 2G\epsilon_x + \lambda\Delta$$

$$\sigma_y = 2G\epsilon_y + \lambda\Delta$$

$$\sigma_z = 2G\epsilon_z + \lambda\Delta$$

$$\tau_{xy} = G\gamma_{xy} \tag{2.76}$$

$$\tau_{yz} = G\gamma_{yz}$$

$$\tau_{zx} = G\gamma_{zx}$$

where
$$\lambda = \nu/E/(1 + \nu)(1 - 2\nu)$$

$$\Delta = \epsilon_x + \epsilon_y + \epsilon_z$$

$$G = E/2(1 + \nu)$$

In *cartesian tensor form* Eq. (2.76) is written

$$\sigma_{ij} = 2G\epsilon_{ij} + \lambda\Delta\delta_{ij} \qquad i,j = x,y,z \tag{2.77}$$

and in general tensor form

$$\sigma_{ij} = 2G\epsilon_{ij} + \lambda\Delta g_{ij} \tag{2.78}$$

where g_{ij} is the "covariant metric tensor" and these coefficients (stress modulus) are often referred to as "Lamé's constants," and $\Delta = g^{mn}\epsilon_{mn}$.

2.5 STRESS-LEVEL EVALUATION

2.5.1 Introduction[1,6]

The detailed elastic and plastic behavior, yield and failure criterion, etc., are repeatable and simply describable for a simple loading state, as in a tensile or torsional specimen. Under any complex loading state, however, no single stress or strain component can be used to describe the *stress state* uniquely; that is, the *yield, flow,* or *rupture* criterion must be obtained by some combination of all the stress and/or strain components, their derivatives, and loading history. In elastic theory the "yield criterion" is related to an "equivalent stress," or "equivalent strain." It is conventional to treat the *stress* criteria.

An "equivalent stress" is defined in terms of the "stress components" such that plastic flow will commence in the body at any position at which this equivalent stress just exceeds the one-dimensional yield-stress value, for the material under consideration. That is, yielding commences when

$$\sigma_{\text{equivalent}} \geq \sigma_E$$

The "elastic safety factor" at a point is defined as the ratio of the one-dimensional yield stress to the equivalent stress at that position, i.e.,

$$n_i = \sigma_E/\sigma_{\text{equivalent}} \tag{2.79}$$

and the elastic safety factor for the entire structure under any specific loading state is taken as the lowest safety factor of consequence that exists anywhere in the structure.

The "margin of safety," defined as $n - 1$, is another measure of the proximity of any structure to yielding. When $n > 1$ the structure has a *positive* margin of safety and will not yield. When $n = 1$, the margin of safety is zero and the structure just yields. When $n < 1$, the margin of safety is *negative* and the structure is considered unsafe. Note that highly localized yielding is often permitted in ductile materials if it is of such nature as to redistribute stresses without failure, building up a "residual stress state" which allows all subsequent loadings to be accomplished with elastic-stress states. This is the basic concept used in the "autofrettage process" in the strengthening of gun tubes, and it also explains why ductile materials often have low "notch sensitivity" (see Sec. 9).

2.5.2 Effective Stress[1,6]

The concept of effective stress is very closely connected with yield criteria. Geometrically it can be shown that a unique surface can be constructed in stress space in terms of principal stresses (σ_1, σ_2, σ_3) such that all nonyielding states of stress lie within that surface, and yielding states lie on or outside the surface. For *ductile materials* the yield surface may be taken as an infinitely long right cylinder having a center axis defined by $\sigma_1 = \sigma_2 = \sigma_3$. Because the yield criterion is represented by a right cylinder it is adequate to define the yield curve, which is the intersection with a plane

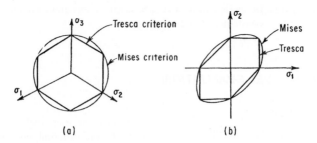

FIG. 2.16 Tresca and Mises criteria. (*a*) General yield criteria. (*b*) One principal stress constant.

normal to the axis of the cylinder. Many yield criteria exist; among these, the "Mises criterion" takes this curve as a circle, and the "Tresca criterion" as a regular hexagon (see Fig. 2.16). The former is often referred to by names such as "Hencky," "Hencky-Mises," or "distortion-energy criterion" and the latter by "shear criterion."

Considering any one of the principal stresses as constant, the "yield locus" can be represented by the intersection of the plane $\sigma_i = $ constant with the cylindrical yield surface, which is represented as an ellipse for the Mises criterion and an elongated hexagon for the Tresca criterion (Fig. 2.16*b*).

In general the yield criteria indicate, upon experimental evidence for a ductile material, that yielding is essentially independent of "hydrostatic compression" or tension for the loadings usually considered in engineering problems. That is, yielding depends only on the "deviatoric" stress component. In general, this can be represented as

$$\sigma_k' = \sigma_k - \tfrac{1}{3}(\sigma_1 + \sigma_2 + \sigma_3) \tag{2.80}$$

where σ' is the principal deviatoric stress component and σ the actual principal stress.

In tensor form

$$\sigma_{ij}' = \sigma_{ij} - \tfrac{1}{3}\sigma_{kk}\delta_{ij} \tag{2.81}$$

The analytical representation of the yield criteria can be shown to be a function of the "deviatoric stress-tensor invariants." In component form these can be expressed as follows, where σ_0 is the yield stress in simple tension.

Mises:

$$\sigma_0 = (1/\sqrt{2})\sqrt{(\sigma_1 - \sigma_2)^2 + (\sigma_2 - \sigma_3)^2 + (\sigma_3 - \alpha_1)^2} \tag{2.82}$$

where σ_1, σ_2, σ_3 are principal stresses.

$$\sigma_0 = (1/\sqrt{2})\sqrt{(\sigma_x - \sigma_y)^2 + (\sigma_y - \sigma_z)^2 + (\sigma_z - \sigma_x)^2 + 6(\tau_{xy}^2 + \tau_{yz}^2 + \tau_{zx}^2)} \tag{2.83}$$

or, in tensor form,

$$\tfrac{2}{3}\sigma_0^2 = \sigma_{ij}'\sigma_{ij}' \tag{2.84}$$

Tresca:

$$\sigma_0 = \sigma_1 - \sigma_3 \tag{2.85}$$

where $\sigma_1 > \sigma_2 > \sigma_3$, or, in general symmetric terms,

$$[(\sigma_1 - \sigma_3)^2 - \sigma_0^2][(\sigma_2 - \sigma_1)^2 - \sigma_0^2][(\sigma_3 - \sigma_2)^2 - \sigma_0^2] = 0 \tag{2.86}$$

2.6 FORMULATION OF GENERAL MECHANICS-OF-MATERIAL PROBLEM

2.6.1 Introduction[2,4,5]

Generally the mechanics-of-material problem is stated as follows: Given a prescribed structural configuration, and surface tractions and/or displacements, find the stresses and/or displacements at any, or all, positions in the body. Additionally it is often desired to use the derived stress information to determine the *maximum load-carrying capacity* of the structure, prior to yielding. This is usually referred to as the problem of *analysis*. Alternatively the problem may be inverted and stated: Given a set of surface tractions and/or displacements, find the geometrical configuration for a constraint such as minimum weight, subject to the yield criterion (or some other general stress or strain limitation). This latter is referred to as the *design* problem.

2.6.2 Classical Formulation[2,4,5]

The classical formulation of the equation for the problem of mechanics of materials is as follows: It is necessary to evaluate the six stress components σ_{ij}, six strain components ϵ_{ij}, and three displacement quantities u_i which satisfy the three equilibrium equations, six strain-displacement relationships, and six stress-strain relationships, all subject to the appropriate stress and/or displacement boundary conditions.

Based on the above discussion and the previous derivations, the most general three-dimensional formulation in cartesian coordinates is:

$$\left.\begin{array}{l} \partial\sigma_x/\partial x + \partial\tau_{xy}/\partial y + \partial\tau_{xz}/\partial z = 0 \\[4pt] \partial\sigma_y/\partial y + \partial\tau_{yz}/\partial z + \partial\tau_{yx}/\partial x = 0 \\[4pt] \partial\sigma_z/\partial z + \partial\tau_{zx}/\partial x + \partial\tau_{zy}/\partial y = 0 \end{array}\right\} \quad \text{(equilibrium)} \qquad (2.87)$$

$$\left.\begin{array}{ll} \epsilon_x = \partial u/\partial x & \gamma_{xy} = \partial v/\partial x + \partial u/\partial y \\[4pt] \epsilon_y = \partial v/\partial y & \gamma_{yz} = \partial w/\partial y + \partial v/\partial z \\[4pt] \epsilon_z = \partial w/\partial z & \gamma_{zx} = \partial u/\partial z + \partial w/\partial x \end{array}\right\} \quad \text{(strain-displacement)} \qquad (2.47)$$

$$\left.\begin{array}{l} \epsilon_x = (1/E)[\sigma_x - v(\sigma_y + \sigma_z)] \\[4pt] \epsilon_y = (1/E)[\sigma_y - v(\sigma_z + \sigma_x)] \\[4pt] \epsilon_z = (1/E)[\sigma_z - v(\sigma_x + \sigma_y)] \\[4pt] \gamma_{xy} = (1/G)\tau_{xy} \\[4pt] \gamma_{yz} = (1/G)\tau_{yz} \\[4pt] \gamma_{zx} = (1/G)\tau_{xz} \end{array}\right\} \quad \text{(stress-strain relationships)} \qquad (2.88)$$

In cartesian tensor form these appear as:

$$\sigma_{ij,j} = 0 \qquad\qquad\qquad \text{(equilibrium)} \qquad (2.89)$$

$$2\epsilon_{ij} = u_{i,j} + u_{j,i} \qquad i,j \to x,y,z \qquad \text{(strain-displacement)} \qquad (2.48)$$

$$\epsilon_{ij} = [(1 + v)/E]\sigma_{ij} - (v/E)\delta_{ij}\sigma_{kk} \quad \text{(stress-strain)} \quad (2.90)$$

All are subject to appropriate boundary conditions.

If the boundary conditions are on displacements, then we can define the displacement field, the six components of strain, and the six components of stress uniquely, using the fifteen equations shown above. If the boundary conditions are on stresses, then the solution process yields six strain components from which three unique displacement components must be determined. In order to assure uniqueness, three constraints must be placed on the strain field. These constraints are provided by the compatibility relationships:

$$\partial^2\epsilon_x/\partial y^2 + \partial^2\epsilon_y/\partial x^2 = \partial^2\gamma_{xy}/\partial x\,\partial y$$

$$\partial^2\epsilon_y/\partial z^2 + \partial^2\epsilon_z/\partial y^2 = \partial^2\gamma_{yz}/\partial y\,\partial z$$

$$\partial^2\epsilon_z/\partial x^2 + \partial^2\epsilon_x/\partial z^2 = \partial^2\gamma_{zx}/\partial z\,\partial x$$

$$2(\partial^2\epsilon_x/\partial y\,\partial z) = (\partial/\partial x)(-\partial\gamma_{yz}/\partial x + \partial\gamma_{zx}/\partial y + \partial\gamma_{xy}/\partial z)$$

$$2(\partial^2\epsilon_y/\partial z\,\partial x) = (\partial/\partial y)(\partial\gamma_{yz}/\partial x - \partial\gamma_{zx}/\partial y + \partial\gamma_{xy}/\partial z)$$

$$2(\partial^2\epsilon_z/\partial x\,\partial y) = (\partial/\partial z)(\partial\gamma_{yz}/\partial x + \partial\gamma_{zx}/\partial y - \partial\gamma_{xy}/\partial z)$$

(compatibility) $\quad (2.91)$

In cartesian tensor form:

$$\epsilon_{ij,kl} + \epsilon_{kl,ij} - \epsilon_{ik,jl} - \epsilon_{jl,ik} = 0 \quad \text{(compatibility)} \quad (2.92)$$

Of the six compatibility equations listed, only three are independent. Therefore, the system can be uniquely solved for the displacement field.

It is possible to simplify the above sets of equations considerably by combining and eliminating many of the unknowns. One such reduction is obtained by eliminating stress and strain:

$$\nabla^2 u + [1/(1 - 2v)](\partial\Delta/\partial x) = 0$$

$$\nabla^2 v + [1/(1 - 2v)](\partial\Delta/\partial y) = 0 \quad (2.93)$$

$$\nabla^2 w + [1/(1 - 2v)](\partial\Delta/\partial z) = 0$$

where ∇^2 is the laplacian operator which in cartesian coordinates is $\partial^2/\partial x^2 + \partial^2/\partial y^2 + \partial^2/\partial z^2$; and Δ is the dilatation, which in cartesian coordinates is $\partial u/\partial x + \partial v/\partial y + \partial w/\partial z$.

Using the above general principles, it is possible to formulate completely many of the technical problems of mechanics of materials which appear under special classifications such as "beam theory" and "shell theory." These formulations and their solutions will be treated under "Special Applications."

2.6.3 Energy Formulations[2,4,5]

Alternative useful approaches exist for the problem of mechanics of materials. These are referred to as "energy," "extremum," or "variational" formulations. From a strictly formalistic point of view these could be obtained by establishing the analogous integral equations, subject to various restrictions, such that they reduce to a minimum. This is not the usual approach; instead energy functions U, W are established so that the stress-strain laws are replaced by

$$\sigma_x = \partial U/\partial \epsilon_x \qquad \tau_{xy} = \partial U/\partial \gamma_{xy}$$

$$\sigma_y = \partial U/\partial \epsilon_y \qquad \tau_{yz} = \partial U/\partial \gamma_{yz}$$

$$\sigma_z = \partial U/\partial \epsilon_z \qquad \tau_{zx} = \partial U/\partial \gamma_{zx}$$

or

$$\sigma_{ij} = \partial U/\partial \epsilon_{ij}$$

and

$$\epsilon_x = \partial W/\partial \sigma_x \qquad \gamma_{xy} = \partial W/\partial \tau_{xy}$$

$$\epsilon_y = \partial W/\partial \sigma_y \qquad \gamma_{yz} = \partial W/\partial \tau_{yz}$$

$$\epsilon_z = \partial W/\partial \sigma_z \qquad \gamma_{zx} = \partial W/\partial \tau_{zx}$$

or

$$\epsilon_{ij} = \partial W/\partial \sigma_{ij}$$

The energy functions are given by

$$U = \tfrac{1}{2}[2G(\epsilon_x^2 + \epsilon_y^2 + \epsilon_z^2) + \lambda(\epsilon_x + \epsilon_y + \epsilon_z)^2 + G(\gamma_{xy}^2 + \gamma_{yz}^2 + \gamma_{zx}^2)] \qquad (2.94)$$

$$W = \tfrac{1}{2}[(1/E)(\sigma_x^2 + \sigma_y^2 + \sigma_z^2) - (2v/E)(\sigma_x\sigma_y + \sigma_y\sigma_z + \sigma_z\sigma_x)$$

$$+ (1/G)(\tau_{xy}^2 + \tau_{yz}^2 + \tau_{zx}^2)] \qquad (2.95)$$

The variational principle for strains, or theorem of minimum potential energy, is stated as follows: Among all states of strain which satisfy the strain-displacement relationships and displacement boundary conditions the associated stress state, derivable through the stress-strain relationships, which also satisfies the equilibrium equations, is determined by the minimization of Π where

$$\Pi = \int_{\text{volume}} U \, dV - \int_{\text{surface}} (\bar{p}_x u + \bar{p}_y v + \bar{p}_z w) \, dS \qquad (2.96)$$

where $\bar{p}_x, \bar{p}_y, \bar{p}_z$ are the x, y, z components of any prescribed surface stresses.

The analogous variational principle for stresses, or principle of least work, is: Among all the states of stress which satisfy the equilibrium equations and stress boundary conditions, the associated strain state, derivable through the stress-strain relationships, which also satisfies the compatibility equations, is determined by the minimization of I, where

$$I = \int_{\text{volume}} W \, dV - \int_{\text{surface}} (p_x \bar{u} + p_y \bar{v} + p_z \bar{w}) \, dS \qquad (2.97)$$

where $\bar{u}, \bar{v}, \bar{w}$ are the x, y, z components of any prescribed surface displacements and p_x, p_y, p_z are the surface stresses.

In the above theorems $\Pi_{\min}$ and $I_{\min}$ replace the equilibrium and compatibility relationships, respectively. Their most powerful advantage arises in obtaining approximate solutions to problems which are generally intractable by exact techniques. In this, one usually introduces a limited class of assumed stress or displacement functions for minimization, which in themselves satisfy all other requirements imposed in the statement of the respective theorems. Then with the use of these theorems it is possible to find the best solution in that limited class which provides the best minimum to the associated Π or I function. This in reality does not satisfy the missing equilibrium or compatibility equation, but it does it as well as possible for the class of function assumed to describe the stress or strain in the body, within the framework of the principle established above. It has been shown that most reasonable assumptions, regardless of their simplicity, provide useful solutions to most problems of mechanics of materials.

2.6.4   Example: Energy Techniques[2,4,5]

It can be shown that for beams the variational principle for strains reduces to

$$\Pi_{min} = \left\{ \int_0^L [\tfrac{1}{2}EI(y'')^2 - qy] \, dx - \sum P_i y_i \right\}_{min} \tag{2.98}$$

where EI is the flexural rigidity of the beam at any position x, I is the moment of inertia of the beam, y is the deflection of the beam, the y' refers to x derivative of y, q is the distributed loading, the P_i's represent concentrated loads, and L is the span length.

If the minimization is carried out, subject to the restrictions of the variational principle for strains, the beam equation results. However, it is both useful and instructive to utilize the above principle to obtain two approximate solutions to a specific problem and then compare these with the exact solutions obtained by other means.

First a centrally loaded, simple-support beam problem will be examined. The function of minimization becomes

$$\Pi = \int_0^{L/2} EI(y'')^2 \, dx - Py_{L/2} \tag{2.99}$$

Select the class of displacement functions described by

$$y = Ax(\tfrac{3}{4}L^2 - x^2) \qquad 0 \le x \le L/2 \tag{2.100}$$

This satisfies the boundary conditions

$$y(0) = y''(0) = y'(L/2) = 0$$

In this A is an arbitrary parameter to be determined from the minimization of Π.

Properly introducing the value of y, y'' into the expression for Π and integrating, then minimizing Π with respect to the open parameter by setting

$$\partial\Pi/\partial A = 0$$

yields

$$y = (Px/12EI) (\tfrac{3}{4}L^2 - x^2) \qquad 0 \le x \le L/2 \tag{2.101}$$

It is coincidental that this is the exact solution to the above problem.
A second class of deflection function is now selected

$$y = A \sin (\pi x/L) \qquad 0 \le x \le L \tag{2.102}$$

which satisfies the boundary conditions

$$y(0) = y''(0) = y(L) = y''(L) = 0$$

which is intuitively the expected deflection shape. Additionally, $y(L/2) = A$. Introducing the above information into the expression for Π and minimizing as before yields

$$y = (PL^3/EI)[(2/\pi^4) \sin (\pi x/L)] \tag{2.103}$$

The ratio of the approximate to the exact central deflection is 0.9855, which indicates that the approximation is of sufficient accuracy for most applications.

2.7 FORMULATION OF GENERAL THERMOELASTIC PROBLEM[2,9]

A nonuniform temperature distribution or a nonuniform material distribution with uniform temperature change introduces additional stresses and/or strains, even in the absence of external tractions.

Within the confines of the linear theory of elasticity and neglecting small coupling effects between the *temperature-distribution problem* and the *thermoelastic problem* it is possible to solve the general mechanics-of-material problem as the superposition of the previously defined mechanics-of-materials problem and an initially traction-free thermoelastic problem.

Taking the same consistent definition of stress and strain as previously presented it can be shown that the strain-displacement, stress-equilibrium, and compatibility relationships remain unchanged in the thermoelastic problem. However, because a structural material can change its size even in the absence of stress, it is necessary to modify the stress-strain laws to account for the additional strain due to temperature (αT). Thus Hooke's law is modified as follows:

$$\epsilon_x = (1/E)[\sigma_x - \nu(\sigma_y + \sigma_z)] + \alpha T$$

$$\epsilon_y = (1/E)[\sigma_y - \nu(\sigma_z + \sigma_x)] + \alpha T \qquad (2.104)$$

$$\epsilon_z = (1/E)[\sigma_z - \nu(\sigma_x + \sigma_y)] + \alpha T$$

The shear strain-stress relationships remain unchanged. α is the coefficient of thermal expansion and T the temperature rise above the ambient stress-free state. In uniform, nonconstrained structures this ambient base temperature is arbitrary, but in problems associated with nonuniform material or constraint this base temperature is quite important.

Expressed in cartesian tensor form the stress-strain relationships become

$$\epsilon_{ij} = [(1 + \nu)/E]\sigma_{ij} - (\nu/E)\delta_{ij}\sigma_{kk} + \alpha T\delta_{ij} \qquad (2.105)$$

In inverted form the modified stress-strain relationships are

$$\sigma_x = 2G\epsilon_x + \lambda\Delta - (3\lambda + 2G)\alpha T$$

$$\sigma_y = 2G\epsilon_y + \lambda\Delta - (3\lambda + 2G)\alpha T \qquad (2.106)$$

$$\sigma_z = 2G\epsilon_z + \lambda\Delta - (3\lambda + 2G)\alpha T$$

or in cartesian tensor form

$$\sigma_{ij} = 2G\epsilon_{ij} + \lambda\Delta \delta_{ij} - (3\lambda + 2G)\alpha T\delta_{ij} \qquad (2.107)$$

Considering the equilibrium compatibility formulations, it can be shown that the analogous thermoelastic displacement formulations result in

$$(\lambda + G)(\partial\Delta/\partial x) + G\nabla^2 u - (3\lambda + 2G)\alpha(\partial T/\partial x) = 0$$

$$(\lambda + G)(\partial\Delta/\partial y) + G\nabla^2 v - (3\lambda + 2G)\alpha(\partial T/\partial y) = 0 \qquad (2.108)$$

$$(\lambda + G)(\partial\Delta/\partial z) + G\nabla^2 w - (3\lambda + 2G)\alpha(\partial T/\partial z) = 0$$

A useful alternate stress formulation is

$$(1 + \nu)\nabla^2\sigma_x + \frac{\partial^2\Theta}{\partial x^2} + \alpha E\left(\frac{1 + \nu}{1 - \nu}\nabla^2 T + \frac{\partial^2 T}{\partial x^2}\right) = 0$$

$$(1 + \nu)\nabla^2\sigma_y + \frac{\partial^2\Theta}{\partial y^2} + \alpha E\left(\frac{1 + \nu}{1 - \nu}\nabla^2 T + \frac{\partial^2 T}{\partial y^2}\right) = 0$$

$$(1 + \nu)\nabla^2\sigma_z + \frac{\partial^2\Theta}{\partial z^2} + \alpha E\left(\frac{1 + \nu}{1 - \nu}\nabla^2 T + \frac{\partial^2 T}{\partial z^2}\right) = 0 \qquad (2.109)$$

$$(1 + \nu)\nabla^2\tau_{xy} + \partial^2\Theta/\partial x\,\partial y + \alpha E(\partial^2 T/\partial x\,\partial y) = 0$$

$$(1 + \nu)\nabla^2\tau_{yz} + \partial^2\Theta/\partial y\,\partial z + \alpha E(\partial^2 T/\partial y\,\partial z) = 0$$

$$(1 + \nu)\nabla^2\tau_{zx} + \partial^2\Theta/\partial z\,\partial x + \alpha E(\partial^2 T/\partial z\,\partial z) = 0$$

where $\Theta = \sigma_x + \sigma_y + \sigma_z$.

2.8 CLASSIFICATION OF PROBLEM TYPES

In mechanics of materials it is frequently desirable to classify problems in terms of their geometric configurations and/or assumptions that will permit their codification and ease of solution. As a result there exist problems in plane stress or strain, beam theory, curved-beam theory, plates, shells, etc. Although the defining equations can be obtained directly from the general theory together with the associated assumptions, it is often instructive and convenient to obtain them directly from physical considerations. The difference between these two approaches marks one of the principal distinguishing differences between the *theory of elasticity* and *mechanics of materials*.

2.9 BEAM THEORY

2.9.1 Mechanics of Materials Approach[1]

The following assumptions are basic in the development of elementary beam theory:

1. Beam sections, originally plane, remain plane and normal to the "neutral axis."
2. The beam is originally straight and all bending displacements are small.
3. The beam cross section is symmetrical with respect to the loading plane, an assumption that is usually removed in the general theory.

4. The beam material obeys Hooke's law, and the moduli of elasticity in tension and compression are equal.

Consider the beam portion loaded as shown in Fig. 2.17a.

For static equilibrium, the internal actions required at section B which are supplied by the immediately adjacent section to the right must consist of a vertical shearing force V and an internal moment M, as shown in Fig. 2.17b.

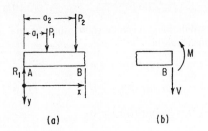

(a)   (b)

FIG. 2.17 Internal reactions due to externally applied loads. (*a*) External loading of beam segment. (*b*) Internal moment and shear.

The evaluation of the shear V is accomplished by noting, from equilibrium $\Sigma F_y = 0$,

$$V = R - P_1 - P_2 \quad \text{(for this example)} \tag{2.110}$$

The algebraic sum of all the shearing forces at one side of the section is called the shearing force at that section. The moment M is obtained from $\Sigma M = 0$:

$$M = R_1 x - P_1(x - a_1) - P_2(x - a_2) \tag{2.111}$$

The algebraic sum of the moments of all external loads to one side of the section is called the bending moment at the section.

Note the sign conventions employed thus far:

1. Shearing force is positive if the right portion of the beam tends to shear downward with respect to the left.
2. Bending moment is positive if it produces bending of the beam concave upward.
3. Loading w is positive if it acts in the positive direction of the y axis.

In Fig. 2.18a a portion of one of the beams previously discussed is shown with the bending moment M applied to the element.

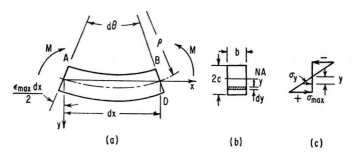

FIG. 2.18 Beam bending with externally applied load. (*a*) Beam element. (*b*) Cross section. (*c*) Bending-stress pattern at section *B–D*.

Equilibrium conditions require that the sum of the normal stresses σ on a cross section must equal zero, a condition satisfied only if the "neutral axis," defined as the plane or axis of zero normal stress, *is also the centroidal axis* of the cross section.

$$\int_{-c}^{+c} \sigma b \, dy = \frac{\sigma}{y} \int_{-c}^{+c} by \, dy = 0 \tag{2.112}$$

where $\sigma/y = (\epsilon/y) E = (\epsilon_{max}/y_{max})E = \text{const}$

Further, if the moments of the stresses acting on the element dy of the figure are summed over the height of the beam,

$$M = \int_{-c}^{+c} \sigma by \, dy = \frac{\sigma}{y} \int_{-c}^{+c} by^2 \, dy = \frac{\sigma}{y} I \tag{2.113}$$

where y = distance from neutral axis to point on cross section being investigated, and

$$I = \int_{-c}^{+c} by^2 \, dy$$

is the area moment of inertia about the centroidal axis of the cross section. Equation (2.113) defines the flexural stress in a beam subject to moment M:

$$\sigma = My/I \tag{2.114}$$

Thus
$$\sigma_{max} = Mc/I \tag{2.115}$$

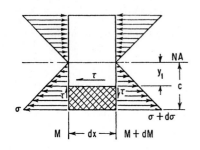

To develop the equations for shear stress τ, the general case of the element of the beam subjected to a varying bending moment is taken as in Fig. 2.19.

Applying axial-equilibrium conditions to the shaded area of Fig. 2.19 yields the following general expression for the horizontal shear stress at the lower surface of the shaded area:

$$\tau = \frac{dM}{dx} \frac{1}{Ib} \int_{y_1}^{c} y\, dA \tag{2.116}$$

FIG. 2.19   Shear-stress diagram for beam subjected to varying bending moment.

or, in familiar terms,

$$\tau = \frac{V}{Ib} \int_{y_1}^{c} y\, dA = \frac{V}{Ib} Q \tag{2.117}$$

where Q = moment of area of cross section about neutral axis for the shaded area above the surface under investigation
V = net vertical shearing force
b = width of beam at surface under investigation

Equilibrium considerations of a small element at the surface where τ is computed will reveal that this value represents both the vertical and horizontal shear.

For a rectangular beam, the vertical shear-stress distribution across a section of the beam is parabolic. The maximum value of this stress (which occurs at the neutral axis) is 1.5 times the average value of the stress obtained by dividing the shear force V by the cross-sectional area.

For many typical structural shapes the maximum value of the shear stress is approximately 1.2 times the average shear stress.

To develop the governing equation for bending deformations of beams, consider again Fig. 2.18. From geometry,

$$\frac{(\epsilon/2)\, dx}{y} = \frac{dx/2}{\rho} \tag{2.118}$$

Combining Eqs. (2.118), (2.114), and (2.72) yields

$$1/\rho = M/EI \tag{2.119}$$

Since
$$1/\rho \approx -d^2y/dx^2 = -y'' \tag{2.120}$$

Therefore
$$y'' = -M/EI \quad \text{(Bernoulli-Euler equation)} \tag{2.121}$$

In Fig. 2.20 the element of the beam subjected to an arbitrary load $w(x)$ is shown together with the shears and bending moments as applied by the adjacent cross sections of the beam. Neglecting higher-order terms, moment summation leads to the following result for the moments acting on the element:

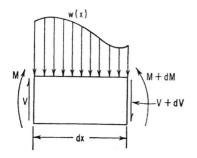

$$dM/dx = V \qquad (2.122)$$

Differentiation of the Bernoulli-Euler equation yields

$$y''' = -V/EI \qquad (2.123)$$

In similar manner, the summation of transverse forces in equilibrium yields

$$dV/dx = -w(x) \qquad (2.124)$$

FIG. 2.20 Shear and bending moments for a beam with load $w(x)$ applied.

or

$$y^{IV} = \frac{w(x)}{EI} \qquad (2.125)$$

where due attention has been given to the proper sign convention.

See Table 2.1 for typical shear, moment, and deflection formulas for beams.

2.9.2 Energy Considerations

The total strain energy of bending is

$$U_b = \int_0^L \frac{M^2}{2EI}\, dx \qquad (2.126)$$

The strain energy due to shear is

$$U_s = \int_0^L \frac{V^2}{2GA}\, dx \qquad (2.127)$$

In calculating the deflections by the energy techniques, shear-strain contributions need not be included unless the beam is short and deep.

The deflections can then be obtained by the application of Castigliano's theorem, of which a general statement is: The partial derivative of the total strain energy of any structure with respect to any one generalized load is equal to the generalized deflection at the point of application of the load, and is in the direction of the load. The generalized loads can be forces or moments and the associated generalized deflections are displacements or rotations:

$$Y_a = \partial U / \partial P_a \qquad (2.128)$$

$$\theta_a = \partial U / \partial M_a \qquad (2.129)$$

where U = total strain energy of bending of the beam
  P_a = load at point a
  M_a = moment at point a
  Y_a = deflection of beam at point a
  θ_a = rotation of beam at point a

Thus

$$Y_a = \frac{\partial U}{\partial P_a} = \frac{\partial}{\partial P_a} \int_0^L M^2 \frac{dx}{2EI} = \int_0^L \frac{M}{EI} \frac{\partial M}{\partial P_a}\, dx \qquad (2.130)$$

TABLE 2.1 Shear, Moment, and Deflection Formulas for Beams[1,12]

Notation: W = load (lb); w = unit load (lb/linear in). V is positive when upward; y is positive when upward. Constraining moments, applied couples, loads, and reactions are positive when acting as shown. All forces are in pounds, all moments in inch-pounds, all deflections and dimensions in inches. θ is in radians and $\tan \theta = \theta$.

Loading, support, and reference number	Reactions R_1 and R_2, vertical shear V	Deflection y, maximum deflection, and end slope θ
End supports Uniform load (1)	$R_1 = +\frac{1}{2}W \qquad R_2 = +\frac{1}{2}W$ $V = \frac{1}{2}W\left(1 - \frac{2z}{l}\right)$	$y = -\frac{1}{24}\frac{Wz}{EIl}(l^3 - 2lz^2 + z^3)$ Max $y = -\frac{5}{384}\frac{Wl^3}{EI}$ at $z = \frac{1}{2}l$ $\theta = -\frac{1}{24}\frac{Wl^2}{EI}$ at $A \qquad \theta = +\frac{1}{24}\frac{Wl^2}{EI}$ at B
End supports Intermediate load (2)	$R_1 = +W\frac{b}{l} \qquad R_2 = +W\frac{a}{l}$ $(A \text{ to } B) \; V = +W\frac{b}{l}$ $(B \text{ to } C) \; V = -W\frac{a}{l}$	$(A \text{ to } B) \; y = -\frac{Wbz}{6EIl}[2l(l-z) - b^2 - (l-z)^2]$ $(B \text{ to } C) \; y = -\frac{Wa(l-z)}{6EIl}[2lb - b^2 - (l-z)^2]$ Max $y = -\frac{Wab}{27EIl}(a+2b)\sqrt{3a(a+2b)}$ at $z = \sqrt{\frac{1}{3}a(a+2b)}$ when $a > b$ $\theta = -\frac{1}{6}\frac{W}{EI}\left(bl - \frac{b^3}{l}\right)$ at A; $\quad \theta = +\frac{1}{6}\frac{W}{EI}\left(2bl + \frac{b^3}{l} - 3b^2\right)$ at C

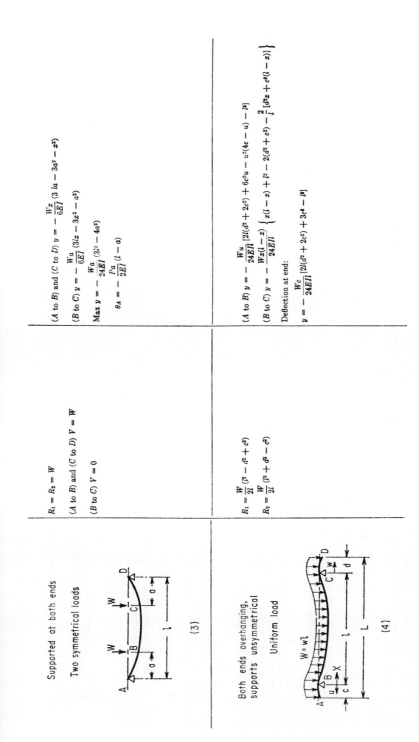

Supported at both ends

Two symmetrical loads

(3)

$R_1 = R_2 = W$

$(A$ to $B)$ and $(C$ to $D)$ $V = W$

$(B$ to $C)$ $V = 0$

$(A$ to $B)$ and $(C$ to $D)$ $y = -\frac{Wz}{6EI}(3la - 3a^2 - z^2)$

$(B$ to $C)$ $y = -\frac{Wa}{6EI}(3lz - 3z^2 - a^2)$

Max $y = -\frac{Wa}{24EI}(3l^2 - 4a^2)$

$\theta_A = -\frac{Pa}{2EI}(l - a)$

Both ends overhanging, supports unsymmetrical

Uniform load

(4)

$W = wl$

$R_1 = \frac{W}{2l}(l^2 - d^2 + c^2)$

$R_2 = \frac{W}{2l}(l^2 + d^2 - c^2)$

$(A$ to $B)$ $y = -\frac{Wu}{24EIl}[2l(d^2 + 2c^2) + 6c^2u - u^2(4c - u) - l^3]$

$(B$ to $C)$ $y = -\frac{Wz(l - x)}{24EIl}\left\{z(l - x) + l^2 - 2(d^2 + c^2) - \frac{2}{l}[d^2z + c^2(l - z)]\right\}$

Deflection at end:

$y = -\frac{Wc}{24EIl}[2l(d^2 + 2c^2) + 3c^4 - l^3]$

2.31

TABLE 2.1 Shear, Moment, and Deflection Formulas for Beams[1,12] (Continued)

Loading, support, and reference number	Reactions R_1 and R_2, vertical shear V	Deflection y, maximum deflection, and end slope θ
Both ends overhanging supports, load at any point between (5)	$R_1 = \dfrac{Wb}{l}$ $R_2 = \dfrac{Wa}{l}$	Between supports same as case 1; for overhang $y = \dfrac{Wabu}{6EIl}\,(l+b)$ Max y same as case 1 Max y at end Max $y = \dfrac{Wabc}{6EIl}\,(l+b)$
Both ends overhanging supports, single overhanging load (6)	$R_1 = \dfrac{W(c+l)}{l}$ $R_2 = \dfrac{Wc}{l}$	$(A \text{ to } B)\; y = -\dfrac{Wu}{6EI}\,(3cu - u^2 + 2cl)$ $(B \text{ to } C)\; y = -\dfrac{Wcz}{6EIl}\,(l-x)(2l-x)$ $(C \text{ to } D)\; y = -\dfrac{Wclw}{6EI}$ $y_A = -\dfrac{Wc^2}{3EI}\,(c+l)$ $y_D = -\dfrac{Wcld}{6EI}$ Max $y = \dfrac{Wcl^2}{15.55EI}$ at $x = 0.42265L$

2.32

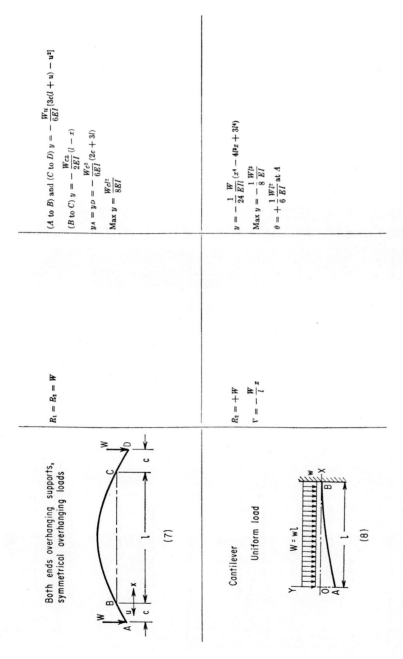

Both ends overhanging supports, symmetrical overhanging loads

(7)

$R_1 = R_2 = W$

(A to B) and (C to D) $y = -\dfrac{Wu}{6EI}[3c(l+u) - u^2]$

(B to C) $y = -\dfrac{Wcx}{2EI}(l - x)$

$y_A = y_D = -\dfrac{Wc^2}{6EI}(2c + 3l)$

Max $y = \dfrac{Wcl^2}{8EI}$

Cantilever

Uniform load

(8)

$W = wl$

$R_2 = +W$

$v = -\dfrac{W}{l}z$

$y = -\dfrac{1}{24}\dfrac{W}{EIl}(z^4 - 4l^3z + 3l^4)$

Max $y = -\dfrac{1}{8}\dfrac{Wl^3}{EI}$

$\theta = +\dfrac{1}{6}\dfrac{Wl^2}{EI}$ at A

2.33

TABLE 2.1 Shear, Moment, and Deflection Formulas for Beams[1,12] *(Continued)*

Loading, support, and reference number	Reactions R_1 and R_2, vertical shear V	Deflection y, maximum deflection, and end slope θ
Cantilever Intermediate load (9)	$R_2 = +W$ $(A \text{ to } B)\ V = 0$ $(B \text{ to } C)\ V = -W$	$(A \text{ to } B)\ y = -\dfrac{1}{6}\dfrac{W}{EI}(-a^3 + 3a^2l - 3a^2z)$ $(B \text{ to } C)\ y = -\dfrac{1}{6}\dfrac{W}{EI}[(z-b)^3 - 3a^2(z-b) + 2a^3]$ $\text{Max } y = -\dfrac{1}{6}\dfrac{W}{EI}(3a^2l - a^3)$ $\theta = +\dfrac{1}{2}\dfrac{Wa^2}{EI}\ (A \text{ to } B)$
One end fixed, one end supported Uniform load (10)	$R_1 = \tfrac{3}{8}W \qquad R_2 = \tfrac{5}{8}W$ $M_2 = \tfrac{1}{8}Wl$ $V = W\left(\dfrac{3}{8} - \dfrac{z}{l}\right)$	$y = \dfrac{1}{48}\dfrac{W}{EIl}(3lz^3 - 2z^4 - l^3z)$ $\text{Max } y = -0.0054\dfrac{Wl^3}{EI}\ \text{at } z = 0.4215l$ $\theta = -\dfrac{1}{48}\dfrac{Wl^2}{EI}\ \text{at } A$

2.34

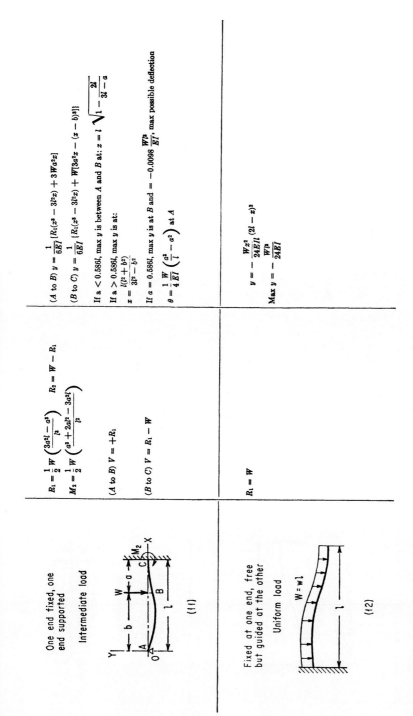

One end fixed, one end supported

Intermediate load

$R_1 = \frac{1}{2} W \left(\dfrac{3a^2l - a^3}{l^3} \right)$ $R_2 = W - R_1$

$M_2 = \frac{1}{2} W \left(\dfrac{a^3 + 2a^2l - 3a^2l}{l^3} \right)$

$(A \text{ to } B)\ V = +R_1$

$(B \text{ to } C)\ V = R_1 - W$

$(A \text{ to } B)\ y = \dfrac{1}{6EI} \left[R_1(z^3 - 3l^2z) + 3Wa^2z \right]$

$(B \text{ to } C)\ y = \dfrac{1}{6EI} \left\{ R_1(z^3 - 3l^2z) + W[3a^2z - (z-b)^3] \right\}$

If $a < 0.586l$, max y is between A and B at: $z = l \sqrt{1 - \dfrac{2l}{3l-a}}$

If $a > 0.586l$, max y is at:
$z = \dfrac{l(l^2 + b^2)}{3l^2 - b^2}$

If $a = 0.586l$, max y is at B and $= -0.0098\ \dfrac{Wl^3}{EI}$, max possible deflection

$\theta = \dfrac{1}{4}\dfrac{W}{EI}\left(\dfrac{a^3}{l} - a^2\right)$ at A

(11)

Fixed at one end, free but guided at the other

Uniform load

$W = wl$

$R_1 = W$

$y = -\dfrac{Wz^2}{24EIl}(2l - z)^2$

$\text{Max } y = -\dfrac{Wl^3}{24EI}$

(12)

TABLE 2.1 Shear, Moment, and Deflection Formulas for Beams[1,12] (*Continued*)

Loading, support, and reference number	Reactions R_1 and R_2, vertical shear V	Deflection y, maximum deflection, and end slope θ
Fixed at one end, free but guided at the other With load (13)	$R_1 = W$	$y = -\dfrac{Wz^2}{12EI}(3l - 2z)$ Max $y = -\dfrac{Wl^3}{12EI}$
Both ends fixed Uniform load (14)	$R_1 = \tfrac{1}{2}W \qquad R_2 = \tfrac{1}{2}W$ $M_1 = \tfrac{1}{12}Wl \qquad M_2 = \tfrac{1}{12}Wl$ $v = \dfrac{1}{2}W\left(1 - \dfrac{2z}{l}\right)$	$y = \dfrac{1}{24}\dfrac{Wz^2}{EIl}(2lz - l^2 - z^2)$ Max $y = -\dfrac{1}{384}\dfrac{Wl^3}{EI}$ at $z = \dfrac{1}{2}l$

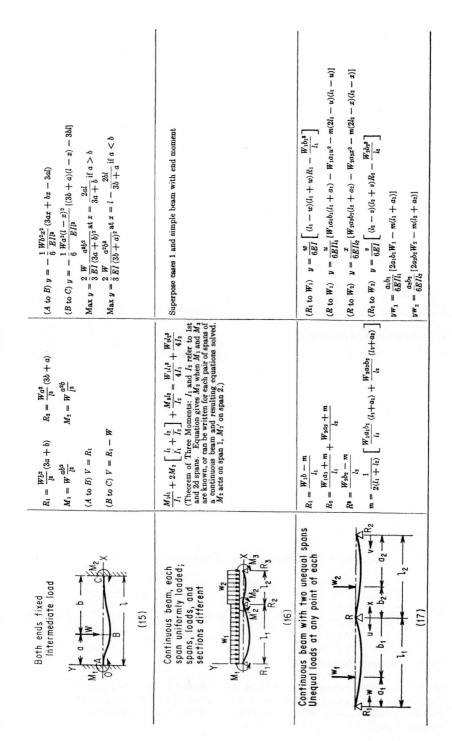

Both ends fixed
Intermediate load

(15)

$$R_1 = \frac{Wb^2}{l^3}(3a+b) \qquad R_2 = \frac{Wa^2}{l^3}(3b+a)$$

$$M_1 = W\frac{ab^2}{l^2} \qquad M_2 = W\frac{a^2b}{l^2}$$

$$(A \text{ to } B)\ V = R_1$$

$$(B \text{ to } C)\ V = R_1 - W$$

$(A \text{ to } B)\ y = -\dfrac{1}{6}\dfrac{Wb^2z^2}{EIl^3}(3az+bz-3al)$

$(B \text{ to } C)\ y = -\dfrac{1}{6}\dfrac{Wa^2(l-x)^2}{EIl^3}[(3b+a)(l-x)-3bl]$

Max $y = \dfrac{2}{3}\dfrac{W}{EI}\dfrac{a^3b^2}{(3a+b)^2}$ at $x = \dfrac{2al}{3a+b}$ if $a > b$

Max $y = \dfrac{2}{3}\dfrac{W}{EI}\dfrac{a^2b^3}{(3b+a)^2}$ at $x = l - \dfrac{2bl}{3b+a}$ if $a < b$

Continuous beam, each span uniformly loaded; spans, loads, and sections different

(16)

$$\frac{M_1l_1}{I_1} + 2M_2\left[\frac{l_1}{I_1}+\frac{l_2}{I_2}\right] + \frac{M_3l_2}{I_2} = \frac{W_1l_1^3}{4I_1} + \frac{W_2l_2^3}{4I_2}$$

(Theorem of Three Moments: I_1 and I_2 refer to 1st and 2d spans. Equation gives M_2 when M_1 and M_3 are known, or can be written for each pair of spans of a continuous beam and resulting equations solved. M_2 acts on span 1, M_2' on span 2.)

Superpose cases 1 and simple beam with end moment

Continuous beam with two unequal spans
Unequal loads at any point of each

(17)

$$R_1 = \frac{W_1b - m}{l_1}$$

$$R_2 = \frac{W_1a_1 + m}{l_1} + \frac{W_2a_2 + m}{l_2}$$

$$R_4 = \frac{W_2b_2 - m}{l_2}$$

$$m = \frac{1}{2(l_1+l_2)}\left[\frac{W_1a_1b_1}{l_1}(l_1+a_1) + \frac{W_2a_2b_2}{l_2}(l_2+a_2)\right]$$

$(R_1 \text{ to } W_1)$ $\quad y = \dfrac{w}{6EI}\left[(l_1-w)(l_1+w)R_1 - \dfrac{W_1b_1^3}{l_1}\right]$

$(R \text{ to } W_1)$ $\quad y = \dfrac{u}{6EIl_1}[W_1a_1b_1(l_1+a_1) - W_1au^2 - m(2l_1-u)(l_1-u)]$

$(R \text{ to } W_2)$ $\quad y = \dfrac{z}{6EIl_2}[W_2a_2b_2(l_2+a_2) - W_2a_2z^2 - m(2l_2-z)(l_2-x)]$

$(R_2 \text{ to } W_2)$ $\quad y = \dfrac{v}{6EI}\left[(l_2-v)(l_2+v)R_4 - \dfrac{W_2b_2^3}{l_2}\right]$

$y_{W_1} = \dfrac{a_1b_1}{6EIl_1}[2a_1b_1W_1 - m(l_1+a_1)]$

$y_{W_2} = \dfrac{a_2b_2}{6EIl_2}[2a_2b_2W_2 - m(l_2+a_2)]$

$$\theta_a = \frac{\partial U}{\partial M_a} = \frac{\partial}{\partial M_a} \int_0^L M^2 \frac{dx}{2EI} = \int_0^L \frac{M}{EI} \frac{\partial M}{\partial M_a} dx \qquad (2.131)$$

An important restriction on the use of this theorem is that the deflection of the beam or structure must be a linear function of the load; i.e., geometrical changes and other nonlinear effects must be neglected.

A second theorem of Castigliano states that

$$P_a = \partial U/\partial Y_a \qquad (2.132)$$

$$M_a = \partial U/\partial \theta_a \qquad (2.133)$$

and is just the inverse of the first theorem. Because it does not have a "linearity" requirement, it is quite useful in special problems.

To illustrate, the deflection y at the center of wire of length $2L$ due to a central load P will be found.

From geometry, the extension δ of each half of the wire is, for small deflections,

$$\delta \approx y^2/2L \qquad (2.134)$$

The strain energy absorbed in the system is

$$U = 2\tfrac{1}{2}(AE/L)\delta^2 = (AE/4L^3)y^4 \qquad (2.135)$$

Then, by the second theorem,

$$P = \partial U/\partial y = (AE/L^3)y^3 \qquad (2.136)$$

or the deflection is

$$y = L \sqrt[3]{P/AE} \qquad (2.137)$$

Among the other useful energy theorems are:

Theorem of Virtual Work. If a beam which is in equilibrium under a system of external loads is given a small deformation ("virtual deformation"), the work done by the load system during this deformation is equal to the increase in internal strain energy.

Principle of Least Work. For beams with statically indeterminate reactions, the partial derivative of the total strain energy with respect to the unknown reactions must be zero.

$$\partial U/\partial P_i = 0 \qquad \partial U/\partial M_i = 0 \qquad (2.138)$$

depending on the type of support. (This follows directly from Castigliano's theorems.) The magnitudes of the reactions thus determined are such as to minimize the strain energy of the system.

2.9.3 Elasticity Approach[2]

In developing the conventional equations for beam theory from the basic equations of elastic theory (i.e., stress equilibrium, strain compatibility, and stress-strain relations) the beam problem is considered a plane-stress problem. The equilibrium equations for plane stress are

$$\partial\sigma_x/\partial x + \partial\tau_{xy}/\partial y = 0 \qquad (2.139)$$

$$\partial\sigma_y/\partial y + \partial\tau_{xy}/\partial x = 0 \qquad (2.140)$$

By using an "Airy stress function" ψ, defined as follows:

$$\sigma_x = \partial^2\psi/\partial y^2 \qquad \sigma_y = \partial^2\psi/\partial x^2 \qquad \tau_{xy} = -\partial^2\psi/\partial x\,\partial y \qquad (2.141)$$

and the compatibility equation for strain, as set forth previously, the governing equations for beams can be developed.

The only compatibility equation not identically satisfied in this case is

$$\partial^2\epsilon_x/\partial y^2 + \partial^2\epsilon_y/\partial x^2 = \partial^2\gamma_{xy}/\partial x\,\partial y \qquad (2.142)$$

Substituting the stress-strain relationships into the compatibility equations and introducing the Airy stress function yields

$$\partial^4\psi/\partial x^4 + 2\partial^4\psi/\partial x^2\,\partial y^2 + \partial^4\psi/\partial y^4 = \nabla^4\psi = 0 \qquad (2.143)$$

which is the "biharmonic" equation where ∇^2 is the Laplace operator.

To illustrate the utility of this equation consider a uniform-thickness cantilever beam (Fig. 2.21) with end load P. The boundary conditions are $\sigma_y = \tau_{xy} = 0$ on the surfaces $y = \pm c$, and the summation of shearing forces must be equal to the external load P at the loaded end,

$$\int_{-c}^{+c} \tau_{xy}b\,dy = P$$

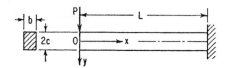

FIG. 2.21 Cantilever beam with end load P.

The solution for σ_x is

$$\sigma_x = \partial^2\psi/\partial y^2 = cxy \qquad (2.144)$$

Introducing $b(2c)^3/12 = I$, the final expressions for the stress components are

$$\sigma_x = -Pxy/I = -My/I$$

$$\sigma_y = 0 \qquad (2.145)$$

$$\tau_{xy} = -P(c^2 - y^2)/2I$$

To extend the theory further to determine the displacements of the beam, the definitions of the strain components are

$$\epsilon_x = \partial u/\partial x = \sigma_x/E = -Pxy/EI$$

$$\epsilon_y = \partial v/\partial y = -\nu\sigma_x/E = \nu Pxy/EI \qquad (2.146)$$

$$\gamma_{xy} = \partial u/\partial y + \partial v/\partial x = [2(1 + \nu)/E]\tau_{xy} = [(1 + \nu)P/EI](c^2 - y^2)$$

Solving explicitly for the u and v subject to the boundary conditions

$$u = v = \partial u/\partial x = 0 \qquad \text{at } x = L \text{ and } y = 0$$

there results

$$v = vPxy^2/2EI + Px^3/6EI - PL^2x/2EI + PL^3/3EI \qquad (2.147)$$

The equation of the deflection curve at $y = 0$ is

$$(v)_{y=0} = (P/6EI)(x^3 - 3L^2x + 2L^3) \qquad (2.148)$$

The curvature of the deflection curve is therefore the Bernoulli-Euler equation

$$1/\rho \approx - (\partial^2v/\partial^2x)_{y=0} = - Px/EI = M/EI = - y'' \qquad (2.149)$$

EXAMPLE 1   The moment at any point x along a simply supported uniformly loaded beam (w lb/ft) of span L is

$$M = wLx/2 - wx^2/2 \qquad (2.150)$$

Integrating Eq. (2.121) and employing the boundary conditions $y(0) = y(L) = 0$, the solution for the elastic or deflection curve becomes

$$y = (wL^4/24EI)(x/L)[1 - 2(x/L)^2 + (x/L)^3] \qquad (2.151)$$

EXAMPLE 2   In order to obtain the general deflection curve, a fictitious load P_a is placed at a distance a from the left support of the previously described uniformly loaded beam.

$$M = -wx^2/2 + wLx/2 + [P_a x(L - a)]/L \qquad 0 < x < a$$
$$M = -wx^2/2 + wLx/2 + [P_a a(L - x)]/L \qquad a < x < L \qquad (2.152)$$

From Castigliano's theorem,

$$y_a = \frac{\partial U}{\partial P_a} = \int_0^L \frac{M}{EI} \frac{\partial M}{\partial P_a} dx \qquad (2.153)$$

and therefore

$$y_a = (1/EI)(\tfrac{1}{24}wa^4 - \tfrac{1}{12}wLa^3 + \tfrac{1}{24}waL^3) \qquad (2.154)$$

The elastic curve of the beam is obtained by substituting x for a in Eq. (2.154), resulting in the same expression as obtained by the double-integration technique.

EXAMPLE 3   It can be shown, considering stresses away from the beam ends, and essentially considering temperature variations only in the direction perpendicular to the beam axis (or in two or more directions by superposition) that the traction-free thermoelastic stress distribution is given by

$$\sigma_x = -\alpha ET + \frac{1}{2c} \int_{-c}^{+c} \alpha ET \, dy + \frac{3y}{2c^3} \int_{-c}^{+c} \alpha ETy \, dy \qquad (2.155)$$

Since this stress distribution results in a net axial force-free, and moment-free, distribution but nonzero bending displacements and slopes, it would be necessary to superpose any additional stresses associated with actual boundary constraints.

Because the stress distribution is independent of any linear temperature gradient, it is always possible to add an arbitrary linear distribution T', such that $T_{tot} = T + T'$ and $\int T_{tot}\, dy = \int T_{tot} y\, dy = 0$.

Thus

$$\sigma_x = - \alpha E T_{tot} \tag{2.156}$$

In general T' can be chosen with sufficient accuracy by *visual examination* of the temperature distribution to make the "total-temperature integral" and its "first moment" equal zero. Thus the simplified formula together with its interpretation presents a useful graphical thermoelastic solution for beam and slab problems.

For the beam with temperature distribution $T = a(c^2 - y^2)$ and a is an arbitrary constant, superimpose a temperature distribution T' such that $A_T = -A_{T'}$, where A_T refers to the area under the temperature distribution curve. Evidently

$$T' = - \tfrac{2}{3}ac^2$$

and

$$T_{tot} = T + T' = a(c^2 - y^2) - \tfrac{2}{3}ac^2$$

and $\int T_{tot}\, dy = \int T_{tot} y\, dy = 0$, evident from the selection of T' and symmetry. The stress is therefore

$$\sigma = - \alpha E T_{tot} = - \alpha E[a(c^2 - y^2) - \tfrac{2}{3}ac^2] \tag{2.157}$$

2.10 CURVED-BEAM THEORY[1]

A "curved beam" (see Fig. 2.22) is defined as a beam in which the line joining the centroid of the cross sections (hereafter referred to as the "center line") is a curve. In the standard developments of the equations for the stresses and deflections for curved beams the following *assumptions* are usually made and represent the *restrictions* on the applicability of curved-beam theory:

The sections of the beam originally plane and normal to the center line of the beam remain so after bending.

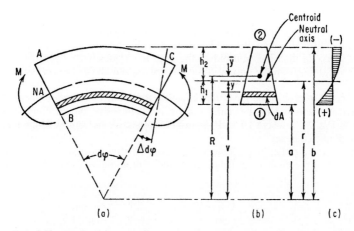

FIG. 2.22 Bending of curved beam element. (*a*) Beam element. (*b*) Cross section. (*c*) Bending-stress pattern at section *C–D*.

All cross sections have an axis of symmetry in the plane of the center line. The beam is subjected to forces and moments acting in the plane of symmetry.

2.10.1 Equilibrium Approach

In Fig. 2.22 a "positive bending" moment is taken as one which tends to *decrease* the *curvature* of the beam. If R denotes the curvature of the beam at the centroid of a section, then it can be shown that the neutral axis is displaced from R a distance toward the center of curvature. As with straight beams, the "neutral axis" is defined as that axis about which the integrated tangential force is zero, when the external traction is restricted to a bending moment. This distance $\bar{y}$ may be computed from

$$\bar{y} = R - \frac{A}{\int dA/v} \tag{2.158}$$

where A = cross-section area of beam, v the distance from the center of curvature to the incremental area; and the integration $\int dA/v$ is carried out over the entire section of the beam.

The flexural stress at any point a distance y from the neutral axis is

$$\sigma = E\epsilon = Ey(\Delta d\phi)/(r - y) \, d\phi \tag{2.159}$$

which shows that the normal stress distribution over a cross section is not linear, as would be the case in simple beam theory, but hyperbolic in shape. Equating the sum of the moments of each of these segments of normal stress to the applied bending moment on the element,

$$\int \sigma y \, dA = \frac{E(\Delta \, d\phi)}{d\phi} \int \frac{y^2 \, dA}{r - y} = M \tag{2.160}$$

yields the resulting expression for the stress

$$\sigma = My/A\bar{y}(r - y) \tag{2.161}$$

with the stresses at the extreme fibers, points 1 and 2, expressed by

$$\sigma_1 = Mh_1/A\bar{y}a \quad \text{(tensile)} \quad \sigma_2 = Mh_2/A\bar{y}b \quad \text{(compressive)} \tag{2.162}$$

The above-described stresses result from pure bending only. If a more general loading condition is given, it must be reduced to the statically equivalent couple and a normal force through the centroid of the section in question.* Then the extensional stresses resulting from the normal force are superposed on the flexural stresses due to the couple.

The development of the governing equation for the displacement of a curved beam, for small deflections with a circular center line, depends on the differential quantities shown in Fig. 2.23, where $1/r$ and $1/r_1$ are the respective curvatures of the undeflected and the deflected beam.

Note that the radial displacement u is taken positive inward in these relations. A comparison of the changes in length $\Delta \, ds$ and central angle $\Delta \, d\phi$ due to deformation leads to

$$1/r_1 - 1/r = u/r^2 + d^2u/ds^2 \tag{2.163}$$

---

*This also applies in the general case when a shearing force also may be acting on this section.

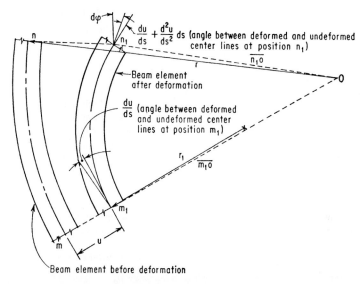

FIG. 2.23 Curved beam with circular center line.

If the thickness of the beam is small compared with the curvature, then

$$1/r_1 - 1/r = -M/EI \qquad (2.164)$$

and the differential equation for the deflection curve of a curved beam, which is entirely analogous to that of simple-beam theory, becomes

$$d^2u/ds^2 + u/r^2 = -M/EI \qquad (2.165)$$

For an infinitely large r this reduces to the Bernoulli-Euler equation for simple beams.

2.10.2 Energy Approach

A second and more powerful approach, which does not require that the center line of the beam be circular, is essentially the application of Castigliano's theorem. The expression for the strain energy in bending of a curved beam is similar to that of simple-beam theory,

$$U = \int_0^s \frac{M^2 \, ds}{2EI} \qquad (2.166)$$

where the integration is over the entire length of the beam. The deflection (or rotation) of the beam at a point under a concentrated load (or moment), is

$$\delta_i = \partial U/\partial P_i \qquad (2.167)$$

The utility of Eqs. (2.166) and (2.167) in calculating the deflection of curved beams can best be shown by example.

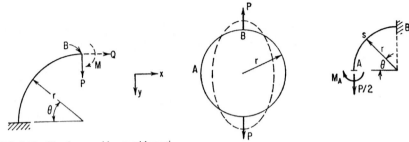

FIG. 2.24   Simple curved beam with vertical load P.

FIG. 2.25   Circular ring.

EXAMPLE 1   The curved beam shown in Fig. 2.24 has a uniform cross section. To determine the horizontal and vertical deflections and rotations, at point B fictitious load Q and M are assumed to act. At any point s on the beam,

$$M_s = -[Pr\cos\theta + Qr(1 - \sin\theta) + M] \qquad (2.168)$$

$$\partial M_s/\partial P = -r\cos\theta \qquad \partial M_s/\partial Q = -r(1 - \sin\theta) \qquad \partial M_s/\partial M = -1 \qquad (2.169)$$

Substituting $Q = 0$, $M = 0$ in the expression M_s we have, for the three deformations with $ds = r\,d\theta$,

$$\delta_y = \left.\frac{\partial U}{\partial P}\right|_{\substack{Q\to 0 \\ M\to 0}} = \int_0^s \frac{M_s}{EI}\frac{\partial M_s}{\partial P}\,ds = \int_0^{\pi/2} \frac{Pr^3}{EI}\cos^2\theta\,d\theta = \frac{\pi}{4}\frac{Pr^3}{EI} \qquad (2.170)$$

$$\delta_x = \left.\frac{\partial U}{\partial Q}\right|_{\substack{Q\to 0 \\ M\to 0}} = \int_0^s \frac{M_s}{EI}\frac{\partial M}{\partial Q}\,ds = \int_0^{\pi/2} \frac{Pr^3}{EI}\cos\theta\,(1 - \sin\theta)\,d\theta = \frac{Pr^3}{2EI} \qquad (2.171)$$

$$\theta = \left.\frac{\partial U}{\partial M}\right|_{\substack{Q\to 0 \\ M\to 0}} = \int_0^s \frac{M_s}{EI}\frac{\partial M_s}{\partial M}\,ds = \int_0^{\pi/2} \frac{Pr^2}{EI}\cos\theta\,d\theta = \frac{Pr^2}{EI} \qquad (2.172)$$

EXAMPLE 2   *Circular Ring-Energy Approach.* The circular ring is subjected to equal and opposite forces P as shown in Fig. 2.25. From symmetry considerations an equivalent model may be constructed where the load on the horizontal section is denoted by moment M_A and force $P/2$. From symmetry, there is no rotation of the horizontal section at point A. Therefore, by Castigliano's theorem,

$$\theta_A = \partial U/\partial M_A = 0 \qquad (2.173)$$

The moment at any point s is given by

$$M_s = M_A - (P/2)r(1 - \cos\theta) \qquad \text{and} \qquad \partial M_s/\partial M_A = 1 \qquad (2.174)$$

Substituting in Eq. (2.173) and imposing the condition of zero rotation,

$$\theta_A = 0 = \frac{\partial U}{\partial M_A} = \int_0^s \frac{M_s}{EI}\frac{\partial M_s}{\partial M_A}\,ds = \int_0^{\pi/2} \frac{1}{EI}\left[M_A - \frac{Pr}{2}(1 - \cos\theta)\right]r\,d\theta \qquad (2.175)$$

which leads to

$$M_A = (Pr/2)(1 - 2/\pi) \qquad (2.176)$$

which yields the moment expression

$$M_s = (Pr/2)(\cos \theta - 2/\pi) \tag{2.177}$$

The total strain energy for the entire ring is four times that of the quadrant considered. To obtain the total increase in the vertical diameter, the following steps are taken:

$$U = 4 \int_0^{\pi/2} \frac{M_s^2}{2EI} r \, d\theta \tag{2.178}$$

$$\delta_y = \frac{\partial U}{\partial P} = \frac{4}{EI} \int_0^{\pi/2} M_s \left(\frac{\partial M_s}{\partial P}\right) r \, d\theta = \frac{Pr^3}{EI} \int_0^{\pi/2} \left(\cos \theta - \frac{2}{\pi}\right)^2 d\theta \tag{2.179}$$

$$\delta_y = \frac{Pr^3}{EI} \left(\frac{\pi}{4} - \frac{2}{\pi}\right) \tag{2.180}$$

Equilibrium Approach: The basic equation for this problem is

$$d^2u/ds^2 + u/r^2 = -M_s/EI \tag{2.181}$$

Substituting Eq. (2.177) yields

$$r^2 \, d^2u/ds^2 + u = (Pr^3/2EI)(2/\pi - \cos \theta) \tag{2.182}$$

The boundary conditions, derived from the symmetry of the ring, are

$$u'(\theta = 0) = u'(\theta = \pi/2) = 0$$

The general solution of Eq. (2.182) is

$$u = A \cos \theta + B \sin \theta + Pr^3/EI\pi - (Pr^3/4EI) \, \theta \sin \theta \tag{2.183}$$

$$A = -Pr^3/4EI \quad \text{and} \quad B = 0 \tag{2.184}$$

Then $u = Pr^3/EI\pi - (Pr^3/4EI) \cos \theta - (Pr^3/4EI)\theta \sin \theta$

The increase in the vertical radius (point B at $\theta = \pi/2$) becomes

$$u(\theta = \pi/2) = (Pr^3/EI)(1/\pi - \pi/8) \tag{2.185}$$

Thus the total increase in the vertical diameter δ_y is

$$\delta_y = -2u(\theta = \pi/2) = (Pr^3/EI)(\pi/4 - 2/\pi) \tag{2.186}$$

which is in agreement with Eq. (2.180).

2.11 THEORY OF COLUMNS[1]

The equilibrium approach for slender symmetrical columns is based on two sets of assumptions:

1. All the basic assumptions inherent in the derivation of the Bernoulli-Euler equation apply to columns also.
2. The transverse deflection of the column at the point of load application is *not* small when compared with the eccentricity of the applied load.

This theory is best described with the aid of a typical column, taken with a built-in support and subjected to an eccentric load, as illustrated in Fig. 2.26.

The moment at any section a distance x from the base is

$$M = -P(\delta + e - y) \tag{2.187}$$

where the negative sign is in accordance with the sign convention for simple beams. Writing the Bernoulli-Euler equation for the bending deflection of this member with the aid of the substitution $p^2 = P/EI$ gives

$$y'' + p^2 y = p^2(\delta + e) \tag{2.188}$$

The general solution of Eq. (2.188) is

$$y = A \sin px + B \cos px + \delta + e \tag{2.189}$$

From the boundary conditions for a built-in end

$$y(0) = y'(0) = 0$$

the equation for the deflection curve is

$$y = (\delta + e)(1 - \cos px) \tag{2.190}$$

The deflection at the end of the column at $x = L$ is seen to be

$$\delta = e(1 - \cos pL)/(\cos pL) \tag{2.191}$$

The complete description of the deflection curve for any point in the column thus becomes

$$y = e(1 - \cos px)/(\cos pL) \tag{2.192}$$

These results can easily be extended to cover a column hinged at both ends by redefining terms as indicated in Fig. 2.27. Thus, from symmetry, the relation of the deflection at the mid-span can be written directly from the previous results:

$$\delta = e\left[1 - \cos\left(\frac{pL}{2}\right)\right] \Big/ \cos\left(\frac{pL}{2}\right) \tag{2.193}$$

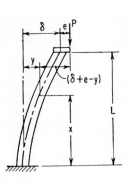

FIG. 2.26 Column subjected to an eccentric load.

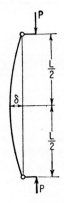

FIG. 2.27 Column hinged at both ends.

In applying these equations note that the deflection is not proportional to the compressive load P. Hence the method of superposing a compressive deflection due to P and a bending deflection due to the couple Pe cannot be used for column action.

In considering column action the concept of "critical load" is of fundamental importance. As the argument of the cosine in the equations for the maximum deflection approaches a value of $\pi/2$, the deflection δ increases without bound and in actual practice the column will fail in a buckling *regardless of the eccentricity e.* Substituting the value $pL = \pi/2$ and $pL/2 = \pi/2$ back into $p^2 = P/EI$ will yield

$$P_{CR} = \pi^2 EI/4L^2 \qquad \text{for the single built-in support} \qquad (2.194)$$

$$P_{CR} = \pi^2 EI/L^2 \qquad \text{for the hinged ends} \qquad (2.195)$$

$$P_{CR} = 4\pi^2 EI/L^2 \qquad \text{for the both ends built-in} \qquad (2.196)$$

The above equations for the critical load (Euler's loads) depend only on the dimensions of the column (I/L^2) and the modulus of the material E. If the moment of inertia is written $I = Ak^2$, where k is the radius of gyration, the critical-load expressions take the form

$$P_{CR} = \frac{C_1 \pi^2 AE}{(L/k)^2} \qquad (2.197)$$

where the dependence is now on the material and a slenderness ratio L/k.

The "critical stress" for a column with hinged ends is given by

$$\sigma_{CR} = \frac{P_{CR}}{A} = \frac{\pi^2 E}{(L/k)^2} \qquad (2.198)$$

A plot of σ_{CR} vs. L/k is hyperbolic in form, as shown in Fig. 2.28.

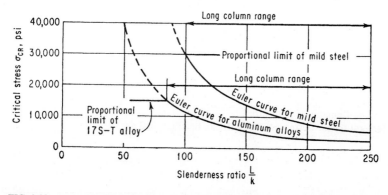

FIG. 2.28   Curve of critical stress vs. slenderness ratio for a column with hinged ends.

The horizontal line in Fig. 2.28 indicates the compressive yield stress of a typical structural steel. For analysis purposes, one uses the compressive yield stress as the design criterion for *small* slenderness ratios and the Euler curve for higher ratios.

Much column design, especially in heavy structural engineering, is accomplished by means of the application of empirical formulas developed as a result of experimental work and practical experience. Several of these formulas are presented below, for hinged bars. Straight-line formulas for structural-steel bars:

$$\sigma_{CR} = P_{CR}/A = 48{,}000 - 210(L/k) \qquad (2.199)$$

Parabolic formula for structural steels:

$$\sigma_{CR} = 40{,}000 - 1.35(L/k)^2 \qquad (2.200)$$

Gordon-Rankine formula for main members with $120 < L/k < 200$:

$$\sigma_{\omega} = \frac{18{,}000}{1 + L^2/18{,}000k^2} \left(1.6 - \frac{L}{200k}\right) \qquad (2.201)$$

where σ_{ω} is a working stress.

2.12   SHAFTS, TORSION, AND COMBINED STRESS[1]

2.12.1   Torsion of Solid Circular Shafts

When a solid circular shaft is subjected to a pure torsional load, the reaction of the shaft for small angles of twist is assumed to be subject to the following restraints:

1. Circular cross sections remain circular and their diameters remain unchanged.
2. The axial distances between adjacent cross sections do not change.
3. A lateral surface element of the shaft is in a state of pure shear.

The shearing stresses acting on an element a distance r from the axis are

$$\tau = Gr\theta \qquad (2.202)$$

where θ = angle of twist per unit length of shaft
G = modulus of rigidity = $E/2(1 + v)$

The shear stress, maximum at the surface is

$$\tau_{max} = \tfrac{1}{2}G\theta D \qquad (2.203)$$

where D = diameter of the shaft.
The total torque acting on the shaft can be expressed as

$$T = \int_A r(\tau\, dA) = \int_A G\theta r^2\, dA = G\theta J \qquad (2.204)$$

$$T = K\theta \qquad (2.205)$$

where J = polar moment of inertia of the circular cross section and K = torsional rigidity = GJ for circular shafts. Therefore, the most useful relations in dealing with circular-shaft torsional problems are

$$\tau_{max} = TD/2J = 16T/\pi D^3 \qquad (2.206)$$

$$\phi = TL/GJ = TL/K \qquad (2.207)$$

where ϕ = total angle of twist for the shaft of length L.

2.12.2 Shafts of Rectangular Cross Section

For a shaft of rectangular cross section,

$$\tau_{max} = (T/ht^2)[3 + 1.8(t/h)] \qquad (h > t) \qquad (2.208)$$

$$\phi = TL/K \qquad (2.209)$$

where $K = \beta ht^3 G$ and the constant β is 0.141, 0.249, and 0.333 for various ratios of h/t of 1.0, 2.5, and ∞, respectively.

2.12.3 Single-Cell Tubular-Section Shaft

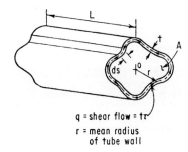

q = shear flow = tr
r = mean radius
of tube wall

FIG. 2.29 Shaft with general tubular shape.

In Fig. 2.29 a general tubular shape is subjected to a pure torque load; the resultant angular rotations will be about point 0. Considering a slice of the tube of length L, the "shear flow" q can be shown to be constant around the tube. In thin-walled-tube problems the shear is expressed in terms of force per unit length and is thought of as "flowing" from *source* to *sink* in much the same manner as in hydrodynamics problems. Thus it can be shown that

$$q = T/2A \qquad (2.210)$$

where A = area enclosed by the median line of the section = πr^2 for a circular tubular section of mean radius r. The total strain energy and total angle of twist (with no warping) for the tube is

$$U = \oint (q^2 L/2tG) \, ds \qquad (2.211)$$

$$\phi = \partial U/\partial T = (TL/4A^2G) \oint ds/t \qquad$$

$$= (L/2AG) \oint (q/t) \, ds \qquad (2.212)$$

The unit angle of twist is

$$\theta = \phi/L = (1/2AG) \oint (q/t) \, ds \qquad (2.213)$$

The torsional rigidity of the tubular section may then be defined as

$$K = \frac{4A^2}{\oint (ds/t)} G \qquad (2.214)$$

For a circular pipe of mean diameter D_m the value of the rigidity becomes

$$K = (\pi D_m^3 t/4)G \qquad (2.215)$$

2.12.4 Combined Stresses

Frequently problems in shafts involve combined torsion and bending. If the weight of the shaft is neglected relative to load P, there are two major components contributing to the maximum stress:

1. Torsional stress τ, which is a maximum at any point on the surface of the shaft
2. Flexural stress σ, which is a maximum at the built-in end on an element most remote from the shaft axis

The direct stresses due to shear are not significant, since they are a maximum at the axis of the shaft where the flexural stress is zero. The loads on the element at the built-in end are $T = Pr$ and $M = -PL$, and the stresses may be evaluated as follows:

$$\tau = TD/2J = PrD/2J \tag{2.216}$$

$$\sigma = Mc/I = \pm PLD/2I \tag{2.217}$$

where r = load offset distance
$\quad\quad D$ = shaft diameter
$\quad\quad L$ = shaft length

The maximum principle stress on the element in tension is

$$\sigma_{max} = \sigma/2 + \tfrac{1}{2}\sqrt{\sigma^2 + 4\tau^2} \tag{2.218}$$

Noting that for a circular shaft the polar moment is twice the area moment about a diameter ($J = 2I$):

$$\sigma_{max} = D/4I(M + \sqrt{M^2 + T^2}) = (PLD/4I)[1 + \sqrt{1 + (r/L)^2}] \tag{2.219}$$

where
$$I = \pi D^4/64$$

The maximum shear stress on the element is

$$\tau_{max} = D/4I \sqrt{M^2 + T^2} = (PLD/4) \sqrt{1 + (r/L)^2} \tag{2.220}$$

EXAMPLE 1 For a hollow circular shaft the loads and stresses are

$$T = Pr \quad\quad \tau = PrD/2J$$

$$M = PL \quad\quad \sigma = PLD/2I$$

$$\sigma_{max} = \frac{16D_0 PL}{\pi(D_0^4 - D_i^4)} \left[1 + \sqrt{1 + \left(\frac{r}{L}\right)^2} \right]$$

$$\tau_{max} = \frac{16D_0 PL}{\pi(D_0^4 - D_i^4)} \left[\sqrt{1 + \left(\frac{r}{L}\right)^2} \right] \tag{2.221}$$

For the angular deflection,

$$K = \pi D^3 tG/4$$

$$\theta = T/K = 4Pr/\pi D^3 tG \tag{2.222}$$

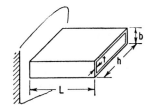

$$\phi = \theta L = 4PrL/\pi D^3 tG$$

where $D = (D_0 + D_i)/2$.

EXAMPLE 2 The shear stresses and angular deflection of a box as shown in Fig. 2.30 will now be investigated.

The shear flow is given by

FIG. 2.30 Rectangular thin-walled section.

$$q = T/2A \tag{2.223}$$

where A = plane area $(b-t)(h-t)$. The shear stress

$$\tau = q/t = T/2t(b-t)(h-t) \tag{2.224}$$

The steps in obtaining the torsional rigidity are

$$\oint ds/t \approx 1/t[2(b - t) + 2(h-t)] \tag{2.225}$$

$$K = \frac{4A^2G}{\oint ds/t} = \frac{2Gt[(b - t)(h - t)]^2}{(b - t) + (h - t)} \tag{2.226}$$

$$\phi = \theta L = \frac{TL}{K} = \frac{TL[(b - t) + (h - t)]}{2Gt[(b - t)(h - t)]^2} \tag{2.227}$$

2.13 PLATE THEORY[3,10]

2.13.1 Fundamental Governing Equation

In deriving the first-order differential equation for a plate under action of a transverse load, the following assumptions are usually made:

1. The plate material is homogeneous, isotropic, and elastic.
2. The least lateral dimension (length or width) of the plate is at least 10 times the thickness h.
3. At the boundary, the edges of the plate are unrestrained in the plane of the plate; thus the reactions at the edges are taken transverse to the plate.
4. The normal to the original middle surface remains normal to the distorted middle surface after bending.
5. Extensional strain in the middle surface is neglected.

The deflected shape of a simply supported plate due to a normal load q is illustrated in Fig. 2.31, which also defines the coordinate system. The positive shears, twists, and moments which act on an element of the plate are depicted in Fig. 2.32.

Application of the equations of equilibrium and Hooke's law to the differential element leads to the following relations:

$$\partial M_x/\partial x + \partial M_{yx}/\partial y = Q_x \qquad M_{xy} = D(1 - v)(\partial^2 w/\partial x\,\partial y) = -M_{yx}$$

$$\partial M_y/\partial y + \partial M_{yx}/\partial x = Q_y \qquad M_x = -D(\partial^2 w/\partial x^2 + v\partial^2 w/\partial y^2)$$

$$\partial Q_x/\partial x + \partial Q_y/\partial y = -q \qquad M_y = -D(\partial^2 w/\partial y^2 + v\partial^2 w/\partial x^2) \tag{2.228}$$

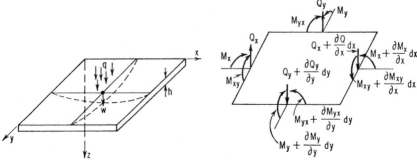

FIG. 2.31 Simply supported plate subjected to normal load q.

FIG. 2.32 Shears, twists, and moments on a plate element.

where $D = Eh^3/12(1 - v^2)$ = plate stiffness and is analogous to the flexural rigidity per unit width (EI) of beam theory.

Properly combining Eqs. (2.228) leads to the basic differential equation of plate theory

$$\nabla^4 w = \partial^4 w/\partial x^4 + 2(\partial^4 w/\partial^2 x\, \partial^2 y) + \partial^4 w/\partial y^4 = q/D \qquad (2.229)$$

This may be compared with the similar equation for beams

$$d^4 y/dx^4 = q/EI \qquad (2.230)$$

The solution of a specific plate problem involves finding a function w which satisfies Eq. (2.229) and the boundary conditions. With w known, the stresses may be evaluated by employing

$$\sigma_{max} = \pm 6M_x/h^2$$

$$\sigma_{max} = \pm 6M_y/h^2 \qquad (2.231)$$

$$\tau_{max} = 6M_{xy}/h^2$$

2.13.2 Boundary Conditions

The usual support conditions and the associated boundary conditions are as follows. Simply supported plate:

$$w = 0 \qquad \text{(zero deflection)}$$

$$M = 0 \qquad \text{(zero moment)}$$

Built-in plate:

$$w = 0 \qquad \text{(zero deflection)}$$

$$w' = 0 \qquad \text{(zero slope)}$$

Free boundary:

$$M = 0 \quad \text{(zero moment)}$$

$$V = 0 \quad \text{(zero reactive shear)}$$

where it is noted that

$$V_x = Q_x - \partial M_{xy}/\partial y = -D[\partial^3 w/\partial x^3 + (2 - \nu)(\partial^3 w/\partial x\, \partial y^2)]$$

$$V_y = Q_y - \partial M_{xy}/\partial x = -D[\partial^3 w/\partial y^3 + (2 - \nu)(\partial^3 w/\partial y\, \partial x^2)]$$

(2.232)

where positive V is directed similar to positive Q.

Note that two boundary conditions are necessary and sufficient to solve the problem of bending for plates with transverse loads.

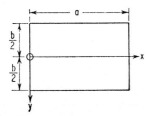

FIG. 2.33   Rectangular plate, simply supported.

EXAMPLE 1   For this problem the loading on a simply supported plate is taken to be distributed uniformly over the surface of that plate, and the coordinate axes are redefined as indicated in Fig. 2.33.

The deflection solution must satisfy Eq. (2.229) and the boundary conditions

$$w = 0 \quad \text{and} \quad \partial^2 w/\partial x^2 = 0 \quad \text{at } x = 0, a$$

$$w = 0 \quad \text{and} \quad \partial^2 w/\partial y^2 = 0 \quad \text{at } y = \pm b/2$$

A series solution of the form

$$w = \sum_{m=1}^{\infty} Y_m \sin \frac{m\pi x}{a} \tag{2.233}$$

is assumed, and after manipulation, the following general expression results for the deflection surface:

$$w = \frac{4qa^4}{\pi^5 D} \sum_{m=1,3,\ldots}^{\infty} \frac{1}{m^5} \left(1 - \frac{\alpha_m \tanh \alpha_m + 2}{2 \cosh \alpha_m} \cosh \frac{2\alpha_m y}{b} \right.$$

$$\left. + \frac{\alpha_m}{2 \cosh \alpha_m} \frac{2y}{b} \sinh \frac{2\alpha_m y}{b} \right) \sin \frac{m\pi x}{a} \tag{2.234}$$

where $\alpha_m = m\pi b/a$. The maximum deflection occurs at $x = a/2$ and $y = 0$:

$$w_{\max} = \frac{4qa^4}{\pi^5 D} \sum_{m=1,3,\ldots}^{\infty} \frac{(-1)^{(m-1)/2}}{m^5} \left(1 - \frac{\alpha_m \tanh \alpha_m + 2}{2 \cosh \alpha_m} \right) = \frac{\alpha q a^4}{D} \tag{2.235}$$

where $\alpha = \alpha(b/a)$. The maximum moments $M_{x\,\max}$ and $M_{y\,\max}$ may be expressed by $\beta q a^2$ and $\beta_1 q a^2$, respectively, where β and β_1 are functions of b/a.

The final group of plate problems deals with edge reactions which are distributed along the edge and the concentrated forces at the corners to keep the corners of the plate from rising, as shown in Fig. 2.34.

Note that the V_x and V_y are negative in sign, whereas R is positive and directed down in accordance with the adopted sign convention for plates.

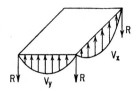

FIG. 2.34 Edge reactions of simply supported plate.

The maximum value of V_x, V_y, and R may be expressed by δqa, $\delta_1 qa$, and ηqa, respectively, where δ, δ_1, and η have the same basic functional relation as the earlier defined coefficients.

Figure 2.35 presents curves of the important plate coefficients.

EXAMPLE 2 The solution for the simple-support uniform-load problem as illustrated above is used as the basis for the solution for a built-in-edge problem. Transposing coordinates as in Fig. 2.36,

$$w = \frac{4qa^4}{\pi^5 D} \sum_{m=1,3,\ldots}^{\infty} \frac{(-1)^{(m-1)/2}}{m^5} \cos \frac{m\pi x}{a} \left(1 - \frac{\alpha_m \tanh \alpha_m + 2}{2 \cosh \alpha_m} \cosh \frac{m\pi y}{a}\right.$$
$$\left. + \frac{1}{2 \cosh \alpha_m} \frac{m\pi y}{a} \sinh \frac{m\pi y}{a}\right) \quad (2.236)$$

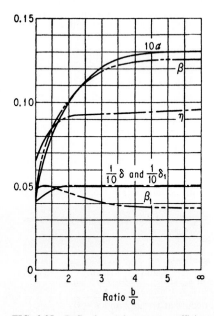

Ratio $\frac{b}{a}$

FIG. 2.35 Deflection and moment coefficients for rectangular plate on simple supports.

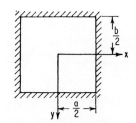

FIG. 2.36 Rectangular plate with built-in edges.

To satisfy the edge restrictions imposed by built-in supports, the deflection of a plate with moments distributed along the edges is superposed on this simple-support solution. These moments are then adjusted to satisfy the built-in boundary condition $\partial w/\partial n = 0$. The procedure finally results in a solution for the maximum deflection at the center of the plate of the form

$$w_{max} = \alpha' qa^4/D \quad (2.237)$$

The coefficient α' can be evaluated as a function of b/a.

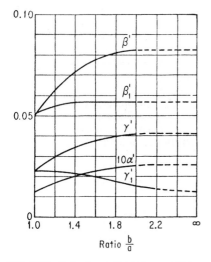

FIG. 2.37 Deflection and moment coefficients for rectangular plate with built-in edges.

Similar expressions for the moments at the center of the plate and at two edges are as follows:

Edge: $\quad M_x = -\beta' qa^2 \quad M_y = -\beta'_1 qa^2$

$$(2.238a)$$

Center: $\quad M_x = \gamma' qa^2 \quad M_y = \gamma'_1 qa^2$

$$(2.238b)$$

Figure 2.37 presents α', β', β'_1, γ', γ'_1, in graphical form.

EXAMPLE 3 We shall use a numerical method to solve the simple-support uniform-load problem. The following expressions define the finite-difference form for the second-order partial derivatives in question (see Fig. 2.38).

$$\frac{\partial^2 w}{\partial x^2}\bigg|_i = \frac{w_{i+1} - 2w_i + w_{i-1}}{h^2} \qquad \frac{\partial^2 w}{\partial y^2}\bigg|_i = \frac{w_{i'+1} - 2w_i + w_{i'-1}}{k^2} \qquad (2.239)$$

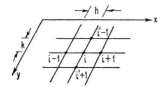 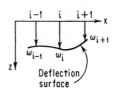

FIG. 2.38 Grid of rectangular plate, simply supported.

For the grid, $h = k$,

$$\nabla^2 w = (1/h^2)(w_{i+1} + w_{i'+1} + w_{i-1} + w_{i'-1} - 4w_i) \qquad (2.240)$$

Utilizing

$$\phi = (\nabla^2 w)D \qquad (2.241)$$

and $\qquad \nabla^2 \phi = (1/h^2)(\phi_{i+1} + \phi_{i'+1} + \phi_{i-1} + \phi_{i'-1} - 4\phi_i) \qquad (2.242)$

The governing equation is

$$\nabla^2 \phi = q \qquad (2.243)$$

Consider now a square plate coarsely divided into four segments. Then evidently $h = k = a/2$ and $w_{i+1} = w_{i'+1} = w_{i-1} = w_{w'-1} = 0$ since these points are on the boundary. Also $\nabla^2 w_{i+1} = \nabla^2 w_{i'+1} = 0$ and, using the last of the above equations, we find

$$\phi_{i+1} = \phi_{i'+1} = \phi_{i-1} = \phi_{i'-1} = 0$$

Therefore, $\qquad\qquad\qquad\qquad \nabla^2 \phi = -4\phi_i/h^2 = q$ $\qquad\qquad\qquad$ (2.244)

Substituting back into the original equation yields

$$\nabla^2 w_i = \phi_i/D = -4w_i/h^2 = -qa^2/16D \qquad\qquad (2.245)$$

$$w_i = (1/256)(qa^4/D) \qquad\qquad (2.246)$$

This is approximately 3.5 percent below the maximum deflection obtained by analytical means:

$$w_{max} = 0.00406(qa^4/D) \qquad\qquad (2.247)$$

Thus, even with this crude grid, the center deflection can be obtained with reasonable accuracy. Since the stress is obtained by combinations of higher derivatives, however, a finer grid is required for a correspondingly accurate stress evaluation.

2.14 SHELL THEORY[3,10]

2.14.1 Membrane Theory: Basic Equation

The general problem of determining stresses in shells may be subdivided into two distinct categories. The first of these is a moment-free "membrane" state of stress, which often predominates over a major portion of the shell. The second, associated with bending effects, is the "discontinuity stress," which affects the shell for a limited segment in the vicinity of a load or profile discontinuity. The final solution often consists of the membrane solution, corrected locally in the regions of the boundaries for the discontinuity effects.

The "membrane solution" for shells in the form of surfaces of revolution, and loaded symmetrically with respect to the axis, often requires the following simplifying assumptions:

1. The thickness h of the shell is small compared with other dimensions of the shell, and the radii of curvature.

2. Linear elements which are normal to the undisturbed middle surface remain straight and normal to the deflected middle surface of the shell.

3. Bending stresses are small and can be neglected; so that only the direct stresses due to strains in the middle surface need be considered.

The basic element for a shell of revolution when subjected to axisymmetric loading is shown in Fig. 2.39. From conditions of static equilibrium the following equations may be derived:

$$(d/d\phi)(N_\varphi r_0) - N_\theta r_1 \cos \phi + Yr_1 r_0 = 0 \qquad\qquad (2.248)$$

$$N_\varphi r_0 + N_\theta r_1 \sin \phi + Zr_1 r_0 = 0 \qquad\qquad (2.249)$$

where Y and Z are the components of the external load parallel to the two-coordinate axis, respectively. Thus, in general, if these components of the load are known and the geometry of the shell is specified, the forces N_ϕ and N_θ can be calculated.

If now the conditions of static equilibrium are applied to a portion of a shell above a parallel circle instead of an element as indicated in Fig. 2.40, then the following equations result:

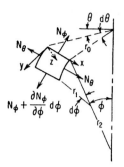

FIG. 2.39   Element of shell of revolution.

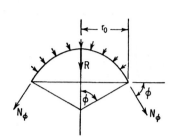

FIG. 2.40   Shell portion above parallel circle.

$$2\pi r_0 N_\phi \sin \phi + R = 0 \tag{2.250}$$

$$N_\phi/r_1 + N_\theta/r_2 = -Z \tag{2.251}$$

where R is the resultant of the total load on that part of the shell in question. Again these two equations suffice for the determination of the two forces N_ϕ and N_θ.

The associated principal stresses for these members are then

$$\sigma_\phi = N_\phi/h \quad \text{and} \quad \sigma_\theta = N_\theta/h \tag{2.252}$$

To complete the consideration of membrane action for shells with axisymmetric loading, the concomitant displacements must be computed. The procedure indicated below applies for symmetrical deformations, in which the displacement (along a meridian) is indicated by v and the displacement in the radial direction is w (with an inward displacement taken as positive).

From geometry and Hooke's law,

$$dv/d\phi - v \cot \phi = f(\phi) \tag{2.253}$$

where $\qquad f(\phi) = (1/Eh)[N_\phi(r_1 + vr_2) - N_\theta(r_2 + vr_1)]$

The solution of Eq. (2.253) is

$$v = \sin \phi \{\int [f(\phi)/\sin \phi] d\phi + C\} \tag{2.254}$$

The constant C is evaluated from the boundary conditions of the problem. The radial displacement is then determined from the following:

$$w = v \cot \phi - (r_2/EH)(N_\theta - vN_\phi) \tag{2.255}$$

A similar analysis on a cylindrical shell without the restriction of symmetric loading will yield the basic equations

$$\partial N_x/\partial x + (1/r)(\partial N_{x\phi}/\partial \phi) = -X \tag{2.256}$$

$$\partial N_{x\phi}/\partial x + (1/r)(\partial N_\phi/\partial \phi) = -Y \tag{2.257}$$

$$N_\phi = -Zr \tag{2.258}$$

where the coordinate axes are redefined in Fig. 2.41 to be consistent with usual practice and Eqs. (2.254) and (2.255) are suitably modified.

2.14.2 Example of Spherical Shell Subjected to Internal Pressure

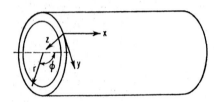

For a hemispherical section of radius a, the forces N_ϕ distributed along the edges are required to maintain equilibrium. These may be determined from the previous equations. Now

$$2\pi a N_\phi \sin \phi + R = 0 \qquad (2.259)$$

FIG. 2.41 Cylindrical shell.

where $\qquad R = -\pi a^2 p \qquad \sin \phi = \sin (\pi/2) = 1$

so that $\qquad N_\phi = pa/2 \qquad (2.260)$

From the other equation of equilibrium,

$$N_\phi/a + N_\theta/a = -Z = p \qquad (2.261)$$

where $\qquad N_\theta = pa/2$

The stresses may now be written

$$\sigma_\phi = \sigma_\theta = pa/2h \qquad (2.262)$$

From loading symmetry, it is concluded that $v = 0$ and the increase in the radial direction is given by

$$w = (-a/Eh)(N_\theta - vN_\phi) \qquad (2.263)$$

$$w = (-pa^2/2Eh)(1 - v) \qquad (2.264)$$

2.14.3 Example of Cylindrical Shell Subjected to Internal Pressure

Using a procedure similar to that outlined above, the following relationships are derived. For open-ended cylinder:

$$\sigma_\phi = pa/h \qquad (2.265)$$

$$w = -pa^2/Eh \qquad (2.266)$$

For closed-ended cylinder:

$$\sigma_\phi = pa/h \qquad (2.267)$$

$$\sigma_x = pa/2h \qquad (2.268)$$

$$w = -(pa^2/Eh)(1 - v/2) \qquad (2.269)$$

2.14.4 Discontinuity Analysis

The membrane solution does not usually satisfy all "edge" conditions, and it is therefore often necessary to superpose a second, "edge-loaded" shell in order to obtain a *complete* solution.

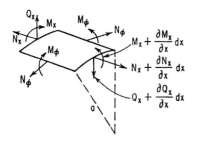

FIG. 2.42 Cylindrical-shell element.

To develop the general equations for the cylindrical shell with axisymmetric load, consider an element as shown in Fig. 2.42, where *bending moments are assumed acting*. The coordinate system is as defined in Fig. 2.41. All forces and moments are shown in their positive directions. Referring to Fig. 2.42, Q_x is the shear force per unit length, M_x is the axial moment per unit length, M_ϕ is the circumferential moment per unit length, and N_x, N_ϕ are the normal forces defined in the discussion of membrane theory.

Applying the equations of equilibrium and Hooke's law and expressing the curvature as a function of moment (as was done in developing the plate equations), the following basic shell differential equation is obtained:

$$d^4w/dx^4 + 4\beta^4w = Z/D \qquad (2.270)$$

where $D = Eh^3/12(1 - v^2)$
$\beta^4 = 3(1 - v^2)/a^2h^2$
w = radial deflection of shell (positive inward)

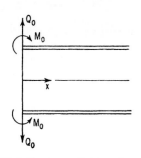

FIG. 2.43 Long-cylinder section.

The solution of the above equation in any particular case depends on the specific boundary conditions at the ends of the cylinder.

One very useful solution is for the case of a long cylinder, without radial pressure, subjected to uniformly distributed forces and moments along the edge, $x = 0$. The assumed positive directions of these loads are as shown in Fig. 2.43. The resultant expression for the deflection is

$$w = (e^{-\beta x}/2\beta^3D)[\beta M_0(\sin \beta x - \cos \beta x) - Q_0 \cos \beta x] \qquad (2.271)$$

The maximum deflection occurs at the loaded end and is evaluated as

$$w_{x=0} = -(1/2\beta^3D)(\beta M_0 + Q_0) \qquad (2.272)$$

The accompanying slope at the loaded end is

$$w'_{x=0} = (1/2\beta^2D)(2\beta M_0 + Q_0) = (dw/dx)_{x=0} \qquad (2.273)$$

The successive derivatives of the above expression for deflection can be written in the following simplified form:

$$w = -(1/2\beta^3D)[\beta M_0\psi(\beta x) + Q_0\theta(\beta x)]$$

$$w' = (1/2\beta^2D)[2\beta M_0\theta(\beta x) + Q_0\phi(\beta x)] \qquad (2.274)$$

$$w'' = -(1/2\beta D)[2\beta M_0\phi(\beta x) + 2Q_0\xi(\beta x)]$$

$$w''' = (1/D)[2\beta M_0\xi(\beta x) - Q_0\psi(\beta x)]$$

where $\qquad\qquad \phi(\beta x) = e^{-\beta x}(\cos \beta + \sin \beta x)$

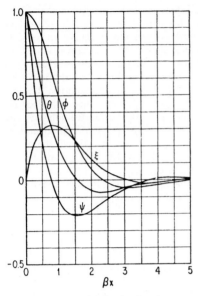

FIG. 2.44   Slope and deflection functions.

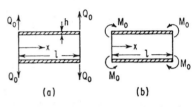

FIG. 2.45   Short shell. (*a*) Bending by shears.
(*b*) Bending by moments.

$$\psi(\beta x) = e^{-\beta x}(\cos \beta x - \sin \beta x)$$

$$\theta(\beta x) = e^{-\beta x} \cos \beta x$$

$$\zeta(\beta x) = e^{-\beta x} \sin \beta x$$

Figure 2.44 is a plot of ϕ, ψ, θ, ζ as a function of βx. Because each function decreases in absolute magnitude with increasing βx, in most engineering applications the effect of edge loads may be neglected at locations for which $\beta x > \pi$.

For the short shell, for which opposite end conditions interact, the following results are obtained. For the case of bending by uniformly distributed shearing forces, as shown in Fig. 2.45*a*, the slope and deflection are given by

$$w_{x=0,l} = -(2Q_0\beta a^2/Eh)\chi_1(\beta l) \tag{2.275}$$

$$w'_{x=0,l} = \pm(2Q_0\beta^2 a^2/Eh)\chi_2(\beta l) \tag{2.276}$$

where $$\chi_1(\beta l) = (\cosh \beta l + \cos \beta l)/(\sinh \beta l + \sin \beta l)$$

$$\chi_2(\beta l) = (\sinh \beta l - \sin \beta l)/(\sinh \beta l + \sin \beta l)$$

For the case of bending by uniformly distributed moments M_0 (as shown in Fig. 2.45*b*), the slope and deflection are given by

$$w_{x=0,l} = -(2M_0\beta^2 a^2/Eh)\chi_2(\beta l) \tag{2.277}$$

$$w'_{x=0,l} = \pm(4M_0\beta^3 a^2/Eh)\chi_3(\beta l) \tag{2.278}$$

where $$\chi_3(\beta l) = (\cosh \beta l - \cos \beta l)/(\sinh \beta + \sin \beta l)$$

Figure 2.46 is a plot of the functions χ_1, χ_2, and χ_3 as a function of βl. The axial bending moment at any location is given by

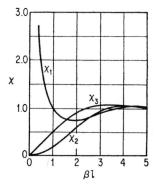

FIG. 2.46 Slope and deflection functions.

$$M_x = -Dw'' \qquad (2.279)$$

and the maximum stress which occurs at the inside and outside surfaces of the shell,

$$\sigma_{x,\text{bending}} = \pm(6M_x/h^2) \qquad (2.280)$$

Likewise the shear force at any point is given by

$$Q_x = -Dw''' \qquad (2.281)$$

and the associated maximum shear stress is $\tau_x = 3Q_x/2h$, which occurs midway between the inner and outer surfaces; the shear stress at the surface is zero.

EXAMPLE Examine the case of a long cylinder subjected to an internal pressure and fixed at the ends as depicted in Fig. 2.47a; axial pressure is taken to be zero.

   The stress and deformation of this shell can be obtained by the superposition of two distinct problems, the membrane and edge-loaded cylinders. The first presupposes free ends and a membrane action as indicated in Fig. 2.47b. The built-in ends resist this membrane deflection at the edges through a system of forces Q_0 and moments M_0 which are required to enforce the boundary conditions of zero deflection and rotation, as shown in Fig. 2.47c.

   The increase in the radius due to membrane action, as a result of pressure, is then obtained from the membrane solution

$$-w_p = \delta = pa^2/Eh \qquad (2.282)$$

The boundary conditions for the edge-loaded problem, based on the actual built-in ends, become

$$w_x = \delta \qquad \text{and} \qquad w'_{x=0} = 0$$

Hence

$$\delta = (1/2\beta^3 D)(\beta M_0 + Q_0) \qquad (2.283)$$

$$0 = (1/2\beta^2 D)(2\beta M_0 + Q_0) \qquad (2.284)$$

Solving Eqs. (2.283) and (2.284), using Eq. (2.282)

$$M_0 = p/2\beta^2 \qquad \text{and} \qquad Q_0 = -p/\beta \qquad (2.285)$$

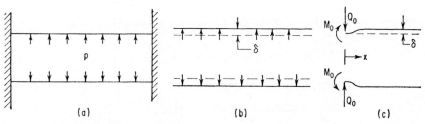

FIG. 2.47 Long cylinder with fixed ends. (a) Action of internal pressure. (b) Membrane action. (c) Discontinuity forces for boundary conditions.

Thus the *complete* solution for the deflection is

$$w = -(1/4\beta^4 D)[p\psi(\beta x) - 2p\theta(\beta x)] - pa^2/Eh \tag{2.286}$$

The axial stresses are given by

$$\sigma_x = \pm(6M_x/h^2) \tag{2.287}$$

where

$$M_x = -Dw'' = (p/2\beta^2)[\phi(\beta x) - 2\xi(\beta x)]$$

The mean circumferential stress can be evaluated from

$$\sigma_{\phi,\text{direct}} = -Ew/a \tag{2.288}$$

and the added component of flexural stress due to the Poisson effect is

$$\sigma_{\phi,\text{bending}} = -v\sigma_{x,\text{bending}} \tag{2.289}$$

so that

$$\sigma_{\phi,\text{total}} = \sigma_{\phi,\text{direct}} + \sigma_{\phi,\text{bending}} \tag{2.290}$$

2.15 CONTACT STRESSES: HERTZIAN THEORY[2]

As discussed in the writings of Hertz (the contact stresses presented here are often termed hertzian stresses), the maximum pressure q due to a compressive force P is given by

$$q = 3P/2\pi a^2 \tag{2.291}$$

and is taken to have a spherical distribution as shown in Fig. 2.48, where

$$a = \sqrt[3]{\frac{3P}{4}\frac{R_1 R_2}{R_1 + R_2}\left(\frac{1 - v_1^2}{E_1} + \frac{1 - v_2^2}{E_2}\right)} \tag{2.292}$$

These expressions may be simplified, if both spheres are composed of identical materials. For Poisson's ratio v of approximately 0.3, which is common to steel, iron, aluminum, and most structural materials, there results

$$a = 1.11\sqrt[3]{(P/E)[R_1 R_2/(R_1 + R_2)]} \tag{2.292a}$$

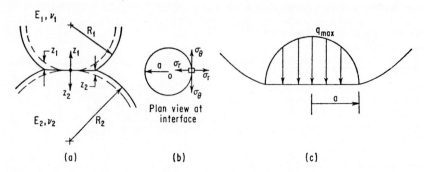

(a)   (b)   (c)

FIG. 2.48 Two spheres in contact. (*a*), (*b*), (*c*) Contact pressure distribution.

and
$$q = 0.388\sqrt[3]{PE^2[(R_1 + R_2)/R_1 R_2]^2}$$
(2.292b)

The general stress levels in the spheres can now be presented based on the above relations. Maximum compressive stress, which occurs at point O, is

$$\sigma_z = -q$$
(2.293)

The maximum tensile stress in the radial direction, which occurs on the periphery of the surface of contact at radius a, is

$$\sigma_r = [(1-2v)/3]q$$
(2.294)

Maximum shear stress, which occurs under point O of Fig. 2.48b, at a depth

$$z_1 = 0.47a$$

is approximately

$$\tau = \frac{1}{3}q$$
(2.295)

and is in a plane inclined to the z axis. This latter stress is usually the governing criterion in the design for bodies in contact, fabricated from ductile materials. A compilation of important contact-stress cases is given in Table 2.2. Other important cases, associated with rolling-element bearings, are discussed in Sec. 19.

2.16 FINITE-ELEMENT NUMERICAL ANALYSIS[14,15,16]

2.16.1 Introduction

The section thus far has dealt with exact solutions to sets of equations which predict the deformation and internal stress distribution of particular bodies such as beams, columns, thin plates, and thin shells. These sets of equations are derived from the same concepts (equilibrium, kinematics, material observation) used to derive the general equations of elastic theory [Eqs. (2.47), (2.76), and (2.87)]. The particular equations for beams, plates, and shells are, in fact, derivable from the general equation by imposing the appropriate limitations with respect to thickness, etc.

In dealing with the more complex structural shapes typical of actual machinery it becomes increasingly difficult to derive appropriate sets of equations for which exact solutions may be found. For such structures predictions of deformation and stress can be effected through numerical solutions of the general equations. Three general numerical techniques are widely used to effect numerical solutions in structural mechanics: transfer-matrix techniques, finite-difference techniques, and finite-element techniques.

Transfer-matrix approaches are typically used to effect solutions in one-dimensional structures. These approaches are widely used in the solution of vibrating shafts and turbine foils. Typical transfer-matrix approaches include, among others, the Holzer method for torsional vibration and the Prohl-Miklestad (Sec. 5) procedure for lateral shaft vibration.

Finite-difference methods[17] are widely used to effect deformation and stress solutions to multidimensional structures. In this technique, the differential operators of the governing equations are replaced with difference operators which relate the values of the unknowns at a gridwork of points in the structure. Example 3 of Sec. 2.13.2 presents a simple illustration of a finite-difference method. The well-known relaxation

TABLE 2.2 Contact Stresses[2.5,12]

Sphere on a sphere,
P = total load

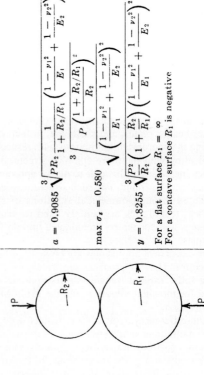

$$a = 0.9085 \sqrt[3]{PR_2 \frac{1}{1+R_2/R_1}\left(\frac{1-\nu_1^2}{E_1}+\frac{1-\nu_2^2}{E_2}\right)}$$

$$\max \sigma_z = 0.580 \sqrt[3]{\frac{P\left(\dfrac{1+R_2/R_1}{R_2}\right)^2}{\left(\dfrac{1-\nu_1^2}{E_1}+\dfrac{1-\nu_2^2}{E_2}\right)^2}}$$

$$y = 0.8255 \sqrt[3]{\frac{P^2}{R_2}\left(1+\frac{R_2}{R_1}\right)\left(\frac{1-\nu_1^2}{E_1}+\frac{1-\nu_2^2}{E_2}\right)^2}$$

For a flat surface $R_1 = \infty$
For a concave surface R_1 is negative

y is the decrease in center-to-center distance between the two spheres.

2.64

Cylinder on cylinder axes, parallel, P = load/linear in.

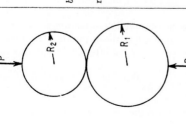

$$b = 1.13 \sqrt{PR_2 \frac{1}{1 + R_2/R_1} \left(\frac{1 - \nu_1^2}{E_1} + \frac{1 - \nu_2^2}{E_2} \right)}$$

$$\max \sigma_z = 0.564 \sqrt{\frac{(P/R_2)(1 + R_2/R_1)}{\dfrac{1 - \nu_1^2}{E_1} + \dfrac{1 - \nu_2^2}{E_2}}}$$

General case of two bodies in contact, P = total pressure

At point of contact minimum and maximum radii of curvature are R_1 and R_1' for body 1, R_2 and R_2' for body 2. Then $1/R_1$ and $1/R_1'$ are *principal curvatures* of body 1, and $1/R_2$ and $1/R_2'$ of body 2, and in each body the principal curvatures are mutually perpendicular. The plane containing curvature $1/R_1$ in body 1 makes with the plane containing curvature $1/R_2$ in body 2 the angle ϕ. Then

$$\max \sigma_z = 1.5P/\pi cd, \quad c = \alpha \sqrt[3]{P\delta/K}, \quad d = \beta \sqrt[3]{P\delta/K}, \text{ and } y = \lambda \sqrt[3]{P^2/K^2\delta}, \text{ where } \delta = \frac{4}{1/R_1 + 1/R_2 + 1/R_1' + 1/R_2'}$$

$$\text{and } K = \frac{8}{3} \frac{E_1 E_2}{E_2(1 - \nu_1^2) + E_1(1 - \nu_2^2)}$$

α and β are given by the following table, where

$$\theta = \arccos \tfrac{1}{4}\delta \sqrt{(1/R_1 - 1/R_1')^2 + (1/R_2 - 1/R_2')^2 + 2(1/R_1 - 1/R_1')(1/R_2 - 1/R_2')\cos 2\phi}$$

θ	0°	10°	20°	30°	35°	40°	45°	50°	55°	60°	65°	70°	75°	80°	85°	90°
α	∞	6.612	3.778	2.731	2.397	2.136	1.926	1.754	1.611	1.486	1.378	1.284	1.202	1.128	1.061	1.00
β	0	0.319	0.408	0.493	0.530	0.567	0.604	0.641	0.678	0.717	0.759	0.802	0.846	0.893	0.944	1.00
λ	—	0.851	1.220	1.453	1.550	1.637	1.709	1.772	1.828	1.875	1.912	1.944	1.967	1.985	1.996	2.00

y is the decrease in center-to-center distance of the two cylinders. b is the half-width of the contact surface.

method for the solution of the governing equation of multidimensional heat-transfer analysis is an example of a finite-difference solution.

Since the 1960s finite-element methods have become the preeminent tool for the numerical solution of deformation and stress problems in structural mechanics. This popularity arises from the ease with which the most general of structural geometries can be considered. Finite-element analysis replaces the exact structure to be considered with a set of simple structural elements (blocks, plates, shells, etc.) interconnected at a finite set of node points. The set of governing equations for this approximate structure can be solved exactly.

Finite-element analysis deals with the spatial approximation of complex structural shapes. It can be used directly to yield solutions in static elasticity or combined with other numerical techniques to obtain the response of structures with nonlinear material properties (plasticity, creep, relaxation), undergoing finite deformation, or subject to shock and vibration excitation. The finite-element technique has also found a wide application in the analysis of heat transfer and fluid flow in complex multidimensional applications.

Many users of the finite-element method do so through the application of large-scale, general-purpose computer codes.[18–20] These codes are widely available, highly user-oriented, and simple to use. They are also easy to misuse. The consequences of misuse are excess expense and, more important, invalid predictions of the state of stress and deformation of the structure due to the applied loading.

The discussion herein introduces the process of finite-element analysis to enable the prospective user to have a suitable understanding of the calculations being performed. The discussion is limited to the *stiffness* approach, which is the most widely used basis of finite elements. Further, for ease of understanding, the presentation deals with structures of two dimensions. The generalization to three dimensions follows directly.

2.16.2 The Concept of Stiffness

The governing equations of the theory of elasticity relate the loads applied externally to a body to the resulting deformation of that body, using the stress equilibrium equations to relate external forces to internal forces, i.e., stresses. The stress-strain relations relate internal forces to internal strains. The strain-displacement equations relate internal strains to observed deformations. The whole solution process can be stated by the relationship

$$F = k\delta \tag{2.296}$$

where F = externally applied forces
δ = observed deformation
k = stiffness of the structure

Thus, all the material and geometric information for the structure is contained in the stiffness term.

Numerical procedures involve relating the observed deformation at a discrete number of points of the body to the forces applied at these points. The relationship between force and displacement is then expressed most effectively in terms of matrix notation.

$$\{F\} = [k]\{\delta\} \tag{2.297}$$

Thus F becomes the vector of applied forces, δ the vector of displacement response, and k the stiffness matrix of the structure.

Stiffness of Simple Discrete Elements. For simple structures, relationships of the type of Eq. (2.297) can be derived directly.

EXAMPLE 1 The governing equation for the simple spring in Fig. 2.49 is

$$F_1 = k(\delta_1 - \delta_2)$$
$$F_2 = k(\delta_2 - \delta_1)$$

which in matrix notation becomes

$$\left\{ \begin{array}{c} F_1 \\ F_2 \end{array} \right\} = \left[\begin{array}{cc} k & -k \\ -k & k \end{array} \right] \left\{ \begin{array}{c} \delta_1 \\ \delta_2 \end{array} \right\} \tag{2.298}$$

where k is the stiffness of the spring, F_1 and F_2 the applied forces at nodes 1 and 2, and δ_1 and δ_2 the resulting displacements at nodes and 1 and 2.

EXAMPLE 2 The truss element in Fig. 2.50 is limited to stretching-compression response under its applied loads. The general governing equation is

$$F = (AE/L)\, \Delta L$$

where A is the cross-sectional area, E is Young's modulus, and L is the length. ΔL is the change in L under the action of the forces. A series of relationships of the form

$$F_i = k_{ij}\, \delta_j$$

may be developed where F_i is the force at node i and δ_j is the displacement of node j. The resulting set of relationships is

$$\left\{ \begin{array}{c} F_{1x} \\ F_{1y} \\ F_{2x} \\ F_{2y} \end{array} \right\} = \frac{AE}{L} \left[\begin{array}{cccc} \cos^2 \alpha & \cos \alpha \sin \alpha & -\cos^2 \alpha & -\cos \alpha \sin \alpha \\ \cos \alpha \sin \alpha & \sin^2 \alpha & -\cos \alpha \sin \alpha & -\sin^2 \alpha \\ -\cos^2 \alpha & -\cos \alpha \sin \alpha & \cos^2 \alpha & \cos \alpha \sin \alpha \\ -\cos \alpha \sin \alpha & -\sin^2 \alpha & \cos \alpha \sin \alpha & \sin^2 \alpha \end{array} \right] \left\{ \begin{array}{c} \delta_{1x} \\ \delta_{1y} \\ \delta_{2x} \\ \delta_{2y} \end{array} \right\} \tag{2.299}$$

which is the form of Eq. (2.297). In general terms, Eq. (2.299) has the form

$$\left\{ \begin{array}{c} F_{1x} \\ F_{1y} \\ F_{2x} \\ F_{2y} \end{array} \right\} = \left[\begin{array}{cccc} k_{11} & k_{12} & k_{13} & k_{14} \\ k_{21} & k_{22} & k_{23} & k_{24} \\ k_{31} & k_{32} & k_{33} & k_{34} \\ k_{41} & k_{42} & k_{43} & k_{44} \end{array} \right] \left\{ \begin{array}{c} \delta_{1x} \\ \delta_{1y} \\ \delta_{2x} \\ \delta_{2y} \end{array} \right\} \tag{2.300}$$

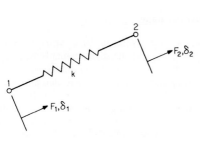

FIG. 2.49 Simple spring finite element.

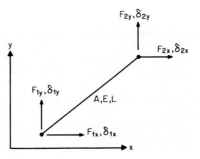

FIG. 2.50 Tension-compression finite element.

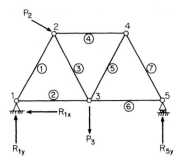

FIG. 2.51   Finite-element model of a simple truss.

Stiffness of a Complex Structure.   The simple structures of these examples often comprise the elements of a more complex structure. Thus, the governing equation of the truss of Fig. 2.51 can be developed by combining the relationships of the individual truss elements from Example 2. One simple procedure is to insert the stiffness contribution from each row and column of each truss element to the stiffness of the appropriate row and column of the complex structure. For the truss of Fig. 2.51 the resulting relationship is shown in Eq. (2.301).

$$
\begin{Bmatrix} R_{1x} \\ R_{1y} \\ P_{2x} \\ P_{2y} \\ 0 \\ P_{3y} \\ 0 \\ 0 \\ 0 \\ R_{5y} \end{Bmatrix}
=
\begin{bmatrix}
k_{11}^1+k_{11}^2 & k_{12}^1+k_{12}^2 & k_{13}^1 & k_{14}^1 & k_{13}^2 & k_{14}^2 & 0 & 0 & 0 & 0 \\[4pt]
k_{21}^1+k_{21}^2 & k_{22}^1+k_{22}^2 & k_{23}^1 & k_{24}^1 & k_{23}^2 & k_{24}^2 & 0 & 0 & 0 & 0 \\[4pt]
k_{31}^1 & k_{32}^1 & k_{33}^1+k_{11}^3+k_{11}^{14} & k_{34}^1+k_{12}^3+k_{12}^4 & k_{13}^3 & k_{14}^3 & k_{13}^4 & k_{14}^4 & 0 & 0 \\[4pt]
k_{41}^1 & k_{42}^1 & k_{43}^1+k_{21}^3+k_{21}^4 & k_{44}^1+k_{22}^3+k_{22}^4 & k_{23}^3 & k_{24}^3 & k_{23}^4 & k_{24}^4 & 0 & 0 \\[4pt]
k_{31}^2 & k_{32}^2 & k_{31}^3 & k_{32}^3 & k_{33}^2+k_{33}^3+k_{11}^5+k_{11}^6 & k_{34}^2+k_{34}^3+k_{12}^5+k_{12}^6 & k_{13}^5 & k_{14}^5 & k_{13}^6 & k_{14}^6 \\[4pt]
k_{41}^2 & k_{42}^2 & k_{41}^3 & k_{42}^3 & k_{43}^2+k_{43}^3+k_{21}^5+k_{21}^6 & k_{44}^2+k_{44}^3+k_{22}^5+k_{22}^6 & k_{23}^5 & k_{24}^5 & k_{23}^6 & k_{24}^6 \\[4pt]
0 & 0 & k_{31}^4 & k_{41}^4 & k_{31}^5 & k_{32}^5 & k_{33}^4+k_{33}^5+k_{11}^7 & k_{34}^4+k_{34}^5+k_{12}^7 & k_{13}^7 & k_{14}^7 \\[4pt]
0 & 0 & k_{41}^4 & k_{42}^4 & k_{41}^5 & k_{42}^5 & k_{43}^4+k_{43}^5+k_{21}^7 & k_{44}^4+k_{44}^5+k_{22}^7 & k_{23}^7 & k_{24}^7 \\[4pt]
0 & 0 & 0 & 0 & k_{31}^6 & k_{32}^6 & k_{31}^7 & k_{32}^7 & k_{33}^6+k_{33}^7 & k_{34}^6+k_{34}^7 \\[4pt]
0 & 0 & 0 & 0 & k_{11}^6 & k_{92}^6 & k_{11}^7 & k_{42}^7 & k_{43}^6+k_{43}^7 & k_{44}^6+k_{44}^7
\end{bmatrix}
\begin{Bmatrix} 0 \\ 0 \\ \delta_{2x} \\ \delta_{2y} \\ \delta_{3x} \\ \delta_{3y} \\ \delta_{4x} \\ \delta_{4y} \\ \delta_{5x} \\ 0 \end{Bmatrix}
$$

(2.301)

2.16.3   Basic Procedure of Finite-Element Analysis

Equation (2.301) comprises 10 linear algebraic equations in 10 unknowns. The unknowns include the forces of reaction R_{1x}, R_{1y}, R_{5y} and the displacements δ_{2x}, δ_{2y}, δ_{3x}, δ_{3y}, δ_{4x}, δ_{4y}, δ_{5x}. Many procedures for the solution of sets of simultaneous, linear algebraic equations are available. One well-known approach is Gauss-Jordan elimination.[17]

Once the nodal displacements are known, the forces acting on each truss element can be determined by solution of the set of equations (2.300) applicable to that element.

The procedure used to determine the displacements and internal forces of the truss of Figure 2.51 illustrates the procedure used to determine the response to applied load inherent in the stiffness finite-element procedure. These steps include:

1. Divide the structure into an appropriate number of discrete (or finite) elements connected only at a finite set of points in the structure.

2. Develop a load-deflection relationship of the form of Eq. (2.300) for each finite element.

3. Sum up the load-deflection relationships for each element to obtain the load-deflection relationship for the entire structure, as in Eq. (2.301).

4. Obtain the deformation pattern for the entire structure using conventional procedures.

5. Determine the internal force distribution for each element from the known deformations using the element force-deflection relationships.

The key steps in finite-element analysis are the discretization of the structure and the development of load-deflection relationships for the finite element. The subsequent assembly of the structure load-deflection relationship, the solution of the resulting set of simultaneous algebraic equations, and the subsequent determination of internal forces are straightforward mechanical procedures. Thus, it remains to illustrate the approximations associated with developing finite elements to the analysis of complex structures.

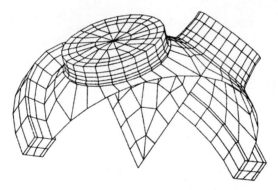

FIG. 2.52   Finite-element model of pressure-vessel head. (*Courtesy of Imo Industries Inc.*)

The truss represents a simplified structure relative to those for which solutions are usually required. A structure such as the pressure-vessel head modeled in Fig. 2.52 is more typical of the component analysis associated with finite-element modeling.

Finite Elements by the Direct Approach.   The direct approach to the development of finite elements requires that a complete set of relationships between the internal and externally applied forces be known a priori. For many structural analyses this is not readily available; i.e., the available equilibrium equations are not sufficient. Therefore, the applicability of the direct approach is limited.

The procedure for development of the load-deflection relationships includes:

1. Define the internal displacement field of the element in terms of the nodal displacements. This requires the assumption of a relationship. Usually polynomial expansions are used.

2. Relate the internal displacement field to the internal force field through the strain-displacement and stress-strain equations.

3. Relate the internal force field to the external forces through the force-equilibrium relations.

4. Combine the results of steps 1–4 to obtain a relationship of the form of Eq. (2.300).

EXAMPLE 1   The beam of Fig. 2.53 has length L, Young's modulus E, and area moment of inertia I. At nodes 1 and 2 it is acted upon by external forces and moments F_1, F_2, $\overline{M}_1$, $\overline{M}_2$. As a result, the nodal displacements and rotations are w_1, w_2, θ_1, θ_2.

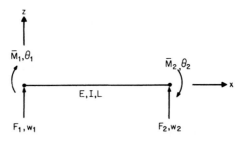

FIG. 2.53   Beam finite element.

Within the beam the deformation pattern is characterized by lateral deflection $w(x)$ and rotation $\theta(x)$, where

$$\theta = -\frac{dw}{dx}$$

Assume that the internal displacement field is governed by the polynomial

$$w(x) = \alpha_1 x^3 + \alpha_2 x^2 + \alpha_3 x + \alpha_4 \tag{2.302}$$

The α's are determined from the boundary conditions on $w(x)$, namely

$$w(0) = w_1$$

$$\theta(0) = \theta_1 = -\frac{dw}{dx}\bigg|_{x=0}$$

$$w(L) = w_2$$

$$\theta(L) = \theta_2 = -\frac{dw}{dx}\bigg|_{x=1}$$

The number of terms in the polynomial expansion for $w(x)$ is, in general, limited to the number of nodal degrees of freedom. With the α's known, Eq. (2.300) becomes

$$w(x) = \frac{1}{L^3}[x^3 \; x^2 \; x \; 1] \begin{bmatrix} 2 & -L & -2 & -L \\ -3L & 2L^2 & 3L & L^2 \\ 0 & -L^3 & 0 & 0 \\ L^3 & 0 & 0 & 0 \end{bmatrix} \tag{2.303}$$

For the special case of a beam, the internal moments are related to the internal displacement field by the Bernoulli-Euler equation

$$M(x) = EI\,(d^2w/dx^2) \tag{2.304}$$

Further, at the nodes

$$M(0) = M_1 = \overline{M}_1 \tag{2.305}$$

$$M(L) = M_2 = -\overline{M}_2 \tag{2.306}$$

For the beam to be in equilibrium under the applied forces and moments it is necessary that

$$F_2L = \overline{M}_1 + \overline{M}_2 \tag{2.307}$$

$$F_1L = -\overline{M}_2 - \overline{M}_1 \tag{2.308}$$

Therefore

$$F_2L = M_1 - M_2 \tag{2.309}$$

$$F_1L = M_2 - M_1 \tag{2.310}$$

Expressing Eqs. (2.305), (2.306), (2.309), (2.310), in matrix format

$$
\begin{Bmatrix} F_1 \\ M_1 \\ F_2 \\ M_2 \end{Bmatrix} =
\begin{bmatrix} -1/L & 1/L \\ 1 & 0 \\ 1/L & -1/L \\ 0 & -1 \end{bmatrix}
\begin{Bmatrix} M_1 \\ M_2 \end{Bmatrix} \tag{2.311}
$$

Combining Eqs. (2.303), (2.304), and (2.311) yields

$$
\begin{Bmatrix} F_r \\ M_1 \\ F_2 \\ M_1 \end{Bmatrix} = \frac{2EI}{L^3}
\begin{bmatrix} 6 & -3L & -6 & -3L \\ -3L & 2L^2 & 3L & L^2 \\ -6 & 3L & 6 & 3L \\ -3L & L^2 & 3L & 2L^2 \end{bmatrix}
\begin{Bmatrix} w_1 \\ \theta_1 \\ w_2 \\ \theta_2 \end{Bmatrix} \tag{2.312}
$$

which is the required load/deflection relationship.

Finite Elements by Energy Minimization. The principle of stationary potential energy states that, for equilibrium to be ensured, the total potential energy must be stationary with respect to variations of admissible displacement fields. An "admissible displacement field" is one which satisfies the natural boundary conditions of the structure, typically those boundary conditions that constrain displacements and slopes. The exact displacement field will result in the minimum value of potential energy.

This energy principle allows the development of a general load-deflection relationship which, in turn, allows the development of a wide variety of finite elements directly from the assumed displacement field. The total potential energy is, in general, defined by

$$\Pi(u, v, w) = U(u, v, w) - V(u, v, w) \tag{2.313}$$

where Π = total potential energy
U = strain energy of deformation
V = work done by applied loads
u, v, w = components of displacement field within the element

For Π to be stationary it is necessary that

$$\frac{\partial \Pi}{\partial u_i} = 0 \qquad \frac{\partial \Pi}{\partial v_i} = 0 \qquad \frac{\partial \Pi}{\partial w_i} = 0 \qquad i = 1, r \qquad (2.314)$$

where the subscript i denotes the ith node of the finite element, and r is the number of nodes. Further, the energy over the volume of the element is

$$U = \int_{\text{vol}} U_0 \, dv \qquad (2.315)$$

$$V = \{\delta\}^T \{F\} \qquad (2.316)$$

where U_0 = strain energy of a unit volume of material
    $\{F\}$ = matrix of nodal forces on the element
    $\{\delta\}$ = matrix of nodal displacements

If we further express the stress-strain and strain-displacement equations [Eqs. (2.76) and (2.47)] in matrix format:

$$\{\sigma\} = [D]\{\epsilon\} \qquad (2.317)$$

$$\{\epsilon\} = [B]\{\delta\} \qquad (2.318)$$

where the element $[B]$ are differential operators, then

$$U_0 = \tfrac{1}{2}\{\epsilon\}[D]\{\epsilon\} \qquad (2.319)$$

Combining Eqs. (2.313) to (2.319) and performing the indicated operations leads to a relationship of the form

$$\{F\} = \int_{\text{vol}} [B]^T [D][B] \, dv \, \{\delta\} \qquad (2.320)$$

Equation (2.320) constitutes a general load-deflection relationship which can be particularized to define a wide variety of finite elements.

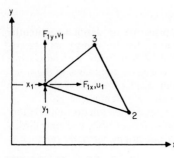

FIG. 2.54   Planar finite element.

EXAMPLE 1   The displacement field within the triangular element in Fig. 2.54 is assumed to be

$$u = \alpha_1 + \alpha_2 x + \alpha_3 y$$
$$v = \alpha_4 + \alpha_5 x + \alpha_6 y \qquad (2.321)$$

The six α's may be determined in terms of the six nodal displacement components as was done for the beam element, whence

$$\begin{Bmatrix} u \\ v \end{Bmatrix} = \begin{bmatrix} N_1 & 0 & N_2 & 0 & N_3 & 0 \\ 0 & N_1 & 0 & N_2 & 0 & N_3 \end{bmatrix} \{\delta\} \qquad (2.322)$$

where $\{\delta\}^T = \{u_1 \, v_1 \, u_1 \, v_2 \, u_3 \, v_3\}^T \qquad (2.323)$

$$N_i = (a_i + b_i x + c_i y)/2\Delta \qquad (2.324)$$

The factors a_i, b_i, c_i, and Δ are constants which evolve from the algebraic manipulations. Continuing, for the two-dimensional case

$$\{\epsilon\} = \left\{\begin{array}{c} \epsilon_x \\ \epsilon_y \\ \epsilon_z \end{array}\right\} = \left\{\begin{array}{c} \partial u/\partial x \\ \partial v/\partial y \\ \partial u/\partial y + \partial v/\partial x \end{array}\right\} \tag{2.325}$$

Substituting Eq. (2.322) into Eq. (2.325) yields

$$\{\epsilon\} = [B]\{\delta\} \tag{2.326}$$

where

$$[B] = \frac{1}{2\Delta}\begin{bmatrix} b_1 & 0 & b_2 & 0 & b_3 & 0 \\ 0 & c_1 & 0 & c_2 & 0 & c_3 \\ c_1 & b_1 & c_2 & b_2 & c_3 & b_3 \end{bmatrix} \tag{2.327}$$

Finally, for the element of Fig. 2.58

$$\{\sigma\} = \left\{\begin{array}{c} \sigma_x \\ \sigma_y \\ \tau_{xy} \end{array}\right\} = [D]\left\{\begin{array}{c} \epsilon_x \\ \epsilon_y \\ \gamma_{xy} \end{array}\right\} \tag{2.328}$$

where, for plane strain

$$[D] = \frac{E(1 - v)}{(1 + v)(1 - 2v)}\begin{bmatrix} 1 & v/(1 - v) & 0 \\ v/(1 - v) & 1 & 0 \\ 0 & 0 & (1 - 2v)/2(1 - v) \end{bmatrix} \tag{2.329}$$

Therefore, all the terms in Eq. (2.320) have been defined and so the load-deflection relationship for this element is established.

Since all the terms under the integral in Eq. (2.320) are constants, the integral may be evaluated exactly. Note that the resulting matrix equation contains six simultaneous algebraic equations, corresponding to the six degrees of freedom associated with the triangular element of Fig. 2.54.

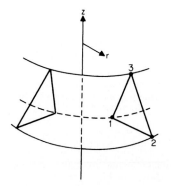

FIG. 2.55 Axisymmetric finite element.

EXAMPLE 2 The displacement function for the axisymmetric element of Fig. 2.55 is

$$u = \alpha_1 + \alpha_2 r + \alpha_3 z$$
$$v = \alpha_4 + \alpha_5 r + \alpha_6 z \tag{2.330}$$

Following the same procedure as in Example 1 we find

$$\{\epsilon\} = \left\{\begin{array}{c} \epsilon_z \\ \epsilon_r \\ \epsilon_\theta \\ \gamma_{rz} \end{array}\right\} = \left\{\begin{array}{c} \partial v/\partial z \\ \partial u/\partial r \\ u/r \\ \partial u/\partial z + \partial v/\partial r \end{array}\right\} \tag{2.331}$$

whence

$$[B] = \begin{bmatrix} 0 & c_1 & 0 & c_2 & 0 & c_3 \\ b_1 & 0 & b_2 & 0 & b_3 & 0 \\ e_1 & 0 & e_2 & 0 & e_3 & 0 \\ c_1 & b_1 & c_2 & b_2 & c_3 & b_3 \end{bmatrix} \tag{2.332}$$

where $e_i = a_i/r + b_i + c_i(z/r)$. Further,

$$[D] = \frac{E(1 - v)}{(1 + v)(1 - 2v)} \begin{bmatrix} 1 & v/(1 - v) & v/(1 - v) & 0 \\ v/(1 - v) & 1 & v/(1 - v) & 0 \\ v/(1 - v) & v/(1 - v) & 1 & 0 \\ 0 & 0 & 0 & (1 - 2v)/2(1 - v) \end{bmatrix} \quad (2.333)$$

The integral in Eq. (2.320) now has the form

$$2\pi \int_{\text{vol}} [B]^T[D][B] r \, dr \, dz$$

However, $[B]$ is no longer a constant array, i.e., $[B] = [B(r,z)]$ so that integration is a complex process. For many elements, the integrand is sufficiently complex that the integration must be carried out numerically. This numerical integration is a wholly different problem from the numerical analysis that is the finite-element method.

The three-dimensional analog to the triangle element of Example 1 is a four-node tetrahedron. A basic feature of these elements is that the strain field within the element is constant. Thus, to model a structure in which the strains vary considerably throughout the body, a large number of elements are required. Constant-strain elements are most useful for modeling thick-walled bodies in which the main action is stretching. Analysis of more flexible bodies in which bending is significant requires elements in which the strain can vary. These higher-order elements contain higher-order terms in the polynomial displacement expressions, e.g., Eq. (2.321).

Higher-Order Elements. The key ingredient in the development of a finite element is the selection of the shape function, that function which relates the internal-element displacement field to the nodal displacement field, e.g., Eq. (2.322). The remainder of the development is a mechanical process.

The shape function may be selected directly to establish some desired element characteristics or it may evolve from the selection of the displacement function as in the elements developed above. If the displacement function approach is used then the size of the polynomial is limited by the number of nodal degrees of freedom of the element, since the $\mathcal{L}$'s must be uniquely expressed in terms of the nodal degrees of freedom. Thus, in the examples above, the beam element is limited to a cubic polynomial, the triangular plane elements to linear polynomials. Higher-order polynomials require the insertion of additional nodes in the elements or of additional degrees of freedom at the existing nodes. Some typical higher-order elements involving additional nodes are shown in Fig. 2.56. An element involving additional nodal degrees of freedom is shown in Fig. 2.57. This latter type is commonly used to model shell- and plate-type structures.

A widely used class of elements in which the shape function is chosen directly is the isoparametric elements. The key fea-

FIG. 2.56 Higher-order finite elements.

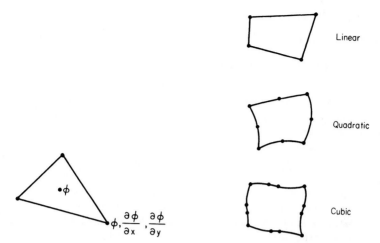

FIG. 2.57   Shell-type finite element.

FIG. 2.58   Isoparametric finite elements.

ture of isoparametric elements is that the elements can have curved sides (Fig. 2.58). This feature allows the element to follow the flow of the structure more readily so that significantly fewer elements are needed to achieve a successful model.

2.16.4   Nature of the Solution

Unless the displacement function used constitutes the exact solution, the equilibrium equation applied within the finite element, or to the total structure, will not be satisfied; i.e., only the exact solution satisfies the equilibrium equations. Further, equilibrium is not satisfied across element boundaries. For example, two adjacent constant-strain (and hence constant-stress) elements cannot correctly represent a continuously varying strain field. Given the approximate nature of the solution, it is appropriate to question whether the response to applied load is at least approximately correct. It can, in fact, be shown that, subject to certain conditions on the finite elements, that the solution will converge to the exact solution with increasing grid refinement. Thus, if questions of accuracy in the analysis of a structure exist, one need only subdivide the critical areas into successively finer element grids. The solutions from these refined analyses will converge toward the correct answer.

The conditions on the elements to assure convergence can be satisfied if the displacement functions used are continuous polynomials of at least the first order within the element and if the elements are compatible. Compatibility requires that at least the nodal variables vary continuously along the boundary between adjacent elements, e.g., the displacement along edge 1-2 of the triangle element of Fig. 2.54 must be the same as along the edge of any other similar element attached to nodes 1 and 2. For the triangular element, since the displacements along edges are straight lines, compatibility is assured.

The above discussion does not preclude the successful use of nonpolynomial displacement functions or nonconverging elements or incompatible elements. However, such elements must be used with great care.

2.16.5   Finite-Element Modeling Guidelines

General rules for finite-element modeling do not exist. However, some reasonable guidelines have evolved to aid the analyst in developing a model which will yield accurate results with a reasonable effort. The more important of these guidelines include:

1. If at all possible, use converging, compatible elements.
2. Grids can be relatively coarse in regions where the state of strain varies slowly. In regions where strains change rapidly, e.g., strain concentrations and structural discontinuities, the grid should be refined.
3. Quadrilateral elements should be used wherever possible in place of triangular elements.
4. Accurate determination of forces and displacements can be accomplished with a more coarse grid than needed for accurate determination of strains and stresses.
5. Prediction of modes of vibration requires a more refined grid than that needed for prediction of natural frequencies.
6. Higher-order elements are generally preferable to constant strain elements.
7. Aspect ratios of multisided two- or three-dimensional elements should be kept below 5.
8. When the accuracy of the solution from a grid is in doubt, the grid should be refined in the critical regions and the analysis rerun.

2.16.6   Generalizations of the Applications

The finite-element method has applications in mechanics of materials beyond the static, linear elastic, isothermal, small-strain class of analyses discussed herein. The generalizations can be classified as related to generalizations of the stress-strain equations and generalizations of the equilibrium equations.

Generalizations of the Stress-Strain Relations.   A more general statement of the stress-strain relations of linear elasticity (Eq. 2.74) is

$$\epsilon_x - \epsilon_x^0 - \alpha(T - T_0) = (1/E)\{(\sigma_x - \sigma_x^0) - \nu[(\sigma_y - \sigma_y^0) + (\sigma_z - \sigma_z^0)]\}$$

$$\epsilon_y - \epsilon_y^0 - \alpha(T - T_0) = (1/E)\{(\sigma_y - \sigma_y^0) - \nu[(\sigma_z - \sigma_z^0) + (\sigma_x - \sigma_x^0)]\}$$

$$\epsilon_z - \epsilon_z^0 - \alpha(T - T_0) = (1/E)\{(\sigma_z - \sigma_z^0) - \nu[(\sigma_x - \sigma_x^0) + (\sigma_y - \sigma_y^0)]\}$$

$$\gamma_{xy} - \gamma_{xy}^0 = (1/G)(\tau_{xy} - \tau_{xy}^0) \tag{2.334}$$

$$\gamma_{yz} - \gamma_{yz}^0 = (1/G)(\tau_{yz} - \tau_{yz}^0)$$

$$\gamma_{zx} - \gamma_{zx}^0 = (1/G)(\tau_{zx} - \tau_{zx}^0)$$

The strain terms with a superscript 0 represent a possible general state of initial strain. The stress terms with a superscript 0 represent a possible general state of initial stress. The strain terms $\alpha(T - T_0)$ represent a possible state of temperature-induced strain. Inclusion of these terms in the development of the finite-element results in a set of additional terms in the load-deflection relationship, which takes on the form

$$\{F\}_{\sigma 0} + \{F\}_{\epsilon 0} + \{F\}_{\alpha T} + \{F\} = \int_{\text{vol}} [B]^T[D][B]\, d\bar{v}\, \{\delta\} \tag{2.335}$$

where
$$\{F\}_{\sigma 0} = \int_{\text{vol}} [B]^T \{\sigma^0\} \, dv$$

$$\{F\}_{\epsilon 0} = -\int_{\text{vol}} [B]^T [D] \{\epsilon^0\} \, d\bar{v}$$

and similarly for $\{F\}_{\alpha T}$.
Nonlinear stress-strain relationships, i.e.,

$$[\sigma] = F[\epsilon] \tag{2.336}$$

are generally incorporated into finite-element analysis in terms of the incremental plasticity formulation (see Refs. 6 to 8). Solutions are effected by applying load to the structure in additive increments. For each load increment a modified linear analysis is performed. Thus the numerical analysis in the space defined by the finite-element model is supplemented by a numerical analysis in the load dimension to yield an analysis of the total problem.

Similarly, creep problems, for which the stress-strain relation is of the form

$$[\sigma] = f([\epsilon], [\partial \epsilon / \partial t]) \tag{2.337}$$

are solved using a numerical analysis in the time domain to supplement the finite-element models in space.

Generalizations of the Equilibrium Equations.   The equilibrium equations, with the addition of body force terms, such as gravitational or inertia, have the form

$$\partial \sigma_x / \partial x + \partial \tau_{xy} / \partial y + \partial \tau_{xz} / \partial z + \bar{F}_x = 0$$
$$\partial \sigma_y / \partial y + \partial \tau_{yz} / \partial z + \partial \tau_{yx} / \partial x + \bar{F}_y = 0 \tag{2.338}$$
$$\partial \sigma_z / \partial z + \partial \tau_{zx} / \partial x + \partial \tau_{zy} / \partial y + \bar{F}_z = 0$$

With these terms, the load-deflection relationship now has the form

$$\{\bar{F}\}_{BF} + \{F\} = \int_{\text{vol}} [B]^T [D] [B] \, d\bar{v} \, \{\delta\} \tag{2.339}$$

where
$$\{\bar{F}\}_{BF} = -\int_{\text{vol}} [N]^T \{\bar{F}\} \, d\bar{v}$$

$[N]^T$ = shape-function matrix, analogous to Eq. (2.322)

If $\{\bar{F}\}_{BF}$ represents an acceleration force per unit volume, then

$$\{\bar{F}\}_{BF} = \rho [N](\partial^2 / \partial t^2)\{\delta\} \tag{2.340}$$

where ρ is the mass per unit volume.
If we further define

$$[M] = \rho \int_{\text{vol}} [N]^T [N] \, d\bar{v}$$

$$[k] = \int_{\text{vol}} [B]^T [D] [B] \, d\bar{v} \tag{2.341}$$

then Eq. (2.297) takes the form

$$[M]\{\ddot{\delta}\} + [k]\{\delta\} = \{F\} \tag{2.342}$$

which is the matrix statement of the general vibration problem discussed in Sec. 5. Therefore, all the solution techniques noted therein are applicable to the spatial finite-element model. A damping force vector can also be developed for Eq. (2.342).

2.16.7 Finite-Element Codes

Structural analysis by the finite-element method contains two major engineering steps: the design of the grid and the use of an appropriate finite element. Finite elements have been developed to represent a broad range of structural configurations, including constant strain and higher-order two- and three-dimensional solids, shells, plates, beams, bars, springs, masses, damping elements, contact elements, fracture mechanics elements, and many others. Elements have been designed for static and dynamic analysis, linear and nonlinear material models, linear and nonlinear deformations.

The finite element depends upon the selection of an appropriate shape or displacement function. The remainder of the analysis is a mechanical process. The element stiffness matrix calculations, including any numerical integrations required, the assembly of the structural load-deflection relationship, the solution of the structure equations for loads and deflections, and the back substitution into the individual element relationships to obtain stress and strain fields require a huge number of calculations but no engineering judgment. The calculation procedure is clearly suited to the "number crunching" digital computer. Effective use of the computer for models of any substantial size requires that efficient computer-oriented numerical integration and simultaneous equation solvers be incorporated into the solution process.

To this end many large, general-purpose, finite-element-based computer codes have been developed[18,19,20] and are available in the marketplace. These codes feature large element libraries, extremely efficient solution algorithms, and a broad range of applications. The code developers strive to make these codes "user friendly" to minimize the effort required to assemble the computer input once the engineering decisions of grid design and element selection from the element library have been made.

Many special-purpose codes with unique finite elements are available to solve problems beyond the range of the general-purpose codes. Beyond the contents of the marketplace, the creation of a finite-element program for any particular application is a relatively simple process once the required finite element has been designed.

REFERENCES

1. Timoshenko, S.: "Strength of Materials," 3d ed., Parts I and II, D. Van Nostrand Company, Princeton, N.J., 1955.

2. Timoshenko, S., and J. N. Goodier: "The Theory of Elasticity," 2d ed., McGraw-Hill Book Company, Inc., New York, 1951.

3. Timoshenko, S., and S. Woinowsky-Krieger: "Theory of Plates and Shells," 2d ed., McGraw-Hill Book Company, Inc., New York, 1959.

4. Sokolnikoff, I. S.: "Mathematical Theory of Elasticity," 2d ed., McGraw-Hill Book Company, Inc., New York, 1956.

5. Love, A. E. H.: "A Treatise on the Mathematical Theory of Elasticity," 4th ed., Dover Publications, Inc., New York, 1944.

6. Prager, W.: "An Introduction to Plasticity," Addison-Wesley Publishing Company, Inc., Reading, Mass., 1959.

7. Mendelson, A.: "Plasticity: Theory and Application," The Macmillan Company, Inc., New York, 1968.

8. Hodge, P. G.: "Plastic Analysis of Structures," McGraw-Hill Book Company, Inc., New York, 1959.

9. Boley, B. A., and J. H. Weiner: "Theory of Thermal Stresses," John Wiley & Sons, Inc., New York, 1960.

10. Flugge, W.: "Stresses in Shells," Springer-Verlag OHG, Berlin, 1960.

11. Hult, J. A. H.: "Creep in Engineering Structures," Blaisdell Publishing Company, Waltham, Mass., 1966.

12. Roark, R. J.: "Formulas for Stress and Strain," 3d ed., McGraw-Hill Book Company, Inc., New York, 1954.

13. McConnell, A. J.: "Applications of Tensor Analysis," Dover Publications, Inc., New York, 1957.

14. Cook, R. D.: "Concepts and Applications of Finite Element Analysis," 2d ed., John Wiley & Sons, Inc., New York, 1981.

15. Zienkiewicz, O. C.: "The Finite Element Method," 3d ed., McGraw-Hill Book Company, (U.K.) Ltd., London, 1977.

16. Gallagher, R. H.: "Finite Element Analysis: Fundamentals," Prentice-Hall, Inc., Englewood Cliffs, N.J., 1975.

17. Hildebrand, F. B.: "Methods of Applied Mathematics," Prentice-Hall, Inc., Englewood Cliffs, N.J., 1952.

18. "ANSYS Engineering Analysis System," Swanson Analysis Systems, Inc., Houston, Pa.

19. "The NASTRAN Theoretical Manual," NASA-SP-221(03), National Aeronautics and Space Administration, Washington, D.C.

20. "The MARC Finite Element Code," MARC Analysis Research Corporation, Palo Alto, Calif.

SECTION 3
KINEMATICS OF MECHANISMS

Ferdinand Freudenstein, Ph.D.
Stevens Professor of Mechanical Engineering
Columbia University
New York, N.Y.

George N. Sandor, Eng.Sc.D., P.E.
Research Professor Emeritus of Mechanical Engineering
Center for Intelligent Machines
University of Florida
Gainesville, Fla.

3.1 DESIGN USE OF THE MECHANISMS SECTION

The design process involves intuition, invention, synthesis, and analysis. Although no arbitrary rules can be given, the following design procedure is suggested:

1. Define the problem in terms of inputs, outputs, their time-displacement curves, sequencing, and interlocks.

2. Select a suitable mechanism, either from experience or with the help of the several available compilations of mechanisms, mechanical movements, and components (Sec. 3.8).

3. To aid systematic selection consider the creation of mechanisms by the separation of structure and function and, if necessary, modify the initial selection (Secs. 3.2 and 3.6).

4. Develop a first approximation to the mechanism proportions from known design requirements, layouts, geometry, velocity and acceleration analysis, and path-curvature considerations (Secs. 3.3 and 3.4).

5. Obtain a more precise dimensional synthesis, such as outlined in Sec. 3.5, possibly with the aid of computer programs, charts, diagrams, tables, and atlases (Secs. 3.5, 3.6, 3.7, and 3.9).

6. Complete the design by the methods outlined in Sec. 3.6 and check end results. Note that cams, power, and precision gearing are treated in Secs. 14, 29, and 30, respectively.

3.2 BASIC CONCEPTS

3.2.1 Kinematic Elements

Mechanisms are often studied as though made up of rigid-body members, or "links," connected to each other by rigid "kinematic elements" or "element pairs." The nature and arrangement of the kinematic links and elements determine the kinematic properties of the mechanism.

If two mating elements are in surface contact, they are said to form a "lower pair"; element pairs with line or point contact form "higher pairs." Three types of lower pairs permit relative motion of one degree of freedom ($f = 1$), turning pairs, sliding pairs, and screw pairs. These and examples of higher pairs are shown in Fig. 3.1. Examples of element pairs whose relative motion possesses up to five degrees of freedom are shown in Fig. 3.2.

A link is called "binary," "ternary," or "n-nary" according to the number of element

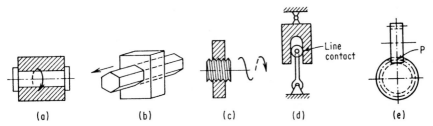

FIG. 3.1 Examples of kinematic-element pairs: lower pairs *a, b, c,* and higher pairs *d* and *e.* (*a*) Turning or revolute pair. (*b*) Sliding or prismatic pair. (*c*) Screw pair. (*d*) Roller in slot. (*e*) Helical gears at right angles.

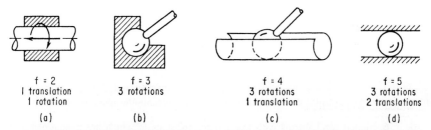

FIG. 3.2 Examples of elements pairs with $f > 1$. (*a*) Turn slide or cylindrical pair. (*b*) Ball joint or spherical pair. (*c*) Ball joint in cylindrical slide. (*d*) Ball between two planes. (Translational freedoms are in mutually perpendicular directions. Rotational freedoms are about mutually perpendicular axes.)

pairs connected to it, i.e., 2, 3, or n. A ternary link, pivoted as in Fig. 3.3*a* and *b,* is often called a "rocker" or a "bell crank," according to whether α is obtuse or acute.

A ternary link having three parallel turning-pair connections with coplanar axes, one of which is fixed, is called a "lever" when used to overcome a weight or resistance (Fig. 3.3*c, d,* and *e*). A link without fixed elements is called a "floating link." Mechanisms consisting of a chain of rigid links (one of which, the "frame," is consid-

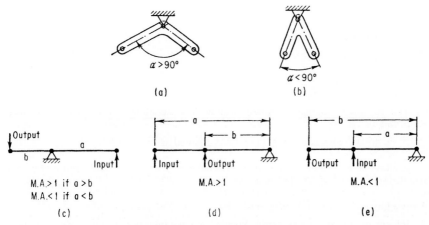

FIG. 3.3 Links and levers. (*a*) Rocker (ternary link). (*b*) Bell crank (ternary link). (*c*) First-class lever. (*d*) Second-class lever. (*e*) Third-class lever.

ered fixed) are said to be closed by "pair closure" if all element pairs are constrained by material boundaries. All others, such as may involve springs or body forces for chain closure, are said to be closed by means of "force closure." In the latter, nonrigid elements may be included in the chain.

3.2.2 Degrees of Freedom[6,9,10,13,94,111,154,242,368]

Let F = degree of freedom of mechanism
   l = total number of links, including fixed link
   j = total number of joints
   f_i = degree of freedom of relative motion between element pairs of ith joint

Then, *in general,*

$$F = \lambda(l - j - 1) + \sum_{i=1}^{j} f_i \tag{3.1}$$

where λ is an integer whose value is determined as follows:
   $\lambda = 3$: Plane mechanisms with turning pairs, or turning and sliding pairs; spatial mechanisms with turning pairs only (motion on sphere); spatial mechanisms with rectilinear sliding pairs only.
   $\lambda = 6$: Spatial mechanisms with lower pairs, the axes of which are nonparallel and nonintersecting; note exceptions such as listed under $\lambda = 2$ and $\lambda = 3$. (See also Ref. 10.)
   $\lambda = 2$: Plane mechanisms with sliding pairs only; spatial mechanisms with "curved" sliding pairs only (motion on a sphere); three-link coaxial screw mechanisms.

Although included under Eq. (3.1), the motions on a sphere are usually referred to as special cases. For a comprehensive discussion and formulas including screw chains and other combinations of elements, see Ref. 13. The freedom of a mechanism with higher pairs should be determined from an equivalent lower-pair mechanism whenever feasible (see Sec. 3.2).

Mechanism Characteristics Depending on Degree of Freedom Only. For plane mechanisms with turning pairs only and one degree of freedom,

$$2j - 3l + 4 = 0 \tag{3.2}$$

except in special cases. Furthermore, if this equation is valid, then the following are true:

1. The number of links is even.
2. The minimum number of binary links is four.
3. The maximum number of joints in a single link cannot exceed one-half the number of links.
4. If one joint connects m links, the joint is counted as $(m - 1)$-fold.

In addition, for nondegenerate plane mechanisms with turning and sliding pairs and one degree of freedom, the following are true.

1. If a link has only sliding elements, they cannot all be parallel.
2. Except for the three-link chain, binary links having sliding pairs only cannot, in general, be directly connected.

3. No closed nonrigid loop can contain less than two turning pairs.

For plane mechanisms, having any combination of higher and/or lower pairs, and with one degree of freedom, the following hold.

1. The number of links may be odd.
2. The maximum number of elements in a link may exceed one-half the number of links, but an upper bound can be determined.[154,368]
3. If a link has only higher-pair connections, it must possess at least three elements.

For constrained spatial mechanisms in which Eq. (3.1) applies with $\lambda = 6$, the sum of the degrees of freedom of all joints must add up to 7 whenever the number of links is equal to the number of joints.

Special Cases. F can exceed the value predicted by Eq. (3.1) in certain special cases. These occur, generally, when a sufficient number of links are parallel in plane motion (Fig. 3.4a) or, in spatial motions, when the axes of the joints intersect (Fig. 3.4b—motion on a sphere, considered special in the sense that $\lambda \neq 6$).

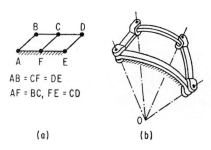

AB = CF = DE
AF = BC, FE = CD

(a)   (b)

FIG. 3.4 Special cases that are exceptions to Eq. (3.1). (a) Parallelogram motion, $F = 1$. (b) Spherical four-bar mechanism, $F = 1$; axes of four turning joints intersect at O.

The existence of these special cases or "critical forms" can sometimes also be detected by multigeneration effects involving pantographs, inversors, or mechanisms derived from these (see Sec. 3.6 and Ref. 154). In the general case, the critical form is associated with the singularity of the functional matrix of the differential displacement equations of the coordinates;[130] this singularity is usually difficult to ascertain, however, especially when higher pairs are involved. Known cases are summarized in Ref. 154. For two-degree-of-freedom systems, additional results are listed in Refs. 111 and 242.

3.2.3 Creation of Mechanisms According to the Separation of Kinematic Structure and Function[54,74,110,132,133]

Basically this is an unbiased procedure for creating mechanisms according to the following sequence of steps:

1. Determine the basic characteristics of the desired motion (degree of freedom, plane or spatial) and of the mechanism (number of moving links, number of independent loops).
2. Find the corresponding kinematic chains from tables, such as in Ref. 133.
3. Find corresponding mechanisms by selecting joint types and fixed link in as many inequivalent ways as possible and sketch each mechanism.
4. Determine functional requirements and, if possible, their relationship to kinematic structure.
5. Eliminate mechanisms which do not meet functional requirements. Consider remaining mechanisms in greater detail and evaluate for potential use.

The method is described in greater detail in Refs. 110 and 133, which show applications to casement window linkages, constant-velocity shaft couplings, other mechanisms, and patent evaluation.

3.2.4 Kinematic Inversion

Kinematic inversion refers to the process of considering different links as the frame in a given kinematic chain. Thereby different and possibly useful mechanisms can be obtained. The slider crank, the turning-block and the swinging-block mechanisms are mutual inversions, as are also drag-link and "crank-and-rocker" mechanisms.

3.2.5 Pin Enlargement

Another method for developing different mechanisms from a base configuration involves enlarging the joints, illustrated in Fig. 3.5.

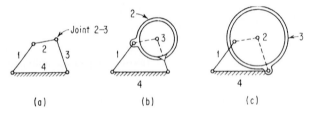

FIG. 3.5 Pin enlargement. (*a*) Base configuration. (*b*) Enlarged pin at joint 2–3; pin part of link 3. (*c*) Enlarged pin at joint 2–3; pin takes place of link 2.

3.2.6 Mechanical Advantage

Neglecting friction and dynamic effects, the instantaneous power input and output of a mechanism must be equal and, in the absence of branching (one input, one output, connected by a single "path"), equal to the "power flow" through any other point of the mechanism.

In a single-degree-of-freedom mechanism without branches, the power flow at any point J is the product of the force F_j at J, and the velocity V_j at J in the direction of the force. Hence, for any point in such a mechanism,

$$F_j V_j = \text{constant} \tag{3.3}$$

neglecting friction and dynamic effects. For the point of input P and the point of output Q of such a mechanism, the mechanical advantage is defined as

$$MA = F_Q/F_P \tag{3.4}$$

3.2.7 Velocity Ratio

The "linear velocity ratio" for the motion of two points P and Q representing the input and output members or "terminals" of a mechanism is defined as V_Q/V_P. If input and output terminals or links P and Q rotate, the "angular velocity ratio" is defined as ω_Q/ω_P, where ω designates the angular velocity of the link. If T_Q and T_P refer to torque output and input in single-branch rotary mechanisms, the power-flow equation, in the absence of friction, becomes

$$T_P\omega_P = T_Q\omega_Q \tag{3.5}$$

3.2.8 Conservation of Energy

Neglecting friction and dynamic effects, the product of the mechanical advantage and the linear velocity ratio is unity for all points in a single-degree-of-freedom mechanism without branch points, since $F_Q V_Q/F_P V_P = 1$.

3.2.9 Toggle

Toggle mechanisms are characterized by sudden snap or overcenter action, such as in Fig. 3.6a and b, schematics of a crushing mechanism and a light switch. The mechanical advantage, as in Fig. 3.6a, can become very high. Hence toggles are often used in such operations as clamping, crushing, and coining.

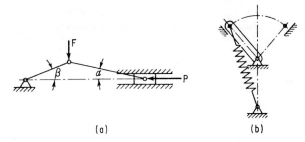

(a)   (b)

FIG. 3.6 Toggle actions. (a) $P/F = (\tan \alpha + \tan \beta)^{-1}$ (neglecting friction). (b) Schematic of a light switch.

3.2.10 Transmission Angle[15,159,160,166–168,176,205] (see Secs. 3.6 and 3.9)

The transmission angle μ is used as a geometrical indication of the ease of motion of a mechanism under static conditions, excluding friction. It is defined by the ratio

$$\tan \mu = \frac{\text{force component tending to move driven link}}{\text{force component tending to apply pressure on driven-link bearing or guide}} \tag{3.6}$$

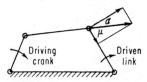

FIG. 3.7 Transmissional angle μ and pressure angle α (also called the deviation angle) in a four-link mechanism.

where μ is the transmission angle

In four-link mechanisms, μ is the angle between the coupler and the driven link (or the supplement of this angle) (Fig. 3.7) and has been used in optimizing linkage proportions (Secs. 3.6 and 3.9). Its ideal value is 90°; in practice it may deviate from this value by 30° and possibly more.

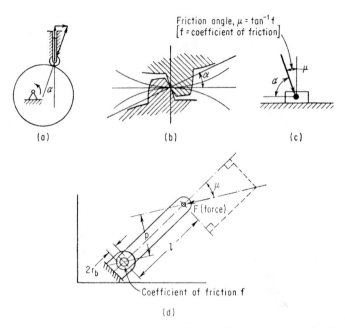

FIG. 3.8 Pressure angle (*a*) Cam and follower. (*b*) Gear teeth in mesh. (*c*) Link in sliding motion; condition of locking by friction $(\alpha + \mu) \geq 90°$. (*d*) Conditions for locking by friction of a rotating link: $\sin \mu \leq f r_b / l$.

3.2.11 Pressure Angle

In cam and gear systems, it is customary to refer to the complement of the transmission angle, called the pressure angle α, defined by the ratio

$$\tan \alpha = \frac{\text{force component tending to put pressure on follower bearing or guide}}{\text{force component tending to move follower}} \tag{3.7}$$

The ideal value of the pressure angle is zero; in practice it is frequently held to within 30° (Fig. 3.8). To ensure movability of the output member the ultimate criterion is to preserve a sufficiently large value of the ratio of driving force (or torque) to friction force (or torque) on the driven link. For a link in pure sliding (Fig. 3.8*c*), the motion will lock if the pressure angle and the friction angle add up to or exceed 90°.

A mechanism, the output link of which is shown in Fig. 3.8*d*, will lock if the ratio of *p*, the distance of the line of action of the force *F* from the fixed pivot axis, to the bearing radius r_b is less than or equal to the coefficient of friction, *f*, i.e., if the line of action of the force *F* cuts the "friction circle" of radius $f r_b$, concentric with the bearing.[171]

3.2.12 Kinematic Equivalence[159,182,288,290,347,376] (see Sec. 3.6)

"Kinematic equivalence," when applied to two mechanisms, refers to equivalence in motion, the precise nature of which must be defined in each case.

The motion of joint *C* in Fig. 3.9*a* and *b* is entirely equivalent if the quadrilaterals

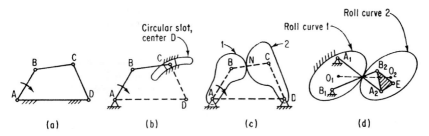

FIG. 3.9 Kinematic equivalence: (*a*), (*b*), (*c*) for four-bar motion; (*d*) illustrates rolling motion and an equivalent mechanism. When O_1 and O_2 are fixed, curves are in rolling contact; when roll curve 1 is fixed and rolling contact is maintained, O_2 generates circle with center O_1.

ABCD are identical; the motion of *C* as a function of the rotation of link *AB* is also equivalent throughout the range allowed by the slot. In Fig. 3.9*c*, *B* and *C* are the centers of curvature of the contacting surfaces at *N*; *ABCD* is one equivalent four-bar mechanism in the sense that, if *AB* is integral with body 1, the angular velocity and angular acceleration of link *CD* and body 2 are the same in the position shown, but not necessarily elsewhere.

Equivalence is used in design to obtain alternate mechanisms, which may be mechanically more desirable than the original. If, as in Fig. 4.9*d*, A_1A_2 and B_1B_2 are conjugate point pairs (see Sec. 3.4), with A_1B_1 fixed on roll curve 1, which is in rolling contact with roll curve 2 (A_2B_2 are fixed on roll curve 2), then the path of *E* on link A_2B_2 and of the coincident point on the body of roll curve 2 will have the same path tangent and path curvature in the position shown, but not generally elsewhere.

3.2.13 The Instant Center

At any instant in the plane motion of a link, the velocities of all points on the link are proportional to their distance from a particular point *P*, called the instant center. The velocity of each point is perpendicular to the line joining that point to *P* (Fig. 3.10).

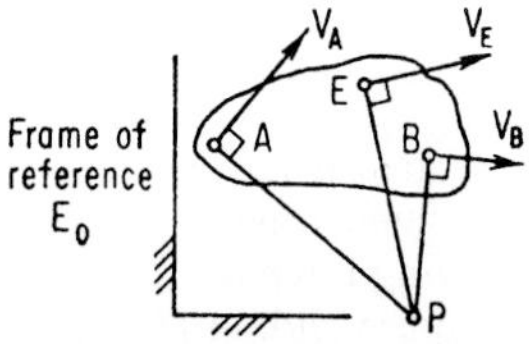

Regarded as a point on the link, *P* has an instantaneous velocity of zero. In pure rectilinear translation, *P* is at infinity.

The instant center is defined in terms of velocities and is not the center of path curvature for the points on the moving link in the instant shown, except in special cases, e.g., points on common tangent between centrodes (see Sec. 3.4).

FIG. 3.10 Instant center, *P*. $V_E/V_B = EP/BP$, $V_E \perp EP$, etc.

An extension of this concept to the "instantaneous screw axis" in spatial motions has been described.[38]

3.2.14 Centrodes, Polodes, Pole Curves

Relative plane motion of two links can be obtained from the pure rolling of two curves, the "fixed" and "movable centrodes" ("polodes" and "pole curves," respectively), which can be constructed as illustrated in the following example.

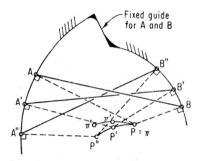

FIG. 3.11 Construction of fixed and movable centroides. Link AB in plane motion, guided at both ends; $PP' = P\pi'$; $\pi'\,\pi'' = P'P''$, etc.

As shown in Fig. 3.11, the intersections of path normals locate successive instant centers P, P', P'', ..., whose locus constitutes the fixed centrode. The movable centrode can be obtained either by inversion (i.e., keeping AB fixed, moving the guide, and constructing the centrode as before) or by "direct construction": superposing triangles $A'B'P'$, $A''B''P''$, ..., on AB so that A' covers A and B' covers B, etc. The new locations thus found for P', P'', ..., marked π', π'', ..., then constitute points on the movable centrode, which rolls without slip on the fixed centrode and carries AB with it, duplicating the original motion. Thus, for the motion of AB, the centrode-rolling motion is kinematically equivalent to the original guided motion.

In the antiparallel equal-crank linkage, with the shortest link fixed, the centrodes for the coupler motion are identical ellipses with foci at the link pivots (Fig. 3.12); if the longer link AB were held fixed, the centrodes for the coupler motion of CD would be identical hyperbolas with foci at A, B, and C, D, respectively.

In the elliptic trammel motion (Fig. 3.13) the centrodes are two circles, the smaller rolling inside the larger, twice its size. Known as "cardanic motion," it is used in press drives, resolvers, and straight-line guidance.

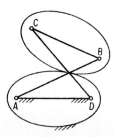

FIG. 3.12 Antiparallel equal-crank linkage; rolling ellipses, foci at A, D, B, C; $AD < AB$.

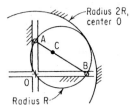

FIG. 3.13 Cardanic motion of the elliptic trammel, so called because any point C of AB describes an ellipse; midpoint of AB describes circle, center O (point C need not be collinear with AB).

Apart from their use in kinematic analysis, the centrodes are used to obtain alternate, kinematically equivalent mechanisms, and sometimes to guide the original mechanism past the "in-line" or "dead-center" positions.[207]

3.2.15 The Theorem of Three Centers

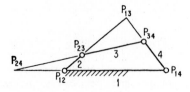

FIG. 3.14 Instant centers in four-bar motion.

Also known as Kennedy's or the Aronhold-Kennedy theorem, this theorem states that, for any three bodies i, j, k in plane motion, the relative instant centers P_{ij}, P_{jk}, P_{ki} are collinear; here P_{ij}, for instance, refers to the instant center of the motion of link i relative to link j, or vice versa. Figure 3.14 illustrates the theorem with respect to four-bar motion. It is used in determining the location of instant centers and in planar path curvature investigations.

3.2.16 Function, Path, and Motion Generation

In "function generation" the input and output motions of a mechanism are linear analogs of the variables of a function $F(x, y, \ldots) = 0$. The number of degrees of freedom of the mechanism is equal to the number of independent variables.

For example, let ϕ and ψ, the linear or rotary motions of the input and output links or "terminals," be linear analogs of x and y, where $y = f(x)$ within the range $x_0 \leq x \leq x_{n+1}$, $y_0 \leq y \leq y_{n+1}$. Let the *input* values ϕ_0, ϕ_j, ϕ_{n+1} and the *output* values ψ_0, ψ_j, ψ_{n+1} correspond to the values x_0, x_j, x_{n+1} and y_0, y_j, y_{n+1}, of x and y, respectively, where the subscripts 0, j, and $(n + 1)$ designate starting, jth intermediate, and terminal values. Scale factors r_ϕ, r_ψ are defined by

$$r_\phi = (x_{n+1} - x_0)/(\phi_{n+1} - \phi_0) \qquad r_\psi = (y_{n+1} - y_0)/(\psi_{n+1} - \psi_0)$$

(it is assumed that $y_0 \neq y_{n+1}$), such that $y - y_j = r_\psi(\psi - \psi_j)$, $x - x_j = r_\phi(\phi - \phi_j)$, whence

$$d\psi/d\phi = (r_\phi/r_\psi)(dy/dx), \quad d^2\psi/d\phi^2 = (r_\phi^2/r_\psi)(d^2y/dx^2),\ \text{and generally,}$$

$$d^n\psi/d\phi^n = (r_\phi^n/r_\psi)(d^ny/dx^n)$$

In "path generation" a point of a floating link traces a prescribed path with reference to the frame. In "motion generation" a mechanism is designed to conduct a floating link through a prescribed sequence of positions (Ref. 382). Positions along the path or specification of the prescribed motion may or may not be coordinated with input displacements.

3.3 PRELIMINARY DESIGN ANALYSIS: DISPLACEMENTS, VELOCITIES, AND ACCELERATIONS (Refs. 41, 58, 61, 62, 96, 116, 117, 129, 145, 172, 181, 194, 212, 263, 278, 298, 302, 309, 361, 384, 428, 487; see also Sec. 3.9)

Displacements in mechanisms are obtained graphically (from scale drawings) or analytically or both. Velocities and accelerations can be conveniently analyzed graphically by the "vector-polygon" method or analytically (in case of plane motion) via complex numbers. In all cases, the "vector equation of closure" is utilized, expressing the fact that the mechanism forms a closed kinematic chain.

3.3.1 Velocity Analysis: Vector-Polygon Method

The method is illustrated using a point D on the connecting rod of a slider-crank mechanism (Fig. 3.15). The vector-velocity equation for C is

$$\mathbf{V}_C = \mathbf{V}_B + \mathbf{V}_{C/B}^n + \mathbf{V}_{C/B}^t = \text{a vector parallel to line AX}$$

where $\mathbf{V}_C$ = velocity of C (Fig. 3.15)
$\mathbf{V}_B$ = velocity of B

$V_{C/B}^n$ = normal component of velocity of C relative to B = component of relative velocity along BC = zero (owing to the rigidity of the connecting rod)

$V_{C/B}^t$ = tangential component of velocity of C relative to B, value (ω_{BC}) BC, perpendicular to BC

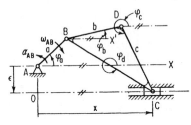

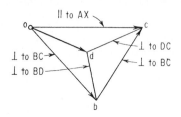

FIG. 3.15 Offset slider-crank mechanism.

FIG. 3.16 Velocity polygon for slider-crank mechanism of Fig. 3.15.

The velocity equation is now "drawn" by means of a vector polygon as follows:

1. Choose an arbitrary origin o (Fig. 3.16).
2. Label terminals of velocity vectors with lowercase letters, such that absolute velocities start at o and terminate with the letter corresponding to the point whose velocity is designated. Thus $V_B = \mathbf{ob}$, $V_c = \mathbf{oc}$, to a certain scale.
3. Draw $\mathbf{ob} = (\omega_{AB})\, \overline{AB}/k_v$, where k_v is the velocity scale factor, say, inches per inch per second.
4. Draw $\mathbf{bc} \perp BC$ and $\mathbf{oc} \| AX$ to determine intersection c.
5. Then $V_C = (\mathbf{oc})/k_v$; absolute velocities always start at o.
6. Relative velocities $V_{C/B}$, etc., connect the terminals of absolute velocities. Thus $V_{C/B} = (\mathbf{bc})/k_v$. Note the reversal of order in C/B and $\mathbf{bc}$.
7. To determine the velocity of D, one way is to write the appropriate velocity-vector equation and draw it on the polygon: $V_D = V_C + V_{D/C}^n + V_{D/C}^t$; the second is to utilize the "principle of the velocity image." This principle states that Δbcd in the velocity polygon is similar to ΔBCD in the mechanism, and the sense $b \rightarrow c \rightarrow d$ is the same as that of $B \rightarrow C \rightarrow D$. This "image construction" applies to any three points on a rigid link in plane motion. It has been used in Fig. 3.16 to locate d, whence $V_D = (\mathbf{od})/k_v$.
8. The angular velocity ω_{BC} of the coupler can now be determined from

$$|\omega_{BC}| = \frac{|V_{B/C}|}{BC} = \frac{|(cb)/k_v|}{BC}$$

The sense of ω_{BC} is determined by imagining B fixed and observing the sense of $V_{C/B}$. Here ω_{BC} is counterclockwise.
9. Note that to determine the velocity of D it is easier to proceed in steps, to determine the velocity of C first and thereafter to use the image-construction method.

3.3.2 Velocity Analysis: Complex-Number Method

Using the slider crank of Fig. 3.15 once more as an illustration with x axis along the center line of the guide, and recalling that $i^2 = -1$, we write the complex-number equations as follows, with the equivalent vector equation below each:

Displacement: $\quad\quad\quad ae^{i\varphi_a} + be^{i\varphi_b} + ce^{i\varphi_c} = x \quad\quad\quad\quad (3.8)$

$$\mathbf{AB} + \mathbf{BD} + \mathbf{DC} = \mathbf{AC}$$

Velocity: $\quad iae^{i\varphi_a}\omega_{AB} + (ibe^{i\varphi_b} + ice^{i\varphi_c})\omega_{BC} = dx/dt \quad (t = \text{time}) \quad (3.9)$

$$\mathbf{V}_B + \mathbf{V}_{D/B} + \mathbf{V}_{C/D} = \mathbf{V}_C$$

Note that $\omega_{AB} = d\varphi_a/dt$ is positive when counterclockwise and negative when clockwise; in this problem ω_{AB} is negative.

The complex conjugate of Eq. (3.9)

$$-iae^{-i\varphi_a}\omega_{AB} - (ibe^{-i\varphi_b} + ice^{-i\varphi_c})\omega_{BC} = dx/dt \quad\quad (3.10)$$

From Eqs. (3.9) and (3.10), regarded as simultaneous equations:

$$\frac{\omega_{BC}}{\omega_{AB}} = \frac{ia(e^{i\varphi_a} + e^{-i\varphi_a})}{-ib(e^{i\varphi_b} + e^{-i\varphi_b}) - ic(e^{i\varphi_c} + e^{-i\varphi_c})} = \frac{a\cos\varphi_a}{-(b\cos\varphi_b + c\cos\varphi_c)}$$

$$\mathbf{V}_D = \mathbf{V}_B + \mathbf{V}_{D/B} = iae^{i\varphi_a}\omega_{AB} + ibe^{i\varphi_b}\omega_{BC}$$

The quantities φ_a, φ_b, φ_c are obtained from a scale drawing or by trigonometry.

Both the vector-polygon and the complex-number methods can be readily extended to accelerations, and the latter also to the higher accelerations.

3.3.3 Acceleration Analysis: Vector-Polygon Method

We continue with the slider crank of Fig. 3.15. After solving for the velocities via the velocity polygon, write out and "draw" the acceleration equations. Again proceed in order of increasing difficulty: from B to C to D, and determine first the acceleration of point C:

$$\mathbf{A}_C = \mathbf{A}_C^n = \mathbf{A}_C^t = \mathbf{A}_B^n + \mathbf{A}_B^t + \mathbf{A}_{C/B}^n + \mathbf{A}_{C/B}^t$$

where $\mathbf{A}_C^n$ = acceleration normal to path of C (equal to zero in this case)
$\mathbf{A}_C^t$ = acceleration parallel to path of C
$\mathbf{A}_B^n$ = acceleration normal to path of B, value $\omega_{AB}^2(\overline{AB})$, direction B to A
$\mathbf{A}_B^t$ = acceleration parallel to path of B, value $\alpha_{AB}(AB)$, $\perp AB$, sense determined by that of α_{AB} (where $\alpha_{AB} = d\omega_{AB}/dt$)
$\mathbf{A}_{C/B}^n$ = acceleration component of C relative to B, in the direction C to B, value $(BC)\omega_{BC}^2$
$\mathbf{A}_{C/B}^t$ = acceleration component of C relative to B, $\perp \overline{BC}$, value $\alpha_{BC}(\overline{BC})$. Since α_{BC} is unknown, so is the magnitude and sense of $\mathbf{A}_{C/B}^t$

The acceleration polygon is now drawn as follows (Fig. 3.17):

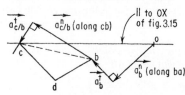

FIG. 3.17 Acceleration polygon for slider crank of Fig. 3.15. $\triangle bcd \approx \triangle BCD$ of Fig. 3.15.

1. Choose an arbitrary origin o, as before.

2. Draw each acceleration of scale k_a (inch per inch per second squared), and label the appropriate vector terminals with the lowercase letter corresponding to the point whose acceleration is designated, e.g., $\mathbf{A}_B = (ob)/k_a$. Draw $\mathbf{A}_B^n$, $\mathbf{A}_B^t$, and $\mathbf{A}_{C/B}^n$.

3. Knowing the direction of $A_{C/B}^t$ ($\perp \overline{BC}$), and also of A_C (along the slide), locate c at the intersection of a line through o, parallel to AX, and the line representing $A_{C/B}^t$. $A_C = (\mathbf{oc})/k_a$.

4. The acceleration of D is obtained using the "principle of the acceleration image," which states that, for any three points on a rigid body, such as link BCD, in plane motion, Δbcd and ΔBCD are similar, and the sense $b \to c \to d$ is the same as that of $B \to C \to D$. $A_D = (\mathbf{od})/k_a$.

5. Relative accelerations can also be found from the polygon. For instance, $A_{C/D} = (\mathbf{dc})/k_a$; note reversal of order of the letters C and D.

6. The angular acceleration α_{BC} of the connecting rod can now be determined from $|\alpha_{BC}| = |A_{C/B}^t|/BC$. Its sense is determined by that of $A_{C/B}^t$.

7. The acceleration of D can also be obtained by direct drawing of the equation $A_D = A_C + A_{D/C}$.

3.3.4 Acceleration Analysis: Complex-Number Method
(see Fig. 3.15)

Differentiating Eq. (3.9), obtain the acceleration equation of the slider-crank mechanism:

$$ae^{i\varphi_a}(i\alpha_{AB} - \omega_{AB}^2) + (be^{i\varphi_b} + ce^{i\varphi_c})(i\alpha_{BC} - \omega_{BC}^2) = d^2x/dt^2 \qquad (3.11)$$

This is equivalent to the vector equation

$$A_B^t + A_B^n + A_{D/B}^t + A_{D/B}^n + A_{C/D}^t + A_{C/D}^n = A_C^n + A_C^t$$

Combining Eq. (3.11) and its complex conjugate, eliminate d^2x/dt^2 and solve for α_{BC}. Substitute the value of α_{BC} in the following equation for A_D:

$$A_D + A_B + A_{D/B} = ae^{i\varphi_a}(i\alpha_{AB} - \omega_{AB}^2) + be^{i\varphi_b}(i\alpha_{BC} - \omega_{BC}^2)$$

The above complex-number approach also lends itself to the analysis of motions involving Coriolis acceleration. The latter is encountered in the determination of the relative acceleration of two instantaneously coincident points on different links.[106,171,384] The general complex-number method is discussed more fully in Ref. 381. An alternate approach, using the acceleration center, is described in Sec. 3.4. The accelerations in certain specific mechanisms are discussed in Sec. 3.9.

3.3.5 Higher Accelerations (see also Sec. 3.4)

The second acceleration (time derivative of acceleration), also known as "shock," "jerk," or "pulse," is significant in the design of high-speed mechanisms and has been investigated in several ways.[41,61,62,106,298,381,384,487] It can be determined by direct differentiation of the complex-number acceleration equation.[381] The following are the basic equations.

Shock of B Relative to A (where A and B represent two points on one link whose angular velocity is ω_p; $\alpha_p = d\omega_p/dt$).[298]   Component along AB:

$$-3\alpha_p\omega_p AB$$

Component perpendicular to AB:

$$AB(d\alpha_p/dt - \omega_p^3)$$

in direction of $\boldsymbol{\omega}_p \times \mathbf{AB}$.

Absolute Shock.[298]   Component along path tangent (in direction of $\boldsymbol{\omega}_p \times \mathbf{AB}$):

$$d^2v/dt^2 - v^3/\rho^2$$

where v = velocity of B and ρ = radius of curvature of path of B.
Component directed toward the center of curvature:

$$\frac{v}{\rho}\left[3\,\frac{dv}{dt} - \frac{v}{\rho}\,\frac{d\rho}{dt}\right]$$

Absolute Shock with Reference to Rolling Centrodes (Fig. 3.18, Sec. 3.4) [l, m as in Eq. (3.22)].   Component along *AP*:

$$-3\omega_p^3\delta\left[\delta\sin\psi\left(\frac{1}{m}+\frac{1}{g}\right) + \delta\cos\psi\left(\frac{1}{l}-\frac{1}{\delta}\right) - \frac{r}{g}\right] \qquad g = \frac{\omega_p^2\delta}{\alpha_p}$$

Component perpendicular to *AP* in direction of $\boldsymbol{\omega}_p \times \mathbf{PA}$:

$$r\left(\frac{d\alpha_p}{dt} - \omega_p^3\right) + 3\delta^2\omega_p^3\left[\cos\psi\left(\frac{1}{m}+\frac{1}{g}\right) - \sin\psi\left(\frac{1}{l}-\frac{1}{g}\right)\right]$$

3.3.6   Accelerations in Complex Mechanisms

When the number of real unknowns in the complex-number or vector equations is greater than two, several methods can be used.[106,145,309] These are applicable to mechanisms with more than four links.

3.3.7   Finite Differences in Velocity and Acceleration Analysis[212,375,419,428]

When the time-displacement curve of a point in a mechanism is known, the calculus of finite differences can be used for the calculation of velocities and accelerations. The data can be numerical or analytical. The method is useful also in ascertaining the existence of local fluctuations in velocities and accelerations, such as occur in cam-follower systems, for instance.

Let a time-displacement curve be subdivided into equal time intervals Δt and define the ith, the general interval, as $t_i \le t \le t_{i+1}$, such that $\Delta t = t_{i+1} - t_i$. The "central-difference" formulas then give the following approximate values for velocities dy/dt, accelerations d^2y/dt^2, and shock d^3y/dt^3, where y_i denotes the displacement y at the time $t = t_i$:

Velocity at $t = t_i + \tfrac{1}{2}\Delta t$: $\dfrac{dy}{dt} = \dfrac{y_{i+1} - y_i}{\Delta t}$ (3.12)

Acceleration at $t = t_i$: $\dfrac{d^2y}{dt^2} = \dfrac{y_{i+1} - 2y_i + y_{i-1}}{(\Delta t)^2}$ (3.13)

Shock at $t = t_i + \tfrac{1}{2}\Delta t$: $\dfrac{d^3y}{dt^3} = \dfrac{y_{i+2} - 3y_{i+1} + 3y_i - y_{i-1}}{(\Delta t)^3}$ (3.14)

If the values of the displacements y_i are known with absolute precision (no error), the values for velocities, accelerations, and shock in the above equations become increasingly accurate as Δt approaches zero, provided the curve is smooth. If, however, the displacements y_i are known only within a given tolerance, say $\pm \epsilon y$, then the accuracy of the computations will be high only if the interval Δt is sufficiently small and, in addition, if

$$2\epsilon y/\Delta t \ll dy/dt \qquad \text{for velocities}$$

$$4\epsilon y/(\Delta t)^2 \ll d^2y/dt^2 \qquad \text{for accelerations}$$

$$8\epsilon y/(\Delta t)^3 \ll d^3y/dt^3 \qquad \text{for shock}$$

and provided also that these requirements are mutually compatible.

Further estimates of errors resulting from the use of Eqs. (3.12), (3.13), and (3.14), as well as alternate formulations involving "forward" and "backward" differences, are found in texts on numerical mathematics (e.g., Ref. 193, pp. 94–97 and 110–112, with a discussion of truncation and round-off errors).

The above equations are particularly useful when the displacement-time curve is given in the form of a numerical table, as frequently happens in checking an existing design and in redesigning.

Some current computer programs in displacement, velocity, and acceleration analysis are listed in Ref. 129; the kinematic properties of specific mechanisms, including spatial mechanisms, are summarized in Sec. 3.9.[129]

3.4 PRELIMINARY DESIGN ANALYSIS: PATH CURVATURE

The following principles apply to the analysis of a mechanism in a given position, as well as to synthesis when motion characteristics are prescribed in the vicinity of a particular position. The technique can be used to obtain a quick "first approximation" to mechanism proportions which can be refined at a later stage.

3.4.1 Polar-Coordinate Convention

Angles are measured counterclockwise from a directed line segment, the "pole tangent" PT, origin at P (see Fig. 3.18); the polar coordinates (r, ψ) of a point A are either $r = |PA|, \psi = \angle TPA$ or $r = -|PA|, \psi = \angle TPA \pm 180°$. For example, in Fig. 3.18 r is positive, but r_c is negative.

3.4.2 The Euler-Savary Equation (Fig. 3.18)

PT = common tangent of fixed and moving centrodes at point of contact P (the instant center).

PN = principal normal at P; $\angle TPN = 90°$.

PA = line or ray through P.

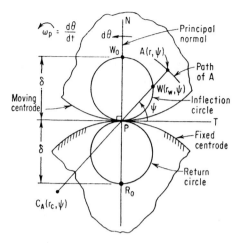

FIG. 3.18 Notation for the Euler-Savary equation.

$C_A(r_c, \psi)$ = center of curvature of path of $A(r, \psi)$ in position shown. A and C_A are called "conjugate points."

θ = angle of rotation of moving centrode, positive counterclockwise.

s = arc length along fixed centrode, measured from P, positive toward T.

The Euler-Savary equation is valid under the following assumptions:

1. During an infinitesimal displacement from the position shown, $d\theta/ds$ is finite and different from zero.
2. Point A does not coincide with P.
3. AP is finite.

Under these conditions, the curvature of the path of A in the position shown can be determined from the following "Euler-Savary" equations:

$$[(1/r) - (1/r_c)] \sin \psi = -d\theta/ds = -\omega_p/v_p \qquad (3.15)$$

where ω_p = angular velocity of moving centrode
$= d\theta/dt$, t = time

v_p = corresponding velocity of point of contact between centrodes along the fixed centrode
$= ds/dt$

Let r_w = polar coordinate of point W on ray PA, such that radius of curvature of path of W is infinite in the position shown; then W is called the "inflection point" on ray PA, and

$$1/r - 1/r_c = 1/r_w \qquad (3.16)$$

The locus of all inflection points W in the moving centrode is the "inflection circle," tangent to PT at P, of diameter $PW_0 = \delta = -ds/d\theta$, where W_0, the "inflection pole," is the inflection point on the principal normal ray. Hence,

$$[(1/r) - (1/r_c)] \sin \psi = 1/\delta \qquad (3.17)$$

The centers of path curvature of all points at infinity in the moving centrode are on the "return circle," also of diameter δ, and obtained as the reflection of the inflection circle about line PT. The reflection of W_0 is known as the "return pole" R_0. For the pole velocity (the time rate change of the position of P along the fixed centrode as the motion progresses, also called the "pole transfer velocity"[421f]) we have

$$v_p = ds/dt = - \omega_p \delta \qquad (3.18)$$

The curvatures of the paths of all points on a given ray are concave toward the inflection point on that ray.

For the diameter of the inflection and return circles we have

$$\delta = r_p r_\pi / (r_\pi - r_p) \qquad (3.19)$$

where r_p and r_π are the polar coordinates of the centers of curvature of the moving and fixed centrodes, respectively, at P. Let $\rho = r_c - r$ be the instantaneous value of the radius of curvature of the path of A, and $w = AW$, then

$$r^2 = \rho w \qquad (3.20)$$

which is known as the "quadratic form" of the Euler-Savary equation.

Conjugate points in the planes of the moving and fixed centrodes are related by a "quadratic transformation."[32] When the above assumptions 1, 2, and 3, establishing the validity of the Euler-Savary equations, are not satisfied, see Ref. 281; for a further curvature theorem, useful in relative motions, see Ref. 23. For a computer-compatible complex-number treatment of path curvature theory, see Ref. 421f, Chap. 4.

EXAMPLE Cylinder of radius 2 in, rolling inside a fixed cylinder of radius 3 in, common tangent horizontal, both cylinders above the tangent, $\delta = 6$ in, $W_0(6, 90°)$. For point $A_1(\sqrt{2}, 45°)$, $r_{c1} = 1.5\sqrt{2}$, $C_{A1}(1.5\sqrt{2}, 45°)$, $\rho_{A1} = 0.5\sqrt{2}$, $r_{w1} = 3\sqrt{2}$, $v_p = -6\omega_p$. For point $A_2(-\sqrt{2}, 135°)$, $r_{c2} = -0.75\sqrt{2}$, $C_{A2}(-0.75\sqrt{2}, 135°)$, $\rho_{A2} = 0.25\sqrt{2}$, $r_{w2} = 3\sqrt{2}$.

Complex-number forms of the Euler-Savary equation[393,421f] and related expressions are independent of the choice of the x, iy coordinate system. They correlate the following complex vectors on any one ray (see Fig. 3.18): $\mathbf{a} = \mathbf{PA}$, $\mathbf{w} = \mathbf{PW}$, $\mathbf{c} = \mathbf{PC}_A$ and $\rho = \mathbf{C}_A\mathbf{A}$, each expressed explicitly in terms of the others:

1. If points P, A, and W are known, find C_A by

$$\rho = (a^2/|\mathbf{a} - \mathbf{w}|)\, e^{i \arg (\mathbf{a} - \mathbf{w})}$$

where $a = |\mathbf{a}|$.

2. If points P, A, and C_A are known, find W by

$$\mathbf{w} = \mathbf{a} - (a/\rho)^2 \rho$$

where $\rho = |\rho|$.

3. If points P, W, and C_A are known, find A by

$$\mathbf{a} = \mathbf{wc}/(\mathbf{w} + \mathbf{c})$$

4. If points A, C_A, and W are known, find P by

$$\mathbf{a} = |(|\mathbf{WA}|\rho)^{1/2}|(\pm e^{i \arg \rho})$$

Note that the last equation yields two possible locations for P, symmetric about A. This is borne out also by Bobillier's construction (see Ref. 421f, Fig. 4.29, p. 329).

5. The vector diameter of the inflection circle, $\delta = \mathbf{PW}_0$, in complex notation:

$$\delta = -\mathbf{r}_p \mathbf{r}_\pi / (\mathbf{r}_p - \mathbf{r}_\pi) \tag{3.19a}$$

where $\mathbf{r}_p = \mathbf{O}_p \mathbf{P}$, $\mathbf{r}_\pi = \mathbf{O}_\pi \mathbf{P}$ and O_p and O_π are the centers of curvature of the fixed and moving centrodes, respectively.

6. The pole velocity in complex vector form is

$$\mathbf{v}_p = i\omega_\pi \delta \tag{3.18a}$$

where ω_π is the angular velocity of the moving centrode.

7. If points P, A, and W_0 are known:

$$\mathbf{w} = \cos (\arg \mathbf{a} - \arg \delta)\delta e^{i(\arg \mathbf{a} - \arg \delta)}$$

With the data of the above example, letting PT be the positive x axis and PN the positive iy axis, we have $\mathbf{r}_p = -i2$, $\mathbf{r}_\pi = -i3$; $\delta = -(-i2)(-i3)/(-i3 + i2) = i6$, which is the same as the vector locating the inflection pole W_0, $\mathbf{w}_0 = \mathbf{PW}_0 = i6$. For point A_1,

$$\mathbf{a}_1 = \sqrt{2}e^{i45°} \qquad \mathbf{w}_1 = \cos (45° - 90°)i6e^{i(45°-90°)} = 3\sqrt{2}e^{i45°}$$

$$\boldsymbol{\rho}_{A1} = (2/|\sqrt{2}e^{i45°} - 3\sqrt{2}e^{i45°}|) \exp[i \arg (\sqrt{2}e^{i45°} - 3\sqrt{2}e^{i45°})] = (\sqrt{2}/2)e^{i(-135°)}$$

$$\mathbf{c}_{A1} = \mathbf{a}_1 - \boldsymbol{\rho}_{A1} = \sqrt{2}e^{i45°} - (\sqrt{2}/2)e^{i(-135°)} = (3\sqrt{2}/2)e^{i45°}$$

$$\mathbf{v}_p = i\omega_\pi i6 = -\omega 6$$

For point A_2,

$$\mathbf{a}_2 = \sqrt{2}e^{i(-45°)} \qquad \mathbf{w}_2 = \cos (-45°-90°)i6e^{i(-45°-90°)} = 3\sqrt{2}e^{i135°}$$

$$\boldsymbol{\rho}_{A2} = (2/|\sqrt{2}e^{i(-45°)} - 3\sqrt{2}e^{i135°}|) \exp[i \arg (\sqrt{2}e^{i(-45°)} - 3\sqrt{2}e^{i135°})] = (\sqrt{2}/4)e^{i(-45°)}$$

and $\mathbf{C}_{A2} = \mathbf{a}_2 - \boldsymbol{\rho}_{A2} = (\sqrt{2} - \sqrt{2}/4)e^{i(-45°)} = (3\sqrt{2}/4)e^{i(-45°)}$

Note that these are equal to the previous results and are readily programmed in a digital computer.

Graphical constructions paralleling the four forms of the Euler-Savary equation are given in Refs. 394 and 421f, p. 3.27.

3.4.3 Generating Curves and Envelopes[368]

Let g-g be a smooth curve attached to the moving centrode and e-e be the curve in the fixed centrode enveloping the successive positions of g-g during the rolling of the centrodes. Then g-g is called a "generating curve" and e-e its "envelope" (Fig. 3.19).

If C_g is the center of curvature of g-g and C_e that of e-e (at M):

1. C_e, P, M, and C_g are collinear (M being the point of contact between g-g and e-e).
2. C_e and C_g are conjugate points; i.e., if C_g is considered a point of the moving cen-

trode, the center of curvature of its path lies at C_e; interchanging the fixed and moving centrodes will invert this relationship.

3. Aronhold's first theorem: The return circle is the locus of the centers of curvature of all envelopes whose generating curves are straight lines.

4. If a straight line in the moving plane always passes through a fixed point by sliding through it and rotating about it, that point is on the return circle.

5. Aronhold's second theorem: The inflection circle is the locus of the centers of curvature for all generating curves whose envelopes are straight lines.

EXAMPLE (utilizing 4 above): In the swinging-block mechanism of Fig. 3.20, point C is on the return circle, and the center of curvature of the path of C as a point of link BD is therefore at C_c, halfway between C and P. Thus $ABCC_c$ constitutes a four-bar mechanism, with C_c as a fixed pivot, equivalent to the original mechanism in the position shown with reference to path tangents and path curvatures of points in the plane of link BD.

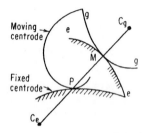

FIG. 3.19   Generating curve and envelope.

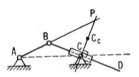

FIG. 3.20   Swinging-block mechanism: $CC_c = C_c P$.

3.4.4   Bobillier's Theorem

Consider two separate rays, 1 and 2 (Fig. 3.21), with a pair of distinct conjugate points on each, A_1, C_1, and A_2, C_2. Let Q_{A1A2} be the intersection of A_1A_2 and C_1C_2. Then the line through PQ_{A1A2} is called the "collineation axis," unique for the pair of rays 1 and 2, regardless of the choice of conjugate point pairs on these rays. Bobillier's theorem states that the angle between the common tangent of the centrodes and one ray is equal to the angle between the other ray and the collineation axis, both angles being described in the same sense.[368] Also see Ref. 421f, p. 3.31.

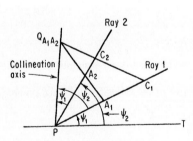

FIG. 3.21   Bobillier's construction.

The collineation axis is parallel to the line joining the inflection points on the two rays.

Bobillier's construction for determining the curvature of point-path trajectories is illustrated for two types of mechanisms in Figs. 3.22 and 3.23.

Another method for finding centers of path curvature is Hartmann's construction, described in Refs. 83 and 421f, pp. 332–336.

Occasionally, especially in the design of linkages with a dwell (temporary rest of output link), one may also use the "sextic of constant curvature," known also as the ρ curve,[32,421f] the locus of all points in the moving centrode whose paths at a given instant have the same numerical value of the radius of curvature.

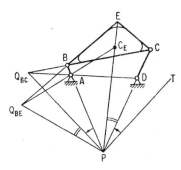

FIG. 3.22 Bobillier's construction for the center of curvature C_E of path of E on coupler of four-bar mechanism in position shown.

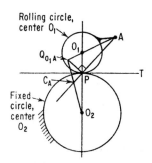

FIG. 3.23 Bobillier's construction for cycloidal motion. Determination of C_A, the center of curvature of the path of A, attached to the rolling circle (in position shown).

The equation of the ρ curve in the cartesian coordinate system in which PT is the positive x axis and PN the positive y axis is

$$(x^2 + y^2)^3 - \rho^2(x^2 + y^2 - \delta y)^2 = 0 \qquad (3.20a)$$

where ρ is the magnitude of the radius of path curvature and δ is that of the inflection circle diameter.

3.4.5 The Cubic of Stationary Curvature (the k_u Curve)[421f]

The "k_u curve" is defined as the locus of all points in the moving centrode whose rate of change of path curvature in a given position is zero: $d\rho/ds = 0$. Paths of points on this curve possess "four-point contact" with their osculating circles. Under the same assumptions as in Sec. 3.4.1, the following is the equation of the k_u curve:

$$(\sin \psi \cos \psi)/r = (\sin \psi)/m + (\cos \psi)/l \qquad (3.21)$$

where (r, ψ) = polar coordinates of a point on the k_u curve

$$m = -3\delta/(d\delta/ds) \qquad (3.22)$$

$$l = 3r_p r_\pi/(2r_\pi - r_p)$$

In cartesian coordinates (x and y axes PT and PN),

$$(x^2 + y^2)(mx + yl) - lmxy = 0 \qquad (3.23)$$

The locus of the centers of curvature of all points on the k_u curve is known as the "cubic of centers of stationary curvature,"[421f] or the "k_a curve." Its equation is

$$(x^2 + y^2)(mx + l^*y) - l^*mxy = 0 \qquad (3.24)$$

where

$$1/l - 1/l^* = 1/\delta \qquad (3.25)$$

The construction and properties of these curves are discussed in Refs. 26, 256, and 421f.

The intersection of the cubic of stationary curvature and the inflection circle yields the "Ball point" $U(r_u, \psi_u)$, which describes an approximate straight line; i.e., its path possesses four-point contact with its tangent (Ref. 421f, pp. 354–356). The coordinates of the Ball point are

$$\psi_u = \tan^{-1} \frac{2r_p - r_\pi}{(r_\pi - r_p)(d\delta/ds)} \qquad (3.26)$$

$$r_u = \delta \sin \psi_u \qquad (3.27)$$

In the case of a circle rolling inside or outside a fixed circle, the Ball point coincides with the inflection pole.

Technical applications of the cubic of stationary curvature, other than design analysis in general, include the generation of n-sided polygons,[32] the design of intermittent-motion mechanisms such as the type described in Ref. 426, and approximate straight-line generation. In many of these cases the curves degenerate into circles and straight lines.[32] Special analyses include the "Cardan positions of a plane" (osculating circle of moving centrode inside that of the fixed centrode, one-half its size; stationary inflection-circle diameter)[49,126] and dwell mechanisms. The latter utilize the "q_1 curve" (locus of points having equal radii of path curvature in two distinct positions of the moving centrode) and its conjugate, the "q_m curve." See also Ref. 395a.

3.4.6 Five and Six Infinitesimally Separated Positions of a Plane (Ref. 421f, pp. 241–245)

In the case of five infinitesimal positions, there are in general four points in the moving plane, called the "Burmester points," whose paths have "five-point contact" with their osculating circles. These points may be all real or pairwise imaginary. Their application to four-bar motion is outlined in Refs. 32, 411, 469, and 489, and related computer programs are listed in Ref. 129, the last also summarizing the applicable results of six-position theory, insofar as they pertain to four-bar motion. Burmester points and points on the cubic of stationary curvature have been used in a variety of six-link dwell mechanisms.[32,159]

3.4.7 Application of Curvature Theory to Accelerations (Ref. 421f, p. 313)

1. The acceleration $\mathbf{A}_p$ of the instant center (as a point of the moving centrode) is given by $\mathbf{A}_p = \omega^2(\mathbf{PW}_0)$; it is the only point of the moving centrode whose acceleration is independent of the angular acceleration α_p.

2. The inflection circle (also called the "de la Hire circle" in this connection) is the locus of points having zero acceleration normal to their paths.

3. The locus of all points on the moving centrode, whose tangential acceleration (i.e., acceleration along path) is zero, is another circle, the "Bresse circle," tangent to the principal normal at P, with diameter equal to $-\omega^2\delta/\alpha_p$ where α_p is the angular acceleration of the moving centrode, the positive sense of which is the same as that of θ. In complex vector form the diameter of the Bresse circle is $i\omega_p^2\delta/\alpha_p$ (Ref. 421f, pp. 336–338).

4. The intersection of these circles, other than P, determines the point F, with zero total acceleration, known as the "acceleration center." It is located at the intersection of the inflection circle and a ray of angle γ, where

$$\gamma = \angle\, W_0PF = \tan^{-1}(\alpha_p/\omega_p^2) \qquad 0 \leq |\gamma| \leq 90°$$

measured in the direction of the angular acceleration (Ref. 421f, p. 337).

5. The acceleration $\mathbf{A}_B$ of any point B in the moving system is proportional to its distance from the acceleration center:

$$\mathbf{A}_B = (\mathbf{B\Gamma})(d^{-i\gamma})|(\omega_p^4 + \alpha_p^2)^{1/2}| \tag{3.28}$$

6. The acceleration vector $\mathbf{A}_B$ of any point B makes an angle γ with the line joining it to the acceleration center [see Eq. (3.28)], where γ is measured from $\mathbf{A}_B$ in the direction of angular acceleration (Ref. 421f, p. 340).

7. When the acceleration vectors of two points (V, U) on one link, other than the pole, are known, the location of the acceleration center can be determined from item 6 and the equation

$$|\tan \gamma| = \frac{|A_{U/V}^t|}{A_{U/V}^n}$$

8. The concept of acceleration centers and images can be extended also to the higher accelerations[41] (see also Sec. 3.3).

3.4.8 Examples of Mechanism Design and Analysis Based on Path Curvature

1. Mechanism used in guiding the grinding tool in large gear generators (Fig. 3.24): The radius of path curvature ρ_m of M at the instant shown: $\rho_m = (W_1 W_2)/(2 \tan^3 \theta)$, at which instant M is on the cubic of stationary curvature belonging to link $W_1 W_2$; ρ_m is arbitrarily large if θ is sufficiently small.

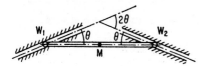

FIG. 3.24 Mechanism used in guiding the grinding tool in large gear generators. (Due to A. H. Candee, Rochester, N.Y.) $MW_1 = MW_2$; link $W_1 W_2$ constrained by straight-line guides for W_1 and W_2.

2. Machining of radii on tensile test specimens[175,488] (Fig. 3.25): C lies on cubic of stationary curvature; AB is the diameter of the inflection circle for the motion of link ABC; radius of curvature of path of C in the position shown:

$$\rho_c = (AC)^2/(BC)$$

3. Pendulum with large period of oscillation, yet limited size[283,434] (Fig. 3.26), as used

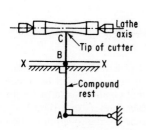

FIG. 3.25 Machining of radii on tensile test specimens. B guided along $X - X$.

FIG. 3.26 Pendulum with large period of oscillation.

in recording ship's vibrations: $AB = a$, $AC = b$, $CS = s$, r_t = radius of gyration of the heavy mass S about its center of gravity. If the mass other than S and friction are negligible, the length l of the equivalent simple pendulum is given by

$$l = \frac{r_t^2 + s^2}{(b/a)(b - a)} - s$$

where the distance CW is equal to $(b/a)(b - a)$. The location of S is slightly below the inflection point W, in order for the oscillation to be stable and slow.

4. Modified geneva drive in high-speed bread wrapper[377] (Fig. 3.27): The driving pin of the geneva motion can be located at or near the Ball point of the pinion motion; the path of the Ball point, approximately square, can be used to give better kinematic characteristics to a four-station geneva than the regular crankpin design, by reducing peak velocities and accelerations.

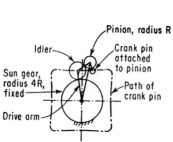

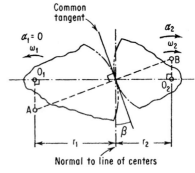

FIG. 3.27 Modified geneva drive in high-speed bread wrapper.

FIG. 3.28 Angular acceleration diagram for noncircular gears.

5. Angular acceleration of noncircular gears (obtainable from equivalent linkage O_1ABO_2) (Ref. 116, discussion by A. H. Candee; Fig. 3.28):

Let ω_1 = angular velocity of left gear, assumed constant, counterclockwise
  ω_2 = angular velocity of right gear, clockwise
  α_2 = clockwise angular acceleration of right gear

Then $\alpha_2 = [r_1(r_1 + r_2)/r_2^2]\,(\tan \beta)\omega_1^2$

3.5 DIMENSIONAL SYNTHESIS: PATH, FUNCTION, AND MOTION GENERATION[106,421f]

In the design of automatic machinery, it is often required to guide a part through a sequence of prescribed positions. Such motions can be mechanized by dimensional synthesis based on the kinematic geometry of distinct positions of a plane. In plane motion, a "kinematic plane," hereafter called a "plane," refers to a rigid body, arbitrary in extent. The position of a plane is determined by the location of two of its points, A and B, designated as A_i, B_i in the ith position.

3.5.1 Two Positions of a Plane

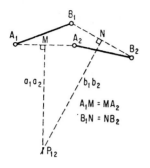

FIG. 3.29 Two positions of a plane. Pole $P_{12} = a_1 a_2 \times b_1 b_2$.

According to "Chasles's theorem," the motion from $A_1 B_1$ to $A_2 B_2$ (Fig. 3.29) can be considered *as though* it were a rotation about a point P_{12}, called the pole, which is the intersection of the perpendicular bisectors $a_1 a_2$, $b_1 b_2$ of $A_1 A_2$ and $B_1 B_2$, respectively. A_1, A_2, ..., are called "corresponding positions" of point A; B_1, B_2, ..., those of point B; $A_1 B_1$, $A_2 B_2$, ..., those of the plane AB.

A similar construction applies to the "relative motion of two planes" (Fig. 3.30) AB and CD (positions $A_i B_i$ and $C_i D_i$, $i = 1, 2$). The "relative pole" Q_{12} is constructed by transferring the figure $A_2 B_2 C_2 D_2$ as a rigid body to bring A_2 and B_2 into coincidence with A_1 and B_1, respectively, and denoting the new positions of C_2, D_2, by C_2^1, D_2^1, respectively. Then Q_{12} is obtained from $C_1 D_1$ and $C_2^1 D_2^1$ as in Fig. 3.29.

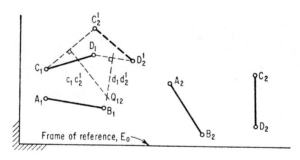

FIG. 3.30 Relative motion of two planes, AB and CD. Relative pole, $Q_{12} = c_1 c_2^1 \times d_1 d_2^1$.

1. The motion of $A_1 B_1$ to $A_2 B_2$ in Fig. 3.29 can be carried out by four-link mechanisms in which A and B are coupler-hinge pivots and the fixed-link pivots A_0, B_0 are located on the perpendicular bisectors $a_1 a_2$, $b_1 b_2$, respectively.
2. To construct a four-bar mechanism $A_0 A B B_0$ when the corresponding angles of rotation of the two cranks are prescribed (in Fig. 3.31 the construction is illustrated with ϕ_{12} clockwise for $A_0 A$ and ψ_{12} clockwise for $B_0 B$):
    a. From line $A_0 B_0 X$, lay off angles $\frac{1}{2}\phi_{12}$ and $\frac{1}{2}\psi_{12}$ opposite to desired direction of rotation of the cranks, locating Q_{12} as shown.
    b. Draw any two straight lines L_1 and L_2 through Q_{12}, such that

$$\angle L_1 Q_{12} L_2 = \angle A_0 Q_{12} B_0$$

in magnitude and sense.

c. A_1 can be located on L_1, B_1 and L_2, and when $A_0 A_1$ rotates clockwise by ϕ_{12}, $B_0 B_1$ will rotate clockwise by ψ_{12}. Care must be taken, however, to ensure that the mechanism will not lock in an intermediate position.

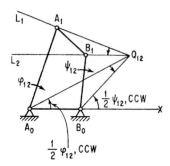

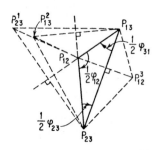

FIG. 3.31 Construction of four-bar mechanism $A_0A_1B_1B_0$ in position 1, for prescribed rotations ϕ_{12} vs. ψ_{12}, both clockwise in this case.

FIG. 3.32 Pole triangle for three positions of a plane. Pole triangle $P_{12}P_{23}P_{13}$ for three positions of a plane; image poles P_{12}^3, P_{23}^1, P_{31}^2; subtended angles $\frac{1}{2}\phi_{23}$, $\frac{1}{2}\phi_{31}$.

3.5.2 Three Positions of a Plane $(A_iB_i,\ i = 1, 2, 3)$[420]

In this case there are three poles P_{12}, P_{23}, P_{31} and three associated rotations ϕ_{12}, ϕ_{23}, ϕ_{31}, where $\phi_{ij} = \angle\ A_iP_{ij}A_j = \angle\ B_iB_{ij}B_j$. The three poles form the vertices of the *pole triangle* (Fig. 3.32). Note that $P_{ij} = P_{ji}$, and $\phi_{ij} = -\ \phi_{ji}$.

Theorem of the Pole Triangle. The internal angles of the pole triangle, corresponding to three distinct positions of a plane, are equal to the corresponding halves of the associated angles of rotation ϕ_{ij} which are connected by the equation

$$\tfrac{1}{2}\phi_{12} + \tfrac{1}{2}\phi_{23} + \tfrac{1}{2}\phi_{31} = 180° \qquad \angle\ \tfrac{1}{2}\phi_{ij} = \angle\ P_{ik}P_{ij}P_{jk}$$

Further developments, especially those involving subtention of equal angles, are found in the literature.[32]

For any three corresponding points A_1, A_2, A_3, the center M of the circle passing through these points is called a "center point." If P_{ij} is considered as though fixed to link A_iB_i (or A_jB_j) and A_iB_i (or A_jB_j) is transferred to position k (A_kB_k), then P_{ij} moves to a new position P_{ij}^k, known as the "image pole," because it is the image of P_{ij} reflected about the line joining $P_{ik}P_{jk}$. $\Delta P_{ik}P_{jk}P_{ij}^k$ is called an "image-pole triangle" (Fig. 3.32). For "circle-point" and "center-point circles" for three finite positions of a moving plane, see Ref. 106, pp. 436–446 and Ref. 421*f*, pp. 114–122.

3.5.3 Four Positions of a Plane $(A_iB_i,\ i = 1, 2, 3, 4)$

With four distinct positions, there are six poles P_{12}, P_{13}, P_{14}, P_{23}, P_{24}, P_{34} and four pole triangles $(P_{12}P_{23}P_{13})$, $(P_{12}P_{24}P_{14})$, $(P_{13}P_{34}P_{14})$, $(P_{23}P_{34}P_{24})$.

Any two poles whose subscripts are all different are called "complementary poles." For example, $P_{23}P_{14}$, or generally $P_{ij}P_{kl}$, where i, j, k, l represents any permutation of the numbers 1, 2, 3, 4. Two complementary-pole pairs constitute the two diagonals of a "complementary-pole quadrilateral," of which there are three: $(P_{12}P_{24}P_{33}P_{13})$, $(P_{13}P_{32}P_{24}P_{14})$, and $(P_{14}P_{43}P_{32}P_{12})$.

Also associated with four positions are six further points $\prod_{ik}$ found by intersections of opposite sides of complementary-pole quadrilaterals, or their extensions, as follows: $\prod_{ik} = P_{il}P_{kl} \times P_{ij}P_{kj}$.

3.5.4　The Center-Point Curve or Pole Curve[32,67,127,421f]

For three positions, a center point corresponds to *any* set of corresponding points; for four corresponding points to have a common center point, point A_1 can no longer be located arbitrarily in plane AB. However, a curve exists in the frame of reference called the "center-point curve" or "pole curve," which is the locus of centers of circles, each of which passes through *four* corresponding points of the plane AB. The center-point curve may be obtained from any complementary-pole quadrilateral; if associated with positions i, j, k, l, the center-point curve will be denoted by m_{ijkl}. Using complex numbers, let $\mathbf{OP}_{13} = \mathbf{a}$, $\mathbf{OP}_{23} = \mathbf{b}$, $\mathbf{OP}_{14} = \mathbf{c}$, $\mathbf{OP}_{24} = \mathbf{d}$, and $\mathbf{OM} = \mathbf{z} = x + iy$, where $\mathbf{OM}$ represents the vector from an arbitrary origin O to a point M on the center-point curve. The equation of the center-point[127] curve is given by

$$\frac{(\bar{\mathbf{z}} - \bar{\mathbf{a}})(\mathbf{z} - \mathbf{b})}{(\mathbf{z} - \mathbf{a})(\bar{\mathbf{z}} - \bar{\mathbf{b}})} = \frac{(\bar{\mathbf{z}} - \bar{\mathbf{c}})(\mathbf{z} - \mathbf{d})}{(\mathbf{z} - \mathbf{c})(\bar{\mathbf{z}} - \bar{\mathbf{d}})} = e^{2i\varphi} \tag{3.29}$$

where $\varphi = \angle\, P_{16}MP_{23} = \angle\, P_{14}MP_{24}$. In cartesian coordinates with origin at P_{12}, this curve is given in Ref. 16 by the following equation:

$$(x^2 + y^2)(j_2x - j_1y) + (j_1k_2 - j_2k_1 - j_3)x^2 + (j_1k_2 - j_2k_1 + j_3)y^2 + 2j_4xy$$

$$+ (-j_1k_3 + j_2k_4 + j_3k_1 - j_4k_2)x + (j_1k_4 + j_2k_3 - j_3k_2 - j_4k_1)y = 0 \tag{3.30}$$

where

$$k_1 = x_{13} + x_{24}$$

$$k_2 = y_{13} + y_{24}$$

$$k_3 = x_{13}y_{24} + y_{13}x_{24}$$

$$k_4 = x_{13}x_{24} - y_{13}y_{24}$$

$$j_1 = x_{23} + x_{14} - k_1 \tag{3.31}$$

$$j_2 = y_{23} + y_{14} - k_2$$

$$j_3 = x_{23}y_{14} + x_{14}y_{23} - k_3$$

$$j_4 = x_{23}x_{14} - y_{23}y_{14} - k_4$$

and (x_{ij}, y_{ij}) are the cartesian coordinates of pole P_{ij}. Equation (3.30) represents a third-degree algebraic curve, passing through the six poles P_{ij} and the six points Π_{ij}. Furthermore, any point M on the center-point curve subtends equal angles, or angles differing by two right angles, at opposite sides $(P_{ij}P_{jl})$ and $(P_{ik}P_{kl})$ of a complementary-pole quadrilateral, provided the sense of rotation of subtended angles is preserved:

$$\angle\, P_{ij}MP_{jl} = \angle\, P_{ik}MP_{kl} \cdots \tag{3.32}$$

Construction of the Center-Point Curve m_{ijkr}[32]　　When the four positions of a plane are known $(A_i, B_i, i = 1, 2, 3, 4)$, the poles P_{ij} are constructed first; thereafter, the center-point curve is found as follows:
　　A chord $P_{ij}P_{jk}$ of a circle, center O, radius

$$R = \overline{P_{ij}P_{jk}}/2 \sin \theta$$

(Fig. 3.33) subtends the angle θ (mod π) at any point on its circumference. For any value of θ, $-180° \le \theta \le 180°$, two corresponding circles can be drawn following Fig.

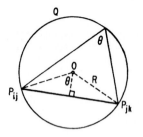

FIG. 3.33   Subtention of equal angles.

3.33, using as chords the opposite sides $P_{ij}P_{jk}$ and $P_{il}P_{kl}$ of a complementary-pole quadrilateral; intersections of such corresponding circles are points (M) on the center-point curve, provided Eq. (3.32) is satisfied.

As a checF, it is useful to keep in mind the following angular equalities:

$$\tfrac{1}{2} \angle A_i MA_l = \angle P_{ij}MP_{jl} = \angle P_{ik}MP_{kl}$$

Also see Ref. 421f, p. 189.

Use of the Center-Point Curve.   Given four positions of a plane $A_i B_i$ ($i = 1, 2, 3, 4$) in a coplanar motion-transfer process, we can mechanize the motion by selecting points on the center-point curve as fixed pivots.

EXAMPLE[91]   A stacker conveyor for corrugated boxes is based on the design shown schematically in Fig. 3.34. The path of C should be as nearly vertical as possible; if A_0, A_1, AC, $C_1C_2C_3C_4$ are chosen to suit the specifications, B_0 should be chosen on the center-point curve determined from A_iC_i, $i = 1, 2, 3, 4$; B_1 is then readily determined by inversion, i.e., by drawing the motion of B_0 relative to A_1C_1 and locating B_1 at the center of the circle thus described by B_0 (also see next paragraphs).

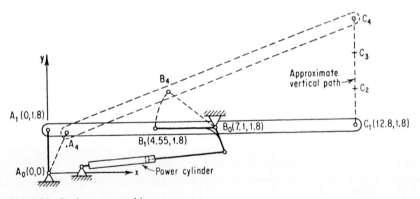

FIG. 3.34   Stacker conveyor drive.

3.5.5   The Circle-Point Curve

The circle-point curve is the kinematical inverse of the center-point curve. It is the locus of all points K in the moving plane whose four corresponding positions lie on one circle. If the circle-point curve is to be determined for positions i of the plane AB, Eqs. (3.29), (3.30), and (3.31) would remain unchanged, except that P_{jk}, P_{kl}, and P_{jl} would be replaced by the image poles P_{jk}^i, P_{kl}^i, and P_{jl}^i, respectively.

The center-point curve lies in the frame or reference plane; the circle-point curve lies in the moving plane. In the above example, point B_1 is on the circle-point curve for plane AC in position 1. The example can be solved also by selecting B_1 on the circle-point curve in A_1C_1; B_0 is then the center of the circle through $B_1B_2B_3B_4$. A computer program for the center-point and circle-point curves (also called "Burmester curves") is outlined in Refs. 383 and 421f, p. 184.

SPECIAL CASE   If the corresponding points $A_1A_2A_3$ lie on a straight line, A_1 must lie on the circle through $P_{12}P_{13}P_{23}{}^1$; for four corresponding points $A_1A_2A_3A_4$ on one straight line, A_1 is located at intersection, other than P_{12}, of circles through $P_{12}P_{13}P_{23}^1$ and $P_{12}P_{14}P_{24}^1$, respectively. Applied to straight-line guidance in slider-crank and four-bar drives in Ref. 251; see also Refs. 32 and 421f, pp. 491–494.

3.5.6   Five Positions of a Plane (A_iB_i, $i = 1, 2, 3, 4, 5$)

In order to obtain accurate motions, it is desirable to specify as many positions as possible; at the same time the design process becomes more involved, and the number of "solutions" becomes more restricted. Frequently four or five positions are the most that can be economically prescribed.

Associated with five positions of a plane are four sets of points $K_u^{(i)}$ ($u = 1, 2, 3, 4$ and i is the position index as before) whose corresponding five positions lie on one circle; to each of these circles, moreover, corresponds a center point M_u. These circle points $K_u^{(1)}$ and corresponding center points M_u are called "Burmester point pairs." These four point pairs may be all real or pairwise imaginary (all real, two-point pairs real and two point pairs imaginary, or all point pairs imaginary).[127,421f] Note the difference, for historical reasons, between the above definition and that given in Sec. 3.4.6 for infinitesimal motion. The location of the center points, M_u, can be obtained as the intersections of two center-point curves, such as m_{1234} and m_{1235}.

A complex-number derivation of their location,[127,421f] as well as a computer program for simultaneous determination of the coordinates of both M_u and $K_u^{(i)}$ is available.[108,127,380,421f]

An algebraic equation for the coordinates (x_u, y_u) of M_u is given in Ref. 16 as follows. Origin at P_{12}, coordinates of P_{ij} are x_{ij}, y_{ij}.

$$x_u = \frac{(u - \tan \tfrac{1}{2}\theta_{12})[l_1(k_2 - k_3 u) - l_2(e_2 - e_3 u)]}{p_1 u^2 + p_2 u + p_3}$$

(3.33)

$$y_u = \frac{(u - \tan \tfrac{1}{2}\theta_{12})[l_1(k_4 + k_1 u) - l_2(e_4 + e_1 u)]}{p_1 u^2 + p_2 u + p_3}$$

where

$$\tan \tfrac{1}{2}\theta_{12} = (x_{13}y_{23} - x_{23}y_{13})(x_{13}x_{23} + y_{13}y_{23})$$

(3.34)

and u is a root of

$$m_4 u^4 + m_3 u^3 + m_2 u^2 + m_1 u + m_0 = 0$$

wherein

$$m_0 = p_3(q_1 + l_3 p_3)$$

$$m_1 = p_2(q_1 + 2l_3 p_3)p_2 + q_2 p_3 - q_3 \tan \tfrac{1}{2}\theta_{12}$$

$$m_2 = q_0 p_3 + q_2 p_2 + q_1 p_1 + l_3(p_2^2 + 2p_1 p_3) - q_5 \tan \tfrac{1}{2}\theta_{12} + q_3$$

(3.35)

$$m_3 = q_0 p_2 + p_1(q_2 + 2l_3 p_2) - q_4 \tan \tfrac{1}{2}\theta_{12} + q_5$$

$$m_4 = p_1(q_0 + l_3 p_1) + q_4$$

$$
\begin{array}{ll}
q_0 = d_1 h_3 - d_3 h_1 & \qquad h_1 = k_1 l_1 - e_1 l_2 \\
q_1 = d_2 h_4 - d_4 h_2 & \qquad h_2 = k_2 l_1 - e_2 l_2 \\
q_2 = -d_1 h_2 + d_2 h_1 - d_3 h_4 + d_4 h_3 & \qquad h_3 = k_3 l_1 - e_3 l_2 \\
q_3 = h_2^2 + h_4^2 & \qquad h_4 = k_4 l_1 - e_4 l_2
\end{array}
$$

(3.36)

$$q_4 = h_1^2 + h_3^2 \qquad\qquad p_1 = k_3 e_1 - k_1 e_3$$

$$q_5 = 2(h_1 h_4 - h_2 h_3) \qquad p_2 = k_3 e_4 + k_1 e_2 - k_2 e_1 - k_4 e_3$$

$$d_1 = x_{15} + x_{25} \qquad\qquad p_3 = k_4 e_2 - k_2 e_4$$

$$d_2 = x_{15} - x_{25} \qquad\qquad e_1 = d_1 - x_{13} - x_{23}$$

$$d_3 = y_{15} + y_{25} \qquad\qquad e_2 = d_2 - x_{13} + x_{23}$$

$$d_4 = y_{15} - y_{25} \qquad\qquad e_3 = d_3 - y_{13} - y_{23} \qquad (3.36)$$

$$k_1 = d_1 - x_{14} - x_{24} \qquad e_4 = d_4 - y_{13} + y_{23}$$

$$k_2 = d_2 - x_{14} + x_{24} \qquad l_1 = x_{13} x_{23} + y_{13} y_{23} - l_3$$

$$k_3 = d_3 - y_{14} - y_{24} \qquad l_2 = x_{14} x_{24} + y_{14} y_{24} - l_3$$

$$k_4 = d_4 - y_{14} + y_{24} \qquad l_3 = x_{15} x_{25} + y_{15} y_{25}$$

The Burmester point pairs are discussed in Refs. 16, 67, and 127 and extensions of the theory in Refs. 382, 400, and 421*f*, pp. 211–230. It is suggested that, except in special cases, their determination warrants programmed computation.[108,421f]

Use of the Burmester Point Pairs.   As in the example of Sec. 3.5.4, the Burmester point pairs frequently serve as convenient pivot points in the design of linked mechanisms. Thus, in the stacker of Sec. 3.5.4, five positions of C_j could have been specified in order to obtain a more accurately vertical path for C; the choice of locations of B_0 and B_1 would then have been limited to at most two Burmester point pairs (since $A_0 A_1$ and $C_1 C_0^\infty$, prescribed, are also Burmester point pairs).

3.5.7   Point-Position Reduction[2,159,194,421f]

"Point-position reduction" refers to a construction for simplifying design procedures involving several positions of a plane. For five positions, graphical methods would involve the construction of two center-point curves or their equivalent. In point-position reduction, a fixed-pivot location, for instance, would be chosen so that one or more poles coincide with it. In the relative motion of the fixed pivot with reference to the moving plane, therefore, one or more of the corresponding positions coincide, thereby reducing the problem to four or fewer positions of the pivot point; the center-point curves, therefore, may not have to be drawn. The reduction in complexity of construction is accompanied, however, by increased restrictions in the choice of mechanism proportions. An exhaustive discussion of this useful tool is found in Ref. 159.

3.5.8   Complex-Number Methods[106,123,371,372,380,381,421f,435]

Burmester-point theory has been applied to function generation as well as to path generation and combined path and function generation.[106,127,380,421f] The most general approach to path and function generation in plane motion utilizes complex numbers. The vector closure equations are used for each independent loop of the mechanism for every prescribed position and are differentiated once or several times if velocities, accelerations, and higher rates of change are prescribed. The equations are then solved for the

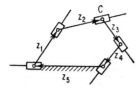

unknown mechanism proportions. This method has been applied to four-bar path and function generators[106,123,127,371,380,384,421f] (the former with prescribed crank rotations), as well as to a variety of other mechanisms. The so-called "path-increment" and "path-increment-ratio" techniques (see below) simplify the mathematics insofar as this is possible. In addition to path and function specification, these methods can take into account prescribed

FIG. 3.35 Mechanism derived from a bar-slider chain.

transmission angles, mechanical advantages, velocity ratios, accelerations, etc., and combinations of these.

Consider, for instance, a chain of links connected by turning-sliding joints (Fig. 3.35). Each bar slider is represented by the vector $\mathbf{z}_j = r_j e^{i\theta_j}$. In this case the closure equation for the position shown, and its derivatives are as follows.

Closure:
$$\sum_{j=1}^{5} \mathbf{z}_j = 0$$

Velocity:
$$\frac{d}{dt} \sum_{j=1}^{5} \mathbf{z}_j = 0 \qquad \text{or} \qquad \sum_{j=1}^{5} \lambda_j \mathbf{z}_j = 0$$

where
$$\lambda_j = (1/r_j)(dr_j/dt) + i(d\theta_j/dt) \qquad (t = \text{time})$$

Acceleration:
$$\frac{d}{dt} \sum_{j=1}^{5} \lambda_j \mathbf{z}_j = 0 \qquad \text{or} \qquad \sum_{j=1}^{5} \lambda_j \mu_j \mathbf{z}_j = 0$$

where
$$\mu_j = \lambda_j + (1/\lambda_j)(d\lambda_j/dt)$$

Similar equations hold for other positions. After suitable constraints are applied on the bar-slider chain (i.e., on r_j, θ_j) in accordance with the properties of the particular type of mechanism under consideration, the equations are solved for the $\mathbf{z}_j$ vectors, i.e., for the "initial" mechanism configuration.

If the path of a point such as C in Fig. 3.35 (although not necessarily a joint in the actual mechanism represented by the schematic or "general" chain) is specified for a number of positions by means of vectors δ_1, δ_2, ..., δ_k, the "path increments" measured from the initial position are $(\delta_j - \delta_1)$, $j = 2, 3, ..., k$. Similarly, the "path increment ratios" are $(\delta_j - \delta_1)/(\delta_2 - \delta_1)$, $j = 3, 4, ..., k$. By working with these quantities, only moving links or their ratios are involved in the computations. The solution of these equations of synthesis usually involves the prior solution of nonlinear "compatibility equations," obtained from matrix considerations. Additional details are covered in the above-mentioned references. A number of related computer programs for the synthesis of linked mechanisms are described in Refs. 129 and 421f. Numerical methods suitable for such syntheses are described in Ref. 372.

3.6 DESIGN REFINEMENT

After the mechanism is selected and its approximate dimensions determined, it may be necessary to refine the design by means of relatively small changes in the proportions, based on more precise design considerations. Equivalent mechanisms and cognates (see Sec. 3.6.6) may also present improvements.

3.6.1 Optimization of Proportions for Generating Prescribed Motions with Minimum Error

Whenever mechanisms possess a limited number of independent dimensions, only a finite number of independent conditions can be imposed on their motion. Thus, if a path is to be generated by a point on a linkage (rather than, say, a cam follower), it is not possible—except in special cases—to generate the curve exactly. A desired path (or function) and the actual, or generated, path (or function) may coincide at several points, called "precision points"; between these, the curves differ.

The minimum distance from a point on the ideal path to the actual path is called the "structural error in path generation." The "structural error in function generation" is defined as the error in the ordinate (dependent variable y) for a given value of the abscissa (independent variable x). Structural errors exist independent of manufacturing tolerances and elastic deformations and are thus inherent in the design. The combined effect of these errors should not exceed the maximum tolerable error.

The structural error can be minimized by the application of the fundamental theorem of P. L. Chebyshev[16,42] phrased nonrigorously for mechanisms as follows:

If n independent, adjustable proportions (parameters) are involved in the design of a mechanism, which is to generate a prescribed path or function, then the largest absolute value of the structural error is minimized when there are n precision points so spaced that the $n + 1$ maximum values of the structural error between each pair of adjacent precision points—as well as between terminals and the nearest precision points—are numerically equal with successive alterations in sign.

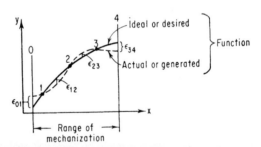

FIG. 3.36 Precision points 1, 2, and 3 and "regions" 01, 12, 23, and 34, in function generation.

In Fig. 3.36 (applied to function generation) the maximum structural error in each "region," such as 01, 12, 23, and 34, is shown as ϵ_{01}, ϵ_{12}, ϵ_{23}, and ϵ_{34}, respectively, which represent vertical distances between ideal and generated functions having three precision points. In general, the mechanism proportions and the structural error will vary with the choice of precision points. The spacing of precision points which yields least maximum structural error is called "optimal spacing." Other definitions and concepts, useful in this connection, are the following:

n-point approximation: Generated path (or function) has n precision points.

nth-order approximation: Limiting case of n-point approximation, as the spacing between precision points approaches zero. In the limit, one precision point is retained, at which point, however, the first $n - 1$ derivatives, or rates of change of the generated path (or function), have the same values as those of the ideal path (or function).

The following paragraphs apply both to function generation and to planar path generation, provided (in the latter case) that x is interpreted as the arc length along the ideal curve and the structural error ϵ_{ij} refers to the distance between generated and ideal curves.

Chebyshev Spacing.[122]   For an n-point approximation to $y = f(x)$, within the range $x_0 \leq x \leq x_{n+1}$, Chebyshev spacing of the n precision points x_j is given by

$$x_j = \tfrac{1}{2}(x_0 + x_{n+1}) - \tfrac{1}{2}(x_{n+1} - x_0) \cos\{[2j - 1)\pi]/2n\} \qquad j = 1, 2, \ldots, n$$

Though not generally optimum for finite ranges, Chebyshev spacing often represents a good first approximation to optimal spacing.

The process of respacing the precision points, so as to minimize the maximum structural error, is carried out numerically[122] unless an algebraic solution is feasible.[42,404]

Respacing of Precision Points to Reduce Structural Error via Successive Approximations.   Let $x_{ij}^{(1)} = x_j^{(1)} - x_i^{(1)}$, where $j = i + 1$, and let $x_i^{(1)}(i = 1, 2, \ldots, n)$ represent precision-point locations in a first approximation as indicated by the superscript (1). Let $\epsilon_{ij}^{(1)}$ represent the maximum structural error between points $x_i^{(1)}, x_j^{(1)}$, in the first approximation with terminal values x_0, x_{n+1}. Then a second spacing $x_{ij}^{(2)} = x_j^{(2)} - x_i^{(2)}$ is sought for which $\epsilon_{ij}^{(2)}$ values are intended to be closer to optimum (i.e., more nearly equal); it is obtained from

$$x_{ij}^{(2)} = \frac{x_{ij}^{(1)}(x_{n+1} - x_0)}{[\epsilon_{ij}^{(1)}]^m \sum\limits_{i=0}^{n} \{x_{ij}^{(1)}/[\epsilon_{ij}^{(1)}]^m\}} \tag{3.37}$$

The value of the exponent m generally lies between 1 and 3. Errors can be minimized also according to other criteria, for instance, according to least squares.[249] Also see Ref. 404.

Estimate of Least Possible Maximum Structural Error.   In the case of an n-point Chebyshev spacing in the range $x_0 \leq x \leq x_n + 1$ with maximum structural errors ϵ_{ij} ($j = i + 1; i = 0, 1, \ldots, n$),

$$\epsilon_{\text{opt(estimate)}}^2 = (1/2n)[\epsilon_{01}^2 + \epsilon_{n(n+1)}^2] + (1/n)[\epsilon_{12}^2 + \epsilon_{23}^2 + \cdots + \epsilon_{(n-1)n}^2] \tag{3.38}$$

In other spacings different estimates should be used; in the absence of more refined evaluations, the root-mean-square value of the prevailing errors can be used in the general case. These estimates may show whether a refinement of precision-point spacing is worthwhile.

Chebyshev Polynomials.   Concerning the effects of increasing the number of precision points or changing the range, some degree of information may be gained from an examination of the "Chebyshev polynomials." The Chebyshev polynomial $T_n(t)$ is that nth-degree polynomial in t (with leading coefficient unity) which deviates least from zero within the interval $\alpha \leq t \leq \beta$. It can be obtained from the following differential-equation identity by equating to zero coefficients of like powers of t:

$$2[t^2 - (\alpha + \beta)t + \alpha\beta] \, T_n''(t) + [2t - (\alpha + \beta)]T_n'(t) - 2n^2 T_n(t) \equiv 0$$

where the primes refer to differentiation with respect to t. The maximum deviation from zero, L_n, is given by

$$L_n = (\alpha - \beta)^n/2^{2n-1}$$

For the interval $-1 \leq t \leq 1$, for instance,

$$T_n(t) = (1/2^{n-1}) \cos (n \cos^{-1} t) \qquad L_n = 1/2^{n-1}$$

$$T_1(t) = t$$

$$T_2(t) = t^2 - \tfrac{1}{2}$$

$$T_3(t) = t^3 - (\tfrac{3}{4})t$$

$$T_4(t) = t^4 - t^2 + \tfrac{1}{8}$$

. . . .

Chebyshev polynomials can be used directly in algebraic synthesis, provided the motion and proportions of the mechanism can be suitably expressed in terms of such polynomials.[42]

Adjusting the Dimensions of a Mechanism for Given Respacing of Precision Points. Once the respacing of the precision points is known, it is possible to recompute the mechanism dimensions by a linear computation[122,174,249,446] provided the changes in the dimensions are sufficiently small.

Let $f(x)$ = ideal or desired functional relationship.

$\quad g(x) = g(x, p_0^{(1)}, p_1^{(1)}, \ldots, p_{n-1}^{(1)})$
$\qquad$ = generated functional relationship in terms of mechanism parameters or pro-
$\qquad$ portions $p_j^{(1)}$, where $p_j^{(k)}$ refers to the jth parameter in the kth approximation.

$\epsilon^{(1)}(x_i^{(2)})$ = value of structural error at $x_i^{(2)}$ in the first approximation, where $x_i^{(2)}$ is a new or
$\qquad$ respaced location of a precision point, such that ideally $\epsilon^{(2)}(x_i^{(2)}) = 0$ (where ϵ
$\qquad$ $= f - g$).

Then the new values of the parameters $p_j^{(2)}$ can be computed from the equations

$$\epsilon^{(1)}(x_i^{(2)}) = \sum_{j=0}^{n-1} \frac{\partial g(x_i^{(2)})}{\partial p_j^{(1)}} (p_j^{(2)} - p_j^{(1)}) \qquad i = 1, 2, \ldots, n \qquad (3.39)$$

These are n linear equations, one each at the n "precision" points $x_i^{(2)}$ in the n unknowns $p_j^{(2)}$. The convergence of this procedure depends on the appropriateness of neglecting higher-order terms in Eq. (3.39); this, in turn, depends on the functional relationship and the mechanism and cannot in general be predicted. For related investigations, see Refs. 131 and 184; for respacing via automatic computation and for accuracy obtainable in four-bar function generators, see Ref. 122, and in geared five-bar function generators, see Ref. 397.

3.6.2 Tolerances and Precision[17,147,158,174,228,243,482]

After the structural error is minimized, the effects of manufacturing errors still remain.

The accuracy of a motion is frequently expressed as a percentage defined as the maximum output error divided by total output travel (range).

For a general discussion of the various types of errors, see Ref. 482.

Machining errors may cause changes in link dimensions, as well as clearances and backlash. Correct tolerancing requires the investigation of both. If the errors in link dimensions are small compared with the link lengths, their effect on displacements, velocities, and accelerations can be determined by a linear computation, using only first-order terms.

The effects of clearances in the joints and of backlash are more complicated and, in addition to kinematic effects, are likely to affect adversely the dynamic behavior of the mechanism.[147] The kinematic effect manifests itself as an uncertainty in displacements, velocities, accelerations, etc., which, in the absence of load reversal, can be computed as though due to a change in link length, equivalent to the clearance or backlash involved. The dynamic effects of clearances in machinery have been investigated in Ref. 98 to 100.

Since the effect of tolerances will depend on the mechanism and on the "location" of the tolerance in the mechanism, each tolerance should be specified in accordance with the magnitude of its effect on the pertinent kinematic behavior.

3.6.3 Harmonic Analysis (see also Sec. 3.9 and bibliography in Ref. 493)

It is sometimes desirable to express the motion of a machine part as a Fourier series in terms of driving motion, in order to analyze dynamic characteristics and to ensure satisfactory performance at high speeds. Harmonic analysis, for example, is used in computing the inertia forces in slider-crank mechanisms in internal-combustion engines[39,367] and also in other mechanisms.[128,286,289,493]

Generally, two types of investigations arise:

1. Determination of the "harmonics" in the motion of a given mechanism as a check on inertial loads and critical speeds
2. Proportioning to minimize higher harmonics[128]

3.6.4 Transmission Angles (see also Sec. 3.2.10)[134–136,155,467]

In mechanisms with varying transmission angles μ, the optimum design involves the minimization of the deviation ν of the transmission angle from its ideal value. Such a design maximizes the force tending to turn the driven link while minimizing frictional resistance, assuming quasi-static operation.

In plane crank-and-rocker linkages, the minimization of the maximum deviation of the transmission angle from 90° has been worked out for given rocker swing angle ψ and corresponding crank rotation ϕ. In the special case of centric crank-and-rocker linkages ($\phi = 180°$) the solution is relatively simple: $a^2 + b^2 = c^2 + d^2$ (where a, b, c, and d denote the lengths of crank, coupler, rocker, and fixed link, respectively). This yields $\sin \nu = (ab/cd)$, $\mu_{max} = 90° + \nu$, $\mu_{min} = 90° - \nu$. The solution for the general case (ϕ arbitrary), including additional size constraints, can be found in Refs. 134 to 136, 155, and 371, and depends on the solution of a cubic equation.

3.6.5 Design Charts

To save labor in the design process, charts and atlases are useful when available. Among these are Refs. 199 and 210 in four-link motion; the VDI-Richtlinien Duesseldorf (obtainable through Beuth-Vertrieb Gmbh, Berlin), such as 2131, 2132 on the offset turning block and the offset slider crank, and 2125, 2126, 2130, 2136 on the offset slider-crank and crank-and-rocker mechanisms; 2123, 2124 on four-bar mechanisms; 2137 on the in-line swinging block; and data sheets in the technical press.

3.6.6 Equivalent and "Substitute" Mechanisms[106,112,421g]

Kinematic equivalence is explained in Sec. 3.2.12. Ways of obtaining equivalent mechanisms include (1) pin enlargement, (2) kinematic inversion, (3) use of centrodes, (4) use of curvature constructions, (5) use of pantograph devices, (6) use of multigeneration properties, (7) substitution of tapes, racks, and chains for rigid links[84,103,159,189,285] and other ways depending on the inventiveness of the designer.* Of these, (5) and (6) require additional explanation.

The "pantograph" can be used to reproduce a given motion, unchanged, enlarged, reduced, or rotated. It is based on "Sylvester's plagiograph," shown in Fig. 3.37.

$AODC$ is a parallelogram linkage with point O fixed with two similar triangles ACC_1, DBC, attached as shown. Points B and C_1 will trace similar curves, altered in the ratio $OC_1/OB = AC_1/AC$ and rotated relative to each other by an amount equal to the angle α. The ordinary pantograph is the special case obtained when B, D, C, and C, A, C_1 are collinear. It is used in engraving machines and other motion-copying devices.

Roberts' theorem[32,182,288,347,421f] states that there are three different but related four-bar mechanisms generating the same coupler curve (Fig. 3.38): the "original" $ABCDE$, the "right cognate" $LKGDE$, and the "left cognate" $LHFAE$. Similarly, slider-crank mechanisms have one cognate each.[182]

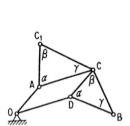

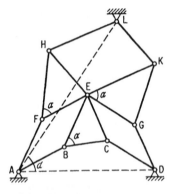

FIG. 3.38 Roberts' theorem. $\triangle BEC \approx \triangle FHE \approx \triangle EKG \approx \triangle ALD \approx \triangle AHC \approx \triangle BKD \approx FLG$; $AFEB$, $EGDC$, $HLKE$ are parallelograms.

FIG. 3.37 Sylvester's plagiograph or skew pantograph.

If the "original" linkage has poor proportions, a cognate may be preferable. When Grashof's inequality is obeyed (Sec. 3.9) and the original is a double rocker, the cognates are crank-and-rocker mechanisms; if the original is a drag link, so are the cognates; if the original does not obey Grashof's inequality, neither do the cognates, and all three are either double rockers or folding linkages. Several well-known straight-line guidance devices (Watt and Evans mechanisms) are cognates.

Geared five-bar mechanisms (Refs. 95, 119, 120, 347, 372, 391, 397, 421f) may also be used to generate the coupler curve of a four-bar mechanism, possibly with better transmission angles and proportions, as, for instance, in the drive of a deep-draw press. The gear ratio in this case is 1:1 (Fig. 3.39), where $ABCDE$ is the four-bar linkage and $AFEGD$ is the five-bar mechanism with links AF and GD geared to each other by 1:1 gearing. The path of E is identical in both mechanisms.

---

*Investigation of enumeration of mechanisms based on degree-of-freedom requirements are found in Refs. 106, 159, and 162 to 165 with application to clamping devices, tools, jigs, fixtures, and vise jaws.

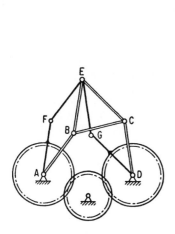

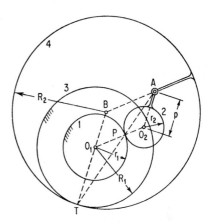

FIG. 3.39 Four-bar linkage $ABCDE$ and equivalent 1:1 geared five-bar mechanism $AFEGD$; $AFEB$ and $DGEC$ are parallelograms.

FIG. 3.40 Double generation of a cycloidal path. For the case shown O_1O_2 and AO_2 rotate in the same direction. $R_2/R_1 = r_2/r_1 + 1$; $R_1 = p(r_1/r_2)$; $R_2 = p[1 + (r_1/r_2)]$. Radius ratios are considered positive or negative depending on whether gearing is internal or external.

In Fig. 3.38 each cognate has one such derived geared five-bar mechanism (as in Fig. 3.39), thus giving a choice of six different mechanisms for the generation of any one coupler curve.

Double Generation of Cycloidal Curves.[315,385,386] A given cycloidal motion can be obtained by two different pairs of rolling circles (Fig. 3.40). Circle 2 rolls on fixed circle 1 and point A, attached to circle 2, describes a cycloidal curve. If O_1, O_2 are centers of circles 1 and 2, P their point of contact, and O_1O_2AB a parallelogram, circle 3, which is also fixed, has center O_1 and radius O_1T, where T is the intersection of extensions of O_1B and AP; circle 4 has center B, radius BT, and rolls on circle 3. If point A is now rigidly attached to circle 4, its path will be the same as before. Dimensional relationships are given in the caption of Fig. 3.40. For analysis of cycloidal motions, see Refs. 385, 386, and 492.

Equivalent mechanisms obtained by multigeneration theory may yield patentable devices by producing "unexpected" results, which constitutes one criterion of patentability. In one application, cycloidal path generation has been used in a speed reducer.[48,426,495] Another form of "cycloidal equivalence" involves adding an idler gear to convert from, say, internal to external gearing; applied to resolver mechanism in Ref. 357.

3.6.7 Computer-Aided Mechanisms Design and Optimization (Refs. 64–66, 78, 79, 82, 106, 108, 109, 185, 186, 200, 217, 218, 246, 247, 317, 322–324, 366, 371, 380, 383, 412–414, 421a, 430, 431, 445, 457, 465, 484, 485)

General mechanisms texts with emphasis on computer-aided design include Refs. 106, 186, 323, 421f, 431, 445. Computer codes having both kinematic analysis and synthesis capability in linkage design include KINSYN[217,218] and LINCAGES.[108] Both codes also include interactive computer graphics features. Codes which can perform both kinematic and dynamic analysis for a large class of mechanisms include DRAM and

ADAMS,[64–66,457] DYMAC,[322,324] IMP,[430,465] kinetoelastodynamic codes,[109,421f] codes for the sensitivity analysis and optimization of mechanisms with intermittent-motion elements,[185,186,200] heuristic codes,[78,79,246,247] and many others.[82,317,322,366,484,485]

The variety of computational techniques is as large as the variety of mechanisms. For specific mechanisms, such as cams and gears, specialized codes are available.

In general, computer codes are capable of analyzing both simple and complex mechanisms. As far as synthesis is concerned the situation is complicated by the non-linearity of the motion parameters in many mechanisms and by the impossibility of limiting most motions to small displacements. For the simpler mechanisms synthesis codes are available. For more complex mechanisms parameter variation of analysis codes or heuristic methods are probably the most powerful currently available tools. The subject remains under intensive development, especially with regard to interactive computer graphics [for example, CADSPAM, computer-aided design of spatial mechanisms (Ref. 421a)].

3.6.8 Balancing of Linkages

At high speeds the inertia forces associated with the moving links cause shaking forces and moments to be transmitted to the frame. Balancing can reduce or eliminate these. An introduction is found in Ref. 421f. (See also Refs. to BAL 25–30, 191, 214, 219, 258–260, 379a, 452, 452a, 452b, 459, 460.)

3.6.9 Kinetoelastodynamics of Linkage Mechanisms

Load and inertia forces may cause cyclic link deformations at high speeds, which change the motion of the mechanism and cannot be neglected. An introduction and copious list of references are found in Ref. 421f. (See also Refs. 53, 107, 202, 416, 417.)

3.7 THREE-DIMENSIONAL MECHANISMS[5,21,35,421f]
(Sec. 3.9)

Three-dimensional mechanisms are also called "spatial mechanisms." Points on these mechanisms move on three-dimensional curves. The basic three-dimensional mechanisms are the "spherical four-bar mechanisms" (Fig. 3.41) and the "offset" or "spatial four-bar mechanism" (Fig. 3.42).

The spherical four-bar mechanism of Fig. 3.41 consists of links AB, BC, CD, and DA, each on a great circle of the sphere with center O; turning joints at A, B, C, and D, whose axes intersect at O; lengths of links measured by great-circle arcs or angles α_i subtended at O. Input θ_2, output θ_1; single degree of freedom, although $\Sigma f_i = 4$ (see Sec. 3.2.2).

Figure 3.42 shows a spatial four-bar mechanism; turning joint at D, turn-slide (also called cylindrical) joints at B, C, and D; a_{ij} denote minimum distances between axes of joints; input θ_2 at D; output at A consists of translation s and rotation θ_1; $\Sigma f_i = 7$; freedom, $F = 1$.

Three-dimensional mechanisms used in practice are usually special cases of the above two mechanisms. Among these are Hooke's joint (a spherical four-bar, with $\alpha_2 = \alpha_3 = \alpha_4 = 90°$, $90° < \alpha_1 \leq 180°$), the wobble plates ($\alpha_4 < 90°$, $\alpha_2 = \alpha_3 = \alpha_1 = 90°$), the space crank,[332] the spherical slider crank,[304] and other mechanisms, whose analysis is outlined in Sec. 3.9.

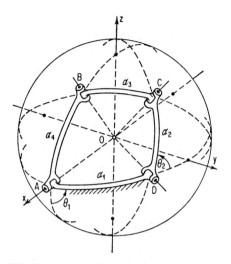

FIG. 3.41   Spherical four-bar mechanism.   **FIG. 3.42**   Offset or spatial four-bar mechanism.

The analysis and synthesis of spatial mechanisms require special mathematical tools to reduce their complexity. The analysis of displacements, velocities, and accelerations of the general spatial chain (Fig. 3.42) is conveniently accomplished with the aid of dual vectors,[421c] numbers, matrices, quaternions, tensors, and Cayley-Klein parameters.[85,87,494] The spherical four-bar (Fig. 3.41) can be analyzed the same way, or by spherical trigonometry.[34] A computer program by J. Denavit and R. S. Hartenberg[129] is available for the analysis and synthesis of a spatial four-link mechanism whose terminal axes are nonparallel and nonintersecting, and whose two moving pivots are ball joints. See also Ref. 474 for additional spatial computer programs. For the simpler problems, for verification of computations and for visualization, graphical layouts are useful.[21,33,36,38,462]

Applications of three-dimensional mechanisms involve these motions:

1. Combined translation and rotation (e.g., door openers to lift and slide simultaneously[3])

2. Compound motions, such as in paint shakers, mixers, dough-kneading machines and filing[8,35,36,304]

3. Motions in shaft couplings, such as universal and constant-velocity joints[4,6,21,35,262] (see Sec. 3.9)

4. Motions around corners and in limited space, such as in aircraft, certain wobble-plate engines, and lawn mowers[60,310,332]

5. Complex motions, such as in aircraft landing gear, remote-control handling devices,[71,270] and pick-and-place devices in automatic assembly machines

When the motion is constrained ($F = 1$), but $\Sigma f_i < 7$ (such as in the mechanism shown in Fig. 3.41), any elastic deformation will tend to cause binding. This is not the case when $\Sigma f_i = 7$, as in Fig. 3.42, for instance. Under light-load, low-speed conditions, however, the former may represent no handicap.[10] The "degenerate" cases, usually associated with parallel or intersecting axes, are discussed more fully in Refs. 3, 10, 143, and 490.

In the analysis of displacements and velocities, extensions of the ideas used in plane kinematic analysis have led to the notions of the "instantaneous screw axis,"[38] valid for displacements and velocities; to spatial Euler-Savary equations; and to concepts involving line geometry.[224]

Care must be taken in designing spatial mechanisms to avoid binding and low mechanical advantages.

3.8 CLASSIFICATION AND SELECTION OF MECHANISMS

In this section, mechanisms and their components are grouped into three categories:

A. Basic mechanism components, such as those adapted for latching, fastening, etc.

B. Basic mechanisms: the building blocks in most mechanism complexes.

C. Groups or assemblies of mechanisms, characterized by one or more displacement-time schedules, sequencing, interlocks, etc.; these consist of combinations from categories A and B and constitute important mechanism units or independent portions of entire machines.

Among the major collections of mechanisms and mechanical movements are the following:

1M. Barber, T. W.: "The Engineer's Sketch-Book," Chemical Publishing Company, Inc., New York, 1940.

2M. Beggs, J. S.: "Mechanism," McGraw-Hill Book Company, Inc., New York, 1955.

3M. Hain, K.: "Die Feinwerktechnik," Fachbuch-Verlag, Dr. Pfanneberg & Co., Giessen, Germany, 1953.

4M. "Ingenious Mechanisms for Designers and Inventors," vols. 1–2, F. D. Jones, ed.: vol. 3, H. L. Horton, ed.; The Industrial Press, New York, 1930–1951.

5M. Rauh, K.: Praktische Getriebelehre," Springer-Verlag OHG, Berlin, vol. I, 1951; vol. II, 1954.

There are, in addition, numerous others, as well as more special compilations, the vast amount of information in the technical press, the AWF publications,[468] and (as a useful reference in depth), the Engineering Index. For some mechanisms, especially the more elementary types involving fewer than six links, a systematic enumeration of kinematic chains based on degrees of freedom may be worthwhile,[162–165,421c] particularly if questions of patentability are involved. Mechanisms are derived from the kinematic chains by holding one link fixed and possibly by using equivalent and substitute mechanisms (Sec. 3.6.6). The present state of the art is summarized in Ref. 159.

In the following list of mechanisms and components, each item is classified according to category (A, B, or C) and is accompanied by references, denoting one or more of the above five sources, or those at the end of this section. In using this listing, it is to be remembered that a mechanism used in one application may frequently be employed in a completely different one, and sometimes combinations of several mechanisms may be useful.

The categories A, B, C, or their combinations are approximate in some cases, since it is often difficult to determine a precise classification.

Adjustments, fine (A, 1M)(A, 2M)(A, 3M)

Adjustments, to a moving mechanism (A, 2M)(AB, 1M); see also Transfer, power

Airplane instruments and linkages (C, 3M)[342]

Analog computing mechanisms;[306] see also Computing mechanisms

Anchoring devices (A, 1M)

Automatic machinery, special-purpose;[161] automatic handling[363]

Ball bearings, guides and slides (A, 3M)

Ball-and-socket joints (A, 1M); see also Joints

Band drives (B, 3M); see also Tapes

Bearings (A, 3M)(A, 1M); jewel;[245] for oscillating motion[56]

Belt gearing (B, 1M)

Bolts (A, 1M)

Brakes (B, 1M)

Business machines, bookkeeping and records (C, 3M)

Calculating devices (C, 3M);[277] see also Mathematical instruments

Cameras (C, 3M); see also Photographic devices

Cam-link mechanisms (BC, 1M)(BC, 5M)[421]

Cams and cam drives (BC, 4M)(BC, 5M)(BC, 1M)[375]

Carriages and cars (BC, 1M)

Centrifugal devices (BC, 1M)

Chain drives (B, 1M)(B, 5M)[265]

Chucks, clamps, grips, holders (A, 1M)

Circular-motion devices (B, 1M)

Clock mechanisms;[18] see also Escapements (Ref. 106 and 421*f*, pp. 37–39)

Clutches, overrunning (BC, 1M)(C, 4M); see also Couplings and clutches

Computing mechanisms (BC, 5M)[80,271,338,446]

Couplings and clutches (B, 1M)(B, 3M)(B, 5M);[139,140,225,261,336] see also Joints

Covers and doors (A, 1M)

Cranes (AC, 1M)[77]

Crank and eccentric gear devices (BC, 1M)

Crushing and grinding devices (BC, 1M)

Curve-drawing devices (BC, 1M); see also Writing instruments and Mathematical instruments

Cushioning devices (AC, 1M)

Cutting devices (A, 1M)[329,354]

Derailleurs or deraillers (see Speed-changing mechanisms) (Refs. 106 and 421*f*, pp. 27)

Detents (A, 3M)

Differential motions (C, 1M)(C, 4M)[188]

Differentials (B, 5M)[4]

Dovetail slides (A, 3M)

Drilling and boring devices (AC, 1M)

Driving mechanisms for reciprocating parts (C, 3M)

Duplicating and copying devices (C, 3M)

Dwell linkages (C, 4M)

Ejecting mechanisms for power presses (C, 4M)

Elliptic motions (B, 1M)

Energy storage, instruments and mechanisms involving[434]

Energy transfer mechanism, special-purpose[267]

Engines, rotary (BC, 1M)

Engines, types of (C, 1M)

Escapements (B, 2M); see also Ratchets and Clock mechanisms

Expansion and contraction devices (AC, 1M)

Fasteners[244,423]

Feed gears (BC, 1M)

Feeding, magazine and attachments (C, 4M)[170]

Feeding mechanisms, automatic (C, 4M)(C, 5M)

Filtering devices (AB, 1M)[230]

Flexure pivots[103,141,269]

Flight-control linkages[365]

Four-bar chains, mechanisms and devices (B, 5M)[106,112,421f]

Frames, machine (A, 1M)

Friction gearing (BC, 1M)

Fuses (see Escapements)

Gears (B, 3M)(B, 1M)

Gear mechanisms (BC, 1M)[106]

Genevas; see Intermittent motions

Geodetic instruments (C, 3M)

Governing and speed-regulating devices (BC, 1M)[461]

Guidance, devices for (BC, 5M)

Guides (A, 1M)(A, 3M)

Handles (A, 1M)

Harmonic drives[101]

High-speed design;[47] special application[339]

Hinges (A, 1M); see also Joints

Hooks (A, 1M)

Hoppers, for automatic machinery (C, 4M), and hopper-feeding devices[234,235,236,421f]

Hydraulic converters (BC, 1M)

Hydraulic and link devices[89,90,92,168,337]

Hydraulic transmissions (C, 4M)

Impact devices (BC, 1M)

Indexing mechanisms (B, 2M); see also Sec. 3.9 and Intermittent motions

Indicating devices (AC, 1M); speed (C, 1M)

Injectors, jets, nozzles (A, 1M)

Integrators, mechanical[358]

Interlocks (C, 4M)

Intermittent motions, general;[44,357,468] see also Sec. 3.9

Intermittent motions from gears and cams (C, 4M)[269]

Intermittent motions, geneva types (BC, 4M)(BC, 5M);[211,357] see also Indexing mechanisms

Intermittent motions from ratchet gearing (C, 1M)(C, 4M)

Joints, all types (A, 1M);[239] see also Couplings and clutches and Hinges

Joints, ball-and-socket[69]

Joints, to couple two sliding members (B, 2M)

Joints, intersecting shafts (B, 2M)

Joints, parallel shafts (B, 2M)

Joints, screwed or bolted (A, 3M)

Joints, skew shafts (B, 2M)

Joints, soldered, welded, riveted (A, 3M)

Joints, special-purpose, three-dimensional[272]

Keys (A, 1M)

Knife edges (A, 3M)[141]

Landing gear, aircraft[71]

Levers (A, 1M)

Limit switches[498]

Link mechanisms (BC, 5M)

Links and connecting rods (A, 1M)

Locking devices (A, 1M)(A, 3M)(A, 5M)

Lubrication devices (A, 1M)

Machine shop, measuring devices (C, 3M)

Mathematical instruments (C, 3M); see also Curve-drawing devices and Calculating devices

Measuring devices (AC, 1M)

Mechanical advantage, mechanisms with high value of (BC, 1M)

Mechanisms, accurate;[482] general[21,96,153,159,176,180,193,263,278]

Medical instruments (C, 3M)

Meteorological instruments (C, 3M)

Miscellaneous mechanical movements (BC, 5M)(C, 4M)

Mixing devices (A, 1M)

Models, kinematic, construction of[51,183]

Noncircular gearing (Sec. 3.9)

Optical instruments (C, 3M)

Oscillating motions (B, 2M)

Overload-relief mechanisms (C, 4M)

Packaging techniques, special-purpose[197]

Packings (A, 1M)

Photographic devices (C, 3M);[478,479,486] see also Cameras

Piping (A, 1M)

Pivots (A, 1M)

Pneumatic devices[57]

Press fits (A, 3M)

Pressure-applying devices (AB, 1M)

Prosthetic devices[311,344,345]

Pulleys (AB, 1M)

Pumping devices (BC, 1M)

Pyrotechnic devices (C, 3M)

Quick-return motions (BC, 1M)(C, 4M)

Raising and lowering, including hydraulics (BC, 1M)[88,340]

Ratchets, detents, latches (AB, 2M)(B, 5M);[18,362,468] see also Escapements

Ratchet motions (BC, 1M)[468]

Reciprocating mechanisms (BC, 1M)(B, 2M)(BC, 4M)

Recording mechanisms, illustrations of;[55,203] recording systems[206]

Reducers, speed; cycloidal;[48,331,495] general[308]

Releasing devices and circuit breakers[84,325,458]

Remote-handling robots;[270] qualitative description[364]

Reversing mechanisms, general (BC, 1M)(C, 4M)

Reversing mechanisms for rotating parts (BC, 1M)(C, 4M)

Robots and manipulators (See Sec. 5.9.10)[421f]

Rope drives (BC, 1M)

Safety devices, automatic (A, 1M)(C, 4M)[20,481]

Screening and sifting (A, 1M)

Screw mechanisms (BC, 1M) (See Ref. 468, no. 6071)

Screws (B, 5M)

Seals, hermetic;[59] O-ring;[209] with gaskets;[46,113,346] multistage[422]

Self-adjusting links and slides (C, 4M)

Separating and concentrating devices (BC, 1M)

Sewing machines (C, 3M)

Shafts (A, 1M)(A, 3M); flexible [198,241]

Ship instruments (C, 3M)

Slider-crank mechanisms (B, 5M)

Slides (A, 1M)(A, 3M)

Snap actions (A, 2M)

Sound, devices using (B, 1M)

Spacecraft, mechanical design of[497]

Spanners (A, 1M)

Spatial body guidance (Refs. 421c, 421d, 421e, 421f)

Spatial function generators with higher pairs (Ref. 421b)

Writing instruments (C, 3M); see also Curve-drawing devices and Mathematical instruments

3.9 KINEMATIC PROPERTIES OF MECHANISMS

For a more complete literature survey, see Refs. 1 to 6 cited in Ref. 124, and the Engineering Index, currently available computer programs are listed in Ref. 129.

3.9.1 The General Slider-Crank Chain (Fig. 3.43)

Nomenclature

A = crankshaft axis
B = crankpin axis
C = wrist-pin axis
FD = guide
AF = e = offset, $\perp FD$
AB = r = crank
BC = l = connecting rod
x = displacement of C in direction of guide, measured from F
t = time

Block at C = slider
s = stroke
θ = crank angle
ϕ = angle between connecting rod and slide, pressure angle
τ = $\angle ABC$
η = $\angle BGP$ = auxiliary angle
PG = collineation axis; CP $\perp FD$

The following mechanisms are derivable from the general slider-crank chain:

1. The *slider-crank mechanism*; guide fixed; if $e \neq 0$, called "offset," if $e = 0$, called "in-line"; $\lambda = r/l$; in case of the in-line slider crank, if $\lambda < 1$, AB rotates; if $\lambda > 1$, AB oscillates.
2. *Swinging-block mechanism*; connecting rod fixed; "offset" or "in-line" as in 1.
3. *Turning-block mechanism*; crank fixed; exact kinematic equivalent of 2; see Fig. 3.44.
4. The *standard geneva mechanism* is derivable from the special case, $e = 0$ (see Fig. 3.45b).
5. Several *variations* of the *geneva mechanism* and other pin-and-slot or block-and-slot drives.

3.9.2 The Offset Slider-Crank Mechanism (see Fig. 3.43 with AFD stationary)

Let $\qquad\qquad \lambda = r/l \qquad \epsilon = e/l$ $\qquad\qquad$ (3.40)

where l is the length of the connecting rod, then

$$s = l[(1 + \lambda)^2 - \epsilon^2]^{1/2} - l[(1 - \lambda)^2 - \epsilon^2]^{1/2} \qquad (3.41)$$

$$\sin \phi = \epsilon + \lambda \sin \theta \qquad (3.42)$$

$$\tau = \pi - \phi - \theta \qquad (3.43)$$

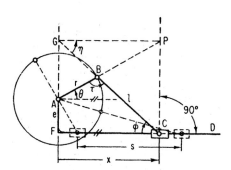

FIG. 3.43 General slider-crank chain.

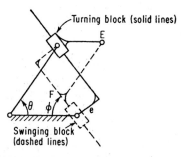

FIG. 3.44 Kinematic equivalence of the swinging-block and turning-block mechanisms, shown by redundant connection *EF*.

and
$$x = r \cos \theta + l \cos \phi \tag{3.44}$$

Let the angular velocity of the crank be $d\theta/dt = \omega$; then the slider velocity is given by

$$dx/dt = r\omega \left[- \sin (\theta + \phi)/\cos \phi \right] \tag{3.45}$$

Extreme value of dx/dt occurs when the auxiliary angle $\eta = 90°$.[116]
   Slider acceleration (ω = constant):

$$\frac{d^2x}{dt^2} = r\omega^2 \left[\frac{-\cos(\theta + \phi)}{\cos \phi} - \lambda \frac{\cos^2 \theta}{\cos^3 \phi} \right] \tag{3.46}$$

Slider shock (ω = constant):

$$\frac{d^3x}{dt^3} = r\omega^3 \left[\frac{\sin(\theta + \phi)}{\cos \phi} + \frac{3\lambda \cos \theta}{\cos^5 \phi} (\sin \theta \cos^2 \phi - \lambda \sin \phi \cos^2 \theta) \right] \tag{3.47}$$

For the angular motion of the connecting rod, let the angular velocity ratio,

$$m_1 = d\phi/d\theta = \lambda(\cos \theta/\cos \phi) \tag{3.48}$$

Then the angular velocity of the connecting rod

$$d\phi/dt = m_1\omega \tag{3.49}$$

Let
$$m_2 = d^2\phi/d\theta^2 = m_1(m_1 \tan \phi - \tan \theta) \tag{3.50}$$

Then the angular acceleration of the connecting rod, at constant ω, is given by

$$d^2\phi/dt^2 = m_2\omega^2 \tag{3.51}$$

In addition, let

$$m_3 = d^3\phi/d\theta^3 = 2m_1m_2 \tan \phi - m_2 \tan \theta + m_1^3 \sec^2 \phi - m_1 \sec^2 \theta \tag{3.52}$$

Then the angular shock of the connecting rod, at constant ω, becomes

$$d^3\phi/dt^3 = m_3\omega^3$$

In general, the $(n-1)$th angular acceleration of the connecting rod, at constant, ω, is given by

$$d^n\phi/dt^n = m_n\omega^n \tag{3.53}$$

where $m_n = dm_{n-1}/d\theta$. In a similar manner, the general expression for the $(n-1)$th linear acceleration of the slider, at constant ω, takes the form

$$d^n x/dt^n = r\omega^n M_n \tag{3.54}$$

where $M_n = dM_{n-1}/d\theta$,

$$M_1 = -\sin(\theta + \phi)/\cos\phi \tag{3.55}$$

and M_2 and M_3 are the bracketed expressions in Eqs. (3.46) and (3.47), respectively.

Kinematic characteristics are governed by Eqs. (3.40) to (3.55). Examples for path and function generation, Ref. 432. Harmonic analyses, Refs. 39 and 296. Coupler curves, Ref. 104. Cognates, Ref. 182. Offset slider-crank mechanism can be used to reduce the friction of the slider in the guide during the "working" stroke; transmission-angle charts, Ref. 467.

Amplitudes of the harmonics are slightly higher than for the in-line slider-crank with the same λ value.

For a nearly constant slider velocity $(1/\omega)(dx/dt) = k$ over a portion of the motion cycle, the proportions[42]

$$12k = 3e \pm \sqrt{9e^2 - 8(l^2 - 9r^2)}$$

may be useful.

3.9.3 The In-Line Slider-Crank Mechanism
$(e = 0)$[32,39,73,104,129,182,296,472,467]

If $\omega \neq$ constant, see Ref. 21. In general, see Eqs. (3.44) to (3.55).

Equations (3.56) and (3.57) give approximate values when $\lambda < 1$, and with $\omega =$ constant. (For nomenclature refer to Fig. 3.43, with $e = 0$, and guide fixed.)

Slider velocity: $dx/dt = r\omega(-\sin\theta - \tfrac{1}{2}\lambda\sin 2\theta)$ (3.56)

Slider acceleration: $d^2x/dt^2 = r\omega^2(-\cos\theta - \lambda\cos 2\theta)$ (3.57)

Extreme Values. $(dx/dt)_{max}$ occurs when the auxiliary angle $\eta = 90°$. For a prescribed extreme value, $(1/r\omega)(dx/dt)_{max}$, λ is obtainable from Eq. (22) of Ref. 116.

At extended dead center: $d^2x/dt^2 = -r\omega^2(1 + \lambda)$ (3.58)

At folded dead center: $d^2x/dt^2 = r\omega^2(1 - \lambda)$ (3.59)

Equations (3.58) and (3.59) yield exact extreme values whenever $0.264 < \lambda < 0.88$.[472]

Computations.   See computer programs in Ref. 129, and also Kent's "Mechanical Engineers Handbook," 1956 ed., Sec. II, Power, Sec. 14, pp. 14-61 to 14-63, for displacements, velocities, and accelerations vs. λ and θ; similar tables, including also kinematics of connecting rod, are found in Ref. 73 for $0.2 \le \lambda \le 0.7$ in increments of 0.1.

Harmonic Analysis[39]

$$x/r = A_0 + \cos\theta + \tfrac{1}{4}A_2\cos 2\theta - \tfrac{1}{16}A_4\cos 4\theta + \tfrac{1}{36}A_6\cos 6\theta - \cdots \tag{3.60}$$

If ω = constant,

$$-(1/r\omega^2)(d^2x/dt^2) = \cos\theta + A_2\cos 2\theta - A_4\cos 4\theta + A_6\cos 6\theta - \cdots \quad (3.61)$$

where A_j are given in Table 3.1[39]

TABLE 3.1   Values of A^*_j

l/r	A_2	A_4	A_6
2 5	0.4173	0.0182	0.0009
3.0	0.3431	0.0101	0.0003
3.5	0.2918	0.0062	0.0001
4.0	0.2540	0.0041	0.0001
4.5	0.2250	0.0028	0.0000
5.0	0.2020	0.0021	
5.5	0.1833	0.0015	
6.0	0.1678	0.0012	

*Biezeno, C. B., and R. Grammel, "Engineering Dynamics," (Trans. M. P. White), vol. 4, Blackie & Son, Ltd., London and Glasgow, 1954.

For harmonic analysis of $\phi(\theta)$, and for inclusion of terms for $\omega \neq$ constant, see Ref. 39; coupler curves (described by a point in the plane of the connecting rod) in Refs. 32 and 104; "cognate" slider-crank mechanism (i.e., one, a point of which describes the same coupler curve as the original slider-crank mechanism), Ref. 182; straight-line coupler-curve guidance, see VDI—Richtlinien No. 2136.

3.9.4 Miscellaneous Mechanisms Based on the Slider-Crank Chain[19,39,42,104,289,291,292,297,299,301,349,367,432,436]

1. In-line swinging-block mechanism
2. In-line turning-block mechanism
3. External geneva motion
4. Shaper drive
5. Offset swinging-block mechanism
6. Offset turning-block mechanism
7. Elliptic slider-crank drive

For "in-line swinging-block" and "in-line turning block" mechanisms, see Fig. 3.45a and b. The following applies to both mechanisms. $\lambda = r/a$; θ is considered as input, with ω_{AB} = constant.

Displacement:
$$\phi = \tan^{-1}\frac{\lambda\sin\theta}{1 - \lambda\cos\theta} \quad (3.62)$$

Angular velocities (positive clockwise):
$$\frac{d\phi}{dt} = \omega_{AB}\frac{\lambda\cos\theta - \lambda^2}{1 + \lambda^2 - 2\lambda\cos\theta} \quad (3.63)$$

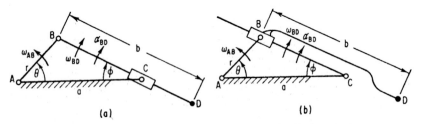

FIG. 3.45 (a) In-line swinging-block mechanism. (b) In-line turning-block mechanism.

$$\frac{1}{\omega_{AB}}\left|\frac{d\phi}{dt}\right|_{min,\theta=180°} = \frac{-\lambda}{\lambda+1} \tag{3.64}$$

$$\frac{1}{\omega_{AB}}\left(\frac{d\phi}{dt}\right)_{max,\theta=0°} = \frac{\lambda}{1-\lambda} \tag{3.65}$$

Angular acceleration: $$\alpha_{BD} = \frac{d^2\phi}{dt^2} = \frac{(\lambda^3-\lambda)\sin\theta}{(1+\lambda^2-2\lambda\cos\theta)^2}\,\omega_{AB}^2 \tag{3.66}$$

Extreme value of α_{BD} occurs when $\theta = \theta_{max}$, where

$$\cos\theta_{max} = -G + (G^2+2)^{1/2} \tag{3.67}$$

and $$G = \tfrac{1}{4}(\lambda + 1/\lambda) \tag{3.68}$$

Angular velocity ratio ω_{BD}/ω_{AB} and the ratio $\alpha_{BD}/\omega_{AB}^2$ are found from Eqs. (3.63) and (3.66), respectively, where $\omega_{BD} = d\phi/dt$. See also Sec. 3.9.7.

Straight-Line Guidance.[42,467]   Point D (see Fig. 3.45a and b) will generate a close point-approximation to a straight line for a portion of its (bread-shaped) path, when

$$b = 3a - r + \sqrt{8a(a-r)} \tag{3.69}$$

Approximate Circular Arc (for a portion of motion cycle).[42]   Point D (Fig. 3.45a and b) will generate an approximately circular arc whose center is at a distance c to the right of A (along AC) when

$$[b(a+c) - c(a-r)]^2 = 4bc(c-a)(a+r)$$

with $b > 0$ and $|c| > a > r$.

Proportions can be used in intermittent drive by attachment of two additional links (VDI-Berichte, vol. 29, 1958, p. 28).[42,473]

Harmonic Analysis[39,286,289,301] (see Fig. 3.45a and b).   Case 1, $\lambda < 1$:

$$\phi = \sum_{n=1}^{\infty} \frac{\lambda^n \sin n\theta}{n} \tag{3.70}$$

$$\frac{\omega_{BD}}{\omega_{AB}} = \frac{d\phi}{d\theta} = \sum_{n=1}^{\infty} \lambda^n \cos n\theta \tag{3.71}$$

Case 2, $\lambda > 1$:

$$\phi = \pi - \theta - \sum_{n=1}^{\infty} \frac{\sin n\theta}{n\lambda^n} \qquad (3.72)$$

$$\frac{d\phi}{d\theta} = -1 - \sum_{n=1}^{\infty} \frac{\cos n\theta}{\lambda^n} \qquad (3.73)$$

Note that in case 1 AB rotates and BC oscillates, while in case 2 both links perform full rotations.

External Geneva Motion. Equations (3.62) to (3.68) apply. For more extensive data, including tables of third derivatives and various numerical values, see Intermittent-Motion Mechanisms, Sec. 3.9.7, and Refs. 252, 253, and 352.

An analysis of the "shaper drive" involving the turning-block mechanism is described in Ref. 289, part 2; see also Ref. 42.

Offset Swinging-Block and Offset Turning-Block Mechanisms.[301] (See Fig. 3.44.) Synthesis of offset turning-block mechanisms for path and function generation described in Ref. 432; see also Ref. 436 for velocities and accelerations; extreme values of angular velocity ratio $d\phi/d\theta = q$ are related by the equation $q_{max}^{-1} + q_{min}^{-1} = -2$; these occur for the same position of the driving link or for those whose crank angles add up to two right angles, depending on whether the driving link swings (oscillates) or rotates, respectively[436]; for graphical analysis of accelerations involving relative motion between two (instantaneously) coincident points on two moving links, use Coriolis's acceleration, or complex numbers in analytical approach.[106]

For the "elliptic slider-crank drive" see Refs. 293, 295, and 297.

3.9.5 Four-Bar Linkages (Plane) (Refs. 2, 12, 16, 32, 42, 58, 61, 62, 104, 106, 115, 118, 122, 123, 125, 127–129, 159, 160, 166, 170, 173, 176, 180, 194, 199, 201, 205, 209, 210, 223, 249, 279, 284, 298, 300, 303, 307, 315, 380, 421*f*, 432, 435, 448, 453, 467, 471, 473, 476, 483, 489, 491)

See Fig. 3.46. Four-bar mechanism, $ABCD$, E on coupler; AB = crank b; BC = coupler c; CD = crank or link d; AD = fixed link a; AB is assumed to be the driving link.

Grashof's Inequality. Length of longest link + length of shortest link < sum of lengths of two intermediate links.

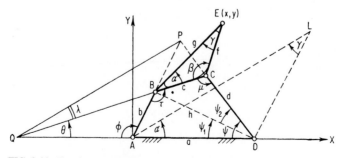

FIG. 3.46 Four-bar mechanism.

Types of Mechanisms

1. If Grashof's inequality is satisfied and b or d is the shortest link, the linkage is a "crank and rocker"; the shortest link is the "crank," and the opposite link is the "rocker."

2. If Grashof's inequality is satisfied and the fixed link is the shortest link, the linkage is a "drag linkage"; both cranks can make complete rotations.

3. All other cases except 4: the linkage is a "double-rocker" mechanism (cranks can only oscillate); this will be the case, for instance, whenever the coupler is the smallest link.

4. Special cases: where the equal sign applies in Grashof's inequality. These involve "folding" linkages and "branch positions," at which the motion is not positive. Example: parallelogram linkage; antiparallel equal-crank linkage ($AB = CD$, $BC = AD$, but AB is not parallel to CD).[207]

Angular Displacement. In Fig. 3.46, $\psi = \psi_1 + \psi_2$ (a minus sign would occur in front of ψ_2 when a mechanism lies entirely on one side of diagonal BD).

$$\psi = \cos^{-1} \frac{h^2 + a^2 - b^2}{2ah} + \cos^{-1} \frac{h^2 + d^2 - c^2}{2hd} \tag{3.74}$$

$$h^2 = a^2 + b^2 + 2ab \cos \phi \tag{3.75}$$

For alternative equation between $\tan \frac{1}{2}\psi$ and $\tan \frac{1}{2}\phi$ (useful for automatic computation) see Ref. 87.

The general closure equation:[115]

$$R_1 \cos \phi - R_2 \cos \psi + R_3 = \cos (\phi - \psi) \tag{3.76}$$

where $\qquad R_1 = a/d \qquad R_2 = a/b \qquad R_3 = (a^2 + b^2 - c^2 + d^2)/2bd \tag{3.77}$

The ϕ, τ equation:

$$p_1 \cos \phi - p_2 \cos \tau + p_3 = \cos (\phi - \tau) \tag{3.78}$$

where $\qquad p_1 = b/c \qquad p_2 = b/a \qquad p_3 = (a^2 + b^2 + c^2 - d^2)/2ac \tag{3.79}$

$$\theta = \tau - \phi = \angle AQB \tag{3.80}$$

Extreme rocker-angle values in a crank and rocker:

$$\psi_{max} = \cos^{-1} \{[a^2 + d^2 - (b + c)^2]/2ad\} \tag{3.81}$$

$$\psi_{min} = \cos^{-1} \{[a^2 + d^2 - (c - b)^2]/2ad\} \tag{3.82}$$

Total range to rocker: $\qquad \Delta\psi = \psi_{max} - \psi_{min}$

To determine inclination of the coupler $\angle AQB = \theta$, determine length of AC:

$$\overline{AC}^2 = a^2 + d^2 - 2ad \cos \psi = k^2 \quad \text{(say)} \tag{3.83}$$

Then compute $\angle ABC = \tau$ from

$$\cos \tau = (b^2 + c^2 - k^2)/2bc \tag{3.84}$$

and use Eq. (3.80). See also Refs. 307 and 448 for angular displacements; for extreme positions see Ref. 284; for geometrical construction of proportions for given ranges and extreme positions, see Ref. 173. For analysis of complex numbers see Ref. 106.

Velocities.

Angular velocity ratio:

$$\omega_{CD}/\omega_{AB} = QA/QD \tag{3.85}$$

Velocity ratio:

$$V_C/V_B = PC/PB \qquad P = AB \times CD \qquad V_p \ (P \text{ on coupler}) = 0 \tag{3.86}$$

Velocity ratio of tracer point E:

$$V_E/V_B = PE/PB \tag{3.87}$$

Angular velocity ratio of coupler to input link:

$$\omega_{BC}/\omega_{AB} = BA/BP \tag{3.88}$$

When cranks are parallel, B and C have the same linear velocity, and $\omega_{BC} = 0$. When coupler and fixed link are parallel, $\omega_{CD}/\omega_{AB} = 1$. At an extreme value of angular velocity ratio, $\lambda = 90°$.[116,433] When ω_{BC}/ω_{AB} is at a maximum or minimum, $QP \perp CD$.

Angular velocity ratio of output to input link is also obtainable by differentiation of Eq. (3.76):

$$m_1 = \frac{\omega_{CD}}{\omega_{AB}} = \frac{d\psi}{d\phi} = \frac{\sin (\phi - \psi) - R_1 \sin \phi}{\sin (\phi - \psi) - R_2 \sin \psi} \tag{3.89}$$

Accelerations (ω_{AB} = constant, t = time)

$$m_2 = \frac{d^2\psi}{d\phi^2} = \frac{1}{\omega_{AB}^2} \frac{d^2\psi}{dt^2} = \frac{(1 - m_1)^2 \cos (\phi - \psi) - R_1 \cos \phi + m_1^2 R_2 \cos \psi}{\sin (\phi - \psi) - R_2 \sin \psi} \tag{3.90}$$

Alternate formulation:

$$\frac{1}{\omega_{AB}^2} \frac{d^2\psi}{dt^2} = m_1(1 - m_1) \cot \lambda \tag{3.91}$$

(useful when $m_1 \neq 1, 0$, and $\lambda \neq 0°, 180°$).

On extreme values, see Ref. 116; velocities, accelerations, and point-path curvature are discussed via complex numbers in Refs. 58, 106, and 427f; computer programs are in Ref. 129.

Second Acceleration or Shock (ω_{AB} = const)

$$\frac{1}{\omega_{AB}^3} \frac{d^3\psi}{dt^3} = \frac{d^3\psi}{d\phi^3} = m_3$$

$$= [R_1 \sin \phi - (m_1^3 \sin \psi - 3m_1 m_2 \cos \psi)R_2 - 3m_2(1 - m_1) \cos (\phi - \psi)$$

$$- (1 - m_1)^3 \sin (\phi - \psi)]/[\sin (\phi - \psi) - R_2 \sin \psi] \tag{3.92}$$

Coupler Motion. Angular velocity of coupler:

$$d\theta/dt = (n_1 - 1)\omega_{AB}$$

where

$$n_1 = 1 + \frac{d\theta}{d\phi} = \frac{\sin(\phi - \tau) - p_1 \sin \phi}{\sin(\phi - \tau) - p_2 \sin \tau} \tag{3.93}$$

where p_1 and p_2 are as before [Eq. (3.79)]. Let

$$n_2 = \frac{d^2\theta}{d\phi^2} = \frac{(1 - n_1)^2 \cos(\phi - \tau) - p_1 \cos \phi + n_1^2 p_2 \cos \tau}{\sin(\phi - \tau) - p_2 \sin \tau} \tag{3.94}$$

Then the angular acceleration of the coupler:

$$d^2\theta/dt^2 = n_2 \omega_{AB}^2 \qquad \omega_{AB} = \text{const} \tag{3.95}$$

If

$$n_3 = \frac{d^3\theta}{d\phi^3}$$
$$= [p_1 \sin \phi - (n_1^3 \sin \tau - 3n_1 n_2 \cos \tau) p_2 - 3n_2(1 - n_1) \cos(\phi - \tau)$$
$$- (1 - n_1)^3 \sin(\phi - \tau)]/[\sin(\phi - \tau) - p_2 \sin \tau] \tag{3.96}$$

The angular shock of the coupler $d^3\theta/dt^3$ at $\omega_{AB} = \text{const}$ is given by

$$d^3\theta/dt^3 = n_3 \omega_{AB}^3 \tag{3.97}$$

See also Refs. 61 and 62 for angular acceleration and shock of coupler; for shock of points on the coupler see Ref. 298.

Harmonic Analysis (ψ vs. ϕ).   Literature survey in Ref. 493. General equations for crank and rocker in Ref. 125. Formulas for special crank-and-rocker mechanisms designed to minimize higher harmonics:[128] Choose $0° << \nu << 90°$, and let

$$AB = \tan \tfrac{1}{2}\nu \qquad BC = (1/\sqrt{2}) \sec \tfrac{1}{2}\nu = CD \qquad AD = 1$$

$$\mu_{max} = 90° + \nu \qquad \mu_{min} = 90° - \nu$$

$$\psi = \text{const} + \sum_{m=1}^{\infty} \frac{(-\tan \tfrac{1}{2}\nu)^m}{-m} \sin m\phi - \frac{C_0}{4} \sin \nu \cos \phi$$
$$- \sum_{m=1}^{\infty} \frac{\sin \nu}{4m} (C_{m-1} - C_{m+1}) \cos m\phi$$

where letting

$$\sin \nu = p \qquad a_2 = \frac{1}{4}p^2 \qquad a_4 = \frac{3}{64}p^4 \qquad a_6 = \frac{5}{512}p^6 \qquad a_8 = \frac{35}{128^2}p^8$$

$$C_m (m \text{ odd}) = 0$$

and
$$C_0 = 1 + C_2 - C_4 + C_6 - C_8 + \cdots \qquad C_2 = a_2 + 4C_4 - 9C_6 + 16C_8 + \cdots$$
$$C_4 = a_4 + 6C_6 - 20C_8 + \cdots \qquad C_6 = a_6 + 8C_8 + \cdots$$
$$C_8 = a_s + \cdots$$

For numerical tables see Ref. 128. For four-bar linkages with adjacent equal links (driven crank = coupler), as in Ref. 128; see also Refs. 45, 300, 303.

Three-Point Function Synthesis.   To find mechanism proportions when (ϕ_i, ψ_i) are prescribed for $i = 1, 2, 3$ (see Fig. 3.46).

$$a = 1 \qquad b = \frac{w_2 w_3 - w_1 w_4}{w_1 w_6 - w_2 w_5} \qquad d = \frac{w_2 w_3 - w_1 w_4}{w_3 w_6 - w_4 w_5}$$

$$c^2 = 1 + b^2 + d^2 - 2bd \cos (\phi_i - \psi_i) - 2d \cos \psi_i + 2b \cos \phi_i \qquad i = 1, 2, 3$$

where

$$w_1 = \cos \phi_1 - \cos \phi_2 \qquad\qquad w_2 = \cos \phi_1 - \cos \phi_3$$

$$w_3 = \cos \psi_1 - \cos \psi_2 \qquad\qquad w_4 = \cos \psi_1 - \cos \psi_3$$

$$w_5 = \cos (\phi_1 - \psi_1) - \cos (\phi_2 - \psi_2) \qquad w_6 = \cos (\phi_1 - \psi_1) - \cos (\phi_3 - \psi_3)$$

Four- and Five-Point Synthesis. For maximum accuracy, use five points; for greater flexibility in choice of proportions and transmission-angle control, choose four points.

Four-Point Path and Function Generation. Path generation together with pre-scribed crank rotations in Refs. 106, 123, 371, and 421*f*. Function generation in Ref. 106, 381, and 421*f*.

Five-Point Path and Function Generation. See Refs. 123, 380, and 421*f*; the latter reference usable for five-point path vs. prescribed crank rotations, for Burmester point-pair determinations pertaining to five distinct positions of a plane, and for func-tion generation with the aid of Ref. 127; additional references include 381, 435, and others at beginning of section; minimization of structural error in Refs. 16, 122, 249, and 421*f*, the latter with least squares; see Refs. 118, 122, and 194 for minimum-error function generators such as log x, sin x, tan x, e^x, x^n, tanh x; infinitesimal motions, Burmester points in Refs. 421*f*, 469, and 489.

General. Atlases for path generation (Ref. 199) and for function generation via "trace deviation" (Refs. 210 and 471); point-position-reduction discussed in Refs. 2, 106, 159, 194, 421*f*, and Sec. 3.5.7; nine-point path generation in Ref. 372.

Coupler Curve.[32,104,315] Traced by point E, in cartesian system with origin at A, and x and y axes as in Fig. 3.46:

$$U = f[(x - a) \cos \gamma + y \sin \gamma](x^2 + y^2 + g^2 - b^2) - gx[(x - a)^2 + y^2 + f^2 - d^2]$$

$$V = f[(x - a) \sin \gamma - y \cos \gamma](x^2 + y^2 + g^2 - b^2) + gy[(x - a)^2 + y^2 + f^2 - d^2]$$

$$W = 2gf \sin \gamma[x(x-a)+y^2-ay \cot \gamma]$$

With these $\qquad\qquad\qquad U^2 + V^2 = W^2 \qquad\qquad\qquad$ (3.98)

Equation (3.98) is a tricircular, trinodal, sextic, algebraic curve. Any intersection of this curve with circle through *ADL* (Fig. 3.46) is a double point, in special cases a cusp; coupler curves may possess up to three real double points or cusps (excluding curves traced by points on folding linkages); construction of coupler curves with cusps and application to instrument design (dwells, noiseless motion reversal, etc.) described in Refs. 32, 159, 279, theory in Ref. 63; detailed discussion of curves, including Watt straight-line motion and equality of two adjacent links, in Ref. 104; instant center (at intersection of cranks, produced if necessary) describes a cusp.

Radius of Path Curvature R for Point E (Fig. 3.46). In this case, as in other linkages, analytical determination of R is readily performed *parametrically*. Parametric equations of the coupler curve:

$$x = x(\phi) = -b \cos \phi + g \cos (\theta + \alpha) \tag{3.99}$$

$$y = y(\phi) = b \sin \phi + g \sin (\theta + \alpha) \tag{3.100}$$

where $\theta = \tau - \phi$, $\tau = \tau(\phi)$ is obtainable from Eqs. (3.74), (3.75), (3.83), and (3.84) and $\cos \alpha = (g^2 - f^2 + c^2)/2gc$.

$$x' = dx/d\phi = b \sin \phi - g(n_1 - 1) \sin (\theta + \alpha) \tag{3.101}$$

$$y' = dy/d\phi = b \cos \phi + g(n_1 - 1) \cos (\theta + \alpha) \tag{3.102}$$

$$x'' = d^2x/d\phi^2 = b \cos \phi - gn_2 \sin (\theta + \alpha) - g(n_1 - 1)^2 \cos (\theta + \alpha) \tag{3.103}$$

$$y'' = d^2y/d\phi^2 = -b \sin \phi + gn_2 \cos (\theta + \alpha) - g(n_1 - 1)^2 \sin (\theta + \alpha) \tag{3.104}$$

where n_1 and n_2 are given in Eqs. (3.93) and (3.94).

$$R = \frac{(x'^2 + y'^2)^{3/2}}{x'y'' - y'x''} \tag{3.105}$$

Equivalent or "Cognate" Four-Bar Linkages. For Roberts' theorem, see Sec. 3.6, Fig. 3.38. Proportions of the cognates are as follows (Figs. 3.38 and 3.46).
Left cognate:

$$\mathbf{AF} = \mathbf{BC}z \qquad \mathbf{HF} = \mathbf{AB}z \qquad \mathbf{HL} = \mathbf{CD}z \qquad \mathbf{AL} = \mathbf{AD}z$$

where $z = (g/c)e^{i\alpha}$ $\qquad \alpha = \angle\ CBE$

and where **AF,** etc., represent the complex-number form of the vector $\overrightarrow{AF}$, etc.
Right cognate:

$$\mathbf{GD} = \mathbf{BC}u \qquad \mathbf{GK} = \mathbf{CD}u \qquad \mathbf{LK} = \mathbf{AB}u \qquad \mathbf{LD} = \mathbf{AD}u$$

where $u = (f/c)e^{-i\beta}$ $\qquad \beta = \angle\ ECB$

The same construction can be studied systematically with the "Cayley diagram."

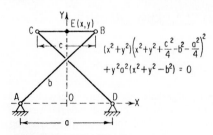

FIG. 3.47 Equal-crank linkage showing equation of symmetric coupler curve generated by point E, midway between C and B.

Symmetrical Coupler Curves. Coupler curves with an axis of symmetry are obtained when $BC = CD = EC$ (Fig. 3.46); also by cognates of such linkages; used by K. Hunt for path of driving pin in geneva motions;[201] also for dwells and straight-line guidance (see Sec. 3.8, 5M). Symmetrical coupler-curve equation[42] for equal-crank linkage, traced by midpoint E of coupler in Fig. 3.47.

Transmission Angles. Angle μ (Fig. 3.46) should be as close to 90° as possible; nontrivial extreme values occur when AB and AD are parallel or antiparallel ($\phi = 0°$, 180°). Generally

$$\cos \mu = [(c^2 + d^2 - a^2 - b^2)/2cd] - (2ab/2cd) \cos \phi \qquad (3.106)$$

$\cos \mu_{max}$ occurs when $\phi = 180°$, $\cos \phi = -1$; $\cos \mu_{min}$ when $\phi = 0°$; $\cos \phi = 1$. Good crank-and-rocker proportions are given in Sec. 3.6:

$$b^2 + a^2 = c^2 + d^2 \qquad |\mu_{min} - 90°| = |\mu_{max} - 90°| = \nu \qquad \sin \nu = ab/cd$$

A computer program for path generation with optimum transmission angles and proportions is described in Ref. 371. Charts for optimum transmission-angle designs are as follows: drag links in Ref. 160; double rockers in Ref. 166; general, in Refs. 170 and 205. See also VDI charts in Ref. 467.

Approximately Constant Angular-Velocity Ratio of Cranks over a Portion of Crank Rotation (see also Ref. 42).   In Fig. 3.46, if $d = 1$, a three-point approximation is obtained when

$$a^2 = c^2 \frac{(1 - 2m_1)(m_1 - 2)}{9m_1} + b^2 \frac{(1 - m_1)(2 - m_1)}{m_1(m_1 + 1)} + \frac{(1 - m_1)(1 - 2m_1)}{(m_1 + 1)}$$

where the angular-velocity ratio m_1 is given by Eq. (3.89). Useful only for limited crank rotations, possibly involving connection of distant shafts, high loads.

Straight-Line Mechanisms.   Survey in Refs. 72, 208; modern and special applications in Refs. 223, 473, 476, theory and classical straight-line mechanisms in Ref. 42; see also below; order-approximation theory in Refs. 421*f* and 453.

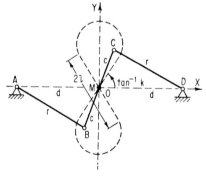

FIG. 3.48   Watt straight-line mechanism.

Fifth-Order Approximate Straight Line via a Watt Mechanism.[42]   "Straight" line of length $2l$, generated by M on coupler, such that $y \cong kx$ (Fig. 3.48).

Choose k, l, r; let $\beta = l(1 + k^2)^{-1/2}$; then maximum error from straight-line path $\cong 0.038(1 + k^2)^3\beta^6$. To compute d and c:

$$(d^2 - c^2) = [r^4 + 6(7 - 4\sqrt{3})l^4 + 3(3 - 2\sqrt{3})l^2r^2]^{1/2}$$

$$p_2 = 3(3 - 2\sqrt{3})\beta$$

$$4k^2d^2 = 2(1 + k^2)(d^2 - c^2 + r^2) + p_2(1 + k^2)^2$$

For less than fifth-order approximation, proportions can be simpler: $AB = CD$, $BM = MC$.

Sixth-Order Straight Line via a Chebyshev Mechanism.[42,491] M will describe an approximate horizontal straight line in the position shown in Fig. 3.49, when

FIG. 3.49   Chebyshev straight-line mechanism. $BN = NC = \frac{1}{2}AD = b$; $AB = CD = r$; $NM = C$ (+ downward). In general, $\phi = 90° - \theta$, where θ is the angle between axis of symmetry and crank in symmetry position.

$$a/r = (2 \cos^2\phi \, \cos 2\phi)/\cos 3\phi$$

$$b/r = -\sin^2 2\phi/\cos 3\phi$$

$$c/r = (\cos^2 \phi \, \cos 2\phi \, \tan 3\phi)/\cos 3\phi$$

when $\phi = 60°$, $NM = 0$.

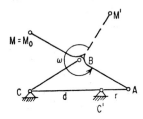

FIG. 3.50 Lambda mechanism.

Lambda Mechanism[12,42] ***and a Related Motion.***   The four-bar lambda mechanism of Fig. 3.50 consists of crank $AC' = r$, fixed link $CC' = d$, coupler AB, driven link BC, with generating point M at the straight-line extension of the coupler, where $BC = MB = BA = 1$. M generates a symmetrical curve. In a related mechanism, $M'B = BA$, $\omega = \angle \, M'BA$ as shown, and M' generates another symmetrical coupler curve.

Case 1.   Either coupler curve of M contained between two concentric circles, center O_1, $O_1 M_0 C$ collinear. $M = M_0$ when $AC'C$ are collinear as shown.

Let ψ'' be a parameter, $0 \le \psi'' \le 45°$. Then a six-point approximate circle is generated by M with least maximum structural error when

$$r = 2 \sin \psi'' \sin 2\psi'' \sqrt{2 \cos 2\psi''}/\sin 3\psi''$$

$$d = \sin 2\psi''/\sin 3\psi''$$

$$O_1 C = 2 \cos^2 \psi''/\sin 3\psi''$$

Radius R of generated circle (at precision points):

$$R = r \cot \psi''$$

Maximum radial (structural) error:

$$2 \cos 2\psi''/\sin 3\psi''$$

For table of numerical values see Ref. 12.

Case 2.   Entire coupler curve contained between two straight lines (six-point approximation of straight line with least maximum structural error). In the equations above, M' generates this curve when $\angle \, M'BA = \omega = \pi + 2\psi''$. Maximum deviation from straight line:[12]

$$2 \sin 2\psi'' \sqrt{2 \cos^3 2\psi''}/\sin 3\psi''$$

Case 3.   Six-point straight line for a portion of the coupler curve of M':

$$r = \tfrac{1}{4} \qquad d = \tfrac{3}{4}$$

Case 4.   Approximate circle for a *portion* of the coupler curve of M. Any proportions for r and d give reasonably good approximation to *some* circle because of symmetry. Exact proportions are shown in Refs. 12 and 42.

Balancing of Four-Bar Linkages for High-Speed Operation.[421f.448]   Make links as light as possible; if necessary, counterbalance cranks, including appropriate fraction of coupler on each.

Multilink Planar Mechanisms. Geared five-bar mechanisms;[95,119,120,421f] six-link mechanisms.[348,421f]

Design Charts.[467] See also section on transmission angles.

3.9.6 Three-Dimensional Mechanisms (Refs. 33–37, 60, 75, 85, 86, 204, 224, 226, 250, 262, 294, 295, 302, 304, 305, 310, 327, 332, 361, 421f, 462–464, 466, 493, 495)

Spherical Four-Bar Mechanisms[85] (θ_1, output vs. θ_2, input) (Fig. 3.51)

$$A \sin \theta_1 + B \cos \theta_1 = C$$

where $A = \sin \alpha_2 \sin \alpha_4 \sin \theta_2$ (3.107)
        $B = -\sin \alpha_4(\sin \alpha_1 \cos \alpha_2 + \cos \alpha_1 \sin \alpha_2 \cos \theta_2)$ (3.108)
        $C = \cos \alpha_3 - \cos \alpha_4(\cos \alpha_1 \cos \alpha_2 - \sin \alpha_1 \sin \alpha_2 \cos \theta_2)$ (3.109)

Other relations given in Ref. 85. Convenient equations between $\tan \frac{1}{2}\theta_1$ and $\tan \frac{1}{2}\theta_2$ given in Ref. 87.

Maximum angular velocity ratio $d\theta_1/d\theta_2$ occurs when $\lambda = 90°$.

Types of mechanisms: Assume α_i ($i = 1, 2, 3, 4$) $< 180°$ and apply Grashof's rule (p. 3.51) to equivalent mechanism with identical axes of turning joints, such that all links except possibly the coupler, $<90°$.

Harmonic analysis: see Ref. 493.

Special Cases of the Spherical Four-Bar Mechanism

Hooke's Joint. $\alpha_2 = \alpha_3 = \alpha_4 = 90°$, $90° < \alpha_1 \le 180°$; in practice, if $\alpha_1 = 180° - \beta$, then $0 \le \beta \le 37\frac{1}{2}°$. If angles θ, φ($\theta = \theta_1 - 90°$, $\varphi = 180° - \theta_2$) are measured from a starting position (shown in Fig. 3.52) in which the planes *ABO* and *OCD* are perpendicular, then

$$\tan \varphi/\tan \theta = \cos \beta \qquad \beta = 180° - \alpha_1$$

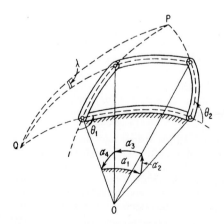

FIG. 3.51 Spherical four-bar mechanism.

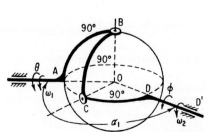

FIG. 3.52 Hooke's joint.

Angular velocity ratio:

$$\omega_2/\omega_1 = \cos\beta/(1 - \sin^2\theta\,\sin^2\beta)$$

Maximum value, $(\cos\beta)^{-1}$, occurs at $\theta = 90°$, $270°$; minimum value, $\cos\beta$, occurs at $\theta = 0°$, $180°$.

A graph of θ vs. φ will show two "waves" per revolution.

If ω_1 = constant, the angular-acceleration ratio of shaft DD' is given by

$$\frac{1}{\omega_1^2}\frac{d^2\varphi}{dt^2} = \frac{\cos\beta\,\sin^2\beta\,\sin 2\theta}{(1 - \sin^2\beta\,\sin^2\theta)^2} \qquad \text{maximum at } \theta = \theta_{max}$$

where
$$\cos 2\theta_{max} = G - (G^2 + 2)^{1/2} \qquad (3.110)$$

and
$$G = (2 - \sin^2\beta)/2\sin^2\beta \qquad (3.111)$$

For $d^2\varphi/dt^2$ as a function of β, see Ref. 264.

Harmonic Analysis of Hooke's Joint.[264,493,495]   If $\alpha_1 = 180° - \beta$, the amplitude ϵ_m of the mth harmonic, in the expression $\varphi(\theta)$, is given by $\epsilon_m = 0$, m odd; and $\epsilon_m = (2/m)(\tan\tfrac{1}{2}\beta)^m$, m even. Two Hooke's joints in series[96,327] can be used to transmit constant (1:1) angular velocity ratio between two intersecting or nonintersecting shafts 1 and 3, provided that the angles between shafts 1 and 3 and the intermediate shaft are the same (Fig. 3.53) and that when fork 1 lies in the plane of shafts 1 and 2, fork 2 lies in the plane of shafts 2 and 3; thus in case shafts 1 and 3 intersect, forks 1 and 2 are coplanar; see also Refs.

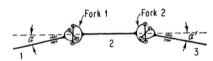

FIG. 3.53   Two Hooke's joints in series.

85 and 327 when $\alpha \neq \alpha'$, which may arise due to misalignment or the effect of manufacturing tolerances, or may be intentional for use as a vibration-excitation drive.

Other Special Cases of the Spherical Four-Bar Mechanism

Wobble-plate mechanism: $\alpha_4 < 90°$, $\alpha_2 = \alpha_3 = \alpha_1 = 90°$; Ref. 35 gives displacements, velocities, and equation of "coupler curve." See also Refs. 36, 294, and 493, the latter giving harmonic analysis.

Two angles 90°, two angles arbitrary[36,295,305]

Fixed link = 90°: Reference 332 includes applications to universal joints, gives velocities and accelerations ("space crank").

Spherical Four-Bar Mechanisms with Dwell.   See Ref. 35.

Spherical Slider-Crank Mechanism.   Three turning pairs, one moving joint a turn slide, input pair and adjacent pair at right angles: see Ref. 304 for displacements; if input and output axes intersect at right angles, obtain "skewed Hooke's joint."[294,304]

Spatial Four-Bar Mechanisms (see Fig. 3.42).   To any spatial four-bar mechanism a corresponding spherical four-bar can be assigned as follows: Through O (Fig. 3.41) draw four radii, parallel to the axes of joints A, B, C, and D in Fig. 3.42 to intersect the surface of a sphere in four points corresponding to the joints of the spherical four-bar. The rotations of the spatial four-bar (Fig. 3.42) are the same as those of the corresponding spherical four-bar and are independent of the offsets a_{ij}, the minimum distances between the (noninteresting) axes i and j ($ij = 12, 23, 34$, and 41) of the spatial four-bar (Fig. 3.42). The input and output angles θ_2, θ_1, of the spatial four-bar, can be measured as in Fig. 3.41 for the corresponding spherical four-bar; for s, the sliding at the output joint, measured from A to Q in Fig. 3.42, the general displacement equations[85,86] are (for rotations see Fig. 3.51 and accompanying equations)

$$s = \frac{A_1 \sin \theta_1 + B_1 \cos \theta_1 - C_1}{-A \cos \theta_1 + B \sin \theta_1} \tag{3.112}$$

where A and B are given in Eqs. (3.107) and (3.108) and

$$A_1 = (a_2 \cos \alpha_2 \sin \alpha_4 + a_4 \sin \alpha_2 \cos \alpha_4) \sin \theta_2 + s_2 \sin \alpha_2 \sin \alpha_4 \cos \theta_2 \tag{3.113}$$

$$B_1 = -a_4 \cos \alpha_4(\sin \alpha_1 \cos \alpha_2 + \cos \alpha_1 \sin \alpha_2 \cos \theta_2)$$

$$-a_1 \sin \alpha_4(\cos \alpha_1 \cos \alpha_2 - \sin \alpha_1 \sin \alpha_2 \cos \theta_2)$$

$$+ a_2 \sin \alpha_4(\sin \alpha_1 \sin \alpha_2 - \cos \alpha_1 \cos \alpha_2 \cos \theta_2)$$

$$+ s_2 \sin \alpha_4 \cos \alpha_1 \sin \alpha_2 \sin \theta_2 \tag{3.114}$$

$$C_1 = -a_3 \sin \alpha_3 + a_4 \sin \alpha_4(\cos \alpha_1 \cos \alpha_2 - \sin \alpha_1 \sin \alpha_2 \cos \theta_2)$$

$$+ a_1 \cos \alpha_4(\sin \alpha_1 \cos \alpha_2 + \cos \alpha_1 \sin \alpha_2 \cos \theta_2)$$

$$+ a_2 \cos \alpha_4(\cos \alpha_1 \sin \alpha_2 + \sin \alpha_1 \cos \alpha_2 \cos \theta_2)$$

$$- s_2 \cos \alpha_4 \sin \alpha_1 \sin \alpha_2 \sin \theta_2 \tag{3.115}$$

where $a_1 = a_{12}$ of Fig. 3.42, and similarly $a_2 = a_{23}$, $a_3 = a_{34}$, $a_4 = a_{41}$, $\alpha_1 =$ angle between axes 1 and 2, $\alpha_2 =$ angle between axes 2 and 3, $\alpha_3 =$ angle between axes 3 and 4, and $\alpha_4 =$ angle between axes 4 and 1 (Figs. 3.41 and 3.42). For complete nomenclature and sign convention, see Ref. 494. These equations are used principally in special cases, in which they simplify.

Special Cases of the Spatial Four-Bar and Related Three-Dimensional Four-Bar Mechanisms[8]

1. Spatial four-bar with two ball joints on coupler and two turning joints: displacements and velocities,[60,201,310,361] synthesis for function generation,[86,250,361] computer programs for displacements and synthesis according to Ref. 86 are listed in Ref. 129; forces and torques are listed in Ref. 60.
2. Spatial four-bar: three angles 90°.[34,35]
3. Spatial four-bar with one ball joint and two turn slides (three links).[35]
4. Spatial four-bar with one ball joint, one turn slide, two turning joints.[462]
5. The "3-D crank slide," one ball joint, one turn slide, intersecting axes; used for agitators.[304]
6. "Degenerate" mechanisms, wherein $F = 1$, $\Sigma f_i < 7$; conditions for,[490] practical constructions in;[3] see also Refs. 143, 194, and 464.
7. Spherical geneva.[37,352]
8. Spatial five-link mechanisms.[6,93]
9. Spatial six-link mechanisms.[9,93]

Harmonic Analysis. Rotations θ_1 vs. θ_2 in Ref. 493, which also includes special cases, such as Hooke's joint, wobble plates, and spherical-crank drive.

8. Special applications.[35,36,294,304,332]

9. Miscellaneous special shaft couplings;[35,294] Cayley-Klein parameters and dual vectors;[85] screw axes and graphical methods.[36,151,196,462]

3.9.7 Intermittent-Motion Mechanisms (Refs. 11, 16, 21, 22, 44, 148, 149, 195, 201, 204, 213, 214, 227, 252, 253, 307, 351, 352, 355, 377, 424, 425, 426, 429, 442, 450, 451, 466, 468)

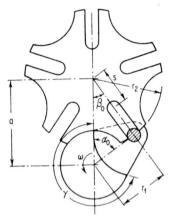

FIG. 3.54   External geneva mechanism in starting position.

The external geneva[252,253,352] is an intermittent-motion mechanism. In Fig. 3.54,

a = center distance

α = angle of driver, radians ($|\alpha| = \alpha_0$)

β = angle of driven or geneva wheel, radians [α and β measured from center-line in the direction of motions ($|\beta| = \beta_0$)]

r_1 = radius to center of driving pin

γ = locking angle of driver, radians

n = number of equally spaced slots in geneva (≥ 3) at start: $|\alpha| = \alpha_0$, $|\beta| = \beta_0 = \pi/2 - \alpha_0$ so that driving pin can enter slot tangentially to reduce shock

r_2 = radius of geneva = $a \cos \beta_0$

r_2' = outside radius of geneva wheel, with correction for finite pin diameter[351]

= $r_2 \sqrt{1 + r_p^2/r_2^2}$; r_p = pin radius

ϵ = gear ratio = $\dfrac{\text{angle moved by driver during motion}}{\text{angle moved by geneva during motion}} = \dfrac{n-2}{2}$

μ = radius ratio = $r_2/r_1 = \cot \beta_0$

s = distance of center of semicircular end of slot from center of geneva $\leq a(1 - \sin \beta_0)$

γ = angle of locking action = $(\pi/n)(n + 2)$ rad; note that classical locking action shown is subject to play in practice; better constructions in 4M, Sec. 3.8

ν = ratio, time of motion of geneva wheel to time for one revolution of driver = $(n - 2)/2n$ ($< \frac{1}{2}$)

r_1 = $a \sin \beta_0$, $\beta_0 = \pi/n$

Let ω = angular velocity of driving wheel, assumed constant, t = time.

Displacement (β vs. α).   Let $r_1/a = \lambda$; then

$$\beta = \tan^{-1} [\lambda \sin \alpha/(1 - \lambda \cos \alpha)] \qquad (3.116)$$

Velocities

$$\frac{1}{\omega} \frac{d\beta}{dt} = \frac{\lambda \cos \alpha - \lambda^2}{1 - 2\lambda \cos \alpha + \lambda^2} \qquad (3.117)$$

$$\frac{1}{\omega} \left(\frac{d\beta}{dt} \right)_{max} = \frac{\lambda}{1 - \lambda} \qquad \text{at } \alpha = 0 \qquad (3.118)$$

Accelerations

$$\frac{1}{\omega^2}\frac{d^2\beta}{dt^2} = \frac{(\lambda^3 - \lambda)\sin\alpha}{(1 + \lambda^2 - 2\lambda\cos\alpha)^2} \qquad (3.119)$$

$$(1/\omega^2)(d^2\beta/dt^2)_{initial\,\alpha\,=\,-\alpha 0} = \tan\beta_0 = r_1/r_2 \qquad (3.120)$$

Maximum acceleration occurs at $\alpha = \alpha_{max}$, where

$$\cos\alpha_{max} = -G + (G^2 + 2)^{1/2} \qquad G = \tfrac{1}{4}(\lambda + 1/\lambda) \qquad (3.121)$$

Second Acceleration or Shock

$$\frac{1}{\omega^3}\frac{d^3\beta}{dt^3} = \frac{\lambda(\lambda^2 - 1)[2\lambda\cos^2\alpha + (1 + \lambda^2)\cos\alpha - 4\lambda]}{(1 + \lambda^2 - 2\lambda\cos\alpha)^3} \qquad (3.122)$$

$$\frac{1}{\omega^3}\frac{d^3\beta}{dt^3}_{(\alpha\,=\,0)} = \frac{\lambda(\lambda + 1)}{(\lambda - 1)^3} \qquad (3.123)$$

Starting Shock $(|\beta| = \beta_0)$

$$\frac{1}{\omega^3}\frac{d^3\beta}{dt^3} = \frac{3\lambda^2}{1 - \lambda^2}$$

Design Procedure (ω = const)

1. Select number of stations ($n \geq 3$).
2. Select center distance a.
3. Compute: $r_1 = a\sin\beta_0$; $r'^2_2 = (a\cos\beta_0)^2 + r_p^2$; $s \leq a(1 - \sin\beta_0)$;

$$|\alpha_0| = (\pi/2)[(n - 2)/n] \qquad \rho|\beta_0| = \pi/2 - \alpha_0 \qquad rad$$

$$\gamma = (\pi/n)(n + 2) \qquad rad$$

4. Determine kinematic characteristics from tables, including maximum velocity, acceleration, and shock: $d^n\beta/dt^n = \omega^n\,d^n\beta/d\alpha^n$, taking the last fraction from the tables.

Check for resulting forces, stresses, and vibrations. See Tables 3.2 and 3.3.

TABLE 3.2 External Geneva Characteristics[253*]

n	β_0	α_0	ϵ	(r_1/a)	(r_2/a)	μ	s_{min}/a	γ	ν	$(d\beta/d\alpha)_{max}$
3	60°	30°	0.5	0.8660	0.5000	0.5774	0.13397	300°	0.1667	6.46
4	45°	45°	1	0.7071	0.7071	1.0000	0.2929	270°	0.2500	2.41
5	36°	54°	1.5	0.5878	0.8090	1.3764	0.4122	252°	0.3000	1.43
6	30°	60°	2	0.5000	0.8660	1.7320	0.5000	240°	0.3333	1.00
7	25°43′	64°17′	2.5	0.4339	0.9009	2.0765	0.5661	231°26′	0.3571	0.766
8	22°30′	67°30′	3	0.3827	0.9239	2.4142	0.6173	225°	0.3750	0.620
9	20°	70°	3.5	0.3420	0.9397	2.7475	0.6580	220°	0.3889	0.520
10	18°	72°	4	0.3090	0.9511	3.0777	0.6910	216°	0.4000	0.447
∞	0°	90°	∞	0	1	∞	1	180°	0.5000	0

*O. Lichtwitz, Getriebe fuer Aussetzende Bewegung, Springer-Verlag OHG, Berlin, 1953.

TABLE 3.3 External Geneva Characteristics[253]*

n	$(d^2\beta/d\alpha^2)_{initial}$ $\alpha_{-} -\alpha_0$	α_{max}	$(d^2\beta/d\alpha^2)_{max}$	$(d^3\beta/d\alpha^3)_{\alpha_{-}0}$
3	1.732	4°46′	31.44	−672
4	1.000	11°24′	5.409	− 48.04
5	0.7265	17°34′	2.299	− 13.32
6	0.5774	22°54′	1.350	− 6.000
7	0.4816	27°33′	0.9284	− 3.429
8	0.4142	31°38′	0.6998	− 2.249
9	0.3640	35°16′	0.5591	− 1.611
10	0.3249	38°30′	0.4648	− 1.236
∞	0	90°	0	0

*O. Lichtwitz, Getriebe fuer Aussetzende Bewegung, Springer-Verlag OHG, Berlin, 1953.

Modifications of Standard External Geneva. More than one driving pin;[252,253] pins not equally spaced;[16,252,253] designs for all small indexing mechanisms;[355] for high-speed indexing;[213] mounting driving pin on a planet pinion to reduce peak loading;[11,253,377] pin guided on four-bar coupler[201] (see also Sec. 3.8, Ref. 5M); double rollers and different entrance and exit slots, especially for starwheels;[227,466] eccentric gear drive for pin.[204]

Internal Genevas.[252,253,352] Used when $v > \frac{1}{2}$; better kinematic characteristics, but more expensive.

Star Wheels.[227,252,253,466] Both internal and external are used; permits considerable freedom in choice of v, which can equal unity, in contrast to genevas. Kinematic properties of external star wheels are better or worse than of external geneva with same n, according as the number of stations (or shoes), n, is less than six or greater than five, respectively.

Special Intermittent and/or Dwell Linkages. The three-gear drive[21,114,195,215,424,442] cardioid drive (slotted link driven by pin on planetary pinion);[352,425,426] link-gear (and/or) -cam mechanisms to produce dwell, reversal, or intermittent motions[22,449,450,451] include link-dwell mechanisms;[148] eccentric-gear mechanisms.[149] These special motions may be required when control of rest, reversal, and kinematic characteristics exceeds that possible with the standard genevas.

3.9.8 Noncircular Cylindrical Gearing and Rolling-Contact Mechanisms
(Refs. 16, 43, 76, 117, 255, 256, 292, 314, 318, 319, 326, 341, 350, 356, 429, 438, 483)

Most of the data for this article are based on Refs. 43 and 318. Noncircular gears can be used for producing positive unidirectional motion; if the pitch curves are closed curves, unlimited rotations may be possible; only externally meshed, plane spur-type gearing will be considered; point of contact between pitch surfaces must lie on the line of centers.

A pair of roll curves may serve as pitch curves for noncircular gears (see Table 3.4): C = center distance; β = angle between the common normal to roll-curves at contact and the line of centers. Angular velocities ω_1 and ω_2 measured in opposite directions; polar-coordinate equations of curves; $R_1 = R_1(\theta_1)$, $R_2 = R_2(\theta_2)$, such that the points $R_1(\theta_1 = 0)$, $R_2(\theta_2 = 0)$ are in mutual contact, where θ_1 and θ_2 are respective oppositely directed rotations from a starting position. Centers O_1, O_2 and contact point Q are collinear.

Twin Rolling Curves.   Mating or pure rolling of two identical curves, e.g., two ellipses when pinned at foci.

Mirror Rolling Curves.   Curve mates with mirror image.

Theorems

1. Every mating curve to a mirror (twin) rolling curve is itself a mirror (twin) rolling curve.
2. All mating curves to a given mirror (twin) rolling curve will mate with each other.
3. A closed roll curve can generally mate with an entire set of different closed roll curves at varying center distances, depending upon the value of the "average gear ratio."

Average Gear Ratio.   For mating closed roll curves: ratio of the total number of teeth on each gear.

Rolling Ellipses and Derived Forms.   If an ellipse, pivoted at the focus, mates with a roll curve so that the average gear ratio n (ratio of number of teeth on mating curve to number of teeth on ellipse) is integral, the mating curve is called an "nth-order ellipse." The case $n = 1$ represents an identical (twin) ellipse; second-order ellipses are oval-shaped and appear similar to ordinary ellipses; third-order ellipses appear pear-shaped with three lobes; fourth-order ellipses appear nearly square; nth-order ellipses appear approximately like n-sided polygons. Equations for several of these are found in Ref. 43. Characteristics for five noncircular gear systems are given in Table 3.4.[43]

Design Data.[43]   Data are usually given in one of three ways:

1. Given $R_1 = R_1(\theta_1)$, C. Find R_2 in parametric form: $R_2 = R_2(\theta_1)$; $\theta_2 = \theta_2(\theta_1)$.

$$\theta_2 = -\theta_1 + C \int_0^{\theta_1} \frac{d\theta_1}{C - R_1(\theta_1)} \qquad R_2 = C - R_1(\theta_1)$$

2. Given $\theta_2 = f(\theta_1)$, C. Find $R_1 = R_1(\theta_1)$, $R_2 = R_2(\theta_2)$.

$$R_1 = \frac{(df/d\theta_1)C}{1 + df/d\theta_1} \qquad R_2 = C - R_1$$

3. Given $\omega_2/\omega_1 = g(\theta_1)$, C. Find $R_1 = R_1(\theta_1)$, $R_2 = R_2(\theta_2)$.

$$R_1 = \frac{Cg(\theta_1)}{1 + g(\theta_1)} \qquad R_2 = C - R_1 \qquad \theta_2 = \int_0^{\theta_1} g(\theta_1)\, d\theta_1$$

4. *Checking for closed curves:* Let $R_1 = R_1(\theta_1)$ be a single-turn closed curve; then $R_2 = R_2(\theta_2)$ will be a single-turn closed curve also, if and only if C is determined from

$$4\pi = C \int_0^{2\pi} \frac{d\theta_1}{C - R_1(\theta_1)}$$

TABLE 3.4 Characteristics of Five Noncircular Gear Systems[43]

Type	Comments	Basic equations	Velocity equation
Two ellipses pivoted at foci	Gears are identical, easy to manufacture. Used for quick-return mechanisms, printing presses, automatic machinery.	$$R = \frac{b^2}{a(1 + \epsilon \cos \theta)}$$ assuming $\theta = 0$ at $R = R_{min}$ ϵ = eccentricity $\quad = \sqrt{1 - (b/a)^2}$ $a = $ ½ major axis $b = $ ½ minor axis	$$\omega_2 = \omega_1 \left[\frac{r^2 + 1 + (r^2-1)\cos\theta_2}{2r} \right]$$ where $r = \dfrac{R_{max}}{R_{min}}$
Second-order ellipses rotated about their geometric centers	Gears are identical. Well-known geometry. Better balance than true elliptical. Two complete speed cycles in one revolution.	$$R = \frac{2ab}{(a+b) - (a-b)\cos 2\theta}$$ assuming $\theta = 0$ at $R = R_{min}$ $C = a + b$ $a = $ max radius $b = $ min radius	$$\omega_2 = \omega_1 \left[\frac{r + 1 + (r^2-1)\cos 2\theta_2}{2r} \right]$$ where $r = \dfrac{a}{b}$

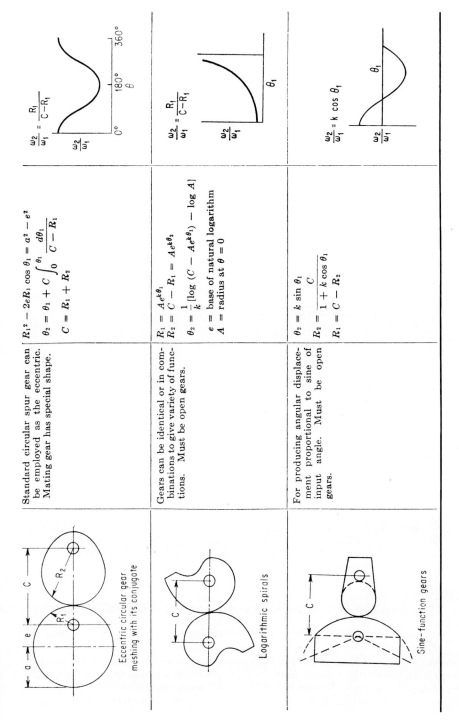

Eccentric circular gear meshing with its conjugate	Standard circular spur gear can be employed as the eccentric. Mating gear has special shape.	$R_1^2 - 2eR_1 \cos \theta_1 = a^2 - e^2$ $\theta_2 = \theta_1 + C \int_0^{\theta_1} \dfrac{d\theta_1}{C - R_1}$ $C = R_1 + R_2$	$\dfrac{\omega_2}{\omega_1} = \dfrac{R_1}{C - R_1}$
Logarithmic spirals	Gears can be identical or in combinations to give variety of functions. Must be open gears.	$R_1 = Ae^{k\theta_1}$ $R_2 = C - R_1 = Ae^{k\theta_2}$ $\theta_2 = \dfrac{1}{k}[\log (C - Ae^{k\theta_1}) - \log A]$ $e =$ base of natural logarithm $A =$ radius at $\theta = 0$	$\dfrac{\omega_2}{\omega_1} = \dfrac{R_1}{C - R_1}$
Sine-function gears	For producing angular displacement proportional to sine of input angle. Must be open gears.	$\theta_2 = k \sin \theta_1$ $R_2 = \dfrac{C}{1 + k \cos \theta_1}$ $R_1 = C - R_2$	$\dfrac{\omega_2}{\omega_1} = k \cos \theta_1$

3.67

When the average gear ratio is not unity, see Ref. 43.

5. *Checking angle* β, also called the angle of obliquity:

$$\beta = \tan^{-1} \frac{1}{R_i} \frac{dR_i}{d\theta_i} \qquad i = 1 \text{ or } 2.$$

Values of β between 0 and 45° are generally considered reasonable.

6. *Checking for tooth undercut:* Let ρ = radius of curvature of pitch curve (roll curve)

$$\rho = \frac{[R_i^2 + (dR_i/d\theta_i)^2]^{3/2}}{R_i^2 - R_i(d^2R_i/d\theta_i^2) + 2(dR_i/d\theta_i)^2} \qquad i = 1 \text{ or } 2$$

Number of teeth:

$$T_{min} = \begin{cases} 32 & \text{for } 14\tfrac{1}{2}° \text{ pressure angle cutting tool} \\ 18 & \text{for } 20° \text{ pressure angle cutting tool} \end{cases}$$

Condition to avoid undercut in noncircular gears:

$$\rho > \frac{T_{min}}{2 \times \text{diametral pitch}}$$

7. *Determining length S of roll curves:*

$$S = \int_0^{2\pi} \left[R^2 + \left(\frac{dR}{d\theta} \right)^2 \right]^{1/2} d\theta$$

This is best computed automatically by numerical integration or determined graphically by large-scale layout.

8. *Check on number of teeth:* For closed single-turn curves,

$$\text{Number of teeth} = (S \times \text{diametral pitch})/\pi$$

Diametral pitch should be integral but may vary by a few percentage points. For symmetrical twin curves, use odd number of teeth for proper meshing following identical machining.

Manufacturing information in Ref. 43.

Special Topics in Noncircular Gearing.   Survey;[341] elliptic gears;[292,350,356] noncircular cams and rolling-contact mechanisms, such as in shears and recording instruments;[117,256,314] noncircular bevel gears;[319] algebraic properties of roll curves;[483] miscellaneous.[255,326,438]

3.9.9   Gear-Linked-Cam Combinations and Miscellaneous Mechanisms[396]

Two-gear drives;[287,391,421f,474,475] straight-type mechanisms in which rack on slide drives output gear (Refs. 285, 330, 390); mechanical analog computing mechanisms;[21,40,137,216,306,328,338,446] three-link screw mechanisms;[359] ratchets;[18,362,370] function generators with two four-bars in series;[232,248] two-degree-of-freedom computing mechanisms;[328] gear-train calculations;[21,24,178,280,282,360] the harmonic drive;[68,126,316] design of variable-speed drives;[31] rubber-covered rollers;[382] eccentric-gear drives.[152]

3.9.10 Robots and Manipulators[97,102,156,187,240,320,373,421f,444,454,470]

Robots are used for production, assembly, materials handling, and other purposes. Mechanically most robots consist of computer-controlled, joint-actuated, open kinematic chains terminating in an "end effector," such as a gripper, hand, or a tool adaptor, which is used for motion transfer. The gripper may have many degrees of freedom, just as does the human hand. The mechanical design of robot mechanisms can include both rigid and elastic elements and involves the determination of kinematic structure, ranges of motion, useful work space, dexterity, kinematics, joint actuation, mechanical advantage, dynamics, power requirements, optimization, and integration with the electronic and computer portions of the robot. A general survey can be found in Heer[187] and Roth,[373] while more specialized investigation can be found in Refs. 97, 102, 156, 240, 320, 421f, 444, 454, and 470. The subject is extensive and continuously expanding.

3.9.11 Hard Automation Mechanisms[421c,421f]

For highly repetitive spatial automation tasks, robotic devices with their multiple programmed inputs are greatly "over-qualified." For these tasks, single-input, purely mechanical spatial mechanisms can be more economical and efficient. For design-synthesis of these see Refs. 421c and 421f.

REFERENCES

1. Alban, C. F.: "Thermostatic Bimetals," *Machine Design,* vol. 18, pp. 124–128, Dec. 1946.

2. Allen, C. W.: "Points Position Reduction," *Trans. Fifth Conf. Mechanisms,* Purdue University, West Lafayette, Ind., pp. 181–193, October 1958.

3. Altmann, F. G.: "Ausgewaehlte Raumgetriebe," *VDI-Berichte,* vol. 12, pp. 69–76, 1956.

4. Altmann, F. G.: "Koppelgetriebe fuer gleichfoermige Uebersetzung," *Z. VDI,* vol. 92, no. 33, pp. 909–916, 1950.

5. Altmann, F. G.: "Raumgetriebe," *Feinwerktechnik,* vol. 60, no. 3, pp. 1–10, 1956.

6. Altmann, F. G.: "Raümliche fuenfgliedrige Koppelgetriebe," *Konstruktion,* vol. 6, no. 7, pp. 254–259, 1954.

7. Altmann, F. G.: "Reibgetriebe mit stufenlos einstellbarer Übersetzung," *Maschinenbautechnik,* supplement *Getriebetechnik,* vol. 10, no. 6, pp. 313–322, 1961.

8. Altmann, F. G.: "Sonderformen raümlicher Koppelgetriebe und Grenzen ihrer Verwendbarkeit," *VDI-Berichte,* Tagungsheft no. 1, *VDI,* Duesseldorf, pp. 51–68, 1953; also in *Konstruktion,* vol. 4, no. 4, pp. 97–106, 1952.

9. Altmann, F. G.: "Ueber raümliche sechsgliedrige Koppelgetriebe," *Z. VDI,* vol. 96, no. 8, pp. 245–249, 1954.

10. Altmann, F. G.: "Zur Zahlsynthese der raümlichen Koppelgetriebe," *Z. VDI,* vol. 93, no. 9, pp. 205–208, March 1951.

11. Arnesen, L.: "Planetary Gears," *Machine Design,* vol. 31, pp. 135–139, Sept. 3, 1959.

12. Artobolevskii, I. I., S. Sh. Blokh, V. V. Dobrovolskii, and N. I. Levitskii: "The Scientific Works of P. L. Chebichev II—Theory of Mechanisms" (Russian), Akad. Nauk, Moscow, p. 192, 1945.

13. Artobolevskii, I. I.: "Theory of Mechanisms and Machines" (Russian). State Publishing

House of Technical-Theoretical Literature, Moscow-Leningrad, 1940 ed., 762 pp.; 1953 ed., 712 pp.

14. Artobolevskii, I. I. (ed.): "Collected Works (of L. V. Assur) on the Study of Plane, Pivoted Mechanisms with Lower Pairs," Akad. Nauk, Moscow, 1952.

15. Artobolevskii, I. I.: "Das dynamische Laufkriterium bei Getrieben," *Maschinenbautechnik (Getriebetechnik), vol. 7, no. 12, pp. 663–667, 1958.

16. Artobolevskii, I. I., N. I. Levitskii, and S. A. Cherkudinov: "Synthesis of Plane Mechanisms" (Russian), Publishing House of Physical-Mathematical Literature, Moscow. 1959.

17. "Transactions of Second U.S.S.R. Conference on General Problems in the Theory of Mechanisms and Machines" (Russian), Board of Editors: I. I. Artobolevskii, S. I. Artobolevskii, G. G. Baranov, A. P. Bessonov, V. A. Gavrilenko, A. E. Kobrinskii, N. I. Levitskii, and L. N. Reshetov, Publishing House of Scientific Technical Studies on Machine Design, Moscow, 1960. 4 Volumes of 1958 Conference: A. Analysis and Synthesis of Mechanisms. B. Theory of Machine Transmissions. C. Dynamics of Machines. D. Theory of Machines in Automatic Operations and Theory of Precision in Machinery and Instrumentation.

18. Assmus, F.: "Technische Laufwerke," Springer-Verlag OHG, Berlin, 1958.

19. Bammert, K., and A. Schmidt: "Die Kinematik der mittelbaren Pleuelanlenkung in Fourier-Reihen," *Ing.-Arch.,* vol. 15, pp. 27–51, 1944.

20. Bareiss, R. A., and P. A. Brand, "Which Type of Torque-limiting Device," *Prod. Eng.,* vol. 29, pp. 50–53, Aug. 4, 1958.

21. Beggs, J. S.: "Mechanism," McGraw-Hill Book Company, Inc., New York, 1955.

22. Beggs, J. S., and R. S. Beggs: "Cams and Gears Join to Stop Shock Loads," *Prod. Eng.,* vol. 28, pp. 84–85, Sept, 16, 1957.

23. Beggs, J. S.: "A Theorem in Plane Kinematics," *J. Appl. Mechanics,* vol. 25, pp. 145–146, March 1958.

24. Benson, A.: "Gear-train Ratios," *Machine Design,* vol. 30, pp. 167–172, Sept. 18, 1958.

25. Berkof, R. S.: "Complete Force and Moment Balancing of in-line Four-Bar Linkages," *Mechanism and Machine Theory,* vol. 8, pp. 397–410, 1973.

26. Berkof, R. S.: "Force Balancing of a Six-Bar Linkage," *Proceedings,* Fifth World Congress on Theory of Machines and Mechanisms, ASME, pp. 1082–1085, 1979.

27. Berkof, R. S.: "The Input Torque in Linkages," *Mechanism and Machine Theory,* vol. 14, pp. 61–73, 1979.

28. Berkof, R. S., and G. G. Lowen: "A New Method for Completely Force Balancing Simple Linkages," *J. Eng. Ind., Trans. ASME,* vol. 91B, pp. 21–26, 1969.

29. Berkof, R. S.: "On the Optimization of Mass Distribution in Mechanisms," Ph.D. thesis, City University of New York, 1969.

30. Berkof, R. S., and G. G. Lowen: "Theory of Shaking Moment Optimization of Force-Balanced Four-Bar Linkages," *J. Eng. Ind., Trans. ASME,* vol. 93B, pp. 53–60, 1971.

31. Berthold, H.: "Stufenlos verstellbare mechanische Getriebe," VEB Verlag Technik, Berlin, 1954.

32. Beyer, R.: "Kinematische Getriebesynthese," Springer-Verlag OHG, Berlin, 1953.

33. Beyer, R., and E. Schoerner: "Raumkinematische Grundlagen," Johann Ambrosius Barth, Munich, 1953.

34. Beyer, R.: "Zur Geometrie und Synthese eigentlicher Raumkurbelgetriebe," *VDI-Berichte,* vol. 5, pp. 5–10, 1955.

35. Beyer, R.: "Zur Synthese und Analyse von Raumkurbelgetrieben," *VDI-Berichte,* vol. 12, pp. 5–20, 1956.

36. Beyer, R.: "Wissenschaftliche Hilfsmittel und Verfahren zur Untersuchung räumlicher Gelenkgetriebe," *VDI Zeitschrift,* vol. 99, pt. I, no. 6, pp. 224–231, February, 1957; pt. 2, no. 7, pp. 285–290, March 1957.

37. Beyer, R.: "Räumliche Malteserkreutzgetriebe," *VDI-Forschungsheft*, no. 461, pp. 32–36, 1957 (see Ref. 120).

38. Beyer, R.: Space Mechanisms, *Trans. Fifth Conf. Mechanisms*, Purdue University, West Lafayette, Ind., pp. 141–163, Oct. 13–14, 1958.

39. Biezeno, C. B., and R. Grammel: "Engineering Dynamics" (Trans. M. P. White), vol. 4, Blackie & Son, Ltd., London and Glasgow, 1954.

40. Billings, J. H.: "Review of Fundamental Computer Mechanisms," *Machine Design*, vol. 27, pp. 213–216, March 1955.

41. Blaschke, W., and H. R. Mueller: "Ebene Kinematik," R. Oldenbourg KG, Munich, 1956.

42. Blokh, S. Sh.: "Angenaeherte Synthese von Mechanismen," VEB Verlag Technik, Berlin, 1951.

43. Bloomfield, B.: "Non-circular Gears," *Prod. Eng.*, vol. 31, pp. 59–66, Mar. 14, 1960.

44. Bogardus, F. J.: "A Survey of Intermittent-motion Mechanisms," *Machine Design*, vol. 28, pp. 124–131, Sept. 20, 1956.

45. Bogdan, R. C., and C. Pelecudi: "Contributions to the Kinematics of Chebichev's Dyad," *Rev. Mecan. Appliquee*, vol. V, no. 2, pp. 229–240, 1960.

46. Bolz, W.: "Gaskets in Design," *Machine Design*, vol. 17, pp. 151–156, March 1945.

47. Bolz, R. W., J. W. Greve, K. R. Harnar, and R. E. Denega: "High Speeds in Design," *Machine Design*, vol. 22, pp. 147–194, April 1950.

48. Botstiber, D. W., and L. Kingston: "Cycloid Speed Reducer," *Machine Design*, vol. 28, pp. 65–69, June 28, 1956.

49. Bottema, O.: "On Cardan Positions for the Plane Motion of a Rigid Body," *Koninkl. Ned. Akad. Watenschap. Proc.*, pp. 643–651, June 1949.

50. Bottema, O.: "On Gruebler's Formulae for Mechanisms," *Appl. Sci. Research*, vol. A2, pp. 162–164, 1950.

51. Brandenberger, H.: "Kinematische Getriebemodelle," Schweizer, Druck & Verlagshaus, AG, Zurich, 1955.

52. Breunich, T. R.: "Sensing Small Motions," *Prod. Eng.*, vol. 21, pp. 113–116, July 1950.

53. Broniarek, C. A., and G. N. Sandor: "Dynamic Stability of an Elastic Parallelogram Linkage," *J. Nonlinear Vibration Problems*, Polish Academy of Science, Institute of Basic Technical Problems, Warsaw, vol. 12, pp. 315–325, 1971.

54. Buchsbaum, F., and F. Freudenstein: "Synthesis of Kinematic Structure of Geared Kinematic Chains and Other Mechanisms," *J. Mechanisms*, vol. 5, pp. 357–392, 1970.

55. Buffenmyer, W. L.: "Pressure Gages," *Machine Design*, vol. 31, p. 119, July 23, 1959.

56. Bus, R. R.: "Bearings for Intermittent Oscillatory Motions," *Machine Design*, vol. 22, pp. 117–119, January 1950.

57. Campbell, J. A.: "Pneumatic Power—3," *Machine Design*, vol. 23, pp. 149–154, July 1951.

58. Capitaine, D.: "Zur Analyse and Synthese ebener Vier und Siebengelenkgetriebe in komplexer Behandlungsweise," doctoral dissertation, Technische Hochschule, Munich, 1955–1960.

59. Carnetti, B., and A. J. Mei: "Hermetic Motor Pumps for Sealed Systems," *Mech. Eng.*, vol. 77, pp. 488–494, June 1955.

60. Carter, B. G.: "Analytical Treatment of Linked Levers and Allied Mechanisms," *J. Royal Aeronautical Society*, vol. 54, pp. 247–252, October 1950.

61. Carter, W. J.: "Kinematic Analysis and Synthesis Using Collineation-Axis Equations," *Trans. ASME*, vol. 79, pp. 1305–1312, 1957.

62. Carter, W. J.: "Second Acceleration in Four-Bar Mechanisms as Related to Rotopole Motion," *J. Appl. Mechanics*, vol. 25, pp. 293–294, June 1958.

63. Cayley, A.: "On Three-Bar Motion," *Proc. London Math. Soc.*, vol. 7, pp. 135–166, 1875–1876.

64. "DRAM and ADAMS" (computer codes), Chace, M. A.: Mechanical Dynamics Inc., 55 South Forest Avenue, Ann Arbor, Mich. 48104.

65. Chace, M. A., and Y. O. Bayazitoglu: "Development and Application of a Generalized d'Alembert Force for Multifreedome Mechanical Systems," *J. Eng. Ind., Trans ASME,* vol. 93B, pp. 317–327, 1971.

66. Chace, M. A., and P. N. Sheth: "Adaptation of Computer Techniques to the Design of Mechanical Dynamic Machinery," ASME Paper 73-DET-58, 1973.

67. Cherkudinov, S. A.: "Synthesis of Plane, Hinged, Link Mechanisms" (Russian), Akad, Nauk, Moscow, 1959.

68. Chironis, N.: "Harmonic Drive," *Prod. Eng.,* vol. 31, pp. 47–51, Feb. 8, 1960.

69. Chow, H.: "Linkage Joints," *Prod. Eng.,* vol. 29, pp. 87–91, Dec. 8, 1958.

70. Conklin, R. M., and H. M. Morgan: "Transducers," *Prod. Eng.,* vol. 25, pp. 158–162, December 1954.

71. Conway, H. G.: "Landing-Gear Design," Chapman & Hall, Ltd., London, 1958.

72. Conway, H. G.: "Straight-Line Linkages," *Machine Design,* vol. 22, p. 90, 1950.

73. Creech, M. D.: "Dynamic Analysis of Slider-Crank Mechanisms," *Prod. Eng.,* vol. 33, pp. 58–65, Oct. 29, 1962.

74. Crossley, F. R. E.: "The Permutations or Kinematic Chains of Eight Members or Less from the Graph-Theoretic Viewpoint," *Developments in Theoretical and Applied Mechanics,* vol. 2, Pergamon Press, Oxford, pp. 467–486, 1965.

75. Crossley, F. R. E., and F. W. Keator: "Three Dimensional Mechanisms," *Machine Design,* vol. 27, no. 8, pp. 175–179, 1955; also, no. 9, pp. 204–209, 1955.

76. Cunningham, F. W.: "Non-circular Gears," *Machine Design,* vol. 31, pp. 161–164, Feb. 19, 1959.

77. "Skoda Harbor Cranes for the Port of Madras, India," *Czechoslovak Heavy Industry;* pp. 21–28, February 1961.

78. Datseris, P.: "Principles of a Heuristic Optimization Technique as an Aid in Design: An Overview," ASME Paper 80-DET-44, 1980.

79. Datseris, P., and F. Freudenstein: "Optimum Synthesis of Mechanisms Using Heuristics for Decomposition and Search," *J. Mechanical Design, ASME Trans.,* vol. 101, pp. 380–385, 1979.

80. Deimel, R. F., and W. A. Black: "Ball Integrator Averages Sextant Readings," *Machine Design,* vol. 18, pp. 116–120, March 1946.

81. Deist, D. H.: "Air Springs," *Machine Design,* vol. 30, pp. 141–146, Jan. 6, 1958.

82. Dix, R. C., and T. J. Lehman: "Simulation of the Dynamics of Machinery," *J. Eng. Ind., Trans. ASME,* vol. 94B, pp. 433–438, 1972.

83. de Jonge, A. E. R.: "A Brief Account of Modern Kinematics," *Trans. ASME,* vol. 65, pp. 663–683, 1943.

84. de Jonge, A. E. R.: "Kinematic Synthesis of Mechanisms," *Mech. Eng.,* vol. 62, pp. 537–542, 1940 (see Refs. 167 and 168).

85. Denavit, J.: "Displacement Analysis of Mechanisms Based on 2×2 Matrices of Dual Numbers" (English), *VDI-Berichte,* vol. 29, pp. 81–88, 1958.

86. Denavit, J., and R. S. Hartenberg: "Approximate Synthesis of Spatial Linkages," *J. Appl. Mechanics,* vol. 27, pp. 201–206, Mar. 1960.

87. Dimentberg, F. M.: "Determination of the Motions of Spatial Mechanisms," Akad. Nauk, Moscow, p. 142, 1950 (see Ref. 78).

88. *Design News:* "Double-Throw Crankshaft Imparts Vertical Motion to Worktable," p. 26, Apr. 14, 1958.

89. *Design News:* "Sky-Lift"-Tractor and Loader with Vertical Lift Attachment," pp. 24–25, Aug. 3, 1959.

90. *Design News:* "Hydraulic Link Senses Changes in Force to Control Movement in Lower Limb," pp. 26–27, Oct. 26, 1959.

91. *Design News:* "Four-Bar Linkage Moves Stacker Tip in Vertical Straight Line," pp. 32–33, Oct. 26, 1959.

92. *Design News:* "Linkage Balances Forces in Thrust Reverser on Jet Star," pp. 30–31, Feb. 15, 1960.

93. Dimentberg, F. M.: "A General Method for the Investigation of Finite Displacements of Spatial Mechanisms and Certain Cases of Passive Constraints" (Russian), Akad. Nauk, Moscow, *Trudi Sem. Teor. Mash. Mekh.,* vol. V, no. 17, pp. 5–39, 1948 (see Ref. 72).

94. Dobrovolskii, V. V.: "Theory of Mechanisms" (Russian), State Publishing House of Technical Scientific Literature, Machine Construction Division, Moscow, p. 464, 1951.

95. Dobrovolskii, V. V.: "Trajectories of Five-Link Mechanisms," *Trans. Moscow Machine-Tool Construction Inst.,* vol. 1, 1937.

96. Doughtie, V. L., and W. H. James: "Elements of Mechanism," John Wiley & Sons, Inc., New York, 1954.

97. Dubowsky, D., and T. D. DesForges: "Robotic Manipulator Control Systems with Invariant Dynamic Characteristics," *Proc. Fifth World Congress on the Theory of Machines and Mechanisms,* Montreal, pp. 101–111, 1979.

98. Dubowsky, S.: "On Predicting the Dynamic Effects of Clearances in Planar Mechanisms," *J. Eng. Ind., Trans. ASME,* vol. 96B, pp. 317–323, 1974.

99. Dubowsky, S., and F. Freudenstein: "Dynamic Analysis of Mechanical Systems and Clearances I, II," *J. Eng. Ind., Trans. ASME,* vol. 93B, pp. 305–316, 1971.

100. Dubowsky, S., and T. N. Gardner: "Dynamic Interactions of Link Elasticity and Clearance Connections in Planar Mechanical Systems," *J. Eng. Ind., Trans. ASME,* vol. 97B, pp. 652–661, 1975.

101. Dudley, D. W.: "Gear Handbook," McGraw-Hill Book Company, Inc., New York, 1962.

102. Duffy, J.: "Mechanisms and Robot Manipulators," Edward Arnold, London, 1981.

103. Eastman, F. S.: "Flexure Pivots to Replace Knife-Edges and Ball Bearings," *Univ. Wash. Eng. Expt. Sta. Bull. 86,* November 1935.

104 Ebner, F.: "Leitfaden der Technischen Wichtigen Kurven," B. G. Teubner Verlagsgesellschaft, GmbH, Leipzig, 1906.

105. Ellis, A. H., and J. H. Howard: "What to Consider When Selecting a Metallic Bellows," *Prod. Eng.,* vol. 21, pp. 86–89, July 1950.

106. Erdman, A. G., and G. N. Sandor: "Mechanism Design, Analysis and Synthesis," vol. 1, Prentice-Hall, Inc., Englewood Cliffs, N.J., 1984.

107. Erdman, A. G., and G. N. Sandor: "Kineto-Elastodynamics—A Frontier in Mechanism Design," *Mechanical Engineering News,* ASEE, vol. 7, pp. 27–28, November 1970.

108. Erdman, A. G.: "LINCAGES," Department of Mechanical Engineering, University of Minnesota, 111 Church Street S.E., Minneapolis, Minn. 55455.

109. Erdman, A. G., G. N. Sandor, and R. G. Oakberg: "A General Method for Kineto-Elastodynamic Analysis and Synthesis of Mechanisms," *J. Eng. Ind., Trans. ASME,* vol. 94B, pp. 1193–1205, 1972.

110. Erdman, A. G., and J. Bowen: "Type and Dimensional Synthesis of Casement Window Mechanisms," *Mech. Eng.,* vol. 103, no. 12, pp. 46–55, 1981.

111. Federhofer, K.: "Graphische Kinematik und Kinetostatik," Springer-Verlag OHG, Berlin, 1932.

112. Franke, R.: "Vom Aufbau der Getriebe," vol. I, Beuth-Vertrieb, GmbH, Berlin, 1948; vol. II, VDI Verlag, Duesseldorf, Germany, 1951.

113. Nonmetallic Gaskets, based on studies by E. C. Frazier, *Machine Design,* vol. 26, pp. 157–188, November 1954.

114. Freudenstein, F.: "Design of Four-Link Mechanisms," doctoral dissertation, Columbia University, University Microfilms, Ann Arbor, Mich., 1954.

115. Freudenstein, F.: "Approximate Synthesis of Four-Bar Linkages," *Trans. ASME,* vol. 77, pp. 853–861, August 1955.

116. Freudenstein, F.: "On the Maximum and Minimum Velocities and the Accelerations in Four-Link Mechanisms," *Trans. ASME,* vol. 78, pp. 779–787, 1956.

117. Freudenstein, F.: "Ungleichfoermigkeitsanalyse der Grundtypen ebener Getriebe," *VDI-Forschungsheft,* vol. 23, no. 461, ser. B, pp. 6–10, 1957.

118. Freudenstein, F.: "Four-Bar Function Generators," *Machine Design,* vol. 30, pp. 119–123, Nov. 27, 1958.

119. Freudenstein, F., and E. J. F. Primrose: "Geared Five-Bar Motion I—Gear Ratio Minus One," *J. Appl. Mech.,* vol. 30: *Trans. ASME,* vol. 85, ser. E, pp. 161–169, June 1963.

120. Freudenstein, F., and E. J. F. Primrose: "Geared Five-Bar Motion II—Arbitrary Commensurate Gear Ratio," *ibid.,* pp. 170–175 (see Ref. 268a).

121. Freudenstein, F., and B. Roth: "Numerical Solution of Systems of Nonlinear Equations," *J. Ass. Computing Machinery,* October 1963.

122. Freudenstein, F.: "Structural Error Analysis in Plane Kinematic Synthesis," *J. Eng. Ind., Trans. ASME,* vol. 81B, pp. 15–22, February 1959.

123. Freudenstein, F., and G. N. Sandor: "Synthesis of Path-Generating Mechanisms by Means of a Programmed Digital Computer," *J. Eng. Ind., Trans. ASME,* vol. 81B, pp. 159–168, 1959.

124. Freudenstein, F.: "Trends in the Kinematics of Mechanisms," *Appl. Mechanics Revs.,* vol. 12, no. 9, September 1959, survey article.

125. Freudenstein, F.: "Harmonic Analysis of Crank-and-Rocker Mechanisms with Application," *J. Appl. Mech.,* vol. 26, pp. 673–675, December 1959.

126. Freudenstein, F.: "The Cardan Positions of a Plane," *Trans. Sixth Conf. Mechanisms,* Purdue University, West Lafayette, Ind., pp. 129–133, October 1960.

127. Freudenstein, F., and G. N. Sandor: "On the Burmester Points of a Plane," *J. Appl. Mechanics,* vol. 28, pp. 41–49, March 1961; discussion, September 1961, pp. 473–475.

128. Freudenstein, F., and K. Mohan: "Harmonic Analysis," *Prod. Eng.,* vol. 32, pp. 47–50, Mar. 6, 1961.

129. Freudenstein, F.: "Automatic Computation in Mechanisms and Mechanical Networks and a Note on Curvature Theory," presented at the International Conference on Mechanisms at Yale University, March, 1961; Shoestring Press, Inc., New Haven, Conn., 1961, pp. 43–62.

130. Freudenstein, F.: "On the Variety of Motions Generated by Mechanisms," *J. Eng. Ind., Trans. ASME,* vol. 81B, pp. 156–160, February 1962.

131. Freudenstein, F.: "Bi-variate, Rectangular, Optimum-Interval Interpolation," *Mathematics of Computation,* vol. 15, no. 75, pp. 288–291, July 1961.

132. Freudenstein, F., and L. Dobrjanskyj: "On a Theory for the Type Synthesis of Mechanisms," *Proc. Eleventh International Congress of Applied Mechanics,* Springer-Verlag, Berlin, pp. 420–428, 1965.

133. Freudenstein, F., and E. R. Maki: "The Creation of Mechanisms According to Kinematic Structure and Function," *J. Environment and Planning B,* vol. 6, pp. 375–391, 1979.

134. Freudenstein, F., and E. J. F. Primrose: "The Classical Transmission Angle Problem," *Proc. Conf. Mechanisms,* Institution of Mechanical Engineers (London), pp. 105–110, 1973.

135. Freudenstein, F.: "Optimum Force Transmission from a Four-Bar Linkage," *Prod. Eng.,* vol. 49, no. 1, pp. 45–47, 1978.

136. Freudenstein, F., and M. S. Chew: "Optimization of Crank-and-Rocker Linkages with Size and Transmission Constraints," *J. Mech. Design, Trans. ASME,* vol. 101, no. 1, pp. 51–57, 1979.

137. Fry, M.: "Designing Computing Mechanisms," *Machine Design,* vol. 17–18, 1945–1946; I, August, pp. 103–108; II, September, pp. 113–120; III, October, pp. 123–128; IV, November,

pp. 141–145; V, December, pp. 123–126; VI, January, pp. 115–118; VII, February, pp. 137–140.

138. Fry, M.: "When Will a Toggle Snap Open," *Machine Design,* vol. 21, p. 126, August 1949.

139. Gagne, A. F., Jr.: "One-Way Clutches," *Machine Design,* vol. 22, pp. 120–128, Apr. 1950.

140. Gagne, A. F., Jr.: "Clutches," *Machine Design,* vol. 24, pp. 123–158, August 1952.

141. Geary, P. J.: "Torsion Devices," Part 3, Survey of Instrument Parts, British Scientific Instrument Research Association, Research Rep. R. 249, 1960, South Hill, Chislehurst, Kent, England; see also Rep. 1, on Flexture Devices; 2 on Knife-Edge Bearings, all by the same author.

142. *General Motors Eng. J.:* "Auto Window Regulator Mechanism," pp. 40–42, January, February, March, 1961.

143. Goldberg, M.: "New Five-Bar and Six-Bar Linkages in Three Dimensions," *Trans. ASME,* vol. 65, pp. 649–661, 1943.

144. Goodman, T. P.: "Toggle Linkage Applications in Different Mechanisms," *Prod. Eng.,* vol. 22, no. 11, pp. 172–173, 1951.

145. Goodman, T. P.: "An Indirect Method for Determining Accelerations in Complex Mechanisms," *Trans. ASME,* vol. 80, pp. 1676–1682, November 1958.

146. Goodman, T. P.: "Four Cornerstones of Kinematic Design," *Trans. Sixth Conf. Mechanisms,* Purdue University, West Lafayette, Ind., pp. 4–30, Oct. 10–11, 1961; also published in *Machine Design,* 1960–1961.

147. Goodman, T. P.: "Dynamic Effects of Backlash," *Trans. Seventh Conf. Mechanisms,* Purdue University, West Lafayette, Ind., pp. 128–138, 1962.

148. Grodzenskaya, L. S.: "Computational Methods of Designing Linked Mechanisms with Dwell" (Russian), *Trudi Inst. Mashinoved.,* Akad. Nauk, Moscow, vol. 19, no. 76, pp. 34–45, 1959; see also vol. 71, pp. 69–90, 1958.

149. Grodzinski, P.: "Eccentric Gear Mechanisms," *Machine Design,* vol. 25, pp. 141–150, May 1953.

150. Grodzinski, P.: "Straight-Line Motion," *Machine Design,* vol. 23, pp. 125–127, June 1951.

151. Grodzinski, P., and E. M. Ewen: "Link Mechanisms in Modern Kinematics," *Proc. Inst. Mech. Eng. (London),* vol. 168, no. 37, pp. 877–896, 1954.

152. Grodzinski, P.: "Applying Eccentric Gearing," *Machine Design,* vol. 26, pp. 147–151, July 1954.

153. Grodzinski, P.: "A Practical Theory of Mechanisms," Emmett & Co., Ltd., Manchester, England, 1947.

154. Gruebler, M.: "Getriebelehre," Springer-Verlag OHG, Berlin, 1917/21.

155. Gupta, K. C.: "Design of Four-Bar Function Generators with Minimax Transmission Angle," *J. Eng. Ind., Trans. ASME,* vol. 99B, pp. 360–366, 1977.

156. Gupta, K. C., and B. Roth: "Design Consideration for Manipulator Workspace," ASME paper 81-DET-79, 1981.

157. Hagedorn, L.: "Getriebetechnik und Ihre praktische Anwendung," *Konstruktion,* vol. 10, no. 1, pp. 1–10, 1958.

158. Hain, K.: "Der Einfluss der Toleranzen bei Gelenkrechengetrieben," *Die Messtechnik,* vol. 20, pp. 1–6, 1944.

159. Hain, K.: "Angewandte Getriebelehre," 2d ed., VDI Verlag, Duesseldorf, 1961, English translation "Applied Kinematics," McGraw-Hill Book Company, Inc., New York, 1967.

160. Hain, K.: "Uebertragungsguenstige unsymmetrische Doppelkurbelgetriebe," *VDI-Forschungsheft,* no. 461, supplement to *Forsch. Gebiete Ingenieurw.,* series B, vol. 23, pp. 23–25, 1957.

161. Hain, K.: "Selbsttaetige Getriebegruppen zur Automatisierung von Arbeitsvorgaengen," *Feinwerktechnik,* vol. 61, no. 9, pp. 327–329, September 1957.

162. Hain, K.: "Achtgliedrige kinematische Ketten mit dem Freiheitsgrad F = 1 fuer gegebene Kraeftever-haeltnisse," *Das Industrieblatt,* vol. 62, no. 6, pp. 331–337, June 1962.

163. Hain, K.: Beispiele zur Systematik von Spannvorrichtungen aus sechsgliedrigen kinematis-chen Ketten mit dem Freiheitsgrad F = 1," *Das Industrieblatt,* vol. 61, no. 12, pp. 779–784, December 1961.

164. Hain, K.: "Die Entwicklung von Spannvorrichtungen mit mehrenren Spannstellen aus kine-matischen Ketten," *Das Industrieblatt,* vol. 59, no. 11, pp. 559–564, November 1959.

165. Hain, K.: "Entwurf viergliedriger kraftverstaerkender Zangen fuer gegebene Kraeftever-haeltnisse," *Das Industrieblatt,* pp. 70–73, February 1962.

166. Hain, K.: "Der Entwurf Uebertragungsguenstiger Kurbelgetriebe mit Hilfe von Kurventafeln," *VDI-Bericht,* Getriebetechnik und Ihre praktische Anwendung, vol. 29, pp. 121–128, 1958.

167. Hain, K.: "Drag Link Mechanisms," *Machine Design,* vol. 30, pp. 104–113, June 26, 1958.

168. Hain, K.: "Hydraulische Schubkolbenantriebe fuer schwierige Bewegungen," *Oelhydraulik & Pneumatik,* vol. 2, no. 6, pp. 193–199, September 1958.

169. Hain, K., and G. Marx: "How to Replace Gears by Mechanisms," *Trans. ASME,* vol. 81, pp. 126–130, May 1959.

170. Hain, K.: "Mechanisms, a 9-Step Refresher Course," (trans. F. R. E. Crossley), *Prod. Eng.,* vol. 32, 1961 (Jan. 2, 1961–Feb. 27, 1961, in 9 parts).

171. Ham, C. W., E. J. Crane, and W. L. Rogers: "Mechanics of Machinery," 4th ed., McGraw-Hill Book Company, Inc., New York, 1958.

172. Hall, A. S., and E. S. Ault: "How Acceleration Analysis Can Be Improved," *Machine Design,* vol. 15, part I, pp. 100–102, February, 1943; part II, pp. 90–92, March 1943.

173. Hall, A. S.: "Mechanism Properties," *Machine Design,* vol. 20, pp. 111–115, February 1948.

174. Hall, A. S., and D. C. Tao: "Linkage Design—A Note on One Method," *Trans. ASME,* vol. 76, no. 4, pp. 633–637, 1954.

175. Hall, A. S.: "A Novel Linkage Design Technique," *Machine Design,* vol. 31, pp. 144–151, July 9, 1959.

176. Hall, A. S.: "Kinematics and Linkage Design," Prentice-Hall, Inc., Englewood Cliffs, N.J., 1961.

177. Hannula, F. W.: "Designing Non-circular Surfaces," *Machine Design,* vol. 23, pp. 111–114, 190, 192, July 1951.

178. Handy, H. W.: "Compound Change Gear and Indexing Problems," The Machinery Publishing Co., Ltd., London.

179. Harnar, R. R.: "Automatic Drives," *Machine Design,* vol. 22, pp. 136–141, April 1950.

180. Harrisberger, L.: "Mechanization of Motion," John Wiley & Sons, Inc., New York, 1961.

181. Hartenberg, R. S.: "Complex Numbers and Four-Bar Linkages," *Machine Design,* vol. 30, pp. 156–163, Mar. 20, 1958.

182. Hartenberg, R. S., and J. Denavit: "Cognate Linkages," *Machine Design,* vol. 31, pp. 149–152, Apr. 16, 1959.

183. Hartenberg, R. S.: "Die Modellsprache in der Getriebetechnik," *VDI-Berichte,* vol. 29, pp. 109–113, 1958.

184. Hastings, C., Jr., J. T. Hayward, and J. P. Wong, Jr.: "Approximations for Digital Computers," Princeton University Press, Princeton, N.J., 1955.

185. Haug, E. J., R. Wehage, and N. C. Barman: "Design Sensitivity Analysis of Planar Mechanisms and Machine Dynamics," *J. Mechanical Design, Trans. ASME,* vol. 103, pp. 560–570, 1981.

186. Haug, E. J., and J. S. Arora: "Applied Optimal Design," John Wiley & Sons, Inc., New York, 1979.

187. Heer, E.: "Robots and Manipulators," *Mechanical Engineering,* vol. 103, no. 11, pp. 42–49, 1981.

188. Heidler, G. R.: "Spring-Loaded Differential Drive (for tensioning)," *Machine Design,* vol. 30, p. 140, Apr. 3, 1958.

189. Hekeler, C. B.: "Flexible Metal Tapes," *Prod. Eng.,* vol. 32, pp. 65–69, Feb. 20, 1961.

190. Herst, R.: "Servomechanisms (types of), *Electrical Manufacturing,* pp. 90–95, May 1950.

191. Hertrich, F. R.: "How to Balance High-Speed Mechanisms with Minimum-Inertia Counterweights," *Machine Design,* pp. 160–164, Mar. 14, 1963.

192. Hildebrand, S.: "Moderne Schreibmaschinenantriebe und Ihre Bewegungsvorgänge," *Getriebetechnik, VDI-Berichte,* vol. 5, pp. 21–29, 1955.

193. Hildebrand, F. B.: "Introduction to Numerical Analysis," McGraw-Hill Book Company, Inc., New York, 1956.

194. Hinkle, R. T.: "Kinematics of Machines," 2d ed., Prentice-Hall, Inc., Englewood Cliffs, N.J., 1960.

195. Hirschhorn, J.: "New Equations Locate Dwell Position of Three-Gear Drive," *Prod. Eng.,* vol. 30, pp. 80–81, June 8, 1959.

196. Hohenberg, F.: "Konstruktive Geometrie in der Technik," 2d ed., Springer-Verlag OHG, Vienna, 1961.

197. Horsteiner, M.: "Getriebetechnische Fragen bei der Faltschachtel-Fertigung," *Getriebetechnik, VDI-Berichte,* vol. 5, pp. 69–74, 1955.

198. Hotchkiss, C., Jr.: "Flexible Shafts," *Prod. Eng.,* vol. 26, pp. 168–177, February 1955.

199. Hrones, J. A., and G. L. Nelson: "Analysis of the Four-Bar Linkage," The Technology Press of the Massachusetts Institute of Technology, Cambridge, Mass., and John Wiley & Sons, Inc., New York, 1951.

200. Huang, R. C., E. J. Haug, and J. G. Andrews: "Sensitivity Analysis and Optimal Design of Mechanical Systems with Intermittent Motion," *J. Mechanical Design, Trans. ASME,* vol. 100, pp. 492–499, 1978.

201. Hunt, K. H.: "Mechanisms and Motion," John Wiley & Sons, Inc., New York, pp. 108, 1959; see also *Proc. Inst. Mech. Eng.* (London), vol. 174, no. 21, pp. 643–668, 1960.

202. Imam, I., G. N. Sandor, W. T. McKie, and C. W. Bobbitt: "Dynamic Analysis of a Spring Mechanism," *Proc. Second OSU Applied Mechanisms Conference,* Stillwater, Okla., pp. 31-1–31-10, Oct. 7–9, 1971.

203. *Instruments and Control Systems:* "Compensation Practice," p. 1185, August 1959.

204. Jahr, W., and P. Knechtel: "Grundzuege der Getriebelehre," Fachbuch Verlag, Leipzig, vol. 1, 1955; vol. II, 1956.

205. Jensen, P. W.: "Four-Bar Mechanisms," *Machine Design,* vol. 33, pp. 173–176, June 22, 1961.

206. Jones, H. B., Jr.: "Recording Systems," *Prod. Eng.,* vol. 26, pp. 180–185, March 1955.

207. de Jonge, A. E. R.: "Analytical Determination of Poles in the Coincidence Position of Links in Four-Bar Mechanisms Required for Valves Correctly Apportioning Three Fluids in a Chemical Apparatus," *J. Eng. Ind., Trans. ASME,* vol. 84B, pp. 359–372, August 1962.

208. de Jonge, A. E. R.: "The Correlation of Hinged Four-Bar Straight-Line Motion Devices by Means of the Roberts Theorem and a New Proof of the Latter," *Ann. N.Y. Acad. Sci.,* vol. 84, pp. 75–145, 1960 (see Refs. 68 and 69).

209. Johnson, C.: "Dynamic Sealing with O-rings," *Machine Design,* vol. 27, pp. 183–188, August 1955.

210. Johnson, H. L.: "Synthesis of the Four-Bar Linkage," M.S. dissertation, Georgia Institute of Technology, Atlanta, June 1958.

211. Johnson, R. C.: "Geneva Mechanisms," *Machine Design,* vol. 28, pp. 107–111, Mar. 22, 1956.

212. Johnson, R. C.: "Method of Finite Differences in Cam Design—Accuracy—Applications," *Machine Design,* vol. 29, pp. 159–161, Nov. 14, 1957.

213. Johnson, R. C.: "Development of a High-Speed Indexing Mechanism," *Machine Design,* vol. 30, pp. 134–138, Sept. 4, 1958.

214. Kamenskii, V. A., "On the Question of the Balancing of Plane Linkages," *Mechanisms,* vol. 3, pp. 303–322, 1968.

215. Kaplan, J., and H. North: "Cyclic Three-Gear Drives," *Machine Design,* vol. 31, pp. 185–188, Mar. 19, 1959.

216. Karplus, W. J., and W. J. Soroka: "Analog Methods in Computation and Simulation," 2d ed., McGraw-Hill Book Company, Inc., New York, 1959.

217. Kaufman, R. E.: "KINSYN: An Interactive Kinematic Design System," *Trans. Third World Congress on the Theory of Machines and Mechanisms,* Dubrovnik, Yugoslavia; September 1971.

218. Kaufman, R. E.: "KINSYN II: A Human Engineered Computer System for Kinematic Design and a New Least Squares Synthesis Operator," *J. Mechanisms and Machine Theory,* vol. 8, no. 4, pp. 469–478, 1973.

219. Kaufman, R. E., and G. N. Sandor: "Complete Force Balancing of Spatial Linkages," *J. Eng. Ind., Trans. ASME,* vol. 93B, pp. 620–626, 1971.

220. Kaufman, R. E., and G. N. Sandor: "The Bicycloidal Crank: A New Four-Link Mechanism," *J. Eng. Ind., Trans. ASME,* vol. 91B, pp. 91–96, 1969.

221. Kaufman, R. E., and G. N. Sandor: "Complete Force Balancing of Spatial Linkages," *J. Eng. Ind., Trans. ASME,* vol. 93B, pp.620–626, 1971.

222. Kaufman, R. E., and G. N. Sandor: "Operators for Kinematic Synthesis of Mechanisms by Stretch-Rotation Techniques," ASME Paper No. 70-Mech-79, *Proc. Eleventh ASME Conf. on Mechanisms,* Columbus, Ohio, Nov. 3–5, 1970.

223. Kearny, W. R., and M. G. Wright: "Straight-Line Mechanisms," *Trans. Second Conf. Mechanisms,* Purdue University, West Lafayette, Ind., pp. 209–216, Oct. 1954.

224. Keler, M. K.: "Analyse und Synthese der Raumkurbelgetriebe mittels Raumliniengeometrie und dualer Groessen," *Forsch. Gebiete Ingenieurw.,* vol. 75, pp. 26–63, 1959.

225. Kinsman, F. W.: "Controlled-Acceleration Single-Revolution Drives," *Trans. Seventh Conf. Mechanisms,* Purdue University, West Lafayette, Ind., pp. 229–233, 1962.

226. Kislitsin, S. G.: "General Tensor Methods in the Theory of Space Mechanisms" (Russian), Akad. Nauk, Moscow, *Trudi Sem. Teor. Mash. Mekh.,* vol. 14, no. 54, 1954.

227. Kist, K. E.: "Modified Starwheels," *Trans. Third Conf. Mechanisms,* Purdue University, West Lafayette, Ind., pp. 16–20, May 1956.

228. Kobrinskii, A. E.: "On the Kinetostatic Calculation of Mechanisms with Passive Constraints and with Play," Akad. Nauk, Moscow, *Trudi Sem. Teor. Mash. Mekh.,* vol. V, no. 20, pp. 5–53, 1948.

229. Kobrinskii, A. E., M. G. Breido, V. S. Gurfinkel, E. P. Polyan, Y. L. Slavitskii, A. Y. Sysin, M. L. Zetlin, and Y. S. Yacobson: "On the Investigation of Creating a Bioelectric Control System," Akad. Nauk, Moscow, *Trudi Inst. Mashinoved,* vol. 20, no. 77, pp. 39–50, 1959.

230. Kovacs, J. P., and R. Wolk: "Filters," *Machine Design,* vol. 27, pp. 167–178, Jan. 1955.

231. Kramer, S. N., and G. N. Sandor: "Finite Kinematic Synthesis of a Cycloidal-Crank Mechanism for Function Generation," *J. Eng. Ind., Trans. ASME,* vol. 92B, pp. 531–536, 1970.

232. Kramer, S. N., and G. N. Sandor: "Kinematic Synthesis of Watt's Mechanism," ASME Paper No. 70-Mech-50, *Proc. Eleventh Conf. Mechanisms,* Columbus, Ohio, Nov. 3–5, 1970.

233. Kramer, S. N., and G. N. Sandor: "Selective Precision Synthesis—A General Method of Optimization for Planar Mechanisms," *J. Eng. Ind., Trans. ASME,* vol. 97B, pp. 689–701, 1975.

234. Kraus, C. E.: "Chuting and Orientation in Automatic Handling," *Machine Design,* vol. 21, pp. 95–98, Sept. 1949.

235. Kraus, C. E., "Automatic Positioning and Inserting Devices," *Machine Design,* vol. 21, pp. 125–128, Oct. 1949.

236. Kraus, C. E., "Elements of Automatic Handling," *Machine Design,* vol. 23, pp. 142–145, Nov. 1951.

237. Kraus, R.: "Getriebelehre," vol. I, VEB Verlag Technik, Berlin, 1954.

238. Kuhlenkamp, A.: "Linkage Layouts," *Prod. Eng.,* vol. 26, no. 8, pp. 165–170, Aug. 1955.

239. Kuhn, H. S.: "Rotating Joints," *Prod. Eng.,* vol. 27, pp. 200–204, Aug. 1956.

240. Kumar, A., and K. J. Waldron: "The Workspace of a Mechanical Manipulator," *J. Mechanical Design, Trans. ASME,* vol. 103, pp. 665–672, 1981.

241. Kupfrian, W. J.: "Flexible Shafts, *Machine Design,* vol. 26, pp. 164–173, October 1954.

242. Kutzbach, K.: "Mechanische Leitungsyerzweigung, ihre Gesetze und Anwendungen, Maschinenbau," vol. 8, pp. 710–716, 1929; see also *Z. VDI,* vol. 77, p. 1168, 1933.

243. Langosch, O.: "Der Einfluss der Toleranzen auf die Genauigkeit von periodischen Getrieben," *Konstruktion,* vol. 12, no. 1, p. 35, 1960.

244. Laughner, V. H., and A. D. Hargan: "Handbook of Fastening and Joining Metal Parts," McGraw-Hill Book Company, Inc., New York, 1956.

245. Lawson, A. C.: "Jewel Bearing Systems," *Machine Design,* vol. 26, pp. 132–137, April 1954.

246. Lee, T. W., and F. Freudenstein: "Heuristic Combinatorial Optimization in the Kinematic Design of Mechanisms I, II," *J. Eng. Ind., Trans. ASME,* vol. 98B, pp. 1277–1284, 1976.

247. Lee, T. W.: "Optimization of High-Speed Geneva Mechanisms," *J. Mechanical Design, Trans. ASME,* vol. 103, pp. 621–630, 1981.

248. Levitskii, N. I.: "On the Synthesis of Plane, Hinged, Six-Link Mechanisms," pp. 98–104. See item *A,* Ref. 17, in this section of book.

249. Levitskii, N. I.: "Design of Plane Mechanisms with Lower Pairs" (Russian), Akad. Nauk, Moscow-Leningrad, 1950.

250. Levitskii, N. I., and Sh. Shakvasian: "Synthesis of Spatial Four-Link Mechanisms with Lower Paris," Akad. Nauk, Moscow, *Trudi Sem. Teor. Mash. Mekh.,* vol. 14, no. 54, pp. 5–24, 1954.

251. Lichtenheldt, W.: "Konstruktionslehre der Getriebe," Akademie Verlag, Berlin, 1964.

252. Lichtwitz, O.: "Mechanisms for Intermittent Motion," *Machine Design,* vol. 23–24, 1951–1952; I, December, pp. 134–148; II, January, pp. 127–141; III, February, pp. 146–155; IV, March, pp. 147–155.

253. Lichtwitz, O.: "Getriebe fuer Aussetzende Bewegung," Springer-Verlag OHG, Berlin, 1953.

254. Lincoln, C. W.: "A Summary of Major Developments in the Steering Mechanisms of American Automobiles," *General Motors Eng. J.,* pp. 2–7, March–April 1955.

255. Litvin, F. L.: "Design of Non-circular Gears and Their Application to Machine Design," Akad. Nauk, Moscow, *Trudi Inst. Machinoved,* vol. 14, no. 55, pp. 20–48, 1954.

256. Lockenvitz, A. E., J. B. Oliphint, W. C. Wilde, and J. M. Young: "Non-circular Cams and Gears," *Machine Design,* vol. 24, pp. 141–145, May 1952.

257. Loerch, R., A. G. Erdman, G. N. Sandor, and A. Midha: "Synthesis of Four-Bar Linkages with Specified Ground Pivots," *Proceedings, Fourth OSU Applied Mechanisms Conference,* pp. 10-1–10-6, Nov. 2–5, 1975.

258. Lowen, G. G., and R. S. Berkof: "Determination of Force-Balanced Four-Bar Linkages with Optimum Shaking Moment Characteristics," *J. Eng. Ind., Trans. ASME,* vol. 93B, pp. 39–46, 1971.

259. Lowen, G. G., and R. S. Berkof: "Survey of Investigations into the Balancing of Linkages," *J. Mechanisms,* vol. 3, pp. 221–231, 1968.

260. Lowen, G. G., F. R. Tepper, and R. S. Berkof: "The Qualitative Influence of Complete Force Balancing on the Forces and Moments of Certain Families of Four-Bar Linkages," *Mechanism and Machine Theory,* vol. 9, pp. 299–323, 1974.

261. Lundquist, I.: "Miniature Mechanical Clutches," *Machine Design,* vol. 28, pp. 124–133, Oct. 18, 1956.

262. Mabie, H. H.: "Constant Velocity Universal Joints," *Machine Design,* vol. 20, pp. 101–105, May 1948.

263. Mabie, H. H., and F. W. Ocvirk: "Mechanisms and Dynamics of Machinery," John Wiley & Sons, Inc., New York, 1957.

264. *Machine Design,* "Velocity and Acceleration Analysis of Universal Joints," vol. 14 (data sheet), pp. 93–94, November 1942.

265. *Machine Design,* "Chain Drive," vol. 23, pp. 137–138, June 1951.

266. *Machine Design,* "Constant Force Action" (manual typewriters), vol. 28, p. 98, Sept. 20, 1956.

267. *Machine Design,* vol. 30, p. 134, Feb. 20, 1958.

268. *Machine Design,* "Constant-Force Output," vol. 30, p. 115, Apr. 17, 1958.

269. *Machine Design,* "Taut-Band Suspension," vol. 30, p. 152, July 24, 1958, "Intermittent-Motion Gear Drive," vol. 30, p. 151, July 24, 1958.

270. *Machine Design,* "Almost Human Engineering," vol. 31, pp. 22–26, Apr. 30, 1959.

271. *Machine Design,* "Aircraft All-Mechanical Computer, vol. 31, pp. 186–187, May 14, 1959.

272. *Machine Design,* "Virtual Hinge Pivot," vol. 33, p. 119, July 6, 1961.

273. Mansfield, J. H.: "Woodworking Machinery," *Mech. Eng.,* vol. 74, pp. 983–995, December 1952.

274. Marich, F.: "Mechanical Timers," *Prod. Eng.,* vol. 32, pp. 54–56, July 17, 1961.

275. Marker, R. C.: "Determining Toggle-Jaw Force," *Machine Design,* vol. 22, pp. 104–107, March 1950.

276. Martin, G. H., and M. F. Spotts: "An Application of Complex Geometry to Relative Velocities and Accelerations in Mechanisms," *Trans. ASME,* vol. 79, pp. 687–693, April 1957.

277. Mathi, W. E., and C. K. Studley, Jr.: "Developing a Counting Mechanism," *Machine Design,* vol. 22, pp. 117–122, June 1950.

278. Maxwell, R. L.: "Kinematics and Dynamics of Machinery," Prentice-Hall, Inc., Englewood Cliffs, N.J., 1960.

279. Mayer, A. E.: "Koppelkurven mit drei Spitzen und Spezielle Koppelkurvenbueschel," *Z. Math. Phys.,* vol. 43, p. 389, 1937.

280. McComb, G. T., and W. N. Matson: "Four Ways to Select Change Gears—And the Faster Fifth Way," *Prod. Eng.,* vol. 31, pp. 64–67, Feb. 15, 1960.

281. Mehmke, R.: "Ueber die Bewegung eines Starren ebenen Systems in seiner Ebene," *Z. Math. Phys.,* vol. 35, pp. 1–23, 65–81, 1890.

282. Merritt, H. E.: "Gear Trains," Sir Isaac Pitman & Sons, Ltd., London, 1947.

283. Meyer zur Capellen, W.: "Getriebependel," *Z. Instrumentenk.,* vol. 55, 1935; pt. I, October, pp. 393–408; pt. II, November, pp. 437–447.

284. Meyer zur Capellen, W.: "Die Totlagen des ebenen, Gelenkvierecks in analytischer Darstellung," *Forsch. Ing. Wes.,* vol. 22, no. 2, pp. 42–50, 1956.

285. Meyer zur Capellen, W.: "Der einfache Zahnstangen-Kurbeltrieb und das entsprechende Bandgetriebe," *Werkstatt u. Betrieb,* vol. 89, no. 2, pp. 67–74, 1956.

286. Meyer zur Capellen, W.: "Harmonische Analyse bei der Kurbelschleife," *Z. angew., Math. u. Mechanik,* vol. 36, pp. 151–152, March–April 1956.

287. Meyer zur Capellen, W.: "Kinematik des Einfachen Koppelraedertriebes," *Werkstatt u. Betrieb,* vol. 89, no. 5, pp. 263–266, 1956.

288. Meyer zur Capellen, W.: "Bemerkung zum Satz von Roberts über die Dreifache Erzeugung der Koppelkurve," *Konstruktion,* vol. 8, no. 7, pp. 268–270, 1956.

289. Meyer zur Capellen, W.: "Kinematik und Dynamik der Kurbelschleife," *Werkstatt u. Betrieb,* pt. I, vol. 89, no. 10, pp. 581–584, 1956; pt. II, vol. 89, no. 12, pp. 677–683, 1956.

290. Meyer zur Capellen, W.: "Ueber gleichwertige periodische Getriebe," *Fette, Seifen, Anstrichmittel,* vol. 59, no. 4, pp. 257–266, 1957.

291. Meyer zur Capellen, W.: "Die Kurbelschleife Zweiter Art," *Werkstatt u. Betrieb,* vol. 90, no. 5, pp. 306–308, 1957.

292. Meyer zur Capellen, W.: "Die elliptische Zahnraeder und die Kurbelschleife," *Werkstatt u. Betrieb,* vol. 91, no. 1, pp. 41–45, 1958.

293. Meyer zur Capellen, W.: "Die harmonische Analyse bei elliptischen Kurbelschleifen," *Z. angew, Math. Mechanik*, vol. 38, no. 1/2, pp. 43–55, 1958.

294. Meyer zur Capellen, W.: "Das Kreuzgelenk als periodisches Getriebe," *Werkstatt u. Betrieb*, vol. 91, no. 7, pp. 435–444, 1958.

295. Meyer zur Capellen, W.: "Ueber elliptische Kurbelschleifen," *Werkstatt u. Betrieb*, vol. 91, no. 12, pp. 723–729, 1958.

296. Meyer zur Capellen, W.: "Bewegungsverhaeltnisse an der geschraenkten Schubkurbel," *Forchungs berichte des Landes Nordrhein-Westfalen, West-Deutscher Verlag*, Cologne, no. 449, 1958.

297. Meyer zur Capellen, W.: "Eine Getriebergruppe mit stationaerem Geschwindigkeitsverlauf" (Elliptic Slidercrank Drive) (in German) *Forschungsberichte des Landes Nordrhein-Westfalen*, West-Deutscher Verlag, Cologne, no. 606, 1958.

298. Meyer zur Capellen, W.: "Die Beschleunigungsaenderung," *Ing-Arch.*, vol. 27, pt. I, no. 1, pp. 53–65; pt. II, no. 2, pp. 73–87, 1959.

299. Meyer zur Capellen, W.: "Die geschraenkte Kurbelschleife zweiter Art," *Werkstatt u. Betrieb*, vol. 92, no. 10, pp. 773–777, 1959.

300. Meyer zur Capellen, W.: "Harmonische Analyse bei Kurbeltrieben," *Forschungsberichte des Landes Nordrhein-Westfalen*, West-Deutscher Verlag, Cologne, pt. I, no. 676, 1959; pt. II, no. 803, 1960.

301. Meyer zur Capellen, W.: "Die geschraenkte Kurbelschleife," *Forschungsberichte des Landes Nordrhein-Westfalen*, West-Deutscher Verlag, Cologne, pt. I, no. 718, 1959; pt. II, no. 804, 1960.

302. Meyer zur Capellen, W.: "Die Extrema der Uebersetzungen in ebenen und sphaerischen Kurbeltrieben," *Ing. Arch.*, vol. 27, no. 5, pp. 352–364, 1960.

303. Meyer zur Capellen, W.: "Die gleichschenklige zentrische Kurbelschwinge," *Z. Prakt. Metallbearbeitung*, vol. 54, no. 7, pp. 305–310, 1960.

304. Meyer zur Capellen, W.: "Three-Dimensional Drives," *Prod. Eng.*, vol. 31, pp. 76–80, June 30, 1960.

305. Meyer zur Capellen, W.: "Kinematik der sphaerischen Schubkurbel," *Forschungsberichte des Landes Nordrhein-Westfalen*, West-Deutscher Verlag, Cologne, no. 873, 1960.

306. Michalec, G. W.: "Analog Computing Mechanisms," *Machine Design*, vol. 31, pp. 157–179, Mar. 19, 1959.

307. Miller, H.: "Analysis of Quadric-Chain Mechanisms," *Prod. Eng.*, vol. 22, pp. 109–113, February 1951.

308. Miller, W. S.: "Packaged Speed Reduces and Gearmotors," *Machine Design*, vol. 29, pp. 121–149, Mar. 21, 1957.

309. Modrey, J.: "Analysis of Complex Kinematic Chains with Influence Coefficients," *J. Appl. Mech.*, vol. 26; *Trans. ASME*, vol. 81, E., pp. 184–188, June 1959; discussion, *J. Appl. Mech.*, vol. 27, pp. 215–216, March, 1960.

310. Moore, J. W., and M. V. Braunagel: "Space Linkages," *Trans. Seventh Conf. Mech.*, Purdue University, West Lafayette, Ind., pp. 114–122, 1962.

311. Moreinis, I. Sh.: "Biomechanical Studies of Some Aspects of Walking on a Prosthetic Device" (Russian), Akad. Nauk, Moscow, *Trudi Inst. Machinoved.*, vol. 21, no. 81–82, pp. 119–131, 1960.

312. Morgan, P.: "Mechanisms for Moppets," *Machine Design*, vol. 34, pp. 105–109, Dec. 20, 1962.

313. Moroshkin, Y. F.: "General Analytical Theory of Mechanisms" (Russian), Akad. Nauk, Moscow, *Trudi Sem. Teor. Mash. Mekh.*, vol. 14, no. 54, pp. 25–50, 1954.

314. Morrison, R. A.: "Rolling-Surface Mechanisms," *Machine Design*, vol. 30, pp. 119–123, Dec. 11, 1958.

315. Mueller, R.: "Einfuehrung in die theoretische Kinematik," Springer-Verlag OHG, Berlin, 1932.

316. Musser, C. W.: "Mechanics Is Not a Closed Book," *Trans. Sixth Conf. Mechanisms,* Purdue University, West Lafayette, Ind., October, 1960, pp. 31–43.

317. Oldham, K., and J. N. Fawcett: "Computer-Aided Synthesis of Linkage—A Motorcycle Design Study," *Proc. Institution of Mechanical Engineers* (London), vol. 190, no. 63/76, pp. 713–720, 1976.

318. Olsson, V.: "Non-circular Cylindrical Gears," *Acta Polytech., Mech. Eng.* Ser., vol. 2, no. 10, Stockholm, 1959.

319. Olsson, V.: "Non-circular Bevel Gears," *Acta Polytech., Mech. Eng. Ser., 5,* Stockholm, 1953.

320. Orlandea, N., and T. Berenyi: "Dynamic Continuous Path Synthesis of Industrial Robots Using ADAMS Computer Program," *J. Mechanical Design, Trans. ASME,* vol. 103, pp. 602–607, 1981.

321. Paul, B.: "A Unified Criterion for the Degree of Constraint of Plane Kinematic Chains," *J. Appl. Mech.,* vol. 27; *Trans. ASME,* ser. E., vol. 82, pp. 196–200, March 1960.

322. Paul, B.: "Analytical Dynamics of Mechanisms—A Computer Oriented Overview," *J. Mechanisms and Machine Theory,* vol. 10, no. 6, pp. 481–508, 1975.

323. Paul, B.: "Kinematics and Dynamics of Planar Machinery," Prentice-Hall, Inc., Englewood Cliffs, N.J., 1979.

324. Paul, B., and D. Krajcinovic: "Computer Analysis of Machines with Planar Motion, I, II," *J. Appl. Mech.,* vol. 37, pp. 697–712, 1979.

325. Peek, H. L.: "Trip-Free Mechanisms," *Mech. Eng.,* vol. 81, pp. 193–199, March 1959.

326. Peyrebrune, H. E.: "Application and Design of Non-circular Gears," *Trans. First Conf. Mechanisms,* Purdue University, West Lafayette, Ind., pp. 13–21, 1953.

327. Philipp, R. E.: "Kinematics of a General Arrangement of Two Hooke's Joints," ASME paper 60-WA-37, 1960.

328. Pike, E. W., and T. R. Silverberg: "Designing Mechanical Computers," *Machine Design,* vol. 24, pt. I, pp. 131–137, July 1952; pt. II, pp. 159–163, August 1952.

329. Pollitt, E. P.: "High-Speed Web-Cutting," *Machine Design,* vol. 27, pp. 155–160, December 1955.

330. Pollitt, E. P.: "Motion Characteristics of Slider-Crank Linkages," *Machine Design,* vol. 30, pp. 136–142, May 15, 1958.

331. Pollitt, E. P.: "Some Applications of the Cycloid in Machine Design," *Trans. ASME, J. Eng. Ind.,* ser. B, vol. 82, no. 4, pp. 407–414, November 1960.

332. Predale, J. O., and A. B. Hulse, Jr.: "The Space Crank," *Prod. Eng.,* vol. 30, pp. 50–53, Mar. 2, 1959.

333. Primrose, E. J. F., and F. Freudenstein: "Geared Five-Bar Motion II—Arbitrary Commensurate Gear Ratio," *J. Appl. Mech.,* vol. 30; *Trans. ASME,* ser. E., vol. 85, pp. 170–175, June 1963.

334. Primrose, E. J. F., F. Freudenstein, and G. N. Sandor: "Finite Burmester Theory in Plane Kinematics," *J. Appl. Mech., Trans. ASME,* vol. 31E, pp. 683–693, 1964.

335. Procopi, J.: "Control Valves," *Machine Design,* vol. 22, pp. 153–155, September 1950.

336. Proctor, J.: "Selecting Clutches for Mechanical Drives," *Prod. Eng.,* vol. 32, pp. 43–58, June 19, 1961.

337. *Prod. Eng.:* "Mechanisms Actuated by Air or Hydraulic Cylinders," vol. 20, pp. 128–129, December 1949.

338. *Prod. Eng.:* "Computing Mechanisms," vol. 27, I, p. 200, Mar.; II, pp. 180–181, April 1956.

339. *Prod. Eng.:* "High-Speed Electrostatic Clutch," vol. 28, pp. 189–191, February 1957.

340. *Prod. Eng.:* "Linkage Keeps Table Flat," vol. 29, p. 63, Feb. 3, 1958.

341. *Prod. Eng.:* vol. 30, pp. 64–65, Mar. 30, 1959.

342. *Prod. Eng.:* "Down to Earth with a Four-Bar Linkage," vol. 31, p. 71, June 22, 1959.

343. *Prod. Eng.:* "Design Work Sheets," no. 14.

344. Radcliffe, C. W.: "Prosthetic Mechanisms for Leg Amputees," *Trans. Sixth Conf. Mechanisms,* Purdue University, West Lafayette, Ind., pp. 143–151, October 1960.

345. Radcliffe, C. W.: "Biomechanical Design of a Lower-Extremity Prosthesis," ASME Paper 60-WA-305, 1960.

346. Rainey, R. S.: "Which Shaft Seal," *Prod. Eng.,* vol. 21, pp. 142–147, May 1950.

347. Rankers, H.: "Vier genau gleichwertige Gelenkgetriebe für die gleiche Koppelkurve," *Des Industrieblatt,* pp. 17–21, January 1959.

348. Rankers, H.: "Anwendungen von sechsgliedrigen Kurbelgetrieben mit Antrieb an einem Koppelpunkt," *Das Industrieblatt,* pp. 78–83, February 1962.

349. Rankers, H.: "Bewegungsverhaeltnisse an der Schubkurbel mit angeschlossenem Kreuzschieber," *Das Industrieblatt,* pp. 790–796, December 1961.

350. Rantsch, E. J.: "Elliptic Gears Depend on Accurate Layout," *Machine Design,* vol. 9, pp. 43–44, March 1937.

351. Rappaport, S.: "A Neglected Design Detail," *Machine Design,* vol. 20, p. 140, September 1948.

352. Rappaport, S.: "Kinematics of Intermittent Mechanisms," *Prod. Eng.,* vols. 20–21, 1949–1951, I, "The External Geneva Wheel," July, pp. 110–112; II, "The Internal Geneva Wheel," August, pp. 109–112; III, "The Spherical Geneva Wheel," October, pp. 137–139; IV, "The Three-Gear Drive," January, 1950, pp. 120–123; V, "The Cardioid Drive," 1950, pp. 133–134.

353. Rappaport, S.: "Crank-and-Slot Drive," *Prod. Eng.,* vol. 21, pp. 136–138, July 1950.

354. Rappaport, S.: "Shearing Moving Webs," *Machine Design,* vol. 28, pp. 101–104, May 3, 1956.

355. Rappaport, S.: "Small Indexing Mechanisms," *Machine Design,* vol. 29, pp. 161–163, Apr. 18, 1957.

356. Rappaport, S.: "Elliptical Gears for Cyclic Speed Variation," *Prod. Eng.,* vol. 31, pp. 68–70, Mar. 28, 1960.

357. Rappaport, S.: "Intermittent Motions and Special Mechanisms," *Trans. Conf. Mechanisms,* Yale University, Shoestring Press, New Haven, Conn., pp. 91–122, 1961.

358. Rappaport, S.: "Review of Mechanical Integrators," *Trans. Seventh Conf. Mechanics,* Purdue University, West Lafayette, Ind., pp. 234–240, 1962.

359. Rasche, W. H.: "Design Formulas for Three-Link Screw Mechanisms," *Machine Design,* vol. 17, pp. 147–149, August 1945.

360. Rasche, W. H.: "Gear Train Design," *Virginia Polytechnic Inst. Eng. Experiment Station Bull.,* 14, 1933.

361. Raven, F. H.: "Velocity and Acceleration Analysis of Plane and Space Mechanisms by Means of Independent-Position Equations," *J. Appl. Mech.,* vol. 25, March 1958; *Trans. ASME,* vol. 80, pp. 1–6.

362. Reuleaux, F.: "The Constructor" (trans. H. H. Suplee), D. Van Nostrand Company, Inc., Princeton, N.J., 1983.

363. Richardson, I. H.: "Trend Toward Automation in Automatic Weighing and Bulk Materials Handling," *Mech. Eng.,* vol. 75, pp. 865–870, November 1953.

364. Ring, F.: "Remote Control Handling Devices," *Mech. Eng.,* vol. 78, pp. 828–831, September 1956.

365. Roemer, R. L.: Flight-Control Linkages, *Mech. Eng.,* vol. 80, pp. 56–60, June 1958.

366. Roger, R. J., and G. C. Andrews: "Simulating Planar Systems Using a Simplified Vector-Network Method," *J. Mechanisms and Machine Theory,* vol. 16, no. 6, pp. 509–519, 1975.

367. Root, R. E., Jr., "Dynamics of Engine and Shaft," John Wiley & Sons, Inc., New York, 1932.

368. Rosenauer, N., and A. H. Willis: "Kinematics of Mechanisms," Associated General Publications, Pty. Ltd., Sydney, Australia, 1953.

369. Rosenauer, N.: "Some Fundamentals of Space Mechanisms," *Mathematical Gazette,* vol. 40, no. 334, pp. 256–259, December 1956.

370. Rossner, E. E.: "Ratchet Layout," *Prod. Eng.,* vol. 29, pp. 89–91, Jan. 20, 1958.

371. Roth, B., F. Freudenstein, and G. N. Sandor: "Synthesis of Four-Link Path Generating Mechanisms with Optimum Transmission Characteristics," *Trans. Seventh Conf. Mechanics,* Purdue University, West Lafayette, Ind., pp. 44–48, October 1962 (available from *Machine Design,* Penton Bldg., Cleveland).

372. Roth, B., and F. Freudenstein: "Synthesis of Path Generating Mechanisms by Numerical Methods," *J. Eng. Ind., Trans. ASME,* vol. 85B, pp. 298–306, August 1963.

373. Roth, B.: "Robots," *Appl. Mechanics Rev.,* vol. 31, no. 11, pp. 1511–1519, 1978.

374. Roth, G. L.: "Modifying Valve Characteristics," *Prod. Eng.,* Annual Handbook, pp. 16–18, 1958.

375. Rothbart, H. A.: "Cams," John Wiley & Sons, Inc., New York, 1956.

376. Rothbart, H. A.: "Equivalent Mechanisms for Cams," *Machine Design,* vol. 30, pp. 175–180, Mar. 20, 1958.

377. Rumsey, R. D.: "Redesigned for Higher Speed," *Machine Design,* vol. 23, pp. 123–129, April 1951.

378. Rzeppa, A. H.: "Universal Joint Drives," *Machine Design,* vol. 25, pp. 162–170, April 1953.

379. Saari, O.: "Universal Joints," *Machine Design,* vol. 26, pp. 175–178, October 1954.

379a. Sadler J. P., and R. W. Mayne: "Balancing of Mechanisms by Nonlinear Programming," *Proc. Third Applied Mechanisms Conf.,* 1973.

380. Sandor, G. N., and F. Freudenstein: "Kinematic Synthesis of Path-Generating Mechanisms by Means of the IBM 650 Computer," Program 9.5.003, IBM Library, Applied Programming Publications, IBM, 590 Madison Avenue, New York, N.Y. 10022, 1958.

381. Sandor, G. N.: "A General Complex-Number Method for Plane Kinematic Synthesis with Applications," doctoral dissertation, Columbia University, University Microfilms, Ann Arbor, Mich., 1959, Library of Congress Card No. Mic. 59-2596.

382. Sandor, G. N.: "On the Kinematics of Rubber-Covered Cylinders Rolling on a Hard Surface," ASME Paper 61-SA-67, Abstr., *Mech. Engrg.,* vol. 83, no. 10, p. 84, October 1961.

383. Sandor, G. N.: "On Computer-Aided Graphical Kinematics Synthesis," Technical Seminar Series, Rep. 4, Princeton University, Dept. of Graphics and Engineering Drawing, Princeton, N.J., 1962.

384. Sandor, G. N.: "On the Loop Equations in Kinematics," *Trans. Seventh Conf. Mechanisms,* Purdue University, West Lafayette, Ind., pp. 49–56, 1962.

385. Sandor, G. N.: "On the Existence of a Cycloidal Burmester Theory in Planar Kinematics," *J. Appl. Mech.,* vol. 31, *Trans. ASME,* vol. 86E, 1964.

386. Sandor, G. N.: "On Infinitesimal Cycloidal Kinematic Theory of Planar Motion," *J. Appl. Mech., Trans. ASME,* vol. 33E, pp. 927–933, 1966.

387. Sandor, G. N., and F. Freudenstein: "Higher-Order Plane Motion Theories in Kinematic Synthesis," *J. Eng. Ind., Trans. ASME,* vol. 89B, pp. 223–230, 1967.

388. Sandor, G. N.: "Principles of a General Quaternion-Operator Method of Spatial Kinematic Synthesis," *J. Appl. Mech., Trans. ASME,* vol. 35E, pp. 40–46, 1968.

389. Sandor, G. N., and K. E. Bisshopp: "On a General Method of Spatial Kinematic Synthesis by Means of a Stretch-Rotation Tensor," *J. Eng. Ind., Trans. ASME,* vol. 91B, pp. 115–122, 1969.

390. Sandor, G. N., with D. R. Wilt: "Synthesis of a Geared Four-Link Mechanism," *Proc. Second International Congress Theory of Machines and Mechanics,* vol. 2, Zakopane, Poland, 1969, pp. 222–232, Sept. 24–27; *J. Mechanisms,* vol. 4, pp. 291–302, 1969.

391. Sandor, G. N., et al.: "Kinematic Synthesis of Geared Linkages," *J. Mechanisms,* vol. 5, pp. 58–87, 1970, *Rev. Rumanian Sci. Tech.-Mech. Appl.,* vol. 15, Bucharest, pp. 841–869, 1970.

392. Sandor, G. N., with A. V. M. Rao: "Extension of Freudenstein's Equation to Geared Linkages," *J. Eng. Ind., Trans. ASME,* vol. 93B, pp. 201–210, 1971.

393. Sandor, G. N., A. G. Erdman, L. Hunt, and E. Raghavacharyulu: "New Complex-Number Forms of the Euler-Savary Equation in a Computer-Oriented Treatment of Planar Path-Curvature Theory for Higher-Pair Rolling contact, *J. Mech. Design, Trans. ASME,* vol. 104, pp. 227–232, 1982.

394. Sandor, G. N., A. G. Erdman, and E. Raghavacharyulu: "A Note on Bobillier Constructions," in preparation.

395. Sandor, G. N., with A. V. Mohan Rao and Steven N. Kramer: "Geared Six-Bar Design," *Proc. Second OSU Applied Mechanisms Conference,* Stillwater, Okla, pp. 25-1–25-13, Oct. 7–9, 1971.

395a. Sandor, G. N., A. G. Erdman, L. Hunt, and E. Raghavacharyulu: "New Complex-Number Form of the Cubic of Stationary Curvature in a Computer-Oriented Treatment of Planar Path-Curvature Theory for Higher-Pair Rolling Contact," *J. Mech. Design, Trans. ASME,* vol. 104, pp. 233–238, 1982.

396. Sandor, G. N., with A. V. M. Rao: "Extension of Freudenstein's Equation to Geared Linkages," *J. Eng. Ind., Trans. ASME,* vol. 93B, pp. 201–210, 1971.

397. Sandor, G. N., with A. G. Erdman: "Kinematic Synthesis of a Geared Five-Bar Function Generator," *J. Eng. Ind., Trans. ASME,* vol. 93B, pp. 11–16, 1971.

398. Sandor, G. N., with A. D. Dimarogonas and A. G. Erdman: "Synthesis of a Geared *N*-bar Linkage," *J. Eng. Ind., Trans. ASME,* vol. 93B, pp. 157–164, 1971.

399. Sandor, G. N., with A. V. M. Rao, and G. A. Erdman: "A General Complex-Number Method of Synthesis and Analysis of Mechanisms Containing Prismatic and Revolute Pairs," *Proc. Third World Congress on the Theory of Machines and Mechanisms,* vol. D, Dubrovnik, Yugoslavia, pp. 237–249, Sept. 13–19, 1971.

400. Sandor, G. N. et al.: "Synthesis of Multi-Loop, Dual-Purpose Planar Mechanisms Utilizing Burmester Theory," *Proc. Second OSU Applied Mechanisms Conf.,* Stillwater, Okla., pp. 7-1–7-23, Oct. 7–9, 1971.

401. Sandor, G. N., with A. D. Dimarogonas: "A General Method for Analysis of Mechanical Systems," *Proc. Third World Congress for the Theory of Machines and Mechanisms,* Dubrovnik, Yugoslavia, pp. 121–132, Sept. 13–19, 1971.

402. Sandor, G. N., with A. V. M. Rao: "Closed Form Synthesis of Four-Bar Path Generators by Linear Superposition," *Proc. Third World Congress on the Theory of Machines and Mechanisms,* Dubrovnik, Yugoslavia, pp. 383–394, Sept. 13–19, 1971.

403. Sandor, G. N., with A. V. M. Rao: "Closed Form Synthesis of Four-Bar Function Generators by Linear Superposition," *Proc. Third World Congress on the Theory of Machines and Mechanisms,* Dubrovnik, Yugoslavia, pp. 395–405, Sept. 13–19, 1971.

404. Sandor, G. N., with R. S. Rose: "Direct Analytic Synthesis of Four-Bar Function Generators with Optimal Structural Error," *J. Eng. Ind., Trans. ASME,* vol. 95B, pp. 563–571, 1973.

405. Sandor, G. N., with A. V. M. Rao, and J. C. Kopanias: "Closed-form Synthesis of Planar Single-Loop Mechanisms for Coordination of Diagonally Opposite Angles," *Proc. Mechanisms 1972 Conf.,* London, Sept. 5–7, 1972.

406. Sandor, G. N., A. V. M. Rao, D. Kohli, and A. H. Soni: "Closed Form Synthesis of Spatial Function Generating Mechanisms for the Maximum Number of Precision Points," *J. Eng. Ind., Trans. ASME,* vol. 95B, pp. 725–736, 1973.

407. Sandor, G. N., with J. F. McGovern: "Kinematic Synthesis of Adjustable Mechanisms," Part 1, "Function Generation," *J. Eng. Ind., Trans. ASME,* vol. 95B, pp. 417–422, 1973.

408. Sandor, G. N., with J. F. McGovern: "Kinematic Synthesis of Adjustable Mechanisms," Part 2, "Path Generation," *J. Eng. Ind., Trans. ASME,* vol. 95B, pp. 423–429, 1973.

409. Sandor, G. N., with Dan Perju: "Contributions to the Kinematic Synthesis of Adjustable

Mechanisms," *Trans. International Symposium on Linkages and Computer Design Methods,* vol. A-46, Bucharest, pp. 636–650, June 7–13, 1973.

410. Sandor, G. N., with A. V. M. Rao: "Synthesis of Function Generating Mechanisms with Scale Factors as Unknown Design Parameters," *Trans. International Symposium on Linkages and Computer Design Methods,* vol. A-44, Bucharest, Romania, pp. 602–623, 1973.

411. Sandor, G. N., J. F. McGovern, and C. Z. Smith: "The Design of Four-Bar Path Generating Linkages by Fifth-Order Path Approximation in the Vicinity of a Single Point," *Proc. Mechanisms 73 Conf.,* University of Newcastle upon Tyne, England, Sept. 11, 1973: *Proc. Institution of Mechanical Engineers* (London), pp. 65–77.

412. Sandor, G. N., with R. Alizade and I. G. Novrusbekov: "Optimization of Four-Bar Function Generating Mechanisms Using Penalty Functions with Inequality and Equality Constraints," *Mechanism and Machine Theory,* vol. 10, no. 4, pp. 327–336, 1975.

413. Sandor, G. N., with R. I. Alizade and A. V. M. Rao: "Optimum Synthesis of Four-Bar and Offset Slider-Crank Planar and Spatial Mechanisms Using the Penalty Function Approach with Inequality and Equality Constraints," *J. Eng. Ind., Trans. ASME,* vol. 97B, pp. 785–790, 1975.

414. Sandor, G. N., with R. I. Alizade and A. V. M. Rao: "Optimum Synthesis of Two-Degree-of-Freedom Planar and Spatial Function Generating Mechanisms Using the Penalty Function Approach," *J. Eng. Ind., Trans. ASME,* vol. 97B, pp. 629–634, 1975.

415. Sandor, G. N., with I. Imam: "High-Speed Mechanism Design—A General Analytical Approach," *J. Eng. Ind., Trans. ASME,* vol. 97B, pp. 609–629, 1975.

416. Sandor, G. N., with D. Kohli: "Elastodynamics of Planar Linkages Including Torsional Vibrations of Input and Output Shafts and Elastic Deflections at Supports," *Proc. Fourth World Congress on the Theory of Machines and Mechanisms,* vol. 2, University of Newcastle upon Tyne, England, pp. 247–252, Sept. 8–13, 1975.

417. Sandor, G. N., with D. Kohli: "Lumped-Parameter Approach for Kineto-Elastodynamic Analysis of Elastic Spatial Mechanisms," *Proc. Fourth World Congress on the Theory of Machines and Mechanisms,* vol. 2, University of Newcastle upon Tyne, England, pp. 253–258, Sept. 8–13, 1975.

418. Sandor, G. N., with S. Dhande: "Analytical Design of Cam-Type Angular-Motion Compensators," *J. Eng., Ind., Trans. ASME,* vol. 99B, pp. 381–387, 1977.

419. Sandor, G. N., with C. F. Reinholtz, and S. G. Dhande: "Kinematic Analysis of Planar Higher-Pair Mechanisms," *Mechanism and Machine Theory,* vol. 13, pp. 619–629, 1978.

420. Sandor, G. N., with R. J. Loerch and A. G. Erdman: "On the Existence of Circle-Point and Center-Point Circles for Three-Precision-Point Dyad Synthesis," *J. Mechanical Design, Trans. ASME,* Oct. 1979, pp. 554–562.

421. Sandor, G. N., with R. Pryor: "On the Classification and Enumeration of Six-Link and Eight-Link Cam-Modulated Linkages," Paper No. USA-66, *Proc. Fifth World Congress on the Theory of Machines and Mechanisms,* Montreal, July, 1979.

421a. Sandor, G. N., et al.: "Computer Aided Design of Spatial Mechanisms," (CADSPAM), Interactive Computer Package for Dimensional Synthesis of Function, Path and Motion Generator Spatial Mechanisms, in preparation.

421b. Sandor, G. N., D. Kohli, and M. Hernandez: "Closed Form Analytic Synthesis of R-Sp-R Three-Link Function Generator for Multiply Separated Positions," ASME Paper No. 82-DET-75, May 1982.

421c. Sandor, G. N., D. Kohli, C. F. Reinholtz, and A. Ghosal: "Closed-Form Analytic Synthesis of a Five-Link Spatial Motion Generator," *Proc. Seventh Applied Mechanisms Conf.,* Kansas City, Mo., pp. XXVI-1–XXVI-7, 1981, "Mechanism and Machine Theory," vol. 19, no. 1, pp. 97–105, 1984.

421d. Sandor, G. N., D. Kohli, and Zhuang Xirong: "Synthesis of a Five-Link Spatial Motion Generator for Four Prescribed Finite Positions," submitted to the ASME Design Engineering Division, 1982.

421*e*. Sandor, G. N., D. Kohli, and Zhuang Xirong: "Synthesis of RSSR-SRR Spatial Motion Generator Mechanism with Prescribed Crank Rotations for Three and Four Finite Positions," submitted to the ASME Design Engineering Division, 1982.

421*f*. Sandor, G. N., and A. G. Erdman: "Advanced Mechanism Design Analysis and Synthesis," vol. 2, Prentice-Hall, Inc., Englewood Cliffs, N.J., 1984.

421*g*. Sandor, G. N., A. G. Erdman, E. Raghavacharyulu, and C. F. Reinholtz: "On the Equivalence of Higher and Lower Pair Planar Mechanisms," *Proc. Sixth World Congress on the Theory of Machines and Mechanisms,* Delhi, India, Dec. 15–20, 1983.

422. Saxon, A. F.: "Multistage Sealing," *Machine Design,* vol. 25, pp. 170–172, March, 1953.

423. Soled, J.: "Industrial Fasteners," *Machine Design,* vol. 28, pp. 105–136, Aug. 23, 1956.

424. Schashkin, A. S.: "Study of an Epicyclic Mechanism with Dwell," Ref. 17*B*, pp. 117–132.

425. Schmidt, E. H.: "Cyclic Variations in Speed," *Machine Design,* vol. 19, pp. 108–111, March 1947.

426. Schmidt, E. H.: "Cycloidal-Crank Mechanisms," *Machine Design,* vol. 31, pp. 111–114, Apr. 2, 1959.

427. Schulze, E. F. C.: "Designing Snap-Action Toggles," *Prod. Eng.,* vol. 26, pp. 168–170, November 1955.

428. Shaffer, B. W., and I. Krause: "Refinement of Finite Difference Calculations in Kinematic Analysis," *Trans. ASME,* vol. 82B, no. 4, pp. 377–381, November 1960.

429. Sheppard, W. H.: "Rolling Curves and Non-circular Gears," *Mech. World,* pp. 5–11, January 1960.

430. Sheth, P. N., and J. J. Uicker: "IMP—A Computer-Aided Design Analysis System for Mechanisms and Linkages," *J. Eng. Ind., Trans. ASME,* vol. 94B, pp. 454–464, 1972.

431. Shigley, J., and J. J. Uicker: "Theory of Machines and Mechanisms," McGraw-Hill Book Company, Inc., New York, 1980.

432. Sieker, K. H.: "Kurbelgetriebe-Rechnerische Verfahren," *VDI-Bildungswerk,* no. 077 (probably 1960–1961).

433. Sieker, K. H.: "Extremwerte der Winkelgeschwindigkeiten in Symmetrischen Doppelkurbeln," *Konstruktion,* vol. 13, no. 9, pp. 351–353, 1961.

434. Sieker, K. H.: "Getriebe mit Energiespeichern," C. F. Winterische Verlagshandlung, Fussen, 1954.

435. Sieker, K. H.: "Zur algebraischen Mass-Synthese ebener Kurbelgetriebe," *Ing. Arch.,* vol. 24, pt. I, no. 3, pp. 188–215, pt. II, no. 4, pp. 233–257, 1956.

436. Sieker, K. H.: "Winkelgeschwindigkeiten und Winkelbeschleunigungen in Kurbelschleifen," *Feinverktechnik,* vol. 64, no. 6, pp. 1–9, 1960.

437. Simonis, F. W.: "Stufenlos verstellbare Getriebe," Werkstattbuecher no. 96, Springer-Verlag OHG, Berlin, 1949.

438. Sloan, W. W.: "Utilizing Irregular Gears for Inertia Control," *Trans. First Conf. Mechanisms,* Purdue University, West Lafayette, Ind., pp. 21–24, 1953.

439. Soni, A. H., et al.: "Linkage Design Handbook," ASME, New York, 1977.

440. Spector, L. F.: "Flexible Couplings," *Machine Design,* vol. 30, pp. 101–128, Oct. 30, 1958.

441. Spector, L. F.: "Mechanical Adjustable-Speed Drives," *Machine Design,* vol. 27, I, April, pp. 163–196; II, June, pp. 178–189, 1955.

442. Spotts, M. F.: "Kinematic Properties of the Three-Gear Drive," *J. Franklin Inst.,* vol. 268, no. 6, pp. 464–473, December 1959.

443. Strasser, F.: "Ten Universal Shaft-Couplings," *Prod. Eng.,* vol. 29, pp. 80–81, Aug. 18, 1958.

444. Sugimoto, K., and J. Duffy: "Determination of Extreme Distances of a Robot Hand—I: A General Theory," *J. Mechanical Design, Trans. ASME,* vol. 103, pp. 631–636, 1981.

445. Suh, C. H., and C. W. Radcliffe: "Kinematics and Mechanisms Design," John Wiley & Sons, Inc., New York, 1978.

446. Svoboda, A.: "Computing Mechanisms and Linkages," MIT Radiation Laboratory Series, vol. 27, McGraw-Hill Book Company, Inc., New York, 1948.

447. Taborek, J. J.: "Mechanics of Vehicles 3, Steering Forces and Stability," *Machine Design,* vol. 29, pp. 92–100, June 27, 1957.

448. Talbourdet, G. J.: "Mathematical Solution of Four-Bar Linkages," *Machine Design,* vol. 13, I, II, no. 5, pp. 65–68; III, no. 6, pp. 81–82; IV, no. 7, pp. 73–77, 1941.

449. Talbourdet, G. J.: "Intermittent Mechanisms (data sheets)," *Machine Design,* vol. 20, pt. I, September, pp. 159–162; pt. II, October, pp. 135–138, 1948.

450. Talbourdet, G. J.: "Motion Analysis of Several Intermittent Variable-Speed Drives," *Trans. ASME,* vol. 71, pp. 83–96, 1949.

451. Talbourdet, G. J.: "Intermittent Mechanisms," *Machine Design,* vol. 22, pt. I, September, pp. 141–146, pt. II, October, pp. 121–125, 1950.

452. Tepper, F. R., and G. G. Lowen: "On the Distribution of the RMS Shaking Moment of Unbalanced Planar Mechanisms: Theory of Isomomental Ellipses," ASME technical paper 72-Mech-4, 1972.

452*a.* Tepper, F. R., and G. G. Lowen: "General Theorems Concerning Full Force Balancing of Planar Linkages by Internal Mass Redistribution," *J. Eng. Ind., Trans. ASME,* vol. 94B, pp. 789–796, 1972.

452*b.* Tepper, F. R., and G. G. Lowen: "A New Criterion for Evaluating the RMS Shaking Moment in Unbalanced Planar Mechanisms," *Proc. Third Applied Mechanisms Conf.,* 1973.

453. Tesar, D.: "Translations of Papers (by R. Mueller) on Geometrical Theory of Motion Applied to Approximate Straight-Line Motion," Kansas State Univ. Eng. Exp. Sta., Spec. Rept. 21, 1962.

454. Tesar, D., M. J. Ohanian, and E. T. Duga: "Summary Report of the Nuclear Reactor Maintenance Technology Assessment," *Proc. Workshop on Machines, Mechanisms and Robotics,* sponsored by the National Science Foundation and the Army Research Office, University of Florida, Gainesville, 1980.

455. Tesar, D., and G. Matthews: "The Dynamic Synthesis, Analysis and Design of Modeled Cam Systems," Lexington Books, D. C. Heath and Company, Lexington, Mass., 1976.

456. Thearle, E. L.: "A Non-reversing Coupling," *Machine Design,* vol. 23, pp. 181–184, April 1951.

457. Threlfall, D. C.: "The Inclusion of Coulomb Friction in Mechanisms Programs with Particular Reference to DRAM," *J. Mechanisms and Machine Theory,* vol. 13, no. 4, pp. 475–483, 1978.

458. Thumin, C.: "Designing Quick-Acting Latch Releases," *Machine Design,* vol. 19, pp. 110–115, September 1947.

459. Timoshenko, S., and D. G. Young: "Advanced Dynamics," McGraw-Hill Book Company, Inc., New York, 1948.

460. Timoshenko, S., and D. G. Young: "Engineering Mechanics," McGraw-Hill Book Company, Inc., New York, pp. 400–403, 1940.

461. Tolle, M.: "Regelung der Kraftmaschinen," 3d ed., Springer-Verlag OHG, Berlin, 1921.

462. Trinkl, F.: "Analytische und zeichnerische Verfahren zur Untersuchung eigentlicher Raumkurbelgetriebe," *Konstruktion,* vol. 11, no. 9, pp. 349–359, 1959.

463. Uhing, J.: "Einfache Raumgetriebe, für ungleichfoermige Dreh-und Schwingbewegung," *Konstruktion,* vol. 9, no. 1, pp. 18–21, 1957.

464. Uicker, J. J., Jr.: "Displacement Analysis of Spatial Mechanisms by an Iterative Method Based on 4×4 Matrices," M.S. dissertation, Northwestern University, Evanston, Ill., 1963.

465. Uicker, J. J.: "IMP" (computer code), Department of Mechanical Engineering, University of Wisconsin, Madison.

466. Vandeman, J. E., and J. R. Wood: "Modifying Starwheel Mechanisms," *Machine Design,* vol. 25, pp. 255–261, April 1953.

467. VDI Richtlinien. VDI Duesseldorf; for transmission-angle charts, refer to (*a*) four-bars, *VDI* 2123, 2124, Aug. 1959; (*b*) slider cranks, *VDI* 2125, Aug. 1959. For straight-line generation, refer to (*a*) in-line swinging-blocks, *VDI* 2137, Aug. 1959; (*b*) in-line slider-cranks, *VDI* 2136, Aug. 1959. (*c*) Planar four-bar, *VDI* 2130-2135, August, 1959.

468. "Sperrgetriebe," AWF-VDMA-VDI Getriebehefte, Ausschuss f. Wirtschaftliche Fertigung, Berlin, no. 6061 pub. 1955, nos. 6062, 6071 pub. 1956, no. 6063 pub. 1957.

469. Veldkamp, G. R.: "Curvature Theory in Plane Kinematics," J. B. Wolters, Groningen, 1963.

470. Vertut, J.: "Contributions to Analyze Manipulator Morphology, Coverage and Dexterity," vol. 1, "On the Theory and Practice of Manipulators," Springer-Verlag, New York, pp. 227–289, 1974.

471. Vidosic, J. P., and H. L. Johnson: "Synthesis of Four-Bar Function Generators," *Trans. Sixth Conf. Mechanisms,* Purdue University, West Lafayette, Ind., pp. 82–86, October 1960.

472. Vogel, W. F.: "Crank Mechanism Motions," *Prod. Eng.,* vol. 12, pt. I, June, pp. 301–305; II, July, pp. 374–379; III, pp. 423–428; August, IV, September, pp. 490–493, 1941.

473. Volmer, J.: "Konstruktion eines Gelenkgetriebes fuer eine Geradfuehrung," *VDI-Berichte,* vol. 12, pp. 175–183, 1956.

474. Volmer, J.: "Systematik," Kinematik des Zweiradgetriebes," *Maschinenbautechnik,* vol. 5, no. 11, pp. 583–589, 1956.

475. Volmer, J.: "Raederkurbelgetriebe," *VDI-Forschungsheft,* no. 461, pp. 52–55, 1957.

476. Volmer, J.: "Gelenkgetriebe zur Geradfuehrung einer Ebene" *Z. Prakt. Metallbearbeitung,* vol. 53, no. 5, pp. 169–174, 1959.

477. Wallace, W. B., Jr.: "Pressure Switches," *Machine Design,* vol. 29, pp. 106–114, Aug. 22, 1957.

478. Weise, H.: "Bewegungsverhaeltnisse an Filmschaltgetriebe, Getriebetechnik," *VDI-Berichte,* vol. 5, pp. 99–106, 1955.

479. Weise, H.: "Getriebe in Photographischen und kinematographischen Geraeten," *VDI-Berichte,* vol. 12, pp. 131–137, 1956.

480. Westinghouse Electric Corp., Pittsburgh, Pa., "Stability of a Lifting Rig," "Engineering Problems," vol. II, approx. 1960.

481. Weyth, N. C., and A. F. Gagne: "Mechanical Torque-Limiting Devices," *Machine Design,* vol. 18, pp. 127–130, May 1946.

482. Whitehead, T. N.: "Instruments and Accurate Mechanisms," Dover Publications, Inc., New York, 1954.

483. Wieleitner, H.: "Spezielle ebene Kurven," G. J. Goeschen Verlag, Leipzig, 1908.

484. Williams, R. J., and A. Seireg: "Interactive Modeling and Analysis of Open or Closed Loop Dynamic Systems with Redundant Actuators," *J. Mechanical Design, Trans. ASME,* vol. 101, pp. 407–416, 1979.

485. Winfrey, R. C.: "Dynamic Analysis of Elastic Link Mechanisms by Reduction of Coordinates," *Trans. ASME,* 94B, *J. Eng. Ind.,* 1972, pp. 577–582.

486. Wittel, O., and D. C. Haefele: "A Non-intermittent Film Projector," *Mech. Eng.,* vol. 79, pp. 345–347, April 1957.

487. Wolford, J. C., and A. S. Hall: "Second-Acceleration Analysis of Plane Mechanisms," ASME paper 57-A-52, 1957.

488. Wolford, J. C., and D. C. Haack: "Applying the Inflection Circle Concept," *Trans. Fifth Conf. Mechanisms,* Purdue University, West Lafayette, Ind., pp. 232–239, 1958.

489. Wolford, J. C.: "An Analytical Method for Locating the Burmester Points for Five Infinitesimally Separated Positions of the Coupler Plane of a Four-Bar Mechanism," *J. Appl. Mech.,* vol. 27, *Trans. ASME,* vol. 82E, pp. 182–186, March 1960.

490. Woerle, H.: "Sonderformen zwangläufiger viergelenkiger Raumkurbelgetriebe," *VDI-Berichte,* vol. 12, pp. 21–28, 1956.

491. Wunderlich, W.: "Zur angenaeherten Geradfuehrung durch symmetrische Gelenkvierecke," *Z. angew Math. Mechanik*, vol. 36, no. 3/4, pp. 103–110, 1956.

492. Wunderlich, W.: "Hoehere Radlinien,: *Osterr. Ing.-Arch.*, vol. 1, pp. 277–296, 1947.

493. Yang, A. T.: "Harmonic Analysis of Spherical Four-Bar Mechanisms," *J. Appl. Mech.*, vol. 29, no. 4, *Trans. ASME*, vol. 84E, pp. 683–688, December 1962.

494. Yang, A. T.: "Application of Quaternion Algebra and Dual Numbers to the Analysis of Spatial Mechanisms," doctoral dissertation, Columbia University, New York, 1963. See also *J. Appl. Mech.*, 1964.

495. Yudin, V. A.: "General Theory of Planetary-Cycloidal Speed Reducer," Akad. Nauk, Moscow, *Trudi Sem. Teor. Mash. Mekh.*, vol. 4, no. 13, pp. 42–77, 1948.

496. "Computer Explosion in Truck Engineering," a collection of seven papers, SAE Publ. SP240, December 1962.

497. "Mechanical Design of Spacecraft," Jet Propulsion Laboratory, Seminar Proceedings, Pasadena, Calif., August 1962.

498. "Guide to Limit Switches," *Prod. Eng.*, vol. 33, pp. 84–101, Nov. 12, 1962.

499. "Mechanical Power Amplifier," *Machine Design*, vol. 32, p. 104, Dec. 22, 1960.

SECTION 4
CAM MECHANISMS

Harold A. Rothbart, D.Eng.

Dean Emeritus
College of Science and Engineering
Fairleigh Dickinson University
Teaneck, N.J.

NOTATIONS AND DEFINITIONS

A = cross-sectional area, in^2

A = acceleration, in/s^2

a, b = distance, in

B = follower bearing length, in

C = constant

c = damping constant, lb·s/in

E = modulus of elasticity, lb/in^2

f = external forces, lb

f = friction force, lb

F = force normal to cam profile, lb

F_{st} = static friction

g = gravitational constant

h = maximum displacement of follower, in

I = moment of inertia, $lb/(in·s^2)$

k = spring constant, lb/in

K = constant

l = length of link, in

L = length of contact between cam and roller followers, in

m = equivalent mass of follower, $lb·s^2/in$

n = integer

N_1, N_2 = forces normal to translating follower stem, lb

r = a number

r = radius from cam center to center of curvature, in

r_c = radius from cam center to center of roller follower, in

R_a = radius of prime circle (smallest circle to the roller center), in

R_p = radius of pitch circle (circle drawn to roller center having maximum pressure angle), in

R_r = radius of follower roller, in

t = time for cam to rotate angle θ, s

$T =$ time for cam to rotate angle θ_0, s

$T =$ torques lb in

$T_p =$ period

$u =$ follower overhang, in

$x =$ displacement, in

$\dot{x} =$ velocity, in/s

$\ddot{x} =$ acceleration in/s^2

$W =$ safe endurance load, lb

$y =$ cam function (follower) displacement, in

$\dot{y} =$ cam function (follower) velocity, in/s

$\ddot{y} =$ cam function (follower) acceleration, in/s^2

$\alpha, \beta, \gamma =$ angles, rad

$\psi =$ pressure angle, rad

$\psi_n =$ maximum pressure angle, rad

$\theta_0 = \omega_t -$ cam angle rotation for rise h, rad

$\theta = \omega_t -$ cam angle rotation for displacement y, rad

$\mu =$ coefficient of friction

$\rho =$ radius of curvature of pitch curve (path of roller center), in

$\rho_c =$ radius of curvature of cam surface at point of maximum stress (positive for concave and negative for convex), in

$\tau =$ angle between radius of curvature and follower motion, rad

$\omega =$ cam angular velocity, rad/s

4.1 INTRODUCTION[28,29,45,48]

A "cam" is a mechanical member for transmitting a desired motion to a follower by direct contact. The driver is called a cam and the driven member is called the "follower." Either may remain stationary, translate, oscillate, or rotate. The motion is given by $y = f(\theta)$.

Kinematically speaking, in its general form the plane cam mechanism (Fig. 4.1)

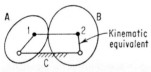

consists of two shaped members A and B connected by a fixed third body C. Either body A or body B may be the driver with the other the follower. We may at each instant replace these shaped bodies by an equivalent mechanism. These are pin-jointed at the instantaneous centers of curvature, 1 and 2, of the contacting surfaces. In general, at any other instant the points 1 and 2 are shifted and the links of the equivalent mechanism have different lengths. Figure 4.2 shows the two most popular cams.

FIG. 4.1 Basic cam mechanism and its kinematic equivalent (points 1 and 2 are centers of curvature) of the contact point. (*Courtesy of John Wiley & Sons, Inc.*[1])

Cam synthesis may be accomplished by (1) shaping the cam body to some known curve, such as involute, spirals, parabolas, or circular arcs; (2) mathematically controlling and establishing the follower motion and forming the cam by tabulated data of this action; (3) establishing the cam contour in parametric form; and (4) layout of cam profile by eye. This last method is acceptable only for low speeds in which a smooth "bumpless" curve may fulfill the requirements. But as either loads, mass, speed, or elasticity of the members increases, a detailed study must be made of both the dynamic aspects of the cam curve and the accuracy of cam fabrication. References 2–5, 16, 27, 31, 32, 40, 41, 45, 47, 48, and 54 discuss the exact synthesis of cam-follower design kinematics. Reference 20 shows cam-computing mechanisms of all sorts.

The roller follower is most frequently employed to distribute and reduce wear between the cam and follower. Obviously the cam and follower must be constrained

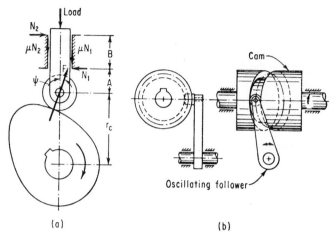

FIG. 4.2   Popular cams. (*a*) Radial cam—translating roller follower (open cam). (*b*) Cylindrical cam—oscillating roller follower (closed cam). (*Courtesy of John Wiley & Sons, Inc.*[1])

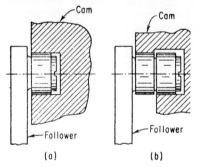

FIG. 4.3   Some roller and groove designs. (*a*) Single roller. (*b*) Double roller, undercut groove. (*Courtesy of John Wiley & Sons, Inc.*[1])

at all speeds. A preloaded compression spring (with an open cam) or a positive drive is used. Positive-drive action is accomplished by either a cam groove or a conjugate follower or followers in contact with opposite sides of a single or double cam. Figure 4.3*a* shows a single roller in a groove. Backlash and sliding become serious at high speed and may necessitate the double roller in a relieved groove (Fig. 4.3*b*). Tapered rollers have been used.

4.2   PRESSURE ANGLE[1,6]

The pressure angle is the angle between the normal to the cam profile and the instantaneous direction of the follower. It is related to the follower force distribution. With flat-faced followers, the pressure-angle force distribution rarely is of concern. With oscillating roller followers, the distance from the cam center to the follower pivot center and the length of the follower arm are pertinent to this study. A radial translating roller follower having rigid members has a theoretical limiting pressure angle based on the forces shown in Fig. 4.2*a*.

$$\psi_m < \arctan\left[B/\mu(2A + B)\right] \qquad (4.1)$$

To be practical, the pressure angle is generally limited to $30°$ for the translating roller follower because of the elasticity of the follower stem and the clearance between the stem in its guide bearing and the friction. We can show that the pressure angle $\tan \psi = \dot{y}/r_c\omega$. Symmetrical cam curves have the point of maximum pressure angle approximately at the midpoint of the rise

$$\tan \psi_m = \dot{y}_m/R_p\omega \tag{4.2}$$

4.3 CURVATURE[1,7]

The cam curvature is pertinent in establishing the cam-follower surface stresses, life, and undercutting. Undercutting is a phenomenon in which inadequate cam curvature yields incorrect follower movement. For a translating follower the curvature of pitch curve (path of roller center) at any point

$$\rho = \frac{[(R_a + y)^2 + (\dot{y}/\omega)^2]^{3/2}}{(R_a + y)^2 + 2(\dot{y}/\omega)^2 - (R_a + y)(\ddot{y}/\omega^2)} \tag{4.3}$$

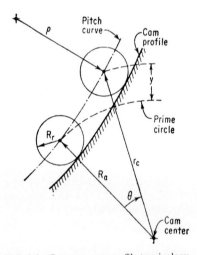

FIG. 4.4   Curvature cam profile terminology.

If ρ is positive, we have a convex cam profile, and if ρ is negative, we have a concave cam profile (Fig. 4.4). For a convex curved cam to prevent undercutting, which generally occurs in the point of maximum negative acceleration, $\rho > R_r$. In all cams the minimum radius of curvature should be determined to establish that the cam-follower system does not have excessive stresses or undercutting.

4.4 COMPLEX NOTATION[17,19]

Complex numbers can be employed to obtain the curvature, profile, cutter location, etc., for any kind of cam and follower. Use of a high-speed computer will aid in its use. Let us take the disk cam with an oscillating roller follower as an example (Fig. 4.5). The angular displacement of the follower from initial position β_0

$$\beta = \beta_0 + f(\theta)$$

The relations to the roller or cutter center

$$\mathbf{r}_c = r_c e^{i\theta_c} = re^{i\alpha} + \rho e^{i\varphi} = a + ib + le^{i\beta} \tag{4.4}$$

The equation of the cam contour measured from the initial position

$$r_c e^{i(\theta + \theta_c)} = (a + ib + le^{i\beta})e^{i\theta} \tag{4.5}$$

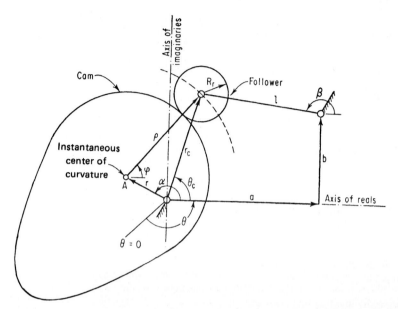

FIG. 4.5   Complex notation for oscillating roller follower.

Solving for r_c and θ_c yields the information necessary to construct the cam profile. Separating Eq. (4.4) into real and imaginary parts

$$r \cos \alpha + \rho \cos \varphi = a + l \cos \beta \qquad (4.6)$$

$$r \sin \alpha + \rho \sin \varphi = b + l \sin \beta \qquad (4.7)$$

Differentiating, we obtain the radius of curvature

$$\rho = \frac{(E^2 + D^2)^{3/2}}{(E^2 + D^2)[1 + f'(\theta)] - (aE + bD)f'(\theta) + (a \sin \beta - b \cos \beta)lf''(\theta)} \qquad (4.8)$$

where
$$E = [1 + f'(\theta)]l \cos \beta + a \qquad (4.9)$$

$$D = [1 + f'(\theta)]l \sin \beta + b \qquad (4.10)$$

to avoid undercutting $\rho > R_r$.

4.5 BASIC DWELL-RISE-DWELL CURVES

The most popular method in establishing the cam shape is by first choosing a simple curve which is easy to construct and analyze. Conventionally such curves are of two families: the simple polynomial, i.e., parabolic and cubic, and the trigonometric, i.e., simple harmonic, cycloidal, etc. The polynomial family

$$y = C_0 + C_1\theta + C_2\theta^2 + C_3\theta^3 + \cdots + C_n\theta^n \tag{4.11}$$

of which the basic curve is generally with a low integer

$$y = C_n\theta^n \tag{4.12}$$

The trigonometric family

$$y = C_0 + C_1\sin\omega t + C_2\cos\omega t + C_3\sin^2\omega t + \cdots \tag{4.13}$$

is reduced to the basic curves using only the lowest-order terms. The trigonometric are better since they yield smaller cams, lower follower side thrust, cheaper manufacturing costs, and easier layout and duplication. A summary of equations and curves is shown in Fig. 4.6 and Tables 4.1 and 4.2.

The parabolic curve is constructed (Fig. 4.6a) by dividing any line OB into odd incremental spacings, i.e., 1, 3, 5, 5, 3, 1, the same number as the abscissa spacing. This line is projected to rise h and then to spacings. The simple harmonic curve is constructed by dividing the h diameter into equal parts and projecting.

4.6   CAM LAYOUT

The fundamental basis for cam layout is one of inversion in which the cam profile is developed by fixing the cam moving the follower to its respective relative positions.

EXAMPLE   A radial cam rotating at 180 rpm is driving a ¾-in-diameter translating roller follower. Construct the cam profile with the pressure angle limited to 20° on the rise. The motion is as follows: (1) rise of 1 in with simple harmonic motion in 150° of cam rotation, (2) dwell for 60°, (3) fall of 1 in with simple harmonic motion in 120° of cam rotation, (4) dwell for the remaining 30°.

solution   The maximum pressure angle for the rise action occurs at the transition point where $\theta = 75°$. From Eq. (4.8) we find $R_p = 1.64$ in which 1¾ in is chosen.

The simple harmonic displacement diagram is constructed in Fig. 4.7, and the cam angle of 150° is divided into equal parts, i.e., 6 parts at 25° on either side of the line of action A3″. This gives radial lines A0″, A1″, A2″, etc. From the cam center A, arcs are swung from points 0′, 1′, 2′, 3′, etc., intersecting respective radial lines, locating points 0, 1, 2, 3, etc. Next we draw the smooth pitch curve, roller circles, and cam profile tangent to the rollers. We continue in the same manner to complete the cam. Computer-aided design of cam profile is shown in Refs. 38 and 50.

4.7   ADVANCED CURVES[15]

Special curve shapes are obtained by (1) employing polynomial or trigonometric relations previously mentioned, (2) numerical procedure by the method of finite differ-

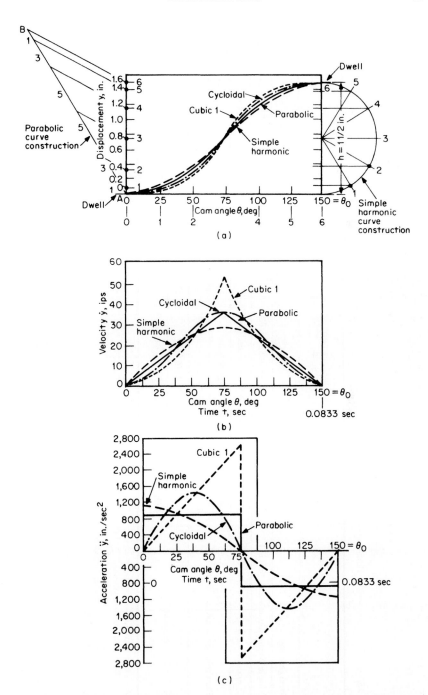

FIG. 4.6 Comparison of basic curves—dwell-rise-dwell cam. (Follower has 1½-in rise in 150° of cam rotation at 300 r/min.) (*a*) Displacement. (*b*) Velocity. (*c*) Acceleration. (*Courtesy of John Wiley & Sons, Inc.*[1])

TABLE 4.1 Characteristic Equations of Basic Curves*

Curves	Displacement y, in.	Velocity $\dot{y}$, ips	Acceleration $\ddot{y}$, in./sec²
Straight line	$h\theta/\theta_0$	$\omega h/\theta_0$	0
Simple harmonic motion (SHM)	$h/2[1 - \cos(\pi\theta/\theta_0)]$	$(h\pi\omega/2\theta_0)\sin(\pi\theta/\theta_0)$	$(h/2)(\pi\omega/\theta_0)^2\cos(\pi\theta/\theta_0)$
Cycloidal	$h/\pi[\pi\theta/\theta_0 - \tfrac{1}{2}\sin(2\pi\theta/\theta_0)]$	$h\omega/\theta_0[1 - \cos(2\pi\theta/\theta_0)]$	$2h\pi\omega^2/\theta_0^2\,\sin(2\pi\theta/\theta_0)$
Parabolic or constant acceleration $\theta/\theta_0 \leq 0.5$	$2h(\theta/\theta_0)^2$	$4h\omega\theta/\theta_0^2$	$4h(\omega^2/\theta_0^2)$
$\theta/\theta_0 \geq 0.5$	$h[1 - 2(1 - \theta/\theta_0)^2]$	$(4h\omega/\theta_0)(1 - \theta/\theta_0)$	$-4h(\omega^2/\theta_0^2)$
Cubic 1 or constant pulse 1 $\theta/\theta_0 \leq 0.5$	$4h(\theta/\theta_0)^3$	$(12h\omega/\theta_0)(\theta/\theta_0)^2$	$(24h\omega^2/\theta_0^2)(\theta/\theta_0)$
$\theta/\theta_0 \geq 0.5$	$h[1 - 4(1 - \theta/\theta_0)^3]$	$(12h\omega/\theta_0)(1 - \theta/\theta_0)^2$	$-(24h\omega^2/\theta_0^2)(1 - \theta/\theta_0)$

where h = maximum rise of follower, in
    θ_0 = cam angle of rotation to give rise h, rad
    ω = cam angular velocity, rad/s
    θ = cam-angle rotation for follower displacement y, rad

*Courtesy of John Wiley & Sons, Inc.[1]

TABLE 4.2 Comparison of Symmetrical Curves*

		Displacement y	Velocity	Acceleration	Pulse $\dddot{y}$	Comments
Polynomial	Parabolic		2, $\theta_0=1$	4, 4	∞, ∞, $\theta_0=1$	Backlash serious
	Cubic No. 1		3, $\theta_0=1$	12, 12	∞, 24	Not suggested
	Cubic No. 2		$1\frac{1}{2}$	6.4, 6.4	∞, ∞, 12	Not suggested for high speeds
	2–3 polynomial		$1\frac{1}{2}$	6, 6	∞, ∞, 8	
	3–4–5 polynomial	Rise, Dwell, h=1 total rise, $\theta_0=1$ Total angle	1.9	5.8, 5.8	60, 30	Similar to the cycloidal
	4–5–6–7 polynomial		2.4	7.3, 7.3	42, 52	Doubtful high-speed improvement over the 3–4–5 polynomial
Trigonometric	Simple harmonic motion		1.6	4.9, 4.9	∞, ∞, 15.5	
	Cycloidal		2	6.3, 6.3	40, 40	Excellent for high speeds
	Double harmonic		2	5.5, 9.9	∞, 20, 40	Best as a dwell-rise-return-dwell cam
Combination	Trapezoidal acceleration		2	5.3, 5.3, $\frac{\beta}{8}$	44, 44	Excellent for high speeds; may have machining difficulty
	Modified trapezoidal acceleration		2	4.9, 4.9, $\frac{\beta}{8}$	61, 61	Excellent vibratory and easier fabrication may prove better than cycloidal

*Courtesy of John Wiley & Sons, Inc.[1]

ences, (3) blending portions of curves, and (4) trial and error employment of a high-speed computer starting with the fourth-derivative (d^4y/dt^4) curve.

Rise or fall portions of any curve are each composed of a positive acceleration period and a negative acceleration period. For any smooth curve, we know that $\int_0^{\theta/\omega} \ddot{y}\, dt = 0$. Therefore, for any rise or fall curve, the area under the positive-acceleration portion equals the area under the negative-acceleration portion for each curve. Reference 37 shows a method for controlling the harmonics of a cam profile based on a minimum squared error performance index over a prescribed speed range.

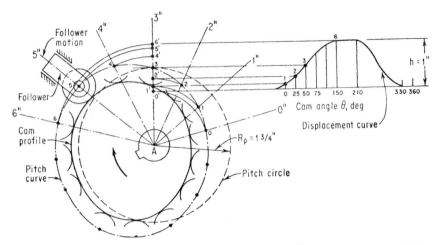

FIG. 4.7   Radial cam—translating roller. Follower example (scale ⅝ in = 1 in). (*Courtesy of John Wiley & Sons, Inc.*[1])

4.7.1   Polynomial and Trigonometric

Equations (4.11) and (4.13) show the general expressions. The symmetrical dwell-rise-dwell cam curve can be developed from a Fourier series with only odd multiples of the fundamental frequency[14]

$$\ddot{y} = \sum_{k\ odd}^{n} C_k \sin \omega_k t = C_1 \sin \omega_1 t + C_3 \sin \omega_3 t + C_5 \sin \omega_5 t + \cdots \qquad (4.14)$$

where $\omega_k = 2\pi k/T$ and C_r's are coefficients.

If $k = 1$, it yields the well-known cycloidal curve; if $k = \infty$, we obtain the parabolic curve.

For $n > 1$ we can progressively flatten curves for any shape desired.

4.7.2   Finite Differences[1,8,38,39]

In the application of finite differences the displacement y is known for pivotal points equally spaced by Δt. The derivation of formulas by Taylor-series expansion of $y(t + \Delta t)$ about t is

$$y(t + \Delta t) = y(t) + \Delta t\, \dot{y}(t) + [(\Delta t)^2/2!]\ddot{y}(t) + [(\Delta t)^3/3!]\dddot{y}(t) + \cdots \qquad (4.15)$$

For simplicity we take a three-point approximation which consists of passing a parabola through the three pivotal points. At any interval i

$$y_{i+1} \simeq y_i + \Delta t\, \dot{y}_i + (\Delta t)^2/2!\ddot{y}_i$$

$$y_{i-1} \simeq y_i - \Delta t\, \dot{y}_i + (\Delta t)^2/2!\ddot{y}_i$$

Solving for the velocity and acceleration, respectively,

$$\dot{y}_i \simeq (y_{i+1} - y_{i-1})/2\Delta t \qquad (4.16)$$

$$\ddot{y}_i \simeq (y_{i+1} + y_{i-1} - 2y_i)/(\Delta t)^2 \qquad (4.17)$$

Equations (4.16) and (4.17) can be used to determine the numerical relationships between y, $\dot{y}$, $\ddot{y}$ characteristics curves for incremental steps to include the cam accuracy if desired.

4.7.3 Blending Portions of Curves[1,33,34]

The primary condition for these combination curves is that (1) there must be no discontinuities in the acceleration curve and (2) the acceleration curve has a smooth shape. Therefore, weaker harmonics result. Table 4.2 gives a comparison of all the basic curves and some combination curves.

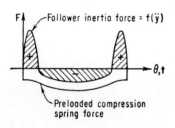

As an illustration of an advanced curve choice, automobile and aircraft designers are employing an unsymmetrical cam acceleration curve in which the maximum positive acceleration is much greater than the maximum negative acceleration (Fig. 4.8). This is a high-speed cam giving minimum spring size, maximum cam radius of curvature, and longer surface life.

FIG. 4.8 Typical high-speed cam curve used for automotive or aircraft engine valve system.

EXAMPLE Derive the relationship for the excellent dwell-rise-fall-dwell cam curve shown in Fig. 4.9 having equal maximum acceleration values. Portions I and III are harmonic; portions II and IV are horizontal straight lines.

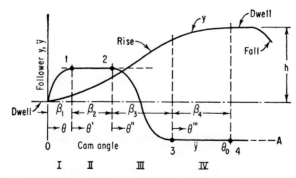

FIG. 4.9 Example of symmetrical dwell-rise-fall-dwell cam curve.

solution Let the θ's and β's be the angles for each portion shown. Note for velocity and acceleration one should multiply values by ω and ω^2, respectively, and the boundary conditions are $y(0) = 0$, $\dot{y}(0) = 0$, and $\dot{y}(\beta_4) = 0$ and $y(\beta_4) = $ total rise h.

Portion I

$$\ddot{y} = A \sin (\pi\theta/2\beta_1)$$

$$\dot{y} = (2\beta_1 A/\pi)[1 - \cos (\pi\theta/2\beta_1)]$$

$$y = - A(2\beta_1/\pi)^2 \sin (\pi\theta/2\beta_1) + (2A\beta_1/\pi)\theta$$

$$\dot{y}_1 = 2A\beta_1/\pi$$

$$y_1 = (2A\beta_1^2/\pi)(1 - 2/\pi)$$

Portion II

$$\ddot{y} = \dot{A}$$

$$\dot{y} = A\theta' + \dot{y}_1$$

$$y = A(\theta')^2/2 + \dot{y}_1\theta' + y_1$$

$$\dot{y}_2 = A\beta_2 + 2A\beta_1/\pi$$

$$y_2 = A\beta_2^2/2 + 2A\beta_1\beta_2/\pi + 2A\beta_1^2/\pi(1 - 2/\pi)$$

Portion III

$$\ddot{y} = A \cos (\pi\theta''/\beta_3)$$

$$\dot{y} = A\beta_3/\pi \sin (\pi\theta''/\beta_3) + \dot{y}_2$$

$$y = - A(\beta_3/\pi)^2[1 - \cos(\pi\theta''/\beta_3)] + \dot{y}_2\theta'' + y_2$$

$$\dot{y}_3 = \dot{y}_2$$

$$y_3 = 2A(\beta_3/\pi)^2 + A\beta_2\beta_3 + 2A\beta_1\beta_3/\pi + A\beta_2^2/2 + 2A\beta_1\beta_2/\pi + (2A\beta_1^2/\pi)(1 - 2/\pi)$$

Portion IV

$$\ddot{y} = - A$$

$$\dot{y} = - A\theta''' + \dot{y}_3$$

$$y = - [A(\theta''')^2/2] + \dot{y}_3\theta''' + y_3$$

$$\dot{y}_4 = - A\beta_4 + \dot{y}_3 = - A\beta_4 + A\beta_2 + 2A\beta_1/\pi = 0$$

$$\beta_4 = \beta_2 + 2\beta_1/\pi$$

Total Rise

$$h = y_4 = - (A\beta_4^2/2) + (A\beta_2 + 2A\beta_1/\pi)\beta_4 + y_3$$

Substituting

$$A = \frac{h}{[- \beta_4^2/2 + \beta_2\beta_4 + 2\beta_1\beta_4/\pi + (\beta_3/\pi)^2 + \beta_2\beta_3 + 2\beta_1\beta_3/\pi + \beta_2^2/2 + 2\beta_1\beta_2/\pi}$$
$$+ (2\beta_1^2/\pi)(1 - 2/\pi)]$$

For a given total rise h for angle θ_0 and any two of the β angles given one can solve for all angles and all values of the derivative curves.

4.8 CAM DYNAMICS

4.8.1 Introduction

Classical cam design and analysis, which treats the cam mechanism as a single-degree-of-freedom rigid body, is in general inadequate to describe the dynamics of cam mechanisms which are inherently compliant. A measure of system compliance is the extent of the deviation between the dynamic response and the intended kinematic response, which is the vibration level.

This deviation tends to increase (1) as any of the input harmonics of cam motion approaches the fundamental frequency of the mechanism, (2) with the increase of the maximum value of the cam function third-time derivative (jerk), and (3) as the cam surface irregularities become more severe. The deleterious effects of vibration are well known. Vibration causes motion perturbations, excessive component stresses, noise, chatter, wear, and cam surface erosion. This wear and cam surface erosion feeds back to further exacerbate the problem. References 1, 9 to 14, 18, 26, 29, 30, 49, and 51 to 54 give more on the subject.

Models.   Modeling[21–23] transforms the system into a set of tenable mathematical equations which describe the system in sufficient detail for the accuracy required. Various tools of analysis are at the disposal of the designer-analyst, ranging in complexity from classical linear analysis for single-degree-of-freedom systems to complex multiple-degree-of-freedom nonlinear computer programs.

4.8.2 Lumped-Model Rigid Camshaft System

The cam mechanism consists of a camshaft, cam, follower train including one or more connecting links, and springs terminating in a load mass or force. For lumped systems, links or springs are divided into two or more ideal mass points interconnected with weightless springs and dampers. Ball bearings are modeled as point masses and/or springs due to radial or axial compliance of the balls and axial spring preloading.

Springs.   The spring force generally follows the law

$$F = kx \pm \sum_{n=1} a_n x^{2n+1}$$

where terms under the Σ sign are nonlinear. Rarely does one go beyond the cubic term $a_1 x^3$ when defining a nonlinear spring.

Dampers.   Damping take the general form

$$C(\dot{x}) = C\dot{x}|\dot{x}|^{r-1}$$

$$\text{For } r = \begin{cases} 0 & C(\dot{x}) \text{ is coulomb damping} \\ 1 & C(\dot{x}) \text{ is viscous damping} \\ 2 & C(\dot{x}) \text{ is quadratic damping} \end{cases}$$

"Stiction" is accounted for as the breakaway force, $\mathscr{F}$, in one element i sliding along another element j.

$$F_{st} = 0 \qquad \text{for } \dot{x}_i \neq \dot{x}_j$$

$$\leq \mathscr{F} \qquad \text{for } \dot{x}_i = \dot{x}_j$$

$$= \mathscr{F} \qquad \text{at breakaway}$$

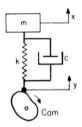

FIG. 4.10   Single-degree-of-freedom system.

Damping in cam-follower systems ranges from 5 to 15 percent of critical damping.

Single-Degree-of-Freedom System.   In its simplest form, the model is the one-degree-of-freedom system which lumps the follower train and mass load into a single equivalent mass with equivalent springs and dampers connected to the cam, as shown in Fig. 4.10.

$$m\ddot{x} + c(\dot{x} - \dot{y}) + k(x - y) = 0 \qquad (4.18)$$

The vibration form of Eq. (4.18) is

$$m(\ddot{x} - \ddot{y}) + c(\dot{x} - \dot{y}) + k(x - y) = -m\ddot{y} \qquad (4.18a)$$

Substituting $z = x - y$ yields

$$m\ddot{z} + c\dot{z} + kz = m\ddot{y} \qquad (4.19)$$

which places in evidence the prominent role of cam function acceleration y on the vibrating system, similar to a system response to a shock input at its foundation.

The general solution to Eq. (4.19) consists of the complementary or transient solution plus the particular solution. The complementary solution is the solution to the homogeneous equation

$$m\ddot{z} + c\dot{z} + kz = 0$$

$$z + 2\zeta\omega_n z + \omega_n^2 = 0$$

$$\text{where } \omega_n = (k/m)^{1/2}$$

$$\zeta = \tfrac{1}{2}\, c(1/km)^{1/2}$$

and is

$$z_c = Ae^{-\zeta\omega_n t} \sin{[(1 - \zeta^2)^{1/2}\omega_n]t} + Be^{-\zeta\omega_n t} \cos{[(1 - \zeta^2)^{1/2}\omega_n]t}$$

where A and B are determined in conjunction with the particular solution and the initial conditions. One form of the particular solution is derivable from the impulse response.

$$z_p = \frac{1}{(1 - \zeta^2)^{1/2}\omega_n} \int_0^t \ddot{y}\, e^{-\zeta\omega 0(t - \tau)} \sin\,[(1 - \zeta^2)^{1/2}\omega_n](t - \tau)d\tau$$

For finite $\ddot{y}$, the case here, $z_p(0) = \dot{z}_p(0) = 0$, so that the general solution to Eq. (4.19) is

$$z = z_p + \left\{ \left[\frac{\dot{z}(0) + z(0)\zeta\omega_n}{(1 - \zeta^2)^{1/2}\omega_n}\right] \sin\,[(1 - \zeta^2)^{1/2}\omega_n]t \right.$$
$$\left. + z(0) \cos\,[(1 - \zeta^2)^{1/2}\omega_n]t \right\} e^{-\zeta\omega_n t} \quad (4.20)$$

where the initial conditions $z(0)$ and $\dot{z}(0)$ are subsumed in z_c. Accordingly, for a system initially at rest, the z_p solution becomes the complete solution. However, the difficulty with this solution is that $\ddot{y}$ is periodic and integration would have to be performed over many cycles depending on the amount of damping present. Solutions would be ana-

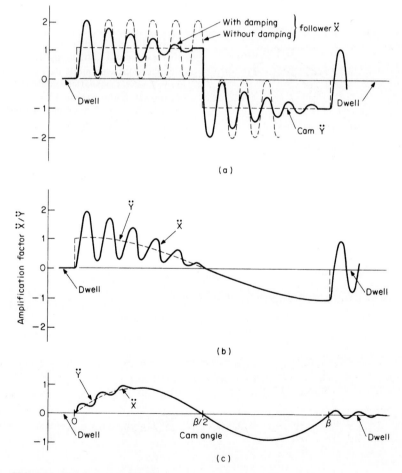

FIG. 4.11 Single-degree-of-freedom response of basic dwell-rise-dwell cam acceleration curves (*a*) Parabolic. (*b*) Simple harmonic. (*c*) Cycloidal.[1]

lyzed first to determine the input "transient" peak response. Finally the "steady-state" response would be reached when

$$z(nt) = z(n + 1)T \quad \text{and} \quad \ddot{z}(nt) = \ddot{z}(n + 1)T$$

Typical transient-response curves for some one-degree-of-freedom systems are shown in Fig. 4.11, where it is noted that discontinuities in the acceleration cause large amplification factors. For improved accuracy, the model may be developed in more detail, leading to the more general multiple-degree-of-freedom system. Reference 25 discusses matrix methods in modeling.

4.8.3   Lumped-Model Flexible Camshaft System: Spring Windup[24]

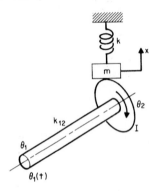

The system takes account of camshaft flexibility and leads to a set of nonlinear equations owing to the fact that the driving-cam function is no longer a known function of time.

As an example, consider the case shown in Fig. 4.12, where $\theta_1(t)$, the input shaft rotating at a known angular rate, is not equal to the cam angle θ_2, or $\theta_1 \neq \theta_2$. This is due to camshaft flexibility, k_{12} is the camshaft spring rate and the cam-follower rise is x. A convenient method of formulation is derivable from Lagrange's equations (see Sec. 5) as follows:

FIG. 4.12   Flexible camshaft system.

$$T = \tfrac{1}{2} M\dot{x}^2 + \tfrac{1}{2} I\dot{\theta}_2^2$$

$$U = \tfrac{1}{2}k_{12}(\theta_2 - \theta_1)^2 + \tfrac{1}{2}Kx^2$$

$$\mathcal{L} = T - V$$

The follower is constrained by the cam function.

$$x = x(\theta_2)$$

where $\theta_1 = \theta_1(t)$, by the Lagrange multiplier method

$$\lambda_1 \, dx - \lambda_1 x' \, d\theta_2 = 0; \, x' = dx(\theta_2)/d\theta_2$$

$$\lambda_2 \, d\theta_1 - \lambda_2 \theta_1' \, dt = 0; \, \theta' = d\theta_1/dt$$

$$\left(\frac{d}{dt}\frac{\partial L}{\partial \dot{x}} - \frac{\partial L}{\partial \dot{x}}\right) dx + \lambda_1 dx = 0 \rightarrow m\ddot{x} + kx + \lambda_1 = 0 \tag{4.24}$$

$$\left(\frac{d}{dt}\frac{\partial L}{\partial \dot{\theta}_2} - \frac{\partial L}{\partial \theta_2}\right) d\theta_2 - \lambda_1 x' d\theta_2 = 0 \rightarrow I\ddot{\theta}_2 + k_{12}(\theta_2 - \theta_1) - \lambda_1 x' = 0 \tag{4.25}$$

$$\left(\frac{d}{dt}\frac{\partial L}{\partial \dot{\theta}_1} - \frac{\partial L}{\partial \theta_1}\right) d\theta_1 + \lambda_2 = 0 \rightarrow k_{12}(\theta_1 - \theta_2) + \lambda_2 = 0 \tag{4.26}$$

where λ_1 = contact force on cam
$\lambda_1 x'$ = torque on cam due to contact force
λ_2 = torque on shaft

Elimination of λ_1 between Eqs. (4.24) and (4.25) and taking two time derivatives of $x(\theta_2)$ yields

$$m\ddot{x}x' + kxx' + I\ddot{\theta}_2 + k_{12}(\theta_2 - \theta_1) = 0$$

$$\ddot{x} = x''\dot{\theta}_2^2 + \ddot{\theta}_2 x'$$

whence

$$(I + mx'^2)\ddot{\theta}_2 + mx'x''\dot{\theta}_2^2 + kxx' + k_{12}\theta_2 = k_{12}\theta_1(t) \tag{4.27}$$

characterized as the spring-windup equation with one degree of freedom.

4.9 CAM PROFILE ACCURACY[1,8,43]

A surface may appear smooth to the eye and yet have poor dynamic properties, since the determination of whether a cam is satisfactory is based on the acceleration curve. The acceleration being the second derivative of lift, the large magnification of error will appear on it. Two methods for establishing the accuracy of the cam fabrication and its dynamic effect are (1) employing the mathematical numerical method of finite differences, and (2) utilizing electronic instrumentation of an accelerometer pickup (located on the cam follower), and cathode-ray oscilloscope.

Reference 42 contains a review of cam manufacturing methods, including a treatment of accuracy. Reference 44 applies stochastic and probabilistic techniques to the study of error in cams. Reference 36 uses a FORTRAN program to compute values to minimize errors in cam fabrication.

Many types of errors were investigated on measured shapes. All cams employed the basic cycloidal curve. In Fig. 4.13 we see a cam investigated at 300 r/min. The

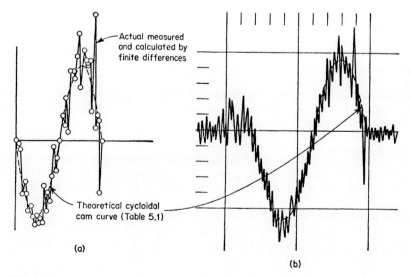

FIG. 4.13 Accuracy investigation of cam turning 300 r/min; in 45°, rise = ¾ in. (a) Finite-difference calculation. (b) Experimental with accelerometer and oscilloscope.

greatest error in fabrication exists at about the 40° cam angle Correlation is observed between the finite-difference and the experimental results. The theoretical maximum acceleration was calculated from the cycloidal formula to be 19.5g and the maximum acceleration of cam with errors is about 34g. On a broad average the maximum acceleration difference for the error only is 12g per 0.001-in error at a speed of 300 r/min. Figure 4.14 shows the dimension of a high-speed cam for a textile loom.

4.10 CAM LOADS[1,35]

Pertinent aspects of cam-follower design are the load and life of the cam and follower contacting surfaces having the parameters of surface roughness, waviness and stresses, prior history of machining, moduli of elasticity, friction (the amount of rolling and sliding), choice of materials, elastohydrodynamic lubrication, corrosion, heat treatment, materials in combination, the area and radii of contacting surfaces, the number of performance cycles, and the contaminants in the environment that may collect on the surfaces.

When the roller follower in contact with the cam is subjected to significant changes in rolling speed, skidding occurs. This skidding action may produce scuffing wear on the concave side of the cam surface at the point of maximum acceleration. Note that proper lubrication is necessary to prevent excessive wear, welding, and scoring of the materials, i.e., generally the cam surface.

Surface fatigue is the predominant cause of failure in cams. In Eq. (4.28) the usual choice of a hardened-steel roller follower is shown in combination with different cam materials with a percent sliding and 10^8 cycles. The test data based on hertzian equations and surface fatigue give the surface endurance load

$$w = \frac{KL}{1/R_r + 1/\rho_c} \tag{4.28}$$

where L is contact length, in inches.

Table 4.3 shows the values of K for various suggested cam materials in contact with a cam roller made of tool steel hardened to Rockwell C 60 to 62. The data shown should be modified depending on experience, the amount of sliding, and the number of cycles during performance. Steels in combination are usually chosen for heavier loads. Although class 20 gray irons have a low endurance limit, classes 30, 45, and others show advantages over free-machining steel. Austempering of Meehanite, or class 30, may be a good choice because of the five-grain structure, excessive dispersion of graphite, and conversion of released austenite. Sometimes bronze or phenolics are selected. Failure of the phenolics and other polymers was by material flows and surface cracking rather than by fatigue.

Alignment between cam and follower is critical. Dynamic deflection of the roller support system can seriously shorten the life of the materials in combination. This is caused by a smaller area of surfaces in contact showing wear on the edges of the cam and follower. Also harmful effects of stress concentrations should be avoided by eliminating sharp cam or follower edges by using generous fillet radii. Reference 46 demonstrates the use of sequential random search techniques for determining cam design parameters in order to minimize contact stress.

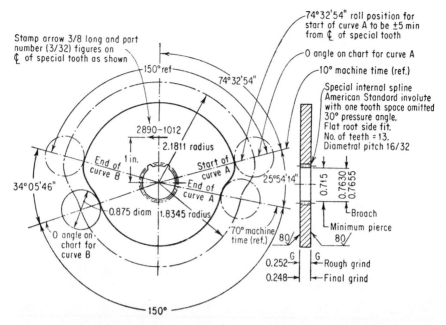

One required, No. A4140 steel ³⁄₁₆ in. thick. Blank and pierce allow ³⁄₆₄ in. on contour for finishing cam to be flat within ±0.010 after blanking. Heat-treat before machining and after blanking to 30 to 34 Rockwell C.

Induction-harden cam periphery 50 to 55 Rockwell C. Tolerance on cam curve must not vary in excess of ±0.001 from true curve, profile surface 32 μin.

Cam profile (roller center)

Points for curve *A*		Points for curve *B*	
Angle	Radius	Angle	Radius
0° 0′ 0″	2.1811	0° 0′ 0″	1.8345
0° 30′ 0″	2.1809	0° 22′ 30″	1.8345
1° 0′ 0″	2.1801	0° 45′ 2″	1.8345
1° 30′ 0″	2.1788	1° 07′ 38″	1.8346
2° 0′ 0″	2.1769	1° 30′ 18″	1.8349
2° 30′ 0″	2.1746	1° 53′ 5″	1.8353
3° 0′ 0″	2.1707	2° 16′ 1″	1.8359
Etc.	Etc.	Etc.	Etc.
23° 34′ 14″	1.8388	32° 35′ 28″	2.1807
24° 24′ 14″	1.8369	32° 58′ 8″	2.1810
24° 54′ 14″	1.8355	33° 20′ 44″	2.1810
25° 24′ 14″	1.8347	33° 43′ 15″	2.1811
25° 54′ 14″	1.8345	34° 35′ 46″	2.1811

150° from end of curve *A* to start of curve *B* at 1.8345 rad	150° from end of curve *B* to start of curve *A* at 2.1811 rad

FIG. 4.14 Cam for loom. (*Courtesy of Warner & Swasey Co., Cleveland, Ohio*)

TABLE 4.3 Cam Load Constant K

Material run against tool steel roll-hardened to Rockwell C 60–62

Material	K
Steel	
1020, 130–150 Bhn	1,700
1020, carburized, 0.045-in minimum depth, Rockwell C 50–60	10,000
4150, heat-treated, 220–300 Bhn, phosphate-coated	7,000
4340, heat-treated, 270–300 Bhn	5,000
4340, induction-hardened, 0.045-in minimum depth, Rockwell C 50–58	9,000
6150, 270–300 Bhn	2,000
Gray iron	
Class 20, 140–160 Bhn	700
Class 30, heat-treated (austempered), 255–300 Bhn, phosphate-coated (Meehanite)	2,500
Class 35, 225–245 Bhn	2,000
Class 45, 220–240 Bhn	1,000
Nodular iron	
Grade 80-60-03, 170–241 Bhn	2,000
Grade 100-70-03, heat-treated, 240–260 Bhn	3,600
Bronze (phosphorous), sand casting, 65–75 Bhn	400
Zinc, die casting, 70 Bhn	200
Phenolic-cotton base	900
Phenolic-linen base, laminate	700
Resin-acetal	600
Polyethylene	400

REFERENCES

1. Rothbart, H. A.: "Cams," John Wiley & Sons, Inc., New York, 1956.

2. Beyer, R.: "Kinematische Getriebesynthese," Springer-Verlag OHG, Berlin, 1953.

3. Artobolevskii, I. I.: "Theory of Mechanisms and Machines" (in Russian), Chief State Publishing House for Technical Theoretical Literature, Moscow, 1953.

4. Chen, F. Y.: "Kinematic Synthesis of Cam Profiles for Prescribed Acceleration by a Finite Integration Method," *Trans. ASME, J. Eng. Ind.*, vol. 95B, p. 769, 1973.

5. Jansen, B.: "Dynamik der Kurvengetriebe," VDI-Berichte 127, 1969.

6. Kloomok, M., and R. V. Muffley: "Determination of Pressure Angles for Swinging-Follower Cam Systems," *Trans. ASME*, vol. 70, p. 473, 1948.

7. Kloomok, M., and R. V. Muffley: "Determination of Radius of Curvature for Radial and Swinging Follower Cam Systems," ASME Paper 55-SA-89, 1955.

8. Johnson, R. C.: "Method of Finite Differences for Cam Design," *Mach. Des.*, vol. 27, p. 195, November 1955.

9. Jehle, F., and W. R. Spiller: "Idiosyncrasies of Valve Mechanisms and Their Causes," *Trans. SAE*, vol. 24, p. 197, 1929.

10. Turkish, M. C.: "Valve Gear Design," Eaton Mfg. Co., Detroit, Mich., 1946.

11. Dudley, W. M.: "New Methods in Valve Cam Design," *Trans. SAE*, vol. 2, p. 19, January 1948.

12. Mitchell, D. B.: "Tests on Dynamic Response of Cam-Follower Systems," *Mech. Eng.*, vol. 72, p. 467, June 1950.

13. Hrones, J. A.: "Analysis of Dynamic Forces in a Cam-Driven System," *Trans. ASME*, vol. 70, p. 473, 1948.

14. Freudenstein, F.: "On the Dynamics of High-Speed Cam Profiles," *Int. J. Mech. Sci.*, vol. 1, pp. 342–349, 1960.

15. Rothbart, H. A.: "Cam Dynamics," *Trans. 1st Int. Conf. Mechanisms,* Yale University, New Haven, Conn., March 27, 1961.

16. Beggs, J. S.: "Mechanism," McGraw-Hill Book Company, Inc., New York, pp. 139–141, 1955.

17. Raven, F. H.: "Analytical Design of Disk Cams and Three-Dimensional Cams by Independent Position Equations," *Trans. ASME,* vol. 26, series E, no. 1, pp. 18–24, March 1959.

18. Rothbart, H. A.: "Limitations on Cam Pressure Angles," *Prod. Eng.,* vol. 29, no. 19, p. 193, 1957.

19. Hinkle, R.: "Kinematics of Machines," Prentice-Hall, Inc., Englewood Cliffs, N.J., pp. 148–161, 1960.

20. Rothbart, H.: "Mechanical Control Cams," *Control Eng.,* vol. 7, no. 6, pp. 118–122, June 1960; vol. 8, no. 11, pp. 97–101, November 1961.

21. Chen, F. Y.: "A Survey of the State of the Art of Cam System Dynamics," *Mech. Mach. Theory,* vol. 12, p. 201, 1977.

22. Chen, F. Y., and N. Polanich: "Dynamics of High Speed Cam Driven Mechanisms," *Trans. ASME, J. Eng. Ind.,* p. 769, August 1975.

23. Mathew, G. K., and O. Tesar: "The Design of Modeled Cam Systems," *Trans. ASME, J. Eng. Ind.,* p. 1175, November 1975.

24. Seckallas, L. E., and M. Savage: "The Characterization of Cam Drive System Windup," *Trans. ASME, J. Mech. Des.,* vol. 102, p. 278, April 1980.

25. Winfrey, R. C.: "Elastic Link Mechanisms Dynamics," *Trans. ASME, J. Eng. Ind.,* vol. 93, no. 1, p. 268, February 1971.

26. Neklutin, C. N.: "Vibration Analysis of Cams," *Trans. 2d Conf. Mechanisms,* Penton Publishing Co., Cleveland, Ohio, p. 6, 1954.

27. Tesar, D.: "The Dynamic Synthesis, Analysis and Design of Modeled Cam Systems," Lexington Books, Lexington, Mass., 1976.

28. Chakraborty, J.: "Kinematics and Geometry of Planar and Spatial Cam Mechanisms," John Wiley & Sons, Inc., New York, 1977.

29. Barkan, P., and McGarrity, R. V.: "A Spring-Actuated Cam-Follower System," *Trans. ASME, J. Eng. Ind.,* vol. 67, ser. B, no. 1, p. 279, 1965.

30. Winfrey, R. C., R. V. Anderson, and C. W. Gnikla: "Analysis of Elastic Machinery with Clearances," *Trans. ASME, J. Eng. Ind.,* vol. 95, no. 2, p. 695, 1973.

31. Huey, C. O., Jr., and M. W. Dixon: "The Cam-Link Mechanism for Structural Error-Free Path and Friction Generation," *Mech. Mach. Theory,* vol. 9, p. 367, 1974.

32. Duca, C. D., and I. Simionescu: "The Exact Synthesis of Single-Disk Cams with Two Oscillating Rigidly Connected Roller Followers," *Mech. Mach. Theory,* vol. 15, p. 213, 1980.

33. Pagel, P. A.: "Custom Cams from Building Blocks," *Mach. Des.,* vol. 50, p. 86, May 11, 1978.

34. Weber, T., Jr.: "Simplifying Complex Cam Design," *Mach. Des.,* vol. 51, p. 115, March 22, 1979.

35. Morrison, R. A." "Load/Life Curve for Gear and Cam Materials," *Mach. Des.,* vol. 40, p. 102, August 1, 1968.

36. Giordana, F., V. Rognoni, and G. Ruggieri: "On the Influence of Measurement Errors in the Kinematic Analysis of Cams," *Mech. Mach. Theory,* vol. 14, p. 337, 1979.

37. Weiderrich, J. L.: "Dynamic Synthesis of Cams Using Finite Trigonometric Series," ASME Paper No. 74-Det-2, 1974.

38. Miske, C. R.: "An Introduction to Computer Aided Design," Prentice-Hall, Inc., Englewood Cliffs, N.J., 1968.

39. Shipley, J. E.: "Kinematic Analysis of Mechanisms," 2d ed., McGraw-Hill Book Company, Inc., New York, 1969.

40. Jackowski, C. S., and J. F. Dubil: "Single Disk Cams with Positively Controlled Oscillating Followers," *Trans. ASME, J. Appl. Mech.,* vol. 34, no. 2, p. 157, 1967.

41. Wunderlick, W.: "Contributions to the Geometry of Cam Mechanisms with Oscillating Followers," *Trans. ASME, J. Appl. Mech.,* vol. 38, no. 1, p. 1, 1971.

42. Grant, B., and A. H. Soni: "A Survey of Cam Manufacture Methods," *Trans. ASME, J. Mech. Des.,* vol. 21, p. 445, July 1979.

43. Kim, H. R., and W. R. Newcombe: "Stochastic Error Analysis in Cam Mechanisms," *Mech. Mach. Theory,* vol. 13, p. 631, 1978.

44. Tuttle, S. B.: "Error Analysis of Mechanisms," *Mach. Des.,* vol. 38, p. 152, June 1966.

45. Neklutin, C. N.: "Mechanisms and Cams for Automatic Machines," Elsevier North Holland, Inc., New York, 1967.

46. Chen, F. Y., and A. H. Soni: "On Harrisberger's Adjustable Trapezoidal Motion Program for Cam Design," *Proc. Appl. Mech. Conf.,* Paper No. 41, Oklahoma State University, Oklahoma City, 1971.

47. Sandler, B. Z.: "Suboptimal Mechanism Synthesis," *Israel J. Tech.,* vol. 12, p. 160, 1974.

48. Levitskii, N. I.: "Cam Mechanisms" (in Russian), *Mach. Tech. Publ.,* Moscow, 1964.

49. Koster, M. P.: "Vibration of Cam Mechanisms," Macmillan Publishers, Ltd., London, 1974.

50. Lafuente, J. M.: "Interactive Graphics in Data Processing Cam Design on Graphic Console," *IBM Syst. J.,* vol. 304, p. 335, 1968.

51. Hart, F. D., and C. F. Forowski: "Coupled Effects of Preloads and Damping on Dynamic Cam-Follower Separation," ASME Paper No. 64-Mech. 18, 1964.

52. Chen, F. Y.: "Analysis and Design of Cam-Driven Mechanisms with Nonlinearities," *Trans. ASME, J. Eng. Ind.,* vol. 78, no. 2, p. 685, August 1976.

53. Baumgarten, J. R.: "Preload Force Necessary to Prevent Separation of Follower from Cam," *Trans. 7th Conf. Mechanisms,* Purdue University, West Lafayette, Ind., 1962.

54. Chen, F. Y., "Mechanics and Design of Cam Mechanisms," Pergamon Press, New York, 1982.

55. Kim, H. R., and W. R. Newcombe: "The Effect of Cam Profile Errors and System Flexibility on Cam Mechanism Output," *Trans. ASME, Mechanisms and Machine Theory,* vol. 17, no. 1, p. 57, 1982.

56. Soni, A. H., and B. Grant: "Cam Design Survey," *Trans. ASME, Design Technology Transfer,* p. 177, 1974.

SECTION 5
MECHANICAL VIBRATIONS

Eric E. Ungar, Eng.Sc.D.

Chief Consulting Engineer
Bolt, Beranek and Newman, Inc.
Cambridge, Mass.

INTRODUCTION

The field of dynamics deals essentially with the interrelation between the motions of objects and the forces causing them. The words "shock" and "vibration" imply particular forces and motions: hence, this chapter concerns itself essentially with a subfield of dynamics. However, oscillatory phenomena occur also in nonmechanical systems, e.g., electric circuits, and many of the methods and some of the nomenclature used for mechanical systems are derived from nonmechanical systems.

Mechanical vibrations may be caused by forces whose magnitudes and/or directions and/or points of application vary with time. Typical forces may be due to rotating unbalanced masses, to impacts, to sinusoidal pressures (as in a sound field), or to random pressures (as in a turbulent boundary layer). In some cases the resulting vibrations may be of no consequence; in others they may be disastrous. Vibrations may be undesirable because they can result in deflections of sufficient magnitude to lead to malfunction, in high stresses which may lead to decreased life by increasing material fatigue, in unwanted noise, or in human discomfort.

Section 5.1 serves to delineate the concepts, phenomena, and analytical methods associated with the motions of systems having a single degree of freedom and to introduce the nomenclature and ideas discussed in the subsequent sections. Section 5.2 deals similarly with systems having a finite number of degrees of freedom, and Sec. 5.3 with continuous systems (having an infinite number of degrees of freedom). Mechanical shocks are discussed in Sec. 5.4, and in Sec. 5.5 appears additional information concerning design considerations, vibration-control techniques, and rotating machinery, as well as charts and tables of natural frequencies, spring constants, and material properties. The appended references substantiate and amplify the presented material.

Complete coverage of mechanical vibrations and associated fields is clearly impossible within the allotted space. However, it was attempted here to present enough information so that an engineer who is not a specialist in this field can solve the most prevalent problems with a minimum amount of reference to other publications.

5.1 SYSTEMS WITH A SINGLE DEGREE OF FREEDOM

A system with a single degree of freedom is one whose configuration at any instant can be described by a single number. A mass constrained to move without rotation along a given path is an example of such a system; its position is completely specified when one specifies its distance from a reference point, as measured along the path.

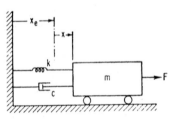

FIG. 5.1 A system with a single degree of freedom.

Single-degree-of-freedom systems can be analyzed more readily than more complicated ones; therefore, actual systems are often approximated by systems with a single degree of freedom, and many concepts are derived from such simple systems and then enlarged to apply also to systems with many degrees of freedom.

Figure 5.1 may serve as a model for all single-degree-of-freedom systems. This model consists of a pure inertia component (mass m supported on rollers which are devoid of friction and inertia), a pure restoring component (massless spring k), a pure energy-dissipation component (massless dashpot c), and a driving component (external force F). The inertia component limits acceleration. The restoring component opposes system deformation from equilibrium and tends to return the system to its equilibrium configuration in absence of other forces.

5.1.1 Linear Single-Degree-of-Freedom Systems[8,17,32,40,61,63]

If the spring supplies a restoring force proportional to its elongation and the dashpot provides a force which opposes motion of the mass proportionally to its velocity, then the system response is proportional to the excitation, and the system is said to be linear. If the position x_e indicated in Fig. 5.1 corresponds to the equilibrium position of the mass and if x denotes displacement from equilibrium, then the spring force may be written as $-kx$ and the dashpot force as $-c\, dx/dt$ (where the displacement x and all forces are taken as positive in the same coordinate direction). The equation of motion of the system then is

$$m\, d^2x/dt^2 + c\, dx/dt + kx = F \qquad (5.1)$$

Free Vibrations. In absence of a driving force F and of damping c, i.e., with $F = c = 0$, Eq. (5.1) has a general solution which may be expressed in any of the following ways:

$$x = A \cos (\omega_n t - \phi) = (A \cos \phi) \cos \omega_n t + (A \sin \phi) \sin \omega_n t$$

$$= A \sin (\omega_n t - \phi + \pi/2) = A \operatorname{Re}\{e^{i(\omega_n t - \phi)}\} \qquad (5.2)$$

A and ϕ are constants which may in general be evaluated from initial conditions. A is the maximum displacement of the mass from its equilibrium position and is called the "displacement amplitude"; ϕ is called the "phase angle." The quantities ω_n and f_n, given by

$$\omega_n = \sqrt{k/m} \qquad f_n = \omega_n/2\pi$$

are known as the "undamped natural frequencies"; the first is in terms of "circular frequency" and is expressed in radians per unit time, the second is in terms of cyclic frequency and is expressed in cycles per unit time.

If damping is present, $c \neq 0$, one may recognize three separate cases depending on the value of the damping factor $\zeta = c/c_c$, where c_c denotes the critical damping coefficient (the smallest value of c for which the motion of the system will not be oscillatory). The critical damping coefficient and the damping factor are given by

$$c_c = 2\sqrt{km} = 2m\omega_n \qquad \zeta = \frac{c}{c_c} = \frac{c}{2\sqrt{km}} = \frac{c}{2m\omega_n}$$

The following general solutions of Eq. (5.1) apply when $F = 0$:

$$c > c_c (\zeta > 1): \qquad x = Be^{(-\zeta + \sqrt{\zeta^2 - 1})\omega_n t} + Ce^{(-\zeta - \sqrt{\zeta^2 - 1})\omega_n t} \qquad (5.3a)$$

$$c = c_c (\zeta = 1): \qquad x = (B + Ct)e^{-\omega_n t} \qquad (5.3b)$$

$$c < c_c (\zeta < 1): \qquad x = Be^{-\zeta \omega_n t} \cos (\omega_d t + \phi) = B \operatorname{Re} \{e^{i(\omega_N t + \phi)}\} \qquad (5.3c)$$

where $\qquad \omega_d \equiv \sqrt{\omega_n 1 - \zeta^2} \qquad i\omega_N = -\zeta \omega_n + i\omega_d$

denote, respectively, the "undamped natural frequency" and the "complex natural frequency," and the B, C, ϕ are constants that must be evaluated from initial conditions in each case.

Equation (5.3a) represents an extremely highly damped system; it contains two decaying exponential terms. Equation (5.3c) applies to a lightly damped system and is essentially a sinusoid with exponentially decaying amplitude. Equation (5.3b) pertains to a critically damped system and may be considered as the dividing line between highly and lightly damped systems.

Figure 5.2 compares the motions of systems (initially displaced from equilibrium by an amount x_0 and released with zero velocity) having several values of the damping factor ζ.

In all cases where $c > 0$ the displacement x approaches zero with increasing time. The damped natural frequency ω_d is generally only slightly lower than the undamped natural frequency ω_n; for $\zeta \leq 0.5$, $\omega_d \geq 0.87\, \omega n$.

The static deflection x_{st} of the spring k due to the weight mg of the mass m (where g denotes the acceleration of gravity) is related to natural frequency as

$$x_{st} = mg/k = g/\omega_n^2 = g/(2\pi f_n)^2$$

This relation provides a quick means for computing the undamped natural frequency (or for approximating the damped natural frequency) of a system from its static deflection. For x in inches or centimeters and f_n in hertz (cycles per second) it becomes

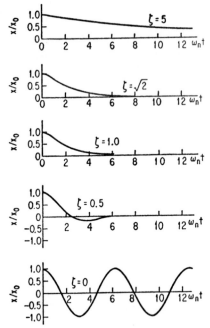

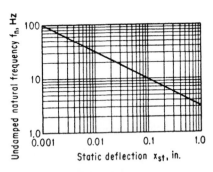

FIG. 5.2 Free motions of linear single-degree-of-freedom systems with various amounts of damping.

FIG. 5.3 Relation between natural frequency and static deflection of linear undamped single-degree-of-freedom system.

$$f_n^2 \text{ (Hz)} = 9.80/x_{st} \text{ (in)} = 24.9/x_{st} \text{ (cm)}$$

which is plotted in Fig. 5.3.

Forced Vibrations. The previous section dealt with cases where the forcing function F of Eq. (5.1) was zero. The solutions obtained were the so-called "general solution of the homogeneous equation" corresponding to Eq. (5.1). Since these solutions vanish with increasing time (for $c > 0$), they are sometimes also called the "transient solutions." For $F \neq 0$ the solutions of Eq. (5.1) are made up of the aforementioned general solution (which incorporates constants of integration that depend on the initial conditions) plus a "particular integral" of Eq. (5.1). The particular integrals contain no constants of integration and do not depend on initial conditions, but do depend on the excitation. They do not tend to zero with increasing time unless the excitation tends to zero and hence are often called the "steady-state" portion of the solution.

The complete solution of Eq. (5.1) may be expressed as the sum of the general (transient) solution of the homogeneous equation and a particular (steady-state) solution of the nonhomogeneous equation. The associated general solutions have already been discussed; hence the present discussion will be concerned primarily with the steady-state solutions.

The steady-state solutions corresponding to a given excitation $F(t)$ may be obtained from the differential equation (5.1) by use of various standard mathematical techniques[12,20] without a great deal of difficulty. Table 5.1 gives the steady-state responses x_{ss} to some common forcing functions $F(t)$.

TABLE 5.1   Steady-State Responses of Linear Single-
Degree-of-Freedom Systems to Several Forcing Functions

$F(t)$	x_{ss}
1	$1/k$
t	$t/k - c/k^2$
t^2	$t^2/k - 2ct/k^2 + 2c^2/k^3 - 2m/k^2$
$e^{\pm\alpha t}$*	$h(\pm\alpha)$
$te^{\pm\alpha t}$*	$h(\pm\alpha)e^{\pm\alpha t} - h^2(\pm\alpha)(c \pm 2m\alpha)te^{\pm\alpha t}$
$e^{i\omega t}$	$(k - m\omega^2 + ic\omega)^{-1}$
$\sin\beta t$†	$g(\beta)[(k - m\beta^2)\sin\beta t - c\beta\cos\beta t]$
$\cos\beta t$†	$g(\beta)[c\beta\sin\beta t + (k - m\beta^2)\cos\beta t]$
$e^{\pm\alpha t}\sin\beta t$‡	$(A\cos\beta t + B\sin\beta t)e^{-\alpha t}/D$
$e^{\pm\alpha t}\cos\beta t$‡	$(A\sin\beta t - B\cos\beta t)e^{-\alpha t}/D$

*$h(\pm\alpha) = (m\alpha^2 \pm c\alpha + k)^{-1}$
†$g(\beta) = [(k - m\beta^2)^2 + (c\beta)^2]^{-1}$
‡$A = k + m(\alpha^2 - \beta^2) \pm \alpha c$
$\quad B = \beta(c \pm 2\alpha m)$
$\quad D = A^2 + B^2$

Superposition.   Since the governing differential equation is linear, the response corresponding to a sum of excitations is equal to the sum of the individual responses; or, if

$$F(t) = A_1F_1(t) + A_2F_2(t) + A_3F_3(t) + \cdots$$

where $A_1, A_2, \ldots$ are constants, and if $x_{ss1}, x_{ss2}, \ldots$ are solutions corresponding, respectively, to $F_1(t), F_2(t), \ldots$, then the steady-state response to $F(t)$ is

$$x_{ss} = A_1x_{ss1} + A_2x_{ss2} + A_3x_{ss3} + \cdots$$

Superposition permits one to determine the response of a linear system to any time-dependent force $F(t)$ if one knows the system's impulse response $h(t)$. This impulse response is the response of the system to a Dirac function $\delta(t)$ of force; also $h(t) = u(t)$, where $u(t)$ is the system response to a unit step function of force [$F(t) = 0$ for $t < 0$, $F(t) = 1$ for $t > 0$]. In the determination of $h(t)$ and $u(t)$ the system is taken as at rest and at equilibrium at $t = 0$.

The motion of the system may be found from

$$x_{ss}(t) = \int_0^t F(\tau)h(t - \tau)\,d\tau \tag{5.4}$$

in conjunction with the proper "transient" solution expression, the constants in which must be adjusted to agree with specified initial conditions. For single-degree-of-freedom systems,

$$h(t) = \frac{\omega_n}{2k\sqrt{\zeta^2-1}}\ [e^{[-\zeta+(\zeta^2-1)^{1/2}]\omega_n t} + e^{[-\zeta-(\zeta^2-1)^{1/2}]\omega_n t}] \qquad \text{for } \zeta > 1$$

$$h(t) = \frac{\omega_n^2 t}{k}\ e_0^{-\omega_n t} \qquad \text{for } \zeta = 1$$

$$h(t) = \frac{\omega_n}{k\sqrt{1-\zeta^2}}\ e^{-\zeta\omega_n t}\sin\omega_\alpha t \qquad \text{for } \zeta < 1$$

$$h(t) = \frac{\omega_n}{k}\ \sin\omega_n t \qquad \text{for } \zeta = 0$$

Sinusoidal (Harmonic) Excitation. With an excitation

$$F(t) = F_0 \sin \omega t$$

one obtains a response which may be expressed as

$$x_{ss} = X_0 \sin (\omega t - \phi)$$

where
$$X_0 = \frac{F_0}{\sqrt{(k-m\omega^2)^2 + (c\omega)^2}} \qquad \tan \phi = \frac{c\omega}{k-m\omega^2} \qquad (5.5)$$

The ratio

$$H_s(\omega) = \frac{X_0}{F_0/k} = \left\{ \left[1 - \left(\frac{\omega}{\omega_n} \right)^2 \right]^2 + \left(2\zeta \frac{\omega}{\omega_n} \right)^2 \right\}^{-1/2} \qquad (5.6)$$

is called the frequency response or the magnification factor. As the latter name implies, this ratio compares the displacement amplitude X_0 with the displacement F_0/k that a force F_0 would produce if it were applied statically. $H_s(\omega)$ is plotted in Fig. 5.4.

Complex notation is convenient for representing general sinusoids.[*] Corresponding to a sinusoidal force

$$F(t) = F_0 e^{i\omega t}$$

one obtains a displacement

$$x_{ss} = X_0 e^{i\omega t}$$

where
$$\frac{X_0}{F_0/k} = H(\omega) = \left[1 - \left(\frac{\omega}{\omega_n} \right)^2 + 2i\zeta \left(\frac{\omega}{\omega_n} \right) \right]^{-1} \qquad (5.7)$$

$H(\omega)$ is called the complex frequency response, or the complex magnification factor[†] and is related to that of Eq. (5.6) as

$$H_s(\omega) = |H(\omega)|$$

From the model of Fig. 5.1 one may determine that the force F_{TR} exerted on the wall at any instant is given by

$$F_{TR} = kx + c\dot{x}$$

The ratio of the amplitude of this transmitted force to the amplitude of the sinusoidal applied force is called the transmissibility TR_s and obeys

$$TR_s = \frac{F_{TR}}{F_0} = \sqrt{\frac{1 + [2\zeta(\omega/\omega_n)]^2}{[1 - (\omega/\omega_n)^2]^2 + [2\zeta(\omega/\omega_n)]^2}} \qquad (5.8)$$

---

[*]In complex notation[40] it is usually implied, though it may not be explicitly stated, that only the *real parts* of excitations and responses represent the physical situation. Thus the complex form $Ae^{i\omega t}$ (where the coefficient $A = a + ib$ is also complex in general) implies the oscillation given by

$$\text{Re}\{Ae^{i\omega t}\} = \text{Re}\{(a + ib)(\cos \omega t + i \sin \omega t)\} = a \cos \omega t - b \sin \omega t$$

[†]An alternate formulation in terms of mechanical impedance is discussed in Sec. 5.2.6.

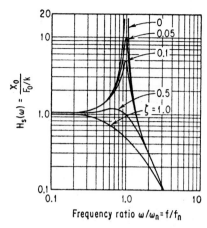

FIG. 5.4 Frequency response (magnification factor) of linear single-degree-of-freedom system.

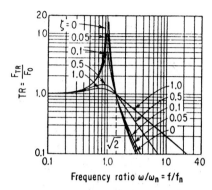

FIG. 5.5 Transmissibility of linear single-degree-of-freedom system.

Transmissibility $TR_s(\omega)$ is plotted in Fig. 5.5.
In complex notation

$$TR = \frac{1 + 2i\zeta(\omega/\omega_n)}{1 - (\omega/\omega_n)^2 + 2i\zeta(\omega/\omega_n)} \qquad TR_s(\omega) = |TR(\omega)|$$

It is evident that

$$|H(\omega)| \approx |TR| \approx \begin{cases} 1 & \text{for } \omega \ll \omega_n \\ (\omega_n/\omega)^2 & \text{for } \omega \gg \omega_n \end{cases}$$

Increased damping ζ always reduces the frequency response H. For $\omega/\omega_n < \sqrt{2}$ increased damping also decreases TR, but for $\omega/\omega_n > \sqrt{2}$ increased damping increases TR.[*]

The frequencies at which the maximum transmissibility and amplification factor occur for a given damping ratio are shown in Fig. 5.6; the magnitudes of these maxima are shown in Fig. 5.7. For small damping ($\zeta < 0.3$, which applies to many practi-

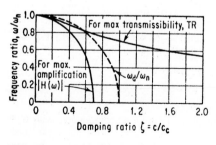

FIG. 5.6 Frequencies at which magnification and transmissibility maxima occur for given damping ratio.

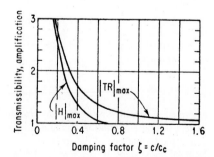

FIG. 5.7 Maximum values of magnification and transmissibility.

---

[*]It is important to note that these remarks apply only for the type of damping represented by a viscous dashpot model; different relations generally apply for other damping mechanisms.[49]

cal problems), the maximum transmissibility $|TR|_{max}$ and maximum amplification factor $|H|_{max}$ both occur at $\omega_d \approx \omega_n$, and

$$|TR|_{max} \approx |H|_{max} \approx (2\zeta)^{-1}$$

The quantity $(2\zeta)^{-1}$ is often given the symbol Q, termed the "quality factor" of the system. The frequency at which the greatest amplification occurs is called the resonance frequency; the system is then said to be in resonance. For lightly damped systems the resonance frequency is practically equal to the natural frequency, and often no distinction is made between the two. Thus, for lightly damped systems, resonance (i.e., maximum amplification) occurs essentially when the exciting frequency ω is equal to the natural frequency ω_n.

Equation (5.6) shows that

$$X_0/F_0 \approx 1/k \text{ (system is stiffness-controlled) for } \omega \ll \omega_n$$

$$\approx 1/2k\zeta \text{ (system is damping-controlled) for } \omega \approx \omega_n \, (\zeta \ll 1)$$

$$\approx 1/m\omega^2 \text{ (system is mass-controlled) for } \omega \gg \omega_n$$

General Periodic Excitation.   Any periodic excitation may be expressed in terms of a Fourier series (i.e., a series of sinusoids) and any aperiodic excitation may be expressed in terms of a Fourier integral, which is an extension of the Fourier-series concept. In view of the superposition principle applicable to linear systems the response can then be obtained in terms of a corresponding series or integral.

A periodic excitation with period T may be expanded in a Fourier series as

$$F(t) = \frac{A_0}{2} + \sum_{r=1}^{\infty} (A_r \cos r\omega_0 t + B_r \sin r\omega_0 t) = \sum_{r=-\infty}^{\infty} C_r e^{ir\omega_0 t} \tag{5.9}$$

where the period T and fundamental frequency ω_0 are related by

$$\omega_0 T = 2\pi$$

The Fourier coefficients A_r, B_r, C_r may be computed from

$$A_r = \frac{2}{T} \int_t^{t+T} F(t) \cos (r\omega_0 t) \, dt \qquad B_r = \frac{2}{T} \int_t^{t+T} F(t) \sin (r\omega_0 t) \, dt$$

$$C_r = \frac{1}{2}(A_r - iB_r) = \frac{1}{T} \int_t^{t+T} F(t) e^{-ir\omega_0 t} \, dt \tag{5.10}$$

Superposition permits the steady-state response to the excitation given by Eq. (5.9) to be expressed as

$$x_{ss} = \frac{1}{k} \sum_{r=-\infty}^{\infty} H_r C_r e^{ir\omega_0 t} \tag{5.11}$$

where H_r is obtained by setting $\omega = r\omega_0$ in Eq. (5.7).

If a periodic excitation contains a large number of harmonic components with $C_r \neq 0$, it is likely that one of the frequencies $r\omega_0$ will come very close to the natural frequency ω_n of the system. If $r_0\omega_0 \approx \omega_n$, $C_{r0} \neq 0$, then $H_{r0}C_{r0}$ will be much greater than the other components of the response (particularly in a very lightly damped system), and

$$x_{ss}k \approx H_{r0}C_{r0}e^{i\omega_n t} + H_{-r0}C_{-r0}e^{-i\omega_n t} + A_0/2$$

$$\approx (1/2\zeta)(A_{r0} \sin \omega_n t - B_{r0} \cos \omega_n t) + A_0/2$$

General Nonperiodic Excitation.[2,3,11,16]   The response of linear systems to any well-behaved* forcing function may be determined from the impulse response as discussed in conjunction with Eq. (5.4) or by application of Fourier integrals. The latter may be visualized as generalizations of Fourier series applicable for functions with infinite period.

A "well-behaved"* forcing function $F(t)$ may be expressed as[†]

$$F(t) = \frac{1}{2\pi} \int_{-\infty}^{\infty} \Phi(\omega)e^{i\omega t}\, d\omega \tag{5.12}$$

where

$$\Phi(\omega) = \int_{-\infty}^{\infty} F(t)e^{-i\omega t}\, dt \tag{5.13}$$

[These are analogous to Eqs. (5.9) and (5.10)]. With the ratio $H(\omega)$ of displacement to force as given by Eq. (5.7), the displacement-response transform then is

$$X(\omega) = (1/k)H(\omega)\Phi(\omega)$$

and, analogously to Eq. (5.11), one finds the displacement given by

$$x_{ss}(t) = (1/2\pi) \int_{-\infty}^{\infty} X(\omega)e^{i\omega t}\, d\omega = (1/2\pi k) \int_{-\infty}^{\infty} H(\omega)\Phi(\omega)e^{i\omega t}\, d\omega$$

$$= (1/2\pi k) \int_{-\infty}^{\infty} H(\omega) \left[\int_{-\infty}^{\infty} F(t)e^{-i\omega t}\, dt \right] e^{i\omega t}\, d\omega$$

One may expect the components of the excitation with frequencies nearest the natural frequency of a system to make the most significant contributions to the response. For lightly damped systems one may assume that these most significant components are contained in a small frequency band containing the natural frequency. Usually one uses a "resonance bandwidth" $\Delta\omega = 2\zeta\omega_n$, thus effectively assuming that the most significant components are those with frequencies between $\omega_n(1 - \zeta)$ and $\omega_n(1 + \zeta)$. (At these two limiting frequencies, commonly called the half-power points, the rate of energy dissipation is one-half of that at resonance. The amplitude of the response at these frequencies is $1/\sqrt{2} \approx 0.707$ times the amplitude at resonance.) Noting that the largest values of the complex amplification factor $H(\omega)$ occur for $\omega \approx \pm \omega_d \approx \pm \omega_n$, one may write

$$2\pi x_{ss}k \approx - i\omega_n e^{-\zeta\omega_n t}[\Phi(\omega_n)e^{i\omega_n t} + \Phi(-\omega_n)e^{-i\omega_n t}]$$

Random Vibrations: Mean Values, Spectra, Spectral Densities.[2,3,11,16]   In many cases one is interested only in some mean value as a characterization of response. The time average of a variable $y(t)$ may be defined as

$$\bar{y} = \lim_{\tau \to \infty} (1/\tau) \int_0^{\tau} y(t)\, dt \tag{5.14}$$

where it is assumed that the limit exists. For periodic $y(t)$ one may take τ equal to a period and omit the limiting process.

---

*"Well-behaved" means that $|F(t)|$ is integrable and $F(t)$ has bounded variation.

[†]Other commonly used forms of the integral transforms can be obtained by substituting $j = - i$. Since $j^2 = i^2 = - 1$, all the developments still hold. Fourier transforms are also variously defined as regards the coefficients. For example, instead of $1/2\pi$ in Eq. (5.12), there often appears a $1/\sqrt{2\pi}$; then a $1/\sqrt{2\pi}$ factor is added in Eq. (5.13) also. In all cases the product of the coefficients for a complete cycle of transformations is $1/2\pi$.

The mean-square value of $y(t)$ thus is given by

$$\overline{y^2} = \lim_{t \to \infty} (1/\tau) \int_0^\tau y^2(t)\, dt$$

and the root-mean-square value by $y_{\mathrm{rms}} = (\overline{y^2})^{1/2}$. For a sinusoid $x = \mathrm{Re}\{Ae^{i\omega t}\}$ one finds $\overline{x^2} = \frac{1}{2}|A|^2 = \frac{1}{2}AA^*$ where $\underline{A}^*$ is the complex conjugate of A.

The mean-square response $\overline{x^2}$ of a single-degree-of-freedom system with frequency response $H(\omega)$ [Eq. (5.7)] to a sinusoidal excitation of the form $F(t) = \mathrm{Re}\{F_0 e^{i\omega t}\}$ is given by

$$k^2\overline{x^2} = H(\omega)F_0 H^*(\omega)F_0^*/2 = |H(\omega)|^2\overline{F^2}$$

Similarly, the mean-square value of a general periodic function $F(t)$, expressed in Fourier-series form as

$$F(t) = \sum_{r=-\infty}^{\infty} C_r e^{ir\omega_0 t}$$

is

$$\overline{F^2} = \sum_{r=-\infty}^{\infty} \frac{C_r C_r^*}{2} = \frac{1}{2}\sum_{r=-\infty}^{\infty} |C_r|^2 \tag{5.15}$$

The mean-square displacement of a single-degree-of-freedom system in response to the aforementioned periodic excitation is given by

$$k^2\overline{x^2} = \frac{1}{2}\sum_{r=-\infty}^{\infty} |C_r|^2|H(r\omega_0)|^2$$

where convergence of all the foregoing infinite series is assumed.

If one were to plot the cumulative value of (the sum representing) the mean-square value of a periodic variable as a function of frequency, starting from zero, one would obtain a diagram somewhat like Fig. 5.8. This graph shows how much each frequency (or "spectral component") adds to the total mean-square value. Such a graph[*] is called the spectrum (or possibly more properly the integrated spectrum) of $F(t)$. It is generally of relatively little interest for periodic functions, but is extremely useful for aperiodic (including random) functions.

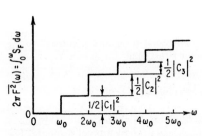

FIG. 5.8 (Integrated) spectrum of periodic function $F(t)$.

The derivative of the (integrated) spectrum with respect to ω is called the "mean-square spectral density" (or power spectral density) of F. Thus the power spectral density S_F of F is defined as[*]

$$S_F(\omega) = 2\pi d(\overline{F^2})/d\omega \tag{5.16}$$

where $\overline{F^2}$ is interpreted as a function of ω as in Fig. 5.8. The mean-square value of F is related to power spectral density as

$$\overline{F^2} = (1/2\pi) \int_0^\infty S_F(\omega)\, d\omega \tag{5.17}$$

---

[*]The factor 2π appearing in Fig. 5.8 and Eqs. (5.16) and (5.17) is a matter of definition. Different constants are sometimes used in the literature, and one must use care in comparing results from different sources.

This integral over all frequencies is analogous to the infinite sum of Eq. (5.15). From Fig. 5.8 and Eq. (5.17) one may visualize that power spectral density is a convenient means for expressing the contributions to the mean-square value in any frequency range.

For nonperiodic functions one obtains contributions to the mean-square value over a continuum of frequencies instead of at discrete frequencies, as in Fig. 5.8. The (integrated) spectrum and the power spectral density then are continuous curves. The relations governing the mean responses to nonperiodic excitation can be obtained by the same limiting processes which permit one to proceed from the Fourier series to Fourier integrals. However, the results are presented here in a slightly more general form so that they can be applied also to systems with random excitation.[*]

Response to Random Excitation: Autocorrelation Functions. For stationary ergodic random processes[†] whose sample functions are $F(t)$ or for completely specified functions $F(t)$ one may define an autocorrelation function $R_F(\tau)$ as

$$R_F(\tau) = \lim_{T \to \infty} (1/2T) \int_{-T}^{T} F(t)F(t + \tau)\, dt \qquad (5.18)$$

This function has the properties

$$R_F(0) = \overline{F^2} \geq R_F(\tau) \qquad R_F(-\tau) = R_F(\tau)$$

For many physical random processes the values of F observed at widely separated intervals are uncorrelated, that is

$$\lim_{\tau \to \infty} R_F(\tau) = (\overline{F})^2$$

or R_F approaches the square of the mean value (not the mean-square value!) of F for large time separation τ. (Many authors define variables measured from a mean value; if such variables are uncorrelated, their $R_F \to 0$ for large τ.) One may generally find some value of τ beyond which R_F does not differ "significantly" from $(\overline{F})^2$. This value of τ is known as the "scale" of the correlation.

The power spectral density of F is given by[‡,§]

---

[*]In the previous discussion the excitation was described as some known function of time and the responses were computed as other completely defined time functions; in each case the values at each instant were specified or could be found. Often the stimuli cannot be defined so precisely; only some statistical information about them may be available. Then, of course, one may only obtain some similar statistical information about the responses.

[†]A "random process" is a mathematical model useful for representing randomly varying physical quantities. Such a process is determined not by its values at various instants but by certain average and spectral properties. One sacrifices precision in the description of the variable for the sake of tractability.

One may envision a large number of sample functions (such as force vs. time records obtained on aircraft landing gears, with time datum at the instant of landing). One may compute an average value of these functions at any given time instant; such an average is called a "statistical average" and generally varies with the instant selected. On the other hand, one may also compute the time average of any given sample function over a long interval. The statistical average will be equal to the time average of almost every sample function, provided that the sample process is both stationary and ergodic.[3]

A random process is stationary essentially if the statistical average of the sample functions is independent of time, i.e., if the ensemble appears unchanged if the time origin is changed. Ergodicity essentially requires that almost every sample function be "typical" of the entire group. General mathematical results are to a large degree available only for stationary ergodic random processes; hence the following discussion is limited to such processes.

[‡]Definitions involving different numerical coefficients are also in general use.

[§]For real $F(t)$ one may multiply the coefficient shown here by 2 and replace the lower limit of integration by zero.

$$S_F(\omega) = 2 \int_{-\infty}^{\infty} R_F(\tau) e^{-i\omega\tau} \, d\tau \tag{5.19}$$

and is equal to twice* the Fourier transform of the autocorrelation function R_F. Inversion of this transform gives*,†

$$R_F(\tau) = \frac{1}{4\pi} \int_{-\infty}^{\infty} S_F(\omega) e^{i\omega\tau} \, d\tau$$

whence*,†

$$R_F(0) = \overline{F^2} = \frac{1}{4\pi} \int_{-\infty}^{\infty} S_F(\omega) \, d\omega \tag{5.20}$$

The last of these relations agrees with Eq. (5.17).

System Response to Random Excitation.   For a system with a complex frequency response $H(\omega)$ as given by Eq. (5.7) one finds that the power spectral density S_x of the response is related to the power spectral density S_F of the exciting force according to

$$k^2 S_x(\omega) = |H(\omega)|^2 S_F(\omega) \tag{5.21}$$

In order to compute the mean-square response of a system to aperiodic (or stationary ergodic random) excitation one may proceed as follows:

1. Calculate $R_F(\tau)$ from Eq. (5.18).
2. Find $S_F(\omega)$ from Eq. (5.19).
3. Determine $S_x(\omega)$ from Eq. (5.21).
4. Find $\overline{x^2}$ from Eq. (5.20) (with F subscripts replaced by x).

"White noise" is a term commonly applied to functions whose power spectral density is constant for all frequencies. Although such functions are not realizable physically, it is possible to obtain power spectra that remain virtually constant over a frequency region of interest in a particular problem (particularly in the neighborhood of the resonance of the system considered, where the response contributes most to the total).

The mean-square displacement of a single-degree-of-freedom system to a (real) white-noise excitation F, having the power spectral density $S_F(\omega) = S_0$, is given by

$$\overline{x^2} = \omega_n S_0 / 8 \zeta k^2 = S_0 / 4ck$$

Probability Distributions of Excitation and Response.[2,3,11,16]   For most practical purposes it is sufficient to define the probability of an event as the fraction of the number of "trials" in which the event occurs, provided that a large number of trials are made. (In throwing an unbiased die a large number of times one expects to obtain a given number, say 2, one-sixth of the time. The probability of the number 2 here is ⅙.)

If a given variable can assume a continuum of values (unlike the die for which the variable, i.e., the number of spots, can assume only a finite number of discrete values) it makes generally little sense to speak of the probability of any given value. Instead, one may profitably apply the concepts of probability distribution and probability density functions. Consider a continuous random variable x and a certain value x_0 of that

---

*Definitions involving different numerical coefficients are also in general use.
†For real $F(t)$ one may multiply the coefficient shown here by 2 and replace the lower limit of integration by zero.

variable. The probability distribution function P_{dis} then is defined as a function expressing the probability P that the variable $x \leq x_0$. Symbolically,

$$P_{dis}(x_0) = P(x \leq x_0)$$

The probability density function P_{dens} is defined by

$$P_{dens}(x_0) = dP_{dis}(x_0)/dx_0$$

so that the probability of x occurring between x_0 and $x_0 + dx_0$ is

$$P(x_0 < x \leq x_0 + dx_0) = P_{dens}(x_0)dx_0$$

Among the most widely studied distributions are the gaussian (or normal) distribution, for which

$$P_{dens}(x) = [1/\sigma\sqrt{2\pi}]e^{-(x-M)2/2\sigma2} \qquad (5.22)$$

and the Rayleigh distribution, for which

$$P_{dens}(x) = (x/\sigma^2)e^{-x2/2\sigma2} \qquad (5.23)$$

In the foregoing, M denotes the statistical mean value, defined by

$$M = \int_{-\infty}^{\infty} xP_{dens}(x)\, dx$$

and σ^2 denotes the variance of x and is defined by

$$\sigma^2 = \int_{-\infty}^{\infty} (x-M)^2 P_{dens}(x)\, dx$$

σ is called the "standard deviation" of the distribution. For a stationary ergodic random process

$$\sigma^2 + M^2 = R_x(0) = \overline{x^2}$$

The gaussian distribution is by far the most important, since it represents many physical conditions relatively well and permits mathematical analysis to be carried out relatively simply. With some qualifications, the "central limit theorem" states that any random process, each of whose sample functions is constructed from the sum of a large number of sample functions selected independently from some other random process, will tend to become gaussian as the number of sample functions added tends to infinity. A stationary gaussian random process with zero mean is completely characterized by either its autocorrelation function or its power spectral density. If the excitation of a linear system is a gaussian random process, then so is the system response.

For a gaussian random process $x(t)$ with autocorrelation function $R_x(\tau)$ and power spectral density $S_x(\omega)$ one may find the *average* number of times N_0 that $x(t)$ passes through zero in unit time from

$$(\pi N_0)^2 = -\left.\frac{d^2R_x}{d\tau^2}\right|_{\tau=0} [R_x(0)]^{-1} = \left[\int_0^\infty \omega^2 S_x(\omega)\, d\omega\right]\left[\int_0^\infty S_x(\omega)\, d\omega\right]^{-1}$$

The quantity N_0 gives an indication of the "apparent frequency" of $x(t)$. A sinusoid crosses zero twice per cycle and has $f = N_0/2$. This relation may be taken as the definition of apparent frequency for a random process.

The average number of times N_α that the aforementioned gaussian $x(t)$ crosses the value $x = \alpha$ per unit time is given by

$$N_\alpha = N_0 e^{-\alpha^2/2R_x(0)}$$

The average number of times per unit time that $x(t)$ passes through α with positive slope is half the foregoing value. The average number of peaks* of $x(t)$ occurring per unit time between $x = \alpha$ and $x = \alpha + d\alpha$ is

$$N_{\alpha,\alpha+d\alpha} = \frac{N_0 \alpha\, d\alpha}{2R_x(0)} e^{-\alpha^2/2R_x(0)}$$

For a linear single-degree-of-freedom system with natural circular frequency ω_n subject to white noise of power spectral density $S_F(\omega) = S_0$, one finds

$$N_0 = \omega_n/\pi \qquad f_{\text{apparent}} = \omega_n/2\pi = f_n$$

The displacement vs. time curve representing the response of a lightly damped system to broadband excitation has the appearance of a sinusoid with the system natural frequency, but with randomly varying amplitude and phase. The average number of peaks per unit time occurring between α and $\alpha + d\alpha$ in such an oscillation is given by

$$(2\pi/\omega_n)N_{\alpha,\alpha+d\alpha} = [\alpha\, d\alpha/R_x(0)]e^{-\alpha^2/2R_x(0)}$$

The term on the right-hand side is, except for the $d\alpha$, the Rayleigh probability density of Eq. (5.23).

5.1.2 Nonlinear Single-Degree-of-Freedom Systems[21,34,60]

The previous discussion dealt with systems whose equations of motion can be expressed as linear differential equations (with constant coefficients), for which solutions can always be found. The present section deals with systems having equations of motion for which solutions cannot be found so readily. Approximate analytical solutions can occasionally be found, but these generally require insight and/or a considerable amount of algebraic manipulation. Numerical or analog computations or graphical methods appear to be the only ones of general applicability.

Practical Solution of General Equations of Motion.   After one sets up the equations of motion of a system one wishes to analyze, one should determine whether solutions of these are available by referring to texts on differential equations and compendia such as Ref. 27. (The latter reference also describes methods of general utility for obtaining approximate solutions, such as that involving series expansion of the variables.) If these approaches fail, one is generally reduced to the use of numerical or graphical methods. A wide range of computer-based methods is available.[44,45] In the following pages two generally useful methods are outlined. Methods and results applicable to some special cases are discussed in subsequent sections.

A Numerical Method.   The equation of motion of a single-degree-of-freedom system can generally be expressed in the form

$$m\ddot{x} + G(x, \dot{x}, t) = 0 \qquad \text{or} \qquad \ddot{x} + f(x, \dot{x}, t) = 0 \qquad (5.24)$$

---

*Actually average excess of peaks over troughs, but for $\alpha \gg x_{\text{rms}}$ the probability of troughs in the interval becomes very small.

where f includes all nonlinear and nonconstant coefficient effects. (f may occasionally also depend on higher time derivatives. These are not considered here, but the method discussed here may be readily extended to account for them.) It is assumed that f is a known function, given in graphical, tabular, or analytic form.

In order to integrate Eq. (5.24) numerically as simply as possible, one assumes that f remains virtually constant in a small time interval Δt. Then one may proceed by the following steps:

1. *a.* Determine $f_0 = f(x_0, \dot{x}_0, 0)$, the initial value of f, from the specified initial displacement x_0 and initial velocity $\dot{x}_0$. Then the initial acceleration is $\ddot{x}_0 = -f_0$.
   b. Calculate the velocity $\dot{x}_1$ at the end of a conveniently chosen small time interval Δt_{0-1}, and the average velocity $\dot{x}_{0-1}$ during the interval from

$$\dot{x}_1 = \dot{x}_0 + \ddot{x}_0\,\Delta t_{0-1} \qquad \overline{\dot{x}}_{0-1} = \tfrac{1}{2}(\dot{x}_0 + \dot{x}_1)$$

   c. Calculate the displacement x_1 at the end of the interval Δt_{0-1} and the average displacement $\overline{x}_{0-1}$ during the interval from

$$x_1 = x_0 + \overline{\dot{x}}_{0-1}\,\Delta t_{0-1} \qquad \overline{x}_{0-1} = \tfrac{1}{2}(x_0 + x_1)$$

   d. Compute a better approximation* to the average f and $\ddot{x}$ during the interval Δt_{0-1} by using $x = \overline{x}_{0-1}$, $\dot{x} = \overline{\dot{x}}_{0-1}$, $t = \tfrac{1}{2}\Delta t_{0-1}$ in the determination of f.
2. Repeat steps 1*b* to 1*d*, beginning with the new approximation of f, until no further changes in f occur (to the desired accuracy).
3. Select a second time interval Δt_{1-2} (not necessarily of the same magnitude as Δt_{0-1}) and continue to
   a. Find $\ddot{x}_1 = -f_1 = -f(x_1, \dot{x}_1, t_1)$.
   b. Calculate the velocity $\dot{x}_2$ at the end of the interval Δt_{1-2}, and the average velocity $\dot{x}_{1-2}$ during the interval, from

$$\dot{x}_2 = \dot{x}_1 + \ddot{x}_1\,\Delta t_{1-2} \qquad \overline{\dot{x}}_{1-2} = \tfrac{1}{2}(\dot{x}_1 + \dot{x}_2)$$

   c. Similarly, find the final and average displacements for the Δt_{1-2} interval from

$$x_2 = x_1 + \overline{\dot{x}}_{1-2}\,\Delta t_{1-2} \qquad \overline{x}_{1-2} = \tfrac{1}{2}(x_1 + x_2)$$

   d. Compute a better approximation to the average f and $\ddot{x}$ during the interval Δt_{1-2} by using $x = \overline{x}_{1-2}$, $\dot{x} = \dot{x}_{1-2}$, $t = \Delta t_{0-1} + \tfrac{1}{2}\Delta t_{1-2}$ in the determination of f.
4. Repeat steps 3*b* to 3*d*, starting with the better value of f, until no changes in f occur to within the desired accuracy.
5. One may then continue by essentially repeating steps 3 and 4 for additional time intervals until one has determined the motion for the desired total time of interest.

Generally, the smaller the time intervals selected, the greater will be the accuracy of the results (regardless of the f-averaging method used). Use of smaller time intervals naturally leads to a considerable increase in computational effort. If high accuracy is required, one may generally benefit by employing one of the many available more sophisticated numerical-integration schemes.[12,44,45] In many practical instances the labor of carrying out the required calculations by "hand" becomes prohibitive, however, and use of a digital computer is indicated.

---

*It should be noted that evaluation of f at the average values of the variables involved is only one of many possible ways of obtaining an average f for the interval considered. Other averages, for example, can be obtained from $\sqrt{f_0 f_1}$, $\tfrac{1}{2}(f_0 + f_1)$. One can rarely predict which average will produce the most accurate results in a given case.

A Semigraphical Method. One may avoid some of the tedium of the foregoing numerical-solution method and gain some insight into a problem by using the "phase-plane delta" method[25,32] discussed here. This method, like the foregoing numerical one, is essentially a stepwise integration for small time increments. It is based on rewriting the equation of motion (5.24) as

$$\dot{x} + \omega_0^2(x + \delta) = 0 \qquad \delta = -x + f(x, \dot{x}, t)/\omega_0^2 \tag{5.25}$$

where ω_0 is any convenient constant circular frequency. (Any value may be chosen for ω_0, but it is usually useful to select one with some physical meaning, e.g., $\omega_0 = \sqrt{k_0/m_0}$, where k_0 and m_0 are values of stiffness and mass for small, x, $\dot{x}$ and t.) If one introduces into Eq. (5.25) a reduced velocity v given by

$$v = \dot{x}/\omega_0$$

and assumes that $\delta(x, \dot{x}, t)$ remains essentially constant in a short time interval one may integrate the resulting equation to obtain

$$v^2 + (x + \delta)^2 = R^2 = \text{const}$$

Thus for small time increments the solutions of (5.25) are represented in the xv plane (Fig. 5.9) by short arcs of circles whose centers are at $x = -\delta$, $v = 0$.

The angle $\Delta\theta$ subtended by the aforementioned circular arc is related to the time interval Δt according to

$$\Delta t \approx \Delta x/v\omega_0 \approx \Delta\theta/\omega_0 \tag{5.26}$$

On the basis of the foregoing discussion one may thus proceed as follows:

1. Calculate $\delta(x_0, \dot{x}_0, t_0)$ from Eq. (5.25) using the given initial conditions.
2. Locate the circle center $(-\delta, 0)$ and the initial point (x_0, v_0) on the xv plane; draw a small clockwise arc.
3. At the end of this arc is the point x_1, v_1 corresponding to the end of the first time interval.
4. Measure or calculate (in radians) the angle $\Delta\theta$ subtended by the arc; calculate the length of the time increment from Eq. (5.26).
5. Calculate $\delta(x_1, \dot{x}_1, t_1)$, and continue as before.
6. Repeat this process until the desired information is obtained.
7. Plots, such as those of x, $\dot{x}$ or $\ddot{x}$, against time, may then be readily obtained from the xv curve and the computed time information.

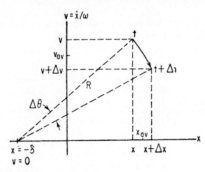

If increased accuracy is desired, particularly where δ changes rapidly, δ should be evaluated from average conditions (x_{av}, $\dot{x}_{av}$, t_{av}) during the time increment instead of conditions at the beginning of this increment (see Fig. 5.9). If δ depends on only one variable, a plot of δ against this variable may generally be used to advantage, particularly if it is superposed onto the vx plane.

Mathematical-Approximation Methods. An analytical expression is usually prefer-

FIG. 5.9   Phase-plane delta method.

able to a series of numerical solutions, since it generally permits greater insight into a given problem. If exact analytical solutions cannot be found, approximate ones may be the next best approach.

Series expansion of the dependent in terms of the independent variable is often a useful expedient. Power series and Fourier series are most commonly used, but occasionally series of other functions may be employed. The approach consists essentially of writing the dependent variable in terms of a series with unknown coefficients, substituting this into the differential equation, and then solving for the coefficients. However, in many cases these solutions may be difficult, or the series may converge slowly or not at all.

Other methods attempt to obtain solutions by separating the governing equations into a linear part (for which a simple solution can be found) and a nonlinear part. The solution of the linear part is then applied to the nonlinear part in some way so as to give a first correction to the solution. The correction process is then repeated until a second better approximation is obtained, and the process is continued. Such methods include:

1. Perturbation,[21,40] which is particularly useful where the nonlinearities (deviations from linearity) are small
2. Reversion,[21] which is a special treatment of the perturbation method
3. Variation of parameters,[21,60] useful where nonlinearities do not result in additive terms
4. Averaging methods, based on error minimization
   a. Galerkin's method[21]
   b. Ritz method[21]

Conservative Systems: The Phase Plane.[21,26,60]   A conservative system is one whose equation of motion can be written

$$m\ddot{x} + f(x) = 0 \tag{5.27}$$

In such systems (which may be visualized as masses attached to springs of variable stiffness) the total energy E remains constant; that is,

$$E = V(\dot{x}) + U(x)$$

where $$V(\dot{x}) = \tfrac{1}{2}m\dot{x}^2 \qquad U(x) = \int_{x_0}^{x} f(x)\, dx$$

where $V(\dot{x})$ and $U(x)$ are, respectively, the kinetic and the potential energies and x_0 is a convenient reference value.

The velocity-displacement ($\dot{x}$ vs. x) plane is called the "phase plane"; a curve in it is called a "phase trajectory." The equation of a phase trajectory of a conservative system with a given total energy E is

$$\dot{x}^2 = (2/m)[E - U(x)] \tag{5.28}$$

The time interval $(t-t_1)$ in which a change of displacement from x_1 to x occurs, is given by

$$t - t_1 = \int_{x_1}^{x} \frac{dx}{\dot{x}} = \int_{x_1}^{x} \frac{dx}{\sqrt{(2/m)[E - U(x)]}} \tag{5.29}$$

Some understanding of the geometry of phase trajectories may be obtained with the aid of Fig. 5.10, which shows the dependence of phase trajectories on total energy for a hypothetical potential energy function $U(x)$. For $E = E_1$ the motion is periodic; zero

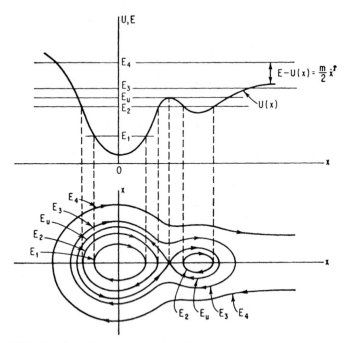

FIG. 5.10 Dependence of phase trajectories on energy.

velocity and velocity reversal occur where $U(x) = E_1$. With $E = E_2$ periodic oscillations are possible about two points; the initial conditions applicable in a given case dictate which type of oscillation occurs in that case. For $E = E_3$ only a single periodic motion is possible; for $E = E_4$ the motion is aperiodic. For $E = E_u$ there exists an instability at x_u; there the mass may move either in the increasing or decreasing x direction. (The arrows on the phase trajectories point in the direction of increasing time.)

For a linear undamped system $f(x) = kx$, $U(x) = \frac{1}{2}kx^2$, and phase trajectories are ellipses with semiaxes $(2E/k)^{1/2}$, $(2E/m)^{1/2}$.

The following facts may be summarized for conservative systems:

1. Oscillatory motions occur about minima in $U(x)$.

2. Phase trajectories are symmetric about the x axis and cross the x axis perpendicularly.

3. If $f(x)$ is single-valued, the phase trajectories for different energies E do not intersect.

4. All finite motions are periodic.

For nonconservative systems the phase trajectories tend to cross the constant-energy trajectories for the corresponding conservative systems. For damped systems the trajectories tend toward lower energy; i.e., they spiral in to a point of stability. For excited systems the trajectories spiral outward, either toward a "limit-cycle" trajectory or indefinitely.[21,60]

The period T of an oscillation of a conservative system occurring with maximum displacement (amplitude) x_{max} may be computed from

$$T = \sqrt{2m} \int_{x_{\min}}^{x_{\max}} \left[\int_{x}^{x_{\max}} f(x)\, dx \right]^{-1/2} dx \tag{5.30}$$

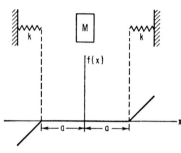

FIG. 5.11 Spring-mass system with clearance.

where $x_{\max}$ and $x_{\min}$ are the largest and smallest values of x (algebraically) for which zero velocity $\dot{x}$ occurs. The frequency f may then be obtained from $f = 1/T$.

For a linear spring-mass system with clearance, as shown in Fig. 5.11, the frequency is given by[17,32]

$$f = \sqrt{\frac{k}{m}} \left\{ 2\left[\pi + \frac{2}{(x_{\max}/a) - 1} \right] \right\}^{-1}$$

where $x_{\max}$ is the maximum excursion of the mass from its middle position.

For a system governed by Eq. (5.27) with $f(x) = kx|x|^{b-1}$ [or $f(x) = kx^b$, if b is odd], the frequency is given by[35]

$$f = \sqrt{\frac{(b+1)k}{8\pi m}} \, x_{max}^{b-1} \frac{\Gamma[1/(b+1) + \tfrac{1}{2}]}{\Gamma[1/(b+1)]}$$

in terms of the gamma function Γ, values of which are available in many tables.

Steady-State Periodic Responses. In many cases, particularly in steady-state analyses of periodically forced systems, periodic oscillatory solutions are of primary interest. A number of mathematical approaches are available to deal with these problems. Most of these, including the well-known methods of Stoker[60] and Schwesinger,[21,54] are based on the idea of "harmonic balance."[21] They essentially assume a Fourier expansion of the solution and then require the coefficients to be adjusted so that relevant conditions on the lowest few harmonic components are satisfied.

For example, in order to find a steady-state periodic solution of

$$m\ddot{x} + g(\dot{x}) + f(x) = F \sin(\omega t + \phi)$$

one may substitute an assumed displacement

$$x = A_1 \sin \omega t + A_2 \sin 2\omega t + \cdots + A_n \sin n\omega t$$

and impose certain restrictions on the error ϵ,

$$\epsilon(t) = m\ddot{x} + g(\dot{x}) + f(x) - F \sin(\omega t + \phi)$$

In Schwesinger's method the mean-square value of the error, $\overline{\epsilon^2} = \int_0^{2\pi} \epsilon^2(t)d(\omega t)$, is minimized, and values of F and ϕ are calculated from this minimization corresponding to an assumed A_1.

Systems and Nonlinear Springs. The restoring forces of many systems (particularly with small amounts of nonlinearity) may be approximated so that the equation of motion may be written as

$$\ddot{x} + 2\zeta\omega_0\dot{x} + \omega_0^2 x + (a/m)x^3 = (F/m)\cos\omega t \tag{5.31}$$

in the presence of viscous damping and a sinusoidal force. ζ is the damping factor and ω_0 the natural frequency of a corresponding undamped linear system (i.e., for $a = 0$).

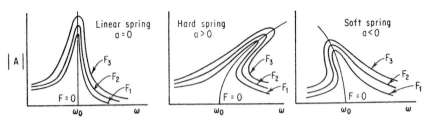

FIG. 5.12   Comparison of frequency responses of linear and nonlinear systems.

For $a > 0$ the spring becomes stiffer with increasing deflection and is called "hard"; for $a < 0$ the spring becomes less stiff and is called "soft."

Figure 5.12 compares the responses of linear and nonlinear lightly damped spring systems. The responses are essentially of the form $x = A \cos \omega t$; curves of response amplitude A vs. forcing frequency ω are sketched for several values of forcing amplitude F, for constant damping ζ. For a linear system the frequency of free oscillations ($F = 0$) is independent of amplitude; for a hard system it increases; for a soft system it decreases with increasing amplitude. The response curves of the nonlinear spring systems may be visualized as "bent-over" forms of the corresponding curves for the linear systems.

As apparent from Fig. 5.12, the response curves of the nonlinear spring systems are triple-valued for some frequencies. This fact leads to "jump" phenomena, as sketched in Figs. 5.13 and 5.14. If a given force amplitude is maintained as forcing frequency is changed slowly, then the response amplitude follows the usual response curve until point 1 of Fig. 5.13 is reached. The hatched regions between points 1 and 3 correspond to unstable conditions; an increase in ω above point 1 causes the amplitude to jump to that corresponding to point 2. A similar conditions occurs when frequency is slowly decreased; the jump then occurs between points 3 and 4.

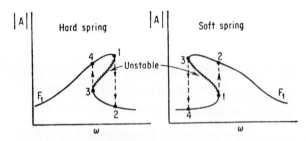

FIG. 5.13   Jump phenomena with variable frequency and constant force.

As also evident from Fig. 5.12, a curve of response amplitude vs. force amplitude at constant frequency is also triple-valued in some regions of frequency. Thus amplitude jumps occur also when one changes the forcing amplitude slowly at constant frequency ω_c. This condition is sketched in Fig. 5.14. For a hard spring this can occur only at frequencies above ω_0, for soft springs below ω_0. The equations characterizing a lightly damped nonlinear spring system and its jumps are summarized in Fig. 5.15.

The previous discussion deals with the system response as if it were a pure sinusoid $x = A \cos \omega t$. However, in nonlinear systems there occur also harmonic components (at frequencies $n\omega$, where n is an integer) and subharmonic components (fre-

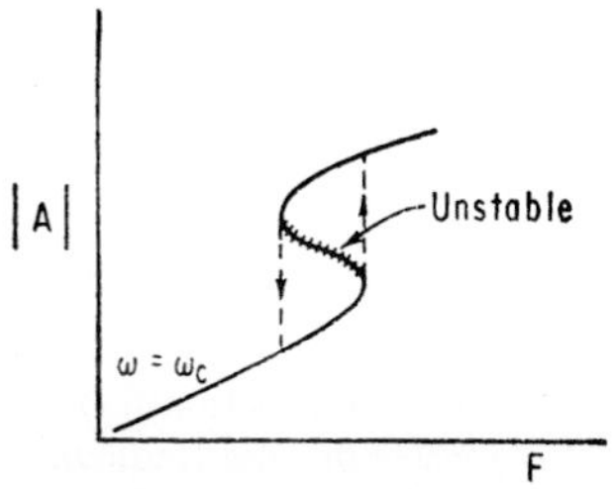

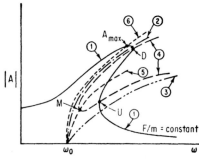

FIG. 5.15 Characteristics of responses of nonlinear springs. Equation of motion: $m\ddot{x} + c\dot{x} + kx + ax^3 = F \cos \omega t$, where ω_0 = undamped natural frequency for linear system (with $a = 0$); ζ = damping ratio for linear system = $c/2m\omega_0$. (1) Response curve for forced vibrations, F/m constant: $[A(\omega_0^2 - \omega^2) + \frac{3}{4}(a/m)A^3]^2 + [2\zeta\omega_0\omega A]^2 = (F/m)^2$. (2) Response curve for undamped free vibrations (approximate locus of downward jump points D, and of A_{max}): $\omega^2 = \omega_0^2 + \frac{3}{4}(a/m)A^2$. (3) Locus of upward jump points U with zero damping; approximate locus of same with finite damping: $\omega^2 = \omega_0^2 + \frac{9}{8}(a/m)A^2$. (4) Locus of upward and downward jump points U and D with finite damping: $[\omega_0^2 - \omega^2 + \frac{3}{4}(a/m)A^2][\omega_0^2 - \omega^2 + \frac{9}{4}(a/m)A^2] + (2\zeta\omega_0\omega)^2 = 0$. (5) Locus of points M below which no jumps occur: $\omega^2 = \omega_0^2 + \frac{9}{8}(a/m)A^2$. (6) Locus of maximum amplitudes A_{max}: $\omega^2 - \frac{3}{4}(a/m)A^2 = \omega_0^2(1 - 2\zeta^2)$.

$\omega_c > \omega_0$ for hard spring

$\omega_c < \omega_0$ for soft spring

FIG. 5.14 Jump phenomena with variable force and constant frequency.

quencies ω/n). In addition, components occur at frequencies which are integral multiples of the subharmonic frequencies. The various harmonic and subharmonic components tend to be small for small amounts of nonlinearity, and damping tends to limit the occurrence of subharmonics. The amplitude of the component with frequency 3ω (the lowest harmonic above the fundamental with finite amplitude for an undamped system) is given by $aA^3/36m\omega^2$, where A is the amplitude of the fundamental response, in view of Eq. (5.31). For more complete discussions see Ref. 21.

Graphical Determination of Response Amplitudes. A relatively easily applied method for approximating response amplitudes was developed by Martienssen[37] and improved by Mahalingam.[34] It is based on the often observed fact that the response to sinusoidal excitation is essentially sinusoidal. The method is here first explained for a linear system, then illustrated for nonlinear ones.

In order to obtain the steady-state response of a linear system one substitutes an assumed trial solution

$$x = A \cos (\omega t - \phi)$$

into the equation of motion

$$m\ddot{x} + c\dot{x} + kx = P \cos \omega t$$

By equating coefficients of corresponding terms on the two sides of the resulting equation one obtains a pair of equations which may be solved to yield

$$\omega_n^2 A = \omega^2 A + \frac{P}{m}\cos\phi \qquad \tan\phi = \frac{c\omega/m}{(\omega_n/\omega)^2 - 1} \qquad (5.32)$$

where ω_n is the undamped natural frequency. One may plot the functions

$$y_1(A) = A\omega_n^2(A) \qquad y_2(A) = A\omega^2 + (P/m)\cos\theta$$

and determine for which value of A these two functions intersect. This value of A then is the desired amplitude. Since P and m are given constants, y_2 plots as a straight line with slope ω^2 and y intercept $(P/m)\cos\phi$. For a linear system one may compute $\tan\phi$ directly from Eq. (5.32), but for a nonlinear system ω_n^2 depends on the amplitude A and direct computation of $\tan\phi$ is not generally possible. Use of a method of successive approximations is then indicated.

Figure 5.16a shows application of this method to a linear system. After calculating ϕ from Eq. (5.32) one may find the y intercept $(P/m)\cos\phi$. For a given frequency ω one may then draw a line of slope ω^2 through that intercept to represent the function y_2. For a linear system y_1 is a straight line with slope ω_n^2 and passing through the origin. The amplitude of the steady-state oscillation may then be determined as the value A_0 of A where the two lines intersect.

Figure 5.16b shows a diagram analogous to Fig. 5.16a, but for an arbitrary nonlinear system. The function $y_1 = A\omega_n^2$ is not a straight line in general since ω_n generally is a function of A. This function may be determined from the restoring function $f(x)$ by use of Eq. (5.30). The possible amplitudes corresponding to a given driving frequency ω and force amplitude P are determined here, as before, by the intersection of the y_1 and y_2 curves. As shown in the figure, more than one amplitude may correspond to a given frequency—a condition often encountered in nonlinear systems.

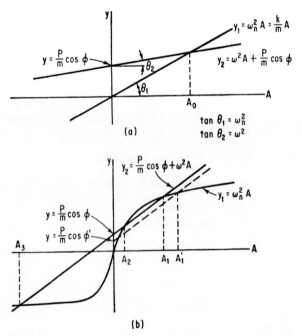

FIG. 5.16 Graphical determination of amplitude. (a) Linear system. (b) Nonlinear system.

In applying the previously outlined method to nonlinear systems one generally cannot find the correct value of φ at once from Eq. (5.32), since ω_n depends on the amplitude A, as has been pointed out. Instead one may assume any value of φ, such as φ′, and determine a first approximation A_1' to A_1. Using the approximate amplitude one may then determine better values of $\omega_n(A)$ and φ from Eq. (5.32) and then use these better values to obtain a better approximation to A_1. This process may be repeated until φ and A_1 have been found to the desired degree of accuracy. A separate iteration process of this sort is generally required for each of the possible amplitudes. (Different values of φ correspond to the different amplitudes A_1, A_2, A_3.)

Systems with Nonlinear Damping. The governing equation in this case may be written

$$m\ddot{x} + C(x, \dot{x}) + kx = F \sin \omega t \qquad (5.33)$$

where $C(x, \dot{x})$ represents the effect of damping. Exact or reasonably good approximate solutions are available only for relatively few cases. However, for many practical cases where the damping is not too great, the system response is essentially sinusoidal. One may then use an "equivalent" viscous-damping term $c_e \dot{x}$ instead of $C(x, \dot{x})$, so that the equivalent damping results in the same amount of energy dissipation per cycle as does the original nonlinear damping. For a response of the form $x = A \sin \omega t$ the equivalent viscous damping may be computed from

$$c_e = \frac{1}{\pi \omega A} \int_0^{2\pi} C(A \sin \omega t, A\omega \cos \omega t) \, d(\omega t) \qquad (5.34)$$

In contrast to the usual linear case, c_e is generally a function of frequency and amplitude. Once c_e has been found, the response amplitude may be computed from Eq. (5.6); successive approximations must be used if c_e is amplitude-dependent.

For "dry" or Coulomb friction, where the friction force is $\pm \mu$ (constant in magnitude but always directed opposite to the velocity) the equivalent viscous damping c_e and response amplitude A are given by

$$c_e = \frac{4\mu}{\pi A \omega} \qquad A = \frac{F}{k(1 - \omega^2/\omega_n^2)} \sqrt{1 - \left(\frac{4}{\pi}\frac{\mu}{F}\right)^2}$$

Further details on Coulomb damped systems appear in Refs. 17 and 63.

If complex notation is used, the damping effect may be expressed in terms of an imaginary stiffness term, and Eq. (5.33) may alternatively be written

$$m\ddot{x} + k(1 + i\gamma)x = Fe^{i\omega t}$$

where γ is known as the "structural damping factor."[18,51,68] For a response given by $x = Ae^{i\omega t}$ one finds

$$|A| = F/|-m\omega^2 + k(1 + i\gamma)| \qquad (5.35)$$

This reduces identically to the linear case with viscous damping c, if γ is defined so that $k\gamma = \omega c$. The steady-state behavior of a system with any reasonable type of damping may be represented by this complex stiffness concept, provided that γ is prescribed with the proper frequency and amplitude dependence. The foregoing relation and Eq. (5.34) may be used to find the aforementioned proper dependences for a given resisting force $C(x, \dot{x})$. The case of constant γ corresponds to "structural damping" (widely used in aircraft flutter calculations) and represents a damping force proportional to displacement but in phase with velocity. The damping factor γ is particularly useful for describing the damping action of rubberlike materials, for which the damping is virtually independent of amplitude (but not of frequency),[55] since then the response is explicitly given by Eq. (5.35).

Response to Random Excitation.   For quantitative results the reader is referred to Refs. 3, 11, and 16.

Qualitatively, the response of a linear system to random excitation is essentially a sinusoid at the system's natural frequency. The amplitudes (i.e., the envelope of this sinusoid) vary slowly and have a Rayleigh distribution. Compared with a linear system a system with a "hard" spring has a higher natural frequency, a lower probability of large excursions, and waves with flattened peaks. ("Soft" spring systems exhibit opposite characteristics.) The effects of the nonlinearities on frequency and on the wave shape are generally very small.

Self-Excited Systems.   If the damping coefficient c of a linear system is negative, the system tends to oscillate with ever-increasing amplitude. Positive damping extracts energy from the system; negative damping contributes energy to it. A system (such as one with negative damping) for which the energy-contributing forces are controlled by the system motion is called self-excited. A source of energy must be available if a system is to be self-excited. The steady-state amplitude of a self-excited oscillation may generally be determined from energy considerations, i.e., by requiring the total energy dissipated per cycle to equal the total energy supplied per cycle.

The chatter of cutting tools, screeching of hinges or locomotive wheels, and chatter of clutches are due to self-excited oscillations associated with friction forces which decrease with increasing relative velocity. The larger friction forces at lower relative velocities add energy to the system; smaller friction forces at higher velocities (more slippage) remove energy. While the oscillations build up, the energy added is greater than that removed; at steady state in each cycle the added energy is equal to the extracted energy.

More detailed discussions of self-excited systems may be found in Refs. 21 and 60.

5.2 SYSTEMS WITH A FINITE NUMBER OF DEGREES OF FREEDOM

The instantaneous configurations of many physical systems can be specified by means of a finite number of coordinates. Continuous systems, which have an infinite number of degrees of freedom, can be approximated for many purposes by systems with only a finite number of degrees of freedom, by "lumping" of stiffnesses, masses, and distributed forces. This concept, which is extremely useful for the analysis of practical problems, has been the basis for numerous computer-based "finite-element" and "modal-analysis" methods.[24,44,45]

5.2.1 Systematic Determination of Equations of Motion

Generalized Coordinates: Constraints.[53,64]   A set of n quantities q_i ($i = 1, 2, ..., n$) which at any time t completely specify the configuration of a system are called "generalized coordinates" of the system. The quantities may or may not be usual space coordinates.

If one selects more generalized coordinates than the minimum number necessary to describe a given system fully, then one finds some interdependence of the selected coordinates dictated by the geometry of the system. This interdependence may be expressed as

$$G(q_1, q_2, \ldots, q_n; \dot{q}_1, \dot{q}_2, \ldots, \dot{q}_n; t) = 0 \qquad (5.36)$$

Relations like those of Eq. (5.36) are known as "equations of constraint,"; if no such equations can be formulated for a given set of generalized coordinates, the set is known as "kinematically independent."

The constraints of a set of generalized coordinates are said to be integrable if all equations like (5.36) either contain no derivatives $\dot{q}_i$ or if such $\dot{q}_i$ that do appear can be eliminated by integration. If to the set of n generalized coordinates there correspond m constraints, all of which are integrable, then one may find a new set of $(n - m)$ generalized coordinates which are subject to no constraints. This new system is called "holonomic," and $(n - m)$ is the number of degrees of freedom of the system (that is, the smallest number of quantities necessary to describe the system configuration at any time). In practice one may often be able to select a holonomic system of generalized coordinates by inspection. Henceforth the discussion will be limited to holonomic systems.

Lagrangian Equations of Motion. The equations governing the motion of any holonomic system may be obtained by application of Lagrange's equation

$$(d/dt)(\partial T/\partial \dot{q}_i) - \partial T/\partial q_i + \partial U/\partial q_i + \partial F/\partial \dot{q}_i = Q_i \qquad i = 1, 2, \ldots, n \qquad (5.37)$$

where T denotes the kinetic energy, U the potential energy of the entire dynamic system, F is a dissipation function, and Q_i the generalized force associated with the generalized coordinate q_i.

The generalized force Q_i may be obtained from

$$\delta W = Q_i \delta q_i \qquad (5.38)$$

where δW is the total work done on the system by all external forces not contributing to U when the single coordinate q_i is changed to $q_i + \delta q_i$. The potential energy U accounts only for forces which are "conservative" (that is, for those forces for which the work done in a displacement of the system is a function of only the initial and final configurations). The dissipation function F represents half the rate at which energy is lost from the system; it accounts for the dissipative forces that appear in the equations of motion.

Lagrange's equations can be applied to nonlinear as well as linear systems, but little can be said about solving the resulting equations of motion if they are nonlinear, except that small oscillations of nonlinear systems about equilibrium can always be approximated by linear equations. Methods of solving linear sets of equations of motion are available and are discussed subsequently.

5.2.2 Matrix Methods for Linear Systems—Formalism

The kinetic energy T, potential energy U, and dissipation function F of any *linear* holonomic system with n degrees of freedom may be written as

$$T = \tfrac{1}{2} \sum_{i=1}^{n} \sum_{j=1}^{n} a_{ij} \dot{q}_i \dot{q}_j \qquad a_{ij} = a_{ji}$$

$$U = \tfrac{1}{2} \sum_{i=1}^{n} \sum_{j=1}^{n} c_{ij} q_i q_j \qquad c_{ij} = c_{ji} \qquad (5.39)$$

$$F = \tfrac{1}{2} \sum_{i=1}^{n} \sum_{j=1}^{n} b_{ij} \dot{q}_i \dot{q}_j \qquad b_{ij} = b_{ji}$$

The equations of motion may then readily be determined by use of Eq. (5.37). They may be expressed in matrix form as

$$A\{\ddot{q}\} + B\{\dot{q}\} + C\{q\} = \{Q(t)\} \tag{5.40}$$

where A, B, C are symmetric square matrices with n rows and n columns whose elements are the coefficients appearing in Eqs. (5.39) and $\{q\}$, $\{\dot{q}\}$, $\{\ddot{q}\}$, $\{Q(t)\}$ are n-dimensional column vectors.

(*Note:* A is called the inertia matrix, B the damping matrix, C the elastic or the stiffness matrix. For example,

$$A = \begin{bmatrix} a_{11} & a_{12} & \cdots & a_{1n} \\ a_{21} & a_{22} & \cdots & a_{2n} \\ \cdot & \cdot & & \cdot \\ \cdot & \cdot & & \cdot \\ a_{n1} & a_{n2} & \cdots & a_{nn} \end{bmatrix} \quad \{q\} = \begin{bmatrix} q_1 \\ q_2 \\ \cdot \\ \cdot \\ q_n \end{bmatrix} \quad \{\dot{q}\} = \begin{bmatrix} \dot{q}_1 \\ \dot{q}_2 \\ \cdot \\ \cdot \\ \dot{q}_n \end{bmatrix} \quad \{Q\} = \begin{bmatrix} Q_1 \\ Q_2 \\ \cdot \\ \cdot \\ Q_n \end{bmatrix}$$

The elements of $\{q\}$ are the coordinates of q_i, the elements of $\{\dot{q}\}$ are the first time derivatives of q_i (i.e., the generalized velocities $\dot{q}_i$); those of $\{\ddot{q}\}$ are the generalized accelerations $\ddot{q}_i$, those of $\{Q\}$ are the generalized forces Q_i.)

Free Vibrations: General System. If all the generalized forces Q_i are zero, then Eq. (5.40) reduces to a set of homogeneous linear differential equations. To solve it one may postulate a time dependence given by

$$\{q\} = \{r\}e^{st} \qquad s = \sigma + i\omega$$

which, when introduced into Eq. (5.40), results in

$$(s^2 A + sB + C)\{r\} = 0 \tag{5.41}$$

One may generally find n nontrivial solutions $\{r^{(j)}\}$, $j = 1, 2, \dots, n$, each corresponding to a specific value $s_{(j)}$ of s. The general solution of the homogeneous equation may then be expressed as

$$\{q\} = \sum_{j=1}^{n} \alpha_j \{r^{(j)}\} e^{s(j)t} = R\{p\} \tag{5.42}$$

in terms of complex constants α_j and the newly defined

$$R = \begin{bmatrix} r_1^{(1)} & r_1^{(2)} & \cdots & r_1^{(n)} \\ r_2^{(1)} & r_2^{(2)} & \cdots & r_2^{(n)} \\ \cdot & \cdot & & \cdot \\ \cdot & \cdot & & \cdot \\ r_n^{(1)} & r_n^{(2)} & \cdots & r_n^{(n)} \end{bmatrix} \quad \{p(t)\} = \begin{bmatrix} \alpha_1 & e^{s(1)t} \\ \alpha_2 & e^{s(2)t} \\ \cdot & \cdot \\ \cdot & \cdot \\ \alpha_n & e^{s(n)t} \end{bmatrix} \tag{5.43}$$

Initial conditions, e.g., $\{q(0)\}$, $\{\dot{q}(0)\}$ may be introduced into Eq. (5.42) to evaluate the constants α_j.

Forced Vibrations: General System. The forced motion may be described in terms of the sum of two motions, one satisfying the homogeneous equation (with all $Q_i = 0$) and including all the constants of integration, the other satisfying the complete Eq. (5.40) and containing no integration constants. (The constants must be evaluated so that the total solution satisfies the prescribed initial conditions.) The latter constant-free solution is often called the "steady-state solution."

Steady-State Solution for Periodic Generalized Forces.   If the generalized forces are harmonic with frequency ω_0 one may set

$$\{Q\}=\{\overline{Q}\}e^{i\omega_0 t} \qquad \{q\}=\{\overline{q}\}e^{i\omega_0 t} \tag{5.44}$$

in Eq. (5.40) and obtain

$$(-\omega_0^2 A + i\omega_0 B + C)\{\overline{q}\}=\{\overline{Q}\}$$

from which $\{\overline{q}\}$ may be determined. Equation (5.44) then gives the steady-state solutions.

If the Q_i are periodic with period T, but not harmonic, one may expand them and the components q_i of the steady-state solution in Fourier series:

$$\{Q\}=\sum_{N=-\infty}^{\infty} \{\overline{Q}^{(N)}\}e^{iN\omega_0 t} \qquad \{q\}=\sum_{N=-\infty}^{\infty} \{\overline{q}^{(N)}\}e^{iN\omega_0 t} \qquad \omega_0=\frac{T}{2\pi}$$

The Fourier components $\{\overline{q}^{(N)}\}$ may be evaluated from

$$(-N^2\omega_0^2 A + iN\omega_0 B + C)\{\overline{q}^{(N)}\} = \{\overline{Q}^{(N)}\}$$

Steady-State Solution for Aperiodic Generalized Forces.   For general $\{Q(t)\}$ one may write the solutions of Eq. (5.40) as

$$q_i(t) = \sum_{j=1}^{n} \int_0^t h_i^{(j)}(t - \tau)Q_j(\tau)\,d\tau \tag{5.45}$$

where $h_i^{(j)}(t)$ is the response of coordinate q_i to a unit impulse acting in place of Q_j. The "impulse response function" $h_i^{(j)}$ may be found from

$$h_i^{(j)}(t) = (d/dt)u_i^{(j)}(t)$$

where $u_i^{(j)}$ is the response of q_i to a unit step function acting in place of Q_j. (A unit step function is zero for $t < 0$, unity for $t > 0$.)

If one defines a square matrix $[H(t)]$ whose elements are $h_i^{(j)}$, one may write Eq. (5.45) alternatively as

$$\{q(t)\}=\int_0^t [H(t-\tau)]\{Q(\tau)\}\,d\tau$$

where $\qquad [H(t)] = \dfrac{1}{2\pi i} \displaystyle\int_{c+i\infty}^{c-i\infty} [T(s)]e^{st}\,ds \qquad [T(s)] = (s^2A + sB + C)^{-1}$

Undamped Systems.   For undamped systems all elements of the B matrix of Eq. (5.40) are zero, and Eq. (5.41) may be rewritten in the classical eigenvalue form

$$E\{r\} = \omega^2\{r\} \qquad E = A^{-1}C \tag{5.46}$$

The eigenvalues $\omega_{(j)}$, i.e., the values for which nonzero solutions $r^{(j)}$ exist, are real. They are the natural frequencies; the corresponding solution vectors $r^{(j)}$ describe the mode shapes.

One may then find a set of principal coordinates ψ_i, in terms of which the equations of motion are uncoupled and may be written in the following forms:

$$\{\ddot{\psi}\} + \Omega\{\psi\} = R^{-1}A^{-1}\{Q\} \quad \text{or} \quad M\{\ddot{\psi}\} + K\{\psi\} = \Phi(t)$$

Here
$$M = \bar{R}AR \quad K = \bar{R}CR \quad \{\Phi(t)\} = \bar{R}\{Q(t)\}$$

Ω is the diagonal matrix of natural frequencies

$$\Omega = \begin{bmatrix} \omega_{(1)}^2 & 0 & \cdots & 0 \\ 0 & \omega_{(2)}^2 & & 0 \\ \cdot & \cdot & & \cdot \\ \cdot & \cdot & & \cdot \\ \cdot & \cdot & & \cdot \\ 0 & 0 & \cdots & \omega_{(n)}^2 \end{bmatrix}$$

R is given by Eq. (5.43), and $\bar{R}$ denotes the transpose of R.

The principal coordinates ψ_i are related to the original coordinates q_i by

$$\{q\} = R\{\psi\} \qquad \{\psi\} = R^{-1}\{q\} \tag{5.47}$$

The response of the system to any forcing function $\{Q(t)\}$ may be determined in terms of the principal coordinates from

$$\psi_j(t) = m_{jj}k_{jj} \int_0^t \Phi_j(\tau) \sin[\omega_{(j)}(t - \tau)] \, d\tau + \psi_j(0) \cos[\omega_{(j)}t] + [1/\omega_{(j)}]\dot{\psi}_j(0) \sin[\omega_{(j)}t] \tag{5.48}$$

The response in terms of the original coordinates q_i may then be obtained by substitution of the results of Eq. (5.48) into Eq. (5.47).

5.2.3 Matrix Iteration Solution of Positive-Definite Undamped Systems

Positive-Definite Systems: Influence Coefficient and Dynamic Matrixes. A system is "positive-definite" if its potential energy U, as given by Eq. (5.39), is greater than zero for any $\{q\} \neq \{0\}$. Systems connected to a fixed frame are positive-definite; systems capable of motion (changes in the coordinates q_i) without increasing U are called "semidefinite."[64] The latter motions occur without energy storage in the elastic elements and are called "rigid-body" motions or "zero modes." (They imply zero natural frequency.)

Rigid-body motions are generally of no interest in vibration study. They may be eliminated by proper choice of the generalized coordinates or by introducing additional relations (constraints) among an arbitrarily chosen system of generalized coordinates by applying conservation-of-momentum concepts. Thus any system of generalized coordinates can be reduced to a positive-definite one.

For positive-definite linear systems C^{-1}, the inverse of the elastic matrix, is known as the influence coefficient matrix D. The elements of D are the influence coefficients; the typical element d_{ij} is the change in coordinate q_i due to a unit generalized force Q_j (applied statically), with all other Q's equal to zero. Since these influence coefficients can be determined from statics, one generally need not find C at all. It should be noted that for systems that are not positive-definite one cannot compute the influence coefficients from statics alone.

For iteration purposes it is useful to rewrite Eq. (5.46), the system equation of free sinusoidal motion, as

$$G\{r\} = (1/\omega^2)\{r\} \tag{5.49}$$

where $\qquad\qquad \{q\}=\{r\}e^{i\omega t} \qquad G = C^{-1}A = DA = E^{-1}$ $\qquad\qquad$ (5.50)

The matrix G is called the "dynamic matrix" and is defined, as above, as the product of the influence coefficient matrix D and the inertia matrix A.

Iteration for Lower Modes. In order to solve Eq. (5.49), which is a standard eigenvalue matrix equation, numerically for the lowest mode one may proceed as follows: Assume any vector $\{r_{(1)}\}$; then compute $G\{r_{(1)}\} = \alpha_{(1)}\{r_{(2)}\}$, where $\alpha_{(1)}$ is a constant chosen so that one element (say, the first) of $\{r_{(2)}\}$ is equal to the corresponding element of $\{r_{(1)}\}$. Then find $G\{r_{(2)}\} = \alpha_{(2)}\{r_{(3)}\}$, with $\alpha_{(2)}$ chosen like $\alpha_{(1)}$ before. Repeat this process until $\{r_{(n+1)}\} \approx \{r_{(n)}\}$ to the desired degree of accuracy. The corresponding constant $\alpha_{(n)}$ which satisfies $G\{r_{(n)}\} = \alpha(n)\{r_{(n+1)}\}$ then yields to lowest natural frequency ω_1 of the system and $\{r_{(n)}\}$ describes the shape of the corresponding (first) mode $\{r^{(1)}\}$. In view of Eq. (5.49)

$$\omega_1^2 = 1/\alpha_{(n)}$$

The second mode $\{r^{(2)}\}$ must satisfy the orthogonality relation

$$\{\overline{r^{(2)}}\}A\{r^{(1)}\} = 0 \qquad \text{or} \qquad \sum_{i=1}^{n}\sum_{j=1}^{n} r_i^{(2)}a_{ij}r_j^{(1)} = 0 \qquad\qquad (5.51)$$

In order to obtain a vector that satisfies Eq. (5.51) from an arbitrary vector $\{r\}$ one may select $(n-1)$ components of $\{r_{(2)}\}$ as equal to the corresponding components of $\{r\}$ and then compute the nth from Eq. (5.51). This process may be expressed as

$$\{r^{(2)}\} = S_1\{r\}$$

where S_1 is called the "first sweeping matrix." S_1 is equal to the identity matrix in n dimensions, except for one row which describes the interrelation Eq. (5.51). If A is diagonal one may take, for example,

$$S_1 = \begin{bmatrix} 0 & -\dfrac{a_{22}r_2^{(1)}}{a_{11}r_1^{(1)}} & -\dfrac{a_{33}r_3^{(1)}}{a_{11}r_1^{(1)}} & \cdots & \dfrac{a_{nn}r_n^{(1)}}{a_{11}r_1^{(1)}} \\ 0 & 1 & 0 & \cdots & 0 \\ 0 & 0 & 1 & \cdots & 0 \\ 0 & \cdot & \cdot & & \cdot \\ \vdots & \cdot & \cdot & & \cdot \\ & \cdot & \cdot & & \cdot \\ 0 & 0 & 0 & \cdots & 1 \end{bmatrix}$$

To obtain the second lowest mode shape $\{r^{(2)}\}$ and the second lowest natural frequency ω_2 one may form $H_1 = GS_1$, and solve

$$H_1\{r\} = (1/\omega^2)\{r\} \qquad\qquad (5.52)$$

by iteration. Since Eq. (5.52) is of the same form as Eq. (5.49), one may proceed here as previously discussed, i.e., by assuming a trial vector $\{r_{(1)}\}$, forming $H_1\{r_{(1)}\} = \alpha_{(1)}\{r_{(2)}\}$, so that one element of $\{r_{(2)}\}$ is equal to the corresponding element of $\{r_{(1)}\}$, then forming $H_1\{r_{(2)}\} = \alpha_{(2)}\{r_{(3)}\}$, etc. This process converges to $\{r^{(2)}\}$ and $\alpha = 1/\omega_2^2$.

The third mode $\{r^{(3)}\}$ similarly must satisfy

$$\{\overline{r^{(3)}}\}A\{r^{(1)}\} = 0 \qquad \{\overline{r^{(3)}}\}A\{r^{(2)}\} = 0$$

or

$$\sum_{i=1}^{n}\sum_{j=1}^{n} r_i^{(3)} a_{ij} r_j^{(1)} = 0 \qquad \sum_{i=1}^{n}\sum_{j=1}^{n} r_i^{(3)} a_{ij} r_j^{(2)} = 0 \qquad (5.53)$$

One may thus select $n - 2$ components of $\{r^{(3)}\}$ as equal to the corresponding components of an arbitrary vector $\{r\}$ and adjust the remaining two components to satisfy Eq. (5.53). The matrix S_2 expressing this operation, or

$$\{r^{(3)}\} = S_2\{r\}$$

is called the "second sweeping matrix." For diagonal A one possible form of S_2 is

$$S_2 = \begin{bmatrix} 0 & -\dfrac{a_{22}r_2^{(1)}}{a_{11}r_1^{(1)}} & -\dfrac{a_{33}r_3^{(1)}}{a_{11}r_1^{(1)}} & -\dfrac{a_{nn}r_n^{(1)}}{a_{11}r_1^{(1)}} \\ 0 & -\dfrac{a_{22}r_2^{(2)}}{a_{11}r_1^{(2)}} & -\dfrac{a_{33}r_3^{(2)}}{a_{11}r_1^{(2)}} & -\dfrac{a_{nn}r_n^{(2)}}{a_{11}r_1^{(2)}} \\ 0 & 0 & 1 & 0 \\ 0 & \cdot & \cdot & \cdot \\ \cdot & \cdot & \cdot & \cdot \\ \cdot & \cdot & \cdot & \cdot \\ 0 & 0 & \cdots & 1 \end{bmatrix}$$

Then one may form $H_2 = GS_2$ and solve

$$H_2\{r\} = (1/\omega^2)\{r\}$$

by iteration. The process here converges to $\{r^{(3)}\}$ and $\alpha = 1/\omega_3^2$.

Higher modes may be treated similarly; each mode must be orthogonal to all the lower ones, so that $p - 1$ relations like Eq. (5.51) must be utilized to find the $(p - 1)$st sweeping matrix. Iteration on $H_{(p-1)} = GS_{(p-1)}$ then converges to the pth mode.

Iteration for the Higher Modes. The previously outlined process begins with the lowest natural frequency and works toward the highest. It is not very useful for the highest few modes because of the tedium and of the accumulation of rounding off errors. Results for the higher modes can be obtained more simply and accurately by starting with the highest frequency and working toward lower ones.

The highest mode may be obtained by solving Eq. (5.46) directly by iteration. This is accomplished by assuming any trial vector $\{r_{(1)}\}$, forming $E\{r_{(1)}\} = \beta_{(1)}\{r_{(2)}\}$ with $\beta_{(1)}$ chosen so that one element of the result $\{r_{(2)}\}$ is equal to the corresponding element of $\{r_{(1)}\}$. Then one may form $E\{r_{(2)}\} = \beta_{(2)}\{r_{(3)}\}$ similarly, and continue until $\{r_{(p+1)}\} \approx \{r_{(p)}\}$ to within the required accuracy. Then $\{r_{(p)}\} = \{r^{(n)}\}$ and $\beta_{(p)} = \omega_n^2$.

The next-to-highest $[(n - 1)$st$]$ mode may be found by writing

$$\{r^{(n-1)}\} = T_1\{r\}$$

where T_1 is a sweeping matrix that "sweeps out" the nth mode. T_1 is equal to the identity matrix, except for one row, which expresses the orthogonality relation

$$\{\overline{r^{(n-1)}}\}A\{r^{(n)}\}=0 \qquad \text{or} \qquad \sum_{i=1}^{n}\sum_{j=1}^{n} r_i^{(n-1)} a_{ij} r_j^{(n)} = 0$$

Iterative solution of

$$J_1\{r\}=\omega^2\{r\} \qquad \text{where} \qquad J_1 = ET_1$$

then converges to $\{r^{(n-1)}\}$ and $\omega_{(n-1)}^2$.

The next lower modes may be obtained similarly, using other sweeping matrixes embodying additional orthogonality relations in complete analogy to the iteration for lower modes described above.

5.2.4 Approximate Natural Frequencies of Conservative Systems

A conservative system is one which executes free oscillations without dissipating energy. The potential energy $\hat{U}$ that the system has at an instant when its velocity (and hence its kinetic energy) is zero must therefore be exactly equal to the kinetic energy $\hat{T}$ of the system when it occupies its equilibrium position (zero potential energy) during its oscillation.

For sinusoidal oscillations the (generalized) coordinates obey $q_j = \bar{q}_j \sin \omega t$ where $\bar{q}_j$ is a constant (i.e., the amplitude of q_j). For linear holonomic systems, in view of Eq. (5.39),

$$\hat{T} = \tfrac{1}{2}\omega^2 \sum_{i=1}^{n} \sum_{j=1}^{n} a_{ij} \bar{q}_i \bar{q}_j \qquad \hat{U} = \tfrac{1}{2} \sum_{i=1}^{n} \sum_{j=1}^{n} c_{ij} \bar{q}_i \bar{q}_j$$

Rayleigh's Quotient. Rayleigh's quotient RQ, defined as[64]

$$RQ = \frac{\displaystyle\sum_{i=1}^{n}\sum_{j=1}^{n} c_{ij} \bar{q}_i \bar{q}_j}{\displaystyle\sum_{i=1}^{n}\sum_{j=1}^{n} a_{ij} \bar{q}_i \bar{q}_j} \tag{5.54}$$

is a function of $\{\bar{q}_1, \bar{q}_2, ..., \bar{q}_n\}$. However, multiplication of each $\bar{q}_i$ by the same number does not change the value of RQ. Rayleigh's quotient has the following properties:

1. The value of RQ one obtains with any $\{\bar{q}_1, \bar{q}_2, ..., \bar{q}_n\}$ always equals or exceeds the square of the lowest natural frequency of the system; $RQ \geq \omega_1^2$.

2. $RQ = \omega_n^2$ if $\{\bar{q}_1, \bar{q}_2, ..., \bar{q}_n\}$ corresponds to the nth mode shape (eigenvector) of the system, but even fairly rough approximations to the eigenvector generally result in good approximations to ω_n^2.

For systems whose influence coefficients d_{ij} are known, one may substitute an arbitrary vector $\{\bar{q}\} = \{\bar{q}_1, \bar{q}_2, ..., \bar{q}_n\}$ into the right-hand side of

$$\{\bar{q}\}_1 = \alpha G\{\bar{q}\}$$

where $G = DA$ is the dynamic matrix of Eq. (5.50) and α is an arbitrary constant. [This relation follows directly from Eq. (5.49).] The resulting vector $\{\bar{q}\}_1$, when substituted into Eq. (5.54), results in a value of RQ which is nearer to ω_1^2 than the value obtained by direct substitution of the arbitrary vector $\{\bar{q}\}$. Often one obtains good results rapidly if one assumes $\{\bar{q}\}$ initially so as to correspond to the deflection of the system due to gravity (i.e., the "static" deflection).

Rayleigh-Ritz Procedure. An alternative method useful for obtaining improved approximations to ω_1^2 from Rayleigh's quotient is the so-called Rayleigh-Ritz procedure. It consists of computing RQ from Eq. (5.54) for a trial vector $\{\bar{q}\}$ made up of a linear combination of arbitrarily selected vectors

$$\{\bar{q}\} = \alpha_1 \{\bar{q}\}_1 + \alpha_2 \{\bar{q}\}_2 + \cdots$$

then minimizing RQ with respect to the coefficients α of the selected vectors. The resulting minimum value of RQ is approximately equal to ω_1^2. That is, RQ evaluated so that $\partial RQ/\partial\alpha_1 = \partial RQ/\partial\alpha_2 = \cdots = 0$ is approximately equal to ω_1^2:

$$RQ \approx \omega_1^2$$

Dunkerley's Equation.[61]   This equation states that

$$1/\omega_1^2 + 1/\omega_2^2 + \cdots + 1/\omega_n^2 = 1/\Omega_1^2 + 1/\Omega_2^2 + \cdots + 1/\Omega_n^2 \qquad (5.55)$$

where ω_i denotes the ith natural frequency of an n-degree-of-freedom system, and Ω_i denotes the natural frequency that the ith inertia element would have if all others were removed from the system. Usually $\omega_n^2 \gg \cdots \gg \omega_2^2 \gg \omega_1^2$, so that the left-hand side of Eq. (5.55) is approximately equal to $1/\omega_1^2$ and Eq. (5.55) may be used directly for estimation of the fundamental frequency ω_1. In many cases the Ω_i are obtainable almost by inspection, or by use of Table 5.8.

5.2.5 Chain Systems

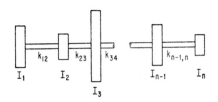

A chain system is one in which the inertia elements are arranged in series, so that each is directly connected only to the one preceding and the one following it. Shafts carrying a number of disks (or other rotational inertia elements) are the most common example and are discussed in more detail subsequently. Translational chain systems, as sketched in Fig. 5.17, may be treated completely analogously and hence will not be discussed separately.

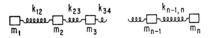

FIG. 5.17   Rotational and translational chain systems.

Sinusoidal Steady-State Forced Motion. If the torque acting on the sth disk is $T_s e^{i\omega t}$, where T_s is a known complex number, then the equations of motion of the system may be written as

$$(k_{12} - I_1\omega^2)\theta_1 - k_{12}\theta_2 = T_1$$

$$-k_{12}\theta_1 + (k_{12} + k_{23} - I_2\omega^2)\theta_2 - k_{23}\theta_3 = T_2$$

$$-k_{23}\theta_2 + (k_{23} + k_{34} - I_3\omega^2)\theta_3 - k_{34}\theta_4 = T_3 \qquad (5.56)$$

$$\cdots\cdots\cdots\cdots\cdots\cdots\cdots\cdots\cdots\cdots\cdots$$

$$-k_{n-1,n}\theta_{n-1} + (k_{n-1,n} - I_n\omega^2)\theta_n = T_n$$

where $\theta_s e^{i\omega t}$ describes the angular motion of the sth disk, as measured from equilibrium. (Damping in the system may be taken into account by assigning complex values to the k's, as in the last portion of Sec. 5.1.2.)

One may solve this set of equations simply by using each equation in turn to eliminate one of the θ's, so that one may finally solve for the last remaining θ, then obtain the others by substitution of the determined value into the given equations. This procedure becomes prohibitively tedious if more than a few disks are involved.

If all T's and k's are real (i.e., if the driving torques are in phase and if damping is neglected), one may assume a real value for θ_1, then calculate the corresponding value of θ_2 from the first of Eqs. (5.56). Then one may find θ_3 from the second equation, θ_4 from the third, and so on. Finally, one may compute T_n from the last (nth) equation and compare it with the given value. This process may be repeated with different initially assumed θ_1 values until the computed value of T_n comes out sufficiently close to the specified one. After a few computations one may often make good use of a plot of computed T_n vs. assumed θ_1 for determining by interpolation or extrapolation a good approximation to the correct value of θ_1.

If all T's are zero, except T_n, one may proceed as before. But, since θ_1 is proportional to T_n in this case, the correct value of θ_1 may be computed directly after a single complete calculation by use of the proportionality

$$\theta_{1,\text{correct}} = \theta_{1,\text{assumed}} \, (T_{n,\text{specified}}/T_{n,\text{calculated}})$$

The latter approach may be used also for damped systems (i.e., with complex k's).

Natural Frequencies: Holzer's Method.   Free oscillations of the system considered obey Eqs. (5.56), but with all $T_s = 0$. To obtain the natural frequencies one may proceed by assuming a value of ω and setting $\theta_1 = 1$, then calculating θ_2 from the first equation, thereafter θ_3 from the second, etc. Finally one may compute T_n from the last equation. If this T_n comes out zero, as required, the assumed frequency is a natural frequency. By repeating this calculation for a number of assumed values of ω one may arrive at a plot of T_n vs. ω, which will aid in the estimation of subsequent trial values of ω. (Natural frequencies are obtained where this curve crosses the ω axis.) One should keep in mind that a system composed of n disks has n natural frequencies. The mode shape (i.e., a set of values of θ's that satisfy the equation of motion) is also obtained in the course of the calculations.

Convenient tabular calculation methods (Holzer tables) may be set up on the basis of the equations of motion Eq. (5.56) rewritten in the following form:

$$I_1\theta_1\omega^2 = k_{12}(\theta_1 - \theta_2)$$

$$(I_1\theta_1 + I_2\theta_2)\omega^2 = k_{23}(\theta_2 - \theta_3)$$

$$\dots\dots\dots\dots\dots\dots\dots$$

$$(I_1\theta_1 + I_2\theta_2 + \cdots + I_s\theta_s)\omega^2 = k_{s,s+1}(\theta_s - \theta_{s+1}) \qquad (5.57)$$

$$\dots\dots\dots\dots\dots\dots\dots\dots\dots$$

$$\sum_{s=1}^{n} (I_s\theta_s)\omega^2 = 0$$

Tabular formats are given in a number of texts.[17,61,63,64] Methods for obtaining good first trial values for the lowest natural frequency are also discussed in Refs. 17 and 61. Branched systems, e.g., where several shafts are interconnected by gears, may also be treated by this method.[9,61,63] Damped systems (complex values of k) may also be treated with no added difficulty in principle.[61]

5.2.6 Mechanical Circuits[20,30,50]

Mechanical-circuit theory is developed in direct analogy to electric-circuit theory in order to permit the highly developed electrical-network-analysis methods to be applied to mechanical systems. A mechanical system is considered as made up of a number of mechanical-circuit elements (e.g., masses, springs, force generators) connected in series or parallel, in much the same way that an electrical network is considered to be made up of a number of interconnected electrical elements (e.g., resistances, capacitances, voltage sources). From known behavior of the elements one may then, by proper combination according to established rules, determine the system responses to given excitations.

Mechanical-circuit concepts are useful for determination of the equations of motion (which may then be solved by classical or transform techniques[20,30]) for analyzing and visualizing the effects of system interconnections, for dealing with electromechanical systems, and for the construction of electrical analogs by means of which one may evaluate the responses by measurement.

Mechanical Impedance and Mobility. As in electric-circuit theory, the sinusoidal steady state is assumed, and complex notation is used in basic mechanical-impedance analysis. That is, forces F and relative velocities V are expressed as

$$F = F_0 e^{i\omega t} \qquad V = V_0 e^{i\omega t}$$

where F_0 and V_0 are complex in the most general case.

Mechanical impedance Z and mobility Y are complex quantities defined by

$$Z = F_0/V_0 \qquad Y = 1/Z = V_0/F_0 \tag{5.58}$$

If the force and velocity refer to an element or system, the corresponding impedance is called the "impedance of the element" or system; if F and V refer to quantities at the same point of a mechanical network, then Z is called the "driving-point impedance" at that point. If F and V refer to different points, the corresponding Z is called the "transfer impedance" between those points.

Mechanical impedances (and mobilities) of elements can be combined exactly like electrical impedances (and admittances), and the driving-point impedances of composites can easily be obtained. From a knowledge of the elemental impedances and of how they combine, one may calculate (and often estimate quickly) the behavior of composite systems.

Basic Impedances, Combination Laws, Analogies. Table 5.2 shows the basic mechanical-circuit elements and their impedances and summarizes the impedances of some simple systems.

In Table 5.3 are indicated the combination laws for mechanical impedances and mobilities. Table 5.4 is a summary of analogies between translational and rotational mechanical systems and electrical networks. The impedances of some distributed mechanical systems are discussed in Sec. 5.3.2 and summarized in Tables 5.6 and 5.7.

Systems which are a combination of rotational and translational elements or electromechanical systems may be treated as either all-mechanical systems of a single type or as all-electrical systems by suitable substitution of analogous elements and variables.[20] In systems where both rotation and translation of a single mass occur, this single mass may have to be represented by two or more mass elements, and the concept of mutual mass (analogous to mutual inductance) may have to be introduced.[20]

Some Results from Electrical-Network Theory:[20,58] Resonances. Resonances occur at those frequencies for which the system impedances are minimum; antiresonances occur when the impedances are maximum.

Force and Velocity Sources. An ideal velocity generator supplies a prescribed relative velocity amplitude regardless of the force amplitude. A force generator supplies a prescribed force amplitude regardless of the velocity. When a velocity source is "turned off," $V = 0$, it acts like a rigid connection between its terminals. When a force generator is turned off, $F = 0$, it acts like no connection between the terminals.

TABLE 5.2 Mechanical Impedances of Simple Systems

SYSTEM	DIAGRAM	IMPEDANCE Z	FREQUENCY DEPENDENCE
MASS		$i\omega m$	$m_1 < m_2 < m_3$, $m = m_1$
VISCOUS DASHPOT		c	$c_1 < c_2 < c_3$, $c = c_1$
SPRING		$k/i\omega$	$k_1 < k_2 < k_3$, $k = k_1$
DRIVEN MASS ON SPRING		$\dfrac{k}{i\omega} + i\omega m$	$\omega_0 = \omega_n = \sqrt{k/m}$
DAMPED MASS		$c + i\omega m$	$\omega_0 = c/m$
SPRING AND DASHPOT IN PARALLEL		$c + k/i\omega$	$\omega_0 = k/c$
DAMPED SINGLE DEGREE OF FREEDOM SYSTEM		$i\omega m + c + \dfrac{k}{i\omega}$	$Z_0^2 = c_c^2 - 2mk = c_c^2(\zeta - 1/2)$, $\omega_0 = \sqrt{k/m} = \omega_n$
MASS ON DRIVEN SPRING		$\dfrac{1}{\dfrac{i\omega}{k} + \dfrac{1}{i\omega m}}$	$\omega_0 = \omega_n = \sqrt{k/m}$
MASS DRIVEN THROUGH DASHPOT		$\dfrac{1}{\dfrac{1}{c} + \dfrac{1}{i\omega m}}$	$\omega_0 = c/m$
SPRING AND DASHPOT IN SERIES		$\dfrac{1}{\dfrac{1}{c} + \dfrac{i\omega}{k}}$	$\omega_0 = k/c$
DAMPED MASS DRIVEN THROUGH SPRING		$\dfrac{1}{\dfrac{1}{c+i\omega m} + \dfrac{i\omega}{k}}$ $= \dfrac{c_c\left[\zeta + \dfrac{i}{2}\dfrac{\omega}{\omega_n}\right]}{1 - \left(\dfrac{\omega}{\omega_n}\right)^2 + 2i\zeta\dfrac{\omega}{\omega_n}}$	$|Z_0| = \dfrac{c_c}{2}\sqrt{1 + \dfrac{1}{4\zeta^2}}$, $\omega_0 = c/m$, $\omega_n = \sqrt{k/m}$

Reciprocity. The transfer impedance Z_{ij} (the force in the jth branch divided by the relative velocity of a generator in the ith branch) is equal to the transfer impedance Z_{ji} (force in the ith branch divided by relative velocity of generator in jth branch).

Foster's Reactance Theorem. For a general undamped system the driving-point impedance can be written $Z = if(\omega)$, where $f(\omega)$ is a real function of (real) ω. The function $f(\omega)$ always has $df/d\omega > 0$ and has a pole or zero at $\omega = 0$ and at $\omega = \infty$. All

TABLE 5.3 Combination of Impedances and Mobilities

Elements or subsystems:

$$Z_1 = 1/Y_1 = F_1/V_1 \qquad Z_2 = 1/Y_2 = F_2/V_2 \qquad \cdots \qquad Z_n = 1/Y_n = F_n/V_n$$
All F, V, Z, Y are complex quantities

Combination in	Parallel	Series
Diagram		
Connection identified by	All components having same V	Same F acting on all components
$Z = F/V =$	$Z_1 + Z_2 + \cdots + Z_n$	$(1/Z_1 + 1/Z_2 + \cdots + 1/Z_n)^{-1}$
$Y = V/F =$	$(1/Y_1 + 1/Y_2 + \cdots + 1/Y_n)^{-1}$	$Y_1 + Y_2 + \cdots + Y_n$
Z's combine like electrical resistances in	Series	Parallel
Y's combine like electrical resistances in	Parallel	Series

poles and zeros are simple (not repeated), poles and zeros alternate (i.e., there is always a zero between two poles), and $f(\omega)$ is determined within a multiplicative factor by its poles and zeros.

Thévenin's and Norton's Theorems. Consider any two terminals of a linear system. Then, as far as the effects of the system at these terminals are concerned, the system may be replaced by (see Fig. 5.18)

1. (Thévenin's equivalent) A series combination of an impedance Z_i and a velocity source V_{oc}
2. (Norton's equivalent) A parallel combination of an impedance Z_i and a force source F_b

Z_i is called the "internal impedance" of the system and is the driving-point impedance obtained at the terminals considered when all sources in the system are turned off (velocity generators replaced by rigid links, force generators replaced by disconnections). V_{oc} is the "open-circuit" velocity, i.e., the velocity occurring between the terminals considered (with all generators active). F_b is the "blocked force," i.e., the force transmitted through a rigid link inserted between the terminals considered, with all generators active.

TABLE 5.4 Mechanical-Electrical Analogies (Lumped Systems)

Mechanical System		Electrical Systems	
Translational	Rotational	Force–voltage (classical) analog	Force–current (mobility) analog
F Force	T Torque	V Voltage	I Current
u Velocity	Ω Angular velocity	I Current	V Voltage
x Displacement	θ Angular displacement	q Charge	ϕ Magnetic flux
m Mass	J Moment of inertia	L Inductance	C Capacitance
c Damping	c_r Rotational damping	R Resistance	$G = \dfrac{1}{R}$ Conductance
k Spring rate	k_r Torsional spring rate	$S = \dfrac{1}{C}$ Elastance	$\Gamma = \dfrac{1}{L}$ Inverse inductance
$Z = \dfrac{F}{u}$ Mechanical impedance	$Z_r = \dfrac{T}{\Omega}$ Rotational impedance	$Z = \dfrac{V}{I}$ Impedance	$Y = \dfrac{I}{V}$ Admittance
$Y = \dfrac{u}{F}$ Mobility	$Y_r = \dfrac{\Omega}{T}$ Rotational mobility	$Y = \dfrac{I}{V}$ Admittance	$Z = \dfrac{V}{I}$ Impedance

5.3 CONTINUOUS LINEAR SYSTEMS

5.3.1 Free Vibrations

The equations that govern the deflection $u(x, y, t)$ of many continuous linear systems (e.g., bars, shafts, strings, membranes, plates) in absence of external forces may be expressed as

$$\mathfrak{M}\ddot{u} + \mathfrak{K}u = 0 \tag{5.59}$$

where $\mathfrak{M}$ and $\mathfrak{K}$ are linear differential operators involving the coordinate variables only. Table 5.5 lists $\mathfrak{M}$ and $\mathfrak{K}$ for a number of common systems. Solutions of Eq. (5.59) may be expressed in terms of series composed of terms of the form $\phi(x,y)e^{i\omega t}$, where ϕ satisfies

$$(\mathfrak{K} - \mathfrak{M}\omega^2)\phi = 0 \tag{5.60}$$

in addition to the boundary conditions of a given problem. In solving the differential equa-

TABLE 5.4 Mechanical-Electrical Analogies (Lumped Systems) (*Continued*)

Mechanical Systems		Electrical Systems	
Translational	Rotational	Force-voltage (classical) analog	Force-current (mobility) analog
Kinetic energy = $\frac{1}{2}mu^2$	Kinetic energy = $\frac{1}{2}I\Omega^2$	Magnetic energy = $\frac{1}{2}LI^2$	Electrical energy = $\frac{1}{2}CV^2$
Potential energy = $\frac{1}{2}kx^2$	Potential energy = $\frac{1}{2}k_r\theta^2$	Electrical energy = $\frac{1}{2}\frac{1}{C}q^2$	Magnetic energy = $\frac{1}{2}\frac{1}{L}\phi^2$
Power loss = u^2c	Power loss = $\Omega^2 c_r$	Power loss = I^2R	Power loss = V^2G
$m\ddot{u}+cu+k\int u\,dt = F(t)$	$J\dot{\Omega}+c_r\Omega+k_r\int\Omega\,dt = T(t)$	$L\dot{I}+RI+\frac{1}{C}\int I\,dt = V(t)$	$C\dot{V}+RV+\frac{1}{L}\int V\,dt = I(t)$
Elements connected to same node have same velocity u	Elements connected to same node have same angular velocity Ω	Elements in same loop have same current I	Elements connected to same node have same voltage v
Element connected to ground has one end at zero velocity	Element connected to ground has one end at zero velocity	Element in one loop has only one loop current through it	Element connected to ground has one end at zero (reference) voltage
Elements between two nodes	Elements between two nodes	Elements in two loops	Elements between two nodes
Elements in parallel	Elements in parallel	Elements in series	Elements in parallel

tion (5.60) by use of standard methods and introducing the boundary conditions applicable in a given case one finds that solutions ϕ that are not identically zero exist only for certain frequencies. These frequencies are called the "natural frequencies of the system"; the equation that the natural frequencies must satisfy for a given system is called the "frequency equation of the system"; the functions ϕ that satisfy Eq. (5.60) in conjunction with the natural frequencies are called the "eigenfunctions" or "mode shapes" of the system. One-dimensional systems (strings, bars) have an infinite number of natural frequencies ω_n and eigenfunctions ω_n; $n = 1, 2, \ldots$. Two-dimensional systems (membranes, plates) have a doubly infinite set of natural frequencies ω_{mn}, and eigenfunctions ϕ_{mn}; $m, n = 1, 2, \ldots$. Table 5.6 lists eigenfunctions and frequency equations for some common systems.

The eigenfunctions of flexural systems where all edges are either free, built-in, or pinned are "orthogonal," that is,

$$\int_0^L \mu(x)\phi_n(x)\phi_{n'}(x)\,dx = \begin{cases} \overline{\mu L\Phi}_n & \text{for } n = n' \\ 0 & \text{for } n \neq n' \end{cases} \tag{5.61}$$

TABLE 5.4   Mechanical-Electrical Analogies (Lumped Systems) (*Continued*)

Mechanical Systems		Electrical Systems	
Translational	Rotational	Force-voltage (classical) analog	Force-current (mobility) analog
Newton's law: sum of forces on node = 0	Newton's law: sum of torques on node = 0	Kirchhoff's voltage law: sum of voltages around loop = 0	Kirchhoff's current law: sum of currents into node = 0
Ideal lever	Ideal gear train	Ideal voltage transformer	Ideal current transformer
$\dfrac{u_1}{u_2} = \dfrac{F_2}{F_1} = \dfrac{d_1}{d_2}$	$\dfrac{\Omega_1}{\Omega_2} = \dfrac{T_2}{T_1} = \dfrac{n_2}{n_1}$	$\dfrac{I_1}{I_2} = \dfrac{V_2}{V_1} = \dfrac{N_2}{N_1}$	$\dfrac{V_1}{V_2} = \dfrac{I_2}{I_1} = \dfrac{N_1}{N_2}$

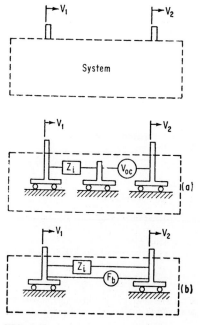

FIG. 5.18   (*a*) Thévenin's and (*b*) Norton's equivalent networks.

TABLE 5.5 Operators [See Eq. (5.59)] for Some Elastic Systems[64]

System	Deflection	$\mathfrak{M} = \mu$	$\mathfrak{K}$
String..............	Lateral	$\rho A(x)$	$T\dfrac{\partial^2}{\partial x^2}$
Bar................	Longitudinal	$\rho A(x)$	$-\dfrac{\partial}{\partial x}\left(EA\dfrac{\partial}{\partial x}\right)$
Bar................	Torsional (angular)	$\rho J(x)$	$-\dfrac{\partial}{\partial x}\left(KG\dfrac{\partial}{\partial x}\right)$
Bar................	Lateral (flexure)	$\rho A(x)$	$\dfrac{\partial^2}{\partial x^2}\left(EI\dfrac{\partial^2}{\partial x^2}\right)$
Membrane.........	Lateral	$\rho h(x,y)$	$-S\nabla^2$
Plate..............	Lateral	$\rho h(x,y)$	$\nabla^2(D\nabla^2)$

Symbols:
ρ = density
A = cross-sectional area
I = centroidal moment of inertia of A
J = polar moment of inertia of A
h = plate thickness
T = tensile force

S = tension/unit length
E = Young's modulus
G = shear modulus
K = torsional constant of A
    ($= J$ for circular sections)
$D = Eh^3/12(1 - \nu^2)$
ν = Poisson's ratio

TABLE 5.6a Modal Properties for Some One-Dimensional Systems

System	Wave velocity c	Boundary conditions $x = 0$	Boundary conditions $x = L$	Mode shape† $\phi_n(x)$	Natural frequency $\omega_n = 2\pi f_n$	$\dfrac{\lambda}{L} = \dfrac{2\pi}{Lk_n}$
String................	$\sqrt{T/\rho A}$	Fixed	Fixed	$\sin\dfrac{n\pi x}{L}$	$\dfrac{n\pi}{L}c$	$\dfrac{2}{n}$
Uniform shaft* in torsion.	$\sqrt{\dfrac{GK}{\rho J}}$	Clamped	Free	$\sin\dfrac{(2n-1)\pi x}{2L}$	$\dfrac{(2n-1)\pi}{2L}c$	$\dfrac{4}{2n-1}$
		Free	Free	$\cos\dfrac{n\pi x}{L}$	$\dfrac{n\pi}{L}c$	$\dfrac{2}{n}$
Uniform bar* in longitudinal vibration	$\sqrt{E/\rho}$	Clamped	Clamped	$\sin\dfrac{n\pi x}{L}$	$\dfrac{n\pi}{L}c$	$\dfrac{2}{n}$

Symbols:
A = cross-sectional area
ρ = material density
T = tensile force
E = Young's modulus

G = shear modulus
J = polar moment of inertia of A
K = torsional constant of A
    ($= J$ for circular sections)
GK = torsional rigidity

λ_n = wavelength
k_n = wave number
n = mode number

*The same modal properties, but with different values of c, apply for uniform shafts vibrating torsionally and uniform bars vibrating longitudinally.

†$\Phi_n = (1/L)\displaystyle\int_0^L \phi_n^2\, dx = \frac{1}{2}$.

TABLE 5.6b Modal Properties for Flexural Vibrations of Uniform Beams [6,7]

$$\text{Natural frequency} \quad \omega_n = 2\pi f_n = k_n^2 c_{LT} = k_n^2 \sqrt{EI/\rho A}$$
$$\text{Wavelength} \quad \lambda_n = 2\pi/k_n$$
$$\text{Wave velocity} \quad c_n = \omega_n/k_n = k_n c_{LT} = k_n \sqrt{EI/\rho A} = \sqrt{\omega_n c_{LT}}$$
$$\text{For mode shapes } \phi_n(x) \text{ listed below, } \Phi_n = (1/L)\int_0^L \phi_n^2 \, dx = 1$$

Boundary conditions		Mode shape $\phi_n(x)$	σ_n	Frequency equation	Roots of frequency equation					
$x=0$	$x=L$				k_1L	k_2L	k_3L	k_4L	k_5L	$k_nL,\, n>5$
Pinned	Pinned	$\sqrt{2}\,\sin(k_n x)$		$\sin(k_n L) = 0$	π	2π	3π	4π	5π	$n\pi$
Clamped	Clamped	$\cosh(k_n x) - \cos(k_n x)$ $-\sigma_n[\sinh(k_n x) - \sin(k_n x)]$	$\dfrac{\cosh(k_n L) - \cos(k_n L)}{\sinh(k_n L) - \sin(k_n L)}$	$\cos(k_n L)\cosh(k_n L) = 1$	4.73	7.85	11.00	14.14	17.29	$\approx \dfrac{2n+1}{2}\pi$
Free	Free	$\cosh(k_n x) + \cos(k_n x)$ $-\sigma_n[\sinh(k_n x) + \sin(k_n x)]$								
Clamped	Free	$\cosh(k_n x) - \cos(k_n x)$ $-\sigma_n[\sinh(k_n x) - \sin(k_n x)]$	$\dfrac{\sinh(k_n L) - \sin(k_n L)}{\cosh(k_n L) + \cos(k_n L)}$	$\cos(k_n L)\cosh(k_n L)$ $= -1$	1.875	4.69	7.85	11.00	14.14	$\approx \dfrac{2n-1}{2}\pi$
Clamped	Pinned	$\cosh(k_n x) - \cos(k_n x)$ $-\sigma_n[\sinh(k_n x) - \sin(k_n x)]$	$\cot(k_n L)$	$\tan(k_n L) = \tanh(k_n L)$	3.93	7.07	10.21	13.35	16.49	$\approx \dfrac{4n+1}{4}\pi$
Free	Pinned	$\cosh(k_n x) + \cos(k_n x)$ $-\sigma_n[\sinh(k_n x) - \sin(k_n x)]$								

Symbols:

k_n = wave number
E = Young's modulus
ρ = material density

$c_L = \sqrt{E/\rho}$ = longitudinal wave velocity
A = cross-section area
I = moment of inertia of A

$r = \sqrt{I/A}$ = radius of gyration
L = beam length

5.41

TABLE 5.6c   Modal Properties for Some Plates[40]

Frequency	$\omega_{mn} = 2\pi f_{mn} = k_{mn}^2\,\sqrt{D/\rho h} \approx k_{mn}^2 c_L r$	$D = Eh^3/12(1 - \nu^2)$
Wavelength	$\lambda_{mn} = 2\pi/k_{mn}$	$r = h/\sqrt{12}$
Wave velocity	$c_{mn} = \omega_{mn}/k_{mn} = k_{mn}\,\sqrt{D/\rho h} \approx k_{mn}c_L r$	$c_L = \sqrt{E/\rho}$

Rectangular, on simple supports; $0 < x < a,\ 0 < y < b$:

$$\phi_{mn}(x,y) = \sin\frac{m\pi x}{a}\sin\frac{n\pi y}{b} \qquad \phi_{mn} = \tfrac{1}{4}$$

$$k_{mn} = \pi\,\sqrt{m^2/a^2 + n^2/b^2}$$

Circular,* clamped at circumference; $0 < r < a,\ 0 < \theta < 2\pi$:

$$\phi_{mn_e}(r,\theta) = \cos\,(m\theta)F_{mn}(r)$$
$$\phi_{mn_o}(r,\theta) = \sin\,(m\theta)F_{mn}(r)$$

$$F_{mn}(r) = J_m(k_{mn}r) - \frac{J_m(k_{mn}a)}{I_m(k_{mn}a)}\,I_m(k_{mn}r)$$

Same orthogonality holds as for membranes (see Table 5.6d)
Frequency equation:

$$I_m(k_{mn}a)\left[\frac{d}{dr}J_m(k_{mn}r)\right]_{r=a} = J_m(k_{mn}a)\left[\frac{d}{dr}I_m(k_{mn}r)\right]_{r=a}$$

Table of $k_{mn}a/\pi$ for Clamped Circular Plate

				n		
		1	**2**	**3**	**>3**
	0	1.015	2.007	3.000	
m	1	1.468	2.483	3.490	$\approx n + \dfrac{m}{2}$
	2	1.879	2.992	4.000	

*For circular plates one finds two sets of ϕ_{mn}, one even (subscript e), one odd (subscript o).

$$\iint_{A_S} \mu(x,y)\phi_{mn}(x,y)\phi_{m'n'}(x,y)\,dA_s = \begin{cases} \bar{\mu}A_s\Phi_{mn} & \text{for } m = m',\, n = n' \\ 0 & \text{otherwise} \end{cases} \tag{5.61}$$

where L is the total length of a (one-dimensional) system, A_s is the total surface area of a (two-dimensional) system, μ is a weighting function, and $\bar{\mu}$ is the mean of value of μ for the system. (The weighting functions of the common systems tabulated in Table 5.5 are equal to the mass operator $\mathfrak{M}$.)

$$\bar{\mu}L = \int_0^L \mu(x)\,dx \qquad \text{for one-dimensional systems}$$

$$\bar{\mu}A_s = \iint_{A_s} \mu(x,y)\,dA_s \qquad \text{for two-dimensional systems}$$

For uniform systems $\mu = \bar{\mu}$ is a constant and may be canceled from Eqs. (5.61).*

*Note that for the systems of Table 5.6 $\mathfrak{M} = \mu$ is a multiplicative factor, not an operator in the more general sense. However, for other systems $\mathfrak{M}$ may be a more general operator (e.g., for lateral vibrations of beams when rotatory inertia is not neglected) and may differ from μ. The distinction between $\mathfrak{M}$ and μ is maintained throughout the subsequent discussion to permit application of the results to systems that are more complicated than those of Table 5.6.

TABLE 5.6d Modal Properties for Some Membranes[40]

Wave velocity $c = \sqrt{S/\rho h}$

Rectangular; $0 < x < a$, $0 < y < b$:

$$\phi_{mn}(x,y) = \sin\frac{m\pi x}{a} \sin\frac{n\pi y}{b} \qquad \phi_{mn} = \tfrac{1}{4}$$

$$\omega_{mn} = 2\pi f_{mn} = \pi c \sqrt{(m/a)^2 + (n/b)^2} \qquad k_{mn} = 2\pi/\lambda_{mn} = \omega_{mn}/c$$

Circular,* $0 < r < a$, $0 < \theta < 2\pi$:

$$\phi_{mn_e}(r,\theta) = \cos\,(m\theta) J_n\,(\omega_{mn}r/c)$$
$$\phi_{mn_o}(r,\theta) = \sin\,(m\theta) J_n\,(\omega_{mn}r/c)$$

$$\iint_{A_s} \phi_{mn_j}\phi_{m'n'_{j'}}\, dA_s = \begin{cases} \pi a^2 \Phi_{mn} & \text{for } m' = m,\ n' = n,\ j' = j \\ 0 & \text{otherwise} \end{cases}$$

$$\Phi_{mn} = \begin{cases} [J_1(\omega_{mn}a/c)]^2 & \text{for } m = 0 \\ \tfrac{1}{2}[J_{m-1}(\omega_{mn}a/c)]^2 & \text{for } m > 0 \end{cases}$$

Frequency equation:

$$J_m(\omega a/c) = 0$$

Table of $\omega_{mn}a/\pi c$ for Circular Membrane

		n			
		1	2	3	>3
m	0	0.7655	1.7571	2.7546	$\approx n + \dfrac{m}{2} - \dfrac{1}{4}$
	1	1.2197	2.2330	3.2383	
	2	1.6347	2.6793	3.6987	

Table of Φ_{mn} for Circular Membrane

		n			
		1	2	3	4
m	0	0.2695	0.1241	0.07371	0.05404
	1	0.08112	0.04503	0.02534	0.02385
	2	0.05770	0.03682	0.02701	0.02134

Symbols for Table 5.6c and d:

S = tension force/unit length
μ = mass/unit area
D = flexural rigidity = $Eb^3/12(1 - \nu^2)$
k_{mn} = wave number
E = Young's modulus
ν = Poisson's ratio

c_L = longitudinal wave velocity = $\sqrt{E/\rho}$
r = radius of gyration of section
b = plate thickness
ρ = material density
J_m = mth-order Bessel function
I_m = mth-order hyperbolic Bessel function

*For circular membranes one finds two set of ϕ_{mn}, one even (subscript e), one odd (subscript o).

The free motions (or "transient" responses) of two-dimensional* systems with initial displacement $u(x, y, 0)$ from equilibrium and initial velocity $\dot{u}(x, y, 0)$ are given by

$$u_t(x,y,t) = \sum_{m=1}^{\infty}\sum_{n=1}^{\infty} \phi_{mn}(B_{mn} \cos \omega_{mn}t + C_{mn} \sin \omega_{mn}t)$$

$$B_{mn} = \frac{1}{\mu A_s \Phi_{mn}} \iint_{A_s} u(x,y,0)\mu(x,y)\phi_{mn}(x,y)\, dA_s \qquad (5.62)$$

$$C_{mn} = \frac{1}{\mu A_s \Phi_{mn}\omega_{mn}} \iint_{A_s} \dot{u}(x,y,0)\mu(x,y)\phi_{mn}(x,y)\, dA_s$$

5.3.2 Forced Vibrations

Forced vibrations of distributed systems are governed by

$$\mathfrak{M}\ddot{u} + \mathfrak{R}u = F(x,y,t)$$

where the right-hand side represents the applied load distribution in space and time.

General Response: Modal Displacements, Forcing Functions, Masses. The displacement $u(x, y, t)$ of a system[†] may, in general, be expressed in a modal series as

$$u(x,y,t) = \sum_{m=1}^{\infty}\sum_{n=1}^{\infty} U_{mn}(t)\phi_{mn}(x,y) \qquad (5.63)$$

where the modal displacement $U_{mn}(t)$ is given by

$$U_{mn}(t) = \frac{1}{\omega_{mn}M_{mn}} \int_0^t G_{mn}(\tau) \sin \omega_{mn}(t - \tau)\, d\tau \qquad (5.64)$$

The modal forcing function $G_{mn}(t)$ and the modal mass M_{mn} are given by

$$G_{mn}(t) = \iint_{A_s} F(x,y,t)\phi_{mn}(x,y)\, dA_s \qquad M_{mn} = \iint_{A_s} \phi_{mn}(x,y)\mathfrak{M}\phi_{mn}(x,y)\, dA_s \quad (5.65)$$

For the special case of a uniform system with $\mathfrak{M} = \mu = \overline{\mu} = $ constant,

$$M_{mn} = \mu A_s \Phi_{mn}$$

where Φ_{mn} is defined as in Eq. (5.61).

The "steady-state" response given by (5.63) and (5.64) must be combined with the "transient" response given by (5.62) if one desires the complete solution.

Sinusoidal Response: Input Impedance. For a sinusoidal forcing function $F(x, y, t) = e^{i\omega t}F_0(x, y)$ one finds a steady-state response given by

$$u_{ss}(x,y,t) = e^{i\omega t}\sum_{m=1}^{\infty}\sum_{n=1}^{\infty} U_{mn}\phi_{mn}(x,y)$$

---

*These equations apply also for one-dimensional systems if all m subscripts and y dependences are deleted, if A_s is replaced by L, and if the double integrations over A_s are replaced by a single integration for Q to L.
[†]These equations apply also for one-dimensional systems if all m subscripts and y dependences are deleted, if A_s is replaced by L, and if the double integrations over A_s are replaced by a single integration from 0 to L.

with $\qquad U_{mn} = \dfrac{G_{mn}}{(\omega_{mn}^2 - \omega^2)M_{mn}} \qquad G_{mn} = \iint_{A_s} F_0(x,y)\phi_{mn}(x,y)\,dA_s$

and M_{mn} given by (5.65).

For a point force $F_1 e^{i\omega t}$ applied at $x = x_0$, $y = y_0$, one finds $G_{mn} = F_1 \phi_{mn}(x_0, y_0)$. The input impedance $Z(x_0, y_0, \omega)$ of the system at $x = x_0$, $y = y_0$ for frequency ω then may be found from

$$\frac{1}{i\omega Z(x_0,y_0,\omega)} = \sum_{m=1}^{\infty} \sum_{n=1}^{\infty} \frac{\phi_{mn}^2(x_0,y_0)}{(\omega_{mn}^2 - \omega^2)M_{mn}} \tag{5.66}$$

Point-Impulse Response: Alternate Formulation of General Response. The response of a system to a point force applied at $x = x_0$, $y = y_0$ and varying like a Dirac impulse function (of unit magnitude) with time is given by

$$u_\delta(x,y; x_0,y_0; t) = \sum_{m=1}^{\infty} \sum_{n=1}^{\infty} \frac{\phi_{mn}(x_0,y_0)\phi_{mn}(x,y)}{\omega_{mn} M_{mn}} \sin \omega_{mn} t$$

An alternate expression for the steady-state response of a system to a general distributed force $F(x, y, t)$ may then be written as

$$u_{ss}(x,y,t) = \iint_{A_s} \left[\int_0^t F(x_0,y_0, \tau) u_\delta(x,y; x_0,y_0; t-\tau)\, d\tau \right] dx_0\, dy_0$$

This expression is entirely analogous to the result one obtains by combining Eqs. (5.63) and (5.64).

5.3.3 Approximation Methods

Finite-Difference Equations, Finite-Element Approximations. One of the most widely applicable numerical methods, particularly if digital-computation equipment is available, consists of replacing the applicable differential equations by finite-difference equations[12,64] which may then be solved numerically.

A second method consists of replacing the continuous (infinite-degrees-of-freedom) system by one made up of a finite number of suitably interconnected elements (masses, springs, dashpots), then applying methods developed for systems with a finite number of degrees of freedom, as outlined in Sec. 5.2. These "finite-element" approximations may be obtained, for example, by dividing a beam or plate to be analyzed into arbitrary segments and assuming the mass of each segment concentrated at its center of gravity or "lumping point." If one then establishes the influence coefficients between the various lumping points, one has enough information to apply directly the methods outlined in Sec. 5.2.[24] In concept, one replaces the structure between lumping points by equivalent springs to obtain a new system analogous to the continuous one; a beam is replaced by a linear array, a plate by a two-dimensional network of masses interconnected by springs. One may then proceed by determining the equations of motion of the new systems and by solving these as discussed in Sec. 5.2. A variety of corresponding computer codes has become available; e.g., see Refs. 44 and 45.

Fundamental Frequencies:* Rayleigh's Quotient. RQ for an arbitrary deflection function $u(x, y)$ of a two-dimensional continuous system is given by[64]

---

*These equations apply also for one-dimensional systems if all m subscripts and y dependences are deleted, if A_s is replaced by L, and if the double integrations over A_s are replaced by a single integration from 0 to L.

$$RQ = \frac{\iint_{A_s} u\mathfrak{N}(u)\, dA_s}{\iint_{A_s} u\mathfrak{M}(u)\, dA_s}$$

For $u = \phi_{mn}$ (the mn mode shape) RQ takes on the value ω_{mn}^2. For any function u that satisfies the boundary conditions of the given system

$$RQ \geq \omega_{11}^2$$

Thus RQ produces an estimate of the fundamental frequency ω_{11} which is always too high.

To obtain a better estimate one may use the Rayleigh-Ritz procedure. In this procedure one forms RQ for a linear combination of any convenient number of functions $u_i(x, y)$ that satisfy the boundary conditions of the problem; that is, one forms RQ from

$$u = \alpha_1 u_1 + \alpha_2 u_2 + \cdots$$

where the α are constants. A good approximation to ω_{11}^2 is then obtained, in general, by minimizing RQ with respect to the various α's.

$$RQ \ (\text{evaluated so that } \partial RQ/\partial\alpha_1 = \partial RQ/\partial\alpha_2 = \cdots = 0) \approx \omega_{11}^2$$

Special Methods for Lateral Vibrations of Nonuniform Beams. In addition to the foregoing methods a number of others are available that have been developed specially for dealing with the vibrations of beams and shafts. In all these methods the beam mass is replaced by a number of masses concentrated at lumping points.

The Stodola method[53,59,61,63] is related to both the Rayleigh and matrix-iteration methods. In it one may proceed as follows:

1. Assume a deflection curve that satisfies the boundary conditions. (Usually the static-deflection curve gives good results.)

2. Determine a first approximation to ω_1 using Rayleigh's quotient, from

$$\omega_1^2 \approx RQ = \frac{g\Sigma m_i u_i}{\Sigma m_i u_i^2}$$

where g denotes the acceleration of gravity and u_i the assumed deflection of the mass m_i.

3. Calculate the deflection of the beam as if inertia forces $(-m_i u_i \omega_1^2)$ were applied statically. (This is usually done best by graphical or numerical means.) Use these new deflections instead of the original u_i in the foregoing equation. Repeat this process until no further changes in RQ result to the degree of accuracy desired. Then $RQ = \omega_1^2$ to within the desired accuracy.

The Myklestad method[41] is essentially the same as Holzer's method, but considerably simpler to use for flexural vibrations. Extensions of Myklestad's original method also apply to coupled bending-torsion vibrations and to vibrations in centrifugal fields.[41] Some simplifications of Myklestad's method have been developed by Thomson.[61] Because of the details necessary for a sufficient discussion of these procedures the reader is referred to the original sources.

5.3.4 Systems of Infinite Extent

Truly infinite or semi-infinite systems do not occur in reality. However, as far as the local response to a local excitation is concerned, finite systems behave like infinite ones if the ends are far (many wavelengths) removed from the excitation and if there is enough dissipation in the system or at the ends so that little effect of reflected waves is felt near the driving point.

The velocities with which waves travel in infinite systems are listed under the heading of "wave velocity" in Table 5.6, but in infinite systems the frequencies and wave numbers are not restricted, as they are in finite ones. In all cases the wave velocity c is related to wavelength λ, wave number k, frequency f (cycles/time), and circular frequency ω (radians/time) as

$$c = f\lambda = \omega\lambda/2\pi = \omega/k$$

Input impedances of infinite structures are useful for estimation of the responses of mechanical systems that are composed of or connected to one or more structures, if the responses of the latter may be approximated by those of corresponding infinite structures in the light of the first paragraph of this article. These impedances may be used precisely like previously discussed impedances of systems with only a few degrees of freedom.

If a force $F_0 e^{i\omega t}$ gives rise to a velocity $V_0 e^{i\omega t}$ at its point of application (where F_0 and V_0 may be complex in general), then the driving-point impedance is defined as $Z = F_0/V_0$. Similarly, the driving-point moment impedance is defined by $Z_M = M_0/\Omega_0$, where M_0 denotes the amplitude of a driving moment $M_0 e^{i\omega t}$ and where $\Omega_0 e^{i\omega t}$ is the angular velocity at the driving point. Table 5.7 lists the driving-point impedances of some infinite and semi-infinite systems.

5.4 MECHANICAL SHOCKS

By a mechanical shock one generally means a relatively suddenly applied transient force or support acceleration. The responses of mechanical systems to shocks may be computed by direct application of the previously discussed methods for determination of transient responses, provided that the forcing functions are known. If the forcing functions are not known precisely, one may approximate them by some idealized functions or else describe them in some rough way, for example, in terms of the subsequently discussed shock spectra.

5.4.1 Idealized Forcing Functions

Among the most widely studied idealized shocks are those associated with sudden support displacements or velocity changes, or with suddenly applied forces. The responses of systems with one or two degrees of freedom to idealized shocks have been studied in considerable detail, since for such systems solutions may be obtained relatively simply by analytical or analog means.

Sudden Support Displacement. Sudden (vertical) support displacements occur, for example, when an automobile hits a sudden change in level of the roadbed. The responses of simple systems to such shocks have been studied in considerable detail.

TABLE 5.7 Driving-Point Impedances of Some Infinite Uniform Systems*

	Beams		
	Extent		
Loading	**Infinite**	**Semi-infinite**	
Axial force...........	$Z = 2A\rho c_L$	$Z = A\rho c_L$	
Lateral force.........	$Z = 2mc_B(1 + i)$	$Z = \dfrac{mc_B}{2}(1 + i)$	
Moment.............	$Z_M = \dfrac{2mc_B}{k^2}(1 + i)$	$Z_M = \dfrac{mc_B}{2k^2}(1 + i)$	
Torsion.............	$Z_M = 2J\rho c_T$	$Z_M = J\rho c_T$	

Symbols:

A = cross-section area
ρ = material density
E = Young's modulus
J = polar moment of inertia of A
K = torsional constant
G = shear modulus
$r = \sqrt{I/A}$

$c_L = \sqrt{E/\rho}$ = longitudinal wave velocity
$c_T = \sqrt{GK/\rho J}$ = torsional wave velocity
$c_B = \sqrt{\omega r c_L}$ = flexural wave velocity
$k = \sqrt{\omega/r c_L}$ = flexural wave number
$\dfrac{c_B}{k^2} = \sqrt{\dfrac{(r c_L)^3}{\omega}}$

Plates

Infinite isotropic: $\quad Z = 8\sqrt{D\rho h} \approx 2.3h^2 \sqrt{\dfrac{E\rho}{1 - \nu^2}}$

Infinite orthotropic: $\quad Z \approx 8\sqrt[4]{D_x D_y}\sqrt{\rho h}$

Isotropic, infinite in x direction, on simple supports at $y = 0, L$, forced at $(0, y_0)$:

$$\dfrac{1}{Z} = \dfrac{i}{2\sqrt{D\rho h}}\sum_{n=1}^{\infty}[(n^2\pi^2 - K)^{-1/2} - (n^2\pi^2 + K)^{-1/2}]\sin^2\left(\dfrac{n\pi y_0}{L}\right)$$

$$K = (kL)^2 = L^2\omega\sqrt{\dfrac{\rho h}{D}}$$

Symbols:

ρ = density of plate material
h = plate thickness
E = Young's modulus of plate material
E_x, E_y = Young's modulus in principal directions
ν = Poisson's ratio of plate material
ν_x, ν_y = Poisson's ratio in principal directions

$k = \sqrt[4]{\omega^2\rho h/D}$
$D = Eh^3/12(1 - \nu^2)$
$D_x = E_x h^3/12(1 - \nu_x^2)$
$D_y = E_y h^3/12(1 - \nu_y^2)$

*Systems are assumed undamped. For finite systems see Eq. (5.66) and Table 5.6.

Families of curves describing system responses to a certain type of rounded step-function displacement are given in Refs. 19, 48, and 52. It is found among other things that the responses are generally of an oscillatory nature, except for peaks in acceleration that occur during the "rise" time of the displacement.

Sudden Support Velocity Change. Velocity shocks, i.e., those associated with instantaneous velocity changes, have been studied considerably. They approximate a number of physical situations where impact occurs; for example, when a piece of equipment is dropped the velocity of its outer parts changes from some finite value to zero at the instant these parts hit the ground. Reference 39 discusses in detail how one may compute the velocity shock responses of single-degree-of-freedom systems with linear and various nonlinear springs. It also presents charts for the computation of the important parameters of responses of simple systems attached to the aforementioned velocity-shock-excited systems. That is, if one system is mounted inside another, one may first compute the motion of the outer system in response to a velocity shock, then the response of the inner system to this motion (assuming the motion of the inner system has little effect on the shock response of the outer). This permits one to estimate, for example, what happens to an electronic component (inner system) when a chassis containing it (outer system) is dropped. Damping effects are generally neglected in Ref. 39, but detailed curves for lightly damped linear two-degree-of-freedom systems appear in Refs. 43 and 46. Some discussion appears also in Ref. 14.

Suddenly Applied Forces. If a force is suddenly applied to a linear system with a single degree of freedom, then this force causes at most twice the displacement and twice the stress that this same force would cause if it were applied statically. This rule of thumb can lead to considerable error for systems with more than one degree of freedom.[46] One generally does well to carry out the necessary calculations in detail for such systems.

5.4.2 Shock Spectra

The shock-spectrum concept is useful for describing shocks and the responses of simple systems exposed to them, particularly where the shocks cannot be described precisely and where they cannot be reasonably approximated by one of the idealized shocks of Sec. 5.4.1.

Physical Interpretation. Assume that a given (support acceleration) shock is applied to an undamped linear single-degree-of-freedom system whose natural frequency is f_1. The maximum displacement of the system (relative to its supports) in response to this shock then gives one point on a plot of displacement vs. frequency. Repetition with the same shock applied to systems with different natural frequencies gives more points which, when joined in a curve, make up the "displacement shock spectrum" corresponding to the given shock. If instead of the maximum displacement one had noted the maximum velocity or acceleration for each test system, one would have obtained the velocity or acceleration shock spectrum.

Definitions.[1,19] If a force $F(t)$ is applied to an undamped single-degree-of-freedom system* with natural frequency ω and mass m, then the displacement, velocity, and

---

*The shock spectra defined here are essentially descriptions of the shock in the frequency domain. Shock spectra are related to the Fourier transforms, the latter being lower bounds to the former.[66] Inclusion of damping appears unnecessary for purposes of describing the shock, but most authors[1,19] include damping in their definitions since they tend to be more concerned with descriptions of system responses to shocks than with descriptions of shocks. The term "shock spectra of a system" is often applied to descriptions of responses of specified system to specified shocks.[62]

acceleration of the system (assumed to be initially at rest and at equilibrium) are given by

$$x(t, \omega) = \frac{1}{m\omega} \int_0^t F(\tau) \sin \omega(t - \tau) \, d\tau$$

$$\dot{x}(t, \omega) = \frac{1}{m} \int_0^t F(\tau) \cos \omega(t - \tau) \, d\tau$$

$$\ddot{x}(t, \omega) = F(t)/m - \omega^2 x(t, \omega)$$

The various shock spectra of the force shock $F(t)$ are then defined as follows:

$$D'_+(\omega) = \text{timewise maximum of } x(t, \omega)$$
$$= \text{positive-displacement spectrum}$$
$$D'_-(\omega) = \text{timewise minimum of } x(t, \omega)$$
$$= \text{negative-displacement spectrum}$$
$$D'(\omega) = \text{timewise maximum of } |x(t, \omega)|$$
$$= \text{displacement spectrum}$$

Similarly, for example,

$$V'(\omega) = \text{timewise maximum of } |\dot{x}(t, \omega)|$$
$$= \text{velocity spectrum}$$
$$A'(\omega) = \text{timewise maximum of } |\ddot{x}(t, \omega)|$$
$$= \text{acceleration spectrum}$$

Similarly, if the previously discussed test system is exposed to a support acceleration $\ddot{s}(t)$, then the displacement, velocity, and acceleration of the system relative to its supports are given by

$$y(t, \omega) = -\frac{1}{\omega} \int_0^t \ddot{s}(\tau) \sin \omega(t - \tau) \, d\tau = x - s$$

$$\dot{y}(t, \omega) = - \int_0^t \ddot{s}(\tau) \cos \omega(t - \tau) \, d\tau = \dot{x} - \dot{s}$$

$$\ddot{y}(t, \omega) = - \ddot{s}(t) = \omega^2 y(t, \omega) = \ddot{x} - \ddot{s}$$

where $y = x - s$ denotes the relative displacement, x the absolute displacement of the mass, and s the displacement of the support.

The various shock spectra of the acceleration shock $\ddot{s}(t)$ are defined as follows:

$$D(\omega) = \text{timewise maximum of } |y(t, \omega)|$$
$$= \text{relative-displacement spectrum}$$
$$V(\omega) = \text{timewise maximum of } |\dot{y}(t, \omega)|$$
$$= \text{relative-velocity spectrum}$$
$$A(\omega) = \text{timewise maximum of } |\ddot{y}(t, \omega)|$$
$$= \text{relative-acceleration spectrum}$$
$$D_a(\omega) = \text{timewise maximum of } |x(t, \omega)|$$
$$= \text{absolute-displacement spectrum}$$

$$V_a(\omega) = \text{timewise maximum of } |\dot{x}(t, \omega)|$$
$$= \text{absolute-velocity spectrum}$$
$$A_a(\omega) = \text{timewise maximum of } |\ddot{x}(t, \omega)|$$
$$= \text{absolute-acceleration spectrum}$$
$$\tilde{V}(\omega) = \omega\, D(\omega) = \text{pseudo-velocity spectrpum}$$

For any $s(t)$ it is true that

$$A_a(\omega) = \omega^2\, D(\omega)$$

but $\tilde{V}(\omega) \neq V(\omega)$, $\tilde{V}(\omega) \neq V_a(\omega)$ in general.

One often distinguishes also between system responses during the action of a shock and responses after the shock action has ceased. The spectra associated with system motion during the shock action are called "primary spectra"; those associated with system motion after the shock are called "residual spectra."

Shock spectra are generally reduced to dimensionless "amplification spectra," e.g., by dividing $D'(\omega)$ by the static displacement due to the maximum value of $F(t)$, or $A(\omega)$ by the maximum value of $\ddot{s}(t)$. Frequency is also usually reduced to dimensionless form by division by some suitable frequencylike parameter.

Simple Shocks. A simple shock is one like that sketched in Fig. 5.19; it is generally nonoscillatory, has a unique absolute maximum reached within a finite "rise time" t_m, and a finite duration t_0. Spectra of simple shocks are usually presented in terms of ft_m or ωt_m (and/or ft_0 or ωt_0) or simple multiples of these dimensionless quantities.

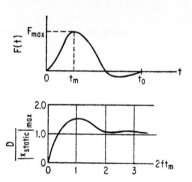

Shock spectra may, by virtue of their definitions, be used directly to determine the maximum responses of single-degree-of-freedom systems to the shocks to which the spectra pertain.

The main utility of the shock-spectrum concept, however, is due to the fact that the amplification spectra of roughly similar shocks, when presented in terms of the foregoing dimensionless-frequency parameters, tend to coincide very nearly; i.e., amplification spectra are relatively insensitive to details of the pulse form.

FIG. 5.19 Typical simple shock and dimensionless amplification spectrum.

(Good coincidence of spectra of similar shocks is obtained for small values of f if the spectra are plotted against ft_0, for large values of f if spectra are plotted against ft_m.)

The effect of damping is to reduce system responses in general. The greatest reduction usually occurs near the peaks of the amplification spectra; near relative minima of these spectra the reduction due to damping is generally small.

The amplification spectra of simple shocks have the following properties:

They pass through the origin.

They do not exceed 2.0.

They approach 1.0 for large ft_m.

They generally reach their maxima between $2\ ft_m = 0.7$ and 2.0, but most often near $2\ ft_m = 1$.

Simple shocks occur in drop tests, aircraft landing impact, hammer impact, gun recoil, explosion blasts, and ground shock due to explosive detonations. Details of many spectra appear in Refs. 1, 19, 64, and 72.

5.5 DESIGN CONSIDERATIONS

5.5.1 Design Approach

Since vibration is generally considered an undesirable side effect, it seldom controls the primary design of a machine or structure. Items usually are designed first to fulfill their main function, then analyzed from a vibration viewpoint in regard to possible equipment damage or malfunction, structural fatigue failure, noise, or human discomfort or annoyance.

The most severe effects of vibration generally occur at resonance; therefore, one usually is concerned first with determination of the resonance frequencies of the preliminary design. (Damping is usually neglected in the pertinent calculations for all but the simplest systems, unless a prominent damping effect is anticipated.) If resonance frequencies are found to lie within the intended range of driving frequencies, one should attempt a redesign to shift the resonances out of the driving-frequency range.

If resonances cannot be avoided reasonably, the designer must determine the severity of these resonances. Damping must then be considered in the pertinent calculations, since it is primarily damping that limits response at resonance. If resonant responses are too severe, one must reduce the excitation and/or incorporate increased damping in the system or structure.[69]

Shifting of Resonances. If resonances are found to occur within the range of excitation frequencies, one should try to redesign the system or structure to change its resonance frequencies. Added stiffness with little addition of mass results in shifting of the resonances to higher frequencies. Added mass with little addition of stiffness results in lowering of the resonance frequencies. Addition of damping generally has little effect on the resonance frequencies.

In cases where the vibrating system cannot be modified satisfactorily, one may avoid resonance effects by not operating at excitation frequencies where resonances are excited. This may be accomplished by automatic controls (e.g., speed controls on a machine) or by prescribing limitations on use of the system (e.g., "red lines" on engine tachometers to show operating speeds to be avoided).

Evaluation of Severity of Resonances. The displacement amplitude X_0 of a linear single-degree-of-freedom system (whose natural frequency is ω_n), excited at resonance by a sinusoidal force $F_0 \sin \omega_n t$, is given by

$$kX_0/F_0 = 1/2\zeta = Q = 1/\eta$$

where ζ is the ratio of damping to critical damping, Q the "quality factor," and η the loss factor of the system. The velocity amplitude V_0 and acceleration amplitude A_0 are given by

$$A_0 = \omega_n V_0 = \omega_n^2 X_0$$

The maximum force exerted by the system's spring (of stiffness k) is

$$F_s = kX_0$$

The maximum spring stress may readily be calculated from the foregoing spring force. Multiple-degree-of-freedom systems generally require detailed analysis in accordance with methods outlined in Sec. 5.2. The previous expressions pertaining to single-degree-of-freedom systems hold also for systems with a number of degrees of freedom if X_0 is taken as the amplitude of a generalized (principal) coordinate that is independent of the other coordinates, and if F_0 is taken as the corresponding generalized force. Then $\eta = 2\zeta = Q^{-1}$ must describe the effective damping for that coordinate.

Stresses in the various members may be determined from the mode shapes (vectors) corresponding to the resonant mode; i.e., from the maximum displacements or forces to which the elements are subjected.

For uniform distributed systems* the modal displacement U_{mn} at resonance of the m, n mode due to a modal excitation $G_{mn} \sin \omega_{mn} t$ is given by

$$|U_{mn}| = |G_{mn}|/M_{mn}\omega_{mn}^2 \eta(\omega_{mn})$$

and is related to modal velocity and acceleration amplitudes $\dot{U}_{mn}, \ddot{U}_{mn}$, according to

$$|\ddot{U}_{m,n}| = \omega_{mn}^2 |U_{mn}| = \omega_{mn}|\dot{U}_{mn}|$$

$\eta(\omega)$ denotes the system loss factor, which varies with frequency depending on the damping mechanism present. For beams and plates in flexure[64] η is related to the usual viscous-damping coefficient c and flexural rigidity D as[†]

$$\eta(\omega) = c\omega/D$$

The maximum stress σ_{max} and maximum strain ϵ_{max} that occur at resonance in bending of beams and plates may be approximated by

$$\sigma_{max} \approx (C/r)|U_{mn}|\omega_{mn}\rho c_{Ln}g_{mn} \approx \epsilon_{max}\rho c_L^2$$

where $c_L = \sqrt{E/\rho}$ denotes the velocity of sound in the material (E is Young's modulus, ρ the material density) and C denotes the distance from the neutral to the outermost fiber, r the radius of gyration of the cross section. For beams with rectangular cross sections and for plates, $C/r = \sqrt{3}$. For plates $M_{mn} = \rho h$; for beams $M_m = \rho A$, where h denotes plate thickness, A beam cross-section area.

The factor g_{mn} accounts for boundary conditions and takes on the following approximate values:

Boundary conditions on pair of opposite edges	g_{mn}
Pinned-pinned or clamped-free	1.00
Clamped-clamped or clamped-pinned	1.33
Free-free or pinned-free	0.80

For conservative design if the boundary conditions are now known one should use $g_{mn} = 1.33$. Otherwise, one should use the largest value of g_{mn} pertinent to the existing boundary conditions.

Reduction of Severity of Resonances.   If redesign to avoid resonances is not feasible, one has only two means for reducing resonant amplitudes: (1) reduction of modal excitation and (2) increase of damping.

---

*The summary here is presented for two-dimensional systems (plates), and double subscripts are used for the various modal parameters. However, the identical expressions apply also for one-dimensional systems (beams, strings) if the n subscripts are deleted. See also Sec. 5.3.

[†]For plates $D = Eh^3/12(1 - \nu^2)$; for beams $D = EI$.

Excitation reduction may take the form of running a machine at reduced power, isolating the resonating system from the source of excitation, or shielding the system from exciting pressures. Increased damping may be obtained by addition of energy-dissipating devices or structures. For example, one might use metals with high internal damping for the primary structure, or else attach coatings or sandwich media with large energy-dissipation capacities to a primary structure of common materials.[51,65] Alternately, one might rely on structural joints or shaft bearings to absorb energy by friction, or else attempt to extract energy by means of viscous friction or acoustic radiation in fluids in contact with the resonant system. Simple dashpots, localized friction pads, or magnetically actuated eddy-current-damping devices may also be used, particularly for systems with a few degrees of freedom; however, such localized devices may not be effective for higher modes of distributed systems.

5.5.2 Source-Path-Receiver Concept

If one is called upon to analyze or modify an existing or projected system from the vibration viewpoint, one may find it useful to examine the system in regard to

1. Sources of vibration (e.g., reciprocating engines, unbalanced rotating masses, fluctuating air pressures)
2. Paths connecting sources to critical items (e.g., substructures, vibration mounts)
3. Receivers of vibratory energy; i.e., critical items that malfunction if exposed to too much vibration (e.g., electronic components)

The problem of limiting the effects of vibration may then be attacked by any or all of the following means:

1. Vibration elimination at the source (e.g., designing machines in opposed pairs so that inertia forces cancel, balancing all rotating items, smoothing or deflecting unsteady air flows)
2. Modification of paths (e.g., changing substructures so as to transmit less vibration, introducing vibration mounts)
3. Decreasing vibration sensitivity of critical items (e.g., changing orientation of components with respect to excitation, using more rugged components, using more fatigue-resistant materials, designing more damping into components in order to decrease the effects of internal resonances)

Vibration Reduction at the Source: Vibration Absorbers.   The strength of vibration sources may often be reduced by proper design. Such design may include balancing, use of the lightest possible reciprocating parts, arranging components so that inertia forces cancel, or attaching vibration absorbers.

A vibration absorber* is essentially a mass m attached by means of a spring k to a primary mass M in order to reduce the response of M to a sinusoidal force acting directly on M (or to a sinusoidal support acceleration when the M is attached to the support by a spring K). If an absorber for which $k/m = \omega_0^2$ is attached to M, then M will experience no excursion if the excitation occurs at a frequency ω_0. (The displacement amplitude of m will be F/k, where F is the force amplitude.)

For driving frequencies very near ω_0, attachment of the vibration absorber results in small amplitudes for M, but for other frequencies it generally results in amplitudes

---

*The discussion is presented here only for translational systems. It may be extended by analogy to apply also to rotational systems.

that are greater than those obtained without an absorber. Vibration absorbers are very frequency-sensitive and hence should be used only where there exists essentially a single relatively accurately known driving frequency. The addition of damping to an absorber[17,63] extends to some extent the frequency range over which the absorber results in vibration reduction of the primary mass, but increased damping also results in more motion of M at the optimum frequency ω_0.

Path Modification: Vibration Mounts. The prime means for modifying paths traversed by vibratory energy consists of the addition of vibration mounts. If a mass M is excited by a vibratory force of frequency ω, one may reduce the vibratory force that the mass exerts on a rigid support to which it is attached by inserting a spring of stiffness k between mass and support such that $k/M < \omega^2/2$. Further reduction in k/M reduces the transmitted force further; added damping with k/M held constant increases the transmitted force.

If a mass M is attached by means of a spring k to a support that is vibrating at frequency ω, then M can be made to vibrate less than the support if k is chosen so that $k/M < \omega^2/2$. Further reduction of k/M further reduces the motion of the mass; addition of damping increases this motion.

Discussions of vibration and shock mounting are to be found in Refs. 14, 32, 55, and 56. Data on commercial mounts may be found in manufacturers' literature.

Modification of Critical Items. Critical items may often be made less sensitive to vibration by redesigning them so that all their internal-component resonance frequencies fall outside the excitation frequency range. Occasionally this can be done merely by reorienting a critical item with respect to the direction of excitation; more often it requires stiffening the components (or possibly adding mass). If fatigue rather than malfunction is a problem, one may obtain improved parts by careful redesign to eliminate stress concentrations and/or by using materials with greater fatigue resistance. If internal resonances cannot be avoided, one may reduce their effect by designing damping into the resonant components.

5.5.3 Rotating Machinery

A considerable amount of information has been amassed in relation to reciprocating and turbine machines. Comprehensive treatments of the associated torsional vibrations may be found in Refs. 42 and 72; less detailed discussions of these appear in standard texts.[17,32,61,63] A few of the most important items pertaining to rotating machinery are outlined subsequently.

Vibrations of rotating machines are caused by the following factors; reduction of these generally serves to reduce vibrations:

1. Unbalance of rotating components
2. Reciprocating components
3. Whirling of shafts
4. Gas forces
5. Instabilities, such as those due to slip-stick phenomena

Balancing. A rotor mounted on frictionless bearings so that its axis of rotation is horizontal remains motionless (if subject only to gravity) in any angular position, if the rotor is *statically* balanced. Static balance implies an even mass distribution around the rotational axis. However, even statically balanced rotors may be dynami-

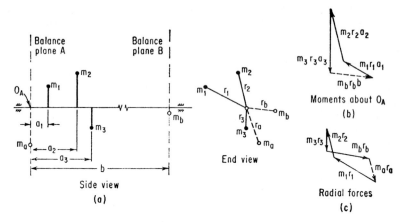

FIG. 5.20   Balancing of rigid rotors.

cally unbalanced. Dynamic unbalance occurs if the centrifugal forces set up during rotation result in a nonzero couple. (Static balance assures only that the centrifugal forces result in zero net radial force.)

Rigid rotors can always be balanced statically by addition of a single weight in any arbitrarily chosen plane, or dynamically (and statically) by addition of two weights in two arbitrarily chosen planes. The procedure, illustrated in Fig. 5.20, is as follows:[9]

1. Divide the rotor into a convenient number of sections by passing planes perpendicular to the rotational axis.

2. Determine the mass and center-of-gravity position for each section.

3. Draw a diagram like Fig. 5.20a where each section is represented by its mass located at the center-of-gravity position.

4. Select balance planes, i.e., planes in which weights are to be attached for balance.

5. Draw a diagram (Fig. 5.20b) of centrifugal force moments* about point Q_A (where rotational axis intersects balance plane A). Each moment is represented by a vector of length $m_i r_i a_i$ (to some suitable scale) parallel to r_i in the end view. The vector required to close the diagram then is $m_b r_b b$; its direction gives the direction of r_b in the end view; $m_b r_b$ may be calculated and either m_b or r_b may be selected arbitrarily.

6. Draw a diagram (Fig. 5.20c) of centrifugal forces* $m_i r_i$, including $m_b r_b$. Each force vector is drawn parallel to r_i in the end view; the vector required to close the diagram is $m_a r_a$ and defines the mass (and its location) to be added in plane A. Again, either m_a or r_a may be selected arbitrarily; r_a in the end view must be parallel to the $m_a r_a$ vector, however.

Balancing of flexible rotors or of rotors on flexible shafts (particularly when operating above critical speeds) can generally be accomplished for only one particular speed or for none at all.[17,61]

Balancing machines, their principles and use, and field balancing procedures are discussed in Refs. 17, 41, 61, and 63, among others.

---

*The common factor ω^2 is omitted from the diagrams.

Balancing of reciprocating engines is treated in Ref. 17 and in a number of texts on dynamics of machines, such as Ref. 23.

Whirling of Shafts: Critical Speeds. Rotating shafts become unstable at certain speeds, and large vibrations are likely to develop. These speeds are known as "critical speeds." At a critical speed the number of revolutions per second is generally very nearly equal to a natural frequency (in cycles per second) of the shaft considered as a nonrotating beam vibrating laterally. Thus critical speeds of shafts may be found by any of the means for calculating the natural frequencies of lateral vibrations of beams (see Sec. 5.3.3).

"Whirling," i.e., violent vibration at critical speeds, occurs in vertical as well as horizontal shafts. In nonvertical shafts gravity effects may introduce additional "critical speeds of second order," as discussed in Ref. 63. Complications may also occur where disks of large inertia are mounted on flexible shafts. Gyroscopic effects due to thin disks generally tend to stiffen the system and thus to increase the critical speeds above those calculated from static flexural vibrations. Thick disks, however, may result in lowering of the critical speeds.[17] Similarly, flexibility in the bearings results in softening of the system and in lowering of the critical speeds.

Turbine Disks. Turbine disks and blades vibrate essentially like disks and beams, and may be treated by previously outlined standard procedures if the rotational speeds are low. However, centrifugal forces result in considerable stiffening effects at high rotational speeds, so that the natural frequencies of rotating disks and blades tend to be considerably higher (and in different ratio to each other) than those of nonrotating assemblies. A brief discussion of these effects and analytical methods may be found in Ref. 63, a more comprehensive one in Ref. 59.

5.5.4 Damping Devices

In this subsection will be presented various devices for the damping or dissipation of mechanical energy. For applications of elastomers employed as damping attachments see Sec. 40, "Dampers and Elastomers," of the Second Edition of the *Mechanical Design and Systems Handbook* (H. Rothbart, ed., McGraw-Hill, New York, 1985).

Electrodynamic Damping. Electrodynamic damping is obtained when a short-circuited electrical conductor is made to move in a magnetic field. An induced current appears in the conductor. The conductor experiences a force proportional to but opposite to the velocity (see Fig. 5.21). The damping force is expressed by

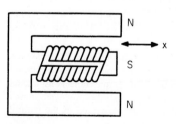

FIG. 5.21 Electrodynamic damper with induced current.

$$F = H^2 A\ell \times 10^{-9}/\rho$$

$$= H^2 V \times 10^{-9}/\rho \qquad (5.67)$$

where F = force
H = average magnetic field strength
A = cross-sectional area of coil wire
ℓ = length of wire
V = volume of wire
ρ = specific resistance

The damping constant ($f = c\dot{x}$) is therefore

$$c = H^2 V \times 10^{-9}/\rho\dot{x} \qquad (5.68)$$

Hydromechanical Damping

Incompressible Fluids (Fig. 5.22). A general form of the damping force may be written

$$F = k\dot{x}^n \tag{5.69}$$

where $1 < n < 2$. In terms of the pressure drop the equation becomes

$$\Delta p = f(\ell/d)(\dot{x}^2/2g) \tag{5.70}$$

where f = a dimensionless friction coefficient which is a function of Reynolds number
ℓ = length of duct
d = diameter of duct

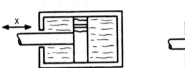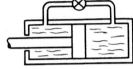

FIG. 5.22   Hydromechanical damping with incompressible fluid.

At low speeds, in which fully developed laminar flow is achieved, Eq. (5.70) assumes the Poiseuille form

$$\Delta p = (32\eta\ell/d^2)\dot{x} \tag{5.71}$$

where η is the viscosity of the liquid.

Compressible Fluids.   The principle of damping in this case is similar to that for incompressible fluids. However, the fluid also acts as a spring. The total behavior can be imagined as a spring in parallel with a damper. Assuming adiabatic compression and ideal gas behavior, the effective spring constant of the gas can be evaluated from

$$k = p\gamma A^2/V \tag{5.72}$$

where p = gas pressure
$\gamma = C_p/C_v$
C_p = specific heat at constant pressure
C_v = specific heat at constant volume
V = volume
A = cross-sectional area of duct

Untuned Viscous Shear Damper (Fig. 5.23).   This damper (also viscous-fluid damper, untuned damped vibration absorber) consists of an annular mass enclosed by an annular ring. A viscous fluid fills the volume between mass and casing. Since there is no elastic connection between the casing, which is attached to the torsionally vibrating system, and the mass, the damper is untuned. Reduction in amplitudes and a lowering of the natural frequency are obtained without introduction of additional resonances.

Reduction of a complex system to an equivalent two-mass system yields a useful design approximation. With $J_F = \infty$, the system and its equivalent are shown in Fig. 5.24. The impedance equations for steady state can be written

$$(-\omega^2 J_D + j\omega c)\theta_1 - (j\omega c)\theta_2 = 0 \tag{5.73}$$

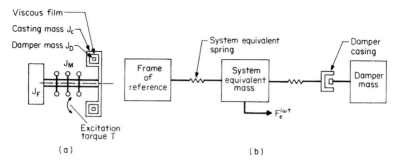

FIG. 5.23 Damping devices. (*a*) Torsional system. (*b*) Translational system.

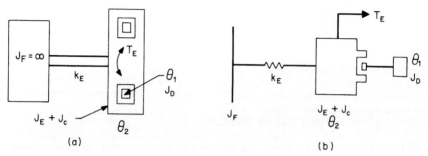

FIG. 5.24 (*a*) Complex system. (*b*) Equivalent two-mass system.

$$-(j\omega c)\theta_1 + [K_2 - \omega^2(J_E + J_c) + j\omega c]\theta_2 = T_E \qquad (5.74)$$

where J_D = mass moment of inertia of damper
J_E = equivalent mass moment of inertia of torsionally vibrating systems
J_c = mass moment of inertia of damper casing
ω = circular natural frequency of vibrating system with damper
θ_1 = steady-state vibration amplitude at damper mass
θ_2 = steady-state vibration amplitude at damper casing

Figure 5.25 is a plot of θ_2 as a function of ω for $c = 0$ and $c = \infty$. For optimum

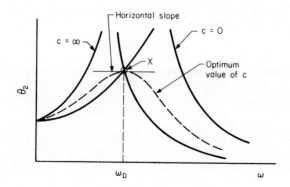

damping, the curve of θ versus ω has a horizontal slope at point X. Setting $\mu = J_D/J_E$ and solving for conditions at point X, the following useful formulas result:

$$\omega_D/\omega_2 = \sqrt{2/(2 + \mu)} \tag{5.75}$$

$$\theta_2/(T_0/K_E) = M = (2 + \mu)/\mu \tag{5.76}$$

$$c/(2J_D\omega_2) = \gamma = 1/\sqrt{2(1 + \mu)(2 + \mu)} \tag{5.77}$$

$$|\theta_{SH}/\theta_2| = |\theta_{SH}/(T_0/K_E)|/M = \sqrt{(1 + \mu)/(2 + \mu)} \tag{5.78}$$

where ω_2 = circular natural frequency of torsionally vibrating system without damper
  ω_D = circular frequency corresponding to point X in Fig. 5.25
  c = optimum value of coefficient of viscous damping
  T_0 = excitation torque
  $\theta_{SH} = |\theta_1 - \theta_2|$
  M = dynamic magnifier

Knowledge of such items as θ_2, T_0, K_E, and ω_2 permits the determination of μ, from which J_D and c may be calculated.

In reciprocating engines μ tends to fall in the range 0.4 to 1.0. Also, J_c/J_D usually lies between 0.35 and 0.8. The dimensions are so proportioned that sufficient surface is provided for proper heat dissipation. Determination of the necessary fluid viscosity is based upon a modified value of c which accounts for varying shear rates on the lateral and peripheral surfaces of the damper.

Slipping-Torque-Type Dampers.   In this type of damper, relative motion between a shaft-fixed hub and a damping mass occurs only when the relative acceleration of the two exceeds a predetermined value. An effective change of natural frequency occurs as the damping mass "locks" and "unlocks" from the hub during each oscillation. Dissipation of energy by damping occurs during the intervals of relative motion.

An analysis of the input and dissipated energy can be made by assuming continual slip, sinusoidal motion of the hub, and a linear time variation of the damper mass. The maximum energy dissipated is

$$U_{D,\max} = (4/\pi)\omega^2\theta_1^2 J_R \tag{5.79}$$

where ω = circular natural frequency of vibrating system without damper
  θ_1 = permissible oscillation amplitude at damper hub
  J_R = moment of inertia of damper

Energy input $U_{in} = T_0\theta_1$, where T_0 is the peak value excitation torque at the hub. Equating input and dissipated energy yields the useful design equation

$$J_R = \pi^2 T_0/4\omega^2\theta_1 \tag{5.80}$$

The Sandner Damper (Pumping-Chamber Type).   A rim and side plates, which act at the damper mass, and a hub are arranged to form internal cavities at certain points of the interfaces. The cavities act as pumping chambers. Oil passes through radial passages starting at the hub, moves to the pumping chambers, returns to the hub to pass through spring-loaded relief valves, and then is discharged into some convenient space. The slipping torque is accurately set by adjustment of the relief valves.

The damper moment of inertia is determined from Eq. (5.80). Maximum torque at which slipping occurs is calculated from

$$T_{R,\max} = (\sqrt{2}/\pi)\omega^2\theta_1 J_R \tag{5.81}$$

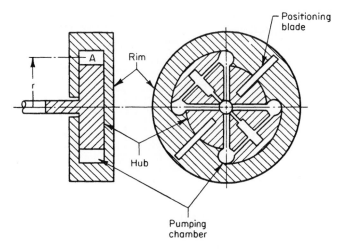

FIG. 5.26   The Sandner damper (pumping-chamber type).

Determination of the relief-valve spring pressure can be calculated from

$$p = T_{R,\max}/2rA$$

where A is the cross-sectional area of one of the pumping chambers and r is the mean radius to the area A, as shown in Fig. 5.26.

The Sandner Damper (Gear-Wheel Type).   In this type of damper a rim is cut with gear teeth on its inner surface. Pinions mesh with this gear and are enclosed in special recesses in the hub. Passages connect opposite sides of each recess to centrally located relief valves.

The rim tends to rotate the pinions because of its inertia torque. Actual rotation takes place beyond a certain critical value as predetermined by the relief-valve adjusted pressure. Oil is passed from one gear chamber to the next as the pinions rotate, as shown in Fig. 5.27. Calculations for this damper are the same as for the pumping-chamber type. However, the pressure is computed as $p = T_{R,\max}/8rA$, where A is the effect area of the gear-pump recess.

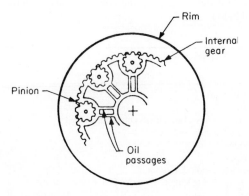

FIG. 5.27   The Sandner damper (gear-wheel type).

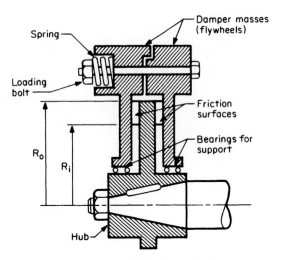

FIG. 5.28   The Lanchester damper (semi-dry-friction type).

The Lanchester Damper (Semi-Dry-Friction Type).   In this type of damper, a hub fixed coaxially to a vibrating shaft, carries friction plates on an annulus near its rim. The damper mass, consisting of two flywheels and loading bolts, presses against the friction surfaces. The damper mass lies coaxial with the shaft but is coupled to it only through the friction surfaces, as shown in Fig. 5.28.

The moment of inertia J_{2R} of the two flywheels is calculated by Eq. (5.80). Maximum torque is computed from $T_{R,\text{max}} = (\sqrt{2}/\pi)\omega^2\theta_1 J_{2R}$ and is equated to the friction torque T_f. Spring load can then be computed from

$$T_f = 4/3\,\mu P(R_o^3 - R_i^3)/(R_o^2 - R_i^2) \tag{5.82}$$

where μ is the coefficient of friction, R_i and R_o are the inner and outer radii of the friction surfaces, respectively, and P is the total spring load.

5.5.5 Charts and Tables

Information on natural frequencies, spring constants, and material properties appears in the tables listed below.

Characteristic	Table
Natural frequencies	
Simple translational systems	5.8*a*
Simple torsional systems	5.8*b*
Beams, bars, shafts, (uniform section unless noted otherwise):	
Uniform and variable section, in flexure	5.8*c*
Flexure; mode shapes	5.6*b*
Torsion; mode shapes	5.6*a*
Longitudinal vibration; mode shapes	5.6*a*
On multiple evenly spaced supports	5.8*d*
Free-free, on elastic foundation	5.8*h*

(Continued)

Characteristic	Table
Natural frequencies (*continued*)	
Strings, uniform; mode shapes	5.6*a*
Membranes, uniform:	
Circular	5.8*g*
Rectangular and circular; mode shapes	5.6*d*
Plates, uniform:	
Cantilever, various shapes	5.8*e*
Circular, various boundary conditions	5.8*f*
Rectangular, on simple supports; mode shapes	5.6*c*
Circular, clamped; mode shapes	5.6*c*
Cylindrical shells	5.8*h*
Rings	5.8*i*
Mass free to rotate and translate in plane	5.8*i*
Mass on spring of finite mass	5.8*i*
Spring constants	
Combinations of springs	5.9*a*
Round-wire helical springs	5.9*b*
Beams with force or moment inputs	5.9*c*
Torsion springs	5.9*d*
Shafts in torsion	5.9*d,e*
Plates loaded at centers	5.9*f*
Miscellaneous	
Longitudinal wavespeed and K_m (for Table 5.8) for engineering materials	5.10

Tables 5.8, 5.9, and 5.10 begin on pages 5.64, 5.74, and 5.79, respectively.

TABLE 5.8a   Natural Frequencies of Simple Translational Systems[7,17]

System	$\omega^2 =$	Remarks
k,m M	$\dfrac{k}{M+0\cdot33m}$	
k,m M	$\dfrac{k}{M+0\cdot23m}$	Point mass on cantilever beam
k,m M	$\dfrac{k}{M+0\cdot375m}$	Point mass at center of clamped beam
k,m M	$\dfrac{k}{M+0\cdot50m}$	Point mass at center of simply supported beam
M_1 k M_2	$k\left[\dfrac{1}{M_1}+\dfrac{1}{M_2}\right]$	
M_1 k_{12} M_2 k_{23} M_3		$2B = \dfrac{k_{12}}{M_1}+\dfrac{k_{23}}{M_3}+\dfrac{k_{12}+k_{23}}{M_2}$ $C = \dfrac{k_{12}+k_{23}}{M_1\,M_2\,M_3}(M_1+M_2+M_3)$
k_{01} M_1 k_{12} M_2	$B\pm\sqrt{B^2-C}$	$2B = \dfrac{k_{01}+k_{12}}{M_1}+\dfrac{k_{12}}{M_2}$ $C = \dfrac{k_{01}\,k_{12}}{M_1\,M_2}$
k_{01} M_1 k_{12} M_2 k_{20}		$2B = \dfrac{k_{01}+k_{12}}{M_1}+\dfrac{k_{12}+k_{20}}{M_2}$ $C = \dfrac{1}{M_1\,M_2}(k_{01}k_{12}+k_{12}k_{20}+k_{20}k_{01})$
d_1 d_2 k_1 k_2 M_1 M_2	$\dfrac{\dfrac{1}{M_1}+\dfrac{1}{R^2M_2}}{\dfrac{1}{k_1}+\dfrac{1}{R^2k_2}}$	Inertia of lever negligible $R = \dfrac{d_2}{d_1}$
d_1 d_2 k M	$\dfrac{k}{M}\left(\dfrac{d_1}{d_1+d_2}\right)^2$	Inertia of lever negligible

(See table 5.9c for k — referring to the first four rows)

Symbols:
ω = circular natural frequency = $2\pi f$
M = mass
m = total mass of spring element
k = spring constant

TABLE 5.8b Natural Frequencies of Simple Torsional Systems[7.17]

System	$\omega^2 =$	Remarks
k, I_s ... I	$\dfrac{k}{I + I_s/3}$	
I_1 ... k ... I_2	$k\left(\dfrac{1}{I_1} + \dfrac{1}{I_2}\right)$	
I_1 ... k_{12} ... I_2 ... k_{23} ... I_3		$2B = \dfrac{k_{12}}{I_1} + \dfrac{k_{23}}{I_3} + \dfrac{k_{12} + k_{23}}{I_2}$ $C = \dfrac{k_{12}\,k_{23}}{I_1\,I_2\,I_3}(I_1 + I_2 + I_3)$
k_{01} ... I_1 ... k_{12} ... I_2	$B \pm \sqrt{B^2 - C}$	$2B = \dfrac{k_{01} + k_{12}}{I_1} + \dfrac{k_{12}}{I_2}$ $C = \dfrac{k_{01}\,k_{12}}{I_1\,I_2}$
k_{01} ... I_1 ... k_{12} ... I_2 ... k_{20}		$2B = \dfrac{k_{01} + k_{12}}{I_1} + \dfrac{k_{12} + k_{20}}{I_2}$ $C = \dfrac{1}{I_1\,I_2}(k_{01}k_{12} + k_{12}k_{20} + k_{20}k_{01})$
I_1 ... k_1 ... G_1 ... G_2 ... k_2 ... I_2	$\dfrac{\dfrac{1}{I_1} + \dfrac{1}{R^2 I_2}}{\dfrac{1}{k_1} + \dfrac{1}{R^2 k_2}}$	Inertia of gears G_1, G_2 assumed negligible $R = \dfrac{\text{number of teeth on gear 1}}{\text{number of teeth on gear 2}}$ $= \dfrac{\text{rpm of shaft 2}}{\text{rpm of shaft 1}}$

Symbols:
ω = circular natural frequency = $2\pi f$
k = torsional spring constant; see Table 5.9d
I = polar mass moment of inertia
I_s = polar mass moment of inertia of entire shaft

TABLE 5.8c Natural Frequencies of Beams in Flexure[7,33]

$$f_n = C_n \frac{r}{L^2} \times 10^4 \times K_m$$

f_n = nth natural frequency, hertz
C_n = frequency constant listed in these tables
r = radius of gyration of cross section = $\sqrt{I/A}$, inches*
L = beam length, inches*
K_m = material constant (Table 5.10) = 1.00 for steel

Uniform-section beams		n				
		1	2	3	4	5
Clamped–Clamped	Free–Free	71.95	198.29	388.73	642.60	959.94
Clamped –Free		11.30	70.85	198.30	388.73	642.60
Clamped –Hinged	Free–Hinged	49.57	160.65	335.17	573.20	874.65
Clamped–Guided	Free-Guided	17.98	97.18	239.98	446.25	715.98
Hinged–Hinged	Guided–Guided	31.73	126.93	285.60	507.73	793.33
Hinged–Guided		7.93	71.40	198.33	388.73	642.60

*For r and L in centimeters, use C_n values listed in table multiplied by 2.54.

TABLE 5.8c Natural Frequencies of Beams in Flexure (*Continued*)

Variable-section beams (Use maximum r to calculate f_n)	Shape		n		
	b/b_0	h/h_0	1	2	3
	1	x/L	17.09	48.89	96.57
	x/L	x/L	26.08	68.08	123.64
	$\sqrt{x/L}$	x/L	22.30	58.18	109.90
	exp x/L	1	15.23	77.78	206.07
	1	x/L	21.21* 35.05†	56.97*	
	x/L	x/L	32.73* 49.50†	76.57*	
	$\sqrt{x/L}$	x/L	25.66* 42.02†	66.06*	

*Symmetric mode.
†Antisymmetric mode.

TABLE 5.8d   Natural Frequencies of Uniform Beams on Multiple Equally Spaced Supports[7,33]

$$f_n = C_n \frac{r}{L^2} \times 10^4 \times K_m$$

f_n = nth natural frequency, hertz
C_n = frequency constant listed in these tables
r = radius of gyration of cross section = $\sqrt{I/A}$, inches*
L = span length, inches*
K_m = material constant (Table 5.10) = 1.00 for steel

	Number of spans	n				
		1	2	3	4	5
Ends simply supported	1	31.73	126.94	285.61	507.76	793.37
	2	31.73	49.59	126.94	160.66	285.61
	3	31.73	40.52	59.56	126.94	143.98
	4	31.73	37.02	49.59	63.99	126.94
	5	31.73	34.99	44.19	55.29	66.72
	6	31.73	34.32	40.52	49.59	59.56
	7	31.73	33.67	38.40	45.70	53.63
	8	31.73	33.02	37.02	42.70	49.59
	9	31.73	33.02	35.66	40.52	46.46
	10	31.73	33.02	34.99	39.10	44.19
	11	31.73	32.37	34.32	37.70	41.97
	12	31.73	32.37	34.32	37.02	40.52
Ends clamped	1	72.36	198.34	388.75	642.63	959.98
	2	49.59	72.36	160.66	198.34	335.20
	3	40.52	59.56	72.36	143.98	178.25
	4	37.02	49.59	63.99	72.36	137.30
	5	34.99	44.19	55.29	66.72	72.36
	6	34.32	40.52	49.59	59.56	67.65
	7	33.67	38.40	45.70	53.63	62.20
	8	33.02	37.02	42.70	49.59	56.98
	9	33.02	35.66	40.52	46.46	52.81
	10	33.02	34.99	39.10	44.19	49.59
	11	32.37	34.32	37.70	41.97	47.23
	12	32.37	34.32	37.02	40.52	44.94
Ends clamped-supported	1	49.59	160.66	335.2	573.21	874.69
	2	37.02	63.99	137.30	185.85	301.05
	3	34.32	49.59	67.65	132.07	160.66
	4	33.02	42.70	56.98	69.51	129 49
	5	33.02	39.10	49.59	61.31	70.45
	6	32.37	37.02	44.94	54.46	63.99
	7	32.37	35.66	41.97	49.59	57.84
	8	32.37	34.99	39.81	45.70	53.63
	9	31.73	34.32	38.40	43.44	49.59
	10	31.73	33.67	37.02	41.24	46.46
	11	31.73	33.67	36.33	39.81	44.19
	12	31.73	33.02	35.66	39.10	42.70

*For r and L in centimeters, use C_n values multiplied by 2.54.

TABLE 5.8e Natural Frequencies of Cantilever Plates[7,33]

$$f_n = C_n \frac{h}{a^2} \times 10^4 \times K_m$$

f_n = nth natural frequency, hertz
C_n = frequency constant listed in table
h = plate thickness, inches*
a = plate dimension, as shown inches*
K_m = material constant (Table 5.10) = 1.00 for steel

Boundary conditions F = free C = clamped S = simply supported	n				
	1	2	3	4	5
$a/b = \frac{1}{2}$	3.41	5.23	9.98	21.36	24.18
$a/b = 1$	3.40	8.32	26.71	20.86	30.32
$a/b = 2$	3.38	14.52	91.92	21.02	47.39
$a/b = 5$	3.36	33.79	548.60	20.94	103.03
$\theta = 15°$	3.50	8.63			
$\theta = 30°$	3.85	9.91			
$\theta = 45°$	4.69	13.38			
$a/b = 2$	6.7	28.6	56.6	137	
$a/b = 4$	6.6	28.5	83.5	240	
$a/b = 8$	6.6	28.4	146.1	457	
$a/b = 14$	6.6	28.4	246	790	
$a/b = 2$	5.5	23.6			
$a/b = 4$	6.2	26.7			
$a/b = 7$	6.4	28.1			

*For h and a in centimeters, use given C_n values multiplied by 2.54.

TABLE 5.8f Natural Frequencies of Circular Plates[7,33]

$$f = C\frac{h}{r^2} \times 10^4 \times K_m$$

f = natural frequency, hertz
C = frequency constant listed in table
h = plate thickness, inches*
r = plate radius, inches*
K_m = material constant (Table 5.10) = 1.00 for steel

Boundary conditions	Number of nodal circles	Number of nodal diameters			
		0	1	2	3
Clamped at circumference	0	9.94	20.67	33.91	49.61
	1	38.66	59.12	82.22	107.9
	2	86.61	116.8	149.5	185.0
	3	153.8	193.5	235.9	281.1
Free	0			5.11	11.89
	1	8.83	19.94	34.27	51.43
	2	37.47	58.19	81.6	108.2
	3	85.35	115.68	149.7	186.1
Clamped at center; free at circumference	0	3.65			
	1	20.33			
	2	59.49			
	3	117.2			
Simply supported at circumference	0	4.84	13.55	24.93	
	1	28.93	47.16	68.18	
	2	72.13	99.93	130.6	
	3	134.4	171.9	212.1	
Clamped at center; simply supported at circumference	0	14.4			
	1	48.0			
Clamped at center and at circumference	0	22.1			
	1	60.2			

Simply supported at radius a; free at circumference	Fundamental mode					
a/r	0	0.2	0.4	0.6	0.8	1.0
C	3.64	4.4	6.5	8.6	7.2	4.84

*For h and r in centimeters, use given C_n values multiplied by 2.54.

TABLE 5.8*g* Natural Frequencies of Circular Membranes[33]

$$f = C\sqrt{T/h\rho r^2}$$

f = natural frequency, hertz
C = frequency constant listed in table
r = membrane radius, inches*
h = membrane thickness, inches*
T = tension at circumference, lb_f/in*
ρ = material density, lb_m/in^3 or $lb_f \cdot s^2/in^4$*

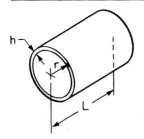

Number of nodal circles	Number of nodal diameters			
	1	2	3	4
1	0.383	0.610	0.817	1.015
2	0.887	1.116	1.340	1.553
3	1.377	1.619	1.849	2.071
4	1.877	2.120	2.355	2.582

*Any set of consistant units may be used that causes all units except *s* to cancel in the square root.

TABLE 5.8*h* Natural Frequencies of Cylindrical Shells[7]

f_{ij} = natural frequency for mode shape ij, hertz
h = shell thickness, inches*
L = length, inches*
i = number of circumferential waves in mode shape
j = number of axial half-waves in mode shape
c_L = Longitudinal wavespeed in shell material, in/s* (See Table 5.10)
v = Poisson's ratio, dimensionless

$$f_{ij} = \frac{\lambda_{ij}}{2\pi r}\frac{c_L}{\sqrt{1 - v^2}}$$

*Any other consistent set of units may be used. For example, h, r, L in centimeters and c_L in centimeters per second.

TABLE 5.8h Natural Frequencies of Cylindrical Shells (*Continued*)

Infinitely long	
Axial modes:	$\lambda_{ij} = i\sqrt{\dfrac{1-\nu}{2}},\qquad i = 1, 2, 3, \ldots$
Radial (extensional) modes:	$\lambda_{ij} = \sqrt{1 + i^2},\qquad i = 0, 1, 2, \ldots$
	Note: $i = 0$ corresponds to breathing mode
Radial-circumferential flexural modes:	$\lambda_{ij} = \dfrac{h/r}{\sqrt{12}}\dfrac{i(i^2 - 1)}{\sqrt{i^2 + 1}},\qquad i = 2, 3, 4, \ldots$

Finite length, simply supported edges without axial constraint	
Torsional modes:	$\lambda_{ij} = \dfrac{j\pi r}{L}\sqrt{\dfrac{1-\nu}{2}}$
Axial modes†:	$\lambda_{ij} = \dfrac{j\pi r}{L}\sqrt{1 - \nu^2}$ $\begin{aligned} i &= 0 \\ j &= 1, 2, 3 \end{aligned}$
Radial modes†:	$\lambda_{ij} = 1$
Bending modes†:	$\lambda_{ij} = \left(\dfrac{j\pi r}{L}\right)^2\sqrt{\dfrac{1-\nu}{2}}$ $\begin{aligned} i &= 1 \\ j &= 1, 2, 3 \end{aligned}$
Radial-axial modes:	$\lambda_{ij} = \dfrac{\sqrt{(1 - \nu^2)\,\beta_j^4 + \dfrac{h^2}{12r^2}(i^2 + \beta_j^2)^4}}{i^2 + \beta_j^2}$ $\beta_j = j\pi r/L$ $\begin{aligned} i &= 2, 3, 4, \ldots \\ j &= 1, 2, 3, \ldots \end{aligned}$

†Values given here apply for long shells, for which $L > 8jr$.

TABLE 5.8i Natural Frequencies of Miscellaneous Systems

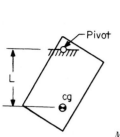

Simple Point-Mass Pendulum[7]

$$\omega = 2\pi f = \sqrt{g/L}$$

Pendulum with Distributed Mass[7]

$$\omega = 2\pi f = \sqrt{MLg/I_0}$$

M = mass
I_0 = mass moment of inertia about pivot
g = acceleration of gravity
L = distance from pivot to center of gravity

Mass Free to Rotate and Translate in Plane[63]

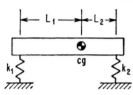

Mass = M, moment of inertia about c.g. = r^2M

$$2\omega^2 = a + c \pm \sqrt{(a - c)^2 + (2b/r)^2}$$

$$a = \frac{k_1 + k_2}{M} \qquad b = \frac{L_2 k_2 - L_1 k_1}{M}$$

$$c = \frac{L_1^2 k_1 + L_2^2 k_2}{r^2 M}$$

Thin Rings[17]

Extension (pure radial mode): $\quad \omega = \dfrac{1}{a} c_L$

Bending in its own plane: $\quad \omega_n = \dfrac{n(n^2 - 1)}{\sqrt{n^2 + 1}} \dfrac{r}{a^2} c_L$

$$n = 2, 3, \ldots$$
$$r = \sqrt{I/A}$$

I = moment of inertia of ring section area A for bending in its own plane
For out-of-plane bending and partial rings, see Den Hartog[17]

Free-free Beam on Elastic Foundation[63]

Same mode shapes as without support, but frequencies increased:

$$\omega^2_{n, \text{ on support}} = \omega^2_{n, \text{ unsupported}} + \frac{k'}{\rho A}$$

$$\frac{k'}{\rho A} = \frac{\text{support spring constant/unit length}}{\text{beam mass/unit length}} = \omega^2_{\text{rigid beam on elastic foundation}}$$

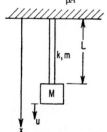

Mass on Rod or Spring of Static Stiffness k, Spring Total Mass m Uniformly Distributed[63]

Frequency equation:

$$(M/m)(\omega\sqrt{m/k}) = \cot(\omega\sqrt{m/k})$$
$$\phi_n(x) = \sin(\omega_n x/L \sqrt{m/k})$$

Weighting function $\mu(x) = m/L + M\delta(x - L)$
[see Sec. 5.3]

M/m	$\omega_1 \sqrt{m/k}$	$\omega_2 \sqrt{m/k}$	$\omega_3 \sqrt{m/k}$	$\omega_4 \sqrt{m/k}$	$\omega_5 \sqrt{m/k}$
0*	$\pi/2$	$3\pi/2$	$5\pi/2$	$7\pi/2$	$9\pi/2$
$\frac{1}{2}$	1.077	3.644	6.579	9.630	12.722
1	0.960	3.435	6.437	9.426	12.645
∞ †	0	π	2π	3π	4π

*Corresponds to spring free at $x = L$.
†Corresponds to spring clamped at $x = L$.

TABLE 5.9a Combination of Spring Constants

	Series combination	$1/k_{\text{total}} = 1/k_1 + 1/k_2 + \cdots + 1/k_n$
	Parallel combination	$k_{\text{total}} = k_1 + k_2 + \cdots + k_n$

TABLE 5.9b Spring Constants of Round-Wire Helical Springs[17]

	Axial loading	$\dfrac{F}{\delta} = k = \dfrac{Gd^4}{8nD^3}$
	Torsion	$\dfrac{T}{\phi} = k_r = \dfrac{Ed^4}{64nD}$
	Bending	$\dfrac{M}{\theta} = k_b = \dfrac{2}{2 + \nu} k_r$

Symbols:

F = axial force	G	= shear modulus
δ = axial deflection	E	= elastic modulus
T = torque	ν	= Poisson's ratio
ϕ = torsion angle	d	= wire diameter
M = bending moment	D	= mean coil diameter
θ = flexure angle	n	= number of coils

TABLE 5.9c Spring Constants of Beams[69]

	Translational (force F)	Rotational (moment M)
	$k = \dfrac{3EI}{L^3}$	$k_r = \dfrac{EI}{L}$
	$k = \dfrac{12EI}{L^3}$	
	$k = \dfrac{3EIL}{(ab)^2}$ <hr> For $a = b$ $k = \dfrac{48EI}{L^3}$	$k_r = \dfrac{3EIL}{L^2 - 3ab}$ <hr> For $a = b$ $k_r = \dfrac{12EI}{L}$
	$k = \dfrac{12EIL^3}{a^3 b^2(3L + b)}$ <hr> For $a = b$ $k = \dfrac{768EI}{7L^3}$	$k_r = \dfrac{4EIL^3/b}{4L^3 - 3b(L + a)^2}$ <hr> For $a = b$ $k_r = \dfrac{64EI}{5L}$
	$k = \dfrac{3EIL^3}{(ab)^3}$ <hr> For $a = b$ $k = \dfrac{192EI}{L^3}$	$k_r = \dfrac{EIL^3}{ab(L^2 - 3ab)}$ <hr> For $a = b$ $k_r = \dfrac{16EI}{L}$
		$k_r = \dfrac{3EI}{L}$
		$k_r = \dfrac{4EI}{L}$

Symbols:
E = modulus of elasticity
I = centroidal moment of inertia
  of cross section
F = force
M = bending moment

$k = F/\delta$, translational spring constant
$k_r = M/\theta$, rotational spring constant
δ = lateral deflection at F
θ = flexural angle at M

	Spiral spring[17] $k_r = T/\phi = EI/L$ E = modulus of elasticity L = total spring length I = moment of inertia of cross section
	Helical spring See Table 5.9b
	Uniform shaft $k_r = GJ/L$ G = shear modulus L = length J = torsional constant (Table 5.9e)
	Stepped shaft $1/k_r = 1/k_{r_1} + 1/k_{r_2} + \cdots 1/k_{r_n}$ $k_{r_j} = k_r$ of jth uniform part by itself

TABLE 5.9e Torsional Constants J of Common Sections[47]

Circle		$J = \dfrac{\pi}{32}\,(D^4 - d^4)$
Ellipse		$J = \dfrac{\pi a^3 b^3}{a^2 + b^2}$
Square		$J = 0.1406\,a^4$
Rectangle		$J = ab^3\left[\dfrac{16}{3} - 3.36\,\dfrac{b}{a}\left(1 - \dfrac{b^4}{12a^4}\right)\right]$
Equilateral triangle		$J = \dfrac{\sqrt{3}\,a^4}{80}$
Any solid compact section without reentrant angles		$J \approx \dfrac{A^4}{40I}$
Any thin closed tube of uniform thickness t		$J \approx \dfrac{4A^2 t}{U}$ $\quad A$ = cross-section area U = mean circumferential length (length of dotted lines shown) I = polar moment of inertia of section about its centroid
Any thin open tube of uniform thickness t		$J \approx \dfrac{Ut^3}{3}$

TABLE 5.9f   Spring Constants of Centrally Loaded Plates[47]

Circular	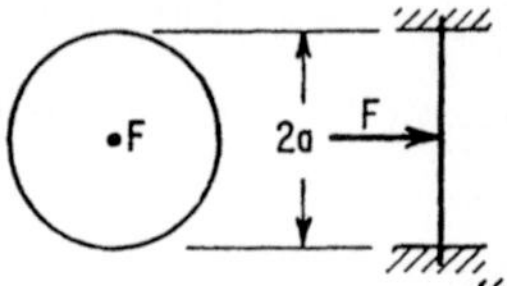	Clamped	$k = \dfrac{16\pi D}{a^2}$
	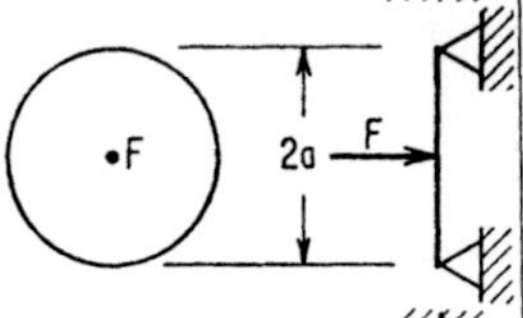	Simply supported	$k = \dfrac{16\pi D}{a^2}\left(\dfrac{1+\nu}{3+\nu}\right)$
	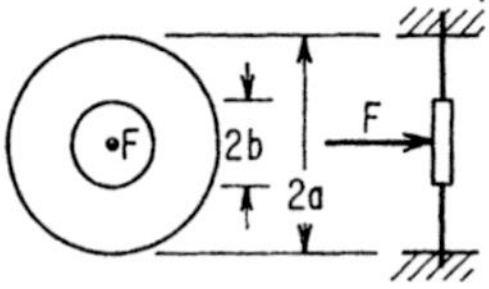	Clamped, with concentric rigid insert	$k = \dfrac{(16\pi D/a^2)(1-c^2)}{(1-c^2)^2 - 4c^2(\ln c)^2}$ $c = b/a$
Square	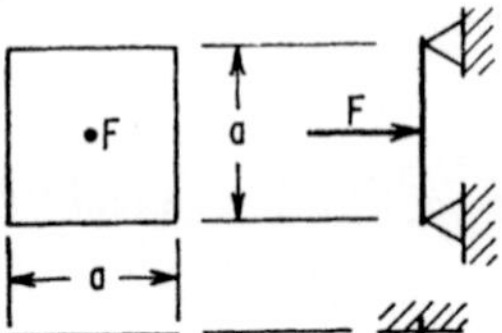	All edges simply supported	$k = 86.1D/a^2$
	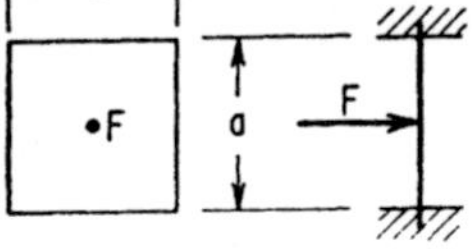	All edges clamped	$k = 192.2D/a^2$
Rectangular	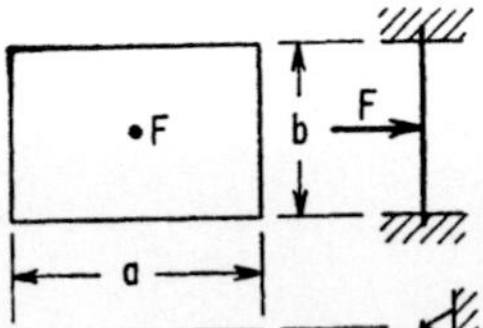	All edges clamped	$k = D\alpha/b^2$
	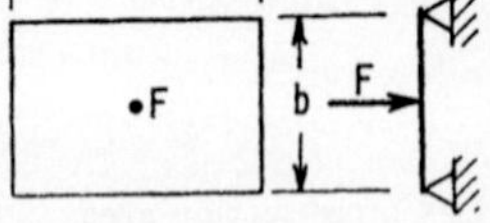	All edges simply supported	$k = 59.2(1 + 0.462c^4)D/b^2$ $c = b/a < 1$
Equilateral triangular	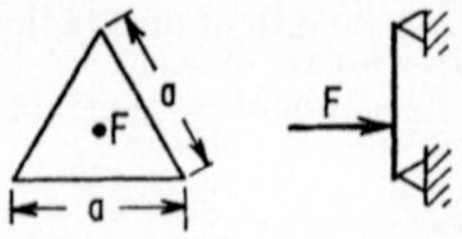	All edges simply supported	$k = 175D/a^2$

For the "All edges clamped" rectangular case:

a/b	4	2	1
α	167	147	192

Symbols:
k = force/deflection, spring constant
$D = \dfrac{Eh^3}{12(1-\nu^2)}$

E = modulus of elasticity
ν = Poisson's ratio
h = plate thickness

TABLE 5.10 Longitudinal Wavespeed and K_m for Engineering Materials

Material	Temperature, °C	Longitudinal wavespeed $c_L = \sqrt{E/\rho}$ 10^5 in/s*	$K_m = \dfrac{(c_L)_{material}}{(c_L)_{structural\ steel}}$
Metals			
Aluminum	20	2.00	1.04
Beryllium	20	4.96	2.57
Brass, bronze	20	1.38–1.57	0.715–0.813
Copper	20	1.34	0.694
Copper	100	1.21	0.627
Copper	200	1.16	0.601
Cupro-nickel	20	1.47–1.61	0.762–0.834
Iron, cast	20	1.04–1.64	0.539–0.850
Lead	20	0.49	0.25
Magnesium	20	2.00	1.04
Monel metal	20	1.76	0.912
Nickel	20	1.24–1.90	0.642–0.984
Silver	20–100	1.03	0.534
Tin	20	0.98	0.51
Titanium	20	1.06–2.00	1.02–1.04
Titanium	90	1.90–1.96	0.984–1.02
Titanium	200	1.84–1.88	0.953–0.974
Titanium	325	1.68–1.75	0.870–0.907
Titanium	400	1.57–1.68	0.813–0.870
Zinc	20	1.46	0.756
Steel, typical	20	2.00	1.04
Steel, structural	−200	2.00	1.04
	−100	1.97	1.02
	25	1.93	1.00
	150	1.92	0.994
	300	1.86	0.963
	400	1.77	0.917
Steel, stainless	−200	1.98–2.04	1.03–1.06
	−100	1.94–2.02	1.005–1.047
	25	1.92–2.06	0.995–1.07
	150	1.89–2.02	0.979–1.05
	300	1.84–1.98	0.953–0.979
	400	1.76–1.94	0.912–1.005
	600	1.70–1.87	0.881–0.969
	800	1.57–1.76	0.813–0.912
Aluminum Alloys	−200	2.08–2.13	1.078–1.104
	−100	2.01–2.06	1.04–1.07
	25	1.97–2.02	1.02–1.05
	200	1.79–1.87	0.927–0.969
Plastics†			
Cellulose acetate	25	0.40	0.21
Methyl methacrylate	25	0.60	0.31
Nylon 6, 6	25	0.54	0.78
Vinyl chloride	25	0.26	0.13
Polyethylene	25	0.41	0.21
Polypropylene	25	0.42	0.22
Polystyrene	25	0.68	0.35

TABLE 5.10 Longitudinal Wavespeed and K_m for Engineering Materials (*Continued*)

Material	Temperature, °C	Longitudinal wavespeed $c_L = \sqrt{E/\rho}$ 10^5 in/s*	$K_m = \dfrac{(c_L)_{material}}{(c_L)_{structural\ steel}}$
Teflon	25	0.17	0.088
Epoxy resin	25	0.61	0.32
Phenolic resin	25	0.57	0.30
Polyester resin	25	0.61	0.32
Glass-fiber–epoxy laminate	25	1.72	0.89
Glass-fiber–phenolic laminate	25	1.44	0.75
Glass-fiber–polyester laminate	25	1.25	0.65
Other			
Concrete	25	1.44–1.80	0.75–0.93
Cork	25	0.20	0.104
Glass	25	1.97–2.36	1.02–1.22
Granite	25	2.36	1.22
Marble	25	1.50	0.78
Plywood	25	0.84	0.44
Woods, along fibers		1.32–1.92	0.68–0.99
Woods, across fibers		0.48–0.54	0.25–0.28

*To convert to centimeters per second, multiply by 2.54
†Values given here are typical. Wide variations may occur with composition, temperature, and frequency.

REFERENCES

1. Barton, M. V. (ed.): "Shock and Structural Response," American Society of Mechanical Engineers, New York, 1960.

2. Bendat, J. S.: "Principles and Applications of Random Noise Theory," John Wiley & Sons, Inc., New York, 1958.

3. Bendat, J. S., and A. G. Piersol: "Engineering Applications of Correlation and Spectral Analysis," John Wiley & Sons, Inc., New York, 1980.

4. Beranek, L. L. (ed.): "Noise and Vibration Control," McGraw-Hill Book Company, Inc., New York, 1971.

5. Biezeno, C. B., and R. Grammel: "Engineering Dynamics," vol. III, "Steam Turbines," Blackie & Son, Ltd., London, 1954.

6. Bishop, R. E. D., and D. C. Johnson: "Vibration Analysis Tables," Cambridge University Press, Cambridge, England, 1956.

7. Blevins, R. D.: "Formulas for Natural Frequency and Mode Shape," Van Nostrand Reinhold Company, Inc., New York, 1979.

8. Burton, R.: "Vibration and Impact," Addison-Wesley Publishing Company, Inc., Reading, Mass., 1958.

9. Church, A. H., and R. Plunkett: "Balancing Flexible Rotors," *Trans. ASME (Ser. B)*, vol. 83, pp. 383–389, November, 1961.

10. Craig, R. R., Jr.: "Structural Dynamics: An Introduction to Computer Methods," John Wiley & Sons, Inc., New York, 1981.

11. Crandall, S. H., and W. D. Mark: "Random Vibration in Mechanical Systems," Academic Press, Inc., New York, 1963.

12. Crandall, S. H.: "Engineering Analysis," McGraw-Hill Book Company, Inc., New York, 1956.

13. Crede, C. E.: "Theory of Vibration Isolation," chap. 30, "Shock and Vibration Handbook," C. M. Harris, and C. E. Crede, eds., McGraw-Hill Book Company, Inc., New York, 1976.

14. Crede, C. E.: "Vibration and Shock Isolation," John Wiley & Sons, Inc., New York, 1951.

15. Cremer, L., M. Heckl, and E. E. Ungar: "Structure-Borne Sound," Springer-Verlag, New York, 1973.

16. Davenport, W. B., Jr.: "Probability and Random Processes," McGraw-Hill Book Company, Inc., New York, 1970.

17. Den Hartog, J. P.: "Mechanical Vibrations," 4th ed., McGraw-Hill Book Company, Inc., New York, 1956.

18. Fung, Y. C.: "An Introduction to the Theory of Aeroelasticity," John Wiley & Sons, Inc., New York, 1955.

19. Fung, Y. C., and M. V. Barton: "Some Shock Spectra Characteristics and Uses," *J. Appl. Mech.*, pp. 365–372, September, 1958.

20. Gardner, M. F., and J. L. Barnes: "Transients in Linear Systems," vol. I, John Wiley & Sons, Inc., New York, 1942.

21. Hagedorn, P.: "Non-Linear Oscillations," Clarendon Press, Oxford, England, 1981.

22. Himelblau, H., Jr., and S. Rubin: "Vibration of a Resiliently Supported Rigid Body," chap. 3, "Shock and Vibration Handbook," C. M. Harris and C. E. Crede, eds., McGraw-Hill Book Company, Inc., New York, 1976.

23. Holowenko, R.: "Dynamics of Machinery," John Wiley & Sons, Inc., New York, 1955.

24. Hueber, K. H.: "The Finite Element Method for Engineers," John Wiley & Sons, Inc., New York, 1975.

25. Jacobsen, L. S., and R. S. Ayre, "Engineering Vibrations," McGraw-Hill Book Company, Inc., New York, 1958.

26. Junger, M. C., and D. Feit: "Sound, Structures, and their Interaction," The MIT Press, Cambridge, Mass., 1972.

27. Kamke, E.: "Differentialgleichungen, Lösungsmethoden and Lösungen," 3d ed., Chelsea Publishing Company, New York, 1948.

28. Leissa, A. W.: "Vibration of Plates," NASA SP-160, U.S. Government Printing Office, Washington, D.C., 1969.

29. Leissa, A. W.: "Vibration of Shells," NASA SP-288, U.S. Government Printing Office, Washington, D.C., 1973.

30. LePage, W. R., and S. Seely: "General Network Analysis," McGraw-Hill Book Company, Inc., New York, 1952.

31. Lowe, R., and R. D. Cavanaugh: "Correlation of Shock Spectra and Pulse Shape with Shock Environment," *Environ. Eng.*, February 1959.

32. Macduff, J. N., and J. R. Curreri: "Vibration Control," McGraw-Hill Book Company, Inc., New York, 1958.

33. Macduff, J. N., and R. P. Felgar: "Vibration Design Charts," *Trans. ASME*, vol. 79, pp. 1459–1475, 1957.

34. Mahalingam, S.: "Forced Vibration of Systems with Non-linear Non-symmetrical Characteristics," *J. Appl. Mech.*, vol. 24, pp. 435–439, September 1957.

35. Major, A.: "Dynamics in Civil Engineering," Akadémiai Kiadó, Budapest, 1980.

36. Marguerre, K., and H. Wölfel: "Mechanics of Vibration," Sijthoff & Noordhoff, Alphen aan den Rijn, The Netherlands, 1979.

37. Martienssen, O.: "Über neue Resonanzerscheinungen in Wechselstromkreisen," *Physik. Z.*, vol. 11, pp. 448–460, 1910.

38. Meirovitch, L.: "Analytical Methods in Vibration," The Macmillan Company, Inc., New York, 1967.

39. Mindlin, R. D.: "Dynamics of Package Cushioning," *Bell System Tech. J.*, vol. 24, nos. 3, 4, pp. 353–461, July–October 1945.

40. Morse, P. M.: "Vibration and Sound," 2d ed., McGraw-Hill Book Company, Inc., New York, 1948.

41. Myklestad, N. O.: "Fundamentals of Vibration Analysis," McGraw-Hill Book Company, Inc., New York, 1956.

42. Nestroides, E. J. (ed.): "Bicera: Handbook on Torsional Vibrations," Cambridge University Press, New York, 1958 (British Internal Combustion Engine Research Association).

43. Ostergren, S. M.: "Shock Response of a Two-Degree-of-Freedom System," Rome Air Development Center Rep. RADC TN 58-251.

44. Perrone, N., and W. Pilkey, and B. Pilkey, (eds.): "Structural Mechanics Software Series," University of Virginia Press, Charlottesville, vols. I–III, 1977–1980.

45. Pilkey, W., and B. Pilkey, (eds.): "Shock and Vibration Computer Programs," The Shock and Vibration Information Center, Naval Research Laboratory, Washington, D.C., 1975.

46. Pistiner, J. S., and H. Reisman: "Dynamic Amplification Factor of a Two-Degree-of-Freedom System," *J. Environ. Sci.,* pp. 4–8, October 1960.

47. Plunkett, R. (ed.): "Mechanical Impedance Methods for Mechanical Vibrations," American Society of Mechanical Engineers, New York, 1958.

48. Richart, F. E., Jr., J. R. Hall, Jr., and R. D. Woods: "Vibration of Soils and Foundations," Prentice-Hall, Inc., Englewood Cliffs, N.J., 1970.

49. Roark, R. J.: "Formulas for Stress and Strain," 3d ed., McGraw-Hill Book Company, Inc., New York, 1954.

50. Rubin, S.: "Concepts in Shock Data Analysis," chap. 23, "Shock and Vibration Handbook," C. M. Harris and C. E. Crede, eds., McGraw-Hill Book Company, Inc., New York, 1976.

51. Ruzicka, J. (ed.): "Structural Damping," American Society of Mechanical Engineers, New York, 1959.

52. Ruzicka, J. E., and T. F. Derby: "Influence of Damping in Vibration Isolation," SVM-7, Shock and Vibration Information Center, Washington, D.C., 1971.

53. Scanlan, R. H., and R. Rosenbaum: "Introduction to the Study of Aircraft Vibration and Flutter," The Macmillan Company, Inc., New York, 1951.

54. Schwesinger, G.: "On One-term Approximations of Forced Nonharmonic Vibrations," *J. Appl. Mech.,* vol. 17, no. 2, pp. 202–208, June 1950.

55. Snowdon, J. C.: "Vibration and Shock in Damped Mechanical Systems," John Wiley & Sons, Inc., New York, 1968.

56. Snowdon, J. C.: "Vibration Isolation: Use and Characterization," NBS Handbook 128, U.S. Department of Commerce, National Bureau of Standards, Washington, D.C., May 1979.

57. Snowdon, J. C., and E. E. Ungar (eds.): "Isolation of Mechanical Vibration, Impact, and Noise," AMD-vol. 1, American Society of Mechanical Engineers, New York, 1973.

58. Skudrzyk, E.: "Simple and Complex Vibratory Systems," The Pennsylvania State University Press, University Park, Pa., 1968.

59. Stodola, A. (Transl. L. C. Lowenstein): "Steam and Gas Turbines," McGraw-Hill Book Company, Inc., New York, 1927.

60. Stoker, J. J.: "Nonlinear Vibrations," Interscience Publishers, Inc., New York, 1950.

61. Thomson, W. T.: "Theory of Vibration with Applications," 2d ed., Prentice-Hall, Inc., Englewood Cliffs, N.J., 1981.

62. Thomson, W. T.: "Shock Spectra of a Nonlinear System," *J. Appl. Mech.,* vol. 27, pp. 528–534, September 1960.

63. Timoshenko, S.: "Vibration Problems in Engineering," 2d ed., Van Nostrand Company, Inc., Princeton, N.J., 1937.

64. Tong, K. N.: "Theory of Mechanical Vibration," John Wiley & Sons, Inc., New York, 1960.

65. Torvik, P. J. (ed.): "Damping Applications for Vibration Control," AMD-vol. 38, American Society of Mechanical Engineers, New York, 1980.

66. Trent, H. M.: "Physical Equivalents of Spectral Notions," *J. Acoust. Soc. Am.,* vol. 32, pp. 348–351, March 1960.

67. Ungar, E. E.: "Maximum Stresses in Beams and Plates Vibrating at Resonance," *Trans. ASME, Ser. B,* vol. 84, pp. 149–155, February 1962.

68. Ungar, E. E.: "Damping of Panels," chap. 14, "Noise and Vibration Control," L. L. Beranek, ed., McGraw-Hill Book Company, Inc., New York, 1971.

69. Ungar, E. E., and R. Cohen: "Vibration Control Techniques," chap. 20, "Handbook of Noise Control," C. M. Harris, ed., McGraw-Hill Book Company, Inc., New York, 1979.

70. Vigness, I.: "Fundamental Nature of Shock and Vibrations," *Elec. Mfg.,* vol. 63, pp. 89–108, June 1959.

71. Warburton, G. B.: "The Dynamical Behavior of Structures," Pergamon Press, Oxford, 1964.

72. Wilson, W. K.: "Practical Solution of Torsional Vibration Problems," John Wiley & Sons, Inc., New York, 1956.

SECTION 6
MACHINE SYSTEMS

Burton Paul, Ph.D.

Asa Whitney Professor of Dynamical Engineering
Department of Mechanical Engineering and Applied Mechanics
University of Pennsylvania
Philadelphia, Pa.

6.1 INTRODUCTION

The following material is a survey of the theory of machine systems based on analytical mechanics and digital computation. Much of the material to be presented is described in greater detail in Refs. 1, 36, and 64 to 66, but some of it is an extension of previously published works. For a discussion of the subject based on the more traditional approach of vector dynamics and graphical constructions see Refs. 51, 52, and 83.

The systems which we shall consider comprise resistant bodies interconnected in such a way that specified input forces and motions are transformed in a predictable way to produce desired output forces and motions. The term "resistant" body includes both rigid bodies and components such as cables or fluid columns which momentarily serve the same function as rigid bodies. This description of a mechanical system is essentially the definition of the term "machine" given by Reuleaux.[75] When discussing the purely kinematic aspects of such a system, the assemblage of "links" (i.e., resistant bodies) is frequently referred to as a "kinematic chain"; when one of the links of a kinematic chain is fixed, the chain is said to form a "mechanism." Various types of "kinematic pairs" or interconnections (e.g., revolutes, sliders, cams, rolling pairs) between pairs of links are described in Sec. 3. Two or more such kinematic pairs may be superposed at an idealized point, or "joint," of the system.

The organization of this section follows that classification[93] of mechanics which divides the subject into "kinematics" (the study of motion irrespective of its cause) and "dynamics" (the study of motion due to forces); dynamics is then subdivided into the categories "statics" (forces in equilibrium) and "kinetics" (forces not in equilibrium).

A Notation. Boldface type is used to represent ordinary vectors, matrices, and column vectors, with their respective scalar components indicated as in the following examples:

$$\mathbf{r} = x\mathbf{i} + y\mathbf{i} + z\mathbf{k} \equiv (x, y, z)$$

$$\mathbf{C} = [c_{ij}] = \begin{bmatrix} c_{11} & \cdots & c_{1N} \\ \cdots & \cdots & \cdots \\ c_{M1} & \cdots & c_{MN} \end{bmatrix}$$

$$\mathbf{x} \equiv \{x_1, x_2, ..., x_N\} \equiv [x_1, x_2, ..., x_N]^T$$

We shall occasionally refer to a matrix such as $\mathbf{C}$ as "an $M \times N$ matrix." Note the use of a superscript T to indicate the transpose of a matrix. In what follows, we shall make frequent use of the rule for transposition of a matrix product, i.e., $(\mathbf{AB})^T = \mathbf{B}^T\mathbf{A}^T$.

If $F(\psi) = F(\psi_1, \psi_2, ..., \psi_M)$, we will use subscript commas to denote the partial derivatives of F with respect to its arguments, as follows:

$$F_{,i} = \partial F(\psi)/\partial \psi_i \qquad F_{,ij} = \partial^2 F(\psi)/\partial \psi_i \partial \psi_j$$

Differentiation with respect to time t will be denoted by an overdot:

$$\dot{x} = dx/dt$$

6.2 ANALYTICAL KINEMATICS

6.2.1 Lagrangian Coordinates

The cartesian coordinates (x^i, y^i, z^i) of a particle P^i with respect to a fixed right-handed reference frame are called the "global" coordinates of the particle. Occasionally, we shall represent the global coordinates by the alternative notation (x_1^i, x_2^i, x_3^i), abbreviated as

$$\mathbf{x}^i \equiv (x_1^i, x_2^i, x_3^i) = (x^i, y^i, z^i) \tag{6.1}$$

When all particles move parallel to the global xy plane, the motion is said to be "planar."

Since the number of particles in a mechanical system is usually infinite, it is convenient to represent the global position vector $\mathbf{x}^k \equiv (x^k, y^k, z^k)$ of an arbitrary particle P^k in terms of a set of M variables $\psi_1 < \psi_2, ..., \psi_M$, called "lagrangian coordinates." By

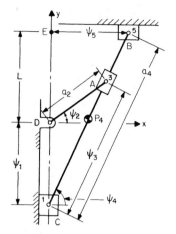

FIG. 6.1 Example of multiloop planar mechanism.

introducing a suitable number of lagrangian coordinates, it is always possible to express the global coordinates of every particle in the system via "transformation equations" of the form

$$\mathbf{x}^k = \mathbf{x}^k(\psi_1, ..., \psi_M) \qquad (6.2a)$$

or $\quad x_i^k = x_i^k(\psi_1, ..., \psi_M)$

$$i = 1, 2, 3 \quad (6.2b)$$

where the time t does not appear explicitly. For example, in Fig. 6.1 the global (x,y) coordinates of the midpoint P^4 of bar BC are given by

$$x_1^4 = x^4 = \tfrac{1}{2} a_4 \cos \psi_4$$
$$\qquad\qquad\qquad\qquad (6.3)$$
$$x_2^{\,4} = y^4 = -\psi_1 + \tfrac{1}{2} a_4 \sin \psi_4$$

In general, the lagrangian coordinates ψ_i need not be independent, but may be related by "equations of constraint" of the form

$$f_i(\psi_1, \psi_2, ..., \psi_M) = 0 \qquad i = 1, 2, ..., N_s \qquad (6.4)$$

or $\qquad g_k(\psi_1, \psi_2, ..., \psi_M, t) = 0 \qquad k = 1, 2, ..., N_t \qquad (6.5)$

Equation (6.4) represents "spatial" (or "scleronomic") constraints because only the space variables ψ_i appear as arguments; there are N_s spatial constraints. On the other hand, time t does appear explicitly in Eq. (6.5), which is therefore said to represent "temporal" (or "rheonomic") constraints; there are N_t such constraints. For example, closure of the two *independent** loops ECBE and DCAD requires that the following *spatial* constraints be satisfied:

$$f_1(\psi) = a_4 \cos \psi_4 - \psi_5 = 0$$

$$f_2(\psi) = a_4 \sin \psi_4 - L - \psi_1 = 0$$

$$\qquad\qquad\qquad\qquad\qquad (6.6)$$

$$f_3(\psi) = \psi_3 \cos \psi_4 - a_2 \cos \psi_2 = 0$$

$$f_4(\psi) = \psi_3 \sin \psi_4 - a_2 \sin \psi_2 - \psi_1 = 0$$

Other equations of spatial constraint described by Eq. (6.4) might arise from the action of gears (see Sec. 7.30 of Ref. 66), from cams (see Sec. 7.40 of Ref. 66), or from kinematic pairs in three-dimensional mechanisms (see p. 752 of Ref. 62).

As an example of a temporal constraint consider the function

---

*The equations of constraint corresponding to loop closure will be independent if each loop used is a "simple closed circuit"; i.e., it encloses exactly one polygon with nonintersecting sides. See Chap. 8 of Ref. 66 for criteria of loop independence.

$$g(\psi, t) = \psi_1 - \beta - \omega t - \alpha t^2/2 = 0 \tag{6.7}$$

where α, β, and ω are constants. This constraint specifies that the variable ψ_1 increases with constant acceleration ($\ddot{\psi}_1 = \alpha$), starting from a reference state (at $t = 0$) where its initial value is β and its initial "velocity" is $\dot{\psi}_1 = \omega$. The special case where $\alpha = 0$ corresponds to a link being driven by a constant-speed motor.

A second example is given by the constraint

$$g(\psi, t) = \psi_1 + \psi_2 + \psi_3 - 3\beta - 3\omega t = 0 \tag{6.8}$$

which requires that the average velocity $\frac{1}{3}(\dot{\psi}_1 + \dot{\psi}_2 + \dot{\psi}_3)$ remain fixed at a constant value ω. Such a constraint could conceivably be imposed by a feedback control system.

If the total number of constraints, given by

$$N_c = N_s + N_t \tag{6.9}$$

just equals the number M of lagrangian variables, the M equations, Eqs. (6.4) and (6.5), which can be written in the form

$$F_i(\psi_1, \psi_2, ..., \psi_M, t) = 0 \qquad i = 1, 2, ..., M \tag{6.10}$$

can be solved for the M unknown values of ψ_j (except for certain *singular* states; see Sec. 8.2 of Ref. 66). Because Eqs. (6.10) are most often highly nonlinear, it is usually necessary to solve them by a numerical procedure, such as the Newton-Raphson algorithm. Details of the technique are given in Chap. 9 of Ref. 66.

To find the lagrangian velocities, it is necessary to differentiate Eqs. (6.10) with respect to time t. The differentiation yields

$$\sum_{j=1}^{M} \frac{\partial F_i}{\partial \psi_j} \dot{\psi}_j + \frac{\partial F_i}{\partial t} = 0 \qquad i = 1, ..., M \tag{6.11}$$

In matrix terms this may be expressed as

$$\mathbf{G}\,\dot{\psi} = \mu \tag{6.12}$$

where $$\mathbf{G} = [\partial F_i/\partial \psi_j] \qquad \mu = -\{\partial F_i/\partial t\} \tag{6.13}$$

In the nonsingular case, Eq. (6.12) can be solved for the $\dot{\psi}_j$.

To find lagrangian accelerations, we may differentiate Eq. (6.11) to obtain

$$\sum_{j=1}^{M} (F_{i,j}\ddot{\psi}_j + \sum_{k=1}^{M} F_{i,jk}\dot{\psi}_k\dot{\psi}_j) + \sum_{j=1}^{M} \left(\frac{\partial F_i}{\partial t}\right)_{,j}\dot{\psi}_j + \frac{\partial^2 F_i}{\partial t^2} \tag{6.14}$$

where use has been made of the comma notation for denoting partial derivatives.

In matrix notation Eq. (6.14) can be expressed as

$$\mathbf{G}\ddot{\psi} = \mathbf{b} \tag{6.15}$$

where $\mathbf{G}$ has been previously defined and

$$\mathbf{b} = \dot{\mu} - \dot{\mathbf{G}}\dot{\psi} \tag{6.16}$$

or
$$b_i = -\frac{\partial^2 F_i}{\partial t^2} - 2\sum_{j=1}^{M}\left(\frac{\partial F_i}{\partial t}\right)_{,j}\dot{\psi}_j - \sum_{j=1}^{M}\sum_{k=1}^{M} F_{i,jk}\dot{\psi}_k\dot{\psi}_j \qquad (6.17)$$

To find velocities of the point of interest $\mathbf{x}^k$, we differentiate Eq. (6.2b) to obtain

$$\dot{x}_i^k = \sum_{j=1}^{M} x_{i,j}^k \dot{\psi}_j \qquad (6.18)$$

or
$$\dot{\mathbf{x}}^k = \mathbf{P}^k\dot{\boldsymbol{\psi}} \qquad (6.19)$$

where the $3 \times M$ matrix $\mathbf{P}^k$ is defined by

$$\mathbf{P}^k = [x_{i,j}^k] \qquad (6.20)$$

The corresponding acceleration is given by

$$\ddot{x}_i^k = \sum_{j=1}^{M} x_{i,j}^k \ddot{\psi}_j + \sum_{j=1}^{M}\sum_{k=1}^{M} x_{i,jk}^k \dot{\psi}_k\dot{\psi}_j \qquad (6.21)$$

or
$$\ddot{\mathbf{x}}^k = \mathbf{P}^k\ddot{\boldsymbol{\psi}} + \mathbf{p}^k \qquad (6.22)$$

where the components of the vector $\mathbf{p}^k$ are given by

$$p_i^k = \sum_{j=1}^{M}\sum_{k=1}^{M} x_{i,jk}^k \dot{\psi}_k\dot{\psi}_j \qquad (6.23)$$

6.2.2 Degree of Freedom

For a system described by M lagrangian coordinates, the most general form for the equations of spatial constraint is

$$f_i(\psi_1, \dots, \psi_M) = 0 \qquad i = 1, \dots, N_s \qquad (6.24)$$

The "rate form" of these equations is

$$\sum_{j=1}^{M} \frac{\partial f_i}{\partial \psi_j}\dot{\psi}_j = 0 \qquad \text{or} \qquad \mathbf{D}\dot{\boldsymbol{\psi}} = \mathbf{0} \qquad (6.25)$$

where
$$\mathbf{D} = [D_{ij}] = [\partial f_i/\partial \psi_j] \qquad (6.26)$$

is an $N_s \times M$ matrix. If the rank of matrix $\mathbf{D}$ is r, we know from a theorem of linear algebra[9] that we may arbitrarily assign any values to $M - r$ of the ψ_j and the others will be uniquely determined by Eq. (6.25). In kinematics terms, we say that the degree of freedom (DOF) or mobility* of the system is

$$F = M - r \qquad (6.27)$$

---

*See Chap. 8 of Ref. 66 for a discussion of mobility criteria.

This result is due to Freudenstein.[24]

In the nonsingular case, $r = N_s$ and the DOF is

$$F = M - N_s \tag{6.28}$$

For example, consider the mechanism of Fig. 6.1, where $M = 5$ and $N_s = 4$ [see Eq. (6.6)]. From Eq. (6.28) we see that this mechanism has one DOF.

If any temporal constraints are also present, we should refer to F as the "unreduced DOF," because the imposition of N_t temporal constraints will lower the degree of freedom still further to the "reduced DOF";

$$F_R = F - N_t = M - N_s - N_t \tag{6.29}$$

For example, if we drive AD of Fig. 6.1 with uniform angular velocity ω, we are in essence imposing the single temporal constraint

$$g_1(\psi, t) = \psi_2 - \omega t = 0 \tag{6.30}$$

Hence the reduced degree of freedom is

$$F_R = 1 - 1 = 0$$

In short, the system is "kinematically determinate." No matter what external forces are applied, the *motion* is always the same. Of course, the internal forces at the joints of the mechanism are dependent on the applied forces, as will be discussed in the following sections on statics and kinetics of machine systems.

6.2.3 Generalized Coordinates

For purposes of kinematics, it is sufficient to work directly with the lagrangian coordinates ψ_i. However, in problems of dynamics, it is useful to express the terms ψ_i in terms of a smaller number of variables q_r, called "generalized coordinates" (also called "primary coordinates"). In problems of kinematics, we may think of the q_r as "driving variables." For example, if a motor drives link DA of Fig. 6.1, $\psi_2 = q_1$ is a good choice for this problem.

In a system with F DOF, we are free to choose any subset of F lagrangian coordinates to serve as generalized coordinates. This choice may be expressed by the matrix relationship

$$\mathbf{T}\psi = \mathbf{q} \tag{6.31}$$

where $\mathbf{T}$ is an $F \times M$ matrix whose elements are defined as

$$T_{ij} = \begin{cases} 1 & \text{if } q_i = \psi_j \\ 0 & \text{otherwise} \end{cases} \tag{6.32a} \tag{6.32b}$$

For example, if $M = 5$ and $F = 2$, we might select ψ_2 and ψ_5 as the primary variables q_1 and q_2. Then Eq. (6.31) becomes

$$\begin{bmatrix} 0 & 1 & 0 & 0 & 0 \\ 0 & 0 & 0 & 0 & 1 \end{bmatrix} \begin{bmatrix} \psi_1 \\ \psi_2 \\ \psi_3 \\ \psi_4 \\ \psi_5 \end{bmatrix} = \begin{bmatrix} q_1 \\ q_2 \end{bmatrix}$$

If desired, the matrix $\mathbf{T}$ may be chosen so that any q_i is some desired linear combination of the ψ_j, but the simpler choice embodied in Eqs. (6.32a) and (6.32b) should suffice for most purposes.

The F equations represented by Eq. (6.31), together with the $M - F$ spatial constraint equations represented by

$$f_i(\psi_1, \ldots, \psi_M) = 0 \qquad i = 1, \ldots, N_s \tag{6.33}$$

constitute a set of M equations in the M variables ψ_k. We can represent the differentiated form of these equations by

$$\sum_{j=1}^{M} f_{i,j} \dot{\psi}_j = 0 \qquad i = 1, \ldots, N_s \tag{6.34}$$

$$\sum_{j=1}^{M} T_{kj} \dot{\psi}_j = \dot{q}_k \qquad k = 1, \ldots, F \tag{6.35}$$

or

$$\mathbf{A}\dot{\psi} = \mathbf{B}\dot{\mathbf{q}} \tag{6.36}$$

where

$$A_{ij} = f_{i,j}, \qquad B_{ij} = 0 \qquad\qquad i = 1, \ldots, N_s \tag{6.37a}$$

$$A_{ij} = T_{ij}, \quad \left. \begin{array}{l} B_{ij} = 1 \text{ if } i = j + N_s \\ B_{ij} = 0 \text{ if } i \neq j + N_s \end{array} \right\}_n \quad i = N_s + 1, \ldots, M \tag{6.37b}$$

In general, the square matrix $\mathbf{A}$ will be nonsingular for some choice of driving variables, i.e., for some arrangement of the ones and zeros in the matrix $\mathbf{T}$. It may be necessary to redefine the initial choice of the $\mathbf{T}$ matrix during the course of time in order for $\mathbf{A}$ to remain nonsingular at certain critical configurations. For example, in a slider-crank mechanism, we could not use the displacement of the slider as a generalized coordinate when the slider is at either of its extreme positions (i.e., in the so-called dead-center positions).

When $\mathbf{A}$ is nonsingular, we may assign independent numerical values to all the q_i, and solve the $M \times M$ system of nonlinear algebraic equations represented by Eqs. (6.31) and (6.33) for $\psi_1, \ldots, \psi_M$ by means of a numerical technique such as the Newton-Raphson method. In order for real solutions to exist, it is necessary that the q_i lie within geometrically meaningful ranges. Then one may find the lagrangian velocities $\dot{\psi}_k$ from Eq. (6.36) in the form

$$\dot{\psi} = \mathbf{A}^{-1}\mathbf{B}\dot{\mathbf{q}} \equiv \mathbf{C}\dot{\mathbf{q}} \qquad \text{or} \qquad \dot{\psi}_j = \sum_{r=1}^{F} C_{jr}\dot{q}_r \tag{6.38}$$

where

$$\mathbf{C} \equiv \mathbf{A}^{-1}\mathbf{B} \tag{6.39}$$

To find the accelerations associated with given values of $\ddot{q}_r$, it is only necessary to solve the differentiated form of Eq. (6.36), i.e.,

$$\mathbf{A}\ddot{\boldsymbol{\psi}} = \mathbf{B}\ddot{\mathbf{q}} + \mathbf{v} \qquad (6.40)$$

where

$$\mathbf{v} \equiv -\dot{\mathbf{A}}\,\dot{\boldsymbol{\psi}} \qquad (6.41)$$

and

$$A_{ij} \equiv \begin{cases} \displaystyle\sum_{k=1} f_{i,jk}\dot{\psi}_k & i = 1, \ldots, N_s \qquad (6.42) \\ 0 & i = N_s + 1, \ldots, M \qquad (6.43) \end{cases}$$

Note that $\dot{\mathbf{B}} \equiv \mathbf{0}$, since all B_{ij} are constants by definition.
From Eq. (6.40), it follows that

$$\ddot{\boldsymbol{\psi}} = \mathbf{C}\ddot{\mathbf{q}} + \mathbf{e} \qquad (6.44)$$

where

$$\mathbf{e} = \mathbf{A}^{-1}\mathbf{v} \qquad (6.45)$$

From previous definitions, it may be verified that each component of the vector $\mathbf{e}$ is a quadratic form in the generalized velocities $\dot{q}_r$. Hence, we see from Eq. (6.44) that any component of lagrangian acceleration is expressible as a linear combination of the generalized accelerations $\ddot{q}_r$ plus a quadratic form in the generalized velocities $\dot{q}_r$. The quadratic velocity terms are manifestations of centripetal and Coriolis accelerations (also called compound centripetal accelerations). We will accordingly refer to them as "centripetal" terms for brevity.

6.2.4   Points of Interest

To express the velocity of point $\mathbf{x}^k$ in terms of $\dot{\mathbf{q}}$ we merely substitute Eq. (6.38) into Eq. (6.19) and find

$$\dot{\mathbf{x}}_k = \mathbf{U}^k\dot{\mathbf{q}} \qquad \text{or} \qquad \dot{x}_i^k = \sum_{j=1}^{M} U_{ij}^k \dot{q}_j \qquad (6.46)$$

where

$$\mathbf{U}^k = \mathbf{P}^k\mathbf{C} \qquad \text{or} \qquad U_{ij}{}^k = \sum_{n=1}^{M} P_{in}^k C_{nj} \qquad (6.47)$$

Similarly, the acceleration for the same point of interest is found by substituting Eq. (6.44) into Eq. (6.22) to yield

$$\ddot{\mathbf{x}}^k = \mathbf{U}^k\ddot{\mathbf{q}} + \mathbf{v}^k \qquad (6.48)$$

where

$$\mathbf{v}^k \equiv \mathbf{P}^k\mathbf{e} + \mathbf{p}^k \qquad (6.49)$$

From previous definitions, it is readily verified that each component of the vector $\mathbf{v}^k$ is a quadratic form in $\dot{q}_r$.
Upon multiplying each term in Eq. (6.46) by an infinitesimal time increment δt we find that any set of displacements which satisfies the spatial constraint equations represented by Eq. (6.33) must satisfy the differential relation

$$\delta x_i^k = \sum_{j=1}^{F} U_{ij}^{\ k} \delta q_j \qquad (6.50)$$

This result will be useful in our later discussion of virtual displacements.

6.2.5 Angular Position, Velocity, and Acceleration

Let $(\xi,\ \eta,\ \zeta)$ be a set of right-handed orthogonal axes ("body axes") fixed in a rigid body, and let $(x,\ y,\ z)$ be a right-handed orthogonal set of axes parallel to directions fixed in inertial space ("space axes"). Both sets of axes have a common origin. The angular orientation of the body relative to fixed space is defined by the three Euler angles Θ, Ψ, Φ shown in Fig. 6.2.

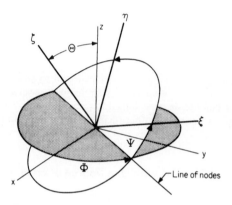

FIG. 6.2   Euler angles.

With this definition of the Euler angles,[15,29,53,61] the body is brought from its initial position (where $\xi,\ \eta,\ \zeta$ coincide with $x,\ y,\ z$) to its final position by the following sequence:

1. Rotation by Φ (precession angle) about the z axis, thereby carrying the ξ axis into the position marked "line of nodes"
2. Rotation by Θ (nutation angle) about the line of nodes, thereby carrying the ζ axis into its final position shown
3. Rotation by Ψ (spin angle) about the ζ axis, thereby carrying the ξ and η axes into their final position as shown

It may be shown[15] that the angular velocity components ω_x, ω_y, ω_z relative to the fixed axes $x,\ y,\ z$ are given by

$$\begin{bmatrix} \omega_x \\ \omega_y \\ \omega_z \end{bmatrix} = \begin{bmatrix} \cos\Phi & \sin\Theta\sin\Phi & 0 \\ \sin\Phi & -\sin\Theta\cos\Phi & 0 \\ 0 & \cos\Theta & 1 \end{bmatrix} \begin{bmatrix} \dot{\Theta} \\ \dot{\Psi} \\ \dot{\Phi} \end{bmatrix} \qquad (6.51)$$

$$\boldsymbol{\omega} = \mathbf{R}\{\dot{\Theta},\ \dot{\psi},\ \dot{\Phi}\} \qquad (6.52)$$

where $\mathbf{R}$ is the square matrix shown in Eq. (6.51).

Now the Euler angles for body k can be defined as a subset of the lagrangian coordinates as follows:

$$\{\Theta, \Psi, \Phi\}^k \equiv \mathbf{E}^k \boldsymbol{\psi} \tag{6.53}$$

where the components of the $3 \times M$ matrices $\mathbf{E}^k$ are all zeros or ones.
Upon noting from Eq. (6.38) that $\boldsymbol{\dot{\psi}} = \mathbf{C}\dot{\mathbf{q}}$, we can write Eq. (6.52) in the form

$$\boldsymbol{\omega}^k = \mathbf{R}^k \mathbf{E}^k \mathbf{C}\dot{\mathbf{q}} = \mathbf{W}^k \dot{\mathbf{q}} \tag{6.54}$$

where $$\mathbf{W}^k \equiv \mathbf{R}^k \mathbf{E}^k \mathbf{C} \tag{6.55}$$

Later we will be interested in studying infinitesimal rotations $(\delta\theta_1, \delta\theta_2, \delta\theta_3)^k$ of body k about the x, y, z axes. From Eq. (6.54) it follows that

$$\delta\boldsymbol{\theta}^k = \mathbf{W}^k \delta\mathbf{q} \qquad \text{or} \qquad \delta\theta_i^k = \sum_{r=1}^{F} W_{ir}^k \delta q_r \tag{6.56}$$

If desired, the angular accelerations are readily found by differentiation of Eq. (6.51) or (6.54).

It is worth noting that use of Euler angles can lead to singular states when Θ is zero or π, because in such cases there is no clear distinction between the angles Φ and Ψ (see Fig 6.2). For this reason, it is sometimes convenient to work with other measures of angular position, such as the four (dependent) "Euler parameters."[29,41,57,63,78]

6.2.6 Kinematics of Systems with One Degree of Freedom

The special case of $F = 1$ is sufficiently important to introduce a notational simplification which is possible because the vector of generalized coordinates $\mathbf{q}$ has only the single component $q_1 \equiv q$. Similarly, we may drop the second subscript in all the matrices which operate on q. For example, Eq. (6.36) becomes

$$\mathbf{A}\boldsymbol{\dot{\psi}} = \mathbf{b}\dot{q} \tag{6.57}$$

where $$b_i = B_{i1} \tag{6.58}$$

In like manner we may rewrite Eqs. (6.38) and (6.44) as

$$\dot{\psi}_i = c_i \dot{q} \tag{6.59}$$

and $$\ddot{\psi}_i = c_i \ddot{q} + e_i = c_i \ddot{q} + c_i' \dot{q}^2 \tag{6.60}$$

where $$c_i' \equiv -\sum_{j=1}^{M} \frac{\partial c_i}{\partial \psi_j} c_j \tag{6.61}$$

Similarly, the velocities and accelerations, for point of interest k, can be written in the compact forms

$$\dot{\mathbf{x}}^k = \mathbf{u}^k \dot{q} \tag{6.62}$$

and
$$\ddot{\mathbf{x}}^k = \mathbf{u}^k\ddot{q} + \mathbf{v}^k = \mathbf{u}^k\ddot{q} + \mathbf{u}'^k\dot{q}^2 \qquad (6.63)$$

where
$$\mathbf{v}_k \equiv \dot{q}\sum_{j=1}^{M}\frac{\partial\mathbf{u}^k}{\partial\psi_j}\dot{\psi}_j = \dot{q}^2\sum_{j=1}^{M}\frac{\partial\mathbf{u}^k}{\partial\psi_j}c_j$$

and
$$\mathbf{u}'^k \equiv \sum_{j=1}^{M}\frac{\partial\mathbf{u}^k}{\partial\psi_j}c_j$$

Finally, we note that the components of the angular velocity of any rigid-body member k of the system are given by Eq. (6.54) in the simplified form

$$\omega_i^k = w_i^k\dot{q} \qquad (6.64)$$

and the corresponding angular accelerations are given by

$$\omega_i^k = w_i^k\ddot{q} + \dot{w}_i^k\dot{q} \qquad (6.65)$$

where
$$w_i^k \equiv W_{i1}^k \qquad (6.66)$$

6.2.7 Computer Programs for Kinematics Analysis

The foregoing analysis forms the basis of a digital computer program called KINMAC (kinematics of machinery) developed by the author and his colleagues.[1,36,69] This FORTRAN program calculates displacements, velocities, and accelerations for up to 30 points of interest in planar mechanisms with up to 10 DOF, up to 10 loops, and up to 10 user-supplied auxiliary constraint relations (in addition to the automatically formulated loop-closure constraint relations). The program also contains built-in subroutines for use in the modeling of cams and temporal constraint relations. A less sophisticated student-oriented program for kinematics analysis, called KINAL (kinematics analysis), is described and listed in full in Ref. 66.

In Ref. 89, Suh and Radcliffe give listings for a package of FORTRAN subroutines called LINCPAC which can be used to solve a variety of fundamental problems in planar kinematics. They also give listings for a package called SPAPAC to be similarly used for problems of spatial kinematics.

Other programs have been developed primarily for dynamics analysis, but they can function in a so-called kinematic mode; among these are IMP and DRAM which will be discussed further in connection with programs for dynamics analysis. Computer programs which have been developed primarily for the problem of kinematic synthesis, but have some application to analysis and simulation, include R. Kaufman's KINSYN[80] and A. Erdman's LINCAGES[22]; these programs are described in the survey article by Kaufman.[42]

6.3 STATICS OF MACHINE SYSTEMS

6.3.1 Direct Use of Newton's Laws (Vectorial Mechanics)

The most direct way of formulating the equations of statics for a machine is to write equations of equilibrium for each body of the system. For example, Fig. 6.3 shows the

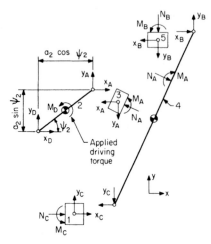

FIG. 6.3 Free-body diagrams for the mechanism of Fig. 6.1.

set of five free-body diagrams associated with the mechanism of Fig. 6.1. The internal reaction forces at pin joints (e.g., X_A, Y_A) and the normal forces and moments at smooth slider contacts (e.g., N_A, M_A) appear as external loadings on the various bodies of the system. A driving torque M_D, acting on link 2, is assumed to be a known function of position. Thus, there exists a total of 14 unknown internal force components, namely, X_A, Y_A, N_A, M_A, X_B, Y_B, N_B M_B, X_C Y_C N_C M_C X_D, and Y_D. In addition, there exist five unknown position variables ψ_1, ..., ψ_5. In order to find the 19 unknown quantities, we can write three equations of equilibrium for each of the five "free bodies." A typical set of such equations, for link 2, is

$$X_D + X_A = 0 \tag{6.67a}$$

$$Y_D + Y_A = 0 \tag{6.67b}$$

$$M_D + (X_D - X_A)(a_2/2)\sin\psi_2 + (Y_A - Y_D)(a_2/2)\cos\psi_2 = 0 \tag{6.67c}$$

In all, we can write a total of 15 such equations of equilibrium for the five links. We can eliminate the 14 internal reactions from these equations to yield one master equation involving the five position variables ψ_1, ..., ψ_5. In addition, we have the four equations of constraint in the form of Eqs. (6.6). Thus all the equations mentioned above may be boiled down to five equations for the five unknown lagrangian coordinates ψ_i. These equations are nonlinear in the ψ_i, but a numerical solution is always possible (see Sec. 9.22 of Ref. 66). Having found the ψ_i, one may solve for the 14 unknown reactions from the equilibrium equations which are linear in the reactions.

If one wishes to find only the configuration of a mechanism under the influence of a set of static external forces, the above procedure is unnecessarily cumbersome compared with the method of virtual work (discussed below).

However, if the configuration of the system is known (all ψ_i given), then the 15 equations of equilibrium are convenient for finding the 14 internal reaction components plus the required applied torque M_D. The details of such a solution are given in Sec. 11.52 of Ref. 66 for arbitrary loads on all the links of the example mechanism.

6.3.2 Terminology of Analytical Mechanics

We now define some key terms and concepts from the subject of analytical mechanics. For a more detailed discussion of the subject see any of the standard works, e.g., Refs. 29, 30, 46, 61, and 102 or Refs. 53, 64, and 66.

1. *Virtual displacements:* Any set of infinitesimal displacements which satisfy all *instantaneous* constraints

This means that if there exist F degrees of freedom, the displacements of a typical point P^k must satisfy the differential relationship

$$\delta x_i^k = \sum_{j=1}^{F} U_{ij}^k \delta q_j \tag{6.68}$$

given by Eq. (6.50).

If, in addition, there exist N_t temporal constraints of the form

$$g_k(\psi_1, \ldots, \psi_M, t) = 0 \qquad k = 1, 2, \ldots, N_t \tag{6.69}$$

they can be expressed in the "rate" form

$$\sum_{j=1}^{M} g_{k,j}\dot\psi_j + \frac{\partial g_k}{\partial t} = 0 \tag{6.70}$$

However, from Eq. (6.38),

$$\dot\psi_j = \sum_{r=1}^{F} C_{jr}\dot q_r$$

Therefore,

$$\sum_{r=1}^{F} \alpha_{kr}\dot q_r + \beta_k = 0 \qquad k = 1, \ldots, N_t \tag{6.71}$$

where

$$\alpha_{kr} = \sum_{j=1}^{M} g_{k,j}C_{jr} \tag{6.72a}$$

$$\beta_k = \partial g_k/\partial t \tag{6.72b}$$

Since $\dot q_r = dq_r/dt$, Eq. (6.71) can be expressed in the form

$$\sum_{r=1}^{F} \alpha_{kr}dq_r + \beta_k dT = 0 \tag{6.73}$$

The virtual displacements δq_r, however, must, *by definition*, satisfy

$$\sum_{r=1}^{F} \alpha_{kr}\delta q_r = 0 \qquad k = 1, \ldots, N_t \tag{6.74}$$

We say that Eq. (6.73) represents the *actual* constraints and Eq. (6.74) represents the *instantaneous* constraints. Any infinitesimals dq_r which satisfy Eq. (6.73) are called "possible displacements," in distinction to the "virtual displacements" which satisfy Eq. (6.74).

In problems of statics, such temporal constraints are not normally present, but they could easily arise in problems of dynamics. Equations of constraint of the form in Eq. (6.77) sometimes arise which cannot be integrated to give a function of the form $g_k(\psi, t) = 0$. Such nonintegrable equations of constraint are said to be "nonholonomic."

2. *Virtual work* (δW): Work done by specified forces on virtual displacements

3. *Active forces:* Those forces which produce nonzero net virtual work

4. *Ideal mechanical systems:* Systems in which constraints are maintained by forces which do no virtual work (e.g., frictionless contacts, rigid links)

5. *Generalized forces:* Terms which multiply the virtual generalized displacements (δq_r) in the expression for virtual work of the form $\delta W = \Sigma_{r=1}^{E} Q_r \delta q_r$

6. *Equilibrium:* State in which the resultant force on each particle of the system vanishes

7. *Generalized equilibrium:* State in which all the generalized forces vanish

6.3.3 The Principle of Virtual Work

The principle of virtual work may be viewed as a basic *postulate* from which all the well-known laws of vector statics may be deduced, or it may be proved directly from the laws of vector statics. The following two-part statement of principle is proved the latter way in Sec. 11.30 of Ref. 66.

1. If an ideal mechanical system is in equilibrium, the net virtual work of all the active forces vanishes for every set of virtual displacements.

2. If the net virtual work of all the active forces vanishes, for every set of virtual displacements, an ideal mechanical system is in a state of generalized equilibrium.

It is pointed out in Ref. 66 that no distinction need be made between generalized equilibrium and equilibrium for systems which are at rest (relative to an inertial frame) before the active forces are applied.

For a system which consists of n rigid bodies, we may express the resultant external force acting on a typical body by $\mathbf{X}^p$ acting through a reference point fixed in the body which momentarily occupies the point $\mathbf{x}^p$ measured in the global coordinate frame. Similarly the active torque $\mathbf{M}^p$ is assumed to act on the same body. The virtual displacements of the reference points are denoted by $\delta\mathbf{x}^p$, and the corresponding rotations of each rigid body may be expressed by the infinitesimal rotation vector

$$\delta\boldsymbol{\theta}^p = (\delta\theta_1^p, \delta\theta_2^p, \delta\theta_3^p) \tag{6.75}$$

where $\delta\theta_i^p$ is the rotation component about an axis, through the point $\mathbf{x}^p$, parallel to the global x_i axis. The virtual work of all the active forces is therefore given by

$$\delta W = \sum_{p=1}^{n} \sum_{i=1}^{3} (X_i^p \, \delta x_i^p + M_i^p \, \delta\theta_i^p) \tag{6.76}$$

From Eqs. (6.50), and (6.56) we know that

$$\delta x_i^p = \sum_{j=1}^{F} U_{ij}^p \, \delta q_j \tag{6.77}$$

$$\delta\theta_i^p = \sum_{j=1}^{F} W_{ij}^p \, \delta q_j \tag{6.78}$$

Upon substitution of Eqs. (6.77) and (6.78) into Eq. (6.76) we find that

$$\delta W = \sum_{j=1}^{F} Q_j \, \delta q_j \tag{6.79}$$

where $\qquad Q_j \equiv \sum_{p=1}^{n} \sum_{i=1}^{3} (U_{ij}^p X_i^p + W_{ij}^p M_i^p) \qquad j = 1, \ldots, F \tag{6.80}$

Equation (6.80) is an explicit formula for the calculation of generalized forces, although they can usually be calculated in more direct, less formal, ways as described in Sec. 11.22 of Ref. 66.

From part 1 of the statement of the principle of virtual work, it follows that if an ideal system is in equilibrium, then for all possible choices of δq_i

$$\delta W = \sum_{j=1}^{F} Q_j \, \delta q_j = 0 \tag{6.81}$$

In the absence of temporal constraints, the δq_j are all-independent, and it follows from Eq. (6.81) that for such cases

$$Q_j = 0 \qquad j = 1, 2, \ldots, F \tag{6.82}$$

Thus, the vanishing of all generalized forces is a necessary condition for equilibrium and a sufficient condition (by definition) for generalized equilibrium.

For examples of the method of virtual work applied to the statics of machine systems, see Sec. 11.30 of Ref. 66.

For those cases where the generalized coordinates q_r are not independent, but are related by constraint relations expressed in the form of Eq. (6.71) (arising either from temporal or nonholonomic constraints), we cannot reduce Eq. (6.81) to the simple form of Eq. (6.82). For such cases we can use the method of Lagrange multipliers.

6.3.4 Lagrange Multipliers

When the F generalized coordinates are not independent, but are related by equations of constraint such as Eq. (6.74), i.e.,

$$\sum_{j=1}^{F} \alpha_{ij} \, \delta q_j \qquad i = 1, \ldots, N_t \tag{6.83}$$

we say that the generalized coordinates form a "redundant," or "superfluous," set.

However, we may choose to think of all the q_i as independent coordinates if we conceptually remove the physical constraints (e.g., hinge pins or slider guides) and consider the reactions produced by the constraints as external forces. In other terms, we must supplement the generalized forces Q_i of the active loads by a set of forces Q'_i, associated with the reactions produced by the (now deleted) constraints. Since the reactions in an ideal system, by definition, produce no work on any displacement compatible with the constraints, it is necessary that the supplemental forces Q'_i satisfy the condition

$$\sum_{j=1}^{F} Q'_j \, \delta q_j = 0 \tag{6.84}$$

where the displacements δq_j satisfy the constraint equation represented by Eq. (6.83).

It may be shown (Ref. 100, p. 562) that in order for Eqs. (6.83) and (6.84) to be satisfied, it is necessary that

$$Q'_j = \sum_{i=1}^{N_t} \lambda_i \alpha_{ij} \tag{6.85}$$

where λ_i are unknown functions of time called "lagrangian multipliers." Accordingly, Eq. (6.85) provides the desired supplements to the generalized forces, and the equations of equilibrium assume the more general form

$$Q_j + Q_j' = Q_j + \sum_{i=1}^{N_t} \lambda_i \alpha_{ij} = 0 \qquad j = 1, ..., F \tag{6.86}$$

If we consider that Q_j and α_{ij} are known functions of the q_r, then Eqs. (6.86) and (6.69) provide for (holonomic constraints) $F + N_t$ equations for determining the unknowns $q_1, ..., q_F$ and $\lambda_1, ..., \lambda_{N_t}$. However, in practice, it is likely that Q_j and α_j are expressed as functions of $\psi_1, ..., \psi_M$. Then it is necessary to utilize the M additional equations consisting of Eqs. (6.24) and (6.31). For nonholonomic constraints, equations of the form of Eq. (6.69) are not available, and one must utilize instead differential equations of the form of Eq. (6.71). Such problems must usually be solved by a process of numerical integration. Numerical integration will be discussed in Sec. 6.5.

6.3.5 Conservative Systems

If the work done on an ideal mechanical system by a set of forces is a function $W(q_1,..., q_F)$ of the generalized coordinates only, the system is said to be conservative. Such forces could arise from elastic springs, gravity, and other field effects.

The increment in W due to a virtual displacement δq_i is

$$\delta \overline{W} (\partial \overline{W}/\partial q_1) \partial q_1 + \cdots + (\partial \overline{W}/\partial q_F) \partial q_F \tag{6.87}$$

By tradition, one defines the "potential energy" or "potential" of the system as

$$V(q) \equiv - \overline{W}(q) \tag{6.88}$$

Therefore

$$\delta \overline{W} = - (\partial V/\partial q_1) \partial q_1 - \cdots - (\partial V/\partial q_F) \delta q_F$$

If, in addition, a set of active forces (not related to $\overline{W}$) acts on the system and produces the generalized forces Q_r^*, the net work done during a virtual displacement is

$$\delta W = (Q_1^* - \partial V/\partial q_1) \partial q_1 - \cdots - (Q_F^* - \partial V/\partial q_F) \partial q_F = 0 \tag{6.89}$$

where the right-hand zero is a consequence of the principle of virtual work. From the definition of Q_r (Sec. 6.3.2) it follows that

$$Q_r = Q_r^* - \partial V/\partial q_r \tag{6.90}$$

If all the δq_r are independent (i.e., no temporal or nonholonomic constraints exist) then Eq. (6.89) predicts that

$$Q_r \equiv Q_r^* - \partial V/\partial q_r = 0 \qquad r = 1, ..., F \tag{6.91}$$

If the δq_r are related, as in Eq. (6.83),

$$\sum_{j=1}^{F} \alpha_{ij} \delta q_j = 0 \qquad i = 1, 2, ..., N_t \tag{6.92}$$

it follows from the discussion leading to Eq. (6.86) that

$$Q_j^* - \frac{\partial V}{\partial q_j} + \sum_{i=1}^{N_t} \lambda_i \alpha_{ij} = 0 \qquad j = 1, ..., F \tag{6.93}$$

It is only necessary to define V within an additive constant, since only the derivative of V enters into the equilibrium equations.

Some examples of the form of V for some common cases follow:

1. A linear spring with free length s_f, current length s, and stiffness K has potential

$$V(s) = \frac{1}{2} K(s - s_e)^2 \tag{6.94}$$

2. A system of mass particles close to the surface of the earth (where the acceleration of gravity g is essentially constant) has potential

$$V(\bar{z}) = W\bar{z} \tag{6.95}$$

where $\bar{z}$ is the elevation of center of gravity of the system vertically above an arbitrary datum plane, and W is the total weight of all particles.

3. A mass particle is located at a distance r from the center of the earth. If its weight is W_e when the particle is at a distance r_e from the earth's center (e.g., at the surface of the earth), its potential in general is

$$V(r) = -W_e r_e^2/r \tag{6.96}$$

In all these examples, V has been expressed in terms of a lagrangian variable. More generally, V is of the form

$$V = V(\psi_1, ..., \psi_M) \tag{6.97}$$

The required derivatives of V are of the form

$$\frac{\partial V(q)}{\partial q_r} = \sum_{j=1}^{M} \frac{\partial V(\psi)}{\partial \psi_j} \frac{\partial \psi_j}{\partial q_r} = \sum_{j=1}^{M} \frac{\partial V(\psi)}{\partial \psi_r} C_{jr} \tag{6.98}$$

and C_{jr} are components of the matrix $\mathbf{C}$ defined by Eq. (6.39). Illustrations of the use of potential functions in problems of statics of machines are given in Sec. 11.42 of Ref. 66.

6.3.6 Reactions at Joints

By Vector Statics. Perhaps the most straightforward way to compute reactions is to establish the three equations of equilibrium for each link, as exemplified by Eqs. (6.67). The equations are linear in the desired reactions and are readily solved by the same linear equation subroutine that is needed for other purposes in a numerical analysis. This method has been described in depth for planar mechanisms in Secs. 11.50 and 11.51 of Ref. 66, and has been used in the author's computer programs STATMAC[70] and DYMAC.[68] It was also used in Ref. 72 and in the computer programs MEDUSA[18] and VECNET.[2]

By Principle of Virtual Work. In order for the internal reactions at a specific joint to appear in the expression for virtual work, it is necessary to imagine that the joint is "broken" and then to introduce, as active forces, the unknown joint reactions needed to maintain closure. This technique is briefly described in Ref. 64, and in more detail by Denavit et al.[16] and by Uicker,[97] who combines the method of virtual work with Lagrange's equations of second type (see Sec. 6.4.2).

By Lagrange Multipliers.   As shown by Chace et al.[10] and Smith,[85] it is possible to find certain crucial joint reactions, simultaneously with the solution of the general dynamics problem, by the method of Lagrange multipliers. A brief description of this method is also given in Ref. 64.

6.3.7   Effects of Friction in Joints

Up to this point, it has been assumed that all internal reactions are workless. If the friction at the joints is velocity-dependent, we can place a damper (not necessarily linear) at the joints; the associated pairs of velocity-dependent force or torque reactions are then included in the active forces $\mathbf{X}^p$ and $\mathbf{M}^p$ as defined in Sec. 6.3.3. These forces, which appear linearly in the generalized forces Q [see Eq. (6.80)], result in velocity-dependent generalized forces.

However, when Coulomb friction exists at the joints, the associated active forces become functions of the unknown joint reactions. These forces complicate the solution of the equations, but they can be accounted for in a number of approximate ways. A detailed treatment of frictional effects in statics of planar mechanisms is given in Secs. 11.60 to 11.63 of Ref. 66. Frictional effects in the dynamics of planar machines are discussed in Sec. 12.70 of Ref. 66. Greenwood[30] (p. 271) utilizes Lagrange multipliers to solve a problem of frictional effects between three bodies moving in a plane with translational motion only.

6.4   KINETICS OF MACHINE SYSTEMS

6.4.1   Lagrange's Form of d'Alembert's Principle

We now assume that active forces $\mathbf{X}^i \equiv (X_1, X_2, X_3)^i$ are applied to the mass center $\mathbf{x}^i \equiv (x_1, x_2, x_3)^i$ of link i and an active torque $\mathbf{M}^i \equiv (M_1, M_2, M_3)^i$ is exerted about the mass center. Any set of active forces which does not pass through the mass center may be replaced by a statically equipollent set of forces which does. In general, $\mathbf{X}^i$ and $\mathbf{M}^i$ may depend explicitly upon position, velocity, and time. Massless springs may be modeled by pairs of equal, but opposite, collinear forces which act on the links at the terminal points of the springs. The spring forces may be any specified linear or nonlinear function of the distance between the spring's terminals. Dampers may be modeled by similar pairs of equal and opposite collinear forces which are dependent on the rate of relative extension of the line joining the damper terminals. Examples of such springs and dampers are given in Sec. 14.43 of Ref. 66.

A typical link is subjected to inertia forces $-m\ddot{\mathbf{x}}$ as well as the real forces $\mathbf{X}$. Thus the virtual work due to *both* these "forces" is given by

$$\delta W_F = \delta \mathbf{x}^T (\mathbf{X} - m\ddot{\mathbf{x}}) \tag{6.99}$$

From Eqs. (6.48) and (6.50) we observe that

$$\ddot{\mathbf{x}} = \mathbf{U}\ddot{\mathbf{q}} + \mathbf{v} \tag{6.100a}$$

$$\delta\mathbf{x} = \mathbf{U}\,\delta\mathbf{q} \qquad \therefore \delta\mathbf{x}^T = \delta\mathbf{q}^T\mathbf{U}^T \tag{6.100b}$$

where superscript T denotes a matrix transpose. Therefore the virtual work due to "forces" on the rigid body is

$$\delta W_F = \delta\mathbf{q}^T[\mathbf{U}^T\mathbf{X} - m\mathbf{U}^T\mathbf{v} - m\mathbf{U}^T\mathbf{U}\ddot{\mathbf{q}}] \tag{6.101}$$

To this must be added the virtual work due to the real and inertial torques which act about the mass center. To help calculate these, we introduce the following notation:

$$
\begin{aligned}
G &= \text{The mass center of a rigid body} \\
(\xi_1, \xi_2, \xi_3) &= \text{Local orthogonal reference axes through } G \text{ aligned with the principal} \\
&\quad \text{axes of inertia of the body} \\
\overline{J}_1, \overline{J}_2, \overline{J}_3 &= \text{Principal moments of inertia of the body, referred to axes } \xi_1, \xi_2, \xi_3 \\
\overline{M}_1, \overline{M}_2, \overline{M}_3 &= \text{Components of external torque in directions } \xi_1, \xi_2, \xi_3 \\
\overline{\omega}_1, \overline{\omega}_2, \overline{\omega}_3 &= \text{Components of angular velocity in directions } \xi_1, \xi_2, \xi_3 \\
\delta\overline{\theta}_1, \delta\overline{\theta}_2, \delta\overline{\theta}_3 &= \text{Infinitesimal rotation components about axes } \xi_1, \xi_2, \xi_3 \\
D_{ij} &= \text{Direction cosine between (local) axis } \xi_i \text{ and (global) axis } x_j
\end{aligned}
$$

Note that an overbar is used on vector components referred to the local (ξ) axes. Corresponding symbols without overbars refer to global (x) axes. The two systems are related by matrix relationships of the type

$$\overline{\mathbf{M}} = \mathbf{D}\mathbf{M} \tag{6.102}$$

$$\overline{\boldsymbol{\omega}} = \mathbf{D}\boldsymbol{\omega} \tag{6.103}$$

$$\delta\overline{\boldsymbol{\theta}} = \mathbf{D}\,\delta\boldsymbol{\theta} \tag{6.104}$$

It may be shown[29,53] that the direction cosines are related to the Euler angles as follows:

$$\mathbf{D} \equiv [D_{ij}] = \begin{bmatrix} C\Psi C\Phi - C\Theta S\Phi S\Psi & C\Psi S\Phi + C\Theta C\Phi S\Psi & S\Psi S\Theta \\ -S\Psi C\Psi - C\Theta S\Phi C\Psi & -S\Psi S\Phi + C\Theta C\Phi C\Psi & C\Psi S\Theta \\ S\Theta S\Phi & -S\Theta C\Phi & C\Theta \end{bmatrix} \tag{6.105}$$

where

$$C\Psi = \cos\Phi \qquad C\Phi = \cos\Phi \qquad C\Theta = \cos\Theta$$

$$S\Psi = \sin\Psi \qquad S\Phi = \sin\Phi \qquad S\Theta = \sin\Theta \tag{6.106}$$

and $\mathbf{D}$ is an orthogonal matrix, i.e.,

$$\mathbf{D}^T = \mathbf{D}^{-1} \tag{6.107}$$

We now take note of Euler's equation of motion for rigid bodies expressed in the form[53,94]

$$\overline{M}_i + \overline{T}_i = 0 \qquad i = 1, 2, 3 \tag{6.108}$$

where the *inertia torque* components $\overline{T}_i$ are given by

$$\overline{T}_1 = -\overline{J}_1\dot{\overline{\omega}}_1 + (\overline{J}_2 - \overline{J}_3)\overline{\omega}_2\overline{\omega}_3 \tag{6.109a}$$

$$\overline{T}_2 = -\overline{J}_2\dot{\overline{\omega}}_2 + (\overline{J}_3 - \overline{J}_1)\overline{\omega}_3\overline{\omega}_1 \tag{6.109b}$$

$$\overline{T}_3 = -\overline{J}_3\dot{\overline{\omega}}_3 + (\overline{J}_1 - \overline{J}_2)\overline{\omega}_1\overline{\omega}_2 \tag{6.109c}$$

In matrix form, we can write

$$\overline{\mathbf{T}} = -\overline{\mathbf{J}}\dot{\overline{\boldsymbol{\omega}}} + \boldsymbol{\tau} \tag{6.110}$$

where

$$\overline{\mathbf{J}} = \begin{bmatrix} \overline{J}_1 & 0 & 0 \\ 0 & \overline{J}_2 & \\ 0 & 0 & \overline{J}_3 \end{bmatrix} \tag{6.111}$$

$$\boldsymbol{\tau} = \begin{Bmatrix} (\overline{J}_2 - \overline{J}_3)\overline{\omega}_2\overline{\omega}_3 \\ (\overline{J}_3 - \overline{J}_1)\overline{\omega}_3\overline{\omega}_1 \\ (\overline{J}_1 - \overline{J}_2)\overline{\omega}_1\overline{\omega}_2 \end{Bmatrix} \tag{6.112}$$

From Eq. (6.54) we recall that

$$\boldsymbol{\omega} = \mathbf{W}\dot{\mathbf{q}} \tag{6.113}$$

Hence, the local velocity components are given by Eq. (6.103) in the form

$$\overline{\boldsymbol{\omega}} = \mathbf{D}\boldsymbol{\omega} = \mathbf{DW}\dot{\mathbf{q}} \tag{6.114}$$

The corresponding acceleration components are therefore

$$\dot{\overline{\boldsymbol{\omega}}} = \overline{\mathbf{W}}\ddot{\mathbf{q}} + \overline{\mathbf{V}} \tag{6.115}$$

where

$$\overline{\mathbf{W}} = \mathbf{DW} \tag{6.116}$$

$$\overline{\mathbf{V}} = (\mathbf{D}\dot{\mathbf{W}} + \dot{\mathbf{D}}\mathbf{W})\dot{\mathbf{q}} \tag{6.117}$$

Thus the inertia torque given by Eq. (6.110) can be expressed as

$$\overline{\mathbf{T}} = -\overline{\mathbf{J}}\,\overline{\mathbf{W}}\,\ddot{\mathbf{q}} - \overline{\mathbf{J}}\,\overline{\mathbf{V}} + \boldsymbol{\tau} \tag{6.118}$$

In accordance with d'Alembert's principle, we may therefore express the virtual work of the active and inertia torques in the form

$$\delta W_M = \sum_{i=1}^{3} \delta\overline{\theta}_i(\overline{M}_i + \overline{T}_i) = \delta\overline{\boldsymbol{\theta}}^T(\overline{\mathbf{M}} + \overline{\mathbf{T}}) \tag{6.119}$$

Upon utilization of Eqs. (6.104) and (6.118) we find

$$\delta W_M = (\mathbf{D}\delta\boldsymbol{\theta})^T(\overline{\mathbf{M}} + \boldsymbol{\tau} - \overline{\mathbf{J}}\,\overline{\mathbf{V}} - \overline{\mathbf{J}}\,\overline{\mathbf{W}}\ddot{\mathbf{q}}) \tag{6.120}$$

Since $\delta\boldsymbol{\theta} = \mathbf{W}\,\delta\mathbf{q}$ and $\overline{\mathbf{W}} = \mathbf{DW}$, we can write

$$(\mathbf{D}\delta\boldsymbol{\theta})^T = (\mathbf{DW}\delta\mathbf{q})^T = (\overline{\mathbf{W}}\delta\mathbf{q})^T = \delta\mathbf{q}^T\overline{\mathbf{W}}^T \tag{6.121}$$

Now we can write Eq. (6.120) in the form

$$\delta W_M = \delta\mathbf{q}^T[\overline{\mathbf{W}}^T(\overline{\mathbf{M}} + \boldsymbol{\tau} - \overline{\mathbf{J}}\,\overline{\mathbf{V}}) - \overline{\mathbf{W}}^T\overline{\mathbf{J}}\,\overline{\mathbf{W}}\ddot{\mathbf{q}}] \tag{6.122}$$

Adding δW_M to δW_F given by Eq. (6.101), the virtual work of all the real and inertial forces and torques is seen to take the form

$$\delta W_M + \delta W_F = \delta\mathbf{q}^T[(\mathbf{U}^T\mathbf{X} + \overline{\mathbf{W}}^T\overline{\mathbf{M}}) - m\mathbf{U}^T\mathbf{v} + \overline{\mathbf{W}}^T(\boldsymbol{\tau} - \overline{\mathbf{J}}\,\overline{\mathbf{V}}) - (m\mathbf{U}^T\mathbf{U} + \overline{\mathbf{W}}^T\overline{\mathbf{J}}\,\overline{\mathbf{W}})\ddot{\mathbf{q}}] \tag{6.123}$$

If we use a superscript k to denote the virtual work done on the kth rigid body in the system, we can express the virtual work of all the real and inertial forces of the complete system by

$$\delta W = \sum_{k=1}^{B} (\delta W_M + \delta W_F)^k = \delta\mathbf{q}^T(\mathbf{Q} + \mathbf{Q}^\dagger - \mathbf{I}\ddot{\mathbf{q}}) = 0 \tag{6.124}$$

where we have utilized the following definitions:

$$\mathbf{I} \equiv \sum_{k=1}^{B} (m\mathbf{U}^T\mathbf{U} + \overline{\mathbf{W}}^T\overline{\mathbf{J}}\,\overline{\mathbf{W}})^k \tag{6.125}$$

$$\mathbf{Q}^\dagger = \sum_{k=1}^{B} [-m\mathbf{U}^T\mathbf{v} + \overline{\mathbf{W}}^T(\boldsymbol{\tau} - \overline{\mathbf{J}}\,\overline{\mathbf{V}})]^k \tag{6.126}$$

$$\mathbf{Q} = \sum_{k=1}^{B} (\mathbf{U}^T\mathbf{X} + \overline{\mathbf{W}}^T\overline{\mathbf{M}})^k = \sum_{k=1}^{B} (\mathbf{U}^T\mathbf{X} + \mathbf{W}^T\mathbf{M})^k \tag{6.127}$$

B is the number of rigid bodies present. It is worth noting that the definition of $\mathbf{I}$ in Eq. (6.125) implies that $\mathbf{I}$ is symmetric, and that the rightmost expression in Eq. (6.127) follows from Eqs. (6.102), (6.107), and (6.116).

We have set $\delta W = 0$ in Eq. (6.124) in accordance with the generalized version of the principle of virtual work which includes the inertial forces. This form of the principle is referred to as Lagrange's form of d'Alembert's principle.[30,64]

When there are no constraints relating the generalized coordinates q_i, the rightmost factor in Eq. (6.124) must vanish, and we have the equations of motion in their final form:

$$\mathbf{I}\ddot{\mathbf{q}} = \mathbf{Q} + \mathbf{Q}^\dagger \tag{6.128a}$$

or

$$\sum_{j=1}^{F} I_{ij}\ddot{q}_j = Q_i + Q_i^\dagger \qquad i = 1, \dots, F \tag{6.128b}$$

Now consider the case where the generalized velocities are related by N_t constraint equations of the form

$$\sum_{j=1}^{F} \alpha_{ij}\dot{q}_j + \beta_i = 0 \qquad i = 1, 2, \ldots, N_t \qquad (6.129)$$

where α_{ij} and β_i may all be functions of the lagrangian coordinates ψ_r and of the time t. These relationships may be nonintegrable (nonholonomic) or derivable from holonomic constraints of the form

$$g_i(\psi_1, \ldots, \psi_M, t) = 0 \qquad (6.130)$$

In either case, the corresponding virtual displacements must satisfy (by definition) the equation

$$\sum_{j=1}^{F} \alpha_{ij}\delta q_j = 0 \qquad (6.131)$$

Using the same argument which led to Eq. (6.85), we conclude that the generalized forces must be supplemented by the "constraint" forces

$$Q'_j = \sum_{i=1}^{N_t} \lambda_i \alpha_{ij} \qquad j = 1, 2, \ldots, F \qquad (6.132a)$$

or

$$\mathbf{Q}' = \boldsymbol{\alpha}^T \boldsymbol{\lambda} \qquad (6.132b)$$

where the Lagrange multipliers $\lambda_1, \ldots, \lambda_{N_t}$ are unknown functions of time to be determined. Therefore the equations of motion, Eqs. (6.128), may be written in the more general form

$$\mathbf{I}\ddot{\mathbf{q}} = \mathbf{Q} + \mathbf{Q}^\dagger + \boldsymbol{\alpha}^T \boldsymbol{\lambda} \qquad (6.133a)$$

or

$$\sum_{j=1}^{F} I_{ij}\ddot{q}_j = Q_i + Q_i^\dagger + \sum_{j=1}^{N_t} \alpha_{ji}\lambda_j \qquad i = 1, \ldots, F \qquad (6.133b)$$

These F equations together with the N_t constraint equations in the form of Eq. (6.129) suffice to find the unknowns $q_1, \ldots, q_F$ and $\lambda_1, \ldots, \lambda_{N_t}$.

From previous definitions, it may be verified that the terms $Q_i^\dagger$ are all quadratic forms in the generalized velocities $\dot{q}_r$, and therefore reflect the effects of the so-called centrifugal and Coriolis (inertia) forces. However, no use is made of this fact in the numerical methods discussed below.

Equations (6.133) form the basis of the author's general-purpose computer program DYMAC (*dynamics of machinery*).[1,66,68]

6.4.2 Equations of Motion from Lagrange's Equations

The kinetic energy T of a system of B rigid bodies is given by

$$T = \frac{1}{2}\sum_{k=1}^{B} (m^k \dot{\mathbf{x}}^T \dot{\mathbf{x}} + \overline{\boldsymbol{\omega}}^T \overline{J}\overline{\boldsymbol{\omega}}) \qquad (6.134)$$

where all terms have been defined above. Upon recalling that $\dot{\mathbf{x}} \equiv \mathbf{U}\dot{\mathbf{q}}$ and $\overline{\boldsymbol{\omega}} = \overline{W}\dot{\mathbf{q}}$, it follows that

$$T = \tfrac{1}{2}\, \dot{\mathbf{q}}^T \mathbf{I}^k \dot{\mathbf{q}} = \frac{1}{2} \sum_{i,j=1}^{F} I_{ij}\dot{q}_i \dot{q}_j \tag{6.135}$$

where the generalized inertia coefficients I_{ij} are precisely those defined by Eq. (6.125). Since the I_{ij} are functions of position only, they may be considered implicit functions of the generalized coordinates q, i.e.,

$$I_{ij}[\psi(q)] \tag{6.136}$$

Until further notice we will assume that I_{ij} is known explicitly in terms of the generalized coordinates [i.e., $I_{ij} = I_{ij}(q)$] and that the q_r are all independent.

Now we may make use of the well-known Lagrange equations (of second type) for independent generalized coordinates*:

$$\frac{d}{dt}\frac{\partial T}{\partial \dot{q}^r} - \frac{\partial T}{\partial q^r} = Q_r = Q_r^* - \frac{\partial V}{\partial q_r} \tag{6.137}$$

where $\mathbf{Q}$ is the generalized force corresponding to $\mathbf{q}$, and $\mathbf{Q}^*$ and $\mathbf{V}$ are as defined in Sec. 6.3.5.

Making use of Eq. (6.135) and the fact that $I_{ij} = I_{ji}$, we find that

$$\frac{\partial T}{\partial \dot{q}_r} = \sum_{j=1}^{F} I_{rj}\dot{q}_j \tag{6.138}$$

and

$$\frac{\partial T}{\partial q_r} = \sum_{i,j=1}^{F} \tfrac{1}{2} I_r^{ij}\dot{q}_i \dot{q}_j \tag{6.139}$$

where

$$I_r^{ij} \equiv \frac{\partial I_{ij}(q)}{\partial q_r} \tag{6.140}$$

Equation (6.138) leads to the further result

$$\frac{d}{dt}\left(\frac{\partial T}{\partial \dot{q}_r}\right) = \sum_{j=1}^{F} I_{rj}\ddot{q}_j + \sum_{j=1}^{F} \dot{I}_{rj}\dot{q}_j \tag{6.141}$$

We may represent $\dot{I}_{rj}$ in the form

$$\dot{I}_{rj} = \sum_{i=1}^{F} \frac{\partial I_{rj}(q)}{\partial q_i}\dot{q}_i = \sum_{i=1}^{F} I_i^{rj}\dot{q}_i \tag{6.142}$$

Upon substitution of Eqs. (6.138) to (6.142) into Lagrange's equation (6.137), we find

$$\sum_{j=1}^{F} I_{rj}\ddot{q}_j + \sum_{i=1}^{F}\sum_{j=1}^{F} G_r^{ij}\dot{q}_i\dot{q}_j = Q_r \qquad r = 1, 2, \ldots, F \tag{6.143}$$

where

$$G_r^{ij} \equiv I_i^{rj} - \tfrac{1}{2} I_r^{ij} \tag{6.144}$$

Equation (6.143) represents a set of second-order differential equations of motion in

---

*These equations may be found in any text on analytical mechanics or variational mechanics, e.g., Refs. 29, 30, 40, 46, 61, and 102.

the generalized coordinates. However, it can be cast in a somewhat more symmetric form by a rearrangement of the coefficients G_r^{ij}. Toward this end, we note that an interchange of the dummy subscripts i and j on the left-hand side of the equation

$$\sum_{i,j=1}^{F} I_i^{rj}\dot{q}_i\dot{q}_j \equiv \sum_{i=1}^{F}\sum_{j=1}^{F} I_j^{ri}\dot{q}_j\dot{q}_i \tag{6.145}$$

results identically in the right-hand side. Hence the double sum in Eq. (6.143) can be expressed in the form

$$\sum_{i=1}^{F}\sum_{j=1}^{F} G_r^{ij}\dot{q}_i\dot{q}_j = \sum_{i=1}^{F}\sum_{j=1}^{F} \begin{bmatrix} ij \\ r \end{bmatrix}\dot{q}_i\dot{q}_j \tag{6.146}$$

where

$$\begin{bmatrix} ij \\ r \end{bmatrix} \equiv \tfrac{1}{2}\,(I_i^{rj}+I_j^{ri} - I_r^{ij}) \tag{6.147}$$

is the "Christoffel symbol of the first kind" associated with the matrix of inertia coefficients I_{ij}.

Upon substitution of Eq. (6.146) into Eq. (6.143) we find

$$\sum_{j=1}^{F} I_{rj}\ddot{q}_j + \sum_{i=1}^{F}\sum_{j=1}^{F} \begin{bmatrix} ij \\ r \end{bmatrix}\dot{q}_i\dot{q}_j = Q_r \tag{6.148}$$

These equations are called the "explicit form of Lagrange's equations" by Whittaker.[102]

Although they are a particularly good starting point for numerical integration of the equations of motion, these "explicit" equations have not been widely used in the engineering literature. They are mentioned by Beyer,[6] who found it necessary to evaluate the coefficients I_{rj} semigraphically and to evaluate coefficients of the type I_r^{ij} by an approximate finite difference formula.

Upon comparing Eqs. (6.128) and (6.148), we see that the terms Q_r† defined by Eq. (6.126) play precisely the same role, in the equations of motion, as the double sum in Eq. (6.148). The great advantage of the form involving $Q_r^†$ is that these terms can be found without knowing explicit expressions for I_{ij} in terms of q_i, as would be required for the calculation of the Christoffel symbols via Eq. (6.140). In order to integrate Eq. (6.148) numerically, it is necessary to solve it explicitly for the acceleration $\ddot{q}_i$. This may be done by any elimination procedure or by inversion of the matrix $[I_{rs}]$.

If we denote by J_{pr} the elements of the matrix $[I_{pr}]^{-1}$, we have

$$\sum_{r=1}^{F} J_{pr}I_{rj} \equiv \delta_{pj} \tag{6.149}$$

where $\delta_{pj} = 1$ if $p = j$ and $\delta_{pj} = 0$ otherwise. Upon multiplying Eq. (6.148) through by J_{pr} and summing over r, we find[64]

$$\ddot{q}_p = \sum_{r=1}^{F} J_{pr}Q_r - \sum_{i=1}^{F}\sum_{j=1}^{F} \begin{Bmatrix} p \\ ij \end{Bmatrix}\dot{q}_i\dot{q}_j \tag{6.150}$$

where

$$\begin{Bmatrix} p \\ ij \end{Bmatrix} \equiv \sum_{r=1}^{F} J_{pr}\begin{bmatrix} ij \\ r \end{bmatrix} \tag{6.151}$$

is the Christoffel symbol of second kind.

When the generalized velocities are related by constraints of the form of Eq. (6.129), Lagrange multipliers may be introduced to generate the supplemental forces of Eqs. (6.132). Then it is only necessary to add the term

$$Q'_r = \sum_{i=1}^{N_t} \lambda_i \alpha_{ir} \tag{6.152}$$

to the right-hand side of Eq. (6.148).

Lagrange's equations (with multipliers) are used in several general-purpose computer programs, such as DRAM,[12,85] ADAMS,[58,60] DADS,[32,101] AAPD,[47] IMP,[81,82] and IMPUM.[81] Lagrange multipliers are not necessary for treating open-loop systems such as robots[34,73,98] or other systems with treelike topology[105] as covered by the program UCIN.[37,38]

6.4.3 Kinetics of Single-Freedom Systems

For systems with one DOF there is a single generalized coordinate, and we may write, for brevity,

$$q_1 \equiv q \qquad Q_1 \equiv Q \qquad Q_1^\dagger \equiv Q^\dagger \qquad I_{11} \equiv I \tag{6.153}$$

Then the equations of motion, Eqs. (6.128), reduce to the single differential equation

$$I\ddot{q} = Q + Q^\dagger \tag{6.154}$$

This result also follows directly from the principle of power balance, which states that the power of the active forces $(Q\dot{q})$ equals the rate of increase of the kinetic energy (T). From Eq. (6.135) we see that

$$T \equiv \tfrac{1}{2} I \dot{q}^2 \tag{6.155}$$

Therefore, the power balance principle is

$$dT/dt = (d/dt)\tfrac{1}{2} I\dot{q}^2 = Q\dot{q} \tag{6.156}$$

$$I\dot{q}\ddot{q} + \tfrac{1}{2}(dI/dq)(dq/dt)\dot{q}^2 = Q\dot{q} \tag{6.157}$$

Hence
$$I\ddot{q} = Q - C\dot{q}^2 \tag{6.158}$$

where
$$C \equiv \tfrac{1}{2} dI/dq \tag{6.159}$$

Equation (6.158) is clearly of the form of Eq. (6.154) with

$$Q^\dagger \equiv -C\dot{q}^2 \tag{6.160}$$

Using the simplified notation of Sec. 6.2.6, we can express I, via Eq. (6.125), in the form

$$I = \sum_{k=1}^{B} \sum_{i=1}^{3} (mu_i^2 + \bar{J}_1 \bar{w}_i^2)^k \tag{6.161}$$

Therefore
$$C = \frac{1}{2}\frac{dI}{dq} = \sum_{k=1}^{B} \sum_{i=1}^{3} (mu_i u_i' + \bar{J}_i \bar{w}_i \bar{w}_i')^k \tag{6.162}$$

where

$$u_i' = \sum_{j=1}^{M} \frac{\partial u_i(\psi)}{\partial \psi_j} \frac{\partial \psi_j}{\partial q} = \sum_{j=1}^{M} \frac{\partial u_i(\psi)}{\partial \psi_j} c_j \tag{6.163}$$

$$\overline{w}_i' = \sum_{j=1}^{M} \frac{\partial \overline{w}_i(\psi)}{\partial \psi_j} c_j \tag{6.164}$$

Similarly, the generalized force given by Eq. (6.127) can be expressed in the form

$$Q = \sum_{k=1}^{B} \sum_{i=1}^{3} (u_i X_i + w_i M_i)^k \tag{6.165}$$

For *planar motions,* in which all particles move in paths parallel to the global (x, y) axes, two of the Euler angles shown in Fig. 6.2 vanish (say $\Theta = \Phi = 0$). It then follows from Eq. (6.105) that **D** is a unit matrix, and from Eq. (6.116) that there is no distinction between the angular velocity coefficients w_i and $\overline{w}_i$.

Numerous examples of the application of these equations to planar single-freedom systems are given in Chap. 12 of Ref. 66. For related approaches to the use of the generalized equation of motion, (6.158), see Refs. 21 and 84.

For *conservative systems* (see Sec. 6.3.5) with one DOF, where all generalized forces are derived from a potential function $V(q)$, the generalized forces can be expressed in the form $Q = -dV/dq$. Therefore, the power balance theorem can be expressed in the form

$$\int_{T_0}^{T} dT = \int_{q_0}^{q} Q \, dq = -\int_{V_0}^{V} dV \tag{6.166}$$

where q_0 is some reference state where the potential energy is V_0 and the kinetic energy is T_0.

The integrated form of this equation is

$$T - T_0 = -(V - V_0) \tag{6.167}$$

or

$$T + V = T_0 + V_0 \equiv E \tag{6.168}$$

where E (a constant) is the total energy of the system. Equation (6.168) is called the "energy integral" or the "first integral" of the equation of motion.

Recalling Eq. (6.155), we can write Eq. (6.168) in the form

$$\frac{dq}{dt} = \dot{q} = \pm (2T/I)^{1/2} = \pm \{2[E - V(q)]/I\}^{1/2} \tag{6.169}$$

Finally, a relation can be found between t and q by means of a simple quadrature expression (e.g., by Simpson's rule) of Eq. (6.169) written in the form $dt = dq/\dot{q}$; i.e.,

$$t - t_0 = \int_{q_0}^{q} \frac{dq}{\dot{q}} = \pm \int_{q_0}^{q} \left\{ \frac{I(q)}{2[E - V(q)]} \right\}^{1/2} dq \tag{6.170}$$

where t_0 is the time at which $q = q_0$. The correct choice of $+$ or $-$ in Eqs. (6.169) and (6.170) is what makes $\dot{q}$ consistent with the given "initial" velocity at $t = t_0$; thereafter, a reversal of sign occurs whenever $\dot{q}$ becomes zero. Examples of the use of these equations are given in Sec. 12.50 of Ref. 66, where it is shown, among other results, that a stable oscillatory motion can occur in the neighborhood of equilibrium points (where $dV/dq = 0$) only if V is a local minimum at this point.

Many slightly different versions of the method of power balance have been published.[3,21,28,33,74] A brief review of the differences among these versions is given in Ref. 64.

6.4.4 Kinetostatics and Internal Reactions

If all the external forces acting on the links of a machine are given, the consequent motion can be found by integrating the generalized differential equation of motion, Eq. (6.133), or any of its equivalent forms. This is the *general problem of kinetics.*

However, one is frequently required to find those forces which must act on a system or subsystem (e.g., a single link) in order to produce a *given motion.* This problem can be solved, as we shall see, by using the methods of statics and is therefore called a problem of "kinetostatics."

Suppose, for example, that the general problem of kinetics has already been solved for a given machine (typically by a numerical solution of the differential equations as explained in Sec. 6.5). This means that the positions, velocities, and accelerations of all points are known at a given instant of time, and the corresponding inertia force $-m\ddot{x}$ is known. Similarly, one may now calculate the inertia torque $\bar{T}$, whose components, referred to principal central axes, are given by Eq. (6.109) or (6.118) for each link. We may then use the matrix D [defined by Eq. (6.105)] to express the components of inertia torque T_i along the global (x_i) axes as follows:

$$T = D^T \bar{T} \tag{6.171}$$

Furthermore, let X' and M' be the net force and moment (about the mass center) exerted on the link by all the other links in contact with the subject link. For example, if the subject link k has its mass center at x^k and is in contact with link n which exerts the fore X'^{kn} and moment M'^{kn} at point x^{kn}, we can write

$$X'^k = \sum_n X'^{kn} \tag{6.172}$$

$$M'^k = \sum_n [M'^{kn} + (x^n - x^k) \times X'^{kn}] \tag{6.173}$$

where the summation is made over all links which contact link k. Using this notation, we may write, for each link k, the following vector equations of "dynamic equilibrium" for forces and moments respectively:

$$X^k - m^k\ddot{x}^k + \sum_n X'^{kn} = 0 \tag{6.174a}$$

$$M^k + T^k + \sum_n [M'^{kn} + (x^n - x^k) \times X'^{kn}] = 0 \tag{6.174b}$$

Two such *vector equations* of kinetostatics can be written for each body in the system ($k = 1, 2, ..., B$), leading to a set of $6B$ scalar equations which are linear in the unknown reaction components $X_i'^{kn}$ and $M_i'^{kn}$. If one expresses Newton's law of action and reaction in the form

$$X_i'^{kn} = -X_i'^{kn} \qquad M_i'^{kn} = -M_i'^{kn}$$

there will be a sufficient number of linear equations to solve for all the unknown reaction force and moment components.[1,64]

When temporal constraints are present, the applied forces $\mathbf{X}^k$ and moments $\mathbf{M}^k$ are not all known a priori, and some of the scalar quantities X_i^k and M_i^k (or some linear combination of them) must be treated as unknowns in Eqs. (6.174a) and (6.174b). For example, if a motor exerts an unknown torque on link 7 about the local axis ξ_2^7 in order to maintain a constant angular speed $\overline{\omega}_2^7$ about that axis, the required *control torque* $\overline{T}_2^7$ is unknown a priori, and the following terms will appear in Eq. (6.174b) as unknown contributions to $\mathbf{M}^7$:

$$\begin{bmatrix} \Delta M_1^7 \\ \Delta M_2^7 \\ \Delta M_3^7 \end{bmatrix} = \begin{bmatrix} D_{11}^7 & D_{21}^7 & D_{31}^7 \\ D_{12}^7 & D_{22}^7 & D_{32}^7 \\ D_{13}^7 & D_{23}^7 & D_{33}^7 \end{bmatrix} \begin{bmatrix} 0 \\ \overline{T}_2^7 \\ 0 \end{bmatrix} = \begin{bmatrix} D_{21}\overline{T}_2^7 \\ D_{22}\overline{T}_2^7 \\ D_{23}\overline{T}_2^7 \end{bmatrix}$$

A systematic procedure for the automatic generation and solution of the appropriate linear equations, when such *control forces* must be accounted for, is given in Ref. 1.

6.4.5 Balancing of Machinery

From elementary mechanics, it is known that the net force acting on a system of particles with fixed total mass equals the total mass times the acceleration of the center of mass. If this force is provided by the foundation of a machine system, the equal but opposite force felt by the foundation is called the "shaking force." For a system of B rigid bodies, the shaking force is precisely the resultant inertia force $\mathbf{F} = -\Sigma m^k \ddot{\mathbf{x}}^k$.

Similarly, a net "shaking couple" $\mathbf{T}$ is transmitted to the foundation where $\mathbf{T}$ is the resultant of all the inertia torques defined by Eqs. (6.109) for an individual link.

The problem of *machine balancing* is usually posed as one of *kinetostatics*, wherein the acceleration of every mass particle in the system is given (usually in the steady-state condition of a machine), and one seeks to determine the size and location of *balancing weights* which will reduce $|\mathbf{F}|$ and $|\mathbf{T}|$ to acceptable levels. The subject is usually discussed under the two categories of *rotor balancing* and *linkage balancing*. Because of the large body of literature on these subjects, we shall only present a brief qualitative discussion of each, which follows closely that in Ref. 67.

Rotor Balancing.[8,44,45,48,54,66,67,104] A rotating shaft supported by coaxial bearings, together with any attached mass (e.g., a turbine disk or motor armature), is called a "rotor." If the center of mass (CM) of a rotor is not located exactly on the bearing axis, a net centrifugal force will be transmitted, via the bearings, to the foundation. The horizontal and vertical components of this periodic shaking force can travel through the foundation to create serious vibration problems in neighboring components. The magnitude of the shaking force is $F = me\omega^2$, where m = rotor mass, ω = angular speed, e = distance (called "eccentricity") of the CM from the rotation axis. The product me, called the "unbalance," depends only on the mass distribution and is usually nonzero because of unavoidable manufacturing tolerances, thermal distortion, etc. When the CM lies exactly on the axis of rotation ($e = 0$), no net shaking force occurs and the rotor is said to be in "static balance." Static balance is achieved in practice by adding or subtracting balancing weights at any convenient radius until the rotor shows no tendency to turn about its axis, starting from any given initial orientation. Static balancing is adequate for relatively thin disks or short rotors (e.g., automobile wheels). However, a long rotor (e.g., in a turbogenerator) may be in static balance and still exert considerable forces upon individual bearing supports. For example, suppose that a long shaft is supported on bearings which are coaxial with the ζ axis, where ξ, η, ζ are body-fixed axes through the CM of a rotor, and ζ coincides with the fixed (global) z axis. If the body rotates with constant angular speed $|\omega| \equiv \omega_3$ about the ζ axis, the Euler equations (see Ref. 94 or Sec. 13.10 of Ref. 66) show that the inertial torques about the rotating ξ and η axes are

$$\overline{T}_1 = \overline{I}_{\zeta\eta}\omega^2$$

$$\overline{T}_2 = -\overline{I}_{\zeta\xi}\omega^2$$

$$\overline{T}_3 = 0$$

where $\overline{I}_{\zeta\eta}$ and $\overline{I}_{\zeta\xi}$ are the (constant) products of inertia referred to the (ξ, η, ζ) axes. From these equations it is clear that the resultant inertia torque rotates with the shaft and produces harmonically fluctuating shaking forces on the foundation. However, if $\overline{I}_{\zeta\eta}$ and $\overline{I}_{\zeta\xi}$ can be made to vanish, the shaking couple will likewise vanish. It is shown in the references cited that any *rigid* shaft may be *dynamically balanced* (i.e., the net shaking force *and* the shaking couple can be simultaneously eliminated) by adding or subtracting a definite amount of mass at any convenient radius in each of two arbitrary transverse cross sections of the rotor. The "balancing planes" selected for this purpose are usually located near the ends of the rotor where suitable shoulders or "balancing rings" have been machined to permit the convenient addition of mass (e.g., lead weights or calibrated bolts) or the removal of mass (e.g., by drilling or grinding). Long rotors, running at high speeds, may undergo appreciable deformations. For such *flexible rotors* it is necessary to utilize more than two balancing planes.

Balancing Machines and Procedures. The most common types of rotor *balancing machines* consist of bearings held in pedestals which support the rotor, which is spun at constant speed (e.g., by a belt drive, or compressed airstream). Electromechanical transducers sense the unbalanced forces (or associated vibrations) transmitted to the pedestals and electric circuits automatically perform the calculations necessary to predict the location and amount of balancing weight to be added or subtracted in preselected balancing planes.

For very large rotors, or for rotors which drive several auxiliary devices, commercial balancing machines may not be convenient. The *field balancing* procedures for such installations may involve the use of accelerometers on the bearing housings, along with vibration meters and phase discriminators (possibly utilizing stroboscopy) to determine the proper location and amount of balance weight. For an example used in field balancing of flexible shafts, see Ref. 26.

Committees of the International Standards Organization (ISO) and the American National Standards Institute (ANSI) have formulated recommendations for the allowable quality grade $G = e\omega$ for various classes of machines.[54]

Balancing of Linkage Machinery. The literature in this field pertaining to the slider-crank mechanism is huge, because of its relevance to the ubiquitous internal combustion engine.[8,14,17,35,39,43,66,79,91] For this relatively simple mechanism (even with multiple cranks attached to the same crankshaft), the theory is well understood and readily applied in practice. However, for more general linkages, the theory and reduction to practice has not yet been as thoroughly developed. The state of the field up to the 1960s has been reviewed periodically by Lowen, Tepper, and Berkof,[4,49,50] and practical techniques for balancing the shaking forces in four-bar and six-bar mechanisms have been described.[5,66,88,95,96,99]

It should be noted that the addition of the balancing weights needed to achieve net *force* balance will tend to increase the shaking moment, bearing forces, and required input torque. The size of the balancing weights can be reduced if one is willing to accept a *partial force balance.*[92,102] Because of the many variables involved in the linkage balancing problem, a number of investigations of the problem have been made from the point of view of optimization theory.[13,96,103]

6.4.6   Effects of Flexibility and Joint Clearance

In the above discussion, it has been assumed that all the links are perfectly rigid and that no clearance (backlash) exists at the joints. A great deal of literature has appeared on the influence of both these departures from the commonly assumed ideal conditions. In view of the extent of this literature, we will point specifically only to surveys of the topics and to recent representative papers which contain additional references.

Publications on link flexibility (up to 1971) have been reviewed by Erdman and Sandor,[23] and more recent work in this area has been described in several papers[7,55,56,90] dealing with the finite element method of analyzing link flexibility. A straightforward, practical, and accurate approach to this problem has been described by Amin,[1] who showed how to model flexible links so that they may be properly included in a dynamic situation by the general-purpose computer program DYMAC.[68]

A number of papers dealing with the effects of looseness and compliance of joints have been published by Dubowsky et al.,[19] and the entire field has been surveyed by Haines.[31] A later discussion of the topic was given by Ehle and Haug.[20]

6.5 NUMERICAL METHODS FOR SOLVING THE DIFFERENTIAL EQUATIONS OF MOTION

A system of n ordinary differential equations is said to be in "standard form" if the derivatives of the unknown functions $w_1, w_2, ..., w_n$, with respect to the independent variable t are given in the form

$$dw_i/dt = f_i(w_1, w_2, ..., w_n, t) \qquad i = 1, ..., n \qquad (6.175)$$

If the initial values $w_1(0), w_2(0), ...,$ are all given we are faced with an *initial value problem,* in standard form, of order n. Fortunately, a number of well-tested and efficient computer subroutines are available for the solution of the problem posed. One major class of such programs is based on the so-called one-step or direct methods such as the Runge-Kutta methods. These methods have the advantage of being self-starting, are free of iterative processes, and readily permit arbitrary spacing Δt between successive times where results are needed.

A second major class of numerical integration routines belong to the so-called predictor-corrector or multistep methods. These are iterative procedures whose main advantage over the one-step procedures is that they permit more direct control over the size of the numerical errors developed. Most predictor-corrector methods suffer, however, from the drawback of starting difficulties and an inability to change the step size t once started. However, "almost automatic" starting procedures for multistep methods have been devised.[27] These use variable-order integration rules for starting or changing interval size. For a review of the theory, and a FORTRAN listing, of the Runge-Kutta algorithm, see Appendix G of Ref. 66. For examples of the use of other types of differential equation solvers see Refs. 10 and 101.

6.5.1 Determining Initial Values of Lagrangian Coordinates

We will assume that we are dealing with a system described by M lagrangian coordinates $\psi_1, \ldots, \psi_M$ which are related by spatial and temporal constraint equations of the form

$$f_i(\psi_1, \ldots, \psi_M) = 0 \qquad i = 1, 2, \ldots, N_s \qquad (6.176)$$

$$g_k(\psi_1, \ldots, \psi_M, t) = 0 \qquad k = 1, 2, \ldots, N_t \qquad (6.177)$$

In order for this system of $N_s + N_t$ equations in M unknowns to have a solution at time $t = 0$, it is necessary that F_R of the ψ_i be given, where

$$F_R \equiv M - N_s - N_t$$

is the reduced degree of freedom discussed in Sec. 6.2.2. Assuming that F_R of the ψ_i are given at $t = 0$, we may solve the $N_s + N_t$ equations for the unknown initial coordinates, using a numerical method, such as the Newton-Raphson algorithm (see Chap. 9 and Appendix F of Ref. 66). At this point, all values of ψ_i will be known.

6.5.2 Initial Values of Lagrangian Velocities

When there are no temporal constraints ($N_t = 0$), we must specify the F independent initial velocities $\dot{q}_1, \ldots, \dot{q}_F$ at time $t = 0$. Then all the lagrangian velocities at $t = 0$ follow from Eq. (6.38) in the form

$$\dot{\psi}_j = \sum_{k=1}^{F} C_{jk} \dot{q}_k \qquad j = 1, \ldots, M \qquad (6.178)$$

When there exist N_t temporal relations expressed in the form of Eq. (6.177), we can express them in their rate form [see Eq. (6.71)]

$$\sum_{k=1}^{F} \alpha_{jk} \dot{q}_k + \beta_j = 0 \qquad j = 1, \ldots, N_t \qquad (6.179)$$

These temporal constraints permit us to specify only F_R ($= F - N_t$) generalized velocities, rather than F of them. With N values of $\dot{q}_k$ left unspecified, we may solve Eqs. (6.178) and (6.179) for all the $M + N_t$ unknown values of $\dot{\psi}_j$ and $\dot{q}_j$. Thus, all values of both ψ_i and $\dot{\psi}_i$ may be found at time $t = 0$, and the numerical integration can proceed.

6.5.3 Numerical Integration

Let us introduce the notation

$$\{\psi_1, \psi_2, \ldots, \psi_M, \dot{q}_1, \ldots, \dot{q}_F\} \equiv \{w_1, w_2, \ldots, w_M, w_{M+1}, \ldots, w_{M+F}\} \qquad (6.180)$$

Now we may rewrite the governing equations (6.178), (6.133), and (6.179), respectively, as

$$\dot{w}_j = \sum_{k=1}^{F} C_{jk} w_{M+K} \qquad\qquad j = 1, 2, \ldots, M \qquad (6.181)$$

$$\sum_{k=1}^{F} I_{jk} \dot{w}_{M+K} = Q_j + Q_j^{\dagger} + \sum_{m=1}^{N_t} \alpha_{mj} \lambda_m \qquad j = 1, \ldots, F \qquad (6.182)$$

$$\sum_{k=1}^{F} \alpha_{jk} \dot{w}_{M+k} = -\beta_j \qquad\qquad j = 1, \ldots, N_t \qquad (6.183)$$

We now note that Q_j, $Q_j\dagger$, and β_j are all known functions of $w_1, \ldots, w_{M+F}$ whenever the state $(\psi, \dot{\psi})$ is known (e.g., at $t = 0$). Then it is clear that we may solve the $F + N_t$ governing equations [Eqs. (6.182) and (6.183)] for the $F + N_t$ unknowns $\dot{w}_{M+1}, \ldots, \dot{w}_{M+F}$ and $\lambda_1, \ldots, \lambda_{N_t}$. Then we may solve Eq. (6.181) for the M remaining velocities $\dot{w}_1, \dot{w}_2, \ldots, \dot{w}_M$. In short, Eqs. (6.181) and (6.183) enable us to calculate all the $M + F$ derivatives $\dot{w}_i$ from a knowledge of the w_i. That is, they constitute a system of $M + F$ first-order differential equations in standard form. Any of the well-known subroutines for such initial-value problems may be utilized to march out the solution for all the ψ_i and $\dot{\psi}_i$ at user-specified time steps. The lagrangian accelerations may be computed at any time step from Eq. (6.44), and the displacements, velocities, and accelerations for any point of interest may be found from Eqs. (6.2), (6.46), and (6.48).

6.6 SURVEY OF MACHINE DYNAMICS LITERATURE AND COMPUTER PROGRAMS

In this section, we shall give a brief review of the large body of literature dealing with our subject that has appeared since the publication of the Second Edition of the *Mechanical Design and Systems Handbook* (1985). Much of the essential historical background on the development of multibody dynamics, including the areas of aerospace and biomechanics, is given by Roberson and Schwertassek[A29] and Huston.[A15] A compact summary of the field's literature is given in Paul[A25] under the headings of "Open-Loop Mechanisms" ("Satellite Controls," "Human Body Models"), "Closed-Loop Mechanisms," "Elastodynamics," and "Machine Balancing."

6.6.1 Computer Programs

A detailed discussion, by their respective authors, of 20 general-purpose multibody dynamics computer programs will be found in Schiehlen,[A33] from which the list in Table 6.1 has been taken.

6.6.2 Survey Papers

Like most rapidly growing scientific fields, that of multibody dynamics has been surveyed on several occasions. One of the earliest such surveys (in the area of theory of machines) was given by Paul.[64] An excellent review of the state of the art in the early 1980s is provided by the proceedings of a NATO sponsored symposium (Haug[A12]). A more recent review was given by Schwertassek and Roberson[A34] who present an in-

TABLE 6.1 General-Purpose Multibody Dynamics Computer Programs

Program	Principal contributors	Country
ADAMS	M. A. Chace, N. Orlandea, R. R. Ryan	United States
AUTODYN	P. Maes, J. C. Samin, P. Y. Willems	Belgium
AUTOLEV	D. A. Levinson, T. R. Kane	United States
CAMS	L. Lilov, B. Bekjarov, M. Lorer	Bulgaria
COMPAMM	J. M. Jimenez, A. Avello, A. Garcia-Alonso, J. Garcia de Jalon	Spain
DADS	E. J. Haug, R. L. Smith	United States
DISCOS	H. Frisch	United States
DYMAC	B. Paul, A. Amin, G. Hud	United States
DYNOCOMBS	R. L. Huston, T. P. King, J. W. Kamman	United States
DYSPAM	B. Paul, R. Schaffa	United States
MEDYNA	O. Wallrapp, C. Führer	Germany
MESA VERDE	J. Wittenburg, U. Wolz, A. Schmidt	Germany
NBOD	H. Frisch	United States
NEWEUL	E. Kreuzer, W. Schiehlen	Germany
NUBEMM	E. Pankiewicz	Germany
PLEXUS	A. Barraco, E. Cuny	France
ROBOTRAN	(see entry under AUTODYN)	Belgium
SIMPACK	W. Rulka	Germany
SPACAR	J. B. Jonker, J. P. Meijaard	Netherlands
SYM	M. Vukobratovic, N. Kirkanski, A. Timcenko, M. Kirkanski	Yugoslavia

depth comparison of the various dynamic formulations used. Many of the books mentioned below also contain extensive literature surveys.

6.6.3 Books on Multibody Dynamics

Suh and Radcliffe[89] broke the mold of the textbook literature by introducing the use of computers for *kinematics* of mechanisms. Their book deals primarily with spatial *kinematics,* although it does contain a relatively brief discussion of *dynamic* analyses for particular mechanisms, utilizing the Newton-Euler formulation. The first book to contain a systematic treatment of the *dynamics* of machinery, formulated in a manner best suited to solution by computer, was Paul.[66] Although this book was focussed on planar problems, the fundamental tools of analytical dynamics (especially the d'Alembert-Lagrange principle) which it expounds are applicable to spatial problems as well. Erdman and Sandor[A6] and Sandor and Erdman[A31] each contain a relatively brief discussion of planar dynamics, but from the more traditional point of view, except for the inclusion of some problems of elastokinetics (contributed by A. Midha). More recently there has appeared a spate of books which treat multibody dynamics from a modern, computer-oriented viewpoint, including Roberson and Schwertassek,[A29] Nikravesh,[A22] Shabana,[A35] Haug,[A13] Schiehlen,[A33] and Huston.[A15] Reference A29 contains a particularly comprehensive list of references, and useful discussions of the historical record.

6.6.4 Books on Dynamics of Robots

There now exists a substantial number of publications which deal with the dynamics of robots, and manipulators. Perhaps the first book to focus on three-dimensional problems of the kinematics and dynamics of robots was that of R. P. Paul.[73] A good idea of the state of the art at that time may be obtained from the collection edited by Brady et al.[A2] This book contains reprints of significant papers, as well as tutorial articles on dynamics, feedback control, trajectory planning, compliance, and task planning. Among the huge number of books on robotics that have since been published, the following are representative books that have significant material on the kinematics, statics, and dynamics of robots: Asada and Slotine,[A2] Fu, Gonzalez, and Lee,[A9] Wolovich,[A40] and Yoshikawa.[A41]

REFERENCES

1. Amin, A.: "Automatic Formulation and Solution Techniques in Dynamics of Machinery," Ph.D. Dissertation, University of Pennsylvania, Philadelphia, 1979.

2. Andrews, G. C., and H. K. Kesavan: "The Vector Network Model: A New Approach to Vector Dynamics," *Mechanism and Machine Theory,* vol. 10, pp. 57–75, 1975.

3. Benedict, C. E., and D. Tesar: "Dynamic Response Analysis of Quasi-rigid Mechanical Systems Using Kinematic Influence Coefficients," *J. Mechanisms,* vol. 6, pp. 383–403, 1971.

4. Berkof, R. S., G. Lowen, and F. R. Tepper: Balancing of Linkages, *Shock & Vibration Dig.,* vol. 9, no. 6, pp. 3–10, 1977.

5. Berkof, R. S., and G. Lowen: A New Method for Completely Force Balancing Simple Linkages, *J. Eng. Ind., Trans. ASME,* ser. B, vol. 91, pp. 21–26, 1969.

6. Beyer, R.: "Kinematisch-getriebedynamisches Praktikum," Springer, Berlin, 1960.

7. Bhagat, B. M., and K. D. Wilmert: "Finite Element Vibration Analysis of Planar Mechanisms," *Mechanism and Machine Theory,* vol. 11, pp. 47–71, 1976.

8. Biezeno, C. B., and R. Grammel: "Engineering Dynamics, Vol. IV—Internal-Combustion Engines," (transl. by M. P. White), Blackie & Son, Ltd., Glasgow, 1954.

9. Bocher, M.: "Introduction to Higher Algebra," The Macmillan Company, New York, pp. 46, 47, 1907.

10. Chace, M. A., and Y. O. Bayazitoglu: "Development and Application of a Generalized d'Alembert Force for Multi-Freedom Mechanical Systems," *J. Eng. Ind., Trans. ASME,* ser. B, vol. 93, pp. 317–327, 1971.

11. Chace, M. A., and P. N. Sheth: "Adaptation of Computer Techniques to the Design of Mechanical Dynamic Machinery," ASME Paper 73-DET-58, ASME, New York, 1973.

12. Chace, M. A., and D. A. Smith: "DAMN—Digital Computer Program for the Dynamic Analysis of Generalized Mechanical Systems," *SAE Trans.,* vol. 80, pp. 969–983, 1971.

13. Conte, F. L., G. R. George, R. W. Mayne, and J. P. Sadler: "Optimum Mechanism Design Combining Kinematic and Dynamic-Force Considerations," *J. Eng. Ind., Trans. ASME,* ser. B, vol. 97, pp. 662–670, 1975.

14. Crandall, S. H.: "Rotating and Reciprocating Machines," in "Handbook of Engineering Mechanics," W. Flugge, ed., McGraw-Hill Book Company, Inc., New York, chap. 58, 1962.

15. Crandall, S. H., D. C. Karnopp, E. F. Kurtz, Jr., and D. C. Pridmore-Brown: "Dynamics of Mechanical and Electromechanical Systems," McGraw-Hill Book Company, Inc., New York, 1968.

16. Denavit, J., R. S. Hartenberg, R. Razi, and J. J. Uicker, Jr.: "Velocity, Acceleration, and Static-Force Analysis of Spatial Linkages," *J. Appl. Mech.,* vol. 32, *Trans. ASME,* ser. E, vol. 87, pp. 903–910, 1965.

17. Den Hartog, J. P.: "Mechanical Vibrations," 4th ed., McGraw-Hill Book Company, Inc., New York, 1956.

18. Dix, R. C., and T. J. Lehman: "Simulation of the Dynamics of Machinery," *J. Eng. Ind. Trans. ASME,* ser. B, vol. 94, pp. 433–438, 1972.

19. Dubowsky, S., and T. N. Gardner: "Design and Analysis of Multilink Flexible Mechanisms with Multiple Clearance Connections," *J. Eng. Ind. Trans. ASME,* ser. B, vol. 99, pp. 88–96, 1977.

20. Ehle, P. E., and E. J. Haug: "A Logical Function Method for Dynamic Design Sensitivity Analysis of Mechanical Systems with Intermittent Motion," *J. Mechanical Design, Trans. ASME,* vol. 104, pp. 90–100, 1982.

21. Eksergian, R.: "Dynamical Analysis of Machines" (a series of 15 installments), appearing in *J. Franklin Inst.,* vols. 209, 210, 211, 1930–1931.

22. Erdman, A. G., and J. E. Gustavson: "LINCAGES: Linkage Interaction Computer Analysis and Graphically Enhanced Synthesis Package," ASME Paper 77-DET-5, 1977.

23. Erdman, A. G., and G. N. Sandor: "Kineto-Elastodynamics—A Review of the Art and Trends," *Mechanism and Machine Theory,* vol. 7, pp. 19–33, 1972.

24. Freudenstein, F.: "On the Variety of Motions Generated by Mechanisms," *J. Eng. Ind., Trans. ASME,* ser. B, vol. 84, pp. 156–160, 1962.

25. Freudenstein, F., and G. N. Sandor: "Kinematics of Mechanisms," sec. 5 of "Mechanical Design and Systems Handbook," 2d ed., H. A. Rothbart, ed., McGraw-Hill Book Company, Inc., New York, 1985.

26. Fujisaw, F., et al.: "Experimental Investigation of Multi-Span Rotor Balancing Using Least Squares Method," *J. Mechanical Design, Trans. ASME,* vol. 102, pp. 589–596, 1980.

27. Gear, C. W.: "Numerical Initial Value Problems in Ordinary Differential Equations," Prentice-Hall, Inc., Englewood Cliffs, N.J., 1971.

28. Givens, E. J., and J. C. Wolford: "Dynamic Characteristics of Spatial Mechanisms," *J. Eng. Ind., Trans. ASME,* ser. B, vol. 91, pp. 228–234, 1969.

29. Goldstein, H.: "Classical Mechanics," 2d ed., Addison-Wesley Publishing Company, Inc., Reading Mass., 1980.

30. Greenwood, D. T.: "Classical Dynamics," Prentice-Hall, Inc., Englewood Cliffs, N.J., 1977.

31. Haines, R. S.: "Survey: 2-Dimensional Motion and Impact at Revolute Joints," *Mechanism and Machine Theory,* vol. 15, pp. 361–370, 1980.

32. Haug, E. J., R. A. Wehage, and N. C. Barman: "Dynamic Analysis and Design of Constrained Mechanical Systems," *J. of Mech. Design, Trans. ASME,* vol. 104, pp. 778–784, 1982.

33. Hirschhorn, J.: "Kinematics and Dynamics of Plane Mechanisms," McGraw-Hill Book Company, Inc., New York, 1962.

34. Hollerback, J. M.: "A Recursive Lagrangian Formulation of Manipulator Dynamics and a Comparative Study of Dynamics Formulation," *IEEE Trans. on Systems, Man and Cybernetics,* SMC-10, 11, pp. 730–736, November 1980.

35. Holowenko, A. R.: "Dynamics of Machinery," John Wiley & Sons, Inc., New York, 1955.

36. Hud, G. C.: "Dynamics of Inertia Variant Machinery," Ph.D. Dissertation, University of Pennsylvania, Philadelphia, 1976.

37. Huston, R. L.: "Multibody Dynamics Including the Effects of Flexibility and Compliance," *Computers and Structures,* vol. 14, pp. 443–451, 1981.

38. Huston, R. L., C. E. Passerello, and M. W. Harlow: Dynamics of Multirigid-Body Systems, *J. Appl. Mech.,* vol. 45, pp. 889–894, 1978.

39. Judge, A. W.: "Automobile and Aircraft Engines," vol. 1, "The Mechanics of Petrol and Diesel Engines," Pitman Publishing Corporation, New York, 1947.

40. Kane, T. R.: "Dynamics," Holt, Rinehart and Winston, New York, 1968.

41. Kane, T. R., and C. F. Wang: "On the Derivation of Equations of Motion," *J. Society for Industrial and Applied Math,* vol. 13, pp. 487–492, 1965.

42. Kaufman, R. E.: "Kinematic and Dynamic Design of Mechanisms," in "Shock and Vibration

Computer Programs Reviews and Summaries," W. & B. Pilkey, ed., Shock and Vibration Information Center, U.S. Naval Research Laboratory, Code 8404, Washington, D.C., 1975.

43. Kozesnik, J.: "Dynamics of Machines," Ervin P. Noordhoff, Ltd., Groningen, Netherlands, 1962.

44. Kroon, R. P.: "Balancing of Rotating Apparatus—I," *J. Appl. Mech.,* vol. 10, *Trans. ASME,* vol. 65, pp. A225–A228, 1943.

45. Kroon, R. P.: "Balancing of Rotating Apparatus—II," *J. Appl. Mech.,* vol. 11, *Trans. ASME,* vol. 66, A47–A50, 1944.

46. Lanczos, C.: "The Variational Principles of Mechanics," 3d ed., University of Toronto Press, Toronto, 1966.

47. Langrana, N. A., and D. L. Bartel: "An Automated Method for Dynamic Analysis of Spatial Linkages for Biomechanical Applications," *J. Eng. Ind., Trans. ASME,* ser. B, vol. 97, pp. 556–574, 1975.

48. Loewy, R. G., and V. J. Piarulli: "Dynamics of Rotating Shafts," Shock and Vibration Information Center, NRL, Washington, D.C., 1969.

49. Lowen, G. G., and R. S. Berkof: "Survey of Investigations into the Balancing of Linkages," *J. Mechanisms,* vol. 3, pp. 221–231, 1968.

50. Lowen, G. G., F. R. Tepper, and R. S. Berkof: "Balancing of Linkages—An Update," *Mechanisms and Machine Theory,* vol. 18, pp. 213–220, 1983.

51. Mabie, H. H., and F. W. Ocvirk: "Mechanisms and Dynamics of Machinery," John Wiley & Sons, Inc., New York, 1975.

52. Martin, G. H.: "Kinematics and Dynamics of Machines," 2d ed., McGraw-Hill Book Company, Inc., New York, 1982.

53. Mitchell, T.: "Mechanics," sec. 4, "Mechanical Design and Systems Handbook," 2d ed., H. A. Rothbart, ed., McGraw-Hill Book Company, Inc., New York, 1985.

54. Muster, D., and D. G. Stadelbauer: "Balancing of Rotating Machinery," in "Shock and Vibration Handbook," 2d ed., C. M. Harris and C. E. Crede, eds., McGraw-Hill Book Company, Inc., New York, 1976, chap. 39.

55. Nath, P. K., and A. Ghosh: "Kineto-Elastodynamic Analysis of Mechanisms by Finite Element Method," *Mechanism and Machine Theory,* vol. 15, pp. 179–197, 1980.

56. Nath, P. K., and A. Ghosh: "Steady State Response of Mechanisms with Elastic Links by Finite Element Methods," *Mechanisms and Machine Theory,* vol. 15, pp. 179–197, 1980.

57. Nikravesh, P. E., and I. S. Chung: "Application of Euler Parameters to the Dynamic Analysis of Three-Dimensional Constrained Systems," *J. Mechanical Design, Trans. ASME,* vol. 104, 1982.

58. Orlandea, N.: "Node-Analogous, Sparsity-Oriented Methods for Simulation of Mechanical Systems," Ph.D. Dissertation, University of Michigan, Ann Arbor, 1973.

59. Orlandea, N., and M. A. Chace: "Simulation of a Vehicle Suspension with the ADAMS Computer Program," SAE Paper No. 770053, Society of Automotive Engineers, Warrendale, Pa., 1977.

60. Orlandea, N., M. A. Chace, and D. A. Calahan: "A Sparsity-Oriented Approach to the Dynamic Analysis and Design of Mechanical Systems—Parts I and II," *J. Eng. Ind., Trans. ASME,* ser. B, vol. 99, pp. 733–779, 780–784, 1977.

61. Pars, L. A.: "A Treatise on Analytical Dynamics," John Wiley & Sons, Inc., New York, 1968.

62. Paul, B.: A Unified Criterion for the Degree of Constraint of Plane Kinematic Chains," *J. Appl. Mech.,* vol. 27, *Trans. ASME* vol. 82, ser. E, pp. 196–200, 1960. See discussion, same volume, pp. 751–753.

63. Paul, B.: "On the Composition of Finite Rotations," *Amer. Math. Monthly,* vol. 70, no. 8, pp. 859–862, 1963.

64. Paul, B.: "Analytical Dynamics of Mechanisms—A Computer-Oriented Overview," *Mechanism and Machine Theory,* vol. 10, pp. 481–507, 1975.

65. Paul, B.: "Dynamic Analysis of Machinery Via Program DYMAC." SAE Paper 770049, Society of Automotive Engineers, Warrendale, Pa., 1977.

66. Paul, B.: "Kinematics and Dynamics of Planar Machinery," Prentice-Hall, Inc., Englewood Cliffs, N.J., 1979.

67. Paul, B.: "Shaft Balancing," in "Encyclopedia of Science and Technology," McGraw-Hill Book Company, Inc., New York, 1993.

68. Paul, B., a   nd A. Amin: "User's Manual for Program DYMAC-G4 (DYnamics of MAChinery—General Version 4)," available from B. Paul, Philadelphia, Pa., 1979.

69. Paul, B., and A. Amin: "User's Manual for Program KINMAC (KINematics of MAChinery)," available from B. Paul, Philadelphia, Pa., 1979.

70. Paul, B., and A. Amin: "User's Manual for Program STATMAC (STATics of MAChinery)," available from B. Paul, Philadelphia, Pa., 1979.

71. Paul, B., and D. Krajcinovic: "Computer Analysis of Machines with Planar Motion—Part 1: Kinematics," *J. Appl. Mech.,* vol. 37, *Trans. ASME,* ser. E, vol. 92, pp. 697–702, 1970a.

72. Paul, B., and D. Krajcinovic: "Computer Analysis of Machines with Planar Motion—Part 2: Dynamics," *J. Appl. Mech.,* vol. 37, *Trans. ASME,* ser. E, vol. 92, pp. 703–712, 1970b.

73. Paul, R. P.: "Robot Manipulators: Mathematics, Programming, and Control," The MIT Press, Cambridge, Mass., 1981.

74. Quinn, B. E.: "Energy Method for Determining Dynamic Characteristics of Mechanisms," *J. Appl. Mech.,* vol. 16, *Trans. ASME,* vol. 71, pp. 283–288, 1949.

75. Reuleaux, F.: "The Kinematics of Machinery" (transl. and annotated by A. B. W. Kennedy), reprinted by Dover Publications, Inc., New York, 1963, original date 1876.

76. Roger, R. J., and G. C. Andrews: "Simulating Planar Systems Using a Simplified Vector-Network Method," *Mechanism and Machine Theory,* vol. 10, pp. 509–519, 1975.

77. Rogers, R. J., and G. C. Andrews: "Dynamic Simulation of Planar Mechanical Systems with Lubricated Bearing Clearances Using Vector Network Methods," *J. Eng. Ind., Trans. ASME,* ser. B, vol. 99, pp. 131–137, 1977.

78. Rongved, L., and H. J. Fletcher: "Rotational Coordinates," *J. Franklin Institute,* vol. 277, pp. 414–421, 1964.

79. Root, R. E.: "Dynamics of Engine and Shaft," John Wiley & Sons, Inc., New York, 1932.

80. Rubel, A. J., and R. E. Kaufman: "KINSYN III: A New Human Engineered System for Interactive Computer-Aided Design of Planar Linkages," ASME Paper No. 76-DET-48, 1976.

81. Sheth, P. N.: "A Digital Computer Based Simulation Procedure for Multiple Degree of Freedom Mechanical Systems with Geometric Constraints," Ph.D. thesis, University of Wisconsin, Madison, 1972.

82. Sheth, P. N., and J. J. Uicker, Jr.: "IMP (Integrated Mechanisms Program), A Computer-Aided Design Analysis System for Mechanisms and Linkage," *J. Eng. Ind., Trans. ASME,* ser. B, vol. 94, pp. 454–464, 1972.

83. Shigley, J. E., and J. J. Uicker, Jr.: "Theory of Machines and Mechanisms," McGraw-Hill Book Company, Inc., New York, 1980.

84. Skreiner, M.: "Dynamic Analysis Used to Complete the Design of a Mechanism," *J. Mechanisms,* vol. 5, pp. 105–109, 1970.

85. Smith, D. A.: "Reaction Forces and Impact in Generalized Two-Dimensional Mechanical Dynamic Systems," Ph.D. dissertation, Mech. Eng., University of Michigan, Ann Arbor, 1971.

86. Smith, D. A.: "Reaction Force Analysis in Generalized Machine Systems," *J. Eng. Ind., Trans. ASME,* ser. B, vol. 95, pp. 617–623, 1973.

87. Smith, D. A., M. A. Chace, and A. C. Rubens: "The Automatic Generation of a Mathematical Model for Machinery Systems," ASME Paper 72-Mech-31, ASME, New York, 1972.

88. Stevensen, E. N., Jr.: "Balancing of Machines," *Trans. ASME, of J. Eng. Ind.* vol. 95, ser. B, pp. 650–656, 1973.

89. Suh, C. H., and C. W. Radcliffe: "Kinematics and Mechanisms Design," John Wiley & Sons, Inc., New York, 1978.

90. Sunoda, W. H., and S. Dubowski: "The Application of Finite Element Methods to the Dynamic Analysis of Flexible Spatial and Co-planar Linkage Systems," *J. Mech. Des., Trans. ASME,* vol. 103, pp. 643–651, 1981.

91. Taylor, C. F.: "The Internal Combustion Engine in Theory and Practice," The MIT Press, Cambridge, Mass., vol. I, 1966; vol. II, 1968.

92. Tepper, F. R., and G. G. Lowen: "Shaking Force Optimization of Four-Bar Linkages with Adjustable Constraints on Ground Bearing Forces," *J. Eng. Ind., Trans. ASME,* ser. B, vol. 97, pp. 643–651, 1975.

93. Thomson, W. (Lord Kelvin), and P. G. Tait: "Treatise on Natural Philosophy, Part I," Cambridge University Press, Cambridge, 1879, p. vi.

94. Timoshenko, S. P., and D. H. Young: "Advanced Dynamics," McGraw-Hill Book Company, Inc., New York, 1948.

95. Tricamo, S. J., and G. G. Lowen: "A New Concept for Force Balancing Machines for Planar Linkages, Part I: Theory," *J. Mechanical Design, Trans. ASME,* vol. 103, pp. 637–642, 1981.

96. Tricamo, S. J., and G. G. Lowen: "A New Concept for Force Balancing Machines for Planar Linkages, Part 2: Application to Four-Bar Linkages and Experiment," *J. Mechanical Design, Trans. ASME,* vol. 103, pp. 784–792, 1981.

97. Uicker, J. J. Jr.: "Dynamic Force Analysis of Spatial Linkages," *J. Appl. Mech.,* vol. 34, *Trans. ASME,* ser. E, vol. 89, pp. 418–424, 1967.

98. Vukobratovic, M., and V. Potkonjak: "Dynamics of Manipulation Robots," Springer Verlag, New York, 1982.

99. Walker, M. J., and K. Oldham: "A General Theory of Force Balancing Using Counterweights, Mechanism and Machine Theory," vol. 13, pp. 175–185, 1978.

100. Webster, A. G.: "Dynamics of Particles and of Rigid Elastic and Fluid Bodies," Dover Publication reprint, New York, 1956.

101. Wehage, R. A., and E. J. Haug: "Generalized Coordinate Partitioning for Dimension Reduction in Analysis of Constrained Dynamic System," *J. Mechanical Design, Trans. ASME,* vol. 104, 1982, pp. 247–255.

102. Whittaker, E. T.: "A Treatise on the Analytical Dynamics of Particles and Rigid Bodies," 4th ed., Dover Publications, Inc., New York, 1944.

103. Wiederrich, J. L., and B. Roth: "Momentum Balancing of Four-Bar Linkages," *J. Eng. Ind., Trans. ASME,* ser. B, vol. 98, pp. 1289–1295, 1976.

104. Wilcox, J. B.: "Dynamic Balancing of Rotating Machinery," Sir Isaac Pitman & Sons, Ltd., London, 1967.

105. Wittenburg, J.: "Dynamics of Systems of Rigid Bodies," B. G. Teubner, Stuttgart, 1977.

SUPPLEMENTARY REFERENCES

A1. Asada, H., and J. E. Slotine: "Robot Analysis and Control," John Wiley & Sons, Inc., New York, 1986.

A2. Brady, M., J. M. Hollerback, T. J. Johnson, Lozano-Pérez, and M. T. Mason (eds.): "Robot Motion: Planning and Control," The MIT Press, Cambridge, Mass., 1982.

A3. Denavit, J., and R. S. Hartenberg: "A Kinematic Notation for Lower-Pair Mechanisms Based on Matrices," *Trans. ASME, J. of Appl. Mech.,* vol. 22, pp. 215–221, 1955.

A4. Desloge, E. A.: "Relationship between Kane's Equations and the Gibbs-Appell Equations," *J. Guidance, Control and Dynamics,* vol. 1, pp. 120–122, 1987.

A5. Erdman, A. G. (ed.): "Modern Kinematics: Developments in the Last Forty Years," John Wiley & Sons, Inc., New York, 1993.

A6. Erdman, A. G., and G. N. Sando: "Mechanism Design Analysis and Synthesis," 2d ed., vol. I, Prentice-Hall, Inc., Englewood Cliffs, N.J., 1991.

A7. Fletcher, H., L. Rongved, and E. Yu: "Dynamics Analysis of a Two-Body Gravitationally Oriented Satellite," *Bell System Tech. J.,* vol. 42 (no. 5, September), pp. 2239–2266, 1963.

A8. Freudenstein, F., J. P. Macey, and E. R. Maki: "Optimum Balancing of Combined Pitching and Yawing Movement in High Speed Machinery," *Trans. ASME, J. Mechanical Design,* vol. 103, pp. 571–577, 1981.

A9. Fu, K., R. Gonzalez, and C. Lee: "Robotics: Control, Sensing, Vision, and Intelligence," McGraw-Hill Book Company, Inc., New York, 1987.

A10. Fujisaw, F. E. A.: "Experimental Investigation of Multi-Span Rotor Balancing Using Least Squares Method," *Trans. ASME, J. Mechanical Design,* vol. 102, pp. 589–596, 1980.

A11. Gabriele, G. A. (ed.), et al.: "Optimization in Mechanisms," chap. 11, Ref. A5, 1992.

A12. Haug, E. J. (ed.): "Computer Aided Analysis and Optimization of Mechanical System Dynamics," NATO ASI Series F, vol. 9, Springer-Verlag, Berlin, 1984.

A13. Haug, E. J.: "Computer Aided Kinematics and Dynamics of Mechanical Systems," vol. I: "Basic Methods," Allyn and Bacon, Boston, 1989.

A14. Hollerback, J. M.: "A Recursive Lagrangian Formulation of Manipulator and a Comparative Study of Dynamics Formulation," *IEEE Transactions on Systems, Man and Cybernetics,* SMC-10,11, 1980.

A15. Huston, R.: "Multibody Dynamics," Butterworth-Heinemann, Stoneham, Mass., 1990.

A16. Huston, R. L., and C. Passerello: "On the Dynamics of a Human Body Model," *J. Biomechanics,* vol. 4, pp. 369–378, 1976.

A17. Huston, R. L., and C. Passerello: "On Multi-Rigid-Body System Dynamics," *Computers and Structures,* vol. 10, pp. 439–446, 1979.

A18. Kane, T. R., and D. A. Levinson: "Dynamics: Theory and Application," McGraw-Hill Book Company, Inc., New York, 1985.

A19. Lo, C-K.: "A General Purpose Comprehensive Dynamics Simulation of Serial Robot Manipulators," Ph.D. dissertation, University of Pennsylvania, Philadelphia, 1990.

A20. Luh, J. Y. S., M. W. Walker, and R. P. C. Paul: "On-Line Computational Scheme for Mechanical Manipulators," *Trans. ASME, J. Dynamic Systems, Measurement and Control,* vol. 102, pp. 69–76, 1980.

A21. Midha, A. M. (ed.), et al.: "Elastic Mechanisms," chap. 9, Ref. A5.

A22. Nikravesh, P. E.: "Computer-Aided Analysis of Mechanical Systems," Prentice-Hall, Inc., Englewood Cliffs, N.J., 1988.

A23. Norton, R. L. (ed.), et al.: "Cams and Cam Followers," chap. 7, Ref. A5.

A24. Paul, B.: "DYMAC (DYnamics of MAChinery)," pp. 305–322, Ref. A33.

A25. Paul, B.: "Dynamics of Multibody Mechanical Systems, A Forty Year History," chap. 8, Ref. A5.

A26. Paul, B., and R. Schaffa: "DYSPAM (DYnamics of SPatial Mechanisms)," pp. 323–340, Ref. A33, 1990. (*Note:* The name of coauthor R. Schaffa was omitted from the text of Ref. A33 due to an editorial error.)

A27. Paul, B., J. W. West, and E. Y. Yu: "A Passive Gravitational Attitude Control System for Satellites," *Bell System Tech. J.,* vol. 42, pp. 2195–2238, 1963.

A28. Pisano, A., and F. Freudenstein: "An Experimental and Analytical Investigation of the Dynamic Response of High-Speed Cam-Follower Systems—Parts I, II," *Trans. ASME, J. Mechanisms, Transmissions and Automation in Design,* vol. 105, pp. 692–698, 699–704, 1983.

A29. Roberson, R. E., and R. Schwertassek: "Dynamics of Multibody Systems" Springer-Verlag, Berlin, New York, 1988.

A30. Roberson, R., and J. Wittenburg: "A Dynamical Formalism for an Arbitrary Number of Interconnected Rigid Bodies, with Reference to the Problem of Satellite Attitude Control," vol. 1, book 3, paper 46D, in *Proc. 3d Congr. Int. Fed. Autom. Control,* Butterworth, 1967, London, 1966.

A31. Sandor, G. N., and A. G. Erdman: "Advanced Mechanism Design, Analysis and Synthesis," vol. II, Prentice-Hall, Inc., Englewood Cliffs, N.J., 1984.

A32. Schaffa, R.: "Dynamics of Spatial Mechanisms and Robots," Ph.D. dissertation, University of Pennsylvania, Philadelphia, 1984.

A33. Schiehlen, W. (ed.): "Multibody Systems Handbook," Springer-Verlag, Berlin, New York, 1990.

A34. Schwertassek, R., and R. E. Roberson: "A perspective on computer oriented multibody dynamic formalisms and their implementations," pp. 261–273 in G. Bianchi and W. Schiehlen, eds., *Proc. IUTAM/IFToMM Symposium on Dynamics of Multibody Systems,* (September 1985, Udine, Italy), Springer-Verlag, Berlin, 1986.

A35. Shabana, A. A.: "Dynamics of Multibody Systems," John Wiley & Sons, Inc., New York, 1989.

A36. Sommerfeld, A.: "Mechanics," Academic Press, Inc., New York, 1964, p. 187.

A37. Vance, J. M.: "Rotordynamics of Turbo Machinery," John Wiley & Sons, Inc., New York, 1988.

A38. Walker, M. W., and D. E. Orin: "Efficient Dynamic Computer Simulation of Robot Mechanisms," *Trans. ASME, J. Dynamic Systems, Measurement, Control,* vol. 104, pp. 205–211, 1982.

A39. Wittenburg, J.: "Dynamics of Systems of Rigid Bodies," B. G. Teubner, Stuttgart, 1977.

A40. Wolovich, W.: "Robotics: Basic Analysis and Design," Holt, Rinehart and Winston, New York, 1987.

A41. Yoshikawa, T.: "Foundations of Robotics," The MIT Press, Cambridge, Mass., 1990.

DESIGN

SECTION 7
STATIC AND FATIGUE DESIGN

Steven M. Tipton, Ph.D., P.E.

Associate Professor of Mechanical Engineering
University of Tulsa
Tulsa, Okla.

SYMBOLS

α = "characteristic length" (empirical curve-fit parameter)

a = crack length

a_f = final crack length

a_i = initial crack length

A = cross-sectional area

A_f = Forman coefficient

A_p = Paris coefficient

A_w = Walker coefficient

b = fatigue strength exponent

b' = baseline fatigue exponent

c = fatigue ductility exponent

$C = 2\tau_{xy,a}/\sigma_{x,a}$ (during axial-torsional fatigue loading)

C' = baseline fatigue coefficient

CM = Coulomb Mohr Theory

d = diameter of tensile test specimen gauge section

DAM_i = cumulative fatigue damage for a particular (ith) cycle

Δ = range of (e.g., stress, strain, etc.) = maximum $-$ minimum

e = nominal axial strain

e_{offset} = offset plastic strain at yield

e_u = engineering strain at ultimate tensile strength

E = modulus of elasticity

E_c = Mohr's circle center

ε = normal strain

ε_a = normal strain amplitude

ε_f = true fracture ductility

ε'_f = fatigue ductility coefficient

ε_u = true strain at ultimate tensile strength

FS = factor of safety

G = elastic shear modulus

γ = shear strain

K = monotonic strength coefficient

K = stress intensity factor

K' = cyclic strength coefficient

K_c = fracture toughness

K_f = fatigue notch factor

K_{Ic} = plane-strain fracture toughness

K_t = elastic stress-concentration factor

MM = modified Mohr Theory

MN = maximum normal stress theory

n = monotonic strain-hardening exponent

n' = cyclic strain-hardening exponent

n_f = Forman exponent

n_p = Paris exponent

n_w and m_w = Walker exponents

N = number of cycles to failure in a fatigue test

N_T = transition fatigue life

P = axial load

ϕ = phase angle between σ_x and τ_{xy} stresses (during axial-torsional fatigue loading)

r = notch root radius

R = cyclic load ratio (minimum load over maximum load)

R = Mohr's circle radius

RA = reduction in area

S = stress amplitude during a fatigue test

S_e = endurance limit

$S_{eq,a}$ = equivalent stress amplitude (multiaxial to uniaxial)

$S_{eq,m}$ = equivalent mean stress (multiaxial to uniaxial)

S_{nom} = nominal stress

$S_u = S_{ut}$ = ultimate tensile strength

S_{uc} = ultimate compressive strength

S_y = yield strength

SALT = equivalent stress amplitude (based on Tresca for axial-torsional loading)

SEQA = equivalent stress amplitude (based on von Mises for axial-torsional loading)

σ = normal stress

σ_a = normal stress amplitude

σ_{eq} = equivalent axial stress

σ_f = true fracture strength

σ'_f = fatigue strength coefficient

σ_m = mean stress during a fatigue cycle

σ_{norm} = normal stress acting on plane of maximum shear stress

σ_{notch} = elastically calculated notch stress

σ_A = maximum principal stress

σ_B = minimum principal stress

σ_u = true ultimate tensile strength

t = thickness of fracture mechanics specimen

τ = shear stress

τ_{max} = maximum shear stress

θ_σ = orientation of maximum principal stress

θ_τ = orientation of maximum shear stress

ν = Poisson's ratio

NOTATIONS

1, 2, 3 = subscripts designating principal stress

I, II, III = subscripts denoting crack loading mode

a,eff = subscript denoting effective stress amplitude (mean stress to fully reversed)

f = subscript referring to final dimensions of a tension test specimen

bend = subscript denoting bending loading

max = subscript denotes maximum or peak during fatigue cycle

min = subscript denoting minimum or valley during fatigue cycle

o = subscript referring to original dimensions of a tension test specimen

Tr = subscript referring to Tresca criterion

vM = subscript referring to von Mises criterion

x, y, z = orthogonal coordinate axes labels

7.1 INTRODUCTION

The design of a component implies a design framework and a design process. A typical design framework requires consideration of the following factors: component function and performance, producibility and cost, safety, reliability, packaging, and operability and maintainability.

The designer should assess the consequences of failure and the normal and abnormal conditions, loads, and environments to which the component may be subjected during its operating life. On the basis of the requirements specified in the design framework, a design process is established which may include the following elements: conceptual design and synthesis, analysis and gathering of relevant data, optimization, design and test of prototypes, further optimization and revision, final design, and monitoring of component performance in the field.

Requirements for a successful design include consideration of data on the past performance of similar components, a good definition of the mechanical and thermal loads (monotonic and cyclic), a definition of the behavior of candidate materials as a function of temperature (with and without stress raisers), load and corrosive environments, a definition of the residual stresses and imperfections owing to processing, and an appreciation of the data which may be missing in the trade-offs among parameters such as cost, safety, and reliability. Designs are typically analyzed to examine the potential for fracture, excessive deformation (under load, creep), wear, corrosion, buckling, and jamming (due to deformation, thermal expansion, and wear). These may be caused by steady, cyclic, or shock loads, and temperatures under a number of environmental conditions and as a function of time. Reference 92 lists the following failures: ductile and brittle fractures, fatigue failures, distortion failures, wear failures, fretting failures, liquid-erosion failures, corrosion failures, stress-corrosion cracking, liquid-metal embrittlement, hydrogen-damage failures, corrosion-fatigue failures, and elevated-temperature failures.

In addition, property changes owing to other considerations, such as radiation, should be considered, as appropriate. The designer needs to decide early in the design process whether a component or system will be designed for infinite life, finite specified life, a fail-safe or damage-tolerant criterion, a required code, or a combination of the above.[3]

In the performance of design trade-offs, in addition to the standard computerized tools of stress analysis, such as the finite-element method, depending upon the complexity of the mathematical formulation of the design constraints and the function to be optimized, the mathematical programming tools of operations research may apply. Mathematical programming can be used to define the most desirable (optimum) behavior of a component as a function of other constraints. In addition, on a systems level, by assigning relative weights to requirements, such as safety, cost, and life, design parameters can be optimized. Techniques such as linear programming, nonlinear programming, and dynamic programming may find greater application in the future in the area of mechanical design.[93]

Numerous factors dictate the overall engineering specifications for mechanical design. This chapter concentrates on philosophies and methodologies for the design of components that must satisfy quantitative strength and endurance specifications. Only deterministic approaches are presented for statically and dynamically loaded components.

Although mechanical components can be susceptible to many modes of failure, approaches in this chapter concentrate on the comparison of the state of stress and/or strain in a component with the strength of candidate materials. For instance, buckling, vibration, wear, impact, corrosion, and other environmental factors are not considered. Means of calculating stress-strain states for complex geometries associated with real mechanical components are vast and wide-ranging in complexity. This topic will be addressed in a general sense only. Although some of the methodologies are presented in terms of general three-dimensional states of stress, the majority of the examples and approaches will be presented in terms of two-dimensional surface stress states. Stresses are generally maximum on the surface, constituting the vast majority of situations of concern to mechanical designers. [Notable exceptions are contact problems,[1-6] components which are surfaced processed (e.g., induction hardened or nitrided[7]), or components with substantial internal defects, such as pores or inclusions.]

In general, the approaches in this chapter are focused on isotropic metallic components, although they can also apply to homogeneous nonmetallics (such as glass, ceramics, or polymers). Complex failure mechanisms and material anisotropy associated with composite materials warrant the separate treatment of these topics.

Typically, prototype testing is relied upon as the ultimate measure of the structural integrity of an engineering component. However, costs associated with expensive and time consuming prototype testing iterations are becoming more and more intolerable. This increases the importance of modeling durability in everyday design situations. In this way, data from prototype tests can provide valuable feedback to enhance the reliability of analytical models for the next iteration and for future designs.

7.2 ESTIMATION OF STRESSES AND STRAINS IN ENGINEERING COMPONENTS

When loads are imposed on an engineering component, stresses and strains develop throughout. Many analytical techniques are available for estimating the state of stress and strain in a component. A comprehensive treatment of this subject is beyond the scope of this chapter. However, the topic is overviewed for engineering design situations.

7.2.1 Definition of Stress and Strain

An engineering definition of "stress" is the force acting over an infinitesimal area. "Strain" refers to the localized deformation associated with stress. There are several important practical aspects of stress in an engineering component:

1. A state of stress-strain must be associated with a particular location on a component.

2. A state of stress-strain is described by stress-strain components, acting over planes.

3. A well-defined coordinate system must be established to properly analyze stress-strain.

4. Stress components are either normal (pulling planes of atoms apart) or shear (sliding planes of atoms across each other).

5. A stress state can be uniaxial, but strains are usually multiaxial (due to the effect described by Poisson's ratio).

The most general three-dimensional state of stress can be represented by Fig. 7.1a. For most engineering analyses, designers are interested in a two-dimensional state of stress, as depicted in Fig. 7.1b. Each side of the square two-dimensional element in Fig. 7.1b represents an infinitesimal area that intersects the surface at 90°.

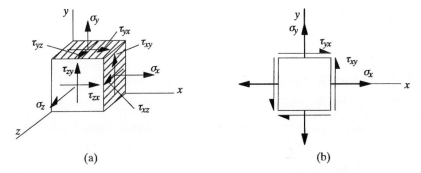

(a)   (b)

FIG. 7.1 The most general (a) three-dimensional and (b) two-dimensional stress states.

By slicing a section of the element in Fig. 7.1b, as shown in Fig. 7.2, and analytically establishing static equilibrium, an expression for the normal stress σ and the shear stress τ acting on any plane of orientation θ can be derived. This expression forms a circle when plotted on axes of *shear stress* versus *normal stress*. This circle is referred to as "Mohr's circle."

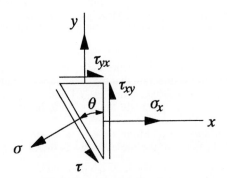

FIG. 7.2 Shear and normal stresses on a plane rotated θ from its original orientation.

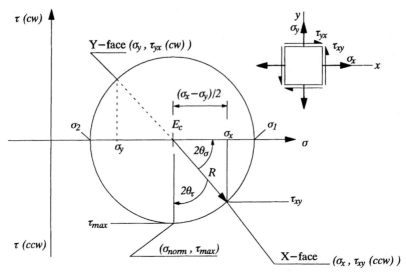

FIG. 7.3   Mohr's circle for a generic state of surface stress.

Mohr's circle is one of the most powerful analytical tools available to a design ana-lyst. Here, the application of Mohr's circle is emphasized for two-dimensional stress states. From this understanding, it is a relatively simple step to extend the analysis to most three-dimensional engineering situations.

Consider the stress state depicted in Fig. 7.1*b* to lie in the surface of an engineering component. To draw the Mohr's circle for this situation (Fig. 7.3), three simple steps are required:

1. Draw the shear-normal axes [τ(cw) positive vertical axis, σ tensile along horizon-tal axis].

2. Define the center of the circle E_c (which always lies on the σ axis):

$$E_c = (\sigma_x + \sigma_y)/2 \qquad (7.1)$$

3. Use the point represented by the "X-face" of the stress element to define a point on the circle (σ_x, τ_{xy}). The X-face on the Mohr's circle refers to the plane whose nor-mal lies in the X direction (or the plane with a normal and shear stress of σ_x and τ_{xy}, respectively).

That's all there is to it. The sense of the shear stress [clockwise (cw) or counter-clockwise (ccw)] refers to the direction that the shear stress attempts to rotate the ele-ment under consideration. For instance, in Figs. 7.1 and 7.2, τ_{xy} is ccw and τ_{yx} is cw. This is apparent in Fig. 7.3, a schematic Mohr's circle for this generic surface ele-ment.

The interpretation and use of Mohr's circle is as simple as its construction. Referring to Fig. 7.3, the radius of the circle R is given by Eq. (7.2).

$$R = \sqrt{\left(\frac{\sigma_x - \sigma_y}{2}\right)^2 + \tau_{xy}^2} \qquad (7.2)$$

This could suggest an alternate step 3: that is, define the radius and draw the circle with the center and radius. The two approaches are equivalent. From the circle, the following important items can be composed: (1) the principal stresses, (2) the maximum shear stress, (3) the orientation of the principal stress planes, (4) the orientation of the maximum shear planes, and (5) the stress normal to and shear stress acting over a plane of *any* orientation.

1. Principal Stresses. It is apparent that

$$\sigma_1 = E_c + R \qquad (7.3)$$

and

$$\sigma_2 = E_c - R \qquad (7.4)$$

2. Maximum Shear Stress. The maximum in-plane shear stress at this location,

$$\tau_{max} = R \qquad (7.5)$$

3. Orientation of Principal Stress Planes. Remember only one rule: A rotation of 2θ around the Mohr's circle corresponds to a rotation of θ for the actual stress element. This means that the principal stresses are acting on faces of an element oriented as shown in Fig. 7.4. In this figure, a counterclockwise rotation from the X-face to σ_1 of $2\theta_\sigma$, means a ccw rotation of θ_σ on the surface of the component, where θ_σ is given by Eq. (7.6):

$$\theta_\sigma = 0.5 \tan^{-1}\left[\frac{2\,\tau_{xy}}{(\sigma_x - \sigma_y)}\right] \qquad (7.6)$$

In Figs. 7.3 and 7.4, since the "X-face" refers to the plane whose normal lies in the x direction, it is associated with the x axis and serves as a reference point on the Mohr's circle for considering normal and shear stresses on any other plane.

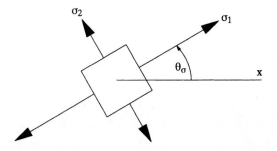

FIG. 7.4 Orientation of the maximum principal stress plane.

4. Orientation of the Maximum Shear Planes. Notice from Fig. 7.3 that the maximum shear stress is the radius of the circle $\tau_{max} = R$. The orientation of the plane of maximum shear is thus defined by rotating through an angle $2\theta_\tau$ around the Mohr's circle, clockwise from the X-face reference point. This means that the plane oriented at an angle θ_τ (cw) from the x axis will feel the maximum shear stress, as shown in Fig. 7.5. Notice that the sum of θ_τ and θ_σ on the Mohr's circle is 90°; this will always be the case. Therefore, the planes feeling the maximum principal (normal) stress and maximum shear stress always lie 45° apart, or

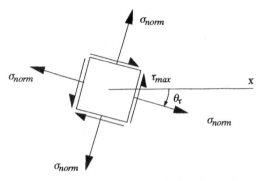

FIG. 7.5 Orientation of the planes feeling the maximum shear stress.

$$\theta_\tau = 45° - \theta_\sigma \tag{7.7}$$

EXAMPLE 1 Suppose a state of stress is given by $\sigma_x = 30$ ksi, $\tau_{xy} = 14$ ksi (ccw) and $\sigma_y = -12$ ksi. If a seam runs through the material 30° from the vertical, as shown, compute the stress normal to the seam and the shear stress acting on the seam.

solution Construct the Mohr's circle by computing the center and radius:

$$E_c = \frac{[30 + (-12)]}{2} = 9 \text{ ksi} \qquad R = \sqrt{\left(\frac{[30 - (-12)]}{2}\right)^2 + 14^2} = 25.24 \text{ ksi}$$

The normal stress σ and shear stress τ acting on the seam are obtained from inspection of the Mohr's circle and shown below:

$$\sigma = E_c + R \cos(33.69° + 60°) = 7.38 \text{ ksi}$$

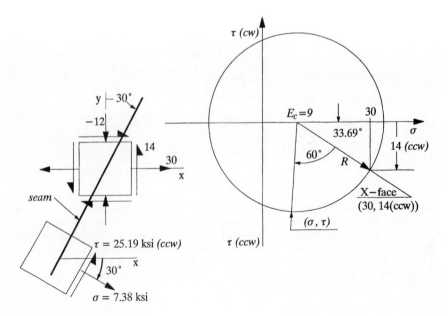

$$\tau = R \sin(33.69° + 60°) = 25.19 \text{ ksi (ccw)}$$

The normal stress σ_{norm} is equal on each face of the maximum shear stress element and $\sigma_{norm} = E_c$, the Mohr's circle center. (This is always the case since the Mohr's circle is always centered on the normal stress axis.)

5. Stress Normal to and Shear Stress on a Plane of Any Orientation. Remember that *the Mohr's circle is a collection of (σ, τ) points that represent the normal stress σ and the shear stress τ acting on a plane at any orientation in the material.* The X-face reference point on the Mohr's circle is the point representing a plane whose stresses are (σ_x, τ_{xy}). Moving an angle 2θ in either sense from the X-face around the Mohr's circle corresponds to a plane whose normal is oriented an angle θ in the same sense from the x axis. (See Example 1.)

More formal definitions for three-dimensional tensoral stress and strain are available.[5,6,8–13] In the majority of engineering design situations, bulk plasticity is avoided. Therefore, the relation between stress and strain components is predominantly elastic, as given by the generalized Hooke's law (with ε and γ referring to normal and shear strain, respectively) in Eqs. (7.8) to (7.13):

$$\varepsilon_x = \frac{1}{E}[\sigma_x - v(\sigma_y + \sigma_z)] \tag{7.8}$$

$$\varepsilon_y = \frac{1}{E}[\sigma_y - v(\sigma_z + \sigma_x)] \tag{7.9}$$

$$\varepsilon_z = \frac{1}{E}[\sigma_z - v(\sigma_x + \sigma_y)] \tag{7.10}$$

$$\gamma_{xy} = \frac{\tau_{xy}}{G} \tag{7.11}$$

$$\gamma_{yz} = \frac{\tau_{yz}}{G} \tag{7.12}$$

$$\gamma_{zx} = \frac{\tau_{zx}}{G} \tag{7.13}$$

where E is the modulus of elasticity, v is Poisson's ratio, and G is the shear modulus, expressed as Eq. (7.14):

$$G = \frac{E}{2(1 + v)} \tag{7.14}$$

7.2.2 Experimental

Experimental stress analysis should probably be referred to as experimental strain analysis. Nearly all commercially available techniques are based on the detection of local states of strain, from which stresses are computed. For elastic situations, stress components are related to strain components by the generalized Hooke's law as shown in Eqs. (7.15) to (7.20):

$$\sigma_x = \frac{E}{1 + v}\varepsilon_x + \frac{vE}{(1 + v)(1 - 2v)}(\varepsilon_x + \varepsilon_y + \varepsilon_z) \tag{7.15}$$

$$\sigma_y = \frac{E}{1 + v}\varepsilon_y + \frac{vE}{(1 + v)(1 - 2v)}(\varepsilon_x + \varepsilon_y + \varepsilon_z) \tag{7.16}$$

$$\sigma_z = \frac{E}{1 + v}\varepsilon_z + \frac{vE}{(1 + v)(1 - 2v)}(\varepsilon_x + \varepsilon_y + \varepsilon_z) \tag{7.17}$$

$$\tau_{xy} = G\gamma_{xy} \tag{7.18}$$

$$\tau_{yz} = G\gamma_{yz} \tag{7.19}$$

$$\tau_{zx} = G\gamma_{zx} \tag{7.20}$$

For strains measured on a stress-free surface where $\sigma_z = 0$; the in-plane normal stress relations simplify to Eqs. (7.21) and (7.22).

$$\sigma_x = \frac{E}{1 - v^2}[\varepsilon_x + v\varepsilon_y] \tag{7.21}$$

$$\sigma_y = \frac{E}{1 - v^2}[\varepsilon_y + v\varepsilon_x] \tag{7.22}$$

Several techniques exist for measuring local states of strain, including electro-mechanical extensometers, photoelasticity, brittle coatings, moiré methods, and holography.[14,15] Other, more sophisticated approaches such as X-ray and neutron diffraction, can provide measurements of stress distributions below the surface. However, the vast majority of experimental strain data are recorded with electrical resistance strain gauges. Strain gauges are mounted directly to a carefully prepared surface using an adhesive. Instrumentation measures the change in resistance of the gauge as it deforms with the material adhered to its gauge section, and a strain is computed from the resistance change. Gauges are readily available in sizes from 0.015 to 0.5 inches in gauge length and can be applied in the roots of notches and other stress concentrations to measure severe strains that can be highly localized.

As implied by Eqs. (7.15) to (7.22), it can be important to measure strains in more than one direction. This is particularly true when the direction of principal stress is unknown. In these situations it is necessary to utilize three-axis rosettes (a pattern of three gauges in one, each oriented along a different direction). If the principal stress directions are known but not the magnitudes, two-axis (biaxial) rosettes can be oriented along principal stress directions and stresses computed with Eqs. (7.21) and (7.22) replacing σ_x and σ_y with σ_1 and σ_2, respectively. These equations can be used to show that severe errors can result in calculated stresses if a biaxial stress state is assumed to be uniaxial. (See Example 2.)

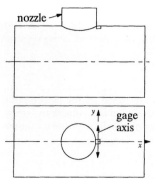

EXAMPLE 2   This example demonstrates how stresses can be underestimated if strain is measured only along a single direction in a biaxial stress field. Compute the hoop stress at the base of the nozzle shown if (1) a hoop strain of 0.0023 is the only measurement taken and (2) an axial strain measurement of +0.0018 is also taken.

solution   For a steel vessel ($E = 30,000$ ksi and $v = 0.3$), if the axial stress is neglected, the hoop stress is calculated to be

$$\sigma_y = E\varepsilon_y = 69 \text{ ksi}$$

However, if the axial strain measurement of +0.0018 is used with Eq. (7.22), then the hoop stress is given by

$$\sigma_x = \frac{E}{1 - v^2}[0.0023 + 0.3(0.0018)] = 93.63 \text{ ksi}$$

In this example, measuring only the hoop strain caused the hoop stress to be underestimated by over 26 percent.

Obviously, in order to measure strains, prototype parts must be available, which is generally not the case in the early design stages. However, rapid prototyping techniques, such as computer numerically controlled machining equipment and stereolithography, can greatly facilitate prototype development. Data from strain-gauge testing of components in the final developmental stages should be compared to preliminary design estimates in order to provide feedback to the analysis.

7.2.3 Strength of Materials

Concise solutions have been developed for pressure vessels; beams in bending, tension and torsion; curved beams; etc.[1-6] These are usually based on considering a section through the point of interest, establishing static equilibrium with externally applied forces, and making assumptions about the distribution of stress or strain throughout the cross section.

Example 3 illustrates the use of traditional bending- and torsional-stress relations, showing how they can be used to improve the efficiency of an experimental strain measurement.

EXAMPLE 3 A steel component is welded to a solid base and loaded as shown. Identify the region of maximum stress. Show where to mount and how to orient a single-axis strain gauge to pick up the maximum signal. Compute the maximum principal stress and strain in the structure for a value of $P = 400,000$ lb.

solution The critical section is the cross section defined by $x = 0$, and the maximum stress can be expected at the origin of the coordinate system shown on page 7.14. The cross section feels bending about the centroidal y axis M_y, and a torque T. (Transverse shear is neglected since it is zero at the point of maximum stress.)

$$\sigma_x = \frac{M_y c}{I_y} = \frac{(1,800,000 \text{ in·lb})(2.5 \text{ in})}{(11 \text{ in})(5 \text{ in})^3/12} = 39.27 \text{ ksi}$$

$$\tau_{xy} = \frac{4,760,000 \text{ in·lb}}{(11 \text{ in})(5 \text{ in})^2}[3 + 1.8(\tfrac{5}{11})] = 66.09 \text{ ksi (from Ref. 1)}$$

Constructing a Mohr's circle (as in Fig. 7.3) the orientations of the principal stresses and their magnitudes are given by

$$\theta = 36.7° \text{ cw}$$

$$\sigma_1 = 88.58 \text{ ksi}$$

$$\sigma_2 = -49.3 \text{ ksi}$$

and Eqs. (7.21) and (7.22) yield

$$\varepsilon_1 = 0.003446$$

$$\varepsilon_2 = -0.002530$$

7.2.4 Elastic Stress-Concentration Factors

Most mechanical components are not smooth. Practical components typically include holes, keyways, notches, bends, fillets, steps, or other structural discontinuities. Stresses tend to become "concentrated" in such regions such that these stresses are sig-

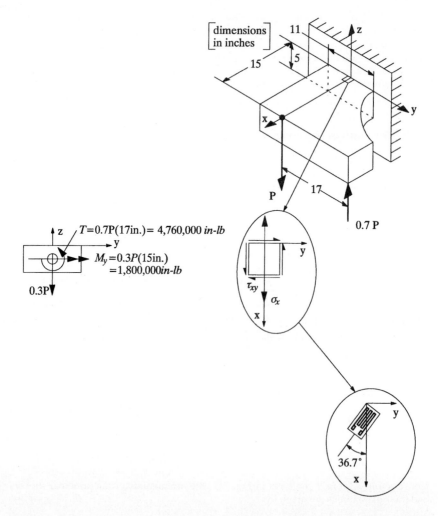

nificantly greater than the nominally calculated stresses, based on the external forces and cross-sectional area. For instance, stress distributions are shown in Fig. 7.6 for a uniform rectangular plate in tension and through the identical cross section of a plate with a filleted step. Notice that the maximum elastically calculated stress at the root of the fillet, σ_{notch}, is greater than the nominal stress, S_{nom}. From this, the *stress-concentration factor K_t* is defined as the maximum stress divided by the nominal stress:

$$K_t = \frac{\sigma_{notch}}{S_{nom}} \tag{7.23}$$

Stress-concentration factors are found by a number of techniques including experimental, finite-element analysis, boundary-element analysis, closed-form elasticity solutions, and others. Fortunately, researchers have tabulated K_t values for many generalized geometries.[16,17]

Reference 16, from Peterson, is a compendium of design charts. Reference 17, from Roark, provides useful empirical formulas that can be programmed into spread-

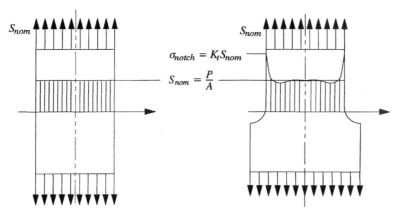

FIG. 7.6   Definition of elastic stress-concentration factor.

sheets or computer routines for design optimization. There are several important points to remember about stress concentration factors:

1. They only apply to elastic states of stress-strain.
2. They are tabulated for a particular mode of loading (axial, bending, torsional, etc.).
3. Since they are elastic, they can be superimposed (i.e., computed separately, then added).

It has been shown that using the full value of stress-concentration factors for evaluating strength can be overly conservative, especially for static design situations. Shigley and Mischke[1] state that one can usually neglect the stress-concentration factor due to the fact that localized yielding can work-harden the material in the notch vicinity and relieve the stresses. On the other hand, neglecting stress-concentration factors can be nonconservative, especially if very brittle material or fatigue loading is involved. As safe design practice, stress-concentration factors should be considered for preliminary analysis. If the resulting solution is unacceptable from a weight, size, or cost standpoint, then reasonable reductions in K_t can be considered based on the potential for the material to deform and locally work-harden. Such decisions can be based on experimental data generated with the material of interest and notches with similar values of K_t. This type of testing can supplement analysis prior to the availability of fully designed prototype parts. Data from testing such as this should be carefully documented, since it can be used as a basis for future design decisions.

7.2.5   Finite-Element Analysis

The use of computers is increasing as rapidly in engineering design as in any other profession. Finite-element analysis (FEA),[18–21] coupled with increasing computational capabilities, is providing increasing analytical power for use on everyday design situations. Commercially available software packages are enabling designers to evaluate states of stress in situations involving complex geometry and loading combinations. However, three important points should be considered when stresses are computed from FEA: (1) Elastic analysis can be straightforward, but the potential for error is great if the analyst has not assured mesh convergence, especially for sharp geometric discontinuities or contact problems. (2) The specification of boundary conditions is critical to obtaining valid results that correlate with the physical stress-strain state in the component being modeled. (3) Elastic-plastic analysis is not yet simplified for everyday design use, particularly for cyclic loading conditions.[22]

These three points are intended to remind the designer not to accept FEA results without an adequate awareness of the assumptions used to implement the analysis (most importantly, mesh density, boundary conditions, and material modeling). Most robust commercial FEA codes provide error estimates associated with their solutions. In high-stress-concentration regions, these errors can be substantial and a locally refined mesh could be called for (often not a simple task). A designer must be careful to avoid the tendency to simply marvel at appealing and colorful FEA output without fully understanding that the results are only as valid as the assumptions used to build the analytical model.

7.3 STRUCTURAL INTEGRITY DESIGN PHILOSOPHIES

An engineer must routinely assure that designs will endure anticipated loading histories with no significant change in geometry or loss in load-carrying capability. Anticipating service-load histories can require experience and/or testing. Techniques for load estimation are as diverse as any other aspect of the design process.[23]

The *design* or *allowable* stress is generally defined as the tension or compressive stress (yield point or ultimate) depending on the type of loading divided by the safety factor. In fatigue the appropriate safety factor is used based on the number of cycles. Also when wear, creep, or deflections are to be limited to a prescribed value during the life of the machine element, the design stress can be based upon values different from above.

The magnitude of the design factor of safety, a number greater than unity, depends upon the application and the uncertainties associated with a particular design. In the determination of the factor of safety, the following should be considered:

1. The possibility that failure of the machine element may cause injury or loss of human life

2. The possibility that failure may result in costly repairs

3. The uncertainty of the loads encountered in service

4. The uncertainty of material properties

5. The assumptions made in the analysis and the uncertainties in the determination of the stress-concentration factors and stresses induced by sudden impact and repeated loads

6. The knowledge of the environmental conditions to which the part will be subjected

7. The knowledge of stresses which will be introduced during fabrication (e.g., residual stresses), assembly, and shipping of the part

8. The extent to which the part can be weakened by corrosion

Many other factors obviously exist. Typical values of design safety factors range from 1.0 (against yield) in the case of aircraft, to 3 in typical machine-design applications, to approximately 10 in the case of some pressure vessels. It is to be noted that these safety factors allow us to compute the allowable stresses given and are not in lieu of the stress-concentration factors which are used to compute stresses in service.

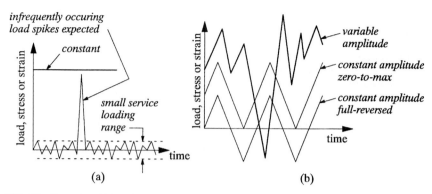

FIG. 7.7 Examples of (*a*) static and (*b*) dynamic, or fatigue, loading.

If the uncertainties are great enough to cause severe weight, volume, or economic penalties, testing and/or more thorough analyses should be performed rather than relying upon very large factors of safety. Factors of safety to be used with standard, commercially available design elements should be those recommended for them by reliable manufacturers and/or by established codes for design of machines.

In probabilistic approaches to design, in terms of stress (or actual load) and strength (or load capability) the safety factor is related to reliability. When a failure may cause injury or otherwise be disastrous, the probability density curves representing the strength of the part and the stress to be sustained should not overlap, and the factor of safety equals the ratio of the mean strength to the mean stress. If the tails of the two curves overlap, a possibility for failure exists.

The "true factor of safety," which may be defined in terms of load, stress, deflection, creep, wear, etc., is the ratio of the magnitude of any of the above parameters resulting in damage to its actual value in service. For example:

$$\text{True factor of safety} = \frac{\text{maximum load part can sustain without damage}}{\text{maximum load part sustains in service}}$$

The true factor of safety is determined after a part is built and tested under service conditions.

In this chapter, loading is classified as "static" or "dynamic." Static loading could be formally defined as loading that remains constant over the life of the component, as depicted in Fig. 7.7*a*. Under this type of loading, the primary concern is avoiding failure by yielding or fracture. Also shown in Fig. 7.7*a* is another type of loading that can be considered "static," from a structural integrity viewpoint. This refers to situations where only a few, *infrequently occurring* load spikes can be expected in service. Dynamic loading fluctuates significantly during the life of the component, as shown in Fig. 7.7*b*. Although peak stresses can remain well below levels associated with yielding, this type of loading can lead to failure by fatigue.

7.3.1 Static Loading

Loading on a mechanical component is rarely steady. However, in many cases, safety factors and service load ratings are used in order to keep in-service load fluctuations small relative to the maximum load the component can sustain. Often this is assured by proof loading a component as the final step in its manufacturing process. Proof loads usually exceed the rated service load by a factor of 2 to 3.5. Examples of this include chains, other lifting hardware, and pressure vessels. Proof loading not only ensures the structural integrity of the part, but can also serve to impart residual stresses that increase the functional elastic limit and increase fatigue life.[24–26]

When a component is designed for static strength, it must be assured that service loads indeed remain well below the strength of the component. Unfortunately, the user, more than the designer, often dictates the maximum load level a component will experience. Users tend to push designs over the limit at every available opportunity. For safety-critical components, designers should consider mechanisms to ensure that loadings do not exceed safe operating levels. For instance, rupture disks are effective "weak links" in the design of pressure systems. Should the operating pressure be exceeded, the rupture disk fails by design, into a discharge tank. Another example would be the use of redundant, or backup, elements. When a specimen begins to deform under too great a load, it gains support when it encounters a backup element, thus avoiding complete fracture (and possibly alerting the end user).

Care must be taken when the loading on a component is classified as "static" for design purposes. The approach is only safe when the static limit is rarely seen in service, as depicted by the load spike in Fig. 7.7a. For example, suppose a nozzle discharges under a constant internal pressure. There is a tendency to utilize that pressure for static design (the constant loading line in Fig. 7.7a). However, if the nozzle discharges for 30 minutes, drains, then repeats, on a regular basis, then fatigue could be important.

7.3.2 Fatigue Loading

Under fatigue loading, cracks develop in high-stress regions which were initially free of any macroscopic defect. A component can endure numerous cycles of loading before the crack is detectable. Once this occurs, a dominant crack usually propagates progressively to fracture. The relative life spent in developing a crack of "engineering size" (usually defined as 1–2 mm in surface length) and then propagating the crack to fracture can define the fatigue design philosophy, as overviewed below.

Infinite-Life Design.   This philosophy is based on the concept of the *fatigue limit,* or the stress amplitude below which fatigue will not occur. For high-cycle components like valve springs, turbo machinery, and other high-speed rotating equipment, this is still a very widely utilized concept. However, the approach is going out of style due to cost- and weight-reduction requirements. There are also problems pertaining to the definition of a fatigue limit for a particular material, since numerous factors have proven influential. These factors include heat treatment, surface condition, residual stresses, temperature, environment, etc. (Furthermore, aluminum and other nonferrous alloys do not exhibit a fatigue limit.) One final cautionary note: Intermittent overloads can reduce or eliminate the fatigue limit.[13]

Finite-Life (Safe-Life) Design.   Instead of designing a component to never fail, parts are designed for a specified life deemed "safe," or unlikely to occur during the rated life of the machine, except in cases of abusive loading. For instance, even a safety-critical automobile suspension component might be designed to sustain only 1000 of the most severe impact loads corresponding to the worst high-speed curb strike on the proving ground. However, the vast majority of automobiles will experience nowhere near this many occurrences. Pressure vessels are sometimes designed to lives on the order of a few hundred cycles, corresponding to cleanout cycles that will occur only a few times annually. Ball joints in automobiles and landing-gear parts in aircraft are other examples of finite-life design situations.

Fail-Safe Design. This strategy, developed primarily in the aircraft industry, should be implemented whenever possible. The approach invokes measures to ensure that, if cracks initiate, they will grow in a controlled manner. Then, measures are taken to ensure that catastrophic failure is avoided, including the use of redundant elements, backup elements, crack-arrest holes positioned at strategic locations, and the use of multiple load paths. This approach is referred to as "leak before burst" in the pressure-vessel industry.

Damage-Tolerant Design. Also developed in the aircraft industry, this philosophy assumes that cracks exist *before* a component is put into service. For instance, cracks are assumed to exist underneath rivet heads or behind a seam, anywhere that they might be concealed during routine inspection. Then, the behavior of the crack is predicted from flight-loading spectra anticipated for the aircraft. Analyses of many key locations on an aircraft are used to schedule maintenance and inspections.

The four methods discussed above cover most design situations, but which of these is utilized depends on the design criteria. It is typical for more than one (sometimes all) of the strategies to be utilized in a single design. For instance, in the design of an aircraft, the fail-safe approach is routinely applied to wings, fuselages, and control surfaces. However, a landing gear and a rotor in the jet engine are designed for finite life.

To provide some size scale to the issue of fatigue crack development, refer to Fig. 7.8.

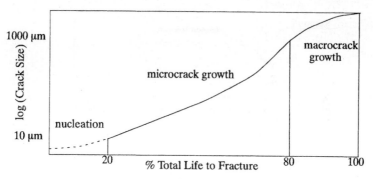

FIG. 7.8 Size scale associated with fatigue crack development.

Crack development can be divided into three separate regimes: nucleation, microcrack growth, and macrocrack growth. The first two regimes are often referred to together as Region I, or *crack initiation*. This is still a very unclear area. Although numerous theories exist, experimental verification is difficult. Obtaining repeatable data in this region has proven difficult and active research is underway. The macrocrack growth regime is referred to as Region II. This region is associated with linear elastic fracture mechanics (LEFM).

Region I: This region involves the formation of surface cracks of the order of 1 mm in length. It is considered to be controlled by the maximum shear stress fluctuation, $\Delta\tau_{max}$, since cracks on this small scale tend to originate on planes experiencing maximum shear stress. Two kinds of maximum shear planes are shown in Fig. 7.9. One intersects the surface at 45° (Sec. A-A) and the other at 90° (Sec. B-B). Note that both are inclined 45° to the applied stress axis. The enlarged views in Fig. 7.9 represent the intersection of slip bands with the free surface. "Slip bands" refer to multiple parallel planes, each accommodating massive dislocation movement, and associated plastic slip. The cross sections depict discontinuities (called intrusions and extrusions) created on the free surface that eventually lead to a macroscopic crack.

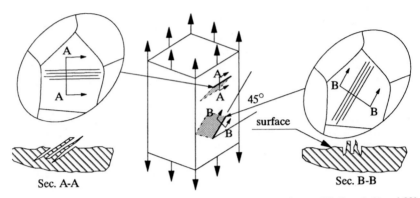

FIG. 7.9 Schematic microscopic shear cracks intersecting the surface at 45° (Sec. A-A) and 90° (Sec. B-B).

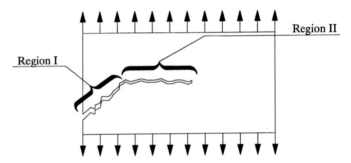

FIG. 7.10 Macroscopic cracks typically propagate perpendicular to maximum principal stress.

Region II: The surface steps created by the slip-band intersection are forced open and decohesion of slip planes forms small cracks oriented 45° to the loading axis. Upon growth, cracks typically turn to become oriented 90° to the principal stress direction. This is illustrated in Fig. 7.10. The crack is forced open and the crack tip blunts. This causes *striations* to form, which result in the *beach marks* that characterize fatigue fracture surfaces. This region is controlled by the range of principal stress ($\Delta\sigma_1$) acting normal to the plane of the crack. Generally, once this stage is reached, fracture mechanics are used to describe subsequent behavior.

7.4 STATIC STRENGTH ANALYSIS

In this section, a state of stress such as that depicted in Fig. 7.1*b*, is considered to be known. Based on this state of stress, the structural integrity of a component is assessed by comparing the stress state to the strength of the material. Although the methodology shown here is applicable to any three-dimensional stress state, surface stress states are emphasized since these comprise the vast majority of engineering design situations. Approaches are described in only a cursory manner, with more detailed references given. Emphasis is given to the application of the approaches to design situations.

7.4.1 Monotonic Tensile Data

The tensile test provides the input data for conducting static strength analysis. A sample of material (usually round or rectangular in cross section) is pulled apart under a monotonically increasing tensile load until failure occurs. Guidelines for conducting tensile tests are found in the American Society for Testing and Materials (ASTM) Specification E-8, "Standard Test Methods of Tension Testing of Metallic Materials."[27] A stress-strain curve from a tensile test is illustrated in Fig. 7.11, with a list of important points corresponding to "properties" measured by the test.

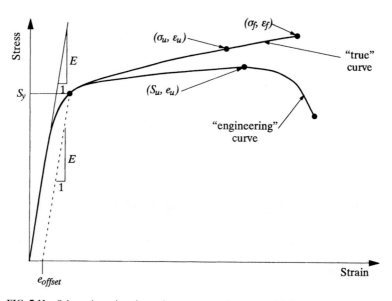

FIG. 7.11 Schematic engineering and true stress-strain curves, with list of properties.

Another important parameter is the *reduction in area*. It is determined from a measurement of the minimum diameter of the broken specimen, and the relation

$$\text{RA} = \frac{A_o - A_f}{A_o} = \frac{d_o^2 - d_f^2}{d_o^2} \tag{7.24}$$

where A_o and d_o are the initial specimen cross-sectional area and diameter, respectively, and the f subscript refers to those dimensions at fracture.

The engineering stress is computed by dividing the applied load by the original gauge section cross-sectional area A_o. Engineering strain is computed by dividing the change in gauge-section length by the initial gauge-section length. The calculation of "true" stress and strain quantities accounts for the fact that, as the loading increases, the cross-sectional area decreases and the gauge length increases. However, the need to distinguish between the two is rare, for everyday engineering design. The two curves are virtually identical up to plastic strains on the order of 10 percent.

A relation was developed by Ramberg and Osgood[28] to describe the stress-strain curve for many metallic engineering alloys. This relation is usually expressed as

$$\varepsilon = \frac{\sigma}{E} + \left(\frac{\sigma}{K}\right)^{1/n} \tag{7.25}$$

where K = strength coefficient
     n = strain-hardening exponent

For situations involving large plasticity (such as forming operations) the approximate log-log linear (power log) relation between stress and plastic strain can sometimes be quite inaccurate over a wide range of plastic strains. Therefore, it can be important to specify the plastic strain range over which n is defined. A designer interested in moderate plastic strain in a notch might be concerned with the range 0.002 to 0.02. However, a manufacturing engineer interested in a forming operation might need more accurate stress-strain information over a range from 0.05 to 0.15. The ASTM Specification E-646, "Tensile Strain-Hardening Exponents (n-Values) of Metallic Sheet Materials,"[29] deals with this issue specifically.

Data from compression tests for engineering materials can be equally important for conducting a static strength assessment. ASTM Specification E-9, "Standard Test Methods of Compression Testing of Metallic Materials at Room Temperature,"[30] describes this type of testing. Data from such a test can be important when attempting to classify a material as ductile or brittle. Failure of "brittle" materials in tension is usually associated with internal stress risers, such as voids or inclusions. Under compression such stress concentrations are less influential and the strength of a brittle material can considerably exceed its own tensile strength (for instance, by a factor of over 4 for some cast irons).

7.4.2 Multiaxial Yielding Theories (Ductile Materials)

Ductile materials are considered to be able to exhibit notable plasticity in a tensile test prior to fracture. No rigorous definition of "ductile" exists. Generally, however, a material is considered ductile if the percent reduction in area is greater than 15 to 20 percent, and the ultimate tensile strength exceeds the yield strength by a notable amount. Another important indicator used to classify a material as ductile is the relation between magnitudes of the tensile and compressive yield strengths. Ductile materials tend to yield in compression at nearly the same stress level as they do in tension, whereas brittle materials are typically quite a bit stronger in compression.

For the design of ductile machine components, two theories are typically utilized: (1) the Tresca criterion (maximum shear stress) and (2) von Mises' criterion (equivalently, the octahedral shear-stress or distortion-energy theory). These approaches can be depicted as safe operating envelopes on axes of minimum versus maximum principal stress (Fig. 7.12). Notice that the Tresca approach is smaller and therefore more conservative than the von Mises.

The Tresca (Maximum Shear-Stress Theory) Criterion.   This approach is based on the premise that yielding will occur when the maximum shear stress under multiaxial loading, τ_{max}, is equal to the maximum shear stress imposed during a tensile test at yield. In other words, yielding occurs when

$$\tau_{max} = \frac{\sigma_A - \sigma_B}{2} = \frac{S_y}{2} \tag{7.26}$$

where σ_A and σ_B are the maximum and minimum principal stresses, respectively. This approach can be restated in terms of an "equivalent stress,"

$$\sigma_{eq,Tr} = \sigma_A - \sigma_B \tag{7.27}$$

which is directly comparable to the axial yield strength of the material. In this way,

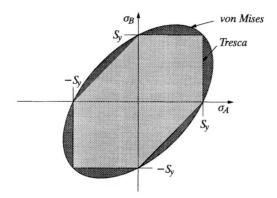

FIG. 7.12   Safe operating regions for von Mises (octahedral shear stress) and Tresca (maximum shear stress) criteria.

the "factor of safety" for the stress state is straightforwardly defined from the Tresca criterion by

$$FS_{Tr} = \frac{S_y}{\sigma_{eq,Tr}} \qquad (7.28)$$

The von Mises Criterion.   The von Mises criterion refers to any of several approaches shown to be essentially identical. These include the distortion energy, octahedral shear stress, and the Mises-Henkey theories. In terms of an equivalent stress, the von Mises approach is given by

$$\sigma_{eq,vM} = \frac{1}{\sqrt{2}} \sqrt{(\sigma_1 - \sigma_2)^2 + (\sigma_2 - \sigma_3)^2 + (\sigma_3 - \sigma_1)^2} \qquad (7.29)$$

Conceptually, the approach can be considered a root-mean-square average of the principal shear stresses, with a scaling factor to assure that the equivalent stress is equal to σ_1 for a uniaxial stress state.

The factor of safety for the von Mises approach is thus given by

$$FS_{vM} = \frac{S_y}{\sigma_{eq,vM}} \qquad (7.30)$$

Experiments have shown that the von Mises criterion is more accurate in terms of describing data trends, but the Tresca approach is a more conservative design option.[13]

7.4.3   Multiaxial Failure Theories (Brittle Materials)

In this section, the use of three design criteria is demonstrated. These approaches are referred to (in order of decreasing conservatism) as the Coulomb-Mohr, modified Mohr, and the maximum normal fracture criteria.[13] Each can be considered to define safe operating envelopes on axes of minimum versus maximum principal stress (Fig. 7.13). The most notable difference between Figs. 7.12 and 7.13 is the typically greater compressive strength S_{uc} exhibited by a brittle material relative to its tensile strength

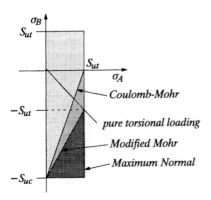

FIG. 7.13 Safe operating regions for the Coulomb-Mohr, modified Mohr, and maximum normal failure theories for brittle materials.

S_{ut}. Also, notice how only the first and fourth quadrants of principal stress space are depicted in Fig. 7.13. This is because the vast majority of all engineering stress states of concern to mechanical designers lie in these quadrants, with the vast majority located in the fourth. (With the exception of the deepest points in the ocean, it is difficult to imagine practical engineering states of surface stress that do not reside in or along the fourth quadrant.)

The three theories are described below, followed by the presentation of a static strength design algorithm. For all three theories, the factor of safety for a state of stress is defined as the ratio of the radial distance to the boundary (through the state of stress) to the radial distance defined by the state of stress. This is depicted in Fig. 7.14.

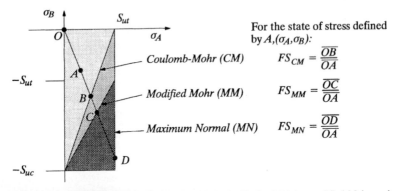

For the state of stress defined by $A,(\sigma_A,\sigma_B)$:

$$FS_{CM} = \frac{\overline{OB}}{\overline{OA}}$$

$$FS_{MM} = \frac{\overline{OC}}{\overline{OA}}$$

$$FS_{MN} = \frac{\overline{OD}}{\overline{OA}}$$

FIG. 7.14 Definition of factor of safety based on the Coulomb-Mohr, modified Mohr, and maximum normal failure theories for brittle materials.

The Coulomb-Mohr Fracture Theory. The Coulomb-Mohr theory is based on the concept that certain combinations of shear stress and stress normal to the plane of maximum shear are responsible for failure. This is manifested in the fourth quadrant of principal stress space by the line from S_{ut} on the tensile stress axis to $-S_{uc}$ on the compressive strength axis.

As is apparent from Figs. 7.13 and 7.14, the Coulomb-Mohr theory is the most conservative design approach. Experimental results have indicated that the approach is typically conservative for design applications.

The Modified Mohr Fracture Theory. This theory is based on empirical observations that the maximum principal stress tends to define failure under torsional loading (or along a line 45° through the fourth quadrant of principal stress space). However, when significant compression accompanies torsion (stress states below the 45° torsion line), the maximum normal stress theory becomes nonconservative. Therefore, the line in the fourth quadrant defined by $(S_{ut},-S_{ut})$ and $(0,-S_{uc})$ is used as the boundary for the modified Mohr theory. Since the approach is formulated from empirical observations, it tends to correlate well with data.

The Maximum Normal Fracture Theory. Conceptually, this is the simplest of the theories in this section. If the magnitude of the maximum (or minimum) principal stress in the material exceeds the material's tensile (or compressive) strength, failure is predicted. Unfortunately, experiments have shown it to be nonconservative for situations involving substantial compression (states of stress in the fourth quadrant, below the line of pure torsion).

7.4.4 Summary Design Algorithm

In practice, the designation of a material as ductile or brittle and the selection of an appropriate failure criterion can be subjective. Major factors include whether or not compressive strength data are available, and whether or not compression constitutes a major portion of the loading. In situations where the choices are not clear, it is advisable to conduct analyses based on limiting assumptions, implementing all potential approaches to bound a solution. To assist in this, an algorithm is presented in the form of a flowchart in Table 7.1 that can be easily coded into a computer program or applied using a computer spreadsheet. The design engineer is responsible for supplying the correct input information (including the classification of the material as ductile or brittle) and for interpreting the output. Several techniques are used to evaluate strength and the designer must decide which is the most appropriate. Output from such a routine is presented in Example 4.

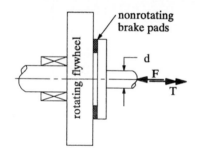

EXAMPLE 4 A round shaft is to be used to apply brake pads to the side of a large flywheel. The shaft is to experience a compressive load of $F = 22,000$ lb, and corresponding torsional load of $T = 23,100$ in·lb. Specify the diameter of the shaft d (to the nearest one-eighth inch) for a safety factor of at least 2.0 using the following materials:

1. ASTM #40 cast iron ($S_{ut} = 42.5$ ksi, $S_{uc} = 140$ ksi)
2. 1020 steel ($S_y = 65$ ksi)
3. Q&T 4340 steel ($S_y = 240$ ksi)

solution For each material, the three failure theories were used from Table 7.1. Diameters (in inches) and safety factors (in parentheses) estimated for each material are presented in tabular form, below.

	1. #40 iron	2. 1020	3. 4340
Tr	N/A	2.0 (2.15)	1.375 (2.62)
vM	N/A	1.875 (2.04)	1.25 (2.27)
CM	1.875 (2.02)	N/A	N/A
MM	1.75 (2.06)	N/A	N/A
MN	1.75 (2.38)	N/A	N/A

TABLE 7.1   Static Strength Analysis

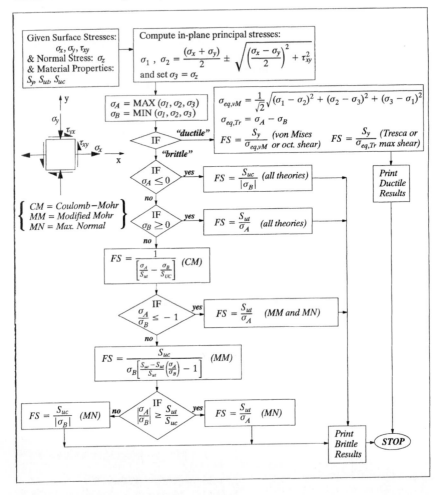

7.5 FATIGUE STRENGTH ANALYSIS

The subject of fatigue analysis is considered in this section from the point of view of an engineering designer. Although this subject is still actively researched, a great deal of solid engineering methodology has been developed. The fatigue design strategy to be described in this section is outlined below.

Crack initiation is defined as the occurrence of a crack of engineering size, usually 1 to 2 mm in surface length. The basis of this definition is illustrated in Fig. 7.15. To obtain baseline fatigue data (stress-life or strain-life), tests are usually conducted on small specimens, 0.25 in (6 mm) in diameter. Usually, "failure" in these tests can be associated with complete fracture of the specimen. *It is assumed that a component experiencing a localized stress-strain history equivalent to the axial specimen will develop a crack of approximately the same size in approximately the same number of cycles.* This concept is often referred to as the local strain approach.[13,23,31,32]

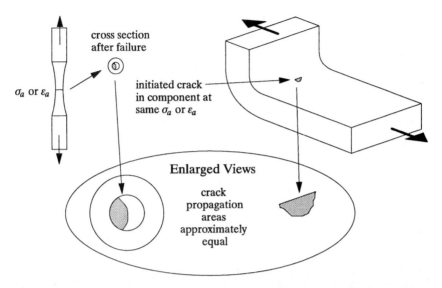

FIG. 7.15 Similitude between failure in a baseline test specimen and crack initiation in an actual engineering component.

In the remainder of this section, "failure" will refer to the occurrence of an engineering-sized crack, roughly the same size as that found in an axial specimen upon its failure in a standard fatigue test. The propagation of the crack due to subsequent fatigue loading is considered separately using fracture-mechanics techniques for damage-tolerant design.

7.5.1 Stress-Life Approaches (Constant-Amplitude Loading)

In this section, the *stress-life (S-N) approach* to fatigue design is overviewed. This is one of the earliest fatigue design approaches to be developed and can still be a useful tool. Its success is based on the fact that, for predominantly elastic loading, the state of stress in a component can often be characterized quite accurately. *As long as the state of fluctuating stress can be accurately estimated, the S-N approach can do a good job of predicting fatigue.* However, fatigue cracks usually develop at structural discontinuities, or notches. In these regions, localized cyclic plastic strains can develop and the task of estimating the state of stress becomes far more difficult. Without a reliable knowledge of the stress state, the utility of the *S-N* approach becomes limited and a strain-based approach (described later) becomes more useful.

Stress-Life Curve.   Baseline data are generated by imposing fully reversed fluctuating stress in a standard specimen, as shown below in Fig. 7.16. This can be done via axial loading or rotating-bending.
   Fully reversed loading refers to the fact that $\sigma_{max} = -\sigma_{min}$ (or, the alternating stress, $\sigma_a = \sigma_{max}$). Tests are conducted by applying loading as shown in Fig. 7.16 until the specimen "fails," usually by fracturing into two separate pieces. Typically, the gauge section ranges in size from 0.25 to 0.5 inch in diameter (6 to 12 mm). To generate *S-N* data for fatigue design purposes, a number of specimens must be tested at varying stress levels. Applicable ASTM guidelines are listed below.

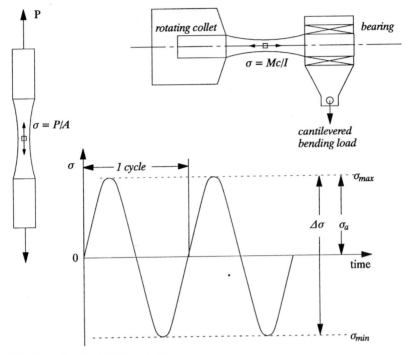

FIG. 7.16   Baseline *S-N* fatigue testing.

- Details for conducting S-N tests are presented in the ASTM E-466-82, "Standard Practice for Conducting Constant Amplitude Axial Stress-Life Tests of Metallic Materials."[33]
- Data from fatigue tests are analyzed according to the ASTM E-739, "Standard Practice for Statistical Analysis of Linear or Linearized Stress-Life (*S-N*) and Strain Life (ε-*N*) Fatigue Data."[34]
- Data are typically presented according to ASTM E-468-82, "Standard Practice for Presentation of Constant Amplitude Fatigue Test Results for Metallic Materials."[35]

There are some fundamental differences between baseline data obtained from rotating-bending and axial testing. The stress amplitude for rotating-bending is computed elastically, even though severe plastic deformation occurs at higher load levels. Therefore, the quantity

$$S = \frac{Mc}{I} \tag{7.31}$$

is actually only a parameter with units of stress, indicating the severity of bending. A plasticity analysis would be required to estimate the actual stress at the specimen surface. And even for high-cycle tests (lower load levels), there is a bending-stress gradient as depicted in Fig. 7.17. For this reason, bending tests are less severe than axial tests and can make the material *appear* stronger. This is due to two factors, both related to the bending versus axial stress distribution: (1) Physically, more of the gauge section is subjected to the maximum stress in an axial test than in a bending test. This

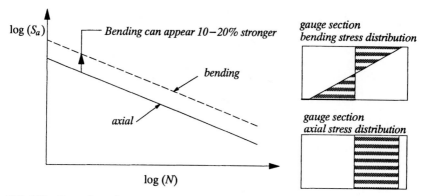

FIG. 7.17 Comparison of stress-life data from axial and rotating-bending test.

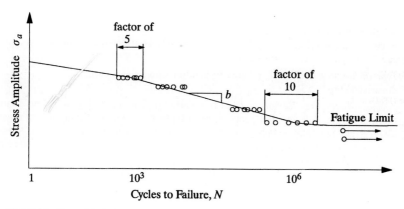

FIG. 7.18 Stress-life fatigue data.

increases the likelihood that a critically sized material defect or properly oriented slip system will experience the most severe stress fluctuation. (2) The bending stress distribution is less severe from the standpoint of crack propagation during microcrack development.

A major benefit to the use of a rotating-bending machine is speed. Motors are used to drive the specimen at a very high rpm, generating data very quickly (e.g., 10,000 rpm). A schematic set of S-N data from such a machine is shown in Fig. 7.18 to illustrate some more fatigue data trends. In Fig. 7.18 and all subsequent S-N plots, axes are logarithmic.

Scatter can plague fatigue data. Factors of 10 or more are not unusual in the high-cycle regime. Scatter is very dependent on cleanliness of material (pores, inclusions, and other microstructural defects). Statistical guidelines from ASTM E-739[34] can be very useful in understanding and utilizing fatigue data.

One of the most utilized features of the S-N curve limit is the fatigue limit. It is important to remember that aluminum and other nonferrous metals do not exhibit a fatigue limit. (Fatigue limits are quoted in the literature for aluminum as the stress amplitude corresponding to a very large number of cycles, such as $5 \cdot 10^7$ to $5 \cdot 10^8$.) For ferrous alloys, fatigue limits can be affected by many factors, as outlined below.

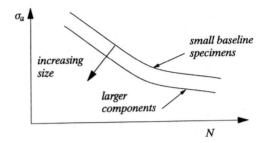

FIG. 7.19 Increasing component size decreases fatigue strength, relative to data generated with small specimens.

- *Size effects.* When a component is considerably larger than the specimen used to generate the baseline fatigue data, a greater volume of material is subjected to a particular stress amplitude. This increases the statistical probability that a microscopic flaw, defect, or slip system will exist that is susceptible to fatigue-crack development. For this reason larger components often fail sooner than smaller specimens, as depicted in Fig. 7.19. This discrepancy is affected by other factors, such as inhomogeneity of microstructure.

- *Type of loading.* Differences between bending and axial loading have already been discussed (Fig. 7.17).

- *Surface processing.* Besides surface roughness in general, plating, nitriding, induction hardening, rolling, shot peening, or any other surface modification can drastically affect the fatigue behavior of a part. Generally, processes improve fatigue resistance if they increase hardness, impose residual compressive surface stresses, and/or reduce surface roughness.

- *Grain size.* This is particularly important for high-cycle fatigue. Typically, smaller grain size means longer fatigue lives. (This is not surprising, since smaller grain size usually means higher yield strength.)

- *Material processing.* The "cleaner" the material, the better its fatigue resistance. For instance, vacuum-melt steel exhibits fatigue lives longer by 50 percent relative to furnace-melt steels. Wrought metals show better fatigue resistance than cast metals (Fig. 7.20). Crack nucleation and microcrack propagation time is avoided since microscopic defects such as inclusions or pores act as instant crack growth sites. (The same can be true for powdered-metal parts.)

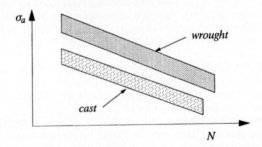

FIG. 7.20 Wrought material is generally more fatigue resistant than cast material.

- *Temperature and environment.* Both of these factors can exhibit profound negative synergism with fatigue mechanisms. (The sum of the two effects can be more than a simple superposition.) When these variables are important, loading frequency and waveform must be considered influential. This is not the case ordinarily (when environmental concerns are not considered influential).

- *Intermittent overloads.* Suppose a component operates in service at a stress level below the fatigue limit, but experiences occasional overloads. Even though the overloads are infrequent and cause no macroscopic plasticity, they can serve to reduce or eliminate the endurance limit.[13]

Numerous attempts have been made to quantify the effects just described.[1-3,31,32] These are generally presented as empirical factors used to reduce the endurance limit. These factors tend to reduce the high-cycle, finite portion of the S-N curve as well. Several useful design charts for these factors appear in Sec. 21.

Finally, designers are frequently forced to evaluate the endurance of a part for which S-N data are not available. Therefore, several textbooks have suggested empirical approaches to estimate the S-N curve from monotonic tensile data.[1,3,12] A comprehensive overview of many of these can be found in Dowling.[13] For example, data have suggested the following relation between the endurance limit and ultimate tensile strength:[1] For wrought steels,

$$S_e = 0.5\, S_u \qquad \text{for } S_u \leq 200 \text{ ksi}$$
$$S_e = 100 \text{ ksi} \qquad \text{for } S_u > 200 \text{ ksi}$$

$$(7.32)$$

For cast iron,

$$S_e = 0.45\, S_u \qquad \text{for } S_u \leq 88 \text{ ksi}$$
$$S_e = 40 \text{ ksi} \qquad \text{for } S_u > 88 \text{ ksi}$$

$$(7.33)$$

S-N Finite-Life Prediction.   Many factors have caused infinite-life design to become impractical, weight and cost being the primary motivators. It has become more common for designers to anticipate typical service-load histories and design for adequate service lives, building in a reasonable allowance for occasional abusive loading. This can result in components without unreasonably high safety factors that are therefore lighter and less expensive. The methodology to be presented here is intended primarily for use in high-cycle fatigue situations ($N > 10^5$ cycles), although it can be useful in other situations so long as stresses can be accurately determined.

For fatigue design based on finite life, the sloping portion of the curve from $10^3 \leq N \leq 10^6$ in Fig. 7.18 must be known from testing or estimated. If data are available, the log-log linear portion of the curve can be characterized by a power law relation,

$$S_a = C'(N)^{b'} \qquad (7.34)$$

where C' and b' are curve-fit parameters used to relate the stress amplitude S_a and number of cycles to failure N. In the absence of fatigue data, the following procedure can be used to estimate these parameters.

- Assume the fatigue limit occurs at a life of 10^6 cycles. [If no fatigue limit data are available, estimate S_e from Eq. (7.32) or (7.33).]

- Assume a stress amplitude of $0.9S_u$ corresponding to a life of 1000 cycles, S_{1000}.

This results in a curve as shown in Fig. 7.21. The coefficient and exponent in Eq. (7.34) are therefore given by

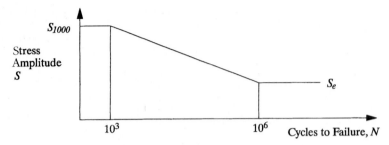

FIG. 7.21   Curve used to approximate *S-N* data.

$$C' = \frac{(S_{1000})^2}{S_e}$$ (7.35)

$$b' = -\tfrac{1}{3}\log\left(\frac{S_{1000}}{S_e}\right)$$ (7.36)

If S_{1000} is assumed to be $0.9S_u$ and S_e to be $0.5S_u$, then $C' = 1.62S_u$ and $b' = -0.0851$.

An equivalent way to express an *S-N* relation is through the use of the following axial fatigue parameters:

$$S = \sigma_f' (2N)^b$$ (7.37)

where σ_f' = fatigue strength coefficient
       b = fatigue strength exponent

This relation is referred to in the literature as the Basquin relation, and its parameters will be discussed in Sec. 7.5.2.

Notice the factor of two that appears in Eq. (7.37). The quantity $2N$ is considered the number of *stress reversals* to failure, since there are two reversals for every cycle (see Fig. 7.22). This is a consequence of some early work on variable-amplitude loading that was taking place while the concept of a "fatigue strength coefficient and exponent" was being developed to characterize fatigue data. At the time, it was felt that considering stress reversals instead of cycles could expedite cumulative fatigue damage analysis. This later proved not to be the case, and consequently, the factor of two must now be accounted for somewhat meaninglessly. This situation is discussed further in Bannantine.[32]

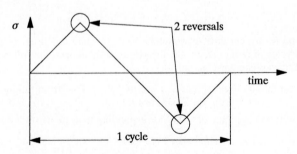

FIG. 7.22   Number of reversals = 2 (number of cycles).

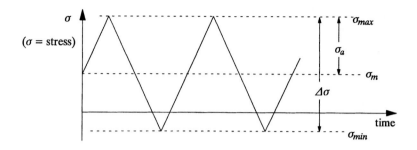

FIG. 7.23 Constant-amplitude loading with a mean stress.

The remainder of this section will be devoted to specifying how to use experimental or estimated baseline *S-N* data (from constant-amplitude, fully reversed specimen loading) on more complex uniaxial stress histories.

Mean Stress Effects. Baseline data are fully reversed ($R = -1$) but actual engineering components are often subjected to loading with *nonzero* mean stress as depicted in Fig. 7.23. From this figure, several parameters are defined, including the stress ratio,

$$R = \frac{\sigma_{min}}{\sigma_{max}} \tag{7.38}$$

stress range,

$$\Delta\sigma = \sigma_{max} - \sigma_{min} \tag{7.39}$$

stress amplitude,

$$\sigma_a = \frac{\sigma_{max} - \sigma_{min}}{2} \tag{7.40}$$

and mean stress,

$$\sigma_m = \frac{\sigma_{max} + \sigma_{min}}{2} \tag{7.41}$$

Mean stresses can act to shorten or lengthen fatigue life, depending on (1) whether the mean stress is positive or negative and (2) whether the loading is predominantly elastic or plastic. This is depicted schematically in Fig. 7.24.

Tensile mean stresses superimpose with applied loading to decrease fatigue life while compressive mean stress decreases the applied loading to increase fatigue life.

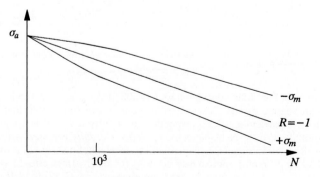

FIG. 7.24 Mean stress effect on *S-N* curve.

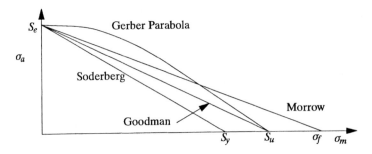

FIG. 7.25   Mean stress constant-life plots (endurance limit).

Mean stress relaxation can occur at higher load levels, diminishing the effect of mean stress at lower lives. This is particularly true when a component has a notch, as discussed later.

Mean stress data are often presented as plots of stress amplitude versus mean stress, corresponding to a particular life. For instance, Fig. 7.25 shows plots of several empirical relations to account for mean stress that have been suggested from testing at endurance-limit load levels. The curves represent combinations of mean stress and stress amplitude (σ_m and σ_a) that correspond to the fatigue limit S_e. Data sets have indeed been shown to lie in the vicinity of these lines and occasionally suggest that particular relations do a better job than others. However, in practice, none of these has been universally agreed upon as superior. In general, the Soderberg line has been determined to be too conservative for practical design use. The Goodman line and Gerber parabola are often more accurate than the Morrow relation. Another popular parameter was proposed by Smith, Watson, and Topper (SWT).[36] This relation is shown schematically with a Goodman line in Fig. 7.26.

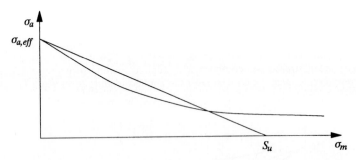

FIG. 7.26   Constant-life plots for Goodman and Smith-Watson-Topper relations.

The curves in Fig. 7.25, considered to describe the *fatigue limit,* can be extended to the finite-life regime by considering combinations of stress amplitude and mean stress that result in a particular life corresponding to a fully reversed test conducted at a stress amplitude of $\sigma_{a,\text{eff}}$. Schematically, this *effective, fully reversed stress amplitude* concept is depicted in Fig. 7.27. The effective stress amplitude, $\sigma_{a,\text{eff}}$, provides a conceptually straightforward approach to account for mean stress effect based on fully reversed baseline data. Relations for $\sigma_{a,\text{eff}}$ are given in Eqs. (7.42) to (7.46) for the criteria illustrated in Figs. 7.25 and 7.26:

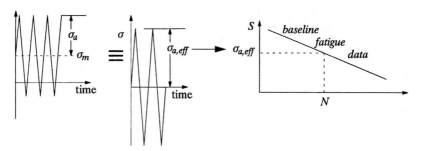

FIG. 7.27 Effective stress-amplitude concept.

Goodman:
$$\sigma_{a,\text{eff}} = \sigma_a\left(\frac{S_u}{S_u - \sigma_m}\right) \tag{7.42}$$

Soderberg:
$$\sigma_{a,\text{eff}} = \sigma_a\left(\frac{S_y}{S_y - \sigma_m}\right) \tag{7.43}$$

Morrow:
$$\sigma_{a,\text{eff}} = \sigma_a\left(\frac{\sigma_f}{\sigma_f - \sigma_m}\right) \tag{7.44}$$

Gerber:
$$\sigma_{a,\text{eff}} = \sigma_a\left(\frac{S_u^2}{S_u^2 - \sigma_m^2}\right) \tag{7.45}$$

Smith-Watson-Topper:
$$\sigma_{a,\text{eff}} = \sqrt{\sigma_a(\sigma_a + \sigma_m)} \tag{7.46}$$

Equations (7.42) to (7.45) are illustrated again in Fig. 7.28. These curves differ from those in Fig. 7.25. Each curve is based on the same input point, that is, the state of stress in the engineering component defined by σ_a and σ_m. But, each implies a different $\sigma_{a,\text{eff}}$ corresponding to the applied stress state. These curves make it apparent that Soderberg is the most conservative from a designer's perspective, since it specifies the highest effective stress, and Gerber is the least conservative.

One final note should be made before leaving mean stress effects. The relations illustrated so far have been discussed primarily in the context of positive mean stress.

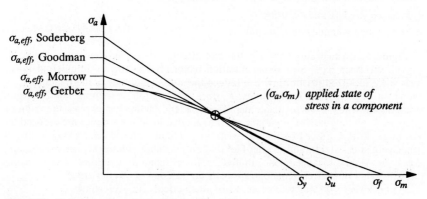

FIG. 7.28 Definition of $\sigma_{a,\text{eff}}$ from relations in Fig. 7.27.

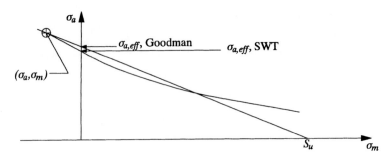

FIG. 7.29   Extension of Goodman and SWT relations into compressive mean stress regime.

The fact that tensile mean stresses have a deleterious effect on fatigue is modeled by the $\sigma_{a,\text{eff}}$ concept. (For example, increasing σ_m increases $\sigma_{a,\text{eff}}$ and decreases estimated life.) However, there are valuable data[37] that demonstrate the beneficial effect of compressive mean stress on fatigue. Therefore, a compressive mean stress should decrease $\sigma_{a,\text{eff}}$. (This is not the case for the Gerber parabola.) The Goodman and SWT relations have been shown to do a good job for small compressive stresses, as illustrated in Fig. 7.29. ("Small" is defined as having a magnitude of less than about $0.5S_y$. A more comprehensive treatment of compressive mean stress effects can be found in Ref. 31.)

The fact that compression can enhance fatigue life can be taken advantage of through material processes such as shot peening, proof loading, carburizing, nitriding, and induction hardening. All of these processes impose large compressive residual stresses at the surface of the material, reducing effective stress amplitudes and increasing fatigue life. (The latter three also considerably harden the surface layer.) Thread rolling and hole stretching are other processes that enhance fatigue resistance by inducing residual surface compression.

Notches.   Figure 7.6 illustrated the concept of an elastic stress-concentration factor K_t defined as the maximum elastic stress at the notch root, divided by the nominal stress (based on net section area). Since the notch stresses increase according to K_t, it would be convenient, analytically, if fatigue strengths were reduced proportionally. However, the effect that a notch has on fatigue is dependent on

- Notch severity (magnitude of K_t)
- Material strength and ductility
- The applied nominal stress magnitude

Figure 7.30 illustrates how a notch can affect a set of fatigue data, relative to smooth-specimen data. Stress-concentration factor effects tend to diminish at lower lives since localized plastic flow can reduce the stress amplitude at the notch root, as shown in Fig. 7.31. At longer lives, K_t does a better job describing notch fatigue strength, but tends to overestimate the effect. Several factors can explain the reduced effect of K_t on fatigue. These include (1) the fact that localized stresses are reduced by yielding, (2) the effect of subsurface stress gradient (microcracks growing into a decreasing stress field), and (3) the fact that only a small volume of material experiences the extreme localized concentrated stresses. From a design point of view, using the full value of a stress-concentration factor to compute notch stresses ($\sigma_{\text{notch}} = K_t S_{\text{nom}}$) is a very safe way to operate, since notch effects are overestimated.

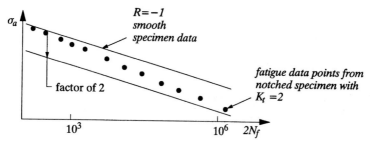

FIG. 7.30 Notch effect on S-N behavior.

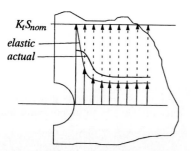

FIG. 7.31 Illustration explaining how elastically calculated K_t can overestimate the effect of a notch on fatigue behavior. Localized yielding and subsurface gradients are apparent.

Fatigue Notch Factors. Recognizing that K_t overestimates fatigue-strength reduction, the concept of an empirical fatigue notch factor (also called a fatigue-strength reduction factor) was developed. The fatigue notch factor K_f is defined from a comparison of fatigue data generated with smooth and notched specimens, as shown in Fig. 7.32.

Unfortunately, the use of fatigue notch factors in design is not straightforward. In many instances, K_f has been shown to vary with life, as is apparent in Fig. 7.32. Furthermore, it can only be reliably determined empirically (by experiment) for the material, geometry, and surface processing of interest.

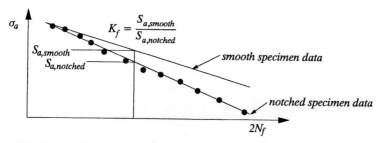

FIG. 7.32 Definition of fatigue notch factor.

To quantify the fatigue-strength reduction associated with a notch, a notch-sensitivity factor was developed as defined in Eq. (7.47):

$$q = \frac{K_f - 1}{K_t - 1} \tag{7.47}$$

where q varies from 0 to 1:

$q = 0$ no notch effect ($K_f = 1$)

$q = 1$ full elastic effect ($K_f = K_t$)

Some researchers have attempted to formulate empirical relations for K_f based on K_t for fatigue-limit load levels. One approach, proposed by Peterson, is given by

$$K_f = 1 + \frac{K_t - 1}{1 + (a/r)} \qquad (7.48)$$

where r = the notch root radius
      a = a "characteristic length" (empirical curve-fit parameter)

$$\alpha \cong \left(\frac{300}{S_{ut}\,(\text{ksi})}\right)^{1.8} \times 10^{-3} \text{ in} \qquad (7.49)$$

For steels, a rule of thumb assessment of the parameter α is often cited:

Annealed steel	Quenched and tempered	Highly hardened
$\alpha \approx 0.010$ in	$\alpha \approx 0.0025$ in	$\alpha \approx 0.001$ in

Consistent with this is the general assessment that harder materials are more notch sensitive than softer materials. Example 5 uses Eq. (7.48) to illustrate this point. It should be remembered that no such empirical relations have been proposed for aluminum or other nonferrous materials.

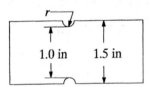

EXAMPLE 5   This example illustrates the effects of tensile strength and notch severity on the estimated values of the fatigue notch factor. Use Eqs. (7.48) and (7.49) to compute K_f for the two different steels and three different notch root radii.

solution   The values of K_f are tabulated below for two steels and three values of r.

	Material A	Material B
	$S_u = 68$ ksi	$S_u = 180$ ksi
	$\alpha = 0.015$ in	$\alpha = 0.0025$ in

		Material A	Material B
r (in)	K_t	$K_f(S_u = 68)$	$K_f(S_u = 180)$
0.2	2.05	1.98	2.03
0.05	3.5	2.92	3.38
0.01	6.0	3.0	5.0

The K_f relations and the example shown above are valid only for fully reversed loading ($R = -1$). Mean stresses can affect notched components differently than smooth ones. To accurately analyze a particular situation, empirical data are usually necessary.

A great deal of data have been generated in the aerospace industry and are published in the form of plots[38,39] such as the one shown in Fig. 7.33. These plots provide direct information on the combined effects of the mean stress and the stress-concentration factor.

A final important trend is noted by those who have studied fatigue notch factors:

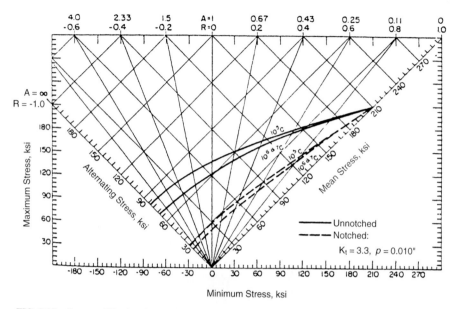

FIG. 7.33 Constant-life plots from MIL Handbook 5 (AISI 4340); S_u = 208 ksi.

there appears to be an upper limit to K_f of about 5 or 6 for very sharp notches.[32] Two possible explanations for this are: (1) the notch tip blunts, reducing K_f, or (2) the notch constitutes a crack and removes the initiation life of the component.

A safe, recommended approach suitable for design is outlined in Table 7.2 for uniaxial loading situations. If more detailed data are available, they can be incorporated into the approach as outlined below:

- Use a measured rather than estimated K_f over the entire range of life.
- Estimate the variation of K_f with life experimentally, or from a source such as that found in "MIL Handbook 5," Fig. 7.33.
- If estimates are unduly conservative, use only the nominal mean stresses ($\sigma_{m,\text{notch}}$ = $S_{m,\text{ax}} + S_{m,\text{bend}}$) to compute the notch mean stress.
- Use only nominal stress amplitudes, and modify baseline S-N data using approaches detailed in Refs. 13 and 32.

7.5.2 Strain-Life Approaches (Constant-Amplitude Loading)

Cyclic Stress-Strain Relation. A standard, low-cycle fatigue specimen is fabricated and tested according to ASTM E-606, "Standard Recommended Practice for Constant-Amplitude Low-Cycle Fatigue Testing,"[40] in strain control. A typical specimen is depicted in Fig. 7.34, along with a stabilized cyclic stress-strain loop.

When fatigue testing is conducted, several specimens (ideally, at least 20) are tested at varying strain amplitudes. At each strain amplitude, a different stabilized loop forms, as depicted in Fig. 7.35. From these loops, the cyclic stress-strain curve may be

TABLE 7.2 Elastic Uniaxial Stress-Life Design Approach

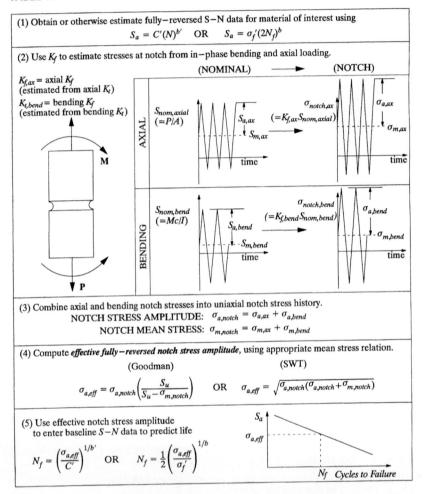

(1) Obtain or otherwise estimate fully−reversed S−N data for material of interest using

$$S_a = C'(N)^{b'} \quad \text{OR} \quad S_a = \sigma_f'(2N_f)^b$$

(2) Use K_f to estimate stresses at notch from in−phase bending and axial loading.

(NOMINAL) ⟶ (NOTCH)

$K_{f,ax}$ = axial K_f
(estimated from axial K_t)

$K_{t,bend}$ = bending K_f
(estimated from bending K_t)

AXIAL

$S_{nom,axial}$
$(=P/A)$

$S_{a,ax}$
$S_{m,ax}$
time

$\sigma_{notch,ax}$
$(=K_{f,ax}S_{nom,axial})$
$\sigma_{a,ax}$
$\sigma_{m,ax}$
time

BENDING

$S_{nom,bend}$
$(=Mc/I)$

$S_{a,bend}$
$S_{m,bend}$
time

$\sigma_{notch,bend}$
$(=K_{f,bend}S_{nom,bend})$
$\sigma_{a,bend}$
$\sigma_{m,bend}$
time

(3) Combine axial and bending notch stresses into uniaxial notch stress history.
NOTCH STRESS AMPLITUDE: $\sigma_{a,notch} = \sigma_{a,ax} + \sigma_{a,bend}$
NOTCH MEAN STRESS: $\sigma_{m,notch} = \sigma_{m,ax} + \sigma_{m,bend}$

(4) Compute *effective fully−reversed notch stress amplitude*, using appropriate mean stress relation.

(Goodman)

$$\sigma_{a,eff} = \sigma_{a,notch}\left(\frac{S_u}{S_u - \sigma_{m,notch}}\right) \quad \text{OR} \quad \sigma_{a,eff} = \sqrt{\sigma_{a,notch}(\sigma_{a,notch} + \sigma_{m,notch})}$$

(SWT)

(5) Use effective notch stress amplitude
to enter baseline $S-N$ data to predict life

$$N_f = \left(\frac{\sigma_{a,eff}}{C'}\right)^{1/b'} \quad \text{OR} \quad N_f = \frac{1}{2}\left(\frac{\sigma_{a,eff}}{\sigma_f'}\right)^{1/b}$$

S_a
$\sigma_{a,eff}$
N_f Cycles to Failure

defined from the locus of the tips of the stabilized hysteresis loops, and expressed using a Ramberg-Osgood[28] relation:

$$\varepsilon_a = \frac{\sigma_a}{E} + \left(\frac{\sigma_a}{K'}\right)^{1/n'} \tag{7.50}$$

where K' = cyclic strength coefficient
n' = cyclic strain-hardening exponent

In this relation, ε_a and σ_a represent strain and stress *amplitudes*, respectively. The curve therefore represents a relation between stress and elastic-plastic strain amplitudes that form during fully reversed strain-controlled testing.

The formation of the stabilized hysteresis loops depicted in Figs. 7.34 and 7.35 usually requires a substantial number of cycles, during which transient softening or hardening may occur. Such behavior is depicted in Fig. 7.36. This can cause the cyclic stress-strain relation to lie below or above the monotonic curve. If the cyclic curve is

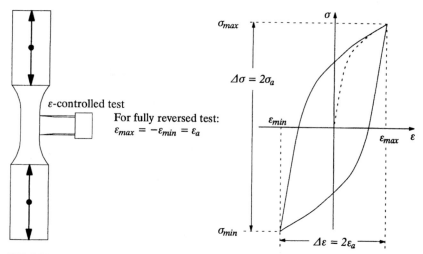

FIG. 7.34 Stabilized stress-strain hysteresis loop from typical ASTM E-606 fatigue test.

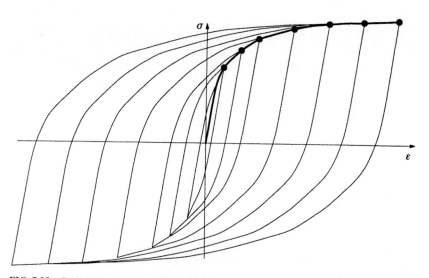

FIG. 7.35 Cyclic stress-strain curve from stable hysteresis loops.

below the monotonic curve, the material can be called a "cyclic softening material." If the cyclic curve is above the monotonic curve, the material "cyclically hardens." Mixed behavior is also observed, depending on the strain amplitude. Examples of each situation are shown in Fig. 7.37.

The transient stress behavior during a typical strain-controlled test is depicted differently in Fig. 7.38 for a cyclically softening material. This figure is a plot of peak and valley stress components at each reversal point throughout the life of the material. There are several noteworthy features to this plot. First, cyclic stabilization is shown to occur within about 10–20 percent of the total life. This depends on the material

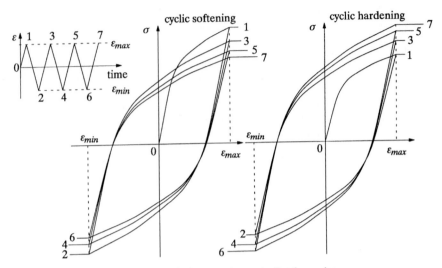

FIG. 7.36 Transient softening and hardening occurring on the first few cycles.

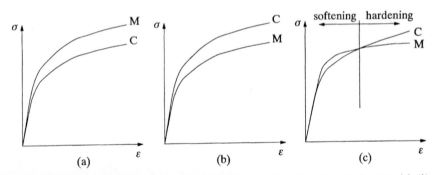

FIG. 7.37 Monotonic (M) and cyclic (C) stress-strain curves for (a) cyclic softening material, (b) cyclic hardening material, and (c) mixed transient behavior.

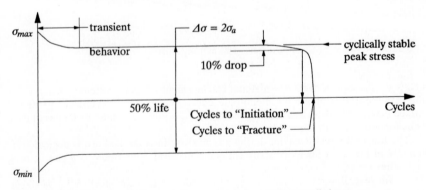

FIG. 7.38 Typical peak and valley stresses versus cycle for a strain-controlled test.

being tested, and the strain amplitude. At larger applied strains (lower lives), stabilization can be less pronounced (that is, the maximum stress can change gradually over the entire test). For this reason, the stabilized stress amplitude is usually defined as the stress amplitude at the "half-life" of the specimen (at near 50 percent of its total life). Near the end of the life of the specimen, notice in Fig. 7.38 how the peak tensile stress drops just prior to final fracture. This results from the decrease in specimen stiffness associated with crack formation. Therefore, this drop in the maximum stress is typically used to define the "crack initiation life" of the specimen, as opposed to the total number of cycles to fracture. (Notice how the compressive valley stress is maintained throughout the test, since the crack faces can sustain the compressive loading.) Some testing laboratories use a peak load drop of 10 percent from the half-life value to define initiation, others use a larger value, such as 50 percent, while others simply use the life to fracture. The discrepancy this causes is usually considered negligible, since the life of a specimen after a discernable crack ("engineering-sized," on the order of 1 to 2 mm in surface length) has formed is generally a small percentage of the life to fracture. However, the subjectivity associated with reducing low-cycle fatigue data is apparent, especially in the low-cycle regime. It is advised that stress-versus-time data be obtained and reviewed by the engineer when low-cycle fatigue testing is conducted.

The definition of the cyclic stress-strain curve requires the testing of several specimens. This is referred to as "companion specimen" testing. Attempts have been made to define the curve from a single test called the "incremental step test."[41] It should be noted that this technique can only approximate the curve and not enough data exist to assess its general reliability.

Refer again to Fig. 7.35, and recall that the dark cyclic stress-strain curve [Eq. (7.50)] is defined by the tips of the hysteresis loops. The light curves (referred to as hysteresis curves) can be approximated well by scaling the dark curve, geometrically, by a factor of two. This is referred to as Massing's hypothesis.[42] To demonstrate this, refer to Fig. 7.39.

In Fig. 7.39, the dark curve from o to a is the cyclic stress-strain curve. The light

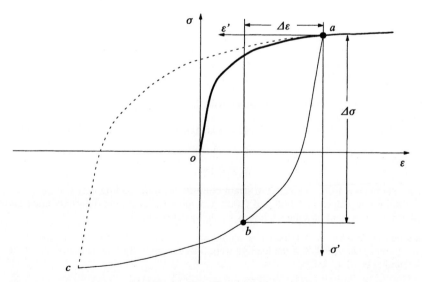

FIG. 7.39 Demonstration of Massing's hypothesis: Solid curve a-b-c is equal to dark curve o-a, geometrically doubled by a factor of two. The dashed curve c-a is obtained by rotating the solid curve, a-b-c, by 180°.

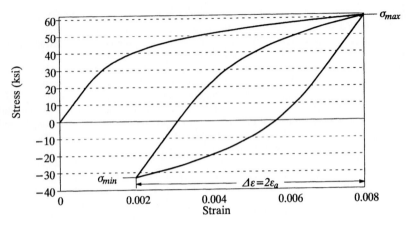

FIG. 7.40 Cyclically stable hysteresis loop computed for Example 6.

curve from a to b to c is the reverse-loading hysteresis curve. The expression for the hysteresis curve on σ'-ε' axes is given by

$$\Delta\varepsilon = \frac{\Delta\sigma}{E} + 2\left(\frac{\Delta\sigma}{2K'}\right)^{1/n'} \tag{7.51}$$

Notice that Eq. (7.51) is given in terms of stress and strain *ranges* (denoted by "Δ"). This is illustrated for point b along the curve a-b-c in Fig. 7.40. For fully reversed loading, the reversal at c would be followed by a stress-strain path along the dashed line from c back to a.

This is important, since it can be used to reveal the location (on σ-ε axes) where a stable hysteresis loop will form during constant-amplitude strain-controlled loading that is *not* fully reversed. The ability to estimate the form of the hysteresis curve provides a mechanism to estimate the path-dependent plasticity behavior of the material under axial loading. The use of this approach is demonstrated in Example 6 for constant-amplitude loading. Life prediction for this example will be discussed later (as will the use of this approach with variable-amplitude loading).

EXAMPLE 6 Consider an axial specimen with the following properties:

$$E = 30{,}000 \text{ ksi}$$
$$K' = 156.88 \text{ ksi}$$
$$n' = 0.184$$

The specimen is to be subjected to strain-controlled cyclic loading between maximum and minimum values of 0.008 and 0.002. Compute the corresponding maximum and minimum stress and plot the cyclically stable hysteresis loop.

solution Using Eq. (7.50), the stress amplitude is computed that would correspond to a strain amplitude of 0.008, if the loading were fully reversed. By trial and error, this value is found to be 61.13 ksi:

$$0.008 = \frac{61.13 \text{ ksi}}{E} + \left(\frac{61.13 \text{ ksi}}{K'}\right)^{1/n'}$$

Now, the stress range $\Delta\sigma$ corresponding to a strain range, $\Delta\varepsilon$, of 0.006 is computed. This corresponds to $\varepsilon_{max} - \varepsilon_{min} = 0.008 - 0.002 = 0.006$. Equation (7.51) is used to define a value of $\Delta\sigma = 94.05$ ksi:

$$0.006 = \frac{94.05 \text{ ksi}}{E} + 2\left(\frac{94.05 \text{ ksi}}{2K'}\right)^{1/n'}$$

From these values, the stress is estimated to fluctuate from a maximum of 61.13 ksi to a minimum of $(61.13 - 94.05 =) -32.92$ ksi. The corresponding stable hysteresis loop is shown in Fig. 7.40.

Strain-Life Relation. As discussed in the preceding section, strain-controlled companion specimen fatigue testing per ASTM E-606[40] results in strain-amplitude versus cycles-to-failure data (defined as complete specimen fracture or the formation of detectable cracks). As shown in Fig. 7.41, the cyclically stable total strain amplitude can be divided into elastic and plastic components. This can be expressed as

$$\varepsilon_a = \varepsilon_a^e + \varepsilon_a^p \tag{7.52}$$

where the superscripts e and p represent elastic and plastic components, respectively. In 1910 Basquin[43] is credited with the observation that log-log plots of stress amplitude (and, therefore, elastic strain amplitude) versus life data behaved linearly. Manson[44] and Coffin,[45] working independently, later observed that log-log plots of plastic strain versus life were also linear. These two observations were combined into the now familiar form

$$\varepsilon_a = \frac{\sigma_f'}{E}(2N_f)^b + \varepsilon_f'(2N_f)^c \tag{7.53}$$

$$\varepsilon_a = \varepsilon_a^e + \varepsilon_a^p$$

where $\sigma_f' =$ fatigue strength coefficient
$\ b =$ fatigue strength exponent

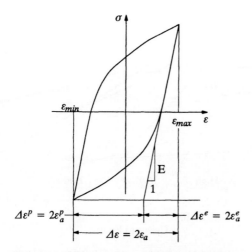

FIG. 7.41 Definition of elastic and plastic strain amplitude from total strain amplitude.

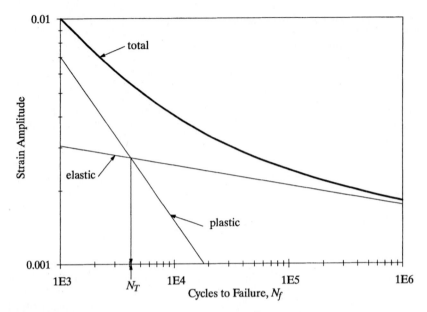

FIG. 7.42   Strain-life relation for a medium-strength steel.[49]

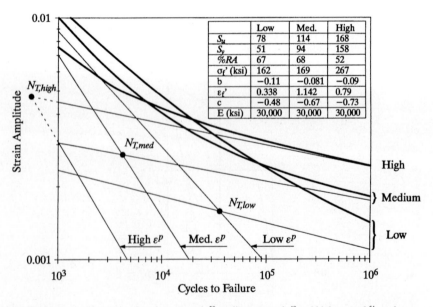

	Low	Med.	High
S_u	78	114	168
S_y	51	94	158
%RA	67	68	52
σ_f' (ksi)	162	169	267
b	−0.11	−0.081	−0.09
ε_f'	0.338	1.142	0.79
c	−0.48	−0.67	−0.73
E (ksi)	30,000	30,000	30,000

FIG. 7.43   Strain-life relation for a low-strength,[50] medium-strength,[49] and high-strength[51] steel.

ε_f' = fatigue ductility coefficient
c = fatigue ductility exponent
E = modulus of elasticity
N_f = number of cycles to failure

Equation (7.53) serves as the foundation of local strain-based fatigue analysis.[13,23,31,32,46–48] It is plotted in Fig. 7.42 for a medium-strength steel.[49] The "transition life," N_t, is also depicted in Fig. 7.42, as the life at which the elastic and plastic strain amplitudes are equal. The transition life can provide an indication of whether straining over a life regime of interest is more elastic or plastic. In general, the higher the tensile strength of the materials, the lower the transition life, and elastic strains tend to dominate for a greater portion of the overall life regime. To demonstrate this, Fig. 7.43 shows the same strain-life curve from Fig. 7.42, replotted with curves for a low-strength steel[50] and a high-strength steel.[51]

Inspection of Fig. 7.43 is interesting from the standpoint of selecting a material for maximum fatigue resistance. For example, in the higher-cycle life regime ($>10^5$ cycles) the higher-strength material provides the greatest fatigue strength. But for a component that must operate in the lower life regime ($<10^4$ cycles) the selection of the lowest-strength steel might be warranted. The medium-strength material could represent a more suitable selection if a combination of high strength and fatigue resistance is required.

As a consequence of Eqs. (7.52) and (7.53), and the Ramberg-Osgood formulation of the cyclic stress-strain curve [Eq. (7.50)], it is apparent that the cyclic strength coefficient and cyclic strain-hardening exponent may be expressed in terms of the fatigue strength and ductility coefficients and exponents as follows:

$$K' = \frac{\sigma_f'}{(\varepsilon_f')^{n'}} \tag{7.54}$$

$$n' = \frac{b}{c} \tag{7.55}$$

However, experience has shown that parameters formed this way can result in a stress-strain curve that *does not* correlate well with actual stress versus strain-amplitude data points. For this reason, use of Eqs. (7.54) and (7.55) is not recommended if K' and n' may be fit directly to data.

Values for cyclic stress-strain and fatigue parameters can be found in Refs. 52 through 54. Unfortunately, tabulated data are available for only a fraction of available engineering alloys and heat treatments. Although such data are becoming more and more available, situations routinely arise where a designer must estimate fatigue properties without the availability of fatigue testing data.

Several attempts have been made to correlate fatigue parameters with monotonic tensile properties,[46,55–57] but no clearly superior approach has been identified. Table 7.3 summarizes several approaches.

These relations should be used *only with a great deal of caution*. The universal slopes[55] approach, the Socie et al.[57] approach, and a more complicated four-point correlation approach were compared to data in Ref. 58. The four-point correlation method provided the best approximation but requires the true fracture stress, a quantity not generally reported during tensile testing. The potential for any of the approaches to provide poor life estimates is great. Considering the differences in deformation and failure mechanisms (on a microscopic level) between monotonic and cyclic loading, limited correlation is not surprising.

Some final notes about the limitations of strain-life data are in order. Equation (7.53) is based on data generated with polished specimens. (ASTM E-606 recom-

TABLE 7.3 Approximate Relations between Tensile Properties and Fatigue Parameters

	Universal slopes[55]	Modified universal slopes[56]	Socie et al.[57]
σ'_f	$1.9018\,S_u$	$0.6227E\left[\dfrac{S_u}{E}\right]^{0.832}$	$S_u + 345\ \text{MPa}$
b	-0.12	-0.09	$-\dfrac{1}{6}\log\left(\dfrac{2(S_u + 345\ \text{MPa})}{S_u}\right)$
ε'_f	$0.7579\left[\ln\left(\dfrac{1}{1-\text{RA}}\right)\right]^{0.6}$	$0.01961\left[\ln\left(\dfrac{1}{1-\text{RA}}\right)\right]^{0.155}\left[\dfrac{S_u}{E}\right]^{-0.53}$	$\ln\left(\dfrac{1}{1-\text{RA}}\right)$
c	-0.6	-0.56	-0.6

mends that specimens be ground and longitudinally lapped to a surface finish of 8 μin or better, with no circumferential grinding marks observable at 20 × magnification.) In the higher life regime, strain-controlled fatigue data are affected by the same factors that can affect stress-life data (surface finish, inclusions, defects, size, etc.). Although it is recognized in the literature that adjustments are necessary to account for surface effects, no accepted methodology to accomplish this exists. It is suggested[23] that the elastic portion of Eq. (7.53) can be modified, by adjusting the fatigue strength exponent b. However, this often affects the total strain-life curve into the lower-cycle regime as well, as is apparent in Fig. 7.44.

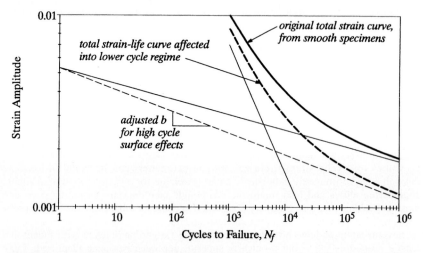

FIG. 7.44 Modifying the strain-life relation by adjusting b to account for surface roughness effects can influence the curve over the entire life regime.

Along these same lines, the strain-life equation itself has no mechanism to incorporate a fatigue limit. When lives in the vicinity of the endurance limit are being considered ($\geq 10^6-10^7$ cycles), the use of stress-based approaches should be considered.

Finally, it can be useful to obtain as much information about a particular data set as possible. Recall that "failure" in strain-controlled tests can be defined as complete fracture, or a drop in the cyclically stable maximum stress (typically 10 to 50 percent). This can be significant, especially if there is a substantial difference between the size of the component being analyzed and the size of the specimens used to generate the data. Published data sets are often based on a very small number of data, generated over a fairly narrow range. There is scatter associated with all fatigue data, and generating more data could change associated strain-life curves drastically. Furthermore, additional uncertainty is associated with utilizing the strain-life relation outside the range over which data were collected. Knowing the range of experimental lives could be very worthwhile towards assessing the reliability of a fatigue life estimate.

Mean Stress Effects.   The standard strain-life relation, Eq. (7.53), is formulated from fully reversed ($R = -1$) strain-controlled test data. However, engineering components are often subjected to loading that induces mean stresses (strains). Although the effect of mean stress on fatigue has been discussed, mean strain, in general, has little effect. For example, consider the stress-strain response during a mean strain axial fatigue test as depicted in Fig. 7.45. The mean stress in the specimen is shown to relax to a cyclically stable response by the seventh reversal. (This response is somewhat exaggerated, as many more cycles are usually required.) This should not be confused with cyclic softening, since cyclically stable materials can exhibit relaxation.

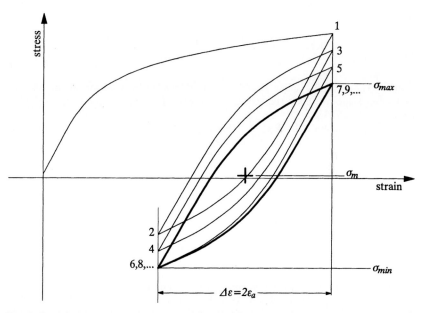

FIG. 7.45   Stress relaxation during a mean strain axial fatigue test.

Mean stress relaxation is the main reason that mean stresses tend to have more of an effect on high-cycle fatigue than on low-cycle fatigue. For high-cycle situations, loading is typically elastic and mean stresses relax little (if any).

Several approaches have been proposed to account for mean stresses with strain-based analysis.[13,31,32] The most effective of these is the Smith-Watson-Topper[36] relation, already discussed in terms of stresses. This approach is based on empirical observations that the product of the stabilized maximum stress and the strain range (or amplitude) during a cycle is proportional to life [Eq. (7.56)]. From fully reversed data, the maximum stress can be approximated in terms of life as Eq. (7.57).

$$\sigma_{max}\varepsilon_a \propto N_f \tag{7.56}$$

$$\sigma_{max} = \sigma_f'(2N_f)^b \tag{7.57}$$

Equations (7.53), (7.56), and (7.57) are combined to form the local-strain-based expression for the SWT relation,

$$\sigma_{max}\varepsilon_a = \frac{(\sigma_f')^2}{E}(2N_f)^{2b} + \sigma_f'\varepsilon_f'(2N_f)^{b+c} \tag{7.58}$$

It should be kept in mind that this expression is strictly empirically based. For extreme cases (e.g., high mean stress) outside the range of data for which the expression was established, the validity of this approach has not been assessed. For purposes of bounding a life estimate, an approximation may be made based on the strain amplitude only, neglecting the mean stress effect [i.e., using Eq. (7.53)]. This is demonstrated in Example 7.

EXAMPLE 7 Compute the life corresponding to the stable hysteresis loop from Example 6 (Fig. 7.40). The low-cycle fatigue parameters for the specimen are given as

$$\sigma_f' = 110 \text{ ksi} \qquad b = -0.105$$
$$\varepsilon_f' = 0.55 \qquad c = -0.625$$

solution The cyclically stable hysteresis loop fluctuated between maximum and minimum strain values of 0.008 and 0.002, with stresses ranging between $\sigma_{max} = 61.13$ ksi and $\sigma_{min} = -32.92$ ksi.

For the strain amplitude of $\varepsilon_a = 0.003$ [$= (0.008-0.002)/2$], and maximum stress of $\sigma_{max} = 61.13$ ksi, Eq. (7.58) appears as

$$(61.13)(0.003) = \frac{(\sigma_f')^2}{E}(2N_f)^{2b} + \sigma_f'\varepsilon_f'(2N_f)^{b+c}$$

This equation is solved iteratively for a life of $N_f = 2618$ cycles. If Eq. (7.53) is solved with no mean stress effect, an upper-bound life estimate of $N_f = 5590$ results.

Notches. Many mechanical components operate with elastically calculated stresses that exceed the yield strength of the material at a notch. However, once yielding occurs the elastic stress concentration factor K_t cannot be expected to accurately predict localized stresses. Since the material has yielded, local stress values are less than the elastically computed quantities. Similarly, strains computed elastically are exceeded in notches. This situation is depicted in Fig. 7.46. Equations (7.59) and (7.60) define notch stress- and strain-concentration factors, where σ_n and ε_n are the actual notch stress and strain, respectively, and S and e are elastically calculated nominal notch stress and strain, respectively.

$$K_\sigma = \frac{\sigma_n}{S} \tag{7.59}$$

$$K_\varepsilon = \frac{\varepsilon_n}{e} \tag{7.60}$$

To conduct a strain-based fatigue life estimate of a notch, the state of stress and strain *at the notch root* must be estimated. To achieve this, three approaches could be considered:

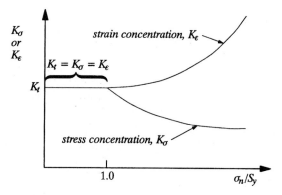

FIG. 7.46 Stress and strain concentration at a notch.

1. Place a strain gauge in the notch root.
2. Conduct elastic-plastic finite-element analysis.
3. Estimate notch root stress-strain from nominal, elastically calculated values.

The first two approaches sound straightforward enough, but both are extremely time consuming and complex. Notch geometries do not usually facilitate the placement of a strain gauge. Furthermore, a part must exist. This is not usually the case early in the design process. The application of finite-element analysis is getting easier for elastic situations, but is still a formidable task for elastic-plastic situations, especially under cyclic loading.[22] Mesh density requirements are great and codes are simply not set up to easily accept cyclic input loading.

For this reason, item 3 in the above list is emphasized here. Specifically, the application of Neuber's rule[59] is discussed. Neuber originally noticed that the geometric mean of the stress- and strain-concentration factors remains approximately constant as plasticity occurs in a notch. This means that the product of notch root stress and strain can be determined from elastically calculated quantities. From this observation, Neuber's rule is stated as

$$K_t^2 Se = \sigma_n \varepsilon_n \tag{7.61}$$

If nominal stress is elastic, the left-hand side of the equation above can be restated as

$$\frac{(K_t S)^2}{E} = \sigma_n \varepsilon_n \tag{7.62}$$

Notice that the left-hand side of this equation represents the applied loading (S) and notch geometry (K_t). The right-hand side is simply the product of the actual stress and strain in the notch root. If the left side is considered as known input, one more relation is needed to solve for the two unknowns. This relation is the cyclic stress-strain curve:

$$\varepsilon_n = \frac{\sigma_n}{E} + \left(\frac{\sigma_n}{K'}\right)^{1/n'} \tag{7.63}$$

Figure 7.47 illustrates the approach for elastic nominal stress-strain. Neuber's rule is represented by the hyperbola. The product of stress and strain is a constant, defined by the geometry and applied loading. The cyclic stress-strain curve provides the second

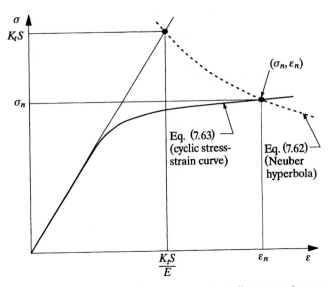

FIG. 7.47 Neuber's rule specifies the point on the cyclic stress-strain curve corresponding to the notch root.

relation necessary to obtain the two unknowns, the stress and strain in the notch root, represented by the intersection of the two curves in Fig. 7.47.

In some instances, a component can be loaded such that the nominal stress S or nominal stress range ΔS is itself large enough to cause plastic straining. In this instance, the material's cyclic stress-strain curve should be used to compute the nominal strain, e, according to

$$e = \frac{S}{E} + \left(\frac{S}{K'}\right)^{1/n'} \tag{7.64}$$

This is demonstrated in Example 8 for applied loading with a mean stress.

EXAMPLE 8 A notched steel bar has an elastic stress-concentration factor of $K_t = 2.42$. The material properties for the steel are given by:

$$E = 30{,}000 \text{ ksi} \qquad K' = 154 \text{ ksi} \qquad n' = 0.123$$
$$\sigma_f' = 169 \text{ ksi} \qquad \varepsilon_f' = 1.14$$
$$b = -0.081 \qquad c = -0.67$$

For applied loading represented by the nominal stress history depicted on page 7.53, compute the cycles to crack initiation for constant-amplitude loading from 50 ksi to −30 ksi.

solution To conduct the analysis, consider loading from (o) to (a), (a) to (b), and finally (b) to (c). Subsequent loadings are repeated (a)-(b)-(c) cycles. First, consider loading from (o) to (a): The nominal stress changes from 0 to 50 ksi, or $\Delta S = 50 - 0$ ksi. For this initial loading segment, Neuber's rule is expressed as

$$K_t^2 \Delta S \Delta e = \Delta \sigma \Delta \varepsilon$$

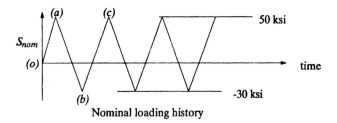

Nominal loading history

On the left side of the Neuber equation, the nominal stress and strain ranges are related by

$$\Delta e = \frac{\Delta S}{E} + \left(\frac{\Delta S}{K'}\right)^{1/n'}$$

On the right side of the Neuber equation, the notch strain and stress ranges are related by

$$\Delta \varepsilon = \frac{\Delta \sigma}{E} + \left(\frac{\Delta \sigma}{K'}\right)^{1/n'}$$

Combining these relations,

$$K_t^2(\Delta S)\left[\frac{\Delta S}{E} + \left(\frac{\Delta S}{K'}\right)^{1/n'}\right] = \Delta \sigma\left[\frac{\Delta \sigma}{E} + \left(\frac{\Delta \sigma}{K'}\right)^{(1/n')}\right]$$

$$(2.42)^2(50)(0.001773) = \Delta \sigma\left[\frac{\Delta \sigma}{E} + \left(\frac{\Delta \sigma}{K'}\right)^{(1/n')}\right]$$

Solving iteratively, $\Delta \sigma = 78.19$ ksi. This corresponds to a strain of

$$\Delta \varepsilon = \frac{78.19}{E} + \left(\frac{78.19}{K'}\right)^{1/n'} = 0.00665$$

Therefore, at point (a), the notch root stress and strain are defined (78.19 ksi, 0.00665).

Next, to get from (a) to (b) the approach is similar, but the stress and strain changes occur along the hysteresis curve, Eq. (7.51), rather than along the cyclic stress-strain curve, Eq. (7.50). The nominal stress changes from 50 to −30 ksi: $\Delta S = 50-(-30) = 80$ ksi. For this segment, Neuber's rule is expressed as:

$$K_t^2 \Delta S \Delta e = \Delta \sigma \Delta \varepsilon$$

On the left side of the Neuber equation, the nominal strain change is related to the nominal stress change by

$$\Delta e = \frac{\Delta S}{E} + 2\left(\frac{\Delta S}{2K'}\right)^{1/n'}$$

On the right side of the Neuber equation, the notch strain range is related to the notch stress by

$$\Delta \varepsilon = \frac{\Delta \sigma}{E} + 2\left(\frac{\Delta \sigma}{2K'}\right)^{1/n'}$$

Combining these relations,

$$K_t^2(\Delta S)\left[\frac{\Delta S}{E} + 2\left(\frac{\Delta S}{2K'}\right)^{1/n'}\right] = \Delta \sigma\left[\frac{\Delta \sigma}{E} + 2\left(\frac{\Delta \sigma}{2K'}\right)^{(1/n')}\right]$$

$$(2.42)^2(80)(0.002701) = \Delta\sigma\left[\frac{\Delta\sigma}{E} + \left(\frac{\Delta\sigma}{2K'}\right)^{(1/n')}\right]$$

Solving iteratively, $\Delta\sigma = 143.55$ ksi.

The strain change corresponding to this stress change is given by

$$\Delta\varepsilon = \frac{143.55}{E} + 2\left(\frac{143.55}{2K'}\right)^{1/n'} = 0.008817$$

Therefore, at point (b), the stress is given by $78.19 - 143.55 = -65.36$ ksi, and the strain is given by $0.00665 - 0.008817 = -0.00217$.

Finally, to get from (b) to (c), notice that the nominal stress change is the same as in the proceeding step. The hysteresis loop for the cycle is shown below.

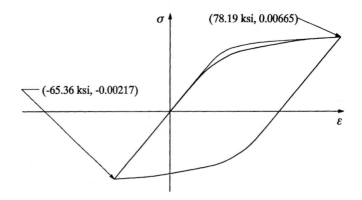

From this, the strain amplitude is given by $\Delta\varepsilon/2 = 0.004409$ and the maximum stress is 78.19 ksi. The SWT estimation of life for this notch root stress-strain response is 7318 cycles (14,636 reversals):

$$(78.19)(0.004409) = \frac{(\sigma_f')^2}{E}[2(7318)]^{2b} + \sigma_f'\varepsilon_f'[2(7318)]^{b+c}$$

7.5.3 Variable-Amplitude Loading

Thus far, the fatigue-life prediction approaches presented in this section have only been discussed in the context of constant-amplitude loading. However, engineering components are seldom subjected to constant-amplitude loading for their entire life. More often, load fluctuations occur with means and amplitudes that vary. In this section, a methodology is presented for applying the approaches presented earlier to variable-amplitude fatigue-life prediction.

Cumulative Damage.   In order to compute the life of a component subjected to irregular loading, the concept of "fatigue damage" was developed, whereby each cycle is considered to expend a finite fraction of the overall life. The most widely utilized damage concept is that of linear damage (also called Miner's rule or the Palmgren-Miner rule).

Linear damage is best explained by considering a single "cycle" of loading (from a minimum to maximum and back to a minimum). If this cycle were applied under constant amplitude to a new component, the techniques described previously in this chapter could be used to compute the number of times N that cycle could be applied before a crack could be expected to develop. With the linear damage concept, each cycle of loading is considered to impose an amount of fatigue "damage" equal to $1/N$. (Equivalently, each cycle is considered to expend $1/N$th of the component life.) If this damage is counted for each cycle of loading, then failure would occur when the cumulative damage reached a value of unity. At this point, 100 percent of the fatigue life has been expended. In other words, damage imposed by the ith cycle of loading DAM_i is expressed by

$$DAM_i = \frac{1}{N_i} \qquad (7.65)$$

where N_i is the number of cycles a new component would be computed to sustain under constant-amplitude loading. Failure is considered to occur when

$$\sum^{i=N_f} DAM_i = 1 \qquad (7.66)$$

Other, more sophisticated nonlinear cumulative-damage approaches have been proposed.[12] Also, several effects, such as sequence of loading or overloads, have been shown to bias linear damage summation in either the conservative or nonconservative directions. However, given all the uncertainties associated with fatigue-life estimation, linear damage theory has been successfully applied to a wide range of engineering design situations, especially when loading is highly irregular and in the absence of major overloads.

Cycle Counting.   In order to sum damage for variable-amplitude load histories, it is necessary to define "cycles" from those histories. For certain situations, this can be a fairly straightforward problem. For instance, consider that n_1 cycles of constant-amplitude loading at an effective stress amplitude of $(\sigma_a)_1$ are applied to a component (Fig. 7.48). Following this, the peak and valley (maximum and minimum) values of stress change, such that an effective stress amplitude of $(\sigma_a)_2$ is now being applied. If this loading continues for n_2 cycles, how many cycles (n_3) could a third range of loading

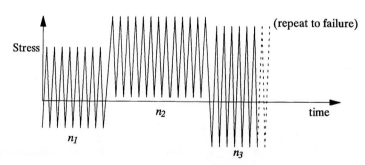

FIG. 7.48   Block loading problem.

be applied until failure is expected? For each effective stress amplitude, a corresponding fatigue life of N_1, N_2, and N_3 can be calculated. Therefore, n_3 can be computed from Eq. (7.66), written for this problem as

$$\sum DAM_i = \frac{n_1}{N_1} + \frac{n_2}{N_2} + \frac{n_3}{N_3} = 1 \qquad (7.67)$$

For situations where loading is more irregular, numerous cycle-counting procedures have been proposed. Several approaches are presented in ASTM E-1049, "Recommended Practices for Cycle Counting in Fatigue Analysis."[60] Of these, rainflow cycle counting is the most popular and widely used approach.

Rainflow Cycle Counting. The concept of rainflow cycle counting has been described by two completely different approaches (referred to as the "falling rain" and "ASTM" approaches) that yield the same results. In each of these, it is necessary to define a "block" of irregular loading. This block could correspond to measurements or estimates over a representative period of service operation (one day, one job, one proving-ground lap, etc.). Furthermore, the block must be reorganized such that the largest peak in the history occurs as the first reversal. Cycles occurring prior to this are simply moved to the end such that the overall peak-valley content of the block remains the same. The "falling rain" approach is described first by considering the stress history, shown in Fig. 7.49. It must be visualized that "gravity" acts along the time axis, as shown in the figure. With this basis, the approach is described below.

1. Define a block (or history) with the highest peak occurring first.
2. Assuming "gravity" acts along the time axis, start a "flow of water droplets" at each peak and valley, *in succession.* The droplets flow along the profile and over the edge.
3. Flow originating at a peak (or valley) stops (⊣) if it "sees" a higher peak (or a deeper valley) than the one from which it started.
4. Flow also stops (>) to avoid collision with rain from a previous flow.

Stress "ranges" (valley-to-peak) are defined by the amount of vertical travel incurred by a particular flow. For instance, the rainflows in Fig. 7.49 define the eight ranges shown at the top of page 7.57.

To predict life, damage imposed by each of the four rainflow counted ranges is computed appropriately. This is demonstrated in Example 9.

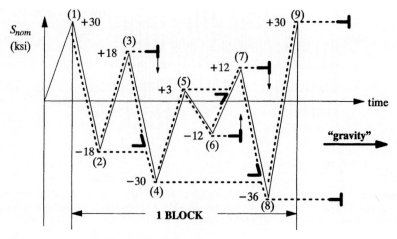

FIG. 7.49   Rainflow cycle counting ("falling rain" analogy).

Ranges

Each of the ranges corresponds to one-half cycle, or one-half of a range pair

$\left\{\begin{array}{l}\text{1. (1) to (8)} \\ \text{2. (2) to (3)} \\ \text{3. (3) to (2)} \\ \text{4. (4) to (7)} \\ \text{5. (5) to (6)} \\ \text{6. (6) to (5)} \\ \text{7. (7) to (4)} \\ \text{8. (8) to (9)}\end{array}\right.$

Therefore, the four "cycles" or range pairs counted by the routine are

1. (1) to (8) & (8) to (9)
2. (2) to (3) & (3) to (2)
3. (4) to (7) & (7) to (4)
4. (5) to (6) & (6) to (5)

(as indicated by the dashed lines connecting the ranges)

EXAMPLE 9 For a forged 1040 steel component with a stress concentration factor of 1.8, conduct an elastic analysis and compute the number of times the nominal stress-loading block shown in Fig. 7.49 can be sustained by the component. The material properties for the material are given by

$$\sigma'_f = 223 \text{ ksi}$$
$$b = -0.14$$

solution The rainflow-cycle-counted nominal and notch stresses are tabulated below. The effective stress amplitude is based on the SWT parameter [Eq. (7.46)].

Range pair	S_{max} (ksi)	S_{min} (ksi)	σ_{max} (ksi)	σ_{min} (ksi)	σ_a (ksi)	$\sigma_{eff,a}$ (ksi)	N (cycles)	D (cyc^{-1})
(1–8)(8–9)	30	−36	54	−64.8	59.4	56.64	8922	1.12E−4
(2–3)(3–2)	18	−18	32.4	−32.4	32.4	32.4	4.82E5	2.08E−6
(4–7)(7–4)	12	−30	21.6	−54	37.8	28.57	1.18E6	8.46E−7
(5–6)(6–5)	3	−12	5.4	−21.6	13.5	8.54	6.61E9	1.5E−10

$$\Sigma D = 0.000115$$

The life of the component, in number of blocks to failure, is computed as shown below:

$$\text{Life} = \frac{1}{\Sigma D} = 8696 \text{ blocks}$$

The ASTM rainflow approach is conceptually quite different from the falling rain approach, but both give identical answers. The ASTM procedure is extremely beneficial analytically, since it can be implemented using a few lines of computer code. A BASIC listing to implement rainflow cycle counting is presented below, as Table 7.4.

Notch Strain Analysis. The estimated life in Example 9 is somewhat low and likely not in the elastic regime, indicating that cyclic plasticity may be occurring in the notch root. To analyze such a situation, a computer program can be written to implement a Neuber analysis (similar to Example 8) for every nominal stress range in the block (RANGE in Table 7.4). This is demonstrated in Example 10.

EXAMPLE 10 The nominal stress history from Fig. 7.49 is to be applied to a forged 1045 steel shaft with a K_t value of 1.8 and again for a K_t value of 3.0. Compute the notch stress-strain response. Rainflow-cycle-count the strain history and estimate the life of each component. The properties are given below.

TABLE 7.4 BASIC Computer Listing for ASTM Rainflow Approach

Nominal stress reversals are stored in an array, Stress(NPTS), where NPTS=number of reversals in the history. Since the history must begin and end with the maximum peak, NPTS will always be an odd integer.

```
        Event(1) = Stress(1)
        N = 1 : 'Counter N initialized
        J = 1 : 'Counter J initialized
1       N = N + 1
        J = J + 1
        IF (J = NPTS + 1) THEN GOTO 3
        Event(N) = Stress(J)
2       IF (N < 3) GOTO 1
        X = ABS(Event(N) - Event(N-1))
        Y = ABS(Event(N-1) - Event(N-2))
        IF (X < Y) THEN GOTO 1
        RANGE = Y
        Xmean = (Event(N-1) + Event(N-2)/2
        Xamplitude = RANGE/2
        N=N-2
        Event(N)=Event(N+2)
        GOTO 2
3       'Finished
        END
```

$$K' = 171.4 \text{ ksi} \qquad \sigma'_f = 223 \text{ ksi}$$

$$n' = 0.18 \qquad b = -0.14$$

$$E = 30{,}000 \text{ ksi} \qquad \varepsilon'_f = 0.61$$

$$c = -0.57$$

solution The rainflow-cycle-counted strain amplitude and maximum stress for each range pair in the stress history are tabulated below for each shaft. The estimated number of cycles corresponding to each range pair are computed using the SWT parameter, Eq. (7.58).

Range pair	For $K_t = 1.8$			For $K_t = 3.0$		
	ε_a	σ_{max} (ksi)	N (cycles)	ε_a	σ_{max} (ksi)	N (cycles)
(1–8)(8–9)	0.002619	46.25	35,760	0.005737	59.60	4,018
(2–3)(3–2)	0.001352	25.08	9.31E5	0.002149	43.87	6.75E4
(4–7)(7–4)	0.001124	29.80	9.66E5	0.002693	39.88	3.19E4
(5–6)(6–5)	0.00045	11.74	4.39E8	0.000756	26.59	9.40E6

The life in number of blocks for each shaft is computed as shown below. For $K_t = 1.8$,

$$\text{Life} = \frac{1}{\sum_{1/N}} = 33{,}248 \text{ blocks}$$

For $K_t = 3.0$,

$$\text{Life} = \frac{1}{\sum_{1/N}} = 3842 \text{ blocks}$$

The mean stress effect on life predictions can be assessed by assuming the strain range to be fully reversed (neglecting mean stress). To do this, the N for each range pair is computed using Eq. (7.53). This results in life predictions of 31,230 blocks for $K_t =$ 1.8 and 3387 blocks for $K_t = 3.0$.

In Example 9, life predictions are based on notch root stresses and strains estimated from the applied nominal stress ranges and Neuber's rule. The estimated notch root stress-strain history can be plotted as a set of repeating hysteresis loops, such as presented in Fig. 7.50. This illustrates an important physical feature associated with rain-

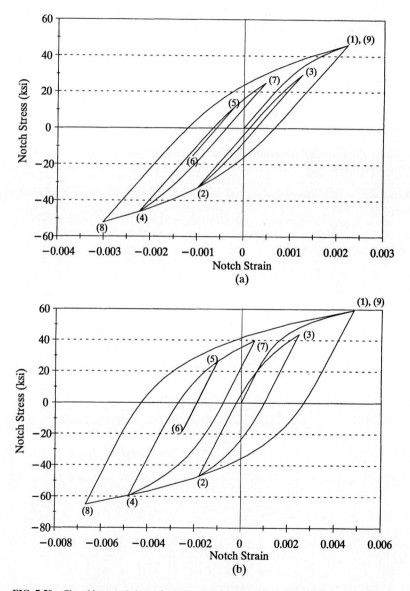

FIG. 7.50 Closed hysteresis loops from Example 10 for (*a*) $K_t = 1.8$ and (*b*) $K_t = 3.0$.

flow cycle counting, in that each event defined by the routine corresponds to a closed hysteresis loop shown in the figure. Comparing Fig. 7.50 and the nominal stress history, Fig. 7.49, notice that each closed hysteresis loop can be directly identified with each rainflow-cycle-counted range pair. Also note that, for this example load history, the largest strain range imposes the vast majority of damage. In general, this may not be the case.

7.6 DAMAGE-TOLERANT DESIGN

The structural integrity assessment methodologies presented in preceding sections of this chapter assume the material to be free of substantial flaws or defects. Stresses and strains are computed and compared to the strength of homogeneous engineering material. Damage-tolerant design recognizes that flaws, specifically cracks, can and do exist, even before a component is placed into service. Therefore, the focus of this design philosophy is on estimating behavior of a crack in an engineering material under service loading, and *fracture mechanics* provide the analytical tools.

Macroscopic cracks are assumed to exist in regions where detection may be difficult or impossible (e.g., under a flange or rivet head), and the behavior of the crack is predicted under anticipated service loading. The estimated behavior is used to schedule inspection and maintenance in order to assure that defects do not propagate to a catastrophic size.

Only a brief overview is presented in this section. This important method, used extensively in the aerospace industry, is covered comprehensively with example applications in several references.[9,13,31,32,61–65] The discussion here is limited to linear elastic approaches, although research on elastic-plastic fracture mechanics is very active, particularly regarding the behavior of very small cracks in locally plastic strain fields.[66–68]

7.6.1 Stress-Intensity Factor

An "ideal" crack in an engineering structure has been modeled analytically as a notch with a root radius of zero (Fig. 7.51). When a stress is applied perpendicular to the

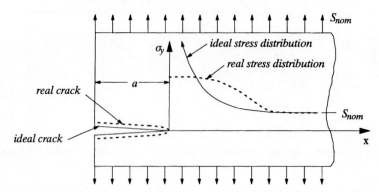

FIG. 7.51 Schematic crack opening and stress distribution from the tip of an edge crack of length a, in a theoretical and real engineering material.

ideal crack, stresses approach infinity at the crack tip. This is evident in Fig. 7.51 and Eq. (7.68) for the stress along the x axis:

$$\sigma_y = \frac{K_I}{\sqrt{2\pi x}} \qquad (7.68)$$

The parameter K_I is referred to as the "stress-intensity factor" and it lies at the center of all fracture mechanics analyses.

The stress-intensity factor can be considered a quantitative measure of the severity of a crack in material attempting to sustain a particular state of stress. It also can be thought of as the rate at which the stresses approach infinity at the tip of a theoretical crack. Since real material cannot sustain infinite stress, intense localized deformation occurs, causing the crack tip to blunt and stresses to redistribute, as depicted in Fig. 7.51. Even though crack-tip deformation is plastic, as long as the size of the plastic zone at the crack tip is small relative to crack dimensions, the elastically calculated K_I has been successfully used to describe the strength of an engineering component. Therefore, the term "linear elastic fracture mechanics" (LEFM) is typically applied to analyses based on K_I.

The significance of the subscript I is shown in Fig. 7.52. Since cracks are considered analytically as planar defects, the subscript refers to the mode in which the two crack faces are displaced. Mode I is the normal loading mode (crack faces are pulled apart) while II and III are shear loading modes (crack faces slide relative to each other). The discussion in the remainder of this section is directly in terms of mode I analyses. However, the extension of the approaches to the shear modes is straightforward.

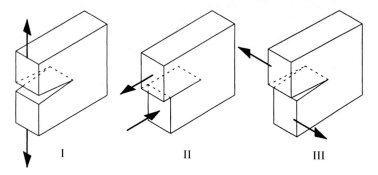

FIG. 7.52 Crack loading modes: (*left*) I, normal; (*center*) II, sliding; (*right*) III, tearing.

Stress-intensity factors can be computed for any geometry and loading combination using finite-element analysis, but not without considerable analytical effort. More typically in design, stress-intensity factors are found from a tabulated reference.[69–72] They are usually expressed as in Eq. (7.69), in terms of the gross nominal stress S_{nom}, crack length, and geometry for a particular type of loading (e.g., bending or tension). Since they are elastically calculated, they can be superimposed.

$$K_I = f(a, \text{geometry})S_{nom}\sqrt{a} \qquad (7.69)$$

7.6.2 Static Loading

The premise of damage-tolerant design is that engineering materials with macroscopic cracks can sustain stresses without failure, up to a point. Obviously, the larger a crack in a component, the less load it can sustain. One of the important features of K_I is its ability to characterize combinations of loading and crack size that correspond to unstable crack propagation (or the strength of the cracked component). Experiments have shown that cracks in engineering materials propagate catastrophically at a certain critical value of the stress-intensity factor, K_c. The critical value is referred to as the *fracture toughness* (not directly associated with the impact strength or area under a stress-strain curve) and is considered a property of the material. In other words, failure is predicted when

$$K_I = f(a,\text{geometry})S_{\text{nom}}\sqrt{a} = K_c \qquad (7.70)$$

Like other "material properties," K_c is influenced by numerous factors including environment, rate of loading, etc. But an important distinction for K_c arises from its observed dependence on specimen size. As the thickness t of the specimen increases along the dimension of the crack front, increasing constraint develops, making plastic deformation more difficult and causing the material to behave in a more brittle manner. Therefore, as shown in Fig. 7.53, as the thickness increases, K_c decreases asymptotically to a value referred to as K_{IC}, the *plane-strain fracture toughness*. Specifications for the experimental determination of K_{IC} are found in ASTM E-399, "Standard Test Method for Plane-Strain Fracture Toughness of Metallic Materials,"[71] and tabulated values can be found in Refs. 73 through 80 for many engineering alloys.

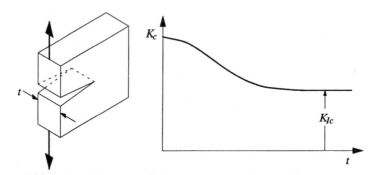

FIG. 7.53 As specimen thickness increases, fracture toughness K_c decreases asymptotically to the value K_{Ic}, the plane-strain fracture toughness.

7.6.3 Fatigue Loading

Cracks in engineering components tend to propagate progressively under fatigue loading. The damage-tolerant design philosophy takes advantage of this phenomenon to schedule periodic inspection and maintenance of components to assure against the growth of a crack to the catastrophic size associated with Eq. (7.70).

Paris and Erdogan[81] are credited with first making the observation that the stress-intensity factor range can be used effectively to correlate crack propagation rates for a particular material under a wide variety of crack geometry and loading combinations. As shown schematically in Fig. 7.54, suppose three separate specimens are subjected to the $R = 0$ loading at different applied loading levels. A stress-intensity factor range is computed from each nominal stress range. Cracks grow through each specimen at different rates, as shown in Fig. 7.54b. However, if the crack propagation rate, da/dN, is plotted versus stress-intensity factor range ΔK_I (on log-log coordinates), then the data from all three tests collapse to the single curve, Fig. 7.54c. This curve is consid-

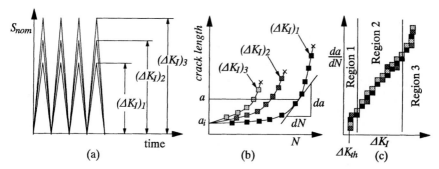

FIG. 7.54 Schematic data from three separate specimens: (*a*) Different nominal stress ranges lead to three different ΔK ranges for a given crack length; (*b*) crack length versus cycles of applied loading; (*c*) crack growth rate versus ΔK for all three tests.

ered to be a characteristic of the material. Guidelines for conducting this type of testing are specified in ASTM E-647, "Standard Test Method for Measurement of Fatigue Crack Growth Rates."[82]

The shape associated with the crack growth rate curve (Fig. 7.54*c*) has been referred to as *sigmoidal,* having three distinct regions. Region 1 is associated with another property called the threshold stress-intensity factor range, ΔK_{th}. This is analogous to the endurance limit for local stress-based fatigue design. Applied loading cycles below ΔK_{th} are considered not to result in any crack advancement. Values of ΔK_{th} for various materials can be found in Refs. 73 through 80. At the upper region of the curve (called Region 3), peak values of K_I are approaching K_c. At this point crack propagation is rapid and growing beyond the limits of LEFM validity. The middle portion of the curve, Region 2, is important since it encompasses a substantial portion of the propagation life. Data in this regime usually appear somewhat linear on logarithmic coordinates, leading to the Paris relation,

$$\frac{da}{dN} = A_p (\Delta K)^{n_p} \tag{7.71}$$

Life prediction based on LEFM is based upon the integration of Eq. (7.71), or any other relation describing the data in Fig. 7.54*c*. This is illustrated in Eq. (7.72) for constant-amplitude loading:

$$N = \int_0^N dN = \int_{a_i}^{a_f} \frac{da}{\Delta K} = \int_{a_i}^{a_f} \frac{da}{f(\text{geometry},a)\sqrt{a}} \tag{7.72}$$

As implied in Eq. (7.72) and Fig. 7.54*b*, a fracture mechanics analysis of crack propagation must begin with the assumption of an initial crack of length a_i. [The final crack length, a_f in Eq. (7.72), can be directly calculated from the maximum anticipated loading and K_c.] The definition of the initial crack size for testing purposes is well defined (ASTM E-647, Ref. 82) but for design purposes can be less objective. Sometimes, cracks are considered to exist where they may be difficult to detect, such as underneath a seam. However, for macroscopically smooth surfaces, the assumption of very small initial crack lengths (cracks that would be difficult to detect without the aid of a microscope, on the order of 0.001 in) can significantly affect the analysis, since small cracks can result in extremely low estimated propagation rates. Very small crack lengths, on the order of typical surface roughness values or surface scratches, are not considered to lie within the valid domain of LEFM.[66-68] The threshold stress intensity factor can be used to define valid initial crack lengths.

Mean stress effects on crack propagation are accounted for with empirical relations usually defined in terms of the stress ratio R [Eq. (7.38)]. Two widely referenced equations were developed by Foreman,[83]

$$\frac{da}{dN} = \frac{A_f(\Delta K)^{n_f}}{(1 - R)K_c - \Delta K}$$

(7.73)

and Walker,[84]

$$\frac{da}{dN} = A_w\left(\frac{\Delta K}{(1 - R)^{1-m_w}}\right)^{n_w}$$

(7.74)

Constants in Eqs. (7.73) and (7.74) are empirically defined by comparison to constant-amplitude data generated over a range of load ratios. The relations can then be used to estimate crack growth during variable-amplitude load histories, usually by assuming that only certain segments in the history will result in incremental crack advancements. For example, damaging segments can be defined as only those that are both *tensile* and *increasing*. Maximum and minimum stresses for a particular segment are used to compute ΔK and R for use in Eq. (7.73) or (7.74), which is thus considered to compute the amount of crack growth resulting from that segment. References 9, 13, 31, 32, and 61 through 65 provide more detailed explanations of the implementation and use of fracture mechanics, including coverage of more advanced topics such as crack closure and sequence effects.

7.7 MULTIAXIAL LOADING

Fatigue under multiaxial loading is an extremely complex phenomenon. Evidence of this is the fact that even though the topic has been actively researched for more than a century, new theories on multiaxial fatigue are still emerging.[85] However, recent work has led to advances in understanding of the mechanisms of multiaxial fatigue and addressed some of the problems associated with implementing those advances for practical design problems.

In this section, only multiaxial fatigue-life prediction approaches are presented that are considered somewhat established. Although such approaches only exist for fairly simplistic multiaxial loading (e.g., constant amplitude, proportional loading, high-cycle regime), they still cover a substantial number of design situations.

7.7.1 Proportional Loading

Loading is defined as proportional when the ratios of principal stresses remains fixed with time. A consequence of this is that principal stress directions do not rotate. The simplest example would be a situation where the three components of surface stress are in phase and fully reversed, as shown in Fig. 7.55a. Figure 7.55b depicts a situation where stresses fluctuate in phase about mean values. In this case, whether or not the loading is purely proportional depends on the ratios of the mean stresses. The mean stresses can cause a slight oscillation of the principal stress axes over a cycle. In either case, a modified Sines approach has been successfully applied to this type of loading in the high-cycle regime.

Sines[37] observed that *mean torsional stresses do not influence fatigue behavior,* while mean tensile stresses decrease life and mean compression improves life. Sines' approach can be summarized as follows:

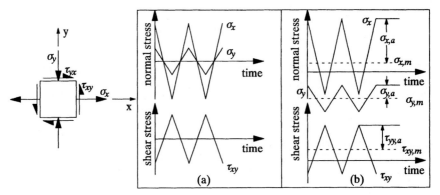

FIG. 7.55 (*a*) Pure proportional loading, fully reversed stress components. (*b*) Proportional stress amplitudes, with mean stresses.

- Compute the amplitude and mean of each normal stress:

$$\sigma_{x,a} = (\sigma_{x,\max} - \sigma_{x,\min})/2$$

$$\sigma_{x,m} = (\sigma_{x,\max} + \sigma_{x,\min})/2$$

$$\sigma_{y,a} = (\sigma_{y,\max} - \sigma_{y,\min})/2$$

$$\sigma_{y,m} = (\sigma_{y,\max} + \sigma_{y,\min})/2$$

$$\sigma_{z,a} = (\sigma_{z,\max} - \sigma_{z,\min})/2$$

$$\sigma_{z,m} = (\sigma_{z,\max} + \sigma_{z,\min})/2$$

- Compute the amplitude of the shear stress:

$$\tau_{xy,a} = (\tau_{xy,\max} - \tau_{xy,\min})/2$$

- Compute the principal stress amplitude from the amplitudes of the normal and shear applied stresses:

$$\sigma_{1,a} = \left(\frac{\sigma_{x,a} + \sigma_{y,a}}{2}\right) + \sqrt{\left(\frac{\sigma_{x,a} - \sigma_{y,a}}{2}\right)^2 + \tau_{xy,a}^2} \qquad (7.75)$$

$$\sigma_{2,a} = \left(\frac{\sigma_{x,a} + \sigma_{y,a}}{2}\right) - \sqrt{\left(\frac{\sigma_{x,a} - \sigma_{y,a}}{2}\right)^2 + \tau_{xy,a}^2} \qquad (7.76)$$

$$\sigma_{3,a} = \sigma_{z,a} \qquad (7.77)$$

- Calculate the equivalent stress amplitude according to a von Mises and an equivalent mean stress as given by

$$S_{eq,a} = \frac{1}{\sqrt{2}} \sqrt{(\sigma_{1,a} - \sigma_{2,a})^2 + (\sigma_{2,a} - \sigma_{3,a})^2 + (\sigma_{3,a} - \sigma_{1,a})^2} \qquad (7.78)$$

$$S_{eq,m} = \sigma_{x,m} + \sigma_{y,m} + \sigma_{z,m} \qquad (7.79)$$

- Use $S_{eq,a}$ and $S_{eq,m}$ as an amplitude and mean stress (in place of σ_a and σ_{mean}) to form an effective, fully reversed stress amplitude, such as described in Fig. 7.27 or given in Eqs. (7.42) to (7.46).

Note that $\tau_{xy,a}$ influences neither $S_{eq,a}$ nor $S_{eq,m}$, and thus does not affect the life estimate. If a pressure-free surface element is being considered, then $\sigma_{3,a} = 0$, and Eq. (7.77) can be simplified to

$$S_{eq,a} = \sqrt{\sigma_{x,a}^2 - (\sigma_{x,a})(\sigma_{y,a}) + \sigma_{y,a}^2 + 3\tau_{xy,a}^2} \qquad (7.80)$$

eliminating the intermediate principal stress-amplitude calculations.

An important point must be kept in mind when implementing this approach: *it can be crucial to keep track of the algebraic sign of the amplitudes of the normal stress components $\sigma_{x,a}$, $\sigma_{y,a}$, and $\sigma_{z,a}$, as well as the principal stresses $S_{1,a}$ and $S_{2,a}$.* Generally, "amplitudes" are considered positive. However, considering amplitude as always positive here can lead to nonconservative life estimates! When two normal stresses *peak* at the same time, then both amplitudes should be considered positive (or the two should have the same algebraic sense). When one is at a *valley* while the other is at a *peak,* then the amplitude of the valley signal is negative while amplitude of the peak signal is positive (or the two should have the opposite algebraic sense). Example 11 serves to illustrate that point.

EXAMPLE 11 Estimate the fatigue life for two materials experiencing the Case A and B stress histories below. The only difference between the two stress histories is the phase relation of the normal stress in the y direction. Properties for the two materials (1045 steels) are as follows: (1) $S_u = 220$ ksi, $\sigma_f' = 843$ ksi, $b = -0.1538$; (2) $S_u = 137$ ksi, $\sigma_f' = 421$ ksi, $b = -0.1607$.

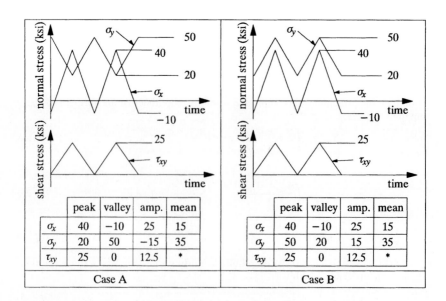

	peak	valley	amp.	mean
σ_x	40	−10	25	15
σ_y	20	50	−15	35
τ_{xy}	25	0	12.5	*

Case A

	peak	valley	amp.	mean
σ_x	40	−10	25	15
σ_y	50	20	15	35
τ_{xy}	25	0	12.5	*

Case B

solution The peak, valley, amplitude, and mean stresses are tabulated above for each load history. For Case A, notice the negative amplitude for $\sigma_{y,a}$. For Case A, $S_{eq,a} = 41.16$ ksi. For Case B, $S_{eq,a} = 30.7$ ksi. For both cases, $S_{eq,m} = 50$ ksi.

Goodman:

$$\sigma_{a,\text{eff}} = S_{\text{eq},a}\left(\frac{S_u}{S_u - S_{\text{eq},m}}\right)$$

SWT:

$$\sigma_{a,\text{eff}} = \sqrt{S_{\text{eq},a}(S_{\text{eq},a} + S_{\text{eq},m})}$$

Using the equivalent amplitude and mean stresses, the Goodman and SWT parameters were used to compute effective stress amplitudes and corresponding lives. Results are tabulated below.

	Case A		Case B	
	$\sigma_{a,\text{eff}}$ (ksi)	N (cycles)	$\sigma_{a,\text{eff}}$ (ksi)	N (cycles)
Material 1				
Goodman	53.26	31.4×10^6	39.76	211×10^6
SWT	61.25	12.7×10^6	49.80	48×10^6
Material 2				
Goodman	64.81	5.7×10^4	48.38	3.52×10^5
SWT	61.25	8.1×10^4	49.80	2.94×10^5

Several important observations can be made from Example 11. Estimated lives are significantly shorter for Case A relative to Case B. This is because the von Mises equivalent stress amplitude reflects the fact that principal shear stresses fluctuate more in Case A than they do in Case B. Also, whether or not the Goodman approach is more or less conservative than the SWT approach depends on the material and load history. For Case B, the SWT prediction is more conservative than Goodman for either material. But, for Case A, SWT is more conservative for the harder material (1), but less so for the softer material (2).

7.7.2 Nonproportional Loading

For nonproportional loading no consensus exists on the most suitable design approach. The ASME Boiler and Pressure Vessel Code[86] presents a generalized procedure, given in Table 7.5. Based on this procedure, an equivalent stress parameter, SEQA, has been derived[87] for combined, constant-amplitude, out-of-phase bending (or axial) and torsional stress:

$$\text{SALT} = \frac{\sigma_{x,a}}{\sqrt{2}} \sqrt{1 + C^2 + \sqrt{1 + 2C^2\cos(2\phi) + C^4}} \tag{7.81}$$

where $\sigma_{x,a}$ and $\tau_{xy,a}$ = elastically calculated notch bending (or axial) and torsional shear-stress amplitudes, respectively, $C = 2\tau_{xy,a}/\sigma_{x,a}$, and ϕ = the phase angle between bending and torsion. This parameter reduces to the Tresca (maximum shear) equivalent stress amplitude for in-phase loading. A similar relation can be defined based on the von Mises criterion:[87]

$$\text{SEQA} = \frac{\sigma}{\sqrt{2}} \sqrt{1 + \tfrac{3}{4}C^2 + \sqrt{1 + \tfrac{3}{2}C^2\cos(2\phi) + \tfrac{9}{16}C^4}} \tag{7.82}$$

For bending-torsion cases, these relations can simplify analysis. For a more random load history, the applications of the ASME approach is demonstrated in Example 12.

EXAMPLE 12 Use the procedure outlined in ASME Code Case NB-3216.2 to compute the Tresca-based S_{alt} for the bending-torsion operating cycle shown on page 7.69. Also,

TABLE 7.5 Excerpt from ASME Boiler and Pressure Vessel Code, Sec. III: Multiaxial Fatigue Evaluation

---

NB-3216.1 Constant Principal Stress Direction.

For any case in which the directions of the principal stresses at the point being considered do not change during the cycle, the steps stipulated in the following subparagraphs shall be taken to determine the alternating stress intensity.

(a) Principal Stresses—Consider the values of the three principal stresses at the point versus time for the complete stress cycle taking into account both the gross and local structural discontinuities and the thermal effects which vary during the cycle. These are designated as σ_1, σ_2 and σ_3 for later identification.

(b) Stress Differences—Determine the stress differences $S_{12} = \sigma_1 - \sigma_2$, $S_{23} = \sigma_2 - \sigma_3$ and $S_{31} = \sigma_3 - \sigma_1$ versus time for the complete cycle. In what follows, the symbol S_{ij} is used to represent any one of these stress differences.

(c) Alternating Stress Intensity—Determine the extremes of the range through which each stress difference (S_{ij}) fluctuates and find the absolute magnitude of this range for each S_{ij}. Call this magnitude $S_{r\,ij}$ and let $S_{alt\,ij} = 0.5\ S_{r\,ij}$. The alternating stress intensity S_{alt}, is the largest of the $S_{alt\,ij}$'s.

NB-3216.2 Varying Principal Stress Direction.

For any case in which the directions of the principal stresses at the point being considered do change during the stress cycle, it is necessary to use the more general procedure of the following subparagraphs.

(a) Consider the values of the six stress components, σ_t, σ_l, σ_r, τ_{tl}, τ_{lr}, τ_{rt} versus time for the complete stress cycle, taking into account both the gross and local structural discontinuities and the thermal effects which vary during the cycle. (*The subscripts t, l and r represent the tangential, longitudinal and radial directions, respectively.*)

(b) Choose a point in time when the conditions are one of the extremes for the cycle (either maximum or minimum, algebraically) and identify the stress components at this time by the subscript i. In most cases, it will be possible to choose at least one time during the cycle when the conditions are known to be extreme. In some cases it may be necessary to try different points in time to find the one which results in the largest value of alternating stress intensity.

(c) Subtract each of the six stress components σ_{ti}, σ_{li}, σ_{ri}, τ_{tli}, τ_{lri}, τ_{rti}, from the corresponding stress components σ_t, σ_l, σ_r, τ_{tl}, τ_{lr}, τ_{rt} at each point in time during the cycle and call the resulting components σ_t', σ_l', σ_r', τ_{tl}', τ_{lr}', τ_{rt}'

(d) At each point in time during the cycle, calculate the principal stresses, σ_1', σ_2' and σ_3', derived from the six stress components, σ_t', σ_l', σ_r', τ_{tl}', τ_{lr}', τ_{rt}'. Note that the directions of the principal stresses may change during the cycle but the principal stress retains its identity as it rotates.

(e) Determine the stress differences $S_{12}' = \sigma_1' - \sigma_2'$, $S_{23}' = \sigma_2' - \sigma_3'$ and $S_{31}' = \sigma_3' - \sigma_1'$ versus time for the complete cycle and find the largest absolute magnitude of any stress difference at any tie. The alternating stress intensity, S_{alt}, is one-half of this magnitude.

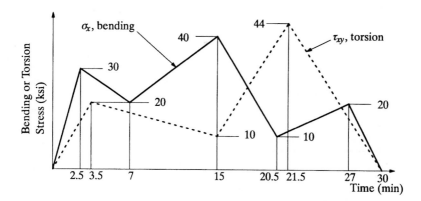

modify the final step (e) of the approach to compute an equivalent stress amplitude based on the von Mises criterion and compute that quantity for the same load history.

solution   The difficult step for implementing the procedure shown in Table 7.5 is step b, selecting the critical time in the load history. In this example, either $t = 0$ ($\sigma_{xi} = 0$ and $\tau_{xyi} = 0$) or $t = 21.5$ min ($\sigma_{xi} = 11.54$ ksi and $\tau_{xyi} = 44$ ksi) give identical results. The Tresca-based equivalent stress amplitude is given by

$$\text{SALT} = 44.38 \text{ ksi}$$

Step e in the procedure can be modified to define a von Mises equivalent stress amplitude:

$$\text{SEQA} = \frac{1}{\sqrt{2}} \sqrt{(\sigma_1{'} - \sigma_2{'})^2 + (\sigma_2{'} - \sigma_3{'})^2 + (\sigma_3{'} - \sigma_1{'})^2}$$

This results in

$$\text{SEQA} = 38.54 \text{ ksi}$$

Although the ASME approach demonstrated in Example 12 is straightforward in its implementation, it has the potential to make very nonconservative life estimates for long loading histories. In fact, predictions made by the approach have been shown to be nonconservative, even for relatively simple constant-amplitude, out-of-phase bending, and torsional loading.[87,88] Active research in multiaxial fatigue-life prediction is still underway to address practical design concerns such as notches[89] and variable-amplitude loading.[90,91]

REFERENCES

1. Shigley, J. E. and C. R. Mischke: "Mechanical Engineering Design," 5th ed., McGraw-Hill Book Company, New York, 1989.

2. Spotts, M. F.: "Design of Machine Elements," 6th ed., Prentice-Hall, Englewood Cliffs, N.J., 1985.

3. Juvinall, R. C.: "Fundamentals of Machine Component Design," John Wiley and Sons, New York, 1983.

4. Slocum, A. H.: "Precision Machine Design," Prentice-Hall, Englewood Cliffs, N.J., 1992.

5. Timoshenko, S. P., and J. N. Goodier: "Theory of Elasticity," 3d ed., McGraw-Hill Book Company, New York, 1970.

6. Boresi, A. P., and O. M. Sidebottom: "Advanced Mechanics of Materials," 4th ed., John Wiley and Sons, New York, 1985.

7. Landgraf, R. W., and R. A. Chernenkoff: "Residual Stress Effects on Fatigue of Surface Processed Steels," "Analytical and Experimental Methods for Residual Stress Effects in Fatigue," ASTM STP 1004, ASTM, Philadelphia, 1986.

8. Fung, Y. C.: "A First Course in Continuum Mechanics," 2d ed., Prentice-Hall, Englewood Cliffs, N.J., 1977.

9. Hertzberg, R. W.: "Deformation and Fracture Mechanics of Engineering Materials," 3d ed., John Wiley and Sons, New York, 1989.

10. Dieter, G. E.: "Mechanical Metallurgy," 3d ed., McGraw-Hill Book Company, New York, 1986.

11. Beer, F. P., and E. R. Johnson: "Mechanics of Materials," 2d ed., McGraw-Hill Book Company, New York, 1992.

12. Collins, J. A.: "Failure of Materials in Mechanical Design," John Wiley and Sons, New York, 1981.

13. Dowling, N. E.: "Mechanical Behavior of Materials," Prentice-Hall, Englewood Cliffs, N.J., 1993.

14. A. S. Cobayashi (ed.): "Manual on Experimental Stress Analysis," Prentice-Hall, Englewood Cliffs, N.J., 1987.

15. Dally, J. W., and W. F. Riley: "Experimental Stress Analysis," 3d ed., McGraw-Hill Book Company, New York, 1991.

16. Peterson, R. E.: "Stress Concentration Factors," John Wiley and Sons, New York, 1974.

17. Young, Y. C.: "Roark's Formulas for Stress and Strain," 6th ed., McGraw-Hill Book Company, New York, 1989.

18. Zienkiewicz, O. C.: "The Finite Element Method," 3d ed., McGraw-Hill Book Company, New York, 1987.

19. Cook, R. D., D. S. Malkus, and M. E. Plesha: "Concepts and Applications of Finite Element Analysis," 3d ed., John Wiley and Sons, New York, 1989.

20. Reddy, J. W.: "An Introduction to the Finite Element Method," McGraw-Hill Book Company, New York, 1984.

21. Bickford, W. B.: "A First Course in the Finite Element Analysis," Richard D. Irwin, Homewood, Ill., 1988.

22. Sorem, J. R., Jr., and S. M. Tipton: "The Use of Finite Element Codes for Cyclic Stress-Strain Analysis," "Fatigue Design," European Structural Integrity Society, vol. 16, J. Solin, G. Marquis, A. Siljander and S. Sipilä, eds., Mechanical Engineering Publications, London, 1993, pp. 187–200.

23. "Fatigue Design Handbook," Society of Automotive Engineers, Warrendale, Pa., 1990.

24. Tipton, S. M., and G. J. Shoup: "The Effect of Proof Loading on the Fatigue Behavior of Open Link Chain," *Trans. ASME, J. Fatigue and Fracture Eng. Mat. Technol.*, vol. 114, no. 1, pp. 27–33, January 1992.

25. Shoup, G. J., S. M. Tipton, and J. R. Sorem: "The Effect of Proof Loading on the Fatigue Behavior of Studded Chain," *Int. J. Fatigue,* vol. 14, no. 1, pp. 35–40, January 1992.

26. Tipton, S. M., and J. R. Sorem: "A Reversed Plasticity Criterion for Specifying Optimal Proof Load Levels," "Advances in Fatigue Lifetime Predictive Techniques," vol. 2, ASTM STP 1211, M. R. Mitchell and R. W. Landgraf, eds., American Society for Testing and Materials STP, Philadelphia, 1993, pp. 186–202.

27. "Standard Test Methods of Tension Testing of Metallic Materials," Specification E-8, American Society for Testing and Materials, Philadelphia.

28. Ramberg, W., and W. Osgood: "Description of Stress-Strain Curves by Three Parameters," NACA TN 902, 1943.

29. "Tensile Strain-Hardening Exponents (*n*-Values) of Metallic Sheet Materials," Specification E-646, American Society for Testing and Materials, Philadelphia.

30. "Standard Test Methods of Compression Testing of Metallic Materials at Room Temperature," Specification E-9, American Society for Testing and Materials, Philadelphia.

31. Fuchs, H. O., and R. I. Stephens: "Metal Fatigue in Engineering," John Wiley and Sons, New York, 1980.

32. Bannantine, J. A., J. J. Comer, and J. H. Handrock: "Fundamentals of Metal Fatigue Analysis," Prentice-Hall, Englewood Cliffs, N.J., 1990.

33. "Standard Practice for Conducting Constant Amplitude Axial Stress-Life Tests of Metallic Materials," Specification E-466, American Society for Testing and Materials, Philadelphia.

34. "Standard Practice for Statistical Analysis of Linear or Linearized Stress-Life (*S-N*) and Strain Life (ε-*N*) Fatigue Data," Specification E-739, American Society for Testing and Materials, Philadelphia.

35. "Standard Practice for Presentation of Constant Amplitude Fatigue Test Results for Metallic Materials," Specification E-468, American Society for Testing and Materials, Philadelphia.

36. Smith, K. N., P. Watson, and T. H. Topper: "A Stress-Strain Function for the Fatigue of Metals," *J. Materials,* vol. 5, no. 4, pp. 767–778, December 1970.

37. Sines, G.: "Failure of Metals under Combined Repeated Stresses with Superimposed Static Stresses," NACA Tech. Note 3495, 1955.

38. "MIL Handbook 5," Department of Defense, Washington, D.C.

39. Grover, H. J.: "Fatigue of Aircraft Structures," NAVAIR 01-1A-13, 1966.

40. "Standard Recommended Practice for Constant-Amplitude Low-Cycle Fatigue Testing," Specification E-606, American Society for Testing and Materials, Philadelphia.

41. Martin, J. F.: "Cyclic Stress-Strain Behavior and Fatigue Resistance of Two Structural Steels," Fracture Control Program Report No. 9, University of Illinois at Urbana-Champaign, 1973.

42. Massing, G.: "Eigenspannungen und Verfestigung beim Messing (Residual Stress and Strain Hardening of Brass)," *Proc. Second International Congress for Applied Mechanics,* Zurich, September 1926.

43. Basquin, O. H.: "The Exponential Law of Endurance Tests," *American Society for Testing and Materials Proceedings,* vol. 10, pp. 625–630, 1910.

44. Manson, S. S.: "Behavior of Materials under Conditions of Thermal Stress," *Heat Transfer Symposium,* University of Michigan Engineering Research Institute, pp. 9–75, 1953 (also published as NACA TN 2933, 1953).

45. Coffin, L. F.: "A Study of the Effects of Thermal Stresses on a Ductile Metal," *Trans. ASME,* vol. 76, pp. 931–950, 1954.

46. Mitchell, M. R.: "Fundamentals of Modern Fatigue Analysis for Design," "Fatigue and Microstructure," Society of Automotive Engineers, pp. 385–437, 1977.

47. Socie, D. F.: "Fatigue Life Prediction Using Local Stress-Strain Concepts," *Experimental Mechanics,* vol. 17, no. 2, 1977.

48. Socie, D. F.: "Fatigue Life Estimation Techniques," Electro General Corporation Technical Report 145A-579, Minnetonka, Minn.

49. Socie, D. F., N. E. Dowling, and P. Kurath: "Fatigue Life Estimation of a Notched Member," ASTM STP 833, *Fracture Mechanics; Fifteenth Symposium,* R. J. Sanford, ed., American Society for Testing and Materials, Philadelphia, 1984, pp. 274–289.

50. Sehitoglu, H.: "Fatigue of Low Carbon Steels as Influenced by Repeated Strain Aging," Fracture Control Program Report No. 40, University of Illinois at Urbana-Champaign, June 1981.

51. Kurath, P.: "Investigation into a Non-arbitrary Fatigue Crack Size Concept," Theoretical and Applied Mechanics, Report No. 429, University of Illinois at Urbana-Champaign, October 1978.

52. "Technical Report on Fatigue Properties—SAE J1099," SAE Informational Report, Society of Automotive Engineers, Warrendale, Pa., 1975.

53. Boller, C., and T. Seeger: "Materials Data for Cyclic Loading," Parts A to E, Elsevier Science Publishers B. V., 1987.

54. Boardman, B. E.: "Crack Initiation Fatigue—Data, Analysis, Trends, and Estimations," SAE Technical Paper 820682, 1982.

55. Manson, S. S.: "Fatigue: A Complex Subject—Some Simple Approximations," *Proc. Soc. Experimental Stress Analysis,* vol. 12, no. 2, 1965.

56. Muralidharan, U., and S. S. Manson: "A Modified Universal Slopes Equation for Estimation of Fatigue Characteristics of Metals," *Trans. ASME, J. Eng. Mat. Technol.,* vol. 110, pp. 55–58, January 1988.

57. Socie, D. F., M. R. Mitchell, and E. M. Caulfield: "Fundamentals of Modern Fatigue Analysis," Fracture Control Program Report No. 26, University of Illinois at Urbana-Champaign, 1977.

58. Ong, J. H.: "An Evaluation of Existing Methods for the Prediction of Axial Fatigue Life from Tensile Data," *Int. J. Fatigue,* vol. 15, no. 1, pp. 13–19, 1993.

59. Neuber, H.: "Theory of Stress Concentration for Shear Strained Prismatical Bodies with Arbitrary Stress-Strain Law," *Trans. ASME, J. Appl. Mech.,* vol. 12, pp. 544–550, 1961.

60. "Recommended Practices for Cycle Counting in Fatigue Analysis," Specification E-1049, American Society for Testing and Materials, Philadelphia.

61. Broek, D.: "Elementary Engineering Fracture Mechanics," 4th ed., Kluwer Academic Publishers, Dordrecht, The Netherlands, 1986.

62. Broek, D.: "The Practical Use of Fracture Mechanics," Kluwer Academic Publishers, Dordrecht, The Netherlands, 1988.

63. Barsom, J. M., and S. T. Rolfe: "Fracture and Fatigue Control in Structures," 2d ed., Prentice-Hall, Englewood Cliffs, N.J., 1987.

64. Anderson, T. L.: "Fracture Mechanics: Fundamentals and Applications," CRC Press, Boca Raton, Fla., 1991.

65. Ewalds, H. L., and R. J. H. Wanhill: "Fracture Mechanics," Edward Arnold Publishers, London, 1984.

66. Newman, J. C., Jr.: "A Review of Modelling Small-Crack Behavior and Fatigue Life Predictions for Aluminum Alloys," *Fatigue and Fracture Eng. Mat. Struct.,* vol. 17, no. 4, pp. 429–439, 1994.

67. Miller, K. J.: "Materials Science Perspective of Metal Fatigue Resistance," *Mat. Sci. Technol.,* vol. 9, no. 6, pp. 453–462, 1993.

68. Ahmad, H. Y., and J. R. Yates: "An Elastic-Plastic Model for Fatigue Crack Growth at Notches," *Fatigue and Fracture Eng. Mat. Struct.,* vol. 17, no. 4, pp. 439–447, 1994.

69. Y. Murakami (ed.): "Stress Intensity Factors Handbook," Pergamon Press, Oxford, 1987.

70. Rooke, D. P., and D. J. Cartwright: "Compendium of Stress Intensity Factors," Her Majesty's Stationery Office, London, 1976.

71. "Standard Test Method for Plane-Strain Fracture Toughness of Metallic Materials," Specification E-399, American Society for Testing and Materials, Philadelphia.

72. Tada, H., P. C. Paris, and G. R. Irwin: "The Stress Analysis of Cracks Handbook," 2d ed., Paris Productions, Inc., 226 Woodbourne Dr., St. Louis, Mo., 1985.

73. J. P. Gallagher (ed.): "Damage Tolerant Design Handbook," Metals and Ceramics Information Center, Battelle Columbus Laboratories, Columbus, Ohio, 1983.

74. "Aerospace Structural Metals Handbook," Metals and Ceramics Information Center, Battelle Columbus Laboratories, Columbus, Ohio, 1991.

75. Hudson, C. M., and S. K. Seward: "A Compendium of Sources of Fracture Toughness and Fatigue Crack Growth Rate Data for Metallic Alloys," *Int. J. Fracture,* vol. 14, pp. R151–R184, 1978.

76. Hudson, C. M., and S. K. Seward: "A Compendium of Sources of Fracture Toughness and Fatigue Crack Growth Data for Metallic Alloys—Part II," *Int. J. Fracture,* vol. 20, pp. R59–R117, 1982.

77. Hudson, C. M., and S. K. Seward: "A Compendium of Sources of Fracture Toughness and Fatigue Crack Growth Data for Metallic Alloys—Part III," *Int. J. Fracture,* vol. 39, pp. R43–R63, 1989.

78. Hudson, C. M., and J. J. Ferraino: "A Compendium of Sources of Fracture Toughness and Fatigue Crack Growth Data for Metallic Alloys—Part IV," *Int. J. Fracture,* vol. 48, no. 2, pp. R19–R43, March 1991.

79. Taylor, D.: "A Compendium of Fatigue Threshold and Growth Rates," EMAS, Great Britain, 1985.

80. "Data Book on Fatigue Growth Rates of Metallic Materials," JSMS, Japan, 1983.

81. Paris, P. C., and R. Erdogan: "A Critical Analysis of Crack Propagation Laws," *Trans. ASME, J. Basic Eng.,* vol. 85, p. 528, 1963.

82. "Standard Test Method for Measurement of Fatigue Crack Growth Rates," Specification E-647, American Society for Testing and Materials, Philadelphia.

83. Forman, R. G., V. E. Kearney, and R. M. Engle: "Numerical Analysis of Crack Propagation in Cyclic-Loaded Structures," *Trans. ASME, J. Basic Eng.,* ser. D., vol. 89, pp. 459–464, 1967.

84. Walker, K.: "The Effect of Stress Ratio during Crack Propagation and Fatigue for 2024-T3 and 7075-T6 Aluminum Alloy Specimens," NASA TN-D 5390, National Aeronautics and Space Administration, 1969.

85. Miller, K. J.: "Metal Fatigue—Past, Current and Future," *Proc. Institution of Mechanical Engineers,* vol. 205, 1991.

86. "ASME Boiler and Pressure Vessel Code," Sec. III, Div. I, Subsection NA, American Society of Mechanical Engineers, New York, 1974.

87. Tipton, S. M.: "Fatigue Behavior under Multiaxial Loading in the Presence of a Notch: Methodologies for the Prediction of Life to Crack Initiation and Life Spent in Crack Propagation," Ph.D. dissertation, Stanford University, Stanford, Calif., 1985.

88. Tipton, S. M., and D. V. Nelson: "Fatigue Life Predictions for a Notched Shaft in Combined Bending and Torsion," "Multiaxial Fatigue," ASTM STP 853, K. J. Miller and M. W. Brown, eds., American Society for Testing and Materials, Philadelphia, 1985, pp. 514–550.

89. Tipton, S. M., and D. V. Nelson: "Developments in Life Prediction of Notched Components Experiencing Multiaxial Fatigue," "Material Durability/Life Prediction Modelling," PVP-vol. 290, S. Y. Zamirk and G. R. Halford, eds., ASME, 1994, pp. 35–48.

90. Bannantine, J. A.: "A Variable Amplitude Multiaxial Fatigue Life Prediction Method," Ph.D. dissertation, The University of Illinois, Champaign, 1989.

91. Chu, C. C.: "Programming of a Multiaxial Stress-Strain Model for Fatigue Analysis," SAE Paper 920662, Society of Automotive Engineers, Warrendale, Pa., 1992.

92. "Failure Analysis and Prevention," "Metals Handbook," 8th ed., vol. 10, American Society for Metals, Metals Park, Ohio, 1978.

SECTION 8
RELIABILITY

Kailash C. Kapur, Ph.D., P.E.

Professor and Director of Industrial Engineering
University of Washington
Seattle, Wash.

Leonard R. Lamberson, Ph.D., P.E.

Professor and Dean
College of Engineering and Applied Sciences
Western Michigan University
Kalamazoo, Mich.

8.1 INTRODUCTION

"Reliability" is concerned with failure-free performance of a system and the duration of time over which this failure-free performance is maintained. Reliability is established at the design stage, and subsequent testing and production will not raise the reliability without basic design changes. Reliability is difficult to conceptualize as part of the usual design calculations; yet it is largely determined by these calculations.

The term often used to describe the overall capability of a system to perform its intended function is "system effectiveness." System effectiveness is defined as the probability that the system can successfully meet an operational demand within a given time when operating under specified conditions. For consumer products the term also reflects customer satisfaction or overall total quality and the degree to which it can be achieved. Effectiveness is influenced by the way the system is designed, manufactured, used, and maintained, and thus is a function of all life-cycle activities.

The effectiveness of a system is a function of several attributes, such as design ade-

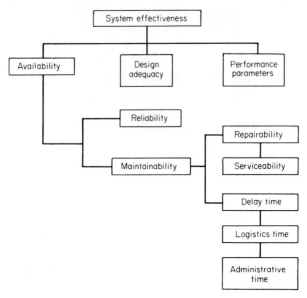

FIG. 8.1   Components of system effectiveness.

quacy, performance measures, safety, reliability, quality, and maintainability. Availability, reliability, and maintainability are time-oriented qualities of a system. Figure 8.1 gives the relationships among some of the components of system effectiveness.

Recently, the term "assurance sciences"[2,4] has been used to address the overall effectiveness of any system. The assurance sciences are (engineering) disciplines which help us with the attainment of product integrity (that is, the product does what it says it is supposed to do). The term "product assurance" is also used by some companies. Product assurance activities are performed throughout the life cycle of a product.

8.2   RELIABILITY MEASURES

Various components of system effectiveness are defined in this section.

8.2.1   Reliability

Reliability is defined as the probability that a given system will perform its intended function satisfactorily for a specified period of time under specified operating conditions. Thus, reliability is related to the probability of successful performance for any system. It is clear that we must define what successful performance for any system is or what we mean by the failure of the system; otherwise, it is not possible to predict when a system will fail to perform its intended function. The time to failure or "life" of a system cannot be deterministically defined and, hence, it is a random variable. Thus, we must quantify reliability by assigning a probability function to the time-to-failure random variable.

Let T be the time-to-failure random variable (rv). Then reliability at time t, $R(t)$, is the probability that the system will not fail by time t, or

$$R(t) = P(T > t) \tag{8.1}$$

$$= 1 - P(T \leq t)$$

$$= 1 - F(t)$$

$$= 1 - \int_0^t f(\tau)\, d\tau$$

where $f(t)$ and $F(t)$ are the probability density function (pdf) and cumulative distribution function (cdf), respectively, for the rv T. For example, if T is exponentially distributed, then

$$f(t) = \lambda e^{-\lambda t} \qquad t \geq 0$$

$$\lambda > 0$$

$$F(t) = \int_0^t \lambda e^{-\lambda \tau}\, d\tau = 1 - e^{-\lambda t} \qquad t \geq 0$$

$$R(t) = e^{-\lambda t} \qquad t \geq 0$$

Another measure that is frequently used as an indirect indicator of system reliability is called the "mean time to failure" (MTTF), which is the expected or mean value of the time-to-failure random variable. Thus, the MTTF is theoretically defined as

$$\text{MTTF} = E(T) = \int_0^\infty t f(t)\, dt = \int_0^\infty R(t)\, dt \tag{8.2}$$

Sometimes the term "mean time between failures" (MTBF) is also used to denote $E(T)$. The problem with only using the MTTF as an indicator of system reliability is that it only uniquely determines reliability if the underlying time-to-failure distribution is exponential. If the failure distribution is other than exponential, the MTTF does not give us the whole picture and can produce erroneous comparisons.

If we have a large population of the items whose reliability we are interested in studying, then for replacement and maintenance purposes we are interested in the rate at which the items in the population, which have survived at any point in time, will fail. This is called the "failure rate" or "hazard rate" and is given by the following relationship:

$$h(t) = f(t)/R(t) \tag{8.3}$$

The failure rate for most components follows the curve shown in Fig. 8.2, which is called the "life-characteristic curve" or "bathtub curve." Shown in Fig. 8.2 are three types of failures, namely quality failures, stress-related failures, and wear-out failures. The sum total of these failures gives the overall failure rate.

The failure-rate curve in Fig. 8.2 has three distinct periods. The initial decreasing failure rate is termed "infant mortality" and is due to the early failure of substandard products. Latent material defects, poor assembly methods, and poor quality control can contribute to an initially high failure rate. A short period of in-plant product testing, termed "burn-in," is used by manufacturers to eliminate these early failures from the consumer market. The flat, middle portion of the failure-rate curve represents the design failure rate for the specific product as used by the consumer market. During the useful-life portion, the failure rate is relatively constant. It might be decreased by redesign or restricting usage. Finally, as products age they reach a wear-out phase characterized by an increasing failure rate. See Refs. 9, 16, and 17 for further details.

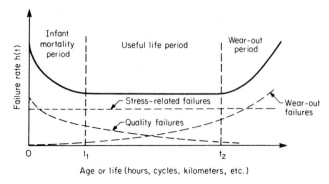

FIG. 8.2   Failure rate vs. life-characteristic curve.

To help the reader understand the notion of failure rate or hazard rate, basic mathematical relations are given below.

The "hazard rate" is defined as the limit of the instantaneous failure rate given no failure up to time t and is given by[8,11]

$$h(t) = \lim_{\Delta t \to 0} \frac{P(t < T \le t + \Delta t | T > t)}{\Delta t} = \lim_{\Delta t \to 0} \frac{R(t) - R(t + \Delta t)}{\Delta t \, R(t)} \qquad (8.4)$$

$$= \frac{1}{R(t)} \left[-\frac{d}{dt} R(t) \right] = \frac{f(t)}{R(t)}$$

Also,
$$f(t) = h(t) \exp\left[-\int_0^t h(\tau) \, d\tau \right] \qquad (8.5)$$

and thus
$$R(t) = \exp\left[-\int_0^t h(\tau) \, d\tau \right] \qquad (8.6)$$

The term "mission reliability" refers to the reliability of a product for a specified time period t^*, i.e.,

$$\text{Mission reliability} = R(t^*) = P(T > t^*) \qquad (8.7)$$

Another term that is used to characterize the reliability of a system is the "B life," which is defined as

$$F(B_\alpha) = \frac{\alpha}{100}$$

Thus B_{10} is the 10th percentile for the life of a system.

8.2.2   Maintainability

"Maintainability" is a system design parameter which has a great impact on system effectiveness. Failures will occur even in highly reliable systems, and the ability of a system to recover is often as important to system effectiveness as is its reliability. Maintainability is defined as the probability that a failed system will be restored to a satisfactory operating condition within a specified interval of downtime. The ease of fault detection, isolation, and repair are all influenced by system design and are the main factors contributing to maintainability. Also contributing are the supply of spare parts, the supporting repair organization, and the preventive maintenance practices. Good maintainability may somewhat offset low reliability. Hence, if the desired reliability cannot be met because of performance constraints, improvements in the system maintainability should be considered.

Let T be the time-to-repair random variable. Then the maintainability function $M(t)$ is given by

$$M(t) = P(T \leq t) \tag{8.8}$$

If the repair time T follows the exponential distribution with mean time to repair (MTTR) of $1/\mu$, where μ is the repair rate, then

$$M(t) = 1 - \exp(-t/\text{MTTR}) \tag{8.9}$$

In addition to the maintainability function, we also use median (50th percentile) and M_{max} (90th or 95th percentile) for the time-to-repair rv as measures of maintainability. The maintenance ratio (MR), which is defined as the total maintenance labor-hours per system operating hour, is also used as a measure of maintainability.

MTTR, which is the mean of the distribution of system repair time, can be evaluated by

$$\text{MTTR} = \frac{\sum_i^n = 1 \, \lambda_i t_i}{\sum_i^n = 1 \, \lambda_i} \tag{8.10}$$

where n = number of components in the system
   λ_i = failure rate of the ith repairable component
   t_i = time required to repair the system when the ith component fails

In addition, other quantities, such as mean active corrective maintenance time and mean active preventive maintenance time, are used to measure maintainability.

Serviceability affects maintainability, as shown in Fig. 8.1, and is very important from the customer viewpoint. Serviceability is a characteristic of design, just as is reliability, and must be considered at the design stage. Serviceability is defined as the degree of ease (or difficulty) with which a system can be repaired. This measure specifically considers fault detection, isolation, and repair. Repairability considers only the actual repair time and is defined as the probability that a failed system will be restored to operation within a specified interval of active repair time. Access covers, plug-in modules, or other features to allow easy removal and replacement of failed components improve the repairability and serviceability.

Maintainability and serviceability can be improved by consideration of some of the following factors during system design:

Accessibility: Provide sufficient working space around subsystems to allow ease of entry for maintenance personnel and tools in order to diagnose, troubleshoot, and complete maintenance activities.

Built-in test/diagnostics: Diagnostic devices indicating the status of equipment should be built-in to support maintainability processes. These devices should indicate the specific failure component for replacement.

Captive hardware: Attach/detach hardware should provide for quick and easy replacement of components, panels, brackets, and chassis. The hardware should be attached so that it is not lost during maintenance activities.

Color coding: Common color coding can greatly speed up maintenance activities. Color coding includes such things as lubrication information, orientation, timing marks, size, etc.

Modularity: Designs should be divided into physically and functionally distinct units for rapid removal and replacement to minimize on-line downtime.

Standardization: Design equipment with subsystems/components that are commercial standards, readily available, and common to the largest extent possible.

Standardized maintenance tools: Equipment should be maintainable with commonly available tools. This decreases tooling expense and training of maintenance personnel.

8.2.3 Availability

Availability is a function of both reliability and maintainability and answers the question, "Is a system available to perform its intended function when it is needed?" Availability measures are time-related, and some of the time elements are (1) storage, free, or off-time; (2) operating time; (3) standby time; and (4) downtime consisting of corrective and preventive maintenance as well as logistics and administrative delay time. "Inherent availability" A_I considers only operating time and corrective maintenance time and is given by

$$A_I = \text{MTBF}/(\text{MTBF} + \text{MTTR}) \tag{8.11}$$

where MTBF = mean time between failures
 MTTR = mean time to repair

Inherent availability is used to measure the design of a system. Achieved availability A_a is used for development and initial production of a system and is given by

$$A_a = \text{OT}/(\text{OT} + \text{TPM} + \text{TCM}) \tag{8.12}$$

where OT = operating time
 TPM = total preventive maintenance time
 TCM = total corrective maintenance time

Operational availability (A_O) covers all segments of time and is given by

$$A_O = (\text{OT} + \text{ST})/(\text{OT} + \text{ST} + \text{TPM} + \text{TCM} + \text{TALDT}) \tag{8.13}$$

where ST = standby time
 TALDT = total administrative and logistics delay time

8.2.4 Durability

Durability is a special case of reliability. It is defined as the probability that an item will successfully survive its projected life, overhaul point, or rebuild point (whichever is the more appropriate durability measure for the item) without a durability failure.

8.3 LIFE DISTRIBUTIONS IN RELIABILITY

In this section we summarize the pdf, cdf, reliability function, hazard-rate function, and MTBF for some of the well-known distributions that are used in reliability and maintainability.

Exponential Distribution

$$f(t) = \lambda e^{-\lambda t} \qquad t \geq 0 \qquad (8.14)$$

$$F(t) = 1 - e^{-\lambda t} \qquad t \geq 0 \qquad (8.15)$$

$$R(t) = e^{-\lambda t} \qquad t \geq 0 \qquad (8.16)$$

$$h(t) = \lambda \qquad (8.17)$$

$$\text{MTBF} = \theta = 1/\lambda \qquad (8.18)$$

Thus, the failure rate for the exponential distribution is always constant.

Normal Distribution

$$f(t) = (1/\sigma\sqrt{2\pi}) \exp\{-\tfrac{1}{2}[(t-\mu)/\sigma]^2\} \qquad -\infty < t < \infty \qquad (8.19)$$

$$F(t) = \Phi[(t-\mu)/\sigma] \qquad (8.20)$$

$$R(t) = 1 - \Phi[(t-\mu)/\sigma] \qquad (8.21)$$

$$h(t) = \frac{\phi[(t-\mu)/\sigma]}{\sigma R(t)} \qquad (8.22)$$

$$\text{MTBF} = \mu \qquad (8.23)$$

$\Phi(Z)$ is the cumulative distribution function and $\phi(Z)$ is the probability density function, respectively, for the standard normal variate Z. The failure rate for the normal distribution is a monotonically increasing function.

Log-Normal Distribution

$$f(t) = (1/\sigma t\sqrt{2\pi}) \exp[-\tfrac{1}{2}[(\ln t - \mu)/\sigma]^2] \qquad t \geq 0 \qquad (8.24)$$

$$F(t) = \Phi[(\ln t - \mu)/\sigma] \qquad (8.25)$$

$$R(t) = 1 - \Phi[(\ln t - \mu)/\sigma] \qquad (8.26)$$

$$h(t) = \frac{\phi[(\ln t - \mu)/\sigma]}{t\sigma R(t)} \qquad (8.27)$$

$$\text{MTBF} = \exp(\mu + \sigma^2/2) \qquad (8.28)$$

The failure rate for the log-normal distribution is neither always increasing nor always decreasing but is increasing on the average.[1] It takes different shapes depending on the parameters μ and σ.

Weibull Distribution

$$f(t) = [\beta(t-\delta)^{\beta-1}/(\theta-\delta)^\beta] \exp\{-[(t-\delta)/(\theta-\delta)]^\beta\} \qquad t \geq \delta \geq 0 \qquad (8.29)$$

$$F(t) = 1 - \exp\{-[(t-\delta)/(\theta-\delta)]^\beta\} \qquad (8.30)$$

$$R(t) = \exp\{-[(t-\delta)/(\theta-\delta)]^\beta\} \qquad (8.31)$$

$$h(t) = \beta(t-\delta)^{\beta-1}/(\theta-\delta)^\beta \qquad (8.32)$$

$$\text{MTBF} = \theta\Gamma(1 + 1/\beta) \tag{8.33}$$

The failure rate for the Weibull distribution is decreasing when $\beta < 1$, is constant when $\beta = 1$ (same as the exponential distribution), and is increasing when $\beta > 1$.

Gamma Distribution

$$f(t) = [\lambda^\eta/\Gamma(\eta)]t^{\eta-1}e^{-\lambda t} \qquad t \geq 0 \tag{8.34}$$

$$F(t) = \sum_{k=n}^{\infty} \frac{(\lambda t)^k \exp(-\lambda t)}{k!} \qquad \text{when } \eta \text{ is integer} \tag{8.35}$$

$$R(t) = \sum_{k=0}^{\eta-1} \frac{(\lambda t)^k \exp(-\lambda t)}{k!} \qquad \text{when } \eta \text{ is integer} \tag{8.36}$$

$$h(t) = f(t)/R(t) \qquad \text{[using Eqs. (8.34) and (8.36)]} \tag{8.37}$$

$$\text{MTBF} = \eta/\lambda \tag{8.38}$$

The failure rate for the gamma distribution is decreasing when $\eta < 1$, is constant when $\eta = 1$, and is increasing when $\eta > 1$.

8.4 RELIABILITY ACTIVITIES DURING THE PRODUCT DESIGN CYCLE

The reliability of a product is determined by the design, so the most economical and effective reliability activities are those that occur early in the design cycle. Such activities must be based largely on experience, since no hard data are available for a quantitative reliability assessment. Some of the procedures used to stimulate reliability improvements throughout the design cycle will now be covered.

8.4.1 Design Review

A design review is a formalized, documented, and systematic review of the design by senior company personnel. The review from a reliability, maintainability, and serviceability standpoint must be a technical review aimed at ferreting out weak points in the design and concerned with design improvements that will ensure early product maturity and improve reliability. Reference 7 gives more on this subject.

8.4.2 Failure Mode, Effects, and Criticality Analysis

For failure prevention during system design, two techniques are presented: failure mode, effects, and criticality analysis and fault-tree analysis. Failure mode, effects, and criticality analysis (FMECA) represents a "bottom-up" approach while fault-tree analysis (FTA) is a "top-down" approach. Both represent qualitative as well as quantitative approaches for assessing the consequences of failure and determining means for prevention or mitigation of the failure effect.

Failure mode and effects analysis (FMEA) is a procedure by which potential failure modes in a system are identified and analyzed. Since design is usually approached from an optimistic viewpoint, the FMEA procedure assumes a pessimistic viewpoint to identify potential product weaknesses. The terms FMEA; design FMA (DFMA); and failure mode, effects, and criticality analysis (FMECA) are also used. The main purpose of the technique is to identify and eliminate failure modes early in the design cycle where they are most economically dealt with. A documented FMECA procedure can be found in MIL-STD-1629A (1980).[14] The essential steps will be included here.

FMEA starts with the selection of a subsystem or component and then identifies and documents all potential failure modes. The effect of each failure mode is traced to other higher-level systems. A documented worksheet is used to record the following:

Function: A concise definition of the functions that the component must perform.

Failure mode: A particular way in which the component can fail to perform a function.

Failure mechanism: A physical or chemical process or hardware deficiency causing the failure mode.

Failure cause: The agent activating the failure mechanism; e.g., saltwater seepage owing to an inadequate seal might produce corrosion as a failure mechanism, though the failure cause was inadequate seal.

Identification of effects of higher-level systems: This determines whether the failure mode is localized, causes higher-level damage, or creates an unsafe condition.

Criticality rating: A measure of severity, probability of failure occurrence, and detectability used to assign priority design actions.

8.4.3 Fault-Tree Analysis

Fault-tree analysis is a technique used for systems safety and reliability analysis.[6] The analysis proceeds from a designated "top event" to basic failure causes called "primary events." A fault tree is a model that graphically portrays the combination of events leading up to the undesirable top event.

8.4.4 Reliability Design Guidelines

We will now discuss some basic principles of reliability in design that are useful for the designer. Each concept is briefly discussed in terms of its role in the design of reliable systems.

Simplicity. Simplification of system configuration contributes to reliability improvement by reducing the number of failure modes. A common approach is called component integration, which is the use of a single part to perform multiple functions.

Use of Proven Components and Preferred Designs. (1) If working within time and cost constraints, use proven components because this minimizes analysis and testing to verify reliability. (2) Mechanical and fluid system design concepts can be categorized and proven configurations given first preference.

Stress-Strength Design.[8] The designer should use various sources of data on strength of materials and strength degradation with time related to fatigue. The traditional and common uses of safety factors do not address reliability, and new techniques such as the probabilistic design approach should be used. The designer would use derating factors, proper reliability or safety margins, and develop stress-strength testing to determine stress and strength distributions. The probabilistic design approach is explained in the following section.

Redundancy. Redundancy sometimes may be the only cost-effective way to design a reliable system from less reliable components.

Local Environmental Control. A severe local environment sometimes prevents the achievement of required component reliability. In that case, the environment should be modified to achieve high reliability. Some typical environmental problems are (1) shock and vibration, (2) heat, and (3) corrosion.

Identification and Elimination of Critical Failure Modes. This is accomplished through FMECA and also by fault-tree analysis.

Self-Healing. A design approach which has possibilities for future development is the use of self-healing devices. Automatic sensing and switching devices represent a form of self-healing.

Detection of Impending Failures. Achieved reliability in the field can be improved by the introduction of methods and/or devices for detecting impending failures. Some of the examples are (1) screening of parts and components, (2) periodic maintenance schedules, and (3) monitoring of operations.

Preventive Maintenance. Preventive-maintenance procedures can enhance the achieved reliability, but the procedures are sometimes difficult to implement. Hence, effective preventive-maintenance procedures must be considered at the design stage.

Tolerance Evaluation. In a complex system, it is necessary to consider the expected range of manufacturing process tolerances, operational environmental, and all stresses, as well as the effect of time. Some of the tolerance evaluation methods are worst-case tolerance analysis, statistical tolerance analysis, and marginal checking.

Human Engineering. Human activities and limitations may be very important to system reliability. The design engineer must consider factors which directly refer to human aspects, such as (1) human factors, (2) human-machine interface, (3) evaluation of the person in the system, and (4) human reliability.

8.4.5 Probabilistic Approach to Design

Reliability is basically a design parameter and must be incorporated in the system at the design stage. One way to quantify reliability during design and to design for reliability is the probabilistic approach.[10] The design variables and parameters are random variables and, hence, the design methodology must consider them as random variables. The reliability of any system is a function of the reliabilities of its components. In order to analyze the reliability of the system, we have to first understand how to compute the reliabilities of the components. The basic idea in reliability analysis from the probabilistic design methodology viewpoint is that a given component has certain strengths which, if exceeded, will result in the failure of the component. The factors which determine the strengths of the component are random variables as well as the factors which determine the stresses or loading acting on the component. "Stress" is used to indicate any agency that tends to induce failures, while "strength" indicates any agency resisting failure. "Failure" itself is taken to mean failure to function as intended; it is said to have occurred when the actual stress exceeds the actual strength for the first time.

Let $f(x)$ and $g(y)$ be the pdf's for the stress random variable X and the strength random variable Y, respectively, for a certain mode of failure. Also, let $F(x)$ and $G(y)$ be the cumulative distribution functions for the random variables X and Y, respectively. Then the reliability R of the component for the failure mode under consideration, with the assumption that the stress and the strength are independent random variables, is given by

$$R = P(Y > X) \tag{8.39}$$

$$= \int_{-\infty}^{\infty} g(y)\left[\int_{-\infty}^{y} f(x)\,dx\right]dy \tag{8.40}$$

$$= \int_{-\infty}^{\infty} g(y)F(y)dy \tag{8.41}$$

$$= \int_{-\infty}^{\infty} f(x)\left[\int_{x}^{\infty} g(y)\,dy\right]dx \tag{8.42}$$

$$= \int_{-\infty}^{\infty} f(x)[1 - G(x)]dx \tag{8.43}$$

For example, suppose the stress rv X is normally distributed with a mean value of μ_X and standard deviation of σ_X, and the strength random variable is also normally distributed with parameters μ_Y and σ_Y. Then the reliability R is given by

$$R = \Phi[(\mu_Y - \mu_X)/\sqrt{\sigma_Y^2 + \sigma_X^2}] \tag{8.44}$$

where $\Phi(\ )$ is the cumulative distribution function for the standard normal variable.

EXAMPLE

$$\mu_Y = 40{,}000 \qquad \mu_X = 30{,}000 \qquad \sigma_Y = 4000 \qquad \sigma_X = 3000$$

Then, factor of safety $= 40{,}000/30{,}000 = 1.33$ and

$$R = \Phi[(40{,}000 - 30{,}000)/\sqrt{(4000)^2 + (3000)^2}] = \Phi(2) = 0.97725$$

If we change μ_X to 20,000 by increasing the factor of safety to 2, we have

$$R = \Phi[(40{,}000 - 20{,}000)/\sqrt{(4000)^2 + (3000)^2}] = \Phi(4) = 0.99997$$

There are four basic ways in which the designer can increase reliability:

1. Increase mean strength: this can be achieved by increasing size, weight, using stronger material, etc.
2. Decrease average stress: this can be done by controlling loads or using higher dimensions.
3. Decrease stress variations: this variation is harder to control but can be effectively truncated by putting limitations on use conditions.
4. Decrease strength variation: the inherent part-to-part variation can be reduced by improving the basic process, controlling the process, and utilizing tests to eliminate the less desirable parts.

8.5   SYSTEM RELIABILITY ANALYSIS

A system reliability model is used as the basis for reliability analysis and apportion-ment. The analysis is usually based on a block diagram that represents system success according to the definition of system reliability. This section will consider the reliabil-ity block diagram and its analysis using both the static and dynamic approaches. Reliability allocation will also be considered.

8.5.1   Reliability Block Diagram

The reliability block diagram (RBD) is constructed from an engineering analysis of the system considering modes of failure and understanding the grouping of similar parts for analysis purposes.

8.5.2   Static Reliability Models

System reliability with static models uses the RBD with a component reliability R_i assigned to each block. The system reliability R_s is then calculated. In the following formulations, n is the number of components.

Series Configuration.   In a series configuration, all subsystems must survive for the system to operate successfully. For the series configuration as shown in Fig. 8.3a, the system reliability is

$$R_s = \prod_{i=1}^{n} R_i \tag{8.45}$$

This is sometimes called the product rule for system reliability.

Parallel Configuration.   In a parallel configuration (Fig. 8.3b) all subsystems must fail before the system is considered to be in a failed state. The system reliability is cal-culated by

$$R_s = 1 - \prod_{i=1}^{n} (1 - R_i) \tag{8.46}$$

A system model may consist of both series and parallel subsystems. The sys-tem reliability can be calculated by re-peated applications of the above series and parallel equations.

8.5.3   Dynamic Reliability Models

Dynamic reliability models are an exten-sion of the static models where time-dependent reliability functions are used for each subsystem. The following nota-tion is defined:

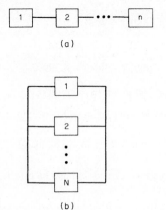

FIG. 8.3   System reliability block diagrams. (a) Series system. (b) Parallel system.

$R_i(t)$ = the reliability function for the ith subsystem
$R_s(t)$ = the system reliability function

These will be used to reformulate series and parallel systems.

Series Configuration.   The system reliability is given by

$$R_s(t) = \prod_{i=1}^{n} R_i(t) \qquad (8.47)$$

where n = the number of subsystems. The failure rate $h_s(t)$ for the series system is given by

$$h_s(t) = \sum_{i=1}^{n} h_i(t) \qquad (8.48)$$

where $h_i(t)$ is the failure rate for the ith subsystem.
  If all subsystems have an exponentially distributed time to failure, then

$$h_s(t) = \sum_{i=1}^{n} \lambda_i \qquad (8.49)$$

where λ_i is the failure rate for the ith subsystem.

Parallel Configuration.   The system reliability for a parallel configuration where all parallel subsystems are activated when the system is turned on is given by

$$R_s(t) = 1 - \prod_{i=1}^{n} [1 - R_i(t)] \qquad (8.50)$$

A standby parallel redundant system is depicted in Fig. 8.4. In this system the standby unit is activated by the switching mechanism S. For a system with two units (i.e., one standby unit), the system reliability is

$$R_s^{\,2}(t) = R_1(t) + \int_0^t f_1(t_1)R_2(t - t_1)dt_1 \qquad (8.51)$$

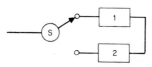

where it is assumed that the switch cannot fail. Here $f_1(t)$ is the pdf for device 1 and $R_2(t)$ is the reliability function for device 2.

FIG. 8.4   Standby redundant configuration.

For the special case where each subsystem is identical with an exponential time to failure, the system reliability is

$$R_s^{\,2}(t) = e^{-\lambda t}(1 + \lambda t) \qquad t \geq 0 \qquad (8.52)$$

If the switch has a reliability of P_s, then the dynamic model for the two-unit system is

$$R_s^{\,2}(t)' = R_1(t) + P_s \int_0^t f_1(t_1)R_2(t - t_1)dt_1 \qquad (8.53)$$

For identical, exponential time-to-failure subsystems

$$R_s^2(t)' = e^{-\lambda t}[1 + P_s \lambda t] \qquad t \geq 0 \qquad (8.54)$$

Many other dynamic model situations are possible. (See Ref. 8.)

8.6 RELIABILITY TESTING

Reliability can be assessed through product testing. The reliability numbers that one obtains from testing are dependent on test conditions, so obviously if one wants to assess product performance in the field by testing, then one must take great care to duplicate field conditions and usage.

It must be remembered that reliability numbers are always relative to the practical implementation of the definition of reliability with respect to what constitutes a failure and the specified environmental and usage conditions. Products that deteriorate over time rather than fail catastrophically are more difficult to quantify. What is usually done is that some degree of loss of function (or degradation) is set as failure.

The data resulting from testing can be used to (1) estimate the reliability (point estimation), (2) set a confidence limit on reliability, and (3) test the hypothesis that a reliability goal has been met. The following will cover the various approaches using the exponential and Weibull distributions as time-to-failure models. The binomial distribution will also be covered for success-failure testing. This coverage should be sufficient for most situations that one might encounter in practice.

8.6.1 Exponential Distribution

The exponential distribution is a very popular and easy-to-use model to represent time to failure. Selection of the exponential distribution as an appropriate model implies that the failure rate is constant over the range of predictions. For certain failure situations and over certain portions of product life, the assumption of a constant failure rate may be appropriate.

Statistical Properties. The pdf for an exponentially distributed time-to-failure random variable T is given by

$$f(t,\lambda) = e^{-\lambda t} \qquad t \geq 0 \qquad (8.55)$$

where the parameter λ is the failure rate. The reciprocal of the failure rate ($\theta = 1/\lambda$) is the mean or expected life. For products that are repairable, the parameter θ is referred to as the mean time between failures (MTBF) and for nonrepairable products θ is called the mean time to failure (MTTF). The parameter λ must be known (or estimated) for any specific application situation.

The reliability function is given by

$$R(t) = e^{-\lambda t} \qquad t \geq 0 \qquad (8.56)$$

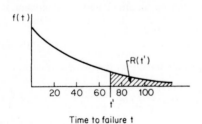

Time to failure t

FIG. 8.5   The exponential distribution.

The relationship between $f(t)$ and $R(t)$ is illustrated in Fig. 8.5. For any value of t the quantity $R(t)$ provides the chance of survival beyond time t. If the reliability function is evaluated at the MTBF, it should be noted that $R(\theta) = 0.368$, or there is only a 36.8 percent chance of surviving the mean life.

Point Estimation. The estimator for the mean life parameter θ is given by

$$\hat{\theta} = T/r \qquad (8.57)$$

where T = the total accumulated test time considering both failed and unfailed items and r = the total number of failures. Then the estimator for λ is $\hat{\lambda} = 1/\hat{\theta}$.

The reliability is estimated by

$$\hat{R}(t) = e^{-\hat{\lambda}t} \qquad t \geq 0 \qquad (8.58)$$

and if one wants to estimate a time for a given reliability level R, this is obtained by

$$\hat{t} = \hat{\theta} \ln{(1/R)} \qquad (8.59)$$

Confidence-Interval Estimates. The confidence intervals for the mean life or for reliability depend on the testing situation. A life-testing situation where n devices are placed on test and it is agreed to terminate the test at the time of the rth failure ($r < n$) is called failure censored and is termed type II life testing. A time-censored life test is one where the total accumulated time (T) is specified. This is termed type I life testing. In type I testing T is specified and r occurs as a result of testing, whereas in a type II situation r is specified and T results from testing. The confidence limits are slightly different for each situation.

Failure-Censored Life Tests. In this situation the number of failures r at which the test will be terminated is specified with n items placed on test. The $100(1 - \alpha)$ percent two-sided confidence interval for θ on the mean life is given by

$$2T/\chi^2_{\alpha/2,2r} \leq \theta \leq 2T/\chi^2_{1-\alpha/2,2r} \qquad (8.60)$$

The quantities $\chi^2_{\alpha,v}$ are the $1 - \alpha$ percentiles of a chi-square distribution with v degrees of freedom.

The one-sided lower $100(1 - \alpha)$ percent confidence limit is

$$2T/\chi^2_{\alpha,2r} \leq \theta \qquad (8.61)$$

The confidence limits on reliability for any specified time can be found by substituting the above limits on θ into the reliability function.

Time-Censored Life Tests. In this situation the accumulated test time T is specified for the test and the test produces r failures. The $100(1 - \alpha)$ percent two-sided confidence interval on the mean life is given by

$$2T/\chi^2_{\alpha/2,2(r+1)} \leq \theta \leq 2T/\chi^2_{1-\alpha/2,2r} \qquad (8.62)$$

If only the one-sided $100(1 - \alpha)$ percent lower confidence interval is desired, it is given by

$$2T/\chi^2_{\alpha,2(r+1)} \leq \theta \qquad (8.63)$$

Hypothesis Testing. Hypothesis testing is another approach to statistical decision making. Whereas confidence limits offer some degree of protection against meager statistical knowledge, the hypothesis-testing approach is easily misused. To use this approach one should be very familiar with the inherent errors involved in applying the hypothesis-testing procedure.

Consider a type II testing situation where n items are placed on test and the test is terminated at the time of the rth failure. An MTBF goal θ_g is to be verified. The hypotheses for this situation are

$$H_0: \theta \leq \theta_g$$
$$H_1: \theta > \theta_g$$

(8.64)

If H_0 is rejected, the conclusion is that the goal has been met.

The statistic calculated from test data is

$$\chi_c^2 = 2T/\theta_g$$

(8.65)

and the decision criteria are to reject H_0 and assume that the goal has been met if $\chi_c^2 > \chi_{\alpha,2r}^2$. Here the level of significance is taken as α.

The probability of accepting a design, that is, concluding that the goal has been met where the design has a true MTBF of θ_1, is

$$P_1 = P[\chi_{2r}^2 \geq (\theta_g/\theta_1)\chi_{\alpha,2r}^2]$$

(8.66)

and can be looked up in χ^2 tables.

8.6.2 Weibull Distribution

The Weibull distribution is considerably more versatile than the exponential distribution and can be expected to fit many different failure patterns. However, when applying a distribution the failure pattern should be carefully studied and the mixture of failure modes noted. The selection and application of a distribution should then be based on this study and on any knowledge of the underlying physical failure phenomena.

Graphical procedures for the Weibull distribution are attractive in that they provide practitioners with a visual representation of the situation. Although the graphical approach will be covered in the following, it should be recognized that there are better statistical estimation procedures; however, these procedures require the use of a computer, so the availability of computer programs will also be covered.

Statistical Properties. The reliability function for the three-parameter Weibull distribution is given by

$$R(t) = \exp\{-[(t - \delta)/(\theta - \delta)]^\beta\} \qquad t \geq \delta$$

(8.67)

where δ = the minimum life ($\delta \geq 0$)
θ = the characteristic life ($\theta > \delta$)
β = the Weibull slope or shape parameter ($\beta > 0$)

The two-parameter Weibull distribution has a minimum life of zero, and the reliability function is

$$R(t) = \exp[-(t/\theta)^\beta] \qquad t \geq 0$$

(8.68)

where θ and β are as previously defined ($\theta > 0$). The term characteristic life resulted from the fact that $R(\theta) = 0.368$; or, there is a 36.8 percent chance of surviving the characteristic life for any Weibull distribution.

The hazard function for the Weibull distribution is given by

$$h(t) = (\beta/\theta^\beta)t^{\beta-1} \qquad t \geq 0 \tag{8.69}$$

It can be seen that the hazard function will decrease for $\beta < 1$, increase for $\beta > 1$, and remain constant for $\beta = 1$.

The expected or mean life for the two-parameter Weibull distribution is given by

$$\mu = \theta\Gamma(1 + 1/\beta) \tag{8.70}$$

where $\Gamma(\ )$ is a gamma function as found in gamma tables. The standard deviation for the Weibull distribution is

$$\sigma = \theta\sqrt{\Gamma(1 + 2/\beta) - \Gamma^2(1 + 1/\beta)} \tag{8.71}$$

Graphical Estimation. The Weibull distribution is very amenable to graphical estimation. This procedure will now be illustrated.

The cumulative distribution for the two-parameter Weibull distribution is given by

$$F(t) = 1 - \exp[-(t/\theta)^\beta] \tag{8.72}$$

then by rearranging and taking logarithms one can obtain

$$\ln(\ln\{1/[1 - F(t)]\}) = \beta \ln t - \beta \ln \theta \tag{8.73}$$

Weibull paper is scaled such that t_j and $P_j = F(t_j)$ can be plotted directly and a straight line fitted to the data. A convenient way to assign values of P_j for plotting is to calculate

$$P_j = (j - 0.3)/(n + 0.4) \tag{8.74}$$

where j is the order of magnitude of the observation and n is the sample size.[8]

EXAMPLE The design for an aluminum flexible drive hub on computer disk packs is under study. The failure mode of interest is fatigue. Data from 12 hubs placed on an accelerated life test follow:

j	Cycles to failure	P_j
1	93,000	5.6
2	147,000	14
3	192,000	22
4	214,000	30
5	260,000	38
6	278,000	46
7	297,000	54
8	319,000	62
9	349,000	70
10	388,000	78
11	460,000	86
12	510,000	94.4

The p_j values were calculated using Eq. (8.74).

The plotted data with a visually fitted line are shown on Weibull paper in Fig. 8.6. The slope of this line provides an estimate of β, which in this case is about 2.3. Most commercially available Weibull papers will have a special scale for estimating β.

The characteristic life can be estimated by recalling that $R(\theta) = 0.368$; or, then

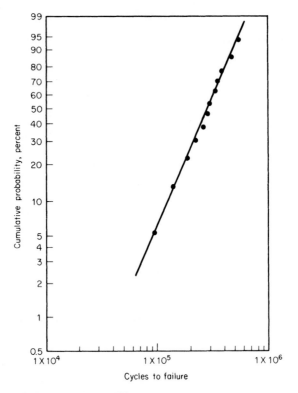

FIG. 8.6 Weibull probability paper.

$F(\theta) = 0.632$. So, one can locate 63.2 percent on the cumulative probability scale, project across to the plotted line, and then project down to the time-to-failure axis. In Fig. 8.6, the characteristic life is about 330,000 cycles.

Weibull paper offers a quick and convenient method for analyzing a failure situation. The population line plotted on the paper can be used to estimate either percent failure at a given time or the time at which a given percentage will fail. Also, a concave plot is indicative of a nonzero minimum life.

8.6.3 Success-Failure Testing

Success-failure testing describes a situation where a product (component, subsystem, etc.) is subjected to a test for a specified length of time T (or cycles, stress reversals, miles, etc.). The product either survives to time T (i.e., survives the test) or fails prior to time T.

Testing of this type can frequently be found in engineering laboratories where a test "bogy" has been established and new designs are tested against this bogy. The bogy will specify a set number of cycles in a certain test environment and at predetermined stress levels.

The probability model for this testing situation is the binomial distribution given by

$$p(y) = \binom{n}{y} R^y (1 - R)^{n-y} \qquad y = 0, 1, 2,..., n \qquad (8.75)$$

where R = the probability of surviving the test

n = the number of items placed on test

y = the number of survivors

The value R is the reliability which is the probability of surviving the test. Also,

$$\binom{n}{y} = \frac{n!}{y!(n-y)!} \tag{8.76}$$

Procedures for estimating product reliability R based on this testing situation will now be covered.

Point Estimate. The point estimate of reliability is simply calculated as

$$\hat{R} = y/n \tag{8.77}$$

Confidence Limit Estimate. The $100(1 - \alpha)$ percent lower confidence limit on the reliability R is calculated by

$$R_L = \frac{y}{y + (n - y + 1)F_{\alpha,2(n-y+1),2y}} \tag{8.78}$$

where $F_{\alpha,2(n-y+1),2y}$ is obtained from F tables. Here again n = the number of items placed on test and y = the number of survivors.

Success Testing. In receiving inspection and sometimes in engineering test labs, one encounters a situation where a no-failure ($r = 0$) test is specified. The concern is usually on ensuring that a reliability level has been achieved at a specified confidence level. A special adaptation of the confidence-limit formula can be derived for this situation.

For the special case where $r = 0$ (i.e., no failures), the lower $100(1 - \alpha)$ percent confidence limit on the reliability is

$$R_L = \alpha^{1/n} \tag{8.79}$$

where α = the level of significance and n = the sample size (i.e., number placed on test). Then with $100(1 - \alpha)$ percent confidence, we can say that

$$R_L \leq R$$

where R is the true reliability.

If we let $C = 1 - \alpha$ be the desired confidence level (i.e., 0.80, 0.90, etc.), then the necessary sample size to demonstrate a desired reliability level R is

$$n = \ln(1 - C)/\ln R \tag{8.80}$$

For example, if $R = 0.80$ is to be demonstrated with 90 percent confidence, we have

$$n = \ln(0.10)/\ln(0.80) = 11$$

Thus, we place 11 items on test and allow no failures. This is frequently referred to as success testing.

8.7 SOURCES OF FAILURE DATA

Relatively few sources for failure-rate data are available. There are some data on electronic components, particularly in military applications. However, on mechanical components practically no good data are commercially available. The use of any existing data bank to obtain failure-rate data on any particular design should be done with great caution and skepticism. The applicability of any past history of failure rate to a current design depends on the degrees of similarity in the design, the environment, and the definition of failure. With these words of caution three sources of failure-rate information are covered in the following sections.

8.7.1 MIL-HDBK 217E[12]

This handbook is concerned with reliability prediction for electronic systems. The handbook contains two methods of reliability prediction: the part-stress-analysis and the parts-count methods. These methods vary in complexity and in the degree of information needed to apply them. The part-stress-analysis method requires the greatest amount of information and is applicable during the later design stages when actual hardware and circuits are being designed. To apply this method, a detailed parts list including part stresses must be available.

The parts-count method requires less information and is relatively easy to apply. The information needed to apply this method is (1) generic part type (resistor, capacitor, etc.) and quantity, (2) part quality levels, and (3) equipment environment. Tables are provided in the handbook to determine the factors in a failure-rate model that can be used to predict the overall failure rate. The parts-count method is obviously easier to apply and, one would assume, less accurate than the stress-analysis method.

8.7.2 Nonelectronic Parts Reliability Data[15]

The *Nonelectronic Parts Reliability Data Handbook* (1981) was prepared under the supervision of the Reliability Analysis Center at Griffiss AFB. It is intended to complement MIL-HDBK-217C in that it has information on some mechanical components. Specifically, the handbook provides failure-rate and failure-mode information for mechanical, electrical, pneumatic, hydraulic, and rotating parts. Again, little is known about the environment, design specifics, etc., that produced the failure-rate data in the handbook.

8.7.3 The Government/Industry Data Exchange Program[13]

The Government/Industry Data Exchange Program (GIDEP, 1974) is a cooperative venture between government and industry participants that provides a means to exchange certain types of technical data. Participants in GIDEP are provided with access to various data banks. The Failure Experience Data Bank contains failure information generated when significant problems are identified on systems in the field. These data are reported to a central data bank by the participants.

The Reliability-Maintainability Data Bank contains failure-rate and failure-mode data on parts, components, and systems based on field operations. The data are reported by the participants. Here again the problem is one of identifying the exact operating conditions that caused failure.

8.8 ELEMENTS OF A RELIABILITY PROGRAM

Management and control of system reliability must be based on a recognition of the system's life cycle beginning at concept, extending through design and production, continuing with the usage of the system, and ending with the removal of the system from the inventory. We wish to achieve a high level of operational reliability in the field, and in order to achieve this objective, various reliability tasks have to be completed throughout the life cycle of the system, starting with the concept stage. Some of the tasks are given below:

1. *Tasks during design and development:*
   a. Initiate reliability activities during conceptual phase, such as reliability and cost trade-off studies, reliability models, and programs.
   b. Develop safety margins from reliability viewpoint.
   c. Predict component reliability from the data bank on the failure rates.
   d. Prepare parts program.
   e. Compute system reliability from component reliability.
   f. Perform worst-case and parts-tolerance analyses.
   g. Determine amount of redundancy needed to achieve a reliability goal.
   h. Determine reliability allocation.
   i. Provide input to human engineering.
   j. Interact with value engineering.
   k. Evaluate design changes.
   l. Perform trade-off analysis.
   m. Compare two or more designs.
   n. Prepare design specifications.
   o. Prepare guidelines for design review.
   p. Perform failure-mode analysis.
   q. Work with cost-reduction programs.

2. *Tasks during development and testing:*
   a. Establish reliability growth curves for the development-and-testing phase.
   b. Develop guidelines for the amount of testing.
   c. Develop bathtub curve based on the failure-rate data and the test data.
   d. Participate in the development of failure definition/scoring criteria document.
   e. Participate in scoring of the test failure.

3. *Tasks during manufacturing:*
   a. Establish guidelines for manufacturing processes.
   b. Provide input to quality control.
   c. Develop input to guidelines to evaluate suppliers and vendors.
   d. Develop product burn-in or debugging time.

4. *Tasks during field usage and maintenance:*
   a. Establish warranty cost and help reduce it.
   b. Optimize the length of warranty.
   c. Reduce inventory costs.
   d. Develop maintenance procedures, both corrective and preventive.
   e. Provide input to the spare parts allocation models.
   f. Participate in the collection and analysis of the field data.
   g. Participate in the feedback process to report and correct the field failures.

8.8.1 Reliability Data-Collection Systems

The ultimate testing of a product occurs in the field under varied customer usages and environments. In order to effectively and knowledgeably guide future design efforts, field information must be accurately collected, summarized, and disseminated to the design function. This information can guide future design approaches, material selection, component or vendor selection, and other activities in the evolution of future products.

Rather than attempt to obtain data from an inaccurate warranty cost-reporting system, it is better to effectively and accurately track a small, representative sample of products. Usually some form of stratified random sampling[3] is needed to include different environmental extremes and different customer usage modes. Each failure should be analyzed and categorized by a qualified technician. For consistency in reporting, a failure-modes dictionary should be developed. This dictionary should have descriptions of the various modes of failure to be used in the reporting procedures. The purpose of the dictionary is to promote consistent reporting among the technicians who analyze the failures.

A computerized data-base-management system (DBMS) must be set up to manage the data. Routine report-generating formats with easy to understand descriptors of reliability should be preestablished. Ultimately, a design engineer should be able to query the DBMS from a terminal to review past product performance.

REFERENCES

1. Barlow, R. E., and F. Proschan: "Statistical Theory of Reliability and Life Testing," Holt, Rinehart and Winston, New York, 1975.

2. Carrubba, E. R., R. D. Gordon, and A. C. Spann: "Assuring Product Integrity," Lexington Books, Lexington, Mass., 1975.

3. Cocharn, W. G.: "Sampling Techniques," 2d ed., John Wiley & Sons, Inc., New York, 1963.

4. Halpern, S.: "The Assurance Sciences," Prentice-Hall, Inc., Englewood Cliffs, N.J., 1978.

5. Haugen, E. G.: "Probabilistic Approach to Design," John Wiley & Sons, Inc., New York, 1968.

6. Henley, E. J., and H. Kumamoto: "Reliability Engineering and Risk Assessment," Prentice-Hall, Inc., Englewood Cliffs, N.J., 1981.

7. Juran, J. M., and F. M. Gryna, Jr.: "Quality Planning and Analysis," 2d ed., McGraw-Hill Book Company, Inc., New York, 1980.

8. Kapur, K. C., and L. R. Lamberson: "Reliability in Engineering Design," John Wiley & Sons, Inc., New York, 1977.

9. Kapur, K. C.: "Reliability and Maintainability," in "Industrial Engineering Handbook," G. Salvendi, ed., John Wiley & Sons, Inc., New York, 1982.

10. Kececioglu, D., and D. Cormier: "Designing a Specified Reliability Directly into a Component," *Proc. 3rd Annu. Aerospace Reliability and Maintainability Conference*, pp. 520–530, 1968.

11. Mann, N. R., R. E. Schafer, and N. D. Singpurwalla: "Methods for Statistical Analysis of Reliability and Life Data," John Wiley & Sons, Inc., New York, 1974.

12. MIL-HDBK-217E: U.S. Department of Defense, "Military Standardization Handbook," Reliability Prediction of Electronic Equipment, April 9, 1979.

13. MIL-STD-1556A: U.S. Department of Defense, Government-Industry Data Exchange Program (GIDEP), U.S. Air Force, February 29, 1976.

14. MIL-STD-1629A: U.S. Department of Defense, Military Standard, "Procedures for Performing a Failure Mode, Effects and Criticality Analysis," November 24, 1980.

15. NPRD-Z: "Nonelectronic Parts Reliability Data," Reliability Analysis Center, Rome Air Development Center, Griffiss Air Force Base, N.Y., 1981.

16. "Quality Assurance-Reliability Handbook," AMC Pamphlet No. 702-3, U.S. Army Materiel Command, Va., October 1968.

17. RDG-376: "Reliability Design Handbook," Reliability Analysis Center, Rome Air Development Center, Griffiss Air Force Base, N.Y., March 1976.

18. Weibull, W.: "A Statistical Distribution Function of Wide Applicability," *J. Appl. Mech.,* pp. 293–296, 1951.

19. Wingo, D. R.: "Solution of the Three-Parameter Weibull Equations by Constrained Modified Quasilinearization (Progressively Censored Samples)," *IEEE Trans. Rel.,* vol. R-22, pp. 96–102, 1973.

SECTION 9
PROPERTIES OF ENGINEERING MATERIALS

Theodore Gela, D.Eng.Sc.
Professor Emeritus of Metallurgy
Stevens Institute of Technology
Hoboken, N.J.

9.1 MATERIAL-SELECTION CRITERIA IN ENGINEERING DESIGN

The selection of materials for engineering components and devices depends upon knowledge of material properties and behavior in particular environmental states. Although a criterion for the choice of material in critically designed parts relates to the performance in a field test, it is usual in preliminary design to use appropriate data obtained from standardized tests. The following considerations are important in material selection:

1. Elastic properties: stiffness and rigidity

2. Plastic properties: yield conditions, stress-strain relations, and hysteresis

3. Time-dependent properties: elastic phenomenon (damping capacity), creep, relaxation, and strain-rate effect

4. Fracture phenomena: crack propagation, fatigue, and ductile-to-brittle transition

5. Thermal properties: thermal expansion, thermal conductivity, and specific heat

6. Chemical interactions with environment: oxidation, corrosion, and diffusion

It is good design practice to analyze the conditions under which test data were obtained and to use the data most pertinent to anticipated service conditions.

The challenge that an advancing technology imposes on the engineer, in specifying treatments to meet stringent material requirements, implies a need for a basic approach which relates properties to structure in metals. As a consequence of the mechanical, thermal, and metallurgical treatments of metals, it is advantageous to explore, for example, the nature of induced internal stresses as well as the processes of stress relief. Better material performance may ensue when particular treatments can be specified to alter the structure in metals so that the likelihood of premature failure in service is lessened. Some of the following concepts are both basic and important:

1. Lattice structure of metals: imperfections, anisotropy, and deformation mechanisms

2. Phase relations in alloys: equilibrium diagrams

3. Kinetic reactions in the solid state: heat treatment by nucleation and by diffusionless processes, precipitation hardening, diffusion, and oxidation

4. Surface treatments: chemical and structural changes in carburizing, nitriding, and localized heating

5. Metallurgical bonds: welded and brazed joints

9.2 STRENGTH PROPERTIES: TENSILE TEST AT ROOM TEMPERATURE

The yield strength determined by a specified offset, 0.2 percent strain, from a stress-strain diagram is an important and widely used property for the design of statically loaded members exhibiting elastic behavior. This property is derived from a test in which the following conditions are normally controlled: surface condition of standard specimen is specified; load is axial; the strain rate is low, i.e., about 10^{-3} in/(in·s); and grain size is known. Appropriate safety factors are applied to the yield strength to allow for uncertainties in the calculated stress and stress-concentration factors and for possible overloads in service. Since relatively small safety factors are used in critically stressed aircraft materials, a proof stress at 0.01 percent strain offset is used because this more nearly approaches the proportional limit for elastic behavior in the material. A typical stress-strain plot from a tensile test is shown in Fig. 9.1, indicating the elastic and plastic behaviors. In order to effect more meaningful comparisons in design strength properties among materials having different specific gravities, the strength property can be divided by the specific gravity, giving units of psi per pound per cubic inch.

The modulus of elasticity is a measure of the stiffness or rigidity in a material. Values of the modulus normally are not exactly determined quantities, and typical values are commonly reported for a given material. When a material is selected on the basis of a high modulus, the tendency toward whip and vibration in shaft or rod applications is reduced. These effects can lead to uneven wear. Furthermore the modulus assumes particular importance in the design of springs and diaphragms, which necessitate a definite degree of motion for a definite load. In this connection, selection of a high-modulus material can lead to a thinner cross section.

The ultimate tensile strength and the ductility, percent elongation in inches per inch or percent reduction in area at fracture are other properties frequently reported from tensile tests. These serve as qualitative measures reflecting the ability of a material in deforming plastically after being stressed beyond the elastic region. The strength properties and ductility of a material subjected to different treatments can vary widely. This is illustrated in Fig. 9.2. When the yield strength is raised by treatment to a high

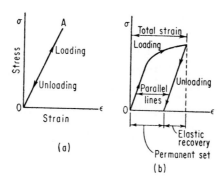

FIG. 9.1   Portions of tensile stress σ-strain ϵ curves in metals.[1] (*a*) Elastic behavior. (*b*) Elastic and plastic behaviors.

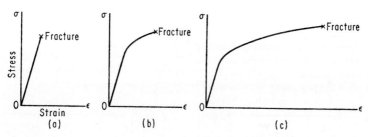

FIG. 9.2   The effects of treatments on tensile characteristics of a metal.[1] (*a*) Perfectly brittle (embrittled)—all elastic behavior. (*b*) Low ductility (hardened)—elastic plus plastic behaviors. (*c*) Ductile (softened)—elastic plus much plastic behaviors.

value, i.e., greater than two-thirds of the tensile strength, special concern should be given to the likelihood of tensile failures by small overloads in service. Members subjected solely to compressive stress may be made from high-yield-strength materials which result in weight reduction.

When failures are examined in statically loaded tensile specimens of circular section, they can exhibit a cup-and-cone fracture characteristic of a ductile material or on the other extreme a brittle fracture in which little or no necking down is apparent. Upon loading the specimen to the plastic region, axial, tangential, and radial stresses are induced. In a ductile material the initial crack forms in the center where the triaxial stresses become equally large, while at the surface the radial component is small and the deformation is principally by biaxial shear. On the other hand, an embrittled material exhibits no such tendency for shear and the fracture is normal to the loading axis. Some types of failures in round tensile specimens are shown in Fig. 9.3.

The properties of some wrought metals presented in Table 9.1 serve to show the significant differences relating to alloy content and treatment. Section 9.17 gives more information.

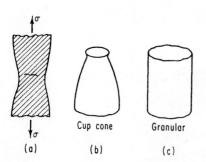

FIG. 9.3   Typical tensile-test fractures.[1] (*a*) Initial crack formation. (*b*) Ductile material. (*c*) Brittle material.

TABLE 9.1   Room-Temperature Tensile Properties for Some Wrought Metals

Metal	Condition	Ultimate tensile σ_{ult}, psi	Yield strength σ_y, psi	% elonga- tion	Modulus of elas- ticity, psi	Density, lb/cu in.
Aluminum (pure)..	Annealed	13,000	5,000	35	10,000,000	0.098
7075T6...........	Heat-treated	76,000	67,000	11	10,400,000	0.101
Copper (pure).....	Annealed	32,000	10,000	45	17,000,000	0.321
Cu-Be(2%).......	Heat-treated	200,000	150,000	2	19,000,000	0.297
Magnesium	Annealed	33,000	18,000	17	6,500,000	0.064
Mg-Al(8.5)A780A.	Strain relieved	44,000	31,000	18	6,500,000	0.065
Nickel (pure).....	Annealed	65,000	20,000	40	30,000,000	0.321
K Monel.........	Heat-treated	190,000	140,000	5	....	0.306
Titanium (pure)...	Annealed	59,000	40,000	28	15,000,000	0.163
Ti-Al(6)-V(14)α-β.	Heat-treated	170,000	150,000	7	15,000,000	0.163

The tensile properties of metals are dependent upon the rate of straining, as shown for aluminum and copper in Fig. 9.4, and are significantly affected by the temperature, as shown in Fig. 9.5. For high-temperature applications it is important to base design on different criteria, notably the stress-rupture and creep characteristics in metals, both of which are also time-dependent phenomena. The use of metals at low temperatures requires a consideration of the possibility of brittleness, which can be measured in the impact test.

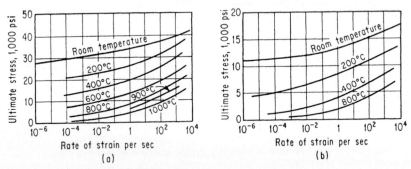

FIG. 9.4   Effects of strain rates and temperatures on tensile-strength properties of copper and aluminum.[1] (a) Copper. (b) Aluminum.

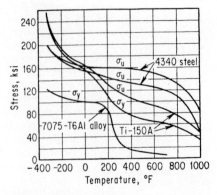

FIG. 9.5   Effects of temperatures on tensile properties. σ_u = ultimate tensile strength; σ_y = yield strength.

9.3 ATOMIC ARRANGEMENTS IN PURE METALS: CRYSTALLINITY

The basic structure of materials provides information upon which properties and behavior of metals may be generalized so that selection can be based on fundamental considerations. A regular and periodic array of atoms (in common metals whose atomic diameters are about one hundred-millionth of an inch) in space, in which a unit cell is the basic structure, is a fundamental characteristic of crystalline solids. Studies of these structures in metals lead to some important considerations of the behaviors in response to externally applied forces, temperature changes, as well as applied electrical and magnetic fields.

The body-centered cubic (bcc) cell shown in Fig. 9.6a is the atomic arrangement characteristic of αFe, W, Mo, Ta, βTi, V, and Nb. It is among this class of metals that transitions from ductile to brittle behavior as a function of temperature are significant to investigate. This structure represents an atomic packing density where about 66 percent of the volume is populated by atoms while the remainder is free space. The elements Al, Cu, γFe, Ni, Pb, Ag, Au, and Pt have a closer packing of atoms in space constituting a face-centered cubic (fcc) cell shown in Fig. 9.6b. Characteristic of these are ductility properties which in many cases extend to very low temperatures. Another structure, common to Mg, Cd, Zn, αTi, and Be, is the hexagonal close-packed (hcp) cell in Fig. 9.6c. These metals are somewhat more difficult to deform plastically than the materials in the two other structures cited above.

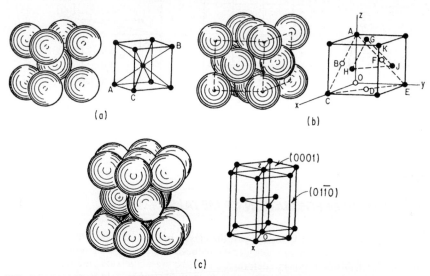

FIG. 9.6 Cell structure. (a) Body-centered cubic (bcc) unit cell structure. (b) Face-centered cubic (fcc) unit cell structure. (c) Hexagonal close-packed (hcp) unit cell structure.

It is apparent, from the atomic arrays represented in these structures, that the closest approach of atoms can vary markedly in different crystallographic directions. Properties in materials are anisotropic when they show significant variations in different directions. Such tendencies are dependent on the particular structure and can be especially pronounced in single crystals (one orientation of the lattices in space). Some examples of these are given in Table 9.2. When materials are processed so that

TABLE 9.2   Examples of Anisotropic Properties in Single Crystals

Property	Material and structure	Properties relation
Elastic module E in tension	αFe (bcc)	$E_{[AB]} \sim 2.2E_{[AC]}$
Elastic module G in shear	Ag (fcc)	$G_{[OC]} \sim 2.3G_{[OK]}$
Magnetization	αFe (bcc)	Ease of magnetization
Thermal expansion coefficient—α	Zn (hcp)	$\alpha_{[OZ]} \sim 4\alpha_{[OA]}$

their final grain size is large (each grain represents one orientation of the lattices) or that the grains are preferentially oriented, as in extrusions, drawn wire, rolled sheet, sometimes in forgings and castings, special evaluation of anisotropy should be made. In the event that directional properties influence design considerations, particular attention must be given to metallurgical treatments which may control the degree of anisotropy. The magnetic anisotropy in a single crystal of iron is shown in Fig. 9.7.

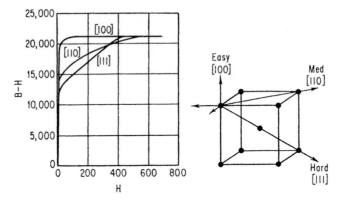

FIG. 9.7   Magnetic anisotropy in a single crystal of iron[2]: $I = (B - H)/4\pi$, where I = intensity of magnetization; B = magnetic induction, gauss; H = field strength, oersteds.

9.4 PLASTIC DEFORMATION OF METALS

When metals are externally loaded past the elastic limit, so that permanent changes in shape occur, it is important to consider the induced internal stresses, property changes, and the mechanisms of plastic deformation. These are matters of practical consideration in the following: materials that are to be strengthened by cold work, machining of cold-worked metals, flow of metals in deep-drawing and impact extrusion operations, forgings where the grain flow patterns may affect the internal soundness, localized surface deformation to enhance fatigue properties, and cold working of some magnetic materials. Experimental studies provide the key by which important phenomena are revealed as a result of the plastic-deformation process. These studies indicate some treatments that may be employed to minimize unfavorable internal-stress distributions and undesirable grain-orientation distributions.

    Plastic deformation in metals occurs by a glide or slip process along densely packed planes fixed by the particular lattice structure in a metal. Therefore, an applied

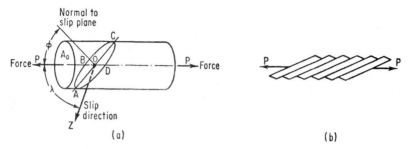

FIG. 9.8 Slip deformation in single crystals. (*a*) Resolved shear stress = P/A_0 cos ϕ cos λ. *ABCD* is plane of slip. *OZ* is slip direction. (*b*) Sketch of single crystal after yielding.

load is resolved as a shear stress, on those particular glide elements (planes and directions) requiring the least amount of deformation work on the system. An example of this deformation process is shown in Fig. 9.8. Face-centered cubic (fcc) structured metals, such as Cu, Al, and Ni, are more ductile than the hexagonal structured metals, such as Mg, Cd, and Zn, at room temperature because in the fcc structure there are four times as many possible slip systems as in a hexagonal structure. Slip is initiated at much lower stresses in metals than theoretical calculations based on a perfect array of atoms would indicate. In real crystals there are inherent structural imperfections termed dislocations (atomic misfits) as shown in Fig. 9.9, which account for the observed yielding phenomenon in metals. In addition, dislocations are made mobile by mechanical and thermal excitations and they can interact to result in strain hardening of metals by cold work. Strength properties can be increased while the ductility is

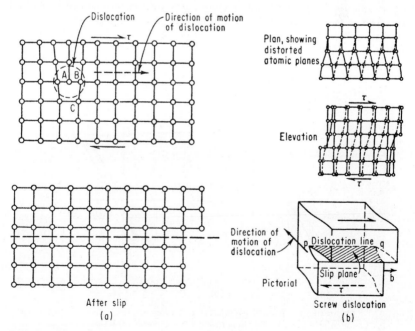

FIG. 9.9 Edge and screw dislocations as types of imperfections in metals.[2] (*a*) Edge dislocation. (*b*) Screw dislocation.

decreased in those metals which are amenable to plastic deformation. Cold working of pure metals and single-phase alloys provides the principal mechanism by which these may be hardened.

The yielding phenomenon is more nonhomogeneous in polycrystalline metals than in single crystals. Plastic deformation in polycrystalline metals initially occurs only in those grains in which the lattice axes are suitably oriented relative to the applied load axis, so that the critically resolved shear stress is exceeded. Other grains rotate and are dependent on the orientation relations of the slip systems and load application; these may deform by differing amounts. As matters of practical considerations the following effects result from plastic deformation:

1. Materials become strain-hardened and the resistance to further strain hardening increases.

2. The tensile and yield strengths increase with increasing deformation, while the ductility properties decrease.

3. Macroscopic internal stresses are induced in which parts of the cross section are in tension while other regions have compressive elastic stresses.

4. Microscopic internal stresses are induced along slip bands and grain boundaries.

5. The grain orientations change with cold work so that some materials may exhibit different mechanical and physical properties in different directions.

The Bauschinger effect in metals is related to the differences in the tensile and compressive yield-strength values, as shown at σ_T and σ_C in Fig. 9.10 when a ductile metal undergoes stress reversal. This change in polycrystalline metals is the result of the nonuniform character of deformation and the different pattern of induced macrostresses. These grains, in which the induced macrostresses are compressive, will yield at lower values upon the application of a reversed compressive stress because they are already part way toward yielding. This effect is encountered in cold-rolled metals where there is lateral contraction together with longitudinal elongation; this accounts for the decreased yield strength in the lateral direction compared with the increased longitudinal yield strength.

The control of metal flow is important in deep-drawing operations performed on sheet metal. It is desirable to achieve a uniform flow in all directions. Cold-rolling

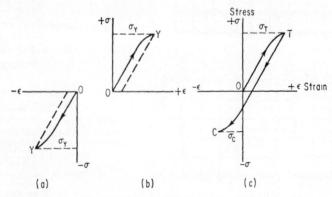

FIG. 9.10   The Bauschinger effect. (*a*) Compression. (*b*) Tension. The application of a compressive stress (*a*) or a tensile stress (*b*) results in the same value of yield strength *y*. (*c*) Stress reversal. A reversal of stress $O \rightarrow T \rightarrow C$ results in different values of tensile and compressive yield strengths; $\sigma_T \neq \sigma_C$.

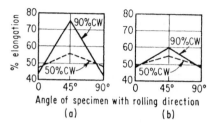

FIG. 9.11 Directionality in ductility in cold-worked and annealed copper sheet.[1] (*a*) Annealed at 1470°F. (*b*) Annealed at 750°F. The variation in ductility with direction for copper sheet is dependent on both the annealing temperature and the amount of cold work (percent CW) prior to annealing.

sheet metal produces a structure in which the grains have a preferred orientation. This characteristic can persist, even though the metal is annealed (recrystallized), resulting in directional properties as shown in Fig. 9.11. A further consequence of this directionality, associated with the deep-drawing operation, is illustrated in Fig. 9.12. The important factors, involved with the control of earing tendencies, are the fabrication practices of the amount of cold work in rolling and duration and temperatures of annealing. When grain textural problems of this kind are encountered, they can be studied by x-ray diffraction techniques and reasonably controlled by the use of optimum cold-working and annealing schedules.

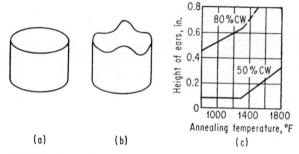

FIG. 9.12 The earing tendencies in cup deep drawn from sheet. (*a*) Uniform flow, nonearing. (*b*) Eared cup, the result of nonuniform flow. (*c*) Height of ears in deep-drawn copper cups related to annealing temperatures and amount of cold work.

9.5 PROPERTY CHANGES RESULTING FROM COLD-WORKING METALS

Cold-working metals by rolling, drawing, swaging, and extrusion is employed to strengthen them and/or to change their shape by plastic deformation. It is used principally on ductile metals which are pure, single-phase alloys and for other alloys which will not crack upon deformation. The increase in tensile strength accompanied by the decrease in ductility characteristic of this process is shown in Fig. 9.13. It is to be noted, especially from the yield-strength curve, that the largest rates of change occur during the initial amounts of cold reduction.

The variations in the macrostresses induced in a cold-drawn bar, illustrated in Fig. 9.14*a*, show that tensile stresses predominate at the surface. The equilibrium state of macrostresses throughout the cross section is altered by removing the surface layers in machining, the result of which may be warping in the machined part. It may be possible, however, to stress-relieve cold-worked metals, which generally have better machinability than softened (annealed) metals, by heating below the recrystallization

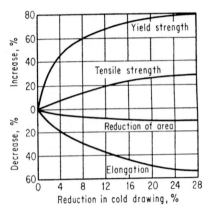

FIG. 9.13 Effect of cold drawing on the tensile properties of steel bars of up to 1-in cross section having tensile strength of 110,000 lb/in^2 or less before cold drawing.[3]

temperature. A typical alteration in the stress distribution, shown in Fig. 9.14b, is achieved so that the warping tendencies on machining are reduced, without decreasing the cold-worked strength properties. This stress-relieving treatment may also inhibit season cracking in cold-worked brasses subjected to corrosive environments containing amines. Since stressed regions in a metal are more anodic (i.e., go into solution more readily) than unstressed regions, it is often important to consider the relieving of stresses so that the designed member is not so likely to be subjected to localized corrosive attack.

Changes in electrical resistivity, elastic springback, and thermoelectric force resulting from cold work can be altered

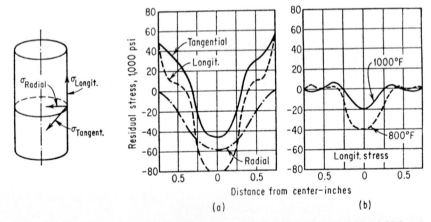

FIG. 9.14 Residual stress.[3] (a) In a cold-drawn steel bar 1½ in in diameter 20 percent cold-drawn, 0.45 percent C steel. (b) After stress-relieving bar.

by a stress-relieval treatment, in a temperature range from A to B, as shown in Fig. 9.15. However, the grain flow pattern (preferred orientation) produced by cold working can be changed only by heating the metal to a temperature at which recrystallized stress-free grains will form.

Residual tensile stresses at the surface of a metal promote crack nucleation in the fatigue of metal parts. The use of a localized surface deformation treatment by shot peening, which induces compressive stresses in the surface fibers, offers the likelihood of improvement in fatigue and corrosion properties in alloys. Shot-peening a forging flash line in high-strength aluminum alloys used in aircraft may also lessen the tendency toward stress-corrosion cracking. The effectiveness of this localized surface-hardening treatment is dependent on both the nature of surface discontinuities formed by shot-peening and the magnitude of compressive stresses induced at the surface.

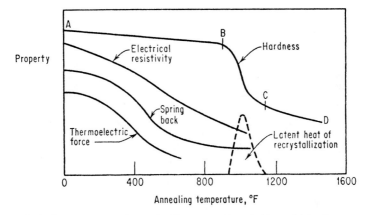

FIG. 9.15   The property changes in 95 percent cold-worked iron with heating temperatures (1 h). The temperature intervals[4]—$A \rightarrow B$, stress relieval; $B \rightarrow C$, recrystallization; and $C \rightarrow D$, grain growth—signify the important phenomena occurring.

9.6 THE ANNEALING PROCESS

Metals are annealed in order to induce softening for further deformation, to relieve
residual stresses, to alter the microstructure, and, in some cases after electroplating, to
expel by diffusion gases entrapped in the lattice. The process of annealing, the attain-
ing of a strain-free recrystallized grain structure, is dependent mainly on the temperature, time, and the amount of prior cold work. The temperature indicated at C in Fig.
9.15 results in the complete annealing after 1 h of the 95 percent cold-worked iron.
Heating beyond this temperature causes grains to grow by coalescence, so that the sur-
face-to-volume ratio of the grains decreases together with decreasing the internal energy of the system. As the amount of cold work (from the originally annealed state)
decreases, the recrystallization temperature increases and the recrystallized grain size
increases. When a metal is cold-worked slightly (less than 10 percent) and subsequently annealed, an undesirable roughened surface forms because of the abnormally large
grain size (orange-peel effect) produced. These aspects of grain-size control in the
annealing process enter in material specifications.

The annealing of iron-carbon-base alloys (steels) is accomplished by heating alloys
of eutectoid and hypoeutectoid compositions (0.8 percent C and less in plain carbon
steels) to the single-phase region; austenite, as shown in Fig. 9.16, above the transition
line GS; and for hypereutectoid alloys (0.8 to 2.0 percent C) between the transition
lines SK and SE in Fig. 9.16; followed by a furnace cool at a rate of about 25°F per
hour to below the eutectoid temperature SK. In the annealed condition, a desirable distribution of the equilibrium phases is thereby produced. A control of the microstructure is manifested by this process in steels. Grain-size effects are principally controlled by the high-temperature treatment, grain sizes increasing with increasing
temperatures, and in some cases minor impurity additions such as vanadium inhibit
grain coarsening to higher temperatures. These factors of grain-size control enter into
the considerations of hardening steels by heat treatment.

The control of the atmosphere in the annealing furnace is desirable in order to prevent gas-metal attack. Moisture-free neutral atmospheres are used for steels which
oxidize readily. When copper and its alloys contain oxygen, as oxide, it is necessary to
keep the hydrogen content in the atmosphere to a minimum. At temperatures lower

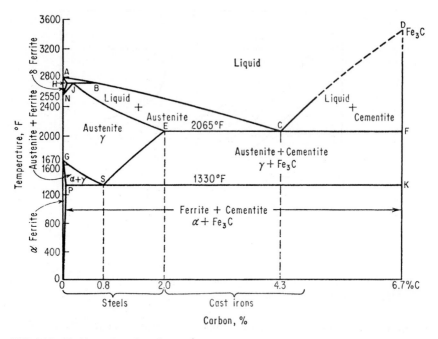

FIG. 9.16   The iron-carbon phase diagram.[4]

than 900°F, the hydrogen should not exceed 1 percent, and as the temperature is increased the hydrogen content should be reduced in order to prevent hydrogen embrittlement. In nickel and its alloys the atmosphere must be free from sulfur and slightly reducing by containing 2 percent or more of CO. Some aluminum alloys containing magnesium are affected by high-temperature oxidation in annealing (and heat treatment) and therefore require atmosphere control.

It is a characteristic property that strengths in all metals decrease with increasing temperatures. The coalescence of precipitate particles is one factor involved, so that material specifications for high-temperature use are concerned with alloy compositions that form particles having lower solubility and lower mobility. A second factor is concerned with the mobility of dislocations which increases at higher temperatures. Since strain hardening is reasoned to be due to the interaction of dislocations, then by the proper additions of solid-solution alloying elements that impede dislocations, resistance to softening will increase at the high temperature. The recrystallization temperature of iron is raised by the addition of 1 atomic percent of Mn, Cr, V, W, Cb, Ta in the same order in which the atomic size of the alloying elements differs from that of iron. The practical implications of these basic atomic considerations are important in selecting metals for high-temperature service.

9.7 THE PHASE DIAGRAM AS AN AID TO ALLOY SELECTION

Phase diagrams, which are determined experimentally and are based upon thermodynamic principles, are temperature-composition representations of slowly cooled alloys (annealed state). They are useful for predicting property changes with composition and selecting feasible fabrication processes. Phase diagrams also indicate the possible response of alloys to hardening by heat treatment. Shown on these diagrams are first-order phase transitions and the phases present. In two-phase regions, the compositions of each phase are shown on the phase boundary lines and the relative amounts of each phase present can be determined by a simple lever relation at a given temperature.

The particular phases that are formed in a system are governed principally by the physical interactions of valence electrons in the atoms and secondarily by atomic-size factors. When two different atoms in the solid state exist on, or where one is in, an atomic lattice, the phase is a solid solution (e.g., γ austenite phase in Fig. 9.16) analogous to a miscible liquid solution. When the atoms are strongly electropositive and strongly electronegative to one another, an intermetallic compound is formed (e.g., Fe_3C, cementite). The two atoms are electronically indifferent to one another and a phase mixture issues (e.g., $\alpha + Fe_3C$) analogous to the immiscibility of water and oil.

The thermodynamic criteria for a first-order phase change, indicated by the solid lines on the phase diagram, are that, at the transition temperature, (1) the change in Gibbs's free energy for the system is zero, (2) there is a discontinuity in entropy (a latent heat of transformation and a discontinuous change in specific heat), and (3) there is a discontinuous change in volume (a dilational effect).

In the selection of alloys for sound castings, particular attention is given that part of the system where the liquidus line ($ABCD$) goes through a minimum. For alloys between the composition limits of $E{\rightarrow}F$, a eutectic reaction occurs at 2065°F such that

$$\text{liq } C = \gamma + Fe_3C$$

It is for this reason in the iron-carbon system that cast irons are classified as having carbon contents greater than 2 percent. For purposes of controlling grain size, obtaining sound castings free from internal porosities (blowholes) and internal shrinkage cavities, and possessing good mold-filling characteristics, alloys and low-melting solutions are chosen near the eutectic composition (i.e., at C). Aluminum-silicon die-casting alloys have a composition of about 11 percent silicon near the eutectic composition. Special considerations need be given to the properties and structures in cast irons because the Fe_3C phase is thermodynamically unstable and decomposition to graphite (in gray cast irons) may result.

The predominant phase-diagram characteristic in steels is the eutectoid reaction, in the solid state, along GSE where $\gamma_s = \alpha + Fe_3C$ (pearlite) at 1330°F. Steels are therefore classified as alloys in the Fe-C system having a carbon content less than 2.0 percent C; and furthermore, according to their applications, compositions are designated as hypoeutectoid (C < 0.8 percent), eutectoid (C = 0.8 percent), and hypereutectoid (C < 2 percent > 0.8 percent). Since the slowly cooled room-temperature structures of steels contain a mechanical aggregate of the ferrite and Fe_3C-cementite phases, the property relations vary linearly as shown in Fig. 9.17. The ductility decreases with increasing carbon contents.

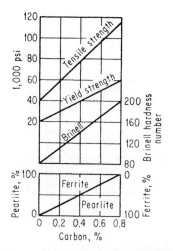

FIG. 9.17 Relation of mechanical properties and structure to carbon content of slowly cooled carbon steels.[4]

Some important characteristics of the equilibrium phases in steels are listed below:

Phase	Characteristics
α ferrite	Low C solubility (less than 0.03%) bcc, ductile, and ferromagnetic below 1440°F
Fe₃C, cementite	Intermetallic compound, orthorhombic, hard, brittle, and fixed composition at 6.7% C
γ austenite	Can dissolve up to 2% C in solid solution, fcc, nonmagnetic, and in this region annealing, hardening, forging, normalizing, and carburizing processes take place

Low-carbon alloys can be readily worked by rolling, drawing, and stamping because of the predominant ductility of the ferrite. Wires for suspension cables having a carbon content of about 0.7 percent are drawn at about 1100°F (patenting) because of the greater difficulty, in room-temperature deformation, caused by the presence of a relatively large amount of the brittle Fe₃C phase.

Extensive substitutional solid-solution alloys form in binary systems when they have similar chemical characteristics and atomic diameters in addition to having the same lattice structure. Such alloys include copper-nickel (monel metal being a commercially useful one), chromium-molybdenum, copper-gold, and silver-gold (jewelry alloys). The phase diagram and the equilibrium-property changes for this system are shown in Fig. 9.18a. Each pure element is strengthened by the addition of the other, whereby the strongest alloy is at an equal atom concentration. There are no first-order phase changes up to the start of melting (the solidus line EHG), so that these are not hardened by heat treatment but only by cold work. The electrical conductivity decreas-

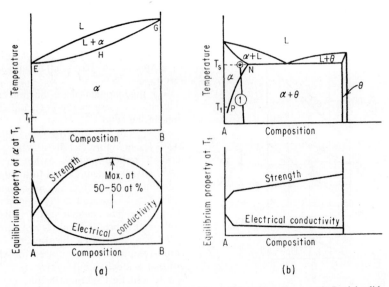

FIG. 9.18 Binary systems.[4] (a) Complete solid-solubility phase diagram. (b) Partial solid-solubility part of phase diagram. α is a substitutional solid solution, a phase with two different atoms on the same lattice. In the AlCu system θ is an intermediate phase (precipitant) having a composition nominally of CuAl₂.

es from each end of the composition axis. Because of the presence of but one phase, these alloys are selected for their resistance to electrochemical corrosion. High-temperature-service metals are alloys which have essentially a single-phase solid solution with minor additions of other elements to achieve specific effects.

Another important system is one in which there are present regions of partial solid solubility as shown in Fig. 9.18b together with equilibrium-property changes. An important consideration in the selection of alloys containing two or more phases is that galvanic-corrosion attack may occur when there exists a difference in the electromotive potential between the phases in the environmental electrolyte. Sacrificial galvanic protection of the base metal in which the coating is more anodic than the base metal is used in zinc-plating iron-base alloys (galvanizing alloys). The intimate mechanical mixture of phases which are electrochemically different may result in pitting corrosion, or even more seriously, intergranular corrosion may result if the alloy is improperly treated by causing localized precipitation at grain boundaries.

Heat treatment by a precipitation-hardening process is indeed an important strengthening mechanism in particular alloys such as the aircraft aluminum-base, copper-beryllium, magnesium-aluminum, and alpha-beta titanium alloys (Ti, Al, and V). In these alloys a distinctive feature is that the solvus line NP in Fig. 9.18b shows decreasing solid solubility with decreasing temperature. This in general is a necessary, but not necessarily sufficient, condition for hardening by precipitation since other thermodynamic conditions as well as coherency relations between the precipitated phases must prevail. The sequence of steps for this process is as follows: An alloy is solution heat-treated to a temperature T_s, rapidly quenched so that a metastable supersaturated solid solution is attained, and then aged at experimentally determined temperature-time aging treatments to achieve desired mechanical properties. This is the principal hardening process for those particular nonferrous alloys (including Inconels) which can respond to a precipitation-hardening process.

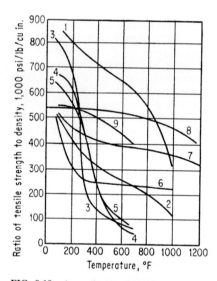

FIG. 9.19 Approximate comparison of materials on a strength-weight basis from room temperature to 1000°F.[6] (1) T_1, 8Mn; (2) 9990 T_1; (3) 75S, T6Al; (4) 24S, T4Al; (5) AZ31A Mg; (6) annealed stainless steel; (7) half-hard stainless steel; (8) Inconel X; (9) glass-cloth laminate.

The engineer is frequently concerned with the strength-to-density ratio of materials and its variation with temperature. A number of materials are compared on this basis in Fig. 9.19 in which the alloys designated by curves 1, 3, 4, 5, and 8 are heat-treatable nonferrous alloys.

9.8 HEAT-TREATMENT CONSIDERATIONS FOR STEEL PARTS

The heat-treating process for steel involves heating to the austenite region where the carbon is soluble, cooling at specific rates, and tempering to relieve some of the stress which results from the transformation. Some important considerations involved in specifying heat-treated parts are strength properties, warping tendencies, mass effects (hardenability), fatigue and impact properties, induced transformation stresses, and the use of surface-hardening processes for enhanced wear resistance. Temperature and time factors affect the structures issuing from the decomposition of austenite; for a eutectoid steel (0.8 percent C) they are as follows:

Decomposition product from γ	Structure	Mechanism	Temperature range, °F
Pearlite	Equilibrium ferrite + Fe$_3$C	Nucleation; growth	1300–1000
Bainite	Nonequilibrium ferrite + carbide	Nucleation; growth	1000–450
Martensite	Supersaturated tetragonal lattice	Diffusionless	M_s (≤450)

The tensile strength of a slowly cooled (annealed) eutectoid steel containing a coarse pearlite structure is about 120,000 lb/in^2. To form bainite, the steel must be cooled rapidly enough to escape pearlite transformation and must be kept at an intermediate temperature range to completion, from which a product having a tensile strength of about 250,000 lb/in^2 can be formed. Martensite, the hardest and most brittle product, forms independently of time by quenching rapidly enough to escape higher-temperature transformation products. The carbon atoms are trapped in the martensite, causing its lattice to be highly strained internally; its tensile strength is in excess of 300,000 lb/in^2. Isothermal transformation characteristics of all steels show the temperature-time and transformation products as in Fig. 9.20, where the lines indicate the start and end of transformation. On the temperature-time coordinates, involved cooling curves can be superimposed which show that, for a 1-in round water-quenched specimen, mixed products will be present. The outside will be martensite and the middle sections will contain pearlite. Alloying elements are added to steels principally to retard pearlite transformation either so that less drastic quenching media can be used or to ensure more uniform hardness throughout. This retardation is shown in Fig. 9.20b for an SAE 4340 steel containing alloying additions of Ni, Cr, or Mo and 0.4 percent C.

The carbon content in steels is the most significant element upon which selection for the maximum attainable hardness of the martensite is based. This relation is shown in Fig. 9.21. Since the atomic rearrangements involved in the transformation from the fcc austenite to the body-centered-tetragonal martensite result in a volumetric expansion, on cooling, of about 1 percent (for a eutectoid steel) nonuniform stress patterns can be induced on transformation. As cooling starts at the surface, by the normal process of heat transfer, parts of a member can be expanding, because of transformation, while further inward normal contraction occurs on the cooling austenite. The danger of cracking and distortion (warping) as a consequence of the steep thermal gradients and the transformation involved in hardening steels can be eliminated by using good design and the selection of the proper alloys. Where section size, time factors, and alloy content (as it affects transformation curves) permit, improved practices by martempering shown by *EFGH* in Fig. 9.20, followed by tempering or austempering shown by *EFK*, may be feasible and are worthy of investigation for the particular alloy used.

Uniform mass distribution and the elimination of sharp corners (potential stress raisers) by the use of generous fillets are recommended. Some design features pertinent to the elimination of quench cracks and the minimization of distortion by warping are illustrated by Fig. 9.22 in which *a*, *c*, and *e* represent poor designs in comparison with the suggested improvements apparent in *b*, *d*, and *f*.

Steels are tempered to relieve stresses, to impart ductility, and to produce a desirable microstructure by a reheating process of the quenched member. The tempering

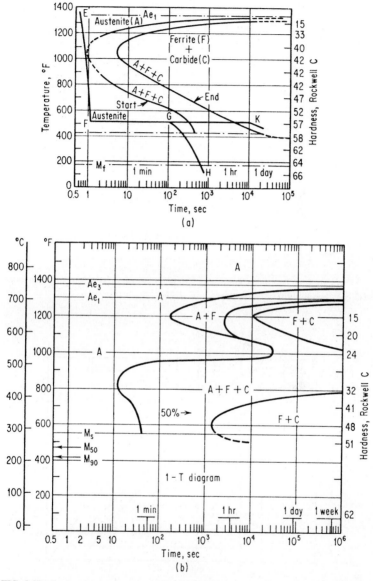

FIG. 9.20   Isothermal transformation diagram.[5] (a) Eutectoid carbon steel. (b) SAE 4340 steel. M_s = start of martensite temperature; M_f = finish of martensite temperature; $EFGH$ = martempering (follow by tempering); EFK = austempering. A, austenite; F, ferrite; C, carbide; M_s and M_f, temperatures for start and finish of martensite transformation; M_{50} and M_{90}, temperatures for 50% and 90% of martensite transformation.

process is dependent on the temperature, time, and alloy content of the steel. Different alloys soften at different rates according to the constitutionally dependent diffusional structure. The response to tempering for 1 h for three different steels of the same carbon content is shown in Fig. 9.23. In addition, the tempering characteristics of a high-

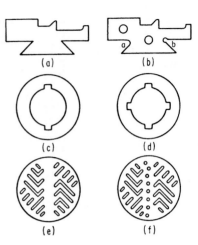

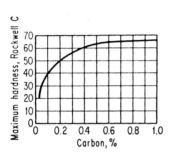

FIG. 9.21   Relation of maximum attainable hardness of quenched steels to carbon content.[7]

FIG. 9.22   Examples of good (*b*, *d*, and *f*) and bad (*a*, *c*, and *e*) designs for heat-treated parts.[8] (1) *b* is better than *a* because of fillets and more uniform mass distribution. (2) In *c* cracks may form at keyways. (3) Warping may be more pronounced in *e* than in *f*, which are blanking dies.[8]

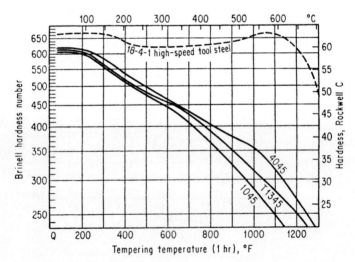

FIG. 9.23   Effect of tempering temperature on the hardnesses of SAE 1045, T1345, and 4045 steels. In the high-speed tool steel 18-4-1, secondary hardening occurs at about 1050°F.[9]

speed tool steel, 18 percent W, 4 percent Cr, 1 percent V, and 0.9 percent C, are shown to illustrate the secondary hardening at about 1050°F. The pronounced tendency for high-carbon steels to retain austenite on transformation normally has deleterious effects on dimensional stability and fatigue performance. In high-speed tool steel, the

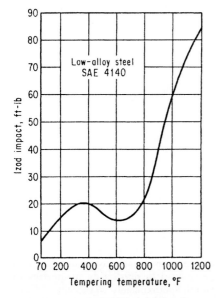

FIG. 9.24 In the tempering of this 4140 steel the notched-bar impact properties decrease in the range of 450 to 650°F.[9]

secondary hardening is due to the transformation of part of the retained austenite to newly transformed martensite. The structure contains tempered and untempered martensite with perhaps some retained austenite. Multiple tempering treatments on this type of steel produce a more uniform product.

In low-alloy steels where the carbon content is above 0.25 percent, there may be a tempering-temperature interval at about 450° to 650°F, during which the notch impact strength goes through a minimum. This is shown in Fig. 9.24 and is associated with the formation of an embrittling carbide network (ϵ carbide) about the martensite subgrain boundaries. Tempering is therefore carried out up to 400°F where the parts are to be used principally for wear resistance, or in the range of 800 to 1100°F where greater toughness is required. In the nomenclature of structural steels adopted by the Society of Automotive Engineers and the American Iron and Steel Institute, the first two numbers designate the type of steel according to the principal alloying elements and the last two numbers designate the carbon content:

Series designation	Types
10xx	Nonsulfurized carbon steels
11xx	Resulfurized carbon steels (free-machining)
12xx	Rephosphorized and resulfurized carbon steels (free-machining)
13xx	Manganese 1.75%
23xx*	Nickel 3.50%
25xx*	Nickel 5.00%
31xx	Nickel 1.25%, chromium 0.65%
33xx	Nickel 3.50%, chromium 1.55%
40xx	Molybdenum 0.20 or 0.25%
41xx	Chromium 0.50 or 0.95%, molybdenum 0.12 or 0.20%
43xx	Nickel 1.80%, chromium 0.50 or 0.80%, molybdenum 0.25%
44xx	Molybdenum 0.40%
45xx	Molybdenum 0.52%
46xx	Nickel 1.80%, molybdenum 0.25%
47xx	Nickel 1.05%, chromium 0.45%, molybdenum 0.20 or 0.35%
48xx	Nickel 3.50%, molybdenum 0.25%
50xx	Chromium 0.25, 0.40, or 0.50%
50xxx	Carbon 1.00%, chromium 0.50%
51xx	Chromium 0.80, 0.90, 0.95, or 1.00%
51xxx	Carbon 1.00%, chromium 1.05%
52xxx	Carbon 1.00%, chromium 1.45%
61xx	Chromium 0.60, 0.80, or 0.95%, vanadium 0.12%, 0.10% min, or 0.15% min

(Continued)

Series designation	Types
81xx	Nickel 0.30%, chromium 0.40%, molybdenum 0.12%
86xx	Nickel 0.55%, chromium 0.50%, molybdenum 0.20%
87xx	Nickel 0.55%, chromium 0.50%, molybdenum 0.25%
88xx	Nickel 0.55%, chromium 0.50%, molybdenum 0.35%
92xx	Manganese 0.85%, silicon 2.00%, chromium 0 or 0.35%
93xx	Nickel 3.25%, chromium 1.20%, molybdenum 0.12%
94xx	Nickel 0.45%, chromium 0.40%, molybdenum 0.12%
98xx	Nickel 1.00%, chromium 0.80%, molybdenum 0.25%

*Not included in the current list of standard steels.

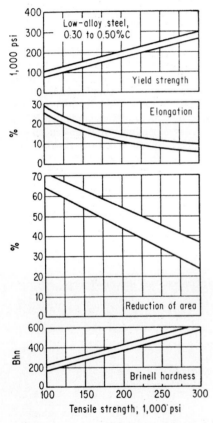

The most probable properties of tempered martensite for low-alloy steels fall within narrow bands even though there are differences in sources and treatments. The relations for these shown in Fig. 9.25 are useful in predicting properties to within approximately 10 percent.

Structural steels may be specified by hardenability requirements, the H designation, rather than stringent specification of the chemistry. Hardenability, determined by the standardized Jominy end-quench test, is a measurement related to the variation in hardness with mass, in quenched steels. Since different structures are formed as a function of the cooling rate and the transformation is affected by the nature of the alloying elements, it is necessary to know whether the particular steel is shallow (A) or deeply hardenable (C), as in Fig. 9.26. The hardenability of a particular steel is a useful criterion in selection because it is related to the mechanical properties pertinent to the section size.

The selection of through-hardened steel based upon carbon content is indicated on the next page for some typical applications.

FIG. 9.25 The most probable properties of tempered martensite for a variety of low-alloy steels.[4]

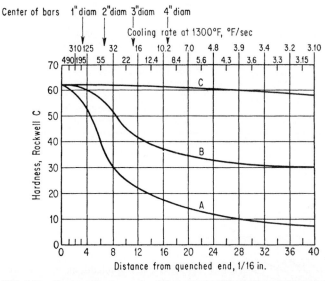

FIG. 9.26 Hardenability curves for different steels with the same carbon content.[7]

Carbon range	Requirement	Approx. tensile strength level, lb/in^2	Applications
Medium, 0.3 to 0.5%	Strength and toughness	150,000	Shafts, bolts, forgings, nuts
Intermediate, 0.5 to 0.7%	Strength	225,000	Springs
High, 0.8 to 1.0%	Wear resistance	300,000	Bearings, rollers, bushings

9.9 SURFACE-HARDENING TREATMENTS

The combination of high surface wear resistance and a tough-ductile core is particularly desirable in gears, shafts, and bearings. Various types of surface-hardening treatments and processes can achieve these characteristics in steels; the most important of these are the following:

Base metal	Process
Low C, to 0.3%	Carburization: A carbon diffusion in the γ-phase region, with controlled hydrocarbon atmosphere or in a box filled with carbon. The case depth is dependent on temperature-time factors. Heat treatment follows process.
Medium C, 0.4 to 0.5%	Localized surface heating by induction or a controlled flame to above the Ac$_3$ temperature; quenched and tempered.

(Continued)

Base metal	Process
Nitriding (nitralloys, stainless steels)	Formation of nitrides (in heat-treated parts) in ammonia atmosphere at 950 to 1000°F, held for long times. A thin and very hard surface forms and there may be dimensional changes.
Low C, 0.2%	Cyaniding: Parts placed in molten salt baths at heat-treating temperatures; some limited carburization and nitriding occur for cases not exceeding 0.020 in. Parts are quenched and tempered.

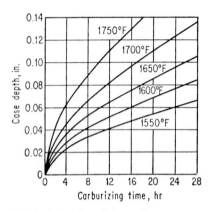

FIG. 9.27 Relation of time and temperature to carbon penetration in gas carburizing.[10]

The carbon penetration in carburization is determined by temperature-time-distance relations issuing from the solution of diffusional equations where D, the diffusion coefficient, is independent of concentrations. These relations, shown in Fig. 9.27, permit the selection of a treatment to provide specific case depths. Typical applications are as follows:

Case depth, in	Applications (automotive)
0.020 or more	Push rods, light-load gears, water pump shafts
0.020–0.040	Valve rocker arms, steering-arm bushings, brake and clutch pedal shifts
0.040–0.060	Ring gears, transmission gears, piston pins, roller bearings
0.060 or more	Camshafts

The heat treatments used on carburized parts depend upon grain-size requirements, minimization of retained austenite in the microstructure, amount of undissolved carbide network, and core-strength requirements. As a result of carburization, the surface fiber stresses are compressive. This leads to better fatigue properties. This treatment, which alters the surface chemistry by diffusion of up to 1 percent carbon in a low-carbon steel, gives better wear resistance because the surface hardness is treated for values above Rockwell C 60, while the low-carbon-content core has ductile properties to be capable of the transmission of torsional or bending loads.

Selection of the nitriding process requires careful consideration of cost because of the long times involved in the case formation. A very hard case having a hardness of about Rockwell C 70 ensures excellent wear resistance. Nitrided parts have good corrosion resistance and improved fatigue properties. Nitriding follows the finish-machining and -grinding operations, and many parts can be nitrided without great likelihood of distortion. Long service (several hundred hours) at 500°F has been attained in nitrided gears made from a chromium-base hot-work steel H11. Some typical nitrided steels and their applications are as follows:

Steel	Nitriding treatment		Case hardness, in	Case depth, Rockwell C	Applications
	h	°F			
4140	48	975	0.025–0.035	53–58	Gears, shafts, splines
4340	48	975	0.025–0.035	50–55	Gears, drive shafts
Nitralloy 135M	48	975	0.020–0.025	65–70	Valve stems, seals, dynamic faceplates
H11	70	960 + 980	0.015–0.020	67–72	High-temperature power gears, shafts, pistons

9.10 PRESTRESSING

Concrete has a tensile strength which is a small fraction of its compressive strength and it can be prestressed (pretensioned) with high carbon, 0.7 to 0.85 percent C, steel wire, or strands. These have a tensile strength of about 260,000 lb/in^2, a yield strength of at least 80 percent of the tensile strength, and an adequate ductility. This wire or strand is pretensioned and when the concrete is cured the tensioning is relaxed. As a result the concrete is in compression, having more favorable load-carrying characteristics. When the steel is put in tension after concrete is cured, by different techniques, this is called "posttensioning."

When SAE 1010 steel was prestressed by roll-induced tension, different effects were observed. By prestraining this steel to 0.9 of its true fracture strain the fatigue life was increased. Prestraining to 0.95 of the true fracture strain, on the other hand, resulted in a decreased fatigue life. Generalizations on the effects of prestressing can create misleading results. Involved are the nature and amount of prestressing, type of materials, and effects on property changes. Plain carbon steels, when prestressed within narrow limits, may have their torsional fracture strains increased. On the other hand, prestressing 304 stainless steel resulted in stress corrosion cracking in a corrosive environment. Transition temperatures on impact can be increased (an undesirable effect) in prestressed low-carbon steels. Shot-peening gear teeth, which can result in improved fatigue properties because of the induced surface compressive stresses, can be a beneficial effect of prestressing.

9.11 SOME PRACTICAL CONSIDERATIONS OF INDUCED RESIDUAL STRESSES IN ALLOYS

The presence of residual stresses, especially at surfaces, may affect the fatigue and bending properties significantly. These stresses can be induced in varying degrees by processes involving thermal changes as in heat treatments, welding, and exposures to steep temperature gradients, as well as mechanical effects such as plastic deformation, machining, grinding, and pretensioning. Effects of these when superimposed on externally applied stresses may reduce fatigue properties, or these properties may be improved by specific surface treatments. Comprehensive studies have been made to quantitatively measure the type—tensile or compressive—and the magnitude of residual surface elastic stresses so that the design engineer and user of structural members can optimize such material behavior.

When surface elastic stresses are present in crystalline material, they may be measured nondestructively by x-ray diffraction techniques, a review of which is given by D. P. Koistenen et al., in the SAE publication "TR-182." A two-exposure method for locating the shifts in x-ray diffraction peaks at high Bragg angles θ can result in stress measurements accurate to within ± 5000 lb/in^2, using elasticity relations and the fact that elastic atom displacements enable x-ray measurements to act as sensitive nondestructive strain gauges. The relation which is used is

$$\sigma_\phi = \frac{E(\cot \theta)(2\theta_n - 2\theta_i)}{2(1 + v)\sin^2\psi} \frac{1}{57.3} \quad \text{lb/in}^2$$

where σ_ϕ = surface elastic stress, lb/in^2
         E = Young's modulus, lb/in^2
         θ_n, θ_i = measured Bragg angle at normal (n) and inclined (i) incidence
         v = Poisson's ratio
         ψ = angle between normal and inclined incident x-ray beam

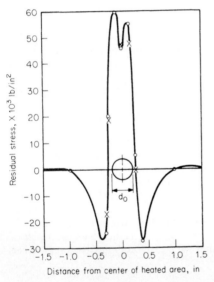

FIG. 9.28   Induced residual stress pattern by localized spot welding.[33]

Typically the back-reflection x-rays used will penetrate the surface by about 0.0002 in. Subsurface stresses may be measured in thick specimens by removing surface layers by electropolishing. These measurements are used for design of gears, bearings, welded plates, aircraft structures, and generally where fatigue and bending properties in use are important. Furthermore, improvements in surface properties can be enhanced by surface treatments which will induce residual compressive stresses.

The effect of localized heating (to about 1300°F) in spot welding a constrained steel specimen had a residual stress pattern as shown in Fig. 9.28.

Surface finishing in heat-treated steel gears, bearings, and dies often entails grinding in the hardened condition because of the greater dimensional tolerances that are desired. Metal removal by the abrasive grinding wheels can cause different degrees of surface residual stresses in the metals because of the frictional heat generated. Very abusive grinding may cause cracks. Examples of the stress patterns and fatigue properties in grinding a hardened aircraft alloy AISI 4340 are shown in Figs. 9.29 and 9.30. The significant decrease in fatigue strength of over 40 percent focuses special attention on the importance of controlling and specifying grinding and metal removal practices in hardened alloys. In addition to this, residual stress patterns may be induced due to the presence of retained austenite in hardened high-carbon steels.

Surface treatments which will induce compressive stresses in surface layers can improve fatigue properties substantially. These include mechanical deformation of surfaces by shot peening and roller burnishing, as well as microstructural effects in carburizing or nitriding steels. The beneficial effects of surface treatments by hardened 4340 steel are shown in Fig. 9.31.

To recover some fatigue properties in decarburized steel surfaces it may be helpful

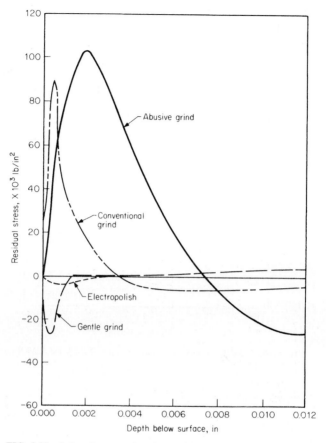

FIG. 9.29 Induced stresses in grinding ASAI 4340 steel heat-treated to Rockwell C 50.

to consider shot peening these. Skin rolling of sheet or tubular products may also induce compressive surface stresses.

9.12 NOTCHED IMPACT PROPERTIES: CRITERIA FOR MATERIAL SELECTION

When materials are subject to high deformation rates and are particularly sensitive to stress concentrations at sharp notches, criteria must be established to indicate safe operating-temperature ranges. The impact test (Izod or Charpy V-notch) performed on notched specimens conducted over a prescribed temperature range indicates the likelihood of ductile (shear-type) or brittle (cleavage-type) failure. In this test the velocity of the striking head at the instant of impact is about 18 fps, so the strain rates are several orders of magnitude greater than in a tensile test. The energy absorbed in fracturing a standard notched specimen is measured by the differences in potential energy from free fall of the hammer to the elevation after fracturing it. The typical effect of temperature upon impact energy for a metal which shows ductile and brittle characteristics is shown

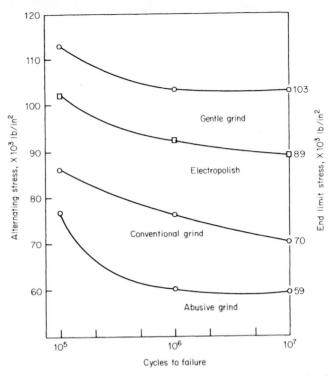

FIG. 9.30 Effect of severity of surface grinding of hardened 4340 Rockwell C (50) steel.[34,35]

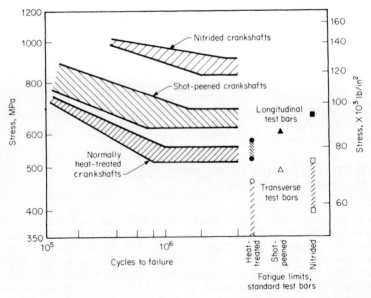

FIG. 9.31 Improved fatigue strength by shot-peening and nitriding 4340 hardened crankshafts.[36]

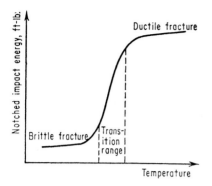

FIG. 9.32   The ductile-to-brittle transition in impact.

in Fig. 9.32. Interest centers on the transition temperature range and the material-sensitive factors, such as composition, microstructure, and embrittling treatments. ASTM specifications for structural steel for ship plates specify a minimum impact energy at a given temperature, as, for example, 15 ft·lb at 40°F. It is desirable to use materials at impact energy levels and at service temperatures where crack propagation does not proceed. Some impact characteristics for construction steels are shown in Fig. 9.33.

It is generally characteristic of pure metals which are fcc in lattice structure to possess toughness (have no brittle fracture tendencies) at very low temperatures. Body-centered-cubic pure metals, as well as hexagonal metals, do show ductile-to-brittle behavior. Tantalum, a bcc metal, is a possible exception and is ductile in impact even at cryogenic temperatures. Alloyed metals do not follow any general pattern of behavior; some specific impact values for these are as follows:

Material	Charpy V notch at 80°F	Impact strength at −100°F, ft·lb
Cu-Be (2%) HT	5.4	5.5
Phosphor bronze (5% Sn):		
Annealed	167	193
Spring temper	46	44
Nilvar Fe-Ni (36%):		
Annealed	218	162
Hard	97	77
2024-T6 aluminum aircraft alloy, HT aged	12	12
7079-T6 aluminum aircraft alloy, HT aged	4.5	3.5
Mg-Al-Zn extruded	7.0	3.0

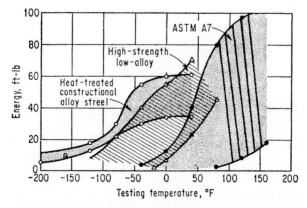

FIG. 9.33   Influence of testing temperature on notch toughness, comparing carbon steel of structural quality (ASTM A7) with high-strength, low-alloy and heat-treated constructional alloy steel.[4] Charpy V notch.

Austenitic stainless steels are ductile and do not exhibit transition in impact down to very low temperatures.

Some brittle service failures in steel structures have occurred in welded ships, gas-transmission pipes, pressure vessels, bridges, turbine generator rotors, and storage tanks. Serious consideration must then be given the effect of stress raisers, service temperatures, tempering embrittling structures in steels, grain-size effects, as well as the effects of minor impurity elements in materials.

9.13 FATIGUE CHARACTERISTICS FOR MATERIALS SPECIFICATIONS

Most fatigue failures observed in service as well as under controlled laboratory tests are principally the result of poor design and machining practice. The introduction of potential stress raisers by inadequate fillets, sharp undercuts, and toolmarks at the surfaces of critically cyclically stressed parts may give rise to crack nucleation and propagation so that ultimate failure occurs. Particular attention should be given to material fatigue properties where rotating and vibrating members experience surface fibers under reversals of stress.

In a fatigue test, a highly polished round standard specimen is subjected to cyclic loading; the number of stress reversals to failure is recorded. For sheets, the standard specimen is cantilever supported. Failure due to tensile stresses usually starts at the surface. Typical of a fatigue fracture is its conchoidal appearance, where there is a smooth region in which the severed sections rubbed against each other and where the crack progressed to a depth where the load could no longer be sustained. From the fatigue data a curve of stress vs. number of cycles to failure is plotted. Note that there can be considerable statistical fluctuation in the results (about ± 15 percent variations in stress).

Two characteristics can be observed in fatigue curves with respect to the endurance limit shown by E_a and E_b in Fig. 9.34:

E_a, curve a, the asymptotic stress value, typical in most materials

E_b, curve b, a stress value taken at an arbitrary number of cycles; e.g., 500,000,000, typical in Al and Mg alloys

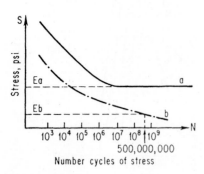

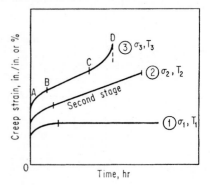

FIG. 9.34 Fatigue curves. (*a*) Most materials have an endurance limit E_a (asymptotic stress). (*b*) Endurance limit E_b (nonasymptotic set at arbitrary value of N).

FIG. 9.35 Typical creep curves. At constant temperature, $\sigma_3 > \sigma_2 > \sigma_1$. At constant stress $T_3 > T_2 > T_1$.

For design specifications, the endurance limit represents a safe working stress for fatigue. The endurance ratio is defined as the ratio of the endurance limit to the ultimate tensile strength. These values are strongly dependent upon the presence of notches on the surface and a corrosive environment, and on surface-hardening treatments. In corrosion, the pits formed act as stress raisers leading generally to greatly reduced endurance ratios. References 15 through 20 provide useful information on corrosion. A poorly machined surface or a rolled sheet with surface scratches evidences low endurance ratios, as do parts in service with sharp undercuts and insufficiently filleted changes in section.

Improvements in fatigue properties are brought about by those surface-hardening treatments which produce induced compressive stresses as in steels, nitrided (about 160,000 lb/in^2 compressive stress) or carburized (about 35,000 lb/in^2 compressive stress).

Metallurgical factors related to poorer fatigue properties are the presence of retained austenite in hardened steels, the presence of flakes or sharp inclusions in the microstructure, and treatments which induce preferential corrosive grain-boundary attack. When parts are quenched or formed so that surface tensile stresses are present, stress-relieval treatments are advisable.

9.14 SHEAR AND TORSIONAL PROPERTIES

When rivets, bolts, and fastening pins are used in structural components, their shear and torsional properties are evaluated mainly for ductile materials. The stress which produces shear fracture is reported as the ultimate shear strength or sometimes as shear strength. Some typical and averaged ratios of strength are given below:

Material	Brinell hardness	Ultimate shear strength/ ultimate tensile strength
Ductile cast irons	170–300	0.9–1.0
Aluminum aircraft alloys	120	0.65
Brasses	100	0.7

In the elastic region the shear or torsional properties are related by

$$G = \tau/\gamma$$

where G = modulus of rigidity in shear
τ = shear stress
γ = shear strain

Furthermore, one can calculate with reasonably good accuracy the relation of elastic moduli by

$$G = E/2(1 + \nu)$$

where E = Young's modulus of rigidity in tension
ν = Poisson's ratio

This relation presumes isotropic elastic behavior.

The elastic shear strength, determined at a fixed offset on the shear stress-strain diagram (i.e., 0.04 percent offset), has been related to the tensile yield strength (0.2

percent offset) such that its ratio is about 0.6 for ductile cast irons and 0.55 for aluminum alloys.

9.15 MATERIALS FOR HIGH-TEMPERATURE APPLICATIONS

9.15.1 Introduction

Selection of materials to withstand stress at high temperatures is based upon experimentally determined temperature stress-time properties. Some useful engineering design criteria follow.

1. Dimensional change, occurring by plastic flow, when metals are stressed at high temperatures for prolonged periods of time, as measured by creep tests
2. Stresses that lead to fracture, after certain set time periods, as determined by stress-rupture tests, where the stresses and deformation rates are higher than in a creep test
3. The effect of environmental exposure on the oxidation or scaling tendencies
4. Considerations of such properties as density, melting point, emissivity, ability to be coated and laminated, elastic modulus, and the temperature dependence on thermal conductivity and thermal expansion

Furthermore, the microstructural changes occurring in alloys used at high temperatures are correlated with property changes in order to account for the significant discontinuities which occur with exposure time. As a result of these evaluations, special alloys that have been (or are being) developed are recommended for use in different temperature ranges extending to about 2800°F (refractory range). Vacuum or electron-beam melting and special welding techniques are of special interest here in fabricating parts.

9.15.2 Creep and Stress: Rupture Properties

In a creep test, the specimen is heated in a temperature-controlled furnace, an axial load is applied, and the deformation is recorded as a function of time, for periods of 1000 to 3000 h. Typical changes in creep strain with time, for different conditions of stress and temperature, are shown in Fig. 9.35. Plastic flow creep, associated with the movement of dislocations by climb sliding of grain boundaries and the diffusion of vacancies, is characterized by:

1. *OA*, elastic extension on application of load
2. *AB*, first stage of creep with changing rate of creep strain
3. *BC*, second stage of creep, in which strain rate is linear and essentially constant
4. *CD*, third stage of increasing creep rate leading to fracture

Increasing stress at a constant temperature or increasing temperature at constant stress results in the transfers from curve 1 to curves 2 and 3 in Fig. 9.35.

The engineering design considerations for dimensional stability are based upon

1. Stresses resulting in a second-stage creep rate of 0.0001 percent per hour (1 percent per 10,000 h or 1 percent per 1.1 years)

2. A second-stage creep rate of 0.00001 percent per hour (1 percent per 100,000 h or 1 percent per 11 years), where weight is of secondary importance relative to long service life, as in stationary turbines

The time at which a stress can be sustained to failure is measured in a stress-rupture test and is normally reported as rupture values for 10, 100, 1000, and 10,000 h or more. Because of the higher stresses applied in stress-rupture tests, shown in Fig. 9.36, some extrapolation of data may be possible and some degree of uncertainty may ensue. Discontinuous changes at points *a* in the stress-rupture data shown in Fig. 9.37 are associated with a change from transgranular to intergranular fracture, and further microstructural changes can occur at increasing times. A composite picture of various high-temperature test results is given in Fig. 9.38 for a type 316 austenitic stainless steel.

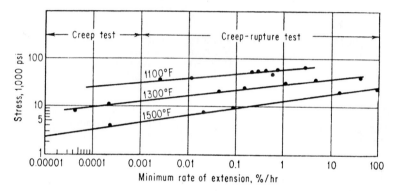

FIG. 9.36   Correlation of creep and rupture test data for type 316 stainless steel (18 Cr, 8 Ni, and Mo).[12]

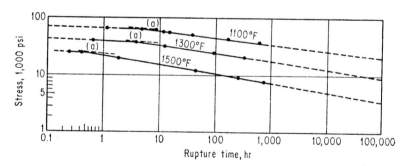

FIG. 9.37   Stress vs. rupture time for type 316 stainless steel.[12] The structural character associated with point (*a*), on each of the three relations, is that the mode of fracture changes from transgranular to intergranular.

9.15.3 Heat-Resistant Superalloys: Thermal Fatigue

The materials especially developed to function at high temperatures for sustained stress application in aircraft (jet engines) are referred to as "superalloys." These are nickel-base alloys designated commercially as Nimonic Hastelloy, Inconel Waspaloy, and René or cobalt-base alloys designated as S-816, HS25, and L605. Significant

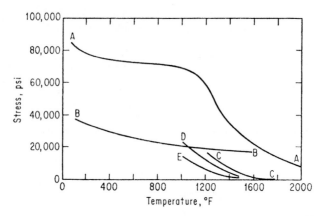

FIG. 9.38   Properties of type 316 stainless steel (18 Cr, 8 Ni, and Mo).[12] A = short-time tensile strength; B = short-time yield strength, 0.2 percent offset; C = stress of rupture, 10,000 h; D = stress for creep rate, 0.0001 percent per hour; E = stress for creep rate, 0.00001 percent per hour.

improvements have been made in engineering design of gas turbine blades (coolant vents) and material processing techniques (directional solidification, single crystal, and protective coating). The developed progress in the past several decades[40] led to increased engine efficiencies and increased time between overhauls from 100 h to more than 10,000 h.

Thermal fatigue is an important high-temperature design property which is related to fracture occurring with cyclical stress applications. This type of fatigue failure, resulting from thermal stresses (constrained expansion and metal structural changes), is also termed low-cycle fatigue. At sustained temperatures above 800°F the *S–N* curves generally do not have an endurance limit (asymptotic stress) and the fatigue stress continually decreases with cycles to failure.

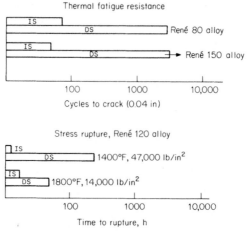

FIG. 9.39   Improved thermal fatigue resistance and stress-rupture properties of directionally solidified (DS) and isotropic superalloy (IS).[37,38]

Processing some superalloy turbine blades by directional solidification (DS) or by eliminating grain-boundary effects in single crystals results in significant improvements in thermal fatigue resistance and stress-rupture lives, as shown in Figure 9.39. DS is controlled by casting with stationary and movable heat sinks so that all grain boundaries are made to be parallel to the applied stress direction. Crystal growth directions are controlled to take advantage of the favorable anisotropic high-strength property values as well as minimizing deleterious grain-boundary reactions.

Materials selections for aircraft gas turbine systems are typically as follows:

Component	Materials used
Turbofin compressor blades	Titanium alloys: Ti-6Al-4V, Ti-6Al-2Sn-4Mo
Combustor (burner)	Sheet alloys, shaped and joined Hastelloy X, Inconel 617
Gas turbine blades	Cast Waspaloy, René, PNA 1422 (special processes—directional or single-crystal solidification)

Thermal shock failures, as in engine valves, can occur as a result of steep temperature gradients, leading to constrained thermal dilational changes when encountering high stress. These restrained stresses can exceed the breaking point of the material at the operating temperatures. Some superalloy properties useful for design are given in Table 9.3.

TABLE 9.3 Superalloy Properties[39]

Alloy	Rupture strength, after 1000 h, lb/in²			Characteristics and applications
	1200°F	1500°F	1800°F	
Hastelloy	34,000	10,000	2,100	Jet engine sheet parts, good oxidation resistance
Inconel 617	47,000	14,000	3,600	Gas turbine aircraft engine parts
René 41	10,200	29,000	—	Jet engine blade, parts
Waspaloy	88,000	25,500	—	Jet engine blades
In 102	52,000	7,400	—	Superheater, jet engine part
TAZ 8A	12,000	61,000	14,000	Good thermal fatigue and oxidation resistance
René	—	5,700	10,000	Compressor disk alloy
S816	46,000	18,000	—	Gas turbine blades, bolts, springs
Haynes 150	—	5,800	1,700	Resists thermal shock, high-temperature corrosion

9.16 MATERIALS FOR LOW-TEMPERATURE APPLICATION

Materials for low-temperature application are of increasing importance because of the technological advances in cryogenics. The most important mechanical properties are usually strength and stiffness, which generally increase as the temperature is decreased. The temperature dependence on ductility is a particularly important criterion in design, because some materials exhibit a transition from ductile to brittle behavior with decreasing temperature. Factors related to this transition are microstructure, stress concentrations present in notches, and the effects of rapidly applied strain rates in materials. Mechanical design can also influence the tendency for brittle failure at low temperature, and for this reason, it is essential that sharp notches (which can result from surface-finishing operations) be eliminated and that corners at changes of section be adequately filleted.

Low-temperature tests on metals are made by measuring the tensile and fatigue properties on unnotched and notched specimens and the notched impact strength. Metals exhibiting brittle characteristics at room temperature, by having low values of percent elongation and percent reduction in area in a tensile test as well as low impact strength, can be expected to be brittle at low temperatures also. Magnesium alloys, some high-strength aluminum alloys in the heat-treated condition, copper-beryllium heat-treated alloys, and tungsten and its alloys all exhibit this behavior. At best, applications of these at low temperatures can be made only provided that they adequately fulfill design requirements at room temperature.

When metals exhibit transitions in ductile-to-brittle behavior, low-temperature applications should be limited to the ductile region, or where experience based on field tests is reliable, a minimum value of impact strength should be specified. The failure, by breaking in two, of 19 out of 250 welded transport ships in World War II, caused by the brittleness of ship plates at ambient temperatures, focused considerable attention on this property. It was further revealed in tests that these materials had Charpy V-notch impact strengths of about 11 ft·lb at this temperature. Design specifications for applications of these materials are now based on higher impact values. For temperatures extending from subatmospheric temperatures to liquid-nitrogen temperatures ($-320°F$), transitions are reported for ferritic and martensitic steels, cast steels, some titanium alloys, and some copper alloys.

Design for low-temperature applications of metals need not be particularly concerned with the Charpy V-notch impact values provided they can sustain some shear deformation and that tensile or torsion loads are slowly applied. Many parts are used successfully in polar regions, being based on material design considerations within the elastic limit. When severe service requirements are expected in use, relative to rapid rates of applied strain on notch-sensitive metals, particular attention is placed on selecting materials which have transition temperatures below that of the environment. Some important factors related to the ductile-to-brittle transition in impact are the composition, microstructure, and changes occurring by heat treatment, preferred directions of grain orientation, grain size, and surface condition.

The transition temperatures in steels are generally raised by increasing carbon content, by the presence of more than 0.05 percent sulfur, and significantly by phosphorus at a rate of 13°F per 0.01 percent P. Manganese up to 1.5 percent decreases the transition temperature and high nickel additions are effective, so that in the austenitic stainless steels the behavior is ductile down to liquid-nitrogen temperatures. In high-strength medium-alloy steels it is desirable, from the standpoint of lowering the transition range, that the structure be composed of a uniformly tempered martensite, rather than containing mixed products of martensite and bainite or martensite and pearlite. This can be controlled by heat treatment. The preferred orientation that can be induced in rolled and forged metals can affect notched impact properties, so that specimens made from the longitudinal or rolling direction have higher impact strengths than those taken from the transverse direction. Transgranular fracture is normally characteristic of low-temperature behavior of metals. The metallurgical factors leading to intergranular fracture, due to the segregation of embrittling constituents at grain boundaries, cause concern in design for low-temperature applications. In addition to the control of these factors for enhanced low-temperature use, it is important to minimize or eliminate notch-producing effects and stress concentration, by specifying proper fabrication methods and providing adequate controls on these, as by surface inspection.

Examples of some low-temperature properties of the refractory metals, all of which have bcc structures, are shown in Fig. 9.40. The high ductility of tantalum at very low temperatures is a distinctive feature in this class that makes it attractive for use as a cryogenic (as well as a high-temperature) material. Based on the increase of yield-strength-to-density ratio with decreasing temperatures shown in Fig. 9.41, the

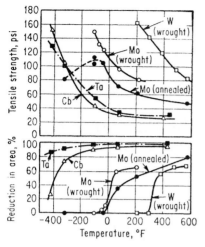

FIG. 9.40 Strength and ductility of refractory metals at low temperatures.[13]

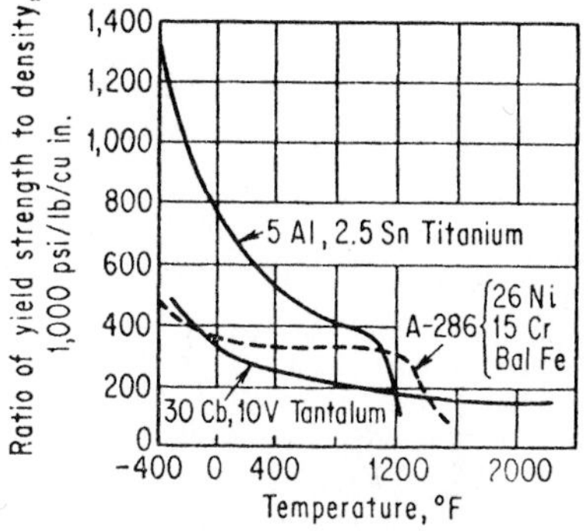

FIG. 9.41 Yield-strength-to-density ratios related to temperature for some alloys of interest in cryogenic applications.[13]

three alloys of a titanium-base Al (5) Sn (2.5), an austenitic iron-base Ni (26), Cr (15) alloy A286, as well as the tantalum-base Cb (30), 10 V alloy, are also useful for cryogenic use.

Comparisons of the magnitude of property changes obtained by testing at room temperature and −100°F for some materials of commercial interest are shown in Table 9.4.

9.17 RADIATION DAMAGE[31]

A close relationship exists between the structure and the properties of materials. Modification and control of these properties are available through the use of various metallurgical processes, among them nuclear radiation. Nuclear radiation is a process whereby an atomic nucleus undergoes a change in its properties brought about by interatomic collisions.

The energy transfer which occurs when neutrons enter a metal may be estimated by simple mechanics, the quantity of energy transferred being dependent upon the atomic mass. The initial atomic collision, or primary "knock-on" as it is called, has enough energy to displace approximately 1000 further atoms, or so-called secondary knock-ons. Each primary or secondary knock-on must leave behind it a resulting vacancy in the lattice. The primaries make very frequent collisions because of their slower movement, and the faster neutrons produce clusters of "damage," in the order of 100 to 1000 Å in size, which are well separated from one another.

Several uncertainties exist about these clusters of damage, and because of this it is more logical to speak of radiation damage than of point defects, although much of the damage in metals consists of point defects. Aside from displacement collisions, replacement collisions are also possible in which moving atoms replace lattice atoms. The latter type of collision consumes less energy than the former.

Another effect, important to the life of the material, is that of transmutation, or the conversion of one element into another. Due to the behavior of complex alloys, the cumulative effect of transmutation over long periods of time will quite often be of

TABLE 9.4 Low-Temperature Test Properties

Tensile Test Properties

Material	Tensile strength, 1,000 psi		Yield strength, 1,000 psi		Elongation, %		Modulus of elasticity, 100 psi	
	80°F	−100°F	80°F	−100°F	80°F	−100°F	80°F	−100°F
Beryllium copper 2% Be —heat-treated sheet....	189	194	154	170	3	3	19.1	19.1
Phosphor bronze 5% Sn; spring temp..........	98	107	90	97	7	11	16.5	17.1
Molybdenum sheet partly recrystallized.........	97	141	75	130	19	3.5	48.7	47.5
Tungsten wire, drawn....	214	...	196	...	2.6	....	56	
Tantalum sheet, annealed	55	71	36	66	27	25	28.2	28.3
Nilvar sheet 36 Ni: 74 Fe, rolled...............	76	101	62	76	27	30	21.8	19.7

Sheet Fatigue Properties

Material	Fatigue strength at 2 × 10⁷ cycles 1,000 psi		Endurance ratios	
	80°F	−100°F	80°F	−100°F
Beryllium copper..........	31	45	0.16	0.23
Molybdenum............. ...	46	65	0.59	0.47
Tantalum.................. .	35	41	0.64	0.58
Nilvar.......	26	30	0.33	0.30

Charpy V-notch Impact Strength

Material	Impact strength, ft-lb	
	80°F	−100°F
Beryllium copper...............	5.4	5.4
Phosphor bronze.....	46	44
Nilvar.................	97	77

importance. U^{235}, the outstanding example of this phenomenon, has enough energy after the capture of a slow neutron to displace one or more atoms.

Moving charged particles may also donate energy to the valence electrons. In metals this energy degenerates into heat, while in nonconductors the electrons remain in excited states and will sometimes produce changes in properties.

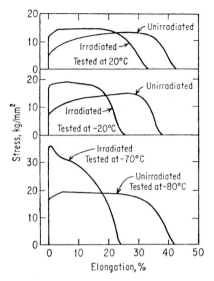

FIG. 9.42 Effect of irradiation on stress-strain curves of Fe single crystals tested at different temperatures. Irradiation dose 8×10^{17} thermal n/cm². (*Courtesy of D. McLean.*[31])

Figure 9.42 illustrates the effect or irradiation on the stress-strain curve of iron crystals at various temperatures.[32] In metals other than iron irradiation tends to produce a ferrous-type yield point and has the effect of hardening a metal. This hardening may be classified as a friction force and a locking force on the dislocations. Some factors of irradiation hardening are:

1. It differs from the usual alloy hardening in that it is less marked in cold-worked than annealed metals.

2. Annealing at intermediate temperatures may increase the hardening.

3. Alloys may exhibit additional effects, due, for example, to accelerated phase changes and aging.

The most noticeable effect of irradiation is the rise in transition temperatures of metals which are susceptible to cold brittleness. Yet another consequence of irradiation is the development of internal cracks produced by growth stresses. At high enough temperatures gas atoms can be diffused and may set up large pressures within the cracks.

Some other effects of irradiation are swelling, phase changes which may result in greater stability, radiation growth, and creep. The reader is referred to Ref. 31 for an analysis of these phenomena.

9.18 PRACTICAL REFERENCE DATA

Tables 9.5 through 9.9 give various properties of commonly used materials. Figure 9.43 provides a hardness conversion graph for steel. References 21 through 30 yield more information.

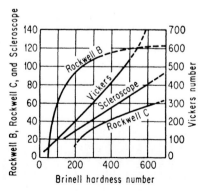

FIG. 9.43 Hardness conversion curves for steel.

TABLE 9.5 Physical Properties of Metallic Elements*

Element	Symbol	Melting point, °F	Boiling point, °F	Specific heat,ᵃ cal/g/°C	Thermal conductivity,ᵃ Btu/hr/sq ft/°F/ft	Density,ᵃ g/cm³	Modulus of elasticity in tension, million psi	Coefficient of linear thermal expansion,ᵃ μ in./in./°F	Electrical resistivity, microohm-cm	Crystal structure
Aluminum	Al	1220	4442	0.215	128.	2.70	9	13.1	2.65	f.c.c.
Antimony	Sb	1167	2516	0.049	10.8	6.62	11.3	4.7	39.	Rhomb.
Arsenic	As	1503 (28 atm)	1135ᵇ	0.082	...	5.72	...	2.6	33.3	Rhomb.
Barium	Ba	1317	2980	0.068	...	3.5	...	...	...	b.c.c.
Beryllium	Be	2332	5020	0.45	84.4	1.85	42	6.4	4	h.c.p.
Bismuth	Bi	520	2840	0.029	4.8	9.80	4.6	7.4	107	Rhomb.
Boron	B	3690	...	0.309	...	2.34	...	4.6	10¹²	Orthorhomb.
Cadmium	Cd	610	1409	0.055	53.	8.65	8.	16.55	6.83	h.c.p.
Calcium	Ca	1540	2625	0.149	72.3	1.55	3.5	12.4	3.91	f.c.c.
Carbon (graphite)	C	6740ᵇ	8730	0.165	13.8	2.25	0.7	0.3 to 2.4	1375	Hexag.
Cerium	Ce	1479	6280	0.045	6.6	6.77	6	4.4	75	f.c.c.
Cesium	Cs	84	1273	0.048	...	1.90	...	54	20	b.c.c.
Chromium	Cr	3407	4829	0.11	40.3	7.19	36	3.4	12.9	b.c.c.
Cobalt	Co	2723	5250	0.099	41.5	8.85	30	7.66	6.24	h.c.p.
Columbium	Cb	4474	8901	0.065	31.5	8.57	16	4.06	12.5	b.c.c.
Copper	Cu	1981	4703	0.092	226.	8.96	...	9.2	1.67	f.c.c.
Gallium	Ga	86	4059	0.079	19.4	5.91	...	10	17.4	Orthorhomb.
Germaniumᶜ	Ge	1719	5125	0.073	33.7	5.32	11.6	3.19	46 × 10⁶	Diam. cubic
Gold	Au	1954	5380	0.031	171.	19.32	1.57	7.9	2.35	f.c.c.
Indium	In	313	3632	0.057	13.8	7.31		18	8.37	f.c.tetr.
Iridium	Ir	4449	9570	0.031	33.7	22.50	76.5	3.8	5.3	f.c.c.
Iron	Fe	2798	5430	0.11	43.3	7.87	28.5	6.53	9.71	b.c.c.
Lanthanum	La	1688	6280	0.048	8.	6.19	10.5	2.77	57	Hexag.
Lead	Pb	621	3137	0.031	20.	11.36	2	16.3	20.6	f.c.c.
Lithium	Li	357	2426	0.79	41.	0.534	...	31	8.55	b.c.c.
Magnesium	Mg	1202	2025	0.245	88.5	1.74	6.35	15.05	4.45	h.c.p.
Manganese	Mn	2273	3900	0.115	...	7.43	23	12.22	185	Complex cubic
Mercury	Hg	−37	675	0.033	4.7	13.55	...	...	98.4	Rhomb.

Element	Symbol	Melting point, °F	Boiling point, °F	Specific heat,[a] cal/g/°C	Thermal conductivity,[a] Btu/hr/sq ft/°F/ft	Density,[a] g/cm³	Modulus of elasticity in tension, million psi	Coefficient of linear thermal expansion,[a] μ in./in./°F	Electrical resistivity, microhm-cm	Crystal structure
Molybdenum	Mo	4730	10040	0.066	82.	10.22	47	2.7	5.2	b.c.c.
Nickel	Ni	2647	4950	0.105	53.	8.90	30	7.39	6.84	f.c.c.
Osmium	Os	4900	9950	0.031		22.57	81	2.6	9.5	h.c.p.
Palladium	Pd	2826	7200	0.058	40.5	12.02	16.3	6.53	10.8	f.c.c.
Phosphorus (white)	P	112	536	0.177		1.83		70	10¹⁷	Cubic
Platinum	Pt	3217	8185	0.0314	39.8	21.45	21.3	4.9	10.6	f.c.c.
Plutonium	Pu	1184	6000	0.033	4.8	19.00	14	30.55	141.4	Monoclinic
Potassium	K	147	1400	0.177	58.	0.86		46	6.15	b.c.c.
Rhenium	Re	5755	10650	0.033	41.	21.04	66.7	3.7	19.3	h.c.p.
Rhodium	Rh	3571	8130	0.059	50.6	12.44	42.5	4.6	4.51	f.c.c.
Rubidium	Rb	102	1270	0.080		1.53		50	12.5	b.c.c.
Ruthenium	Ru	4530	8850	0.057		12.20	60	5.1	7.6	h.c.p.
Selenium[c]	Se	423	1265	0.084	48.2	4.79	8.4	21	12	Hexag.
Silicon[c]	Si	2570	4860	0.162		2.33	16.35	1.6 to 4.1	10⁵	Diam. cubic
Silver	Ag	1761	4010	0.056	242.	10.49	11	10.9	1.59	f.c.c.
Sodium	Na	208	1638	0.295	77.2	0.971		39	4.2	b.c.c.
Strontium	Sr	1414	2520	0.176		2.60			23	f.c.c.
Tantalum	Ta	5425	9800	0.034	31.3	16.60	27	3.6	12.45	b.c.c.
Tellurium	Te	841	1814	0.047	3.3	6.24	6	9.3	4.4 × 10⁵	Hexag.
Thallium	Tl	577	2655	0.031	22.5	11.85		16	18	h.c.p.
Thorium	Th	3182	7000	0.034	21.7	11.66		6.9	13	f.c.c.
Tin	Sn	449	4120	0.054	36.2	7.30	6.3	13	11	Tetrag.
Titanium	Ti	3035	5900	0.124	9.8	4.51	16.8	4.67	42	h.c.p.
Tungsten	W	6170	10706	0.033	96.	19.30	50	2.55	5.65	b.c.c.
Uranium	U	2070	6904	0.028	17.1	19.07	24	3.8 to 7.8	30	Orthorhomb.
Vanadium	V	3450	6150	0.119	16.9	6.1	19	4.6	26	b.c.c.
Yttrium	Y	2748	5490	0.071	8.5	4.47	17		57	h.c.p.
Zinc	Zn	787	1663	0.092	65.	7.13		22	5.92	h.c.p.
Zirconium	Zr	3366	6470	0.067	9.6	6.49	13.7	3.2	40	h.c.p.

*Courtesy of "Metals Handbook," vol. 1, 8th ed., American Society of Metals, Cleveland, 1961.

[a]Near 68°F (20°C).
[b]Sublimes; triple point at 2028 atm.
[c]Semiconductor.

TABLE 9.6 Typical Mechanical Properties of Cast Iron, Cast Steel, and Other Metals (Room Temperature)*

Metal	Modulus of elasticity E, million psi	Ultimate tensile strength σ_u, thousand psi	Yield strength σ_y, thousand psi	Endurance limit σ_{end}, thousand psi	Hardness, Brinell
Gray cast iron, ASTM 20, med. sec.	12	22	......	10	180
Gray cast iron, ASTM 50, med. sec.	19	53	......	25	240
Nodular ductile cast iron:					
Type 60-45-10	22–25	60–80	45–60	35	140–190
Type 120-90-02	22–25	120–150	90–125	52	240–325
Austenitic	18.5	58–68	32–38	32	140–200
Malleable cast iron, ferritic 32510	25	50	32.5	28	110–156
Malleable cast iron, pearlitic 60003	28	80–100	60–80	39–40	197–269
Ingot iron, hot rolled	29.8	44	23	28	83
Ingot iron, cold drawn	29.8	73	69	33	142
Wrought iron, hot rolled longit	29.5	48	27	23	97–105
Cast carbon steel, normalized 70000	30	70	38	31	140
Cast steel, low alloy, 100,000 norm. and temp	29–30	100	68	45	209
Cast steel, low alloy, 200,000 quench. and temp	29–30	200	170	85	400
Wrought plain C steel:					
C1020 hot rolled	29–30	66	44	32	143
C1045 hard. and temp. 1000°F	29–30	118	88	......	277
C1095 hard. and temp. 700°F	29–30	180	118	......	375
Low-alloy steels:					
Wrought 1330, HT and temp. 1000°F	29–30	122	100	......	248
Wrought 2317, HT and temp. 1000°F	29–30	107	72	......	222
Wrought 4340, HT and temp. 800°F	29–30	220	200	......	445
Wrought 6150, HT and temp. 1000°F	29–30	187	179	......	444
Wrought 8750, HT and temp. 800°F	29–30	214	194	......	423
Ultra high strength steel H11, HT	30	295–311	241–247	132	
300M HT and temp. 500°F		289	242	116	
4340 HT and temp. 400°F	30	287	270	107	
25 Ni Maraging	24	319	284		
Austenitic stainless steel 302, cold worked	28	110	75	34	240
Ferritic Stainless Steel 430, cold worked	29	75–90	45–80		
Martensitic Stainless Steel 410, HT	29	90–190	60–145	40	180–390
Martensitic Stainless Steel 440A, HT	29	260	240	......	510
Nitriding Steels, 135 Mod., hard. and temp. (core properties)	29–30	145–159	125–141	45–90	285–320
Nitiding Steels 5Ni-2A, hard. and temp	29–30	206	202	90	
Structural Steel	30	50–65	30–40	......	120
Aluminum Alloys, cast:					
195 SHT and aged	10.1	36	24	8	75
220 SHT	9.5	48	26	8	75
142 SHT and aged	10.3	28–47	25–42	9.5	75–110
355 SHT and aged	10.2	35–42	25–27	9–10	80–90
A13 as cast	10.3	39	21	19	

TABLE 9.6   Typical Mechanical Properties of Cast Iron, Cast Steel, and Other Metals (Room Temperature)* (*Continued*)

Metal	Modulus of elasticity E, million psi	Ultimate tensile strength σ_u, thousand psi	Yield strength σ_y, thousand psi	Endurance limit σ_{end}, thousand psi	Hardness, Brinell
Aluminum Alloys, wrought:					
EC ann	10	12	4		
ECH 19, hard	10	27	24	7	
3003 H 18, hard	10	29	27	10	55
2024 H T (T3)	10.6	70	50	20	120
5052 H 38, hard	10.2	42	37	20	77
7075 HT (T6)	10.4	83	73	23	150
7079 HT (T6)	10.3	78	68	23	145
Copper alloys, cast:					
Leaded red brass BB11-4A	9–14.8	33–46	17–24	. . . . .	55
Leaded tin bronze BB11-2A	12–16	36–48	16–21	. . . . .	60–72
Yellow brass BB11-7A	12–14	60–78	25–40	. . . . .	80–95
Aluminum bronze BB11-9BHT	15	90	40	. . . . .	180
Copper alloys, wrought:					
Oxygen-free 102 ann	17	32–35	10	11	
Hard		50–55	45	13	
Beryllium copper, 172 HT	19	165–183	150–170	35–40	
Cartridge Brass, 260 hard	16	76	63	21	
Muntz metal, 280 ann	15	54	21		
Admiralty, 442 ann	16	53	22		
Manganese bronze, 675 hard	15	84	60		
Phosphor bronze, 521 spring	16	112	70		
Silicon bronze, 647 HT	18	100	88		
Cupro-Nickel, 715 hard	22	80	73		
Magnesium Alloys, cast:					
AZ63A, aged	6.5	30–40	14–19	11–15	55–73
AZ92A, aged	6.5	26–40	16–21	11–15	80–84
HK31A, T6	6.5	31	16	9–11	55
Magnesium Alloys, wrought:					
AZ61AF, forged	6.5	43	26	17–22	55
ZE-10A-H24	6.5	34–38	19–28	20–24	
HM31A-T5	6.5	42	33	12–14	
Nickel-alloy castings 210	21.5	45–60	20–30	. . . . .	80–125
Monel 411 cast	19	65–90	32–45	. . . . .	125–150
Inconel 610 cast	23	70–95	30–45	. . . . .	190
Nickel alloys, wrought:					
200 Spring	30	90–130	70–115		
Duranickel, 301 Spring	30	155–190			
K Monel, K500 Spring	26	145–165	130–180		
Titanium alloys, wrought, unalloyed	15–16	60–110	40–95	60–70	
5Al-2.5 Sn	16–17	115–140	110–135	95	
13V-11 Cr-3Al HT	14.5–16	190–240	170–220	50–55	
Zinc, wrought, comm. rolled	. . . . .	25–31	. . . . .	4.1	
Zirconium, wrought:	14	64	53		
Reactor grade	14	49	29		
Zircaloy 2	13.8	68	61		
Pure metals, wrought:					
Beryllium, ann	44	60–90	45–55		
Hafnium, ann	20	77	32		
Thorium, ann	10	34	26		
Vanadium, ann	20	72	64		
Uranium, ann	30	90	25		

TABLE 9.6 Typical Mechanical Properties of Cast Iron, Cast Steel, and Other Metals (Room Temperature)* *(Continued)*

Metal	Modulus of elasticity E, million psi	Ultimate tensile strength σ_u, thousand psi	Yield strength σ_y, thousand psi	Endurance limit σ_{end}, thousand psi	Hardness, Brinell
Precious metals:					
Gold, ann.	12	19	. . . . . .	46	25
Silver, ann.	11	22	8	. . . . . .	25–35
Platinum, ann.	21	17–26	2–5.5	. . . . . .	38–52
Palladium, ann.	17	30	5	. . . . . .	46
Rhodium, ann.	42	73	. . . . . .	. . . . . .	55–156
Osmium, cast.	80	. . . . . .	. . . . . .	. . . . . .	350
Iridium, ann.	74	. . . . . .	. . . . . .	. . . . . .	170

Babbitt has a compressive elastic limit of 1.3 to 2.5 $\times$ 10^3 lb/in^2 and a Brinell hardness of 20.
Compressive yield strength of all metals, except those cold-worked = tensile yield strength.
Poisson's ratio is in the range 0.25 to 0.35 for metals.
Yield strength is determined at 0.2 percent permanent deformation.
Modulus of elasticity in shear for metals is approximately 0.4 of the modulus of elasticity in tension E.
Compressive yield strength of cast iron 80,000 to 150,000 $\times$ 10^3 lb/in^2.
*From *Materials in Design Engineering*, Materials Selector Issue, vol. 56, no. 5, 1962; courtesy of Reinhold Publishing Corporation, New York.

TABLE 9.7 Mechanical Properties and Applications of Steels

Carbon steels*

AISI	Condition	Tensile strength, lb/in²	Yield strength, lb/in²	Elongation in 2 in, %	Brinell hardness	Machinability	Comments and uses
1020	Annealed	55,000	30,000	25	111	65	Weldable, structural beams, column: case hardener, gears, camshafts, kingpins (hard surface and soft core)
1040	Annealed	76,000	42,000	18	149	60	Not readily weldable. Hardened bolts (shallow). case-fused axles, shafts, chain pins, hobbing splints
1070	Annealed	102,000	56,000	12	212	55	Hardened springs, hand tools, saw material
1095	Annealed	120,000	66,000	10	248	45	Hardened springs, knives, ore screens

Alloy steels (quenched and tempered)†

AISI	Tempering temperature, °F	Tensile strength, lb/in²	Yield strength, lb/in²	Elongation, % in 2 in	Brinell hardness	Machinability (soft)	Comments and uses
1141	400	237,000	176,000	6	461	70	Free-machining (sulfur); gears, transmission, and spline shafts works
	1000	130,000	111,000	18	262		
1345	400	270,000	240,000	10	530	45	Steering knuckle forgings
	1000	140,000	124,000	16	310		
4140	400	257,000	238,000	8	510	65	Can be nitrided, transmission shafts, tractor pins, pinions
	1000	138,000	121,000	18	285		
4340	400	272,000	243,000	10	520	50	Drums for helicoptor transmission (nitrided)
	1000	170,000	156,000	13	360		
6150	400	280,000	245,000	8	538	55	High-stress, high-temperature springs, die-casting dies, plastic molds
	1000	168,000	155,000	13	345		
8650	400	281,000	243,000	10	525	60	Springs, gears
	1000	170,000	153,000	15	340		
9310	400	174,000	150,000	15	390	50	Carburized aircraft gears, high shock resistance, surface C not to exceed 0.9 percent to avoid retained austenite (poorer wear)

TABLE 9.7 Mechanical Properties and Applications of Steels (*Continued*)

AISI	Typical composition			Mechanical properties at 70°F, annealed			Maximum continuous service temperature, °F (air)	Characteristics and applications
	Cr	Ni	C	Tensile strength, lb/in²	Yield strength, lb/in²	Elongation, %		
205	17	1.8	0.20	120,000	69,000	58	—	Spinning and drawing operations, nonmagnetic cryogenic parts
304	18	8	0.08	85,000	35,000	60	1650	Chemical and food processing, cryogenic hypodermic needles
316	17	12	0.08	84,000	32,000	50	1650	Higher corrosion resistance, high creep strength, chemical and pulp equipment
317	18§	10§	0.08§	90,000	35,000	45	1650	Stabilized for weldments, exhaust manifold aircraft, fire walls, pressure vessels
403	12	—	0.15	70,000	45,000	25	1300	Martensitic, turbine quality, jet engine rings
440C	17	—	1.2	110,000	65,000	14	1400	Martensitic, highest hardness, balls, bearings, cases, and nozzles
S15500	17	3.5	0.07	160,000	145,000	15	—	Martensitic, maraging high strength, cams, gears, shafting

Tool Steels¶

AISI	Type	Typical composition					Applications
		C	W	Nb	Cr	V	
W1	Water hardening	0.6 1.4	—	—	—	—	Low C, blanking tools, forging dies; high C, lathe tools, reamers, taps
S1	Shock resisting	0.5	2.5	—	1.5	—	Pneumatic tools, swaging dies, shear blades, master hobs
O7	Oil hardening	1.2	1.75	•	0.75	—	Mandrels, slitters, drills, taps, blanking dies
A2	Age hardening	1.0	—	1.0	5.0	—	Thread rolling dies, extrusion dies, coining dies
D2	High C–high Cr	1.5	—	1.0	12.0	—	Blanking dies, roll thread dies, shear blades, slitters

AISI	Type	Typical composition					Applications
		C	W	Mo	Cr	V	
D3	High C–high Cr	2.25	—	—	12.0	—	More wear-resistant than D2
H12	Hot work	0.35	1.5	1.5	5.0	0.4	Extrusion dies, forging die inserts, punches
T1	High speed	0.75	18	—	4	1	Drills, taps, tool cutters, high hot hardness
M4	High speed	1.3	5.5	4.5	4	4	Lathe and planer tools, brooches, swaging dies
P2	Mold dies	0.07	—	0.2	2	0.5 Ni	Plastic molding dies (hobbed and carburized)

*From "Metals Handbook," vol. 1, 8th ed., American Society for Metals, 1961. Machinability values from "SAE Handbook," 1979; refer to a rating of 100 for AISI 1212 (a resulfurized and rephosphorized) free-machining steel in cold-drawn condition. These plain carbon steels have a low or shallow hardenability when heat-treated.
†From "Metal Progress Databook," 1981; "ASM Metals Handbook," vol. 1, 8th ed., American Society for Metals, 1961. These alloyed steels have greater hardenability than the plain carbon steels when heat-treated.
‡"Metal Progress Databook," 1981.
§Plus Ti (5 × C).
¶From "Metal Progress Databook," 1981.

TABLE 9.8 Typical Properties of Refractory Ceramics and Cermets and Other Materials

Type	Specific gravity	Melting point, °F	Mean specific heat, Btu/lb/ °F	Mean coefficient of thermal expansion, μ in./in./ °F	Mean thermal conductivity, Btu/hr/ sq ft/ °F/ft	Hardness, Mohs scale	Maximum service temperature (oxidiz.), °F
Alumina (99+) (Al$_2$O$_3$).........	3.85	3725	0.23	4.3	10.7	9	3540
Beryllia (BeO)................	3.0	4620	0.29	5.3	9.5	9	4350
Magnesia (MgO)...............	3.6	5070	0.26	7.8	1.47	6	4350
Thoria (ThO$_2$)...................	....	6000	....	5.28	0.0	7	4890
Zirconia (ZrO$_2$)...............	5.5–6	4710	0.16	3.06	0.53	7–8	4530
Quartzite (SiO$_2$)...............	2.65	2552	0.26	0.28	0.8		
Silicon carbide (Dens)(SiC)......	3.2	......	0.33	2.17	25	9–10	3000
Boron carbide..................	2.5	......	....	1.73	16	....	1000
Titanium carbide (TiC).........	6.5	......	....	4.3–7.5			
Tungsten carbide (WC).........	14.3	......	....	2.5–3.9	26–50		
Boron nitride.................	2.05–2.15	4930	....	5.5	10–20		
Graphite.....................	2.25	......	0.18	1.0–1.3	70–120	1.2	

TABLE 9.9 Typical Properties of Plastics at Room Temperature

Type	Specific gravity	Coefficient of thermal expansion, $10^{-5}/$ °F	Thermal conductivity, Btu/hr/ sq ft/ °F/ft	Volume resistivity, ohm-cm	Dielectric strength [a], volts/ mil	Modulus of elasticity in tension, 10^5 psi	Tensile strength, 10^3 psi
Acrylic, general purpose, type I.................	1.17–1.19	4.5	0.12	$>10^{15}$	450–530	3.5–4.5	6–9
Cellulose acetate, type I (med.)..................	1.24–1.34	4.4–9.0	0.1–0.19	10^{12}	250–600	....	2.7–6.5
Epoxy, general purpose.....	1.12–2.4	1.7–5.0	0.1–0.8	10^{13}	350–550	....	2–12
Nylon 6....................	1.13–1.14	4.6–5.4	0.1–0.14	10^{14}	420–485	2.5–3.4	10.2–12
Phenolic, type I (mech.)....	1.31	3.3–4.4	....	1.7×10^{12}	350–400	4–5	6–9
Polyester, Allyl type........	1.30–1.45	2.8–5.6	0.12	$>10^{13}$	330–500	2–3	4.5–7
Silicone cast (mineral)...	1.80–2.0	2.8–3.2	0.09	$>10^{13}$	350–400	....	4.2
Polystyrene, general purpose	1.04–1.07	3.3–4.8	0.06–0.09	10^{18}	>500	4–5	5–8
Polyethylene, low density...	0.92	8.9–11	0.19	10^{18}	480	0.22	1.4–2
Polyethylene, medium density....................	0.93	8.3–16.7	0.19	$>10^{15}$	480	....	2
Polyethylene, high density..	0.96	8.3–16.7	0.19	$>10^{15}$	480	....	4.4
Polypropylene.............	0.89–0.91	6.2	0.08	10^{16}	769–820	1.4–1.7	5

[a] Short time.

REFERENCES

1. Richards, C. W.: "Engineering Materials Science," Wadsworth Publishing Co., San Francisco, 1961.
2. Barrett, C. S.: "Structure of Metals," 2d ed., McGraw-Hill Book Company, Inc., New York, 1952.
3. Sachs and Van Horn: "Practical Metallurgy," American Society for Metallurgy, 1940.

4. "Metals Handbook," vol. 1, 8th ed., American Society for Metals, Cleveland, 1961.

5. "Heat Treatment and Properties of Iron and Steel," *Natl. Bur. Stand. (U.S.) Monograph* 18, 1960.

6. "Metals Handbook," 1954 Supplement, American Society for Metals, Cleveland.

7. "Heat Treatment and Properties of Iron and Steel," *Natl. Bur. Stands (U.S.) Monograph* 18, 1960.

8. Palmer, F. R., and G. V. Luersson: "Tool Steel Simplified," Carpenter Steel Co., 1948.

9. "Suiting the Heat Treatment to the Job," United States Steel Co.

10. "Metals Handbook," American Society for Metals, Cleveland, 1939 ed.

11. "Three Keys to Satisfaction," Climax Molybdenum Co., New York.

12. "Steels for Elevated Temperature Service," United States Steel Co.

13. *Metal Prog.,* vol. 80, nos. 4 and 5, October and November, 1961.

14. Norton, J. T., and D. Rosenthal: *Welding J.,* vol. 2, pp. 295–307, 1945.

15. "ASME Handbook, Metals Engineering—Design," McGraw-Hill Book Company, Inc., New York, 1953.

16. "Symposium on Corrosion Fundamentals," A series of lectures presented at the University of Tennessee Corrosion Conference at Knoxville, The University of Tennessee Press, Knoxville, 1956.

17. Evans, Ulich R.: "The Corrosion and Oxidation of Metals," St. Martin's Press, Inc., New York, 1960.

18. Burns, R. M., and W. W. Bradley: "Protective Coatings for Metals," Reinhold Publishing Corporation, New York, 1955.

19. Bresle, Ake: "Recent Advances in Stress Corrosion," Royal Swedish Academy of Engineering Sciences, Stockholm, Sweden, 1961.

20. "ASME Handbook, Metals Engineering—Design," McGraw-Hill Book Company, Inc., New York, 1953.

Some suggested references recommended for the selections and properties of engineering materials are the following:

21. "Metals Handbook," vol. 1, 8th ed., American Society for Metals, Cleveland, 1961.

22. *Metals Prog.,* vol. 66, no. 1-A, July 15, 1954.

23. *Metals Prog.,* vol. 68, no. 2-A, Aug. 15, 1955.

24. Dumond, T. C.: "Engineering Materials Manual," Reinhold Publishing Corporation, New York, 1951.

25. "Steels for Elevated Temperature Service," United States Steel Co.

26. "Three Keys to Satisfaction," Climax Molybdenum Co., New York.

27. Zwikker, C.: "Physical Properties of Solid Materials," Interscience Publishers, Inc., New York, 1954.

28. Teed, P. L.: "The Properties of Metallic Materials at Low Temperatures," John Wiley & Sons, Inc., New York, 1950.

29. Hoyt, S. L.: "Metals and Alloys Data Book," Reinhold Publishing Corporation, New York, 1943.

30. *Materials in Design Engineering,* Materials Selector Issue, vol. 56, no. 5, Reinhold Publishing Corporation, New York, 1962.

31. McLean, D.: "Mechanical Properties of Metals," John Wiley & Sons, Inc., New York, 1962, pp. 363–382.

32. Edmonson, B.: *Proc. Roy. Soc. (London), Ser. A,* vol. 264, p. 176, 1961.

33. Norton, J. T., and D. Rosenthal: "X-ray Diffraction Measurements," *Welding J.,* vol. 24, pp. 295–307, 1945.

34. Field, Metal: "Machining of High Strength Steels with Emphasis on Surface Integrity," Air Force Machine Data Center, Cincinnati, 1970.

35. Suh, N. P., and A. P. L. Turner: "Elements of Mechanical Behavior of Solids," McGraw-Hill Book Co., Inc., New York, pp. 489–490, 1975.

36. "Metals Handbook," vol. 1, 9th ed. *American Society for Metals,* Cleveland, p. 674, 1978.

37. Woodford, D. A., and D. F. Mawbray: *Mater. Sci. Eng.,* vol. 16, pp. 5–43, 1974.

38. Wright, P. K., and A. F. Anderson: *Met. Tech. G.E.* pp. 31–35, Spring 1981.

39. "Metal Progress Databook," American Society for Metals, Metals Park, Ohio, 1980.

SECTION 10
DIMENSIONS AND TOLERANCES

Earlwood T. Fortini, M.S.

Director of Engineering
Cordell Engineering, Inc.
Peabody, Mass.

10.1 INTRODUCTION

Designers and detailers of mechanical products face two related tasks. One is determining ideal values for dimensions so that parts made to these ideal dimensions assemble and operate in the best possible way; the other is assigning tolerances so that even when not made to ideal dimensions, parts still assemble and operate satisfactorily. This section concerns the second task.

Parts cannot be made to ideal dimensions because of inaccuracies inherent in manufacturing processes. To allow for manufacturing inaccuracies dimensions must include *tolerances*. Tolerance size and manufacturing cost are related, as in Fig. 10.1; tolerance values, therefore, should be assigned based on cost.

This article concerns methods for relating dimensions and tolerances to conditions affecting assembly or operation. Determining value for conditions controlling assembly or operation is one of the principal activities of product designers. This activity may be accomplished on the basis of theory, experiment, or experience.

After a mechanical product is in use, dimensional values degrade because of wear, corrosion, changes in temperature, and so forth. Product designers must accommodate for these changes since dimensional values assigned to product drawings apply to the conditions of assembly or operation that will occur in newly manufactured products.

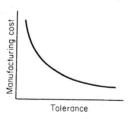

FIG. 10.1 Relationship between manufacturing cost and tolerance.

10.2 DIMENSIONING PRACTICES

Relationships between dimensions, tolerances, and conditions affecting assembly and operation depend upon assumptions concerning dimensioning practices. A dimensioning practice consists of a combination of basis of dimensioning and a dimensioning system.

"Basis of dimensioning" concerns expected values of dimensions or tolerances. The two common bases of dimensioning are the worst-limit and statistical. For the "worst-limit" basis, it is expected that limit values of dimensions or tolerances will combine to give extreme conditions of assembly or operation. For the "statistical" basis, it is expected that dimensions or tolerances will combine according to the laws of chance so that extreme conditions of assembly or operation are not likely. The statistical basis allows the use of greater tolerance values than permitted by the worst-limit basis, hence results in lower manufacturing costs.

"Dimensioning system" concerns drafting conventions for specifying dimensions and tolerances. The two dimensioning systems in common use are the toleranced dimension and true position. In the "toleranced dimension" system, a dimension has definite maximum and minimum limits, and therefore tolerance. In the "true position" system, an untoleranced dimension locates a position and the position is toleranced; the position tolerance has a basic value that may increase as the actual sizes of data or positioned features depart from maximum material conditions.

The set of drawings representing a mechanical product, and indeed, individual drawings of the part, will ordinarily include a mixture of practices. Depending upon its functions, a part may involve several dimensioning problems. Each problem will require careful study before deciding which dimensioning practice applies best to its solution.

10.3 TERMINOLOGY AND SYMBOLS

The term "dimension" can apply to a broad variety of physical values that influence the assembly and functioning of mechanical parts. Thus, the shear modulus of elasticity of a helical compression spring can be called a dimension since it affects the amount of force exerted by the deflected spring. When applied to the dimensioning of product drawings, the term dimension is usually associated to the specification of size and length.

A dimension is "functional" if it has an effect on assembly or operation; otherwise it is "nonfunctional." Only functional dimensions are of concern in this article.

Toleranced dimensions are denoted by the symbol x. The symbol for a specific dimension includes a subscript, for instance x_a. The minimum limit of dimension x_a is denoted by a and the maximum limit by A. Similarly, b is the minimum limit of x_b and B is the maximum limit, and so forth. Subscripts can be suitably assigned to designate dimensions of length, angle, or more general physical values such as shear modulus of elasticity.

"Tolerance," the difference between the limits of a dimension, is denoted by the symbol t. For dimension X_a, $t_a = A - a$; for x_b, $t_b = B - b$; and so forth. A tolerance always has a positive value.

A "dimension condition," the result of a combination of dimensions, will affect the assembly or operation of a mechanical product. For instance, the dimension condition of a helical compression spring is a force resulting from assembly dimensions, number of coils, wire diameter, shear modulus of elasticity, and so forth. The symbol y denotes dimension conditions. Assembly fits, the most common dimension conditions, are denoted by subscript w. Dimension conditions for other than assembly fits are denoted by subscript z or by other subscripts; for instance, subscript F stands for force. Numbers can be added to distinguish between different dimension conditions; for instance, y_{w1}, y_{w2}, y_{z4},....

The minimum limit of dimension condition y_{w1} is denoted by w_1 and its maximum limit by W_1; z_1 is the minimum limit of y_{z1}, and Z_1 the maximum limit, etc.

When the statistical basis of dimensioning applies, limits of dimension conditions have probable values. Subscript p denotes this fact. Thus w_{p1} is the symbol for the probable minimum limit of dimension condition y_{wp1}.

"Variation of a dimension condition," the difference between limits of dimension condition, is denoted by the symbol v. For dimension condition y_{w1}, $v_{w1} = W1 - w_1$; for y_{z4}, $v_{z4} = Z_4 - z_4$; etc.

Table 10.1 summarizes the symbols used for dimensions, tolerances, dimension conditions, and variations of dimension conditions.

TABLE 10.1   Symbols for Dimensions, Tolerances, Dimension Conditions, and Variations of Dimension Conditions

x	General symbol for a dimension
x_a, x_b, x_c, $\ldots$	Symbols of particular dimensions
a, b, c, $\ldots$	Minimum limits for x_a, x_b, x_c, $\ldots$
A, B, C, $\ldots$	Maximum limits for x_a, x_b, x_c, $\ldots$
t	General symbol for tolerance
t_a, t_b, t_c, $\ldots$	Tolerances for x_a, x_b, x_c, $\ldots$
y	General symbol for a dimension condition
y_{w1}, y_{w2}, y_{z1}, $\ldots$	Symbols for particular dimension conditions; w denotes assembly fits
w_1, w_2, z_1, $\ldots$	Minimum limits for y_{w1}, y_{w2}, y_{z1}, $\ldots$
W_1, W_2, Z_1, $\ldots$	Maximum limits for y_{w1}, y_{w2}, y_{z1}, $\ldots$
W_{p1}, w_{p1}, $\ldots$	Probable limits for y_{wp1}, $\ldots$
v	General symbol for variation of dimension condition
v_{w1}, v_{w2}, v_{z1}, $\ldots$	Variation of dimension condition for y_{w1}, y_{w2}, y_{z1}, $\ldots$

10.4   GENERAL DIMENSION AND TOLERANCE RELATIONSHIPS

A mechanical product, depending upon the number of parts it contains, will have from a few to a great many dimension conditions affecting assembly and operation. For the simple case, a dimension condition can be represented by a single equation having the form

$$y = f(x_a, x_b, \ldots) \qquad (10.1)$$

in which y is not related to any other dimension, and dimensions x_a, x_b, $\ldots$, are not included in any other dimension condition. There are two complex cases.

1. Dimension conditions y_1 and y_2, for example, may be related by equations having the forms

$$y_1 = f(x_a, x_b, \ldots, y_2) \qquad (10.2a)$$

$$y_2 = f(x_g, x_h, \dots) \tag{10.2b}$$

2. The same dimension may be present in two or more equations:

$$y_3 = f(x_a, x_b, \dots) \tag{10.3a}$$

$$y_4 = f(x_b, x_g, \dots) \tag{10.3b}$$

Equations (10.1) through (10.3) are called "equations of dimensional condition." For each equation of dimension condition, a "tolerance equation" can be derived relating the variation of dimension condition to tolerances. A tolerance equation derived from Eq. (10.1) can have the general form:

$$v = f(x_a, t_a, x_b, t_b, \dots) \tag{10.4}$$

In some tolerance equations any or all of the terms $x_a, x_b, \dots$, may not be present.

Before deriving an equation of dimension condition, a choice must be made concerning the basis of dimensioning. In addition to the worst-limit basis, two forms of the statistical basis are considered here: the equal-likelihood and normal distributions.

The "worst-limit" basis assumes that limit values of dimensions combine to give extreme possible value of dimension conditions. Tolerance equations can be derived by operating on equations of dimension condition with the worst-limit basis equation:

$$v = \left| \frac{\partial f}{\partial x_a} \right| t_a + \left| \frac{\partial f}{\partial x_b} \right| t_b + \cdots \tag{10.5}$$

The worst-limit basis applies when no risk can be permitted that will allow values of dimension conditions to occur outside of design limits.

"The statistical basis of dimensioning with dimensions having equal-likelihood distributions" assumes that any value of a dimension between limits will have the same chance of occurring as any other value. Tolerance equations can be derived by operating on equations of dimension condition with the statistical basis equation with equal-likelihood distribution:

$$v = \frac{u}{\sqrt{3}} \left[\left(\frac{\partial f}{\partial x_a} t_a \right)^2 + \left(\frac{\partial f}{\partial x_b} t_b \right)^2 + \cdots \right]^{1/2} \tag{10.6}$$

Calculations will give one form of statistical tolerancing. With statistical tolerancing there is a risk that values of the dimension condition will occur outside of design limits because of larger tolerance values than would be calculated using worst-limit tolerancing; this risk can be kept small. Larger tolerance values usually lead to lower manufacturing costs.

The equal-likelihood distribution is also called the rectangular distribution. In practice this distribution will approximate a variety of actual distributions. A typical actual distribution is shown in Fig. 10.2 along with the equal-likelihood distribution it is

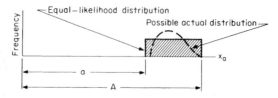

FIG. 10.2   Possible actual distribution for dimension x_a conservatively represented by an equal-likelihood distribution.

assumed to approximate. The equal-likelihood distribution usually conservatively represents actual distributions since dimensional values tend to cluster around the mean.

Tolerance equations derived by the application of the statistical basis equation with normal distribution apply only if the number of tolerances is sufficiently great and if the equation is not dominated by a few larger tolerances.

$$v = \left[\left(\frac{\partial f}{\partial x_a} t_a \right)^2 + \left(\frac{\partial f}{\partial x_b} t_b \right)^2 + \cdots \right]^{1/2} \tag{10.7}$$

The distribution of the dimension condition will tend to become normal as the number of dimensions in the equation of dimension condition increases. This fact is a consequence of the central limit theorem. Figure 10.3 shows distributions of dimension conditions for sums of two, three, and four dimensions having equal-likelihood distributions. Probabilities associated with the standardized normal variable u in Eq. (10.6) should be taken as relative values only (see Table 10.2). The larger the value of u, the less the risk will be that values of the dimension condition will occur outside of probable limits. The risk will also decrease if actual distributions are like that shown in Fig. 10.2. If actual distributions of dimensions fall outside of assumed equal-likelihood distributions, the risk that dimension conditions will occur outside of probable limits will increase.

The "normal distribution" applies when parts are made under statistical quality control so that actual distributions can be conservatively represented by ideal normal distributions. Tolerance equations based on the normal distribution can be derived by operating on equations of dimension condition with Eq. (10.7)

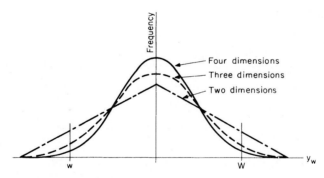

FIG. 10.3 Distributions of dimension conditions for sums of two, three, and four dimensions having equal-likelihood distributions.

TABLE 10.2 Relationship of Standardized Normal Variable u and Risk that a Dimension Condition Will Fall Outside of Design Limits

$\pm u$	Risk, %
2.0	4.56
2.5	1.24
3.0	0.26
3.5	0.04

The standardized normal variable u is implicit in Eq. (10.7), since each tolerance is equal to $2u\sigma$, where σ is the allowable standard deviation due to the combination of manufacturing and inspection inaccuracies. The probability that dimension conditions will occur within design limits is the same probability that dimension values will occur within limits specified on drawings.

If the actual standard deviation of a dimension is less than the allowable standard deviation, the chance that the value of the dimension condition will occur within its probable limits will increase. Conversely, if the actual standard deviation of a dimension is more than the allowable standard deviation, the chance that the value of the dimension conditions will occur within its probable limits will decrease.

The following example illustrates the formulation of equations of dimension condition and tolerance equations.

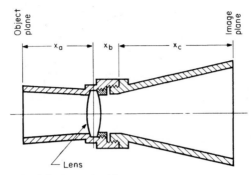

FIG. 10.4   Lens assembly.

EXAMPLE 1   For the lens assembly shown in Fig. 10.4, optical magnification y_m is a dimension condition related to dimensions x_a, x_b, and x_c by equation of dimension condition

$$y_m = (x_b + x_c)/x_a \qquad (10.8)$$

Assuming the worst-limit basis of dimensioning, the tolerance equation derived by the application of Eq. (10.5) is

$$v_m = M - m = [(b + c)/a^2]t_a + (1/a)t_b + (1/a)t_c \qquad (10.9)$$

10.5   THE ASSIGNMENT OF VALUES TO DIMENSIONS AND TOLERANCES

Values assigned to dimensions and tolerances for the two preceding examples must satisfy both an equation of dimension condition and a tolerance equation.

A dimension with known value at the outset of a calculation is called a "restricted dimension"; the tolerance of such a dimension is called a "restricted tolerance." A dimension with unknown value at the outset of a calculation is called an "unrestricted dimension" and its tolerance an "unrestricted tolerance."

For simple cases, only one equation of dimension condition and its corresponding tolerance equation are involved in a dimension problem. Mean values can be assumed for the dimension condition and all unrestricted dimensions except one; the mean value of the remaining dimension can then be calculated using the equation of dimension condition.

After substituting into a tolerance equation, mean values of dimensions, limits of dimension condition, and restricted tolerances, a tolerance equation, derived by the application of Eq. (10.5), can take the form

$$K = k_a t_a + k_b t_b + \cdots \tag{10.13}$$

Compare, for example, Eq. (10.13) with Eq. (10.9). If derived by the application of Eq. (10.6) or (10.7), then a tolerance equation can take the form

$$K^2 = (k_a t_a)^2 + (k_b t_b)^2 + \cdots \tag{10.14}$$

Compare, for example, Eq. (10.14) with Eq. (10.11).

In the foregoing equations K, k_a, k_b, ... will be numerical values.

If a tolerance equation includes only a single unrestricted tolerance, the calculations of a value for that tolerance can be simply undertaken. However, when a tolerance equation includes more than one unrestricted tolerance, additional constraints are required in order to calculate tolerance values.

An ideal assignment of tolerance values will result in lowest manufacturing cost. Achieving this ideal requires cost data as well as the skills and facilities required for calculations. Various methods for optimum tolerance assignment have been described. However, for many problems neither cost data nor the time and effort required for optimal tolerance calculations are available.

The difficulty factors method offers a simple means for assigning tolerance values and provides an approach to optimum tolerance assignment. The relative difficulty of producing dimensions on manufactured parts to within tolerance is estimated by assigning "unit factors" corresponding to difficulties or costs associated with manufacturing processes, materials from which parts are made, shapes of the parts, and sizes of dimensions. The "difficulty factor" for a particular unrestricted tolerance is equal to the sum of its unit factors. Difficulty factors can be generated from a worksheet as in Fig. 10.5. ΣR denotes the sum of difficulty factors.

Tolerances	Unit factors				Difficulty factors
	Process	Material	Shape	Size	
t_a	1.3	1.7	1.2	1.0	$R_a = 5.2$
t_c	1.5	1.3	1.2	1.0	$R_c = 5.0$
t_D	2.6	1.5	2.0	1.0	$R_D = 7.1$
t_L	3.6	1.5	2.0	1.0	$R_L = 8.1$
				Sum of difficulty factors:	$\Sigma R = 25.4$

FIG. 10.5 Worksheet for calculating difficulty factors and the sum of difficulty factors for the unrestricted tolerances of the plunger assembly.

Values for unrestricted tolerances in equations having the form of Eq. (10.13) may be calculated using the equation

$$t_i = KR_i / k_i \Sigma R \tag{10.15}$$

Values for unrestricted tolerances in equations having the form of Eq. (10.14) may be calculated using the equation

$$t_i = (K/k_i)(R_i/\Sigma R)^{1/2} \tag{10.16}$$

EXAMPLE 2   The dimensioning problem started in Example 1 is continued here. Suppose that the design requires that the minimum limit of magnification m be 1.996, and that the maximum limit of magnification M be 2.004; the mean value of magnification $\overline{m}$ is therefore 2.000. From the design layout for the lens assembly, values are scaled: $\overline{a} = 4.000$ and $b = 2.000$. From Eq. (10.8), $\overline{c} = 6.00$. Substituting values for M, m, $\overline{a}$, b, and $\overline{c}$ into Eq. (10.9) gives

$$0.008 = 0.500t_a + 0.250t_b + 0.250t_c \qquad (10.17)$$

This equation has the form of Eq. (10.13). Let it be assumed that after assigning unit factors, the difficulty factors for unrestricted tolerances are $R_a = 6.2$, $R_b = 6.2$, and $R_c = 5.2$ so that $\Sigma R = 17.6$. Using Eq. (10.15), $t_a = 0.0056$, $t_b = 0.0113$, and $t_c = 0.0095$. The problem is solved since dimensions X_a, X_b, X_c now have restricted values.

For complex cases, more than one equation of dimension condition and more than one tolerance equation are involved in a dimension problem. For each equation of dimension condition, mean values can be selected for the dimension condition and all dimensions except one; the mean value of the remaining dimension can then be calculated. The order for the solution of equations of dimension condition requires that each equation have at least one unrestricted dimension that will be determined by calculation.

An order of solution for tolerance equations must be followed that will result in the most restrictive values being assigned to common tolerances. Value can be calculated using Eq. (10.15) or (10.16). The least value of each common tolerance is then adopted. These least values can be substituted back into tolerance equations and a new set of equations obtained which will not have common tolerances.

10.6   INTERNATIONAL TOLERANCE GRADES

The last step in a dimension calculation is the evaluation of tolerances to determine whether or not tolerance values are reasonable. The difficulty in manufacturing a part to tolerance depends not only on the value of the tolerance but, in the case of a length dimension, on length itself. Every toleranced length dimension within the usual ranges of practice has an associated international tolerance grade that indicates whether or not tolerance values are reasonable.

Table 10.3 lists international tolerance grades for combinations of tolerance values and ranges of dimension sizes. The table conforms to standards of the International Organization for Standardization. Both English and metric unit versions of the table are included in American National Standards.[8,9]

Notation such as IT5, IT6, etc., denotes international tolerance grades. Notation IT6, for instance, reads international tolerance grade 6. Length dimensions having the same tolerance grade and produced by the same manufacturing process are expected to be equally difficult to manufacture. Thus, the feature of a part having a dimension with limits 1.735/1.731 is expected to incur the same difficulty in manufacturing as a similar dimension with limits 5.506/5.500; both dimensions have tolerances with grade IT10.

Table 10.4 lists typical manufactured parts and manufacturing processes associated with international tolerance grades. IT1 through IT6 apply to gage work; IT5 through IT8 apply to functional dimensions where a high order of precision is necessary; IT9 through IT12 apply to less accurate functional dimensions. Except for very coarse work, functional dimensions usually do not have international tolerance grades higher than 12.

TABLE 10.3 Tolerance Values for International Tolerance Grades

Range of sizes*		International tolerance grades†							
Over	To	IT1	IT2	IT3	IT4	IT5	IT6	IT7	IT8
0.04	0.12	0.06	0.08	0.12	0.15	0.20	0.25	0.4	0.6
0.12	0.24	0.06	0.08	0.12	0.15	0.20	0.30	0.5	0.7
0.24	0.40	0.06	0.08	0.12	0.15	0.25	0.40	0.6	0.9
0.40	0.71	0.06	0.08	0.12	0.20	0.30	0.4	0.7	1.0
0.71	1.19	0.06	0.08	0.16	0.25	0.40	0.5	0.8	1.2
1.19	1.97	0.08	0.12	0.16	0.30	0.40	0.6	1.0	1.6
1.97	3.15	0.08	0.12	0.20	0.3	0.5	0.7	1.2	1.8
3.15	4.73	0.12	0.16	0.24	0.4	0.6	0.9	1.4	2.2
4.73	7.09	0.16	0.20	0.32	0.5	0.7	1.0	1.6	2.5
7.09	9.85	0.20	0.28	0.40	0.6	0.8	1.2	1.8	2.8
9.85	12.41	0.24	0.32	0.48	0.6	0.9	1.2	2.0	3.0
12.41	15.75	0.28	0.36	0.50	0.7	1.0	1.4	2.2	3.5
15.75	19.69	0.32	0.40	0.60	0.8	1.0	1.6	2.5	4.0

Range of sizes		International tolerance grades							
Over	To	IT9	IT10	IT11	IT12	IT13	IT14	IT15	IT16
0.04	0.12	1.0	1.6	2.5	4	6	10	16	25
0.12	0.24	1.2	1.8	3.0	5	7	12	18	30
0.24	0.40	1.4	2.2	3.5	6	9	14	22	35
0.40	0.71	1.6	2.8	4	7	10	16	28	40
0.71	1.19	2.0	3.5	5	8	12	20	35	50
1.19	1.97	2.5	4.0	6	10	16	25	40	60
1.97	3.15	3.0	4.5	7	12	18	30	45	70
3.15	4.73	3.5	5.0	9	14	22	35	50	90
4.73	7.09	4.0	6.0	10	16	25	40	60	100
7.09	9.85	4.5	7	12	18	28	45	70	120
9.85	12.41	5.0	8	12	20	30	50	80	120
12.41	15.75	6.0	9	14	22	35	60	90	140
15.75	19.69	6	10	16	25	40	60	100	160

*In inches.
†Tolerances in thousandths of an inch.

TABLE 10.4 International Tolerance Grades Associated with Typical Manufactured Parts and Manufacturing Processes

Grade	Manufactured parts or manufacturing process
IT1	Master gages
IT2	High-quality gages
IT3	Good-quality gages
IT4	Medium-quality gages
IT5	Ball bearings, machine lapping, fine grinding
IT6	Grinding, fine honing
IT7	High-quality turning, broaching, and honing
IT8	Center lathe turning and boring, reaming
IT9	Worn capstan or automatic lathe work
IT10	Milling, slotting, extruding
IT11	Drilling, rough turning
IT12	Light press work, precision tube drawing
IT13	Press work, tube rolling
IT14	Die casting, rubber molding
IT15	Stamping
IT16	Sand casting, flame cutting

10.7 ASSEMBLY FITS

Three types of assembly-fit problems occur often in mechanical design: length fits, cylindrical fits, and multiple-hole assemblies. Figure 10.6 shows a simple assembly which includes a length fit, four cylindrical fits, and two multiple-hole assemblies.

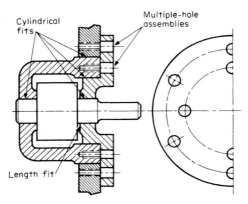

FIG. 10.6 Assembly fits.

10.7.1 Length Fits

A length fit can be simple or complex. A simple length fit involves a relationship of dimensions between parallel surfaces and a condition of fit; no dimension in the relationship is part of another dimensional relationship. For most length fits the condition of fit is clearance. For a complex length fit, one or more length dimensions may be other than a perpendicular distance between parallel surfaces; one or more dimensions may be part of other dimensional relationships.

Equations of dimension condition for length fits can be derived from "dimension loops." A simple dimension loop is shown in Fig. 10.7. The dimension condition and dimensions are represented by vectors. Vectors are either positive or negative, depending on their directions. Vectors of dimension condition are, by definition, in the positive direction; dimension vectors in the same direction are also positive. Vectors in the direction opposite to the vector of dimension condition are negative.

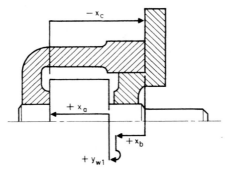

FIG. 10.7 Dimension loop for a length fit.

The general equation of dimension condition for the dimension loop of Fig. 10.7 is

$$y_{w1} + x_a + x_b = x_c \tag{10.18a}$$

This equation, however, is not useful in calculations. Three useful equations can be written. If all values are mean,

$$\overline{w}_1 + \overline{a} + \overline{b} = \overline{c} \tag{10.18b}$$

When the dimension condition is at its maximum limit,

$$W_1 + a + b = C \tag{10.18c}$$

When the dimension condition is at its minimum limit,

$$w_1 + A + B = c \tag{10.18d}$$

Only one of these last three equations need be used in a calculation.
Equations (10.18b), (10.18c), and (10.18d) are generalized as follows:

$$\Sigma(+\overline{x}) + \overline{w} = \Sigma(-\overline{x}) \tag{10.19a}$$

$$\Sigma(+x_{max}) + w = \Sigma(-x_{min}) \tag{10.19b}$$

$$\Sigma(+x_{min}) + W = \Sigma(-x_{max}) \tag{10.19c}$$

Equation (10.19a) states that the sum of mean values of dimensions whose vectors are in the positive direction, plus the mean value of the dimension condition, is equal to the sum of mean values of dimensions whose vectors are in the negative direction. Equation (10.19b) states that the sum of maximum values of dimensions in the positive direction, plus the minimum value of the dimension condition, is equal to the sum of the minimum values of dimensions whose vectors are in the negative direction. Equation (10.19c) states that the sum of the minimum values of dimensions in the positive direction, plus the maximum value of the dimension condition, is equal to the sum of the maximum values of dimensions whose vectors are in the negative direction. Equation 10.20 defines the mean value of the dimension condition:

$$\overline{w} = (W + w)/2 = (W_p + w_p)/2 \tag{10.20}$$

Equations of dimension condition for length fits have the form $y_w = \pm x_a \pm x_b \cdots$. Therefore, tolerance equations for length fits derived by application of the equations in Table 10.2 all have simpler forms than Eqs. (10.13) and (10.14), since coefficients k are unity. It is advantageous, therefore, to use special forms of Eqs. (10.15) and (10.16) applicable to length fits. These special forms are as follows.
Worst-limit basis:

$$t_i = \frac{\Sigma t_u R_i}{\Sigma R} \tag{10.21}$$

$$\Sigma t_u = (W - w) - \Sigma t_r \tag{10.22}$$

Statistical basis with equal-likelihood distribution:

$$t_i = \left(\frac{\Sigma t_u^2 R_i}{\Sigma R}\right)^{1/2} \tag{10.23}$$

$$\Sigma t_u^2 = (3/u^2)(W_p - w_p)^2 - \Sigma t_r^2 \tag{10.24}$$

Statistical basis with normal distribution:

$$t_i = \left(\frac{\Sigma t_u^2 R_i}{\Sigma R}\right)^{1/2} \tag{10.23}$$

$$\Sigma t_u^2 = (W_p - w_p)^2 - \Sigma t_r^2 \tag{10.25}$$

Σt_u denotes the sum of unrestricted tolerances and Σt_r the sum of restricted tolerances. Σt_u^2 denotes the sum of the squares of unrestricted tolerances, and Σt_r^2 the sum of the squares of restricted tolerances.

EXAMPLE 3 In this example, mean values of dimensions and values of tolerances are assigned or calculated for the dimensions in Figure 10.7, assuming the statistical basis of dimensioning with normally distributed dimensions. The designer has specified probable limits of dimension condition $W_{p1} = 0.022$ and $w_{p1} = 0.004$. The design layout was scaled to obtain $\bar{a} = 1.000$ and $\bar{b} = 0.470$. The value of $\bar{c}$ calculated from Eq. (10.18b) is 1.483. Difficulty factors are $R_a = 3.4$, $R_b = 3.6$, and $R_c = 4.5$. From Eqs. (10.23) and (10.25), tolerances are calculated giving $t_a = 0.010$, $t_b = 0.010$, and $t_c = 0.011$.

A complex length fit may also involve angular surfaces such as those shown in the dovetail slide of Fig. 10.8a. In problems of this sort, dimension loops must be written

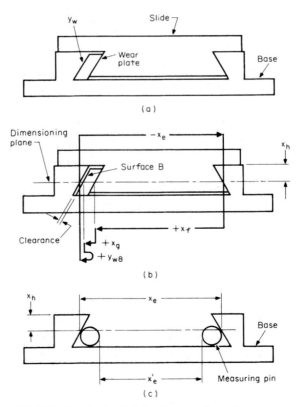

FIG. 10.8 Complex length fit involving angular surfaces.

for specific dimensioning planes, as in Fig. 10.8*b*. Dimensioning planes may be located to be coincident with the use of measuring pins as in Fig. 10.8*c*.

10.7.2 Cylindrical Fits

Cylindrical fits consist of two mating elements: a cylindrical hole and a cylindrical external surface. It is conventional to call the cylindrical external surface a "shaft" element. Hole and shaft elements are depicted in Fig. 10.9. Maximum and minimum diameters of hole and shaft elements are denoted by symbols $x_{H,\max}$, $x_{H,\min}$, $x_{S,\max}$, and $x_{S,\min}$; corresponding hole and shaft tolerances are denoted by t_H and t_S.

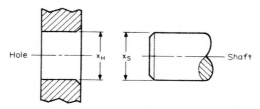

FIG. 10.9 Hole and shaft elements of a cylindrical fit.

Cylindrical fits serve a variety of functions: alignment, transmission of torque, rotation, sliding, etc. Designers must determine values for limits of dimension conditions on the basis of experience, theory, or recommended practices. Dimension limits can be derived from the following equations of dimension condition and tolerance equations.

Equations of dimension condition:

$$\bar{w} = \bar{x}_H - \bar{x}_S \tag{10.26a}$$

$$W = x_{H,\max} - x_{S,\min} \tag{10.26b}$$

$$w = x_{H,\min} - x_{S,\max} \tag{10.26c}$$

Tolerance equations for the worst-limit basis of dimensioning:

$$W - w = t_H + t_S \tag{10.27}$$

$$t_H = \frac{R_H(W - w)}{\Sigma R} \tag{10.28}$$

$$t_S = \frac{R_S(W - w)}{\Sigma R} \tag{10.29}$$

Tolerance equations for statistical basis of dimensioning with normal distribution:

$$(W_p - w_p)^2 = t_H^2 + t_S^2 \tag{10.30}$$

$$t_H = (W_p - w_p)\left(\frac{R_H}{\Sigma R}\right)^{1/2} \tag{10.31}$$

$$t_S = (W_p - w_p)\left(\frac{R_H}{\Sigma R}\right)^{1/2} \tag{10.32}$$

$$\Sigma R = R_H + R_S \tag{10.33}$$

Values of W and w in Eqs. (10.26b) and (10.26c) can also be probable limits. Tolerance equations apply only to the worst-limit basis of dimensioning and to the statistical basis with normally distributed dimensions. Since only two dimensions control a cylindrical fit, the statistical basis with equal-likelihood dimensions is not applicable.

Every cylindrical fit has two possible limits of fit. One limit occurs when both hole and shaft diameters are at maximum material conditions; the other limit occurs when both diameters are at minimum material conditions. A condition of fit is "interference" if its numerical value is negative, "line contact" if its numerical value is zero, and "clearance" if its numerical value is positive. Five categories of cylindrical fits can be designated, depending on the values of minimum and maximum conditions of fit. These five categories are listed in Table 10.5.

TABLE 10.5   Categories of Cylindrical Fits

Description of fit	Value of minimum condition of fit W	Value of maximum condition of fit W
Clearance	Positive	Positive
Line clearance	Zero	Positive
Transition	Negative	Positive
Line interference	Negative	Zero
Interference	Negative	Negative

EXAMPLE 4   A designer has specified a cylindrical fit with probable limits $W_p = 0.010$ and $w_p = 0.002$, with $x_{H,\min} = 2.000$. Difficulty factor $R_H = 1.4$ applies to the hole element and $R_s = 1.0$ to the shaft element. Tolerance values calculated using Eqs. (10.31) and (10.32) are $t_H = 0.006$ and $t_s = 0.005$. Mean hole diameter $\bar{x}_H = 2.0000 + 0.0030 = 2.0030$. Since $\bar{w} = 0.0060$, from Eq. (10.26a), $\bar{x}_s = 2.0030 - 0.0060 = 1.9970$. Mean values and tolerances are now known for both x_H and x_S, so the problem is solved.

Because cylindrical fits are so common in mechanical products, many standards have been developed for assigning dimensions to hole and shaft elements. Most current national standards are derived from standards of the International Standards Organization. The International System of Cylindrical Fits is particularly useful, since it permits selecting fits that are common worldwide. Metric and inch versions are both included in standards of the American National Standards Institute.[8,9]

The International System of Cylindrical Fits includes the following:

1. A scheme of notation for specifying fits
2. The system of tolerance grades discussed in Sec. 10.6
3. Tables of deviations for calculating limits of hole and shaft diameters
4. Recommendations for preferred classes of fits with tables to assist in fit selections and the calculation of hole and shaft diameters

A standard cylindrical fit is specified by a notation such as 3.5000 H9d8, where 3.5000 is the basic diameter, the size from which deviations are taken when calculating limits of hole and shaft diameters. H9 in the notation denotes a class of deviation for calculating hole limits; d8 denotes a class of deviations for calculating shaft limits. In the hole deviation notation H9, the 9 indicates an international tolerance grade of 9 (IT9); similarly, the shaft deviation notation d8 indicates an international tolerance grade of 8 (IT8).

Letters assigned to hole and shaft deviations indicate relative positions of diameter

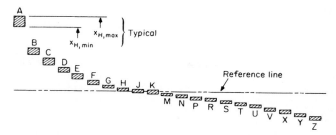

FIG. 10.10 Fit diagrams showing general positions of limits for holes.

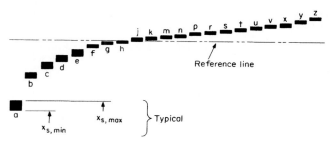

FIG. 10.11 Fit diagrams showing general positions of limits for shafts.

limits. Figure 10.10 shows the general positions of limits for holes; Fig. 10.11 shows the general positions of limits for shafts. Many combinations of hole deviations and shaft deviations are possible, for instance, g8D7. However, most design requirements are satisfied by a limited number of preferred basic hole fits. In a basic hole fit, $x_{H,\text{basic}} = x_{H,\text{min}}$. Figure 10.12 shows relationships between hole and shaft limits $x_{H,\text{basic}}$ and deviations δ. Tables for selecting preferred basic hole fits and deviations for calculating hole and shaft limits can be found in standards.[8,9] Table 10.6 lists preferred classes of basic hole fits. However, each engineering department should select a limited number of fits suitable for its products.

The International System of Cylindrical Fits applies also to standard metric ball bearings. Since outside and bore diameters are restricted, symbols designating fits are abbreviated. Figure 10.13 shows typical fit symbols on a design layout. Symbol 35H7

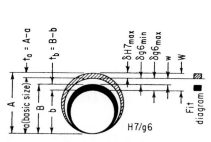

FIG. 10.12 Elements of a typical basic hole fit. $\delta H7_{\text{max}}$, $\delta g6_{\text{max}}$, and $\delta g6_{\text{min}}$ are deviations used to calculate diameters A, B, and b.

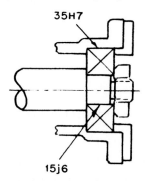

FIG. 10.13 Design layout annotated with symbols for ball bearing fits.

TABLE 10.6 Preferred Classes of Basic Hole Fits

Running and sliding fits

These fits have clearance for both w and W. Fits within a class are intended to give similar running performance, with suitable lubrication, throughout the range of sizes.

Close sliding fits, H5g4. Require very accurate manufacture; intended for the accurate location of parts which must assemble without perceptible play. Clearance increases slowly with diameter so that accurate location is maintained at the sacrifice of free relative motion.

Sliding fits, H6g5. Similar to close sliding fits except that tolerances are one grade coarser, minimum clearances are the same, but maximum clearances are slightly greater. Parts made to this fit move and turn freely but are not intended to run freely. Larger sizes may seize with small temperature changes. Clearances increase slowly with diameter so that accurate location is maintained at the sacrifice of free relative motion.

Precision running fits, H7f6. Give about the closest fits that can be expected to run freely. Intended for precision work at slow speeds and light journal pressures. Not suitable where temperature differences are likely.

Close running fits, H8f7. Intended for use on machinery with moderate surface speeds and journal pressures, and where accurate location and minimum play are desired.

Medium running fits, H8e7 and H9e8. Suitable for higher running speeds and journal pressures than with previous fits. The H9e8 fits are the same as H8e7 fits except that tolerance grades are coarser, resulting in slightly greater maximum clearances.

Free running fits, H9d8. Intended for use where accuracy is not essential, where large temperature variations are anticipated, or for both of these conditions

Loose running fits, H10c9. Intended where large tolerances are permissible.

Locational clearance fits

Fits in this category are intended for normally stationary parts.

Snug locational fits, H6h5, H7h6, H8h7, and H10h9. Line clearance fits, the minimum condition of fit w is zero and the maximum condition of fit W is clearance. These fits are both basic hole and basic shaft.

Medium locational fits, H7g6, H9f8, and H10e9. Apply where accuracy of location is necessary but where greater ease of assembly is probable than with snug locational fits.

Loose locational fits, H10d9 and H11c10. Used where freedom of assembly is most important.

denotes a fit for a bearing with 35-mm outside diameter; symbol 15j6 denotes a fit for a bearing with a 15-mm bore. H7 represents a class of hole deviations and j6 a class of shaft deviations. Tables 10.7 and 10.8 list fit selection recommendations using symbols for hole and shaft deviations. Complete tables for selecting and dimensioning fits can be found in bearing manufacturer catalogs.

The term "inner ring rotation" in Tables 10.7 and 10.8 means that the inner ring rotates relative to the load so that all points on the inner raceway are subject to loading. "Outer ring rotation" means that the outer ring rotates relative to the load so that all points on the outer raceway are subject to loading. "Indeterminate rotation" means that both inner ring and outer ring rotation can occur.

10.7.3 Multiple-Hole Assemblies

A multiple-hole assembly consists of a hole or group of holes in one part that must align with a matching hole or group of holes in a mating part. Fasteners assembled through aligned holes may be bolts, cap screws, studs, dowels, and so forth. Holes must be large enough so that fasteners can assemble despite the misalignment allowed by hole position tolerances; however, holes cannot be so large that bearing surfaces of screws, nuts, washers, and so forth, are not adequately supported. Equations for calculating tentative maximum hole size H' are as follows.

TABLE 10.6 Preferred Classes of Basic Hole Fits (*Continued*)

Locational transition fits

Values of the minimum condition of fit w are interference; values of the maximum condition of fit W are clearance. Fits of these types are used where very little clearance is necessary but where some interference can be tolerated. All locational transition fits have relatively small tolerance grades.
 	
   Wringing fits, H7k6 and H8k7. Apply where less freedom of rotation is wanted than is probable with push fits. With H7k6 the mean condition of fit is practically zero, so that the chance of clearance or interference is equally divided. With H8k7 the mean condition of fit has slightly more clearance than H7k6 but is still close to zero.
 	
   Tight fits, H7n6 and H7n7. Fits usually result in light interferences with only a small chance of clearance. These fits can be used where a slight clearance is acceptable, or where random selective assembly can be relied upon for the few clearance fits that are likely to occur.

Locational interference fits

Fits in this category are used where accuracy of location is of prime importance, and for parts requiring rigidity and alignment with no special needs for bore pressure. Locational interference fits should not be used where frictional loads must be transmitted by virtue of interference.
 	
   Light interference fits, H6n5 and H7p6. Give no interference, or very little, in the least material conditions of the hole and shaft, and light interference in the maximum material conditions.
 	
   Medium interference fits, H7r6. Give a small amount of interference at the least material conditions of the hole and shaft.

Force or shrink fits

Fits in these classes maintain constant bore pressure throughout the range of sizes. Interference varies almost directly with diameter. The difference between minimum and maximum interference is kept small so as to keep pressure variations within reasonable limits. Stresses associated with force or shrink fits depend on many factors, so that fits of these types should not be used without thorough theoretical and practical consideration. The use of shrink-fitting methods (either heating the hole member, cooling the internal member, or both) is often necessary for good assembly practice.
 	
   Medium drive fits, H7s6. Suitable for ordinary steel parts and for shrink fits involving thin sections. Fits have about the most interference possible with high-grade cast-iron hole members.
 	
   Heavy drive fits, H7t6. Intended for heavy steel parts or shrink fits in medium sections. The smallest diameter to which this type of fit is applied is about 1 in.
 	
   Force fits, H7u6 and H8x7. Can only be applied to parts capable of high stresses.

For relatively large bearing areas,

$$H' = (f + 3b)/4 \qquad\qquad (10.34)$$

For relatively small bearing areas,

$$H' = (f + 2b)/3 \qquad\qquad (10.35)$$

where H' = tentative maximum hole size
 f = minimum diameter of bearing surface
 b = minimum fastener diameter

The selection of a final hole size may depend on the choice of preferred punch and drill sizes. H denotes a final maximum hole size, and h the final minimum hole size. Hole tolerance $t_b = H - h$ must be sufficiently large to permit economical manufacture.

EXAMPLE 5 Using Eq. (10.35) for relatively small bearing areas, calculate a tentative hole diameter for a multiple-hole assembly having a ⅜-16 UNC-2A American Standard hexagonal washer-head machine screw. From standard sources, minimum body diameter

TABLE 10.7 Housing Fit Selection for Ball Bearings

		Conditions	Fit class	Remarks
Housing not split radially	Outer ring rotation	Heavy loads on bearings in thin walled housings	P7	The outer ring is not displaceable axially
		Normal and heavy loads	N7	
		Light and variable loads	M7	
Housing split or not split radially	Indeterminate rotation	Heavy shock loads	K7	The outer ring is, as a rule, not displaceable axially
		Heavy and normal loads, axial displacement of outer ring not required		
		Normal and light loads, axial displacement of outer ring desirable	J7	The outer ring is, as a rule, displaceable axially
		Shock loads, temporary complete unloading		
	Inner ring rotation	All loads	H7	The outer ring is easily displaced axially
		Normal and light loads, loads under simple operating conditions	H8	
		Heat supplied through the shaft	G7	
		Applications with larger differential temperature between bearing outer ring and housing	F7	
Housing not split radially	Applications requiring particular accuracy	Very accurate running and small deflections under variable loads	P6 N6 M6	The outer ring is not displaceable axially
		Very accurate running under light loads and *indeterminate load direction*	K6	The outer ring is, as a rule, not displaceable axially
		Very accurate running, axial displacement of outer ring desirable	J6	The outer ring is displaceable axially

b is 0.366, and minimum fastener bearing diameter f is 0.720. Substituting values into Eq. (10.35) gives $H' = 0.484$.

Hole patterns of multiple-hole assemblies may be located with toleranced dimensions; true position dimensioning, however, is better suited. True position dimensioning is particularly applicable if parts are to be inspected using functional gages. True position dimensioning is defined in Ref. 10.

Equations for calculating true position tolerances are as follows:

Type TPA:

$$t_{p1} + t_{p2} = h_1 + h_2 - 2(B + w) \quad \text{(general)} \quad (10.36)$$

$$t_p = h - B - w \quad \text{(simplified)} \quad (10.37)$$

TABLE 10.8 Shaft Fit Selection for Ball Bearings

		Conditions	Examples	Shaft diameters, mm	Fit class
Inner ring rotation		The inner ring to be easily displaced on the shaft	Wheel on nonrotating shaft	All	g6
		The inner ring does not need to be easily displaced on shaft	Tension pulleys and rope sheaves		h6
Outer ring rotation or indeterminate rotation	Light load		Electrical apparatus, machine tools, pumps, ventilators, industrial trucks	Less than 18	h5
				Over 18 to 100	j6
				Over 100 to 200	k6
	Normal load		Applications in general, electrical motors, turbines, pumps, combustion engines, gear transmissions, woodworking machines	Less than 18	j5
				Over 18 to 100	k5
				Over 100 to 140	m5
				Over 140 to 200	m6
				Over 200 to 250	n6

Type TPB:

$$t_{p1} + t_{p2} = h - B - 2w \qquad \text{(general)} \qquad (10.38)$$

$$t_p = (h - B)/2 - w \qquad \text{(simplified)} \qquad (10.39)$$

where t_p = positional tolerance
   h = minimum clearance hole diameter
   B = maximum fastener diameter
   w = minimum assembly clearance

Type TPA equations apply when both mating hole patterns contain clearance holes. Equation (10.36) must be used when positional tolerances t_p and/or minimum clearance hole sizes h are different for each hole pattern; Eq. (10.39) can be used when h and t_p are the same for both hole patterns. Type TPB equations apply when one mating hole pattern contains nonclearance (usually threaded) holes. Equation (10.40) must be used when t_p is not the same for both patterns; Eq. (10.39) can be used when t_p is the same for both hole patterns.

EXAMPLE 6 Both hole patterns for a multiple-hole assembly are to have clearance holes whose minimum diameter is 0.460; a ⅜-16 UNC-2A machine screw is to assemble through aligned holes. Positional tolerances are to be the same for both hole patterns. For a minimum assembly clearance of 0.010, the positional tolerance calculated with Eq. (10.37) is 0.075.

10.8 ERRORS IN MECHANISM AND MECHANISM TRAINS

A "mechanism," for purpose of this discussion, is an assembly of mechanical parts that converts or transmits motion from one or more input terminals to one or more output terminals. A typical mechanism is a gear set consisting of a driving and a driven gear. Other mechanisms are couplings, belt and pulley pairs, screw and nut pairs, cam and follower pairs, Scotch yokes, etc.

A mechanism error can be attributed to many sources. In a gear set, for instance, the mechanism error may be due to eccentricities between bearing surfaces of supporting shafts, eccentricities and play in bearings, runout of gear teeth, minimum backlash, and torsional deflections.

Mechanism errors in high-speed machinery can cause early wear, noise, and fracture. This article, however, is not concerned with dynamic effects but only with errors in position or motion.

Each mechanism in a mechanism train can be the source of more than one type of error. Error types are classified as follows:

Structural error, the error resulting from the use of an approximate instead of an exact mechanism. For example, gear ratio of 6.13:1 may be an ideal requirement, but because of design limitations the approximate ratio of 6:1 must be used instead. The difference between ideal and approximate ratios results in a structural error. Structural errors of variable-aspect mechanisms will not necessarily be uniform throughout the operating range of the mechanism, as shown in Fig. 10.14a.

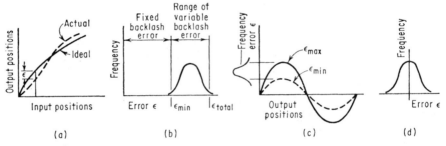

FIG. 10.14   Examples of common types of mechanism errors. (*a*) Structural. (*b*) Backlash. (*c*) Cyclic. (*d*) Random.

Minimum backlash error.   Many mechanisms have clearances necessary for assembly and operation, for example, clearance between gear teeth, or the tongue and groove of an Oldham's coupling. The smallest necessary clearance will result in a minimum backlash error. For many mechanical systems, backlash is the major source of error and special means are often used to reduce or eliminate it entirely.

Variable backlash error, due to the additional clearance caused by tolerances. If the tolerances controlling the amount of additional clearance are normally distributed, then values of additional backlash will also be normally distributed. Figure 10.14b shows total backlash as a combination of fixed and probable values.

Systematic or cyclic error, usually associated with the eccentricity of rotating elements, the wobble of gear faces, etc. The amplitude of cyclic errors may be normally distributed as shown in Fig. 10.14c.

Random error, produced by errors in the location or profile of driving surfaces. Errors of this type are generally assumed to be normally distributed as shown in Fig. 10.14d.

Transient or short-period error results from vibrations, shock, foreign particles lodged between driving surfaces, etc. The character of these errors may be complex and unpredictable.

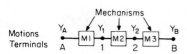

FIG. 10.15 Block diagram representing a simple train of three mechanisms.

A simple train of three mechanisms is represented by the diagram shown in Fig. 10.15. Input motion at terminal A is successively modified by mechanism 1, mechanism 2, and mechanism 3. Mechanism functions $Y_1 = f(Y_A)$, $Y_2 = f(Y_1)$, and $Y_B = f(Y_2)$ must be written connecting output and input terminals of each mechanism. Error components $\epsilon_{A.1}$, $\epsilon_{A.2}$,..., can be associated with input motion Y_A; error components $\epsilon_{1.1}$, $\epsilon_{1.2}$,..., with mechanism 1; etc. As an error component propagates through the train, it will be modified by each succeeding mechanism and contribute to the total output error at terminal B. Each error component is modified by a train propagation factor. The "train propagation factor" for errors originating at terminal A is

$$M_A = \left|(dY_1/dY_A)(dY_2/dY_1)(dY_B/dY_2)\right| \qquad (10.40)$$

The train propagation factor for errors originating at terminal 1 is

$$M_1 = \left|(dY_2/dY_1)(dY_B/dY_2)\right| \qquad (10.41)$$

And, finally, the train propagation factor for errors originating at terminal 2 is

$$M_2 = \left|dY_B/dY_2\right| \qquad (10.42)$$

Depending on their natures, error components will combine either as worst-limit or as normally distributed values. For the mechanism train represented by Fig. 10.15, if $\epsilon_{A.1}$, $\epsilon_{1.1}$, $\epsilon_{2.1}$, and $\epsilon_{3.1}$ are expected to combine as worst-limits, the resulting error at terminal B is

$$\epsilon_{B.1} = M_A\epsilon_{A.1} + M_1\epsilon_{1.1} + M_3\epsilon_{2.1} + \epsilon_{3.1} \qquad (10.43)$$

If errors $\epsilon_{A.2}$, $\epsilon_{1.2}$, $\epsilon_{2.2}$, and $\epsilon_{3.2}$ are normally distributed, then

$$\epsilon_{B.j} = [(M_A\epsilon_{A.2})^2 + (M_1\epsilon_{1.2})^2 + (M_2\epsilon_{2.2})^2 + \epsilon_{3.2}^2]^{1/2} \qquad (10.44)$$

The total error at terminal B is a combination of worst-limit components $\epsilon_{B.1}$, $\epsilon_{B.2}$, ..., and normally distributed components $\epsilon_{B.j}$ $\epsilon_{B.k}$, ...:

$$\epsilon_{B.total} = \epsilon_{B.1} + \epsilon_{B.2} + \cdots + (\epsilon_{B.j}^2 + \epsilon_{B.k}^2 + \cdots)^{1/2} \qquad (10.45)$$

A mechanism train may branch as shown in Fig. 10.16a. When calculating errors at output Terminals B and C, the trains may be treated as two simple trains as shown in Fig. 10.16b and c.

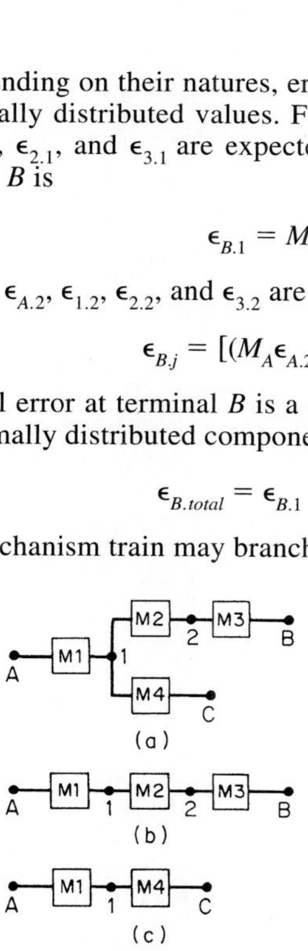

A mechanism train may contain a "compound mechanism"; that is, a mechanism with more than one input and/or one output terminal. A gear differential is a familiar compound mechanism. The diagram in Fig. 10.17 represents a mechanism train having a compound mechanism. Mechanism 4 has a mechanism function $Y_4 = f(Y_2, Y_3)$. In calculating train propagation factors for mechanism trains having a compound mechanism the partial derivative of the mechanism func-

FIG. 10.16 Block diagram for (a) a branched mechanism train and (b and c) equivalent simple trains.

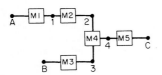

FIG. 10.17 Block diagram for a mechanism train having a compound mechanism.

tion must be taken with respect to the appropriate terminal. Thus the train propagation factor for the mechanism error originating at terminal 1 is

$$M_1 = |(dY_2/dY_1)(\partial Y_4/\partial Y_2)(dY_c/dY_4)| \quad (10.46)$$

EXAMPLE 7 Figure 10.18a shows a lead screw mechanism for positioning the lens on a process camera; the mechanism train is represented by the diagram at Fig. 10.18b. Input turns to the counter θ_A are transmitted through the Oldham coupling M_1, modified by gear set M_2, and transformed to linear motion by the screw and nut pair M_3. Mechanism functions are as follows:

$$\theta_1 = \theta_A \qquad \theta_2 = (N_1/N_2)\theta_1 \qquad S_B = p\theta_2 \qquad (10.47)$$

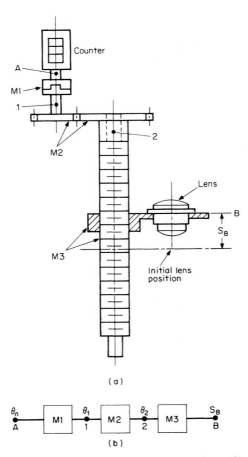

(a)

(b)

FIG. 10.18 (a) Lead screw mechanism. (b) Mechanism train.

in which N_1 and N_2 are numbers of teeth in the driving and driven gears of mechanism 2, and p is the pitch of the lead screw. If $N_1/N_2 = 0.33$ and $p = 0.25$, then the effect of errors originating at terminals A, 1, 2, and B, assuming combination of worst-limit values, is

$$\epsilon_B = 0.083\epsilon_A + 0.083\epsilon_1, + 0.25\epsilon_2 + \epsilon_3 \qquad (10.48)$$

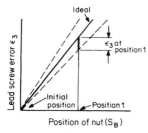

Position of nut (S_B)

FIG. 10.19 Lead screw error as a function of position.

Input error ϵ_A is due to incorrect setting of the counter. Errors in the Oldham's coupling M_2 are due to fixed and variable backlash. Errors in the gear set are cyclic and are due to runout, fixed backlash, and variable backlash. Errors in the lead screw are due to play and lead error. Lead error will increase from some starting position as shown in Fig. 10.19.

REFERENCES

1. Speckhart, F. H.: "Calculation of Tolerance Based on a Minimum Cost Approach," ASME Paper 71-Vibr-114, 1971.

2. Spotts, M. F.: "Allocation of Tolerances to Minimize Cost of Assembly," ASME Paper 72-WA/DE-6, 1972.

3. Smathers, E. W., and P. F. Oswald: "Optimization of Component Functional Dimensions and Tolerances," ASME Paper 72-DE-18, 1972.

4. Wilde, D., and E. Prentice: "Minimum Exponential Cost Allocation of Sure-Fit Tolerances," ASME Paper 75-DET-93, 1975.

5. Huang, J., and P. F. Oswald: "A Method for Optimal Tolerance Selection," ASME Paper 76-WA/DE-23, 1976.

6. Mitchael, W., and J. N. Siddall: "The Optimization Problem with Optimal Tolerance Assignment and Full Acceptance," ASME Paper 80-DET-47, 1980.

7. Fortini, E. T.: "Dimensioning for Interchangeable Manufacture," Industrial Press, Inc., New York, 1967.

8. "Preferred Limits and Fits for Cylindrical Parts," B4.1, American National Standards Institute, New York.

9. "Preferred Metric Limits and Fits," B4.2, American National Standards Institute, New York.

10. "Dimensioning and Tolerancing," Y14.5, American National Standards Institute, New York.

SECTION 11

STANDARDS FOR MECHANICAL ELEMENTS*

John J. Viegas, M.M.E., M.B.A.

Consulting Engineer
Saddle River, N. J.

Note: This section shows inch dimension notation with proper standard references indicated. Equivalent metric values are obtainable from the same respective references.

11.1 *SCREW-THREAD SYSTEMS*

11.1.1 Introduction

The most commonly known threads used in this country today are the following:

1. United States thread or Sellers' profile thread introduced by Sellers in 1864
2. N thread
3. American National screw-thread system developed in 1933
4. Unified system—called the Unified and American Standard—accepted in 1948; recommended for all new designs
5. British Standard Whitworth (B.S.W.)
6. British Standard Fine (B.S.F.)
7. Whitworth truncated or American truncated
8. Whitworth—developed as a war standard in 1944
9. British Association Standard (B.A.)
10. International metric standard—uses the metric system—European
11. Unified miniature screw thread
12. Microscopic-objective screw thread

11.1.2 Classification

Screws for transmitting power are commonly termed "power screws." Screw fastenings are generally categorized as to function, i.e., machine screws, cap screws, setscrews, bolts.

11.1.3 Unified Thread Standardization and Unification

Essentially, only one screw-thread form is now being used for fasteners in this country; it is the *unified thread form*[1] (Fig. 11.1) and is based on earlier standards. This new standard is additional to earlier American standards.[2] Its outstanding characteristic is general interchangeability of threads.

Set up by the standardization committees from Canada, Great Britain, and the United States in 1948, the unified system is herein referred to as the *Unified and American Standard* and is identified by the letters UN. Briefly, the standard comprises the following categories:

Diameter-Pitch Combinations. (*a*) Coarse series (UNC and NC), (*b*) fine series (UNF and NF), (*c*) extra-fine series (UNEF and NEF), (*d*) miniature (UNM), (*e*) constant-pitch series (4UN, 6UN, 8UN, 12UN, 16UN, 20UN, 28UN, 32UN).

Tolerance Classes. External (1A, 2A, 3A, 2, 3); internal (1B, 2B, 3B, 2, 3).

11.1.4 Unified Thread-Form Terminology* (Fig. 11.2)

Major diameter D_m of a threaded screw or nut is the largest diameter of the screw thread.

---

*See Fig. 11.1 for thread form, Fig. 11.3 for dimension symbols.

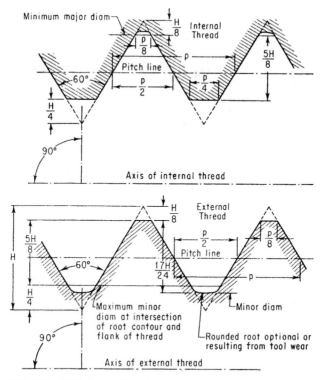

FIG. 11.1   Unified thread form.[1]

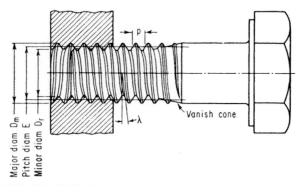

FIG. 11.2   Unified thread form terminology.

Minor diameter D_r is the smallest diameter of the screw thread.

Pitch diameter E, also called "effective diameter," is the diameter of an imaginary cylinder whose surface would pass through the thread profiles at such points as to make the width of the groove equal to one-half the basic pitch. On a perfect thread, this occurs at the point where the widths of thread and groove are equal.

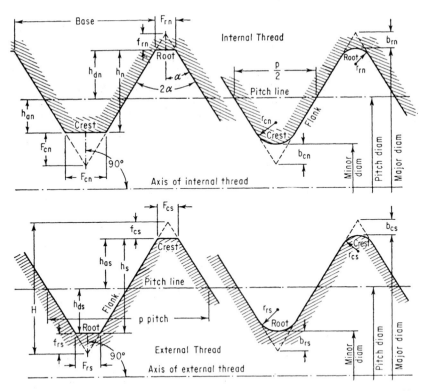

FIG. 11.3   Unified thread symbol application. n = threads per inch; p = pitch; $F_{cn} = p/4 = 0.250$; $F_{rn} = F_{ca} = 0.125$; $H = 0.866025$ = height of sharp V thread; $f_{rn} = f_{cs} = H/8 = 0.10825p$ = truncation of internal thread root and external thread crest; $b_{rs} = H/6 = 0.14434p$ = truncation of external thread root; $\frac{3}{16}H = 0.16238p$ = half addendum of external thread; $f_{cn} = H/4 = 0.21651p$ = truncation of internal thread crest; $h_{as} = \frac{3}{8}H = 0.32476p$ = addendum of external thread; $h_n = h_e = \frac{5}{8}H = 0.54127p$ = height of internal thread and depth of thread engagement; $h_s = \frac{17}{24}H = 0.61343$ = height of external thread; $h_b = 2h_{as} = \frac{3}{4}H = 0.649519p$ = twice the external thread addendum; $\frac{11}{12}H = 0.79386p$ = difference between maximum major pitch diameters of internal thread; $2h_n = 1\frac{1}{4}H = 1.08253p$ = double height of internal thread; $1\frac{5}{12}H = 1.22687p$ = double height of external thread.

Pitch p is the distance from a point on a screw thread to a corresponding point on the next thread measured parallel to the axis.

Lead angle λ on a straight thread is the angle made by the helix of the thread at the pitch line with a plane perpendicular to the axis.

11.1.5   Diameter-Pitch Selection[3]

Coarse-Thread Series (UNC), Table 11.1.   Used where rapid assembly or disassembly is required. Applicable with cast and malleable iron, soft metals, and plastics.

Fine-Thread Series (UNF), Table 11.2.   For application, where thread engagement or wall thickness is limited. Normally not recommended for use in soft metals or plastics.

TABLE 11.1 Unified and American Coarse-Thread Series UNC and NC[1]

Sizes	Basic major diam D_m, in	Threads per inch n	Basic pitch diam E, in	Minor diam external threads D_{rb}, in	Minor diam internal threads D_{rn}, in	Lead angle at basic pitch diam λ Deg	Min	Section at minor diam at $D - 2b_b$, in²	Tensile stress area, in²
1 (0.073)	0.0730	64	0.0629	0.0538	0.0561	4	31	0.00218	0.00263
2 (0.086)	0.0860	56	0.0744	0.0641	0.0667	4	22	0.00310	0.00370
3 (0.099)	0.0990	48	0.0855	0.0734	0.0764	4	26	0.00406	0.00487
4 (0.112)	**0.1120**	**40**	**0.0958**	**0.0813**	**0.0849**	**4**	**45**	0.00496	0.00604
5 (0.125)	0.1250	40	0.1088	0.0943	0.0979	4	11	0.00672	0.00796
6 (0.138)	**0.1380**	**32**	**0.1177**	**0.0997**	**0.1042**	**4**	**50**	0.00745	0.00909
8 (0.164)	**0.1640**	**32**	**0.1437**	**0.1257**	**0.1302**	**3**	**58**	0.01196	0.0140
10 (0.190)	**0.1900**	**24**	**0.1629**	**0.1389**	**0.1449**	**4**	**39**	0.01450	0.0175
12 (0.216)	0.2160	24	0.1889	0.1649	0.1709	4	1	0.0206	0.0242
¼	**0.2500**	**20**	**0.2175**	**0.1887**	**0.1959**	**4**	**11**	0.0269	0.0318
⁵⁄₁₆	**0.3125**	**18**	**0.2764**	**0.2443**	**0.2524**	**3**	**40**	0.0454	0.0524
⅜	**0.3750**	**16**	**0.3344**	**0.2983**	**0.3073**	**3**	**24**	0.0678	0.0775
⁷⁄₁₆	**0.4375**	**14**	**0.3911**	**0.3499**	**0.3602**	**3**	**20**	0.0933	0.1063
½	**0.5000**	**13**	**0.4500**	**0.4056**	**0.4167**	**3**	**7**	0.1257	0.1419
⁹⁄₁₆	**0.5625**	**12**	**0.5084**	**0.4603**	**0.4723**	**2**	**59**	0.162	0.182
⅝	**0.6250**	**11**	**0.5660**	**0.5135**	**0.5266**	**2**	**56**	0.202	0.226
¾	**0.7500**	**10**	**0.6850**	**0.6273**	**0.6417**	**2**	**40**	0.302	0.334
⅞	**0.8750**	**9**	**0.8028**	**0.7387**	**0.7547**	**2**	**31**	0.419	0.462
1	**1.0000**	**8**	**0.9188**	**0.8466**	**0.8647**	**2**	**29**	0.551	0.606
1⅛	**1.1250**	**7**	**1.0322**	**0.9497**	**0.9704**	**2**	**31**	0.693	0.763
1¼	**1.2500**	**7**	**1.1572**	**1.0747**	**1.0954**	**2**	**15**	0.890	0.969
1⅜	**1.3750**	**6**	**1.2667**	**1.1705**	**1.1946**	**2**	**24**	1.054	1.155
1½	**1.5000**	**6**	**1.3917**	**1.2955**	**1.3196**	**2**	**11**	1.294	1.405
1¾	**1.7500**	**5**	**1.6201**	**1.5046**	**1.5335**	**2**	**15**	1.74	1.90
2	**2.0000**	**4½**	**1.8557**	**1.7274**	**1.7594**	**2**	**11**	2.30	2.50
2¼	**2.2500**	**4½**	**2.1057**	**1.9774**	**2.0094**	**1**	**55**	3.02	3.25
2½	**2.5000**	**4**	**2.3376**	**2.1933**	**2.2294**	**1**	**57**	3.72	4.00
2¾	**2.7500**	**4**	**2.5876**	**2.4433**	**2.4794**	**1**	**46**	4.62	4.93
3	**3.0000**	**4**	**2.8376**	**2.6933**	**2.7294**	**1**	**36**	5.62	5.97
3¼	**3.2500**	**4**	**3.0876**	**2.9433**	**2.9794**	**1**	**29**	6.72	7.10
3½	**3.5000**	**4**	**3.3376**	**3.1933**	**3.2294**	**1**	**22**	7.92	8.33
3¾	**3.7500**	**4**	**3.5876**	**3.4433**	**3.4794**	**1**	**16**	9.21	9.66
4	**4.0000**	**4**	**3.8376**	**3.6933**	**3.7294**	**1**	**11**	10.61	11.08

Note: Bold type indicates unified threads, UNC.

Extra-Fine-Thread Series (UNEF), Table 11.3.   Extensively used in aircraft, missile, and other allied fields where requirements of threaded thin walls, minimum thread depth of nuts and coupling flanges, and maximum number of threads required within a given length of thread engagement are important.

Miniature[4] Threads (UNM).   Primarily used on instrument parts and miniature mechanisms.

Uniform-Pitch Series.   4UN, 6UN, 8UN, 12UN, 16UN, 20UN, 28UN, and 32UN; see Ref. 1.

TABLE 11.2 Unified and American Fine-Thread Series, UNF and NF[1]

Sizes	Basic major diam D_m, in.	Threads per inch n	Basic pitch diam E, in.	Minor diam external threads D_{rb}, in.	Minor diam internal threads D_{rn}, in.	Lead angle at basic pitch diam λ Deg	Min	Section at minor diam at $D - 2h_b$, in.^{2}	Tensile stress area, in.^{2}
0 (0.060)	0.0600	80	0.0519	0.0447	0.0465	4	23	0.00151	0.00180
1 (0.073)	0.0730	72	0.0640	0.0560	0.0580	3	57	0.00237	0.00278
2 (0.086)	0.0860	64	0.0759	0.0668	0.0691	3	45	0.00339	0.00394
3 (0.099)	0.0990	56	0.0874	0.0771	0.0797	3	43	0.00451	0.00523
4 (0.112)	0.1120	48	0.0985	0.0864	0.0894	3	51	0.00566	0.00661
5 (0.125)	0.1250	44	0.1102	0.0971	0.1004	3	45	0.00716	0.00830
6 (0.138)	0.1380	40	0.1218	0.1073	0.1109	3	44	0.00874	0.01015
8 (0.164)	0.1640	36	0.1460	0.1299	0.1339	3	28	0.01285	0.01474
10 (0.190)	**0.1900**	**32**	**0.1697**	**0.1517**	**0.1562**	**3**	**21**	0.0175	0.0200
12 (0.216)	0.2160	28	0.1928	0.1722	0.1773	3	22	0.0226	0.0258
¼	**0.2500**	**28**	**0.2268**	**0.2062**	**0.2113**	**2**	**52**	0.0326	0.0364
⁵⁄₁₆	**0.3125**	**24**	**0.2854**	**0.2614**	**0.2674**	**2**	**40**	0.0524	0.0580
⅜	**0.3750**	**24**	**0.3479**	**0.3239**	**0.3299**	**2**	**11**	0.0809	0.0878
⁷⁄₁₆	**0.4375**	**20**	**0.4050**	**0.3762**	**0.3834**	**2**	**15**	0.1090	0.1187
½	**0.5000**	**20**	**0.4675**	**0.4387**	**0.4459**	**1**	**57**	0.1486	0.1599
⁹⁄₁₆	**0.5625**	**18**	**0.5264**	**0.4943**	**0.5024**	**1**	**55**	0.189	0.203
⅝	**0.6250**	**18**	**0.5889**	**0.5568**	**0.5649**	**1**	**43**	0.240	0.256
¾	**0.7500**	**16**	**0.7094**	**0.6733**	**0.6823**	**1**	**36**	0.351	0.373
⅞	**0.8750**	**14**	**0.8286**	**0.7874**	**0.7977**	**1**	**34**	0.480	0.509
1	**1.0000**	**12**	**0.9459**	**0.8978**	**0.9098**	**1**	**36**	0.625	0.663
1⅛	**1.1250**	**12**	**1.0709**	**1.0228**	**1.0348**	**1**	**25**	0.812	0.856
1¼	**1.2500**	**12**	**1.1959**	**1.1478**	**1.1598**	**1**	**16**	1.024	1.073
1⅜	**1.3750**	**12**	**1.3209**	**1.2728**	**1.2848**	**1**	**9**	1.260	1.315
1½	**1.5000**	**12**	**1.4459**	**1.3978**	**1.4098**	**1**	**3**	1.521	1.581

Note: Bold type indicates unified threads, UNF.

11.1.6 Unified Thread Class Selection

The required assembly fit is obtained by selecting the thread class for each component. The class dictates the amount of manufacturing tolerance and allowance permitted (Fig. 11.4). The letter A designates tolerances and allowances applicable to externally threaded fasteners. The letter B identifies internally threaded numbers.

Classes 1A/1B. Loose fit, quick and easy assembly.

Classes 2A/2B. Commercially manufactured in the majority of threaded fasteners. The allowance of this class minimizes galling, minimizes seizure in high-temperature applications, and allows for plating.

Classes 3A/3B. For use where close fit and accuracy of lead and angle of thread are required. Since there is no allowance intended, highly accurate tools are used in its production, together with controlled inspection methods. See Fig. 11.4, for example, showing the physical relationship of the tolerance as designated by the different classes.[5]

TABLE 11.3 Unified and American Extra-Fine-Thread Series, UNEF and NEF[1]

Sizes	Basic major diam D_m, in.	Threads per inch n	Basic pitch diam E, in.	Minor diam external threads D_{rb}, in.	Minor diam internal threads D_{rn}, in.	Lead angle at basic pitch diam λ Deg	Min	Section at minor diam at $D - 2h_b$, in.²	Tensile stress area, in.²
12 (0.216)	0.2160	32	0.1957	0.1777	0.1822	2	55	0.0242	0.0270
¼	0.2500	32	0.2297	0.2117	0.2162	2	29	0.0344	0.0379
5⁄16	0.3125	32	0.2922	0.2742	0.2787	1	57	0.0581	0.0625
3⁄8	0.3750	32	0.3547	0.3367	0.3412	1	36	0.0878	0.0932
7⁄16	**0.4375**	**28**	**0.4143**	**0.3937**	**0.3988**	**1**	**34**	0.1201	0.1274
½	**0.5000**	**28**	**0.4768**	**0.4562**	**0.4613**	**1**	**22**	0.162	0.170
9⁄16	0.5625	24	0.5354	0.5114	0.5174	1	25	0.203	0.214
5⁄8	0.6250	24	0.5979	0.5739	0.5799	1	16	0.256	0.268
11⁄16	0.6875	24	0.6604	0.6364	0.6424	1	9	0.315	0.329
¾	**0.7500**	**20**	**0.7175**	**0.6887**	**0.6959**	**1**	**16**	0.369	0.386
13⁄16	**0.8125**	**20**	**0.7800**	**0.7512**	**0.7584**	**1**	**10**	0.439	0.458
7⁄8	**0.8750**	**20**	**0.8425**	**0.8137**	**0.8209**	**1**	**5**	0.515	0.536
15⁄16	**0.9375**	**20**	**0.9050**	**0.8762**	**0.8834**	**1**	**0**	0.598	0.620
1	**1.0000**	**20**	**0.9675**	**0.9387**	**0.9459**	**0**	**57**	0.687	0.711
1 1⁄16	1.0625	18	1.0264	0.9943	1.0024	0	59	0.770	0.799
1 1⁄8	1.1250	18	1.0889	1.0568	1.0649	0	56	0.871	0.901
1 3⁄16	1.1875	18	1.1514	1.1193	1.1274	0	53	0.977	1.009
1 ¼	1.2500	18	1.2139	1.1818	1.1899	0	50	1.090	1.123
1 5⁄16	1.3125	18	1.2764	1.2443	1.2524	0	48	1.208	1.244
1 3⁄8	1.3750	18	1.3389	1.3068	1.3149	0	45	1.333	1.370
1 7⁄16	1.4375	18	1.4014	1.3693	1.3774	0	43	1.464	1.503
1 ½	1.5000	18	1.4639	1.4318	1.4399	0	42	1.60	1.64
1 9⁄16	1.5625	18	1.5264	1.4943	1.5024	0	40	1.74	1.79
1 5⁄8	1.6250	18	1.5889	1.5568	1.5649	0	38	1.89	1.94
1 11⁄16	1.6875	18	1.6514	1.6193	1.6274	0	37	2.05	2.10
1 ¾	**1.7500**	**16**	**1.7094**	**1.6733**	**1.6823**	**0**	**40**	2.19	2.24
2	**2.0000**	**16**	**1.9594**	**1.9233**	**1.9323**	**0**	**35**	2.89	2.95

Note: Bold type indicates unified threads, UNEF.

11.1.7 Unified Screw Thread Designation (Fig. 11.5)

The method of designating a screw thread is by the use of the initial letters of the thread series preceded by the nominal size (diameter in inches or the screw number) and the number of threads per inch followed by the thread class and, if a left-handed thread is required, the symbol L.H. following the class. For right-handed threads, no symbol is necessary. Thus, ¼-20UNC-2A designates the following: ¼ in. nominal diameter having 20 threads to the inch is a unified series, therefore interchangeable with the thread system of Great Britain and Canada as well as the former American Standard threads in the United States and is in the coarse series. 2A signifies it is an external thread by the letter A and requires a commercial thread which provides sufficient clearance for plating if so desired.

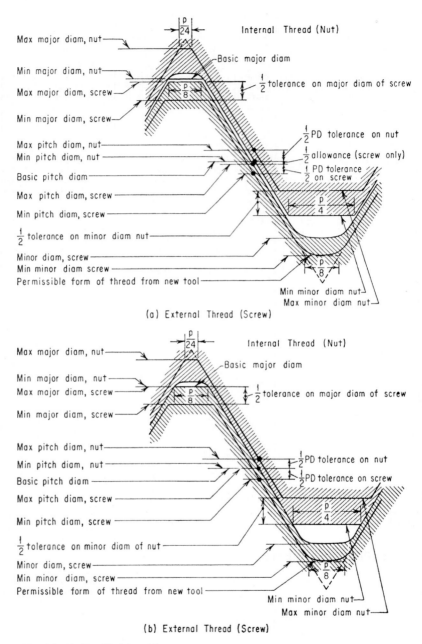

FIG. 11.4   Basic unified thread form.

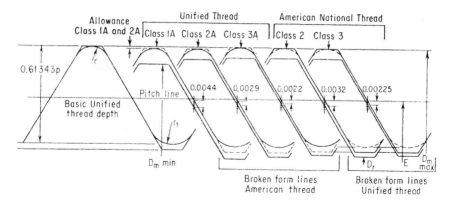

FIG. 11.5   Unified thread and American National pitch diameter tolerance figures.[5] (Various classes of thread for one-half of ¾-10 screws are shown.) Pitch-diameter tolerance figures shown are one-half of ¾-10 screws for the various classes of thread.

11.1.8   Unified Thread Limiting Dimensions: Allowances and Tolerances[6]
(Fig. 11.4)

Allowance.   The minimum clearance or maximum interference.

Tolerance.   The total permissible variation or the difference between maximum and minimum limits.

11.1.9   American National Thread Standard[2]

Also called American Standard or National Standard, this standard was formerly known as the United States thread or Sellers profile and provided the foundation to its successor, the Unified and American Standard. It is interchangeable with the Unified and American Standard.

The American Standard consisted of the following *three thread series:*

1. Coarse series (Table 11.4); general engineering work for rapid assembly.

2. Fine series (Table 11.5).

3. Special-pitch series: 8, 12, 16; see Ref. 2.

Four classes are provided:

   Class 1: Used where clearance is not objectionable

   Class 2: Used in commercial items and most commonly specified

   Class 3: Used in high-quality equipment

   Class 4: For unusual requirements, selective fit

11.1.10   American National Thread Form[2]

This screw thread is substantially identical with the unified thread form (Fig. 11.1). The formulas for its proportions are given in Fig 11.3.

TABLE 11.4   American National Coarse-Thread Series[1]

Size	Threads per in. n	Major diam D	Pitch diam E	Minor diam K	Class 1	Class 2	Classes 3 and 4	Max minor screw diam	Max minor nut diam	Minor nut diam tolerance	Class 1	Class 2	Class 3	Class 4
1	64	0.0730	0.0629	0.0527	0.0052	0.0692	0.0038	0.0538	0.0623	0.0062	0.0026	0.0019	0.0014	
2	56	0.0860	0.0744	0.0628	0.0056	0.0820	0.0040	0.0641	0.0737	0.0070	0.0028	0.0020	0.0015	
3	48	0.0990	0.0855	0.0719	0.0062	0.0946	0.0044	0.0734	0.0841	0.0077	0.0031	0.0022	0.0016	
4	40	0.1120	0.0958	0.0795	0.0068	0.1072	0.0048	0.0813	0.0938	0.0089	0.0034	0.0024	0.0017	
5	40	0.1250	0.1088	0.0925	0.0068	0.1202	0.0048	0.0943	0.1062	0.0083	0.0034	0.0024	0.0017	
6	32	0.1380	0.1177	0.0974	0.0076	0.1326	0.0054	0.0997	0.1145	0.0103	0.0038	0.0027	0.0019	
8	32	0.1640	0.1437	0.1234	0.0076	0.1586	0.0054	0.1257	0.1384	0.0028	0.0038	0.0027	0.0019	
10	24	0.1900	0.1629	0.1359	0.0092	0.1834	0.0066	0.1389	0.1559	0.0110	0.0046	0.0033	0.0024	
12	24	0.2160	0.1889	0.1619	0.0092	0.2094	0.0066	0.1649	0.1801	0.0092	0.0046	0.0033	0.0024	
¼	20	0.2500	0.2175	0.1850	0.0102	0.2428	0.0072	0.1887	0.2060	0.0101	0.0051	0.0036	0.0026	0.0013
⁵⁄₁₆	18	0.3125	0.2764	0.2403	0.0114	0.3043	0.0082	0.2443	0.2630	0.0106	0.0057	0.0041	0.0030	0.0015
⅜	16	0.3750	0.3344	0.2938	0.0126	0.3660	0.0090	0.2983	0.3184	0.0111	0.0036	0.0045	0.0032	0.0016
⁷⁄₁₆	14	0.4375	0.3911	0.3447	0.0140	0.4277	0.0098	0.3499	0.3721	0.0119	0.0070	0.0049	0.0036	0.0018
½	13	0.5000	0.4500	0.4001	0.0148	0.4896	0.0104	0.4056	0.4290	0.0123	0.0074	0.0052	0.0037	0.0019
⁹⁄₁₆	12	0.5625	0.5084	0.4542	0.0158	0.5513	0.0112	0.4603	0.4850	0.0127	0.0079	0.0056	0.0040	0.0020
⅝	11	0.6250	0.5660	0.5069	0.0170	0.6132	0.0118	0.5135	0.5397	0.0131	0.0085	0.0059	0.0042	0.0021
¾	10	0.7500	0.6850	0.6201	0.0184	0.7372	0.0128	0.6273	0.6553	0.0136	0.0092	0.0064	0.0045	0.0023
⅞	9	0.8750	0.8028	0.7307	0.0200	0.8610	0.0140	0.7387	0.7689	0.0142	0.0100	0.0070	0.0049	0.0024
1	8	1.0000	0.9188	0.8376	0.0222	0.9848	0.0152	0.8466	0.8795	0.0148	0.0111	0.0076	0.0054	0.0027
1⅛	7	1.1250	1.0322	0.9394	0.0248	1.1080	0.0170	0.9497	0.9858	0.0154	0.0124	0.0085	0.0059	0.0030
1¼	7	1.2500	1.1572	1.0644	0.0248	1.2330	0.0170	1.0747	1.1108	0.0154	0.0124	0.0085	0.0059	0.0030
1⅜	6	1.3750	1.2667	1.1585	0.0290	1.3548	0.0202	1.1705	1.2126	0.0180	0.0145	0.0101	0.0071	0.0036
1½	6	1.5000	1.3917	1.2835	0.0290	1.4798	0.0202	1.2955	1.3376	0.0180	0.0145	0.0101	0.0071	0.0036
1¾	5	1.7500	1.6201	1.4902	0.0338	1.7268	0.0232	1.5046	1.5551	0.0216	0.0169	0.0116	0.0082	0.0041
2	4½	2.0000	1.8557	1.7113	0.0368	1.9746	0.0254	1.7274	1.7835	0.0241	0.0184	0.0127	0.0089	0.0044
2¼	4½	2.2500	2.1057	1.9613	0.0368	2.2246	0.0254	1.9774	2.0335	0.0241	0.0184	0.0127	0.0089	0.0044
2½	4	2.5000	2.3376	2.1752	0.0408	2.4720	0.0280	2.1933	2.2564	0.0270	0.0204	0.0140	0.0097	0.0048
2¾	4	2.7500	2.5876	2.4252	0.0408	2.7220	0.0280	2.4433	2.5064	0.0270	0.0204	0.0140	0.0097	0.0048
3	4	3.0000	2.8376	2.6752	0.0408	2.9720	0.0280	2.6933	2.7564	0.0270	0.0204	0.0140	0.0097	0.0048
3¼	4	3.2500	3.0876	2.9252	0.0408	3.2220	0.0280	2.9433	3.0064	0.0270	0.0204	0.0140	0.0097	0.0048
3½	4	3.5000	3.3376	3.1752	0.0408	3.4720	0.0280	3.1933	3.2564	0.0270	0.0204	0.0140	0.0097	0.0048
3¾	4	3.7500	3.5876	3.4252	0.0408	3.7220	0.0280	3.4433	3.5064	0.0270	0.0204	0.0140	0.0097	0.0048
4	4	4.0000	3.3876	3.6752	0.0408	3.9720	0.0280	3.6933	3.7564	0.0270	0.0204	0.0140	0.0097	0.0048

11.1.11   British Association Thread (Fig. 11.6)

This type is recommended by the British Standards for screws below ¼-in diameter. This thread has an included angle of $47\frac{1}{2}°$. It is used in watchmaking and instrument work. Basic sizes of bolts are listed in Table 11.6.

11.1.12   International Metric Thread Standard

This thread form (Fig. 11.7 and data of Table 11.7) is similar to the American National Standard in that the included angle is 60°. A rounded root profile is recommended.

TABLE 11.5 American National Fine-Thread Series[17]

		Screw basic diam				Major screw diam tolerances		Max minor screw diam		Max minor nut diam		Nut pitch diam tolerances		
Size	Threads per in., n	Major diam, classes 2 and 3, D_m	Major diam, class 4, D_m	Pitch diam, classes 2 and 3	Pitch diam, class 4	Classes 2 and 3	Class 4	Classes 2 and 3	Class 4	Classes 2 and 3	Class 4	Class 2	Class 3	Class 4
0	80	0.0600	......	0.0519	......	0.0034	......	0.0447	......	0.0514	......	0.0017	0.0013	
1	72	0.0730	......	0.0640	......	0.0036	......	0.0560	......	0.0634	......	0.0018	0.0013	
2	64	0.0860	......	0.0759	......	0.0038	......	0.0668	......	0.0746	......	0.0019	0.0014	
3	56	0.0990	......	0.0874	......	0.0040	......	0.0771	......	0.0856	......	0.0020	0.0015	
4	48	0.1120	......	0.0985	......	0.0044	......	0.0864	......	0.0960	......	0.0022	0.0016	
5	44	0.1250	......	0.1102	......	0.0046	......	0.0971	......	0.1068	......	0.0023	0.0016	
6	40	0.1380	......	0.1218	......	0.0048	......	0.1073	......	0.1179	......	0.0024	0.0017	
8	36	0.1640	......	0.1460	......	0.0050	......	0.1299	......	0.1402	......	0.0025	0.0018	
10	32	0.1900	......	0.1697	......	0.0054	......	0.1517	......	0.1624	......	0.0027	0.0019	
12	28	0.2160	......	0.1928	......	0.0062	......	0.1722	......	0.1835	......	0.0031	0.0022	
¼	28	0.2500	0.2500	0.2268	0.2270	0.0062	0.0062	0.2062	0.2062	0.2173	0.2173	0.0031	0.0022	0.0011
⁵⁄₁₆	24	0.3125	0.3125	0.2854	0.2857	0.0066	0.0066	0.2614	0.2614	0.2739	0.2739	0.0033	0.0024	0.0012
⅜	24	0.3750	0.3750	0.3479	0.3482	0.0066	0.0066	0.3239	0.3238	0.3364	0.3364	0.0033	0.0024	0.0012
⁷⁄₁₆	20	0.4375	0.4375	0.4050	0.4053	0.0072	0.0072	0.3762	0.3672	0.3906	0.3906	0.0036	0.0026	0.0013
½	20	0.5000	0.5000	0.4675	0.4678	0.0072	0.0072	0.4387	0.4387	0.4531	0.4531	0.0036	0.0026	0.0013
⁹⁄₁₆	18	0.5625	0.5625	0.5264	0.5267	0.0082	0.0082	0.4943	0.4943	0.5100	0.5100	0.0041	0.0030	0.0015
⅝	18	0.6250	0.6250	0.5889	0.5892	0.0082	0.0082	0.5568	0.5568	0.5725	0.5725	0.0041	0.0030	0.0015
¾	16	0.7500	0.7500	0.7094	0.7098	0.0090	0.0090	0.6733	0.6733	0.6903	0.6903	0.0045	0.0032	0.0016
⅞	14	0.8750	0.8750	0.8286	0.8290	0.0098	0.0098	0.7874	0.7874	0.8062	0.8062	0.0049	0.0036	0.0018
1	14	1.0000	1.0000	0.9536	0.9540	0.0098	0.0008	0.9124	0.9124	0.9312	0.9312	0.0049	0.0036	0.0018
1⅛	12	1.1250	1.1250	1.0709	1.0714	0.0112	0.0112	1.0228	1.0228	1.0438	1.0438	0.0056	0.0040	0.0020
1¼	12	1.2500	1.2500	1.1959	1.1964	0.0112	0.0112	1.1478	1.1478	1.1688	1.1688	0.0056	0.0040	0.0020
1⅜	12	1.3750	1.3750	1.3209	1.3214	0.0112	0.0112	1.2728	1.2728	1.2938	1.2938	0.0056	0.0040	0.0020
1½	12	1.5000	1.5000	1.4459	1.4464	0.0112	0.0112	1.3978	1.3978	1.4188	1.4188	0.0056	0.0040	0.0020

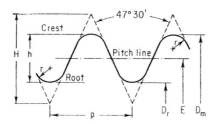

FIG. 11.6 British Association thread.[18] $E = D - h = D - 0.60p$; $H = 1.13634p$; $h = 0.60p$; $r = 0.1818p$.

11.1.13 Löwenherz Thread[8]

Based on the metric system, this thread is used for the fine threads of measuring instruments (see Fig. 11.8). Crest and root flats resemble the Unified and American

TABLE 11.6 British Association Threads[18]

No.	Metric dimensions					Approx. in equivalents						
	Pitch p, mm	Depth of thread 0.6p, mm	Major diam, mm, bolt and nut D_m	Effective* diam, mm, bolt and nut $D_m - h$	Minor diam, mm, bolt and nut $D_m - 2h$	Approx pitch, in.	Approx threads per in.	Depth of thread, in.	Major diam, in., bolt and nut	Effective* diam, in., bolt and nut	Minor diam, in., bolt and nut	Approx area at bottom of thread, in.[2]
0	1.0000	0.600	6.00	5.400	4.80	0.03937	25.4	0.0236	0.2362	0.2126	0.1890	0.0281
1	0.9000	0.540	5.30	4.760	4.22	0.03543	28.2	0.0213	0.2087	0.1874	0.1661	0.0217
2	0.8100	0.485	4.70	4.215	3.73	0.03189	31.3	0.0191	0.1850	0.1659	0.1468	0.0169
3	0.7300	0.440	4.10	3.660	3.22	0.02874	34.8	0.0173	0.1614	0.1441	0.1268	0.0126
4	0.6600	0.395	3.60	3.205	2.81	0.02598	38.5	0.0156	0.1417	0.1262	0.1106	0.0096
5	0.5900	0.355	3.20	2.845	2.49	0.02323	43.1	0.0140	0.1260	0.1120	0.0980	0.0075
6	0.5300	0.320	2.80	2.480	2.16	0.02087	47.9	0.0126	0.1102	0.0976	0.0850	0.0057
7	0.4800	0.290	2.50	2.210	1.92	0.01890	52.9	0.0114	0.0984	0.0870	0.0756	0.0045
8	0.4300	0.260	2.20	1.940	1.68	0.01693	59.1	0.0102	0.0866	0.0764	0.0661	0.0034
9	0.3900	0.235	1.90	1.665	1.43	0.01535	65.1	0.0093	0.0748	0.0656	0.0563	0.0025
10	0.3500	0.210	1.70	1.490	1.28	0.01378	72.6	0.0083	0.0669	0.0587	0.0504	0.0020
11	0.3100	0.185	1.50	1.315	1.13	0.01220	81.9	0.0073	0.0591	0.0518	0.0445	0.0016
12	0.2800	0.170	1.30	1.130	0.96	0.01102	90.7	0.0067	0.0512	0.0445	0.0378	0.0011
13	0.2500	0.150	1.20	1.050	0.90	0.00984	102.0	0.0059	0.0472	0.0413	0.0354	0.0010
14	0.2300	0.140	1.00	0.860	0.72	0.00905	110.0	0.0055	0.0394	0.0339	0.0283	0.0006
15	0.2100	0.125	0.90	0.775	0.65	0.00827	121.0	0.0049	0.0354	0.0305	0.0256	0.0005
16	0.1900	0.115	0.79	0.675	0.56	0.00748	134.0	0.0045	0.0311	0.0266	0.0220	0.0004
17	0.1700	0.100	0.70	0.600	0.50	0.00669	149.0	0.0039	0.0276	0.0236	0.0197	0.0003
18	0.1500	0.090	0.62	0.530	0.44	0.00590	169.0	0.0035	0.0244	0.0209	0.0173	0.0002
19	0.1400	0.085	0.54	0.455	0.37	0.00551	182.0	0.0033	0.0213	0.0179	0.0146	0.0002
20	0.1200	0.070	0.48	0.410	0.34	0.00472	213.0	0.0028	0.0189	0.0161	0.0134	0.0001
21	0.1100	0.065	0.42	0.355	0.29	0.00433	232.0	0.0026	0.0165	0.0140	0.0114	0.0001
22	0.1000	0.060	0.37	0.310	0.25	0.00394	256.0	0.0024	0.0146	0.0122	0.0098	0.00008
23	0.0900	0.055	0.33	0.275	0.22	0.00354	286.0	0.0022	0.0130	0.0108	0.0087	0.00006
24	0.0800	0.050	0.29	0.240	0.19	0.00315	322.0	0.0020	0.0114	0.0094	0.0075	0.00004
25	0.0700	0.040	0.25	0.210	0.17	0.00276	357.0	0.0016	0.0098	0.0083	0.0067	0.00004

*Basic pitch diameter.

11.12

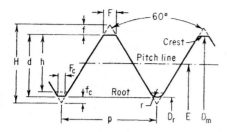

FIG. 11.7 International 60° metric thread. $E = D - h = D - 0.64952p$; $F = 0.1250p$; $F_e = 0.0625p$; $G = 0.10825p$; $G_e = 0.05412p$; $H = 0.86603p$; $h = 0.64952p$; $d = 0.70364p$; $r = 0.054p$ (optional); $f = 0.125H$.

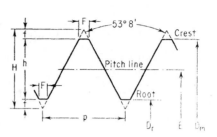

FIG. 11.8 Löwenherz thread.[8] $E = D - h = D - 0.75p$; $F = 0.125p$; $H = p$; $h = 0.75p$.

Standard. Since both the German (DIN) and the French metric thread systems carry diameters down to 1 mm with other details varying slightly, the Löwenherz has been superseded.

11.1.14 Dardelet Thread[8] (Fig. 11.9)

This type is a self-locking thread with an included angle of 29°. When the nut is tightened, a wedging action due to the 6° angle results, locking both bolt and nut. Note the inclined crests on the nut and inclined flats at the roots of the bolt.

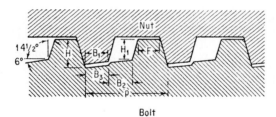

FIG. 11.9 Dardelet relieved thread (unlocked).[8]

11.1.15 Lok-Thread[9]

This has an included thread angle of 60° (Fig. 11.10) and is seen to be similar to the Unified and American Standard. The flat at the root of the screw is a 6° inclined spiral. Locking is obtained by the interference between the crest of the internal thread and the tapered flats of the external thread.

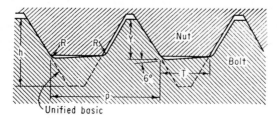

FIG. 11.10 Lok-Thread.[9]

TABLE 11.7 British Standard for International Metric-Isometric Threads[19]

		Millimeter sizes			Equivalent sizes, in.			
Diam of screw D	Pitch p	Depth of thread 0.6495p (h)	Pitch diam $D_m - h$ (E)	Minor diam* $D_m - 2h$ (Dr)	Depth of thread 0.6495p (h)	Major diam D_m	Pitch diam $D_m - h$ (E)	Minor diam $D - 2h$ (Dr)
3	0.50	0.32475	2.67525	2.350	0.01279	0.11811	0.10532	0.09253
3.5	0.60	0.38970	3.11030	2.721	0.01534	0.13780	0.12246	0.10712
4	0.70	0.45465	3.54535	3.091	0.01790	0.15748	0.13958	0.12168
4.5	0.75	0.48712	4.01288	3.526	0.01918	0.17716	0.15798	0.13880
5	0.80	0.51960	4.48040	3.961	0.02046	0.19685	0.17639	0.15593
5.5	0.90	0.58455	4.91545	4.331	0.02301	0.21654	0.19353	0.17052
6	1.00	0.64950	5.35050	4.70	0.02557	0.23622	0.21065	0.18508
7	1.00	0.64950	6.35050	5.70	0.02557	0.27559	0.25002	0.22445
8	1.25	0.81188	7.18813	6.38	0.03196	0.31496	0.28300	0.25104
9	1.25	0.81188	8.18813	7.38	0.03196	0.35433	0.32237	0.29041
10	1.5	0.97425	9.026	8.05	0.03836	0.39370	0.35534	0.31698
11	1.5	0.97425	10.026	9.05	0.03836	0.43307	0.39471	0.35635
12	1.5	0.97425	11.026	10.05	0.03836	0.47244	0.43408	0.39572
12	1.75	1.13662	10.863	9.73	0.04475	0.47244	0.42769	0.38294
14	2.0	1.2990	12.701	11.40	0.05114	0.55118	0.50004	0.44890
16	2.0	1.2990	14.701	13.40	0.05114	0.62992	0.57878	0.52764
18	2.5	1.62375	16.376	14.75	0.06393	0.70866	0.64473	0.58080
20	2.5	1.62375	18.376	16.75	0.06393	0.78740	0.72347	0.65954
22	2.5	1.62375	20.376	18.75	0.06393	0.86614	0.80221	0.73828
22	3.0	1.9485	20.052	18.10	0.07671	0.86614	0.78943	0.71272
24	3.0	1.9485	22.052	20.10	0.07671	0.94488	0.86817	0.79146
26	3.0	1.9485	24.052	22.10	0.07671	1.02362	0.94691	0.87020
27	3.0	1.9485	25.052	23.10	0.07671	1.06299	0.98628	0.90957
28	3.0	1.9485	26.052	24.10	0.07671	1.10236	1.02565	0.94894
30	3.5	2.27325	27.727	25.45	0.08950	1.18110	1.09160	1.00210
32	3.5	2.27325	29.727	27.45	0.08950	1.25984	1.17034	1.08084
33	3.5	2.27325	30.727	28.45	0.08950	1.29921	1.20971	1.12021
34	3.5	2.27325	31.727	29.45	0.08950	1.33858	1.24908	1.15958
36	4.0	2.5980	33.420	30.80	0.10228	1.41732	1.31504	1.21276
38	4.0	2.5980	35.402	32.80	0.10228	1.49606	1.39378	1.29150
39	4.0	2.5980	36.402	33.80	0.10228	1.53543	1.43315	1.33087
40	4.0	2.5980	37.402	34.80	0.10228	1.57480	1.47252	1.37024
42	4.5	2.92275	39.077	36.15	0.11507	1.65354	1.53847	1.42340
44	4.5	2.92275	41.077	38.15	0.11507	1.73228	1.61721	1.50214
45	4.5	2.92275	42.077	39.15	0.11507	1.77165	1.65658	1.54151
46	4.5	2.92275	43.077	40.15	0.11057	1.81102	1.69595	1.58088
48	5.0	3.2475	44.752	41.50	0.12785	1.88976	1.76191	1.63406
50	5.0	3.2475	46.752	43.50	0.12785	1.96850	1.84065	1.71280
52	5.0	3.2475	48.752	45.50	0.12785	2.04724	1.91939	1.79154
56	5.5	3.57225	52.428	48.86	0.14064	2.20472	2.06408	1.92344
60	5.5	3.57225	56.428	52.86	0.14064	2.36220	2.22156	2.08092
64	6.0	3.8970	60.103	56.21	0.15342	2.51968	2.36626	2.21284
68	6.0	3.8970	64.103	60.21	0.15342	2.67716	2.52374	2.37032
72	6.5	4.22175	67.778	63.56	0.16621	2.83464	2.66843	2.50222
76	6.5	4.22175	71.778	67.56	0.16621	2.99212	2.82591	2.65970
80	7.0	4.5465	75.454	70.91	0.17900	3.14960	2.97060	2.79160

*Minor diameter varies with tolerance class and fit.

Recommended Thread Depths. For cap screws and bolts, 70 percent thread:

$$\text{Thread depth } d = 0.7x \text{ (Unified and American depth)}$$

$$\text{Width of root } T = 0.5p \qquad R = 0.137p$$

For studs, 60 percent thread:

$$\text{Thread depth } d = 0.6x \text{ (Unified and American depth)}$$

$$\text{Width of root } T = 0.425p \qquad R = 0.137p$$

11.1.16   Aero-Thread[10] (Fig. 11.11)

This type is particularly applicable where soft and light internally threaded parts such as aluminum or magnesium alloys are used, as in the aircraft field, with a high-strength steel screw. Wear of the light-alloy thread is prevented by a spring-shaped insert, usually made of phosphor bronze.

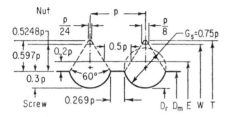

FIG. 11.11   Aero-Thread.[10] D = major diameter of screw; also, minor diameter of tapped hole. E = $D + 0.4p$ = pitch diameter. $K = D - 0.6p$ = minor diameter. $W = D + 1.0495p$ = major diameter, insert. $G_s = 0.75p$ = diameter of thread-form circle. T = major diameter of tap $(D + 1.194p)$; also minimum $(D + 1.122p)$. (*Aircraft Screw Products Co., Inc.*)

11.1.17   Microscopic Objective Screw Thread[11]

Based on the Whitworth thread, except truncated, this thread is used for mounting the microscope objective to the nosepiece. This thread is recommended for other optical assemblies of microscopes, such as photomicrographic equipment. It has become universally accepted as a standard for microscopic-objectives and nosepiece threads.[12]

For data on hose-coupling threads, fire-hose-coupling threads, and threads for electric sockets and lamp bases, see Refs. 13, 14, and 15, respectively.

11.2   BOLTS, NUTS, AND SCREWS

11.2.1   Introduction

Screw fasteners are generally classified according to function, style, and application, such as bolts, nuts, machine screws, and cap screws.

Proportions of these fasteners have been standardized by the American Standards Association and are summarized herein.

11.2.2 Bolts and Nuts

Bolts, used in through holes with mating nut, are shown with leading dimensions in Tables 11.8 through 11.10.

TABLE 11.8 Regular Unfinished Square Bolts[21]

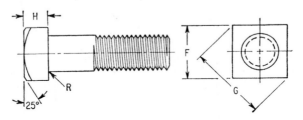

Nominal size or basic major diam of thread	Body diam, max	Width across flats F		Width across corners G		Height H			Radius of fillet R
		Max (basic)	Min	Max	Min	Nom	Max	Min	Max
1/4	0.2500	0.280	3/8	0.3750	0.362	0.530	0.498	11/64	0.188	0.156	0.031
5/16	0.3125	0.342	1/2	0.5000	0.484	0.707	0.665	13/64	0.220	0.186	0.031
3/8	0.3750	0.405	9/16	0.5625	0.544	0.795	0.747	1/4	0.268	0.232	0.031
7/16	0.4375	0.468	5/8	0.6250	0.603	0.884	0.828	19/64	0.316	0.278	0.031
1/2	0.5000	0.530	3/4	0.7500	0.725	1.061	0.995	21/64	0.348	0.308	0.031
5/8	0.6250	0.675	15/16	0.9375	0.906	1.326	1.244	27/64	0.444	0.400	0.062
3/4	0.7500	0.800	1 1/8	1.1250	1.088	1.591	1.494	1/2	0.524	0.476	0.062
7/8	0.8750	0.938	1 5/16	1.3125	1.269	1.856	1.742	19/32	0.620	0.568	0.062
1	1.0000	1.063	1 1/2	1.5000	1.450	2.121	1.991	21/32	0.684	0.628	0.062
1 1/8	1.1250	1.188	1 11/16	1.6875	1.631	2.386	2.239	3/4	0.780	0.720	0.125
1 1/4	1.2500	1.313	1 7/8	1.8750	1.812	2.652	2.489	27/32	0.876	0.812	0.125
1 3/8	1.3750	1.469	2 1/16	2.0625	1.994	2.917	2.738	29/32	0.940	0.872	0.125
1 1/2	1.5000	1.594	2 1/4	2.2500	2.175	3.182	2.986	1	1.036	0.964	0.125
1 5/8	1.6250	1.719	2 7/16	2.4375	2.356	3.447	3.235	1 3/32	1.132	1.056	0.125

Note: Bolt is not finished on any surface.

Minimum thread length shall be twice the diameter plus 1/4 in for lengths up to and including 6 in and twice the diameter plus 1/2 in for lengths over 6 in.

Thread shall be coarse-thread series, class 2A.

TABLE 11.9 Hexagon Bolts[21]

Dimensions of Regular Semifinished Hexagon Bolts

Dimensions of Regular Hexagon Bolts

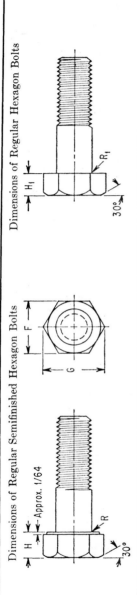

Nominal size or basic major diam of thread	Body diam, max	Width across flats F Max (basic)	Width across flats F Min	Width across corners G Max	Width across corners G Min	Semifinished height H Nom	Semifinished height H Max	Semifinished height H Min	Radius of fillet R Max	Radius of fillet R Min	Regular height H_1 Nom	Regular height H_1 Max	Regular height H_1 Min	Radius of fillet R_1 Max
1/4	0.2500	7/16 0.4375	0.425	0.505	0.484	5/32	0.163	0.150	0.031	0.016	11/64	0.188	0.150	0.031
5/16	0.3125	1/2 0.5000	0.484	0.577	0.552	13/64	0.211	0.195	0.031	0.016	7/32	0.235	0.195	0.031
3/8	0.3750	9/16 0.5625	0.544	0.650	0.620	15/64	0.243	0.226	0.031	0.016	1/4	0.268	0.226	0.031
7/16	0.4375	5/8 0.6250	0.603	0.722	0.687	9/32	0.291	0.272	0.031	0.016	19/64	0.316	0.272	0.031
1/2	0.5000	3/4 0.7500	0.725	0.866	0.826	5/16	0.323	0.302	0.031	0.016	13/32	0.364	0.302	0.031
5/8	0.6250	15/16 0.9375	0.906	1.083	1.033	25/64	0.403	0.378	0.031	0.016	27/64	0.444	0.378	0.062
3/4	0.7500	1 1/8 1.1250	1.088	1.299	1.240	15/32	0.483	0.455	0.047	0.031	1/2	0.524	0.455	0.062
7/8	0.8750	1 5/16 1.3125	1.269	1.516	1.447	35/64	0.563	0.531	0.047	0.031	37/64	0.604	0.531	0.062
1	1.0000	1 1/2 1.5000	1.450	1.732	1.653	39/64	0.627	0.591	0.047	0.031	43/64	0.700	0.591	0.062
1 1/8	1.1250	1 11/16 1.6875	1.631	1.949	1.859	11/16	0.718	0.658	0.062	0.047	3/4	0.780	0.658	0.125
1 1/4	1.2500	1 7/8 1.8750	1.812	2.165	2.066	25/32	0.813	0.749	0.062	0.047	27/32	0.876	0.749	0.125
1 3/8	1.3750	2 1/16 2.0625	1.994	2.382	2.273	27/32	0.878	0.810	0.062	0.047	29/32	0.940	0.810	0.125

11.17

TABLE 11.9 Hexagon Bolts[21] (*Continued*)

Nominal size or basic major diam of thread		Body diam, max	Width across flats *F*			Width across corners *G*		Semifinished height *H*			Radius of fillet *R*		Regular height H_1			Radius of fillet R_1
			Max (basic)		Min	Max	Min	Nom	Max	Min	Max	Min	Nom	Max	Min	Max
1½	1.5000	1.594	2¼	2.2500	2.175	2.598	2.480	15/16	0.974	0.902	0.062	0.047	1	1.036	0.902	0.125
1⅝	1.6250	1.719	2 7/16	2.4375	2.356	2.815	2.686	1	1.038	0.962	0.062	0.047	1 1/16	1.100	0.962	0.125
1¾	1.7500	1.844	2⅝	2.6250	2.538	3.031	2.893	1 3/32	1.134	1.054	0.062	0.047	1 5/32	1.196	1.054	0.125
1⅞	1.8750	1.969	2 13/16	2.8125	2.719	3.248	3.100	1 5/32	1.198	1.114	0.062	0.047	1 7/32	1.260	1.114	0.125
2	2.0000	2.094	3	3.0000	2.900	3.464	3.306	1 7/32	1.263	1.175	0.062	0.047	1 11/32	1.388	1.175	0.125
2¼	2.2500	2.375	3⅜	3.3750	3.262	3.897	3.719	1⅜	1.423	1.327	0.062	0.047	1½	1.548	1.327	0.188
2½	2.5000	2.625	3¾	3.7500	3.625	4.330	4.133	1 17/32	1.583	1.479	0.062	0.047	1 21/32	1.708	1.479	0.188
2¾	2.7500	2.875	4⅛	4.1250	3.988	4.763	4.546	1 11/16	1.744	1.632	0.062	0.047	1 13/16	1.869	1.632	0.188
3	3.0000	3.125	4½	4.5000	4.350	5.196	4.959	1⅞	1.935	1.815	0.062	0.047	2	2.060	1.815	0.188
3¼	3.2500	3.438	4⅞	4.8750	4.712	5.629	5.372	2	2.064	1.936	0.062	0.047	2 3/16	2.251	1.936	0.188
3½	3.5000	3.688	5¼	5.2500	5.075	6.062	5.786	2⅛	2.193	2.057	0.062	0.047	2 5/16	2.380	2.057	0.188
3¾	3.7500	3.938	5⅝	5.6250	5.437	6.495	6.193	2 5/16	2.385	2.241	0.062	0.047	2½	2.572	2.241	0.188
4	4.0000	4.188	6	6.0000	5.800	6.928	6.612	2½	2.576	2.424	0.062	0.047	2 11/16	2.764	2.424	0.188

TABLE 11.10   Finished Hexagon Bolts[21]

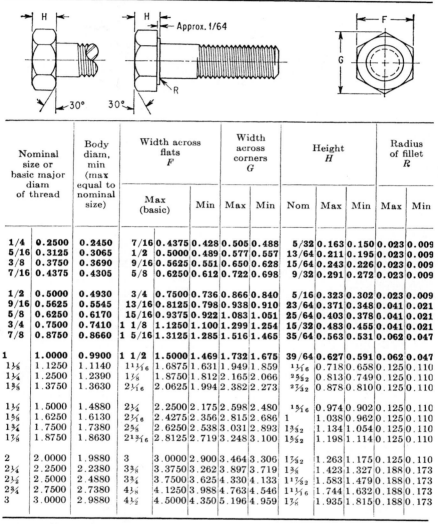

Nominal size or basic major diam of thread		Body diam, min (max equal to nominal size)	Width across flats F		Width across corners G		Height H			Radius of fillet R	
			Max (basic)	Min	Max	Min	Nom	Max	Min	Max	Min
1/4	0.2500	0.2450	7/16 0.4375	0.428	0.505	0.488	5/32	0.163	0.150	0.023	0.009
5/16	0.3125	0.3065	1/2 0.5000	0.489	0.577	0.557	13/64	0.211	0.195	0.023	0.009
3/8	0.3750	0.3690	9/16 0.5625	0.551	0.650	0.628	15/64	0.243	0.226	0.023	0.009
7/16	0.4375	0.4305	5/8 0.6250	0.612	0.722	0.698	9/32	0.291	0.272	0.023	0.009
1/2	0.5000	0.4930	3/4 0.7500	0.736	0.866	0.840	5/16	0.323	0.302	0.023	0.009
9/16	0.5625	0.5545	13/16 0.8125	0.798	0.938	0.910	23/64	0.371	0.348	0.041	0.021
5/8	0.6250	0.6170	15/16 0.9375	0.922	1.083	1.051	25/64	0.403	0.378	0.041	0.021
3/4	0.7500	0.7410	1 1/8 1.1250	1.100	1.299	1.254	15/32	0.483	0.455	0.041	0.021
7/8	0.8750	0.8660	1 5/16 1.3125	1.285	1.516	1.465	35/64	0.563	0.531	0.062	0.047
1	1.0000	0.9900	1 1/2 1.5000	1.469	1.732	1.675	39/64	0.627	0.591	0.062	0.047
1 1/8	1.1250	1.1140	1 11/16 1.6875	1.631	1.949	1.859	11/16	0.718	0.658	0.125	0.110
1 1/4	1.2500	1.2390	1 7/8 1.8750	1.812	2.165	2.066	25/32	0.813	0.749	0.125	0.110
1 3/8	1.3750	1.3630	2 1/16 2.0625	1.994	2.382	2.273	27/32	0.878	0.810	0.125	0.110
1 1/2	1.5000	1.4880	2 1/4 2.2500	2.175	2.598	2.480	15/16	0.974	0.902	0.125	0.110
1 5/8	1.6250	1.6130	2 7/16 2.4275	2.356	2.815	2.686	1	1.038	0.962	0.125	0.110
1 3/4	1.7500	1.7380	2 5/8 2.6250	2.538	3.031	2.893	1 3/32	1.134	1.054	0.125	0.110
1 7/8	1.8750	1.8630	2 13/16 2.8125	2.719	3.248	3.100	1 5/32	1.198	1.114	0.125	0.110
2	2.0000	1.9880	3 3.0000	2.900	3.464	3.306	1 7/32	1.263	1.175	0.125	0.110
2 1/4	2.2500	2.2380	3 3/8 3.3750	3.262	3.897	3.719	1 3/8	1.423	1.327	0.188	0.173
2 1/2	2.5000	2.4880	3 3/4 3.7500	3.625	4.330	4.133	1 17/32	1.583	1.479	0.188	0.173
2 3/4	2.7500	2.7380	4 1/8 4.1250	3.988	4.763	4.546	1 11/16	1.744	1.632	0.188	0.173
3	3.0000	2.9880	4 1/2 4.5000	4.350	5.196	4.959	1 7/8	1.935	1.815	0.188	0.173

Note: Bold type indicates unified thread.

Terms Applied to Square and Hexagon Bolts and Nuts

Series Selection.   Square and hexagon-head bolts and nuts[21] are supplied in two series of sizes, known as "regular" series and "heavy" series. Both series refer to the bolt-head or nut proportions. The heavy series is used where greater bearing surface is required or a greater wrench bearing surface is essential. Regular-series bolt heads are for general use.

Finish Selection. Unless otherwise specified, only the thread portion is finished. Note in Tables 11.8 to 11.10 the unfinished bolts do not have a washer face beneath the bolt heads.

Semifinished bolt heads and nuts (Table 11.9) are machined or otherwise formed or treated on the bearing surface so as to provide a washer face for bolt heads and either a washer face or a circular bearing surface by chamfering the corners for nuts.

Finished bolt heads and nuts (Table 11.10) refers to quality of manufacture and closeness of tolerance and does not indicate that surfaces are necessarily machined.

Nut styles are shown with leading dimensions in Tables 11.11 to 11.14.

TABLE 11.11 Regular Square Nuts[21]

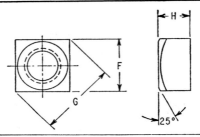

Nominal size or basic major diam of thread		Width across flats F			Width across corners G		Thickness H		
		Max (basic)	Min		Max	Min	Nom	Max	Min
¼	0.2500	7/16	0.4375	0.425	0.619	0.584	7/32	0.235	0.203
5/16	0.3125	9/16	0.5625	0.547	0.795	0.751	17/64	0.283	0.249
3/8	0.3750	5/8	0.6250	0.606	0.884	0.832	21/64	0.346	0.310
7/16	0.4375	3/4	0.7500	0.728	1.061	1.000	3/8	0.394	0.356
½	0.5000	13/16	0.8125	0.788	1.149	1.082	7/16	0.458	0.418
5/8	0.6250	1	1.0000	0.969	1.414	1.330	35/64	0.569	0.525
¾	0.7500	1⅛	1.1250	1.088	1.591	1.494	21/32	0.680	0.632
7/8	0.8750	1 5/16	1.3125	1.269	1.856	1.742	49/64	0.792	0.740
1	1.0000	1½	1.5000	1.450	2.121	1.991	7/8	0.903	0.847
1⅛	1.1250	1 11/16	1.6875	1.631	2.386	2.239	1	1.030	0.970
1¼	1.2500	1 7/8	1.8750	1.812	2.652	2.489	1 3/32	1.126	1.062
1⅜	1.3750	2 1/16	2.0625	1.994	2.917	2.738	1 13/64	1.237	1.169
1½	1.5000	2¼	2.2500	2.175	3.182	2.986	1 5/16	1.348	1.276
1⅝	1.6250	2 7/16	2.4375	2.356	3.447	3.235	1 27/64	1.460	1.384

Stud Bolts. To date there is no American Standard for stud bolts; however, the Industrial Fasteners Institute[31] has published a recommended standard. This standard has four basic classes of studs.

Class 1: Tap end

Class 2: Double end

Class 3: Bolt studs

Class 4: Continuous thread

TABLE 11.12 Hexagon and Hexagon Jam Nuts[21]

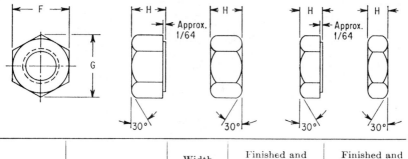

Nominal size or basic major diam of thread		Width across flats F		Width across corners G		Finished and regular semi-finished hexagon nuts thickness H			Finished and regular semi-finished jam nuts thickness H		
		Max (basic)	Min	Max	Min	Nom	Max	Min	Nom	Max	Min
1/4	**0.2500**	7/16 0.4375	0.428	0.505	0.488	7/32	0.226	0.212	5/32	0.163	0.150
5/16	**0.3125**	1/2 0.5000	0.489	0.577	0.557	17/64	0.273	0.258	3/16	0.195	0.180
3/8	**0.3750**	9/16 0.5625	0.551	0.650	0.628	21/64	0.337	0.320	7/32	0.227	0.210
7/16	**0.4375**	11/16 0.6875	0.675	0.794	0.768	3/8	0.385	0.365	1/4	0.260	0.240
1/2	**0.5000**	3/4 0.7500	0.736	0.866	0.840	7/16	0.448	0.427	5/16	0.323	0.302
9/16	**0.5625**	7/8 0.8750	0.861	1.010	0.982	31/64	0.496	0.473	6/16	0.324	0.301
5/8	**0.6250**	15/16 0.9375	0.922	1.083	1.051	35/64	0.559	0.535	3/8	0.387	0.363
3/4	**0.7500**	1 1/8 1.1250	1.088	1.299	1.240	41/64	0.665	0.617	27/64	0.446	0.398
7/8	**0.8750**	1 5/16 1.3125	1.269	1.516	1.447	3/4	0.776	0.724	31/64	0.510	0.458
1	**1.0000**	1 1/2 1.5000	1.450	1.732	1.653	55/64	0.887	0.831	35/64	0.575	0.519
1⅛	**1.1250**	1 11/16 1.6875	1.631	1.949	1.859	31/32	0.999	0.939	39/64	0.639	0.579
1¼	**1.2500**	1 7/8 1.8750	1.812	2.165	2.066	1 1/16	1.094	1.030	23/32	0.751	0.687
1⅜	**1.3750**	2 1/16 2.0625	1.994	2.382	2.273	1 11/64	1.206	1.138	25/32	0.815	0.747
1½	**1.5000**	2 1/4 2.2500	2.175	2.598	2.480	1 9/32	1.317	1.245	27/32	0.880	0.808
1⅝	**1.6250**	2 7/16 2.4375	2.356	2.815	2.686	1 25/64	1.429	1.353	29/32	0.944	0.868
1¾	**1.7500**	2 5/8 2.6250	2.538	3.031	2.893	1 1/2	1.540	1.460	31/32	1.009	0.929
1⅞	**1.8750**	2 13/16 2.8125	2.719	3.248	3.100	1 39/64	1.651	1.567	1 1/32	1.073	0.989
2	**2.0000**	3 3.0000	2.900	3.464	3.306	1 23/32	1.763	1.675	1 3/32	1.138	1.050
2¼	**2.2500**	3 3/8 3.3750	3.262	3.897	3.719	1 59/64	1.970	1.874	1 13/64	1.251	1.155
2½	**2.5000**	3 3/4 3.7500	3.625	4.330	4.133	2 9/64	2.193	2.089	1 29/64	1.505	1.401
2¾	**2.7500**	4 1/8 4.1250	3.988	4.763	4.546	2 23/64	2.415	2.303	1 37/64	1.634	1.522
3	**3.0000**	4 1/2 4.5000	4.350	5.196	4.959	2 37/64	2.638	2.518	1 45/64	1.763	1.643

Note: Bold type indicates unified threads.

The first two classes are then subdivided into four types:

Type a: Unfinished stud

Type b: Optional, full or undersized body

Type c: Full-sized body

Type d: Close-tolerance body

For stud proportions in general use today, see Ref. 21.

TABLE 11.13   Finished Hexagon Slotted Nuts[21]

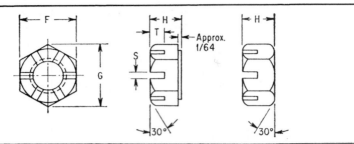

Nominal size or basic major diam of thread		Width across flats F		Width across corners G		Thickness H			Slot	
		Max (basic)	Min	Max	Min	Nom	Max	Min	Width S	Depth T
1/4	**0.2500**	7/16	0.4375	0.428	0.505	0.488	7/32	0.226	0.212	0.078	0.094
5/16	**0.3125**	1/2	0.5000	0.489	0.577	0.557	17/64	0.273	0.258	0.094	0.094
3/8	**0.3750**	9/16	0.5625	0.551	0.650	0.628	21/64	0.337	0.320	0.125	0.125
7/16	**0.4375**	11/16	0.6875	0.675	0.794	0.768	3/8	0.385	0.365	0.125	0.156
1/2	**0.5000**	3/4	0.7500	0.736	0.866	0.840	7/16	0.448	0.427	0.156	0.156
9/16	**0.5625**	7/8	0.8750	0.861	1.010	0.982	31/64	0.496	0.473	0.156	0.188
5/8	**0.6250**	15/16	0.9375	0.922	1.083	1.051	35/64	0.559	0.535	0.188	0.219
3/4	**0.7500**	1 1/8	1.1250	1.088	1.299	1.240	41/64	0.665	0.617	0.188	0.250
7/8	**0.8750**	1 5/16	1.3125	1.269	1.516	1.447	3/4	0.776	0.724	0.188	0.250
1	**1.0000**	1 1/2	1.5000	1.450	1.732	1.653	55/64	0.887	0.831	0.250	0.281
1⅛	1.1250	1¹¹⁄₁₆	1.6875	1.631	1.949	1.859	³¹⁄₃₂	0.999	0.939	0.250	0.344
1¼	1.2500	1⅞	1.8750	1.812	2.165	2.066	1¹⁄₁₆	1.094	1.030	0.312	0.375
1⅜	1.3750	2¹⁄₁₆	2.0625	1.994	2.382	2.273	1¹¹⁄₆₄	1.206	1.138	0.312	0.375
1½	1.5000	2¼	2.2500	2.175	2.598	2.480	1⁹⁄₃₂	1.317	1.245	0.375	0.438
1⅝	1.6250	2⁷⁄₁₆	2.4375	2.356	2.815	2.686	1²⁵⁄₆₄	1.429	1.353	0.375	0.438
1¾	1.7500	2⅝	2.6250	2.538	3.031	2.893	1½	1.540	1.460	0.438	0.500
1⅞	1.8750	2¹³⁄₁₆	2.8125	2.719	3.248	3.100	1³⁹⁄₆₄	1.651	1.567	0.438	0.562
2	2.0000	3	3.0000	2.900	3.464	3.306	1²³⁄₃₂	1.763	1.675	0.438	0.562
2¼	2.2500	3⅜	3.3750	3.262	3.897	3.719	1⁵⁹⁄₆₄	1.970	1.874	0.438	0.562
2½	2.5000	3¾	3.7500	3.625	4.330	4.133	2⁹⁄₆₄	2.193	2.089	0.562	0.688
2¾	2.7500	4⅛	4.1250	3.988	4.763	4.546	2²³⁄₆₄	2.415	2.303	0.562	0.688
3	3.0000	4½	4.5000	4.350	5.196	4.959	2³⁷⁄₆₄	2.638	2.518	0.625	0.750

Note: Bold type indicates unified threads.

Round-Head Bolts.   Carriage bolts, used where the head bears against wood, are illustrated in Table 11.15. Threads are Unified National coarse class 2A. Minimum thread length $= 2D + \frac{1}{4}$ for bolts up to 6 in long, $2D + \frac{1}{2}$ for bolts over 6 in long.

Styles and leading dimensions of countersunk, buttonhead, and step bolts are given in Table 11.16.

Elevator bolts[23] and plow bolts[24] are similar to carriage bolts.[23]

Track bolts[25] are specials, designed to meet the requirements of the railways.

TABLE 11.14   Regular Hexagon and Hexagon Jam Nuts

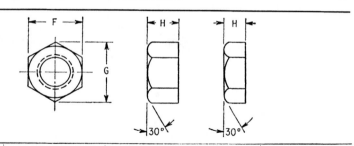

Nominal size or basic major diam of thread		Width across flats F		Width across corners G		Thickness regular nuts H			Thickness regular jam nuts H		
		Max (basic)	Min	Max	Min	Nom	Max	Min	Nom	Max	Min
1/4	0.2500	7/16	0.4375	0.425	0.505	0.484	7/32	0.235	0.203	5/32	0.172	0.140
5/16	0.3125	9/16	0.5625	0.547	0.650	0.624	17/64	0.283	0.249	3/16	0.204	0.170
3/8	0.3750	5/8	0.6250	0.606	0.722	0.691	21/64	0.346	0.310	7/32	0.237	0.201
7/16	0.4375	3/4	0.7500	0.728	0.866	0.830	3/8	0.394	0.356	1/4	0.269	0.231
1/2	0.5000	13/16	0.8125	0.788	0.938	0.898	7/16	0.458	0.418	5/16	0.332	0.292
9/16	0.5625	7/8	0.8750	0.847	1.010	0.966	1/2	0.521	0.479	11/32	0.365	0.323
5/8	0.6250	1	1.0000	0.969	1.155	1.104	35/64	0.569	0.525	3/8	0.397	0.353
3/4	0.7500	1 1/8	1.1250	1.088	1.299	1.240	21/32	0.680	0.632	7/16	0.462	0.414
7/8	0.8750	1 5/16	1.3125	1.269	1.516	1.447	49/64	0.792	0.740	1/2	0.526	0.474
1	1.0000	1 1/2	1.5000	1.450	1.732	1.653	7/8	0.903	0.847	9/16	0.590	0.534
1 1/8	1.1250	1 11/16	1.6875	1.631	1.949	1.859	1	1.030	0.970	5/8	0.655	0.595
1 1/4	1.2500	1 7/8	1.8750	1.812	2.165	2.066	1 3/32	1.126	1.062	3/4	0.782	0.718
1 3/8	1.3750	2 1/16	2.0625	1.994	2.382	2.273	1 11/64	1.237	1.169	13/16	0.846	0.778
1 1/2	1.5000	2 1/4	2.2500	2.175	2.598	2.480	1 5/16	1.348	1.276	7/8	0.911	0.839

11.2.3   Machine Screws (Tables 11.17 to 11.19)

Machine screws are classified according to head style. From the tables note that the No. 0 series has a body diameter of 0.060 in and successive numbers add 0.013 in to the diameter; e.g.,

$$\text{Body diameter of No. 6 screw} = 0.060 + 0.013(6) = 0.060 + 0.078 = 0.138$$

Machine screws are regularly supplied with plain sheared ends, not pointed. On machine screws up to 2 in long, complete threads extend to within two threads of the bearing surface of the head; long screws have a minimum complete thread length of $1\frac{3}{4}$ in.

Screw threads of machine screws are Unified or American National coarse- or fine-thread series, class 2 or 2A, supplied with a naturally bright finish, not heat-treated.

Machine-Screw Heads.   Four head styles are available, namely, screwdriver slot (Tables 11.17 to 11.19), two styles of cross recess,[26] and the "clutch head,"[26] not shown.

TABLE 11.15 Carriage Bolts

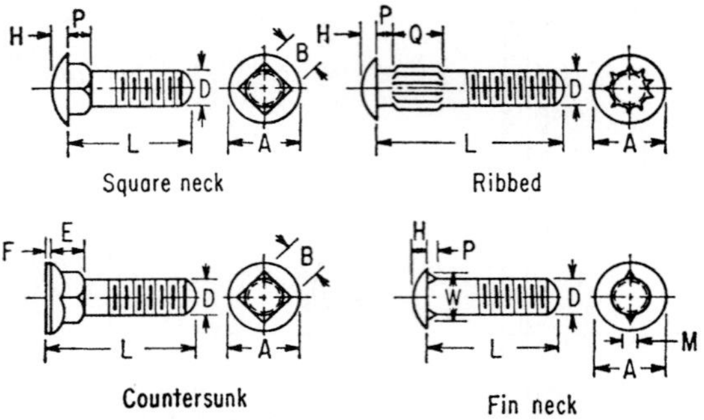

Nominal diam of bolt *D*	Countersunk				Fin neck				
	Diam of head, max, *A*	Feed thickness *F*	Depth of square and countersink, max, *E*	Width of square, max, *B*	Diam of head, max, *A*	Height of head, max, *H*	Depth of fins, max, *P*	Distance across fins, max, *W*	Thickness of fins, max, *M*
No. 10	0.520	0.016	0.250	0.199	0.469	0.114	0.088	0.395	0.098
1/4	0.645	0.016	0.312	0.260	0.594	0.145	0.104	0.458	0.114
5/16	0.770	0.031	0.375	0.324	0.719	0.176	0.135	0.551	0.145
3/8	0.895	0.031	0.437	0.388	0.844	0.208	0.151	0.645	0.161
7/16	1.020	0.031	0.500	0.452	0.969	0.239	0.182	0.739	0.192
1/2	1.145	0.031	0.562	0.515	1.094	0.270	0.198	0.833	0.208
5/8	1.400	0.031	0.687	0.642					
3/4	1.650	0.047	0.812	0.768					

Nominal diam of bolt *D*	Diam of head, max, *A*	Height of head, max, *H*	Square neck		Ribbed neck					
			Depth of square, max, *P*	Width of square, max, *B*	Ribs below head *P*		Length of ribs *Q*			Number of ribs
					$L \leqq 7/8$	$L \geqq 1$	$L \leqq 7/8$	$L = 1$, $L = 1\frac{1}{8}$	$L \geqq 1\frac{1}{4}$	
No. 10	0.469	0.114	0.125	0.199	0.031	0.063	0.188	0.313	0.500	9
1/4	0.594	0.145	0.156	0.260	0.031	0.063	0.188	0.313	0.500	10
5/16	0.719	0.176	0.187	0.324	0.031	0.063	0.188	0.313	0.500	12
3/8	0.844	0.208	0.219	0.388	0.031	0.063	0.188	0.313	0.500	12
7/16	0.969	0.239	0.250	0.452	0.031	0.063	0.188	0.313	0.500	14
1/2	1.094	0.270	0.281	0.515	0.031	0.063	0.188	0.313	0.500	16
5/8	1.344	0.344	0.344	0.642	0.094	0.094	0.188	0.313	0.500	19
3/4	1.594	0.406	0.406	0.768	0.094	0.094	0.188	0.313	0.500	22
7/8	1.844	0.469	0.469	0.895						
1	2.094	0.531	0.531	1.022						

TABLE 11.16 Countersunk, Buttonhead, and Step Bolts

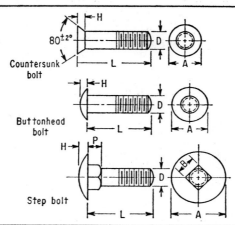

Countersunk bolt			Buttonhead		Step bolt			
Nominal diam of bolt D		Head diam, max, A	Head depth H	Head diam, max, A	Head height, max, H	Head diam, max, A	Head height, max, H	Depth of square, max, P	Width of square, max, B
10	24	. . . . .	. . . . .	0.469	0.114	0.656	0.114	0.125	0.199
¼	20	0.493	0.140	0.594	0.145	0.844	0.145	0.156	0.260
⁵⁄₁₆	18	0.618	0.176	0.719	0.176	1.031	0.176	0.187	0.324
⅜	16	0.740	0.210	0.844	0.208	1.219	0.208	0.219	0.388
⁷⁄₁₆	14	0.803	0.210	0.969	0.239	1.406	0:239	0.250	0.452
½	13	0.935	0.250	1.094	0.270	1.594	0.270	0.281	0.515
⅝	11	1.169	0.313	1.344	0.344				
¾	10	1.402	0.375	1.594	0.406				
⅞	9	1.637	0.438	1.844	0.469				
1	8	1.869	0.500	2.094	0.531				
1⅛	7	2.104	0.563						
1¼	7	2.337	0.625						
1⅜	6	2.571	0.688						
1½	6	2.804	0.750						

TABLE 11.17   Machine-Screw Heads[26,33]

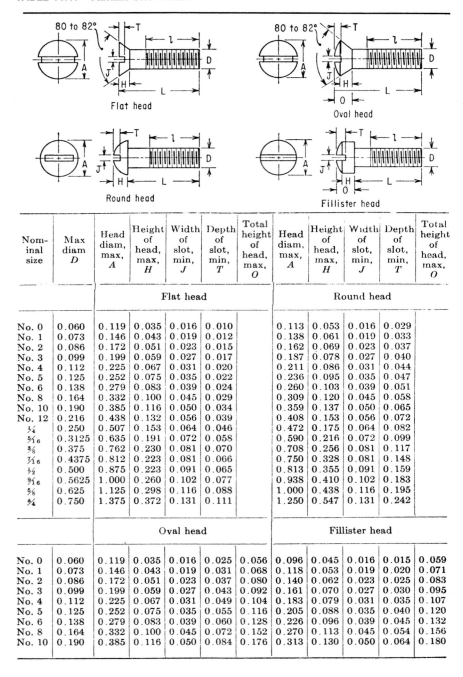

Nom-inal size	Max diam D	Head diam, max, A	Height of head, max, H	Width of slot, min, J	Depth of slot, min, T	Total height of head, max, O	Head diam, max, A	Height of head, max, H	Width of slot, min, J	Depth of slot, min, T	Total height of head, max, O
			Flat head						Round head		
No. 0	0.060	0.119	0.035	0.016	0.010		0.113	0.053	0.016	0.029	
No. 1	0.073	0.146	0.043	0.019	0.012		0.138	0.061	0.019	0.033	
No. 2	0.086	0.172	0.051	0.023	0.015		0.162	0.069	0.023	0.037	
No. 3	0.099	0.199	0.059	0.027	0.017		0.187	0.078	0.027	0.040	
No. 4	0.112	0.225	0.067	0.031	0.020		0.211	0.086	0.031	0.044	
No. 5	0.125	0.252	0.075	0.035	0.022		0.236	0.095	0.035	0.047	
No. 6	0.138	0.279	0.083	0.039	0.024		0.260	0.103	0.039	0.051	
No. 8	0.164	0.332	0.100	0.045	0.029		0.309	0.120	0.045	0.058	
No. 10	0.190	0.385	0.116	0.050	0.034		0.359	0.137	0.050	0.065	
No. 12	0.216	0.438	0.132	0.056	0.039		0.408	0.153	0.056	0.072	
¼	0.250	0.507	0.153	0.064	0.046		0.472	0.175	0.064	0.082	
⁵⁄₁₆	0.3125	0.635	0.191	0.072	0.058		0.590	0.216	0.072	0.099	
⅜	0.375	0.762	0.230	0.081	0.070		0.708	0.256	0.081	0.117	
⁷⁄₁₆	0.4375	0.812	0.223	0.081	0.066		0.750	0.328	0.081	0.148	
½	0.500	0.875	0.223	0.091	0.065		0.813	0.355	0.091	0.159	
⁹⁄₁₆	0.5625	1.000	0.260	0.102	0.077		0.938	0.410	0.102	0.183	
⅝	0.625	1.125	0.298	0.116	0.088		1.000	0.438	0.116	0.195	
¾	0.750	1.375	0.372	0.131	0.111		1.250	0.547	0.131	0.242	
			Oval head						Fillister head		
No. 0	0.060	0.119	0.035	0.016	0.025	0.056	0.096	0.045	0.016	0.015	0.059
No. 1	0.073	0.146	0.043	0.019	0.031	0.068	0.118	0.053	0.019	0.020	0.071
No. 2	0.086	0.172	0.051	0.023	0.037	0.080	0.140	0.062	0.023	0.025	0.083
No. 3	0.099	0.199	0.059	0.027	0.043	0.092	0.161	0.070	0.027	0.030	0.095
No. 4	0.112	0.225	0.067	0.031	0.049	0.104	0.183	0.079	0.031	0.035	0.107
No. 5	0.125	0.252	0.075	0.035	0.055	0.116	0.205	0.088	0.035	0.040	0.120
No. 6	0.138	0.279	0.083	0.039	0.060	0.128	0.226	0.096	0.039	0.045	0.132
No. 8	0.164	0.332	0.100	0.045	0.072	0.152	0.270	0.113	0.045	0.054	0.156
No. 10	0.190	0.385	0.116	0.050	0.084	0.176	0.313	0.130	0.050	0.064	0.180

TABLE 11.17 Machine-Screw Heads[26,33] (*Continued*)

Nom-inal size	Max diam D	Head diam, max, A	Height of head, max, H	Width of slot, min. J	Depth of slot, min. T	Total height of head, max, O	Head diam, max, A	Height of head, max, H	Width of slot, min, J	Depth of slot, min, T	Total height of head, max, O
				Oval head					Fillister head		
No. 12	0.216	0.438	0.132	0.056	0.096	0.200	0.357	0.148	0.056	0.074	0.205
¼	0.250	0.507	0.153	0.064	0.112	0.232	0.414	0.170	0.064	0.087	0.237
⁵⁄₁₆	0.3125	0.635	0.191	0.072	0.141	0.290	0.518	0.211	0.072	0.110	0.295
⅜	0.375	0.762	0.230	0.081	0.170	0.347	0.622	0.253	0.081	0.133	0.355
⁷⁄₁₆	0.4375	0.812	0.223	0.081	0.174	0.345	0.625	0.265	0.081	0.135	0.368
½	0.500	0.875	0.223	0.091	0.176	0.354	0.750	0.297	0.091	0.151	0.412
⁹⁄₁₆	0.5625	1.000	0.260	0.102	0.207	0.410	0.812	0.336	0.102	0.172	0.466
⅝	0.625	1.125	0.298	0.116	0.235	0.467	0.875	0.375	0.116	0.193	0.521
¾	0.750	1.375	0.372	0.131	0.293	0.578	1.000	0.441	0.131	0.226	0.612

Note: Edges of head on flat- and oval-head machine screws may be rounded.

Radius of fillet at base of flat- and oval-head machine screws shall not exceed twice the pitch of the screw thread.

Radius of fillet at base of round- and fillister-head machine screws shall not exceed one-half the pitch of the screw thread.

All four types of screws in this table may be furnished with cross-recessed heads.

Fillister-head machine screws in sizes No. 2 to ¾ in, inclusive, may be furnished with a drilled hole through the head along a diameter at right angles to the slot but not breaking through the slot.

11.2.4 Cap Screws[28]

Slotted-head styles (Table 11.20) are round, flat, or fillister, in screw diameters from ¼ to 1½ in.

Socket-Head Cap Screw. Hexagon and fluted dimensions are given in Table 11.21. Cap screws have chamfered points. Threads on cap screws are Unified and American coarse, fine, or eight-thread series, class 2A for plain (uncoated screws).

Minimum thread length for fluted or socket-head cap screw, National coarse, is $l = 2D + ¼$ in or $½L$, whichever is greater. Thread length for fine threads is $l = 1½D + ½$ or $⅜L$, whichever is greater. Screw thread is class 3A; body is made of high-grade alloy steel, hardened by oil quenching and tempered to Rockwell C 36 to 43.

Hexagon-head cap screws are shown and dimensions given in Table 11.10.

11.2.5 Shoulder Screws[26]

Sometimes termed *stripper bolts*, these are used as pivots for oscillating or rotating parts.

TABLE 11.18 Machine-Screw Heads—Pan, Hexagon, Truss, and 100° Flat Heads[26,33]

Pan head · Crown on recessed pan head · Truss head · Trimmed head Hexagon head · Upset head · 100° flat head

Nominal size	Max diam D	Head diam, max, A	Height of slotted head, max, H	Width of slot, min, J	Depth of slot, min, T	Radius R	Height of recessed head, max, O	Head diam, max, A	Height of head, max, H	Width of slot, min, J	Depth of slot, min, T
			Pan head					Hexagon head			
No. 2	0.086	0.167	0.053	0.023	0.023	0.035	0.062	0.125	0.050		
No. 3	0.099	0.193	0.060	0.027	0.027	0.037	0.071	0.187	0.055		
No. 4	0.112	0.219	0.068	0.031	0.030	0.042	0.080	0.187	0.060	0.031	0.025
No. 5	0.125	0.245	0.075	0.035	0.032	0.044	0.089	0.187	0.070	0.035	0.030
No. 6	0.138	0.270	0.082	0.039	0.038	0.046	0.097	0.250	0.080	0.039	0.033
No. 8	0.164	0.322	0.096	0.045	0.043	0.052	0.115	0.250	0.110	0.045	0.052
No. 10	0.190	0.373	0.110	0.050	0.050	0.061	0.133	0.312	0.120	0.050	0.057
No. 12	0.216	0.425	0.125	0.056	0.060	0.078	0.151	0.312	0.155	0.056	0.077
¼	0.250	0.492	0.144	0.064	0.070	0.087	0.175	0.375	0.190	0.064	0.083
⁵⁄₁₆	0.3125	0.615	0.178	0.072	0.092	0.099	0.218	0.500	0.230	0.072	0.100
⅜	0.375	0.740	0.212	0.081	0.113	0.143	0.261	0.562	0.295	0.081	0.131
			Truss head					100° flat head			
No. 2	0.086	0.194	0.053	0.023	0.022	0.129					
No. 3	0.099	0.226	0.061	0.027	0.026	0.151					
No. 4	0.112	0.257	0.069	0.031	0.030	0.169		0.225	0.048	0.031	0.017
No. 5	0.125	0.289	0.078	0.035	0.034	0.191					
No. 6	0.138	0.321	0.086	0.039	0.037	0.211		0.279	0.060	0.039	0.022
No. 8	0.164	0.384	0.102	0.045	0.045	0.254		0.332	0.072	0.045	0.027
No. 10	0.190	0.448	0.118	0.050	0.053	0.283		0.385	0.083	0.050	0.031
No. 12	0.216	0.511	0.134	0.056	0.061	0.336					
¼	0.250	0.573	0.150	0.064	0.070	0.375		0.507	0.110	0.064	0.042
⁵⁄₁₆	0.3125	0.698	0.183	0.072	0.085	0.457		0.635	0.138	0.072	0.053
⅜	0.375	0.823	0.215	0.081	0.100	0.538		0.762	0.165	0.081	0.064
⁷⁄₁₆	0.4375	0.948	0.248	0.081	0.116	0.619					
½	0.500	1.073	0.280	0.091	0.131	0.701					
⁹⁄₁₆	0.5625	1.198	0.312	0.102	0.146	0.783					
⅝	0.625	1.323	0.345	0.116	0.162	0.863					
¾	0.750	1.573	0.410	0.131	0.182	1.024					

Note: Radius of fillet at base of truss- and pan-head machine screws shall not exceed one-half the pitch of the screw thread.

Truss-, pan-, and 100° flat-head machine screws may be furnished with cross-recessed heads.

Hexagon-head machine screws are usually not slotted; the slot is optional. Also optional is an upset-head type for hexagon-head machine screws of sizes 4, 5, 8, 12, and ¼ in.

TABLE 11.19   Machine-Screw Heads—Binding Head[26,33]

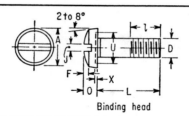

Binding head

Nom-inal size	Max diam D	Head diam, max, A	Total height of head, max, O	Width of slot, min, J	Depth of slot, min, T	Height of oval, max, F	Diam of under-cut,* min, U	Depth of under-cut, min, X
No. 2	0.086	0.181	0.046	0.023	0.024	0.018	0.124	0.005
No. 3	0.099	0.208	0.054	0.027	0.029	0.022	0.143	0.006
No. 4	0.112	0.235	0.063	0.031	0.034	0.025	0.161	0.007
No. 5	0.125	0.263	0.071	0.035	0.039	0.029	0.180	0.009
No. 6	0.138	0.290	0.080	0.039	0.044	0.032	0.199	0.010
No. 8	0.164	0.344	0.097	0.045	0.054	0.039	0.236	0.012
No. 10	0.190	0.399	0.114	0.050	0.064	0.045	0.274	0.015
No. 12	0.216	0.454	0.130	0.056	0.074	0.052	0.311	0.018
¼	0.250	0.513	0.153	0.064	0.088	0.061	0.360	0.021
⁵⁄₁₆	0.3125	0.641	0.193	0.072	0.112	0.077	0.450	0.027
⅜	0.375	0.769	0.234	0.081	0.136	0.094	0.540	0.034

*Use of undercut is optional.
Note: Binding-head machine screws may be furnished with cross-recessed heads.

11.2.6   Setscrews

These are made with either of three head styles:

1. Square head[28]
2. Headless,[28] with a slot for a screwdriver
3. Socket head,[29] or fluted key

Square-head setscrews (Table 11.22) are threaded the entire length of the body. Point styles are illustrated and dimensions are given in Table 11.23.

Headless setscrews are threaded the entire length (Table 11.24). Threads are coarse or fine, class 2 or 2A. Slotted-type setscrews are regularly furnished case-hardened.

Dimensions of keys for *hexagon and fluted sockets* are given in Table 11.25.

A setscrew is often employed to provide frictional resistance in order to prevent relative motion, e.g., between a shaft and a hub. Setscrews are made with heads and headless. Headed setscrews have square heads. A setscrew with an unprotected head on a moving machine member represents a safety hazard. Headless setscrews are manufactured to accommodate either a hexagonal- or spline-shaped wrench (hollow setscrews) or a screwdriver blade. The disadvantage of the cone and cup types is associated with the shaft burr which may form, making removal of parts difficult. For this reason, a flat or conical seat may be machined into the shaft. Setscrews with dog and cone points and sometimes oval points are spotted in holes. Flat (used on hardened

TABLE 11.20 Slotted-Head Cap Screws[28]

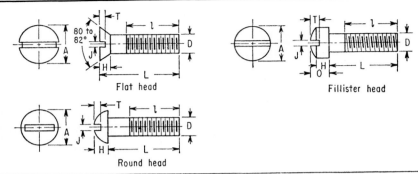

Flat head

Fillister head

Round head

Nominal size (body diam, max) D	Width of slot, min, J	Fillister head					Flat head			Round head		
		Head diam, max, A	Height of head, max, H	Total height of head, max, O	Depth of slot, min, T		Head diam, max, A	Height of head, average, H	Depth of slot min, T	Head diam, max, A	Height of head max, H	Depth of slot, min, T
¼	0.064	0.375	0.172	0.216	0.077		0.500	0.140	0.046	0.437	0.191	0.097
5/16	0.072	0.437	0.203	0.253	0.090		0.625	0.176	0.057	0.562	0.246	0.126
3/8	0.081	0.562	0.250	0.314	0.113		0.750	0.210	0.069	0.625	0.273	0.138
7/16	0.081	0.625	0.297	0.368	0.133		0.8125	0.210	0.069	0.750	0.328	0.167
½	0.091	0.750	0.328	0.412	0.148		0.875	0.210	0.069	0.812	0.355	0.179
9/16	0.102	0.812	0.375	0.466	0.169		1.000	0.245	0.080	0.937	0.410	0.208
5/8	0.116	0.875	0.422	0.521	0.190		1.125	0.281	1.092	1.000	0.438	0.220
¾	0.131	1.000	0.500	0.612	0.233		1.375	0.352	0.115	0.125	0.547	0 277
7/8	0.147	1.125	0.594	0.720	0.264		1.625	0.423	0.139			
1	0.166	1.312	0.656	0.802	0.292		1.875	0.494	0.162			
1⅛	0.178	.....	.....	.....	.....		2.062	0.529	0.173			
1¼	0.193	.....	.....	.....	.....		2.312	0.600	0.197			
1⅜	0.208	.....	.....	.....	.....		2.562	0.665	0.220			
1½	0.240	.....	.....	.....	.....		2.812	0.742	0.244			

steels) and cup points (used with softer materials) are generally used without spotting. Setscrews usually have coarse threads and hardened points (which may be knurled to prevent loosening under vibration).

11.2.7 Tapping Screws[30,31]

Tapping screws tap their own mating threads in holes in materials such as metals and plastics. Tapping screws are of two types, thread-forming (displaces the material), type A, B, BP, and C, U, or thread-cutting (actual removal of material), type D, F, G, T, BF, BG, and BT.

Thread-Forming Screws

Type A. Spaced thread with gimlet point provides a strong joint in light sheet metal up to 18 gage.

TABLE 11.21 Socket-Head Cap Screws[29]

Nominal size	Body diam, max, D	Head diam, max, A	Head height, max, H	Head side height, max, S	Hexagon* Socket width across flats, min, J	Fluted socket Number of flutes	Socket diam minor, min, K	Socket diam major, min, M	Width of socket land, max, N
No. 0	0.060	0.096							
No. 1	0.073	0.118							
No. 2	0.0860	0.140	0.086	0.0803	1/16	6	0.063	0.073	0.016
No. 3	0.0990	0.161	0.099	0.0923	5/64	6	0.080	0.097	0.022
No. 4	0.1120	0.183	0.112	0.1043	5/64	6	0.080	0.097	0.022
No. 5	0.1250	0.205	0.125	0.1163	3/32	6	0.096	0.113	0.025
No. 6	0.1380	0.226	0.138	0.1284	3/32	6	0.096	0.113	0.025
No. 8	0.1640	0.270	0.164	0.1522	1/8	6	0.126	0.147	0.032
No. 10	0.1900	5/16	0.190	0.1765	5/32	6	0.161	0.186	0.039
No. 12	0.2160	11/32	0.216	0.2005	5/32	6	0.161	0.186	0.039
1/4	0.2500	3/8	1/4	0.2317	3/16	6	0.188	0.219	0.050
5/16	0.3125	7/16	5/16	0.2894	7/32	6	0.219	0.254	0.060
3/8	0.3750	9/16	3/8	0.3469	5/16	6	0.316	0.377	0.092
7/16	0.4375	5/8	7/16	0.4046	5/16	6	0.316	0.377	0.092
1/2	0.5000	3/4	1/2	0.4620	3/8	6	0.383	0.460	0.112
9/16	0.5625	13/16	9/16	0.5196	3/8	6	0.383	0.460	0.112
5/8	0.6250	7/8	5/8	0.5771	1/2	6	0.506	0.601	0.138
3/4	0.7500	1	3/4	0.6920	9/16	6	0.531	0.627	0.149
7/8	0.8750	1 1/8	7/8	0.8069	9/16	6	0.600	0.705	0.168
1	1.0000	1 5/16	1	0.9220	5/8	6	0.681	0.797	0.189
1 1/8	1.1250	1 1/2	1 1/8	1.0372	3/4	6	0.824	0.966	0.231
1 1/4	1.2500	1 3/4	1 1/4	1.1516	3/4	6	0.824	0.966	0.231
1 3/8	1.3750	1 7/8	1 3/8	1.2675	3/4	6	0.824	0.966	0.231
1 1/2	1.5000	2	1 1/2	1.3821	1	6	1.003	1.271	0.298

*Maximum socket depth T should not exceed three-fourths of minimum head height H.
Note: Head chamfer angle E is 28 to 32°, the edge between flat and chamfer being slightly rounded.
Screw point chamfer angle 35 to 40°, the chamfer extending to the bottom of the thread. Edge between flat and chamfer is slightly rounded.

Type B. Spaced thread with blunt point. Used for light and heavy sheet metal, nonferrous castings, plastics, or soft metals. Drives faster than any screw except type A and U.

Type BP. Spaced-thread screw same as type B but has a cone point. Used in assemblies where holes are misaligned.

Type C. Used where a machine screw is preferred. Threads are the same pitch as standard machine screws. Useful when chips from machine-screw thread-cutting screws are objectionable. The finer pitch and more engaged thread surface offers an increased functional resistance to loosening.

TABLE 11.22 Square-Head Setscrews[28]

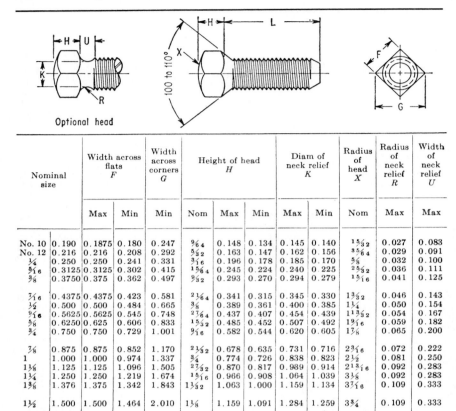

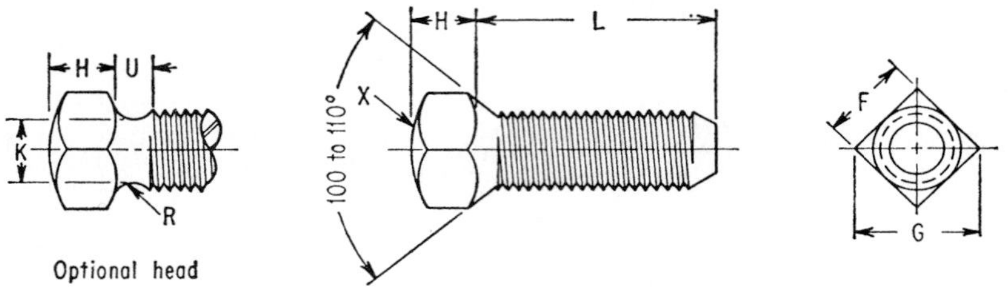
H
U
K
R
Optional head
100 to 110°
X
H
L
F
G

Nominal size		Width across flats F		Width across corners G	Height of head H			Diam of neck relief K		Radius of head X	Radius of neck relief R	Width of neck relief U
		Max	Min	Min	Nom	Max	Min	Max	Min	Nom	Max	Max
No. 10	0.190	0.1875	0.180	0.247	9/64	0.148	0.134	0.145	0.140	15/32	0.027	0.083
No. 12	0.216	0.216	0.208	0.292	5/32	0.163	0.147	0.162	0.156	35/64	0.029	0.091
1/4	0.250	0.250	0.241	0.331	3/16	0.196	0.178	0.185	0.170	5/8	0.032	0.100
5/16	0.3125	0.3125	0.302	0.415	15/64	0.245	0.224	0.240	0.225	25/32	0.036	0.111
3/8	0.3750	0.375	0.362	0.497	9/32	0.293	0.270	0.294	0.279	15/16	0.041	0.125
7/16	0.4375	0.4375	0.423	0.581	21/64	0.341	0.315	0.345	0.330	1 3/32	0.046	0.143
1/2	0.500	0.500	0.484	0.665	3/8	0.389	0.361	0.400	0.385	1 1/4	0.050	0.154
9/16	0.5625	0.5625	0.545	0.748	27/64	0.437	0.407	0.454	0.439	1 13/32	0.054	0.167
5/8	0.6250	0.625	0.606	0.833	15/32	0.485	0.452	0.507	0.492	1 9/16	0.059	0.182
3/4	0.750	0.750	0.729	1.001	9/16	0.582	0.544	0.620	0.605	1 7/8	0.065	0.200
7/8	0.875	0.875	0.852	1.170	21/32	0.678	0.635	0.731	0.716	2 3/16	0.072	0.222
1	1.000	1.000	0.974	1.337	3/4	0.774	0.726	0.838	0.823	2 1/2	0.081	0.250
1 1/8	1.125	1.125	1.096	1.505	27/32	0.870	0.817	0.989	0.914	2 13/16	0.092	0.283
1 1/4	1.250	1.250	1.219	1.674	15/16	0.966	0.908	1.064	1.039	3 1/8	0.092	0.283
1 3/8	1.376	1.375	1.342	1.843	1 1/32	1.063	1.000	1.159	1.134	3 7/16	0.109	0.333
1 1/2	1.500	1.500	1.464	2.010	1 1/8	1.159	1.091	1.284	1.259	3 3/4	0.109	0.333

Type U. Multiple-thread, blunt point, metallic-drive screw threads have a large helix angle, resulting in good holding power, and must be hammered or forced into the work. Used for permanent fastening, since it is difficult to remove.

Thread-Cutting Screws

Types D, F, G, T. Threads approximate machine-screw threads, with blunt point. Front end has tapered entering threads for easy starting. Type T has same usage as D, except wide flute provides more chip clearance. Type G is used where less driving torque than type D is required. Type F replaces type C where a low driving torque is required; however, chip space of the flutes is not suitable for deep penetration. The above types are suitable for use with aluminum, zinc, die castings, carbon- and stainless-steel sheets and shapes, cast iron, brass, and plastics.

Types BF, BG, and BT. Spaced threads blunt point same as type B. Used in plastics and die castings.

TABLE 11.23 Square-Head Setscrew Points[28]

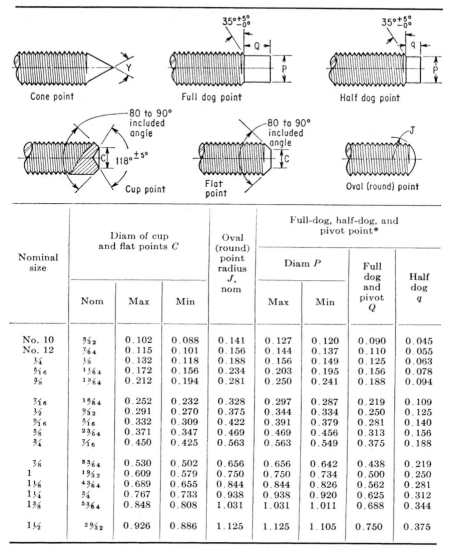

Nominal size	Diam of cup and flat points C			Oval (round) point radius J, nom	Full-dog, half-dog, and pivot point*			
					Diam P		Full dog and pivot Q	Half dog q
	Nom	Max	Min		Max	Min		
No. 10	$\frac{3}{32}$	0.102	0.088	0.141	0.127	0.120	0.090	0.045
No. 12	$\frac{7}{64}$	0.115	0.101	0.156	0.144	0.137	0.110	0.055
$\frac{1}{4}$	$\frac{1}{8}$	0.132	0.118	0.188	0.156	0.149	0.125	0.063
$\frac{5}{16}$	$\frac{11}{64}$	0.172	0.156	0.234	0.203	0.195	0.156	0.078
$\frac{3}{8}$	$\frac{13}{64}$	0.212	0.194	0.281	0.250	0.241	0.188	0.094
$\frac{7}{16}$	$\frac{15}{64}$	0.252	0.232	0.328	0.297	0.287	0.219	0.109
$\frac{1}{2}$	$\frac{9}{32}$	0.291	0.270	0.375	0.344	0.334	0.250	0.125
$\frac{9}{16}$	$\frac{5}{16}$	0.332	0.309	0.422	0.391	0.379	0.281	0.140
$\frac{5}{8}$	$\frac{23}{64}$	0.371	0.347	0.469	0.469	0.456	0.313	0.156
$\frac{3}{4}$	$\frac{7}{16}$	0.450	0.425	0.563	0.563	0.549	0.375	0.188
$\frac{7}{8}$	$\frac{33}{64}$	0.530	0.502	0.656	0.656	0.642	0.438	0.219
1	$\frac{19}{32}$	0.609	0.579	0.750	0.750	0.734	0.500	0.250
$1\frac{1}{8}$	$\frac{43}{64}$	0.689	0.655	0.844	0.844	0.826	0.562	0.281
$1\frac{1}{4}$	$\frac{3}{4}$	0.767	0.733	0.938	0.938	0.920	0.625	0.312
$1\frac{3}{8}$	$\frac{53}{64}$	0.848	0.808	1.031	1.031	1.011	0.688	0.344
$1\frac{1}{2}$	$\frac{29}{32}$	0.926	0.886	1.125	1.125	1.105	0.750	0.375

*Pivot points are similar to full-dog point except that the point is rounded by a radius equal to J.

Where usable length of thread is less than the nominal diameter, half-dog point shall be used.

When length equals nominal diameter or less, $Y = 118° \pm 2°$; when length exceeds nominal diameter, $Y = 90° \pm 2°$

Note: All dimensions are given in inches.

TABLE 11.24   Slotted Headless Setscrews[28]

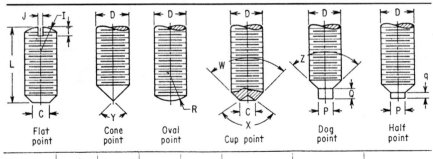

| Flat point | Cone point | Oval point | Cup point | Dog point | Half point |

Nominal size D	Radius of headless crown I	Width of slot J	Depth of slot T	Oval-point radius R	Diam of cup and flat points C		Diam of dog point P		Length of dog point*	
					Max	Min	Max	Min	Fill Q	Half q
5	0.125	0.125	0.023	0.031	0.094	0.067	0.057	0.083	0.078	0.060	0.030
6	0.138	0.138	0.025	0.035	0.109	0.074	0.064	0.092	0.087	0.070	0.035
8	0.164	0.164	0.029	0.041	0.125	0.087	0.076	0.109	0.103	0.080	0.040
10	0.190	0.190	0.032	0.048	0.141	0.102	0.088	0.127	0.120	0.090	0.045
12	0.216	0.216	0.036	0.054	0.156	0.115	0.101	0.144	0.137	0.110	0.055
¼	0.250	0.250	0.045	0.063	0.188	0.132	0.118	0.156	0.149	0.125	0.063
⁵⁄₁₆	0.3125	0.313	0.051	0.078	0.234	0.172	0.156	0.203	0.195	0.156	0.078
³⁄₈	0.375	0.375	0.064	0.094	0.281	0.212	0.194	0.250	0.241	0.188	0.094
⁷⁄₁₆	0.4375	0.438	0.072	0.109	0.328	0.252	0.232	0.297	0.287	0.219	0.109
½	0.500	0.500	0.081	0.125	0.375	0.291	0.270	0.344	0.344	0.250	0.135
⁹⁄₁₆	0.5625	0.563	0.091	0.141	0.422	0.332	0.309	0.391	0.379	0.281	0.140
⁵⁄₈	0.625	0.625	0.102	0.156	0.469	0.371	0.347	0.469	0.456	0.313	0.156
¾	0.750	0.750	0.129	0.188	0.563	0.450	0.425	0.563	0.549	0.375	0.188

*Where usable length of thread is less than the nominal diameter, half-dog point shall be used.

When L (length of screw) equals nominal diameter or less = $118° \pm 2°$; when L exceeds nominal diameter, $Y = 90° \pm 2°$.

Point angles, $W = 80$ to $90°$; $X = 118° \pm 5°$; $Z = 100$ to $110°$.

Allowable eccentricity of dog-point axis with respect to axis of screw shall not exceed 3 percent of nominal screw diameter with maximum 0.005 in.

Note: All dimensions given in inches.

11.3 POWER-TRANSMITTING SCREW THREADS[38]

11.3.1 Types

Acme screw thread

Stub Acme or 29° stub Acme, form 1 and form 2 stub

60° stub

Modified square screw thread

Buttress

11.3.2 Acme Screw Thread[34] (Fig. 11.12)

Developed for the purpose of producing transverse motions, Acme screw thread is standardized for two applications, namely, (1) general purpose and (2) centralizing.

1. General-purpose thread has clearance on all diameters for free movement. Class 2G is preferred for general-purpose assemblies. Classes 3G and 4G are used where backlash and end play are undesirable. Table 11.26 lists the recommended diameter-pitch Acme threads.

2. Centralizing threads have a limited clearance at the major diameter, resulting in controlled concentricity of screw and nut. External threads must have the crest corners clarified 45° × $p/15$ maximum.

Classes. For classes 2C, 3C, 4C the fillet at the minor diameter is equal to $0.1p$ or less. For classes 5C, 6C the minimum fillet is $0.07p$ and maximum fillet is $0.1p$. The major diameter of all classes may have a fillet not greater than $0.06p$. Class 2C provides the maximum end play or backlash.

11.3.3 Stub Acme Screw Thread[35] (Fig. 11.13)

The stub Acme, or 29° stub, thread is used where a coarse-pitch thread of shallow depth is required. Like the Acme thread, it has an angle of 29° between flanks. The stub or height is $0.3p$. Only one class is provided, class 2G (general-purpose) (Table 11.26).

Alternate stub Acme threads[35] (Fig. 11.14) are provided where design requirements necessitate their use.

11.3.4 60° Stub Acme Thread*,[36]

The angle between the flanks of the thread is 60°. The threads are truncated top and bottom. Basic dimensions are given in Table 11.27.

---

*In accordance with standard practice, this thread is designated as follows, for example, "1⅛-9 special form, 60° or 10° thread," followed by all limits of size.

TABLE 11.25 Fluted- and Hexagon-Socket Headless Setscrews[28]

Flat point   Oval point   Cup point   Cone point   Full dog point   Half dog point

Screw size nominal diam D	Cup- and flat-point diameter C		Oval-point radius R	Cone-point angle Y		Dog point				Key engagement† min	Hexagon type		Fluted type*					
				118 ± 2° for these lengths and under	90 ± 2° for these lengths and over	Diam P		Full Q	Half q		Socket width across flats J		Socket diam, minor J		Socket diam, major M		Socket land width N	
	Max	Min				Max	Min				Max	Min	Max	Min	Max	Min	Max	Min
No. 0	0.033	0.027	3⁄64	1⁄16	5⁄64	0.040	0.037	0.030	0.015	0.022	0.0285	0.028	0.026	0.0255	0.035	0.034	0.012	0.0115
No. 1	0.040	0.033	0.055	5⁄64	3⁄32	0.049	0.045	0.037	0.019	0.028	0.0355	0.035	0.026	0.0255	0.035	0.034	0.012	0.0115
No. 2	0.047	0.039	1⁄16	3⁄32	7⁄64	0.057	0.053	0.043	0.022	0.028	0.0355	0.035	0.038	0.0375	0.050	0.049	0.017	0.016
No. 3	0.054	0.045	5⁄64	7⁄64	1⁄8	0.066	0.062	0.050	0.025	0.040	0.051	0.050	0.038	0.0375	0.050	0.049	0.017	0.016
No. 4	0.061	0.051	0.084	1⁄8	5⁄32	0.075	0.070	0.056	0.028	0.040	0.051	0.050	0.051	0.050	0.062	0.061	0.014	0.013

Fluted and hexagon socket types

Nominal Size																		
No. 5	0.067	0.057	3/32	1/8	3/16	0.083	0.078	0.06	0.03	0.050	0.0635	1/16	0.053	0.052	0.071	0.070	0.022	0.021
No. 6	0.074	0.064	7/64	1/8	3/16	0.092	0.087	0.07	0.035	0.050	0.0635	1/16	0.056	0.055	0.079	0.078	0.023	0.022
No. 8	0.087	0.076	1/8	3/16	1/4	0.109	0.103	0.08	0.04	0.062	0.0791	5/64	0.082	0.080	0.098	0.097	0.022	0.021
No. 10	0.102	0.088	9/64	3/16	1/4	0.127	0.120	0.09	0.045	0.075	0.0947	3/32	0.098	0.096	0.115	0.113	0.025	0.023
No. 12	0.115	0.101	5/32	3/16	1/4	0.144	0.137	0.11	0.055	0.075	0.0947	3/32	0.098	0.096	0.115	0.113	0.025	0.023
1/4	0.132	0.118	3/16	1/4	5/16	5/32	0.149	1/8	1/16	0.100	0.1270	1/8	0.128	0.126	0.149	0.147	0.032	0.030
5/16	0.172	0.156	13/64	5/16	3/8	13/64	0.195	5/32	5/64	0.125	0.1582	5/32	0.163	0.161	0.188	0.186	0.039	0.037
3/8	0.212	0.194	9/32	3/8	7/16	1/4	0.241	3/16	3/32	0.150	0.1895	3/16	0.190	0.188	0.221	0.219	0.050	0.048
7/16	0.252	0.232	23/64	7/16	1/2	19/64	0.287	7/32	7/64	0.175	0.2207	7/32	0.221	0.219	0.256	0.254	0.060	0.058
1/2	0.291	0.270	3/8	1/2	9/16	11/32	0.334	1/4	1/8	0.200	0.2520	1/4	0.254	0.252	0.298	0.296	0.068	0.066
9/16	0.332	0.309	27/64	9/16	5/8	25/64	0.379	9/32	9/64	0.200	0.2520	1/4	0.254	0.252	0.298	0.296	0.068	0.066
5/8	0.371	0.347	15/32	5/8	3/4	15/32	0.456	5/16	5/32	0.250	0.3155	5/16	0.319	0.316	0.380	0.377	0.092	0.089
3/4	0.450	0.425	9/16	3/4	7/8	9/16	0.549	3/8	3/16	0.300	0.3780	3/8	0.386	0.383	0.463	0.460	0.112	0.109
7/8	0.530	0.502	23/32	7/8	1	21/32	0.642	7/16	7/32	0.400	0.5030	1/2	0.509	0.506	0.604	0.601	0.138	0.134
1	0.609	0.579	3/4	1	1 1/8	3/4	0.734	1/2	1/4	0.450	0.5655	9/16	0.535	0.531	0.631	0.627	0.149	0.145
1 1/8	0.689	0.655	27/32	1 1/8	1 1/4	27/32	0.826	9/16	9/32	0.450	0.5655	9/16	0.604	0.600	0.709	0.705	0.168	0.164
1 1/4	0.767	0.733	15/16	1 1/4	1 1/2	15/16	0.920	5/8	5/16	0.500	0.6290	5/8	0.685	0.681	0.801	0.797	0.189	0.185
1 3/8	0.848	0.808	1 1/32	1 3/8	1 5/8	1 1/32	1.011	11/16	11/32	0.500	0.6290	5/8	0.744	0.740	0.869	0.865	0.207	0.203
1 1/2	0.926	0.886	1 1/8	1 1/2	1 3/4	1 1/8	1.105	3/4	3/8	0.600	0.7540	3/4	0.828	0.824	0.970	0.966	0.231	0.227
1 3/4	1.086	1.039	1 5/16	1 3/4	2	1 5/16	1.289	7/8	7/16	0.800	1.0040	1	1.007	1.003	1.275	1.271	0.298	0.294
2	1.244	1.193	1 1/2	2	2 1/4	1 1/2	1.474	1	1/2	0.800	1.0040	1	1.007	1.003	1.275	1.271	0.298	0.294

*The number of flutes for setscrews Nos. 0, 1, 2, 3, 5, and 6 is four. The number of flutes for Nos. 4, 8, and larger is six.

†These dimensions apply to cup- and flat-point screws one diameter in length or longer. For screws shorter than one diameter in length, and for other types of points, socket to be as deep as practicable.

Note: All dimensions are given in inches.

11.37

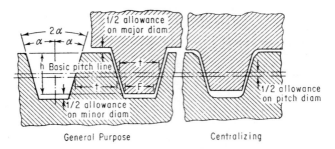

FIG. 11.12   Acme thread forms. $\alpha = 14°30'$; $h = 0.5p$; $t = 0.5p$; $F = 0.3707p$; $p = $ pitch, in.

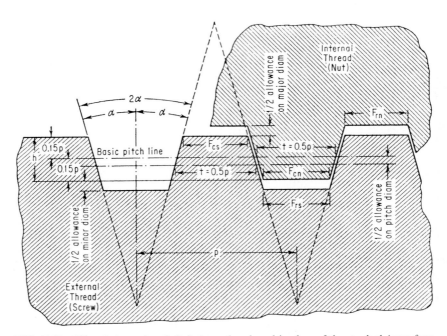

FIG. 11.13   29° stub Acme thread. Stub Acme thread used in place of the standard Acme form when shallow depth is required. $2\alpha = 29°$; $\alpha = 14°30'$; $p = $ pitch; $n = $ number of threads per inch; $N = $ number of turns per inch; $h = 0.3p$, basic height of thread; $F_{cn} = 0.4224p = $ basic width of flat of crest of internal thread; $F_{cs} = 0.4224p = $ basic width of flat of crest of external thread; $F_{rn} = 0.4224p - 0.259 \times$ major-diameter allowance on internal thread; $F_{ra} = 0.4224p - 0.259 \times$ (minor-diameter allowance on external thread $-$ pitch-diameter allowance on external thread).

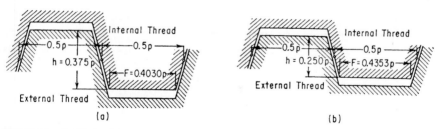

FIG. 11.14 Modified stub Acme thread with basic height of 0.250p. (a) Modified stub Acme thread with basic height of 0.375p (form 1). (b) Modified stub Acme thread with basic height of 0.250p (form 2).

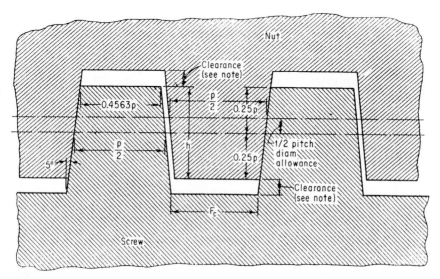

FIG. 11.15 10° modified square thread. *Note:* A clearance should be added to h to produce extra depth, thus avoiding interference with threads of mating parts at major and minor diameters. The amount of this clearance must be determined from the application. p = pitch; $h = 0.5p$ = basic depth of thread; $H = 0.5p$ + clearance = total depth of thread; $t = 0.5p$ = thickness of thread; $F_c =$ $0.4563p - (0.17 \times$ clearance) = flat at root of thread; $F = 0.4563p$ = basic width of flat at crest of thread.

11.3.5 Modified Square Thread[36]

Illustrated in Fig. 11.15, the angle of 10° results in a thread which is the equivalent of a square thread, yet is economical to produce.

11.3.6 Buttress Thread[37] (Fig. 11.16)

Used in specially designed components involving high stresses in one direction only, along the thread axis. As the thrust side of the thread is made very nearly perpendicular to the thread axis, the radial component of the thrust is reduced to a minimum, making this form of thread particularly suitable for tubular members which have to be

TABLE 11.26   Acme and Stub Acme Thread Series[39]

| | | | Acme threads | | | | Stub Acme threads | |
| | | | General-purpose (all classes) and centralizing classes 2C, 3C, and 4C | | Centralizing classes 5C and 6C | | | |
Nominal size	Threads per in. $1/p$	Basic height of thread h	Basic major diam D_m	Helix angle at basic pitch diam λ	Basic major diam D_m'	Helix angle at basic pitch diam λ	Basic height of thread h'	Helix angle at basic pitch diam
¼	16	0.03125	0.2500	5° 12′	......	.....	0.01875	4° 54′
⁵⁄₁₆	14	0.03571	0.3125	4° 42′	......	.....	0.02143	4° 28′
⅜	12	0.04167	0.3750	4° 33′	......	.....	0.02500	4° 20′
⁷⁄₁₆	12	0.04167	0.4375	3° 50′	......	.....	0.02500	3° 41′
½	10	0.05000	0.5000	4° 3′	0.4823	4° 13′	0.03000	3° 52′
⅝	8	0.06250	0.6250	4° 3′	0.6052	4° 12′	0.03750	3° 52′
¾	6	0.08333	0.7500	4° 33′	0.7284	4° 42′	0.05000	4° 20′
⅞	6	0.08333	0.8750	3° 50′	0.8516	3° 57′	0.05000	3° 41′
1	5	0.10000	1.0000	4° 3′	0.9750	4° 10′	0.06000	3° 52′
1⅛	5	0.10000	1.1250	3° 33′	1.0985	3° 39′	0.06000	3° 25′
1¼	5	0.10000	1.2500	3° 10′	1.2220	3° 15′	0.06000	3° 4′
1⅜	4	0.12500	1.3750	3° 39′	1.3457	3° 44′	0.07500	3° 30′
1½	4	0.12500	1.5000	3° 19′	1.4694	3° 23′	0.07500	3° 12′
1¾	4	0.12500	1.7500	2° 48′	1.7169	2° 52′	0.07500	2° 43′
2	4	0.12500	2.0000	2° 26′	1.9646	2° 29′	0.07500	2° 22′
2¼	3	0.16667	2.2500	2° 55′	2.2125	2° 58′	0.10000	2° 50′
2½	3	0.16667	2.5000	2° 36′	2.4605	2° 39′	0.10000	2° 32′
2¾	3	0.16667	2.7500	2° 21′	2.7085	2° 23′	0.10000	2° 18′
3	2	0.25000	3.0000	3° 19′	2.9567	3° 22′	0.15000	3° 12′
3½	2	0.25000	3.5000	2° 48′	3.4532	2° 51′	0.15000	2° 43′
4	2	0.25000	4.0000	2° 26′	3.9500	2° 28′	0.15000	2° 22′
4½	2	0.25000	4.5000	2° 8′	4.4470	2° 10′	0.15000	2° 6′
5	2	0.25000	5.0000	1° 55′	4.9441	1° 56′	0.15000	1° 53′

Note: For general-purpose and centralizing classes 2C, 3C, and 4C, basic pitch diameter = $D_m - h$; basic minor diameter = $D_m - 2h$, $h = p/2$.

For centralizing classes 5C and 6C, basic pitch diameter = $D_m - h$; basic minor diameter = $D_m - 2h$.

For stub Acme, basic pitch diameter = $D_m - h'$; basic minor diameter = $D_m - 2h'$.

screwed together. Applications include breech mechanisms of large guns and airplane-propeller hubs.

Pressure flank angle measured in an axial plane is 7° from the normal to the axis, and the trailing flank angle is 45°. There is no standard series, inasmuch as this thread applies mainly to specially designed components. However, it is recommended that the nominal major diameters be selected from the following geometric series:[36]

TABLE 11.27   60° Stub Threads[35]

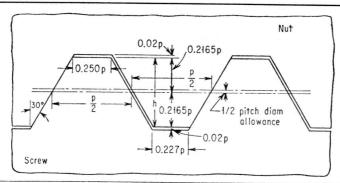

Threads per in.	Pitch p, in.	Depth of thread (basic) $h = 0.433p$, in.	Total depth of thread $(h + 0.02p)$, in.*	Thread thickness (basic) $t = 0.5p$, in.	Width of flat, in.	
					Crest of screw (basic) $F = 0.250p$	Root of screw $F_c = 0.227p$
16	0.06250	0.0271	0.0283	0.0313	0.0156	0.0142
14	0.07143	0.0309	0.0324	0.0357	0.0179	0.0162
12	0.08333	0.0361	0.0378	0.0417	0.0208	0.0189
10	0.10000	0.0433	0.0453	0.0500	0.0250	0.0227
9	0.11111	0.0481	0.0503	0.0556	0.0278	0.0252
8	0.12500	0.0541	0.0566	0.0625	0.0313	0.0284
7	0.14286	0.0619	0.0647	0.0714	0.0357	0.0324
6	0.16667	0.0722	0.0755	0.0833	0.0417	0.0378
5	0.20000	0.0866	0.0906	0.1000	0.0500	0.0454
4	0.25000	0.1083	0.1133	0.1250	0.0625	0.0567

*A clearance of at least $0.02p$ is added to h to produce extra depth, thus avoiding interference with threads of mating parts at minor or major diameters.

Inches				
½	1⅛	2½	5½	12
⁹⁄₁₆	1¼	2¾	6	14
⅝	1⅜	3	7	16
¹¹⁄₁₆	1½	3½	8	18
¾	1¾	4	9	20
⅞	2	4½	10	22
1	2¼	5	11	24

It is recommended that the pitches of buttress threads be selected from the following geometric (10) series:

Internal Thread (Nut)

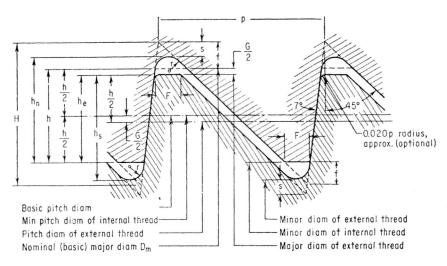

Basic pitch diam
Min pitch diam of internal thread
Pitch diam of external thread
Nominal (basic) major diam D_m

Minor diam of external thread
Minor diam of internal thread
Major diam of external thread

External Thread (Screw)

FIG. 11.16 Form of buttress thread having 7° pressure flank and 45° clearance flank. $D =$ nominal major diameter; $H = 0.89064p =$ height of sharp V thread; $h = 0.6p =$ basic height of thread; $r = 0.07141p =$ root radius; $s = 0.08261p =$ root truncation; $h_e = h - G/2 =$ depth of engagement; $f = 0.14532p =$ crest truncation; $F = 0.16316p =$ crest width; $D_n = D + 0.12542p =$ major diameter of internal thread (nut); $K_s = D - 1.32542p - G =$ minor diameter of external thread (screw); $h_n = 0.66271p =$ height of thread of internal thread (nut); $h_s = 0.66271p =$ height of thread of external thread (screw).

Threads per inch		
20	6	2
16	5	1½
12	4	1¼
10	3	1
8	2½	

The following suggestions are made regarding suitable associations of diameters and pitches:

Diam range, in	Associated pitches, threads per in
From ½ to ¹¹⁄₁₆	20, 16, 12
Over ¹¹⁄₁₆ to 1	16, 12, 10
Over 1 to 1½	16, 12, 10, 8, 6
Over 1½ to 2½	16, 12, 10, 8, 6, 5, 4
Over 2½ to 4	16, 12, 10, 8, 6, 5, 4
Over 4 to 6	12, 10, 8, 6, 5, 4, 3
Over 6 to 10	10, 8, 6, 5, 4, 3, 2½, 2
Over 10 to 16	10, 8, 6, 5, 4, 3, 2½, 2, 1½, 1¼
Over 16 to 24	8, 6, 5, 4, 3, 2½, 2, 1½, 1¼, 1

11.4 WASHERS

11.4.1 Usage

Washers[40] under the heads of screws or bolts are used for the following purposes: to prevent the loosening of the associated screw, bolt, or nut; to distribute the compressive stress over areas larger than that of the bolt head or nut; to distribute thrust loads; between rotating parts to reduce wear; and to prevent damage to finishes by bolt head or nut.

11.4.2 Plain Washers

Type A washer (Table 11.28) is selected by the choosing of the desired inside diameter, outside diameter, and thickness. Type B (Table 11.29) is selected by the nominal screw or bolt size and appropriate thickness: narrow, regular, or wide. Metal washers are normally made from SAE 1060 steel or equivalent, heat-treated to a hardness of Rockwell C 45 to 53.

11.4.3 Helical-Spring Lock Washers

These have the standard dimensions listed in Table 11.30. Spring lock washers are made from carbon steel and stainless steel, types 302 and 420, aluminum zinc alloy, phosphorus or silicon bronze, and K monel. Consult Ref. 41 for specifications on plating and heat treatment of the various materials. Spring lock washers come in three series: regular, heavy, and extra duty. Designation is made by the nominal size and the series, e.g., $\frac{3}{8}$-in regular.

11.4.4 Tooth Lock Washers[41,42]

Tables 11.31 and 11.32 illustrate the four basic series: internal-light and heavy, external, countersunk external, and internal-external. Designation is made by the nominal size, description, and type, e.g., $\frac{3}{8}$-in internal tooth, type A.

11.5 PINS

For more on pins, see Sec. 21.

11.5.1 Dowel Pins

Dowel pins are used primarily to preserve alignment or to retain parts in a fixed position. Dowel-pin size is governed by its application. The general rule is to use dowel pins of the same size as the screws used in fastening the work. The length of the dowel pin should be approximately $1\frac{1}{2}$ to 2 times its diameter. Good practice, when using hardened dowel pins in soft parts, calls for a reamed hole 0.001 in smaller than the pin. Tables 11.33 to 11.35 list pin dimensions.

TABLE 11.28 Type A Plain Washers[40]

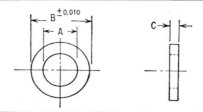

ID*	OD†	Thickness			ID*	OD†	Thickness		
		Nom	Max	Min			Nom	Max	Min
5/64	3/16	0.020	0.025	0.016	5/8	1 1/2	0.109	0.132	0.086
3/32	7/32	0.020	0.025	0.016	5/8	2 1/8	0.134	0.160	0.108
3/32	1/4	0.020	0.025	0.016	21/32	1 5/16	0.095	0.121	0.074
1/8	1/4	0.022	0.028	0.017	11/16	1 1/2	0.134	0.160	0.108
1/8	5/16	0.032	0.040	0.025	11/16	1 3/4	0.134	0.160	0.108
5/32	3/16	0.035	0.048	0.027	11/16	2 3/8	0.165	0.192	0.136
5/32	3/8	0.049	0.065	0.036	13/16	1 1/2	0.134	0.160	0.108
11/64	13/32	0.049	0.065	0.036	13/16	1 3/4	0.148	0.177	0.122
3/16	3/8	0.049	0.065	0.036	13/16	2	0.148	0.177	0.122
3/16	7/16	0.049	0.065	0.036	13/16	2 7/8	0.165	0.192	0.136
13/64	15/32	0.049	0.065	0.036	15/16	1 3/4	0.134	0.160	0.108
7/32	7/16	0.049	0.065	0.036	15/16	2	0.165	0.192	0.136
7/32	1/2	0.049	0.065	0.036	15/16	2 1/4	0.165	0.192	0.136
15/64	17/32	0.049	0.065	0.036	15/16	3 3/8	0.180	0.213	0.153
1/4	1/2	0.049	0.065	0.036	1 1/16	2	0.134	0.160	0.108
1/4*	5/16	0.049	0.065	0.036	1 1/16	2 1/4	0.165	0.192	0.136
1/4*	9/16	0.065	0.080	0.051	1 1/16	2 1/2	0.165	0.192	0.136
17/64	5/8	0.049	0.065	0.036	1 1/16	3 3/8	0.238	0.280	0.210
9/32	5/8	0.065	0.080	0.051	1 3/16	2 1/2	0.165	0.192	0.136
5/16	3/4	0.065	0.080	0.051	1 1/4	2 3/4	0.165	0.192	0.136
5/16	7/8	0.065	0.080	0.051	1 5/16	2 3/4	0.165	0.192	0.136
11/32	11/16	0.065	0.080	0.051	1 3/8	3	0.165	0.192	0.136
3/8	3/4	0.065	0.080	0.051	1 7/16	3	0.180	0.213	0.153
3/8	7/8	0.083	0.104	0.064	1 1/2	3 1/4	0.180	0.213	0.153
3/8	1 1/8	0.065	0.080	0.051	1 9/16	3 1/4	0.180	0.213	0.153
13/32	13/16	0.065	0.080	0.051	1 5/8	3 1/2	0.180	0.213	0.153
7/16	7/8	0.083	0.104	0.064	1 11/16	3 1/2	0.180	0.213	0.153
7/16	1	0.083	0.104	0.064	1 3/4	3 3/4	0.180	0.213	0.153
7/16	1 3/8	0.083	0.104	0.064	1 13/16	3 3/4	0.180	0.213	0.153
15/32	59/64	0.065	0.080	0.051	1 7/8	4	0.180	0.213	0.153
1/2	1 1/8	0.083	0.104	0.064	1 15/16	4	0.180	0.213	0.153
1/2	1 1/4	0.083	0.104	0.064	2	4 1/4	0.180	0.213	0.153
1/2	1 3/8	0.083	0.104	0.064	2 1/16	4 1/4	0.180	0.213	0.153
17/32	1 1/16	0.095	0.121	0.074	2 1/8	4 1/2	0.180	0.213	0.153
9/16	1 1/4	0.109	0.132	0.086	2 3/8	4 3/4	0.220	0.248	0.193
9/16	1 3/8	0.109	0.132	0.086	2 5/8	5	0.238	0.280	0.210
9/16	1 7/8	0.109	0.132	0.086	2 7/8	5 1/4	0.259	0.310	0.228
19/32	1 5/16	0.095	0.121	0.074	3 1/8	5 1/2	0.284	0.327	0.249
5/8	1 3/8	0.109	0.132	0.086					

*Tolerance is ±0.005 in for inside diameters up to 7/32 in, inclusive, and also for the two 1/4-in inside diameters marked with an asterisk (*). Tolerance for all other inside diameters larger than 7/32 in is ±0.010 in.

†Tolerance is ±0.010 in for all outside diameters.

Note: All dimensions are given in inches.

TABLE 11.29   Type B Plain Washers[40]

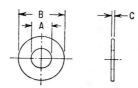

Screw or bolt size	Series	ID A Nom	OD B Nom	Thickness C Nom	Max	Min
0.060 (No. 0)	Narrow		3/16	0.025	0.028	0.022
	Regular	0.068	3/16	0.025	0.028	0.022
	Wide		1/4	0.025	0.028	0.022
0.073 (No. 1)	Narrow		5/32	0.025	0.028	0.022
	Regular	0.084	7/32	0.025	0.028	0.022
	Wide		9/32	0.032	0.036	0.028
0.086 (No. 2)	Narrow		3/16	0.025	0.028	0.022
	Regular	0.094	1/4	0.032	0.036	0.028
	Wide		11/32	0.032	0.036	0.028
0.099 (No. 3)	Narrow		7/32	0.025	0.028	0.022
	Regular	0.109	5/16	0.032	0.036	0.028
	Wide		13/32	0.040	0.045	0.036
0.112 (No. 4)	Narrow		1/4	0.032	0.036	0.028
	Regular	0.125	3/8	0.040	0.045	0.036
	Wide		7/16	0.040	0.045	0.036
0.125 (No. 5)	Narrow		9/32	0.032	0.036	0.028
	Regular	0.141	13/32	0.040	0.045	0.036
	Wide		1/2	0.040	0.045	0.036
0.138 (No. 6)	Narrow		5/16	0.032	0.036	0.028
	Regular	0.156	7/16	0.040	0.045	0.036
	Wide		9/16	0.040	0.045	0.036
0.164 (No. 8)	Narrow		3/8	0.040	0.045	0.036
	Regular	0.188	1/2	0.040	0.045	0.036
	Wide		5/8	0.063	0.071	0.056
0.190 (No. 10)	Narrow		13/32	0.040	0.045	0.036
	Regular	0.208	9/16	0.040	0.045	0.036
	Wide		47/64	0.063	0.071	0.056
0.216 (No. 12)	Narrow		7/16	0.040	0.045	0.036
	Regular	0.240	5/8	0.063	0.071	0.056
	Wide		3/4	0.063	0.071	0.056
1/4	Narrow		1/2	0.063	0.071	0.056
	Regular	0.281	47/64	0.063	0.071	0.056
	Wide		1	0.063	0.071	0.056
5/16	Narrow		5/8	0.063	0.071	0.056
	Regular	0.344	7/8	0.063	0.071	0.056
	Wide		1 1/8	0.063	0.071	0.056
3/8	Narrow		47/64	0.063	0.071	0.056
	Regular	0.406	1	0.063	0.071	0.056
	Wide		1 1/4	0.100	0.112	0.090
7/16	Narrow		7/8	0.063	0.071	0.056
	Regular	0.480	1 1/8	0.063	0.071	0.056
	Wide		1 15/32	0.100	0.112	0.090
1/2	Narrow		1	0.063	0.071	0.056
	Regular	0.540	1 1/4	0.100	0.112	0.090
	Wide		1 3/4	0.100	0.112	0.090
9/16	Narrow		1 1/8	0.063	0.071	0.056
	Regular	0.604	1 15/32	0.100	0.112	0.090

TABLE 11.29 Type B Plain Washers[40] (*Continued*)

Screw or bolt size	Series	ID *A*	OD *B*	Thickness *C*		
		Nom	Nom	Nom	Max	Min
	Wide		2	0.100	0.112	0.090
⅝	Narrow		1¼	0.100	0.112	0.090
	Regular	0.666	1¾	0.100	0.112	0.090
	Wide		2¼	0.160	0.174	0.146
¾	Narrow		1⅜	0.100	0.112	0.090
	Regular	0.812	2	0.100	0.112	0.090
	Wide		2½	0.160	0.174	0.146
⅞	Narrow		1 15/32	0.100	0.112	0.090
	Regular	0.938	2¼	0.160	0.174	0.146
	Wide		2¾	0.160	0.174	0.146
1	Narrow		1¾	0.100	0.112	0.090
	Regular	1 1/16	2½	0.160	0.174	0.146
	Wide		3	0.160	0.174	0.146
1⅛	Narrow		2	0.100	0.112	0.090
	Regular	1 3/16	2¾	0.160	0.174	0.146
	Wide		3¼	0.160	0.174	0.146
1¼	Narrow		2¼	0.160	0.174	0.146
	Regular	1 5/16	3	0.160	0.174	0.146
	Wide		3½	0.250	0.266	0.234
1⅜	Narrow		2½	0.160	0.174	0.146
	Regular	1 7/16	3¼	0.160	0.174	0.146
	Wide		3¾	0.250	0.266	0.234
1½	Narrow		2¾	0.160	0.174	0.146
	Regular	1 9/16	3½	0.250	0.266	0.234
	Wide		4	0.250	0.266	0.234
1⅝	Narrow		3	0.160	0.174	0.146
	Regular	1¾	3¾	0.250	0.266	0.234
	Wide		4¼	0.250	0.266	0.234
1¾	Narrow		3¼	0.160	0.174	0.146
	Regular	1⅞	4	0.250	0.266	0.234
	Wide		4½	0.250	0.266	0.234
1⅞	Narrow		3½	0.250	0.266	0.234
	Regular	2	4¼	0.250	0.266	0.234
	Wide		4¾	0.250	0.266	0.234
2	Narrow		3¾	0.250	0.266	0.234
	Regular	2⅛	4½	0.250	0.266	0.234
	Wide		5	0.250	0.266	0.234
2¼	Narrow		4	0.250	0.266	0.234
	Regular	2⅜	5	0.250	0.266	0.234
	Wide		5½	0.375	0.393	0.357
2½	Narrow		4½	0.250	0.266	0.234
	Regular	2⅝	5½	0.375	0.393	0.357
	Wide		6	0.375	0.393	0.357
2¾	Narrow		5	0.250	0.266	0.234
	Regular	2⅞	6	0.375	0.393	0.357
	Wide		6½	0.375	0.393	0.357
3	Narrow		5½	0.375	0.393	0.357
	Regular	3⅛	6½	0.375	0.393	0.357
	Wide		7	0.375	0.393	0.357

Note: All dimensions are given in inches.

The inside diameter has a tolerance of −0.005 in for sizes 0.060 to 0.190 in, inclusive; −0.010 for sizes 0.216 to $\frac{7}{8}$ in, inclusive; and ±0.010 in for all other sizes. The outside diameter has a tolerance of ±0.005 for narrow series sizes 0.060 to $\frac{5}{16}$ in, inclusive; regular series sizes 0.060 to 0.216 in, inclusive; and wide series sizes 0.060 to 0.164 in, inclusive. All other outside diameters have a tolerance of ±0.010 in.

11.5.2 Taper Pins

These are made in standard sizes with a taper of $\frac{1}{4}$ in/ft and in length increments of $\frac{1}{4}$ in. For standard dimensions see Table 11.36.

11.5.3 Grooved Pins

These are resistant to loosening from shock or vibration. Longitudinal grooves pressed into the cylindrical body deform the pin stock outward. When forced into a drilled hole of proper size, a locking fit is obtained. The various types of grooved pins are:

Type A: The full-length taper grooved pin replaces the taper dowel pin in many applications.

Type B: The half-length taper grooved pin, often used as a dowel pin or stop, has tapering grooves and body on only half of the pin length.

Type C: The full-length constant-diameter pin has straight grooves along its length and a pilot at one end to facilitate easy assembly. Particularly applicable where longitudinal stress due to vibration and shock is severe.

Type F: The full-length constant-diameter pin is similar to type C except it has a pilot at both ends. This pin has applications similar to type C.

Type D: The half-length reverse-taper grooved pin has applications similar to those of type B. Note the pin diameter is constant but stepped.

Type E: The center-grooved half-length pin has oval grooves and body. Used as a fulcrum bolt, T handle, and hinge.

For dimensions, see Ref. 43.

11.5.4 Spring-Type Straight Pins (Roll Pins)[43,44]

Driven into a hole, it exerts continuous spring pressure against the sides of the hole, preventing loosening by vibration. No reaming is required. Construction of the pin permits using a screw through the pin in some applications. It can be used as a replacement for stop pins, setscrews, taper pins, hinge pins, dowel pins, grooved pins, or clevis pins. One important advantage of this type of pin is that it is easily removed and replaced with little loss in holding power. The characteristics are shown in Table 11.37.

11.5.5 Cotter Pins

These are used with threaded members as a positive locking device (see Table 11.38).

11.6 KEYS AND SPLINES

11.6.1 Introduction

Keys and splines are used to prevent relative circumferential movement between rotating parts and respective shafts. Except for extremely heavy loads, the key is sized according to the shaft diameter rather than to the torsional loads developed.

TABLE 11.30 American National Standard Helical Spring Lock Washers[41]

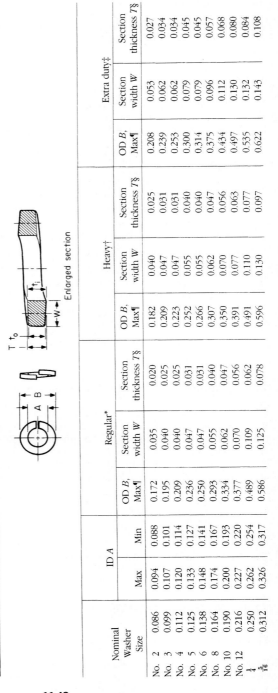

Enlarged section

Nominal Washer Size		ID A		Regular*			Heavy†			Extra duty‡		
		Max	Min	OD B, Max¶	Section width W	Section thickness T§	OD B, Max¶	Section width W	Section thickness T§	OD B, Max¶	Section width W	Section thickness T§
No. 2	0.086	0.094	0.088	0.172	0.035	0.020	0.182	0.040	0.025	0.208	0.053	0.027
No. 3	0.099	0.107	0.101	0.195	0.040	0.025	0.209	0.047	0.031	0.239	0.062	0.034
No. 4	0.112	0.120	0.114	0.209	0.040	0.025	0.223	0.047	0.031	0.253	0.062	0.034
No. 5	0.125	0.133	0.127	0.236	0.047	0.031	0.252	0.055	0.040	0.300	0.079	0.045
No. 6	0.138	0.148	0.141	0.250	0.047	0.031	0.266	0.055	0.040	0.314	0.079	0.045
No. 8	0.164	0.174	0.167	0.293	0.055	0.040	0.307	0.062	0.047	0.375	0.096	0.057
No. 10	0.190	0.200	0.193	0.334	0.062	0.047	0.350	0.070	0.056	0.434	0.112	0.068
No. 12	0.216	0.227	0.220	0.377	0.070	0.056	0.391	0.077	0.063	0.497	0.130	0.080
¼	0.250	0.262	0.254	0.489	0.109	0.062	0.491	0.110	0.077	0.535	0.132	0.084
5/16	0.312	0.326	0.317	0.586	0.125	0.078	0.596	0.130	0.097	0.622	0.143	0.108

3/8	0.375	0.390	0.380	0.683	0.141	0.094	0.691	0.145	0.115	0.741	0.170	0.123
7/16	0.438	0.455	0.443	0.779	0.156	0.109	0.787	0.160	0.133	0.839	0.186	0.143
1/2	0.500	0.518	0.506	0.873	0.171	0.125	0.883	0.176	0.151	0.939	0.204	0.162
9/16	0.562	0.582	0.570	0.971	0.188	0.141	0.981	0.193	0.170	1.041	0.223	0.182
5/8	0.625	0.650	0.635	1.079	0.203	0.156	1.093	0.210	0.189	1.157	0.242	0.202
11/16	0.688	0.713	0.698	1.176	0.219	0.172	1.192	0.227	0.207	1.258	0.260	0.221
3/4	0.750	0.775	0.760	1.271	0.234	0.188	1.291	0.244	0.226	1.361	0.279	0.241
13/16	0.812	0.843	0.824	1.367	0.250	0.203	1.391	0.262	0.246	1.463	0.298	0.261
7/8	0.875	0.905	0.887	1.464	0.266	0.219	1.494	0.281	0.266	1.576	0.322	0.285
15/16	0.938	0.970	0.950	1.560	0.281	0.234	1.594	0.298	0.284	1.688	0.345	0.308
1	1.000	1.042	1.017	1.661	0.297	0.250	1.705	0.319	0.306	1.799	0.366	0.330
1 1/16	1.062	1.107	1.080	1.756	0.312	0.266	1.808	0.338	0.326	1.910	0.389	0.352
1 1/8	1.125	1.172	1.144	1.853	0.328	0.281	1.909	0.356	0.345	2.019	0.411	0.375
1 3/16	1.188	1.237	1.208	1.950	0.344	0.297	2.008	0.373	0.364	2.124	0.341	0.396
1 1/4	1.250	1.302	1.271	2.045	0.359	0.312	2.113	0.393	0.384	2.231	0.452	0.417
1 5/16	1.312	1.366	1.334	2.141	0.375	0.328	2.211	0.410	0.403	2.335	0.472	0.438
1 3/8	1.375	1.432	1.398	2.239	0.391	0.344	2.311	0.427	0.422	2.439	0.491	0.458
1 7/16	1.438	1.497	1.462	2.334	0.406	0.359	2.406	0.442	0.440	2.540	0.509	0.478
1 1/2	1.500	1.561	1.525	2.430	0.422	0.375	2.502	0.458	0.458	2.638	0.526	0.496

*Formerly designated "medium helical spring lock washers."

†Not recommended for new applications.

‡Formerly designated "extra heavy helical spring lock washers."

¶The maximum outside diameters specified allow for the commercial tolerances on cold-drawn wire.

§T = mean section thickness = $(t_i + t_0)/2$.

Note: All dimensions are given in inches.

11.49

TABLE 11.31 American National Standard Internal and External Tooth Lock Washers

Internal tooth — Type A, Type B
External tooth — Type A, Type B
Countersunk external tooth — Type A (80°–82°), Type B (80°–82°)

Internal tooth lock washers

Size		No. 2	No. 3	No. 4	No. 5	No. 6	No. 8	No. 10	No. 12	1/4	5/16	3/8	7/16	1/2	9/16	5/8	11/16	3/4	13/16	7/8	1	1 1/8	1 1/4
A	Max	0.095	0.109	0.123	0.136	0.150	0.176	0.204	0.231	0.267	0.332	0.398	0.464	0.530	0.596	0.663	0.728	0.795	0.861	0.927	1.060	1.192	1.325
	Min	0.089	0.102	0.115	0.129	0.141	0.168	0.195	0.221	0.256	0.320	0.384	0.448	0.512	0.576	0.640	0.704	0.769	0.832	0.894	1.019	1.144	1.275
B	Max	0.200	0.232	0.270	0.280	0.295	0.340	0.381	0.410	0.478	0.610	0.692	0.789	0.900	0.985	1.071	1.166	1.245	1.315	1.410	1.637	1.830	1.975
	Min	0.175	0.215	0.255	0.245	0.275	0.325	0.365	0.394	0.460	0.594	0.670	0.740	0.867	0.957	1.045	1.130	1.220	1.290	1.364	1.590	1.799	1.921
C	Max	0.015	0.019	0.019	0.019	0.021	0.023	0.025	0.025	0.028	0.034	0.040	0.040	0.045	0.045	0.050	0.050	0.055	0.055	0.060	0.060	0.067	0.067
	Min	0.010	0.012	0.015	0.017	0.017	0.018	0.020	0.020	0.023	0.028	0.032	0.032	0.037	0.037	0.042	0.042	0.047	0.047	0.052	0.059	0.059	0.059

External tooth lock washers

Size		No. 2	No. 3	No. 4	No. 5	No. 6	No. 8	No. 10	No. 12	1/4	5/16	3/8	7/16	1/2	9/16	5/8	11/16	3/4	13/16	7/8	1
A	Max	—	—	0.109	0.136	0.150	0.176	0.204	0.231	0.267	0.332	0.398	0.464	0.530	0.596	0.663	0.728	0.795	0.861	0.927	1.060
	Min	—	—	0.102	0.129	0.141	0.168	0.195	0.221	0.256	0.320	0.384	0.448	0.513	0.576	0.641	0.704	0.768	0.833	0.897	1.025
B	Max	—	—	0.235	0.285	0.320	0.381	0.410	0.475	0.510	0.610	0.694	0.760	0.900	0.985	1.070	1.155	1.260	1.315	1.410	1.620
	Min	—	—	0.220	0.270	0.305	0.365	0.395	0.460	0.494	0.588	0.670	0.740	0.880	0.960	1.045	1.130	1.220	1.290	1.380	1.590
C	Max	—	—	0.015	0.019	0.022	0.023	0.025	0.028	0.028	0.034	0.040	0.040	0.045	0.045	0.050	0.050	0.055	0.055	0.060	0.067
	Min	—	—	0.012	0.014	0.016	0.018	0.020	0.023	0.023	0.028	0.032	0.032	0.037	0.037	0.042	0.042	0.047	0.047	0.052	0.059

Heavy internal tooth lock washers

Size		1/4	5/16	3/8	7/16	1/2	9/16	5/8	3/4	7/8
A	Max	0.267	0.332	0.398	0.464	0.530	0.596	0.663	0.795	0.927
	Min	0.256	0.320	0.384	0.448	0.512	0.576	0.640	0.768	0.894
B	Max	0.536	0.607	0.748	0.858	0.924	1.034	1.135	1.265	1.447
	Min	0.500	0.590	0.700	0.800	0.880	0.990	1.100	1.240	1.400
C	Max	0.045	0.050	0.050	0.067	0.067	0.067	0.067	0.084	0.084
	Min	0.035	0.040	0.042	0.050	0.055	0.055	0.059	0.070	0.075

Countersunk external tooth lock washers*

Size		No. 4	No. 6	No. 8	No. 10	No. 12	No. 16	1/4	5/16	3/8	7/16	1/2
A	Max	0.123	0.150	0.177	0.205	0.231	0.287	0.267	0.333	0.398	0.463	0.529
	Min	0.113	0.140	0.167	0.195	0.220	0.273	0.255	0.318	0.383	0.448	0.512
C	Max	0.019	0.021	0.021	0.025	0.025	0.028	0.025	0.028	0.034	0.045	0.045
	Min	0.015	0.017	0.017	0.020	0.020	0.023	0.020	0.023	0.028	0.037	0.037
D	Max	0.065	0.092	0.105	0.099	0.128	0.147	0.128	0.192	0.255	0.270	0.304
	Min	0.050	0.082	0.088	0.083	0.118	0.137	0.113	0.165	0.242	0.260	0.294

*Starting with No. 4, approximate ODs, are 0.213, 0.289, 0.322, 0.354, 0.421, 0.454, 0.505, 0.599, 0.765, 0.867, and 0.976.

TABLE 11.32 American National Standard Internal-External Tooth Lock Washers

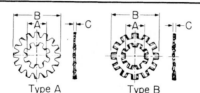

Type A   Type B

Size	A Inside diameter Max	Min	B Outside diameter Max	Min	C Thickness Max	Min	Size	A Inside diameter Max	Min	B Outside diameter Max	Min	C Thickness Max	Min
No. 4	0.123	0.115	0.475	0.460	0.021	0.016	5/16	0.332	0.320	0.900	0.865	0.040	0.032
	0.123	0.115	0.510	0.495	0.021	0.017		0.332	0.320	0.985	0.965	0.045	0.037
			0.610	0.580				0.332	0.320	1.070	1.045	0.050	0.042
										1.155	1.130		
No. 6	0.150	0.141	0.510	0.495	0.028	0.023	3/8	0.398	0.384	0.985	0.965	0.045	0.037
			0.610	0.580				0.398	0.384	1.070	1.045	0.050	0.042
			0.690	0.670						1.155	1.130		
										1.260	1.220		
No. 8	0.176	0.168	0.610	0.580	0.034	0.028	7/16	0.464	0.448	1.070	1.045	0.050	0.042
			0.690	0.670						1.155	1.130		
			0.760	0.740				0.464	0.448	1.260	1.220	0.055	0.047
										1.315	1.290		
No. 10	0.204	0.195	0.610	0.580	0.034	0.028	1/2	0.530	0.512	1.260	1.220	0.055	0.047
	0.204	0.195	0.690	0.670	0.040	0.032				1.315	1.290		
			0.760	0.740				0.530	0.512	1.410	1.380	0.060	0.052
			0.900	0.880				0.530	0.512	1.620	1.590	0.067	0.059
No. 12	0.231	0.221	0.690	0.670	0.040	0.032	9/16	0.596	0.576	1.315	1.290	0.055	0.047
			0.760	0.725				0.596	0.576	1.430	1.380	0.060	0.052
			0.900	0.880				0.596	0.576	1.620	1.590	0.067	0.059
	0.231	0.221	0.985	0.965	0.045	0.037				1.830	1.797		
1/4	0.267	0.256	0.760	0.725	0.040	0.032	5/8	0.663	0.640	1.410	1.380	0.060	0.052
			0.900	0.880						1.620	1.590		
	0.267	0.256	0.985	0.965	0.045	0.037		0.663	0.640	1.830	1.797	0.067	0.059
			1.070	1.045						1.975	1.935		

Note: All dimensions are given in inches except whole numbers under "Size."

TABLE 11.33 Hardened and Ground Dowel Pins[43]

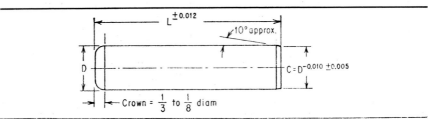

Length L	\(\frac{1}{8}\)	\(\frac{3}{16}\)	\(\frac{1}{4}\)	\(\frac{5}{16}\)	\(\frac{3}{8}\)	\(\frac{7}{16}\)	\(\frac{1}{2}\)	\(\frac{5}{8}\)	\(\frac{3}{4}\)	\(\frac{7}{8}\)
					Nominal diam. D					
Diam. standard pins ±0.0001	0.1252	0.1877	0.2502	0.3127	0.3752	0.4377	0.5002	0.6252	0.7502	0.8752
Diam. oversize pins ±0.0001	0.1260	0.1885	0.2510	0.3135	0.3760	0.4385	0.5010	0.6260	0.7510	0.8760
$\frac{1}{2}$	X	X	X	X						
$\frac{5}{8}$	X	X	X	X						
$\frac{3}{4}$	X	X	X	X	X					
$\frac{7}{8}$	X	X	X	X	X	X				
1	X	X	X	X	X	X				
$1\frac{1}{4}$		X	X	X	X	X	X	X		
$1\frac{1}{2}$		X	X	X	X	X	X	X	X	
$1\frac{3}{4}$		X	X	X	X	X	X	X	X	
2		X	X	X	X	X	X	X	X	X
$2\frac{1}{4}$				X	X	X	X			
$2\frac{1}{2}$			X	X	X		X	X	X	X
3							X	X	X	X
$3\frac{1}{2}$							X	X		
4							X	X	X	X
$4\frac{1}{2}$								X	X	X
5									X	X
$5\frac{1}{2}$									X	X

Note: The letter X indicates range of diameters for corresponding length.

All dimensions are given in inches.

These pins are extensively used in the tool and machine industry, and a machine reamer of nominal size may be used to produce the holes into which these pins tap or press fit. They must be straight and free from any defects.

TABLE 11.34 Dimensions of Ground Dowel Pins, Not Hardened[42]

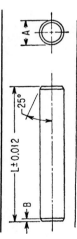

Nominal diam	Diam A Max	Diam A Min	Chamfer B
0.062	0.0600	0.0595	0.010
0.094	0.0912	0.0907	0.010
0.109	0.1068	0.1063	0.010
0.125	0.1223	0.1218	0.010
0.156	0.1535	0.1530	3/64
0.188	0.1847	0.1842	3/64
0.219	0.2159	0.2154	3/64
0.250	0.2470	0.2465	3/64
0.312	0.3094	0.3089	1/32
0.375	0.3717	0.3712	1/32
0.438	0.4341	0.4336	1/32
0.500	0.4964	0.4999	1/32
0.625	0.6211	0.6206	3/64
0.750	0.7458	0.7453	3/64
0.875	0.8705	0.8700	1/16
1.000	0.9952	0.9947	1/16

Note: All dimensions are given in inches.

Maximum diameters are graduated from 0.0005 on $\frac{1}{16}$-in pins to 0.0028 on 1-in pins under the minimum commercial bar stock sizes.

TABLE 11.35 Dimensions of Straight Pins[43]

Chamfered end

Square end

Nominal diam	Diam A Max	Diam A Min	Chamfer B
0.062	0.0625	0.0605	0.015
0.094	0.0937	0.0917	0.015
0.109	0.1094	0.1074	0.015
0.125	0.1250	0.1230	0.015
0.156	0.1562	0.1542	0.015
0.188	0.1875	0.1855	0.015
0.219	0.2187	0.2167	0.015
0.250	0.2500	0.2480	0.015
0.312	0.3125	0.3095	0.030
0.375	0.3750	0.3720	0.030
0.438	0.4375	0.4345	0.030
0.500	0.500	0.4970	0.030

Note: All dimensions are given in inches.

These pins must be straight and free from burrs or any other defects that will affect their serviceability.

TABLE 11.36 Dimensions of Taper Pins[43]

Number	7/0	6/0	5/0	4/0	3/0	2/0	0	1	2	3	4	5	6	7	8	9	10
Size (large end)	0.0625	0.0780	0.0940	0.1090	0.1250	0.1410	0.1560	0.1720	0.1930	0.2190	0.2500	0.2890	0.3410	0.4090	0.4920	0.5910	0.7060
Length, L																	
0.375	X	X	X	X	X	X	X										
0.500	X	X	X	X	X	X	X	X	X								
0.625		X	X	X	X	X	X	X	X								
0.750			X	X	X	X	X	X	X	X							
0.875			X	X		X	X	X	X	X	X						
1.000							X	X	X	X	X	X					
1.250							X	X	X	X	X	X	X				
1.500								X	X	X	X	X	X	X			
1.750								X	X	X	X	X	X	X	X		
2.000									X	X	X	X	X	X	X		
2.250									X	X	X	X	X	X	X	X	X
2.500										X	X	X	X	X	X	X	X
2.750										X	X	X	X	X	X	X	X
3.000										X	X	X	X	X	X	X	X
3.250											X		X	X	X	X	X
3.500											X		X	X	X	X	X
3.750											X		X	X	X	X	X
4.000													X		X	X	X
4.250															X	X	X
4.500															X	X	X
4.750																X	X
5.000																X	X
5.250																X	X
5.500																X	X
5.750																X	X
6.000																	X

Note: All dimensions are given in inches.

Standard reamers are available for pins given above the line.

Pin Nos. 11 (size 0.8600), 12 (size 1.032), 13 (size 1.241), and 14 (1.523) are special sizes—hence their lengths are special.

To find small diameter of pin, multiply the length by 0.02083 and subtract the result from the large diameter.

Types	Commercial type	Precision type
Sizes	7/0 to 14	7/0 to 10
Tolerance on diameter	+0.0013, −0.0007	(+0.0013, −0.0007)
Taper	¼ in./ft	¼ in./ft
	(±0.030)	(±0.030)
Length tolerance	(±0.030)	0.0005 up to 1 in. long
Concavity tolerance	None	0.001 1½ to 2 in. long

11.54

TABLE 11.37 Spring-Type Straight Pins[43,44]

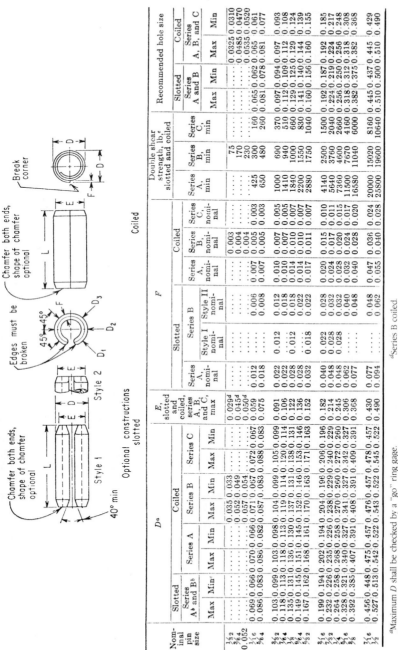

Style 1 — Chamfer both ends, shape of chamfer optional

Style 2 — Chamfer both ends, shape of chamfer optional; Edges must be broken; 45° 45°; 40° min

Optional constructions slotted

Coiled — Chamfer both ends, shape of chamfer optional; Break corner

Nominal pin size	D[a] Slotted Series A[b] and B[b] Max[a]	Slotted Min[c]	D Coiled Series A Max	Coiled Series A Min	D Coiled Series B Max	Coiled Series B Min	D Coiled Series C Max	Coiled Series C Min	E slotted and coiled, series A, B, and C, max	F Slotted Series A nom	F Slotted Series B Style I nom	F Slotted Series B Style II nom	F Coiled Series A nom	F Coiled Series B nom	F Coiled Series C nom	Double shear strength, lb,[e] Series A min	Series B min	Series C min	Rec. hole size Slotted Series A and B Max	Slotted Min	Rec. hole size Coiled Series A, B, and C Max	Coiled Min
1/32	.....	.....	.....	.....	0.035	0.033	.....	.....	0.029[d]	.....	.....	.....	.....	.....	.....	.....	75	.....	.....	.....	0.0325	0.0310
3/64	.....	.....	.....	.....	0.052	0.049	.....	.....	0.045[d]	.....	.....	0.006	.....	0.003	.....	.....	170	.....	.....	.....	0.0485	0.0470
0.052	.....	.....	.....	.....	0.057	0.054	.....	.....	0.050[d]	.....	.....	0.008	.....	0.004	.....	.....	230	.....	.....	.....	0.0535	0.0520
1/16	0.069	0.066	0.070	0.066	0.071	0.067	0.072	0.067	0.059	0.012	.....	.....	0.007	0.004	0.003	425	300	160	0.065	0.062	0.065	0.061
5/64	0.086	0.083	0.086	0.082	0.087	0.083	0.088	0.083	0.075	0.018	.....	.....	0.007	0.005	0.003	650	480	260	0.081	0.078	0.081	0.077
3/32	0.103	0.099	0.103	0.098	0.104	0.099	0.105	0.099	0.091	0.022	0.012	0.012	0.010	0.007	0.005	1000	690	370	0.097	0.094	0.097	0.093
7/64	0.118	0.113	0.118	0.113	0.119	0.114	0.120	0.114	0.106	0.022	.....	0.018	0.010	0.007	0.005	1410	940	510	0.112	0.109	0.112	0.108
1/8	0.135	0.131	0.136	0.130	0.137	0.131	0.138	0.131	0.122	0.028	0.012	0.018	0.014	0.010	0.007	1840	1000	660	0.129	0.125	0.129	0.124
9/64	0.149	0.145	0.151	0.145	0.152	0.146	0.153	0.146	0.136	0.028	.....	0.022	0.014	0.010	0.007	2200	1550	830	0.141	0.140	0.144	0.139
5/32	0.167	0.162	0.168	0.161	0.170	0.163	0.171	0.163	0.152	0.032	0.018	0.022	0.017	0.011	0.010	2880	1750	1040	0.160	0.156	0.160	0.155
3/16	0.199	0.194	0.202	0.194	0.204	0.196	0.206	0.196	0.182	0.040	0.022	0.028	0.020	0.015	0.011	4140	2500	1500	0.192	0.187	0.192	0.185
7/32	0.232	0.226	0.235	0.226	0.238	0.229	0.240	0.229	0.214	0.048	0.028	0.032	0.024	0.017	0.015	5640	3760	2040	0.224	0.219	0.224	0.217
1/4	0.264	0.258	0.268	0.258	0.270	0.260	0.272	0.260	0.245	0.048	0.028	0.032	0.028	0.020	0.017	7360	4600	2660	0.256	0.250	0.256	0.248
5/16	0.328	0.321	0.340	0.321	0.341	0.327	0.342	0.327	0.306	0.062	.....	.....	0.032	0.024	0.020	11500	7670	4160	0.318	0.312	0.318	0.308
3/8	0.392	0.385	0.407	0.391	0.408	0.391	0.409	0.391	0.368	0.077	.....	.....	0.040	0.028	0.024	16580	11040	6000	0.382	0.375	0.382	0.368
7/16	0.456	0.448	0.475	0.457	0.476	0.457	0.478	0.457	0.430	0.077	.....	0.048	0.047	0.036	0.028	20000	15020	8160	0.445	0.437	0.445	0.429
1/2	0.527	0.513	0.542	0.522	0.543	0.522	0.545	0.522	0.490	0.094	.....	0.062	0.055	0.040	0.028	25800	19600	10640	0.510	0.500	0.510	0.490

[a] Maximum D shall be checked by a "go" ring gage.
[b] Series designation applies to stock thickness, A being heaviest.
[c] Minimum D shall be the average of the D_1, D_2, and D_3 diameters.
[d] Series B coiled
[e] Applies to pins made from SAE 1070 to 1095 steel and SAE 51410 or AISI 420 corrosion-resistant steel. SAE 30302 stainless steel has a minimum shear strength equal to 85 percent of values shown for coiled pins.

TABLE 11.38 American National Standard Cotter Pins[43]

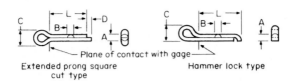

Extended prong square cut type

Hammer lock type

Plane of contact with gage

Nom size	Diam A^* and width B max	Wire width B min	Head diam C min	Prong length D min	Hole size	Nom size	Diam A^* and width B max	Wire width B min	Head diam C min	Prong length D min	Hole size
1/32	0.032	0.022	0.06	0.01	0.047	3/16	0.176	0.137	0.38	0.09	0.203
3/64	0.048	0.035	0.09	0.02	0.062	7/32	0.207	0.161	0.44	0.10	0.234
1/16	0.060	0.044	0.12	0.03	0.078	1/4	0.225	0.176	0.50	0.11	0.266
5/64	0.076	0.057	0.16	0.04	0.094	5/16	0.280	0.220	0.62	0.14	0.312
3/32	0.090	0.069	0.19	0.04	0.109	3/8	0.335	0.263	0.75	0.16	0.375
7/64	0.104	0.080	0.22	0.05	0.125	7/16	0.406	0.320	0.88	0.20	0.438
1/8	0.120	0.093	0.25	0.06	0.141	1/2	0.473	0.373	1.00	0.23	0.500
9/64	0.134	0.104	0.28	0.06	0.156	5/8	0.598	0.472	1.25	0.30	0.625
5/32	0.150	0.116	0.31	0.07	0.172	3/4	0.723	0.572	1.50	0.36	0.750

All dimensions are given in inches.
*Tolerances are − 0.004 in for the 1/32- to 1/16-in sizes, inclusive; − 0.005 in for the 5/32- to 1/8-in sizes, inclusive; −0.006 inch for the 3/8- to 1/2-in sizes, inclusive; and −0.008 inch for the 5/8- and 3/4-in sizes.

11.6.2 Sunk and Feather Keys

The "sunk" key (Table 11.39) is fitted in a groove in the shaft and projects into a keyway in the hub. "Feather keys" permit longitudinal movement between the shaft and the hub.

Some of the key forms most commonly used are the following.

1. *Plain parallel key:*[45] Square keys have sides equal to approximately one-fourth the shaft diameter. Both square and flat keys should fit tightly on all four sides to prevent any tendency to rock. Standard dimensions of square and flat keys are listed in Table 11.39.

2. *Plain taper key:*[46] One surface of this square- or rectangular-section key is tapered, resulting in a wedging action. Refer to Table 11.40 for standard dimensions.

3. *Gib-head key:*[47,48] Used where the small end of the key is not accessible. The gib provides a means for removal as well as a driving head when assembling; refer to Table 11.41.

11.6.3. Woodruff Key[48]

This key cannot move axially but can adjust itself to the taper, if any, in the hub keyway. The key is tightly fitted in a keyway formed in the shaft. Table 11.42 lists the standard key dimensions and Table 11.43 the keyseat (shaft) data.

Refer to references for the following key types: round end,[49] pin,[50] beveled or Barth,[51] Kennedy,[55] and Lewis.[51]

TABLE 11.39 Square and Rectangular Plain Parallel Stock Keys[48]

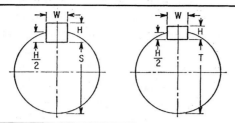

Shaft diam	Square key $W \times H$	Rectangular key $W \times H$	Tolerance*,† on W and H (minus)	Bottom of keyseat to opposite side of shaft	
				Square key S	Rectangular key T
$\frac{1}{2}$	$\frac{1}{8} \times \frac{1}{8}$	$\frac{1}{8} \times \frac{3}{32}$	0.0020	0.430	0.445
$\frac{9}{16}$	$\frac{1}{8} \times \frac{1}{8}$	$\frac{1}{8} \times \frac{3}{32}$	0.0020	0.493	0.509
$\frac{5}{8}$	$\frac{3}{16} \times \frac{3}{16}$	$\frac{3}{16} \times \frac{1}{8}$	0.0020	0.517	0.548
$\frac{11}{16}$	$\frac{3}{16} \times \frac{3}{16}$	$\frac{3}{16} \times \frac{1}{8}$	0.0020	0.581	0.612
$\frac{3}{4}$	$\frac{3}{16} \times \frac{3}{16}$	$\frac{3}{16} \times \frac{1}{8}$	0.0020	0.644	0.676
$\frac{13}{16}$	$\frac{3}{16} \times \frac{3}{16}$	$\frac{3}{16} \times \frac{1}{8}$	0.0020	0.708	0.739
$\frac{7}{8}$	$\frac{3}{16} \times \frac{3}{16}$	$\frac{3}{16} \times \frac{1}{8}$	0.0020	0.771	0.802
$\frac{15}{16}$	$\frac{1}{4} \times \frac{1}{4}$	$\frac{1}{4} \times \frac{3}{16}$	0.0020	0.796	0.827
1	$\frac{1}{4} \times \frac{1}{4}$	$\frac{1}{4} \times \frac{3}{16}$	0.0020	0.859	0.890
$1\frac{1}{16}$	$\frac{1}{4} \times \frac{1}{4}$	$\frac{1}{4} \times \frac{3}{16}$	0.0020	0.923	0.954
$1\frac{1}{8}$	$\frac{1}{4} \times \frac{1}{4}$	$\frac{1}{4} \times \frac{3}{16}$	0.0020	0.986	1.017
$1\frac{3}{16}$	$\frac{1}{4} \times \frac{1}{4}$	$\frac{1}{4} \times \frac{3}{16}$	0.0020	1.049	1.081
$1\frac{1}{4}$	$\frac{1}{4} \times \frac{1}{4}$	$\frac{1}{4} \times \frac{3}{16}$	0.0020	1.112	1.144
$1\frac{5}{16}$	$\frac{5}{16} \times \frac{5}{16}$	$\frac{5}{16} \times \frac{1}{4}$	0.0020	1.137	1.169
$1\frac{3}{8}$	$\frac{5}{16} \times \frac{5}{16}$	$\frac{5}{16} \times \frac{1}{4}$	0.0020	1.201	1.232
$1\frac{7}{16}$	$\frac{3}{8} \times \frac{3}{8}$	$\frac{3}{8} \times \frac{1}{4}$	0.0020	1.225	1.288
$1\frac{1}{2}$	$\frac{3}{8} \times \frac{3}{8}$	$\frac{3}{8} \times \frac{1}{4}$	0.0020	1.289	1.351
$1\frac{9}{16}$	$\frac{3}{8} \times \frac{3}{8}$	$\frac{3}{8} \times \frac{1}{4}$	0.0020	1.352	1.415
$1\frac{5}{8}$	$\frac{3}{8} \times \frac{3}{8}$	$\frac{3}{8} \times \frac{1}{4}$	0.0020	1.416	1.478
$1\frac{11}{16}$	$\frac{3}{8} \times \frac{3}{8}$	$\frac{3}{8} \times \frac{1}{4}$	0.0020	1.479	1.542
$1\frac{3}{4}$	$\frac{3}{8} \times \frac{3}{8}$	$\frac{3}{8} \times \frac{1}{4}$	0.0020	1.542	1.605
$1\frac{13}{16}$	$\frac{1}{2} \times \frac{1}{2}$	$\frac{1}{2} \times \frac{3}{8}$	0.0025	1.527	1.590
$1\frac{7}{8}$	$\frac{1}{2} \times \frac{1}{2}$	$\frac{1}{2} \times \frac{3}{8}$	0.0025	1.591	1.654
$1\frac{15}{16}$	$\frac{1}{2} \times \frac{1}{2}$	$\frac{1}{2} \times \frac{3}{8}$	0.0025	1.655	1.717
2	$\frac{1}{2} \times \frac{1}{2}$	$\frac{1}{2} \times \frac{3}{8}$	0.0025	1.718	1.781
$2\frac{1}{16}$	$\frac{1}{2} \times \frac{1}{2}$	$\frac{1}{2} \times \frac{3}{8}$	0.0025	1.782	1.843
$2\frac{1}{8}$	$\frac{1}{2} \times \frac{1}{2}$	$\frac{1}{2} \times \frac{3}{8}$	0.0025	1.845	1.908
$2\frac{3}{16}$	$\frac{1}{2} \times \frac{1}{2}$	$\frac{1}{2} \times \frac{3}{8}$	0.0025	1.909	1.971
$2\frac{1}{4}$	$\frac{1}{2} \times \frac{1}{2}$	$\frac{1}{2} \times \frac{3}{8}$	0.0025	1.972	2.034
$2\frac{5}{16}$	$\frac{5}{8} \times \frac{5}{8}$	$\frac{5}{8} \times \frac{7}{16}$	0.0025	1.957	2.051
$2\frac{3}{8}$	$\frac{5}{8} \times \frac{5}{8}$	$\frac{5}{8} \times \frac{7}{16}$	0.0025	2.021	2.114
$2\frac{7}{16}$	$\frac{5}{8} \times \frac{5}{8}$	$\frac{5}{8} \times \frac{7}{16}$	0.0025	2.084	2.178
$2\frac{1}{2}$	$\frac{5}{8} \times \frac{5}{8}$	$\frac{5}{8} \times \frac{7}{16}$	0.0025	2.148	2.242
$2\frac{5}{8}$	$\frac{5}{8} \times \frac{5}{8}$	$\frac{5}{8} \times \frac{7}{16}$	0.0025	2.275	2.368
$2\frac{3}{4}$	$\frac{5}{8} \times \frac{5}{8}$	$\frac{5}{8} \times \frac{7}{16}$	0.0025	2.402	2.495

TABLE 11.39   Square and Rectangular Plain Parallel Stock Keys[48] (*Continued*)

Shaft diam	Square key $W \times H$	Rectangular key $W \times H$	Tolerance*,† on W and H (minus)	Bottom of keyseat to opposite side of shaft — Square key S	Bottom of keyseat to opposite side of shaft — Rectangular key T
$2\frac{7}{8}$	$\frac{3}{4} \times \frac{3}{4}$	$\frac{3}{4} \times \frac{1}{2}$	0.0025	2.450	2.575
$2\frac{15}{16}$	$\frac{3}{4} \times \frac{3}{4}$	$\frac{3}{4} \times \frac{1}{2}$	0.0025	2.514	2.639
3	$\frac{3}{4} \times \frac{3}{4}$	$\frac{3}{4} \times \frac{1}{2}$	0.0025	2.577	2.702
$3\frac{1}{8}$	$\frac{3}{4} \times \frac{3}{4}$	$\frac{3}{4} \times \frac{1}{2}$	0.0025	2.704	2.829
$3\frac{1}{4}$	$\frac{3}{4} \times \frac{3}{4}$	$\frac{3}{4} \times \frac{1}{2}$	0.0025	2.831	2.956
$3\frac{3}{8}$	$\frac{7}{8} \times \frac{7}{8}$	$\frac{7}{8} \times \frac{5}{8}$	0.0030	2.880	3.005
$3\frac{7}{16}$	$\frac{7}{8} \times \frac{7}{8}$	$\frac{7}{8} \times \frac{5}{8}$	0.0030	2.944	3.069
$3\frac{1}{2}$	$\frac{7}{8} \times \frac{7}{8}$	$\frac{7}{8} \times \frac{5}{8}$	0.0030	3.007	3.132
$3\frac{5}{8}$	$\frac{7}{8} \times \frac{7}{8}$	$\frac{7}{8} \times \frac{5}{8}$	0.0030	3.140	3.259
$3\frac{3}{4}$	$\frac{7}{8} \times \frac{7}{8}$	$\frac{7}{8} \times \frac{5}{8}$	0.0030	3.261	3.386
$3\frac{7}{8}$	1×1	$1 \times \frac{3}{4}$	0.0030	3.309	3.434
$3\frac{15}{16}$	1×1	$1 \times \frac{3}{4}$	0.0030	3.373	3.498
4	1×1	$1 \times \frac{3}{4}$	0.0030	3.437	3.562
$4\frac{1}{4}$	1×1	$1 \times \frac{3}{4}$	0.0030	3.690	3.815
$4\frac{7}{16}$	1×1	$1 \times \frac{3}{4}$	0.0030	3.881	4.006
$4\frac{1}{2}$	1×1	$1 \times \frac{3}{4}$	0.0030	3.944	4.069
$4\frac{3}{4}$	$1\frac{1}{4} \times 1\frac{1}{4}$	$1\frac{1}{4} \times \frac{7}{8}$	0.0030	4.042	4.229
$4\frac{15}{16}$	$1\frac{1}{4} \times 1\frac{1}{4}$	$1\frac{1}{4} \times \frac{7}{8}$	0.0030	4.232	4.420
5	$1\frac{1}{4} \times 1\frac{1}{4}$	$1\frac{1}{4} \times \frac{7}{8}$	0.0030	4.296	4.483
$5\frac{1}{4}$	$1\frac{1}{4} \times 1\frac{1}{4}$	$1\frac{1}{4} \times \frac{7}{8}$	0.0030	4.550	4.733
$5\frac{7}{16}$	$1\frac{1}{4} \times 1\frac{1}{4}$	$1\frac{1}{4} \times \frac{7}{8}$	0.0030	4.740	4.927
$5\frac{1}{2}$	$1\frac{1}{4} \times 1\frac{1}{4}$	$1\frac{1}{4} \times \frac{7}{8}$	0.0030	4.803	4.991
$5\frac{3}{4}$	$1\frac{1}{2} \times 1\frac{1}{2}$	$1\frac{1}{2} \times 1$	0.0030	4.900	5.150
$5\frac{15}{16}$	$1\frac{1}{2} \times 1\frac{1}{2}$	$1\frac{1}{2} \times 1$	0.0030	5.091	5.341
6	$1\frac{1}{2} \times 1\frac{1}{2}$	$1\frac{1}{2} \times 1$	0.0030	5.155	5.405

*Stock keys are applicable to the general run of work, and the tolerances have been set accordingly. It is understood that these keys are to be cut from cold-finished stock and are to be used without machining. They are not intended to cover the finer applications where a closer fit may be required.

†These tolerances are *negative* and represent the maximum allowable variation *below* the exact nominal size. For example, the standard stock square key for a 2-in shaft has a maximum size of 0.500 × 0.500 in and a minimum size of 0.4975 × 0.4975 in.

11.6.4   Splines[53,55]

Splines are multiple keys cut in a shaft and hub to prevent relative motion and are used in preference to multiple keys for the transmission of power. Types of splines commonly used are the parallel or square side and the more recent involute spline.[53] Table 11.44 lists dimensions for the square-side splines.

11.6.5   Involute Splines

The involute spline is similar in form to involute gears and has pressure angles of 30°, 37.5°, and 45° and a stub tooth with a height equal to one-half that used in gear practice. Table 11.45 shows the basic dimension for splines to indicate the relative tooth sizes for various spline pitches.

TABLE 11.40   Square and Flat Plain Taper Stock Keys[45,46]

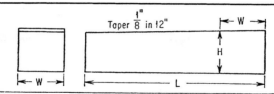

Shaft diam, incl	Square type		Flat type		Tolerance*	
	Max width W	Height at large end† H	Max width W	Height at large end† H	On width (minus)	On height (plus)
$\frac{1}{2}$ – $\frac{9}{16}$	$\frac{1}{8}$	$\frac{1}{8}$	$\frac{1}{8}$	$\frac{3}{32}$	0.0020	0.0020
$\frac{5}{8}$ – $\frac{7}{8}$	$\frac{3}{16}$	$\frac{3}{16}$	$\frac{3}{16}$	$\frac{1}{8}$	0.0020	0.0020
$\frac{15}{16}$–$1\frac{1}{4}$	$\frac{1}{4}$	$\frac{1}{4}$	$\frac{1}{4}$	$\frac{3}{16}$	0.0020	0.0020
$1\frac{5}{16}$–$1\frac{3}{8}$	$\frac{5}{16}$	$\frac{5}{16}$	$\frac{5}{16}$	$\frac{1}{4}$	0.0020	0.0020
$1\frac{7}{16}$–$1\frac{3}{4}$	$\frac{3}{8}$	$\frac{3}{8}$	$\frac{3}{8}$	$\frac{1}{4}$	0.0020	0.0020
$1\frac{13}{16}$–$2\frac{1}{4}$	$\frac{1}{2}$	$\frac{1}{2}$	$\frac{1}{2}$	$\frac{3}{8}$	0.0025	0.0025
$2\frac{5}{16}$–$2\frac{3}{4}$	$\frac{5}{8}$	$\frac{5}{8}$	$\frac{5}{8}$	$\frac{7}{16}$	0.0025	0.0025
$2\frac{7}{8}$–$3\frac{1}{4}$	$\frac{3}{4}$	$\frac{3}{4}$	$\frac{3}{4}$	$\frac{1}{2}$	0.0025	0.0025
$3\frac{3}{8}$–$3\frac{3}{4}$	$\frac{7}{8}$	$\frac{7}{8}$	$\frac{7}{8}$	$\frac{5}{8}$	0.0030	0.0030
$3\frac{7}{8}$–$4\frac{1}{2}$	1	1	1	$\frac{3}{4}$	0.0030	0.0030
$4\frac{3}{4}$ –$5\frac{1}{2}$	$1\frac{1}{4}$	$1\frac{1}{4}$	$1\frac{1}{4}$	$\frac{7}{8}$	0.0030	0.0030
$5\frac{3}{4}$ –6	$1\frac{1}{2}$	$1\frac{1}{2}$	$1\frac{1}{2}$	1	0.0030	0.0030

*Not intended to cover the finer applications where a closer fit may be required.
†This height of the key is measured at the distance W, equal to the width of the key, from the large end.
Note: All dimensions given in inches.

TABLE 11.41   Gib-Head Keys[47,48]

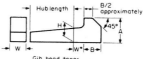

Nominal key size width W	Square		Rectangular			Nominal key size width W	Square		Rectangular		
	H	A	B	H	A	B		H	A	B	H	A	B
$\frac{1}{8}$	$\frac{1}{8}$	$\frac{1}{4}$	$\frac{1}{4}$	$\frac{3}{32}$	$\frac{3}{16}$	$\frac{1}{8}$	1	1	$1\frac{1}{8}$	$1\frac{1}{8}$	$\frac{3}{4}$	$1\frac{1}{4}$	$\frac{7}{8}$
$\frac{3}{16}$	$\frac{3}{16}$	$\frac{5}{16}$	$\frac{5}{16}$	$\frac{1}{8}$	$\frac{1}{4}$	$\frac{1}{4}$	$1\frac{1}{4}$	$1\frac{1}{4}$	2	$1\frac{7}{16}$	$\frac{7}{8}$	$1\frac{3}{8}$	1
$\frac{1}{4}$	$\frac{1}{4}$	$\frac{7}{16}$	$\frac{3}{8}$	$\frac{3}{16}$	$\frac{5}{16}$	$\frac{5}{16}$	$1\frac{1}{2}$	$1\frac{1}{2}$	$2\frac{3}{8}$	$1\frac{3}{4}$	1	$1\frac{5}{8}$	$1\frac{1}{8}$
$\frac{5}{16}$	$\frac{5}{16}$	$\frac{1}{2}$	$\frac{7}{16}$	$\frac{1}{4}$	$\frac{7}{16}$	$\frac{3}{8}$	$1\frac{3}{4}$	$1\frac{3}{4}$	$2\frac{3}{4}$	2	$1\frac{1}{4}$	$2\frac{3}{8}$	$1\frac{3}{8}$
$\frac{3}{8}$	$\frac{3}{8}$	$\frac{5}{8}$	$\frac{1}{2}$	$\frac{1}{4}$	$\frac{7}{16}$	$\frac{3}{8}$	2	2	$3\frac{1}{2}$	$2\frac{1}{4}$	$1\frac{1}{2}$	$2\frac{5}{8}$	$1\frac{3}{4}$
$\frac{1}{2}$	$\frac{1}{2}$	$\frac{7}{8}$	$\frac{5}{8}$	$\frac{3}{8}$	$\frac{5}{8}$	$\frac{1}{2}$	$2\frac{1}{2}$	$2\frac{1}{2}$	4	3	$1\frac{3}{4}$	$2\frac{3}{4}$	2
$\frac{5}{8}$	$\frac{5}{8}$	1	$\frac{3}{4}$	$\frac{7}{16}$	$\frac{3}{4}$	$\frac{9}{16}$	3	3	5	$3\frac{1}{2}$	2	$3\frac{1}{2}$	$2\frac{1}{4}$
$\frac{3}{4}$	$\frac{3}{4}$	$1\frac{1}{4}$	$\frac{7}{8}$	$\frac{1}{2}$	$\frac{7}{8}$	$\frac{3}{4}$	$3\frac{1}{2}$	$3\frac{1}{2}$	6	4	$2\frac{1}{2}$	4	3
$\frac{7}{8}$	$\frac{7}{8}$	$1\frac{3}{8}$	1	$\frac{5}{8}$	1	$\frac{3}{4}$							

*For locating position of dimension H. Tolerance does not apply.
For larger sizes the following relationships are suggested as guides for establishing A and B: $A = 1.8H$ and $B = 1.2H$.
Note: All dimensions are given in inches.

TABLE 11.42   ANSI Standard Woodruff Keys[48]

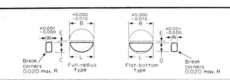

Key no.	Nominal key size $W \times B$	Actual length L +0.000 −0.010	Height of key				Distance below center E
			C		D		
			Max	Min	Max	Min	
202	$\frac{1}{16} \times \frac{1}{4}$	0.248	0.109	0.104	0.109	0.104	$\frac{1}{64}$
202.5	$\frac{1}{16} \times \frac{5}{16}$	0.311	0.140	0.135	0.140	0.135	$\frac{1}{64}$
302.5	$\frac{3}{32} \times \frac{5}{16}$	0.311	0.140	0.135	0.140	0.135	$\frac{1}{64}$
203	$\frac{1}{16} \times \frac{3}{8}$	0.374	0.172	0.167	0.172	0.167	$\frac{1}{64}$
303	$\frac{3}{32} \times \frac{3}{8}$	0.374	0.172	0.167	0.172	0.167	$\frac{1}{64}$
403	$\frac{1}{8} \times \frac{3}{8}$	0.374	0.172	0.167	0.172	0.167	$\frac{1}{64}$
204	$\frac{1}{16} \times \frac{1}{2}$	0.491	0.203	0.198	0.194	0.188	$\frac{3}{64}$
304	$\frac{3}{32} \times \frac{1}{2}$	0.491	0.203	0.198	0.194	0.188	$\frac{3}{64}$
404	$\frac{1}{8} \times \frac{1}{2}$	0.491	0.203	0.198	0.194	0.188	$\frac{3}{64}$
305	$\frac{3}{32} \times \frac{5}{8}$	0.612	0.250	0.245	0.240	0.234	$\frac{1}{16}$
405	$\frac{1}{8} \times \frac{5}{8}$	0.612	0.250	0.245	0.240	0.234	$\frac{1}{16}$
505	$\frac{5}{32} \times \frac{5}{8}$	0.612	0.250	0.245	0.240	0.234	$\frac{1}{16}$
605	$\frac{3}{16} \times \frac{5}{8}$	0.612	0.250	0.245	0.240	0.234	$\frac{1}{16}$
406	$\frac{1}{8} \times \frac{3}{4}$	0.740	0.313	0.308	0.303	0.297	$\frac{1}{16}$
506	$\frac{5}{32} \times \frac{3}{4}$	0.740	0.313	0.308	0.303	0.297	$\frac{1}{16}$
606	$\frac{3}{16} \times \frac{3}{4}$	0.740	0.313	0.308	0.303	0.297	$\frac{1}{16}$
806	$\frac{1}{4} \times \frac{3}{4}$	0.740	0.313	0.308	0.303	0.297	$\frac{1}{16}$
507	$\frac{5}{32} \times \frac{7}{8}$	0.866	0.375	0.370	0.365	0.359	$\frac{1}{16}$

11.7   WIRE AND SHEET-METAL GAGES

In the United States today, two wire gages are most commonly used.

11.7.1   United States Steel Wire Gage[56]

Also called "steel wire gage," this is used for practically all steel wire in the United States. See column 1, Table 11.46.

11.7.2   American Wire Gage

Also called Brown & Sharp gage, this is used to identify copper and aluminum wire as well as sheets of copper, aluminum, and other nonferrous metals. See column 2, Table 11.46.

Other gages in use are listed in Table 11.46 but are not recommended for new design specifications.

TABLE 11.42 ANSI Standard Woodruff Keys[48] (*Continued*)

Key no.	Nominal key size $W \times B$	Actual length L +0.000 −0.010	Height of key C Max	C Min	D Max	D Min	Distance below center E
607	$\frac{3}{16} \times \frac{7}{8}$	0.866	0.375	0.370	0.365	0.359	$\frac{1}{16}$
707	$\frac{7}{32} \times \frac{7}{8}$	0.866	0.375	0.370	0.365	0.359	$\frac{1}{16}$
807	$\frac{1}{4} \times \frac{7}{8}$	0.866	0.375	0.370	0.365	0.359	$\frac{1}{16}$
608	$\frac{3}{16} \times 1$	0.992	0.438	0.433	0.428	0.422	$\frac{1}{16}$
708	$\frac{7}{32} \times 1$	0.992	0.438	0.433	0.428	0.422	$\frac{1}{16}$
808	$\frac{1}{4} \times 1$	0.992	0.438	0.433	0.428	0.422	$\frac{1}{16}$
1008	$\frac{5}{16} \times 1$	0.992	0.438	0.433	0.428	0.422	$\frac{1}{16}$
1208	$\frac{3}{8} \times 1$	0.992	0.438	0.433	0.428	0.422	$\frac{1}{16}$
609	$\frac{3}{16} \times 1\frac{1}{8}$	1.114	0.484	0.479	0.475	0.469	$\frac{5}{64}$
709	$\frac{7}{32} \times 1\frac{1}{8}$	1.114	0.484	0.479	0.475	0.469	$\frac{5}{64}$
809	$\frac{1}{4} \times 1\frac{1}{8}$	1.114	0.484	0.479	0.475	0.469	$\frac{5}{64}$
1009	$\frac{5}{16} \times 1\frac{1}{8}$	1.114	0.484	0.479	0.475	0.469	$\frac{5}{64}$
610	$\frac{3}{16} \times 1\frac{1}{4}$	1.240	0.547	0.542	0.537	0.531	$\frac{5}{64}$
710	$\frac{7}{32} \times 1\frac{1}{4}$	1.240	0.547	0.542	0.537	0.531	$\frac{5}{64}$
810	$\frac{1}{4} \times 1\frac{1}{4}$	1.240	0.547	0.542	0.537	0.531	$\frac{5}{64}$
1010	$\frac{5}{16} \times 1\frac{1}{4}$	1.240	0.547	0.542	0.537	0.531	$\frac{5}{64}$
1210	$\frac{3}{8} \times 1\frac{1}{4}$	1.240	0.547	0.542	0.537	0.531	$\frac{5}{64}$
811	$\frac{1}{4} \times 1\frac{3}{8}$	1.362	0.594	0.589	0.584	0.578	$\frac{3}{32}$
1011	$\frac{5}{16} \times 1\frac{3}{8}$	1.362	0.594	0.589	0.584	0.578	$\frac{3}{32}$
1211	$\frac{3}{8} \times 1\frac{3}{8}$	1.362	0.594	0.589	0.584	0.578	$\frac{3}{32}$
812	$\frac{1}{4} \times 1\frac{1}{2}$	1.484	0.641	0.636	0.631	0.625	$\frac{7}{64}$
1012	$\frac{5}{16} \times 1\frac{1}{2}$	1.484	0.641	0.636	0.631	0.625	$\frac{7}{64}$
1212	$\frac{3}{8} \times 1\frac{1}{2}$	1.484	0.641	0.636	0.631	0.625	$\frac{7}{64}$
617-1	$\frac{3}{16} \times 2\frac{1}{8}$	1.380	0.406	0.401	0.396	0.390	$\frac{21}{32}$
817-1	$\frac{1}{4} \times 2\frac{1}{8}$	1.380	0.406	0.401	0.396	0.390	$\frac{21}{32}$
1017-1	$\frac{5}{16} \times 2\frac{1}{8}$	1.380	0.406	0.401	0.396	0.390	$\frac{21}{32}$
1217-1	$\frac{3}{8} \times 2\frac{1}{8}$	1.380	0.406	0.401	0.396	0.390	$\frac{21}{32}$
617	$\frac{3}{16} \times 2\frac{1}{8}$	1.723	0.531	0.526	0.521	0.515	$\frac{17}{32}$
817	$\frac{1}{4} \times 2\frac{1}{8}$	1.723	0.531	0.526	0.521	0.515	$\frac{17}{32}$
1017	$\frac{5}{16} \times 2\frac{1}{8}$	1.723	0.531	0.526	0.521	0.515	$\frac{17}{32}$
1217	$\frac{3}{8} \times 2\frac{1}{8}$	1.723	0.531	0.526	0.521	0.515	$\frac{17}{32}$
822-1	$\frac{1}{4} \times 2\frac{3}{4}$	2.000	0.594	0.589	0.584	0.578	$\frac{25}{32}$
1022-1	$\frac{5}{16} \times 2\frac{3}{4}$	2.000	0.594	0.589	0.584	0.578	$\frac{25}{32}$
1222-1	$\frac{3}{8} \times 2\frac{3}{4}$	2.000	0.594	0.589	0.584	0.578	$\frac{25}{32}$
1422-1	$\frac{7}{16} \times 2\frac{3}{4}$	2.000	0.594	0.589	0.584	0.578	$\frac{25}{32}$
1622-1	$\frac{1}{2} \times 2\frac{3}{4}$	2.000	0.594	0.589	0.584	0.578	$\frac{25}{32}$
822	$\frac{1}{4} \times 2\frac{3}{4}$	2.317	0.750	0.745	0.740	0.734	$\frac{5}{8}$
1022	$\frac{5}{16} \times 2\frac{3}{4}$	2.317	0.750	0.745	0.740	0.734	$\frac{5}{8}$
1222	$\frac{3}{8} \times 2\frac{3}{4}$	2.317	0.750	0.745	0.740	0.734	$\frac{5}{8}$
1422	$\frac{7}{16} \times 2\frac{3}{4}$	2.317	0.750	0.745	0.740	0.734	$\frac{5}{8}$
1622	$\frac{1}{2} \times 2\frac{3}{4}$	2.317	0.750	0.745	0.740	0.734	$\frac{5}{8}$
1228	$\frac{3}{8} \times 3\frac{1}{2}$	2.880	0.938	0.933	0.928	0.922	$\frac{13}{16}$
1428	$\frac{7}{16} \times 3\frac{1}{2}$	2.880	0.938	0.933	0.928	0.922	$\frac{13}{16}$
1628	$\frac{1}{2} \times 3\frac{1}{2}$	2.880	0.938	0.933	0.928	0.922	$\frac{13}{16}$
1828	$\frac{9}{16} \times 3\frac{1}{2}$	2.880	0.938	0.933	0.928	0.922	$\frac{13}{16}$
2028	$\frac{5}{8} \times 3\frac{1}{2}$	2.880	0.938	0.933	0.928	0.922	$\frac{13}{16}$
2228	$\frac{11}{16} \times 3\frac{1}{2}$	2.880	0.938	0.933	0.928	0.922	$\frac{13}{16}$
2428	$\frac{3}{4} \times 3\frac{1}{2}$	2.880	0.938	0.933	0.928	0.922	$\frac{13}{16}$

The key numbers with the −1 designation, while representing the nominal key size have a shorter length L and due to a greater distance below center E are less in height than the keys of the same number without the −1 designation.

TABLE 11.43   ANSI Standard Keyseat Dimensions for Woodruff Keys[48,52]

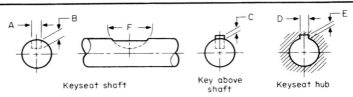

Keyseat shaft   Key above shaft   Keyseat hub

Key no.	Nominal size key	Keyseat (Shaft)					Key above shaft	Keyseat (Hub)	
		Width A^*		Depth B	Diameter F		Height C	Width D	Depth E
		Min	Max	$+0.005$ -0.000	Min	Max	$+0.005$ -0.005	$+0.002$ -0.000	$+0.005$ -0.000
202	$\frac{1}{16} \times \frac{1}{4}$	0.0615	0.0630	0.0728	0.250	0.268	0.0312	0.0635	0.0372
202.5	$\frac{1}{16} \times \frac{5}{16}$	0.0615	0.0630	0.1038	0.312	0.330	0.0312	0.0635	0.0372
302.5	$\frac{3}{32} \times \frac{5}{16}$	0.0928	0.0943	0.0882	0.312	0.330	0.0469	0.0948	0.0529
203	$\frac{1}{16} \times \frac{3}{8}$	0.0615	0.0630	0.1358	0.375	0.393	0.0312	0.0635	0.0372
303	$\frac{3}{32} \times \frac{3}{8}$	0.0928	0.0943	0.1202	0.375	0.393	0.0469	0.0948	0.0529
403	$\frac{1}{8} \times \frac{3}{8}$	0.1240	0.1255	0.1045	0.375	0.393	0.0625	0.1260	0.0685
204	$\frac{1}{16} \times \frac{1}{2}$	0.0615	0.0630	0.1668	0.500	0.518	0.0312	0.0635	0.0372
304	$\frac{3}{32} \times \frac{1}{2}$	0.0928	0.0943	0.1511	0.500	0.518	0.0469	0.0948	0.0529
404	$\frac{1}{8} \times \frac{1}{2}$	0.1240	0.1255	0.1355	0.500	0.518	0.0625	0.1260	0.0685
305	$\frac{3}{32} \times \frac{5}{8}$	0.0928	0.0943	0.1981	0.625	0.643	0.0469	0.0948	0.0529
405	$\frac{1}{8} \times \frac{5}{8}$	0.1240	0.1255	0.1825	0.625	0.643	0.0625	0.1260	0.0685
505	$\frac{5}{32} \times \frac{5}{8}$	0.1553	0.1568	0.1669	0.625	0.643	0.0781	0.1573	0.0841
605	$\frac{3}{16} \times \frac{5}{8}$	0.1863	0.1880	0.1513	0.625	0.643	0.0937	0.1885	0.0997
406	$\frac{1}{8} \times \frac{3}{4}$	0.1240	0.1255	0.2455	0.750	0.768	0.0625	0.1260	0.0685
506	$\frac{5}{32} \times \frac{3}{4}$	0.1553	0.1568	0.2299	0.750	0.768	0.0781	0.1573	0.0841
606	$\frac{3}{16} \times \frac{3}{4}$	0.1863	0.1880	0.2143	0.750	0.768	0.0937	0.1885	0.0997
806	$\frac{1}{4} \times \frac{3}{4}$	0.2487	0.2505	0.1830	0.750	0.768	0.1250	0.2510	0.1310
507	$\frac{5}{32} \times \frac{7}{8}$	0.1553	0.1568	0.2919	0.875	0.895	0.0781	0.1573	0.0841
607	$\frac{3}{16} \times \frac{7}{8}$	0.1863	0.1880	0.2763	0.875	0.895	0.0937	0.1885	0.0997
707	$\frac{7}{32} \times \frac{7}{8}$	0.2175	0.2193	0.2607	0.875	0.895	0.1093	0.2198	0.1153
807	$\frac{1}{4} \times \frac{7}{8}$	0.2487	0.2505	0.2450	0.875	0.895	0.1250	0.2510	0.1310
608	$\frac{3}{16} \times 1$	0.1863	0.1880	0.3393	1.000	1.020	0.0937	0.1885	0.0997
708	$\frac{7}{32} \times 1$	0.2175	0.2193	0.3237	1.000	1.020	0.1093	0.2198	0.1153
808	$\frac{1}{4} \times 1$	0.2487	0.2505	0.3080	1.000	1.020	0.1250	0.2510	0.1310
1008	$\frac{5}{16} \times 1$	0.3111	0.3130	0.2768	1.000	1.020	0.1562	0.3135	0.1622
1208	$\frac{3}{8} \times 1$	0.3735	0.3755	0.2455	1.000	1.020	0.1875	0.3760	0.1935
609	$\frac{3}{16} \times 1\frac{1}{8}$	0.1863	0.1880	0.3853	1.125	1.145	0.0937	0.1885	0.0997
709	$\frac{7}{32} \times 1\frac{1}{8}$	0.2175	0.2193	0.3697	1.125	1.145	0.1093	0.2198	0.1153
809	$\frac{1}{4} \times 1\frac{1}{8}$	0.2487	0.2505	0.3540	1.125	1.145	0.1250	0.2510	0.1310
1009	$\frac{5}{16} \times 1\frac{1}{8}$	0.3111	0.3130	0.3228	1.125	1.145	0.1562	0.3135	0.1622
610	$\frac{3}{16} \times 1\frac{1}{4}$	0.1863	0.1880	0.4483	1.250	1.273	0.0937	0.1885	0.0997
710	$\frac{7}{32} \times 1\frac{1}{4}$	0.2175	0.2193	0.4327	1.250	1.273	0.1093	0.2198	0.1153
810	$\frac{1}{4} \times 1\frac{1}{4}$	0.2487	0.2505	0.4170	1.250	1.273	0.1250	0.2510	0.1310
1010	$\frac{5}{16} \times 1\frac{1}{4}$	0.3111	0.3130	0.3858	1.250	1.273	0.1562	0.3135	0.1622
1210	$\frac{3}{8} \times 1\frac{1}{4}$	0.3735	0.3755	0.3545	1.250	1.273	0.1875	0.3760	0.1935
811	$\frac{1}{4} \times 1\frac{3}{8}$	0.2487	0.2505	0.4640	1.375	1.398	0.1250	0.2510	0.1310
1011	$\frac{5}{16} \times 1\frac{3}{8}$	0.3111	0.3130	0.4328	1.375	1.398	0.1562	0.3135	0.1622
1211	$\frac{3}{8} \times 1\frac{3}{8}$	0.3735	0.3755	0.4015	1.375	1.398	0.1875	0.3760	0.1935

*See footnote at end of table.
Note: All dimensions are given in inches.

TABLE 11.43 ANSI Standard Keyseat Dimensions for Woodruff Keys[48,52] (*Continued*)

Key no.	Nominal size key	Keyseat (Shaft) Width A^* Min	Max	Depth B +0.005 −0.000	Diameter F Min	Max	Key above shaft Height C +0.005 −0.005	Keyseat (Hub) Width D +0.002 −0.000	Depth E +0.005 −0.000
812	$\frac{1}{4} \times 1\frac{1}{2}$	0.2487	0.2505	0.5110	1.500	1.523	0.1250	0.2510	0.1310
1012	$\frac{5}{16} \times 1\frac{1}{2}$	0.3111	0.3130	0.4798	1.500	1.523	0.1562	0.3135	0.1622
1212	$\frac{3}{8} \times 1\frac{1}{2}$	0.3735	0.3755	0.4485	1.500	1.523	0.1875	0.3760	0.1935
617-1	$\frac{3}{16} \times 2\frac{1}{8}$	0.1863	0.1880	0.3073	2.125	2.160	0.0937	0.1885	0.0997
817-1	$\frac{1}{4} \times 2\frac{1}{8}$	0.2487	0.2505	0.2760	2.125	2.160	0.1250	0.2510	0.1310
1017-1	$\frac{5}{16} \times 2\frac{1}{8}$	0.3111	0.3130	0.2448	2.125	2.160	0.1562	0.3135	0.1622
1217-1	$\frac{3}{8} \times 2\frac{1}{8}$	0.3735	0.3755	0.2135	2.125	2.160	0.1875	0.3760	0.1935
617	$\frac{3}{16} \times 2\frac{1}{8}$	0.1863	0.1880	0.4323	2.125	2.160	0.0937	0.1885	0.0997
817	$\frac{1}{4} \times 2\frac{1}{8}$	0.2487	0.2505	0.4010	2.125	2.160	0.1250	0.2510	0.1310
1017	$\frac{5}{16} \times 2\frac{1}{8}$	0.3111	0.3130	0.3698	2.125	2.160	0.1562	0.3135	0.1622
1217	$\frac{3}{8} \times 2\frac{1}{8}$	0.3735	0.3755	0.3385	2.125	2.160	0.1875	0.3760	0.1935
822-1	$\frac{1}{4} \times 2\frac{3}{4}$	0.2487	0.2505	0.4640	2.750	2.785	0.1250	0.2510	0.1310
1022-1	$\frac{5}{16} \times 2\frac{3}{4}$	0.3111	0.3130	0.4328	2.750	2.785	0.1562	0.3135	0.1622
1222-1	$\frac{3}{8} \times 2\frac{3}{4}$	0.3735	0.3755	0.4015	2.750	2.785	0.1875	0.3760	0.1935
1422-1	$\frac{7}{16} \times 2\frac{3}{4}$	0.4360	0.4380	0.3703	2.750	2.785	0.2187	0.4385	0.2247
1622-1	$\frac{1}{2} \times 2\frac{3}{4}$	0.4985	0.5005	0.3390	2.750	2.785	0.2500	0.5010	0.2560
822	$\frac{1}{4} \times 2\frac{3}{4}$	0.2487	0.2505	0.6200	2.750	2.785	0.1250	0.2510	0.1310
1022	$\frac{5}{16} \times 2\frac{3}{4}$	0.3111	0.3130	0.5888	2.750	2.785	0.1562	0.3135	0.1622
1222	$\frac{3}{8} \times 2\frac{3}{4}$	0.3735	0.3755	0.5575	2.750	2.785	0.1875	0.3760	0.1935
1422	$\frac{7}{16} \times 2\frac{3}{4}$	0.4360	0.4380	0.5263	2.750	2.785	0.2187	0.4385	0.2247
1622	$\frac{1}{2} \times 2\frac{3}{4}$	0.4985	0.5005	0.4950	2.750	2.785	0.2500	0.5010	0.2560
1228	$\frac{3}{8} \times 3\frac{1}{2}$	0.3735	0.3755	0.7455	3.500	3.535	0.1875	0.3760	0.1935
1428	$\frac{7}{16} \times 3\frac{1}{2}$	0.4360	0.4380	0.7143	3.500	3.535	0.2187	0.4385	0.2247
1628	$\frac{1}{2} \times 3\frac{1}{2}$	0.4985	0.5005	0.6830	3.500	3.535	0.2500	0.5010	0.2560
1828	$\frac{9}{16} \times 3\frac{1}{2}$	0.5610	0.5630	0.6518	3.500	3.535	0.2812	0.5635	0.2872
2028	$\frac{5}{8} \times 3\frac{1}{2}$	0.6235	0.6255	0.6205	3.500	3.535	0.3125	0.6260	0.3185
2228	$\frac{11}{16} \times 3\frac{1}{2}$	0.6860	0.6880	0.5893	3.500	3.535	0.3437	0.6885	0.3497
2428	$\frac{3}{4} \times 3\frac{1}{2}$	0.7485	0.7505	0.5580	3.500	3.535	0.3750	0.7510	0.3810

*These width A values were set with the maximum keyseat (shaft) width as that figure which will receive a key with the greatest amount of looseness consistent with assuring the key's sticking in the keyseat (shaft). Minimum keyseat width is that figure permitting largest shaft distortion acceptable when assembling maximum key in minimum keyseat.

Dimensions A, B, C, D are taken at side intersection.

Note: All dimensions are given in inches.

11.7.3 Sheet-Metal Gage

The United States Standard and the Manufacturers Standard gage for sheet metal are listed in columns 6 and 7. The gage values listed in Table 11.46 are currently used and accepted. Where tolerances are important, however, it is recommended that the decimal equivalent in parentheses follow the gage number.

11.7.4 Twist-Drill Gage

Twist-drill gage[57] is given by decimal-size drill diameters ranging from 0.0135 to 0.2280, inclusive (Table 11.47).

TABLE 11.44 Dimensions of Spline Fittings for All Fits[53,54]

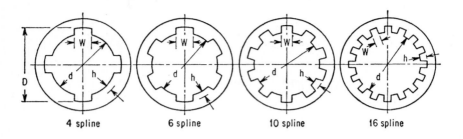

4 spline   6 spline   10 spline   16 spline

Nominal diam	4-spline		6-spline		10-spline		16-spline	
	D max[1]	W max[2]	D max[1]	W max[2]	D max[1]	W max[2]	D max[1]	W max[2]
¾	0.750	0.181	0.750	0.188	0.750	0.117		
⅞	0.875	0.211	0.875	0.219	0.875	0.137		
1	1.000	0.241	1.000	0.250	1.000	0.156		
1⅛	1.125	0.271	1.125	0.281	1.125	0.176		
1¼	1.250	0.301	1.250	0.313	1.250	0.195		
1⅜	1.375	0.331	1.375	0.344	1.375	0.215		
1½	1.500	0.361	1.500	0.375	1.500	0.234		
1⅝	1.625	0.391	1.625	0.406	1.625	0.254		
1¾	1.750	0.422	1.750	0.438	1.750	0.273		
2	2.000	0.482	2.000	0.500	2.000	0.312	2.000	0.196
2¼	2.250	0.542	2.250	0.563	2.250	0.351		
2½	2.500	0.602	2.500	0.625	2.500	0.390	2.500	0.245
3	3.000	0.723	3.000	0.750	3.000	0.468	3.000	0.294
3½	.....	.....	.....	.....	3.500	0.546	3.500	0.343
4	.....	.....	.....	.....	4.000	0.624	4.000	0.392
4½	.....	.....	.....	.....	4.500	0.702	4.500	0.441
5	.....	.....	.....	.....	5.000	0.780	5.000	0.490
5½	.....	.....	.....	.....	5.500	0.858	5.500	0.539
6	.....	.....	.....	.....	6.000	0.936	6.000	0.588

[1]Tolerance allowed of −0.001 in for shafts ¾ to 1¾ in, inclusive; of −0.002 in for shafts 2 to 3 in, inclusive; −0.003 in. for shafts 3½ to 6 in, inclusive, for 4-, 6-, and 10-spline fittings; tolerance of −0.003 in allowed for all sizes of 16-spline fittings.

[2]Tolerance allowed of −0.002 in for shafts ¾ to 1¾ in, inclusive; of −0.003 in for shafts 2 to 6 in, inclusive, for 4-, 6-, and 10-spline fittings; tolerance of −0.003 allowed for all sizes of 16-spline fittings.

TABLE 11.45 Basic Dimensions for Involute Splines

30° φ teeth*	Spline pitch P/Ps	Circular pitch P	Min effective space width (basic) S_v min		
			30° φ	37.5° φ	45° φ
	2.5/5	1.2566	0.6283	0.6683	—
	3/6	1.0472	0.5236	0.5569	—
	4/8	0.7854	0.3927	0.4177	—
	5/10	0.6283	0.3142	0.3342	—
	6/12	0.5236	0.2618	0.2785	—
	8/16	0.3927	0.1963	0.2088	—
	10/20	0.3142	0.1571	0.1671	0.1771
	12/24	0.2618	0.1309	0.1392	0.1476
	16/32	0.1963	0.0982	0.1044	0.1107
	20/40	0.1571	0.0785	0.0835	0.0885

TABLE 11.45 Basic Dimensions for Involute Splines (*Continued*)

30° φ teeth*	Spline pitch P/Ps	Circular pitch P	Min effective space width (basic) S_y min		
			30° φ	37.5° φ	45° φ
	24/48	0.1309	0.0654	0.0696	0.0738
	32/64	0.0982	0.0491	0.0522	0.0553
	40/80	0.0785	0.0393	0.0418	0.0443
	48/96	0.0654	0.0327	0.0348	0.0369
	64/128	0.0491	—	—	0.0277
	80/160	0.0393	—	—	0.0221
	128/256	0.0246	—	—	0.0138

*Shown to illustrate the relative tooth sizes for the various spline pitches.

11.8 STEEL PIPE AND FITTINGS

11.8.1 Pipe-Thread Types

1. Taper pipe threads for general use (NPT)
2. Internal straight threads in pipe couplings (NPSC)
3. Taper pipe threads for railing joints (NPTR)
4. Straight pipe threads for mechanical joints (NPSM, NPSL, NPSH)
5. Dryseal pressure-tight joints
6. Aeronautical

11.8.2 American Standard Pipe-Thread Form

The basic form known as the "American Standard taper pipe-thread form" is shown in Fig. 11.17. American Standard taper pipe threads are designated by specifying in sequence the nominal size, number of threads per inch, and the symbols for thread series and form, e.g., (1) ⅜-18 NPT LH, (2) ⅛-27 NPSC.

TABLE 11.46 Wire and Sheet-Metal Gages

No. of wire gage	Steel wire gage (U.S.)* (1)	American Wire or Brown & Sharpe gage (2)	Birmingham or Stub's iron wire gage (3)	Stub's steel wire gage (4)	No. of wire gage	Stub's steel wire gage	Music or piano wire gage (5)	U.S. Standard (6)	Manufacturers Standard for sheet steel nominal (7)
7/0	0.4900	......	......	.....	51	0.066	.....	0.500	
6/0	0.4615	0.5800	......	.....	52	0.063	0.004	0.469	
5/0	0.4305	0.5165	0.5000	.....	53	0.058	0.005	0.438	
4/0	0.3938	0.4600	0.4540	.....	54	0.055	0.006	0.406	
3/0	0.3625	0.4096	0.4250	.....	55	0.050	0.007	0.375	
2/0	0.3310	0.3648	0.3800	.....	56	0.045	0.008	0.344	
1/0	0.3065	0.3249	0.3400	.....	57	0.042	0.009	0.312	
1	0.2830	0.2893	0.3000	0.227	58	0.041	0.010	0.281	
2	0.2625	0.2576	0.2840	0.219	59	0.040	0.011	0.266	
3	0.2437	0.2294	0.2590	0.212	60	0.039	0.012	0.250	0.2391
4	0.2253	0.2043	0.2380	0.207	61	0.038	0.013	0.234	0.2242
5	0.2070	0.1819	0.2200	0.204	62	0.037	0.014	0.219	0.2092
6	0.1920	0.1620	0.2030	0.201	63	0.036	0.016	0.203	0.1943
7	0.1770	0.1443	0.1800	0.199	64	0.035	0.018	0.188	0.1793
8	0.1620	0.1285	0.1650	0.197	65	0.033	0.020	0.172	0.1644
9	0.1483	0.1144	0.1480	0.194	66	0.032	0.022	0.156	0.1495
10	0.1350	0.1019	0.1340	0.191	67	0.031	0.024	0.141	0.1345
11	0.1205	0.0907	0.1200	0.188	68	0.030	0.026	0.125	0.1196
12	0.1055	0.0808	0.1090	0.185	69	0.029	0.029	0.109	0.1046
13	0.0915	0.0720	0.0950	0.182	70	0.027	0.031	0.0938	0.0897
14	0.0800	0.0641	0.0830	0.180	71	0.026	0.033	0.0781	0.0747
15	0.0720	0.0571	0.0720	0.178	72	0.024	0.035	0.0703	0.0673
16	0.0625	0.0508	0.0650	0.175	73	0.023	0.037	0.0625	0.0598
17	0.0540	0.0453	0.0580	0.172	74	0.022	0.039	0.0562	0.0538
18	0.0475	0.0403	0.0490	0.168	75	0.020	0.041	0.0500	0.0478
19	0.0410	0.0359	0.0420	0.164	76	0.018	0.043	0.0438	0.0418
20	0.0348	0.0320	0.0350	0.161	77	0.016	0.045	0.0375	0.0359
21	0.0317	0.0285	0.0320	0.157	78	0.015	0.047	0.0344	0.0329
22	0.0286	0.0253	0.0280	0.155	79	0.014	0.049	0.0312	0.0299
23	0.0258	0.0226	0.0250	0.153	80	0.013	0.051	0.0281	0.0269
24	0.0230	0.0201	0.0220	0.151	...	.....	0.055	0.0250	0.0239
25	0.0204	0.0179	0.0200	0.148	...	.....	0.059	0.0219	0.0209
26	0.0181	0.0159	0.0180	0.146	...	.....	0.063	0.0188	0.0179
27	0.0173	0.0142	0.0160	0.143	...	.....	0.067	0.0172	0.0164
28	0.0162	0.0126	0.0140	0.139	...	.....	0.071	0.0156	0.0149
29	0.0150	0.0113	0.0130	0.134	...	.....	0.075	0.0141	0.0135
30	0.0140	0.0100	0.0120	0.127	...	.....	0.080	0.0125	0.0120
31	0.0132	0.00893	0.0100	0.120	...	.....	0.085	0.0109	0.0105
32	0.0128	0.00795	0.0090	0.115	...	.....	0.090	0.0102	0.0097
33	0.0118	0.00708	0.0080	0.112	...	.....	0.095	0.00938	0.0090
34	0.0104	0.00630	0.0070	0.110	...	.....	0.100	0.00859	0.0082
35	0.0095	0.00561	0.0050	0.108	...	.....	0.106	0.00781	0.0075
36	0.0090	0.00500	0.0040	0.106	...	.....	0.112	0.00703	0.0067
37	0.0085	0.00445	......	0.103	...	.....	0.118	0.00664	0.0064
38	0.0080	0.00396	......	0.101	...	.....	0.124	0.00625	0.0060
39	0.0075	0.00353	......	0.099	...	.....	0.130		
40	0.0070	0.00314	......	0.097	...	.....	0.138		
41	0.0066	0.00280	......	0.095	...	.....	0.146		
42	0.0062	0.00249	......	0.092	...	.....	0.154		
43	0.0060	0.00222	......	0.088	...	.....	0.162		
44	0.0058	0.00198	......	0.085	...	.....	0.170		
45	0.0055	0.00176	......	0.081	...	.....	0.180		
46	0.0052	0.00157	......	0.079					
47	0.0050	0.00140	......	0.077					
48	0.0048	0.00124	......	0.075					
49	0.0046	0.00111	......	0.072					
50	0.0044	0.00099	......	0.069					

*Also known as Washburn and Moen, American Steel and Wire Co., and Roebling wire gages. A greater selection of sizes is available and is specified by what are known as split gage numbers. They can be recognized by the $\frac{1}{4}$ and $\frac{1}{2}$ fractions which follow the gage number; i.e., $4\frac{1}{4}$, $4\frac{1}{2}$, $4\frac{3}{4}$. The decimal equivalents of split gage numbers are in the Steel Products Manual entitled "Wire and Rods, Carbon Steel" published by the American Iron and Steel Institute, New York, N.Y.

TABLE 11.47   Twist-Drill and Steel Wire Gage

No.	Size, in.	No.	Size, in.	No.	Size, in.	No.	Size, in.	No.	Size, in.	No.	Size, in.
1	0.2280	14	0.1820	27	0.1440	40	0.0980	53	0.0595	67	0.0320
2	0.2210	15	0.1800	28	0.1405	41	0.0960	54	0.0550	68	0.0310
3	0.2130	16	0.1770	29	0.1360	42	0.0935	55	0.0520	69	0.0292
4	0.2090	17	0.1730	30	0.1285	43	0.0890	56	0.0465	70	0.0280
5	0.2055	18	0.1695	31	0.1200	44	0.0860	57	0.0430	71	0.0260
6	0.2040	19	0.1660	32	0.1160	45	0.0820	58	0.0420	72	0.0250
7	0.2010	20	0.1610	33	0.1130	46	0.0810	59	0.0410	73	0.0240
8	0.1990	21	0.1590	34	0.1110	47	0.0785	60	0.0400	74	0.0225
9	0.1960	22	0.1570	35	0.1100	48	0.0760	61	0.0390	75	0.0210
10	0.1935	23	0.1540	36	0.1065	49	0.0730	62	0.0380	76	0.0200
11	0.1910	24	0.1520	37	0.1040	50	0.0700	63	0.0370	77	0.0180
12	0.1890	25	0.1495	38	0.1015	51	0.0670	64	0.0360	78	0.0160
13	0.1850	26	0.1470	39	0.0995	52	0.0635	65	0.0350	79	0.0145
								66	0.0330	80	0.0135

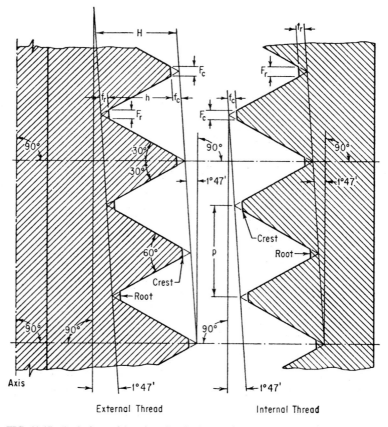

FIG. 11.17   Basic form of American Standard taper pipe thread. $H = 0.866025p$ = height of 60° sharp V thread; $h = 0.800000p$ = height of thread on product; $p = 1/n$ = pitch (measured parallel to axis); n = number of threads per inch; f_c = depth of truncation at crest; f_r = depth of truncation at root; F_c = width of flat at crest; F_r = width of flat at root.

11.8.3   Pipe-Thread Symbols

N = American (National) Standard

P = pipe

T = taper

C = coupling

S = straight

M = mechanical

L = locknut

H = hose coupling

R = rail fitting

Unless otherwise specified right-hand threads are obtained.

11.8.4   Taper Pipe-Thread Formula

The taper of the thread is 1:16 or 0.750 in/ft measured on the diameter along the axis. Basic taper pipe-thread dimensions, obtained by substituting into the following formulas, are given in Table 11.48.

$$L_2 = (0.80D + 6.8)(1/n) = (0.80D + 6.8)p$$

where L_2 = length of engagement

D = outside diameter of pipe

n = threads per inch

This formula determines directly the length of effective thread engagement, which includes two usable threads slightly imperfect.

The basic pitch diameters of the taper thread are determined by the following formulas:*

$$E_0 = D - (0.05D + 1.1)(1/n)$$

$$E_1 = E_0 + 0.0625L_1$$

where D = outside diameter of pipe

E_0 = pitch diameter of threaded end of pipe

E_1 = pitch diameter of thread at large end of internal thread

L_1 = normal engagement by hand between external and internal threads

n = threads per inch

It should be noted that, in special applications as in high-pressure work, longer thread engagement is used; thus E_1, the pitch diameter shown in Table 11.48, is maintained, and the pitch diameter E_0 at the end of the pipe is made smaller.

11.8.5   Pipe-Coupling Threads

The internal straight threads in pipe couplings[59] are straight or parallel threads of the same thread form as the American taper pipe thread. This thread is normally specified where a tight low-pressure joint is required with an American Standard external taper pipe thread.

---

*For the ⅛-27 and ¼-18 sizes, E_1 approx = $D - (0.05D + 0.827)p$.

TABLE 11.48   American Standard Taper Pipe Thread, NPT[58,62]

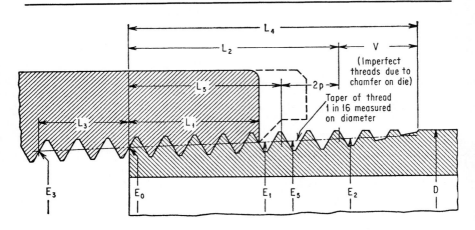

Nominal pipe size	OD of pipe D	Threads per in. n	Pitch of thread p	Pitch diam at beginning of external thread E_0	Handtight engagement			Effective thread, external		
					Length L_1		Diam E_1	Length L_2		Diam E_2
					In.	Threads		In.	Threads	
$\frac{1}{16}$	0.3125	27	0.03704	0.27118	0.160	4.32	0.28118	0.2611	7.05	0.28750
$\frac{1}{8}$	0.405	27	0.03704	0.36351	0.1615	4.36	0.37360	0.2639	7.12	0.38000
$\frac{1}{4}$	0.540	18	0.05556	0.47739	0.2278	4.10	0.49163	0.4018	7.23	0.50250
$\frac{3}{8}$	0.675	18	0.05556	0.61201	0.240	4.32	0.62701	0.4078	7.34	0.63750
$\frac{1}{2}$	0.840	14	0.07143	0.75843	0.320	4.48	0.77843	0.5337	7.47	0.79179
$\frac{3}{4}$	1.050	14	0.07143	0.96768	0.339	4.75	0.98887	0.5457	7.64	1.00179
1	1.315	$11\frac{1}{2}$	0.08696	1.21363	0.400	4.60	1.23863	0.6828	7.85	1.25630
$1\frac{1}{4}$	1.660	$11\frac{1}{2}$	0.08696	1.55713	0.420	4.83	1.58338	0.7068	8.13	1.60130
$1\frac{1}{2}$	1.900	$11\frac{1}{2}$	0.08696	1.79609	0.420	4.83	1.82234	0.7235	8.32	1.84130
2	2.375	$11\frac{1}{2}$	0.08696	2.26902	0.436	5.01	2.29627	0.7565	8.70	2.31630
$2\frac{1}{2}$	2.875	8	0.12500	2.71953	0.682	5.46	2.76216	1.1375	9.10	2.79062
3	3.500	8	0.12500	3.34062	0.766	6.13	3.38850	1.2000	9.60	3.41562
$3\frac{1}{2}$	4.000	8	0.12500	3.83750	0.821	6.57	3.88881	1.2500	10.00	3.91562
4	4.500	8	0.12500	4.33438	0.844	6.75	4.38712	1.3000	10.40	4.41562
5	5.563	8	0.12500	5.39073	0.937	7.50	5.44929	1.4063	11.25	5.47862
6	6.625	8	0.12500	6.44609	0.958	7.66	6.50597	1.5125	12.10	6.54062
8	8.625	8	0.12500	8.43359	1.063	8.50	8.50003	1.7125	13.70	8.54062
10	10.750	8	0.12500	10.54531	1.210	9.68	10.62094	1.9250	15.40	10.66562
12	12.750	8	0.12500	12.53281	1.360	10.88	12.61781	2.1250	17.00	12.66562
14 OD	14.000	8	0.12500	13.77500	1.562	12.50	13.87262	2.2500	18.90	13.91562
16 OD	16.000	8	0.12500	15.76250	1.812	14.50	15.87575	2.4500	19.60	15.91562
18 OD	18.000	8	0.12500	17.75000	2.000	16.00	17.87500	2.6500	21.20	17.91562
20 OD	20.000	8	0.12500	19.73750	2.125	17.00	19.87031	2.8500	22.80	19.91562
24 OD	24.000	8	0.12500	23.71250	2.375	19.00	23.86094	3.2500	26.00	23.91562

TABLE 11.48   American Standard Taper Pipe Thread, NPT[58,62] (*Continued*)

Nominal pipe size	Wrench makeup length for internal thread			Vanish thread V		Overall length external thread L_4	Nominal perfect external threads		Height of thread h	Increase in diam per thread 0.0625/n
	Length L_3		Diam E_3				Length L_5*	Diam E_5		
	In.	Threads		In.	Threads					
$\frac{1}{16}$	0.1111	3	0.26424	0.1285	3.47	0.3896	0.1870	0.28287	0.02963	0.00231
$\frac{1}{8}$	0.1111	3	0.35656	0.1285	3.47	0.3924	0.1898	0.37537	0.02963	0.00231
$\frac{1}{4}$	0.1667	3	0.46697	0.1928	3.47	0.5946	0.2907	0.49556	0.04444	0.00347
$\frac{3}{8}$	0.1667	3	0.60160	0.1928	3.47	0.6006	0.2967	0.63056	0.04444	0.00347
$\frac{1}{2}$	0.2143	3	0.74504	0.2478	3.47	0.7815	0.3909	0.78286	0.05714	0.00446
$\frac{3}{4}$	0.2143	3	0.95429	0.2478	3.47	0.7935	0.4029	0.99286	0.05714	0.00446
1	0.2609	3	1.19733	0.3017	3.47	0.9845	0.5089	1.24543	0.06957	0.00543
$1\frac{1}{4}$	0.2609	3	1.54083	0.3017	3.47	1.0085	0.5329	1.59043	0.06957	0.00543
$1\frac{1}{2}$	0.2609	3	1.77978	0.3017	3.47	1.0252	0.5496	1.83043	0.06957	0.00543
2	0.2609	3	2.25272	0.3107	3.47	1.0582	0.5826	2.30543	0.06957	0.00543
$2\frac{1}{2}$	0.2500	2	2.70391	0.4337	3.47	1.5712	0.8875	2.77500	0.100000	0.00781
3	0.2500	2	3.32500	0.4337	3.47	1.6337	0.9500	3.40000	0.100000	0.00781
$3\frac{1}{2}$	0.2500	2	3.82188	0.4337	3.47	1.6837	1.0000	3.90000	0.100000	0.00781
4	0.2500	2	4.31875	0.4337	3.47	1.7337	1.0500	4.40000	0.100000	0.00781
5	0.2500	2	5.37511	0.4337	3.47	1.8400	1.1563	5.46300	0.100000	0.00781
6	0.2500	2	6.43047	0.4337	3.47	1.9462	1.2625	6.52500	0.100000	0.00781
8	0.2500	2	8.41797	0.4337	3.47	2.1462	1.4625	8.52500	0.100000	0.00781
10	0.2500	2	10.52969	0.4337	3.47	2.3587	1.6750	10.65000	0.100000	0.00781
12	0.2500	2	12.51719	0.4337	3.47	2.5587	1.8750	12.65000	0.100000	0.00781
14 OD	0.2500	2	13.75938	0.4337	3.47	2.6837	2.0000	13.90000	0.100000	0.00781
16 OD	0.2500	2	15.74688	0.4337	3.47	2.8837	2.2000	15.90000	0.100000	0.00781
18 OD	0.2500	2	17.73438	0.4337	3.47	3.0837	2.4000	17.90000	0.100000	0.00781
20 OD	0.2500	2	19.72188	0.4337	3.47	3.2837	2.6000	19.90000	0.100000	0.00781
24 OD	0.2500	2	23.69688	0.4337	3.47	3.6837	3.0000	23.90000	0.100000	0.00781

*The length L_5 from the end of the pipe determines the plane beyond which the thread form is imperfect at the crest. The next two threads are perfect at the root. At this plane the cone formed by the crests of the thread intersects the cylinder forming the external surface of the pipe.

11.8.6   Railing Joints

These require a more rigid joint obtained by using a shortened external thread, permitting the use of the larger end of the pipe thread. The form of thread is the same as the form of the American Standard taper pipe thread (Fig. 11.17). Straight pipe threads[59] are based on the pitch diameter of the American Standard pipe thread at the gaging notch, E_1 of Table 11.49.

11.8.7   Dryseal Pipe Thread[60]

Dryseal threads are used for both external and internal threads where a pressure-tight joint is required and the use of a sealer is objectionable. The external thread is tapered while the internal threads may be tapered or straight. The crest and root truncations are modified American Standard pipe threads, in most cases, to produce a pressure-tight joint.

TABLE 11.49   Dimensions of Welded and Seamless Steel Pipe[63,64]

Nom diam, in.	Schedule	OD, in.	Wall thickness, in.	ID, in.	Internal area, in.^{2}
$\frac{1}{8}$	40 (S)	0.405	0.068	0.269	0.0568
	80 (XS)		0.095	0.215	0.363
$\frac{1}{4}$	40 (S)	0.540	0.088	0.364	0.1041
	80 (XS)		0.119	0.302	0.0716
$\frac{3}{8}$	40 (S)	0.675	0.091	0.493	0.1909
	80 (XS)		0.126	0.423	0.1405
$\frac{1}{2}$	40 (S)	0.840	0.109	0.622	0.3039
	80 (XS)		0.147	0.546	0.2341
	160		0.187	0.466	0.1706
	(XXS)		0.294	0.252	0.0499
$\frac{3}{4}$	40 (S)	1.050	0.113	0.824	0.5333
	80 (XS)		0.154	0.742	0.4324
	160		0.219	0.612	0.2942
	(XXS)		0.308	0.434	0.1479
1	40 (S)	1.315	0.133	1.049	0.8643
	80 (XS)		0.179	0.957	0.7193
	160		0.250	0.815	0.5217
	(XXS)		0.358	0.599	0.2818
$1\frac{1}{4}$	40 (S)	1.660	0.140	1.380	1.496
	80 (XS)		0.191	1.278	1.283
	160		0.250	1.160	1.057
	(XXS)		0.382	0.896	0.6305
$1\frac{1}{2}$	40 (S)	1.900	0.145	1.610	2.036
	80 (XS)		0.200	1.500	1.767
	160		0.281	1.338	1.406
	(XXS)		0.400	1.100	0.9503
2	40 (S)	2.375	0.154	2.067	3.356
	80 (XS)		0.218	1.939	2.953
	160		0.344	1.687	2.235
	(XXS)		0.436	1.503	1.774
$2\frac{1}{2}$	40 (S)	2.875	0.203	2.469	4.788
	80 (XS)		0.276	2.323	4.233
	160		0.375	2.125	3.547
	(XXS)		0.552	1.771	2.464
3	40 (S)	3.500	0.216	3.068	7.393
	80 (XS)		0.300	2.900	6.605
	160		0.438	2.624	5.408
	(XXS)		0.600	2.300	4.155
$3\frac{1}{2}$	40 (S)	4.000	0.226	3.548	9.887
	80 (XS)		0.318	3.364	8.888
4	40 (S)	4.500	0.237	4.026	12.73
	80 (XS)		0.337	3.826	11.50
	120		0.438	3.624	10.32
	160		0.531	3.438	9.283
	(XXS)		0.674	3.152	7.803
5	40 (S)	5.563	0.258	5.047	20.02
	80 (XS)		0.375	4.813	18.19
	120		0.500	4.563	16.35
	160		0.625	4.313	14.61
	(XXS)		0.750	4.063	12.97
6	40 (S)	6.625	0.280	6.065	28.89
	80 (XS)		0.432	5.761	26.07
	120		0.562	5.501	23.77
	160 (XXS)		0.719	5.187	21.13
			0.864	4.897	18.83

TABLE 11.49 Dimensions of Welded and Seamless Steel Pipe[63,64] (*Continued*)

Nom diam, in.	Schedule	OD, in.	Wall thickness, in.	ID, in.	Internal area, in.^{2}
8	20	8.625	0.250	8.125	51.85
	30		0.277	8.071	51.16
	40 (S)		0.322	7.981	50.03
	60		0.406	7.813	47.94
	80 (XS)		0.500	7.625	45.66
	100		0.594	7.437	43.44
	120		0.719	7.187	40.57
	140		0.812	7.001	38.50
	(XXS)		0.875	6.875	37.12
	160		0.906	6.813	36.46
10	20	10.75	0.250	10.250	82.52
	30		0.307	10.136	80.69
	40 (S)		0.365	10.020	78.85
	60 (XS)		0.500	9.750	74.66
	80		0.594	9.562	71.81
	100		0.719	9.312	68.11
	120		0.844	9.062	64.50
	140 (XXS)		1.000	8.750	60.13
	160		1.125	8.500	56.75
12	20	12.75	0.250	12.250	117.86
	30		0.330	12.090	114.80
	(S)		0.375	12.000	113.10
	40		0.406	11.938	111.93
	(XS)		0.500	11.750	108.43
	60		0.562	11.626	106.16
	80		0.688	11.374	101.61
	100		0.844	11.062	96.11
	120 (XXS)		1.000	10.750	90.76
	140		1.125	10.500	86.59
	160		1.312	10.126	80.53
14 OD	10	14.00	0.250	13.500	143.14
	20		0.312	13.376	140.52
	30 (S)		0.375	13.250	137.89
	40		0.438	13.124	135.28
	(XS)		0.500	13.000	132.67
	60		0.594	12.812	128.92
	80		0.750	12.500	122.72
	100		0.938	12.124	115.45
	120		1.094	11.812	109.58
	140		1.250	11.500	103.87
	160		1.406	11.188	98.31
16 OD	10	16.00	0.250	15.500	188.69
	20		0.312	15.376	185.69
	30 (S)		0.375	15.250	182.65
	40 (XS)		0.500	15.000	176.72
	60		0.656	14.688	169.44
	80		0.844	14.312	160.88
	100		1.031	13.938	152.58
	120		1.219	13.562	144.46
	140		1.438	13.124	135.28
	160		1.594	12.812	128.92
18 OD	10	18.00	0.250	17.500	240.53
	20		0.312	17.376	237.13
	(S)		0.375	17.250	233.71
	30		0.438	17.124	230.00

TABLE 11.49   Dimensions of Welded and Seamless Steel Pipe[63,64] (*Continued*)

Nom diam, in.	Schedule	OD, in.	Wall thickness, in.	ID, in.	Internal area, in.[2]
18 OD	(XS)		0.500	17.000	226.98
	40		0.562	16.876	223.68
	60		0.750	16.500	213.83
	80		0.938	16.124	204.19
	100		1.156	15.688	193.30
	120		1.375	15.250	182.65
	140		1.562	14.876	173.81
	160		1.781	14.438	163.72
20 OD	10	20.00	0.250	19.500	298.65
	20 (S)		0.375	19.250	291.04
	30 (XS)		0.500	19.000	283.53
	40		0.594	18.812	277.95
	60		0.812	18.376	265.21
	80		1.031	17.938	252.72
	100		1.281	17.438	238.83
	120		1.500	17.000	226.98
	140		1.750	16.500	213.83
	160		1.969	16.062	202.62
24 OD	10	24.00	0.250	23.500	433.74
	20 (S)		0.375	23.250	424.56
	(XS)		0.500	23.000	415.48
	30		0.562	22.876	411.01
	40		0.688	22.624	402.00
	60		0.969	22.062	382.28
	80		1.219	21.562	365.15
	100		1.531	20.938	344.32
	120		1.812	20.376	326.92
	140		2.062	19.876	310.28
	160		2.344	19.312	292.92
30 OD	10	30.00	0.312	29.376	677.76
	(S)		0.375	29.250	671.62
	20 (XS)		0.500	29.000	660.52
	30		0.625	28.750	649.18

Note:
   S = wall thickness formerly designated "standard weight."
   XS = wall thickness formerly designated "extra strong."
   XXS = wall thickness formerly designated "double extra strong."

11.8.8 Aeronautical Taper Pipe Thread[62]

As given in Military Specifications MILP-7105, these are identical in all respects with American Standard taper pipe threads except for E_3 and L_3 dimensions in the $2\frac{1}{2}$ and 3 sizes. The importance of the Aeronautical pipe-thread specifications lies in the gaging system, which provides a closer dimensional control than gages used for American Standard taper pipe threads or American Standard Dryseal pipe threads.

11.8.9 Pipe Designation

Commercial sizes of wrought iron and steel pipe[61] are known by their nominal inside diameter from $\frac{1}{8}$ to 12 in. Above 12-in diameter, the outside diameter is referred to. All classes of pipe of a given nominal size have the same outside diameter; the extra thickness is reflected on the inside diameters.

The former designation, standard, extra strong, and double extra weight pipe, has been replaced by a simpler and more versatile system. The new standard uses pipe-thickness schedules.[61] It gives a larger selection of wall thicknesses with the flexibility of adding new wall thicknesses (Table 11.49).

11.8.10 Pipe Wall Thickness[62]

Minimum wall thickness is determined by the following formula:

$$t_m = PD/2S + C \qquad \text{or} \qquad (D/2)[1 - (S - P)(S + P)] + C$$

where t_m = minimum pipe wall thickness, in*
P = maximum internal pressure, lb/in^2 gage
D = outside diameter of pipe, in
S = allowable stress, lb/in^2
C = thickness allowance for corrosion, threading
$$C = \begin{cases} 0.15 & \text{for cast iron} \\ 0.05 & \text{for threaded steel or wrought iron } \frac{3}{8} \text{ in and smaller} \\ 0.8/n & \text{for threaded steel or wrought iron } \frac{1}{2} \text{ in or larger} \end{cases}$$

11.8.11 Pipe-Fitting Dimensions

For dimensions of fittings refer to Tables 11.50 to 11.55.

---

*For pipe sizes 4 in or greater, $t_m = 10$ percent $\times$ ID.

TABLE 11.50 General Dimensions of Class 125 Flanged Fittings Straight Sizes[62,64,65]

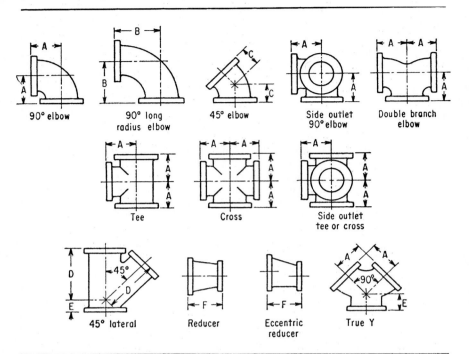

Nominal pipe size	ID of fittings	Center to face 90° elbow, tees, crosses, true Y, and double-branch elbow A	Center to face 90° long radius elbow B	Center to face 45° elbow C	Center to face lateral D	Short center to face true Y and lateral E	Face to face reducer F	Diam of flange	Thickness of flange, min	Wall thickness
1	1	3½	5	1¾	5¾	1¾	....	4¼	7/16	5/16
1¼	1¼	3¾	5½	2	6¼	1¾	....	4⅝	½	5/16
1½	1½	4	6	2¼	7	2	....	5	9/16	5/16
2	2	4½	6½	2½	8	2½	5	6	⅝	5/16
2½	2½	5	7	3	9½	2½	5½	7	11/16	5/16
3	3	5½	7¾	3	10	3	6	7½	¾	⅜
3½	3½	6	8½	3½	11½	3	6½	8½	13/16	7/16
4	4	6½	9	4	12	3	7	9	15/16	½
5	5	7½	10¼	4½	13½	3½	8	10	15/16	½
6	6	8	11½	5	14½	3½	9	11	1	9/16
8	8	9	14	5½	17½	4½	11	13½	1⅛	⅝
10	10	11	16½	6½	20½	5	12	16	1 3/16	¾
12	12	12	19	7½	24½	5½	14	19	1¼	13/16
14	14	14	21½	7½	27	6	16	21	1⅜	⅞
16	16	15	24	8	30	6½	18	23½	1 7/16	1
18	18	16½	26½	8½	32	7	19	25	1 9/16	1 1/16
20	20	18	29	9½	35	8	20	27½	1 11/16	1⅛
24	24	22	34	11	40½	9	24	32	1⅞	1¼
30	30	25	41½	15	49	10	30	38¾	2⅛	1 7/16
36	36	28	49	18	....	....	36	46	2⅜	1⅝
42	42	31	56½	21	....	....	42	53	2⅝	1 13/16
48	48	34	64	24	....	....	48	59½	2¾	2

TABLE 11.51 Center to Contact Surface of American Standard Steel Flanged Fittings[62,64,65]

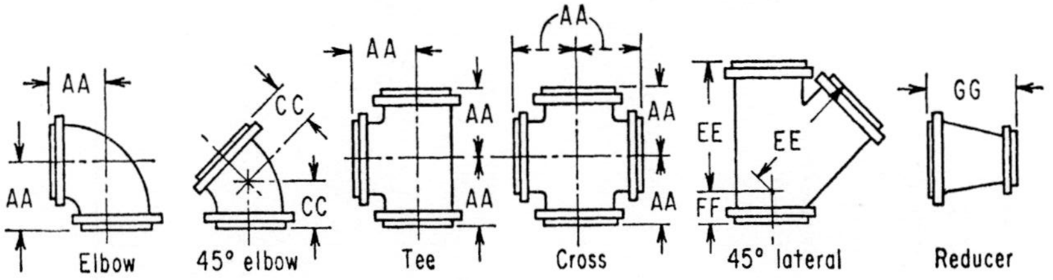
AA
AA
Elbow
CC
CC
45° elbow
AA
AA
AA
Tee
AA
AA
AA
Cross
EE
EE
FF
45° lateral
GG
Reducer

Nom pipe size	400-lb standard					600-lb standard					900-lb standard					1,500-lb standard				
	AA	*CC*	*EE*	*FF*	*GG*	*AA*	*CC*	*EE*	*FF*	*GG*	*AA*	*CC*	*EE*	*FF*	*GG*	*AA*	*CC*	*EE*	*FF*	*GG*
½	For sizes below 4 in., use the dimensions of 600-lb fittings					3¼	2	5¾	1¾	5	For sizes below 3 in., use the dimensions of 1,500-lb fittings					4¼	3			
¾						3¾	2¼	6¾	2	5						4½	3¼			
1						4¼	2¼	7¼	2¼	5						5	3½	9	2½	5
1¼	....	....	....	....	....	4½	2¾	8	2½	5	....	....	....	....	....	5½	4	10	3	5¾
1½	....	....	....	....	....	4¾	3	9	2¾	5	....	....	....	....	....	6	4¼	11	3½	6¼
2	....	....	....	....	....	5¾	4¼	10¼	3½	6	....	....	....	....	....	7¼	4¾	13¼	4	7¼
2½	....	....	....	....	....	6½	4½	11½	3½	6¾	....	....	....	....	....	8¼	5¼	15¼	4½	8¼
3	....	....	....	....	....	7	5	12¾	4	7¼	7½	5½	14½	4½	7¾	9¼	5¾	17¼	5	9¼
3½	....	....	....	....	....	7½	5½	14	4½	7¾										
4	8	5½	16	4½	8¼	8½	6	16½	4½	8¾	9	6½	17½	5½	9¼	10¾	7¼	19¼	6	10¾
5	9	6	16¾	5	9¼	10	7	19½	6	10¼	11	7½	21	6½	11¼	13¼	8¾	23¼	7½	13¾
6	9¾	6¼	18¾	5¼	10	11	7½	21	6½	11¼	12	8	22½	6½	12¼	13⅞	9⅜	24⅞	8⅛	14½
8	11¾	6¾	22¼	5¾	12	13	8½	24½	7	13¼	14½	9	27½	7½	14¾	16⅜	10⅞	29⅞	9⅛	17
10	13¼	7¾	25¾	6¼	13¼	15½	9½	29½	8	15¾	16½	10	31½	8½	16¾	19	12	36	10¼	20¼
12	15	8¾	29¾	6½	15¼	16½	10	31½	8½	16¾	19	11	34½	9	17¾	22¼	13¼	40¾	12	23
14 OD	16¼	9¼	32¾	7	16½	17½	10¾	34½	9	17¾	20¼	11½	36½	9½	19	24¾	14¼	44	12½	25¾
16 OD	17¾	10¼	36¼	8	18½	19½	11¾	38½	10	19¾	22¼	12½	40¾	10½	21	27¼	16¼	48¼	14¾	28¼
18 OD	19¼	10¾	39¼	8½	19½	21½	12¼	42	10½	21¾	24	13¼	45½	12	24½	30¼	17¾	53¼	16½	31½
20 OD	20¾	11¼	42¾	9	21	23½	13	45½	11	23¾	26	14½	50¼	13	26½	32¾	18¾	57¾	17¾	34
24 OD	24½	12¾	50¼	10½	24½	27½	14¾	53	13	27¾	30½	18	60	15½	30½	38¼	20¾	67¼	20½	39¾

Note: All dimensions in inches.

TABLE 11.52   Dimensions of American 150-lb Standard Malleable-Iron Screwed Fittings (Straight Sizes)[64]

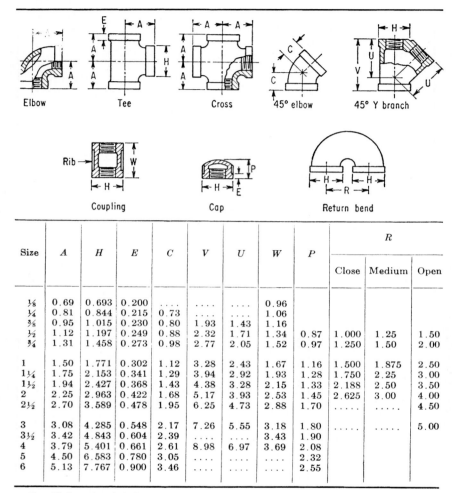

Elbow   Tee   Cross   45° elbow   45° Y branch

Rib → Coupling   Cap   Return bend

Size	A	H	E	C	V	U	W	P	R		
									Close	Medium	Open
⅛	0.69	0.693	0.200	....	....	....	0.96				
¼	0.81	0.844	0.215	0.73	....	....	1.06				
⅜	0.95	1.015	0.230	0.80	1.93	1.43	1.16				
½	1.12	1.197	0.249	0.88	2.32	1.71	1.34	0.87	1.000	1.25	1.50
¾	1.31	1.458	0.273	0.98	2.77	2.05	1.52	0.97	1.250	1.50	2.00
1	1.50	1.771	0.302	1.12	3.28	2.43	1.67	1.16	1.500	1.875	2.50
1¼	1.75	2.153	0.341	1.29	3.94	2.92	1.93	1.28	1.750	2.25	3.00
1½	1.94	2.427	0.368	1.43	4.38	3.28	2.15	1.33	2.188	2.50	3.50
2	2.25	2.963	0.422	1.68	5.17	3.93	2.53	1.45	2.625	3.00	4.00
2½	2.70	3.589	0.478	1.95	6.25	4.73	2.88	1.70	....	....	4.50
3	3.08	4.285	0.548	2.17	7.26	5.55	3.18	1.80	....	....	5.00
3½	3.42	4.843	0.604	2.39	....	....	3.43	1.90			
4	3.79	5.401	0.661	2.61	8.98	6.97	3.69	2.08			
5	4.50	6.583	0.780	3.05	....	....	....	2.32			
6	5.13	7.767	0.900	3.46	....	....	....	2.55			

Note: All dimensions in inches.

TABLE 11.53 Dimensions of American 125- and 250-lb Standard Cast-Iron Screwed Fittings (Straight Sizes)*[64]

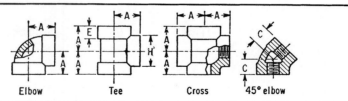

Elbow   Tee   Cross   45° elbow

Size	125 lb				250 lb			
	A	*H*	*E*	*C*	*A*	*H*	*E*	*C*
¼	0.81	0.93	0.38	0.73	0.94	1.17	0.49	0.81
⅜	0.95	1.12	0.44	0.80	1.06	1.36	0.55	0.88
½	1.12	1.34	0.50	0.88	1.25	1.59	0.60	1.00
¾	1.13	1.63	0.56	0.98	1.44	1.88	0.68	1.13
1	1.50	1.95	0.62	1.12	1.63	2.24	0.76	1.31
1¼	1.75	2.39	0.69	1.29	1.94	2.73	0.88	1.50
1½	1.94	2.68	0.75	1.43	2.13	3.07	0.97	1.69
2	2.25	3.28	0.84	1.68	2.50	3.74	1.12	2.00
2½	2.70	3.86	0.94	1.95	2.94	4.60	1.30	2.25
3	3.08	4.62	1.00	2.17	3.38	5.36	1.40	2.50
3½	3.42	5.20	1.06	2.39	3.75	5.98	1.49	2.63
4	3.79	5.79	1.12	2.61	4.13	6.61	1.57	2.81
5	4.50	7.05	1.18	3.05	4.88	7.92	1.74	3.19
6	5.13	8.28	1.28	3.46	5.63	9.24	1.91	3.50
8	6.56	10.63	1.47	4.28	7.00	11.73	2.24	4.31
10	8.08	13.12	1.68	5.16	8.63	14.37	2.58	5.19
12	9.50	15.47	1.88	5.97	10.00	16.84	2.91	6.00

*This applies to elbows and tees only.
Note: All dimensions in inches.

TABLE 11.54 Dimensions of American Standard Cast-Iron Screwed Drainage Fittings*[64]

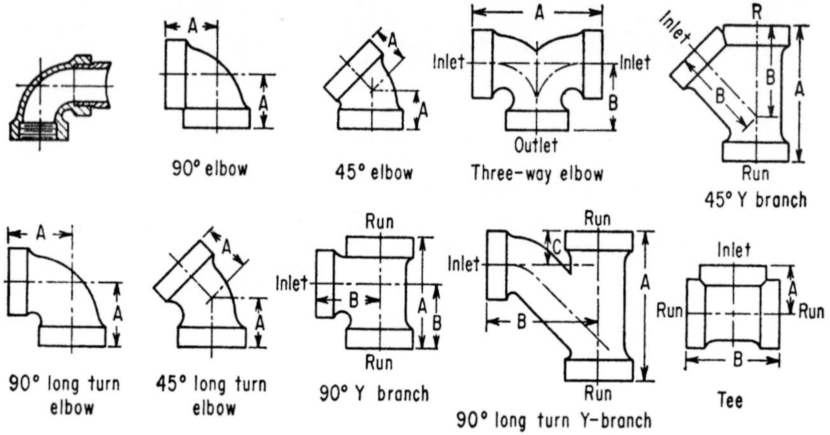
A
90° elbow
A
45° elbow
A
Inlet
Inlet
B
Outlet
Three-way elbow
R
Inlet
B
B
A
Run
45° Y branch
A
90° long turn elbow
A
45° long turn elbow
Run
Inlet
B
A
B
Run
90° Y branch
Run
Inlet
C
A
B
Run
90° long turn Y-branch
Inlet
Run
A
Run
B
Tee

Size, in.	90° elbows	45° elbows	90° long-turn elbows	45° long-turn elbows	Three-way elbows		Tees		90° Y branches		90° long-turn Y branches			45° Y branches	
	A	*A*	*A*	*A*	*A*	*B*	*A*	*B*	*A*	*B*	*A*	*B*	*C*	*A*	*B*
1 1/4	1 3/4	1 3/8	2 1/4	1 3/4	4 1/2	2 1/4	1 3/4	3 1/2	3 3/4	2 1/4	4 3/4	3 5/8	1 3/16	5	3 1/4
1 1/2	2 3/16	1 7/16	2 1/2	1 7/8	5 1/4	2 5/8	2 3/16	4 3/8	4 1/4	2 1/2	5 3/8	4 1/8	1 1/4	5 1/2	3 5/8
2	2 5/8	1 3/4	3 1/16	2 1/4	6 1/4	3 1/8	2 5/16	4 5/8	5 3/16	3 1/16	7 1/16	5 7/16	1 5/8	6 7/16	4 5/16
2 1/2	2 13/16	2 1/16	3 11/16	2 5/8	7 3/8	3 11/16	2 13/16	5 5/8	6 5/16	3 11/16	8 1/4	6 1/4	2	7 7/8	5 3/8
3	3 3/16	2 3/8	4 1/4	2 7/8	8 5/8	4 5/16	3 3/16	6 3/8	7 1/4	4 1/4	9 13/16	7 1/2	2 5/16	9	6 3/16
4	3 11/16	2 3/4	5 3/16	3 7/16	10 3/8	5 3/16	4	8	8 3/4	5 3/16	13 3/4	9 7/8	2 7/8	10 7/8	7 11/16
5	4 1/2	3 3/16	6 1/8	4 1/8	12 1/4	6 1/8	4 5/8	9 1/4	10 5/16	6 1/8	15 3/4	12 1/4	3 1/2	12 15/16	9 3/16
6	5 3/16	3 1/2	7 1/8	4 7/8	14 1/4	7 1/8	5 3/16	11 3/8	11 15/16	7 1/8	18 11/16	14 9/16	4 1/8	14 3/4	10 3/4
8	6 1/2	4 3/16	9	6	....	....	6 1/2	13	15 1/16	9	24 9/16	19 5/16	5 1/4	18 13/16	13 9/16

*Crane Co.
Note: All dimensions in inches.

TABLE 11.55 Soldered-Joint Fittings—Dimensions of Elbows, Tees, and Crosses[64]

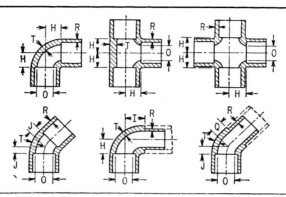

Nominal size	Cast brass							Wrought metal (T and R)
	H	I	J	Q	O	T	R	
$\frac{1}{4}$	$\frac{1}{4}$	$\frac{3}{8}$	$\frac{3}{16}$	$\frac{1}{4}$	0.31	0.08	0.048	0.030
$\frac{3}{8}$	$\frac{5}{16}$	$\frac{7}{16}$	$\frac{3}{16}$	$\frac{5}{16}$	0.43	0.08	0.048	0.035
$\frac{1}{2}$	$\frac{7}{16}$	$\frac{9}{16}$	$\frac{3}{16}$	$\frac{9}{16}$	0.54	0.09	0.054	0.040
$\frac{3}{4}$	$\frac{9}{16}$	$1\frac{1}{16}$	$\frac{1}{4}$	$\frac{3}{8}$	0.78	0.10	0.060	0.045
1	$\frac{3}{4}$	$\frac{7}{8}$	$\frac{5}{16}$	$\frac{7}{16}$	1.02	0.11	0.066	0.050
$1\frac{1}{4}$	$\frac{7}{8}$	1	$\frac{7}{16}$	$\frac{9}{16}$	1.26	0.12	0.072	0.055
$1\frac{1}{2}$	1	$1\frac{1}{8}$	$\frac{1}{2}$	$\frac{5}{8}$	1.50	0.13	0.078	0.060
2	$1\frac{1}{4}$	$1\frac{3}{8}$	$\frac{9}{16}$	$\frac{3}{4}$	1.98	0.15	0.090	0.070
$2\frac{1}{2}$	$1\frac{1}{2}$	$1\frac{5}{8}$	$\frac{5}{8}$	$\frac{7}{8}$	2.46	0.17	0.102	0.080
3	$1\frac{3}{4}$	$1\frac{7}{8}$	$\frac{3}{4}$	1	2.94	0.19	0.114	0.090
$3\frac{1}{2}$	2	$2\frac{1}{8}$	$\frac{7}{8}$	$1\frac{1}{8}$	3.42	0.20	0.120	0.100
4	$2\frac{1}{4}$	$2\frac{3}{8}$	$1\frac{5}{16}$	$1\frac{1}{4}$	3.90	0.22	0.132	0.110
5	$3\frac{1}{8}$	....	$1\frac{7}{16}$	....	4.87	0.28	0.168	0.125
6	$3\frac{5}{8}$	....	$1\frac{5}{8}$	....	5.84	0.34	0.204	0.140

11.9 STRUCTURAL SECTIONS[66,67]

The most economical form of beam cross section for vertical loading only is a beam of I section. Beams designed to resist bending should have the greatest depth of material in the direction of force applied, as indicated by the formula for section modulus.

This section includes pertinent data of standard sections useful to the designer as well as the detailer of fabricated steel; see Tables 11.56 to 11.62.

TABLE 11.56 Wide-Flange Beams and Columns

Nominal size, in.	Weight /ft, lb	Area of section, in.²	Depth of section, in.	Flange Width, in.	Flange Thickness, in.	Web thickness, in.	Neutral axis perpendicular to web at center I, in.⁴	S, in.³	r, in.	Neutral axis parallel to web at center I, in.⁴	S, in.³	r, in.
36 × 16½	300	88.17	36.72	16.65	1.680	0.945	20,290	1,105	15.17	1,225	147.1	3.73
	280	82.32	36.50	16.59	1.570	0.885	18,819	1,031	15.12	1,127	135.9	3.70
	260	76.56	36.24	16.55	1.440	0.845	17,233	951	15.00	1,020	123.3	3.65
	245	72.03	36.06	16.51	1.350	0.802	16,092	892	14.95	944	114.4	3.62
	230	67.73	35.88	16.47	1.260	0.765	14,988	835	14.88	870	105.7	3.59
36 × 12	194	57.11	36.48	12.11	1.260	0.770	12,103	663	14.56	355	58.7	2.49
	182	53.54	36.32	12.07	1.180	0.725	11,281	621	14.52	327	54.3	2.47
	170	49.98	36.16	12.02	1.100	0.680	10,470	579	14.47	300	50.0	2.45
	160	47.09	36.00	12.00	1.020	0.653	9,738	541	14.38	275	45.9	2.42
	150	44.16	35.84	11.97	0.940	0.625	9,012	502	14.29	250	41.8	2.38
33 × 15¾	240	70.52	33.50	15.86	1.400	0.830	13,585	811	13.88	874	110.2	3.52
	220	64.73	33.25	15.81	1.275	0.775	12,312	740	13.79	782	99.0	3.48
	200	58.79	33.00	15.75	1.150	0.715	11,048	669	13.71	691	87.8	3.43
33 × 11½	152	44.71	33.50	11.56	1.055	0.635	8,147	486	13.50	256	44.3	2.39
	141	41.51	33.31	11.53	0.960	0.605	7,442	446	13.39	229	39.8	2.35
	130	38.26	33.10	11.51	0.855	0.580	6,699	404	13.23	201	35.0	2.29
30 × 15	210	61.78	30.38	15.10	1.315	0.775	9,872	649	12.64	707	93.7	3.38
	190	55.90	30.12	15.04	1.185	0.710	8,825	586	12.57	624	83.1	3.34
	172	50.65	29.88	14.98	1.065	0.655	7,891	528	12.48	550	73.4	3.30
30 × 10½	132	38.83	30.30	10.55	1.000	0.615	5,753	379	12.17	185	35.1	2.18
	124	36.45	30.16	10.52	0.930	0.585	5,347	354	12.11	169	32.3	2.16
	116	34.13	30.00	10.50	0.850	0.564	4,919	327	12.00	153	29.2	2.12
	108	31.77	29.82	10.48	0.760	0.548	4,461	299	11.85	135	25.8	2.06
27 × 14	177	52.10	27.31	14.09	1.190	0.725	6,728	492.8	11.36	518.9	73.7	3.16
	160	47.04	27.08	14.02	1.075	0.658	6,018	444.0	11.31	458.0	65.3	3.12
	145	42.68	26.88	13.96	0.975	0.600	5,414	402.9	11.26	406.9	58.3	3.09
27 × 10	114	33 53	27.28	10.07	0.932	0.570	4,080	299.2	11.03	149.6	29.7	2.11
	102	30.01	27.07	10.02	0.827	0.518	3,604	266.3	10.96	129.5	25.9	2 08
	94	27.65	26.91	9.99	0.747	0.490	3,266	242.8	10.87	115.1	23.0	2.04
24 × 14	160	47.04	24.72	14.09	1.135	0.656	5,110	413.5	10.42	492.6	69.9	3.23
	145	42.62	24.49	14.04	1.020	0.608	4,561	372.5	10.34	434.3	61.8	3.19
	130	38.21	24.25	14.00	0.900	0.565	4,009	330.7	10.24	375.2	53.6	3.13
24 × 12	120	35.29	24.31	12.08	0.930	0.556	3,635	299.1	10.15	254.0	42.0	2.68
	110	32.36	24.16	12.04	0.855	0.510	3,315	274.4	10.12	229.1	38.0	2.66
	100	29.43	24.00	12.00	0.775	0.468	2,987	248.9	10.08	203.5	33.9	2.63
24 × 9	94	27.63	24.29	9.06	0.872	0.516	2,683	220.9	9.85	102.2	22.6	1.92
	84	24.71	24.09	9.02	0.772	0.470	2,364	196.3	9.78	88.3	19.6	1.89
	76	22.37	23.91	8.98	0.682	0.440	2,096	175.4	9.68	76.5	17.0	1.85
21 × 13	142	41.76	21.46	13.13	1.095	0.659	3,403	317.2	9.03	385.9	58.8	3.04
	127	37.34	21.24	13.06	0.985	0.588	3,017	284.1	8.99	338.6	51.8	3.01
	112	32.93	21.00	13.00	0.865	0.527	2,620	249.6	8.92	289.7	44.6	2.96
21 × 9	96	28.21	21.14	9.03	0.935	0.575	2,088	197.6	8.60	109.3	24.2	1.97
	82	24.10	20.86	8.96	0.795	0.499	1,752	168.0	8.53	89.6	20.0	1.93
21 × 8¼	73	21.46	21.24	8.29	0.740	0.455	1,600	150.7	8.64	66.2	16.0	1.76
	68	20.02	21.13	8.27	0.685	0.430	1,478	139.9	8.59	60.4	14.6	1.74
	62	18.23	20.99	8.24	0.615	0.400	1,326	126.4	8.53	53.1	12.9	1.71
18 × 11¾	114	33.51	18.48	11.83	0.991	0.595	2,033	220.1	7.79	255.6	43.2	2.76
	105	30.86	18.32	11.79	0.911	0.554	1,852	202.2	7.75	231.0	39.2	2.73
	96	28.22	18.16	11 75	0.831	0.512	1,674	184.4	7.70	206.8	35.2	2.71

TABLE 11.56 Wide-Flange Beams and Columns (*Continued*)

Nominal size, in.	Weight /ft, lb	Area of section, in.²	Depth of section, in.	Flange Width, in.	Flange Thickness, in.	Web thickness, in.	Neutral axis perpendicular to web at center I, in.⁴	S, in.³	r, in.	Neutral axis parallel to web at center I, in.⁴	S, in.³	r, in.
18 × 8¾	85	24.97	18.32	8.83	0.911	0.526	1,429	156.1	7.57	99.4	22.5	2.00
	77	22.63	18.16	8.78	0.831	0.475	1,286	141.7	7.54	88.6	20.2	1.98
	70	20.56	18.00	8.75	0.751	0.438	1,153	128.2	7.49	78.5	17.9	1.95
	64	18.80	17.87	8.71	0.686	0.403	1,045	117.0	7.46	70.3	16.1	1.93
18 × 7½	60	17.64	18.25	7.56	0.695	0.416	984	107.8	7.47	47.1	12.5	1.63
	55	16.19	18.12	7.53	0.630	0.390	889	98.2	7.41	42.0	11.1	1.61
	50	14.71	18.00	7.50	0.570	0.358	800	89.0	7.38	37.2	9.9	1.59
16 × 11½	96	28.22	16.32	11.53	0.875	0.535	1,355	166.1	6.93	207.2	35.9	2.71
	88	25.86	16.16	11.50	0.795	0.504	1,222	151.3	6.87	185.2	32.2	2.67
16 × 8½	78	22.92	16.32	8.58	0.875	0.529	1,042	127.8	6.74	87.5	20.4	1.95
	71	20.86	16.16	8.54	0.795	0.486	936	115.9	6.70	77.9	18.2	1.93
	64	18.80	16.00	8.50	0.715	0.443	833	104.2	6.66	68.4	16.1	1.91
	58	17.04	15.86	8.46	0.645	0.407	746	94.1	6.62	60.5	14.3	1.88
16 × 7	50	14.70	16.25	7.07	0.628	0.380	655	80.7	6.68	34.8	9.8	1.54
	45	13.24	16.12	7.03	0.563	0.346	583	72.4	6.64	30.5	8.7	1.52
	40	11.77	16.00	7.00	0.503	0.307	515	64.4	6.62	26.5	7.6	1.50
	36	10.59	15.85	6.99	0.428	0.299	466	56.3	6.49	22.1	6.3	1.45
14 × 16	426	125.2	18.69	16.69	3.033	1.875	6,610	707.4	7.26	2,359	282.7	4.34
	398	116.9	18.31	16.59	2.843	1.770	6,013	656.9	7.17	2,169	261.6	4.31
	370	108.8	17.94	16.47	2.658	1.655	5,454	608.1	7.08	1,986	241.1	4.27
	342	100.6	17.56	16.36	2.468	1.545	4,911	559.4	6.99	1,806	220.8	4.24
	314	92.30	17.19	16.23	2.283	1.415	4,399	511.9	6.90	1,631	201.0	4.20
	287	84.37	16.81	16.13	2.093	1.310	3,912	465.5	6.81	1,466	181.8	4.17
	264	77.63	16.50	16.02	1.938	1.205	3,526	427.4	6.74	1,331	166.1	4.14
	246	72.33	16.25	15.94	1.813	1.125	3,228	397.4	6.68	1,226	153.9	4.12
	237	69.69	16.12	15.91	1.748	1.090	3,080	382.2	6.65	1,174	147.7	4.11
	228	67.06	16.00	15.86	1.688	1.045	2,942	367.8	6.62	1,124	141.8	4.10
	219	64.36	15.82	15.80	1.623	1.005	2,798	352.6	6.59	1,073	135.6	4.08
	211	62.07	15.75	15.80	1.563	0.980	2,671	339.2	6.56	1,028	130.2	4.07
	202	59.39	15.63	15.75	1.503	0.930	2,538	324.9	6.54	979	124.4	4.06
14 × 16	193	56.73	15.50	15.71	1.438	0.890	2,402	310.0	6.51	930	118.4	4.05
	184	54.07	15.38	15.66	1.378	0.840	2,274	295.8	6.49	882	112.7	4.04
	176	51.73	15.25	15.64	1.313	0.820	2,149	281.9	6.45	837	107.1	4.02
	167	49.09	15.12	15.60	1.248	0.780	2,020	267.3	6.42	790	101.3	4.01
	158	46.47	15.00	15.55	1.188	0.730	1,900	253.4	6.40	745	95.8	4.00
	150	44.08	14.88	15.51	1.128	0.695	1,786	240.2	6.37	702	90.6	3.99
	142	41.85	14.75	15.50	1.063	0.680	1,672	226.7	6.32	660	85.2	3.97
14 × 14½	320	94.12	16.81	16.71	2.093	1.890	4,141	492.8	6.63	1,635	195.7	4.17
	136	39.98	14.75	14.74	1.063	0.660	1,593	216.0	6.31	567	77.0	3.77
	127	37.33	14.62	14.69	0.998	0.610	1,476	202.0	6.29	527	71.8	3.76
	119	34.99	14.50	14.65	0.938	0.570	1,373	189.4	6.26	491	67.1	3.75
	111	32.65	14.37	14.62	0.873	0.540	1,266	176.3	6.23	454	62.2	3.73
	103	30.26	14.25	14.57	0.813	0.495	1,165	163.6	6.21	419	57.6	3.72
	95	27.94	14.12	14.54	0.748	0.465	1,063	150.6	6.17	383	52.8	3.71
	87	25.56	14.00	14.50	0.688	0.420	966	138.1	6.15	349	48.2	3.70
14 × 12	84	24.71	14.18	12.02	0.778	0.451	928	130.9	6.13	225.5	37.5	3.02
	78	22.94	14.06	12.00	0.718	0.428	851	121.1	6.09	206.9	34.5	3.00
14 × 10	74	21.76	14.19	10.07	0.783	0.450	796	112.3	6.05	133.5	26.5	2.48
	68	20.00	14.06	10.04	0.718	0.418	724	103.0	6.02	121.2	24.1	2.46
	61	17.94	13.91	10.00	0.643	0.378	641	92.2	5.98	107.3	21.5	2.45
14 × 8	53	15.59	13.94	8.06	0.658	0.370	542	77.8	5.90	57.5	14.3	1.92
	48	14.11	13.81	8.03	0.593	0.339	484	70.2	5.86	51.3	12.8	1.91
	43	12.65	13.68	8.00	0.528	0.308	429	62.7	5.82	45.1	11.3	1.89
14 × 6¾	38	11.17	14.12	6.77	0.513	0.313	385	54.6	5.87	24.6	7.3	1.49
	34	10.00	14.00	6.75	0.453	0.287	339	48.5	5.83	21.3	6.3	1.46
	30	8.81	13.86	6.73	0.383	0.270	289	41.8	5.73	17.5	5.2	1.41
12 × 12	190	55.86	14.38	12.67	1.736	1.060	1,892	263.2	5.82	589.7	93.1	3.25
	161	47.38	13.88	12.51	1.486	0.905	1,541	222.2	5.70	486.2	77.7	3.20
	133	39.11	13.38	12.36	1.236	0.755	1,221	182.5	5.59	389.9	63.1	3.16

TABLE 11.56 Wide-Flange Beams and Columns (*Continued*)

Nominal size, in.	Weight /ft, lb	Area of section, in.²	Depth of section, in.	Flange Width, in.	Flange Thickness, in.	Web thickness, in.	I, in.⁴	S, in.³	r, in.	I, in.⁴	S, in.³	r, in.
12 × 12	120	35.31	13.12	12.32	1.106	0.710	1,071	163.4	5.51	345.1	56.0	3.13
(cont'd)	106	31.19	12.88	12.23	0.986	0.620	930	144.5	5.46	300.9	49.2	3.11
	99	29.02	12.75	12.19	0.921	0.580	858	134.7	5.43	278.2	45.7	3.09
	92	27.06	12.62	12.15	0.856	0.545	788	125.0	5.40	256.4	42.2	3.08
	85	24.98	12.50	12.10	0.796	0.495	723	115.7	5.38	235.5	38.9	3.07
	79	23.22	12.38	12.08	0.736	0.470	663	107.1	5.34	216.4	35.8	3.05
	72	21.16	12.25	12.04	0.671	0.430	597	97.5	5.31	195.3	32.4	3.04
	65	19.11	12.12	12.00	0.606	0.390	533	88.0	5.28	174.6	29.1	3.02
12 × 10	58	17.06	12.19	10.01	0.641	0.359	476	78.1	5.28	107.4	21.4	2.51
	53	15.59	12.06	10.00	0.576	0.345	426	70.7	5.23	96.1	19.2	2.48
12 × 8	50	14.71	12.19	8.07	0.641	0.371	394	64.7	5.18	56.4	14.0	1.96
	45	13.24	12.06	8.04	0.576	0.336	350	58.2	5.15	50.0	12.4	1.94
	40	11.77	11.94	8.00	0.516	0.294	310	51.9	5.13	44.1	11.0	1.94
12 × 6½	36	10.59	12.24	6.56	0.540	0.305	280	45.9	5.15	23.7	7.2	1.50
	31	9.12	12.09	6.52	0.465	0.265	238	39.4	5.11	19.8	6.1	1.47
	27	7.97	11.95	6.50	0.400	0.240	204	34.1	5.06	16.6	5.1	1.44
10 × 10	112	32.92	11.38	10.41	1.248	0.755	718.7	126.3	4.67	235.4	45.2	2.67
	100	29.43	11.12	10.34	1.118	0.685	625.0	112.4	4.61	206.6	39.9	2.65
	89	26.19	10.88	10.27	0.998	0.615	542.4	99.7	4.55	180.6	35.2	2.63
	77	22.67	10.62	10.19	0.868	0.535	457.2	86.1	4.49	153.4	30.1	2.60
	72	21.18	10.50	10.17	0.808	0.510	420.7	80.1	4.46	141.8	27.9	2.59
	66	19.41	10.38	10.11	0.748	0.457	382.5	73.7	4.44	129.2	25.5	2.58
	60	17.66	10.25	10.07	0.683	0.415	343.7	67.1	4.41	116.5	23.1	2.57
	54	15.88	10.12	10.02	0.618	0.368	305.7	60.4	4.39	103.9	20.7	2.56
	49	14.40	10.00	10.00	0.558	0.340	272.9	54.6	4.35	93.0	18.6	2.54
10 × 8	45	13.24	10.12	8.02	0.618	0.350	248.6	49.1	4.33	53.2	13.3	2.00
	39	11.48	9.94	7.99	0.528	0.318	209.7	42.2	4.27	44.9	11.2	1.98
	33	9.71	9.75	7.96	0.433	0.292	170.9	35.0	4.20	36.5	9.2	1.94
10 × 5¾	29	8.53	10.22	5.79	0.500	0.289	157.3	30.8	4.29	15.2	5.2	1.34
	25	7.35	10.08	5.76	0.430	0.252	133.2	26.4	4.26	12.7	4.4	1.31
	21	6.19	9.90	5.75	0.340	0.240	106.3	21.5	4.14	9.7	3.4	1.25
8 × 8	67	19.70	9.00	8.28	0.933	0.575	271.8	60.4	3.71	88.6	21.4	2.12
	58	17.06	8.75	8.22	0.808	0.510	227.3	52.0	3.65	74.9	18.2	2.10
	48	14.11	8.50	8.11	0.683	0.405	183.7	43.2	3.61	60.9	15.0	2.08
	40	11.76	8.25	8.07	0.558	0.365	146.3	35.5	3.53	49.0	12.1	2.04
	35	10.30	8.12	8.02	0.493	0.315	126.5	31.1	3.50	42.5	10.6	2.03
	31	9.12	8.00	8.00	0.433	0.228	109.7	27.4	3.47	37.0	9.2	2.01
8 × 6½	28	8.23	8.06	6.54	0.463	0.285	97.8	24.3	3.45	21.6	6.6	1.62
	24	7.06	7.93	6.50	0.398	0.245	8.25	20.8	3.42	18.2	5.6	1.61
8 × 5¼	20	5.88	8.14	5.27	0.378	0.248	69.2	17.0	3.43	8.5	3.2	1.20
	17	5.00	8.00	5.25	0.308	0.230	56.4	14.1	3.36	6.7	2.6	1.16

Note: Flanges of wide-flange beams and columns are not tapered, have constant thickness.
Lightweight beams for each nominal size, and beams with depth in even inches, are most usually stocked.
Designation of wide-flange beams is made by giving nominal depth and weight, thus, 8WF40.

TABLE 11.57 American Standard I Beams

Depth of beam, in.	Weight ft, lb	Area of section, in.	Width of flange, in.	Thick-ness of web, in.	Neutral axis perpendicular to web at center			Neutral axis coincident with center line of web		
					I, in.^{4}	r, in.	z in.^{3}	I, in.^{4}	r, in.	z, in.^{3}
24	120.0	35.13	8.048	0.798	3010.8	9.26	250.9	84.9	1.56	21.1
	105.9	30.98	7.875	0.625	2811.5	9.53	234.3	78.9	1.60	20.0
	100.0	29.25	7.247	0.747	2371.8	9.05	197.6	48.4	1.29	13.4
	90.0	26.30	7.124	0.624	2230.1	9.21	185.8	45.5	1.32	12.8
	79.9	23.33	7.000	0.500	2087.2	9.46	173.9	42.9	1.36	12.2
20	95.0	27.74	7.200	0.800	1599.7	7.59	160.0	50.5	1.35	14.0
	85.0	24.80	7.053	0.653	1501.7	7.78	150.2	47.0	1.38	13.3
	75.0	21.90	6.391	0.641	1263.5	7.60	126.3	30.1	1.17	9.4
	65.4	19.08	6.250	0.500	1169.5	7.83	116.9	27.9	1.21	8.9
18	70.0	20.46	6.251	0.711	917.5	6.70	101.9	24.5	1.09	7.8
	54.7	15.94	6.000	0.460	795.5	7.07	88.4	21.2	1.15	7.1
15	50.0	14.59	5.640	0.550	481.1	5.74	64.2	16.0	1.05	5.7
	42.9	12.49	5.500	0.410	441.8	5.95	58.9	14.6	1.08	5.3
12	50.0	14.57	5.477	0.687	301.6	4.55	50.3	16.0	1.05	5.8
	40.8	11.84	5.250	0.460	268.9	4.77	44.8	13.8	1.08	5.3
	35.0	10.20	5.078	0.428	227.0	4.72	37.8	10.0	0.99	3.9
	31.8	9.26	5.000	0.350	215.8	4.83	36.0	9.5	1.01	3.8
10	35.0	10.22	4.944	0.594	145.8	3.78	29.2	8.5	0.91	3.4
	25.4	7.38	4.660	0.310	122.1	4.07	24.4	6.9	0.97	3.0
8	23.0	6.71	4.171	0.441	64.2	3.09	16.0	4.4	0.81	2.1
	18.4	5.34	4.000	0.270	56.9	3.26	14.2	3.8	0.84	1.9
7	20.0	5.83	3.860	0.450	41.9	2.68	12.0	3.1	0.74	1.6
	15.3	4.43	3.660	0.250	36.2	2.86	10.4	2.7	0.78	1.5
6	17.25	5.02	3.565	0.465	26.0	2.28	8.7	2.3	0.68	1.3
	12.5	3.61	3.330	0.230	21.8	2.46	7.3	1.8	0.72	1.1
5	14.75	4.29	3.284	0.494	15.0	1.87	6.0	1.7	0.63	1.0
	10.0	2.87	3.000	0.210	12.1	2.05	4.8	1.2	0.65	0.82
4	9.5	2.76	2.796	0.326	6.7	1.56	3.3	0.91	0.58	0.65
	7.7	2.21	2.660	0.190	6.0	1.64	3.0	0.77	0.59	0.58
3	7.5	2.17	2.509	0.349	2.9	1.15	1.9	0.59	0.52	0.47
	5.7	1.64	2.330	0.170	2.5	1.23	1.7	0.46	0.53	0.40

TABLE 11.58   American Standard Channels

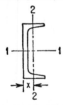

Depth of channel, in.	Weight ft, lb	Area of section, in.²	Width of flange, in.	Thick-ness of web, in.	Axis 1-1			Axis 2-2	x, in.
					I, in.⁴	r, in.	S, in.³	r, in.	
15	50.0	14.64	3.716	0.716	401.4	5.24	53.6	0.87	0.80
	40.0	11.70	3.520	0.520	346.3	5.44	46.2	0.89	0.78
	33.9	9.90	3.400	0.400	312.6	5.62	41.7	0.91	0.79
12	30.0	8.79	3.170	0.510	161.2	4.28	26.9	0.77	0.68
	25.0	7.32	3.047	0.387	143.5	4.43	23.9	0.79	0.68
	20.7	6.03	2.940	0.280	128.1	4.61	21.4	0.81	0.70
10	30.0	8.80	3.033	0.673	103.0	3.42	20.6	0.67	0.65
	25.0	7.33	2.886	0.526	90.7	3.52	18.1	0.68	0.62
	20.0	5.86	2.739	0.379	78.5	3.66	15.7	0.70	0.61
	15.3	4.47	2.600	0.240	66.9	3.87	13.4	0.72	0.64
9	20.0	5.86	2.648	0.448	60.6	3.22	13.5	0.65	0.59
	15.0	4.39	2.485	0.285	50.7	3.40	11.3	0.67	0.59
	13.4	3.89	2.430	0.230	47.3	3.49	10.5	0.67	0.61
8	18.75	5.49	2.527	0.487	43.7	2.82	10.9	0.60	0.57
	13.75	4.02	2.343	0.303	35.8	2.99	9.0	0.62	0.56
	11.5	3.36	2.260	0.220	32.3	3.10	8.1	0.63	0.58
7	14.75	4.32	2.299	0.419	27.1	2.51	7.7	0.57	0.53
	12.25	3.58	2.194	0.314	24.1	2.59	6.9	0.58	0.53
	9.8	2.85	2.090	0.210	21.1	2.72	6.0	0.59	0.55
6	13.0	3.81	2.157	0.437	17.3	2.13	5.8	0.53	0.52
	10.5	3.07	2.034	0.314	15.1	2.22	5.0	0.53	0.50
	8.2	2.39	1.920	0.200	13.0	2.34	4.3	0.54	0.52
5	9.0	2.63	1.885	0.325	8.8	1.83	3.5	0.49	0.48
	6.7	1.95	1.750	0.190	7.4	1.95	3.0	0.50	0.49
4	7.25	2.12	1.720	0.320	4.5	1.47	2.3	0.46	0.46
	5.4	1.56	1.580	0.180	3.8	1.56	1.9	0.45	0.46
3	6.0	1.75	1.596	0.356	2.1	1.08	1.4	0.42	0.46
	5.0	1.46	1.498	0.258	1.8	1.12	1.2	0.41	0.44
	4.1	1.19	1.410	0.170	1.6	1.17	1.1	0.41	0.44

TABLE 11.59 Standard Angles, Equal Legs

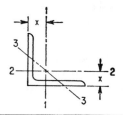

Size, in.			Weight / ft, lb	Area of section, in.²	Axis 1-1 and axis 2-2				Axis 3-3, r min, in.
					I, in.⁴	r, in.	S, in.³	x, in.	
8	× 8	× 1⅛	56.9	16.73	98.0	2.42	17.5	2.41	1.56
		1	51.0	15.00	89.0	2.44	15.8	2.37	1.56
		⅞	45.0	13.23	79.6	2.45	14.0	2.32	1.57
		¾	38.9	11.44	69.7	2.47	12.2	2.28	1.57
		⅝	32.7	9.61	59.4	2.49	10.3	2.23	1.58
		½	26.4	7.75	48.6	2.50	8.4	2.19	1.59
6	× 6	× 1	37.4	11.00	35.5	1.80	8.6	1.86	1.17
		⅞	33.1	9.73	31.9	1.81	7.6	1.82	1.17
		¾	28.7	8.44	28.2	1.83	6.7	1.78	1.17
		⅝	24.2	7.11	24.2	1.84	5.7	1.73	1.18
		½	19.6	5.75	19.9	1.86	4.6	1.68	1.18
		⅜	14.9	4.36	15.4	1.88	3.5	1.64	1.19
5	× 5	× ⅞	27.2	7.98	17.8	1.49	5.2	1.57	0.97
		¾	23.6	6.94	15.7	1.51	4.5	1.52	0.97
		⅝	20.0	5.86	13.6	1.52	3.9	1.48	0.98
		½	16.2	4.75	11.3	1.54	3.2	1.43	0.98
		⅜	12.3	3.61	8.7	1.56	2.4	1.39	0.99
4	× 4	× ¾	18.5	5.44	7.7	1.19	2.8	1.27	0.78
		⅝	15.7	4.61	6.7	1.20	2.4	1.23	0.78
		½	12.8	3.75	5.6	1.22	2.0	1.18	0.78
		⅜	9.8	2.86	4.4	1.23	1.5	1.14	0.79
		¼	6.6	1.94	3.0	1.25	1.1	1.09	0.80
3½ × 3½ ×		½	11.1	3.25	3.6	1.06	1.5	1.06	0.68
		⅜	8.5	2.48	2.9	1.07	1.2	1.01	0.69
		¼	5.8	1.69	2.0	1.09	0.79	0.97	0.69
3	× 3	× ½	9.4	2.75	2.2	0.90	1.1	0.93	0.58
		⅜	7.2	2.11	1.8	0.91	0.83	0.89	0.58
		¼	4.9	1.44	1.2	0.93	0.58	0.84	0.59
2½ × 2½ ×		½	7.7	2.25	1.2	0.74	0.72	0.81	0.49
		⅜	5.9	1.73	0.98	0.75	0.57	0.76	0.49
		¼	4.1	1.19	0.70	0.77	0.39	0.72	0.49
2	× 2	× ⅜	4.7	1.36	0.48	0.59	0.35	0.64	0.39
		¼	3.19	0.94	0.35	0.61	0.25	0.59	0.39
		⅛	1.65	0.48	0.19	0.63	0.13	0.55	0.40
1¾ × 1¾ ×		¼	2.77	0.81	0.23	0.53	0.19	0.53	0.34
		⅛	1.44	0.42	0.13	0.55	0.10	0.48	0.35
1½ × 1½ ×		¼	2.34	0.69	0.14	0.45	0.13	0.47	0.29
		⅛	1.23	0.36	0.08	0.47	0.07	0.42	0.30
1¼ × 1¼ ×		¼	1.92	0.56	0.08	0.37	0.09	0.40	0.24
		⅛	1.01	0.30	0.04	0.38	0.05	0.36	0.25
1	× 1	× ¼	1.49	0.44	0.04	0.29	0.06	0.34	0.20
		⅛	0.80	0.23	0.02	0.30	0.03	0.30	0.20

TABLE 11.60   Selected Standard Angles, Unequal Legs

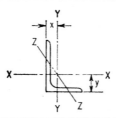

Size, in.	Thickness, in.	Weight/ ft. lb	Area of section, in.²	Axis X-X I, in.⁴	S, in.³	r, in.	y, in.	Axis Y-Y I, in.⁴	S, in.³	r, in.	x, in.	Axis Z-Z r, in.
8 × 6	1	44.2	13.00	80.8	15.1	2.49	2.65	38.8	8.9	1.73	1.65	1.28
	¾	33.8	9.94	63.4	11.7	2.53	2.56	30.7	6.9	1.76	1.56	1.29
	½	23.0	6.75	44.3	8.0	2.56	2.47	21.7	4.8	1.79	1.47	1.30
	7⁄16	20.2	5.93	39.2	7.1	2.57	2.45	19.3	4.2	1.80	1.45	1.31
8 × 4	1	37.4	11.00	69.6	14.1	2.52	3.05	11.6	3.9	1.03	1.05	0.85
	¾	28.7	8.44	54.9	10.9	2.55	2.95	9.4	3.1	1.05	0.95	0.85
	½	19.6	5.75	38.5	7.5	2.59	2.86	6.7	2.2	1.08	0.86	0.86
	7⁄16	17.2	5.06	34.1	6.6	2.60	2.83	6.0	1.9	1.09	0.83	0.87
7 × 4	⅞	30.2	8.86	42.9	9.7	2.20	2.55	10.2	3.5	1.07	1.05	0.86
	¾	26.2	7.69	37.8	8.4	2.22	2.51	9.1	3.0	1.09	1.01	0.86
	½	17.9	5.25	26.7	5.8	2.25	2.42	6.5	2.1	1.11	0.92	0.87
	⅜	13.6	3.98	20.6	4.4	2.27	2.37	5.1	1.6	1.13	0.87	0.88
6 × 4	⅞	27.2	7.98	27.7	7.2	1.86	2.12	9.8	3.4	1.11	1.12	0.86
	¾	23.6	6.94	24.5	6.3	1.88	2.08	8.7	3.0	1.12	1.08	0.86
	½	16.2	4.75	17.4	4.3	1.91	1.99	6.3	2.1	1.15	0.99	0.87
	⅜	12.3	3.61	13.5	3.3	1.93	1.94	4.9	1.6	1.17	0.94	0.88
	5⁄16	10.3	3.03	11.4	2.8	1.94	1.92	4.2	1.4	1.17	0.92	0.88
6 × 3½	½	15.3	4.50	16.6	4.2	1.92	2.08	4.3	1.6	0.97	0.83	0.76
	⅜	11.7	3.42	12.9	3.2	1.94	2.04	3.3	1.2	0.99	0.79	0.77
	¼	7.9	2.31	8.9	2.2	1.96	1.99	2.3	0.85	1.01	0.74	0.78
5 × 3½	¾	19.8	5.81	13.9	4.3	1.55	1.75	5.6	2.2	0.98	1.00	0.75
	½	13.6	4.00	10.0	3.0	1.58	1.66	4.1	1.6	1.01	0.91	0.75
	¼	7.0	2.06	5.4	1.6	1.61	1.56	2.2	0.83	1.04	0.81	0.76
5 × 3	½	12.8	3.75	9.5	2.9	1.59	1.75	2.6	1.1	0.83	0.75	0.65
	⅜	9.8	2.86	7.4	2.2	1.61	1.70	2.0	0.89	0.84	0.70	0.65
	¼	6.6	1.94	5.1	1.5	1.62	1.66	1.4	0.61	0.86	0.66	0.66
4 × 3½	⅝	14.7	4.30	6.4	2.4	1.22	1.29	4.5	1.8	1.03	1.04	0.72
	½	11.9	3.50	5.3	1.9	1.23	1.25	3.8	1.5	1.04	1.00	0.72
	⅜	9.1	2.67	4.2	1.5	1.25	1.21	3.0	1.2	1.06	0.96	0.73
	¼	6.2	1.81	2.9	1.0	1.27	1.16	2.1	0.81	1.07	0.91	0.73
4 × 3	⅝	13.6	3.98	6.0	2.3	1.23	1.37	2.9	1.4	0.85	0.87	0.64
	½	11.1	3.25	5.1	1.9	1.25	1.33	2.4	1.1	0.86	0.83	0.64
	¼	5.8	1.69	2.8	1.0	1.28	1.24	1.4	0.60	0.90	0.74	0.65
3½ × 3	½	10.2	3.00	3.5	1.5	1.07	1.13	2.3	1.1	0.88	0.88	0.62
	¼	5.4	1.56	1.9	0.78	1.11	1.04	1.3	0.59	0.91	0.79	0.63
3½ × 2½	½	9.4	2.75	3.2	1.4	1.09	1.20	1.4	0.76	0.70	0.70	0.53
	¼	4.9	1.44	1.8	0.75	1.12	1.11	0.78	0.41	0.74	0.61	0.54
3 × 2½	½	8.5	2.50	2.1	1.0	0.91	1.00	1.3	0.74	0.72	0.75	0.52
	⅜	6.6	1.92	1.7	0.81	0.93	0.96	1.0	0.58	0.74	0.71	0.52
	¼	4.5	1.31	1.2	0.56	0.95	0.91	0.74	0.40	0.75	0.66	0.53
3 × 2	½	7.7	2.25	1.9	1.0	0.92	1.08	0.67	0.47	0.55	0.58	0.43
	3⁄16	3.07	0.90	0.84	0.41	0.97	0.97	0.31	0.20	0.58	0.47	0.44
2½ × 2	⅜	5.3	1.55	0.91	0.55	0.77	0.83	0.51	0.36	0.58	0.58	0.42
	3⁄16	2.75	0.81	0.51	0.29	0.79	0.76	0.29	0.20	0.60	0.51	0.43
2 × 1½	¼	2.77	0.81	0.32	0.24	0.62	0.66	0.15	0.14	0.43	0.41	0.32
	⅛	1.44	0.42	0.17	0.13	0.64	0.62	0.09	0.08	0.45	0.37	0.33
1¾ × 1¼	¼	2.44	0.69	0.20	0.18	0.54	0.60	0.09	0.10	0.35	0.35	0.27
	⅛	1.23	0.36	0.11	0.09	0.56	0.56	0.05	0.05	0.37	0.31	0.27

TABLE 11.61 Carnegie Zees

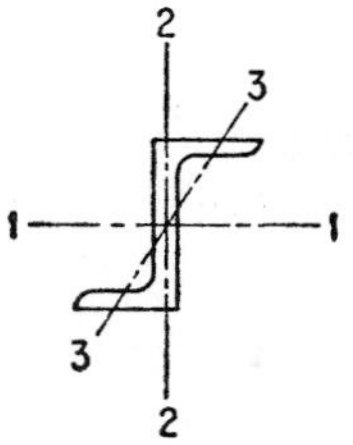

Size			Weight/ft, lb	Area of section, in.^{2}	Axis 1-1			Axis 2-2			Axis 3-3
Depth, in.	Flanges, in.	Thickness, in.			I, in.^{4}	r, in.	S, in.^{3}	I, in.^{4}	r, in.	S, in.^{3}	r min, in.
6 1/8	3 5/8	1/2	21.1	6.19	34.4	2.36	11.2	12.9	1.44	3.8	0.84
6	3 1/2	3/8	15.7	4.59	25.3	2.35	8.4	9.1	1.41	2.8	0.83
5	3 1/4	1/2	17.9	5.25	19.2	1.91	7.7	9.1	1.31	3.0	0.74
5 1/8	3 3/8	7/16	16.4	4.81	19.1	1.99	7.4	9.2	1.38	2.9	0.77
5 1/16	3 5/16	3/8	14.0	4.10	16.2	1.99	6.4	7.7	1.37	2.5	0.76
5	3 1/4	5/16	11.6	3.40	13.4	1.98	5.3	6.2	1.35	2.0	0.75
4 1/16	3 1/8	1/2	15.9	4.66	11.2	1.55	5.5	8.0	1.31	2.8	0.67
4 1/8	3 3/16	3/8	12.5	3.66	9.6	1.62	4.7	6.8	1.36	2.3	0.69
4 1/16	3 1/8	5/16	10.3	3.03	7.9	1.62	3.9	5.5	1.34	1.8	0.68
4	3 1/16	1/4	8.2	2.41	6.3	1.62	3.1	4.2	1.33	1.4	0.67
3	2 11/16	1/2	12.6	3.69	4.6	1.12	3.1	4.9	1.15	2.0	0.53
3	2 11/16	3/8	9.8	2.86	3.9	1.16	2.6	3.9	1.17	1.6	0.54
3	2 11/16	1/4	6.7	1.97	2.9	1.21	1.9	2.8	1.19	1.1	0.55

TABLE 11.62 Gages for Angles, Inches

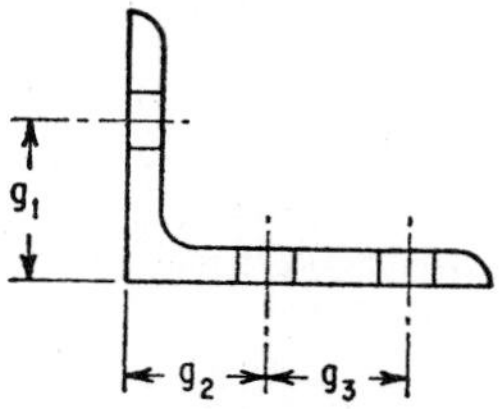

Leg	8	7	6	5	4	3 1/2	3	2 1/2	2	1 3/4	1 1/2	1 3/8	1 1/4	1
g_1	4 1/2	4	3 1/2	3	2 1/2	2	1 3/4	1 3/8	1 1/8	1	7/8	7/8	3/4	5/8
g_2	3	2 1/2	2 1/4	2										
g_3	3	3	2 1/2	1 3/4										
Max rivet	1 1/8	1	7/8	7/8	7/8	7/8	7/8	3/4	5/8	1/2	3/8	3/8	3/8	1/4

REFERENCES*

Screw-Thread Systems

1. ANSI B1.1, "Unified Screw Threads," 1974.
2. ANSI B1.1, "American National Screw Threads, Unified inch terminology," Appendix A, 1974.
3. ANSI B1.1, "Standard Thread Series," Appendix C, 1960.
4. ANSI B1.1, "Unified Miniature Screw Threads," 1958.
5. ANSI "Thread Elements and Formulas," Jones & Lamson Machine Company, Springfield, Vt.
6. ANSI B1.1, "Allowance and Tolerance Tables," Appendix E, 1974.
7. "British Standards for Workshop Practice," Handbook 2, B.S.I. Report 84, 1940.
8. "Machinery's Handbook," Screw Thread Systems, The Industrial Press, New York, 1979.
9. Trade name, patented by Lock Thread Corp.
10. Trade name, patented by Aircraft Screw Products Company.
11. ANSI B1.11 (R1978), "Microscopic Objective Threads," 1958.
12. "Screw Thread Standards for Federal Services," Part III, p. 36, U.S. Department of Commerce, National Bureau of Standards Handbook H28, 1957.
13. ANSI B2.4 (R1974), "Hose Coupling Screw Threads," 1966.
14. ANSI B26, "Fire Hose Coupling Screw Threads," 1953.
15. "Screw Thread Standards for Federal Service," U.S. Department of Commerce, National Bureau of Standards Handbook H28, 1957.
16. Kent, W.: "Mechanical Engineers' Handbook," 12th ed., Carmichael, Design and Production, John Wiley & Sons, Inc., New York, 1951.
17. Marks, Lionel S.: "Mechanical Engineers' Handbook" (Baumeister rev.), McGraw-Hill Book Company, Inc., New York, 1978.
18. British Standard 93, 1951.
19. British Standard 3643, Part II, 1966.
20. British Standard 1095, 1943.
21. ANSI B18.2, "Bolts, Nuts, Screws, Square and Hexagon Bolts and Nuts," 1960.
22. "Machinery Handbook," Screw Thread Systems, The Industrial Press, New York, 1979.
23. ANSI B18.5, "Round Head Bolts," 1978.
24. ANSI B18.9 (R1977), "Plow Bolts," 1958.
25. ANSI B18.10 (R1975), "Track Bolts and Nuts," 1963.
26. ANSI B18.6.3 (R1977), "Machine Screws," 1972.
27. Kent, "Mechanical Engineers' Handbook," 12th ed., Carmichael, Design and Production, John Wiley & Sons, Inc., New York, 1951.
28. ANSI B18.6.2 (R1977), "Hexagon, Slotted Cap Screws; Square, Slotted Headless Set Screws," 1972.
29. ANSI B18.3, "Socket Head Cap and Set Screws," 1976.
30. ANSI B18.6.4 (R1975), "Slotted and Recessed Head Tapping Screws and Metallic Drive Screws," 1966.
31. Parker Kalon Corp., Clifton, N.J.

*These standard references have been prepared in both inch and metric versions.

32. "Bolt, Nut and Rivet Standards," Industrial Fasteners Institute, Cleveland, Ohio.

33. "SAE Handbook," Society of Automotive Engineers, 1957.

Power Screws

34. ANSI B1.5-1977, "Acme Screw Threads," 1977.

35. ANSI B1.8, "Stub Acme Screw Threads," 1977.

36. U.S. Department of Commerce, National Bureau of Standards Handbook H28, p. 46, part III, 1957.

37. U.S. Department of Commerce, National Bureau of Standards Handbook H28, part III, 1957.

38. Laughner, V. H., and A. D. Hargan: "Handbook of Fastening and Joining Metal Parts," McGraw-Hill Book Company, Inc., New York, 1956.

39. Kent, W.: "Mechanical Engineers' Handbook," 12th ed., Power, John Wiley & Sons, Inc., New York, 1951.

Washers

40. ANSI B18.22.H965 (R1975), "Plain Washers," 1975.

41. ANSI B18.21.1, "Lock Washers," 1972.

42. "Machinery's Handbook," The Industrial Press, New York, 1979.

Pins

43. ANSI B18.8.1-1972, ANSI B18.8.2-1978, "Machine Pins."

44. Spring Type Straight Pin, trade name, "Rollpin," patented by Elastic Stop Nut Corp.

Keys and Splines

45. Laughner, H., and A. D. Hargan: "Handbook of Fastening and Joining Metal Parts," table 11.1, McGraw-Hill Book Company, Inc., New York, 1956.

46. Ref. 45, table 11.2.

47. Ref. 45, table 11.3.

48. ANSI B17.1-1967 (R1973), B17.2-1967 (R1978), "Woodruff Keys, Keyslots and Cutters."

49. Ref. 45, table 11.4.

50. Phelan, Richard M.: "Fundamentals of Mechanical Design," McGraw-Hill Book Company, Inc., New York, 1957.

51. Kent, W.: "Mechanical Engineers' Handbook," chap. 11, p. 15, 12th ed., John Wiley & Sons, Inc., New York, 1951.

52. "Machinery's Handbook," The Industrial Press, New York, 1979.

53. ANSI B921, "Involute Splines," 1970.

54. LeGrande, R. E.: "The New American Machinists' Handbook," McGraw-Hill Book Company, Inc., New York, 1955.

55. "SAE Handbook," Society of Automotive Engineers, Warrendale, Pa., 1957.

Wire and Sheet Metal Gages

56. "Machinery's Handbook," The Industrial Press, New York, 1979.
57. ANSI B94.2, "Twist Drills," 1964.

Pipe

58. ANSI B2.1, "Pipe Threads," 1960.
59. ANSI B33.1, "Hose Coupling Screw Threads," 1970.
60. ANSI B1.20.3, "Dryseal Pipe Threads," 1976.
61. ANSI B36.10, "Wrought Steel and Wrought Iron Pipe," 1959.
62. "Machinery's Handbook," The Industrial Press, New York, 1979.
63. "American Petroleum Institute Handbook," Washington, D.C., 1967.
64. ANSI B36.10, "Welded & Seamless Wrought Steel Pipe," 1979.
65. ANSI B16.5, "Steel Pipe Flanges and Flanged Fittings," 1977.

Structural Sections

66. AISC Handbook, "Manual of Steel Construction," 6th ed., American Institute of Steel Construction, New York, 1975.
67. "Bethlehem Steel Handbook," Bethlehem, Pa., 1965.

General References

68. Air Force and Navy: Air Material Command, Wright Patterson Air Force Base, Dayton, Ohio.
69. Federal: Business Service Center, General Services Administration, Washington, D.C.
70. IFI: Industrial Fasteners Institute, 1517 Terminal Tower, Cleveland, Ohio.
71. MIL: Commanding Officer, Naval Aviation Supply Depot, 700 Robbins Ave., Philadelphia, Pa. Att: C D S.
72. NAS: National Aircraft Standards Committee, National Standards Association, 610 Washington Loan & Trust Bldg., Washington, D.C.
73. Machine Design, "The Fastener's Book," Sec. 2, Penton Publishing Company, Cleveland, Ohio, Sept. 29, 1960.

SECTION 12

BOLTS AND BOLTED JOINTS*

M. O. M. Osman, Dr. Tech. Sc.

Professor of Mechanical Engineering
Concordia University
Montreal, Canada

12.1 GENERAL

The joining of removable machine and structural elements is commonly accomplished by means of threaded bolts and screws. Proper tolerancing of hole patterns between mating elements is important for good assembly. Using the method of true position tolerancing, optimum clearance hole sizes can be determined based on standard drill sizes and hole tolerances.[31]

External threads, such as those found on screws, bolts, and pipes, are usually formed by cutting or rolling a helicoidal groove on a screw blank, bolt, or rod. Internal threads are generally produced by cutting. Cold hardening causes rolled threads to be generally stronger than threads formed by cutting.[24]

Section 12 deals with proportions of threaded fasteners and thread strength. References 14 and 15 concern the physical and mechanical properties of threaded fasteners.

A fine thread is especially suited to joints subject to vibration. It offers greater strength at the thread root, may be used to provide fine adjustment, and is recommended for tapping hard nuts. The strength of fine threads is greater for sizes 1 in and under, increasing as the size decreases.

---

*Preliminary note: The reader is referred to Sec. 11 of this handbook for a compilation of standards on the subject matter of this section.

A coarse thread is desirable where the hole is tapped in a weaker material as, for example, cast iron, aluminum, and magnesium alloys. It is stronger in sizes 1 in and over and is especially suited for tapping in brittle materials.

Threads are scored by repeated unscrewing and tightening of steel nuts. After 50 tightenings, the torsional resistance is approximately 100 percent greater than for the first tightening. After 200 tightenings, it is approximately 150 percent greater.

Three basic forms of standard bolts are shown in Fig. 12.1. The through bolt (Fig. 12.1a) is least expensive but requires that both ends of the bolt be accessible. The tap bolt and stud (Fig. 12.1b and c, respectively) require strong threaded sections. An important disadvantage of the tap bolt is that wear may occur in the tapped material, which is not easily replaced.

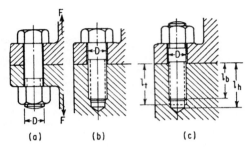

FIG. 12.1   Bolts. (a) Through. (b) Tap. (c) Stud.

The required length of a tapped hole depends upon the material. For a steel bolt or stud it is common practice to use the following dimensions. Referring to Fig. 12.1c:

$$\text{Length of threaded bolt portion} = l_b = C_1D \qquad \text{in}$$

$$\text{Length of tap} = l_t = C_1D + \frac{1}{8} \text{ to } \frac{1}{4} \text{ in}$$

$$\text{Length of blind hole} = l_h = C_1D + \frac{1}{4} \text{ to } \frac{1}{2} \text{ in}$$

where C_1 = 1.0 for steel, 1.5 for cast iron and bronze, and 2.0 for aluminum.

An oversize tap drill should be used for softer metals or the tap will roll the material, forming a deeper thread.

12.2 HOLE TOLERANCE

True position tolerances for hole patterns that must match with mating holes for good assembly are shown in Tables 12.1 and 12.2.[31] These tolerances are based on standard drill sizes and hole tolerances. Rectangular toleranced hole pattern calculations and charts must take into consideration the difference in a two-hole pattern versus three or more patterns for a given size fastener.

By using the method of true position tolerancing, the same size hole can be used for one or any number of holes.

These tables list the clearance holes for fixed condition and floating condition. A fixed condition is where one mating member has studs, pins, and countersunk holes, while the other matching member has clearance holes, Fig. 12.1b and c. A floating condition is one in which both members have clearance holes, or one member could have a floating nut, the amount of float being the same as the amount of clearance in the clearance hole, Fig. 12.1a.

TABLE 12.1 Clearance Hole Sizes for Mating Multiple Hole Patterns in a Floating Condition Based on True Position Tolerances

Hole size = 2TP + S*

Maximum screw or bolt body diameter S	True position TP (diameter) and equivalent rectangular tolerance							
	0.005 (±0.0013)	0.010 (±0.0035)	0.015 (±0.0035)	0.020 (±0.007)	0.025 (±0.009)	0.030 (±0.0106)	0.035 (±0.012)	0.040 (±0.014)
No. 4—0.112	0.125	0.136	0.144	0.152	0.166†	0.173†	0.182†	0.193†
No. 6—0.138	0.149	0.159	0.169	0.180	0.189	0.199†	0.209†	0.219†
No. 8—0.164	0.177	0.185	0.196	0.204	0.219	0.228	0.231†	0.246†
No. 10—0.190	0.201	0.213	0.221	0.234	0.242	0.250	0.261	0.272†
$\frac{1}{4}$—0.250	0.261	0.272	0.281	0.290	0.302	0.312	0.323	0.332
$\frac{5}{16}$—0.312	0.323	0.332	0.344	0.358	0.368	0.375	0.386	0.397
$\frac{3}{8}$—0.375	0.386	0.397	0.406	0.422	0.437	0.437	0.453	0.469
$\frac{7}{16}$—0.437	0.453	0.469	0.469	0.484	0.500	0.500	0.516	0.531
$\frac{1}{2}$—0.500	0.516	0.531	0.531	0.547	0.562	0.562	0.578	0.594
$\frac{9}{16}$—0.562	0.578	0.594	0.591	0.609	0.625	0.625	0.641	0.656
$\frac{5}{8}$—0.625	0.641	0.656	0.656	0.672	0.687	0.687	0.703	0.719
$\frac{3}{4}$—0.750	0.766	0.781	0.781	0.797	0.812	0.812	0.828	0.844

*Hole size given is the next largest drill size to the required hole.
†Requires washers to assure proper seating of the screw head.

12.3

TABLE 12.2 Clearance Hole Sizes for Mating Multiple Hole Patterns in a Floating Condition Based on True Position Tolerances

Hole size = TP + S*

Maximum screw or bolt body diameter S	True position TP (diameter) and equivalent rectangular tolerance								
	0.015 (±0.0053)	0.020 (±0.007)	0.025 (±0.009)	0.030 (±0.0106)	0.035 (±0.012)	0.040 (±0.014)	0.045 (±0.016)	0.050 (±0.0176)
No. 4—0.112	0.128	0.136	0.140	0.144	0.147	0.152	0.157	0.166
No. 6—0.138	0.154	0.159	0.165	0.169	0.173	0.180	0.185	0.189
No. 8—0.164	0.180	0.185	0.189	0.196	0.199	0.204	0.209	0.219
No. 10—0.190	0.205	0.213	0.219	0.224	0.228	0.234	0.238	0.242
$\frac{1}{4}$—0.250	0.265	0.272	0.277	0.281	0.290	0.290	0.295	0.302
$\frac{5}{16}$—0.312	0.328	0.332	0.339	0.344	0.348	0.358	0.358	0.368
$\frac{3}{8}$—0.375	0.391	0.397	0.401	0.406	0.413	0.422	0.422	0.437
$\frac{7}{16}$—0.437	0.453	0.469	0.469	0.469	0.481	0.484	0.484	0.500
$\frac{1}{2}$—0.500	0.516	0.531	0.531	0.531	0.547	0.547	0.547	0.562
$\frac{9}{16}$—0.562	0.578	0.594	0.591	0.591	0.609	0.603	0.609	0.625
$\frac{5}{8}$—0.625	0.611	0.656	0.656	0.656	0.672	0.672	0.672	0.697
$\frac{3}{4}$—0.750	0.766	0.781	0.781	0.781	0.797	0.797	0.797	0.812

*Hole size given is the next largest drill size to the required hole.

12.4

Since we are only dealing with holes for fixed and floating conditions, it is advisable that the design engineer study the method of true position tolerancing as outlined in USASI Y14.5,[32] since this standard has become mandatory for all military drawings.

For a better understanding of the method of true position tolerancing, the book *Fundamentals of Position Tolerance*[33] is highly recommended.

12.3 MATERIALS AND STRENGTH

12.3.1 General

In addition to strength, materials used for threaded fasteners must often meet specifications related to corrosion resistance, magnetic properties, electrical conductivity, thermal conductivity, and cost.

The majority of bolts, however, are made of steel. Standard specifications of steels used for threaded fasteners cover a very broad range of mechanical properties. Table 12.3 lists properties of various SAE grades of steel. Commercial screws and bolts are made of low-carbon steels: SAE 1010, 1018, 1025, 1030, and 1040. Cap screws have been made of C1335, 3135, 4047, and 8635, and studs of 1035 to 1045. Heat-treatable alloy steels are the 4700, 8600, and 8700 series. Automatic screw machines use the free-cutting 1100 series. AISI 430 has been popular as a corrosion-resistant metal. High-strength bolts with hardened washers having the same dimensions as standard (ASTM A325) are being used in large structures.

The pulling of a hardened-steel threaded mandrel through a nut provides a stripping test for nuts known as "proof strength." This strength is slightly less than the yield strength.

References 20 and 21 compare riveted and bolted joints and indicate that, with a significant clearance hole, the clamping force is the most important factor in the fatigue strength of the joint. They also indicate that, because of a somewhat better controlled preload, bolted joints have a stronger fatigue strength than riveted ones. In some cases the joint strength may be increased by application of high-strength materials for bolts or rivets.

Standard aircraft bolts have an ultimate tensile and shear strength of 160,000 and 95,000 lb/in^2, respectively. High-strength aircraft bolts have been developed having strengths of 230,000 and 130,000 lb/in^2, respectively. Titanium bolts[19] are stronger and 43 percent lighter in weight than high-strength aircraft bolts.

Table 12.4 presents properties of nonferrous fastener materials.[15]

12.3.2 High Temperature

Three factors affecting the performance of a bolted assembly at high working temperature are the rupture strength of the locknut, relaxation occurring in the locknut and bolt, and thermal expansion.[17] The first two are predicted by test, and the third can be avoided by selection of materials having similar expansion rates. For high temperatures the following are recommended material choices:

Temperature range	Recommended materials
60–400°F	Aluminum alloy 7075-T6, and steel SAE 4140 and 8740
400–800°F	Alloy steels, 300–400 series, stainless steels, and titanium
Up to 900°F	Martensitic chromium steels, and 17-7PH
900°F	Superalloy 8A and 286 Iconel

TABLE 12.3   Properties of ASTM and SAE Grades of Steel Bolts and Cap Screws

SAE grade	Steel designation	Bolt size diam, in.	Yield strength σ_y, psi	Ultimate tensile strength (min) σ_u, psi	Hardness (Brinell)	Comments
0	Low-carbon	All sizes	....	....	....	General use
1	ASTM A307 low-carbon commercial	All sizes	....	55,000	207 max	Cold or hot heading
2	Low-carbon bright finish 0.28C, 0.04P, max 0.05S	Up to ½ incl.	55,000	69,000	241 max	Cold-headed product. Automotive applications
		Over ½ to ¾ incl.	52,000	64,000		
3	Medium-carbon 0.28 to 0.55C, max 0.04P, max 0.05S. May be aged to 700°F to suit	Up to ½ incl.	85,000	110,000	207–269	Cold worked by heading or roll threading. Used for high fatigue strength
		Over ½ to ⅝ incl.	80,000	100,000		
4	Commercial	Over ¾ to 1½ incl.	28,000	55,000	207 max	
5	ASTM A325 medium-carbon 0.28 to 0.55C, max 0.04P, max 0.05S, quenched and tempered at 800°F	Up to ¾ incl.	85,000	120,000	241–302	Used where high preload is necessary
		Over ¾ to 1 incl.	78,000	115,000	235–302	
		Over 1 up 1½ incl.	74,000	105,000	223–285	
6	Special medium-carbon, oil-quenched and tempered at 800°F min	Up to ⅝ incl.	110,000	140,000	285–331	Used where higher strength than grade 5 is required
		Over ⅝ to ¾ incl.	105,000	133,000	269–331	
7	Medium-carbon fine-grain alloy 0.28 to 0.55C, max 0.04P, max 0.05S, oil-quenched and tempered at 800°F min	Up to 1½ incl.	105,000	133,000	269–331	Roll-threaded after heat-treatment for improved fatigue strength
8	Same as grade 7 but higher strength	Up to 1½ incl.	120,000	150,000	302–352	

Notch effects cannot be predicted except by test. High-temperature bolts have all thread dimensions reduced by 0.003 in from standard dimensions to eliminate galling of the bolt thread in the nut, and the shape of the nut is an important factor in high-temperature service. Above 500°F bolts are not reusable.

TABLE 12.4   Nonferrous Fastener Materials[15]

Material	Tensile strength, psi	Yield strength, 0.5% elongation, psi	Rockwell hardness
Aluminum:			
2024-T4	55,000 min 60,000 avg	50,000	54–63 B
2011-T3	55,000 avg 13,000 min	48,000	50–60 B
1100	16,000 avg	5,000	20–25 F
Brasses:			
Yellow brass	60,000 min 72,000 avg	43,000	57–65 B
Free-cutting brass	50,000 min 55,000 avg	25,000	53–59 B
Commercial bronze	45,000 min 50,000 avg	40,000	40–50 B
Naval bronze, composition A	55,000 min 65,000 avg	33,000	45–50 B
Naval bronze, composition B	50,000 min 60,000 avg	31,000	45–50 B
Copper	35,000–45,000	10,000–37,000	40–80 F
Silicon bronze:			
High-silicon, type A	70,000 min 80,000 avg	38,000	74–80 B
Low-silicon, type B	70,000 min 76,000 avg	35,000	67–75 B
Silicon-aluminum	80,000 min 85,000 avg	42,000	76–82 B
Nickel and high nickel:			
Monel	82,000 min 97,000 avg	60,000	90 B
Nickel	68,000–82,000	20,000–65,000	75–86 B
Inconel	80,000–120,000	25,000–70,000	72 B to 24 C
Stainless steel (AISI):			
Type 302	90,000–124,000	35,000–116,000	89 B to 25 C
Type 303	90,000–124,000	35,000–116,000	89 B to 25 C
Type 304	85,000–112,000	30,000–92,000	87–98 B
Type 305	80,000–110,000	30,000–95,000	84–97 B
Type 309	100,000–120,000	40,000–100,000	94 B to 23 C
Type 310	100,000–120,000	40,000–100,000	94 B to 23 C
Type 316	90,000–115,000	30,000–95,000	89–99 B
Type 317	90,000–115,000	30,000–95,000	89–99 B
Type 321	90,000–112,000	30,000–92,000	89–98 B
Type 347	90,000–112,000	35,000–95,000	89–98 B
Type 410	75,000–190,000	40,000–140,000	81 B to 42 C
Type 416	75,000–190,000	40,000–140,000	81 B to 42 C
Type 430	70,000–90,000	40,000–183,000	77–90 B

12.4   BOLTS SUBJECTED TO STEADY LOADS

If a bolt is loaded by a steady force (time-invariant), the influence of local stress and stress concentration in the threaded portion of the bolt may be disregarded if the stress is less than the ultimate stress. In a ductile material, a readjustment of local stress will occur as a consequence of material yielding.

12.4.1 Axial Load Only

Consider a bolt loaded by an external load only, e.g., the threaded portion of a hook used on a hoist. Only an axial load acts; i.e., no twisting moment exists in the bolt because of the initial turning of the nut. The stress area

$$A_s = \pi D_s^2/4 = F_e/\sigma_d \tag{12.1}$$

where $D_s = (D_m + D_r)/2$ = diameter corresponding to stress area, in
$\quad D_m$ = major or nominal diameter, in
$\quad D_r$ = root diameter, in
$\quad \sigma_d = \sigma_y/N$ = allowable nominal bolt stress, lb/in^2
$\quad \sigma_y$ = yield strength of bolt material, lb/in^2 (see Table 12.3 for carbon steel)
$\quad N$ = factor of safety = 2 to 3 for carbon steels = 1.5 to 3 for alloy steels
$\quad F_e$ = external bolt load, lb

The strength contributed by the threads is small in comparison with other unknowns. The value of N depends on the degree of accuracy with which the magnitudes of load and mechanical properties of bolt material are known, as well as upon the quality of the work. For inferior work the allowable stress in Eq. (12.1) must be decreased. The allowable stress should be lowered for screws having a nominal diameter of $\frac{1}{2}$ in or smaller. The allowable nominal stress should be also lowered when a number of bolts in the joint work together and an uneven load distribution among the individual bolts is possible. For additional simplicity and safety, the root area may be used in lieu of the stress area.

12.4.2 Tension Bolt Stresses

The permissible magnitude of the external load per bolt may be reasonably expressed by

$$F_e = C_2(A_s)^{1.42} = \sigma_w A_s \tag{12.2}$$

where F_e = external load per bolt, lb
$\quad C_2$ = an empirical constant
$\quad \sigma_w = C_2(A_s)^{0.42}$ = working stress, lb/in^2
$\quad A_s$ = bolt stress area, in^2

Equation (12.2) applies for bolts of $\frac{3}{4}$ in diameter and over made of steel containing 0.08 to 0.25 percent carbon.

For bolts 2 in and smaller, $C_2 = 5000$ for carbon steel of ultimate strength 60,000 lb/in^2, increasing in proportion to the ultimate strength to 15,000 for alloy-steel bolts. For bronze bolts, $C_2 = 1000$.

For bolts 2 in and larger, C_2 is increased by 40 percent.

ASME Boiler and Pressure Vessel Code[2] recommends a method by which bolt stress is approximately determined by the bolt material regardless of bolt size, for assemblies where leak prevention as well as strength must be considered. The code lists the values for permissible stresses for common bolt materials at temperatures from -20 to $+1100°F$. The maximum axial load acting on the bolt is assumed to be greater than the externally applied load plus the initial tightening force necessary to prevent leakage under the operating conditions. This axial load is assumed to be also larger than the initial tightening force required to "seat" the contact surfaces of the joint without the external load. The code contains data concerning the properties of gasket materials necessary for calculating these initial tightening forces.

12.4.3   Bolts Subjected to Initial Tightening

When the nut shown in Fig. 12.1 is tightened, the friction between the nut and screw threads causes a twisting moment in the bolt which results in principal tensile and shear stresses.

Because of the many factors influencing this moment, it is convenient in practice to use the following simple relationship for the twisting moment M:

$$M = C_3 F_i D_m \qquad \text{in·lb} \tag{12.3}$$

where C_3 = an experimentally determined coefficient
$\quad\quad F_i$ = tightening-up load, lb
$\quad\quad D_m$ = bolt nominal diameter, in

Lacking experimental data, the following average values of C_3 may be used for conventional bolt threads:[3]

Well-lubricated, smooth surface:	$C_3 = 0.10$
Unlubricated, smooth surface:	$C_3 = 0.12$
Well-lubricated, surfaces not smooth:	$C_3 = 0.13$–0.15
Unlubricated, surfaces not smooth:	$C_3 = 0.18$–0.20

The values of C_3 given are for unplated steel bolts with the threads cut (not rolled or ground).

Wrench torque overcomes frictional resistance between the nut and bolt, and between the nut and supporting surface; therefore, the wrench torque may be 50 to 100 percent greater than moment M.

In particular cases where the bolts are highly stressed, it is desirable to remove the torsional moment by turning the nut back through a small angle after it has been tightened. If this is done after the nut is locked (by a pin driven through the nut and bolt), the initial bolt tension is unaffected.

In general, the influence of the torsional stress on the bolt strength may be taken into account by decreasing the allowable nominal stress σ_d by 25 to 30 percent. Equation (12.1) then takes the form

$$A_s = \pi D_s^2/4 = F_i/0.75\sigma_d \tag{12.4}$$

where F_i = initial tightening induced by the nut, lb.

In metal-to-metal surfaces, the tightening force should be somewhat greater than F_e. If $F_e > F_i$, then F_e should be used in Eq. (12.4).

Equation (12.4) should be employed only for screws with nominal diameters ¾ in or larger. For diameters less than ¾ in, the tightening moment depends to such a great extent upon the judgment and experience of the mechanic that the calculation of the actual combined stress in the thread is almost impossible.

The tightening moment may be established by means of a torque wrench, by measurement of the actual elongation of a through bolt, or by inducing a predetermined strain through temperature control. These methods may be employed to achieve the desirable initial tightening and to predict, if only approximately, the nominal stresses. The approximations arise in the unknown friction between the nut and the supporting surface.

Special care must be exercised if consistent results are to be achieved when using the torque wrench. Threads must be in a good condition and the fit must be such that the nut may be screwed up snug by finger pressure alone. Good graphite grease is suggested. Where faulty lubrication occurs, the variation in bolt tension may be as high as 10:1.[23]

12.4.4 Combined Initial-Tightening Load and Applied Load

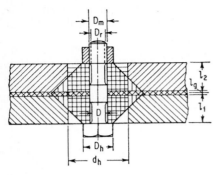

FIG. 12.2 Assembly of two plates and a gasket.

In the discussion which follows, reference is made to Fig. 12.2, in which a gasketed bolted joint is shown. The bolt has been tightened to load F_i, and an external load F_e is applied to the joint. In order to size the bolt properly, it is necessary to calculate the resultant bolt load.

Before application of the external load, the bolt is in tension because of the tightening load F_i, F_e therefore causes an increase in the bolt deformation. The gasket (and other connected parts) undergo a decrease in deformation because of the compression relief associated with F_e.

The change of bolt (and gasket) length δ associated with the application of external load F_e (considering only bolt and gasket) is

$$\delta = \frac{F_e}{E_g A_g / l_g + E_b A_b / l_b} \tag{12.5}$$

where E_g = modulus of elasticity of gasket, lb/in²
E_b = modulus of elasticity of bolt, lb/in²
l_g = gasket thickness, in
l_b = head nut to bolt length, in
A_g = gasket load-carrying area, in²
A_b = bolt-cross-sectional area, in³

The resultant bolt load is therefore

$$F_t = \frac{\delta E_b A_b}{l_b} + F_i = F_e \frac{E_b A_b / l_b}{E_g A_g / l_g + E_b A_b / l_b} + F_i$$

$$= F_e \frac{k_b}{k_b k_g} + F_i = K F_e + F_i \tag{12.6}$$

where k_g = gasket stiffness, lb/in
k_b = bolt stiffness, lb/in

The magnitude of K in Eq. (12.6) lies between zero and unity. Where soft gasketing material is used, the gasket stiffness may become so small in comparison with the bolt stiffness as to cause K to approach unity. A hard thin gasket of large area results in a high gasket stiffness with corresponding diminution of K. The gasket elasticity approaches zero as its thickness diminishes, and therefore k_g is infinite in the absence of a gasket. In this case, $K = 0$.

Equation (12.6) may be generalized to include any connected members

$$F_t = F_e(k_b / k_m k_b) + F_i \tag{12.7}$$

where k_m is the resultant stiffness of the connected members (including gasket, if any), defined by

$$1/k_m = 1/k_1 + 1/k_2 + 1/k_3 + \cdots + 1/k_n \tag{12.8}$$

where $k_1, k_2, k_3, \ldots, k_n$ are the stiffness constants EA/l of the connected members.

If the stiffness constant of a flange is to be evaluated, the value of A used should be the effective cross-sectional area of the flange portions undergoing deformation. These cross-sections can be represented by compression cones shown in Fig. 12.2, which intersect the bearing surfaces under the nut and head at approximately 45° angles. The stiffness of a double cone can be determined approximately by replacing the cones with a hollow cylinder of effective cross-sectional area.

$$A_e = (\pi/4)(d_h^2 - D_m^2) \tag{12.9}$$

where d_h = diameter of the compression cylinder = $D_h + (l_1 + l_2)/2$, in
l_1, l_2 = flange thicknesses, in
D_h = Diameter of bearing surface of nut or head, in

The resultant compression of the connected members is

$$F_g = F_i - F_e[k_m/(k_b + k_m)] \tag{12.10}$$

If F_e becomes so large as to cause F_g to equal zero, i.e., the joint is unloaded, the joint members separate, and the bolt must carry the entire load F_e.

Representative values of K are given in Table 12.5.

When the assembly of machine members is complex in form (e.g., the bolted assembly of a connecting rod), the stiffness constants must be determined experimentally.

TABLE 12.5 Stiffness of Bolted Assembly

Type of joint	Ratio	
	$\dfrac{k_g}{k_b + k_g}$	$\dfrac{k_b}{k_b + k_g}$
Soft packing with studs..........................	0.00	1.00
Soft packing with through bolts....................	0.25	0.75
Asbestos gasket.................................	0.40	0.60
Soft-copper gasket with long through bolts..........	0.50	0.50
Hard-copper gasket with long through bolts..........	0.75	0.25
Very rigid metal-to-metal joint with long through bolts	1.00	0.00

12.4.5 Bolts Subjected to External Shear Load

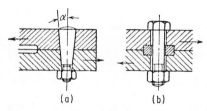

FIG. 12.3 Bolted assemblies for transverse force. (*a*) Taper bolt. (*b*) Cylindrical key.

When small transverse forces act on the bolted assembly, fitted bolts in reamed holes may be used. For larger forces, taper bolts (Fig. 12.3*a*) with tan $\alpha = \frac{1}{20}$ to $\frac{1}{10}$ are used. However, it is usually better to relieve the transverse force by the use of special members, such as the cylindrical key in Fig. 12.3*b*. Preloading bolts so that the frictional resistance between the joints exceeds the shear load is another commonly used method. In this case the bolt load is equal to the initial tightening load only, since the local stresses due to the pressure between the hole wall and the bolt shank do not exist. Dowel pins in reamed holes are also used to take shear. In addition, a knurled bolt body has been successfully used.

12.4.6 Bolted Joints with Eccentric Load

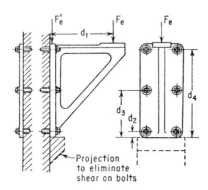

FIG. 12.4   Bracket subjected to eccentric load.

In Fig. 12.4 a cast-iron bracket, which is connected by six bolts to a wall, is shown. An external force F_e applied at the distance d_1 from the wall may be replaced by moment $M_e = F_e d_1$ and force $F_e'(F_e' = F_e)$. The moment tends to rotate the bracket clockwise about its lower edge, while the force F_e' tends to move the bracket downward along the wall surface. The projection shown prevents downward motion of the bracket so that no direct shear load acts on the bolts.

When the bracket flange is heavy and the bolts are relatively long, as in the case presented in the figure, it is reasonable to assume that the wall and the bracket flange are very rigid when compared with the bolts. In this case, as long as the individual bolt load is less than its initial tightening force, the load on each bolt is equal to the tightening load. In practice, all bolts are of the same size and are tightened with approximately the same initial force

$$F_i = C_4 F_e d_1/2(d_2 + d_3 + d_4) \qquad (12.11)$$

where C_4 is a coefficient larger than unity to assure that the moment produced by bolt tightening is larger than the external moment M_e, and the bracket does not separate from the wall. This coefficient C_4 may be from 1.2 to 2 depending upon the accuracy with which the external force is known, the uniformity of initial tightening among the individual bolts, and the type of external load F_e. Larger values for C_4 should be used for variable external loads. If the initial tightening force is determined from Eq. (12.11), the bolt size can be selected by using Eq. (12.4).

Bolts in a bolted connection of this kind should not be subjected to direct shear. If there is no projection in the wall preventing the bracket from sliding, the frictional force between the wall and the bracket surfaces must be larger than the external force which tends to move the bracket along the wall. Thus

$$F_i = C_5 F_e/\mu n \qquad (12.12)$$

where μ = coefficient of friction between wall and bracket surfaces
   n = number of bolts in connection
   C_5 = a coefficient larger than unity to provide a frictional force larger than external force

This coefficient C_5 may be between 1.5 and 3 depending upon the accuracy with which the values of external loads and coefficient of friction are known, and the uniformity of tightening of bolts. The larger values for C_5 should be used in cases when F_e is variable or when vibrations may occur.

For selecting the proper size of bolts, Eq. (12.4) is again applicable. The larger force found from Eqs. (12.11) and (12.12) must be substituted in Eq. (12.4).

Reference 19 discusses other kinds of eccentric loads.

12.5   BOLT DESIGN FOR REPEATED LOADING

12.5.1   General

Under conditions such that the stiffness of the connected parts (e.g., flanges, gaskets) is very high compared with the stiffness of the bolt, and the initial tightening force is large enough to prevent the connected parts from being separated from each other under an applied external load, the axial load on the bolt becomes practically steady and equal to the initial tightening force. Thus the bolt may be designed by using the methods for Sec. 12.2, even if the bolt connection is actually subjected to periodically changing loads. However, these methods become unsatisfactory if the compressive stiffness of the connected members is of the same order as the tensile stiffness of the bolt and a variable external force exists.

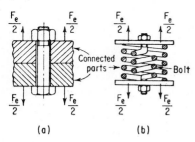

The tightened bolted assembly can be represented schematically as a system comprised of two springs (Fig. 12.5). One of these springs represents the connected parts (including gasket) and is compressed.

FIG. 12.5   Bolted joint (F_e = external load). (*a*) Bolted assembly. (*b*) Elastic equivalent.

12.5.2   Bolt Loads

The additional symbols defined below pertain to the discussion which follows.

F_b =   portion of F_e taken by bolt, lb

$F_{i(c)}$ =   critical tightening load, the smallest magnitude of F_i necessary to prevent separation of the connected parts when F_e is applied, lb

F_{max} =   maximum load applied to bolted connection under cyclic loading, lb

F_{min} =   minimum load applied to bolted connection under cyclic loading, lb

R = F_{min}/F_{max} = range ratio of cyclic load

S_e =   corrected endurance limit, lb/in²

S_e' =   endurance limit of the rotating beam specimen, lb/in²

c =   stiffness coefficient

d_b =   bolt diameter, in

n =   factor of safety

λ_1 =   surface factor

λ_2 =   reliability factor

λ_3 =   modifying factor for stress concentration

σ =   nominal bolt stress, lb/in²

σ_a =   stress amplitude, lb/in²

σ_m =   steady (mean) component of stress lb/in²

The axial load on the bolt

$$F_b = F_i + cF_e \qquad (12.13)$$

where $$c = k_b/(k_b + k_m) \tag{12.14}$$

The critical tightening load $(F_b \rightarrow F_e)$

$$F_{i(c)} = F_e[k_m/(k_b + k_m)] = cF_e(k_m/k_b) \tag{12.15}$$

In order to avoid separation conditions and to ensure against leakage, the initial tightening force

$$F_i = C_6 F_{i(c)} = C_6 cF_e(k_m/k_b) \tag{12.16}$$

where C_6 usually falls between 1.2 and 1.5 but can be greater depending upon the type of assembly and upon the accuracy with which F_e, k_m, and k_b can be determined.

The axial force acting on the bolt due to cyclic loading

$$F_{b,min} = F_i + cF_{e,min} = C_6 cF_{e,max}(k_m/k_b) + cF_{e,min}$$

$$F_{b,max} = F_i + cF_{e,max} = C_6 cF_{e,max}(k_m/k_b) + cF_{e,max} \tag{12.17}$$

The range ratio of cycle of the load on the bolt

$$R = F_{b,min}/F_{b,max} = (C_6 F_{e,max}k_m + F_{e,min}k_b)/F_{e,max}(C_6 k_m + k_b) \tag{12.18}$$

If external force varies from zero to F_e, then $F_{b,min} = F_i$ and Eq. (12.18) reduces to

$$R = C_6 k_m/(C_6 k_m + k_b) \tag{12.19}$$

12.5.3   Design of Bolted Connection Using Fatigue Analysis

The main objective in fatigue design is to keep the number of load applications that will cause failure above the number expected during the lifetime of the part. Fatigue life increases with the increase in stress ratio γ ($\gamma = \sigma_{min}/\sigma_{max}$). The tightening pressure imposed upon the fastener is an important factor contributing to the fatigue strength of the bolted joint.

Whenever the stress in the bolt exceeds the elastic limit, plastic elongation of the bolt will occur, resulting in loosening of the bolted joint. Therefore, if it is possible, the peak stresses should be kept below the elastic limit at all times. Loosening is, however, unavoidable at extremely high values of external loads, and the amount of loosening can be decreased by tightening larger bolts to higher values of initial tension.

It can easily be shown that the bolt is subjected to tensile stress fluctuations. Therefore, the following define a general tensile stress condition:

$$\sigma_a = (\sigma_{max} - \sigma_{min})/2 \qquad \sigma_m = (\sigma_{max} + \sigma_{min})/2 \tag{12.20}$$

The stresses in the bolt are related to the bolt diameter and external load by the following:

$$\sigma_{ba} = (2c/\pi d_b^2)(F_{e,max} - f_{e,min}) \tag{12.21}$$

$$\sigma_{bm} = 4F_i/\pi d_b^2 + (2c/\pi d_b^2)(F_{e,max} + F_{e,min}) \tag{12.22}$$

The yield strength S_y and ultimate tensile strength S_{ut} of the bolt material may be obtained from Tables 12.3 and 12.4.

The Soderberg criteria[34] can be used in the case of a cyclic stress situation.

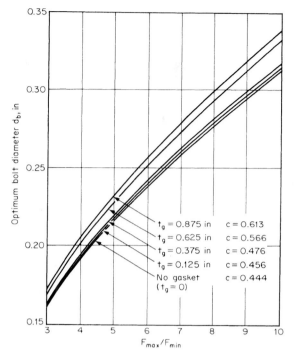

FIG. 12.6 Optimum bolt diameter d_b versus the ratio F_{max}/F_{min} for various gasket thicknesses. Gasket material, copper ($E_g = 17 \times 10^6$ lb/in²); member material, steel ($E_m = 30 \times 10^6$ lb/in²). Bolt material, steel ($F_{min} = 200$ lb).

Considering Soderberg's line for a factor of a safety n, as shown in Fig. 12.6, one can write

$$S_e/n = S_a[1 + (S_e/S_y)(1/\tan \phi)] \tag{12.23}$$

The fatigue or endurance limit S_e for the bolt may be calculated using the following relationship:

$$S_e = \lambda_1 \lambda_2 \lambda_3 S_e' \tag{12.24}$$

Simplifying Eq. (12.23) using Eqs. (12.21) and (12.22) gives the following expression for the optimum bolt diameter.[35]

$$d_b = \{(2nc/\pi S_e)(F_{e,max} - F_{e,min}) + (n/\pi S_y)[4F_i + 2c(F_{e,max} + F_{e,min})]\}^{1/2} \tag{12.25}$$

The stiffness coefficient c is given by Eq. (12.14) in the case of no gasket. However, when a gasket is used, the following equation can be used:

$$c = k_b/(k_b + k_g) \tag{12.26}$$

When gasket is employed in the design, gasket stiffness K_g usually is taken only for determining the stiffness coefficient c. This is due to the fact that most of the deformation takes places within the more elastic gasket material. In cases where member stiffness K_m is comparable to gasket stiffness K_g, the combined effect should be taken. This results in a stiffness coefficient c given by

$$c = K_b(K_m + K_g)/(K_bK_m + K_bK_g + K_mK_g) \qquad (12.27)$$

Figure 12.6[35] illustrates the solution of Eq. (12.25) for a sample case.

12.5.4 Nut Loosening by Vibration[22]

In a bolted joint, the nut is under compressive loading while the bolt is loaded in tension. Increasing the bolt load results in an increase in longitudinal strain and a small decrease in bolt diameter because of the Poisson's ratio effect. Corresponding to the increased tensile bolt load is an increased compressive nut load which causes a slight increase in the nut diameter. The decreased bolt diameter and increased nut diameter (nut flattening) cause a radial slippage of the engaged thread. It is not purely radial because of the circumferential component of the normal thread pressure.

The net effect of each vibration cycle is a small amount of loosening called "fric-

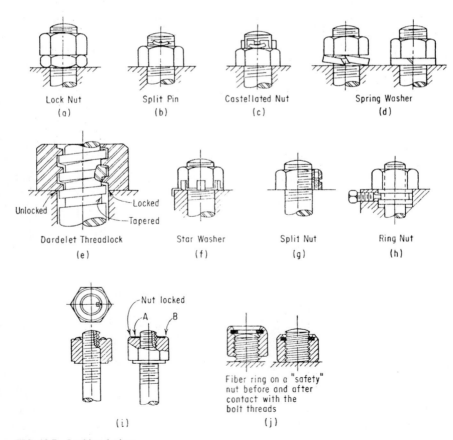

FIG. 12.7 Locking devices.

tional ratchet." Where the applied load variation and a relatively soft gasket material result in large bolt-load variation, special care must be exercised to prevent nut loosening. To reduce the possibility of nut loosening by vibration, the amount of stretch in tightening should be as great as possible.

Since threaded fasteners lose their effectiveness and purpose when they become loosened by vibration, continuous use, or other reasons, auxiliary locking devices have been established. These devices are of two types, friction and positive engagement. Figure 12.7 shows such devices. *a, d, e, i,* and *j* rely on friction and *b, c, f, g,* and *h* are of positive engagement. The most positive device is *c,* where a cotter pin is placed through the bolt and nut. The most recent positive fastening device is a self-aligning nut that reduces the stress concentration which would be present in the initial thread as shown in Fig. 12.7*c.* Thread inserts have some advantages over other types. This is discussed in Ref. 27. For a discussion on locknut devices as shown in Fig. 12.7, Ref. 27 is recommended. For a guide to locknut performance of critical application, aircraft specifications AN-N-5 and AN-N-10 are recommended.

The trend to higher-strength fasteners is discussed in Ref. 28. Tensile strengths of over 100,000 lb/in^2 are not uncommon.

12.6 IMPROVING BOLT PERFORMANCE

In some cases the reliability of a bolted assembly can be increased considerably, without changing the basic diameter of the bolt or increasing the size or weight of the entire assembly. This can be accomplished by (1) decreasing the stress-concentration factor in the threads and other critical cross sections of the bolt, (2) decreasing the longitudinal stiffness of the bolt, (3) improving the load distribution among the individual threads of the nut, (4) reducing bending stresses in the bolt arising from the deformation of the assembled members, and (5) controlling the bolt preload. The stress-concentration factor can be decreased by using rolled threads and by increasing the fillet radii at the junction between the shank and the head and at other places where the diameter changes. Longitudinal bolt stiffness can be decreased by decreasing the shank cross-sectional area. This method is limited because of the accompanying increase in shank stress. Theoretically, the highest bolt reliability is achieved when the safety factor at the critical section in the threaded portion of the bolt and that of the critical section of the shank are equal. For cases in which the external load acting on the assembly changes from zero to maximum and the bolt is unloaded from the torsional moment, the safety factors in the two critical cross sections become equal when[3,6]

$$D = D_r \sqrt{K''_{-1}/K'_{-1}} \qquad (12.28)$$

where D = shank diameter, in.

The axial load tends to extend the bolt, including the threaded portion, inside the nut. At the same time, the nut is subjected to compression. Consequently, load F tends to increase the pitch of the bolt threads and decrease the pitch of the nut threads. Since the nut and bolt threads are engaged, they must deform equally under load, however. As a result the load distribution among individual threads becomes uneven and the thread closest to the bearing nut surface becomes overloaded. Analytical investigations[5] show that, if six or more threads in the nut are engaged with the bolt threads, and if the nut is supported in a conventional way, about 35 percent of the total load is taken by the thread closest to the bearing nut surface. Figure 12.8 shows an "improved" form of the nut[5,7] with a nearly even load distribution among the threads

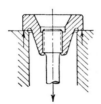

FIG. 12.8   Improved nut.

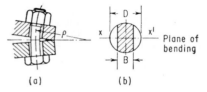

FIG. 12.9   Bolts in bending. (*a*) Deformation in assembly. (*b*) Bolt shank cross section for reducing bending moment in the bolt.

and with "overlapping" threads used to decrease the effect of the stress concentration in the first thread. The stress-concentration factor K_{-1}' is 25 to 30 percent lower for this design than for the conventional nut form. Under nonuniform thread loading, the nut expands as a result of radial forces, which tends to make the load distribution more uniform. A soft nut relative to the bolt also results in a more uniform load distribution because the threads deform plastically, transferring some of their load to less loaded threads. Nuts are generally overdesigned, permitting this choice.

Repeated bending stresses occur in the bolt because of deformation of bolted members under variable external load. The bending stiffness of these members is generally much higher than the bending stiffness of the bolts, and the bending deformation (the magnitude of the radius of the curvature of the bolt axis) of a bolt therefore depends primarily upon the stiffness of the assembled members (Fig. 12.9). Consequently, the bending moment and therefore the stress in the critical cross section of the bolt (usually in the threaded portion) depend upon the moment of inertia of the shank cross section, and it is highly desirable to decrease this moment of inertia as much as possible. The actual radius of curvature cannot be readily calculated in most cases. It is therefore difficult to consider the nominal bending stress when determining the safety factor analytically. However, its influence on the strength of the assembly may be rather significant.[30] The cross section of the shank shown in Fig. 12.9*b* is a good design because the bending stresses in the bolt cross sections and the longitudinal stiffness of the bolt shank are reduced if the bending moment acts in one definite plane $(x - x')$ relative to the bolt.[3,6]

In general, if the load producing the deflection of an assembly is a variable load, any precaution which tends to decrease the bending stresses in the critical sections of the bolt may considerably increase the reliability of a joint. Control of preload is quite significant in bolt reliability. Preloading should be maintained throughout the life of the assembly. An assembly tends to relax or reduce its initial preload and torsional and shear stress in service. This is due to surface high points and flow of metal at these points of high stress and the crushing of bolted material (especially when hard fasteners are used on softer bolted materials). Adequate base area is necessary. As mating surfaces adjust themselves, relaxation reaches a stable value lower than the initial preload. Since the amount of relaxation is fixed, this factor becomes more important in short stiff bolts than in longer, more resilient ones.

Preloading can generally reduce the problem of stress changes in bolts so that fatigue is not significant. When excessive amplitude of bolt stress change exists, preloading may not prevent loosening. In such cases locking devices are necessary. Increasing the thread length somewhat increases the fatigue strength and makes a bolt more resistant to loosening.

All locking devices used to combat the loosening effect of vibrations and shock loads are based on increasing frictional drag between contacting surfaces. Liquid threadlocking agents have been applied on thread surfaces to develop different locking strengths. Thermosetting plastics[16] have also been used.

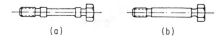

(a)   (b)

FIG. 12.10   Bolts designed for shock and creep.
(a) Shock. (b) Creep.

Figure 12.10 shows bolts designed for shock and creep. Both have a smooth change in cross section. Figure 12.10a for shock design has short lengths of normal shank diameter provided for location in boltholes. Also, the shank of the bolt is reduced to the core diameter of the thread to provide better resilience. Figure 12.10b shows a bolt designed for creep. The large shank diameter results in lower stresses and thus lower creep rates than the previous bolt.

12.7   FINITE-ELEMENT MODELING OF BOLTED JOINTS FOR COMPUTER-AIDED DESIGN

There is an increasing interest in the analysis of bolted joints with complicated geometry support conditions, load applications, and material properties. This has led to a need for more precise information on the behavior of bolted connections in steel components.[39]

Bolted connections are among the most difficult to analyze of all the components of machine elements, to the extent that analysis is usually made on very simplified configurations under idealized support and load conditions. Then the resulting design coefficients are adjusted to reflect test findings, themselves often limited in scope and number.

Since the middle sixties, the finite-element technique has been successfully used in analyzing the problems of elasticity and elasto-plasticity;[36,37] however, only recently has it been used in modeling the behavior of bolted connections.[38] Due to the variety and complexity of finite-element techniques, no effort will be made to derive the algorithms. For additional details reference may be made to publications cited in Ref. 38 and also Sec. 2.

REFERENCES

1. Seaton, A. E., and H. M. Routhwaite: "Marine Engineer's Pocket Book," 16th ed. Charles Griffin & Company, Ltd., London, 1914.

2. ASME Boiler and Pressure Vessel Code, Sec. VIII, 1952 ed.

3. Radzimovsky, E. I.: "Bolt Design for Repeated Loading," *Machine Design,* November 1952.

4. Rotscher, F.: "Maschinenelemente," Springer-Verlag OHG, Berlin, 1929.

5. Maduschka, L.: *Forsch Gebiete Ingenieurw,* vol. 7, no. 6, 1936.

6. Radzimovsky, E. I.: "Schraubenverbindungen bei veränderlicher Belastung," Manu Verlag, Augsburg, 1949.

7. Wiegand, H., and B. Haas: "Berechnung und Gestaltung Schraubenverbindungen," Springer-Verlag OHG, Berlin, 1940.

8. Almen: "On the Strength of Highly Stressed Dynamically Loaded Bolts and Studs," *Diesel Power,* vol. 24, August 1946.

9. Dolan, T. J., and J. H. McClow: The Influence of Bolt Tension and Eccentric Tensile Loads on the Behavior of a Bolt Joint, *Proc. SESA,* vol. 8, no. 1, 1950.

10. Field, J. E.: "Fatigue Strength of Screw Threads," *Engineer,* vol. 198, July 1954.

11. Boomsma, M.: "Loosening and Fatigue Strength of Bolted Joints, *Engineer,,* vol. 200, August 1955.

12. Stewart, W. S.: "Properties of Preloaded Steel Bolts," *Prod. Eng.,* vol. 24, November 1953.

13. Boomsma, M.: "Strength Calculation of Bolted Joints," *Engineer,* vol. 203, May 1957.

14. "ASME Handbook, Metals Engineering—Design," p. 126, McGraw-Hill Book Company, Inc., New York, 1953.

15. "Machine Fasteners Book," Penton Publishing Company, Cleveland, Ohio, 1960.

16. Krieble, R. H.: "New Developments in Thread Locking Agents," *Machine Design,* vol. 29, no. 19, pp. 129–134, Sept. 19, 1957.

17. Selwones, R. S., and R. A. Degen: "High Temperature Design of Bolted Assemblies," *Prod. Eng.,* vol. 28, no. 12, pp. 79–83, Sept. 30, 1957.

18. Viglione, J.: "Strength of Titanium Bolts," *Prod. Eng.,* vol. 26, no. 4, pp. 129–133, April 1955.

19. Shigley, J. E.: "Machine Design," McGraw-Hill Book Company, Inc., New York, 1956.

20. Baron F., and E. A. Larson: "Comparative Behavior of Bolted and Riveted Joints." *Trans. ASME,* ser. E. 26, pp. 285–290, June 1959.

21. Carter, J. W., K. E. Lenzen, and L. T. Wyly: "Fatigue in Riveted and Bolted Single Lap Joints," *Proc. ASCE,* vol. 85, no. ST3, pp 7–28, March 1959.

22. Goodier, J. N., and R. J. Sweeney: "Loosening by Vibration of Threaded Fastenings," *Mech. Eng.,* vol. 67, pp. 798–802, December 1945.

23. Almen, J. O.: "Tightening Is a Vital Factor in Bolt Endurance," *Machine Design,* vol. 16, pp. 158–162, February 1944.

24. Almen, J. O.: "Fatigue Durability of Prestressed Screw Thread," *Prod. Eng.,* vol. 22, pp. 153–156, April 1951.

25. Ollis, R., Jr.: "Self Aligning Nuts," *Machine Design,* pp. 176–179, June 21, 1962.

26. Wolfe, P. C.: "Threaded Inserts," *Elec. Mfg.,* pp. 120–123, January 1954.

27. Feroni, C. C.: "Fundamentals of Selecting Lock Nuts," *Prod. Eng.,* pp. 177–179, December 1953.

28. Stewart, W. C.: "Mechanical Fasteners," *Machine Design,* pp. 220–224, September 1954.

29. Vallance, A., and V. L. Doughtie: "Design of Machine Elements," 3d ed., McGraw-Hill Book Company, Inc., New York, 1951.

30. Radzimovsky, E. I., and Kasuba R.: "Bending Stresses in the Bolts of a Bolted Assembly," *Experimental Mechanics,* vol. 2, no. 9, pp. 264–270, September 1962.

31. Trucks, H. E.: "Designing for Economical Production," Society of Manufacturing Engineers, Dearborn, Mich., 1974.

32. "Dimensioning and Tolerancing for Engineering Drawings," USASI Y14.5-1974, American Society of Mechanical Engineers, New York.

33. Liggett, John V.: "Fundamentals of Position Tolerance," Society of Manufacturing Engineers, 1970.

34. Shigley, J. E.: "Mechanical Engineering Design," 2d ed., McGraw-Hill Book Company, Inc., New York, 1972.

35. Osman, M. O. M., Mansour, W. M., and Dukkipati, R. V.: "On the Design of Bolted Connections with Gaskets Subjected to Fatigue Loading," ASME Transactions, *Journal of Engineering for Industry,* vol. 99, no. 2, May 1977.

36. Zienkiewicz, O. V.: "The Finite Element Method in Engineering Science," McGraw-Hill Book Company, Inc., New York, 1971.

37. Marcal, P. V., and King, I. P.: "Elasto-Plastic Analysis of Two-Dimensional System by Finite-Element Method," *Int. J. Mech. Sci.,* vol. 9, pp. 143–155, 1968.

38. Krishnamurthy, N.: "Modelling and Prediction of Steel Bolted Connection Behaviour," *Computer and Structures,* vol. 2, pp. 75–82, 1980.

39. Ito, Y., M. Korzumi, and M. Masuko: "One Proposal to the Computing Procedure of CAD Considering a Bolted Joint," *Bull. of the ASME,* vol. 20, no. 149, November 1977.

SECTION 13
RIVETS AND RIVETED JOINTS

Glenn L. Haugen, B.S.

Programmer Analyst
City of Los Angeles
Los Angeles, Calif.

13.1   INTRODUCTION

A rivet is a fastening device which is secured by distortion or upsetting of the shank and ends. This definition is used to differentiate rivets from threaded fasteners, nails, and screws, but it also includes upset pins with enlarged ends. Rivets have been made of many materials; soft steel and aluminum alloy are the most common.

Figure 13.1 shows some types of cold-formed fastening devices. Figures 13.1*a* and *b* show a *solid rivet* and a *tubular rivet,* respectively. Tubular rivets can be used to advantage in thin material which would buckle under the action of solid rivets of small diameter. Since the strength of the tubular shank may be limited by the buckling strength of the thin wall, the strength of such rivets should be determined by test rather than computation. In driving, a mandrel should be forced into the rivet to expand it (called "setting") into close contact with the pieces being joined before the ends are flared to form the heads. Split or bifurcated rivets and eyelets are similar. An eyelet has a through hole and is applied where great strength is not required. Figure 13.1*c* shows a staked joint which is often used as a connection to thin parts. "Metal stitching" is the stapling of sheet metal and other softer materials with hard wire.

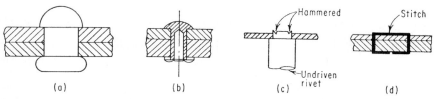

FIG. 13.1   Types of rivets and fasteners. (*a*) Solid rivet. (*b*) Tubular rivet. (*c*) Staking. (*d*) Metal stitching.

Production has been as high as 300 stitches per minute with $\frac{1}{16}$-in steel plate. Figure 13.1*d* shows a flat-clinch type, and Ref. 23 provides strength and pitch information.

For those special cases in which both sides of the work are not accessible, the rivet must be inserted and driven from the same side. Several varieties of "blind rivets" have been developed. The manufacturer's recommendations for hole size and grip length should be strictly followed. Many of the rivets, in addition to their blind-driving characteristics, offer the additional advantage that they can be driven rapidly by only one worker. Blind rivets are usually of a tubular nature. Driving may be accomplished by drawing a present mandrel either completely or partially through the tube, thus upsetting the end. In the latter case, the connection can be made pressure-tight. An explosive force, detonated by heat, may also be used.

13.2   GENERAL COMMENTS

13.2.1   Head Forms and Driving Pressure

The shapes and proportions of many of the heads commonly used in various applications are shown in Fig. 13.2.

The tinners' rivets are similar to flat-head rivets.[22] The 100° flat-top countersunk head is used almost exclusively in aircraft and the cone point, No. 9, is described in a tentative specification of the American Institute of Bolt, Nut and Rivet Manufacturers and is used extensively for structural-size rivets of high-strength materials. The annular point, No. 10, was developed by the builders of the aluminum bridge at Arvida, Canada, and is covered by a British standard for large aluminum-alloy rivets. It shows considerable advantage for upsetting high-strength materials.

The effort required to form the driven head depends on the shape and diameter of the head, the material, and the temperature at the time of driving.

The force required for cold forming buttonheads on mild-steel rivets can be estimated on the basis of a stress of 150,000 lb/in^2 and the diameter of the head. For example, the forming of a buttonhead on a $\frac{3}{4}$-in-diameter rivet requires a force of about 200,000 lb. For hot driving (1600 to 1900°F), the driving pressures are only about 70 percent as great. The pressures required to form heads on rivets of other materials can be found in the same way.

The high pressure required to develop some head forms may result in bulging or even fracturing of the edge of the workpiece or buckling of the workpiece between rivets, especially in the case of thin material. The feasibility of some combinations of conditions should be evaluated by test. Undersized hammers should be avoided because they tend to peen the rivet rather than upset it. Special squeeze riveters have been built capable of cold-driving 3-in-diameter steel rivets. Since the force varies roughly as the square of the diameter of the rivet head, a small driven head such as a buttonhead may have an advantage over a mushroom head. The cone-point head can be formed with from 20 to 50 percent of the force required to form the buttonhead.

In many cases, the malleability of the rivet can be increased and the driving force reduced by hot driving. Care must be taken to control the temperature properly, so that there will be no adverse effect on the strength or other characteristics of the material. Rivets of materials subject to natural age hardening, such as aluminum alloy 2024, require less force if driven immediately after quenching or if refrigerated between the quenching and driving operations to arrest the age hardening.

Tests have shown that rather small heads are sufficient to develop the full tensile strength of the rivet shank, and heads of even insignificant size are sufficient to devel-

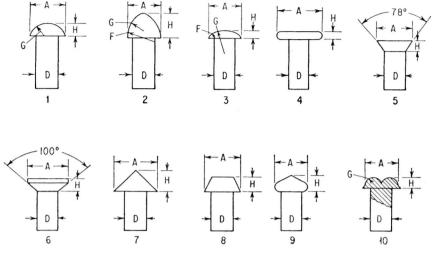

FIG. 13.2 Common types of heads for rivets.

Head type	Head diam A	Head depth H	Head radius G	Edge radius F
1. Standard buttonhead [11,21,22]	1.75D	0.75D	0.885D	
2. High buttonhead [21]	1.5D + 0.031	0.75D + 0.125	0.75D − 0.281	0.75D + 0.281
3. Mushroom head [24]	2D	0.625D	1.634D	0.5D
4. Tinners' rivet	2.25D	0.3D		
5. 78° flat-top countersunk head	1.81D	0.5D		
6. 100° flat-top countersunk head [24]				
Driven heads:				
7. Steeple point [11]	2D	D		
8. Pan point [21]	1.75D	0.7D		
9. Cone point	1.5D	0.75D		
10. Annular point	1.5D	0.48D	0.375D	

op the shear strength under double-shear loading. Therefore, where ease of driving is important and where large, fully formed heads of the button or brazier type are not required for the sake of appearance or other reasons, heads of smaller size and simpler shape should be satisfactory. Even mildly upset pins will be satisfactory for many applications.[1] Obviously, if the rivet must resist tensile loads, the heads must be of such diameter that they will not pull through the holes, and of such height that a ring will not be sheared off.

Shear cracks in heads resulting from overdriving should be avoided, but with some materials, cracks may develop in driven heads not yet completely formed. Tests indicate that these cracks do not adversely affect the static tensile strength or the shear strength under either static or cyclic loading. The only valid objection to such head cracks would seem to be a poor appearance.[2]

Reliability was not impaired according to tests of structural-steel rivets subjected to tensile loading, even though some of the specimens were intentionally the product of malpractices and would not pass ordinary inspection.[3] Even the rivets heated to such a high temperature that they were considered badly burned developed tensile strengths equal to that of the rods from which they were made. On the other hand, in the case of heat-treatable materials, the strictest attention must be paid to the control of the temperature and time at temperature.

13.2.2 Flush-Riveting Dimpled Sheets

Where a flush surface of thin parts must be maintained, as in the skin of high-performance aircraft, the sheets are sometimes dimpled, as shown in Fig. 13.3, to accommodate the countersunk head of the rivet and to increase the strength of the joint. In air-

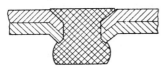

craft riveting, as opposed to most structural riveting, the countersunk head is the manufactured one and the driven head is usually flat. In order to avoid cracked dimples, it may be necessary to increase the temperature of the sheets to be dimpled.[4]

FIG. 13.3   Riveted joints in dimpled sheet.

13.2.3 Folds under Heads

Since the portion of the shank adjacent to the manufactured head is restrained against swelling, the metal in a thoroughly upset shank folds down against the head to form a flat reentrant angle. The fold has the appearance and many of the characteristics of a crack. Although such folds seem to have no effect on the strength of the rivet in double shear, it is probable that they promote rivet failures of the type in which the heads "pop off."

The formation of folds under the manufactured heads can be eliminated by (1) use of small clearance in the holes, (2) avoidance of overdriving, (3) use of a small fillet or chamfer at the junction of the head and shank, and (4) use of a rivet set which bears heavily at the axis of the head.

13.2.4 Hole Clearance

A factor which may have considerable influence upon the strength and behavior of a rivet is the degree to which the shank is upset to fill the hole. For practical reasons, the holes must be larger than the rivet shank and the greater the clearance the greater is the possibility of bending the shank and producing eccentric heads. The best clearance is the smallest one that permits easy insertion of the rivet. For cold-driven rivets, the hole diameter need not be more than about 4 percent greater than the nominal diameter of the rivet, while for hot-driven rivets in structural work a clearance of $\frac{1}{16}$ in is commonly used. Table 13.1 gives the cross-sectional areas of rivets expanded 4 percent over the nominal diameter, of rivets filling holes with a clearance of $\frac{1}{16}$ in, and of standard rivet-hole sizes for aircraft work.[5] The lengths of undriven rivets depend on the rivet diameter, kind of head, plate thickness being riveted, and hole diameter.

13.2.5 Minimum Spacing

The minimum rivet spacing and clearance in corners are controlled by the driving tools available. Small rivets impose few restrictions on spacing because special tools

TABLE 13.1 Cross-Sectional Areas of Driven Rivets

Nominal diam d, in.	Area for clearance of 4%, in.^{2}	Area for clearance of $\frac{1}{16}$ in., in.^{2}	Military standard[5]		
			Drill size	Diam, in.	Area, in.^{2}
$\frac{1}{8}$	0.0133	0.0276	30	0.1285	0.0130
$\frac{5}{32}$	0.0207	0.0376	21	0.159	0.0200
$\frac{3}{16}$	0.0298	0.0491	11	0.191	0.0287
$\frac{7}{32}$	0.0406	0.0621			
$\frac{1}{4}$	0.0530	0.0767	F	0.257	0.0519
$\frac{5}{16}$	0.0828	0.1104	P	0.323	0.0819
$\frac{3}{8}$	0.119	0.150	W	0.386	0.117
$\frac{7}{16}$	0.162	0.196			
$\frac{1}{2}$	0.212	0.249			
$\frac{9}{16}$	0.268	0.307			
$\frac{5}{8}$	0.331	0.371			
$\frac{11}{16}$	0.401	0.442			
$\frac{3}{4}$	0.477	0.518			
$\frac{7}{8}$	0.649	0.690			
1	0.848	0.887			
$1\frac{1}{8}$	1.074	1.108			
$1\frac{1}{4}$	1.325	1.353			

can be readily developed to accommodate the relatively small driving loads. The controlling dimensions usually used in structural engineering are shown in Fig. 13.4.

13.2.6 Initial Tension

The initial tension in structural-steel rivets with well-driven heads may be more than 70 percent[3] of the yield-point strength of the rod if the heads are formed while the rivet is at a cherry-red temperature. The initial tension in rivets with countersunk or flat heads is somewhat less, and that in cold-driven rivets is very low.

13.3 STRENGTH OF RIVETED JOINTS

13.3.1 Static Strength

Under conditions approaching the ultimate load of a joint in which the rivets are subjected to double shear, as indicated in Fig. 13.5, experience and test results indicate that the load can be assumed uniformly distributed among the rivets. Failure may occur by one or a combination of processes; i.e., the rivets may shear or the plates may fail in tension or in bearing. The bearing failure of the plates may take the form of bulging in the region adjacent to the rivet or splitting out to the edge of the plate, or shearing of a portion of the plate to the edge on longitudinal planes tangent to the rivet at the ends of the transverse diameter, as shown in Fig. 13.6.

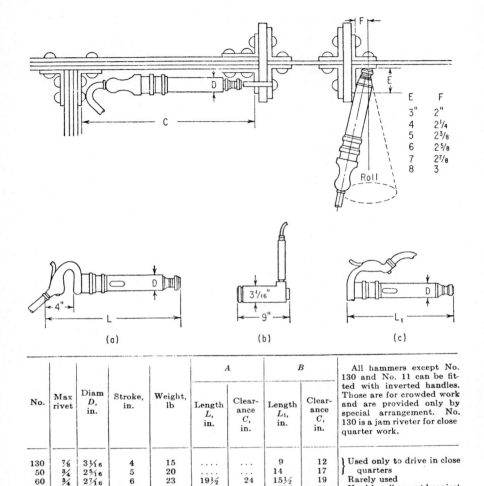

E	F
3"	2"
4	2¼
5	2³⁄₈
6	2⁵⁄₈
7	2⁷⁄₈
8	3

(a)   (b)   (c)

No.	Max rivet	Diam D, in.	Stroke, in.	Weight, lb	A Length L, in.	A Clearance C, in.	B Length L_1, in.	B Clearance C, in.	All hammers except No. 130 and No. 11 can be fitted with inverted handles. Those are for crowded work and are provided only by special arrangement. No. 130 is a jam riveter for close quarter work.
130	⅞	3³⁄₁₆	4	15	....	...	9	12	⎱ Used only to drive in close
50	¾	2⁹⁄₁₆	5	20	....	...	14	17	⎰ quarters
60	¾	2⁷⁄₁₆	6	23	19½	24	15½	19	Rarely used
80	1	2⁷⁄₁₆	8	25	21½	26	17½	21	Used for all except heaviest riveting
90	1¼	2⁷⁄₁₆	9	26	23¾	28	19¾	23	⎱ Used for heaviest riveting
11	1½	2⁷⁄₁₆	11	32	26½	31	.....	...	⎰

FIG. 13.4   Erection clearances for inserting and driving rivets. If hammer can be rolled, easier driving and more symmetrical heads are obtained. To permit this, distance F must be as given here and field rivets must have a perfect stagger with shop rivets. (*a*) Standard open-handle riveter. (*b*) Jam riveter No. 130. (*c*) Inverted-handle riveter. (*Reproduced by permission of AISC Manual.*)

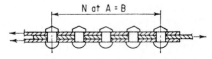

FIG. 13.5   Rivets stressed in double shear.

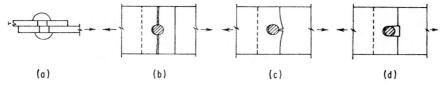

| (a) | (b) | (c) | (d) |

FIG. 13.6   Possible modes of failure of a riveted joint. (*a*) Shear in rivet. (*b*) Tension in plate. (*c*) Tearing of plate behind hole. (*d*) Shearing of plate behind hole.

The strength of the joint is essentially equal to the smallest of the following quantities:

$$P_s = nNA_R\tau_u \qquad (13.1)$$

$$P_t = W_n t\sigma_u \qquad (13.2)$$

$$P_b = NDt\beta_u \qquad (13.3)$$

where P_s = shear strength of all rivets, lb
  P_t = tensile strength of plates, lb
  P_b = bearing strength of plates, lb
   n = number of planes on which shearing failure must occur, 1 for single shear or 2 for double shear
   N = number of rivets in the connection
  A_r = cross-sectional area of the driven rivet shank, in²
  W_n = least net width of the joint, in
   t = thickness of the plate, in
   D = diameter of the driven rivet shank, in
  τ_u = shear strength of rivet material, lb/in²
  σ_u = tensile strength of plate material, lb/in²
  β_u = bearing strength of plate material, lb/in² (function of the edge distance and of the ratio of the rivet diameter to the thickness of the material being joined)

When the three values of strength computed by Eqs. (13.1), (13.2), and (13.3) are equal, the joint is said to have balanced design and is equally likely to fail by each of the three modes. The efficiency of the joint is the ratio of the smallest of these three loads to the computed strength of the gross section of the members being joined. Ultimate tensile properties of steel rivet materials are given in Table 13.2*a*. Ultimate shear strengths of aluminum-alloy rivets are given in Table 13.2*b*. Values of strength of a number of structural materials are given in Table 13.3.

Much effort has been expended in developing a rivet pattern which leads to a balanced design and, hence, maximum efficiency of the materials. Most of the effort has concerned the subject of net width of the joint, W_n. In joints with a single row of rivets and those with chain riveting (Fig. 13.7), the net width is simply the gross width less the product of the hole diameter and the number of holes per row. In joints with staggered riveting or with some rivets omitted from the first row (Fig. 13.8), the spacing of the rows must be considered in determining the net width.

TABLE 13.2a Strength of Steel Rivet Materials

ASTM specification	Description of material	Ultimate tensile strength σ_u, lb/in^2	Yield point σ_y, lb/in^2	Elongation in 8 in, %
A31	Boiler rivet steel and rivets, grade A	45,000–55,000	23,000	27
	Boiler rivet steel and rivets, grade B	58,000–68,000	29,000	22
A131	Structural steel for ships	55,000–65,000	30,000	23
A502	Steel structural rivets, grade 1	52,000–62,000	28,000	24
	Steel structural rivets, grade 2	68,000–82,000	38,000	20
	Steel structural rivets, grade 3	68,000–82,000	50,000	20

Note: For zinc coatings (hot-dip) on iron and steel hardware see ASTM A153.

TABLE 13.2b Shear Strengths of Aluminum-Alloy Rivets

ASTM specification	Description of material	Ultimate tensile strength σ_u, lb/in^2
A31	Boiler rivet steel and rivets, grade A	45,000–55,000
	Boiler rivet steel and rivets, grade B	58,000–68,000
A131	Structural steel for ships	55,000–65,000
A502	Steel structural rivets, grade 1	52,000–62,000
	Steel structural rivets, grade 2	68,000–82,000
	Steel structural rivets, grade 3	68,000–82,000

Note: For zinc coatings (hot-dip) on iron and steel hardware see ASTM A

TABLE 13.3a Tensile Properties of Structural Steels (¾-Inch-Thick Plate)

Material	ASTM specification	Ultimate strength, lb/in^2	Yield point, lb/in^2	Elongation in 8 in, %
Carbon, structural	A36	58,000–80,000	36,000	20
High-strength low-alloy	A242	70,000	50,000	18
Carbon-manganese-silicon	A537	70,000–100,000	50,000	18
Columbiam-vanadium, grade 60	A572	75,000	60,000	15

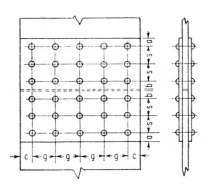

FIG. 13.7 Joint with chain riveting.

TABLE 13.3*b*   Minimum Tensile Properties of Aluminum-Alloy Plate (ASTM B209)

Alloy	Temper	Thickness, in	Ultimate tensile strength σ_u, lb/in^2	Yield strength σ_y, lb/in^2	Elongation in 2 in, %
1100	H14	0.250–0.499	16,000	14,000	6
		0.500–1.000	16,000	14,000	10
3003	H14	0.250–0.499	20,000	17,000	8
		0.500–1.000	20,000	17,000	10
5456	H321	0.500–1.500	44,000	31,000	12
Alclad					
2014	T42	0.250–0.499	57,000	34,000	15
		0.500–1.000	58,000*	34,000*	15
	T62	0.250–0.499	64,000	57,000	8
		0.500–1.000	67,000*	59,000	6
2024	T42	0.250–0.499	64,000	38,000	12
		0.500–1.000	62,000	38,000	8
Alclad					
2024	T42	0.250–0.499	60,000	36,000	12
		0.500–1.000	61,000*	38,000*	8
6061	T651	0.250–0.499	42,000	35,000	10
		0.500–1.000	42,000	35,000	9
		1.001–2.000	42,000	35,000	8
Alclad					
6061	T651	0.250–0.499	38,000	32,000	10
		0.500–1.000	42,000*	35,000	9
		1.001–2.000	42,000*	35,000*	8
7075	T651	0.250–0.499	78,000	67,000	9

*The tension test specimen from plate 0.500 in and thicker is machined from the core and does not include the cladding alloy.

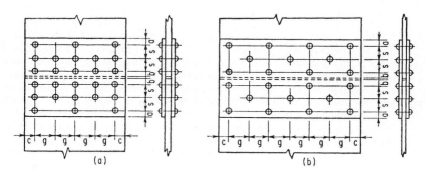

FIG. 13.8   Joints with rivets omitted from first row.

The rules for determining the net width, as given in several of the specifications for structures, are as follows:[6]

1. In the case of a chain of holes extending across a part in any diagonal or zigzag line, the net width of the part shall be obtained by deducting from the gross width the sum of the diameters of all holes in the chain, and adding, for each gage space in the chain, the quantity

$$b = s^2/4g \qquad (13.4)$$

where s = longitudinal spacing (pitch), in, of any two successive holes
g = transverse spacing (gage), in, of same two holes

The critical net section of the part is obtained from the chain which gives the least net width.

2. For angles, the gross width shall be the sum of the widths of the legs less the thickness. The gage for holes in opposite legs shall be the sum of the gages from the back of the angle less the thickness. The gage distances commonly used are shown in Table 13.4.

TABLE 13.4   Usual Gages for Angles

All figures in inches

	Leg	8	7	6	5	4	$3\frac{1}{2}$	3	$2\frac{1}{2}$	2	$1\frac{3}{4}$	$1\frac{1}{2}$	$1\frac{3}{8}$	$1\frac{1}{4}$	1
	g	$4\frac{1}{2}$	4	$3\frac{1}{2}$	3	$2\frac{1}{2}$	2	$1\frac{3}{4}$	$1\frac{3}{8}$	$1\frac{1}{8}$	1	$\frac{7}{8}$	$\frac{7}{8}$	$\frac{3}{4}$	$\frac{5}{8}$
	g_1	3	$2\frac{1}{2}$	$2\frac{1}{4}$	2										
	g_2	3	3	$2\frac{1}{2}$	$1\frac{3}{4}$										

3. In computing the net area, the diameter of a rivet hole shall be taken as $\frac{1}{8}$ in greater than the nominal diameter of the rivet.

An analysis of much of the test data available in 1934, on the static strength of riveted joints in steel members, together with the results of an extensive series of tests undertaken in connection with the design of the San Francisco–Oakland Bay Bridge, concluded:[7] (1) Nothing is gained by an attempt to detail a tension member with a critical net area greater than about 75 percent of the gross area. (2) There is no justification for elaborate formulas to calculate the effect of rivet stagger on net section. (3) Joints should be as compact as practicable (optimum results will probably be obtained by full-row riveting with a gage of about 4.5 rivet diameters and a pitch of 3.5 to 4 rivet diameters. (4) Statements 1 to 3 lead to the suggestion that the allowable loads on riveted tension members be based on, and expressed in terms of, stress in the gross section; and that the net section be not less than 75 percent of the gross section. (5) The practice of assuming equal shear per rivet, regardless of length of joint, is satisfactory. (6) Except in comparatively heavy structures where reduction in size or weight of splice is important, there is little reason for using manganese-steel rivets rather than carbon-steel rivets. Because of the greater slip which occurs in joints employing manganese-steel rivets, carbon-steel rivets are preferable in members subject to stress reversals.

In his discussion of Ref. 7, Professor W. M. Wilson suggested that the effective net width of a riveted tension member is given by the equation

$$W_n = 0.85(W_g - ND)(1 + D/g) \qquad (13.5)$$

where W_n = effective net width, in
W_g = gross width, in
N = number of rivet holes to be deducted
D = nominal diameter of the rivet plus $\frac{1}{8}$ in, in

g = transverse distance between rivets or two times the edge distance,*
whichever is greater, in

His analysis of data involving 25 rivet patterns showed that this formula yielded net widths within 7 percent of the value deduced from the test results, whereas one difference of 15 percent was obtained from the rule involving Eq. (13.4).

In another discussion of Ref. 7, Hartmann and Holt present an analysis of data for 98 steel specimens and 39 aluminum-alloy specimens and suggest the addition of a statement similar to the following:

"The transverse distance between rivets in the outside row shall not exceed eight times the nominal diameter of the rivet, and the edge distance for the end rivet in the outside row shall not exceed four times the nominal diameter of the rivet."

This discussion of net width is applicable to joints which are transverse to the load line. The use of diagonal joints, however (Fig. 13.9), could result in structures of greater efficiency, as in the case of spirally wound pipe. The number of rivets to be considered in determining the net width may be relatively small. Results of tests on such a joint which developed an efficiency of 86 percent are reported in Ref. 8.

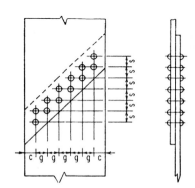

FIG. 13.9 Diagonal rivet pattern.

For high-performance structures, there may be some question as to the value of τ_u to be used in Eq. (13.1). Reference 8 reports an average shear strength of 41,250 lb/in² for 12 undriven steel rivets tested in single shear and only 37,750 lb/in² for 12 other rivets from the same lot tested in double shear. It is to be expected that the strength of hot-driven rivets would be a little higher and that of cold-driven rivets considerably higher than the strengths of these undriven rivets. References 9 and 10 employ a safety factor for steel rivets in shear of about 3 based on the ultimate strength. References 15 and 16 for aluminum-alloy rivets in shear use a factor of safety of about 2½.

For general purposes, a value of shear strength equal to about 60 percent of the tensile strength can be used; however, Ref. 11 sets the allowable stress for shear equal to 80 percent of the allowable stress in tension.

When a rivet is used in conjunction with a relatively thin plate, the plate seems to cut into the rivet and cause an apparent lowering of the shear strength of the rivet. The associated reduction in shear strength depends upon whether the rivet is in single or double shear. Table 13.5 shows the percent reduction in shear strength of aluminum-alloy rivets resulting from the use of thin plates and shapes. These reductions are applicable to the average shear strengths. This type of behavior is not recognized in the specifications of steel structures.

A small amount of data on aluminum-alloy rivets at elevated temperatures[12] indicates that the shear strength is affected by changes in temperature in the same way that the tensile strength of the material is affected.

The bearing strength of parts being joined ,/., is greatly influenced by the distance

---

*In a discussion of test specimens or narrow joints, this is not to be confused with the use of the expression in the discussion of bearing strength.

TABLE 13.5 Percentage Reduction in Shear Strength of Aluminum-Alloy Rivets Resulting from Their Use in Thin Plates and Shapes[15,16]

Ratio* D/t	Loss in double shear†	Ratio* D/t	Loss in double shear†	Ratio* D/t	Loss in		Ratio* D/t	Loss in	
					Single shear	Double shear		Single shear	Double shear
1.5	0	2.2	9.1	2.9	0	18.2	3.5	2.0	26.0
1.6	1.3	2.3	10.4	3.0	0	19.5	3.6	2.4	27.3
1.7	2.6	2.4	11.7	3.1	0.4	20.8	3.7	2.8	28.6
1.8	3.9	2.5	13.0	3.2	0.8	22.1	3.8	3.2	29.9
1.9	5.2	2.6	14.3	3.3	1.2	23.4	3.9	3.6	31.2
2.0	6.5	2.7	15.6	3.4	1.6	24.7	4.0	4.0	32.5
2.1	7.8	2.8	16.9						

*Ratio of the rivet diameter D to the plate thickness t. The thickness used is that of the thinnest plate in a single-shear joint or of the middle plate in a double-shear joint.
†The percentage loss of strength in single shear is zero for D/t less than 3.0.

(parallel to the load line) between the center of the loaded rivet (or pin) and the free edge of the plate (or specimen) when that distance is less than about two times the diameter of the rivet. The ASME Boiler and Pressure Vessel Code[11] permits the use of allowable stress values for bearing equal to 1.60 times the allowable stress value for tension, when the edge distance e is at least $2D$. Tests on a wide variety of aluminum alloys[13] indicate the following relations:

$$\text{Bearing strength} = \begin{cases} (1.5 \text{ to } 1.7) \times \text{tensile strength} & \text{for } e = 1.5D \\ (1.8 \text{ to } 2.4) \times \text{tensile strength} & \text{for } e = 2.0D \\ (2.4 \text{ to } 3.4) \times \text{tensile strength} & \text{for } e = 4.0D \end{cases}$$

In Ref. 13, bearing yield strength is defined as the stress corresponding to a deformation of the hole equal to 2.0 percent of the initial diameter, and the following ratios between bearing and tensile yield strengths for aluminum alloys are given:

$$\text{Bearing yield strength} = \begin{cases} (1.4 \text{ to } 2.1) \times (\text{tensile yield strength}) & \text{for } e = 1.5D \\ (1.5 \text{ to } 2.5) \times (\text{tensile yield strength}) & \text{for } e = 2.0D \\ (1.6 \text{ to } 2.6) \times (\text{tensile yield strength}) & \text{for } e = 4.0D \end{cases}$$

13.3.2 Distribution of Load under Working Conditions

In the preceding discussion, it has been considered that, at failure, the load is uniformly distributed among the rivets. As pointed out in a discussion of Ref. 14 by A. E. R. de Jonge, all the analyses made for determining the distribution of load among the rivets under working conditions lead to the conclusion that these loads are not uniformly distributed and that the end rivets resist a relatively large percentage of the load.

The analysis developed in Ref. 14 leads to the load-distribution factors given in Table 13.6 for certain special cases based on rivets and plates of the same material. In a discussion of Ref. 14, Hill and Holt present experimental and calculated distribution factors for double-strap butt joints between aluminum-alloy and steel plates given in Table 13.7. In every case involving three or more rows of rivets, the end rivets support more than their share of working loads.

TABLE 13.6 Values of Distribution Coefficients C

Diameter of rivets = ⅞ in

No. of rivets	Width of strip, in.	Pitch of rivets, in.	Thickness of plate t, in.		Stress factor k, in./kip	Distribution coefficients				% over-stress
			t_1	t_2		C	C_1	C_2	C_3	
					Lap Joints					
3	3	2.5	0.5		1/8200	0.333	0.371	0.257	......	11.4
	3	5.0	0.5		1/8200	0.333	0.395	0.210	......	18.6
	3	2.5	1.0		1/6400	0.333	0.351	0.298	......	5.4
	3	5.0	1.0		1/6400	0.333	0.365	0.270	......	9.6
4	3	2.5	0.5		1/8200	0.250	0.327	0.173	......	30.8
	3	5.0	0.5		1/8200	0.250	0.368	0.132	......	47.1
	3	2.5	1.0		1/6400	0.250	0.288	0.212	......	15.2
	3	5.0	1.0		1/6400	0.250	0.316	0.184	......	26.4
5	3	2.5	0.5		1/8200	0.200	0.312	0.140	0.096	56.0
	3	5.0	0.5		1/8200	0.200	0.3618	0.1095	0.0574	80.9
	3	2.5	1.0		1/6400	0.200	0.2570	0.1705	0.1450	28.5
	3	5.0	1.0		1/6400	0.200	0.2955	0.1495	0.110	47.7
6	3	2.5	0.5		1/8200	0.1667	0.305	0.128	0.067	83.0
	3	5.0	0.5		1/8200	0.1667	0.3596	0.1036	0.0368	115.6
	3	2.5	1.0		1/6400	0.1667	0.241	0.149	0.110	44.5
	3	5.0	1.0		1/6400	0.1667	0.286	0.135	0.079	71.5
					Butt Joints					
2	3	2.5	0.5	0.375	1/6180	0.250	0.268	0.232	......	7.2
	3	5.0	0.5	0.375	1/6180	0.250	0.277	0.223	......	10.8
	*	*	0.75	0.375	1/7150	0.250	0.250	0.250	......	0.0
	3	2.5	0.75	0.5	1/7645	0.250	0.262	0.238	......	4.8
	3	5.0	0.75	0.5	1/7645	0.250	0.268	0.232	......	7.2
	*	*	1.00	0.5	1/8200	0.250	0.250	0.250	......	0.0
3	3	2.5	0.75	0.5	1/7645	0.1667	0.2052	0.1252	0.1696	23.1
	3	5.0	0.75	0.5	1/7645	0.1667	0.2237	0.1003	0.1760	34.2
	3	2.5	1.00	0.5	1/8200	0.1667	0.1860	0.1280	0.1860	11.6
	3	5.0	1.0	0.5	1/8200	0.1667	0.1981	0.1039	0.1981	18.8
	3	2.5	1.0	0.75	1/7200	0.1667	0.2019	0.1363	0.1618	21.1
	3	5.0	1.0	0.75	1/7200	0.1667	0.2211	0.1152	0.1637	32.6

*Any value.

13.3.3 Fatigue Strength

The nonuniform distribution of load among rivets is given special consideration in structures subject to fatigue, which requires an understanding of the entire previous history of the joint, including the surface condition of the plates, rivet-driving temperature, and previous loading.

Tests on joints in aluminum-alloy plates indicate that the shear fatigue strengths of driven rivets are a much higher percentage of the ultimate strength than are the tensile fatigue strengths of rolled plates with either open holes or holes filled with idle rivets. Thus the designer of riveted joints subject to cyclic loading is especially concerned with the plates or shapes, rather than with the rivets.

TABLE 13.7   Percentage of Total Load Carried by Each Rivet in a Large Butt Joint at Normal Working Load

Specimen No.	Thickness of main plates, in.	Thickness of cover plates, in.	Kind of rivets	Determination	Aluminum main plate rivets Nos.			Steel main plate rivets Nos.				
					1 and 5	2 and 4	3	1	2	3	4	5
1	¾	⅜	Steel	Experimental	34	13	6	17	8.5	12.5	27.5	34.5
				Calculated	31.5	14	9	17	9	10	19.5	44.5
2	1	½	Steel	Experimental	29.5	14	13	23	14.5	16	22	24
					25.5*	19	11	16	17	16	19	32
					25.5†	18	13	18.5	9.5	13.5	23.5	35
				Calculated	28	16	12	17	12	13	20	38
3	1	½	Aluminum	Experimental	24.5	19	13	19.5	10	19	21.5	30
				Calculated	24.5	18	15	17	15	16	21	31

*Specimen straightened before test.
†Specimen loaded in compression.

For members subject to reversal of stress, except for wind loads, Ref. 6 provides the following rule:

To the net total compressive stress, and to the net total tensile stress, add arithmetically 50 percent of the smaller of these two; proportion the connected material and the connecting rivets, bolts, pins, or welds for each of the two increased stresses thus separately obtained at the unit stresses prescribed (for rivets, 15,000 lb/in² shear, 44,000 lb/in² bearing under double shear, 35,000 lb/in² bearing under single shear, supplementary provisions of June 1960). If 100,000 reversals are expected throughout the life of the building, the stresses in the connected material and in the connecting rivets, bolts, pins, or welds shall not exceed 75 percent of those specified. Sharp notches, copes, and other sudden changes of cross section shall be particularly avoided in and adjacent to such connections.

References 9 and 10 provide that the connections of the members subject to alternating (cyclic) loadings occurring in succession during one passage of the live load shall be proportioned for the sum of the net alternating stresses.

References 15 and 16 provide the following rule for members subjected to cyclic loading:

Riveted members designed in accordance with the other requirements of the specifications and constructed so as to be free of severe reentrant corners and other unusual stress raisers require no further consideration of cyclic loadings for numbers of cycles less than 300,000 cycles for structures of 6061-T6 and 100,000 cycles for structures of 2014-T6.

When greater numbers of repetitions of some particular loading cycle are expected during the life of a structure, the calculated net section tensile stresses for the loading in question shall not exceed the values given by the curves in Figs. 13.10 and 13.11.

The following points are worthy of note:

1. The most severe combination of loadings for which a structure is designed (dead load, maximum live load, maximum impact, maximum wind, etc.) rarely occurs in actual service and is of little or no interest from the standpoint of fatigue.
2. The loading of most interest from the fatigue standpoint is the steady dead load with a superimposed, repetitive applied live load under normal operating conditions.

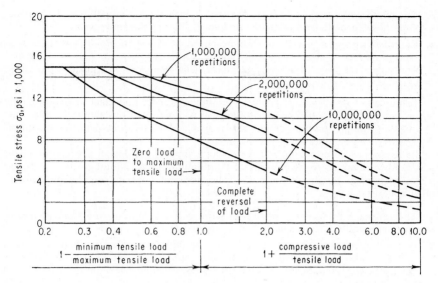

FIG. 13.10 Allowable tensile stresses on net section for various numbers of repetitions of load application, aluminum alloy 6061-T6.[15]

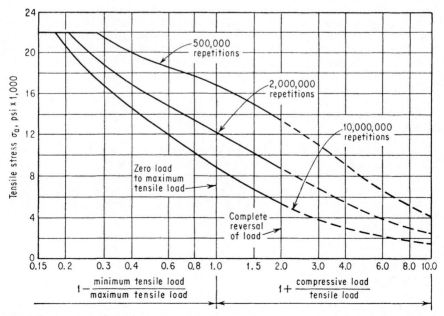

FIG. 13.11 Allowable tensile stresses on net section for various numbers of repetitions of load application, aluminum alloy 2014-T6.[16]

3. The number of cycles of load encountered in structures is usually small compared with those encountered in fatigue problems involving machine parts.

4. Careful attention to details of design and fabrication will greatly enhance resistance to fatigue. A structural fatigue failure is usually at a point of stress concentration where the state of stress could have been improved with little or no added expense.

Test results[17] show that the geometry of a joint is more important in governing the fatigue strength than is the choice of alloy. Stress concentrations should be alleviated by avoiding sharp reentrant corners and by using generous fillets and smooth transitions. Secondary bending should be minimized.

Reference 18 describes fatigue tests which support the design procedures in the ASCE Specifications for Structures of Aluminum Alloys. Although the specimens were of structural aluminum alloys, the following principles are equally applicable to other materials:

1. The fatigue strengths of specimens with well-driven idle rivets are generally greater than those of specimens with open holes.

2. Because of their unsymmetrical geometry, lap joints or single-strap butt joints have lower fatigue strengths than double-strap butt joints. Increasing the stiffness against flexing increases the fatigue strength of unsymmetrical joints.

The fatigue strength of a butt joint of balanced static design with a computed gross-area efficiency of 76 percent was found to be less than that of a simple butt joint with a computed efficiency of only 68 percent.[18] This difference can be attributed to the fact that the computed stress concentration was greater in the joint of balanced design than in the simple joint. Thus the differences in stress-concentration factors associated with different rivet spacings must be recognized. The fatigue strength of a joint will be controlled by the concentrated stress and not by the average stress.

Reference 18 points out that the ASCE design curves (Figs. 13.10 and 13.11) are based on a stress-concentration factor equal to 2.4. Fatigue strengths of joints with larger values of the stress-concentration factor are obtained by multiplying the allowable fatigue stress derived from the design curves by the ratio of 2.4 to the stress-concentration factor of the joint.

13.3.4 Fretting

Under cyclic loading, the small relative movement of the parts of a joint may develop fretting or galling of the surfaces. Ordinarily fretting acts as a very serious stress raiser. Applications of paint or sealing compounds on the faying surfaces may reduce the frictional forces between parts, but at the same time they may serve a useful purpose in reducing the tendency of the surfaces to fret. The small amount of data available indicates that such treatments of the faying surfaces have no deleterious effect on the fatigue strengths of joints.

13.3.5 Creep

Although the high local stresses may be reduced by creep deformations, the fatigue life of a joint may still be reduced by the interaction of the mechanisms of fatigue and

creep. The creep strength and stress-rupture strength are not affected by the stress concentration and interaction of fatigue mechanisms.[19] It is improbable that highly localized creep deformations in a joint would affect the behavior of the structure as a
whole.

13.3.6 Slip

Wilson and Moore[20] describe a series of tests designed to determine whether serious
error is introduced into computations for stresses in steel frames by the assumption
that the joints are perfectly rigid. They found that connections of the type used for the
specimens shown in Fig. 13.12 were so rigid that they could be considered perfectly
rigid. The connections of the type used for the specimens shown in Fig. 13.13 cannot
be considered perfectly rigid.

Additional material related to structural specifications may be found in Refs. 10,
11, 15, 16, and 23.

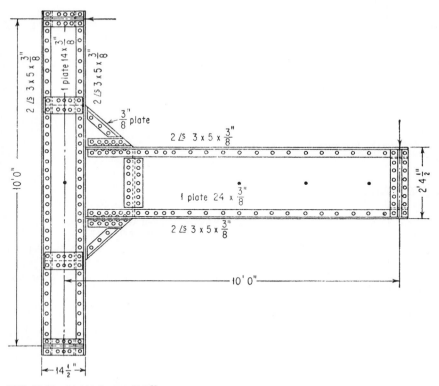

FIG. 13.12 Rigid joint (no slip).[20]

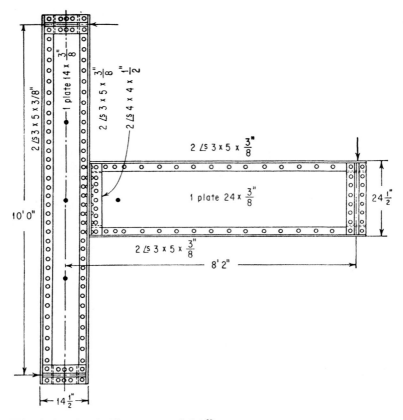

FIG. 13.13 Joint (significant amount of slip).[20]

REFERENCES

1. Moisseiff, L. S., E. C. Hartmann, and R. L. Moore: "Riveted and Pin-Connected Joints of Steel and Aluminum Alloys," *Trans. ASCE,* vol. 109, pp. 1359–1399, 1944.

2. Hartmann, E. C., C. F. Wescoat, and M. W. Brennecke: "Prescriptions for Head Cracks in 24S-T Rivets," *Aviation,* November 1943.

3. Wilson, W. M., and W. A. Oliver: "Tension Tests of Rivets," *Univ. Illinois Eng. Expt. Sta. Bull.* 210, 1930.

4. Finch, D. M., and J. E. Dorn: Dimpling Technique Developed for High Strength Aluminum Alloys, OPRD Research Project, February 1945.

5. "Metallic Materials and Elements for Aerospace Vehicle Structures," MIL-HDBK 5D, chap. 8, U.S. Government Printing Office, Washington, D.C.

6. "AISC Manual," American Institute of Steel Construction, New York, 5th ed., 1956; 7th ed., 1978.

7. Davis, R. E., G. B. Woodruff, and H. E. Davis: "Tension Tests of Large Riveted Joints," *Trans. ASCE,* vol. 105, pp. 1193–1299, 1940. Discussion by W. M. Wilson, pp. 1264–1275; discussion by E. C. Hartmann and Marshall Holt, pp. 1291–1295.

8. Wilson, W. M., J. Mather, and C. O. Harris: "Tests of Joints in Wide Plates," *Univ. Illinois Eng. Expt. Sta. Bull.* 239.

9. American Railway Engineering Association Specifications for Steel Railway Bridges, 1981–1982.

10. American Association of State Highway Officials Standard Specifications for Highway Bridges, 1980.

11. Sec. VIII, ASME Boiler and Pressure Vessel Code.

12. Dewalt, W. J., and K. O. Bogardus: "Static Shear Strength of 2117-T4 Aluminum-Alloy Rivets at Elevated Temperatures," *NACA Res. Mem.* 55130, January 1956.

13. Finley, E. M.: "Bearing Strengths of Some 75S-T6 and 14S-T6 Aluminum-Alloy Hand Forgings," *NACA Tech. Note* 2883, January 1953.

14. Hrennikoff, A.: "Work of Rivets in Riveted Joints," *Trans. ASCE,* vol. 99, pp. 437–449, 1934. Discussion by A. E. R. de Jonge, pp. 474–484; discussion by H. N. Hill and Marshall Holt, pp. 464–469.

15. ASCE Committee for Lightweight Structures: Specifications for Structures of Aluminum Alloy 6061-T6, *Proc. ASCE,* Paper 970.

16. ASCE Committee for Lightweight Structures: Specifications for Structures of Aluminum Alloy 2014-T6, *Proc. ASCE,* Paper 971.

17. Hartmann, E. C., Marshall Holt, and I. D. Eaton: "Static and Fatigue Strengths of High-Strength Aluminum-Alloy Bolted Joints, *NACA Tech. Note* 2276, February 1951.

18. Holt, M., I. D. Eaton, and R. B. Matthiesen: "Fatigue Tests of Riveted or Bolted Aluminum Alloy Joints," *J. Struct. Div. ASCE,* Paper 1148.

19. Mordfin, L., H. Nixon, and G. E. Greene: "Investigations of Creep Behavior of Structural Joints under Cyclic Loads and Temperatures," *NASA Tech. Note* D-181.

20. Wilson, W. M., and Herbert F. Moore: "Tests to Determine the Rigidity of Riveted Joints in Steel Structures," *Univ. Illinois Eng. Expt. Sta. Bull.* 104.

21. "Large Rivets (½ in. Nominal Diameter and Larger)," ANS/ASME B18.1.2, American Society of Mechanical Engineers, New York, 1972.

22. "Small Solid Rivets," ANS/ASME B18.1.1, American Society of Mechanical Engineers, New York, 1972.

23. Fisher, J. W., and J. H. A. Strunk: "Guide to Design Criteria for Bolted and Riveted Joints," John Wiley & Sons, Inc., New York, 1974.

24. "Rivet, Solid Countersunk 100°, Precision Head, Aluminum and Titanium Columbian Alloy," Department of Defense Military Standard 20426, draft dated March 1983.

SECTION 14
SHRINK- AND PRESS-FITTED ASSEMBLIES

Stanley J. Becker, Ph.D.

Adjunct Professor of Mechanical Engineering
Villanova University
Philadelphia, Pa.

14.1 INTRODUCTION

"Shrink" and "press fits" are permanent connections which operate by interference of materials. They differ in method of assembling. A press fit or force fit is obtained by forcing a shaft into a smaller hole. A shrink fit is produced by heating the member having the hole and allowing it to cool to the shaft and ambient temperature. The same effect is sometimes utilized when the shaft is initially subcooled. Both methods may be used for holding a hub and shaft together. The holding power (torsional and axial) of a shrink joint is considerably greater than that of a press joint with the same interference. Fits are classified in Sec. 19. However, the shrink fit may also be applied to compound cylinders, designed to withstand high internal pressures with a minimum use of material. In the earlier portion of this section the theoretical relationships of the compounded-cylinders design will be emphasized. Joint designs represent simpler cases and utilize the fundamentals shown.

14.2 SHRINK FITTING

The elastic design of cylinders which sustain an internal pressure greater than one-tenth of the allowable tensile stress requires the use of the thick-cylinder formulas originally developed by Lamé.[1] The designer may frequently find formulas in use for thick cylinders or for shrink fitting which appear to be modifications of the original Lamé formulas. Often these are Lamé's formulas combined with an end condition of closure or nonclosure and a yield condition to provide some equivalent, but not real, stress value. For a clear understanding of design, it is best to give separate consideration to stress, end closure, and yield condition.

14.2.1 Nomenclature

$\sigma_t, \sigma_r, \sigma_z =$ tangent, radial, and axial normal stresses, respectively, positive if tension

$\epsilon_t, \epsilon_r, \epsilon_z =$ tangent, radial, and axial normal strains, respectively, positive if extension

$p_j =$ pressure acting at typical radius j

$p_j' =$ passive interference pressure at radius j after first subassembly

$p_j'' =$ passive interference pressure at radius j after second subassembly

$a, b, \ldots =$ typical boundary radii

$\delta_j =$ diametrical interference at typical radius j

$\Delta_1\delta_j =$ change in measured δ_j after first subassembly

$\Delta_2\delta_j =$ change in measured δ_j after second subassembly

$\nu =$ Poisson's ratio

$E =$ Young's modulus at operating temperature

$E' =$ Young's modulus at ambient temperature

$\alpha =$ thermal coefficient of expansion

$\Delta T =$ temperature change

$u =$ radial outward deflection at radius r

$r =$ arbitrary radius

$C_1, C_2 =$ constants of integration

$\tau_{fg} =$ shear yield stress in cylinder with inside boundary radius f and outside boundary radius g. Shear yield stress is taken at 57.7 percent of tensile yield stress

14.2.2 Basic Equations

Lamé's equations can be developed directly from a fundamental law, a mathematical relationship, and two assumed physical relationships.

Fundamental Law.   Equilibrium of forces on an elementary volume of the cylinder (Fig. 14.1). Since every radius exhibits the same force conditions, equilibrium leads to the stress equation

$$\sigma_t - \sigma_r - r\, d\sigma_r/dr = 0 \qquad (14.1)$$

Mathematical Relationship—Compatibility (Fig. 14.2).   By definition $\epsilon_r = du/dr$. If a circle of radius r is expanded to radius $r + u$, then its new length is $2\pi(r + u)$ while its original length was $2\pi r$. Thus its length has increased by $2\pi u$ and the tangent strain along the circle is $2\pi u/2\pi r = u/r$. Since ϵ_r and ϵ_t are both defined by the same quantity u, they must be related (i.e., compatible). The compatibility relationship is

$$\epsilon_t - \epsilon_r + r\, d\epsilon_t/dr = 0 \qquad (14.2)$$

The *assumed physical relationships* hold reasonably well for most materials and most cylinders.

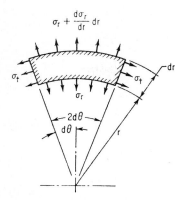

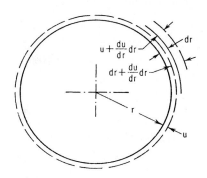

FIG. 14.1 Elementary cylindrical volume under radially symmetric forces.

FIG. 14.2 Strains under radial expansion. $\epsilon_r = ([dr + (du/dr)dr] - dr)/dr = du/dr.$ $\epsilon_t = [2\pi(r +$

1. *Hooke's law:* This linear relationship between stress and strain is based upon experimental evidence provided by simple tension and compression tests on many specimens. Within the elastic limit, it is a good approximation for some materials such as mild steel, and a poor approximation for others. In either case, it is the simplest relationship available for analysis which approximates the true state of affairs.

Taking into account the effect of lateral contraction of the material, the three principal strains can be expressed in terms of the three principal stresses in the following form of Hooke's law:

$$E\epsilon_r = \sigma_r - \nu(\sigma_t + \sigma_z) \tag{14.3}$$

$$E\epsilon_t = \sigma_t - \nu(\sigma_z + \sigma_r) \tag{14.4}$$

$$E\epsilon_z = \sigma_z - \nu(\sigma_r + \sigma_t) \tag{14.5}$$

2. *Generalized plane strain:* The second assumed relationship is generalized plain strain; i.e., ϵ_z is a constant all across any given cross section of the cylinder. Its value depends on the axial and radial load and the particular cross section in question. This relationship is very nearly true for most cylinders at distances from an end closure or a very sudden change in section greater than about three times the mean value of average radius and wall thickness.

14.2.3 Lamé's Equations for Cylinders

Solving Eq. (14.5) for σ_z and applying the result to Eqs. (14.3) and (14.4) yields

$$E\epsilon_r = (1 - \nu^2)\sigma_r - \nu(1 + \nu)\sigma_t - \nu E\epsilon_z \tag{14.6}$$

and $$E\epsilon_t = (1 - \nu^2)\sigma_t - \nu(1 + \nu)\sigma_r - \nu E\epsilon_z \tag{14.7}$$

By inserting Eqs. (14.6) and (14.7) into the compatibility equation [Eq. (14.2)] and using the generalized plane strain relationship that ϵ_z is independent of r and therefore has no derivative with respect to r, the following stress equation results:

$$(1 - \nu^2)r \, d\sigma_t/dr - \nu(1 + \nu)r \, d\sigma_r/dr = (1 + \nu)(\sigma_r - \sigma_t) \tag{14.8}$$

From the equilibrium equation,

$$\sigma_r - \sigma_t = -r\, d\sigma_t/dr \qquad \text{and} \qquad d\sigma_r/dr = r\, d^2\sigma_t/dr^2 + 2\, d\sigma_t/dr$$

Inserting these into Eq. (14.8) and simplifying the result,

$$d^2\sigma_t/dr^2 + (3/r)(d\sigma_t/dr) = 0$$

or, in another form,

$$(1/r^3)[(d/dr)(r^3\, d\sigma_t/dr)] = 0 \tag{14.9}$$

This integrates immediately to

$$\sigma_r = C_1 + C_2/r^2$$

and therefore

$$\sigma_t = C_1 - C_2/r^2 \tag{14.10}$$

where C_1 and C_2 are arbitrary constants. Use is made of the following boundary conditions: When $r = a$, $\sigma_r = -p_a$; when $r = b$, $\sigma_r = -p_b$. Therefore,

$$C_1 = \frac{p_a a^2 - p_b b^2}{b^2 - a^2} \qquad C_2 = -\frac{a^2 b^2}{b^2 - a^2}(p_a - p_b) \tag{14.11}$$

In the case of the solid shaft $C_2 = 0$, since σ_r cannot be infinite at $r = 0$, and therefore $C_1 = -p_b$.

Hence, for the hollow cylinder,

$$\sigma_r = \frac{p_a a^2 - p_b b^2}{b^2 - a^2} - \frac{p_a - p_b}{r^2}\frac{a^2 b^2}{b^2 - a^2} \tag{14.12}$$

$$\sigma_t = \frac{p_a a^2 - p_b b^2}{b^2 - a^2} + \frac{p_a - p_b}{r^2}\frac{a^2 b^2}{b^2 - a^2} \tag{14.13}$$

and, for the solid shaft,

$$\sigma_r = \sigma_t = -p_b \tag{14.14}$$

Since $\epsilon_t = u/r$, from Eq. (14.7),

$$u = (r/E)[(1 - v^2)\sigma_t - v(1 + v)\sigma_r - vE\epsilon_z] \tag{14.15}$$

or

$$u = \frac{(1 + v)(1 - 2v)}{E}\frac{p_a a^2 - p_b b^2}{b^2 - a^2}r + \frac{1 + v}{E}\frac{p_a - p_b}{r}\frac{a^2 b^2}{b^2 - a^2} - vr\epsilon_z \tag{14.16}$$

From Eq. (14.10), it follows that $\sigma_r + \sigma_t = 2C_1$, a constant. From Eq. (14.5), since ϵ_z is also constant for a given cross section, it follows that σ_z is constant for a section. This has not been previously assumed. Hence σ_z is known once the axial force is known; that is, σ_z = axial force per unit cross section. It follows from $\epsilon_t = u/r$ [Eq. (14.4) and Eqs. (14.10) and (14.11)] that

$$u = \frac{1 - v}{E}\frac{p_a a^2 - p_b b^2}{b^2 - a^2}r + \frac{1+v}{E}\frac{p_a - p_b}{r}\frac{a^2 b^2}{b^2 - a^2} - \frac{vr}{E}\sigma_z \tag{14.17}$$

Equation (14.16) will often prove more useful than the usual alternative form, Eq. (14.17).

14.2.4 The Compound Cylinder (Fig. 14.3)

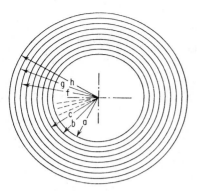

FIG. 14.3 The compound cylinder.

From the form of Eqs. (14.12) and (14.13), it can be seen that the largest magnitude of stress difference, $\sigma_t - \sigma_r$, in the thick hollow cylinder appears at the bore of the cylinder and decreases rapidly with increasing radius. Because of this unequal distribution of stress along a radius, the material in a thick hollow cylinder is not efficiently utilized in resisting internal pressure. The situation can be improved by proper prestressing of the material.

One method of prestressing a thick cylinder against an internal pressure is by constructing the cylinder of two or more concentric cylinders with an interference fit between the component cylinders. This places an external pressure on the innermost cylinder which tends to balance the internal pressure load on this component. At the same time, internal pressure acts on the outermost cylinder before any pressure load is put on the assembly, thus increasing the maximum internal pressure on the outermost component. Such an assemblage of concentric cylinders is known as a *compound cylinder.*

If two adjacent components with the same elastic modulus of such a compound cylinder are considered, with inside radius f, interference radius g, and outer radius h, and with pressures acting at these radii with values p_f, p_g, and p_h, respectively, the outward radial deflection at the outer radius of the inner cylinder from Eq. (14.16) is

$$u = \frac{(1+v)(1-2v)}{E} \frac{p_f f^2 - p_g g^2}{g^2 - f^2} g + \frac{1+v}{E} \frac{p_f - p_g}{g} \frac{f^2 g^2}{g^2 - f^2} - vg\epsilon_z$$

The outward radial deflection at the inner radius of the outer cylinder is

$$u = \frac{(1+v)(1-2v)}{E} \frac{p_g g^2 - p_h h^2}{h^2 - g^2} g + \frac{1+v}{E} \frac{p_g - p_h}{g} \frac{g^2 h^2}{h^2 - g^2} - vg\epsilon_z$$

The difference between these two deflections must equal one-half the diametral interference. The diametral interference at g is therefore

$$\delta_g = \frac{4g(1-v^2)}{E} \left[\frac{p_g - p_h}{1 - (g/h)^2} - \left(\frac{f}{g}\right)^2 \frac{p_f - p_g}{1 - (f/g)^2} \right] \tag{14.18}$$

δ_g is not changed by changes in pressure loads at any internal or external radius of this subassembly or any other larger subassembly or the entire assembly of the compound cylinder.[2] Therefore, the values of p_f, p_g, and p_h in Eq. (14.18) can be either the passive interference load or the active external load plus interference load. In particular, for the two-component assembly with radii a, b, and c (in order of magnitude), the passive interference pressure at b is

$$p_b = \frac{E\delta_b}{4b(1-v^2)} \frac{[1 - (a/b)^2][1 - (b/c)^2]}{1 - (a/c)^2} \tag{14.19}$$

The case of a cylindrical sleeve of outside radius b, placed over a solid shaft of radius a, with diametral interference δ_a, will be considered separately. For this case Eqs. (14.14) and (14.15) result, for the solid shaft, in

$$u = (r/E)[(1 + v)(1 - 2v)p_a + vE\epsilon_z]$$

which when evaluated at $r = a$, deducting the value from that of Eq. (14.16) at $r = a$, results in one-half of δ_a. Thus

$$\delta_a = \frac{4a(1 - v^2)}{E} \frac{p_a - p_b}{1 - (a/b)^2} \tag{14.20}$$

or

$$p_a - p_b = [E/4a(1 - v^2)]\{\delta_a[1 - (a/b)^2]\} \tag{14.21}$$

14.2.5 Balanced Design for the Compound Cylinder

A criterion for the elastic design of a thick-walled cylinder is quite properly chosen as that of yielding. This criterion is quite separate from that of rupture and should not be confused with the latter. Other design criteria which may be considered are fatigue and creep.

Since yielding does not occur all at once in the cylinder throughout its wall thickness, the yield criterion may be considered in various stages.[9] For basic elastic design, initial yield at the inside wall of the cylinder is a necessary criterion. A balanced design of the compound cylinder under such a criterion is one in which every cylindrical component yields simultaneously upon application of internal pressure on the whole assemblage.

The yield criterion requires a yield condition for the materials used. To this end the Tresca (also called the maximum shear) yield condition is selected, modified so that it will match, for the case of plane strain (zero third principal strain), the Von Mises (also called the distortion energy or octahedral shear stress) yield condition. To use this condition, it is required that the yield shear stress is taken to be 57.7 percent of the tensile yield and that the maximum difference of principal stress at any point in the material is $\sigma_t - \sigma_r$.

Stress geometry requires that the maximum shear stress at a point be one-half of the maximum difference of principal stresses at the point. The modified Tresca condition thus cannot satisfy one-dimensional tensile or compressive stress test data but because of its construction is quite satisfactory for design of thick cylindrical pressure vessels.

For a typical cylindrical component with inner radius f and outer radius g, using Eqs. (14.10) and (14.11), the maximum shear stress, which must occur at the smallest radius, is

$$\tau_{fg} = \left(\frac{\sigma_t - \sigma_r}{2}\right)_{r=f} = \frac{C_2}{f^2} = \frac{p_f - p_g}{1 - (f/g)^2} \tag{14.22}$$

From Eqs. (14.18) and (14.22), it follows that the diametral interference necessary to achieve the balanced design is given by

$$\delta_g = [4g(1 - v^2)/E][\tau_{gh} - (f/g)^2 \tau_{fg}] \tag{14.23}$$

The pressure at f, from Eq. (14.22),

$$p_f = \tau_{fg}[1 - (f/g)^2] + \tau_{gh}[1 - (g/h)^2] + p_h \tag{14.24}$$

It is desired to adjust g so that p_f will be maximized with all other quantities unaltered. This requires that $\partial p_f / \partial g = 0$, or that

$$\tau_{fg}(f/g)^2 = \tau_{gh}(g/h)^2 \tag{14.25}$$

or

$$g = (\tau_{fg}/\tau_{gh})^{1/4}(fh)^{1/2} \tag{14.26}$$

Since $\partial^2 p_f / \partial g^2$ is readily shown to be negative for any radius g, Eqs. (14.25) and (14.26) are a maximizing condition for p_f.

14.2.6 Corrections for Operating Temperature and Assembly Order

The general method of economical balanced design of compound cylinders can be achieved by repeated and successive use of Eqs. (14.22), (14.23), and (14.25). However, the values of interference given by Eqs. (14.18) and (14.23) are those of the completely disassembled compound cylinder. Since it is usually preferable to manufacture the interferences at various stages of fabrication, proper account must be made of this modification. Also the operating temperature may not be the same as the fabrication temperature and proper account must be made of this effect on the interference values. (The optimum design radii are so slightly affected by the latter change that it is not worthwhile to alter the basic radii for operating-temperature considerations.)

In placing larger cylinders over smaller subassemblies, the growth of the outside diameter of the last subassembly due to that assembly must be added to the free-state (disassembled) interference at the same diameter in order to maintain the correct interference pressures.[2] Likewise, in placing smaller cylinders inside subassemblies, the contraction of the inside diameter of the last subassembly should be added to the free-state (disassembled) interference at the same diameter in order to maintain the correct interference pressure.

If the operating temperature is ΔT above ambient or fabrication temperature, and if α is the coefficient of expansion of the material used, then, at ambient, all dimensions including the interferences are smaller by a factor of $1 - \alpha \Delta T$. The pressures and stresses should be multiplied by $(E'/E)(1 - \alpha \Delta T)$ in going from operating to ambient conditions, where E' is Young's modulus at ambient conditions while E is Young's modulus at operating conditions.

Assume that the calculated interferences of disassembly have been corrected to ambient temperature and measured at this temperature. Suppose that a four-cylinder assembly is to be manufactured with successive radii $e < f < g < h < j$ and suppose that the middle subassembly $f < g < h$ is to be assembled first with design interference at g as given by Eq. (14.18) or (14.23) and corrected to ambient conditions. From Eq. (14.16), the additional diametral interference necessary at f at the ambient temperature to maintain the proper operating interference pressure is

$$\Delta_1 \delta_f = \frac{4f(1 - \nu^2)}{1 - (f/g)^2} \frac{p'_g}{E'} \tag{14.27}$$

where p'_g is the passive interference pressure at g at ambient temperature caused by the first subassembly [and calculated from Eq. (14.18)].

Similarly, the additional diametral interference at h to be accounted for is

$$\Delta_1 \delta_h = \frac{(g/h)^2 4h(1 - \nu^2)}{1 - (g/h)^2} \frac{p'_g}{E'} \tag{14.28}$$

at ambient temperature.

Suppose that the first subassembly is followed by slipping the subassembly over cylinder *ef* with corrected interference. This will cause a passive interference pressure at *f* at ambient conditions of p''_f and the total corrected diametral interference at *h* at ambient conditions will be

$$\Delta_2 \delta_h = \frac{4(1 - \nu^2)}{E'} \left[\frac{hp'_g (g/h)^2}{1 - (g/h)^2} + \frac{hp''_f (f/h)^2}{1 - (f/h)^2} \right] \tag{14.29}$$

However, after the second subassembly, the ambient interference pressure at *g*, using Eq. (14.12) and superposition, is

$$p''_g = p'_g + p''_f \frac{(f/g)^2 - (f/h)^2}{1 - (f/h)^2} \tag{14.30}$$

If Eq. (14.30) is solved for p'_g and the result applied to Eq. (14.29), it is easily shown that the correction to the interference at *h* is

$$\Delta_2 \delta_h = \frac{(g/h)^2 4h(1 - \nu^2)}{1 - (g/h)^2} \frac{p''_g}{E'} \tag{14.31}$$

which has the same form as Eq. (14.28) with p''_g substituted for p'_g.

The passive interference pressure of the first subassembly at ambient temperature, from Eq. (14.18) with $p'_f = p'_h = 0$, is

$$p'_g = \frac{E'}{1 - \nu^2} (1 - \alpha \Delta T) \frac{\delta_g}{4g} \frac{[1 - (f/g)^2][1 - (g/h)^2]}{1 - (f/h)^2} \tag{14.32}$$

The factor $1 - \alpha \Delta T$ is used here because it is applied to δ_g in actual measurement but not to *g*, which is kept at nominal value as measured.

For passive interference pressure at operating condition of this same subassembly at *g*, the factor $E'(1 - \alpha \Delta T)$ in Eq. (14.32) is changed to *E*.

For the next subassembly, with $p''_e = p''_h = 0$, the passive interference pressure at *f* at ambient temperature is

$$p''_f = \frac{E'}{1 - \nu^2} (1 - \alpha \Delta T) \frac{\delta_f + \Delta_1 \delta_f}{4f} \frac{[1 - (e/f)^2][1 - (f/h)^2]}{1 - (e/h)^2} \tag{14.33}$$

where $(1 - \alpha \Delta T) \Delta_1 \delta_f$ is the correction to be made to δ_f for the previous subassembly and is given by Eq. (14.27).

The method of calculation can be extended for any order of assemblage.

14.2.7   Unbalanced Design, Creep, Fatigue, Partial Yielding, Autofrettage

The analysis of the balanced design ignores certain inelastic effects which may make an unbalanced design preferable. For example, a design which causes a particular component of an assembly to operate above its fatigue limit for the anticipated number of cycles during a lifetime, even though below the yield point, would not be preferred to one utilizing more material in an unbalanced design, but below the fatigue limit. Furthermore, the effect of creep may destroy the balance of an original design, thereby reducing its strength. The latter may be especially true where heavy shrink stresses are involved. Even if the assembly is fully loaded over a short portion of its lifetime and unloaded during a major portion of its lifetime, creep properties of the materials involved may govern the design. In this connection, it is noted that creep acts to lower the design interference pressures, thereby decreasing the support and strength of the innermost cylindrical components. The design interference pressure may therefore be governed by permissible creep limits rather than by the yield strength.

It is not the purpose here to enter into a full discussion of fatigue and creep properties of simple or compound cylinders. Work on these subjects is not extensive and much research is in progress. Rather, reference is made to a few representative published articles.[3,4,5,6]

Because the balanced design is based upon initial yield criteria, it is not directly related to either full yielding or bursting of such pressure vessels. Some theoretical work has been done on these subjects for simple cylinders (Refs. 7 and 8, for example) but little on compound cylinders. Full yielding is treated incidental to partial yielding of compound cylinders under restricted conditions in Ref. 9, and the possibility of increasing the strength of such compound cylinders by a measured amount of partial yielding (autofrettaging) has been investigated, all as an extension of the previous theory. It has been found that the ideal autofrettage pressure, to be used as a test pressure, is given by the formula [Eq. (10) of Ref. 9]

$$p_a = (\tau_{ab+} - \tau_{ab-})[1 - (a/l)^2] \qquad (14.34)$$

provided that certain conditions, given below, are met. In this formula, p_a is the internal pressure that, when released, will cause the compound cylinder to be on the verge of reversed yielding at the inside radius a, *and no other place,* provided that the Tresca yield condition, as given by Eq. (14.22), is satisfied on load application and that the materials exhibit no strain hardening. l is the outside radius of the assembly, τ_{ab+} is the shear yield stress for the inner cylinder under load (57.7 percent of tensile yield), and τ_{ab-} is the corresponding shear yield stress on reversed yielding. τ_{ab-} is negative to τ_{ab+} and usually $\tau_{ab-} = -\tau_{ab+}$; however, the formula permits different conditions of yielding on stress reversal.

14.2.8   Unequal Tube Lengths

The development previously presented is essentially one-dimensional, in that all effects depend solely on a radius location as the one independent space coordinate. Recently more attention has been given to the effects of variations along the axial length of the cylinders. In particular, attention has been focused on the problem of pressure and stress concentrations at compound cylinder tube ends which are short of the ends of the rest of the cylinder. The same problem occurs where a cylindrical hub is shrunk onto a solid shaft of greater length (Fig. 14.4). The nature of the problem is indicated in Ref. 10, where the problem is treated by means of thin-shell theory (with application to thin shells). The solution indicates that the interference pressure becomes a concentrated force per unit circumference at such short ends. Concentrated forces are permitted in thin-shell theory just as in beam theory.

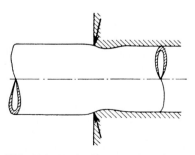

FIG. 14.4   Pinch effect at a discontinuity of an interference fit.

In Refs. 11 and 12, the effect of unequal tube lengths is studied in greater detail without restriction to thin-shell theory. There is no contradiction in the results from that of the previous reference.

Because of this concentration of forces, design modifications have been suggested to alleviate the condition. Essentially, these efforts have been to make the ends of the shorter tubes more flexible.

14.3 PRESS AND SHRINK JOINTS

14.3.1 Description

The term "joints" applied to a mechanism denotes a connection between two members, capable of power transmission. A press joint or a shrink joint refers specifically to a connection between a shaft and tube, or between two tubes, or between two shafts connected by a tube. Such a joint is an interference fit depending solely upon the resulting friction for its ability to transmit power. There are two design questions associated with this joint:

1. Is there enough friction present to transmit the necessary power without slippage?
2. Is there enough strength in the structural members to carry both the stresses caused by the interference and the additional stresses caused by power transmission?

The first question, which is basic to the design, is easily answered. The second, somewhat more complex, can be answered most easily if the effects of interference and power transmission are considered separately and then superposed, together with other effects to be discussed shortly.

14.3.2 Power vs. Friction

There are two types of power that can be transmitted by a press or shrink joint: torsional and axial. The holding ability is

$$T = 2\pi f p_b L b^2 \qquad \text{for torsion} \qquad (14.35)$$

and

$$F = 2\pi f p_b L b \qquad \text{for axial force} \qquad (14.36)$$

where T = torque
 F = force
 L = interference length
 b = interference radius
 f = coefficient of friction
 p_b = interference pressure

Equation (14.36) is approximate since it assumes a constant value of p_b. The axial force, however, will cause p_b to vary along the length of the joint because of the Poisson effect of the axial stresses. This is of no practical consequence, however, because of the uncertainty in the coefficient of friction.

Table 14.1, from Ref. 16 via 15, is a guide to the selection of coefficients of friction.

Equations (14.35) and (14.36), together with a knowledge of the coefficient of friction, are sufficient to answer the first design question, provided that b and p_b are known. These can be selected, essentially, on the basis of Lamé's equations and the theory of the compound cylinder. The stresses caused by power transmission should

TABLE 14.1   Coefficients of Friction for Pressed and Shrunk Assemblies

	Coefficient of friction		
	Min	Max	Avg
a. Assembly of 62 press fits; lower values due to hub yielding as a result of too large fit allowance.........	0.030	0.250	0.086
b. Pressing four steel axles of $5\frac{1}{4}$ in. diam and 175 Brinell hardness into a cast-steel spider....................	0.100	0.140	0.115
Pressing off the pressed-on spider................	0.120	0.150	0.137
Pressing off the shrunk-on spider................	0.160	0.170	0.165
c. Pressing 15 axles of $6\frac{3}{16}$ in. diam into gears, both of which were steel having a hardness of 170 Brinell.....	0.075	0.250	0.150
d. Pressing steel gears off steel shafts having tapered fits, where the mean diameter was $6\frac{1}{2}$ in.:			
When the parts were pressed together.............	0.150	0.210	0.177
When the parts were shrunk together.............	0.210	0.230	0.220

be superposed on the interference stresses. (Also stresses caused by discontinuity, as indicated in Refs. 10, 11, and 12, should be included.)

14.3.3   Interference, Interference Pressure, and Interference Stresses

In the case of joint interference, unlike that of the compound cylinder, the condition of the interference is essentially plane stress. Equations (14.12) and (14.13) remain valid since they are independent of the value of σ_z or ϵ_z, but Eq. (14.17) with σ_z set equal to zero will be found more useful than Eq. (14.16). If the joint members all have the same elastic modulus, then Eqs. (14.18) to (14.21) remain valid with the substitution of unity for the factor $1 - v^2$ wherever it appears so that plane stress conditions are maintained. In the present notation, considering only interference pressures, these equations become

$$\delta_b = \frac{4b}{E} p_b \frac{1 - (a/c)^2}{[1 - (a/b)^2][1 - (b/c)^2]} \qquad \text{for a hollow shaft} \qquad (14.18a)$$

$$p_b = \frac{E\delta_b}{4b} \frac{[1 - (a/b)^2][1 - (b/c)^2]}{1 - (a/c)^2} \qquad \text{for a hollow shaft} \qquad (14.19a)$$

$$\delta_b = \frac{4b}{E} p_b \frac{1}{1 - (b/c)^2} \qquad \text{for a solid shaft} \qquad (14.20a)$$

$$p_b = (E\delta_b/4b)[1 - (b/c)^2] \qquad \text{for a solid shaft} \qquad (14.21a)$$

Where the mating parts of a press or shrink joint are made of materials with different elastic moduli, direct application of Eq. (14.17) with $\sigma_z = 0$ results in

$$\delta_b = 2p_b b \left[\frac{1}{E_{ab}} \left(\frac{b^2 + a^2}{b^2 - a^2} - v_{ab} \right) + \frac{1}{E_{bc}} \left(\frac{c^2 + b^2}{c^2 - b^2} + v_{bc} \right) \right] \qquad \text{for a hollow shaft} \quad (14.37)$$

where E_{ab} and v_{ab} are Young's modulus and Poisson's ratio, respectively, in the shaft with inside radius a and interference radius b, and E_{bc} and v_{bc} are Young's modulus and Poisson's ratio, respectively, in the hub with interference radius b and outside radius c.

Hence, the interference pressure is

$$p_b = \left(\frac{\delta_b}{2b}\right)\left[\frac{1}{E_{ab}}\left(\frac{b^2+a^2}{b^2-a^2}-v_{ab}\right)+\frac{1}{E_{bc}}\left(\frac{c^2+b^2}{c^2-b^2}+v_{bc}\right)\right]^{-1} \qquad \text{with a hollow shaft}$$
$$(14.37a)$$

Similarly, for a solid shaft, using Eq. (14.5) with $\sigma_z = 0$, Eqs. (14.14) and (14.16) for the solid shaft, and Eq. (14.17) for the hub, with $\sigma_z = 0$, the interference relations become the same as given by Eqs. (14.37) and (14.37a), but with $a = 0$. Thus for a solid shaft

$$\delta_b = 2p_b b\left[\frac{1-v_{ab}}{E_{ab}}+\frac{1}{E_{bc}}\left(\frac{c^2+b^2}{c^2-b^2}+v_{bc}\right)\right] \qquad (14.37b)$$

The stresses caused by the interference are found by application of Eqs. (14.12) to (14.14):

$$\sigma_r = -\frac{p_b}{1-(a/b)^2}\left[1-\left(\frac{a}{r}\right)^2\right] \qquad \text{for a hollow shaft} \qquad (14.12a)$$

$$\sigma_t = -\frac{p_b}{1-(a/b)^2}\left[1+\left(\frac{a}{r}\right)^2\right] \qquad \text{for a hollow shaft} \qquad (14.13a)$$

$$\sigma_\tau = -\frac{p_b}{1-(b/c)^2}\left[\left(\frac{b}{r}\right)^2-\left(\frac{b}{c}\right)^2\right] \qquad \text{for the hub} \qquad (14.12b)$$

$$\sigma_t = \frac{p_b}{1-(b/c)^2}\left[\left(\frac{b}{r}\right)^2+\left(\frac{b}{c}\right)^2\right] \qquad \text{for the hub} \qquad (14.13b)$$

$$\sigma_t = \sigma_r = -p_b \qquad \text{for the solid shaft} \qquad (14.14)$$

14.3.4 Stresses Due to Power Transmission in Combination with Interferences

Superposed upon the interference stresses and pressures must be those stresses caused by power transmission, bending, and centrifugal action. The latter is discussed in Sec. 24. The true stress picture associated with power transmission is quite complex because load transfer from the shaft to the hub does not occur at one position but rather all along the length of the interference. However, it is conservative and simplifying to assume this transfer occurs at one position along the length of the interference.

Employing this assumption in the case of only axial force and interference, appeal is made to St. Venant's principle, so that at an axial distance of about $c - b$ in the hub from the point of load transfer, and still within the zone of effect of the interference, the axial shear stresses have become essentially dispersed to give a resultant uniform axial normal stress of $F/\pi(c^2 - b^2)$ in the hub. This stress is simply treated as the third principal stress, the other two being σ_r and σ_t as determined by the interference fit. Similarly for the shaft, the axial principal stress is taken as $F/\pi(b^2 - a^2)$. Such axial forces may affect the interference fit, although this effect is not calculated. Of course, other sources of axial stress, such as shaft bending and discontinuity effects, must be superposed.

A similar treatment and use of St. Venant's principle are used for torsional power transmission, but in this case the effect of torsion is to rotate the directions of the principal stresses so that σ_t and σ_z are no longer principal values. Furthermore, since maximum torsional shear stress occurs at the outside while maximum tangential interference stress is at the inside surface of either mating part, it can be shown that it is possible for the maximum principal stress to occur at any point along the radius of either mating part (with the sole exception of the solid shaft), depending on the geometry, interference, and torque. The formulas to be used are Eqs. (14.13a), (14.13b), (14.14), and the following:

$$\tau = 2Tr/\pi(b^4 - a^4) \qquad \text{in the shaft} \qquad (14.38)$$

$$\tau = 2Tr/\pi(c^4 - b^4) \qquad \text{in the hub} \qquad (14.39)$$

and principal stresses at any point are net σ_r (including centrifugal action) and

$$\tfrac{1}{2}(\sigma_t + \sigma_z) \pm \sqrt{[(\sigma_t - \sigma_z)/2]^2 + \tau^2} \qquad (14.40)$$

where σ_t = total tangent stress due to all sources (pressure, discontinuity effects, centrifugal) and σ_z = total axial stress due to all sources (axial force, bending of shaft as a beam, discontinuities).

For further aid in design, assistance in the design of power transmitting joints can be obtained from the charts of Ref. 15. It should be remembered, however, that the total stress in these components is made up of stress due to interference, power transmission, bending, discontinuity effects (as outlined in Refs. 10, 11, and 12), and centrifugal action, if any.

14.3.5 Centrifugal Action on Shafts and Disks

The study of centrifugal effects on rotating parts appears to be a subject quite different from that of shrunk or press fits. However, in the case of shafts or flat disks, the mathematical treatment is so similar to that which has preceded that it is quite natural to discuss the theory in this section.

Equation of Motion. The same conditions exist as were used in writing Eq. (14.1), except that the volume element now has a constant acceleration of $\omega^2\rho = v^2/\rho$ toward the center of rotation, where ω = angular velocity, v = tangent linear velocity, and ρ = radius of rotation.

It is assumed that the center of rotation coincides with the geometric center of the shaft or disk, so that $\rho = r$. In this case Newton's second law of motion ($F = ma$) becomes

$$\sigma_t - \sigma_r - r\,d\sigma_r/dr = (\gamma/g)\omega^2 r^2 \qquad (14.41)$$

where γ = weight density of the rotating part and g = acceleration of gravity.

Hooke's Law under Plane Stress. Since most problems of this type occur under conditions of essentially plane stress or one on which a known axial stress can be superposed upon a system with zero axial stress, Eqs. (14.3) and (14.4) can be used in the simplified form

$$E\epsilon_r = \sigma_r - \nu\sigma_t \qquad (14.42)$$

and
$$E\epsilon_t = \sigma_t - \nu\sigma_r \qquad (14.43)$$

while Eq. (14.5) loses all utility.

Compatibility.   Equation (14.2) remains valid. In a development quite similar to that of Eq. (14.9), the following differential equation can be constructed:

$$(1/r^3)(d/dr)(r^3\, d\sigma_r/dr) = -(3 + \nu)(\gamma/g)\omega^2 \qquad (14.44)$$

Note that this equation is linear in σ_r. It is therefore permissible to separate the static effects of pressure entirely from the dynamic system and superimpose them later, since the static effects are simply part of the homogeneous solution to this equation. If conditions of generalized plane strain had been assumed ($\epsilon_z = 0$) instead of plane stress ($\sigma_z = 0$), the resulting equation of motion would have contained the factor $(3 - 2\nu)/(1 - \nu)$ in place of $3 + \nu$ on the right-hand side, and otherwise would be identical to Eq. (14.44).

The general solution to Eq. (14.44) (homogeneous solution plus particular integral) is

$$\sigma_r = C_1 + C_2/r^2 - [(3 + \nu)/8](\gamma/g)\omega^2 r^2 \qquad (14.45)$$

For the solid shaft or disk, C_2 must be zero for stress to be finite, and if c is the outside radius of shaft, disk, or hub of a press- or shrink-fitted assembly, then $\sigma_r = 0$ at this radius. Thus for a system with a solid shaft, and with the same elastic constants for hub and shaft, the stresses due solely to centrifugal action are given by (with $r \le c$)

$$\sigma_r = [(3 + \nu)/8](\gamma/g)\omega^2(c^2 - r^2) \qquad \text{with solid shaft} \qquad (14.46)$$

and, from Eq. (14.41),

$$\sigma_t = (\gamma\omega^2/8g)[(3+\nu)c^2 - (1 + 3\nu)r^2] \qquad \text{with solid shaft} \qquad (14.47)$$

If the shaft is hollow with inside radius a, then C_2 is not zero and the boundary conditions for centrifugal stresses are $\sigma_r = 0$ at both $r = a$ and $r = c$. This results in stresses due solely to centrifugal action which are given by ($a \le r \le c$)

$$\sigma_r = [(3 + \nu)/8](\gamma/g)\omega^2(c^2 + a^2 - r^2 - a^2c^2/r^2) \qquad (14.48)$$

and, from Eq. (14.41),

$$\sigma_t = (\gamma\omega^2/8g)[(3 + \nu)(c^2 + a^2 + a^2c^2/r^2) - (1 + 3\nu)r^2] \qquad (14.49)$$

For generalized plane strain, the factor $3 + \nu$ in Eqs. (14.46) to (14.49) is replaced by $(3 - 2\nu)/(1 - \nu)$ and the factor $1 + 3\nu$ is replaced by $(1 + 2\nu)/(1 - \nu)$.

This system of stresses is to be superposed upon the stresses due to interference and other causes. In particular, note that σ_r at the interference radius b is always positive and therefore tensile and thus reduces the interference pressure.

Further discussion of this subject will be found in Refs. 14 and 15. The former contains a description of an interesting and useful method for extending the analysis of the flat disk to one of variable thickness. Essentially, the method consists of approximating the variable thickness by a set of concentric rings, each of uniform thickness, and then imposing the condition that the radial force (not stress) per unit circumference shall be continuous across the mating boundaries. The same technique can be applied, where necessary, to the compound cylindrical vessel.

The radial deflections because of centrifugal action can be found from Eq. (14.43) [or Eq. (14.7) for the case of generalized plane strain] since $\epsilon_t = u/r$. Thus, for centrifugal action alone,

$$u = (r/E)(\gamma/8g)\omega^2[(3 + \nu)(1 - \nu)(c^2 + a^2) + (3 + \nu)(1 + \nu)a^2c^2/r^2) - (1 - \nu^2)r^2] \tag{14.50}$$

For generalized plane strain, the factor $3 + \nu$ in Eq. (14.50) is replaced by $(3 - 2\nu)/(1 - \nu)$ and $1 - \nu^2$ is replaced by $1 - [2\nu^2/(1 - \nu)]$, while $-\nu r\epsilon_z$ is added to the total radial deflection.

14.3.6 Design

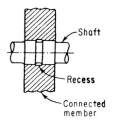

FIG. 14.5 A press-fit design.

Figure 14.5 shows an excellent design for shrink or press fit. Internal recessing provides better symmetry of the shrink-fit stresses and prevents a fulcrum in imperfect fits. Diameters of shaft are reduced at ends of the hub, reducing the stress concentration and shrinkage allowance at these points. If a key is added to the fit, the shaft seat diameter may be further increased and a more gradual step down to the shaft diameter is provided.[17] The reader is referred to Sec. 15 for filament- and fiber-reinforced composite cylinders.

REFERENCES

1. Lamé and Clapeyron: Mémoire sur l'équilibre interieur des corps solides homogénes. *Memoires présentés par divers savans,* vol. 4, 1833.
2. Becker, S. J., and L. Mollick: "The Theory of the Ideal Design of a Compound Vessel," *Trans. ASME,* vol. 82, ser. B, pp. 136–142, May 1960.
3. Morrison, J. L. M., B. Crossland, and J. S. C. Perry: "Fatigue under Triaxial Stress, Development of a Testing Machine and Preliminary Results," *Proc. Inst. Mech. Engrs. (London),* vol. 170, pp. 697–712, 1956.
4. Davis, E. A.: "Relaxation of a Cylinder on a Rigid Shaft," *J. Appl. Mech.* vol. 27, ser. E, no. 1, pp. 41–44, March 1960.
5. Davis, E. A.: "Relaxation of Stress in a Heat-Exchanger Tube of Ideal Material," *Trans. ASME,* vol. 75, no. 1, pp. 381–385, April 1952.
6. Finnie, Iain: "Steady State Creep of a Thick-Walled Cylinder under Combined Axial Load and Internal Pressure," *Trans. ASME,* vol. 82, ser. D, no. 3, pp. 689–694, September 1960.
7. Svensson, N. L.: "The Bursting Pressure of Cylindrical and Spherical Vessels," *J. Appl. Mech.* vol. 25, pp. 89–96, 1958.
8. Marin, J., and F. P. J. Rimrott: "Design of Thick-Walled Pressure Vessels Based on the Plastic Range," *Welding Research Council Bulletin Series, Bull.* 41, July 1958 (Welding Research Council of the Engineering Foundation; other pertinent references are listed in the bibliography of this bulletin).
9. Becker, S. J.: "An Analysis of the Yielded Compound Cylinder," *Trans. ASME,* vol. 83, ser. B, pp. 43–49, February 1961.
10. Friedrich, C. M., and S. J. Becker: "A Thin Cylinder under Semi-Infinite Outward Restraint,"

Bettis Tech. Rev., WAPD-BT-9, pp. 78–83, August 1958 (available from the Office of Technical Services, Department of Commerce, Washington, D.C.).

11. Sparenberg, J. A.: "On a Shrink-Fit Problem," *Appl. Sci. Res.,* Sec. A, vol. 7, pp. 109–120, 1958.

12. Severn, R. T.: "Shrink-fit Stresses between Tubes Having a Finite Interval of Contact," *Quart. J. Mech. Appl. Math.,* vol. 12, pp. 82–88, February 1959.

13. "Manual of Standard and Recommended Practice Wheel and Axle Manual," Association of American Railroads.

14. Timoshenko, S.: "Strength of Materials," part II, D. Van Nostrand Company, Inc., Princeton, N.J., 1958.

15. Horger, O. J.: "ASME Handbook, Metals Engineering—Design," Press-Fitted Assembly, pp. 178–189, McGraw-Hill Book Company, Inc., New York, 1953.

16. Baugher, J. W.: Transmission of Torque by Means of Press and Shrink Fits, *Trans. ASME,* vol. 53, paper MSP-53-10, pp. 85–92, 1931.

17. Baugher, J. W.: Internal Stress and Fracture of Metals, *Metallurgia,* March 1946.

SECTION 15
COMPOSITES

Keith T. Kedward, Ph.D.

Professor of Mechanical Engineering
University of California—Santa Barbara
Santa Barbara, Calif.

SYMBOLS

$[A], A_{ij}$ = in-plane stiffness matrix and a component thereof (laminate)

$[B], B_{ij}$ = in-plane/flexure coupling matrix and a component thereof (laminate)

$[D], D_{ij}$ = flexural stiffness matrix and a component thereof (laminate)

E_{fL}, E_{fT} = elastic modulus of fiber reinforcement in axial (L), transverse (T) directions

E_L, E_T, E_N = elastic modulus of unidirectional layer in fiber direction (L), transverse direction (T), and normal to the plane (N)

E_x, E_y, E_z = elastic modulus of multidirectional laminate along structural axes x, y, z

E_θ, E_r = elastic modulus of composite curved beam in circumferential (θ) and radial (r) directions

E_m = elastic modulus of matrix material

F_L^{tu}, F_L^{cu} = ultimate tensile strength (tu) and compressive strength (cu) in fiber direction (L) for unidirectional layer

F_T^{tu}, F_T^{cu} = ultimate tensile strength (tu) and compressive strength (cu) transverse to fiber direction (T) for unidirectional layer

F_{LT}^{su} = ultimate shear strength (su) referred to L,T directions for unidirectional layer

F_x^{tu}, F_x^{cu}
F_y^{tu}, F_y^{cu}
F_{xy}^{su} = corresponding ultimate strengths for a multidirectional laminate referred to structural (x,y) axes

F^{isu} = ultimate, interlaminar shear strength

G_{LT} = shear modulus for unidirectional layer (in-plane)

G_{xy} = corresponding shear modulus for multidirectional laminate referred to structural, x,y axes

$\mathcal{G}_{Ic}, \mathcal{G}_{IIc}$ = critical strain energy release rates for Mode I, Mode II crack propagation

M = bending moment applied to curved beam

M_x, M_y, M_{xy} = normal moment resultants referred to structural x,y axes, per unit width

N_x, N_y, N_{xy} = normal force resultants re-ferred to structural x,y axes, per unit width

$[Q], Q_{ij}$ = stiffness matrix for a unidirectional layer and a component thereof

$[\overline{Q}], \overline{Q}_{ij}$ = stiffness matrix for an off-axis layer and a component thereof

R_i, R_o, R_m = radii for curved beam to inner surface, outer surface, and midplane, respectively

r = arbitrary radius location for curved beam

$[S], S_{ij}$ = compliance matrix for a unidirectional layer and a component thereof

$[T]$ = transformation matrix

t = thickness dimension for curved beam

u, v, w = displacement components along structural x,y,z axes

u_0, v_0, w_0 = displacement components along structural $x,y,z = 0$ axes

z = distance from midplane of laminate to arbitrary thickness location

α_{fL}, α_{fT} = coefficient of thermal expansion of fiber reinforcement in axial (L) and transverse (T) directions

α_L, α_T = coefficient of thermal expansion of unidirectional layer in fiber direction (L) and transverse direction (T)

α_x, α_y = coefficient of thermal expansion of multidirectional laminate along structural axes x,y

ΔT = change of temperature referenced to stress-free state

ε = strain components

K = curvature components

θ = angle of orientation of fiber direction (L) referenced from structural axis (x) in unidirectional layer

ν_{LT} = major Poisson's ratio characterizing strain produced transverse to fiber direction (T) for an applied stress in the fiber direction (L)

ν_{xy} = major Poisson's ratio characterizing strain produced in y direction for an applied stress in the x direction (multidirectional laminate)

ρ = mass density

σ = direct strain components

τ = shear stress component

ω = angular velocity

15.1 INTRODUCTION

Composite materials are simply a combination of two or more different materials that may provide superior and unique mechanical and physical properties. The most attractive composite systems effectively combine the most desirable properties of their constituents and simultaneously suppresses the least desirable properties. For example, a glass-fiber-reinforced plastic combines the high strength of thin glass fibers with the

ductility and environmental resistance of an epoxy resin; the inherent damage susceptibility of the fiber surface is thereby suppressed, whereas the low stiffness and
strength of the resin are enhanced.

The opportunity to develop superior products for aerospace, automotive, and recre-
ational applications has sustained the interest in advanced composites. Currently composites are being considered on a broader basis; for applications that include civil
engineering structures such as bridges and freeway pillar reinforcement and for bio-
medical products such as prosthetic devices. The recent trend toward affordable composite structures with a somewhat decreased emphasis on performance will have a
major impact on the wider exploitation of composites in engineering.

Composites typically comprise a high-strength synthetic fiber embedded within a
protective matrix. The most mature and widely used composite systems are *polymer
matrix composites* (PMCs), and this system will provide the major focus for this section. Contemporary PMCs typically use a ceramic type of reinforcing fiber such as
carbon, Kevlar, or glass in a resin matrix wherein the fibers make up approximately 60
volume percent of the PMC. Metal or ceramic matrices can be substituted for the resin
matrix to provide a higher-temperature capability. These specialized systems are
termed *metal matrix composites* (MMCs) and *ceramic matrix composites* (CMCs), and
a general qualitative comparison of the relative merits of all three categories is summarized in Table 15.1.

15.1.1 Short-Fiber/Particulate Composites

The fibrous reinforcing constituent of composites may consist of thin continuous
fibers or of relatively short fiber segments, or whiskers. However, reinforcing effectiveness is realized by using segments of relatively high aspect ratio (length-to-diameter ratio). Nevertheless, as a reinforcement for PMCs, these short-fiber or whisker systems are structurally less efficient and very susceptible to long-term and/or cyclic
loadings. On the other hand, the substantially lower cost and reduced anisotropy on
the macroscopic scale render these composite systems appropriate in structurally less
demanding industrial applications.

15.1.2 Continuous-Fiber Composites

Continuous-fiber reinforcements are generally required for structural or high-performance applications. The specific strength (strength-to-density ratio) and specific stiffness (elastic modulus-to-density ratio) of continuous-fiber-reinforced PMCs, for
example, can be vastly superior to conventional metal alloys as illustrated in Fig. 15.1.
These types of composite can also be designed to provide other attractive properties
such as high thermal or electrical conductivity and low coefficient of thermal expansion. Also, depending on how the fibers are oriented or interwoven within the matrix,
these composites can be "tailored" to provide the desired structural properties for a
specific structural component. Thus designing for, and with, anisotropy is a unique
aspect of contemporary composites in that the design engineer must simultaneously
design the structure and the "material" of construction. Of course anisotropy brings
problems as well as unique opportunities. With reference to Fig. 15.1 it should be
appreciated that the vertical bars representing the conventional metals signify the
potential variation in specific strength that may be brought about by changes in alloy
constituents and heat treatment. The angled bars for the continuous-fiber composites
represent the range of specific properties from the unidirectional, all 0° fiber orienta-

TABLE 15.1 Composite Design Comparisons

	PMC	CMC	MMC
Material usage	Not recommended (handling and general damage-tolerance concerns)	Not recommended (handling and general damage-tolerance concerns)	Can prove feasible for dominant axial loads
Nonlinear effects	Usually not important	Nonlinearity due to matrix breakdown and interface failure	Nonlinearities due to plasticity and failures in matrix/interface
Temperature capability	Poor	Good	Moderate
Fatigue	Not major issue (if design avoids out-of-plane loads)	Potential concern	Potential concern
Stiffness, strength, anisotropy	Can lead to significant problems due to coupling, joining, etc.	Moderate concern due to shear strength, interface variability, and residual stress	Not major issue

15.4

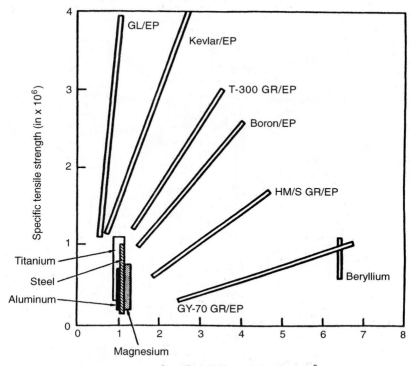

FIG. 15.1 A weight-efficiency comparison.

tion at the upper end to the pseudo-isotropic laminate with equal proportions of fibers in the 0°, +45°, −45°, and 90° orientations at the lower end. In the case of the composites, the variations in between the upper and lower ends of the bars is achieved by "tailoring" in the form of laminate design, which will be treated in Sec. 15.3.

15.1.3 Overview of Composites Design Methodology

Many aspects of the design procedure for composite structures parallel that for conventional metallic structures. However, the approach of the effective composite designer must be based on an even more fundamental foundation since the range of options open to the composites designer is more extensive. The necessity for simultaneous design of the structure and the material of construction is the basic reason for this situation. It is consequential that the composites design function is demanding of a high degree of interaction among many disciplines. Subsequently, the designer of a new composite component must actively promote his or her creation by explaining the advantages it affords, and must strive to gain acceptance through "certification" of the integrity of the product.

An appreciation of the complete design procedure can be gained from a step-by-step description such as that summarized in Fig. 15.2.

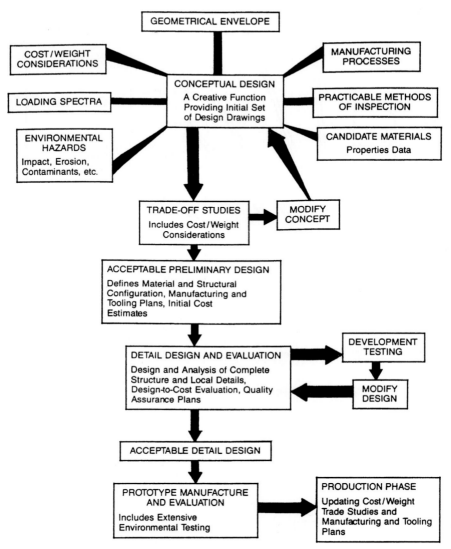

FIG. 15.2   An overview of composites design methodology.

Conceptual Design Considerations.   The starting point is the concept design activity in which consideration is obviously given to the defined geometric envelope, loading, and environmental conditions. Closely interrelated aspects of cost and manufacturing processes also form important, if not key, considerations in the conceptual design phase. A further consideration, which may not be obvious at the outset, is that of potential methods and techniques of inspection that are essential elements involved in the achievement of product acceptance. The selection of materials represents a focal point of the activity and is particularly challenging in view of the bewildering array of material options. This wide range of options adds a new dimension to the design process relative to conventional design with metallic materials. As a consequence, the composites designer is forced to proceed with two or often many more candidate material choices. The first concept is committed to paper, and these design drawings signify completion of one of the most creative phases of the design procedure.

Trade-Off Studies. On the basis of the conceptual design drawings, a structures analysis team is assigned to perform the necessary trade-off studies incorporating cost and weight considerations. It is imperative that the team work closely with the designer and with representatives from design-to-cost, manufacturing technology, quality assurance, materials and processes, and testing functions. It is difficult to clearly separate the concept design and trade-off study activities because the procedure is iterative and interactive. Co-location of the design team is invaluable in all phases of composite design and product development. The outcome of the trade-off study is a preliminary material configuration and a structural configuration supplemented by sufficiently definitive manufacturing and tooling plans to permit cost estimates. With this information the designer enters into the detail design and evaluation phases.

Detail Design and Evaluation. Close attention to design-to-cost and the associated fabrication, tooling, and quality assurance aspects must be exercised in this phase. Both the nonrecurring, or acquisition, costs associated with tooling and facilities and the recurring costs of labor and materials production should be considered. The premium that results from the introduction of vehicle weight savings also enters the picture; a realistic estimate of this value in dollars per pound must be assigned to aid the designer. Structurally, design details (e.g., the joint concepts selected) are considered, and appropriate methods of structural analysis are used to size the component. It is imperative that the development of the transverse or secondary stresses is identified and carefully analyzed in detail design. Differences between composite designs and designs based on conventional materials are emphasized in this activity. Typically, structural analyses in this phase utilize "global" and "local" finite-element models; substructuring techniques are invaluable in facilitating cost-effective and often iterative design studies.

Depending on the extent of innovative features composing the design, experimental evaluation (development testing) or specific features may be included to substantiate the predictions from structural analysis and to provide assurance that the design concept is sound.

Prototype Manufacture and Evaluation. From the detail design and layout drawings created in the previous phase, the necessary tooling and facilities are established and a prototype structure produced that is subjected to representative environmental testing. Because the process may be iterative, prototype tooling and facilities may be of a temporary form. The most successful composite designs emerge from an effort that involves the responsible design engineer or engineering team on an intimate basis throughout the complete cycle.

After successful completion of the prototype evaluation, the design process moves into the production phase, which commences with a reevaluation of the design, examining closely opportunities for reducing part count and identifying and scrutinizing high-cost details to effect cost reductions where possible. Repetition of cost-weight trade-off studies may be necessary so as to reflect high-volume production aspects. Most assuredly, some considerable modification to the manufacturing and tooling plans will be introduced.

15.2 COMPOSITE PROPERTIES

The class of composites which forms the focus of this section is PMCs with continuous-fiber reinforcement. In this type of composite the properties of an arbitrary laminated composite architecture are derived from the elastic and strength properties of a unidirectional layer. The unidirectional layer properties can be derived from the constituent properties of the fiber and matrix that ranges between 50 percent and 65 percent by volume of the fiber reinforcement phase, typically. For our purposes nominal value of 60 percent by volume of fiber will be adopted.

Fiber reinforcements most commonly encountered in contemporary composites include carbon or graphite fibers, Kevlar fibers, and glass fibers, all of which can be obtained in similar diameters, i.e., 0.003–0.005 in. Both the carbon/graphite and Kevlar fibers are inherently anisotropic in themselves, although it is the axial properties that dominate the behavior of unidirectional and, generally, multidirectional arrays or laminates. Typical fiber properties are presented in Table 15.2, where the degree of individual fiber anisotropy is indicated.

The properties of polymer matrices range over a much smaller spectrum as indicated in Table 15.3, and the relatively low stiffness and strength properties rarely dominate the composite behavior, with certain exceptions. Most notable exceptions are the interlaminar shear strength and the thickness-direction tensile strength, wherein the fiber-to-matrix interface may play an important role. For these reasons greatest attention will be placed on the macroscopic composite properties that are of most direct interest to the mechanical or structural engineer. Typical values for such properties are provided in Table 15.4 for three widely used composites. A widely used carbon fiber/epoxy composite is chosen to illustrate typical properties and degrees of anisotropy in elastic, strength, and thermal properties as shown in Table 15.5. Engineers responsible for design and structural evaluation should take particular note of both the degree of anisotropy in strength and stiffness properties. Usually the matrix-dominated properties such as the shear and transverse tensile strengths are very

TABLE 15.2 Typical Fiber Properties

Fiber	Density (lb/in^3)	Axial elastic modulus (10^6 lb/in^2)	Transverse elastic modulus (10^6 lb/in^2)	Tensile strength (10^3 lb/in^2)
E-Glass	0.091	10.5	10.5	500
S-Glass	0.090	12.4	12.4	600
Kevlar 49	0.052	18.0	1.3	400
AS4 Carbon	0.064	35.0	2.0	350

TABLE 15.3 Typical Properties for Polymer Matrices

Polymer	Density (lb/in^3)	Elastic modulus (10^6 lb/in^2)	Tensile strength (10^6 lb/in^2)	Poisson's ratio
HERCULES 3501-6, Epoxy	0.044	0.62	12.0	0.34
NARMCO 5208, Epoxy	0.044	0.50	11.0	0.35
EPON 828, Epoxy	0.044	0.47	13.0	0.35

TABLE 15.4 Properties for Typical Continuous-Fiber-Reinforced Composites and Structural Metals

| Property | Unidirectional composite (60% fiber/40% resin, by volume) | | | | Metals | |
	E-Glass/resin resin	Kevlar/ resin	HS carbon/ epoxy	UHM Gr./ epoxy	7075.T6 aluminum	4130 steel
Elastic:						
Density (lb/in^3)	0.070	0.047	0.058	0.060	0.100	0.284
E_L (msi)	6.5	11.0	19.5	40.0	10.3	30.0
E_T (msi)	1.8	1.0	1.5	1.2	10.3	30.0
G_{LT} (msi)	0.7	0.4	0.9	0.65	4.0	12.0
ν_{LT}	0.32	0.33	0.30	0.28	0.30	0.28
Strength:						
F_L^{tu} (ksi)	180	220	200	100	79	100
F_T^{tu} (ksi)	6	4.5	7	5	77	100
F_L^{cu} (ksi)	120	45	170	90	70	130
F_T^{cu} (ksi)	20	20	20	20	70	130
F_{LT}^{su} (ksi)	8	4	10	9	47	60

TABLE 15.5 Typical Unidirectional Properties for Graphite/Epoxy System

Stiffness properties		Strength properties		Thermal properties	
E_L (msi)	20.0	F_L^{tu} (ksi)	240.0	α_L ($\mu\varepsilon$/°F)	-0.3
E_T (msi)	1.4	F_L^{cu} (ksi)	200.0	α_T ($\mu\varepsilon$/°F)	17.0
G_{LT} (msi)	0.8	F_T^{tu} (ksi)	7.0	K_L (btu in/h ft^2 °F)	40.0
ν_{LT}	0.28	F_T^{cu} (ksi)	20.0	K_T (btu in/h ft^2 °F)	4.5
		F_{LT}^{su} (ksi)	10.0		
$\dfrac{\nu_{LT}}{E_L} = \dfrac{\nu_{TL}}{E_T}$		F^{isu} (ksi)	9.0		

low and the avoidance of matrix-dominated failure modes represents a major challenge for the structural designer. It is also worthy of note that compression strength in the fiber direction, $F_L{}^{cu}$, is significantly lower than the equivalent tensile strength $F_L{}^{tu}$ due to a microfiber instability mechanism. In fact, the ratio of these two strengths $(F_L{}^{cu}/F_L{}^{tu})$ may be much lower for some other systems, e.g., Kevlar/epoxy and more recent high strain-to-failure carbon fibers. The lower compression strengths relative to the tensile strengths are also influenced by the fiber diameter and the matrix properties that are themselves affected by moisture, temperature, interface integrity, and porosity.

15.2.1 In Situ Properties

An important fundamental aspect of multidirectional composite laminates is the manner in which the individual unidirectional layer or laminar properties translate into laminate properties. For all the thermoelastic properties this translation is straightforward as described in Sec. 15.3 and is accomplished by the usual rules for transformation of stress and strain. However, the strength properties tend to be modified by the mutual constraint imposed by adjacent layers and therefore are a function of the individual layer thickness. The result is a need to modify the basic unidirectional properties, one of the most significant being the ultimate transverse strain to failure intension of individual layers. Unidirectional-layer compressive strength and the associated ultimate strain to failure is also influenced to a significant degree by the mutual support offered by an adjacent transverse or angled layer. As a consequence, correction factors are sometimes introduced to compensate, but more routinely tests are conducted on the actual laminate configuration designed to establish reliable allowables for use in design.

15.3 LAMINATE DESIGN AND ANALYSIS

For the simultaneous design of material and structure that is the basic philosophy for composite structures design, laminated-plate theory (LPT), and the associated computer codes, represents the fundamental tool for the composite designer. The anatomy of a composite laminate, indicating the translation from the constituent fiber and matrix properties to those of a built-up laminate, is illustrated in Fig. 15.3. Values contained in this figure compare with those presented in Tables 15.2 through 15.4. Figure 15.3 also illustrates the use of an alternative form of material, a fabric laminate that can provide similar, but slightly inferior, properties in a reduced thickness. The ability to produce a single layer comprised of equal proportions of fibers woven into 0° and 90°

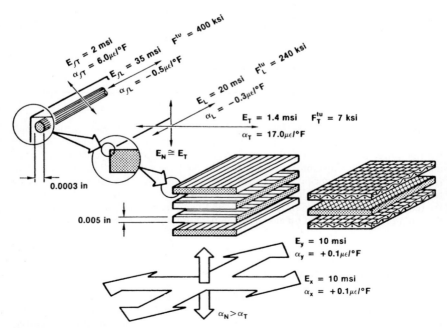

FIG. 15.3 The anatomy of a composite laminate.

orientations offered by this "textile system" therefore represents a valuable composite form. A state of plane stress and, for bending, that plane sections remain plane is assumed in the following theoretical development.

15.3.1 Laminate Configuration Notation

A method for specifying a given multidirectional laminate configuration has been established and is now routinely used on engineering drawings and documents. The following items essentially explain this laminate orientation notation:

1. Each layer or lamina is denoted by the angle representing the orientation (in degrees) between its fiber orientation and the reference structural (x-axis) of the laminate.
2. Individual adjacent angles, if different, are separated by a slash.
3. Layers are listed in sequence starting with first layer laid up, adjacent to the tool surface.
4. Adjacent layers of the same angle are denoted by a numerical subscript.
5. The total laminate is contained between square brackets with a subscript indicating that it is the total laminate (subscript "T") or one-half of a symmetric laminate* (subscript "S").
6. Positive angles are assumed clockwise looking toward the layup tool surface and adjacent layers of equal and opposite signs are specified with + or − signs as appropriate.
7. Symmetrical laminates with an odd number of layers are denoted as symmetric laminates with an even number of plies but with the center layer overlined.

The notations for some commonly used laminate configurations are illustrated in Fig. 15.4a through d.

In essence lamination theory is involved in the transformation of the individual stiffnesses of each layer in the principal directions to the direction of orientation in the laminate thereby providing the stiffness characterization for the specified laminate configuration. Subsequently, application of a given system of loads is broken down into individual layer contributions and referred back to the principal directions in each layer. A failure criterion is then used to assess the margin of safety arising in each layer. The complete process is illustrated in Fig. 15.5.

The development of the necessary relationships starts from the basic Hooke's law that is expressed in matrix form as

$$\{ \ \} = [\qquad\qquad]\{ \ \}$$

$$= \begin{bmatrix} S_{11} & S_{12} & 0 \\ S_{12} & S_{22} & 0 \\ 0 & 0 & S_{LL} \end{bmatrix} \{ \ \}$$

or in compact form as

---

*The definition of a symmetric laminate will be given in a following subsection.

(a)

Ply	Angle
8	0
7	−45
6	90
5	+45
4	+45
3	90
2	−45
1	0

or

Ply	Angle
6	9
5	60
4	120
3	120
2	60
1	0

Notation $[0/\pm 45/90]_s$ $[0/60/120]_s$

Required
(i) When properties required to be equal in all directions in plane of laminate:
Tensile strength
Compressive strength
Modulus
Coefficient of thermal expansion (CTE)
(ii) In some cases of combined tension or compression and shear.

(b)

Ply	Angle
11,12	0
10	45
9	90
8	−45
6,7	0
5	−45
4	90
3	45
1,2	0

or

Ply	Angle
20	90
19	60
16,17,18	0
15	120
12,13,14	0
11	60
10	60
7,8,9	0
6	120
3,4,5	0
2	60
1	90

Notation $[0_3/\pm 45/90]_s$ $[0_6/60_2/120/90]_s$

Used
(i) When load or stiffness requirement is highly directional.
(ii) To attain low CTE in one direction with low modulus fibers.

(c)

Ply	Angle
8	90
7	+θ
6	0
5	−θ
4	−θ
3	0
2	+θ
1	90

Laminate "balanced" about median plane

Notation $[90/\theta/0/-\theta]_s$

(i) Required to prevent warpage of flat laminate.
(ii) Can be departed from in contained circular parts: cylinders, conus, etc.
(iii) Small deviations near median plane possible; verify by test.

(d)

Ply	Angle
10	90
8 9	+θ
7	0
6	−θ
5	−θ
4	0
2 3	+θ
1	90

Balanced but more +θ than −θ

Notation $[90/+\theta_2/0/-\theta]_s$

(i) Will not warp, but 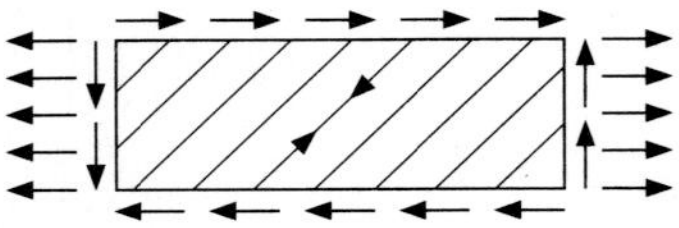
(ii) Applied axial load induces shear due to strain in excess +θ plies.

FIG. 15.4 Examples of laminates and conventional notations: (*a*) pseudo (quasi) isotropic laminates; (*b*) directional laminates; (*c*) a symmetrical laminate; (*d*) a symmetrical, but unbalanced, laminate.

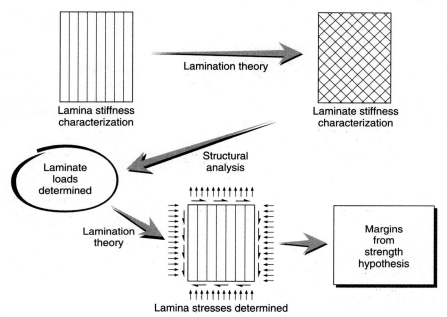

FIG. 15.5   Procedure for strength determination.

$$\{\varepsilon_i\} = [S_{ij}]\{\sigma_j\} \tag{15.1}$$

Here the elastic compliances S_{ij} are expressed in terms of the principal elastic constants E_L, E_T, G_{LT}, ν_{TL}.

The inverse relationship is expressed as

$$\left\{ \quad \right\} = \left[\qquad\qquad\qquad\qquad\qquad\qquad \right]\left\{ \quad \right\}$$

or as

$$\{\sigma_i\} = [Q_{ij}]\{\varepsilon_j\} \tag{15.2}$$

Following the established procedures, involving transformation rules, the components of both stress and strain may be related for an arbitrary angle θ from the fiber direction (Refs. 1–6):

$$\{\varepsilon'\} = [T^{-1}][S][T]\{\sigma'\}$$

$$= [S]\{\sigma'\} \tag{15.3}$$

and

$$\{\sigma'\} = [T^{-1}][Q][T]\{\varepsilon'\}$$

$$= [\overline{Q}]\{\varepsilon'\} \tag{15.4}$$

Several of the relationships represented by Eqs. (15.1) through (15.4) may be involved in the transformations for each individual layer forming a general laminate.

They are introduced into the laminate, which may be subjected to membrane and bending loads through the conventional strain-displacement equations,[1] to obtain

$$\{\sigma'\} = [\overline{Q}]_k\{\varepsilon_0'\} + z[\overline{Q}]_k\{K!\} \tag{15.5}$$

This equation now facilitates the ability to predict the stresses at any point z and in any layer of the laminate if the midplane strains $\{\varepsilon_o\}$ and curvatures $\{K'\}$ are known.

At this juncture the force resultants (N_x, N_y, N_{xy}) or force per unit width of the plate and the moment resultants (M_x, M_y, M_{xy}) or moment per unit width of the plate may be utilized to obtain the laminate constituent equations, i.e.,

$$\left\{ \quad \right\} = \int_{-h/2}^{h/2} \left\{ \quad \right\} dz \tag{15.6}$$

and

$$\left\{ \begin{array}{c} M_x \\ M_y \\ M_{xy} \end{array} \right\} = \int_{-h/2}^{h/2} \left\{ \begin{array}{c} \sigma_x \\ \sigma_y \\ \tau_{xy} \end{array} \right\} z\,dz \tag{15.7}$$

Introducing Eq. (15.5) into Eqs. (15.6) and (15.7) and separating the continuous integral into a sum of discrete integrals across each layer gives

$$\{N\} = \sum_{k=1}^{n} \left\{ \int_{h_{K-1}}^{h_K} [\overline{Q}]_K\{\varepsilon_o'\}dz + \int_{k_{K-1}}^{H_K} [\overline{Q}]_K^{\{K'\}} z\,dz \right\} \tag{15.8}$$

and

$$\{M\} = \sum_{k=1}^{n} \left\{ \int_{h_{K-1}}^{k_K} [\overline{Q}]_K\{\varepsilon_o'\}z\,dz + \int_{k_{K-1}}^{H_K} [\overline{Q}]_K\{K'\}z^2\,dz \right\} \tag{15.9}$$

These equations may be expressed in more compact notation as

$$\{N\} = [A]\{\varepsilon_o'\} + [B]\{K'\} \tag{15.10}$$

$$\{M\} = [B]\{\varepsilon_o'\} + [D]\{K'\} \tag{15.11}$$

where

$$[A] = [A_{ij}] = \sum_{k=1}^{n} [\overline{Q}_{ij}]_k(h_k - h_{k-1}) \tag{15.12}$$

$$[B] = [B_{ij}] = \tfrac{1}{2} \sum_{k=1}^{n} [\overline{Q}_{ij}]_k(h_k^2 - h_{k-1}^2) \tag{15.13}$$

$$[D] = [D_{ij}] = \tfrac{1}{3} \sum_{k=1}^{n} [\overline{Q}_{ij}]_k(h_k^3 - h_{k-1}^3) \tag{15.14}$$

Or, again, these expressions can be reduced to

$$\left\{ \quad \right\} = \left[\qquad\qquad \right]\left\{ \quad \right\} \tag{15.15}$$

This constitutive relation for a laminated plate embodies all the coupling effects that will be discussed from a physical viewpoint in the following section. The submatrix [B] will be shown to represent some unique behavioral responses inherent in the class of unsymmetric laminates.

The inverse relationship of Eq. (15.15) is expressed as $\tag{15.16}$

For the case of a laminate that is *symmetric,* or more precisely, *midplane symmetric,* the submatrix [B] in Eq. (15.15) will be zero. Similarly submatrices [B´] and [C´] will be zero, in Eq. (15.16) for the symmetric laminate.

15.3.2 Failure Criteria

Although much debate and development has occurred with regard to the most appropriate failure criteria for composite laminates, the most widely adopted approach in composite applications is the *maximum-strain criterion.* The application of this relatively simple criterion requires an experimental database for the ultimate strains for each of the three fundamental loading directions for the individual "orthotropic" layer comprising the laminate. The "three fundamental loading directions" refer to axial loading in the fiber direction, axial loading transverse to the fiber direction, and in-plane shear associated with the former directions. However, it should be acknowledged that the ultimate strain values may be markedly different for tension and compression both in the fiber direction and transverse to it. Thus a total of the following *five* ultimate strains are required to facilitate application of the maximum strain centers:

1. ε_L^{tu}, the ultimate *tensile* strain in the fiber direction (L)
2. ε_L^{cu}, the ultimate *compressive* strain in the fiber direction (L)
3. ε_T^{tu}, the ultimate *tensile* strain transverse to the fiber direction (T)
4. ε_T^{cu}, the ultimate *compressive* strain transverse to the fiber direction (T)
5. γ_{LT}^{su}, the ultimate *shear* strain referred to the L,T directions

In connection with the actual values used for strains 1 through 5 the reader is referred to Sec. 15.2.1, which explains how the individual layer properties must be adjusted to represent the strength, or ultimate strain values of a given layer that is contained within a multidirectional laminate. The prudent approach in engineering development work is to identify special laminate configurations that may be used to establish representative in situ properties for the range of potential candidate laminates for application to a specific design.

15.3.3 Coupling, Balance, and Symmetry

From the above theoretical development, the mathematical relationships define all the coupling relationships arising in the arbitrary laminate. However, a discussion of the physical aspects of such coupling phenomena and the laminate designs that may be invoked to suppress these responses is helpful to the structural engineer.

Extension-Shear Coupling. First, the in-plane coupling between extension and shear or vice versa arises in the case of any off-axis layer: e.g.,

$$\gamma_{xy} = S_{16}\,\sigma_x \qquad \text{or} \qquad \varepsilon_x = S_{16}\,\tau_{xy} \qquad (15.17)$$

or, for the inverse situation,

$$\sigma_x = Q_{16}\,\gamma_{xy} \qquad \text{or} \qquad \tau_{xy} = Q_{16}\,\varepsilon_x \qquad (15.18)$$

From a physical point of view, the shear deformation induced by an axial tensile stress is caused by the tendency for the layer to contract along the diagonals by unequal amounts due to differences in the Poisson's ratio in these two directions. Alternatively, considering the special case of a $+45°$ layer, the axial stress may be resolved into planes at $+45°$ and $-45°$ to the direction of applied stress. The resulting strains due to equal resolved stress components along these directions obviously produce very different strains.

Intuitively it is easily rationalized that the use of a $[\pm\theta]_T$ laminate will result in the mutual suppression of the tension-induced shear deformation in each individual layer. In the general case equal numbers of layers in the off-axis ($+\theta$ and $-\theta$) layers will suppress this coupling; the resulting laminate is termed a "balanced" laminate.

Extension-Torsion Coupling. For this the previous balanced laminate $[\pm\theta]_T$ is considered. The spatial separation in the thickness direction results in equal and opposite deformations in shear deformation induced by an axial tensile stress. This deformation situation therefore results in twisting of the laminate, a condition that is illustrated in Fig. 15.6. From a simplistic viewpoint the illustration presented in Fig. 15.7 provides a type of designers' guide to coupling evaluations which facilitates rational judgments in laminate design. All the responses indicated in these two figures can be confirmed by use of the lamination theory discussed above. Suppression of the twisting deformation is achieved by use of a symmetric laminate in which the off-axis layers below the central plane are "mirrored" by an identical off-axis layer at the same distance above the central plane (see Fig. 15.7).

Extension-Bending Coupling (Related through B_{11}, B_{12}, and B_{13} Matrix Components). The simplest form of laminate exhibiting a coupling between in-plane extension (or compression) and bending deformation is the $[0, 90]_T$ unsymmetric laminate.

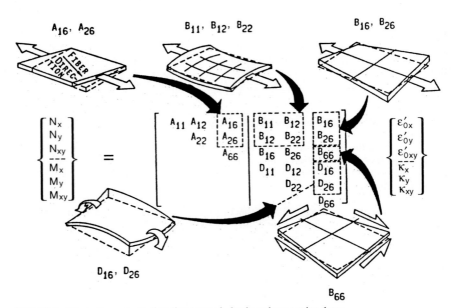

FIG. 15.6 Illustrations of coupling phenomena in laminated composite plates.

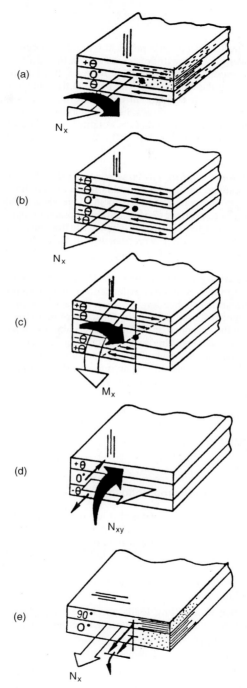

FIG. 15.7 Designer's guide to coupling evaluation. (*a*) Unsymmetric laminate under axial tension, N_x, illustrating B_{16} coupling phenomenon. (*b*) Symmetric laminate under axial tension, N_x; $B_{16} = 0$. (*c*) Symmetric laminate under bending moment, M_x, illustrating D_{16} coupling phenomenon. (*d*) Unsymmetric laminate under in-plane shear, N_{xy}, illustrating B_{66} coupling phenomenon. (*e*) Unsymmetric laminate under axial tension, N_x, illustrating B_{11} coupling phenomenon.

This response can be rationalized, on a physical basis, by recognizing that the neutral plane for this two-layer laminate will be located within the stiffest 0° layer giving rise to a bending moment produced by the in-plane forces applied at the midplane and the associated effect between the two planes. The bending deformation created in the primary loading direction will also create an anticlastic bending response via the mismatch in Poisson's ratio of the two layers, hence the existence of the B_{12} matrix component. For this case it is clearly seen that the coupling would be suppressed by use of a four-layer symmetric laminate, i.e., $[0, 90]_s$, or a three-layer symmetric laminate such as $[0, \overline{90}]_s$.

In-Plane Shear-Torsion Coupling (Related through the B_{66} Matrix Component).
To visualize the mechanism associated with this mode of coupling consider a $[\pm 45]_T$ unsymmetric, two-layer laminate subjected to in-plane shear loads. By recognizing that the in-plane shear is equivalent to a biaxial tension and compression loading with the tensile direction in the lower layer aligned with the fiber direction, and in the upper layer, transverse to the fiber direction, it will be realized that the plate will assume a torsional deformation (see Fig. 15.6).

Bending-Torsion Coupling (Related through D_{16} and D_{66} Matrix Components).
For this mode of coupling a four-layer balanced symmetric laminate, i.e., $[\pm \theta]_s$, is considered. The application of a bending moment, and an associated strain gradient, to this laminate will induce different degrees of shear coupling to the outer and inner layers. As a consequence the response of the outer layers will dominate due to the higher strain levels in these layers, resulting in a net torsional deformation as illustrated in Fig. 15.6. For qualitative assessment of this mode of coupling the magnitude of the shear responses can be considered to exert an internal couple on the laminated plate as illustrated in Fig. 15.7. A similar rationale can be used to design a laminate that would not exhibit this coupling. For example, an eight-layer laminate of the configuration

$$[(\pm \theta)_s/(\mp \theta_s)]_T \qquad \text{or} \qquad [\pm \theta, \mp \theta, \mp \theta, \pm \theta]_T$$

will exhibit no bending-torsion coupling.

15.3.4   General Laminate Design Philosophy

The recommended approach for laminates that are required to support biaxial loads is conveyed in the family of laminates represented in the set of carpet plots found in Figs. 15.8 through 15.14. Even for highly directional loading, a *nominal percentage of layers,* approximately 10 percent, should be included for the following reasons:

1. Providing restraints that inhibit development of microcracks that typically form in directions parallel to fibers
2. Improved resistance to handling loads and enhanced damage tolerance (this is especially relevant for relatively thin laminates, i.e., less than 0.20 in thick)
3. More manageable values of the major Poisson's ratio (v_{xy}), particularly where interfaces with other materials or laminates with values in the 0.30 range exist
4. Compatibility with thermal expansion coefficient with respect to adjacent structure

Other commonly adopted and recommended practices include laminate designs that minimize the subtended angle between adjacent layers and use of the minimum practi-

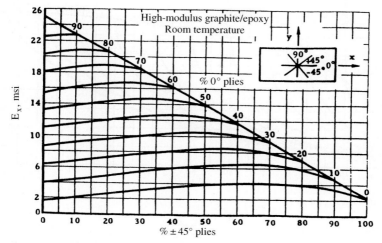

FIG. 15.8   Carpet plot for elastic modulus, E_x.

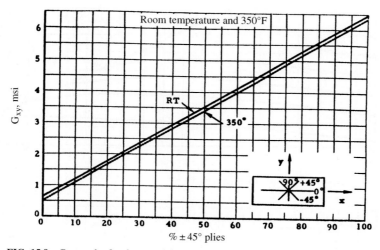

FIG. 15.9   Carpet plot for shear modulus, G_{xy}.

cable number of layers of the same orientation in one group. To illustrate the former, a laminate configuration of $[0, +45, 0, -45, 90]_s$ is preferred over a laminate such as $[0, +45, -45, 0, 90]_s$ even though the in-plane thermoelastic properties would be identical for these two laminates. For the latter the length of transverse microcracks tends to be limited by the existence of the layer boundaries; hence a $[0, +45, 0, -45, 0, 90]_s$ laminate is preferred over a $[0_3, +45, -45, 90]_s$ laminate.

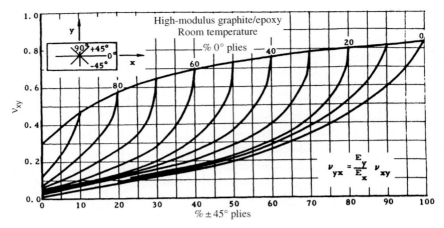

FIG. 15.10   Carpet plot for Poisson's ratio, ν_{xy}.

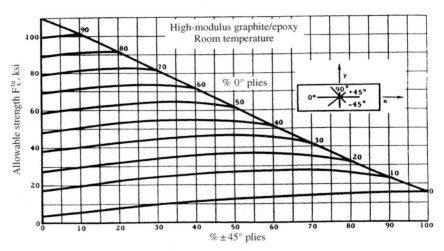

FIG. 15.11   Carpet plot for ultimate tensile strength, F_x^{tu}.

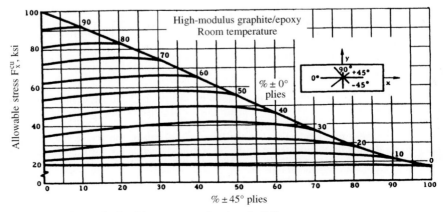

FIG. 15.12   Carpet plot for ultimate compressive strength, F_x^{cu}.

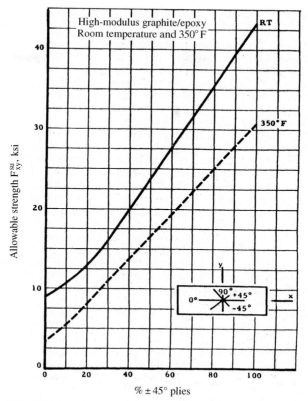

FIG. 15.13 Carpet plot for ultimate shear strength, F_{xy}^{su}.

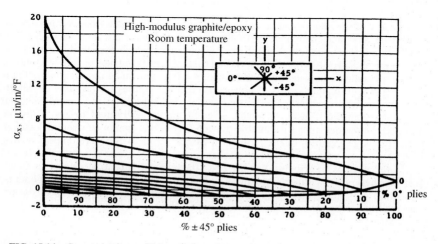

FIG. 15.14 Carpet plot for coefficient of thermal expansion, α_x.

15.4 APPLICATIONS AND DESIGN APPROACHES

In view of the significantly higher cost of most contemporary composite structures, it is strongly recommended that component design evaluations should be based on the full potential benefits of the characteristics of composites for the total system. For example, the opportunities for reduced weight, represented by the higher specific stiffness and specific strengths illustrated in Fig. 15.1, may translate into higher-speed processing capabilities, reduced maintenance and handling costs, lower assembly costs, as well as significant savings in supporting structure.

For example, consider the system benefits that arise in design of a segmented helicopter drive shaft for which the reduction in the number of shaft segments permitted by use of the composite culminates in reduction of the number of bearings, bearing support frames, and associated lubrication systems. The ability of composite drive shafts to span greater distances due to higher specific stiffness has also been exploited in the drive system used to power large cooling-tower fans. Applications that can benefit from the economics of part consolidation have been demonstrated in aerospace products and are now being realized in commercial ground-transportation applications such as monocoque car bodies and bicycle frames. Currently, the availability of a wide range of both reinforcing fiber "preforms" and polymer matrix processing approaches such as resin transfer molding (RTM), resin film infusion (RFI), advanced tow placement (ATP), filament and tape winding, roll wrapping, and pultrusions are providing increased opportunities for net shape manufacturing and hence reduced manufacturing costs.

A requirement for extreme dimensional stability in concert with high stiffness that has been demonstrated over many years in aerospace for precision optical benches, telescope structures, antennae, and wave-guide "plumbing" is also emerging in the form of automated precision machining, tape-laying, and assembly and measurement equipment. Opportunities for enhanced life-cycle advantages that accrue from reductions in corrosion, wear, and fatigue represent a continuing area of interest in numerous commercial applications.

15.4.1 Design Approaches, Issues, and Limitations

In the design of the majority of complex hardware products made from composites the engineering designer encounters difficulties associated with the low "secondary" or "matrix-dominated" properties of these nonisotropic material forms. By recourse to Tables 15.4 and 15.5 it will be observed that transverse (in-plane) tensile strength of the unidirectional composite laminate is merely a few percent of the longitudinal tensile strength. Consequently, it is of no surprise that the "through-thickness" or "short-transverse" tensile strength of a multidirectional laminate is of the same order, but even lower than the transverse tensile strength of the individual layers. Thus the importance of the designer's awareness of such limitations cannot be overemphasized. In fact, the large majority of the failures in composite hardware development testing has arisen due to underestimated or unrecognized out-of-plane loading effects and interrelated regions of structural joints and attachments. Due to the many common adverse experiences with delaminations induced by out-of-plane load components, this section will be devoted to the identification of the numerous sources of out-of-plane load development and the candidate approaches to eliminate or minimize their influence.

First, a general overview of many of the common problems created for the engineering designer that are consequences of low matrix-dominated, elastic, and strength

TABLE 15.6 General Overview of Problems Created by the Low "Secondary" (Matrix-Dominated) Properties of Advanced Composites

Controlling property	Problem	
F^{isu}	Failure induced by shear in beams under flexural loading	Premature crippling failure in compression
	Premature torsional failures	Failure of adherends in structural bonded joints
F_T^{tu}	Failure induced by transverse tensile fracture of curved beams in flexure	Failure of laminate due to free-edge effects, e.g., cutouts, ply drops
	Shock waves during normal impacts	
G_{LT}	Reduction in flexural and torsional stiffness	
	Reduction in resonant frequencies of plate and beam members	
	Reduction of elastic buckling capability	
	Interpretation of experimental stress analysis data	
α_T	Distortion at fillets due to high expansion coefficient (through thickness)	
α_T, F_T^{tu}	Failure due to thermal stresses in thick-walled composite cylinders	

properties are summarized in Table 15.6. Several of the most common sources will now be discussed in more detail. Figure 15.15a illustrates these major sources, which may be broadly categorized as follows:

Category A—Curved sections: Curved segments, rings, and hollow cylinders and spherical vessels that are representative of angle bracket design details, curved frames, and internally or externally pressurized vessels

Category B—Tapers and transitions: Local changes of section that are representative of laminate layer terminations, doublers, and stiffener terminations as well as the end details of bonded and bolted joints

As mentioned, commonplace structural details of both categories have contributed to numerous unanticipated failures in composite hardware components. In some cases such failures can propagate catastrophically after initiation and may therefore be a serious safety threat. Other instances have arisen where initial failures may self-arrest, resulting in benign failures but with some degree of local stiffness degradation. Subsequent load redistribution may, however, precipitate eventual catastrophic failure depending on load spectrum characteristics.

A further illustration of a built-up composite structure in the form of a torque box is presented in Fig. 15.15b. Several of the aforementioned sources of delamination potentially induced by out-of-plane load components are evident from a cursory review of this figure.

15.4.2 Category A—Curved Sections

A range of detailed structural hardware configurations that contain out-of-plane load paths induced by curvature and miscellaneous test methods are illustrated in Fig. 15.16.

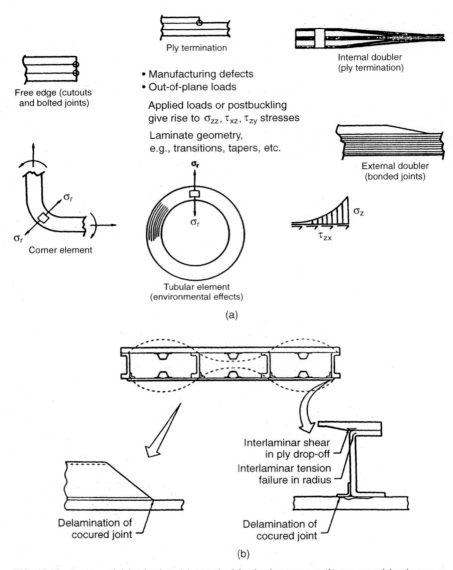

Ply termination

• Manufacturing defects
• Out-of-plane loads

Free edge (cutouts and bolted joints)

Internal doubler (ply termination)

Applied loads or postbuckling give rise to σ_{zz}, τ_{xz}, τ_{zy} stresses

Laminate geometry, e.g., transitions, tapers, etc.

External doubler (bonded joints)

σ_r

Corner element

σ_r

σ_r

σ_r

σ_z

τ_{zx}

Tubular element (environmental effects)

(a)

Interlaminar shear in ply drop-off
Interlaminar tension failure in radius

Delamination of cocured joint

Delamination of cocured joint

(b)

FIG. 15.15 Sources of delamination: (*a*) generic delamination sources; (*b*) sources arising in torque box assembly.

A broad awareness of the sources of out-of-plane loads is critically important to the composite designer to assist him or her in developing a carefully calibrated methodology that is simple to apply whilst providing adequate prediction of critical loading states. Premature structural failures arising in development hardware require that the designer make a determined effort to initiate such an approach dealing with the flexure of simply curved composite shapes. In such work a combination of finite-element

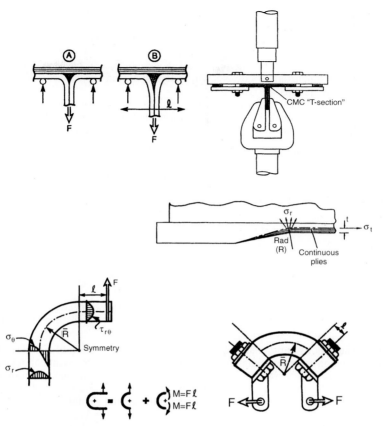

FIG. 15.16   Miscellaneous test methods and sources of delamination.

methods, classical elasticity theory, and simplified strength-of-materials approaches should be applied and compared. The following simplified expressions represent approximate approaches for estimating the interlaminar tensile stress developed in the presence of an opening bending moment M per unit width:

$$\sigma_r(\text{max}) = 3M/2tR_m \tag{15.19}$$

and

$$\sigma_r(\text{max}) = 3M/2t\sqrt{R_iR_o} \tag{15.20}$$

The analytical work should be extended through the selection of carefully conceived experimental methods and by modifications and enhancements to the simplified methods as appropriate.

One specific enhancement deals with effect of width of the specimen or component in relationship to plane-stress, plane-strain, or generalized plane-strain solution approaches. It is widely acknowledged that the restraint of anticlastic curvature that inherently exists in a relatively wide curved segment will result in the development of a variation in stress components for finite-width curved beams. The development of a simplified approach, based on a classical elasticity formulation, that indicates the peak value of interlaminar tensile stress relative to the nominal values determined by plane-stress or plane-strain approaches was described in Ref. 10 and elsewhere. This may be derived from an adaptation of the equations used to predict the variation in interlaminar shear stress across the width of simple finite-width orthotropic beams under transverse shear loading.[11] Again the theoretical development can be "calibrated" using two- and three-dimensional finite-element analyses as well as judiciously configured experimental evaluations.

The above treatment can be extended to the state of thermal/residual deformation and stress-prediction methodology. For this a refinement of the simple expressions, developed in Ref. 10, can be used:

1. Rotation of 90° segment (i.e., "springback")

$$\gamma = \frac{\pi}{2}(\alpha_\theta - \alpha_T)\Delta T \tag{15.21}$$

2. Maximum radial tensile stress

$$\sigma_r(\max) = \frac{E_\theta}{8}\left(\frac{t}{R_m}\right)^2(\alpha_\theta - \alpha_r)\Delta T \tag{15.22}$$

where E_θ, α_θ = the circumferential elastic modulus and coefficient of thermal expansion, respectively

E_r, α_r = the radial (through-thickness) elastic modulus and coefficient of thermal expression, respectively

R_m, t = the mean radius and thickness of the section, respectively

ΔT = the temperature change relative to some selected zero stress or zero distortion state

On the subject of multiaxial stress conditions and the selection of a failure criterion, an evaluation of both a simple maximum-stress–based approach and an interaction approach can be used. However, the observation is offered that failure initiation due to transverse tensile stresses generally develops in a region where the in-plane stresses are extremely small; consequently an experimental program should focus on establishing confidence in the postulate that an interaction formulation will be unnecessary.

15.4.3 Category B—Tapers and Transitions

An extensive background in the design and analysis of composite bolted and bonded joints is invaluable in also developing approaches for changes of cross section including ply drops and tapered sections (see Table 15.7). The generic similarity between the

TABLE 15.7 Relative Advantages of Mechanically Fastened vs. Adhesively Bonded Joints

Advantages of mechanical joints	Advantages of bonded joints
Tolerant to the effects of environment and fatigue	High joint efficiency index (relative strength/weight of the loaded joint region)
Ease of inspection	Low part count
Capability for repeated assembly	No strength degradation of basic laminate by use of cutouts
High reliability	Low cost potential
No special surface preparation required	Potential corrosion problems minimized

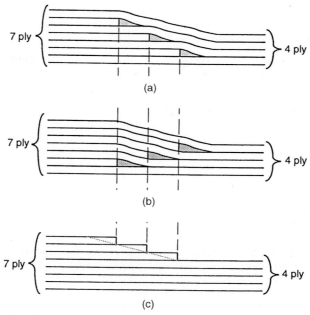

FIG. 15.17 Generic tapered laminate designs. (*a*) External plies terminated first; (*b*) internal plies terminated first; (*c*) tapered external doubler.

detailed structural features permits application of the numerous mechanics and design insights and techniques developed for structural composite joints. Again, a combination of analyses from simplified strength-of-materials approaches, classical elasticity approaches, and by two-dimensional and three-dimensional finite-element techniques can be adopted.

To address a common dilemma in design that is illustrated in Fig. 15.17, which shows three generic laminate-stacking designs for thickness transitions, is beyond the scope of this section. Note that there is currently no available design-analysis approach to determine which of these three configurations would result in the highest structural integrity, and the designer must resort to expensive testing programs as part of a "building block" philosophy.

It is widely recognized that detailed three-dimensional, or even two-dimensional, finite-element modeling down to the ply level in complex regions can be both prohibitively time consuming, and, most important, often does not portray the as-fabricated detail in a representative manner. Consequently, a manageable and physically rationalized analysis should be directed at representative fabricated details.

Delamination initiation and growth prediction represent an important component in predicting and assessing the possible existence of self-arrest mechanisms. Conventional use of G_{Ic} and G_{IIc} energy-release-rate methods can be incorporated with judicious use of experimental and analysis techniques and results.

Ultimate objectives directed toward providing a combined theoretical and experimental background and attendant physical insights to support design guidelines and rules of thumb, e.g., 20-to-1 scarf angle ratios, $30\times$ thickness overlap lengths or ply-drop rates of three plies per one-quarter inch of axial length in the major loading direction are recommended. Such rules of thumb have to be clearly associated with the specific class of composite, possibly indexed according to stiffness anisotropy. For example, one would anticipate a significant difference in the rules of thumb for glass/polyester vis-à-vis high-modulus graphite/epoxy.

REFERENCES

1. Mallick, P. K.: "Fiber-Reinforced Composites: Materials, Manufacturing, and Design," 2d ed., Marcel-Dekker, 1993.

2. Halpin, J. C.: "Primer on Composite Materials—Analysis," Technomic, 1984.

3. Whitney, J. M.: "Structural Analysis of Laminated Anisotropic Plates," Technomic, 1987.

4. Tsai, S. W., and H. T. Hahn: "Introduction to Composite Materials," Technomic, 1980.

5. "Engineered Materials Handbook," vol. 1: "Composites," ASM International, Material Park, Ohio, 1987.

6. Kelly, A. (ed.): "Concise Encyclopedia of Composite Materials," Pergamon Press, Oxford, 1988.

7. Dharan, C. K. H.: "Design of Automotive Components with Advanced Composites," *Proc. 2d International Conference on Composite Materials, ICCM2,* The Metallurgical Society of AIME, New York, 1978.

8. Linsenmann, D. R.: "Hybridization for Cost-Effective Design," *Proc. ASME Winter Annual Meeting on Composite Materials in the Automobile Industry,* 1978.

9. "Advanced Composite Design Guide," 3d ed., Air Force Materials Laboratory, Wright Patterson Air Force Base, Ohio, 1973.

10. Kedward, K. T., R. S. Wilson, and S. K. McLear: "Flexure of Simply Curved Composite Shapes," *Composite J.,* vol. 20, no. 6, pp. 527–536, November 1989.

11. Kedward K. T.: "On the Short Beam Test Method," *J. Fiber Sci. and Technol.,* 1972.

12. Everett, R. A.: "The Role of Peel Stresses in Cyclic Debonding," NASA TM-84504, 1982.

13. Hart-Smith, L. J.: "Advances in the Analysis and Design of Adhesive-Bonded Joints in Composite Aerospace Structures," *Proc. 19th SAMPE Symposium,* pp. 722–737, April 1974.

14. Williams, J. H., S. S. Lee, and T. K. Wang: "Nondestructive Evaluation of Strength and Separation Modes in Adhesively Bonded Automotive Glass Fiber Composite Single Lap Joints," *J. Composite Materials,* vol. 21, pp. 14–35, 1987.

15. Hart-Smith, L. J.: "Analysis and Design of Advanced Composite Bonded Joints," NASA CR-2218, 1974.

16. Goland, M., and E. Reissner: "The Stresses in Cemented Joints," *J. Appl. Mech.,* vol. 11, pp. 17–27, 1944.

17. Johnson, W. S., and S. Mall: "A Fracture Mechanics Approach for Designing Adhesively Bonded Joints," NASA TM-85694, 1983.

18. Mall, S., and W. S. Johnson: "Characterization of Mode I and Mixed-Mode Failure of Adhesive Bonds between Composite Adherends," ASTM STP 893, J. M. Whitney, ed., pp. 322–334, 1986.

19. Eisenmann, J. R.: "Bolted Joint Static Strength Model for Composite Materials," *3d Conf. on Fibrous Composites in Flight Vehicle Design,* Pt. II, NASA TMX-3377, Williamsburg, Virginia, pp. 563–602, November 4–6, 1975.

20. Waddoups, M. E., J. R. Eisenmann, and B. E. Kaminski: "Macroscopic Fracture Mechanics of Advanced Composite Materials," *J. Composite Materials,* pp. 446–454, 1971.

21. Konish, H. J., Jr., and T. A. Cruse: "Determination of Fracture Strength in Orthotropic Graphite/Epoxy Laminates," *Symposium on Composite Reliability,* Las Vegas, Nevada, ASTM STP 580, pp. 490–503, April 15–16, 1974.

22. Lekhnitskii, S. G.: *Anisotropic Plates,* Gordon & Breach, New York, 1968.

23. Waszczak, J. P., and T. A. Cruse: "A Synthesis Procedure for Mechanically-Fastened Joints in Advanced Composite Materials," AFML-TR-73-145, vol. II, 1973.

24. Waszczak, J. P.: "Mode I and Mode II Stress Intensity Factors for Bolt Bearing Specimens Containing Symmetric Radial Cracks in Anisotropic Materials," Dept. of Mech. Eng., Carnegie-Mellon Univ., Report No. SM-72-28, September 1972.

25. Hart-Smith, L. J.: "Bolted Joints in Graphite-Epoxy Composites," McDonnell Douglas Corporation, NASA CR-144899, 1977.

26. Hyer, M. W., E. C. Klang, and D. E. Cooper: "The Effects of Pin Elasticity, Clearance and Friction on the Stresses in a Pin-Loaded Orthotropic Plate," *J. Composite Materials,* vol. 21, pp. 190–206, 1987.

27. Eisenmann, J. R., and B. E. Kaminski: "Fracture Control for Composite Structures," *Engineering Fracture Mechanics,* vol. 4, pp. 907–913, 1972.

SECTION 16

WELDING, BRAZING, SOLDERING, AND ADHESIVE BONDING

Phyllis Zych Budka, M.S.
President
Technical Communications Unlimited
Schenectady, N.Y.

Ernest F. Nippes, Ph.D.
Professor Emeritus of Metallurgical Engineering
Rensselaer Polytechnic Institute
Troy, N.Y.

16.1 WELDING

16.1.1 Introduction

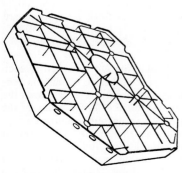

FIG. 16.1 Welded base structure.

Advances in welding, brazing, soldering, and adhesive bonding during the past decade have permitted consideration of these fastener methods in machine design. By use of these fasteners, cast or machined components assembled by bolting, screwing, riveting, or doweling can often be replaced by standard structural shapes cut to size. A welded base structure is shown in Fig. 16.1. As shown, extensive use can be made of standard steel shapes in lieu of special castings. But the use of such construction often requires new perspectives of designers and engineers.[1,2,9] It is misleading to consider welding as merely one of the various methods of fastening component parts together. In order to

16.1

employ welding successfully as a method of joining machine components, the machine must be designed for welding.[44-46,48-53] To duplicate a design fabricated by other fastener methods is unsatisfactory and will be wasteful in both the weight (size, thickness, reinforcement) and fabrication costs. With welded construction it is often possible to use steel plates, bars, angles, channels, etc., and steel castings as required to meet the actual stress and service requirements, in lieu of the use of ordinary cast-iron castings which have greater toughness. Such standard steel shapes are usually three to six times as strong as standard iron castings. One major design consideration that must be recognized in using the welding processes is that welding is positive in action and that the joints so made are *rigid* joints. This contrasts with other joining processes where slippage or movement between parts always occurs. Thus a design that functions satisfactorily when riveted may be completely unsatisfactory when welded.

Discussed herein are the design parameters affecting welding processes and the results that can be expected from the use of these processes. Intentionally omitted are the process details of the various fasteners under discussion. Process information is found in the specialized handbooks devoted to those subjects.[3-9] Process descriptions herein have been limited to those factors required to clarify the design factors presented.

16.1.2 The Joining Process

Of the more than 60 different welding processes (Ref. 55) available, most machine fastener requirements can be met by three, namely, arc and resistance welding and brazing. Within these three categories, shielded metal arc, gas-metal arc, flux-cored arc, gas tungsten arc, submerged arc, shielded or unshielded stud welding, seam, spot, and projection welding, and furnace or induction brazing are most widely used. All these processes are characterized by the fact that the metal components being joined have either been molten during the joining process or have been at a temperature approaching the melting point of the base material. The heat associated with these processes has an effect on design and must be considered as discussed later. Brazing is a group of welding processes in which the parts are heated to a suitable temperature and the filler metal used has a melting temperature about 450°C (840°F) and below the solidus of the base metal. The filler metal is distributed between the closely fitted surfaces of the joint by capillary attraction. Braze welding is differentiated from brazing because the filler metal is deposited in a groove or fillet exactly at the point where it is to be used and capillary action is not a factor. Brazing is arbitrarily distinguished from soldering by the filler metal melting temperature.[9] Soldering, in which the filler metal has a liquidus below 840°F (450°C) and below the solidus of the base metal, is generally limited to nonstructural applications. Dip and torch methods are the standard methods. Soldering is a low-temperature operation [generally below 750°F (400°C) and most usually below 500°F (260°C)] which has little if any significant effect on the properties of the material being joined.

Filler metals[3,4,9,11] used in welding, brazing, and soldering are specified to each process and its variables, often by the filler metal producer. Adhesive bonding is generally the lowest-temperature joining process and as such rarely, if ever, raises the temperature of the surfaces being joined to much above 300°F (150°C). Many of the adhesives, specifically designed for fastener use, are room-temperature curing. Advances in materials for adhesive bonding indicate that this method of joining of machine parts is in its infancy and will become more significant in future years.

The design factor dictating the selection of fusion welding over all other welding processes is the type of load the joint carries. A tension joint dictates the use of a fusion weld; such joints as shear, compression, and scarf permit joining by any of the methods under discussion in this chapter.

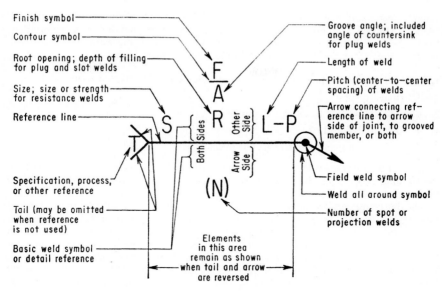

FIG. 16.2 Standard location of elements of a welding symbol.

Fillet	Plug or slot	Arc-spot or arc-seam	Groove								Back or backing	Melt-thru	Surfacing	Flange	
			Square	V	Bevel	U	J	Flare-V	Flare-bevel					Edge	Corner
△	▽	▬	‖	∨	∨	∪	⊔	⋎	Ⅰⴾ	⌒	◢	⌣	⅃∟	⅃∟

Basic Arc and Gas Welding Symbols

(a)

Type of weld			
Resistance–spot	Projection	Resistance–seam	Flash or upset
✳	✕	✕✕✕	Ⅰ

Basic Resistance Welding Symbols

(b)

Weld all around	Field weld	Contour	
		Flush	Convex
○	●	—	⌒

General Welding Symbols

(c)

FIG. 16.3 Welding symbols.

The design processes for welded *construction* are standard, using known mathematical formulas. But the design process for the *joint* involves many factors unique to the processes and must be clearly and carefully thought through.

The designers' specifications as to the type of weld (fusion or resistance) desired are expressed on the drawing through the use of the American Welding Society (AWS) weld symbol. In its basic form it appears as a broken arrow with symbol elements as shown (Figs. 16.2 and 16.3). The symbols used are shown in Fig. 16.3. Welding should not be specified on drawings except through the use of the AWS welding symbols. They are precise, eliminate discussions with manufacturing and inspection personnel, and assure the design engineer that the joint will be fabricated as wanted. Working stresses need not be reduced to take into consideration defective materials or work when the desired parameters are specified by means of the welding symbol. The abnormalities of welded construction that must be considered are:

1. Undesirable mechanical properties in the weld- and heat-affected zone
2. Thermal stresses
3. Limited knowledge of fatigue characteristics
4. Warpage and distortion
5. Formation of notches and stress concentrators

These factors are related to the fact that the metal in the joint has been molten during the welding process and has not received mechanical work subsequent to solidification.

On the positive side, welding is the most economical method of joining tension members. At least a 20 percent reduction in net cross section can be realized in welded construction by the elimination of rivet or other fastener holes. Further reductions in weight can be experienced in the final design because welded construction does not require heavy end-connection details. Ease of fabrication should also be considered, especially in machines where such items as pressure vessels are used. With any other type of construction a secondary sealing operation, which at best is highly unreliable, would be required. Generally, fusion-welding equipment is the lowest-cost joining equipment that can be used for assembly purposes.

In addition to stresses caused by weld-metal solidification, welded joints are subjected to the forces produced by external loads. The stress distributions and concentrations caused by external loads on three common types of joints are shown in Fig. 16.4.

Note that the single lap joint is the poorest design. In a lap joint the weld metal is strained at each end by shearing moments (tending to cause the two parts to slide over one another) and by the bending moment developed (which tends to tear the weld metal). Thus it can be seen that the strength of a lap joint is proportional to the width of the joint but does not increase linearly with increasing overlap. It is affected also by the thickness of the members. The tension joint shows a nonuniform stress across the length of the joint, with the edges again carrying the major load. The scarf joint shows the most uniform stress distribution, both within the weld metal and in the adjoining base metal. Uneven concentrations of stress will cause the joint to fail under load at a value lower than that calculated for the joint. These theoretical facts, including the presence of notches and other stress concentrators, must be considered in many of the joint designs later discussed.

16.1.3 Welding Processes, Selection Considerations

Spot, roll, seam, flash, and upset are the most common resistance-welding processes used in machine construction. All are characterized by the fact that the heat for coalescence is obtained from electrical resistance at the interface of the pieces being joined. For process and equipment details see Ref. 9.

In oxyfuel gas welding, the heat is generated by the combustion of a fuel gas, principally acetylene with air or oxygen.

In arc welding, heat is generated by the resistance to flow of an electric current across a gaseous gap. Filler metal may or may not be added to the joints. Shielded metal arc (SMA), gas-metal arc (GMA or MIG), and gas-tungsten arc (GTA or TIG) are the most commonly used processes.

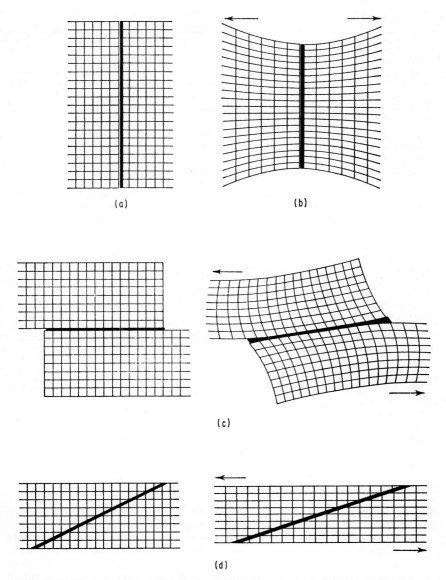

FIG. 16.4 Schematic of stress distribution within joints (exaggerated scale). (*a*) Butt joint is depicted here in two dimensions as it might look before (left) and after (right) being subjected to an applied load. (*b*) Arrows indicate the direction of stress; the constriction in the center, the nonuniform distribution of stress across the area of the joint. (*c*) Lap-joint adhesive under an applied load is strained chiefly at both ends. One component of the stress results in a shearing motion, which tends to slide the two portions of the joint over each other; a second component acts to force the joint open. (*d*) Scarf joint is far stronger than either a butt joint or a lap joint (see illustrations above). This is a type of joint in which the two adherent materials are tapered off so that they stretch uniformly throughout the length of the joint when subjected to an applied load.

TABLE 16.1 Welding Processes Compared

Process	Advantages	Disadvantages

Resistance welding
  Spot
  Roll
  Seam
  Flash
 	
  Upset
  Projection
	Very flexible
	Economical
	Joins three or more pieces with
	  one weld
	Properties of joint independent
	  of process
	Can join any combination of
	  metals that alloy together
		Poor in fatigue
		Requires expensive equipment
		Must be able to reach both sides
		  of the joint

Oxyfuel gas welding	Good control of weld metal
	Heat input anneals weld metal
	  and heat-affected zone
	Inexpensive equipment
		High cost
		Low production rate
		Higher heat input; steep temper-
		  ature gradient
		More warpage, distortion, etc.
		Material thicknesses must be
		  nearly equal

Arc welding
  Shielded metal arc
  Gas-tungsten arc
  Gas-metal arc
  Flux-cored arc
  Submerged arc
	High production rate
	Minimum distortion
	Low annealing of base metal
		Stress concentrations at edge of
		  welds, undercutting, over-
		  laps, etc., likely

The advantages and disadvantages associated with these processes are shown in Table 16.1.

Selection of the actual process depends upon (1) the alloy being welded, (2) metal thickness, (3) desired properties in the joint, (4) production rate, (5) quality standards, (6) number of units to be fabricated, (7) amount and type of tooling available.

16.1.4 Resistance Welding

Material Weldability. The basic theory and uses of resistance welding are given in Ref. 9. The characteristics of the spot weld formed when joining different materials together are shown in Fig. 16.5. This table shows that it is physically possible to weld a wide variety of metals and alloys to one another. In actual practice, spot properties, corrosion resistance, and other factors must be taken into account. Recommendations for a variety of current aluminum, magnesium, steel, and nickel alloys are shown in Figs. 16.6 and 16.7.

When two or more sheets of unequal thickness are to be welded, the maximum ratio of the thicknesses of the two sheets is given in Fig. 16.8. For welding three or more sheets, the minimum thickness of a thinner outer sheet is given in Fig. 16.9. Combinations of more than three thicknesses can be welded, but the welding technique needs to be developed for each application and therefore should be avoided if possible,

Spot Pattern. The following spot patterns are most commonly used:

	Aluminum	Stainless steel	Brass	Copper	Galvanized iron	Steel	Lead	Monel	Nickel	Nichrome	Tinplate	Zinc	Phosphor bronze	Nickel silver	Terneplate
Aluminum	B	E	D	E	C	D	E	D	D	D	C	C	C	F	C
Stainless steel	F	A	B	E	B	A	F	C	C	C	B	F	D	D	A
Brass	D	B	C	D	D	D	F	C	C	C	D	E	C	C	D
Copper	E	E	D	F	E	E	E	D	D	D	E	E	C	C	D
Galvanized iron	C	B	D	E	B	B	D	C	D	C	B	C	D	E	A
Steel	D	A	D	E	B	A	C	C	C	C	B	F	C	D	B
Lead	E	F	F	E	D	E	C	E	E	E	—	F	E	E	D
Monel	D	C	C	D	C	C	A	A	B	B	C	C	E	B	D
Nickel	D	C	C	D	C	C	B	B	A	B	C	F	C	B	C
Nichrome	D	C	C	D	C	C	F	A	B	A	C	F	D	A	C
Tinplate	C	B	D	E	A	B	—	C	C	A	C	C	D	D	C
Zinc	C	F	E	B	C	F	F	C	F	C	C	D	D	F	C
Phosphor bronze	C	D	C	C	D	C	E	C	C	D	D	D	B	B	D
Nickel silver	F	D	C	C	E	D	E	B	D	B	D	F	B	A	D
Terneplate	C	A	D	D	B	B	D	D	C	C	C	C	D	D	B

FIG. 16.5 Relative weldability ratings of selected metal and alloy combinations used for resistance spot-welding operations by J. G. Bralla.[56] (A, excellent; B, good; C, fair; D, poor; E, very poor; F, impractical.)

Basic alloy designation	1100-0, 1100-H14, 5052-0, 5052-H34 aluminum	2024-0 bare aluminum	2024-0 clad aluminum	2024-T3 clad aluminum	6061 (all tempers) aluminum	7075-T6 bare aluminum	7075-T6 clad aluminum	2014 clad aluminum	AZ31A magnesium
1100-0, 1100-H14, 5052-0, 5052-H34, aluminum......	G	P	F	N	P	P	N	P	
2024-0 bare aluminum........	P	N	F	F	P	N	F	N	
2024-0 clad aluminum........	F	F	G	N	P	F	N	F	
2024-T3 clad aluminum.......	N	F	N	G	P	F	G	F	
6061 (all tempers) aluminum..	P	P	P	P	G	P	P	P	
7075-T6 bare aluminum......	P	N	F	F	P	N	F	N	
7075-T6 clad aluminum......	N	F	N	G	P	F	G	F	
2014 clad aluminum..........	P	N	F	F	P	N	F	G	
AZ31A magnesium...........									G

Code:
G = good
F = fair
P = poor
N = not recommended

FIG. 16.6   Weldability of various aluminum and magnesium alloys (resistance spot welding).

	Inconel	Monel	K Monel	AISI 1010 through 1035	AISI 1040 and above	AISI 4130	AISI 4135	AISI 8630	18-8 stainless steel*
Inconel	G	N	N	N	N	N	N	N	N
Monel	N	F	P	N	N	N	N	N	N
K Monel	N	P	F	N	N	N	N	N	N
AISI 1010 through 1035	N	N	N	G	F	F	F	F	P
AISI 1040 and above	N	N	N	F	F	F	F	F	P
AISI 4130	N	N	N	F	F	G	F	G	P
AISI 4135	N	N	N	F	F	F	G	F	P
AISI 8630	N	N	N	F	F	G	F	G	P
18-8 stainless steel*	N	N	N	P	P	P	P	P	G

G = good; F = fair; P = poor; N = not recommended.

*18-8 welded to carbon or alloy steel may produce a brittle nugget unless postheating is used.

Note: In spot welding carbon or low-alloy steels with a carbon content above 0.2%, postweld tempering may be required.

FIG. 16.7   Weldability of various steel and nickel-base alloys (resistance spot welding).

Steel and nickel-base alloys		All other metallic materials	
Thinner sheet	Max ratio	Thinner sheet	Max ratio
0.020 and under	1:2	0.020 and under	1:1.5
0.021 and over	1:3	0.021 and over	1:2.5

FIG. 16.8   Resistance-welding thickness limitations, two sheets.

Steel and nickel-base alloys		All other metallic materials	
Thinner outer sheet, in	% of total thickness	Thinner outer sheet, in	% of total thickness
0.020 and under	33⅓	0.020 and under	40
0.021 and over	25	0.021 and over	33⅓

FIG. 16.9   Resistance-welding thickness limitations, three or more sheets.

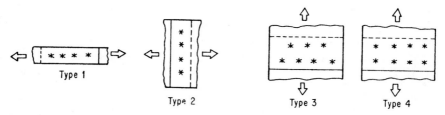

FIG. 16.10   Typical spot patterns.

1. Single rows with the spacing according to structural requirements. A Type 1 joint is not as reliable as a Type 2 joint because of the possibility of progressive failure (see Fig. 16.10). More than three welds in tandem are of no structural value for loading in a Type 1 welded joint.

2. Double rows either staggered or in line. Type 4 joints are somewhat more efficient for the loadings indicated in Fig. 16.10 than Type 3 joints, provided the minimum weld spacing is maintained.

Spot Location.   Spots may be located as desired provided the minimum spacing shown in Fig. 16.11 is maintained. The values shown in Fig. 16.11 are based on a control thickness which is equal to one-half the total thickness of the combinations being considered. Closer spot spacings than the minimum shown in the table may cause partial short circuiting of the welding current through the adjacent spots, producing a weld of smaller diameter and lower shear strength.

Loading of Spot-Welded Joints.   Spot-welded joints should be designed to transmit shear or compression loads only. The following limitations must be observed in spot-welded joints:

1. Avoid welds in tension or those subject to undue vibration (fatigue failures must be considered).

2. Avoid designs where considerable "tension-field" action is developed even though the tension load is below the static strength of the joint (fatigue failures must be considered).

3. Single isolated welds should not be used in structural assemblies since spot welds provide little resistance to torsional stresses.

4. A minimum of two spot welds should always be used.

The shear strength of single spot welds in a joint consisting of two equal thicknesses of material is shown in Table 16.2. As noted, single spot welds cannot be used in actual designs. When two or more spot welds are used, the shear strengths shown in Table 16.2 must be reduced by at least 10 percent to allow for shunting action (flow of

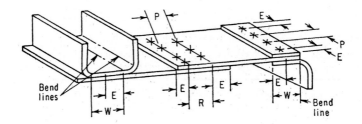

Min values for spot location

Control thickness ($\frac{1}{2}$ total thickness)	Steel and nickel-base alloys					Other metallic materials				
	P min	E min	E min^a	R min	W min	P min	E min	E min^a	R min	W min
0.010	0.25	0.18	0.12	0.22	0.36	0.31	0.18	0.12	0.27	0.36
0.015	0.30	0.18	0.12	0.26	0.36	0.31	0.18	0.12	0.27	0.36
0.020	0.38	0.22	0.12	0.33	0.44	0.38	0.18	0.12	0.33	0.36
0.025	0.44	0.25	0.14	0.38	0.50	0.38	0.22	0.14	0.33	0.44
0.030	0.50	0.25	0.14	0.43	0.50	0.38	0.25	0.16	0.33	0.50
0.035	0.50	0.25	0.16	0.43	0.50	0.40	0.25	0.16	0.34	0.50
0.040	0.62	0.25	0.18	0.54	0.50	0.40	0.25	0.18	0.38	0.50
0.045	0.75	0.25	0.20	0.65	0.50	0.46	0.28	0.20	0.40	0.56
0.050	0.88	0.31	0.20	0.76	0.62	0.50	0.30	0.20	0.43	0.60
0.055	0.94	0.31	0.22	0.81	0.62	0.50	0.33	0.22	0.43	0.66
0.060	0.98	0.31	0.22	0.85	0.62	0.50	0.36	0.22	0.43	0.72
0.065	1.03	0.31	0.22	0.89	0.62	0.50	0.38	0.22	0.43	0.76
0.070	1.12	0.31	0.24	0.97	0.62	0.54	0.38	0.24	0.47	0.76
0.075	1.20	0.31	0.24	1.04	0.62	0.58	0.38	0.25	0.50	0.76
0.080	1.25	0.38	0.24	1.08	0.76	0.62	0.38	0.25	0.54	0.76
0.085	1.30	0.38	0.25	1.12	0.76	0.62	0.40	0.25	0.54	0.80
0.090	1.34	0.38	0.25	1.16	0.76	0.62	0.44	0.25	0.54	0.88
0.095	1.38	0.38	0.25	1.19	0.76	0.66	0.44	0.25	0.57	0.88
0.100	1.44	0.38	0.25	1.25	0.76	0.73	0.44	0.28	0.63	0.88
0.105	1.48	0.38	0.28	1.28	0.76	0.78	0.44	0.28	0.67	0.88
0.110	1.54	0.40	0.28	1.33	0.80	0.84	0.46	0.28	0.73	0.92
0.115	1.60	0.40	0.28	1.38	0.80	0.90	0.46	0.28	0.78	0.92
0.120	1.64	0.40	0.28	1.42	0.80	0.94	0.48	0.28	0.81	0.96
0.125	1.68	0.44	0.28	1.45	0.88	1.00	0.50	0.28	0.87	1.00

^{a} Nonstructural only.

FIG. 16.11   Design limitations and spot locations for resistance welding. $P = pitch\ distance$, the center-to-center distance between any two welds; $E = edge\ distance$, the distance from the center of any weld to an edge or a bend line; $W = width\ distance$, the distance between two edges, or the distance between an edge and a bend line, or the distance between two bend lines; $R = row\ distance$, the center-line-to-centerline distance between adjacent rows of welds.

a portion of the current through the already made spot weld which reduces the size of the second and subsequent welds). In large-assembly fixture-type spot-welding machines, the actual shear strength is determined on the basis of tests run on actual parts. In general, the test values will approach those shown in Table 16.2.

TABLE 16.2 Single-Spot Shear Strength

Alloy thickness	Low-carbon steel* <70,000 psi	Low-carbon steel* >70,000 psi	Stainless steel, tensile strength			Magnesium AZ-31	Nickel	Inconel
			70/90,000 psi	90/150,000 psi	<150,000 psi			
0.010	130	180	150	170	210	100	135	175
0.021	320	440	370	470	500	176	350	545
0.031	570	800	680	800	930	272	760	920
0.040	920	1,200	1,000	1,270	1,400	344		
0.050	1,350		1,450	1,700	2,000	432		
0.062	1,850		1,950	2,400	2,900	544	2,400	2,750
0.078	2,700		2,700	3,400	4,000	660		
0.094	3,450		3,550	4,200	5,300	770	3,600	4,400
0.109	4,150		4,200	5,000	6,400	925		
0.125	5,000		5,000	6,000	7,600	1,075	5,600	6,400

*These steels may contain a low content of alloying elements.

Joint Design and Equipment Limitations.   It is of the utmost importance to provide
for accessibility of welding apparatus in all resistance-welding applications. The high
force and current involved require heavy machine structures, and the need for water
cooling the electrodes further complicates the accessibility problem. Wherever possible, the design should be such that simplified electrode components can be used with a
minimum of offsetting. Figure 16.12 illustrates typical machine-joint designs, classified as to desirability.

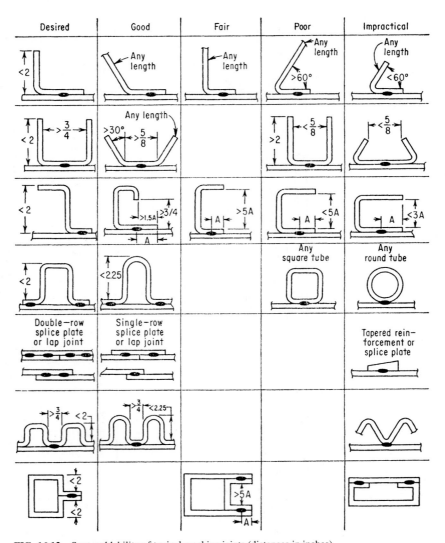

FIG. 16.12   Spot weldability of typical machine joints (distances in inches).

16.1.5 Oxyfuel Gas Fusion Welding

General Design Considerations.   Welding may be employed in the fabrication of practically all types of structures. Oxyfuel gas welding is more expensive than arc welding and may cause more buckling and warping when used to weld very thin flat sections unless excessive clamping is employed. Oxyfuel gas welding is usually specified for metals between 0.010 and $\frac{1}{16}$ in. thick, stainless steel frequently being an exception. Arc welding is used in applications where these limitations apply, and in general where high production rates are involved since arc welding is a lower-production-cost process. Reference 12 summarizes all the recommended welding practices (for automotive welding design), showing specifics in welding-process selection.

The following design considerations apply:

1. Metals being joined must have approximately the same melting point.

2. Welded joints have the same strength as the annealed material of the parts welded if the weld metal is of the same analysis as the base metal. When tensile tested to failure, the joint will usually break at the edge of the weld in the base-metal area annealed by the welding heat rather than at the weld. Welded fittings may be heat-treated after welding to remove cooling strains and overheated spots in the fitting. Welded alloy steels should be heat-treated after welding if possible.

3. Materials of equal thicknesses are best suited for welding. Where the thickness ratio is 1:1.5 or less, the materials are considered to be of equal heating value and can be welded easily. Materials where the thickness ratio is greater than 1:1.5 are considered to be of unequal heating value and are difficult to weld. A maximum ratio of 1:3 is normally considered the limit. An exception to this exists where one side of the joint has a greater thickness but smaller mass. For instance, a $\frac{1}{8}$-in continuous sheet can be welded to a 1-in cube because the block, once heated, will not dissipate the heat so rapidly as the continuous sheet containing the greater area.

4. Joints prepared for welding should have the edges beveled on one or both pieces to allow space for filling in the welding material, which should be of nearly the same analysis as the welded pieces (see Table 16.3).

TABLE 16.3   Weld-Rod Types for Various Base Alloys

Base metal	Weld rod
Steel castings	Steel
Steel pipe	Steel
Steel plate	Steel
Steel sheet	Steel or bronze
High-carbon steel	Steel
Wrought iron	Steel
Galvanized iron	Steel
Cast iron, ductile	Nickel-base alloys (approx. 50% Fe, 50% Ni)
Cast iron, gray	Nickel-base alloys, cast iron, or bronze
Cast iron, malleable	Nickel-base alloys or bronze
Cast-iron pipe	Nickel-base alloys or cast iron
Chrome-nickel steel castings	Base metal or 25-12 chrome-nickel stainless steel
Chrome-nickel steel sheet	Columbium stainless steel or base metal
Chromium steel	Columbium stainless steel or base metal

5. Single-plate gussets, welded on one side only, should have the weld carried around the end to provide positive anchoring of the gusset end.

6. Welded joints should not be designed to take bending loads unless supported by auxiliary means such as pinning, riveting, telescoping, or other positive support.

7. Avoid the convergence of more than six members to form a welded cluster. The repeated application of heat on the small area where all members meet weakens the base metal adjacent to the remelted weld metal.

8. Do not use tin-lead solder in the proximity of a joint. Repair of the joint might cause contamination of the weld metal with subsequent embrittlement and cracking of the affected area.

9. Do not weld a brazed joint. The reverse practice of brazing a welded joint is permissible.

10. Avoid a joint which requires welding at the inside junction of two members forming an acute angle.

11. Avoid joints in parts which are heat-treated before welding, especially to a tensile strength in excess of 125,000 lb/in^2.

12. Avoid washers welded to lugs or other parts of fittings.

13. Do not weld parts previously spot-welded because of the probability of inclusion of flux or flux-removal fluids in the spot-welded seam.

14. Avoid designs which require welding into deep pockets, which are difficult to handle because of arc blow. Arc blow may be defined as the swerving of the arc from its normal path because of magnetic forces. Maximum arc blow usually occurs at corners where three plates intersect at right angles to one another. Corners should be left open if possible.

15. Avoid welding parts having widely different melting points.

16. Avoid hydrogen-induced cracking (a.k.a.: delayed cracking, cold cracking, underbead cracking, toe cracking, hydrogen-assisted cracking).

Hydrogen-Induced Cracking (HIC).[57-59] Plain-carbon, and low-, medium-, and high-alloy steels, when exposed to hydrogen environments during arc welding, may experience cracking in the weld or heat-affected zones. This hydrogen-induced cracking, which occurs at low temperatures [below 400°F (200°C)], may be delayed in some cases for several days. The incidence of cracking increases with increasing hardness and strength of the steel and with increasing carbon and alloy contents.

Hydrogen-induced cracking (HIC) requires the simultaneous presence of four factors:

1. *A metallurgical factor:* A crack-sensitive microstructure must be present. The greater the hardness of the microstructure, the greater is the sensitivity to HIC. From a practical point of view, HIC is rarely encountered with a hardness less than Rockwell C 30 (Vickers or Brinell Hardness 300).

The use of higher values of energy input (higher arc current and lower travel speed) and the use of higher preheat and interpass temperatures, will result in slower cooling rates in the weld and heat-affected zones; these zones will have softer and more HIC-resistant metallurgical structures. Additionally, the use of these welding parameters will result in much slower cooling rates at low temperatures [such as 400°F (200°C) or below] and will provide longer time for hydrogen to diffuse from the weldment.

2. *A chemical factor:* A critical value of hydrogen must be present; this level varies inversely with the yield strength of the steel. The principal source of hydrogen

in arc welding results from the dissociation of hydrogen-containing compounds, such as water and hydrocarbons, by the high temperatures of the arc. The level of hydrogen can be controlled by appropriate choice of welding process variables (discussed above in item 1) and by attention to the care and storage of welding consumable ("low-hydrogen practice," described below).

3. *A mechanical factor:* A critical stress level of yield-strength magnitude must be present for HIC to occur. Unfortunately, the residual stresses, which occur as a result of welding, have yield-strength magnitudes.

4. *A temperature factor:* HIC occurs only within the temperature range from about $-100°C$ ($-148°F$) to $200°C$ ($392°F$). Unfortunately, most structures must serve within this temperature range.

Low-hydrogen practice includes the following:

1. Use low-hydrogen shielded metal-arc welding (SMAW) electrodes which are supplied in a dried condition and are enclosed in hermetically sealed containers.
2. Store low-hydrogen SMAW electrodes and submerged arc welding (SAW) fluxes in ovens at a holding temperature of 290 to 300°F (143 to 149°C) to avoid moisture pickup from the ambient atmosphere.
3. Limit the time of exposure of electrodes to ambient atmosphere after removal from the holding ovens to minimize moisture pickup prior to welding. Follow manufacturer's recommendations for atmospheric exposure and for reconditioning electrodes [usually heating for 1 h at 800°F (425°C) to remove moisture].
4. Remove hydrocarbons (oil, grease, paint) and moisture from the work surface prior to welding.

Detail Fusion-Welding Design Considerations. Three factors influence the length of fusion welds, namely, strength requirements, design of the parts, and distortion of the parts and possible resultant cracking of the weld. Depending upon these factors, welds may be either continuous, intermittent, or tack welds. Tack welds are generally used to hold material in place prior to normal welding; tack welds do not provide structural support.

Continuous welds are used whenever strength requirements are high, or where a liquid- or gastight joint is required. They are costly because of the postwelding straightening operation usually required to eliminate distortion caused by the heat of the welding operation. Intermittent welds are used on long joints where strength and rigidity requirements are not exacting enough to warrant the extra cost and weight of a continuous weld.

There are five fundamental types of welded joints, namely, butt, lap, corner, edge or flange, and tee. The more common types of welding applied to these joints are given in Fig. 16.13. Recommendations on structural efficiency are given in Fig. 16.14.

In all types of butt welds, the strength characteristics may be improved by welding both sides. However, welding both sides is not recommended on steel or aluminum alloys in gages less than 0.125 in, or on any gage of magnesium alloy.[30]

Fittings should be made simply, with the smallest practicable number of component parts to reduce welding to a minimum. Single-piece forgings or castings are preferable to built-up welded fittings.

When welding sheet or tube to fittings, the thinner stock is frequently burned away before the thicker material is brought to the fusing point. Furthermore, finish machining of a heavy fitting included in a welded assembly cannot be completed because excessive warpage or shrinkage has left insufficient material for the final operation. It is therefore necessary to rough-machine all heavy fittings prior to oxyfuel gas welding to eliminate excess material, reducing the area to be welded to a thickness not to

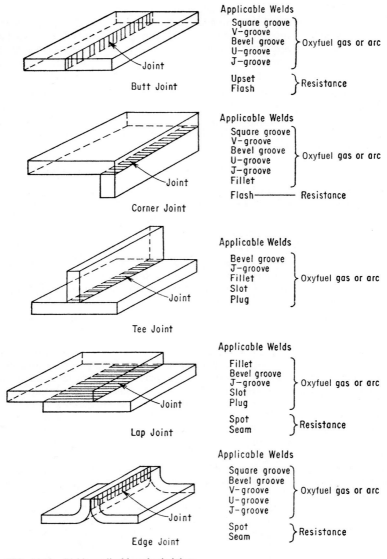

FIG. 16.13   Welds applicable to basic joints.

exceed three times the thickness of the attaching members. When such precautions are exercised, the heavy fitting may be machined to within ⅓₂ in of its final thickness before welding.

The joints shown in Fig. 16.15 represent good practice because equal gages of material are welded. The fitting in the area to be joined is first beveled at 45°.

A weld bead around a welded fitting helps eliminate warpage and cracking by permitting expansion and contraction (see Fig. 16.16).

Type of weld	Illustration	Sheet gage	Efficiency in			Fatigue resist- ance	Application
			Shear	Tension	Com- pression		
Square butt welded one side	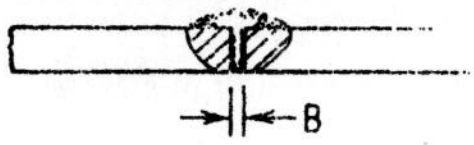	Up to 0.124	High	High	High	Fair	
Single V welded one side	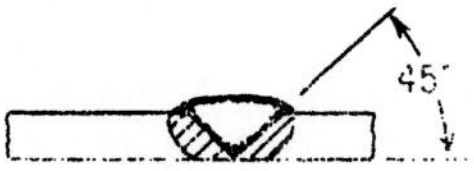	0.125 and over	High	High	High	Good	All general appli- cations. The choice depends upon the mate- rial, gage, load- ing, fatigue re- quirements, etc.
Single bevel welded one side		0.125 and over	High	High	High	Good	
Double V welded both sides	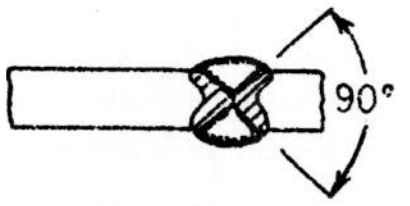	0.375 and over	High	High	High	Good	
Open-square groove corner	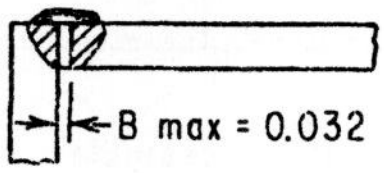	Up to 0.065	Medium	Low	High	Poor	
Single V corner fillet	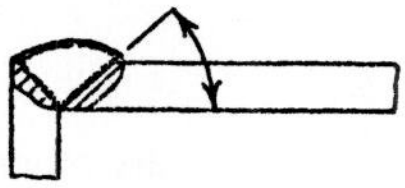	0.065 and over	Medium	Low	....	Poor	Closed structures and heavy-gage tanks
Outside or inside sin- gle fillet	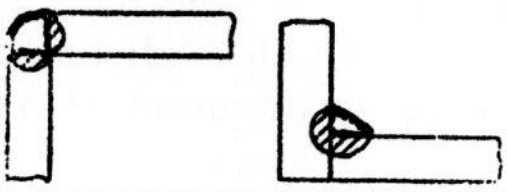	Up to 0.094	High	Low	....	Poor	
Single fillet tee	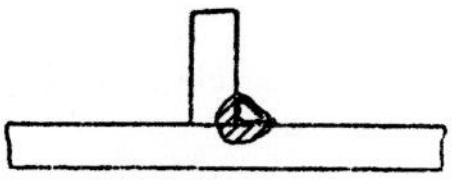	Any gage	Medium	Low	Low	Poor	
Double fillet tee	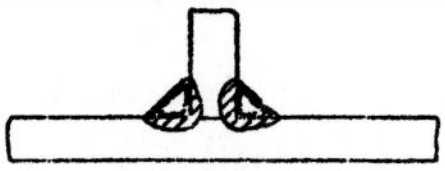	Any gage	High	High	High	Good	General purposes
Edge fillet weld		Any gage	Medium	Low	....	Fair	
Double-lap fillet		Up to 0.250	High	High	....	Fair	Closed structure and heavy-gage tanks
Flange weld		Up to 0.081	Medium	Low	....	Poor	Tanks and closed nonstructural parts

Flange height	Material gage
4 × material gage	0.040 or less
3 × material gage	0.041–0.081

Root Opening B
0.064 gage and less, no gap
0.064–0.124 gage, gap = ½ sheet thickness

FIG. 16.14 Structural efficiency of various types of welds (dimensions in inches).

FIG. 16.15   Preferred fitting joints.

FIG. 16.16   Method of stress relief around a fitting.

If the members to be joined are of the same diameter, they should be welded at a 30° angle scarf joint (Fig. 16.17). In a scarfed joint, the plane of the joint is inclined to the main axis of the members. A tubular liner should be used, extending at least 1 in or $1\frac{1}{4}D$, whichever is greater, beyond the welded joint. This not only prevents reduction of diameter when the joint is under tension but also serves to align the tube ends for welding. All liners must fit snugly inside the tubular members.

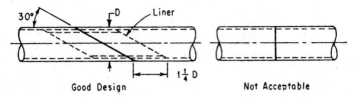

FIG. 16.17   Tubular joint splices, equal-size tubes.

Tubes of different diameters should be telescoped one inside the other. The tubing may be swaged or a sleeve may be used to ensure a snug fit. The female tube (and sleeve, if used) must be scarfed or fishmouthed (see Fig. 16.18). A fishmouthed joint is double-scarfed; as a result, as shown in Fig. 16.18, the maximum overlap is reduced for a constant value of D.

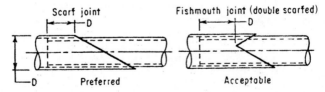

FIG. 16.18   Tubular joint splices, unequal-size tubes. Fishmouth joint is double-scarfed.

Truss joints generally involve one continuous member, to which are welded vertical and/or diagonal members. The axes of several members must intersect at a common point to avoid eccentric loading. The fixity and tensile strength of such joints are increased by the use of gussets (see Fig. 16.19). They also permit gradual introduction of the stress into the joint.

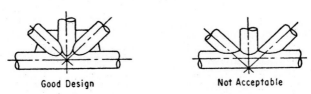

FIG. 16.19   Tubular truss joints, acceptable and unacceptable.

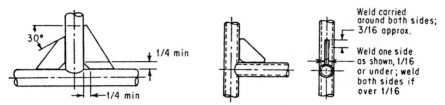

FIG. 16.20   Double-plate gusset reinforcement, tubular joints.

FIG. 16.21   Single-plate gusset reinforcement, tubular joints.

Because of stress concentration, welds in slotted tubes should not end diametrically opposite one another. A minimum diagonal of 30° across the tube should be maintained (see Fig. 16.20).

Single-plate gussets ¹⁄₁₆ in thick and under should be welded on one side only and have the weld carried around the end to provide positive anchoring of the gusset plate (see Fig. 16.21). Gussets over ¹⁄₁₆ in thick should be welded on both sides.

Splicing of tubes can often be done conveniently near a truss joint, which tends to strengthen the splice.

Lugs may be welded to any point on a member if the load on the lug is of low magnitude. High-stress lugs should be located at truss points. Washers or bushings may be welded to lugs to increase bearing area when weight is critical, but this practice *must not* be followed on highly stressed fittings.

End fittings designed for tubular members may be forgings flash-welded to the tubes where large quantities are involved. For highly stressed members of limited quantity, the fitting may be a forging or high-tensile steel casting jointed to the tube by a fishmouth weld or similar method. Abrupt changes in section must be avoided because of the probability of high stress concentrations.

Figure 16.22 illustrates both tubular terminal, end-fitting designs.

Figure 16.23 illustrates methods of welding a tube to a sheet or two sheets together. In the latter case, except for capping a tube or for a low-stressed joint, the tube should be inserted into a hole punched in the sheet.

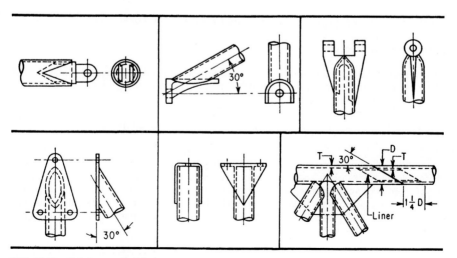

FIG. 16.22   Tubular terminal design.

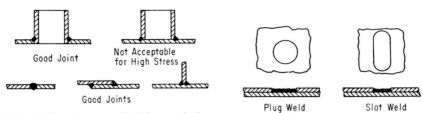

FIG. 16.23 Recommended sheet and plate joints.

FIG. 16.24 Recommended plug and slot welds.

Plug or slot welds may be employed for attaching two sheets or plates at specific points. A certain degree of local warping usually accompanies this process (see Fig. 16.24). Plug welds, often used in lieu of spot welds to assemble plate stock too thick to be spot-welded with available machines, may be made with oxyfuel gas or arc welding.

Weldability of Materials. The relative welding qualities of similar and dissimilar metals and types of welding which may be employed involve complex metallurgical considerations. Figure 16.25 presents weldability data for a limited number of combinations. More information may be found in Refs. 4 and 37; a more complete discussion of weldability of steels is given in Ref. 13 and of nonferrous metals in Ref. 14.

16.1.6 Arc Welding

The methods used for most steel applications for thicknesses over 0.062 in involve shielded metal arc (SMA), gas-metal arc (GMA), flux-cored arc (FCA) and submerged arc (SA) welding. For aluminum alloys over 0.091 in, the preferred processes are gas-metal arc and plasma arc welding. For thinner materials, particularly for aluminum alloys, the preferred process is gas-tungsten arc welding (GTAW). Joint recommendations are given in Fig. 16.26.

Design Considerations for Magnesium and High-Nickel Alloys. In addition to the general and detail design considerations listed above, the following specific considerations should be observed.

Magnesium Alloys. Gas-tungsten arc welding should be employed on sheets and plates with thicknesses ranging from 0.030 to 1.00 in. A backing plate is desirable to minimize "drop-through" and distortion due to heat dissipation.

All magnesium joints should be stress-relieved.

Desirable joint designs for magnesium alloys are illustrated in Fig. 16.27.

High-Nickel Alloys. For Monel, nickel, and Inconel, filler metal should be the same as the base alloy or electrodes which have been commercially developed for welding these alloys. Such joints are equal in corrosion resistance and strength to the parent metal, requiring no aftertreatment in heavy gages. Fabricated assemblies in thin gages should be stress-relieved after welding. The high coefficient of thermal expansion makes it necessary to use heavier clamping pressure to keep parts from buckling than is used with low-alloy steel.

Basic alloy designation	Alum. 1100 and 3003	Alum. 5052	Alum. 5053	Alum. 6061	Magnesium AZ61X	Tool steel M1	AISI 1010, 1020, 1025	AISI 2330	AISI 4130	AISI 4140	AISI 4340	Stainless steel 18-8
1100 and 3003	B E 1											
5052	B E 1	B E 1										
5053	B E 1	B E 1	B E 1									
6061	B E 1	B E 1	B E 1	B E 1								
AZ61X					E 1							
M1					E 1	E 1						
1010, 1020, 1025							A B C D 1					
2330							A B C D 2	A B C D 2				
4130							A B C D 4	A B C D 4	A B C D 4			
4140							A B C D 4	A B C D 5	A B C D 5	A B C D E 5		
4340							A B C D 5	A B C D 5	A B C D 5	A B C D E 5	A B C D E 5	
18-8							A B C D 1	A B C D 1	A B C D 2	A B C D E 2	A B C D E 2	A B C D E 1

A = oxyfuel gas; B = gas-metal arc; C = shielded metallic arc; D = flux-cored arc; E = gas-tungsten arc.

1 = good; 2 = fair; 3 = poor; 4 = heat-treat after welding; 5 = special methods, such as preheating and controlled cooling, required.

The necessity for heat-treatment after welding should be judged on the basis of the specific design. Welding may be done after heat-treatment, provided the reduced strength in the weld area is taken into consideration.

Oxyfuel gas welding may be used on any steel fitting where deemed expedient.

FIG. 16.25 Weldability of various alloys.

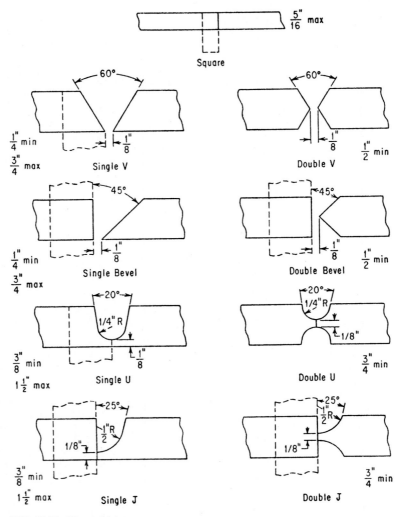

FIG. 16.26 Types of joints.

16.1.7 Distortion and Heat Flow

Distortion associated with a given material and welding process is a function of joint design. This phenomenon is due to the contractions and unequal cooling rates which exist between the materials being welded and the materials being used for the weld. The effect of unequal contractions and cooling rates upon two welded structures is shown in Fig. 16.28.

If a complete mathematical approach is desired, a general theory of welding deformation and stress is treated in Ref. 16. If further information is desired on residual welding stresses, Ref. 17 is recommended.

The following rules should be observed to minimize distortion:

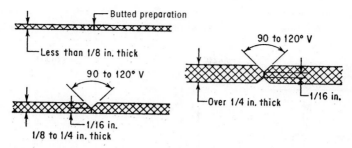

FIG. 16.27   Recommended joint chamfer for magnesium alloys.

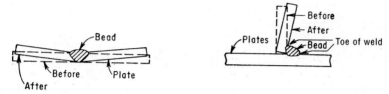

FIG. 16.28   Distortion after cooling. (*Note:* The toe of the weld is the junction between the face of the weld and the base metal.)

Use U or J joints in preference to V or bevel.

Use double joints (front and back) where thickness warrants instead of a single joint.

In both cases the prescribed method uses less weld metal; this reduces heat input and therefore distortion. Distortion is also related to the joint design. See Fig. 16.28.

Preheating before welding serves to reduce distortion by reducing thermal gradients. Metal thickness and joint type affect the flow of heat away from the weld bead and therefore affect resulting distortion. This effect is shown in Fig. 16.29,[9] in a simplistic way. For a butt weld in ¼-in plate, heat flow occurs in two directions; the thermal severity number has been arbitrarily assigned a value of 2. With a change in thickness to 1 in, the heat-flow path has been doubled; thus, the thermal severity number has been doubled to a value of 4. For a tee joint, heat-flow occurs in three directions; thus, with ¼-in plate the thermal severity number is 3, and with a ½-in plate, is 6.

16.1.8   Unfavorable Factors

The major undesirable factors to be considered in selecting welding as opposed to other joining methods are

1. The possibility of generating inferior welds as a result of improper welding conditions and poor workmanship

2. Incomplete knowledge of the fatigue characteristics of welded joints, including the influence of notches and stress concentrations

3. Undesirable mechanical properties in the weld and heat-affected zones

4. Thermal stresses, including warpage and distortion

Heat flow	Sheet thickness, in.		Thermal severity no.
Two way			
	Both	$\frac{1}{4}$	2
		$\frac{1}{4}-\frac{1}{2}$	3
		$\frac{1}{4}-\frac{3}{4}$	4
	Both	$\frac{1}{2}$	4
		$\frac{1}{2}-1$	6
		1	8
		$1-2$	12
Three way			
	Both	$\frac{1}{4}$	3
		$\frac{1}{2}$	6
		1	12
		2	24
		$\frac{1}{4} + \frac{1}{2} + \frac{1}{2}$	5
		$\frac{1}{2} + 1 + 1$	10
		$2 + 1 + 1$	16
Four way			
	Both	$\frac{1}{4}$	4
		$\frac{1}{2}$	8
		1	16
		2	32
		$\frac{1}{4} + \frac{1}{2} + \frac{1}{2} + \frac{1}{2}$	7
		$\frac{1}{2} + \frac{1}{2} + 1 + 1$	12

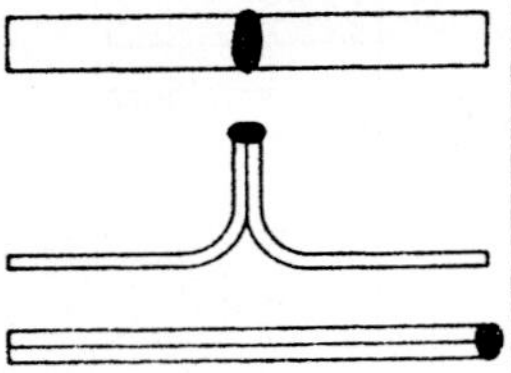

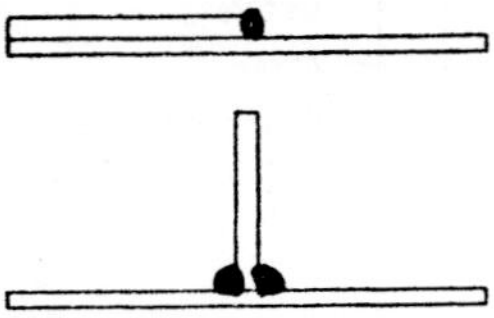

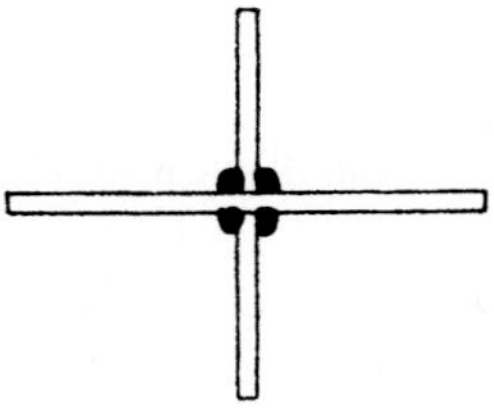

FIG. 16.29   Heat-flow paths in typical joints and nominal thermal severity.

A nondestructive test is often specified to evaluate the physical soundness of the weld. Such methods include x-ray, liquid penetrant, fluorescent-penetrant, magnetic-particle, and ultrasonic inspection. Exact standards must be specified for each joint and should be based upon tests on the actual part.

Fatigue failures show after some time has passed and may not be recognized as a limiting factor until many units are in service. Parameters for welded joints which will experience fatigue are given by the AWS and AISC codes. These codes were developed for structures but are adaptable to machines and equipment which also experience fatigue.

A general rule for the design of weldments for fatigue resistance is similar to that for any fatigue-prone part: abrupt section changes should be avoided. For this reason, simple butt welds are better in fatigue than lap and fillet joints. Butt welds without reinforcement will have greater fatigue life than those with reinforcement because section change is absent. Transverse attachments also lower fatigue resistance.

Designs for welded joints which will experience fatigue should avoid excessive reinforcements, undercuts, lack of penetration, and rough welds. Other design considerations include the following: (1) Welded joints should not be made in a region which will experience flexure; (2) fatigue resistance is generally higher in the rolling direc-

TABLE 16.4   Relative Fatigue Strength of Various Welded Joints

Type of joint	Notes	Endurance limit, ksi (1 ksi = 1000 lb/in²)	
		Tension/ compression	Tension only
	Weld machine flush	±18.0	14.0 ± 14.0
	As welded	±14.0	11.5 ± 11.5
	Incomplete root* fusion	±10.0	8.0 ± 8.0
	Complete root* fusion	±11.0	9.5 ± 9.5
	Incomplete root* fusion	±7.4	7.0 ± 7.0
	No joint preparation	±4.5	4.0 ± 4.0

*The root of a weld occurs at a point in the cross section at which the bottom of the weld intersects the base-metal surfaces. With incomplete root fusion, i.e., lack of penetration, the depth of the weld metal does not extend into the root of the joint.

tion of sheet metal; and (3) eccentric loads, biaxial and triaxial stresses, and restrained internal sections should be avoided. Detailed information on fatigue is found in Refs. 18 to 33. Typical data are given in Table 16.4.

The fatigue strength is a function of both the type of joint and the loading. Low working stresses are necessary because the weld joint results in stress concentration. Joint-design[4,5,7,9,34,35] changes to eliminate the effects of stress raisers in brazed joints are shown in Fig. 16.30. These changes include thickening the thinner member at the point of stress concentration, reshaping members, changing joint type, and adding reinforcement. Accepted values of the stress-concentration factor K are as follows: butt weld machined, 1.2; fillet weld at toe, 1.5; fillet weld at end, 2.7; and tee butt joint, sharp corners, 2.0. More desirable values of the stress-concentration factor can be obtained with improved work quality. The data in Table 16.4 together with the information mentioned above show that the smooth butt welds should be used wherever possible.

The effect of combined loads (static plus dynamic) must be considered in some machine components as shown in Fig. 16.30. The equation for the straight line relating

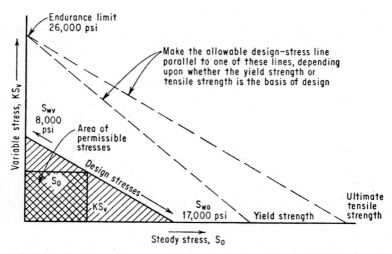

FIG. 16.30 Variable and steady stress relationships for designing to avoid fatigue failures.

the steady stress S_0 and the variable stress KS_V and joining the variable working stress $S_{u'v}$ and the steady working stress $S_{u'0}$ is

$$KS_V/S_{u'v} + S_0/S_{u'0} = 1$$

By substituting into the above formula and using the preceding formulas to solve for the length of weld needed to carry the stress calculated, a determination can be made as to whether there is sufficient joint area to produce the length of weld required for safe operation.

The subject of corrosion[4,9,35,38] and its relation to welded metals is complex. The combination of base metal and weld metal of a different composition in the presence of an electrolyte can lead to corrosion. Also, localized corrosion can occur if the metal anode surface or volume is small in comparison with that of the cathode.

16.1.9 Stress Calculations[39,40]

The basic formulas for determining shear and normal stresses in welded joints are given in Fig. 16.31. Reference 11 gives data for allowable shear values for various weld-metal strengths. Strength level varies between 60,000 and 100,000 lb/in^2 with an allowable shear stress between 18,000 and 33,000 lb/in^2.

16.2 BRAZING AND SOLDERING

16.2.1 Design

AWS has arbitrarily defined the difference between soldering and brazing. If the joining filler metal has a liquidus above 840°F (450°C) and if the liquidus of the filler metal is below the solidus of the metals being joined, the process is brazing. In the

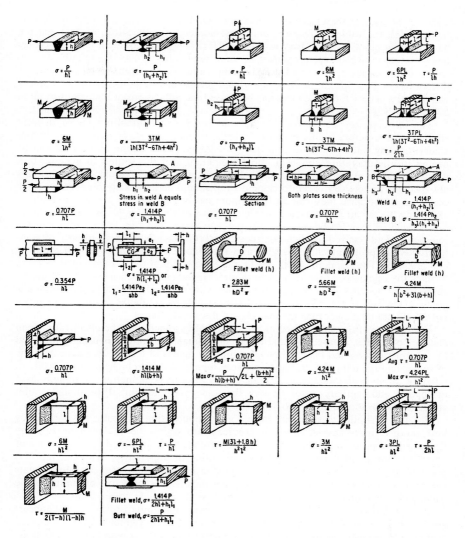

FIG. 16.31 Stress calculations. σ = normal stress, lb/in^2; τ = shear stress, lb/in^2; M = bending moment, in·lb; P = external load, lb; L = linear distance, in; h = size of weld, in; l = length of weld, in.

soldering process, the filler metal has a liquidus temperature which does not exceed 840°F (450°C) and which does *not* exceed the solidus temperature of the base metal. The important consideration is not what the process is called, but how the joint is designed and what the allowable stresses are.

The major difference in design between soldering and brazing is that soldering is a sealing process whereas brazing is a structural joining process. Soldering is used extensively for making electrical connections, hermetic seals, joining of lead or copper pipe, filling casting or sheet-metal imperfections, and general repair work. In such applications the mechanical load is usually carried by a mechanical joint (e.g., twisted wires in the case of the electrical connection and crimped stovepipe joint in the hermetic joint), the solder providing continuity. Solders, usually alloys of lead and tin, have melting points ranging from 450 to 600°F (232 to 316°C). Such lead-tin solders have very low mechanical properties at room temperature (tensile strength of 5800 lb/in^2) which drop off very rapidly as the temperature is increased. To overcome these difficulties, a few special solders have been formulated, e.g., silver-tin-lead solder which melts at 589°F (309°C) and has been used in applications at temperatures up to 572°F (300°C).

The basic design of the joint to be soldered is the controlling factor in joint strength. All structural joints must be crimped, riveted, bolted, or spot-welded before soldering. Any thickness or section can be soldered provided the base metal can be heated to a temperature above the flow temperature of the solder. In general, butt welds should be avoided because of the low tensile strength of solders. If such joints are absolutely necessary, a reinforcement strap should be used. Lap joints are in more common use because they can be fitted more easily and achieve greater strength by extending the lap area. The area of the overlap is a function of application. Long overlaps should be avoided because it is difficult to assure full flow of the solder in the joint area. Voids caused by such poor design practice can be determined only by nondestructive inspection techniques such as x-ray (which is not usually used in commercial practice). The standard practice is to make the width of the overlap equal to $1\frac{1}{2}$ to 3 times the thickness of the thinner base-metal members.[37] Such joints should not be subjected to bending loads because the solder will fail rapidly under the tension loads applied (see Fig. 16.4 for loading distribution). In overlap tubular members, the overlap should not be greater than the diameter of the smaller tube.

For solder to flow, the surface of the metal being joined must be chemically clean and free of oxides, grease, dirt, and other contaminants. Such dirt is usually removed by solvent cleaning and then chemical etching by means of a flux. To assure the flow of solder into the joint, the clearance between members should be held within 0.001 to 0.010 in. Joints with 0.003 in of clearance offer the best combination of strength, ease of deposit, and capillary action to draw the solder into the joint. Typical solder-joint applications are shown in Fig. 16.32.

Brazing is similar to soldering in that good joint design is similar for both. They both require joints having capillary crevices into which the solder or braze alloy flows. Advances in braze alloys and heating methods have increased the importance of brazing to the point where brazing rivals welding in importance. With brazing it is possible to produce joints of low cost, high ductility, and high strength. Most metals can be joined either to themselves or to other metals. The principal exception is the exclusion of aluminum and magnesium from dissimilar-metal combinations.

From a design point of view, brazing offers minimum joint distortion, good appearance as brazed, absence of weld spatter or other deposits (which require removal), adaptability to mechanization, and ability to perform operations with lower labor skill.

The disadvantages include the following: joint strength may be lost at lower temperatures than welded joints; the color of the braze alloy may not match the color of the alloy being joined; the braze alloy may have lower corrosion resistance than the base alloy and therefore need protection (paint, etc.) if the machine is to be used in a corrosion area.

The braze joints used are similar in principle to those specified for soldering except that thicker base-metal sections are often employed. Braze joints include lap, scarf, butt, tee, and combinations thereof. Typical joints are shown in Fig. 16.33.

The lap joint is preferred when the joint must be leakproof or is subjected to high loads. Here, also, the strength of the joint can be varied by increasing or decreasing the area of the lap. The lap joint is preferred because it permits the easy assembly of members and the maintenance of proper capillary clearances through inexpensive jig-

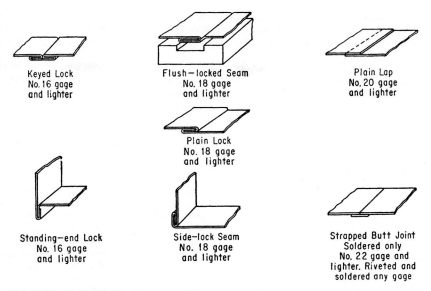

FIG. 16.32 Typical designs for soldered joints. Nominal thicknesses: 16 gage, 0.060 in; 18 gage, 0.048 in; 20 gage, 0.036 in; 22 gage, 0.030 in.

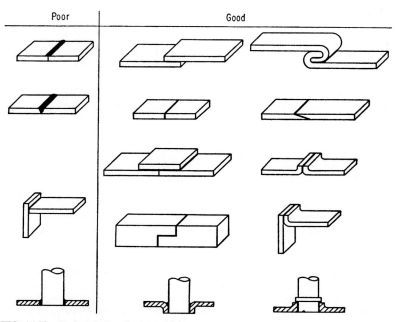

FIG. 16.33 Typical designs for brazed joints.

ging. For tubular lap joints, correct fit tolerances must be achieved through control of the diametric tolerances. To assure correctness of fit, the machining of one or both members to a tolerance closer than that obtained with "as-drawn" tubing must usually be specified.

Square butt joints can be successfully used if care is taken to specify that the joint components are machined square to provide uniform joint clearance between members. Square butt joints are usually specified for applications where double-lap joints cannot be used. The scarf butt joint is a variation of the butt joint offering a stronger, more ductile, but more costly joint.

16.2.2 Joint Clearances

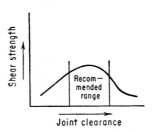

FIG. 16.34   Shear strength vs. joint clearance.

The strength of a brazed joint is a function of joint clearance, joint type, braze alloy, and the alloys being joined. To achieve a maximum joint strength, the thickness of the braze alloy, in a joint as brazed, should be thin enough to assure failure in the base alloy. The relationship of shear strength (of braze alloy) to joint clearance is shown in Fig. 16.34.

As related to specific braze alloys, recommended joint clearances (at brazing temperature) are shown in Table 16.5.

If the joint clearance is too small or too large, the braze alloy will not flow into the joint and it will not develop full strength.

TABLE 16.5   Joint Clearances (at Brazing Temperature) Recommended for Maximum Joint Shear Strength per AWS B2.2 Specification

AWS filler-metal classification	Joint clearance
B Al Si	0.002–0.010 for laps under ¼ in.
	0.010–0.025 for laps over ¼ in.
B Cu P	0.001–0.005
B Ag	0.002–0.005
B Cu	0.000–0.002
B Cu Zn	0.002–0.005
B Mg	0.004–0.010
B Ni	0.000–0.005*
B Au	0.002–0.005

*A large number of Ni-base filler metals are available; joint clearances vary among these brazing filler metals.

Composition and uses of braze alloys are given in AWS Specification B2.2. For machine-component assembly, the copper- and silver-base alloys are usually used. A shear strength of 15,000 lb/in^2 can be used in calculating joint strength with either copper braze or silver braze when the joint does not exceed 0.010 in of thickness.[41] For thicker materials a shear strength of 12,000 lb/in^2 can be used safely. Where copper braze is used, brazed assemblies may be heat-treated after brazing without affecting the characteristics of the joint.

Brazing may be used where the gages are too thin for welding or where it is desirable to avoid forging and machining costs by simultaneously brazing a number of simple fabricated details into an assembly.

Braze welding is similar to brazing in that the base metal is not melted but is joined to another piece of metal by an alloy of a lower melting point. The main difference is that in braze welding the alloy is not drawn into the joint by capillary action. Braze welding is used extensively for repair work and some fabrication on such metals as cast iron, malleable iron, wrought iron, and steel. It is also used, but to a lesser extent, on copper, nickel, and high-melting-point brasses and bronzes. The strength of a braze-welded joint is dependent on the quality of the bond between the filler metal and the base metal. It is also dependent on the quality of the filler metal after it is deposited. The tensile strength of a braze-welded joint on steel or cast iron is approximately 50,000 lb/in^2.

16.3 ADHESIVE BONDING

"Adhesive bonding" is the process of joining materials chemically through the formation of interatomic or intermolecular bonds. Adhesives can be used to join a wide variety of materials (metal to metal, metal to ceramic, metal to polymer, etc.) for both structural and nonstructural use.

The application of adhesive technology to the aerospace industry is responsible for significant recent progress in the theoretical understanding of adhesive bonding and its broadened use as a material-joining method. Theories of adhesion and adhesive joint strength are given in the references (Refs. 60–64).

Some possible advantages of bonding over other fastening methods are:[43,44]

Uniform stress distribution with resultant increased service life

Reduced weight

Fatigue resistance

Ability to join thick or thin materials

May not require high temperature for joining

Ability to join dissimilar materials

No perforation required

Can provide leakproof joints

Can incorporate vibration-damping and insulation properties

Can provide corrosion resistance

May reduce production costs

Some of the significant limitations of bonding metallic materials are: (1) The maximum operating temperature of a bonded machine tool is determined by the adhesive; (2) the best synthetic-resin adhesives available have a maximum operating temperature of about 500°F (260°C); (3) tooling costs for bonding may be excessive when only a few assemblies are made; (4) structural adhesives generally have high shear and tensile strength, but their resistance to peel and cleavage stresses is relatively poor. Information on standard test procedures for structural adhesives is found in Refs. 62 and 63.

From an application standpoint, adhesives are generally classified as structural and nonstructural. Structural adhesives are those which produce high-strength joints used

TABLE 16.6 Structural Adhesives Classification

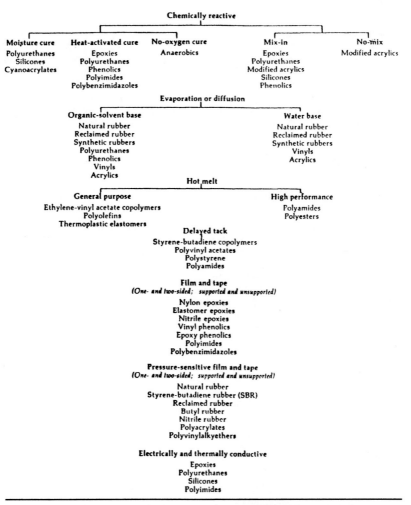

Chemically reactive

Moisture cure	Heat-activated cure	No-oxygen cure	Mix-in	No-mix
Polyurethanes	Epoxies	Anaerobics	Epoxies	Modified acrylics
Silicones	Polyurethanes		Polyurethanes	
Cyanoacrylates	Phenolics		Modified acrylics	
	Polyimides		Silicones	
	Polybenzimidazoles		Phenolics	

Evaporation or diffusion

Organic-solvent base	Water base
Natural rubber	Natural rubber
Reclaimed rubber	Reclaimed rubber
Synthetic rubbers	Synthetic rubbers
Polyurethanes	Vinyls
Phenolics	Acrylics
Vinyls	
Acrylics	

Hot melt

General purpose	High performance
Ethylene-vinyl acetate copolymers	Polyamides
Polyolefins	Polyesters
Thermoplastic elastomers	

Delayed tack
Styrene-butadiene copolymers
Polyvinyl acetates
Polystyrene
Polyamides

Film and tape
(One- and two-sided; supported and unsupported)
Nylon epoxies
Elastomer epoxies
Nitrile epoxies
Vinyl phenolics
Epoxy phenolics
Polyimides
Polybenzimidazoles

Pressure-sensitive film and tape
(One- and two-sided; supported and unsupported)
Natural rubber
Styrene-butadiene rubber (SBR)
Reclaimed rubber
Butyl rubber
Nitrile rubber
Polyacrylates
Polyvinylalkyethers

Electrically and thermally conductive
Epoxies
Polyurethanes
Silicones
Polyimides

From M. M. Gauthier, "Sorting Out Structural Adhesives: Part 1," *Advanced Materials & Processes*, vol. 138, no. 1, pp. 26–35, ASM International, Materials Park, Ohio, 1990. Reprinted with permission.

in high-stress areas. The nonstructural adhesives are used in areas where stresses are low because of either small applied loads or large bond areas in relation to the load being applied. Chemically, adhesives are divided into thermosetting and thermoplastic resins, elastomeric materials, and organic and inorganic materials. A comprehensive list of structural adhesives is given in Table 16.6.[65]

Typical properties and characteristics of structural-adhesives are given in Table 16.7.[65] Table 16.8 gives a summary of adhesive types used to join metals.[60]

Surface treatment of metals to be bonded is an important factor in obtaining strong joints. Reference 66 provides information on surface-preparation techniques for adhe-

TABLE 16.7 Typical Properties and Characteristics of Structural Adhesives

	Epoxy	Polyurethane	Modified acrylic	Cyanoacrylate	Anaerobic
Impact resistance	Poor	Excellent	Good	Poor	Fair
Tensile shear strength, MPa (10^3 psi)	15.4 (2.2)	15.4 (2.2)	25.9 (3.7)	18.9 (2.7)	17.5 (2.5)
T-peel strength, N/m (lbf/in)	<525 (3)	14,000 (80)	5,250 (30)	<525 (3)	1,750 (10)
Substrates bonded	Most	Most smooth, nonporous	Most smooth, nonporous	Most metals or plastics	Metals, glass, thermosets
Service temperature range, °C (°F)	−55 to 120 (−70 to 250)	−160 to 80 (−250 to 175)	−70 to 120 (−100 to 250)	−55 to 80 (−70 to 175)	−55 to 150 (−70 to 300)
Heat cure or mixing required	Yes	Yes	No	No	No
Solvent resistance	Excellent	Good	Good	Good	Excellent
Moisture resistance	Excellent	Fair	Good	Poor	Good
Gap limitation, mm (in)	None	None	0.75 (0.03)	0.25 (0.01)	0.60 (0.025)
Odor	Mild	Mild	Strong	Moderate	Mild
Toxicity	Moderate	Moderate	Moderate	Low	Low
Flammability	Low	Low	High	Low	Low

From M. M. Gauthier, "Sorting Out Structural Adhesives: Part 1," *Advanced Materials & Processes*, vol. 138, no. 1, pp. 26–35. ASM International, Materials Park, Ohio, 1990. Reprinted with permission.

16.33

TABLE 16.8 Summary of Adhesive Types vs. Adherends

Adhesive	Metal																
	Al	Ag	Au	Be	Brass/bronze	Cd	Cu	Mg	Ni	Pb	Sn	Steel	Stainless steel	Ti	U	W	Zn
Acrylics	X				X		X					X	X				
Second-generation acrylics			X														
Anaerobics	X				X	X											
Cyanoacrylates		X					X	X		X							X
Epoxies	X				X		X	X	X	X	X	X	X	X	X	X	X
Epoxy-phenolics	X	X	X	X				X	X	X		X	X	X			
Modified epoxies	X	X	X														
Modified phenolics	X																
Neoprene-phenolics	X				X		X	X									
Nitrile-epoxies	X			X		X	X							X			X
Nitrile-phenolics				X	X	X	X	X	X			X	X	X			
Nitrile-rubbers																X	
Nylon-epoxy				X	X		X	X	X								
Polyacrylates									X	X	X						
Polyamides									X								
Polybenzimidazoles				X					X			X	X				
Polyimides				X					X			X	X	X			
Polyurethanes					X		X	X		X							
Polyvinyl acetates								X									
Polyvinyl alkyl ether			X							X	X						X
Silicones	X	X			X		X	X		X	X						
Styrene-butadiene											X						
Vinyl-plasticols	X						X	X									
Vinyl-phenolics								X	X								

From R. W. Messler, Jr., *Joining of Advanced Materials*, Butterworth-Heinemann, Stoneham, Mass., 1993. Reprinted with permission.

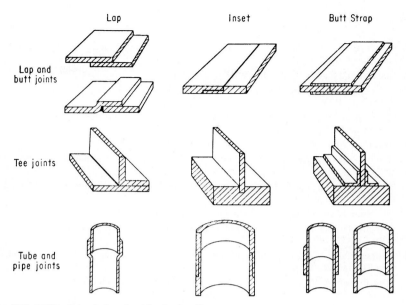

FIG. 16.35   Joint designs for adhesive bonding.

sive bonding. In designing joints to be bonded, the high shear strength and relatively low peel strength of adhesively fastened metal systems must be taken into account. Lap joints are the most common type of adhesive joint. Recommended joint designs for adhesive bonding are shown in Fig. 16.35. The thickness of the adhesive is a contributing factor in bonded joint strength. Generally, the bond strength of the more rigid adhesives used for structural applications decreases as the thickness of the bond line increases.

At higher temperatures the variation in strength with bond thickness becomes less pronounced. For flexible adhesives, bond strength may increase as the thickness of the adhesive bond line increases, which is opposite to the effect obtained in brazed and soldered joints.

Most high-temperature adhesives, which consist of cross-linked networks of large molecules, have no melting points. Their limiting thermal factor is strength reduction due to oxidation. Oxidation starts a progressive breaking (scission) of the molecular bonds of the long-chain molecules. This scission results in loss of strength, elongation, and toughness—leading to failure of the joint.

Oxidative degradation of an adhesive is affected by any catalytic reaction between the adhesive and the substrates. As an example, under identical conditions an adhesive used for bonding stainless steel may oxidize more than when bonding aluminum, but an adhesive with a different chemical makeup may show exactly the opposite effect.

High temperature curing agents can extend the service temperature of epoxy resins to as high as 500°F (260°C). These curing agents increase the number of cross-links between the epoxy molecules.

Polyimides are specifically designed for high-temperature service and can withstand continuous exposure at 500°F (260°C); they are unmatched by any other commercially available adhesive.

REFERENCES

1. Snyder, G. L.: "Design Considerations for Welded Machining Parts," *J. Am. Welding Soc.,* vol. 25, February, pp. 105–118, 1946.
2. "Ideas for Designing Machining Parts," *J. Am. Welding Soc.,* vol. 30, pp. 269–270, 1951.
3. Carey, Howard B.: "Modern Welding Technology," 2d ed., Prentice-Hall, Inc., Englewood Cliffs, N.J., 1989.
4. "Welding, Brazing, and Soldering," in "Metals Handbook," ASM International, vol. 6., 10th ed., Materials Park, Ohio, 1993.
5. "Brazing Handbook," American Welding Society, Miami, Fla., 1991.
6. Phillips, Arthur L., ed.: "Current Welding Processes," American Welding Society, New York, 1966.
7. Phillips, Arthur L., ed.: "Modern Joining Processes," American Welding Society, New York, 1966.
8. "Soldering Manual," American Welding Society, Miami, Fla., 1977.
9. "Welding Handbook," American Welding Society, 8th ed., Miami, Fla. vol. 1 (1987), vol. 2 (1991), vol. 3 (1994).
10. Blodgett, Omer W.: "Design of Weldments," The James F. Lincoln Arc Welding Foundation, Cleveland, Ohio, 1976.
11. "The Procedure Handbook of Arc Welding," The Lincoln Electric Company, 12th ed., Cleveland, Ohio, 1973.
12. "Recommended Practices for Automotive Welding," AWS Automotive Welding Committee, American Welding Society, New York, 1962.
13. Stout, Robert D., and Doty, W. D'Orville: "Weldability of Steels," Welding Research Council, New York, 1971.
14. West, E. G.: "The Welding of Non-Ferrous Metals," Chapman & Hall, Ltd., London, 1951.
15. Fusion Welding of Nickel and High Nickel Alloys, International Nickel Company, *Tech. Bull.* T2.
16. Masubuchi, K.: "Analysis of Welded Structures," Pergamon Press, N.Y., 1980.
17. Gunnert, R.: "Residual Welding Stresses," Almqvist & Wiksell, Stockholm, 1955.
18. Weymueller, Carl R.: "Designing Weldments for Fatigue Resistance," *Welding Design and Fabrication,* vol. 51, no. 5, May 1978.
19. Blodgett, Omer W.: "Weld Failures: They Could Be the Result of Violating Simple Design Principles—Part I and II," *Welding Journal,* vol. 61, nos. 3 and 4, March and April 1982.
20. Harman, C. R.: "Handbook for Welding Design," vol. I, Pitman Publishing Corporation, New York, 1956.
21. Wilson, W. M.: "Fatigue of Structural Joints," *J. Am. Welding Soc.,* vol. 29, no. 3, p. 204, March, 1950.
22. Structural Welding Code—Steel, ANSI/AWS D1.1-94, American Welding Society, Miami, Fla., 1994.
23. "Fatigue Tests of Riveted Joints," *Univ. Illinois Bull.* 302, 1938.
24. "Fatigue Tests of Butt Welds in Structural Steel Plates," *Univ. Illinois Bull.* 310, 1938.
25. "Fatigue Tests of Connection Angles," *Univ. Illinois Bull.* 317, 1939.
26. "Fatigue Tests of Welded Joints in Structural Plates," *Univ. Illinois Bull.* 327, 1941.
27. "Fatigue Tests of Commercial Butt Welds in Structural Steel Plates," *Univ. Illinois Bull.* 344, 1943.
28. "Fatigue Tests of Fillet Welds and Plug-Weld Connections in Steel Structural Members," *Univ. Illinois Bull.* 350, 1944.
29. "Rate of Propagation of Fatigue Cracks in $12 \times \frac{3}{4}$ inch Steel Plates with Severe Geometrical Stress Risers," *Univ. Illinois Bull.* 371, 1947.

30. "Flexural Fatigue Strength of Steel Beams," *Univ. Illinois Bull.* 377, 1948.

31. "Fatigue Strength of Fillet-Weld, Plug-Weld, and Slot Weld Joints Connecting Steel Structural Members," *Univ. Illinois Bull.* 380, 1949.

32. "The Fatigue Strength of Various Details Used for the Repair of Bridge Members," *Univ. Illinois Bull.* 382, 1949.

33. "The Fatigue Strength of Various Types of Butt Welds Connecting Steel Plates," *Univ. Illinois Bull.* 384, 1949.

34. Lindberg, Roy A., and Normam R. Braton: "Welding and Other Joining Processes," Allyn and Bacon, Inc., Boston, Mass., 1976.

35. Linnert, George E.: "Welding Metallurgy," vol. 1 and 2, American Welding Society, New York, 1965.

36. Lucas, Victor A.: "Designer's Guide to Welded Structures: Parts I and II," *Welding Design and Fabrication.* vol. 52, no. 5, May 1979, and vol. 52, no. 6, June 1979.

37. Nippes, E. F. (ed.): "Welding, Brazing and Soldering Metals Handbook," 9th ed., vol. 6, ASM International, Materials Park, Ohio, 1983.

38. Fontana, Mars C., and Norbert D. Greene: "Corrosion Engineering," McGraw-Hill Book Company, Inc., New York, 1967.

39. Jeanings, C. H.: "Welding Design," *Welding J.* vol. 15, no. 10, pp. 58–70, 1936.

40. Blodgett, O. W.: "A Simplified Approach to Calculating Design Stresses," *J. Am. Welding Soc.,* pp. 1182–1192, December 1958.

41. MIL-HOBK-5 (Military Handbook) Armed Forces Supply Center, Washington, D.C., March 1959.

42. Guttman, Werner H.: "Concise Guide to Structural Adhesives," Reinhold Publishing Corporation, New York, 1956.

43. Epstein, G.: "Adhesive Bonding for Metals," Reinhold Publishing Corporation, New York, 1954.

44. Koopman, K. H.: "Elements of Joint Design for Welding," *J. Am. Welding Soc.,* pp. 579–588, June 1958.

45. DeWitt, E. J.: "Welding of Small Subassemblies in a Program of All Welded Machines," *J. Am. Welding Soc.,* pp. 27–31, January 1939.

46. Mikulak, J.: "Design of Welded Machining," *J. Am. Welding Soc.,* vol. 24, pp. 13–21, January 1945.

47. De Brugne, N. A., and R. Houwink: "Adhesion and Adhesives," Elsevier Publishing Company, New York, 1951.

48. Brockenbrough, R. L., and B. G. Johnston: "Steel Design Manual," United States Steel Corporation, Chicago, Ill., 1968.

49. "Designer's Guide for Welded Construction," Lincoln Electric Company, Cleveland, Ohio, 1962.

50. Harman, R. C. (ed.): "Handbook for Welding Design," vol. 1, Sir Isaac Pitman & Sons, Ltd., London, 1967.

51. "Manual of Steel Construction," 6th ed., American Institute of Steel Construction, Inc., New York, 1963.

52. "Manual of Steel Construction," 7th ed., American Institute of Steel Construction, Inc., New York, 1970.

53. Hinkel, I. E.: "Joint Designs Can Be Both Practical and Economical," *Welding Journal,* vol. 60, p. 301, June 1981.

54. "ASM Metals Reference Book," 2d ed., ASM International, Materials Park, Ohio, 1983.

55. Nippes, E. F. (ed.): "Metals Handbook, Desk Edition," sec. 30, "Joining," ASM International, Materials Park, Ohio, 1985.

56. Bralla, J. G. (ed.): "Handbook of Product Design for Manufacturing," McGraw-Hill Book Company, Inc., N.Y., 1986.

57. Groville, B. A. (ed.): "The Principles of Cold Cracking Control in Welds," Dominion Bridge Co., Montreal, Quebec, Canada, 1975.

58. Gibala, R., and R. F. Hehemann (eds.): "Hydrogen Embrittlement and Stress Corrosion Cracking," ASM International, Materials Park, Ohio, 1984.

59. Interrante, C. G., and G. M. Pressouyre (eds.): "Current Solutions to Hydrogen Problems in Steels," ASM International, Materials Park, Ohio, 1982.

60. Messler, R. W., Jr.: "Joining of Advanced Materials," Butterworth-Heinemann, Stoneham, Mass., 1993.

61. Adams, R. D.: "Theoretical Stress Analysis of Adhesively Bonded Joints," in "Joining Fibre-Reinforced Plastics," F. L. Matthews, ed., pp. 185–226, Elsevier, London, 1987.

62. Adams, R. D., and W. C. Wake: "Structural Joints in Engineering," Elsevier, London, 1984.

63. Landrock, A. H.: "Adhesives Technology Handbook," Noyes Publ., Park Ridge, New Jersey, 1985.

64. Lewis, A. F.: "Epoxy Resin Adhesives," in "Epoxy Resins—Chemistry and Technology," C. A. May, ed., pp. 653–718, Marcel Dekker, Inc., N.Y., 1988.

65. Gauthier, M. M.: "Sorting Out Structural Adhesives: Part 1," *Advanced Materials and Processes*, vol. 138, no. 1, pp. 26–35, ASM International, Materials Park, Ohio, 1990.

66. Wegman, R. F.: "Surface Preparation Techniques for Adhesive Bonding," Noyes Publications, Park Ridge, N.J., 1989.

SECTION 17

SEALS

Izhak Etsion, D.Sc.

Professor of Mechanical Engineering
Technion—Israel Institute of Technology
Haifa, Israel

Bruce M. Steinetz, Ph.D., P.E.

Senior Research Engineer
NASA Lewis Research Center
Cleveland, Ohio

17.1 INTRODUCTION

A breakdown of pressure components into specific categories is difficult because of their great variety, but this section will be devoted entirely to seals—static and dynamic. The static seal will include those normally referred to as gaskets as well as other types which serve the same purpose and, in some cases, have additional uses. Dynamic seals are those which function between moving components, although the seal itself or part of the seal may be stationary.

The discussion of these seals will cover their use from a design viewpoint, not the design of the seal so much, but rather the design and characteristics of the component parts in order to best utilize the seal. A number of important factors in design which affect performance are discussed. These should aid in the design of the proper machine component and in the selection of the best seal for the application.

For more information than is supplied here on seal materials, see Ref. 1.

STATIC SEALS

17.2 COMPRESSION TYPE

17.2.1 General

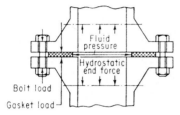

FIG. 17.1   Forces on a gasketed joint.

Compression-type static seals are those which are installed between the surfaces to be sealed and are preloaded to cause them to conform to the sealing surfaces and thereby eliminate or at least reduce leakage. When one of these seals, referred to as gaskets, is in use, certain forces are always present, and it is a balance of these forces, in combination with the gasket and joint characteristics, which determines the resultant sealability. A diagrammatic representation of these forces is shown in Fig. 17.1.

The mechanics of this type of system can be explained by the use of a simple equation:

$$F_b = N_b \sigma_b A_b \tag{17.1}$$

where F_b = total bolt load, lb
N_b = number of bolts
σ_b = bolt stress, lb/in^2
A_b = stress area (mean of pitch and root area) per bolt, in^2

Very often the number of bolts is selected on the basis of aesthetic reasoning, and it is little wonder that some gasketed joints give trouble. A later section presents a method for calculating the number of bolts to use in a particular joint.

In almost all cases in the design of a gasketed joint, the total bolt load required F_b can be calculated. From this, the load per bolt or the product $\sigma_b A_b$ can be calculated by dividing the total bolt load by the number of bolts. If the maximum allowable stress σ_b for the particular bolting material is known, the stress area A_b can be calculated and the size of the required bolt determined. Until quite recently the root diameter of the bolts was generally used as the stress area, but a stress area based upon a diameter that is the mean of the pitch and root diameters has been adopted by ASA (Standard B1.1-1960).

When a joint or joint design is already available, the calculations can be made in the reverse manner. The stress area of the bolts is multiplied by the maximum allowable stress for the bolting material and by the number of bolts to give the total bolt load F_b.

The maximum allowable stress and the yield stress for standard bolting materials

TABLE 17.1 Bolting Materials and Allowable Stresses

Type of metal	ASTM specification	Nominal composition	Max allowable design stress, psi		Min yield stress, psi
Carbon steel.....	A261-BD	0.55 C	16,250		75,000
Carbon steel.....	A307-B	. . . .	7,000		
Carbon steel.....	A325	0.30 C	18,750		77,000, 1–1½ in. 55,000, 1¾–3 in.
Low-alloy steel...	A193-BA	. . . .	16,250		
Low-alloy steel...	A193-BB	. . . .	18,750		
Low-alloy steel...	A193-BC	. . . .	20,000		
Low-alloy steel...	A193-B5	5 Cr, ½ Mo	20,000		75,000– 90,000
Chrome steel.....	A193-B6	12 Cr	20,000		80,000–120,000
Low-alloy steel...	A193-B7	1 Cr, 0.2 Mo	20,000		65,000–115,000
Low-alloy steel...	A193-B7a	1 Cr, 0.6 Mo	20,000		95,000 120,000
Low-alloy steel...	A193-B14	1 Cr, 0.3 Mo-V	20,000		105,000–120,000
Low-alloy steel...	A193-B16	1 Cr, ½ Mo-V	20,000		85,000–105,000
Chrome-nickel steel.........	B8, B8C, B87	18 Cr, 8 Ni	15,000		30,000–100,000
Low-alloy steel...	A354-BB	. . . .	18,750		75,000– 80,000
Low-alloy steel...	A354-BC	. . . .	20,000		95,000–105,000
Low-alloy steel...	A354-BD	. . . .	20,000		120,000
			100°F	400°F	
Aluminum.......	B211-GS11A, T6	. . . .	7,000	2,750	35,000
Aluminum.......	B211-CG42A, T4	. . . .	8,000	3,650	40,000
Aluminum.......	B211-CS41A, T6	. . . .	10,800	2,350	55,000
Copper..........	B-12	Copper	2,000	1,600	10,000
Copper..........	B98-A, C, and D	Copper-silicon	3,000	. . . .	15,000
Copper..........	B98-B	Copper-silicon	2,400–11,000	. . . .	12,000– 55,000
Nickel..........	B160	Low-carbon Nickel	2,000	1,900	10,000
Nickel..........	B164	Nickel-copper	4,900– 8,000	4,000–7,200	25,000– 40,000
Nickel..........	B166	Nickel-chromium Iron	5,900– 6,900	5,300–6,300	30,000– 35,000

are given in Table 17.1. These are established by the American Society of Mechanical Engineers, but other organizations such as the Society of Automotive Engineers may set up different values for the same or special materials. Under no circumstances should the yield stress be exceeded, and many times, especially with larger bolts, even the maximum design stress cannot be utilized.

Two methods of estimating the bolt load actually applied to a bolted joint are by the use of a torque wrench and a table or formula and by the use of an empirical equation. Bolt stresses as a function of torque are given in Table 17.2. A general relationship is

$$F_b = N_b T / 0.2 D_b \tag{17.2}$$

where T = torque, in·lb

D_b = nominal bolt diameter, in

An empirical equation which can be used when a torque wrench is not available is based on the normal pull a mechanic would put on a standard open-end wrench for the required bolt size:

$$F_b = 16,000 D_b N_b \tag{17.3}$$

It should be understood that neither of these equations gives a very accurate determination of the bolt load, but often nothing more is available. Torques, especially, are misleading because the stress developed is so dependent on the friction factors present. Test data of applied torque vs. bolt stress calculated from bolt elongation show a

TABLE 17.2 Torque vs. Bolt Stress and Bolt Load

Nominal diam	Bolt stress, psi											
	7,500		15,000		30,000		45,000		60,000		90,000	
	Torque, ft-lb	Load, lb	Torque, ft-lb	Load, lb	Torque, ft-lb	Load, lb	Torque, ft-lb	Load, lb	Torque, ft-lb	Load, lb	Torque, ft-lb	Load, lb
National Coarse-thread Series												
1/4	1	240	2	480	4	950	6	1,430	8	1,900		
5/16	2	390	4	780	8	1,570	12	2,350	16	3,130		
3/8	3	600	6	1,160	12	2,320	18	3,480	24	4,640		
7/16	5	800	10	1,590	20	3,180	30	4,770	40	6,360		
1/2	8	1,060	15	2,120	30	4,250	45	6,370	60	8,500		
9/16	12	1,360	23	2,720	45	5,450	68	8,170	90	10,900		
5/8	15	1,690	30	3,380	60	6,770	90	10,200	120	13,500		
3/4	25	2,510	50	5,010	100	10,000	150	15,000	200	20,000		
7/8	40	3,460	80	6,920	160	13,800	240	20,800	320	27,700		
1	62	4,540	123	9,080	245	18,200						
1 1/8	98	5,720	195	11,400	390	22,900						
1 1/4	137	7,260	273	14,500	545	29,100						
1 3/8	183	8,650	365	17,300	730	34,600						
1 1/2	219	10,500	437	21,100	875	42,100						
1 3/4	390	14,200	775	28,500	1,550	56,900						
2	563	18,700	1,125	37,500	2,250	74,900						
8-thread Series												
1	…	…	…	…	245	18,200	368	27,200	490	36,300		
1 1/8	…	…	…	…	355	23,700	533	35,500	710	47,400		
1 1/4	…	…	…	…	500	30,000	750	44,900	1,000	59,900		
1 3/8	…	…	…	…	680	37,000	1,020	55,400	1,360	73,900		
1 1/2	…	…	…	…	800	44,700	1,200	67,000	1,600	89,400		

17.4

TABLE 17.2 Torque vs. Bolt Stress and Bolt Load (*Continued*)

Nominal diam	Bolt stress, psi											
	7,500		15,000		30,000		45,000		60,000		90,000	
	Torque, ft-lb	Load, lb	Torque, ft-lb	Load, lb	Torque, ft-lb	Load, lb	Torque, ft-lb	Load, lb	Torque, ft-lb	Load, lb	Torque, ft-lb	Load, lb
8-thread Series (Continued)												
1⅝	......	......	......	......	1,100	53,200	1,650	79,800	2,200	106,300		
1¾	......	......	......	......	1,500	62,400	2,250	93,600	3,000	124,800		
1⅞	......	......	......	......	2,000	72,300	3,000	108,500	4,000	144,600		
2	......	......	......	......	2,200	83,000	3,300	124,500	4,400	166,000		
2¼	......	......	......	......	3,180	106,600	4,770	159,800	6,360	213,100		
2½	......	......	......	......	4,400	133,100	6,600	199,600	8,800	266,100		
2¾	......	......	......	......	5,920	162,500	8,880	243,700	11,840	325,000		
3	......	......	......	......	7,720	194,900	11,580	292,300	15,440	389,700		
National Fine-thread Series												
¼	......	......	......	......	6	1,090	9	1,630	12	2,120	18	2,930
5⁄16	......	......	......	......	12	1,740	18	2,600	24	3,480	36	5,210
⅜	......	......	......	......	22	2,630	31	3,940	44	5,260	66	7,880
7⁄16	......	......	......	......	32	3,560	50	5,070	64	7,100	96	10,700
½	......	......	......	......	50	4,790	76	7,100	100	9,600	150	14,200
9⁄16	......	......	......	......	72	6,080	109	9,120	144	12,200	216	18,200
⅝	......	......	......	......	98	7,670	148	11,500	196	15,300	294	23,000
¾	......	......	......	......	170	11,200	255	16,800	340	22,300	510	33,500
⅞	......	......	......	......	270	15,300	404	22,800	540	30,500	810	45,800
1	......	......	......	......	390	19,900	587	29,800	780	39,700	1,170	59,600
1⅛	......	......	......	......	574	25,600	860	38,500	1,148	51,300	1,722	76,900
1¼	......	......	......	......	790	32,200	1,180	48,300	1,580	64,300	2,370	96,500
1⅜	......	......	......	......	1,044	42,400	1,565	59,000	2,088	78,800	3,132	118,200
1½	......	......	......	......	1,358	47,400	2,060	71,000	2,716	94,800	4,074	142,200

band variation of about ± 40 percent. This is largely attributed to the variation in the torque readings. The only accurate way to measure bolt load is by the measurement of bolt elongation using an extensometer or strain gages.

As can be seen from Fig. 17.1, the bolt load is applied to the gasket during installation, and therefore it is equal to the gasket load at that time. The gasket stress is this value divided by the gasket area which is in contact with the sealing faces. When a gasketed joint is placed in service so that an internal pressure is present, a new force, the hydrostatic end force, comes into play. This force can be calculated from the known pressure and the area over which it is effective. A number of experiments with different types of gaskets of different size have conclusively shown that the effective area at the leakage pressure is essentially the area based on the mid-diameter of the gasket.

$$F_h = P_i A_m \tag{17.4}$$

where F_h = hydrostatic end load, lb
$\quad P_i$ = internal pressure, lb/in^2
$\quad A_m = (\pi/4)D_m^{\,2}$ = effective hydrostatic pressure area, in^2
$\quad D_m$ = mid-diameter of gasket, in

The effect of this hydrostatic end force is highly dependent upon the particular assembly. This is one of the reasons a gasketed joint design should include an ample factor of safety. In a rigid assembly where a constant load is maintained on the joint, the hydrostatic end force balances some of the initial load. When the two become equal even the best gasketed joint will leak.

In a rigid assembly where a constant load is maintained on the gasket rather than on the joint (constant strain), the hydrostatic end load adds directly to the initial load. Most gasketed assemblies are somewhere between these two extremes, but no one has yet been able to calculate exactly where, because of other factors such as flange rotation, gasket stress relaxation, and bolt creep.

A factor in all seal applications which most designers do not usually consider is the amount of leakage which can be tolerated. This will vary from a relatively large amount which can be expected in certain oven or furnace seals to the molecular-diffusion magnitude of leakage which is almost too much in certain radioactive or toxic-fluid applications. The time element also becomes important in establishing a criterion of leakage because it may change radically, especially if there are temperature and pressure cycles involved.

17.2.2 ASME Code

The American Society of Mechanical Engineers' Code for Unfired Vessels[2] presents the most commonly used design methods for gasketed joints. Although these design calculations have been used for many years, it is doubtful whether the intricacies required to obtain some of the values are worthwhile because of the lack of real data and the number of other factors of unknown value which enter into any static-sealing problem.

The code makes use of two basic equations to calculate bolt load, with the larger calculated load being used for design:

$$W_{m1} = H + H_p = 0.785G^2P + (2b \times 3.14GmP) \tag{17.5}$$

$$W_{m2} = H_y = 3.14bGy \tag{17.6}$$

where W_{m1} = required bolt load for maximum operating or working conditions, lb
$\qquad W_{m2}$ = required initial bolt load at atmospheric-temperature conditions without internal pressure, lb
$\qquad H$ = total hydrostatic end force, lb $(0.785G^2P)$
$\qquad H_p$ = total joint-contact-surface compression load, lb
$\qquad H_y$ = total joint-contact-surface seating load, lb
$\qquad G$ = diameter at location of gasket load reaction; generally defined as follows: When $b_0 < \frac{1}{4}$ in, G = mean diameter of gasket contact face, in; when $b_0 > \frac{1}{4}$ in, G = outside diameter of gasket contact face less $2b$, in
$\qquad P$ = maximum allowable working pressure, lb/in^2
$\qquad b$ = effective gasket or joint-contact-surface seating width, in
$\qquad 2b$ = effective gasket or joint-contact-surface pressure width, in
$\qquad b_0$ = basic gasket seating width per table;* the table defines b_0 in terms of flange finish and type of gasket, usually from one-half to one-fourth gasket contact width
$\qquad m$ = gasket factor per table;* the table shows m for different types and thicknesses of gaskets ranging from 0.5 to 6.5
$\qquad y$ = gasket or joint-contact-surface unit seating load, lb/in^3, per table,* which shows values from 0 to 26,000 lb/in^2

In Eq. (17.5) use is made of an m value which provides a margin of safety to be applied when hydrostatic end force becomes a determining factor. This value is very difficult to determine experimentally since it is not a constant. Equation (17.6) assumes that a certain unit stress is required on a gasket to make it conform to the sealing surfaces and be effective. The y or yield-stress values given in the code were apparently based on experience and judgment but are very difficult to obtain experimentally. Other factors considered by the code relate to the surface finish and the effective gasket width to be used with these finishes. No design criteria are given for the spacing of the bolts, an item of great importance.

17.2.3 Simplified Design

A much simpler method of calculation has been found to be very effective.[3] This is based on the stress required to seat the gasket and the hydrostatic end force developed in the joint. Basically, the equations do the same thing as the code but they are simplified by using the full gasket contact width regardless of the flange width or the surface finish of the sealing faces.

The method of calculation depends upon whether a joint is being designed or whether a gasket is being determined for a joint already in existence.

In all cases it is necessary to select the type of gasket required based on temperature, pressure, fluid, flange finish, etc. These are described in numerous articles and in manufacturers' catalogs and literature. After the desired gasket has been selected, the minimum seating stress, as given in Table 17.3, is used to calculate the total bolt load required by Eq. (17.7).

$$F_b = S_g A_g \qquad (17.7)$$

where S_g = gasket seating stress, lb/in^2
$\qquad A_g$ = gasket contact area, in^2

---

*Tables UA-47.1 and 47.2 in ASME Code for Unfired Pressure Vessels.[2]

TABLE 17.3 Minimum Seating Stresses for Gaskets

Material	Thickness	Minimum seating stresses, psi
Rubber sheet:		
75 durometer....................	$\frac{1}{32}$ in. and up	200
60 durometer....................	$\frac{1}{32}$ in. and up	175
Compressed asbestos................	$\frac{1}{64}$ in.	3,500
	$\frac{1}{32}$ in.	2,500
	$\frac{1}{16}$ in.	1,800
	$\frac{1}{8}$ in.	1,400
Rubberized cloth..................	2 ply	2,500
	3 ply	2,100
	4 ply	1,800
Vegetable-fiber sheets..............	All	750
Teflon:		
Virgin.........................	$\frac{1}{64}$ in.	14,000
	$\frac{1}{32}$ in.	6,500
	$\frac{1}{16}$ in.	3,700
	$\frac{1}{8}$ in.	1,600
Glass-filled....................	$\frac{1}{64}$ in.	14,000
	$\frac{1}{32}$ in.	11,000
	$\frac{1}{16}$ in.	6,000
	$\frac{1}{8}$ in.	3,000
Asbestos cloth, impregnated..........	$\frac{3}{32}$ in.	1,600
Flat metals:		
Aluminum......................	All	20,000
Copper........................		45,000
Carbon steel...................		70,000
Monel.........................		85,000
Stainless......................		95,000
Corrugated sheet metal:		
Lead..........................	All	500*
Aluminum.....................		1,000
Copper........................		3,500
Monel........................		4,500
Stainless......................		6,000
Profile solid metal:		
Aluminum.....................	All	25,000†
Copper........................		35,000
Carbon steel...................		55,000
Monel........................		65,000
Stainless......		75,000
Corrugated sheet with asbestos:		
Aluminum......................	All	2,000*
Copper........................		2,500
Carbon steel...................		3,000
Monel.........................		3,500
Stainless......................		4,000
Plain metal jacketed:		
Lead..........................	All	500
Aluminum.....................		2,500
Copper........................		4,000
Carbon steel...................		6,000
Monel.......		7,500
Stainless......................		10,000
Spiral wound...	All	3,000–30,000

*Based on total projected area.
†Based on actual contact area (0.010-in-width profiles).

This equation will be satisfactory for the lower pressures in the smaller sizes since hydrostatic end force is not a major factor. No limits can be set to specify these values of size and pressure, so another calculation is required:

$$F_b \geq KP_t A_m \tag{17.8}$$

where P_t = test pressure, lb/in²
    A_m = effective hydrostatic area based on gasket mid-diameter, in²
    K = safety factor

Safety factors are based upon the joint conditions and the operating conditions but not on surface finish or the type of gasket. For minimum-weight applications where all installation factors (bolt lubrication, tension, parallel surfaces, etc.) are carefully controlled, low-temperature applications, and applications where adequate proof pressure is applied, the K factor is 1.2 to 1.4. For most normal designs where weight is not a major factor, vibration is moderate, and temperatures do not exceed 700 to 800°F, the K factor is 1.5 to 2.5. Use the high end of the range where bolts are not lubricated. For cases of extreme fluctuations in pressure, temperature, or vibration; where no test pressure is applied; or where uniform bolt tension is difficult to ensure; the K factor is 2.6 to 4.0.

In essence, these factors are similar to the m values in the code. As noted, the total bolt load F_b calculated by Eq. (17.7) should always be equal to or greater than the value given by Eq. (17.8). If it is not, it should be increased. In most cases this can be done safely without changing the gasket because most gaskets can withstand at least four times the minimum gasket seating stress without sustaining any damage. However, in some cases it may be necessary to change the gasket dimensions or even to change the type of gasket in order to meet the requirement of Eq. (17.8).

17.2.4 Rule of Thumb

A rule of thumb for determining the range of usage of different types of gasket materials is based on the product of temperature in degrees Fahrenheit and pressure in pounds per square inch with a maximum temperature being specified. Values are given in Table 17.4.

TABLE 17.4 *PT* Values for Gasket Materials

Material	Max $P_i T$*	Max temp, °F
Rubber	15,000	300
Vegetable fiber	40,000	250
Rubberized cloth	125,000	400
Compressed asbestos	250,000	850
Metals	>250,000	Depends on metal

*P_i = pressure, lb/in²; T = temperature, °F.

17.2.5 Flange Width

There are no known methods of calculating the optimum flange width for gaskets, and the widths are usually picked on a basis of common-sense judgment based upon the following factors:

1. Most gaskets have a minimum flange width based upon their strength and fabrication limitations. These can range from a minimum of about ½ in for a rubber or solid metal O ring to ⅜ in for a corrugated, jacketed metal gasket.
2. The maximum flange width will usually be limited by the bolt load available or by the practicability of manufacture.
3. The optimum flange width is that width which will give an effective seal with a minimum bolt load. It can be arrived at by proper consideration of the seating stress, hydrostatic end force, and compression characteristics of the gasket.

17.2.6 Flange Rotation

Flange rotation is a factor which is present in all flanged joints, but it becomes important only in large sizes, thin flanges, and at relatively high pressures. Generally, if the proper gasket is selected and the joint is well designed, flange rotation is not important because it has very little effect on maintaining a seal. When it does become a problem, some means to prevent excessive rotation such as a full-face gasket or a limiting device outside the bolts are usually incorporated into the gasket. The rotation is caused by the bolt load being applied a great distance from the centerline of the flange and the effect of the uneven hydrostatic stresses over the flange width.

17.2.7 Gasket Thickness

Gasket thickness is another factor which is difficult to specify. Generally, the thinnest gasket consistent with the requirements should be used. These requirements are determined by the limits of manufacture, flange surface finish, flange flatness, flange parallellism, and compressibility required. Rubber and compressed-sheet gaskets are often used in ½-in thicknesses when the flanges have a good finish and are parallel; they are seldom used above ⅛-in thickness. Some gaskets like the spiral-wound asbestos-steel types are not generally available in thicknesses less than ⅛ in.

17.2.8 Compressibility and Recovery

For years compressibility and recovery values have been specified for gasket materials without any real correlations between these values and the performance of the material. Actually, compression and recovery values were for quality-control purposes in specifications for materials which had been found to work in a certain application. When a new material came along, the compressibility and recovery probably had to be in a certain range for it to get further consideration. This further study consisted of other tests more related to performance or even an actual application test. Many materials were then discarded despite their successful compressibility and recovery results. It is probable that as many good materials are discarded because of their compression and recovery values as are obtained after additional application tests.

Knowledge of gasket compressibility is most important in two particular types of use. First, if the mating surfaces have a major deficiency in surface finish, flatness, or parallelism, a certain amount of compressibility is necessary in order to obtain contact between the gasket and flanges over their entire area. Second, many joints require a definite spacing (e.g., metal-to-metal connections and parts with critically aligned components), and the gasket must therefore have enough compressibility to allow the joint to be made up metal-to-metal with the available bolt load. It must not be so com-

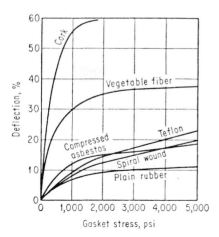

FIG. 17.2 Compression of gasket materials.

pressible that the seating stress is not obtained on the gasket before the load shifts to the metal-to-metal joint.

Recovery of a gasket material can be related to sealability in long-time service or where flange separation can occur. The recovery as usually measured, however, has no real value because it is taken from full to zero (or essentially zero) load and any joint would undoubtedly leak long before this recovery could occur. If recovery were a straight-line function, these figures might still be of some use, but usually most of the recovery occurs at the very low loads which cannot be allowed in actual service.

Figure 17.2 gives some typical compression curves for a few types of gasket materials. It should be understood that different formulations of the same types can give widely different values.

17.2.9   Sealability

Sealability is the primary property of a gasket material. Since there is no general, all-inclusive definition of sealability, there is no general method of measuring it. Sealability is usually empirically determined; that is, an actual or simulated joint is set up under a certain loading and tested for sealability under certain conditions of temperature, pressure, time, and fluid. Many times a compression machine is used to provide a more accurate load although it does not duplicate the load behavior in an actual joint. A standardized sealability test procedure (D2025-62T) has been published by ASTM and SAE.

Probably one of the most important properties of a gasket material is stress relaxation. Although there is undoubtedly a correlation between recovery and stress relaxation, most of the measured recovery is in the area of very low stress, whereas the stress relaxation is usually measured at stresses 50 percent of the full stress or higher. Although data are very limited, Fig. 17.3 shows the type of plot obtained with two different types of materials. ASTM has recently adopted standard D2139-62T for this characteristic.

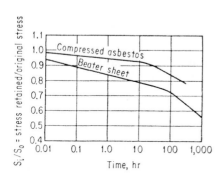

FIG. 17.3 Stress relaxation of gasket materials.

17.2.10   Bolt Spacing

Bolt spacing is very important in gasketed joints, but little attention is paid to it and there are no good rules. Generally, the total load required is calculated, and either the number of bolts is selected from an appearance or manufacturing viewpoint or the size of bolt is determined by appearance or flange width. This situation led Roberts[4] to develop the following approximate technique.

Based upon the assumption that a flange is analogous to an elastic beam with point loading representing the bolt loads, equations were derived to show the load midway between the two bolts. It was further assumed that this load should be 95 percent of the load at the bolts, and a curve was derived plotting the relationship between the ratio of bolt spacing to flange thickness and a function S defined as follows:

$$S = tb_f E_f/db_g E_g \tag{17.9}$$

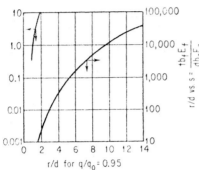

FIG. 17.4   Plot for determining flange bolt spacing.

where t = gasket thickness, in
b_f = flange width, in
b_g = gasket width, in
d = flange thickness, in
r = bolt spacing, in
E_g = modulus of elasticity of gasket
E_f = modulus of elasticity of flange material

This plot is shown in Fig. 17.4 where q is the load midway between two bolts and q_0 is the load at the bolts. When the factor r/d has been determined, the value of r is calculated and the number of bolts obtained by dividing the spacing into the bolt-circle circumference.

17.2.11   Confined vs. Unconfined

The primary example of the confined gasket is the static O-ring seal. Here it is very necessary to confine the O ring because it is usually a relatively thick gasket with high elongation. However, it is desirable also to confine many other types of gaskets in order to:

1. Prevent blowout
2. Allow easier centering
3. Ensure controlled compression to prevent damage to the gasket and misalignment of parts

The major disadvantages of confined gaskets are:

1. It is more difficult to produce the desired surface finish.
2. Gasket width must be reduced for the same overall flange width.
3. Flange thickness may have to be increased.
4. Incorrect gasket thickness will result in too much or too little load on the gasket.
5. Bolt elasticity has no effect on maintaining tightness of joint if stress relaxation occurs in the gasket.

Although the disadvantages outnumber the desirable characteristics, there are many times when the confined gasket joint is very advantageous.

17.3 PRESSURE-ACTUATED TYPES

17.3.1 Theory and Design

One of the troublesome aspects in the use of standard flat-faced gaskets is the difficulty of sealing high pressures. This is caused by the hydrostatic end force which tends to separate the flanges, thereby decreasing the flange load on the gasket. The same net effect occurs if the gasket or the joint tends to relax during service.

At high pressures a compensating action can be obtained if the pressure can be utilized to force the gasket into sealing contact with the flanges. This can be accomplished in several ways, depending upon whether the gasket is sealing in a radial or axial direction.

Probably the most common type of gasket which may be pressure-actuated is the O ring. When an O ring is installed in a groove between two flanges, a certain amount of compression or gasket load is supplied to give at least an initial seal. Let us assume that the flanges are brought together metal-to-metal. As pressure is applied to this joint, it causes the O ring to expand outward against the outside diameter of the groove and at the line between the flanges. With an increase in internal pressure the flanges may tend to separate, but the O ring is forced into this opening and thereby seals it off. This will continue until the pressure becomes so high or the opening between the flanges becomes so great that the O ring is extruded out between the flanges and failure occurs.

The same principle applies to other types of gaskets such as the delta, lens, and new ring joint, where radial movement of the gasket causes it to seal the opening between the two flanges.

The other type of pressure-actuated gasket is represented by the Bridgman joint, bellows gasket, perforated metal O ring, and other specialty types such as the Skinner, Cadillac, and Bar-X seals. In the case of the Bridgman joint the method of sealing is the same as for the O ring, delta, etc., but the forces are applied in an axial direction. However, the sealing is accomplished by sealing off a space between two mating parts by wedging the gasket in between.

With the other types which seal by the action of pressure in the axial direction, the sealing is conventional except that the internal pressure gives the ultimate gasket load. The primary factor in design is to provide as large an area as possible for the pressure actuation and as small an area as possible for the sealing area. This results in a high stress at the sealing surface and thereby promotes good sealing ability. The design and mechanics of pressure-actuated joints are given in Fig. 17.5a and b. Additional discussion of other types of gaskets and other devices for static sealing can be found in Refs. 7 and 8.

17.4 DIAPHRAGM SEALS

17.4.1 Theory and Design

Diaphragm seals form a continuous membrane across the fluid path and therefore prevent leakage past the plane of the diaphragm. They are used exclusively in valves and pumps where relative motion between two components is quite small and in a reciprocating action. In addition to creating a seal across the flow path of the fluid, the diaphragm serves as a gasket around its perimeter and center which serve as the points of attachment to the equipment.

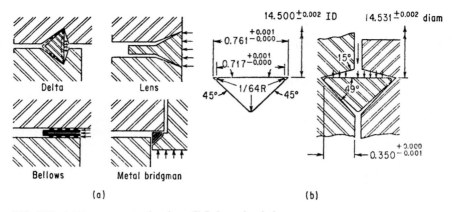

FIG. 17.5 (*a*) Pressure-actuated gaskets. (*b*) Delta gasket design.

In a valve, the diaphragm center is attached to the valve stem or valve-operating mechanism so that it flexes and allows reciprocating motion of the valve-stem assembly as the valve is opened or closed. At the extremes of motion the diaphragm is extended close to its maximum without causing excessive tensile stress. The action in a diaphragm pump is essentially the same. The outer periphery of the diaphragm is attached between the bonnet and the body of the valve or between the halves of the pump as with a standard gasket.

The design of a diaphragm seal is relatively simple since the major requirement is that the diaphragm material have the flex life, ease of flexing, and pressure resistance required. The gasket requirements must also be met, and the basic design criteria for gaskets can be applied.

Although many diaphragm materials are used, the most common are rubber, rubberized cloth with and without wire reinforcement, Teflon, polyethylene, and beryllium bronze.

DYNAMIC SEALS

17.5   GENERAL

Dynamic seals are required to seal four types of shaft motion: rotary, reciprocating, helical, and swinging-rotary. There is a wide variety of materials and designs from which to choose. Before the selection of the proper packing can be made, many factors must be considered: cost, equipment requirements necessary for the installation, service conditions, and performance level required. The types of dynamic seals or packing are classified as follows: (1) compression types, sometimes referred to as mechanical packings or jam type; (2) automatic or pressure-actuated, almost always molded or machined; (3) oil seals; (4) mechanical seals, including face seals; (5) labyrinth or positive-clearance seal; and (6) piston rings, both metallic and nonmetallic.

The selection of the type of dynamic seal required for a particular application can be guided by estimating the severity of the pressure and velocity conditions in terms of a *PV* factor, which is defined as the product of the fluid pressure in pounds per square inch and the surface velocity in feet per minute. The *PV* parameter is an

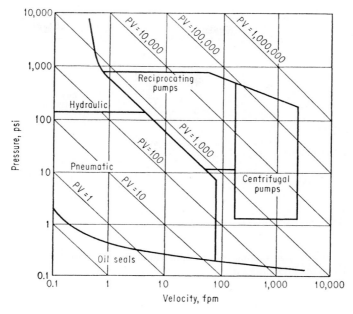

FIG. 17.6 *PV* application ranges for dynamic seals.

approximate measure of the frictional heat generated between the packing and the shaft. Figure 17.6 is a plot of the *PV* ranges for various common dynamic-seal applications.

17.6 COMPRESSION TYPES

17.6.1 General

There are many types of compression packings and some variations in installation, but all function by restricting the axial flow of the medium being sealed to an acceptable rate. Whenever there is considerable relative motion with this type of packing, there must always be some leakage to provide lubrication and prevent temperature buildup. Figure 17.7 illustrates the basic installation of a compression packing.

The packing may be considered as providing an adjustable labyrinth which expands radially against the shaft because of a compressive force applied at one end, either from applied gland load or from the pressure of the fluid being sealed. When fluid flow or leakage occurs, there is a drop in fluid pressure from 100 percent at the packing ring at the bottom of the stuffing box to almost 0 percent at the packing ring next to the gland. The leakage rate is dependent upon the ability of the packing to maintain its sealing force against the shaft under the operating conditions. With fluctuating fluid pressures, compression packings do not function satisfactorily without gland adjustment. Although the packing can be seated effectively for a high fluid pressure, an abrupt pressure drop will cause the leakage to be reduced and the packing will

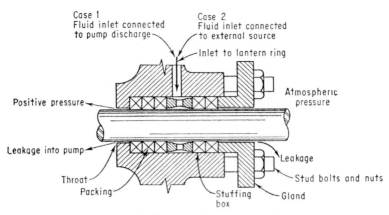

FIG. 17.7 Typical stuffing box with lantern ring.

overheat; or if effective sealing is accomplished at the low pressure, the leakage will be excessive at the high pressure.

17.6.2 Design

The stuffing box shown in Fig. 17.8a and b and Table 17.5 for rotating and reciprocating shaft dimensions is based upon the recommendations of the Mechanical Packings Association.[5] These dimensions are intended for soft packings to approximately 1500 lb/in^2 and are independent of shaft speed. The shaft clearance at the throat can vary between 0.008 and 0.030 in, dependent upon the pressure sealed. Gland clearance at the inside diameter and outside diameter should be held to a minimum to prevent extrusion of the packing because of the relative motion. The inside diameter clearance is usually between 0.020 and 0.075 in; clearance at the outside diameter can be held at 0.010 to 0.046 in. These clearances provide gland alignment and should not permit shaft interference at the inside diameter. The recommended dimensions will also accommodate molded-type packings.

Soft-compression-type packings are usually made in straight lengths which are cut to form rings that fit the stuffing-box dimensions. The ratio of shaft to bore dimensions should conform to the dimensions given in order to minimize uneven density of the packing from inside to outside diameter. If the width W of the packing is too large, the inside diameter of the formed ring will bulge and the outside diameter will neck down, resulting in high shaft friction while still tending to leak along the bore.

In the following discussion numerous references are made to the excellent work of Denny and Turnbull.[6]

17.6.3. Number of Rings, Stress Decay, and Pressure Drop

When a packing is installed, it is seated by tightening the gland stud bolts and thus applying an axial force on the packing. The axial stress results in a radial stress which determines the sealing effectiveness of the packing, the magnitude of the radial stress depending upon the elasticity of the packing. In a packing set, the greatest axial stress

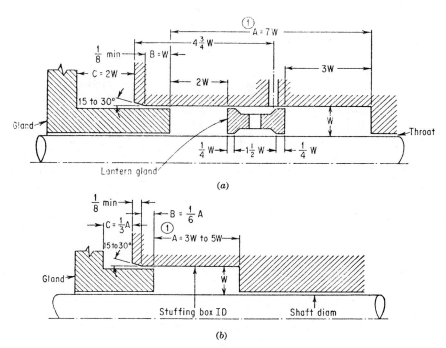

FIG. 17.8 Stuffing-box dimensions for rotating shafts and reciprocating rods. (*a*) Stuffing-box dimensions for rotating shafts 1. A depth dimension of 7W is used where lantern gland is used. (*b*) Stuffing-box dimensions for reciprocating rods 1. A = total depth of packing.

TABLE 17.5 Packing Width vs. Shaft Diameter

Rotary shafts	
Shaft diam, in.	Packing width, in. W
$\frac{5}{8}$ to and including $1\frac{1}{8}$	$\frac{5}{16}$
Over $1\frac{1}{8}$ to and including $1\frac{7}{8}$	$\frac{3}{8}$
Over $1\frac{7}{8}$ to and including 3	$\frac{1}{2}$
Over 3 to and including $4\frac{3}{4}$	$\frac{5}{8}$
Over $4\frac{3}{4}$ to and including 12	$\frac{3}{4}$

Reciprocating rods		
Rod diam, in.	Packing width, in.	Dash No.
$\frac{1}{4}$ to and including $1\frac{1}{4}$	$\frac{1}{4}$	8-24
Over $1\frac{1}{4}$ to and including $2\frac{1}{2}$	$\frac{5}{16}$	25-35
Over $2\frac{1}{2}$ to and including $3\frac{3}{4}$	$\frac{3}{8}$	36-46
Over $3\frac{3}{4}$ to and including $5\frac{1}{2}$	$\frac{7}{16}$	49-55
Over $5\frac{1}{2}$ to and including 15	$\frac{1}{2}$	56-80

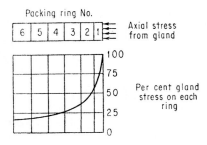

FIG. 17.9   Stress decay through a set of packing.

exists on the ring next to the gland. Each succeeding ring is subject to less axial stress, as shown in Fig. 17.9 for a six-ring set.

This loss in stress transmission through the packing set is principally caused by the friction between the stuffing-box bore and the packing outside diameter and, to some extent, the friction between the shaft and the inside diameter of the packing. The rougher the bore, the greater the proportion of gland force taken up by the last ring installed. There are no fixed axial-stress values required to seat compression packing, but the following estimates may be useful as a guide:

Type of packing	"Estimated" minimum axial stress, lb/in², required to seat packings
Teflon-impregnated braided asbestos	200
Plaited asbestos, lubricated	250
Plastic	200
Braided vegetable fiber, lubricated	160
Braided metallic	400–500

The fluid pressure distribution along a set of packing can be illustrated as shown in Fig. 17.10.

On the basis of the foregoing, it might appear that a packing set would require two or possibly three rings only because the two rings next to the gland do approximately 75 percent of the sealing. However, additional rings are required to compensate for packing wear, to help throttle pressure, to aid in spreading out the shaft wear, and in some cases, to protect the sealing rings from abrasive media. Suggested number of packing rings to be used are: for a stuffing box without lantern ring, 5 or 6 rings; for a stuffing box with lantern ring, 3 rings inside, 2 outside; for smothering-type gland (used with hazardous fluids), 6, 8, or 10 rings depending on service. Greater numbers are often used in high-pressure installations.

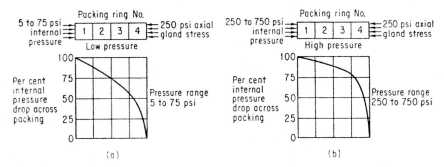

FIG. 17.10   Fluid pressure drop vs. packing length. (*a*) Low pressure. (*b*) High pressure.

17.6.4 Power Requirements

The power requirement of a compression packing is influenced by many variables such as packing coefficient of friction, packing length, initial gland-compression load on packing, fluid pressure, shaft diameter, shaft speed, shaft finish, leakage rate, and lubrication.

The packing coefficient of friction as determined by the pull required to move a piece of packing under load across a steel plate is shown in Fig. 17.11 for various packings and is useful in calculating the starting torque for an installation. Once a pump is started and leakage is established, there is a rapid decline in shaft torque, as shown in Fig. 17.12. This indicates the need for high-starting-torque motors for compression-packing applications. The effect of packing length on shaft torque is dependent on whether the applied gland stress is greater or less than the fluid pressure. For *low fluid pressures* where the fluid pressure gradient is relatively uniform, the friction torque is distributed almost uniformly along the length of the packing. As a result, the torque increases almost linearly with the number of packing rings, as shown in Fig. 17.13. This torque decays with time and packing wear until the axial force caused by gland load becomes less than the axial force due to fluid pressure. For *high fluid pressures* the sealing is automatically maintained by the last 10 to 20 percent of packing length. Thus most of the friction torque will occur there, and increasing the total number of rings will not result in a marked increase in torque (Fig. 17.13). This torque will tend to remain constant until the packing no longer seals effectively.

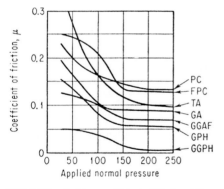

FIG. 17.11 Packing coefficient of friction. TA = Teflon-impregnated plaited asbestos; GA = graphite-plaited asbestos; GGAF = graphite-greasy asbestos fiber; GPH = graphite-plaited hemp; GGPH = graphite-greasy plaited hemp; PC = plain plaited cotton; FPC = Teflon-impregnated plaited cotton.

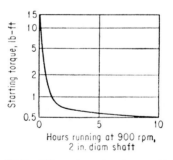

Hours running at 900 rpm, 2 in. diam shaft

FIG. 17.12 Effect of running time on friction torque. Four rings of Teflon-impregnated plaited cotton. Applied gland pressure, 15 lb/in². Water pressure, 75 lb/in². Curve calculated from coefficient of friction.

The initial compression load on the packing directly affects the initial starting torque and will be a determining factor with respect to friction torque as long as the gland load exceeds the fluid-pressure load on the packing. When the fluid-pressure load is greater, the gland load does not affect friction torque, which increases in proportion to fluid pressure. Figure 17.14 confirms these effects.

Shaft speed and diameter directly determine the amount of work done and are therefore basic in calculating the power requirements. Shaft finish also affects the

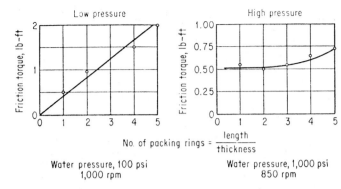

FIG. 17.13 Variation of friction torque with number of packing rings. Graphite-greasy plaited hemp.

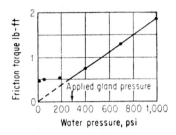

FIG. 17.14 Effect of water pressure on friction torque. Average of two glands. Teflon-impregnated cotton, 850 r/min.

power requirements because the coefficient of friction is a function of surface finish. The effect of shaft speed on torque is complex and little information is available at high speeds, but the data plotted in Fig. 17.15 give an indication of results at slower speeds. Data at 3500 r/min show the average friction torque was 1.3 ft·lb for ⅝-in-square plastic packing in a 300-h test against water at 300°F and 150 lb/in² with a 3¼-in shaft. The leakage rate has also been found to be independent of shaft speed.

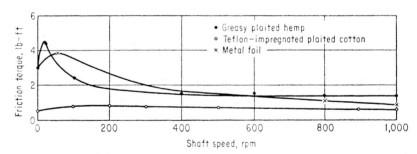

FIG. 17.15 Typical torque-vs.-shaft-speed values. Three rings. Applied gland pressure, 200 lb/in²; water pressure, 75 lb/in².

The leakage rate is a determining factor in providing lubrication and influencing the power requirements as shown in Fig. 17.16. In laboratory tests, a 5-hp electric motor driving a 2-in shaft was readily stalled when packing leakage dried up. Figure 17.17 shows how the rate of leakage is controlled by gland pressure when it exceeds fluid pressure.

In addition to the lubrication provided by the leakage of the sealed fluid, a lubricating fluid such as oil or grease may be introduced through the lantern gland and this more desirable lubricant can reduce the power requirements.

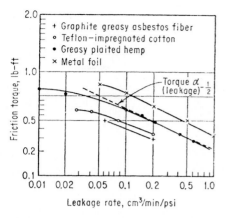

FIG. 17.16 Friction torque vs. leakage rate. Three rings of packing, 1000 r/min.

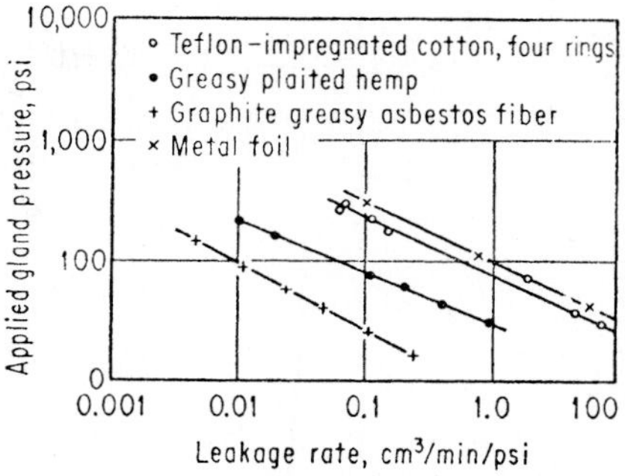

FIG. 17.17 Effect of gland load on leakage rate. Shaft speed, 1000 r/min.

17.6.5 *PV* Factor

The *PV* factor (pressure times shaft linear velocity) referred to previously is a useful guide in determining the selection of a compression packing. This factor may be converted to heat input if a packing coefficient of friction is estimated. A packing selection can then be made on the basis of the heat resistance of the packing materials available. The temperature condition of the application must also be considered in making this selection. The heat generated in a stuffing box can be expressed as

$$Q = PLdV\mu/J \tag{17.10}$$

where Q = quantity of heat generated, Btu/min
P = fluid pressure to be sealed, lb/in^2
L = depth of packing, in
d = shaft diameter, in
V = shaft velocity, ft/min = $\pi d/12$ (r/min)
μ = packing coefficient of friction
J = mechanical equivalent of heat, a constant of 778 ft·lb/Btu

The chart in Fig. 17.18, in which the *PV* factor is automatically established by the revolutions per minute, shaft diameter in inches, and fluid pressure in pounds per square inch, will be helpful in the selection of a compression packing material.

17.6.6 Wear

One of the advantages of compression packings, as compared with lip seals or mechanical seals, is the ease with which packing wear can be compensated for with gland take-up. The amount of take-up available is established by gland clearance and the compressibility of the packing. Packing life can be increased and shaft wear decreased by the use of hardened shafting with a 20-rms or better finish. Hard chrome-plating shafts and finishing to 10 rms improves shaft and packing wear. It also reduces the possibility of shaft corrosion, which often is the cause of premature packing fail-

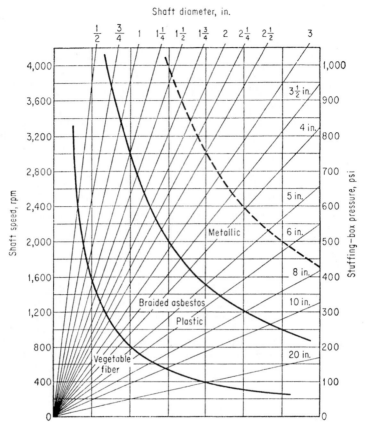

FIG. 17.18 *PV* chart.

ure. Another common practice is to provide shaft sleeves in the stuffing-box area that may be replaced when worn so that new packings will operate against optimum shaft conditions. It is good practice to have a shaft hardness in excess of 400 Bhn; for metallic packings a shaft hardness of 500 Bhn or better is recommended.

17.6.7 Shaft Finish

In addition to reducing wear, a good shaft finish of 16 to 20 rms will improve the sealing performance of packings. This is especially important in reciprocating service where any shaft imperfections would establish a leakage channel across the sealing surface of the packing.

17.6.8 Misalignment and Shaft Deflection

Misalignment between shaft and bore makes the sealing problem more difficult, and it should be held to an acceptably low value. If a shaft runs true but is located eccentri-

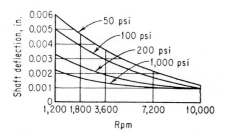

FIG. 17.19 Maximum shaft deflection vs. shaft speed and fluid pressure.

cally to the bore, most compression pack-ings can accommodate this quite readily up to about 0.010 in. The effect is most harmful at high pressures because exces-sive clearance is available on one side of the shaft, and extrusion of the packing may occur. The misalignment resulting in a runout condition or eccentric motion of the shaft about its centerline is the most difficult for the packing to seal against. This results in a continuous pounding of the packing during each revolution or stroke. The packing has to be extremely resilient to withstand this action and still seal effectively. Runout may be caused by faulty installation, out-of-round shaft, imbalance, or inadequate bearing support and should usually be held to 0.003 in. Figure 17.19 shows some curves for allowable runout vs. shaft speed and fluid pressure.

17.6.9 Lubrication—Lantern Rings

A lantern ring is an annular member used to establish a liquid seal around the shaft and to provide a means of supplying lubrication to the stuffing-box packing. The lubricant may be the fluid being pumped or fluid from another source, i.e., integrally lubricated stuffing box or externally lubricated stuffing box. A typical stuffing box with lantern ring was shown in Fig. 17.7. For an integrally lubricated stuffing box, it is essential that the fluid being pumped be a clear liquid with lubricating qualities and a temperature below its atmospheric boiling point. This type of arrangement is neces-sary when the pump has vacuum on its suction, in order to prevent the packing from running dry and to prevent the contamination of the fluid by another lubricant. Since the fluid is withdrawn from the discharge of the pump, its pressure is always higher than at the stuffing box and the action is automatic.

When the pump fluid is a poor lubricant, contains abrasive material, or cannot be allowed to leak out because of its highest toxicity, fire hazard, high temperature, or corrosiveness, an externally lubricated lantern ring is used. A lubricant should be selected that will contaminate the fluid being pumped as little as possible. This arrangement can also be used to provide a cooling action in the stuffing box.

17.6.10 Beveled Boxes

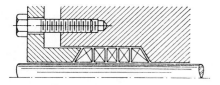

FIG. 17.20 Beveled-end stuffing box.

Although square-end stuffing boxes are generally used for compression packings, some stuffing boxes are made with beveled ends, as illustrated in Fig. 17.20. The theory is that the beveled ends tend to force the packing against the shaft where the sealing takes place. However, this configuration also tends to force the packing into the shaft and follower clear-ance, thus promoting extrusion, and

results in greater shaft and packing wear.

17.6.11 Spring Followers

Spring followers are sometimes used to provide a follow-up on gland load to compensate for the decay of gland stress because of packing wear and loss of impregnation. Laboratory experiments have shown that, although this technique prevents increase of leakage rate with time, the friction torque remains very high, and wear of the packing becomes excessive. Spring followers should be used only in special low-speed applications.

17.6.12 Dynamic Seals for Reciprocating Service

Compression packings are used in reciprocating service. Many of the installation recommendations for rotary application, such as shaft finish, runout, and lubrication, apply equally in this case. In reciprocating service the packing is almost invariably subjected to fluctuating pressure, from atmospheric to pump output pressure. As a result, leakage is intermittent and shaft friction varies during a pressure stroke. Because the gland load on the packing is adjusted to seal maximum pump pressure, during the low-pressure part of the stroke the packing is not lubricated by the fluid being pumped. If heavy leakage were permitted, packing erosion would result in premature failure; hence it is usually necessary to keep a positive gland load on the packing all the time. Outside lubrication such as an oil drip on a horizontal shaft is commonly employed to ensure adequate lubrication. Temperature buildup caused by the packing is usually not a problem because packing contact is not concentrated and speeds are relatively slow. The maximum shaft speed recommended is 250 ft/min.

17.7 AUTOMATIC OR PRESSURE-ACTUATED

17.7.1 Theory

In automatic or pressure-actuated packings, fluid pressure automatically supplies the seating force required by the packing to effect a seal. There are two general classifications: (1) lip packings such as cups, U cups, and V rings; and (2) squeeze types such as O rings. They all rely upon built-in interference to establish an initial low-pressure seal; as fluid pressure increases, the packing is forced tighter against the surfaces to be sealed, as illustrated in Fig. 17.21.

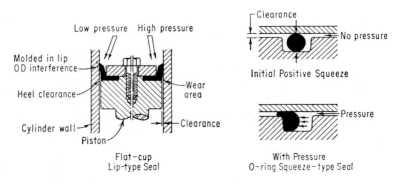

FIG. 17.21 Typical pressure-actuated packing installations.

Pressure-actuated packings are generally used to seal reciprocating motion, but occasionally for slow-speed rotary motion. These packings seal very effectively so that there is a greater temperature rise than with compression packings where a certain amount of leakage is allowed. In reciprocating service the heat tends to dissipate because the packing is continuously in contact with a new surface which has been lubricated and cooled. This is not the case in rotary service where the heat developed within the packing under pressure will destroy the packing unless some heat-removing means is provided.

Most pressure-actuated packings are installed with no provision for packing take-up. Therefore, in order to ensure long life, wear must be held to a minimum; one of the best methods to accomplish this is to have a good shaft finish of 16 rms or better. This will also reduce frictional heat, which shortens seal life. In hydraulic service, the fluid will usually supply sufficient lubrication, but in pneumatic service external lubrication usually must be provided. In any installation the clearance between the sealed members is critical. The packing under pressure, if improperly supported, tends to extrude into the clearance and wear away. The higher the pressure, the more critical the clearance becomes; therefore, it is always advisable to hold clearance to a minimum. Backup washers are necessary in some applications. These packings should never be installed over sharp edges or come into contact with them during operation.

17.7.2 Lip-Type Packings

These packings are generally made of leather, rubber, or fabric-reinforced rubber with the choice of material depending on the service conditions. A properly selected and installed lip-type pressure-actuated packing will seal with almost no leakage. They are sensitive to pressure and will wipe lubrication from the shaft more effectively as pressure is increased. Friction on a pressure stroke varies proportionally with the pressure. On the return stroke the packing relaxes and friction drops to a low level.

The oldest type of lip packing is a flat cup with a single lip on the outer circumference. This is sometimes classified as unbalanced packing. When the lip is on the inner circumference, it is referred to as a hat or flange gasket. The flat cup is usually used to seal a piston or other driving or driven members, while the hat gasket is almost always stationary and seals a shaft or ram. Both types seal across the base as a static seal or gasket, while the lip is pressure-actuated. Typical installations are shown in Fig. 17.22*a* and *b*.

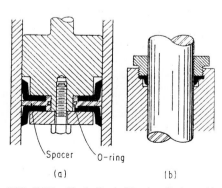

(a)   (b)

FIG. 17.22 Single-lip packing installations. (*a*) Piston assembly. (*b*) Internally threaded design, hat gasket shaft assembly.

Figure 17.22*a* illustrates an application where the cup flange is sealed by a known amount of compression as determined by the dimensions of the spacer boss. The spacer also prevents transfer of load between cups. There are numerous designs for flat-cup installations, each of which has certain functional or cost advantages. For example, an all-compound or homogeneous flat cup is not clamped across the base but is permitted to float and depends upon fluid pressure to create a base seal. This results in a simplified design of minimum no-pressure friction. The design should be such that the cup base is not overstressed, or the base of the lip or heel will be forced into contact with the cylinder walls to create high friction and cause rapid wear. Overcompression also causes the lip to tilt inward, resulting in loss of lip interference. Clearance must be provided at the base of the cup to prevent binding action that might result from swelling of the packing. The hat gasket (Fig. 17.22b) has application where space is limited and functions best up to about 5-in inside diameter. Single-lip packings may be installed with a spring expander or contractor against the sealing lip in order to maintain contact with the cylinder wall or shaft.

FIG. 17.23   U-cup packing installation.

The U cup and V ring are considered balanced packings because they have a sealing lip on the inside and outside diameters. They are completely automatic, seal with low friction, and require no gland adjustment. They can be used to seal either a shaft or single- and double-acting pistons.

Figure 17.23 illustrates installations of a U cup with a pedestal ring used to center the packing and hold it in position. The tongue should allow clearance for the lip, both at the sides and at the tip. The corners on the tongue should be rounded so as not to cut the cup. Some compression of the cup base is desirable, but this should not be excessive or lip distortion will occur. About $\frac{1}{64}$-in compression is usually satisfactory. Table 17.6 gives recommended flange widths for various sizes of U cups.

TABLE 17.6   Homogeneous U-Cup Packing—Nominal Sizes

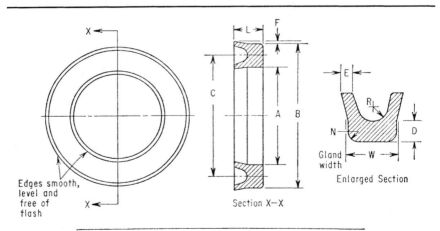

Dash No.	Dimensions, width and length, in.	Nominal size, in.	
		ID	OD
8–24	$\frac{1}{4}$	$\frac{1}{4}$–$1\frac{1}{4}$	$\frac{3}{4}$–$1\frac{3}{4}$
25–35	$\frac{5}{16}$	$1\frac{1}{4}$–$2\frac{1}{2}$	$1\frac{3}{4}$–$3\frac{1}{8}$
36–40	$\frac{3}{8}$	$2\frac{1}{2}$–3	$3\frac{1}{4}$–$3\frac{3}{4}$

17.7.3 V Rings

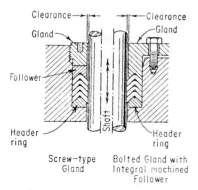

Screw-type Gland

Bolted Gland with Integral machined Follower

FIG. 17.24 V-ring packing installation.

The most popular of the multiple-lip seals or nested sets are V rings, although there are other types which can be used to advantage in many applications. Properly installed, V rings will seal considerable pressure with long life and low friction. They are a balanced packing that may be used to seal a shaft and single- or double-acting piston. Typical installations are as shown in Fig. 17.24.

Table 17.7 illustrates the importance of good design and realistic operating conditions relative to the performance of a set of typical all-compound lip packing as obtained in laboratory tests. The standard test was run using four rings of 3 in inside diameter by 3¾ in outside diameter packing, 20-rms finish shaft, 8000 lb/in² hydraulic fluid at 160°F, 0.005-in follower to shaft clearance, and sixty 1-ft strokes per minute with pressure cycled on alternate strokes.

TABLE 17.7 Lip-Packing Performance vs. Equipment and Environment Variables

Test variable	Leakage rate, %	Life,* %
None, standard test...........................	100	100
0.015-in. follower clearance......................	400	24
0.015-in. follower clearance and 32 rms shaft.......	360	21
0.015-in. follower clearance and 60 rms shaft.	620	8
32 rms shaft...................................	166	60
120 rms shaft.................................	500	6
0.010-in. undersize shaft.......................	300	24
6,000 psi..................................	515	4
2-ring set..................................	220	31
3-ring set......	101	79
200°F.....	180	11
250°F.......	286	8

*Life based on number of cycles before leakage rate became excessive.

In some installations, such as control valves, spring loading is used (with the spring located under the male adapter) to maintain the lip interference of a V ring set. This is especially effective with an inelastic low-friction material such as Teflon. The spring load can be calculated to provide the exact value necessary to effect a seal without creating high friction. This will compensate for some lip wear. The design of V rings is fairly well established. They are made to standard dimensions, as are the header and follower adapter rings required to make them function. A tolerance of ± 0.010 in is permitted in the height of each V ring and adapter, so that in a set of four V rings the accumulated variance could be 0.060 in. The design of the gland and stuffing box should allow for the maximum stack height if a nonadjustable gland is to be used, and the tolerance should be compensated for by the use of shims. V rings should always be snugly held by the gland, with leather V rings requiring slight compression. Tables 17.8 through 17.10 provide data on V rings.

TABLE 17.8 Recommended Number of V Rings vs. Pressure

Pressure, psi	No. of V-rings*		
	Leather	Homogeneous	Fabricated
Up to 500	3	3	3
500– 1,500	4	4	4
1,500– 3,000	4	5	4
3,000– 5,000	5	5	5
5,000–10,000	5	. . .	6
10,000 and over	6		

*For solid rings.

TABLE 17.9 Homogeneous, Leather, and Fabricated V Ring Packings—Nominal Sizes

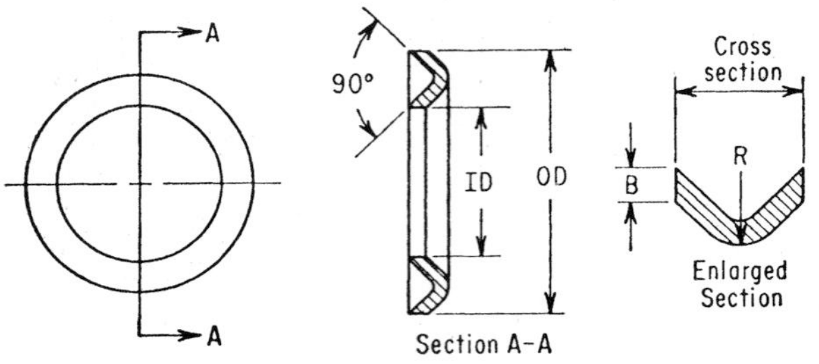

Dash No.	Gross section *W*, in.	Nominal size		*B*, in. −0.010 in.	Stack height, in., 3-ring set*
		ID	OD		
8-24	1/4	1/4–1 1/4	3/4–1 3/4	0.083	5/8
25-35	5/16	1 1/4–2 1/2	1 7/8–3 1/8	0.140	55/64
36-46	3/8	2 1/2–3 3/4	3 1/4–4 1/2	0.156	31/32
49-55	7/16	4 –5 1/2	4 7/8–6 3/8	0.197	1 5/32
56-80	1/2	5 1/2–15	6 1/2–16	0.197	1 7/32

*Includes adapters.

17.7.4 Squeeze-Type Packings

This style of packing is the simplest sealing device, and it will seal effectively over a wide range of temperatures, pressures, and fluids. The O ring is the simplest and most popular of the squeeze-type seals; its design and installation are fairly well standardized. The other squeeze types are D ring, T ring, delta ring, and X ring, each designed to overcome some of the limitations of the O ring, such as spiral failure, friction, or

TABLE 17.10 Dimensions for V-Ring Adapters

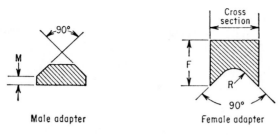

Male adapter   Female adapter

Dash No.	Cross section	Dimensions, in.		
		F	*M*	*R*
8-24	1/4	1/4	1/8	1/16
25-35	5/16	5/16	1/8	7/64
36-46	3/8	3/8	1/8	1/8
49-55	7/16	7/16	1/8	5/32
56-80	1/2	1/2	1/8	9/32

extrusion. Squeeze-type seals are generally installed in rectangular grooves with 5 to 10 percent initial compression to establish initial sealing. Fluid pressure forces the confined ring against one side of the groove and mating part, sealing off the clearance between. Since the functioning of all squeeze-type packings is basically similar, this discussion will relate to the more commonly used O ring.

O rings are primarily used to seal reciprocating motion, but they are being more widely used to seal helical motion (valve stems) and, with special design, to seal low-speed and low-pressure rotary motion. The three types of packing grooves are:

1. *Rectangular:* Most common for dynamic applications.
2. *V:* Recommended for low temperature where they increase squeeze. High friction and wear are an objection.
3. *Undercut or dovetail:* Used for slow-speed reciprocating motion. Reduce friction and tendency to blow out. Have concentrated wear and high cost. Useful for valve-seat seals.

Figure 17.21 illustrates the operation of O rings.

Antiextrusion backup rings are used mainly to raise the pressure range by preventing extrusion of the O ring into the clearance. They are sometimes employed to eliminate excessive clearance beyond the values recommended in Table 17.11.

The softer O rings will seal more effectively and have less friction at zero pressure. Operating friction at pressure is influenced by the amount of O-ring surface making contact, as illustrated by Fig. 17.25.

There are two types of recommended groove designs for O rings. One is for military aircraft, MIL-R-5514-D; and the other, for industrial applications, is based upon the military specification MIL-R-5514.

Tables of standard size and installation data are available from government specifications and O-ring manufacturers.

TABLE 17.11 Recommended Clearances for O Rings

Operating pressure, psi	Clearance, in.		
	For 70 Shore A	For 80 Shore A	For 90 Shore A
0	0.010	0.010	0.010
250	0.010	0.010	0.010
500	0.008	0.010	0.010
1,000	0.005	0.008	0.010
1,500	0.003	0.005	0.008
2,000	. . . . .	0.004	0.005
3,000	. . . . .	0.003	0.004
5,000	. . . . .	. . . . .	0.003

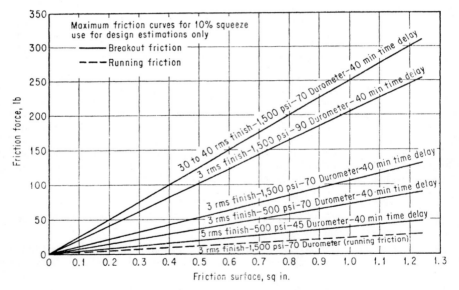

FIG. 17.25 O-ring friction. Maximum friction curves for 10 percent squeeze. Use for design estimations only. Solid line shows breakout friction. Dashed line shows running friction.

Limitations on surface-finish requirements are given below.

For military applications:
1. Diameter over which packing must slide, 16 μin maximum
2. Groove root diameter, 32 μin maximum
3. O ring groove sides without backup rings, 32 μin maximum
4. O ring groove sides with backup rings, 62 μin maximum

For commercial applications:

1. Diameter over which O ring must slide, 16 μin maximum
2. O ring groove root diameter, 64 to 125 μin
3. O ring groove side, with or without backup rings, 64 to 125 μin

If possible, toolmarks should run in the same direction as O-ring movement to promote better sealing and reduce wear. Honed and hardened steel that has been hard-nickel plated is best. Hard chrome plating is good but increases friction. Aluminum alloy, bronze, brass, Monel, and soft stainless steels are not recommended for long life. They have limited low-pressure application.

The design of O rings for rotary applications up to 600 ft/min has been established with good lubrication and cooling. For successful rotary usage O rings should be installed so that they are in circumferential compression rather than in tension (developed in stretching around a shaft) in order to eliminate the "Joule effect" common to some rubbers. Design tables are available that give about 5 percent peripheral compression by having the O-ring inside diameter larger than the shaft it seals and the groove width equal to the O-ring cross section. Shaft speed should be limited as follows:

Shaft diameter, in	Maximum speed, ft/min
$\frac{1}{8}$–$\frac{9}{32}$	350
$\frac{3}{8}$–$1\frac{1}{16}$	450
$\frac{3}{4}$–$1\frac{1}{4}$	600

Additional information on various sealing devices being used in this area can be found in Refs. 7 and 8.

17.8 OIL SEALS

17.8.1 Theory

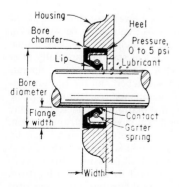

FIG. 17.26 Oil-seal installation.

An oil seal, sometimes referred to as a "radial positive-contact seal," is generally used to seal off lubricant flow along a rotating shaft as illustrated in Fig. 17.26.

The oil seal has two essential elements, a rigid heel and a flexible lip. The heel is a force fit into the bore to seal off outside-diameter leakage and prevent seal rotation, and the flexible lip makes sealing contact with the rotating shaft. The seal lips can be either springless or spring-loaded, and both usually have some interference fit on the shaft in the as-molded condition. In the spring-loaded type the spring provides the additional lip-contact pressure for sealing and to compensate for wear and compression set. The spring load is critical and may be applied either by a garter spring or by a finger spring to provide a lip pressure sufficient to seal. The pressure must not be so great as to cause a rupture of the lubricant film under the lip. Seal lips are pressure-actuated, and any increase of internal pressure will result in tighter sealing. Oil-seal lips must almost always be lubricated, and usually a thin film of the fluid being sealed forms between the seal lip and the rotating shaft. There are many seal designs, e.g., single lip, double lip, metal cased, molded heel section, rubber-coated heel, integral wiper or exclusion lip, outside lip; and many materials: rubber, leather, Teflon, various metals, etc. The selection of the best seal is dictated by the service conditions, cost, sealing effectiveness, and desired seal life. The metal-rubber combination is the most widely used. The seal manufacturer can recommend the best type of seal. Seals of special design are required for installations where fluid pressure exceeds 10 lb/in². Oil seals also have limited application for reciprocating and oscillating motion.

17.8.2   Design

The determination of flange width and seal width (bore depth) is important in order that seals of standard dimension can be used without going to the expense of securing special designs required by poorly dimensioned recesses. A square cross section is best for designing any seal. Table 17.12 presents suggested oil-seal cross-sectional dimensions as a function of shaft diameters where seal flange and seal width are equal.

TABLE 17.12   Oil-Seal Dimensions

Shaft diam, in.	Seal flange and width, in.
To ½	¼
$\frac{9}{16}$– 1	$\frac{5}{16}$
1$\frac{1}{16}$– 2	⅜
2$\frac{1}{16}$– 5	½
5$\frac{1}{16}$– 8	¾
8$\frac{1}{16}$–16	¾
16$\frac{1}{16}$–24	¾ and 1
Over 24	1 and over

17.8.3   Power or Torque Requirements

TABLE 17.13   Oil-Seal Torque

Shaft Diameter vs. Seal Torque	
Shaft diam., in.	Torque, oz-in.
⅜	8–10
⅞	10–20
1	15–30
2	60–90
3	90–120

Seal* Torque vs. Pressure	
Pressure, psi	Torque, oz-in.
Atm	10
5	17
10	23
15	50

*$\frac{7}{8}$-in-diameter shaft at 1750 r/min.

Generally the power absorption of an oil seal is not great enough to be significant in a normal application. The peak torque usually occurs at startup, especially after a lengthy shutdown period. Once operating speed is reached and the lip lubricating film established, the torque will be low. For installations where torque may be a factor, such as in instrumentation, seals with minimum lip load can be designed. Torque will be appreciable if there is any fluid pressure; this usually requires special design consideration. Typical oil-seal torque values are presented in Table 17.13.

The horsepower requirement may be calculated as follows:

$$\text{hp} = 9.917 \times 10^{-7} NT \qquad (17.11)$$

where N = shaft revolutions per minute
$\quad T$ = seal torque, oz·in

The following torque limits for oil seals established by Naval Ordinance, MIL Specification 13882, are a useful guide in estimating power requirements:

	Maximum torque, oz·in
A limitation (low-friction seals)	$4.8\pi D^2$
B limitations (normal-friction seals)	$10.4\pi D^2$

Note: D = shaft diameter, in.

17.8.4 Performance Variables

Among the important factors which influence seal performance are:

1. Shaft speed in terms of feet per minute rubbing speed.
2. Temperature. For a discussion dealing with the problems of dynamic O-ring seals and glands operating at high temperatures see Ref. 18.
3. Shaft finish.
4. Fluid pressure.
5. Severity of fluid sealed and lubricity.
6. Shaft concentricity and runout.

An oil seal can give satisfactory performance with some of the factors at a relatively severe level but will have short life if all factors are adverse. The shaft speed limit for leather seals is about 2000 ft/min, and for rubber-lip seals about 3000 ft/min (although when conditions are favorable the latter has been increased to 4000 ft/min or higher). The temperature limit is dependent upon the heat resistance of the lip material.

A finely finished shaft is required for maximum seal life; 10 to 20 μin root-mean-square (rms) is the general recommendation, but 10 to 16 μin rms is preferred. Laboratory tests show that sealing effectiveness is not improved with better than a 10-μin rms finish, and seal life has actually been shortened with a 2-μin rms finish. It has been suggested that, with a superfinish, a lubricating film cannot be maintained. Tool finish marks sometimes create a spiral lead for leakage; if they do exist, the direction of shaft rotation should be in the direction that would direct the flow inward.

A shaft hardness of from Rockwell C 40 to 50 will promote good performance by limiting shaft wear. Soft metal shafts such as brass or aluminum should be provided with a hardened sleeve or ring to contact the seal lip. Since seal lips are readily pressure-actuated, a means of preventing pressure buildup should be provided, especially at high speeds, unless the seal is designed for a pressure application. If the fluid being sealed is a poor lubricant, such as air, auxiliary lubrication must be provided.

The built-in seal lip interference plus spring load will compensate to some extent for a shaft that is off-center but true-running. Runout, however, whether from an out-of-round shaft or shaft whip, in combination with high speed, will tend to throw the lip off the shaft and permit leakage. Table 17.14 will act as a guide to indicate maximum misalignment and runout.

A design guide for helicopter transmission seals (Ref. 9) discusses the various

TABLE 17.14   Shaft Speed vs. Shaft Misalignment for Oil Seals

Shaft speed, rpm	Shaft to bore misalignment (static measurements) in., max	Shaft runout and whip (dynamic measurement) T.I.R.,* in., max
0– 800	0.010	0.010
800–2,000	0.005	0.005
2,000–4,000	0.005	0.003

*T.I.R. = total indicator reading.

design parameters needed to be considered for speeds, pressures, and temperatures for acceptable seal lives. This design guide discusses recommended practices for conventional lip seals, hydrodynamic lip seals, circumferential seals, and face seals in the transmission application.

17.9   MECHANICAL FACE SEALS*

17.9.1   Introduction

Mechanical face seals (sometimes also termed "end face seals") are used in a host of applications to seal fluids of various types. These fluids range from liquid lubricants and highly toxic chemicals to various gases. The applications range from automotive water pumps to high-speed gas turbines. The function of a mechanical seal is to permit a rotating shaft to penetrate an enclosure (transmission box, submarine hull, pump housing, etc.) while maintaining separation of the environments on the inside and outside of the enclosure. Thus, leakage and loss of sealed fluids is restricted, and entry of foreign bodies into the operating medium is prevented.

Mechanical seals were first introduced at the turn of the century.[19] Since then, more and more they are replacing the traditional soft packings and stuffing boxes, having the advantages of smaller leakage loss, reduced maintenance, and longer life.

17.9.2   Fundamentals

Basically a mechanical face seal (Fig. 17.27) has two main components: One is attached to the shaft and rotates with it; the other is attached to the housing and is non-rotating. One of the components (termed "primary seal ring" or "seal head") is flexibly mounted to provide angular and axial freedom of motion. This flexibility allows accommodation of axial and angular displacements that arise from various causes, e.g., shaft vibration, face runout, thermal growth. Figure 17.27a shows a mechanical face seal in which the seal seat (sometimes termed "mating ring") is rigidly mounted to the shaft and the primary seal ring is flexibly mounted to the housing. In Figure 17.27b the primary seal ring is flexibly mounted to the shaft and is the rotating component.

Axial loading from the sealed pressure and the springs tend to force the flexibly mounted component into alignment with the rigidly mounted one. Relative sliding

---

*By Izhak Etsion.

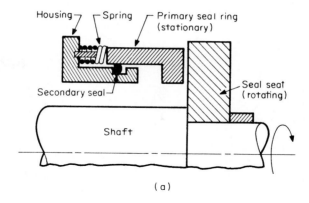

(a)

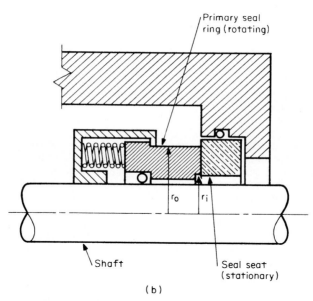

(b)

FIG. 17.27 Mechanical seal. (*a*) Rotating seat. (*b*) Stationary seat. (*c*) Arrangement of components.

motion takes place between the faces of these two components where the primary sealing occurs. Secondary seals prevent leakage between the seal components and the shaft and housing. In order to avoid wear and to achieve long life, the two faces of the primary seal must be separated by a film of the sealed fluid (Fig. 17.28). This film must be very thin to keep the leakage rate within acceptable limits. In most mechanical seals currently in use where speeds and pressures are low to moderate, the separation between seal faces is only a fraction of 1 μm. Moreover, this separation is due mainly to surface macroroughness and is thus not complete; therefore occasional contact between the stationary and rotating faces occurs (Fig. 17.28*b*). Such seals are termed contacting seals. As pressures and speeds in more demanding applications increase, any contact between seal faces should be avoided; the required separation

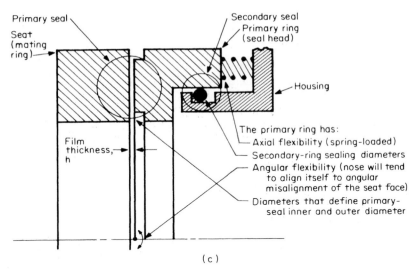

Primary seal

Seat (mating ring)

Secondary seal
Primary ring (seal head)

Housing

Film thickness, h

The primary ring has:
- Axial flexibility (spring-loaded)
- Secondary-ring sealing diameters
- Angular flexibility (nose will tend to align itself to angular misalignment of the seat face)
- Diameters that define primary-seal inner and outer diameter

(c)

FIG. 17.27 (*Continued*)

between the faces is then of the order of a few micrometers (Fig. 17.28a). These seals are termed "noncontacting seals."

The amount of face separation in a mechanical seal depends on a balance between all axial forces acting on the flexibly mounted seal head. These forces include the previously mentioned axial loading from the sealed pressure and springs as well as an opening force due to the pressure generated in the lubricating film between the faces. Mechanical contact contributes yet another force in cases of contacting seals. The mechanism of pressure generation in the lubricating film is essentially the same as in thrust bearings. The film pressure for liquid lubricating films is given by the Reynolds equation

$$\frac{\partial}{\partial r}\left(rh^3 \frac{\partial p}{\partial r}\right) + \frac{1}{r}\frac{\partial}{\partial \theta}\left(h^3 \frac{\partial p}{\partial \theta}\right) = 6\mu\omega r \frac{\partial h}{\partial \theta} + 12\mu r \frac{\partial h}{\partial t} \qquad (17.12)$$

The radial extent of the seal interface is much smaller than its circumferential extent ($r_i/r_o > 0.8$); hence, the second term on the left-hand side of Eq. (17.12) can be neglected without losing the accuracy of its solution. The Reynolds equation for mechanical seals thus becomes

$$\frac{\partial}{\partial r}\left(h^3 \frac{\partial p}{\partial r}\right) = 6\mu\left(\omega \frac{\partial h}{\partial \theta} + 2\frac{\partial h}{\partial t}\right) \qquad (17.13)$$

where $h(r, \theta, t)$, the local film thickness between the mating faces, depends on the face geometry and on the system's dynamics.

The boundary conditions of Eq. (17.13) are $p = p_i$ at $r = r_i$ and $p = p_o$ at $r = r_o$. The pressure p has two components: p_s, the hydrostatic component contributed by the pressure differential across the seal, and p_d, a hydrodynamic component contributed by the relative velocity between the mating faces.

The two pressure components are given by

$$p_s = p_o - (p_o - p_i)[h_i^2/(h_o^2 - h_i^2)][(h_o/h)^2 - 1] \qquad (17.14a)$$

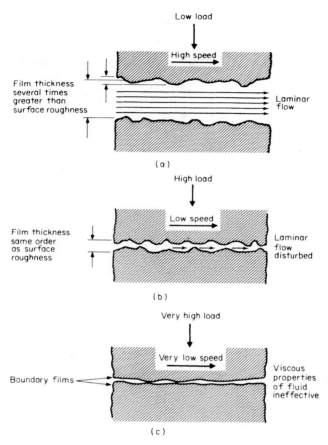

FIG. 17.28 Seal lubrication models. (*a*) Full-film lubrication, noncontacting. (*b*) Thin-film (mixed) lubrication, contacting. (*c*) Boundary lubrication, contacting.

and
$$p_d = -3\mu\left(\omega\,\frac{\partial h}{\partial\theta} + 2\,\frac{\partial h}{\partial t}\right)\frac{(r_o - r)(r - r_i)}{h_m h^2} \qquad (17.14b)$$

The total pressure $p(r, \theta, t)$ is given by

$$p = p_s + p_d \qquad (17.15)$$

In Eqs. (17.14) the indices i and o correspond to the inner and outer radii of the seal interface r_i and r_o, respectively; h_m is average film thickness $(h_i + h_o/2)$; μ is fluid viscosity; ω is shaft speed; and t is time. As can be seen the pressure p is very much affected by the geometry and time variation of the film thickness h in the primary seal. Figure 17.29 illustrates some of the possible primary seal geometries.

The opening force acting to separate the mating faces is given by

$$F_o = \int_{r_i}^{r_o}\int_0^{2\pi} pr\,d\theta\,dr \qquad (17.16)$$

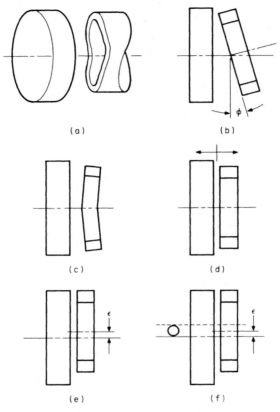

FIG. 17.29 Possible primary seal geometries. (*a*) Waviness. (*b*) Angular misalignment. (*c*) Coning. (*d*) Axial vibration. (*e*) Parallel misalignment. (*f*) Shaft whirl.

and the average fluid film pressure is

$$P_f = F_o/\pi(r_o^2 - r_i^2) \tag{17.17}$$

Referring to Fig. 17.30*a*, the contribution of the axial loading from the sealed pressure p_h to the closing force is

$$F_c = \pi(r_o^2 - r_b^2)p_h \tag{17.18}$$

where r_b is the radial location of the secondary seal. The ratio between these closing and opening forces is

$$F_c/F_o = [(r_o^2 - r_b^2)/(r_o^2 - r_i^2)](p_h/p_f) \tag{17.19}$$

The ratio $B = (r_o^2 - r_b^2)/(r_o^2 - r_i^2)$ can be introduced giving

$$F_c/F_o = B(p_h/p_f) \tag{17.20}$$

The parameter B which presents the ratio between the area on which the sealed pres-

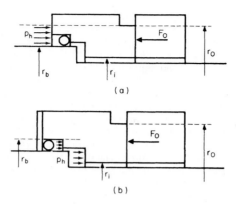

FIG. 17.30 Force balance. (*a*) Sealed pressure at outside diameter. (*b*) Sealed pressure at inside diameter.

sure p_h and the area on which the average film pressure p_f are acting is termed the balance ratio of the seal. In the case of flat parallel faces with constant film thickness h, the pressure drops linearly across the sealing dam. In this case $p_f = 0.5p_h$ and Eq. (17.20) becomes

$$F_c/F_o = 2B$$

For an equilibrium of the closing and opening forces, a balance ratio B of 0.5 is required. If $b > 0.5$, the closing force is larger than the opening force, whereas the opposite is true when $B < 0.5$. In contacting seals it is important to maintain a closing force component F_c larger than the opening force F_o in order to provide mechanical contact between the mating faces. Hence, balance ratios larger than 0.5 are used. Too high a balance ratio may, however, result in excessive friction and wear. It has become, therefore, a common practice to choose for contacting seals balance ratio ranging from 0.65 to 0.75 dictated by the operating conditions.

When the sealed pressure is at the inside diameter of the primary seal ring (Fig. 17.30*b*), the balance ratio $B = (r_b^2 - r_i^2)/(r_o^2 - r_i^2)$.

The actual average film pressure p_f is affected by various phenomena. Hence, for the actual force balance one should use Eq. (17.19) in the form

$$F_c/F_o = B(p_h/p_f)$$

with p_f obtained from Eqs. (17.14) to (17.17).

Equations (17.13) and (17.14) are quite accurate for the model shown in Fig. 17.28*a*. However, surface roughness and wear,[20] thermoelastic effects,[21] and phase change[22] play an important role under various conditions of seal operation and have to be considered under these conditions.

Basic Configurations. There are four basic configurations of seal arrangement. The flexibly mounted component can be either rotating or stationary, and the sealed fluid can be either at the outside diameter or at the inside diameter of the primary ring. The most common configuration, especially in pumps, is the rotating flexibly mounted ring as in Fig. 17.27*b*. The advantage of this configuration is the relative ease of assembly and routine inspection, as well as a better heat transfer between the seal and the sealed fluid. The stationary flexibly mounted ring, Fig. 17.27*a,* is found mainly in applications where rotating speed is high. Its main advantage is the absence of centrifugal forces on the flexible support. The location of the sealed fluid at the outside diameter of the seal has a number of advantages compared with the location at the inside diameter: centrifugal effect opposing the radial inward flow of the sealed fluid, centrifuging solid particles away from the primary seal, compression rather than tensile stresses acting on the seal rings due to the sealed pressure, and better static and dynamic stability.

On the other hand, when sealing corrosive fluids, better protection of metal parts can be achieved when the sealed fluid is at the inside diameter of the seal. Either a single- or a multiple-seal arrangement can be used. Figure 17.31 shows a single-seal arrangement where the seal can be mounted inside or outside the housing. The inside-mounted seal is normally preferred for better operation. The entire primary ring is surrounded by fluid, and the sealed pressure is usually at the outer diameter. Double seals (Fig. 17.32) consist of two single seals mounted back to back (Fig. 17.32a) or face to face with a common seat (Fig. 17.32b). A secondary fluid is circulated in the double-seal cavity at a pressure slightly above the pressure of the process fluid. Double seals are often used when the process fluid is abrasive or when sealing toxic or flammable fluids. When the process fluid is a gas, circulation of liquid in the double-seal cavity provides lubrication to the mating faces. The secondary buffer fluid also removes heat from the two seals. In this case a heat exchanger is normally added in the fluid cycle to prevent accumulation of heat.

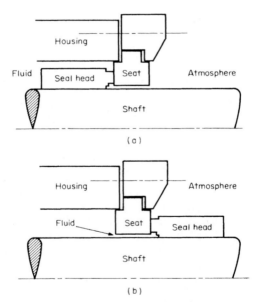

FIG. 17.31 Single-seal arrangement. (a) Inside-mounted. (b) Outside mounted.

Figure 17.32c shows a tandem seal arrangement. This arrangement consists of two seals of the same design mounted in the same direction. The buffer zone pressure is not necessarily higher than the process fluid pressure. However, the buffer fluid prevents process fluid from leaking to the atmosphere. Tandem arrangement can be used with more than two seals to seal very high pressures. In this case, each seal takes only a portion of the total pressure differential. An example is the high-pressure seals of primary coolant pumps for nuclear reactors.[23]

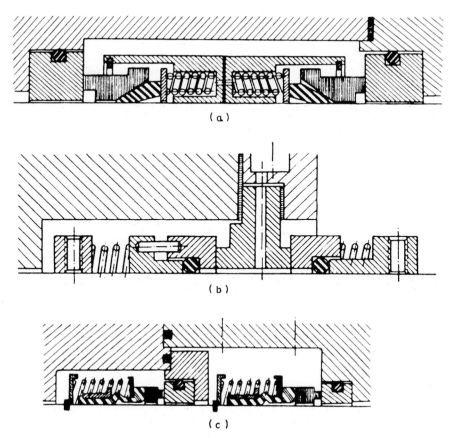

FIG. 17.32 Double-seal arrangement. (*a*) Back to back. (*b*) Face to face. (*c*) Tandem.

The Primary Seal. The primary seal consists of the two mating rings in relative motion. These rings are often referred to as the "seal head," or primary ring (for the flexibly mounted ring), and the "seal seat." Carbon-graphite materials of various types, used in combination with some other compatible hard material, are found in the majority of mechanical seal applications. Table 17.15 lists various head-seat combinations for several fluids and environments.

Mating rings of the primary seal are normally lapped to obtain smooth and flat surfaces. Surface roughness of 0.15 to 0.5 μm and flatness of 2 to 3 helium light bands are common requirements. However, in spite of the great care taken to lap the rings, the surface profile changes during the operation. Mechanical and thermal distortions are the main reasons for these changes, and the result is often waviness and coning as shown in Fig. 17.29*a* and *c*.

The Secondary Seal. The secondary seal provides sealing between the primary seal rings and housing or shaft, Fig. 17.33. The most commonly used secondary sealing consists of elastomer rings of various cross sections such as O rings (Fig. 17.33*a*), V rings, U cup rings, etc. Another type of secondary sealing consists of bellows. This group is subdivided into elastomer bellows (Fig. 17.33*b*), and metal bellows of the corrugated (Fig. 17.33*c*) or welded (Fig. 17.33*d*) type.

TABLE 17.15 Seal Materials and Environment Combination

Ennvironment	Seal head material	Seat material
Water	Carbon graphite (also carbon graphite containing various metals: copper, lead, babbit, etc.)	Bronze
		Ni-resist
		Nickel cast iron (for constant operation only)
		Ceramic
		Stellite (hard facing on 316 stainless or any stabilized stainless)
		Tungsten carbide
		Malcomized 316 stainless
		Carbon-filled PTFE
		Glass-filled PTFE
		Chrome plate on various parent materials (must be thick enough)
		Ceramic facing on stainless
	Tungsten carbide	Tungsten carbide
	Carbon containing various metals	Stainless steel (series 400, hardened to Rockwell C 50 or higher)
Caustics	Carbon graphite	Carbon-filled PTFE
		Stellite-faced stainless steel
	Carbon graphite (nonmetallic)	Hard-faced 316 stainless steel
	Carbon graphite (metallic, for dilute solutions)	Stellite-faced stainless steel
Salt Solution	Carbon graphite	Stainless steel
		Ceramic
		Monel
	Ceramic	Ceramic-faced stainless
		Ceramic
	Carbon babbit	Phosphor bronze
Sea Water	Carbon babbit	Aluminum bronze
	Stellite on stainless	Aluminum bronze
	Tungsten carbide	Tungsten carbide
	Bronze	Laminated plastic
	Carbon graphite	Stellite
Acids	Carbon graphite	Hard-faced 316 stainless
		Carpenter 20 stainless
		Stellite
		Chromium boride
		Ceramic
		Masteloy A, B, C
		Carbon-filled PTFE (for nonoxidizing acids)
	Ceramic	Stellite (attacked by many mineral acids)
		Glass-filled PTFE (oxidizing acids)
		PTFE

*This is a general recommendation, as 316 stainless is not hardenable.
†For high-temperature capabilities (maximum approximately 700°F).
‡Sodium and fluorine compounds and radioactivity may adversely affect PTFE.

TABLE 17.15   Seal Materials and Environment Combination (*Continued*)

Environment	Seal head material	Seat material
Gasoline	Carbon graphite	Cast iron Carbon-filled PTFE Glass-filled PTFE Ni-resist Nitralloy Ceramic Stellite facing on stainless steel Stainless steel, 400 series
Oil	Carbon graphite	Bronze (for few applications) Ni-resist Cast iron Ceramic Stellite (hard facing on 316 stainless steel, especially for high pressures and high velocity) Tungsten carbide Malcomized 316 stainless Carbon-filled PTFE Glass-filled PTFE Sintered iron or bronze Nitralloy, hardened Tool steel, hardened SAE-1040 steel Stainless steel (400 series hardened to Rockwell C 50*) Bronze Bronze
	Cast iron Graphite molybdenum	
Oxidizing fluids	Boron carbide	Boron carbide
Slurry	Carbon	Ceramic Tungsten carbide
	Bronze Tungsten	Tungsten carbide Tungsten
Heat transfer fluids (bithenyl class)	Carbon graphite	Ni-resist Stellite Ceramics Chrome carbide plate Tungsten carbide† Ceramic†
	Tungsten carbide	Tungsten carbide†
Gas (air, CO_2, H_2, He, N_2, O_2)	Carbon graphite	Tool steels Chrome plate Tungsten carbide, plate and solids Chrome carbide plate Ceramics 300 stainless steel 400 stainless steel 440-C
	Glass-filled PTFE‡ Carbon filled PTFE‡ (not for N_2 service)	4140, 4340 Tool steels (hardened) Chrome oxide

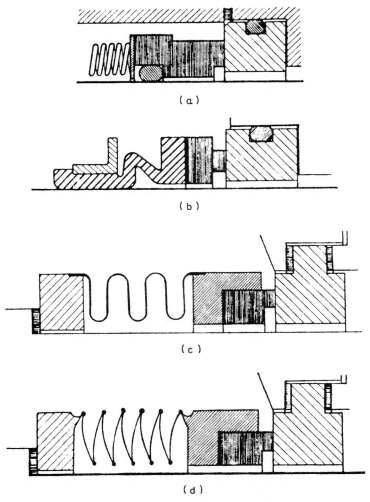

FIG. 17.33 Various secondary seals. (*a*) O rings. (*b*) Elastomeric bellows. (*c*) Corrugated metal bellows. (*d*) Welded metal bellows.

Elastomers have severe temperature limitations as well as limited chemical compatibility. Table 17.16 presents properties of the most common elastomers and their compatibility with various fluids.

Metal bellows seal assemblies are usually found in applications involving extreme temperature or highly reactive media incompatible with elastomers. The welded bellows require less space and have better fatigue life and higher pressure capabilities than the corrugated bellows.

Metal bellows eliminate the need for additional springs since they function as both secondary seal and the face loading elements. Another advantage of metal bellows over elastomeric rings is the elimination of sliding friction in the secondary seal as the seal head tracks misalignment of the seat. The absence of such friction eliminates abrasion and fretting corrosion and provides better tracking capability and face alignment.

TABLE 17.16 The Most Important Elastomers and Their Properties

Elastomer	Composition	Working temperature range, °C	Tensile strength, bar	Elongation, %	Hardness, °Shore	Water	Steam	Hydraulic fluids, non-flammable (ester based)	Mineral fats and oils	Vegetable and animal fats and oils	Ozone	Hydro-carbons Aliphatic	Hydro-carbons Aromatic	Hydro-carbons Halogenated	Alcohols	Ketones	Esters	Dilute acids	Concentrated acids	Dilute alkalis	Concentrated alkalis	Saline solutions
Natural rubber	Rubber, K. W. Coil	−30 to 120	50 to 280	1000	30 to 98	x	x	—	—	—	—	—	—	—	x	x	—	o	—	x	o	x
S.B.R.	Refining-type polymerisate Butadiene-styrene copolymer	−30 to 130	50 to 240	700	40 to 95	x	x	—	—	—	o	—	—	—	x	x	—	x	o	x	x	x
Nitrile N	Butadiene-acrylonitrile copolymer	−30 to 130	50 to 240	700	40 to 95	o	o	—	x	x	o	x	o	—	x	—	—	o	—	o	o	x
Neoprene	Chlorinated-butadiene polymerisate	−40 to 140	50 to 270	800	40 to 90	x	x	—	o	o	x	o	—	—	x	—	—	x	o	x	x	x
Butyl	Isobutylene-isoprene copolymer	−50 to 150	40 to 170	900	40 to 95	x	x	o	o	o	x	—	o	—	x	o	o	x	o	x	x	x
Hypalon	Chloro-sulfonated polyethylene	−40 to 140	40 to 200	600	40 to 80	o	o	—	o	o	x	—	—	—	x	—	—	x	o	x	x	x
Silicone rubber	Polycondensates of dialkylsiloxanes	−100 to 200	20 to 80	500	65 to 80	o	—	—	x	x	x	o	o	o	x	o	x	x	o	x	o	o
Thiokol	Alkylpolysulfide	−40 to 80	10 to 60	200	70 to 85	x	—	x	x	x	x	x	o	o	x	o	x	x	o	x	x	x
Polyacrylic	Polyacrylate	−30 to 120	20 to 70	700	70 to 95	o	—	x	x	x	x	x	—	—	o	—	—	—	—	o	—	o
Vulcollan	Polyurethane	−30 to 80	200 to 320	600	70 to 95	o	—	—	x	x	x	x	—	—	o	—	—	—	—	—	—	—
Adiprene	Polyurethane	−40 to 120	80 to 300	700	60 to 90	x	o	—	x	x	x	x	—	—	o	—	—	—	—	o	—	—
Kel-F	Copolymer of chlorotriethylene and vinylidene fluoride	−50 to 180	30 to 120	700		x	x	—	—	o	x	o	—	—	x	—	—	x	x	x	x	x
Viton	Vinylidene fluoride–hexafluoropropylene copolymer	−60 to 200	80 to 160	300	60 to 95	x	o	o	x	x	x	x	x	o	x	—	—	x	o	o	—	x
PTFE	Polytetrafluoroethylene	−200 to 280	140 to 310	200	55D	x	x	x	x	x	x	x	x	x	x	x	x	x	x	x	x	x
E.P.R.	Ethylene-propylene	−55 to 200	50 to 160	400	70 to 95	x	x	x	—	—	x	—	x	—	x	x	o	x	o	x	x	x
F.S.R.	Fluoro-silicone rubber	−60 to 230	55 to 85	400	40 to 80	o	o	o	x	x	x	x	x	o	x	—	o	o	—	x	o	x

Note: x, stable; o, stable under certain conditions; —, unstable.

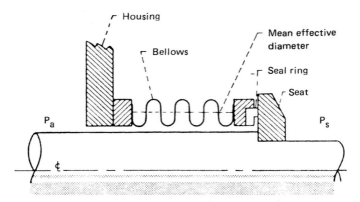

FIG. 17.34 Mean effective diameter in metal bellows seal.

Metal bellows are not suited, however, to withstand high pressures and are limited to a maximum pressure differential of about 6 MPa. Another disadvantage of metal bellows is the changes in their mean effective diameter (Fig. 17.34) as a result of changes in the sealed pressure. This makes impossible the design of a metal bellows seal with a constant balance ratio for the full range of pressure conditions.

Loading Elements.   Axial loading of the mating faces is essential for proper sealing. Figure 17.35 illustrates the various springs that may be found in conventional mechanical seals to accomplish this function.

The single spring (Fig. 17.35a) has a relatively large wire cross section, and it is therefore less affected by corrosion degradation. One disadvantage is that a new spring is required for any shaft size. Another is that the single spring is rigid and does not provide uniform load distribution around the entire seal-head circumference. This may cause misalignment between the mating faces and increased leakage.

Multiple springs (Fig. 17.35b) make seal design independent of the shaft diameter. Loading can be easily varied by varying the number of springs. More uniform loading is achieved around the head circumference. The disadvantage of the multiple springs is the greater effect corrosion may have on the spring load. However, with the use of stainless steel springs this danger is greatly minimized.

Wave springs (Fig. 17.35c) are used where only a small spring load is required and axial space is limited. Their main advantage is light weight and small axial space requirement. Their disadvantages are high susceptibility to degradation of the spring loading due to corrosion, low available working length, and nonuniform load distribution around the seal circumference.

17.9.3   Environmental Control

Proper seal operation is very much dependent on the environmental conditions around the mating faces.

Figure 17.36 shows different seal arrangements that are used to control the environment of the seal faces. Figure 17.36a illustrates bypass cooling where a small amount of clean process liquid is being diverted and passed over the seal area. When additional cooling is required or when the process liquid is not very clean, the seal area can be flushed from an independent source (Fig. 17.36b).

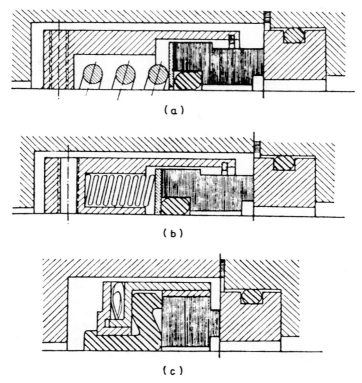

FIG. 17.35 Various spring arrangements. (*a*) Single spring. (*b*) Multiple springs. (*c*) Wave springs.

Some liquids, like sugar and salt solutions, evaporate when they come into contact with the atmosphere, leaving abrasive crystals behind. Figure 17.36*c* shows a quench arrangement in which a barrier fluid is used to isolate the atmosphere from the seal interface. When the abrasive crystals are in solution only above a certain temperature, the seal environment can be heated—by applying electric heating elements, for example. Double and tandem seals are often used for environmental control, as illustrated in Fig. 17.37.

17.9.4 Seal Wear

Changes in the face geometry of the mating rings due to wear have significant effects on the pressure distribution in the fluid film of the seal interface. This in turn affects the seal balance, the leakage across the seal, and the susceptibility of lubrication failure and destruction wear. The mating rings in a mechanical seal usually operate in unidirectional sliding, but oscillation and impacts due to the flexibility of the system are also common. The materials used for the mating rings have to be good bearing materials and at the same time also chemically compatible with the sealed fluids. All these factors contribute to the seal wear and greatly affect its life. A number of wear mechanisms occur in mechanical seals. They are: adhesive wear, abrasive wear, corrosive

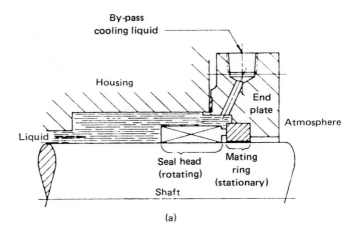

(a)

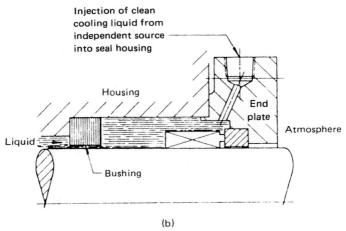

(b)

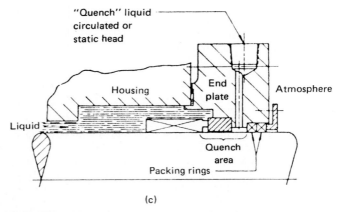

(c)

FIG. 17.36 Single-seal arrangements for environmental control. (*a*) With bypass. (*b*) With injection. (*c*) With quench.

17.48

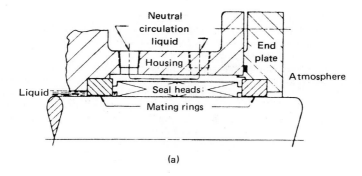

(a)

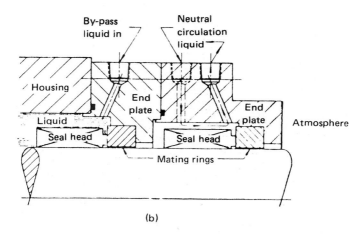

(b)

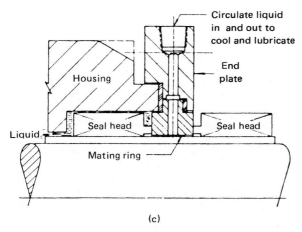

(c)

FIG. 17.37 Multiple-seal arrangements. (*a*) Double seals. (*b*) Tandem seals. (*c*) Opposed double seals.

wear, pitting, and fretting.[28] The dominant type of wear in mechanical seals is the adhesive wear. Adhesion between the mating surfaces is caused by physical and chemical bonds. With relative motion, shear occurs in the direction of sliding along the weakest shear plane, and a transfer-film process develops. Abrasive wear results from the operating environment of the seal when contaminating particles enter into the seal interface. Corrosive wear is common in applications where the sealed fluids are chemically active. Pitting is usually related to fatigue wear and will happen in dynamically unstable seals where impact of the mating faces occurs. Fretting normally occurs on the secondary sealing surfaces due to reciprocating axial motion of the flexibly mounted element as it tries to follow displacements of the mating rings.

A PV factor is often used to evaluate the severity of application by giving some measure of wear. The PV factor is the product of the actual face pressure P in bars (10^5 N/m^2) and the mean face velocity in meters per second. A wear coefficient can be defined as

$$K = WH/Fd \qquad (17.21)$$

where W = volumetric wear
H = material hardness
F = imposed load
d = sliding distance

Substituting

$$d = Vt \qquad (17.22)$$

$$F = PA \qquad (17.23)$$

and $$W = bA \qquad (17.24)$$

into Eq. (17.21) where A = contact area and b = linear wear, we obtain the dimensionless wear coefficient in the form

$$K = (b/t)(H/PV) \qquad (17.25)$$

The PV factor also gives the friction power loss of a given seal in the form

$$L = PVfA \qquad (17.26)$$

where L = power loss and f = friction coefficient.

Tests to evaluate PV limits for various seal materials are performed by seal manufacturers, seal users, and seal material suppliers. In these tests wear rates are measured at various levels of PV values, and a PV limit is defined for a uniform wear rate corresponding to a typical life span of about two years. No standard test procedure has been established for the PV limits, and the PV value is therefore an imperfect relationship but, if used in the proper context, it can be a useful reference tool.

Table 17.17 gives order of magnitude values for wear coefficient K of various seal materials, their friction coefficient f, and PV limits. The data correspond to tests with water at about 50°C being the sealed fluid. The friction coefficient is obtained from tests at $PV = 35$ bar·m/s. Temperature has a marked influence on material wear, and therefore the information in Table 17.17 should be used for comparison only.

17.9.5 Noncontacting Mechanical Seals

Noncontacting mechanical seals are used in applications where high-pressure, high-speed, and long-life requirements exceed the capacity of conventional contacting

TABLE 17.17 Wear Coefficient K (Order of Magnitude), PV Limit, and Friction Coefficient f for Seal Face Materials

Sliding Materials		Wear coefficient K	PV limit, bar $\cdot$ m/s	Friction coefficient f
Rotating	Stationary			
Carbon-graphite (resin-filled)	NI-resist cast iron	10^{-6}	35	0.07
Carbon-graphite (resin-filled)	Ceramic (85% Al_2O_3)	10^{-7}		
Carbon-graphite (babbitt-filled)		10^{-7}		
Carbon-graphite (bronze-filled)	Tungsten carbide (6% cobalt)	10^{-8}	175	
Tungsten carbide (6% cobalt)		10^{-8}	42	0.08
Silicon carbide converted carbon	Silicon carbide converted carbon	10^{-9}	175	0.05

seals. In noncontacting mechanical seals, face separation of the order of a few micrometers is maintained by hydrostatic or hydrodynamic effects similar to those in various thrust bearings. The noncontacting mechanical seals can be divided into self-energized hydrostatic, externally pressurized hydrostatic, and hydrodynamic seals.

The self-energized hydrostatic seals (Fig. 17.38) utilize the sealed pressure to induce opening force and maintain controlled face separation. A minimum pressure differential across the sealing dam is required to overcome the closing force and separate the mating faces; therefore, transient rubbing contact usually occurs when the sealed pressure falls below the minimum required value.

A common feature of the three configurations of self-energized hydrostatic seals shown in Fig. 17.38 is the variation in the average film pressure p_f with the change in face separation. The upper limit of the pressure distribution shown in Fig. 17.38 corresponds to the case of contact between the mating faces, namely, zero gap. The lower limit corresponds to the case where the gap is very large. In this case, the geometry of the interface has no effect on the pressure drop which becomes linear as in any conventional flat face seal. The equilibrium pressure profile shown in the three examples of Fig. 17.38 corresponds to a certain gap which is the desired clearance between the mating faces. Any change in the design clearance results in an increase or decrease of the opening force in a stabilizing sense. That is, if the gap decreases the opening force increases and vice versa. Of the three configurations in Fig. 17.38, the conical seal (Fig. 17.38c) is the most popular one. The depth of the recessed step (Fig. 17.38b) or the taper (Fig. 17.38c) must be very small, of the order of a few micrometers, for adequate fluid film stiffness and dynamic stability. The recessed pads with orifice compensation design often suffer from orifices clogging by contaminants in the sealed fluid.

A design which overcomes the transient rubbing contact of the self-energized hydrostatic seals is shown in Fig. 17.39. Here an external pressure source is used to pressurize the interface of the seal independent of the sealed pressure. However, the additional complexity of the pressurizing system, the danger of clogged orifices, and poor dynamic stability of the seal are disadvantages of this design.

The hydrodynamic seals, Fig. 17.40, utilize lift pads to generate hydrodynamic

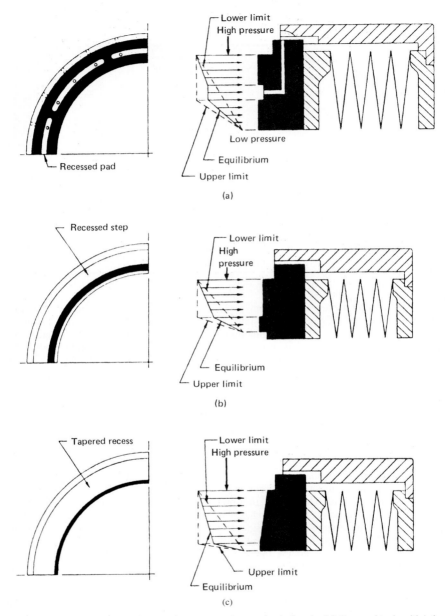

FIG. 17.38 Self-energized hydrostatic noncontacting mechanical seals. (*a*) Recessed pads with orifice compensation. (*b*) Recessed step. (*c*) Convergent tapered face.

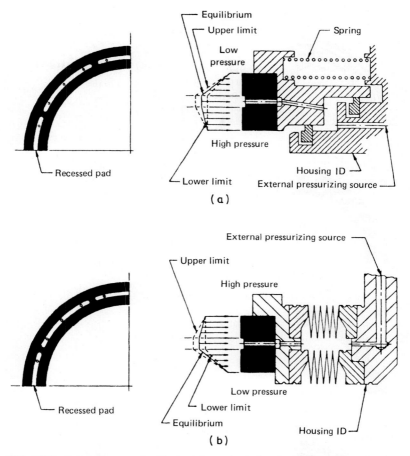

FIG. 17.39 Externally pressurized hydrostatic mechanical seals. (*a*) Piston-ring secondary element. (*b*) Bellows secondary element.

fluid film pressure by the rotational speed. The hydrodynamic lift is independent of the sealed pressure; it is proportional to the rotational speed and to the fluid viscosity. Therefore, a minimum speed is required to develop sufficient lift force for face separation, and rubbing contact generally occurs during the start and stop transients. Hydrodynamic seals require relatively wide face to incorporate the lifting pads in addition to the seal dam. This increases the friction drag and the potential danger of thermal and mechanical distortions in hydrodynamic seals as compared with hydrostatic seals.

A common disadvantage of all the noncontacting seals is higher leakage rates as compared with conventional contacting seals.

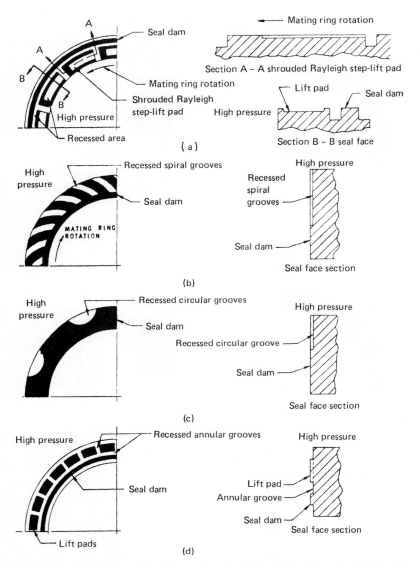

FIG. 17.40 Various types of hydrodynamic noncontacting mechanical seals. (*a*) Shrouded Rayleigh step. (*b*) Spiral grooves. (*c*) Circular grooves. (*d*) Annular grooves.

17.9.6 Mechanical Seal Dynamics

A mechanical seal is actually a complex dynamic system. The flexibly mounted seal head (Fig. 17.27) has five degrees of freedom: It can move axially, tilt about two perpendicular diameters, and move radially in two perpendicular directions. Circumferential rotation is the only excluded motion due to antirotation devices. The leakage across the seal is very sensitive to the face separation; hence small variations in the relative position between the primary seal components may cause excessive leakage. In noncontacting seals rubbing contact has to be avoided to prevent failure due to friction and wear. Seal dynamics is therefore a key factor for safe operation of mechanical seals and particularly of high-speed noncontacting seals.

Wobble of the seal seat due to inevitable runout is transmitted to the primary seal ring via the fluid film. The primary seal ring tracks the seat wobble, but its motion is affected by the stiffness and damping of both the flexible support and the fluid film, as well as by its own inertia. Thus, the primary seal ring does not perfectly track the wobbling seat but has a different amplitude and a certain phase shift. The result is a relative misalignment between the mating faces, as shown in Fig. 17.41 which corresponds to a noncontacting coned face seal.

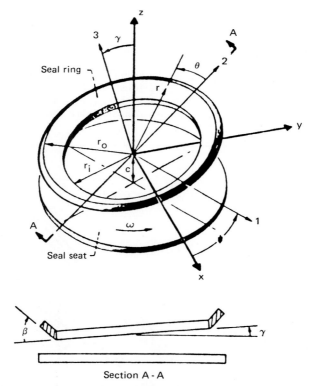

FIG. 17.41 Relative misalignment between mating rings in noncontacting seal.

The general approach in analyzing seal dynamics is to solve the equations of motion of the primary seal ring in its three major degrees of freedom. These are the axial one along the z axis and the two orthogonal tilts about the x and y axes (Fig. 17.41). The various parameters involved in the dynamic analysis are shown schematically in Fig. 17.42. The seat displacement represents runout of amplitude y_o and frequency ω. The seat motion is transmitted to the mass m of the seal ring via the fluid film having stiffness and damping coefficients K_f and C_f respectively. The seal ring is flexibly mounted on its support consisting of springs and secondary seal with stiffness

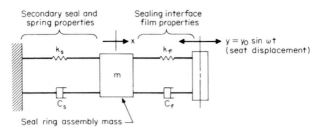

FIG. 17.42 Lumped-parameter model for analyzing seal dynamics.

and damping coefficients K_s and C_s, respectively. The outcome is the displacement $x = x(t)$ of the seal ring. The equations of motion are usually nonlinear coupled equations, and their solution requires a transient numerical procedure. Examples of such solutions can be found in Ref. 33 for the seal shown in Fig. 17.38c and in Ref. 35 for the seal of Fig. 17.40a. Stability maps can be found similar to that shown in Fig. 17.43. Three modes of seal operation are identified: a stable mode in which the flexibly mounted primary ring tracks synchronously the seat runout; an unstable mode in which seal failure results because of uncontrolled vibrations; and a transition mode in

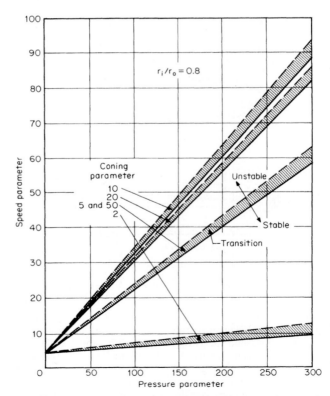

FIG. 17.43 Stability maps for noncontacting self-energized hydrostatic seal with coned face.

which approximately half-frequency whirl of the primary ring is superimposed on the basic synchronous tracking of the stable mode.

An expression for the angular stability threshold of noncontacting mechanical seals can be found in the form

$$\omega_{cr}^2 = 4[(K_f + K_s)/I](1 + C_s/C_f) \qquad (17.27)$$

where I = moment of inertia of the primary seal ring
K_f, K_s = angular stiffness coefficients
C_f, C_s = damping coefficients

For stable operation the shaft speed ω must be less than the critical speed ω_r. Axial stability in mechanical seals is less critical than the angular one and hence is not of concern.

A general solution for the seal ring displacement in the stable mode of operation has the form

$$x = x_o \sin(\omega t - \phi) \qquad (17.28)$$

where x_o is the amplitude of the wobble motion of the seal ring and ϕ is a phase shift between the seat and ring displacements. The relative misalignment between seat and ring (Fig. 17.41) depends on the transmissibility x_o/y_o and the phase shift ϕ. In some stable seals the combination of transmissibility and phase shift may be such that the relative misalignment is large enough to cause contact between the mating faces. These seals will fail due to excessive leakage and wear although they are dynamically stable. Uniform circumferential load distribution from the springs, low damping of the flexible support, and low mass inertia of the seal ring reduce the danger of failure in dynamically stable seals.

17.10 LABYRINTH SEALS

17.10.1 Theory

The basic theory of a labyrinth seal is that a pressure drop will accompany the flow of a fluid through a passage. The longer the passage and the more turbulent the flow, the greater the drop in pressure will be. Since the primary purpose is to limit leakage as much as possible, the cross-sectional flow area should be a minimum. Labyrinth seals are always used between moving parts, although in essence a static seal which allows some leakage is also a labyrinth seal. Similarly, any dynamic packing which allows leakage acts as a labyrinth seal, and this especially applies to compression-type packings. One manner to differentiate between a true labyrinth seal and these pseudo-labyrinth seals is to define a labyrinth seal as one "designed to function between two moving parts by having a fixed clearance between the parts so that the restricted flow of fluid through this clearance creates the required pressure drop with a tolerable loss of fluid." This definition also excludes the segmented carbon rings used in many installations.

17.10.2 Design

There are innumerable labyrinth-seal designs, depending upon fluid, temperature, pressure, type of motion, speed, allowable leakage, space, and material requirements.

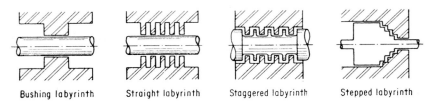

| Bushing labyrinth | Straight labyrinth | Staggered labyrinth | Stepped labyrinth |

FIG. 17.44   Labyrinth seals.

The simplest type, of course, is a close-fitting bushing around the shaft. This could be used under the mildest conditions where the laminar fluid flow results in the small pressure drop required, without excessive leakage or temperature rise.

The most common types of labyrinth seal are those which use blades or steps to promote turbulent flow and increase the length of seal. The more common types are shown in Fig. 17.44. The primary design objective is the calculation of the length and clearance of the seal and the spacing between the steps for the staggered type which results in a leakage rate below the maximum allowable.

17.10.3   Bushing Labyrinth

The bushing labyrinth presents the simplest type, but even so, complications arise when the relative motion and possible eccentricity are considered. Standard fluid-flow theories are the basis for the calculations and were used by Tao and Donovan[10] in their work in which a general equation relating the friction factor to the Reynolds number was derived:

$$F = C/(N_{Re})n \qquad (17.29)$$

For laminar flow through a small annulus with no relative motion between adjoining surfaces, the values of C and n were found to be 170 and 1.03, respectively. For turbulent flow, the values were 0.316 and 0.21, and in both cases the equation applies for both concentric and eccentric parts.

When relative motion is considered, a number of variations must be made. By the application of these to a specific system Tao and Donovan developed nomographs for laminar and turbulent flows showing the relationship between pressure drop, radial clearance, sleeve length, shaft radius, and shaft speed for water at 68°F with concentric motion. By using a relationship between eccentricity and the ratio of eccentric to concentric flow, it is a simple matter to calculate the flow with eccentric motion. Reproductions of the nomographs are shown in Fig. 17.45a and b. The axial velocity is read directly from the charts, and when multiplied by the annulus area, the amount of flow with concentric motion is obtained. This value multiplied by the ratio of eccentric to concentric flow from Fig. 17.46 will give the eccentric-flow value.

17.10.4   Straight Labyrinth

For a straight labyrinth seal the method presented by Heffner[11] gives the best correlation. Heffner states that the leak rate of gases through labyrinth seals is analogous to the flow rate through a single orifice, and a similar type of equation can be used:

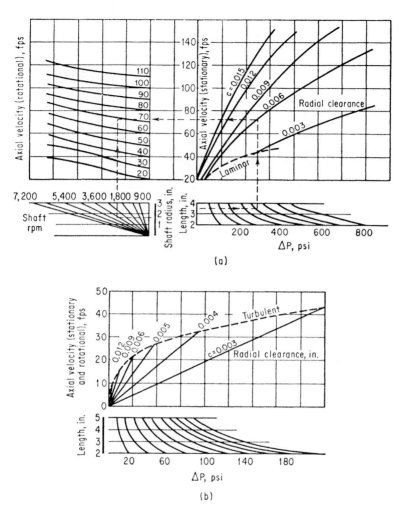

FIG. 17.45 Leakage past the labyrinth seal. (*a*) For turbulent flow of water at 68°F. (*b*) For laminar flow of water at 68°F.

$$M = \alpha A \phi P (g/RT)^{1/2} \tag{17.30}$$

where M = leakage rate, lb/s
   α = contraction coefficient
   A = flow area, in^2
   ϕ = pressure-ratio function
   P = absolute pressure at seal inlet, lb/in^2
   g = gravitational constant, lbm·in/lbf·s^2
   R = gas constant, lb·in/lbm·°R
   T = absolute temperature, °R

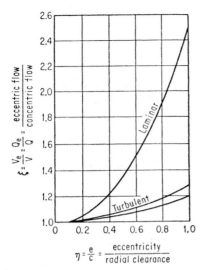

$$\eta = \frac{e}{c} = \frac{\text{eccentricity}}{\text{radial clearance}}$$

FIG. 17.46   Flow ratio vs. eccentricity ratio.

In order to use this equation effectively, proper expressions must be found for α and ϕ. α can be determined experimentally for a single-blade seal using the equation

$$M = \alpha AP(g/RT)^{1/2}[2K/(K-1)]^{1/2}r^{1/K}$$
$$\times (1 - r^{K-1/K})^{1/2} \qquad (17.31)$$

where K = gas specific-heat ratio
     r = pressure ratio P_a/P
     P_a = absolute pressure at seal outlet, lb/in^2

The correlation between single-blade data and multiple-blade seals can be made by means of Reynolds number:

$$N_{\text{Re}} = (W/\alpha)[2Ph/\mu(T)^{1/2}]\alpha \qquad (17.32)$$

where h = radial clearance, in

The pressure-ratio function is defined by the following equation from Egli:[12]

$$\phi = \left[\frac{1 - r^2}{N + (2/K)\ln(1/r)}\right]^{1/2} \qquad (17.33)$$

where N = number of blades

If W is now defined as the specific leakage rate in lbm·°R$^{1/2}$·s·lbf, or $W = G(T^{1/2})/P$, the basic leakage-rate equation can be rewritten as

$$W/\alpha = \phi(g/R)^{1/2} \qquad (17.34a)$$

Heffner's experimental data did not quite correlate with Eq. (17.34a); so it is rewritten as

$$W/\alpha = F(\phi) \qquad (17.34b)$$

A plot of this equation based upon experimental data will allow the determination of W/α for a specific value of ϕ calculated from Eq. (17.33). The value for W/α is then substituted in Eq. (17.32), which had previously been plotted for the experimental data. A value of α which will fit the experimental plot can then be determined and the specific leakage rate W can be calculated with W/α and α known.

Although the above summary of Heffner's work is probably not in enough detail to work out a design of a labyrinth seal, it does give an indication of the method and the complexity of the problem.

Equations (17.35) and (17.36), which will give at least an estimate of the number of rings, or the leakage rate for labyrinth seals, are given by Peickii and Christensen:[13]

$$N = \frac{(40P - 2600)(W/A)}{540(W/A) - P} \qquad (17.35)$$

where N = number of rings
     W = permissible leakage, lb/s
     A = cross-sectional flow area $C\pi D$, in^2

C = clearance, in
D = diameter, in
P = absolute pressure, lb/in^2

It is necessary to calculate N for the inlet and outlet pressures for a constant W/A and then subtract the two values to obtain the number of rings required to control the leakage at a definite rate with the specified inlet and outlet pressures.

The leakage rate can be calculated from the following equation if the number of rings is known:

$$W = 25KA\left\{\frac{P_1/V_1 - [1 - (P_2/P_1)^2]}{N - \ln(P_2/P_1)}\right\}^{1/2} \tag{17.36}$$

where W = flow rate, lb/h
V_1 = initial specific volume, ft^3/lb
K = experimental coefficient

For interlocking labyrinths $K \cong 55$ and is independent of clearance in the usual range. For noninterlocking types, K varies with the ratio of labyrinth spacing to radial clearance such that $K = 100$ for a ratio of 5 and $K = 60$ for a ratio of 50.

Additional labyrinth-seal theory and data may be found in Refs. 16 and 17.

17.11 BRUSH SEALS[36–41]

Brush seals are often a replacement for labyrinth seals. Brush seals can reduce flow leakage to 10 to 50 percent of that of the labyrinth seal, depending on the time in service. Brush seals can tolerate transient shaft movement without losing their sealing capacity. The heat generated by the brush seal is considerably lower than the heat generated by windage in a labyrinth seal. Also, repair procedures are very simple: brush elements can be easily replaced by new ones and the worn rotor coating can be relatively easily resprayed.

Figure 17.47 shows a double-element seal arrangement. The seal consists of many wire bristles sandwiched between a backing ring and a sideplate. The wire bristles protrude radially inward, forming a brush bore that fits around a mating shaft. The outer diameter of the sideplate, backing ring, and bristle ends are welded together and

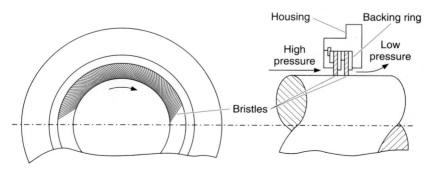

FIG. 17.47   Brush seal arrangement (double element).

then machined to form an outside diameter which can be fitted into a housing. For high-pressure sealing application (pressure drop more than 100 psi), two or three brush-seal elements can be placed together to prevent excessive bristle wear.

When making contact with the mating rotor, the bristles are angled rather than radial as shown in Fig. 17.47. This allows the bristles to bend without buckling while accommodating relative radial movement between the static brush and its mating rotor. The clearance between the backing ring, on the low-pressure side of each element, and the rotating shaft is kept to a minimum for better sealing.

Brush wire bristles range in diameter from 0.002 to 0.0028 in and have a density range from 2300 to 4500 wires per inch. The most commonly used material for brush seals is the cobalt-base alloy Haynes 25. The usual selection of the shaft counterface coating materials against which the brush will ride are ceramic coatings, chromium carbide, and aluminum oxides. Selecting the correct mating wire and shaft surface finish for a given application can reduce frictional heating and extend seal life through reduced oxidation and wear.

17.12   ROPE SEALS*

Advanced seal technology may employ rope seals to satisfy the extreme operating, high-temperature conditions found in supersonic vehicles. In this application, for slow-speed interface panel sealing, braided-rope seal made from high-temperature fiber materials allows the seals to operate at temperatures above 1500°F. With a coolant to purge the seals, the braided design is the most effective method for high-heat-flux environments. A hybrid braided-rope seal is fabricated with a dense, flow-resistant, lightweight core of ceramic fibers. These fibers are overbraided with a durable sheath made of fine-diameter super-alloy wires. The design employs many linear feet of rope to seal the gaps between movable surfaces (see Refs. 42 through 44).

REFERENCES

1. "Review and Bibliography on Aspects of Fluid Sealing," British Hydromechanics Research Association (BHRA) Cranfield, Bedford, England, 1972, and "Second Review and Bibliography on Aspects of Fluid Sealing" 1975.

2. "Rules for Construction of Unfired Pressure Vessels," American Society of Mechanical Engineers, New York, N.Y., 1952.

3. Whalen, J. J.: "How to Select the Right Gasket Material," *Prod. Eng.,* pp. 52–56, Oct. 3, 1960.

4. Roberts, I.: "Gaskets and Bolted Joints," *J. Appl. Mech.,* vol. 17, pp. 169–179, June 1950.

5. "Handbook of Mechanical Packings and Gasket Materials," Mechanical Packing Association, Washington, D.C., 1960.

6. Denny, D. F., and D. E. Turnbull: "Sealing Characteristics of Stuffing-box Seals for Rotating Shafts," Proceedings of the Institution of Mechanical Engineers, London, 174, no. 6, pp. 171–179, 1960.

7. Buchter, H. H.: "Industrial Sealing Technology," Wiley Interscience Publication, John Wiley & Sons, Inc., New York, 1979.

*By Bruce M. Steinetz.

8. McKillop, G. R.: "Compression Packings," Machine Design, Penton Publishing Company, Cleveland, Ohio, p. 64, Sept. 13, 1973.

9. Hayden, T. S., and Keller, C. H.: "Design Guide for Helicopter Transmission Seals," Sikorsky Aircraft NAS3-15684, NASA CR-120997, 1973.

10. Tao, L. N., and W. E. Donovan: "Through-Flow in Concentric and Eccentric Annuli of Fine Clearance with and without Relative Motion of Boundaries," *Trans. ASME,* vol. 77, no. 8, pp. 1291–1299, November 1955.

11. Heffner, F. E.: "A General Method for Correlating Labyrinth Seal Leak-Rate Data," *J. Basic Eng.,* pp. 265–275, June 1960.

12. Egli, A.: "The Leakage of Steam through Labyrinth Seals," *Trans. ASME,* vol. 57, pp. 115–122, 1935.

13. Peickii, V. L., and D. A. Christensen: "How to Choose a Dynamic Seal," *Prod. Eng.,* pp. 57–69, Mar. 20, 1961.

14. "The Seals Book," *Machine Design,* Jan. 19, 1961.

15. Williams, B. G.: "Mechanical Seals—A Survey of Present Day Practice—part 2," *Chem. & Process Eng.,* pp. 124–126, April 1955.

16. Zabriskie, W., and B. Sternlicht: "Labyrinth-Seal Leakage Analysis," *J. Basic Eng.,* vol. 81, ser. D, no. 3, September 1959.

17. Kearton, W. J.: "The Flow of Air through Radial Labyrinth Glands," *Proc. Inst. Mech. Engrs. (London),* vol. 169, 1955.

18. McCuistion, T. J.: "O Ring and Gland Design for High Temperature Seals," *Prod. Eng.,* pp. 151–155, January 1956.

19. Wilkinson, J.: U.S. Patent 1,063,633, June 1913.

20. Young, L. A., and Lebeck, A. O.: "Experimental Evaluation of a Mixed Friction Hydrostatic Mechanical Face Seal Model Considering Radial Taper, Thermal Taper and Wear," *Trans. ASME, J. Lub. Technol.,* vol. 104, no. 4, pp. 439–448, October 1982.

21. Burton, R. A., ed.: "Thermal Deformation in Frictionally Heated Systems," Elsevier Sequoia, 1980.

22. Hughes, W. F., and Chao, N. H.: "Phase Change in Liquid-Face Seals II—Isothermal and Adiabatic Bounds with Real Fluids," *Trans. ASME J. Lub. Technol.,* vol. 102, no. 3, pp. 350–359, July 1980.

23. Martinon, A. R.: "Design, Development and Testing of Large-Diameter, High Pressure Seals for Nuclear-Reactor, Primary-Coolant Pumps—A Challenge to the Pump Manufacturer," *Lubrication Engineering,* vol. 36, no. 6, pp. 325–340, 1980.

24. Ludwig, L. P., and Greiner, H. F.: "Designing Mechanical Face Seals for Improved Performance. Part 1," Basic Configurations, *Mechanical Engineering,* pp. 38–46, November 1978, and Part 2 Lubrication, *Mechanical Engineering,* pp. 18–23, December 1978.

25. Mayer, E.: "Mechanical Seals," Newnes-Butterworth, 1977.

26. Engineered Fluid Sealing, Crane Packing Co., Morton Grove, Ill., 1976.

27. Hugobuchter, H.: "Industrial Sealing Technology," John Wiley & Sons, Inc., New York, 1979.

28. Johnson, R. L., and Schoenherr, K.: Seal Wear, "Wear Control Handbook," New York, ASME, pp. 727–753, 1980.

29. Abar, J. W.: Failure of Mechanical Face Seals, "Metals Handbook," ASME, New York, 8th ed., vol. 10, pp. 437–439, 1975.

30. Austin, R. M., and Nau, B. S.: "The Seal Users Handbook," BHRA Fluid Engineering, 1974.

31. Neale, M. J., ed.: "Tribology Handbook," Butterworth, 1973.

32. Liquid Rocket Engine Turbopump Rotating-Shaft Seals, NASA SP-8121, February 1978.

33. Etsion, I.: "Dynamic Analysis of Noncontacting Face Seals," *Trans. ASME, J. Lub. Technol.,* vol. 194, no. 4, pp. 460–468, October 1982.

34. Shapiro, W., and Colsher, R.: "Steady State and Dynamic Analysis of a Jet Engine Gas-Lubricated Shaft Seal," *ASLE Trans.,* vol. 17, no. 3, pp. 190–200, July 1974.

35. Etsion, I.: "A Review of Mechanical Face Seal Dynamics," *The Shock and Vibration Digest,* vol. 14, no. 3, pp. 9–14, March 1982.

36. Atkinson, E., and B. Bristol: "Effects of Material Choices on Brush Seal Performance," *Lubrication Engineering,* vol. 48, no. 9, pp. 740–746, September 1992.

37. Steinetz, B. M., and R. C. Hendricks: "Engine Seals," chap. 9 in "Tribology for Aerospace Applications," E. V. Zaretsky, ed., to be published by Society of Tribologist and Lubrication Engineers, 1994.

38. Ferguson, J. G.: "Brushes as High Performance Gas Turbine Seals," ASME Paper 88-GT-182, presented at the Gas Turbine and Aero Engine Congress, Amsterdam, The Netherlands, June 6–9, 1988.

39. Flower, R.: "Brush Seal Development System," AIAA Paper AIAA 90-2143, presented at AIAA/SAE/ASME/ASEE 26th Joint Propulsion Conference, July 1990.

40. Holle, G. F., and M. R. Krishnan: "Gas Turbine Brush Seal Applications," AIAA Paper AIAA 90-2142, presented at AIAA/SAE/ASME/ASEE 26th Joint Propulsion Conference, July 1990.

41. Chupp, R. E., and Lt. P. Nelson: "Evaluation of Brush Seals for Limited-Life Engines," AIAA Paper AIAA 90-2140, presented at AIAA/SAE/ASME/ASEE 26th Propulsion Conference, July 1990.

42. Steinetz, B. M., and R. C. Hendricks: "Engine Seal Technology Requirements to Meet NASA's Advanced Subsonic Technology Goals," AIAA Paper AIAA-94-2698, presented at AIAA/SAE/ASME/ASEE 30th Joint Propulsion Conference, Indianapolis, Ind., June 1994.

43. Steinetz, B. M., L. Kren, and Z. Cai: "High Temperature Flow and Durability Assessments of Hypersonic Engine Seals," to be published as NASA TP-3483, 1994.

44. Munson, J., "Testing of a High Performance Compressor Discharge Seal," AIAA Paper 93-1997, presented at AIAA/SAE/ASME/ASEE 29th Joint Propulsion Conference, June 1993.

SECTION 18

FLUID-FILM BEARINGS*

David P. Fleming, Ph.D., P.E.

Senior Research Engineer
NASA Lewis Research Center
Cleveland, Ohio

SYMBOLS

A_R = bearing characteristic number [Eq. (18.12)]

A = area, m² (in²)

a = restrictor area, m² (in²)

b_1 = distance from inlet to step, m (in)

b_2 = distance from step to outlet or exhaust, m (in)

C_D = orifice discharge coefficient

c = bearing radial clearance, m (in)

D = bearing diameter, m (in)

DN = shaft bore diameter, mm, multiplied by speed, rpm

d = orifice diameter, m (in)

e = bearing eccentricity, m (in)

F_μ = friction force per unit length, N/m (lbf/in)

H = power loss or pumping loss, W (in-lbf/s)

h = film thickness, m (in)

h_0 = minimum film thickness, m (in)

h_1, h_2 = clearances in b_1 and b_2 regions, respectively, m (in)

K = thrust bearing parameter, $6\mu\omega(R_2/h_2)^2$

k = ratio of specific heat at constant pressure to that at constant volume

L = bearing length, m (in)

*Portions of this section are reprinted with permission from Brewe, David E., Fleming, David P., Dimofte, Florin, and Hendricks, Robert C., "Fluid and Gas Film Lubrication," in *Tribology for Aerospace Applications,* E. V. Zaretsky, ed., Society of Tribologists and Lubrication Engineers, Park Ridge, Ill., 1995.

18.1

l_c = length of element or flow element in direction of flow, m (in)

M = mass flow rate, kg/s (lbm/s)

m = dimensionless ratio, 2000 c/R

N = shaft speed, rpm

P = load per unit area, Pa (psi)

p = pressure, Pa (psi)

p_i = intermediate (downstream) pressure

p_s = supply pressure

Q = oil flow rate, m³/s (in³/s)

R = rotor radius, m (in)

R_1 = inner bearing (pad) radius, m (in)

R_2 = outer bearing (pad) radius, m (in)

Re = Reynolds number

$\mathcal{R}$ = gas constant

s = minimum clearance region in step bearing, m (in)

T = temperature

t = time, s

U = velocity of moving bearing surface, m/s (in/s)

W = load, N (lbf)

x, y, z = coordinates (x, direction of motion; y, normal to direction of motion; z, normal to bearing surfaces)

Z = absolute viscosity, cP

β = helix angle; angular extent of pad

γ = pitch angle

Δ = step or recess depth, m (in)

ε = eccentricity ratio, e/c, dimensionless

η = y/R

θ = angular coordinate

Λ = journal bearing compressibility number, $6\mu\omega R^2/p_a c^2$

Λ_s = bearing compressibility number, $6\mu\omega R^2/p_a s^2$

Λ_Δ = bearing compressibility number $6\mu\omega R^2/p_a \Delta^2$

Λ_0 = orifice-restrictor bearing parameter (Fig. 18.18)

μ = viscosity, N·s/m² (lbf·s/in²)

ν = kinematic viscosity, m²/s (in²/s)

ρ = mass density, kg/m³ (lbf·s²/in⁴)

ϕ = attitude angle, degrees

ω = angular velocity, rad/s

Subscripts

a = atmospheric or ambient

cp = center of pressure

p = pitch line

s = supply

0 = initial or reference

18.1 FLUID-FILM BEARING TYPES

Fluid-film bearings have been successfully used for millennia, and may well be the most used machine element in our civilization. Today large numbers of fluid-film bearings are used in a myriad of applications from small electric motors, to automobile and aircraft piston engines, to large steam turbines for electric power generation. Properly designed and lubricated, these bearings are very reliable and offer lives measured in decades for some applications. The advantages and limitations of fluid-film bearings are listed in Table 18.1.

TABLE 18.1 Advantages and Limitations of Fluid-Film Bearings

		Bearing type			
Condition	General comments	Hydrodynamic fluid-film bearings	Hydrostatic fluid-film bearings	Self-acting gas bearings	Externally pressurized gas bearings
High temperature	Attention to differential expansions and their effect on axial clearance is necessary	Attention to oxidation resistance of lubricant is necessary		Excellent	Excellent
Low temperature	Attention to differential expansions and starting torques is necessary	Lubricant may impose limitations; consideration of starting torque is necessary	Lubricant may impose limitations	Excellent; thorough drying of gas is necessary	
External vibration	Attention to possibility of fretting damage is necessary (except for hydrostatic bearings)	Satisfactory	Excellent	Normally satisfactory	Excellent
Space requirements	—	Small radial extent but total space requirement depends on lubricant feed system		Small radial extent	Small radial total space requirement depends on gas feed system
Dirt or dust	—	Satisfactory; filtration of lubricant is important		Sealing is important	Satisfactory
Vacuum		Lubricant may impose limitations		Not normally applicable	Not applicable when vacuum has to be maintained
Wetness and humidity	Attention to possibility of metallic corrosion is necessary	Satisfactory		Satisfactory	Satisfactory
Radiation	—	Lubricant may impose limitations		Excellent	Excellent
Low starting torque	—	Satisfactory	Excellent	Satisfactory	Excellent
Low running torque	—	Satisfactory	Satisfactory	Excellent	Excellent
Accuracy of radial location	—	Good	Excellent	Good	Excellent
Life	—	Theoretically infinite but affected by filtration and number of start-stops	Theoretically infinite	Theoretically infinite but affected by number of start-stops	Theoretically infinite

Fluid-film bearings may be classified by the intended function: journal bearings support a shaft in the radial direction, and thrust bearings restrain a shaft in the axial direction. Bearings may also be classified according to the method of fluid pressure development: in hydrodynamic (self-acting) bearings, the pressure is generated by the motion of the bearing acting on a wedge-shaped film; in hydrostatic (externally pressurized) bearings the film pressure is supplied from outside the bearing by a high-

TABLE 18.1 Advantages and Limitations of Fluid-Film Bearings (*Continued*)

Condition	General comments	Bearing type			
		Hydrodynamic fluid-film bearings	Hydrostatic fluid-film bearings	Self-acting gas bearings	Externally pressurized gas bearings
Combination of axial and radial load capacity	—	A journal bearing surface must be provided to carry radial loads and a thrust face to carry axial loads			
Silent running	—	Excellent	Excellent except for possible pump noise	Excellent	Excellent except for possible compressor noise
Simplicity of lubrication	—	Self-contained assemblies can be used with certain limits of load, speed, and diameter; beyond this, oil circulation is necessary	Auxiliary high pressure is necessary	Excellent	Pressurized supply of dry, clean gas is necessary
Availability of standard parts	—	Good	Poor	Poor	Poor
Prevention of contamination of product and surroundings	—	Normally satisfactory, but attention to sealing is necessary, except where process liquid can be used as lubricant		Excellent	Excellent
Tolerance to manufacturing and assembly inaccuracies	—	Poor	Satisfactory	Poor	Satisfactory
Type of motion					
Frequent start-stops	—	Good	Excellent	Poor	Excellent
Unidirectional	—	Suitable	Suitable	Suitable	Suitable
Bidirectional	—	Some types are suitable	Suitable	Some types are suitable	Suitable
Oscillatory	—	Unsuitable	Suitable	Unsuitable	Suitable
Running costs	—	Depends on complexity of lubrication system	Cost of lubricant supply has to be considered	Nil	Cost of gas supply has to be considered

Source: From Ref. 55.

pressure lubricant source. Finally, bearings may be classified by the type of lubricant: oil or another incompressible lubricant, or a compressible lubricant (e.g., air or liquid hydrogen).

The load capacity of any fluid-film bearing is determined by the fluid pressure acting on the effective area of the bearing. Substantial pressure can be generated within a hydrodynamic bearing by the relative motion of bearing surfaces. However, hydrostatic bearings often have greater load capacity than hydrodynamic bearings; also, with external pressurization a load can be supported with no relative motion of the bearing surfaces. This is ideal for providing near frictionless positioning bearings. On the other side, a source of high-pressure lubricant is required; the lubricant supply pressure must generally be about twice the load pressure supported.

18.1.1 Journal Bearings

In its simplest form a journal bearing may be a bushing or a reamed hole in a machine, supporting a rotating shaft as shown in Fig. 18.1. If the bearing extends completely around the shaft, it is called a full journal bearing. If the angle of wrap is less than

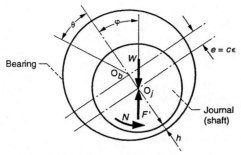

(a) Bearing operating configuration.

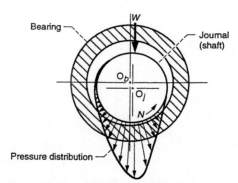

(b) Circumferential pressure distribution.

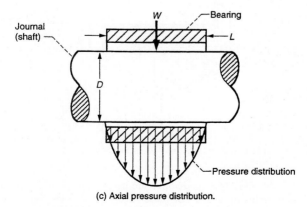

(c) Axial pressure distribution.

FIG. 18.1 Hydrodynamic journal bearing. (*From Bisson and Anderson, Ref. 11.*)

360°, it is called a partial journal bearing; this may be used when the load always acts in one direction.

The following paragraphs describe oil- or grease-lubricated hydrodynamic journal bearings commonly used in machinery. The lubricant may be supplied by several means.

Sintered materials impregnated with oil (e.g., Oilite bronze) are often used for bushings; these bearings usually operate in the boundary lubrication regime rather than in the hydrodynamic regime.

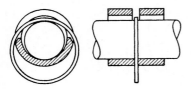

FIG. 18.2   Ring-oiled bearing. (*From Wilcock and Booser, Ref. 56.*)

For horizontal shafts, an oil ring (Fig. 18.2) which is rotated by contact with the shaft transfers oil from an oil reservoir beneath the shaft to the top of the shaft. Grooves may be used to distribute the oil along the bearing.

In wick-oiled bearings (Fig. 18.3) an oil-saturated wick rubbing against the shaft supplies oil to the bearing. Seals retain the oil within the bearing, and slingers return oil to a porous packing material which transmits oil to the wick.

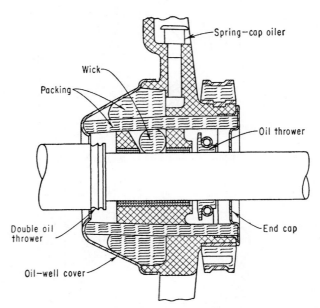

FIG. 18.3   A typical wick-oiled bearing installation for fractional-horsepower electric motors. (*From Wilcock and Booser, Ref. 56.*)

The lubrication schemes described above are used primarily for bearings operating at low speeds with moderate loads. For more severe conditions, the lubricant is pressure-fed to the bearing via holes or grooves (Fig. 18.4). This is not the same as hydrostatic lubrication; lubricant supply pressures for hydrodynamic bearings are typically

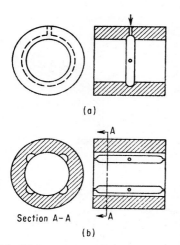

FIG. 18.4 Pressure-fed bearings. (*a*) Circumferential-groove. (*b*) Multiple-groove. (*From Wilcock and Booser, Ref. 56.*)

1 to 5 bar (15 to 75 psig), which is considerably less than the load pressures supported. A pressure-fed system may include oil supply and return lines, a storage tank, filter, cooler, and pressure regulators.

Other lubrication schemes include hand oiling from an oil can, cup oiling, bath lubrication (immersion), splash lubrication, and, for very low speeds, grease lubrication.

Grooves are often used to distribute the oil within the bearing. Generally, (1) grooves should not connect high-pressure areas to low-pressure areas of the oil film, (2) the lubricant should be introduced in a low-pressure region, and (3) a minimum amount of grooving should be used. In particular, grooves should be used sparingly in the high-pressure region of the bearing to avoid a reduction in load capacity.

18.1.2 Thrust Bearings

The simplest form of thrust bearing consists of two opposed washers with some means of supplying lubricant to the interface. However, lubrication theory shows that such an arrangement does not develop any hydrodynamic load capacity; the loads carried in practice are assumed to be due to imperfections in the flat surfaces. Thus for any substantial load capacity, one of the surfaces must be contoured in some manner that allows a converging film to occur. Commonly used fixed-geometry thrust bearings make use of tapered lands, Rayleigh steps, and spiral grooves (Fig. 18.5). Types shown in Fig. 18.5a through 18.5d are used as several discrete pads, while spiral and herringbone grooves are placed on continuous rings. Pivoted pads are also commonly used in thrust bearings (Fig. 18.6).

18.1.3 Hydrodynamic Bearings

Pressure in a hydrodynamic bearing is generated by relative motion of the bearing surfaces. Full hydrodynamic lubrication occurs only when the lubricant film thickness is greater than the asperity heights of the mating surfaces. The parameter which indicates whether hydrodynamic lubrication can occur in journal bearings is the grouping (viscosity × speed ÷ load pressure), often denoted *ZN/P*. Figure 18.7 shows the relationship between this grouping and the friction coefficient. For small values of *ZN/P*, the thin-film boundary region from A to B, rubbing between surface asperities causes most of the friction. In the transition region from B to C, viscous forces begin to prevail and rubbing between asperities diminishes. The location of the minimum friction coefficient region depends on the surface finish of the bearing materials; a smoother surface can operate at lower values of *ZN/P* without asperity contact. In the region from C to D, true hydrodynamic lubrication is established; there is no asperity contact, and the friction coefficient does not depend on surface conditions. With no surface-to-surface contact, bearing life is theoretically infinite; this is the region for which bearings are usually designed.

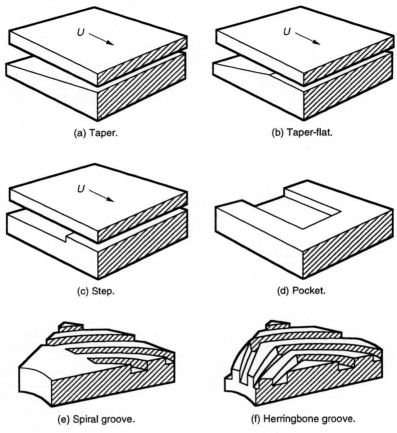

(a) Taper.   (b) Taper-flat.

(c) Step.   (d) Pocket.

(e) Spiral groove.   (f) Herringbone groove.

FIG. 18.5   Contoured shapes for thrust bearings. (*From Ausman, Ref. 37.*)

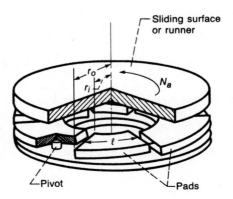

FIG. 18.6   Pivoted-pad thrust bearing. (*From Raimondi and Boyd, Ref. 57.*)

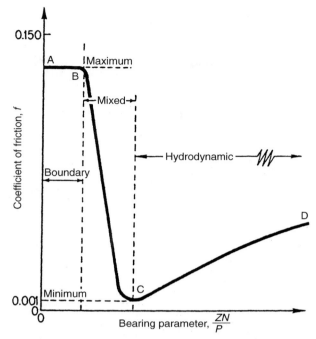

FIG. 18.7   Variation of coefficient of friction with *ZN/R*. (*From Rippel, Ref. 10.*)

Consider the tapered-land thrust bearing (Fig. 18.5*a*). Pressure is generated because more fluid enters the leading edge than can leave the smaller exit area at the trailing edge. This fluid pressure between the bearing surfaces causes them to remain separated. The pressure can be increased and the film thickness maintained to accommodate a higher load by increasing either the fluid viscosity or the relative speed of the runner.[2]

In journal bearings (Fig. 18.1) the shaft reaches an eccentric position under load, and a fluid film, tapered in the direction of motion, is created. Fluid pressures are generated in the film by the resistance to outflow of the oil dragged into the converging space by the relative motion of the journal and the bearing. Such a hydrodynamic journal bearing has circumferential flow of fluid that remains in the bearing and side flow of fluid that is lost from the bearing. The side flow removes heated lubricant from the bearing.[2]

In dynamically loaded bearings the relative normal motion of the shaft to the bearing through the line of centers (Fig. 18.8) creates an additional contribution to the load capacity. The pressures depicted in this figure are a result of the squeezing action caused by the normal approach of shaft and bearing. Conversely, negative pressures are created if the shaft and the bearing are receding from each other. Negative pressure can release dissolved gases from the liquid and/or vaporize the liquid to cause cavitation.

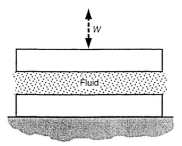

(a) Plane surfaces.

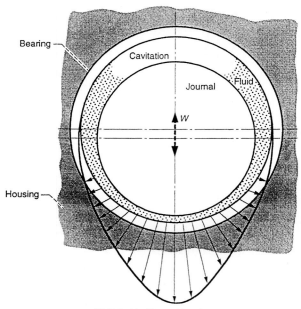

(b) Cylindrical journal bearing.

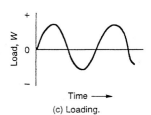

(c) Loading.

FIG. 18.8   Dynamically loaded squeeze-film bearing.

18.1.4 Hydrostatic Bearings

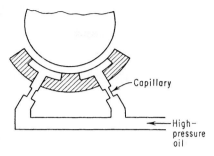

FIG. 18.9   Hydrostatic bearing.

The load capacity of hydrostatic bearings (Fig. 18.9) depends principally on some source of pressure external to the bearings—usually a pump. More than one bearing is normally supplied by the same pump, and some form of hydraulic restriction or flow control is required between the pressure source and the bearing film.[2]

In operation the loaded member is raised by the pressure in the recess acting on the bearing surface until flow through the restriction is equal to flow from the recess. The integrated product of pressure and bearing area is then equal to the applied load. In this way bearing clearance changes will accommodate load changes. Such hydrostatic bearings can be designed to have extreme accuracy with very little shaft motion, even with large load changes. Because hydrostatic bearings do not rely on relative motion to support loads, they are ideally suited for force-measuring equipment, precision machine tools, and laboratory instruments. Their high stiffness (resistance to clearance change under load) and good damping, coupled with extremely good rotational accuracy, result in very accurate use as machine-tool spindles.[2]

18.1.5 Gas-Lubricated Bearings

Because gases have viscosity, they can be used in hydrodynamic and hydrostatic bearings to support loads in essentially the same manner as any other fluid. Although they offer the advantage of low friction over wide ranges of speed and temperature, they have their own particular problems. Relative to liquid lubricants, gases have low viscosity, are compressible, and offer little, if any, value as a boundary lubricant. Because their viscosity is low, hydrodynamic gas-lubricated bearings operate with thin films and thus require careful control of bearing geometry and surface finish. The compressibility of gases limits the maximum pressure that can be generated within a bearing. From a practical standpoint maximum unit pressure of hydrodynamic gas bearing operating with 1 atm ambient pressure is about 350 kPa (50 psi), based on the projected area of the bearing. Hydrodynamic gas bearings have been run above 500,000 rpm with small-diameter spindles.[2]

The load capacity of hydrostatic gas bearings is limited primarily by the supply pressure and the need to prevent a bearing instability known as pneumatic hammer. This instability occurs often enough that an experimental development program is often needed, in addition to a design analysis, for many gas-bearing applications.[2]

18.2   FLUID-FILM PRESSURE DEVELOPMENT

Osborne Reynolds[3] first developed the equation which has been named for him relating viscosity, film thickness, pressure, and sliding velocity. For compressible or incompressible lubricants, the Reynolds equation is[4]

$$\frac{\partial}{\partial x}\left(\frac{\rho h^3}{12\mu}\frac{\partial p}{\partial x}\right) + \frac{\partial}{\partial y}\left(\frac{\rho h^3}{12\mu}\frac{\partial p}{\partial y}\right) = \frac{\partial}{\partial x}\left[\frac{\rho h(U_1 + U_2)}{2}\right] + \frac{\partial(\rho h)}{\partial t} \tag{18.1}$$

In deriving the Reynolds equation the following assumptions were made:

1. The flow is laminar.
2. Inertia forces are small relative to viscous shear forces and are neglected. (Reynolds number is low.)
3. No external forces (e.g., gravity) act on the fluid.
4. No slip exists at the fluid-solid interface.
5. The film thickness is much smaller than any radius of curvature or transverse bearing dimension.
6. The pressure across the film is constant, i.e., $\partial p/\partial z = 0$.

An additional assumption made in Eq. (18.1) is that the sliding velocities of the two bearing surfaces are collinear. This last assumption can easily be eliminated.[4] The Reynolds equation can be applied to determine the film pressure if the assumptions above hold. Equation (18.1) is in cartesian coordinates; appropriate transformations can be made to suit the bearing geometry if needed. Solving for the pressures requires knowledge of the boundary conditions suited to the application. The pressures can then be expressed as a function of viscosity, film shape, and relative motion of the bearing surfaces. Quite often, only one surface of the bearing is moving; in this case $U_1 = U$ and $U_2 = 0$.

No exact solution of the generalized form of the Reynolds equation has been obtained. Approximate solutions for specific conditions have been obtained with infinite series and relaxation techniques, and accurate numerical solutions have been obtained with computer codes. An exact solution is possible for two specialized cases: (1) the bearing is much wider in the transverse direction than in the direction of motion (the so-called infinite bearing) or (2) for journal bearings, the bearing is short enough that the pressure-induced circumferential flow is negligible (the short-bearing solution).

In addition to the load capacity obtained by solving the Reynolds equation, an expression for the power loss in a bearing is needed to completely define the bearing characteristics. The friction force F_μ of the bearing can be calculated from the following equation:

$$F_\mu = \pm \iint \frac{\mu U}{h} \, dx \, dy - \frac{1}{2}\iint (p - p_a)\frac{\partial h}{\partial x} \, dx \, dy \tag{18.2}$$

The power loss is then

$$H = F_\mu U \tag{18.3}$$

18.3   JOURNAL BEARINGS

18.3.1   Steady-State Loading

Steady-state operation occurs when the bearing and shaft centers remain fixed relative to each other during operation. Thus the transient term in Eq. (18.1) is zero. To obtain the pressure, the film shape must first be defined and appropriate boundary conditions applied upon integration of the Reynolds equation. From Fig. 18.1 the film thickness h can be expressed as

$$h = c(1 + \epsilon \cos \theta) \tag{18.4}$$

where the *eccentricity ratio* $\epsilon = e/c$. The film is convergent for $0 \leq \theta \leq \pi$ and divergent for $\pi \leq \theta \leq 2\pi$. Pressure builds up in the convergent region because fluid is being convected into a constricting space. This pressure buildup reaches a maximum near the minimum film thickness and within the convergent region.

As mentioned above, the steady-state Reynolds equation has been solved analytically for only a few cases. Solutions have been obtained by using the infinitely long bearing approximation[5,6] and the finite-short-bearing approximation.[7]

Long-Bearing Solution. The infinitely long bearing assumes that side flow is negligible (i.e., $\partial p/\partial y = 0$.) With this assumption and the assumption of constant density and zero normal velocity, the Reynolds equation, Eq. (18.1), can be written in polar coordinates as

$$\frac{d}{d\theta}\left[h^3 \frac{dp}{d\theta} \right] = 6\mu R^2 \omega \frac{dh}{d\theta} \tag{18.5}$$

where $\theta = x/R$. The Sommerfeld[5] noncavitating boundary conditions are

$$p = 0 \text{ at } \theta = 0, 2\pi \tag{18.6}$$

Equation (18.5) can be integrated twice over the circumference of the bearing; the integration constants are determined from the boundary conditions. The resulting pressure solution turns out to be antisymmetric about $\theta = \pi$ and predicts negative pressures in one half of the bearing equal in magnitude to the positive pressures in the other half. For various reasons, negative pressures rarely exist in bearings; therefore it is common to simply ignore any pressures less than zero. This yields the so-called *half-Sommerfeld solution*. To determine the load carried by the bearing, the positive pressure is integrated around the bearing circumference to obtain load components parallel to and perpendicular to the line of centers. These components are then combined to obtain the total load. Results are[4]

$$\frac{W}{\mu \omega RL}\left(\frac{c}{R}\right)^2 = \frac{6\epsilon[\pi^2 - \epsilon^2(\pi^2 - 4)]^{1/2}}{(2 + \epsilon^2)(1 - \epsilon^2)} \tag{18.7}$$

$$\phi = \tan^{-1}\left[\frac{\pi}{2\epsilon}(1 - \epsilon^2)^{1/2}\right] \tag{18.8}$$

where ϕ, known as the *attitude angle,* is the angle between the bearing load and the line of centers (see Fig. 18.1).

There are valid objections to merely ignoring the negative pressures, and considerable work has been done to obtain better solutions. However, Dowson and Taylor[8] state that the half-Sommerfeld solution may not be in error by more than 15 percent when compared with more precise solutions. A great deal of work has been done to accurately determine cavitation boundaries; Ref. 1 presents a summary of this work.

Short-Bearing Solution. Most journal bearings have their length shorter than their diameter; for this case the long-bearing solution is not accurate. DuBois and Ocvirk[7] derived a solution of the Reynolds equation applicable for bearings whose length is much shorter than their diameter; consequently the pressure-induced circumferential flow can be neglected relative to the other terms. The Reynolds equation becomes

$$\frac{\partial}{\partial y}\left(h^3 \frac{\partial p}{\partial y}\right) = 6\mu\omega \frac{\partial h}{\partial \theta} \tag{18.9}$$

This is readily integrated to determine the film pressure; which in turn is integrated to determine the bearing load. Again utilizing only positive pressures, the results are[4]

$$\frac{W}{\mu\omega RL}\left(\frac{c}{R}\right)^2 = \left(\frac{L}{R}\right)^2 \frac{\epsilon}{4(1-\epsilon^2)^2}[16\epsilon^2 + \pi^2(1-\epsilon^2)]^{1/2} \tag{18.10}$$

$$\phi = \tan^{-1}\left[\frac{\pi(1-\epsilon^2)^{1/2}}{4\epsilon}\right] \tag{18.11}$$

The results for both long and short bearings indicate that the load capacity increases without limit as the eccentricity ratio ε approaches 1. However, bearing surfaces are not perfectly smooth, and asperity contact will occur at an eccentricity ratio below 1; also, heat generation may be excessive when film thickness is very small. Hamrock[4] notes that most journal bearings operate with maximum eccentricity ratios less than 0.8.

The short-bearing solution is considered usable for $L/R \le 1$.[4] Longer bearings require use of approximations or numerical methods. The design of a journal bearing for the steady state usually involves optimizing clearance, bearing length, lubricant inlet slots, and oil flow rate. Hydrodynamic equation solutions provide pressure distributions in the bearing and the resulting minimum oil-film thickness for given loads. However, manipulating the equations is tedious and time consuming unless computer codes are used;[9] such codes are now commercially available for personal computers from several vendors. Traditionally, journal bearings and other types of fluid-film bearings have been designed by using several types of graphs developed from theory, such as that by Rippel[10] shown in Fig. 18.10. Such methods are adequate for most steady-state applications. To use this figure, one first calculates the bearing *character-istic number* A_R:

$$A_R = 1.45 \times 10^{-7} \frac{m^2 W}{D^2 \mu N} \tag{18.12}$$

where $m = 2000\,c/D$, i.e., 1000 times the clearance ratio. Any consistent units may be used in this equation except for N, which is in rpm; the numerical multiplier in Eq. (18.12) appears because as originally presented A_R was defined with inconsistent units. The feasibility of hydrodynamic operation can be estimated by inspecting the magnitude of A_R. If within the bearing size limitations the value of A_R falls between 0.0005 and 0.50, full film lubrication is feasible.[10]

When the bearing characteristic number is known, the L/D ratio and the eccentricity ratio ε can be determined from Fig. 18.10. The minimum film thickness can be obtained from Eq. (18.4) for $\theta = \pi$, that is,

$$h_0 = c(1 - \varepsilon) \tag{18.13}$$

By using the recommended eccentricity ratio line, bearing dimensions can be found for various operating conditions.

Bearing power loss may be estimated from[11]

$$H = \frac{\pi}{4} \frac{\mu\omega^2 L D^3}{c\sqrt{1-\epsilon^2}} \tag{18.14}$$

EXAMPLE 1   Select the proper length bearing to carry a 2000-N radial load at 3000 rpm. The diameter of the machine shaft is 50 mm; it is desirable to use this as the bearing journal. The available lubricant is an SAE 20 oil which has a viscosity of 0.007 Pa·s at the expected operating temperature of 90°C. Also calculate the bearing power loss.

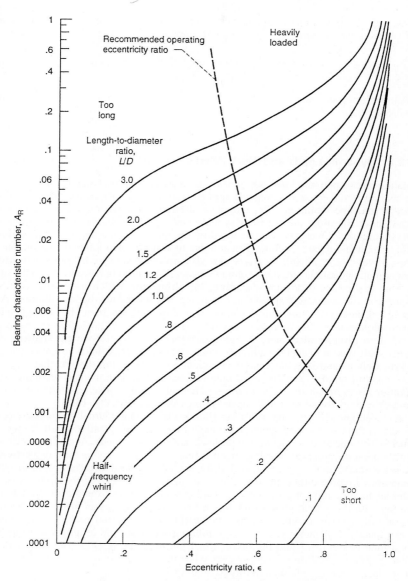

FIG. 18.10   Design chart for hydrodynamic journal bearings. (*From Rippel, Ref. 10.*)

solution   A typical bearing clearance ratio c/R is 0.001; try this value. The bearing radial clearance is then 0.025 mm and the parameter $m = 1$. From Eq. (18.12),

$$A = \frac{1.45 \times 10^7(2000)}{(0.05)^2(0.007)(3000)} = 0.0055 \tag{18.15}$$

Entering Fig. 18.10 on the recommended operating eccentricity ratio line we find L/D = 0.5 and ε = 0.65; thus the proposed bearing diameter and clearance are feasible. The bearing length is then 25 mm and the minimum film thickness, by Eq. (18.13), 0.009 mm. The power loss is calculated from Eq. (18.14); note that in this formula the shaft speed must be expressed in radians per second.

$$H = \frac{\pi}{4} \frac{0.007(3000\pi/30)^2(0.025)(0.05)^3}{25 \times 10^{-6}\sqrt{1 - 0.65^2}} = 90 \text{ W} \tag{18.16}$$

Although not addressed here, Rippel[10] also gives information to calculate oil flow requirements.

18.3.2   Dynamic Loads and Shaft Whirl

Rotating and/or periodic loads are typical examples of bearing dynamic loads. Even if the bearing is subjected to a constant applied load, it can experience dynamic loading through machine vibrations. Depending on the ratio of vibration frequency to shaft speed, these vibrations are either subsynchronous, synchronous, or supersynchronous. Subsynchronous vibrations can be caused by the journal interacting with the working fluid (e.g., half-frequency whirl). Friction due to seal rubs and internal friction due to stress reversals in shrink-fit parts can also produce subsynchronous vibration. Synchronous vibrations are most commonly caused by an imbalance in the shaft, such as from blade loss in a turbine engine. Misalignment, which normally creates a response at twice the shaft speed, is an example of a supersynchronous vibration. Any number of these vibration sources can occur in a rotating machine, thus complicating the procedures necessary to achieve smooth operation.

Modern rotating machinery is being designed with lighter-weight shafting and higher shaft speeds to achieve a higher power density. These conditions are conducive to synchronous and subsynchronous vibration. Several bearing designs that incorporate antiwhirl features are discussed below. Another effective method of suppressing vibrations is to mount the rotor on rolling-element bearings in series with nonrotating journal bearings (squeeze-film dampers) to provide damping (Fig. 18.8).

It is well known that plain, full-fluid-film, hydrodynamic journal bearings are readily susceptible to the self-generated instability commonly called "half-frequency whirl" when operated at relatively light loads and/or high speeds. According to Bisson and Anderson,[11] journal-bearing instabilities occur most frequently in vertically mounted shafts and, in general, in lightly loaded bearings. Efforts to inhibit whirl sometimes involve either creating artificial loads in the bearing to supplement the external load or reducing the bearing load capacity. As an example of the latter, a bearing with three axial grooves inhibits whirl better than does an ungrooved bearing because the grooves essentially make the bearing a 120° partial bearing with a lower load capacity. Still more effective in inhibiting whirl are the elliptical bearing, the three-lobe bearing,[12,13] and the pivoted-pad journal bearing (Fig. 18.11). All these bearings provide built-in oil wedges even when there is no external load. Pinkus[14] discusses these bearings and their ability to inhibit whirl.

Early designs used a centrally lobed bearing having an equally distributed converg-

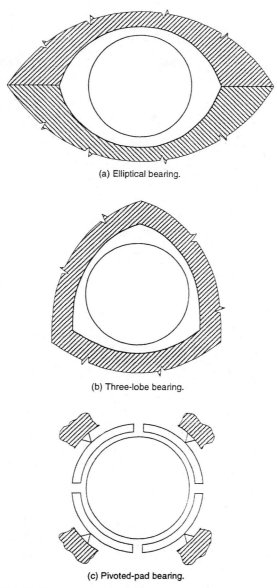

(a) Elliptical bearing.

(b) Three-lobe bearing.

(c) Pivoted-pad bearing.

FIG. 18.11   Three types of whirl-resistant journal bearings. (*From Bisson and Anderson, Ref. 11.*)

ing-diverging film. However, only the converging wedge region of the lobe generates pressure to support a load. Chadbourne et al.[15] showed that tilting the lobe toward the shaft at the trailing edge results in a larger converging wedge region, greater load capacity, and improved stability.

Bearings designed to include steps and grooves so that the films are discontinuous are also known to be more stable.[16] Good examples are the herringbone-groove bearing[17,18] and the Rayleigh step bearing[19]—both fixed-geometry bearings.

Pivoted-pad bearings are somewhat similar to lobe bearings but are more complex and have a variable geometry. These bearings are exceptionally stable, but they may be subject to pivot surface damage.[20] Schuller and Anderson tested pivoted-pad bearings in liquid sodium and found they were more stable than any of the fixed-geometry bearings tested.[20]

Schuller et al.[21–24] investigated the stability characteristics of several fixed-geometry journal-bearing configurations, using water as the lubricant, to provide design information for space power applications using liquid sodium. Their results should indicate the relative stability of various bearing types for other lubricants as well.

The three-tilted-lobe bearing with grooves (offset factor of 1.0) performed best. In order of diminishing stability the other bearings ranked as follows: herringbone-groove journal mated with a plain bearing; one-segment, three-pad, shrouded Rayleigh step bearing; three-tilted-lobe journal with grooves (offset factor of 1.0) mated with a plain bearing; and three-central-lobe bearing with grooves. In the tests in which the journals had tilted lobes the shaft speed could be increased beyond the point of initial fractional-frequency whirl without the whirl growing excessively. In some cases bearing distress was not noticed until a shaft speed of twice the initial whirl onset speed was reached.

18.3.3 Cavitation

Machine elements in relative motion separated by a lubricating fluid are often subjected to conditions that cause the fluid to cavitate. Cavitation is defined as the disruption of a continuous liquid phase by the emergence of gas or vapor and can be a result of either (1) dissolved gas coming out of solution and/or entrained gas from the surrounding environment or (2) evaporation (flashing) of the fluid. Both types of cavitation are commonly observed in journal bearings and squeeze-film dampers. Jakobsson and Floberg[25] showed that the occurrence of cavitation in journal bearings affects power loss, coefficient of friction, bearing torque, and load capacity.

Dowson and Taylor,[8] in an excellent review of cavitation, point out that cavitation need not have a deleterious effect on the load capacity of fluid-film bearings. However, imposing higher loads and speeds together with more complicated loading cycles can result in a bearing experiencing "cavitation erosion-damage,"[26] such as may be found in heavily loaded diesel engine bearings. This damage occurs under conditions of vapor cavitation (as opposed to gas cavitation). Very large implosive forces are created and confined to a small area as the vapor bubble collapses. The net effect is that metal is hammered out of the bearing surfaces by a fatigue process, resulting in erosion damage to the bearing surfaces. This phenomenon arises under conditions of dynamic loading and is frequently observed in main and connecting-rod bearings in compression-ignition engines.

18.4 THRUST BEARINGS

Except for very light loading, for which two opposed flat surfaces suffice, thrust bearings require contouring of one of the surfaces in order to ensure generation of a fluid film. The most common thrust bearings use several discrete sector-shaped pads, which may be of the flat, tapered land, or stepped type (Fig. 18.5). The pads may be fixed, or for flat pads, supported on a point or line pivot. A film may also be generated by means of spiral grooving on one of the surfaces.

18.4.1 Flat Sector Pads

An analysis of incompressibly lubricated flat sector pad bearings was made by Etsion,[27] his was the first analysis to accurately account for the film thickness distribution in a sector pad. Figure 18.12 shows the pad with arbitrary roll and pitch angles. The incompressible Reynolds equation for this configuration is

$$\frac{\partial}{\partial r}\left(\frac{rh^3}{\mu}\frac{\partial p}{\partial r}\right) + \frac{1}{r}\frac{\partial}{\partial \theta}\left(\frac{h^3}{\mu}\frac{\partial p}{\partial \theta}\right) = 6\omega r\frac{\partial h}{\partial \theta} \qquad (18.17)$$

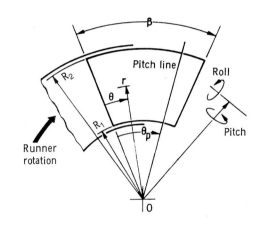

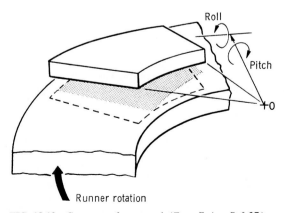

FIG. 18.12   Geometry of sector pad. (*From Etsion, Ref. 27.*)

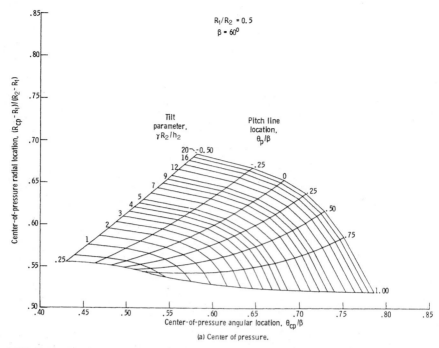

FIG. 18.13   Design charts for flat, sector-shaped pad with ratio of inner to outer radius r_i/r_o of 0.5 and angular extent β of 60°. (*From Etsion, Ref. 27.*)

Etsion realized that the pitch and roll of a sector pad about any point can be transformed to a corresponding pure pitch about a certain radial line called the *pitch line,* which may or may not be within the confines of the pad. The film thickness h can be conveniently expressed as

$$h = h_p + \gamma r \sin(\theta_p - \theta) \tag{18.18}$$

where h_p is the clearance along the pitch line, γ the pitch angle, and θ_p the angle measured from the pitch line. Results were obtained numerically and presented for a range of the four pad parameters: radius ratio R_1/R_2, pad angle β, pitch line location θ_p/β, and tilt angle $\gamma R_2/h_2$. A significant finding of the work was that load capacity is maximized by having a uniform minimum film thickness along the trailing edge of the pad. Examples of Etsion's results are shown in Fig. 18.13. Etsion's results as shown in this figure may be used to design a bearing, as the following example shows.

EXAMPLE 2   Design a fixed-pad thrust bearing to most efficiently carry a 4000-N load at 3000 rpm, using as the lubricant an oil with a viscosity of 0.007 Pa·s (SAE 20 at 90°C). Use 60° sector pads with a radius ratio of 0.5. Minimum film thickness is to be 0.02 mm.

solution   The maximum number of 60° sectors that can be used, allowing some space between sectors, is 5. Thus each pad must carry $W = 4000/5 = 800$ N. The most efficient design has the minimum film thickness uniform along the trailing edge of the pad; thus $\theta_p/\beta = 1.0$. From Fig. 18.13b, the optimum unit load W/KA equals 0.0038 for a tilt para-

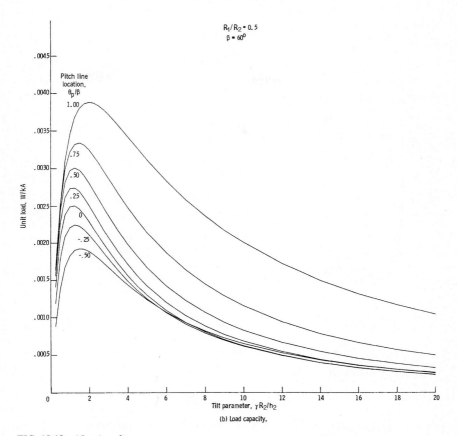

$R_1/R_2 = 0.5$
$\beta = 60^0$

Pitch line
location,
θ_p/β

1.00

.75
.50
.25
0
-.25
-.50

Unit load, W/kA

Tilt parameter, $\gamma R_2/h_2$

(b) Load capacity.

FIG. 18.13 (*Continued*)

meter $\gamma R_2/h_2$ of approximately 3. Inserting the definitions of K and A, and solving for pad radius, we obtain

$$R_2^4 = \frac{Wh_2^2}{(0.0038)6\mu\omega\beta[1 - (R_1/R_2)^2]} \tag{18.19}$$

Working out the numbers, pad sector outer radius $R_2 = 53$ mm, and tilt angle $\gamma = 1.1$ mrad. The power loss is obtained from Fig. 18.13c: $H/W\omega h_2 = 15$, leading to $H = 75$ W per pad or 375 W for the bearing assembly.

EXAMPLE 3 For the same conditions as Example 2, design a bearing whose pads are supported on point pivots. Determine the effect of reducing the speed to 1500 rpm.

solution At the design speed of 3000 rpm, the bearing will operate exactly the same as the fixed-pad design; the only additional information to be determined is the pivot location. This is simply read off Fig. 18.13a as $(R_{cp} - R_1)/(R_2 - R_1) = 0.52$ and $\theta_p/\beta = 0.65$. In other words, the pivot should be located 52 percent of the distance from the inner to the outer radius and 65 percent of the distance from the leading to the trailing edge of the

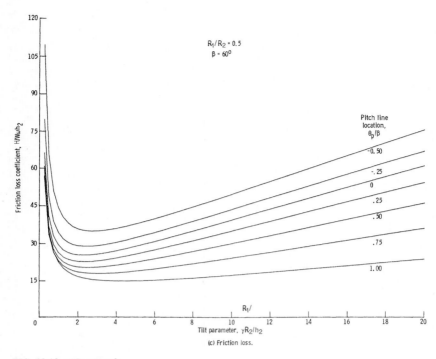

FIG. 18.13   (*Continued*)

pad. For a given bearing the pivot location is fixed; thus if the speed drops to 1500 rpm, by Fig. 18.13*a*, the tilt parameter and pitch-line location are unaltered. From Fig. 18.13*b* and *c*, the load parameter and friction coefficient are also unaltered. Looking at the definition of the load parameter, we see that all quantities are fixed save speed and trailing-edge clearance; h_2 is proportional to the square root of the speed ω. Therefore if the speed is halved, the clearance drops to $1/\sqrt{2}$ of its previous value, or 0.014 mm. For constant friction coefficient and load, power loss will drop to $1/(2\sqrt{2})$ of its former value, or 133 W for the bearing assembly.

The center of pressure on a fixed pad changes with operating conditions. Determining changes in minimum film thickness and power loss is thus somewhat more complex than for a pivoted-pad bearing; Etsion[27] outlines the procedure.

Figure 18.13 and the similar figures published by Etsion[27,28] can also be used to design thrust bearings with line pivots. In this case, for optimum performance, the pivot line must be parallel to the trailing edge of the pad rather than radial.

In order to allow bidirectional operation, the pivot may be placed midway between the leading and trailing edges ($\theta_p/\beta = 0.5$). Such a bearing with a radial line pivot will have zero theoretical load capacity.[27] A point-pivoted bearing will carry a load if the pivot is properly placed in the radial direction; however, load capacity will be considerably less than for the optimum angular pivot location.

18.5 HYDROSTATIC BEARINGS

Hydrodynamic bearings require relative motion of the bearing surfaces to build up the load-supporting pressures. When a bearing must support a load with little or no relative motion between the surfaces, an externally pressurized or hydrostatic bearing must be used. Hydrostatic bearings may also be required in applications where touching or rubbing of the bearing surfaces cannot be permitted at startup and shutdown. For example, under extreme temperature or corrosive conditions, it may be necessary to use bearing materials with poor sliding friction properties. Surface contact may then result in galling and surface welding, making the bearing inoperative. In such circumstances a hydrostatic bearing or a combination hydrostatic-hydrodynamic bearing will prevent surface contact at startup and shutdown. Hydrostatic bearings are also used where low-friction supports are required for stationary assemblies. They offer the advantages of high stiffness and high load capacity independent of rotation, but require high-pressure pumps and high lubricant flows.[11]

18.5.1 Liquid Lubricant

The basic principles of hydrostatic bearings can be illustrated by examination of the circular thrust pad shown in Fig. 18.14a. The Reynolds equation for this configuration is

$$\frac{\partial}{\partial r}\left(rh^3\frac{\partial p}{\partial r}\right) = 0 \tag{18.20}$$

This equation is easy to solve for load capacity and flow rate: results are

$$W = \frac{\pi(p_i - p_a)}{2}\left(\frac{R_2^2 - R_1^2}{\ln(R_2/R_1)}\right) \tag{18.21}$$

$$Q = \frac{\pi h^3(p_i - p_a)}{6\mu \ln(R_2/R_1)} \tag{18.22}$$

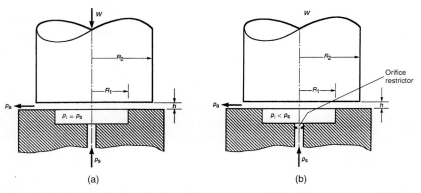

FIG. 18.14 Hydrostatic thrust bearing: (a) uncompensated; (b) orifice compensated. (*From Bisson and Anderson, Ref. 11.*)

It is immediately apparent from these results that the load supported does not depend on either clearance or fluid viscosity, although the flow rate is affected by viscosity. Lack of load dependence on clearance means that the bearing has no stiffness; this is not usually a desirable situation. A hydrostatic bearing is given stiffness by placing a flow restrictor in the lubricant supply line, for example an orifice as shown in Fig. 18.14*b*. The pressure in the bearing recess drops from p_s to the intermediate pressure p_i. Now, if the clearance decreases for some reason, the resistance to flow out of the bearing increases, dropping the flow rate. A lower flow rate results in a smaller pressure drop across the orifice, increasing the intermediate pressure p_i (the supply pressure p_s is constant); this in turn increases the bearing load. Thus stiffness has been created; the bearing load increases as the clearance decreases.

This principle of compensation is also necessary for the proper functioning of journal bearings. When load is applied, the journal becomes eccentric. The pressure increases in the side with the reduced clearance and decreases in the side with the increased clearance. This difference in pressures supports the load.

Commonly used types of compensators include capillary tubes, orifices, and flow-control valves. The pressure drop across capillaries and orifices varies with flow rate. Flow-control valves maintain constant flow regardless of the pressure drop.

The flow through a capillary tube is given by the Hagen-Poiseuille law:

$$Q = \frac{\pi \, \Delta p \, d^4}{128 \, l_c \mu} \tag{18.23}$$

The velocity of discharge from a sharp-edged orifice is obtained from Bernoulli's equation. The corresponding volume flow rate is

$$Q = C_D a \sqrt{\frac{2\Delta p}{\rho}} \tag{18.24}$$

The orifice discharge coefficient C_D is typically about 0.6.

A flow-control valve maintains a constant flow rate regardless of the pressure drop across it, so that

$$Q = \text{constant} \tag{18.25}$$

With the type of compensating element and bearing geometry known, the operating film thickness under load can be determined. For the simple bearing (Fig. 18.14) with capillary compensation, flow into the pad [Eq. (18.23)] is equated with flow out of the pad [Eq. (18.22)]:

$$\frac{\pi(p_s - p_i)d^4}{128 \, l_c \mu} = \frac{\pi h^3(p_i - p_a)}{6\mu \, \ln(R_2/R_1)} \tag{18.26}$$

The required pad pressure to support a given load W is obtained from Eq. (18.21). With the pad pressure determined, the supply pressure is then a known function of the film thickness h and the dimensions of the capillary. These variables can be adjusted to suit the particular installation.

In a journal bearing the zero-load film thickness is fixed at half the total bearing clearance. The compensating elements will determine the zero-load pad pressures in terms of the supply pressure p_s. The compensators are usually designed so that the pressure drop through them is approximately the same as that through the bearing at zero load.

In addition to the pressure and the flow, knowing the pumping power requirements for a hydrostatic bearing is important. The power required is

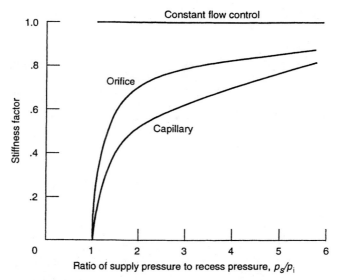

FIG. 18.15 Hydrostatic bearing stiffness at constant load and film thickness for three types of compensation. (*From Malanoski and Loeb, Ref. 29.*)

$$H = (p_s - p_a)Q \qquad (18.27)$$

In many installations the bearing deflection under load, or stiffness, is important. Malanoski and Loeb[29] worked out "stiffness factors" for orifice, capillary, and flow-control-valve compensation based on bearing operation at a given load and film thickness (Fig. 18.15). Under these conditions the bearing compensated by use of a flow-control valve was always stiffer than the orifice-compensated bearing, which was always stiffer than the capillary-compensated bearing. For orifice or capillary compensation, stiffness increases monotonically with pressure ratio.

H. C. Rippel[30] has consolidated the published work and experimental results of most contributors to this technology into design charts and specifications that facilitate the efficient design of hydrostatic bearings for specific applications.

18.5.2 Gaseous Lubricant

Although some characteristics of gas-lubricated bearings are similar to those of liquid-lubricated bearings, the compressibility of gases produces marked differences in some respects, especially for flow through orifices.

For both compressible and incompressible lubricants, a restrictor must be placed in the pressurized supply line in order for the bearing to have any significant stiffness. For gas-lubricated bearings three types of restrictors are commonly used:[31] laminar restrictors, such as capillary tubes; fixed orifices; inherent (variable orifice) restrictors. Laminar restrictors operate the same way with gas lubricants as with liquid lubricants. Fixed-orifice and inherent restrictors are illustrated in Fig. 18.16. Although orifices are mechanically the same for gas and liquid lubricants, they operate quite differently for a gas lubricant. When the lubricant velocity in the orifice is less than the speed of sound, the mass flow rate is given by

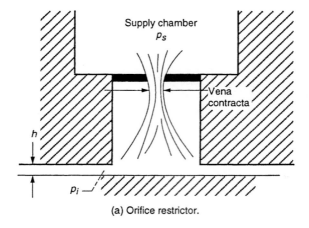

(a) Orifice restrictor.

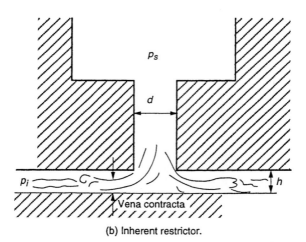

(b) Inherent restrictor.

FIG. 18.16 Orifice and inherent restrictors. (*From Vohr, Ref. 31.*)

$$M = C_D a p_s \left[\frac{2k}{(k-1)\mathfrak{R}T} \left[\left(\frac{p_i}{p_s} \right)^{2/k} - \left(\frac{p_i}{p_s} \right)^{k+1/k} \right] \right]^{1/2} \tag{18.28}$$

For constant supply pressure p_s the mass flow rate increases as the downstream pressure p_i decreases until sonic velocity is reached in the orifice. The flow is then said to be choked, and further decreases in downstream pressure do not affect the flow rate. Under choked conditions the mass flow rate is given by

$$M = C_D a p_s \left[\frac{2k}{(k-1)\mathfrak{R}T} \left[\left(\frac{p_c}{p_s} \right)^{2/k} - \left(\frac{p_c}{p_s} \right)^{k+1/k} \right] \right]^{1/2} \tag{18.29}$$

where $p_c/p_s = [2/(k+1)]^{k/(k-1)}$ is the pressure ratio when the velocity becomes sonic. Both relationships assume isentropic flow. The restrictor area a is $\pi d^2/4$ for an orifice restrictor and πdh for an inherent restrictor.

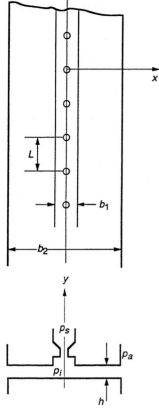

FIG. 18.17 Infinitely long strip. (*From Cheng, Ref. 32.*)

For an inherent restrictor the resistance to gas flow decreases as the clearance increases—qualitatively the same relationship as for flow through the lubricant film. However, the film resistance is inversely proportional to the cube of the clearance and thus changes much more rapidly with clearance than does the orifice resistance. This rapid change results in effective compensation, although an inherently compensated bearing is less stiff than a fixed-orifice-compensated bearing. Inherently compensated bearings have the advantage of being much less susceptible to pneumatic hammer.

The basic characteristics of a hydrostatic bearing may be illustrated by analysis of the infinitely long strip shown in Fig. 18.17. For this bearing the Reynolds equation [Eq. (18.1)] becomes

$$\frac{d^2p^2}{dx^2} = 0 \qquad (18.30)$$

The solution is easily obtained as

$$p = p_i\left[1 - \left(1 - \frac{p_a^2}{p_i^2}\right)\frac{x - b_1}{b_2}\right]^{1/2} \qquad (18.31)$$

resulting in the mass flow[32]

$$M = \frac{h^3 L}{12\mu b_2 \mathcal{R}T}(p_i^2 - p_a^2) \qquad (18.32)$$

where L is the spacing between two orifices. The problem is now to match this mass flow with that through the orifices in order to determine the intermediate pressure p_i. Matching is usually done numerically through an iterative procedure. With p_i known, the bearing load capacity is calculated from

$$W = L\left[p_i b_1 + \int_{b_1}^{b_1+b_2} p\,dx\right] \qquad (18.33)$$

Tang and Gross[33] calculated results for a large number of cases; examples of their results appear in Fig. 18.18 for mass flow, Fig. 18.19 for load capacity, and Fig. 18.20 for stiffness. Not surprisingly, mass flow, load capacity, and stiffness all increase as supply pressure p_s increases. Stiffness information is useful when the bearing displacement under load must be minimized. As Fig. 18.20 shows, stiffness varies widely as the bearing parameter Λ_0 is changed; there is a definite optimum that maximizes stiffness.

As shown for the journal bearing of Fig. 18.21, external pressurization introduces high-pressure gas into the interior of the bearing. When the journal is displaced from its central position, the film pressure increases on the low-clearance side and decreases on the high-clearance side. The result of these changes is a force that pushes the journal back toward the central position. In addition to increasing the load capacity (and also stability) of the bearing, external pressurization provides load capacity even without journal rotation. This aspect finds use in zero-friction positioning mechanisms

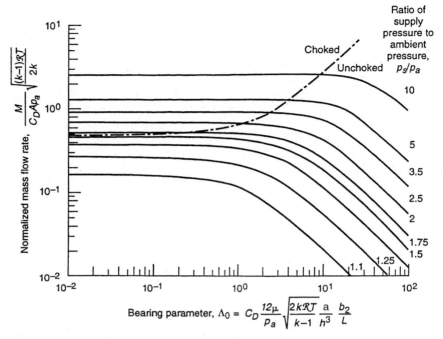

FIG. 18.18   Mass flow rate. (*From Tang and Gross, Ref. 33.*)

and for starting bearings without rubbing contact.[34] When the bearing does rotate, however, load capacity increases above the zero-speed value. Cunningham et al.[35] measured load capacity increases approaching 50 percent for a supply pressure ratio of 2.3 but, as expected, lesser increases at higher supply pressures.

   External pressurization improves stability as well as enhancing load capacity. Fleming et al.[36] calculated and measured the stability of an unloaded bearing. The analysis showed that trapped gas volume around the bearing orifices (orifice recess volume) is very detrimental to stability. If the volume is large enough, the bearing may be unstable at zero speed, a phenomenon known as pneumatic hammer.

18.6   GAS-LUBRICATED HYDRODYNAMIC BEARINGS

Gas-lubricated bearings are like liquid-lubricated bearings in many respects. Gases are viscous and can act as lubricants, although the bearing generally must have much closer clearances because gases are less viscous than liquids. The compressibility of gases complicates the analysis by making the Reynolds equation nonlinear. On the other hand, certain aspects of gas bearings make them simpler to analyze than liquid bearings. Perhaps the most significant simplification is that the gas film is continuous (or complete) throughout the bearing. That is, there is no film breakdown due to cavitation as occurs in the low-pressure region of liquid-lubricated bearings. Further simplifications are that the gas lubricating film remains more nearly isothermal and more nearly laminar than does a liquid lubricating film.[37]

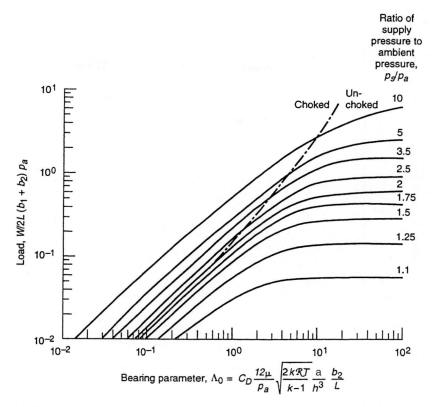

FIG. 18.19   Load capacity for longitudinal flow [$b/(b_1 + b_2) = 0.1$]. (*From Tang and Gross, Ref. 33.*)

One of the assumptions in the derivation of the Reynolds equation [Eq. (18.1)] was that the flow is laminar. Transition from laminar to turbulent flow is largely dependent on Reynolds number, where

$$\text{Re} = Uh/\nu \tag{18.34}$$

and ν is the kinematic viscosity of the lubricant. The critical Reynolds number above which turbulence occurs is somewhat vague, but it is generally believed to be in the range $500 < \text{Re}_{cr} < 2000$. Since the kinematic viscosity of gases is much less than that of oils, it is advisable to verify whether the laminar-flow assumption also applies to gas lubrication. A typical 25-mm (1-in) diameter, air-lubricated journal bearing operating with a radial clearance of 0.025 mm (10^{-3} in) at a speed of 50,000 rpm would have a Reynolds number less than 100. Clearly, then, most gas-lubricated bearings will operate in the laminar-flow region. Therefore, the basic Reynolds equation, which neglects fluid inertia terms, is valid.

It is common practice to assume that the bearing gas film is isothermal.[37] If additionally the perfect gas law applies ($p = \rho \mathcal{R} T$), then the Reynolds equation [Eq. (18.1)] becomes

$$\frac{\partial}{\partial x}\left(\frac{ph^3}{12\mu}\frac{\partial p}{\partial x}\right) + \frac{\partial}{\partial y}\left(\frac{ph^3}{12\mu}\frac{\partial p}{\partial y}\right) = \frac{\partial}{\partial x}\left[\frac{ph(U_1 + U_2)}{2}\right] + \frac{\partial(ph)}{\partial t} \tag{18.35}$$

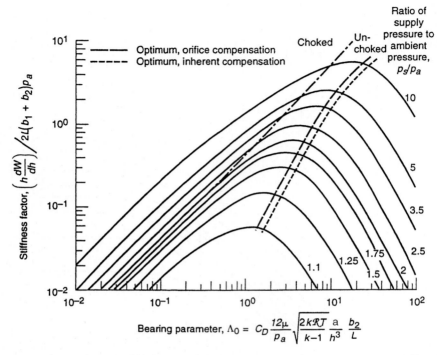

FIG. 18.20   Stiffness factor for longitudinal flow $[b/(b_1 + b_2) = 0.1]$. (*From Tang and Gross, Ref. 33.*)

In contrast to the Reynolds equation for liquid lubrication, Eq. (18.32) is nonlinear in the pressure p. Approximate methods of solution usually rely on some means of linearization, one of which will be described in the next section.

Mechanical Technology Inc.[38] has published a manual for gas-lubricated bearings that can facilitate their design for specific applications. Extensive design curves are presented for hydrodynamic and hydrostatic thrust and journal bearings.

18.6.1   Journal Bearings

A gas-lubricated journal bearing is similar to its liquid-lubricated counterpart shown in Fig. 18.1. Eq. (18.35) for steady state is[37]

$$\frac{\partial}{\partial x}\left(ph^3\frac{\partial p}{\partial x}\right) + \frac{\partial}{\partial y}\left(ph^3\frac{\partial p}{\partial y}\right) = 6\mu U\frac{\partial}{\partial x}(ph) \qquad (18.36)$$

Ausman[39] developed the perturbation method for linearizing Eq. (18.36) in order to obtain an approximate solution. The general concept is to assume that the pressure may be expressed as

$$p(\theta,\eta) = p_a + p_1(\theta,\eta) \qquad (18.37)$$

where p_1/p_a is of order $\varepsilon << 1$. Substitute this and Eq. (18.4) into Eq. (18.36) and

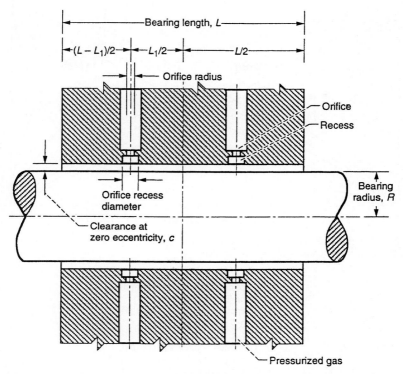

FIG. 18.21 Externally pressurized journal bearing. (*From Fleming et al., Ref. 36.*)

neglect all terms of order ε^2 or higher. The first-order (in ε) perturbation equation is then

$$\frac{\partial^2 p_1}{\partial \theta^2} + \frac{\partial^2 p_1}{\partial \eta^2} = \Lambda \left(\frac{\partial p_1}{\partial \theta} - \epsilon p_a \sin\theta \right) \qquad (18.38)$$

where

$$\Lambda = \frac{6\mu\omega R^2}{p_a c^2} \qquad (18.39)$$

The dimensionless parameter Λ is called the bearing compressibility number. It or some similar dimensionless number is a fundamental parameter in hydrodynamic gas lubrication.[37]

Equation (18.38) can be solved for the first-order perturbation pressure,[39] which in turn can be integrated in the usual fashion to obtain load components parallel and perpendicular to the line of centers.

Figure 18.22 is a graph of total load and attitude angle for several L/D ratios. The first-order perturbation solution yields a load linearly related to eccentricity ratio ε. This relationship is a consequence of the linearization and is valid only for small eccentricity. Actually, the load increases much more sharply at high values of ε as indicated in Fig. 18.23.[37] The *linearized ph* solution was developed by Ausman[40] to yield more accurate results at high eccentricities. However, the basic perturbation solution is reasonably accurate up to $\varepsilon = 0.5$, and is conservative for higher eccentricities.

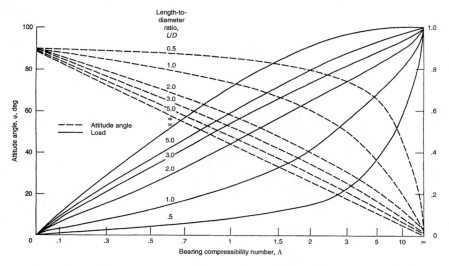

FIG. 18.22 Design chart for radially loaded, self-acting, gas-lubricated journal bearings (isothermal first-order perturbation solution). (*From Ausman, Ref. 37.*)

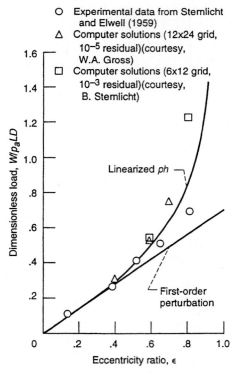

FIG. 18.23 Experimental and predicted load as a function of eccentricity ratio for finite-length, self-acting, gas-lubricated journal bearing. Bearing compressibility number, Λ, 1.3; length-to-diameter ratio, L/D, 1.5. (*From Ausman, Ref. 40.*)

The bearing friction torque may be calculated by the same formula as for liquid-lubricated bearings, Eq. (18.14).

18.6.2 Helical-Groove (Herringbone) Journal Bearings

A variation of the plain journal bearing is the helical-groove (herringbone) journal bearing shown in Fig. 18.24 where shallow grooves (on the order of the bearing clearance) are placed in the shaft. The terms "helical groove" and "herringbone groove" are often used interchangeably. Vohr and Chow[17] were the first to predict that the herringbone-groove bearing geometry can suppress half-frequency (shaft) whirl and increase bearing load capacity. Additionally, they reported that friction torque would be lower than that for an equivalent plain, cylindrical journal bearing. These predictions were verified independently by the theoretical work of Hirs[41] and experimentally by Malanoski and Loeb[29] and by Cunningham et al.,[42,43] who ran a herringbone-groove journal bearing up to 76,000 rpm.

Herringbone-groove journal bearings provide the high load capacity and stability of hydrostatic bearings without requiring a source of pressurized gas. The grooved journal acts as a pump, pumping gas from the ends of the bearing toward the center and thereby increasing the pressure toward the center of the bearing. Of course, there is no load capacity at zero speed, but at high speed the herringbone-groove bearing's load capacity and stability may exceed those of the hydrostatic bearing.

Hamrock and Fleming[45,46] calculated groove parameters to maximize either the radial load capacity or the stability of helical-groove bearings. (Radial load capacity was chosen rather than total load capacity because the tangential component of total load capacity was considered to be destabilizing.)

18.6.3 Thrust Bearings

Other than pivoted-pad bearings, hydrodynamic gas-lubricated thrust bearings usually are made by contouring the surface of one of the two opposing disks. As for the liquid-lubricated bearing this is necessary because two opposing flat disks would produce little or no load capacity. Various contours may be used as indicated in Fig. 18.5. Either the fixed disk or the rotating disk may be contoured. Of these types of thrust bearing the gas-lubricated, spiral-groove bearing is probably best.[37]

The step bearing is perhaps the easiest to analyze because the bearing clearance h is constant throughout each of two regions. The Reynolds equation can be linearized by a simple perturbation method and solved in each region separately. The two solutions are then matched at the step. For constant bearing clearance, the steady-state Reynolds equation is

$$\frac{\partial^2 p^2}{\partial^2 x} + \frac{\partial^2 p^2}{\partial^2 y} = \frac{12\mu U}{h^2}\frac{\partial p}{\partial x} \qquad (18.40)$$

The linearized form of the equation can be written directly by substituting

$$p = p_a + p_1 \qquad (18.41)$$

Assuming p_1 is small compared to p_a and neglecting all but first-order terms in p_1 we obtain

$$\frac{\partial^2 p_1}{\partial x^2} + \frac{\partial^2 p_1}{\partial y^2} = \frac{6\mu U}{p_a h^2}\frac{\partial p_1}{\partial x} \qquad (18.42)$$

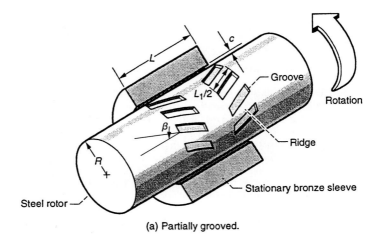

(a) Partially grooved.

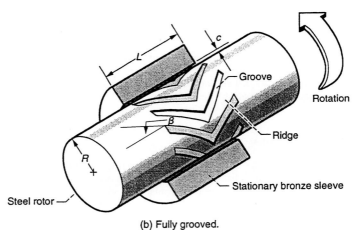

(b) Fully grooved.

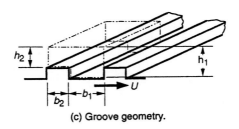

(c) Groove geometry.

FIG. 18.24   Herringbone-groove bearings.

Ausman[46] solved Eq. (18.42) in polar coordinates. He obtained results as functions of the parameter Λ_Δ where

$$\Lambda_\Delta = \frac{6\mu\omega R^2}{p_a \Delta^2}$$

(18.43)

in which Δ is the height of the step separating the two regions.

Table 18.2 gives the optimum number of pads and step location for a given value of Λ_Δ or $(\Lambda_\Delta)_1$ and the ratio of inner to outer radius. From this table the unit load (load per unit area) P can be calculated for the given values of $(\Lambda_\Delta)_1$ when the minimum clearance s is equal to the depth of the step Δ. For $s \neq \Delta$ and values of Λ_Δ not listed in Table 18.2, Fig. 18.25 is used to adjust the tabulated load capacity. In this figure

$$\Lambda_s = \frac{6\mu\omega R^2}{p_a s^2}$$

(18.44)

Ausman[37] notes that a pocket bearing (Fig. 18.5d) is likely to perform better than a step bearing, but that the step-bearing analysis can be used to conservatively design a pocket bearing.

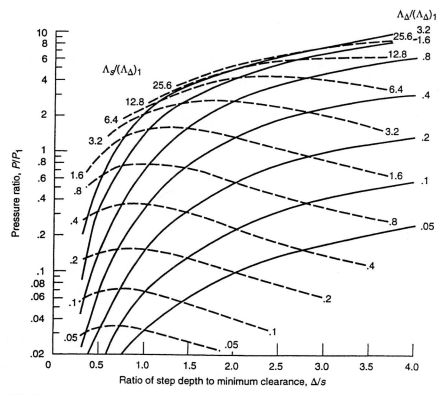

FIG. 18.25   Design chart for stepped-sector, thrust pad bearing. (*From Ausman, Ref. 37.*)

TABLE 18.2 Optimum Number of Pads and Location of Step for Maximum Load at Clearance Ratio of 2*

Ratio of inner to outer radius	$\Lambda_\Delta = \Lambda_s$	Number of pads, N	Ratio of leading sector angle to total pad angle per pad	Ratio of unit load to ambient pressure, P/P_a
0.20	10	4	0.45	0.064
	20	4	.39	.141
	40	3	.26	.286
	80	3	.16	.470
	160	3	.10	.638
0.30	10	5	0.45	0.059
	20	5	.39	.131
	40	4	.29	.270
	80	4	.20	.457
	160	4	.12	.622
0.40	10	6	0.46	0.053
	20	6	.39	.119
	40	5	.31	.248
	80	5	.23	.431
	160	5	.14	.602
0.50	10	8	0.48	0.046
	20	7	.39	.103
	40	7	.33	.219
	80	6	.23	.397
	160	6	.14	.572
0.60	10	11	0.49	0.038
	20	10	.42	.084
	40	9	.36	.184
	80	8	.25	.349
	160	7	.17	.530
0.70	10	15	0.50	0.029
	20	14	.45	.063
	40	12	.36	.141
	80	11	.28	.284
	160	9	.17	.466
0.75	10	18	0.51	0.024
	20	17	.46	.052
	40	15	.37	.116
	80	13	.28	.243
	160	11	.19	.421
0.80	10	23	0.53	0.019
	20	21	.48	.041
	40	19	.40	.091
	80	17	.31	.194
	160	14	.22	.363

*Ratio of maximum clearance (leading sector) to minimum clearance (trailing sector), $(s + \Delta)/s$, 2; angular width of groove between pads, 2°.

Source: From Ref. 3.

Hamrock[47] determined optimum pad geometries by using a linearized analysis and rectangular approximation for a sector-shaped pad. He obtained results to maximize either load capacity or stiffness for a wide range of compressibility numbers. For low compressibility numbers (near incompressible lubrication), maximum load capacity and maximum stiffness were obtained with nearly the same geometry (nearly square pads with the step occupying about half of the pad length and a step film thickness about one and a half times that in the ridge region). At higher compressibility numbers longer pads (relative to pad width) were called for, with the step region becoming relatively shorter and the film thickness ratio increasing. Required changes from optimum incompressible geometry were greater to maximize stiffness than to maximize load.

Etsion and Fleming[48] developed an improved analysis and presented design curves for a wide range of flat sector bearing geometries and operating conditions. Among the conclusions of this work were (1) as for liquid-lubricated sectors, the bearing load capacity is maximized when the minimum clearance is at the trailing edge of the pad and is uniform along the trailing edge, (2) for a given outside diameter the unit load is higher for a smaller inside diameter (in addition to the load capacity increase due to the larger area), and (3) the rectangular slider approximation seriously overestimates the load capacity, by 60 percent for some of the cases studied.

18.6.4 Foil Bearings

A radical departure from conventional rigid-geometry bearings is the foil bearing which has been used in small, high-speed, lightweight turbomachinery ranging from air units, such as supercharger and auxiliary air-handling units, to Brayton-cycle systems and cryogenic turbomachinery for space power.[49–53] Foil bearings accommodate distortions and misalignments well, are very stable, and have excellent load capacity. On the debit side, analytical tools and design charts are not readily available, and starting torques tend to be high.

The materials that have been used for foil bearings are Inconel, beryllium copper, and various stainless steels. Solid-lubricant coatings and other coatings extend foil life.

The so-called bending-dominated foil bearing has seen the greatest application. In this bearing, the foil is not tensioned, but one or more foil segments are supported on a corrugated bump foil (Fig. 18.26) or other foil segments (Fig. 18.27). Figure 18.26, for the first type of bearing, shows the shaft, top foil, and bump foil, which fits between the top foil and the housing. Figure 18.27 shows an alternative design for thrust and journal bearings called overlapping foils. In the journal bearing (Fig. 18.27b) several arc-shaped foils overlap around the shaft circumference. Part of each foil fits into a groove in the bearing housing. Each foil has a defined radius of curvature and thickness.

In general, foil thickness varies from 0.1 to 0.3 mm (0.004 to 0.011 in) depending on the bearing and the fit. The bearing slides on and over the shaft as a single unit. The radial clearance ratio c/R is usually from 0.001 to 0.02—approximately 100 times greater than for conventional bearings with film thicknesses of 100 to 200 μm. Bearing stiffness is a combination of that generated by the convergent fluid-film wedge and the mechanical design of the leaves and foil bumps. The leaves and foil bumps are usually separated by a distance of 0.02 to 0.08 mm (0.001 to 0.003 in), which results in a nonlinear stiffness. Damping is provided both by the fluid film and the friction characteristics of the foil and/or bumps.

Foil bearings have been run at speeds of 4.2 million DN and average pressures of 0.7 MPa (100 psi).[50] Bearing load capacity increases as speed increases. However, at

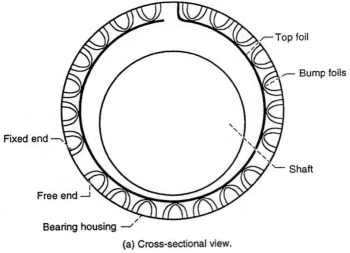

(a) Cross-sectional view.

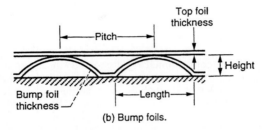

(b) Bump foils.

FIG. 18.26 Bending-dominated foil journal bearing design. (*From Heshmat et al., Ref. 59.*)

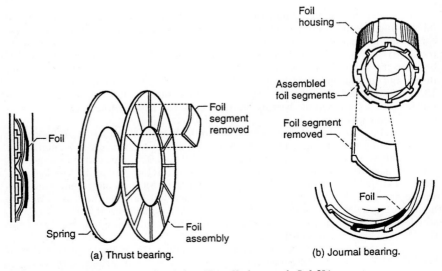

(a) Thrust bearing.

(b) Journal bearing.

FIG. 18.27 Overlapping foil bearing design. (*From Heshmat et al., Ref. 59.*)

low speeds the shaft will rub the bearing foils. Also, at low speeds and high loads the shaft can have significant runout.

18.7 MATERIALS

E. R. Booser[54] has summarized the materials requirements for fluid-film bearing applications. Bearings operating with full film lubrication theoretically never touch the shaft. However, bearings do start and stop, and some are required to operate with boundary lubrication, or no lubrication at all, for short periods. Hence, successful use of a fluid-film bearing material depends on matching its properties to the demands of the application. Operating factors to be considered include load, speed, contamination, life required, temperature, and the type and quantity of lubrication available. No bearing material is equally good with respect to all requirements. Proper selection is a matter of judgment based on the following factors:[54]

1. *Compatibility,* a measure of the antiweld and antiscoring characteristics of a bearing material when operated with a given mating material. All journal bearings experience some metal-to-metal contact, if only during starting. During metal-to-metal contact the sliding surfaces rub on microscopic high spots. The friction developed at these points can produce localized welding, causing a seizing damage that resembles scoring. Therefore, a good bearing material will not weld easily to the shaft material. Liquid cadmium, for instance, will not wet a steel shaft.

2. *Conformability,* the ability to compensate for misalignment and to conform to other geometric errors. Good conformability is characteristic of soft metals with a low modulus of elasticity.

3. *Embeddability,* the ability to absorb dirt and other foreign particles to avoid scoring and wear. Usually, in metallic bearing materials, good embeddability is found in materials with good conformability. This is not true in nonmetallic materials. Carbon graphite has a low modulus and good conformability, but its hardness gives it a poor ability to absorb dirt.

4. *Load capacity:* Largely, this is an experience factor related equally to design and lubrication, as well as to the strength and other performance properties of the material. It is a measure of the maximum hydrodynamic pressure that a material can be expected to endure.

5. *Fatigue resistance:* This factor is particularly important in applications where the load changes direction, such as in reciprocating automotive and aircraft engines. When a bearing material experiences a fatigue failure, cracks begin just below the surface of the material at the points of maximum shear stress. In babbitt these cracks propagate at a right angle to the surface down to the bond with the backing material. Then they progress along the backing material and join other similar fatigue cracks. When sufficiently weakened, the bearing material begins to flake loose. Low-modulus materials are often used as thin overlays on strip steel to improve fatigue resistance.

6. *Corrosion resistance:* This factor is important where uninhibited industrial lubricating oils are used. Oxidation products from some lubricating oils can attack and corrode cadmium, lead, zinc, and some copper-bearing alloys. Corrosion effects can be minimized by such modifications as applying a thin flash of indium over cadmium or lead alloys or by adding tin to lead babbitt. Both tin and aluminum alloys are usually unaffected by oxidized oils.

7. *Cost factors:* Overall costs are frequently the deciding factor in selecting a bearing material. Some materials cost more than others. Frequently, however, a thin layer of a high-cost material can supply the necessary bearing characteristics while a low-cost substrate provides the necessary structural strength.

Performance characteristics of common fluid-film bearing alloys are indicated in Table 18.3. For instance, the babbitts and cadmium-base materials are rated highest on compatibility for operation against steel shafts. Babbitts, which are soft, low-melting-point alloys, are best in conformability and in embedding dirt. The harder, higher-strength materials, such as the bronzes, aluminum, and silver, carry heavier loads and offer better fatigue strength.[54]

18.7.1 Babbitts

Tin- and lead-base babbitts are probably the best known of all bearing materials. With their ability to embed dirt and excellent compatibility properties under boundary lubrication conditions, babbitt bearings are used in a wide range of applications. In small bushings for fractional-horsepower motors and in automotive engine bearings, the babbitt is generally used as a thin coating over a steel strip. For larger bearings in heavy-duty equipment a thicker babbitt is cast on a rigid backing of steel or cast iron. After machining, the babbitt layer is usually 3 to 6 mm (⅛ to ¼ in) thick.[54]

Relative to other bearing materials, babbitts generally have lower load capacity and fatigue strength, are a little higher in cost, and require a more complicated design. Also, their strength decreases rapidly with increasing temperature. These shortcomings can often be avoided by using an intermediate layer of high-strength, fatigue-resisting material between a steel backing and a thin babbitt surface layer. Such composite bearings frequently eliminate any need for using alternative materials with poorer compatibility characteristics.[54]

Specifications established by the Society of Automotive Engineers (SAE) for babbitt compositions are given in Table 18.4. Grades 11 and 12 are tin-base babbitts and grades 13 to 15 are lead base. The American Society for Testing and Materials (ASTM) has adopted similar specifications (B23). ASTM grades 1 to 5 are tin base, and grades 6 to 19 are lead base.[54]

18.7.2 Bronzes and Copper Alloys

Dozens of copper alloys are available as bearing materials. Most of these can be grouped into four classes: copper-lead, leaded bronze, tin-bronze, and aluminum-bronze. Characteristics of typical copper alloy cast bearing materials are listed in Table 18.5. The alloys at the top of this listing have higher lead content and correspondingly better compatibility properties. Thus, copper-lead and leaded bronzes are generally the best bearing alloys despite their lower physical properties. Load capacity, strength, hardness, wear resistance, and fatigue strength increase as other alloying elements (such as tin, aluminum, and iron) are added. Costs vary widely. Simple bronze bushings are usually cheaper than babbitt bearings; copper-lead bearings are more expensive. Copper alloy bearings provide greater load capacity than the softer babbitts, better high-temperature operation, and greater wear resistance, but poorer antiscoring properties.[54]

TABLE 18.3 Performance Characteristics of Fluid-Film Bearing Alloys

Alloy	Brinell hardness			Load capacity		Maximum operating temperature		Compat-ibility*	Conform-ability and embedd-ability*	Corrosion resistance*	Fatigue strength*
	At room temperature	At 149°C (300°F)	Minimum shaft	MPa	psi	°C	°F				
Tin-base babbitt	6–12	20–30	130–170	5.5–10.3	800–1500	149	300	1	1	1	5
Lead-base babbitt	6–12	15–20	130–170	5.5–8.3	800–1200	149	300	1	1	3	5
Three-component bearings, babbitt surfaced	——	——	200–300	13.8–34.5+	2000–5000+	149	300	1	2	2	2
Cadmium base	15	30–40	200–250	10.3–13.8	1500–2000	260	500	1	2	5	4
Copper-lead	20–23	20–30	200	10.3–17.2	1500–2500	177	350	2	2	5	3
Lead bronze	40–60	40–70	300	20.7–27.6	3000–4000	232	450	3	4	4	2
Tin bronze	60–70	60–80	300–400	27.6+	4000+	260+	500+	5	5	2	1
Aluminum alloy	40–45	45–50	200–300	34.5+	5000+	121	250	4	3	1	2
Silver (overplated)	25	25	300–400	34.5+	5000+	260	500	2	3	1	1

*Arbitrary scale with 1 being the best material and 5 the worst.

Source: From Ref. 54.

TABLE 18.4 Composition of Babbitts*

(Percent composition)

	Tin base		Lead base		
Element	SAE 11	SAE 12 (ASTM 2)	SAE 13	SAE 14 (ASTM 7)	SAE 15 (ASTM 15)
Tin	86.0 min	88.25 min	5.0 to 7.0	9.25 to 10.75	0.9 to 1.25
Antimony	6.0 to 7.5	7.0 to 8.0	9.0 to 11.0	14.0 to 16.0	14.0 to 15.5
Lead	0.50	0.50	Remainder	Remainder	Remainder
Copper	5.0 to 6.5	3.0 to 4.0	0.50	0.50	0.50
Iron	0.08	0.08	——	——	——
Arsenic	0.10	0.10	0.25	0.60	0.80 to 1.20
Bismuth	0.08	0.08	0.10	0.10	0.10
Zinc	0.005	0.005	0.005	0.005	0.005
Aluminum	0.005	0.005	0.005	0.005	0.005
Cadmium	——	——	0.05	0.05	0.02
Others, total	0.20	0.20	0.20	0.20	0.20

*All values not given as ranges are maximum except as shown.
Source: From Ref. 54.

18.7.3 Aluminum

With their high load capacity, good fatigue strength, high thermal conductivity, excellent corrosion resistance, and low cost, aluminum alloys are used extensively in connecting-rod and main bearings in internal combustion engines, in roll-neck bearings in steel mills, and in reciprocating compressors and aircraft equipment. Aluminum has not replaced babbitt for industrial equipment operating under a steady unidirectional load. Aluminum alloys require hardened shafts and improved surface finish. They have relatively poor compatibility characteristics and lack embeddability and conformability. For automotive use, antiscoring characteristics and embeddability are considerably improved by using a thin lead babbitt or an electrodeposited lead-tin overlay.[54]

The most common aluminum bearing materials is the SAE 770 listed in Table 18.6. In this casting alloy 6.5 percent tin is included for improved compatibility. The small amounts of copper and nickel increase strength, hardness, and scuff resistance. A wrought alloy is available of almost the same composition with the addition of 1.5 percent silicon (SAE 780). This alloy can be used to form bimetal bearings of SAE 780 sheet, 0.76 mm (0.030 in) thick, bonded by hot rolling to nickel-plated, low-carbon steel.[54]

A second major type of aluminum alloy, such as SAE 781, contains 1 to 3 percent cadmium for improved bearing properties. Varying amounts of silicon, copper, and nickel may be used. This type of alloy is used in making steel-backed bearings with a 0.01-mm (0.0005-in) overlay containing approximately 90 percent lead and 10 percent tin. This thin layer of babbitt improves the surface properties of composite bearings during break-in.[54]

A third type of aluminum bearing alloy is a reticular 20 to 30 percent tin alloy containing up to 3 percent copper.[54]

18.7.4 Gas-Lubricated Bearing Materials

The following categories were developed for selecting gas-lubricated bearing materials and coatings for sliding compatibility:

TABLE 18.5 Typical Copper Alloy Cast Bearing Materials

Alloy number	Material	Nominal composition, percent				Hardness (Brinell)	Tensile strength		Maximum operating temperature		Maximum load	
		Cu	Sn	Pb	Zn		MPa	psi	°C	°F	MPa	psi
SAE 490	Copper-lead	65	—	35	—	25	55	8 000	180	350	14	2000
SAE 48	Copper-lead	70	—	30	—	28	60	8 500	180	350	14	2000
AMS 4840	High-lead tin bronze	70	5	25	—	48	170	25 000	200+	400+	21+	3000+
SAE 67	Semiplastic bronze	78	6	16	—	55	210	30 000	230	450	21+	3000+
SAE 40	Leaded red brass	85	5	5	5	60	240	35 000	230	450	24	3500
SAE 660	Bronze bearings	83	7	7	3	60	240	35 000	230+	450+	28	4000
SAE 64	Phosphor bronze	80	10	10	—	63	240	35 000	230+	450+	28	4000
SAE 62	Gun metal	88	10	—	2	65	310	45 000	260+	500+	28	4000
SAE 620	Navy G	88	8	—	4	68	280	40 000	260	500	28+	4000+
SAE 63	Leaded gun metal	88	10	2	—	70	280	40 000	260	500	28+	4000+
ASTM B148-52-9c	Aluminum bronze	85	*	—	—	195	620	90 000	260+	500+	31+	4500+

*4 Fe, 11 Al.

Source: From Ref. 54.

TABLE 18.6 Composition of Aluminum Bearing Alloys

Material	Bearing type	Nominal composition, percent					
		Al	Sn	Cu	Ni	Si	Cd
SAE 770	Cast, solid	91.5	6.5	1	1	—	—
SAE 780	Rolled, bimetal, or trimetal	90.5	6.5	1	.5	1.5	—
SAE 781	Wrought, trimetal	95	—	—	—	4	1
Reticular tin-aluminum	Trimetal	79	20	1	—	—	—

Source: From Ref. 54.

1. Hard, wear-resistant material combinations, such as solid members or coatings of ceramic or carbide material, nitrided or malcomized alloys, and hardened tool steels.

2. Self-lubricating materials mated against a hard, lapped shaft. These materials can be carbon graphite, or sintered bronze on a thin steel backing with an impregnated surface of PTFE and lead.

3. Soft phase materials melted against a hard, lapped shaft. These materials are leaded bronze and powder compresses containing an infiltrated phase.

4. Solid lubricant films melted against a hard, lapped shaft. These can be resin- or ceramic-bonded or applied in any other acceptable manner.

For operation in an inert environment consideration should be given to the following material combinations: hard material combinations, particularly oxides such as chromium oxide; self-lubricating materials which depend on PTFE for sliding properties; and solid lubricant films containing MoS_2.

Results of tests with different material combinations are summarized in Table 18.7. Table 18.8 contains suggested material combinations for use at various temperatures.

TABLE 18.7 Gas-Lubricated Bearing Material Test Results from Large Bearing Evaluations

Material combination		Behavior in start-stop tests	Behavior in high-speed rub tests
Pad	Shaft		
Chromium oxide coating	Chromium oxide coating	Excellent; negligible wear	Excellent; material transfer
Nickel-bonded tungsten carbide coating	Aluminum oxide coating	Excellent; negligible wear	Poor; material transfer and scoring
Resin-bonded to MoS_2 film	Hardened 416 stainless steel	Good: MoS_2 coating almost worn away after 1000 start-stops	Fair; coating beginning to wear through and scoring commencing
Sintered-bronze facing impregnated with PTFE and lead	Molybdenum coating	Fair; would not lift off if stress was greater than 14 kPa (2 psi)	Excellent; smooth, polished surfaces
Cobalt-bonded tungsten carbide coating	Hardened AISI M-50 steel	Poor; metal transfer and scoring after about 350 start-stops	Poor; material transfer and scoring
Nickel-bonded tungsten carbide coating	Nickel-bonded tungsten carbide coating	Fair; wear debris formed streaks on shaft and friction increased substantially	Not run, but probably poor
Nitrided Nitralloy	Nitrided Nitralloy	Poor; tendency for pickup and scoring	Not run, but probably poor
Carbon graphite	Hard, lapped chromium plate	Excellent; for best results	Not run, but carbon-graphite can score chromium plate

TABLE 18.8 Gas Bearing Material Combinations for Use at Various Temperatures

Temperature range °C	°F	Typical process gas	Bearing materials	Journal materials	Problem areas	State of the art
Cryogenic		Helium, nitrogen	Fused PTFE film on metal substrates	Fused teflon film on metal substrates	Wear life of PTFE film under start-stop conditions not known; durability of film during high-speed rubs not known	Basic friction and wear tests indicate feasibility of obtaining at least 100 starts and stops at low stresses (14 to 21 kPa; 2 to 3 psi).
-54 to 149	-65 to 300	Air	Carbon graphites	Hard chromium plate; nitrided steels	Temperature limitation due to design considerations, such as designing for differential thermal expansion; bearing may score shaft due to high-speed rub	Proven to be capable of more than 1000 start-stops at stresses to 28 kPa (4 psi)
121 to 177	250 to 350	Steam	Silver-impregnated carbon graphite	Hard chromium-plated stainless steel	Same as above	Excellent start-stop capability
121 to 232	250 to 450	Steam	Phosphor bronze (solid or sprayed coating)	Stellite or Stellite hard facing	Sliding compatibility	Used successfully in externally pressurized bearings
-54 to 260	-65 to 500	Air, inert gases	DU bonded to metallic substrate	Hard chromium plate; nitrided surfaces; hardened steel (40+ Rockwell C) and electrolyzed steel	Relatively poor resistance to start-stops at stress levels > 14 kPa (2 psi)	Excellent resistance to damage during high-speed rubs

TABLE 18.8 Gas Bearing Material Combinations for Use at Various Temperatures (*Continued*)

Temperature range °C	°F	Typical process gas	Bearing materials	Journal materials	Problem areas	State of the art
−54 to 260	−65 to 500	Air, inert gases	Resin-bonded solid lubricants bonded to metallic substrate	Same as above	Require careful preparation and run-in	Capable of 1000 start-stops at stress levels to 21 kPa (3 psi); wears rapidly during high-speed rubs
−54 to 482	−65 to 900	Air	Plasma-sprayed, metal-bonded carbide coating*	Plasma-sprayed Al_2O_3 coating*	Poor for high-speed rubs	Excellent start-stop behavior; negligible wear after 1000 start-stops at 28-kPa (4-psi) stress
−54 to 482	−65 to 900	Inert or reactive gases, air	Plasma-sprayed chromium oxide*	Plasma-sprayed chromium oxide*	Nonconductive nature of coating makes grooving difficult	Excellent start-stop behavior at 28 KPa (4 psi); very promising for resistance to damage during high-speed rubs
Above 482	900	Air, inert or reactive gases	Hot-pressed Al_2O_3	Same as bearing	Brittle materials	Very limited experience
			Cemented Al_2O_3	Same as bearing	Difficult to fabricate	
			Nickel-bonded titanium carbide†	Al_2O_3	Thermal shock	

*Must be used on corrosion-resistant substrate for operation in corrosive envionment.
†Oxidizes in are above 538°C (1000 °F).

18.7.5 CONTAMINATION

Contaminants in both fluid-film and gas-film systems can significantly affect bearing performance, life, and reliability. Contaminant particles can brinell the shaft (journal) and bearing surfaces, causing stress raisers that can act as nuclei for surface pitting (spalling). Hard particles act as an abrasive medium and wear the softer surface. The particles can also act to clog lubricant orifices and jets, resulting in lubricant starvation of the bearing surfaces.

Even in systems that are initially clean, multiple startups and shutdowns cause metal-to-metal contact, resulting in wear debris. This debris with time can cause damage. As a result good filtration is required in the lubrication system for long life and reliability.

REFERENCES

1. Brewe, David E., David P. Fleming, Florin Dimofte, and Robert C. Hendricks: "Fluid and Gas Film Lubrication," in *Tribology for Aerospace Applications,* E. V. Zaretsky, ed., Society of Tribologists and Lubrication Engineers, Park Ridge, Ill., 1995.

2. DeHart, A. O.: "Basic Bearing Types," *Machine Design,* vol. 36, no. 6, pp. 15–17, 1966.

3. Reynolds, O.: "On the Theory of Lubrication and Its Application to Mr. Beauchamp Tower's Experiments, Including an Experimental Determination of the Viscosity of Olive Oil," *Phil. Trans. Roy. Soc., London,* vol. 177, pp. 157–234, 1886.

4. Hamrock, Bernard J.: *Fundamentals of Fluid Film Lubrication,* McGraw-Hill Book Company, Inc., New York, 1994.

5. Sommerfeld, A.: "The Hydrodynamic Theory of Lubrication Friction," *Mathematics and Physics,* vol. 50, nos. 1–2, pp. 97–155, 1904.

6. Harrison, W. J.: "The Hydrodynamical Theory of Lubrication with Special Reference to Air as a Lubricant," *Proc. Cambridge Phil. Soc.,* vol. 22, no. 3, pp. 39–54, 1913.

7. DuBois, G. B., and F. Ocvirk: "Analytical Derivation and Experimental Evaluation of Short-Bearing Approximation for Full Journal Bearings," NACA Report 1157, 1953.

8. Dowson, D., and C. M. Taylor: "Cavitation in Bearings—Lubricating Films," *Annu. Rev. Fluid Mechanics,* vol. 11, M. Van Dyke, J. V. Wehausen, and J. L. Lumley, eds., Annual Reviews, Palo Alto, Calif., pp. 35–66, 1979.

9. Glaeser, W. A.: "Plain Bearings," *Machine Design,* vol. 42, no. 15, pp. 8–14, 1970.

10. Rippel, H. C.: "Cast Bronze Bearing Design Manual," 2d ed., Cast Bronze Bearing Institute, Inc., Cleveland, Ohio, 1965.

11. Bisson, E. E., and W. J. Anderson, eds.: "Advanced Bearing Technology," NASA SP-38, pp. 63–138, 1964.

12. Marsh, H.: "The Stability of Aerodynamic Gas Bearings. Part II: Noncircular Bearing," Ph.D thesis, St. John's College, Cambridge, England, 1964.

13. Lund, J. W.: "Rotor-Bearing Dynamics Design Technology. Part 7: The Three-Lobe Bearing and Floating Ring Bearing," Report MTI-67TR47, Mechanical Technology Inc., Latham, N.Y., 1968.

14. Pinkus, O.: "Sleeve Bearing Design," *Product Eng.,* vol. 26, no. 8, pp. 134–139, 1955.

15. Chadbourne, L. E., F. X. Dobler, and A. D. Rottler: "SNAP 50/SPUR Nuclear Mechanical Power Unit, Experimental Research and Development Program. Part 2: Bearing," Report APS-5249-R, AFAPL-TR-67-34, PT. 2, AiResearch, Phoenix, Ariz., 1966.

16. Sternlicht, B., and L. W. Winn: "Geometry Effects on the Threshold of Half-Frequency Whirl in Self-Acting, Gas-Lubricated Journal Bearings," *J. Basic Eng.,* vol. 86, no. 2, pp. 313–320, 1964.

17. Vohr, J. H., and C. Y. Chow: "Characteristics of Herringbone-Grooved, Gas-Lubricated Journal Bearings," *J. Basic Eng.,* vol. 87, no. 3, pp. 568–578, 1965.

18. Malanoski, S. B.: "Experiments on an Ultrastable Gas Journal Bearing," *J. Lubr. Technol.,* vol. 89, no. 4, pp. 433–438, 1967.

19. Boeker, G. F., and B. Sternlicht: "Investigation of Translatory Fluid Whirl in Vertical Machines," *Trans. ASME,* vol. 78, pp. 13–19, 1956.

20. Schuller, F. T., and W. J. Anderson: "Experiments on the Stability of Water-Lubricated Three-Sector Hydrodynamic Journal Bearings at Zero Load," NASA TN D-5752, 1970.

21. Schuller, F. T.: "Experiments on the Stability of Various Water-Lubricated Fixed Geometry Hydrodynamic Journal Bearings at Zero Load," *J. Lubr. Technol.,* vol. 95, no. 4, pp. 434–446, 1973.

22. Schuller, F. T.: "Designs of Various Fixed-Geometry Water-Lubricated Hydrodynamic Journal Bearings for Maximum Stability," NASA SP-333, 1973.

23. Schuller, F. T.: "Effect of Number of Lobes and Length-Diameter Ratio on Stability of Tilted-Lobe Hydrodynamic Journal Bearings at Zero Load," NASA TN D-7902, 1975.

24. Schuller, F. T.: "Stability Experiments with Hydrodynamic Tilted-Lobe Journal Bearings of Various Numbers of Lobes and Length-to-Diameter Ratios," *ASLE Trans.,* vol. 20, no. 4, pp. 271–281, 1977.

25. Jakobsson, B., and L. Floberg: "The Finite Journal Bearing Considering Vaporization," *Trans. Chalmers University of Technology,* Gothenburg, Sweden, p. 190, 1957.

26. Wilson, R. W.: "Cavitation Damage in Plain Bearings," *Cavitation and Related Phenomena in Lubrication: Proc. of First Leeds-Lyon Symp. Tribology,* D. Dowson, M. Godet, and C. Taylor, eds., Mechanical Engineering Publications, New York, pp., 177–184, 1974.

27. Etsion, I.: "Design Charts for Arbitrarily Pivoted, Liquid-Lubricated, Flat-Sector-Pad Thrust Bearing," *J. Lubr. Technol.,* vol. 100, no. 2, pp. 279–286, 1978.

28. Etsion, I.: "Design Charts for Arbitrarily Pivoted, Liquid-Lubricated, Flat-Sector-Pad Thrust Bearing," NASA TN D-8344, 1977.

29. Malanoski, S. B., and A. M. Loeb: "The Effect of the Method of Compensation of Hydrostatic Bearing Stiffness," ASME Paper 60-Lub-12, 1960.

30. Rippel, H. C.: "Cast Bronze Hydrostatic Bearing Design Manual," 2d ed., Cast Bronze Bearing Institute, Inc., Cleveland, Ohio, 1965.

31. Vohr, J. H.: "Restrictor Flows," "Design of Gas Bearings, Vol. 1: Design Notes," N. F. Rieger, ed., Mechanical Technology Inc., Latham, N.Y., 1966.

32. Cheng, H. S.: "Basic Relations," "Design of Gas Bearings, Vol. 1: Design Notes," N. F. Rieger, ed., Mechanical Technology Inc., Latham, N.Y., 1966.

33. Tang, I. C., and W. A. Gross: "Analysis and Design of Externally Pressurized Gas Bearings," *ASLE Trans.,* vol. 5, no. 1, pp. 261–284, 1962.

34. Davis, J. E.: "Design and Fabrication of the Brayton Rotating Unit," NASA CR-1870, 1972.

35. Cunningham, R. E., D. P. Fleming, and W. J. Anderson, "Steady-State Experiments on Rotating Externally Pressurized Air-Lubricated Journal Bearings," *J. Lubr. Technol.,* vol. 92, no. 2, pp. 336–345, 1970.

36. Fleming, D. P., R. E. Cunningham, and W. J. Anderson: "Zero-Load Stability of Rotating Externally Pressurized Gas-Lubricated Journal Bearings," *J. Lubr. Technol,* vol. 92, no. 2, pp. 325–335, 1970.

37. Ausman, J. S.: "Gas-Lubricated Bearings," "Advanced Bearing Technology," E. E. Bisson and W. J. Anderson, eds., NASA SP-38, pp. 109–138, 1964.

38. "Design of Gas Bearings, Vol 1: Design Notes," N. F. Rieger, ed., Mechanical Technology Inc., Latham, N.Y., 1966.

39. Ausman, J. S.: "Theory and Design of Self-Acting Gas Lubricated Journal Bearings Including Misalignment Effects," *First International Symp. Gas-Lubricated Bearings,* ACR-49, Office Naval Research, pp. 161–192, 1959.

40. Ausman, J. S.: "An Improved Analytical Solution for Self-Acting Gas-Lubricated Journal Bearings of Finite Length, *J. Basic Eng.,* vol. 83, no. 2, pp. 188–194, 1961.

41. Hirs, G. G.: "The Load Capacity and Stability Characteristics of Hydrodynamic Grooved Journal Bearings," ASLE Paper 64LC-24, 1964.

42. Cunningham, R. E., D. P. Fleming, and W. J. Anderson: "Experimental Stability Studies of the Herringbone-Grooved Gas-Lubricated Journal Bearings," *J. Lubr. Techol.,* vol. 91, no. 1, pp. 52–59, 1969.

43. Cunningham, R. E., D. P. Fleming, and W. J. Anderson: "Experimental Load Capacity and Power Loss of Herringbone-Grooved Gas-Lubricated Journal Bearings," *J. Lubr. Technol.,* vol. 93, no. 3, pp. 415–422, 1971.

44. Hamrock, B. J., and D. P. Fleming: "Optimization of Self-Acting Herringbone Journal Bearings for Maximum Radial Load Capacity," NASA TN D-6351, 1971.

45. Fleming, D. P., and B. J. Hamrock: "Optimization of Self-Acting Herringbone Journal Bearings for Maximum Stability," *Proc. of Sixth International Gas Bearing Symp.,* N. G. Coles, ed., BHRA Fluid Engineering, Paper No. C1, 1974.

46. Ausman, J. S.: "An Approximate Analytical Solution for Self-Acting Gas Lubrication of Stepped Sector Thrust Bearings," *ASLE Trans.,* vol. 4, pp. 304–313, 1961.

47. Hamrock, B. J.: "Optimization of Self-Acting Step Thrust Bearings for Load Capacity and Stiffness," *ASLE Trans.,* vol. 15, pp. 159–170, 1972.

48. Etsion, I., and D. P. Fleming: "An Accurate Solution of the Gas Lubricated, Flat Sector Thrust Bearing," *J. Lubr. Technol.,* vol, 99, no. 1, pp. 82–88, 1977.

49. O'Connor, L.: Fluid-Film Foil Bearings Control Engine Heat," *Mech. Eng.,* vol. 115, no. 5, pp. 71–75, 1993.

50. Heshmat, H.: "The Advancement in Performance of Aerodynamic Foil Bearings: High Speed and Load Capacity," paper presented at STLE/ASME Tribology Conference, New Orleans, October 24–27, 1993.

51. Saville, M., A. Gu, and R. Capaldi: "Liquid Hydrogen Turbopump Foil Bearing," AIAA Paper 91-2108, 1991.

52. Gilbrech, R. J., A. Gu, T. Rigney, M. Saville, and I. Rossoni: "Liquid Hydrogen Foil-Bearing Turbopump," AIAA Paper 93-2537, 1993.

53. Genge, G. G., M. Saville, and A. Gu: "Foil Bearing Performance in Liquid Nitrogen and Liquid Oxygen," AIAA Paper 93-2536, 1993.

54. Booser, E. R.: "Plain Bearing Materials," *Machine Design,* vol. 42, no. 15, pp. 14–20, 1970.

55. "General Guide to the Choice of Thrust Bearing Type," Engineering Sciences Data Unit, Item 67033, Institution of Mechanical Engineers, London, 1967.

56. Wilcock, D. F., and E. R. Booser: "Bearing Design and Application," McGraw-Hill Book Company, Inc., New York, 1957.

57. Raimondi, A. A., and J. Boyd: "Applying Bearing Theory to the Analysis and Design of Pad-Type Bearings," *ASME Trans.,* vol. 77, pp. 287–309, 1955.

58. Maleev, V. L., and J. B. Hartman: "Machine Design," International Textbook Co., Scranton, Pa., 1954.

59. Heshmat, H., J. A. Walowit, and O. Pinkus: "Analysis of Gas-Lubricated Foil Journal Bearings," *J. Lubr. Techol.,* vol. 105, no. 4, pp. 647–655, 1983.

SECTION 19
ROLLING-ELEMENT BEARINGS

Bernard J. Hamrock, Ph.D.
Professor of Mechanical Engineering
Ohio State University
Columbus, Ohio

William J. Anderson, M.S.M.E.
Vice President
NASTEC Inc.
Cleveland, Ohio

SYMBOLS

a_1, a_2, a_3 = life factors

B = total conformity of bearing

b = semiminor axis of roller contact, m

C = dynamic load capacity, N

$c_1, ..., c_4$ = constants

D = distance between race curvature centers, m

$\tilde{D}$ = material factor

D_x = diameter of contact ellipse along x axis, m

D_y = diameter of contact ellipse along y axis, m

d = rolling-element diameter, m

d_a = overall diameter of bearing (Fig. 19.5), m

d_b = bore diameter, mm

d_e = pitch diameter, m

d_i = inner-race diameter, m

d_o = outer-race diameter, m

E = modulus of elasticity, N/m^2

E' = effective elastic modulus, $2/[(1 - v_a^2)/E_a + (1 - v_b^2)/E_b]$, N/m^2

$\tilde{E}$ = metallurgical processing factor

$\mathscr{E}$ = elliptic integral of second kind

$\overline{\mathscr{E}}$ = approximate elliptic integral of second kind

e = percentage of error

$\tilde{F}$ = lubrication factor

F = applied load, N

F' = load per unit length, N/m

F_e = bearing equivalent load, N

F_r = applied radial load, N

F_t = applied thrust load, N

$\mathscr{F}$ = elliptic integral of first kind

$\overline{\mathscr{F}}$ = approximate elliptic integral of first kind

f = race conformity ratio

f_a = rms surface finish of rolling element

f_b = rms surface finish of race

f_c = coefficient dependent on materials and bearing type (Table 19.14)

G = dimensionless materials parameter, $\xi E'$

$\tilde{G}$ = speed-effect factor

H = dimensionless film thickness, h/R_x

$\tilde{H}$ = misalignment factor

H_{min} = dimensionless minimum film thickness

h = film thickness, m

J = number of stress cycles

K = load-deflection constant

K_1 = load-deflection constant for roller bearing

$K_{1.5}$ = load-deflection constant for ball bearing

k = ellipticity parameter, D_y/D_x

$\bar{k}$ = approximate ellipticity parameter

L = fatigue life

L_A = adjusted fatigue life

L_{10} = fatigue life where 90 percent of bearing population will endure

L_{50} = fatigue life where 50 percent of bearing population will endure

l = bearing length, m

l_r = roller effective length, m

l_t = roller length, m

l_v = length dimension in stress volume, m

M = probability of failure

m = number of rows of rolling elements

N = rotational speed, rpm

n = number of rolling elements

P = dimensionless pressure, p/E'

P_d = diametral clearance, m

P_e = free endplay, m

p = pressure, N/m^2

q = constant, $\pi/2 - 1$

R = curvature sum, m

R = reliability factor, percent

R_x = effective radius in x direction, m

R_y = effective radius in y direction, m

r = race curvature radius, m

r_a = ball radius, m

r_c = roller corner radius, m

r_y = radius of roller in y direction, m

S = probability of survival

s = shoulder height, m

T = tangential force, N

U = dimensionless speed parameter, $u\eta_0/E'R_x$

u = mean surface velocity in direction of motion, $(u_a + u_b)/2$, m/s

V = stressed volume, m^3

v = elementary volume, m^3

W = dimensionless load parameter, $F/E'R_x^2$

X,Y = factors for calculation of equivalent load

x, y, z = coordinate system

Z = constant defined by Eq. (19.28)

Z_0 = depth of maximum shear stress, m

α = radius ratio, R_y/R_x

β = contact angle, degrees

β' = iterated value of contact angle, degrees

β_f = free or initial contact angle, degrees

Γ = curvature difference

δ = total elastic deformation, m

$\bar{\delta}$ = approximate total elastic deformation, m

ξ = pressure-viscosity coefficient of lubrication, m²/N

η = absolute viscosity at gage pressure, N·s/m²

η_0 = viscosity at atmospheric pressure, N·s/m²

θ = angle used to define shoulder height

Λ = film parameter (ratio of film thickness to composite surface roughness)

μ = coefficient of rolling friction

ν = Poisson's ratio

ρ = lubricant density, N·s²/m⁴

ρ_0 = density at atmospheric pressure, N·s²/m⁴

σ_{max} = maximum hertzian stress, N/m²

γ_0 = maximum shear stress, N/m²

ψ = angular location

ψ_l = limiting value of ψ

ω = angular velocity, rad/s

ω_B = angular velocity of rolling-element–race contact, rad/s

ω_b = angular velocity of rolling element about its own center, rad/s

ω_c = angular velocity of rolling element about shaft center, rad/s

Subscripts

a = solid a

b = solid b

i = inner race

o = outer race

x, y, z = coordinate system

Superscript

$(^-)$ = approximate

19.1 INTRODUCTION

Ball bearings are used in many kinds of machines and devices with rotating parts. The designer is often confronted with decisions on whether a rolling-element or hydrodynamic bearing should be used in a particular application. The following characteristics make ball bearings *more desirable* than hydrodynamic bearings in many situations: (1) low starting and good operating friction, (2) the ability to support combined radial and thrust loads, (3) less sensitivity to interruptions in lubrication, (4) no self-excited instabilities, and (5) good low-temperature starting. Within reasonable limits, changes in load, speed, and operating temperature have but little effect on the satisfactory performance of ball bearings.

The following characteristics make ball bearings *less desirable* than hydrodynamic bearings: (1) finite fatigue life subject to wide fluctuations, (2) larger space required in the radial direction, (3) low damping capacity, (4) higher noise level, (5) more severe alignment requirements, and (6) higher cost. Each type of bearing has its particular strong points, and care should be taken in choosing the most appropriate type of bearing for a given application.

Useful guidance on the important issue of bearing selection has been presented by the Engineering Science Data Unit.[17,18] These Engineering Science Data Unit documents provide an excellent guide to the selection of the type of journal or thrust bearing most likely to give the required performance when considering the load, speed, and geometry of the bearing. The following types of bearings were considered:

1. Rubbing bearings, where the two bearing surfaces rub together [e.g., unlubricated bushings made from materials based on nylon, polytetrafluoroethylene (PTFE), and carbon]

2. Oil-impregnated porous metal bearings, where a porous metal bushing is impregnated with lubricant and thus gives a self-lubricating effect (as in sintered-iron and sintered-bronze bearings)

3. Rolling-element bearings, where relative motion is facilitated by interposing rolling elements between stationary and moving components (as in ball, roller, and needle bearings)

4. Hydrodynamic film bearings, where the surfaces in relative motion are kept apart by pressures generated hydrodynamically in the lubricant film

Figure 19.1 gives a guide to the typical load that can be carried at various speeds, for a nominal life of 10,000 h at room temperature, by journal bearings of various types of shafts of the diameters quoted. The heavy curves indicate the preferred type of journal bearing for a particular load, speed, and diameter and thus divide the graph into distinct regions. From Fig. 19.1 it is observed that rolling-element bearings are preferred at lower speeds and hydrodynamic oil-film bearings are preferred at higher speeds. Rubbing bearings and oil-impregnated porous metal bearings are not preferred for any of the speeds, loads, or shaft diameters considered. Also, as the shaft diameter is increased, the transitional point at which hydrodynamic bearings are preferred over rolling-element bearings moves to the left.

The applied load and speed are usually known, and this enables a preliminary assessment to be made of the type of journal bearing most likely to be suitable for a particular application. In many cases, the shaft diameter will already have been determined by other considerations, and Fig. 19.1 can be used to find the type of journal bearing that will give adequate load capacity at the required speed. These curves are based on good engineering practice and commercially available parts. Higher loads and speeds or smaller shaft diameters are possible with exceptionally high engineering standards or specially produced materials. Except for rolling-element bearings, the curves are drawn for bearings with a width equal to the diameter. A medium-viscosity mineral-oil lubricant is assumed for the hydrodynamic bearings.

Similarly, Fig. 19.2 gives a guide to the typical maximum load that can be carried at various speeds for a nominal life of 10,000 h at room temperature by thrust bearings of various diameters. The heavy curves again indicate the preferred type of bearing for a particular load, speed, and diameter and thus divide the graph into major regions. As with the journal bearing results (Fig. 19.1), the hydrodynamic bearing is preferred at higher speeds and the rolling-element bearing is preferred at lower speeds. A difference between Figs. 19.1 and 19.2 is that at very low speeds there is a portion of the latter figure in which the rubbing bearing is preferred. Also, as the shaft diameter is increased, the transitional point at which hydrodynamic bearings are preferred over rolling-element bearings moves to the left. Note also from this figure that oil-impregnated porous metal bearings are not preferred for any of the speeds, loads, or shaft diameters considered.

19.2   BEARING TYPES

A great variety of both design and size ranges of ball and roller bearings is available to the designer. It is the intent of this section to not duplicate the complete descriptions

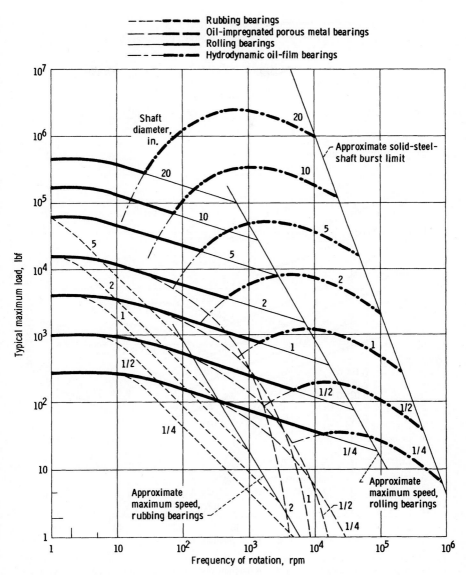

FIG. 19.1 General guide to journal bearing type. (Except for roller bearings, curves are drawn for bearings with width equal to diameter. A medium-viscosity mineral-oil lubricant is assumed for hydrodynamic bearings.) (*From Engineering Science Data Unit.*[17])

given in manufacturers' catalogs, but rather to present a guide to representative bearing types along with the approximate range of sizes available. Tables 19.1 to 19.9 illustrate some of the more widely used bearing types. In addition, there are numerous types of specialty bearings available; space does not permit a complete cataloging of all available bearings. Size ranges are given in metric units. Traditionally, most rolling-element bearings have been manufactured to metric dimensions, predating the efforts toward a metric standard.

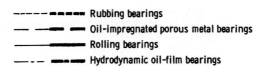

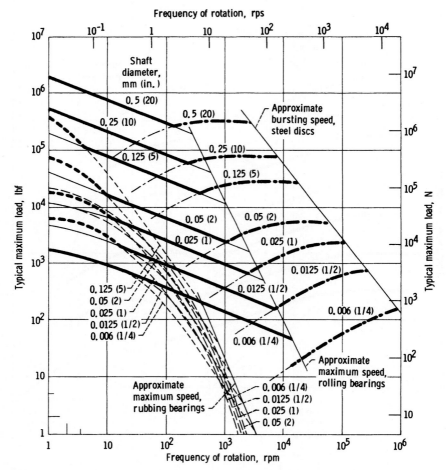

FIG. 19.2 General guide to thrust bearing type. Except for roller bearings, curves are drawn for typical ratios of inside to outside diameter. A medium-viscosity mineral-oil lubricant is assumed for hydrodynamic bearings. (*From Engineering Science Data Unit.*[18])

In addition to available bearing types and approximate size ranges, Tables 19.1 to 19.9 also list approximate relative load-carrying capabilities, both radial and thrust, and, where relevant, approximate tolerances to misalignment.

Rolling bearings are an assembly of several parts—an inner race, an outer race, a set of balls or rollers, and a cage or separator. The cage or separator maintains even spacing of the rolling elements. A cageless bearing, in which the annulus is packed with the maximum rolling-element complement, is called a "full-complement bearing." Full-complement bearings have high load capacity but lower speed limits than bearings equipped with cages. Tapered-roller bearings are an assembly of a cup, a cone, a set of tapered rollers, and a cage.

19.2.1 Ball Bearings

Ball bearings are used in greater quantity than any other type of rolling bearing. For an application where the load is primarily radial with some thrust load present, one of the types in Table 19.1 can be chosen. A Conrad, or deep-groove, bearing has a ball complement limited by the number of balls that can be packed into the annulus between the inner and outer races with the inner race resting against the inside diameter of the outer race. A stamped and riveted two-piece cage, piloted on the ball set, or a machined two-piece cage, ball piloted or race piloted, is almost always used in a Conrad bearing. The only exception is a one-piece cage with open-sided pockets that

TABLE 19.1 Representative Radial Ball Bearings

Type	Approx. range of bore sizes, mm		Relative capacity		Limiting speed factor	Tolerance to misalignment
	Min.	Max.	Radial	Thrust		
Conrad or deep groove	3	1060	1.00	0.7 (2 directions)	1.0	±0°15′
Maximum capacity or filling notch	10	130	1.2–1.4	0.2 (2 directions)	1.0	±0°3′
Magneto or counterbored outer	3	200	0.9–1.3	0.5–0.9 (1 direction)	1.0	±0°5′
Airframe or aircraft control	4.826	31.75	High static capacity	0.5 (2 directions)	0.2	0°
Self-aligning, internal	5	120	0.7	0.2 (1 direction)	1.0	±2°30′
Self-aligning, external	—	—	1.0	0.7 (2 directions)	1.0	High
Double row, maximum	6	110	1.5	0.2 (2 directions)	1.0	±0°3′
Double row, deep groove	6	110	1.5	1.4 (2 directions)	1.0	0°

TABLE 19.2 Representative Angular-Contact Ball Bearings

Type		Approx. range of bore sizes, mm		Relative capacity		Limiting speed factor	Tolerance to misalignment
		Min.	Max.	Radial	Thrust		
One-directional thrust		10	320	1.00–1.15*	1.5–2.3* (1 direction)	1.1–3.0*	±0°2'
Duplex, back-to-back		10	320	1.85	1.5 (2 directions)	3.0	0°
Duplex, face-to-face		10	320	1.85	1.5 (2 directions)	3.0	0°
Duplex, tandem		10	320	1.85	2.4 (1 direction)	3.0	0°
Two-directional or split ring		10	110	1.15	1.5 (2 directions)	3.0	±0°2'
Double row		10	140	1.5	1.85 (2 directions)	0.8	0°
Double row, maximum		10	110	1.65	0.5 (in 1 direction) 1.5 (in other direction)	0.7	0°

*Depends on contact angle.

is snapped into place. A filling-notch bearing has both inner and outer races notched so that a ball complement limited only by the annular space between the races can be used. It has low thrust capacity because of the filling notch.

The self-aligning internal bearing shown in Table 19.1 has an outer-race ball path ground in a spherical shape so that it can accept high levels of misalignment. The self-aligning external bearing has a multipiece outer race with a spherical interface. It too can accept high misalignment and has higher capacity than the self-aligning internal bearing. However, the external self-aligning bearing is somewhat less self-aligning than its internal counterpart because of friction in the multipiece outer race.

Representative angular-contact ball bearings are illustrated in Table 19.2. An angular-contact ball bearing has a two-shouldered ball groove in one race and a single-shouldered ball groove in the other race. Thus it is capable of supporting only a unidirectional thrust load. The cutaway shoulder allows assembly of the bearing by snapping over the ball set after it is positioned in the cage and outer race. This also permits use of a one-piece, machined, race-piloted cage that can be balanced for high-speed operation. Typical contact angles vary from 15 to 25°.

Angular-contact ball bearings are used in duplex pairs mounted either back to back or face to face as shown in Table 19.2. Duplex bearing pairs are manufactured so that they "preload" each other when clamped together in the housing and on the shaft. The use of preloading provides stiffer shaft support and helps prevent bearing skidding at light loads. Proper levels of preload can be obtained from the manufacturer. A duplex pair can support bidirectional thrust load. The back-to-back arrangement offers more resistance to moment or overturning loads than does the face-to-face arrangement.

Where thrust loads exceed the capability of a simple bearing, two bearings can be used in tandem, with both bearings supporting part of the thrust load. Three or more bearings are occasionally used in tandem, but this is discouraged because of the difficulty in achieving good load sharing. Even slight differences in operating temperature will cause a maldistribution of load sharing.

The split-ring bearing shown in Table 19.2 offers several advantages. The split ring (usually the inner) has its ball groove ground as a circular arc with a shim between the ring halves. The shim is then removed when the bearing is assembled so that the split-ring ball groove has the shape of a gothic arch. This reduces the axial play for a given radial play and results in more accurate axial positioning of the shaft. The bearing can support bidirectional thrust loads but must not be operated for prolonged periods of time at predominantly radial loads. This results in three-point ball-race contact and relatively high frictional losses. As with the conventional angular-contact bearing, a one-piece precision-machined cage is used.

Ball thrust bearings (Table 19.3), which have a 90° contact angle, are used almost exclusively for machinery with vertically oriented shafts. The flat-race bearing allows eccentricity of the fixed and rotating members. An additional bearing must be used for radial positioning. It has low load capacity because of the very small ball-race contacts and consequent high hertzian stress. Grooved-race bearings have higher load capacities and are capable of supporting low-magnitude radial loads. All of the pure thrust ball bearings have modest speed capability because of the 90° contact angle and the consequent high level of ball spinning and frictional losses.

19.2.2 Roller Bearings

Cylindrical-roller bearings (Table 19.4) provide purely radial load support in most applications. An N- or U-type bearing will allow free axial movement of the shaft relative to the housing to accommodate differences in thermal growth. An F- or J-type bearing will support a light thrust load in one direction; and a T-type bearing, a light bidirectional thrust load.

TABLE 19.3 Representative Thrust Ball Bearings

Type		Approx. range of bore sizes, mm		Relative capacity		Limiting speed factor	Tolerance to misalignment
		Min.	Max.	Radial	Thrust		
One directional, flat race		6.45	88.9	0	0.7 (1 direction)	0.10	0° (accepts eccentricity)
One directional, grooved race		6.45	1180	0	1.5 (1 direction)	0.30	0°
Two directional, grooved race		15	220	0	1.5 (2 directions)	0.30	0°

Cylindrical-roller bearings have moderately high radial load capacity as well as high speed capability. Their speed capability exceeds that of either spherical or tapered-roller bearings. A commonly used bearing combination for support of a high-speed rotor is an angular-contact ball bearing or duplex pair and a cylindrical-roller bearing.

As explained in Sec. 19.3 on bearing geometry, the rollers in cylindrical-roller bearings are seldom pure cylinders. They are crowned or made slightly barrel shaped to relieve stress concentrations of the roller ends when any misalignment of the shaft and housing is present.

Cylindrical-roller bearings may be equipped with one- or two-piece cages, usually race piloted. For greater load capacity, full-complement bearings can be used, but at a significant sacrifice in speed capability.

Spherical-roller bearings (Tables 19.5 to 19.7) are made as either single- or double-row bearings. The more popular bearing design uses barrel-shaped rollers. An alternative design employs hourglass-shaped rollers. Spherical-roller bearings combine very high radial load capacity with modest thrust load capacity (with the exception of the thrust type) and excellent tolerance to misalignment. They find widespread use in heavy-duty rolling mill and industrial gear drives, where all of these bearing characteristics are requisite.

Tapered-roller bearings (Table 19.8) are also made as single- or double-row bearings with combinations of one- or two-piece cups and cones. A four-row bearing assembly with two- or three-piece cups and cones is also available. Bearings are made with either a standard angle for applications in which moderate thrust loads are present or with a steep angle for high thrust capacity. Standard and special cages are available to suit the application requirements.

Single-row tapered-roller bearings must be used in pairs because a radially loaded bearing generates a thrust reaction that must be taken by a second bearing. Tapered-roller bearings are normally set up with spacers designed so that they operate with some internal play. Manufacturers' engineering journals should be consulted for proper setup procedures.

Needle-roller bearings (Table 19.9) are characterized by compactness in the radial direction and are frequently used without an inner race. In the latter case, the shaft is hardened and ground to serve as the inner race. Drawn cups, both open and closed end, are frequently used for grease retention. Drawn cups are thin walled and require substantial support from the housing. Heavy-duty roller bearings have relatively rigid

TABLE 19.4 Representative Cylindrical Roller Bearings

Type		Approx. range of bore sizes, mm		Relative capacity		Limiting speed factor	Tolerance to misalignment
		Min.	Max.	Radial	Thrust		
Separable outer ring, nonlocating (RN, RIN)		10	320	1.55	0	1.20	±0°5′
Separable inner ring, nonlocating (RU, RIU)		12	500	1.55	0	1.20	±0°5′
Separable outer ring, one-direction locating (RF, RIF)		40	177.8	1.55	Locating (1 direction)	1.15	±0°5′
Separable inner ring, one-direction locating (RJ, RIJ)		12	320	1.55	Locating (1 direction)	1.15	±0°5′
Self-contained, two-direction locating		12	100	1.35	Locating (2 directions)	1.15	±0°5′
Separable inner ring, two-direction locating (RT, RIT)		20	320	1.55	Locating (2 directions)	1.15	±0°5′
Nonlocating, full complement (RK, RIK)		17	75	2.10	0	0.20	±0°5′
Double row, separable outer ring, nonlocating (RD)		30	1060	1.85	0	1.00	0°
Double row, separable inner ring, nonlocating		70	1060	1.85	0	1.00	0°

TABLE 19.5 Representative Spherical Roller Bearings

Type	Approx. range of bore sizes, mm		Relative capacity		Limiting speed factor	Tolerance to misalignment
	Min.	Max.	Radial	Thrust		
Single row, barrel or convex	20	320	2.10	0.20	0.50	± 2°
Double row, barrel or convex	25	1250	2.40	0.70	0.50	± 1°30′
Thrust	85	360	0.10* 0.10†	1.80* 2.40†	0.35–0.50	± 3°
Double row, concave	50	130	2.40	0.70	0.50	± 1°30′

*Symmetric rollers.
†Asymmetric rollers.

TABLE 19.6 Standardized Double-Row Spherical Roller Bearings

Type	Roller design	Retainer design	Roller guidance	Roller-race contact
SLB	Symmetric	Machined, roller piloted	Retainer pockets	Modified line, both races
SC	Symmetric	Stamped, race piloted	Floating guide ring	Modified line, both races
SD	Asymmetric	Machined, race piloted	Inner-ring center rib	Line contact, outer; point contact, inner

races and are more akin to cylindrical roller bearings with high length-to-diameter ratio rollers.

Needle-roller bearings are more speed limited than cylindrical roller bearings because of roller skewing at high speeds. A high percentage of needle-roller bearings are full-complement bearings. Relative to a caged needle bearing, these have higher load capacity but lower speed capability.

There are many types of specialty bearings available other than those discussed here. Aircraft bearings for control systems, thin-section bearings, and fractured-ring bearings are some of the more widely used bearings among the many types manufactured. A complete coverage of all bearing types is beyond the scope of this chapter.

Angular-contact ball bearings and cylindrical-roller bearings are generally consid-

TABLE 19.7 Characteristics of Spherical Roller Bearings

Series	Types	Approx. range of bore sizes, mm		Approx. relative capacity*		Limiting speed factor
		Min.	Max.	Radial	Thrust	
202	Single-row barrel	20	320	1.0	0.11	0.5
203	Single-row barrel	20	240	1.7	0.18	0.5
204	Single-row barrel	25	110	2.1	0.22	0.4
212	SLB	35	75	1.0	0.26	0.6
213	SLB	30	70	1.7	0.53	0.6
22, 22K	SLB, SC, SD	30	320	1.7	0.46	0.6
23, 23K	SLB, SC, SD	40	280	2.7	1.0	0.6
30, 30K	SLB, SC, SD	120	1250	1.2	0.29	0.7
31, 31K	SLB, SC, SD	110	1250	1.7	0.54	0.6
32, 32K	SLB, SC, SD	100	850	2.1	0.78	0.6
39, 39K	SD	120	1250	0.7	0.18	0.7
40, 40K	SD	180	250	1.5	—	0.7

*Load capacities are comparative within the various series of spherical roller bearings only. For a given envelope size, a spherical roller bearing has a radial capacity approximately equal to that of a cylindrical roller bearing.

ered to have the highest speed capabilities. Speed limits of roller bearings are discussed in conjunction with lubrication methods. The lubrication system employed has as great an influence on bearing limiting speed as does the bearing design.

19.3 KINEMATICS

The relative motions of the separator, the balls or rollers, and the races of rolling-element bearings are important to understanding their performance. The relative velocities in a ball bearing are somewhat more complex than those in roller bearings, the latter being analogous to the specialized case of a zero- or fixed-value-contact-angle ball bearing. For that reason, the ball bearing is used as an example here to develop approximate expressions for relative velocities. These are useful for rapid but reasonably accurate calculation of elastohydrodynamic film thickness, which can be used with surface roughnesses to calculate the lubrication life factor.

The precise calculation of relative velocities in a ball bearing in which speed or centrifugal force effects, contact deformations, and elastohydrodynamic traction effects are considered requires a large computer to numerically solve the relevant equations. The reader is referred to the growing body of computer codes discussed in Sec. 19.12 for precise calculations of bearing performance. Such a treatment is beyond the scope of this section. However, approximate expressions that yield answers with accuracies satisfactory for many situations are available.

When a ball bearing operates at high speeds, the centrifugal force acting on the ball creates a divergency of the inner- and outer-race contact angles, as shown in Fig. 19.3, in order to maintain force equilibrium on the ball. For the most general case of rolling and spinning at both inner- and outer-race contacts, the rolling and spinning velocities of the ball are as shown in Fig. 19.4.

The equations for ball and separator angular velocity for all combinations of inner- and outer-race rotation were developed in Ref. 40. Without introducing additional relationships to describe the elastohydrodynamic conditions at both ball-race contacts,

TABLE 19.8   Principal Types of Tapered Roller Bearings

Type	Subtype	Approx. range of bore sizes, mm
Single row		8–1690
TS	TST—tapered bore	24–430
	TSS—steep angle	16–1270
	TS—pin cage	—
	TSE, TSK—keyway cones	12–380
	TSF, TSSF—flanged cup	8–1070
	TSG—steering gear (without cone)	—
Two-row, double cone, single cups		30–1200
TDI	TDIK, TDIT, TDITP—tapered bore	30–860
	TDIE, TDIKE—slotted double cone	24–690
	TDIS—steep angle	55–520
Two-row, double cup, single cones, adjustable	TDO	8–1830
TDO	TDOS—steep angle	20–1430
Two-row, double cup, single cones, nonadjustable	TNA	20–60
TNA	TNASW—slotted cones	30–260
	TNASWE—extended cone rib	20–305
	TNASWH—slotted cones, sealed	8–70
	TNADA, TNHDADX—self-aligning cup AD	—
Four-row, cup adjusted		70–1500
TQO	TQO, TOOT—tapered bore	250–1500
Four-row, cup adjusted		
TQI	TQIT—tapered bore	—

however, the ball spin-axis orientation angle θ cannot be obtained. As mentioned, this requires a lengthy numerical solution except for the two extreme cases of outer- or inner-race control. These are illustrated in Fig. 19.5.

Race control assumes that pure rolling occurs at the controlling race, with all of the ball spin occurring at the outer-race contact. The orientation of the ball rotational axis is then easily determinable from bearing geometry. Race control probably occurs only in dry bearings or dry-film-lubricated bearings where Coulomb friction conditions exist in the ball-race contact ellipses. The moment-resisting spin will always be greater at one of the race contacts. Pure rolling will occur at the race contact with the higher magnitude of moment-resisting spin. This is usually the inner race at low speeds and the outer race at high speeds.

In oil-lubricated bearings in which elastohydrodynamic films exist in both ball-race contacts, rolling with spin occurs at both contacts. Therefore, precise ball motions

TABLE 19.9 Representative Needle Roller Bearings

Type	Bore sizes, mm		Relative load capacity		Limiting speed factor	Misalignment tolerance
	Min.	Max.	Dynamic	Static		
Drawn cup, needle	3	185	High	Moderate	0.3	Low
Drawn cup, needle, grease retained	4	25	High	Moderate	0.3	Low
Drawn cup, roller	5	70	Moderate	Moderate	0.9	Moderate
Heavy-duty roller	16	235	Very high	Moderate	1.0	Moderate
Caged roller	12	100	Very high	High	1.0	Moderate
Cam follower	12	150	Moderate to high	Moderate to high	0.3–0.9	Low
Needle thrust	6	105	Very high	Very high	0.7	Low

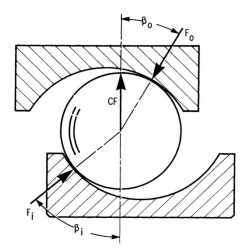

FIG. 19.3 Contact angles in a ball bearing at appreciable speeds.

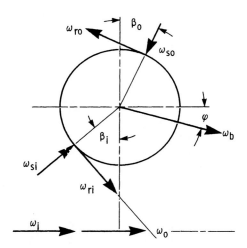

FIG. 19.4 Angular velocities of a ball.

can only be determined through use of a computer analysis. We can approximate the situation with a reasonable degree of accuracy, however, by assuming that the ball rolling axis is normal to the line drawn through the centers of the two ball-race contacts.

The angular velocity of the separator or ball set ω_c about the shaft axis can be shown to be (Ref. 4),

$$\omega_c = \frac{(v_i + v_o)/2}{d_e/2} = \frac{1}{2}\left[\omega_i\left(1 - \frac{d\cos\beta}{d_e}\right) + \omega_o\left(1 + \frac{d\cos\beta}{d_e}\right)\right] \qquad (19.1)$$

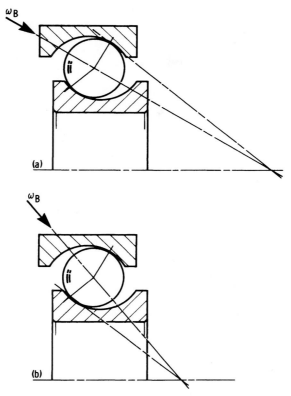

FIG. 19.5 Ball spin-axis orientations for outer- and inner-race control. (*a*) Outer-race control. (*b*) Inner-race control.

where v_i and v_o are the linear velocities of the inner and outer contacts. The angular velocity of a ball about its own axis ω_b is

$$\omega_b = \frac{v_i - v_o}{d_e/2} = \frac{d_e}{2d}\left[\omega_i\left(1 - \frac{d\cos\beta}{d_e}\right) - \omega_o\left(1 + \frac{d\cos\beta}{d_e}\right)\right] \qquad (19.2)$$

To calculate the velocities of the ball-race contacts, which are required for calculating elastohydrodynamic film thicknesses, it is convenient to use a coordinate system that rotates at ω_c. This fixes the ball-race contacts relative to the observer. In the rotating coordinate system, the angular velocities of the inner and outer races become

$$\omega_{ir} = \omega_i - \omega_c = [(\omega_i - \omega_o)/2]\,[1 + (d\cos\beta)/d_e]$$

$$\omega_{or} = \omega_o - \omega_c = [(\omega_o - \omega_i)/2]\,[1 - (d\cos\beta)/d_e]$$

The surface velocities entering the *ball–inner-race contact* for pure rolling are

$$u_{ai} = u_{bi} = [(d_e - d\cos\beta)/2]\omega_{ir} \qquad (19.3)$$

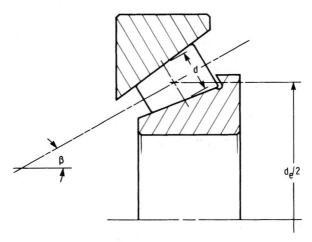

FIG. 19.6   Simplified geometry for tapered-roller bearing.

or $\qquad u_{ai} = u_{bi} = [d_e(\omega_i - \omega_o)/4][1 - (d^2 \cos^2 \beta)/d_e^{\,2}]$ $\qquad$ (19.4)

and those at the *ball–outer-race contact* are

$$u_{ao} = u_{bo} = [(d_e + d \cos \beta)/2]\omega_{or}$$

or $\qquad u_{ao} = u_{bo} = [d_e(\omega_o - \omega_i)/4][(1 - d^2 \cos^2 \beta)/d_e^{\,2}]$ $\qquad$ (19.5)

For a cylindrical roller bearing, $\beta = 0°$ and Eqs. (19.1), (19.2), (19.4), and (19.5) become, if d is roller diameter,

$$\omega_c = \tfrac{1}{2}[\omega_i(1 - d/d_e) + \omega_o(1 + d/d_e)]$$

$$\omega_R = (d_e/2d)[\omega_i(1 - d/d_e) + \omega_o(1 + d/d_e)]$$

$$u_{ai} = u_{bi} = [d_e(\omega_i - \omega_o)/4](1 - d^2/d_e^{\,2})$$

$$u_{ao} = u_{bo} = [d_e(\omega_o - \omega_i)/4](1 - d^2/d_e^{\,2})$$

$\qquad$ (19.6)

For a tapered-roller bearing, equations directly analogous to those for a ball bearing can be used if d is the average diameter of the tapered roller, d_e is the diameter at which the geometric center of the rollers is located, and β is the angle as shown in Fig. 19.6.

19.4 *MATERIALS AND MANUFACTURING PROCESSES*

Until about 1955, rolling-element bearing materials technology did not receive much attention from materials scientists. Bearing materials were restricted to SAE 52100 and some carburizing grades, such as AISI 4320 and AISI 9310, which seemed to be adequate for most bearing applications, despite the limitation in temperature of about 176°C (350°F) for 52100 steel. A minimum acceptable hardness of Rockwell C 58 should be specified. Experiments indicate that fatigue life increases with increasing hardness.

The advent of the aircraft gas turbine engine, with its need for advanced rolling-element bearings, provided the major impetus for advancements in rolling-element bearing materials technology. Increased temperatures, higher speeds and loads, and the need for greater durability and reliability all served as incentives for the development and evaluation of a broad range of new materials and processing methods. The combined research efforts of bearing manufacturers, engine manufacturers, and government agencies over the past three decades have resulted in startling advances in rolling-element bearing life and reliability and in performance. The discussion here is brief in scope. For a comprehensive treatment of the research status of current bearing technology and current bearing designs, refer to Ref. 10.

19.4.1 Ferrous Alloys

The need for higher temperature capability led to the evaluation of a number of available molybdenum and tungsten alloy tool steels as bearing materials (Table 19.10). These alloys have excellent high-temperature hardness retention. Such alloys melted and cast in an air environment, however, were generally deficient in fatigue resistance because of the presence of nonmetallic inclusions. Vacuum processing techniques can reduce or eliminate these inclusions. Techniques used include vacuum induction melting (VIM) and vacuum arc remelting (VAR). These have been extensively explored, not only with the tool steels now used as bearing materials, but with SAE 52100 and some of the carburizing steels as well. Table 19.10 lists a fairly complete array of ferrous alloys, both fully developed and experimental, from which present-day bearings are fabricated. AISI M-50, usually VIM-VAR or consumable electrode vacuum melted (CEVM) processed, has become a very widely used quality bearing material. It is usable at temperatures to 315°C (600°F), and it is usually assigned a materials life fac-

TABLE 19.10 Typical Chemical Compositions of Selected Bearing Steels[10]

Alloying element percent by weight

Designation	C	P (max)	S (max)	Mn	Si	Cr	V	W	Mo	Co	Cb	Ni
SAE 52100	1.00	0.025	0.025	0.35	0.30	1.45						
MHT*	1.03	0.025	0.025	0.35	0.35	1.50						
AISI M-1	0.80	0.030	0.030	0.30	0.30	4.00	1.00	1.50	8.00			
AISI M-2	0.83			0.30	0.30	3.85	1.90	6.15	5.00			
AISI M-10	0.85			0.25	0.30	4.00	2.00	—	8.00			
AISI M-50	0.80			0.30	0.25	4.00	1.00	—	4.25			
T-1 (18-4-1)	0.70			0.30	0.25	4.00	1.00	18.0				
T15	1.52	0.010	0.004	0.26	0.25	4.70	4.90	12.5	0.20	5.10		
44OC	1.03	0.018	0.014	0.48	0.41	17.30	0.14	—	0.50			
AMS 5749	1.15	0.012	0.004	0.50	0.30	14.50	1.20	—	4.00			
Vasco matrix II	0.53	0.014	0.013	0.12	0.21	4.13	1.08	1.40	4.80	7.81	—	0.10
CRB-7	1.10	0.016	0.003	0.43	0.31	14.00	1.03	—	2.02	—	0.32	
AISI 9310†	0.10	0.006	0.001	0.54	0.28	1.18	—	—	0.11	—	—	3.15
CBS 600†	0.19	0.007	0.014	0.61	1.05	1.50	—	—	0.94	—	—	0.18
CBS 1000M†	0.14	0.018	0.019	0.48	0.43	1.12	—	—	4.77	—	—	2.94
Vasco X-2†	0.14	0.011	0.011	0.24	0.94	4.76	0.45	1.40	1.40	0.03	—	0.10

*Also contains 1.36 percent Al.
†Carburizing grades.

tor of 3 to 5 (Sec. 9). T-1 tool steel has also come into fairly wide use, mostly in Europe, in bearings. Its hot hardness retention is slightly superior to that of M-50 and approximately equal to that of M-1 and M-2. These alloys retain adequate hardness to about 400°C (750°F).

Surface-hardened or carburized steels are used in many bearings where, because of shock loads or cyclic bending stresses, the fracture toughness of the through-hardened steels is inadequate. Some of the newer materials being developed, such as CBS 1000 and Vasco X-2 have hot hardness retention comparable to that of the tool steels (Fig. 19.7). They too are available as ultraclean, vacuum-processed materials and should offer adequate resistance to fatigue. Carburized steels may become of increasing importance in ultrahigh-speed applications. Bearings with through-hardened steel races are currently limited to $d_b N \approx 2.5 \times 10^6$ (where d_b is bore diameter in millimeters and N is rotational speed in revolutions per minute) because, at higher $d_b N$ values, fatigue cracks propagate through the rotating race as a result of the excessive hoop stress present.[9]

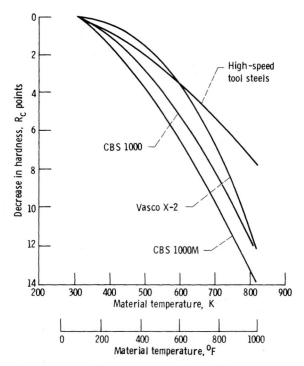

FIG. 19.7 Hot hardness of CBS 1000, CBS 1000M, Vasco X-2, and high-speed tool steels. (*From Anderson and Zaretsky.*[3])

In applications where the bearings are not lubricated with conventional oils and protected from corrosion at all times, a corrosion-resistant alloy should be used. Dry-film-lubricated bearings, bearings cooled by liquefied cryogenic gases, and bearings exposed to corrosive environments such as very high humidity and salt water are applications where corrosion-resistant alloys should be considered. Of the alloys listed in Table 19.10, both 440C and AMS 5749 are readily available in vacuum-melted heats.

In addition to improved melting practice, forging and forming methods that result in improved resistance to fatigue have been developed. Experiments indicate that fiber or grain flow parallel to the stressed surface is superior to fiber flow that intersects the stressed surface.[7,63] Forming methods that result in more parallel grain flow are now being used in the manufacture of many bearings, especially those used in high-load applications.

19.4.2 Ceramics

Experimental bearings have been made from a variety of ceramics including alumina, silicon carbide, titanium carbide, and silicon nitride. The use of ceramics as bearing materials for specialized applications will probably continue to grow for several reasons. These include:

1. *High-temperature capability:* Because ceramics can exhibit elastic behavior to temperatures beyond 1000°C (1821°F), they are an obvious choice for extreme-temperature applications.
2. *Corrosion resistance:* Ceramics are essentially chemically inert and able to function in many environments hostile to ferrous alloys.
3. *Low density:* This can be translated into improved bearing capacity at high speeds, where centrifugal effects predominate.
4. *Low coefficient of thermal expansion:* Under conditions of severe thermal gradients, ceramic bearings exhibit less drastic changes in geometry and internal play than do ferrous alloy bearings.

At the present time, silicon nitride is being actively developed as a bearing material.[15,53] Silicon nitride bearings have exhibited fatigue lives comparable to and, in some instances, superior to that of high-quality vacuum-melted M-50. Two problems remain: (1) quality control and precise nondestructive inspection techniques to determine acceptability, and (2) cost. Improved hot isostatic compaction, metrology, and finishing techniques are all being actively pursued. However, silicon nitride bearings must still be considered experimental.

19.5 SEPARATORS

Ball- and roller-bearing separators, sometimes called "cages" or "retainers," are bearing components that, although never carrying load, are capable of exerting a vital influence on the efficiency of the bearing. In a bearing without a separator, the rolling elements contact each other during operation and in so doing experience severe sliding and friction. The primary function of a separator is to maintain the proper distance between the rolling element and to ensure proper load distribution and balance within the bearing. Another function of the separator is to maintain control of the rolling elements in such a manner as to produce the least possible friction through sliding contact. Furthermore, a separator is necessary for several types of bearings to prevent the rolling elements from falling out of the bearing during handling. Most separator troubles occur from improper mounting, misaligned bearings, or improper (inadequate or excessive) clearance in the rolling-element pocket.

The materials used for separators vary according to the type of bearing and the application. In ball bearings and some sizes of roller bearings, the most common type of separator is made from two strips of carbon steel that are pressed and riveted together. Called ribbon separators, they are the least expensive to manufacture and are entirely suitable for many applications. They are also lightweight and usually require little space.

The design and construction of angular-contact ball bearings allow the use of a one-piece separator. The simplicity and inherent strength of one-piece separators permit their fabrication from many desirable materials. Reinforced phenolic and bronze are the two most commonly used materials. Bronze separators offer strength and low-friction characteristics and can be operated at temperatures to 230°C (450°F). Machined, silver-plated ferrous alloy separators are used in many demanding applications. Because reinforced cotton-base phenolic separators combine the advantages of low weight, strength, and nongalling properties, they are used for such high-speed applications as gyro bearings. In high-speed bearings, lightness and strength are particularly desirable, since the stresses increase with speed but may be greatly minimized by reduction of separator weight. A limitation of phenolic separators, however, is that they have an allowable maximum temperature of about 135°C (275°F).

19.6 CONTACT STRESSES AND DEFORMATIONS

The loads carried by rolling-element bearings are transmitted through the rolling element from one race to the other. The magnitude of the load carried by an individual rolling element depends on the internal geometry of the bearing and the location of the rolling element at any instant. A load-deflection relationship for the rolling-element–race contact is developed in this section. The deformation within the contact is a function of, among other things, the ellipticity parameter and the elliptic integrals of the first and second kinds. Simplified expressions that allow quick calculations of the stresses and deformations to be made easily from a knowledge of the applied load, the material properties, and the geometry of the contacting elements are presented in this section.

19.6.1 Elliptical Contacts

When two elastic solids are brought together under a load, a contact area develops, the shape and size of which depend on the applied load, the elastic properties of the materials, and the curvatures of the surfaces. The shape of the contact area may be elliptical, with D_y being the diameter in the y direction (transverse direction) and D_x being the diameter in the x direction (direction of motion). For the special case where $r_{ax} = r_{ay}$ and $r_{bx} = r_{by}$, the resulting contact is a circle rather than an ellipse. Where r_{ay} and r_{by} are both infinity, the initial line contact develops into a rectangle when load is applied.

The contact ellipses obtained with either a radial or a thrust load for the ball–inner-race and ball–outer-race contacts in a ball bearing are shown in Fig. 19.8. Inasmuch as the size and shape of these contact areas are highly significant in the successful operation of rolling elements, it is important to understand their characteristics.

The ellipticity parameter k is defined as the elliptical-contact diameter in the y direction (transverse direction) divided by the elliptical-contact diameter in the x direction (direction of motion), or $k = D_y/D_x$. If the equation $1/r_{ax} + 1/r_{bx} > 1/r_{ay} + 1/r_{by}$ is satisfied and $\alpha \geq 1$, the contact ellipse will be oriented so that its major diameter will be transverse to the direction of motion, and consequently $k \geq 1$. Otherwise, the major diameter would lie along the direction of motion with both $\alpha \leq 1$ and $k \leq 1$. Figure 19.9 shows the ellipticity parameter and the elliptic integrals of the first and

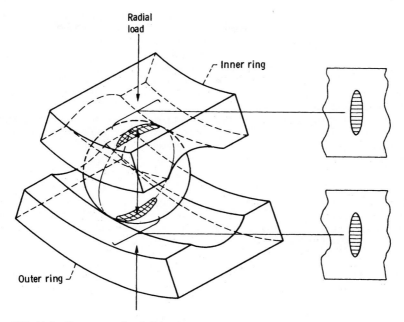

FIG. 19.8 Contact areas in a ball bearing.

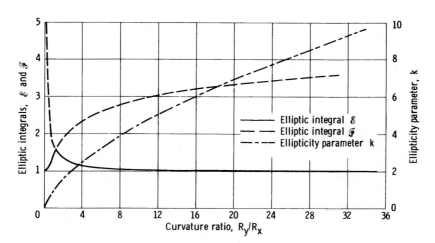

FIG. 19.9 Ellipticity parameter and elliptic integrals of the first and second kinds as a function of the curvature ratio.

second kinds for a range of the curvature ratio ($\alpha = R_y/R_x$) usually encountered in concentrated contacts.

Simplified Solutions for $\alpha \geq 1$. The classical hertzian solution requires the calculation of the ellipticity parameter k and the complete elliptic integrals of the first and second kinds $\mathcal{F}$ and $\mathcal{E}$. This entails finding a solution to a transcendental equation relating k, $\mathcal{F}$, and $\mathcal{E}$ to the geometry of the contacting solids. This is usually accomplished by some iterative numerical procedure, as described in Ref. 25, or with the aid of charts, as shown in Ref. 37. Reference 26 provides a shortcut to the classical hertzian solution for the local stress and deformation of two elastic bodies in contact. The shortcut is accomplished by using simplified forms of the ellipticity parameter and the complete elliptic integrals, expressing them as functions of the geometry. The results are summarized here.

Table 19.11 shows various values of radius-of-curvature ratios and corresponding values of k, $\mathcal{F}$, and $\mathcal{E}$ obtained from the numerical procedure given in Ref. 25. For the set of pairs of data $[(k_i, \alpha_i), i = 1, 2,..., 26]$, a power fit using linear regression by the method of least squares resulted in the following equation:

$$\bar{k} = \alpha^{2/\pi} \qquad \text{for } \alpha \geq 1 \qquad (19.7)$$

The asymptotic behavior of $\mathcal{E}$ and $\mathcal{F}$ ($\alpha \rightarrow 1$ implies $\mathcal{E} \rightarrow \mathcal{F} \rightarrow \pi/2$, and $\alpha \rightarrow \infty$ implies $\mathcal{F} \rightarrow \infty$ and $\mathcal{E} \rightarrow 1$) was suggestive of the type of functional dependence that $\mathcal{E}$ and $\mathcal{F}$ might follow. As a result, an inverse and a logarithmic curve fit were tried for $\mathcal{E}$ and $\mathcal{F}$, respectively. The following expressions provided excellent curve fits:

TABLE 19.11 Comparison of Numerically Determined Values with Curve-Fit Values for Geometrically Dependent Variables

($R_x = 1.0$ cm)

Radius-of-curvature ratio α	Elliptical parameter k	Ellipticity $\bar{k}$	Percent error e	Complete elliptic integral of first kind $\mathcal{F}$	$\bar{\mathcal{F}}$	Percent error e	Complete elliptic integral of second kind $\mathcal{E}$	$\bar{\mathcal{E}}$	Percent error e
1.00	1.0000	1.0000	0	1.5708	1.5708	0	1.5708	1.5708	0
1.25	1.1604	1.1526	0.66	1.6897	1.6892	−0.50	1.4643	1.4566	0.52
1.50	1.3101	1.2945	1.19	1.7898	1.8022	−0.70	1.3911	1.3805	0.76
1.75	1.4514	1.4280	1.61	1.8761	1.8902	−0.75	1.3378	1.3262	0.87
2.00	1.5858	1.5547	1.96	1.9521	1.9664	−0.73	1.2972	1.2854	0.91
3.00	2.0720	2.0125	2.87	2.1883	2.1979	−0.44	1.2002	1.1903	0.83
4.00	2.5007	2.4170	3.35	2.3595	2.3621	−0.11	1.1506	1.1427	0.69
5.00	2.8902	2.7860	3.61	2.4937	2.4895	0.17	1.1205	1.1142	0.57
6.00	3.2505	3.1289	3.74	2.6040	2.5935	0.40	1.1004	1.0951	0.48
7.00	3.5878	3.4515	3.80	2.6975	2.6815	0.59	1.0859	1.0815	0.40
8.00	3.9065	3.7577	3.81	2.7786	2.7577	0.75	1.0751	1.0713	0.35
9.00	4.2096	4.0503	3.78	2.8502	2.8250	0.88	1.0666	1.0634	0.30
10.00	4.4994	4.3313	3.74	2.9142	2.8851	1.00	1.0599	1.0571	0.26
15.00	5.7996	5.6069	3.32	3.1603	3.1165	1.38	1.0397	1.0381	0.15
20.00	6.9287	6.7338	2.81	3.3342	3.2807	1.60	1.0296	1.0285	0.10
25.00	7.9440	7.7617	2.29	3.4685	3.4081	1.74	1.0236	1.0228	0.07
30.00	8.8762	8.7170	1.79	3.5779	3.5122	1.84	1.0196	1.0190	0.05
35.00	9.7442	9.6158	1.32	3.6700	3.6002	1.90	1.0167	1.0163	0.04
40.00	10.5605	10.4689	0.87	3.7496	3.6764	1.95	1.0146	1.0143	0.03
45.00	11.3340	11.2841	0.44	3.8196	3.7436	1.99	1.0129	1.0127	0.02
50.00	12.0711	12.0670	0.03	3.8821	3.8038	2.02	1.0116	1.0114	0.02
60.00	13.4557	13.5521	−0.72	3.9898	3.9078	2.06	1.0096	1.0095	0.01
70.00	14.7430	14.9495	−1.40	4.0806	3.9958	2.08	1.0082	1.0082	0.01
80.00	15.9522	16.2759	−2.03	4.1590	4.0720	2.09	1.0072	1.0071	0.01
90.00	17.0969	17.5432	−2.61	4.2280	4.1393	2.10	1.0064	1.0063	0
100.00	18.1871	18.7603	−3.15	4.2895	4.1994	2.10	1.0057	1.0057	0

$$\overline{\mathscr{E}} = 1 + q/\alpha \qquad \text{for } \alpha \geq 1 \tag{19.8}$$

$$\overline{\mathscr{F}} = \pi/2 + q \ln \alpha \qquad \text{for } \alpha \geq 1 \tag{19.9}$$

where
$$q = \pi/2 - 1$$

Values of $\overline{k}$, $\overline{\mathscr{E}}$, and $\overline{\mathscr{F}}$ are presented in Table 19.11 and compared with the numerically determined values of k, $\mathscr{E}$, and $\mathscr{F}$. Table 19.11 also gives the percentage of error determined as

$$e = (\overline{z} - z)100/z \qquad z = \{k, \mathscr{E}, \mathscr{F}\} \qquad \overline{z} = \{\overline{k}, \overline{\mathscr{E}}, \overline{\mathscr{F}}\}$$

When the ellipticity parameter k [Eq. (19.7)], the elliptic integrals of the first and second kinds [Eqs. (19.9) and (19.8)], the normal applied load F, Poisson's ratio v, and the modulus of elasticity E of the contacting solids are known, we can write the major and minor axes of the contact ellipse and the maximum deformation at the center of the contact, from the analysis of Ref. 34, as

$$D_y = 2(6\overline{k}^2\overline{\mathscr{E}} FR/\pi E')^{1/3} \tag{19.10}$$

$$D_x = 2(6\overline{\mathscr{E}} FR/\pi \overline{k} E')^{1/3} \tag{19.11}$$

$$\delta = \overline{\mathscr{F}}[(9/2\overline{\mathscr{E}} R)(F/\pi \overline{k} E')^2]^{1/3} \tag{19.12}$$

$$E' = \frac{2}{(1 - v_a^2)/E_a + (1 - v_b^2)/E_b} \tag{19.13}$$

In these equations, D_y and D_x are proportional to $F^{1/3}$ and δ is proportional to $F^{2/3}$.

The maximum hertzian stress at the center of the contact can also be determined by using Eqs. (19.10) and (19.11).

$$\sigma_{max} = 6F/\pi D_x D_y$$

Simplified Solutions for $\alpha \leq 1$. Table 19.12 gives the simplified equations for $\alpha < 1$ as well as for $\alpha \geq 1$. It is important to make the proper evaluation of α since it has a great significance in the outcome of the simplified equations. Figure 19.10 shows three diverse situations in which the simplified equations can be usefully applied. The locomotive wheel on a rail, Fig. 19.10a, illustrates an example in which the ellipticity parameter k and the radius ratio α are less than 1. The ball rolling against a flat plate, Fig. 19.10b, provides pure circular contact (i.e., $\alpha = k = 1.0$). Figure 19.10c shows how the contact ellipse is formed in the ball–outer-ring contact of a ball bearing. Here the semimajor axis is normal to the direction of rolling and consequently α and k are greater than 1. Table 19.13 uses this figure to show how the degree of conformity affects the contact parameters.

19.6.2 Rectangular Contacts

For this situation, the contact ellipse discussed in the preceding section is of infinite length in the transverse direction ($D_y \rightarrow \infty$). This type of contact is exemplified by a cylinder loaded against a plate, a groove, or another, parallel, cylinder or by a roller loaded against an inner or outer ring. In these situations, the contact semiwidth is given by

$$b = R_x \sqrt{8W/\pi}$$

where the dimensionless load

TABLE 19.12   Simplified Equations

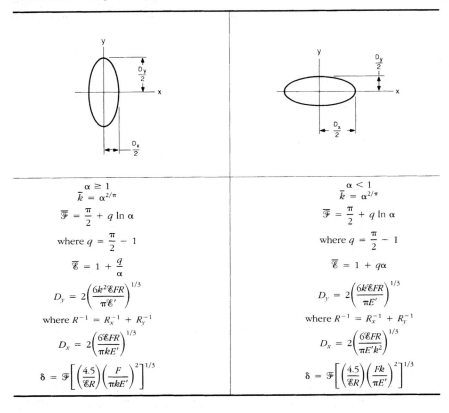

$$W = F'/E'R_x$$

and F' is the load per unit length along the contact. The maximum deformation for a rectangular contact can be written as (Ref. 19)

$$\delta = (2WR_x/\pi)[\tfrac{2}{3} + \ln(4r_{ax}/b) + \ln(4r_{bx}/b)] \qquad (19.14)$$

The maximum hertzian stress in a rectangular contact can be written as

$$\sigma_{max} = E'\sqrt{W/2\pi}$$

19.7   STATIC LOAD DISTRIBUTION

With the simple analytical expression for deformation in terms of load defined in Sec. 19.6, it is possible to consider how the bearing load is distributed among the rolling elements. Most rolling-element bearing applications involve steady-state rotation of either the inner or outer race or both; however, the speeds of rotation are usually not so great as to cause ball or roller centrifugal forces or gyroscopic moments of significant magnitudes. In analyzing the loading distribution on the rolling elements, it is usually satisfactory to ignore these effects in most applications. In this section the load-deflection relationships for ball and roller bearings are given, along with radial and thrust load distributions of statically loaded rolling elements.

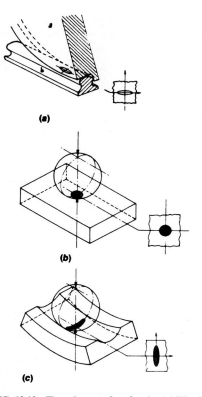

(a)

(b)

(c)

FIG. 19.10 Three degrees of conformity. (*a*) Wheel on rail. (*b*) Ball on plane. (*c*)Ball–outer-race contact.

19.7.1 Load-Deflection Relationships

For an elliptical contact, the load-deflection relationship given in Eq. (19.12) can be written as

$$F = K_{1.5}\delta^{3/2} \qquad (19.15)$$

where $\quad K_{1.5} = \pi k E' \sqrt{2\mathscr{E}R/9\mathscr{F}^3} \quad (19.16)$

Similarly, for a rectangular contact, Eq. (19.14) gives

$$f = K_1\delta$$

where $\quad K_1 = \dfrac{\pi/E'/2}{\tfrac{2}{3} + \ln{(4r_{ax}/b)} + \ln{(4r_{bx}/b)}}$

$$(19.17)$$

In general then,

$$F = K_j\delta^j \qquad (19.18)$$

TABLE 19.13 Practical Applications for Differing Conformities

($E' = 2.197 \times 10^7$ N/cm²)

Contact parameters	Wheel on rail	Ball on plane	Ball–outer-race contact
F, N	1.00×10^5	222.4111	222.4111
r_{ax}, cm	50.1900	0.6350	0.6350
r_{ay}, cm	∞	0.6350	0.6350
r_{bx}, cm	∞	∞	-3.8900
r_{by}, cm	30.0000	∞	-0.6600
α	0.5977	1.0000	22.0905
k	0.7099	1.0000	7.3649
$\bar{k}$	0.7206	1.0000	7.1738
$\mathscr{E}$	1.3526	1.5708	1.0267
$\bar{\mathscr{E}}$	1.3412	1.5708	1.0258
$\mathscr{F}$	1.8508	1.5708	3.3941
$\bar{\mathscr{F}}$	1.8645	1.5708	3.3375
D_y, cm	1.0783	0.0426	0.1842
$\bar{D}_y$, cm	1.0807	0.0426	0.1810
D_x, cm	1.5190	0.0426	0.0250
$\bar{D}_x$, cm	1.4997	0.0426	0.0252
δ, cm	0.0106	7.13×10^{-4}	3.56×10^{-4}
$\bar{\delta}$, cm	0.0108	7.13×10^{-4}	3.57×10^{-4}
σ_{max}, N/cm²	1.66×10^5	2.34×10^5	9.22×10^4
$\bar{\sigma}_{max}$, N/cm²	1.1784×10^5	2.34×10^5	9.30×10^4

in which $j = 1.5$ for ball bearings and 1.0 for roller bearings. The total normal approach between two races separated by a rolling element is the sum of the deformations under load between the rolling element and both races. Therefore,

$$\delta = \delta_o + \delta_i \tag{19.19}$$

where

$$\delta_o = [F/(K_j)_o]^{1/j} \tag{19.20}$$

$$\delta_i = [F/(K_j)_i]^{1/j} \tag{19.21}$$

Substituting Eqs. (19.19) through (19.21) into Eq. (19.18) gives

$$K_j = \{[1/(K_j)_o]^{1/j} + [1/(K_j)_i]^{1/j}\}^{-j}$$

Recall that $(K_j)_o$ and $(K_j)_i$ are defined by Eq. (19.16) or (19.17) for an elliptical or rectangular contact, respectively. From these equations, we observe that $(K_j)_o$ and $(K_j)_i$ are functions of only the geometry of the contact and the material properties. The radial and thrust load analyses are presented in the following two sections and are directly applicable for radially loaded ball and roller bearings and thrust-loaded ball bearings.

19.7.2 Radially Loaded Ball and Roller Bearings

A radially loaded rolling element with radial clearance P_d is shown in Fig. 19.11. In the concentric position shown in Fig. 19.11a, a uniform radial clearance between the rolling element and the races of $P_d/2$ is evident. The application of a small radial load to the shaft causes the inner ring to move a distance $P_d/2$ before contact is made between a rolling element located on the load line and the inner and outer races. At any angle, there will still be a radial clearance c that, if P_d is small compared with the radius of the tracks, can be expressed with adequate accuracy by $c = (1 - \cos \psi)P_d/2$. On the load line where $\psi = 0$, the clearance is zero, but where $\psi = 90°$, the clearance retains its initial value of $P_d/2$.

The application of further load will cause elastic deformation of the balls and the elimination of clearance around an arc $2\psi_c$. If the interference or total elastic compression on the load line is δ_{max}, the corresponding elastic compression of the ball δ_ψ along a radius at angle ψ to the load line will be given by

$$\delta_\psi = (\delta_{max} \cos \psi - c) = (\delta_{max} + P_d/2) \cos \psi - P_d/2$$

This assumes that the races are rigid. Now, it is clear from Fig. 19.11c that $(\delta_{max} + P_d/2)$ represents the total relative radial displacement of the inner and outer races. Hence,

$$\delta_\psi = \delta \cos \psi_d - P/2 \tag{19.22}$$

The relationship between load and elastic compression along the radius at angle ψ to the load vector is given by Eq. (19.18) as

$$F_\psi = K_j \delta_\psi{}^j$$

Substituting Eq. (19.22) into this equation gives $F_\psi = K_j(\delta \cos \psi - P_d/2)^j$.

For static equilibrium, the applied load must equal the sum of the components of the rolling-element loads parallel to the direction of the applied load. $F_r = \Sigma F_\psi \cos \psi$. Therefore,

$$F_r = K_j\Sigma(\delta \cos \psi - P_d/2)^j \cos \psi \tag{19.23}$$

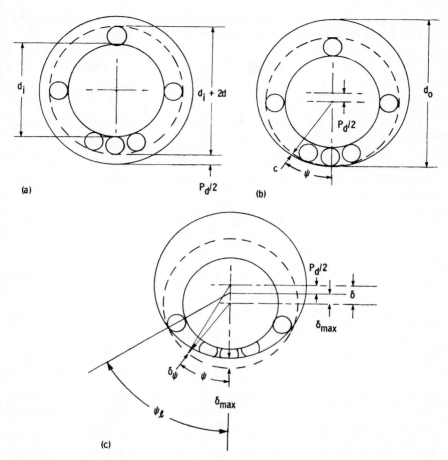

FIG. 19.11 Radially loaded rolling-element bearing. (*a*) Concentric arrangement. (*b*) Initial contact. (*c*) Interference.

The angular extent of the bearing arc $2\psi_p$ in which the rolling elements are loaded, is obtained by setting the root expression in Eq. (19.23) equal to zero and solving for ψ:

$$\psi_l = \arccos(P_d/2\delta)$$

The summation in Eq. (19.23) applies only to the angular extent of the loaded region. This equation can be written for a *roller bearing* as

$$F_r = [\psi_l - (P_d/2\delta) \sin \psi_l](nK_1\delta/2\pi) \tag{19.24}$$

and similarly in integral form for a *ball bearing* as

$$F_r = \frac{n}{\pi} k_{1.5}\delta^{3/2} \int_0^{\psi l} \left(\cos \psi - \frac{P_d}{2\delta}\right)^{3/2} \cos \psi \, d\psi$$

The integral in the equation can be reduced to a standard elliptic integral by the hypergeometric series and the beta function. If the integral is numerically evaluated directly, the following approximate expression is derived:

$$\int_0^{\psi_l} \left(\cos \psi - \frac{P_d}{2\delta}\right)^{3/2} \cos \psi \, d\psi = 2.491 \left\{\left[1 + \left(\frac{P_d/2\delta - 1}{1.23}\right)^2\right]^{1/2} - 1\right\}$$

This approximate expression fits the exact numerical solution to within ± 2 percent for a complete range of $P_d/2\delta$.

The load carried by the most heavily loaded ball is obtained by substituting $\psi = 0°$ in Eq. (19.23) and dropping the summation sign

$$F_{max} = K_j \delta^j (1 - P_d/2\delta)^j \tag{19.25}$$

Dividing the maximum ball load, Eq. (19.25), by the total radial load for a *roller bearing*, Eq. (19.24), gives

$$F_r = \frac{[\psi_l - (P_d/2\delta) \sin \psi_l](nF_{max}/2\pi)}{1 - P_d/2\delta} \tag{19.26}$$

and similarly for a *ball bearing*

$$F_r = nF_{max}/Z \tag{19.27}$$

where

$$Z = \frac{\pi(1 - P_d/2\delta)^{3/2}}{2.491\{[1 + ((1 - P_d/2\delta)/1.23))^2]^{1/2} - 1\}} \tag{19.28}$$

For *roller bearings* when the diametral clearance P_d is zero, Eq. (19.26) gives

$$F_r = nF_{max}/4 \tag{19.29}$$

For *ball bearings* when the diametral clearance P_d is zero, the value of Z in Eq. (19.27) becomes 4.37. This is the value derived for ball bearings of zero diametral clearance.[57] The approach used was to evaluate the finite summation for various numbers of balls. The celebrated Stribeck equation for static load-carrying capacity was then derived by writing the more conservative value of 5 for the theoretical value of 4.37:

$$F_r = nF_{max}/5 \tag{19.30}$$

In using Eq. (19.30), it should be remembered that Z was considered to be a constant and that the effects of clearance and applied load on load distribution were not taken into account. However, these effects were considered in obtaining Eq. (19.27).

19.7.3 Thrust-Loaded Ball Bearings

The "static-thrust-load capacity" of a ball bearing may be defined as the maximum thrust load that the bearing can endure before the contact ellipse approaches a race shoulder, as shown in Fig. 19.12, or the load at which the allowable mean compressive stress is reached, whichever is smaller. Both the limiting shoulder height and the mean compressive stress must be calculated to find the static-thrust-load capacity.

The contact ellipse in a bearing race under a load is shown in Fig. 19.12. Each ball is subjected to an identical thrust component F/n, where F_t is the total thrust load. The initial contact angle before the application of a thrust load is denoted by β_f. Under load, the normal ball thrust load F acts at the contact angle β and is written as

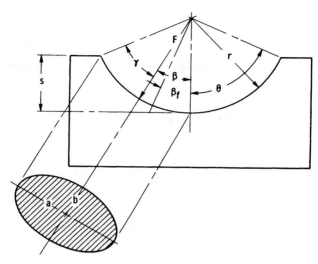

FIG. 19.12 Contact ellipse in bearing race.

$$F = F_t/(n \sin \beta) \tag{19.31}$$

A cross section through an angular-contact bearing under a thrust load F_t is shown in Fig. 19.13. From this figure, the contact angle after the thrust load has been applied can be written as

$$\beta = \arccos\left(\frac{D - P_d/2}{D + \delta}\right) \tag{19.32}$$

The initial contact angle is given in the equation $\cos \beta_f = (D - P_d/2)/D$. Using this equation and rearranging terms in Eq. (19.32) gives, solely from geometry (Fig. 19.13).

$$\delta = D(\cos \beta_f/\cos \beta - 1)$$

$$\delta = \delta_o + \delta_i$$

$$\delta = [F/(K_j)_o]^{1/j} + [F/(K_j)_i]^{1/j}$$

$$K_j = \{[1/(K_j)_o]^{1/j} + [1/(K_j)_i]^{1/j}\}^{-j}$$

$$= \{[4.5\overline{\mathscr{F}}_o^3/\pi k_o E_o'(R_o\overline{\mathscr{E}}_o)^{1/2}]^{2/3} + [4.5\overline{\mathscr{F}}_i^3/\pi k_i E_i(R_i\overline{\mathscr{E}}_i)^{1/2}]^{2/3}\}^{-1} \tag{19.33}$$

$$F = K_j D^{3/2}(\cos \beta_f/\cos \beta - 1)^{3/2} \tag{19.34}$$

where $\qquad K_{1.5} = \pi k E'(R\overline{\mathscr{E}}/4.5\overline{\mathscr{F}}^3)^{1/2} \tag{19.35}$

and k, $\overline{\mathscr{E}}$, and $\overline{\mathscr{F}}$ are given by Eqs. (19.7), (19.8), and (19.9), respectively.
From Eqs. (19.31) and (19.34), we can write

$$F_t/(n \sin \beta) = F$$

$$F_t/nK_j D^{3/2} = (\sin \beta)(\cos \beta_f/\cos \beta - 1)^{3/2} \tag{19.36}$$

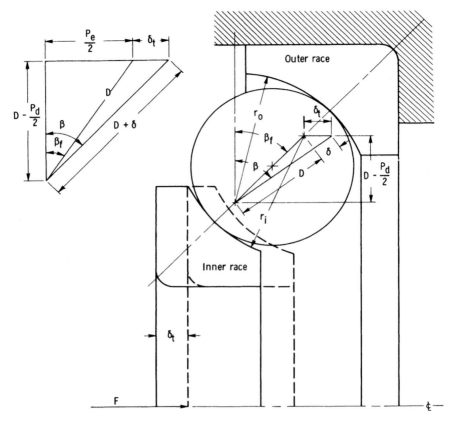

FIG. 19.13 Angular-contact ball bearing under thrust load.

This equation can be solved numerically by the Newton-Raphson method. The iterative equation to be satisfied is

$$\beta' - \beta = \frac{F/nK_{1.5}D^{3/2} - (\sin \beta)(\cos \beta_f/\cos \beta - 1)^{3/2}}{(\cos \beta)(\cos \beta_f/\cos \beta - 1)^{3/2} + \frac{3}{2}(\cos \beta_f)(\tan^2 \beta)(\cos \beta_f/\cos \beta - 1)^{1/2}} \quad (19.37)$$

which converges and is satisfied when $\beta' - \beta$ becomes essentially zero.

When a thrust load is applied, the shoulder height is limited to the distance by which the pressure-contact ellipse can approach the shoulder. As long as the following inequality is satisfied, the pressure-contact ellipse will not exceed the shoulder-height limit $\theta > \beta + \arcsin (D_y/fd)$. The angle used to define the shoulder height θ may be written as $\theta = \arccos (1 - s/fd)$. The axial deflection δ_t corresponding to a thrust load may be written as

$$\delta_t = (D + \delta) \sin \beta - D \sin \beta_f \quad (19.38)$$

Substituting Eq. (19.33) into Eq. (19.38) gives

$$\delta_t = [D \sin (\beta - \beta_f)]/(\cos \beta)$$

Having determined β from Eq. (19.37) and β_f from Eq. (19.19), we can easily evaluate the relationship for δ_r.

19.7.4 Preloading

In Sec. 19.2.1, the use of angular-contact bearings as duplex pairs preloaded against each other is discussed. As shown in Table 19.2, duplex bearing pairs are used in either back-to-back or face-to-face arrangements. Such bearings are usually preloaded against each other by providing what is called "stickout" in the manufacture of the baring. This is illustrated in Fig. 19.14 for a bearing pair used in a back-to-back arrangement. The magnitude of the stickout and the bearing design determine the level of preload on each bearing when the bearings are clamped together as in Fig. 19.14. The magnitude of pre-load and the load-deflection characteristics for a given bearing pair can be calculated by using Eqs. (19.15), (19.31), (19.33), (19.34), (19.35), and (19.36).

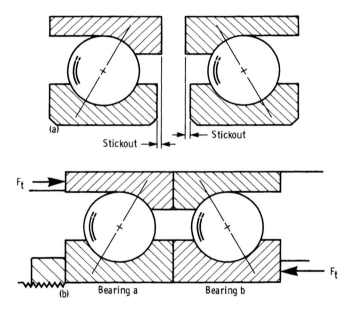

FIG. 19.14 Angular-contact bearings in back-to-back arrangement, shown individually as manufactured and as mounted with preload. (*a*) Separated. (*b*) Mounted and preloaded.

The relationship of initial preload, system load, and final load for bearings *a* and *b* is shown in Fig. 19.15. The load-deflection curve follows the relationship $\delta = KF^{2/3}$. When a system thrust load F_t is imposed on the bearing pairs, the magnitude of load on bearing *b* increases while that on bearing *a* decreases until the difference equals the system load. The physical situation demands that the *change* in each bearing deflection be the same (Δa and Δb in Fig. 19.15). The increments in bearing load, however, are *not* the same. This is important because it always requires a system thrust load far greater than twice the preload before one bearing becomes unloaded. Prevention of bearing unloading, which can result in skidding and early failure, is an objective of preloading.

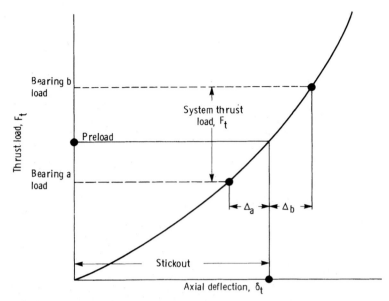

FIG. 19.15 Thrust load–axial deflection curve for a typical ball bearing.

19.8 ROLLING FRICTION AND FRICTION IN BEARING

19.8.1 Rolling Friction

The concepts of rolling friction are important, generally, in understanding the behavior of machine elements in rolling contact and particularly because rolling friction influences the overall behavior of rolling-element bearings. The theories of Refs. 33 and 50 attempted to explain rolling friction in terms of the energy required to overcome the interfacial slip that occurs because of the curved shape of the contact area. As shown in Fig. 19.16, the ball rolls about the y axis and makes contact with the groove from a to b. If the groove is fixed, then for zero slip over the contact area, no point within the area should have a velocity in the direction of rolling. The surface of the contact area is curved, however, so that points a and b are at different radii from the y axis than are points c and d. For a rigid ball, points a and b must have different velocities with respect to the y axis than do points c and d, because the velocity of any point on the ball relative to the y axis equals the angular velocity times the radius of the y axis. Slip must occur at various points over the contact area unless the body is so elastic that yielding can take place in the contact area to prevent this interfacial slip. The theories assumed that this interfacial slip took place and that the forces required to make a ball roll were those required to overcome the friction due to this interfacial slip. In the contact area, rolling without slip will occur at a specific radius from the y axis. Where the radius is greater than this radius to the rolling point, slip will occur in one direction; where it is less than this radius to the rolling point, slip will occur in the other direction. In Fig. 19.16, the lines to points c and d represent the approximate location of the rolling bands, and the arrows shown in the three portions of the contact area represent the directions of interfacial slip when the ball is rolling into the paper.

The location of the two rolling bands relative to the axes of the contact ellipse can

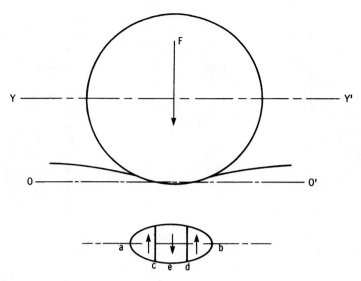

FIG. 19.16 Differential slip due to curvature of contact ellipse.

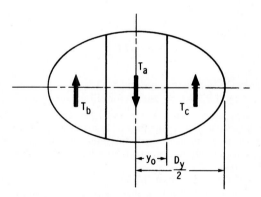

FIG. 19.17 Friction forces in contact ellipse.

be obtained by summing the forces acting on the ball in the direction of rolling. In Fig. 19.17, these are $2T_b - T_a = \mu F$, where μ is the coefficient of rolling friction. If a hertzian ellipsoidal pressure distribution is assumed, the location of the rolling bands can be determined (Ref. 11). For bearing steels $y_o/D_y \simeq 0.174$, where D_y is the diameter of the contact ellipse along the y axis [Eq. (19.10)].

In actuality, all materials are elastic so that areas of no slip as well as areas of microslip exist within the contact, as pointed out in Ref. 36. Differential strains in the materials in contact will cause slip unless it is prevented by friction. In high-conformity contacts, slip is likely to occur over most of the contact region, as shown in Fig. 19.18. The limiting value of friction force T is $T = 0.2\mu D_y^2 F/r_{ax}^2$.

A portion of the elastic energy of compression in rolling is always lost because of hysteresis. The effect of hysteresis losses on rolling resistance has been studied in Ref. 58, in which the following expression was developed:

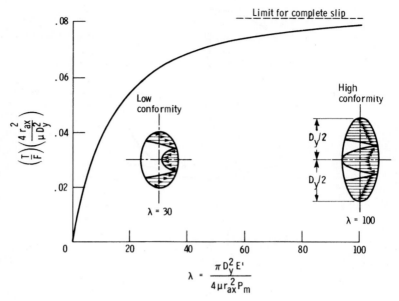

FIG. 19.18 Frictional resistance of ball in conforming groove.

$$T = c_4 \lambda F D_y / r_{ax}$$

where $\lambda = \pi D_y^2 E'/4\mu r_{ax}^2 P_m$

$\quad p_m$ = mean pressure within the contact

$\quad c_4$ = $\frac{1}{3}$ for rectangular contacts and $\frac{3}{32}$ for elliptical contacts

Two hard-steel surfaces show a value λ of about 1 percent.

Of far greater importance in contributing to frictional losses in rolling-element bearings, especially at high speeds, is ball spinning. Spinning as well as rolling takes place in one or both of the ball–race contacts of a ball bearing. The situation is shown schematically in Fig. 19.4.

High speeds cause a divergency of the contact angles β_i and β_o. In Fig. 19.5b the ball is rolling with inner-race control, so that approximately pure rolling takes place at the inner-race contact. The rolling and spinning vectors ω_r and ω_s at the outer-race contact are shown in Fig. 19.4. The higher the ratio ω_s/ω_r, the higher the friction losses.

19.8.2 Friction Losses in Rolling-Element Bearings

Some of the factors that affect the magnitude of friction losses in rolling bearings are (1) bearing size, (2) bearing type, (3) bearing design, (4) load (magnitude and type, either thrust or radial), (5) speed, (6) oil viscosity, and (7) oil flow. In a specific bearing, friction losses consist of the following:

1. Sliding friction losses in the contacts between the rolling elements and the races. These losses include differential slip and slip due to ball spinning. They are complicated by the presence of elastohydrodynamic lubrication films. The shearing of

elastohydrodynamic lubrication films, which are extremely thin and contain oil whose viscosity is increased by orders of magnitude above its atmospheric pressure value, accounts for a significant fraction of the friction losses in rolling-element bearings.

2. Hysteresis losses due to the damping capacity of the race and ball materials.

3. Sliding friction losses between the separator and its locating race surface and between the separator pockets and the rolling elements.

4. Shearing of oil films between the bearing parts and oil churning losses caused by excess lubricant within the bearing.

5. Flinging of oil off the rotating parts of the bearing.

Because of the dominant role played in bearing frictional losses by the method of lubrication, there are no quick and easy formulas for computing rolling-element bearing power loss. Coefficients of friction for a particular bearing can vary by a factor of 5, depending on lubrication. A flood-lubricated bearing may consume five times the power of one that is merely wetted by oil–air mist lubrication.

As a rough but useful guide, we can use the friction coefficients given in Ref. 46. These values were computed at a bearing load that will give a life of 1×10^9 revolutions for the respective bearings. The friction coefficients for several different bearings are shown below.

Bearing type	Friction coefficient
Self-aligning ball	0.0010
Cylindrical roller, with flange-guided short rollers	0.0011
Thrust ball	0.0013
Single-row, deep-groove ball	0.0015
Tapered and spherical roller, with flange-guided rollers	0.0018
Needle roller	0.0045

All of the friction coefficients are referenced to the bearing bore.

More accurate estimates of bearing power loss and temperature rise can be obtained by using one or more of the available computer codes that represent the basis for current design methodology for rolling-element bearings. These methodologies are discussed in Refs. 5 and 49, and are briefly reviewed later in this chapter.

19.9 LUBRICANTS AND LUBRICATION SYSTEMS

A liquid lubricant has several functions in a rolling-element bearing. It provides the medium for the establishment of separating films between the bearing parts (elastohydrodynamic between the races and rolling elements and hydrodynamic between the cage or separator and its locating surface). It serves as a coolant if circulated through the bearing either to an external heat exchanger or simply brought into contact with the bearing housing and machine casing. A circulating lubricant also serves to flush out wear debris and carry it to a filter where it can be removed from the system. Finally, it provides corrosion protection. The different methods of providing liquid lubricant to a bearing are discussed here individually.

19.9.1 Solid Lubrication

An increasing number of rolling-element bearings are lubricated with solid-film lubricants, usually in applications such as extreme temperature or the vacuum of space where conventional liquid lubricants are not suitable.

Success in cryogenic applications where the bearing is cooled by the cryogenic fluid (liquid oxygen or hydrogen) has been achieved with transfer films of polytetrafluoroethylene (Ref. 52). Bonded films of soft metals, such as silver, gold, and lead, applied by ion plating as very thin films (0.2 to 0.3 μm) have also been used (Ref. 60). Silver, and lead in particular, have found use in bearings used to support the rotating anode in x-ray tubes. Very thin films are required in rolling-element bearings in order not to significantly alter the bearing internal geometry and to retain the basic mechanical properties of the substrate materials in the hertzian contacts.

19.9.2 Liquid Lubrication

The great majority of rolling-element bearings are lubricated by liquids. The liquid in greases can be used or liquid can be supplied to the bearing from either noncirculating or circulating systems.

Greases. The most common and probably least expensive mode of lubrication is grease lubrication. In the strictest sense, a grease is not a liquid, but the liquid or fluid constituent in the grease is the lubricant. Greases consist of a fluid phase of either a petroleum oil or a synthetic oil and a thickener. The most common thickeners are sodium-, calcium-, or lithium-based soaps, although thickeners of inorganic materials such as bentonite clay have been used in synthetic greases. Some discussion of the characteristics and temperature limits of greases is given in Refs. 11 and 44.

Greases are usually retained within the bearing by shields or seals that are an integral part of the assembled bearing. Since there is no recirculating fluid, grease-lubricated bearings must reject their heat by conduction and convection. They are therefore limited in operating speeds to maximum d_bN values of 0.25×10^6 to 0.4×10^6.

The proper grease for a particular application depends on the temperature, speed, and ambient pressure environment to which the bearing is exposed. Reference 44 presents a comprehensive discussion useful in the selection of a grease; the bearing manufacturer can also recommend the most suitable grease and bearing type.

Nonrecirculating Liquid Lubrication Systems. At low to moderate speeds, where the use of grease lubrication is not suitable, other methods of supplying lubricant to the bearing can be used. These include splash or bath lubrication, wick, oil-ring, and oil–air mist lubrication. Felt wicks can be used to transport oil by capillary action from a nearby reservoir. Oil rings, which are driven by frictional contact with the rotating shaft, run partially immersed in an oil reservoir and feed oil mechanically to the shaft, which is adjacent to the bearing. The bearing may itself be partially immersed in an oil reservoir to splash-lubricate itself. All of these methods require very modest ambient temperatures and thermal conditions as well as speed conditions equivalent to a maximum d_bN of about 0.5×10^6. The machinery must also remain in a fixed-gravity orientation.

Oil–air mist lubrication supplies atomized oil in an airstream to the bearing, where a reclassifier increases the droplet size, allowing it to condense on the bearing surfaces. Feed rates are low and a portion of the oil flow escapes with the feed air to the atmosphere. Commercial oil–air mist generators are available for systems ranging from a single bearing to hundreds of bearings. Bearing friction losses and heat generation with mist lubrication are low, but ambient temperatures and cooling requirements must be moderate because oil–air mist systems provide minimal cooling. Many bearings, especially small-bore bearings, are successfully operated at high speeds (d_bN values to greater than 10^6) with oil–air mist lubrication.

Jet Lubrication.   In applications where speed or heat-rejection requirements are too high, jet lubrication is frequently used to lubricate and control bearing temperatures. A number of variables are critical to achieving not only satisfactory but near-optimal performance and bearing operation. These include the placement of the nozzles, the number of nozzles, the jet velocity, the lubricant flow rate, and the scavenging of the lubricant from the bearing and immediate vicinity. The importance of proper jet lubricating system design is shown in Ref. 43 and is summarized in Fig. 19.19.

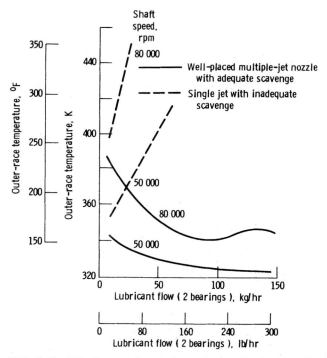

FIG. 19.19   Effectiveness of proper jet lubrication. Test bearings, 20-mm-bore angular-contact ball bearings; thrust load, 222 N (50 lbf). (*From Matt and Gianotti.*[43])

Proper placement of the jets should take advantage of any natural pumping ability of the bearings. Figure 19.20 (Ref. 47) illustrates jet lubrication of ball and tapered-roller bearings. Centrifugal forces aid in moving the oil through the bearing to cool and lubricate the elements. Directing jets at the radial gaps between the cage and the races achieves maximum penetration of oil into the interior of the bearing. References 6, 45, 48, and 64 present useful data on the influence of jet placement and velocity on the lubrication of several types of bearings.

Under-Race Lubrication.   As bearing speeds increase, centrifugal effects become more predominant, making it increasingly difficult to effectively lubricate and cool a

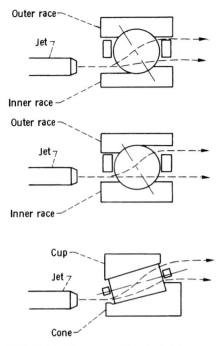

FIG. 19.20 Placement of jets for ball bearings with relieved rings and tapered-roller bearings.

bearing. The jetted oil is thrown off the sides of the bearing rather than penetrating to the interior. At extremely high $d_b N$ values (2.4×10^6 and higher), jet lubrication becomes ineffective. Increasing the flow rate only adds to heat generation through increased churning losses. Reference 12 describes an under-race oiling system used in a turbofan engine for both ball and cylindrical roller bearings. Figure 19.21 illustrates the technique. The lubricant is directed radially under the bearings. Centrifugal effects assist in pumping oil out through the bearings through suitable slots and holes, which are made a part of both the shaft and mounting system and the bearings themselves. Holes and slots are provided within the bearing to feed oil directly to the ball-race and cage-race contacts.

This lubrication technique has been thoroughly tested for large-bore ball and roller bearings up to $d_b N = 3 \times 10^6$. Pertinent data are reported in Refs. 13, 51, 54, and 55.

19.10 *ELASTOHYDRODYNAMIC LUBRICATION*

Elastohydrodynamic lubrication is a form of fluid-film lubrication where elastic deformation of the bearing surfaces becomes significant. It is usually associated with highly stressed machine components such as rolling-element bearings. Historically, elastohydrodynamic lubrication may be viewed as one of the major developments in the field of lubrication in the twentieth century. It not only revealed the existence of a previously unsuspected regime of lubrication in highly stressed nonconforming machine ele-

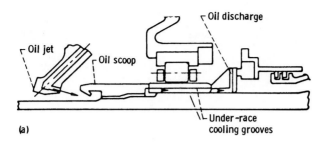

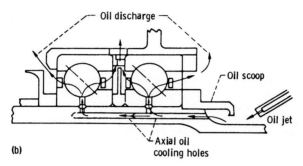

FIG. 19.21   Under-race oiling system for main shaft bearings on turbofan engine. (*a*) Cylindrical-roller bearing. (*b*) Ball thrust bearing. (*From Brown, 1970.*[12])

ments, but it also brought order to the complete spectrum of lubrication regimes, ranging from boundary to hydrodynamic.

19.10.1   Relevant Equations

The relevant equations used in elastohydrodynamic lubrication are given here.

Lubrication Equation (Reynolds Equation)

$$\frac{\partial}{\partial x}\left(\frac{\rho h^3}{\eta}\frac{\partial p}{\partial x}\right) + \frac{\partial}{\partial y}\left(\frac{\rho h^3}{\eta}\frac{\partial p}{\partial y}\right) = 12u\frac{\partial}{\partial x}(\rho h)$$

where $u = (u_a + u_b)/2$.

Viscosity Variation

$$\eta = \eta_0 e^{\xi p}$$

where η_0 = coefficient of absolute or dynamic viscosity at atmospheric pressure and ξ = pressure-viscosity coefficient of the fluid.

Density Variation (for Mineral Oils)

$$\rho = \rho_0\left(1 + \frac{0.6p}{1 + 1.17p}\right)$$

where ρ_0 = density of atmospheric conditions.

Elasticity Equation

$$\delta = \frac{2}{E'} \iint \frac{p(x,y)\, dx\, dy}{\sqrt{(x - x_1)^2 + (y - y_1)^2}}$$

where

$$E' = \frac{2}{(1 - v_a^2)/E_a + (1 - v_b^2)/E_b}$$

Film Thickness Equation

$$h = h_0 + x^2/2R_x + y^2/2R_y + \delta(x,y)$$

where

$$1/R_x = 1/r_{ax} + 1/r_{bx} \qquad (19.39)$$

$$1/R_y = 1/r_{ay} + 1/r_{by} \qquad (19.40)$$

The elastohydrodynamic lubrication solution therefore requires the calculation of the pressure distribution within the conjunction, at the same time allowing for the effects that this pressure will have on the properties of the fluid and on the geometry of the elastic solids. The solution will also provide the shape of the lubricant film, particularly the minimum clearance between the solids. A detailed description of the elasticity model one could use is given in Ref. 16, and the complete elastohydrodynamic lubrication theory is given in Ref. 29.

19.10.2   Dimensionless Grouping

The variables resulting from the elastohydrodynamic lubrication theory are

E' = effective elastic modulus, N/m^2

F = normal applied load, N

h = film thickness, m

R_x = effective radius in x (motion) direction, m

R_y = effective radius in y (transverse) direction, m

u = mean surface velocity in x direction, m/s

ξ = pressure-viscosity coefficient of fluid, m^2/N

η_0 = atmospheric viscosity, N·s/m^2

From these variables the following five dimensionless groupings can be established.

Dimensionless Film Thickness

$$H = h/R_x$$

Ellipticity Parameter

$$k = D_y/D_x = (R_y/R_x)^{2/\pi}$$

Dimensionless Load Parameter

$$W = F/E'R_x^2 \qquad (19.41)$$

Dimensionless Speed Parameter

$$U = \eta_0 u / E' R_x \qquad (19.42)$$

Dimensionless Materials Parameter

$$G = \xi E' \qquad (19.43)$$

The dimensionless minimum film thickness can thus be written as a function of the other four parameters:

$$H = f(k, U, W, G)$$

The most important practical aspect of elastohydrodynamic lubrication theory is the determination of the minimum film thickness within a conjunction. That is, maintaining a fluid-film thickness of adequate magnitude is extremely important to the operation of such machine elements as rolling-element bearings. Specifically, elastohydrodynamic film thickness influences fatigue life, as discussed in Sec. 19.11.

19.10.3 Minimum-Film-Thickness Formula

By using the numerical procedures outlined in Ref. 27, the influence of the ellipticity parameter and the dimensionless speed, load, and materials parameters on minimum film thickness has been investigated in Ref. 28. The ellipticity parameter k was varied from 1 (a ball-on-plate configuration) to 8 (a configuration approaching a rectangular contact). The dimensionless speed parameter U was varied over a range of nearly two orders of magnitude, and the dimensionless load parameter W over a range of one order of magnitude. Situations equivalent to using solid materials of bronze, steel, and silicon nitride and lubricants of paraffinic and naphthenic oils were considered in the investigation of the role of the dimensionless materials parameter G. Thirty-four cases were used in generating the minimum-film-thickness formula given here:

$$H_{min} = 3.63 \, U^{0.68} G^{0.49} W^{-0.073} (1 - e^{-0.68k}) \qquad (19.44)$$

In this equation, the most dominant exponent occurs on the speed parameter, and the exponent on the load parameter is very small and negative. The materials parameter also carries a significant exponent, although the range of this variable in engineering situations is limited.

19.10.4 Pressure and Film-Thickness Plots

A representative contour plot of dimensionless pressure is shown in Fig. 19.22 for $k = 1.25$, $U = 0.168 \times 10^{-11}$, $W = 0.111 \times 10^{-6}$, and $G = 4522$. In this figure and in Fig. 19.23, the $+$ symbol indicates the center of the hertzian contact zone. The dimensionless representation of the x and y coordinates causes the actual hertzian contact ellipse to be a circle, regardless of the value of the ellipticity parameter. The hertzian contact circle is shown by asterisks. There is a key in this figure that shows the contour labels and each corresponding value of dimensionless pressure. The inlet region is to the left and the exit region is to the right. The pressure gradient at the exit end of the conjunction is much larger than that in the inlet region. In Fig. 19.22 a pressure spike is visible at the exit of the contact.

Contour plots of the film thickness are shown in Fig. 19.23 for $k = 1.25$, $U = 0.168 \times 10^{-11}$, $W = 0.111 \times 10^{-6}$, and $G = 4522$. In this figure two minimum regions

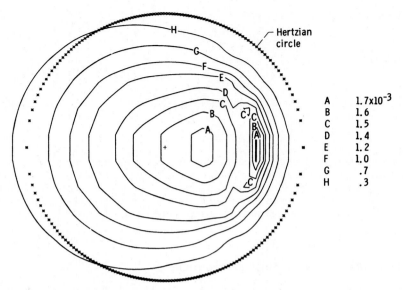

FIG. 19.22   Contour plot of dimensionless pressure. $k = 1.25$; $U = 0.168 \times 10^{-11}$; $W = 0.111 \times 10^{-6}$; $G = 4522$.

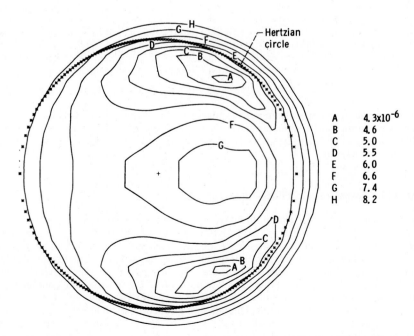

FIG. 19.23   Contour plot of dimensionless film thickness. $k = 1.25$; $U = 0.168 \times 10^{-11}$; $W = 0.111 \times 10^{-6}$; $G = 4522$.

occur in well-defined side lobes that follow, and are close to, the edge of the hertzian contact circle. These results produce all the essential features of previously reported experimental observations based on optical interferometry.[14]

19.11 BEARING LIFE

19.11.1 Lundberg-Palmgren Theory

Calculations of bearing life are based upon the assumption that a bearing will fail from rolling-element fatigue.

Rolling fatigue is a material failure caused by the application of repeated stresses to a small volume of material. It is a unique failure type. It is essentially a process by which the first failure occurs at the weakest point. A typical spall is shown in Fig. 19.24. We can surmise that on a microscale there will be a wide dispersion in material strength or resistance to fatigue because of inhomogeneities in the material. Because bearing materials are complex alloys, we would not expect them to be homogeneous or equally resistant to failure at all points. Therefore, the fatigue process can be expected to be one in which a group of supposedly identical specimens exhibit wide variations in failure time when stressed in the same way. For this reason, it is necessary to treat the fatigue process statistically.

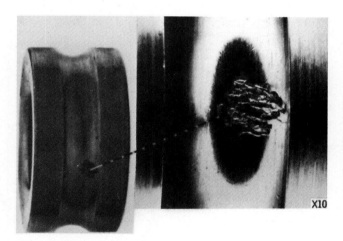

FIG. 19.24   Typical fatigue spall.

To be able to predict how long a particular bearing will run under a specific load, we must have the following two essential pieces of information: (1) an accurate, quantitative estimate of the life dispersion or scatter, and (2) the life at a given survival rate or reliability level. This translates into an expression for the "load capacity," or the ability of the bearing to endure a given load for a stipulated number of stress cycles or revolutions. If a group of supposedly identical bearings is tested at a specific load and speed, there will be a wide scatter in bearing lives, as shown in Fig. 19.25. This distribution is exponential and is typical for most rolling-element bearings. Using Weibull analysis (Ref. 62), the distribution shown in Fig. 19.25 can be plotted linearly on a

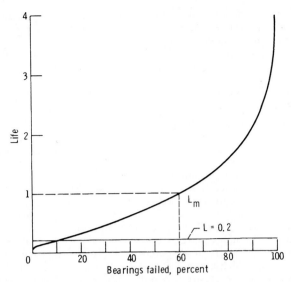

FIG. 19.25 Distribution of bearing fatigue failures.

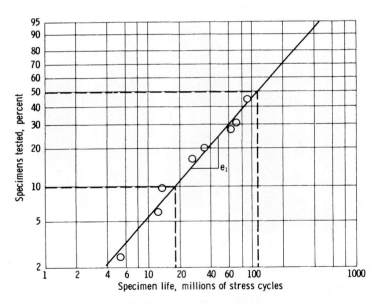

FIG. 19.26 Typical Weibull plot of bearing fatigue failures.

Weibull plot shown in Fig. 19.26. The formula for this distribution can be expressed as a Weibull function,

$$\ln \ln (1/S) = e_1 \ln (L/A) \qquad (19.45)$$

where S is the probability of survival expressed as a fraction, e is the Weibull slope, L

is the life or time at a probability of survival S, and A is the characteristic life or time at a probability of survival of 0.368. The slope e_1 for most bearings ranges from 1 to 2 and can be assumed for purposes of most analysis to be 1.1.

Lundberg and Palmgren (Refs. 41 and 42) using Weibull analysis developed the Lundberg-Palmgren theory on which bearing life and ratings are based, where

$$\ln (1/S) = \frac{\tau_0^{c_1} L^{e_1}}{Z_0^{c_3}} V \tag{19.46}$$

and

$$V = D_y Z_0 l_v \tag{19.47}$$

Then Eq. (19.46) becomes

$$\ln (1/S) = \frac{\tau_0^{c_1} L^{e_1}}{Z_0^{c_3-1}} D_y l_v \tag{19.48}$$

The concept of a bearing dynamic-load capacity C is introduced by Lundberg and Palmgren. This is the theoretical load placed on a bearing which will produce 1 million inner-race revolutions at a probability of survival of 90 percent, or $S = 0.9$. This S of 0.9 is also referred to as the 10 percent or L_{10} life where 10 percent of a group of bearings can be expected to fail before reaching a given time. By substituting the appropriate values in Eq. (19.48) this value of C can be calculated.

Factors on which dynamic load capacity and bearing life depend are:

1. Size of rolling element
2. Number of rolling elements per row
3. Number of rows of rolling elements
4. Conformity between rolling elements and races
5. Contact angle under load
6. Material properties

For different size and types of bearings, values of C can be obtained using AFBMA and ISO ratings (Refs. 1 and 2) and bearing manufacturers' catalogs. Knowing C, the L_{10} life of any other load F_e can be calculated using the following equation from Lundberg-Palmgren:

$$L = (C/F_e)^m \tag{19.49}$$

where C = basic dynamic capacity or load rating
F_e = equivalent bearing load
m = 3 for ball bearings and 10/3 for roller bearings
L = life, millions of race revolutions

The equivalent bearing load F_e can be calculated from the following equation:

$$F_e = XF_r + YF_1 \tag{19.50}$$

Factors X and Y are given in bearing manufacturers' catalogs for specific bearings.

19.11.2 ANSI/AFBMA Standards

The ANSI/AFBMA standards (Refs. 1 and 2) became effective in 1950 for ball bearings and in 1953 for roller bearings. By the early 1960s many bearing manufacturers began to incorporate life factors into their bearing catalog ratings. These life factors are not recognizable to the casual bearing user. Therefore, life estimates obtained from different bearing manufacturers' catalogs can be significantly different.

The Lundberg-Palmgren life theory incorporated into its equations a material factor that reflected bearing technology and life existing prior to 1950. This material factor was based on groups of bearings that were laboratory life-tested over several decades. When these equations were incorporated into the ANSI/AFBMA standards, the material factor used by Lundberg and Palmgren was also incorporated into the standards.

Research in steel metallurgy and processing, lubrication and lubricants, and bearing-manufacturing process has resulted in significant improvement in bearing life over that obtained in the early 1950s. To account for these technology advancements, the ANSI/AFBMA and ISO standards combine three factors into the following formula to adjust for life.

$$L_{na} = a_1 a_2 a_3 L_{10} = a_1 a_2 a_3 \left(\frac{C}{F_e}\right)^m \tag{19.51}$$

where L_{na} = adjusted life, millions of revolutions
  C = dynamic load capacity, lb
  e = equivalent bearing load, lb
  a_1 = adjustment factor for reliability
  a_2 = adjustment factor for materials and processing
  a_3 = adjustment factor for operating conditions

Zaretsky, in "STLE Life Factors for Rolling Bearings,"[65] lists the variables affecting bearing life and gives specific values as they may apply to specific applications. These variables are listed in Table 19.14. (Reference 65 can be obtained from The Society of Tribologist and Lubrication Engineers, Park Ridge, Ill.)

The value for reliability is obtained from the Weibull equation (19.45) and is in the ANSI/AFBMA standards:

$$a_1 = 4.48 \left(\ln \frac{100}{R}\right)^{2/3} \tag{19.52}$$

where R = reliability factor, percent. Research has shown that at values of S equal 0.999, $a_1 = 0.053$. At values of S above 0.999, no failure due to fatigue would be expected. At values of S equal 0.9, $a_1 = 1.65$.

Material and processing variables affect life and are reflected in life factor a_2. For over a century, AISI 52100 steel has been the predominant material for rolling-element bearings. In fact, the basic dynamic capacity as defined by AFBMA in 1949 is based on an air-melted 52100 steel, hardened to at least Rockwell C 58. Since that time, as discussed in Sec. 19.4.1, better control of air-melting processes and the introduction of vacuum remelting processes have resulted in more homogeneous steels with fewer impurities. Such steels have extended rolling-element bearing fatigue lives to several times the AFBMA or catalog life. Life improvements of 3 to 8 times are not uncommon. Other steel compositions, such as AISI M-1 and AISI M-50, chosen for their higher temperature capabilities and resistance to corrosion, also have shown greater resistance to fatigue pitting when vacuum melting techniques are employed. Case-hardened materials, such as AISI 4620, AISI 4118, and AISI 8620, used primarily for roller bearings, have the advantage of a tough, ductile steel core with a hard, fatigue-resistant surface.

Life factor a_3 reflects the effect of lubricant on bearing life. Until approximately 1960, the role of the lubricant between surfaces in rolling contact was not fully appreciated. Metal-to-metal contact was presumed to occur in all applications, with atten-

TABLE 19.14 Summary of Life Factors (LF) for Rolling-Element Bearings (From Zaretsky, Ref. 65)

a_1 (reliability)	a_2 (materials and processing)	a_3 (operating conditions)

Probability of failure	Bearing steel*
	 Melting process†
	 Metalworking
	 Material hardness
	 Residual stress
	 Carbide factor‡
	 Nonferrous materials
	 Ceramic hybrid bearings
	 Bearing refurbishment
	 and restoration
	 Coatings and surface treatments
		Load:
		   Heavy load and ring distortion
		   Roller bearing in housing with
		   rigid legs
		   Dish or sag of horizontally
		   mounted large-diameter ball
		   bearing
		   Misalignment
		   Housing clearance
		   Axially loaded cylindrical
		   bearings
		   Rotordynamics
		   Hoop stresses
		   Speed
		 Temperature:
		   Steel
		   Ceramic
		 Lubrication:
		   Lubricant film thickness
		   Starvation
		   Surface finish
		   Grease
		   Additives
		   Traction
		   Water
		   Filtration

*To be used with melting-process LF.
†To be used with bearing-steel LF.
‡To be used in place of bearing-steel LF but with melting-process LF.

dant required boundary lubrication. The development of elastohydrodynamic lubrication theory showed that lubricant films of thicknesses of the order of microinches and tens of microinches occur in rolling contact. Since surface finishes are of the same order of magnitude as the lubricant film thicknesses, the significance of rolling-element bearing surface roughnesses to bearing performance became apparent. Reference 59 first reported on the importance on bearing life of the ratio of elastohydrodynamic lubrication film thickness to surface roughness. Figure 19.27 shows life as a percentage of calculated L_{10} as a function of Λ, where

$$\Lambda = h_{min}/\sqrt{f_a^2 + f_b^2} \tag{19.53}$$

Figure 19.28, from Ref. 8, presents a curve of the recommended $\tilde{F}$ factor as a function of the Λ parameter. A mean of the curves presented in Ref. 59 for ball bearings and in Ref. 56 for roller bearings is recommended for use. A formula for calculating the minimum film thickness h_{min} is given in Eq. (19.44).

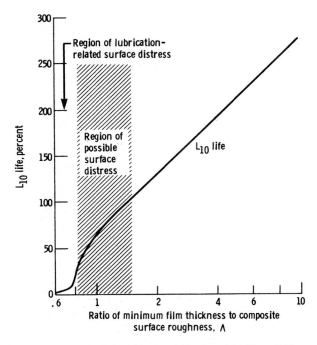

FIG. 19.27   Group fatigue life L_{10} as function of Λ. (*From Tallian, 1967.*[54])

19.12   DYNAMIC ANALYSES AND COMPUTER CODES

As has been stated, a precise analysis of rolling-bearing kinematics, stresses, deflections, and life can be accomplished only by using a large-scale computer code. Presented here is a very brief discussion of some of the significant analyses that have led to the formulation of modern bearing analytical computer codes. The reader is referred to COSMIC, University of Georgia, Athens, Ga., for the public availability of program listings.

19.12.1   Quasi-Static Analyses

Rolling-element bearing analysis began with the work of Jones on ball bearings.[38,39] This work was done before there was a general awareness of elastohydrodynamic lubrication and assumed Coulomb friction in the race contacts. This led to the commonly known "race control" theory, which assumes that pure rolling (except for Heathcote interfacial slip) can occur at one of the ball-race contacts. All of the spinning required for dynamic equilibrium of the balls would then take place at the other, or "noncontrolling," race contact. Jones' analysis proved to be quite effective for predicting fatigue life but less useful for predicting cage slip, which usually occurs at

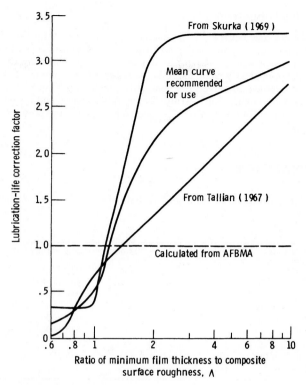

FIG. 19.28 Lubrication-life correction factor as function of Λ. (*From Bamberger.*[8])

high speeds and light loads. Reference 32 extended the analysis, retaining the assumption of Coulomb friction, but allowing a frictional resistance to gyroscopic moments at the noncontrolling as well as the controlling race contact. Reference 32 is adequate for predicting bearing performance under conditions of dry-film lubrication or whenever there is a complete absence of any elastohydrodynamic film.

Reference 31 first incorporated elastohydrodynamic relationships into a ball-bearing analysis. A revised version of Harris' computer program,[32] called SHABERTH, was developed that incorporated actual traction data from a disk machine. The program SHABERTH has been expanded until today it encompasses ball, cylindrical roller, and tapered-roller bearings.

Reference 30 first introduced elastohydrodynamics into a cylindrical roller-bearing analysis. The initial analysis has been augmented with more precise viscosity-pressure and temperature relationships and traction data for the lubricant. This has evolved into the program CYBEAN, and more recently has been incorporated into SHABERTH. Parallel efforts by Harris' associates have resulted in SPHERBEAN, a program that can be used to predict the performance of spherical roller bearings. These analyses can range from relatively simple force balance and life analyses through a complete thermal analysis of a shaft bearing system in several steps of varying complexity.

19.12.2 Dynamic Analyses

The work of Jones and Harris[32,38,39] is categorized as quasi-static because it applies only when steady-state conditions prevail. Under highly transient conditions, such as accelerations or decelerations, only a true dynamic analysis will suffice. Reference 61 made the first attempt at a dynamic analysis to explain cage dynamics in gyro-spin-axis ball bearings. Reference 20 solved the generalized differential equations of motion of the ball in an angular-contact ball bearing. The work was continued for both cylindrical roller bearings[21,22] and for ball bearings[23,24] and is available in the program DREB.

19.13 APPLICATION

Consider a single-row, radial, deep-groove ball bearing with the following dimensions:

Dimensions	Minimum elastohydrodynamic film thickness
Inner-race diameter d_i, m	0.052291
Outer-race diameter d_o, m	0.077706
Ball diameter d, m	0.012700
Number of balls in complete bearing, n	9
Inner-groove radius r_i, m	0.006604
Outer-groove radius r_o, m	0.006604
Contact angle β, degrees	0
RMS surface finish of balls, f_b, μm	0.0625
RMS surface finish of races, f_a, μm	0.0175

A bearing of this kind might well experience the following operation conditions:

Operating condition	Result
Radial load F_r, N	8900
Inner-race angular velocity ω_i, rad/s	400
Outer-race angular velocity ω_o, rad/s	0
Lubricant viscosity at atmospheric pressure and effective operating temperature of bearing, η_0, N·s/m^2	0.04
Viscosity-pressure coefficient ξ, m^2/N	2.3×10^{-8}
Modulus of elasticity for both balls and races, E, N/m^2	2×10^{11}
Poisson's ratio for both balls and races, ν	0.3

The essential features of the geometry of the inner and outer conjunctions can be ascertained as follows:

Pitch Diameter

$$d_e = 0.5(d_o + d_i) = 0.065 \text{ m}$$

Diametral Clearance

$$P_d = d_o - d_i - 2d = 1.5 \times 10^{-5} \text{ m}$$

Race Conformity

$$f_i = f_o = r/d = 0.52$$

Equivalent Radius

$$R_{x,i} = d(d_e - d)/2d_e = 0.00511 \text{ m}$$

$$R_{x,o} = d(d_e + d)/2d_e = 0.00759 \text{ m}$$

$$R_{y,i} = f_i d/(2f_i - 1) = 0.165 \text{ m}$$

$$R_{y,o} = f_o d/(2f_o - 1) = 0.165 \text{ m}$$

The curvature sum

$$1/R_i = 1/R_{x,i} + 1/R_{y,i} = 201.76$$

gives $R_i = 4.956 \times 10^{-3}$ m, and the curvature sum

$$1/R_o = 1/R_{x,o} + 1/R_{y,o} = 137.81$$

gives $R_o = 7.256 \times 10^{-3}$ m. Also, $\alpha_i = R_{y,i}/R_{x,i} = 32.35$ and $\alpha_o = R_{y,o}/R_{x,o} = 21.74$. The nature of the hertzian contact conditions can now be assessed.

Ellipticity Parameters.   From Eq. (19.7),

$$\bar{k}_i = \alpha_i^{2/\pi} = 9.42 \qquad \bar{k}_o = \alpha_o^{2/\pi} = 7.09$$

Elliptic Integrals

$$q = \pi/2 - 1$$

From Eq. (19.8),

$$\bar{\mathscr{E}}_i = 1 + q/\alpha_i = 1.0188 \qquad \bar{\mathscr{E}}_o = 1 + q/\alpha_o = 1.0278$$

From Eq. (19.9),

$$\bar{\mathscr{F}}_i = \pi/2 + q \ln \alpha_i = 3.6205 \qquad \bar{\mathscr{F}}_o = \pi/2 + q \ln \alpha_o = 3.3823$$

The effective elastic modulus E' is given by

$$E' = \frac{2}{(1 - v_a^2)/E_a + 1 - v_b^2/E_b} = 2.198 \times 10^{11} \text{ N/m}^2$$

To determine the load carried by the most heavily loaded ball in the bearing, it is necessary to adopt an iterative procedure based on the calculation of local static compression and the analysis presented in Sec. 19.7. Reference 57 found that the value of Z was about 4.37 by using the expression [Eq. (19.27)]

$$F_{max} = ZF_r/n$$

where F_{max} = load on most heavily loaded ball
F_r = radial load on bearing
n = number of balls

However, it is customary to adopt a value of $Z = 5$ in simple calculations in order to produce a conservative design, and this value will be used to begin the iterative procedure.

Stage 1.   Assume $Z = 5$. Then

$$F_{max} = 5F_r/9 = \tfrac{5}{9}(8900) = 4944 \text{ N}$$

From Eq. (19.12) the maximum local elastic compression is

$$\delta_i = \overline{\mathscr{F}}_i[(9/2\overline{\mathscr{E}}_i R_i)(F_{max}/\pi \overline{k}_i E')^2]^{1/3} = 2.902 \times 10^{-5} \text{ m}$$

$$\delta_o = \overline{\mathscr{F}}_o[(9/2\overline{\mathscr{E}}_o R_o)(F_{max}/\pi \overline{k}_o E')^2]^{1/3} = 2.877 \times 10^{-5} \text{ m}$$

The sum of the local compressions on the inner and outer races is

$$\delta = \delta_i + \delta_o = 5.779 \times 10^{-5} \text{ m}$$

A better value for Z can now be obtained from

$$Z = \frac{\pi(1 - P_d/2\delta)^{3/2}}{2.491(\{1 + [(1 - P_d/2\delta)/1.23]^2\}^{1/2} - 1)}$$

since $P_d/2\delta = (1.5 \times 10^{-5})/(5.779 \times 10^{-5}) = 0.1298$. Thus,

$$Z = 4.551$$

Stage 2

$$Z = 4.551$$

$$F_{max} = (4.551)(8900)/9 = 4500 \text{ N}$$

$$\delta_i = 2.725 \times 10^{-5} \text{ m}$$

$$\delta_o = 2.702 \times 10^{-5} \text{ m}$$

$$\delta = 5.427 \times 10^{-5} \text{ m}$$

$$P_d/2\delta = 0.1382$$

Thus, $Z = 4.565$.

Stage 3

$$Z = 4.565$$

$$F_{max} = (4.565)(8900)/9 = 4514 \text{ N}$$

$$\delta_i = 2.731 \times 10^{-5} \text{ m}$$

$$\delta_o = 2.708 \times 10^{-5} \text{ m}$$

$$\delta = 5.439 \times 10^{-5} \text{ m}$$

$$P_d/2\delta = 0.1379$$

and hence $Z = 4.564$.

This value is very close to the previous value of 4.565, from stage 2, and a further iteration confirms its accuracy.

Stage 4

$$Z = 4.564$$

$$F_{max} = (4.564)(8900)/9 = 4513 \text{ N}$$

$$\delta_i = 2.731 \times 10^{-5} \text{ m}$$

$$\delta_o = 2.707 \times 10^{-5} \text{ m}$$

$$\delta = 5.438 \times 10^{-5} \text{ m}$$

$$P_d/2\delta = 0.1379$$

and hence $Z = 4.564$.

The load on the most heavily loaded ball is thus 4513 N.

Elastohydrodynamic Minimum Film Thickness.   For pure rolling, from Eq. (19.3),

$$u = |\omega_o - \omega_i|(d_e^2 - d^2)/4d_e = 6.252 \text{ m/s}$$

The dimensionless load, speed, and materials parameters for the inner- and outer-race conjunctions thus become [from Eqs. (19.41) to (19.43), respectively]

$$W_i = F/E'(R_{x,i})^2 = 4513/(2.198 \times 10^{11})(5.11)^2 \times 10^{-6} = 7.863 \times 10^{-4}$$

$$U_i = \eta_0 u/E'R_{x,i} = (0.04)(6.252)/(2.198 \times 10^{11})(5.11 \times 10^{-3}) = 2.227 \times 10^{-10}$$

$$G_i = \xi E' = (2.3 \times 10^{-8})(2.198 \times 10^{11}) = 5055$$

$$W_o = F/E'(R_{x,o})^2 = 4513/(2.198 \times 10^{11})(7.59)^2(10^{-6}) = 3.564 \times 10^{-4}$$

$$U_o = \eta_0 u/E'R_{x,o} = (0.04)(6.252)/(2.198 \times 10^{11}) \times (7.59 \times 10^{-3}) = 1.499 \times 10^{-10}$$

$$G_o = \xi E' = (2.3 \times 10^{-8})(2.198 \times 10^{11}) = 5055$$

The dimensionless minimum elastohydrodynamic film thickness in a fully flooded elliptical contact is given by Eq. (19.44):

$$H_{min} = h_{min}/R_x = 3.63 U^{0.68} G^{0.49} W^{-0.073}(1 - e^{-0.68k})$$

Ball–Inner-Race Conjunction.   From Eq. (19.44),

$$(h_{min})_i/R_{x,i} = (3.63)(2.732 \times 10^{-7})(65.29)(1.685)(0.9983) = 1.09 \times 10^{-4}$$

Thus, $\qquad\qquad (h_{min})_i = (1.09 \times 10^{-4})R_{x,i} = 0.557 \text{ } \mu\text{m}$

The lubrication factor Λ was found to play a significant role in determining the fatigue life of rolling-element bearings. In this case, from Eq. (19.53),

$$\Lambda_i = (h_{min})_i/\sqrt{f_a^2 + f_b^2} = (0.557 \times 10^{-6})/\{[(0.175)^2 + (0.0625)^2]^{1/2} \times 10^{-6}\} = 3.00 \quad (19.53)$$

Ball–Outer-Race Conjunction.   From Eq. (19.44),

$$(H_{min})_o = (h_{min})_o/R_{x,o} = 3.63 U_o^{0.68} G_o^{0.49} W^{-0.073}(1 - e^{-0.68k_o})$$

$$= (3.63)(2.087 \times 10^{-7})(65.29)(1.785)(0.9919) = 0.876 \times 10^{-4}$$

Thus, $(h_{min})_o = (0.876 \times 10^{-4})R_{x,o} = 0.665 \ \mu m$

In this case, the lubrication factor Λ is given by Eq. (19.53):

$$\Lambda = (0.665 \times 10^{-6})/\{[(0.175)^2 + (0.0625)^2]^{1/2} \times 10^{-6}\} = 3.58$$

Once again, it is evident that the smaller minimum film thickness occurs between the most heavily loaded ball and the inner race. However, in this case the minimum elastohydrodynamic film thickness is about 3 times the composite surface roughness, and the bearing lubrication can be deemed to be entirely satisfactory. Indeed, it is clear from Fig. 19.28 that very little improvement in the lubrication factor $\bar{F}$ and thus in the fatigue life of the bearing could be achieved by further improving the minimum film thickness and hence Λ.

REFERENCES

1. "Load Ratings and Fatigue Life for Ball Bearings," ANSI/AFBMA 9-1990, The Anti-Friction Bearing Manufactures Association, Washington, D.C., 1990.

2. "Load Ratings and Fatigue Life for Roller Bearings," ANSI/AFBMA 11-1990, The Anti-Friction Bearing Manufacturers Association, Washington, D.C., 1990.

3. Anderson, N. E., and E. V. Zaretsky: "Short-Term Hot Hardness Characteristics of Five Case-Hardened Steels," NASA TN D-8031, National Aeronautics and Space Administration, 1975.

4. Anderson, W. J.: "Elastohydrodynamic Lubrication Theory as a Design Parameter for Rolling Element Bearings," ASME Paper 70-DE-19, American Society of Mechanical Engineers, New York, 1970.

5. Anderson, W. J.: "Practical Impact of Elastohydrodynamic Lubrication," in Proc 5th Leeds-Lyon Symposium in Tribology on Elastohydrodynamics and Related Topics," D. Dowson, C. M. Taylor, M. Godet, and D. Berthe, eds., Mechanical Engineering Publication, Bury St. Edmunds, Suffolk, England, p. 217, 1979.

6. Anderson, W. J., E. J. Macks, and Z. N. Nemeth: "Comparison of Performance of Experimental and Conventional Cage Designs and Materials for 75-Millimeter-Bore Cylindrical Roller Bearings at High Speeds," NACA TR-1177, National Advisory Committee for Aeronautics, 1954.

7. Bamberger, E. N.: "Effect of Materials—Metallurgy Viewpoint," in "Interdisciplinary Approach to the Lubrication of Concentrated Contacts," P. M. Ku, ed., NASA SP-237, National Aeronautics and Space Administration, 1970.

8. Bamberger, E. N.: "Life Adjustment Factors for Ball and Roller Bearings—An Engineering Design Guide," American Society for Mechanical Engineers, New York, 1971.

9. Bamberger, E. N., E. V. Zaretsky, and H. Signer: "Endurance and Failure Characteristics of Main-Shaft Jet Engine Bearings at 3×10^6 DN," ASME Trans., J. Lubr. Technol., ser. F, vol. 98, no. 4, p. 510, 1976.

10. Bamberger, E. N., et al.: "Materials for Rolling Element Bearings," in "Bearing Design—Historical Aspects, Present Technology and Future Problems," W. J. Anderson, ed., American Society of Mechanical Engineers, New York, p. 1, 1980.

11. Bisson, E. E., and Anderson, W. J.: "Advanced Bearing Technology," NASA SP-38, National Aeronautics and Space Administration, 1964.

12. Brown, P. F.: "Bearing and Dampers for Advanced Jet Engines," SAE Paper 700318, Society of Automotive Engineers, 1970.

13. Brown, P. F., L. C. Dobek, F. C. Hsing, and J. R. Miner: "Mainshaft High Speed Cylindrical Roller Bearings for Gas Engines," PWA-FR-8615, Pratt & Whitney Group, West Palm Beach, Fla., 1977.

14. Cameron, A., and R. Gohar: "Theoretical and Experimental Studies of the Oil Film in Lubricated Point Contact," *Proc. Roy. Soc. London,* ser. A, vol. 291, p. 520, 1976.

15. Cundill, R. T., and F. Giordano: "Lightweight Materials for Rolling Elements in Aircraft Bearings," in "Problems in Bearing and Lubrication," AGARD Conf. Preprint No. 323, Advisory Group for Aeronautical Research and Development (NATO), Paris, p. 6–1, 1982.

16. Dowson, D., and B. J. Hamrock: "Numerical Evaluation of the Surface Deformation of Elastic Solids Subjected to a Hertzian Contact Stress," *ASLE Trans.,* vol. 19, no. 4, p. 279, 1976.

17. "General Guide to the Choice of Journal Bearing Type," Engineering Science Data Unit, item 65007, Institution of Mechanical Engineers, London, 1965.

18. "General Guide to the Choice of Thrust Bearing Type," Engineering Science Data Unit, item 67033, Institution of Mechanical Engineers, London, 1967.

19. "Contact Stresses," Engineering Science Data Unit, item 78035, Institution of Mechanical Engineers, London, 1978.

20. Gupta, P. K.: "Transient Ball Motion and Skid in Ball Bearings," *Trans. ASME, J. Lubr. Technol.,* vol. 97, no. 2, p. 261, 1975.

21. Gupta, P. K.: "Dynamics of Rolling Element Bearings—Part I, Cylindrical Roller Bearings Analysis," *Trans. ASME, J. Lubr. Technol.,* vol. 101, no. 3, p. 293, 1979.

22. Gupta, P. K.: "Dynamics of Rolling Element Bearings—Part II, Cylindrical Roller Bearing Results," *Trans. ASME, J. Lubr. Technol.,* vol. 101, no. 3, p. 305, 1979.

23. Gupta, P. K.: "Dynamics of Rolling Element Bearings—Part III, Ball Bearing Analysis," *Trans. ASME, J. Lubr. Technol.,* vol. 101, no. 3, p. 312, 1979.

24. Gupta, P. K.: "Dynamics of Rolling Element Bearings—Part IV, Ball Bearing Results," *Trans. ASME, J. Lubr. Technol.,* vol. 101, no. 3, p. 319, 1979.

25. Hamrock, B. J., and W. J. Anderson: "Analysis of an Arched Outer-Race Ball Bearing Considering Centrifugal Forces," *Trans. ASME, J. Lubr. Technol.,* vol. 95, no. 3, p. 265, 1973.

26. Hamrock, B. J., and D. Brewe: "Simplified Solution for Stresses and Deformations," *Trans. ASME, J. Lubr. Technol.,* vol. 105, p. 171, 1983.

27. Hamrock, B. J., and D. Dowson: "Isothermal Elastohydrodynamic Lubrication of Point Contacts—Part I, Theoretical Formulation," *ASME J. Lubr. Technol.,* vol. 98, no. 22, p. 223, 1976.

28. Hamrock, B. J., and D. Dowson: "Isothermal Elastohydrodynamic Lubrication of Point Contacts—Part III, Fully Flooded Results," *ASME J. Lubr. Technol.,* vol. 99, no. 2, p. 264, 1977.

29. Hamrock, B. J., and D. Dowson: "Ball Bearing Lubrication—The Elastohydrodynamics of Elliptical Contacts," John Wiley & Sons, Inc., New York, 1981.

30. Harris, T. A.: "An Analytical Method to Predict Skidding in High Speed Roller Bearings," *ASLE Trans.,* vol. 9, p. 229, 1966.

31. Harris, T. A.: "An Analytical Method to Predict Skidding in Thrust-Loaded, Angular-Contact Ball Bearings," *Trans. ASME, J. Lubr. Technol.,* vol. 93, no. 1, p. 17, 1971.

32. Harris, T. A.: "Ball Motion in Thrust-Loaded, Angular Contact Bearings with Coulomb Friction," *Trans. ASME, J. Lubr. Technol.,* vol. 93, no. 1, p. 32, 1971.

33. Heathcote, H. L.: "The Ball Bearing in the Making, Under Test and in Service," *Proc. Inst. Automot. Engrs.,* vol. 15, p. 569, London, 1921.

34. Hertz, H.: "The Contact of Elastic Solids," *J. Reine Angew, Math.,* vol. 92, p. 156, 1881.

35. "Rolling Bearings, Dynamic Load Ratings and Rating Life," ISO/TC4/JC8, rev. of ISOR281, International Organization for Standardization, Technical Committee ISO/TC4, 1976.

36. Johnson, K. L., and D. Tabor: "Rolling Friction," *Proc. Inst. Mech. Engs.,* vol. 182, part 3A, p. 168, 1967/1968.

37. Jones, A. B.: "Analysis of Stresses and Deflections," New Departure Engineering Data, General Motors Corp., Bristol, Conn., 1946.

38. Jones, A. B.: "Ball Motion and Sliding Friction in Ball Bearings," *Trans. ASME, J. Basic Eng.,* vol. 81, no. 1, p. 1, 1959.

39. Jones, A. B.: "A General Theory for Elastically Constrained Ball and Radial Roller Bearings Under Arbitrary Load and Speed Conditions," *Trans. ASME, J. Basic Eng.,* vol. 82, no. 2, p. 309, 1960.

40. Jones, A. B.: "The Mathematical Theory of Rolling Element Bearings," in "Mechanical Design and Systems Handbook," 1st ed., H. A. Rothbart, ed., McGraw-Hill Book Company, Inc., New York, p. 13-1, 1964.

41. Lundberg, G., and A. Palmgren: "Dynamic Capacity of Rolling Bearings," *Acta Polytech.* (Mechanical Engineering Series), vol. I, no. 3, 1947.

42. Lundberg, G., and A. Palmgren: "Dynamic Capacity of Rolling Bearings," *Acta Polytech.* (Mechanical Engineering Series), vol. II, no. 4, 1952.

43. Matt, R. J., and R. J. Gianotti: "Performance of High Speed Ball Bearings with Jet Oil Lubrication," *Lub. Eng.,* vol. 22, no. 8, p. 316, 1966.

44. McCarthy, P. R.: "Greases," in "Interdisciplinary Approach to Liquid Lubricant Technology," P. M. Ku, ed., NASA SP-318, p. 137, National Aeronautics and Space Administration, 1973.

45. Miyakawa, Y., K. Seki, and M. Yokoyama: "Study on the Performance of Ball Bearings at High *DN* Values," NAL-TR-284, National Aerospace Lab., Tokyo, (NASA TTF-15017, 1973), 1972.

46. Palmgren, A.: "Ball and Roller Bearing Engineering," 3d ed., SKF Industries, Inc., Philadelphia, Pa., 1959.

47. Parker, R. J.: "Lubrication of Rolling Element Bearings," in "Bearing Design Historical Aspects, Present Technology and Future Problems," W. J. Anderson, ed., American Society of Mechanical Engineers, New York, p. 87, 1980.

48. Parker, R. J., and H. R. Signer: "Lubrication of High-Speed, Large Bore Tapered-Roller Bearings," *Trans. ASME, J. Lubr. Technol.,* vol. 100, no. 1, p. 31, 1978.

49. Pirvics, J.: "Numerical Analysis Techniques and Design Methodology for Rolling Element Bearing Load Support Systems," in "Bearing Design Historical Aspects, Present Technology and Future Problems," W. J. Anderson, ed., American Society of Mechanical Engineers, New York, p. 47, 1980.

50. Reynolds, O.: "On Rolling Friction," *Phil. Trans. Roy. Soc. London,* part 1, vol. 166, p. 155, 1875.

51. Schuller, F. T.: "Operating Characteristics of a Large-Bore Roller Bearing to Speed of 3×10^6 *DN*," NASA TP-1413, 1979.

52. Scibbe, H. W.: "Bearing and Seal for Cryogenic Fluids," SAE Paper 680550, 1968.

53. Sibley, L. W.: "Silicon Nitride Bearing Elements for High Speed High Temperature Applications," in "Problems in Bearing Lubrication," AGARD Conf., no. 323, Advisory Group for Aeronautical Research and Development (NATO), Paris, 1982.

54. Signer, H. R., E. N. Bamberger, and E. V. Zaretsky: "Parametric Study of the Lubrication of Thrust Loaded 120-mm Bore Ball Bearings to 3 Million *DN*," *Trans. ASME, J. Lubr. Technol.,* vol. 96, no. 3, p. 515, 1974.

55. Signer, H. R., and F. T. Schuller: "Lubrication of 35-Millimeter-Bore Ball Bearings of Several Designs at Speeds to 2.5 Million *DN*," in "Problems in Bearing and Lubrication," AGARD Conf., no. 323, Advisory Group for Research and Development (NATO), Paris, France, p. 8-1, 1982.

56. Skurka, J. C.: "Elastohydrodynamic Lubrication of Roller Bearings," *Trans. ASME, J. Lubr. Technol.,* vol. 92, no. 2, p. 281, 1970.

57. Stribeck, R.: "Kugellager fur beliebige Belastungen," *Z. Ver, dt. Ing.,* vol. 45, no. 3, p. 73, 1901.

58. Tabor, D.: "The Mechanism of Rolling Friction. II. The Elastic Range," *Proc. Roy. Soc. London,* part A, vol. 229, p. 198, 1955.

59. Tallian, T. E.: "On Competing Failure Modes in Rolling Contact," *Trans. ASLE,* vol. 10, p. 418, 1967.

60. Todd, M. J., and R. H. Bentall: "Lead Film Lubrication in Vacuum," *Proc. 2d ASLE Int Conf. Solid Lubrication,* American Society of Lubrication Engineers, p. 148, 1978.

61. Walters, C. T.: "The Dynamics of Ball Bearings," *Trans. ASME, J. Lubr. Technol.,* vol. 93, no. 1, p. 1, 1975.

62. Weibull, W.: "A Statistical Representation of Fatigue Failures in Solids," *Trans. Roy. Inst. Technol.,* Stockholm, vol. 27, 1949.

63. Zaretsky, E. V., and W. J. Anderson: "Material Properties and Processing Variables and Their Effect on Rolling-Element Fatigue," NASA TM X-52226, National Aeronautics and Space Administration, 1966.

64. Zaretsky, E. V., H. Signer, and E. N. Bamberger: "Operating Limitations of High-Speed Jet-Lubricated Ball Bearings," *Trans. ASME, J. Lubr. Technol.,* vol. 8, no. 1, p. 32, 1976.

65. Zaretsky, E. V.: "STLE Life Factors for Rolling Bearings," Society of Tribologist and Lubrication Engineers, Park Ridge, Ill., 1992.

SECTION 20
POWER SCREWS*

John E. Johnson

Manager, Mechanical Model Shops
T.R.W. Corp.
Redondo Beach, Calif.

Power screws are used to translate rotary motion into uniform longitudinal motion. Common applications include jacks, valves, machine tools, and presses. Screws also permit very accurate position adjustment because they provide a high reduction ratio from rotational to longitudinal displacement.[8–10]

20.1 THREADS

20.1.1 V Threads

The efficiency of a power screw depends upon the profile angle of the thread: the larger the angle, the lower the efficiency. V threads are therefore not well suited for transmission of great loads. They are generally used only where accuracy of adjustment and low production cost are required and the power demands are quite small.

20.1.2 Square Threads

The square thread (Fig. 20.1) has the greatest efficiency (zero profile angle). However, it is costly to manufacture, because it cannot be cut with dies, and it is difficult to engage with a moving split nut as is sometimes required.

---

*Preliminary note: The reader is referred to Sec. 11 of this handbook for a compilation of standards on the subject matter of this section.

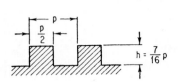

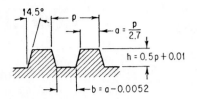

FIG. 20.1 Square thread.

FIG. 20.2 Acme thread, profile angle $\beta = 14.5°$.

20.1.3 Acme Threads

The Acme thread (Fig. 20.2) is often used in order to overcome the difficulties associated with the square thread.

While its efficiency is lower than that of a square thread, it has the advantage that lost motion resulting from manufacturing tolerances or wear can be taken out by using a split nut.

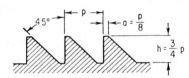

FIG. 20.3 Buttress thread.

20.1.4 Buttress Threads

Where unidirectional power transmission is required and the nut returns with little or no load, the buttress thread (Fig. 20.3) can be used. Because its square face is used for power transmission, it has the square-thread efficiency but slightly lower manufacturing cost.

20.1.5 Multiple Threads

Two or more parallel threads can be used to reduce the ratio of screw rotation to nut displacement. This reduces mechanical advantage but increases efficiency because of increased helix angle.

20.2 FORCES

The pitch of a thread p is the distance from a point on one thread to the corresponding point on an adjacent thread regardless of whether the screw has a single or multiple thread. The displacement d_s of the nut or screw resulting from one full turn of either is the lead l. Thus for a multiple thread of m threads,

$$l = mp \tag{20.1}$$

and the displacement for n revolutions of either screw or nut

$$d_s = nl = nmp \tag{20.2}$$

The angle of the thread at its mean diameter with respect to a normal to the screw axis

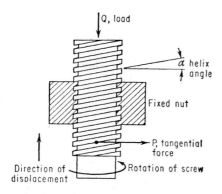

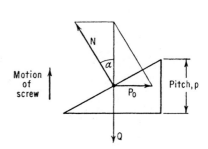

FIG. 20.4 Force P required to overcome load Q.

FIG. 20.5 Vector solution for force P_0 required to overcome load Q, neglecting friction.

is the helix angle α. To determine the force P required to overcome a certain load Q it is necessary to observe the relation of the load direction with respect to the displacement direction. If the load opposes the direction of motion (Fig. 20.4) the force required to overcome it, neglecting friction, is given by

$$P_0 = Q \tan \alpha \qquad \text{(see Fig. 20.5)} \tag{20.3}$$

However, friction displaces the normal N (Fig. 20.6) by angle ϕ, and P_1 is given by

$$P_1 = Q \tan (\alpha + \phi) = Q(\tan \alpha + \tan \phi)/(1 - \tan \alpha \tan \phi) \tag{20.4}$$

Replacing $\tan \phi$ with μ, the coefficient of friction,

$$P_1 = Q(\tan \alpha + \mu)/(1 - \mu \tan \alpha) \tag{20.5}$$

Because of the thread profile angle β, the resultant R should be replaced by $R/\cos \beta$. This affects only the friction terms since friction gave rise to R to begin with; hence these must be divided by $\cos \beta$. Thus

$$P_1 = Q(\mu \sec \beta + \tan \alpha)/(1 - \mu \sec \beta \tan \alpha) \tag{20.6}$$

When a load is applied to the screw in Fig. 20.4 but P is removed, friction alone keeps the screw from turning and moving in the direction of Q. Hence, when motion in the direction of the load is required, the applied force need only be large enough to overcome the friction (Fig. 20.7).

$$P_2 = Q \tan (\phi - \alpha) \tag{20.7}$$

For a profile angle β

$$P_2 = Q(\mu \sec \beta - \tan \alpha)/(1 + \mu \sec \beta \tan \alpha) \tag{20.8}$$

When motion of the screw or nut may be caused by the load applied to either, the

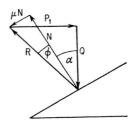

FIG. 20.6 Vector solution for force P_1 required to overcome load Q, including frictional force μN.

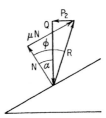

FIG. 20.7 Vector solution for force P_2 required to overcome the frictional force when motion is in the direction of the load.

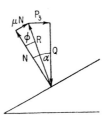

FIG. 20.8 Force P_3 required to prevent motion of screw or nut due to an applied load.

screw and nut are not self-locking and a tangential force will be required to prevent motion (Fig. 20.8). This force is given by

$$P_3 = Q(\tan \alpha - \mu \sec \beta)/(1 + \mu \sec \beta \tan \alpha) \quad (20.9)$$

When $P_3 = 0$ the friction force will just cancel the tangential force produced by the load, and the screw will be self-locking. Setting $P_3 = 0$, the helix angle α at which the screw is self-locking is given by

$$\tan \alpha = \mu \sec \beta \quad (20.10)$$

Note that, for the helix angle at the equilibrium point when the screw is just self-locking, force P_1, to move against the load, will be

$$P_1 = Q \tan 2\alpha \quad (20.11)$$

The torque required to overcome the load Q on the screw is given by

$$T_1 = \tfrac{1}{2} D_s P_1 \qquad \text{for motion against the load} \quad (20.12)$$

$$T_2 = \tfrac{1}{2} D_s P_2 \qquad \text{for motion with the load} \quad (20.13)$$

where D_s is the mean diameter of the screw.

20.3 FRICTION

From Eq. (20.10), it can be seen that the helix angle at which a screw will be self-locking depends upon the coefficient of friction as well as the thread profile. For an average coefficient of friction $\mu = 0.150$, the helix angle must be at least 9° for square

TABLE 20.1 Coefficient of Friction μ

Screw	Nut			
	Steel	Brass	Bronze	Cast iron
Steel, dry.................	0.15–0.25	0.15–0.23	0.15–0.19	0.15–0.25
Steel, machine oil..........	0.11–0.17	0.10–0.16	0.103–0.15	0.11–0.17
Bronze....................	0.08–0.12	0.04–0.06	....	0.06–0.09

and buttress and 10° for Acme threads. These values allow for a small ($\frac{1}{2}$°) margin of safety so that the screw will not keep turning under load for kinetic coefficient of friction slightly less than 0.150 if the applied force P is removed. Coefficients of friction are given in Table 20.1.[2]

The effect of friction in bearings and thrust collars, which must always be used either on the nut or the screw depending upon application, was not included in the preceding considerations. The force necessary to overcome these frictional forces must be determined separately and added to Eqs. (20.6), (20.7), and (20.8).

Where two surfaces are in sliding contact, good design practice requires that nut and screw be made of different materials in order to reduce both wear and friction. Because the nut is usually smaller and easier to replace than the screw, it is made of the softer material, usually high-grade bronze or brass where loads are light.

Workmanship also has an important effect upon friction. A 30- to 50-μin finish will result in a coefficient of friction about one-third lower than a finish of 100 to 125 μin.[3]

20.4 DIFFERENTIAL AND COMPOUND SCREWS

The displacement of the nut or screw depends upon the pitch. From Eqs. (20.1) and (20.2),

$$d_s = nmp \qquad (20.2)$$

where n is the number of turns of the screw and m the number of threads. If small displacement per turn is required, m must equal 1 and $d = p$. However, very small displacements require a very small pitch, and this results in a weak thread. This difficulty can be overcome to some extent by using a differential screw.

When a fast motion is required, a multiple thread of $m = 2$, 3, or more can be used. However, machining is expensive because each thread must be separately cut. Also, since the helix angle on multiple-thread screws is quite large, the screw will not be self-locking. Here, the remedy may be the use of a compound screw.

20.4.1 Differential Screw

The differential screw has two threads in series. Both are of the same hand but different pitch. Every revolution of the screw will move the two nuts toward or away from each other by an amount equal to the difference in pitch. For the arrangement in Fig. 20.9 the nuts C and F will separate, and with the coarser pitch at the fixed nut

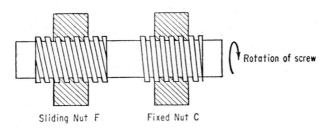

Sliding Nut F   Fixed Nut C

FIG. 20.9   Differential screw.

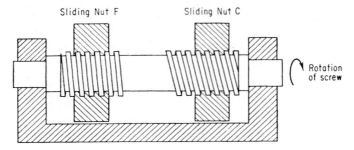

FIG. 20.10   Differential screw.

$$d_s = p_C - p_F \tag{20.14}$$

If the fixed nut has the finer pitch, then, for the same sense of rotation for the screw,

$$- d_s = p_F - p_C \tag{20.15}$$

that is, the nuts will move toward each other. Another arrangement (Fig. 20.10) shows the nuts F and C approaching each other. Their relative displacement

$$d_R = p_C - p_F \tag{20.16}$$

If the rotation of the screw is reversed, the nuts will separate:

$$- d_R = p_F - p_C \tag{20.17}$$

20.4.2   Compound Screw

If as in the arrangements in Fig. 20.11 the threads are of opposing hands, the result will be a compound screw. The displacement will then be the sum of the two pitches.

$$d_s = p_C + p_F \tag{20.18}$$

Note that for the arrangement in Fig. 20.10 interchange of coarse and fine pitch between the nuts will not result in a reversal of the displacement direction as it would

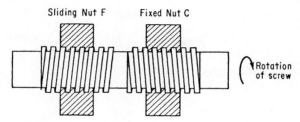

FIG. 20.11   Compound screw.

for a differential screw. Also note that a differential screw must always have threads of different pitch, whereas a compound screw may have both threads of the same pitch, but they must always be of opposite hand.

20.5 EFFICIENCY

The efficiency of a screw is the ratio of the force required for motion against load without friction to that required when friction is present:

$$e_1 = P_0/P_1 = (\tan \alpha)(1 - \mu \sec \beta \tan \alpha)/(\tan \alpha + \mu \sec \beta) \qquad (20.19)$$

$$e_1 = (\cos \beta - \mu \tan \alpha)/(\cos \beta + \mu \cot \alpha) \qquad (20.20)$$

Thus the efficiency of a square thread ($\beta = 0$),

$$e_{\text{sq. thread}} = (1 - \mu \tan \alpha)/(1 + \mu \cot \alpha) \qquad (20.21)$$

From Eq. (20.20) it is seen that max e occurs when $\alpha = 45°$. The maximum attainable efficiency

$$e_{\text{max}} = (\cos \beta - \mu)/(\cos \beta + \mu) \qquad (20.22)$$

To obtain the greatest efficiency, μ should be as small as possible.

When the motion occurs in the direction of the load, it is meaningless to speak of efficiency unless it is proposed to convert an axial load into a tangential force with a non-self-locking screw, a most unlikely design application. When the screw is self-locking, efficiency must be considered negative because a tangential force is required to do what the axial load cannot accomplish by itself. Zero efficiency then represents the equilibrium condition given by Eqs. (20.9) and (20.10). However, the efficiencies are required to consider differential and compound screws, and mathematical expression can be derived provided the foregoing is kept in mind. When the screw is not self-locking the efficiency is given by

$$e_3 = P_3/P_0 = (\cos \beta - \mu \cot \alpha)/(\cos \beta + \mu \tan \alpha) \qquad (20.23)$$

The 100 percent efficient case occurs when P_3 equals P_0, to keep the nut or screw from overrunning. Zero efficiency represents the equilibrium case when screw and nut are just self-locking. For the latter, setting Eq. (20.23) equal to zero,

$$\cos \beta = \mu \cot \alpha \tag{20.24}$$

or
$$\tan \alpha = \mu \sec \beta \tag{20.25}$$

For a self-locking nut, the negative efficiency is given by

$$e_2 = 1 - \frac{P_0}{P_0 + P_2} = 1 - \frac{\cos \beta + \mu \tan \alpha}{\mu (\tan \alpha + \cot \alpha)} \tag{20.26}$$

For the equilibrium case of self-locking, when $e = 0$,

$$\tan \alpha = \mu \sec \beta \tag{20.27}$$

For a differential screw, note that, regardless of rotational direction, one thread will always cause displacement against the load, while the other causes displacement with the load. For the screw in Fig. 20.9, nut F may move with the load with respect to the screw thread, and the nut at C against the load. Reversal of rotation reverses the motions as well. Therefore, the efficiency for a differential screw must be checked for both directions of rotation. The efficiency is given by

$$e_4 = e_1 e_2 \tag{20.28}$$

when both threads are self-locking. Care must be taken to substitute the proper values of α, μ, and β for each thread into the respective efficiency relationships. When only one thread is self-locking,

$$e_5 = e_1 e_3 \tag{20.29}$$

for one direction of motion, and e_4 from Eq. (20.28) for the other. When neither thread is self-locking, Eq. (20.29) applies for either direction.

While the efficiency for a differential screw is quite low for self-locking threads, it can be improved by making one or both threads overrunning. With proper design the tangential force P_3 resulting from a load Q on one thread is not large enough to cause motion against the load on the other thread; so that, even if either thread is separately not self-locking, the combined result in a differential screw produces the desired self-locking feature.

Since the threads are of opposite hand in a compound screw, the displacement with respect to the load is the same for both threads. The efficiency is therefore the product of the efficiency of each thread.

$$e_5 = e_{1,\text{thread1}} e_{1,\text{thread2}} \tag{20.30}$$

20.6 DESIGN CONSIDERATIONS

20.6.1 Thread Bearing Pressure

Design of power screws, as with all other screws, is based upon the assumption that the load is uniformly distributed over all threads. This assumption is not true. When the screw is in tension and the nut in compression, only the first threads will carry the load. The other threads merely maintain contact with one another. This is especially true when the threads are new and not worn. After the threads have undergone some plastic and elastic deformation, some of the load will be carried by the other threads too, but the load distribution nevertheless is far from uniform.

When both screw and nut are in tension or compression, the load is distributed among all threads in engagement, but the distribution again is not uniform. It depends upon the total number of threads in engagement. For three threads, the distribution for a screw in tension is $\frac{2}{3}Q$ from the first to the second thread and $\frac{1}{3}Q$ from the second to the third thread. The distribution in the nut is the same in reverse order.[4] This condition can be alleviated by using a nut of variable cross section. The outside of the nut is parabolic; uniform pressures produced in such nuts greatly reduce wear.[5] To simplify design, however, the assumption of uniformly distributed load is usually made, allowance being made by selecting low values of bearing pressure. The required number of threads is then given by

$$Q = 0.785(D^2 - d^2)p_b N \qquad (20.31)$$

where D is the outside diameter of the screw, d the inside diameter of the nut (root diameter of the screw), and N the number of threads required. Values for p_b may be taken from Table 20.2)[4,6]

TABLE 20.2 Safe Bearing Pressures

Screw	Nut	Safe bearing pressure p_b, psi	Speed range
Steel...........	Bronze	2,500–3,500	Low speed
Steel...........	Bronze	1,600–2,500	Not over 10 fpm
	Cast iron	1,800–2,500	Not over 8 fpm
Steel...........	Bronze	800–1,400	20–40 fpm
	Cast iron	600–1,000	20–40 fpm
Steel...........	Bronze	150–240	50 fpm and over

20.6.2 Tensile and Compressive Stresses

The screw tensile stress is based upon the cross-sectional area at the root diameter. A factor of 4 to 5 is recommended for stress concentrations in the absence of fillets at the root of the thread.

Compressive stresses also cause stress concentrations, but these are not so dangerous and may be neglected.

If the screw is longer than 6 times its root diameter, it should be treated as a short column.

20.6.3 Shear Stresses

The shear produced at the thread or root in a direction parallel to the screw axis is generally not dangerous.

The number of threads in the nut, where shear occurs at the major diameter, is usually controlled by wear considerations.

20.6.4 Deflection

An applied turning moment produces torsional shear, the effect of which should be considered. It must also be borne in mind that torsional, bending, and axial deflections are not negligible. A torsional deflection of $1°$ causes a pitch variation of 0.00005 in. For a precision screw, where the pitch tolerance may be ± 0.0002, this represents a 25 percent deviation. When the screw is a beam, the bending deflection will cause the threads on the compression side to move together and those on the tension side to separate. When the effect of all three deflections is added, the resulting deflection may be quite large (0.008 to 0.015 in is not unusual on heavily loaded screws) and must not be neglected lest undue wear result.[3]

20.6.5 Collar Friction

The frictional forces associated with thrust collars must be determined separately and added to the frictional forces arising from thread loads in order to determine the torque required to turn the screw or nut. Since power screws almost always move with higher than negligible speed, the equations used to determine the static-friction forces at the head of a screw or nut fastening should not be used. A useful relationship is $pv = 20,000$, where p is the allowable pressure, lb/in^2, and v the velocity at the mean diameter of the collar, ft/min. The bearing pressure thus determined should not exceed the value associated with load Q over the thrust-collar area. Coefficients of collar friction are found in Table 20.3.[2]

TABLE 20.3   Coefficients of Friction for Thrust Collars

Material	Running	Starting
Soft steel on cast iron...........	0.12	0.17
Hard steel on cast iron..........	0.09	0.15
Soft steel on bronze.............	0.08	0.10
Hard steel on bronze............	0.06	0.08

20.6.6 Efficiency

Advantageous use of Eq. (20.10) should be made in order to obtain the most efficient thread for a self-locking screw. The helix angle so derived will always be the most efficient. When the screw can be overrunning, the most efficient thread will have a helix angle as close to $45°$ as possible.

20.7   ROLLING-ELEMENT BEARING POWER SCREWS[11,13-15]

To reduce friction, special threads using a ball or roller between the screw and nut are employed. The efficiency of these devices is about 90 percent. The most popular kinds are *ball screws* and *roller screws*. In addition there are *reversing screws, linear ball mechanisms, ball splines, ball plates,* and *ball bushings.* All replace sliding action

FIG. 20.12   Ball screw. (*Courtesy Warner Electric/Dana, South Beloit, Ill.*)

with lower rolling-element friction. *Ball screws* (Fig. 20.12) have semicircular grooves and the nut has rows of bearing balls. The nut has grooves so that the ball at the end of the nut returns to the starting thread in the nut. *Roller screws* are similar to ball screws except that they utilize threaded rollers and have an efficiency slightly less than ball screws. *Reversing screws* are ball screws that provide a steady reciprocating motion with a constant rotation input. The lead angle of these screws is between 15 and 30°, used in fishing reels. Another application of reversing screws is the rotating-output screwdriver with a lead angle between 50 and 80°. *Linear ball mechanisms*[12] are not power screws. They are mechanisms that permit axial movement by use of rolling elements.

It is suggested to utilize manufacturers' catalogs for details of size and dynamic performance. Proper lubrication is essential: oil bath or oil mist is desired as lubricant. A second choice is to use sodium- or lithium-based greases. The life in distance traveled of the bearing is $L = (C/F_e)^3$ where F_e is the equivalent load on the bearing and C is the dynamic constant given in the manufacturer catalog. Rolling-element bearing power screws do not handle shock loads very well. For more on rolling-element bearings see Sec. 19. References 10 and 11 show design techniques to approach the problems of thermal expansion and internal friction.

REFERENCES

1. *Engineering Forum,* vol. 21, July 1960, Eaton Manufacturing Co., Cleveland, Ohio.

2. Ham, C. W., and D. G. Ryan: *Univ. Illinois Bull.,* vol. 29, no. 81, June 1932.

3. Hieber, G. E.: "Power Screws," *Mach. Des.,* November 1953.

4. Maleev, V. L., and J. B. Hartman: "Machine Design," 3d ed., International Textbook Co., Scranton, Pa., p. 385, 1954.

5. Timoshenko, S., and J. M. Lessels: "Applied Elasticity," Westinghouse Press, East Pittsburgh, Pa., 1925.

6. Vallance, A., and V. L. Doughtie: "Design of Machine Members," 3d ed., McGraw-Hill Book Company, Inc., New York, p. 169, 1951.

7. Faires, V. M.: "Designs of Machine Elements," 3d ed., The Macmillan Co., New York, 1955.

8. Niemann, G.: "Maschinen Elemente," Springer-Verlag OHG, Berlin, p. 161, 1960.

9. Spotts, M. F.: "Design of Machine Elements," 2d ed., Prentice-Hall, Inc., Englewood Cliffs, N.J., 1953.

10. Ninomiya, M.: "Maintaining Ball Screw Precision," *Mach. Des.,* vol. 51, no. 8, p. 105, Apr. 12, 1979.

11. "Ball Screws," ANSI B5.48, American Society of Mechanical Engineers, 1977.

12. Linear Motion Systems, Catalog No. 100-IDE, THK Co., Ltd., Tokyo, 1993.

13. Ball Bearing Screws Master Catalog, P-978, Warner Electric Corp., Walterboro, S.C., 1993.

14. Advanced Linear Motion Systems, Form T1-001B, Thomson Industries Inc., Port Washington, N.Y., 1992.

15. "Linear Actuator Technology Guide," TS9105248, Thomson Saginaw, Port Washington, N.Y., 1993.

SECTION 21

SHAFTS, COUPLINGS, KEYS, ETC.*

Stuart H. Loewenthal, P.E., M.S.

Staff Technical Consultant
Lockheed Missile and Space Corp.
Sunnyvale, Calif.

21.1 SHAFTING

21.1.1 Introduction

The term "shaft" applies to rotating machine members used for power or torque transmission. The shaft is subject to torsion, bending, and occasionally axial loading. Stationary and rotating members, called axles, carry rotating elements, and are subjected primarily to cyclic bending. Crankshafts are irregularly shaped (see Fig. 21.1) and often carry large dynamic loads.

According to their position within the system, shafts are referred to as "transmission" or "line shafts." These are relatively long shafts which transmit torque from motor to machine. "Countershafts" are short shafts between driven motor and driven machine or line shaft. "Head shafts" or "stub shafts" are shafts directly connected to the motor.

The transmission of motion or power through an angle can be accomplished without

---

Preliminary note: The reader is referred to Sec. 11 of this handbook for a compilation of standards on the subject matter of this section.

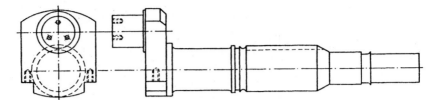

FIG. 21.1   One-piece crankshaft with side crank.

gear trains, chains, or belts by using flexible shafting. Such shafting is fabricated by building up on a single central wire one or more superimposed layers of coiled wire.[1]

Regardless of design requirement, care must be taken to reduce the stress concentration in notches, keyways, etc. Proper consideration of notch sensitivity may improve the strength more significantly than material consideration. Equally important to the design is the proper consideration of factors known to influence the fatigue strength of the shaft, such as surface condition, size, temperature, residual stress, and corrosive environment.

High-speed shafts require not only higher shaft stiffness but also stiff bearing supports, machine housing, etc. High-speed shafts must be carefully checked for static and dynamic unbalance and for first- and second-order critical speeds. The design of shafts in some cases, such as those for turbopumps, is dictated by shaft dynamics rather than fatigue strength considerations.[2]

The lengths of journals, clutches, pulleys, and hubs should be viewed critically because these very strongly influence the overall assembly length. Pulleys, gear coupling, etc., should be placed as close as possible to the bearing supports in order to reduce the bending stresses.

Shafts designed for fatigue or static strength have dimensions selected relative to the working stress of the shaft material, torque, bending loads to be sustained, and any stress concentrations or other factors influencing fatigue strength present. Shafts designed for rigidity have one or more dimensions exceeding those determined by strength criteria in order to meet deflection requirements on axial twist, lateral deflection, or some combination.

21.1.2   Static or Steady Loading

The complexity of stresses is a result of torque transmitted to the shaft, bending of the shaft due to its weight or load, and axial forces imparted to the shaft. The three basic types or cases will be considered.

Case 1 considers *pure torque*. For a shaft transmitting power P_o at a rotational speed n, the transmitted torque T can be found from

$$T \text{ (in·lb)} = 63,025 \, \frac{P_o \text{ (hp)}}{n \text{ (r/min)}} \tag{21.1}$$

or

$$T \text{ (N·mm)} = 9,550,000 \, \frac{P_o \text{ (kW)}}{n \text{ (r/min)}}$$

The relation between nominal shear stress τ_{nom}, torque T, and polar section modulus Z_p is given by

$$\tau_{nom} \text{(in·lb)} = \frac{T \text{(in·lb)}}{Z_p \text{(in}^3)} \tag{21.2}$$

or
$$\tau_{nom} \text{(N·mm)} = \frac{T \text{(N·mm)}}{Z_p \text{(mm}^3)}$$

For a circular shaft the nominal shear stress

$$\tau_{nom} = \frac{16T}{\pi d_0^3} B \tag{21.3}$$

where $B = 1$. For a hollow circular shaft

$B = 1/(1 - \alpha^4)$ d_o = outside diameter
$\alpha = d_i/d_o$ d_i = inside diameter

Shear stress in noncircular shafts may be calculated from relations found in Table 21.1 and Sec. 2. Torsion and design of crankshafts are found in Ref. 4.

Case 2 considers *simple bending.* For a given bending moment M (lb·in), transverse section modulus Z (in^3), and nominal stress in bendingσ_{nom},

$$M = \sigma_{nom}Z$$

For a circular cross section

$$\sigma_{nom} = (32/\pi d_0^3)BM \tag{21.4}$$

where B for a solid circular shaft is equal to 1.

Case 3 considers *combined torsion, bending,* and *axial loading* in a circular shaft. When a shaft is subjected to torsion and bending, the induced stresses are larger than the direct stress due to T and M alone. The effective stress σ_{eff} must be considered:

$$\sigma_{eff} = (32/\pi d_0^3)B\sqrt{[M + (Fd_o/8)(1 + \alpha^2)]^2 + 3/4(T)^2} \tag{21.5}$$

where σ_{eff} = effective normal stress and F = axial force.

In terms of static or steady-state stress failure, a ductile metal is usually considered to have failed when it has suffered elastic failure, that is, when marked plastic deformation has begun. For ductile metals under static or steady loading, local yielding due to stress concentrations, such as small holes, notches, or fillets, is generally not troublesome. This is true provided that the volume of the material affected and its location do not seriously reduce the strength of the member as a whole. Local plastic yielding of certain highly stressed elements permits some degree of stress relieving to occur, passing part of the stress on to adjacent elements within the member. Under these circumstances, the local stress concentration factor can be neglected. Failure will generally occur when the effective stress σ_{eff} exceeds the yield strength of the bulk material, σ_y. Thus for *ductile metals,* where local yielding at stress concentration is acceptable, the shaft diameter (or outside diameter) can be found from

$$d_0^3 = (FS/\sigma_y)(32/\pi)B\sqrt{[M + (Fd_o/8)(1 + \alpha^2)]^2 + 3/4(T)^2} \tag{21.6}$$

where FS = factor of safety.

Note: The value to use for FS is based on judgment. It depends on the consequences of failure, i.e., cost, time, safety. Some factors to consider when selecting a

TABLE 21.1 Shaft Data

Cross section of shaft	Equivalent length of circular bar of diameter D	Maximum shear stress
Solid circle, diameter D_1	$L = L_1 D^4 \left[\dfrac{1}{D_1^4}\right]$	$\tau = \dfrac{5 \cdot 1\,T}{D_1^3}$ at periphery
Hollow circle, d_1, D_1	$L = L_1 D^4 \left[\dfrac{1}{D_1^4 - d_1^4}\right]$	$\tau = \dfrac{5 \cdot 1 D_1\,T}{(D_1^4 - d_1^2)}$ at periphery
Slit hollow circle, d_1, D_1	$L = L_1 D^4 \left[\dfrac{1 \cdot 5}{(D_1 + d_1)(D_1 - d_1)^3}\right]$	$\tau = \dfrac{7 \cdot 6\,T}{(D_1 + d_1)(D_1 - d_1)^2}$
Ellipse, a, b	$L = L_1 D^4 \left[\dfrac{(a^2 + b^2)}{2 a^3 b^3}\right]$	$\tau = \dfrac{5 \cdot 1\,T}{a b^2}$ at X
Rectangle, $a > b$	$L = L_1 D^4 \dfrac{(a^2 + b^2)}{3 \cdot 1 a^3 \cdot b^3}$	$\tau = \left[\dfrac{15a + 9b}{5 a^2 \cdot b^2}\right] T$ at X
Square, a	$L = L_1 D^4 \left[\dfrac{1}{1 \cdot 43 a^4}\right]$	$\tau = \dfrac{4 \cdot 8\,T}{a^3}$ at X
Triangle, $0.866a$, a	$L = L_1 D^4 \left[\dfrac{4.53}{a^4}\right]$	$\tau = \dfrac{20\,T}{a^3}$ at X
Hexagon, D_1	$L = L_1 D^4 \left[\dfrac{1}{1 \cdot 18 D_1^4}\right]$	$\tau = \dfrac{5 \cdot 3\,T}{D_1^3}$ at X
Octagon, D_1	$L = L_1 D^4 \left[\dfrac{1}{1 \cdot 10 D_1^4}\right]$	$\tau = \dfrac{5 \cdot 4\,T}{D_1^3}$ at X

value for FS are how well the actual loads, operating environment, and material strength properties are known, as well as possible inaccuracies of the calculation method. Values typically range from 1.3 to 6, depending on the confidence in the prediction technique and the criticality of the application. Unless experience or special circumstances dictate, the use of FS values of less than 1.5 is not normally recommended.

TABLE 21.1 Shaft Data (*Continued*)

Cross section of shaft	Equivalent length of circular bar of diameter D	Maximum shear stress
	$L = L_1 D^4 \left[\dfrac{1 \cdot 16}{D_1^4}\right]$	$\tau = \dfrac{8 \cdot 1 T}{D_1^3}$ at X

| | $L = L_1 D^4 \left[\dfrac{K}{D_1^4}\right]$ | $\tau = \dfrac{K_1 N}{D_1^3}$ |

	r	K	r	K_1 At A	K_1 At B
	$D_1/100$	1.28	$D_1/100$	31.5	13.4
	$D_1/50$	1.27	$D_1/50$	19.7	13.0
	$D_1/25$	1.26	$D_1/25$	13.5	12.6
	$D_1/15$	1.24	$D_1/15$	11.0	11.7

| | $L = L_1 D^4 \left[\dfrac{K}{D_1^4}\right]$ | $\tau = \dfrac{K_1 T}{D_1^3}$ |

	r	K	r	K_1 At A	K_1 At B
	$D_1/100$	1.29	$D_1/100$	13.0	8.7
	$D_1/50$	1.29	$D_1/50$	9.7	8.6
	$D_1/25$	1.28	$D_1/25$	8.6	8.5
	$D_1/15$	1.26	$D_1/15$	8.3	8.4

| | $L = L_1 D^4 \left[\dfrac{K}{D_1^4}\right]$ | $\tau = \dfrac{K_1 T}{D_1^3}$ |

	r	K	r	K_1
	$D_1/100$	1.86	$D_1/100$	20.0
	$D_1/75$	1.79	$D_1/75$	15.5
	$D_1/50$	1.72	$D_1/50$	14.6
	$D_1/40$	1.64	$D_1/40$	13.8
				at X

L = length of solid circular shaft of diameter D having the same torsional rigidity as length L_1 of the actual shaft

T = torque transmitted by shaft

In the case of *brittle metals,* such as cast iron, cracking rather than local yielding at stress concentration points is likely to occur. These cracks can propagate through the member leading to shaft fracture. For the design of shafts made of metals with low ductility or fracture toughness, it is customary to apply theoretical stress concentration factors K_t to account for this effect. Because brittle metals experience fracture failure, rather than plastic deformation, it is also customary to use the ultimate tensile strength of the material σ_u as the limiting strength factor. The diameter of shafts made from brittle metals can be found from

$$d_0^3 = (FS/\sigma_u)(32/\pi)B\sqrt{[(K_t)_b M + (K_t)_a(Fd_o/8)(1 + \alpha^2)]^2 + 3/4[(K_t)_t T]^2} \quad (21.7)$$

where $(K_t)_b$ = theoretical stress concentration factors in bending
$\quad\quad (K_t)_a$ = theoretical stress concentration factors in axial loading
$\quad\quad (K_t)_t$ = theoretical stress concentration factor in torsion

Typical values of K_t for shafts with fillets and holes can be found in Figs. 21.2 and 21.3. Factors for numerous other cases appear in Ref. 5.

It is worth noting that applying the full value of K_t in the design of shafts made of some brittle metals may not be justified from strictly a strength standpoint. However, it is usually prudent to use full K_t values in view of a brittle metal's poor resistance to shock loading. A second point to be made is that surface-hardened ductile steel shafts can exhibit brittle behavior at stress concentration points. Although cracks may stop at the softer core interface, some reduction in strength will still occur. This reduction in strength should be taken into account when selecting FS values.

Some shafts can fail under heavy transverse shear loading. For short, solid shafts having only transverse shear loading, the shaft diameter is given by

$$d_o = \sqrt{\frac{1.7\,V}{\tau_y/FS}} \quad\quad\quad (21.8)$$

where τ_y = shear yield strength = $0.577\sigma_y$ for most steels and V = maximum transverse shear load.

Shafts having a high degree of hollowness—or, in other words, tubes—may fail from buckling rather than from stress. (Reference 6 should be consulted.)

21.1.3   Fluctuating Loads

Ductile machine elements subjected to repeat fluctuating stresses above their endurance strength but below their yield strength will eventually fail from fatigue. The insidious nature of fatigue is that it occurs without visual warning at bulk operating stresses below plastic deformation. Shafts sized to avoid fatigue will usually be strong enough to avoid elastic failure, unless severe transient or shock overloads occur.

Failure from fatigue is statistical in nature, inasmuch as the fatigue life of a particular specimen cannot be precisely predicted but rather the likelihood of failure based on a large population of specimens. For a group of specimens or parts made to the same specification, the key fatigue variables would be the effective operating stress, the number of stress cycles, and volume of material under stress. Since the effective stresses are usually the highest at points along the surface where discontinuities occur, such as keyways, splines, and fillets, these are the points from which fatigue cracks are most likely to emanate. However, each volume of material under stress carries with it a finite probability of failure. The product of these elemental probabilities (the "weakest link" criterion) yields the likelihood of failure for the entire part for a given number of loading cycles.

At present there is no unified, statistical failure theory to predict shafting fatigue. However, reasonably accurate life estimates can be derived from general design equations coupled with bench-type fatigue data and material static properties. Fatigue test data are often obtained on either a rotating-beam tester under the conditions of reversed bending or on an axial fatigue tester. The data generated from these machines are usually plotted in the form of stress/life (S/N) diagrams. On these diagrams the bending stress at which the specimens did not fail after some high number of stress cycles, usually 10^6 to 10^7 cycles for steel, is commonly referred to as the "fatigue

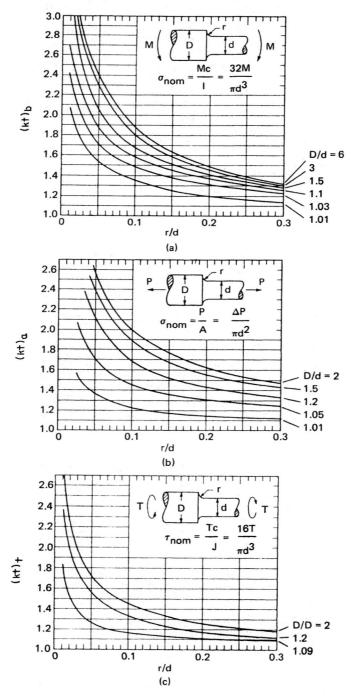

FIG. 21.2 Theoretical stress concentration factor for shaft with fillet (*a*) bending, (*b*) axial load, (*c*) torsion.[8]

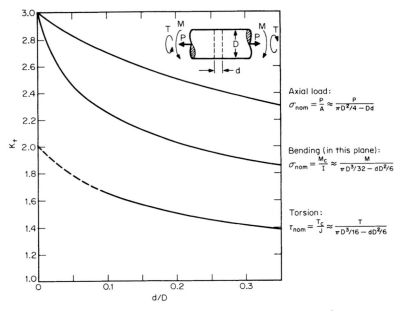

FIG. 21.3   Theoretical stress concentration factors for shaft with radial hole.[8]

limit" σ_f. For mild steels it is the stress at which the *S/N* curve becomes nearly horizontal. This seems to imply that operating stresses below the fatigue limit will lead to "infinite" service life. However, this is misleading, since no part can have a 100 percent probability of survival. In fact, fatigue-limit values determined from *S/N* diagrams normally represent the mean value of the failure distribution due to test-data scatter. Statistical corrections must be applied for designs requiring high reliabilities as will be discussed. Furthermore, many high-strength steels, nonferrous materials, and even mild steel in a corrosive environment *do not* exhibit a distinct fatigue limit. In view of this, it is best to consider that the fatigue limit represents a point of *very long life* (> 10^6 cycles).

Up until now the design equations given are for shafts with steady or quasi-steady loading. However, most shafts rotate with gear, sprocket, or pulley radial loads, thereby producing fluctuating stresses. Fluctuating stresses can be broken into two components: an "alternating stress" component σ_a superimposed on a mean stress component σ_m. Under the special circumstance when the mean stress component σ_m is zero, the alternating stress σ_a is said to be a "fully reversing stress" σ_r.

The formulas that follow are for solid circular shafts. These formulas can be applied, with modification, to hollow shafting, provided that the wall is not extremely thin ($\alpha < 0.9$).

Simple Loading.   The loading is considered to be "simple" when only one kind of stress exists, that is, only fluctuating bending or torsion or tension.

For simple loading several failure relations have been proposed, but the "modified Goodman line" is, perhaps, the most widely used. It is given by

$$\sigma_a/\sigma_f + \sigma_m/\sigma_u = 1 \tag{21.9}$$

where σ_a and σ_m are, respectively, the alternating and mean components of the simple stress, σ_u is the ultimate tensile or torsional strength, and σ_f is the fatigue limit of the shaft, that is, the fatigue limit of the shaft material after it has been corrected by certain application factors known to affect fatigue strength, such as those due to surface finish, size, and stress concentration. These fatigue-modifying factors are addressed later in this chapter.

Fluctuating Bending. For the case of a fluctuating bending moment load comprised of an alternating bending moment M_a superimposed on a steady bending moment M_m, the appropriate solid shaft diameter for long life (at least 10^6 cycles) can be found from

$$d^3 = [32(\text{FS})/\pi](M_a/\sigma_f + M_m/\sigma_u) \tag{21.10}$$

where $\qquad M_m = (M_{max} + M_{min})/2 \qquad$ and $\qquad M_a = (M_{max} - M_{min})/2 \tag{21.11}$

Fluctuating Torsion. For the case of a fluctuating torsional load, composed of an alternating torque T_a superimposed on a steady torque T_m, the shaft diameter can be found from

$$d^3 = [16(\text{FS})/\pi](T_a/\tau_f + T_m/\tau_u) \tag{21.12}$$

where τ_f is the torsional fatigue limit and τ_u is the ultimate shear strength, and where expressions similar to Eq. (21.11) apply for T_a and T_m. From the maximum distortion-energy theory of failure, the shear strength and torsional fatigue limit properties of a material are approximately related to the tensile properties by

$$\tau_f = \sigma_f/\sqrt{3} \qquad \tau_u = \sigma_u/\sqrt{3}$$

Making the above substitutions into Eq. (21.12) gives

$$d^3 = (16\sqrt{3}/\pi)(\text{FS})(T_a/\sigma_{ft} + T_m/\sigma_u) \tag{21.13}$$

where σ_{ft} = fatigue limit of the shaft, considering a *torsional* rather than bending fatigue stress concentration factors.

It is instructive to note that there is a large amount of fatigue data[7] on various metals, including steel, iron, aluminum, and copper, which shows that the torsional fatigue strength of smooth, notch-free members is unaffected by the presence of a mean torsional stress component up to and slightly beyond the torsional yield strength of the material. However, when significant stress raisers are present, the more common case, test data cited by Ref. 8 indicate that the torsional fatigue strength properties of the shaft member will be reduced by the presence of a mean or steady torsional stress component in accordance with Eq. (21.9). This is attributed to the observation that the state of stress at the point of stress concentration deviates from that of pure shear. Thus, Eq. (21.13) is recommended except when the shaft to be designed is substantially free from points of stress concentration, in which case the following may be used:

$$d^3 = (16\sqrt{3}/\pi)(\text{FS})(T_a/\sigma_{ft}) \tag{21.14}$$

To avoid possible yielding failure, the minimum shaft diameter should be no smaller than

$$d^3 = (16\sqrt{3}/\pi)(\text{FS})[(T_a + T_m)/\sigma_y] \tag{21.15}$$

21.1.4 Fatigue under Combined Stresses

For applications where a simple fluctuating stress of the same kind is acting, for example, an alternating bending stress superimposed on a steady bending stress, the Goodman failure line method just described provides an acceptable design. However, most power transmission shafting is subjected to a combination of reversed bending stress (a rotating shaft with constant moment loading) and steady or nearly steady torsional stress. Although a large body of test data has been generated for the simple stress condition, such as pure tensile, flexural, or torsional stress, little information has been published for the combined bending and torsional stress condition. However, some cyclic bending and steady torsional fatigue test data[9] for alloy steel shows a reduction in reversed-bending fatigue strength with mean torsion stress according to the elliptical relation

$$(\sigma_r/\sigma_f)^2 + (\tau_m/\tau_y)^2 = 1 \tag{21.16}$$

Reversed Bending with Steady Torsion. From the failure relation given in Eq. (21.16), the following formula may be used to size solid or hollow shafts under reversed bending M_r and steady torsional T_m loading with negligible axial loading:

$$d_o^3 = [32(FS)/\pi]B\sqrt{(M_r/\sigma_f)^2 + \tfrac{3}{4}(T_m/\sigma_y)^2} \tag{21.17}$$

Equation (21.17) is the basic shaft design equation for the proposed ASME Standard B106.1M, "Design of Transmission Shafting."[9] It can also be derived theoretically from the distortion-energy failure theory as applied to fatigue loading.

Fluctuating Bending Combined with Fluctuating Torsion. In the general case when both the bending and torsional moments acting as the shaft are fluctuating, the safe shaft diameter, according to the distortion-energy theory, can be found from

$$d_o^3 = [32(FS)/\pi]B\sqrt{(M_m/\sigma_u + M_a/\sigma_f)^2 + \tfrac{3}{4}(T_m/\sigma_y + T_a/\sigma_{ft})^2} \tag{21.18}$$

21.1.5 Fatigue Life–Modifying Factors

In Eqs. (21.10), (21.13), (21.17), and (21.18), the fatigue limit of the shaft to be designed is *almost always different* from the fatigue limit of the highly polished, notch-free fatigue test specimen, commonly listed in material property tables. A number of service factors that are known to affect fatigue strength have been identified. These factors can be used to modify the uncorrected fatigue limit of the test specimen, σ_f^*, as follows:

$$\sigma_f = k_a k_b k_c k_d (k_e)_b k_f k_g k_h k_i \sigma_f^* \tag{21.19}$$

or

$$\sigma_{ft} = k_a k_b k_c k_d (k_e)_t k_f k_g k_h k_i \sigma_f^*$$

where σ_f = corrected bending fatigue limit of shaft
$\quad\sigma_{ft}$ = corrected bending fatigue limit of shaft considering $(k_e)_t$ rather than $(k_e)_b$
$\quad\sigma_f^*$ = bending or tensile fatigue limit of polished, unnotched test specimen without mean stress
$\quad k_a$ = surface factor
$\quad k_b$ = size factor
$\quad k_c$ = reliability factor
$\quad k_d$ = temperature factor

$(k_e)_b$ = fatigue stress concentration factor in bending
$(k_e)_t$ = stress concentration factor in torsion
k_f = press-fitted collar factor
k_g = residual stress factor
k_h = corrosion factor
k_i = miscellaneous effects factor

Table 21.2 lists representative σ_f^* values for selected steels along with other material fatigue properties. In the event that actual test data are unavailable, a rough rule of thumb for steel is

$$\sigma_f^* \simeq 0.5\sigma_u \qquad\qquad \text{for } \sigma_u \le 200{,}000 \text{ lb/in}^2 \text{ (1400 MPa)}$$

$$\sigma_f^* \simeq 100{,}000 \text{ lb/in}^2 \text{ (700 MPa)} \qquad \text{for } \sigma_u > 200{,}000 \text{ lb/in}^2 \text{ (1400 MPa)}$$

(21.20)

It is important to remember that σ_f^* values in Table 21.2 do not represent final design values, but must be corrected by the factors given in Eq. (21.19).

Surface Factor k_a. Since the shaft surface is the most likely place for fatigue cracks to start, the surface condition significantly affects the fatigue limit as shown in Fig. 21.4. This figure is based on a compilation of test data from several investigations for a variety of ferrous metals and alloys.[10] The figure shows that the endurance characteristics of the higher tensile strength steels are more adversely affected by poorer surface condition. Furthermore, it shows that surface decarburization that often accompanies forging can cause a severe reduction in fatigue strength. Most, if not all, of the strength reduction due to surface condition can be recovered by cold rolling, shot peening, and other means of inducing residual compressive stress into the surface, as is discussed later.

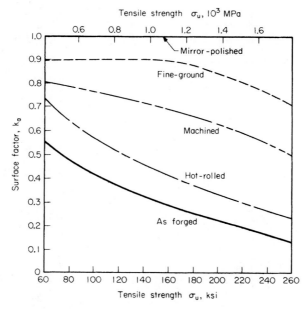

FIG. 21.4 Surface factor k_a as a function of surface condition and strength.[10]

TABLE 21.2 Representative Strength and Fatigue Properties of Selected Steels[19]

(Based on test specimen data without mean stresses)

SAE specification	Bhn	Process description	Ultimate strength σ_u, 10^3 lb/in² (MPa)	Yield strength σ_y, 10^3 lb/in² (MPa)	Fatigue strength coefficient σ'_f, 10^3 lb/in² (MPa)	Fatigue strength exponent b	Fatigue limit at 10^6 cycles, σ^*_f, 10^3 lb/in² (MPa)
1005–1009	125	CD sheet	60 (414)	58 (400)	78 (538)	0.073	35 (244)
1005–1009	90	HR sheet	50 (345)	38 (262)	93 (641)	0.109	29 (202)
1015	80	Normalized	60 (414)	33 (228)	120 (827)	0.11	27 (186)
1018	126	CD bar	64 (441)	54 (372)	—	—	—
1020	108	HR plate	64 (441)	38 (262)	130 (896)	0.12	30 (208)
1040	170	CD bar	85 (586)	71 (480)	—	—	—
1040	225	As forged	90 (621)	50 (345)	223 (1538)	0.14	33 (233)
1045	225	Q&T	105 (724)	92 (634)	178 (1227)	0.095	47 (323)
1045	390	Q&T	195 (1344)	185 (1276)	230 (1586)	0.074	79 (547)
1045	500	Q&T	265 (1827)	245 (1689)	330 (2275)	0.08	104 (715)
1045	595	Q&T	325 (2241)	270 (1862)	395 (2723)	0.081	122 (843)
1050	197	CD bar	100 (690)	84 (579)	—	—	—
1144	305	Drawn at temperature	150 (1034)	148 (1020)	230 (1586)	0.09	66 (454)
1541F	290	Q&T forging	138 (951)	129 (889)	185 (1276)	0.076	63 (435)
4130	258	Q&T	130 (896)	113 (779)	185 (1276)	0.083	59 (404)
4130	365	Q&T	207 (1427)	197 (1358)	246 (1696)	0.081	77 (532)
4140	310	Q&T drawn at temperature	156 (1076)	140 (965)	265 (1827)	0.08	90 (619)
4142	310	Drawn at temperature	154 (1062)	152 (1048)	210 (1448)	0.10	53 (366)
4142	380	Q&T	205 (1413)	200 (1379)	265 (1827)	0.08	83 (574)
4142	450	Q&T and deformed	280 (1931)	270 (1862)	305 (2103)	0.09	83 (572)
4340	243	HR and annealed	120 (827)	92 (634)	174 (1200)	0.095	49 (337)
4340	409	Q&T	213 (1469)	199 (1372)	290 (1999)	0.091	80 (550)
4340	350	Q&T	180 (1241)	170 (1172)	240 (1655)	0.076	82 (567)
5160	430	Q&T	242 (1669)	222 (1531)	280 (1931)	0.071	103 (709)

Note: Values listed are typical. Specific values should be obtained from the steel producer. CD = cold drawn. HR = hot rolled. QT = quenched and tempered.

21.12

From Ref. 10, the "ground-surface" category includes all types of surface finishing which do not affect the fatigue limit by more than 10 percent. Polished, ground, honed, lapped, or superfinished shafts are included in this ground category as well as commercial shafts that are turned, ground and polished, or turned and polished. The "machined-surface" category includes shafts that are either rough or finished machined with roughness ranging between 62 anad 250 μin. The "hot-rolled" category covers surface conditions encountered on hot-rolled shafts which have slight irregularities, some including oxide and scale defects with partial surface decarburization. The "as-forged" category includes shafts with larger surface irregularities, included oxide, and scale defects, with total surface decarburization.

Size Factor k_b. There is considerable experimental evidence that the bending and torsional fatigue strength of large shafts can be significantly less than that of the small test specimens (typically 0.30 in in diameter) which are used to generate fatigue data.[11,12] The size effect is attributed to the greater volume of material under stress and thus, the greater likelihood of encountering a potential fatigue-initiating defect in the material's metallurgical structure. Also the heat treatment of large parts may produce a metallurgical structure that is not as uniform and does not have as fine a grain structure as that obtained with smaller parts. Another factor is that smaller shafts have a higher stress gradient; that is, the rate of stress change with depth is greater.[11]

Figure 21.5 shows the effect of size on bending fatigue strength, of unnotched, polished-steel test specimens up to 2 in diameter from many investigations. The curve appearing in this figure is due to Kuguel's theory[11] which relates the reduction in fatigue strength to the increase in the volume of material under 95 percent of the peak stress. The Kuguel expression can be written as

$$k_b = (d/0.3)^{-0.068} \tag{21.21}$$

where d is the diameter of the shaft in inches.

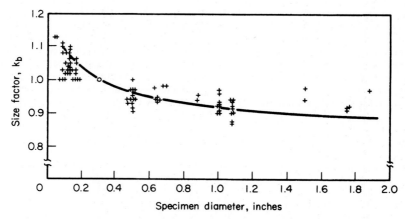

FIG. 21.5 Size factor k_b, Eq. (21.21) (data for unnotched, polished steel specimens having σ_y = 50 to 165 ksi).[8,11]

For larger shafts, that is, above 2 in in diameter, there is insufficient data for establishing a definitive formula. The few relevant tests for 6-in- and 8.5-in-diameter, plain carbon steel specimens in rotating bending have shown a considerable reduction in fatigue strength.[12] The following expression, in the absence of actual data for the shaft to be designed, should provide a reasonable estimate of the size factor for shafts between 2 and 10 in in diameter:

$$k_b = d^{-0.19} \tag{21.22}$$

Reliability Factor k_c. Even under well-controlled test conditions, it is clear that the unavoidable variability in the preparation of test specimens and their metallurgical structure will cause a variability in their measured fatigue strengths. Fatigue-limit data published in standard design references usually represent an average value of endurance for the sample of test specimens. Most designs require a much higher survival rate than 50 percent, that is, the probability that at least half of the population will not fail in service. Consequently, fatigue-limit values must be *reduced* by some amount to *increase* reliability. The amount of this reduction is dependent on the failure distribution curve and required reliability. In the absence of test data, a good rule of thumb for determining k_c is to assume a normal or gaussian failure distribution with a standard deviation of 8 percent for most steels. These values are given in Table 21.3.

TABLE 21.3 Reliability Factor k_c

Shaft nominal reliability	k_c
0.5	1.0
0.90	0.897
0.99	0.814
0.999	0.753
0.9999	0.702

As a cautionary note, correction factors for very high levels of reliability are highly sensitive to the type of failure distribution assumed. Accordingly, k_c values for reliability levels of 99 percent and above are to be used as guidelines.

Temperature Factor k_d. Test data[7,13] indicate that the fatigue limit of carbon and alloy steel is relatively unaffected by operating temperatures between approximately -100 and $+400°F$ (see Table 21.4). For this temperature range $k_d = 1$ is recommended. At low temperatures (to $-200°F$) carbon and alloy steel both possess significantly greater bending fatigue strength. As the temperature is increased to approximately 700°F, carbon steels actually show a small improvement in fatigue strength relative to room-temperature values, while the fatigue strength of alloy steel (SAE 4340) slightly decreases. At elevated temperatures, above 800°F, the fatigue resistance of both types of steel drops sharply as the effects of creep and loss of material strength properties become more pronounced. For applications outside the normal temperature range, the fatigue properties for the shaft material should be ascertained from published or user-generated test data. Reference 7 should provide some guidance on k_d values for typical steels.

TABLE 21.4 Temperature Factor k_d: Fatigue Properties as Related to Room-Temperature Properties, 70°F (23°C)[7,13]

Steel (condition)	Temperature, °F (°C)								
	-200 (-129)	-100 (-73)	0 (-18)	$+70$ ($+23$)	200 (93)	400 (204)	600 (316)	800 (427)	1000 (538)
SAE 1035	1.7	1.3	1.1	1.0	1.0	1.2	1.4	1.3	0.8
SAE 1060	1.5	1.2	1.1	1.0	1.0	1.1	1.2	1.0	0.2
SAE 4340	1.3	1.1	1.0	1.0	0.9	0.9	0.9	0.8	0.6
SAE 4340 (notched)				1.0			0.9	0.9	0.8
0.17% carbon				1.0	1.0	1.0	1.4	1.2	0.6
SAE 4340				1.0			1.0	1.0	0.5
Carbon steel		1.3		1.0					
Carbon steel (notched)		1.1		1.0					
Alloy steel (notched)		1.1		1.0					

Fatigue Stress Concentration Factor k_e. Experience has shown that a shaft fatigue failure almost always occurs at a notch, hole, keyway, shoulder, or other discontinuity where the effective stresses have been amplified. The effect of a stress concentration on the fatigue limit of the shaft is represented by the fatigue stress concentration factor k_e, where

$$k_e = \frac{\text{fatigue limit of the notched specimen}}{\text{fatigue limit of a specimen free of notches}} = \frac{1}{K_e} \qquad (21.23)$$

and where K_e = fatigue-strength reduction factor.

Experimental data[5] indicate that low-strength steels are significantly less sensitive in fatigue to notches or other stress raisers than high-strength steels, as shown in Fig. 21.6. The notch sensitivity q of material can be used to relate the fatigue-strength reduction factor K_e to the theoretical (static) stress concentration factor K_t as follows:

$$K_e = 1 + q(K_t - 1) \qquad (21.24)$$

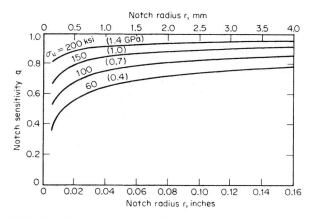

FIG. 21.6 Notch-sensitivity q charts for steel subjected to reversed bending or reversed axial loads.[5,8,69]

The appropriate theoretical stress concentration factor, K_t, to be used in Eq. (21.24) is the one that corresponds to the type of cyclic loading that is present. Thus for a cyclic bending stress, the bending fatigue stress factor is

$$(k_e)_b = \{1 + q[(K_t)_b - 1]\}^{-1} \qquad (21.25)$$

Similarly, for cyclic torsion,

$$(k_e)_t = \{1 + q[(K_t)_t - 1]\}^{-1} \qquad (21.26)$$

Representative values of $(K_t)_b$ and $(K_t)_t$ for shafts with fillets and holes can be found in Figs. 21.2 and 21.3. Typical design values of $(k_e)_b$ and $(k_e)_t$ for steel shafts with keyways appear in Table 21.5.

Press-Fitted Collar Factor k_f. A common method of attaching gears, bearings, couplings, pulleys, and wheels to shafts or axles is through the use of an interference fit. The change in section creates a point of stress concentration at the face of the collar.

TABLE 21.5 Fatigue-Stress-Concentration Factors $(k_e)_b$ and $(k_e)_t$ for Keyways in Solid Round Steel Shafts[14,*]

Steel	Profiled keyway		Sled-runner keyway	
	Bending	Torsion	Bending	Torsion
Annealed (< 200 bhn)	0.63	0.77	0.77	0.77
Quenched and drown (> 200 Bhn)	0.50	0.63	0.63	0.63

*Courtesy McGraw-Hill Book Company.
Note: Nominal stresses should be based on the section modulus for the total shaft section.

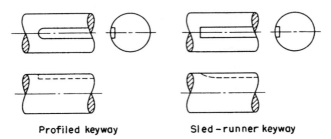

Profiled keyway   **Sled-runner keyway**

This stress concentration coupled with the fretting action of the collar as the shaft flexes is responsible for many shaft failures in service. A limited amount of fatigue test data has been generated for steel shafts having press-fitted, plain (without grooves or tapers) collars in pure bending.[5] Based on these data from several sources,[5] typical fatigue-life reductions range from about 50 to 70 percent. Therefore, the approximate range of press-fitted collar factors is $k_f = 0.3$ to 0.5.

Larger shafts having diameters greater than about 3 in tend to have k_f values less than 0.4 when the collars are loaded. Smaller shafts with unloaded collars tend to have k_f values greater than 0.4. The effect of interference pressure over a wide range between collar and shaft has been found to be small, except for very light fits (less than about 4000 lb/in²) which reduce the penalty to fatigue strength.[5] Surface treatments producing favorable compressive residual stresses and hardening, such as cold rolling, peening, induction, or flame hardening, can often fully restore fatigue strength[7] ($k_f = 1$). Stepping the shaft seat with a generous shoulder fillet radius or providing stress relieving grooves on the bore of the collar can also provide substantial strength improvements.

Residual Stress Factor k_g. The introduction of residual stress through various mechanical or thermal processes can have significant harmful or beneficial effects on fatigue strength. Residual stresses have the same effect on fatigue strength as mean stresses of the same kind and magnitude. Thus residual tensile stresses behave as static tensile loads that reduce strength, while residual compressive stresses behave as static compressive stresses which are beneficial to fatigue strength. Table 21.6 lists the most common manufacturing processes and the type of residual stress they are likely to produce. The extent that the residual stresses from these processes reduce or benefit fatigue strength is dependent on several factors including the severity of the loading cycle and the yield strength of the material in question. Since the maximum residual stress (either compressive or tensile) that can be produced in a part can be no greater than the yield strength of the material minus the applied stress, harder, higher-strength materials can benefit more or be harmed more by residual stress.[15] This coupled with an increase in notch sensitivity makes it important to stress-relieve welded parts made from stronger steels and increases the need to cold-work critical areas. For low cycle

TABLE 21.6 Manufacturing Processes That Produce Residual Stresses

Beneficial residual compressive stress	Harmful residual tensile stress
Prestressing or overstraining	Cold straightening
Shot or hammer peening	Grinding or machining
Sand or grit blasting	Electrodischarge machining (EDM)
Cold surface rolling	Welding
Coining	Flame cutting
Tumbling	Chrome, nickel, or zinc plating
Burnishing	
Flame or induction hardening	
Carburizing or nitriding	

fatigue operations it usually does not pay to shot peen or cold-roll mild steel parts with relatively low yield strengths since much of the beneficial residual compressive stress can be "washed out" with the first application of a large stress.

Cold working of parts or the other means listed in Table 21.6 to instill residual compressive stress is most often applied to minimize or eliminate the damaging effect of a notch, fillet, or other defects producing high stress concentration or residual tensile stresses. Cold-working processes not only generate favorable compressive stresses but also work harden the surface of the part leading to increased fatigue strength. The following discussion reviews some of the processes which generate residual stresses and their likely effect on fatigue.

Prestressing is commonly employed in the manufacture of springs and torsion bars to produce a surface with high residual compressive stress. When applied to notched steel tensile specimens having k_t values of 2.5 and 3.2, pretest stretching was found to eliminate the notch effect completely.[15] The stretching caused the notch to locally yield in tension. When the stretching load was released, the "spring back" from the surrounding, unyielded material created large residual compressive stresses in the notch region.

Shot peening is also used in the manufacture of springs and for shafts that have been plated, welded, or cold-straightened. Normally, peening is only required at the point of stress concentrations. Keyways, splines, fillets, grooves, holes, etc. can be successfully treated. Peening is also useful in minimizing the adverse effects of corrosion fatigue and fretting fatigue. It is effective in restoring the fatigue strength of rough-forged shafts having decarburized surfaces and those that have been heavily machined or ground. Gentle grinding or lapping operations after peening for improved surface finish will have little detrimental effect as long as the layer removed is less than about 10 percent of the induced compressive layer.[7] Generally speaking, if the depth and magnitude of the compressive stress is sufficiently great, the peening process can virtually negate the effect of a notch or other stress concentration. Accordingly, $k_g \approx 0.5/(k_d k_e)$ to $1/(k_d k_e)$, depending on the application. References 7, 15, and 16 should be consulted for more detail.

Surface rolling can be even more effective than shot peening, since it is possible to produce a larger and deeper layer of compressive stress and also achieve a higher degree of work hardening. Furthermore, the surface finish remains undimpled. With heavy cold rolling, the fatigue strength of even an unnotched shaft can be improved up to 80 percent according to test data appearing in Ref. 7. However, like peening, cold rolling is normally applied at points of high stress concentration and where residual tensile stresses are present. Surface rolling of railroad wheel axles and crank pins is commonplace. The fatigue strength of crankshafts has been shown to increase by 60 to 80 percent after rolling the fillets with steel balls.[7]

Hardening processes such as *flame* and *induction hardening,* as well as *case carburizing* and *nitriding,* can considerably strengthen both unnotched and notched parts. This arises from the generation of large residual compressive stresses in combination with an intrinsically stronger, hardened surface layer. These hardening techniques are particularly effective in combating corrosion fatigue and fretting fatigue. Flame hardening of notched and unnotched carbon-steel rotating-beam specimens will typically increase fatigue strength from 40 to 190 percent, while axles with diameters from 2 to 9.5 in having press-fitted wheels showed 46 to 246 percent improvement.[7] Induction hardening of unnotched carbon- and alloy-steel specimens showed fatigue strength improvements from 19 to 54 percent.[7] Case carburizing plain rotating-beam specimens made from various steels caused a 32 to 105 percent increase in fatigue strength, while the notched specimens (0.5-mm-radius notch) showed an improved strength of 82 to 230 percent.[7] Similarly, plain nitrided specimens experienced a 10 to 36 percent benefit, while notched nitrided specimens improved 50 to 300 percent.[7]

The presence of residual tensile stresses due to the processes appearing in Table 21.6 require a derating in fatigue strength. *Cold-straightening* operations tend to introduce tensile residual stress in areas where the material was originally overstrained in compression. References 7 and 14 report fatigue strength reductions ranging from 20 to 50 percent as the result of cold straightening and that such reductions can be avoided if hammer or shot peening is applied during straightening.

While nickel and chrome plating are effective in increasing wear resistance and in improving resistance to corrosion or corrosion fatigue, the resulting residual tensile stresses generated in the plated layer can cause up to a 60 percent reduction in the fatigue limits according to published data.[7] Much of the loss in fatigue strength can be restored if nitriding, shot peening, or surface rolling is performed prior to plating. Shot peening after plating may be even more effective than before plating.

Corrosion Fatigue Factor k_b. The formation of pits and crevices on the surface of shafts due to corrosion, particularly under stress, can cause a major loss in fatigue strength. Exposed shafts on outdoor and marine equipment as well as those in contact with corrosive chemicals are particularly vulnerable. Corrosion fatigue cracks can even be generated in stainless steel parts where there may be no visible signs of rusting. Furthermore, designs strictly based on the fatigue limit may be inadequate for lives much beyond 10^6 to 10^7 cycles in a corrosive environment. Metals fatigue-tested even in a mildly corrosive liquid such as fresh water rarely show a distinct fatigue limit.[7] For example, the *S/N* curve for mild carbon steel tested in a saltwater spray shows a very steep downward slope, even beyond 10^8 cycles. Corrosion fatigue strength has also been found to decrease with an increase in the rate of cycling so both the cycling rate and number of stress cycles should be specified when quoting fatigue strengths of meals in a corrosive environment. Reference 7 contains a wealth of information on the corrosive fatigue strength of metals. Typically, the bending fatigue strength of chromium steels at 10^7 cycles ranges from about 60 to 80 percent of the air-tested fatigue limit when tested in a saltwater spray. In fresh water, carbon and low-alloy steels have a fatigue limit that varies from approximately 80 percent of the air limit for 40,000 lb/in^2 tensile strength steels, 40 percent of the air limit for $\sigma_u = 80,000$ lb/in^2 and 20 percent of the air limit for $\sigma_u = 180,000$ lb/in^2 when cycled in bending at 1450 cycles/min at 2×10^7 cycles.[7] In salt water under the same test conditions, the fatigue limit of carbon steels is about 50 percent, 30 percent, and 15 percent of the air limit for steels having σ_u values of 40,000, 80,000, and 180,000 lb/in^2, respectively. Surface treatments such as galvanizing, sherardizing, zinc or cadmium plating, surface rolling, or nitriding can normally restore the fatigue strength of carbon steels tested in fresh water or salt spray to approximately 60 to 90 percent of the normal fatigue limit in air.[7]

Miscellaneous Effects Factor k_r.   Since fatigue failures nearly always occur at or near the surface of the shaft, where the stresses are the greatest, surface condition strongly influences fatigue life. A number of factors that have not been previously discussed but are known to affect the fatigue strength of a part are listed below:

1. Fretting corrosion
2. Thermal fatigue
3. Electrochemical environment
4. Radiation
5. Shock or vibration loading
6. Ultrahigh-speed cycling
7. Welding
8. Surface decarburization

Although only limited quantitative data have been published for these factors,[7,8,15] they should nonetheless be considered and accounted for if applicable. Some of these factors can have a considerable effect on the shaft's endurance characteristics. In the absence of published data, it is advisable to conduct fatigue tests that closely simulate the shaft condition and its operating environment.

21.1.6 Variable-Amplitude Loading

The analysis presented thus far, for simplicity, assumes that the nominal loads acting on the shaft are essentially of constant amplitude and that the shaft life is to exceed 10^6 or 10^7 cycles. However, most shafts in service are generally exposed to a spectrum of loading. Occasionally, shafts are designed for lives that are less than 10^6 cycles for purposes of economy. Both of these requirements complicate the method of analysis and increase the uncertainty of the prediction. Under these conditions, prototype component fatigue testing under simulated loading becomes even more important.

Short-Life Design.   Local yielding of notches, fillets, and other points of stress concentration is to be expected for shafts designed for short service lives, less than about 1000 cycles. Since fatigue cracks inevitably originate at these discontinuities, the plastic fatigue behavior of the material dictates its service life. Most materials have been observed to either cyclically harden or soften, depending upon their initial state, when subjected to cyclic plastic strain. Therefore, the cyclic fatigue properties of the material, which can be significantly different than its static or monotonic strength properties, needs to be considered in the analysis. For low-cycle, short-life designs, the plastic notch strain analysis, discussed in detail in Refs. 15, 17, and 18, is considered to be the most accurate design approach. This method, used widely in the automotive industry, predicts the time to crack formation based on an experimentally determined relationship between local plastic and elastic strain and the number of reversals to failure. The aforementioned references should be consulted for details of this method.

Intermediate- and Long-Life Designs.   For intermediate- and long-life designs both total strain/life and nominal stress/life (*S/N* curve) methods have been successfully applied. Although the nominal stress/life approach, adopted here, is much older and more widely known, both methods have provided reasonable fatigue-life predictions. Both methods should be applied to the same problem for comparison whenever possible. However, only the nominal stress-life method will be outlined here.

Obviously, the key to accurate fatigue-life prediction is obtaining a good definition of stress/life (S/N) characteristics of the shaft material. Mean bending and/or torsional stress effects should be taken into account if present. Furthermore, a good definition of the loading history is also required. Even when the requirements are met, the accuracy of the prediction is, at best, approximate. As an example, an extensive cumulative fatigue-damage test program was conducted by the SAE to assess the validity of various fatigue-life prediction methods.[17] Numerous simple-geometry, notched steel-plate specimens were fatigue-tested in uniaxial tension. Tests were conducted under constant-amplitude loading and also under a variable-amplitude loading that closely simulated the service loading history. The test specimens' material fatigue properties and the actual force-time history were very well defined. Under these well-controlled conditions, predicted mean life from the best available method was within a factor of $\frac{1}{3}$ to 3 times the true experimental value for about 80 percent of the test specimens, while some of the other methods were considerably less accurate.[17] Under less ideal conditions, such as when the loading history and material properties are not as well known or when a multiaxial stress state is imposed, a predictive accuracy within a factor of 10 of the true fatigue life would not be unacceptable with today's state of knowledge.

The following is a greatly simplified approach to estimate the required shaft diameter for a limited number of stress cycles under a variable-amplitude loading history. It assumes that the loading history can be broken into blocks of constant-amplitude loading and that the sum of the resulting fatigue damage at each block loading equals one at the time of failure in accordance with the Palmgren-Miner linear damage rule. Great care must be exercised in reducing a complex, irregular loading history into a series of constant-amplitude events in order to preserve the fidelity of the prediction. Reference 17 discusses the merits of several cycle-counting schemes that are commonly used in practice for prediction purposes.

A shortcoming of Miner's rule is that it assumes that damage occurs at a linear rate without regard to the sequence of loading. There is ample experimental evidence that a virgin material will have shorter fatigue life, that is, Miner's sums less than one, when first exposed to high cyclic stress before low cyclic stress.[7,18] This "overstressing" is thought to create submicroscopic cracks in the material structure that can accelerate the damage rate. On the other hand, test specimens exposed first to stresses just below the fatigue limit are often stronger in fatigue than when new. This "coaxing" or "understraining" effect which can produce Miner's sums much larger than 1 is believed due to a beneficial strain-aging phenomenon. While Miner's sums at the time of failure can range from 0.25 to 4, depending on loading sequence and magnitude, the experimental range shrinks to approximately 0.6 to 1.6 when the loading is in a more random manner.[18] This is often acceptable for failure estimates. More complicated cumulative damage theories have been devised to account for "sequencing" effects. In fact, Ref. 18 discusses seven different ones, but none of them have been shown to be completely reliable for all practical shaft loading histories. In most cases, Miner's rule serves almost as well, and, because of its simplicity, it is still preferred by many.

In order to determine the upper shaft size for a given number of stress cycles under a variable-amplitude loading situation, it is necessary to construct an S/N curve for the shaft under the proper mean loading condition. If an experimentally determined S/N curve for the shaft material is available, then, of course, it is to be used *after being corrected* for the fatigue life modifying factors identified in Eq. (21.19). However, if actual test data are not available, it is still possible to generate a reasonable estimate of the S/N characteristics as shown in Fig. 21.7. In Fig. 21.7, a straight line connects the true fatigue strength coefficient σ'_f at 1 cycle with the shaft's corrected fatigue limit σ_f at 10^6 stress cycles (or 10^7 cycles if applicable) on log-log coordinates. The coefficient σ'_f is the true stress (considering necking) required to cause fracture on the first applied stress cycle. This method assumes that the fracture strength of the material in the outer fibers of the shaft is unaffected by the presence of mean bending or torsional

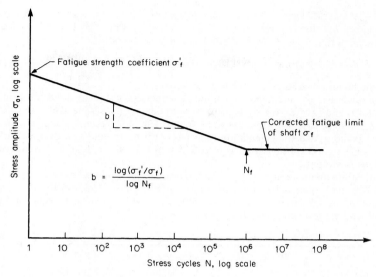

FIG. 21.7 Generalized S/N curve constructed from σ'_f and σ_f.

stresses or the presence of a notch.[15] In the case of axial loading, this assumption is not correct since the whole section rather than the outer fibers must support the mean load. Typical values of σ'_f for a number of different steel compositions can be found in Table 21.2 taken from Ref. 19. For steels in the low and intermediate hardness range (less than 500 Bhn)[20] not listed in Table 21.2:

$$\sigma'_f \simeq \sigma_u + 50,000 \text{ lb/in}^2 \tag{21.27}$$

where σ_u = ultimate tensile strength.

The "fatigue limit of the shaft" σ_f can be found from the fatigue limit of polished, unnotched specimen σ^*_f corrected by the k factors of Eq. (21.19). If a mean bending moment M_m and/or mean torsional load T_m are present, then the specimen fatigue life σ^*_f is altered according to Eq. (21.18) approximately as follows:

$$\sigma^*_{fm} = \sigma^*_f [\sqrt{1 - 77.8(T_m/d^3\sigma_y)^2} - 10.2(M_m/d^3\sigma_u)] \tag{21.28}$$

where σ^*_{fm} = test specimen fatigue limit with M_m or $T_m \neq 0$.

Assuming that the shaft is exposed to a series of alternating bending moments of constant amplitude M_a, acting for n_1 loading cycles, M_{a2} for n_2 cycles, M_{a3} for n_3 cycles, etc., then according to Miner's rule,

$$n_1/N_1 + n_2/N_2 + n_3/N_3 + \cdots = 1 \tag{21.29}$$

where N_1 is the number of cycles to failure at bending moment M_{a1}, N_2 is the cycles to failure at M_{a2}, etc.

From the straight line on the log-log S/N plot of Fig. 21.7, it is clear that

$$\sigma_{a1}/\sigma_f = (N_f/N_1)^b \qquad \sigma_{a2}/\sigma_f = (N_f/N_2)^b \qquad \sigma_{a3}/\sigma_f = (N_f/N_3)^b \tag{21.30}$$

where σ_{ai} is the alternating bending stress at bending moment M_a, N_f is the number of stress cycles corresponding to the fatigue limit σ_f (usually 10^6 to 10^7 cycles), and b is the slope of the *S/N* curve taken as a *positive* value where $b = \log(\sigma_f'/\sigma_f)/6$ for $N_f = 10^6$ cycles, or $b = \log(\sigma_f'/\sigma_f)/7$ for $N_f = 10^7$ cycles. Substituting Eq. (21.30) back into Eq. (21.27), noting $\sigma_{ai} = 32\,M_{ai}/\pi d^3$, and simplifying yields

$$d^3 = [32(FS)/\pi\sigma_f][(n_1/N_f)(M_{a1})^{1/b} + (n_2/N_f)(M_{a2})^{1/b} + (n_3/N_f)(M_{a3})^{1/b}]^b \quad (21.31)$$

where the factor of safety term FS has been introduced.

Note that since σ_f, and hence b, can be dependent on the shaft diameter d through Eqs. (21.19) and (21.28), it may be necessary to take an initial guess of d to calculate σ_f and b so that d from Eq. (21.31) can be found. The most recently found d should be used to update σ_f and the calculation repeated until the change in d becomes acceptably small. [A good starting point is to calculate d from Eq. (21.31) assuming that no mean load is present.]

To illustrate application of this method, consider that a shaft is to be designed from SAE 1045 steel, quenched and tempered (225 Bhn, $\sigma_u = 105{,}000$ lb/in² and $\sigma_y = 92{,}000$ lb/in², from Table 21.2) for 100,000 cycles under a steady torque of 25,000 in[cdt]lb and the following variable bending moment schedule.

M_{ai}, in·lb	Percent time	Number of cycles n_i	Fraction of N_f, n_i/N_f
15,000	15	15,000	0.015
10,000	35	35,000	0.035
5,000	50	50,000	0.050
	100	100,000	

The fatigue limit of a smooth 1045 steel specimen without mean stress σ_f^* is listed as 47,000 lb/in² in Table 21.2. (Note this is somewhat smaller than the approximation $0.5\sigma_u$ or 52,500 lb/in².)

Starting with an initial shaft diameter guess of $d = 2.0$ in, the effect of the mean torque of 25,000 in·lb on σ_f^* can be found from Eq. (21.28) as follows:

$$\sigma_{fm}^* = 47{,}000\sqrt{1 - 77.8[25{,}000/(2^3)(92{,}000)]^2} = 44{,}840 \text{ lb/in}^2$$

Assume that in this example the product of all the k factors described by Eq. (21.19) is equal to 0.4. Then the shaft's corrected bending fatigue limit is

$$\sigma_f = 0.4(44{,}840) = 17{,}900 \text{ lb/in}^2$$

For this material σ_f' is given as 178,000 lb/in², so the *S/N* curve slope is

$$b = \log(178{,}000/17{,}900)/6 = 0.166$$

or $1/b = 6.0$

Finally, selecting FS = 2.0, the required shaft diameter d can be found from Eq. (21.31) to be

$$d^3 = [32(2)/\pi\ 17{,}900][0.015(15{,}000)^{6.0} + 0.035(10{,}000)^{6.0} +$$
$$0.05(5000)^{6.0}]^{0.166} = 8.44 \text{ in}^3$$

or $d = 2.04$ in.

It is instructive to note that if the calculation were repeated considering that only the maximum bending moment of 15,000 in·lb acted 15 percent of the time and that if the shaft ran unloaded the rest of the time, that is, $M_{a_2} = M_{a_3} = 0$, then $d = 2.01$ in. The insignificant reduction in shaft diameter from ignoring the lower loads clearly illustrates the dominant effect that peak loads have on fatigue life. This is also apparent from Fig. 21.7 where life is inversely proportional to the $1/b$ power of stress amplitude. The exponent $1/b$ typically ranges from about 5 for heavily notched shafts to about 14 for some polished, unnotched steel test specimens without mean stresses (see Table 21.2). Even at a modest $1/b$ value of 6, 64 times more fatigue damage is caused by doubling the alternating bending moment, hence the stress amplitude. This underscores the necessity of paying close attention to overload conditions in both shaft and structural-element fatigue designs.

21.1.7 Rigidity

Gears, bearings, couplings, and other drive-line components perform best when they are maintained in correct alignment. Operating positioning errors due to shaft, bearing, and housing deflections under load can lead to increased operating temperatures, and vibratory loads, as well as reduced service life. Shaft stiffness rather than strength can sometimes dictate shaft size, particularly for high-speed, precision machinery where accurate alignment becomes more important. Dynamic rather than static deflections and runouts determine operating characteristics, so shaft vibration should also be considered. Information on dynamic load deflections can be found in Secs. 2, 5, and 6.

The permissible values of shaft deflection are normally dictated by the sensitivity of selected transmission components to positional errors and the quality of service desired. For example, small, high-speed precision gears will obviously require greater positioning accuracy than large, low-speed gears intended for normal industrial service. Thus it is good practice to determine the effects of shaft deflection on the performance of the specific components rather than to use generalized rules of thumb. However, certain general rules can often be helpful for preliminary design purposes.

In the case of gears, positional errors due to shaft, bearing, and housing deflections cause changes in the operating center distance and backlash, and cause a maldistribution of tooth loads, lowering gear mesh capacity. Normally, involute gears can tolerate somewhat greater changes in operating center distance, that is, parallel deflections between driver and driven gear shafts, then changes in parallelism (misalignment) between shafts. According to Ref. 21, the deviation between the operating and ideal center distance of involute toothed gears for proper meshing action should be less than either 2 percent or one quarter of the working depth, whichever is smaller. For spur and straight bevel gears, working depth = 2 × pitch diameter/number of teeth. Permissible misalignment values for gears can be more restrictive, depending on quality of service according to Table 21.7. In the case of bevel gears having 6- to 15-in diameter, some manufacturers recommend that either pinion or gear deflections should not generally exceed 0.003 in. These are general maximum limits which can usually

TABLE 21.7 Representative Maximum Permissible Misalignment Values for Geared Shafts[22]

Gear-train quality	Misalignment, inches per inch of shaft length between bearings
Commercial	0.005
Precision	0.003
High precision	0.0015
Ultra precision	0.001

be improved upon by mounting gears close to bearings and using a straddle mount rather than an overhung mount design when possible.

Shaft misalignment through plain and rolling-element bearings can cause stress concentration to develop, leading to shorter life. For journal bearings, the shaft angle through the bearing should never cause contact between the shaft and the bearing. Thus the shaft angle in radians should never exceed the bearing diametral clearance/bearing width. Spherical and deep-grooved ball bearings have greater tolerance to misalignment than cylindrical and tapered roller bearings, according to Table 21.8.

TABLE 21.8   Limits of Rolling-Element Bearing Misalignment Based on Manufacturer's General Experience[23]

Bearing type	Allowable misalignment, rad (in/in)
Cylindrical and tapered roller	0.001
Spherical	0.0087
Deep-grooved ball	0.0035–0.0047

Flexible shaft couplings are designed to accept some degree of shaft misalignment and offset. Acceptable limits for gear-type couplings are normally established by maximum permissible sliding velocities, while foil and diaphragm types are limited by flexible-element fatigue. Coupling manufacturers' product literature should be consulted for recommended values.

In some applications where there is a need to synchronize the position of several machine elements mounted on the same shaft, the maximum angular deflection of the shaft between elements may need to be limited. Permissible values for shaft twist commonly quoted in the literature are 1° twist for a shaft length of $20d$ or 10 ft, whichever is more stringent. For a solid circular shaft in torsion, the shaft twist in degrees θ due to torque T is given by

$$\theta = 584TL/d^4G \qquad (21.32)$$

where d = shaft diameter
G = elastic shear modulus, 11.5×10^6 lb/in^2 for steel
L = shaft length of circular shaft or equivalent length for noncircular shaft from Table 21.1

21.1.8   Shaft Materials

Power-transmission shafts and axles are most commonly machined from plain carbon (AISI/SAE 1040, 1045, and 1050) or alloy (AISI/SAE 4140, 4145, 4150, 4340, and 8620) steel bar stock. The bar stock may be either hot rolled or cold finished (cold drawn and machined). Cold drawing not only improves mechanical strength but also improves machinability, surface finish, and dimensional accuracy. Hot-rolled shafts are often quenched and tempered for greater strength and then finished (turned and polished or turned, ground and polished) for improved finish and dimensional accuracy. Table 21.9 shows normal surface finish ranges.

In general, standard commercial-line shafting up to 3 in in diameter is normally cold drawn and is sufficiently straight that no turning is normally needed. However, a straightening operation is usually required if keyways are cut into the shaft, since the keyways relieve the compression on the shaft surface generated during cold drawing.

TABLE 21.9 Normal Range of Surface Finish for Shaft Finishing Operations[24]

Finishing operation	Normal range of surface finish, 10^{-6} in, AA*
Cold drawn	50–125
Turned and polished	15–40
Cold drawn, ground, and polished	8–20
Turned, ground, and polished	8–20

*AA ≃ arithmetic average surface roughness. RMS (root-mean-square) roughness ≃ 1.11 AA roughness.

Generally, turned bars can be machined or key-seated with practically no danger of warpage. The diameter of machinery shafting is normally available in $\frac{1}{16}$-in increments from $\frac{1}{2}$ to $2\frac{1}{2}$ in in $\frac{1}{8}$-in increments to 4 in, and in $\frac{1}{4}$-in increments to 5 in, with standard stock lengths of 16, 20, and 24 ft. Transmission shafting diameters are graduated in $\frac{1}{4}$-in increments from $1\frac{5}{16}$ to $2\frac{7}{16}$, and $\frac{1}{2}$-in increments to $5\frac{15}{16}$ in.

Crankshafts, and hollow, large-diameter, or flanged shafts are often forged or partially forged and then ground or machined to final dimensions. Quite often, crankshafts, forged shafts, and those of large section require a higher carbon steel or alloy steel to preserve strength. In the case of crankshafts, steel grades 1046, 1052, 1078, 4340, and 5145 have been successfully used.

For some light-duty applications, low-strength, plain carbon steels, such as 1018 and 1022, can be used. However, the cost savings expected by switching to a lower grade of steel can sometimes be negated by the need to increase the size of the shaft and the components attached to it. In many cases, particularly weight- and size-sensitive applications, it may be more cost-effective to switch to a higher-strength alloy steel, such as 4140 or 4340.

Case-carburized steels, such as 8620, 1060, and the 4000 series, are frequently used for splined and geared shafts. Case-hardened shafts are also specified for bearing application where the balls or rollers contact the shaft directly. General design information on shafting steels can be found in Ref. 13.

It should be noted that it is pointless to specify a more expensive, higher-strength steel for applications where shaft rigidity is the deciding factor. This is because both low- and high-strength steels have essentially equal elastic modulus properties. Only an increase in shaft section or a decrease in the bearing support span can lessen the shaft deflection under a fixed load.

21.1.9 Shaft Vibration

As the speed of a rotating shaft is increased, a critical speed (system natural frequency) may be approached which may cause a significant increase in shaft deflections and vibrations. Operation at or near a system critical speed can lead to high vibration amplitudes which, if left unchecked, can lead to substantial damage to the machine. The subject of rotor dynamics is much too diverse (e.g., see Refs. 25 to 32) to be covered here in detail. Therefore, the following discussion will be limited to some of the more basic design considerations with reference to more sophisticated approaches.

The avoidance of potentially dangerous vibrations through careful modeling of the shaft-bearing system should be undertaken early in the design process. Although a shaft may be sufficiently strong to meet service-life requirements, it may not be sufficiently stiff (or soft) to avoid rotor dynamics problems. Modern high-speed rotating machinery can be designed to operate at speeds substantially greater than the fundamental critical frequency (so-called supercritical shafting), but proper attention to balancing and provisions for damping are usually required.[25] However, most rotating machinery is designed to operate at speeds below the first critical frequency, and only this will be addressed here.

As a first approximation, the first critical or fundamental frequency of a shaft-bearing system with concentrated body masses can be found from the Rayleigh-Ritz equation[26] as follows:

$$\omega_c = \left[\frac{g \sum_1^m W_i \delta_i}{\sum_1^m W_i \delta_i^2} \right]^{1/2} \tag{21.33}$$

where ω_c = first critical frequency, rad/s
W_i = weight of ith body
δ_i = static deflection of ith body
m = total number of bodies
g = gravitational constant, 9.8 m/s² (386 in/s²)

It should be emphasized that the proper deflections to use in Eq. (21.33) are those due to the weight of the body and not to any external forces.

Equation (21.33) considers the shaft to be weightless; however, the shaft's weight can also be taken into account by dividing the shaft into a number of discrete elements. Then the weight of each shaft element, acting at its center of gravity, and its deflection can be readily included into the summation of Eq. (21.33). This technique can also be used to estimate the first natural frequency of the shaft by itself.

In applying Eq. (21.33) a judgment must be made whether the bearings will freely accommodate shaft misalignment (simply supported shaft) or resist misalignment (fixed support) or lie somewhere in between. Clearly shafts supported by spherical or self-aligning bearings can be treated as simply supported while those supported by cylindrical or tapered roller bearings are considered fixed. Furthermore, bearings supporting long, flexible shafts tend to behave more like a fixed support than when supporting short, rigid shafts which have relatively small misalignment. When in doubt, the best course to follow is to calculate shaft deflections both ways to "bracket" the critical speed estimate.

Up until now, the support bearings have been considered as laterally rigid. However, rolling-element and fluid-film bearings are not infinitely stiff and will contribute to shaft deflections. The effect of bearing lateral flexibility on the shaft system's critical speed can be approximately accounted for by calculating the additional deflection at each load point δ_{bi} due to support-bearing deflection. The total deflection at each load point δ_i to be incorporated into Eq. (21.33) will then equal the sum of that due to static shaft bending δ_{si} plus δ_{bi} or $\delta_i = \delta_{si} + \delta_{bi}$.

It is clear from Eq. (21.33) that the addition of bearing deflection will cause a reduction in the critical speed relative to a shaft supported on completely rigid bearings. Normally this reduction in speed is relatively small (less than about 20 percent) since the support bearings and housing are usually significantly stiffer than the shaft. It also follows that the critical speed can be raised by reducing total shaft deflections such as by selecting bearings with increased stiffness or adding preload to them, or by reducing support-bearing span, increasing local shaft section diameter, or reducing transmission component weight.

The above method is approximate in that it ignores inertia effects, assumes that bearings have linear deflection characteristics, and neglects damping and unbalance forces. If the predicted critical speed is within about 20 percent of the operating speed, it is desirable to perform a more detailed analysis. General critical-speed computer codes for shafts (Refs. 28 and 29) can be used in place of or in addition to Eq. (21.33).

Reference 27 is a comprehensive document which treats the above topics and many others from a practical design standpoint. Hysteretic effects due to press-fitted components and splines, damping effects due to fluid-film bearings, gyroscopic effects due to turbine and compressor disks, and balancing methods are some of the other topics addressed. Multiplane balancing techniques are extremely important for minimizing shaft excursions for high-speed flexible rotors. Names and sources of a number of rotor dynamics codes also appear in Ref. 27. A general critical-speed computer program which calculates the critical speed and mode shapes for a multispan shaft with any combination of couplings and bearings up to nine elements can be found in Ref. 29. Reference 30 is a good source of information for ball and cylindrical roller-bearing stiffness values to be used for input data to critical speed and rotor dynamics response-type computer codes. Considerable practical information on torsional vibration problems appears in Ref. 31, and simplified analytical techniques appear in Sec. 5. Reference 32 provides a modern treatment of general vibration problems with discussion of modal analysis and application.

21.2 KEYS AND KEYWAYS

21.2.1 Fundamentals

The primary function of a key is to transmit torque between a shaft and the mating machine element, such as a flywheel, a gear, a sprocket, or a coupling between shafts of a motor and the machine it drives. Keys prevent relative rotational motion of the joined elements and in most cases also prevent relative axial motion, except in the cases of feather and spline keys. Section 11 shows standards.

The distribution of forces on the surfaces of a key is complex and depends upon the key fit in the shaft and hub. The force distribution is not uniform over the key surfaces. Because of uncertainties in analysis, a sufficiently high factor of safety should be employed in calculating key dimensions. This factor of safety may be taken as 1.5 for a steady torque, and it should be increased up to 2.5 for a strong fluctuating torque. Practical considerations require that the minimum hub length be 1.2 shaft diameters for good grip and to prevent rocking on the shaft. This length is also recommended as the minimum length of the key. A light press fit between the shaft and the hub should be used in order to prevent rocking and eccentricity (except with feather keys). Of the keys shown, parallel and tapered keys are the most popular. Recommended nominal dimensions for various shaft diameters are shown in Sec. 11.

Class 1 fit for parallel keys lists clearances from 0.004 in for keys ½ in wide and less to 0.017 in for keys from 6 to 7 in wide. Class 2 fit requires these clearances be reduced to a 0.001-in interference fit to 0.002-in clearance for keys 1.25 in wide or narrower, with no clearance to exceed 0.002 in. Class 3 fit is a recognized, but untabulated, interference fit.[33]

21.2.2 Prestressed Keys

The standard key taper is $\frac{1}{8}$ in/ft. The key seat in the shaft is plain and the seat in the hub is tapered and slightly more shallow to provide for fitting.

In the most usual case of a "top-fitting" key (see Fig. 21.8a) compressive stresses exist on the upper and lower surfaces (ab and cd) and a clearance may exist as shown. For "top and side" fitting keys, an additional side fit of the key is required.

The mechanism of torque transmission is indicated in Fig. 21.9.[17] The equations

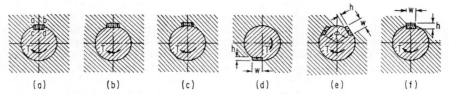

FIG. 21.8 Sunk key installations. (*a*) Top fitting key. (*b*) Flat. (*c*) Saddle. (*d*) Single tangential. (*e*) Double tangential. (*f*) Parallel.

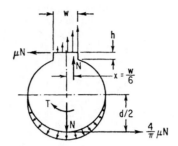

FIG. 21.9 Top fitting key—torque transmission.

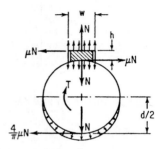

FIG. 21.10 Saddle key—torque transmission.

which follow are based upon torque transmission by friction and it is therefore advisable also to check shaft bearing stress and key shear stress. The normal force

$$N = WL\sigma/2 \tag{21.34}$$

The force necessary to drive the key home

$$Q = N(\tan \alpha + 2\mu) \tag{21.35}$$

The transmitted torque

$$T \leq (WL\sigma_d/4)(W/3 + 1.44\mu d) \tag{21.36}$$

where L = effective key length
σ_d = design stress
α = taper
μ = coefficient of friction = 0.1 for greased keys, 0.15 for dry keys

Flat keys (Fig. 21.8*b*) and *saddle keys* (Fig. 21.8*c*) are used for transmission of small steady torques in cases where shaft weakening cannot be tolerated. Under excessive load, they tend to loosen, rotate, and damage the shaft.

The torque transmission occurs because of friction between key and hub, key and shaft, and shaft and hub (Fig. 21.10). With $N = WL\sigma$, considering the equilibrium condition for moments of forces μN and $(4/\pi)\mu N$, the transmitted torque is given by

$$T \leq 1.14Nd \tag{21.37}$$

or

$$T \leq 1.14L\sigma_d\mu \tag{21.38}$$

With $d \approx 3W_1$ and $\mu \approx 0.15$,

$$T \le 0.5W^2 L\sigma_d \tag{21.39}$$

With *tangential keys* (Fig. 21.8*d*) torque is transmitted by means of compressive stress alone. For each key connection, two mating taper keys are used as shown. The key seats in the shaft and in the hub are *not* tapered. A single tangential key transmits torque in one direction only. For intermittent torque transmission double keying of shafts is used (Fig. 21.8*d*); $\phi = 120°$ to $135°$. Tangential keys are suitable for transmission of large and oscillating torques such as those associated with flywheels. They are more effective in torque transmission than square keys of equal cross section. However, they weaken the shaft more severely than do square keys. With

$$T = N(d/2) \tag{21.40}$$

and

$$N = hL\sigma \tag{21.41}$$

the transmitted torque

$$T \le \tfrac{1}{2}dhL\sigma_d \tag{21.42}$$

21.2.3 Nonprestressed Keys

A *parallel key* is shown in Fig. 21.8*f*. These keys are made square ($w = h$) up to $w = 1\tfrac{1}{2}$ in and $w = \tfrac{1}{4}d_{shaft}$. For larger key width w, the key height h is usually smaller than

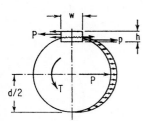

FIG. 21.11 Parallel key—torque transmission.

w. The depth of the key seat in the shaft and in the hub usually is equal to $\tfrac{1}{2}h$. Upper and lower surfaces of the key are parallel. The key is fitted on the sides of the key seat and may have clearance on the top. Torque is transmitted through the sides and no axial forces can be resisted. Neglecting the friction between contacting surfaces, and considering the transmission of torque by means of compressive and shear stresses, the transmitted torque (Fig. 21.11) is

$$T \le dwL\tau_d/2 \quad \text{for shear} \tag{21.43}$$

and

$$T \le \tfrac{1}{4}dhL\sigma_d \quad \text{for compression} \tag{21.44}$$

or, for $d \simeq 4w$ and $h \simeq \tfrac{3}{2}w$,

$$T \le 2w^2 L\tau_d \quad \text{for shear} \tag{21.45}$$

$$T \le \tfrac{3}{2}w^2 L\sigma_d \quad \text{for compression} \tag{21.46}$$

Woodruff keys (Fig. 21.12) are used for light or medium torque transmission, especially in machine-tool and automotive engineering, usually in shafts up to $2\tfrac{1}{2}$ in in diameter. They offer the advantage of permitting easy removal of pulleys from shafts and of preventing tipping of the keys. They adjust themselves easily to a tapered-hub key seat and may also be used for angular location of parts on tapered shaft ends. Because of the relatively deep key seats, a disadvantage associated with Woodruff keys is shaft weakening. Two keys should therefore be used for longer hubs. Woodruff keys should not be used as sliding keys.

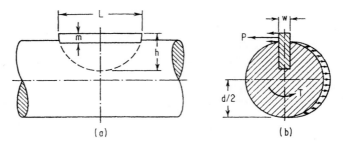

FIG. 21.12 Woodruff key. (*a*) Installation. (*b*) Torque transmission.

The torque transmitted can be calculated (Fig. 21.12*b*) from

$$T \leq \frac{1}{2}d\tau_b \qquad \text{for shear} \tag{21.47}$$

and
$$T \leq \frac{1}{2}dm\sigma_d \qquad \text{for compression} \tag{21.48}$$

Sliding and *guiding* keys (feather keys) (Fig. 21.13) prevent only relative rotational motion between hub and shaft, but permit axial motion. The tight-fit key is fastened either to the shaft or hub, usually by means of a countersunk screw or riveted pin. The pressure *p* on the bearing faces of the key should be less than 1000 lb/in² if the hub is to slide under load.

"Roll pin keys," Fig. 21.14, are used for economy in lightly loaded shafts (see Sec. 11).

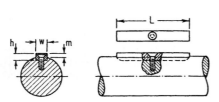

FIG. 21.13 Sliding and guiding keys.

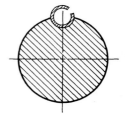

FIG. 21.14 Roll pin key.

21.2.4 Heavy-Duty Keys

The Barth key (Fig. 21.15*a*) utilizes the torque to force the key into the key seat, thereby inducing a compressive stress. This key does not require a tight fit.

Other heavy-duty keys are the round or pin key (Nordberg key) (Fig. 21.15*c*), the Lewis key (Fig. 21.15*c*), and the Kennedy key (Fig. 21.15*d*).

21.2.5 Effect of Elasticity

With a tapered key and fitted key, the shaft has a tendency to twist in the hub during torque fluctuations. The torque distribution is approximately the same as with the fitted key; where the torque is a maximum, the shaft will move slightly in the hub, which

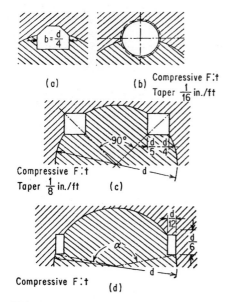

FIG. 21.15 Heavy-duty keys (*a*) Barth. (*b*) Round or pin. (*c*) Lewis. (*d*) Kennedy.

produces a gradual wear of the surfaces pressed together. The wear will therefore extend the motion farther into the hub until the shaft becomes loose. A good fit on the sides will limit the torsional motion of the shaft and will lengthen the life of this key connection. If the torque fluctuates strongly, the key will ultimately loosen. Therefore, a taper rectangular key is not satisfactory for heavy and fluctuating torque. A nonlinear-spring effect exists at all keyed joints.

21.2.6 Stress Concentration

Stress concentration in keyways is especially pronounced at the keyway ends, and fatigue cracks usually start there. The size of the fillet radius r strongly influences the value of the stress-concentration factor K_t, and rectangular keyways should be avoided, especially if the shaft is already highly stressed. For keyways in bending, Hetenyi[35] finds in photoelastic tests that

$$K_t = 1.38 \qquad \text{for sled-runner keyways (Fig. 21.16}a)$$

$$K_t = 1.79 \qquad \text{for profiled keyways (Fig. 21.16}b)$$

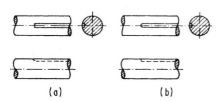

FIG. 21.16 Keyways. (*a*) Sled runners. (*b*) Profiled.

According to Peterson[36,37] these values may be somewhat low, especially for heat-treated alloy steels. For keyways in torsion, the stress-concentration factor can be calculated from the following expression due to Neuber:[38]

$$K_{ts} = 1 + \sqrt{t/rC} \qquad (21.49)$$

where $C = 2(R - T)3R$.

Equation (21.49) is shown graphically in Fig. 21.17*a,* where the results of a mathematical analysis (due to Leven, see Ref. 36) are shown for comparison. Figure 21.17*b* shows stress-concentration values found by the plaster-model method for hollow shafts in torsion.[22] References 40 to 43 report stress concentration and crack propagation of actually loaded keys in their seats. They show that with a clearance of 0.002 in or less, the force distribution on the keyway wall was not uniform. Also, stress concentrations can, in some cases, be higher than shown here.

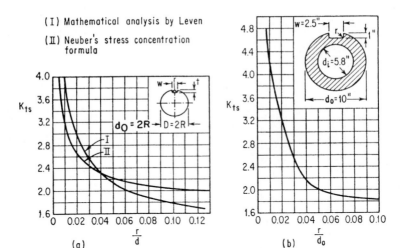

FIG. 21.17   Stress concentration values for shafts. (*a*) Solid. (*b*) Hollow.

21.3 SPLINES

21.3.1 Introduction

"Splines" may be considered multiple keys, integral with the shaft. Compared with keys, splines have the advantage of self-centering. For a given torque transmission, a splined shaft is weakened less than a keyed shaft. Thus a splined shaft can transmit higher torque than a keyed shaft of comparable dimension. See also Sec. 11.

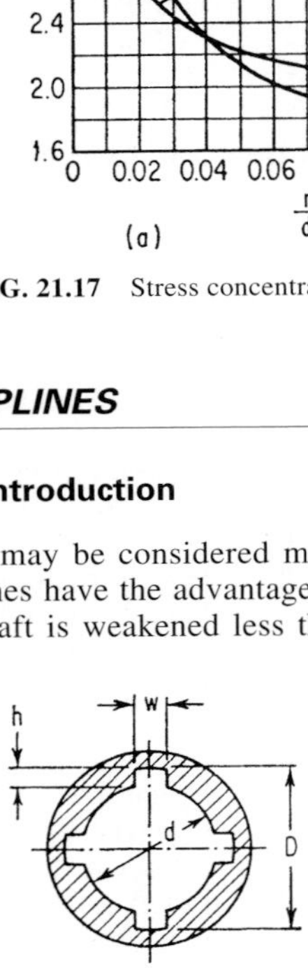

   Parallel-side splines[44,45] are available as four-, six-, ten-, and sixteen-spline fits (see Fig. 21.18 for a four-spline fit).

   Involute splines[44,45] have the same general form as involute gear teeth, except that the pressure angle is 30° and the depth of tooth is half the depth of a standard gear tooth. Involute splines are stronger; there is no danger of excessive stress concentration, and a better surface quality can be achieved for manufacturing and accuracy. ASA standards provide 15 diametral pitches, from ½ to ⁴⁸⁄₁₆, with the number of teeth varying within each group from 6 to 50. Splines are made with *flat root* (Fig. 21.19*a*) and with *fillet root* (Fig. 21.19*b*). *Ball-bearing* splines are discussed in Ref. 46.

FIG. 21.18   Parallel splines—four-spline fit.

   Three types of fits are provided (see Fig. 21.19) by varying the *outside diameter* of the external spline, the *tooth thickness* (used mainly for fillet root splines), and the *root diameter* of the internal spline. Within each type of fit the following three classes may be discerned:

1. Sliding fit

2. Close fit: close on either the outside diameter, inside diameter, or tooth sides

3. Press fit: with interference on the outside diameter, inside diameter, or tooth sides

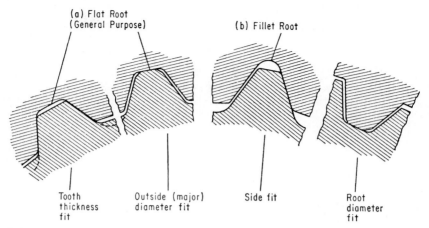

FIG. 21.19 Involute splines.

21.3.2 Stresses

A splined shaft is weaker in torsion than in bending. The stiffness of a splined shaft is usually based upon root diameter d, so that the equivalent length

$$L_6 = L(d_e^4/d^4) \qquad (21.50)$$

The strength of a straight splined shaft is based upon the equivalent shaft diameter $d_e = D - h^4$. The torque-carrying capacity for parallel-side splines is described in Ref. 45.

$$T = i(D_m/2)Lps \qquad (21.51)$$

where i = number of splines
$D_m = (D + d)/2$, mean diameter, in
L = effective spline length, in
s = contact distance, in = $\frac{1}{2}(D - d) - 2h$
p = permissible side pressure, lb/in^2

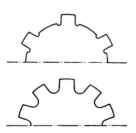

FIG. 21.20 Stress-relieving fillets in spline roots.

p is limited by material properties and stress concentration; suggested value, p = 4 × permissible nominal shear stress.

Maximum shear stress may be calculated by utilizing Neuber's formula; see Eq. (21.52). Stress relieving of the fillets at the spline roots (Fig. 21.20) is advisable.

21.3.3 Design

The following equations relate to parallel-side and involute splines. For relations specifically for involute splines see Ref. 71 and Table 21.10.

For splines in shear, the force P acting on a spline at the mean spline diameter D_m is given by

TABLE 21.10   Basic Formulas for Involute Splines

Flat and fillet roots
$D = N/P_d$
$D_o = (N + 1)/P_d$ (external)
$D_i = (N + 1)/P_d$ (inside-diameter fits only)
$p = \pi/P_d$
$t = p/2$
$a = 0.500P_d$ *
$b = 0.600/P_d + 0.002$ *

	Flat roots only	
Pitch	From 2.5/5 through $\frac{12}{24}$	From $\frac{16}{32}$ through $\frac{48}{96}$
D_0 (internal)	$(N + 1.8)/P_d$	$(N + 1.8)/P_d$
D_i (external)	$(N - 1.8)/P_d$	$(N - 2)/P_d$
b (internal	$0.900/P_d$	$0.900/P_d$
b (external)	$0.900/P_d$	$1.000/P_d$

*For outside-diameter fits, internal spline $a = b$; for inside-diameter fits, external spline $a = b$.

Note:

D = pitch diameter	b = dedendum
D_o = outside diameter	P_d = diametral pitch
D_i = inside diameter	p = circular pitch
a = addendum	t = circular tooth thickness
N = number of teeth	

$$P = 2T/D_m = 4T/(D + d) \tag{21.52}$$

and
$$P = Lbik\tau_d \tag{21.53}$$

where b = spline width measured at the base, in
k = coefficient accounting for nonuniform load distribution; suggested value,[19] $k = 0.75$

For splines in bending, given the bending moment

$$M_b = P[(D - d)/2] \tag{21.54}$$

and the rectangular section modulus

$$Z = Lb^2k/6 \tag{21.55}$$

the bending stress

$$\sigma = 3P(D - d)/Lb^2ik = 12T(D - d)/(D + d)Lbik \tag{21.56}$$

For bearing stresses with the nominal bearing stress twice the nominal stress in compression σ,

$$P = (D - d)Lik\sigma \tag{21.57}$$

21.3.4   Involute Serrations

Involute serrations are involute splines with a pressure angle of 45° and a much finer pitch. All the definitions related to involute splines apply to involute serrations, and the same symbols are used for both. Involute serrations are used in a pitch range of $^{50}/_{120}$ to $^{128}/_{256}$ and a pitch-diameter range from 0.1 to 10.0 in.

Involute serrations are mainly used for close fits, with no provision for sliding. Other classes of fits may be employed, however. Compared with involute splines, the teeth of involute serrations are shallower, stronger, have less radial depth of contact, and frequently offer manufacturing advantages. Under the same load conditions, the contact pressure, radial forces, and resistance to sliding are greater in serrations than in splines. Finer pitches in the case of serrations result in a greater number of teeth and therefore provide a wider range of index positions.

21.3.5 Miscellaneous

Another type of form-locked shaft connection, the Tri-Lobe spline shaft connection, is shown in Fig. 21.21.

21.4 *SNAP RINGS (RETAINING RINGS)*[47]

Snap rings are used to prevent axial motion of mating concentric parts, e.g., shafts and bearing sleeves. Snap rings may be installed in expansion (Fig. 21.22a, external snap ring expanded on a shaft) or in contraction (Fig. 21.22b, internal snap ring contracted in a groove on a hollow shaft).

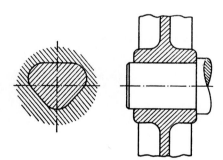

FIG. 21.21 Tri-Lobe spiral shaft.

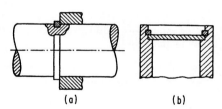

(a)   (b)

FIG. 21.22 Snap rings. (*a*) External. (*b*) Internal.

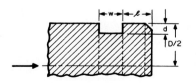

FIG. 21.23 Snap-ring groove.

Usually SAE 1060–1090 spring steel is used for snap rings, with beryllium copper alloy 25, stainless steel type PH 15-7 MO or equivalent, and aluminum Alclad 7075-T6 and no. 3003 used for special conditions.[47]

Successful use of snap rings depends upon groove location and dimensions such that both the shaft and the ring can carry the design load (see Fig. 21.23). It is recommended that

$$l = 3d$$

and groove depth

$$d = Pfs/\pi CD\sigma_u \qquad (21.58)$$

where P = axial load
   fs = safety factor
   D = shaft diameter
   σ_u = ultimate tensile stress of the shaft
   C = load factor dependent upon ring geometry (shown in Fig. 21.24)

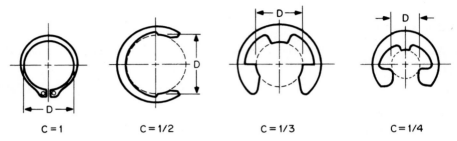

| $C = 1$ | $C = 1/2$ | $C = 1/3$ | $C = 1/4$ |

FIG. 21.24   Snap-ring load factors.

Ring thickness t should not be less than

$$t = Pfs/\pi CD\tau_u \qquad (21.59)$$

where τ_u = ultimate shear stress of the ring material. For carbon spring steel τ_u = 120,000 lb/in^2, for beryllium copper (alloy 25 CDA no. 172) τ_u = 75,000 lb/in^2, and for stainless steel (PH 15-7) τ_u = 110,000 lb/in^2.

Common groove dimensions and tolerances for standard external snap rings are given in Table 21.11.

TABLE 21.11   External Snap Rings

Groove dimensions and tolerances, in

Shaft diameter, D	Groove		Width W	Tolerance $\pm$
	Diameter $D - 2d$	Tolerance $\pm$		
0.25	0.230	±0.0015	0.029	±0.003
0.50	0.468	±0.0020	0.039	±0.003
1.00	0.940	±0.0030	0.046	±0.003
2.00	1.886	±0.0050	0.068	±0.004
5.00	4.790	±0.0060	0.120	±0.005
10.00	9.575	±0.008	0.209	±0.008

Each play due to manufacturing tolerances or wear may be compensated by beveled snap rings, which provide an essentially rigid shoulder, or by bowed snap rings, which allow some flexibility at the shoulder.

21.5   COUPLINGS

21.5.1   Introduction

In contrast to clutches, couplings do not permit, during shaft rotation, disengagement of the two shafts which they connect. The selection of couplings is based upon the following considerations:

Loading: Maximum load, steady and vibratory load, torque transmitted.

Misalignment: Maximum parallel and angular misalignment, and often ability to compensate for axial displacement.

Stiffness: Especially in the case of flexible couplings, the maximum permissible deflection across the coupling due to static and dynamic loads, and also the ability to provide damping and detuning effects.

Most couplings have not been standardized, and specific information regarding a particular coupling is therefore best obtained from the manufacturer. It is current practice for manufacturers to quote a single value of coupling stiffness. For couplings with variable stiffness, the stiffness characteristics should be experimentally determined.[48]

In the design of couplings, it is important to consider:

Compactness: The total length of the assembly.

Weight: A factor in static and dynamic loading of the shafting system.

Static and dynamic balancing: The latter is especially recommended for high-speed application.

Installation and maintenance: Require accessibility without shafting-system disassembly.

21.5.2   Rigid Couplings

While rigid coupling offer the advantage of design simplicity, they do not provide compensation for misalignment. This type of coupling may be employed in low-rotational-speed applications and in cases where only small misalignments are anticipated. *Sleeve-type* couplings (Fig. 21.25*a*) are fitted over both shaft ends, connected by means of keys, taper pins, or setscrews. Sleeve couplings are generally used with small-diameter shafts, except in some marine installations utilizing heavy-duty couplings in which special hydraulic sleeve-expanding devices are provided to disconnect the sleeves. The difficulties associated with disconnecting sleeve couplings can be overcome for small-diameter shafts by employing *compression* couplings. The *ring compression* coupling, shown in Fig. 21.25*b,* represents a simple and inexpensive solution and is in use for shaft diameters up to $5\frac{5}{16}$ in. Likewise, *flange compression* couplings (keyed or keyless) have the advantage of simple adjustability and low cost (Fig. 21.25*c*). The number and size of the bolts used depend upon the size of the coupling. Outside dimensions are the same as those for flange-faced couplings (discussed later). Generally in compression couplings $L = 4d$; other dimensions can be calculated from strength considerations.

Clamp-type couplings (see Fig. 21.25*d*) are split longitudinally, and both parts of the coupling are bolted together. Torque is generally transmitted through a key. Flange-type couplings (see Fig. 21.25*e*) transmit torque by virtue of friction between flange faces, and shearing stress in the bolts or pins. Where friction is employed, bolts

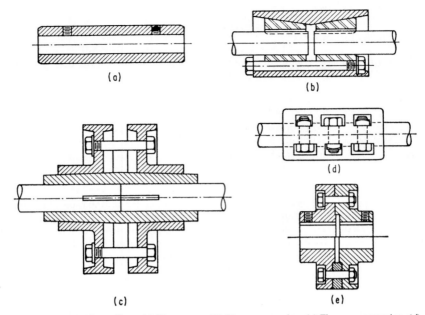

FIG. 21.25 Rigid couplings. (*a*) Sleeve type. (*b*) Ring compression. (*c*) Flange compression. (*d*) Clamp type. (*e*) Flange type.

are required to provide the normal force associated with sufficient friction. Should the available frictional torque be exceeded by the load torque, slipping will occur and the coupling will rely upon bolt shear stress. Flange-faced couplings are generally used for heavy loads and in permanent installations. They have the advantage of being shorter and lighter than many other heavy-duty couplings. Flange couplings are not standardized, except for integrally forged hydroelectric unit couplings which have been standardized for optimum weight and strength.

In the following equations which relate to flange-faced couplings the following symbols are used:

d = shaft diameter

d_b = bolt diameter

D = diameter of bolt circle

D_h = hub diameter

D_m = mean diameter at which F acts

F = tangential force acting at the bolts

F_0 = compressive force in each of the bolts

i = number of bolts

L = hub length

T_{fr} = friction torque at the flange interface

T = transmitted torque

Z_0 = polar sectional modulus of the shaft

τ = allowable shear stress

μ = coefficient of friction

σ = compressive stress

Case 1 considers *frictional contact.* The transmitted torque

$$T = F_0 i D_m/2 \leq T_{fr} \qquad (21.60)$$

Furthermore, with $T = Z_0\tau$, the necessary compression force can be found to be

$$F_0 = 2Z_0\tau/\mu i D_m \qquad (21.61)$$

Here it is advisable to check the strength of bolts in compression:

$$\sigma = 4F_0/d^2$$

Case 2 considers *bolts in shear.* The tangential force can be expressed as

$$F = 2T/D \qquad (21.61a)$$

or $$F = i\pi d_b^2\tau/4 \qquad (21.61b)$$

and the shear stress in bolts

$$\tau = 8T/i\pi d_b^2 D \le \tau_{design} \qquad (21.62)$$

Since there are no standards available, empirical values may be used for preliminary calculations.

$$d_b = 0.5d/\sqrt{i}$$

$$D = 2(d + 1)$$

$$D_h = 1.5d + 1$$

$$i = 0.5d_b + 3$$

$$L = 1.25d + 0.75$$

21.5.3 Flexible Couplings[49–52]

Flexible couplings provide compensation for misalignment of shaft ends by virtue of angular, lateral, and longitudinal (axial) flexibility. Generally, flexibility can be achieved by:

Loose fitting of hard-coupling members (for low-speed application only)

Close fitting of hard-coupling members and provision for lubricated sliding motion between these parts

Employing resilient metallic or nonmetallic members

Couplings in which flexibility is obtained by the first two means are frequently considered rigid or only kinematically flexible. Such couplings are not capable of compensating for dynamic disturbances. By way of contrast, couplings employing resilient materials not only possess kinematic flexibility, but also are able to withstand shock and vibration during rotation and also to change the dynamic characteristics of the shafting system (detuning action).

For the dynamics of high-speed couplings see Refs. 53 to 59.

Flexible Couplings Employing Rigid Members. The *double-slider (Oldham)* coupling (Fig. 21.26) may be used for shafts with lateral misalignment amounting to as much as 5 percent of the shaft diameter with considerable axial play and with relatively small angular misalignment (1°). The tongues of the centerpiece are perpendicular to one another and are free to slide in the grooves in both hubs. Lubrication and cleanliness of the sliding surfaces are important.

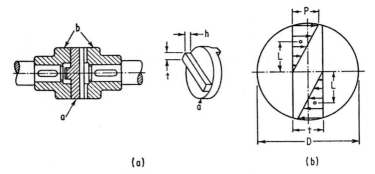

(a)   **(b)**

FIG. 21.26   Double slider (Oldham) coupling. (*a*) Installation. (*b*) Pressure distribution.

The design of the Oldham coupling is based upon an allowable pressure of the sliding surfaces $p = 1000$ lb/in². The pressure distribution is shown in Fig. 21.26*b*. The
force F due to the total pressure on each side of the tongue

$$F = \tfrac{1}{4}pDh \qquad (21.63)$$

from which the torque transmitted by both sides of the tongue

$$T = 2FL = \tfrac{1}{6}pD^2h \qquad (21.64)$$

and the horsepower transmitted at a rotational speed of n, revolutions per minute

$$P = Tn/63{,}030 = pD^2hn/378{,}180 \qquad (21.65)$$

where h = axial dimension of the contact area (approximately the height of the
tongue)
D = outside diameter of the centerpiece
F = force due to the total pressure on each side of the tongue
L = distance of the pressure-area centroid from the center line (Fig. 21.26*a*)

Suggested values for preliminary calculations: $D \approx (3$ to $4)d_{\text{shaft}}$ and the width of the
tongue $t = 0.45d_{\text{shaft}}$.

Gear-tooth couplings (Fig. 21.27) are double-engagement type and may be used in
the case of angular and parallel misalign-
ment. The use of gears improves the
accuracy of the fits and results in
smoother and quieter operation. The coupling consists of two externally geared
hubs (*a*) and an internally geared sleeve
(*b*). The sleeve gear meshes with the hub
gear and the coupling acts as a spline
drive. Clearance between shaft ends
should be provided.

*Flexible Couplings Employing Resilient
Members.*   The ability of a coupling to
compensate for fluctuating torsional load
depends primarily upon the dynamic
characteristics of the resilient member

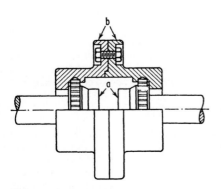

FIG. 21.27   Gear-tooth coupling.

employed. The design and selection of the coupling will therefore be based upon strength considerations as well as the effective spring constant of the coupling elements. Applications of spring elements as resilient members are described in Refs. 48 and 67. Discussion of a number of resilient elements follows:

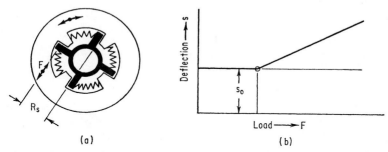

FIG. 21.28   Peripheral-spring coupling. (*a*) Installation. (*b*) Torsion characteristic of non-prestressed spring coupling.

Helical springs (Fig. 21.28) are employed in *peripheral-spring* couplings, the springs being inserted between the hubs of the coupling (against hardened spring seats). A representative torsional characteristic of nonpreloaded spring couplings is given in Fig. 21.28*b*. Because of a certain amount of backlash, the coupling is capable of detuning action. For *calculation for strength* the effective load F can be calculated from

$$F = (T/R_s)\gamma \tag{21.66}$$

where T = transmitted torque
  R_s = effective spring radius
  γ = load factor ($\gamma > 1$)

The *torsional spring constant* of the coupling

$$k_s = T/\theta = kR_s^2 i \tag{21.67}$$

where i = number of active springs
  k_s = compression stiffness of each spring
  R_s = effective spring radius
  θ = relative angular displacement of the hub

Spoked couplings (Fig. 21.29) permit angular and transverse misalignment. The spokes are usually fixed at one end and may be fixed, pivoted, or free at the other end. Loading of a spoke is similar to the loading of a cantilever beam. Angular displacement changes the effective length of the spoke, which, in turn, changes the spring constant and the effect force $F = T/i(R + L)$. Both can be controlled e.g., by varying the slope α in Fig. 21.29*b*. A typical load-deflection curve is shown in Fig. 21.29*c*.

In *Bibby* couplings (Fig. 21.30) each of the hubs is provided with teeth, and both hubs are coupled together by a continuous-strip spring. The teeth are flared and the stiffness of the coupling spring changes with the deflection or load. These couplings allow axial, transverse, and angular misalignment and have a detuning ability in case of vibratory disturbances. Generally, sections of the coupling spring are subjected to bending; shear may prevail under impact loading. The force F may be calculated for n rpm from

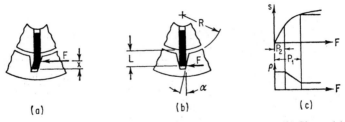

FIG. 21.29 Spoked coupling. (*a*) Angular misalignment. (*b*) Slope. (*c*) Deflection.

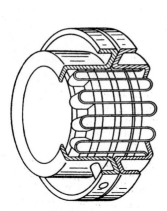

FIG. 21.30 Bibby coupling.

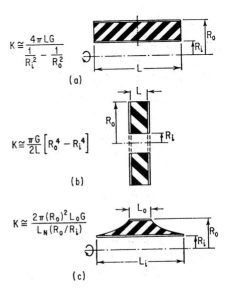

$$K \cong \frac{4\pi L G}{\dfrac{1}{R_i^2} - \dfrac{1}{R_o^2}}$$

(*a*)

$$K \cong \frac{\pi G}{2L}\left[R_o^4 - R_i^4\right]$$

(*b*)

$$K \cong \frac{2\pi (R_o)^2 L_o G}{L_N (R_o/R_i)}$$

(*c*)

FIG. 21.31 Rubber-annulus coupling. (*a*) Solid. (*b*) Hollow. (*c*) Hyperbolic contour.

$$F = 63{,}000 P_o \gamma / R_s i \text{ in} \tag{21.68}$$

where P_o = power
γ = loading coefficient (1 to 3, in special cases up to 6)
R_s = effective spring radius
i = number of springs

Resilient-Material Flexible Couplings.[60] These couplings offer the advantages of simplicity of design and low cost. Torque is transmitted through resilient elements (elastomers) generally exposed to shear and usually permitting larger misalignments.

Because of strength considerations, the application of rubber elements is limited mainly to lighter loads. Owing to the nonlinear characteristics of rubber elements, this type of flexible coupling can also be used for detuning purposes. Approximate formulas for spring constants of some simply shaped rubber elements are given in Fig. 21.31.

21.5.4 Hydraulic Couplings

Hydraulic couplings operate on hydrokinetic principles. These couplings consist of two radially vaned rotors, the first of which (impeller) is connected to the prime mover, and the second (runner) to the output shaft. When the impeller rotates, the liquid (oil) located between the vanes of the impeller flows radially outward under the influence of centrifugal force, as indicated by the arrows in Fig. 21.32a, and into the runner. From here it returns to the impeller, thus closing the circuit of the vortex of liquid. The drag produced on the runner causes it to rotate with the impeller. The impeller acts as a hydrodynamic pump, the runner as a turbine.

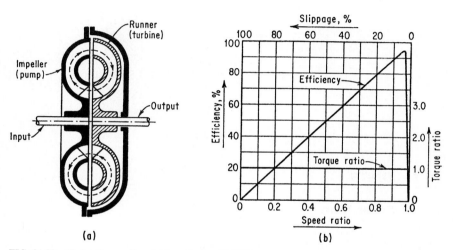

FIG. 21.32 Hydraulic coupling. (*a*) Installation. (*b*) Efficiency.

The torque ratio of a fluid coupling is 1:1, which means that the coupling does not aid the driver in overcoming high torques. Any speed difference at the output side of the coupling is due to slip between the runner and the impeller. This slip increases nearly in proportion to the torque transmitted (Fig. 21.32b). Slip of 1 to 3 percent, which corresponds to transmission efficiency of 97 to 99 percent, is usual. Usually a mineral oil (viscosity of 180 to 200 SSU at 130°F) is used as the driving liquid.

Two different designs with respect to the filling of hydraulic couplings are:

Constant-filling design with a constant volume of fluid sealed in the rotating parts (application mainly to electric motors and internal-combustion engines).

Variable-filling design in which the fluid volume can be controlled by levers (up to 300 hp, approximately) or by pumps (up to 3000 hp). Applications include fans, centrifugal and reciprocating pumps, agitators, and mixers.

With regard to torsional vibrations, it is recommended that a hydraulic coupling be considered as having zero torsional stiffness. For calculations which include damping effects, the viscous friction within the coupling should be considered and the damping values for the coupling should be determined experimentally or obtained from the manufacturer.

21.6 PINS

Pins are employed when parts are to be accurately aligned. Figure 21.33 shows installation of dowel, tapered pins, and tapered dowel pins. These are in sequence of
increasing strength requirements. Cylindrical dowel pins are preferred in tool and gage
work. For more stringent requirements regarding accuracy of alignment and for parts
which must be taken apart frequently, tapered dowel pins (with a threaded larger-
diameter end) are advisable.

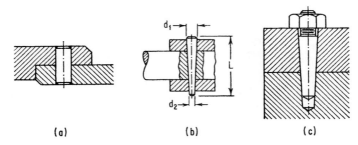

FIG. 21.33   Pins. (*a*) Dowel. (*b*) Taper. (*c*) Taper dowel.

21.7 JOINTS

21.7.1 Cotter Joints

A "cotter" is a form of key. In cotter joints, the connected parts are subjected to either
shear or compression; the cotter itself is subjected to shear at two cross sections.
    Prestressed cotter joints (Fig. 21.34*a*) can withstand alternating loads. Only unidi-
rectional loads can be resisted by a nonprestressed cotter joint (Fig. 21.34*b*).

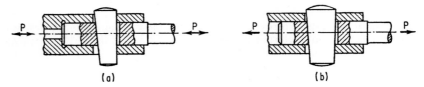

FIG. 21.34   Cotter joints. (*a*) Prestressed. (*b*) Nonprestressed.

    Taper on the cotter should be properly selected in order to prevent loosening. For
fluctuating loads, a taper of $\frac{1}{2}$ in/ft is normally sufficient. The taper may be increased
if additional locking of the cotter is provided, e.g., by a setscrew. Additional locking
devices are especially advisable for vibration applications. For machine tapers and
self-holding key drives, see Sec. 11.

Geometry and Force Analysis.   The equation of equilibrium for a symmetrical non-
prestressed cotter (Fig. 21.35) is written

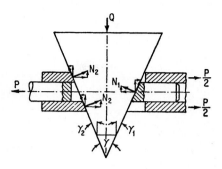

FIG. 21.35 Cotter-pin force analysis.

$$Q = 2N_2(\sin \alpha_2 + \mu_2 \cos \alpha_2) + N_1(\sin \alpha_1 + \mu_1 \cos \alpha_1) \quad (21.69)$$

Since

$$N_1 = P/(\cos \alpha_1 - \mu_1 \sin \alpha_1) \quad (21.70)$$

$$N_2 = P/2(\cos \alpha_2 - \mu_2 \sin \alpha_2) \quad (21.71)$$

$$Q = P\left(\frac{\sin \alpha_2 + \mu_2 \cos \alpha_2}{\cos \alpha_2 - \mu_2 \sin \alpha_2} + \frac{\sin \alpha_1 + \mu_1 \cos \alpha_1}{\cos \alpha_1 - \mu_1 \sin \alpha_1}\right) \quad (21.72)$$

For $\mu_1 = \mu_2 = \mu$ and $\mu = \tan \rho$,

$$Q = P[\tan (\alpha_2 \pm \rho) + \tan (\alpha_1 \pm \rho)] \quad (21.73)$$

The positive sign applies if the driven motion of the cotter and Q have the same direction, the negative sign if the directions are opposite.

The condition for self-locking, $Q \leq 0$, for a one-sided cotter ($\alpha_2 = 0$) yields

$$\alpha_1 = \alpha_2 \leq 2\rho \qquad \mu \simeq 0.1 \qquad \tan \alpha \simeq 0.2$$

Stresses. (Refer to Fig. 21.36.)

Rod in Tension

$$P = (\pi d_1^2/4)\sigma_d \quad (21.74)$$

or

$$d_1 \geq \sqrt{4P/\pi\sigma_d} = 1.13 \sqrt{P/\sigma_d} \quad (21.75)$$

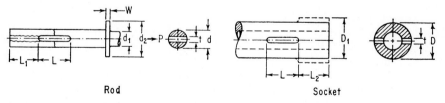

Rod   Socket

FIG. 21.36 Stressed area of rod and socket.

Crushing of the Cotter Bearing Surface, at the Rod End

$$P \leq td\sigma_b \quad (21.76)$$

(where the ultimate strength of the bearing surface $\sigma_b = 2\sigma$) and $t \simeq (\frac{1}{3} \text{ to } \frac{1}{4})d$, the rod diameter can be calculated as

$$d \geq \sqrt{(3 \text{ to } 4)P/2\sigma_d} \quad (21.77)$$

Rod End Tearing across the Cotter Hole

$$P = (\pi d^2/4 - td)\sigma \qquad\qquad (21.78)$$

Shearing Rod End

$$P = 2L_1 d_{\tau_d} \qquad\qquad (21.79)$$

Socket End Tearing across the Cotter Hole

$$P = [(\pi/4)(D^2 - d^2) - (D - d)t]\sigma \qquad\qquad (21.80)$$

or

$$\sigma = \frac{P}{(D - d)[(\pi/4)(D + d) - t]} \leq \sigma_d \qquad\qquad (21.81)$$

It is advisable to check whether an increase in D is required.

Crushing of the Bearing Surface in the Socket End

$$P = [(D - d)t]\sigma_{\text{crushing}} \qquad\qquad (21.81)$$

Note that $\sigma_{\text{crushing}} = 2\sigma_a$. Check for possible need to increase D.

Shearing Socket End

$$P \leq 2[L_2(D - d)\tau_d] \qquad\qquad (21.82)$$

Double Shearing of the Cotter

$$P \leq 2th_{\text{cotter}}\tau_d \qquad\qquad (21.83)$$

Crushing the Collar of the Rod

$$P = 2\pi(d_2^2 - d^2)\sigma_d \qquad\qquad (21.84)$$

Shearing of the Collar of the Rod

$$P \leq d^2 \pi w \tau_d \qquad\qquad (21.85)$$

Other Dimensions.   Considering the cotter loading condition, and assuming $d_1 = \frac{3}{4}d$, the following dimensions can be derived:

$$L_{\text{cotter}} = (1.3 \text{ to } 1.5)d \qquad \text{and} \qquad h_{\text{cotter}} = 1.3d$$

Further recommended dimensions:

$$L_1 = L_2 = (0.7 \text{ to } 1.0)L_{\text{cotter}}$$

$$D = (0.8 \text{ to } 0.9)D_1$$

$$D_1 = 2d$$

$$w = \frac{1}{4}d$$

$$d_2 = 1.3d$$

21.7.2 Friction Joints

Connections between hub and shaft can be achieved by means of friction in a cylindrical or conical fit. Typical examples of friction hub-shaft fits are shown in Fig. 21.37.

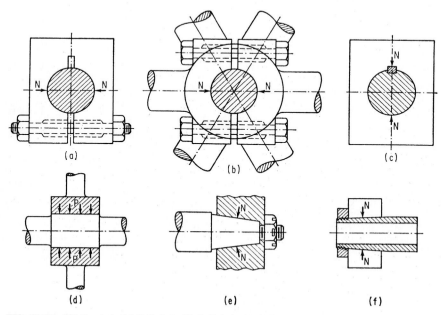

FIG. 21.37 Friction hub. (*a*) Split hub. (*b*) Split hub. (*c*) Solid hub with taper key. (*d*) Press fit. (*e*) Conical fit with taper fit. (*f*) conical fit with nut.

Assuming a uniform pressure distribution at the hub-shaft interface, the retaining force may be expressed

$$R = \Sigma N\mu \geq F_t = 2T/d \qquad (21.86)$$

where $N = p(\pi dL)$ and μ = coefficient of friction; $\mu = 0.13$ for steel on cast iron, 0.33 for steel on cast iron or steel in shrink joints in which cold welding occurs, and as much as 0.65 for steel and cast iron if the surfaces are treated with carborundum.[45]

Calculations of forces and stress are identical to pressure-vessel calculations. The bursting force F is given by

$$F = dLp \qquad (21.87)$$

and $$F = F/\mu\pi = 2T/\pi\mu d \qquad (21.88)$$

The average normal stress in section *a-a* (hub stress)

$$\sigma_{av} = F/2(D - d)L \qquad (21.89)$$

If *taper* keys are used in order to originate the normal force N, the retaining friction force R can be expressed as

$$R = 2F_k u \qquad (21.90)$$

or, from Eq. (21.86),

$$R \geq F_t = 2T/d \tag{21.91}$$

where the force of the key F_k can be calculated from the force Q necessary to drive home the key:

$$Q = F_k(\tan \alpha + 2\mu) \tag{21.92}$$

and, from Eq. (21.90),

$$Q = (T/d\mu)(\tan \alpha + 2\mu) \tag{21.93}$$

The driving force Q is limited by key deformation. Thus

$$Q \leq bh\sigma_{d,key} \tag{21.94}$$

The preceding equations are based upon the assumption that the normal force N is originated by the key at least two points of the shaft circumference.

21.7.3 Universal Joints[61–65]

Universal joints are used to transmit torque between shafts with intersecting axes. As a rule, the angle between the driving and the driven shafts should be kept as small as possible (5 to 15°, seldom more than 30°), but an exception can be made where low speeds and torques are involved. The angle between shafts is held below 11.5° for speeds of 1500 r/min and below 3.5° for speeds over 1500 r/min. A Hooke (or Cardan) joint is shown in Fig. 21.38 with the hubs (a) and the inner yoke (b) made of cast iron, and the pins (c) made of steel. All universal joints provide oscillatory angular velocity (accelerations) output when driven by a constant speed input. Uniformity of angular motion can be restored if the two shafts are connected by an intermediate shaft and two universal joints, provided that both shafts make the same angle with the intermediate shaft, and the forks on the intermediate shaft are in the same plane (for the most common case of parallel driving and driven shafts). If the intermediate shaft can telescope, and driving and driven shafts can be moved longitudinally and laterally, and independently of one another, e.g., many machine-tool drives.

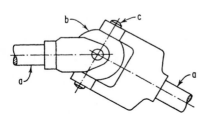

FIG. 21.38   Hooke universal joint.

Single-joint velocity drives are discussed in Ref. 46. Other universal joints are ball, roller, slides-block trunion joints, and the Rzeppa[62] universal joint.

21.8 RIGID-BODY CLUTCHES

21.8.1   Jaw Clutch

"Positive-engagement clutches" (Fig. 21.39) with spiral jaws are made for driving in one direction only (positive-engagement clutches with square jaws are designed for

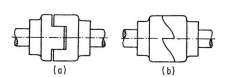

FIG. 21.39 Positive clutch. (*a*) Square jaw. (*b*)
Spiral jaw.

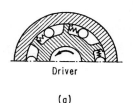
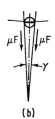

FIG. 21.40 Roller clutch.

driving in both directions). Design of jaw clutches should be based on strength considerations, and it is advisable to check jaws for shearing and for crushing at the sides.
Approximate principal dimensions for cast-iron jaw clutches are shown in Ref. 71.

21.8.2 Unidirectional Clutches

Freewheeling or overrunning clutches are employed when torque must be transmitted
in one direction only or where the driven member is to be permitted to "overrun" the
driver. They can be considered a simple and inexpensive substitute for a differential
drive and are also used in multiple-speed drives where they automatically disengage
the lower-speed unit when the higher-speed unit is engaged. When one of the members
of the overrunning clutch is fixed, the other member can rotate only in one direction
and the clutch becomes "backstopping," e.g., in gear reducers where reverse motion
must be excluded.

The most common designs utilize the wedging principle (Fig. 21.40). Generally,
the construction consists of a shell-shaped outer member and an inner member with
wedge-shaped pockets, or flat cam profiles, around its periphery; the coupling
between the two members is provided by the locking action of rolls. Generally, instantaneous wedging action at any position without backlash or slip is required. The
wedge-shaped pocket may have the disadvantages of expensive manufacturing costs
and of the relatively small number of rolls employed.

For proper locking action, the condition for self-locking, $\alpha \leq 2\phi$, must be satisfied.
Here α = the angle between tangents to the cam contour and to the roller surface at
the point of contact, and $\phi = \tan^{-1}\mu$ (μ = coefficient of friction). To check the
strength of the roller the crushing force

$$F_c = F_t/\tan \alpha = 2T/D \tan \alpha \qquad (21.95)$$

where T = the torque transmitted and D = the effective diameter of the rollers. For
more than three rollers, uniform distribution of load is difficult to achieve without special precautions in regard to accuracy of manufacture.[67]

21.9 SETSCREWS

The setscrew diameter d is related to shaft diameter D by the following empirical relationship:[66]

$$d = D/8 + {}^5\!/_{16} \text{ in} \qquad (21.96)$$

The maximum safe holding force is given by[30]

$$F = 2500d^{2.31} \tag{21.97}$$

A factor of safety of 1.5 to 2 is recommended.[71] If a lower-grade steel is used, the holding power is approximately one-sixth of the values given. The holding power of headless setscrews is still lower.[71]

It should be noted that setscrew holding power is related to the initial torque of the screw, the shaft diameter, the setscrew-point size, the shaft material, the setscrew material, vibration of the shaft, etc. Smaller cup points have been developed with improved results.[66]

REFERENCES

1. "Rotary Motion Flexible Shafts," 7th ed., S. S. White Industrial Products Division of Pennwalt, Piscataway, N.J., 1977.

2. "Liquid Rocket Engine Turbopump Shafts and Couplings," NASA SP-8101, Sept. 1972.

3. Sadowy, M.: "Shafts, Couplings, Keys, etc.," in "Mechanical Design and Systems Handbook," H. A. Rothbart, ed., McGraw-Hill Book Company, Inc., New York, pp. 27-1 to 27-12, 1964.

4. Lowell, C. M.: "A Rational Approach to Crankshaft Design," ASME Paper 55-A-57, 1955.

5. Peterson, R. E.: "Stress Concentration Factors," John Wiley & Sons, Inc., New York, 1974.

6. Roark, R. M., and W. C. Young: "Formulas for Stress and Strain," 5th ed., McGraw-Hill Book Company, Inc., New York, 1975.

7. Forrest, P. G.: "Fatigue of Metals," Pergamon Press, London, 1962.

8. Juvinall, R. C.: "Engineering Considerations of Stress and Strain," McGraw-Hill Book Company, Inc., New York, 1967.

9. ASME Standard B106.M (Draft) "Design of Transmission Shafting," ASME, New York, 1984.

10. Noll, G. C., and C. Lipson: "Allowable Working Stresses," *Proc. Soc. Exp. Stress Anal.,* vol. III, no. 2, pp. 86–101, 1946.

11. Kuguel, R.: "A Relation Between Theoretical Stress Concentration Factor and Fatigue Notch Factor Deduced from the Concept of Highly Stressed Volume," *Proc. ASTM,* vol. 61, pp. 732–748, 1969.

12. Horger, O. J., and H. R. Neifert: "Fatigue Properties of Large Specimens with Related Size and Statistical Effects," Symp. Fatigue with Emphasis on Statistical Approach, Special Technical Publication No. 137, ASTM, 1952.

13. ASM Committee on Fatigue of Steel: "The Selection of Steel for Fatigue Resistance," "Metals Handbook, Properties and Selection of Metals," vol. 1, 8th ed., T. Lyman, ed., American Society for Metals, p. 224, 1961.

14. Lipson, C., G. C. Noll, and L. S. Clock: "Stress and Strength of Manufactured Parts," McGraw-Hill Book Company, Inc., New York, 1950.

15. Fuchs, H. O., and R. I. Stephens: "Metal Fatigue in Engineering," John Wiley & Sons, Inc., New York, 1980.

16. Almen, J. O., and P. H. Black: "Residual Stresses and Fatigue in Metals," McGraw-Hill Book Company, Inc., New York, 1963.

17. Cumulative Fatigue Damage Division of the SAE Fatigue Design and Evaluation Committee: "Fatigue Under Complex Loading: Analysis and Experiments," vol. 6, R. M. Wetzel, ed., Society of Automotive Engineers, 1977.

18. Collins, J. A.: "Failure of Materials in Mechanical Design: Analysis, Prediction and Prevention," John Wiley & Sons, Inc., New York, 1981.

19. "SAE Information Report J1099 on Fatigue Properties," "SAE Handbook," p. 4.44, 1978.

20. J. A. Graham, ed., "Fatigue Design Handbook," Society of Automotive Engineers, New York, 1968.

21. Tuplin, W. A., and G. W. Michalec: "Gearing," "Mechanical Design and Systems Handbook," H. A. Rothbart, McGraw-Hill Book Company, Inc., New York, pp. 32-18–32-19, 1964.

22. Michalec, G. W.: "Precision Gearing: Theory and Practice," John Wiley & Sons, Inc., New York, p. 346, 1966.

23. Bamberger, E. N., et al.,: "Life Adjustment Factors For Ball and Roller Bearings—An Engineering Guide," American Society of Mechanical Engineers, New York, 1971.

24. "SAE Report J935 on High Strength Carbon and Alloy Die Drawn Steel," in "SAE Handbook," Society of Automotive Engineers, New York, p. 304, 1978.

25. Gunter, E. J.: "Dynamic Stability of Rotor-Bearing Systems," NASA SP-113, 1966.

26. Temple, G. A., and W. G. Bickley: "Rayleigh's Principles and Its Application to Engineering Problems," Dover Publications, Inc., New York, 1949.

27. Rieger, N. F.: "Vibration of Rotating Machinery, Part 1: Rotor-Bearing Dynamics," 2d ed., The Vibration Institute, Clarendon Hills, Ill., February 1982.

28. Pan, C. H. T., E. R. Wu, and A. I. Krauter: "Rotor-Bearing Dynamics Technology Design Guide—Part I: Flexible Rotor Dynamics," AFAPL-TR-78-6 Part I., 1980, Air Force Aero Propulsion Laboratory, Wright Patterson AFB, Dayton, Ohio. (Also see Parts II through IX.)

29. Trivisonno, R. J.: "Fortran IV Computer Program for Calculating Critical Speeds of Rotating Shafts," NASA TN-D-7385, Aug. 1973.

30. Jones, A. B., and J. M. McGrew, Jr.: "Rotor-Bearing Dynamics Technology Design Guide. Part II: Ball Bearings and Part IV: Cylindrical Roller Bearings," AFAPL TR-78-6 Part II, February 1978, and Part IV, December 1978, Air Force Aero Propulsion Laboratory, Wright Patterson AFB, Dayton, Ohio.

31. Nestorides, E. J.: "A Handbook on Torsional Vibration," The British Internal Combustion Engine Research Association, Cambridge University Press, Cambridge, England, 1958.

32. Thomson, W. T.: "Theory of Vibration with Applications," Prentice-Hall, Inc., Englewood Cliffs, N.J., 1981.

33. "Keys and Keyseats," ANSI B 17.1—1967, and "Woodruff Keys and Keyseats," ANSI B 17.2—1967.

34. Kotchina, N. I., W. S. Lomiakov, and others: "Machine Elements" (Russian), Gos Nautchno-Techn. Isdat., Moscow, 1958.

35. Hetenyi, M. I.: "The Application of Hardening Resins to Three Dimensional Photoelastic Studies," *J. Appl. Phys.,* vol. 10, 1939.

36. Peterson, R. E.: "Stress Concentration Design Factors," John Wiley & Sons, Inc., New York, 1953.

37. Peterson, R. E.: "Fatigue of Shafts Having Keyways," *ASTM Proc.,* vol. 32, part II, p. 413, 1932.

38. Neuber, H.: "Kerspannungslehre," Springer-Verlag OHG, Berlin, 1939.

39. Seely, S., and A. Dolan: "Stress Concentration at Fillets, Holes, and Keyways as Found by the Plaster-Model Method," *Univ. Illinois Eng. Expt. Sta. Bull.,* no. 276, June 1935.

40. Orthwein, W. C.: "Keyway Stresses When Torsional Loading Is Applied by the Keys," *Exp. Mech.,* vol. 15, pp. 245–248, 1975.

41. Orthwein, W. C.: "A New Key and Keyway Design," *J. Mech. Design, Trans. ASME,* vol. 101, pp. 338–341, 1979; also, "Failure Analysis and Prevention," "Metals Handbook," vol. 10, American Society for Metals, Cleveland, p. 385, 1975.

42. Dorey, S. F.: "Large Scale Torsional and Fatigue Testing of Marine Shafts," *Proc. Inst. Eng.,* pp. 339–406, 1948.

43. Graham, J. A., J. F. Millan, and F. Appl: "Fatigue Design Handbook," vol. 4, "Advances in Eng.," Society of Automotive Engineers, Inc., New York, 1968.

44. Dudley, D. W.: "How to Design Involute Splines," *Prod. Eng.,* Oct. 28, 1957.

45. "SAE Handbook," Society of Automotive Engineers, Inc., New York, 1977.

46. Rowland, D. R.: "Matching Ball Bearing Splines to Torque, Radial-Load, and Life Requirements," *Machine Design,* July 20, 1961.

47. "Technical Manual: Design Data and Engineering Specifications," Waldes Kohinoor, Inc., Long Island City, N.Y., 1972.

48. Nestorides, E. T.: "A Handbook of Torsional Vibrations," Cambridge University Press, New York, 1958.

49. Spector, L. F.: "Design Guide—Flexible Couplings," *Machine Design,* Oct. 30, 1958.

50. Conway, H. G.: "Couplings of Parallel Shafts," Products Design Handbook Issue, *Prod. Eng.,* mid-October, 1955.

51. Pampel, W. P.: "Kupplungen," vol. I, VEB Verlag Technik, Berlin, 1958.

52. Gensheimer, J. R.: "How to Design Flexible Couplings," *Machine Design,* Sept. 14, 1961.

53. "High Speed Gear Coupling Development," Pennsylvania State University Seminar on High Speed Flexible Couplings, University Park, Pa., June 10–13, 1962.

54. Allen, E. A.: "A Coupling Design to Minimize Dynamic Instabilities of High Speed Systems," Pennsylvania State University Seminar on High Speed Couplings, University Park, Pa., June 10–13, 1962.

55. Grundtner, R. R.: "The Torsionally Soft Coupling and Its Application," Pennsylvania State University Seminar on High Speed Flexible Couplings, University Park, Pa., June 10–13, 1962.

56. Thoma, F. A.: "Coupling Induced Shaft Vibration," Pennsylvania State University.

57. Seminar on High Speed Flexible Couplings, University Park, Pa., June 10–13, 1962, "Some Design Problems Associated with a New High Speed Flexible Disk Type Coupling," Pennsylvania State University Seminar on High Speed Flexible Couplings, University Park, Pa., June 10–13, 1962.

58. Ho, J. Y. L.: "Dynamic Balance of Gear Type Flexible Coupling," Pennsylvania State University Seminar on High Speed Flexible Couplings, University Park, Pa., June 10–13, 1962.

59. Rothfus, N. B., and C. B. Gibbons: "Contoured Flexible Diaphragm Type High Speed Couplings," Pennsylvania State University Seminar on Flexible High Speed Couplings, University Park, Pa., June 10–13, 1962.

60. Larsen, P. J.: "Special Report on Engineering Elastomers," *Machine Design,* Jan. 8, 1962.

61. "Universal Joint and Driveshaft Design Manual," Advances in Engineering Series, no. 7, Society of Automotive Engineers, New York, 1979.

62. Rzeppa, A. H.: "University Joint Drives," *Machine Design,* April 1953.

63. Rosenburg, R. M.: "On the Dynamical Behavior of Rotating Shafts Driven by Universal (Hooke) Couplings," ASME Paper 57-A-1, 1957.

64. Saari, O.: "How to Obtain Useful Speed Variations with Universal Joints," *Machine Design,* Oct. 1954.

65. Mabie, H. H.: "Constant Velocity Joints," *Machine Design,* May 1948.

66. Gates, C. S.: "Allenpoint Cup Point," Allen Manufacturing Co., Hartford, Conn.

67. Niemann, A.: "Machinelemente," vol. 2, Springer-Verlag OHG, Berlin, 1960.

68. Pinknery, B. H. D.: *Machinery,* Oct. 15, 1914.

69. Simes, G., and Wassman, J. L.: "Metal Fatigue," McGraw-Hill Book Company, Inc., New York, 1959.

70. Gibson, W. B.: "Fluid Coupling," *Machine Design,* Mar. 31, 1960.

71. Maleev, V., and J. B. Hartman: "Machine Design," International Textbook Co., Scranton, Pa., 1957.

SECTION 22

FRICTION CLUTCHES

Thomas A. Dow, Ph.D.

Professor of Mechanical and Aerospace Engineering
North Carolina State University
Raleigh, N.C.

22.1 INTRODUCTION

The operation of a friction clutch is dependent upon the phenomenon of friction. A more detailed discussion of this subject can be found in Sec. 1 on friction, lubrication, and wear, and in Sec. 23 on friction brakes. The key parameters of a successful clutch design are its operating temperature and its dynamic engagement characteristics. These two parameters are related, since the temperature has an effect on the static/dynamic friction coefficients which influence the engagement and the smoothness of operation.

The following sections describe the common types of friction clutches, the important properties of the friction materials, methods of determining operating temperatures, and the dynamics of friction clutch systems.

22.2 TYPES OF FRICTION CLUTCHES[1,2]

22.2.1 Spring-Loaded Dry Clutch

The spring-loaded dry clutch is widely used in automobiles for spring-applied and manually released applications. The output shaft is connected to a hub assembly faced

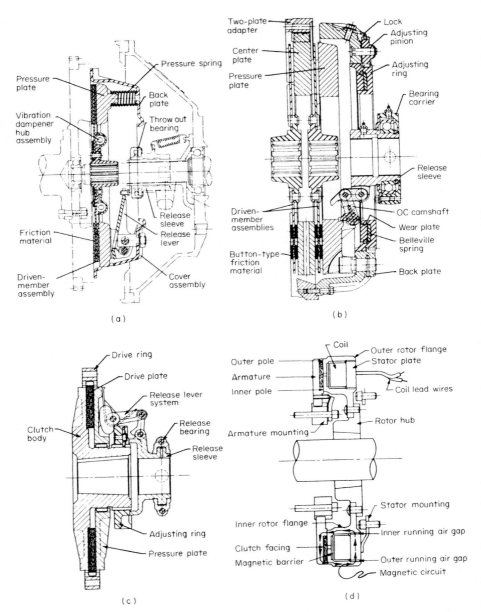

FIG. 22.1   Types of friction clutches.[1] (*a*) Single-plate clutch. (*b*) Twin-plate clutch. (*c*) Power takeoff over-center clutch. (*d*) Single-plate electromagnetic clutch.

with a friction material, usually an asbestos-based material. This output shaft is then connected to the input shaft by clamping the hub assembly between two rotating plates: the flywheel of the prime mover and the pressure plate of the clutch assembly. The clamping force is obtained by coil or Belleville springs which hold the clutch in the applied state. The clutch is released by levers on the rotating plate which compress the springs. A throw-out bearing is necessary to connect the external clutch release mechanism to this rotating assembly.

The parts of such a clutch are shown in Fig. 22.1. The pressure plate must be massive enough to prevent distortion and to provide thermal mass to absorb the heat generated during engagement without overheating the clutch facing. The springs must be designed to provide sufficient clamping pressure over the range of motion caused by clutch-facing wear, and not be excessively influenced by clutch temperature. In some designs, the friction material is isolated from the output shaft by torsional dampers. The purpose of these dampers is to prevent the torsional vibrations of the engine from being transmitted through the clutch at frequencies which would create objectionable gear and driveline noises.

22.2.2 Over-Center Dry Clutch

The over-center dry clutch differs from the spring-loaded design in the method by which clamping force is generated. The lever system incorporates an over-center toggle lever action or cam to develop the clamping force, as shown in Fig. 22.1*b*. The over-center clutch, therefore, will remain in either the released or the engaged position with no external force being applied. This clutch is usually activated by a hand lever and is utilized in many farm and off-road applications.

22.2.3 Over-Center Power Takeoff Dry Clutch

The over-center power takeoff dry clutch is usually used in conjunction with a power takeoff on an engine to attach auxiliary or stationary equipment. This clutch does not use the engine flywheel as part of the friction surface, but contains both clamping faces. The drive is typically through a spline or gear, as shown in Fig. 22.1*c*. Some applications of such clutch systems[3,4] also use the clutch to protect the driveline by slipping when a critical torque is exceeded. The operation of the clutch under these conditions is somewhat different than normal engagement because the overload condition occurs at maximum clamping load.

22.2.4 Electromagnetic Single-Plate Dry Clutch

The electromagnetic single-plate dry clutch utilizes an electromagnet to provide the clamping force, and the single-plate design has only a single friction surface, as compared to the two surfaces of a mechanical single-plate clutch (see Fig. 22.1*d*). Nevertheless, such a clutch can provide fairly high torque capacity for its size. Electromagnetic clutches are attractive for automatically actuated operations such as for air-conditioning compressor drives. The major limitation is the heat capacity, since the clutch operating temperature is limited by the temperature rating of the coil insulation (see Sec. 22.7).

22.2.5 Mechanically Actuated Wet Clutches

The design of mechanically actuated wet clutches is similar to dry clutches, but requires oiltight cover assemblies, proper oil circulation, and adequate release mechanisms. The wet clutch[5] is a much better heat exchanger and relies on the oil to carry much of the heat away during engagement, making it less sensitive than the dry clutch to total heat generation. Since the coefficient of friction of materials operating in oil is lower than under dry conditions, it is necessary to provide higher clamping force, more friction faces, or larger clutch diameter to have the same torque capacity. Clutch assemblies with three or more plates are commonly used to develop sufficient torque in a minimum size. The unit pressure for a wet clutch is typically higher than for a dry clutch, ranging from a high of 200 lb/in^2 in a mechanically actuated wet clutch to seldom greater than 50 lb/in^2 in a dry-clutch design.

22.2.6 Hydraulic Multiple-Disk Clutch

The hydraulic multiple-disk clutch is finding its way into many modern applications. With proper valving it can be controlled either manually or automatically through auxiliary controls. It is widely used in heavy-duty power-shift transmissions for earth moving, construction equipment, farm tractors, and motor trucks. It has also been applied to cooling-fan drives.[6]

The basic designs used for the other clutches apply for this case, but the actuation utilizes hydraulic pressure. The effect of the rotation of the clutch on the hydraulic pressure must be considered when the disengagement mechanisms of such clutches are designed, and balance pistons or return springs are built into most designs.

22.2.7 Electromagnetic Multiple-Disk Clutch

The electromagnetic multiple-disk clutch is an electrically operated oil-bath clutch, using a friction disk pack like the hydraulic clutch. It is used in machine-tool transmissions which require power shifting for automatic sequence operation. However, as with other electromagnetic designs, this clutch has a limited torque capacity when compared to a hydraulically actuated clutch of the same size. The additional requirement of a separate electrical power supply as well as a hydraulic system to circulate the fluid makes this design less attractive for heavy-duty applications. However, the adaptability to automatic control of the electrical clamping force makes this clutch an attractive choice for some applications.

22.2.8 Centrifugal-Force Clutch

Centrifugal force is often utilized to generate the clamping force when automatic clutch engagement and release dependent upon prime-mover speed are desired. This permits load-free start-up of the prime mover and can prevent stalling. The rotating weights of such designs often work against spring load, so that at a predetermined speed the centrifugal-force effects balance the spring preload. Any further increase in speed will result in a gradual clamp force buildup and a soft start of the driven member. Such a gradual start is desirable, but high torque at low speed can cause excessive slip and thermal destruction of the clutch plates and facing.

22.2.9 Wrapped-Spring Clutch

The clamping force for this clutch is produced by twisting a coil spring so that its inside diameter is reduced and it clamps to a rotating central shaft. Such clutches are typically used in low-power applications[7] where they provide a compact, efficient drive system.

22.3 CLUTCH MATERIALS

22.3.1 Dry-Clutch Lining Materials[8]

Friction materials used for clutch facings are similar to those employed for brakes and, at the present time, three different types of materials are used: organic, sintered iron, and ceramic.

By far, the most widely used is the organic material which consists mainly of asbestos yarn, woven with copper or brass wire. The maximum operating pressure is 30 lb/in^2 for these materials, and the friction coefficient ranges from 0.2 to 0.45. A good design estimate for friction coefficient is 0.25. The torque requirement of the clutch will determine the size and the number of plates. These materials have changed very little over the past 25 years, and still perform satisfactorily at low cost in many applications.

Sintered iron and ceramic materials are capable of withstanding operating pressures and temperatures far in excess of organic facings. Organic facing materials are limited to a maximum operating temperature of approximately 450°F, and rapid wear will result from operation over this temperature (see discussion in Sec. 23.1 on brake wear). Ceramic and sintered iron will operate at twice this temperature, but their cost is substantially higher. The ceramic material also has another advantage—light weight—and clutches using such facings have a low moment of inertia. Such clutch plates are made by riveting small buttons of the ceramic facing material to the disk. The number of buttons will depend upon the type of application and the size of the clutch. The result is an assembly which is 10 percent lighter than a similar disk faced with organic material, and 45 percent lighter than one faced with sintered iron. This puts less load on synchronizer rings, and more rapid shifts are possible. The button design also allows more area for heat transfer from the clutch disk which reduces its operating temperature. However, the trade-off of increased life must be compared to the increased initial cost of iron and ceramic materials.

22.3.2 Wet-Clutch Lining Materials

The main application of wet friction clutches began with the GM prototype automatic transmission in 1938. Dry-type friction materials—asbestos-based sintered metal and woven products—were used in this application. Although these materials performed satisfactorily, the desire for lower cost materials with improved performance led to new friction products. The next development was the use of semimetallic linings after World War II. These materials gave higher static and dynamic friction coefficients, but were less durable. These early friction materials for wet clutches were hard and dense because such properties were desirable for dry applications. However, resilient materials are more desirable for wet-clutch designs (natural cork has this property and a high wet friction coefficient in oil of 0.13). Cork was used in automatic transmissions into the 1950s. The next development was Bakelite-type materials with a phenolic binder. These friction materials had enough physical strength that they could also be used as structural elements and thus eliminate the steel carrier. However, higher horsepower applications soon made these materials inadequate and a new type of wet friction material was developed in the late 1950s. This class of friction materials is known as "paper" type because it is manufactured on a paper-making machine. The process uses a water slurry of materials (asbestos, cellulose, and fillers) which are dried on a screen into a continuous sheet of material. This paper is blanked into the appropriate shape, saturated with oil or liquid resins, cured, and bonded to a steel core. "Paper" materials are noted for their low cost, high dynamic friction coefficient, and extremely low static/dynamic coefficient ratios.

Figure 22.2 shows the engagement characteristics (in oil) of the older wet friction materials and the "paper" type. The sintered metal and the semimetallic friction material have low dynamic friction coefficients and sharp friction peaks as the clutch approaches static contact. The static/dynamic friction ratio for these materials is high. On the other hand, the "paper" material exhibits uniform friction and a smooth transition to static contact. These characteristics make the "paper" material among the smoothest and quietest wet friction materials available.

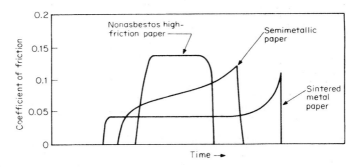

FIG. 22.2   Dynamic engagement characteristics (wet friction trace) of wet friction material.

In the 1960s, "paper" materials were improved with the addition of graphite. This change made the paper more tolerant to surface finish of the mating steel plates, and also allowed higher horsepower transmission because of the high specific heat and conductivity of the graphite. These benefits led to the development of a new family of friction materials known as "graphitics." These materials are superior for high-energy transmission applications.

Another class of materials known as "polymerics" is the extension of rubber compounding technology into the friction materials industry. These materials are a blend of elastomer, fibers (ceramic and organic), fillers, lubricants, friction particles, and a curing system. The material is formed to shape, cured, and bonded to a steel core. The advantages of polymerics are high resilience and durability, high friction coefficient, and good engagement properties. These materials have the highest power-absorbing capabilities of any current clutch material.

Table 22.1 compares the characteristics of different clutch materials operating in oil. The key operating characteristics for smooth, rapid engagement are high dynamic friction and medium-to-high static/dynamic friction ratio. However, these factors must be weighed against the cost and durability of the friction material for a specific application.

TABLE 22.1 Characteristics of Various Types of Friction Materials Currently Available[9]

Characteristic	Molded semi-metallic	Sintered metal	Woven	"Paper"			Graphitic	Polymeric
				Low static	High static	Graphitic		
Dynamic friction	Low	Low to medium	Medium	High	High	High	Medium	Medium to high
Static/dynamic ratio	High	High	Medium	Low	Very high	Medium	Medium	Medium
Resistance to heat	Good	Excellent	Good	Good	Fair	Very good	Excellent	Good
Durability	Good	Excellent	Very good	Good	Fairly good	Very good	Excellent	Excellent
Resistance to mating plate distortion	Fair	Fair	Good	Very good	Very good	Very good	Very good	Very good
Compressive strength	High	High	Medium	Low	Low	Medium	Medium	High
Facing thickness, in	0.030–0.075	0.025–0.060	0.125–0.300	0.020–0.045	0.020–0.045	0.020–0.040	0.020–0.060	0.020–0.100
Dynamic friction	0.06–0.08	0.05–0.07	0.09–0.11	0.13–0.15	0.12–0.14	0.10–0.12	0.09–0.11	0.11–0.13
Static friction:								
Dexron oil	0.12–0.15	0.11–0.14	0.13–0.16	0.13–0.15	0.18–0.21	0.13–0.15	0.12–0.15	0.14–0.17
Engine oil	0.18–0.20	0.17–0.20	0.19–0.21	0.18–0.20	0.25–0.30	0.18–0.20	0.18–0.20	0.17–0.19
Mating surface requirement	Rough to medium	Rough	Medium	Very smooth	Very smooth	Smooth	Medium	Rough to smooth
Optimum grooving configuration	Spiral or spiral with radial	Sunburst or spiral with radial	Multiple parallel large land	Full waffle or multiple parallel	Full waffle or multiple parallel	Full waffle or multiple parallel	Sunburst or multiple parallel	Spiral/radial sunburst full waffle

Table 22.1 also shows the friction coefficient for the different materials as a function of facing thickness, and the roughness requirements[10] of the mating surfaces from the point of view of retaining reasonable clutch facing wear.

The grooving pattern on the plates of a lubricated clutch will have a significant effect on the friction coefficient. The optimum grooving pattern is shown in Table 22.1 for the different friction materials.

The oil used for clutch applications is a highly compounded hydrocarbon lubricant. The additives include viscosity-index improvers, antifoam agents, antioxidants, and extreme-pressure lubricants. These additives modify the frictional characteristics of the clutch materials, and have an important effect on the life of the clutch.[11,12] As shown in Table 22.1, the choice of oil can have a significant impact on the friction coefficient; for example, the static friction of a semi-metallic material is increased by 50 percent when the lubricant is changed from Dexron to engine oil. Therefore, it is essential that both the friction material and the fluid be considered in a new clutch design.

22.4   TORQUE EQUATIONS

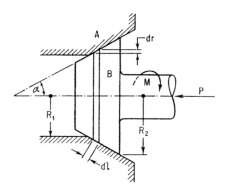

FIG. 22.3   Friction clutch.

Friction clutches are used in power-transmission systems (1) to transmit the desired torque from one point in the mechanism to the other and simultaneously to limit this torque to a desired value, and (2) to couple the prime mover to the load in a manner which allows the acceleration of the load to be controlled.

Consider the arrangement shown in Fig. 22.3, in which it is desired to determine the torque M due to frictional forces existing at the interface of members A and B, and due to the axial force P.

The normal load acting on the differential strip of width dl is $2\pi rp\, dl$, where p is the pressure. Since $dr = dl \sin \alpha$, the axial load supported by the strip

$$dP = 2\pi rp\, dr \tag{22.1}$$

so that
$$P = 2\pi \int_{R_1}^{R_2} pr\, dr \tag{22.2}$$

The frictional force acting on the strip dl

$$\mu\, dP = 2\pi\mu pr\, dl \tag{22.3}$$

where μ = coefficient of friction.

The associated moment

$$dM = 2\pi\mu pr^2 \csc \alpha\, dr \tag{22.4}$$

and the total frictional moment

$$M = 2\pi \csc \alpha \int_{R_1}^{R_2} \mu p r^2 \, dr \qquad (22.5)$$

Before Eq. (22.2) can be integrated, the dependence of μ and p on r and the relationship connecting μ and p must be known.

Variation of μ with r. If friction is independent of sliding velocity, μ is independent of r. If, as is commonly the case in practice, friction is a decreasing function of linear sliding velocity v, the relationship can be written

$$\mu = f(v) = f(\omega_r r) \qquad (22.6)$$

where ω_r is the relative angular velocity between the members A and B of the clutch in Fig. 22.3.

In case μ varies linearly with v, Eq. (22.6) can be written

$$\mu = \mu_0 - mv = \mu_0 - m\omega_r r \qquad (22.7)$$

where μ_0 = static coefficient of friction and m = slope of friction speed characteristic of the interface (depends on material combination).

Variation of p and r. There is, in any practical situation, considerable uncertainty as to the distribution of axial load P over the contact area. Because of the difficulty of obtaining perfect geometrical fit (especially in the case of cone clutches) between the two surfaces, and particularly with relatively hard materials, such as sintered metals and ceramics, contact between the surfaces is far from uniform and pressure distribution highly irregular. When dealing with resilient friction materials such as granulated cork or paper, the pressure is reasonably independent of r, and the assumption of uniform contact is safe.

Relationship between μ and p. As previously indicated, μ may be a function of p. If p is invariant with r, the expression for the clutch frictional moment will not be affected by the variation of μ with p. However, the design of the clutch (i.e., the number of plates required, etc.) will be different for different values of p.

If p varies with r and μ is a function of p as in

$$\mu = \mu_p - ap \qquad (22.8)$$

and if we assume that the wear rate is constant, i.e., the pressure times the surface speed (proportional to radius) is a constant C, one obtains[13]

$$\mu = \mu_p - aC/r \qquad (22.9)$$

Examination of the characteristics of those materials in which μ varies with p shows that Eq. (22.8) and hence Eq. (22.9) can be expected to hold only over a limited range of values of p. A nonlinear representation of μ as a function of p will probably be necessary in most cases.

In subsequent discussion the variation of friction with sliding velocity is of importance. If the variation is linear, Eq. (22.7) may be used, and depending on which of the two assumptions, that of the uniform rate of wear or of the constancy of p over the frictional area, is taken as the basis upon which Eqs. (22.2) and (22.5) are integrated, one obtains two expressions for the frictional moment of the clutch:

1. p invariant with r, $\mu = \mu_0 - mr\omega_r$:

$$M = \frac{2nP\mu_0}{3 \sin \alpha} \frac{R_2^3 - R_1^3}{R_2^2 - R_1^2} - \frac{nmP}{2 \sin \alpha}(R_2^2 + R_1^2)\omega_r \tag{22.10}$$

2. Uniform rate of wear, $pr = C_3$, $\mu = \mu_0 - mr\omega_r$:

$$M = \frac{Pn\mu_0}{2 \sin \alpha}(R_1 + R_2) - \frac{nmP}{3 \sin \alpha} \frac{R_2^3 - R_1^3}{R_2 - R_1}\omega_r \tag{22.11}$$

where n = number of pairs of frictional surfaces in contact.

It is seen that Eqs. (22.10) and (22.11) have the form

$$M = M_0 - \alpha\omega_r \tag{22.12}$$

where M_0 is recognized as the static torque capacity of a friction clutch

$$M_0 = (2nP\mu_0/3 \sin \alpha)/[(R_2^3 - R_1^3)/(R_2^2 - R_1^2)]$$

$$\tag{22.13}$$

or $$M_0 = (nP\mu_0/2 \sin \alpha)(R_1 + R_2)$$

depending on the assumption made in integrating Eqs. (22.2) and (22.5).

Simple calculation will show that for clutches so designed that R_2 is not very much larger than R_1, both equations (22.10) and (22.11) give values of M which, for all practical purposes, are identical. In general, M obtained from Eq. (22.11) has a slightly lower value than that given by Eq. (22.10).

Cone Clutches. Equations (22.10) and (22.11) can be used as they stand for computation of M.

Disk Clutches. In this case, $\alpha = \pi/2$, $\sin \alpha = 1$, and Eqs. (22.10) and (22.11) become, respectively,

$$M = \tfrac{2}{3} nP\mu_0[(R_2^3 - R_1^3)/(R_2^2 - R_1^2)] - \tfrac{1}{2} nmP(R_2^2 + R_1^2)\omega_r \tag{22.14}$$

$$M = \tfrac{1}{2} nP\mu_0(R_1 + R_2) - \tfrac{1}{3} nmP[(R_2^3 - R_1^3)/(R_2 - R_1)]\omega_r \tag{22.15}$$

Multiple-Disk Clutches. In the case of multiple-disk clutches Eqs. (22.10) through (22.13) require correction in that not all the disks in the assembly are subject to the same externally applied axial clamping force P_0.

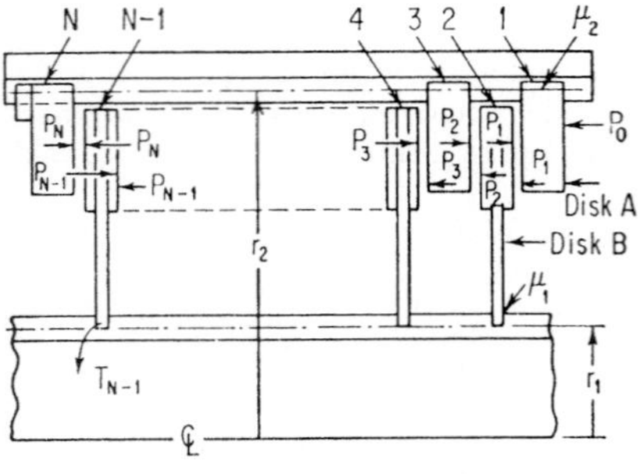

FIG. 22.4 Multiple-disk clutch.

Referring to Fig. 22.4, individual disks are in equilibrium under the action of three axial forces P_N, P_{N-1}, and the force originating at the hub spline because of friction between the disk and the hub as a consequence of torque transmitted by that disk, T_{N-1}/r_1.

Considering disk 1, the following relationship holds between forces acting on this disk when the clutch transmits full rated torque:

$$P_1 = P_0 - T_1\mu_2/r_2 \tag{22.16}$$

where T_1 = frictional moment transmitted by disk 1
μ_2 = coefficient of friction between disk 1 and its housing
r_2 = radius, as indicated in Fig. 22.4

But

$$T_1 = kP_1 \tag{22.17}$$

where, for simplicity, k may be taken from Eq. (22.13).

$$k = \tfrac{1}{2}\mu_0(R_1 + R_2)$$

for $\alpha = \pi/2$ and $n = 1$.
Thus,

$$P_1 = \frac{1}{1 + k\mu_2/r_2} \tag{22.18}$$

Considering now the conditions under which disk 2 remains in balance one obtains

$$P_2 = P_1 - (\mu_1/r_1)(T_1 + T_2)$$

which becomes

$$P_2 = \frac{1 - k\mu_1/r_1}{1 + k\mu_1/r_1} \quad P_1 = \frac{1}{1 + k\mu_2/r_2}\frac{1 - k\mu_1/r_1}{1 + k\mu_1/r_1} P_0 \tag{22.19}$$

where μ_1 = coefficient of friction between disk 2 and the hub
r_1 = radius, as indicated in Fig. 22.4

Extending this procedure to include all disks in the assembly and expressing axial forces in terms of P_0, one can write the equivalent load P

$$\frac{P}{P_0} = \frac{1}{(N-1)(1+a)}\left(1 + \frac{1-b}{1+b}\right)\sum_{n=1}^{j}\left[\frac{(1-b)(1-a)}{(1+b)(1+a)}\right]^{n-1} \tag{22.20}$$

where $a = k\mu_2/r_2$
$b = k\mu_1/r_1$
$j = (N-1)/2$
N = total number of disks (both sets)

With P_0 known, P is computed through Eq. (22.20) and then inserted into Eqs. (22.10) through (22.13) to compute the value of the total frictional moment of the clutch.

22.5 THERMAL PROBLEMS IN CLUTCH DESIGN

When one solid body slides over another, most of the work done against frictional forces opposing the motion will be liberated as heat at the interface. Consequently the temperature of the rubbing surfaces will increase and may reach values high enough to destroy the clutch.

Even under moderate loads and slow speeds, high instantaneous values of surface temperature are reached at points of actual contact of the two surfaces. Thermoelectric methods employed in the measurement of these transient local temperatures indicate values as high as 1000°C, and their duration has been observed to be of the order of 10^{-4} s. The results are similar when the surfaces are lubricated with mineral oil except that peak temperatures are reduced. Also, it has been observed that, when large quantities of heat are liberated at the rubbing interface during a short time interval, high surface temperatures are generated while the bulk of the two bodies in contact remains relatively cool.

Most premature clutch failures can be attributed to excessive surface temperatures generated during slipping under axial load. In the case of metallic clutch plates, high temperatures existing at the rubbing interface may cause the individual plates to be welded together. When nonmetallic or semimetallic plates are used with steel or cast-iron separating plates, the high temperatures can cause excessive wear. Other effects of high surface temperatures are distortion of the shape of the plates, surface cracks in solid metallic plates caused by thermal stresses, appreciable fluctuation in the value of the friction coefficient ("fading"), and in the case of lubricated clutches, oxidation of oil resulting in the formation of deposits on the working surfaces and in grooves.

In order to design a clutch that is satisfactory from the thermal point of view, it is necessary to know or estimate the value of surface temperature considered safe, and the maximum value of surface temperature likely to occur in a given assembly operating under known conditions of loading.

Because the required information is not readily obtained, it is probably best to perform a carefully executed series of tests. Direct measurement of surface temperatures in a clutch assembly presents great difficulties, however, especially when the friction material is a poor electrical conductor. Furthermore, experimentation with different clutch designs working under different conditions is rather expensive and time-consuming.

If a model is created, based on some simplifying assumptions, surface temperature of the clutch plates sliding relative to each other may be easily computed. The calculated values may then be compared with the safe temperature values usually supplied by friction-material manufacturers. Complete correlation cannot be expected if this procedure is followed, but in most situations a good criterion for establishing the size of the total frictional area of the clutch is thus obtained.

The imperfection with which the model represents the actual system should be borne in mind. The results obtained analytically should be regarded as an approximation. Even though the calculated values of the surface temperature of a clutch plate may differ significantly from the actual transient local values, the equations express the manner in which temperature will vary with system parameters and thus will show which of them should be altered to obtain desired change in temperature.

One widely used method of evaluating the thermal capacity of a friction clutch consists of calculating the average rate of energy dissipation per unit area of the clutch frictional surface during the slip period. The answer, usually stated in terms of hp/in^2 (often of the order of 0.3 to 0.5 hp/in^2), is compared with a similar figure computed for some other successfully operating design and is adjusted accordingly by variation of plate area and number of plates in the assembly. This method affords a quick and rough estimation of the extent of thermal loading to which the active surfaces are subjected and can serve as a very approximate basis for comparing two similar assemblies operating under identical conditions. This method completely neglects such factors of utmost importance in heat transfer as conductivity, specific heat, and plate thickness which have large effects on the surface temperature.

The temperature of the clutch plates will be evaluated by solving the Fourier conduction equation, and the necessary assumptions discussed below.

1. The heat flux generated at the rubbing interface is uniformly distributed over the total frictional area of the clutch; i.e., the rate of energy dissipation is a function of time only. This assumption is based upon the premise that the rate of wear of the rubbing surfaces and not the intensity of pressure is uniformly distributed over the area of the plate. This assumption is justified as follows: First, in a properly designed system, the rate of heat dissipation is a function of time—it decreases from an initial maximum value at the beginning of the slip period to zero at the end, i.e., at the time when the two shafts are effectively coupled by the clutch. Secondly, most clutches are so designed that the ratio of the outside diameter to the inside diameter of the plate is not very much greater than unity (about 1.3) and, therefore, the effect of the heat-flux variation with radius (if any) on plate temperature will be small. Finally, the manner in which the thermal parameters affect the surface temperature is not greatly influenced by the flux distribution.

2. Heat transfer from the clutch plates to the surroundings during slipping is negligible.

This assumption is well justified when dealing with slip periods of short duration, of the order of several seconds or less. Also, in most clutches the edge area of the plate is much smaller than the area receiving heat; thus the heat flow in the radial direction is small and can be neglected in comparison with axial direction.

3. The effect of the lubricating oil is neglected. The effect of the lubricant is difficult to account for in the analysis, mainly because, during the time interval in which the plates move closer together under the action of the axial force, lubrication changes from hydrodynamic to boundary while the oil is squeezed out of the space between the plates. The presence of oil generally lowers surface temperature considerably, depending on the type of lubricant. Under boundary lubrication the temperature can be expected to be between 30 to 60 percent lower than in the case of dry friction.

22.5.1 Derivation of Temperature Equation

First to be considered is the form of the expression for the rate of heat dissipation to be used in subsequent analysis. Consider a basic power-transmission system shown in Fig. 22.5.

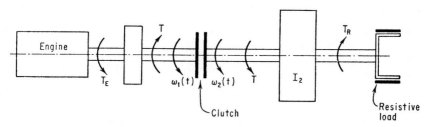

FIG. 22.5   Basic power-transmission system.

T_E is the engine output torque, T the clutch frictional torque, T_R the resistive load torque, and $\omega_1(t)$ and $\omega_2(t)$ are angular velocities of engine and load sides of clutch, respectively. All torques acting on the system will, for simplicity, be assumed constant.

It is easily shown that, when the compliance of the system is neglected,

$$\omega_1(t) = \frac{T_E - T}{I_1} t + \Omega_1 \qquad \text{rad/s} \qquad (22.21)$$

$$\omega_2(t) = \frac{T - T_R}{I_2} t + \Omega_2 \qquad \text{rad/s} \qquad (22.22)$$

where Ω_1 and Ω_2 are initial velocities of the two shafts.

The rate at which energy is dissipated in the clutch during slipping is given by

$$q_0(t) = T[\omega_1(t) - \omega_2(t)] \qquad \text{lb} \cdot \text{ft/s} \qquad (22.23)$$

$$q_0(t) = T(\{[I_2T_E + I_1T_R - T(I_1 + I_2)]/I_1I_2\}t + \Omega_1 - \Omega_2) \qquad (22.24)$$

The duration of the slip period is given by

$$t_0 = I_1I_2(\Omega_1 - \Omega_2)/[T(I_1 + I_2) - (I_2T_E + I_1T_R)] \qquad \text{s} \qquad (22.25)$$

The total amount of energy dissipated during slipping (in one cycle) is obtained by integrating Eq. (22.24) between the limits $t = 0$ and $t = t_0$. Thus

$$Q = T(\Omega_1 - \Omega_2)^2 I_1 I_2/[2T(I_1 + I_2) - (I_2T_E + I_1T_R)] \qquad \text{lb} \cdot \text{ft} \qquad (22.26)$$

In the special case, when $T_R = T_E = 0$, i.e, when the system consists of two flywheels I_1 and I_2 rotating at different speeds in the absence of external torques, Eqs. (22.24), (22.25), and (22.26) reduce to

$$q_0(t) = \{\Omega_1 - \Omega_2 - T[(I_1 + I_2)/I_1I_2]t\} T \qquad \text{lb} \cdot \text{ft/s} \qquad (22.27)$$

$$t_0 = I_1I_2(\Omega_1 - \Omega_2)/T(I_1 + I_2) \qquad \text{s} \qquad (22.28)$$

$$Q = I_1I_2(\Omega_1 - \Omega_2)^2/2(I_1 + I_2) \qquad \text{lb} \cdot \text{ft} \qquad (22.29)$$

There is considerable experimental evidence to suggest that the rate of heat dissipation varies approximately linearly with time and thus has the general form

$$- At + B \qquad (22.30)$$

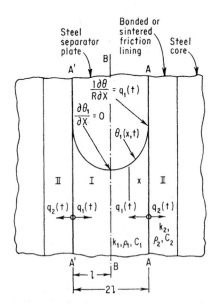

Steel separator plate

Bonded or sintered friction lining

Steel core

$$\frac{1}{R}\frac{\partial\theta}{\partial X} = q_1(t)$$

$$\frac{\partial\theta_1}{\partial X} = 0$$

$\theta_1(x,t)$

$q_2(t) \quad q_1(t) \qquad q_1(t) \quad q_2(t)$

$k_2,$

$k_1, A_1, C_1 \quad P_2, C_2$

FIG. 22.6 Two clutch plates in contact under the action of the axial clamping force.

which is identical with Eqs. (22.24) and (22.27).

Figure 22.6 shows the two clutch plates in contact under the action of the axial clamping force. Surface A-A, where rubbing occurs, can be thought of as a plane source of heat producing $q(t)$, Btu/s·ft². At the rubbing interface the surface temperature of both plates I and II must be the same. The rates of heat flow into I and II will not, in general, be equal but will depend on the thermal properties of the two materials in contact.

While the manner of division of the total flux is yet to be determined, the mathematical form of solution of the heat-conduction equation can be written

$$q_s t = m(-at + b)$$

$$= -\alpha_s t + \beta_s \qquad (22.31)$$

$$q_f(t) = (1 - m)(-at + b)$$

$$= -\alpha_f t + \beta_f \qquad (22.32)$$

where m is as yet undetermined and $q_s(t)$ = heat flux entering steel plate, $q_f(t)$ = heat flux entering friction lining. The heat-diffusion equation written for the steel plate is[14]

$$\partial^2\theta_s/\partial x^2 = C_sR_s(\partial\theta_s/\partial t) \tag{22.33}$$

where $C_s = c_s\rho_s$
$R_s = 1/k_s$
θ_s = temperature of the steel plate over ambient temperature
c_s = specific heat of steel
k_s = thermal conductivity of steel
ρ_s = density of steel

Using Eq. (22.31) together with the boundary conditions,

$$-(1/R_s)(\partial\theta_s/\partial x) = q_s(t) \qquad \text{at } x = 0$$
$$\partial\theta_s/\partial x = 0 \qquad \text{at } x = l \tag{22.34}$$

From the solution of Eq. (22.33), the surface temperature of a clutch plate as a function of time is determined.

$$\theta_s(0,t) = \zeta\left(\frac{t}{l\sqrt{R_sC_s}} + \frac{l^3\sqrt{R_sC_s}}{3} - \frac{2l\sqrt{R_sC_s}}{\pi^2}\sum_{n=1}^{\infty}\frac{1}{n^2}\exp-\frac{n^2\pi^2}{l^2R_sC_s}t\right)$$
$$- \nu\left[\frac{t^2}{2l\sqrt{R_sC_s}} + \frac{l\sqrt{R_sC_s}}{3} - \frac{l^3(C_sR_s)^{3/2}}{45} + \frac{2l^3(R_sC_s)^{3/2}}{\pi^4}\sum_{n=1}^{\infty}\frac{1}{n^4}\exp-\frac{n^2\pi^2}{l^2R_sC_s}t\right] \tag{22.35}$$

where $\zeta = \beta_s\sqrt{R_s/C_s}$ and $\nu = \alpha_s\sqrt{R_s/C_s}$.

For very large values of $l\sqrt{R_sC_s}$ or very small values of t, Eq. (22.35) is not convenient for computation of the surface temperature. It can be shown that, for very large plate thickness, such that reflection from the opposite side can be neglected,

$$\theta_s(0, t)_{l\to\infty} = 2\zeta\sqrt{t/\pi} - \tfrac{4}{3}(\nu t_{3/2}/\sqrt{\pi}) \tag{22.36}$$

In this case, the maximum temperature occurs at the time $t_m = \frac{1}{2}t_0$ and its value is $\sqrt{2}$ times that at $t = t_0$, i.e., the temperature at the end of engagement period.

The value of m in Eqs. (22.31) and (22.32) is determined by considering that the temperature at the rubbing interface ($x = 0$) must be the same for both the steel plate and the friction lining, i.e.,

$$\theta_s(0, t) = \theta_f(0, t) \tag{22.37}$$

The result, obtained by using any one of the temperature expressions, is

$$m = \frac{\sqrt{R_f/C_f}}{\sqrt{R_s/C_s} + \sqrt{R_f/C_f}} \tag{22.38}$$

It can be seen from Eq. (22.38) that the thermal properties and the densities of both mating materials are important in determining the value of the surface temperature.

22.5.2 Application

Equation (22.38) rewritten in more familiar notation

$$m = \sqrt{c_f \rho_f k_f}/(\sqrt{c_f \rho_f k_f} + \sqrt{c_s \rho_s k_s}) \tag{22.39}$$

shows that the total heat flux divides equally between the two contacting plates when the products of their thermal properties and densities are equal.

From Eqs. (22.31), (22.32), (22.35), and (22.39) it can be shown that the temperature at the interface is proportional to $(\sqrt{c_f \rho_f k_f} + \sqrt{c_s \rho_s k_s})^{-1}$. High values of thermal conductivity, specific heat, and high density will decrease the temperature at the rubbing interface. These properties, however, are not the only criteria to be used in the selection of materials when designing a clutch.

Other factors, such as coefficient of friction, fading, crushing strength, wear characteristics, and conformability, must also be considered when designing a friction clutch. Other criteria are discussed in the section on temperature flashes.

In practice, many clutches, especially those of multiple-disk type, are so designed that the thickness of the plate $2l$ does not exceed 0.10 in. When l (0.05 in) is substituted in Eq. (22.35) the constant in the exponent is

$$n^2 \pi^2/l^2 R_s C_s = 71.3 n^2 \qquad \text{for steel}$$

Thus, even with $t = 0.1$ and $n = 1$, $e^{-7.13} = 9 \times 10^{-4}$ and for all practical purposes the contribution of the infinite series in n can be neglected. Thus

$$\theta_s(0, t) = -\frac{vt^2}{2l\sqrt{R_s C_s}} + \left(\frac{\zeta}{l\sqrt{R_s C_s}} - \frac{vl\sqrt{R_s C_s}}{3}\right)t + \left[\frac{\zeta l\sqrt{R_s C_s}}{3} + \frac{vl^3 (R_s C_s)^{3/2}}{45}\right] \tag{22.40}$$

In order to find the maximum surface temperature of the plates, we differentiate Eq. (22.40) and set $\partial\theta_s/\partial t = 0$. The time of incidence of the maximum surface temperature is thus given by

$$t_m = \zeta/v - l^2 R_s C_s/3 \qquad \text{s} \tag{22.41}$$

or, since $\zeta/v = t_0$,

$$t_m = t_0 - l^2 R_s C_s/3 \qquad \text{s} \tag{22.42}$$

Substituting the value of t_m from Eq. (22.41) into Eq. (22.40), the maximum surface temperature is found to be (for small l)

$$\theta_{s,\max} = \tfrac{1}{2}\, \zeta^2/vl\sqrt{R_s C_s} + \tfrac{7}{90}\, vl^3 (R_s C_s)^{3/2} \tag{22.43}$$

A detailed examination of Eq. (22.43) will yield interesting and useful results depending on the nature of the system in which the clutch is used. Thus, if the system consists of two flywheels rotating at different speeds in the absence of external torques, and the purpose of the clutch is to transfer energy from one flywheel to the other (such systems form the basis of various forms of inertia starters), Eq. (22.29) shows that the total amount of energy dissipated in the clutch during the process of coupling is independent of the clutch frictional torque T. On the other hand, the duration of the slip period t_0, according to Eq. (22.28), is inversely proportional to T. The question may be raised, how does the change in t_0, caused by varying T, affect the maximum surface temperature of the plates? It is to be understood that the variation of clutch frictional torque is effected by varying the axial clamping force F only, and that clutch dimensions, number of plates, and their frictional and thermal properties are held constant.

This problem may arise when it is desired to adapt a clutch of certain thermal and mechanical capacity to a system demanding a higher-capacity clutch, and the deficiency is to be compensated for by increased pressure on the frictional surfaces.

It can be shown that the amount of energy absorbed by the steel plate per unit surface area is given by

$$Q_s = \bar{\mu}\,\delta(Q/A) \tag{22.44}$$

where δ = conversion factor from lb·ft to Btu = 1.285×10^{-3}
   A = total frictional area of clutch, ft^2

and $\bar{\mu} = \dfrac{\sqrt{R_f/C_f}}{\sqrt{R_f/C_f} + \sqrt{R_s/C_s}}$

Q was defined in Eq. (22.29).

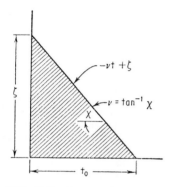

FIG. 22.7   Area number $- vt + \zeta$ plot which is proportional to Q_s.

Also, it is not difficult to verify that Q_s will be proportional to the shaded area in Fig. 22.7, and since Q_s is independent of either T or t_0, we may write

$$\zeta t_0 = K_1 = \text{const} \tag{22.45}$$

where

$$K_1 = 2Q_s\sqrt{R_s/C_s}$$

Now

$$\zeta/t_0 = v \tag{22.46}$$

Then, from Eqs. (22.45) and (22.46),

$$\zeta = K_1/t_0 \tag{22.47}$$

$$v = K_1/t_0^2 \tag{22.48}$$

$$\zeta^2/v = K_1 \tag{22.49}$$

but $t_0 = K_2/T$ where $K_2 = I_1 I_2(\Omega_1 - \Omega_2)/(I_1 + I_2)$. Substituting for t_0 into Eq. (22.48),

$$v = (K_1/K_2)T^2 \tag{22.50}$$

Substitution of ζ^2/v from Eq. (22.49) and v from Eq. (22.50) into Eq. (22.43) results in

$$\theta_{s,\max} = \frac{Q_s}{lC_s} + \frac{7}{45}\frac{Q_s T^2 l^3 R_s^2 (I_1 + I_2)^2}{I_1^2 I_2^2 (\Omega_1 - \Omega_2)^2} \tag{22.51}$$

Although the contribution of the second term in Eq. (22.51) to the value of θ_s max becomes appreciable only at very high values of T, it will be observed that surface temperature is proportional to the square of the clutch frictional torque. Thus, in the case when the clutch is used to start or stop pure inertia loads, too small and *not* too large values of t_0 will tend to damage the clutch.

Another much more frequently used system is one in which external driving and resistive torques, as well as the clutch frictional torques, act on the rotating inertias. In such systems, Eqs. (22.24), (22.25), and (22.26) would be used to calculate the rate of energy dissipation, duration of slip period, and the total amount of energy dissipated per cycle. It will be observed that, in the present case, Q is no longer independent of T and that both $t_0 \to \infty$ and $Q \to \infty$ when $T = (I_2 T_E + I_1 T_R)/(I_1 + I_2)$.

We again wish to examine the effect of varying clutch frictional torque on maximum surface temperature when the change in T is brought about by varying axial force on the clutch plates only.

From Eqs. (22.24) and (22.26), it follows that

$$\alpha_s = \frac{\overline{\mu}\delta}{A} \frac{T^2(I_1 + I_2) - T(I_2 T_E + I_1 T_r)}{I_1 I_2} \qquad \text{Btu/s}^2 \cdot \text{ft}^2 \qquad (22.52)$$

$$\beta_s = (\overline{\mu}\,\delta/A)T(\Omega_1 - \Omega_2) \qquad \text{Btu/s}^2 \cdot \text{ft}^2 \qquad (22.53)$$

then

$$\zeta_s = (\overline{\mu}\,\delta/A)T(\Omega_1 - \Omega_2)\sqrt{R_s/C_s} \qquad (22.54)$$

$$v_s = (\overline{\mu}\,\delta/A)[(T^2 - MT)/P]\sqrt{R_s/C_s} \qquad (22.55)$$

where

$$M = (I_1 T_E + I_1 T_R)/(I_1 + I_2)$$

$$P = I_1 I_2/(I_2 + I_2)$$

Substituting for ζ_s and v_s from Eqs. (22.54) and (22.55) into Eq. (22.43), one obtains

$$\theta_{s,max} = \frac{\delta\overline{\mu}T(\Omega_1 - \Omega_2)^2 P}{2AI(T - M)C_s} + \frac{7\overline{\mu}\,\delta T(T - M)l^3 R_s^2 C_s}{90PA} \qquad (22.56)$$

Equation (22.56) expresses analytically what may, perhaps, have been arrived at by some qualitative speculation, namely, that when driving and resistive torques are present in the system shown in Fig. 22.5, maximum surface temperature will reach very high values when the frictional clutch torque approaches the value of the driving torque. The same equation also shows that, when $T \gg (I_2 T_E + I_1 T_R)/(I_1 + I_2)$ (i.e., when t_0 is very small), the contribution of the second term to the value of θ_s max may become appreciable; so that, in spite of decreased slip period, the surface temperature of clutch plates may begin to increase.

22.5.3 Variation of Maximum Surface Temperature with Plate Thickness

If in the previously discussed system it is desired to lower the value of the maximum surface temperature by properly designing the friction elements, two methods are generally available: (1) increase in the total clutch frictional area A by increasing the number of plates or their size, and (2) increase in the thickness of the steel separator plate.

In applying the first of these methods, Eq. (22.56) shows that the maximum surface temperature is inversely proportional to the total frictional area A. Thus it follows that virtually any desired value of maximum surface temperature may be obtained by assigning proper value to A. The addition of one or more sets of plates, however, will generally be more expensive than the increase in thickness of the steel separator plates in an already designed clutch.

A general criterion for the maximum useful plate thickness when heated on both sides is based upon plotting the nondimensional Fourier number $\mathcal{F} = t_0/l^2 CR$ as a function of plate thickness. It can be shown that any increase beyond the value corresponding approximately to $1/\mathcal{F} = 1.75$ ceases to be significant:

$$l_{max} = \sqrt{1.75 t_0/C_s R_s} = \sqrt{1.75 t_0 k_s/c_s \rho_s} \qquad (22.57)$$

where $l_{max} = \frac{1}{2}$ maximum useful plate thickness when heated on both sides.

In the case when the plate is heated on one side only, Eq. (22.57) gives the actual maximum useful plate thickness.

22.5.4 Optimum Radial Dimensions of a Clutch Plate

The optimum dimensions of a clutch plate are dependent on the criterion used to define the meaning of the word "optimum." If the criterion is formulated on the basis that the ratio of frictional torque to the clutch-plate surface temperature should be a maximum one can obtain the relationship between the inside and outside radii as follows:

$$T \propto R_o + R_i \qquad (22.58)$$

where T = torque
R_o = outside radius of plate
R_i = inside radius of plate

The surface temperature is inversely proportional to the area of the plate

$$\theta_s \propto [1/(R_o^2 - R_i^2)] \qquad (22.59)$$

$$T/\theta_s = (1 - k^2)(1 + k) \qquad (22.60)$$

where $k = R_i/R_o$, and from which T/θ_s is a maximum when $k = \frac{1}{3}$ or

$$R_i = \frac{1}{3} R_o \qquad (22.61)$$

22.5.5 Flash Temperatures

Very high values of transient surface temperature occur at the interface between two sliding bodies, these temperature flashes occurring at points of intimate contact between the two surfaces.[15-17]

The effective or microscopic contact between the two surfaces is confined to surface asperities, and it can be shown that the relationship between the effective and the nominal or macroscopic area of contact is

$$A'/A = p/p_m \qquad (22.62)$$

where p_m = mean yield pressure of the softer material
p = nominal pressure (average pressure)
A' = effective frictional area over which contact actually occurs

If contact occurs at n spots, and if these spots are assumed to be circular in shape and equal in size

$$A' = \pi n r^2 \qquad (22.63)$$

where r = radius of the spot.

When relative motion exists between the surfaces, it has been observed that a number of luminous points appear at the rubbing interface and that the position of these spots changes rapidly as points of intimate contact wear away and new asperities come into contact. Referring to Fig. 22.8, which represents a portion of a

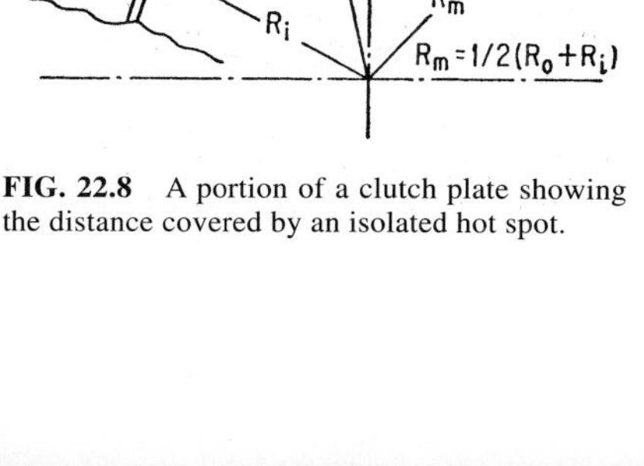

FIG. 22.8 A portion of a clutch plate showing the distance covered by an isolated hot spot.

clutch plate, let S be the distance covered by such an isolated hot spot in the small time interval t_s. Then

$$t_s = S/\omega R_m \qquad (22.64)$$

where ω = angular velocity and R_m = mean radius of the plate.

Also, $$S \acute{\ } r \qquad (22.65)$$

If a clutch with only one frictional surface is considered, the rate at which heat enters each spot is given by

$$q_r = \delta T \omega / \pi n r^2 \qquad (22.66)$$

Because t_s is very small, q_r will remain essentially constant. Also, the effect of heat reflection from the other side of the plate can be neglected. This allows the use of the simplest expression for the surface temperature, which is derived for the case of a semi-infinite solid receiving heat at a constant rate:

$$\theta_f(t) \propto \sqrt{R_s/C_s} q_r \sqrt{t_c} \qquad (22.67)$$

From $T = \mu p A R_m$, $A \propto R2m$, and Eqs. (22.62) through (22.67), the following expression is obtained:

$$\theta_f(t) \propto \sqrt{R_s/C_s}(\mu p_m^{3/4} R_m p^{1/4} \omega^{1/2}/n^{1/4}) \qquad (22.68)$$

Equation (22.68) indicates that, in order to decrease the value of flash temperature occurring at the areas of intimate contact, the following points should be considered when selecting friction material and clutch-plate size:

1. Soft friction materials having good surface conformability are preferable.
2. The coefficient of friction should be low. Since, in order to minimize clutch size, high values of μ are usually preferred, a material having sharp fade characteristics at high temperatures should be used.
3. Contact between the mating plates should be as uniform as possible; i.e., the number of contact spots n should be as large as possible.
4. For a clutch of given nominal frictional area A, the mean radius of the plate should be kept small. Since, for the plate of a given outside diameter, the frictional torque of the clutch is proportional to R_m, a compromise decision regarding the value of R_m should be made which takes into account both flash temperatures and the value of the frictional torque desired.

The foregoing analysis to a certain extent explains why substitution of hard sintered-bronze plates for those coated with such materials as cork or wood flour and asbestos mixtures does not always result in solving the high-temperature problems in a poorly designed clutch. Even though, according to Eq. (22.35), the use of sintered bronze should, because of its higher thermal conductivity, result in lower surface temperature than that obtained with cork lining, this advantage is offset by high values of surface-temperature flashes if the contact between the rubbing plates is not uniform. Therefore, when using metallic friction materials, great care should be taken in the preparation and finish of working surfaces.

22.5.6   Bulk Temperature of Clutch Assembly

When in a given power-transmission system the clutch is cycled continuously, it is necessary to have some means of computing the value of the average temperature at which the assembly will operate. This temperature is based upon radiative and convective heat transfer.

Problems encountered in analysis include the following:

1. Although most clutch assemblies are similar in outside appearance, i.e., cylindrical, the details of their design vary widely and may, in many cases, influence the process of heat transfer considerably (see Sec. 22.2).

2. Heat-transfer data, which could be directly applied to conditions under which the clutch is required to perform, are not available (e.g., shape, large diameter/length ratio, oil-mist atmosphere, end effects).

3. When formulas and methods applicable to simple and idealized cases are used to describe complicated systems errors result, mainly because many of the so-called "constants" rarely remain truly constant, and also some of the effects which can be considered negligible in idealized cases may become quite prominent in an actual system.

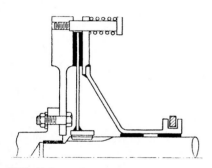

FIG. 22.9   Design where Newton's "law of cooling" may be used.

Two approaches are possible toward the estimation of the bulk temperature of the clutch, depending on its construction. When the design is such as shown in Fig. 22.9, i.e., when the heat generated at the rubbing surfaces is transmitted directly by conduction through a homogeneous material to the outer surfaces and from there transferred to the surroundings by the mechanism of forced convection, Newton's "law of cooling" may be used. This method is strictly correct only when the temperature everywhere within the body, including the surface, is the same and the surface coefficient of heat transfer is a constant independent of temperature.

In practice, the above conditions are approximated if the thermal conductivity of the material in question is high, when temperature does not vary too rapidly, and when the relative size of the body is small.

When a solid body of weight W, specific heat c, having an exposed surface area A cools slowly from an initial temperature θ'', the temperature at any subsequent time t is given by

$$\theta - \theta_a = (\theta'' - \theta_a) \exp(-\alpha t) \qquad (22.69)$$

where θ_a = ambient temperature
    $\alpha = Ah/Wc$
    h = surface coefficient of heat transfer

The temperature rise per cycle is

$$\theta = \theta'' - \theta' = \delta Q/Wc \qquad °F \qquad (22.70)$$

where $\theta' =$ temperature at the beginning of the slip period (or cycle) and $\theta'' =$ temperature at the end of the slip period.

It is assumed that the clutch attains the temperature θ'' instantaneously, which is approximately correct when the time during which it is allowed to cool is large in comparison with the time required for the heat to diffuse throughout the bulk of the plates.

If t_0 is the equal time interval between successive applications of the clutch, the temperatures θ' and θ'' are given by the following equations. First cycle,

$$\theta' = \theta_a \qquad \theta'' = \theta_a + \Delta\theta$$

Second cycle,

$$\theta'_2 = \theta_a + \Delta\theta \exp(-\alpha t_c) \qquad \theta''_2 = \theta_a + \Delta\theta[1 + \exp(-\alpha t_c)]$$

nth cycle,

$$\theta'_n = \theta_a + \Delta\theta[\exp(-\alpha)t_c + \exp(-2\alpha)t_c + \cdots + \exp(-n\alpha t_c)]$$

$$= \theta_a + \Delta\theta \, \frac{\exp[-(n-1)\alpha t_c] - 1}{\exp(-\alpha t_c) - 1} \exp(-\alpha t_c)$$

$$\theta''_n = \theta_a + \Delta\theta[1 + \exp(-\alpha t_c) + \exp(-2\alpha t_c) + \cdots + \exp(-n\alpha t_c)]$$

$$= \theta_a + \Delta\theta \, \frac{\exp(-n\alpha t_c) - 1}{\exp(-\alpha t_c) - 1}$$

The steady-state values of θ'_n and θ''_n are obtained by making $n \to \infty$. Thus

$$\theta'_{ss} = \theta_a + \Delta\theta \exp(-\alpha t_c)/[1 - \exp(-\alpha t_c)] \tag{22.71}$$

$$\theta''_{ss} = \theta_a + \Delta\theta[1 - \exp(-\alpha t_c)]^{-1} \tag{22.72}$$

Before Eqs. (22.71) and (22.72) can be used, the value of h must be determined. This can be approximately obtained by the use of the charts given in Figs. 22.10 and 22.11, taken from Refs. 18 and 19. Strictly speaking, the data contained in Fig. 22.10 pertain to rotating cylinders in which the length/diameter ratio is large, so that end effects are negligible, which is not quite the case when dealing with clutches. Figure 22.11 represents the results of experiments designed to establish the heat transfer in rotating electrical machines. Figure 22.12 gives those properties of air which are needed to evaluate Re and Nu.

The other method of evaluating the average temperature of the clutch assembly may be applied in cases where the preceding exponential development cannot be used. Consider that the clutch shown in Fig. 22.13 (disregard, for this purpose, the presence of cooling oil) is intended to be cycled continuously and at a regular time interval of 60-s duration. The clutch cylinder rotates at a constant speed of 4000 r/min and is completely enclosed in a cylindrical housing made of good heat-conducting material. The ambient temperature outside the housing is 100°F. Assume that the average rate of energy dissipation at the rubbing surfaces during the slip period is 0.5 hp/in^2 per cycle and that $t_0 = 0.8$ s. With $A = 117$ in^2, the total amount of energy liberated per cycle in the form of heat is 33.1 Btu. Thus in the steady state the average rate of heat rejection from the clutch assembly is $q = 33.1/60 = 0.55$ Btu/s.

Because of the high thermal conductivity of the housing material and its small wall thickness, the temperature drop across it will be neglected in the foregoing.

Referring to Fig. 22.14, the heat-dissipating areas of the housing and of the clutch cylinder are not equal:

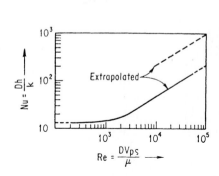

FIG. 22.10 Correlation of Nu (Nusselt number) vs. Re (Reynolds number) for a rotating cylinder without cross flow in air. D = diameter; h = coefficient of heat transfer; k = thermal conductivity of air; V_p = peripheral velocity; s = specific weight of air; μ = viscosity of air.

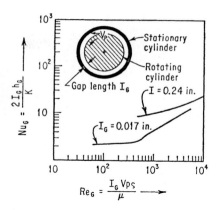

FIG. 22.11 Correlation of Nu_G (gap Nusselt number) vs. Re_G (gap Reynolds number) for a pair of coaxial cylinders. Inner cylinder rotating in air without axial flow (heat transfer from rotor surface to stator surface). Note curves show heat transfer across gap from rotating to stationary cylinder.

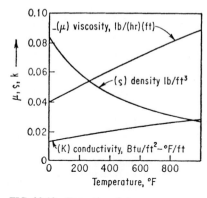

FIG. 22.12 Properties of air.

$$q_{12} = q/A_I = 0.55/0.44 = 1.25 \text{ Btu}/(s\cdot ft^2)$$

$$q_{23} = q/A_{II} = 0.55/0.52 = 1.06 \text{ Btu}/(s\cdot ft^2)$$

where q_{12} = rate of heat flow per square foot from I to II

q_{23} = rate of heat flow per square foot from II to III

A_I and A_{II} = heat-dissipating areas of clutch cylinder and housing, respectively

Now
$$q_{12} = h_g(\theta_I - \theta_{II}) + (E_I - E_{II})\epsilon_c \qquad (22.73)$$

where h_g = gap heat-transfer coefficient, Btu/(ft²·s²·°F)

θ_I = surface temperature (°F) of the clutch cylinder

θ_{II} = temperature of the housing

E_I = rate at which energy is radiated from the surface of I at a temperature θ_I (°F) (black-body emission)

E_{II} = rate at which energy is reflected from II back to I at a temperature θ_{II} (°F) (black-body reflection)

ϵ_c = combined emissivity of I and II (associated with geometry of system)

$$\epsilon_c = \frac{\epsilon_I \epsilon_{II}}{\epsilon_I(A_I/A_{II}') + \epsilon_{II} - \epsilon_I \epsilon_{II}(A_I/A_{II}')}$$

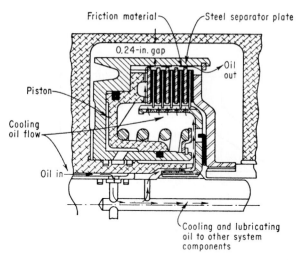

FIG. 22.13   Design where Newton's "law of cooling" cannot be used.

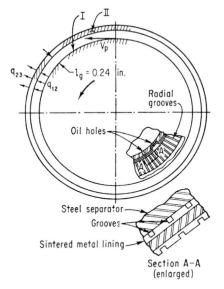

FIG. 22.14   Heat dissipation from outer surface of clutch cylinder for clutch shown in Fig. 22.13. I = clutch cylinder outer surface; II = housing; III = surrounding air.

where ϵ_I = emissivity of I
ϵ_{II} = emissivity of II
A_I = outside area of I
A'_{II} = inside area of II

The rate of heat flow from II to III is given by

$$q_{23} = h_c(\theta_{II} - \theta_{III}) + (E_{II} - E_{III})\epsilon_{II} \quad (22.74)$$

where h_c = coefficient of heat transfer due to natural (or forced) convection, Btu/(ft²·s·°F)
θ_{III} = ambient temperature, °F
E_{III} = rate of radiant-heat reflection from the surroundings

The value of the coefficient h_c may be obtained from Fig. 22.15 computed from the correlation of Nu vs. Pr × Gr for horizontal cylinders in air. The coefficient h_g is obtained from Fig. 22.11. The values of black-body radiant-heat transfer are plotted in Fig. 22.16 in a manner designed to facilitate computation. Values of ϵ_I and ϵ_{II} may be obtained from Ref. 14.

All five quantities, h_g, h_c, E_I, E_{II}, and E_{III}, are, for any physical configuration, functions of temperatures θ_I, θ_{II}, and θ_{III}. The simplest method of solution is outlined below:

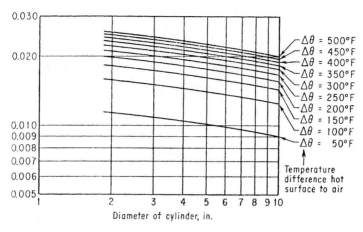

FIG. 22.15   Surface coefficient of heat transfer h_c, Btu/(ft²·s·°F).

1. With θ_{III} known, assume a series of several values of θ_{II} and calculate corresponding values of q_{23}. Plot these values on a temperature scale as a function of θ_{II}. The resulting graph is shown in Fig. 22.17, curve A. The intersection of this curve (point X) with the horizontal line $q_{23} = 1.06$ Btu/(ft²·s) determines the value of θ_{II}.

2. The same procedure is applied in order to determine the surface temperature of the clutch cylinder θ_{I}. The point of intersection of curve B with the horizontal line $q_{12} = 1.25$ Btu/(ft²·s) determines the surface temperature of the rotating clutch cylinder, which in this case is 537°F.

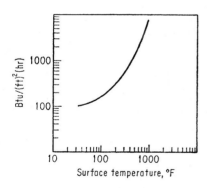

FIG. 22.16   Power radiated [Btu/(ft²·h)] by a blackbody surface.

With this value of the surface temperature of the clutch cylinder, the temperature existing at the rubbing surfaces of the plates during the slip period would be several hundred degrees higher, and thus steps must be taken to provide added cooling of the clutch assembly. In the case of lubricated clutches, such as shown in Fig. 22.13, this is most effectively accomplished by providing sufficient amounts of cooling oil with passages arranged so that the flow is directed at the surfaces where heat is generated. The flow of oil should be continuous, and the clutch so designed that the oil is in direct contact with the plates prior to, during, and after the slip period. In a heavy-duty clutch, sintered-metal friction plates would probably be used and, in this case, radial grooves, such as shown in Fig. 22.14, can be most easily and inexpensively provided.

Heat transfer from the clutch assembly by conduction to other system components (shafts, gears, etc.) has been neglected. This would probably amount to some 10 to 15 percent of the average rate of heat generation; so that the calculated value of $\theta_{\text{I}} = 537°$F may be too high.

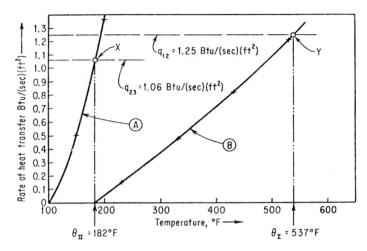

FIG. 22.17   Plot of values of q from assumed values of θ in order to find θ_I and θ_{II}.

22.6   DYNAMICS OF SYSTEMS EMPLOYING FRICTION CLUTCHES

A rigorous dynamic analysis of a mechanical power-transmission system containing a friction clutch presents a very difficult problem in derivation and solution of the differential equations describing the motion of the system.[20–22] This is because these systems are usually characterized by several degrees of freedom and are generally nonlinear. Because simple and general analytical techniques for dealing with nonlinear differential equations of higher order than the second are not available, there will generally be some sacrifice of mathematical rigor and elegance in the analysis.

The problem is to determine the behavior of a system following application of the clutch, given a friction clutch of certain specified characteristics, and given a set of initial conditions.

As a first step in arriving at a mathematical representation of a friction clutch it can be stated that a device whose operation depends upon friction exerts a frictional force or torque on the system only when relative motion exists between contacting members.[23]

This frictional force or torque is, in general, a function of both the magnitude and the direction of the relative velocity, and it may also be made time-dependent by introducing external control.

Thus a friction clutch may be characterized

$$M_c = f(\Delta\dot\theta, t) \tag{22.75}$$

where M_c = frictional torque
  $\Delta\dot\theta$ = relative angular velocity
  t = time

The variation in the friction coefficient with variables such as temperature, humidity, duty cycle, and loading history is slow in comparison with the frequencies encountered in vibrating mechanical systems.

For a clutch in which the torque is not externally controlled,

$$M_c = f(\dot\theta_1 - \dot\theta_2) \tag{22.76}$$

where $\dot\theta_1$ and $\dot\theta_2$ are the velocities of input and output plates, respectively. This relationship could be represented graphically as shown in Fig. 22.18.

It should be recognized that Fig. 22.18, to a different scale, provides the variation of frictional coefficient of the materials used in the construction of the clutch. This information will normally be available in graphical form as a result of testing. Equation (22.76) may be represented analytically:

$$M_c = \frac{\dot\theta_1 - \dot\theta_2}{|(\dot\theta_1 - \dot\theta_2)|} M_c(0) + \sum_{n=1}^{\infty} \frac{1}{n!} (\dot\theta_1 - \dot\theta_2)^n M_c^{(n)}(0) \tag{22.77}$$

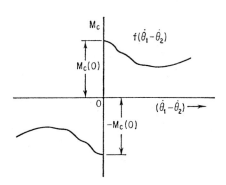

FIG. 22.18 Curve of $M_c = f(\theta_1 - \theta_2)$ for a clutch in which the torque is not externally controlled.

where $M_c(0)$ is the value of the clutch torque when $(\dot\theta_1 - \dot\theta_2) = 0$ (i.e., "static" friction torque) and primes denote derivatives with respect to $(\dot\theta_1 - \dot\theta_2)$. In practice, as many terms of the series (22.77) are used as are necessary to obtain an adequate "fitting" to the curve in Fig. 22.18. It should be noted that Eq. (22.77) is nonlinear not only by virtue of the terms $(\dot\theta_1 - \dot\theta_2)^2$, $(\dot\theta_1 - \dot\theta_2)^3$, ... $(\dot\theta_1 - \dot\theta_2)^n$, but also because the term $M_c(0)$ is a function of the relative velocity $\dot\theta_1 - \dot\theta_2$.

Since the torque-speed curve of a clutch may be characterized over the whole or part of the range by negative slope, it is of interest to see what effect this may have on the behavior of the dynamical system of which the clutch is a part.

Consider a simple mechanical oscillator consisting of mass, m, spring of constant k, and a damper of constant α. The equation of motion is

$$m\ddot{x} + \alpha\dot{x} + kx = 0 \tag{22.78}$$

Multiply Eq. (22.78) by x and integrating from 0 to τ

$$\left[\frac{m(\dot{x})^2}{2}\right]_0^\tau + \left[\frac{kx^2}{2}\right]_0^\tau = -\alpha \int_0^\tau (\dot{x})^2 \, dt \tag{22.79}$$

The sum of the left-hand side of Eq. (22.79) represents the variation of the total energy, kinetic plus potential, during this time. It can be seen that this variation may be either positive or negative depending upon the sign of α. If α is positive, the variation in the total energy is negative, i.e., the system loses energy as $\tau \to \infty$. If α is negative the system gains energy as $\tau \to \infty$. In a system without a source of energy this is not physically realizable. α represents the slope of the friction-speed characteristic of the damper.

A power-transmission system to be considered here consists of three essential elements: (1) a source of energy (in the form of some type of prime mover on the input to the clutch), (2) a clutch controlling the flow of power, and (3) the load on the output side of the clutch.

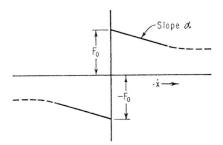

FIG. 22.19 Characteristics of damper of mechanical oscillator.

Since the presence of inertia and compliance is inevitable in mechanical systems, the entire system may become oscillatory following the application of the clutch. If the clutch characteristic has anywhere a portion with negative slope, these oscillations will tend to build up as suggested by Eq. (22.78) until limited by some system nonlinearity. In this case, the nonlinearity which eventually limits the growth of the oscillation and causes the amplitude to decrease to zero in a properly designed system is the term $[(\dot{\theta}_1 - \dot{\theta}_2)/|(\dot{\theta}_1 - \dot{\theta}_2)|] M_c(0)$ in Eq. (22.77).

Consider now a simple mechanical oscillator, letting the damper have a characteristic shown in Fig. 22.19.

This characteristic has been constructed by retaining only the first and the second terms of Eq. (22.77), i.e.,

$$F = (\dot{x}/|\dot{x}|)F_0 - \alpha\dot{x} \tag{22.80}$$

Equation (22.78) is now written

$$\ddot{x} + (1/m)[(\dot{x}/|\dot{x}|)F_0 - \alpha\dot{x}] + \omega^2 x = 0 \tag{22.81}$$

where $\omega = \sqrt{k/m}$ = natural frequency of the oscillator.

If oscillations are admitted they will be of the form $x = A(t) \sin(\omega t + \Phi)$. If the rate of variation of $A(t)$ is small in comparison with ω, we find

$$\frac{d[A(t)]}{dt} = -\frac{1}{2\pi\omega m}\left[\int_0^{\pi/2} (F_0 - \alpha A\omega \cos\psi)\cos\psi\, d\psi\right.$$

$$\left. - \int_{\pi/2}^{3\pi/2} (-F_0 - \alpha A\omega \cos\psi)\cos\psi\, d\psi - \int_{3\pi/2}^{2\pi} (F_0 - \alpha A\omega \cos\psi)\cos\psi\, d\psi\right] \tag{22.82}$$

Integration yields

$$d[A(t)]/dt = \alpha A/2m - 2F_0/\pi\omega m \tag{22.83}$$

Equation (22.83), a linear differential equation of the first order, is solved to obtain

$$A(t) = F_0/\pi\alpha\omega + (A_0 - F_0/\alpha\pi\omega)\exp(\alpha/2m) \tag{22.84}$$

where A_0 = initial amplitude [value of $A(t)$ at $t = 0$].

The damper characteristic given by Eq. (22.80) is a mathematical function used to obtain Eq. (22.84). A more realistic case would be represented by the dashed extension of the curves in Fig. 22.19. Thus Eq. (22.84) should be used with the restriction that $|A_0| < |4F_0/\alpha\pi\omega|$, which implies that the case of $|\alpha\dot{x}| > |F_0|$ is inadmissible.

Examining Eq. (22.84) in the light of restrictions just imposed, it is seen that the amplitude of the oscillation decreases exponentially despite the fact that the damper is characterized by negative slope.

When $F_0 = 0$, the amplitude increases exponentially, e.g.,

$$A(t)|_{F_0 = 0} = A_0 \exp[(\alpha/2m)t] \tag{22.85}$$

which shows that it is the nonlinearity of the fractional characteristics of the damper which has the effect of positive damping.

22.6.1 Transient Analysis of Mechanical Power Transmission Containing a Friction Clutch

A mechanical power-transmission system to be discussed in this section consists of a prime mover, a friction clutch, and a load. It will also be characterized by the presence of a multiplicity of energy-storage elements in the form of inertias, compliances, and dissipative elements. Included will be various nonlinear and linear coupling elements.

Backlash and hysteresis in various coupling elements will not be considered, but otherwise no restriction is made as to the linearity of the components present in the system, with the exception that inertia is assumed to remain invariant.

The following method of analysis is general enough to permit extension to systems employing any number of energy-storage elements and any number of dissipative elements.

Consider a system shown in Fig. 22.20.

$$
\begin{aligned}
I_1 &= \text{moment of inertia of prime mover} \\
I_2 &= \text{moment of inertia of input plate of clutch} \\
I_3 &= \text{moment of inertia of output plate of clutch} \\
I_4, I_5 &= \text{moment of inertia of load side of clutch} \\
M_1 &= \text{driving torque} \\
M_c &= \text{clutch frictional torque} \\
M_2 &= \text{load torque} \\
f_1(\theta_1 - \theta_2), f_2(\theta_2 - \theta_4), f_3(\theta_4 - \theta_5) &= \text{equivalent characteristics of various shafts;} \\
&\quad \text{torque as a function of the angle of twist} \\
C_1, C_2, C_3, C_4, C_5 &= \text{coefficients of viscous resistance corresponding} \\
&\quad \text{to } \dot\theta_1, \dot\theta_2, \dot\theta_3, \dot\theta_4, \text{ and } \dot\theta_5, \text{ respectively.} \\
g(\dot\theta_4 \text{ and } \dot\theta_5) &= \text{characteristic of a damper placed in the system} \\
&\quad \text{as shown in Fig. 22.20, torque as a function of} \\
&\quad \text{relative velocity } \dot\theta_4 - \dot\theta_5
\end{aligned}
$$

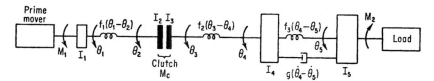

FIG. 22.20 Mechanical power-transmission system with a friction clutch.

Having described the lumped parameters of the transmission system, consider next the characteristics of the prime mover and the load. In the general case, torques M_1 and M_2 can be considered functions of time, angular velocity, and angular displacement.

In practice, the torque pulsations of multicylinder engines are small in comparison with the average value of torque and can be neglected in the present analysis.

If the torque M_1 is considered a function of speed only, it can be represented analytically by an infinite series.

$$
M_1(\dot\theta_1) = M_1(0) + \sum_{n=1}^{\infty} \frac{1}{n!} (\dot\theta_1)^n M_1^{(n)}(0) \tag{22.86}
$$

where $M_1(0)$ is the value of $M_1(\dot\theta_1)$ at $\dot\theta_1 = 0$.

Generally, the load torque M_2 can be considered a function of both speed $\dot{\theta}_s$ and the angular position of the shaft θ_s and can be represented by the sum of two infinite series of the same form

$$M_2(\dot{\theta}_s, \theta_s) = M_2(0) + \sum_{n=1}^{\infty} \frac{1}{n!} (\dot{\theta}_s)^{(n)} M_2^{(n)}(0) + \overline{M}_2(0) + \sum_{r=1}^{\infty} \frac{1}{r!} (\theta_s)^r \overline{M}_2^{(r)}(0) \qquad (22.87)$$

In certain situations it may be more convenient to replace the second term on the right-hand side of Eq. (22.87) with a Fourier series, e.g., if $M_2(\dot{\theta}_s, \theta_s)$ is periodic in θ_s.

In Eq. (22.87) $M_2(0)$ is the value of the load torque at $\dot{\theta}_s = 0$ and $M_2(0)$ is the value of the load torque at $\theta_s = 0$.

For a mechanical system with k degrees of freedom the form of Lagrange's equations will be suited for deriving the equations of motion if the system of Fig. 22.20 is written

$$(d/dt)(\partial T/\partial \dot{q}_k) - \partial T/\partial q_k + \partial V/\partial q_k + \partial F/\partial \dot{q}_k = Q_k \qquad (22.88)$$

where T = total kinetic energy of the system
$\quad\quad V$ = total potential energy of the system
$\quad\quad F$ = dissipation function
$\quad\quad Q_k$ = generalized force
$\quad\quad q_k$ = generalized position coordinate
$\quad\quad \dot{q}_k$ = generalized velocity coordinate
$\quad\quad k$ = 1,2,3, ... = number of independent coordinates

The computation of T is normally quite straightforward and needs no special explanation. In the computation of V should be included all elastic forces whether internal or external that are functions of displacement and all constant forces. All external and internal forces and moments that are functions of velocities must be included in the computation of F. Q_k includes all external forces and moments which are functions of time.

The coordinates used below follow.

Velocities	Positions	k
$\dot{q}_1, \dot{\theta}_1$	q_1, θ_1	1
$\dot{q}_2, \dot{\theta}_2$	q_2, θ_2	2
$\dot{q}_3, \dot{\theta}_3$	q_3, θ_3	3
$\dot{q}_4, \dot{\theta}_4$	q_4, θ_4	4
$\dot{q}_5, \dot{\theta}_5$	q_5, θ_5	5

In the case considered presently $Q_k = 0$, since it was assumed that no time-dependent torques act on the system. The potential energy stored in a single elastic element whose torque-deflection characteristic is given by $f(\theta_1 - \theta_2)$ can be expressed as

$$V = \int_0^{\theta_1 - \theta_2} f(\theta_1 - \theta_2) \, d(\theta_1 - \theta_2) \qquad (22.89)$$

Similarly the dissipation function or the energy dissipated in a damper whose torque-velocity characteristic is $g(\dot{\theta}_1 - \dot{\theta}_2)$ is given by

$$F = \int_0^{\dot{\theta}_1 - \dot{\theta}_2} g(\dot{\theta}_1 - \dot{\theta}_2) \, d(\dot{\theta}_1 - \dot{\theta}_2) \qquad (22.90)$$

T, V, and F may now be established:

$$T = \frac{1}{2} \left[I_1(\dot{\theta}_1)^2 + I_2(\dot{\theta}_2)^2 + I_3(\dot{\theta}_3)^2 + I_4(\dot{\theta}_4)^2 + I_5(\dot{\theta}_5)^2 \right] \tag{22.91}$$

$$V = \int_0^{\theta_1 - \theta_2} f_1(\theta_1 - \theta_2)\, d(\theta_1 - \theta_2) + \int_0^{\theta_3 - \theta_4} f_2(\theta_3 - \theta_4)\, d(\theta_3 - \theta_4)$$

$$+ \int_0^{\theta_4 - \theta_5} f_3(\theta_4 - \theta_5)\, d(\theta_4 - \theta_5) + \int_0^{\theta_5} \left[M_2(0) + \sum_{r=1}^{\infty} \frac{1}{r!} (\theta_5)^r \overline{M}_2^{(r)}(0) \right] d\theta_5 \tag{22.92}$$

Note that the part of the load torque M_2 which is a function of position θ_5 has been included in the computation of V.

In cases dealing with linear springs the first three terms of Eq. (22.92) become $\frac{1}{2}k_1(\theta_1 - \theta_2)^2 + \frac{1}{2}k_2(\theta_3 - \theta_4)^2 + \frac{1}{2}k_3(\theta_4 - \theta_5)^2$, where k_1, k_2, and k_3 are spring constants.

The dissipation function for this system is given by

$$F = \frac{1}{2} \left[C_1(\dot{\theta}_1)^2 + C_2(\dot{\theta}_2)^2 + C_3(\dot{\theta}_3)^2 + C_4(\dot{\theta}_4)^2 + C_5(\dot{\theta}_5) \right]$$

$$+ \int_0^{\dot{\theta}_2 - \dot{\theta}_3} M_c(\dot{\theta}_2 - \dot{\theta}_3)\, d(\dot{\theta}_2 - \dot{\theta}_3) + \int_0^{\dot{\theta}_4 - \dot{\theta}_5} g(\dot{\theta}_4 - \dot{\theta}_5)\, d(\dot{\theta}_4 - \dot{\theta}_5)$$

$$- \int_0^{\dot{\theta}_1} M_1(\dot{\theta}_1)\, d\dot{\theta}_1 + \int_0^{\dot{\theta}_5} M_2(\dot{\theta}_5, \theta_5)\, d\dot{\theta}_5 \tag{22.93}$$

In computing F the driving torque $M_1(\dot{\theta}_1)$ supplies energy to the system. Therefore, the associated dissipation function is negative.

It should be noted that dissipation due to load torque $M_2(\dot{\theta}, \theta_5, \theta_5)$ is obtained by integration with respect to $\dot{\theta}_5$ only.

In order to obtain the dynamical equations of motion for the system, Eqs. (22.91), (22.92), and (22.93) are substituted into Eq. (22.75) and indicated partial differentiation performed in order for $k = 1, 2$, etc.

The result is

$$I_1\ddot{\theta}_1 + \frac{\partial}{\partial \theta_1} \int_0^{\theta_1 - \theta_2} f_1(\theta_1 - \theta_2)\, d(\theta_1 - \theta_2) + C_1\dot{\theta}_1 - M_1(\dot{\theta}_1) = 0$$

$$I_2\ddot{\theta}_2 + \frac{\partial}{\partial \theta_2} \int_0^{\theta_1 - \theta_2} f_1(\theta_1 - \theta_2)\, d(\theta_1 - \theta_2)$$

$$+ \frac{\partial}{\partial \dot{\theta}_2} \int_0^{\dot{\theta}_2 - \dot{\theta}_3} M_c(\dot{\theta}_2 - \dot{\theta}_3)\, d(\dot{\theta}_2 - \dot{\theta}_3) + C_2\dot{\theta}_2 = 0$$

$$I_3\ddot{\theta}_3 + \frac{\partial}{\partial \theta_3} \int_0^{\theta_3 - \theta_4} f_2(\theta_3 - \theta_4)\, d(\theta_3 - \theta_4) \tag{22.94}$$

$$+ \frac{\partial}{\partial \dot{\theta}_3} \int_0^{\dot{\theta}_2 - \dot{\theta}_3} M_c(\dot{\theta}_2 - \dot{\theta}_3)\, d(\dot{\theta}_2 - \dot{\theta}_3) + C_3\dot{\theta}_3 = 0$$

$$I_4\ddot{\theta}_4 + \frac{\partial}{2\theta_4} \int_0^{\theta_3 - \theta_4} f_2(\theta_3 - \theta_4)\, d(\theta_3 - \theta_4) + \frac{\partial}{\partial \theta_4} \int_0^{\theta_4 - \theta_5} f_3(\theta_4 - \theta_5)\, d(\theta_4 - \theta_5)$$

$$+ \frac{\partial}{\partial \dot{\theta}_4} \int_0^{\dot{\theta}_4 - \dot{\theta}_5} g(\dot{\theta}_4 - \dot{\theta}_5)\, d(\dot{\theta}_4 - \dot{\theta}_5) + C_4\dot{\theta}_4 = 0$$

$$I_5\ddot{\theta}_5 + \frac{\partial}{\partial\theta_5}\int_0^{\theta_4-\theta_5} f_3(\theta_4 - \theta_5)\, d(\theta_4 - \theta_5)$$

$$+ \frac{\partial}{\partial\dot{\theta}_k}\int_0^{\dot{\theta}_4-\dot{\theta}_5} g(\dot{\theta}_4 - \dot{\theta}_5)\, d(\dot{\theta}_4 - \dot{\theta}_5) + C_5\dot{\theta}_5 + M_2(\dot{\theta}_5, \theta_5) = 0 \quad (22.94)$$

Equations (22.94) apply only during the transient, i.e., only when $(\dot{\theta}_2 - \dot{\theta}_3) \neq 0$. When $(\dot{\theta}_2 - \dot{\theta}_3) = 0$ and the clutch is engaged permanently a different set of equations must be derived to represent the system.

In deriving the equations of motion, the coefficient of the highest derivative, no matter how small, must not be neglected or the resulting equations will not properly represent the system.

Equations (22.94) represent a starting point in the analysis of power transmissions utilizing friction clutches when the various functions describing system parameters are known. If it is desired to obtain explicit solutions for the various speeds, shaft deflections, and torques existing in the system and their variation in time, because of the complexity of the mathematics recourse must be made to analog or digital computers.

In order to gain some approximate information regarding the behavior of the system, Eqs. (22.94) can be "linearized" and some useful techniques normally applicable to sets of linear differential equations can be applied with the understanding that only a limited amount of information can be obtained in this way.

If the system does not contain any highly nonlinear springs (or couplings), or if these can be represented by means of piecewise linear functions, the functions $f_1(\theta_1 - \theta_2)$, etc., can be considered linear of the form $k_1(\theta_1 - \theta_2)$. The same pertains to various dissipation functions, which can also be made representable by linear functions or combinations of linear functions.

It is often useful to determine, even approximately, the natural frequencies of the system. A method of calculating these is given below and is based upon the assumption that Eqs. (22.94) can be linearized.

Equations (22.94) are rewritten, omitting all dissipation terms and all external torques acting on the system and considering all springs to be linear.

$$I_1\ddot{\theta}_1 + k_1(\theta_1 - \theta_2) = 0$$

$$I_2\ddot{\theta}_2 - k_1(\theta_1 - \theta_2) = 0$$

$$I_3\ddot{\theta}_3 + k_2(\theta_3 - \theta_4) = 0 \quad\quad (22.95)$$

$$I_4\ddot{\theta}_4 - k_2(\theta_3 - \theta_4) + k_3(\theta_4 - \theta_5) = 0$$

$$I_5\ddot{\theta}_5 - k_3(\theta_4 - \theta_5) = 0$$

When the expression for the clutch torque is omitted, as it is in Eqs. (22.95), there is no coupling between the input and output sides of the clutch and the transmission system consists of two dynamically independent subsystems. Since the presence of the dissipation term which acts as a coupling term between the two subsystems will not greatly influence their natural frequencies, they remain essentially independent during the transient. The situation changes after the engagement of the clutch is completed since then $\dot{\theta}_2 = \dot{\theta}_3$ and I_2 and I_3 form one flywheel.

In order to obtain the natural frequencies of the system, solutions of the form $\theta_k = A_k \sin(\omega t + \phi_k)$ are sought, where A_k is the amplitude and ϕ_k is the phase angle of the kth harmonic, with $\omega = 2\pi f$ the natural angular frequency of the system. Substituting θ_k for θ_1 and θ_2 in Eqs. (22.95), a set of equations is obtained

$$(k_1 - \lambda I_1)A_1 - k_1A_2 = 0$$

$$-k_1 A_1 + (k_1 - \lambda I_2)A_2 = 0 \tag{22.96}$$

where $\lambda = \omega^2$, which results in a quadratic equation in λ which, when solved, yields the roots

$$\lambda_1 = \omega_1^2 = 0$$

$$\lambda_2 = \omega_2^2 = k_1(I_1 + I_2)/I_1 I_2$$

The natural frequency of the input side of the clutch is given by

$$f_2 = (1/2\pi)\sqrt{k_1(I_1 + I_2)/I_1 I_2} \tag{22.97}$$

Similarly,

$$f_3 = (1/2\pi)[(a + \sqrt{a^2 - 4b})/2]^{1/2} \tag{22.98}$$

$$f_4 = (1/2\pi)[(a - \sqrt{a^2 - 4b})/2]^{1/2} \tag{22.99}$$

where

$$a = [k_2(I_3 I_5 + I_4 I_5) + k_3(I_3 I_5 + I_3 I_4)]/I_3 I_4 I_5$$

$$b = k_2 k_3(I_3 + I_4 + I_5)/I_3 I_4 I_5$$

Equations (22.94) can be useful in another way to gain more information about the dynamic behavior of the system during the transient following the application of the clutch.

Earlier, a simple mechanical oscillator was analyzed from the standpoint of stability, and in this connection, a question may well be asked how will a more complicated system behave under similar conditions.

For the purpose of the present discussion it is more convenient to take as a model

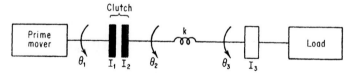

FIG. 22.21   Simplified power-transmission system.

the simpler system shown in Fig. 22.21.

The stiffness of the shaft connecting the prime mover to the input member of the clutch is made very large in comparison with the shaft connecting the clutch and the load. Such a system approximately represents a power transmission consisting of an engine, a clutch, flexible coupling of stiffness k, a gearbox, and a load. It will be assumed that the prime mover, the clutch, and the load have torque-speed characteristics as shown in Fig. 22.22.

Equations (22.77), (22.86), and (22.87) yield the following, respectively: clutch torque, driving torque, load torque

$$M_c = M_c(0)(\dot\theta_1 - \dot\theta_2)/|(\dot\theta_1 - \dot\theta_2)| + \alpha(\dot\theta_1 - \dot\theta_2) \tag{22.100}$$

$$M_1 = M_1(0) + a\dot\theta_1 \tag{22.101}$$

$$M_2 = M_2(0) + b\dot\theta_3 \tag{22.102}$$

where there is no restriction placed on the algebraic sign of α, a, and b.

Furthermore, if the discussion is limited to apply only for values of $0 < (\dot\theta_1 - \dot\theta_2) <$

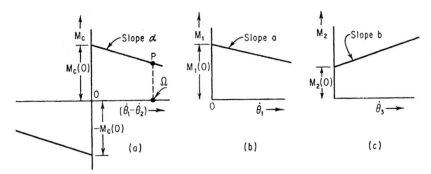

FIG. 22.22 Torque-speed characteristics for clutch, prime mover, and load shown in Fig. 22.21. (a) Clutch characteristics. (b) Prime-mover characteristics. (c) Load characteristics.

Ω where Ω is the value of the relative velocity of clutch plates corresponding to some point P on the torque-speed characteristic of the clutch in Fig. 22.22a, set of equations of motion for the system can be obtained which will be linear in the interval specified. With these restrictions, the expression for the clutch torque becomes

$$M_c = M_c(0) + \alpha(\dot{\theta}_1 - \dot{\theta}_2) \tag{22.103}$$

and the equations of motion are written

$$I_1\ddot{\theta}_1 + c_1\dot{\theta}_1 - [M_1(0) + a\dot{\theta}_1] + [M_c(0) + \alpha(\dot{\theta}_1 - \dot{\theta}_2)] = 0$$
$$I_2\ddot{\theta}_2 + c_2\dot{\theta}_2 + k(\theta_2 - \theta_3) - [M_c(0) + \alpha(\dot{\theta}_1 - \dot{\theta}_2)] = 0 \tag{22.104}$$
$$I_3\ddot{\theta}_3 + c_3\dot{\theta}_3 - k(\theta_2 - \theta_3) + [M_2(0) + b\dot{\theta}_3] = 0$$

Expansion of the determinant resulting from Eq. (22.104) leads to

$$p_2\{I_1I_2I_3p^4 + [I_1(I_2R_3 + I_3R_2) + R_1I_2I_3]p^3$$
$$+ [I_1(I_2k + R_2R_3 + I_3k) + R_1(I_2R_3 + I_3R_2) - \alpha^2I_3]p^2$$
$$+ [I_1k(R_2 + R_3) + (I_2k + R_2R_3 + I_3k) + \alpha R_3]p$$
$$+ k[R_1(R_2 + R_3) + \alpha]\} = 0 \tag{22.105}$$

where
$$p = d/dt$$

$$R_1 = c_1 - a + \alpha \tag{22.106}$$

$$R_2 = c_2 + \alpha \tag{22.107}$$

$$R_3 = c_3 + b \tag{22.108}$$

If the system represented by Eqs. (22.104) is to be stable, the roots of Eq. (22.105) must have all their real parts negative. The presence of two equal roots $p^2 = 0$ merely signifies that the system is capable of rotation in space without executing any oscillations, and is of no consequence in this discussion.

In general, in order to discover whether or not an algebraic equation of the form

$$a_0p^n + a_1p^{n-1} + a_2p^{n-2} + \cdots + a_n = 0 \tag{22.109}$$

possesses roots having positive real parts, an array of coefficients is formed. The coefficients are first written in two rows:

Row 1: $\qquad a_0 \quad a_2 \quad a_4 \quad \cdots$

Row 2: $\qquad a_1 \quad a_3 \quad a_5 \quad \cdots$

By cross multiplication of the first column with each succeeding column in turn a further $n-1$ rows are formed:

Row 3: $\qquad b_1 = (a_1 a_2 - a_0 a_3) \qquad b_2 = (a_1 a_4 - a_0 a_5) \qquad \cdots$

Row 4: $\qquad c_1 = (b_1 a_3 - a_1 b_2) \qquad c_2 = (b_1 a_5 - a_1 b_3) \qquad \cdots$

Equation (22.109) contains no roots with positive real parts if, and only if, all terms in the first column of the array (i.e., a_0, a_2, b_2, c_1, etc.) are positive for positive a_0, or negative for negative a_0. The number of times a change in sign takes place in going from one term to the next in the first column gives the number of roots with positive real parts. This is known as "Routh's stability criterion."

The first conclusion that can be made is that a necessary but not sufficient condition for stability is that all coefficients of p in Eq. (22.109) must be present and that they must all have the same algebraic sign. Since inertia and stiffness in a mechanical system will generally be positive, this statement imposes certain restrictions on the values of R_1, R_2, and R_3 in Eq. (22.105), and consequently on the values of α, a, and b in Eqs. (22.100), (22.101), and (22.102). Normally the values of the coefficients of viscous resistance c_1, c_2, and c_3 will be made as small as possible in an efficient power-transmission system, so that the three other dissipation terms a, α, and b which characterize the prime mover, the clutch, and the load will play an important part in determining the dynamic behavior of the system during the transient following the application of the clutch.

Since it is desirable from the viewpoint of stability that R_1, R_2, and R_3 in Eq. (22.105) be positive the following statements can be made:

1. The output torque of the prime mover should decrease with speed.
2. The torque-speed characteristic of the clutch should be a curve having positive slope.
3. The torque-speed characteristic of the load should be a curve having positive slope, but the value of the load torque at any particular speed within the range of operation should not exceed that of the static clutch torque, $M_c(0)$.

In most cases of practical importance statements 1 and 3 above are satisfied, but the case of a friction clutch exhibiting a torque-speed characteristic with positive slope is rare. In a friction clutch, the choice of friction materials, the preparation of surfaces, lubrication, temperature, and pressure on the frictional surfaces will all influence the torque-speed characteristic, which in some cases can be made, if not increasing with speed, at least of zero slope over most of the operating range.

If the transmission system is such that the vibrations of the drive line are particularly disturbing during the transient caused by the application of the clutch, some form of tuned and damped vibration absorber is necessary if the change in the torque-speed characteristic of the clutch is inadmissible. Drive-line vibration frequency caused by the action of a friction clutch usually falls within the lower end of the audio spectrum and may also have values low enough to be quite unpleasant to a human being. Examples of this are clutch "chatter"[24] in various forms of motor vehicles when starting from rest and the noise occurring during the shift from one geared ratio to the next in automatic transmissions.

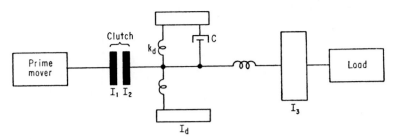

FIG. 22.23   Transmission system shown in Fig. 22.21 with damped vibration absorber added.

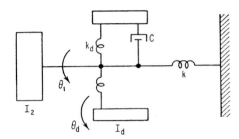

FIG. 22.24   Output end of clutch shown in Fig. 22.23.

In order to develop criteria for the design of a damped vibration absorber to be used with a friction clutch, consider a system shown in Fig. 22.23, which is essentially the same as that shown in Fig. 22.21, the difference being that another oscillating system consisting of a small flywheel has been flexibly attached to the clutch plate and a viscous damper connected between the clutch plate and the absorber inertia.

If, as is usually the case with mechanical power transmission, $I_3 \gg I_2$, the natural frequency of the system on the output side of the clutch is very closely given by $\sqrt{k/I_2}$. The output end of the clutch can thus be represented as in Fig. 22.24.

I_2 = moment of inertia of drive clutch plate(s)

c = viscous damping coefficient of the absorber

k_1 = spring constant of the main drive shaft

α = slope of the torque-speed characteristic of the clutch, equivalent dynamically to viscous damping coefficient

k_d = spring constant of the absorber

The total kinetic energy of the system is given by

$$T = \tfrac{1}{2} I_2(\dot{\theta}_2)^2 + \tfrac{1}{2} I_d(\dot{\theta}_d)^2 \tag{22.110}$$

The total potential energy of the system is given by

$$V = \tfrac{1}{2} k_1(\theta_2)^2 + \tfrac{1}{2} k_d(\theta_2 - \theta_d)^2 \tag{22.111}$$

The dissipation function is

$$F = \frac{1}{2}\,\alpha(\dot{\theta}_2)^2 + \frac{1}{2}\,c(\dot{\theta}_2 - \dot{\theta}_d)^2 \tag{22.112}$$

Substituting Eqs. (22.110) through (22.112) into Eq. (22.88) yields, in the absence of any external forces,

$$I_2\ddot{\theta}_1 + k_1\theta_1 + k_d(\theta_1 - \theta_d) + \alpha\dot{\theta}_1 + c(\dot{\theta}_1 - \dot{\theta}_d) = 0$$

$$I_d\ddot{\theta}_d - k_d(\theta_1 - \theta_d) - c(\dot{\theta}_1 - \dot{\theta}_d) = 0 \tag{22.113}$$

Equations (22.113) describe the motion of the system of Fig. 22.24.

Making α negative in Eqs. (22.113) and expanding the determinant associated with Eqs. (22.113), an equation in p is obtained:

$$I_1 I_d p^4 + [c(I_1 + I_d) - \alpha I_d]p^3 + [k_1 I_d + k_d(I_1 + I_d) - c\alpha]p^2 + (k_1 c - k_d\alpha)p + k_1 k_d = 0 \tag{22.114}$$

Applying Routh's criterion to Eq. (22.114) yields the following array of coefficients:

Row 1: $I_2 I_d$

Row 2: $c(I_2 + I_d) - I_d\alpha$

Row 3: $[c(I_2 + I_d) - I_d\alpha][k_2 I_2 + k_2(I_2 + I_d) - c\alpha] - I_2 I_d(k_1 c - k_2\alpha)$

Row 4: $[c(I_2 + I_d) - I_d\alpha][k_1 I_d + k_2(I_2 + I_d) - c\alpha](k_1 c - k_2\alpha)$
$$- [c(I_2 + I_d) - I_d\alpha]^2 k_1 k_2 - I_2 I_d(k_1 c - k_2\alpha)^2$$

Since $I_2 I_d > 0$, all the subsequent rows must be positive if the system is to be dynamically stable. This affords a means of obtaining a range of values for the viscous damping coefficient of the vibration absorber c which will result in stable operation.

In order for all the terms in the array to be positive, all the coefficients of all powers of p in Eq. (22.114) must be positive. This gives a very rough indication of the range of acceptable values of c.

$$c > I_d\alpha/(I_2 + I_d)$$

$$c > (k_d/k_1)\alpha \tag{22.115}$$

$$c < [k_1 I_d + k_d(I_2 + I_d)]\alpha$$

The vibration absorber must also be proportioned so that the natural frequency of the inertia I_d (the damper flywheel) is the same as that of I_2. This imposes a second condition

$$k_1/I_2 = k_d/I_d \tag{22.116}$$

In cases where I_3 in Fig. 22.23 is not very large in comparison with I_2, for the natural frequencies to be the same,

$$k_d/I_d = k_1 I_2 I_3/(I_2 + I_3) \tag{22.117}$$

Substituting Eq. (22.116) into Eq. (22.115) yields

$$c > I_d\alpha/(I_2 + I_d) \qquad c > (k_d/k_1)\alpha \qquad c < (2I_2 + I_d)k_d/\alpha \tag{22.118}$$

22.6.2 Simplified Approach to Dynamic Problems Involving Friction Clutches

In designing power-transmission systems containing friction clutches situations frequently arise in which it is required to calculate energy dissipation in the clutch during the engagement period or simply to compute the inertial torques acting on the system during a transition from one ratio to the other. These problems can be greatly simplified if the compliance is neglected.

This procedure is justified in cases where the natural frequencies of the system are relatively high and the resulting amplitude of oscillation small in comparison with the changes in velocities which take place as a consequence of the application of the clutch.

Consider a simple two-inertia system shown in Fig. 22.25. Such systems frequently form the basis of various forms of inertia starters.

It is convenient here to use velocities as coordinates rather than angular displacements, because the compliance is neglected.

The coefficient of friction will be assumed to be a constant, independent of speed. Assume also that no external torques act on the system.

In Fig. 22.25, the two flywheels initially rotate at two different angular velocities Ω_1 and Ω_2. Let the clutch be instantaneously applied at time $t = 0$ and let the torque exerted by it on the two flywheels be M_c = constant. This torque will act on the system only as long as there exists a difference in speeds between the two shafts, as shown in Fig. 22.26.

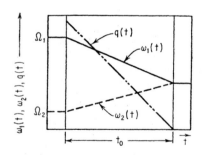

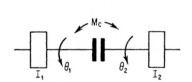

FIG. 22.25 Simple two-inertia system.

FIG. 22.26 Plot showing time t_0, which torque will be exerted, and the rate of energy dissipation.

Equations for speed, rate of heat dissipation, etc., are therefore valid for time $0 < t < t_0$, where t_0 is the time required to couple the two shafts.

The equations of motion for the two sides of the clutch are

$$I_1(d\omega_1/dt) = -M_c \qquad (22.119)$$

$$I_2(d\omega_2/dt) = M_c \qquad (22.120)$$

where $\omega_1 = \dot{\theta}_1$ and $\omega_2 = \dot{\theta}_2$, the angular velocities of I_1 and I_2, respectively. Integrating Eqs. (22.119) and (22.120) and applying the conditions at $t = 0$, $\omega_1(t) = \Omega_1$, and $\omega_2(t) = \Omega_2$ yields

$$\omega_1(t) = -(M_c/I_1)t + \Omega_1 \qquad (22.121)$$

$$\omega_2(t) = (M_c/I_2)t + \Omega_2 \qquad (22.122)$$

The relative velocity of I_1 with respect to I_2 is given by

$$\omega_1(t) = \omega_2(t) = \omega_r(t) = -M_c[(I_1 + I_2)/I_1 I_2]t + \Omega_1 - \Omega_2 \qquad (22.123)$$

The rate at which energy is dissipated in the clutch during the engagement period is given by $q(t) = T\omega_r(t)$.

$$q(t) = M_c\{\Omega_1 - \Omega_2 - M_c[(I_1 + I_2)/I_1 I_2]t\} \qquad \text{lb·in/s} \qquad (22.124)$$

The duration of the engagement period, found by considering that when $t = t_0$, $\omega_1(t) = \omega_2(t)$, is given by

$$t_0 = I_1 I_2 (\Omega_2 - \Omega_2)/M_c(I_1 + I_2) \qquad \text{s} \qquad (22.125)$$

The total energy dissipated is obtained by integration of Eq. (22.124):

$$Q = \int_0^{t_0} q(t)\, dt = \frac{I_1 I_2 (\Omega_1 - \Omega_2)^2}{2(I_1 + I_2)} \qquad \text{lb·in} \qquad (22.126)$$

The efficiency with which energy is transferred from I_1 to I_2 during the clutching operation is

$$\eta = 1 - Q/E_0 \qquad (22.127)$$

where E_0 = kinetic energy of the system at a time $t = 0$. Thus

$$\eta = (I_1 \Omega_2 + I_2 \Omega_2)^2 / (I_1 + I_2)(I_1 \Omega_1^2 + I_2 \Omega_2^2) \qquad (22.128)$$

If $\Omega_2 = 0$, this expression reduces to

$$\eta = I_1/(I_1 + I_2)$$

and if $I_1 = I_2 = I = \Omega_2 \neq 0$

$$\eta = (\Omega_1 + \Omega_2)^2 / 2(\Omega_1^2 + \Omega_2^2)$$

The case where the frictional clutch torque is a function of relative velocity of clutch plates is analyzed next, assuming negligible compliance and the dependence of friction on velocity represented by a piecewise linear function, shown in Fig. 22.27.

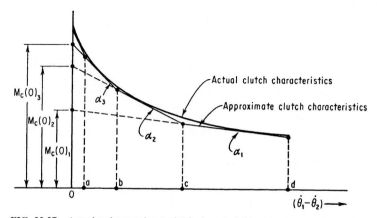

FIG. 22.27   Actual and approximate clutch characteristics.

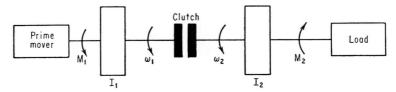

FIG. 22.28   Two-inertia transmission system.

For the system shown in Fig. 22.28, the equations of motion are

$$I_1\dot{\omega}_1 - \alpha_1\omega_1 + \alpha_1\omega_2 = M_1 - M_c(0)_1 \tag{22.129}$$

$$I_2\dot{\omega}_2 - \alpha_1\omega_2 + \alpha_1\omega_1 = M_c(0)_1 - M_2 \tag{22.130}$$

In Eqs. (22.129) and (22.130), α is the slope of the clutch frictional characteristic corresponding to the interval $c < (\omega_1 - \omega_2) < d$, and $M_c(0)$ is the point on the M_c axis obtained by intersection of the line of slope α_1 with the axis.

Solution of Eqs. (22.129) and (22.130) yields

$$\omega_1(t) = [(A_1\beta^2 + B_1\beta + C_1)\exp(\beta t) - \beta(C_1 t + B_1) - C_1]/I_1 I_2\beta^2 \tag{22.131}$$

$$\omega_2(t) = [(A_2\beta^2 + B_2\beta + C_1)\exp(\beta t) - \beta(C_1 t + B_2) - C_1]/I_1 I_2\beta^2 \tag{22.132}$$

where $A_1 = I_1 I_2\omega_1(0)$
$B_1 = I_2[M_1 - M_c(0)_1] - \alpha_1[I_1\omega_1(0) + I_2\omega_2(0)]$
$C_1 = M_2 - M_1$
$\beta = \alpha_1(I_1 + I_2)/I_1 I_2$ (22.133)
$A_2 = I_1 I_2\omega_2(0)$
$B_2 = I_1[M_c(0)_1 - M_2] - \alpha_1[I_1\omega_1(0) + I_2\omega_2(0)]$

In Eqs. (22.133) $\omega_1(0)$ and $\omega_2(0)$ are the values of the angular velocities $\omega_1(t)$ and $\omega_2(t)$ at $t = 0$, i.e., at point (d) in Fig. 22.27.

Equations (22.131) and (22.132) are used to calculate the time which must elapse for the two velocities to reach the value $\omega_1(t) - \omega_2(t) = \omega_r(c)$ (where ω_r denotes relative velocity), the point on the $\omega_1(t) - \omega_2(t)$ axis in Fig. 22.27 at which the slope of the clutch characteristic changes discontinuously. This value of time is given by

$$t_c = \frac{1}{\beta}\ln\frac{\omega_r(c)I_1 I_2\beta + (B_1 + B_2)}{(A_1 - A_2)\beta + (B_1 - B_2)} \tag{22.134}$$

The computation is carried out along the line of slope α_2 with α_2 substituted for α_1, $\omega_1(t_c)$ and $\omega_2(t_c)$ for $\omega_1(0)$ and $\omega_2(0)$, respectively, and $M_c(0)_2$ for $M_c(0)_1$ in Eqs. (22.133). The process is repeated at points (b), (a), and (0), each time using appropriate constants in Eqs. (22.133) and finding the time the system takes to arrive at the appropriate point on the $\omega = \omega_1 - \omega_2$ axis in Fig. 22.27.

The rate at which the energy is dissipated in the clutch when operating along the linear segment of its characteristic is given by

$$q_0(t) = \{M_c(0) - \alpha[\omega_1(t) - \omega_2(t)]\}\,[\omega_1(t) - \omega_2(t)] \tag{22.135}$$

Since the coefficient of friction is in this case a variable depending on the relative speed between clutch plates, it is of interest to examine the conditions under which engagement of the clutch can occur.

To this end, Eqs. (22.131) and (22.132) are differentiated, and by setting $t = 0$, expressions for accelerations of the two

$$\dot{\omega}_1(0) = \{M_1 - [M_2 - \alpha(\Omega_1 - \Omega_2)]\}/I_1 \tag{22.136}$$

$$\dot{\omega}_2(0) = (M_1 - M_2)/I_2 - \{M_1 - [M_2 - \alpha(\Omega_1 - \Omega_2)]\}/I_2 \tag{22.137}$$

Examining Eqs. (22.136) and (22.137), it can be noted that:

1. The most desirable condition for fast and positive engagement occurs when the following inequality is satisfied:

$$M_1 < \{M_c(0)_1 - \alpha_1[\omega_1(0) - \omega_2(0)]\} \tag{22.138}$$

2. When

$$M_1 = \{M_c(0)_1 - \alpha_1[\omega_1(0) - \omega_2(0)]\} \tag{22.139}$$

engagement will be slower but will be accomplished in a finite length of time, depending on the value of $M_1 - M_2$. $M_1 - M_2$ must be positive in order that engagement may be accomplished.

The rate at which $\omega_2(t)$ will begin to increase immediately after the clutch is applied is determined by the quantity $(M_1 - M_2)/I_2$.

The increase of speed of the output side of the clutch decreases the relative velocity of clutch plates, thus increasing clutch torque. Any increase in clutch torque above the value given by Eq. (22.139) will cause the deceleration of input shaft. Thus the effect of the initial increase in speed of the load side of clutch is cumulative and leads to positive engagement.

3. When the prime-mover output torque is greater than the clutch torque at the instant of clutch application, i.e., when

$$M_1 > \{M_c(0)_1 - \alpha_1[\omega_1(0) - \omega_2(0)]\} \tag{22.140}$$

engagement is possible only when the following inequality is satisfied:

$$(M_1 - \{M_c(0)_1 - \alpha_1[\omega_1(0) - \omega_2(0)]\})/I_1$$
$$< (\{M_c(0)_1 - \alpha_1[\omega_1(0) - \omega_2(0)]\} - M_2)/I_2 \tag{22.141}$$

Nevertheless, the analysis provides a basis for continued study of friction clutches and associated control components pertinent to any specific design through the use of modern computing equipment which makes it possible to include in the analysis the dependence of friction on speed, temperature, etc., while simultaneously greatly reducing the time required for computation.

22.7 ELECTROMAGNETIC FRICTION CLUTCHES

A method widely used for controlling friction clutches depends on the electromagnetic force of attraction between two magnetized iron parts. The axial clamping force on the friction disks is thus a function of exciting current, and current may be made time-dependent, thus controlling the torque transmitted through the clutch. Electromagnetic clutches are frequently used in applications requiring close control of acceleration of rotating masses because the exciting current is easily and accurately controlled.

Electromagnetic clutches can be operated dry or in oil and can be of single- and multiple-disk design. The clutch may be so designed that the magnetic flux either (1) passes through the disk stack or (2) does not pass through the stack. The latter case provides more flexibility in design, since the disk stack can then be formed and friction materials selected without regard to the geometry of the magnetic circuit and the magnetic properties of the disk materials. Figure 22.29 shows one example of such a clutch.

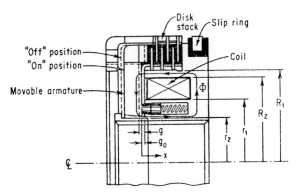

FIG. 22.29   Electromagnetic clutch.

In the following analysis, it will be assumed that effects of saturation, leakage flux, flux fringing, eddy currents, and hysteresis are all negligible, and that the ampere-turns required to maintain the flux in the iron parts of the magnetic circuit are very small compared with the ampere-turns required to maintain the field across the air gap. Let

α = mechanical friction coefficient (armature parts)

d = value of x at which spring force equals zero

E = voltage applied to the coil

N = number of turns in the coil

i = coil current, A

μ = magnetic permeability of air (rationalized mks)

Φ = magnetic flux, Wb

λ = flux linkages, Wb·turn

L = inductance of the coil, H

R = resistance of the coil, Ω

x = displacement of armature, m

g = maximum air-gap length, m

g_0 = minimum air-gap length, m

m = mass of the armature, kg

k = return spring elastance, N/m

R_1, R_2, r_1, r_2 are defined in Fig. 22.29.
Under the assumptions made above (using rationalized mks units),

$$\Phi = \mu A_1 A_2 Ni/(g - x)(A_1 + A_2) \qquad \text{Wb} \qquad (22.142)$$

where $A_1 = \pi(R_1^2 - R_2^2)$ and $A_2 = \pi(r_1^2 - r_2^2)$. Also

$$\lambda = N\Phi \qquad (22.143)$$

The self-inductance of the coil can be defined as

$$L = L_0/(g - x) \qquad H \tag{22.144}$$

where

$$L_0 = \mu A_1 A_2 N^2/(A_1 + A_2) \tag{22.145}$$

The magnetic energy stored in the field existing in the air gap is given by

$$W_m = \tfrac{1}{2} L i^2 \qquad J \tag{22.146}$$

From Eq. (22.146) the electromagnetic force acting on the armature is

$$F_m = \partial W_m/\partial x = -[L_0 i^2/2(g - x)^2] \qquad N \tag{22.147}$$

Equation (22.147) shows that the force, and also the torque transmitted by the clutch, are very sensitive to change in the coil current and to the difference $(g - x)$. Whereas in the case of the hydraulically actuated clutch the position of the actuating piston relative to the main body of the clutch is immaterial when computing torque transmitted, the minimum distance of the armature from the body of the electromagnetic clutch is an important parameter.

The applied coil voltage is given by

$$E = Ri + d\lambda/dt \tag{22.148}$$

However, λ is a function of x and i, and both x and i are functions of time. Therefore,

$$E = R_i + (dL/dx)(dx/dt) + L(di/dt) \tag{22.149}$$

Summing all forces acting on the armature and performing the differentiation indicated by Eq. (22.149) yields

$$M(d^2x/dt^2) + \alpha(dx/dt) - k(d + x) + L_0 i^2/2(g - x)^2 = 0 \tag{22.150}$$

$$- [L_0 i/(g - x)^2](dx/dt) + [L_0/(g - x)](di/dt) + R_i = E \tag{22.151}$$

Equations (22.150) and (22.151) describing dynamic behavior of the clutch system are nonlinear and cannot be solved by employing elementary methods. However, they provide a reasonable basis for studying the response of the system to the control voltage E using an analog computer, in which case the nonlinearity of the magnetic circuit (saturation and hysteresis) can also be taken into account when formulating system equations.

It is difficult to calculate complete response of the system from the time the voltage E is applied across the coil, but when it is realized that the clutch begins to transmit torque only after the armature has been attracted and the clamping force applied to the disk stack, Eqs. (22.150) and (22.151) can be simplified considerably and can be used to examine the factors affecting the torque capacity of the clutch.

After the motion of the armature has ceased $d^2x/dt^2 = dx/dt = 0$ and $g - x = g_0$ and Eqs. (22.150) and (22.151) become, respectively,

$$-k(d + g - g_0) + L_0 i^2/2(g_0)^2 = F_c \tag{22.152}$$

$$L(di/dt) + Ri = E \tag{22.153}$$

where $L = L_0/g_0$ and F_c = net axial clamping force.
Solution of Eq. (22.153) yields

$$i(t) = E/R - (E/R - I_0) \exp[-(R/L)t] \tag{22.154}$$

where I_0 is the value of i at the time $t = 0$, i.e., at the time when $x = g - g_0$ and armature motion ceased.

Substituting for $i(t)$ from Eq. (22.154) into (22.152) the response of the axial clamping force to the sudden application of control voltage E is obtained:

$$F_c = [L_0/2(g_0)^2]\{(E/R)^2 - 2(E/R)[(E/R) - I_0] \exp[-(R/L)t]$$

$$+ [(E/R) - I_0]^2 \exp[(2R/L)t]\} - K(d + g - g_0) \quad (22.155)$$

For sufficiently large values of t, the steady-state value of the clamping force on the disk stack is given by

$$F_c = [L_0/2(g_0)^2](E/R) - K(d + g - g_0) \quad (22.156)$$

To disengage the clutch, the control voltage E is reduced to zero, and a similar procedure leads to the expression for F_c

$$F_c = \frac{L_0}{2(g_0)^2} \left(\frac{E}{R + r}\right)^2 \exp\left[-\frac{2(R + r)}{L} t\right] - K(d + g - g_0) \quad (22.157)$$

Equation (22.157) shows how the clamping force F_c, acting on the disk stack, decreases with time following disappearance of the control voltage E at a time $t = 0$. In this equation, r is any external resistance through which the coil discharges and which may be added to decrease the disengagement time by decreasing the time constant of the system.

If the coefficient of friction of the disk surfaces is a reasonably invariant quantity, Eqs. (22.155), (22.156), and (22.157) also show how the clutch torque varies with time. In practice, these equations may represent clutch torque variation with time to a very good approximation in the case of dry clutches, but when dealing with oil-lubricated clutches additional factors must be considered which pertain to the mechanical-design details of friction disks.

Frictional torque characteristics of magnetic clutches have been studied experimentally and the results of these investigations are discussed in Ref. 25.

The value of frictional torque developed by the clutch and its variation with time depends not only on the magnetizing current but also on the kind of disks and disk material used. Similarly, the decrease of the clutch torque besides being a function of time such as indicated by Eq. (22.157) also depends on adhesion between the disks, which in turn is a function of disk design and material.

REFERENCES

1. Harting, G. R.: "Design and Application of Heavy-Duty Clutches," *SAE Trans.*, vol. 72, pp. 405–432, 1964.

2. Steinhagen, H. G.: "The Plate Clutch," SAE Paper 800978, 1980.

3. Crolla, D. A.: "Friction Materials for Overload Clutches," *Tribology Int.*, vol. 12, no. 4, pp. 155–160, Aug. 1979.

4. Chestney, A. A. W., and D. A. Crolla: "The Effect of Grooves on the Performance of Frictional Materials for Overload Clutches," *Wear*, vol. 53, no. 1, pp. 143–163, 1979.

5. Harting, G. R.: "The Why, Status, and Future of Wet Clutches for Trucks," SAE Paper 700875, 1970.

6. Cummins, G. F.: "A New Wet Clutch Fan Drive System," ASME Paper 74-DE-12, 1974.

7. Mitchell, R. K., T. R. Kosmatka, and O. Braunschweig: "Clutch for Lawn Mower Engine Auxiliary Shaft," SAE Paper 800039, 1980.

8. Hermanns, M. J.: "Heavy-Duty, Long-Life, Dry Clutches," SAE Paper 700874, 1970.

9. Lloyd, F. A., and M. A. Dipino: "Advances in Wet Friction Materials—75 Years of Progress," SAE Paper 800977, 1980.

10. Fish, R. L., and F. A. Lloyd: "Surface Finish Requirements of Spacer Plates for Paper Friction Applications," SAE Preprint 730840, 1973.

11. Fish, R. L.: "Wet Friction Material—Some Modes of Failure and Methods of Correction," SAE Paper 760664, 1976.

12. Anderson, A. E.: "Friction and Wear of Paper Type Wet Friction Elements," SAE Paper 720521, 1972.

13. Morse, I. E., and R. T. Hinkle: "Taking Guesswork out of Disk Clutch Design," *Mach. Des.,* vol. 47, no. 28, pp. 64–65, Nov. 27, 1965.

14. Jakob, M.: "Heat Transfer," vols. 1 and 2, John Wiley & Sons, Inc., New York, 1949, 1957.

15. Blok, H.: "Measurement of Temperatures Flashes on Gear Teeth under Extreme Pressure Conditions," *Proc. Gen. Disc. Lubrication, Inst. Mech. Eng. (London),* vol. 2, pp. 14–20, 1937.

16. Barber, J. R.: "Thermoelastic Instabilities in the Sliding of Conforming Solids," *Proc. Roy. Soc. London,* ser. A, vol. 312, pp. 381–394, 1969.

17. Dow, T. A.: "Thermoelastic Effects in a Thin Sliding Seal—A Review," *Wear,* vol. 59, pp. 31–52, 1980.

18. Gazley, C.: "Heat Transfer Characteristics of the Rotational and Axial Flow between Coaxial Cylinders," ASME Paper 56-A-128, 1958.

19. Kays, W. M., and I. S. Bjorklund: "Heat Transfer from a Rotating Cylinder with and without Crossflow," ASME Paper 56-A-71, 1957.

20. Andronow, A. A., and C. E. Chaikin: "Theory of Oscillations," Princeton University Press, Princeton, N.J., 1949.

21. MacLachlan, N. W.: "Ordinary Nonlinear Differential Equations," 2d ed., Oxford University Press, New York, 1956.

22. Timoshenko, S.: "Vibration Problems in Engineering," 3d ed., D. Van Nostrand Co., Inc., Princeton, N.J., 1955.

23. Bunda, T., et al.: "Friction Behavior of Clutch-Facing Materials: Friction Characteristics in the Low-Velocity Slippage," SAE Paper 720522, 1972.

24. Jarvis, R. P., and R. M. Oldershaw: "Clutch Chatter in Automobile Drivelines," *Proc. Inst. Mech. Eng. (London),* vol. 187, no. 27, pp. 369–379, 1973.

25. Nitsche, C.: "Electromagnetic Multi-Disk Clutches," ASME Paper 58-5A-61, 1957.

SECTION 23
FRICTION BRAKES

Thomas A. Dow, Ph.D.

Professor of Mechanical and Aerospace Engineering
North Carolina State University
Raleigh, N.C.

23.1 FRICTION MATERIALS

23.1.1 Types and Physical Properties

The logical selection of a brake material for a specific application requires a knowledge of friction material formulation. The "formulation of a brake lining" is defined as the specific mixture of materials and the sequence of production steps which together determine the characteristics of the brake lining. There are three basic types of linings[1,2] currently in use.

Organic Linings.   Organic linings are generally compounded of six basic ingredients.

1. *Asbestos:* for heat resistance and high coefficient of friction

2. *Friction modifiers:*[3] for example, the oil of cashew nut shells to give desired friction qualities

3. *Fillers:* for example, rubber chips to control noise

4. *Curing agents:* to produce the desired chemical reactions during manufacture

5. *Other materials:* for example, powdered lead, brass chips, and aluminum powders for improved overall braking performance

6. *Binders:* phenolic resins for holding the ingredients together

Asbestos has characteristics that make it well suited for friction applications: thermal stability, adequate wear resistance, and strength. For these reasons it has found universal acceptance as the basic ingredient in brake linings. However, because occupational exposure to asbestos during lining manufacture has been shown to be a health hazard, the lining industry is looking for replacements. One possibility[4] is Kevlar aramid fiber which has been shown to boost the performance of nonasbestos friction materials to that currently available for asbestos materials.

Semimetallic Linings.[5]   These linings substitute iron, steel, and graphite for part of the asbestos and organic components of an organic lining. The reasons for this substitution are

1. Improved frictional stability and high-temperature performing (fade resistance)
2. Excellent compatibility with rotor and high-temperature wear resistance at temperatures greater than 450°F (230°C)
3. High performance with minimal noise

Such linings were first released for police cars and taxi cabs and their use has spread to some original-equipment passenger cars.

Metallic Linings.   These linings have received attention for special application involving large heat dissipation and high temperatures. Sintered metallic–ceramic friction materials[6] have been successfully applied to jet aircraft brakes,[7] heavy-duty clutch facings, and race and police car brakes.

Two methods are in use for making brake linings—weaving and molding. Both types are basically asbestos with binders to hold the asbestos fibers together. For the woven-type material, asbestos thread and fine copper wire are woven together with the other components, cured, and cut to the appropriate shape. For the organic molded type, asbestos fibers and other compounds are ground and pressed into the final shape. Molded linings are more commonly used.

Molded organic linings may be formed from a dry mix or a wet mix. The dry mix consists of asbestos compounds, filler materials, and powdered resins. These ingredients are mixed, preformed to shape, and heated under pressure to form a hard slatelike board. This board is then cut and drilled for specific pad shapes. The wet mix, on the other hand, consists of asbestos compounds, organic fillers, and liquid resins. High-pressure extruding, screw extruding, calendering, and other processes are used to form the brake pad from this mix. In general, the softer the friction material, the greater the friction coefficient and the shorter the life of the lining. The hard ceramic and metallic linings have comparatively low friction coefficients but have high wear resistance and can survive in adverse operating conditions, for example, at high temperature. Because many linings are riveted to the backing plate, the material must have sufficient strength[8] to remain firmly attached to the backing plate under panic stop conditions. Properties[9,10] of different friction materials are shown in Table 23.1.

Composition brake blocks have a number of ingredients, and quality control[11] of the final product is a continuing problem. Quality control typically consists of 100 percent inspection of the pads for electrical conductivity (shows whether the binder is fully cured), metal content (checks for the presence of metal impurities), and visual imperfections (determines the overall integrity of the pads). Partial inspection is also carried out for specific gravity and hardness.

23.1.2 Brake Friction

The general concepts of friction have been developed over many years.[12–14] As applied to a composition brake lining on a steel or cast-iron brake surface, the main sources of friction are[15,16]

TABLE 23.1   Properties of Friction Materials

Type	Woven lining (flexible)	Molded lining (sheeter molded)	Molded block (rigid molded)
Compressive strength, lb/in²	10,000–15,000	10,000–18,000	10,000–15,000
Shear strength, lb/in²	3500–5000	8000–10,000	4500–5500
Tensile strength, lb/in²	2500–3000	4000–5000	3000–4000
Modulus of rupture, lb/in²	4000–5000	5000–7000	5000–7000
Hardness, Brinell	*	*	20–28
Specific gravity	2.0–2.5†	1.7–2.0	2.1–2.5
Thermal expansion, in/(in.°F) (mean to 350°F)	*	1.5×10^{-5}	1.3×10^{-5}
Thermal conductivity, Btu·in/(h·ft²·°F) (125°F mean)	5.5	3	3.5–4
Principal type of service	Industrial	Automotive (passenger)	Automotive (truck and bus) and railway
Specific heat			
Maximum continuous operating temperature, °F	400–500	500	750
Maximum rubbing speed, ft/min	7500	5000	7500
Maximum continuous operating pressure, lb/in²	50–100	100	150
Average friction coefficient (350°F, 50 lb/in², 600 ft/min)	0.45	0.47	0.40–0.45

*Nonhomogeneous.
†High wire-mesh content.

1. *Adhesion:*[17] As the lining material moves over the drum or rotor surface, the metallic constituents in the lining weld to the rotor and drum material. Shearing of these junctions produces a frictional force.

2. *Shear deformation:* The coefficient of friction increasing with increasing temperature suggests that deformation effects play an important role since the resin softens with increasing temperature. It is believed that the deformation effect is due to the formation of wavelike surface deformation[18] rather than elastic hysteresis loss.

3. *Plowing:* In the process of tangential motion of the surfaces, the roughness protrusions on the disk-drum interlock with particles of ingredients and displace them. When the ultimate strength is exceeded, the polymer structure is disrupted and particles of ingredients are torn away. Long asbestos or steel fibers supply the mechanical strength to prevent excessive loss of material during plowing.

4. *Hysteresis:* The energy losses involved with elastically stressing and unstressing the brake linings produce a small source of brake friction.

5. *Surface films:* Contamination of the surface from decomposed lining materials or fluid-borne contaminants will greatly affect the friction coefficient by reducing adhesion and shear deformation.

The importance of each component of friction discussed[19] above will vary over the life of the lining. The initial operation of the system may involve high plowing losses due to the high original surface roughness. As the roughness is reduced due to wear, the positive effect of increased adhesive bonds will become more important as will the adverse effects of surface-contamination films.

The friction coefficient[20] for a brake material rubbing against cast iron is a function of load, speed, and bulk temperature. A general expression for the friction force F can be written as

$$F = K(T)\, P^{a(T)}\, V^{b(T)} \qquad (23.1)$$

where $K(T)$ = constant, dependent upon temperature
     P = normal load
    $a(T)$ = exponent of load dependent upon temperature (varies between $+0.8$ and $+1.25$)
     V = sliding speed
    $b(T)$ = exponent of speed dependent upon temperature (varies between -0.25 and $+0.25$)

The influence of load, speed, and temperature for one asbestos friction material is shown in Figs. 23.1 and 23.2. Figure 23.1 shows the influence of bulk temperature and normal load on the friction coefficient. Increasing the load causes a slight decrease in friction coefficient (except at low loads and 600°F). As a function of temperature, this friction material shows a peak in friction coefficient at a temperature between 400 and 500°F. At a 100-lb normal load (200 lb/in^2), the friction coefficient varies from 0.25 to 0.4, depending on the lining temperature. An increase in sliding speed results in a decrease in friction coefficient, as shown in Fig. 23.2.

Because of the strong influence of surface temperature on the friction coefficient, considerable care must be exercised in looking at independent influences of speed and load on friction. For the experiments described above, a small specimen (0.7 in × 0.7 in) of brake lining was loaded against the inside of a 11-in-diameter cast-iron drum (SAE standard J661). The temperature was measured by a thermocouple embedded in the drum slightly below its surface. The experimental conditions were such that heat was added to the system to achieve the test temperatures, and the thermocouple was used to control the heating system. This situation was selected to minimize the influence of frictionally generated heat on the surface temperature. However, such experimental conditions may not guarantee identical surface temperatures for different speeds and loads. Since the frictional heat is generated at the surface, even though it is a small part of the total heat input, the frictional heating (due to increased speed or load) may produce a different temperature profile on the lining and be responsible for part of the decrease in the friction coefficient measured in Figs. 23.1 and 23.2.

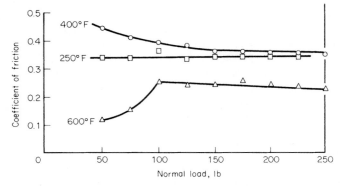

FIG. 23.1 Friction coefficient as function of normal load for asbestos friction material, 300 r/min speed.[20]

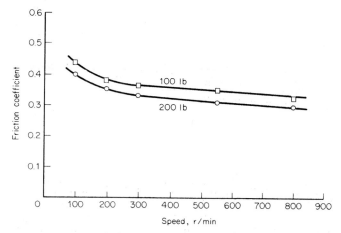

FIG. 23.2 Friction coefficient as function of speed for asbestos friction material, 350°F drum temperature.[20]

Because of these complexities, the determination of the coefficients of Eq. (23.1) must be corroborated by experimental measurements.

23.1.3 Brake Wear

The wear[21-24] of composition brake materials when pressed against a steel or cast-iron rotating surface is a complex mixture of adhesive and oxidative wear and surface fatigue. At low temperatures (below 450°F) the controlling factor for wear is adhesive wear and surface fatigue. However, above 450°F, the dominant wear mechanisms are thermal degradation of the resin binder[25] (phenolic) and mechanical and thermal changes in the asbestos reinforcing fibers.[26]

One effect which makes wear prediction difficult is that during braking, the friction material rubs against a reaction product[27] considerably different from that of the virgin cast iron. The reaction product is a layer on the rotor surface from 50 to 250 μin thick which influences both the friction and wear behavior of the brake. Wear data have been compared for stainless-steel and quartz rotors, with each of two different surface finishes, and the results indicated that surface finish and conductivity of the rotor were more important than the material in determining the transfer film and therefore the friction and wear behavior.

Wear models[28-31] have been developed as a result of extensive dynomometer and field experiments. The results of these experiments show that the wear of organic and semimetallic friction materials can be modeled by a modified Archard wear law, separated into two temperature regimes. At temperatures less than 450°F,

$$\text{Wear} = P^\alpha \, V^\beta \, t \tag{23.2}$$

At temperatures greater than 450°F,

$$\text{Wear} = P^\alpha \, V^\beta \, t e^{E/RT} \tag{23.3}$$

where W = weight loss from lining material
$\quad\quad\;\; P$ = load
$\quad\quad \alpha, \beta$ = material constants
$\quad\quad\;\; V$ = sliding speed
$\quad\quad\;\; t$ = time
$\quad\quad\;\; E$ = activation energy, Btu/mol
$\quad\quad\;\; R$ = universal gas constant = 1.986 Btu/(mol·°R)
$\quad\quad\;\; T$ = absolute temperature, °R = °F + 460

Values of α, ß, and E for selected brake materials[29] are shown in the following table.

Lining material	α	ß	E, Btu/mol
Commercial lining	0.42		38.1
Secondary lining*	0.63	1.45	30.6
Disk pad*	0.28		15.9

*High-temperature materials.

The reason for the two ranges of temperature has to do with the modes of wear which take place under these conditions. At low temperatures (below 450°F) the wear is a result of plowing, cutting, three-body abrasive wear, and adhesive wear. For the high-temperature region, the wear is controlled by thermal processes. As a result the wear of friction materials increases exponentially above some critical temperature. For most of the current composition materials held together with phenolic resins, the critical temperature is approximately 450°F.

The exponential term in Eq. (23.3) includes the influence of pyrolysis by using the activation energy of the brake material and the absolute temperature of the brake surface. Using Eqs. (23.2) and (23.3), the wear of different operating conditions can be compared. For example, changing the operating temperature of secondary lining shown in Table 23.1 from 450 to 600°F will increase the wear by a factor of 3. However, the absolute value of the wear rate cannot be accurately predicted without experimental corroboration.

23.2   VEHICLE BRAKING

A friction brake turns the kinetic energy of the vehicle into heat. But because of the design of most vehicles, this heat is not equally distributed to the wheels. The heat dissipated at each brake will be a function of the static and dynamic weight distribution on the wheels and the design of the brake system. The dynamic loading will depend upon the design of the vehicle (the static weight distribution, the height of the center of gravity, and the wheelbase) and the deceleration. The sum of the forces during braking shows that the deceleration of the vehicle in percentages of the acceleration of gravity g must be less than or equal to the friction coefficient between tire and road. This friction coefficient[32] will depend upon the tire size and construction, the road surface, and the relative slip between tire and road.

If the weight is uniformly distributed right to left, the front and rear wheel loading (L_F and L_R) can be written as

$$L_F = W(F + \mu h/d)$$
$$L_R = W(R - \mu h/d)$$

$$(23.4)$$

where F = static front wheel loading = d_R/d
     R = static rear wheel loading = d_F/d
     d = wheel base
     d_F = distance from center of gravity to front wheel
     d_R = distance from center of gravity to rear wheel
     μ = friction coefficient, equivalent to deceleration expressed as a multiple of g
     h = vertical distance from road to center of gravity

This expression can be used to estimate the change in wheel loading due to friction forces at the road during a stop. For short, high vehicles significant weight transfer will occur, while for long, low vehicles, smaller percentages of weight will be transferred. For example, a typical front-wheel-drive sedan of 2500 lb will have a static weight distribution of 60/40 front to rear. During a 0.8g stop, this distribution will change to about 80/20.

The balance[33] of braking between front and rear is an important design consideration. The brake system could be designed so that the front brakes produced 4 times the torque of the rear; and for the 0.8g stop described above, maximum tire friction at both ends of the car could be utilized. However, if under wet conditions, the friction coefficient between tire and road were reduced to 0.2, the dynamic weight distribution would be 65/35. Under these conditions, the brake system balanced 80 percent front to 20 percent rear would cause the front wheels to slide. If the brake system were balanced for this lower-deceleration dynamic weight distribution, the rear wheels would slide first during maximum deceleration for dry conditions.

To make a decision as to the design of the brake balance, the influence of wheel slide on vehicle control must be considered. The control[34,35] of a vehicle is related to wheel slide in the following way: Locking only the rear wheels produces a condition which results in partial or complete loss of control of the vehicle. Depending upon the vehicle's characteristics, this situation could lead to a spin. Locking only the front wheel produces a stable condition where the vehicle travels in a straight line but nearly all steering control is lost. The result is that for most vehicles it is best to balance the brake system so that the front wheels lock first.

To improve vehicle control during braking, antiskid braking systems have been developed. These systems measure the relative wheel and vehicle speed and modulate the brake pressure to keep each wheel (or each set of wheels) at the limit of adhesion[36,37] without sliding. The maximum friction coefficient for the tire on the road occurs at a low percentage of slip, which is nearly rolling conditions, rather than gross sliding which would occur if the brake were locked. Thus an antiskid system can be designed to produce maximum brake torque, while still retaining full steering control, a situation which would otherwise only exist for the most experienced driver.

23.3 BRAKE TEMPERATURE

The actual temperature at the lining-rotor interface plays a key role in the friction and wear associated with a brake material.[40–42] It is at this interface that the frictional heat is generated and the highest temperatures occur. The surface temperature of the lining material determines (1) the mode of wear and (2) the film present on the surface which influences the friction coefficient (see Sec. 23.1). The equilibrium temperature of the brake will be influenced by the heat input (proportional to the vehicle weight, initial speed, and stopping frequency) and the magnitude of the heat dissipation from the brake. The heat is lost through conduction to the brake assembly as well as convection and radiation to the surroundings.

23.3.1   Heat Input

The instantaneous heat input to the brakes q is equal to the change in kinetic energy of the vehicle:

$$q = \Delta KE = \frac{\partial}{\partial t} KE = \frac{\partial}{\partial t} \left(\frac{1}{2} m v^2 \right) \qquad (23.5)$$

where q = rate of heat input to brakes, Btu/s
    KE = kinetic energy of vehicle, Btu
      m = mass of vehicle = weight $\div$ 32.2 ft/s^2
      v = instantaneous speed of vehicle, ft/s

If we assume that the deceleration of the vehicle is uniform, the change in speed will be a linear function of time, i.e.,

$$v(t) = v(0)(1 - t/t_s) \qquad (23.6)$$

where t_s = stopping time = $\sqrt{2D/ag}$
    D = stopping distance, ft
    a = deceleration rate of vehicle, percent of g
    g = acceleration of gravity, ft/s^2
  $v(0)$ = initial speed of vehicle, ft/s

The heat generation rate can then be written as

$$q = mv(t)\, dv(t)/dt = magv(t) \qquad (23.7)$$

$$= magv(0)(1 - t/t_s) \qquad (23.8)$$

The total heat generated over the stop will be the sum of the instantaneous heat rates or the total change in kinetic energy of the vehicle, which for a stop to rest will be its initial kinetic energy

$$Q = \tfrac{1}{2}\, mv(0)^2 \qquad \text{Btu} \qquad (23.9)$$

For the vehicle discussed, the instantaneous heat input will be a maximum at the beginning of the stop and will have a value of $q(0) = 212$ Btu/s and will linearly decrease over the stopping time. The total heat generated over the stop will be $Q = 390$ Btu.

The design of the brake system will determine the percentage of the total heat generated which will be dissipated at each wheel. If the system is balanced 60/40 front to rear, then the heat input to one front brake will be 30 percent of the total, or 117 Btu.

23.3.2   Average Brake Temperature

The change in the average brake temperature for a single stop can be calculated from

$$\Delta T = Q/M_B c \qquad \text{°F} \qquad (23.10)$$

where Q = heat input to brake, Btu
    M_B = mass of brake, lb
     c = specific heat of brake rotor = 0.1 Btu/(lb·°F) for cast iron

For a single stop from ambient conditions, the bulk of the heat will be absorbed by the rotor or drum, since it has much higher conductivity than the lining material. For a 20-lb cast-iron front disk brake, the average temperature rise for the single stop described previously would be about 60°F. The temperature of the brake after repeated stops will depend upon how much of the heat generated is lost due to conduction, convection, and radiation. Another significant factor will be the residual brake torque. While in itself this torque will not generate high temperatures, it will reduce the heat losses from the brake and greatly change the equilibrium temperature after multiple stops.

23.3.3 Brake Surface Temperature

The surface temperature of the disk or drum and the lining will be significantly higher than the average brake temperature. Peak surface temperatures on disk brakes have been measured in the range of 1200 to 1400°F. It is this temperature which influences the wear of the lining material since the wear particles are generated at the surface. Other effects of high brake surface temperatures are distortion of the drum or disk or surface cracking in these members as a result of plastic deformation from high thermal stresses. However, surface temperature is difficult to predict because of the nature of the contact. The rotor surface and the lining actually make contact at small localized regions (asperities) and the peak temperatures for these contacts are quite high. Some operating conditions result in larger localized "hot spots" or "fire-banding"[43,44] which also produce high localized surface temperatures. Therefore, the problem of predicting the surface temperature becomes one of determining the size and duration of the contacts.

For a linearly decreasing heat input into the surface of a semi-infinite body, the surface temperature rise can be estimated from the expression[45]

$$T_s(t) = [2\,q(0)\,\sqrt{t/A\sqrt{\pi K \rho c}}](1 - 2t/3t_s) \qquad (23.11)$$

where $q(0)$ = initial heat input, Btu/s
  t = time, s
  t_s = total stopping time, s
  K = thermal conductivity of body, Btu/(in·s·°F)
  ρ = density of body, lb/in^3
  c = specific heat of body, Btu/(lb·°F)
  A = surface area, in^2

The maximum value of the temperature occurs after half of the stopping time and can be written as

$$T_{max} = [0.53q(0)/A]\,\sqrt{t_s/K\rho c} \qquad (23.12)$$

The difficulty in applying the above expressions comes in deciding upon an appropriate value of the area over which it acts. The percentage of the heat that flows into the disk or into the pad varies with the operating conditions. Therefore, the decision cannot be made arbitrarily, but requires a complete analysis of the problem.[46–48] However, several approximations can be made which illustrate the range of surface temperatures possible.

Average Surface Temperature.   One estimate of the surface temperature can be made by assuming that the heat uniformly enters the entire swept area of the disk. For the 2500-lb vehicle previously discussed, with ventilated cast-iron disks, the values for Eq. (23.11) will be

$q(0)$ = heat input rate per brake surface
   = 59 Btu/s
t_s = 3.5 s
K = 6.9×10^{-4} Btu/(in·s°F)

ρ = 0.26 lb/in³
c = 0.1 Btu/(lb·°F)
A = 34 in²

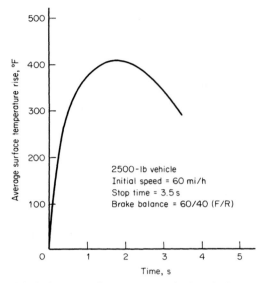

FIG. 23.3 Disk surface-temperature rise for a single stop.

The surface temperature as a function of time is plotted in Fig. 23.3, and the maximum temperature from Eq. (23.12) is 407°F. This temperature is significantly hotter than the bulk temperature rise of the disk calculated from Eq. (23.10) for the same stop.

Peak Surface Temperature.   Real surfaces contact at small isolated regions defined by plastic deformation of the softer material. For brakes, the softer material is the brake lining and its hardness is a function of temperature. Studies[30] of debris generated during braking experiments showed it to be generated at near 900°F. The Meyer's hardness associated with this temperature is 5700 lb/in². Therefore, real area of contact, A_R, between lining material and disk at this temperature can be written as

$$A_R = L/H_M \qquad (23.13)$$

where L = load on brake pad and H_M = Meyer's hardness.
   Typical brake loading for a current 2500-lb automobile will be 5000 lb over a pad area of 12 in². From Eq. (23.13), the real area of contact will be 0.88 in² or about 7 percent of the total pad area. This real area will be divided into small contact regions and part of the heat will enter the disk through each of these regions. If we assume that all of these regions are collected into a circular region of area A_R at the center of the pad, the real area of the brake surface, A_{SR}, swept by the contact in one revolution can be written as

$$A_{SR} = d_A(4\pi A_R)^{1/2} \qquad (23.14)$$

where d_A = average diameter of disk or drum surface.

When the real swept area from Eq. (23.14) is substituted to find the maximum surface temperature from Eq. (23.12), the result is a surface rise temperature of 760°F. If
this temperature rise is added to the ambient conditions (100°F), the temperature will
be similar to that expected by the structure of the wear debris.

Equilibrium Temperature.   The heat dissipated during a stop is transferred to the
disk and then lost to the surroundings by conduction, convection, and radiation. The
heat loss is a function of disk temperature so that there will be an equilibrium temperature for intermittent braking where the energy gained during a single brake application will be lost before the next application. This equilibrium temperature will be
influenced by the weight of the vehicle, the deceleration, the brake design and balance, and the time between stops. The loss of heat from a disk brake due to convection
is significantly improved by the use of a ventilated rotor. Figure 23.4 shows the
results[49] of a series of brake applications at a constant disk speed. The ventilated rotor
reached an equilibrium temperature of 700°F after eight cycles, whereas the solid disk
of the same diameter and mass reached a significantly higher equilibrium temperature
(1100°F) after 20 cycles. In the case of the solid rotor, more of the heat energy was
retained during each cooling cycle, and subsequent brake applications produced a
higher rotor equilibrium temperature. Figure 23.5 shows the definite advantage in
cooling rate of the ventilated disk. This advantage is due to the increased surface and
of the ventilated disk as well as the air-pumping properties of that design. The convection heat transfer coefficient[46] for a ventilated rotor is shown in Fig. 23.6. These values can be used to estimate the heat lost to convection from the rotor. Radiation and
conduction losses must be taken into account.

Brake Fade.   Brake fade has been the main contributor to the nearly total replacement of drum brakes with disk brakes on the front wheels of new cars. It is caused by
two factors. The first cause is the drop in the coefficient of friction at high temperatures. The friction coefficient for an asbestos-based friction material is shown in Fig.

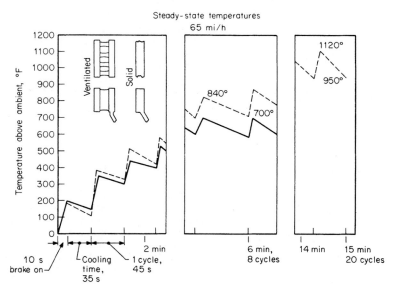

FIG. 23.4   Stabilization temperature curves for solid and ventilated rotors of equal mass
and diameter.[49]

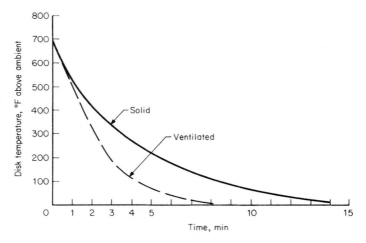

FIG. 23.5 Cooling temperature curves for solid and ventilated rotors of equal mass and diameter.[49] Disks have common initial temperatures. Speed is 65 mi/h.

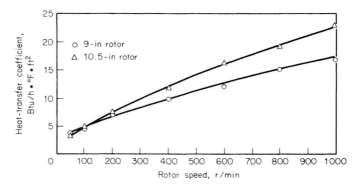

FIG. 23.6 Vent heat-transfer coefficient versus rotor speed in r/min for cast-iron rotors with radial vanes.[46]

23.7 for several unit loads. For all cases, the friction coefficient drops dramatically above 500°F. At this temperature, the phenolic binder and other lining constituents decompose, oxidize, or melt. The result is an increase in wear and a reduction in friction coefficient known as "fade." Fade behavior has been attributed to the formation of a liquid or low-friction solid phase at the interface as a result of the high surface temperature. This is purely a thermal effect which depends upon the lining and will occur for any brake design. The second cause of fade is mechanical loss of contact which occurs in drum brakes. During severe braking, the effective drum radius will

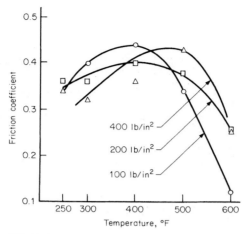

FIG. 23.7 Variation of friction coefficients with temperature for different unit loads.[20]

grow more than the brake shoe radius due to the larger thermal conductivity of the brake drum.[50] The result will be a change in pressure distribution over the brake lining and a reduction in brake torque of 20 percent for the same actuating force.

The performance of a drum brake suffers more under severe operating conditions than the disk brake. Since a drum brake is self-energizing, a reduction in friction coefficient with high temperature will cause a relatively larger reduction in brake torque at constant operating pressures than for the non-self-energizing disk brake. Then the additional mechanical change in the contact geometry of the drum and shoe lower the brake torque even more. Attempts to combat fade on drum brakes by using hard linings that retain friction coefficient at high temperatures were not too successful. The reason was that the fade caused by changes in geometry was more pronounced when unyielding linings were used, even though the fade due to the friction reduction was somewhat alleviated.

The disk brake does not experience mechanical fade so that disk linings can be hard with high friction coefficient at high temperatures. Semimetallic and ceramic linings have these qualities and have been used for disk brakes operating at severe conditions. The disk brake still experiences some fade[51,52] due to friction reduction with temperature, but a modern automobile braking system with ventilated disks will show no fade (negligible increase in pedal effort) for six consecutive 0.5g stops from 60 mi/h.

23.4 SIMPLE BRAKES

23.4.1 Block Brakes

The block brake is the simplest form of brake. Single-lever block brakes are not widely used because the normal braking force results directly in shaft bearing pressure and shaft deflection. Shaft bending is prevented by arranging two brake blocks opposite one another in a double block brake as shown in Fig. 23.8.

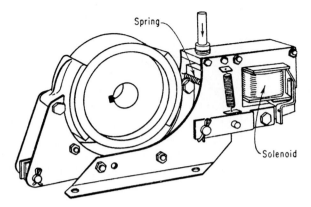

FIG. 23.8   Double block brake.

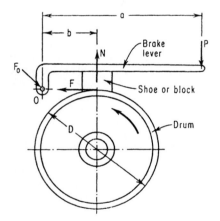

FIG. 23.9   Single-lever block brake (case I).

Single-Lever Block Brake (Case I).
Shoe subtends small angle on brake drum.
Line of action F passes through fulcrum
O as shown in Fig. 23.9. Summing
moments about O and equating to zero
yields

$$\Sigma M_0 = Pa - Nb = Pa - (F/\mu)b = 0$$

or $\qquad F = \mu(a/b)P \qquad (23.15)$

The braking torque

$$T = FD/2 = \mu a\, DP/2b \qquad (23.16)$$

where N = normal reaction between
           drum and shoe
    P = applied load
    F = friction force = μN
    μ = coefficient of friction

The above equations apply to drum rotation in either direction.

Single-Lever Block Brake (Case II).   Shoe subtends small angle on brake drum.
Line of action F passes a distance e below fulcrum O. For counterclockwise drum
rotation,

$$\Sigma M_0 = Pa - (F/\mu)b + Fe = 0$$

or $\qquad\qquad\qquad\qquad F = \dfrac{Pa}{b/\mu - e} \qquad\qquad\qquad\qquad (23.17)$

$$T = \frac{Pa}{b/\mu - e}\, \frac{D}{2} \qquad (23.18)$$

The frictional force in this case helps to apply the brake. The brake is therefore "self-
energizing."

If $e > b/\mu$, the brake is self-locking, and some force P is necessary to disengage the brake. For clockwise drum rotation,

$$F = \frac{Pa}{b/\mu + e} \tag{23.19}$$

Single-Lever Block Brake (Case III). Shoe subtends small angle on brake drum. Line of action F passes a distance e above fulcrum O. For counterclockwise drum rotation,

$$\Sigma M_0 = Pa - (F/\mu)b - Fe = 0 \tag{23.20}$$

$$F = \frac{Pa}{b/\mu + e}$$

For clockwise drum rotation,

$$F = \frac{Pa}{b/\mu - e} \tag{23.21}$$

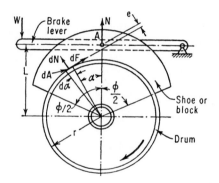

FIG. 23.10 Single-lever block brake with long pivoted shoe (case IV).

Single-Lever Block Brake, Long Pivoted Shoe (Case IV). (See Fig. 23.10.) Assume the distribution of normal pressure p on the shoe to be

$$p = P \cos \alpha$$

where P is a constant. For a face width w, the differential area of the shoe

$$dA = wr \, d\alpha$$

and the normal and frictional forces are, respectively,

$$dN = pwr \, d\alpha = Pwr \cos \alpha \, d\alpha \tag{23.22}$$

$$dF = \mu Pwr \cos \alpha \, d\alpha \tag{23.23}$$

The moment of dF about A (not necessarily the center of the pin) is

$$dM_A = e \, dF = \mu Pwr(L \cos^2 \alpha - r \cos \alpha) \, d\alpha$$

$$M_A = \int dM_A = \mu Pwr(L\alpha/2 + L \sin 2\alpha/4 - r \sin \alpha)_{-\phi/2}^{\phi/2} = 0 \tag{23.24}$$

and

$$L = \frac{4r \sin \phi/2}{\phi + \sin\phi} \tag{23.25}$$

where L is the distance from the drum center to the line of action of F. Point A is the center of pressure. If the pivot pin is located at A, the shoe will not tip. The normal force N is given by

$$N = \int dN = Pwr \int_{-\phi/2}^{\phi/2} \cos \alpha \, d\alpha = 2 \, Pwr \sin \phi/2 \tag{23.26}$$

and the braking torque is approximately

$$T = \int r \, dF = \int \mu pwr^2 \, d\alpha = \mu Pwr^2 \int \cos \alpha \, d\alpha = \mu Nr \tag{23.27}$$

23.4.2   Band Brakes

In a band brake, braking action is obtained by pulling a band tightly against the drum. The braking force F_t is defined as the difference between the tensions F_1 and F_2 at the two ends of the band. Thus

$$F_t = F_1 - F_2 \tag{23.28}$$

Referring to Fig. 23.11, summation of the horizontal and vertical components of the forces acting on a differential element of band length yields, respectively,

$$\mu F_N - dF \cos (d\theta/2) = 0$$

and

$$F_N - (2F + dF) \sin (d\theta/2) = 0$$

where F_N = force between band and drum
F = band tension
μF_N = frictional force
θ = angle of contact

Making the small-angle approximations $\sin (d\theta/2) = d\theta/2$ and $\cos (d\theta/2) = 1$ and eliminating F_N from the equilibrium equations yields

$$\mu F \, d\theta - dF = 0$$

Integrating,

$$\int_{F_2}^{F_1} \frac{dF}{F} = \mu \int_0^{\theta} d\theta$$

Thus

$$F_1/F_2 = e^{\mu\theta} \qquad \text{or} \qquad F_1 = F_t e^{\mu\theta}/(e^{\mu\theta} - 1) \tag{23.29}$$

Figure 23.12 shows a differential band brake. Equating to zero the moments acting on the lever about point A,

$$M_A = Pa + F_1 b_1 - F_2 b_2 = 0$$

and substituting Eq. (23.29), the solution for P is

$$P = F_t (b_2 - e^{\mu\theta} b_1)/a(e^{\mu\theta} - 1) \tag{23.30}$$

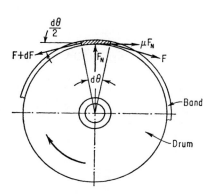

FIG. 23.11 Band tension.

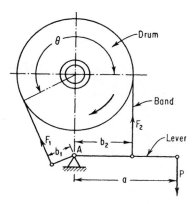

FIG. 23.12 Differential band brake.

Normal operation of the arrangement shown in Fig. 23.12 requires that $b_2 > b_1 e^{\mu\theta}$. If $b_2 = b_1 e^{\mu\theta}$, the brake is self-locking. If $b_2 < b_1 e^{\mu\theta}$, a force must be applied in the opposite direction in order to permit drum rotation. Similar analyses are used for other combinations of direction of P, fulcrum location, and direction of drum rotation.

For $b_1 = 0$ and clockwise drum rotation

$$P = F_t b_2/a(e^{\mu\theta} - 1) \tag{23.31}$$

and the arrangement is termed a simple band brake.

23.5 ANALYSIS OF SELF-ACTUATING DRUM BRAKES

Automotive brakes commonly used in the United States are self-actuating or "self-energizing" brakes. They utilize wedging action to cause the normal force exerted by the lining on the drum to increase nonlinearly with the coefficient of friction. The "caliper" disk brake is not self-actuating. The normal force is independent of the coefficient of friction and depends only on the mechanical design.

Self-actuating brakes are rated by their effectiveness, the ratio of friction force to applied force.* The effectiveness of the drum brake is complicated by its geometry. Since the friction force developed by self-actuating brakes is not proportional to the coefficient of friction, measurement of the coefficient on self-actuating brakes is impractical. The drum brake is particularly unsuited for this purpose, because its geometry and hence its effectiveness for a given coefficient of friction may alter with applied load because of drum and shoe distortion.

---

*The effectiveness is often defined as the ratio of vehicle deceleration to hydraulic pressure. This quantity is directly proportional to the effectiveness as defined above, the proportionality factor varying with type of brake and weight of vehicle.

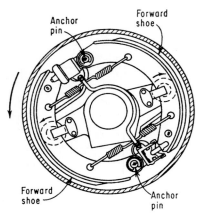

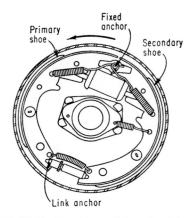

FIG. 23.13   Fixed-anchor internal drum brake.

FIG. 23.14   Movable- or link-anchor internal drum brake (Duo-Serve).

23.5.1   The Internal Drum Brake

Self-actuating drum brakes are of two types: (1) fixed anchor (Fig. 23.13) and (2) movable or link anchor (Fig. 23.14). Each brake contains two shoes. Both shoes may be self-actuating, called "forward shoes," or one shoe may be anti-self-actuating, called the "drag shoe." In passenger car brakes, the load is applied through hydraulic wheel cylinders and is given by the product of the fluid pressure and wheel-cylinder area. In many truck brakes, the load is applied through cams and must be measured.

Several methods are in use for calculating brake effectiveness. They usually assume negligible drum and shoe distortion, an assumption valid for light loads only. They also assume a uniform coefficient of friction over the surface of the lining, valid for most practical purposes. The equations derived herein are based upon a graphical method described by Fazekas.[53]

Accurate methods involve integration of the friction and normal forces on an element of lining. Some methods require the calculation of a definite integral for each type of brake and for each change of dimensions. The present method applies to all cylinder-loaded brakes and avoids special integration. The actuation equation gives the effectiveness, in most cases more quickly and accurately than graphical analysis. Structural and finite-element analyses[54-56] have also been applied to drum brakes to study the role of contact geometry and distortion on the torque characteristics.

23.5.2   Theory of Resultant Forces on Shoe

Three forces act on the shoe (Figs. 23.18 and 23.19), the applied load L, the reaction R of the anchor, and the vector resultant reaction V of the friction and normal forces exerted by the drum through the lining. The force V acts through the so-called "drag point" or "center of pressure" (CP). In most practical cases, the CP lies outside the drum of any internal curved-shoe brake. With negligible distortion, the pressure distribution is harmonic. With moderate distortion, it is practically uniform over the lining.

Moment of Friction Forces about Center of Drum.   Figure 23.15 is a schematic diagram of the circular surface of the lining of angular arc 2ϕ. The origin of coordinates is the center of the brake. Let p_m be the normal (radial) pressure at θ_m, locating the line

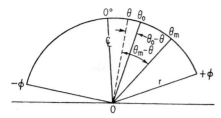

FIG. 23.15 Schematic diagram of the circular surface of the lining of angular arc 2ϕ.

of maximum pressure (LMP), θ_m being measured from the center line of the lining. The pressure at an angle θ from the center line is

$$p = p_m \cos(\theta_m - \theta) \qquad (23.32)$$

The normal force on an element of area of lining is $pwr\, d\theta$, where w is the width of lining and r is the drum radius. The corresponding friction force in the direction of rotation of the drum is $\mu pwr\, d\theta$ and the moment of this force about the center is $\mu pwr^2\, d\theta$, where μ is the coefficient of friction.

Integrating these moments over the lining gives the total friction moment

$$Fr = \int_{-\phi}^{+\phi} \mu p_m wr^2 \cos(\theta_m - \theta)\, d\theta$$

$$Fr = 2\mu p_m wr^2 \cos\theta_m \sin\phi \qquad (23.33)$$

Resultant Normal Force.   The normal forces $pwr\, d\theta$ acting together produce a vector resultant normal force N along a drum radius at some angle θ_0 from the center line of the lining. The component of the pressure p in the direction θ_0 is $p\cos(\theta_0 - \theta)$, so that the component of the normal force in this direction is $pwr\cos(\theta_0 - \theta)\, d\theta$.

Integrating these forces over the lining gives the resultant normal force

$$N = p_m wr \int_{-\phi}^{+\phi} \cos(\theta_m - \theta)\cos(\theta_0 - \theta)\, d\theta$$

$$N = p_m wr[\cos\theta_m \cos\theta_0(\phi + \sin\phi\cos\phi) + \sin\theta_m \sin\theta_0(\phi - \sin\phi\cos\phi)] \quad (23.34)$$

Relation of θ_0 to θ_m.   By definition, the resultant normal force perpendicular to the direction θ_0 is zero. The component of the normal force perpendicular to θ_0 is $pwr\sin(\theta_0 - \theta)\, d\theta$. Integrating these forces over the lining gives

$$\int_{-\phi}^{+\phi} \cos(\theta_m - \theta)\sin(\theta_0 - \theta)\, d\theta = 0$$

$$\cos\theta_m \sin\theta_0(\phi + \sin\phi\cos\phi) - \sin\theta_m \cos\theta_0(\phi - \sin\phi\cos\phi) = 0$$

Transposing and dividing gives

$$\tan\theta_m = \tan\theta_0[(\phi + \sin\phi\cos\phi)/(\phi - \sin\phi\cos\phi)] \qquad (23.35)$$

Center-of Pressure Circle: CP Locus.   The resultant friction force μN acts through the CP at right angles to N. The moment of μN about the center of the drum $\mu Nc = Fr$, where c is the distance of the CP from the center of the drum. Using the values for Fr and N from Eqs. (23.33) and (23.34) together with Eq. (23.35),

$$c = Fr/\mu N = (2r\sin\phi\cos\theta)/(\phi + \sin\phi\cos\phi) \qquad (23.36)$$

This is the polar equation of a circle with the origin on its circumference, its center at point $d/2$, 0, and of diameter

$$d = (2r \sin \phi)/(\phi + \sin \phi \cos \phi) \tag{23.37}$$

The relationship between d/r and ϕ is shown in Fig. 23.16, curve A.

Relation of Center of Pressure to Line of Maximum Pressure. This relationship is given in Eq. (23.35) which, employing Eq. (23.37), may be written

$$\tan \theta_0 = \tan \theta_m[1 - (d/r) \cos \phi] \tag{23.38}$$

The relationship between $\tan \theta_m/\tan \theta_0$ and ϕ is shown in Fig. 23.17.

23.5.3 Effect of Moderate Distortion of Drum and Shoe: Uniform Pressure

Assume the CP is initially at the center line of the lining, i.e., $\theta_0 = 0$, for the usual coefficient of friction. This gives the minimum variation of pressure over the lining and consequently the most uniform heating and wear, and maximum effectiveness. In well-adjusted brakes, the wear is often uniform, i.e., the pressure is uniform over the lining. This means that the CP remains at the center line after moderate distortion. This will not necessarily be the case if the coefficient changes considerably, since, in general, the CP is at the center line for only one coefficient of friction. With uniform pressure p,

$$Fr = \int_{-\phi}^{+\phi} \mu pwr^2 \, d\theta = 2\mu pwr^2\phi$$

The component of p in the direction θ_0 is $p \cos \theta$, since $\theta_0 = 0$. Therefore,

$$N = \int_{-\phi}^{+\phi} pwr \cos \theta \, d\theta = 2pwr \sin \phi$$

The distance of the CP from the center of the drum

$$\begin{aligned}
c &= Fr/\mu N \\
&= 2\mu pwr^2\phi/2\mu pwr \sin \phi \\
&= r\phi/\sin \phi \tag{23.39}
\end{aligned}$$

This is a circle of radius c with its center at the drum center. The function c/r versus 2ϕ is shown plotted in Fig. 23.16, curve B.

23.5.4 The Actuation Equation: Effectiveness

The above relations, together with the known geometry of a given brake, suffice to obtain the actuation equation and hence the effectiveness, usually without graphical construction. Either the LMP (θ_m) or the CP (θ_0) can always be found from the geometry of the brake, and the other found from Fig. 23.17.

Fixed-anchor brakes are the most stable. Each shoe has a separate anchor (Fig. 23.13) which allows rotation about a fixed axis. The actuation equation for the drag shoe, anchored at the leading end, is obtained by reversing the sign of the actuation factor. The anchored end is herein called the "heel" and the loaded end the "toe."

Figure 23.18 is a schematic diagram showing the lining of a forward shoe of an 11-

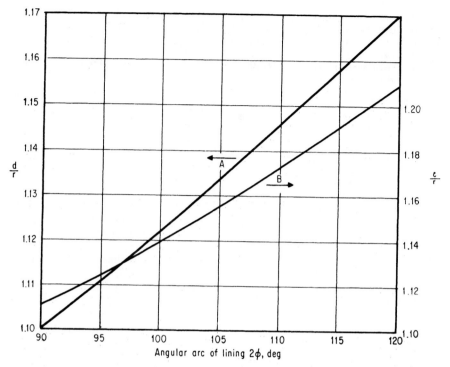

FIG. 23.16 Relationship between d/r, c/r, and ϕ.

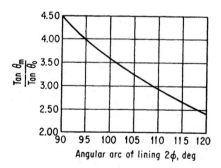

FIG. 23.17 Relationship between $\tan \theta_m/\tan \theta_0$, and ϕ.

in-diameter brake and the angles and dimensions required to obtain the actuation equation. The drum reaction V is drawn through the CP, making the friction angle $17° = \arctan 0.3$ with the normal force N. The anchor reaction R is drawn through the anchor and the intersection of V with the load L (off the diagram). The magnitudes of R and V may be obtained by constructing a force triangle on L.

The angles θ_m and θ_0 are measured from the lining center line. All other lining angles are measured from the anchor line, a line from the center of the brake to the anchor. From Fig. 23.16 for $2\phi = 117°$, the CP-circle diameter $d = 6.4$ in for no drum distortion. The CP circle is drawn as defined by Eq. (23.37), with its center on the center line of the lining at $85\frac{1}{2}°$.

In a fixed-anchor brake, the LMP necessarily lies on a line at right angles to the anchor line.[53] Therefore, $\theta_m = 90 - 85\frac{1}{2}° = 4\frac{1}{2}°$. From Fig. 23.17 $\theta_0 = 1°46'$.

The CP lies on the CP circle at its intersection with the drum radius inclined at $1°46'$ to the center line of the lining. The pressure distribution on this lining is thus very nearly symmetrical.

Since the LMP is fixed in a fixed-anchor brake, the location of the CP is independent of the coefficient of friction and depends only on the length and position of the lining. This explains its greater stability. In link- and sliding-anchor brakes, the CP and the LMP both move as the coefficient of friction varies.

It is convenient to write the actuation equation in terms of the ratio of the braking torque to the applied torque about the anchor:

$$\frac{Fr}{Lr_2} = \frac{\mu Nc}{Na - \mu Nb} = \frac{\mu(c/a)}{1 - \mu(b/a)} \tag{23.40}$$

where μ = coefficient of friction
$\quad a$ = moment arm of the normal force N about the anchor
$\quad b$ = moment arm of the friction force μN about the anchor
$\quad c$ = moment arm of the friction force μN about the center of the brake
$\quad r$ = radius of the brake drum
$\quad r_2$ = moment arm of the load L about the anchor

The ratio b/a is the actuation constant Q. The factor $\mu(b/a)$ is the actuation factor A. The shoe locks when the drum reaction V passes through the anchor, i.e., when $\mu = a/b = \tan \angle (V - N) = \tan 34\frac{1}{2}° = 0.687$, in this case.

In a drag shoe, the algebraic sign of the actuation factor is changed. It cannot lock and its effectiveness is less than if non-self-actuating.

Figure 23.18 illustrates the graphical method. The quantities needed to obtain the actuation equation may be measured on a full-scale drawing, or they may be obtained analytically. From Fig. 23.18,

$$a = r_1 \cos (\theta_m - \theta_0)$$
$$b = c - r_1 \sin (\theta_m - \theta_0) \tag{23.41}$$
$$c = d \cos \theta_0$$

where r_1 is the length of the anchor line. These quantities can all be obtained from Figs. 23.16 and 23.17 and the geometry of the brake and lining, without graphical construction.

The effectiveness of two forward shoes of the 11-in brake, referred to a single cylinder load L, is plotted in Fig. 23.20, curve B. For any coefficient, the effectiveness is a maximum when the CP lies on the center line of the lining. The magnitude and distribution of pressure on the lining can be calculated from Eqs. (23.22), (23.33), (23.40), and (23.41).

The link-anchor brake, exemplified by the Bendix Duo-Servo brake[57] (Fig. 23.14), has one movable-anchor shoe, called the primary, and one fixed-anchor shoe, called the secondary. The load is applied near the toe of the primary and the heel of the secondary through hydraulic cylinders placed back to back. The heel of the primary is connected to the toe of the secondary through a rigid, hinged link. The end of the link attached to the primary is the link anchor which can move toward or away from the drum as the primary rotates about it.

Figure 23.19 is a schematic diagram of the primary in the 12-in Bendix brake. The forces shown are those exerted on the primary shoe and lining. The drum rotation moves the primary out of contact with the fixed anchor. The load L acts at a right angle to the center line of the brake. The reaction R through the link is, for all practical purposes, parallel to L. Therefore, $V = R + L$. The moments about the link anchor $Vr_3 = Lr_2$. These two equations give the magnitude of V and R. The pressure distribution is harmonic, but the LMP is not known a priori as for a fixed anchor.

In the graphical method, the CP is found by drawing a line parallel to L through a point P located on the CP circle at an angular distance $2\theta_\mu$ from the lining center line.

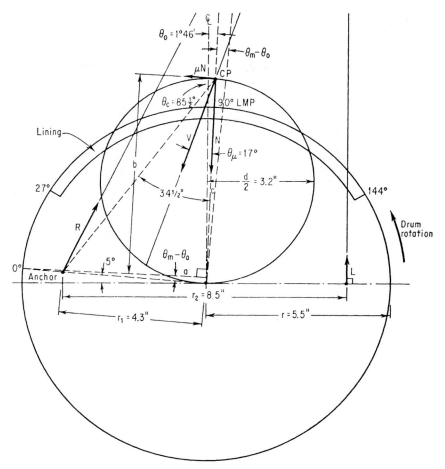

FIG. 23.18 Forces on the fixed-anchor brake.

This line also intersects the CP circle at the CP. Since an inscribed angle is measured by half the intercepted arc, this line makes the angle θ_μ with the drum radius through the CP, thus defining the position and direction of V and N.

Analytically, from Fig. 23.19,

$$\theta_a = 74 - \theta_\mu = 52° \tag{23.42}$$

$$\theta_0 = \theta_c - \theta_a = \theta_c + \theta_\mu - 74° = 18° \tag{23.43}$$

where θ_a = angular position of the CP measured from the anchor line
θ_μ = friction angle, arctan μ
θ_c = angular position of the lining

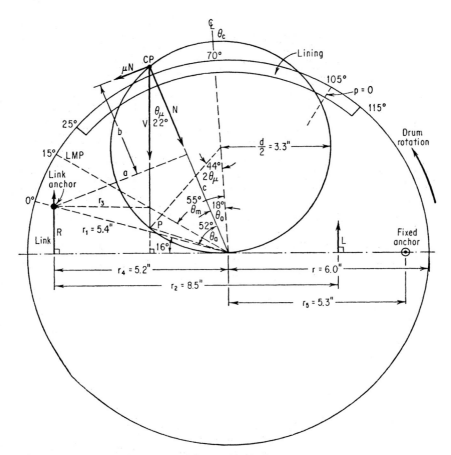

FIG. 23.19 Forces on the primary shoe of link-anchor brake.

The LMP (θ_m) is found from Eq. (23.43) and Figs. 23.15 and 23.16 to be 15° from the anchor line, i.e., 10° beyond the heel of the lining. At 90° from this, or 10° in from the toe, the pressure is zero. The remainder of the toe is lifted off the drum.

The pressure distribution is extremely asymmetrical with the lining in this position. For $\mu = 0.4$, the pressure distribution would be symmetrical if the lining were moved 18° toward the link anchor. In the symmetrical-pressure position, the effectiveness is 10 percent higher than in the position of Fig. 23.17.

The actuation equation is, as before,

$$\frac{Fr}{Lr_2} = \frac{\mu(c/a)}{1 - \mu(b/a)} \tag{23.44}$$

From Fig. 23.19,

$$\begin{aligned}
a &= r_1 \sin \theta_a \\
b &= c - r_1 \cos \theta_a \\
c &= d \cos \theta_0
\end{aligned} \tag{23.45}$$

Since θ_0 and θ_a are both functions of μ, the effectiveness is a complicated function of μ. In general, both the numerator and denominator of the right-hand side of Eq. (23.44) are quadratic functions of μ. This shoe cannot lock, since the drum reaction V cannot pass through the anchor; the quadratic denominator of the right-hand side of Eq. (23.44) has no real roots. The secondary shoe, however, can lock.

In this case, the friction force in the primary can be found by a simpler method without using the actuation equation. Since V is normal to the brake center line, the moment of V about the center of the brake is $Vc \sin \theta_\mu = (Lr_2/r_3)c \sin \theta_\mu$, which, in turn, is equal to Fr.

The secondary is analyzed as is any other fixed-anchor shoe. The drum rotation forces the secondary against the fixed anchor. Two loads are exerted on the secondary, the reaction R of the primary and the applied load L, both opposite in direction to that shown in Fig. 23.19. The total torque exerted by these two loads about the fixed anchor is equal to that exerted by V.

The total torque T applied to the secondary about the fixed anchor is therefore

$$T = L(r_2/r_3)(r_5 + c \sin \theta_\mu) \tag{23.46}$$

where r_2 = moment arm of L about the link anchor
r_3 = moment arm of V about the link anchor = $r_4 - c \sin \theta_\mu$
r_4 = distance from the link anchor to the center of the brake
r_5 = distance from the fixed anchor to the center of the brake

The actuation equation is

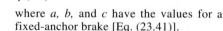

$$\frac{Fr}{T} = \frac{\mu(c/a)}{1 - \mu(b/a)} \tag{23.47}$$

where a, b, and c have the values for a fixed-anchor brake [Eq. (23.41)].

For the values given in Fig. 23.19, $T = 22.8L$, which may be compared with $Lr_2 = 8.5L$ for the primary. The secondary lining commonly has a lower coefficient but greater arc length than the primary. Since this brake is symmetrical, except for the linings, its behavior is identical in reverse, the primary and secondary becoming interchanged.

The nearly threefold greater braking torque exerted by the secondary results in an asymmetrically loaded and distorted drum. The pressure on the secondary is three to four times that on the primary and the rate of wear correspondingly greater. The combined effectiveness of the two shoes of the 12-in Duo-Servo brake is shown in curve C of Fig. 23.20.

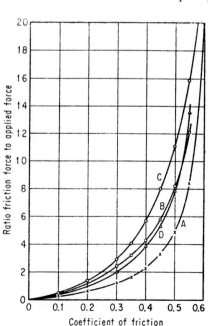

FIG. 23.20 Ratio of friction force to applied force vs. coefficient of friction.

23.5.5 Drum and Shoe Distortion

Distortion, which can greatly alter the magnitude and distribution of pressure on the lining and, to some extent, the effectiveness, depends upon the applied load, the rigidity of the drum and shoe, and the compressibility of the lining.

Moderate distortion produces the desirable condition of uniform pressure, and its effect can be calculated (see Sec. 23.5.3). Severe distortion, caused by the high loads required by a low lining coefficient, markedly increases the pressure on the toe and heel and decreases the pressure on the center of the lining. The changes in pressure are minimized by a flexible shoe and a soft lining which tend to conform to the drum when it distorts. The effect on brake performance must be measured.

Computation of lining pressure for a distorted drum required measurement of lining compressibility and numerical integration of the moment equations. The computed effectiveness for the distorted drum was about 20 percent greater than when undistorted, for all coefficients of friction.

The effect of distortion on the Bendix brake is reported by Winge.[50] He finds that the effectiveness is increased more for lower coefficients than for higher coefficients. The wide variation of effectiveness with coefficient is thus somewhat less than shown in Fig. 23.20, curve C.

23.5.6 Cam Brakes: Truck Brakes

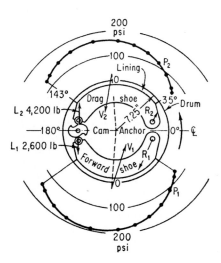

FIG. 23.21 Forces and lining pressure distribution in a Timken brake.

Cam-operated drum brakes have one forward and one drag shoe loaded by a double cam[58] mounted on a single pivot and actuated through a lever arm. The load and hence the effectiveness of cam brakes are difficult to calculate, because the rotation of the drum pushes the drag shoe against the cam, thereby increasing the load, and tends to pull the forward shoe away from the cam, thereby decreasing the load. The amount of this effect varies with the coefficients of friction between lining and drum and between cam and shoe, the clearance in the cam pivot, the compressibility of the lining, and the distortion of the drum and shoes. The usual result is that the effectiveness of the two shoes is about equal.

Figure 23.21 shows the forces and lining pressure distribution in a Timken 14½-in-diameter cam-and-roller-type brake. A similar brake is the cam-and-plate type.

23.6 DISK BRAKES

The disk-brake concept is among the oldest brake designs,[59] the first patent having been issued in 1902. However, because of their lack of self-energization, disk brakes were applied only to aircraft before 1940. After World War II, the development of disk brakes was accelerated because the increases in vehicle weight and speeds demanded a brake with high heat-dissipation capabilities. The disk brake provided a solution to this need. The Crosley Hot Shot of the early 1950s was the first American production car to come equipped with disk brakes. They were applied to heavier passenger cars in 1965 and are standard front brakes for most automobiles at the present time. Rear disk brakes[60,61] have also been applied to many automobiles, but the problems of increased cost, parking brake complications, and reduced heat-dissipation requirements at the rear wheels have limited their application.

Disk brakes have become popular because they overcome some basic disadvantages of the drum brakes they replace. Ironically, the basic mechanical advantage of drum brakes (self-energization), which spurred their development is the source of their main problem. This problem is related to the influence of friction coefficient on the ratio of friction force to applied force, which is defined as the effectiveness of the brake. The drum brake is extremely sensitive to changes in friction coefficient, as shown in Fig. 23.22. Small changes in friction due to moisture, high temperature, or speed require a large change in pedal pressure for the same braking torque. A change in friction coefficient from 0.3 to 0.4 (which is not unlikely, as shown in Sec. 23.1) will require twice the normal force on the brake shoe for the same braking torque. This behavior is not desirable for automobile brakes, where predictable performance is a requirement. The disk brake overcomes this problem because its torque output is linearly proportional to friction coefficient; thus, a change in friction from 0.3 to 0.4 requires only a 33 percent increase in pedal effort. However, because no self-energization occurs, the disk brake requires power assist to provide reasonable pedal effort on vehicles over 3500 lb.

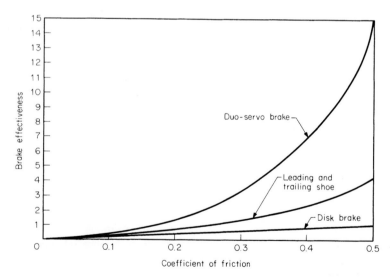

FIG. 23.22   Influence of friction coefficient[59] on brake torque for drum and disk brakes.

Basically, the advantages[59] of the disk brake over the drum brake stem from three factors:

1. Linear relationship between torque output and friction coefficient
2. Higher heat-transfer rates from radiation and convection (ventilated rotors)
3. Lack of mechanical fade (due to distortion of drum brakes)

The disadvantages of disk brakes are higher cost due to the number of castings required and the necessity of power assistance for many applications.

23.6.1 Types of Disk Brakes

Fixed Caliper. Early disk brakes had a caliper rigidly mounted to the vehicle suspension straddling the disk (Fig. 23.23). A pair of hydraulic pistons mounted in the caliper applied the clamping force to the disk from each side. To apply this design to heavier American vehicles in a limited space, a four-piston caliper was developed, but has largely been superseded by the floating-caliper brake.

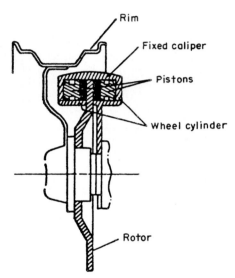

FIG. 23.23 Opposing piston—fixed-caliper disk brake.

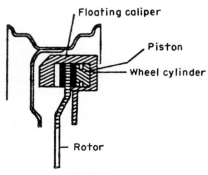

FIG. 23.24 Single piston—floating-caliper disk brake.

Floating Caliper. Most disk brakes now use a floating caliper (Fig. 23.24) which can move perpendicular to the disk surface. This caliper supports the inner and outer linings and contains a single piston, usually inboard of the disk. Upon application of pressure to the piston, a force is exerted on one shoe and the reaction on the caliper forces it to slide and apply the same force to the other shoe. Thus a single cylinder clamps both linings against the disk. The floating caliper is necessary to accommodate the wear of the linings.

23.6.2 Actuation

Because no self-actuation occurs in most disk brakes, a significant mechanical advantage is required to apply the clamping force. Typically for automotive applications, vacuum-boosted hydraulic systems are utilized; however, straight hydraulic actuation is used on lighter vehicles. Disk brakes are also being applied to railroad vehicles and large trucks. These applications typically utilize an air-mechanical or air-over-hydraulic system for actuation.

23.6.3 Design[62,63]

Disk.   The disk size is based upon the space available and the energy-dissipation requirements of the vehicle. The brake duty cycle can be divided into two states: the transient state, as found in rapid deceleration, and the steady state, as experienced in prolonged applications. In the transient state, the heat enters the disk more rapidly than it can be dissipated and the disk becomes a heat reservoir. The mass of the disk must be such that the temperature will not be excessively high. Under steady-state conditions, or repeated transient states, the heat will be lost to the surroundings by radiation and convection. Ventilated disk brakes with properly designed and ducted rotors, can appreciably increase the cooling rate of the disk (see Sec. 23.3).

The disk material must have a high conductivity so that the heat generated at the disk-lining interface is quickly conducted to the disk interior. For the initial brake application, the temperature of the disk will be a function of the vehicle speed, brake balance, and disk material. For subsequent stops, the equilibrium temperature will depend upon the heat losses from the disk and the zero-pressure (residual) brake torque. This latter quantity, while it does not generate large quantities of heat, influences the loss of heat from the disk between stops. Most modern disk brakes use the hydraulic cylinder seals[59] to back the pads away from the disk when the pressure is released. Thus the residual brake torque for a well-maintained system will be quite low. However, corrosion or wear may influence the motion of the pads and increase the residual torque.

Lining.   The face area of the lining material and its wrap on the disk face are important considerations for the design of disk brakes. The face area of the lining is usually specified in the form of maximum horsepower dissipation per square inch of each pad surface. For heavy-duty applications[49] this can be as much as 8 hp/in², but lower values are more commonly used (2 hp/in²) for automotive applications. The wrap of the lining should not exceed one-third of the swept area in order to retain the benefits of heat loss from the disk face outside the lining, which accounts for about 15 percent[64] of the total heat loss.

23.7 BRAKE SQUEAL

Brake squeal has been a problem since the early days of motoring. However, it has been an elusive problem, and the occurrence of squeal, even for the same vehicle, will depend[65] upon the driver and the history of brake operation. The elimination of squeal has generally been tackled empirically on a case-by-case basis, and a number of ways of reducing squeal on existing brake systems have been utilized. Techniques used to reduce squeal include (1) applying boundary lubricants on the lining surface or backing plate to reduce friction coefficient, (2) inserting antisqueal shims between piston and back plates on disk brakes, (3) adding weight to shoes or calipers at strategic locations, or (4) changing the stiffness of the pad. Other solutions have included changes in the disk or drum material, such as the use of pearlitic-gray iron because of its high internal damping.

Theoretical explanations of the problem have been varied, including stick-slip theory,[66-68] variable dynamic friction coefficient,[69] sprag-slip theory,[70] and the theory of geometric coupling.[71] The current understanding of squeal relates its occurrence to a coupling of vibrations within the brake assembly. The most important parameters are the coefficient of friction of the lining material (higher coefficients leading to higher probability of squeal), the geometry of the brakes, and, in particular, the location of the areas of contact between the lining and the disk or drum. (A leading angle of contact, that is, the contact a specific distance ahead of the pivot location, can lead to brake squeal.) The mechanism for squeal is an increase in normal load (and therefore friction) and the release of that load due to geometric changes or deflections. Measurements of accelerations[65,72] of brake components have shown that during squeal the friction material is deflected elastically along the pad surface by the frictional forces. That deflection causes a second deflection with a component vertical to the surface of the pad. This vertical component reduces the friction force; thus, the stored energy returns to the first configuration. The cycle is then repeated. The phase difference between the coupled vibrations necessary to maintain squeal arises from the inertia of the components in the system.

Holographic measurements[65] of a disk brake have shown the modes of vibration present and the main sources of noise. Under squealing conditions, the disk can have from 6 to 10 circumferential modes. However, the magnitude of the disk vibrations is not the main source of vibration. The friction pads and the calipers have higher vibration amplitudes and these components transmit the main part of the noise energy. The pad vibration occurs perpendicular to the disk surface and has several mode shapes, but, in general, the leading edge of the pad has the greatest amplitude of vibration (frequently more than 0.0001 in). This result reinforces the idea of sprag-slip motion as being the key explanation of the squeal behavior.

23.7.1 Sprag-Slip Theory

The prevailing understanding of squeal is based, in part, on a model[70] first published in 1961. As shown in Fig. 23.25, the model consists of a block at A in contact with a plane sliding to the right with speed V. The arm AO' is rigid and connected to a pivot which is mounted in a flexible support. A second pivot exists at O'', also flexibly mounted. An increase in friction force will, because of the geometry, produce an increase in normal load on the contact. The increase in load due to friction force, L_f, can be written from the sum of the moments about O' as

$$L_f = F_F \tan \theta \qquad (23.48)$$

where F_F = friction force and θ = leading angle for front strut. Then the sum of the horizontal forces at the contact yields

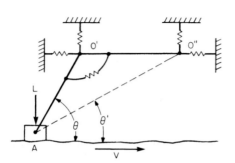

FIG. 23.25 Geometry for sprag-slip model of brake contact.

$$F_F = L\mu/(1 - \mu \tan \theta) \tag{23.49}$$

where L = initial load on block and μ = friction coefficient

The value of friction goes unbounded as the friction coefficient approaches cot θ, which is called the spragging angle. When this condition occurs, there is no slip between the block and the moving surface; A is therefore displaced elastically to the right and θ increases slightly. This motion reduces the pivot angle at O' and the elastic moment at O' becomes large enough such that $AO'O''$ becomes like a rigid triangle pivoting ATO'', which can be replaced by AO''. For this case, the angle θ' is less than the spragging angle and the normal, and friction force is reduced as in Eq. (23.49). This action releases the elastic energy at O' and slip occurs. This "sprag-slip" behavior leads to a variation in the normal load on the brake pad, setting up vibrations of the pad and its support which account for the squeal behavior.

This contact behavior can occur for internal and external drum brakes and disk brakes. For disks, the vibration mode consists of motion of the pad perpendicular to the disk and vibration of the pad and the caliper. For the drum brakes, the vibration involves the shoe and the backing plate motion.

Solutions to squeal problems are as varied as their sources. Since the contact of the pad and disk or drum varies with (1) the life of the pad, (2) the current operating temperature, (3) the pad stiffness (dependent upon pressure), and (4) the thermal distortion of the system, no universal fix for squealing problems is available. However, observations as to the sources of vibration and therefore noise generation have been published. These results were generated using holographic[65] pictures of squealing disk brakes, and the conclusions included the following:

1. Increasing stiffness of brake pad is likely to reduce squeal.

2. Brake pads can be supported at antinodes (points of minimum vibration) to reduce transmitted vibration.

3. The natural frequencies of pads and calipers should be made significantly different.

4. Reduction in friction coefficient will reduce tendency to squeal.

REFERENCES

1. Joshi, M. N.: "Disk Linings—Analytical Study and Selection Criteria," SAE Paper 800782, 1980.

2. Newcomb, T. P., and R. T. Spurr: "Friction Materials for Brakes," *Tribology (London)*, vol. 4, no. 2, pp. 75–81, May 1971.

3. Schiefer, H. M., and G. V. Kubczak: "Controlled Friction Additives for Brake Pads and Clutches," SAE Paper 790717, 1979.

4. Token, H. Y.: "Asbestos Free Brakes and Dry Clutches Reinforced with Kevlar Aramid Fiber," SAE Paper 800667, 1980.

5. Aldrich, F. W.: "Semi-Metallics: A New Type of Friction Material," SAE Paper 710591, 1971.

6. Rhee, S. K.: "Ceramics in Automotive Brake Materials," *Am. Ceram. Soc. Bull.*, vol. 55, no. 6, pp. 585–588, June 1976.

7. Ho, T. L., F. E. Kennedy, and M. B. Peterson: "Evaluation of Aircraft Brake Materials," *ASLE Trans.*, vol. 22, no. 1, pp. 71–78, Jan. 1979.

8. Knapp, R. A., and A. E. Anderson: "Friction Material Riveting Studies: Load-Deflection Characteristics and Finite Element Modeling," SAE Paper 790716, 1979.

9. Unpublished Reports. As manufactured and tested by Johns Manville.

10. Anderson, A. E., and R. A. Knapp: "Brake Lining Mechanical Properties, Laboratory Specimen Studies," SAE Paper 790715 for Passenger Car Meeting, June 1979.

11. Russell, G. R.: "The Manufacture of Disk Brake Linings," SAE Paper 750228, 1975.

12. Bowden, F. P., and D. Tabor: "Friction and Lubrication of Solids," Part 1, 1954 and Part 2, 1964, Oxford University Press, New York.

13. Rabinowicz, E.: "Friction and Wear of Materials," John Wiley & Sons, Inc., New York, 1965.

14. Kragelskii, I. V.: "Friction and Wear," Butterworth, Washington, D.C., p. 301, 1965.

15. Tanaka, K., S. Veda, and N. Noguchi: "Fundamental Studies on the Brake Friction of Resin-Based Friction Materials," Wear, vol. 23, no. 2, pp. 349–365, March 1973.

16. Scieszka, S. F.: "Tribological Phenomena in Steel Composite Brake Material Friction Pairs," Wear, vol. 64, no. 2, pp. 367–378, November 1980.

17. Bros, J., and S. F. Scieszka: "The Investigation of Factors Influencing Dry Friction in Brakes," Wear, vol. 44, no. 2, pp. 271–286, 1977.

18. Tanaka, K.: "Friction and Deformation of Polymers," J. Phys. Soc. Japan, vol. 16, p. 2003, 1961.

19. Rhee, S. K., and P. A. Thesier: "Effects of Surface Roughness of Brake Drums on Coefficient of Friction and Lining Wear," SAE Paper 720449, 1972.

20. Rhee, S. K.: "Friction Coefficient of Automotive Friction Materials—Its Sensitivity to Load, Speed, and Temperature," SAE Trans., Paper 740415, pp. 1575–1580, 1974.

21. Libsch, T. A., and S. K. Rhee: "Microstructural Changes in Semi-Metallic Disk Brake Pads Created by Low Temperature Dynomometer Testing," Wear, vol. 46, no. 2, pp. 203–212, 1978.

22. Rhee, S. K.: "Wear Mechanisms for Asbestos-Reinforced Automotive Friction Materials," Wear, vol. 29, no. 3, pp. 391–393, 1974.

23. Pogosian, A. K., and N. A. Lambarian: "Wear and Thermal Processes in Asbestos-Reinforced Friction Materials," Trans. ASME, J. Lubr. Technol., vol. 101, no. 4, pp. 451–485, October 1979.

24. Weintraub, M. H., A. E. Anderson, and R. L. Gealer: "Wear of Resin Asbestos Friction Materials," Polym. Sci. Technol., vol. 5B, pp. 623–649, 1974.

25. Liu, T., and S. K. Rhee: "High Temperature Wear of Semi-Metallic Disk Brake Pads," Wear, vol. 46, no. 2, pp. 213–218, 1978.

26. Jacko, M. G.: "Physical and Chemical Changes of Organic Disk Pads in Service," Wear, vol. 46, no. 1, pp. 163–175, 1978.

27. Liu, T., and S. K. Rhee: "A Study of Wear Rates and Transfer Films of Friction Materials," Wear, vol. 60, no. 1, pp. 1–12, 1980.

28. Rhee, S. K.: "Wear Mechanism at Low-Temperature for Metal-Reinforced Phenolic Resins," Wear, vol. 23, no. 2, pp. 261–263, 1973.

29. Liu, T., and S. K. Rhee: "High Temperature Wear of Asbestos-Reinforced Friction Materials," Wear, vol. 37, no. 2, pp. 291–297, 1976.

30. Bush, H. D., D. M. Rowson, and S. E. Warren: "The Application of Neutron-Activation Analysis to the Measurement of the Wear of a Frictional Material," Wear, vol. 20, no. 2, pp. 211–225, June 1972.

31. Pavelescu, D., and M. Musat: "Some Relations for Determining the Wear of Composite Brake Materials," Wear, vol. 20, no. 1, pp. 91–97, 1974.

32. Kant, S., et al.: "Prediction of the Coefficient of Friction for Pneumatic Tires on Hard Pavement," Proc. Inst. Mech. Eng. (London), vol. 189, no. 34, pp. 259–266, 1975.

33. Limpert, R., and C. Y. Warner: "Proportional Braking of Solid-Frame Vehicles," SAE Paper 710047, 1971.

34. Campbell, C.: "The Sports Car," Robert Bentley, Inc., Cambridge, Mass., p. 244, 1970.

35. Yim, B., et al.: "Highway Vehicle Stability in Braking Maneuvers," SAE Paper 700515, 1970.

36. Hartley, J.: "Anti-Lock Braking," *Automot. Des. Eng.,* vol. 13, pp. 15–19, June 1974.

37. Clemett, H. R., and J. W. Moules: "Brakes and Skid Resistance," *Highway Res. Rec.,* no. 477, pp. 27–33, 1973.

38. Harned, J. L., L. E. Johnson, and G. Scharpf: "Measurement of Tire Brake Force Characteristics as Related to Anti-Lock Control System Design," SAE Paper 690214, 1969.

39. Bernard, J. E., L. Segel, and R. E. Wild: "Tire Shear Force Generation During Combined Steering and Brake Maneuvers," SAE Paper 770852, 1977.

40. Jacko, M. G., W. M. Spurgeon, and S. B. Catalano: "Thermal Stability and Fade Characteristics of Friction Materials," SAE Paper 680417, 1968.

41. Ho, T., M. B. Peterson, and F. F. Ling: "Effect of Frictional Heating on Brake Materials," *Wear,* vol. 30, no. 1, pp. 73–91, October 1974.

42. Bark, L. S., D. Moran, and S. J. Percival: "Polymer Changes During Frictional Material Performance," *Wear,* vol. 41, no. 2, pp. 309–314, February 1977.

43. Wetenkamp, H. R., and R. M. Kipp: "Hot Spot Heating by Composition Shoes," *Trans. ASME,* ser. B, vol. 98, no. 2, pp. 453–458, May 1976.

44. Santini, J. J., and F. E. Kennedy: "Experimental Investigation of Surface Temperature and Wear in Disk Brakes," *Lubr. Eng.,* vol. 31, no. 8, pp. 402–404 and 413–417, August, 1975.

45. Rowson, D. M.: "The Interfacial Surface Temperature of a Disk Brake," *Wear,* vol. 47, pp. 323–328, 1978.

46. Sisson, A. E.: "Thermal Analysis of Vented Brake Rotors," SAE Paper 780352, pp. 1685–1694, 1978.

47. Limpert, R.: "Thermal Performance of Automotive Disk Brakes," SAE Paper 750873, pp. 2355–2368, 1975.

48. Morgan, S., and R. W. Dennis: "A Theoretical Prediction of Disk Brake Temperatures and a Comparison with Experimental Data," SAE Paper 720090, 1972.

49. Soltis, P. J.: "Straight Air Disk Brakes," SAE Paper 791041, 1979.

50. Winge, J. L.: "Instrumentation and Methods for the Evaluation of Variables in Passenger Car Brakes," SAE Paper 361 B, 1961.

51. Secrist, D. A., and R. W. Hornbeck: "Analysis of Heat Transfer and Fade in Disk Brakes," ASME Paper 75-DE-A, 1975.

52. Newcomb, T. P.: "Stopping Revolutions: Developments in the Braking of Cars from the Earliest Days," *Proc. Inst. Mech. Eng.,* vol. 195, p. 139, 1981.

53. Fazekas, G. A. G.: "Graphical Shoe-Brake Analysis," *Trans. ASME,* vol. 79, p. 1322, August 1957, also "Some Basic Properties of Shoe Brakes," *J. Appl. Mech.,* vol. 25, p. 7, March 1958.

54. Millner, N., and B. Parsons: "Effect of Contact Geometry and Elastic Deformations on the Torque Characteristics of a Drum Brake," *Proc. Inst. Mech. Eng. (London),* vol. 187, no. 26, pp. 317–331, 1973.

55. Petrof, R. C.: "Structural Analysis of Automotive Brake Drums," SAE Paper 760788, 1976.

56. Day, A. J., P. R. Harding, and T. P. Newcomb: "Finite Element Approach to Drum Brake Analysis," *Automot. Eng. (London),* vol. 5, no. 5, pp. 21–22, October/November, 1980.

57. Lupton, C. R.: "Drum Brakes," *Mach. Des.,* vol. 22, p. 146, July 1950.

58. Frazee, I., and E. L. Bedell: "Automotive Brakes and Power Transmission Systems," Chap. 1, American Technical Society, Chicago, Ill., 1956.

59. Lueck, F. E.: "Why Disk Brakes," SAE Paper 1004B, 1965.

60. Ballard, C. E., R. W. Emmons, F. L. Janoski, and K. Goering: "Rear Disk Brake for American Passenger Cars," SAE Paper 741064, 1974.

61. Klein, H. C., and Rowell, J. M.: "Development of 4-wheel Disk Brake Systems in Europe," SAE Paper 741065, 1974.

62. Dike, G.: "On Optimum Design of Disk Brakes," *J. Eng. Inc., Trans. ASME,* ser. B, vol. 96, no. 3, pp. 863–870, August 1974.

63. Cords, F. W., and J. B. Dale: "Designing the Brake System Step-by-Step," SAE Paper 760637, 1976.

64. Airheart, F. B.: "Design Approaches to Truck Disk Brakes," SAE Paper 740604, 1974.

65. Felske, A., G. Hoppe, and H. Matthai: "Oscillation in Squealing Disk Brakes—Analysis of Vibration Modes by Holographic Interferometry," SAE Paper 786333, p. 1577, 1978.

66. Blok, H. "Fundamental Mechanical Aspects of Boundary Lubrication," *Automot. Eng.,* vol. 46, p. 54, 1940.

67. Sinclair, D.: "Frictional Vibrations," *J. Appl. Mech.,* pp. 207–214, 1955.

68. Rabinowicz, E.: "A Study of Stick-Slip Process," *Proc. Symp. Friction and Wear,* Detroit, Mich., 1957.

69. Rabinowicz, E.: *Proc. Symp. Friction and Wear,* Elsevier Publishing Co., Amsterdam, The Netherlands, p. 149, 1959.

70. Spurr, R. T. "A Theory of Brake Squeal," *Proc. Inst. Mech. Eng.,* no. 1, pp. 33–40, 1961/1962.

71. Jarvis, R. P., and B. Mills: "Vibration Induced by Dry Friction," *Proc. Inst. Mech. Eng.,* vol. 178, Part 1, p. 847, 1963/1964.

72. Millner, N.: "An Analysis of Disk Brake Squeal," SAE Paper 780332, p. 1565, 1978.

SECTION 24

BELTS

William H. Baier, Ph.D.

Director of Engineering
The Fitzpatrick Co.
Elmhurst, Ill.

INTRODUCTION

Belt drives, employing flat, V, or other belt cross sections, are used in the transmission of power between shafts which may be located some distance apart. There are basically two types of belting: friction and positive drive.

Friction belting transmits power by means of friction between belt and pulley or sheave; hence, because of slippage, the velocity ratio between shafts is not constant. Positive drives have projections which mesh with toothed or grooved pulleys.

A number of belt materials are currently in use. Among these are leather, natural and synthetic rubber, cotton, canvas, nylon, and metal. Shaft centerline position may be fixed or may vary.

Belting is generally operated at speeds less than 6500 ft/min. Flat woven or synthetic endless belts are used at speeds up to 18,000 ft/min, film belts to 47,000 ft/min.

24.2 FLAT BELTS

24.2.1 Forces

In transmitting power from one shaft to another by means of a flat belt and pulleys, the belt must have an initial tension T_0. When power is being transmitted, the tension T_1 in the tight side exceeds the tension T_2 in the slack side (Fig. 24.1).

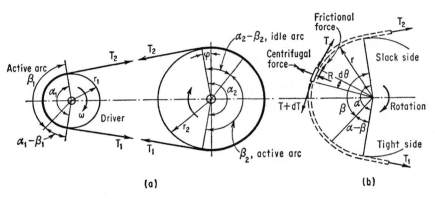

FIG. 24.1   Action of belts on pulleys.

The tension in a belt transmitting power is determined by considering belt deformation under load. For leather belts the stress-strain relationship is not linear. If it is assumed that the total stretch of the belt is the same when it is transmitting power as it is when it is at rest, T_1, T_2, and T_0 are related by the following equations. For vertical (and short horizontal) belts,

$$T_1^{1/2} + T_2^{1/2} = 2T_0^{1/2} \tag{24.1}$$

and for horizontal belts,

$$T_1^{1/2} + T_2^{1/2} = 2T_0^{1/2} + \frac{w^2 E_1 C^2 A^{1/2}}{24} \left(\frac{1}{T_1^2} + \frac{1}{T_2^2} - \frac{2}{T_0^2} \right) \tag{24.2}$$

where w = weight of belt, lb/ft (Sec. 24.3)
  E = elastic constant for leather, determined[1] to be in the range 860 to 900 $(lb/in^2)^{1/2}$

A = the cross-sectional area, in^2
C = center distance, in

The last term in Eq. (24.2) is introduced by the catenary effect resulting from belt sag. For a given T_1 the resulting T_2 is higher for a horizontal belt.

For belts having a linear stress-strain relation the above equations become, for vertical (and short horizontal) belts,

$$T_1 + T_2 = 2T_0 \tag{24.3}$$

and for horizontal belts,

$$T_1 + T_2 = 2T_0 + \frac{w^2 E C^2 A}{24}\left(\frac{1}{T_1^2} + \frac{1}{T_2^2} - \frac{2}{T_0^2}\right) \tag{24.4}$$

where E is the modulus of elasticity of the belt material, lb/in^2.

The horsepower transmitted by the belt is given by

$$hp = (T_1 - T_2)V/33{,}000 \tag{24.5}$$

where V = belt velocity, ft/min.

24.2.2 Action of Belt on Pulley

Because the tension T_1 in the tight side of the belt is greater than that in the slack side T_2, the belt material undergoes a change in strain as it passes around the belt. The belt has a small relative motion, called "creep," with respect to the pulley to compensate for these different strains. Except when the drive delivers maximum power, this creep occurs only on the so-called active arc β of the pulley (see Fig. 24.1). The active arc increases as the effective tension $(T_1 - T_2)$ increases until the active arc equals the entire arc of contact.

The forces acting on the belt are shown in Fig. 24.1b. Summing forces in the tangential and normal directions and substituting and integrating the tangential forces over the active arc, 0 to β, yields

$$(T_1 - wv^2/g)/(T_2 - wv^2/g) = e^{\mu\beta} \tag{24.6}$$

where wv^2/g accounts for centrifugal effects, v = belt velocity, ft/s, and μ = coefficient of friction (Table 24.1). The output of the belt is shown to decrease with increasing belt velocities as a result of centrifugal effects. Equation (24.6) indicates that, at some particular speed, the output drops to zero. Effects other than those considered allow the belt to transmit power even at these speeds and the equation therefore gives conservative results, especially in the case of heavy belts.

Flat belts in line-shaft and machine-belting service are generally operated in the 1000 to 4500 ft/min range. While operation in the 3500 to 4500 ft/min range results in the highest horsepower capacity per dollar expended for belts and pulleys, drive efficiency is less than that obtained when operating at lower speed. Total annual cost of lower-speed (1000 to 3000 ft/min) drives is in most cases lower.

To utilize the weight of the belt in maintaining tension (by increasing the arc of contact) the driving member should rotate so that the tension side of the belt is at the bottom in a horizontal drive. When the drive is vertical, steeply inclined, or horizontal with short centerline distance, no such advantage can be gained from the belt weight.

A number of methods are in use to increase contact pressure or angle of contact to enable the belt to transmit a greater effective tension. One such method utilizes an

TABLE 24.1   Coefficients of Friction μ for Belts and Pulleys*

Belt material	Pulley material					
	Iron, steel	Wood	Paper	Wet iron	Greasy iron	Oily iron
Oak-tanned leather...........	0.25	0.30	0.35	0.20	0.15	0.12
Mineral-tanned leather........	0.40	0.45	0.50	0.35	0.25	0.20
Canvas stitched..............	0.20	0.23	0.25	0.15	0.12	0.10
Balata......................	0.32	0.35	0.40	0.20		
Cotton woven................	0.22	0.25	0.28	0.15	0.12	0.10
Camel hair..................	0.35	0.40	0.45	0.25	0.20	0.15
Rubber, friction.............	0.30	0.32	0.35	0.18		
Rubber, covered.............	0.32	0.35	0.38	0.15		
Rubber on fabric............	0.35	0.38	0.40	0.20		

*Belt dressings or cork inserts tend to raise these values.

idler pulley between driver and driven pulleys. While effective tension is increased, belt life is reduced because of the reversed flexing introduced by the idler.

In a vertical drive, the belt tension at the top of the belt on each side of the pulley is greater than the tension at the bottom, by an amount equal to the weight of the freely hanging belt. It is therefore preferable to have the larger of the two pulleys at the bottom, so that the loss of tension may be offset by an increased contact arc.

24.2.3   Belt Stresses

Figure 24.2 shows the tensile stress acting in an operating belt. Note the location of the maximum stress.

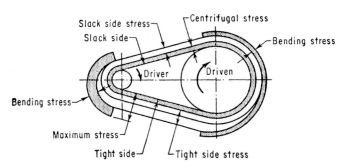

FIG. 24.2   Belt stresses.

24.2.4   Belt Geometry

Open and Crossed Belts.   The length of belt L required for an open drive (Fig. 24.3) is given by the series

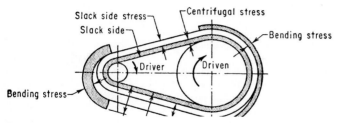

FIG. 24.3   Belt drives.

$$L = 2C + (\pi/2)(D_1 + D_2) + (D_1 - D_2)^2/4C + \cdots \tag{24.7}$$

where C is the center distance, and D_1, D_2 are the diameters of the large and small pulleys, respectively. The length of a crossed belt is given by the series

$$L = 2C + (\pi/2)(D_1 + D_2) + (D_1 + D_2)^2/4C + \cdots \tag{24.8}$$

For rough calculations, the first two terms provide sufficient accuracy. For increased accuracy, D_1 and D_2 should be taken as the pulley diameter plus one belt thickness. For large center distances, it is evident that the same belt may be used for either an open or a crossed drive. Center distance for an open belt is given by

$$4C = b + [b^2 - 2(D_1 - D_2)^2]^{1/2} \tag{24.9}$$

where $b = L - (\pi/2)(D_1 + D_2)$. The above discussion neglects the effects of belt sag.

With drives employing a short center distance and a high speed ratio, the angle of contact on the small pulley is decreased and therefore the capacity of the drive is reduced.

To provide reasonable values of effective tension in drives having angles of contact less than 165° on the small pulley requires high belt tensions. Contact angles less than 155° are never recommended.

Cone Pulleys.   Cone or stepped pulleys (Fig. 24.4) may be used to obtain a variable velocity ratio. Generally, the velocity ratios are chosen so that the driven velocities increase in a geometric ratio from step to step. That is, if the angular velocity of the first pulley is a, that of the second is ka, of the third k^2a, etc. In practice, k is in the range 1.25 to 1.75. The value[4] for machine tools is about 1.2.

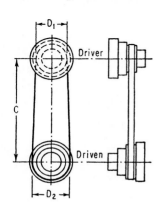

FIG. 24.4   Cone or stepped pulleys.

For crossed belts, for a constant belt length, $D_{11} + D_{21} = D_{1i} + D_{2i}$, where D_{11}, D_{21} are the diameters of the first set of pulleys and D_{1i}, D_{2i} are the diameters of any succeeding pair. In addition, for the velocity ratio n_{11}/n_{2i}, where n_{11} is the angular velocity of the driver shaft and n_{2i} is the angular velocity of the ith driven pulley, the diameters D_{1i} and D_{2i} must satisfy the ratio $D_{2i}/D_{1i} = n_{11}/n_{2i}$. These two relations must be solved simultaneously to determine D_{1i}, D_{2i}.

A more complex relation exists for open belts. If the required diameter ratio is substituted in the open-belt-length equation, the result is

$$L = 2C + (\pi/2)(1 + n_{1i}/n_{2i})D_{1i} + (1 - n_{1i}/n_{2i})^2(D_{1i}^2/4C) + \cdots \qquad (24.10)$$

This equation is solved for D_{1i} and D_{2i} obtained from the required ratio of diameters. Graphical methods for solution of this equation are described in books on mechanisms.

An alternative approximate method for open belts, accurate to four decimal places, follows. Let $m = 1.58114C - D_0$, where D_0 is the diameter of either pulley when $D_2/D_1 = 1$. Then $(D_1 + m)^2 + (k^n D_1 + m)^2 = 5L^2$, where k is the geometric ratio between steps. This equation is solved for D_1 after D_0, C, and k have been chosen. C is the center distance, L the belt length.

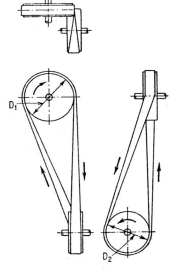

Quarter-Turn Drive. The length of belt L for a quarter-turn drive (Fig. 24.5) is given by

$$L = (\pi/2)(D_1 + D_2) + (C^2 + D_1^2)^{1/2} + (C^2 + D_2^2)^{1/2} \qquad (24.11)$$

Arc of Contact. Arc of contact for an open drive is

$$\theta = \pi \pm \sin^{-1}(D_1 - D_2)/C \qquad (24.12)$$

where the plus sign corresponds to the larger pulley and the minus sign to the smaller pulley. Arc of contact for both pulleys in a crossed drive is

$$\theta = \pi + \sin^{-1}(D_1 + D_2)/C \qquad (24.13)$$

FIG. 24.5   Quarter-turn drive.

24.2.5 Coefficient of Friction

The coefficient of friction between belt and pulley depends upon the materials involved and the condition of the surfaces. Coefficients of friction, for design purposes, for a number of belt and pulley combinations are given in Table 24.1. Barth has shown that the coefficient varies with velocity. Experimental data for leather belts on iron pulleys gave the equation

$$\mu = 0.54 - 140/(500 + V) \qquad (24.14)$$

The coefficient of friction may also be influenced by the cleanliness of the surrounding air, its temperature and humidity, the area of surface contact, the belt-to-pulley-thickness-diameter ratio, and slip. The coefficient rises rapidly after installation of a new belt; changes from 0.25 to 0.65 in a short time have been observed. The variation of the coefficient with belt tension has not been determined. For design purposes, it is ordinarily assumed that the coefficient remains constant.

24.2.6 Creep and Slip

The tension in a belt as it passes around the driver pulley increases from slack-side tension to the tight-side tension. At the driven pulley, the reverse is true. Consequently, the belt elongates as it passes around the driver and shortens at the driven. A greater length of belt leaves the driver than is delivered by the driven pulley; hence the belt must move relative to the surface of both pulleys. This motion, caused by elongation, is called "creep."

At low values of effective tension, creep occurs over only a small arc of the pulley. This arc, called the "active arc of contact," increases with increasing effective tension until the belt creeps over the entire pulley. If the effective tension is further increased, the belt begins to slide on the pulley. The latter motion is called "slip."

A creep velocity of 2 percent of belt velocity is considered reasonable. Slip usually begins at a creep velocity of about 1 to 3 percent of belt velocity. The equation for creep velocity in a leather belt is

$$v = (V_1/2)(\sigma_1^{1/2} - \sigma_2^{1/2})/(830 + \sigma_2^{1/2}) \qquad (24.15)$$

where σ_1 and σ_2 are the unit stresses (lb/in^2) in the tight and slack sides, respectively, and V_1 is the tight-side velocity.

24.2.7 Angle Drives

Open or crossed-belt drives are used to greatest advantage where center distances are large. In many instances, however, space is at a premium and power must be transmitted between shafts whose center lines are not parallel. Shafts which lie at an angle to each other may lie in the same plane, in which case they would intersect if extended, or they may not lie in the same plane. The most common arrangements are parallel and right-angle or quarter-turn shafting.

Parallel shafting on close centers can be connected by means of the nonreversible drive shown in Fig. 24.6. The effect is similar to a crossed drive, the shafts rotating in opposite directions. The drive of Fig. 24.6 is nonreversible because pulley *B* does not deliver the belt in the plane of guide pulley *C*. A reversible drive can be made, however, but the guide pulleys will no longer have a common shaft.

One of the most common angular drives is the quarter turn, in which the driven pulley is mounted at right angles to the driver. As the belt leaves one pulley, it must be turned so as to run onto the other pulley correctly. When rotation is in one direction, guides are usually not required. Guide pulleys are required for reversible operation.

To overcome the belt distortion introduced in quarter-turn drives, the belt is turned through a 180° angle so that, at the joint, the grain side is adjacent to the flesh side. With this construction, wear is equalized on the faces and edges of the belt. An alternative method of combating belt distortion is to make one edge longer than the other. For heavy belts, this method is preferred.

A reversible one-pulley quarter-turn drive is shown in Fig. 24.7. A two-guide-pulley arrangement, not shown, has an advantage in that the belt leaves the pulley in a straighter condition, reducing side stretch. The possibility of the belt's rubbing on itself is eliminated, and tension adjustment is possible if one guide pulley is made movable.

If the shafts intersect in a horizontal plane, the guide pulleys may have to be carried on a vertical axis. In this case, they are called "mule pulleys" and must be heavily crowned to hold the belt. For long belts, crowning will not hold the belt on the pulley and flanges must be provided. The flanges must be made with a reentrant fillet to reduce the tendency for the belt to ride upon the flanges. A guide plate mounted on the shaft adjacent to the pulley may be substituted for the flange. The flange or plate should be regarded as a safety measure; the crowning should be sufficient to keep the belt on the pulley under ordinary circumstances.

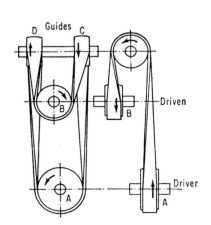

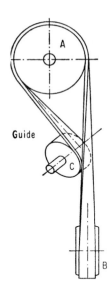

FIG. 24.6   Reversible drive between close parallel shafts.

FIG. 24.7   Reversible quarter-turn drive, one guide pulley.

24.2.8   Flat-Belt Fasteners

Belts may be made endless by splicing and cementing, or the ends may be joined by means of belt fasteners. The strength of a properly made splice is equal to that of the belt, while the use of a fastener decreases belt capacity from 10 to 50 percent dependent upon the fastener used. Fasteners offer a more convenient method of joining belt ends, however, when it becomes necessary to take up stretch or repair a damaged section.

Fasteners are made in four general classes: laces, hooks, plates, and the pressure type in which the belt's ends are held together. Laces are made of rawhide or wire. Hooks are formed from wire and plates are of solid or hinged construction.

Metallic fasteners are fabricated from ordinary steel for normal applications. Stainless steel and monel are used for highly corrosive or abrasive conditions; Everdur (copper, silicon, and manganese alloy) where electrical static protection is required.

24.3   FLAT-BELT MATERIALS

24.3.1   Leather Belting

Leather belting is vegetable-tanned, chrome-tanned, or combination-tanned rawhide, or semi-rawhide. Leather belting comes in the sizes shown in Table 24.2. The weight of leather is about 0.035 lb/in^3. Oak-tanned belts are preferred in dry applications, chrome-tanned and rawhide in damp areas. Waterproof belts can be used in damp or oily applications. Leather belts can be joined by cementing. Such joints, when properly made, have a strength equal to that of the belt. They can also be joined by leather lacing or riveted joints which have a strength one-third to two-thirds of the belt

TABLE 24.2   Horsepower per Inch of Width, Oak-Tanned Leather Belting†

Belt speed, fpm	Single ply		Double ply			Triple ply	
	$^{11}\!/_{64}$ medium	$^{13}\!/_{64}$ heavy	$^{18}\!/_{64}$ light	$^{29}\!/_{64}$ medium	$^{23}\!/_{64}$ heavy	$^{39}\!/_{64}$ medium	$^{34}\!/_{64}$ heavy
1,000	1.8	2.1	2.6	3.1	3.6	4.1	4.5
2,000	3.5	4.1	4.9	6.0	6.9	8.1	8.9
3,000	5.2	5.9	7.2	8.7	10.0	11.6	12.8
4,000	6.4	7.4	9.0	10.9	12.6	14.5	16.0
5,000	7.4	8.4	10.3	12.5	14.3	16.5	18.2
6,000	7.8	8.9	10.9	13.2	15.2	17.6	19.3

	rpm							
Min pulley diam,* in.	0–2,500	2½	3	4	5	8	16	20
	2,500–4,500	3	3½	4½	6	9	18	22
	4,000–6,000	3½	4	5	7	10	20	24

*Belts to 8 in wide. For wider belts add 2 in to minimum pulley diameter for single- and double-ply belts. Add 4 in for triple-ply.
†From National Industrial Leather Association.

strength, and wire-laced joints which have a strength of 85 to 90 percent of the belt strength. Oak-tanned leather can be used up to 110°F, chrome or retanned leather to a somewhat higher temperature. Efficiency is about 98 percent. Belts in which the surface is ribbed or treated to increase friction are available in mineral-tanned or oak-tanned forms. They are not recommended for installations where the belt can slip momentarily. Leather belts are also available as V belts, laminated V belts, block V belts, and round, oval, solid, or plied belting. On step-cone pulleys the narrowest, thickest belt that is applicable should be chosen. Pulley faces should be 0.5 to 2 in wider than the belt.

24.3.2   Nylon-Core Belting

These belts have a nylon-core load-carrying member. Leather or rubberized friction material is used in both surfaces or on the pulley side with nitrile rubber, nylon cloth, or nylon on the other side to provide protection. Lightweight belts are available without facing. Drives range from those which transmit a few inch-ounces of torque to over 1000 hp. Belt speeds of 60,000 ft/min are possible. The high tensile strengths of the core, 40,000 lb/in², results in a thin belt in which the energy required to flex the belt and centrifugal forces are low.

Dehydration and moist heat will result in premature aging. Normal operating temperatures are −20 to 212°F while some plastic-covered belts can stand brief exposures to 300°F. Available friction materials have resistance to oil, gasoline, and static. Nylon core thicknesses of 0.020 to 0.217 in, which corresponds to belt thickness of 0.125 to 0.3125 in, lengths to 75 in, and widths to 36 in are available. Nylon belts can be supplied endless or with ends prepared for hot or cold cementing in the field. Some nylon belts are suitable for joining with fasteners.

Proper initial tension is attained when the belt is elongated as shown in the following table:

Condition	Elongation, %
Normal	1.5–2
Shock loads, heavy starting	2–3
High pulley ratios, dusty pulleys, cone pulleys, variable humidity and temperature	3–4

When the center distance is adjustable, initial tension should be just great enough to transmit the peak load without slipping. Belt pitch length is the measured steel-tape length plus 1 in for 0.020-in belts and 1.75 in for 0.217-in belts.

The thickest suitable belt should be chosen, except where flexibility is important, as in very short drives where the pulleys are almost touching or in multiple-pulley drives. For additional design information, see Refs. 8 to 10 and 14.

24.3.3 Rubber Flat Belts

Such belts are made up of plies of fabric or cord load-carrying members impregnated or laminated with protective or friction-improving materials. Load-carrying materials include prestretched cotton, rayon, nylon, polyester, or steel cables. Because of their distinctive properties, nylon-core belts are discussed in Sec. 24.3.2. Belts in which the load-carrying member consists of plies of rubberized fabric permit the use of metal belt fasteners. When this member consists of cords it must be supplied in endless form or an oil-field-type clamp used. Some belts are spliced in place in the field.

Laminating materials include neoprene, polyvinyl chloride (PVC), rubber, chrome leather, hypalon, and polyurethane. Belts without laminating materials are used where high strength and flexibility are of primary importance. Neoprene or hypalon are used in the place of rubber where oil or grease is present. Neoprene belts can operate at 250°F. Static-resistant belts and dust-jacketed belts, which reduce wear resistance on the edges and provide chemical resistance, are available. A disadvantage of rubberized belts is the decrease in friction that occurs when the rubber lamination wears away.

Initial tension should be in the range of 15- to 25-lb/ply per inch of width, or such that the belt elongates 1 percent for normal conditions and 1.5 percent for severe load. These percentages should be reduced 50 percent under humid conditions. Ultimate tensile strength ranges from 280 to 600 lb/ply or more per inch of width. A take-up of 2 to 4 percent of belt length should be provided for stretch in service or for manufacturing variations. Depending on construction, widths to 60 in are available. Belt speeds in the range 2500 to 3000 ft/min are considered low speed; 3000 to 4000 ft/min, medium speed; and 4000 to 5500 ft/min, high speed. Such belts are limited to 6000 ft/min. Maximum power is about 500 hp. Belts with cord load-carrying members are well suited to serpentine drives.

24.3.4 Cotton and Canvas Belting

Fabric belts of this type are made up of layers of woven material held together by gums, rubbers, or synthetic resins which also serve to protect the fabric. Stitched canvas belts are inexpensive belts which need little protection against oil or moisture. Friction coefficient is low (0.15 to 0.22 against a steel pulley); hence pulley bearing pressure will be high. Resistance to high temperature is good. Bituminous-coated belts resist moisture and are recommended for rough usage. Combination leather and cotton belts consist of a single or double ply of oak or chrome leather cemented to a woven cotton backing.

Initial tension should be 20 to 25 lb/ply per inch of width.

Because of the lower coefficient of friction, the largest possible pulleys which do not produce excessive belt speeds should be used to reduce bearing loads. Cotton and canvas belts are limited to belt velocities of 5000 to 6000 ft/min.

24.3.5 Balata Belting

Balata belts are made from closely woven duck of high tensile strength impregnated with balata gum obtained from South American trees. Balata is tough, stretches very little, is waterproof, resistant to aging, and affected by mineral oil but not by animal oils or humidity. Maximum allowable ambient temperature is 100° to 120°F. Initial tension is ordinarily in the range 22 to 25 lb/ply/per inch of width. Balata belting can be laced by metal fasteners or made endless. Its horsepower capacity is similar to that of canvas stitched or rubber belts. The coefficient of friction remains high even under damp conditions.

The specific weight of balata and other belts in lb/in^3 is as follows: leather ranges from 0.035 to 0.045, while canvas is 0.044, rubber 0.041, balata 0.040, single-woven cotton belt, 0.042, and double-woven cotton belt 0.045.

24.3.6 Steel Belting

Thin steel belts are generally used in installations where belt speed is high. Initial tension is generally high because of the lower coefficient of friction. Stretch is low because of the high elastic modulus of steel. Steel belts provide very high power transmission for a given belt cross section. Belt material is similar to that used for clock springs. Carbon content is high; the material is drawn and rolled, and ground to size. It is available in widths from ½ to 3 in in 0.01-in thickness. Tensile strength is in excess of 300,000 lb/in^2. Belt material has rounded edges to reduce the hazard while running.

Joints may be riveted or silver-soldered. Rivets can be phosphor bronze or cold-drawn iron wire. Joints are about 80 percent efficient, failure occurring by shear. Silver-soldered butt joints, with 60° beveled edges, which can be used with small pulleys, are about 65 percent efficient, rupture occurring in the adjoining belting, which is softened by the joining process. Cover-plate joints are bent to conform to the smallest pulley.

Hampton,[3] using 0.01- by 0.75-in steel belting on cork pulleys, found that horsepower transmitted increased in proportion to slip velocity until the tension in the slack side approached that due to centrifugal effects, at which point slip velocity increased rapidly without a corresponding increase in horsepower transmitted. This breakaway point occurs at higher horsepowers for higher belt speeds because effective belt pull for a given horsepower is lower at higher speeds. The power in the range before slip occurred is given by

$$\text{hp} = 0.17T_1^{0.65}V_s \qquad (24.16)$$

where V_s = slip velocity and T_1 = tight-side tension.

Friction coefficient f rises with slip velocity, but drops with increasing belt speed. For the belts of the previous reference,

$$f = V_s^{1.37} \times 3.2 \times 10^5/(p^{0.5} \times V_d^{1.5}) \tag{24.17}$$

where p = unit pressure on the driver pulley face, lb/in^2, and V_d = belt velocity, ft/min. Drive efficiency is over 98 percent. Slip velocities to approximately 3 ft/min are considered reasonable. Young[4] showed that the horsepower which can be transmitted by a steel belt running on wood pulleys is

$$\mathrm{hp} = 1.51p^{0.81}/1000V_d^{0.78} \tag{24.18}$$

24.3.7   V-Ribbed Belting (Grooved, Poly-V)

These belts are flat belts with a ribbed underside. The ribs, which run in grooves in the sheaves, provide some wedging action. V-ribbed belts require less tension than flat belts, but more than V belts. They are usually more efficient than a flat belt and sometimes a V belt. The thin belt construction, which minimizes flexing, permits the use of small pulleys. They may be used on vertical and turned drives and, with back-bend idlers, on serpentine drives. They are not used in multibelt drives. If used in cross drives, a steel plate, located at the point of belt intersection, must be provided to stop the belt from rubbing on itself. Standard designations for these belts are J, L, and M. Designation H is used on belts which can run on small pulleys and K for automotive applications. Because the ribs bottom in the groove, the drive ratio does not change as the belt wears, as is the case with V belts.

24.4   FLAT-BELT WORKING STRESSES

24.4.1   Design Stresses and Loads

Allowable tension for flat-belt materials is customarily expressed in terms of lb/in width or lb/ply/in width rather than in terms of allowable stress, lb/in^2. Ultimate strength for oak-tanned leather belts is in the range 3000 to 4500 lb/in^3; for chrome-tanned leather, 4000 to 5500 lb/in^2; for rubber, balata, cotton, or canvas belts, 900 to 1500 lb/in^2; and for nylon-core belts, 35,000 lb/in^2.

As belt thickness and number of plies increase, recommended maximum allowable tension, lb/in width, increases. For leather belting particularly, however, recommended maximum allowable stress, lb/in^2, and maximum allowable tension per ply, lb/ply/in width, decrease as a result of increased bending stresses at the pulleys.

Recommended maximum values of allowable tension (lb/in width) for oak-tanned leather belts range from 108 for single-ply medium to 275 for triple-ply heavy; for rubber, canvas, and balata, 32-oz duck, they range from 75 for 3-ply to 300 for 12-ply. Recommended maximum stress for oak-tanned leather belts ranges from 620 lb/in^2 for single-ply medium to 520 lb/in^2 for triple-ply heavy. For rubber, balata, or canvas belts the corresponding value is 500 lb/in^2.

24.4.2   Horsepower Ratings of Flat Belts

The Goodyear Company has developed the following empirical relation relating service life I to d, the small pulley diameter; l, the belt length; S, the belt speed; N, the belt thickness; and T, the tight-side tension (Ref. 2):

$$I = kd^{5.35}l/S^{0.5}N^{6.27}T^{4.12} \qquad (24.19)$$

where k is a constant dependent upon the particular belt characteristics. From Eq. (24.19) it is seen that a reduction in small pulley diameter, an increase in belt thickness, or an increase in tight-side tension have the greatest effect in decreasing belt life. A 50 percent decrease in pulley diameter will decrease drive life to $\frac{1}{32}$ its original value; a 10 percent increase in tension will decrease belt life 60 percent.

The difference in horsepower recommendations among several manufacturers for a given type of belt may result from a difference in choice of service life. In addition some manufacturers base their recommended horsepower for smaller belts on a shorter service life because ordinarily the user does not expect so long a life from such drives.

24.5 FLAT-BELT DESIGN

Design techniques for all types of flat belts are basically similar; however, variations are encountered as outlined in this section. Current design is generally based on empirical data concerning the load-carrying capacity of a given belt. These recommended horsepowers are chosen to provide reasonable life under most circumstances. As may be deduced from Eq. (24.19), belt horsepower capacity is a function of desired service life, pulley diameters, belt length, belt speed, belt thickness, and tight-side tension. In addition, such factors as service conditions and contact angle must be included.

Existing practice is such that the effect of some factors such as belt length is ignored. In the case of leather belts, that of contact angle is neglected, while for nylon belts, no account is taken of service conditions. These omissions are permissible because of their comparatively minor effects or because recommended horsepower values are somewhat conservative.

In the design of a flat-belt drive, the diameters of the driver and driven pulleys, center distance, horsepower, and belt speed are first determined. Then, in accordance with the recommendations of Table 24.2 for leather belting; Table 24.3 for rubber, balata, woven cotton, or canvas stitched belting; and Table 24.4 for nylon belting, a belt of proper thickness is chosen. From these tables, the rated horsepower per inch of width (hp_R) is also obtained.

To account for variations from the normal conditions assumed in Tables 24.2, 24.3, and 24.4 for (hp_R), these values must be modified by service factors to obtain the belt design horsepower per inch of width (hp_D). Recommendations regarding appropriate service factors vary for different types of belting as listed below.

1. Leather:

   F, special operating conditions (Table 24.5)

   M, motor and starting method (Table 24.5)

   P, small pulley diameter (Table 24.5)

   Accordingly, for leather belts,

$$hp_D = hp_R(P/FM) \qquad (24.20)$$

2. Rubber, cotton, canvas, balata:

   K_1, arc of contact (Table 24.6)

   F, special operating conditions (Table 24.5)

   M, motor and starting method (Table 24.5)

TABLE 24.3 Horsepower per Inch of Width: Rubber, Balata, Cotton Duck, and Canvas Stitched Belts

35-oz silver hard duck, ply	Small pulley diam, in.	Belt speed, fpm			
		1,000	2,000	4,000	6,000
4	4	1.07	1.87	2.75	2.31
	8	1.98	3.52	5.72	6.49
	12	2.20	4.18	7.04	8.47
	14+	2.20	4.29	7.59	9.24
6	8	1.87	3.41	5.17	4.84
	12	2.86	5.06	8.36	9.46
	16	3.30	6.05	10.01	11.77
	28+	3.30	6.49	11.33	13.86
8	14	2.86	5.17	8.14	8.36
	20	4.07	7.37	12.21	13.97
	26	4.40	8.58	14.52	17.49
	36+	4.40	8.58	15.18	18.37
10	24	4.18	7.48	11.88	12.76
	30	5.39	9.68	17.16	18.81
	36	5.50	10.67	17.82	21.45
	48+	5.50	10.78	18.92	22.99
32-oz duck:					
4	4	0.9	1.2	2.2	1.8
	8	1.5	2.8	4.5	5.1
	12	1.8	3.5	5.8	7.0
	18+	1.8	3.6	6.3	7.7
6	8	1.4	2.5	3.9	3.5
	12	2.3	4.1	6.6	7.5
	16	2.7	4.9	8.1	9.6
	26+	2.7	5.3	9.4	11.5
8	16	2.5	4.4	7.0	7.4
	22	3.5	6.3	10.3	12.1
	28	3.6	6.9	11.5	13.9
	36+	3.6	7.1	12.5	15.3
10	28	3.7	6.7	10.9	12.2
	34	4.5	8.6	14.4	17.3
	40	4.5	8.9	15.2	18.4
	44+	4.5	8.9	15.7	19.2

Accordingly, for such belts,

$$\text{hp}_D = \text{hp}_R(K_1/FM) \tag{24.21}$$

3. Nylon core:

K_1, arc of contact (Table 24.6)

Accordingly, for nylon-core belts,

$$\text{hp}_D = \text{hp}_R(K_1) \tag{24.22}$$

When the allowable design horsepower per inch of width (hp_D) is established, the necessary belt width t is obtained from

$$t = \text{hp}_T/\text{hp}_D \tag{24.23}$$

where hp_T is the horsepower being transmitted by the belt. For more information on flat-belt design, see Refs. 5–7.

TABLE 24.4 Horsepower Capacity per Inch of Width, Nylon-Core Belts

Nylon-core thick-ness, in., approx	Small pulley diam, in.	Belt speed, fpm									
		1,000	2,000	3,000	4,000	5,000	6,000	7,000	8,000	9,000	10,000
0.020	½	0.15	0.20	0.30	0.40	0.50					
	1	0.20	0.30	0.50	0.62	0.81					
	2	0.30	0.55	0.85	1.12	1.45					
0.030	1	0.55	1.0	1.62	2.02	2.62	3.25	3.37	3.5	3.62	3.7
	4	1.12	2.25	3.12	4.0	5.25	6.0	6.5	6.7	6.8	6.9
	8	1.3	2.37	3.62	4.7	5.7	6.7	7.5	8.0	8.1	8.2
0.060	2	1.5	2.75	4.37	5.02	6.7	8.0	9.2	10.0	10.5	10.6
	4	1.87	4.6	6.0	6.5	10.0	11.7	13.2	13.7	15.2	15.4
	12	2.75	5.0	8.0	10.0	12.5	15.0	17.5	18.7	20.0	20.2
0.090	4	3.75	5.62	9.0	11.7	13.7	17.0	18.7	21.2	22.5	23.0
	8	4.12	6.87	10.5	13.7	17.5	20.0	22.5	25.0	27.6	28.0
	20	4.5	8.12	13.0	16.0	21.2	25.0	27.5	31.2	33.7	35.0
0.120	8	5.5	9.5	14.5	18.7	22.7	27.7	32.5	35.5	40.0	42.0
	15	6.35	11.87	18.7	23.8	29.0	35.0	37.0	40.0	44.0	46.0
	40	7.0	13.0	20.0	25.5	32.5	37.5	42.5	45.0	48.0	50.0

Note: Published data vary. Other recommendations range from 75 to 100 percent of these values.

TABLE 24.5 Correction Factors for Belt Drives

Factor	Condition	Factor F
F, special operating conditions............	Oily, wet, dusty	1.35
	Vertical drive	1.2
	Jerky loads	1.2
	Shock and reversing	1.4
	Method	**Factor M**
M, motor and starting method...........	Squirrel cage, compensator start	1.5
	Squirrel cage, line start	2.0
	Slip ring and high start torque	2.5
	Diam, in.	**Factor P**
P, small pulley diameter....	0–4	0.5
	4.5–8	0.6
	9–12	0.7
	13–16	0.8
	17–30	0.9
	30+	1.0

TABLE 24.6   Arc Correction Factor K_1

Arc of contact,* deg	Factor
180	1.00
174	0.99
168	0.97
162	0.96
156	0.94
150	0.92
144	0.90
138	0.88
132	0.87
126	0.85
120	0.83
114	0.80
108	0.78
102	0.75
96	0.72
90	0.69

*Small sheave.

24.6   FLAT-BELT PULLEYS AND IDLERS

24.6.1   Pulleys

As the belt passes around the pulley, it flexes, and bending stresses are induced. As a result of the reduced radius of curvature, bending stresses are higher for small pulleys. Because of the reduced fatigue stresses, a large pulley results in an increased belt life. Pulleys should not be so large, however, as to cause excessive belt speeds.

Table 24.2 lists minimum pulley sizes for leather belting. If pulley diameters less than those shown are used, belt life will be greatly reduced as a result of the higher bending stresses and greater elongation of the outer belt fiber.

In order to aid in keeping the belt on the pulley face, it is recommended that the face be crowned as shown in Fig. 24.8, which also shows approximate pulley proportions. An alternate form of crowning consists of making the central half of the pulley face cylindrical and incorporating a slight straight or curved taper in the face at either side of the cylindrical portion.

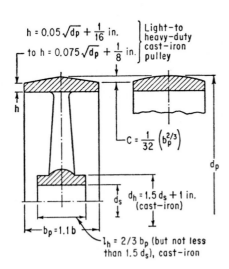

$$h = 0.05\sqrt{d_p} + \tfrac{1}{16} \text{ in.}$$
$$\text{to } h = 0.075\sqrt{d_p} + \tfrac{1}{8} \text{ in.}$$
Light-to heavy-duty cast-iron pulley

$$C = \tfrac{1}{32}\left(b_p^{2/3}\right)$$

$$d_h = 1.5\,d_s + 1 \text{ in.}$$
(cast-iron)

$$b_p = 1.1\,b$$

$$l_h = 2/3\,b_p \text{ (but not less than } 1.5\,d_s), \text{ cast-iron}$$

FIG. 24.8   Crowned pulley proportions.

24.6.2 Idlers

Idlers are always flat-faced and located on the slack side of the belt next to the small pulley, whether the small pulley is driving or driven.

Only where the drive is steady and the small pulley is driven may the idler be fixed.

Spring-loaded idlers used as gravity idlers have the disadvantage of relieving the spring tension on the belt when the slack increases upon application of load.

The clearance between the small pulley and the idler should not exceed 2 in, and the idler should be positioned so that the belt wrap around the small pulley is 225 to 245°.

Idlers should be well balanced.

24.7   V BELTS

24.7.1   Introduction

The cross section of a V belt (Fig. 24.9) is such that belt tension wedges the belt into the sheave groove, resulting in increased frictional forces. When compared with a flat-belt drive, a V-belt drive can operate at a reduced angle of contact, with lower initial tension, is usually more compact, and has less severe pulley alignment requirements. The load-carrying member consists of one or more layers of cord composed of nylon, rayon, cotton, glass, or steel. The cords are often encased in a soft rubber matrix, the cushion section. This section is encased in harder synthetic or natural rubber. The entire belt is enclosed in an abrasion-resistant rubberized canvas cover.

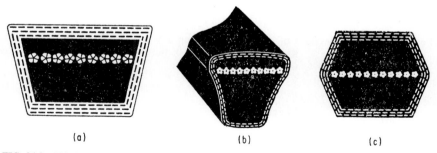

(a)   (b)   (c)

FIG. 24.9   V-belt construction.

V belts may be used for small center-distance drives without idlers, for speed ratios as high as 10:1, and for belt speeds to 15,000 ft/min. Belts operate most efficiently at belt speeds of 1500 to 6000 ft/min, and at temperatures in the range −30 to 140°F. Belts are available for operation at −60 and 185°F. Because they are subject to creep, they should not be used on synchronous drives.

If the belt squeals in service, belt tension should be increased. Belt dressing, oil, or gasoline should not be applied. Soap and water should be used for cleaning. Grooved idlers should be used on the inside surface of the belt. Flat idlers, whose diameter is 1.33 times that of the smallest pulley, should be used on the back surface. A tight-to-slack-side tension ratio of 5:1 is used on 180° arc of contact drives. V belts are made

TABLE 24.7   Properties of V Belts

Belt section	Width a, in	Thickness b, in
Heavy-duty standard and super		
A	0.50	0.31
B	0.66	0.41
C	0.88	0.53
D	1.25	0.75
E	1.50	0.91
Heavy-duty narrow		
3V	0.38	0.32
5V	0.62	0.54
8V	1.0	0.88
Light-duty		
2L	0.25	0.16
3L	0.38	0.22
4L	0.50	0.31
5L	0.66	0.38

Note: The included angle of the sheave is less than that of the belt, so that wedging will increase the effective friction. The difference should not be too great or the force required to pull the belt free will be excessive.

in the standard letter sizes of Table 24.7 and numerous other sizes. For information regarding V belt drives, see Refs. 11 to 13.

24.7.2   Forces

Forces normal to the sides of the grooves act on a V belt as shown in Fig. 24.10. Referring to Fig. 24.10, the normal force on the groove face

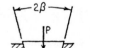

FIG. 24.10   V-belt forces.

$$P_n = P/2 \sin \beta \qquad (24.24)$$

The tractive force

$$F = 2\mu P_n = 2\mu P/2 \sin \beta = \mu_e P \qquad (24.25)$$

where P = the sum of the tight- and slack-side tensions = $T_1 + T_2$
μ = coefficient of friction
μ_c = effective coefficient of friction = $\mu/\sin \beta$

The relation between tight and slack tensions [Eq. (24.6)] is applicable to V belts when μ_e replaces μ in that equation.

24.7.3   Geometry

Center Distance.   Drive center distance is generally larger than the larger sheave diameter, but less than the sum of the larger and smaller sheave diameters. Because the life of the V belt is substantially shortened by slack-side shock and vibration, long center distances, in excess of 2½ to 3 times the large sheave diameter, are usually not recommended for V belts.

Provision should be made to permit a small increase or decrease in the center distance to accommodate the nearest standard size of belt.

Larger center distances may be used with link-type V belts which, because of better balance, are subject to less vibration.

The drive should make provision for the adjustment of center distance above and below the nominal value for belt take-up after stretching and installation, respectively. Take-up allowance should be at least 2½ percent of the belt length for belts up to 200 in long, slightly less for longer belts. The application allowance should be at least equal to the belt thickness, and up to three times this long for heavier sections and longer belts.

In multiple-belt drives, all belts should elongate equally in order to assure an equal distribution of load. When one belt breaks, all the belts should be replaced in order that all the belts will have nearly equal properties.

The adjustable sheave in a multiple-belt drive should permit an increase in center distance of at least 5 percent above the nominal value and should also permit a decrease in this distance by an amount at least equal to the belt thickness to permit installation.

24.7.4   Sheaves

Sheaves are generally made of wood, stamped sheet steel, or cast iron. For higher speeds, cast-steel sheaves are used.

A minimum clearance of ⅛ in is required between the inner surface or base of the belt and the bottom of the sheave groove. This clearance prevents the belt from bottoming as it becomes narrower from wear.

It is important that sheaves be statically balanced for applications up to 6500 ft/min and dynamically balanced for higher speeds.

Sheaves are available which permit the groove width to be adjusted, thus varying the effective pitch diameter of the sheave and permitting moderate changes in the speed ratio.

The diameter of the small sheave should not be less than the values given in Table 24.7.

Sheaves should be as large as possible; however, if possible, the belt speed should not exceed 6500 ft/min.

24.7.5   V-Flat Drives

Where the speed ratio is 3:1 or more and the center distance is short (equal to or slightly less than the diameter of the large sheave), it is possible to develop the required tractive force and maintain efficiency while omitting the grooving from the face of the large sheave. In this case, the belt contact is at its inner surface and the cost of cutting the grooves of the large sheave is eliminated. The maximum tractive power of the belts is obtained when the contact angle on the larger sheave is in the range 240 to 250° and that on the smaller sheave is 110 to 120°.

When power must be transmitted to grooved sheaves from both the top and bottom of the belt, double V belts, made to fit standard V grooves, are used (see Fig. 24.9c).

24.7.6 Quarter-Turn V-Belt Drives

Quarter-turn drives may be used with V belts. The horsepower rating of a V belt used in this arrangement is 75 percent of its rating in a straight drive under the same load conditions.

If the sheaves are relatively small, and take-up can be provided on one of the shafts, an idler may be omitted. The minimum recommended center distance in this case is $6(D + b)$, where D is the large sheave diameter and b is the belt width.

An idler is necessary to keep the belt on the sheaves when a larger speed ratio is used. The center distance between the idler and the small sheave should not be less than $8b$.

24.8 V-BELT TYPES

24.8.1 Heavy Duty Narrow (Industrial Narrow)

This line of three belts, 3V, 5V, and 8V, whose dimensions are given in Table 24.7, serves the same application area as classical V belts. While their constructions are generally similar, the narrower wedge shape of the narrow belt results in a greater power-transmitting capability. A narrow belt drive can transmit up to three times the horsepower of a classical belt of the same size. They have high resistance to oil, temperature, and environmental conditions, and can run at speeds to 6500 ft/min without dynamically balanced pulleys.

24.8.2 Heavy Duty Standard (Classical, Conventional)

This line of five standard cross sections A, B, C, D, and E, whose dimensions are given in Table 24.7, is normally used in multiple belt installations. They are made with one or more layers of load-carrying cords encased in gum rubber. Single-layer cord belts are used in high-speed, short-center-distance drives with small sheaves. Temperature range is from -30 to $145°F$. Standard belts are ordinarily run at speeds of less than 6000 ft/min.

24.8.3 Heavy Duty Super (Heavy Duty Classical)

This line of five belts has a construction similar to conventional belts and the same code designation, as given in Table 24.7. They provide about 30 percent more power transmission capability than a conventional belt of the same size. Some are notched to permit running over small pulleys.

24.8.4 Light Duty (Standard Light, Single-V)

The dimensions of this line of four belts, 2L, 3L, 4L, and 5L, whose construction is similar to and slightly thinner than conventional belts, are given in Table 24.7. They are intended for fractional horsepower applications where one belt is required. Because they have only one load-carrying layer encased in rubber and are usually enclosed in a fabric layer, they can flex around small pulleys. Service is usually intermittent and overall operating life shorter than in industrial applications.

24.8.5   Agricultural Belts

Agricultural V belts are available in nine cross sections. Five sections, HA, HB, HC, HD, and HE, have the same cross sections as classical V belts. Four sections, HAA, HBB, HCC, and HDD, are available in a double-V or hexagonal cross section. To meet agricultural application requirements, construction differs from that of the classical belt. These requirements include intermittent shock loads and the ability to bend in reverse over small-diameter sheaves. As a result, agricultural belts incorporate an undercord which acts to prevent the belt from wedging into the sheave when subjected to shock loads, and a flexible jacket. Many agricultural drives have more than one driven shaft and, in addition, have idlers. Double-V belts are intended for such drives. Agricultural belts are available in the joined configuration for multiple-belt drives. These belts operate in classical V-belt drive sheaves.

24.8.6   Automotive Belts

These belts, intended for automotive accessory applications, where space is limited, have a narrow cross section. They must transmit high horsepower over small-diameter pulleys. They are available in six sizes, having top widths of 0.38 to 1 in. The two smallest sizes, SAE sizes 0.380 and 0.500, where the designation equals the top width in inches, are most widely used. Because several accessories are often driven from a single belt, design problems are complex and similar to those for agricultural belts.

24.8.7   Open-End V Belts

These belts, which have cross sections similar to those of conventional belts, have prepunched holes. They are cut to length and the open ends connected by a fastener. The load-carrying members consists of layers of fabric which hold the fasteners securely. They are limited to low-power applications and to speeds under 4000 ft/min.

24.8.8   Link V Belts

These belts consist of a series of links, having a V cross section, which are fastened together with metallic fasteners. They may be joined together in the field. Such belts run without vibration. They tend to elongate until all links are run in and seated. The higher-power types are designated by a T symbol (TD and TE, for example).

24.8.9   Joined V Belts

These belts consist of two or more heavy duty conventional or narrow V belts bonded together at their upper surfaces. They tend to reduce lateral vibrations and the possibility of the belt turning over.

24.8.10   Double-Angle V Belts

These belts are essentially two heavy duty conventional belts molded back to back with the load-carrying layer at the junction. They may be used on serpentine drives containing reverse bends or where power is to be transmitted from the top and bottom of the belt. The five belts in this series are designated AA, BB, CC, DD, and EE. The symbol DG, used together with the belt symbol, indicates that the belt runs in a deep-groove pulley.

24.8.11   Low-Stretch Cable (Steel Cable) Belts

These belts are similar in construction to standard belts. They have high-strength steel or glass-fiber cords to provide for high power-transmission capability. They are notched for greater flexibility and can run efficiently at speeds to 10,000 ft/min. Their construction provides little shock-absorbing capability and is intolerant of misalignment.

24.8.12   Wide Angle V Belts

These belts have a large V angle that results in reduced wedging action and more dependence on the capacity of the load-carrying members. The high friction coefficient of the polyurethane encasing material compensates for the reduced wedging action. Applications include fractional horsepower office equipment, automotive, and light industrial equipment. They run over pulleys as small as 0.67 inch in diameter and at belt speeds over 10,000 ft/min. There are four belts in this metric series: 3M, 5M, 7M, and 11M.

24.8.13   Flexible Sidewall V Belts

The load-carrying member of these belts is a prestretched aircraft cable around which is molded a polyurethane, nylon, or other plastic V-shaped member. The cross section of the V-shaped member is such that the sidewalls are free to flex to permit the belt to conform to sheaves of different included groove angles. The belt consists of a series of individual V-blocks spaced along the load-carrying member. The individual blocks are free to move relative to each other to permit flexing around smaller pulleys than conventional V belts. The belts are available in four sizes, 2V, 3V, 4V, and 5V, having the same cross section as the corresponding L-series light-duty belt. Published data indicate the belts may be run at speeds to 6500 ft/min and transmit up to about 15 hp.[19]

24.9   V-BELT DESIGN

Current design techniques for V belts are based on empirical data concerning the load-carrying capacity of a given belt. Drives designed in accordance with these recommendations will have reasonable life under most circumstances. These techniques are described in detail in Refs. 11, 20, and in many V belt manufacturers' catalogs. A summary is given in the next paragraph.

In the design of a drive, the pitch diameters of the driving and driven pulley, center distance, given horsepower, belt speed, and speed ratio are first determined. If the driver is an electric motor, the diameter of the smallest pulley must exceed those recommended by manufacturers of V belts and motors. The proper belt section, drive service factor, and degree-of-maintenance service factor are determined from Ref. 20 or manufacturers' catalogs. The horsepower for which the drive is designed is called the transmitted horsepower hp_T. It is determined by multiplying the given horsepower by the appropriate service factors.

The belt length is determined from Eqs. (24.8) or (24.9), where D and d are the resulting pitch diameters. The nearest standard belt length is chosen and the resulting center distance calculated. The arc of contact for the smaller sheave is calculated from Eq. (24.12) and the arc of contact factor K_1 determined from Table 24.6. A length correction factor K_2, the additional horsepower for the design speed ratio, and the nominal horsepower per belt hp_B for a drive having an arc of contact of 180° are determined from Ref. 20 or manufacturers' catalogs. The additional horsepower for the design speed ratio is added to the nominal horsepower belt to obtain the rated horsepower per belt hp_R. This sum is multiplied by K_1 and K_2 to obtain the design horsepower per belt hp_D.

Accordingly, design horsepower for V belts is

$$hp_D = hp_R(K_1K_2) \tag{24.26}$$

where the rated horsepower per belt is based on the rating diameter.

The number of belts required, n, is determined from

$$n = hp_T/hp_D \tag{24.27}$$

where hp_T is the horsepower transmitted by the drive.

24.10 SYNCHRONOUS BELTS (TIMING, POSITIVE)

In these belts, one surface is formed into teeth which mesh with similar teeth in the mating sheave. They do not depend on friction to transmit power; speed is uniform, there being no chordal rise and fall of pitch line. The synchronous belt, shown in Fig. 24.11, has load-carrying members which can be carbon steel, rayon, glass, or stainless steel. Encasing material is usually neoprene. The teeth, molded integrally with the encasing material, can be a moderately hard rubber or neoprene which may be nylon-jacketed. Power may also be transmitted by frictional means from the back of the belt; however, such transmission will not be synchronous. Stock belts can transmit up to 275 hp, special belts to 800 hp. Speeds to 16,000 ft/min are possible, although the larger pitch belts, 0.875- and 1.25-in pitch, should not be run over 8000 ft/min. Stock neoprene belts can be run at temperatures from −30 to over 185°F. Special encasing materials permit operation from −65 to 260°F, and may be static-conductive or resistant to oil and chemicals.

There are two series of synchronous belts, the conventional trapezoidal tooth series and the newer high-torque drive (HTD) curvilinear tooth series, which was developed

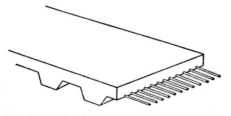

FIG. 24.11 Synchronous belt construction.

TABLE 24.8 Synchronous-Belt Sizes

Conventional belts		HTD belts
Designation	Pitch, in	Pitch, mm
MXL	0.080	3
XL	$\frac{1}{5}$	5
L	$\frac{3}{8}$	8
H	$\frac{1}{2}$	14
XH	$\frac{7}{8}$	
XXH	$1\frac{1}{4}$	

for applications where chains can be used. Conventional belts are available in six pitches; HTD belts, in four pitches, as shown in Table 24.8.

Synchronous belts do not require an initial tension. The two largest conventional pitch belts can be installed slightly slack if shock loads are high. Belts must be taut enough to avoid contact between tight and slack side teeth. Belt width should be approximately equal to the diameter of the smaller pulley. If the belt is too wide, the inherent side thrust of these belts may be excessive. For short drives, with center distance less than 8 times the small pulley diameter, and with both shafts horizontal, one pulley should be flanged. For all drives in which the shafts are vertical and for long horizontal-shaft drives, both pulleys should be flanged. When the drive has more than two pulleys, every other pulley should be flanged. Idlers may contact either face to adjust belt length, for power takeoff, to increase arc of contact, and to prevent belt whip. In the last case, the idler is placed on the slack side on either face. If the idler contacting the inside face of the belt is at least 6 in in diameter, it may be a noncrowned flat pulley. All flat belt idlers should be noncrowned.

To reduce slippage on a high-ratio flat-belt application, a synchronous belt running over a synchronous-belt small pulley and an uncrowned flat-belt large pulley may be used. Synchronous belts are not recommended for quarter-turn, mule, and crossed drives. If they are used, center distance should be large and the belts themselves as narrow as possible. Quarter-turn drives above 2.5:1 ratio should be replaced by a two-step drive consisting of a 1:1 quarter-turn drive, followed by a conventional two-pulley reducing drive. In high-speed drives, the smallest pitch belt that will transmit the power should be used to keep centrifugal forces low.

Conventional belts can be ordered in other pitches, widths, and lengths. Widths to 14 in are available. Belts over 180 in long are available in the two largest pitches; however, they are factory-spliced and have reduced power-transmitting capability. HTD 14-mm belts are available in lengths up to 270 in. Synchronous belts, with teeth on both sides, are available in both systems for use in serpentine drives.

Belts are available in 40DP, 0.0816-in pitch with polyester load-carrying materials and molded polyurethane encasing material. A design recently introduced may be spliced to any length in the field. No pulley side flanges are required because the load member, which is encased in polyurethane, runs in a groove in the pulleys.

24.11 VARIABLE-SPEED BELTS

V belts may be used in variable-speed drives whose speed variation is affected by moving one sheave side wall with respect to the other. The speed is infinitely variable within the specified range. In simple drives, the drive must be stopped to change speed. Drives are available in which speed is adjusted manually or automatically in response to speed or torque requirements without stopping the drive.

Five types of V belt are used in variable-speed drives. Where the range of speed variation is limited, classical V belts are often used. Usually only the motor pulley has variable pitch. Under these conditions, speed variation is limited to 1:1.4. If both pulleys have variable pitch, variation can be 1:1.96. Sections A, B, and C are used in these drives.

Industrial variable-speed belts have thin, wide sections which permit speed variations to 8:1 and horsepowers per belt of 32 or more. There are many cross sections having top widths from 0.875 to 3.0 in. Some are standardized including P, Z, R, T, and W.

Agricultural variable-speed belts are thicker than industrial variable-speed belts to handle the higher power requirements of such applications. There are five standard cross sections having top widths from 1.0 to 2.0 in. If both pulleys have variable pitch, the maximum speed variation is 1:3.67.

Recreational-vehicle belts are used in lightweight, highly responsive variable-speed transmissions. They are subject to high, rapidly changing loads with sidewall temperatures reaching 400°F. Belts are customized using special rubber compounds and tensile reinforcement.

Automotive variable-speed belts have been in limited use for main traction and accessory drives for many years. Development is being accelerated because of the high efficiency of such a drive when compared to existing automatic transmission. Improved materials and configurations are aiding the development of these custom belts.

Permissible speed variation increases as belt thickness decreases, width increases, and as groove angle increases. Because a wide, thin belt required to produce a large speed variation tends to collapse under heavy loading, a compromise between speed variation and load-carrying capacities is necessary.

24.12 ENDLESS BELTS

Endless belts are made in a variety of constructions, shapes, and materials. Load-carrying members are both natural and synthetic yarns. Coating materials include butyl, neoprene, acrylonitriles, and fluorocarbon rubbers; vinyl and acrylic resins; and polyurethanes and polyethylenes whose durometer measurements range from 50 to 90 on the Shore A scale. They can be made in widths to 72 in and lengths to 900 in. While many are used in instrumentation, computers, and fractional-horsepower drives, they can be used to transmit a significant amount of power.

24.12.1 Film Belts

Film belts are thin (0.5 to 14 mils), flat belts available in widths of 0.125 to 2 in and lengths of 4 to 50 in. They are made of polyesters, Mylar or Kapton, or polyimides. Drive velocity may be as high as 47,000 ft/min and pulley diameters as small as 0.050 in. Application temperature may reach 225°F.

The largest permissible pulleys should be used to reduce belt tension. The design should be based on steady-state torque because the belt will slip under starting or shock loading. The length-to-width ratio should exceed 30:1 to promote good tracking. The diameter of the smallest pulley should be at least 100 times the belt thickness. The thinnest belt possible should be used.

Applications for film belts include lightly loaded, high-speed drives in business machines and instruments.

24.12.2 Woven Belts

In woven endless belts, the carcass or load-carrying member is woven as an endless belt in that it has yarns running in both the warp and fill directions. These belts are smooth-operating and are stable in both the lateral and longitudinal directions.

Multi-ply belts are available with the plies either stitched or cemented together. Natural cotton and synthetic yarns such as nylon are used as carcass materials. Typical applications include business machines and portable power tools. Single-ply belts 0.025 in thick, made of Dacron or Kevlar nylon, are run at belt speeds of 20,000 ft/min. Thinner single-ply belts, 0.0010 to 0.0015 in thick, are used in the same low-power, precision drives as film belts. Single-ply elastic and semielastic belts, 0.010 to 0.040 in thick, are used in fixed center distance applications and where their elasticity dampens vibrations induced by the driving motor.

Some woven belts are furnished with one surface rough and the other smooth. The rough side contacts the pulley in oily applications, the smooth side when the application is dry. Carcass materials are impregnated with a number of materials, including black, white, and red neoprene and synthetic resins. Widths to 60 in and lengths to 900 in are available. Tight-to-slack-side tension ratio should be in the range 2 to 4. For a ratio over 4, efficiency decreases.

24.12.3 Round Belts

Round belts are available as all-yarn, yarn-and-elastomer, and all-elastomer types. They may be purchased endless or made endless in the field by splicing. Round belts can operate on a drive in which the pulleys are in several planes. They are available in diameters from 0.03125 to 0.75 in, and lengths are unlimited. Pulley pitch diameter may be as small as 0.5 in.

Various natural and synthetic yarns are used for the carcass. In wound construction, yarn or braid is wound in a continuous helix, with or without a cover. If the winding is in one direction, stretch is about 1 percent; if alternate windings are in opposite directions, about 3 to 5 percent. Some round belts can run in V-belt as well as V-grooved pulleys. A tight-side-to-slack-side ratio of about 3.5 is most efficient. Elastic round belts have helically wound carcasses that can stretch. They are used in two-stage drives and applications similar to those for O-ring belts.

Two types of round belts, elastomeric and cast polyethylene, are treated in Secs. 24.12.4 and 24.12.5.

24.12.4 Elastomeric Round Belts

Elastomeric round belts, called "O-ring belts," are usually used in sound reproduction applications. They should be able to withstand oil, ozone, and weather, and be stretchable. Compounds used are neoprene, urethane, natural rubber, ethylene-propylene-terpolymer (EPT), and ethylene-propylene-diene-monomer (EPDM). Neoprene is a general-purpose material. EPT and EPDM are low-cost materials. Urethane is expensive and has high strength and abrasion resistance. Natural rubber can be run over small pulleys at speeds to 2000 ft/min. Compounding improves its poor aging properties.

In designing belt drives several factors should be considered. The minimum pulley diameter should be 6 times the cross-sectional diameters of the belt. Belt life is about inversely proportional to the fifth power of the pulley diameter. Pulley grooves should be semicircular with a radius of $0.45d$, where d is the O-ring diameter. Belt stretch should be 16 percent to optimize belt life and efficiency. Overall efficiency is about 80 percent. O-ring diameters of less than 0.070 in should not be used. O-ring belts will not run well in V grooves.

24.12.5 Cast Polyurethane Belts

Polyurethane belts are made in flat, round, V, and special cross sections. They are available in solid urethane and endless woven fabric load-carrying members. The urethane can be solid or foam. These belts have high resistance to abrasion, dirt, oil, water, most chemicals, shock, and vibration. Normal operating range is −20 to 150°F, with a maximum of 190°F. The coefficient of friction is high, 0.7 on steel. They can be used in limited space, reversed bends, and serpentine drives. The surface can be ground to ± 0.0005 in. Solid urethane belts may be purchased endless or spliced, hot or cold, to length, in the field.

Because of their high coefficient of friction, urethane belts should not be used on applications which require the belt to slip on the pulley. On cross drives the belt strands must not touch. A steel plate must be placed between the strands at the point of intersection. Belt-tightening drives, such as belt tighteners, must not be used. Urethane V belts can be run in V belt pulleys. Round belts can be run in V belt pulleys or round-belting pulleys.

Pulleys must be carefully aligned. Because of the high coefficient of friction of urethane, the belt will turn over if this is not the case. If pulleys must be misaligned, round belts should be used. A small amount of petroleum jelly on the belt will assist in preventing the belt from climbing out of the pulley grooves.

Flat belts should be operated on crowned pulleys. The crown should be 2° except for a level section of 0.2 of the belt width at the center of the pulley. The pulley width should be 0.125 in more than the belt width. Flat belting cannot be used for converging or diverging applications. They can be used on twist belt drives if both pulley shafts are in parallel planes and the pulley faces have a common centerline. For installations involving multiple flat belts on common shaft, maximum shaft deflection must not exceed 0.2 percent of the shaft length between bearings.

Because of the high hysteresis of urethane, the allowable minimum pulley diameter is a function of belt speed, length, and number of pulley diameters. Urethane belts should be installed with an initial stretch of 5 to 10 percent, with 12 percent as a maximum. Proper belt tension is that which just eliminates slack in the belt adjacent to the drive pulley on the slack side.

24.13 POSITIVE-DRIVE BELTS

Positive-drive belts maintain a positive rotational relationship between shafts by means of projections on the belt which fit into depressions in the pulley. One form of positive drive belt, the synchronous belt, is examined in Sec. 24.10.

24.13.1 Miniature Sprocket Chains and Pulley Belts

Two closely related means of power transmission, miniature sprocket chains and pulley belts, are used in instrumentation, servo, and other low-power installations. They have 0.03125-in-diameter stainless-steel load-carrying members jacketed in polyurethane. The chains, which are similar in appearance to roller chains, are available in single and double configurations, having two and three load-carrying members, respectively, joined by crosspins. The belts have a single load-carrying member and either a single side cog extending to either side of this member or two sets of side cogs at right angles to each other. The latter, referred to as a three-dimensional belt, provides a positive drive in the applications where a round belt is ordinarily used. Because chains are flexible laterally and belts can twist, chain sprockets and belt pulleys can be at an angle to each other. Two pitches are available, 20 pitch (0.15708 in) and 32 pitch (0.0982 in), a standard gear pitch. Sprockets and grooved gears are available in pitch diameters from 0.5 to 4.0 in.

Plastic chains and belts produce low noise even at high speeds and they dampen shock. Wear rate is low and they do not require lubrication. In most drives, the small pulley rotates at 6000 r/min or less; speeds of 15,000 r/min have been attained. Maximum recommended horsepower capacity for pulley belts is 1.5; for single-strand sprocket chains, 2.5; and for double chains, 7.5. Ratios are limited only by available sprockets. No pretensioning is necessary. The belts may be produced in endless form or the ends may be joined in the field, at a loss of strength of about 50 percent, using a soft sintered cast bushing.[17]

24.13.2 Spur-Gear Cable-Drive Chains

This positive drive has two prestretched aircraft standard cables as load-carrying members which are connected by crosspins. Encasing material is polyurethane, nylon, or fire-resistant plastic. It is available in three pitches, 12 (0.2168 CP), 16 (0.1965 CP), and 24 (0.1309 CP), where CP = circular pitch in inches, and mates with standard 14.5 and 20° spur gears of these pitches. Maximum horsepower at 1800 r/min is 12 pitch, 10.2 hp; 16 pitch, 48 hp; and 12 pitch, 123 hp. No lubrication is required.[17]

24.13.3 Spur-Gear Drive Belts (Pinned Drive)

This positive drive has features similar to those of the spur-gear cable-drive chain, but has only one load-carrying member. It is available in two pitches: 0.1 CP and 32 (0.0982 CP). Both are rated at a maximum of 5.10 hp at 1800 r/min. To mate with these belts, standard pulleys with 0.1875-in-wide faces are modified. The modification consists of a groove, 0.0937 in wide by 0.077 in deep, cut in the face for the load-carrying member to ride in.[17]

24.13.4 Metal Belts

These very thin belts have high strength-to-weight ratios and stretch very little. They are made of stainless steel. They are available in thicknesses of 0.002 to 0.030 in, widths to 36 in, and circumferences down to 5 in. They can be obtained plain or in various hole configurations. Standard perforations for 16- and 35-mm tape drives and others specified by the user are available. Holes can be produced mechanically or chemically. Such belts are intended for precision electromechanical drive systems, encoder drives, and XY plotters.[15]

NASA has patented two steel positive drive belts, originally developed for spacecraft instrument drives. In a chain-type version, the load-carrying members are wires wrapped around metal rods which run at right angles to them. These rods serve as link pins. In another design, a steel strip is formed into a series of U-shaped teeth. The steel is flexible enough to flex around a sprocket but resists stretching. Both belts are coated with plastic to reduce noise and wear.[16]

24.13.5 Bead-Chain Belts

The load member of these belts consists of a stainless-steel or Aramid cable covered with polyurethane. At specified intervals, this coating is formed into beads which seat into conical recesses in the pulley. Because the belt is flexible, it can be used with misaligned or skewed pulleys, with guide tubes for the belt, and to provide rotary or oscillatory motion. Because the beads may come out of the sprockets under heavy loading, sprocket chains and pulley belts were developed to deal with this problem.[16]

These belts are used in electronically controlled, fast film transport and positioning. A soft polyurethane web 0.180 by 0.020 in, reinforced by 0.004-in stainless-steel wire, supports the teeth. One tooth system, on an outer pitch line, engages the film, while the other, on an inner pitch line, engages the drive and driven sprockets. The belts are produced in circular form.[18]

24.14 PIVOTED MOTOR BASES

With small-diameter pulleys necessary with high-speed motors and short center-distance drives, it is impractical to maintain belt tension by belt sag, so pivoted motor bases and idler pulley drives are employed. Pivoted motor bases, which automatically maintain correct belt tension, are of two types, gravity and reaction-torque.

The gravity type (Fig. 24.12) utilizes part or all of the weight of the driving motor to maintain belt tension. In horizontal drives, with the driven pulley above the driver, motor weight keeps the belt in tension as it stretches because of the torque transmitted. Vertical drives are also used.

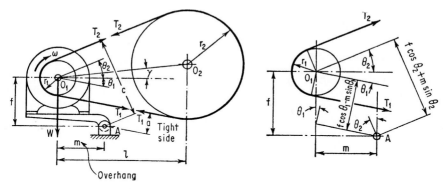

FIG. 24.12 Gravity-type pivoted motor base.

Reaction-torque motor bases are of all-steel construction. The standard base (Fig. 24.13) is designed to function satisfactorily only with the tight side nearest the pivot axis. Special bases are available for reverse-direction drives. Reaction torque on the motor stator tends to rotate the stator about the pivot axis in the opposite direction to the belt pull, thus tightening the belt.

Vertical drives with the motor mounted above the driven pulley are called "down drives." A counterweight is utilized to balance the motor weight.

The spring-type motor base maintains belt tensions by means of springs built into the motor base.

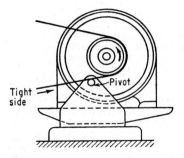

FIG. 24.13 Reaction-torque pivoted motor base.

Belt tension in a gravity-type pivoted motor base drive (Fig. 24.12) is found by taking moments about the pivot A. The resulting moment equation,

$$Wm = T_1a + T_2c$$

is solved simultaneously with the equation for tension ratio, $T_1/T_2 = e^{\mu\beta}$, to find the belt tensions. Maximum power, which can be transmitted when $T_2 = 0$, is

$$\text{hp} = mWNd_1/126{,}000a$$

where N is the motor r/min and all dimensions are in inches. During operation, only about 75 percent of this horsepower can be transmitted by the drive. For ordinary oak-tanned-leather belt drives values of $T_1/T_2 = 5$ at 180° arc of contact and $T_1/T_2 = 3$ at 120° are recommended. For high-capacity belts, these values may be increased to 6.5 and 3.5, respectively. Centrifugal effects in the belt are usually neglected because the motor weight maintains belt pulley contact. From Fig. 24.12, the moment arms a and c are given by $a = f\cos\theta_1 - m\sin\theta_1 - r_1$, and $c = f\cos\theta_1 + m\sin\theta_1 + r_1$.

Belts should be thin and wide to reduce flexing around the small pulley. Pulleys should be as large as possible consistent with reasonable belt speeds, and little, if any, crown should be provided for wider belts. V belts can also be used with pivoted motor bases.

24.15 BELT DYNAMICS

24.15.1 Transverse Vibration

When the belt is excited by forces which have a frequency corresponding to its transverse natural frequency, it may undergo serious flapping vibrations. It is accordingly desirable to determine the natural frequency of the belt to avoid such operation. In the treatment below, it is assumed that the flexural rigidity of the belt is negligible.

The velocity of sound u, which is equal to the velocity of propagation of a disturbance along the belt, is given by

$$u = (Pg/\rho A)^{1/2} \quad \text{in/s} \quad (24.28)$$

where P is the tension in the belt, lb; g the gravitational constant, in/s²; ρ the density of the belt, lb/in³; and A the cross-sectional area, in².

The velocity v of the belt is

$$v = 2\pi(RN/60) \quad \text{in/s} \quad (24.29)$$

where R and N are the radius and r/min of the sheave. The time for a disturbance to travel up and down the belt is $t = 2L_su/(u^2 - v^2)$ s, where L_s is the distance between points of tangency of the belt, so that the frequency of the fundamental mode is

$$F_1 = 60/t = 60(u^2 - v^2)2L_su \quad \text{cycles/min} \quad (24.30)$$

In addition to the fundamental mode, the belt may execute higher modes of vibration whose frequency is given by

$$F_i = mF_1 \tag{24.31}$$

where $m = 1, 2, 3$, etc. It should be noted that the natural frequency of the slack side differs from that of the tight side and that this frequency differs with belt speed.

To account for its flexural rigidity, the belt is assumed to be a simply supported beam in lateral vibration whose natural frequency

$$F_2 = 30(m^2/L_S^2)(EIg/A)^{1/2} \qquad \text{cycles/min} \tag{24.32}$$

where $m = 1, 2, 3$, etc. For a steel flat belt this reduces to

$$F_2 = 5.5 \times 10^6 bm^2/L_S^2 \tag{24.33}$$

where b is the belt thickness.

The approximate natural frequency F of a belt having flexural rigidity is given by

$$F^2 = F_1^2 + F_2^2 \tag{24.34}$$

The effect of flexural rigidity, even for a steel belt, is generally negligible.

Excitation of lateral vibrations in a belt drive can arise as a joint, debris, or thickened area of the belt passes over a pulley. The excitation frequency in this case is obtained by dividing the linear velocity of the belt by the total length of the belt, or

$$F_e = 2\pi R N_R/L_t \qquad \text{cycles/min} \tag{24.35}$$

24.15.2 Torsional Rigidity of Belt Drives

The torsional rigidity of a belt drive is the ratio of the moment applied to one pulley to the angular deflection undergone by the pulley. The torsional rigidity for the open drive may be written

$$C_R = \frac{R^2\beta}{e[L_S + (R + r)/3]} \qquad \text{in·lb/rad} \tag{24.36}$$

$$C_r = \frac{r^2\beta}{e[L_S + (R + r)/3]} \qquad \text{in·lb/rad} \tag{24.37}$$

where C_R, C_r are the torsional rigidities referred to the larger and smaller pulleys, respectively, $e = 1/AE$ where A is the cross-section area, E the modulus of elasticity of the belt, L_S the distance between points of tangency of the belt

$$L_S = [C^2 - (R - r)^2]^{1/2}$$

and R and r are the radii of the large and small pulley, respectively. The empirical factor $(R + r)/3$ is included to account for stretching effects as the belt passes around the pulley and may be omitted in approximate calculations. β is a constant whose value depends on the relation between steady-state and vibratory torques imposed. If the magnitude of the vibratory force is less than the magnitude of the steady-state force in the slack side, $\beta = 2$. If the vibratory torque is large enough that the force in the slack side is negligible throughout the cycle, $\beta = 1$. If the force in the slack side is zero for $\frac{1}{2}$ cycle, $\beta = 1.5$ (approximately). For steady-state conditions, with no vibratory torque, $\beta = 2$.

The belt may be converted to an equivalent length of either drive or driven shafting for frequency calculations. This value is, when referred to the larger pulley,

$$L_E = (\pi/32)(GD^4/\beta EAR^2)L_S \tag{24.38}$$

or, when referred to the smaller pulley,

$$L_e = (\pi/32)(Gd^4/\beta EAr^2)L_S \tag{24.39}$$

where G is the torsional rigidity of the shaft and D and d are the diameters of large and small pulleys, respectively.

EXAMPLE Calculate the torsional rigidity of a steel belt 1.0 in wide and 0.004 in thick; diameter of both pulleys is 6 in and center distance is 7 in. With these values, $e = 1/(1.0 \times 0.004) \times 30 \times 10^6 = 250/30 \times 10^6$.

$$C_r = C_r = \frac{2 \times 9 \times 30 \times 10^6}{250 \times 7 + (3 + 3)/3} = 240,000 \text{ in·lb/rad}$$

The value of E for leather is usually taken as 25,000 lb/in^2.

When calculating the fundamental frequency of a multimass system in which a belt drive is included, it is usually assumed that this mode of vibration has a node in the belt, whose stiffness is low compared with that of other members of the system.

Excitations at this frequency can be caused by sheave unbalance or a bent or misaligned shaft. Excitations at twice this frequency are produced by out-of-round sheaves or by shaft misalignment. Excitations at ½, 1, and 1½ times this frequency arise when the belt is driven by four-stroke-cycle internal-combustion engines.

REFERENCES

1. Barth, C. G.: "The Transmission of Power by Leather Belting," *Trans. ASME,* vol. 31, p. 29, 1909.

2. "Handbook of Power Transmission," "Flat Belting," p. 11, Publication S-5119, The Goodyear Tire and Rubber Co., Akron, Ohio, 1954.

3. Hampton, F. G., C. F. Leh, and W. E. Helmick: "An Experimental Investigation of Steel Belting," *Mech. Eng.,* vol. 42, p. 369, July 1920.

4. Young, G. L., and G. V. D. Marx: "Performance Tests of Steel Belts with Compressed Spruce Pulleys," *Mech. Eng.,* vol. 45, p. 246, 1923.

5. "1981–1982 Power Transmission Design Handbook," Penton IPC, p. C/108.

6. Erickson, W.: "Straight Talk about Belt Drives," *Mach. Des.,* p. 199, April 21, 1977.

7. "Flat Belts," *Mach. Des.,* (Mechanical Drives Reference Issue), p. 22, June 3, 1976.

8. Page-Lon Catalog, Page Belting Company, Concord, N.H.

9. Nycor-M Catalog, L. H. Shingle Co., Worcester, Mass.

10. Foulds Vitalastic Horsepower Tables, I. Foulds & Sons, Inc., Hudson, Mass.

11. Dodge Engineering Catalog, D78, Dodge Division, Reliance Electric Co., Mishawaka, Ind., 1978.

12. Erickson, W.: "New Standards for Power Transmission Belts," *Mach. Des.,* p. 83, January 22, 1981.

13. Hitchcox, A.: "V-belts—Designed to Deliver," *Power Transm.* p. 22, November 1981.

14. Tantastic Nylon Core Belting, Catalog 3M, J. E. Rhoads & Sons, Inc., Wilmington, Del., March 1977.

15. "Drive Bands Data," Metal Belts Inc. Agawam, Mass., 1981.

16. "Durable, Nonslip, Stainless Steel Drive Belts," *Mech. Eng.,* p. 48, July 1979.

17. "Getting in Step With Hybrid Belts," *Des. Eng.,* p. 63, April 1981.

18. Stefanides, E. J.: "Polyurethane Belt Safely Links Servo Drive," *Des. News,* p. 44, December 12, 1978.

19. Catalog Supplement, W. B. Berg, Inc., East Rockaway, N.Y., 1976.

20. "Narrow Multiple V-Belts" (3V, 5V, and 8V Cross Sections), IP-22, Rubber Manufacturers Association, Washington, D.C., 1977.

21. Sun, D. C.: "Performance Analysis of Variable Speed-Ratio Metal V-Belt Drive," *Trans. ASME, J. Mechanisms, Transmissions and Automation in Design,* vol. 110, p. 472, 1988.

SECTION 25
CHAINS

George V. Tordion, Ing. P.

Professor of Mechanical Engineering
Université Laval
Quebec, Canada

25.1 INTRODUCTION (Refs. 4, 6, 8, 9, 12, 13, 15–18, 23–27, 29)

Power-transmission chains are primarily of two kinds, roller and silent. They transmit power in a positive manner through sprockets rotating in the same plane. This fairly old transmission element (sketches of its design have been found in notebooks of Leonardo da Vinci) has been improved in the course of time to a very high standard of precision and quality.

Chain transmission is positive and there is no slip present as in belt drives. Large center distances can be dealt with more easily, with fewer elements and in less space than with gears. Chain drives have high efficiency. No initial tension is necessary and shaft loads are therefore smaller. The only maintenance required, after a careful alignment of elements, is lubrication. Chains as well as sprockets are thoroughly standardized in ASA standard B.29.1. The primary specifications of the standard single-strand roller chain are given in Table 25.1.

25.2 NOMENCLATURE

p = pitch, in

T = chain tension, lb

T_c = chain tension due to centrifugal forces, lb

w = chain weight per unit length, lb/in

$m = w/g$ = chain mass per unit length, lb·s²/in²

D_1, D_2 = sprocket diameters, in

N_1, N_2 = sprocket-teeth numbers

n_1, n_2 = revolutions per minute of the sprockets

$\rho = n_2/n_1$ = speed ratio

V = chain speed, in/s in equations, ft/min in specifications

25.1

TABLE 25.1   American Standard Roller Chain

Pitch, in	ASA No.	Weight, lb/ft	Avg ultimate strength, lb	Roller width, in	Roller diam, in	Side plate	
						Thickness, in	Height, in
$\frac{1}{4}$	25	0.085	875	$\frac{1}{8}$	0.130	0.03	0.23
$\frac{3}{8}$	35	0.22	2,100	$\frac{3}{16}$	0.200	0.050	0.36
$\frac{1}{2}$	41	0.28	2,000	$\frac{1}{4}$	0.306	0.050	0.39
$\frac{1}{2}$	40	0.41	3,700	$\frac{5}{16}$	$\frac{5}{16}$	0.060	0.46
$\frac{5}{8}$	50	0.68	6,100	$\frac{3}{8}$	0.400	0.080	0.59
$\frac{3}{4}$	60	0.96	8,500	$\frac{1}{2}$	$\frac{15}{32}$	0.094	0.68
1	80	1.70	14,500	$\frac{5}{8}$	$\frac{5}{8}$	0.125	$\frac{7}{8}$
$1\frac{1}{4}$	100	2.70	24,000	$\frac{3}{4}$	$\frac{3}{4}$	0.156	$1\frac{5}{32}$
$1\frac{1}{2}$	120	4.00	34,000	1	$\frac{7}{8}$	0.187	$1\frac{13}{32}$
$1\frac{3}{4}$	140	5.20	46,000	1	1	0.218	$1\frac{5}{8}$
2	160	6.80	58,000	$1\frac{1}{4}$	$1\frac{1}{4}$	0.250	$1\frac{7}{8}$
$2\frac{1}{4}$	180	9.10	76,000	$1\frac{13}{32}$	$1\frac{13}{32}$	0.281	$2\frac{1}{4}$
$2\frac{1}{2}$	200	10.80	95,000	$1\frac{1}{2}$	$1\frac{9}{16}$	0.312	$2\frac{5}{16}$
3	240	16.5	130,000	$1\frac{7}{8}$	1.875	0.375	2.8

$V_0 = \sqrt{T/m}$ = velocity of propagation of transverse waves, in/s

$L = L'p$ = chain length, in

$C = C'p$ = center distance of the sprockets, in

a = strand length between seated rollers, in

h = chord length of the strand between seated rollers, in

s = strand maximum sag, in

γ = pressure angle

$\alpha_1 = 180°/N_1$

$\alpha_2 = 180°/N_2$

ω_n = natural circular frequency of vibrations, rad/s

E = Young's modulus, lb/in^2

f = roller width, in

25.3   DESIGN OF ROLLER CHAINS (Fig. 25.1)

The speed ratio is given by a $\rho = n_2/n_1 = N_1/N_2$. For one-step transmission, it is recommended that $\rho < 7$. Values between 7 and 10 may be used at low speeds (< 650 ft/min). The minimum wrap angle of the chain on the smaller sprocket is 120°. A smaller angle may be used on idler sprockets used for adjustment of the chain slack,

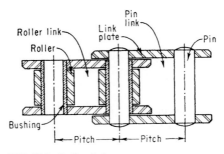

FIG. 25.1   Roller chain.

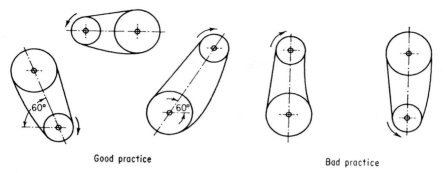

Good practice   **Bad practice**

FIG. 25.2 Chain-drive arrangements.

where the center distance is not adjustable. Horizontal drive is recommended, in which case the power should be transmitted by the upper strand. Preferred inclined-drive arrangements are shown in Fig. 25.2. Vertical drives should be used with idlers to prevent the chain from sagging and to avoid disengagement from the lower sprocket. When running outside the chain, idlers should be located near the smaller sprocket.

25.3.1 Chain Length

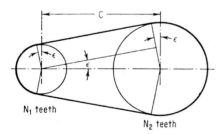

FIG. 25.3 Angle ϵ used in chain-length calculation.

The chain length L' in pitch numbers, for a given distance C' in pitch numbers, is exactly (Fig. 25.3)

$$L' = 2C' \cos \epsilon + (N_1 + N_2)/2 + (N_2 - N_1)\epsilon/\pi \quad (27.1)$$

where

$$\epsilon = \arcsin \frac{1/\sin \alpha_2 - 1/\sin \alpha_1}{2C'} \quad (27.2)$$

Since ϵ is generally small, it is sufficient in most cases to use the approximation $\epsilon = (N_2 - N_1)/(2C')$. Then

$$L' = 2C' + (N_1 + N_2)/2 + S/C' \quad (27.3)$$

where $S = [(N_2 - N_1)/2\pi]^2$. The values of S are tabulated in Table 25.2. The actual chain length is $L = L'p$, and the actual sprocket center distance $C = C'p$. It is recommended that C' lie between 30 and 50 pitches. If the center distance is not given, the designer is free to fix C' and to calculate L' from Eq. (25.3). The nearest larger (preferably even) integer L' should be chosen. In this case, an offset link is avoided. With L' an integer, the center distance becomes exactly

$$C' = e/4 + \sqrt{(e/4)^2 - S/2} \quad (25.4)$$

with $e = L' - (N_1 + N_2)/2$. C' should be decreased by about 1 percent to provide slack in the nondriving chain strand. For horizontal drive, this will result in a chain sag of some 2 percent of the strand length.

The chain speed is $V = \pi n_1 D_1/12$ (ft/min), or approximately $V = n_1 p N_1/12$ (ft/min). The sprocket diameters are $D_i = p/\sin \alpha_i$ (in), $i = 1$ or 2.

TABLE 25.2 Values of S for Different $N_2 - N_1$

$N_2 - N_1$	S	$N_2 - N_1$	S	$N_2 - N_1$	S	$N_2 - N_1$	S
1	0.02533	16	6.4846	31	24.342	46	53.599
2	0.10132	17	7.3205	32	25.938	47	55.955
3	0.22797	18	8.2070	33	27.585	48	58.361
4	0.40528	19	9.1442	34	29.282	49	60.818
5	0.63326	20	10.132	35	31.030	50	63.326
6	0.91189	21	11.171	36	32.828	51	65.884
7	1.2412	22	12.260	37	34.677	52	68.493
8	1.6211	23	13.400	38	36.577	53	71.153
9	2.0518	24	14.590	39	38.527	54	73.863
10	2.5330	25	15.831	40	40.528	55	76.624
11	3.0650	26	17.123	41	42.580	56	79.436
12	3.6476	27	18.466	42	44.683	57	82.298
13	4.2808	28	19.859	43	46.836	58	85.211
14	4.9647	29	21.303	44	49.039	59	88.175
15	5.6993	30	22.797	45	51.294	60	91.189

25.3.2 Ratings

The transmitted horsepower is related to the required horsepower rating by the following:

$$\text{Required hp rating} = \frac{\text{hp transmitted} \times \text{service factor}}{\text{multiple-strand factor}} \tag{25.5}$$

where the service factors and the strand factors are given in Tables 25.3 and 25.4. The ratings are given in extensive tables in Ref. 2. They are partially reproduced in graphical form in Fig. 25.4. For every chain pitch and number of teeth in the small sprocket, the horsepower ratings are given as a function of speed in revolutions per minute. The service life expectancy is approximately 15,000 hr. The lubrication conditions are: type I, manual lubrication; type II, drip lubrication from a lubricator; type III, bath lubrication or disk lubrication; type IV, forced-circulation lubrication.

The rating graphs have the peculiar "tent form." At lower speeds chain fatigue is responsible for the chain failures. From a certain speed the roller impact resistance becomes responsible for roller breakage, and the rated horsepower falls rapidly until, at the ultimate speed, failure is caused by joint galling.[3] Even with the new reliable

TABLE 25.3 Service Factor[2]

Type of driven load	Type of input power		
	Internal-combustion engine with hydraulic drive	Electric motor or turbine	Internal-combustion engine with mechanical drive
Smooth	1.0	1.0	1.2
Moderate shock	1.2	1.3	1.4
Heavy shock	1.4	1.5	1.7

TABLE 25.4   Strand Factor[2]

No. of strands	Multiple-strand factor
2	1.7
3	2.5
4	3.3

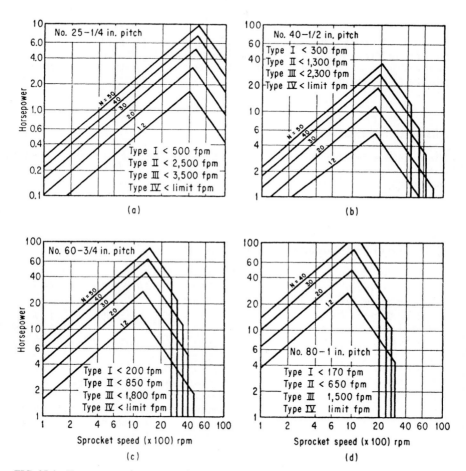

FIG. 25.4   Horsepower ratings vs. sprocket speed.

ratings, careful consideration should be given the application. The design is a compromise between life and cost. A large pitch provides more bearing area than a small one for the same load, but requires fewer teeth on the sprocket. This leads to strong chordal action and thus to large dynamic effects, which, in turn, results in premature wear. A multiple strand with smaller pitch is preferable in such cases. Reference 23 gives more information.

25.3.3 Roller Impact Velocity

The roller impact velocity is given by $\omega p \sin (2\alpha + \gamma)$. How much of the chain mass is involved in the impact is unknown. It is sometimes assumed[1,5] that two links of mass mp are involved. In this case, the kinetic energy is $mp\omega^2 p^2 \sin^2 (2\alpha + \gamma)$. The strain energy is $\frac{1}{2}P\Delta$ (P = impact force, Δ = local deformation). Using the approximation[5] $\Delta = 3P/fE$, the strain energy becomes $3P^2/2fE$ where f is the roller width. If the kinetic energy is completely transformed into strain energy, the impact force

$$P = (2\pi V/N_1)\sqrt{2wpfE/3g}\,\sin(2\alpha+\gamma) \qquad (25.6)$$

A sprocket having a small number of teeth, running at high speeds, leads to roller breakage. Lubricant viscosity also plays an important part in roller fatigue because the rollers must squeeze the lubricant before they are completely seated, and squeezing provides good damping.

25.3.4 Centrifugal Force

The chain loop is subjected to a uniformly distributed additional tension $T_c = mV^2$ caused by centrifugal force. T_c is generally low and does not affect sprocket tooth pressure but it must be added to the useful tension because it affects chain-joint pressure.

25.3.5 Force Distribution on the Sprocket Teeth

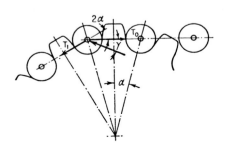

FIG. 25.5   Forces on sprocket teeth.

Referring to Fig. 25.5 and providing that the pressure angle is constant for all seated rollers,

$$T_n = T_0\{(\sin \gamma)/[\sin (2\alpha + \gamma)]\}^n \quad (25.7)$$

where n is the number of seated rollers. For example, for $N = 18$, $n = 6$, and $\gamma = 25°$, $T_6 = 0.045T_0$. Therefore, T_6 is only 4.5 percent of T_0. The tension in the chain decreases very rapidly but never becomes zero unless the pressure angle is zero. For a new chain the pressure angle is given by ASA standard:

$$\gamma = 35° - 120°/N$$

As the chain wears, the pitch of each link is increased, but because wear has a random nature, it is difficult to predict the pressure angles. No actual measurements of γ under running conditions have been made. For elongated chains, the roller engages the top curve of the teeth farther and farther away from the working curve, leading to a dangerous situation. An elongation limit of $\Delta L = 80/N_2$ percent has been proposed, but a maximum of 3 percent is admitted in the new ratings for 15,000 h of service. Increased elongation indicates that the hardened layers of pins and bushings are worn away and the soft cores reached. A very rapid wear increase follows at this stage.

25.4 SYSTEM ANALYSIS

25.4.1 Chordal Action

Chordal action, or polygonal effect, is the variation of the velocity of the driven sprocket which results because the actual instantaneous velocity ratio is not constant but a function of the sprocket angular position $\rho = d\varphi_2/d\varphi_1 = f(\varphi_1)$. Therefore, we have the angular-displacement equation $\dot{\varphi}_2 \int f(\varphi_1) d\dot{\varphi}_1 + $ constant, the velocity equation $\dot{\varphi}_2 = \rho(\varphi_1)\dot{\varphi}_1$, and the acceleration equation $\ddot{\varphi}_2 = \rho(\varphi_1)\ddot{\varphi}_1 + (d\rho/d\varphi_1)\dot{\varphi}_1^2$. From the last equation, it is seen that, even in the case of constant driving speed $\dot{\varphi}_1$, the angular acceleration of the driven sprocket is not zero. This produces dynamic loads, vibrations, and premature chain wear. The analysis of chordal action is made under the assumption that the chain strand acts like a connecting rod in a four-bar linkage. Experiments at low speeds[7] yield excellent agreement with the theoretical results. At higher speeds the chain vibrations are sufficient to alter the static behavior of the chordal action. Nevertheless this action still remains a main source of excitation. Analysis shows[1,7] that the speed variation is maximum for a strand length equal to an *odd multiple of half pitches*. For this case,

$$(\Delta\omega/\omega)_{max} = (N_2/N_1)(\tan \alpha_2/\sin \alpha_1 - \sin \alpha_2/\tan \alpha_1) \qquad (25.8)$$

If the strand length is equal to an *even number of half pitches,* the speed variation is close to a minimum:

$$(\Delta\omega/\omega)_{min} = (N_2/N_1)(\sin \alpha_2/\sin \alpha_1 - \tan \alpha_2/\tan \alpha_1) \qquad (25.9)$$

For $N_1 = N_2$, and an even number of half pitches, the speed variation is zero. Generally, for a fractional number of half pitches the speed variation fluctuates between maximum and minimum (Fig. 25.6). For large N_2/N_1 the values of $\Delta\omega/\omega$ in both cases approach the same limit (Fig. 25.7). There is a considerable advantage in having as many teeth as possible on the small sprocket. Chordal action, in general, may be ignored if the minimum number of sprocket teeth is 15.

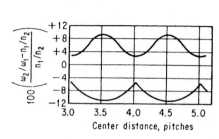

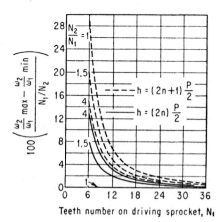

FIG. 25.6 Speed variation vs. center distance in pitches.

FIG. 25.7 Speed fluctuation vs. number of teeth on driving sprocket.

25.4.2 Transverse Vibrations of Chains

The power-transmitting strand may be considered as a tight string with a uniformly distributed mass. The small, transverse, undamped free vibrations $y(x, t)$ are described by the differential equation[10]

$$(V^2 - V_0^2)(\partial^2 y/\partial x^2) + 2V(\partial^2 y/\partial x\,\partial t) + \partial^2 y/\partial t^2 = 0 \tag{25.10}$$

which has the solution

$$y = C_1 \cos\left[\omega t + \omega x/(V_0 - V) + \phi_1\right] + C_2 \cos\left[\omega t - \omega x/(V_0 + V) + \phi_2\right] \tag{25.11}$$

The natural frequency, for the boundary conditions $y = 0$ at $x = 0$ and $x = a$, is

$$\omega_n = (n\pi V_0/a)[1 - (V/V_0)^2] \qquad n = 1, 2, 3,\ldots \tag{25.12}$$

To each ω_n corresponds a mode of vibration. For strings with continuously distributed mass, the number of modes is infinite. For a chain, there is a limit to this number because the strand is composed of a finite number of rollers s. The string approximation is therefore unsuitable for n close to s, and for small s, in general. Considering the chain as a tight string of zero mass with point masses m_0 at every pitch distance, the natural frequencies at zero speed

$$\omega_n = (n\pi V_0/a)(2s/n\pi) \sin\left[n\pi/(2s + 2)\right] \qquad n = 1, 2,\ldots, s \tag{25.13}$$

For $s \to \infty$ Eq. (25.13) becomes identical with Eq. (25.12) with $V = 0$. For $s = 19$ and $s = 44$ the errors are 5 percent and 2.2 percent, respectively. One important source of vibration excitation is the chordal action discussed above. The fundamental frequency of this action is the tooth-engagement circular frequency $2\pi V/p$; harmonics are also possible. Neglecting harmonics, the critical chain speeds

$$V_{\text{crit}} = (aV_0/np)[\sqrt{1 + (pn/a)^2} - 1] \cong nV_0 p/2a \qquad n = 1, 2, 3,\ldots \tag{25.14}$$

Another important source of vibration excitation is the runout of the shaft or sprocket. Its frequency is $2\pi V/Np$. The critical speed attributable to eccentricity is therefore

$$V_{\text{crit}} = (aV_0/npN)[\sqrt{1 + (Npn/a)^2} - 1] \qquad n = 1, 2, 3,\ldots \tag{25.15}$$

The additional tension $T_c = mV^2$ changes the natural frequency of the strand, since $V_0^2 = T/m + mV^2$, and therefore

$$\omega_n = \frac{n\pi}{a} \frac{T/m}{(T/m + V^2)^{1/2}} \tag{25.16}$$

but the critical speeds [Eq. (25.14)] remain unchanged. All critical speeds must be avoided, even when a resonance does not result in infinite vibration amplitudes; damping is always present and reduces resonant amplitudes. For the computation of transverse amplitudes as a function of $V \neq V_{\text{crit}}$, see Ref. 10. Resonance may be remedied by varying tight-strand tension and length, and the possible use of guides or idlers.

The traveling curve of the slack strand under the action of gravity is the catenary of a heavy chain in equilibrium. The catenary is disturbed by the chordal action at the ends. The stability criteria are formulated in Ref. 14, where the problem is reduced, after linearization, to a Mathieu differential equation of the type

$$\ddot{y} + (b - 2q \cos 2t)y = 0$$

The dimensionless parameters

$$b = [(\omega_n/V)^2 - (7.5/a)^2](p/\pi)^2$$

and
$$q = 2(\omega_n^2/g)(p^2/D_1) \qquad (25.17)$$

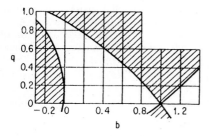

FIG. 25.8 Plot of the dimensionless parameters b and q of the Mathieu differential equation.

are shown in Fig. 25.8. The shaded regions represent unstable running conditions and must be avoided if possible. The slack strand ω_n can be determined either experimentally or from Table 25.5. The given values are valid for a horizontal arrangement of the suspension points (seated rollers) and small s/h, when the catenary of the heavy chain can be approximated by a parabola. The values were experimentally verified to within 5.7 percent.

TABLE 25.5   Values of ω_n

h/a	0.95	0.96	0.97	0.98	0.99	0.998
s/h	0.140	0.125	0.110	0.088	0.060	0.030
$\omega_n^2 a/g$	10.3	14.2	18	20.4	29	91.5

25.4.3 Longitudinal Vibrations of the Tight Strand

The chain is elastic in tension also, and the cushioning of shocks between the driving and driven shafts is one of the advantages of chain transmissions. The elongation of a chain under 50 percent of ultimate load has been measured with great care.[11] A statistical average for all pitches is 0.081 in/ft. Not more than 5 percent of chains of a given pitch will vary from this average by more than 0.014 in/ft. The values of the elastic constant given in Table 25.6 are based upon an assumed linear relationship between load and elongation.

TABLE 25.6   Values of the Elastic Constant K

p, in	$\tfrac{3}{8}$	$\tfrac{1}{2}$	$\tfrac{5}{8}$	$\tfrac{3}{4}$	1	$1\tfrac{1}{4}$	$1\tfrac{1}{2}$	$1\tfrac{3}{4}$	2
$K/10^6$ lb-in./in.	0.163	0.269	0.420	0.604	1.075	1.68	2.41	3.3	4.3

The natural frequency at zero speed is

$$\omega_n = (\pi n/a)\sqrt{K/m} \qquad n = 1, 2, 3,\ldots \qquad (25.18)$$

while the critical speeds due to the tooth-engagement frequency become

$$V_{\text{crit}} = (np/2a)\sqrt{K/m} \qquad n = 1, 2, 3,\ldots \qquad (25.19)$$

25.4.4 Equivalent Stiffness of Chain for Torsional-Vibration Calculations

The arrangement to be considered is that of a chain drive with two sprockets. The torsional stiffness depends upon whether the shaft of the first or second sprocket is taken as a reference value. With reference to the first shaft, the torsional stiffness of the tight strand is $K_1 = (K/a)R_1^2 (\text{lb·in/rad})$. With reference to the second shaft, it is $K_2 = (K/a)R_2^2 = K_1(R_2/R_1)^2 (\text{lb·in/rad})$. In both equations K is the tensile elastic constant of the chain. The equivalent shaft length in terms of a reference diameter d_0 and shear modulus G (lb/in^2) is

$$L_{1,eq} = \pi d_0^4 G/32 K_1 \quad \text{in}$$

or
$$L_{2,eq} = \pi d_0^4 G/32 K_2 \quad \text{in} \qquad (25.20)$$

Further development follows the standard torsional-vibration procedure. For sufficiently high torsional amplitudes, however, the tight strand can become slack during a half cycle of vibration, introducing a nonlinear element into the computations. A theory taking this phenomenon into account has not yet been developed.

25.4.5 Noise

The transverse, longitudinal, and torsional vibrations radiate noise. The most important sources of their excitation are the roller impacts on the sprocket (high-frequency bursts at meshing-frequency rate) and the polygonal action with its very many harmonics, even if their kinematic effect is negligible. The lubrication is an excellent damper of noise. Some qualitative relations are given in Ref. 28.

25.5 SILENT OR INVERTED-TOOTH CHAIN

Silent chains consist of inverted-toothed links alternately assembled on articulating-joint parts. There are two types of chain, (1) flank contact and (2) heavy-duty chordal-action-compensating.

The *flank-contact-type chains*[19,20] are shown in Fig. 25.9a. The chain width may be as great as 16 times the pitch, and speed-reduction ratios as high as 8 are employed. The speed limit is about 6000 r/min for ⅜-in-pitch chain and 900 r/min for 2-in-pitch chain. The minimum and maximum number of sprocket teeth suggested is 17 and 150, respectively. The sprockets have a pitch diameter approximately equal to the outside diameter.

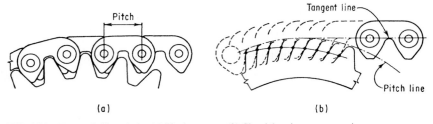

FIG. 25.9 Types of silent chain. (*a*) Flank-contact. (*b*) Chordal-action-compensating.

These chains are subject to the same speed variations and chordal rise and fall due to the polygonal chordal action as roller chains. In general, they are suggested for slow to moderate speed and 20,000 or more hours of life expectancy. They have an approximate ultimate tensile strength of 12,500 × pitch × width.

Heavy-duty chordal-action-compensating-type chains are designed to obtain the optimum in link strength and to eliminate the effects of chordal action. Figure 25.9*b* shows how the chain pitch line is maintained in a constant tangential relation to the sprocket pitch circle. This characteristic is obtained by generating the link profile for conjugate action with the sprocket tooth and/or by special articulating-joint design. These chains are used for a life requirement of less than 20,000 h and/or widely varying load conditions. They have an approximate ultimate tensile strength of 20,000 × pitch × width.

For further information on silent-tooth chains see Refs. 19 to 22.

REFERENCES

1. Binder, R. C.: "Mechanics of the Roller Chain Drive," Prentice-Hall, Inc, Englewood Cliffs, N.J., 1956.

2. "New Horsepower Ratings of American Standard Roller Chains," proposed by ARSCM, Diamond Chain Company, Inc., Indianapolis, Ind., 1960.

3. Frank, J. F., and C. O. Sundberg: "New Roller Chain Horsepower Ratings," *Mach. Des.,* July 6, 1961.

4. Kuntzmann, P.: "Roller Chain Drives" (French), Dunod, Paris, 1961.

5. Niemann, G.: "Machine Design" (German), vol. 2. Springer-Verlag OHG, Berlin, 1960.

6. Jackson and Moreland: "Design Manual for Roller and Silent Chain Drives," prepared for ARSCM, 1955.

7. Bouillon, G., and G. V. Tordion: "On Polygonal Action in Roller Chain Drives," *Trans. ASME, J. Eng. Ind.,* p. 243, May 1965.

8. Morrison, R.: "Polygonal Action in Chain Drives," *Mach. Des.,* vol. 24, no. 9, September 1952.

9. Mahalingham, S.: "Polygonal Action in Chain Drives," *J. Franklin Inst.,* vol. 265, no. 1, January 1958.

10. Mahalingham, S.: "Transverse Vibrations of Power Transmission Chains," *Brit. J. Appl. Phys.,* vol. 8, April 1957.

11. Whitney, L. H., and P. M. MacDonald: "Elastic Elongation of Chains," *Prod. Eng.,* February 1952.

12. Germond, H. S.: "Wear Limits for Roller and Silent Chain Drives," ASME Paper 60-WA-8, 1960.

13. Schakel, R. A., and C. O. Sundberg: "Proven Concepts in Oil Field Roller Chain Drive Selection," ASME Paper 57-PET-24, 1957.

14. Ignatenko, V. V.: "Variable Forces in the Strand of a Chain Drive" (Russian), *Bull. Inst. Higher Educ.,* no 4, 1961.

15. "Roller Chains and Sprockets," Catalog 8 ACME Chain Corp., Holyoke, Mass.

16. "Stock Power Transmission and Conveyor Products," Catalog 760, Diamond Chain Co., Inc., Indianapolis, Ind.

17. Radzimovsky, E. I.: "Eliminating Pulsations in Chain Drives," *Prod. Eng.,* July 1955.

18. Hofmeister, W. F., and H. Klaucke: "Dynamic Check Point Way to Longer Chain Life," *Iron Age,* August 1956.

19. American Standards Association: ASA B29.1, "Transmission Roller Chains and Sprocket Teeth" (SAE SP-69).

20. American Standards Association: ASA B29.3, "Double Pitch Power Transmission Chains and Sprockets" (SAE SP-69).

21. American Standards Association: ASA B29.2, "Inverted Tooth (Silent) Chains and Sprocket Teeth" (SAE SP-68).

22. American Standards Association: ASA B29.9, "Small Pitch Silent Chains and Sprocket Tooth Form (Less Than ⅜ Inch Pitch)" (SAE TR-96).

23. Rudolph, R. O., and P. J. Imse: "Designing Sprocket Teeth," *Mach. Des.*, pp. 102–107, Feb. 1, 1962.

24. Turnbull, S. R., and J. N. Fawcett: "An Approximate Kinematic Analysis of the Roller Chain Drive," *Proc. Inst. Mech. Eng. 4th World Cong. Theory Mach. Mech.*, p. 907, 1975.

25. Shimizu, H., and A. Sueoka: "Nonlinear Free Vibration of Roller Chain Stretched Vertically," *Bull. Japan Soc. Mech. Eng.*, vol. 19, no. 127, p. 22, January 1976.

26. Marshek, K. M.: "On the Analysis of Sprocket Load Distribution," *Mech. Mach. Theory*, vol. 14, pp. 135–139, 1979.

27. Fawcett, J. N., and S. W. Nicol: "A Theoretical Investigation of the Vibration of Roller Chain Drives," *ASME Proc. 5th World Cong. Theory Mach. Mech.*, vol. 2, p. 1482, 1979.

28. Uehara, K., and T. Nakajima: "On the Noise of Roller Chain Drives," *ASME Proc. 5th World Cong. Theory Mach. Mech.*, p. 906, 1979.

29. Fawcett, J. N., and S. W. Nicol: "Vibration of a Roller Chain Drive Operating at Constant Speed and Load," *Proc. Inst. Mech. Eng.*, vol. 194, p. 97, 1980.

30. Lee, T. W.: "Automated Dynamic Analysis of Chain-Driven Mechanical Systems," *Trans. ASME J. Mechanisms, Transmissions and Automation in Design*, vol. 105, p. 362, 1983.

SECTION 26
WIRE ROPE

George A. Costello, Ph.D.
Professor of Theoretical and Applied Mechanics
University of Illinois
Urbana, Ill.

26.1 INTRODUCTION

Wire rope has been applied in various mechanical systems, e.g., hoists, elevators, instruments, and aircraft arresting gear. In some rare cases it has been used as a means for power transmission; for these the rope wrap tension is given in Sec. 23 and the coefficient of friction μ for the wire rope in a cast-iron or steel groove in 0.1. Wire rope manufacturers' catalogs and "Wire Rope Users Manual"[7] contain much valuable information.

26.2 WIRE ROPE DATA

The various components of wire rope are shown in Fig. 26.1. Wire rope is composed of wires of basic classifications 7, 19, or 37, twisted into a strand. Under these classifications other numbers of wires exist. A number of preformed strands (usually six, seven, or eight) are then wound helically about a core of hemp, polypropylene, or steel wire. The core acts as an elastic support for the strands and is intended to prevent excessive contact stresses between the wires.

The strands are laid around the core either to the right or to the left, the result being accordingly designated as right- or left-lay wire rope.

There are two kinds of lay wire rope, *regular lay* and *lang lay*. The wires on regular lay appear to line up with the axis of the rope, while those on lang lay form an angle with the axis.

Lang-lay ropes are generally more resistant to bending fatigue and have a greater wearing surface than regular-lay ropes. Lang-lay ropes are, however, more susceptible to handling abuses, pinching in undersized sheave grooves, and crushing when improperly wound on drums. Also, the twisting moment acting in the strand tends to unwind the strand,[8] thus causing excessive rotation of the rope. Hence lang-lay ropes should always be secured at the ends in order to prevent the rope from unlaying. Figure 26.2 shows several basic cross-sectional constrictions around which standard wire ropes are made.

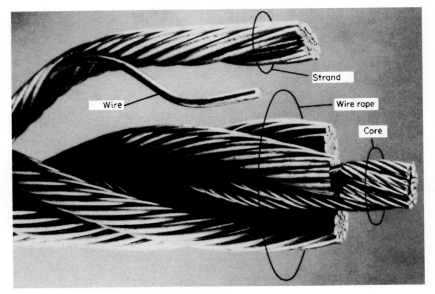

FIG. 26.1   Component parts of a wire rope.

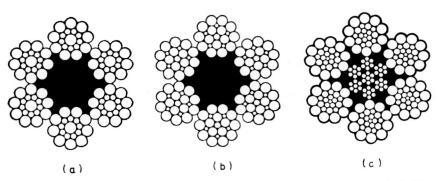

(a)   (b)   (c)

FIG. 26.2   Basic construction around which standard wire ropes are built. (a) 6 × 19 Seale, fiber core. (b) 6 × 21 filler wire (FW), fiber core. (c) 6 × 26 Warrington Seale, independent wire rope core (IWRC).

Although steel is most popular, wire rope is made of many kinds of metal, such as copper, bronze, stainless steel, and wrought iron. Table 26.1 presents the characteristics of 6 × 19 class wire rope. The two most common grades of rope are improved plow steel and extra improved plow steel. The normal strength shown in Table 26.1 is the safe strength of the wire rope when subjected to a static tensile load. Data are shown for two types of rope core, fiber core and independent wire rope core (IWRC).

TABLE 26.1   Wire Rope: 6 × 19 Class

Diameter, in	Nominal strength, tons*			Approximate weight, lb/ft	
	Improved plow steel		Extra improved plow steel		
	Fiber core	IWRC	IWRC	Fiber core	IWRC
3/16	1.55	1.67	—	0.059	0.065
1/4	2.74	2.94	3.40	0.105	0.116
5/16	4.26	4.58	5.27	0.164	0.18
3/8	6.10	6.56	7.55	0.236	0.26
7/16	8.27	8.89	10.2	0.32	0.35
1/2	10.7	11.5	13.3	0.42	0.46
9/16	13.5	14.5	16.8	0.53	0.59
5/8	16.7	17.9	20.6	0.66	0.72
3/4	23.8	25.6	29.4	0.95	1.04
7/8	32.2	34.6	39.8	1.29	1.42
1	41.8	44.9	51.7	1.68	1.85
1 1/8	52.6	56.5	65.0	2.13	2.34
1 1/4	64.6	69.4	79.9	2.63	2.89
1 3/8	77.7	83.5	96	3.18	3.50
1 1/2	92.0	98.9	114	3.78	4.16
1 5/8	107	115	132	4.44	4.88
1 3/4	124	133	153	5.15	5.67
1 7/8	141	152	174	5.91	6.50
2	160	172	198	6.72	7.39
2 1/8	179	192	221	7.59	8.35
2 1/4	200	215	247	8.51	9.36
2 3/8	222	239	274	9.48	10.4
2 1/2	244	262	302	10.5	11.6
2 5/8	268	288	331	11.6	12.8
2 3/4	292	314	361	12.7	14.0
2 7/8	317	341	392	13.9	15.3
3	—	370	425	—	16.6
3 1/8	—	399	458	—	18.0
3 1/4	—	429	492	—	19.5
3 3/8	—	459	529	—	21.0
3 1/2	—	491	564	—	22.6

*Galvanized at 10% lower strengths.

26.3   DESIGN DATA

The design factors are flexibility, wear, strength, core strength, and corrosion resistance. Flexibility and fatigue strength (bending) are obtained by using a large number of small-diameter wires. Large wires give better resistance, and strength depends on the wires and the structure. Corrosion by moisture and stray electric currents is generally inhibited by proper lubrication of the wire rope.

Data for the same type of wire rope produced by different manufacturers can differ considerably.

Minimum factors of safety based on breaking strength are: 5 to 8 in hoists and cranes, 3.5 in guy ropes, 7 to 12 in elevators, and 4 to 8 in mine shafts, depending on the risks. Note that for mine shafts the minimum factor of safety is 8 for depths up to

500 ft and decreases to 4 for depths over 3000 ft. The lower design factors of safety are indicated for deeper shafts (by necessity) because the weight of the rope becomes significant in the system design. The factors of safety are based upon favorable conditions of environment and lubrication of the rope and depend on the loads; acceleration; speed; attachments; number, size, and arrangements of sheaves and drums; corrosion; abrasion; rope length; and accuracy of stress investigation. A factor of safety of 1.2 is used in the design of U.S. Navy arresting gear under controlled design conditions and for limited life. The breaking strength is 80 percent or more of the total strength of the wires in the rope. The modulus of elasticity of wire rope averages 14,000,000 lb/in^2, which is less than half the modulus of elasticity of the wires in the rope because the wires are composed of twisted wires, which act like springs. A friction loss of 5 percent is assumed in hoisting practice, and 3 percent is used for very flexible rope with extra large pulleys.

The basic stresses in a wire rope are tension due to primary loads and dynamic loads and also tension due to bending or wrapping around a drum and sheaves. The fatigue conditions associated with all must certainly be considered. Bending often accounts for the predominant load. Contact stresses between the wires also exist.

Splices and *attachments* will next be considered. The use of spliced rope should be avoided since each splice may have an efficiency as low as 80 percent in the larger rope sizes. Many kinds of attachments have been used. The best is the rope socket (efficiency 100 percent), in which the wires are opened and high-grade zinc poured on (Fig. 26.3*a*). Figure 26.3*b* shows a thimble and clamp which has an approximate efficiency of 75 percent. Woven wire has been used in lieu of clamps. Because of stress concentration, thimble and other splices cannot strengthen filler rope.

Drums and sheaves should be of the largest practical diameter. Where feasible, a grooved drum is desirable and grooves should provide ample clearance between windings. The winding of one layer of rope on another should be avoided since this considerably reduces rope life. Long ropes may run as fast as 6000 ft/min. Reverse bending of wire rope should also be avoided. Figure 26.4 shows recommended sheave and drum proportions. Sheaves for rope are made similar to those of chains. The life of the

FIG. 26.3 Rope attachments. (*a*) Rope socket. (*b*) Thimble and clamps.

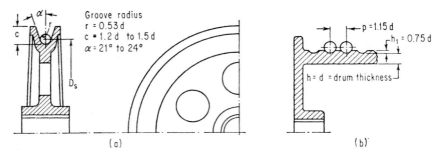

FIG. 26.4 Sheave and drum proportions. (*a*) Sheave (arms may be of elliptical cross section). (*b*) Drum.

TABLE 26.2 Recommended Sheave and Drum Ratios

Construction	Suggested D/d ratio*	Minimum D/d ratio*
6 × 7	72	42
19 × 7 or 18 × 7	51	34
6 × 19 Seale	51	34
6 × 25 B	45	30
6 × 27 H	45	30
6 × 30 G	45	30
6 × 21 filler wire	45	30
6 × 25 filler wire	39	26
6 × 31 Warrington Seale	39	26
6 × 36 Warrington Seale	35	23
8 × 19 Seale	41	27
8 × 25 filler wire	32	21
6 × 41 Warrington Seale	32	21
6 × 42 Tiller	21	14

*D = tread diameter of sheave; d = nominal diameter of rope.

rope can be lengthened by use of an insert at bottom of the sheave groove. The insert may be made of leather, hard rubber, wood, etc. Small drums on hoists are made plain.

The drum thickness should be made equal to the rope diameter and checked for strength. Ropes are subjected to cyclic bending stresses when operating over sheaves resulting in fatigue damage. Larger sheaves yield lower bending and fatigue stresses. Table 26.2 shows the recommended sheave-to-wire diameter ratio for various ropes.

26.4 STRESS ANALYSIS

In spite of the long-time use of wire ropes as a structural element and because of the complexity of the problem, relatively few theoretical papers of any significance have appeared.[1,4] As is the case in many engineering areas, the experimental work preceded the theory. The results of extensive testing of wire rope running over sheaves, for example, have been reported.[2,3] For a more complete list of references, see Refs. 4 and 5.

The theory presented is a consequence of several investigations.[8–14] It is assumed that the wire material is linearly elastic, friction is neglected, and the wire material is stress-free when the rope is in an unloaded state.

The cross section of a loaded strand, shown in Fig. 26.5, initially consists of a straight center wire of radius R_1, surrounded by m_2 helical wires of wire radius R_2. It is assumed that the center wire is of sufficient size to prevent the outer wires from touching each other. This is usually the case in wire rope design since this tends to minimize the effects of friction.

If the outer wires with the elliptical cross sections shown in Fig. 26.5 are initially touching each other, the initial radius r_2 of the helix of these outer wires is given by the relation[8]

$$\frac{r_2}{R_2} = \left[1 + \frac{\tan^2(\pi/2 - \pi/m_2)}{\sin^2 \alpha_2}\right]^{1/2} \tag{26.1}$$

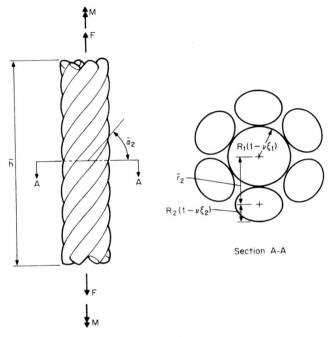

FIG. 26.5   A single-lay wire rope strand.

in which m = number of outer wires and α_2 = initial helix angle of the outer wires. In the usual case $r_2 < R_1 + R_2$, so that the outside wires are not touching each other. The expression

$$\tan \alpha_2 = p_2/2\pi r_2 \tag{26.2}$$

where p_2 = the pitch of the outer wires, is also valid. Since the pitch is generally known, a combination of Eqs. (26.1) and (26.2) yields the initial helix radius r_2, assuming that the outside wires are touching each other. If $r_2 < R_1 + R_2$, as determined by Eqs. (26.1) and (26.2), then the actual initial helical radius is given by

$$r_2 = R_1 + R_2 \tag{26.3}$$

Figure 26.6 shows an undeformed and a deformed configuration of an outer wire. An analysis of these configurations yields

$$\xi_1 = (\bar{h} - h)/h = \xi_2 + \Delta\alpha_2/\tan \alpha_2 \tag{26.4}$$

and

$$\beta_2 = r_2 \frac{\bar{\theta}_2 - \theta_2}{h} = \frac{1 + \nu}{\tan \alpha_2} \xi_2 + \left[\frac{\nu R_1}{(R_1 + R_2)\tan^2 \alpha_2} - 1 \right] \Delta\alpha_2 \tag{26.5}$$

where ξ_1 = axial strain in the center wire and of course the strand
h = original length of the strand
$\bar{h}$ = final length of the strand

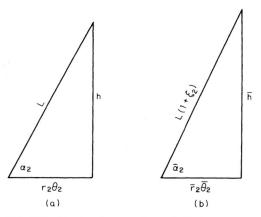

FIG. 26.6 Developed views of outer helical wire center line. (*a*) Initial configuration. (*b*) Final configuration.

ξ_2 = axial strain in the outer wires
$\alpha_2 + \Delta\alpha_2 = \bar{\alpha}_2$ = final helix angle of the outer wires
β_2 = rotational strain of the outer wires
v = Poisson's ratio for the wire material
θ_2 = original total angle the outer wires sweep out in a plane perpendicular to the axis of the strand
$\bar{\theta}_2$ = final total angle the outer wires sweep out in a plane perpendicular to the axis of the strand.

The angle of twist ϕ per unit length of the strand is given by the expression

$$\phi = (\bar{\theta}_2 - \theta_2)/h \tag{26.6}$$

It is assumed in the above that the axial strains ξ_1 and ξ_2 and the increase in the helix angle $\Delta\alpha_2$ are small, an assumption which is valid for most metallic strands.

In Fig. 26.7, the actual loads acting on the outer wires are shown. The equations yielding these loads are

$$\frac{H}{ER_2^3} = \frac{\pi}{4(1+v)} \frac{R_2}{R_1 + R_2}$$

$$\left[\left(1 - 2\sin^2\alpha_2 + \frac{vR_1\cos^2\alpha_2}{R_1 + R_2}\right)\Delta\alpha_2 + v(\sin\alpha_2\cos\alpha_2)\xi_2\right] \tag{26.7}$$

$$\frac{G'}{ER_2^3} = \frac{\pi}{4} \frac{R_2}{R_1 + R_2}$$

$$\left[\left(\frac{vR_1}{R_1 + R_2}\frac{\cos^2\alpha_2}{\tan\alpha_2} - 2\sin\alpha_2\cos\alpha_2\right)\Delta\alpha_2 + v(\cos^2\alpha_2)\xi_2\right] \tag{26.8}$$

$$\frac{T}{ER_2^2} = \pi\xi_2 \tag{26.9}$$

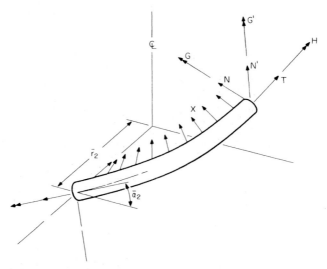

FIG. 26.7   Resultant loads acting on an outer wire.

$$\frac{N'}{ER_2^2} = \frac{H}{ER_2^3}\frac{\cos^2\alpha_2}{r_2/R_2} - \frac{G'}{ER_2^3}\frac{\sin\alpha_2\cos\alpha_2}{r_2/R_2} \tag{26.10}$$

and
$$\frac{X}{ER_2} = \frac{N'}{ER_2^2}\frac{\sin\alpha_2\cos\alpha_2}{r_2/R_2} - \frac{T}{ER_2^3}\frac{\cos^2\alpha_2}{r_2/R_2} \tag{26.11}$$

where H = axial twisting moment in an outer wire
     G' = bending moment in an outer wire
     T = axial force in an outer wire
     N' = shearing force in an outer wire
     X = contact force per unit length acting on the outer wire

Once these loads are known, the stresses in the wire can be computed.

The axial force F and the axial twisting moment M (Fig. 26.5) are given by the expressions

$$F = m_2 ER_2^2\left(\frac{T}{ER_2^2}\sin\alpha_2 + \frac{N'}{ER_2^2}\cos\alpha_2\right) + \pi R_1^2 E\xi_1 \tag{26.12}$$

and
$$M = m_2 ER_2^3\left(\frac{H}{ER_2^3}\sin\alpha_2 + \frac{G'}{ER_2^3}\cos\alpha_2 + \frac{T}{ER_2^2}\frac{r_2}{R_2}\cos\alpha_2 - \frac{N'}{ER_2^2}\frac{r_2}{R_2}\sin\alpha_2\right)$$
$$+ \frac{\pi ER_1^4}{4(1+\nu)}\frac{\beta_2}{r_2} \tag{26.13}$$

where F = the total axial force in the strand and M = the total axial twisting moment in the strand.

Consider, for example, an axially loaded strand in which the ends are not allowed to rotate. Let the strand be subjected to a prescribed axial strain ξ_1 (Fig. 26.6). Since $\phi = \beta_2 = 0$, Eqs. (26.4) and (26.5) can be used to determine ξ_2 and $\Delta\alpha_2$, and hence Eqs. (26.7) through (26.11) yield H, G', T, N', and X. The axial force F and the axial twisting moment M can be determined from Eqs. (26.12) and (26.13).

Let, for example, $m_2 = 6$, $R_1 = 0.03155$ in, $R_2 = 0.02893$ in, $E = 30 \times 10^6$ lb/in^2, $p = 1.30$ in, and $v = 0.29$. In this case, the outer wires do not touch each other. Hence $r_2 = 0.06048$ in and $\alpha_2 = 72.7069°$. For an axial wire strain of $\xi_2 = 0.001$ and $\phi = 0°$, the following values are determined: $\xi_1 = 0.001112$, $\Delta\alpha_2 = 0.000382$ rad, $F = 558.6$ lb, and $M = 7.60$ in·lb.

It is interesting to note that for $\phi = 0$, a load of 558.6 lb produces a stress of $E\xi_1 = 33,350$ lb/in^2 in the center wire. Since the results are linear, a load of 1000 lb, for example, would produce a stress of 59,700 lb/in^2 in the center wire with $\phi = 0$. Also, for the axial load of 558.6 lb and $\phi = 0$, the maximum tensile stress on the cross-sectional area of the outer wires is $E\xi_2 + 4G'R_2/\pi R_2^4 = (30,000 + 2600)$ lb/in^2 = 32,600 lb/in^2, which is the sum of the stresses due to the axial load T and the bending moment G' in the outer wires. This stress is less than the stress in the center wire.

In general, the total axial force and the total axial twisting moment for a rope can be expressed as follows[11] (see Fig. 26.7):

$$F/AE = K_1 \epsilon + K_2 \beta \tag{26.14}$$

and
$$M/ER^3 = K_3 \epsilon + K_4 \beta \tag{26.15}$$

where $A = \Sigma \pi R_i^2$ = metallic cross-sectional area of the rope
R = outside radius of the rope
R_i = radius of an individual wire
ϵ = axial strain of the rope
K_1, K_2, K_3, K_4 = dimensionless constants
β = rotational strain defined by

$$\beta = R\phi \tag{26.16}$$

where ϕ is the angle of twist per unit length of the rope. For example, in the case of a straight wire $K_1 = 1$, $K_2 = 0$, $K_3 = 0$, and $K_4 = \pi/4(1 + v)$. For the straight strand discussed above $K_1 = 0.886$, $K_2 = 0.144$, $K_3 = 0.319$, and $K_4 = 0.118$.

The above results can be extended to more complex cross sections,[11] such as that shown in Fig. 26.8. For example let $R_1 = 0.03155$ in, $R_2 = 0.02893$ in, $R_3 = 0.027723$

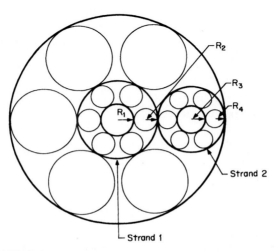

FIG. 26.8 Wire rope structure.

in, $R_4 = 0.02585$ in, $\alpha_2 = 73.707°$, $\alpha_4 = 81.066°$, and $\alpha_2^* = 70.830°$, where α_2^* is the helix angle of strand 2. The following equations result from an application of above theory

$$F/AE = 0.7984\epsilon + 0.1799\beta \tag{26.17}$$

and
$$M/ER^3 = 0.3092\epsilon + 0.0840\beta \tag{26.18}$$

For the 6×21 Seale wire rope with an IWRC, the following equations result:

$$F/AE = 0.7019\epsilon + 0.1232\beta \tag{26.19}$$

and
$$M/ER^3 = 0.2060\epsilon + 0.0403\beta \tag{26.20}$$

It is instructive at this point to express the effective modulus of elasticity of a rope E_e as

$$E_e = K_1E \tag{26.21}$$

which is the modulus of elasticity of the rope when $\phi = 0°$. Strand 1 has an effective modulus of $0.8864E$ and has a helix angle of $73.707°$, while strand 2, with a helix angle of $81.066°$, has an effective modulus of $0.9642E$. When strands 1 and 2 are placed together to form a rope, the effective modulus drops to $0.798E$. When strands 1, 2, and 3 are placed together to form the Seale IWRC rope, the effective modulus becomes $0.702E$. A recent test[11] on a 1.306-in-diameter 6×19 Seale rope with an independent wire rope core yielded an E_e of about 18,400,000 lb/in², while the preceding theory predicts an effective modulus of 19,950,000 lb/in² based on $E = 28,500,000$ lb/in². It should be remembered that the theory assumes that the wires just fit together and that contact deformation is neglected. Contact deformation and improper seating of the wires in the rope tend to lower the effective modulus of elasticity. The theory also assumes that the ends of the rope are held fast so that no relative motion between the wires occurs in the axial direction, a condition which is difficult to maintain even in a zinc socket.

It is interesting to compare, for a Seale IWRC, the relative axial wire strains in the individual wires when the rope is not allowed to rotate. For $\xi_{11} = 1.00$, $\xi_{12} = 0.9010$, $\xi_{21} = 0.8643$, $\xi_{22} = 0.8266$, $\xi_{31} = 0.8580$, $\xi_{32} = 0.8171$, and $\xi_{33} = 0.7408$ where ξ_{ij} is the axial strain of a wire in the ith strand and the jth layer.

Figure 26.9 shows a rope bent over a sheave, where

$\rho = $ radius of curvature of the center-line of the rope

$p = $ normal load per unit length acting on the rope

$M_B = $ bending moment in the rope

$M_T = $ torsional moment in the rope

$F = $ axial force in the rope

$q = $ twisting couple per unit length

The loads p and q are applied to the rope by the sheave in order to keep the rope in equilibrium. Equilibrium of the rope requires that

$$F = p(\rho - R) \tag{26.22}$$

and
$$M_T = pq \tag{26.23}$$

This twisting couple per unit length q must be applied to the rope over a small contact area.[14] This twisting couple per unit length is rather large for large-diameter ropes and

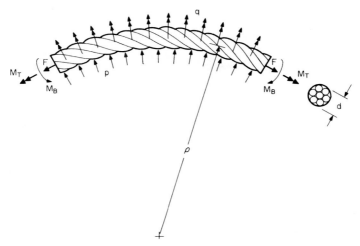

FIG. 26.9   Resultant loads acting on a rope bent over a sheave.

may thus be the main cause of excessive wear due to the shearing stresses acting on the outer wires as the rope passes over the sheave.

There are many parameters that can vary in the construction of a wire rope, e.g., the number of wires and the helix angle. For simplicity, a rope or strand consisting of six helical wires wrapped around a center wire as shown in Fig. 26.5 will be considered first. The bending stiffness of the rope will be approximated by a summation of the bending stiffness of each of the wires in the rope, i.e., by treating the rope as an assemblage of helical springs.[8] This assumption has been confirmed recently by experiments.[15]

The bending stiffness of a helical wire with wire radius R_1 is less than the bending stiffness of a straight wire of wire radius R_2; i.e., the helical wire is more flexible than the straight wire. When the strand is wrapped around the sheave, the center wire receives the larger bending stress because it is initially straight and also because it generally has a larger wire radius than with the outside helical wires.[13]

Consider, for example, strand 1 discussed above, for which $R_1 = 0.03155$ in and $R_2 = 0.02893$ in. Let the strand be bent over a sheave with a radius $\rho = 6$ ft and let the axial force F in the strand be, as previously calculated with $\beta = 0$, 558.6 lb. The twisting moment M_T is 7.60 in·lb. The bending stress σ_B in the center wire is given by the expression

$$\sigma_B = ER_1/\rho \qquad (26.24)$$

and hence for the case under consideration $\sigma_B = (30 \times 10^6)(0.03155)/(6 \times 12)$ lb/in² $= 13,150$ lb/in². The maximum tensile stress occurs in the center wire and is equal to the sum of stresses due to the axial loads and the bending stresses.[14] Hence $\sigma_{max} = (33,350 + 13,150)$ lb/in² $= 46,500$ lb/in². It is interesting to note that increasing the axial loads does not cause an increase in the bending stresses of the center wire since the curvature remains the same.

The above theory has been extended to ropes with complex cross sections.[12] In the case of a 6 × 19 Seale IWRC, however, the maximum tensile stress on a wire cross section may shift from the center wire in strand 1 to the center wire in the outside strand, since the center wire in the outside strand has a larger wire radius than the center wire in strand 1. This can cause larger bending stresses in the center wire of the

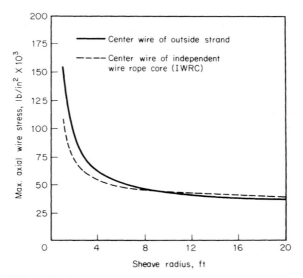

FIG. 26.10 Maximum axial center wire tensile stress as a function of sheave radius of a 1.25-in-diameter 6 × 19 Seale rope (axial force = 18,250 lb).

outside strand and hence the maximum wire stress depends upon the radius of curvature ρ. Figure 26.10 shows a plot of the tensile stress in a Seale IWRC rope for the two wires discussed above as a function of the radius of curvature ρ.

While the previous analytical discussion has been applied to a Seale IWRC wire rope, similar remarks can also be made for other types of ropes. The dimensions and constructions of these ropes as well as the breaking loads can be found in wire rope catalogs of various manufacturers.

In an actual rope there is of course some friction. In the axially loaded rope, friction is felt to be small. However, many ropes fail by a mechanism called "fretting fatigue." As a rope passes over a sheave, wires which are pressed together move relative to one another. This relative motion results in pitting and scarring of the surfaces due to the abrasive action of the debris. Frequently, cracks propagate into the material from the stem of the base of these pits. At present, however, there is no generalized theory of fretting fatigue.[16]

It should be noted that no rope is designed to last forever and hence ropes have to be inspected on a regular basis. This is critical. Among factors which are generally involved in the inspection are abrasion, rope stretch, reduction in rope diameter, corrosion, kinks, heat damage, protruding core, damaged end attachments, fatigue failure, scrubbing, and broken wires. For additional information along these lines, the interested reader is again referred to Ref. 7. A more elaborate discussion of the theory can be found in Ref. 17.

26.5 SPECIAL ROPE DRIVES

Rope drives consisting of manila or cotton rope on multiple-grooved pulleys may often be a most economical form of drive over long distances and for large amounts of power. The velocity should be quite high; 5000 ft/min is a good speed for economy.

Maximum permissible working load in pounds for manila rope for hoisting purposes is $200d^2$ at 400 to 800 ft/min, $400d_r^2$ at 150 to 300 ft/min, and $1000d^2$ below 100 ft/min, where d is rope diameter in inches. Reference 18 shows data for manila rope. Nylon rope is also used. It has from 1½ to 2 times the tensile strength and 3 times the elasticity of manila rope and gives greater resistance to fatigue and surface abrasion. For power transmission the rope wrap tension is shown in Sec. 23. The coefficient of friction for hemp rope on a cast-iron pulley is $\mu = 0.2$ and on a wood pulley $\mu = 0.4$.

REFERENCES

1. Suslov, B. M.: "On the modulus of elasticity of wire ropes," *Wire and Wire Products,* vol. 11, p. 176, 1936.

2. Scoble, W. A.: "First Report of Wire Rope Research Committee," *Proc. Inst. Mech. Engrs.,* vol. 115, p. 835, 1920; second report, vol. 119, p. 1193, 1924; third report, vol. 123, p. 353, 1928; fourth report, vol. 125, p. 553, 1930; fifth report, vol. 130, p. 373, 1935.

3. de Forest, A. V., and L. W. Hopkins: "Testing of rope wire and wire rope," *Proc. ASTM,* vol. 32, p. 398, 1932.

4. Costello, G. A.: "Analytical investigation of wire rope," *Appl. Mech. Rev.,* vol. 31, no. 7, July 1978.

5. Sayenga, D.: "The birth and evolution of the American wire rope industry," *Proc. 1st Annual Wire Rope Symposium,* Washington State University, Pullman, Wash., p. 275, 1980.

6. "Wire Rope Manufacturing," Leschen Wire Rope Company, St. Joseph, Mo., 1979.

7. "Wire Rope Users Manual," Committee of Wire Rope Producers, American Iron and Steel Institute, Washington, 1983.

8. Costello, G. A., and R. E. Miller: "Lay effect of wire rope," *J. Eng. Mech. Div. ASCE,* vol. 105, no EM4, Proc. Paper 14753, p. 597, August 1979.

9. Costello, G. A., and J. W. Phillips: "Effective modulus of twisted wire cables," *J. Eng. Mech. Div. ASCE,* vol. 102, no. EM1, Proc. Paper 11944, p. 171, February 1976.

10. Costello, G. A., and R. E. Miller: "Static response of reduced rotation rope," *J. Eng. Mech. Div. ASCE,* vol. 106, no. EM4, Proc. Paper 15612, p. 623, August 1980.

11. Velinsky, S. A., G. L. Anderson, and G. A. Costello: "Wire rope with complex cross sections," T & AM report no. 448, Dept. Theoretical and Applied Mechanics, University of Illinois, Urbana-Champaign, p. 1, June 1981.

12. Velinsky, S. A.: "Analysis of wire rope with complex cross sections," Ph.D. thesis, Dept. Theoretical and Applied Mechanics, University of Illinois, Urbana-Champaign, 1981.

13. Costello, G. A.: "Large deflections of helical spring due to bending," *J. Eng. Mech. Div., ASCE,* vol. 103, no, EM3, Proc. Paper 12964, p. 479, June 1977.

14. Costello, G. A., and G. J. Butson: "A simplified bending theory for wire rope," *J. Eng. Mech. Div., ASCE,* vol. 108, no. EM2, Apr. 1982.

15. McConnell, K. G., and W. P. Zemke: "The measurement of flexural stiffness of multistranded electrical conductors while under tension," *J. Soc. Experimental Stress Analysis, Experimental Mechanics,* vol. 20, no. 6, p. 198, June 1980.

16. Hoeppner, D. W., and F. L. Gates: "Fretting fatigue considerations in engineering design," *Wear,* 70, p. 155, 1981.

17. Costello, G. A.: *Theory of Wire Rope,* Springer-Verlag, New York, 1990.

18. U.S. Government, Spec. TR 601A, November 26, 1935.

SECTION 27
GEARING

John J. Coy, M.S.M.E., Ph.D., P.E.

Chief of Mechanical System Technology Branch
NASA Lewis Research Center
Cleveland, Ohio

Dennis P. Townsend, B.S.M.E.

Senior Research Engineer
NASA Lewis Research Center
Cleveland, Ohio

Erwin V. Zaretsky, B.S.M.E., J.D., P.E.

Chief Engineer of Structures
NASA Lewis Research Center
Cleveland, Ohio

27.1 SYMBOLS

A = Cone distance, m (in)

a = Semimajor axis of hertzian contact ellipse, m (in)

B = Backlash, m (in)

B_{10} = Bearing life at 90 percent reliability level

b = Semiminor axis of hertzian contact ellipse, m (in)

C = Center distance, m (in)

C_1 to C_7 = Special constants (see Table 27.19)

c = Distance from neutral axis to outermost fiber of beam, m (in)

c = Exponent

D = Diameter of pitch circle, m (in)

D_G = Pitch diameter of gear wheel, m (in)

D_W = Pitch diameter of worm gear, m (in)

d = Impingement depth, m (in)

E = Young's modulus of elasticity, N/m² (lb/in²)

E' = Effective elastic modulus, N/m² (lb/in²)

e = Weibull slope, dimensionless

F = Face width of tooth, m (in)

F_e = Effective face width, m (in)

f = Coefficient of friction

G = Dimensionless materials parameter

G_{10} = Life of gear at 90 percent reliability level, in millions of revolutions or hours

H = Dimensionless film thickness

H_B = Brinell hardness

h = Film thickness, m (in)

h = Exponent

I = Area moment of inertia, m⁴ (in⁴)

J = Geometry factor

k = Ellipticity ratio, dimensionless

K = Load intensity factor, N/m² (lb/in²)

K_a = Application factor

K_L = Life factor

K_m = Load distribution factor

K_R = Reliability factor

K_s = Size factor

K_T = Temperature factor

K_v = Dynamic factor

K_x = Cutter radius factor

L = Life, hours or millions of revolutions or stress cycles

L_W = Lead of worm gear, m (in)

l = Length, m (in)

M = Bending moment, N·m (lb·in)

m_c = Contact ratio

m_g = Gear ratio, N_2/N_1

N = Number of teeth

N_e = Equivalent number of teeth

n = Angular velocity, r/min

O = Centerpoint location

P = Diametral pitch, m⁻¹ (in⁻¹)

P_n = Normal diametral pitch, m⁻¹ (in⁻¹)

P_t = Transverse diametral pitch, m⁻¹ (in⁻¹)

p = Circular pitch, m (in)

p_a = Axial pitch, m (in)

p_b = Base pitch, m (in)

p_n = Normal circular pitch, m (in)

p_t = Transverse circular pitch, m (in)

Q = Power loss, kW (hp)

q = Number of planets

R_c = Rockwell C scale hardness

r = Radius, m (in)

r_a = Addendum circle radius, m (in)

r_b = Base circle radius, m (in)

r_c = Cutter radius, m (in)

r_p = Pitch circle radius, m (in)

r_{av} = Mean pitch cone radius, m (in)

S = Probability of survival

S = Stress, N/m² (lb/in²)

S_{at} = Allowable tensile stress, N/m^2 (lb/in^2)

S_{ay} = Allowable yield stress, N/m^2 (lb/in^2)

S_c = Maximum hertzian compressive stress, N/m^2 (lb/in^2)

S_r = Residual shear stress, N/m^2 (lb/in^2)

S_T = Probability of survival of complete transmission including bearings and gears

s = Distance measured from the pitch point along line of action, m (in)

T = Torque, N·m (lb·in)

T_{10} = Tooth life at 90 percent reliability, in millions of stress cycles or hours

t = Tooth thickness, m (in)

U = Dimensionless speed parameter

u = Integer number

V = Stressed volume, m^3 (in^3)

$\mathbf{v}$ = Velocity vector, m/s (ft/s)

v_s = Sliding velocity, m/s (ft/s)

v_j = Jet velocity, m/s (ft/s)

v_r = Rolling velocity, m/s (ft/s)

W = Dimensionless load parameter

W_N = Load normal to surface, N (lb)

W_r = Radial component of load, N (lb)

W_t = Transverse or tangential component of load, N (lb)

W_x = Axial component of load, N (lb)

w = Semiwidth of path of contact, m (in)

X_1 = Distance from line of centers to impingement point on pinion, m (in)

X_2 = Same as X_1, but for gear

Y = Lewis form factor, dimensionless

Z = Length of line of action, m (in)

z = Depth to critical shear stress, m (in)

α = Pressure-viscosity exponent, m^2/N (in^2/lb)

Γ = Pitch angle, degrees

γ = Angle, degrees

Δc = Change in center distance, m (in)

$\Delta \rho$ = Curvature difference, dimensionless

Δp = Differential pressure between oil and ambient, N/m^2 (lb/in^2)

δ = Tooth deflection, m (in)

ϵ = Involute roll angle, degrees (rad)

ϵ_H = Increment of roll angle for which a single pair of teeth is in contact, degrees (rad)

ϵ_L = Increment of roll angle during one of the periods of two pairs of teeth in contact, degrees (rad)

ϵ_c = Involute roll angle to lowest point of double tooth pair contact

η = Life, millions of stress cycles

θ = Angle of rotation, rad

ξ = Efficiency

Λ = Film thickness parameter, ratio of film thickness to rms surface roughness, dimensionless

λ_w = Lead angle of worm gear, degrees

μ = Lubricant absolute viscosity, N·s/m^2 (lb·s/in^2)

μ_o = Absolute viscosity at standard atmospheric pressure, N·s/m^2 (lb·s/in^2)

ν = Poisson's ratio, dimensionless

ρ = Radius of curvature, m (in)

Σ = Shaft angle, degrees

$\Sigma \rho$ = Curvature sum, m^{-1} (in^{-1})

σ = Surface rms roughness, m (in)

τ = Shear stress, N/m^2 (lb/in^2)

τ_{max} = Maximum shear stress, N/m^2 (lb/in^2)

τ_o = Maximum subsurface orthogonal reversing shear stress, N/m^2 (lb/in^2)

Y = Parameter defined in Table 27.15

ϕ = Pressure angle, degrees

ϕ_n = Normal pressure angle, degrees

ϕ_t = Transverse pressure angle, degrees

ϕ_w' = Angle between load vector and tooth centerline (Fig. 27.36), degrees

ψ = Helix angle, degrees

ψ = Spiral angle for spiral bevel gear, degrees

ω = Angular velocity, rad/s

Subscripts

1	=	Pinion	s =	Sliding, sun
2	=	Gear	t =	Threads
a	=	At addendum, at addendum circle	T =	Total
B	=	Bearing	w =	Windage
G	=	Gear wheel	W =	Worm gear
j	=	Jet	x, y =	Mutually perpendicular planes in which the minimum and maximum (principal) radii of surface curvature lie
m	=	Mesh		
p	=	Planet		
r	=	Rolling, ring		

27.2 INTRODUCTION

Gears are the means by which power is transferred from source to application. Gearing and geared transmissions drive the machines of modern industry. Gears move the wheels and propellers that transport us through the sea, land, and air. A very sizable section of industry and commerce in today's world depends on gearing for its economy, production, and likelihood.

The art and science of gearing has its roots before the common era. Yet many engineers and researchers continue to delve into the regions where improvements are necessary, seeking to quantify, establish, and codify the methods to make gears meet the ever-widening needs of advancing technology.[1] References 135–142 are recommended for further reading on gearing.

27.2.1 Early History of Gearing

The earliest written descriptions of gears are said to have been made by Aristotle in the fourth century B.C.[2] It has been pointed out[3,4] that the passage attributed to Aristotle by some[2] was actually from the writings of his school, in "Mechanical Problems of Aristotle" (ca. 280 B.C.). In the passage in question, there was no mention of gear teeth on the parallel wheels, and they may just as well have been smooth wheels in frictional contact. Therefore, the attribution of gearing to Aristotle is, most likely, incorrect.

The real beginning of gearing was probably with Archimedes who about 250 B.C. invented the endless screw turning a toothed wheel, which was used in engines of war. Archimedes also used gears to simulate astronomical ratios. The Archimedian spiral was continued in the hodometer and dioptra, which were early forms of wagon mileage indicators (odometer) and surveying instruments. These devices were probably "thought" experiments of Heron of Alexandria (ca A.D. 60), who wrote on the subjects of theoretical mechanics and the basic elements of mechanism. The oldest surviving relic containing gears is the Antikythera mechanism, so named because of the Greek island of that name near which the mechanism was discovered in a sunken ship in 1900. Professor Price[3] of Yale University has written an authoritative account of this mechanism. The mechanism is not only the earliest relic of gearing, but it also is an extremely complex arrangement of epicyclic differential gearing. The mechanism is identified as a calendrical computing mechanism for the sun and moon, and has been dated to about 87 B.C.

The art of gearing was carried through the European dark ages after the fall of

Rome, appearing in Islamic instruments such as the geared astrolabes which were used to calculate the positions of the celestial bodies. Perhaps the art was relearned by the clock- and instrument-making artisans of fourteenth-century Europe, or perhaps some crystallizing ideas and mechanisms were imported from the East after the crusades of the eleventh through the thirteenth centuries.

It appears that the English abbot of St. Alban's monastery, born Richard of Wallingford, in A.D. 1330, reinvented the epicyclic gearing concept. He applied it to an astronomical clock, which he began to build at that time and which was completed after his death.

A mechanical clock of a slightly later period was conceived by Giovanni de Dondi (1348–1364). Diagrams of this clock, which did not use differential gearing,[3] appear in the sketchbooks of Leonardo da Vinci, who designed geared mechanisms himself.[3,5] In 1967 two of da Vinci's manuscripts, lost in the National Library in Madrid since 1830,[6] were rediscovered. One of the manuscripts, written between 1493 and 1497 and known as "Codex Madrid I,"[7] contains 382 pages with some 1600 sketches. Included among this display of Leonardo's artistic skill and engineering ability are his studies of gearing. Among these are tooth profile designs and gearing arrangements that were centuries ahead of their "invention."

27.2.2 Beginning of Modern Gear Technology

In the period 1450 to 1750, the mathematics of gear-tooth profiles and theories of geared mechanisms became established, Albrecht Dürer is credited with discovering the epicycloidal shape (ca. 1525). Philip de la Hire is said to have worked out the analysis of epicycloids and recommended the involute curve for gear teeth (ca. 1694). Leonard Euler worked out the law of conjugate action (ca. 1754).[5] Gears designed according to this law have a steady speed ratio.

Since the industrial revolution in the mid-nineteenth century, the art of gearing blossomed, and gear designs steadily became based on more scientific principles. In 1893 Wilfred Lewis published a formula for computing stress in gear teeth.[8] This formula is in wide use today in gear design. In 1899 George B. Grant, the founder of five gear manufacturing companies, published "A Treatise on Gear Wheels."[9] New inventions led to new applications for gearing. For example, in the early part of this century (1910), parallel shaft gears were introduced to reduce the speed of the newly developed reaction steam turbine enough to turn the driving screws of ocean-going vessels. This application achieved an overall increase in efficiency of 25 percent in sea travel.[2]

The need for more accurate and quiet-running gears became obvious with the advent of the automobile. Although the hypoid gear was within our manufacturing capabilities by 1916, it was not used practically until 1926, when it was used in the Packard automobile. The hypoid gear made it possible to lower the drive shaft and gain more usable floor space. By 1937 almost all cars used hypoid-geared rear axles. Special lubricant antiwear additives were formulated in the 1920s which made it practical to use hypoid gearing. In 1931 Earle Buckingham, chairman of an American Society of Mechanical Engineers (ASME) research committee on gearing, published a milestone report on gear-tooth dynamic loading.[10] This led to a better understanding of why faster-running gears sometimes could not carry as much load as slower-running gears.

High-strength alloy steels for gearing were developed during the 1920s and 1930s. Nitriding and case-hardening techniques to increase the surface strength of gearing were introduced in the 1930s. Induction hardening was introduced in 1950. Extremely clean steels produced by vacuum melting processes introduced in 1960 have proved effective in prolonging gear life.

Since the early 1960s there has been increased use of industrial gas turbines for electric power generation. In the range of 1000 to 14,000 hp, epicyclic gear systems have been used successfully. Pitch-line velocities are from 50 to 100 m/s (10,000 to 20,000 ft/min). These gear sets must work reliably for 10,000 to 30,000 h between overhauls.[1]

In 1976 bevel gears produced to drive a compressor test stand ran successfully for 235 h at 2984 kW (4000 hp) and 200 m/s (40,000 ft/min).[11] From all indications these gears could be used in an industrial application if needed. A reasonable maximum pitch-line velocity for commercial spiral-bevel gears with curved teeth is 60 m/s (12,000 ft/min).[12]

Gear system development methods have been advanced in which lightweight, high-speed, highly loaded gears are used in aircraft applications. The problems of strength and dynamic loads, as well as resonant frequencies for such gearing, are now treatable with techniques such as finite-element analysis, siren and impulse testing for mode shapes, and application of damping treatments where required.[13]

The material contained in this section will assist in the design, selection, applications, and evaluation of gear drives. Sizing criteria, lubricating consideration, material selection, and methods to estimate service life and power loss are presented.

27.3 TYPES AND GEOMETRY

References 14 and 15 outline the various gear types including information on proper gear selection. These references classify single-mesh gears according to the arrangement of their shaft axes in a single-mesh gear set. These arrangements are *parallel shafts; intersecting shafts*; and *nonparallel, nonintersecting shafts*. Table 27.1 lists and compares these gear sets for a single mesh.

27.3.1 Parallel Shaft Gear Types

Spur Gears. Spur gears are the most common type of gears. A representative pair of external spur gears is shown in Fig. 27.1. The basic geometry of spur gears is shown in Fig. 27.2. Teeth are straight and parallel to the shaft axis. This is consequently the simplest possible type of gear. The smaller of two gears in mesh is called the "pinion." The larger is customarily designated as the gear. In most applications (for speed reduction) the pinion is the driving element whereas the gear is the driven element.

Most spur gear tooth profiles are cut to conform to an involute curve which ensures conjugate action. "Conjugate action" is defined as a constant angular velocity ratio between two meshing gears. Conjugate action may be obtained with any tooth profile shape for the pinion, provided the mating gear is made with a tooth shape that is conjugate to the pinion tooth shape. The conjugate tooth profiles are such that the common normal at the point of contact between the two teeth will always pass through a fixed point on the line of centers. The fixed point is called the "pitch point."

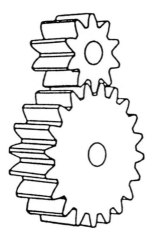

FIG. 27.1 Spur gears.

TABLE 27.1 Comparison of Single-Mesh Gears

Type of gear	Gear ratio range (nominal)	Maximum pitch line velocity, ft/min		Nominal efficiency at rated power, %	Remarks
		Aircraft, high precision	Commercial		
Parallel shafts					
Spur	1:1–5:1	20,000	4,000	97–99.5	
Helical	1:1–10:1	40,000	4,000	97–99.5	15–35° helix angle
Double helical "herringbone"	1:1–20:1	40,000	4,000	97–99.5	
Conformal	1:1–12:1	—	—	97–99.5	
Intersecting shafts					
Straight bevel	1:1–8:1	10,000	1,000	97–99.5	
Spiral bevel	1:1–8:1	25,000	12,000	97–99.5	35° spiral angle most common
Zerol®* bevel	1:1–8:1	10,000	1,000	97–99.5	
Nonintersecting and nonparallel shafts					
Worm	3:1–100:1	10,000	5,000	50–90	
Double enveloping worm	3:1–100:1	10,000	4,000	50–98	
Crossed helical	1:1–100:1	10,000	4,000	50–95	Higher ratios have lower efficiency
Hypoid	1:1–50:1	10,000	4,000	50–98	

*Zerol is a registered trademark of Gleason Works, Rochester, N.Y.

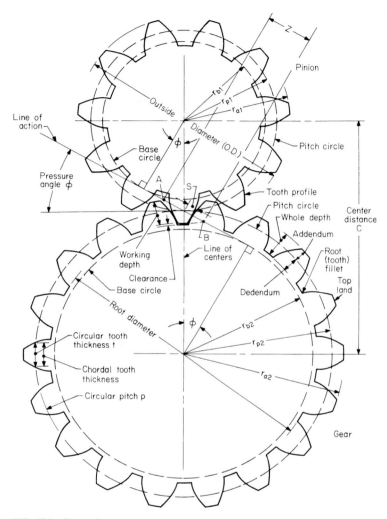

FIG. 27.2   External spur gear geometry.

Referring to Fig. 27.2, the line of action or pressure line is shown as the line which is tangent to both base circles. This is the line along which all points of contact between the two teeth will lie. The pressure angle φ is defined as the acute angle between the line of action and a line perpendicular to the line of centers. While the pressure angle can vary for different applications, most spur gears are cut to operate at pressure angles of 20 or 25°.

The circular pitch p of a spur gear is defined as the distance, on the pitch circle, from a point on a tooth to the corresponding point of the adjacent tooth. The circular pitch

$$p = \frac{2\pi r_p}{N}$$

(27.1)

where r_p is the pitch circle radius and N is the number of teeth. Similarly, the base pitch p_b is the distance on the base circle from a point on one tooth to the corresponding point on an adjacent tooth. Hence, base pitch

$$p_b = \frac{2\pi r_b}{N} \tag{27.2}$$

where r_b is the base circle radius. Also, from geometry,

$$p_b = p \cos \phi \tag{27.3}$$

The diametral pitch P is defined as the number of teeth on the gear divided by the pitch circle diameter in inches (determines the relative sizes of gear teeth):

$$P = \frac{N}{D} \tag{27.4}$$

where D is the pitch circle diameter and

$$P \cdot p = \pi \tag{27.5}$$

For a given size gear diameter, the larger the value for P, the greater the number of gear teeth and the smaller their size. Module is the reciprocal of diametral pitch in concept; but whereas diametral pitch is expressed as number of teeth per inch, the module is generally expressed in millimeters per tooth. Metric gears (in which tooth size is expressed in module) and American standard-inch-diametral pitch gears are not interchangeable.

Referring to Fig. 27.2, the distance between the centers of two gears is

$$C = \frac{D_1 + D_2}{2} \tag{27.6}$$

The space between the teeth must be larger than the mating gear tooth thickness in order to prevent jamming of the gears.

The difference between the tooth thickness and tooth space as measured along the pitch circle is called *backlash*. Backlash may be created by cutting the gear teeth slightly thinner than the space between teeth or by setting the center distance slightly greater between the two gears. In the second case, the operating pressure angle of the gear pair is increased accordingly. The backlash for a gear pair must be sufficient to permit free action under the most severe combination of manufacturing tolerances and operating temperature variations. Backlash should be very small in positioning control systems, and it should be quite generous for single-direction power gearing. Table 27.2 gives recommended backlash amounts. If the center distance between mating

TABLE 27.2 Recommended Backlash for Assembled Gears

P	Backlash, in	P	Backlash, in
1	0.025–0.040	5	0.006–0.009
$1\frac{1}{2}$	0.018–0.027	6	0.005–0.008
2	0.014–0.020	7	0.004–0.007
$2\frac{1}{2}$	0.011–0.016	8–9	0.004–0.006
3	0.009–0.014	10–13	0.003–0.005
4	0.007–0.011	14–32	0.002–0.004

external gears is increased by an amount ΔC, the resulting increase in backlash ΔB is given by the following equation:

$$\Delta B = \Delta C \times 2 \tan \phi \qquad (27.7)$$

For the mesh of internal gears, Eq. 27.7 gives the decrease in backlash for an increase in center distance.

Referring to Fig. 27.2, the distance between points A and B is the length of contact along the line of action, which is defined as the distance between where the gear outside circumference intersects the line of action and where the pinion outside circumference intersects the line of action. This is the total length along the line of action for which there is tooth contact. This distance can be determined by the various radii and the pressure angle as follows:

$$Z = [r_{a1}^2 - (r_{p1} \cos \phi)^2]^{1/2} + [r_{a2}^2 - (r_{p2} \cos \phi)^2]^{1/2} - C \sin \phi \qquad (27.8)$$

The radii of curvature of the teeth when contact is at a distance s along the line of action from the pitch point can be represented for the pinion as

$$\rho_1 = r_{p1} \sin \phi \pm s \qquad (27.9)$$

and for the gear

$$\rho_2 = r_{p2} \sin \phi \pm s \qquad (27.10)$$

Where s is positive for Eq. (27.9), it must be negative for Eq. (27.10) and vice versa.

In order to determine how many teeth are in contact, the contact ratio of the gear mesh must be determined. The contact ratio m_c is defined as

$$m_c = \frac{Z}{P_b} \qquad (27.11)$$

Gears are generally designed with contact ratios between 1.2 and 1.6. A contact ratio of 1.6, for example, means that for 40 percent of the time, there will be one pair of teeth in contact, and for 60 percent of the time, two pairs of teeth will be in contact. A contact ratio of 1.2 means that 80 percent of the time one pair of teeth will be in contact and two pairs of teeth will be in contact 20 percent of the time.

Gears with contact ratios greater than 2 are referred to as *high-contact ratio gears*. For high-contact ratio gears there are never fewer than two pairs of teeth in contact. A contact ratio of 2.2 means that for 80 percent of the time there will be two pairs of teeth in contact and for 20 percent of the time there will be three pairs of teeth in contact. High-contact ratio gears are generally used in select applications where long life is required. Figure 27.3a shows a high-contact-ratio (2.25) modified tooth, and Fig. 27.3b shows a normal-contact-ratio (1.3) involute tooth. Analysis should be cautious in designing high-contact ratio gearing because of higher bending stresses which may occur in the tooth dedendum region. Also, higher sliding in the tooth contact may contribute to distress of the tooth surfaces. In addition, higher dynamic loading may occur with high-contact ratio gearing.[16]

Interference of the gear teeth is an important consideration. The portion of the spur gear below the base circle is sometimes cut as a straight radial line, but never as an involute curve (Fig. 27.4). Hence, if contact should occur below the base circle, nonconjugate action (interference) will occur. The maximum addendum radius of the gear without interference is the value r_a' calculated from the following equation:

$$r_a' = (r_b^2 + c^2 \sin^2 \phi)^{1/2} \qquad (27.12)$$

FIG. 27.3 Comparison of gear contact ratio. (*a*) Modified tooth profile; contact ratio 2.25. (*b*) Conventional involute tooth; contact ratio 1.3.

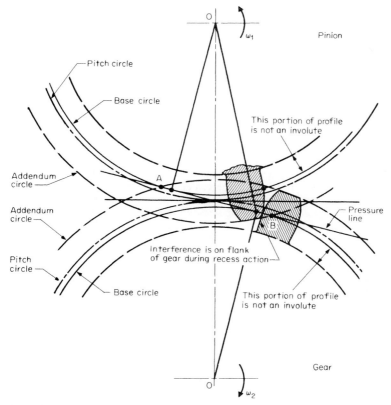

FIG. 27.4 Interference in the action of gear teeth. (Interference is on flank of pinion during approach action and flank of gear during recess action.)

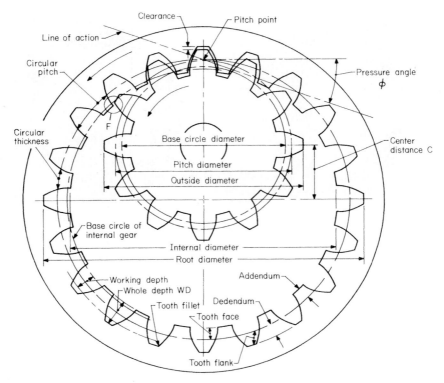

FIG. 27.5   Internal spur gear geometry.

where r'_a is the radial distance from the gear center to the point of tangency of the mating gear's base circle with the line of action. Should interference be indicated, there are several methods to eliminate it. The center distance may be increased, which will also increase the pressure angle and change the contact ratio. Also, the addendum may be shortened, with a corresponding increase in the addendum of the mating member. The method used depends upon the application and experience.

The geometry for *internal gearing* where teeth are cut on the inside of the rim is shown in Fig. 27.5. The pitch relationships previously discussed for external gears apply also to internal gears [Eqs. (27.1) through (27.5)]. A secondary type of involute interference called "tip fouling" may also occur. The geometry should be carefully checked in the region labeled F in Fig. 27.5 to see if this occurs.

Helical Gears.   Figure 27.6 shows a helical gear set. These are cylindrical gears with the teeth cut in the form of a helix; one gear has a right-hand helix and the mating gear a left-hand helix. Helical gears have a greater load-carrying capacity than equivalent-size spur gears. Although these gears produce less vibration than spur gears because of overlapping of the teeth, a high thrust load is produced along the axis of rotation. This results in high rolling-element bearing loads which may reduce the life and reliability of a transmission system. This end thrust increases with helix angle.

In order to overcome the axial-thrust load, double-helical gear sets (herringbone gears) are used (Fig. 27.7). The thrust loads are counterbalanced, so that no resultant axial load is transmitted to the bearings. Space is sometimes provided between the two

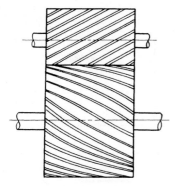

FIG. 27.6 Single-helical gear set.

FIG. 27.7 Double-helical, or herringbone, gear set.

sets of teeth to allow for runout of the tooth-cutting tool, or the gear is assembled in two halves. The space should be small because it reduces the active face width. Double-helical gears are used for the transmission of high torques at high speeds in continuous service.

The geometry of a helical gear is shown in Fig. 27.8. While spur gears have only one diametral and circular pitch, helical gears have two additional pitches. The normal circular pitch p_n is the distance between corresponding points of adjacent teeth as measured in plane BB of Fig. 27.8, which is perpendicular to the helix.

$$p_n = p_t \cos \psi \qquad (27.13)$$

where p_t is the transverse circular pitch and ψ is the helix angle.

The transverse circular pitch p_t is measured in plane AA of Fig. 27.8 perpendicular to the shaft axis. The axial pitch p_a is a similar distance measured in a plane parallel to the shaft axis:

$$p_a = p_t \cot \psi \qquad (27.14)$$

The diametral pitch p_t in the transverse plane is

$$p_t = \frac{N}{D} \qquad (27.15)$$

where N is the number of teeth of the gear and D is the diameter of the pitch circle.

The normal diametral pitch is

$$p_n = \frac{p_t}{\cos \phi} \qquad (27.16)$$

To maintain contact across the entire tooth face, the minimum face width F must be greater than the axial pitch p_a. However, to assure a smooth transfer of load during tooth engagement the face width F should be at least 1.2 to 2 times the axial pitch p_a.

The two pressure angles associated with the helical gear are the transverse pressure angle ϕ_t measured in plane AA (Fig. 27.8) and the normal pressure angle ϕ_n measured in plane BB can be related as follows:

$$\tan \varphi_n = \tan \phi_t \cos \psi \qquad (27.17)$$

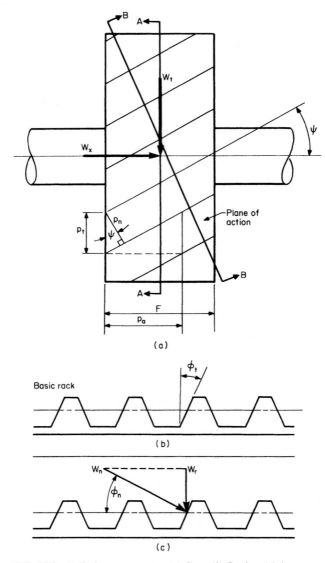

FIG. 27.8 Helical gear geometry. (*a*) Gear. (*b*) Section *AA* (transverse plane). (*c*) Section *BB* (normal plane).

The three components of normal load W_N acting on a helical gear tooth can be written as follows:

Tangential: $$W_t = W_N \cos \phi_n \cos \psi \qquad (27.18)$$

Radial: $$W_r = W_N \sin \varphi_n \qquad (27.19)$$

Axial: $$W_x = W_N \cos \varphi_n \sin \psi \qquad (27.20)$$

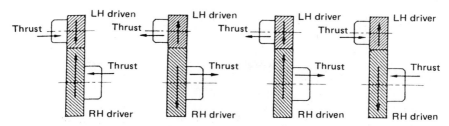

FIG. 27.9 Direction of the axial thrust load for helical gears on parallel shafts. LH indicates left-hand and RH indicates right-hand.

where W_t is the tangential load acting at the pitch circle perpendicular to the axis of rotation. Figure 27.9 shows the direction of the axial-thrust loads for pairs of helical gears, considering whether the driver has a right-hand or left-hand helix and the direction of rotation. These loads must be considered when selecting and sizing the rolling-element bearings to support the shafts and gears.

The plane normal to the gear teeth, *BB* in Fig. 27.8, intersects the pitch cylinder. The gear tooth profile generated in this plane would be a spur gear, having the same properties as the actual helical gear. The number of teeth of the equivalent spur gear in the normal plane is known as either the "virtual" or "equivalent" number of teeth. The equivalent number of teeth is

$$N_e = \frac{N}{\cos^3 \psi} \tag{27.21}$$

Conformal (Wildhaber-Novikov) Gears. In 1923, E. Wildhaber filed a U.S. patent application[17] for helical gearing of a circular arc form. In 1926 he filed a patent application for a method of grinding of this form of gearing.[18] Until Wildhaber, helical gears were made only with screw involute surfaces. However, Wildhaber-type gearing failed to receive the support necessary to develop it into a working system.

In 1956, M. L. Novikov in the U.S.S.R. was granted a patent for a similar form of gearing.[19] Conformal gearing then advanced in the U.S.S.R., but the concept found only limited application. Conformal gearing was also the subject of research in Japan and China but did not find widespread industrial application.

The first and probably only application of conformal gearing to aerospace technology was by Westland Helicopter in Great Britain.[20,21] Westland modified the geometry of the conformal gears as proposed by Novikov and Wildhaber and successfully applied the modified geometry to the transmission of the Lynx helicopter. The major advantage demonstrated on this transmission system is increased gear load capacity without gear tooth fracture. It would be anticipated that the gear system would exhibit improved power-to-weight ratio and/or increased life and reliability. However, these assumptions have neither been analytically shown nor experimentally proved.

The tooth geometry of a conformal gear pair is shown in Fig. 27.10. To achieve the maximum contact area, the radii of two mating surfaces, r_1 and r_2, should be identical. However, this is not practical. Inaccuracies in fabrication will occur on the curvatures and in other dimensions such as the center distance. Such inaccuracies can cause stress concentrations on the edges and tips of the gear teeth. To prevent this edge loading, the radii are made slightly unequal and are struck from different centers. Although this slight mismatch decreases the stress concentrations, the contact surface area is also reduced, which amplifies the hertzian contact stress. The resultant contact stresses are dependent on the helix angle.

As a historical note, in early designs thermal expansion in gearcases and shafts was

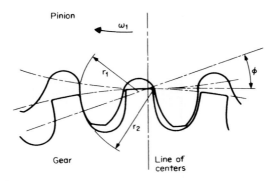

FIG. 27.10 Conformal tooth geometry.

sufficient to affect the gear pair center distances; the contact stresses were thus increased, leading to early failure.[22]

As was discussed for involute gearing, a constant velocity ratio exists for spur gear tooth pairs at every angular position because the teeth "roll" over each other. However, with conformal gearing, there is only one angular position where a tooth is in contact with its opposite number. Immediately before and after that point there is no contact. To achieve a constant velocity ratio, the teeth must run in a helical form across the gear and an overlap must be achieved from one tooth pair to the next as the contact area shifts across the gear, from one side to the other, during meshing.[21] The sweep velocity of the contact area across the face width is often thought of as being a

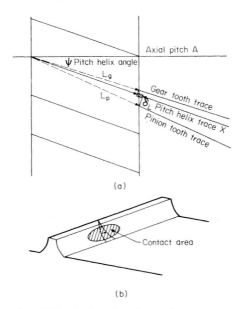

FIG. 27.11 Conformal gearing tooth contact trace diagram. (*a*) Sliding components in rolling action. (*b*) Relative sliding velocities in contact area.

pure rolling action, because no physical translation of the metal occurs in this direction. However, there is a small sliding component acting along the tooth.

If one considers the pitch surfaces of the mating gears, the length of the meshing helices would be the same on both the pinion and the gear. But the conformal system uses an all-addendum pinion and all-dedendum wheel, so when the contact area traverses one axial pitch, it must sweep over a longer distance on the pinion and a shorter distance on the gear. Thus, a small sliding component exists as shown in Fig. 27.11a. The magnitude of the sliding can vary with tooth design. However, since the sliding velocity is dependent on the displacement from the pitch surface, the percentage sliding will vary across the contact area, as shown in Fig. 27.11b. The main sliding velocity which occurs at the tooth contact acts up and down the tooth height. The magnitude of this sliding component can vary up to approximately 15 percent of the pitch helix sweep velocity. Thus, in comparison with involute gears, the slide-roll ratios are significantly lower, although the two motions being compared are in different directions. This lower slide-roll ratio seems to indicate why slightly lower power losses can be attained using conformal gears.[21]

27.3.2 Intersecting Shaft Gear Types

There are several types of gear systems which can be used for power transfer between intersecting shafts. The most common of these are straight-bevel gears and spiral-bevel gears. In addition, there are special bevel gears which accomplish the same or similar results and are made with special geometrical characteristics for economy. Among these are Zerol* gears, Conifex* gears, Formate* gears, Revacycle* gears, and face gears. For the most part, the geometry of spiral-bevel gears is extremely complex and does not lend itself to simplified formulas or analysis.

The gear tooth geometry is dictated by the machine tool being used to generate the gear teeth and the machine tool settings. The following are descriptions of these gears.

Straight-Bevel Gears. Bevel gear arrangements are shown in Fig. 27.12. Figure 27.13 is a drawing of a straight-bevel gear showing the terminology and important

---

*Registered trademarks of Gleason Works, Rochester, N.Y.

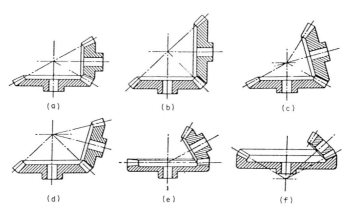

FIG. 27.12 Bevel gear arrangements. (*a*) Usual form. (*b*) Miter gears. (*c, d*) Other forms. (*e*) Crown gear. (*f*) Internal bevel.

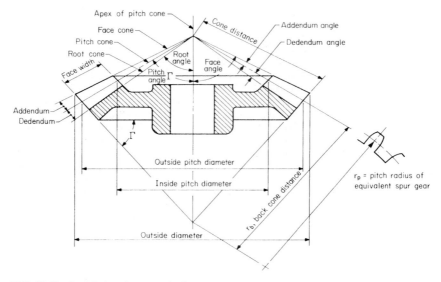

FIG. 27.13   Straight-bevel gear terminology.

physical dimensions. Straight-bevel gears are used generally for relatively low speed applications with pitch line velocities up to 1000 ft/min and where vibration and noise are not important criteria. However, with carefully machined and ground straight-bevel gears it may be possible to achieve speeds up to 15,000 ft/min.

Bevel gears are mounted on intersecting shafts at any desired angle, although 90° is most common. They are designed and manufactured in pairs and as a result are not always interchangeable. One gear is mounted on the cantilevered or outboard end of the shaft. Because of the outboard mounting, the deflection of the shaft where the gear is attached can be rather large. This would result in the teeth at the small end moving out of mesh. The load would thus be unequally distributed, with the larger ends of the teeth taking most of the load. To reduce this effect, the tooth face width is usually made no greater than one-third the cone distance. Bevel gears are usually classified according to their pitch angle. A bevel gear having a pitch angle of 90° and a plane for its pitch surface is known as a *crown gear.*

When the pitch angle of a bevel gear exceeds 90°, it is called an *internal bevel gear.* Internal bevel gears cannot have a pitch angle very much greater than 90° because of problems incurred in manufacturing such gears. These manufacturing diffi-culties are the main reason why internal bevel gears are rarely used.

Bevel gears with pitch angles less than 90° are the type most commonly used. When two meshing bevel gears have a shaft angle of 90° and have the same number of teeth, they are called *miter gears.* In other words, miter gears have a speed ratio of 1. Each of the two gears has a 45° pitch angle.

The relationship (Tregold's approximation) between the actual number of teeth for the bevel gear, the pitch angle, and the virtual or equivalent number of teeth is

$$N_e = \frac{N}{\cos \Gamma} \tag{27.22}$$

where Γ is the pitch angle. The back cone distance becomes the pitch radius for the equivalent spur gear.

Face Gears.   These gears have teeth cut on the flat face of the blank. The face gear meshes at right angles with a spur or helical pinion. When the shafts intersect, it is known as an "on-center" face gear. These gears may also be offset to provide a right-angle nonintersecting shaft drive.

Coniflex Gears.   These are straight-bevel gears whose teeth are crowned in a length-wise direction to accommodate small shaft misalignments.

Formate Gears.   These have the gear member of the pair with nongenerated teeth, usually with straight tooth profiles. The pinion member of the pair has generated teeth that are conjugate to the rotating gear.

Revacycle Gears.   These are straight-bevel gears generated by a special process, with a special tooth form.

Spiral-Bevel Gears.   A spiral-bevel gear pair is shown in Fig. 27.14a. The teeth of spiral-bevel gears are curved and oblique. These gears are suitable for pitch-line velocities up to 12,000 ft/min. Ground teeth extend this limit to 25,000 ıt/min.

There are different types of spiral-bevel gears depending on the method used to generate the gear tooth surfaces. Among these are spiral-bevel gears made by the Gleason method, the Klingelnberg system, and the Oerlikon system. These companies have developed specified and detailed directions for the design of spiral-bevel gears which are related to the respective method of manufacture. However, there are some general considerations which are common for the generation of all types of spiral-bevel gears (Fig. 27.14b) such as the concept of pitch cones, the generating gear, and the conditions of force transmission. The actual tooth gear geometry is dictated by the machine tool settings supplied by the respective manufacture.

The standard spiral-bevel gear has a pressure angle of 20°, although 14½ and 16° angles are used. The usual spiral angle is 35°. Because spiral gears are much stronger than similar-sized straight or Zerol gears, they can be used for large speed reduction ratio applications at a reduced overall installation size. The hand of the spiral should be selected so as to cause the gears to separate from each other during rotation and not to force them together, which might cause jamming.

Thrust loads produced in operation with spiral-bevel gears are greater than those produced with straight-bevel gears. An analogy for the relationship that exists between straight- and spiral-bevel gears is that existing between spur and helical gears. Localized tooth contact is also achieved. This means that some mounting and load deflections can occur without a resultant load concentration at the end of the teeth. The total force W_N acting normal to the pinion tooth and assumed concentrated at the average radius of the pitch cone may be divided into three perpendicular components. These are the transmitted, or tangential, load W_t; the axial, or thrust, component W_x; and the separating, or radial, component W_r. The force W_t may, of course, be comput-ed from the following:

$$W_t = \frac{T}{r_{av}} \tag{27.23}$$

where T is the input torque and r_{av} is the radius of the pitch cone measured at the midpoint of the tooth. The forces W_x and W_r depend upon the hand of the spiral and the direction of rotation. Thus there are four possible cases to consider. For a right-hand pinion spiral with clockwise pinion rotation and for a left-hand spiral with counterclockwise rotation. For gears with 90° between shaft axes, the equations are

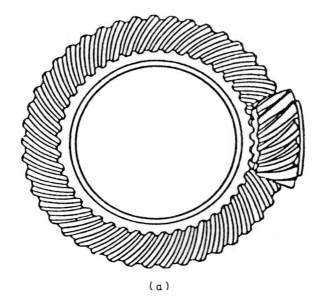

(a)

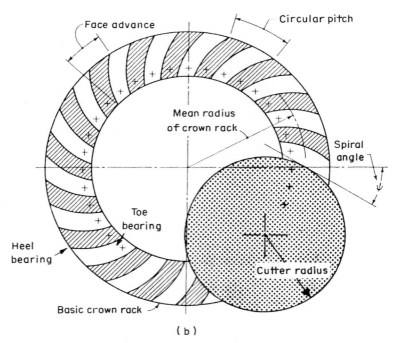

(b)

FIG. 27.14   Spiral-bevel gears. (*a*) Spiral-bevel gear pair. (*b*) Cutting spiral gear teeth on basic crown rack.

$$W_x = \frac{W_t}{\cos \psi} (\tan \phi_n \sin \Gamma - \sin \psi \cos \Gamma) \qquad (27.24)$$

$$W_r = \frac{W_t}{\cos \psi} (\tan \phi_n \cos \Gamma + \sin \psi \sin \Gamma) \qquad (27.25)$$

The other two cases are a left-hand spiral with clockwise rotation and a right-hand spiral with counterclockwise rotation. For these two cases, the equations are

$$W_x = \frac{W_t}{\cos \psi} (\tan \phi_n \sin \Gamma + \sin \psi \cos \Gamma) \qquad (27.26)$$

$$W_r = \frac{W_t}{\cos \psi} (\tan \phi_n \cos \Gamma - \sin \psi \sin \Gamma) \qquad (27.27)$$

where ψ = spiral angle
Γ = pinion pitch angle
ϕ_n = normal pressure angle

and the rotation is observed from the input end of the pinion shaft. Equations (27.24) to (27.27) give the forces exerted by the gear on the pinion. A positive sign for either W_x or W_r indicates that it is directed away from the cone center. The forces exerted by the pinion on the gear are equal and opposite. Of course, the opposite of an axial pinion load is a radial gear load, and the opposite of a radial pinion is an axial gear load.

The applications in which Zerol bevel gears are used are very much the same as those for straight-bevel gears (Fig. 27.15). The minimum suggested number of teeth is 14, one or more teeth should always be in contact, and the basic pressure angle is 20°, although angles of 22½ or 25° are sometimes used to eliminate undercutting.

FIG. 27.15 Zerol gears. (*Courtesy of Gleason Works, Rochester, N.Y.*)

27.3.3 Nonparallel, Nonintersecting Shafts

Nonparallel, nonintersecting shafts lie in parallel planes and may be skewed at any angle between 0 and 90°. Among the gear systems used for the purpose of power-transfer between nonparallel, nonintersecting shafts are hypoid gears, crossed-helical gears, worm gears, Cone-Drive* worm gears, face gears, Spiroid[†] gears, Planoid[†] gears, Helicon[†] gears, and Beveloid[‡] gears. The following are descriptions of these gears.

Hypoid Gears.   A hypoid gear pair is shown in Fig. 27.16. Hypoid gears are very similar to spiral-bevel gears. The main difference is that their pitch surfaces are hyperboloids rather than cones. As a result, their pitch axes do not intersect, the pinion axis being above or below the gear axis. In general, hypoid gears are most desirable for those applications involving large speed reduction ratios, those having nonintersecting shafts, and those requiring great smoothness and quietness of operation. Hypoid gears are almost universally used for automotive applications, allowing the drive shaft to be located underneath the passenger compartment floor. They operate more smoothly and quietly than spiral-bevel gears and are stronger for a given ratio. Because the two supporting shafts do not intersect, bearings can be mounted on both sides of the gear to provide extra rigidity. The pressure angles usually range between 19 and $22\frac{1}{2}°$. The minimum number of teeth suggested is 8 for speed ratios greater than 6:1, and 6 for smaller ratios. High-reduction hypoids permit ratios between 10:1 and 120:1 and even as high as 360:1 in fine pitches.

FIG. 27.16   Hypoid gears.

Crossed-Helical Gears.   A pair of crossed-helical gears, also known as spiral gears, are shown in Fig. 27.17. They are in fact ordinary helical gears used in nonparallel shaft

---

*Registered trademark, Michigan Tool Company, Detroit, Michigan.
†Registered trademark, Illinois Tool Works, Chicago, Illinois.
‡Registered trademark, Vinco Corporation, Detroit, Michigan.

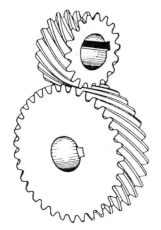

FIG. 27.17 Crossed-helical gears.

applications. They transmit relatively small amounts of power because of sliding action and limited tooth contact area. However, they permit a wide range of speed ratios without change of center distance or gear size. They can be used at angles other than 90°. In order for two helical gears to operate as crossed-helical gears, they must have the same normal diametral pitch P_n and normal pressure angle ϕ_n. The gears do not need to have the same helix angle or be opposite of hand. In most crossed gear applications, the gears have the same hand.

The relationship between the helix angles of the gears and the angle between the shafts on which the gears are mounted is shown in Fig. 27.17 and given by

$$\Sigma = \psi_1 + \psi_2 \tag{27.28}$$

for gears having the same hand and

$$\Sigma = \psi_1 - \psi_2 \tag{27.29}$$

for gears having opposite hands, where Σ is the shaft angle. For crossed-helical gears where the shaft angle is 90°, the helical gears must be of the same hand.

The center distance between the gears is given by the following relation:

$$C = \frac{1}{2P_n}\left(\frac{N_1}{\cos \psi_1} + \frac{N_2}{\cos \psi_2}\right) \tag{27.30}$$

The sliding velocity v_s acts at the pitch point tangentially along the tooth surface (Fig. 27.18), where by the law of cosines

$$v_s = [v_1^2 + v_2^2 - 2v_1 v_2 \cos \Sigma]^{1/2} \tag{27.31}$$

For shaft angles of 90°, the sliding velocity is given by

$$v_s = \frac{v_1}{\cos \psi_2} = \frac{v_2}{\cos \psi_1} \tag{27.32}$$

Referring to Fig. 27.19, the direction of the thrust load for crossed-helical gears may be determined. The forces on the crossed-helical gears are

Transmitted force: $\qquad\qquad W_t = W_N \cos \phi_n \cos \psi \qquad\qquad$ (27.33)

Axial thrust load: $\qquad\qquad W_x = W_N \cos \phi_n \sin \psi \qquad\qquad$ (27.34)

Radial or separating force: $\qquad\quad W_r = W_N \sin \phi_n \qquad\qquad$ (27.35)

where ϕ_n is the normal pressure angle.

The pitch diameter is

$$D = \frac{N}{P_n \cos \psi} \tag{27.36}$$

Since the pitch diameters are not directly related to the tooth numbers, they cannot be used to obtain the angular velocity ratio. This ratio must be obtained from the ratio of the tooth numbers.

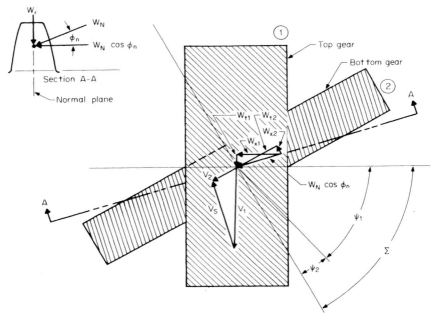

FIG. 27.18 Meshing crossed-helical gears showing the relationship between the helix angles, shaft angle, and load and velocity vectors.

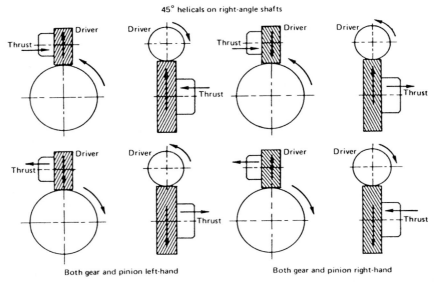

FIG. 27.19 Direction of thrust load for two meshing crossed-helical gears mounted on shafts that are oriented 90° to each other.

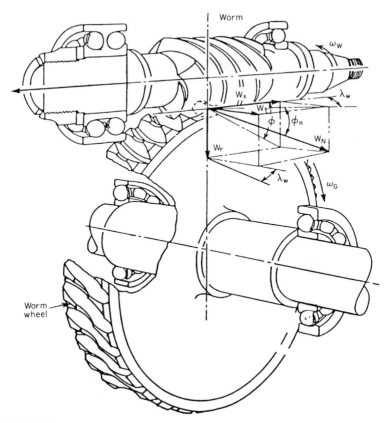

FIG. 27.20   Worm gear set showing load vectors.

Worm Gears.   A worm gear set is depicted in Fig. 27.20 showing forces which are discussed later. Referring to Fig. 27.21, the worm gear set is comprised of a worm which resembles a screw and a worm wheel which is a helical gear. The shafts on which the worm and wheel are mounted are usually 90° apart. Because of the screw action of worm gear drives, they are quiet, vibration-free, and produce a smooth output. On a given center distance, much higher ratios can be obtained through a worm gear set than with other conventional types of gearing. The contact (hertz) stresses are lower with worm gears than with crossed-helical gears. Although there is a large component of sliding motion in a worm gear set (as in crossed-helical gears), a worm gear set will have a higher load capacity than a crossed-helical gear pair.

Worm gear sets may be either single- or double-enveloping. In a single-enveloping set, Fig. 27.20, the worm wheel has its width cut into a concave surface, thus partially enclosing the worm when in mesh. The double-enveloping set, in addition to having the helical gear width cut concavely, has the worm length cut concavely. The result is that both the worm and the gear partially enclose each other. A double-enveloping set will have more teeth in contact and will have area rather than line contact, thus permitting greater load transmission.

All worm gear sets must be carefully mounted in order to assure proper operation.

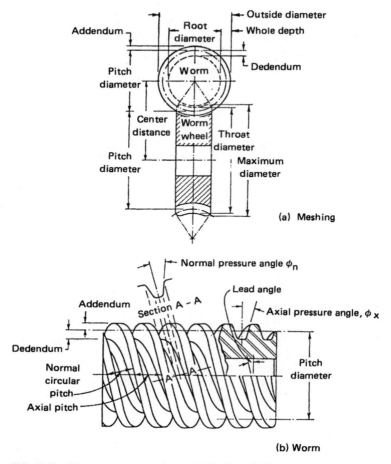

FIG. 27.21   Worm gear set nomenclature. (*a*) Meshing. (*b*) Worm.

The double-enveloping or cone type is much more difficult to mount than the single-enveloping.

Referring to Figs. 27.20 and 27.21, the relationships to define the worm gear set are as follows.

The lead angle of the worm is

$$\lambda_w \text{ arctan } \frac{L_W}{\pi D_W} \qquad (27.37)$$

This is the angle between a tangent to the pitch helix and the plane of rotation. If the lead angle is less than 5°, a rotation worm gear set cannot be driven backwards or, in other words, is self-locking. The pitch diameter is D_W. The lead is the number of teeth or threads multiplied by the axial pitch of the worm, or

$$L_W = N_W \times p_a \qquad (27.38)$$

The lead and helix angles of the worm and worm wheel are complementary:

$$\lambda_w + \psi_w = 90° \tag{27.39}$$

where ψ_w is the helix angle of the worm. Also

$$\lambda_G + \psi_G = 90° \tag{27.40}$$

and for 90° shaft angles,

$$\lambda_w = \psi_G \tag{27.41}$$

The center distance is given by

$$C = \frac{D_w + D_G}{2} \tag{27.42}$$

where D_G is the pitch diameter of the wheel.

The AGMA recommends the following equation be used to check the magnitude of D_w:

$$D_w \geq \frac{C^{0.875}}{2.2} \approx 3P_G \tag{27.43}$$

where P_G is the circular pitch of the worm wheel in inches. The gear ratio is given by

$$m_g = \frac{\omega_w}{\omega_G} = \frac{N_G}{N_W} = \frac{\pi D_G}{L_W} \tag{27.44}$$

Referring again to Fig. 27.20, the forces acting on the tooth surface are as follows if friction is neglected:

W_t = transmitted force which is tangential to worm and axial to the worm wheel
$\quad = W_N \cos \phi_n \sin \lambda_w \tag{27.45}$

W_x = axial thrust load on worm and tangential or transmitted force on worm wheel
$\quad = W_N \cos \phi_n \cos \lambda_w \tag{27.46}$

W_r = the radial or separating force $= W_N \sin \phi_n \tag{27.47}$

The directions of the thrust loads for worm gear sets are given in Fig. 27.22.

27.4 PROCESSING AND MANUFACTURE

27.4.1 Materials

A wide variety of gear materials are available today for the gear designer. Depending on the application, the designer may choose from materials such as wood, plastics, aluminum, magnesium, titanium, bronze, brass, gray cast iron, nodular and malleable iron, and a range of low-, medium-, and high-alloy steels. In addition, there are many different ways of modifying or processing the materials to improve their properties or to reduce the cost of manufacture. These include reinforcements for plastics, coating and processing for aluminum and titanium, and hardening and forging for many of the iron-based (or ferrous) gear materials.

When selecting a gear material for an application, the gear designer will first determine the actual requirements for the gears being considered. The design requirements for a gear in a given application will depend on such things as accuracy, load, speed, material, and noise limitations. The more stringent these requirements are, the more costly the gear will become. For instance, a gear requiring high accuracy because of

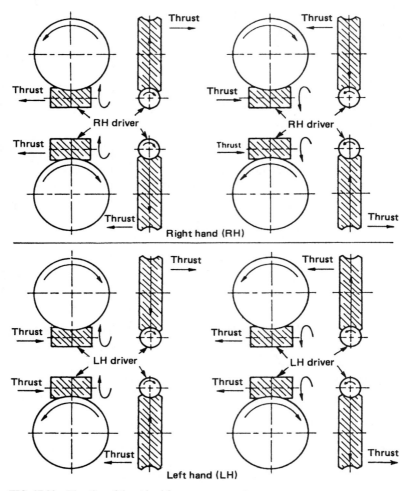

FIG. 27.22   Direction of thrust load for a worm gear set.

speed or noise limitations may require several processing operations in its manufacture. Each operation will add to the cost. Machined gears, which are the most accurate, can be made from materials with good strength characteristics. However, these gears are very expensive. The cost is further increased if hardening and grinding are required, as in applications where noise limitation is a critical requirement. Because of cost, machined gears should be a last choice for a high-production gear application. Figure 27.23 shows the relative costs for high-volume gearing.

Some of the considerations in the choice of a material include allowable bending and hertzian stress, wear resistance, impact strength, water and corrosion resistance, manufacturing cost, size, weight, reliability, and lubrication requirements. Steel, under proper heat treatment, meets most of these qualifications. Where wear is relatively severe, as with worm gearing, a high-quality, chill-cast nickel bronze may be used for rim material. The smaller worm gears may be entirely of nickel bronze.

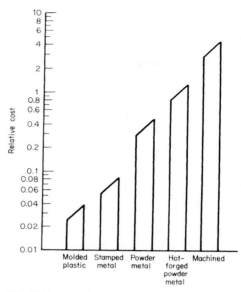

FIG. 27.23 Relative costs of gear materials.

In aircraft applications, such as helicopter, vertical or short take-off and landing (V/STOL), and turboprop aircraft, the dominant factors are reliability and weight. Cost is of secondary importance. Off-the-road vehicles, tanks, and some actuators may be required to operate at extremely high loads with a corresponding reduction in life. These loads may produce bending stresses in excess of 150,000 lb/in^2 and hertzian stresses in excess of 400,000 lb/in^2 for portions of the duty cycle. This may be acceptable because of the relatively short life of the vehicle and a deemphasis on reliability. (As a contrast, aircraft gearing typically operates at maximum bending stresses of 65,000 lb/in^2 and maximum hertzian stresses of 180,000 lb/in^2.)

Plastics. There has always been a need for lightweight, low-cost gear material for light-duty applications. In the past, gears were made from wood or phenolic-resin-impregnated cloth. However, in recent years with the development of many new polymers, many gears are made of various "plastic" materials. Table 27.3 lists plastic

TABLE 27.3 Properties of Plastic Gear Materials

Property	ASTM	Acetate	Nylon	Polyimide
Yield strength, lb/in^2	D 638	10,000	11,800	10,500
Shear strength, lb/in^2	D 732	9510	9600	11,900
Impact strength (Izod)	D 256	1.4	0.9	0.9
Elongation at yield, percent	D 638	15	5	6.5
Modulus of elasticity, lb/in^2	D 790	410 000	410 000	460 000
Coefficient of linear thermal expansion, in/in·°F	D 696	4.5×10^{-5}	4.5×10^{-5}	2.8×10^{-5}
Water absorption (24 h), percent	D 570	0.25	1.5	0.32
Specific gravity	D 792	1.425	1.14	1.43
Temperature limit, °F	—	250	250	600

materials used for molded gears. The most common molded plastic gears are the acetate and nylon resins. These materials are limited in strength, temperature resistance, and accuracy. The nylon and acetate resins have a room-temperature yield strength of approximately 10,000 lb/in². This is reduced to approximately 4000 lb/in² at their upper temperature limit of 250°F. Nylon resin is subject to considerable moisture absorption, which reduces its strength and causes considerable expansion. Larger gears are made with a steel hub that has a plastic tire for better dimensional control. Plastic gears can operate for long periods in adverse environments, such as dirt, water, and corrosive fluids, where other materials would tend to wear excessively. They can also operate without lubrication or can be lubricated by the processed material as in the food industry. The cost of plastic gears can be as low as a few cents per gear for a simple gear on a high-volume production basis. This is probably the most economical gear available. Often a plastic gear is run in combination with a metal gear to give better dimensional control, low wear, and quiet operation.

Polyimide is a more expensive plastic material than nylon or acetate resin, but it has an operating temperature limit of approximately 600°F. This makes the polyimides suitable for many adverse applications that might otherwise require metal gears. Polyimides can be used very effectively in combination with a metal gear without lubrication because of polyimide's good sliding properties.[23–25] However, polyimide gears are more expensive than other plastic gears because they are difficult to mold and the material is more expensive.

Nonferrous Metals. Several grades of wrought and cast aluminum alloys are used for gearing. The wrought alloy 7075-T6 is stronger than 2024-T4 but is also more expensive.

Aluminum does not have good sliding and wear properties. However, it can be anodized with a thin, hard surface layer that will give it better operating characteristics. Anodizing gives aluminum good corrosion protection in salt water applications. But the coating is thin and brittle and may crack under excessive load.

Magnesium is not considered a good gear material because of its low elastic modulus and other poor mechanical properties. Titanium has excellent mechanical properties, approaching those of some heat-treated steels with a density nearly half that of steel. However, because of its very poor sliding properties, producing high friction and wear, it is not generally used as a gear material. Several attempts have been made to develop a wear-resistant coating, such as chromium plating, iron coating, and nitriding for titanium.[26–29] Titanium would be an excellent gear material if a satisfactory coating or alloy could be developed to provide improved sliding and wear properties.

Several alloys of zinc, aluminum, magnesium, brass, and bronze are used as die-cast materials. Many prior die-cast applications now use less expensive plastic gears. The die-cast materials have higher strength properties, are not affected by water, and will operate at higher temperatures than the plastics. As a result, they still have a place in moderate-cost, high-volume applications. Die-cast gears are made from copper or lower-cost zinc or aluminum alloys. The main advantage of die casting is that the requirement for machining is either completely eliminated or drastically reduced. The high fixed cost of the dies makes low production runs uneconomical. Some of the die-cast alloys used for gearing are listed in Refs. 30 and 31.

Copper Alloys. Several copper alloys are used in gearing. Most are the bronze alloys containing varying amounts of tin, zinc, manganese, aluminum, phosphorus, silicon, lead, nickel, and iron. The brass alloys contain primarily copper and zinc with small amounts of aluminum, manganese, tin, and iron. The copper alloys are most often used in combination with an iron or steel gear to give good wear and load capacity, especially in worm gear applications where there is a high sliding component. In these

TABLE 27.4 Properties of Copper Alloy Gear Materials

Material	Modulus of elasticity, 10^6 lb/in^2	Yield strength, lb/in^2	Ultimate strength, lb/in^2
Yellow brass	15	50,000	60,000
Naval brass	15	45,000	70,000
Phosphor bronze	15	40,000	75,000
Aluminum bronze	19	50,000	100,000
Manganese bronze	16	45,000	80,000
Silicon bronze	15	30,000	60,000
Nickel-tin bronze	15	25,000	50,000

cases, the steel worm drives a bronze gear. Bronze gears are also used where corrosion and water are a problem. Several copper alloys are listed in Table 27.4. The bronze alloys are either aluminum bronze, manganese bronze, silicon bronze, or phosphorus bronze. These bronze alloys have yield strengths ranging from 20,000 to 60,000 lb/in^2 and all have good machinability.

Cast Iron. Cast iron is used extensively in gearing because of its low cost, good machinability, and moderate mechanical properties. Many gear applications can use gears made from cast iron because of its good sliding and wear properties, which are in part a result of the free graphite and porosity. There are three basic cast irons distinguished by the structure of graphite in the matrix of ferrite. These are (1) gray cast iron, where the graphite is in flake form; (2) malleable cast iron, where the graphite consists of uniformly dispersed fine, free-carbon particles or nodules; and (3) ductile iron, where the graphite is in the form of tiny balls or spherules. The malleable iron and ductile iron have more shock and impact resistance. The cast irons can be heat-treated to give improved mechanical properties. The bending strength of cast iron[32] ranges from 5000 to 25,000 lb/in^2, and the surface fatigue strength[33] ranges from 50,000 to 115,000 lb/in^2. In many worm gear drives a cast iron gear can be used to replace the bronze gear at a lower cost because of the sliding properties of the cast iron.

Sintered Powder Metals. Sintering of powder metals is one of the more common methods used in high-volume, low-cost production of medium-strength gears with fair dimensional tolerance.[34] In this method a fine metal powder of iron or other material is placed in a high-pressure die and formed into the desired shape and density under very high pressure. The unsintered (green) part has no strength as it comes from the press. It is then sintered in a furnace under a controlled atmosphere to fuse the powder together for increased strength and toughness. Usually, an additive (such as copper in iron) is used in the powder for added strength. The sintering temperature is then set at the melting temperature of the copper to fuse the iron powder together for a stronger bond than would be obtained with the iron powder alone. The parts must be properly sintered to give the desired strength.

There are several materials available for sintered powder-metal gears that give a wide range of properties. Table 27.5 lists properties of some of the more commonly used gear materials, although other materials are available. The cost for volume production of sintered powder-metal gears is an order of magnitude lower than that for machined gears.

A process that has been more recently developed is the hot-forming powder-metal process.[35,36] In this process a powder-metal preform is made and sintered. The sintered powder-metal preformed part is heated to forging temperature and finish-forged. The hot-formed parts have strengths and mechanical properties approaching the ultimate

TABLE 27.5 Properties of Sintered Powder-Metal Alloy Gear Materials

Composition	Ultimate tensile strength, psi	Apparent hardness, Rockwell	Comment
1 to 5Cu-0.6C-94Fe	60,000	B 60	Good impact strength
7Cu-93Fe	32,000	B 35	Good lubricant impregnation
15Cu-0.6C-84Fe	85,000	B 80	Good fatigue strength
0.15C-98Fe	52,000	A 60	Good impact strength
0.5C-96Fe	50,000	B 75	Good impact strength
2.5Mo-0.3C-1.7N-95Fe	130,000	C 35	High strength, good wear
4Ni-1Cu-0.25C-94Fe	120,000	C 40	Carburized and hardened
5Sn-95Cu	20,000	H 52	Bronze alloy
10Sn-87Cu-0.4P	30,000	H 75	Phosphorous-bronze alloy
1.5Be-0.25Co-98Cu	80,000	B 85	Beryllium alloy

mechanical properties of the wrought materials. A wide choice of materials is available for the hot-forming powder-metal process. Since this is a fairly new process, there will undoubtedly be improvements in the materials made from this process and reductions in the cost. Several promising high-temperature, cobalt-base alloy materials are being developed.

Because there are additional processes involved, hot-formed powder-metal parts are more expensive than those formed by the sintered powder-metal process. However, either process is more economical than machining or conventional forging and produces the desired mechanical properties. This makes the hot-forming powder-metal process attractive for high-production parts where high strength is needed, as in the automotive industry.

Accuracy of the powder-metal and hot-formed processes is generally in the AGMA class 8 range. Better accuracy can be obtained with special precautions if die wear is limited. This tends to increase the cost. Figure 27.23 shows the relative costs of some of the materials or processes for high-volume, low-cost gearing.

Hardened Steels. A large variety of iron or steel alloys are used for gearing. The choice of which material to use is based on a combination of operating conditions such as load, speed, lubrication system, and temperature plus the cost of producing the gears. When operating conditions are moderate, such as medium loads with ambient temperatures, a low-alloy steel can be used without the added cost of heat treatment and additional processing. The low-alloy material in the non-heat-treated condition can be used for bending stresses in the 20,000 lb/in^2 range and surface durability hertzian stresses of approximately 85,000 lb/in^2. As the operating conditions become more severe, it becomes necessary to harden the gear teeth for improved strength and to case-harden the gear tooth surface by case-carburizing or case-nitriding for longer pitting fatigue life, better scoring resistance, and better wear resistance. There are several medium-alloy steels that can be hardened to give good load-carrying capacity with bending stresses of 50,000 to 60,000 lb/in^2 and contact stresses of 160,000 to 180,000 lb/in^2. The higher alloy steels are much stronger and must be used in heavy-duty applications. AISI 9310, AISI 8620, Nitralloy N, and Super-Nitralloy are good materials for these applications and can operate with bending stresses of 70,000 lb/in^2 and maximum contact (hertzian) stresses of 200,000 lb/in^2. These high-alloy steels should be case-carburized (AISI 8620 and 9310) or case-nitrided (Nitralloy) for a very hard wear-resistant surface. Gears that are case-carburized will usually require grinding after the hardening operation because of distortion during the heat-treating process. The nitrided materials offer the advantage of much less distortion during

TABLE 27.6 Properties of Steel Alloy Gear Materials

Material	Tensile strength, lb/in²	Yield strength, lb/in²	Elongation in 2 in, percent
Cast iron	45,000	—	—
Ductile iron	80,000	60,000	3
1020	80,000	70,000	20
1040	100,000	60,000	27
1066	120,000	90,000	19
4146	140,000	128,000	18
4340	135,000	120,000	16
8620	170,000	140,000	14
8645	210,000	180,000	13
9310	185,000	160,000	15
440C	110,000	65,000	14
416	160,000	140,000	19
304	110,000	75,000	35
Nitralloy 135M	135,000	100,000	16
Super-Nitralloy	210,000	190,000	15
CBS 600	222,000	180,000	15
CBS 1000M	212,000	174,000	16
Vasco X-2	248,000	225,000	6.8
EX-53	171,000	141,000	16
EX-14	169,000	115,000	19

nitriding and therefore can be used in the as-nitrided condition without additional finishing. This is very helpful for large gears with small cross sections where distortion can be a problem. Since case depth for nitriding is limited to approximately 0.020 in, case crushing can occur if the load is too high. Some of the steel alloys used in the gearing industry are listed in Table 27.6.

Gear surface fatigue strength and bending strength can be improved by shot peening.[37,38] The 10 percent surface fatigue life of the shot-peened gears is 1.6 times that of the standard ground gears.[37]

The low- and medium-alloy steels have a limited operating temperature above which they begin to lose their hardness and strength, usually around 300°F. Above this temperature, the material is tempered, and early bending failures, surface pitting failures, or scoring will occur. To avoid this condition, a material is needed that has a higher tempering temperature and that maintains its hardness at high temperatures. The generally accepted minimum hardness required at operating temperature is Rockwell C 58. In recent years several materials have been developed that maintain a Rockwell C 58 hardness at temperatures from 450 to 600°F.[39] Several materials have shown promise of improved life at normal operating temperature. The hot hardness data indicate that they will also provide good fatigue life at higher operating temperatures.

AISI M-50 has been used successfully for several years as a rolling-element bearing material for temperatures to 600°F.[40-44] It has also been used for lightly loaded accessory gears for aircraft applications at high temperatures. However, the standard AISI M-50 material is generally considered too brittle for more heavily loaded gears. AISI M-50 is considerably better as a gear material when forged with integral teeth. The grain flow from the forging process improves the bending strength and impact resistance of the AISI M-50 considerably.[45]

The M-50 material can also be ausforged with gear teeth to give good bending strength and better pitting life.[46,47] However, around 1400°F, the ausforging temperature is so low that forging gear teeth is very difficult and expensive. As a result, ausforging for gears has considerably limited application.[46] Test results show that the forged and ausforged gear can give lifetimes approximately 3 times those of the standard AISI 9310 gears.[46]

Nitralloy N is a low-alloy nitriding steel that has been used for several years as a gear material. It can be used for applications requiring temperatures of 400 to 450°F. A modified Nitralloy N called Super-Nitralloy, or 5Ni-2A1 Nitralloy, was used in the U.S. supersonic aircraft program for gears. It can be used for gear applications requiring temperatures to 500°F. Surface fatigue data for the Super-Nitralloy gears and other steels are shown in Table 27.7.

TABLE 27.7   Relative Surface Pitting Fatigue Life Compared to VAR AISI 9310 Steel for Aircraft-Quality Gear Steels (Rockwell C 59–62)

Steel	10 percent relative life
VAR AISI 9310	1
VAR AISI 9310 (shot-peened)	1.6
VAR Carpenter EX-53	2.1
CVM CBS 600	1.4
VAR CBS 1000	2.1
CVM VASCO X-2	1.0
CVM Super-Nitralloy (5Ni-2A1)	1.3
VIM-VAR AISI M-50 (forged)	3.2
VIM-VAR AISI M-50 (ausforged)	2.4

Two materials that were developed for case-carburized tapered roller bearings but also show promise as high-temperature gear materials are CBS 1000M and CBS 600.[48,49] These materials are low- to medium-alloy steels that can be carburized and hardened to give a hard case of Rockwell C 60 with a core of Rockwell C 38. Surface fatigue test results for CBS 600 and AISI 9310 are shown in Table 27.7. The CBS 600 has a medium fracture toughness that could cause fracture failures after a surface fatigue spall has occurred.

Two other materials that have recently been developed as advanced gear materials are EX-53 and EX-14. Reference 50 reports that the fracture toughness of EX-53 is excellent at room temperature and improves considerably as temperature increases. The EX-53 surface fatigue results show a 10 percent life that is twice that of the AISI 9310. Vasco X-2 is a high-temperature gear material[51] that is used in advanced CH-47 helicopter transmissions. This material has an operating temperature limit of 600°F and has been shown to have good gear load-carrying capacity when properly heat treated. The material has a high chromium content (4.9 percent) that oxidizes on the surface and can cause soft spots when the material is carburized and hardened. A special process has been developed that eliminates these soft spots when the process is closely followed.[52] Several groups of Vasco X-2 with different heat treatments were surface fatigue-tested in the NASA gear test facility. All groups except that with the special processing gave poor results.[53] Vasco X-2 has a lower fracture toughness than AISI 9310 and is subject to tooth fractures after a fatigue spall.

27.4.2   Metallurgical Processing Variables

Research reported in the literature on gear metallurgical processing variables is not as extensive as that for rolling-element bearings. However, an element of material in a hertzian stress field does not recognize whether it is in a bearing or a gear. It only recognizes the resultant shearing stress acting on it. Consequently, the behavior of the material in a gear will be very similar to that in a rolling-element bearing. The metallurgical processing variables to be considered are

1. Melting practice, such as air, vacuum induction, consumable electrode vacuum melt (CVM), vacuum degassing, electroslag (electroflux) remelt, and vacuum induction melting–vacuum arc remelting (VIM-VAR)
2. Heat treatment which gives hardness, differential hardness, and residual stress
3. Metalworking consisting of thermomechanical working and fiber orientation

The above items can significantly affect gear performance. It should be mentioned that some factors that can also significantly affect gear fatigue life and that have some meaningful documentation are not included. These are trace elements, retained austenite, gas content, inclusion type, and content. While any of these can exercise some effect on gear fatigue life, they are, from a practical standpoint, too difficult to measure or control by normal quality control procedures.

Heat-treatment procedures and cycles, per se, can also affect gear performance. However, at present no controls, as such, are being exercised over heat treatment. The exact thermal cycle is left to the individual producer with the supposition that a certain grain size and hardness range must be met. Hardness is discussed in some detail in reference to gear life. In the case of grain size, lack of any definitive data precludes any meaningful discussion.

Melting Practice. Sufficient data and practical experience exist to indicate the use of vacuum-melted materials and, specifically, CVM can increase gear surface pitting fatigue life beyond that obtainable by air-melted materials.[54-59] Since in essentially all critical applications such as gears for helicopter transmissions and turboprop aircraft, vacuum-melt material is specified, a multiplication factor can be introduced into the life estimation equation.

Life improvements up to 13 times by CVM processing[41-44] and up to 100 times by VIM-VAR processing[59] over air-melted steels are indicated in the literature. However, it is recommended that conservative estimates for life improvements be considered such as a factor of 3 for CVM processing and a factor of 10 for VIM-VAR processing. While these levels may be somewhat conservative, the confidence factor for achieving this level of improvement is high. Consequently, the extra margin of reliability desired in critical gear applications will be inherent in the life calculations.

Data available on other melting techniques such as vacuum induction, vacuum degassing, and electroslag remelting would indicate that the life improvement approaches or equals that of the CVM process. However, it is also important to differentiate between CVM and CVD (carbon vacuum degassing). The CVM process yields cleaner and more homogeneous steels than CVD.

Heat Treatment. Reference 14 discusses the various methods of heat treatment as applied to gears. Heat treatment is accomplished by either furnace hardening, carburizing, induction hardening, or flame hardening. Gears are either case-hardened, through-hardened, nitrided, or precipitation-hardened for the proper combination of toughness and tooth hardness.

The use of high-hardness, heat-treated steels permits smaller gears for a given load. Also, hardening can significantly increase service life without increasing size or weight. But the gear must have at least the accuracies associated with softer gears and, for maximum service life, even greater precision.

Heat-treatment distortion must be minimized if the gear is to have increased service life. Several hardening techniques have proved useful. For moderate service-life increases, the gears are hardened but kept within the range of machinability so that distortion produced by heat treatment can be machined away.

Where maximum durability is required, surface hardening is necessary. Carburizing, nitriding, and induction hardening are generally used. However, precision gearing (quality 10 or better) can only be assured by finishing after hardening. Full-contour induction hardening is an economical and effective method for surface-hardening gears. The extremely high but localized heat allows small sections to come to hardening temperatures while the balance of the gear dissipates heat. Thus, major distortions are eliminated.

While conventional methods such as flame hardening reduce wear by hardening the tooth flank, gear strength is not necessarily improved. In fact, stresses built up at the juncture of the hard and soft material may actually weaken the tooth. Induction hardening provides a hardened tooth contour with a heat-treated core to increase both surface durability and tooth strength. The uniformly hardened tooth surface extends from the flank, around the tooth, to the flank. No stress concentrations are developed to impair gear life.

Nitriding is a satisfactory method of hardening small- and medium-size gears. Distortion is minimal because furnace temperatures are comparatively low. Hardening pattern is uniform but depth of hardness is limited. Best results are achieved when special materials, suited to nitriding, are specified.

Most gear manufacturing specifications do not designate heat treatment, but rather call for material characteristics, i.e., hardness and grain size, which are controlled by the heat-treatment cycle. Hardness is the most influential heat-treatment-induced variable.[60-62] It is recommended that Rockwell C 58 be considered the minimum hardness required for critical gear applications.

A relationship has been proposed in Ref. 42 which approximates the effect of hardness on surface fatigue life:

$$\frac{L_2}{L_1} = e^{0.1(RC_2 - RC_1)} \qquad (27.48)$$

where L_1 and L_2 are 10 percent fatigue lives at gear hardness R_{C_1} and R_{C_2}, respectively. Although this relationship was obtained for AISI 52100, it may be extended to other steels. The life-vs.-hardness curve in Fig. 27.24 represents Eq. (27.48) with the relative life at Rockwell C 60 equal to 1.0. The hardness-vs.-temperature curves in Fig. 27.24 indicate the decrease in hardness with increased temperature for various initial room-temperature material hardnesses.

To use the nomograph (Fig. 27.24), first determine the gear operating temperature. Then follow a horizontal line to the appropriate room-temperature hardness curve. Then move vertically to the life-hardness curve and read the relative life at that point. See the dashed line example for a 300°F operating temperature and Rockwell C 64 at room temperature hardness. Relative life for this example is approximately 1.3.

Another concept to be considered is the effect of differences in hardness between the pinion and the gear.[63,64] Evidence exists to show that hardness differences between the mating components can positively affect system life by inducing compressive residual stresses during operation.[64] Differential hardness (ΔH) is defined as the hardness of the larger of two mating gears minus the hardness of the smaller of the two. It appears that a differential hardness (ΔH) of two points Rockwell C may be an optimum for maximum life. For critical applications, as a practical matter, it would be advisable to match the hardness of the mating gears to assure a ΔH of zero and at the same time assure that the case hardness of the gear teeth at room temperature is the maximum attainable. This will allow for maximum elevated operating temperature and maximum life. While the ΔH effect has been verified experimentally with rolling-element bearings, there is no similar published work with gears.

As previously discussed, residual stresses can be induced by the heat-treatment process, differential hardness, and/or shot peening. There is no analytical method to predict the amount of residual stress in the subsurface region of gear tooth contact.

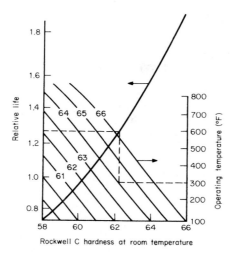

FIG. 27.24 Nomograph for determining relative life at operating temperature as a function of room-temperature hardness.

However, these residual stresses can be measured in test samples by x-ray diffraction methods. The effect of these residual stresses on gear pitting fatigue life can be determined by the following equation[37]:

$$\text{Life} \propto \left(\frac{1}{\tau_{max} - S_{ry}/2} \right)^9 \tag{27.49}$$

where τ_{max} is the maximum shearing stress (45° plane) and S_{ry} is the measured compressive residual stress below the surface at the location of τ_{max}.

Metalworking. Proper grain flow or fiber orientation can significantly affect pitting fatigue life[65,66] and may improve bending strength of gear teeth. Proper fiber orientation can be defined as grain flow parallel to the gear tooth shape. Standard forging of gears with integral gear teeth as opposed to machining teeth in a forged disk is one way of obtaining proper fiber orientation.[46] A controlled-energy-flow forming (CEFF) technique can be used for this purpose. This is a high-velocity metalworking procedure that has been a production process for several years.

AISI M-50 steel is a through-hardened material often used for rolling-element bearings in critical applications. Test gears forged from M-50 steel yielded approximately 3 times the fatigue life of machined vacuum-arc-remelted (VAR) AISI 9310 gears.[46] Despite the excellent fatigue life, M-50 is not recommended for gears because its low fracture toughness makes gears prone to sudden catastrophic tooth fracture after a surface fatigue spall has begun rather than the gradual failure and noisy operation typical of surface pitting. It is expected that forged AISI 9310 (VAR) gears would achieve similar life improvement while retaining the greater reliability of the tougher material.

Ausforging, a thermomechanical fabrication method, has potential for improving strength and life of gear teeth. Rolling-element tests with AISI M-50 steel (not recommended for gears—see above paragraph) show that 75 to 80 percent work (reduction of area) produces maximum benefit.[67-69] The suitability of candidate steels to the ausforging process must be individually evaluated. AISI 9310 is not suitable because of its austenite-to-martensite transformation characteristics. Tests reported in Ref. 46 found no statistically significant difference in lives of ausforged and standard-forged AISI M-50 gears. The lack of improvement of the ausforged gears is attributed to the final machining required of the ausforged gears which removed some material with preferential grain flow. Reference 46 also reported a slightly greater tendency toward tooth fracture. This tendency is attributed to better grain flow obtained in the standard forged gears. The energy required limits the ausforging process to gears no larger than 90 mm (3½ in) in diameter.

27.4.3 Manufacturing

Gears can be formed by various processes that can be classified under the general head of milling, generating, and molding.

Milling.   Almost any tooth form can be milled. However, only spur, helical, and straight-bevel gears are usually milled. Surface finish can be held to 125 μin rms.

Generating.   In the generating process, teeth are formed in a series of passes by a generating tool shaped somewhat like a mating gear. Either hobs or shapers can be used. Hobs resemble worms that have cutting edges ground into their teeth. Hobbing can produce almost any external tooth form except bevel gear teeth, which are generated with face-mill cutters, side-mill cutters, and reciprocating tools. Hobbing closely controls tooth spacing, lead, and profile. Surface finishes as fine as 63 μin rms can be obtained. Shapers are reciprocating pinion or rack-shaped tools. They can produce external and internal spur, helical, herringbone, and face gears. Shaping is limited in the length of cut it can produce. Surface finishes as fine as 63 μin rms are possible.

Molding.   Large-volume production of gears can often be achieved by molding. Injection molding produces light gears of thermoplastic material. Die casting is a similar process using molten metal. Zinc, brass, aluminum, and magnesium gears are made by this process.

Sintering is used in small, heavy-duty gears for instruments and pumps. Iron and brass are the materials most used. Investment casting and shell molding produce medium-duty iron and steel gears for rough applications.

Gear Finishing.   To improve accuracy and finish, gears may be shaved. Shaving removes only a small amount of surface metal. A very hard mating gear with many small cutting edges is run with the gear to be shaved. Surface finish can be as fine as 32 μin rms.

Lapping corrects minute heat-treatment distortion errors in hardened gears. The gear is run in mesh with a gear-shaped lapping tool, or another mating gear. An abrasive lapping compound is used between them. Lapping improves tooth contact but does not increase accuracy of the gear. Finish is on the order of 32 μin rms.

Grinding is the most accurate tooth-finishing process. Profiles can be controlled or altered to improve tooth contact. For example, barreling or crowning the flanks of teeth promotes good center contact where the tooth is strong, and minimizes edge and corner contact where the tooth is unsupported. Surface finishes as fine as 16 μin rms or better can be obtained. However, at surface finishes better than 16 μin rms, tooth errors can be induced which may affect gear tooth dynamic loading and, as a result, nullify the advantage of the improved surface finish.

27.5 STRESSES AND DEFLECTIONS

There are several approaches to determining the stresses and deflections in gears. The one used most commonly for determining the bending stress and deflection is a modified form of the Lewis equation.[8] This approach considers the gear tooth to be a cantilevered beam or plate. Modifying factors based on geometry or type of application are then used to amend the calculated stress and allowed design strength. It is now known that the calculated stress may not accurately represent the true stress. But since a large amount of experience has accrued across the gearing industry, compatible "allowed stress" has been compiled, and the modified Lewis equation is used successfully to design gears. The Lewis method should be thought of as a means of comparing a proposed new gear design for a given application with successful operation in the past of similar designs. If the computed stress number is less than the computed allowable stress number, the design will be satisfactory. The American Gear Manufacturer's Association has published a standard (AGMA 218.01)[70] for calculating the bending strength for spur and helical gears, as well as many other standards for gear design and applications. A second method of calculation of stresses and deflections comes from classical theory on elasticity. Several examples of such methods are found in the literature.[71–73] The methods involve the use of complex variable theory to map the shape of the gear tooth onto a semi-infinite space in which the stress equations are solved. The methods have the advantage of computing accurate stresses in the regions of stress concentration. The methods do, however, seem limited to plane stress problems. Therefore, they will not work for bevel or helical gears where the effect of the wide teeth and distributed tooth load is important.

The most powerful method for determining accurate stress and deflection information is the finite-element method. However, the method is too expensive and cumbersome to use in an everyday design situation. While at first the method would seem to answer all problems of computational accuracy, it does not. Research is continuing in this promising area.[74–79]

The versatility of the finite-element method introduces other questions of how best to use the method. These questions include (1) how many elements there are and the most efficient arrangement of elements in the regions of stress concentration, (2) how to represent the boundary support conditions, and (3) how to choose the aspect ratios of the solid brick element in three-dimensional stress problems. Pre- and postprocessors specifically for gears have been developed.[79] Various experiments have been conducted around specific questions such as the effect of rim thickness on critical stress in lightweight gears.[80,81] It is expected that methods such as the finite element method and the classical theory of elasticity will continue to be used in a research type mode, with research results being used to supply application factors and stress modifying factors for the modified Lewis (AGMA) method. One such approach has been suggested for the effect of the ratio of rim thickness to gear tooth height.[81]

Experimental methods and testing of all proposed gear applications cannot be overemphasized. The foregoing discussion has centered around the bending fatigue strength of the gear teeth, where the stress causing failure is at the gear tooth fillet between the tooth profile and root of the gear. Surface pitting fatigue and scoring types of failure are discussed in Sec. 27.6 and are also likely modes of failure. Experimental testing of gear prototype designs will reveal any weakness in the design which could not be anticipated by conventional design equations, such as where special geometry factors will cause the failure-causing stress to appear at locations other than the tooth fillet.* In critical applications such as aircraft, full-scale testing of every

*In Ref. 82 it is shown that for thin-rimmed internal gears, the maximum tensile stress occurs several teeth away from the loaded tooth. In Ref. 79 it is shown that for thin-rimmed external gears the bending stress in the root can be a higher stress than the fillet stress. These observations emphasize the need for care in applying conventional design equations with a cavalier confidence in their universal applicability.

prototype gearbox is done in ground-based test rigs, and every piece of a flight hard-ware gearbox assembly is tested in a "green run" transmission test rig as a distinct step in the production process for the completed assembly. An experimental verifica-tion of a proposed gear design is essential to a complete design process.

27.5.1 Lewis Equation Approach for Bending Stress Number (Modified by the AGMA)

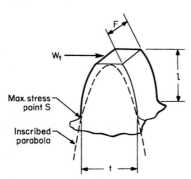

FIG. 27.25   Gear tooth load.

Figure 27.25 shows a gear tooth loaded by the transmitted tangential load W_t, which is assumed to act at the tip of the tooth. The Lewis formula for the stress at the fillet fol-lows the simple strength of materials formula for bending stress in a beam:

$$S = \frac{Mc}{I} = \frac{(W_t l) \times t/2}{\frac{1}{12}Ft^3} \tag{27.50}$$

The values for the tooth thickness at the criti-cal section and the beam length are obtained from inscribing a parabola inside the tooth, tangent to the fillet, with the apex at the point of load application. The procedure is based on the fact that a parabolic-shaped beam is a beam of constant bending stress. For a family of geometrically similar tooth forms, a dimensionless tooth form factor Y is defined as follows:

$$Y = \frac{t^2 P}{6l} \tag{27.51}$$

where P is the diametral pitch.

Using the form factor, the tooth bending stress is calculated by

$$S = \frac{W_t}{F} \frac{P}{Y} \tag{27.52}$$

where W_t is transmitted tangential load and F is face width. The assumptions from which Eq. (27.52) was derived are (1) the radially directed load was neglected; (2) only one tooth carried the full transmitted load; (3) there was uniform line contact between teeth, causing a uniform load distribution across the face width; and (4) the effect of stress concentration at the fillet was neglected.

A more comprehensive equation (defined by the AGMA) for bending stress number is

$$S = \frac{W_t}{F} \frac{P}{J} \frac{K_a K_s K_m}{K_v} \tag{27.53}$$

where S = the bending stress number, lb/in² (MPa)
$\quad K_a$ = application factor
$\quad K_s$ = size factor
$\quad K_m$ = load distribution factor
$\quad K_v$ = dynamic factor or velocity factor
$\quad J$ = geometry factor

The *application factor* K_a is intended to modify the calculated stress number, as a result of the type of service the gear will see. Some of the pertinent application influences include type of load, type of prime mover, acceleration rates, vibration, shock, and momentary overloads. Application factors are established after considerable field experience is gained with a particular type application. It is suggested that designers establish their own values of application factors based on past experience with actual designs that have run successfully. If they are unable to do this, suggested factors from past experience are available in Table 27.8.

TABLE 27.8 Suggested Application Factors K_a (for Speed Increasing and Decreasing Drives)*

Power source	Load on driven machine		
	Uniform	Moderate shock	Heavy shock
Uniform	1.00	1.25	1.75 or higher
Light shock	1.25	1.50	2.00 or higher
Medium shock	1.50	1.75	2.25 or higher

*For speed increasing drives of spur and bevel gears (but not helical and herringbone gears), add $0.01(N_2/N_1)^2$ to the factors where N_1 = number of teeth in pinion, N_2 = number of teeth in gear.
Source: AGMA

The *size factor* K_s reflects the influence of nonhomogeneous materials. On the basis of the weakest link theory, a larger section of material may be expected to be weaker than a small section because of the greater probability of the presence of a "weak" link. The size factor K_s corrects the stress calculation to account for the known fact that larger parts are more prone to fail. At present, a size factor $K_s = 1$ is recommended for spur and helical gears by the AGMA, and for bevel gears the Gleason Works recommends

$$K_s = P^{-0.25} \qquad P < 16$$

$$= 0.5 \qquad P > 16 \tag{27.54}$$

The *load distribution factor* K_m is to account for the effects of possible misalignments in the gear which will cause uneven loading and a magnification of the stress above the uniform case. If possible, keep the ratio of face width to pitch diameter small (less than 2) for less sensitivity to misalignment and uneven load distribution due to load-sensitive deflections. A second rule of thumb is to keep the face width between 3 and 4 times the circumferential pitch. Often gear teeth are crowned or edge-relieved to diminish the effect of misaligned axes on tooth stresses. Normally, the range of this factor can be approximately 1 to 3. Use a factor larger than 2 for gears mounted with less than full face contact. When more precision is desired and enough detail of the proposed design is in hand to make the necessary calculations, the AGMA standards should be consulted. For most preliminary design calculations the data in Table 27.9 for spur and helical gears and Table 27.10 and Fig. 27.26 for bevel gears should be used.

The *dynamic factor* K_v is intended to correct for the effects of the speed of rotation and the degree of precision in the gear accuracy. As a first approximation K_v may be taken from the chart in Fig. 27.27. This chart is intended only to account for the effect of tooth inaccuracies. It should not be used for lightly loaded, lightly damped gears, or resonant conditions. If the gear system approaches torsional resonance, or if the gear blank is near resonance, then computerized numerical methods must be used such as

TABLE 27.9 Spur and Helical Gear Load Distribution Factor K_m

	Face width							
	2-in face and under		6-in face		9-in face		16-in face and over	
Condition of support	Spur	Helical	Spur	Helical	Spur	Helical	Spur	Helical
Accurate mounting, low bearing clearances, minimum elastic deflection, precision gears	1.3	1.2	1.4	1.3	1.5	1.4	1.8	1.7
Less rigid mountings, less accurate gears, contact across full face	1.6	1.5	1.7	1.6	1.8	1.7	2.0	2.0
Accuracy and mounting such that less than full face contact exists			Over 2.0					

Source: AGMA

TABLE 27.10 Bevel Gears Load Distribution Factor K_m

Application	Both members straddle-mounted	One member straddle-mounted	Neither member straddle-mounted
General industrial	1.00 to 1.10	1.10 to 1.25	1.25 to 1.40
Automotive	1.00 to 1.10	1.10 to 1.25	—
Aircraft	1.00 to 1.25	1.10 to 1.40	1.25 to 1.50

Source: Gleason Works, Rochester, N.Y.

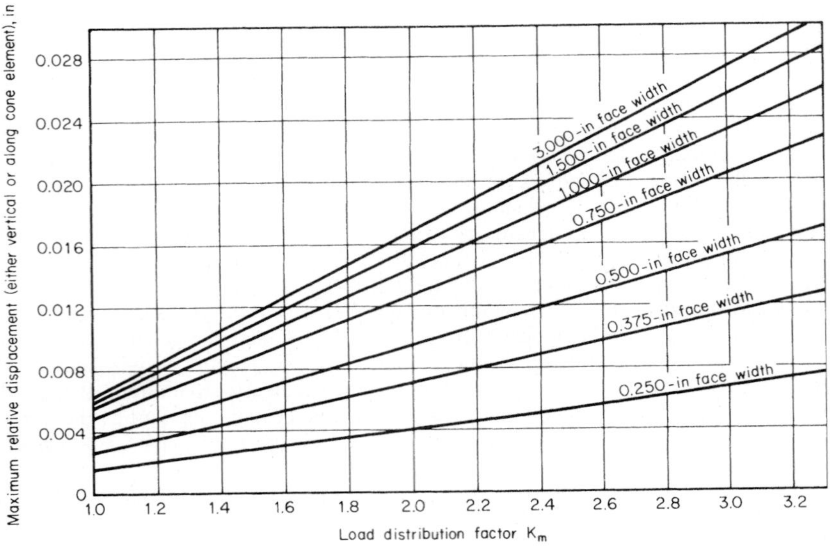

FIG. 27.26 Bevel gear load distribution factor K_m. (*Courtesy of Gleason Works, Rochester, N.Y.*)

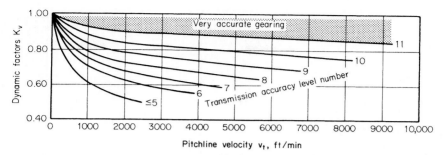

FIG. 27.27 Gear dynamic factor K_v. Transmission accuracy level number specifies the AGMA class for accuracy of manufacture. (*Courtesy of AGMA.*)

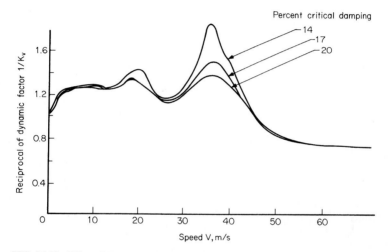

FIG. 27.28 Effect of damping ratio on dynamic load factor K_v ($m_g = 1$, $N = 28$, $P = 8$).

presented in Refs. 83 and 84. The chart in Fig. 27.28 shows the effect of damping on the dynamic factor for a certain specific design. No comprehensive data for a broad range of parameters is yet available. However, the computer program TELSGE (*ther-mal elastohydrodynamic lubrication of spur gears*) and GRDYNSING (*gear dynamic analysis for single mesh*) are available from the NASA computer program library.*

The *geometry factor J* is really just a modification of the form factor Y to account for three more influences. They are stress concentration, load sharing between the teeth, and changing the load application point to the highest point of single tooth contact. The form factor Y can be found from a geometrical layout of the Lewis parabola within the gear tooth outline (Fig. 27.25). The load-sharing ratio may be worked out using a statically indeterminate analysis of the pairs of teeth in contact, considering the flexibilities of the teeth in contact, and stress concentration factors from the work of Dolan and Broghammer[85] may be applied to get J. The J factors have already been

---

*Available from the Computer Software Management and Information Center (COSMIC) Computing and Information Services, University of Georgia, Athens.

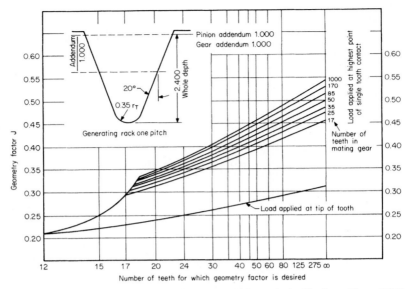

FIG. 27.29 Geometry factor J for 20° spur gear with standard addendum. (*From AGMA 218.01, December 1982.*)

computed for a wide range of standard gears and are available in the AGMA standards. For gear applications, the charts presented in Figs. 27.29 through 29.34 for spur, helical, and bevel gears may be used.

The *transmitted tooth load* W_t is equal to the torque divided by the pitch radius for spur and helical gears. For bevel gears the calculation should use the large-end pitch cone radius.

A *rim thickness factor* K_b which multiplies the stress computed by the AGMA formula [Eq. (27.53)] has been proposed in Ref. 81. The results in Fig. 27.35 show that for thin-rimmed gears, the calculation stress should be adjusted by a factor K_b if the backup ratio is less than 2.

For spiral-bevel gears, *a cutter radius factor* K_x (a stress correction factor) has been given by Gleason to account for the effects of cutter radius curvature, which is a measure of the tooth lengthwise curvature:

$$K_x = 0.2111\left(\frac{r_c}{A}\right)^{0.2788/\log \sin \psi} + 0.7889 \qquad (27.55)$$

where A = gear mean cone distance
ψ = gear mean spiral angle
r_c = cutter radius

Normally (r_c/A) ranges from 0.2 to 1 and ψ from 10 to 50°, giving a range of K_x from 1 to 1.14. Recommended practice is to select the next larger nominal cutter that has a radius given by

$$r_c = 1.1A \sin \psi \qquad (27.56)$$

The cutter radius factor should be entered in the denominator of the stress equation [Eq. (27.53)].

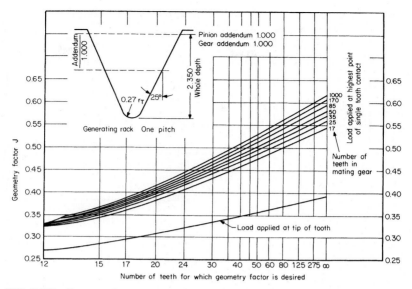

FIG. 27.30 Geometry factor J for 25° spur gear with standard addendum. (*From AGMA 218.01, December 1982.*)

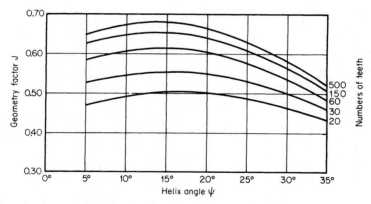

FIG. 27.31 Helical gear geometry factor J for 20° normal pressure angle, standard addendum, full fillet hole. (*From AGMA 218.01, December 1982.*)

27.5.2 Allowable Bending Stress Number

The previous section has presented a consistent method for calculating a bending stress index number. The stresses calculated by Eq. (27.53) may be much less than stresses measured by strain gauges or calculated by other methods such as finite-element methods. However, there is a large body of allowable stress data available in the AGMA standards which is consistent with the calculation procedure of Eq. (27.53). If the calculated stress number S is less than the allowed stress number S_{at}, the design should be satisfactory. The equation is

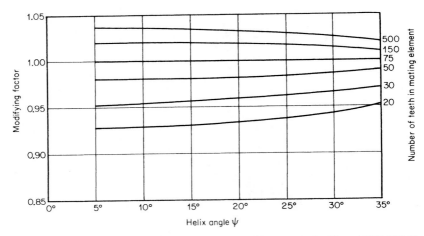

FIG. 27.32   Helical gear *J* factor multipliers, 20° normal pressure angle. (*From AGMA 218.01, December 1982.*)

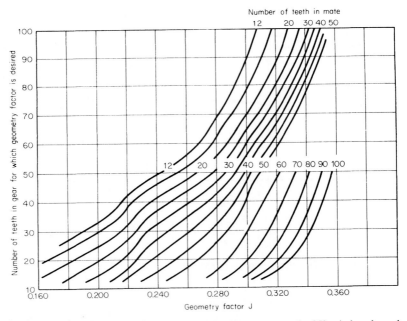

FIG. 27.33   Spiral bevel gear geometry factor for 20° pressure angle, 35° spiral angle, and 90° shaft angle. (*From Gleason publication SD 3103, April 1981.*)

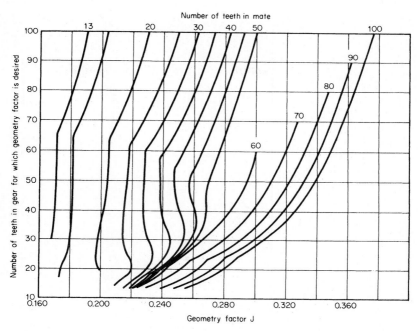

FIG. 27.34 Straight-bevel gear geometry factor for 20° pressure angle and 90° shaft angle. (*From Gleason publication SD 3103E, April 1981.*)

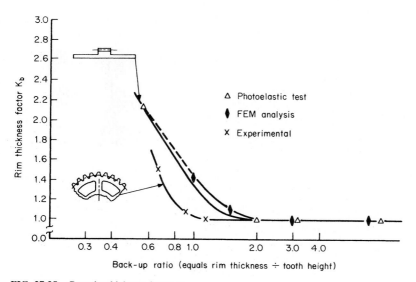

FIG. 27.35 Gear rim thickness factor K_b.

TABLE 27.11 Allowable Bending Stress Number, S_{at} for 10^7 Stress Cycles

Material	AGMA class	Commercial designation	Heat treatment	Minimum hardness surface	Cores	S_{at}, lb/in² Spur and helical	Bevel
Steel	A-1 through A-5	—	Through-hardened and tempered	180 Brinell	—	25–33,000	14,000
				240 Brinell	—	31–41,000	19,000
				300 Brinell	—	36–47,000	
				360 Brinell	—	40–52,000	
				400 Brinell	—	42–56,000	
			Flame or induction-hardened section	50–54 Rockwell C	—	45–55,000	—
				55 Rockwell C min		—	30,000
			Flame or induction-hardened section		—	22,000	13,500
			Carburized and case-hardened	55 Rockwell C	—	55–65,000	27,500
				60 Rockwell C	—	55–70,000	30,000

Cast Iron		AISI 4140	Nitrided	48 Rockwell C	300 Brinell	34–45,000	20,000
		AISI 4340	Nitrided	46 Rockwell C	300 Brinell	36–47,000	
		Nitralloy 135M	Nitrided	60 Rockwell C	300 Brinell	38–48,000	
		2½% chrome	Nitrided	54–60 Rockwell C	350 Brinell	55–65,000	
	20		As cast	—	—	5000	2,700
	30		As cast	175 Brinell	—	8500	4,600
	40		As cast	200 Brinell	—	13,000	7,000
Nodular (ductile) iron	A-7-a	ASTM 60-40-18	Annealed	140 Brinell	—	15,000	8,000
	A-7-c	ASTM 80-55-06	Quenched and tempered	180 Brinell	—	20,000	11,000
	A-7-d	ASTM 100-70-03	Quenched and tempered	230 Brinell	—	26,000	14,000
	A-7-e	ASTM 120-90-02	Quenched and temepred	270 Brinell	—	30,000	18,500
Malleable iron (pearlitic)	A-8-c	45007	—	165 Brinell	—	10,000	—
	A-8-e	50005	—	180 Brinell	—	13,000	—
	A-8-f	53007	—	195 Brinell	—	16,000	—
	A-8-i	80002	—	240 Brinell	—	21,000	—
Bronze	Bronze 2	AGMA 2C	Sand cast Sand cast	Tensile strength min. 40,000 lb/in² (275 MPa)		5,700	3,000
	Al/Br 3	ASTM B-148-52 Alloy 9C	Heat-treated	Tensile strength min. 90,000 lb/in² (620 MPa)		23,600	12,000

Source: AGMA and Gleason Works, Rochester, N.Y.

27.49

TABLE 27.12 Gear Life Factor K_L

Number of cycles	Spur, helical, and herringbone				Bevel gears
	160 Brinell hardness	250 Brinell hardness	450 Brinell hardness	Case-carburized*	Case-carburized*
Up to 1000	1.6	2.4	3.4	2.7	4.6
10,000	1.4	1.9	2.4	2.0	3.1
100,000	1.2	1.4	1.7	1.5	2.1
1 million	1.1	1.1	1.2	1.1	1.4
10 million	1.0	1.0	1.0	1.0	1.0
100 million and over	1.0–0.8	1.0–0.8	1.0–0.8	1.0–0.8	1.0

*Rockwell C 55 min.
Source: AGMA.

$$S \leq S_{at}\left(\frac{K_L}{K_T K_R}\right) \tag{27.57}$$

where the K factors are to modify the allowable stress to account for the effects that alter the basic allowed stress in bending. Table 27.11 shows the basic allowable stress number S_{at} in bending fatigue, which is for 10^7 stress cycles for single-direction bending. Use 70 percent of these values for reversed bending such as in an idler gear. If there are momentary overloads, the design should be checked against the possibility of exceeding the yield strength and causing permanent deformation of the gear teeth. For through-hardened steel the allowable yield strength S_{av} is dependent on Brinell hardness H_B, according to the following empirical equation:

$$S_{ay} = (482H_B - 32,800) \text{ psi} \tag{27.58}$$

The *life factor* K_L may be selected from Table 27.12 and is used to adjust the allowed stress number for the effect of designing to lives other than 10^7 stress cycles.

The *temperature factor* K_T derates the strength to account for loss of basic material strength at elevated temperature. For temperatures up to 250°F the factor is unity. For higher temperatures there is loss of strength due to the tempering effect, and in some materials over 350°F the value of K_T must be greater than unity. One alternative method that is used widely is to form the K_T factor as follows:

$$K_T = \frac{460 + °F}{620} \tag{27.59}$$

where the temperature range is between 160 and 300°F. Operating at temperatures above 300°F is normally not recommended.

The *reliability factor* K_R is used to adjust for desired reliability levels either less or greater than 99 percent, which is the level for the allowed bending strength data in Table 27.11. If actual statistical data on the strength distribution of the gear material are in hand, a suitable K_R factor may be selected. In lieu of this, use the values from Table 27.13.

27.5.3 Other Methods

There is no available methodology that applies to all types of gears for stress and deflection analysis. However, some selected methods that have a limited application are practical for design use and for gaining some insight into the behavior of gears under load.

TABLE 27.13 Reliability Factor K_R

Requirements of application	K_R	Reliability
Fewer than one failure in 10,000	1.50	0.9999
Fewer than one failure in 1000	1.25	0.999
Fewer than one failure in 100	1.00	0.99
Fewer than one failure in 10	0.85*	0.9

*At this value plastic flow might occur rather than pitting.
Source: AGMA.

A method for determining the stress in bending has been successfully applied to spur gears.[86] The stress is calculated by the following equations:

$$S = \frac{W_n \cos \phi_w'}{F}\left(1 + 0.26\,\frac{t_s^{0.7}}{2r}\right)\left[\frac{6l_s'}{t_s^2} + \left(\frac{0.72}{t_s l_s}\right)^{1/2}\left(1 - \frac{t_w}{t_s}\,v\tan\phi_w'\right) - \frac{\tan\phi_w'}{t_s}\right] \quad (27.60)$$

where S = root fillet tensile stress at location indicated in Fig. 27.36
   F = tooth face width measured parallel to gear axis
   W_N = instantaneous force normal to tooth surface transmitted by tooth
   $v \approx \frac{1}{4}$

The remaining quantities in Eq. (27.60) are defined in Fig. 27.36. Angle γ in Fig. 27.36 defines the point where the root fillet tensile stress is calculated by Eq. (27.60).

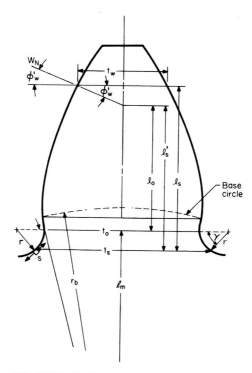

FIG. 27.36 Tooth geometry for evaluation of root bending stress formula parameters.

Reference 86 provides an equation for the value of γ that locates the position of the maximum root stress:

$$\tan \gamma_{i+1} = \frac{(1 + 0.16A_i^{0.7})A_i}{B_i(4 + 0.416A_i^{0.7}) - (\frac{1}{3} + 0.016A_i^{0.7})A_i \tan \phi_W'} \tag{27.61}$$

where

$$A_i = \frac{t_o}{r} + 2(1 - \cos \gamma_i) \tag{27.62}$$

and

$$B_i = \frac{l_o}{r} + \sin \gamma_i \tag{27.63}$$

where t_o and l_o are defined in Fig. 27.36, and subscripts i and $i + 1$ denote that the transcendental equation [Eq. (27.61)] can be solved iteratively for γ with i counting the steps in the iteration procedure.

Once the angle γ is determined, the dimension t_s shown in Fig. 27.36 also is determined by the formula

$$t_s = t_o + 2r(1 - \cos \gamma) \tag{27.64}$$

Deflections of gear teeth at the load contact point are important to determining the correct amount of tip relief necessary to minimize dynamic loads. A finite-element study has been done for a gear with a fully backed-up tooth (thick rim).[84] Figure 27.37

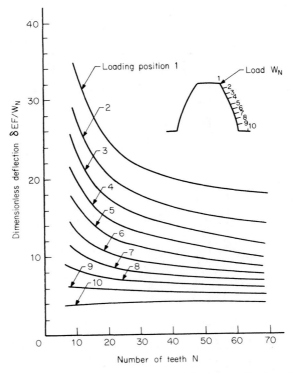

FIG. 27.37 Gear tooth dimensionless deflection as a function of number of teeth and loading position.

shows the results. The ordinate is the dimensionless deflection $\delta EF/W_N$, where δ is deflection, W_N/F is the load per unit of face width, and E is Young's modulus. The dimensionless deflection depends only on the load position and the tooth-gear blank configuration and not on the size of the gear. The load positions are numbered 1 to 10 with the pitch point being load position 5.5. These deflection values are also useful for the construction of a mathematical model of torsional vibration in the gear train.

The methods presented in Ref. 86 allow study of the separate effects contributing to gear tooth deflection. The method is based on beam deflection theory with corrections included to account for hertz deflection, shear deformation, and the flexibility effects of the gear blank itself. The results of the method applied to a 3.5-in pitch diameter gear with 28 teeth, 8 diametral pitch, 0.25-in face width, and solid gear blank are shown in Fig. 27.38. This figure gives an idea of the relative contributions of individual flexibility effects to the total deflection. The hertzian deflection is approximately 25 percent of the total. The gear blank deflection is approximately 40 percent. The beam bending effect of the tooth is approximately 35 percent of the total. The beam bending effect occurs in three parts: the part in the fillet region; the zone of blending between fillet and involute profile (called "undercut" in Fig. 27.38); and the tooth proper (labeled "involute" in the figure). Figure 27.38 shows deflection as a function of load position on the tooth. The discontinuities in slope are the points where the load sharing changes from single tooth pair contact to double pair contact.

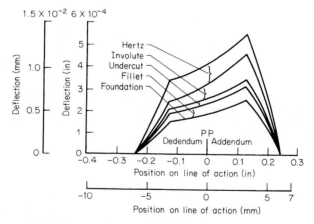

FIG. 27.38 Tooth deflection calculated by the method of Cornell.[87]

27.6 GEAR LIFE PREDICTIONS

Gears in operation are subjected to repeated cycles of contact stress. Even if the gears are properly designed to have acceptable bending stresses to avoid bending fatigue failure, there remains the likelihood of failure by surface contact fatigue. Contact fatigue is unlike bending fatigue in that there is no endurance limit for contact fatigue failure. Eventually a pit will form when the gear has been stressed by enough repeated load cycles. This section shows how to use the method that was developed at NASA for prediction of surface fatigue life for gears.[87-91] The method is based on the assumption that pitting fatigue failures for gears and bearings are similar. A short explanation of the theoretical basis of the life prediction formulas presented in this section follows.

27.6.1 Theory of Gear Tooth Life

The fatigue-life model proposed in 1946 by Lundberg and Palmgren[92] is the common-ly accepted theory to determine the fatigue life of rolling-element bearings. The prob-ability of survival as a function of stress cycles is expressed as

$$\log \frac{1}{S} \propto \frac{\tau_o^c \eta^e V}{z^h} \tag{27.65}$$

Hence, if the probability of survival S is specified, the life η in millions of stress cycles for the required reliability S can be considered a function of the stressed vol-ume V, the maximum critical shearing stress τ_o, and the depth to the critical shearing stress z. As a result, the proportionality can be written as

$$\eta \propto \frac{z^{h/e}}{\tau_o^{c/e} V^{1/e}} \tag{27.66}$$

Equation (27.66) reflects the idea that greater stress shortens life. The depth below the surface z at which the critical stress occurs is also a factor. A microcrack beginning at a point below the surface requires time to work its way to the surface. Therefore, we expect that life varies by power of depth to the critically stressed zone.

The stressed volume V is also an important factor. Pitting initiation occurs near any small stress-raising imperfection in the material. The larger the stressed volume, the greater the likelihood of fatigue failure. By the very nature of the fatigue-failure phe-nomenon, it is the repetitive application of stress that causes cumulative damage to the material. The greater the number of stress cycles, the greater the probability of failure. By experience it has been found that the failure distribution follows the Weibull* model, with $e = 2.5$ for gears. Experiments have determined that exponent $c = 23.25$ and exponent $h = 2.75$. Based on direct application of statistical fundamentals and using the hertzian stress equations, the following formulas enable the straightforward calculation of fatigue life for spur gears. The expression for reliability as a function of stress cycles on a single gear tooth is

$$\log \frac{1}{S} = \left(\frac{\eta}{T_{10}}\right)^e \log\left(\frac{1}{0.9}\right) \tag{27.67}$$

where S is the probability of survival and T_{10} is the life corresponding to a 90 percent reliability. The dispersion or scatter parameter e (Weibull exponent) based on NASA fatigue tests is equal to 2.5. The single tooth life T_{10} is determined from

$$T_{10} = 9.18 \times 10^{18} W_N^{-4.3} F_e^{3.9} \Sigma \rho^{-5} l^{-0.4} \tag{27.68}$$

where T_{10} = millions of stress cycles
  W_N = normal load, lb
  F_e = loaded or effective face width, in
  l = loaded profile length, in
  $\Sigma \rho$ = curvature sum

The curvature sum is given by

---

*The Weibull distribution is defined as $S = \exp[(n/n_c)^e]$ where e is the Weibull exponent and n_c is the characteristic life.

$$\Sigma \rho = \left(\frac{1}{r_{p1}} + \frac{1}{r_{p2}} \right) \frac{1}{\sin \phi} \tag{27.69}$$

where r_{p1} and r_{p2} are the pitch radii in inches and ϕ is the pressure angle. The factor 9.18×10^{18} is an experimentally obtained constant based on NASA gear fatigue tests using aircraft-quality gears of vacuum-arc-remelted (VAR) AISI 9310 steel. If another material with a known fatigue life expectancy relative to the AISI 9310 material is selected, then the life calculated by Eq. (27.68) for T_{10} should be multiplied by an appropriate factor. Table 27.7 lists relative life for several steels based on NASA gear fatigue data. These values can be used as life adjustment factors. A similar treatment of life adjustment factors was made for rolling-element bearings for effects of material, processing, and lubrication variables.[93] As always, the designer should base the life calculations on experimental life data whenever possible.

An alternate procedure for estimating life is given by the American Gear Manufacturers' (AGMA) Standard 218.01.[70] The method is based on allowable stress which may be corrected for certain application and geometry effects. Because of the wide availability of Standard 218.01 the method will not be repeated here. Unlike the Lundberg and Palmgren[92] approach presented herein, the AGMA method does not include stressed volume directly and depth to critical stress sensitivities.

EXAMPLE A gear is to be made of an experimental steel with a known life of three times that of AISI 9310 (based on experimental surface fatigue results with a group of specimens). The gear tooth geometry has a 20° pressure angle, 8 pitch, 28 teeth, face width equals 0.375 in, standard addendum, and meshes with an identical gear. Determine the expected life at the 90 percent reliability level and the 99 percent reliability level for a single tooth, if the mesh is to carry 200 hp at 10,000 r/min.

solution First the necessary geometry must be worked out. The pitch radius is

$$r_{p1} = r_{p2} = \frac{28 \text{ teeth}}{8 \text{ teeth/in of pitch dia.}} \times \frac{1}{2} \times 1.75 \text{ in}$$

The load normal to the gear tooth surface is determined from the power. The torque is

$$T = 63,000 \times \frac{\text{hp}}{\text{r/min}} = 63,000 \times \frac{200}{10,000} = 1260 \text{ lb·in}$$

The normal load is

$$W_N = \frac{T}{r_{p1} \cos \phi} = \frac{1260}{1.75 \cos 20°} = 766 \text{ lb}$$

The curvature sum is

$$\Sigma \rho = \left(\frac{1}{1.75} + \frac{1}{1.75} \right) \frac{1}{\sin 20°} = 3.34 \text{ in}^{-1}$$

The contact ratio is defined as the average number of tooth pairs in contact. It can be calculated by dividing the contact path length by the base pitch. The standard addendum gear tooth has an addendum radius $r_a = r_p + 1/P$. Therefore, the addendum radius is 1.88 in. The length of contact path along the line of action is

$$Z = [r_{a1}^2 - (r_{p1} \cos \phi)^2]^{1/2} + [r_{a2}^2 - (r_{p2} \cos \phi)^2]^{1/2} - (r_{p1} + r_{p2}) \sin \phi$$

$$= 0.604 \text{ in}$$

The base pitch is

$$p_b = 2\pi r_{p1} \frac{\cos \varphi}{N}$$

$$= 2\pi \frac{1.75}{28} \cos 20°$$

$$= 0.369 \text{ in}$$

The contact ratio is 1.64, which means that the gear mesh alternates between having one pair of teeth and two pairs of teeth in contact at any instant. This means that the load is shared for a portion of the time. Since the load is most severe for the single pair of teeth in contact, the length of loaded profile length corresponds to the profile length for single-tooth pair contact. If the contact ratio is above 2.0, then two separate zones of double-tooth pair contact contribute to the loaded profile length. For the current example, the pinion roll angle increment for the single-pair contact (heavy load zone) is

$$\epsilon_H = \frac{2p_b - Z}{r_{p1} \cos \phi}$$

$$= \frac{(2 \times 0.369) - 0.604}{1.75 \cos 20°}$$

$$= 0.081 \text{ rad}$$

The pinion roll angle increment for the double-tooth pair contact (light load zone) is

$$\epsilon_L = \frac{Z - p_b}{r_{p1} \cos \phi}$$

$$= \frac{0.604 - 0.369}{1.75 \cos 20°}$$

$$= 0.143 \text{ rad}$$

The pinion roll angle from the base of the involute to where the load begins (lowest point of double-tooth contact) is

$$\epsilon_c = \frac{(r_{p1} + r_{p2})\sin \varphi - [r_{a2}^2 - (r_{p2} \cos \phi)^2]^{1/2}}{r_{p1} \cos \phi}$$

$$= \frac{(1.75 + 1.75) \sin 20° - [1.88^2 - (1.75 \cos 20°)^2]^{1/2}}{1.75 \cos 20°}$$

$$= 0.174 \text{ rad}$$

The length of involute corresponding to the single-tooth pair contact is

$$l_1 = r_{p1} \cos \phi \, \epsilon_H(\epsilon_c + \epsilon_L + \tfrac{1}{2}\epsilon_H)$$

$$= 1.75 \cos 20°(0.081)(0.174 + 0.143 + 0.081/2)$$

$$= 0.048 \text{ in}$$

Finally the life is obtained as follows:

$$T_{10} = 9.18 \times 10^{18}(766)^{-4.3}(0.375)^{3.9}(3.34)^{-5}(0.048)^{-0.4} = 643 \text{ million stress cycles}$$

In hours, the life is

$$T_{10} = \frac{643 \times 10^6 \, r}{10,000 \, r/min \times 60 \, min/h} = 1072 \, h$$

Conversion to the basis of 99 percent reliability gives

$$T_{10} = \left(\frac{\log 1/0.99}{\log 1/0.9} \right)^{1/2.5} = 419 \, h$$

27.6.2 Life for the Gear

From basic probability notions, the life of the gear is based on the survival probabilities of the individual teeth. For a simple gear having N teeth meshing with another gear, the gear life corresponding to the tooth life at the same reliability level is given by the following relation:

$$G_{10} = T_{10} N^{-1/e} \qquad \text{simple gear} \qquad (27.70)$$

where e = Weibull slope $(2 \cdot 5)$
$\quad G_{10}$ = 10 percent life for gear
$\quad T_{10}$ = 10 percent life for single tooth as determined by previous equations
$\quad N$ = number of teeth.

For the case of an idler gear, where the power is transferred from an input gear to an output gear through the idler gear, both sides of the tooth are loaded once during each revolution of the idler gear. This is equivalent to a simple gear with twice the number of teeth. Therefore, the life of an idler gear with N teeth is

$$G_{10} = T_{10}(2N)^{-1/e} \qquad \text{idler gear} \qquad (27.71)$$

when it meshes with gears of equal face width.

The life of a power-collecting (bull) gear, where there are u number of identical pinions transmitting equal power to the bull gear which has N teeth, is

$$G_{10} = \frac{T_{10}}{u}(N)^{-1/e} \qquad \text{bull gear} \qquad (27.72)$$

27.6.3 Gear System Life

From probability theory, it can be shown that the life of two gears in mesh is given by the following:

$$L = (L_1^{-e} + L_2^{-e})^{-1/e} \qquad (27.73)$$

where L_1 and L_2 are the gear lives.

Equation (27.73) may be generalized in n gears, in which case

$$L = \left(\sum_{i=1}^{n} L_i^{-e} \right)^{-1/e} \qquad (27.74)$$

If the power transmission system has rolling-element bearings, then the system life equations must be solved with different values for the Weibull scatter parameter. For cylindrical or tapered bearings $e = \frac{3}{2}$ and for ball bearings $e = \frac{10}{9}$. The probability distribution for the total power transmission is obtained from the following equation[94]:

$$\log\left(\frac{1}{S_T}\right) = \log\left(\frac{1}{0.9}\right)\left[\left(\frac{L}{L_1}\right)^{e_1} + \left(\frac{L}{L_2}\right)^{e_2} + \left(\frac{L}{L_3}\right)^{e_3} + \cdots\right] \tag{27.75}$$

where L_1, L_2, L_3, etc., are the component lives corresponding to 90 percent reliability. Information on life calculation for bearings may be found in the sections dealing with bearings in this handbook.

EXAMPLE A gearbox has been designed and the component lives have been determined as follows. Find the system life at the 95 percent reliability level.

Bearing	e	B_{10} life	Gear	G_{10} life
No. 1 ball	(10/9)	1200 h	No. 1 pinion	(2.5) 1000 h
No. 2 ball	(10/9)	1500 h		
No. 3 roller	(3/2)	1300 h	No. 2 gear	(2.5) 1900 h
No. 4 roller	(3/2)	1700 h		

solution

$$\log\left(\frac{1}{0.95}\right) = \left(\log\frac{1}{0.9}\right)\left[\left(\frac{L}{1200}\right)^{10/9} + \left(\frac{L}{1500}\right)^{10/9} + \left(\frac{L}{1300}\right)^{3/2}\right.$$
$$\left. + \left(\frac{L}{1700}\right)^{3/2} + \left(\frac{L}{1000}\right)^{2.5} + \left(\frac{L}{1900}\right)^{2.5}\right]$$

Using a programmable calculator, iteration yields the following result for the life at the 95 percent reliability level:

$$L = 249 \text{ h}$$

This result is surprisingly low and emphasizes the need for high-reliability components if the required system reliability and life is to be met.

27.6.4 Helical Gear Life

Life calculation of helical gear teeth is presented in this section. The approach is to think of the helical gear as a modified spur gear, with the tooth profile in the plane normal to the gear axis being preserved but with the spur tooth being sliced like a loaf of bread and wrapped around the base circle along a helix of angle ψ. The helical tooth life may be obtained as a "corrected" spur tooth life where adjustment factors for each of the factors which influence life in the previous equations are given as follows:[88]

Load adjustment:
$$W_{helical} = \frac{W_{spur}\, P_b}{0.95Z \cos\psi} \tag{27.76}$$

Curvature sum adjustment:
$$\Sigma\, \rho_{helical} = (\Sigma\, \rho_{spur})\cos\psi \tag{27.77}$$

Face width in contact adjustment:
$$F_{helical} = F_{spur} \sec\psi \tag{27.78}$$

No adjustment factor for involute length l is recommended. The above adjustment factors are recommended for helical gears with axial contact ratios above 2. For lesser axial contact ratio, the life should be calculated as if the gear were a spur gear.

EXAMPLE A gear has the following properties: standard addendum, transverse pressure angle, 20°; diametral pitch, 8; number of teeth, 28; face width, 1 in; helix angle, 45°; and material, AISI 9310. Calculate the tooth life at 90 percent reliability if 766 lb is transmitted along the pitch line. (The properties in the transverse plane are taken from the example problem for spur gear life.)

solution From the preceding example problem, the transverse base pitch is 0.369 in and the length of the contact path is 0.604 in. The axial pitch p_a is calculated as follows:

$$p_a = \frac{p_b}{\tan \psi} = \frac{0.369}{\tan 45°} = 0.369 \text{ in}$$

The axial contact ratio is 2.71, which allows use of the defined correction factors:

$$W_{helical} = \frac{W_{spur} p_b}{0.95Z \cos \psi}$$

$$= (766 \times 0.369)/(0.95 \times 0.604 \times \cos 45°)$$

$$= 697 \text{ lb}$$

$$\Sigma \rho_{helical} = \Sigma \rho_{spur} \cos \psi$$

$$= 3.34 \cos 45°$$

$$= 2.362 \text{ in}^{-1}$$

$$F_{helical} = F_{spur} \sec \psi$$

$$= 1/\cos 45°$$

$$= 1.414 \text{ in}$$

The above terms are now substituted into the tooth life equation as follows:

$$T_{10} = 9.18 \times 10^{18} W_{helical}^{-4.3} F_{helical}^{3.9} \Sigma \rho_{helical}^{-5} l^{-0.4}$$

$$= 9.18 \times 10^{18}(697)^{-4.3}(1.414)^{3.9}(2.362)^{-5}(0.048)^{-0.4}$$

$$= 9.66 \times 10^5 \text{ million stress cycles}$$

27.6.5 Bevel and Hypoid Gear Life

For bevel and hypoid gears, the contact pattern may be determined from a tooth contact analysis (TCA). The TCA is done with a computer program developed by the Gleason Works.[95] The mathematics is very complicated,[96,97] but the needed parameters for a life analysis are simply the surface curvatures and path of contact (from which the hertzian stress may be calculated). Assuming that the result from a TCA-type analysis is available, the bevel or hypoid tooth life is calculated by the following formulas, which are adapted from Ref. 98. Reference 99 shows an alternate method of assessing surface durability:

$$T_{10} = \left(\frac{3.58 \times 10^{56} z^{7/3}}{\tau^{31/3} V}\right)^{1/2.5} \qquad \text{where } V = wzl \qquad (27.79)$$

TABLE 27.14 Values of Contract Stress Parameters

Δe	a^*	b^*	v
0	1	1	1.2808
0.1075	1.0760	0.9318	1.2302
0.3204	1.2623	0.8114	1.1483
0.4795	1.4556	0.7278	1.0993
0.5916	1.6440	0.6687	1.0701
0.6716	1.8258	0.6245	1.0517
0.7332	2.011	0.5881	1.0389
0.7948	2.265	0.5480	1.0274
0.83495	2.494	0.5186	1.0206
0.87366	2.800	0.4863	1.0146
0.90999	3.233	0.4499	1.00946
0.93657	3.738	0.4166	1.00612
0.95738	4.395	0.3830	1.00376
0.97290	5.267	0.3490	1.00218
0.983797	6.448	0.3150	1.00119
0.990902	8.062	0.2814	1.000608
0.995112	10.222	0.2497	1.000298
0.997300	12.789	0.2232	1.000152
0.9981847	14.839	0.2070	1.000097
0.9989156	17.974	0.18822	1.000055
0.9994785	23.55	0.16442	1.000024
0.9998527	37.38	0.13050	1.000006
1	∞	0	1.000000

$$\Sigma \rho = \frac{1}{\rho_{1x}} + \frac{1}{\rho_{2x}} + \frac{1}{\rho_{1y}} + \frac{1}{\rho_{2y}} \qquad \text{curvature sum} \qquad (27.80)$$

$$\Delta \rho = (1/\rho_{1x} + 1/\rho_{2x}) - (1/\rho_{1y} + 1/\rho_{2y}) \qquad \text{curvature difference} \qquad (27.81)$$

From Table 27.14 we get a^*, b^*, v:

$$a = a^* \left[\frac{3W_N}{2\Sigma\rho} \frac{(1 - v_1^2)}{E_1} + \frac{(1 - v_2^2)}{E_2} \right]^{1/3} \qquad (27.82)$$

$$b = b^* \left[\frac{3W_N}{2\Sigma\rho} \frac{(1 - v_1^2)}{E_1} + \frac{(1 - v_2^2)}{E_2} \right]^{1/3} \qquad (27.83)$$

$$S_c = \frac{3W_N}{2\pi ab} \qquad (27.84)$$

$$\tau_o = \frac{(2v - 1)^{1/2} S_c}{2v(Y + 1)} \qquad (27.85)$$

$$z = \frac{b}{(v + 1)(2v - 1)^{1/2}} \qquad (27.86)$$

EXAMPLE From the results of a TCA study, it is found that the contact path length on the bevel pinion is approximately 0.3 in long. At the central point the principal radii of curvature are as follows:

Pinion	Gear
$\rho_{1x} = 2$ in	$\rho_{2x} = \infty$
$\rho_{1y} = 4$ in	$\rho_{2y} = -5$ in

The load normal to the tooth surface is 700 lb. Estimate the pinion tooth life.

solution

1. Calculate the curvature sum.

$$\Sigma \rho = \frac{1}{\rho_{1x}} + \frac{1}{\rho_{2x}} + \frac{1}{\rho_{1y}} + \frac{1}{\rho_{2y}}$$

$$\Sigma \rho = \frac{1}{2} + \frac{1}{\infty} + \frac{1}{4} - \frac{1}{5}$$

$$= 0.550 \text{ in}^{-1}$$

2. Calculate the curvature difference.

$$\Delta \rho = \frac{\left(\dfrac{1}{\rho_{1x}} + \dfrac{1}{\rho_{2x}}\right) - \left(\dfrac{1}{\rho_{1y}} + \dfrac{1}{\rho_{2y}}\right)}{}$$

$$= \frac{\left(\dfrac{1}{2} + \dfrac{1}{\infty}\right) - \left(\dfrac{1}{4} - \dfrac{1}{5}\right)}{} = 0.818$$

3. From Table 27.14, by interpolation,

$$a^* = 2.397$$
$$b^* = .531$$
$$v = 1.023$$

4. Calculate the dimensions of the hertz ellipse, assuming the properties of steel, $E = 30 \times 10^6$ lb/in^2, $v = 0.25$.

$$a = a^* \left[\frac{3W_N}{2\Sigma\rho} \left(\frac{(1 - v_1^2)}{E_1} + \frac{(1 - v_2^2)}{E_2} \right) \right]^{1/3}$$

$$= 2.397 \left[\frac{3 \times 700}{2 \times 0.550} \left(\frac{1 - 0.25^2}{30 \times 10^6} \right) \times 2 \right]^{1/3} = 0.118 \text{ in}$$

$$b = b^* \left[\frac{3W_N}{2\Sigma\rho} \left(\frac{(1 - v_1^2)}{E_1} + \frac{(1 - v_2^2)}{E_2} \right) \right]^{1/3}$$

$$= 0.531 \left[\frac{3 \times 700}{2 \times 0.550} \left(\frac{1 - 0.25^2}{30 \times 10^6} \right) \times 2 \right]^{1/3} = 0.026 \text{ in}$$

5. Calculate the maximum hertzian contact stress.

$$S_c = \frac{3W_N}{2\pi ab} = \frac{3 \times 700}{2\pi \times 0.118 \times 0.026} = 108,000 \text{ lb/in}^2$$

6. Calculate the maximum reversing orthogonal shear stress and depth under the surface at which it occurs.

$$\tau_o = \frac{(2v - 1)^{1/2}}{2Y(v + 1)} S_c$$

$$= \frac{(2 \times 1.023 - 1)^{1/2}}{2 \times 1.023(1.023 + 1)} \times 108{,}000 = 26{,}700 \text{ psi}$$

$$z = \frac{1}{(v + 1)(2v - 1)^{1/2}} b$$

$$= \frac{1}{(1.023 + 1)[2(1.023) - 1]^{1/2}} \times 0.026 = 0.013 \text{ in}$$

7. Assume that the semiwidth of the hertzian contact ellipse is representative of the width of the contact path across the tooth, and calculate the stressed volume as follows:

$$V = wzl \approx azl$$

$$= 0.118 \times 0.013 \times 0.3$$

$$= 460 \times 10^{-6} \text{ in}^3$$

8. Finally the tooth life at 90 percent reliability is calculated as follows:

$$T_{10} = \left(\frac{3.58 \times 10^{56} z^{7/3}}{\tau_o^{31/3} V} \right)^{1/2.5}$$

$$= \left[\frac{3.58 \times 10^{56} (0.013)^{7/3}}{(26{,}700)^{31/3} 460 \times 10^{-6}} \right]^{1/2.5}$$

$$= 7.94 \times 10^9 \text{ stress cycles}$$

27.7 LUBRICATION

In gear applications, fluid lubrication serves four functions:

1. Provides a separating film between rolling and sliding contacting surfaces, thus preventing wear
2. Acts as a coolant to maintain proper gear temperature
3. Protects the gear from dirt and other contaminants
4. Prevents corrosion of gear surfaces

The first two lubrication functions are not necessarily separable. The degree of surface separation affects friction mode and the magnitude of friction force. This in turn influences frictional heat generation and gear temperatures.

Until the last two decades the role of lubrication between surfaces in rolling and sliding contact was not fully appreciated. Metal-to-metal contact was presumed to occur in all applications with attendant required boundary lubrication. An excessive quantity of lubricant sometimes generated excessive gear tooth temperatures causing thermal failure. The development of the elastohydrodynamic lubrication theory showed that lubricant films of thicknesses of the order of microinches and tens of microinches occur in rolling contact. Since surface finishes are of the same order of magnitude as the lubricant film thickness, the significance of gear tooth finish to gear performance became apparent.

27.7.1 Lubricant Selection

The useful bulk-temperature limits of several classes of fluid lubricants in an oxidative environment are given in Table 27.15.[100] Grease lubricants are listed in Table 27.16.[100] The most commonly used lubricant is mineral oil, both in liquid and grease form (Table 27.17).[100] As a liquid, mineral oil usually has an antiwear or extreme pressure (EP) additive, an antifoam agent, and an oxidation inhibitor. In grease, the antifoam agent is not required.

TABLE 27.15 Lubricants

Lubricant type and specification	Bulk temperature limit in air, °F
Mineral oil MIL-L-6081	200+
Diester MIL-L-7808	300+
Type II ester MIL-L-23699	425
Triester MIL-L-9236B	425
Super-refined and synthetic mineral oils	425
Fluorocarbon*	550
Polyphenyl ether*	600

*Not recommended for gears.

Synthetic lubricants have been developed to overcome some of the harmful effects of lubricant oxidation. However, synthetic lubricants should not be selected over readily available and invariably less-expensive mineral oils if operating conditions do not require them. Incorporation of synthetic lubricants in a new design is usually more easily accomplished than conversion of an existing machine to their use.

The heat-transfer requirements of gears dictate whether a grease lubricant can be used. Grease lubrication permits the use of simplified housing and seals.

27.7.2 Elastohydrodynamic Film Thickness

Elastohydrodynamics is discussed in detail in Sec. 19 for rolling-element bearings. Rolling elements, which gear teeth can be regarded as, are separated by a highly compressed, extremely thin lubricant film. Because of the presence of high pressures in the contact zone, the lubrication process is accompanied by some elastic deformation

TABLE 27.16 Greases

Type	Specification	Recommended temp. range, °F	General composition
Aircraft: High-speed, ball and roller bearing	MIL-G-38220	− 40 to + 400	Thickening agent and fluorocarbon
Aircraft: Synthetic, extreme pressure	MIL-G-23827	− 100 to + 250	Thickening agent, low-temperature synthetic oils, or mixtures EP additive
Aircraft: Synthetic, molybdenum disulfide	MIL-G-21164	− 100 to + 250	Similar to MIL-G-23827 plus MoS₂
Aircraft: General purpose, wide temp. range	MIL-G-81322	− 65 to + 350	Thickening agent and syntehtic hydrocarbon
Helicopter: Oscillating, bearing	MIL-G-25537	− 65 to + 160	Thickening agent and mineral oil
Plug valve: Gasoline- and oil-resistant	MIL-G-6032	+ 32 to + 200	Thickening agent, vegetable oils, glycerols and/or polyesters
Aircraft Fuel- and oil-resistant	MIL-G-27617	− 30 to + 400	Thickening agent and fluorocarbon or fluorosilicone
Ball and roller bearing: Extreme high temperature	MIL-G-25013	− 100 to + 450	Thickening agent and silicone fluid

of the contact surface. Accordingly, this process is referred to as elastohydrodynamic (EHD) lubrication. Ertel[101] and later Grubin[102] were among the first to identify this phenomenon. Hamrock and Dowson[103] have derived a simplified approach to calculating the EHD film thickness as follows:

$$H_{min} = 3.63 U^{0.68} G^{0.49} W^{-0.073} (1 - e^{-0.68k}) \qquad (27.87)$$

In this equation the most dominant exponent occurs on the speed parameter U, and the exponent on the load parameter W is very small and negative. The materials parameter G also carries a significant exponent, although the range of this variable in engineering situations is limited.

The variables related in Eq. (27.87) consist of five dimensionless groupings. These are

Dimensionless film thickness:

$$H = \frac{h}{\rho_x} \qquad (27.88)$$

Ellipticity parameter:

$$k = \left(\frac{\rho_y}{\rho_x}\right)^{2/\pi} \qquad (27.89)$$

Dimensionless load parameter:

$$W = \frac{W_N}{E' \rho_x^2} \qquad (27.90)$$

TABLE 27.17 Mineral Oil Classification and Comparative Viscosities

Category	SAE number	ASTM grade*	AGMA gear oil	Approx. viscosity (CS) 100°F	210°F
	—	32	—	2	—
	—	40	—	3–5.5	—
	—	60	—	8.5–12	—
	—	75	—	12–16	—
	—	105	—	19–24	—
	—	150	—	29–35	—
Extra light	—	215	—	42–51	—
	10W	—	1	45	4.2
	—	315	—	61–75	—
Light	20W	465	2	69	5.7
	—	—	—	90–110	—
Medium	30W	—	3	128	10
	—	700	—	130–166	—
Medium heavy	40	—	—	183	13
	—	—	4	—	—
Heavy	50	—	—	—	17
	—	1000	—	194–237	18
	—	—	5	291–356	—
	—	1500	—	417–525	—
	140	2150	6	—	24
	—	—	—	—	25
	—	3150	—	630–780	—
Super heavy	—	4650	7†	—	29
	—	—	8†	910–1120	36
	250	—	—	—	43
	—	7000	8A†	1370–1670	47
	—	—	9	—	—
	—	—	10	—	97
	—	—	11	—	227
	—	—	—	—	464

*Grade number is equivalent to average Saybolt universal viscosity (SUS) at 100°F.
†Compounded with fatty oil.

Dimensionless speed parameter:

$$U = \frac{\mu_o v}{E' \rho_x} \tag{27.91}$$

Dimensionless materials parameter:

$$G = \alpha E' \tag{27.92}$$

where E' = effective elastic modulus, N/m^2 (lb/in^2)
W_N = normal applied load, N (lb)
h = film thickness, m (in)
ρ_x = effective radius of curvature in x (motion) direction, m in
ρ_y = effective radius of curvature in y (transverse) direction, m in
v = mean surface velocity in x direction, m/s (in/s)
α = pressure-viscosity coefficient of fluid, m^2/N (in^2/lb)
μ_o = atmospheric viscosity, $N \cdot s/m^2$ ($lb \cdot s/in^2$)

$$E' = \frac{2}{(1 - v_1^2)/E_1 + (1 - v_2^2)/E_2} \tag{27.93}$$

$$\frac{1}{\rho_x} = \frac{1}{\rho_{1x}} + \frac{1}{\rho_{2x}} \tag{27.94}$$

$$\frac{1}{\rho_y} = \frac{1}{\rho_{1y}} + \frac{1}{\rho_{2y}}$$

where E_1, E_2 = elastic modulus of bodies 1 and 2
v_1, v_2 = Poisson's ratios of bodies 1 and 2
ρ_{1x}, ρ_{2x} = radius in x (motion) direction of bodies 1 and 2
ρ_{1y}, ρ_{2y} = radius in y (transverse) direction of bodies 1 and 2

For typical steels, E' is 33×10^6 lb/in^2, and, for mineral oils, a typical value of α is 1.5×10^{-4} in^2/lb. Thus, for mineral-oil-lubricated rolling elements, $G = 5000$.

For bodies in sliding and rolling contact with parallel axes of rotation, the tangential surface velocities are

$$v_1 = \omega_1 R_{1x}$$

$$v_2 = \omega_2 R_{2x} \tag{27.95}$$

where R_{1x} and R_{2x} are the radii from the centers of rotation to the contact point.

The geometry of an involute gear contact at distance s along the line of action from the pitch point can be represented by two cylinders rotating at the angular velocity of the respective gears. The distance s is positive when measured toward the pinion (member 1). Equivalent radius, from Eq. (27.94), is

$$\rho_x = \frac{(r_{p1} \sin \phi - s)(r_{p2} \sin \phi + s)}{(r_{p1} + r_{p2})\sin \phi} \tag{27.96}$$

Contact speeds from Eq. (27.95) are

$$v_1 = \omega_1(r_{p1} \sin \phi - s)$$

$$v_2 = \omega_2(r_{p2} \sin \phi + s) \tag{27.97}$$

Surface topography is important to the EHD lubrication process. EHD theory is based on the assumption of perfectly smooth surfaces, that is, no interaction of surface asperities. Actually, of course, this is not the case. An EHD film of several millionths of an inch can be considered adequate for highly loaded rolling elements in a high-temperature environment. However, the calculated film might be less than the combined surface roughness of the contacting elements. If this condition exists, surface asperity contact, surface distress (in the form of surface glazing and pitting), and surface smearing or deformation can occur. Extended operation under these conditions can result in high wear, excessive vibration, and seizure of mating components. A surface-roughness criterion for determining the extent of asperity contact is based upon the ratio of film thickness to a composite surface roughness. The film parameter Λ is

$$\Lambda = \frac{h}{\sigma} \qquad (27.98)$$

where composite roughness σ is

$$\sigma = \left(\sigma_1^2 + \sigma_2^2\right)^{1/2} \qquad (27.99)$$

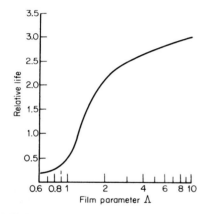

and σ_1 and σ_2 are the rms roughnesses of the two surfaces in contact. Figure 27.39 is a plot of relative life as a function of the film parameter Λ. Although the curve of Fig. 27.39 has been prepared using the rms estimate, an arithmetic average of AA might be used.

In addition to life, the Λ ratio can be used as an indicator of gear performance. At values of Λ less than 1, surface scoring, smearing, or deformation, accompanied by wear, will occur on the gear surface. When Λ is between 1 and 1.5, surface distress can occur accompanied by superficial surface pitting. For values between 1.5 and 3 some surface glazing can occur with eventual gear failure by classical subsurface-originated rolling-element (pitting) fatigue. At values of 3 or greater, minimal wear can be expected with extremely long life. Failure will eventually be by classical subsurface-originated rolling-element fatigue.[93]

FIG. 27.39 Gear relative life as a function of film parameter Λ.

27.7.3 Boundary Lubrication

Extreme-pressure lubricants can significantly increase the load-carrying capacity of gears. The extreme-pressure additives in the lubricating fluid form a film on the surfaces by a chemical reaction, adsorption, and/or chemisorption. These boundary films can be less than 1 μin to several microinches thick.[104] These films may be from the chemical reaction of sulfur[105] or from the chemisorption of iron stearate. Figure 27.40 shows the range of film thicknesses for various types of films.[105] The films can provide separation of the metal surfaces when the lubricant becomes thin enough for the asperities to interact. The boundary film probably provides lubrication by microasperity-elastohydrodynamic lubrication,[104] where the asperities deform under the load. The boundary film prevents contact of the asperities and at the same time provides low-

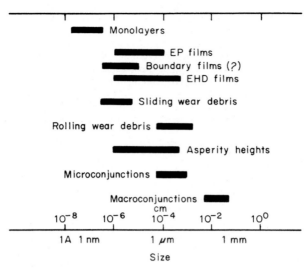

FIG. 27.40 Range of film thickness for various types of lubricant films, with relation to gear surface roughness and wear.

shear-strength properties that prevent shearing of the metal and reduce the friction coefficient over that of the base metal. These boundary films provide lubrication at different temperature conditions, depending on the materials used. For example, some boundary films will melt at a lower temperature than others[105] and will then fail to provide protection of the surfaces. The "failure temperature" is the temperature at which the lubricant film fails. In extreme-pressure lubrication this failure temperature is the temperature at which the boundary film melts. The melting point or thermal stability of surface films appears to be one unifying physical property governing failure temperature for a wide range of materials.[105] It is based on the observation[106] that only a film which is solid can properly interfere with potential asperity contacts. For this reason, many extreme-pressure lubricants contain more than one chemical for protection over a wide temperature range. For instance, Borsoff[107,108] found that phosphorus compounds were superior to chlorine and sulfur at slow speeds, while sulfur was superior at high speeds. He explains this as a result of the increased surface temperature at the higher speeds. It should be remembered that most extreme-pressure additives are chemically reactive and increase their chemical activity as temperature is increased. Horlick[109] found that some metals such as zinc and copper had to be removed from their systems when using certain extreme-pressure additives.

27.7.4 Lubricant Additive Selection

Some of the extreme-pressure additives commonly used for gear oils are those containing one or more compounds of chlorine, phosphorus, sulfur, or lead soaps.[110] Many chlorine-containing compounds have been suggested for extreme-pressure additives, but few have actually been used. Some lubricants are made with chlorine containing molecules where the Cl_3—C linkage is used. For example, either tri(trichloroethyl) or tri(trichlortert butyl) phosphate additives have shown high load-carrying capacity. Other chlorine-containing additives are chlorinated paraffin or petroleum waxes and hexachloroethene.

The phosphorus-containing compounds are perhaps the most commonly used additives for gear oils. Some aircraft lubricants have 3 to 5 percent tricresyl phosphate or tributyl phosphite as either an extreme-pressure or antiwear agent. Other phosphorus extreme-pressure agents used in percentages of 0.1 to 2.0 percent could be dodecyl dihydrogen phosphate; diethyl-, dibutyl-, or dicresyl-phenyl trichloroethyl phosphite; and a phosphate ester containing a pentachlorophenyl radical. Most of the phosphorus compounds in gear oils also have other active elements. The sulfur-containing extreme-pressure additives are believed to form iron sulfide films that prevent wear up to very high loads and speeds. However, they give higher friction coefficients and are therefore usually supplemented by other boundary film forming ingredients that reduce friction. The sulfur compounds should have controlled chemical activity (such as oils containing dibenzyl disulfide of 0.1 or more percent). Other sulfur-containing extreme-pressure additives are diamyl disulfide, dilauryl disulfide, sulfurized oleic acid and sperm oil mixtures, and dibutylxanthic acid disulfide.

Lead soaps have been used in lubricants for many years. They resist the wiping and sliding action in gears, and they help prevent corrosion of steel in the presence of water. Some of the lead soaps used in lubricants are lead oleate, lead fishate, lead-12-hydroxystearate, and lead naphthenate. The lead soap additive most often used is lead naphthenate because of its solubility. Lead soaps are used in concentrations of from 5 to 30 percent.

There are other additive compounds that contain combinations of these elements and most extreme-pressure lubricants contain more than one extreme-pressure additive. Needless to say, the selection of a proper extreme-pressure additive is a complicated process. The word susceptibility is frequently used with reference to additives in oils to indicate the ability of the oil to accept the additive without deleterious effects. Such things as solubility, volatility, stability, compatibility, load-carrying capacity, cost, etc., must be considered.

Many gear oil compounds depend upon the use of proprietary or packaged extreme-pressure additives. As a result, the lubricant manufacturer does not evaluate the additives' effectiveness. Because of this, any selection of extreme-pressure additives should be supported by an evaluation program to determine their effectiveness for a given application. However, a few firms have considerable background in the manufacture and use of extreme-pressure additives for gear lubrication, and their recommendations are usually accepted without question by users of gear oils.

27.7.5 Jet Lubrication

For most noncritical gear applications where cooling the gear teeth is not an important criterion, splash or mist lubrication is more than adequate to provide acceptable lubrication. However, in critical applications such as helicopter transmission systems and turboprop aircraft gear boxes, heat rejection becomes important. Hence, jet lubrication is used. It is important to have the oil penetrate into the dedendum region of the gear teeth in order to provide for cooling and lubrication. Failure to do so can result in premature gear failure primarily by scoring and wear. Methods of oil jet lubrication were analytically determined and experimentally verified by Akin and Townsend.[111-117]

Out-of-Mesh. An analysis along with experimental results was given in Refs. 111 and 114 for out-of-mesh jet lubrication of spur gears. The analysis determines the impingement depth for both gear and pinion when the jet is pointed at the out-of-mesh location and directed at the pitch point of the gear and pinion. Figure 27.41 shows the oil jet as it clears the gear tooth and impinges on the pinion tooth and also gives the nomenclature for the equations. The time of flight for the oil jet to clear the gear tooth and impinge on the pinion tooth must equal the time of rotation of the gears during this time and is given by the following equation:

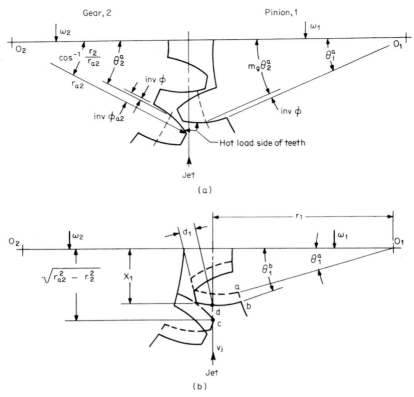

FIG. 27.41 Model for jet lubrication of out-of-mesh gears. (*a*) Gear tooth is at position where oil jet just clears the gear tooth and begins flight toward the pinion tooth. Pinion is shown in position *a*. (*b*) As jet moves from point *c* on the gear to point *d* on the pinion, the pinion rotates from position *a* to *b*.

Time of rotation = time of flight

$$\{\tan^{-1} x_1/r_1 + \text{inv}\{\cos^{-1}[r_{b1}/(r_1{}^2 + x_1{}^2)^{1/2}]\}$$

$$- m_g(\cos^{-1} r_2/r_{a2} - \text{inv } \phi_{a2} + \text{inv } \phi) - \text{inv } \phi\}/w_1 = [(r_{a2}^2 - r_2^2)^{1/2} - x_1]/v_j \qquad (27.100)$$

where inv $\phi = \tan \phi - \phi$.

Equation (27.100) must be solved iteratively for x_1 and substituted into the following equation to determine the pinion impingement depth:

$$d_1 = r_{a1} - (r_1^2 + x_1^2)^{1/2} \qquad (27.101)$$

A similar equation is used to determine the time of flight and rotation of the gears for the impingement on the unloaded side of the gear tooth:

$$[\tan^{-1}(x_2/r_2) + \text{inv}\{\cos^{-1}[r_{b2}/(r_2^2 + x_2^2)^{1/2}\}$$

$$- \{\cos^{-1}(r_1/r_{a1}) - \text{inv } \phi_{a1} + \text{inv } \phi\}/m_g - \text{inv } \phi]/w_2 = [(r_{a1}^2 - r_1^2)^{1/2} - x_2]v_f \qquad (27.102)$$

Solving for x_2 iteratively and substituting in the following equation gives the impingement depth for the gear:

$$d_2 = r_{a2} - (r_2^2 + x_2^2)^{1/2} \tag{27.103}$$

At high gear ratios, it is possible for the gear to shield the pinion entirely so that the pinion would receive no cooling unless the jet direction and/or position is modified. The following three tests may be used to determine the condition for zero impingement depth on the pinion, when

$$\tan^{-1}\frac{x_1}{r_1} < \cos^{-1}\frac{r_1}{r_{a1}} \quad \text{or} \quad x_1 < (r_{a1}^2 - r_1^2)^{1/2}$$

and when $d_1 \le 0$.

The pinion tooth profile is not impinged on directly for speed reducers with ratios above about 1.25, depending upon the jet velocity ratio. Gear impingement depth is also limited by leading-tooth blocking at ratios above about 1.25 to 1.5.

Into-Mesh. In Refs. 116 and 117 it was shown for into-mesh lubrication there is an optimum oil jet velocity to obtain best impingement depth and therefore best lubrication and cooling for the gear and pinion. This optimum oil jet velocity is equal to the gear and pinion pitch line velocity. When the oil jet velocity is greater or less than pitch line velocity, the impingement depth will be less than optimum. Also the oil jet should be pointed at the intersections of the gear and pinion outside diameters and should intersect the pitch diameters for best results. A complete analytical treatment for into-mesh lubrication is given in Refs. 116 and 117. Into-mesh oil jet lubrication will give much better impingement depth than out-of-mesh jet lubrication. However, in many applications, there will be considerable power loss from oil being trapped in the gear mesh. Radial oil jet lubrication at the out-of-mesh location will give the best efficiency, lubrication, and cooling.[112]

Radial Oil Jet Lubrication. The vectorial model for impingement depth using a radially directed oil jet from Ref. 113 is shown in Fig. 27.42. The impingement depth d can be calculated to be

$$d = \frac{1.57 + 2\tan\phi + BP/2}{P\{nD/[2977(\Delta p)^{1/2}] + \tan\phi\}} \tag{27.104}$$

where the oil jet velocity $V_j = 13\sqrt{\Delta p}$ ft/s with Δp in pounds per square inch gauge. If the required d is known, then the ΔP that is required to obtain that impingement depth is

$$\Delta p = \left\{\frac{ndPD}{2977[1.57 + BP/2 + (2 - dP)\tan\phi]}\right\}^2 \tag{27.105}$$

High-speed motion pictures were used to determine the impingement depth experimentally for various oil pressures and gear revolutions per minute. Figure 27.43 is a plot of calculated and experimental impingement depth vs. jet nozzle pressure for gear speeds of 2560 and 4920 r/min. At both speeds there is good agreement between the calculated and experimental impingement depths at higher pressures. However, at the lower pressures, there is considerable difference between the calculated and experimental impingement depths. Most of this difference in impingement depth is due to viscous losses in the nozzle with the very viscous oil used.

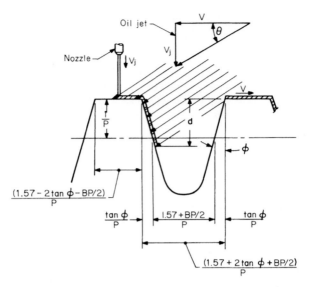

FIG. 27.42   Vectorial model for oil jet penetration depth.

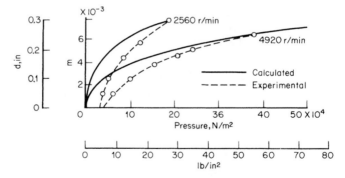

FIG. 27.43   Calculated and experimental impingement depth vs. oil jet pressure at 4920 and 2560 r/min.

27.7.6   Gear Tooth Temperature

A computer program was developed using a finite element analysis to predict gear tooth temperatures.[83,84,118] However, this program does not include the effects of oil jet cooling and oil jet impingement depth. It uses an average surface heat-transfer coefficient for surface temperature calculation based on the best information available at that time.

In order to have a method for predicting gear tooth temperature more accurately, it is necessary to have an analysis that allows for the use of a heat transfer coefficient for oil jet cooling coupled with a coefficient for air–oil mist cooling for that part of the time that each condition exists. Once the analysis can make use of these different coefficients, it can be combined with a method that determines the oil jet impingement depth to give a more complete gear temperature analysis program.[115] However, both the oil jet and air–oil mist heat transfer coefficients are unknowns and must be determined experimentally.

Once the heat generation and the oil jet impingement depth have been calculated, the heat transfer coefficients are either calculated or estimated. Then, a finite element analysis can be used to calculate the temperature profile of the gear teeth as shown in Fig. 27.44. The isotherms in this figure are the temperature differences between the calculated gear tooth temperatures and the inlet cooling oil.

Infrared temperature measurements of the gear tooth during operation verified the accuracy of the calculated values for the gear tooth surface.[112,115]

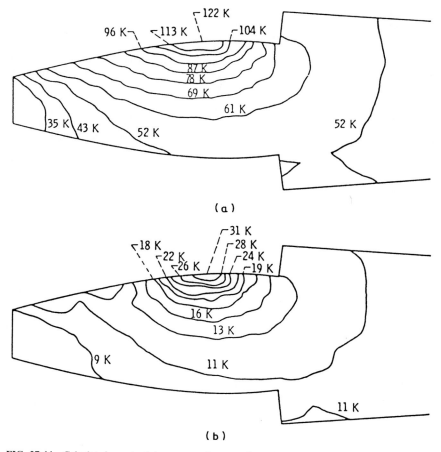

FIG. 27.44 Calculated gear tooth temperature increase above oil-in temperature at a speed of 10,000 r/min and load of 5903 N/cm (3373 lb·in) with 3½-in pitch diameter gear. (*a*) Zero impingement depth. (*b*) 87.5 percent impingement depth.

27.8 POWER-LOSS PREDICTIONS

Figure 27.45 is an experimental plot of efficiency as a function of input torque for a 317-hp, 17:1 ratio helicopter transmission system comprising a spiral-bevel gear input and a three-planet system. The nominal maximum efficiency of this transmission is 98.4 percent. The power loss of 1.6 percent of 317 hp is highly dependent on load and almost independent of speed. Table 27.1 gives maximum efficiency values for various types of gear systems. By multiplying the efficiency for each gear stage, the overall maximum efficiency of the gear system may be estimated. Unfortunately, most transmission systems do not operate at maximum rated torque. As a result, the percent power-loss prediction at part load will generally be high. In addition, gear geometry can significantly influence power losses and, in turn, the efficiency of gear sets.

As an example, consider a set of spur gears with a 4-in (10-cm) pinion. A reduction in gear diametral pitch from 32 to 4 can degrade peak efficiency from 99.8 to 99.4 percent under certain operating conditions. Although this reduced efficiency may at first glance appear to be of little significance, the change does represent a 200 percent increase in power loss.

Although some analytical methods have been developed for predicting gear power losses, most of them do not conveniently account for gear-mesh geometry or operating conditions. Generally, these predictive techniques base power-loss estimates on the friction coefficient at the mesh, among other parameters. Experience shows that the accuracy of these analytical tools often leaves much to be desired, especially for estimating part-load efficiency.

Figure 27.46 shows a typical power-loss breakdown for a gear system. An evaluation of all the power-loss components as a function of torque and speed shows that an unloaded gear set rotating at moderate to high speeds can comprise more than half of the total power losses at full load. Although the sliding loss is dominant at low operating speeds, it becomes only a moderate portion of the total gear-system losses at higher speeds. Both the rolling loss and support ball bearing losses increase with operating speed. Good estimates of these speed-dependent losses are vital for accurately deter-

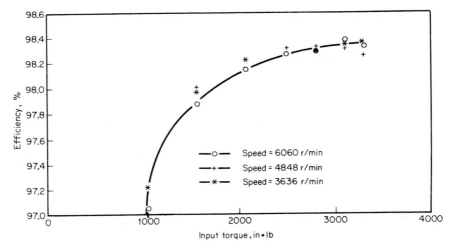

FIG. 27.45 Experimental input efficiency of a three-planet 317-hp helicopter transmission with a spiral-bevel gear input and an input-output ratio of 17:1.

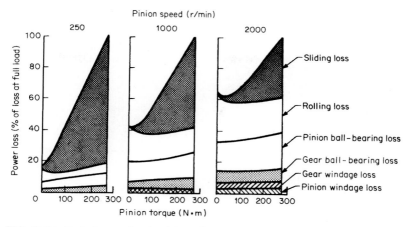

FIG. 27.46 Typical power-loss breakdown for a gear system. Gear set parameters include pitch diameter, 15.2 cm; gear ratio, 1.66; diametral pitch, 8; pinion face width-to-diameter ratio, 0.5; lubricant viscosity, 30 cP.

mining the power consumption of gearboxes. Anderson and Loewenthal[119–122] have developed an analytical method to predict power losses accurately in spur gear sets which have been correlated with experimental data. Their method applies to spur gears which have standard tooth proportions and are jet- or splash-lubricated. Moreover, it considers the individual contributions of sliding, rolling, and windage to power loss at the gear mesh. Some of the power-loss computations required by the method involve the use of mathematical expressions based on average operating conditions where the friction coefficient f, average sliding velocity v_s, average total rolling velocity v_r, and average EHD film thickness h are all evaluated at the point halfway between the pitch point and the start of engagement along the line of action. The mesh losses are determined simply by summing the sliding, rolling, and windage power loss components. Thus,

$$Q_m = Q_s + Q_r + Q_{w1} + Q_{w2} \qquad (27.106)$$

It should be noted that the churning loss of gears running submerged in oil is not accounted for in this analysis.

27.8.1 Sliding Loss

The sliding loss Q_s at the mesh results from frictional forces that develop as the teeth slide across one another. This loss can be estimated as

$$Q_s = C_1 f \overline{W}_N \overline{v}_s \qquad (27.107)$$

Values of C_1 through C_7 are given in Table 27.18. Friction coefficient f, average normal load $\overline{W}_N$, and average sliding velocity $\overline{v}_s$ are

$$f = 0.0127 \log\left(\frac{C_6 \overline{W}_N}{F\mu \, \overline{v}_s \overline{v}_r^2}\right) \qquad (27.108)$$

TABLE 27.18 Constants for Computing Gear-Mesh Power Losses

Constant	Value for SI metric units	Value for U.S. customary units
C_1	2×10^{-3}	3.03×10^{-4}
C_2	9×10^4	1.970
C_3	1.16×10^{-8}	1.67×10^{-14}
C_4	0.019	2.86×10^{-9}
C_5	39.37	1.0
C_6	29.66	45.94
C_7	2.05×10^{-7}	4.34×10^{-3}

$$\overline{W}_N = \frac{T_1}{D_1 \cos \phi} \tag{27.109}$$

$$\overline{v}_s = |\mathbf{v}_2 - \mathbf{v}_1| \quad \text{(in the vector sense)} \tag{27.110}$$

$$= 0.0262 n_1 \left(\frac{1 + m_g}{m_g} \right) Z \ \text{ft/s}$$

It should be noted that the expression for f is based upon test data applied to gear sliding-loss computations involving the elastohydrodynamic (EHD) lubrication regime, where some asperity contact occurs. Such a lubrication regime generally can be considered as the common case for gearset operation. Parameter f should be limited to a range from 0.01 to 0.02.

For computations of $\overline{v}_s$ and f, the contact-length path Z and average total rolling velocity $\overline{v}_r$ are

$$Z = 0.5 \left\{ \left[\left(D_1 + \frac{2}{C_5 P} \right)^2 - (D_1 \cos \phi)^2 \right]^{1/2} \right.$$

$$\left. + \left[\left(D_2 + \frac{2}{C_5 P} \right)^2 - (D_2 \cos \phi)^2 \right]^{1/2} - (D_1 + D_2)\sin \phi \right\} \tag{27.111}$$

$$\overline{v}_r = 0.1047 n_1 \left[D_1 \sin \phi - \frac{Z}{4} \left(\frac{m_g - 1}{m_g} \right) \right] \tag{27.112}$$

27.8.2 Rolling Loss

Rolling loss occurs during the formation of an EHD film when oil is squeezed between the gear teeth, then subsequently pressurized. The rolling loss Q_r is a function of the average EHD film thickness h and the contact ratio m_c, which is the average number of teeth in contact. Therefore,

$$Q_r = C_2 \overline{h} \overline{v}_r F m_c \tag{27.113}$$

where the "central" EHD film thickness $\overline{h}$ for a typical mineral oil lubricating steel gear is

$$\overline{h} = C_7 (\overline{v}_r \mu)^{0.67} \overline{W}_N^{-0.067} \rho_{eq}^{0.464} \tag{27.114}$$

and the contact ratio m_c is

$$m_c = \frac{C_5 ZP}{\pi \cos \phi} \tag{27.115}$$

The expression for $\bar{h}$ in Eq. (27.114) does not consider the thermal and starvation effects occurring at high operating speeds, generally above 40 m/s (8000 ft/min) pitch line velocity. These effects typically tend to constrain the film thickness to some limiting value. For computing h, the equivalent-contact rolling radius is

$$\rho_{eq} = \frac{\left[D_1(\sin \phi) + \dfrac{Z}{2} \right] \left[D_2(\sin \phi) - \dfrac{Z}{2} \right]}{2(D_1 + D_2)\sin \phi} \tag{27.116}$$

27.8.3 Windage Loss

In addition to sliding and rolling losses, power losses due to pinion and gear windage occur. Such losses can be estimated from expressions based on turbine-disk drag test data. Thus, the pinion and gear windage losses are approximated as

$$Q_{w1} = C_3 \left(1 + 4.6 \frac{F}{D_1} \right) n_1^{2.8} D_1^{4.6} (0.028\mu + C_4)^{0.2} \tag{27.117}$$

$$Q_{w2} = C_3 \left(1 + 4.6 \frac{F}{D_2} \right) \left(\frac{n_2}{m_g} \right)^{2.8} D_2^{4.6} (0.028\mu + C_4)^{0.2} \tag{27.118}$$

These windage loss expressions apply for an air-oil atmosphere, commonly found in typical gearboxes. The expressions assume a oil specific gravity of 0.9 and constant values for air density and viscosity at 339 K (150°F). To further account for the oil-rich gearbox atmosphere, both density and viscosity of the atmosphere are corrected to reflect a 34.25:1 air-oil combination, which often is reported for helicopter transmissions lubricated with oil jets. Most of the preceding expressions involve special constants, which are labeled as C_1 through C_7. The appropriate values for these constants can be obtained from Table 27.18.

27.8.4 Other Losses

To determine the total gear-system loss, power losses due to the support bearings also should be considered. Both hydrodynamic and rolling-element bearing losses often can equal or exceed the gear-mesh losses. Bearing losses can be quantified either theoretically or by testing.

Total gear-system loss Q_T is

$$Q_T = Q_m + Q_h \tag{27.119}$$

where Q_m is gear-mesh losses and P_B is total bearing power losses. Given Q_T and input power Q_{in}, gear-system efficiency is predicted to be

$$\zeta = \frac{Q_{in} - Q_T}{Q_{in}} \times 100 \tag{27.120}$$

27.8.5 Optimizing Efficiency

The method described for predicting spur-gear power losses also can be employed in a repetitive manner for a parametric study of the various geometric and operating variables for gears. Typical results of such parametric studies are summarized by the two efficiency maps shown in Figs. 27.47 and 27.48, commonly called *carpet plots,* for light and moderate gear tooth loads. These carpet plots describe the simultaneous effects on gear-mesh efficiency when the diametral pitch, pinion pitch diameter, and pitchline velocity are varied. The computed gear-mesh efficiencies do not include support-bearing losses.

On these plots, the three key variables are represented along orthogonal intersecting planes for three different values of diametral pitch, pinion pitch diameter, and pitchline velocity. The gear-mesh efficiency accompanying any combination of these three gear parameters is represented by an intersecting point for the three planes. Efficiencies at intermediate values can be readily found by interpolation among the planes.

The light and moderate spur gear loads are specified in terms of a contact load factor, usually called a K factor. The K factor is a widely used parameter that normalizes the degree of loading on gears of different size and ratio. Essentially, the K factor describes the load intensity on a gear tooth. It is evaluated as

$$K = \frac{\overline{W}_i(m_g + 1)}{FD_1 m_g} \qquad (27.121)$$

The two carpet plots of Figs. 27.47 and 27.48 were generated for light and moderate loads where $K = 10$ and 300 lb/in^2, respectively. Allowable K factors for spur gears generally range from about 100 lb/in^2 for low-hardness steel gears with generated teeth to 1000 lb/in^2 for aircraft-quality, high-speed gears that are case-hardened and ground. A nominal K factor for a general-purpose industrial drive with 300 Bhn steel gears, carrying a uniform load at a pitchline velocity of 15 m/s (3000 ft/min) or less, typically ranges from 275 to 325 lb/in^2.

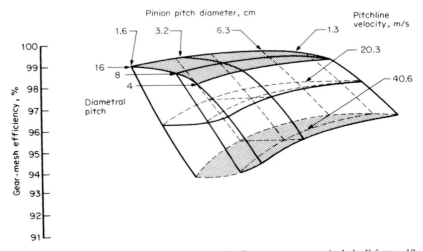

FIG. 27.47 Parametric study for a light gear load. Gear set parameters include K factor, 10 lb/in^2; gear ratio, 1; pinion face width-to-diameter ratio, 0.5; lubricant viscosity, 30 cP.

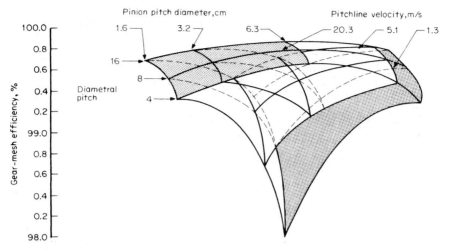

FIG. 27.48 Parametric study for a moderate gear load. Gear set parameters include K factor, 300 lb/in²; gear ratio, 1; pinion face width-to-diameter ratio, 0.5; lubricant viscosity, 30 cP.

The plot for a light load indicates that increasing diametral pitch (resulting in smaller gear teeth) and decreasing pitchline velocity tends to improve gear-mesh efficiency. For a moderate to heavy load, increasing both diametral pitch and pitchline velocity improves efficiency. The reason for this reversal in efficiency characteristics is that speed-dependent losses at light loads dominate, whereas sliding loss is more pronounced at higher loads. Also, at higher speed, the sliding loss is reduced because the sliding-friction coefficient is lower.

Essentially, these two parametric studies indicate that large-diameter, fine-pitch gears tend to be most efficient when operated under appreciable load and high pitchline velocity. Consequently, to attain best efficiency under these operating conditions, the gear geometry should be selected as noted.

27.9 OPTIMAL DESIGN OF SPUR GEAR MESH

When designing gear systems, it is cost-effective to make the design efficient and long-lived with a minimum weight-to-power ratio from some standard point of view. This introduces the concept of optimization in the design, selecting as a point of reference either minimum size (which relates also to minimum weight) or maximum strength. While the designer is seeking to optimize some criterion index, all the basic design requirements (constraints) such as speed ratio, reliability, producibility, acceptable cost, and no failures due to wear, pitting, tooth breakage, or scoring must also be met. The design process must carefully assess all of the possible modes of failure and make adjustments in the allowable design parameters in order to achieve the optimum design. Figure 27.49 presents a qualitative diagram of expected failure modes for operation in various regimes of load and speed. The job of the designer is to anticipate the type of load and speed conditions the gears will see in service and choose a design that will not fail and is optimum in some sense.

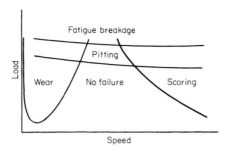

FIG. 27.49 Various failure regimes encountered in gear teeth.

27.9.1 Minimum Size

The index for optimization is often minimum size. A measure of size is center distance. A procedure for optimization where minimum center distance was the objective has been developed.[123,124] In the procedure, the following items were the constraints: (1) allowed maximum hertzian (normal) stress S_{max} equals 200,000 lb/in², which affects scoring and pitting; (2) the allowed bending stress, $S_b = 60,000$ lb/in², which affects tooth breakage due to fatigue; (3) the geometric constraint for no involute interference, which is the contact between the teeth under the base circle, or tip fouling in the case of internal gears; and (4) to help in keeping the misalignment factor to a minimum, a ratio of 0.25 of tooth width to pinion pitch diameter was selected. The results of the optimization study for a gear ratio of 1 are shown in Fig. 27.50. After consideration of all the design parameters it was found that for a given set of allowed stresses, gear ratio, and pressure angle, the design space could be shown on a plane with the vertical axis as the number of teeth on the pinion and the horizontal axis as the diametral pitch. In Fig. 27.50, the region for designs that meet acceptable conditions is shown. But not all the designs resulting from those combinations of number of teeth and pitch are best in approaching a minimum center-distance design.

The combinations of number of teeth and diametral pitch that fall on any straight line emanating from the origin will give a constant-sized design. The slope of the line is directly related to the center distance. The minimum slope line that falls into the region of acceptable designs gives the optimum choice of pitch and tooth number for an optimally small (light) design.

The data in Fig. 27.51 summarize results from a large number of plots similar to those in Fig. 27.50. Figure 27.51 may be used to select trial values of designs that can lead to optimum final designs. The value of these charts is that initial designs may be quickly selected, and then a more detailed study of the design should be made, using the more detailed methods for calculating bending fatigue stress, life, scoring, and efficiency previously described. In Fig. 27.50 the point labeled A is the point nearest optimum on the minimum center distance criterion since that is the point which lies on the minimum slope straight line through the origin and yet contains points in the regions of acceptable designs. The point B is the point that would be chosen in a selection based only on bending strength criteria and kinematic conditions of no involute interference.

In the consideration of the gear and pinion, the hertzian stress is common to both gear and pinion tooth, while the bending strength criterion was applied only to the pinion since it is the weaker element of the pinion and gear teeth.

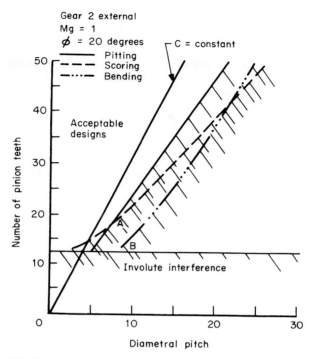

FIG. 27.50 External gear mesh design space ($F/D = 0.25$, $T_1 = 113$ n·m (1000 lb·in), $S_{max} = 1.38$ GPa (200 ksi), $S_b = 414$ MPa (60 ksi), $E = 205$ GPa (30×10^6 lb/in^2), and $\nu = 0.25$).

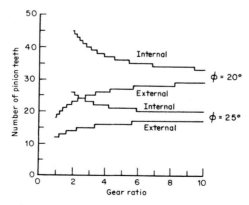

FIG. 27.51 Optimal number of pinion gear teeth for pressure angle ϕ of 20 and 25°.

Other approaches to optimization have included the balanced strength approach, where the pinion was made thicker and with shorter addendum while the gear was made with a longer addendum. The idea was to balance the strength of the pinion tooth with the gear tooth. In Ref. 125 this idea of balanced design was applied to scoring temperature index balance as well as bending strength balance. In Ref. 126 long and short addenda gear sets were examined for their optimum tooth numbers. In Ref. 127 a direct-search computer algorithm was applied to the minimum center distance design problem.

27.9.2 Specific Torque Capacity

Criteria for comparing two designs have been described in the literature as *specific torque capacity (STC).*[21] The STC rating of a single gear is defined as the torque transmitted by that gear divided by the superficial volume of its pitch cylinder. That is, torque per unit volume is equal to

$$\frac{T}{V} = \frac{4T}{\pi D^2 F} \tag{27.122}$$

where D is pitch circle diameter, and F is face width.

For a pair of gears with a 1:1 ratio, the same STC value can be attributed to the pair as that for each gear individually. Thus, for such a gear pair, the average STC value is

$$\frac{T_1/V_1 + T_2/V_2}{2} = \frac{1}{2}\left(\frac{4T_1}{\pi D_1^2 F_1} + \frac{4T_2}{\pi D_2^2 F_2}\right) \tag{27.123}$$

But if T_1 and T_2 are the torque loadings of a pinion and a wheel, then $T_2 = m_g T_1$, where m_g is the gear ratio. Similarly, $D_2 = m_g D_1$, but $F_2 = F_1$. Thus,

$$\text{STC}_{\text{av}} = \frac{2}{\pi}\left(\frac{T_1}{D_1^2 F_1} + \frac{m_g T_1}{m_g^2 D_1^2 F_1}\right)$$

$$\tag{27.124}$$

or
$$\text{STC}_{\text{av}} = \frac{2T_1}{\pi D_1^2 F_1}\frac{1 + m_g}{m_g}$$

But $T = W_t R/2$, where W_t is total tangential load acting at the pitch surface. Thus

$$\text{STC}_{\text{av}} = \frac{W_t}{F_1 D_1 \pi}\frac{1 + m_g}{m_g} \tag{27.125}$$

This expression bears a close resemblance to the AGMA K factor for contact stress, where

$$K = \frac{W_t}{F_1 D_1}\frac{1 + m_g}{m_g} \tag{27.126}$$

Hence, $\text{STC}_{\text{av}} = K/\pi$

Thus, although K has been considered as a surface-loading criterion for a gear pair, it also has a direct relationship to the basic torque capacity of a given volume of gears. It can, therefore, be used as a general comparator of the relative load capacity of gear pairs, without any specific association with tooth surface or root stress conditions. The STC value thus becomes a very useful quantity for comparing the performance of gears based on completely different tooth systems, even where the factors that limit performance are not consistent.

27.10   GEAR TRANSMISSION CONCEPTS

27.10.1   Series Trains

Speed Reducers.   The overall ratio of any reduction gear train is the input shaft speed divided by the output speed. It is also the product of the individual ratios at each mesh, except in planetary arrangements. The ratio is most easily determined by dividing the product of the numbers of teeth of all driven gears by the product of the numbers of teeth of all driving gears. By manipulating numbers, any desired ratio can be obtained, either exactly or with an extremely close approximation.

In multiple-mesh series trains the forces transmitted through the gear teeth are higher at the low-speed end of the train. Therefore, the pitches and face widths of the gears are usually not the same throughout the train. In instrument gears, which transmit negligible power, this variation may not be necessary.[14] Table 27.19 gives several possible speed ratios with multiple meshes.

TABLE 27.19   Gear Speed Ratios with Multiple Meshes

Type of gear	Ratio
Helical, double-helical, herringbone gears:	
Double reduction	10:1–75:1
Triple reduction	75:1–350:1
Spiral bevel (first reduction) with helical gear:	
Double reduction	9:1–50:1
Triple reduction	50:1–350:1
Worm gears:	
Double reduction	100:1–8000:1
Helicals with worm gears:	
Double reduction	50:1–270:1
Planetary gear sets:	
Simple	4:1–10:1
Compound	10:1–125:1

Speed Increasers.   Speed-increasing gear trains require greater care in design, especially at high ratios. Because most gear sets and gear trains are intended for speed reduction, standards and published data in general apply to such drives. It is not safe to assume that these data can be applied without modification to a speed-increasing drive. Efficiency is sometimes lower in an increasing drive, requiring substantial input torques to overcome output load; in extreme instances, self-locking may occur.[14] Traction and hybrid drives should be considered as reasonable alternatives to gear drives for this purpose.

Reverted Trains.   When two sets of parallel-shaft gears are so arranged that the output shaft is concentric with the input, the drive is called a "reverted train." The requirement of equal center distance for the two trains complicates determination of how many teeth should be in each gear to satisfy ratio requirements with standard pitches. Helical gears provide greater design freedom through possible variation of helix angle.[14]

27.10.2 Multispeed Trains

Sliding Gears.   Speed adjustment is effected by sliding gears on one or more inter-mediate parallel shafts (Fig. 27.52). Shifting is generally accomplished by disengaging the input shaft. Sliding-gear transmissions are usually manually shifted by means of a lever or a handwheel. A variety of shaft arrangements and mountings are available.[128]

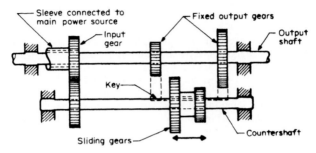

FIG. 27.52   Sliding gear assembly.

Constant Mesh.   Several gears of different sizes mounted rigidly to one shaft mesh with mating gears free to rotate on the other shaft (Fig. 27.53). Speed adjustment is obtained by locking different gears individually on the second shaft by means of splined clutches or sliding couplings.

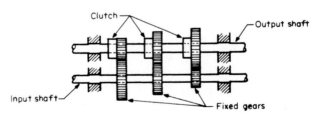

FIG. 27.53   Constant-mesh gear assembly.

Constant-mesh gears are used in numerous applications, among them heavy-duty industrial transmissions. This arrangement can use virtually any type of gearing—for example, spur, helical, herringbone, and bevel gears.

Manually shifted automotive transmissions combine two arrangements. The for-ward speeds are constant-mesh helical gears, whereas reverse uses sliding spur gears.[128]

Idler Gears.   An adjustable speed drive comprising an idler gear consists of one shaft which carries several different-size gears rigidly mounted as shown in Fig. 27.54.

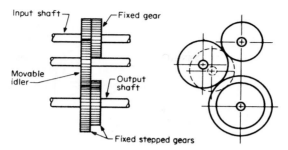

FIG. 27.54   Idler gear assembly.

Speed adjustment is through an adjustable arm which carries the idler gear to connect with fixed or sliding gears on the other shaft.

This arrangement is used to provide stepped speeds in small increments and is frequently found in machine tools. A transmission of this type can be connected to another multispeed train to provide an extremely wide range of speeds with constant-speed input.

Shifting is usually manual. The transmission is disengaged and allowed to come to a stop before the sliding idler gear is moved to the desired setting.[128]

27.10.3   Epicyclic Gearing

An epicyclic gear train such as shown in Fig. 27.55 is a reverted-gear arrangement in which one or more of the gears (planets) moves around the circumference of coaxial gears, which may be fixed or rotating with respect to their own axes. The planet gears have a motion consisting of rotation about their own axes and rotation about the axes of the coaxial gears.[14] The arrangement shown in Fig. 27.55 consists of a central sun gear with external teeth, a ring gear with internal teeth, revolving planet pinions which engage the sun gear and the internal ring gear, and a planet carrier in which the planet pinions are supported.[14] Epicyclic trains may incorporate spur or helical gears, external or internal, or bevel gears arranged in numerous ways either for purposes of fixed ratio or for multispeed applications.

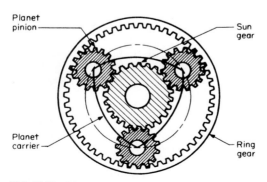

FIG. 27.55   Planetary gears.

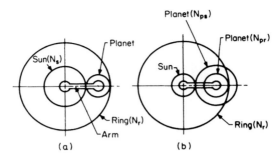

FIG. 27.56 Epicyclic gear trains. (*a*) Simple train. (*b*) Compound train (planet gears keyed to same shaft).

Single Epicyclic Trains. A simple single epicyclic train is shown in Fig. 27.56*a*. The coaxial sun and ring gears are connected by a single intermediate planet gear carried by a planet carrier arm. A compound epicyclic train is shown in Fig. 27.56*b*. The intermediate planet pinions are compound gears. Table 27.19 gives possible speed ratios which can be attained. For single planetary arrangements, to make assembly possible, $(N_r + N_s)/q$ must be a whole number, where N_r and N_s are the numbers of teeth in the ring and sun gears, and q is the number of planets equally spaced around the sun.[14] In order to make assembly possible for a compound planetary arrangement $(N_r N_{ps} - N_s N_{pr})/q$ must equal a whole number,[14] where N_{ps} and N_{pr} are the numbers of teeth in the planet gears in contact with the sun and ring, respectively.

Coupled Epicyclic Trains. Coupled epicyclic trains consist of two or more single epicyclic trains arranged so that two members in one train are common to the adjacent train.[14]

Multispeed Epicyclic Trains. Multispeed epicyclic trains are the most versatile and compact gear arrangement for a given ratio range and torque capacity. Tables 27.20 and 27.21 show the speed ratios possible with simple and compound epicyclic trains when one member is fixed and another is driving. With a suitable arrangement of clutches and brakes, an epicyclic train can be the basis of a change-speed transmission. With the gears locked to each other, an epicyclic train has a 1:1 speed ratio.[14]

Most planetary-gear transmissions are of the "automatic" type, in which speed

TABLE 27.20 Simple Epicyclic Train Ratios (See Fig. 27.56*a*)

Condition	Revolution of		
	Sun	Arm	Ring
Sun fixed	0	1	$1 + \dfrac{N_s}{N_r}$
Arm fixed	1	0	$-\dfrac{N_s}{N_r}$
Ring fixed	$1 + \dfrac{N_r}{N_s}$	1	0

Note: In the equations, N_s and N_r denote either pitch diameter or number of teeth in the respective gears.

TABLE 27.21   Compound Epicyclic Train Ratios (See Fig. 27.56b)

Condition	Revolution of		
	Sun	Arm	Ring
Sun fixed	0	1	$1 + \dfrac{N_s N_{pr}}{N_{ps} N_r}$
Arm fixed	1	0	$-\dfrac{N_s N_{pr}}{N_{ps} N_r}$
Ring fixed	$1 + \dfrac{N_{ps} N_r}{N_s N_{pr}}$	1	0

Note: In the equations, N_r, N_s, N_{pr} and N_{ps} denote either pitch diameter or number of teeth in the respective gears.

changes are carried out automatically at a selected speed or torque level. At the same time it is also the most expensive, because of the clutching and braking elements necessary to control operation of the unit. In addition, practical ratios available with planetary sets are limited.[128]

Bearingless Planetary Trains.   The self-aligning bearingless planetary transmission (Fig. 27.57) covers a variety of planetary-gear configurations, which share the common characteristic that the planet carrier, or spider, is eliminated, as are conventional planet-mounted bearings. The bearings are eliminated by load-balancing the gears, which are separated in the axial direction. All forces and reactions are transmitted through the gear meshes and contained by simple rolling rings. The concept was first demonstrated by Curtis Wright Corporation under sponsorship of the U.S. Army Aviation Research and Development Command.[129,130]

Hybrid Transmission.   Geared planetary transmissions have a limitation on the speed ratio that can be obtained in a single stage without the planets interfering with each other. For example, a four-planet drive would have a maximum speed ratio of 6.8

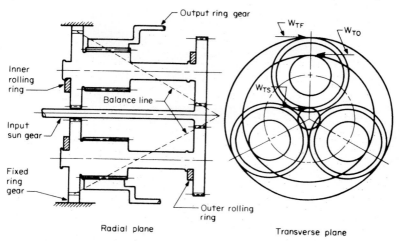

Radial plane   Transverse plane

FIG. 27.57   A 500-hp bearingless planetary transmission.

before the planets interfered with each other. A five-planet drive would be limited to a ratio of 4.8, and so on. A remedy to the speed-ratio and planet number limitations of simple, single-row planetary systems was devised by A. L. Nasvytis.[131] His drive system used the sun and ring roller of the simple planetary traction drive, but he replaced the single row of equal-diameter planet rollers with two or more rows of stepped, or dual diameter, planets. With this new, multiroller arrangement, practical speed ratios of 250:1 could be obtained in a single stage with three planet rows. Furthermore, the number of planets carrying the load in parallel could be greatly increased for a given ratio. This resulted in a significant reduction in individual roller contact loading with a corresponding improvement in torque capacity and fatigue life.

To further reduce the size and the weight of the drive for helicopter transmission applications, pinion gears in contact with a ring gear are incorporated with the second row of rollers (Fig. 27.58). The ring gear is connected through a spider to the output rotor shaft. The number of planet-roller rows and the relative diameter ratios at each contact are variables to be optimized according to the overall speed ratio and the uniformity of contact forces. The traction-gear combination is referred to as the hybrid transmission. The potential advantages of the hybrid transmission over a conventional planetary are higher speed ratios in a single stage, higher power-to-weight ratios, lower noise by replacing gear contacts with traction rollers, and longer life because of load sharing by multiple power paths.

FIG. 27.58   A 500-hp hybrid helicopter transmission.

27.10.4 Split-Torque Transmissions

A means to decrease the weight-to-power ratio of a transmission or to decrease the unit stress of gear teeth is by load sharing through multiple power paths. This concept is referred to as the *split-torque transmission*.[132-134] The concept is illustrated in Fig. 27.59 for single-input and double-input variants. Instead of a planetary-gear arrange-

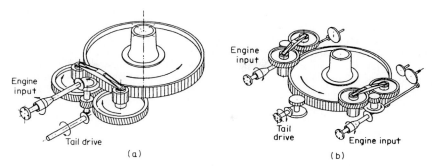

FIG. 27.59 Conceptual sketch of a split-torque transmission. (*a*) Single input. (*b*) Dual input.

ment, the input power is split into two or more power paths and recombined in a bull gear to the output power shaft. This concept appears to offer weight advantages over conventional planetary concepts without high-contact-ratio gearing. The effect of incorporating high-contact-ratio gearing into the split-torque concept is expected to further reduce transmission weight.

27.10.5 Differential Gearing

Differential gearing is an arrangement in which the normal ratio of the unit can be changed by driving into the unit with a second drive. This arrangement, or one having two outputs and one input, is used to vary ratio. It is called the *free type*. Simple differentials may use bevel gears (Fig. 27.60*a*) or spur gears (Fig. 27.60*b*). The bevel-gear type is used in automotive rear-end drives. The input-output speed relationship is

$$2\omega = \omega_1 - \omega_2 \tag{27.127}$$

where ω is the speed of the arm, and ω_1, ω_2 are the two shaft speeds.

Another type of differential (called the *fixed type*) has a large, fixed ratio. Such a drive is an evolution of the compound epicyclic train in Fig. 27.56*b*. If the ring gear is replaced with a sun gear which meshes with planet N_{pr}, the equations in Table 27.21 apply, except that the plus signs change to minus. If the sun and planet gears are made

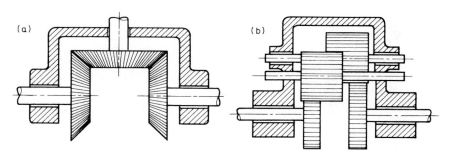

FIG. 27.60 Simple differentials. (*a*) Bevel gears. (*b*) Spur gears.

almost but not exactly equal, the output speed is the small difference between two terms that are almost equal. Such a drive is good for ratios from 10:1 to 3000:1.

Using two ring gears instead of a ring and a sun as in Fig. 27.56b results in a fixed-differential drive suitable for ratios of 15:1 to 100:1.

Any epicyclic gear train can be designed for differential operation. For instance, instead of one of the elements, such as a sun or ring gear, being fixed, two of the elements may be driven independently. Output speed is then the net result of the two inputs.

Compound epicyclic trains can produce several input or output speeds by the addition of extra sun or ring gears meshing with the planet pinions. Thus, the compound epicyclic train in Fig. 27.56b may have another sun gear meshing with N_{Pr} and another ring gear meshing with N_{Ps}. Any one of the four (two suns and two rings) could be fixed, and the others available for input, output, or free.[14]

27.10.6 Closed-Loop Trains

Epicyclic gear trains with more than one planet pinion meshing with the same sun and ring gears have parallel paths through which power can flow. Other gear trains, too, sometimes use multiple-tooth contact to increase the capacity within a given space. Such gearing is sometimes called "locked-train gearing."

Two considerations arise with multiple-tooth contact which are not present in open trains. One is the proper selection of tooth numbers and spacing to ensure assembly. The other is gear accuracy and adjustment to ensure equal distribution of the load to each mesh. Multiple-mesh or locked-train gearing requires careful attention to tooth accuracy and support of the gears. Backlash and backlash tolerances should be as nearly equal as possible at each mesh to ensure equal load distribution.[14]

REFERENCES

1. Coy, J. J.: "Geared Powered Transmission Technology," *Advanced Power Transmission Technology*, G. K. Fischer, ed., NASA CP-2210, pp. 49–77, 1983.

2. "The New Encyclopedia Britannica," 15th ed., Encyclopedia Britannica, Inc., 1974.

3. de Solla Price, Derek: "Gears from the Greeks," Science History Publications, 1975.

4. Drachmann, A. G.: "The Mechanical Technology of Greek and Roman Antiquity," Munksgeard Publisher, Copenhagen, 1963.

5. Dudley, D. W.: "The Evolution of the Gear Art," American Gear Manufacturers Association, 1969.

6. Reti, Ladislao: "Leonardo on Bearings and Gears," *Sci. Am.,* vol. 224, no. 2, pp. 100–110, February 1971.

7. da Vinci, Leonardo (L. Reti, trans.): "The Madrid Codices of Leonardo da Vinci," McGraw-Hill Book Company, Inc., New York, 1974.

8. Lewis, Wilfred: "Investigations of the Strength of Gear Teeth," *Proc. of the Engineers Club of Philadelphia,* Philadelphia, pp. 16–23, 1893.

9. Grant, George B.: "A Treatise on Gear Wheels," 21st ed., Philadelphia Gear Works, Inc., 1980.

10. Buckingham, Earle: "Dynamic Loads on Gear Teeth," *Am. Soc. Mech. Eng.,* 1931.

11. Tucker, A. I.: "Bevel Gears at 203 Meters per Second (40,000 FPM)," ASME paper 77-DET-178, 1977.

12. Haas, L. L.: "Latest Developments in the Manufacture of Large Spiral Bevel and Hypoid Gears," ASME paper 80C2/DET-119, August 1980.

13. Drago, R. J., and F. W. Brown: "The Analytical and Experimental Evolution of Resonant Response in High-Speed, Light-Weight, Highly Loaded Gearing," *J. Mech. Des.,* vol. 103, no. 2, pp. 346–356, April 1981.

14. Crawshaw, S. L., and Kron, H. O.: "Mechanical Drives," *Machine Design,* vol. 39, pp. 18–23, September 21, 1967.

15. "Gears and Gear Drives," *Machine Design,* vol. 55, no. 15, pp. 16–32, June 30, 1983.

16. Kasuba, R.: "A Method for Static and Dynamic Load Analysis of Standard and Modified Spur Gears," *Advanced Power Transmission Technology,* G. K. Fischer, ed., NASA CP-2210, 1983, pp. 403–419.

17. Wildhaber, E.: "Helical Gearing," U.S. Patent 1,601,750, 1926.

18. Wildhaber, E.: "Method of Grinding Gears," U.S. Patent 1,858, 568, 1932.

19. Novikov, M. L.: U.S.S.R. Patent 109,750, 1956.

20. Shotter, B. A.: "The Lynx Transmission and Conformal Gearing," SAE paper 7810141, 1979.

21. Shotter, B. A.: "Experiences with Conformal/W-N Gearing," *Machinery,* pp. 322–326, October 5, 1977.

22. Costomiris, G. H., D. P. Daley, and W. Grube: "Heat Generated in High Power Reduction Gearing," Pratt & Whitney Aircraft Division, PWA-3718, p. 43, June 1969.

23. Fusaro, R. L.: "Effects of Atmosphere and Temperature on Wear, Friction and Transfer of Polyimide Films," *ASLE Trans.,* vol. 21, no. 2, pp. 125–133, April 1978.

24. Fusaro, R. L.: "Tribological Properties at 25°C of Seven Polyimide Films Bonded to 440C High-Temperature Stainless Steel," NASA TP-1944, 1982.

25. Fusaro, R. L.: "Polyimides Tribological Properties and Their Use as Lubricants," NASA TM-82959, 1982.

26. Johansen, K. M.: "Investigation of the Feasibility of Fabricating Bimetallic Coextruded Gears," AFAPL-TR-73-112, (AD-776795), AiResearch Mfg. Co., December 1973.

27. Hirsch, R. A.: "Lightweight Gearbox Development for Propeller Gearbox System Applications Potential Coatings for Titanium Alloy Gears," AFAPL-TR-72-90, (AD-753417), General Motors Corp., December 1972.

28. Delgrosso, E. J., et al.: "Lightweight Gearbox Development for Propeller Gearbox Systems Applications," AFAPL-TR-71-41-PH-1, (AD-729839), Hamilton Standard, August 1971.

29. Manty, B. A., and H. R. Liss: "Wear Resistant Coatings for Titanium Alloys," FR-8400, (AD-A042443), Pratt & Whitney Aircraft, March 1977.

30. Dudley, D. W.: "Gear Handbook, The Design, Manufacture and Applications of Gears," McGraw-Hill Book Company, Inc., New York, 1962.

31. Michalec, G. W.: "Precision Gearing Theory and Practice," John Wiley & Sons, Inc., New York, 1966.

32. "AGMA Standard for Rating the Strength of Spur Gear Teeth," AGMA 220.02, American Gear Manufacturers Association, August 1966.

33. "AGMA Standard for Surface Durability (Pitting) of Spur Gear Teeth," AGMA 210.02, American Gear Manufacturers Association, January 1965.

34. Smith, W. E.: "Ferrous-Based Powder Metallurgy Gears," *Gear Manufacture and Performance,* P. J. Guichelaar, B. S. Levy, and N. M. Parikh, eds., American Society for Metals, pp. 257–269, 1974.

35. Antes, H. W.: "P/M Hot Formed Gears," *Gear Manufacture and Performance,* P. J. Guichelaar, B. S. Levy, and N. M. Parikh, eds., American Society for Metals, pp. 271–292, 1974.

36. Ferguson, B. L., and D. T. Ostberg: "Forging of Powder Metallurgy Gears," TARADCOM-TR-12517, TRW-ER-8037-F, TRW, Inc., May 1980 (AD-A095556).

37. Townsend, D. P., and E. V. Zaretsky: "Effect of Shot Peening on Surface Fatigue Life of Carburized and Hardened AISI 9310 Spur Gears," NASA TP-2047, 1982.

38. Straub, J. C.: "Shot Peening in Gear Design," AGMA Paper 109.13, June 1964.

39. Anderson, N. E., and E. V. Zaretsky: "Short-Term, Hot-Hardness Characteristics of Five Case-Hardened Steels," NASA TN D-8031, 1975.

40. Hingley, C. G., H. E. Southerling, and L. B. Sibley: "Supersonic Transport Lubrication System Investigation" (AL65T038 SAR-1, SKF Industries, Inc.; NASA Contract NAS3-6267), NASA CR-54311, May 1965.

41. Bamberger, E. N., E. V. Zaretsky, and W. J. Anderson: "Fatigue Life of 120-mm-Bore Ball Bearings at 600°F and Fluorocarbon, Polyphenyl Ether and Synthetic Paraffinic Base Lubricants," NASA TN D-4850, 1968.

42. Zaretsky, E. V., W. J. Anderson, and E. N. Bamberger: "Rolling-Element Bearing Life from 400° to 600°F," NASA TN D-5002, 1969.

43. Bamberger, E. N., E. V. Zaretsky, and W. J. Anderson: "Effect of Three Advanced Lubricants on High-Temperature Bearing Life," *J. Lubr. Technol., Trans. ASME,* vol. 92, no. 1, pp. 23–33, January 1970.

44. Bamberger, E. N., and E. V. Zaretsky: "Fatigue Lives at 600°F of 120-Millimeter-Bore Ball Bearings of AISI M-50, AISI M-1 and WB-49 Steels," NASA TN D-6156, 1971.

45. Bamberger, E. N.: "The Development and Production of Thermo-Mechanically Forged Tool Steel Spur Gears," (R73AEG284, General Electric Co.; NASA Contract NAS3-15338), NASA CR-121227, July 1973.

46. Townsend, D. P., E. N. Bamberger, and E. V. Zaretsky: "A Life Study of Ausforged, Standard Forged, and Standard Machined AISI M-50 Spur Gears," *J. Lubr. Technol.,* vol. 98, no. 3, pp. 418–425, July 1976.

47. Bamberger, E. N.: "The Effect of Ausforming on the Rolling Contact Fatigue Life of a Typical Bearing Steel," *J. Lubr. Technol.,* vol. 89, no. 1, pp. 63–75, January 1967.

48. Jatczak, C. F.: "Specialty Carburizing Steels for Elevated Temperature Service," *Met. Prog.,* vol. 113, no. 4, pp. 70–78, April 1978.

49. Townsend, D. P., R. J. Parker, and E. V. Zaretsky: "Evaluation of CBS 600 Carburized Steel as a Gear Material," NASA TP-1390, 1979.

50. Culler, R. A., E. C. Goodman, R. R. Hendrickson, and W. C. Leslie: "Elevated Temperature Fracture Toughness and Fatigue Testing of Steels for Geothermal Applications," TER-RATEK Report No. Tr 81-97, Terra Tek, Inc., October 1981.

51. Roberts, G. A., and J. C. Hamaker: "Strong, Low Carbon Hardenable Alloy Steels," U.S. Patent No. 3,036,912, May 29, 1962.

52. Cunningham, R. J., and W. N. J. Lieberman: "Process for Carburizing High Alloy Steels," U.S. Patent No. 3,885,995, May 27, 1975.

53. Townsend, D. P., and E. V. Zaretsky: "Endurance and Failure Characteristics of Modified Vasco X-2, CBS 600 and AISI 9310 Spur Gears," *J. Mech. Des., Trans. ASME,* vol. 103, no. 2, pp. 506–515, April 1981.

54. Walp, H. O., R. P. Remorenko, and J. V. Porter: "Endurance Tests of Rolling-Contact Bearings of Conventional and High Temperature Steels under Conditions Simulating Aircraft Gas Turbine Applications," (AD-212904) TR 58-392, Wright Air Development Center, Dayton, Ohio, 1959.

55. Baile, G. H., O. G. Gustafsson, H. Mahncke, and L. B. Sibley: "A Synopsis of Active SKF Research and Development Programs of Interest to the Aerospace Industry," AL-63M003, SKF Industries, Inc., King of Prussia, Pa., July 1963.

56. Scott, D.: "Comparative Rolling Contact Fatigue Tests on En 31 Ball Bearing Steels of Recent Manufacture," NFL 69T96, National Engineering Laboratory, Glasgow, Scotland, 1969.

57. "1968 Aero Space Research Report," AL 68Q013, SKF Industries, Inc., King of Prussia, Pa., 1968.

58. Scott, D., and J. Blackwell, J.: "Steel Refining as an Aid to Improved Rolling Bearing Life," NEL Report No. 354 National Eng. Lab.

59. Bamberger, E. N., E. V. Zaretsky, and H. Signer: "Endurance and Failure Characteristic of Main-Shaft Jet Engine Bearing at 3×10^6 DN," *J. Lubr. Technol., Trans. ASME,* vol. 98, no. 4, pp. 580–585, October 1976.

60. Carter, T. L., E. V. Zaretsky, and W. J. Anderson: "Effect of Hardness and Other Mechanical Properties on Rolling-Contact Fatigue Life of Four High-Temperature Bearing Steels," NASA TN D-270, 1960.

61. Jackson, E. G.: "Rolling-Contact Fatigue Evaluations of Bearing Materials and Lubricants," *ASLE Transactions,* vol. 2, no. 1, pp. 121–128, 1959.

62. Baughman, R. A.: "Effect of Hardness, Surface Finish, and Grain Size on Rolling-Contact Fatigue Life of M-50 Bearing Steel," *J. of Basic Engineering,* vol. 82, no. 2, pp. 287–294, June 1960.

63. Zaretsky, E. V., R. J. Parker, W. J. Anderson, and D. W. Reichard: "Bearing Life and Failure Distribution as Affected by Actual Component Differential Hardness," NASA TN D-3101, 1965.

64. Zaretsky, E. V., R. J. Parker, and W. J. Anderson: "Component Hardness Differences and Their Effect on Bearing Fatigue," *J. Lubr. Technol.,* vol. 89, no. 1, January 1967.

65. Anderson, W. J., and T. L. Carter: "Effect of Fiber Orientation, Temperature and Dry Powder Lubricants on Rolling-Contact Fatigue," *ASLE Transactions,* vol. 2, no. 1, pp. 108–120, 1959.

66. Carter, T. L.: "A Study of Some Factors Affecting Rolling Contact Fatigue Life," NASA TR R-60, 1960.

67. Bamberger, E. N.: "The Effect of Ausforming on the Rolling Contact Fatigue Life of a Typical Bearing Steel," *J. Lubr. Technol.,* vol. 89, no. 1, pp. 63–75, January 1967.

68. Bamberger, E. N.: "The Production, Testing, and Evaluation of Ausformed Ball Bearings," Final Engineering Report, AD-637576, June 1966, General Electric Co., Cincinnati, Ohio.

69. Parker, R. J., and E. V. Zaretsky: "Rolling Element Fatigue Life of Ausformed M-50 Steel Balls," NASA TN D-4954, 1968.

70. "AGMA Standard for Rating the Pitting Resistance and Bending Strength of Spur and Helical Involute Gear Teeth," AGMA 218.01, American Gear Manufacturers Association, December 1982.

71. Cardou, A., and G. V. Tordion: "Numerical Implementation of Complex Potentials for Gear Tooth Stress Analysis," *ASME J. Mech. Des.,* vol. 103, no. 2, pp. 460–466, April 1981.

72. Rubenchik, V.: "Boundary-Integral Equation Method Applied to Gear Strength Rating," *ASME Journal of Mechanisms, Transmissions, and Automation in Design,* vol. 105, no. 1, pp. 129–131, March 1983.

73. Alemanni, M., S. Bertoglio, and P. Strona: "B.E.M. in Gear Teeth Stress Analysis: Comparison with Classical Methods," *Proc. of the International Symposium on Gearing and Power Transmissions, Japan Soc. of Mech. Engrs., Tokyo,* vol. 2, pp. 177–182, 1981.

74. Wilcox, L., and W. Coleman: "Application of Finite Elements to the Analysis of Gear Tooth Stresses," *ASME Journal of Engineering for Industry,* vol. 95, no. 4, pp. 1139–1148, November 1973.

75. Wilcox, L. E.: "An Exact Analytical Method for Calculating Stresses in Bevel and Hypoid Gear Teeth," *Proc. of the International Symposium on Gearing and Power Transmissions, Japan Soc. of Mech. Engrs., Tokyo,* vol. 2, pp. 115–121, 1981.

76. Frater, J. L., and R. Kasuba: "Extended Load-Stress Analysis of Spur Gearing: Bending Strength Considerations," *Proc. of the International Symposium on Gearing and Power Transmissions, Japan Soc. of Mech. Engrs., Tokyo,* vol. 2, pp. 147–152, 1981.

77. Coy, J. J., and C. H. Chao: "A Method of Selecting Grid Size to Account for Hertz Deformation in Finite Element Analysis of Spur Gears," *J. Mech. Des., ASME Trans.,* vol. 104, no. 4, pp. 759–766, 1982.

78. Drago, R. J.: "Results of an Experimental Program Utilized to Verify a New Gear Tooth Strength Analysis," AGMA technical paper P229.27, October 1983.

79. Drago, R. J., and R. V. Lutthans: "An Experimental Investigation of the Combined Effects of Rim Thickness and Pitch Diameter on Spur Gear Tooth Root and Fillet Stresses," AGMA technical paper P229.22, October 1981.

80. Chang, S. H., R. L. Huston, and J. J. Coy: "A Finite Element Stress Analysis of Spur Gears Including Fillet Radii and Rim Thickness Effects," *ASME Trans., Journal of Mechanisms, Transmissions, and Automation in Design,* vol. 105, no. 3, pp. 327–330, 1983.

81. Drago, R. J.: "An Improvement in the Conventional Analysis of Gear Tooth Bending Fatigue Strength," AGMA technical paper P229.24, October 1982.

82. Chong, T. H., T. Aida, and H. Fujio: "Bending Stresses of Internal Spur Gear," *Bulletin of the JSME,* vol. 25, no. 202, pp. 679–686, April 1982.

83. Wang, K. L., and H. S. Cheng: "A Numerical Solution to the Dynamic Load, Film Thickness, and Surface Temperatures in Spur Gears, Part I—Analysis," *J. Mech. Des.,* vol. 103, pp. 177–187, January 1981.

84. Wang, K. L., and H. S. Cheng: "A Numerical Solution to the Dynamic Load, Film Thickness, and Surface Temperatures in Spur Gears, Part II—Results," *J. Mech. Des.,* vol. 103, pp. 188–194, January 1981.

85. Dolan, T. J., and E. L. Broghamer: "A Photoelastic Study of the Stresses in Gear Tooth Fillets," University of Illinois Engineering Experiment Station, Bulletin No. 335, 1942.

86. Cornell, R. W.: "Compliance and Stress Sensitivity of Spur Gear Teeth," *J. Mech. Des.,* vol. 103, no. 2, pp. 447–459, April 1981.

87. Coy, J. J., D. P. Townsend, and E. V. Zaretsky: "Analysis of Dynamic Capacity of Low-Contact-Ratio Spur Gears Using Lundberg-Palmgren Theory," NASA TN D-8029, 1975.

88. Coy, J. J., and E. V. Zaretsky: "Life Analysis of Helical Gear Sets Using Lundberg-Palmgren Theory," NASA TN D-8045, 1975.

89. Coy, J. J., D. P. Townsend, and E. V. Zaretsky: "Dynamic Capacity and Surface Fatigue Life for Spur and Helical Gears," *ASME Trans., J. Lubr. Technol.,* vol. 98, no. 2, pp. 267–276, April 1976.

90. Townsend, D. P., J. J. Coy, and E. V. Zaretsky: "Experimental and Analytical Load-Life Relation for AISI 9310 Steel Spur Gears," *J. Mech. Des.,* vol. 100, no. 1, pp. 54–60, January 1978. (Also NASA TM X-73590, 1977.)

91. Coy, J. J., D. P. Townsend, and E. V. Zaretsky: "An Update on the Life Analysis of Spur Gears," *Advanced Power Transmission Technology,* NASA CP 2210, AVRADCOM TR 82-C-16, pp. 421–433, 1983.

92. Lundberg, G., and A. Palmgren: "Dynamic Capacity of Rolling Bearings," *Acta Polytech., Mech. Eng. Ser.,* vol. 1, no. 3, 1947.

93. Bamberger, E. N., et al.: "Life Adjustment Factors for Ball and Roller Bearings," *An Engineering Design Guide,* American Society of Mechanical Engineers, 1971.

94. Savage, M., C. A. Paridon, and J. J. Coy: "Reliability Model for Planetary Gear Trains," *ASME Trans., J. Mechanisms, Transmissions, and Automation in Design,* vol. 105, no. 5, pp. 291–297, 1983.

95. Krenzer, T. J., and R. Knebel: "Computer-Aided Inspection of Bevel and Hypoid Gears," SAE Paper 831266, 1983.

96. Litvin, F. L., and Y. Gutman: "Methods of Synthesis and Analysis for Hypoid Gear-Drives of 'Formate' and 'Helixform,'" *ASME Trans., J. Mech. Des.,* vol. 103, no. 1, pp. 83–113, January 1981.

97. Litvin, F. L., and Y. Gutman: "A Method of Local Synthesis of Gears Grounded on the Connections Between the Principal and Geodetic Curvatures of Surfaces," *ASME Trans., J. Mech. Des.,* vol. 106, no. 1, pp. 114–125, January 1981.

98. Coy, J. J., D. A. Rohn, and S. H. Loewenthal: "Constrained Fatigue Life Optimization for a Nasvytis Multiroller Traction Drive," *ASME Trans., J. Mech. Des.,* vol. 103, no. 2, pp. 423–429, April 1981.

99. Coleman, W.: Bevel and Hypoid Gear Surface Durability: Pitting and Scuffing, "Source Book on Gear Design, Technology and Performance," M. A. H. Howes, ed., Am. Soc., for Metals, Metals Park, Ohio 44073, pp. 243–258, 1980.

100. Anderson, W. J., and E. V. Zaretsky: "Rolling-Element Bearings," *Machine Design,* vol. 42, no. 15, pp. 21–37, June 18, 1970.

101. Ertel, A. M.: "Hydrodynamic Lubrication Based on New Principles," *Prikl. Mat. Melch.,* vol. 3, no. 2, 1939 (in Russian).

102. Grubin, A. N.: "Fundamentals of the Hydrodynamic Theory of Lubrication of Heavily Loaded Cylindrical Surfaces," "Investigation of the Contact of Machine Components," Kh. F. Ketova, ed., translation of Russian Book No. 30, Central Scientific Institute for Technology and Mechanical Engineering, Moscow, 1949, chap. 2. (Available from Dept. of Scientific and Industrial Research, Great Britain, Transl. CTS-235 and Special Libraries Assoc., Transl. R-3554.)

103. Hamrock, B. J., and D. Dowson: "Minimum Film Thickness in Elliptical Contacts for Different Regimes of Fluid-Film Lubrication," NASA TP 1342, October 1978.

104. Fein, R. S.: "Chemistry in Concentrated-Conjunction Lubrication," *Interdisciplinary Approach to the Lubrication of Concentrated Contacts,* SP-237, NASA, Washington, D.C., pp. 489–527, 1970.

105. Godfrey, D.: "Boundary Lubrication," *Interdisciplinary Approach to Friction and Wear,* SP-181, NASA, Washington, D.C., pp. 335–384, 1968.

106. Bowden, F. P., and Tabor, D: "The Friction and Lubrication of Solids," vol. 2, Clarendon Press, Oxford, 1964.

107. Borsoff, V. A.: "Fundamentals of Gear Lubrication," Annual Report, Shell Development Co., Emeryville, Calif. [Work under Contract NOa(s) 53-356-c], June 1955.

108. Borsoff, V. N., and R. Lulwack: "Fundamentals of Gear Lubrication," Final Report, Shell Development Co., Emeryville, Calif. [Work under Contract NOa(s) 53-356-c], June 1957.

109. Horlick, E. J., and D. E. O'D. Thomas: "Recent Experiences in the Lubrication of Naval Gearing," *Gear Lubrication Symposium,* Institute of Petroleum, 1966.

110. Boner, C. J.: "Gear and Transmission Lubricant," Reinhold Publishing Company, New York, 1964.

111. Townsend, D. P., and L. S. Akin: "Study of Lubricant Jet Flow Phenomena in Spur Gears— Out of Mesh Condition," Advanced Power Transmission Technology, NASA CP-2210, G. K. Fischer, ed., pp. 461–476, 1983.

112. Townsend, D. P., and L. S. Akin: "Gear Lubrication and Cooling Experiment and Analysis," Advanced Power Transmission Technology, NASA CP-2210, G. K. Fischer, ed., pp. 477–490, 1983.

113. Akin, L. S., J. J. Mross, and D. P. Townsend: "Study of Lubricant Jet Flow Phenomena in Spur Gears," *ASME Trans., J. Lubr. Technol.,* vol. 97, no. 2, pp. 283–288, 295, April 1975.

114. Townsend, D. P., and L. S. Akin: "Study of Lubricant Jet Flow Phenomena in Spur Gears—Out of Mesh Condition," *ASME Trans., J. Mech. Des.,* vol. 100, no. 1, pp. 61–68, January 1978.

115. Townsend, D. P., and L. S. Akin: "Analytical and Experimental Spur Gear Tooth Temperature as Affected by Operating Variables," *ASME Trans., J. Mech. Des.,* vol. 103, no. 1, pp. 219–226, January 1981.

116. Akin, L. S., and D. P. Townsend: "Into Mesh Lubrication of Spur Gears with Arbitrary Offset Oil Jet. Part 1: For Jet Velocity Less Than or Equal to Gear Velocity," *ASME Trans., J. Mech. Trans. and Automation in Design,* vol. 105, no. 4, pp. 713–718, December 1983.

117. Akin, L. S., and D. P. Townsend: "Into Mesh Lubrication of Spur Gears with Arbitrary Offset Oil Jet. Part 2: For Jet Velocities Equal to or Greater Than Gear Velocity," *ASME Trans., J. Mech. Trans. and Automation in Design,* vol. 105, no. 4, pp. 719–724, December 1983.

118. Patir, N., and H. S. Cheng: "Prediction of Bulk Temperature in Spur Gears Based on Finite Element Temperature Analysis," *ASLE Trans.,* vol. 22, no. 1, pp. 25–36, January 1979.

119. Anderson, N. E., and S. H. Loewenthal: "Spur-Gear-System Efficiency at Part and Full Load," NASA TP-1622, 1980.

120. Anderson, N. E., and S. H. Loewenthal: "Effect of Geometry and Operating Conditions on Spur Gear System Power Loss," *ASME Trans., J. Mech. Des.,* vol. 103, no. 1, pp. 151–159, January 1981.

121. Anderson, N. E., and S. H. Loewenthal: "Design of Spur Gears for Improved Efficiency," *ASME Trans., J. Mech. Des.,* vol. 104, no. 3, pp. 283–292, October 1982.

122. Anderson, N. E., and S. H. Loewenthal: "Selecting Spur-Gear Geometry for Greater Efficiency," *Machine Design,* vol. 55, no. 10, pp. 101–105, May 12, 1983.

123. Savage, M., J. J. Coy, and D. P. Townsend: "Optimal Tooth Numbers for Compact Standard Spur Gear Sets," *ASME Trans., J. Mech. Des.,* vol. 104, no. 4, pp. 749–758, October 1982.

124. Savage, M., J. J. Coy, and D. P. Townsend: "The Optimal Design of Standard Gearsets," Advanced Power Transmission Technology, NASA CP-2210, G. K. Fischer, ed., pp. 435–469, 1983.

125. Lynwander, P.: "Gear Tooth Scoring Design Considerations," AGMA paper P219.10, October 1981.

126. Savage, M., J. J. Coy, and D. P. Townsend: "The Optimal Design of Involute Gear Teeth with Unequal Addenda," NASA TM-82866, AVRADCOM TR 82-C-7, 1982.

127. Carroll, R. K., and G. E. Johnson: "Optimal Design of Compact Spur Gear Sets," ASME paper 83-DET-33, 1983.

128. Wadlington, R. P.: "Gear Drives," *Machine Design,* vol. 39, pp. 24–26, September 21, 1967.

129. DeBruyne, Neil A.: "Design and Development Testing of Free Planet Transmission Concept," USAAMRDL-TR-74-27, Army Air Mobility Research and Development Laboratory, 1974 (AD-782857).

130. Folenta, D. J.: "Design Study of Self-Aligning Bearingless Planetary Gear (SABP)," *Advanced Power Transmission Technology,* G. K. Fischer, ed., NASA CP-2210, 1983, pp. 161–172.

131. Nasvytis, A. L.: "Multiroller Planetary Friction Drives," SAE Paper 660763, October 1966.

132. White, G.: "New Family of High-Ratio Reduction Gears with Multiple Drive Paths," *Proc. Inst. Mech. Eng. (London),* vol. 188, no. 23/74, pp. 281–288, 1974.

133. Lastine, J. L., and G. White: "Advanced Technology VTOL Drive Train Configuration Study," USAAVLABS-TR-69-69, Army Aviation Material Laboratories, 1970 (AD-867905).

134. White, G.: "Helicopter Transmission Arrangements with Split-Torque Gear Trans," *Advanced Power Transmission Technology,* G. K. Fischer, ed., NASA CP-2210, pp. 141–150, 1983.

135. Buckingham, E.: "Analytical Mechanics of Gears," Dover Publications, New York, 1963.

136. Boner, C. J.: "Gear and Transmission Lubricants," Reinhold Publishing, New York, 1964.

137. Deutschman, A., W. J. Michels, and C. E. Wilson: "Machine Design," The Macmillan Company, Inc., New York, 1975, pp. 519–659.

138. Dudley, D. W.: "Practical Gear Design," McGraw-Hill Book Company, Inc., New York, 1954.

139. Dudley, D. W.: "Gear Handbook," McGraw-Hill Book Company, Inc., New York, 1962.

140. Khiralla, T. W.: "On the Geometry of External Involute Spur Gears," T. W. Khiralla, Studio City, Calif., 1976.

141. Michalec, G. W.: "Precision Gearing: Theory and Practice," John Wiley & Sons, Inc., New York, 1966.

142. "Mechanical Drives Reference Issue," *Machine Design,* 3d ed., vol. 39, no. 22, September 21, 1967.

SECTION 28

SPRINGS*

Abilio A. Relvas, B.A., B.S.M.E.

Manager—Technical Assistance
Associated Spring
Barnes Group Inc.
Bristol, Conn.

---

*This section is based on material copyrighted by Associated Spring, Barnes Group Inc.

28.1 INTRODUCTION

A spring is a device which stores energy or provides a force over a distance by elastic deflection. Energy may be stored in a compressed gas or liquid, or in a solid that is bent, twisted, stretched, or compressed. The energy is recoverable by the elastic return of the distorted material. Spring structures are characterized by their ability to withstand relatively large deflections elastically. The energy stored is proportional to the volume of elastically distorted material, and in the case of metal springs, the volume of distorted material is limited by the spring configuration and the stress-carrying capacity (elastic limit) of the most highly strained portion. For this reason, most commercial spring materials come from the group of high-strength materials, including high-carbon steel, cold-rolled and precipitation-hardening stainless and nonferrous alloys, and a few specialized nonmetallics such as laminated fiberglass. Thus spring design is basically an analysis of machine elements which undergo relatively large elastic deflections and are made of materials capable of withstanding high stress levels without yielding.

Symbols.   Following are the letter symbols for equations and their corresponding units. Where a symbol can have more than one interpretation its meaning is made explicit in each section where it is used.

b = breadth or width, in

C = spring index, dimensionless

d = wire or bar diameter, in

D = diameter, in

E = modulus of elasticity, lb/in^2

f = frequency of vibration, Hz (cycles per second)

g = gravitational constant, 386.4 in/s^2

G = modulus of elasticity in shear, lb/in^2

h = height, in

k = rate, lb/in or in·lb/rad

K = factor, used variously

ln = natural logarithm

L = length, in

M = moment of force, in·lb

N = number of active coils in helically wound springs, dimensionless

p = pitch, in

P = force, lb

r = radius, in

R = radius, in

t = thickness, in

T = number of turns, dimensionless

U = energy, in·lb

v = velocity, in/s

V = volume, in^3

δ = linear deflection, in

θ = angular deflection, rad

μ = Poisson's ratio, dimensionless

γ = density, lb/in^3

σ = normal stress, lb/in^2

τ = shear stress, lb/in^2

ϕ = angle, °

The following terms are peculiar to spring engineering. Definitions of general terms are not given.

Rate (see Sec. 28.2): Also referred to as scale, gradient, and load factor.

Spring index: Applies to helical-type springs. For springs wound with circular wire, spring index $C = D/d$; for springs wound with rectangular wire $C = D/t$ if t is in the radial direction, or $C = D/b$ if b is in the radial direction. See Figs. 28.7 and 28.11b.

Active coils: Coils free to deflect under load.

Inactive coils: Coils not free to deflect under load.

Total coils: The sum of active and inactive coils.

Free-length or height: A length or height measurement of a spring in the free (unloaded) condition.

Solid length or height: A length or height measurement of a compression spring when no further deflection is possible.

Solid stress: The stress corresponding to deflection to solid length or height.

Set: A permanent distortion resulting when a spring is stressed beyond the elastic limit.

Initial tension: The force holding coils closed in the free or unloaded condition of helical extension or torsion springs.

Sizes: For economy and availability, preferred sizes for spring steel wire and strip should be used. Sizes are usually specified in thousandths of an inch.

28.2 SPRING CHARACTERISTICS

The load-deflection curve is the most fundamental of all spring properties. For a load-deflection curve of the form $P = f(\delta)$ the force P will cause a linear deflection δ in the direction of P; when the load is a moment of force M, the load-deflection curve has the form $M = f(\theta)$ and M is in the direction of θ. Four types of load-deflection curve are represented in Fig. 28.1a to d.

28.2.1 Rate

By definition, rate is the slope of the load-deflection curve

$$k = dP/d\delta \quad \text{or} \quad k = dM/d\theta \tag{28.1}$$

When the load-deflection curve is linear, rate is constant so that

$$k = \Delta P/\Delta\delta \quad \text{or} \quad k = \Delta M/\Delta\theta \tag{28.2}$$

For a linear load-deflection curve with no initial load, Eq. (28.2) reduces to

$$k = P/\delta \quad \text{or} \quad k = M/\theta \tag{28.3}$$

28.2.2 Operating Range

Many springs operate over only a part of their useful load-deflection curve. The operating range is defined by the limits of load and deflection as shown in Fig. 28.1e.

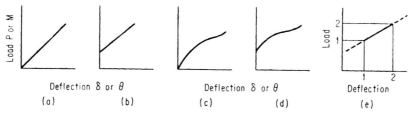

FIG. 28.1 Types of load-deflection curve (a) Linear, no initial load. (b) Linear, initial load. (c) Nonlinear, no initial load. (d) Nonlinear, initial load. (e) Operating range.

28.2.3 Energy Storage

The energy stored by deflecting a spring over its operating range,

$$U = \int_{\delta_1}^{\delta_2} P \, d\delta \qquad \text{or} \qquad U = \int_{\theta_1}^{\theta_2} M \, d\theta \qquad (28.4)$$

For any linear load-deflection curve,

$$U = (P_2 + P_1)(\delta_2 - \delta_1)/2 \qquad \text{or} \qquad U = (M_2 + M_1)(\theta_2 - \theta_1)/2 \qquad (28.5)$$

If in addition to the load-deflection curve being linear, the initial load is zero, Eq. (28.5) reduces to

$$U = P\delta/2 \qquad \text{or} \qquad U = M\theta/2 \qquad (28.6)$$

28.2.4 Stress

Values of maximum allowable load are determined by equations relating stress and load. Calculated stresses differ from true stresses because of the approximate nature of the stress formulas, because these formulas do not take into account residual stresses induced in manufacturing, and because of the difficulty in evaluating stress concentrations. For these reasons, limits for calculated stresses are not always based merely on the permissible stress values for spring materials as determined by standard tests. Rather, the amount by which calculated stresses can approach permissible values for spring materials often depends on factors devised from tests.

Residual stresses exist. They may be either beneficial or detrimental, depending upon the direction of the applied service stresses. Favorable residual stresses may be deliberately added to the system already present from the cold-forming operations.

As a general rule, plastic flow caused by forming in the same direction as in service will result in a favorable residual-stress system, one in which the pertinent residual stresses are opposite in sign to the service stresses. These residual stresses may be both tension and shear, and calculations must be made on a combined-stress basis. For example, a compression spring will be able to support a higher load elastically if it has been previously set, removed, or overloaded into the plastic region by a compression load. If, however, it has been stretched out previously (to meet a blueprint dimension), the residual-stress system is unfavorable, and the elastic limit in compression will be below expectations unless the spring is again stress-relieved. Similarly, if a flat spring is formed by bending pretempered material, the residual stresses will be favorable to loads tending to close up the bend and unfavorable to loads opening the bend. A cold-wound torsion spring is best stressed when its coils tend to wind down onto an arbor and unfavorably stressed when the coils unwind.

For best spring design it is thus necessary to predict the type and direction of the residual stresses and add them to the service stresses. In extreme cases of unfavorable residual stresses, hardening the spring after forming may be necessary. However, it is always best to attempt to design so that residual stresses can be utilized effectively; if not, the spring manufacturer must have the necessary design information so that he can stress-relieve the springs at the highest possible temperature.

The simple spring formulas do not take into account the residual-stress system which is always present in cold-formed springs. Most stress calculations are therefore directed toward determining relative rather than absolute stress figures. Charts of allowable design stresses take into account residual stresses.

Conditions of stress concentrations introduced by notches, holes, and curvature not otherwise included in stress equations should be included in stress calculations.

28.2.5 Energy Storage Capacity

The energy storage capacity (ESC)[2] of a spring is the amount of energy that can be stored per unit volume of active spring material when the spring is loaded so that stress is the maximum allowable. For springs with load-deflection curves of the type shown in Fig. 28.1a and subject to a static load, the resilience

$$U/V = K(\max \sigma)^2/E \qquad \text{or} \qquad U/V = K(\max \tau)^2/G \qquad (28.7)$$

V is the volume of active spring material. Values of K for common spring types are given in Table 28.1; K is an index of volume efficiency.

TABLE 28.1 Values of K

Spring type	K
1. Helical extension and compression, round wire:	
$\qquad$ C-3	1/5.4
$\qquad$ C-6	1/4.7
$\qquad$ C-9	1/4.5
$\qquad$ C-12	1/4.3
2. Helical torsion, round wire	$\frac{1}{8}$
3. Helical torsion, rectangular wire	$\frac{1}{6}$
4. Cantilever leaf, uniform rectangular section	$\frac{1}{18}$
5. Triangular cantilever leaf, rectangular section	$\frac{1}{6}$
6. Spiral spring, rectangular section	$\frac{1}{6}$
7. Torsion bar, circular section	$\frac{1}{4}$
8. Belleville spring	$\frac{1}{40}$ to $\frac{1}{10}$

28.3 SPRINGS IN COMBINATION

Springs are often used in combination because of space limitations, because a combination may be more efficient than a single equivalent spring or because a combination will give a load-deflection curve or dynamic characteristics not possible with a single spring of standard design.

For simplicity, the examples shown are of combinations having the fewest number of springs; these examples can be easily generalized to apply to larger combinations.

28.3.1 Series and Parallel Combinations

Figure 28.2 shows schematically how springs are combined in parallel and series. For the parallel combination,

$$P_{12} = P_1 + P_2 \qquad \delta_{12} = \delta_1 = \delta_2 \qquad k_{12} = k_1 + k_2 \qquad (28.8)$$

For the series combination,

$$P_{12} = P_1 = P_2 \qquad \delta_{12} = \delta_1 + \delta_2 \qquad 1/k_{12} = 1/k_1 + 1/k_2 \qquad (28.9)$$

A parallel-series combination is shown in Fig. 28.2c.

When the operating range of a combination is within the operating range of each component spring, the resulting load-deflection curve will not have any abrupt changes. However, for springs combined as in Fig. 28.3, the load-deflection curve will

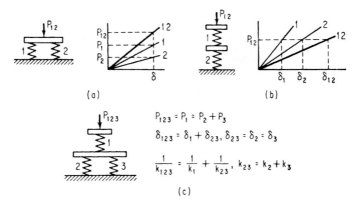

FIG. 28.2 Spring combination. (*a*) Parallel springs, linear load deflection. $P = 0$ where $\delta = 0$. (*b*) Series springs, linear load deflection. $P = 0$ where $\delta = 0$. (*c*) Springs in series-parallel combinations.

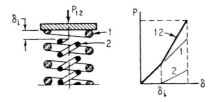

FIG. 28.3 Nest of helical springs. Inner spring shorter than outer results in abrupt change in load-deflection curve. Note that springs are wound opposite in hand to prevent tangling.

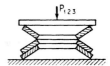

FIG. 28.4 Nest of Belleville springs.

change abruptly when the inner spring begins to deflect. Similarly, the load-deflection curve for the combination of Belleville springs shown in Fig. 28.4 will change abruptly when the single spring has deflected to the flat condition.

28.3.2 Energy Storage

The energy stored within the operating range of a combination of springs may be determined from the load-deflection characteristics of the combination or by adding together the energy stored by the separate springs.

28.4 SPRING MATERIALS AND OPERATING CONDITIONS

In this section properties of some common spring materials are given. Data applicable only to specific types of springs are given in the sections devoted to these springs. For further data consult Ref. 5.

28.4.1 Common Spring Materials

Designations for and properties of typical common spring materials are given in Table 28.2.

28.4.2 Strength of Spring Wire and Strip

Figure 28.5 shows the dependence of tensile strength on wire diameter. Table 28.3 gives recommended working stresses for flat-spring materials in static applications. Data apply to springs subject to static and low-cycle duty at normal temperatures. Recommendations as to the use of the data are given under specific spring types.

28.4.3 Cyclic Loading

Fatigue failures usually start at a surface defect or point of stress concentration. Metals hardened by heat treatment show typical fatigue fractures. Some hard-drawn wires develop long longitudinal cracks, which are often mistaken for metallurgical seams; hence tests of springs made from hard-drawn wire can be misleading if not properly monitored, as a crack can grow over several hundred thousand cycles without external sign of failure.

It is difficult to apply to springs much of the data obtained from the usual reverse-bending and torsional-fatigue tests. There are numerous reasons for this difficulty. The chief reasons are: residual stresses are inherent in springs but not found in test specimens; the hardness at which metals are used in springs makes them more sensitive to stress concentrations than the same metals fatigue-tested at lower hardnesses; most springs are not subject to complete stress reversals as is the case with test specimens.

The difference between stresses at the extremes of the operating range is called the *stress range*. Allowable stress ranges must be evaluated by using Goodman or Soderberg diagrams, *SN* curves, etc. Because of the variables involved in cyclic loading, the number of cycles to failure between similar springs may differ by a factor of 5 or more.

Springs in cyclic service may also fail by setting, particularly when highly stressed, as small amounts of plastic flow undetectable in a single cycle can accumulate over many operations.

28.4.4 Selection of Materials

When fatigue resistance is important and wire sizes are small, use:

1. Music wire where space is limited and stresses are high
2. Stainless steel when corrosive conditions exist or temperatures are as high as 600°F
3. Oil-tempered or hard-drawn wire when space is available and costs must be kept low
4. Chrome-vanadium steel when temperature approaches 400°F
5. Phosphor bronze when electrical conductivity is important

When fatigue resistance is important and wire sizes are above $\frac{1}{8}$ in, use:

TABLE 28.2 Typical Properties of Common Spring Materials

Common name, specification	E, 10^6 lb/in^2	G, 10^6 lb/in^2	Density γ, lb/in^3	Max service temp., °F	Electrical conductivity*	Principal characteristics
			High-carbon steels			
1. Music wire, ASTM A228	30	11.5	0.283	250	7	High strength, excellent fatigue life
2. Hard drawn, ASTM A227	30	11.5	0.283	250	7	General-purpose use, poor fatigue life
3. Valve spring, ASTM A230	30	11.5	0.283	300	7	For excellent fatigue life at average stress range
4. Oil tempered, ASTM A229	30	11.5	0.283	300	7	For moderate stress; low impact or shock
			Alloy steels			
5. Chrome vanadium, ASTM A232	30	11.5	0.283	425	7	High strength, resists shock, good
6. Chrome silicon, ASTM A401	30	11.5	0.283	475	5	fatigue life. ASTM A59 used as less
7. Silicon manganese, ASTM A59	30	11.5	0.283	450	4.5	expensive substitute for ASTM A232

Stainless steels

8. Martensitic, AISI 410, 420	29	11	0.280	500	2.5	Unsatisfactory for subzero applications
9. Austenitic, AISI 301, 302	28	10	0.282	600	2	Good strength at moderate temperatures, low stress relaxation
10. Precipitation hardening, 17-7PH	29.5	11	0.286	600	2	Long life under extreme service conditions
11. High temperature, A286	29	10.4	0.290	950	2	For elevated-temperature applications

Copper-base alloys

12. Spring brass, ASTM B134	16	6	0.308	200	17	Low cost, high conductivity; poor mechanical properties
13. Phosphor bronze, ASTM B159	15	6.3	0.320	200	18	Able to withstand repeated flexures. Popular alloy
14. Silicon bronze, ASTM B99	15	6.4	0.308	200	7	Used as less expensive substitute for phosphor bronze
15. Beryllium copper, ASTM B197	19	6.5	0.297	400	21	High elastic and fatigue strength; hardenable

Nickel-base alloys

16. Inconel 600	31	11	0.307	600	1.5	Good strength, high corrosion resistance
17. Inconel X-750	31	11	0.298	1100	1	Precipitation hardening; for high temperatures
18. Ni-Span C	27	9.6	0.294	200	1.6	Constant modulus over a wide temperature range

*Electrical conductivity given as a percent of the conductivity of the International Annealed Copper Standard.

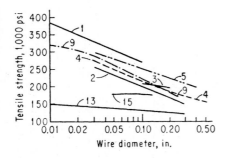

FIG. 28.5 Tensile strength of spring wires vs. diameter. Numbers correspond to listing in Table 28.2. (1) Music wire ASTM A228. (2) Hard-drawn ASTM A227. (3) Valve spring ASTM A230. (4) Oil-tempered ASTM A229. (5) Chrome vanadium ASTM A232. (9) Stainless steel AISI 302. (13) Phosphor bronze ASTM B159. (15) Beryllium copper ASTM B197.

TABLE 28.3 Recommended Working Stresses for Flat-Spring Materials in Static Applications[3]

Material, specifications	1000 lb/in² σ
Stainless steel, AISI 302, hard	130
Stainless steel 17-7 PH, hardened	160
Phosphor bronze (521),* spring temper	70
Beryllium copper (172),* ½ hard	90
Beryllium copper (172),* hard	110
Inconel 600 spring temper	90
Inconel X-750 at 500°F	100
At 700°F	90
At 900°F	75

*Numbers are of the Copper Development Association Inc. standard alloys.

1. Valve-spring wire for critical applications
2. Oil-tempered or hard-drawn wire when space is available and costs must be kept low
3. Chrome-vanadium steel when temperatures approach 400°F
4. Chrome-silicon steel when temperatures are as high as 500°F
5. Stainless steel when corrosive conditions exist
6. Phosphor bronze when electrical conductivity is important

When resistance to set in static load is important, use:

1. Music wire where space is limited and wire size small
2. Oil-tempered wire, lower stress than music wire
3. Hard-drawn wire where low cost is essential
4. Chrome-vanadium steel when temperatures are as high as 400°F
5. Chrome-silicon steel when temperatures are as high as 500°F
6. Stainless steel for corrosion resistance and temperatures approaching 600°F
7. Inconel, A286 for temperatures 500 to 800°F
8. Inconel X for temperatures above 800°F
9. Beryllium copper when electrical conductivity is important and stresses are high

When fatigue resistance or high resistance to setting is important, use (flat springs):

1. AISI 1095 for high-duty applications
2. AISI 1075 for general application
3. Stainless steel for corrosive conditions
4. Phosphor bronze for electrical conductivity
5. Beryllium copper for electrical conductivity and shapes that demand forming while soft followed by hardening

Formability requirements may dictate the need to use annealed materials and hardening and tempering after forming. It would not apply to 300 stainless steel and phosphor bronze. When low cost is essential, use:

1. AISI 1050 for low-cost clips where volume is substantial
2. AISI 1065
3. Brass for decorative purposes

28.4.5 Tolerances

Allowable variations in commercial spring wire sizes are given in Table 28.4. Standard commercial tolerances for coil diameter, free length, and load are given in Tables 28.5 through 28.7. Tolerances on squareness is 3°.

28.4.6 Springs at Reduced and Elevated Temperatures

The mechanical properties of spring metals vary with temperature. For most spring metals E and G decrease almost linearly with temperature within their useful ranges of application. One exception is the family of nickel-base alloys which includes Elinvar and Ni-Span C whose composition is specifically designed to confer a near zero temperature coefficient of spring rate over a limited temperature range around room temperature. This anomalous behavior is caused by a gradual and reversible loss of ferromagnetic effect as the material is raised in temperature toward its Curie point. Once the Curie point is neared (at temperatures above 200°F), the temperature coefficient of rate is comparable with that of other spring alloys.

Reduced Temperature. The body-centered-cubic materials such as spring steel are normally considered to be brittle at subzero temperatures, and the plain-carbon and low-alloy steels are not generally recommended for such service. However, a spring structure is designed to be resilient, and thus the spring configuration may protect the spring material from brittle failure. In special cases of low-temperature application where a large quantity of springs is required and the cost must be low, ordinary spring materials should not be disregarded without a service test. Otherwise, the face-centered-cubic materials, such as copper-base, nickel-base, or austenitic iron-base alloys, should be specified.

Elevated Temperature. Springs used at temperatures above room temperature must be designed with allowance for the rate change associated with modulus decrease, if accuracy of rate is essential.

TABLE 28.4   Allowable Variations in Commercial Spring-Wire Sizes

Material	ASTM Specification No.	Wire diam, in.	Permissible variation, in.
Hard-drawn spring wire	A227	0.028–0.072	±0.001
Chrome-vanadium spring wire	A231	0.073–0.375	±0.002
Oil-tempered wire	A229	0.376 and over	±0.003
Music wire	A228	0.026 and under	±0.0003
		0.027–0.063	±0.0005
		0.064 and over	±0.001
Carbon-steel valve spring wire	A230	0.093–0.148	±0.001
		0.149–0.177	±0.0015
Chrome-vanadium valve spring wire	A232	0.178–0.250	±0.002

TABLE 28.5   Coil-Diameter Tolerances, Helical Compression and Extension Springs, ± in

Wire diameter, in	Spring index D/d						
	4	6	8	10	12	14	16
0.015	0.002	0.002	0.003	0.004	0.005	0.006	0.007
0.023	0.002	0.003	0.004	0.006	0.007	0.008	0.010
0.035	0.002	0.004	0.006	0.007	0.009	0.011	0.013
0.051	0.003	0.005	0.007	0.010	0.012	0.015	0.017
0.076	0.004	0.007	0.010	0.013	0.016	0.019	0.022
0.114	0.006	0.009	0.013	0.018	0.021	0.025	0.029
0.171	0.008	0.012	0.017	0.023	0.028	0.033	0.038
0.250	0.011	0.015	0.021	0.028	0.035	0.042	0.049
0.375	0.016	0.020	0.026	0.037	0.046	0.054	0.064
0.500	0.021	0.030	0.040	0.062	0.080	0.100	0.125

TABLE 28.6   Free-Length Tolerances, Closed-and-Ground Helical Compression Springs, ± in/in of Free Length

Number of active coils per inch	Spring index D/d						
	4	6	8	10	12	14	16
0.5	0.010	0.011	0.012	0.013	0.015	0.016	0.016
1	0.011	0.013	0.015	0.016	0.017	0.018	0.019
2	0.013	0.015	0.017	0.019	0.020	0.022	0.023
4	0.016	0.018	0.021	0.023	0.024	0.026	0.027
8	0.019	0.022	0.024	0.026	0.028	0.030	0.032
12	0.021	0.024	0.027	0.030	0.032	0.034	0.036
16	0.022	0.026	0.029	0.032	0.034	0.036	0.038
20	0.023	0.027	0.031	0.034	0.036	0.038	0.040

For springs less than ½ in long, use the tolerances for ½ in.
For closed ends not ground, multiply above values by 1.7.

TABLE 28.7 Load Tolerances, Helical Compression Springs, ± Percent of Load (Start with Tolerances from Table 28.6 Multiplied by Free Length)

Length tolerance, ± in	Deflection from free length to load, in														
	0.050	0.100	0.150	0.200	0.250	0.300	0.400	0.500	0.750	1.000	1.500	2.000	3.000	4.000	6.000
0.005	12	7	6	5	6.5	5.5	5								
0.010		12	8.5	7	10	8.5	7	6							
0.020		22	15.5	12	14	12	9.5	8	5						
0.030			22	17	18	15.5	12	10	6	5					
0.040				22	22	19	14.5	12	7.5	6	5				
0.050					25	22	17	14	9	7	5.5	5			
0.060						25	19.5	16	10	8	6	5.5			
0.070							22	18	11	9	6.5	6	5		
0.080							25	20	12.5	10	7.5	6	5		
0.090								22	14	11	8	7	5.5		
0.100									15.5	12	8.5	12	8.5		
0.200										22	15.5	17	12	7	5.5
0.300											22	21	15	9.5	7
0.400												25	18.5	12	8.5
0.500														14.5	10.5

First load test at not less than 15% of available deflection. Final load test at not more than 85% of available deflection.

28.13

The type of failure characteristic of springs at elevated temperature is setting or loss of load. The setting of a spring is the result of two types of flow—plastic and anelastic. Plastic flow of the spring material is simply the result of exceeding the elastic limit or yield point at the service temperature. This level limits the allowable stress at any given temperature, regardless of length of time under load. The loss of load caused by plastic flow is not recoverable.

In addition, a spring under load for a period of time will lose more load or set more, by a process called *anelastic flow.* This flow is time- and temperature-dependent and may be recoverable when the load is removed. The net effect in spring service is that a spring will lose load or relax in service at a rate which is stress-, time-, and temperature-dependent. If the spring is unloaded at temperature, the load loss may be partially recoverable. The complete relaxation curve for a spring loaded to a constant deflection at elevated temperature consists of a very rapid first stage characterized by plastic flow caused by the lower yield point at elevated temperature, and a second stage of anelastic flow at a rate that is dependent on the log of the time under load. There has been no evidence of a third stage or that relaxation ever stops, although the rate is decreasing. If the spring is loaded under constant-load (creep) conditions, instead of constant-deflection (relaxation) conditions, the rate of load loss is far higher, particularly at longer test times.[4]

The accurate and efficient design of a high-temperature spring is a difficult undertaking. Assuming that operating temperatures and spring-loading conditions are well known (and they seldom are), the designer must know what relaxation he can accept within his specifications of load loss and time of service, for zero change is close to unobtainable. The more stable the spring must be and the longer the time of service, the lower the design stress must be. On the other hand, if the service is intermittent with periods of no load at temperature, the spring may recover a portion of its load loss in the "rest" periods and relax far less than might be anticipated. Because of these variables, data plots can only indicate the worst service condition, that of constant loading for the times and temperatures indicated.

If the spring requirements for minimum relaxation are severe, heat setting may be used. This is a process for inducing favorable residual stresses during manufacture by preloading the spring at higher than service temperatures. Such a process can contribute markedly to the apparent relaxation resistance of a spring, but since the heat setting is actually an anelastic-flow process, its effects are recoverable. In other words, a heat-set spring will grow and lose its stability if it is unloaded at the service temperature. Furthermore, if the heat-setting operation is improperly carried out, its effects may be transient, and the spring will behave unpredictably.

28.4.7 Constancy of Rate

Some springs are such that their rate is constant over the entire usable range of deflection. For example, a helical extension spring has an essentially constant rate after the initial tension has been exceeded and before the yield point is reached. A torsion spring exhibits a constant rate within the elastic range, because the decrease of diameter as the spring winds down on an arbor is compensated for by the increase in number of coils. In both examples cited, minor deviations from a straight-line rate may be caused by end or loop deflection.

A helical compression spring will show a constant rate only in the central portion of its load-deflection plot. Initially, nonuniform compression of the end coils appears to produce a lower rate than calculated, and as the spring approaches solid, the rate increases. A constant-rate design should utilize only the center 70 percent of the total deflection. If the compression spring must have a constant rate over a major portion of its deflection range, certain special techniques can achieve this at added cost. The closed-end coils may be soldered or brazed together to reduce the effect of early end closure. A more exact solution is to braze an insert into the dead coils.

Some springs are deliberately designed with a nonconstant rate, e.g., variable-pitch compression springs, conical springs, and volute springs. These are useful in shock-absorbing operations because their rate continues to increase with deflection. No helical compression spring can be designed with a decreasing rate.

A few specialized spring designs may have a rate close to zero, a feature desirable in counterbalance springs and springs used in seals. Belleville springs of special configurations, the constant-force spring and a buckling column, fall in this category. Belleville springs can also be designed with a decreasing rate or even a negative rate over a given deflection range, so that snap-through action will occur. Buckling flat springs, which will snap through, are also a common spring application of negative rate.

The torque output of a motor spring decreases with each turn, and the rate of change depends upon the spring design and manufacturing technique. A constant-force spring can be made to have a torque output that is nearly constant over the useful range of the spring.

28.5 HELICAL COMPRESSION SPRINGS

28.5.1 Principal Characteristics

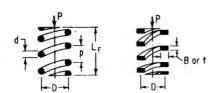

FIG. 28.6 Helical compression springs. L_F is the free length. For rectangular wire $b > t$; $b = t$ if the wire has a square section.

The load-deflection curve is of the type shown in Fig. 28.1a. For springs wound from circular wire (see Fig. 28.6).

$$P = \delta G d^4 / 8 N D^3 \qquad (28.10)$$

$$\tau = 8 K_A P D / \pi d^3 \qquad (28.11)$$

For springs wound from rectangular wire

$$P = K_1 \delta G b t^3 / N D^3 \qquad (28.12)$$

$$\tau = K_A K_2 P D / b t^2 \qquad (28.13)$$

Values for stress factor K_A are given in Table 28.8. Values for the shape factors K_1 and K_2 are given in Table 28.9.[6]

TABLE 28.8 Stress Factors for Helical Springs

C	3	4	5	6	8	10	12	16
K_A	1.58	1.40	1.31	1.25	1.18	1.15	1.12	1.09
K_B	1.33	1.23	1.17	1.14	1.10	1.08	1.07	1.05
K_C	1.29	1.20	1.15	1.12	1.09	1.07	1.05	1.04

TABLE 28.9 Shape Factors K_1 and K_2 for Helical Compression Springs Wound from Rectangular Wire

b/t	1.00	1.50	1.75	2.00	2.50	3.00	4.00	6.00
K_1	0.180	0.250	0.272	0.292	0.317	0.335	0.358	0.381
K_2	2.41	2.16	2.09	2.04	1.94	1.87	1.77	1.67

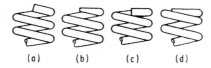

(a)   (b)   (c)   (d)

FIG. 28.7 Types of ends for helical compression springs. (*a*) Open, not ground. (*b*) Open and ground. (*c*) Closed, not ground. (*d*) Closed and ground.

Types of ends for helical compression springs are shown in Fig. 28.7. *Open ends not ground* are satisfactory only if accuracy of load is not important. Springs of this type tangle easily when loose packed for shipment or storage. *Open ends ground* also tangle easily and are generally specified only where it is necessary to obtain as many active coils as possible in a limited space. *Closed ends* not ground are preferred for wire sizes less than 0.020 in diameter or thickness, or if $C > 12$. Closed ends ground are preferred for wire sizes greater than 0.020 in diameter or thickness and if $C < 12$. Springs of this type are usually ground to a bearing surface of 270 to 330. Relationships between types of ends and spring measurements are given in Table 28.10.

TABLE 28.10 Helical Compression Spring Measurements for Common Types of Ends

Type of ends	L_F	N_T	L_S
Open, not ground	$pN + d$	N	$dN + d$
Open and ground	$pN + p$	$N + 1$	$dN + d$
Closed, not ground	$pN + 3d$	$N + 2$	$dN + 3d$
Closed and ground	$pN + 2d$	$N + 2$	$dN + 2d$

L_F = free length, N_T = total number of coils, L_S = solid height.

Table applies to springs wound from rectangular wire if b or t as appropriate is substituted for d. However, rectangular wire keystones in coiling and the axial dimension increase must be allowed for in the solid height calculation, according to

$$t_1 = t(C + 0.5)/C \quad \text{or} \quad b_1 = b(C + 0.5)/C$$

where t_1 or b_1 is the thickness of the inner edge of the material after coiling and t or b the thickness before coiling.

28.5.2 Lateral Loading

Occasionally an application requires lateral as well as axial loading of a helical compression spring. Although the axial rate of a spring is constant for all loads, the lateral rate varies with compression. An adjustable spring-rate system can thus be designed to take advantage of this fact.

The combined stress of a laterally loaded spring

$$\tau = \tau_A(1 + F_L/D + P_L L/PD) \tag{28.14}$$

where τ_A is the axial stress, P the axial load, L the length of the spring under load, and F_L and P_L the lateral deflection and lateral load, respectively.

28.5.3 Buckling and Squareness in Helical Springs

If the free length of a compression spring is more than four times the spring diameter, its stability under load may become critical, and the spring may buckle as a column.

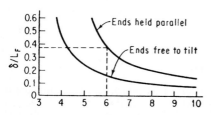

FIG. 28.8 Criterion for stability. Example: A spring having the proportions $L_F/D > 6$ and ends held parallel will buckle if $\delta/L_F > 3.7$.

The lateral stability of a spring depends on the ratio of its free length to mean diameter L_F/D. A slender spring will buckle under load when the ratio of deflection to free length δ/L_F exceeds a critical value. Figure 28.8 can be used to evaluate lateral stability of closed and ground springs. If possible, the spring should be so designed that it will not buckle under service conditions; otherwise the spring must be guided within a tube or on a rod. The friction between spring and guides will interfere somewhat with the accuracy of the load-deflection curve and reduce the spring endurance limit (see Table 28.11).

TABLE 28.11 Factors for Determining Limiting Stress Values for Helical Compression Springs

Material	Factor (before set removed)	Factor (after set removed)
Music wire	0.45	0.65–0.75
Valve spring	0.50	for all materials
Chrome vanadium	0.50	
Oil-tempered	0.50	
Hard-drawn	0.45	
Phosphor bronze	0.35	
Beryllium copper	0.35	
Stainless steel	0.35	

Torsion springs, if long and slender, may also buckle. This may sometimes be prevented by properly clamping the ends or by using initial tension between coils; if the spring still buckles, guides or arbors must be used.

A coiled compression spring does not exert force directly along its axis; the loading is usually eccentric. If the spring must be square under load, as where uniform pressure must be exerted on a valve in guides, special design considerations are used. It is difficult to design a spring which will be square when unloaded and square throughout its loading cycle. Squareness under a specific load is often achieved at the expense of squareness at other deflections. Practical experience in spring design is far ahead of theory in this area at present.

The effect of eccentric loading is to cause an increase in torsional stress on one side of a compression spring and a decrease in stress on the other side.

An approximate relationship for the load eccentricity

$$e = 1.12R(0.504/N + 0.121/N^2 + 2.06/N^3) \tag{28.15}$$

Equation (28.15) indicates that the eccentricity is especially important when there are few active coils since

$$\frac{\text{Stress in eccentrically loaded spring}}{\text{Stress in spring under pure axial load}} = 1 + \frac{e}{R} \tag{28.16}$$

where e is a measure of load displacement from the spring axis and R is the spring mean radius. However, this stress correction is usually ignored unless the number of active turns is two or less. Since the eccentricity varies with load, these calculations are only approximate.

28.5.4 Change in Diameter during Deflection

A helical spring will change in diameter during deflection, which may be significant if the spring is functioning under severe space restrictions. If the spring ends are not allowed to unwind during compression the maximum increase in diameter of a compression spring deflected to solid height is given by

$$\Delta D_{\text{max}} = 0.05(p^2 - d^2)/D \tag{28.17}$$

where pitch p and mean diameter D are those of the spring in the free condition.

28.5.5 Surge and Vibration

When a spring is impact loaded, maximum stress is limited not by total deflection but by the impact velocity of loading v

$$\tau \cong v \sqrt{2\gamma G/g} \tag{28.18}$$

For a steel spring

$$\tau = 131v \text{ (Ref. 7)} \tag{28.19}$$

Surge waves can travel through a spring at a frequency characteristic of the spring. The fundamental frequency of a helical compression spring wound from circular wire

$$f = (Kd/9ND^2) \sqrt{gG/\gamma} \tag{28.20}$$

For springs with one end fixed and one free, $K = \frac{1}{2}$ for springs with both ends fixed, $K = 1$. For a steel spring

$$f = 13,900Kd/ND^2 \tag{28.21}$$

To avoid vibration, the fundamental frequency of a spring should be at least 13 times the frequency of the operating cycle.

28.5.6 Limiting-Stress Values for Static and Low-Cycle Duty

The value of τ calculated by Eq. (28.11) or (28.13) must be less than the permissible limiting-stress value of the spring materials by a suitable margin of safety. Values of permissible limiting stress are determined by multiplying the tensile strength obtained from Fig. 28.5 by the factors given in Table 28.11. Preferred practice is to design for deflection to the solid height.

Set removal increases the load carrying ability of springs in static applications by inducing favorable residual stresses. This allows the use of a higher factor as indicated in Table 28.11. For springs with set removed, the stress correction factor K_A in Eq. (28.11) should be replaced by $1 + 0.5/C$.

28.5.7 Limiting-Stress Values for Cyclic Loading

The value of τ calculated by Eq. (28.11) should not exceed the permissible limiting stress determined by multiplying the tensile strength by the factor given in Table 28.12. Values in Table 28.12 are guidelines and are based on a stress ratio (minimum stress to maximum stress) of zero, no surging, room temperature, and noncorrosive environment. To estimate the life at any other stress ratio, a combined S-N curve and modified Goodman diagram can be used. See Refs. 5 and 6.

Shot peening improves fatigue life, as shown in Table 28.12.

TABLE 28.12 Factors for Determining Limiting-Stress Values for Round Wire Helical Compression Springs in Cyclic Loading

| Fatigue life, cycles | ASTM A228, austenitic stainless steel, and nonferrous | | ASTM A230 and A232 | |
	Not shot-peened	Shot-peened	Not shot-peened	Shot-peened
10^5	0.36	0.42	0.42	0.49
10^6	0.33	0.39	0.40	0.47
10^7	0.30	0.36	0.38	0.46

28.6 HELICAL EXTENSION SPRINGS (Fig. 28.9)

28.6.1 Principal Characteristics

For close-wound springs the load-deflection curve is of the type shown in Fig. 28.1b and

$$P = P_i + \delta G d^4/8ND^3 \qquad \tau = 8K_A PD/\pi d^3 \qquad (28.22)$$

P_i is the initial tension wound into the spring and τ_i the corresponding torsional stress,

$$P_i = \pi \tau_i d^3/8D \qquad (28.23)$$

Recommended values of τ_i depend on the spring index (see Fig. 28.10).

For open-wound springs τ_i and P_i are zero and the load-deflection curve is of the type shown in Fig. 28.1a.

28.6.2 Stresses in Hooks

Critical stresses may occur at sections A and B as shown in Fig. 28.9. At section A the stress is due to bending, and at section B the stress is due to torsion:

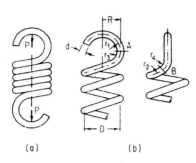

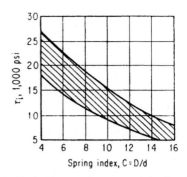

FIG. 28.9 Helical extension springs. (*a*) Close-wound. (*b*) Open-wound.

FIG. 28.10 Ranges of recommended values for stress due to initial tension in close-wound helical extension springs.[9]

$$\sigma = (32PR/\pi d^3)(r_1/r_3) \qquad \tau = (16PR/\pi d^3)(r_2/r_4) \qquad (28.24)$$

Recommended practice is that $r_4 > 2d$. By winding the last few coils with a smaller diameter than the main part of the spring so that R is reduced, the stresses in hooks can be minimized.

28.6.3 Limiting-Stress Values for Static and Low-Cycle Duty

Values τ and σ calculated by Eqs. (28.22) and (28.24) must be less than the permissible limiting stress of the spring by a suitable margin of safety. Values of permissible limiting stress are determined by multiplying the value of tensile strength obtained from Fig. 28.5 by the multiplying factor given in Table 28.13. Recommended values in Table 28.13 are based on set not removed and low-temperature heat treatment.

TABLE 28.13 Factors for Determining Limiting-Stress Values for Helical Extension Springs

Material	Factor for τ		Factor for σ (loop)
	Body	Loop	
Music wire	0.45–0.50	0.40	0.75
Valve spring	0.45–0.50	0.40	0.75
Chrome vanadium	0.45–0.50	0.40	0.75
Oil-tempered	0.45–0.50	0.40	0.75
Hard-drawn	0.45–0.50	0.40	0.75
Phosphor bronze	0.35	0.30	0.55
Beryllium copper	0.35	0.30	0.55
Stainless steel	0.35	0.30	0.55

28.6.4 Limiting-Stress Values for Cyclic Loading

Calculated stresses should not exceed the recommended permissible limiting stresses determined by multiplying the tensile strength by the factors given in Table 28.14. Values in Table 28.14 are based on a stress ratio of 0, no surging, no shot peening, ambient environment, and low-temperature heat treatment.

TABLE 28.14 Factors for Determining Limiting-Stress Values for ASTM A 228 and Type 302 Stainless Steel Helical Extension Springs in Cyclic Loading

Fatigue life, cycles	Factor for τ		Factor for (loop)
	Body	Loop	
10^5	0.36	0.34	0.51
10^6	0.33	0.30	0.47
10^7	0.30	0.28	0.45

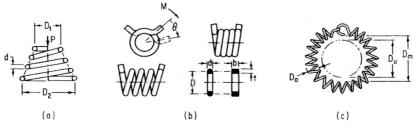

| (a) | (b) | (c) |

FIG. 28.11 Wound springs. (*a*) Conical spring. (*b*) Helical torsion spring. (*c*) Garter spring.

28.7 *CONICAL SPRINGS* (Fig. 28.11*a*)

28.7.1 Principal Characteristics

The load deflection curve shown in Fig. 28.1*a* applies only until the bottom coil closes:

$$P = \delta GD^4/8ND^3 \qquad \tau = 8K_A DP/\pi d^3 \qquad (28.25)$$

A conical spring of uniformly changing diameter can be calculated using the average diameter as the mean diameter, or

$$D = (D_1 + D_2)/2$$

provided the bottom coil does not close. After the bottom coil closes, the conical spring should be calculated on a turn-by-turn basis as if the spring were a series of separate coils. The highest stress at a given load should be calculated by using the mean diameter of the largest active coil at load.

The conical spring can be designed so that each coil nests wholly or partially in the adjacent larger coil so that the solid height of the spring can be as small as one wire diameter. Conical springs can be designed to have a uniform rate by varying the pitch.

28.7.2 Limiting-Stress Values

For static and cyclic loading, limiting-stress values may be determined on the same basis as for helical compression springs.

28.8 HELICAL TORSION SPRINGS (Fig. 28.11b)

28.8.1 Principal Characteristics

For open-wound springs the load-deflection curve is of the type shown in Fig. 28.1a. For springs wound from circular wire

$$M = \theta E d^4 / 67.8ND \qquad \sigma = 32 K_B M / \pi d^3 \qquad (28.26)$$

The factor 67.8 is greater than the theoretical factor of 64 and has been found satisfactory in practice.

For springs wound from rectangular wire,

$$M = \theta E b t^3 / 13.2 \pi ND \qquad \sigma = 6 K_C M / b t^2 \qquad (28.27)$$

The factor 13.2 is greater than the theoretical factor of 12 and has been found satisfactory in practice. Deflection is given in radians to facilitate the calculation of energy storage. However, it is often convenient to express deflection in number of turns $T = \theta/(2\pi)$. Stress factors K_B and K_C at the spring ID are given in Table 28.8. For stress factors at the spring OD see Ref. 5.

For close-wound springs there may be an initial turning moment due to friction between the coils. Friction between coils will affect rate and cause considerable hysteresis in a complete deflection cycle.

28.8.2 Changes of Dimension with Deflection

As a helical torsion spring winds up, mean diameter decreases and the number of coils increases. After T turns the mean diameter $D_T = DN/(T + N)$ and the number of coils $N_T = T + N$; D and N are the mean diameter and number of coils before windup.

28.8.3 Limiting-Stress Values for Static and Low-Cycle Duty

The value of σ calculated by Eq. (28.26) or (28.27) can be as high as 100 percent of the tensile strength if the residual stress is favorable. If it is unfavorable, use 60 to 80 percent of the value for stress-relieved springs.

28.8.4 Limiting-Stress Values for Cyclic Loading

Calculated stresses should not exceed the recommended permissible limiting stresses determined by multiplying the tensile strength by the factors given in Table 28.15.

TABLE 28.15 Factors for Determining Limiting-Stress Values for Helical Torsion Springs in Cyclic Loading

Fatigue life, cycles	ASTM A228 and type 302 stainless steel	ASTM A230 and A232
10^5	0.53	0.55
10^6	0.50	0.53

Values in Table 28.15 are based on no shot peening, no surging, and springs in the "as stress-relieved" condition.

28.9 GARTER SPRINGS (Fig. 28.11c)

28.9.1 Principal Characteristics

Garter springs are special forms of helical extension springs with ends fastened together to form garterlike rings. The inside diameter D_u of an unmounted spring is made to assemble over a larger mounting diameter D_m so that the circumferential load P_c lb exerted by the spring per inch of circumference,

$$P_c = 2\pi K - 2\pi D_u K/D_m + 2P_i/D_m \qquad (28.28)$$

28.9.2 Application

Garter springs are used in mechanical seals for shafting and to hold circular segments in place.

28.10 LEAF SPRINGS

The term "leaf spring" as it is used here applies to springs whose load-deflection and load-stress formulas are based on beam equations. Load-deflection curves are of the type shown in Fig. 28.1a.

28.10.1 Single-Leaf Cantilever Springs (Fig. 28.12)

For springs of either rectangular or tapered section

$$P = \delta Ebt^3/4L^3K \qquad \sigma = 6PL/bt^2 \qquad (28.29)$$

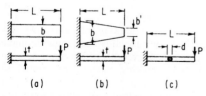

(a)   (b)   (c)

FIG. 28.12 Single-leaf cantilever spring. (a) Uniform cantilever section. (b) Tapering uniform rectangular section. (c) Uniform circular section.

If the rectangular section is uniform as shown in Fig. 28.12a, $K = 1$. A more economical use is made of material where a cantilever leaf spring is tapered as shown in Fig. 28.12b. For tapered springs, values of K are given in Table 28.16.

For cantilever springs made from circular wire

$$P = 3\pi\delta Ed^4/64L^3$$
$$\sigma = 32PL/\pi d^3 \qquad (28.30)$$

TABLE 28.16 Values of K for Tapered Leaf Springs[6]

b'/b	0	0.1	0.2	0.3	0.4	0.5	0.6	0.7	0.8	0.9	1.0
K	1.50	1.39	1.32	1.25	1.20	1.16	1.12	1.09	1.05	1.03	1.00

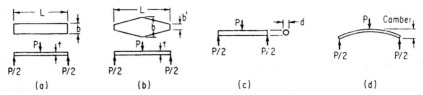

FIG. 28.13 Single-leaf end-supported springs. (*a*) Uniform rectangular section. (*b*) Tapering rectangular section. (*c*) Uniform circular section. (*d*) Elliptical type.

28.10.2 Single-Leaf End-Supported Springs (Fig. 28.13)

For springs of rectangular or tapered section

$$P = 4\delta Ebt^3/L^3K \qquad \sigma = 3PL/2bt^2 \qquad (28.31)$$

If the rectangular spring is uniform as shown in Fig. 28.13*a*, $K = 1$. A more economical use is made of material when an end-supported leaf spring is tapered as shown in Fig. 28.13*b*. For tapered springs, values of K are given in Table 28.16.

For end-supported springs made from round wire

$$P = 3\pi\delta Ed^4/4L^3 \qquad \sigma = 8PL/\pi d^3 \qquad (28.32)$$

Springs initially bowed as shown in Fig. 28.13*d* are called *elliptical springs*. Camber at no load is often made so the spring will be flat under load.

28.10.3 Multiple-Leaf Springs (Fig. 28.14)

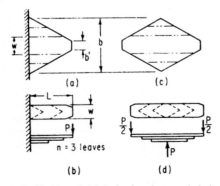

The formulas for multiple-leaf springs are based on those for equivalent single-leaf springs. For example, the single-leaf cantilever spring shown in Fig. 28.14*a* is equivalent to a multiple-leaf cantilever spring for which $b = nw$, n being the number of leaves.

In using equations for equivalent single-leaf springs, allowance must be made for interleaf friction, which may increase the load-carrying capacity by from 2 to 12 percent depending on the number of leaves and conditions of lubrication.

FIG. 28.14 Multiple-leaf springs and their equivalent single-leaf springs.

28.10.4 Limiting-Stress Values for Static and Low-Cycle Duty

Values of calculated stress can be as high as 100 percent of tensile strength of ferrous material and as high as 80 percent of tensile strength for nonferrous material provided residual stress is favorable.

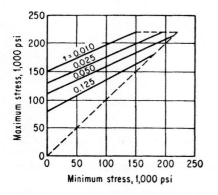

FIG. 28.15 Design chart for cyclically loaded flat steel springs, applies to 0.65 and 0.80 carbon steel at 45 to 48 Rockwell C hardness.[5] Data are for 10^6 cycles.

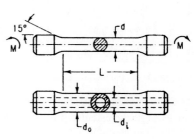

FIG. 28.16 Torsion-bar springs.

28.10.5 Cyclic Loading

Maximum and minimum values of stress for cyclic loading must fall within the permissible limits defined by a Goodman or similar diagram. The design chart shown in Fig. 28.15 applies to carbon-steel springs.

28.11 TORSION-BAR SPRINGS (Fig. 28.16)

28.11.1 Principal Characteristics

The load-deflection curve is of the type shown in Fig. 28.1a. For torsion bars with hollow circular sections,

$$M = \pi\theta G(d_0^4 - d_i^4)/32L \qquad \tau = 16Md_0/\pi(d_0^4 - d_i^4) \qquad (28.33)$$

For torsion bars with solid circular sections, $d_i = 0$ and $d_0 = d$. Torsion-bar ends should be stronger than the body to prevent failure at the ends.

28.11.2 Material

For highly stressed applications, torsion bars are generally made of silicomanganese. Fatigue resistance is increased by shot peening. Presetting increases strength in the direction of preset but should not be used if bars are loaded in both directions.

28.11.3 Application

Torsion-bar springs are used where highly efficient energy-storage devices are required. The most familiar applications are in automotive equipment.

28.12   *SPIRAL TORSION SPRINGS* (Fig. 28.17)

28.12.1   Principal Characteristics

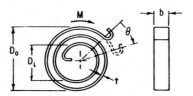

FIG. 28.17   Spiral torsion spring.

The load-deflection characteristic is of the type shown in Fig. 28.1a and

$$M = \theta E b t^3 / 12L \qquad \sigma = 6M/bt^2 \qquad (28.34)$$

The length of active material $L \approx \pi n (D_0 + D_i)/2$; n is the number of coils. Deflection in Eq. (28.34) is expressed in radians to facilitate the calculation of energy storage. The corresponding number of rotations $T = \theta/(2\pi)$.

Stresses at the spring ends may exceed the values of stress calculated by Eq. (28.34) depending on the design of the end and its treatment in manufacture.

28.12.2   Limiting-Stress Values for Static and Low-Cycle Duty

Calculated stress should be less than 80 percent of the tensile strength.

28.12.3   Application

Spiral torsion springs are widely used for locks in automobiles and to maintain pressure on carbon brushes of electric motors and generators.

28.13   *POWER SPRINGS* (Fig. 28.18a)

28.13.1   Principal Characteristics

The power or clock spring is a special form of the spiral torsion spring. However, Eq. (28.34) applies only to maximum torque when the spring is fully wound. Figure

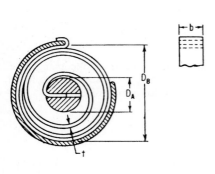

(a)

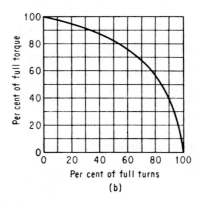

(b)

FIG. 28.18   Power spring. (*a*) Orientation. (*b*) Load-deflection curve.

28.18b indicates the form of a typical load-deflection curve. To deliver the maximum number of turns, the space occupied by the spring coil should be half the space available in the drum so that

$$Lt = (\pi/8)(D_B^2 - D_A^2) \tag{28.35}$$

in which case the number of initial coils n_A wound tight on the arbor, and the number of rotations T that can be made by the arbor before the spring is unwound solid against the drum,

$$n_A = \frac{-D_A + \sqrt{1/2(D_A^2 + D_B^2)}}{2t} \qquad T = \frac{-D_A - D_B + \sqrt{2(D_A^2 + D_B^2)}}{2.55t} \tag{28.36}$$

The arbor diameter D_A should be from 15t to 25t. The ratio of length to thickness L/t is generally between 3000 and 4000. If an extremely large number of operating cycles are desired, L/t should be less than 1000.

28.13.2   Materials and Recommended Working Stresses

Equation (28.34) is used for calculation of maximum stress when the spring is tightly wound on its arbor so that M is a maximum. Power springs are usually made from cold-rolled tempered spring steel AISI 1074 or cold-rolled tempered spring steel AISI 1095. Recommended maximum values of working stress for these materials depend on desired life. They may be as high as 80 percent of tensile strength for conventional power springs and as high as 100 percent of tensile strength for prestressed power springs.

28.14   *CONSTANT-FORCE SPRINGS*

Constant-force springs are made from flat stock that has been given a constant natural curvature R by prestressing. A force is required to increase R_n during extension. The principle of the constant-force spring is applied to extension and motor types.

28.14.1   Extension Type (Fig. 28.19a, b)

In its relaxed condition the spring will form a tightly wound spiral. An extension spring is mounted on a spool of radius $R_2 > R_n$. The force P is virtually constant; it will vary somewhat as the coil unwinds and the radius R decreases:

$$P = (Ebt^3/26.4)(2/R_1R_n - 1/R_1) \qquad R_1 = R_2 + nt \tag{28.37}$$

where n is the number of coils still wound on the roller. R_2 should be about $1.2R_n$. The length of prestressed extension springs should be

$$L = \delta + 10R_2 \tag{28.38}$$

A stress factor K relates spring material, proportions, load, and number of operations[1]

$$P = Kbt$$

Values of K for materials commonly used for constant-force springs are given in Fig. 28.19c. Recommended practice is that $50 < b/t < 200$, $b/t = 100$ being a good median value.

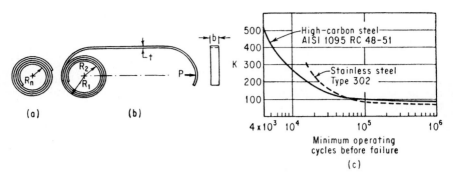

FIG. 28.19 Constant-force spring. (*a*) Unmounted coil. (*b*) Extension type. (*c*) Stress factors.[1]

Constant-force extension springs are used where large deflections and a zero rate are required as in counterbalancing devices.

28.14.2 Motor Type (Fig. 28.20)

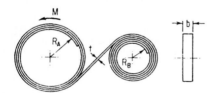

FIG. 28.20 Motor-type constant-force spring.

A constant-force spring reverse-wound on the larger of two spools will wind up on the smaller spool, driving the larger spool with a torque approximated by the equation

$$M = ER_A bt^3/24(1/R_A + 1/R_B)^2 \quad (28.39)$$

Spring load and stress are related by the equation

$$\sigma = 2Et(1/R_n + 1/R_A) \tag{28.40}$$

Recommended practice is that $50 < b/t < 200$, $b/t = 100$ being a good median value.

28.15 BELLEVILLE WASHERS (Fig. 28.21*a*)

28.15.1 Principal Characteristics

The load-deflection curve is of the type shown in Fig. 28.1*c*. Belleville washers give relatively high loads for small deflections. By a suitable choice of spring proportions a wide variation of load-deflection curves can be obtained as shown in Fig. 28.21*b*. Rate is almost constant when $h/t = 0.4$ and close to zero for a large part of the deflection range when $h/t = 1.4$. When $h/t > 2.8$, a Belleville washer becomes unstable and will snap over. The general equations for Belleville washers are[9]

$$P = \frac{4E\delta}{(1 - \mu^2)K_1 D_0^2}\left[(h - \delta)\left(h - \frac{\delta}{2}\right)t + t^3\right] \tag{28.41}$$

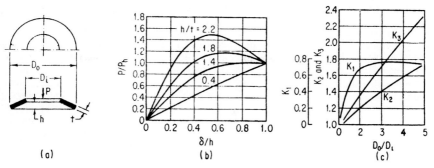

FIG. 28.21 Belleville washer. (*a*) Orientation. (*b*) Load-deflection curve P/P_h is the ratio of any P to the load P_h when deflected to the flat condition. (*c*) Factors K_1, K_2, and K_3 for spring equation.

$$\sigma = \frac{4E\delta}{(1 - \mu^2)K_1 D_0^2}\left[K_2\left(h - \frac{\delta}{2}\right) + K_3 t\right] \tag{28.42}$$

The factors K_1, K_2, and K_3 are all functions of D_0/D_i as shown in Fig. 28.21*c*.

Simpler equations are possible if washers are designed on the basis of selected proportions. The ratio h/t as shown in Fig. 28.21*b* determines load-deflection characteristics. The ratio of diameters D_0/D_i is usually restricted by assembly requirements; one of the ratios 1.2, 1.8, 2.4, or 3.0 will ordinarily be found suitable. If the most efficient use of spring material is a factor, then D_0/D_i should be 1.8. For washers designed to deflect to the flat positions so that σ is a solid stress

$$P = \frac{\sigma^2(1 - \mu^2)K_4 D_0^2}{4E} \qquad t = \left[\frac{\sigma(1 - \mu^2)}{4E}\right]^{1/2} K_5 D_0 \tag{28.43}$$

For steel springs $\mu = 0.3$ and $E = 30 \times 10^6$ lb/in²; hence

$$P = 75.8 \times 10^{-10}K_4\sigma^2 D_0^2 \qquad t = 87.0 \times 10^{-6}K_5\sigma^{1/2}D_0 \tag{28.44}$$

Factors K_4 and K_5 are given in Table 28.17.

TABLE 28.17 Factors K_4 and K_5 for Simplified Belleville-Washer Equations

D_0/D_i	K	$h/t = 0.4$	$h/t = 1.4$	$h/t = 1.8$	$h/t = 2.2$
1.2	K_4	0.475	0.068	0.043	0.029
	K_5	0.770	0.346	0.290	0.248
1.8	K_4	0.681	0.103	0.064	0.043
	K_5	1.024	0.465	0.388	0.336
2.4	K_4	0.598	0.089	0.056	0.038
	K_5	1.039	0.467	0.391	0.338
3.0	K_4	0.480	0.075	0.048	0.033
	K_5	0.990	0.452	0.381	0.328

28.15.2 Materials and Recommended Working Stresses

The value of stress calculated by Eqs. (28.42) and (28.43) is a compressive stress at the upper edge of the inside diameter. This stress usually determines the load-carrying ability of the washer and can be as high as 120 percent of tensile strength when set is not removed. In cyclic service cracks may start at the upper edge of the inside diameter but will usually not progress. Actual fatigue failures start as cracks at the outer diameter in a zone stressed in tension. To determine the stress at these locations see Ref. 5. Stresses in static applications for carbon or alloy steel can be as high as 120 percent of the tensile strength when set is not removed and 275 percent when set is removed. For nonferrous metals and austenitic stainless steel, stresses can be as high as 95 percent of the tensile strength when set is not removed and 160 percent when set is removed.

28.15.3 Application

Belleville washers are used where a high load with a relatively small deflection is required or where their unique load-deflection characteristics can be used to advantage. Belleville washers are used for preloading bearings in spindles, as pressure disks for power brakes, and in buffer assemblies for absorbing impact loads.

28.16 WAVE-WASHER SPRINGS (Fig. 28.22a)

28.16.1 Principal Characteristics

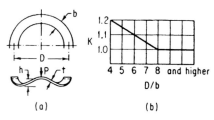

(a) (b)

FIG. 28.22 Wave-washer spring. (a) Orientation. (b) Factor K for wave-washer spring equation.

The load-deflection curve is of the type shown in Fig. 28.1a and

$$P = 13K\delta Ebt^3 n^4/\pi^3 D^3$$
$$\sigma = 3\pi PD/4bt^2 n^2 \tag{28.45}$$

The factor K is plotted in Fig. 28.22b. The number of waves n can be three or more. Usual practice is to have three, four, or six waves.

Allowance must be made in the assembly of wave washers for the slight decrease in diameters from the loaded to the unloaded condition.

28.16.2 Limiting-Stress Values for Static and Low-Cycle Duty

Wave washers are similar to leaf springs; hence the recommendations in Sec. 28.10.4 apply. Maximum stress should be based on deflection to the solid height, in which case

$$\sigma = 39KhEtn^2/4\pi^2 D^2 \tag{28.46}$$

28.16.3 Application

Wave-washer springs are used because of their compact form where a static load or small deflection range is required. A common use for wave washers is to assemble them between parts mounted on a shaft in order to compensate for variations in assembly clearances.

28.17 NONMETALLIC SYSTEMS

28.17.1 Elastomer Springs

An elastomer is rubber or a rubberlike material which can be stretched to at least twice its free length at normal temperatures and will quickly return to its original length when the load is released. The ability to undergo very large deformations makes elastomers ideally suited to applications where energy absorption is important. Because of their high damping properties elastomers are used for resilient mountings in applications requiring vibration isolation. The stress-strain characteristics of elastomers are not linear; hence load-deflection curves are of the type shown in Fig. 28.1c.

Elastomers can be applied to a variety of spring designs. Only the simplest types of elastomer springs are considered here. The load-deflection curves for these simple types are nearly linear for limited deflections. For more complete information on elastomer springs, see Refs. 6, 10, and 11.

Compression Cylinders (Fig. 28.23a).   Principal formulas are

$$P = \pi\delta ED^2/4h \qquad \sigma = 4P/\pi D^2 \tag{28.47}$$

The value of E is not constant but depends on the relative area available for bulging as measured by the form factor.[10]

$$K = [(\pi/4)D^2]/\pi Dh = D/4h$$

The relationship between E and K is given by Fig. 28.24a. Recommended practice is for $\delta < 0.2h$ to avoid creep.

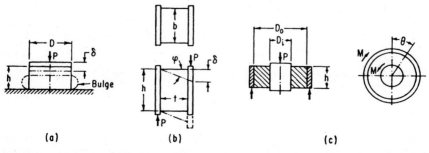

FIG. 28.23 Elastomer springs. (a) Compression cylinder. (b) Cylindrical bushing—force applied for direct shear loading—applied torque. (c) Shear sandwich.

Shear Sandwiches (Fig. 28.23c).   For $\phi < 20°$ the load-deflection curve is nearly linear. Principal formulas are

$$P = \delta bhG/t \qquad \tau = P/bh \tag{28.48}$$

Recommended practice is that $t < h/4$ or $t < b/4$, whichever is the least value.

Cylindrical Bushings (Fig. 28.23b).   Cylindrical bushings can be loaded in either direct shear or torsion. For direct-shear loading

$$\delta = P \ln(D_0/D_i)/2\pi hG \qquad \tau = P/\pi hD \tag{28.49}$$

Equation (28.49) is sufficiently accurate for practical purposes if $P/(\pi hGD_0) < 0.4$. Shear stress calculated by Eq. (28.49) refers to the bond between elastomers and the surface at D_i. For torsion loading

$$\theta = (M/\pi hG)(1/D_i^2 - 1/D_0^2) \qquad \tau = 2M/\pi hD_i^2 \tag{28.50}$$

Equation (28.50) is valid for $\theta < 0.7$ rad (40°). Shear stress calculated by Eq. (28.50) is for the bond between the elastomer and the surface at D_i.

Mechanical Properties and Recommended Working Stresses.   Both E and G depend on hardeners; values for static loading are shown in Fig. 28.24b to have different values for dynamic loading.

For compression cylinders subject to static or low-cycle loading, maximum recommended compressive stress is 700 lb/in². Lower values should be used for cyclic duty.

For shear sandwiches and cylindrical bushings loaded in direct shear, recommended maximum design stress based on the bond strength is 35 lb/in²; for cylindrical bushings loaded in torsion, recommended design stress based on the bond strength is 50 lb/in². Durometer hardness of 30 to 60 is usual for elastomer springs subject to shear loading.

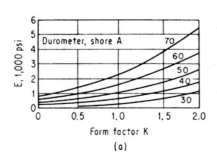

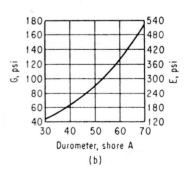

FIG. 28.24   Natural-rubber charts. (*a*) Modulus of elasticity vs. form factor K. (*b*) G and E vs. durometer.

28.17.2   Compressible Liquids

Special spring devices utilize the compressibility of silicone-base liquids, which can be compressed to 90 percent of their volume at pressures maintainable in production units. These compact springs will sustain high loads in a much smaller space than possible with other spring systems, because of their high spring rate. They are costly because of the necessity of excellent pressure seals. They are also sensitive to temperature variations. These springs are applicable where space is at a premium, as in aircraft landing gear, presses, and dies. Operating frequency generally should be kept below 1 (Hz).

Compressible-liquid springs may also be used as shock absorbers with excellent damping qualities attained by utilizing dashpot action.

28.17.3 Compressible Gases

The air springs in use in many shock-absorbing applications utilize gas compressibility.[12] Many applications take advantage of the increasing rate of such a spring as the deflection continues. The spring rate can be changed at will by intentionally altering the gas pressure, which also affords a method of adjusting the loaded height of the spring system. There are problems associated with the design of a leakproof flexible gas container and with increasing temperatures. Generally spring rates are low, deflections high, and considerable space is needed.

REFERENCES

1. Mechanical Drawing Requirements for Springs, Military Standard, MIL-STD-29A, March 1962.

2. Maier, Karl W.: "Springs That Store Energy Best," *Prod. Eng.,* Nov. 10, 1958, pp. 71–75.

3. Carson, Robert W.: "Flat Spring Materials," *Prod. Eng.,* June 11, 1962, pp. 68–80.

4. Crooks, R. D., and W. R. Johnson: "The Performance of Springs at Temperatures above 900°F," *Trans. SAE,* vol. 69, pp. 325–330, 1961.

5. "Design Handbook," Associated Spring: Barnes Group Inc., Bristol, Conn., 1981.

6. Wahl, A. M.: "Mechanical Springs," 2d ed., McGraw-Hill Book Company, Inc., New York, 1963.

7. Maier, Karl W.: "Dynamic Loading of Compression Springs," part II, *Prod. Eng.,* March 1955, pp. 162–174.

8. Chironis, Nicholas P. (ed.).: "Spring Design and Application," McGraw-Hill Book Company, Inc., New York, 1961.

9. Almen, J. O., and A. Laszlo: "The Uniform Section Disc Spring," *Trans. ASME,* vol. 58, no. 4, pp. 305–314, May 1936.

10. Gobel, E. F.: "Berechnung and Gestaltung von Gummifedern," Springer-Verlag OHG, Berlin, 1955.

11. McPherson, A. T., and A. Klemmin (eds.): "Engineering Uses of Rubber," Reinhold Publishing Corporation, New York, 1956.

12. Deist: "Air Springs Cushion the Ride," *Mech. Eng.,* June 1958, p. 61.

SECTION 29
PNEUMATIC COMPONENTS

Richard F. Lytle, B.S.M.E.

Technical Consultant
Fisher Controls Company
Marshalltown, Iowa

29.1 INTRODUCTION

Pneumatic components can be built up into a wide variety of systems offering a high degree of dependability and safety along with relatively low cost. Their desirability is evidenced by the continued popularity of pneumatic systems utilized in the process control field, where the features of economical energy storage and environmental safety are of major importance.

Systems composed of nozzle-flapper amplifiers, mechanical components, pneumatic relays, and volume-restriction networks can exhibit high gain, high power, and a variety of desirable dynamic characteristics.

Most of the advantages of pneumatic systems arise from the inherent compressibility of air, a factor which unfortunately contributes to the time lag of the components. In spite of this drawback, systems can be designed with excellent dynamic characteristics. Intelligent design of pneumatic control systems requires a knowledge of feedback control theory and an understanding of fundamental physical laws to develop practical models of basic elements and components.

This section will illustrate the methods of developing mathematical models of commonly used physical devices for the design and application of pneumatic control systems.

29.2 BASIC RELATIONSHIPS

The flow of compressible fluids through orifices, relay valves, and similar restrictions can be described by a reversible adiabatic process from the inlet to the vena contracta. From this point of minimum area to a downstream point where pressure recovery has occurred the process is irreversible.[4]

Of the various formulas proposed for flow-rate calculations based on this overall loss, the form given by Perry[6] is probably the most useful one for design purposes.

$$W = KYA[2g\omega(P_1 - P_2)]^{1/2} \qquad (29.1)$$

where P_1 = upstream pressure, lb/ft^2 abs
P_2 = downstream pressure, lb/ft^2 abs
W = weight rate of flow, lb/s
K = discharge coefficient, including the velocity-of-approach factor
Y = expansion factor
ω = upstream specific weight, lb/ft^3
g = acceleration of gravity, 32.2 ft/s^2
A = area of restriction, ft^2

For air at 70°F the relationship reduces to

$$W = 0.048\ KYA[P_1(P_1 - P_2)]^{1/2} \qquad (29.2)$$

The equations are usable in both the critical and subcritical regions, P_2 being used as the actual downstream pressure and not limited to $0.53P_1$. In pneumatic components the ratio of restriction area to pipe area is usually small, so that the product KY is a function only of P_2/P_1 and the ratio of specific heats. Figure 29.1 gives values of KY for airflow through a sharp-edged orifice. The slope in the subcritical region is inversely proportional to k, the ratio of specific heats, so that the curve can be used for fluids other than air. The work of Cunningham[5] indicates that k has practically no effect on the slope in the critical region.

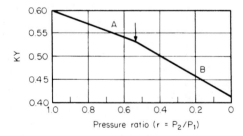

FIG. 29.1 Effect of pressure ratio on KY for airflow through a sharp-edged orifice.[6]

As the configuration of the restriction departs from the "sharp-edged" concept, the KY curve starts at higher values because of an increase in K, and the slope in both flow regions becomes steeper because of less opportunity for radial expansion at the vena contracta. Since the designs of restrictions vary considerably, it is usually necessary to obtain the KY curve experimentally.

The flow of compressible fluids in tubes is characterized by a region of laminar flow and a region of turbulent flow, a transition between the two occurring at Reynolds numbers between 2000 and 4000. In the turbulent range at high Reynolds numbers and low pressure drops, the flow varies nearly with the square root of the pressure drop, but it reaches limiting values as the pressure drop is increased (Ref. 1, Chap. 6). In most pneumatic devices the steady-state flow through tubing is zero, so that an analysis can be made on the basis of small pressure drops. This permits the assumption of laminar flow, which gives a linear relationship between flow and pressure drop.

For laminar flow of air through a capillary tube [Ref. 1, Eq. (5.10)] in the range $N_r < 2000$ $(P_1 - P_2)/P_1 << 1$

$$W = \pi\omega D^4(P_1 - P_2)/128\mu L \qquad (29.3)$$

where W = weight rate of flow, lb/s
$\quad\quad D$ = inside diameter of tube, ft
$\quad\quad \mu$ = viscosity, (lb·s)/ft^2, 0.37×10^{-6} for air at 60°F
$\quad\quad L$ = tube length, ft
$\quad\quad P_1$ = upstream pressure, lb/ft^2 abs
$\quad\quad P_2$ = downstream pressure, lb/ft^2 abs
$\quad\quad \omega$ = specific weight, lb/ft^3 (mean)
$\quad\quad N_r$ = Reynolds number, $\rho U D/\mu$
$\quad\quad \rho$ = density, slugs/ft^3 = (ω/g)
$\quad\quad U$ = velocity, ft/s

Laminar flow of air through a needle valve is derived from Ref. 1, Eq. (1.5), and a force balance; see also Ref. 3. With $b << D_m$ and ϕ small, $N_r < 2000$, and $(P_1 - P_2)/P_1 << 1$

$$W = \pi\omega D_m b^3(P_1 - P_2)/12\mu L \qquad (29.4)$$

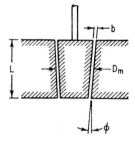

where W, μ, P_1, P_2, and ω are as in Eq. (29.3) and L, b, and D_m are as shown in Fig. 29.2, measured in feet. The length to use in Reynolds number is $2b$. Typical dimensions for valves used in pneumatic circuits in process-control equipment are $L = \frac{1}{8}$ in, $D_m = \frac{1}{16}$ in, $\phi = 1°$, and $b = 0.001$ in, which gives laminar flow over the usual pressure drops encountered.

The rate of pressure change within a volume is derived from conservation of mass, $PV^k =$ constant, $C^2 = kgRT$ [Eq. (6.28), Ref. 1], and the perfect gas law. It is given by

FIG. 29.2 Needle-valve nomenclature.

$$dP/dt = WC^2/gV \qquad (29.5)$$

for a reversible adiabatic process. For PV = constant we obtain $dP/dt = WC^2/gVk$ for an isothermal process

where P = pressure, lb/ft^2
$\quad\quad W$ = net flow into the volume, lb/s
$\quad\quad t$ = time, s
$\quad\quad C$ = velocity of sound (1120 ft/s for air at 60°F)
$\quad\quad k$ = ratio of specific heats C_p/C_v (1.4 for air)
$\quad\quad g$ = acceleration of gravity, 32.2 ft/s^2
$\quad\quad V$ = volume, ft^3

For most applications the isothermal relationship is realistic only at steady state or very low frequencies, probably less than 0.01 hertz (Hz). For pneumatic devices we are usually concerned with frequencies above 0.1 Hz, so that the reversible adiabatic relationship is applicable and will be used throughout this section except as noted. The relationship [Eq. (29.5)] is not affected by the irreversible expansion which occurs at the entrance to the volume considered.

The transfer function relating pressure to flow from Eq. (29.5) is

$$\mathscr{L}P/\mathscr{L}W = C^2/gVS = Z_v \qquad (29.6)$$

In cases where the volume V is free to change with pressure as in Fig. 29.3, a correction can be made for the apparent increase in volume as seen by the flow source. This can be derived from conservation of mass [Eq. (29.5)], PV^k = constant, $C^2 = kgRT$ (Eq. 6.28, Ref.1) and the perfect gas law. In this case the equivalent volume V to use in Eq. (29.6) is

$$V = \overline{V}_m + 1.4\overline{P}A^2/K_s \qquad (29.7)$$

where $\overline{V}_m$ = mean volume, ft^3
   $\overline{P}$ = mean pressure, lb/ft^2 abs
   A = area, ft^2
   K_s = spring rate, lb/ft

This correction for spring loading should not be overlooked in design, since the volume given by Eq. (29.7) is often many times larger than that given by $\overline{V}_m$ alone.

If the output pressure acts on a significant inertia load, the "equivalent-volume" concept is not applicable. In this situation the mean volume should be used, with the load velocity times mean specific weight fed back through the load ram area to the flow-summing point.

29.3 SIMPLE NETWORKS

The majority of pneumatic systems are made up of simple networks involving a volume and either one or two restrictions. The static and dynamic characteristics of these networks depend upon the linearity of the restrictions and whether or not a steady flow is established through the device.

29.3.1 Case 1. Single Linear Restriction, No Steady Flow

The combination shown in Fig. 29.4 is commonly used as a feedback element for integral or derivative action with a needle valve, scratch valve or capillary tube for the restriction. The transfer function, derived from conservation of mass [Eq. (29.5)] and $d\omega/dt = (g/C^2)dP/dt$ (Ref. 2, Chap. 7), is

$$\mathscr{L}P_2/\mathscr{L}P_1 = 1/(\tau S + 1) \qquad (29.8)$$

where τ is $128\mu LgV/\omega\pi D^4C^2$ for the capillary tube and $12\mu LgV/\omega\pi C^2 D_m b^3$ for a needle valve. Nomenclature is from Eqs. (29.3) through (29.5).

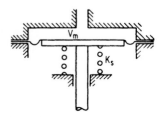

FIG. 29.3   Spring-loaded volume.

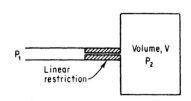

FIG. 29.4   Simple volume-restriction network.

29.3.2 Case 2. Single Nonlinear Restriction, No Steady Flow

If a restriction such as a sharp-edged orifice is used in the network of Fig. 29.4, the system will be nonlinear and cannot be linearized even with the assumption of small perturbations. This is the only case that we shall consider for which it is not possible to write a transfer function. We can, of course, obtain time-domain solutions for a specified disturbance. If, for example, a small step change is made in P_1, we have, from Eqs. (29.1) and (29.5) and also conservation of mass

$$t = [P_1^{1/2} - (P_1 - P_2)^{1/2}]V/\omega KYAC^2$$

with P_1 being the magnitude of the step and P_2 the deviation of the pressure in the volume from its initial value. This solution is plotted in Fig. 29.5. Note that P_2 becomes equal to P_1 in a finite length of time rather than approaching it asymptotically. This is the result of an "apparent time constant," which approaches zero as the pressure difference approaches zero. This feature makes nonlinear restrictions undesirable in networks with no established steady flow.

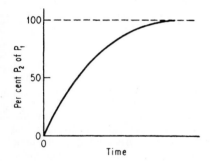

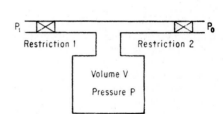

FIG. 29.5 Response of orifice volume system to step input.

FIG. 29.6 Double-restriction network.

29.3.3 Case 3. Two-Linear or Nonlinear Restrictions, Established Steady Flow

Figure 29.6 shows the general case of a single volume with two restrictions. The restrictions may be fixed or variable but in all cases a steady flow exists through the device. The output of the system is P and the input or disturbance may be P_1, P_0, or a change in either restriction. The transfer function for any input X is obtained from the block diagram in Fig. 29.7 [since system flow $\overline{W} = f(P, X)$] as

$$\mathscr{L}P/\mathscr{L}X = (\partial W/\partial X)(\mathscr{L}P/\mathscr{L}W)/[1 + (\mathscr{L}P/\mathscr{L}W)(\partial W_1/\partial P + \partial W_2/\partial P)]$$

where $\partial W/\partial X$ is the change in flow produced by a change in input with P held constant. Likewise, $\partial W_1/\partial P$ and $\partial W_2/\partial P$ are the flow changes produced by a change in P with the input or disturbance held constant.

Letting $G = (\partial W/\partial X)/(\partial W_1/\partial P + \partial W_2/\partial P)$ we obtain the general transfer function

$$\mathscr{L}P/\mathscr{L}X = G/\{[1/Z(\partial W_1/\partial P + \partial W_2/\partial P)] + 1\} \tag{29.9}$$

where G is the steady-state change in P produced by the change in the input X. It is best obtained from the slope dP/dX of the calculated characteristic curve. This curve is normally calculated from static considerations and is therefore an isothermal gain. The

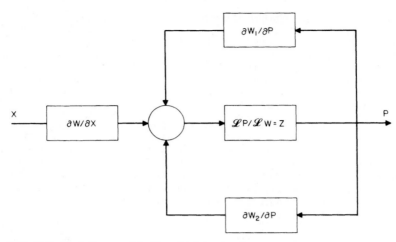

FIG. 29.7   Block diagram of double-restriction network.

reversible adiabatic gain is slightly higher, a maximum difference of 16 percent occurring with critical drop across both restrictions. This reversible adiabatic gain is given by $(\partial W/\partial X)/(\partial W_1/\partial P + \partial W_2/\partial P)$, with these quantities evaluated as shown below.

The partial derivatives $\partial W_1/\partial P$ and $\partial W_2/\partial P$ are functions of the operating point of the system and can be obtained by differentiation of the basic flow equations. Formulas for $\partial W/\partial P$ for various flow regimes have been calculated by Lloyd and Anderson[2] from basic equations and are given in Table 29.1. All symbols in the table with a bar over the variable refer to the steady-state operating conditions about which the deviations occur; W is in pounds per second and pressures are in pounds per square foot absolute; and Z is $\mathscr{L}P/\mathscr{L}W$. For the simple volume termination, $Z = Z_v = C^2/gVS$, giving for a transfer function

$$\mathscr{L}P/\mathscr{L}X = G/(\tau S + 1) \qquad (29.10)$$

with $\tau = gV/C^2 \, (\partial W_1/\partial P + \partial W_2/\partial P)$.

The calculation of Z for a transmission line between the output volume and the load volume is given in Sec. 29.10.

TABLE 29.1   Pressure Sensitivity Equations

Condition		Equation
Upstream pressure variation	Subcritical flow	$\dfrac{\partial W}{\partial P_1} = \dfrac{\bar{W}}{2(\bar{P}_1 - \bar{P}_2)}$
	Critical flow	$\dfrac{\partial W}{\partial P_1} = \dfrac{\bar{W}}{\bar{P}_1}$
Downstream pressure variation	Subcritical flow	$\dfrac{\partial W}{\partial P_2} = -\dfrac{\bar{W}}{2\bar{P}_2}\dfrac{(2\bar{P}_2 - \bar{P}_1)}{(\bar{P}_1 - \bar{P}_2)}$
	Critical flow	$\dfrac{\partial W}{\partial P_2} = 0$

29.4 NOZZLE-FLAPPER SYSTEMS

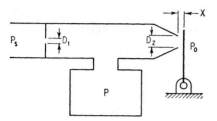

FIG. 29.8 Nozzle-flapper system.

Figure 29.8 shows a common system used to convert mechanical motion into a pneumatic pressure. The gain of the device is high, often above 10,000 lb/in² per inch, and the load on the input element is usually insignificant. Figure 29.9 shows the steady-state characteristics and Fig. 29.10 the time constants of a typical system with $P_s = 35$ lb/in² abs, $D_1 = 0.010$ in, $D_2 = 0.020$ in, and $V = 1$ in³ (no transmission line). The curve of pressure vs. flapper displacement is calculated from

$$X = \frac{D_1^2 Y_1 K_1}{4 D_2 Y_2 K_2} \left[\frac{P_s (P_s - P)}{P(P - P_0)} \right]^{1/2}$$

(29.11)

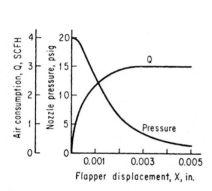

FIG. 29.9 Typical steady-state characteristics of the nozzle flapper.

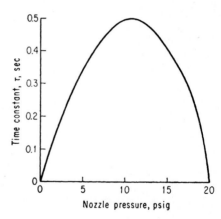

FIG. 29.10 Effect of operating point on nozzle-flapper time constant.

The nomenclature is from Fig. 29.8 with Y as the expansion factor and K as the flow coefficient. The dimensions are in inches with pressures in pounds per square inch absolute. Equation (29.1) gives the flow across the first restriction and, letting $Q = 47,000W$ (volumetric flow, standard cubic feet per hour), general transfer function is given by Eq. (29.9) and that for the volume load by Eq. (29.10).

Since both G and τ vary with flapper position, it is advisable to follow the nozzle flapper with a volume relay with a fixed pressure-to-pressure-gain ratio so that the operation of the nozzle flapper will be restricted to the relatively linear portion of the characteristic curve.

In process control equipment the primary orifice diameter is usually between 0.009 and 0.020 in and the nozzle diameter is 0.020 to 0.060 in. Smaller primary orifice dimensions present problems in plugging and produce excessive time constants, as indicated by Eq. (29.10). Small nozzle diameters result in low gain in the low-pressure range and limit the minimum pressure that can be obtained.

Equation (29.11) is based on the assumption that the minimum pressure is governed by the nozzle-flapper clearance area, not by the nozzle area. A high ratio D_2/D_1 can produce high gain but nozzle-flapper alignment becomes critical. A large primary orifice diameter results in low gain unless a corresponding increase is made in nozzle diameter. The nozzle pressure creates a force load on the flapper system. Excessive flapper forces and high-frequency mechanical flapper oscillation sometimes limit the maximum nozzle size that can be used. These oscillations can be minimized by reducing nozzle pressure or diameter, increasing flapper spring rate, or decreasing the flapper mass. A study of this problem in connection with hydraulic systems has been made by Feng.[7]

29.5 EJECTOR SYSTEMS

Occasionally it becomes desirable to omit a power relay and work a nozzle-flapper system over most of its range. In this situation an ejector system can be used to minimize the nonlinearities of the basic system. Figure 29.11 shows a typical design. The geometry roughly follows recommendations for fluid-handling ejectors[8] except that no attempt is made to recover pressure in the discharge section. These recommendations show discharge throat diameters 2 or 3 times the supply nozzle diameter with the nozzle positioned about one nozzle diameter past the start of the converging section. For proper operation the output connection must be in the general area shown but the nozzle-flapper can be some distance away if necessary.

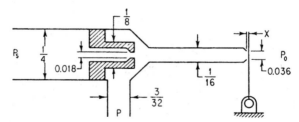

FIG. 29.11   Typical ejector design.

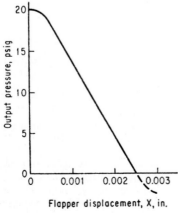

FIG. 29.12   Ejector characteristic.

Figure 29.12 illustrates the excellent linearity that can be obtained by properly positioning the supply restriction in the converging section. Varying the location of this restriction can produce characteristics between those of the basic nozzle-flapper system and the ejector. This becomes necessary when negative load pressures are to be avoided. Analytical relationships for the static and dynamic characteristics of these systems are not available, but laboratory tests indicate time constants on the order of 2 or 3 times those of a basic nozzle-flapper system with the same size restrictions.

29.6 BLEED RELAYS

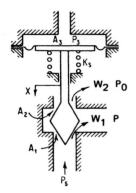

Figure 29.13 shows a typical design for a bleed relay. The input can be a pressure on the diaphragm, as shown, or direct mechanical input. A port size of 0.060 in is typical when the device is used as a first-stage motion-pressure transducer. For a power relay, $\frac{1}{8}$- to $\frac{3}{16}$-in ports are common, with the valve travel being about half the port diameter. The characteristic curve is obtained by first relating the port-area ratio A_2/A_1 to the motion input X or pressure input ($X = A_3P_3/K_s$). The area ratio is then related to the output pressure P by

$$\frac{A_2}{A_1} = \frac{Y_1 K_1}{Y_2 K_2} \left[\frac{P_s(P_s - P)}{P(P - P_0)} \right]^{1/2} \tag{29.12}$$

FIG. 29.13 Bleed relay.

The nomenclature is from Fig. 29.13 with Y as the expansion factor and K as the flow coefficient. Areas are in square inches with pressures in pounds per square inch absolute. The steady-state air consumption is obtained by using Eq. (29.1) across either restriction, with the flow in standard cubic feet per hour given by 47,000 W. Figures 29.14 and 29.15 show the characteristics typical of the $\frac{1}{16}$-in port designs with a supply pressure of 35 lb/in² abs and a volume load of 1 in³. The seat-to-seat distance L is sometimes made adjustable so that the gain can be varied. The magnitude of gain variation obtained by this method is limited by the effect of L on linearity, air consumption, and the time constant.

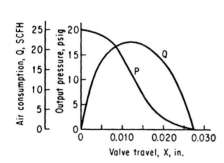

FIG. 29.14 Typical bleed-relay characteristics.

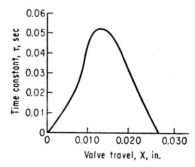

FIG. 29.15 Effect of valve position on bleed-relay time constant.

Where the bleed relay is used without a transmission line, its transfer function, obtained from Eq. (29.10), is

$$\mathscr{L}P/\mathscr{L}X = G/(\tau S + 1) \tag{29.13}$$

with $G = dP/dX$ from the calculated characteristic curve and $\tau = gV/C^2(\partial W_1/\partial P + \partial W_2/\partial P)$. For a pressure input, the time constant is the same and G is dP/dP_3. When the output connects to a transmission line, the general relationship (29.9) must be used, with Z obtained from the characteristics of the line and the load.

The bleed relay has no dead zone and therefore can be designed with a small diaphragm offering a high input impedance (low equivalent volume). The input impedance is also affected by the spring rate [see Eq. (29.7)] so that high gains obtained by low spring rates increase the load on the preceding stage. Since there is no internal pressure feedback, the time constant for a fixed pressure range and load is determined by the port size—increasing the port size decreases the time constant but also increases the steady-state air consumption.

29.7 PRESSURE DIVIDERS

The three-way valve may also be used as a pressure divider, as shown in Figure 29.16. One application is in the feedback path of a controller to vary the closed-loop gain. As in the case of the bleed relay, the dynamic response is characterized by a first-order lag with the time constant equal to $gV/C^2(\partial W_1/\partial P + \partial W_2/\partial P)$. In the subcritical region the gain dP/dP_1 is obtained by plotting the characteristic curve from Eq. (29.12).

$$P = \tfrac{1}{2}\{(-P_1/C_1 - P_0) + [(P_1/C_1 - P_0)^2 + 4P_1^2/C_1]^{1/2}\} \tag{29.14}$$

where $C_1 = (A_2 Y_2 K_2 / A_1 Y_1 K_1)^2$.

Over the usual pressure ranges used in process instrumentation, the ratio $Y_2 K_2 / Y_1 K_1$ remains very close to unity, and in spite of the apparent nonlinearity of the relationship given by Eq. (29.14), the gain is essentially constant. Figure 29.17 is a plot of Eq. (29.14) for an area ratio of 1.11, showing a nearly constant gain of 0.54.

The way in which the time constant and air consumption vary with valve position is the same as for the bleed relay, with τ and Q becoming smaller as P_1 is increased.

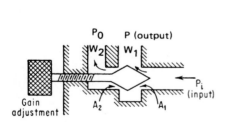

FIG. 29.16 Pressure divider.

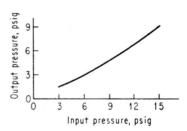

FIG. 29.17 Pressure-divider characteristics.

29.8 NONBLEED RELAYS

Figure 29.18 shows the design of a relay that uses air only during transients. Variations include relays with a spring force rather than a pressure input and different diaphragm arrangements to balance plug forces and minimize dead zone.

A generalized transfer function for the nonbleed relay with a simple volume termination can be approximated by

$$\mathscr{L}P_2/\mathscr{L}P_1 = (A_1/A_2)/(K_s gV/C^2 C_1 A_2)S + 1 \tag{29.15}$$

where K_s = main spring rate, lb/ft
A_1 = input area, ft^2
A_2 = feedback area, ft^2
C_1 = valve gain, (lb/s)/ft

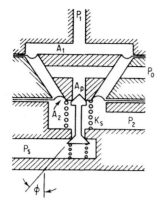

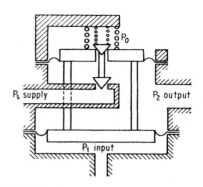

FIG. 29.18   Nonbleed relay.

FIG. 29.19   High-gain nonbleed relay.

The value of C_1 can be obtained from Eq. (29.1) or (29.2) with $A = \pi D \sin \phi$. C_1 will vary with P_2 and will in general be different for exhaust and supply. This results in a transfer function which is different for exhaust or supply, and a detailed nonlinear simulation or experimental means are usually required to properly model the relay.

The relay gain is seen to be dependent only on the area ratio A_1/A_2 with the time constant proportional to $VK_s/A_2D \sin \phi$. The dead zone as a fractional part of the input signal is, from a force balance,

$$\text{Dead zone} = [A_p(P_s - P_0) + F_s]/A_1(\Delta P_1) \qquad (29.16)$$

where ΔP_1 is the input signal range and F_s is the force exerted by the inner valve spring with the system in a neutral position. The effect of the dead zone is minimized by keeping the ratio A_1/A_p as large as practical, and the steady-state error is eliminated by unavoidable (and sometimes intentional) leakage across the valve ports.

Flow-closed valves have a tendency to sustain high-frequency oscillations unless damping is provided. The normal input system provides some damping (see Sec. 29.9), but unless other damping is provided the port size and pressure drop are limited. In the low-pressure range $\frac{1}{8}$-in ports are common; $\frac{1}{2}$-in ports with 150-lb/in^2 drop have been used in conjunction with air or dry-friction dampers.

Because of the internal feedback in these relays they are characterized by relatively low gain (usually less than 5) and small time constants. Variations in the design to produce high gain (Fig. 29.19) give a corresponding increase in the time constant as indicated by the transfer function.

Another variation includes a bypass restriction around the exhaust port of such size that the supply port is throttled but the exhaust port never opens on normal signal amplitudes. The transfer function for small-amplitude signals is given by Eq. (29.9), with feedback added, and for large signals is approximated by the relationships developed for the nonbleed relay. Some designs omit the exhaust end of the inner valve in favor of a small fixed restriction. It should be noted that the input to a nonbleed relay appears as a spring-loaded diaphragm only when the relay time constant is much larger than the time constant of the device driving the relay. Where this is not the case, the flow utilized in moving the relay must be fed back to the flow-summing point in the preceding stage.

29.9 AIR SPRINGS AND DAMPERS

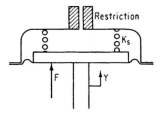

FIG. 29.20 Pneumatic damping device.

Figure 29.20 shows a component used as a damping device in various control mechanisms. With the restriction closed, the rate of the air spring alone, from $PV^k = $ constant and the spring-rate relationship, is

$$K_a = 1.4A^2P/V \qquad (29.17)$$

where K_a = air spring rate, lb/ft
A = diaphragm area, ft^2
P = pressure, lb/ft^2 abs
V = chamber volume, ft^3

The rate is nonlinear, increasing with P. It is linearized by using a mean pressure for P and restricting the analysis to a narrow range of pressures around this mean. At very low frequencies the process becomes isothermal and we would use 1.0 in place of 1.4. Consequently the system shows a higher spring rate at high frequencies than it does at low frequencies. The break point associated with this change could be determined by a heat transfer analysis but is probably less than 0.01 Hz. The equations in this section will neglect this low-frequency change, but it should be kept in mind that for certain configurations the steady-state characteristics will depend upon the isothermal case. An analysis of an air-filled dashpot with a linear restriction is given by Lloyd.[12] Figure 29.21 shows the block diagram used for the linearized analysis.

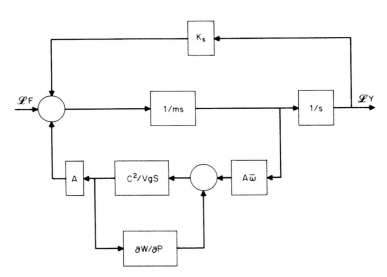

FIG. 29.21 Block diagram of pneumatic damping device.

The transfer function is

$$\frac{\mathscr{L}Y}{\mathscr{L}F} = \frac{(1/K_s)(\tau S + 1)}{\tau(s^3/\omega_n^2) + s^2/\omega_n^2 + [1 + (K_a/K_s)]\tau S + 1} \qquad (29.18)$$

where $\omega_n = \sqrt{K_s/M}$, undamped natural frequency of the spring mass system
  K_a = air spring rate, $A^2 \overline{\omega} C^2/Vg$, lb/ft
  K_s = mechanical spring rate, lb/ft
  $\tau = Vg/C^2(\partial W/\partial P)$

29.10 PNEUMATIC TRANSMISSION LINES

The output of a pneumatic component is often connected to the load volume by means of a length of pipe or tubing. The dynamic characteristics of the line affect the performance of both the component and the overall system. A treatment of transmission-line dynamics is complicated by the fact that the line's resistance, capacitance, and inductance are distributed over its length.

Two transfer functions are of interest: $\mathscr{L}P_2/\mathscr{L}P_1$ for the input/output characteristics and $\mathscr{L}P_1/\mathscr{L}W_1$ (Z_i, driving-point impedance) to describe the loading effects of the line upon the component.

Rohmann and Grogan[10] have treated this subject by analogy with electrical transmission lines. The application of their equations to a typical system is shown in Figs. 29.22 and 29.23. They give, as a low-frequency approximation to their complete solution, the following transfer functions:

$$\frac{P_r}{P_i} = \frac{1}{1 + [(2g_0 + 1)/2]\{RC\ell^2 S[(L/R)S + 1]\}} \tag{29.19}$$

$$Z_i = \frac{1}{(1 + g_0)C\ell S} \frac{1 + [(2g_0 + 1)/2][RC\ell^2 S (L/R\,S + 1)]}{1 + \frac{1}{6}[(3g_0 + 1)/(g_0 + 1)]\{RC\ell^2 S[(L/R\,S + 1)]\}} \tag{29.20}$$

where P_r = Laplace transform of terminal pressure, lb/in^2
  P_i = Laplace transform of input pressure, lb/in^2
  $g_0 = C_r/C\ell$
  R = resistance per unit length, (lb·s/in^5)/in = $(1.2)8\mu/\pi r^4$
  C = capacitance per unit length, (in^5/lb)/in
    = $\pi r^2/np$
  L = inertia per unit length, (lb·s^2/in^5)/in
    = $\delta/\pi r^2$
  ℓ = length of tube, in
  Z_i = driving-point impedance, lb·s/in^5
  C_r = capacitance of receiver, in^5/lb (V/np)
  I_i = volumetric flow rate, in^3/s
  δ = mass density, lb·s^2/in^4
  n = polytropic exponent
  V = receiver volume, in^3
  p = initial system pressure, lb/in^2 abs
  μ = viscosity, lb·s/in^2
  r = tube inside radius, in

The input-output relationship is adequate for the majority of applications since the dynamics of the process and the final control element will restrict the frequency ranges of interest. For lines with large terminal volumes the relationship in Sec. 29.3.1, case 1, offers a further simplification for very low frequencies. In all cases there is a dead time equal to the line length divided by the velocity of sound.

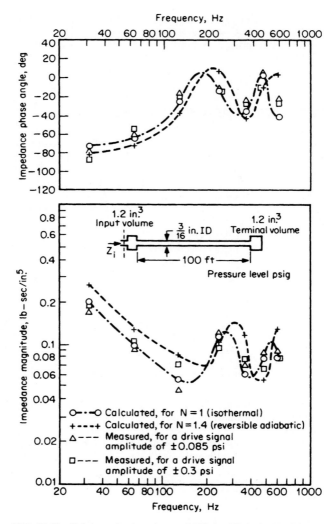

FIG. 29.22 Driving-point impedance of 100 ft of ³⁄₁₆-in inside-diameter copper tubing.

The relationship for driving-point impedance might be used in a preliminary study of a pneumatic component but the complete relationship should be used for accurate work, since the frequencies involved in the component loop may be considerably higher than those of interest in the overall system.

The transient response of pneumatic transmission lines has been investigated by Schuder and Binder.[11] Figure 29.24 indicates the ways in which these lines can respond to step inputs. In practice, most long lines are well damped, and their step response may be approximated by

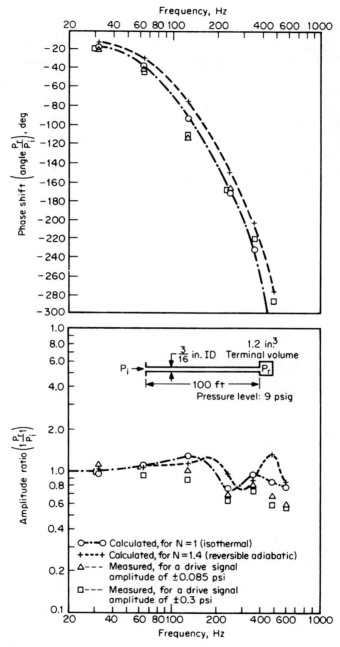

FIG. 29.23 Output-input frequency response of 100 ft of ³⁄₁₆-in inside-diameter copper tubing.

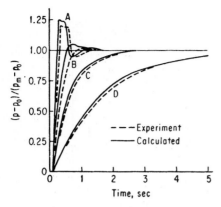

FIG. 29.24 Response of pneumatic transmission lines to step inputs. Curve A: $L = 100$ ft, $Q = 91.2$ in³, OD = ⅜ in. Curve B: $L = 100$ ft, $Q = 294.0$ in³, OD = ⅜ in. Curve C: $L = 100$ ft, $Q = 91.2$ in³, OD = ¼ in. Curve D: $L = 100$ in³, $Q = 294.0$ in³, OD = ¼ in.

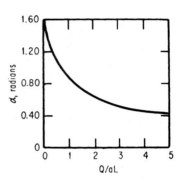

FIG. 29.25 Solution of $\alpha \tan \alpha = aL/Q$.

$$(p - p_0)/(p_m - p_0) = 0 \qquad \text{for } 0 < t < L/c$$

$$(p - p_0)/(p_m - p_0) = 1 - Ge^{-tM} \qquad \text{for } t \geq L/c \tag{29.21}$$

where $G = [1 + R/(\rho\theta)]/\{a[(Q/aL + 1) \sin \alpha + (Q/aL) \cos \alpha]\}$
$M = 1/2 (\theta - R/\rho)$
α = solution of $\alpha \tan \alpha = aL/Q$, rad (from Fig. 29.25)
Q = receiver volume, ft³
a = cross-sectional area of tube, ft²
L = tube length, ft
t = time, s
$\theta = [(R/\rho)^2 - (2 \alpha C/L)^2]^{1/2}$
ρ = air density, slugs/ft³
R = resistance, (lb·s)/ft⁴, $32\mu/d^2$
c = velocity of sound, ft/s
d = tube inside diameter, ft
μ = viscosity, slugs/(ft·s)
p = receiver pressure at time t, lb/ft²
p_0 = initial tubing and receiver pressure, lb/ft²
p_m = step pressure applied at input, lb/ft²

Only a small portion of the work that has been done in this field is covered by Refs. 10 and 11. The additional Refs. 15 through 19 cover other approaches to the problem and contain considerable useful test data.

29.11 DYNAMIC SIMULATION

Simulation of a physical system is an analysis tool which can be used in the evaluation of hardware design prior to actual construction. As computing costs have decreased

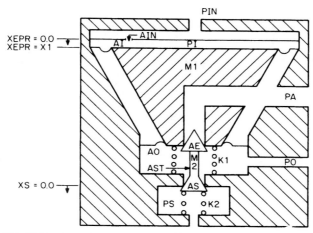

FIG. 29.26 Schematic of nonbleed relay.

and numerical methods improved, digital simulation software packages such as MIMIC, DYSIM, CSMP, DYNAMO, and ACSL began to offer the advantage of analog and hybrid computers and incorporate powerful output printing and plotting routines. Simulation software packages provide a simple method of realistic analysis of pneumatic components, including the nonlinear effects.

The results of an example problem utilizing ACSL (Advanced Computer Simulation Language) are described below. ACSL was attractive for simulating a nonbleed relay because of the nonlinearities present in the model such as dead zone and nonlinear flows. Since ACSL is based on the use of algebraic and differential equations, linearization techniques are not required. Model performance is not limited to small perturbation inputs, and the model yields a nonlinear, more realistic solution.

Figure 29.26 shows a schematic of the simulated nonbleed relay. Variable definitions are described in Table 29.2, and Table 29.3 is a listing of the program. Figures 29.27 through 29.30 show a comparison of simulated relay to actual test results for a sinusoidal and a first-order step input.

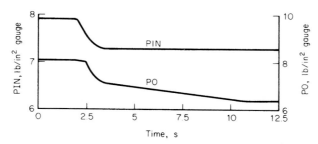

FIG. 29.27 Response of simulated nonbleed relay to first-order step input.

TABLE 29.2   Nonbleed Relay Variable Definitions

AE	EXHAUST PORT AREA, (IN**2)
AEPR1	AEPR IF XE.LT.(.0068), (IN**2)
AEPR2	AEPR IF XE.GE.(.0068), 0.0115 (IN**2)
AEPR	EXHAUST FLOW AREA, (IN**2)
AEZ	EXHAUST PORT LEAKAGE AREA, 1.9E-5 (IN**2)
AI	INPUT DIAPHRAGM AREA, 3.09 (IN**2)
AIN	DAMPING PLATE ORIFICE AREA, 7.67E-4 (IN**2)
AMP	AMPLITUDE OF SINUSOIDAL INPUT PRESSURE, 0.208 (PSIG)
AOAE	A0-AE, 1.15 (IN**2)
AOAST	A0-AST, 1.157 (IN**2)
AS	SUPPLY PORT AREA, (IN**2)
ASAST	AS-AST, 0.012 (IN**2)
ASPR1	ASPR IF XS.LT.(.0045), (IN**2)
ASPR2	ASPR IF XS.GE.(.0045), 0.0122 (IN**2)
ASPR	SUPPLY PORT FLOW AREA, (IN**2)
AST	STEM AREA, (IN**2)
B1	DAMPING COEFFICIENT, (LB/FT/SEC)
B2	DAMPING COEFFICIENT, (LB/FT/SEC)
BEGIN	INITIAL VALUE OF PI, 7.45 (PSIG)
C1	CONSTANT USED IN FLOW CALCULATIONS, 0.048
C5	OUTPUT CAPACITANCE, 5.8E-5 (IN**2)
CIN	INPUT CAPACITANCE, (IN**2)
DR	DAMPING RATIO, 1.0
FREQ	FREQUENCY OF SINUSOIDAL PRESSURE INPUT, 7.85 (RAD/SEC)
K1	BIAS SPRING AND DIAPHRAGM SPRING RATE, 336 (LB/FT)
K2	SPRING RATE OF PLUG RETAINING SPRING, 19.2 (LB/FT)
KY	DISCHARGE COEFFICIENT*EXPANSION FACTOR
KYE	KY FOR EXHAUST PORT
KYS	KY FOR SUPPLY PORT
M1	MASS OF RELAY HEAD, 3.086E-3 (LB*SEC/FT)
M2	MASS OF RELAY PLUG, 1.229E-4 (LB*SEC/FT)
PAABS	ABSOLUTE ATMOSPHERIC PRESSURE, 14.3 (PSIA)
PI	PRESSURE ACTING ON INPUT DIAPHRAGM, (PSIG)
PIABIC	INITIAL VALUE OF PIABS, 21.75 (PSIA)
PIABS	ABSOLUTE PI, (PSIA)
PIE	3.416
PIN	INPUT PRESSURE, (PSIG)
PINABS	ABSOLUTE INPUT PRESSURE, (PSIA)
PO	OUTPUT PRESSURE, (PSIG)
POABIC	INITIAL VALUE OF POABS, 21.26 (PSIA)
POABS	ABSOLUTE RELAY OUTPUT PRESSURE, (PSIA)
POABSH	DUMMY VARIABLE USED TO LIMIT POABS
PR	DAMPING PLATE PRESSURE RATIO
PR1	PR WHEN PI.GT.PIN
PR2	PR WHEN PIN.GT.PI
PRE	POABS/PSABS
PRS	PAABS/POABS
PS	SUPPLY PRESSURE, (PSIG)
PSABS	ABSOLUTE SUPPLY PRESSURE, (PSIA)
PSAS	PS*AS, 0.362 (LB)
SQT	SQRT(P1 * (P1-P2)), (PSIA)
SQTE	EXHAUST PORT SQT, (PSIA)
SQTS	SUPPLY PORT SQT, (PSIA)
TSTP	MAXIMUM SIMULATION TIME, (SEC)
VDEAD	SMALL VOLUME, 0.0143 (IN**3)
W	FLOW THROUGH DAMPING PLATE ORIFICE, (LB/SEC)
WE	EXHAUST PORT FLOW, (LB/SEC)
WS	SUPPLY PORT FLOW, (LB/SEC)
X1	MAXIMUM VALUE OF XEPR, 0.0058 (FT)
XE	RELAY HEAD POSITION RELATIVE TO PLUG, (FT)
XEPDIC	INITIAL CONDITION ON XEPRD, (FT/SEC)
XEPR	RELAY HEAD POSITION REL. TO ZERO PRESSURE START PT., (FT)
XEPRD	TIME DERIVATIVE OF XEPR, (FT/SEC)
XEPRDD	SECOND TIME DERIVATIVE OF XEPR, (FT/SEC**2)
XEPRIC	INITIAL CONDITION ON XEPR, (FT)
XEPRH	DUMMY VARIABLE FOR XEPR, (FT)
X01	BIAS SPRING INITIAL COMPRESSION, .0368 (FT)
X02	PLUG RETAINING SPRING INITIAL COMPRESSION, 0.0268 (FT)
XS	RELAY HEAD/PLUG POSITION REL. TO SUPPLY PORT, (FT)
XSD	TIME DERIVATIVE OF XS, (FT/SEC)
XSDD	SECOND TIME DERIVATIVE OF XS, (FT/SEC**2)
XSDIC	INITIAL CONDITION ON XSD, 0.0 (FT)
XSIC	INITIAL CONDITION ON XS, 0.0 (FT)
XSH	DUMMY VARIABLE FOR XS, (FT)
XSMAX	MAXIMUM VALUE OF XS, 0.0069 (FT)

TABLE 29.3 Nonbleed Relay Program Listing

```
PROGRAM RELAY
     CINTERVAL CINT = .001  $  ALGORITHM IALG = 3   $  NSTEPS NSTP = 1
     B1  =  2 * DR * SQRT(K1*M1)
     B2  =  2 * DR * SQRT ( (K1 +K2) * (M1 + M2) )
     PIN = BEGIN*AMP*HARM(1.0,FREQ,0.0)
     PINABS = PIN + PAABS
     PR1 = PIABS/PINABS
     PR2 = PINABS/PIABS
     PR = RSW(PIN.GT.PI,PR1,PR2)
     KY = RSW(PR.LE.0.53,0.2245*PR+.4143,0.1419*PR+.4554)
PROCEDURAL  (SQT=PINABS,PIABS)
     IF(PINABS.GT.PIABS)SQT=SQRT(PINABS*(PINABS-PIABS))
     IF(PINABS.LE.PIABS)SQT=-SQRT(PIABS*(PIABS-PINABS))
END $ 'OF PROCEDURAL'
     W = AIN * C1 * KY * SQT
     CIN = (VDEAD + (AI*12.0*(XEPR+XS))) * 2.21763E-6
     PIABS = INTEG(W/CIN,PIABIC)
     PI = PIABS-PAABS
     XE = X1 - XEPR
     AEPR1 = 2.35619449*XE-97.94516567*(XE**2)+AEZ
     ASPR1 = 2.940530724*XS-97.94516568*(XS**2)
     AEPR = RSW(XE.LT.(0.0812/12.0),AEPR1,AEPR2)
     ASPR = RSW(XS.LT.(0.0539/12.0),ASPR1,ASPR2)
     PSABS = PS + PAABS
     PRS = POABS/PSABS  $  PRE = PAABS/POABS
     KYS = RSW(PRS.LE.0.53,.2245*PRS+.4130,.1419*PRS+.4554)
     KYE = RSW(PRE.LE.0.53,.2245*PRE+.4130,.1419*PRE+.4554)
     SQTS = SQRT(PSABS*(PSABS-POABS))
     SQTE = SQRT(POABS*(POABS-PAABS))
     WS = ASPR * C1 *KYS * SQTS
     WE = AEPR * C1 *KYE *SQTE
     PO = POABS - PAABS
     DBLINT (XEPRH, XEPRD = XEPRIC,XEPRDD,XEPDIC,0.0,X1)
     XEPR = BOUND(0.0,X1,XEPRH)
     DBLINT(XSH,XSD = XSIC,XSDD,XSDIC,0.0,XSMAX)
     XS = BOUND(0.0,XSMAX,XSH)
     XEPRDD = ((PI*AI)-(PO*AOAE) - (B1*XEPRD) -(K1*...
     (X01 + XEPR)))/M1
     XSDD = ((PI * AI) - (PO * AOAST)-(PSAS -(PO*ASAST))...
     -(B2*XSD)-(K1*(X01+X1+XS)) - (K2*(X02+XS))/(M1+M2)
     POABSH = LIMINT((WS-WE)/C5,POABIC,14.3,34.3)
     POABS = BOUND(14.3,34.3,POABSH)
     TERMT (T.GE.TSTP)
END $ 'OF PROGRAM'
```

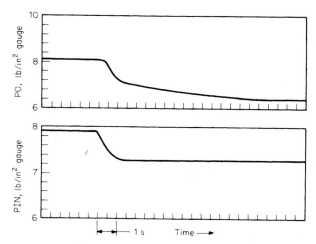

FIG. 29.28 Test results of actual nonbleed relay to first-order step input.

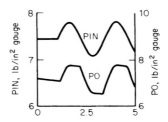

FIG. 29.29   Response of simulated nonbleed relay to sinusoidal input.

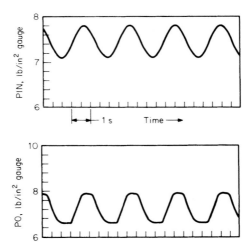

FIG. 29.30   Test results of actual nonbleed relay to sinusoidal input.

REFERENCES

1. Binder, R. C.: "Fluid Mechanics," Prentice-Hall, Inc., Englewood Cliffs, N.J., 1973.

2. Lloyd, S. G., and G. D. Anderson: "Industrial Process Control," Fisher Controls Co., Marshalltown, Iowa, 1971.

3. Blackburn, J. F., G. Reethof, and J. L. Shearer: "Fluid Power Control," John Wiley & Sons, Inc., New York, 1960.

4. Schuder, C. B.: "Thermodynamics of Flow Through Control Valves," *Instrumentation Technology,* June 1973.

5. Cunningham, R. G.: "Orifice Meters with Supercritical Compressible Flow," *Trans. ASME,* vol. 73, p. 625, July 1951.

6. Perry, J. A., Jr.: "Critical Flow through Sharp-Edged Orifices," *Trans. ASME,* vol. 71, p. 757, October 1949.

7. Feng, Tsun-Ying: "Static and Dynamic Characteristics of Flapper-Nozzle Valves," *Trans. ASME, J. Basic Eng.,* vol. 81, p. 275, 1959.

8. Elrod, H. G., Jr.: "Theory of Ejectors," *J. Appl. Mech.,* June 1942.

9. Hayashi, S., T. Matsui, and T. Ito: "Study of Flow and Thrust in Nozzle-Flapper Valves," *J. Fluids Eng., Trans. ASME,* March 1975.

10. Rohmann, C. P., and E. C. Grogan: "On the Dynamics of Pneumatic Transmission Lines," *Trans. ASME,* vol. 79, p. 853, 1957.

11. Schuder, C. B., and R. C. Binder: "The Response of Pneumatic Transmission Lines to Step Inputs," *Trans. ASME, J. Basic Eng.,* vol. 81, 1959.

12. Lloyd, S. G.: "The Dynamics of an Air-Filled Dashpot with Linear Restriction," ISA paper 17.10-4-65.

13. Mitchell and Gauthier Associates: "Advanced Continuous Simulation Language, User Guide/Reference Manual," P.O. Box 685, Concord, Mass. 01742, 1975.

14. Cheng, R. M. H., and O. G. Harris: "Characteristics of a Pneumatic Pressure Relay Valve," *Proceedings 1979 Summer Computer Simulation Conference,* Toronto, Ont., 1979.

15. Moore, E. F., and M. E. Franke: "The Small Signal Response of Annular Pneumatic Transmission Lines," *J. Fluids Eng., ASME,* December 1974.

16. Schuder, C. B., and G. C. Blunck: "The Driving Point Impedance of Fluid Process Lines," *ISA Trans,* vol. 2, no. 1, January 1963.

17. Sidell, R. S. and D. N. Wormley: "An Efficient Simulation Method for Distributed Lumped Fluid Networks," *Trans. ASME,* Series G, *J. Dyn. Sys. Meas. Cont.,* March 1977.

18. Goodson, R. E., and R. G. Leonard: "A Survey of Modeling Techniques for Fluid Line Transient," *Trans. ASME,* Series D, *J. Bas. Eng.,* June 1972.

19. Katz, S.: "Transient Response of Terminated Pneumatic Transmission Lines by Frequency Response Conversion," Fluidics Committee, ASME, 1976.

20. Stenning, A. H.: "An Experimental Study of Two-Dimensional Gas Flow Through Valve-Type Orifices," ASME paper 54-A-45.

21. Tsai, D. H., and E. C. Cassidy: "Dynamic Behavior of a Simple Pneumatic Pressure Reducer," ASME Paper 60-Wa-186.

22. Reethof, G.: "Analysis and Design of a Servomotor Operating on High Pressure Compressed Gas," *Trans. ASME,* vol. 79, p. 875, May 1957.

23. Benedict, R. P.: "The Response of a Pressure-Sensing System," ASME Paper 59-A-289.

24. Shearer, J. L.: "Study of Pneumatic Processes in the Continuous Control of Motion with Compressed Air," *Trans. AMSE,* parts I, II, vol. 78, p. 233, February 1956.

25. Newell, Floyd B.: "Diaphragm Characteristics, Design and Terminology," ASME, New York.

26. Liu, F. F.: "Dynamic Response Behavior of Diaphragms in Relation to Driving Function and Surrounding Media," paper no. 58-A-224, ASME meeting, Nov. 30–Dec. 5, 1958.

27. Matheny, James D.: Bellows Spring Rate for Seven Typical Convolution Shapes, *Machine Design,* Jan. 4, 1962.

28. Merriman, D. B.: "Modeling Friction in a Spring-Mass-Damper System," Technical Report T-79-72, U.S. Army Missile Research and Development Command, Redstone Arsenal, Ala. 35809, 1979.

HYDRAULIC COMPONENTS

Anthony J. Healey, Ph.D.

Consultant
Marine Division, Brown and Root, Inc.
Houston, Tex.

30.1   INTRODUCTION

Hydraulic components find their widest application as the actuation (power-output) elements of power-control systems. The more important advantages of hydraulic systems are summarized as follows:

1. High-pressure hydraulic power can be generated efficiently, with pump efficiencies of 92 percent common.

2. Hydraulic components are comparatively light in weight compared with equivalent mechanical and electrical components because the highly stressed structures of the hydraulic system make very efficient use of structural material. Hydraulic pumps and motors with a power density of less than 1 lb/hp are common. This light weight is made possible by the high pressures now available from commercially

available pumps. Hydraulic systems operating at 3000 lb/in^2 are quite common and higher-pressure systems are available.

3. As seen by the load, the hydraulic actuator is extremely stiff compared with an equivalent pneumatic or electrical system. The hydraulic approach permits the maintenance of load position against significant and varying load forces, with lower loop gain and higher response speed.

4. Hydraulic actuation offers the highest torque (or force) to inertia ratio in comparison with most mechanical, pneumatic, and electrical systems. This property, coupled with the incompressible nature of the medium, results in exceptionally fast response and high power output.

Some of the disadvantages of hydraulic systems follow:

1. Most hydraulic systems use organic-base fluids which present serious fire and explosion hazards because of fluid spillage from leaks in piping or seals as well as breaks. The wide use of hydraulics in the die-casting and machine-tool field has resulted in the search for nonflammable hydraulic fluids, with good lubricity and no toxicity. Although good progress has been made, no fully satisfactory fluid is available at this time. Section 30.2 will discuss some of the fluids and their relative fire-hazard problems.

2. Associated with the fire-hazard problem is the inherent difficulty of preventing leaks in normal usage and the subsequent "messiness" of the hydraulic system. This problem can be avoided at the expense of maintainability by all-welded plumbing.

3. High-speed-of-response hydraulic systems are sensitive to solid contaminants in the fluid which interfere with the smooth and proper function of the control valves and have been known to cause either poor operation or outright failure. Cleanliness in manufacture, assembly, normal operation, and service is an absolute necessity to assure good reliability. Adequate provision for filtering is only one of the necessities of good hydraulic design.

Even a cursory review of the design requirements for such systems as high-speed steel rolling-mill tension controls, jet-engine controls, or radar-antenna drives conveys the need for the building of comprehensive and representative analytical models to assure satisfactory dynamic response and adequate stability throughout widely varying operating conditions and realistic tolerances.

The purpose of this section is to provide the designer with necessary basic building blocks to permit the preparation of an analytical model of the hydraulic system for experimentation and design specification. For illustrations of present design practice of commercially available hydraulic components, see Ref. 38.

30.2 PROPERTIES OF HYDRAULIC FLUIDS

30.2.1 Density and Related Properties

Density ρ is defined as the mass per unit of volume.

Specific weight W is defined as the weight per unit of volume.

Specific gravity σ is the ratio of the density of the substance in question to that of water at 60°F.

The petroleum industry uses a measure of relative density called "API gravity." API gravity in terms of specific gravity is given by Eq. (30.1):

$$\text{Degrees API} = 141.5/(\sigma 60°F/60°F) - 131.5 \tag{30.1}$$

$\sigma 60°F/60°F$ represents the specific gravity of the substance at 60°F relative to water at 60°F.

Specific gravity in terms of degrees API is given by

$$\sigma 60°F/60°F = 141.5/(\text{deg API} + 131.5) \tag{30.2}$$

The density of a liquid is a function of both pressure and temperature. At any one temperature a good approximation is given by Eq. (30.3):

$$\rho = \rho_0(1 + aP - bP^2) \tag{30.3}$$

where P is pressure and ρ_0 is the density at the reference condition. Empirical constants a and b are functions of temperature. Typical values for hydraulic oils at $\sigma 60°F$ are[1]

$$a = 4.38 \times 10^{-6} \, \text{in}^2/\text{lb}$$

$$b = 5.65 \times 10^{-11} \, \text{in}^4/\text{lb}^2$$

The decrease of density with increasing temperature (thermal expansion) is approximated by Eq. (30.4):

$$\rho = \rho_1[1 - \alpha(T - T_1)] \tag{30.4}$$

where α is called the cubical-expansion coefficient. This linear approximation is accurate within 0.5 percent for most hydraulic fluids over temperature ranges of 500°F.

Compressibility K is a most important property to the designer of fast-response hydraulic systems. It is defined as the unit rate of change of density with change in pressure.

For small pressure changes

$$K \triangleq (1/\rho)(d\rho/dP) \tag{30.5}$$

The *bulk modulus* β of a fluid is the reciprocal of compressibility,

$$\beta \triangleq \rho(dP/d\rho) \tag{30.6}$$

From Eq. (30.3), compressibility and bulk modulus are respectively

$$K = (a - 2bP)/(1 + aP - bP^2) \tag{30.7}$$

$$\beta = (1 + aP - bP^2)/(a - 2bP) \tag{30.8}$$

Typical values of β are given in the table of fluid properties (Table 30.1).

The computed values of bulk moduli for a given fluid should be used with great caution in the analysis of hydraulic control systems. The effective compressibility as seen by the flow source must be considered. Entrained air and transmission conduit compliance can drastically reduce the bulk modulus of the pure liquid.

30.2.2 Viscosity

A true fluid is commonly defined as matter which is unable to store energy in shear; it is not elastic in shear. Therefore, any shear force applied to a fluid film will result in a

TABLE 30.1 Properties of Hydraulic Fluids

Type	Light turbine	Phosphate ester	Chlorinated hydrocarbon	SAE 30	Petroleum base MIL-H-5606 A	Phosphate ester
Use	Petroleum industry	Industrial fire resistant	Industrial fire resistant	Mobile petroleum	Aircraft missile	Aircraft missile
Typical commercial	DTE light	Pydraul F9	Aroclor 1242		Esso Univis J43	Skydrol 500A
Specific gravity, 60°F/60°F	0.865	1.20	1.32	0.887	0.848	1.086
Expansion coefficient α, 1/°F	4.21×10^{-4}	4×10^{-4}		4.21×10^{-4}	0.5×10^{-3}	
Viscosity, cs:						
−40°F					2130/−65°F	
0°F		18,000/20°F	2,000/30°F		500	240
100°F	32	47	18	120	14.3	11.5
210°F	5.2	6.3	2.0	12	5.1	3.92
400°F		2.5/300°F			1.9	
550°F						
700°F						
Viscosity-temperature coefficient, λ, 1/°F	1.85×10^{-2}				1.04×10^{-2}	
Pour point, °F	0	−5	2	0	−90	−85
Fire point, °F	445	675	None	495	235	425
Flash point, °F	400	430	348	435	225	360
Autogenous ignition temperature, °F		1100			700	>1100
Bulk modulus $\times 10^{-3}$ psi	≈250	387		≈250	270	387
Thermal conductivity, Btu/(hr)(ft^2)(°F/ft)		0.065/70°F	0.053	0.082	0.0615	0.0365
Specific heat, Btu/(lb) (°F)		0.33	0.285/60°F		0.5/100°F	0.42/68°F

TABLE 30.1 Properties of Hydraulic Fluids (*Continued*)

Type	Halogenated silicone	Silicone ester	Diester MIL-L 7808 C	Poly-phenyl ether	NAK 77	Mercury	JP 6
Use	Aircraft missile	Aircraft missile	Jet-engine lubrication	Aircraft nuclear	Aircraft nuclear		Jet fuel
Typical commercial	G.E. F50	Oronite 8200					
Specific gravity, 60°F/60°F	1.03	0.93	0.93	1.18	0.87	12.76/662°F, 13.55/68°F	0.75 to 0.84
Expansion coefficient α, 1/°F		0.445×10^{-3} 2400/−65°F	0.44×10^{-3}				
Viscosity, cs:							
−40°F	934	630	1920			0.1305	5.22
0°F	287	195	230			0.1087	2.6
100°F	60	32.5	14.2	609	6.7	0.0935	0.97
210°F	21.2	11.3	3.6	6.0	0.5	0.0797	0.53
400°F	6.5	3.8	1.2	1.3	0.4	0.0745	
550°F	3.5	2.2	0.8	0.9	0.3	0.0710	
700°F	2.2			0.5	0.2		
Viscosity-temperature coefficient, λ, 1/°F							
Pour point, °F	−100	<−100	<−75	+5	+10		
Fire point, °F	>650	450	485		NA		
Flash point, °F	>525	390	430		NA		
Autogenous ignition temperature, °F	900	760	485	465	NA		449
Bulk modulus × 10⁻³ psi	150/75°F, 30/700°F	218	290/0°F, 60/440°F			2,600/90°F, 1,400/600°F	170/100°F, 80/200°F, 30/400°F
Thermal conductivity, Btu/(hr)(ft²)(°F/ft)		0.070	0.084/200°F			4.7/50°F, 8.1/600°F	0.087/0°F, 0.082/210°F
Specific heat, Btu/(lb)(°F)	0.37/75°F, 0.51/500°F	0.47/68°F	0.35/65°F, 0.60/400°F			0.033 2240°F(Tc)* −38°F(FP)† 674.4°F(BP)‡	0.43/0°F, 0.53/210°F

*Tc—critical temperature. †FP—freezing point. ‡BP—boiling point.

finite shear rate. A Newtonian fluid is one for which shear rate is proportional to the shear stress. A constant of proportionality μ, the absolute viscosity, is defined by the following expression:

$$\mu = \tau/(du/dx) \tag{30.9}$$

where τ is the shear stress, du the incremental change in velocity resulting from the shear stress, and x the direction of the shear stress.

Most hydraulic fluids behave like Newtonian liquids up to shear rates of about 1,000,000 lb/in²/s. However, some of the highly compounded high-temperature hydraulic fluids may lose as much as 40 percent viscosity as the long-chain molecules are oriented or sheared.[2] This loss is temporary in some cases, with the original viscosity restored after an appreciable lapse of time. In other cases the loss is permanent.

Since several units of viscosity are in use, they should be carefully defined:

Reyn.   A very large inconvenient unit in the English system.

$$1 \text{ Reyn} \equiv 1 \text{ lb s/in}^2$$

Centipoise (cP) (metric system).   One centipoise is the viscosity of a fluid such that a force of 1 dyne will give two parallel surfaces 1 cm² area, 1 cm apart, a velocity of 0.01 cm/s. The centipoise is thus 0.01 dyne-s/cm².

Centistoke (cSt).   Absolute viscosity divided by density is defined as kinematic viscosity. The common units are the stoke and the centistoke corresponding to the poise and the centipoise, respectively, divided by the density in consistent units. The centistoke is thus 0.01 cm²/s.

Saybolt universal second.   The Saybolt viscosimeter is commonly used to determine the viscosity of petroleum products. The time required for 60 mL of the sample to flow through a 0.176-cm-diameter and 1.225-cm-long tube is measured and designated SSU.

The following lists various conversion factors: 1 lb·s/in² = 68,747.2 poises, 1 lb·s/ft² = 478.8 poises, 1 in²/s = 6.4516 stokes, 1 ft²/s = 229.03 stokes, 1 dyne·s/cm² = 1 poise, and 1 cm²/s = 1 stoke.

Effect of Pressure.   The viscosity of liquids increases with pressure in a manner approximated as follows:

$$\log_{10} \mu/\mu_0 = cP \tag{30.10}$$

where c is a constant which for petroleum products at room temperature is approximately[3]

$$c = 7 \times 10^{-4} \text{ in}^2/\text{lb}$$

Effect of Temperature.   The viscosity of hydraulic fluids decreases markedly with increasing temperature. Of the many empirical formulas which have been proposed to describe the variation, Eq. (30.11) has found broadest acceptance.

$$\mu_t = \mu_0 e^{-\lambda(T-T0)} \tag{30.11}$$

where T is the temperature °F, T_0 is the reference temperature, and λ is the temperature coefficient of viscosity.

For most petroleum-base oils an empirical formula called the *Walther formula* is a better approximation.[4] It is valid for relatively pure oils in the temperature range -20

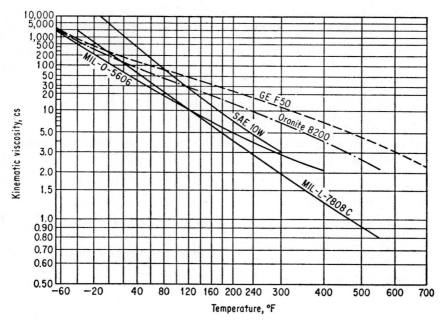

FIG. 30.1 Viscosity-temperature curves for various hydraulic fluids.

to $+160°F$. The ASTM charts[5] are based on this approximation. Figure 30.1 gives the viscosity-temperature characteristics of several typical hydraulic fluids on the ASTM chart. The temperature coefficients of viscosity of several typical hydraulic fluids are given in Table 30.1.

Pour Point.[6] The pour point is the temperature at which a fluid will no longer pour from a standard container when tested according to a standard ASTM procedure. It is considerably below the temperature which can be considered a practical lower limit for use in hydraulic systems. A viscosity in excess of 2500 cSt is not generally practical for hydraulic-system use.

30.2.3 Chemical Properties

Thermal stability relates to the fact that some hydraulic fluids when heated to high temperatures either decompose to form gaseous, liquid, or solid products or polymerize to form gels, varnishes, or even cokes.

Oxidative stability relates to the reaction of the hydraulic fluid with the oxygen of either the atmospheric air, the dissolved air, or other oxidizing agents. Since the products of oxidation are often acidic in nature, corrosion problems may result. Sludges and varnishes which result can clog filters, freeze sleeve valves, and foul orifices. Oxidation problems arise primarily in high-temperature applications.

Hydrolytic stability relates to the reaction of the hydraulic fluid with free water.

Fire Safety. Standard tests have been devised to rate the relative "fire safety" of various fluids.[7]

The *flash point* is the temperature at which sufficient vapors are evolved from the fluid in a heated cup to cause a transient flame when a pilot flame is brought into the test area.

The *fire point* is the temperature at which the flame above a test cup will be self-sustaining.

The *autogenous ignition* temperature is considerably above the fire point and is the temperature at which a liquid droplet will ignite upon contact with heated air.

Compatibility of hydraulic fluid relates to the property of the fluid to either be affected or affect surrounding metallic and nonmetallic materials.

Toxicity. Several of the special hydraulic fluids for use in high-temperature applications contain special additives which may be injurious when inhaled as vapor.

30.2.4 Thermal Properties

The specific heat is the amount of heat which must be supplied to a unit mass to raise its temperature one degree.

For most fluids the specific heat increases with temperature. The specific heats of pure petroleum-base oils are well approximated by a Bureau of Standards equation

$$C = (1/\sqrt{\sigma})(0.388 + 0.00045T) \tag{30.12}$$

where C = specific heat, Btu/lb·°F
σ = specific gravity at 60°F/60°F
T = temperature, °F

The thermal conductivity for petroleum-base oils

$$k = (0.813/\sigma)[1 - 0.0003(T - 32)] \qquad \text{Btu/h·ft}^2\text{·°F·in} \tag{30.13}$$

30.2.5 Surface Properties

Two areas involving surface energies which are of interest to the designer of hydraulic equipment are foaming and boundary lubrication. A foam is an emulsion of gas bubbles in a liquid. Since foam is many times more compressible than the gas-free liquid, the presence of foam in a servosystem will drastically decrease the bulk modulus of the mixture depending on the volumetric proportions of foam and liquid in the chamber. Developers of hydraulic fluids frequently add antifoaming agents.[6]

Friction and wear between metal surfaces in sliding are strongly affected by the molecular structure of the fluid-metal interfaces. Boundary lubrication relates to physicochemical relations which occur in very thin films. "Oiliness" is sometimes defined as a property of a fluid which will give low coefficients of friction to two sliding surfaces.

30.2.6 Choice of Hydraulic Fluid

System performance, both steady and transient, is affected by fluid properties as follows:[3,4]

Viscosity affects damping effects, pipe flow, lubrication, leakage, motor and pump efficiencies.

Density affects orifice flow, acoustic effects, pump and motor efficiency.

Specific heat and thermal conductivity combined with viscosity and density affect temperature rise and heat rejection.

Compressibility is all-important in determining transmission characteristics, stability and response of closed-loop control systems, and pressure pulsations from pumps.

Vapor pressure and gas solubility affect cavitation effects and the associated pump-filling problem as well as the potential effects on compressibility.

Hydraulic-system life and reliability are closely associated with such fluid properties as:

Lubricity. Boundary lubrication affects wear in pumps and motors, and friction in sliding valves.

Thermal stability, where poor performance results in high acidity, gas evolution, solid-particle formation, gum formation, varnish formation, depending upon fluid and temperature level.

Compatibility. Poor performance may result in seal deterioration and other side effects.

Not to be overlooked should be *hydrolytic stability, oxidative stability,* and *flammability* for consideration of *fire safety.*

Certain of the disadvantages of the types of fluids listed in Table 30.1[4] are as follows:

1. Petroleum-base oils have poor viscosity-temperature characteristics, are volatile and highly flammable.
2. Esters of dibasic organic acids have poor low-temperature viscosity, poor thermal stability, and poor compatibility with elastomers and other organic materials.
3. Phosphate esters are good in many respects for aircraft and industrial applications, but thermal stability is poor above 250°F.
4. Polyglycols have poor thermal stability above 400°F and their decomposition is catalyzed by some metals. They are incompatible with even a trace of hydrocarbon, this incompatibility resulting in the formation of gummy precipitates.
5. Fluorocarbons and chlorinated hydrocarbons have poor viscosity-temperature characteristics, do not accept additives, and are heavy.
6. Silicones have poor lubricity and are unresponsive to additives.
7. Polysiloxanes and orthosilicate esters have poor hydraulic stability and are as flammable as petroleum-base oils.
8. Phenyl ethers have poor low-temperature characteristics limiting their use to above 0°F.
9. Liquid metals such as NAK 77 (a eutectic of 77 percent Na, 23 percent K) and mercury have poor lubricity, present serious handling problems. Mercury is highly toxic, NAK 77 highly flammable in air.

30.3 FUNDAMENTAL RELATIONSHIPS IN HYDRAULIC FLOW

30.3.1 Introduction

In the design of such hydraulic components as pumps, motors, valves, and conduits the observation and proper application of certain basic relationships will assure sound fundamental design. These concepts are[4]

1. Conservation of mass (continuity)
2. Conservation of momentum
3. Conservation of energy

Conservation of mass requires that the rate of mass flow into a defined volume equal the rate of mass flow out plus the rate at which mass accumulates within the control volume. Thus (see Fig. 30.2)

$$\int_{A_s} \rho V_n dA_s + d/dt \int_V \rho \, dv = 0 \tag{30.14}$$

where V_n = velocity normal to control surface
A_s = control surface
v = control volume

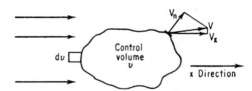

FIG. 30.2 Control volume in a flow field.

Conservation of momentum requires that the net rate of outflow of momentum in a specific direction x plus the rate at which x momentum accumulates within the control volume be equal to the force applied to the control volume in the x direction. Thus

$$\Sigma F_x = d/dt \int_v \rho V_x \, dv + \int_{A_s} \rho V_x V_n \, dA \tag{30.15}$$

EXAMPLE 1[4] How is the pressure difference $P_1 - P_2$ related to the rate of change of flow of a frictionless incompressible fluid in a uniform tube of length L and area A?

Conservation of mass requires that

$$\rho Q_1 = \rho Q_2 \quad \text{or} \quad Q_1 = Q_2 = Q \quad (\rho = \text{const}) \tag{30.16}$$

where Q = volume flow, in³/s.
Conservation of momentum requires that

$$P_1 A - P_2 A = \rho Q V - \rho Q V + (d/dt)(\rho V A L) \tag{30.17}$$

Thus

$$P_1 - P_2 = (\rho L/A)(dQ/dt) \tag{30.18}$$

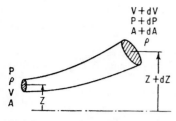

FIG. 30.3 One-dimensional, friction-less, steady, incompressible streamline flow with area change.

EXAMPLE 2 One-dimensional, frictionless, incom-pressible, streamline flow of a fluid (Fig. 30.3).

Conservation of mass demands that

$$V\,dA + A\,dV = 0 \qquad (30.19)$$

Conservation of momentum demands that

$$-gp\,dZ - dP = \rho V^2(dA/A) + \rho V\,dV \qquad (30.20)$$

Combining Eqs. (30.19) and (30.20) yields Euler's equation:

$$V\,dV + dP/\rho + g\,dZ = 0 \qquad (30.21)$$

Since ρ is constant, Euler's equation can be integrated yielding Bernoulli's equation:

$$V^2/2g + P/\rho g + Z = \text{const} \qquad (30.22)$$

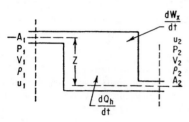

FIG. 30.4 Conservation of energy applied to a simple control volume.

Conservation of energy requires that the increase in internal energy of a sys-tem of fixed identity be equal to the work done on the system plus the heat added to the system. Applied to a control volume fixed in space and not subject to acceler-ation, conservation of energy yields (see Fig. 30.4)

$$dQ_h/dt + dW_x/dt - dE/dt$$
$$+ \int_{A_s} (P/\rho + e)\rho V_n\,dA \qquad (30.23)$$

where Q_h = heat flow to the control volume
W_x = shaft and shear work done on the system
E = total internal energy of fluid inside the control volume
e = total internal energy per unit of mass ($e = u + gZ + V^2/2$)
u = intrinsic internal energy per unit of mass of fluid
Z = height above the reference point, in (gZ is the potential energy per unit of mass)
P = pressure on an element of area at the surface of the control volume

30.3.2 Orifice Flow

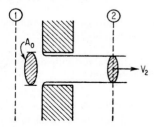

FIG. 30.5 Frictionless flow through a nozzle or orifice.

The design of valves for control and regulation purposes and the design of pumps and motors require the analysis of flow through rounded and sharp-edged orifices.

Consider the case of *frictionless flow through a nozzle or orifice* as illustrated in Fig. 30.5. At condition 1 the velocity is negligible. From Bernoulli's equation and $\rho AV = $ constant,

$$\rho Q = A_2 \sqrt{2\rho(P_1 - P_2)} \qquad (30.24)$$

The discharge coefficient C_d is defined

$$C_d \triangleq A_2/A_0 \qquad (30.25)$$

where A_0 is the nominal area of the constriction. The discharge coefficient varies from 0.6 to 1.0 depending upon the geometry of the upstream passage. With this definition the mass-flow equation becomes

$$\rho Q = C_d A_0 \sqrt{2\rho(P_1 - P_2)} \qquad (30.26)$$

The fact that friction phenomena from viscous effects are neglected requires that the Reynolds number of the flow at section 2 must be in excess of a certain critical value, which varies from 260 for the sharp-edged annular orifice to 100,000 for certain circular orifices in pipes. The Reynolds number, a dimensionless quantity, is defined by

$$\text{Re} \triangleq \rho V D_e/\mu$$

where D_e is the equivalent hydraulic diameter of the constriction, μ is the absolute viscosity, V is the velocity, and ρ is the density. For Re larger than the critical value the discharge coefficient C_d is constant and thus independent of Re. For Re less than the critical value C_d is a function of both geometry and Re, as demonstrated in the following paragraphs.

The Sharp-Edged Circular Orifice. Orifice meters for flow measurement commonly consist of sharp-edged circular orifices in pipes with pressure taps on each side of the orifice plate. Sharp-edged circular orifices are also commonly used in flow-control valves as fixed restrictions. The effect of upstream pipe diameter is commonly included as the ratio of orifice diameter to pipe diameter β. The mass-flow equation thus becomes

$$\rho Q = C_d A_0 \sqrt{2\rho(P_1 - P_2)/(1 - \beta^4)} \qquad (30.27)$$

For a circular orifice

$$\text{Re} = 4\rho Q/\pi\mu D \qquad (20.28)$$

Figure 30.6 presents data for knife-edged orifices[9] with 45° chamfers. As a properly constructed measuring device, the sharp-edged orifice will give a measurement accuracy of about ±0.5 percent. As a control device, the sharp-edged circular orifice can be expected to give the most reproducible results as compared with other orifice configurations (see Fig. 30.7b).

The thick-plate circular orifice, particularly in the smaller sizes and for diameter-to-length ratio of unity, gives a 30 percent spread in discharge coefficient at any one Reynolds number for Reynolds numbers from 100 to 6000. A diameter length ratio below 0.2 is recommended in order to assure reasonable insensitivity to upstream conditions and manufacturing variations.

The quadrant-edge orifice is reported to give excellent results over large ranges of Reynolds number and diameter ratios[10] (see Table 30.2 and Fig. 30.7c).

The Sharp-Edged Annual Orifice. For sliding-type control valves with sharp edges (see Fig. 30.7a) the discharge coefficient C_d is reasonably constant

$$C_d = 0.60 \text{ to } 0.65$$

above the critical value of the Reynolds number of 300. Below the critical value of Re, the discharge coefficient increases to about 0.95, then decreases almost linearly to

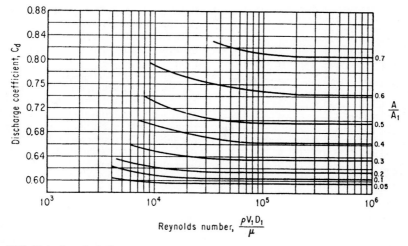

FIG. 30.6   Data for knife-edged orifices.

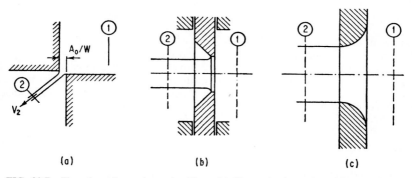

FIG. 30.7   Flow through nozzles and orifices. (*a*) Sharp-edged annular orifice. (*b*) Sharp-edged orifice. (*c*) Quadrant-edged orifice.

TABLE 30.2   Discharge Coefficients for Quadrant-Edged Orifices[10]

β	r/d	$C_d = K(\sqrt{1 - \beta^4})$	Limits of constant C_d within $\pm 0.5\%$	
			$R_{D\mathrm{min}}$	$R_{D\mathrm{max}}$
0.225	0.100	0.77	760	56,000
0.400	0.114	0.78	650	140,000
0.500	0.135	0.801	430	230,000
0.600	0.209	0.838	350	250,000
0.625	0.285	0.859	330	250,000
0.656	0.380	0.890	270	250,000
0.692	0.446	0.890	250	250,000

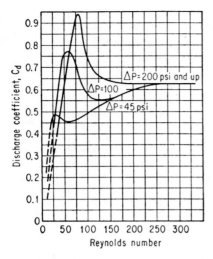

FIG. 30.8 Orifice coefficients for annular sharp-edged orifices as a function of orifice Reynolds number.

Re = 0 (see Fig. 30.8). With decreasing pressure drop the peak value of C_d tends to decrease. This phenomenon becomes pronounced for ΔP below about 150 lb/in².

Poppet-type valves show a similar variation of discharge coefficient with Re. Experiments[11] show general agreement with Fig. 30.8 over pressure-drop ranges of 500 to 200 lb/in² and 45° poppet angles. The geometry of the valve and the valve chamber should be considered in applying any data, as the effect may be large for special cases.

The "Flapper-Valve" Type of Annular Orifice.

The "flapper valve" has been for many years and is today frequently employed in control valves. Figure 30.9a shows the sharp-edged orifice. Figure 30.9b is the more practical chamfered and flat design. Experiments for a two-dimensional glass-sided model gave discharge coefficients of 0.65 for Reynolds numbers in excess of 600 for the sharp-edged orifices as shown in Fig. 30.10. Figure 30.11 presents the results for several flats and chamfers as shown in Fig. 30.9b.[12] Outside chamfer angles α can vary from 30 to 60° without materially affecting the result. The results indicate that internally chamfered orifices of the flapper type can have discharge coefficients up to 0.90, which remain constant for Reynolds numbers in excess of 600, where Re is defined by

$$Re = \rho Vx/\mu \qquad (30.29)$$

where $V = \sqrt{2(\Delta P)/\rho}$ (30.30)

        x = flapper opening, in
   ΔP = pressure drop, lb/in²
    ρ = density of fluid, lb·s/in⁴

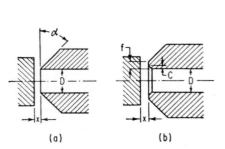

(a)   (b)

FIG. 30.9 Schematic diagrams of flapper orifice configurations.

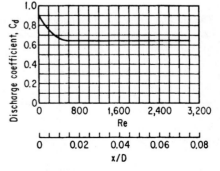

FIG. 30.10 Discharge coefficients as a function of flapper opening for sharp-edged flapper nozzle with 60° outside chamfer.

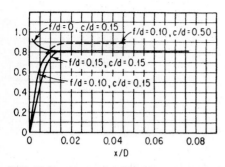

FIG. 30.11 Discharge coefficients as a function of flapper opening for several nozzles with flats and inside chamfers at 45°.

V = average velocity, in/s

since $\qquad Q = C_d A \sqrt{2(\Delta P)/\rho}$ (30.31)

and $\qquad A = \pi D x$ (30.32)

$\qquad \text{Re} = Q/\pi D C_d \mu$ (30.33)

30.3.3 Laminar Flow

In the design of such hydraulic components as certain closely fitted valves, the pistons and valve plates of positive-displacement pumps and motors, and certain seals, the equations of fully developed laminar flow are of interest. The following cases assume constant fluid viscosity and density.[4]

Steady Flow through Circular Pipes (see Fig. 30.12)

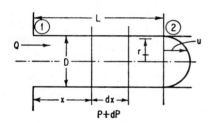

FIG. 30.12 Laminar flow in a round tube.

$dP/dx = -128 \, \mu Q/\pi D^4$ (30.34)

$dP/dx = -32 \, \mu V_{av}/D^2$ (30.35)

$Q = (\pi D^4/128 \, \mu L)(P_1 - P_2)$ (30.36)

$u = 2V_{av}(1 - 4r^2/D^2)$ (30.37)

$V_{av} = 4Q/\pi D^2$ (30.38)

If the friction factor f is defined by

$$\Delta P = f\rho(L/D)(V^2/2)$$ (30.39)

$$f \triangleq 2D/\rho V^2(dP/dx)$$ (30.40)

then for laminar flow the friction factor is given by Eq. (30.41):

$$f = 64/\text{Re}$$ (30.41)

The distance from the entrance of a tube to the length at which fully developed laminar flow is obtained is given by

$$L/D = 0.058 \, \text{Re} \qquad \text{where Re} = \rho VD/\mu < 2000$$ (30.42)

Steady Flow through Stationary Flat Plates (Fig. 30.13)

$$dP/dx = -12 \, \mu Q/wh^3 = -12 \, \mu V_{av}/h^2$$ (30.43)

$$V_{av} = Q/hw$$ (30.44)

$$Q = (wh^3/12 \, \mu L)(P_1 - P_2) \qquad \text{for } w >> h$$ (30.45)

where h is the thickness of the passage, w is the width, and L is the length of the passage.

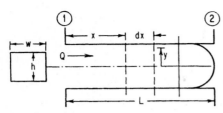

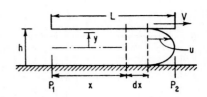

FIG. 30.13 Steady laminar flow through stationary flat plates.

FIG. 30.14 Steady laminar flow through moving flat plates.

Steady Flow between Stationary and Moving Flat Parallel Plates (Fig. 30.14)

$$Q/w = Vh/2 - (h^3/12\,\mu)(dP/dx) \qquad \text{if } w >> h \tag{30.46}$$

where Q is rate of flow through plates relative to fixed surface.

$$dP/dx = (P_1 - P_2)/L \tag{30.47}$$

$$Q/w = Vh/2 - (h^3/12\,\mu)[(P_1 - P_2)/L] \tag{30.48}$$

The force on the upper plate

$$F_u = Lw\{u(V/h) + h/2[(P_1 - P_2)/L]\} \tag{30.49}$$

The force on the lower plate

$$F_1 = Lw\{-uV/h + (h/2)[(P_1 - P_2)/L\} \tag{30.50}$$

Steady Flow through an Eccentric Circular Annulus (Fig. 30.15)

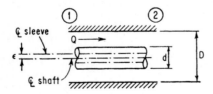

$$Q = (\pi Dh^3/12\,\mu L)(P_1 - P_2)$$
$$[1 + 1.5(\epsilon/h)^3] \tag{30.51}$$

where ϵ is the eccentricity and $h = (D - d)/2$.

FIG. 30.15 Steady laminar flow through an eccentric circular annulus.

30.3.4 Flow with Varying Viscosity and Density

The assumption of constant density and constant viscosity will not fully explain the behavior of such devices as pistons moving in cylinders or valve plates rotating against cylinder blocks. Since work is done on a particle of fluid in viscous shear as it passes through a passage the temperature of the fluid will increase for the adiabatic process. Experiments[4] have verified the following relationships and design curves.

Consider that fluid properties as a function of temperature are defined by

$$\rho = \rho_1(1 - \alpha\Delta T) \tag{30.52}$$

$$E = E_1 + C_v\Delta T \tag{30.53}$$

where C_v is the specific heat, in·lb/lb°F

$$\mu = \mu_1 e^{-\lambda}(\Delta T) \tag{30.54}$$

The temperature gradient is approximated by the following equation, assuming no side leakage:

$$dT/dx = (1/\lambda)[K_2/(1 + K_2 x)] \tag{30.55}$$

where
$$K_2 \triangleq 2Vx\mu_1/h^2 C_v \tag{30.56}$$

The gradient is largest at $x = 0$, the leading edge of the slider, since the fluid there is coolest and therefore most viscous. At $x = 0$,

$$\left. \frac{dT}{dx} \right|_{x=0} = \frac{2V\mu_1}{h^2 C_v} \tag{30.57}$$

The pressure distribution is given by

$$p = P_m - \frac{3\alpha C_v}{2x^2} \left(\ln \frac{1 + K_2 x}{1 + K_2 x_m} \right)^2 \tag{30.58}$$

where the subscript m refers to conditions for the maximum pressure point or where

$$dp/dx = 0 \tag{30.59}$$

Integrating the local pressure p over the area of the pad results in the expression for the load-carrying capacity of the parallel slider

$$F = K_3 \left[\frac{2 + K_2 L}{K_2 L} \ln (1 + K_2 L) - 2 \right] \tag{30.60}$$

For small values of $K_2 L$ (up to approximately 0.1) the simplified force equation may be written

$$F = (K_3/6)[(K_2 L)^2 - (K_2 L)^3] \tag{30.61}$$

where
$$K_3 \triangleq 3\alpha C_v Lw/2\lambda^2 \tag{30.62}$$

For design purposes plots of normalized forces, clearance and temperatures will be found useful. The normalized parameters are defined by

$$F^* \triangleq F/F_1 \tag{30.63}$$

$$h^* = 1/\sqrt{K_2 L} \tag{30.64}$$

$$\Delta T^* = \ln (1 + K_2 L) \tag{30.65}$$

$$\Delta T^* = \lambda \Delta T_L \tag{30.66}$$

Figure 30.16 provides a cross plot of the normalized quantities.

The centering force on a rotating cylinder is of interest in gear and vane pump design. Figure 30.17 schematically shows a typical example, and Fig. 30.18 is the plot of the nondimensional centering force vs. the eccentricity in the clearance space.

30.4 HYDRAULIC POWER GENERATION

Two fundamentally different types of hydraulic pumps are available:

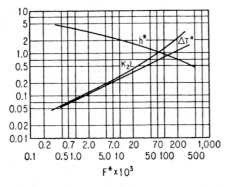

FIG. 30.16 Normalized force vs. normalized clearances; normalized temperature rise and length for the parallel slider.[4]

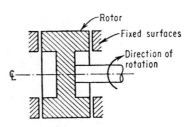

FIG. 30.17 Rotating cylinder between flat side plates.

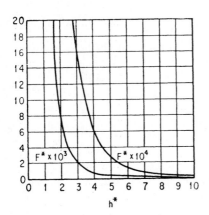

FIG. 30.18 Nondimensional axial centering force vs. eccentricity for the rotating cylinder between side plates.[4]

1. Positive-displacement pumps of the piston, vane, or gear type used to generate relatively high pressures at relatively low flows. Clean fluids of good lubricity and adequate viscosity are commonly used.

2. Hydrokinetic pumps of the axial-flow or centrifugal type used to generate large volume flows at relatively low pressures. Centrifugal pumps will tolerate such low-lubricity fluids as liquid hydrogen, liquid metals, jet fuels, and fluids with considerable contaminant content.

30.4.1 Positive-Displacement Pumps

Gear pumps are used extensively for jet fuels, lubricating oils, and other applications where pressures up to 1500 lb/in² suffice. Gear pumps are fixed-displacement pumps; i.e., delivery per revolution cannot be changed over large ranges with good retention of efficiency.

Vane pumps find broad use in such applications as roadworking machinery, machine-tool application, and many other uses where pressures do not exceed 2000 lb/in². Variable-displacement vane pumps are available but pressures rarely exceed 1500 lb/in².

High-pressure generation of fluid power in the 3000 to 5000 lb/in² range can be accomplished by the *piston* pump. Either axial-piston or radial-piston pumps, fixed and variable delivery, are available.

Figure 30.19 shows a diagram of a bent-axis piston pump. The unit is compact and provides a high power-to-volume ratio. It is capable of operating in configurations such that the angle between the drive shaft and the cylinder block is adjustable.

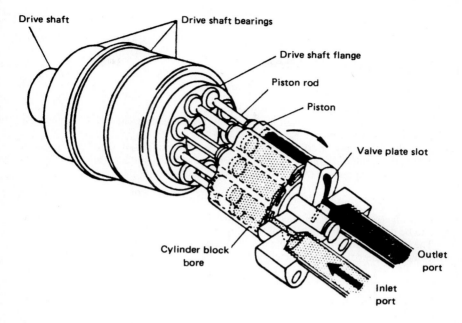

FIG. 30.19   Diagram of a bent-axis piston pump. (*Courtesy John Wiley & Sons, Inc.*[14])

Adjustment of this angle can cause the pump displacement to vary continuously from zero to maximum in either direction.

Efficiency.   The power required to drive a pump of specific delivery and pressure rise and the heat rejected into the fluid during the pumping process are a function of pump efficiency.

The volumetric efficiency η_v of a positive-displacement machine is defined as the ratio of the actual delivery to the displacement or the theoretical delivery:

$$\eta_v = Q_A/Q_T = (Q_T - Q_L)/Q_T = 1 - Q_L/Q_T \tag{30.67}$$

where Q_A = actual delivery
$\quad Q_T$ = theoretical delivery
$\quad Q_L$ = internal leakage

For most well-designed pumps the dominant leakage is laminar in nature, and therefore loss due to cavitation at the inlet, and orifice-type backflow can be assumed negligible. The leakage flow Q_L can be expressed in terms of a nondimensional parameter[4] $\Delta P D_p/2\pi\mu$ and a slip coefficient C_s, where ΔP = pressure drop, D_p = displacement per revolution at zero pressure drop across pump, μ = viscosity of fluid in clearance passages.

The mechanical or torque efficiency η_m is defined as the theoretical torque T_P required to drive a pump at a particular speed N, pressure rise ΔP, and delivery Q_A divided by the actual torque T_A.

$$\eta_m = T_P/T_A = T_P/(T_P + T_F) \tag{30.68}$$

where T_F is the friction drag torque.

The applied torque T_A is

$$T_A = \Delta P D_p/2\pi + C_d D_p \mu N + C_f(\Delta P D_p/2\pi) + T_c \qquad (30.69)$$

The theoretical torque is given by

$$T_p = \Delta P D_p/2\pi \qquad (30.70)$$

C_d is the viscous drag coefficient (dimensionless).
C_f is the drag coefficient resulting from pressure-dependent and speed-independent friction sources such as bearings and seals.
T_c is the remaining frictional drag, which is independent of speed and pressure drop. It is the result of tight seals, poorly lubricated rubbing surfaces, etc.
Substituting expression (30.69) into (30.68) yields

$$\eta_m = 1/[1 + C_d(\mu N/\Delta P) + C_f + T_c/D_p\Delta P] \qquad (30.71)$$

C_d is inversely proportional to the first power of typical pump clearances. T_c and C_f tend to be proportional to the size of the pump.
The overall pump efficiency is equal to the product of volumetric and mechanical efficiencies. The volumetric efficiency is

$$\eta_v = 1 - C_s(\Delta P/\mu N) \qquad (30.72)$$

It can be shown that the slip (or leakage) coefficient C_s is proportional to the cube of the typical clearance of a pump. The pump efficiency is given by

$$\eta_p = \eta_v\eta_m \qquad (30.73)$$

From Eqs. (30.71) and (30.72), Eq. (30.73) becomes

$$\eta_p = \frac{1 - C_s(\Delta P/\mu N)}{1 + C_d(\mu N/\Delta P) + C_f + T_c/D_p\,\Delta P} \qquad (30.74)$$

The efficiency of a pump is thus determined by C_s, C_d, C_p and $\mu N/\Delta P$ if T_c is considered negligible. Table 30.3 gives typical values for several hydraulic pumps.[13]
The volume under compression in the pump as seen by the high-pressure circuit is of interest to the system designer. Typical values for fixed- and variable-delivery pumps are as follows:

Fixed displacement

$$\frac{\text{Volume under compression}}{\text{Displacement per revolution}} = 1.2 \text{ to } 1.5$$

TABLE 30.3   Hydraulic Pump Constants

Hydraulic unit	D, in.^{3}/rev	C_d	C_s	C_f	T_c, in.-lb
Piston pump.........	3.60	16.8×10^4	0.15×10^{-7}	0.045	0
Vane pump..........	2.865	7.3×10^4	0.477×10^{-7}	0.212	0
Spur-gear pump.....	2.965	10.25×10^4	0.48×10^{-7}	0.179	0
Internal-gear pump..	2.965	9.77×10^4	1.02×10^{-7}	0.045	0
Internal-gear pump..	1.800	5.06×10^4	1.6×10^{-7}	0.075	13

Variable delivery

$$\frac{\text{Volume under compression}}{\text{Displacement per revolution}} = 1.8 \text{ to } 3.0$$

30.4.2 Power Supply Using Fixed-Displacement Pumps

A constant-pressure hydraulic power supply usually consists of a pump, a prime mover (an electric motor in stationary supplies) to drive the pump, a reservoir of fluid, a pressure regulator, a relief valve for safety purposes, filters, and perhaps an oil cooler and an accumulator. Such a supply is shown schematically in Fig. 30.20. The accumulator functions to filter pressure pulsations from the pump and to provide additional fluid under pressure to accommodate peak flow demands. Accumulators are often not necessary.

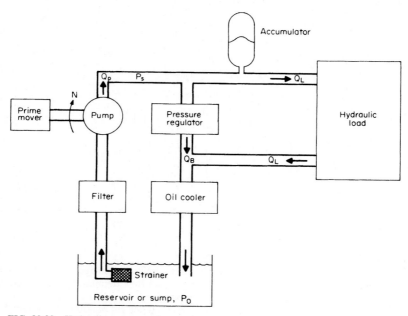

FIG. 30.20   Hydraulic power supply.

The pressure regulator is needed because the fixed-displacement pump, driven at constant speed, delivers a fixed volumetric flow. The pump should be sized for normal peak flow. While the accumulator can make up for short-duration transient flow demands, excess flow must be bypassed when load demands fall. The pressure regulator acts to limit the increase in system pressure when demand drops and provides a safety relief valve function. The disadvantage of this system is its low efficiency, caused by the power loss across the bypass regulator valve. This power loss heats the hydraulic fluid and increases cooling requirements. The hydraulic power delivered by the system is given by

$$\text{hp} = Q_p P_s / 1787 \tag{30.75}$$

where Q_p is in gal/min, P_s is in lb/in^2, hp is horsepower, the power loss as heat is $Q_B P_s$.

This objectionable heat input can be drastically reduced by automatically adjusting the delivered flow from the pump to maintain the desired pressure.

30.4.3   Variable-Delivery Pressure-Compensated Power Supply

A common method of overcoming the low efficiency of the fixed delivery system is to use a variable-delivery pump with the delivery controlled to maintain system pressure over a wide flow range. Efficiency is improved because only the flow required by load demand is generated by the pump. Figure 30.21 shows a diagram of this type of system. The control valve compares the force associated with the actual system pressure P_s with that resulting from the desired pressure K_p as determined by the spring. If P_s is too high, the valve opens so that the control pressure P_c rises, moving the stroke control mechanism to reduce delivery.

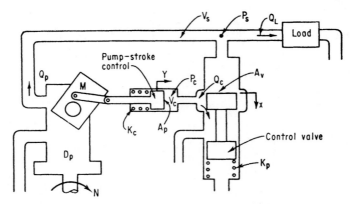

FIG. 30.21   Schematic diagram of pressure-compensated variable-delivery pump circuit.

This system, although efficient, responds more slowly to changes in load demand and has a tendency for supply pressure to drop as load flow demand increases. This is because its operation employs a pressure control function having time lags due to swash-plate and oil-pressure responses. These are damped by making the system less sensitive to supply-pressure variations. The steady-state drop of this control system is typically around 10 percent but can be reduced by increasing the valve area A_v and reducing the stoking piston area A_p.

Figure 30.22 is a normalized curve for the step response of a typical variable-delivery pump with a purely resistive load. The disturbance in each case is a step change in a load-flow demand. Further details concerning the dynamics of this power source are given in Ref. 14.

System Considerations.   The stability of the variable-delivery servo-type pump (either pressure-controlled or electrohydraulically controlled stroke mechanism) is sensitive to many design parameters. The more important causes of instability are:

1. Operation near cutoff can be shown to be the least-damped condition.

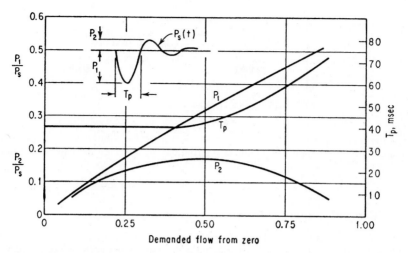

FIG. 30.22 Response characteristics of typical pressure-compensated variable-delivery pumps.

2. Servo-valve dead band, because of the integrator action of the control, will cause hunting.
3. High friction level in the stoking mechanism with high gain control will cause instability.
4. Backlash in the stroking mechanism can induce instability.
5. Piston pumps develop pressure pulsations in the output line which, because of the nonlinear nature of the system, can induce subharmonic oscillations which with low system damping may result in sustained oscillations.
6. The frequency of the pressure pulsations in the output line, if too close to the characteristic frequencies of the servo valve or cam plate system, can be troublesome.

30.5 HYDRAULIC POWER STORAGE

The basic intentional power-storage element in hydraulic systems is the accumulator, which may be either the spring-loaded or gas-loaded ram (Fig. 30.23a and b), the bladder-type accumulator (Fig. 30.24), or the pure liquid volume (surge tank). In addi-

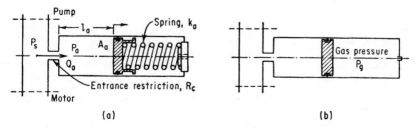

FIG. 30.23 Ram-type accumulator. (a) Spring-loaded type. (b) Air-loaded type.

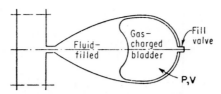

FIG. 30.24 Bladder-type accumulator.

tion to the intentional use of devices for power storage by volume (capacitance storage), there are other storage phenomena (inertial energy storage) that can have significant effects.

30.5.1 Capacitive Energy Storage

The accumulators mentioned above store energy in a volume of pressurized oil. The formulas for energy stored and the mathematical relations between pressure and volume flow are given below.

Spring-Backed Piston. For this type of accumulator

$$C(dP_a/dt) = Q_1 - Q_2 \qquad C = A_a^2/k_a \qquad (30.76)$$

and the energy stored is given by

$$E = (C/2)(P_a^2 - P_0^2) \qquad (30.77)$$

where P_0 is the pressure to overcome the initial spring force.

Gas-Bag Accumulator (see Fig. 30.24). For this type of accumulator,

$$C(dP_a/dt) = Q_1 - Q_2 \qquad C = V_0 P_0^{1/\gamma}/\gamma P_a^{\gamma+1/\gamma} \qquad (30.78)$$

and the energy stored E is given by

$$E = P_0 V_0 \ln (P_a/P_0) \qquad \text{and} \qquad E = (P_a/P_0)^{\gamma-1/\gamma} - 1)P_0 V_0/(\gamma - 1) \qquad (30.79)$$

for isothermal compression and for adiabatic processes ($\gamma = 1.4$ for air), respectively.

30.5.2 Inertial Energy Storage

The kinetic energy stored in a mass of flowing fluid can give rise to significant system effects, of which the water hammer phenomenon is probably the best known. However, for design purposes and particularly in dealing with hydraulic-system dynamic performance, this effect needs to be analyzed.

The kinetic energy stored by a fluid flowing with volumetric rate Q_a is a uniform conduit of area A, where the rate Q_a is unsteady, can be modeled by the equations

$$I(dQ_a/dt) = P_a - P_b \qquad I = \rho L/A \qquad (30.80)$$

and the kinetic energy stored is $\frac{1}{2}IQ_a^2$. $\qquad (30.81)$

The model above assumes the flow velocity to be uniform across the sectional area.

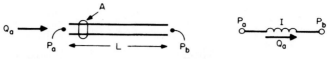

FIG. 30.25   Fluid inertia and electrical equivalent.

When a parabolic profile for fully developed flow is taken into account, the inertial energy is increased by a factor of $\frac{4}{3}$ and $I = \frac{4}{3}(\rho L/A)$.[15] Figure 30.25 illustrates a segment of a fluid conduit and its electrical circuit equivalent.[16]

30.5.3   Capacitive Energy Storage by Hoop Strain in Piping and Oil Compressibility

When a trapped volume of oil is present between hydraulic-system components, elastic energy will be stored both in the fluid and by stretch in the pipe wall. As an example, consider a trapped volume of oil between a piston and an orifice, as indicated in Fig. 30.26.

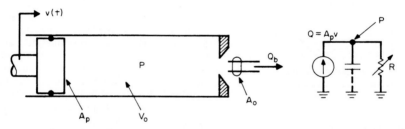

FIG. 30.26   Trapped-volume compressibility effect and electrical circuit equivalent.

Under circumstances of rapid movement of the piston, a significant displaced volume must be present to compress the trapped fluid and to establish an increase in its pressure. Changes in oil pressure are accompanied by changes in density. The equation of state can be written as

$$d\rho/\rho = dP/\beta \qquad (30.82)$$

where β is the isothermal bulk modulus of oil.

The inflow displacement from the piston is $A_p v$ and if it is not equal to the outflow rate Q_b, then the trapped volume V_0 is pressurized at a rate

$$dP/dt = (\beta/V_0)[A_p v(t) - Q_b] \qquad (30.83)$$

By comparison with the formulas above, the capacitance C is given by

$$C = V_0/\beta \qquad (30.84)$$

Its effective value is reduced by the presence of air bubbles and by mechanical stretch in the pipe wall. The equivalent bulk modulus, accounting for radial pipe wall stretch, is

$$\beta_e = \{\beta^{-1} + [E(D_o/D_i - 1)]^{-1}\}^{-1} \qquad (30.85)$$

where D_o = pipe outside diameter
D_i = pipe inside diameter
E = Young's modulus

The compressive energy stored E is given by

$$E = (C/2)P^2 \qquad (30.86)$$

where $C = V_0/\beta_e$. Apart from the amount of energy stored by fluid compressibility, this effect leads to time lags in response of the pressure P to inflow changes. Equation (30.83) will be used many times in the analysis of hydraulic systems. The outflow from the control volume Q_b depends on the overpressure P but this pressure takes a certain time to change. Using the orifice flow equation, the simple piston-driven system shown in Fig. 30.26 has a dynamic system response similar to that of an electrical capacitance-resistance circuit as modeled by Eq. (30.87):

$$\frac{dP}{dt} = -\frac{\beta_e C_d A_0}{V_0}\sqrt{\frac{2P}{\rho}} + \frac{\beta_e A_p}{V_0} v \qquad (30.87)$$

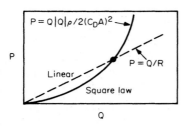

FIG. 30.27 Square-law and linear fluid resistance.[16]

Equation (30.87) is a nonlinear ordinary differential equation with piston velocity v as the input, which is easily solved for the response P by numerical techniques (see Sec. 30.2). However, some insight can be gained by representing the orifice as a linear fluid resistance having an equivalent linear pressure flow characteristic and by considering small piston movements, so that the volume V_0 does not change appreciably (see Fig. 30.27).

An equivalent linear resistance can be roughly modeled by the secant value $R \simeq P_{max}/Q_{max}$, where P_{max} and Q_{max} are estimated maximum values of P and Q. Using this secant model, Eq. (30.87) is linearized and gives

$$dP/dt = -(\beta_e/V_0 R)P + (\beta_e A_p/V_0)v \qquad (30.88)$$

with the solution for a step input in v equal to v_1 and an initial condition $P(0) = 0$, given by

$$P(t) = RA_p v_1[1 - e^{-t/\tau}] \qquad (30.89)$$

and a time constant τ given by

$$\tau = V_0 R/\beta_e \qquad (30.90)$$

The approach used here will be illustrated again later and is useful for modeling pressure responses wherever significant trapped volumes of oil occur, as in hydraulic cylinder chambers or in short transmission lines between valves, pumps, and motors.

30.6 HYDRAULIC POWER TRANSMISSION

30.6.1 Introduction

The connection between the power-generating, power-controlling, and power-utiliza-
tion devices requires the transmission of flows and pressures through the transmission
lines. In hydraulic systems the major proportion of power is transmitted as flow.
Because the change in fluid power requires pressure changes, transmission of pressure
signals becomes an all-important consideration in system design to assure a dynamically stable and responsive system. The problem of hydraulic-power transmission can
thus be divided into two parts: (1) steady flow through pipes, tubes, and fittings; (2)
dynamic response of the hydraulic transmission line.

30.6.2 Steady Flow through Pipes and Fittings

Flow in Tubing. The tubing used in normal hydraulic practice is made of drawn-alu-
minum alloy, copper, or stainless steel of very smooth surface.

For Reynolds numbers below the lower critical value of 2000, the friction factor is
given by

$$f = 64/\text{Re} \tag{30.91}$$

where $\text{Re} = \rho VD/\mu$.

For Reynolds numbers above the upper critical value of 4000, the resistance law
for pipes is given by the Karman-Prandtl equation

$$1/\sqrt{f} = 2.0 \log_{10}(\text{Re} \sqrt{f}) - 0.8 \tag{30.92}$$

This relationship agrees exactly with experiment for ultrasmooth pipes with roughness ratios k/D less than 0.00001.

Figure 30.28 is the Stanton diagram, which presents the friction factor as a function
of Reynolds number.[8] For standard drawn tubing an absolute roughness of

$$k = 0.00006 \text{ in}$$

should be used.

The pressure drop in a tube of length L and diameter D is given by

$$\Delta P = f(L/D)(\rho V^2/2) \tag{30.93}$$

Hydraulic circuits are characterized by a large number of bends in tubing and fittings
of various types such as elbows, tees, unions, contractions, and enlargements. To evaluate the total pressure drop in a complex system, it is convenient to compute an equivalent length L_e for each bend and fitting, using the information in Fig. 30.29.[17] The
total pressure drop in a system is then given by Eq. (30.94). Note should be taken of
the fact that the length of the bend (case F) must be added to the loss from change of
direction of flow.

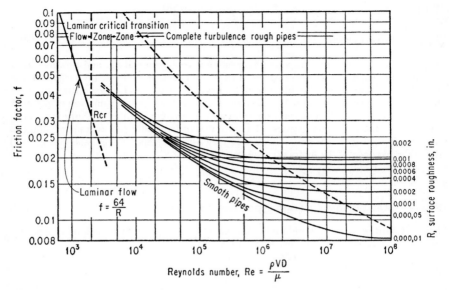

FIG. 30.28   Plot of friction factor vs. Reynolds number of various roughness ratios k/D.

$$\Delta P(\text{system}) = \underbrace{\sum_{i=1}^{i=n} f_i \frac{L_i}{D_i} \frac{\rho V_i^2}{2}}_{\text{tubing}} + \underbrace{\sum_{i=1}^{i=n} f_i \frac{L_{ei}}{D_i} \frac{\rho V_i^2}{2}}_{\text{fittings}} + \underbrace{\sum_{i=1}^{i=n} f_i \frac{L_{ei}}{D_i} \frac{\rho V_i^2}{2}}_{\text{bends}} \qquad (30.94)$$

Since the coefficient of energy loss $L_e f/D$ is inversely proportional to fluid density, the plots must be corrected for wide departures from the reference values of the aircraft hydraulic oil (An-0-366b) at 75°F ($W = 0.0312$ lb/in^3).

In many systems flexible hosing is used. Where actual pressure-drop test data are not available from the manufacturer, Fig. 30.30 can be used as a guide in estimating the effect of roughness ratios on terminal-friction factors. For orifice data, see Sec. 30.3.2.

It is common practice to limit velocities in hydraulic circuits to certain proved ranges. Exceeding the recommended values will mean large pressure losses (and temperature rises). Weight and cost penalties result from velocities which are too low, however. Table 30.4 lists recommended velocities.

30.6.3   Dynamic Response of Hydraulic Transmission Lines

In many instances the flows through the piping of a hydraulic system are not steady. With unsteady flow, fluid mass and compressibility effects can introduce undesirable transients and deterioration of system response. Short lines may either increase the effective mass of the system or increase time lags from fluid compressibility. The effects of medium or long lines are complex, introducing water hammer and other traveling wave phenomena, and are best analyzed by using mathematical models and computer simulation.[18–20] Fluid transmission lines act dynamically as wave guides for conducting pressure and flow signals and are modeled mathematically by wave equations. While fluid mass, friction, and compressibility are really distributed throughout the length of the line, wave equation models are complex, and simpler models using

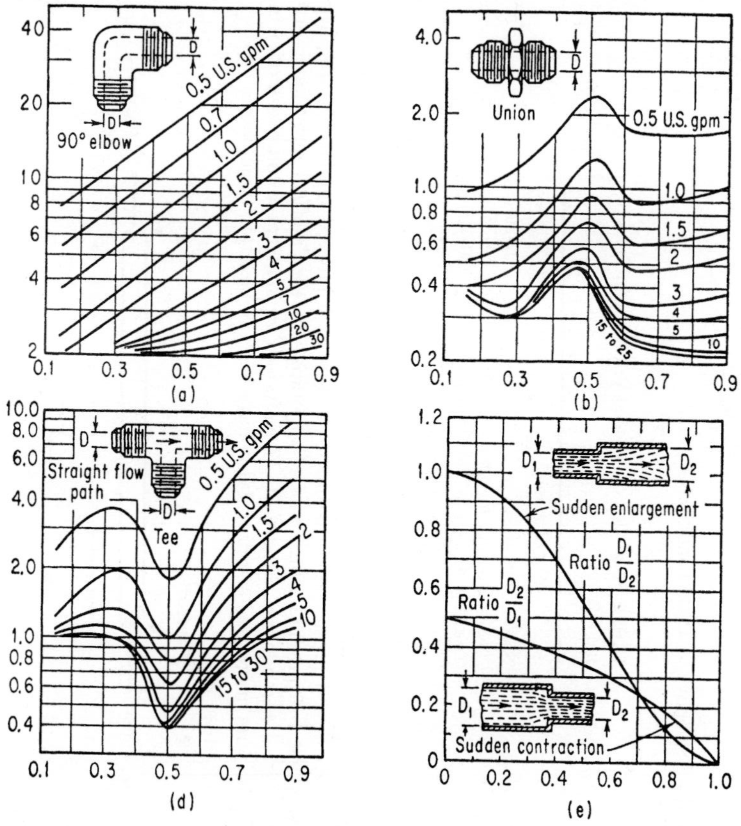

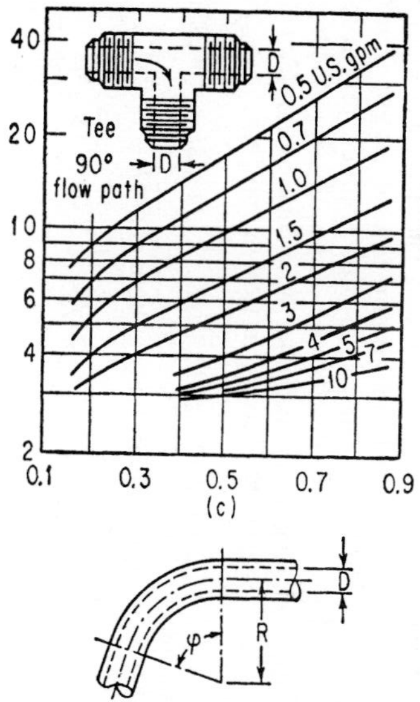

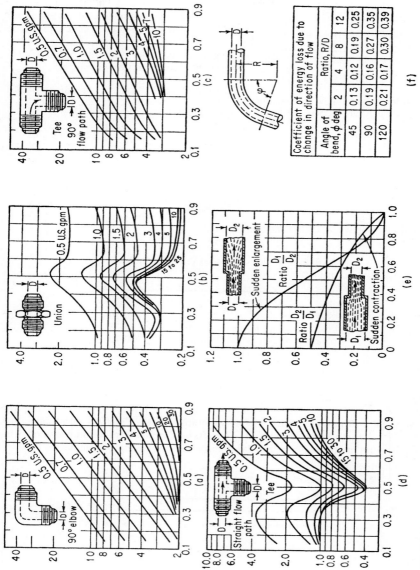

Coefficient of energy loss due to change in direction of flow				
Angle of bend, ϕ deg	Ratio, R/D			
	2	4	8	12
45	0.13	0.12	0.19	0.25
90	0.19	0.16	0.27	0.35
120	0.21	0.17	0.30	0.39

(f)

FIG. 30.29 Plots of coefficient of energy loss $L_c f/D$.

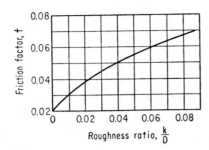

FIG. 30.30 Turbulent-flow friction factors as a function of roughness ratio.

lumped-parameter techniques have been developed. Lumped models can be used effectively, particularly if the length of the line is less than the shortest wavelength of the signals transmitted. The natural frequencies of a transmission line of length L are given by the organ pipe frequencies from classical physics and, depending on whether similar or dissimilar boundary conditions (in terms of pressure or flow) apply, are

$$f = c_0/2L, \; c_0/L, \; 3c_0/2L, \; ..., \; nc_0/2L \quad (30.95)$$

for similar inputs at both ends or

$$f = c_0/4L, \; 3c_0/4L, \; 5c_0/L, \; ..., \; (2n - 1)c_0/4L \quad (30.96)$$

for dissimilar inputs at the ends.

For these calculations, the speed of wave propagation c_0, given by

$$c_0 = (\beta_e/\rho_0)^{1/2} \quad (30.97)$$

generally lies between 35,000 and 50,000 in/s.

TABLE 30.4 Recommended Velocities for Pipes and Fittings

Area of system	Velocity, in/s
Suction lines, $\frac{1}{4}$ to 1 in	20–50
Suction lines, $1\frac{1}{4}$ and up	50–75
Discharge lines, $\frac{1}{4}$ to 2 in	100–200
Discharge through valves	250
Flow in relief and safety valves	1000

Simple Lumped Models. When the line length is short compared with the wavelengths contained in the pressure and flow signals, a lumped model can simplify analysis and will provide a means of incorporating the line's effect into models for other system components. Figure 30.31 shows an example of a lumped model in which the fluid inertia and compressibility have been divided into three lumps.

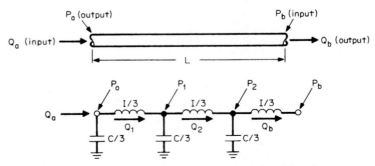

FIG. 30.31 Transmission line model for three lumps with Q_a and P_b as inputs.

Rules to be observed in dealing with lumped models include the need to place a capacitive lump next to a flow input or an inertial lump next to a pressure input. Where the line is driven by similar inputs at each end, there will be an odd number of total elements.

The differential equations of the three-lump model illustrated in Fig. 30.31 require the use of auxiliary pressure P_1 and P_2 and become

$$dP_a/dt = (-Q_1 + Q_a)(3/C) \qquad dQ_1/dt = (P_a - P_1)(3/I)$$

$$dP_1/dt = (Q_1 - Q_2)(3/C) \qquad dQ_2/dt = (P_1 - P_2)(3/I) \qquad (30.98)$$

$$dP_2/dt = (Q_2 - Q_b)(3/C) \qquad dQ_b/dt = (P_2 - P_b)(3/I)$$

where $Q_a(t)$ and $P_b(t)$ are inputs and C and I are found by using Sec. 30.5.

Equations (30.98) can be solved numerically. The use of linear-state transition matrix methods[21] is especially convenient where $Q_a(t)$ and $P_b(t)$ are arbitrary input functions of time and the terminal variables $P_a(t)$ and $Q_b(t)$ are outputs. Such mathematical models as Eqs. (30.98) are simple to generate and easy to use but require many lumps for accurate representation of true wave propagation effects. A rule of thumb[22] is to use 10 lumps per shortest-signal wavelength, where the wavelength λ is related to the signal frequency f by

$$\lambda = c_0/f$$

Effects of Friction. Laminar fluid friction acts to damp out transmission line transients. With hydraulic oil as the fluid, line resonant phenomena are sometimes significantly attenuated. Two models used in analyzing the distributed effects of fluid friction are the constant and the frequency-dependent friction models. The former model applies the laminar steady-flow friction to unsteady flow cases and generally underestimates line damping. It gives constant attenuation per wavelength of travel, while the latter model includes the variation of flow velocity profiles that occur at different wave frequencies and gives attenuation increasing with frequency.

Constant-Friction Model. In the constant-friction model the wave equations are modified to include an additional pressure-gradient term which is proportional to the average flow velocity at a line cross section where the proportionality constant is found from the laminar steady-flow friction (Hagen-Poiseuille) formula. The wave propagation operator for lines of length L and the characteristic impedance are now modified to

$$\Gamma = Ts[1 + (8\pi\nu/A_s)]^{1/2} \quad \text{and} \quad Z_0(s) = (\rho c_0/A)[1 + (8\pi\nu/A_s)]^{1/2} \quad (30.99)$$

and ν is the kinematic viscosity of the oil.

These formulas can be manipulated to produce complex arguments for the hyperbolic transfer functions and require the use of a computer for their evaluation. For frequency response calculations Eq. (30.99) can be used directly, but the model will underpredict the wave attenuation at high frequencies.[27] Also, the simplification afforded by using the finite product expansions for reducing complex hyperbolic transfer functions to rational polynomial and differential equations is now made more difficult.

Frequency-Dependent Friction Model. In the frequency-dependent friction model, account is taken of the radial dependence of fluid velocity and its variation with wave frequency.[28,29] Several references[29–31] dealing with this subject have included the use of round, rectangular, and annular passages as transmission lines. This model has been experimentally verified for both liquid[32] and gaseous[29] media. The

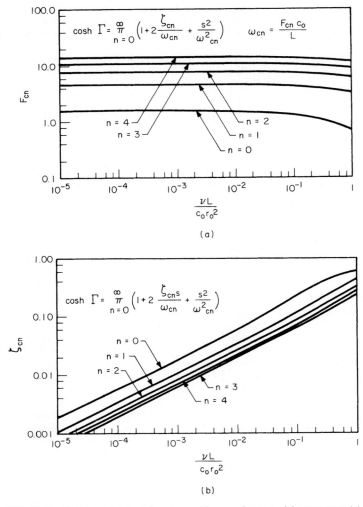

FIG. 30.32 Variation of natural frequency with approximate model parameters: (*a*) F_{cn}; (*b*) ζ_{cn}; (*c*) F_{sn}; (*d*) ζ_{sn}.

expressions for the propagation operator and characteristic impedance become functions of Bessel functions with complex arguments. These require complex computer programs for the evaluation of the line transfer functions.

To simplify the modeling effort, finite product expressions can still be used. With friction, each mode has an appropriate damping factor. The evaluation of mode frequency and damping ratio requires much computation but has been described by Gerlach[24] in terms of the viscous damping number D:

$$D = \nu L / r^2 c_0 \tag{30.100}$$

where r = line radius.

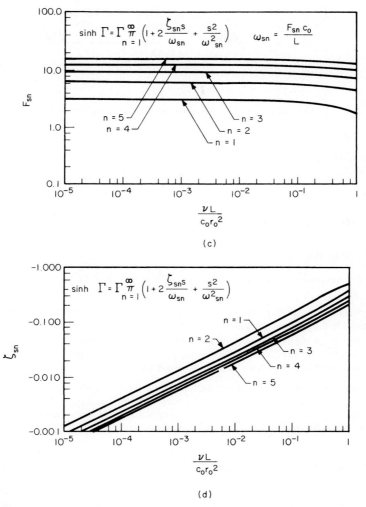

FIG. 30.32 (*Continued*)

Figure 30.32 shows how the natural frequency and damping ratio vary with the damping number for use with the approximations

$$\cosh \Gamma L \simeq \prod_{n=0}^{N} [1 + 2\zeta_n(s/\omega_n) + (s/\omega_n)^2] \qquad (30.101)$$

$$\sinh \Gamma L \simeq (Ts) \prod_{m=1}^{M} [1 + 2\zeta_m(s/\omega_m) + (s/\omega_m)^2]$$

For a 0.25-in-diameter, 40-ft-long line with an oil of 30-cSt viscosity and an effective wave speed of 45,000 in/s

$$D = (0.3)(480)/[(2.54)^2(0.125)^2(45,000)]$$

$$= 0.0327$$

TABLE 30.5 Table of Natural Frequencies and Damping
Ratios for $D = 0.008$ and Transfer Function $1/(\cosh \Gamma)$

Mode	Undamped natural frequencies, rad/s	Damping ratio
1	125	0.13
2	375	.06
3	625	.04

Note: For more accurate numerical values see Table 5 in Ref. 33.

and the mode, natural frequencies, and damping ratios from Fig. 30.32 are as listed in Table 30.5.

30.7 HYDRAULIC POWER CONTROL

30.7.1 Introduction

There are essentially two methods of controlling the flow of hydraulic power to a load:

1. By varying some characteristic of the pump (power generator) so that the rate at which fluid energy is generated is controlled in a definable manner.
2. By throttling the fluid power in a single- or multiple-orifice valve, thereby effecting control by predictable flow restrictions. Some of the fluid power is thus changed into heat, resulting in a fluid-temperature rise.

In addition to the flow characteristics, the flow- and pressure-induced forces on the valve components must be described. Understanding the nature of these forces is important in the design of pressure and flow regulators as well as the specification and analysis of the behavior of the flow-valve actuators such as torque motors.

30.7.2 Flow-Valve Configurations

The hydraulic control valve is an assembly of flow-controlling elements which operate in a prescribed and predictable manner. Several common constructions are described as follows:

The Spool Valve. The spool valve is one of the most common types. Figure 30.33 is the two-way application, Fig. 30.34 the three-way, and Fig. 30.35 the more complex but most versatile four-way valve system. Each system permits the driving of a load.

The *two-way valve* introduces a restriction in the line to the load which reduces a part of the supply pressure and dissipates some of the supply power. Thus the fluid power to the load can be controlled. With a simple restriction, the direction of flow to the load cannot be changed.

The *three-way valve* (Fig. 30.34) is the simplest configuration which permits load reversal. The differential-area double-acting piston exposes its smaller area to the supply pressure and the larger (usually double) area to the valve-controlled pressure. The exhaust port permits fluid to escape to the return line so that the supply pressure can push the actuating piston to the left.

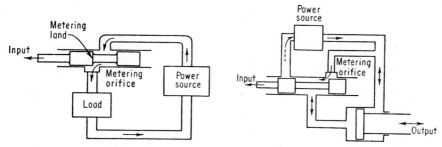

FIG. 30.33   Two-way spool valve.

FIG. 30.34   Three-way spool valve.

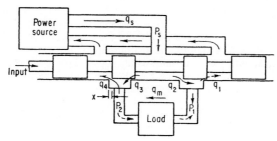

FIG. 30.35   Four-way spool valve, closed-center type.

If the distance between the metering edges on the spool and the sleeve is identical, the valve is said to be of the *closed-center* and zero-lapped type. If the distance between the metering edges on the spool is larger than those on the sleeve, the valve is defined as an *open-center* (or underlapped) valve. With the open-center valve there is a steady flow of fluid from the source to the return line for valve positions in the underlap region. The closed-center and overlapped valve has the reverse geometry with the load sealed from both source and return over the overlap region of the valve. In practice the unavoidable clearance between valve and sleeve will permit some small flow to occur.

The three-way valve permits an interesting variation in the power source. The source can be a *constant-flow source* working into the underlapped valve, with the valve operating primarily in the underlap region and the pressure in the source built up to meet load-force power requirements. Beyond the underlap region the valve functions as a two-way valve, with the load velocity determined by the source flow. The source can be a *constant-pressure source* with a closed-center valve. The flow from the source will vary depending upon load-velocity demands, load direction and leakage.

The most common method of controlling a reversible load is by means of the *four-way valve* (Fig. 30.35). With the valve to the right of its null position, the source flow is throttled through orifice 2 to the load and from the load through orifice 4 to return. The reverse flows from and to the load are attained by moving the spool to the left of the centered position as shown by the dotted arrows. The four-way valve, similar to the three-way valve, is used as either an underlapped valve with a constant-flow source (power steering) or a closed-center valve with a constant-pressure source. If a single hydraulic power supply has to provide power to several valve-motor-load systems, the closed-center constant-pressure design is used with either a variable-delivery pump or a constant-delivery pump and bypassing pressure regulator. Power dissipation, load cycles, and cost consideration dictate which system is used.

The three-way valve has several limitations compared with the four-way valve:

1. It is difficult to apply to a rotary motor.

2. The volume under compression compared with an equivalent four-way system is larger, thus tending to give poorer dynamic response in a closed-loop system.

3. Under certain conditions there is a lower load stiffness.

4. It tends to be more nonlinear in its flow-pressure characteristics.

The prime advantage of the three-way valve is its simplicity. There is only one critical axial port dimension rather than three for the single-spool four-way valve.

The Flapper Valve. Many applications of hydraulic controls require pilot valves (for the first stage of amplification) which are simple and insensitive to contaminants. One mechanization is the popular flapper valve. The ideal flapper valve, illustrated in Fig. 30.36, has sharp edges and behaves essentially like a pure orifice. The ideal flapper valve is too sensitive to nicking of the sealing edges. For this reason the practical designs are equipped with a face of finite width, illustrated in Fig. 30.37. For openings where the clearance is of the same order as the seat width, the orifice law no longer holds. For very small openings the flow will become laminar and thus the flow will vary as the third power of the opening and be inversely proportional to the viscosity. In general the flow-pressure relationships of the flat-faced flapper are not very predictable unless the normal operation opening is much larger than the seat width.

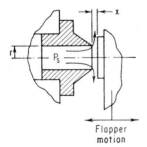

FIG. 30.36   Ideal flapper valve.

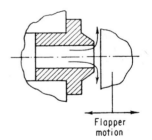

FIG. 30.37   Flat-faced flapper valve.

The Poppet Valve. Probably the most widely used type of seating valve is the poppet valve. Relief valves, pressure and flow regulators, check valves, and shutoff valves use the poppet valve extensively. A typical unbalanced poppet is shown in Fig. 30.38. The inverted and pressure-balanced valve is shown in Fig. 30.39. With a sharp-edged seat it behaves like a pure orifice. Near the cutoff point, similar to the flapper, the valve is highly nonlinear, since a particular design must have a seat of finite width, usually 0.002 to 0.005 in.

Jet-Pipe Valve. The process-control field has long used the jet-pipe valve. It is a hydrokinetic valve in that the jet-velocity impingement on the receiver (Fig. 30.40) creates the pressure by momentum recovery. The flow pattern is highly complex and subject to wide variation depending upon fabrication details.[27] The valve, since it con-

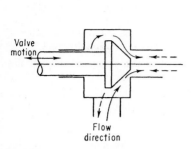

FIG. 30.38   Poppet valve.

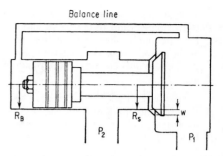

FIG. 30.39   Balanced inverted poppet valve. For approximate balance $R_B = R_s + \frac{1}{2}W$.

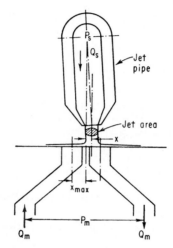

FIG. 30.40   Jet-pipe valve.

tains neither closely fitted sliding surfaces nor small orifices, is remarkably insensitive to dirt. Its behavior is parallel to that of beam deflection fluid amplifiers.

30.7.3   Pressure-Flow Relationships

Introduction.   The pressure-flow-displacement relationships for many of the configurations mentioned in the previous paragraphs will be derived, based upon the following assumptions:

1. The fluid is nonviscous and incompressible. For very small valve openings of flappers and poppet valves and low pressure drops and for flow through clearance spaces in spool valves, these assumptions should be examined as the case demands. The assumption of incompressibility inside the valve, for the flow relationships, results in negligible errors.

2. The fluid source is ideal. This assumption demands a constant-pressure-source (CP) supply fluid to the valve and a constant pressure at the intake regardless of flow demands. The constant-flow source (CQ) similarly provides a constant flow at the valve intake port regardless of pressure demands. The idealized characteristics should thus be corrected for excessive line drops and regulation droops.

3. The geometry of the valve is ideal. It is assumed that the metering edges of the valve are sharp and that the clearances between spool and sleeve are zero.

4. Steady-state conditions prevail inside the valve. For a stable valve and system the assumption is sound. For valve squeal and chatter the characteristics no longer hold. Valve stability will be discussed in Sec. 30.7.4.

The volume flow through an orifice, derived previously, is given by

$$Q = C_d A_0 \sqrt{2(\Delta P)/\rho} \tag{30.102}$$

The discharge coefficient C_d is very nearly constant for Reynolds numbers larger than 260 ($Re = 2Vx\rho/\mu$)

where A_0 = nominal area for annular opening ($A_0 = \pi D_0 x_0$) in^2
D_0 = annular mean diameter, in
x = width of annular slit, in
ΔP = pressure drop

For the sharp-edged orifice it is assumed that $C_d = 0.625$.
Caution should be exercised in interpreting test data. If the valve saturates, meaning that an appreciable portion of the pressure drop occurs in the valve chambers and passages rather than across the orifice, the analysis is not valid. Similarly, if the opening approaches the clearance, the relationships are in error.

The characteristics of the following cases will be given:

1. The symmetrical four-way valve with constant-pressure source: (a) and (b) four arms variable, (c) two arms variable

2. The three-way valve with constant-pressure source: (a) two arms variable, (b) one arm variable

3. The partially underlapped symmetrical case of the four-way valve with the constant-pressure source

4. Asymmetric cases

For details of the derivations the reader is referred to Refs. 4 and 14. The plots of these equations are useful in understanding valve characteristics. The slopes and spacing of the individual curves, the *valve gains,* are of paramount interest to the system designer and are presented in analytical form.

Case 1a. Symmetrical Closed-Center Four-Way Valve with Constant-Pressure Source (Rectangular Ports) (see Fig. 30.35). The dimensionless equation relating flow to pressure to valve displacement is

$$P_m = 1 - 2Q_m^2/y^2 \tag{30.103}$$

where $P_m = (P_1 - P_2)/P_s$ = motor pressure   (30.104)

$Q_m = Q/G \sqrt{P_s}$ = flow to motor   (30.105)

$y = x/x_0$ = displacement where x is valve opening   (30.106)

x_0 = maximum valve opening

$$G = g/y = \text{unit valve conductance} \tag{30.107}$$

with $g = AC_d \sqrt{2/\rho}$ (A is the port area corresponding to x) where G is the conductance for $y = 1$ (the valve fully open).

Equation (30.103) represents a set of parabolas and is plotted in Fig. 30.41a.

The flow sensitivity is given by k_q, the spacing of the curves at constant load pressure.

$$k_q = \partial q_m / \partial x \big|_{pm = \text{const}} \tag{30.108}$$

The pressure sensitivity is given by k_{p0}, the slope of the curves at constant flow to the load,

$$k_{p0} = \partial p_m / \partial x \big|_{qm = \text{const}} \tag{30.109}$$

Thus, near the origin

$$k_{q0} = (G/x_0) \sqrt{P_s/2} \tag{30.110}$$

The valve behaves like an adjustable resistance and to all intents and purposes acts in concert with a constant pressure supply to provide an adjustable flow into and out of the actuator. The flow gain K_q represents the valve flow rate per unit opening and is often critical in determining drive response speed and stability.

Case 1b. Symmetrical Open-Center Four-Way Valve with Constant-Pressure Source. If the distance between the lands on the spool in Fig. 30.34 is less than the distance between the lands on the sleeve by an amount $2x_0$ (x_0 being the so-called underlap or open center) the valve becomes the underlapped valve. For operation in the underlap region the valve characteristics are given by (see Fig. 30.41b)

$$Q_m^2 = 1 + y^2 - 2yP_m - (1 - y^2) \sqrt{1 - P_m^2} \tag{30.111}$$

The curves in Fig. 30.41b are remarkably straight, parallel, and evenly spaced, indicat-

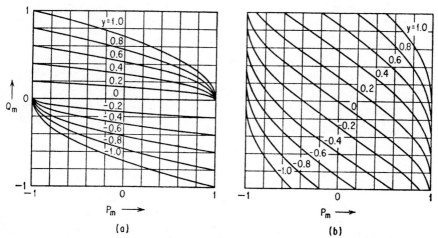

FIG. 30.41 Dimensionless pressure-flow valve characteristics. (*a*) Closed-center four-way valve, constant pressure supply. (*b*) Open-center four-way valve, constant pressure supply.

ing a linear device over most of its operating range. The standby power loss is very high relative to the power supplied to the load. If the operation of the valve is not limited to the underlap region, case 3 results.

For this case, the values of valve gain are

$$k_{q0} = 2(G/x_0)\sqrt{P_s/2} \tag{30.112}$$

$$k_{p0} = 2(P_s/x_0) \tag{30.113}$$

Case 1c. Symmetrical Four-Way Valve with Constant-Pressure Source with the Two Upstream Arms Fixed and the Two Downstream Arms Variable (Twin Flapper-Nozzle System) (see Fig. 30.42). The dimensionless equations relating flow, pressure, and geometry are very complicated for the total valve. A reasonable set of equations based upon two three-way valves is given by the following equations:

$$P_1 = (P_s/\alpha^2)[\alpha + (\alpha - 2)Q_m^2 - 2A_m(1 - y)\sqrt{\alpha - A_m^2}] \tag{30.114}$$

$$P_4 = (P_s/\beta^2)[\beta + (\beta - 2)Q_m^2 - 2Q_m(1 + y)\sqrt{\beta - Q_m^2}] \tag{30.115}$$

$$\alpha \triangleq y^2 + 2 - 2y \tag{30.116}$$

where $y \triangleq x/u$

$$\beta \triangleq y^2 + 2 + 2y \tag{30.117}$$

$$P_m = (P_1 - P_4)/P_s \tag{30.118}$$

$$Q_m = Q_m/G\sqrt{P_s} \tag{30.119}$$

Figure 30.42 is a plot of the valve characteristics.

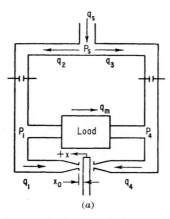

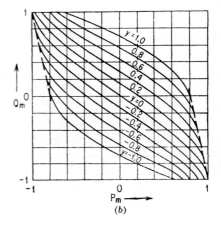

FIG. 30.42 Twin flapper-nozzle valve.

The valve gains at the origin are

$$k_{q0} = (G/x_0)\sqrt{P_s/2} \tag{30.120}$$

$$k_{p0} = P_s/x_0 \tag{30.121}$$

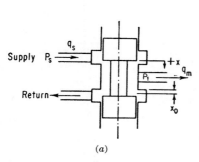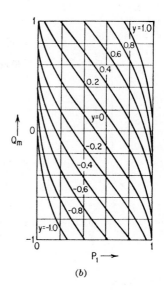

FIG. 30.43   Three-way underlapped valve.

Case 2a. Underlapped Three-Way Valve with Constant-Pressure Source and Both Arms Variable (Fig. 30.43a).   The characteristics here given will be valid within the underlap region (Fig. 30.43b).

$$x = \pm x_0$$

The governing relationships are

$$Q_m = (1 + y) \sqrt{1 - P_1} - (1 - y) \sqrt{P_1} \tag{30.122}$$

where

$$Q_m = Q_m/G \sqrt{P_s} \qquad P_1 = p_1/P_s \qquad y = x/x_0$$

Rearrangement yields

$$P_1 = \frac{(y + 1)^2(y^2 + 1) - 2yQ_m^2 - (1 - y^2)Q_m \sqrt{2(1 - y)^2 - Q_m^2}}{2(y^2 + 1)^2} \tag{30.123}$$

The flow sensitivity is given by

$$(2G/x_0) \sqrt{P_s/2} \tag{30.124}$$

and the pressure sensitivity by

$$P_s/x_0 \tag{30.125}$$

Case 2b. Three-Way Valve with Constant-Pressure Source and the Downstream Arm Variable.   The single-jet flapper is shown in Fig. 30.44a. While its simplicity recommends it, experience has shown that its characteristics are sensitive to variations in supply pressure. The governing equations are

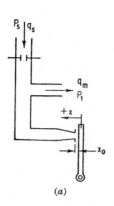

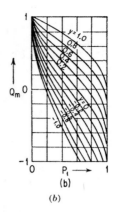

FIG. 30.44   Single-jet flapper-nozzle valve.

$$Q_m = \sqrt{1 - P_1} - (1 - y)\sqrt{P_1} \tag{30.126}$$

$$P_1 = \frac{2y - y^2(1 - Q_m^2) - 2(1 - y)Q_m\sqrt{2 - 2y + y^2 - Q_m^2}}{(2 - 2y + y^2)^2} \tag{30.127}$$

Figure 30.44b gives the valve characteristics.
    The flow sensitivity

$$k_{q0} = (G/x_0)\sqrt{P_s/2} \tag{30.128}$$

The pressure sensitivity

$$k_{p0} = P_s/x_0 \tag{30.129}$$

***Case 3. Partially Underlapped Symmetrical Four-Way Valve with Constant-Pressure
Source.***   Case 1a, the closed-center valve in its ideal form, assumes no leakage from
clearance between spool and sleeve and metering edge rounding. All practical designs
will have clearances in excess of 0.0002 in (radial) and an equivalent radius on the
metering edge. The metering edge radius, in particular, will increase with use because
of erosion from fluid and contaminant flow, and crushing from hard foreign particles.
It is important to evaluate carefully the valve gain near the closed position. k_p is usually less than 10^6 lb/in². The valve characteristics of the actual valve are thus similar to
an open-center valve near the null position. They approach closed-center characteristics for openings two to three times larger than the valve clearance or 0.0004 to 0.0006
in. Depending upon the full travel of the valve, the underlap region may occupy an
appreciable portion of the valve travel. A typical case is illustrated in Fig. 30.45. The
characteristics should be compared with those of Fig. 30.41a and b. For a particular
case the valve characteristics can be constructed by appropriate approximations using
Figs. 30.45 and 30.41.

Case 4. Unsymmetrical Cases.   The nonlinear characteristics of the zero-lapped
valve and particularly its high gain near the null position make the design of stable
systems difficult. The underlapped valve has been shown to result in more stable systems by virtue of its lower and more constant gains as well as inherently higher damping coefficients. The high standby power loss of this valve has led to studies of

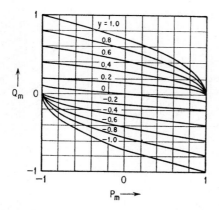

FIG. 30.45 Dimensionless pressure-flow characteristics of the partially underlapped (20 percent) four-way valve with constant pressure supply.

unsymmetrically lapped valves which could combine the good features of both types, namely, low standby power loss and good damping near the null. Another reason for opening the supply pressure to the load ports while keeping the return ports closed at center is the need for cooling flow through the actuator in high-temperature applications. Figure 30.46 shows four cases of a specific valve design in which $P_s = 950$ lb/in² $x_0 = 0.005$ in, and W = port width = 1.0 in with varying unsymmetrical underlap. The term k is the ratio of return-port underlap to pressure-port underlap as shown in Fig. 30.47, thus $k = 1$ represents a symmetrically lapped valve.

The relative effects upon valve gain and standby power loss of the uneven underlap are shown in Figs. 30.48 to 30.50 in which

$$C_1 = \frac{(\partial Q_m/\partial P_m)x = 0 \quad \text{for uneven lap}}{(\partial Q_m/\partial P_m)x = 0 \quad \text{for even lap}} \tag{30.130}$$

$$K_1 = \frac{(\partial Q_m/\partial x)P_m = 0 \quad \text{for uneven lap}}{(\partial Q_m/\partial x)P_m = 0 \quad \text{for even lap}} \tag{30.131}$$

$$\text{HP} = \frac{\text{standby power loss for uneven lap}}{\text{standby power loss for even lap}} \tag{30.132}$$

The Jet-Pipe Valve. The jet-pipe valve differs from those previously described in that it is more akin to a hydrokinetic class of devices (see Fig. 30.40). The momentum of the fluid against the two receiver holes is regulated by controlling the jet direction relative to the receiver block. Pressure recoveries up to 90 percent can be achieved with well-designed receivers. Jet-pipe flow is complex three-dimensional flow which has defied useful analysis to date. Experience indicated that:

1. The receiver holes should be about twice the jet diameter.
2. The jet pipe should be separated from the receiver block by at least two jet diameters. Separation in excess of two jet diameters has little effect on the valve characteristics.

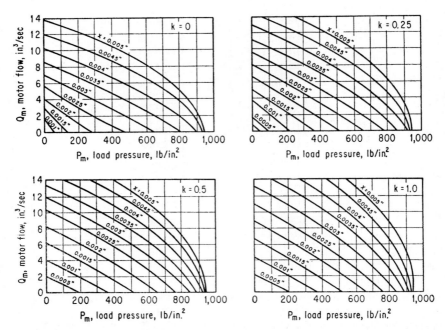

FIG. 30.46 Flow-pressure characteristics of unevenly underlapped four-way valves with constant supply pressure.

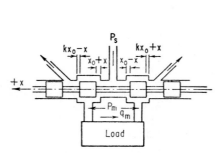

FIG. 30.47 Schematic diagram of unevenly underlapped four-way valve.

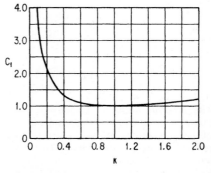

FIG. 30.48 Dimensionless relative valve gain $\delta Q_m/\delta P_m|x = 0$ for the uneven lapped valve relative to the symmetrical valve for various values of asymmetry k.

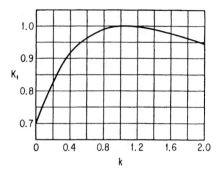

FIG. 30.49 Dimensionless relative valve gain $\delta Q_m/\delta x | P_m = 0$ for the uneven lapped valve relative to the symmetrical valve for various values of asymmetry k.

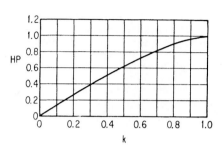

FIG. 30.50 Ratio of standby power loss of unevenly lapped valve over standby power loss of symmetrical valve for various values of asymmetry k.

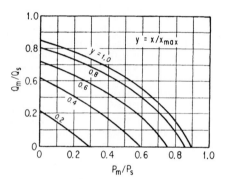

FIG. 30.51 Dimensionless pressure-flow characteristics of the jet-pipe valve.

The flow-pressure characteristics are given by Fig. 30.51 and are seen to be similar to the characteristics of the four-way flapper valve (Fig. 30.42).

30.7.4 Steady-State Forces in Flow-Control Valves

The ability to position the spool valve accurately in its sleeve, the flapper valve relative to its plug, and the jet-pipe valve relative to its receiver is the heart of good fluid-power control-system design. The forces acting on the valve elements can be divided as follows: flow induced, pressure unbalance induced, inertia, viscous drag, and nonviscous drag.

For *spool-type valves,* the flow-induced forces generally represent the major portion of the force to be overcome by the valve actuator. The origin of the force is best explained by studying the pressure distribution on the walls of the spool lands. Figure 30.52 shows that flow either into the spool or out of the spool at the metering edge results in a flow-induced force causing the valve to close.

The force can be shown to be

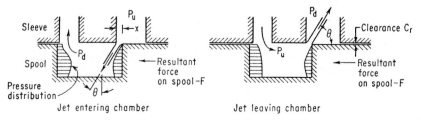

FIG. 30.52 Origin of the flow-induced force on the spool-type valve.

$$F = \sqrt{\rho} \cos \theta Q \sqrt{P_u - P_d} \tag{30.133}$$

where θ = angle between jet and valve-chamber wall
P_u = upstream pressure
P_d = downstream pressure

Angle θ is equal to 69° subject to the following limitations:

1. The fluid is nonviscous and incompressible.
2. The peripheral width of the orifice (into the paper) is large compared with the axial length (two-dimensional flow).
3. Flow is irrotational (potential flow) ahead of the orifice.
4. The upstream chamber angles are 90° (within 10°).
5. Valve clearance is an order of magnitude smaller than valve opening x.
6. Orifice edges are sharp.

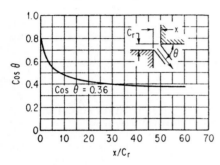

FIG. 30.53 Effect of radial clearance between spool and sleeve on efflux charge θ.

For the case of finite clearance between valve spool and valve sleeve C_r the efflux angle is effected as shown in Fig. 30.53. The effect of rounded corners is similar because for small openings θ will become less than 69°, resulting in an increased flow force.

For petroleum-base fluids with a specific gravity of 0.85, square sharp corners, and clearance small relative to opening, the following useful relations result for a single land:

$$F = 0.0032Q \sqrt{P_u - P_d} \tag{30.134}$$

$$F = 4.8 \times 10^{-5}Q^2/A \tag{30.135}$$

$$F = 0.215A(P_u - P_d) \tag{30.136}$$

These flow forces can be substantially reduced or even reversed with proper design of the valve to direct the efflux momentum. The design considerations are rather complex in that both upstream and downstream chamber designs affect the efflux angle. In principle, the flow is to be guided with retention of jet velocity until the jet leaves at an angle θ of 90°.[4]

Similar fundamental considerations hold for the *poppet valve*. Yet the resulting forces on a poppet valve have been found to be difficult to predict because of the effect of fluid shear forces on the poppet cone in addition to the complex flow pattern in the downstream chamber. Guiding the flow as illustrated in Fig. 30.54 has been found to reduce the flow-induced closing forces and result in improved performance of pressure regulators.

The flapper-nozzle-valve forces are best studied for two configurations, the ideal sharp-edged nozzle flapper and the more common flat-faced nozzle. If the sharp-edged flapper is tightly shut (Fig. 30.55) the force on the flapper

$$F = \pi r^2 P_s \tag{30.137}$$

where r is the radius of the nozzle.

As the flapper opens and orifice-type flow occurs, the total force becomes the stat-

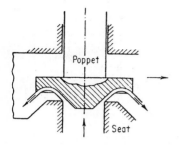

FIG. 30.54 Flow-force-compensated poppet valve.

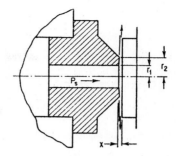

FIG. 30.55 Flat-faced flapper-nozzle valve.

ic pressure P_1 times the nozzle area plus the momentum transferred to the flapper. For small values of x/r, it can be shown that the force varies parabolically with opening x (Ref. 4)

$$F = \pi r^2 P_s [1 + (2C_d x/r)^2] \qquad (30.138)$$

In the limit all the momentum is transferred to the flapper:

$$F = 2\pi r^2 P_s \qquad (30.139)$$

Design realities do not permit either the manufacture or maintenance of a sharp-edged flapper-nozzle unit. Many flapper valves are therefore made with relatively wide flats. For very small values of $x/(r_2 - r_1)$ (see Fig. 30.55), the flow is essentially laminar and proportional to x^3 as given by Eq. (30.140). The limiting value of $x/(r_2 - r_1)$ is determined by the onset of separation. A good estimate can be made by limiting Re at r_1 to less than 2000.

$$q = \frac{\pi x^3 P_s}{6\mu \ln (r_2/r_1)} \qquad (30.140)$$

The force is given by

$$F = \pi(r_2^2 - r_1^2)P_s/(2 \ln r_2/r_1) \qquad (30.141)$$

The pressure distribution across the flat may result in a force considerably larger than would be obtained from the sharp-edged orifice. For values of $x/(r_2 - r_1)$ larger than the critical value, flow separation occurs, making analytical predictions impossible.

30.7.5 Transient Flow Forces in Flow-Control Valves

The rate of change of flow into and out of valve chambers results in forces on the moving valve element. These forces are usually proportional to the valve velocity and thus similar to damping forces. Since the force may either aid or oppose the valve velocity, the damping may be negative or positive. If the valve, as is usually the case, is part of a spring-mass system (torque motor or valve actuator), negative damping will contribute to dynamic instability.

The detailed derivation of the transient flow forces can be found in Ref. 4. The total flow-induced force on a spool with constant pressure drop across the land is given by

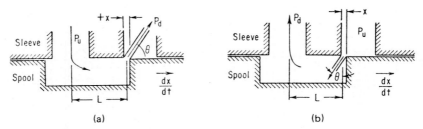

FIG. 30.56   Valve ports. (*a*) Positively damped. (*b*) Negatively damped.

$$F = -[2C_d^2W \cos \theta(P_u - P_d)]x - [C_dW\sqrt{2(P_u - P_d)\rho}\ L](dx/dt) \qquad (30.142)$$

for the configuration of Fig. 30.56, where W is the valve port-area per unit opening distance and x is the distance opened. The first term on the right-hand side is similar to a spring force tending to close the valve; the second term, since it opposes motion in the direction of opening, is like a positive damping force. If the flow is reversed as shown in Fig. 30.56*b*, the spring force still tends to close the valve, but the damping force is now in the direction of motion and tends to open the valve, thus producing a negative damping effect

$$F = -[2C_d^2W \cos \theta(P_u - P_d)]x + [C_dW\sqrt{2(P_u - P_d)\rho}\ L](dx/dt) \qquad (30.143)$$

The porting design can thus be arranged to give positive damping to assure stability.

However, even valves with positively damped ports can be unstable when installed in a specific system. Such instabilities are usually associated with acoustic resonances between the transmission line, the valve, and the load.

Section 30.6.3 discusses various methods of analyzing and simulating the resonant transmission line.

30.8 *HYDRAULIC MOTORS*

30.8.1 Introduction

Hydraulic motors are the output devices of the hydraulic system. They convert the controlled fluid power into mechanical position (either rotary or linear) for the position servo, or controlled power for the power control in the form of torque at angular velocity (or force at linear velocity) (see Fig. 30.57). The input impedance and load sensitivity are the characteristics of importance.

Several types of motors (actuators) are commonly used:

Linear actuators, or ram-type motors, are shown in Fig. 30.58*a, b*, and *c*. Figure 30.58*a* is the single-acting ram, which is rather uncommon because it relies on the load force for return.

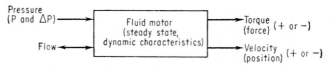

FIG. 30.57   Block diagram of motor.

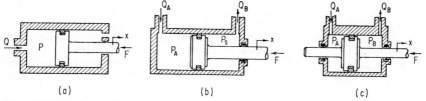

FIG. 30.58   Ram-type motors.

Figure 30.58*b* is the double-acting differential-area ram used commonly with three-way valves such that the small area is constantly at supply pressure and the area ratio is 2:1. For applications with four-way valves the area ratio is usually much less than 2:1 and is determined by the structural requirements of the actuator rod.

Figure 30.58*b* is the double-acting equal-area ram. In each case differential pressures across the ports result in force on the piston which is transmitted through the rod to the load. Pressures vary from a few hundred to 5000 lb/in^2 and up.

Rotary actuators can be of the gear, vane, or multiple-piston (axial or radial) types. Gear and vane motors are usable up to 1000 lb/in^2, with piston motors available for hydraulic pressures up to 5000 lb/in^2.

Hydraulic motors are used for both steady and intermittent duty.

30.8.2   Characteristics of Motors

The steady-state characteristics of motors can be expressed in a manner similar to those of the pump as developed in Sec. 30.3. The torque output of the motor is

$$T_{AM} = \Delta P D_m/2\pi - C_d D_m \mu N_m - C_f(\Delta P D_m/2\pi) - T_c \qquad (30.144)$$

The flow to the motor is given by

$$Q_{AM} = D_m N_m + (2\pi C_s D_m/\mu)\, \Delta P - Q_r \qquad (30.145)$$

Certain piston motors have additional retarding torques and leakage flows proportional to the sum of the port pressures. Appropriate terms can be added to Eqs. (30.144) and (30.145)

where T_{AM} = shaft torque
  ΔP = pressure drop
  D_m = motor ideal displacement
  C_f = drag coefficient dependent on pressure
  T_c' = friction torque dependent on neither speed nor pressure from shaft seals and bearings
  Q_{AM} = flow to the motor, volume per revolution
  N_m = motor speed
  C_s = leakage coefficient
  Q_r = flow-loss rate from cavitation effects

The overall efficiency η_m, torque efficiency η_{mm}, and volumetric efficiency η_{vm} are given by Eqs. (30.146) and (30.147):

$$\eta_m = \frac{1 + C_s(\Delta P/\mu N) - Q_r/2\pi D_m N}{1 - C_d(\mu N/\Delta P) - C_f - T_c/D_m \,\Delta P} \qquad (30.146)$$

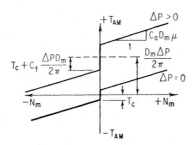

FIG. 30.59 Torque-speed characteristics of positive-displacement motors.

$$\eta_m = \eta_{vm}/\eta_{mm} \tag{30.147}$$

The steady-state characteristics can be shown on torque-speed diagrams (Fig. 30.59) for further clarification.[13]

The dynamic characteristics of motors are described by several design parameters: volume under compression V_m, inertia relative to torque J_m, mechanical compliance of output shaft K_m, friction as a function of speed only C_d, friction as a function of pressure only C_p, and nonviscous friction T_c.

The volume under compression as seen by the load circuit for fixed-displacement gear, vane, and piston motors varies between 1.2 and 1.5 times the displacement (volume per revolution).

The inertia of axial-piston motors is extremely low relative to the torque at the maximum pressure differential at no load. Typical values for an aircraft-type axial-piston motor at $\Delta P = 3000$ lb/in²:

$$T_m/J_m = 100,000 \text{ 1/s}^2$$

$$T_m^2/J_m = 4 \times 10^7 \text{ lb·in/s}^2$$

Vane motors at $\Delta P = 1000$ lb/in²:

$$T_m/J_m = 200,000 \text{ 1/s}^2$$

$$T_m^2/J_m = 7.5 \times 10^7 \text{ lb·in/s}^2$$

Values for gear-type motors are similar in magnitude to vane motors.

The compliance of the output member is a result of the mechanical connections between the pistons and the shaft. Flexibility of connecting rods, universal joints, wobble plates, etc., results in shaft motion with torque change and no flow to the backlash in the output drive. For reversals in direction this phenomenon becomes important and must be considered in the systems simulation.

Friction as a function of speed results in damping in the closed-loop position-controlled system.

Friction as a function of pressure only can be considered as a reduction in the effective displacement of the motor.

The nonviscous friction term T_c resulting from seals is a major obstacle to achieving accurate performance and stability. T_c is strongly dependent upon the average pressure of the ports to the motor. Thus the lower the average pressure, the lower T_c.

A further problem exists in the large difference between starting torque and running torque for some motors. A straight-line approximation for the pressure drop required to start a hydraulic motor as a function of the downstream port pressure is valid:

$$\Delta P = A + BP_d \tag{30.148}$$

B varies from 0.220 to 0.550 depending upon the motor design; A varies from 20 to 50 lb/in² for rotary motors with displacement ranging from 0.065 to 0.375 in³/r. Ram-type motors depending upon the rod-seal design may have considerably higher breakaway pressure differentials ranging into several hundred pounds per square inch.

The no-load running torque can also be represented by a straight-line approximation. The intercept A is the same as for the previous case, but the slope is generally one-quarter to one-third of the starting P/P_d. Certain balanced-vane-type motors of special design exhibit exceedingly low starting pressure drops with P/P_d equal to almost zero. Special attention to this phenomenon in the design of motors can thus result in good starting performance. Figure 30.59 is somewhat of an idealization in that it does not recognize the starting-torque problem. Under slow speed conditions, stick-slip motions can arise, and additional damping is needed to eliminate this problem.[34]

30.9 HYDRAULIC CONTROL SYSTEMS

30.9.1 Introduction

This section describes the configuration of some typical hydraulic drive systems comprised of hydraulic motors (either rotary or ram type) and valves or pumps as control elements. The major advantage in using hydraulic drive systems is that large output powers can be quickly controlled by very small power level signals (from a microcomputer, for example). Valve-controlled hydraulic servo drives are an integral part of modern aircraft flap, wing surface, engine, and landing-gear controls and of controls for steel mills and other heavy industrial equipment.

Most of the drives use the valve to control oil flow rate to the motor. The valve opening sets the speed of the motor. Variations of the basic servomotor drive to provide position and force controls involve feedback control of the primary variable. Dynamic performance and particularly stability are design issues that are solved by using analytical models and computer simulation. In the following section, analytical models and their limitations are described for valve-controlled and pump-controlled drives. The solution of the model differential equations will require a digital or analog computer. Numerical techniques for the solution of ordinary differential equations are well established and will not be discussed here.

30.9.2 Valve-Controlled Drives

The most widely used method of hydraulic control consists of a four-way control valve of either the overlapped or underlapped type as described in Sec. 30.7 and a ram- or rotary-type motor as described in Sec. 30.8. The valve may be actuated by mechanical, electrical, or fluid means. The system is shown in Fig. 30.60. The following assumptions are made in the analysis:[4]

1. The control valve is ideal, edges are sharp, and radial clearances are zero between spool and sleeve.
2. Flow through orifices is simple-type orifice flow with no viscous effects.
3. All flow passages between valve and motor are short so that fluid-mass effects can be neglected.
4. Friction losses in lines are negligible.
5. Leakage flow in the motor is laminar.
6. Supply characteristics are unaffected by valve demands (pressure remains constant or flow remains constant, respectively).

7. Exhaust pressure is zero.
8. Motion of the ram is restricted to small excursions from its center position, and the valve characteristics can be assumed linear for small load-pressure, load-flow, and valve-travel changes.

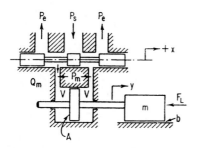

FIG. 30.60 Valve-controlled hydraulic servo-motor with inertia load.

Referring to Fig. 30.60, the steady-state speed of the load $\dot{y}$ depends on valve opening, although it will slow down as the applied mechanical force F_L increases. The steady-state speed vs. valve opening and load relationships can be established by using the pressure-flow valve characteristics of Fig. 30.40, if the load pressure drop P_m is replaced by F_L/AP_s and the valve flow rate Q is replaced by $\dot{y}A$.

The dynamic characteristics of the system are similar to those of a second-order spring mass system. The response of the mass is limited by two phenomena, relating respectively to the compressibility of the oil in the trapped volumes V and the inertial forces induced by acceleration of the mass m.

Using mathematical models based on the concepts described in Sec. 30.5, the equations of motion and continuity applied to the system give

Left-hand chambers: $\qquad (V/\beta_e)(dP_1/dt) = Q - Av \qquad$ (30.149)

Right-hand chambers: $\qquad (V/\beta_e)(dP_2/dt) = Av - Q \qquad$ (30.150)

Mass motion: $\qquad m(dv/dt) = -bv + P_1A - P_2A - F_L \qquad$ (30.151)

Valve flow: $\qquad Q = C_d\,\pi D_0 x\,\sqrt{(2/\rho)(P_s - P_1)} \qquad$ (30.152)

$\qquad\qquad = C_d\pi D_0 x\,\sqrt{2P_2/\rho} \qquad$ (30.153)

and $\qquad\qquad \dfrac{dy}{dt} = v$

Now, defining $(P_1 - P_2) = P_m$ and rearranging, two first-order "state" equations[21] can be established for modeling the drive dynamic response, with a third, Eq. (30.153), used to compute drive displacement from velocity.

$$dP_m/dt = -(2\beta_e A/V)v + C_d\pi D_0 x(t)(P_s - P_m)/\rho \qquad (30.154)$$

$$dv/dt = -(b/m)v + (A/m)P_m - F_L(t)/m \qquad (30.155)$$

$$dy/dt = v \qquad \text{and} \qquad P_m < P_s \qquad (30.156)$$

The system equations, when linearized, give

$$d(\Delta P_m)/dt = -(2A\beta_e/V)\,\Delta v - (\partial Q/\partial P_m)(\Delta P_m) + K_q[\Delta x(t)] \qquad (30.157)$$

$$d(\Delta v)/dt = -(b/m)\,\Delta v + (A/m)\,\Delta P_m - (1/m)\,\Delta F_L(t) \qquad (30.158)$$

$$d(\Delta y)/dt = \Delta v \qquad (30.159)$$

These linearized equations model the response of drive speed and load-pressure differential to input changes from external loads (ΔF_L) and/or changes in control valve opening Δx. This basic model has been found to predict transient and frequency response of hydraulic drives with good agreement within the limits given in the assumptions. The equations of the model have been extended to include cavitation effects and the effects of nonlinearity with the pressure-flow valve characteristics.[11] The linearized equations (30.157) and (30.158) correspond to a second-order system. For changes in valve opening as the input and load velocity as the output, the drive transfer function can be reduced to

$$\Delta v(s)/\Delta x(s) = K_1/[(s/\omega_0)^2 + 2\zeta(s/\omega_0) + 1] \tag{30.160}$$

where $K_1 = K_q A/(K_p b + A^2)$ and $K_p = (\partial Q/\partial P_m)$. For lightly loaded drives, K_p is very small, and K_1 reduces to approximately K_q/A.

The undamped natural frequency ω_0 is given by

$$\omega_0 = \sqrt{2(K_p b + A^2)\beta_e/mV} \tag{30.161}$$

with the damping ratio ζ given by

$$\zeta = (K_p m + bV/2\beta_e)/2 \sqrt{(Vm/2\beta_e)(K_p b + A_2)} \tag{30.162}$$

The basic servomotor as modeled here has a transfer-function block diagram for the "motor-load" dynamics as shown in Fig. 30.61.

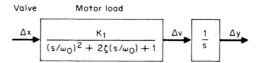

FIG. 30.61 Transfer function of the hydraulic motor drive with inertial load without external load disturbances.

Figure 30.62 shows the transfer function block diagram resulting from the reduction of Eq. (30.157) and (30.158) when load disturbances are included. By careful manipulation of the equations, it can be shown that the "stiffness" of the drive for resisting movement when loads are applied is the hydraulic "spring rate"

$$k = \Delta F_L/\Delta y\big|_{\Delta x = 0} \simeq 2\beta_e A^2/V$$

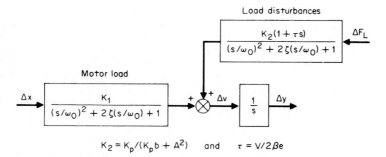

FIG. 30.62 Transfer function of the hydraulic motor drive with inertial load and external load disturbances.

The diagrams enable the design engineer both to understand the behavior of the valve-controlled drive as a mechanical control actuator and to provide formulas to estimate the key design parameters corresponding to stiffness, natural frequency, and damping ratio.

An important result of the simple analysis given here is that these drives are lightly damped. Increased damping can be achieved by increasing the valve pressure-sensitivity coefficient K_p. This can be achieved by such techniques as[34]

1. Using underlapped valves
2. Operating at high steady-load conditions
3. Providing leakage across the main drive piston

The validity of this simplified approach must be carefully examined. If a pressure-compensated variable-delivery pump is used as a fluid power source, as described in Sec. 30.4, the sensitivity of the source-to-flow demand variations from the valve and their effect on supply pressure and flow dynamics must be examined. Of particular interest are the pump damping coefficient and natural frequency. A low damping coefficient (below 0.5) and a pump natural frequency close to the valve and load natural frequency require closer study of power source and drive interactions.[14]

In certain jet-engine applications the ram motor is removed from the valve by as much as 20 ft. If the frequency of the standing wave and its harmonics in the transmission line even approaches the system frequency, the transmission-line dynamics as described in Sec. 30.6.3 must be considered to assure a stable system.

Of prime importance is the full understanding of the nature of the load. Mechanical compliance between ram and inertia and inertia against ground and the nonviscous friction in the load are among the considerations which must be carefully examined.

30.9.3 Electrohydraulic Servo Drives (Single-Stage)

The previous section dealt with the valve-controlled hydraulic drive as a control actuator. The valve opening Δx was considered as the input. In many cases—ranging from aerospace, heavy industry, and robotics to offshore petroleum applications—valve actuation is accomplished by electrical signals. The electrical actuator can be a solenoid for on-off operation, stepper motor for digital control operation, or electrical torque motor for continuous control operation. The advantage of electrical operation lies in convenience, ease of control, and ease of interfacing the mechanical power elements with microprocessor control components.

Control valves can be arranged in single- or two-stage forms. The single-stage control valve shown in Fig. 30.63 shows how a permanent-magnet torque motor can be arranged to operate a four-way valve. The torque motor is well suited to this application because the valve operates with very small movement (< 0.020 in). The motor must be sized to overcome the valve loading arising from the flow forces established in the spool chamber. This load behaves like a spring load for the motor. An input current of a fraction of an ampere can actuate a valve controlling oil flow of a few gallons per minute at pressure levels up to 3000 lb/in^2.

While the simplicity of single-stage operation is an advantage, one of the drawbacks is the potential instability arising from motor-valve interaction. If an undersized motor is used, changes in main-drive load pressure will change the spool-valve flow force, which in turn may cause motor displacements and unstable system interaction.

In order for the single-stage motor to operate with reasonable accuracy as a control device, the electrical "stiffness" of the motor K_a must be higher than the stiffness associated with the spool-valve flow force K_f.[4,14]

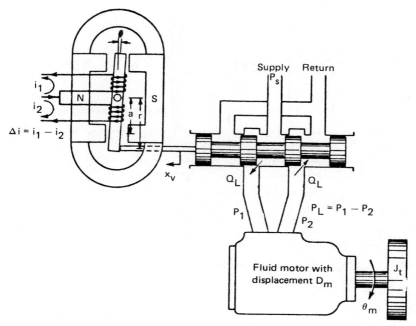

FIG. 30.63 Schematic of a single-stage electrohydraulic servo valve connected to a motor with inertia load. (*Courtesy John Wiley & Sons, Inc.*[14])

K_f can be computed by using the results of Sec. 30.7.4 from

$$F = (0.215 \, \pi D_0 P_s)x \qquad (30.163)$$

so that

$$K_f = 0.215 \, \pi D_0 P_s$$

The mechanical stiffness K_a associated with changes in torque-motor position for a unit change in motor load force at constant motor current is best obtained experimentally.

Another drawback is that the steady-state accuracy of the drive in terms of valve flow for a given electrical input is susceptible to change as system pressure changes. This is an undesirable feature from the control system standpoint. The unstable interaction of the drive and torque motor is discussed in detail in Ref. 35. In some applications, where rapid response is not required and the expense of two-stage operation cannot be justified, a low-power servomotor or synchro with a mechanical gear reduction to the valve could be used to provide motor stiffness.

To overcome torque motor-valve-drive instabilities and provide more accurate and rapid response, a two-stage servo valve concept is commonly used as indicated in Figs. 30.64 and 30.65.

The two designs shown use a torque motor to drive a flapper-nozzle valve pilot stage, which in turn drives the main spool valve stage. The flapper nozzles are two variable arms of a four-arm bridge (Fig. 30.42). The mass of the spool and the compressibility of the oil in the trapped volumes V_{op} combine to give a second-order characteristic for the spool movement response. The effective load on the torque motor is now reduced by the ratio of nozzle area to spool valve piston area.

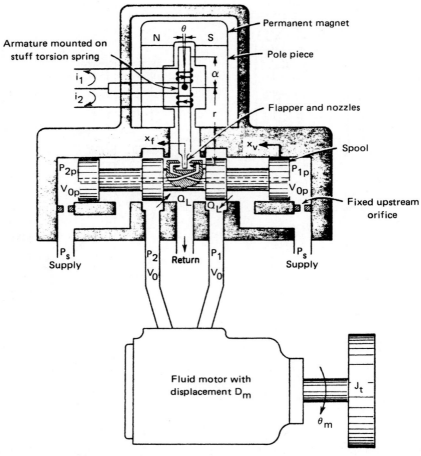

FIG. 30.64 Schematic of a two-stage electrohydraulic servo valve with direct feedback controlling a motor with inertia load. (*Courtesy John Wiley & Sons, Inc.*[35])

In both designs a feedback of main-spool valve position to produce a proportional load on the flapper is used to provide improved steady-state accuracy and reduce susceptibility to changes in conditions. Two-stage electrohydraulic servo valves have found wide military and commercial usage.

Practical considerations relating to the use of high-performance servo valves required a high degree of filtration. Silt and oil contamination affect the accuracy of the valve as a control element. To some degree these problems can be removed by using a high-frequency dither signal added to the control signal. This keeps the valve in motion without affecting the drive. Other factors such as threshold and null shifts with temperature and pressure changes should be considered. Electrical hysteresis in the torque motor can also be significant and may require separate analysis, as shown in Sec. 30.9.4.

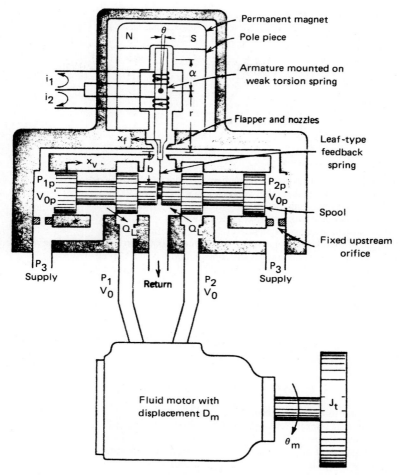

FIG. 30.65   Schematic of a two-stage electrohydraulic servo valve with force feedback controlling a motor with inertia load. (*Courtesy John Wiley & Sons, Inc.*[35])

30.9.4   Pump-Displacement-Controlled Variable-Speed Drives[4]

For high-power applications, a variable-speed output positive-displacement hydraulic transmission is used. This transmission consists of a variable-displacement hydraulic pump driven by a constant-speed power source and transmission lines to and from the fixed-displacement hydraulic motor which drives an inertial load with some viscous damping (see Fig. 30.66).

The following assumptions are made in the simplified analysis:

1. Pressures P_a and P_b are uniform throughout the lines and chambers in the pump and motor. Short lines are thus assumed between pump and motor so that fluid-inertia effects and wave phenomena are neglected.

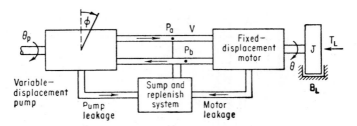

FIG. 30.66   Pump-displacement-controlled servomotor.

2. Leakage in pump and motor is considered to consist of three parts:
   a. Internal laminar leakage between P_a and P_b.
   b. External laminar leakage from P_a to case.
   c. External laminar leakage from P_b to case.
3. Cavitation phenomena are considered negligible.
4. Operation of the system is such that relief valves between lines a and b never open.
5. The replenishing system assures that lines a and b remain filled.
6. Motor friction torque is proportional to $(P_a - P_b)$ and motor speed.
7. Motor-to-load connection is rigid.

For small changes of all variables from the condition of zero motor speed, the system equations are

$$(V/\beta_e)(dP_a/dt) = -D_m\,\Omega - C_e P_a - C_i P_m + (Nk_p)\phi(t) \tag{30.164}$$

$$(V/\beta_e)(dP_b/dt) = D_m\,\Omega - C_e P_b - C_i P_m \tag{30.165}$$

$$J_e\,d\Omega/dt = -B\Omega + D_m P_m - C_{fm} P_m + T_L(t) \tag{30.166}$$

where J_e = polar moment of inertia of load and motor
C_i = sum of internal leakage coefficients of pump and motor
C_e = sum of external leakage coefficients of pump and motor
B = equivalent motor and load viscous friction coefficient
C_{fm} = motor coefficient of friction due to pressure
D_m = motor angular displacement
$\dot\theta_m$ = motor angular speed
$\dot\theta_p$ = pump shaft speed
k_p = displacement coefficient of pump
ϕ = pump stroke angle
T_L = load torque
V = volume under compression in each line and motor
β_e = isothermal equivalent modulus

These equations can be reduced, similarly to Eqs. (30.154) to (30.156), to two first-order "state" equations. These equations can also be incorporated with other system equations if more complex loading or control system inputs are being simulated by use of a computer. For example, other equations, to representing structural responses, could be added and used to generate the load torque T_L.

Rearranging the equations of the drive given above, the block diagram of Fig. 30.67 can be derived. This drive system has its own resonant natural frequency w_0 and damping ratio, given by

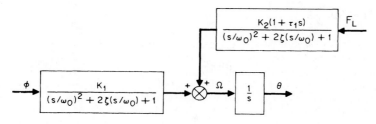

Where:

$$K_1 = \frac{(1 - C_{fm})D_m k_p N}{[(C_i + C_e/2)B + (1 - C_{fm})D_m^2]}$$

$$K_2 = \frac{(C_i + C_e/2)}{[(C_i + C_e/2)B + (1 - C_{fm})D_m^2]}$$

$$\tau_1 = \frac{V}{2\beta_e(C_i + C_e/2)}$$

FIG. 30.67 Block diagram of the transfer-function model of a pump-controlled drive. J_e = polar moment of inertia of load and motor; C_i = sum of internal leakage coefficients of pump and motor; C_e = sum of external leakage coefficients of pump and motor; B = equivalent motor and load viscous friction coefficient; C_{fm} = motor coefficient of friction due to pressure; D_m = motor displacement; Ω = motor angular speed; N = pump shaft speed; k_p = displacement coefficient of pump; ϕ = pump stroke angle; T_L = load torque; β_e = isothermal equivalent modulus; V = volume under compression in each line and motor.

$$w_0 = \sqrt{\frac{(C_i + C_e/2)B + (1 - C_f)D_m^2}{J_e(1/2)V/\beta_e}} \tag{30.167}$$

$$\xi = [(C_i + C_e/2)J_e + \tfrac{1}{2}(V/\beta_e)B]/(w_0 J_e V/\beta_e) \tag{30.168}$$

The estimation of the motor viscous friction coefficient requires special mention. The mechanical seal friction in the motor is closely modeled by coulomb friction. Unless the drive motor changes direction, coulomb friction will not provide any damping and only the motor leakage can be used effectively. With motor directional changes, however, coulomb friction will act to dissipate energy. Coulomb friction can be modeled in some cases as an equivalent viscous friction by using the concept of equivalent energy dissipation for each cycle of oscillation of the motor. For oscillatory displacements, the equivalent viscous rate depends on the velocity amplitude of motor oscillations[36] and can be estimated for this application as

$$B = \frac{4D_m P_c}{\pi |\Omega|} \tag{30.169}$$

where P_c is the pressure drop to overcome coulomb friction. Use of Eq. (30.169) is only recommended when the drive operates in an oscillatory mode in which the magnitude of the speed fluctuations is $|\Omega|$, being approximately equal in each rotation direction. The coulomb equivalent damping will be zero if speed fluctuations occur about a nominal speed in one direction of rotation only.

REFERENCES

1. Dow, R. B., and C. F. Fink: "Computation of Some Physical Properties of Lubricating Oils of High Pressures," *J. Appl. Phys.,* vol. 11, pp. 353–357, May 1940.

2. Klaus, E. E., and M. R. Fenske: "Some Viscosity-Shear Characteristics of Lubricants," *Lubrication Eng.,* March–April 1955, pp. 101–108.

3. U.S. Army Material Command (AMCRD-TV); "Hydraulic Fluids," Engineering Design Handbook, Pamphlet No. AMCP-706-123, Washington, Apr. 1971.

4. Blackburn, J. F., G. Reethof, and J. L. Shearer: "Fluid Power Control," John Wiley & Sons, Inc., New York, 1960.

5. ASTM D 341–43, "Standard Viscosity-Temperature Charts for Liquid Petroleum Products."

6. ASTM D 97–47, "Cloud and Pour Points."

7. ASTM D 92–52, "Flash and Fire Points by Means of Cleveland Open Cup."

8. Shames, I. H.: "Mechanics of Fluids," McGraw-Hill Book Company, Inc., New York, 1962.

9. Grace, H. P., and C. E. Lapple: "Discharge Coefficients of Small-Diameter Orifices and Flow Nozzles," *Trans. ASME,* vol. 73, pp. 639–647, July 1951.

10. Jorissen, A. L.: "Discharge Measurements at Low Reynolds Numbers—Special Devices," *Trans. ASME,* February 1956, pp. 365–368.

11. McCloy, D., and H. R. Martin: "The Control of fluid Power," Halsted Press (John Wiley & Sons, Inc.), New York, 1973, p. 46.

12. Merrill, R. F.: "Effect of Nozzle Geometry on Flapper Valve Characteristics," S. B. Thesis, M. E. Dept., MIT, Jan. 16, 1956.

13. Wilson, W. E.: "Positive-Displacement Pumps and Fluid Motors," Pitman Publishing Co., New York, 1950.

14. Merritt, H. E.: "Hydraulic Control Systems," John Wiley & Sons, Inc., New York, 1967.

15. Shearer, J. L.: "Handbook of Fluid Dynamics," Sec. 21, McGraw-Hill Book Company, Inc., New York, 1961.

16. Shearer, J. L., A. T., Murphy, and H. H. Richardson: "Introduction to System Dynamics," Addison-Wesley, Inc., Reading, Mass., 1967.

17. Sverdrup, N. N.: "Calculating the Energy Losses in Hydraulic Systems," *Prod. Eng.,* p. 146, April 1951.

18. Goodson, R. E., and R. G. Leonard: "A Survey of Modeling Techniques for Fluid Transmission Line Transients," *J. Basic Eng., Trans. ASME,* ser. D, vol. 94, June 1972.

19. Brown, F. T.: "The Transient Response of Fluid Lines," *Journal of Basic Engineering, Trans. ASME,* Series D, Vol. 84, December 1962, p. 547.

20. Oldenburger, R., and R. E. Goodson: "Simplification of Hydraulic Line Dynamics by Use of Infinite Products," *J. Basic Eng. Trans, ASME.,* ser. D, vol. 86, p. 1, January 1965.

21. Schultz, D. G., and J. L. Melsa: "Linear Control Systems," McGraw-Hill Book Company, Inc., New York, 1969.

22. Kirshner, J. M., and S. Katz: "Design Theory of Fluidic Components," Academic Press, Inc., New York, p. 70, 1975.

23. Kaplan, W.: "Advanced Mathematics for Engineers," Addison-Wesley Inc., Reading, Mass., p. 183, 1981.

24. Gerlach, C. R.: "Dynamic Models for Viscous Fluid Transmission Lines," *Proc. 10th Joint Automatic Control Conference,* Boulder, Colo., Published by ASME, 1969 (Also Ph.D., Dissertation, Okla. State Univ., 1966).

25. Franke, M. E., and T. M. Drzewiecki (ed.): "Fluid Transmission Lines," ASME Special Publication, 1981.

26. Doebelin, E. O.: "System Modelling and Response, Theoretical and Experimental Approaches," John Wiley & Sons, Inc., New York, 1980.

27. Kirshner, J. M., and S. Katz: "Design Theory of Fluidic Components," Academic Press Inc., New York, 1975.

28. Nichols, N. B.: "The Linear Properties of Pneumatic Transmission Lines," *Trans. ISA.,* vol. 2, p. 39, 1963.

29. Karam, J. T., and M. E. Franke: "The Frequency Response of Pneumatic Lines," *J. Basic Eng., Trans. ASME,* p. 371, June 1967.

30. Schaedel, H.: "A Theoretical Investigation of Fluidic Transmission Lines with Rectangular Cross Sections," Paper K3, Proc. 3d Cranfield Fluidics Conference, BHRA, 1968.

31. Healey, A. J., and J. A. Nicholson: "Dynamic Response of Annular Fluid Transmission Lines," *J. Fluids Eng., Trans. ASME,* December 1974.

32. D'Souza, A. F., and R. Oldenburger: "Dynamic Response of Fluid Lines," *J. Basic Eng., Trans. ASME,* ser. D, vol. 36, p. 589, 1964.

33. Hullender, D. A., and A. J. Healey: "Rational Polynomial Approximations for Fluid Transmission Line Models," ASME Special Publication, Fluid Transmission Line Dynamics, p. 33, 1981.

34. Royle, J. K.: "Hydraulic Damping Techniques at Low Velocity," *Proc. Conf. Hydraulic Power Transmission and Control,* Inst. Mech. Engrs., London, 1961.

35. Merrit, H.: "Hydraulic Control Systems," John Wiley & Sons, Inc., New York, p. 202, 1967.

36. Harris, C. M., and C. E. Crede: "Shock and Vibration Handbook," McGraw-Hill Book Company, Inc., New York, p. 362, 1972.

37. D'Azzo, J. J., and C. H. Houpis: "Feedback Control Systems Analysis and Synthesis," chap. 18, McGraw-Hill Book Co., Inc., New York, 1966.

38. Keller, R.: "Hydraulic System Analysis," 2d ed., Hydraulics & Pneumatics Publisher, 1974.

SECTION 31
IMPACT DESIGN

Philip Barkan, Ph.D.

Professor Emeritus of Mechanical Engineering
Stanford University
Stanford, Calif.

31.1   INTRODUCTION

Impact involves the study of the physical phenomena which attend the collision of bodies with initial relative velocity. Several books[1-11] cited in the bibliography are recommended to the reader interested in a more rigorous and more extensive treatment than is provided here.

Impact phenomena are especially important to the machine designer since in nearly all systems the highest forces and greatest stresses arise as a consequence of impact. Many serious machine failures arise because impact forces were not properly recognized. Impact can be quite useful, since by means of impact it is possible to achieve many extreme short-duration effects. High force levels, rapid dissipation of energy, large accelerations and decelerations, and enormous power amplification may be so obtained from low-energy sources operating at low force levels.

31.1.1   Nomenclature

$A =$ cross-sectional area
$a =$ rod diameter

$B =$ function defined by Eqs. (31.10) and (31.11)

C = constant of proportionality

c = wave-propagation velocity

d = lateral dimension of a rod

e = coefficient of restitution

E = Young's modulus

F = force

f = wave function

G = shear modulus

g = acceleration due to gravity or wave function

I = moment of inertia

K = net stiffness between centers of gravity of two impacting bodies

L = length of a body, measured in direction of impact

M, m = mass

n = number of a sequence or a constant

p = plastic stress, assumed independent of strain

P = gas pressure or dynamic stress

R, r = radius of curvature

T = kinetic energy

t = time

τ = kinetic energy

U = strain energy

U_s = strain energy density

u = displacement

v = velocity

V = volume

X, x = displacement or position

z = amplitude of bounce

α = deformation or approach distance between centers of gravity of impacting bodies

γ = ratio of specific heat of a gas

δ = deflection of a beam

ϵ = strain

ζ = parameter involving strain rate and temperature

η = natural frequency

θ = absolute temperature

Λ = wavelength

γ = function defined by Eq. (31.22)

v = Poisson's ratio

ρ = mass density

σ = stress

ϕ = force due to impact

Φ = function proportional to energy in a mode of vibration

Subscripts

c = common

e = externally applied

f = at end of impact

I = incident

m = maximum

0 = at beginning of impact

p = elastic limit

R = reflected

T = transmitted

t = at time t during impact

u = ultimate

y = yield

31.1.2   Accuracy of Impact Calculations

Impact calculations suffer from several practical limitations which limit their value to establishing the approximate magnitude of the various phenomena involved and for providing some insight into the manner in which the various parameters affect the final result. In spite of these limitations available techniques provide a useful guide to the design of impacting systems and to the interpretation of final results. For precise data resort to experimental measurements is essential, employing such powerful tools as strain gages, high-speed photography, velocity, and motion transducers.

The limitations to precise analysis include

1. The mathematics required for rigorous solution of impact problems is excessively complex, and approximate methods which omit one or more basic characteristics of the impact phenomenon are invariably employed.

2. The problem of adequately defining a complicated physical system such as a machine with discontinuities in both geometry and materials is formidable.

3. The properties of even conventional engineering materials are modified to some extent by the duration and rate of loading. These effects are neither well defined nor fully understood.

31.1.3 Inadvertent Sources of Impact

Occasionally impact is overlooked in machine designs with unhappy results. It is therefore important for the designer to recognize and cope with possible conditions of severe impact. Typical sources include

1. Clearances as between cams and followers

2. Backlash or bearing clearances in mechanisms undergoing force or motion reversal

3. Mechanisms with kinematic discontinuities so that components with large relative velocities are engaged

4. High-speed systems which are nonlinearly elastic so that abrupt changes in stiffness occur with results similar to impact

31.1.4 Design Factors to Be Considered in Minimizing Impact Effects[30]

1. Minimize the velocity of impact.
2. Minimize the mass of impacting bodies.
3. Design for minimum stiffness in the vicinity of the point of impact.
4. Design critical components for maximum strain-energy storage capacity within the tolerable maximum stress by the following means:
   a. The maximum possible volume of materials should be stressed uniformly to the peak stress.
   b. Other things being equal, select materials whose energy storage capacity per unit volume called "modulus of resilience" $\sigma_y^2/2E$ is the maximum, where σ_y = yield stress and E = Young's modulus.
5. Minimize sensitivity to local stress concentrations by these means:
   a. Employ a ductile material with some capacity for plastic deformation to cope with localized stress concentrations.
   b. Avoid surface irregularities, sharp discontinuities, and internal inhomogeneities in the design and manufacture of critical parts.

31.2 CLASSICAL THEORY OF IMPACT AND THE COEFFICIENT OF RESTITUTION

The simplest form of impact analysis, based on the laws of conservation of momentum and an empirical manifestation of the conservation of energy, can provide information on the net change in velocity of the center of mass of each body involved in the impact, the net impulse, and the energy-exchange processes accompanying the impact. This method can be employed when external forces either are not present during the period of impact or are negligibly small compared with the impact-induced forces.

31.2.1 Central Impact

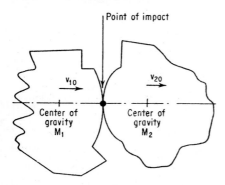

FIG. 31.1 Schematic representation of central impact. Initial impact velocity $= v_{10} - v_{20}$.

In the simplest case two bodies collide on surfaces which are normal to the common line connecting their centers of mass (Fig. 31.1) and have velocity components only along this common line. With these restrictions no rotational or sliding effects occur.

Conservation of momentum requires that momentum immediately before and immediately after the impact be equal.

$$m_1 v_{10} + m_2 v_{20} = m_1 v_{1f} + m_2 v_{2f} \quad (31.1)$$

The second relationship involved can be regarded as a form of the conservation of energy. It recognizes that the total kinetic energy in the bodies after the impact must be equal to or less than the initial kinetic energy. The loss in kinetic energy is best expressed in terms of the coefficient of restitution (e), which Newton defined as

$$e = (v_{1f} - v_{2f})/(v_{10} - v_{20}) \quad (31.2)$$

It can then be shown that the kinetic energy loss is

$$T_{loss} = [(1 - e^2)/2][(m_1 m_2)/(m_1 + m_2)](v_{10} - v_{20})^2 \quad (31.3)$$

The coefficient of restitution is discussed in more detail at a later point in this section.

Simultaneous solution of Eqs. (31.1) and (31.2) results in the following dimensionless expressions for the terminal velocity of each of the impacting bodies:

$$\left. \begin{aligned} \frac{v_{1f} - v_{10}}{v_{10} - v_{20}} &= -\frac{(1 + e)(m_2/m_1)}{m_2/m_1 + 1} \\[2ex] \frac{v_{2f} - v_{20}}{v_{10} - v_{20}} &= \frac{1 + e}{1 + m_2/m_1} \end{aligned} \right\} \quad (31.4)$$

Figure 31.2, which is based on Eq. (31.4), illustrates the effect of mass ratio and coefficient of restitution upon the velocities of two impacting bodies for the special case when the initial velocity of one of the masses is zero.

The net impulses produced by the impact upon the two masses are equal and opposite; their magnitude is obtained from conservation of momentum and Eq. (31.4).

$$\left| \int \phi \, dt \right| = \left| (1 + e) \frac{m_1 m_2}{m_1 + m_2} (v_{10} - v_{20}) \right| \quad (31.5)$$

During the impact the total kinetic energy of the two bodies will decrease and

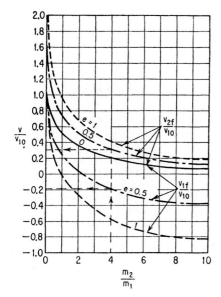

FIG. 31.2   Effect of coefficient of restitution and mass ratio upon velocity change when one mass is initially motionless. *Example:*
Given

$$e = 0.5, \quad m_2/m_1 = 4, \quad v_2 = 0$$

From graph

$$v_{2f}/v_{10} = 0.32 \qquad v_{1f}/v_{10} = -0.18$$

strain energy will be developed within the impacting bodies. The maximum strain energy developed by the impact will occur at that instant when the centers of mass of the two impacting bodies share a common velocity. From conservation of momentum the common velocity is

$$v_c = (m_1 v_{10} + m_2 v_{20})/(m_1 + m_2)$$

and the maximum possible strain energy is dictated by conservation of energy

$$\begin{aligned} U &= \frac{m_1}{2} v_{10}^2 + \frac{m_2}{2} v_{20}^2 - \frac{m_1 + m_2}{2} v_c^2 \\ &= \frac{m_1 m_2}{m_1 + m_2} \frac{(v_{10} - v_{20})^2}{2} \end{aligned} \tag{31.6}$$

31.2.2   Most General Form of Classical Theory of Impact[38]

In its most general form, the calculation of the change in motion resulting from an impact between two bodies which are free of external constraints and external forces will require the designation of six quantities for each body:

Three components of translational velocity of the center of gravity of each body

Three components of angular velocity of each body about axes through the center of gravity of each body

Thus 12 equations are necessary to define the motions of the two bodies. Six of these equations are determined by the condition that the angular momentum of each body about any axis through the point of contact is unchanged, since the impact forces must act through this point. The seventh equation requires that the momentum of the entire system in the direction normal to the surface in contact be constant, since the normal forces acting on the two bodies are equal and opposite. The eighth equation involves the conservation of energy or the coefficient of restitution of the bodies, i.e., Eqs. (31.4) and (31.5). The last four equations consider conditions at the impact surface.

Case 1.   If friction at the impact surface is negligible then two equations are employed for each body to state that the momentum of *each* body is unchanged in directions tangent to the plane of contact.

Case 2.   If friction at the surface is sufficiently high (i.e., perfect roughness) so that no slippage or sliding occurs, then two equations are employed to state that the linear momentum of the system in mutually perpendicular directions parallel to the impact plane is unchanged. The last two equations then require that there be no relative motion between the two bodies in the plane of impact.

Case 3.   If friction is intermediate between these two extreme cases, then as in case 2, the system invariance of linear momentum in the impact plane is still valid and this accounts for two of the equations. The last two equations equate the variation in linear momentum of each body along the direction of slip with the frictional impulse.

31.2.3 The Coefficient of Restitution

The coefficient of restitution e is defined by Eq. (31.2) as the ratio (relative velocity between two bodies immediately after impact)/(relative velocity just prior to impact). However, it is most convenient for purposes of understanding to regard the coefficient of restitution as an energy-loss function, since all impacts are basically processes of energy exchange and energy transformation. When two bodies collide, a portion of the original kinetic energy is converted into strain energy within the impacting bodies. Subsequently, some fraction of the strain energy is reconverted back into the kinetic energy of the impacting bodies. The remainder of the energy is trapped within the impacting bodies in exciting various modes of vibration within the bodies or is dissipated as energy of plastic deformation. Since the velocity of bodies which deform cannot be described in terms of a single value, the coefficient of restitution will, in general, be expressed in terms of the velocities of the centers of gravity of the bodies or in terms of a momentum averaged velocity of each body, which is

$$v_{av} = \frac{\int_0^L [dm(x)/dx]\, v(x)\, dx}{\int_0^L [dm(x)/dx]\, dx} \tag{31.7}$$

The coefficient of restitution is not a basic material property. Its magnitude is dependent upon the geometry of the bodies, the presence or absence of slip at the contact interface, the peak stresses produced, and the duration of contact as well as basic material properties such as E, ρ, and the elastic limit σ_p.

The parameters primarily control the magnitude of the coefficient of restitution. One is the volume of plastically deformed material generated by the impact. The second parameter t_f/t_η is the ratio (duration of the impact period)/(the period of the fundamental natural frequency of the impacting bodies). As the ratio t_f/t_η becomes smaller, the vibrational modes of the bodies "trap" an increasingly larger proportion of the strain energy generated by the impact.

Coefficient of Restitution for Compact Bodies.   For compact bodies such as spheres, the ratio t_f/t_η is quite large, and as a consequence the proportion of kinetic energy lost because of excitation of various modes of vibration is quite small. For such shapes the coefficient of restitution is primarily controlled by plastic deformation near the point of impact, or internal friction within the impacting bodies. The extent of plastic deformation is dependent upon the velocity of impact, and this effect accounts for the basic velocity sensitivity of the coefficient of restitution of spherelike bodies.

Typical coefficient-of-restitution data for the impact of spheres are shown in Fig. 31.3. Tests have shown that for spheres the effect of size upon the coefficient of restitution is rather small. For example, variation of about 100:1 in the weight of cast-iron balls (i.e., 5:1 in diameter) produces only 10 percent variation in e. Bowden and Tabor[45] have derived an approximate relationship based on Hertz equations for deformation of spherical surfaces[46] and conservation of energy which accounts for the velocity variation of the coefficient of restitution of spheres. For spheres the relative velocities before and after impact are related by an equation of the form

$$v_f = k(v_0^2 - \tfrac{3}{8}v_f^2)^{3/8} \tag{31.8}$$

This nonlinear relationship yields values of $e = -v_f/v_0$, which correlate well in shape with experimental data such as Fig. 31.3. k is proportional to

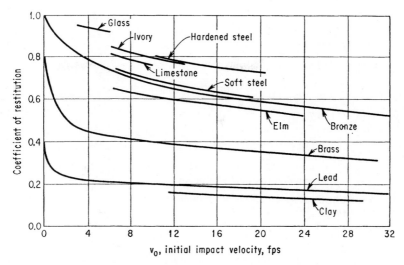

FIG. 31.3 Coefficient of restitution as a function of impact velocity for spheres of the same material.

$$\sigma_p^{5/8} r^{3/8}/m^{1/8}/[(E_1 + E_2)^{1/2}/E_1 E_2]$$

Note that since for spheres $m \propto r^3$, the coefficient of restitution should be very nearly independent of size.

Coefficient of Restitution for Distributed Systems. For bodies such as rods, plates, or beams for which the period of impact tends to be short compared with the period of propagation of stress waves, the amount of energy remaining within the bodies after impact in the form of internal vibration or traveling stress waves is relatively large, and this factor has a predominant influence upon the coefficient of restitution. For such systems even if plastic deformation does not occur, a large proportion of the initial kinetic energy can be lost during the impact, and this is accounted for in trapped strain energy within the bodies.

When a relatively light, compact mass such as a sphere impacts a relatively massive rod, beam, or plate, the duration of impact is very short compared with the basic period of vibration of the resilient member and very long compared with the period of the fundamental mode of vibration of the sphere. The elastic vibration of the sphere then absorbs only a negligible fraction of the initial kinetic energy[31] and it is only necessary to calculate the energy absorbed by the body which the sphere strikes. Zener and Feshbach[21] showed that for such situations the strain energy trapped within the body at the end of the impact can be accurately expressed in terms which do not require very precise estimates of the impact force-time characteristic during the impact. An approximation of this force-time history was made employing the Hertz methods of Sec. 31.3, by assuming that the force history is dictated solely by strain conditions within the immediate region of contact. By normalizing this approximate interaction force in a form which is necessarily quite insensitive to the inaccuracy of the approximation, they calculated the energy remaining within a beam subject to such a short-duration force. Employing this technique, the coefficient of restitution for the impact of a relatively light sphere striking a relatively massive beam or plate was calculated to be

$$e = (1 - B)/(1 + B) \tag{31.9}$$

B is a function whose value is as follows:

1. For a sphere striking a simple supported beam:

$$B = (m/M)Q_r^{1/2}G(Q_1) \tag{31.10}$$

where $Q_1 = 1.087\eta_1 t_f$
$G(Q_1) = $ function defined by Table 31.1
$\quad m = $ mass of striking sphere
$\quad M = $ mass of beam
$\quad t_f = $ approximate duration of impact as calculated by Eq. (31.29)
$\quad \eta_1 = $ fundamental frequency of beam

The results of these calculations are plotted in Fig. 31.4a for a wide range of the parameter $(m/M)Q_1^{1/2}$.

2. Sphere striking a circular plate:

$$B = [3m/16\pi^2(1.087t_f)]/[3(1 - v^2)/\rho E]^{1/2} \tag{31.11}$$

TABLE 31.1 Table of Function $G(Q_1)$ for Calculation of Coefficient of Restitution for Impact of a Sphere on a Simple Hinged Beam[21]

Q_1	$G(Q_1)$	Q_1	$G(Q_1)$	Q_1	$G(Q_1)$
0	0.840	0.25	0.922	1.0	0.500
0.02	0.840	0.30	0.976	1.1	0.378
0.04	0.840	0.35	1.008	1.2	0.278
0.06	0.840	0.40	1.028	1.3	0.205
0.08	0.846	0.45	1.032	1.4	0.123
0.10	0.832	0.50	1.018	1.5	0.078
0.12	0.814	0.55	1.000	1.6	0.038
0.14	0.803	0.60	0.972	1.7	0.010
0.16	0.810	0.7	0.880	1.8	0.002
0.18	0.824	0.8	0.766	1.9	0.000
0.20	0.850	0.9	0.632		

Source: C. Zener and H. Feshback, "A Method of Calculating Energy Losses during Impact," *J. Appl. Mech.*, Vol. 70, June 1948.

Coefficient of Restitution for Longitudinal Impact between Bars. Sears[16] made a detailed theoretical and experimental study of the longitudinal impact of steel bars of various lengths with rounded ends and reported the results shown in Fig. 31.4b. At the relatively low impact velocity employed in this study, losses associated with plastic deformation near the point of impact on the rounded end are relatively small compared with the strain energy trapped as traveling waves within the bodies. This factor accounts for the variation in e with bar length.

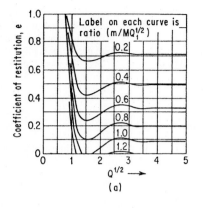
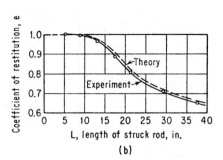

FIG. 31.4 Coefficient of restitution (*a*) of a sphere colliding with a beam pivoted at both ends,[21] (*b*) of originally stationary rod for the longitudinal impact of two ½-in-diameter alloy-steel bars (striker 5⅓ in long, impact velocity 5 in/s).[16]

31.3 SOLUTIONS OF EQUATIONS OF MOTION FOR IMPACT

Since the classical methods provide no information about the transient forces, stresses, and accelerations produced within the impacting bodies and tell nothing about the duration of the impact, it is necessary to solve the more basic equations of motion for such information. Because of the inherent complexity of these general solutions, it is necessary to exploit reasonable simplifications whenever the opportunity arises. Hence several different approaches are employed in this section, each with its pertinent field of application. The types of equations necessary to define the impacting systems adequately, and hence the solutions, depend largely upon the geometry of the bodies.

The first and simplest type of solution assumes that stresses are transmitted instantaneously to all points in the system. This quasi-static property is reasonably valid for one-degree-of-freedom systems in which the duration of impact is long compared with the time required for stress waves to propagate through the body.

Problems which can be solved in this way include collision of bodies in which most of the deformation occurs in the vicinity of the point of impact and the center of mass of the component is well behind the point of impact. These criteria are satisfied by compact bodies with curved surfaces near the region of impact and more complex geometries in which, by virtue of shape or material, the dominant source of flexibility is located in the immediate vicinity of the impact point. For such bodies it is possible to represent one region as having flexibility but negligible mass and a second region as having mass but infinite rigidity. These methods are described in Sec. 31.3.1.

More precise theory, including the effects of both elasticity and inertia, gives rise to the concept of a finite velocity of propagation of stress waves. These considerations become necessary when the forces or stresses vary sufficiently rapidly during the time of propagation of these signals so that substantial deviation from quasi-static equilibrium occurs. Typical problems requiring this more rigorous approach include the longitudinal and transverse impact of long beams and plates. Practical solutions employing this approach are restricted to simple geometries with uniform properties. These methods are covered in Sec. 31.3.3.

Intermediate between bodies which are amenable to solution by these two methods are bodies which can be conveniently represented by multimass lumped systems, such as in Fig. 31.8. Many nonuniform geometries and nonlinear properties can be represented in this way. Although few general solutions are available, particular solutions are readily obtained on automatic computers. These methods are discussed in Sec. 31.3.2.

31.3.1 The Quasi-Static Approach to Central Impact

The methods to be described here are most useful when impact is considered between bodies which satisfy the following criteria:

1. The period or duration of the impact is relatively long compared with the period of the fundamental natural frequency of the impacting bodies.
2. The stress-strain relationship is strain-rate-independent. To a reasonable approximation this can be assumed true for most common metals employed in machine work, *at stress levels below the elastic limit.*
3. Forces produced at the point of impact can be defined in terms of one-degree-of-freedom models.
4. The motion before and after the impact is adequately defined in terms of the translational motion of the center of gravity and rotational motion about the center of gravity of each body.
5. During impact the body deforms essentially as it would if deformed statically by the slow application of comparable forces between the centers of gravity of the two impacting bodies.

These restrictions allow the use of the simplest methods of impact, which are of considerable value in many machine-design applications.

Consider the collision of two masses with the following additional restrictions:

1. The line connecting the centers of gravity of the two bodies passes through the point of impact.
2. The center of the radius defining the curvature of the two surfaces also lies along this axis.

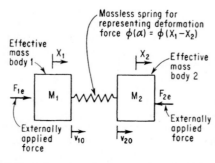

FIG. 31.5   Schematic representation of collision between two bodies satisfying quasi-static criteria. Positive direction is to right.

Unless these latter two conditions are met there will be both sliding at the surface of impact, requiring consideration of friction, and rotational effects.

Figure 31.5 shows a schematic diagram of a collision between two bodies satisfying these criteria. Positive motion is to the right, and external forces F_{e1} and F_{e2} drive the two bodies. The sign of velocities and forces is dictated by their direction. X_1 and X_2 define motions of centers of gravity of the two bodies. The equations of motion of each body during impact are

$$\ddot{X}_1 = F_{1e}/m_1 - \phi/m_1 \qquad (31.12)$$

$$\ddot{X}_2 = F_{2e}/m_2 + \phi/m_2 \qquad (31.13)$$

By subtracting $\ddot{X}_2$ from $\ddot{X}_1$ and defining $X_1 - X_2 = \alpha = $ deformation, the basic equation for this type of impact is obtained.

$$\frac{d^2\alpha}{dt^2} = \frac{d}{d\alpha}\frac{\dot{\alpha}^2}{2} = \frac{F_{1e}}{m_1} - \frac{F_{2e}}{m_2} - \frac{m_1 + m_2}{m_1 m_2}\phi(\alpha) \qquad (31.14)$$

$\phi(\alpha)$ represents the reaction force produced at the surface of the impacting bodies. Even for perfectly elastic, constant-modulus materials $\phi(\alpha)$ will not, in general, be a linear function nor will it even be single-valued. The dissipation of kinetic energy normally attending the impact process necessitates that the force-deformation relationship $\phi(\alpha)$ exhibit a hysteresis effect. Equation (31.15) will be solved for general simple functions relating ϕ and α. The selection of the best relationship for any particular problem is a matter of engineering judgment.

The solution to Eq. (31.14) yields a relationship between α and $\dot{\alpha}$ at any instant t during the impact.

$$\dot{\alpha}_t^2 = \dot{\alpha}_0^2 - 2\frac{m_1 + m_2}{m_1 m_2}\int_0^{\alpha_t}\phi(\alpha)\,d\alpha - \left(\frac{F_{1e}}{m_1} - \frac{F_{2e}}{m_2}\right)\alpha \qquad (31.15)$$

Solution of Eq. (31.15) with the terminal condition $\dot{\alpha}_t = 0$ yields Eq. (31.16) from which the value of α_m can be determined.

$$\frac{m_1 + m_2}{m_1 m_2}\int_0^{\alpha_m}\phi(\alpha)\,d\alpha - \left(\frac{F_{1e}}{m_1} - \frac{F_{2e}}{m_2}\right)\alpha_m = \dot{\alpha}_0^2 \qquad (31.16)$$

Peak force is determined by substituting α_m in the known function $\phi(\alpha)$.

The order of magnitude of the impact duration can be estimated from Eq. (31.15) by the relationship

$$t_f = \int_0^{\alpha_m}\frac{d\alpha}{\dot{\alpha}} + \int_{\alpha_m}^{\alpha_f}\frac{d\alpha}{\dot{\alpha}} \qquad (31.17)$$

Impact-Force and Impact-Period Formulas.   The evaluation of peak force and impact duration depends finally upon the value of $\phi(\alpha)$, which will be a function of the properties of the materials involved, the geometry, and the stress levels produced.

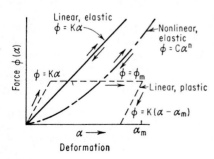

The following compilation represents specific solutions to Eqs. (31.15), (31.16), and (31.17) for several functions $\phi(\alpha)$ which are of general interest. The force-deflection characteristics considered are illustrated in Fig. 31.6.

Linear Elastic Collision.   If the geometry of a structure is such that a single-valued linear force-deflection relationship is valid, then $\phi = K\alpha$, $e = 1$, and solutions can be obtained either from Eq. (31.16) or by direct solution of Eq. (31.14). For compact bodies with high values of Young's modulus, impact force ϕ_m is normally quite large compared with external forces F_{1e} and F_{2e}.

FIG. 31.6 Characteristic forces produced by impact deformation which are analyzed by the methods of Sec. 31.3.1.

When forces F_{1e} and F_{2e} are negligible compared with ϕ_m, the solution to Eq. (31.14) yields an expression for peak force

$$\phi_m = (v_{10} - v_{20})[K(m_1 m_2/(m_1 + m_2)^{1/2}] = \dot{\alpha}_0[K(m_1 m_2)/(m_1 + m_2)] \qquad (31.18)$$

Maximum deceleration of each mass is then

$$(\ddot{X}_1)_m = -\phi_m/m_1 \qquad (\ddot{X}_2)_m = +\phi_m/m_2 \qquad (31.19)$$

and duration of impact

$$t_f = \pi[m_1 m_2/K(m_1 + m_2)]^{1/2} \qquad (31.20)$$

For the special but important case where one of the masses is both infinitely large and stiff, Eqs. (31.18) and (31.20) reduce to

$$\phi_m = v_{10}(Km_1)^{1/2} \qquad (31.18a)$$

$$t_f = \pi(m_1/K)^{1/2} \qquad (31.20a)$$

When external forces F_{1e} and F_{2e} are significant with respect to ϕ_m,

$$\phi_m = K\{(F_{1e}/m_1 - F_{2e}/m_2)\eta^{-2} + [(F_{1e}/m_1 - F_{2e}/m_2)^2\eta^{-4} + \dot{\alpha}_0^2\eta^{-2}]^{1/2}\} \qquad (31.21)$$

where

$$\eta = [K(m_1 + m_2)/m_1 m_2]^{1/2} \qquad (31.21a)$$

Impact duration is determined by the relationship

$$(\sin \eta t)/(1 - \cos \eta t) = -(F_{1e}/m_1 - F_{2e}/m_2)(\dot{\alpha}_0\eta)^{-1} = \lambda \qquad (31.22)$$

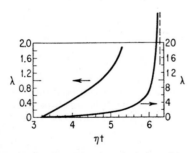

By entering Fig. 31.7, which is a plot of Eq. (31.22), with the argument λ, a value of ηt is obtained from which impact duration can be determined.

Linear Plastic Collision. For quasi-rigid bodies whose force deflection is linear, plastic (e.g., the dashed curve, Fig. 31.6) deformation occurs until a prescribed value of peak force ϕ_m is reached. At force level ϕ_m plastic yielding begins. Further compression occurs plastically at constant force until relative motion between the bodies ceases. Re-

FIG. 31.7 Curve for the calculation of impact duration according to Eq. (31.22).

laxation occurs elastically, resulting in permanent deformation. For this case the coefficient of restitution is

$$e = (\phi_m/\dot{\alpha}_0)[(m_1 + m_2)/Km_1 m_2]^{1/2} \qquad (31.23)$$

When Eq. (31.23) gives values of $e > 1$, this indicates no plastic deformation and hence $e = 1$.

For the general case, the impact duration is

$$t_f = \frac{\dot{\alpha}_0}{\phi_m} \frac{m_1 m_2}{m_1 + m_2}\left[1 - \left(\frac{\phi_m \eta}{K\dot{\alpha}_0}\right)^2\right] + \frac{\pi}{2\eta} + \frac{1}{\eta}\arcsin\frac{\phi_m \eta}{K\dot{\alpha}_0} \qquad (31.24)$$

where η is defined by Eq. (31.21a). In the limit, as the elastic intervals become negligibly short,

$$t_f = (\dot{\alpha}_0/\phi_m)[m_1 m_2/(m_1 + m_2)] \tag{31.25}$$

Nonlinear Elastic Collision. A general form of force-deflection relationship which is useful in representing many nonlinear systems is

$$\phi = C\alpha^n \tag{31.26}$$

(where C and n are constants). For the case where F_1 and F_{2e} are small compared with ϕ_m. Solution of Eqs. (31.12) and (31.13) yields

$$\text{Peak force } \phi_m = \left[\frac{\dot{\alpha}_0^2}{2}\left(\frac{m_1 m_2}{m_1 + m_2}\right)(n + 1)C^{1/n}\right]^{n/(n+1)} \tag{31.27}$$

$$\text{Maximum compression } \alpha_m = \left[\frac{n + 1}{C}\left(\frac{\dot{\alpha}_0^2}{2}\right)\frac{m_1 m_2}{m_1 + m_2}\right]^{1/(n+1)} \tag{31.28}$$

$$\text{Duration of impact } t_f = \frac{2\alpha_m}{\dot{\alpha}_0}\int_0^1 \frac{dz}{(1 - z^{n+1})^{1/2}} \tag{31.29}$$

Impact of Compact Bodies with Spherical Surfaces in the Vicinity of the Impact Region. In the case of impact of two compact bodies such as spheres, most of the deformation is in the immediate vicinity of the point of contact. Hertz[46] derived solutions for this case assuming perfect elasticity and ignoring plastic deformation. For the case of two spheres or comparable compact bodies with radii of curvature R_1, R_2 in the vicinity of the point of impact, solutions may be obtained from Eqs. (31.26) through (31.29), employing the following values for constants C and n:

$$C = \frac{4}{3\pi}\left(\frac{R_1 R_2}{R_1 + R_2}\right)^{1/2}\left(\frac{1 - v_1^2}{\pi E_1} + \frac{1 - v_2^2}{\pi E_2}\right)^{-1} \tag{31.30}$$

$$n = \tfrac{3}{2}$$

With $n = \tfrac{3}{2}$, Eq. (31.29) reduces to the following expression for impact duration:

$$t_f = 2.943(\alpha_m/\dot{\alpha}_0) \tag{31.29a}$$

The case of a sphere impacting a massive plane is readily obtained: In the limit as m_1 and $R_1 \to \infty$, Eqs. (31.27) to (31.29) yield

$$\phi_m = (5\dot{\alpha}_0^2 m_2 C^{2/3}/4)^{3/5} \tag{31.27a}$$

$$\alpha_m = (5\dot{\alpha}_0^2 m_2/4C)^{2/5} \tag{31.28a}$$

where
$$C = (4/3\pi)(R_2)^{1/2}\left(\frac{1 - v_1^2}{\pi E_1} + \frac{1 - v_2^2}{\pi E_2}\right)^{-1} \tag{31.30a}$$

The assumption of perfectly elastic behavior in such cases has been shown not to be completely valid. Goldsmith and Lyman have suggested that for more accurate work the assumption of perfect elasticity should not be used, but rather a quasi-statically measured force-indentation relationship should be employed. Correlations between force-indentation relationships obtained quasi statically and under impact were found to be within 25 percent.

Recognizing that the Hertz assumption of perfectly elastic behavior is not fully valid, completely plastic behavior has been assumed by some investigators of the impact of spherical surfaces.[45,48] Bowden and Tabor[45] analyzed the case for the impact of two masses with equal radii of curvature in the vicinity of the impact zone when the indentation is small compared with the curvature radius R.

Under such conditions the force is

$$\phi = \sigma_y R \alpha$$

$$\phi_m = \dot{\alpha}_0 \left(\sigma_y \pi R \, \frac{m_1 m_2}{m_1 + m_2} \right)^{1/2} \tag{31.31}$$

Impact duration is then

$$t_f = \frac{\pi}{2} \left(\frac{m_1 m_2}{m_1 + m_2} \, \frac{1}{\sigma_y \pi R} \right)^{1/2} \tag{31.32}$$

Neither assumption, perfect elasticity nor perfect plasticity, results in perfect correlation with test. Such discrepancies have been attributed by Bowden and Tabor[45] to strain-rate dependence in the material or they may possibly be due to the presence of shock-wave phenomena as studied by Bell.[50] (See Sec. 31.4.6.)

Post-Impact Behavior.   How bodies behave subsequent to impact is a problem of frequent interest. As in Fig. 31.5, consider a pair of impacting bodies m_1 and m_2 under the influence of forces F_{1e} and F_{2e} which tend to force the two bodies together. For such a system, what is normally regarded as a single impact is, in fact, a series of impacts in rapid succession. This series of impacts is theoretically infinite in number but finite in total duration and involves the dissipation of enough kinetic energy to allow the two bodies to move together. From Eq. (31.3), the kinetic energy to be dissipated by the series of impacts is

$$\Delta T = \tfrac{1}{2} [m_1 m_2/(m_1 + m_2)] \dot{\alpha}_0^2 \tag{31.33}$$

The duration of the nth interval between impacts during which the bodies are separated is

$$t_n = 2e^n \dot{\alpha}_0 (F_{1e}/m_1 - F_{2e}/m_2)^{-1} \qquad n = 1, 2, 3, \dots \tag{31.34}$$

If the duration of impact is negligible compared to the bounce period, then the total time during which bouncing between the members will occur is

$$t_{\text{total}} = \sum^n t_n = \frac{2e}{1 - e} \, \dot{\alpha}_0 \left(\frac{F_{1e}}{m_1} - \frac{F_{2e}}{m_2} \right)^{-1} \tag{31.35}$$

The amplitude of the nth bounce is

$$z_n = (\dot{\alpha}_0^2/2) \, e^{2n} (F_{1e}/m_1 - F_{2e}/m_2)^{-1} = t_n^2/8(F_{1e}/m_1 - F_{2e}/m_2) \tag{31.36}$$

The common velocity of the two bodies after the bouncing has ceased is given by Eq. (31.6) to a first approximation.

31.3.2   Multi-Degree-of-Freedom Systems

Between the single-degree-of-freedom systems which were solved in Sec. 31.3.1 and the continuum systems with an infinite number of degrees of freedom discussed in

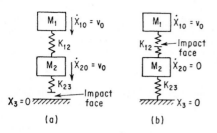

(a)   (b)

FIG. 31.8   Impact of two-degree-of-freedom systems.

Sec. 31.3.3 lies the important class of problems associated with a multiple but limited number of degrees of freedom. For many problems this approach offers a practical and reasonable approximation.

In representing complex mechanisms this approach enjoys the advantage of considerable versatility, particularly when numerical solutions are to be obtained. However, this approach does not lend itself to great generalization because of the large number of variables involved, and specific solutions are normally developed. For linear cases, closed-form solutions are possible but may sometimes be too cumbersome for convenient application. The simplest multimass linear problem is of the type shown in Fig. 31.8, representing either the impact of a two-degree-of-freedom structure against an infinitely massive surface (Fig. 31.8a) or the impact of two one-degree-of-freedom systems, with one mass resiliently mounted to an infinitely massive plane as shown in Fig. 31.8b. The equations of motion for both systems are

$$m_1\ddot{X}_1 + K_{12}(X_1 - X_2) = 0$$

$$m_2\ddot{X}_2 - K_{12}(X_1 - X_2) + K_{23}X_2 = 0$$

(31.37)

The two natural frequencies η_1, η_2 in terms of which the solutions may be expressed are obtained as the real positive roots from the following equations:

$$\eta_1^2 = \frac{H}{2J}\left[1 - \left(1 - \frac{4J}{H_2}\right)^{1/2}\right] \qquad \eta_2^2 = \frac{H}{2J}\left[1 + \left(1 - \frac{4J}{H^2}\right)^{1/2}\right]$$

(31.38)

where $H = (m_1 + m_2)/K_{23} + m_1/K_{12}$ and $J = m_1 m_2/K_{12}K_{23}$

The solutions to Eqs. (31.37) then depend upon the boundary conditions:

1. For the case described by Fig. 31.8a the initial conditions are, at $t = 0$,

$$\ddot{X}_{10} = \ddot{X}_{20} = v_0 \qquad X_1 = X_2 = 0$$

The solutions for displacement, velocity, and acceleration as functions of time are then

$$X_2 = \frac{F_{23}}{K_{23}} = \frac{v_0}{\eta_2^2 - \eta_1^2}\left(\frac{\beta^2 - \eta_1^2}{\eta_1}\sin\eta_1 t + \frac{\eta_2^2 - \beta^2}{\eta_2}\sin\eta_2 t\right)$$

(31.39a)

$$X_1 - X_2 = \frac{F_{12}}{K_{12}} = \frac{v_0}{\eta_2^2 - \eta_1^2}\frac{K_{23}}{M_2}\left(\frac{\sin\eta_1 t}{\eta_1} - \frac{\sin\eta_2 t}{\eta_2}\right)$$

(31.39b)

where $\beta^2 = K_{12}(m_1 + m_2)/m_1 m_2$

(31.39c)

$$\dot{X}_2 = \frac{v_0}{\eta_2^2 - \eta_1^2}[(\beta^2 - \eta_1^2)\cos\eta_1 t + (\eta_2^2 - \beta^2)\cos\eta_2 t]$$

(31.39d)

$$\ddot{X}_2 = \frac{-v_0}{\eta_2^2 - \eta_1^2}[\eta_1(\beta^2 - \eta_1^2)\sin\eta_1 t + \eta_2(\eta_2^2 - \beta^2)\sin\eta_2 t]$$

(31.39e)

$$\dot{X}_1 = \dot{X}_2 + \frac{v_0}{\eta_2^2 - \eta_1^2} \frac{K_{23}}{m_2} [\cos \eta_1 t - \cos \eta_2 t] \tag{31.39f}$$

$$\ddot{X}_1 = \ddot{X}_2 - \frac{v_0}{\eta_2^2 - \eta_1^2} \frac{K_{23}}{m_2} [\eta_1 \sin \eta_1 t - \eta_2 \sin \eta_2 t] \tag{31.39g}$$

The duration of the impact, determined by equating Eq. (31.39a) to zero, is implicitly given by the following relationship:

$$\sin \eta_1 t_f / \sin \eta_2 t_f = - [(\eta_2^2 - \beta^2)/(\beta^2 - \eta_1^2)](\eta_1/\eta_2) \tag{31.39h}$$

where graphical solution for t_f can be made. The terminal velocities $\dot{X}_{1f}$ and $\dot{X}_{2f}$ are found by substituting t_f in Eqs. (31.39c) and (31.39e). The coefficient of restitution is approximately

$$e = - (m_1 \dot{X}_{1f} + m_2 \dot{X}_{2f})/(m_1 + m_2)v_0 \tag{31.39i}$$

2. For the case described by Fig. 31.8b the initial conditions are, at $t = 0$,

$$X_{10} = X_{20} = 0 \qquad \dot{X}_{10} = v_0 \qquad \dot{X}_{20} = 0$$

and the solution to Eq. (31.35) is

$$\frac{F_{23}}{K_{23}} = X_2 = \frac{v_0 \eta_1^2 \eta_2^2 (M_1/K_{23})}{\eta_1^2 - \eta_2^2} \left(\frac{\sin \eta_2 t}{\eta_2} - \frac{\sin \eta_1 t}{\eta_1} \right) \tag{31.39j}$$

$$\frac{F_{12}}{K_{12}} = X_1 - X_2 = \frac{-v_0}{\eta_1^2 - \eta_2^2} \left(\frac{\eta_2^2 - \gamma^2}{\eta_2} \sin \eta_2 t - \frac{\eta_1^2 - \gamma^2}{\eta_1} \sin \eta_1 t \right) \tag{31.39k}$$

where $\gamma^2 = (2K_{12} + K_{23})M_2$

$$\dot{X}_2 = \frac{v_0 \eta_1^2 \eta_2^2}{\eta_1^2 - \eta_2^2} \frac{M_1}{K_{23}} (\cos \eta_2 t - \cos \eta_1 t) \tag{31.39l}$$

$$\ddot{X}_2 = -\frac{v_0 \eta_1^2 \eta_2^2}{\eta_1^2 - \eta_2^2} \frac{M_1}{K_{23}} (\eta_2 \sin \eta_2 t - \eta_1 \sin \eta_1 t) \tag{31.39m}$$

$$\dot{X}_1 = \dot{X}_2 - \frac{v_0}{\eta_1^2 - \eta_2^2} [(\eta_2^2 - \gamma^2) \cos \eta_2 t - (\eta_1^2 - \gamma^2) \cos \eta_1 t] \tag{31.39n}$$

$$\ddot{X}_1 = \ddot{X}_2 + \frac{v_0}{\eta_1^2 - \eta_2^2} [\eta_2(\eta_2^2 - \gamma^2) \sin \eta_2 t - \eta_1(\eta_1^2 - \gamma^2) \sin \eta_1 t] \tag{31.39o}$$

The duration of impact, obtained by equating Eq. (31.39k) to zero, is implicitly defined by the equation

$$\sin \eta_1 t_f / \sin \eta_2 t_f = (\eta_1/\eta_2)(\eta_2^2 - \gamma_2)/(\eta_1^2 - \gamma^2) \tag{31.39p}$$

This solution is useful in evaluating the effect upon peak impact force F_{12} produced by the stiffness of the mounting K_{23} and for evaluating the peak force or shock transmitted to the foundation F_{23}. Further discussion of the problem of impact-induced shock is given in Sec. 5.

31.3.3 Wave Phenomena in Impact

Consideration of the effects of stress-wave propagation during impact becomes necessary when the applied forces or stresses change significantly during the time required
for the propagation of these signals throughout the system. Under such conditions substantial deviation from static equilibrium occurs. For example, at the beginning of
impact remote portions of the body remain unstressed while large stresses develop
near the point of impact and propagate at finite velocity.

As waves travel through a medium they alter both the stress magnitude and the
velocity of the particles composing the medium. In most general terms plane waves
are defined as "dilational" if they produce particle motion along the direction of propagation or "distortional" if they produce particle motion perpendicular to the direction
of propagation. Different types of stress give rise to different waves with different
velocities of propagation. For machine-design work it is convenient to limit consideration to the following types of waves.

1. Longitudinal waves which transmit compressive and tensile stresses are generated
   in such typical applications as pile driving, connecting rods in high-speed
   machines, pump rods in oil-well machinery, and elevator cables.

   Longitudinal waves are a kind of plane dilational wave in that particle motion
   is along the direction of propagation. Although the velocity of propagation of longitudinal waves is somewhat dependent upon the frequency of the waves, the relatively low-frequency waves of engineering interest propagate at the uniform
   velocity

$$c = (E/\rho)^{1/2}$$

2. Torsional waves transmitting shear stresses are encountered typically in (*a*) rapidly
   applied motions on helical springs and (*b*) long drill rods. Torsional waves are one
   example of distortional waves in that particle motion is normal to the direction of
   wave propagation. Plane torsional waves propagate at velocity

$$c = (G/\rho)^{1/2}$$

3. Flexural waves, such as are encountered in the transverse impact of beams, are
   among the most difficult to handle analytically, since they are multidimensional
   and because their velocity of propagation, even in elementary theory, is strongly
   dependent upon the frequency of the wave. Thus the ordinary case, in which the
   total wave is composed of many harmonics or frequency components, is strongly
   influenced by dispersive effects, as are discussed later. More detailed discussion of
   flexural waves is given by Kolsky,[11] Wasley,[2] and Achenbach.[5]

While the velocity of propagation of longitudinal waves is also dependent upon
their frequency or wavelength, the effect is relatively small in the region of most practical importance. With flexural waves, however, the influence of wavelength is quite
large in precisely the range of frequencies which is of interest. Figure 31.9 illustrates
the manner in which the velocity of propagation of longitudinal and flexural waves is
influenced by wavelength in magnesium rods. For small values of the parameter α/Λ,
where α is the transverse dimension of the rod and Λ is the wavelength of a particular
sinusoidal wave, it is seen that the velocity of propagation of longitudinal waves is
essentially constant, whereas long flexural waves are propagated at a velocity

$$c = 2\pi c_0 K/\Lambda$$

where K = radius of gyration of the beam cross section and $c_0 = (E/\rho)^{1/2}$

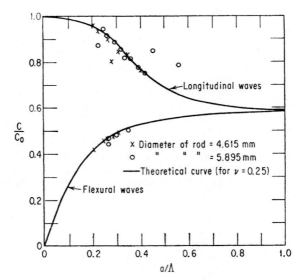

FIG. 31.9 Longitudinal and flexural wave-propagation velocities in magnesium rods as a function of wavelength A and rod diameter a.[11]

Wave Dispersion.   Wave dispersion is one of the most important omissions in the approximate wave theory which is normally employed in engineering work. Because of wave dispersion a stress wave, as it propagates through a system, may become distorted in shape and attenuated in size, because of the following:

1. Internal friction within the propagating medium ordinarily tends more rapidly to damp out the higher-frequency components.

2. Large differences in propagation velocity are characteristic of different types of waves. Hence a combined stress composed of, say, shear and tension becomes distorted and spread out and may separate into completely separate signals.

3. The time-varying impact stress is made of harmonic components of different amplitude and frequency; the velocity of propagation varies with the frequency and hence distorts the wave. This effect exists even for elastic materials with uniform properties and becomes of primary importance in materials exhibiting strain-rate sensitivity.

4. A dilational wave such as a longitudinal wave incident upon a free surface not only gives rise to a reflected dilational wave but also produces a distortional wave as well. Kolsky[11] also shows that, when an elastic wave reaches a slip-free boundary, four waves are generated. Two of these waves are refracted into the second medium and two are reflected back. More detailed considerations of such phenomena, which are prohibitively complex for most engineering applications, are described in Refs. 2, 5, and 11. Thus reflection from boundaries, free surfaces, or discontinuities gives rise to further distortion of the original wave.

In longitudinal impact dispersive effects are usually significant only in the vicinity of the point of impact, within the order of the lateral dimensions of the rod. Within this interval strong radial effects occur and plane-wave theory is not adequate. More modest radial effects occur even in the case of a uniform rod by virtue of Poisson's ratio, which causes lateral change and hence produces shear waves in response to a normal stress wave. For most engineering problems these effects are not predominant and the simple one-dimensional theory of plane waves is adequate. It is possible to regard the waves as essentially plane with substantial simplification of the theory provided that the duration of the pulse is such that $d/t_f c \ll 1$, where d = lateral dimensions of the impacting rod, t_f = period of impact, and c = velocity of propagation, or provided that the excitation does not include high-frequency harmonics of significant amplitude whose wavelengths are less than the transverse dimensions of the bar.

Torsional waves are not dispersive for circular cross sections.[11] Flexural waves, because of their great sensitivity to wavelength, are strongly dispersive.

Longitudinal Impact of Uniform Bars. Consider a bar of uniform cross section composed of an isotropic material, with a known stress-strain relationship $\sigma(\epsilon)$, in which stresses are uniform in planes perpendicular to the boundaries. We ignore the secondary effects of Poisson's ratio which give rise to radial-shear and radial-inertia effects. As a result the equation formed will not involve any dispersive effects and waves will be propagated through the medium without change. It should be pointed out that real systems never undergo the perfectly abrupt discontinuities which are implied in the solution for the plane stress wave. The effects which arise from our assumption of perfect isotropy mean that these methods cannot be accurate when considering variations within the atomic dimensions or crystal-lattice dimensions of the medium. For the gross effects of engineering importance this simplification poses no serious difficulty. Let u equal the displacement of a point originally located at position x along the bar. Then the equation of motion is

$$\partial^2 u/\partial t^2 = (1/\rho)(d\sigma/d\epsilon)(\partial^2 u/\partial x^2) = c^2(\partial^2 u/\partial x^2) \tag{31.40}$$

$$c = \text{velocity of propagation} = \sqrt{(d\sigma/d\epsilon)\rho^{-1}} \tag{31.40a}$$

In the case of linearly elastic waves $d\sigma/d\epsilon = E$, whence

$$c = (E/\rho)^{1/2} \tag{31.40b}$$

Any stress signal imposed at one point will be propagated at velocity c, along the bar. Table 31.2 summarizes propagation velocity for several common materials. For any material whose properties are not strongly strain-rate sensitive and for which the static stress-strain relationship is known, it is possible to determine with good accuracy the velocity of propagation.[49]

TABLE 31.2 Propagation Velocities of Longitudinal and Transverse Waves in Various Materials[22]

Material	Long., c_L, fps	Transv., c_T, fps
Aluminum................	20,900	10,200
Brass.........	14,000	6,700
Glass (window)............	22,300	10,700
Iron.....................	19,500	10,500
Lead....................	7,100	2,300
Lucite...................	8,700	4,200
Polystyrene..............	7,500	3,900
Steel...................	19,500	10,200

St. Venant's general solution to Eq. (31.40) is of the form $u = f(x - ct) + g(x +$

ct). This is commonly called the wave solution since the two functions can be interpreted as two waves, a function $f(x - ct)$ moving in the direction of increasing x, and a function $g(x + ct)$ moving in the direction of decreasing x as in Fig. 31.8. The velocity of any point in the bar is

$$\partial u/\partial t = c(-\partial f/\partial x + \partial g/\partial x)$$

The strain is represented by

$$\epsilon_x = \partial u/\partial x = \partial f/\partial x + \partial g/\partial x \tag{31.41}$$

and the stress

$$\sigma_x = E\epsilon_x = E(\partial f/\partial x + \partial g/\partial x)$$

The functions f and g may be found by applying suitable boundary conditions to define the initial velocity and stress distribution according to Eq. (31.41).

Some Properties of Plane, Linear Longitudinal Waves. The following concepts, helpful in understanding and applying wave phenomena, arise directly from the solution of the linear wave equation.

Particle Velocity and Wave Velocity. The velocity of wave propagation is a property of the medium [Eq. (31.40b)] and is distinct and different from the particle velocity. When a velocity change Δv is imposed on a linearly elastic bar, a stress wave of magnitude

$$\Delta\sigma = \Delta v \, (E\rho)^{1/2} = \Delta v \, \rho c \tag{31.42}$$

travels down the bar at velocity c and modifies the stress and particle velocity in the bar. The direction of the particle velocity is in the same direction as the wave motion for compression waves, and in the opposite direction for tension waves. If the cross-sectional area of the bar is A, the corresponding force is approximately $\Delta F = \Delta\sigma A$. The approximation arises from the fact that real plane waves are not fully uniform but are influenced by the boundaries of the body.[43]

Effect of Stress Wave on Particle Motion. In a compressive stress wave the velocity change of the particles is in the same direction as the stress wave is propagated. In a tension wave, conversely, the velocity change of the particles is in the direction opposite to the direction of wave propagation.

Reflection of Waves. When an incident plane wave σ_I reaches a free end, reflected wave σ_R is reflected back into the medium in order to satisfy the boundary condition that the total stress at a free boundary must be zero. Thus $\sigma_R + \sigma_I = 0$; hence a compression wave reaching a free end is reflected back as a tensile wave of equal and opposite magnitude. Corresponding to the reflected tensile wave, the particle velocity increases in the direction of the original incident compressive wave. An incident plane tension wave reaching a free end is covered by the converse of this description.

When an incident plane wave σ_I reaches a perfectly rigid, fixed end, the boundary condition requires that the total particle velocity must be zero. Thus if particle velocity v_I corresponds to σ_I, then $v_I - V_R = 0$. Accordingly an incident wave is reflected without change in sign or magnitude, and as a result of superposition, the stress at the fixed boundary instantaneously doubles.

When an incident longitudinal plane wave σ_I passes through a discontinuity in the medium, either as a result of a change in material or to a first approximation as a result of a change in geometry, there will, in general, arise a reflected wave σ_R and a transmitted wave σ_T which must satisfy the following boundary conditions:

1. The force across the discontinuity must be continuous; thus

$$(\sigma_I + \sigma_R)A_1 = \sigma_T A_2$$

2. The particle velocity must be continuous; thus

$$v_I + v_R = v_T$$

Combined with Eq. (31.42), these two conditions yield relationships for the magnitudes of these two waves:

$$\frac{\sigma_T}{\sigma_I} = \frac{2A_1\rho_2 c_2}{A_1\rho_1 c_1 + A_2\rho_2 c_2} \qquad \frac{\sigma_R}{\sigma_I} = \frac{A_2\rho_2 c_2 - A_1\rho_1 c_1}{A_2\rho_2 c_2 + A_1\rho_1 c_1} \qquad (31.43)$$

Initial Velocity at an Impact Interface. When two surfaces collide, the initial velocity at the common impact interface of both bodies is assumed to change instantaneously to a value $v = (v_{1i} - v_{2i})/2$. Subsequent reflections of the waves will, of course, modify this value in time.

Energy in Wave. The energy in a linear, longitudinal plane wave is composed of two equal parts: one half is the kinetic energy of the particles; the second half represents the strain energy of deformation. The total energy in a bar which is subjected for a time t to a uniform stress at one end is

$$T + U = (A/E)ct\sigma^2 \qquad (31.44)$$

Interference or Combining of Waves. As long as the deformations within a body remain elastic, the principle of superposition can be used to determine the stress magnitude and particle velocity. Thus when several longitudinal waves of different sign and magnitude coincide in space and time at some point in the system, the net stress and velocity can be determined by simply adding the magnitudes of all the coincident waves with due regard for the sign of each wave, provided that the final stress does not exceed the elastic limit.

Waves moving in opposite directions can move through one another with no alteration, provided that their combined stress magnitude does not exceed the elastic limit.

Figure 31.10 illustrates the superposition resulting from a decaying compression wave incident upon the free end of a bar, resulting in a reflected tension wave. The superposition of the two waves results in a net tensile stress within the body. This phenomenon accounts for scabbing-type failures which are encountered in violent impact of armor plate by projectiles.[4]

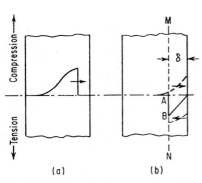

FIG. 31.10 Example of wave reflection and superposition near a free surface in a semiinfinite plate. (*a*) Before reflection. (*b*) Shortly after wave reflection.[4]

31.4 DYNAMIC BEHAVIOR OF MATERIALS

Several properties of engineering materials are modified as a consequence of impact-type rapid loading. For nonmetals such as plastics, rubbers, and other elastomers profound changes can occur as a consequence of rapid loading and these modified properties must be considered in even approximate calculation of dynamic behavior. For metals stressed below their elastic limit these effects are not normally so extreme, but under certain conditions significant deviations from quasi-static behavior can occur. Because of the incomplete state of our knowledge of these properties and the complexity of the applications, it is not generally possible to allow for these effects quantitatively. In general, stress limits for common metals under impact are approximately equivalent to quasi-static stress limits. However, in some cases this assumption may not be conservative. A qualitative understanding of these effects can be valuable as a guide to design and as an aid in the understanding and interpretation of some impact phenomena.

31.4.1 Delayed Yield Phenomenon

Metals which exhibit well-defined static yield points also exhibit the characteristic of delayed yield when subjected to rapidly applied short-duration stresses. It is possible to subject low-carbon steels and iron (which exhibit well-defined yield points) to tensile stresses in excess of their static elastic limit without permanent deformation, provided that the time during which the stress exceeds the elastic limit is less than some well-defined limit. Tests on a 0.17 percent carbon steel produced the relationships shown in Fig. 31.11 between the maximum stress and the cumulative time such a stress could be maintained without permanent deformation.[22] These times are so short at normal temperatures as to preclude their exploitation in repetitive operation, unless the cumulative effects are erased by heat treatment. Temperature sensitivity of this phenomenon is also seen to be significant.

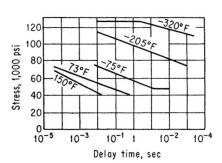

FIG. 31.11 Delay time for the initiation of yielding at rapidly applied stresses on 0.17 percent carbon steel at various temperatures.[22]

Materials such as stainless steels, aluminum, and copper which do not exhibit well-defined yield points apparently do not exhibit the delayed-yield phenomenon. Clark[22] reports that tests on such metals, including 18-8 stainless steel, normalized SAE 4130 steel, quenched and tempered 4130 steel, 24-ST aluminum alloy, and 75-ST aluminum alloy, all deformed plastically during the loading period with no evidence of delayed yield.

31.4.2 Critical Impact Velocity[22]

A characteristic of importance in some applications is the existence of a "critical" impact velocity of metal. Wires or rods subjected to impact develop initial strains which are related to the difference between the velocity of propagation of strain in the metal and the velocity to which a point on the rod is impulsively subjected, as indicated by Eq. (31.42). Because $d\sigma/d\epsilon$ becomes smaller as σ is increased beyond the elastic limit, Eq. (31.40a) shows that high stresses propagate at lower velocities than low stresses. Thus a maximum velocity exists above which the large plastic strains, which are being generated, cannot propagate so rapidly as the end of the rod is being pulled, and fracture of the rod occurs almost immediately.

There exists then a critical velocity at which a tensile specimen will always fail under impact, independent of cross section of the specimen or the mass of the impacting mass, provided that the kinetic energy is at least sufficient to supply the energy absorbed by the necking-down process associated with tensile failures. This critical velocity is defined by Eq. (31.45).

$$v_{cr} = \int_0^{\epsilon_{ult}} \left(\frac{1}{\rho} \frac{d\sigma}{d\epsilon} \right)^{1/2} d\epsilon \tag{31.45}$$

Table 31.3 shows results of impact experiments on critical velocity.

TABLE 31.3   Critical Impact Velocity and Tensile Properties of Common Materials[22]

Material	Ultimate strength, psi		Elongation in 8 in.		Critical velocity, fps	
	Static	Dynamic	Static	Dynamic*	Experimental	Theoretical
Ingot iron, annealed........	37,100	57,400	25.7	16.2	100	†
SAE 1015, annealed.......	50,600	63,500	28.0	30.0	100	†
SAE 1022, cold-rolled.....	84,000	105,000	6.0	15.0	100	95
SAE 1040, annealed.......	78,050	91,800	20.4	20.7	200	†
SAE 1045, quenched and tempered.............	142,900	169,000	5.7	9.2	190	88
SAE 2345, quenched and tempered.............	145,250	175,250	8.4	14.1	200	161
SAE 4140, quenched and tempered.............	134,250	151,000	8.5	14.7	175	132
SAE 5150, quenched and tempered.............	139,000	148,100	8.5	13.3	170	159
Type 302 stainless steel....	93,300	110,800	58.5	46.6	200	490
Copper, annealed.........	29,900	36,700	32.7	43.8	200	231
Copper, cold-rolled........	45,000	60,000	2.5	10.7	50	42
2S aluminum, annealed....	11,600	15,400	23.0	30.0	200	176
2S aluminum, ½ hard.....	17,200	22,100	4.6	7.0	110	36
24S-T aluminum alloy.....	65,150	68,600	11.3	13.5	200	290
Magnesium alloy (Dow J).	43,750	51,360	9.6	10.9	200	303

*Maximum percentage elongation up to the critical velocity.
†Existence of a yield point prevents computation of critical velocity. Values are given in feet per second.

31.4.3   Transition from Ductile to Brittle Behavior

Ductility is normally regarded as a property of materials subject to stresses beyond their elastic limit. As such it might ordinarily be expected not to play too important a role in the behavior of machine components which are intended for long, repetitive life and are consequently designed to operate at stress levels well within their elastic limit. It might ordinarily be concluded then that the elastic limit or yield strength should be the all-important design criteria.

On a gross basis in ordinary machine design, while yield strength is the important criterion, high local stress conditions in materials frequently arise which make ductility an important criterion for repetitive operations. Ductile behavior allows flow of the metal, providing a redistribution of the stress concentrations, and therefore helps to avoid failure due to local stress conditions. For impact applications the question of ductility takes on additional significance since, under impact, bodies are forced to absorb a definite amount of strain energy, for example, as is dictated by Eq. (31.7).

The significant distinctions between brittle and ductile failures are the amount of flow in response to shear stresses and the energy absorbed by the piece in order to produce fracture. Fractures are characterized as being brittle if failures involve relatively little flow of the metal and relatively little energy absorption or as ductile if considerable flow of metal and relatively greater energy are required to failure. Brittle behavior becomes more likely as temperature is reduced, the rate of loading is increased, and transverse or multiaxial stresses are increased, by shape and previous history.

The presence of sharp corners, notches, or structural discontinuities produces complex multiaxial stresses, composed of tensile, compression, and shear components. The net result of these combined stresses is to increase the tensile stress necessary to produce failure.[27] Under such conditions, normally ductile materials can be made to fail in a brittle manner with a substantial reduction in the energy absorbed during failure.

For rapid loading the shape of a piece becomes important for still another reason since stress reflections occur at discontinuities which can give rise to instantaneous doubling of local stresses. The location and character of stress-wave reflections will determine location of peak stresses and determine preferred failure locations.

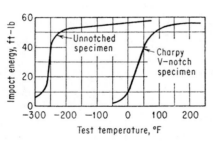

FIG. 31.12 Transition between ductile and brittle fracture in 0.1 percent carbon-steel plate illustrating notch sensitivity.[39]

The temperature at which a particular specimen first demonstrates a change from ductile to brittle failure at a particular rate of loading is termed the "transition temperature." The transition temperature is not a fundamental property of the material, but at least for irons and ferritic steels is particularly sensitive to notch effects, as is evidenced by Fig. 31.12, and is quite sensitive to rate of strain. The existence of a high transition temperature in a metal intended for use at low temperatures should indicate the need for careful evaluation of the particular shape under the particular environmental conditions.

Increasing rate of strain also tends to produce a change from ductile to brittle fracture. Although ductile materials characteristically flow when subjected to shear stresses, with rapidly applied loads of short duration the shear stresses are not able to produce significant flow in the available time. Such materials then may fail in a brittle manner and at higher ultimate stresses, depending upon temperature and rate-of-strain conditions.

The following interesting relationship between temperature and rate of strain, as they influence the transition from ductile to brittle failure, has been proposed.[27]

$$\dot{\epsilon}_c/\dot{\epsilon}_0 = \exp\left[-(Q/R)(\theta_c^{-1} - \theta_0^{-1})\right]$$

where $\dot{\epsilon}_c$ and $\dot{\epsilon}_0$ are the minimum strain rates producing brittle fracture at temperatures θ_c and θ_0, respectively, and Q/R is an empirical constant related to the activation energy of the material.

The ultimate tensile strength σ_u in tension is also influenced by rate of strain and temperature. A parameter ζ derived from fundamental reasoning which appears successfully to relate these factors with ultimate strength is[27]

$$\sigma_u = \zeta^\gamma = \left[\frac{\dot{\epsilon} \exp (Q/R\theta)}{f_0} \right]^\gamma \qquad (31.46)$$

where $f_0 = JQ/Nh$
    Q = heat of activation, cal/g·atom (an empirical factor)
    R = constant = 1.96
    γ = empirical constant
    N = Avogadro's number
    J = mechanical equivalent of heat energy
    h = Planck's constant
    θ = temperature

Figure 31.13 shows a plot for the ultimate strength as a function of ζ for copper at temperatures between room temperature and 400°C. At higher temperatures correlation is not so successful. Comparable data for a typical steel are shown in Fig. 31.14. These results show that only a limited amount of experimental data is necessary to obtain extensive information on the variation of ultimate strength over a wide range of strain rate and temperature.

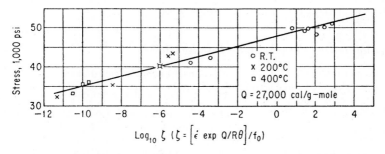

FIG. 31.13 Ultimate tensile strength of copper as a function of a parameter (ζ) including strain rate and temperature.[27]

31.4.4 Effect of Residual Stresses

The importance of residual stresses upon impact behavior is not too well known. Little work has been done in this area and the results are not conclusive. Under normal static-loading conditions, time permits the leveling of high local residual stresses through plastic flow, provided severe restraints such as sharp notches are not acting. Under the conditions of impact loading, superposition of the impact stresses and residual stresses may result because time may be too short to allow plastic flow to attenuate and redistribute the peak residual stresses. Although validity of this theory is not yet proved, Miklowitz[36] has suggested that residual stresses may become important when high rates of loading are combined with low temperatures and notching. For such conditions the use of some method of stress relief is desirable.

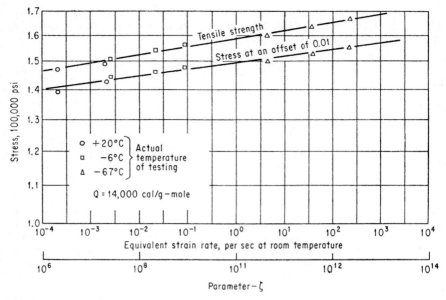

FIG. 31.14   Strain rate and temperature dependence of a typical steel. Ultimate tensile strength and plastic stress at a strain of 0.01 are plotted as functions of parameter.[37]

31.4.5   Influence of Impact upon Fatigue Life[28,29]

The change in fatigue life resulting from impact loading rather than conventional, relatively slowly applied sinusoidal loading has been studied for some alloy steels loaded in fully reversed bending.

The degree of hardness or the heat treatment of the steel was found to have a measurable effect upon fatigue life. Table 31.4 summarizes these data, indicating that for soft, relatively ductile steels the tolerable stress under impact is increased at least 5 percent and for hardened steels the tolerable stresses may be reduced up to 12 percent for 100,000 operations.

TABLE 31.4   Effect of Impact-Type Loading and Heat Treatment upon Fatigue Life of 10-mm-Diameter Bars Loaded in Reverse Bending[28,29] Dynamic Stresses Measured by Resistance-Type Strain Gages

Russian designation	Composition of steel					Stress limit for 100,000 operations					
						Tempered at 600°C			Tempered at 200°C		
	C	S	Mn	Cr	Ni	σ_d	σ_s	σ_d/σ_s	σ_d	σ_s	σ_d/σ
45Kh...........	0.43	0.26	0.67	1.00	0.29	82,000	76,000	1.05	107,000	122,000	0.88
30KhGSA.......	0.3	1.10	0.95	1.02	....	88,000	81,000	1.08	109,000	117,000	0.92

σ_d = impact fatigue strength, lb/in²; σ_s = conventional (quasi-static) fatigue strength, lb/in².

Scale or size effects also are important, however, and these may tend to have greater significance upon fatigue life. Hence generalizations from these data must be made with caution.

31.4.6 Rate Sensitivity of Stress-Strain Relationships

When as a consequence of rapid impact-type loadings metals undergo large irreversible plastic strains, modifications of the normal quasi-static stress-strain relationships are sometimes observed.[33–35,50] Typical relationships for 0.24 percent carbon steel, annealed copper, and annealed aluminum are shown in Figs. 31.15, 31.16, and 31.17.

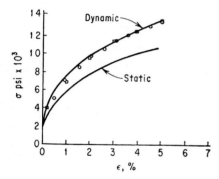

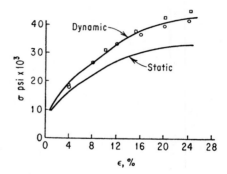

FIG. 31.15 A comparison of Bell's theoretical dynamic stress-strain curve in annealed aluminum, with the experimental (solid line) curve of Ref. 35. The lower solid line is the static stress-strain curve.[50]

FIG. 31.16 A comparison of Bell's theoretical dynamic stress-strain curve in copper, with experimental data (solid line). The lower solid line is the static stress-strain curve.[50]

Bell[50] made a careful study of this behavior in annealed aluminum and copper. He showed that compression impact of long bars at velocities exceeding the critical impact velocity would produce this anomalous behavior, but only within a region restricted to the first one or two diameters near the impact surface. This behavior is associated with the creation of a nondispersive shock front at the impact interface which ultimately develops into dispersive plastic waves as it propagates along the bar. In the region encompassed by the shock wave, modified stress-strain relationships are observed. Beyond the immediate vicinity of the impact zone (distances greater than 1½ to 2 diameters), the normal strain-rate-independent stress-strain relationships apply and can be used to determine the velocity of propagation with good accuracy, employing Eq. (31.38a).

The dynamic stress-strain curves for annealed aluminum and copper do not appear to be a measure of material properties in the same sense as the static-stress-strain curve but rather reflect the mechanics of the development of initial plastic wave fronts. Within the shock zone in the immediate vicinity of the impact point, Bell[50] was able to calculate the dynamically modified properties from the static-stress-strain properties by imposing conservation of mass, momentum, and energy ultimately resulting in the equations

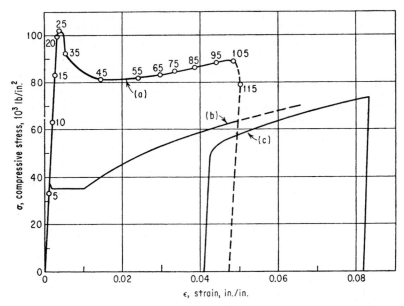

FIG. 31.17 Static and dynamic compressive stress-strain curves for a 0.24 percent carbon steel, annealed at 1652°F. (*a*) Impact loading. (*b*) Static test of annealed specimen. (*c*) Static test of dynamically shocked specimen.

$$\frac{P\epsilon_0}{2} = \frac{\rho v_0^2}{2} = \int_0^{\epsilon_0} \sigma_x \, d\epsilon_x \tag{31.47}$$

By equating the latter two terms and solving first for ϵ_0, the dynamic stress P can be obtained. Calculations employing this simple procedure produced the excellent correlations with experiment shown in Figs. 31.15 and 31.16.

Detailed discussion of shock-wave phenomena in solids may be found in Refs. 42 and 43.

Comparison of dynamic and static-stress-strain relationships for a 0.24 percent carbon steel is shown in Fig. 31.17. This material exhibits both the delayed yield phenomenon and considerable difference between static and dynamic stresses in the plastic region. The elastic modulus appears to be independent of strain rate.

31.4.7 Some Properties of Fluid and Elastomeric Buffers

Buffers are frequently needed in machine design to provide means for mitigating the effects of impact or for rapidly decelerating a rapidly moving body after it has traveled a prescribed distance. Properties of buffers which are particularly important for such applications include: (1) the peak force produced by the impact must be limited to some safe value, (2) kinetic energy must be irreversibly dissipated so that rebound is minimized, and (3) the buffing means must be reusable for successive operations. Means available for providing these important functions include the use of metal springs, fluid dashpots, and elastomers, each with its advantages and limitations. This discussion will be limited to a brief review of these means.

Metal springs are largely insensitive to temperature, rate of loading, and environment. They are economical and simple to design, and their behavior is well understood. The methods of Secs. 31.3.1 and 31.3.2 may be employed in their analysis. The major limitations of metal springs are

1. They do not enjoy a particularly high ratio of energy-absorbing capacity per unit volume.
2. They have only a small energy-dissipating capacity, which means that rebound is a problem unless friction devices, which are inherently more complex, and less reproducible, are employed.

"Fluid dashpots" (Fig. 31.18) utilize the kinetic energy of the moving mass to drive a piston which forces a fluid through a small restricting orifice. The flow of the fluid irreversibly dissipates much of the energy and produces a decelerating force proportional to the fluid pressure produced. Dashpots may be either of the pneumatic type if air or a gas is employed as the fluid medium, or hydraulic if a liquid is employed. "Pneumatic dashpots" are most effective when they can be operated in environments at pressures considerably higher than normal atmospheric pressure. When operating in a normal 14.7 lb/in^2 atmosphere, their energy-absorbing capacity is relatively low per unit volume of dashpot, and they require high compression ratios to achieve even modest peak pressures. Pneumatic dashpots are rather critical in manufacture and in operation, require small restricting orifices, and are highly sensitive to leakage and stroke variations.

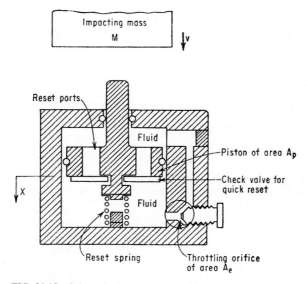

FIG. 31.18   Schematic drawing of fluid dashpot-type buffer.

Because of compressibility effects, pneumatic dashpots pose mathematical difficulties, although the underlying theory is quite well understood. The differential equations of motion of a relatively rigid mass decelerated by a pneumatic dashpot containing an ideal gas are

$$M(d^2X/dt^2) + (P - P_{atm})A_P = F_e$$

$$V = V_0 - A_p X$$

$$[A_e(RT_i 2g)^{1/2}]^{-1}\left(\frac{V}{P\gamma}\frac{dP}{dt} + \frac{dV}{dt}\right) = \left(\frac{\gamma}{\gamma - 1}\right)^{1/2} f\left(\frac{P}{P_{atm}}\right) \tag{31.48}$$

$$f\left(\frac{P}{P_{atm}}\right) = \left\{\left(\frac{P_c}{P_{atm}}\right)^{(\gamma+1)/\gamma}\left(\frac{P_{atm}}{P}\right)^{2/\gamma}\left[\left(\frac{P}{P_c}\right)^{(\gamma-1)/\gamma} - 1\right]\right\}^{1/2}$$

where A_p = piston cross-sectional area
$\quad\quad A_e$ = bleed-orifice effective area
$\quad\quad M$ = mass of body being decelerated
$\quad\quad V$ = volume of gas in dashpot
$\quad\quad P$ = pressure
$\quad\quad \gamma$ = ratio of specific heats of gas
$\quad\quad R$ = gas constant for particular gas employed
$\quad\quad X$ = stroke

and $\quad\quad\quad\quad\quad\quad\quad\quad\quad P_c = [2/(\gamma + 1)]^{\gamma/\gamma-1}P$

when $\quad\quad\quad\quad\quad\quad\quad [2/(\gamma + 1)]^{\gamma/\gamma-1}P > P_{atm}$

or $\quad\quad\quad\quad\quad\quad\quad\quad\quad P_c = P_{atm}$

when $\quad\quad\quad\quad\quad\quad\quad [2/(\gamma + 1)]^{\gamma/\gamma-1}P < P_{atm}$

These equations normally require solution by step-by-step numerical methods or by automatic computers. For crude approximations the limiting values for the behavior of a pneumatic dashpot must lie between the following limits:

1. With negligibly small leakage the following relationship holds approximately

$$PV^\gamma = \text{const}$$

and the energy absorbed by the dashpot is

$$\Delta T = P_0 V_0^\gamma(V_f^{1-\gamma} - V_0^{1-\gamma})$$

2. The performance of a pneumatic dashpot with a large orifice for which the gas pressure and density variations are small may be roughly estimated employing the incompressible-dashpot equations (31.49) and (31.50).

Hydraulic dashpots are extensively used as buffers and enjoy relatively high energy-absorbing capacity since they can readily accommodate generated pressures of 5 to 10,000 lb/in². Analysis is much simpler than with pneumatic dashpots and much tighter control is possible because of the relatively small compressibility of liquids compared with gases. Disadvantages of hydraulic dashpots include their cost, precision, and number of parts required, and the possible danger of gas-bubble formation within the dashpot which may alter the performance. They are relatively temperature-insensitive if designed with a sharp-edged orifice since then the density of the liquid predominates rather than its viscosity. The behavior of a relatively rigid mass impacting a hydraulic dashpot may be determined from the relationship

$$\ln\frac{V_0}{V_x} = \frac{\rho A_p}{M}\int_0^x \left[\left(\frac{A_p}{A_e}\right)^2 - 1\right]dx \tag{31.49}$$

In which the orifice area A_e may be allowed to vary with the stroke of the dashpot. The pressure generated is

$$\Delta P = (\rho/2)[(A_p/A_e)^2 - 1]v^2 \qquad (31.50)$$

Rubber and Elastomeric Materials. For many applications the use of a rubber or similar elastomeric material is a good solution to the buffer problem. These materials excel in low cost, simplicity, reliability, and economy of space. Their use normally results in smaller deflections and hence higher impact forces than can be achieved with fluid dashpots.

Because of the complex properties of elastomers, practical solutions to the impact problem are greatly limited. In addition to sensitivity to rate of strain and temperature, elastomers are characterized by a rather long memory so that their behavior depends heavily upon their previous history, such as the magnitude of strain to which they have been previously subjected and the length of time they are allowed to "recover" following previous operations. In general the effect of frequent operations is to stiffen the elastomer, since the long, complex elastomer molecules are not allowed sufficient time to unwind. This effect also tends to reduce their energy-dissipating properties.

An additional complication arises because of the large strains to which elastomers are normally subjected in impact. Since the volume of the elastomer remains essentially constant, the elastomer must dilate radially, normal to the direction of applied strain. The dilation in turn is strongly influenced by the end conditions. Thus, if rigid surfaces are firmly bonded to the elastomer, a higher effective modulus will be observed than if the engaging surfaces are lubricated, allowing radial dilation to occur at the ends. It has also been found[40] that when elastomer ends are rigidly bonded the response is sensitive to a "shape factor" which is the ratio of unstressed area normal to the direction of applied load divided by the lateral free or "bulge" area. When the ends are unconstrained the elastomer response is not strongly influenced by the shape factor. It is because of the large number of variables which markedly influence elastomer response that few data exist for predicting their impact behavior.

One attempt to determine the impact behavior of elastomers is described in Ref. 41. The following information is drawn from that source.

The strong sensitivity to temperature and to rate of loading of the modulus of elasticity of an elastomer is shown in Figs. 31.19 and 31.20. Stress-strain curves are given in Fig. 31.20 for one sample of a Buna N buffer under dynamic and quasi-static strain rates (strain rate $= v_0/L = \epsilon$, where $L =$ initial length of the rubber buffer). It is evident from these data that quasi-static data cannot be employed to predict dynamic behavior. If the initial slope of the stress-strain curve of Fig. 31.20 is taken as a measure of Young's modulus, it is seen that the modulus varies from 800 lb/in^2 when $\dot{\epsilon} = 0.0053$ to 8660 lb/in^2 when $\dot{\epsilon}$ is in the range 200 to 500. From these results, it is seen that a change in strain rate of 10^5 changes the apparent modulus by about a factor of 10.

The very large temperature sensitivity of Young's modulus as shown in Fig. 31.19 points up an important limitation in the use of elastomers for applications requiring precise response if large variations in ambient temperature can occur.

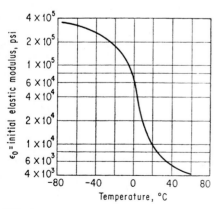

FIG. 31.19 Effect of temperature upon the dynamic modulus of a Buna N elastomer, $40 < \dot{\epsilon}_0 < 400$ per second.[41]

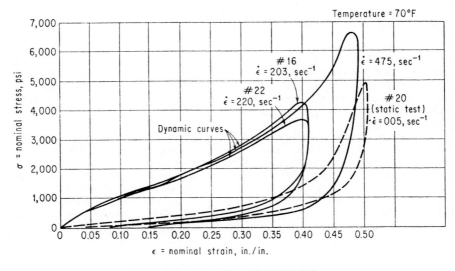

Initial strain rate, s^{-1}	Specific energy input, lb/in^3	Approx. initial modulus, lb/in^2
475	1244	
220	752	8660
203	805	
0.0053	296	800

FIG. 31.20 Comparison of quasi-static and dynamic stress-strain curves for a Buna N elastomer, dynamic data from impact tests.[41]

It was found possible to correlate several facets of compression impact behavior which are of engineering interest by restricting the problem to the case of one Buna N elastomer, with (1) ambient temperature of 70°F, (2) simple cylindrical samples of elastomer, (3) unbonded ends, (4) compression impact, (5) initial strain rates $40 < \dot{\epsilon} < 400$ per second, (6) input energies into the elastomer limited to about 400 in·lb/in^3 of elastomer, and (7) fairly long intervals of the order of minutes between operations.

The following two design parameters were found to be adequate under these conditions to correlate the observed data:

$$U_s = (1/2g)[W_1 W_2/(W_1 + W_2)][v_{10} - v_{20})^2/2AL] \quad \text{in·lb/in}^3$$

$$\lambda = [W_1 W_2/(W_1 + W_2)](L/A) \quad \text{lb/in}$$

where W_1, W_2 are weights of impacting bodies, A is load-bearing area. In terms of these two parameters, with the units specified above, the following empirical relationships were found to hold:

Duration of impact:

$$t_f = (\pi/E_{0}g)^{1/2}\lambda^{1/2}(U_s/100)^{-0.1}$$

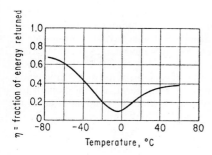

FIG. 31.21 Effect of temperature upon fraction of energy returned as kinetic energy for a Buna N elastomer.[41]

Here E_0 is the initial dynamic modulus as is given in Fig. 31.21.

Fraction of energy returned as kinetic energy after the impact:

$$\eta = 0.5(\lambda/U_s)^{0.2}$$

Maximum stress produced by the impact:

$$\sigma_{max} = 49U_s^{0.7}\lambda^{-0.1}$$

Maximum strain produced by the impact:

$$\epsilon_{max} = 0.026U_s^{0.38}\lambda^{0.12}$$

31.4.8 Impact Properties of Nonmetals

Published information on the dynamic properties of such materials as thermoplastic and thermosetting compounds is extremely meager. In part this may be due to the wide and ever-increasing variety of such materials, their heterogeneous composition, and the extreme complexity of their properties. Fillers of diverse types ranging from wood to sand to fiberglass, combined with a resin binder, produce materials which are non-isotropic and which are sensitive to geometry and molding procedure. In addition, such materials are far more sensitive than metals to environmental conditions such as strain rate, frequency of loading, and temperature. Because of these complications it is particularly difficult to make generalizations. It is well to keep in mind, however, that under impactlike conditions, published quasi-static data may bear little resemblance to the dynamic properties of plastic materials. As one example, the data shown in Fig. 31.22 are of interest. This curve shows the pronounced rate sensitivity of the energy-

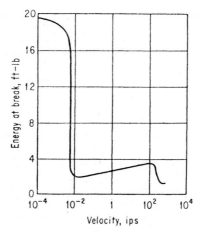

FIG. 31.22 The impact strength of phenol-formaldehyde resin as a function of the rate of application of load.[42]

absorbing capacity of specimens of one phenol-formaldehyde resin. Note that large and abrupt changes in energy absorption occur at impact velocities of 10^{-2} and 10^2 in/s. The trend is evidently strong toward increasingly brittle behavior as strain rate is increased. It should be expected that the apparent Young's modulus of such materials is also rate-sensitive, as is demonstrated for elastomers in Fig. 31.20. In their impact-fatigue properties plastic materials exhibit a pronounced sensitivity to the frequency of loading. As the frequency is increased, there is, in general, a reduction in the fatigue life. This is probably due to the long relaxation time for residual stresses in plastics and to their high internal friction, which gives rise to thermal effects.

31.4.9 Digital Computation of Impact

The principal emphasis in this section has been on relatively simple, one-dimensional analyses which emphasize order-of-magnitude estimates, physical insights, and approximate closed-form solutions. This section also has treated one-dimensional models which more accurately treat the distribution of mass, stiffness, and internal losses, and include multiple degrees of freedom and approximate, nonlinear properties of geometry or materials. For most machine-design problems, such methods, which are readily amenable to computer solutions are to be preferred, particularly when combined with judicious testing and measurement. Usually more elaborate analyses are not justified because of limitations in the accuracy of modeling of structural stiffness, internal damping, and material properties.

There exists, however, in addition to those simple treatments, a class of computer solutions to impact behavior which map in two and three dimensions the detailed deformation response in the vicinity of the impact point. This treatment is more appropriately a subject of interest to theoretical and applied mechanics rather than machine design. Zukas[1] has compiled a useful overview of the state of the art of numerical computation of impact phenomena. Such numerical solutions greatly extend the range of solutions which can be obtained, including detailed treatment through failure and postfailure, oblique impact effects, and situations involving three-dimensional stress states. However, the accuracy of the results has been limited by the extent to which material properties are known. Use of such large-scale computer solutions has been principally of interest in aerospace and military applications. Current effort is focused on three-dimensional simulations in order to determine structural configurations to protect space vehicles against hypervelocity impacts with meteorites and perforation of solids by projectiles.

These elaborate programs usually embody three distinct stages:

1. A "preprocessor" stage in which the geometry, material properties, and initial conditions are specified

2. A main program, typically employing a finite-element analysis combined with conservation laws of mass, momentum, and energy; an equation of state for the determination of pressures; a constitutive (stress-strain) relationship; and a failure criterion; and a postfailure model

3. A "postprocessor" computer program which prepares graphical summaries of the vast quantity of data compiled in stage 2, involving mapping in three dimensions of the distribution of deformation, velocity, temperature, etc.

Zukas[1] emphasizes the costly time-consuming effort entailed in even acquiring mastery of existing programs at the current state of the art. In cases where the material properties are known well enough, such solutions have been found to reproduce observed behavior extremely well. Among the typical problems which have been studied in depth are

1. Hypervelocity impact of a nylon sphere against a steel plate, leading to scabbing failure of the steel plate
2. Ballistic impact of a long bar which strikes an armor plate at an oblique angle
3. Oblique impact of a steel sphere into an aluminum target
4. Impact involving anisotropic media

The interested reader is referred to Zukas[1] for an entree to this advanced technology.

REFERENCES

1. Zukas, J. A., et al.: "Impact Dynamics," John Wiley & Sons, Inc., New York, 1982.
2. Wasley, R. J.: "Stress Wave Propagation in Solids: An Introduction," Marcel Dekker, Inc., New York, 1973.
3. Goldsmith, W.: "Impact," Edward Arnold (Publishers) Ltd., London, 1960.
4. Rhinehart, J. S., and J. Pearson: "Behavior of Metals under Impulsive Loads," American Society for Testing and Materials, Cleveland, 1954.
5. Achenbach, J. D.: "Wave Propagation in Elastic Solids," American Elsevier Publishing Co., Inc., New York, 1975.
6. Johnson, W.: "Impact Strength of Materials," Crane, Russak & Co., Inc., New York, 1972.
7. Lindholm, U. S.: "Mechanical Behavior of Materials under Dynamic Loads," Springer-Verlag, New York, 1968.
8. Huffinston, N. J.: "Behavior of Materials under Dynamic Loading," American Society of Mechanical Engineers, New York, 1965.
9. Rakhmatulin, K., and Y. Dem Yanov: "Strength under High Transient Loads," (Trans. from Russian), Israel Program for Scientific Translations, Jerusalem, 1966.
10. Kinslolo, R.: "High Velocity Impact Phenomenon," Academic Press, Inc., New York, 1970.
11. Kolsky, H.: "Stress Waves in Solids," Oxford University Press, Fair Lawn, N.J., 1953.
12. DeJuhasz, K. J.: "Graphical Analysis of Impact of Bars above the Elastic Range," *J. Franklin Inst.*, vol. 248, nos. 1 and 2, July and August 1949.
13. Marti, W.: "Vibrations in Valve Springs of Internal Combustion Engines," *Sulzer Tech. Rev. Switz.*, no. 2, 1936.
14. Timoshenko, S., and J. N. Goodier: "Theory of Elasticity," 2d ed., McGraw-Hill Book Company, Inc., New York, 1951.
15. Donnell, L. H.: "Longitudinal Wave Transmission and Impact," *Trans. ASME,* vol. 2, p. 153, 1930.
16. Sears, J. E.: "On the Impact of Bars with Rounded Ends," *Trans. Cambridge Phil. Soc.*, vol. 21, 1909–1911.
17. Timoshenko, S.: "Zur Frage nach der Wirkung eines Stosse auf einer Balken, Z.," *Math. Physik,*" vol. 62, p. 193.
18. Lee, E. H.: "The Impact of a Mass Striking a Beam," *J. Appl. Mech.*, vol. 62, p. A-129, December 1940.
19. Krefeld, W. J. et al.: "An Impact Investigation of Beams with Buttwelded Splices under Impact," *Welding J. N. Y. Res. Suppl.*, vol. 12, no. 7, pp. 373S–432S, 1947.
20. Hoppmann, W. H.: "Impact of a Mass on a Damped, Elastically Supported Beam," *J. Appl. Mech.*, vol. 70, p. 125, June 1948.

21. Zener, C., and H. Feshbach: "A Method of Calculating Energy Losses during Impact," *J. Appl. Mech.,* p. A67, June 1939.

22. Clark, D. S.: "The Behavior of Metals under Dynamic Loading," *Metal Prog.,* p. 67, November 1953.

23. Brooks, W. A., and I. W. Wilder: "Effect of Dynamic Loading on Strength of an Inelastic Column," *NACA Tech. Note* 3077, March 1954.

24. Hartz, B. J., and R. W. Clough: "Inelastic Response of Columns to Dynamic Loading," *Proc. ASCE, J. Eng. Mech. Div.,* vol. 83, no. Em2, April 1957.

25. Gerard, G., and H. Becker: "Column Behavior under Conditions of Compressive Stress Wave Propagation," *J. Appl. Phys.,* vol. 22, no. 10, p. 1298, 1951.

26. Coppa, A. P.: "On the Mechanism of Buckling of a Circular Cylindrical Shell under Longitudinal Impact," *Gen. Elec. Rept.* R60SD494, Missile & Space Vehicle Dept., Philadelphia.

27. Zener, C., and J. H. Hollomon: "Plastic Flow and Rupture in Metals," *Trans. ASM,* vol. 33, pp. 188–226, 1944. Morkovin, D.: *ibid.,* p. 221 (discussion).

28. Davidenkov, N. N., and E. I. Belyaeva: "Investigation of Fatigue under Repeated Impact," *Brutcher Trans.,* no. 4719, POB 157, Altadena, Calif. (Russian original in *Metalloved. i Obrabotka Metal.,* no. 11, November 1956).

29. Davidenkov, N. N., and E. I. Belyaeva: Study of Impact Fatigue Strength, *Brutcher Trans.,* no. 4349, POB 157, Altadena, Calif. (Russian original in *Metalloved. i Obrabotka Metal.,* no. 9, September 1958).

30. Welch, W. P.: "ASME Handbook—Metals Engineering—Design," Chap. 8, "Impact Considerations in Design," McGraw-Hill Book Company, Inc., New York, 1953.

31. Rayleigh, J. W. S.: "On the Production of Vibrations by Forces of Relatively Long Duration with Application to the Theory of Collisions," *Phil. Mag.,* ser. 6, vol. 11, 1906.

32. Dohrwend, C. O., D. C. Drucker, and P. Moore: "Transverse Impact Transients," *Proc. SESA,* vol. 1, no. 2, pp. 1–11, 1944.

33. Campbell, J. D., and J. Duby: "The Yield Behavior of Mild Steel in Dynamic Compression," *Proc. Roy. Soc. (London),* vol. A236, 1956.

34. Maiden, C. J., and J. D. Campbell: "The Static and Dynamic Strength of a Carbon Steel at Low Temperatures," *Phil. Mag.,* ser. 8, vol. 3, 1958.

35. Johnson, J. E., D. S. Wood, and D. S. Clark: *J. Appl. Mech.,* p. 523, December 1953.

36. Miklowitz, J.: "The Effect of Residual Stresses on High-speed Impact Resistance," in "Residual Stresses in Metals Construction," W. R. Osgood, ed., Reinhold Publishing Corporation, New York, 1954.

37. Zener, C., and J. H. Hollomon: "Effect of Strain Rate upon Plastic Flow in Steel," *J. Appl. Phys.,* vol. 15, 1944.

38. Whittaker, E. T.: "A Treatise on the Analytical Dynamics of Particles and Rigid Bodies," Cambridge University Press, New York, 1937.

39. Davidenkov, N. N.: "Allowable Working Stresses under Impact," *Trans. ASME,* vol. 56, 1934.

40. Cardillo, R. M., and D. F. Kruse: "Load Bearing Characteristics of Butyl Rubber," ASME paper 61-WA-335.

41. Barkan, P., and M. F. Sirkin: "Impact Behavior of Elastomers," ASME annual meet. paper, 1962; published in *Trans. Soc. of Plastics Engrs.,* pp. 210–219, July 1963.

42. Bradley, J. N.: "Shock Waves in Chemistry and Physics," John Wiley & Sons, Inc., New York, 1962.

43. Rice, M. H., R. G. McQueen, and J. M. Walsh: "Compression of Solids by Strong Shock Waves," in "Solid State Physics," vol. 6, Seitz and Turnbull, eds., Academic Press, Inc., New York, 1958.

44. Maxwell, B.: "Mechanical Properties of Plastic Dielectrics," *Elec. Mfg.,* vol. 58, p. 146, September 1956.

45. Bowden, F. P., and D. Tabor: "The Friction and Lubrication of Solids," Oxford University Press, Fair Lawn, N.J., 1954.

46. Love, A. E. H.: "Mathematical Theory of Elasticity," 4th ed., Dover Publications, Inc., New York, 1927.

47. Goldsmith, W., and P. T. Lyman: "The Penetration of Hard-Steel Spheres into Plane Metal Surfaces," ASME paper 60-WA-15, *J. Appl. Mech.*

48. Crook, A. W.: "A Study of Some Impacts between Metal Bodies by a Piezo-Electric Method," *Proc. Roy. Soc. (London) Series,* vol. A212, p. 377, 1952.

49. Bell, J. F.: Propagation of Large Amplitude Waves in Annealed Aluminum, *J. Appl. Phys.,* vol. 31, no. 2, pp. 277–282, February 1960.

50. Bell, J. F.: Study of Initial Conditions in Constant Velocity Impact, *J. Appl. Phys.,* vol. 31, no. 12, pp. 2188–2195, December 1960.

INDEX

ABOUT THE EDITOR

Harold A. Rothbart is a noted consulting engineer, lecturer, and author on mechanical engineering. He was formerly Dean of the College of Science and Engineering at Fairleigh Dickinson University. Dr. Rothbart has lectured in Europe and Asia for industries and universities on high-speed machinery, education, imagination, invention, and design. In addition, he has published and presented more than 50 papers and lectures on engineering and applied sciences in journals and at conferences throughout the world. Dr. Rothbart is also editor of the Second Edition of *Mechanical Design and Systems Handbook*.